Index of Applications

D1305705

Finite Mathematics and Applied Calculus

Finite Mathematics and Applied Calculus

Sixth Edition

Stefan Waner
Hofstra University

Steven R. Costenoble
Hofstra University

BROOKS/COLE
CENGAGE Learning·

Australia • Brazil • Japan • Korea • Mexico • Singapore • Spain • United Kingdom • United States

BROOKS/COLE
CENGAGE Learning

Finite Mathematics and Applied Calculus, Sixth Edition
Stefan Waner, Steven R. Costenoble

Publisher: Richard Stratton

Development Editor: Jay Campbell

Editorial Assistant: Alex Gontar

Media Editor: Andrew Coppola

Brand Manager: Gordon Lee

Marketing Coordinator: Lindsy Lettre

Marketing Communications Manager:
Linda Yip

Content Project Manager:
Alison Eigel Zade

Senior Art Director: Linda May

Print Buyer: Doug Bertke

Rights Acquisition Specialist:
Shalice Shah-Caldwell

Production Service: MPS Limited

Text Designer: RHDG Design

Cover Designer: Chris Miller

Cover Image: Bathsheba Grossman
(bathsheba.com)

Compositor: MPS Limited

For product information and technology assistance, contact us at **Cengage Learning Customer & Sales Support, 1-800-354-9706**

For permission to use material from this text or product, submit all requests online at **www.cengage.com/permissions**. Further permissions questions can be emailed to **permissionrequest@cengage.com**.

Library of Congress Control Number: 2012947816

Student Edition:
ISBN-13: 978-1-133-60770-0
ISBN-10: 1-133-60770-5

Brooks/Cole
20 Channel Center Street
Boston, MA 02210
USA

Cengage Learning is a leading provider of customized learning solutions with office locations around the globe, including Singapore, the United Kingdom, Australia, Mexico, Brazil and Japan. Locate your local office at **international.cengage.com/region**

Cengage Learning products are represented in Canada by Nelson Education, Ltd.

For your course and learning solutions, visit **www.cengage.com**.

Purchase any of our products at your local college store or at our preferred online store **www.cengagebrain.com**.

Instructors: Please visit **login.cengage.com** and log in to access instructor-specific resources.

Printed in Canada
3 4 5 6 7 16 15

Brief Contents

Contents

Preface

Finite Mathematics and Applied Calculus, sixth edition, is intended for a one- or two-term course for students majoring in business, the social sciences, or the liberal arts. Like the earlier editions, the sixth edition of *Finite Mathematics and Applied Calculus* is designed to address the challenge of generating enthusiasm and mathematical sophistication in an audience that is often underprepared and lacks motivation for traditional mathematics courses. We meet this challenge by focusing on real-life applications and topics of current interest that students can relate to, by presenting mathematical concepts intuitively and thoroughly, and by employing a writing style that is informal, engaging, and occasionally even humorous.

The sixth edition goes further than earlier editions in implementing support for a wide range of instructional paradigms: from traditional face-to-face courses to online distance learning courses, from settings incorporating little or no technology to courses taught in computerized classrooms, and from classes in which a single form of technology is used exclusively to those incorporating several technologies. We fully support three forms of technology in this text: TI-83/84 Plus graphing calculators, spreadsheets, and powerful online utilities we have created for the book. In particular, our comprehensive support for spreadsheet technology, both in the text and online, is highly relevant for students who are studying business and economics, where skill with spreadsheets may be vital to their future careers.

Our Approach to Pedagogy

Real-World Orientation We are confident that you will appreciate the diversity, breadth, and abundance of examples and exercises included in this edition. A large number of these are based on real, referenced data from business, economics, the life sciences, and the social sciences. Examples and exercises based on dated information have generally been replaced by more current versions; applications based on unique or historically interesting data have been kept.

Adapting real data for pedagogical use can be tricky; available data can be numerically complex, intimidating for students, or incomplete. We have modified and streamlined many of the real-world applications, rendering them as tractable as any "made-up" application. At the same time, we have been careful to strike a pedagogically sound balance between applications based on real data and more traditional "generic" applications. Thus, the density and selection of real data–based applications has been tailored to the pedagogical goals and appropriate difficulty level for each section.

Readability We would like students to read this book. We would like students to *enjoy* reading this book. Thus, we have written the book in a conversational and student-oriented style, and have made frequent use of question-and-answer dialogues to encourage the development of the student's mathematical curiosity and intuition. We hope that this text will give the student insight into how a mathematician develops and thinks about mathematical ideas and their applications.

Rigor We feel that mathematical rigor need not be antithetical to the kind of applied focus and conceptual approach that are earmarks of this book. We have worked hard to ensure that we are always mathematically honest without being unnecessarily formal. Sometimes we do this through the question-and-answer dialogues and sometimes through the "Before we go on . . ." discussions that follow examples, but always in manner designed to provoke the interest of the student.

Five Elements of Mathematical Pedagogy to Address Different Learning Styles The "Rule of Four" is a common theme in many texts. Implementing this approach, we discuss many of the central concepts **numerically**, **graphically,** and **algebraically** and clearly delineate these distinctions. The fourth element, **verbal communication** of mathematical concepts, is emphasized through our discussions on translating English sentences into mathematical statements and in our extensive Communication and Reasoning exercises at the end of each section. A fifth element, **interactivity**, is implemented through expanded use of question-and-answer dialogues but is seen most dramatically in the student Website. Using this resource, students can interact with the material in several ways: through interactive tutorials in the form of games, chapter summaries, and chapter review exercises, all in reference to concepts and examples covered in sections and with online utilities that automate a variety of tasks, from graphing to regression and matrix algebra.

Exercise Sets Our comprehensive collection of exercises provides a wealth of material that can be used to challenge students at almost every level of preparation and includes everything from straightforward drill exercises to interesting and rather challenging applications. The exercise sets have been carefully graded to move from straightforward basic exercises and exercises that are similar to examples in the text to more interesting and advanced ones, marked as "more advanced" for easy reference. There are also several much more difficult exercises, designated as "challenging." We have also included, in virtually every section of every chapter, interesting applications based on real data, Communication and Reasoning exercises that help students articulate mathematical concepts and recognize common errors, and exercises ideal for the use of technology.

Many of the scenarios used in application examples and exercises are revisited several times throughout the book. Thus, for instance, students will find themselves using a variety of techniques, from solving systems of equations to linear programming, or graphing through the use of derivatives and elasticity, to analyze the same application. Reusing scenarios and important functions provides unifying threads and shows students the complex texture of real-life problems.

New to This Edition

Content

- Chapter 1 (page 39): We now include, in Section 1.1, careful discussion of the common practice of representing functions as equations and vice versa; for instance, a cost equation like $C = 10x + 50$ can be thought of as defining a cost *function* $C(x) = 10x + 50$. Instead of rejecting this practice, we encourage the student to see this connection between functions and equations and to be able to switch from one interpretation to the other.

  Our discussion of functions and models in Section 1.2 now includes a careful discussion of the algebra of functions presented through the context of important applications rather than as an abstract concept. Thus, the student will see from the outset *why* we want to talk about sums, products, etc. of functions rather than simply *how* to manipulate them.

- Chapter 2 (page 125): The Mathematics of Finance is now Chapter 2 of the text because the discussion of many important topics in finance relate directly to the first discussions of compound interest and other mathematical models in Chapter 1. Note that our discussion of the Mathematics of Finance does not require the use of logarithmic functions to solve for exponents analytically but instead focuses on numerical solution using the technologies we discuss. However, the use of logarithms is presented as an option for students and instructors who prefer to use them.

- Chapter 10 (page 687): Our discussion of limits now discusses extensively when, and why, substitution can be used to obtain a limit. We now also follow the usual convention of allowing only one-sided limits at endpoints of domains. This approach also applies to derivatives, where we now disallow derivatives at endpoints of domains, as is the normal convention.

- Chapter 11 (page 783): The closed-form formula for the derivative of $|x|$, introduced in Section 11.1, is now more fully integrated into the text, as is that for its antiderivative (in Chapter 13). (It is puzzling that it is not standard fare in other calculus books.)

- Chapter 13 (page 945): The sections on antiderivatives and substitution have been reorganized and streamlined and now include discussion of the closed-form antiderivative of $|x|$ and well as new exercises featuring absolute values.

  The definite integral is now introduced in the realistic context of the volume of oil released in an oil spill comparable in size to the *BP* 2011 Gulf oil spill.

- Chapter 15 (page 1081): The discussion of level curves in Section 15.1 is now more extensive and includes added examples and exercises.

- **Case Studies:** A number of the Case Studies at the ends of the chapters have been extensively revised, using updated real data, and continue to reflect topics of current interest, such as subprime mortgages, hybrid car production, the diet problem (in linear programming), spending on housing construction, modeling tax revenues, and pollution control.

Current Topics in the Applications

- We have added and updated numerous real data exercises and examples based on topics that are either of intense current interest or of general interest to contemporary students, including Facebook, XBoxes, iPhones, Androids, iPads, foreclosure rates, the housing crisis, subprime mortgages, travel to Cancun, the *BP* 2011 Gulf oil spill, and the U.S. stock market "flash crash" of May 6, 2010. (Also see the list, in the inside back cover, of the corporations we reference in the applications.)

Exercises

- We have expanded the chapter review exercise sets to be more representative of the material within the chapter. Note that all the applications in the chapter review exercises revolve around the fictitious online bookseller, *OHaganBooks.com,* and the various—often amusing—travails of *OHaganBooks.com* CEO John O'Hagan and his business associate Marjory Duffin.

- We have added many new conceptual Communication and Reasoning exercises, including many dealing with common student errors and misconceptions.

End-of-Chapter Technology Guides

- Our end-of-chapter detailed Technology Guides now discuss the use of spreadsheets in general rather than focusing exclusively on Microsoft® Excel, thus enabling readers to use any of the several alternatives now available, such as Google's online Google Sheets®, Open Office®, and Apple's Numbers®.

Continuing Features

Case Study Each chapter ends with a section entitled "Case Study," an extended application that uses and illustrates the central ideas of the chapter, focusing on the development of mathematical models appropriate to the topics. These applications are ideal for assignment as projects, and to this end we have included groups of exercises at the end of each.

- **Before We Go On** Most examples are followed by supplementary discussions, which may include a check on the answer, a discussion of the feasibility and significance of a solution, or an in-depth look at what the solution means.

- **Quick Examples** Most definition boxes include quick, straightforward examples that a student can use to solidify each new concept.

- **Question-and-Answer Dialogue** We frequently use informal question-and-answer dialogues that anticipate the kinds of questions that may occur to the student and also guide the student through the development of new concepts.

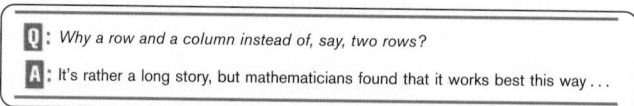

Q: Why a row and a column instead of, say, two rows?

A: It's rather a long story, but mathematicians found that it works best this way . . .

- **Marginal Technology Notes** We give brief marginal technology notes to outline the use of graphing calculator, spreadsheet, and Website technology in appropriate examples. When necessary, the reader is referred to more detailed discussion in the end-of-chapter Technology Guides.

- **End-of-Chapter Technology Guides** We continue to include detailed TI-83/84 Plus and Spreadsheet Guides at the end of each chapter. These Guides are referenced liberally in marginal technology notes at appropriate points in the chapter, so instructors and students can easily use this material or not, as they prefer. Groups of exercises for which the use of technology is suggested or required appear throughout the exercise sets.

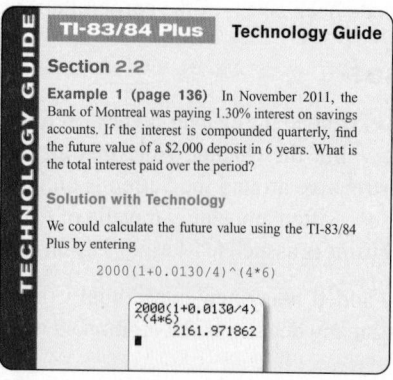

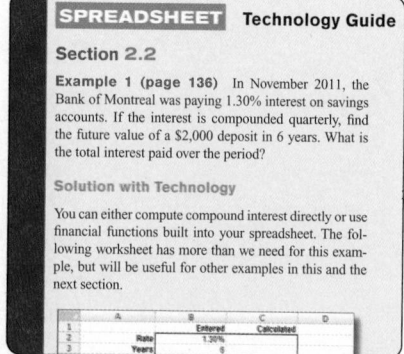

- **Communication and Reasoning Exercises for Writing and Discussion**
  These are exercises designed to broaden the student's grasp of the mathematical concepts and develop modeling skills. They include exercises in which the student is asked to provide his or her own examples to illustrate a point or design an application with a given solution. They also include "fill in the blank" type exercises, exercises that invite discussion and debate, and exercises in which the student must identify common errors. These exercises often have no single correct answer.

Supplemental Material

For Instructors and Students

Enhanced WebAssign®

Content

Exclusively from Cengage Learning, Enhanced WebAssign® combines the exceptional mathematics content in Waner and Costenoble's text with the most powerful online homework solution, WebAssign. Enhanced WebAssign engages students with immediate feedback, rich tutorial content, videos, animations, and an interactive eBook, helping students to develop a deeper conceptual understanding of the subject matter. The interactive eBook contains helpful search, highlighting, and note-taking features.

Instructors can build online assignments by selecting from thousands of text-specific problems, supplemented if desired with problems from any Cengage Learning textbook. Flexible assignment options give instructors the ability to choose how feedback and tutorial content is released to students as well as the ability to release assignments conditionally based on students' prerequisite assignment scores. Increase student engagement, improve course outcomes, and experience the superior service offered through CourseCare. Visit us at http://webassign.net/cengage or www.cengage.com/ewa to learn more.

Service

Your adoption of Enhanced WebAssign® includes CourseCare, Cengage Learning's industry leading service and training program designed to ensure that you have everything that you need to make the most of your use of Enhanced WebAssign. CourseCare provides one-on-one service, from finding the right solutions for your course to training and support. A team of Cengage representatives, including Digital Solutions Managers and Coordinators as well as Service and Training Consultants, assists you every step of the way. For additional information about CourseCare, please visit www.cengage.com/coursecare.

Our Enhanced WebAssign training program provides a comprehensive curriculum of beginner, intermediate, and advanced sessions, designed to get you started and effectively integrate Enhanced WebAssign into your course. We offer a flexible online and recorded training program designed to accommodate your busy schedule. Whether you are using Enhanced WebAssign for the first time or an experienced user, there is a training option to meet your needs.

www.WanerMath.com

The authors' Website, accessible through www.WanerMath.com and linked within Enhanced WebAssign, has been evolving for more than a decade and has been receiving increasingly more recognition. Students, raised in an environment in which computers permeate both work and play, now use the Internet to engage with the material in an active way. The following features of the authors' Website are fully integrated with the text and can be used as a personalized study resource as well as a valuable teaching aid for instructors:

- *Interactive tutorials* on almost all topics, with guided exercises, which also can be used in classroom instruction or in distance learning courses

- *More challenging game versions of tutorials* with randomized questions that complement the traditional interactive tutorials and can be used as in-class quizzes

- *Detailed interactive chapter summaries* that review basic definitions and problem-solving techniques and can act as pre-test study tools

- *Downloadable Excel tutorials* keyed to examples given in the text

- *Online utilities for use in solving many of the technology-based application exercises*. The utilities, for instructor use in class and student use out of class, include a function grapher and evaluator that now also graphs derivatives and offers curve-fitting, regression tools, an interactive Riemann sum grapher with an improved numerical integrator, a matrix algebra tool, linear programming tools, and a line entry calculator that calculates permutations and combinations, expands multinomial expressions, does derivatives and numerical integration, and gives Taylor series.

- *Chapter true-false quizzes* with feedback for many incorrect answers

- *Supplemental topics* including interactive text and exercise sets for selected topics not found in the printed texts

- *Spanish versions* of chapter summaries, tutorials, game tutorials, and utilities

For Students

Student Solutions Manual *by Waner and Costenoble*
ISBN: 9781285085661
The student solutions manual provides worked-out solutions to the odd-numbered exercises in the text, plus problem-solving strategies and additional algebra steps and review for selected problems.

    To access this and other course materials and companion resources, please visit **www.cengagebrain.com**. At the CengageBrain.com home page, search for the ISBN of your title (from the back cover of your book) using the search box at the top of the page. This will take you to the product page where free companion resources can be found.

For Instructors

Complete Solution Manual *by Waner and Costenoble*
ISBN: 9781285085678
The instructor's solutions manual provides worked-out solutions to all of the exercises in the text.

Solution Builder *by Waner and Costenoble*
ISBN: 9781285085685
This time-saving resource offers fully worked instructor solutions to all exercises in the text in customizable online format. Adopting instructors can sign up for access at www.cengage.com/solutionbuilder.

PowerLecture™ with ExamView® computerized testing *by Waner and Costenoble*
ISBN: 9781285085708
This CD-ROM provides the instructor with dynamic media tools for teaching, including Microsoft® PowerPoint® lecture slides, figures from the book, and the Test Bank. You can create, deliver, and customize tests (both print and online) in minutes with ExamView® computerized testing, which includes Test Bank items in electronic format. In addition, you can easily build solution sets for homework or exams by linking to Solution Builder's online solutions manual.

Instructor's Edition
ISBN: 9781133610595

www.WanerMath.com
The Instructor's Resource Page at www.WanerMath.com features an expanded collection of instructor resources, including an updated corrections page, an expanding set of author-created teaching videos for use in distance learning courses, and a utility that automatically updates homework exercise sets from the fifth edition to the sixth.

Acknowledgments

This project would not have been possible without the contributions and suggestions of numerous colleagues, students, and friends. We are particularly grateful to our colleagues at Hofstra and elsewhere who used and gave us useful feedback on previous editions. We are also grateful to everyone at Cengage for their encouragement and guidance throughout the project. Specifically, we would like to thank Richard Stratton and Jay Campbell for their unflagging enthusiasm and Alison Eigel Zade for whipping the book into shape.

We would also like to thank our accuracy checker, Jerrold Grossman, and the numerous reviewers who provided many helpful suggestions that have shaped the development of this book.

Priscilla Chaffe-Stengel, *California State University, Fresno*

Celeste Hernandez, *Richland College*

Erick Hofacker, *University of Wisconsin–River Falls*

Jan Rychtar, *University of North Carolina at Greensboro*

Jack Narayan, *State University of New York at Oswego*

Mary Wagner-Krankel, *St. Mary's University*

Stefan Waner
Steven R. Costenoble

O

Precalculus Review

 **Website**

www.WanerMath.com

- At the Website you will find section-by-section interactive tutorials for further study and practice.

DreamPictures/Taxi/Getty Images

Introduction

In this chapter we review some topics from algebra that you need to know to get the most out of this book. This chapter can be used either as a refresher course or as a reference.

There is one crucial fact you must always keep in mind: The letters used in algebraic expressions stand for numbers. All the rules of algebra are just facts about the arithmetic of numbers. If you are not sure whether some algebraic manipulation you are about to do is legitimate, try it first with numbers. If it doesn't work with numbers, it doesn't work.

0.1 Real Numbers

The **real numbers** are the numbers that can be written in decimal notation, including those that require an infinite decimal expansion. The set of real numbers includes all integers, positive and negative; all fractions; and the irrational numbers, those with decimal expansions that never repeat. Examples of irrational numbers are

$$\sqrt{2} = 1.414213562373\ldots$$

and

$$\pi = 3.141592653589\ldots$$

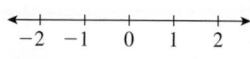

Figure 1

It is very useful to picture the real numbers as points on a line. As shown in Figure 1, larger numbers appear to the right, in the sense that if $a < b$ then the point corresponding to b is to the right of the one corresponding to a.

Intervals

Some subsets of the set of real numbers, called **intervals**, show up quite often and so we have a compact notation for them.

Interval Notation

Here is a list of types of intervals along with examples.

	Interval	*Description*	*Picture*	*Example*
Closed	$[a, b]$	Set of numbers x with $a \leq x \leq b$	a b (includes end points)	$[0, 10]$
Open	(a, b)	Set of numbers x with $a < x < b$	a b (excludes end points)	$(-1, 5)$
Half-Open	$(a, b]$	Set of numbers x with $a < x \leq b$	a b	$(-3, 1]$
	$[a, b)$	Set of numbers x with $a \leq x < b$	a b	$[0, 5)$

Infinite	$[a, +\infty)$	Set of numbers x with $a \le x$		$[10, +\infty)$
	$(a, +\infty)$	Set of numbers x with $a < x$		$(-3, +\infty)$
	$(-\infty, b]$	Set of numbers x with $x \le b$		$(-\infty, -3]$
	$(-\infty, b)$	Set of numbers x with $x < b$		$(-\infty, 10)$
	$(-\infty, +\infty)$	Set of all real numbers		$(-\infty, +\infty)$

Operations

There are five important operations on real numbers: addition, subtraction, multiplication, division, and exponentiation. "Exponentiation" means raising a real number to a power; for instance, $3^2 = 3 \cdot 3 = 9$; $2^3 = 2 \cdot 2 \cdot 2 = 8$.

A note on technology: Most graphing calculators and spreadsheets use an asterisk * for multiplication and a caret sign ^ for exponentiation. Thus, for instance, 3×5 is entered as 3*5, $3x$ as 3*x, and 3^2 as 3^2.

When we write an expression involving two or more operations, like

$$2 \cdot 3 + 4$$

or

$$\frac{2 \cdot 3^2 - 5}{4 - (-1)}$$

we need to agree on the order in which to do the operations. Does $2 \cdot 3 + 4$ mean $(2 \cdot 3) + 4 = 10$ or $2 \cdot (3 + 4) = 14$? We all agree to use the following rules for the order in which we do the operations.

Standard Order of Operations

Parentheses and Fraction Bars First, calculate the values of all expressions inside parentheses or brackets, working from the innermost parentheses out, before using them in other operations. In a fraction, calculate the numerator and denominator separately before doing the division.

Quick Examples

1. $6(2 + [3 - 5] - 4) = 6(2 + (-2) - 4) = 6(-4) = -24$
2. $\dfrac{(4 - 2)}{3(-2 + 1)} = \dfrac{2}{3(-1)} = \dfrac{2}{-3} = -\dfrac{2}{3}$
3. $3/(2 + 4) = \dfrac{3}{2 + 4} = \dfrac{3}{6} = \dfrac{1}{2}$
4. $(x + 4x)/(y + 3y) = 5x/(4y)$

Exponents   Next, perform exponentiation.

Quick Examples

$$\left. \begin{array}{l} \textbf{1.}\ 2 + 4^2 = 2 + 16 = 18 \\ \textbf{2.}\ (2 + 4)^2 = 6^2 = 36 \end{array} \right\} \quad \text{Note the difference.}$$

3. $2\left(\dfrac{3}{4-5}\right)^2 = 2\left(\dfrac{3}{-1}\right)^2 = 2(-3)^2 = 2 \times 9 = 18$

4. $2(1 + 1/10)^2 = 2(1.1)^2 = 2 \times 1.21 = 2.42$

Multiplication and Division   Next, do all multiplications and divisions, from left to right.

Quick Examples

1. $2(3 - 5)/4 \cdot 2 = 2(-2)/4 \cdot 2$	Parentheses first
$= -4/4 \cdot 2$	Left-most product
$= -1 \cdot 2 = -2$	Multiplications and divisions, left to right
2. $2(1 + 1/10)^2 \times 2/10 = 2(1.1)^2 \times 2/10$	Parentheses first
$= 2 \times 1.21 \times 2/10$	Exponent
$= 4.84/10 = 0.484$	Multiplications and divisions, left to right

3. $4\dfrac{2(4-2)}{3(-2 \cdot 5)} = 4\dfrac{2(2)}{3(-10)} = 4\dfrac{4}{-30} = \dfrac{16}{-30} = -\dfrac{8}{15}$

Addition and Subtraction   Last, do all additions and subtractions, from left to right.

Quick Examples

1. $2(3 - 5)^2 + 6 - 1 = 2(-2)^2 + 6 - 1 = 2(4) + 6 - 1$
$\qquad = 8 + 6 - 1 = 13$

2. $\left(\dfrac{1}{2}\right)^2 - (-1)^2 + 4 = \dfrac{1}{4} - 1 + 4 = -\dfrac{3}{4} + 4 = \dfrac{13}{4}$

$$\left. \begin{array}{l} \textbf{3.}\ 3/2 + 4 = 1.5 + 4 = 5.5 \\ \textbf{4.}\ 3/(2 + 4) = 3/6 = 1/2 = 0.5 \end{array} \right\} \quad \text{Note the difference.}$$

5. $4/2^2 + (4/2)^2 = 4/2^2 + 2^2 = 4/4 + 4 = 1 + 4 = 5$

⬛ Entering Formulas

Any good calculator or spreadsheet will respect the standard order of operations. However, we must be careful with division and exponentiation and use parentheses as necessary. The following table gives some examples of simple mathematical expressions and their equivalents in the functional format used in most graphing calculators, spreadsheets, and computer programs.

Mathematical Expression	*Formula*	*Comments*
$\dfrac{2}{3-x}$	`2/(3-x)`	Note the use of parentheses instead of the fraction bar. If we omit the parentheses, we get the expression shown next.
$\dfrac{2}{3}-x$	`2/3-x`	The calculator follows the usual order of operations.
$\dfrac{2}{3\times5}$	`2/(3*5)`	Putting the denominator in parentheses ensures that the multiplication is carried out first. The asterisk is usually used for multiplication in graphing calculators and computers.
$\dfrac{2}{x}\times5$	`(2/x)*5`	Putting the fraction in parentheses ensures that it is calculated first. Some calculators will interpret `2/3*5` as $\dfrac{2}{3\times5}$, but `2/3(5)` as $\dfrac{2}{3}\times5$.
$\dfrac{2-3}{4+5}$	`(2-3)/(4+5)`	Note once again the use of parentheses in place of the fraction bar.
2^3	`2^3`	The caret ^ is commonly used to denote exponentiation.
2^{3-x}	`2^(3-x)`	Be careful to use parentheses to tell the calculator where the exponent ends. Enclose the *entire exponent* in parentheses.
2^3-x	`2^3-x`	Without parentheses, the calculator will follow the usual order of operations: exponentiation and then subtraction.
3×2^{-4}	`3*2^(-4)`	On some calculators, the negation key is separate from the minus key.
$2^{-4\times3}\times5$	`2^(-4*3)*5`	Note once again how parentheses enclose the entire exponent.
$100\left(1+\dfrac{0.05}{12}\right)^{60}$	`100*(1+0.05/12)^60`	This is a typical calculation for compound interest.
$PV\left(1+\dfrac{r}{m}\right)^{mt}$	`PV*(1+r/m)^(m*t)`	This is the compound interest formula. *PV* is understood to be a single number (present value) and not the product of *P* and *V* (or else we would have used `P*V`).
$\dfrac{2^{3-2}\times5}{y-x}$	`2^(3-2)*5/(y-x)` or `(2^(3-2)*5)/(y -x)`	Notice again the use of parentheses to hold the denominator together. We could also have enclosed the numerator in parentheses, although this is optional. (Why?)
$\dfrac{2^y+1}{2-4^{3x}}$	`(2^y+1)/(2-4^(3*x))`	Here, it is necessary to enclose both the numerator and the denominator in parentheses.
$2^y+\dfrac{1}{2}-4^{3x}$	`2^y+1/2-4^(3*x)`	This is the effect of leaving out the parentheses around the numerator and denominator in the previous expression.

Accuracy and Rounding

When we use a calculator or computer, the results of our calculations are often given to far more decimal places than are useful. For example, suppose we are told that a square has an area of 2.0 square feet and we are asked how long its sides are. Each side is the square root of the area, which the calculator tells us is

$$\sqrt{2} \approx 1.414213562$$

However, the measurement of 2.0 square feet is probably accurate to only two digits, so our estimate of the lengths of the sides can be no more accurate than that. Therefore, we round the answer to two digits:

Length of one side ≈ 1.4 feet

The digits that follow 1.4 are meaningless. The following guide makes these ideas more precise.

Significant Digits, Decimal Places, and Rounding

The number of **significant digits** in a decimal representation of a number is the number of digits that are not leading zeros after the decimal point (as in .0005) or trailing zeros before the decimal point (as in 5,400,000). We say that a value is **accurate to _n_ significant digits** if only the first _n_ significant digits are meaningful.

When to Round

After doing a computation in which all the quantities are accurate to no more than _n_ significant digits, round the final result to _n_ significant digits.

Quick Examples

1. 0.00067 has two significant digits.
   The 000 before 67 are leading zeros.

2. 0.000670 has three significant digits.
   The 0 after 67 is significant.

3. 5,400,000 has two or more significant digits.
   We can't say how many of the zeros are trailing.[1]

4. 5,400,001 has 7 significant digits.
   The string of zeros is not trailing.

5. Rounding 63,918 to three significant digits gives 63,900.

6. Rounding 63,958 to three significant digits gives 64,000.

7. $\pi = 3.141592653...$ $\frac{22}{7} = 3.142857142...$ Therefore, $\frac{22}{7}$ is an approximation of π that is accurate to only three significant digits (3.14).

8. $4.02(1 + 0.02)^{1.4} \approx 4.13$
   We rounded to three significant digits.

---

[1]If we obtained 5,400,000 by rounding 5,401,011, then it has three significant digits because the zero after the 4 is significant. On the other hand, if we obtained it by rounding 5,411,234, then it has only two significant digits. The use of scientific notation avoids this ambiguity: 5.40×10^6 (or 5.40 E6 on a calculator or computer) is accurate to three digits and 5.4×10^6 is accurate to two.

One more point, though: If, in a long calculation, you round the intermediate results, your final answer may be even less accurate than you think. As a general rule,

When calculating, don't round intermediate results. Rather, use the most accurate results obtainable or have your calculator or computer store them for you.

When you are done with the calculation, *then* round your answer to the appropriate number of digits of accuracy.

0.1 EXERCISES

Calculate each expression in Exercises 1–24, giving the answer as a whole number or a fraction in lowest terms.

1. $2(4+(-1))(2\cdot-4)$

2. $3+([4-2]\cdot 9)$

3. `20/(3*4)-1`

4. `2-(3*4)/10`

5. $\dfrac{3+([3+(-5)])}{3-2\times 2}$

6. $\dfrac{12-(1-4)}{2(5-1)\cdot 2-1}$

7. `(2-5*(-1))/1-2*(-1)`

8. `2-5*(-1)/(1-2*(-1))`

9. $2\cdot(-1)^2/2$

10. $2+4\cdot 3^2$

11. $2\cdot 4^2+1$

12. $1-3\cdot(-2)^2\times 2$

13. `3^2+2^2+1`

14. `2^(2^2-2)`

15. $\dfrac{3-2(-3)^2}{-6(4-1)^2}$

16. $\dfrac{1-2(1-4)^2}{2(5-1)^2\cdot 2}$

17. `10*(1+1/10)^3`

18. `121/(1+1/10)^2`

19. $3\left(\dfrac{-2\cdot 3^2}{-(4-1)^2}\right)$

20. $-\left(\dfrac{8(1-4)^2}{-9(5-1)^2}\right)$

21. $3\left(1-\left(-\dfrac{1}{2}\right)^2\right)^2+1$

22. $3\left(\dfrac{1}{9}-\left(\dfrac{2}{3}\right)^2\right)^2+1$

23. `(1/2)^2-1/2^2`

24. `2/(1^2)-(2/1)^2`

Convert each expression in Exercises 25–50 into its technology formula equivalent as in the table in the text.

25. $3\times(2-5)$

26. $4+\dfrac{5}{9}$

27. $\dfrac{3}{2-5}$

28. $\dfrac{4-1}{3}$

29. $\dfrac{3-1}{8+6}$

30. $3+\dfrac{3}{2-9}$

31. $3-\dfrac{4+7}{8}$

32. $\dfrac{4\times 2}{\left(\frac{2}{3}\right)}$

33. $\dfrac{2}{3+x}-xy^2$

34. $3+\dfrac{3+x}{xy}$

35. $3.1x^3-4x^{-2}-\dfrac{60}{x^2-1}$

36. $2.1x^{-3}-x^{-1}+\dfrac{x^2-3}{2}$

37. $\dfrac{\left(\frac{2}{3}\right)}{5}$

38. $\dfrac{2}{\left(\frac{3}{5}\right)}$

39. $3^{4-5}\times 6$

40. $\dfrac{2}{3+5^{7-9}}$

41. $3\left(1+\dfrac{4}{100}\right)^{-3}$

42. $3\left(\dfrac{1+4}{100}\right)^{-3}$

43. $3^{2x-1}+4^x-1$

44. $2^{x^2}-(2^{2x})^2$

45. 2^{2x^2-x+1}

46. $2^{2x^2-x}+1$

47. $\dfrac{4e^{-2x}}{2-3e^{-2x}}$

48. $\dfrac{e^{2x}+e^{-2x}}{e^{2x}-e^{-2x}}$

49. $3\left(1-\left(-\dfrac{1}{2}\right)^2\right)^2+1$

50. $3\left(\dfrac{1}{9}-\left(\dfrac{2}{3}\right)^2\right)^2+1$

0.2 Exponents and Radicals

In Section 1 we discussed exponentiation, or "raising to a power"; for example, $2^3=2\cdot 2\cdot 2$. In this section we discuss the algebra of exponentials more fully. First, we look at *integer* exponents: cases in which the powers are positive or negative whole numbers.

Integer Exponents

Positive Integer Exponents

If a is any real number and n is any positive integer, then by a^n we mean the quantity $a \cdot a \cdot \cdots \cdot a$ (n times); thus, $a^1 = a$, $a^2 = a \cdot a$, $a^5 = a \cdot a \cdot a \cdot a \cdot a$. In the expression a^n the number n is called the **exponent**, and the number a is called the **base**.

Quick Examples

$$3^2 = 9 \qquad\qquad 2^3 = 8$$
$$0^{34} = 0 \qquad\qquad (-1)^5 = -1$$
$$10^3 = 1,000 \qquad 10^5 = 100,000$$

Negative Integer Exponents

If a is any real number *other than zero* and n is any positive integer, then we define

$$a^{-n} = \frac{1}{a^n} = \frac{1}{a \cdot a \cdot \cdots \cdot a} \ (n \text{ times})$$

Quick Examples

$$2^{-3} = \frac{1}{2^3} = \frac{1}{8} \qquad\qquad 1^{-27} = \frac{1}{1^{27}} = 1$$

$$x^{-1} = \frac{1}{x^1} = \frac{1}{x} \qquad\qquad (-3)^{-2} = \frac{1}{(-3)^2} = \frac{1}{9}$$

$$y^7 y^{-2} = y^7 \frac{1}{y^2} = y^5 \qquad 0^{-2} \text{ is not defined}$$

Zero Exponent

If a is any real number other than zero, then we define

$$a^0 = 1$$

Quick Examples

$$3^0 = 1 \qquad\qquad\qquad 1,000,000^0 = 1$$

$$0^0 \text{ is not defined}$$

When combining exponential expressions, we use the following identities.

Exponent Identity	Quick Examples
1. $a^m a^n = a^{m+n}$	$2^3 2^2 = 2^{3+2} = 2^5 = 32$ $x^3 x^{-4} = x^{3-4} = x^{-1} = \dfrac{1}{x}$ $\dfrac{x^3}{x^{-2}} = x^3 \dfrac{1}{x^{-2}} = x^3 x^2 = x^5$
2. $\dfrac{a^m}{a^n} = a^{m-n}$ if $a \neq 0$	$\dfrac{4^3}{4^2} = 4^{3-2} = 4^1 = 4$ $\dfrac{x^3}{x^{-2}} = x^{3-(-2)} = x^5$ $\dfrac{3^2}{3^4} = 3^{2-4} = 3^{-2} = \dfrac{1}{9}$
3. $(a^n)^m = a^{nm}$	$(3^2)^2 = 3^4 = 81$ $(2^x)^2 = 2^{2x}$
4. $(ab)^n = a^n b^n$	$(4 \cdot 2)^2 = 4^2 2^2 = 64$ $(-2y)^4 = (-2)^4 y^4 = 16y^4$
5. $\left(\dfrac{a}{b}\right)^n = \dfrac{a^n}{b^n}$ if $b \neq 0$	$\left(\dfrac{4}{3}\right)^2 = \dfrac{4^2}{3^2} = \dfrac{16}{9}$ $\left(\dfrac{x}{-y}\right)^3 = \dfrac{x^3}{(-y)^3} = -\dfrac{x^3}{y^3}$

Caution

- In the first two identities, the bases of the expressions must be the same. For example, the first gives $3^2 3^4 = 3^6$, but does *not* apply to $3^2 4^2$.
- People sometimes invent their own identities, such as $a^m + a^n = a^{m+n}$, which is wrong! (Try it with $a = m = n = 1$.) If you wind up with something like $2^3 + 2^4$, you are stuck with it; there are no identities around to simplify it further. (You can factor out 2^3, but whether or not that is a simplification depends on what you are going to do with the expression next.)

EXAMPLE 1 Combining the Identities

$$\frac{(x^2)^3}{x^3} = \frac{x^6}{x^3} \qquad \text{By (3)}$$

$$= x^{6-3} \qquad \text{By (2)}$$

$$= x^3$$

$$\frac{(x^4 y)^3}{y} = \frac{(x^4)^3 y^3}{y} \qquad \text{By (4)}$$

$$= \frac{x^{12} y^3}{y} \qquad \text{By (3)}$$

$$= x^{12} y^{3-1} \qquad \text{By (2)}$$

$$= x^{12} y^2$$

EXAMPLE 2 **Eliminating Negative Exponents**

Simplify the following and express the answer using no negative exponents.

a. $\dfrac{x^4 y^{-3}}{x^5 y^2}$ **b.** $\left(\dfrac{x^{-1}}{x^2 y}\right)^5$

Solution

a. $\dfrac{x^4 y^{-3}}{x^5 y^2} = x^{4-5} y^{-3-2} = x^{-1} y^{-5} = \dfrac{1}{xy^5}$

b. $\left(\dfrac{x^{-1}}{x^2 y}\right)^5 = \dfrac{(x^{-1})^5}{(x^2 y)^5} = \dfrac{x^{-5}}{x^{10} y^5} = \dfrac{1}{x^{15} y^5}$

Radicals

If a is any non-negative real number, then its **square root** is the non-negative number whose square is a. For example, the square root of 16 is 4, because $4^2 = 16$. We write the square root of n as $\sqrt{n}$. (Roots are also referred to as **radicals**.) It is important to remember that $\sqrt{n}$ is never negative. Thus, for instance, $\sqrt{9}$ is 3, and not -3, even though $(-3)^2 = 9$. If we want to speak of the "negative square root" of 9, we write it as $-\sqrt{9} = -3$. If we want to write both square roots at once, we write $\pm\sqrt{9} = \pm 3$.

The **cube root** of a real number a is the number whose cube is a. The cube root of a is written as $\sqrt[3]{a}$ so that, for example, $\sqrt[3]{8} = 2$ (because $2^3 = 8$). Note that we can take the cube root of any number, positive, negative, or zero. For instance, the cube root of -8 is $\sqrt[3]{-8} = -2$ because $(-2)^3 = -8$. Unlike square roots, the cube root of a number may be negative. In fact, the cube root of a always has the same sign as a.

Higher roots are defined similarly. The **fourth root** of the *non-negative* number a is defined as the non-negative number whose fourth power is a, and written $\sqrt[4]{a}$. The **fifth root** of any number a is the number whose fifth power is a, and so on.

Note We cannot take an even-numbered root of a negative number, but we can take an odd-numbered root of any number. Even roots are always positive, whereas odd roots have the same sign as the number we start with. ∎

EXAMPLE 3 **nth Roots**

$$\sqrt{4} = 2 \qquad \text{Because } 2^2 = 4$$

$$\sqrt{16} = 4 \qquad \text{Because } 4^2 = 16$$

$$\sqrt{1} = 1 \qquad \text{Because } 1^2 = 1$$

$$\text{If } x \geq 0, \text{ then } \sqrt{x^2} = x \qquad \text{Because } x^2 = x^2$$

$$\sqrt{2} \approx 1.414213562 \qquad \sqrt{2} \text{ is not a whole number.}$$

$$\sqrt{1+1} = \sqrt{2} \approx 1.414213562 \qquad \text{First add, then take the square root.}^2$$

$$\sqrt{9+16} = \sqrt{25} = 5 \qquad \text{Contrast with } \sqrt{9} + \sqrt{16} = 3 + 4 = 7.$$

---

[2] In general, $\sqrt{a+b}$ means the square root of the *quantity* $(a+b)$. The radical sign acts as a pair of parentheses or a fraction bar, telling us to evaluate what is inside before taking the root. (See the Caution on the next page.)

$$\frac{1}{\sqrt{2}} = \frac{\sqrt{2}}{2}$$ Multiply top and bottom by $\sqrt{2}$.

$$\sqrt[3]{27} = 3$$ Because $3^3 = 27$

$$\sqrt[3]{-64} = -4$$ Because $(-4)^3 = -64$

$$\sqrt[4]{16} = 2$$ Because $2^4 = 16$

$\sqrt[4]{-16}$ is not defined   Even-numbered root of a negative number

$\sqrt[5]{-1} = -1$, since $(-1)^5 = -1$ Odd-numbered root of a negative number

$\sqrt[n]{-1} = -1$ if n is any odd number

Q: *In the example we saw that $\sqrt{x^2} = x$ if x is non-negative. What happens if x is negative?*

A: If x is negative, then x^2 is positive, and so $\sqrt{x^2}$ is still defined as the non-negative number whose square is x^2. This number must be $|x|$, the **absolute value of x**, which is the non-negative number with the same size as x. For instance, $|-3| = 3$, while $|3| = 3$, and $|0| = 0$. It follows that

$$\sqrt{x^2} = |x|$$

for every real number x, positive or negative. For instance,

$$\sqrt{(-3)^2} = \sqrt{9} = 3 = |-3|$$

and $$\sqrt{3^2} = \sqrt{9} = 3 = |3|.$$

In general, we find that

$$\sqrt[n]{x^n} = x \text{ if } n \text{ is odd, and } \sqrt[n]{x^n} = |x| \text{ if } n \text{ is even.}$$

We use the following identities to evaluate radicals of products and quotients.

Radicals of Products and Quotients

If a and b are any real numbers (non-negative in the case of even-numbered roots), then

$$\sqrt[n]{ab} = \sqrt[n]{a}\,\sqrt[n]{b}$$ Radical of a product = Product of radicals

$$\sqrt[n]{\frac{a}{b}} = \frac{\sqrt[n]{a}}{\sqrt[n]{b}} \quad \text{if } b \neq 0$$ Radical of a quotient = Quotient of radicals

Notes

- The first rule is similar to the rule $(a \cdot b)^2 = a^2 b^2$ for the square of a product, and the second rule is similar to the rule $\left(\frac{a}{b}\right)^2 = \frac{a^2}{b^2}$ for the square of a quotient.

- *Caution* There is no corresponding identity for addition:
  $$\sqrt{a+b} \text{ is } not \text{ equal to } \sqrt{a} + \sqrt{b}$$

(Consider $a = b = 1$, for example.) Equating these expressions is a common error, so be careful! ∎

Quick Examples

1. $\sqrt{9 \cdot 4} = \sqrt{9}\sqrt{4} = 3 \times 2 = 6$ Alternatively, $\sqrt{9 \cdot 4} = \sqrt{36} = 6$

2. $\sqrt{\dfrac{9}{4}} = \dfrac{\sqrt{9}}{\sqrt{4}} = \dfrac{3}{2}$

3. $\dfrac{\sqrt{2}}{\sqrt{5}} = \dfrac{\sqrt{2}\sqrt{5}}{\sqrt{5}\sqrt{5}} = \dfrac{\sqrt{10}}{5}$

4. $\sqrt{4(3 + 13)} = \sqrt{4(16)} = \sqrt{4}\sqrt{16} = 2 \times 4 = 8$

5. $\sqrt[3]{-216} = \sqrt[3]{(-27)8} = \sqrt[3]{-27}\sqrt[3]{8} = (-3)2 = -6$

6. $\sqrt{x^3} = \sqrt{x^2 \cdot x} = \sqrt{x^2}\sqrt{x} = x\sqrt{x}$ if $x \geq 0$

7. $\sqrt{\dfrac{x^2 + y^2}{z^2}} = \dfrac{\sqrt{x^2 + y^2}}{\sqrt{z^2}} = \dfrac{\sqrt{x^2 + y^2}}{|z|}$ We can't simplify the numerator any further.

Rational Exponents

We already know what we mean by expressions such as x^4 and a^{-6}. The next step is to make sense of *rational* exponents: exponents of the form p/q with p and q integers as in $a^{1/2}$ and $3^{-2/3}$.

Q: *What should we mean by* $a^{1/2}$?

A: The overriding concern here is that all the exponent identities should remain true. In this case the identity to look at is the one that says that $(a^m)^n = a^{mn}$. This identity tells us that

$$(a^{1/2})^2 = a^1 = a.$$

That is, $a^{1/2}$, when squared, gives us a. But that must mean that $a^{1/2}$ is the *square root* of a, or

$$a^{1/2} = \sqrt{a}.$$

A similar argument tells us that, if q is any positive whole number, then

$$a^{1/q} = \sqrt[q]{a}, \text{ the } q\text{th root of } a.$$

Notice that if a is negative, this makes sense only for q odd. To avoid this problem, we usually stick to positive a.

Q: *If p and q are integers (q positive), what should we mean by* $a^{p/q}$?

A: By the exponent identities, $a^{p/q}$ should equal both $(a^p)^{1/q}$ and $(a^{1/q})^p$. The first is the qth root of a^p, and the second is the pth power of $a^{1/q}$, which gives us the following.

Conversion Between Rational Exponents and Radicals

If a is any non-negative number, then
$$a^{p/q} = \sqrt[q]{a^p} = \left(\sqrt[q]{a}\right)^p.$$

Using exponents   Using radicals

In particular,
$$a^{1/q} = \sqrt[q]{a}, \text{ the } q\text{th root of } a.$$

Notes

- If a is negative, all of this makes sense only if q is odd.
- All of the exponent identities continue to work when we allow rational exponents p/q. In other words, we are free to use all the exponent identities even though the exponents are not integers. ∎

Quick Examples

1. $4^{3/2} = (\sqrt{4})^3 = 2^3 = 8$
2. $8^{2/3} = (\sqrt[3]{8})^2 = 2^2 = 4$
3. $9^{-3/2} = \dfrac{1}{9^{3/2}} = \dfrac{1}{(\sqrt{9})^3} = \dfrac{1}{3^3} = \dfrac{1}{27}$
4. $\dfrac{\sqrt{3}}{\sqrt[3]{3}} = \dfrac{3^{1/2}}{3^{1/3}} = 3^{1/2-1/3} = 3^{1/6} = \sqrt[6]{3}$
5. $2^2 2^{7/2} = 2^2 2^{3+1/2} = 2^2 2^3 2^{1/2} = 2^5 2^{1/2} = 2^5\sqrt{2}$

EXAMPLE 4 Simplifying Algebraic Expressions

Simplify the following.

a. $\dfrac{(x^3)^{5/3}}{x^3}$ b. $\sqrt[4]{a^6}$ c. $\dfrac{(xy)^{-3}y^{-3/2}}{x^{-2}\sqrt{y}}$

Solution

a. $\dfrac{(x^3)^{5/3}}{x^3} = \dfrac{x^5}{x^3} = x^2$

b. $\sqrt[4]{a^6} = a^{6/4} = a^{3/2} = a \cdot a^{1/2} = a\sqrt{a}$

c. $\dfrac{(xy)^{-3}y^{-3/2}}{x^{-2}\sqrt{y}} = \dfrac{x^{-3}y^{-3}y^{-3/2}}{x^{-2}y^{1/2}} = \dfrac{1}{x^{-2+3}y^{1/2+3+3/2}} = \dfrac{1}{xy^5}$

Converting Between Rational, Radical, and Exponent Form

In calculus we must often convert algebraic expressions involving powers of x, such as $\dfrac{3}{2x^2}$, into expressions in which x does not appear in the denominator, such as $\dfrac{3}{2}x^{-2}$. Also, we must often convert expressions with radicals, such as $\dfrac{1}{\sqrt{1+x^2}}$, into expressions with no radicals and all powers in the numerator, such as $(1 + x^2)^{-1/2}$. In these cases, we are converting from **rational form** or **radical form** to **exponent form**.

Rational Form

An expression is in **rational form** if it is written with positive exponents only.

Quick Examples

1. $\dfrac{2}{3x^2}$ is in rational form.

2. $\dfrac{2x^{-1}}{3}$ is not in rational form because the exponent of x is negative.

3. $\dfrac{x}{6} + \dfrac{6}{x}$ is in rational form.

Radical Form

An expression is in **radical form** if it is written with integer powers and roots only.

Quick Examples

1. $\dfrac{2}{5\sqrt[3]{x}} + \dfrac{2}{x}$ is in radical form.

2. $\dfrac{2x^{-1/3}}{5} + 2x^{-1}$ is not in radical form because $x^{-1/3}$ appears.

3. $\dfrac{1}{\sqrt{1 + x^2}}$ is in radical form, but $(1 + x^2)^{-1/2}$ is not.

Exponent Form

An expression is in **exponent form** if there are no radicals and all powers of unknowns occur in the numerator. We write such expressions as sums or differences of terms of the form

$$\text{Constant} \times (\text{Expression with } x)^p \qquad \text{As in } \tfrac{1}{3}x^{-3/2}$$

Quick Examples

1. $\dfrac{2}{3}x^4 - 3x^{-1/3}$ is in exponent form.

2. $\dfrac{x}{6} + \dfrac{6}{x}$ is not in exponent form because the second expression has x in the denominator.

3. $\sqrt[3]{x}$ is not in exponent form because it has a radical.

4. $(1 + x^2)^{-1/2}$ is in exponent form, but $\dfrac{1}{\sqrt{1 + x^2}}$ is not.

EXAMPLE 5 **Converting from One Form to Another**

Convert the following to rational form:

a. $\dfrac{1}{2}x^{-2} + \dfrac{4}{3}x^{-5}$ **b.** $\dfrac{2}{\sqrt{x}} - \dfrac{2}{x^{-4}}$

Convert the following to radical form:

c. $\dfrac{1}{2}x^{-1/2} + \dfrac{4}{3}x^{-5/4}$ **d.** $\dfrac{(3+x)^{-1/3}}{5}$

Convert the following to exponent form:

e. $\dfrac{3}{4x^2} - \dfrac{x}{6} + \dfrac{6}{x} + \dfrac{4}{3\sqrt{x}}$ **f.** $\dfrac{2}{(x+1)^2} - \dfrac{3}{4\sqrt[5]{2x-1}}$

Solution For (a) and (b), we eliminate negative exponents as we did in Example 2:

a. $\dfrac{1}{2}x^{-2} + \dfrac{4}{3}x^{-5} = \dfrac{1}{2} \cdot \dfrac{1}{x^2} + \dfrac{4}{3} \cdot \dfrac{1}{x^5} = \dfrac{1}{2x^2} + \dfrac{4}{3x^5}$

b. $\dfrac{2}{\sqrt{x}} - \dfrac{2}{x^{-4}} = \dfrac{2}{\sqrt{x}} - 2x^4$

For (c) and (d), we rewrite all terms with fractional exponents as radicals:

c. $\dfrac{1}{2}x^{-1/2} + \dfrac{4}{3}x^{-5/4} = \dfrac{1}{2} \cdot \dfrac{1}{x^{1/2}} + \dfrac{4}{3} \cdot \dfrac{1}{x^{5/4}}$

$= \dfrac{1}{2} \cdot \dfrac{1}{\sqrt{x}} + \dfrac{4}{3} \cdot \dfrac{1}{\sqrt[4]{x^5}} = \dfrac{1}{2\sqrt{x}} + \dfrac{4}{3\sqrt[4]{x^5}}$

d. $\dfrac{(3+x)^{-1/3}}{5} = \dfrac{1}{5(3+x)^{1/3}} = \dfrac{1}{5\sqrt[3]{3+x}}$

For (e) and (f), we eliminate any radicals and move all expressions involving x to the numerator:

e. $\dfrac{3}{4x^2} - \dfrac{x}{6} + \dfrac{6}{x} + \dfrac{4}{3\sqrt{x}} = \dfrac{3}{4}x^{-2} - \dfrac{1}{6}x + 6x^{-1} + \dfrac{4}{3x^{1/2}}$

$= \dfrac{3}{4}x^{-2} - \dfrac{1}{6}x + 6x^{-1} + \dfrac{4}{3}x^{-1/2}$

f. $\dfrac{2}{(x+1)^2} - \dfrac{3}{4\sqrt[5]{2x-1}} = 2(x+1)^{-2} - \dfrac{3}{4(2x-1)^{1/5}}$

$= 2(x+1)^{-2} - \dfrac{3}{4}(2x-1)^{-1/5}$

Solving Equations with Exponents

EXAMPLE 6 **Solving Equations**

Solve the following equations:

a. $x^3 + 8 = 0$ **b.** $x^2 - \dfrac{1}{2} = 0$ **c.** $x^{3/2} - 64 = 0$

Solution

a. Subtracting 8 from both sides gives $x^3 = -8$. Taking the cube root of both sides gives $x = -2$.

b. Adding $\frac{1}{2}$ to both sides gives $x^2 = \frac{1}{2}$. Thus, $x = \pm\sqrt{\frac{1}{2}} = \pm\frac{1}{\sqrt{2}}$.

c. Adding 64 to both sides gives $x^{3/2} = 64$. Taking the reciprocal (2/3) power of both sides gives

$$(x^{3/2})^{2/3} = 64^{2/3}$$
$$x^1 = \left(\sqrt[3]{64}\right)^2 = 4^2 = 16$$

so $x = 16.$

0.2   EXERCISES

Evaluate the expressions in Exercises 1–16.

1. 3^3 **2.** $(-2)^3$ **3.** $-(2 \cdot 3)^2$ **4.** $(4 \cdot 2)^2$

5. $\left(\dfrac{-2}{3}\right)^2$ **6.** $\left(\dfrac{3}{2}\right)^3$ **7.** $(-2)^{-3}$ **8.** -2^{-3}

9. $\left(\dfrac{1}{4}\right)^{-2}$ **10.** $\left(\dfrac{-2}{3}\right)^{-2}$ **11.** $2 \cdot 3^0$ **12.** $3 \cdot (-2)^0$

13. $2^3 \, 2^2$ **14.** $3^2 3$ **15.** $2^2 2^{-1} 2^4 2^{-4}$ **16.** $5^2 5^{-3} 5^2 5^{-2}$

Simplify each expression in Exercises 17–30, expressing your answer in rational form.

17. $x^3 x^2$ **18.** $x^4 x^{-1}$ **19.** $-x^2 x^{-3} y$ **20.** $-xy^{-1}x^{-1}$

21. $\dfrac{x^3}{x^4}$ **22.** $\dfrac{y^5}{y^3}$ **23.** $\dfrac{x^2 y^2}{x^{-1} y}$ **24.** $\dfrac{x^{-1} y}{x^2 y^2}$

25. $\dfrac{(xy^{-1}z^3)^2}{x^2 yz^2}$ **26.** $\dfrac{x^2 yz^2}{(xyz^{-1})^{-1}}$ **27.** $\left(\dfrac{xy^{-2}z}{x^{-1}z}\right)^3$

28. $\left(\dfrac{x^2 y^{-1}z^0}{xyz}\right)^2$ **29.** $\left(\dfrac{x^{-1}y^{-2}z^2}{xy}\right)^{-2}$ **30.** $\left(\dfrac{xy^{-2}}{x^2 y^{-1}z}\right)^{-3}$

Convert the expressions in Exercises 31–36 to rational form.

31. $3x^{-4}$ **32.** $\dfrac{1}{2}x^{-4}$ **33.** $\dfrac{3}{4}x^{-2/3}$

34. $\dfrac{4}{5}y^{-3/4}$ **35.** $1 - \dfrac{0.3}{x^{-2}} - \dfrac{6}{5}x^{-1}$ **36.** $\dfrac{1}{3x^{-4}} + \dfrac{0.1x^{-2}}{3}$

Evaluate the expressions in Exercises 37–56, rounding your answer to four significant digits where necessary.

37. $\sqrt{4}$ **38.** $\sqrt{5}$ **39.** $\sqrt{\dfrac{1}{4}}$

40. $\sqrt{\dfrac{1}{9}}$ **41.** $\sqrt{\dfrac{16}{9}}$ **42.** $\sqrt{\dfrac{9}{4}}$

43. $\dfrac{\sqrt{4}}{5}$ **44.** $\dfrac{6}{\sqrt{25}}$ **45.** $\sqrt{9} + \sqrt{16}$

46. $\sqrt{25} - \sqrt{16}$ **47.** $\sqrt{9 + 16}$ **48.** $\sqrt{25 - 16}$

49. $\sqrt[3]{8 - 27}$ **50.** $\sqrt[4]{81 - 16}$ **51.** $\sqrt[3]{27/8}$

52. $\sqrt[3]{8 \times 64}$ **53.** $\sqrt{(-2)^2}$ **54.** $\sqrt{(-1)^2}$

55. $\sqrt{\dfrac{1}{4}(1 + 15)}$ **56.** $\sqrt{\dfrac{1}{9}(3 + 33)}$

Simplify the expressions in Exercises 57–64, given that x, y, z, a, b, and c are positive real numbers.

57. $\sqrt{a^2 b^2}$ **58.** $\sqrt{\dfrac{a^2}{b^2}}$ **59.** $\sqrt{(x + 9)^2}$

60. $(\sqrt{x + 9})^2$ **61.** $\sqrt[3]{x^3(a^3 + b^3)}$ **62.** $\sqrt[4]{\dfrac{x^4}{a^4 b^4}}$

63. $\sqrt{\dfrac{4xy^3}{x^2 y}}$ **64.** $\sqrt{\dfrac{4(x^2 + y^2)}{c^2}}$

Convert the expressions in Exercises 65–84 to exponent form.

65. $\sqrt{3}$ **66.** $\sqrt{8}$ **67.** $\sqrt{x^3}$

68. $\sqrt[3]{x^2}$ **69.** $\sqrt[3]{xy^2}$ **70.** $\sqrt{x^2 y}$

71. $\dfrac{x^2}{\sqrt{x}}$ **72.** $\dfrac{x}{\sqrt{x}}$ **73.** $\dfrac{3}{5x^2}$

74. $\dfrac{2}{5x^{-3}}$ **75.** $\dfrac{3x^{-1.2}}{2} - \dfrac{1}{3x^{2.1}}$ **76.** $\dfrac{2}{3x^{-1.2}} - \dfrac{x^{2.1}}{3}$

77. $\dfrac{2x}{3} - \dfrac{x^{0.1}}{2} + \dfrac{4}{3x^{1.1}}$ **78.** $\dfrac{4x^2}{3} + \dfrac{x^{3/2}}{6} - \dfrac{2}{3x^2}$

79. $\dfrac{3\sqrt{x}}{4} - \dfrac{5}{3\sqrt{x}} + \dfrac{4}{3x\sqrt{x}}$ **80.** $\dfrac{3}{5\sqrt{x}} - \dfrac{5\sqrt{x}}{8} + \dfrac{7}{2\sqrt[3]{x}}$

81. $\dfrac{3\sqrt[5]{x^2}}{4} - \dfrac{7}{2\sqrt{x^3}}$ **82.** $\dfrac{1}{8x\sqrt{x}} - \dfrac{2}{3\sqrt[5]{x^3}}$

83. $\dfrac{1}{(x^2 + 1)^3} - \dfrac{3}{4\sqrt[3]{(x^2 + 1)}}$ **84.** $\dfrac{2}{3(x^2 + 1)^{-3}} - \dfrac{3\sqrt[3]{(x^2 + 1)^7}}{4}$

Convert the expressions in Exercises 85–96 to radical form.

85. $2^{2/3}$ **86.** $3^{4/5}$ **87.** $x^{4/3}$ **88.** $y^{7/4}$

89. $(x^{1/2}y^{1/3})^{1/5}$ **90.** $x^{-1/3}y^{3/2}$ **91.** $-\dfrac{3}{2}x^{-1/4}$ **92.** $\dfrac{4}{5}x^{3/2}$

93. $0.2x^{-2/3} + \dfrac{3}{7x^{-1/2}}$ 94. $\dfrac{3.1}{x^{-4/3}} - \dfrac{11}{7}x^{-1/7}$

95. $\dfrac{3}{4(1-x)^{5/2}}$ 96. $\dfrac{9}{4(1-x)^{-7/3}}$

Simplify the expressions in Exercises 97–106.

97. $4^{-1/2}4^{7/2}$ 98. $2^{1/a}/2^{2/a}$ 99. $3^{2/3}3^{-1/6}$

100. $2^{1/3}2^{-1}2^{2/3}2^{-1/3}$ 101. $\dfrac{x^{3/2}}{x^{5/2}}$ 102. $\dfrac{y^{5/4}}{y^{3/4}}$

103. $\dfrac{x^{1/2}y^2}{x^{-1/2}y}$ 104. $\dfrac{x^{-1/2}y}{x^2y^{3/2}}$

105. $\left(\dfrac{x}{y}\right)^{1/3}\left(\dfrac{y}{x}\right)^{2/3}$ 106. $\left(\dfrac{x}{y}\right)^{-1/3}\left(\dfrac{y}{x}\right)^{1/3}$

Solve each equation in Exercises 107–120 for x, rounding your answer to four significant digits where necessary.

107. $x^2 - 16 = 0$ 108. $x^2 - 1 = 0$

109. $x^2 - \dfrac{4}{9} = 0$ 110. $x^2 - \dfrac{1}{10} = 0$

111. $x^2 - (1+2x)^2 = 0$ 112. $x^2 - (2-3x)^2 = 0$

113. $x^5 + 32 = 0$ 114. $x^4 - 81 = 0$

115. $x^{1/2} - 4 = 0$ 116. $x^{1/3} - 2 = 0$

117. $1 - \dfrac{1}{x^2} = 0$ 118. $\dfrac{2}{x^3} - \dfrac{6}{x^4} = 0$

119. $(x-4)^{-1/3} = 2$ 120. $(x-4)^{2/3} + 1 = 5$

0.3 Multiplying and Factoring Algebraic Expressions

Multiplying Algebraic Expressions

Distributive Law

The **distributive law** for real numbers states that

$$a(b \pm c) = ab \pm ac$$
$$(a \pm b)c = ac \pm bc$$

for any real numbers a, b, and c.

Quick Examples

1. $2(x-3)$ is *not* equal to $2x - 3$ but is equal to $2x - 2(3) = 2x - 6$.
2. $x(x+1) = x^2 + x$
3. $2x(3x-4) = 6x^2 - 8x$
4. $(x-4)x^2 = x^3 - 4x^2$
5. $(x+2)(x+3) = (x+2)x + (x+2)3$
   $$= (x^2 + 2x) + (3x + 6) = x^2 + 5x + 6$$
6. $(x+2)(x-3) = (x+2)x - (x+2)3$
   $$= (x^2 + 2x) - (3x + 6) = x^2 - x - 6$$

There is a quicker way of expanding expressions like the last two, called the "FOIL" method (First, Outer, Inner, Last). Consider, for instance, the expression $(x+1)(x-2)$. The FOIL method says: Take the product of the first terms: $x \cdot x = x^2$, the product of the outer terms: $x \cdot (-2) = -2x$, the product of the inner terms: $1 \cdot x = x$, and the product of the last terms: $1 \cdot (-2) = -2$, and then add them all up, getting $x^2 - 2x + x - 2 = x^2 - x - 2$.

EXAMPLE 1 FOIL

a. $(x-2)(2x+5) = 2x^2 + 5x - 4x - 10 = 2x^2 + x - 10$

$\qquad\qquad\quad\uparrow\quad\ \ \uparrow\quad\ \uparrow\quad\ \uparrow$
$\qquad\qquad\text{First\ \ Outer\ \ Inner\ \ Last}$

b. $(x^2+1)(x-4) = x^3 - 4x^2 + x - 4$

c. $(a-b)(a+b) = a^2 + ab - ab - b^2 = a^2 - b^2$

d. $(a+b)^2 = (a+b)(a+b) = a^2 + ab + ab + b^2 = a^2 + 2ab + b^2$

e. $(a-b)^2 = (a-b)(a-b) = a^2 - ab - ab + b^2 = a^2 - 2ab + b^2$

The last three are particularly important and are worth memorizing.

Special Formulas

$$(a-b)(a+b) = a^2 - b^2 \qquad \text{Difference of two squares}$$
$$(a+b)^2 = a^2 + 2ab + b^2 \qquad \text{Square of a sum}$$
$$(a-b)^2 = a^2 - 2ab + b^2 \qquad \text{Square of a difference}$$

Quick Examples

1. $(2-x)(2+x) = 4 - x^2$

2. $(1+a)(1-a) = 1 - a^2$

3. $(x+3)^2 = x^2 + 6x + 9$

4. $(4-x)^2 = 16 - 8x + x^2$

Here are some longer examples that require the distributive law.

EXAMPLE 2 Multiplying Algebraic Expressions

a. $(x+1)(x^2 + 3x - 4) = (x+1)x^2 + (x+1)3x - (x+1)4$

$$= (x^3 + x^2) + (3x^2 + 3x) - (4x + 4)$$
$$= x^3 + 4x^2 - x - 4$$

b. $\left(x^2 - \dfrac{1}{x} + 1\right)(2x+5) = \left(x^2 - \dfrac{1}{x} + 1\right)2x + \left(x^2 - \dfrac{1}{x} + 1\right)5$

$$= (2x^3 - 2 + 2x) + \left(5x^2 - \dfrac{5}{x} + 5\right)$$
$$= 2x^3 + 5x^2 + 2x + 3 - \dfrac{5}{x}$$

c. $(x-y)(x-y)(x-y) = (x^2 - 2xy + y^2)(x-y)$

$$= (x^2 - 2xy + y^2)x - (x^2 - 2xy + y^2)y$$
$$= (x^3 - 2x^2y + xy^2) - (x^2y - 2xy^2 + y^3)$$
$$= x^3 - 3x^2y + 3xy^2 - y^3$$

Factoring Algebraic Expressions

We can think of factoring as applying the distributive law in reverse—for example,

$$2x^2 + x = x(2x + 1),$$

which can be checked by using the distributive law. Factoring is an art that you will learn with experience and the help of a few useful techniques.

Factoring Using a Common Factor

To use this technique, locate a **common factor**—a term that occurs as a factor in each of the expressions being added or subtracted (for example, x is a common factor in $2x^2 + x$, because it is a factor of both $2x^2$ and x). Once you have located a common factor, "factor it out" by applying the distributive law.

Quick Examples

1. $2x^3 - x^2 + x$ has x as a common factor, so
$$2x^3 - x^2 + x = x(2x^2 - x + 1)$$

2. $2x^2 + 4x$ has $2x$ as a common factor, so
$$2x^2 + 4x = 2x(x + 2)$$

3. $2x^2y + xy^2 - x^2y^2$ has xy as a common factor, so
$$2x^2y + xy^2 - x^2y^2 = xy(2x + y - xy)$$

4. $(x^2 + 1)(x + 2) - (x^2 + 1)(x + 3)$ has $x^2 + 1$ as a common factor, so
$$(x^2 + 1)(x + 2) - (x^2 + 1)(x + 3) = (x^2 + 1)[(x + 2) - (x + 3)]$$
$$= (x^2 + 1)(x + 2 - x - 3)$$
$$= (x^2 + 1)(-1) = -(x^2 + 1)$$

5. $12x(x^2 - 1)^5(x^3 + 1)^6 + 18x^2(x^2 - 1)^6(x^3 + 1)^5$ has $6x(x^2 - 1)^5(x^3 + 1)^5$ as a common factor, so
$$12x(x^2 - 1)^5(x^3 + 1)^6 + 18x^2(x^2 - 1)^6(x^3 + 1)^5$$
$$= 6x(x^2 - 1)^5(x^3 + 1)^5[2(x^3 + 1) + 3x(x^2 - 1)]$$
$$= 6x(x^2 - 1)^5(x^3 + 1)^5(2x^3 + 2 + 3x^3 - 3x)$$
$$= 6x(x^2 - 1)^5(x^3 + 1)^5(5x^3 - 3x + 2)$$

We would also like to be able to reverse calculations such as $(x + 2)(2x - 5) = 2x^2 - x - 10$. That is, starting with the expression $2x^2 - x - 10$, we would like to **factor** it to get the expression $(x + 2)(2x - 5)$. An expression of the form $ax^2 + bx + c$, where a, b, and c are real numbers, is called a **quadratic** expression in x. Thus, given a quadratic expression $ax^2 + bx + c$, we would like to write it in the form $(dx + e)(fx + g)$ for some real numbers $d, e, f,$ and g. There are some quadratics, such as $x^2 + x + 1$, that cannot be factored in this form at all. Here, we consider only quadratics that do factor, and in such a way that the numbers $d, e, f,$ and g are integers (whole numbers; other cases are discussed in Section 5). The usual technique of factoring such quadratics is a "trial and error" approach.

Factoring Quadratics by Trial and Error

To factor the quadratic $ax^2 + bx + c$, factor ax^2 as $(a_1x)(a_2x)$ (with a_1 positive) and c as c_1c_2, and then check whether or not $ax^2 + bx + c = (a_1x \pm c_1)(a_2x \pm c_2)$. If not, try other factorizations of ax^2 and c.

Quick Examples

1. To factor $x^2 - 6x + 5$, first factor x^2 as $(x)(x)$, and 5 as $(5)(1)$:

$$(x + 5)(x + 1) = x^2 + 6x + 5. \qquad \text{No good}$$
$$(x - 5)(x - 1) = x^2 - 6x + 5. \qquad \text{Desired factorization}$$

2. To factor $x^2 - 4x - 12$, first factor x^2 as $(x)(x)$, and -12 as $(1)(-12)$, $(2)(-6)$, or $(3)(-4)$. Trying them one by one gives

$$(x + 1)(x - 12) = x^2 - 11x - 12. \qquad \text{No good}$$
$$(x - 1)(x + 12) = x^2 + 11x - 12. \qquad \text{No good}$$
$$(x + 2)(x - 6) = x^2 - 4x - 12. \qquad \text{Desired factorization}$$

3. To factor $4x^2 - 25$, we can follow the above procedure, or recognize $4x^2 - 25$ as the difference of two squares:

$$4x^2 - 25 = (2x)^2 - 5^2 = (2x - 5)(2x + 5).$$

Note: Not all quadratic expressions factor. In Section 5 we look at a test that tells us whether or not a given quadratic factors.

Here are examples requiring either a little more work or a little more thought.

EXAMPLE 3 Factoring Quadratics

Factor the following: **a.** $4x^2 - 5x - 6$ **b.** $x^4 - 5x^2 + 6$

Solution

a. Possible factorizations of $4x^2$ are $(2x)(2x)$ or $(x)(4x)$. Possible factorizations of -6 are $(1)(-6)$, $(2)(-3)$. We now systematically try out all the possibilities until we come up with the correct one.

$(2x)(2x)$ and $(1)(-6)$:	$(2x + 1)(2x - 6) = 4x^2 - 10x - 6$	No good
$(2x)(2x)$ and $(2)(-3)$:	$(2x + 2)(2x - 3) = 4x^2 - 2x - 6$	No good
$(x)(4x)$ and $(1)(-6)$:	$(x + 1)(4x - 6) = 4x^2 - 2x - 6$	No good
$(x)(4x)$ and $(2)(-3)$:	$(x + 2)(4x - 3) = 4x^2 + 5x - 6$	Almost!
Change signs:	$(x - 2)(4x + 3) = 4x^2 - 5x - 6$	Correct

b. The expression $x^4 - 5x^2 + 6$ is not a quadratic, you say? Correct. It's a quartic (a fourth degree expression). However, it looks rather like a quadratic. In fact, it is quadratic *in* x^2, meaning that it is

$$(x^2)^2 - 5(x^2) + 6 = y^2 - 5y + 6$$

where $y = x^2$. The quadratic $y^2 - 5y + 6$ factors as

$$y^2 - 5y + 6 = (y - 3)(y - 2)$$

so

$$x^4 - 5x^2 + 6 = (x^2 - 3)(x^2 - 2)$$

This is a sometimes useful technique.

Our last example is here to remind you why we should want to factor polynomials in the first place. We shall return to this in Section 5.

EXAMPLE 4 Solving a Quadratic Equation by Factoring

Solve the equation $3x^2 + 4x - 4 = 0$.

Solution We first factor the left-hand side to get

$$(3x - 2)(x + 2) = 0.$$

Thus, the product of the two quantities $(3x - 2)$ and $(x + 2)$ is zero. Now, if a product of two numbers is zero, one of the two must be zero. In other words, either $3x - 2 = 0$, giving $x = \frac{2}{3}$, or $x + 2 = 0$, giving $x = -2$. Thus, there are two solutions: $x = \frac{2}{3}$ and $x = -2$.

0.3 EXERCISES

Expand each expression in Exercises 1–22.

1. $x(4x + 6)$

2. $(4y - 2)y$

3. $(2x - y)y$

4. $x(3x + y)$

5. $(x + 1)(x - 3)$

6. $(y + 3)(y + 4)$

7. $(2y + 3)(y + 5)$

8. $(2x - 2)(3x - 4)$

9. $(2x - 3)^2$

10. $(3x + 1)^2$

11. $\left(x + \dfrac{1}{x}\right)^2$

12. $\left(y - \dfrac{1}{y}\right)^2$

13. $(2x - 3)(2x + 3)$

14. $(4 + 2x)(4 - 2x)$

15. $\left(y - \dfrac{1}{y}\right)\left(y + \dfrac{1}{y}\right)$

16. $(x - x^2)(x + x^2)$

17. $(x^2 + x - 1)(2x + 4)$

18. $(3x + 1)(2x^2 - x + 1)$

19. $(x^2 - 2x + 1)^2$

20. $(x + y - xy)^2$

21. $(y^3 + 2y^2 + y)(y^2 + 2y - 1)$

22. $(x^3 - 2x^2 + 4)(3x^2 - x + 2)$

In Exercises 23–30, factor each expression and simplify as much as possible.

23. $(x + 1)(x + 2) + (x + 1)(x + 3)$

24. $(x + 1)(x + 2)^2 + (x + 1)^2(x + 2)$

25. $(x^2 + 1)^5(x + 3)^4 + (x^2 + 1)^6(x + 3)^3$

26. $10x(x^2 + 1)^4(x^3 + 1)^5 + 15x^2(x^2 + 1)^5(x^3 + 1)^4$

27. $(x^3 + 1)\sqrt{x + 1} - (x^3 + 1)^2\sqrt{x + 1}$

28. $(x^2 + 1)\sqrt{x + 1} - \sqrt{(x + 1)^3}$

29. $\sqrt{(x + 1)^3} + \sqrt{(x + 1)^5}$

30. $(x^2 + 1)\sqrt[3]{(x + 1)^4} - \sqrt[3]{(x + 1)^7}$

In Exercises 31–48, (a) factor the given expression; (b) set the expression equal to zero and solve for the unknown (x in the odd-numbered exercises and y in the even-numbered exercises).

31. $2x + 3x^2$

32. $y^2 - 4y$

33. $6x^3 - 2x^2$

34. $3y^3 - 9y^2$

35. $x^2 - 8x + 7$

36. $y^2 + 6y + 8$

37. $x^2 + x - 12$

38. $y^2 + y - 6$

39. $2x^2 - 3x - 2$

40. $3y^2 - 8y - 3$

41. $6x^2 + 13x + 6$

42. $6y^2 + 17y + 12$

43. $12x^2 + x - 6$

44. $20y^2 + 7y - 3$

45. $x^2 + 4xy + 4y^2$

46. $4y^2 - 4xy + x^2$

47. $x^4 - 5x^2 + 4$

48. $y^4 + 2y^2 - 3$

0.4   Rational Expressions

Rational Expression

A **rational expression** is an algebraic expression of the form $\dfrac{P}{Q}$, where P and Q are simpler expressions (usually polynomials) and the denominator Q is not zero.

Quick Examples

1. $\dfrac{x^2 - 3x}{x}$ $P = x^2 - 3x,\ Q = x$

2. $\dfrac{x + \frac{1}{x} + 1}{2x^2y + 1}$ $P = x + \dfrac{1}{x} + 1,\ Q = 2x^2y + 1$

3. $3xy - x^2$ $P = 3xy - x^2,\ Q = 1$

Algebra of Rational Expressions

We manipulate rational expressions in the same way that we manipulate fractions, using the following rules:

Algebraic Rule	**Quick Example**
Product: $\dfrac{P}{Q} \cdot \dfrac{R}{S} = \dfrac{PR}{QS}$	$\dfrac{x+1}{x} \cdot \dfrac{x-1}{2x+1} = \dfrac{(x+1)(x-1)}{x(2x+1)} = \dfrac{x^2-1}{2x^2+x}$
Sum: $\dfrac{P}{Q} + \dfrac{R}{S} = \dfrac{PS+RQ}{QS}$	$\dfrac{2x-1}{3x+2} + \dfrac{1}{x} = \dfrac{(2x-1)x + 1(3x+2)}{x(3x+2)}$ $= \dfrac{2x^2+2x+2}{3x^2+2x}$
Difference: $\dfrac{P}{Q} - \dfrac{R}{S} = \dfrac{PS-RQ}{QS}$	$\dfrac{x}{3x+2} - \dfrac{x-4}{x} = \dfrac{x^2 - (x-4)(3x+2)}{x(3x+2)}$ $= \dfrac{-2x^2+10x+8}{3x^2+2x}$
Reciprocal: $\dfrac{1}{\left(\frac{P}{Q}\right)} = \dfrac{Q}{P}$	$\dfrac{1}{\left(\frac{2xy}{3x-1}\right)} = \dfrac{3x-1}{2xy}$
Quotient: $\dfrac{\left(\frac{P}{Q}\right)}{\left(\frac{R}{S}\right)} = \dfrac{P}{Q} \cdot \dfrac{S}{R} = \dfrac{PS}{QR}$	$\dfrac{\left(\frac{x}{x-1}\right)}{\left(\frac{y-1}{y}\right)} = \dfrac{xy}{(x-1)(y-1)} = \dfrac{xy}{xy-x-y+1}$
Cancellation: $\dfrac{P\check{R}}{Q\check{R}} = \dfrac{P}{Q}$	$\dfrac{(x-1)(xy+4)}{(x^2y-8)(x-1)} = \dfrac{xy+4}{x^2y-8}$

Caution Cancellation of summands is *invalid*. For instance,

$$\frac{\cancel{x} + (2xy^2 - y)}{\cancel{x} + 4y} = \frac{(2xy^2 - y)}{4y} \quad \text{✗ WRONG!} \quad \text{Do } not \text{ cancel a summand.}$$

$$\frac{\cancel{x}(2xy^2 - y)}{4\cancel{x}y} = \frac{(2xy^2 - y)}{4y} \quad \text{✓ CORRECT} \quad \text{Do cancel a factor.}$$

Here are some examples that require several algebraic operations.

EXAMPLE 1 Simplifying Rational Expressions

a. $\dfrac{\left(\frac{1}{x+y} - \frac{1}{x}\right)}{y} = \dfrac{\left(\frac{x-(x+y)}{x(x+y)}\right)}{y} = \dfrac{\left(\frac{-y}{x(x+y)}\right)}{y} = \dfrac{-y}{xy(x+y)} = -\dfrac{1}{x(x+y)}$

b. $\dfrac{(x+1)(x+2)^2 - (x+1)^2(x+2)}{(x+2)^4} = \dfrac{(x+1)(x+2)[(x+2)-(x+1)]}{(x+2)^4}$

$$= \dfrac{(x+1)(x+2)(x+2-x-1)}{(x+2)^4} = \dfrac{(x+1)(x+2)}{(x+2)^4} = \dfrac{x+1}{(x+2)^3}$$

c. $\dfrac{2x\sqrt{x+1} - \frac{x^2}{\sqrt{x+1}}}{x+1} = \dfrac{\left(\frac{2x\left(\sqrt{x+1}\right)^2 - x^2}{\sqrt{x+1}}\right)}{x+1} = \dfrac{2x(x+1) - x^2}{(x+1)\sqrt{x+1}}$

$$= \dfrac{2x^2 + 2x - x^2}{(x+1)\sqrt{x+1}} = \dfrac{x^2 + 2x}{\sqrt{(x+1)^3}} = \dfrac{x(x+2)}{\sqrt{(x+1)^3}}$$

0.4 EXERCISES

Rewrite each expression in Exercises 1–16 as a single rational expression, simplified as much as possible.

1. $\dfrac{x-4}{x+1} \cdot \dfrac{2x+1}{x-1}$

2. $\dfrac{2x-3}{x-2} \cdot \dfrac{x+3}{x+1}$

3. $\dfrac{x-4}{x+1} + \dfrac{2x+1}{x-1}$

4. $\dfrac{2x-3}{x-2} + \dfrac{x+3}{x+1}$

5. $\dfrac{x^2}{x+1} - \dfrac{x-1}{x+1}$

6. $\dfrac{x^2-1}{x-2} - \dfrac{1}{x-1}$

7. $\dfrac{1}{\left(\frac{x}{x-1}\right)} + x - 1$

8. $\dfrac{2}{\left(\frac{x-2}{x^2}\right)} - \dfrac{1}{x-2}$

9. $\dfrac{1}{x}\left[\dfrac{x-3}{xy} + \dfrac{1}{y}\right]$

10. $\dfrac{y^2}{x}\left[\dfrac{2x-3}{y} + \dfrac{x}{y}\right]$

11. $\dfrac{(x+1)^2(x+2)^3 - (x+1)^3(x+2)^2}{(x+2)^6}$

12. $\dfrac{6x(x^2+1)^2(x^3+2)^3 - 9x^2(x^2+1)^3(x^3+2)^2}{(x^3+2)^6}$

13. $\dfrac{(x^2-1)\sqrt{x^2+1} - \frac{x^4}{\sqrt{x^2+1}}}{x^2+1}$

14. $\dfrac{x\sqrt{x^3-1} - \frac{3x^4}{\sqrt{x^3-1}}}{x^3-1}$

15. $\dfrac{\frac{1}{(x+y)^2} - \frac{1}{x^2}}{y}$

16. $\dfrac{\frac{1}{(x+y)^3} - \frac{1}{x^3}}{y}$

0.5 Solving Polynomial Equations

Polynomial Equation

A **polynomial equation** in one unknown is an equation that can be written in the form

$$ax^n + bx^{n-1} + \cdots + rx + s = 0$$

where $a, b, \ldots, r$, and s are constants.

We call the largest exponent of x appearing in a nonzero term of a polynomial the **degree** of that polynomial.

Quick Examples

1. $3x + 1 = 0$ has degree 1 because the largest power of x that occurs is $x = x^1$. Degree 1 equations are called **linear** equations.
2. $x^2 - x - 1 = 0$ has degree 2 because the largest power of x that occurs is x^2. Degree 2 equations are also called **quadratic equations**, or just **quadratics**.
3. $x^3 = 2x^2 + 1$ is a degree 3 polynomial (or **cubic**) in disguise. It can be rewritten as $x^3 - 2x^2 - 1 = 0$, which is in the standard form for a degree 3 equation.
4. $x^4 - x = 0$ has degree 4. It is called a **quartic**.

Now comes the question: How do we solve these equations for x? This question was asked by mathematicians as early as 1600 BCE. Let's look at these equations one degree at a time.

Solution of Linear Equations

By definition, a linear equation can be written in the form

$$ax + b = 0. \qquad \text{a and b are fixed numbers with $a \neq 0$.}$$

Solving this is a nice mental exercise: Subtract b from both sides and then divide by a, getting $x = -b/a$. Don't bother memorizing this formula; just go ahead and solve linear equations as they arise. If you feel you need practice, see the exercises at the end of the section.

Solution of Quadratic Equations

By definition, a quadratic equation has the form

$$ax^2 + bx + c = 0. \qquad \text{a, b, and c are fixed numbers and $a \neq 0$.[3]}$$

---

[3] What happens if $a = 0$?

The solutions of this equation are also called the **roots** of $ax^2 + bx + c$. We're assuming that you saw quadratic equations somewhere in high school but may be a little hazy about the details of their solution. There are two ways of solving these equations—one works sometimes, and the other works every time.

Solving Quadratic Equations by Factoring (works sometimes)

If we can factor[4] a quadratic equation $ax^2 + bx + c = 0$, we can solve the equation by setting each factor equal to zero.

Quick Examples

1. $x^2 + 7x + 10 = 0$

   $(x + 5)(x + 2) = 0$ Factor the left-hand side.

   $x + 5 = 0$ or $x + 2 = 0$ If a product is zero, one or both factors is zero.

   Solutions: $x = -5$ and $x = -2$

2. $2x^2 - 5x - 12 = 0$

   $(2x + 3)(x - 4) = 0$ Factor the left-hand side.

   $2x + 3 = 0$ or $x - 4 = 0$

   Solutions: $x = -3/2$ and $x = 4$

Test for Factoring

The quadratic $ax^2 + bx + c$, with a, b, and c being integers (whole numbers), factors into an expression of the form $(rx + s)(tx + u)$ with r, s, t, and u integers precisely when the quantity $b^2 - 4ac$ is a perfect square. (That is, it is the square of an integer.) If this happens, we say that the quadratic **factors over the integers**.

Quick Examples

1. $x^2 + x + 1$ has $a = 1$, $b = 1$, and $c = 1$, so $b^2 - 4ac = -3$, which is not a perfect square. Therefore, this quadratic does not factor over the integers.

2. $2x^2 - 5x - 12$ has $a = 2$, $b = -5$, and $c = -12$, so $b^2 - 4ac = 121$. Because $121 = 11^2$, this quadratic does factor over the integers. (We factored it above.)

Solving Quadratic Equations with the Quadratic Formula (works every time)

The solutions of the general quadratic $ax^2 + bx + c = 0$ $(a \neq 0)$ are given by

$$x = \frac{-b \pm \sqrt{b^2 - 4ac}}{2a}.$$

---

[4]See the section on factoring for a review of how to factor quadratics.

We call the quantity $\Delta = b^2 - 4ac$ the **discriminant** of the quadratic (Δ is the Greek letter delta), and we have the following general rules:

- If Δ is positive, there are two distinct real solutions.
- If Δ is zero, there is only one real solution: $x = -\dfrac{b}{2a}$. (Why?)
- If Δ is negative, there are no real solutions.

Quick Examples

1. $2x^2 - 5x - 12 = 0$ has $a = 2$, $b = -5$, and $c = -12$.

$$x = \frac{-b \pm \sqrt{b^2 - 4ac}}{2a} = \frac{5 \pm \sqrt{25 + 96}}{4} = \frac{5 \pm \sqrt{121}}{4} = \frac{5 \pm 11}{4}$$

$$= \frac{16}{4} \text{ or } -\frac{6}{4} = 4 \text{ or } -3/2 \qquad \text{Δ is positive in this example.}$$

2. $4x^2 = 12x - 9$ can be rewritten as $4x^2 - 12x + 9 = 0$, which has $a = 4$, $b = -12$, and $c = 9$.

$$x = \frac{-b \pm \sqrt{b^2 - 4ac}}{2a} = \frac{12 \pm \sqrt{144 - 144}}{8} = \frac{12 \pm 0}{8} = \frac{12}{8} = \frac{3}{2}$$

$$\text{Δ is zero in this example.}$$

3. $x^2 + 2x - 1 = 0$ has $a = 1$, $b = 2$, and $c = -1$.

$$x = \frac{-b \pm \sqrt{b^2 - 4ac}}{2a} = \frac{-2 \pm \sqrt{8}}{2} = \frac{-2 \pm 2\sqrt{2}}{2} = -1 \pm \sqrt{2}$$

The two solutions are $x = -1 + \sqrt{2} = 0.414\ldots$ and $x = -1 - \sqrt{2} = -2.414\ldots$ $\qquad$ Δ is positive in this example.

4. $x^2 + x + 1 = 0$ has $a = 1$, $b = 1$, and $c = 1$. Because $\Delta = -3$ is negative, there are no real solutions. $\qquad$ Δ is negative in this example.

Q: *This is all very useful, but where does the quadratic formula come from?*

A: To see where it comes from, we will solve a general quadratic equation using "brute force." Start with the general quadratic equation.

$$ax^2 + bx + c = 0.$$

First, divide out the nonzero number a to get

$$x^2 + \frac{bx}{a} + \frac{c}{a} = 0.$$

Now we **complete the square**: Add and subtract the quantity $\dfrac{b^2}{4a^2}$ to get

$$x^2 + \frac{bx}{a} + \frac{b^2}{4a^2} - \frac{b^2}{4a^2} + \frac{c}{a} = 0.$$

We do this to get the first three terms to factor as a perfect square:

$$\left(x + \frac{b}{2a}\right)^2 - \frac{b^2}{4a^2} + \frac{c}{a} = 0.$$

(Check this by multiplying out.) Adding $\dfrac{b^2}{4a^2} - \dfrac{c}{a}$ to both sides gives:

$$\left(x + \frac{b}{2a}\right)^2 = \frac{b^2}{4a^2} - \frac{c}{a} = \frac{b^2 - 4ac}{4a^2}.$$

Taking square roots gives

$$x + \frac{b}{2a} = \frac{\pm\sqrt{b^2 - 4ac}}{2a}.$$

Finally, adding $-\dfrac{b}{2a}$ to both sides yields the result:

$$x = -\frac{b}{2a} + \frac{\pm\sqrt{b^2 - 4ac}}{2a}$$

or

$$x = \frac{-b \pm \sqrt{b^2 - 4ac}}{2a}.$$

Solution of Cubic Equations

By definition, a cubic equation can be written in the form

$$ax^3 + bx^2 + cx + d = 0. \qquad a, b, c, \text{ and } d \text{ are fixed numbers and } a \neq 0.$$

Now we get into something of a bind. Although there is a perfectly respectable formula for the solutions, it is very complicated and involves the use of complex numbers rather heavily.[5] So we discuss instead a much simpler method that *sometimes* works nicely. Here is the method in a nutshell.

Solving Cubics by Finding One Factor

Start with a given cubic equation $ax^3 + bx^2 + cx + d = 0$.

Step 1 By trial and error, find one solution $x = s$. If a, b, c, and d are integers, the only possible *rational* solutions[6] are those of the form $s = \pm$(factor of d)/(factor of a).

Step 2 It will now be possible to factor the cubic as

$$ax^3 + bx^2 + cx + d = (x - s)(ax^2 + ex + f) = 0$$

To find $ax^2 + ex + f$, divide the cubic by $x - s$, using long division.[7]

Step 3 The factored equation says that either $x - s = 0$ or $ax^2 + ex + f = 0$. We already know that s is a solution, and now we see that the other solutions are the roots of the quadratic. Note that this quadratic may or may not have any real solutions, as usual.

---

[5]It was when this formula was discovered in the 16th century that complex numbers were first taken seriously. Although we would like to show you the formula, it is too large to fit in this footnote.

[6]There may be *irrational* solutions, however; for example, $x^3 - 2 = 0$ has the single solution $x = \sqrt[3]{2}$.

[7]Alternatively, use "synthetic division," a shortcut that would take us too far afield to describe.

> ### Quick Example
>
> To solve the cubic $x^3 - x^2 + x - 1 = 0$, we first find a single solution. Here, $a = 1$ and $d = -1$. Because the only factors of ± 1 are ± 1, the only possible rational solutions are $x = \pm 1$. By substitution, we see that $x = 1$ is a solution. Thus, $(x - 1)$ is a factor. Dividing by $(x - 1)$ yields the quotient $(x^2 + 1)$. Thus,
>
> $$x^3 - x^2 + x - 1 = (x - 1)(x^2 + 1) = 0$$
>
> so that either $x - 1 = 0$ or $x^2 + 1 = 0$.
>
> Because the discriminant of the quadratic $x^2 + 1$ is negative, we don't get any real solutions from $x^2 + 1 = 0$, so the only real solution is $x = 1$.

Possible Outcomes When Solving a Cubic Equation

If you consider all the cases, there are three possible outcomes when solving a cubic equation:

1. One real solution (as in the Quick Example above)
2. Two real solutions (try, for example, $x^3 + x^2 - x - 1 = 0$)
3. Three real solutions (see the next example)

EXAMPLE 1 Solving a Cubic

Solve the cubic $2x^3 - 3x^2 - 17x + 30 = 0$.

Solution First we look for a single solution. Here, $a = 2$ and $d = 30$. The factors of a are ± 1 and ± 2, and the factors of d are $\pm 1, \pm 2, \pm 3, \pm 5, \pm 6, \pm 10, \pm 15,$ and ± 30. This gives us a large number of possible ratios: $\pm 1, \pm 2, \pm 3, \pm 5, \pm 6, \pm 10, \pm 15, \pm 30, \pm 1/2, \pm 3/2, \pm 5/2, \pm 15/2$. Undaunted, we first try $x = 1$ and $x = -1$, getting nowhere. So we move on to $x = 2$, and we hit the jackpot, because substituting $x = 2$ gives $16 - 12 - 34 + 30 = 0$. Thus, $(x - 2)$ is a factor. Dividing yields the quotient $2x^2 + x - 15$. Here is the calculation:

$$
\begin{array}{r}
2x^2 + x - 15 \\
x - 2 \enclose{longdiv}{2x^3 - 3x^2 - 17x + 30} \\
\underline{2x^3 - 4x^2} \\
x^2 - 17x \\
\underline{x^2 - 2x} \\
-15x + 30 \\
\underline{-15x + 30} \\
0.
\end{array}
$$

Thus,

$$2x^3 - 3x^2 - 17x + 30 = (x - 2)(2x^2 + x - 15) = 0.$$

Setting the factors equal to zero gives either $x - 2 = 0$ or $2x^2 + x - 15 = 0$. We could solve the quadratic using the quadratic formula, but, luckily, we notice that it factors as

$$2x^2 + x - 15 = (x + 3)(2x - 5).$$

Thus, the solutions are $x = 2$, $x = -3$ and $x = 5/2$.

Solution of Higher-Order Polynomial Equations

Logically speaking, our next step should be a discussion of quartics, then quintics (fifth degree equations), and so on forever. Well, we've got to stop somewhere, and cubics may be as good a place as any. On the other hand, since we've gotten so far, we ought to at least tell you what is known about higher order polynomials.

Quartics Just as in the case of cubics, there is a formula to find the solutions of quartics.[8]

Quintics and Beyond All good things must come to an end, we're afraid. It turns out that there is no "quintic formula." In other words, there is no single algebraic formula or collection of algebraic formulas that gives the solutions to all quintics. This question was settled by the Norwegian mathematician Niels Henrik Abel in 1824 after almost 300 years of controversy about this question. (In fact, several notable mathematicians had previously claimed to have devised formulas for solving the quintic, but these were all shot down by other mathematicians—this being one of the favorite pastimes of practitioners of our art.) The same negative answer applies to polynomial equations of degree 6 and higher. It's not that these equations don't have solutions; it's just that they can't be found using algebraic formulas.[9] However, there are certain special classes of polynomial equations that can be solved with algebraic methods. The way of identifying such equations was discovered around 1829 by the French mathematician Évariste Galois.[10]

---

[8]See, for example, *First Course in the Theory of Equations* by L. E. Dickson (New York: Wiley, 1922), or *Modern Algebra* by B. L. van der Waerden (New York: Frederick Ungar, 1953).

[9]What we mean by an "algebraic formula" is a formula in the coefficients using the operations of addition, subtraction, multiplication, division, and the taking of radicals. Mathematicians call the use of such formulas in solving polynomial equations "solution by radicals." If you were a math major, you would eventually go on to study this under the heading of Galois theory.

[10]Both Abel (1802–1829) and Galois (1811–1832) died young. Abel died of tuberculosis at the age of 26, while Galois was killed in a duel at the age of 20.

0.5 EXERCISES

Solve the equations in Exercises 1–12 for x (mentally, if possible).

1. $x + 1 = 0$

2. $x - 3 = 1$

3. $-x + 5 = 0$

4. $2x + 4 = 1$

5. $4x - 5 = 8$

6. $\frac{3}{4}x + 1 = 0$

7. $7x + 55 = 98$

8. $3x + 1 = x$

9. $x + 1 = 2x + 2$

10. $x + 1 = 3x + 1$

11. $ax + b = c$ $(a \neq 0)$

12. $x - 1 = cx + d$ $(c \neq 1)$

By any method, determine all possible real solutions of each equation in Exercises 13–30. Check your answers by substitution.

13. $2x^2 + 7x - 4 = 0$

14. $x^2 + x + 1 = 0$

15. $x^2 - x + 1 = 0$

16. $2x^2 - 4x + 3 = 0$

17. $2x^2 - 5 = 0$

18. $3x^2 - 1 = 0$

19. $-x^2 - 2x - 1 = 0$

20. $2x^2 - x - 3 = 0$

21. $\frac{1}{2}x^2 - x - \frac{3}{2} = 0$

22. $-\frac{1}{2}x^2 - \frac{1}{2}x + 1 = 0$

23. $x^2 - x = 1$

24. $16x^2 = -24x - 9$

25. $x = 2 - \frac{1}{x}$

26. $x + 4 = \frac{1}{x - 2}$

27. $x^4 - 10x^2 + 9 = 0$

28. $x^4 - 2x^2 + 1 = 0$

29. $x^4 + x^2 - 1 = 0$

30. $x^3 + 2x^2 + x = 0$

Find all possible real solutions of each equation in Exercises 31–44.

31. $x^3 + 6x^2 + 11x + 6 = 0$

32. $x^3 - 6x^2 + 12x - 8 = 0$

33. $x^3 + 4x^2 + 4x + 3 = 0$

34. $y^3 + 64 = 0$

35. $x^3 - 1 = 0$

36. $x^3 - 27 = 0$

37. $y^3 + 3y^2 + 3y + 2 = 0$

38. $y^3 - 2y^2 - 2y - 3 = 0$

39. $x^3 - x^2 - 5x + 5 = 0$

40. $x^3 - x^2 - 3x + 3 = 0$

41. $2x^6 - x^4 - 2x^2 + 1 = 0$

42. $3x^6 - x^4 - 12x^2 + 4 = 0$

43. $(x^2 + 3x + 2)(x^2 - 5x + 6) = 0$

44. $(x^2 - 4x + 4)^2(x^2 + 6x + 5)^3 = 0$

0.6 Solving Miscellaneous Equations

Equations often arise in calculus that are not polynomial equations of low degree. Many of these complicated-looking equations can be solved easily if you remember the following, which we used in the previous section:

Solving an Equation of the Form $P \cdot Q = 0$

If a product is equal to 0, then at least one of the factors must be 0. That is, if $P \cdot Q = 0$, then either $P = 0$ or $Q = 0$.

Quick Examples

1. $x^5 - 4x^3 = 0$

$x^3(x^2 - 4) = 0$ Factor the left-hand side.

Either $x^3 = 0$ or $x^2 - 4 = 0$ Either $P = 0$ or $Q = 0$.

$x = 0, 2$ or -2.   Solve the individual equations.

2. $(x^2 - 1)(x + 2) + (x^2 - 1)(x + 4) = 0$

$(x^2 - 1)[(x + 2) + (x + 4)] = 0$ Factor the left-hand side.

$(x^2 - 1)(2x + 6) = 0$

Either $x^2 - 1 = 0$ or $2x + 6 = 0$ Either $P = 0$ or $Q = 0$.

$x = -3, -1,$ or 1.   Solve the individual equations.

EXAMPLE 1 Solving by Factoring

Solve $12x(x^2 - 4)^5(x^2 + 2)^6 + 12x(x^2 - 4)^6(x^2 + 2)^5 = 0$.

Solution

Again, we start by factoring the left-hand side:

$$12x(x^2 - 4)^5(x^2 + 2)^6 + 12x(x^2 - 4)^6(x^2 + 2)^5$$
$$= 12x(x^2 - 4)^5(x^2 + 2)^5[(x^2 + 2) + (x^2 - 4)]$$
$$= 12x(x^2 - 4)^5(x^2 + 2)^5(2x^2 - 2)$$
$$= 24x(x^2 - 4)^5(x^2 + 2)^5(x^2 - 1).$$

Setting this equal to 0, we get:

$$24x(x^2 - 4)^5(x^2 + 2)^5(x^2 - 1) = 0,$$

which means that at least one of the factors of this product must be zero. Now it certainly cannot be the 24, but it could be the x: $x = 0$ is one solution. It could also be that

$$(x^2 - 4)^5 = 0$$

or

$$x^2 - 4 = 0,$$

which has solutions $x = \pm 2$. Could it be that $(x^2 + 2)^5 = 0$? If so, then $x^2 + 2 = 0$, but this is impossible because $x^2 + 2 \geq 2$, no matter what x is. Finally, it could be that $x^2 - 1 = 0$, which has solutions $x = \pm 1$. This gives us five solutions to the original equation:

$$x = -2, -1, 0, 1, \text{ or } 2.$$

EXAMPLE 2 Solving by Factoring

Solve $(x^2 - 1)(x^2 - 4) = 10$.

Solution Watch out! You may be tempted to say that $x^2 - 1 = 10$ or $x^2 - 4 = 10$, but this does not follow. If two numbers multiply to give you 10, what must they be? There are lots of possibilities: 2 and 5, 1 and 10, $-500{,}000$ and -0.00002 are just a few. The fact that the left-hand side is factored is nearly useless to us if we want to solve this equation. What we will have to do is multiply out, bring the 10 over to the left, and hope that we can factor what we get. Here goes:

$$x^4 - 5x^2 + 4 = 10$$
$$x^4 - 5x^2 - 6 = 0$$
$$(x^2 - 6)(x^2 + 1) = 0$$

(Here we used a sometimes useful trick that we mentioned in Section 3: We treated x^2 like x and x^4 like x^2, so factoring $x^4 - 5x^2 - 6$ is essentially the same as factoring $x^2 - 5x - 6$.) *Now* we are allowed to say that one of the factors must be 0: $x^2 - 6 = 0$ has solutions $x = \pm\sqrt{6} = \pm 2.449\ldots$ and $x^2 + 1 = 0$ has no real solutions. Therefore, we get exactly two solutions, $x = \pm\sqrt{6} = \pm 2.449\ldots$.

To solve equations involving rational expressions, the following rule is very useful.

Solving an Equation of the Form *P/Q* = 0

If $\dfrac{P}{Q} = 0$, then $P = 0$.

How else could a fraction equal 0? If that is not convincing, multiply both sides by Q (which cannot be 0 if the quotient is defined).

Quick Example

$$\frac{(x + 1)(x + 2)^2 - (x + 1)^2(x + 2)}{(x + 2)^4} = 0$$

$(x + 1)(x + 2)^2 - (x + 1)^2(x + 2) = 0$ If $\frac{P}{Q} = 0$, then $P = 0$.

$(x + 1)(x + 2)[(x + 2) - (x + 1)] = 0$ Factor.

$(x + 1)(x + 2)(1) = 0$

Either $x + 1 = 0$ or $x + 2 = 0$,

$x = -1$ or $x = -2$

$x = -1$ $x = -2$ does not make sense in the original equation: it makes the denominator 0. So it is not a solution and $x = -1$ is the only solution.

EXAMPLE 3 Solving a Rational Equation

Solve $1 - \dfrac{1}{x^2} = 0$.

Solution Write 1 as $\frac{1}{1}$, so that we now have a difference of two rational expressions:

$$\frac{1}{1} - \frac{1}{x^2} = 0.$$

To combine these we can put both over a common denominator of x^2, which gives

$$\frac{x^2 - 1}{x^2} = 0.$$

Now we can set the numerator, $x^2 - 1$, equal to zero. Thus,

$$x^2 - 1 = 0$$

so

$$(x - 1)(x + 1) = 0,$$

giving $x = \pm 1$.

➡ **Before we go on...** This equation could also have been solved by writing

$$1 = \frac{1}{x^2}$$

and then multiplying both sides by x^2. ∎

EXAMPLE 4 Another Rational Equation

Solve $\dfrac{2x - 1}{x} + \dfrac{3}{x - 2} = 0$.

Solution We *could* first perform the addition on the left and then set the top equal to 0, but here is another approach. Subtracting the second expression from both sides gives

$$\frac{2x - 1}{x} = \frac{-3}{x - 2}$$

Cross-multiplying [multiplying both sides by both denominators—that is, by $x(x - 2)$] now gives

$$(2x - 1)(x - 2) = -3x$$

so

$$2x^2 - 5x + 2 = -3x.$$

Adding $3x$ to both sides gives the quadratic equation

$$2x^2 - 2x + 1 = 0.$$

The discriminant is $(-2)^2 - 4 \cdot 2 \cdot 1 = -4 < 0$, so we conclude that there is no real solution.

➡ **Before we go on...** Notice that when we said that $(2x - 1)(x - 2) = -3x$, we were *not* allowed to conclude that $2x - 1 = -3x$ or $x - 2 = -3x$. ∎

EXAMPLE 5 A Rational Equation with Radicals

Solve $\dfrac{\left(2x\sqrt{x + 1} - \frac{x^2}{\sqrt{x+1}}\right)}{x + 1} = 0$.

Solution Setting the top equal to 0 gives

$$2x\sqrt{x + 1} - \frac{x^2}{\sqrt{x + 1}} = 0.$$

This still involves fractions. To get rid of the fractions, we could put everything over a common denominator ($\sqrt{x + 1}$) and then set the top equal to 0, or we could multiply the whole equation by that common denominator in the first place to clear fractions. If we do the second, we get

$$2x(x + 1) - x^2 = 0$$
$$2x^2 + 2x - x^2 = 0$$
$$x^2 + 2x = 0.$$

Factoring,

$$x(x + 2) = 0$$

so either $x = 0$ or $x + 2 = 0$, giving us $x = 0$ or $x = -2$. Again, one of these is not really a solution. The problem is that $x = -2$ cannot be substituted into $\sqrt{x + 1}$, because we would then have to take the square root of -1, and we are not allowing ourselves to do that. Therefore, $x = 0$ is the only solution.

0.6   EXERCISES

Solve the following equations:

1. $x^4 - 3x^3 = 0$

2. $x^6 - 9x^4 = 0$

3. $x^4 - 4x^2 = -4$

4. $x^4 - x^2 = 6$

5. $(x+1)(x+2) + (x+1)(x+3) = 0$

6. $(x+1)(x+2)^2 + (x+1)^2(x+2) = 0$

7. $(x^2+1)^5(x+3)^4 + (x^2+1)^6(x+3)^3 = 0$

8. $10x(x^2+1)^4(x^3+1)^5 - 10x^2(x^2+1)^5(x^3+1)^4 = 0$

9. $(x^3+1)\sqrt{x+1} - (x^3+1)^2\sqrt{x+1} = 0$

10. $(x^2+1)\sqrt{x+1} - \sqrt{(x+1)^3} = 0$

11. $\sqrt{(x+1)^3} + \sqrt{(x+1)^5} = 0$

12. $(x^2+1)\sqrt[3]{(x+1)^4} - \sqrt[3]{(x+1)^7} = 0$

13. $(x+1)^2(2x+3) - (x+1)(2x+3)^2 = 0$

14. $(x^2-1)^2(x+2)^3 - (x^2-1)^3(x+2)^2 = 0$

15. $\dfrac{(x+1)^2(x+2)^3 - (x+1)^3(x+2)^2}{(x+2)^6} = 0$

16. $\dfrac{6x(x^2+1)^2(x^2+2)^4 - 8x(x^2+1)^3(x^2+2)^3}{(x^2+2)^8} = 0$

17. $\dfrac{2(x^2-1)\sqrt{x^2+1} - \frac{x^4}{\sqrt{x^2+1}}}{x^2+1} = 0$

18. $\dfrac{4x\sqrt{x^3-1} - \frac{3x^4}{\sqrt{x^3-1}}}{x^3-1} = 0$

19. $x - \dfrac{1}{x} = 0$

20. $1 - \dfrac{4}{x^2} = 0$

21. $\dfrac{1}{x} - \dfrac{9}{x^3} = 0$

22. $\dfrac{1}{x^2} - \dfrac{1}{x+1} = 0$

23. $\dfrac{x-4}{x+1} - \dfrac{x}{x-1} = 0$

24. $\dfrac{2x-3}{x-1} - \dfrac{2x+3}{x+1} = 0$

25. $\dfrac{x+4}{x+1} + \dfrac{x+4}{3x} = 0$

26. $\dfrac{2x-3}{x} - \dfrac{2x-3}{x+1} = 0$

0.7   The Coordinate Plane

Q: *Just what is the xy-plane?*

A: The *xy*-plane is an infinite flat surface with two perpendicular lines, usually labeled the **x-axis** and **y-axis**. These axes are calibrated as shown in Figure 2. (Notice also how the plane is divided into four **quadrants**.)

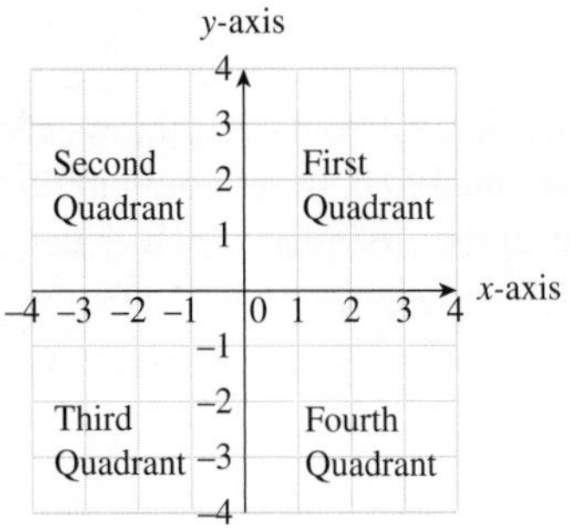

y-axis, Second Quadrant, First Quadrant, x-axis, Third Quadrant, Fourth Quadrant

The *xy*-plane

Figure 2

Thus, the *xy*-plane is nothing more than a very large—in fact, infinitely large—flat surface. The purpose of the axes is to allow us to locate specific positions, or **points**, on the plane, with the use of **coordinates**. (If Captain Picard wants to have himself beamed to a specific location, he must supply its coordinates, or he's in trouble.)

Q: *So how do we use coordinates to locate points?*

A: The rule is simple. Each point in the plane has two coordinates, an **x-coordinate** and a **y-coordinate**. These can be determined in two ways:

1. The *x*-coordinate measures a point's distance to the right or left of the *y*-axis. It is positive if the point is to the right of the axis, negative if it is to the left of the axis, and 0 if it is on the axis. The *y*-coordinate measures a point's distance above or below the *x*-axis. It is positive if the point is above the axis, negative if it is below the axis, and 0 if it is on the axis. Briefly, the *x*-coordinate tells us the *horizontal* position (distance left or right), and the *y*-coordinate tells us the *vertical* position (height).

2. Given a point P, we get its x-coordinate by drawing a vertical line from P and seeing where it intersects the x-axis. Similarly, we get the y-coordinate by extending a horizontal line from P and seeing where it intersects the y-axis.

This way of assigning coordinates to points in the plane is often called the system of **Cartesian** coordinates, in honor of the mathematician and philosopher René Descartes (1596–1650), who was the first to use them extensively.

Here are a few examples to help you review coordinates.

EXAMPLE 1 Coordinates of Points

a. Find the coordinates of the indicated points. (See Figure 3. The grid lines are placed at intervals of one unit.)

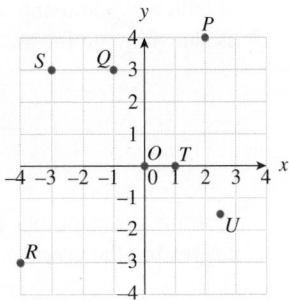

Figure 3

b. Locate the following points in the xy-plane.

$$A(2, 3), \ B(-4, 2), \ C(3, -2.5), \ D(0, -3), \ E(3.5, 0), \ F(-2.5, -1.5)$$

Solution

a. Taking them in alphabetical order, we start with the origin O. This point has height zero and is also zero units to the right of the y-axis, so its coordinates are $(0, 0)$. Turning to P, dropping a vertical line gives $x = 2$ and extending a horizontal line gives $y = 4$. Thus, P has coordinates $(2, 4)$. For practice, determine the coordinates of the remaining points, and check your work against the list that follows:

$$Q(-1, 3), \ R(-4, -3), \ S(-3, 3), \ T(1, 0), \ U(2.5, -1.5)$$

b. In order to locate the given points, we start at the origin $(0, 0)$, and proceed as follows. (See Figure 4.)

To locate A, we move 2 units to the right and 3 up, as shown.

To locate B, we move -4 units to the right (that is, 4 to the *left*) and 2 up, as shown.

To locate C, we move 3 units right and 2.5 down.

We locate the remaining points in a similar way.

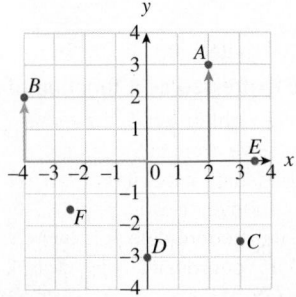

Figure 4

The Graph of an Equation

One of the more surprising developments of mathematics was the realization that equations, which are algebraic objects, can be represented by graphs, which are geometric objects. The kinds of equations that we have in mind are equations in x and y, such as

$$y = 4x - 1, \quad 2x^2 - y = 0, \quad y = 3x^2 + 1, \quad y = \sqrt{x - 1}.$$

The **graph** of an equation in the two variables x and y consists of all points (x, y) in the plane whose coordinates are solutions of the equation.

EXAMPLE 2 Graph of an Equation

Obtain the graph of the equation $y - x^2 = 0$.

Solution We can solve the equation for y to obtain $y = x^2$. Solutions can then be obtained by choosing values for x and then computing y by squaring the value of x, as shown in the following table:

x	-3	-2	-1	0	1	2	3
$y = x^2$	9	4	1	0	1	4	9

Plotting these points (x, y) gives the following picture (left side of Figure 5), suggesting the graph on the right in Figure 5.

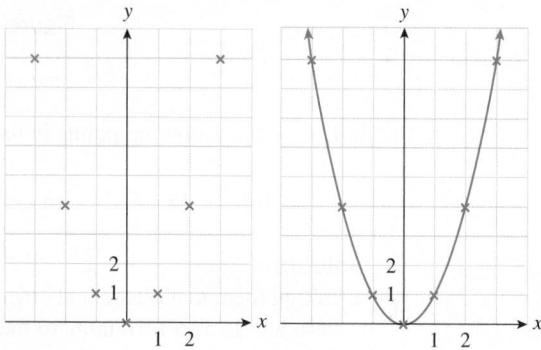

Figure 5

Distance

The distance between two points in the xy-plane can be expressed as a function of their coordinates, as follows:

Distance Formula

The distance between the points $P(x_1, y_1)$ and $Q(x_2, y_2)$ is

$$d = \sqrt{(x_2 - x_1)^2 + (y_2 - y_1)^2} = \sqrt{(\Delta x)^2 + (\Delta y)^2}.$$

Derivation

The distance d is shown in the figure below.

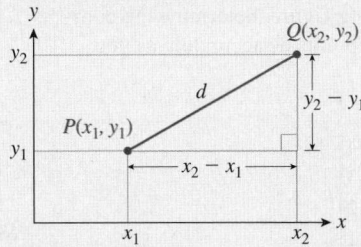

By the Pythagorean theorem applied to the right triangle shown, we get

$$d^2 = (x_2 - x_1)^2 + (y_2 - y_1)^2.$$

Taking square roots (d is a distance, so we take the positive square root), we get the distance formula. Notice that if we switch x_1 with x_2 or y_1 with y_2, we get the same result.

Quick Examples

1. The distance between the points $(3, -2)$ and $(-1, 1)$ is
$$d = \sqrt{(-1-3)^2 + (1+2)^2} = \sqrt{25} = 5.$$

2. The distance from (x, y) to the origin $(0, 0)$ is
$$d = \sqrt{(x-0)^2 + (y-0)^2} = \sqrt{x^2 + y^2}. \qquad \text{Distance to the origin}$$

The set of all points (x, y) whose distance from the origin $(0, 0)$ is a fixed quantity r is a circle centered at the origin with radius r. From the second Quick Example, we get the following equation for the circle centered at the origin with radius r:

$$\sqrt{x^2 + y^2} = r. \qquad \text{Distance from the origin} = r.$$

Squaring both sides gives the following equation:

Equation of the Circle of Radius r Centered at the Origin

$$x^2 + y^2 = r^2$$

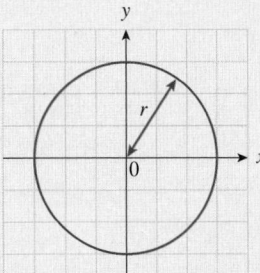

Quick Examples

1. The circle of radius 1 centered at the origin has equation $x^2 + y^2 = 1$.
2. The circle of radius 2 centered at the origin has equation $x^2 + y^2 = 4$.

0.7   EXERCISES

1. Referring to the following figure, determine the coordinates of the indicated points as accurately as you can.

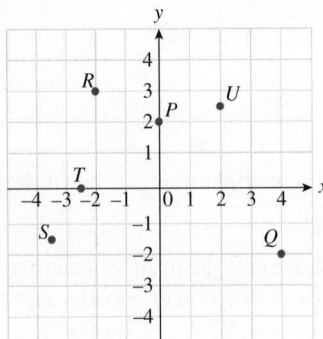

2. Referring to the following figure, determine the coordinates of the indicated points as accurately as you can.

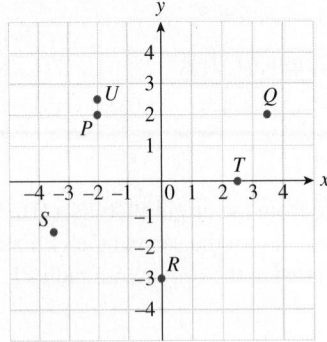

3. Graph the following points.

$P(4, 4), Q(4, -4), R(3, 0), S(4, 0.5), T(0.5, 2.5),$
$U(-2, 0), V(-4, 4)$

4. Graph the following points.

$P(4, -2), Q(2, -4), R(1, -3), S(-4, 2), T(2, -1),$
$U(-2, 0), V(-4, -4)$

Sketch the graphs of the equations in Exercises 5–12.

5. $x + y = 1$ 6. $y - x = -1$

7. $2y - x^2 = 1$ 8. $2y + \sqrt{x} = 1$

9. $xy = 4$ 10. $x^2 y = -1$

11. $xy = x^2 + 1$ 12. $xy = 2x^3 + 1$

In Exercises 13–16, find the distance between the given pairs of points.

13. $(1, -1)$ and $(2, -2)$ 14. $(1, 0)$ and $(6, 1)$

15. $(a, 0)$ and $(0, b)$ 16. (a, a) and (b, b)

17. Find the value of k such that $(1, k)$ is equidistant from $(0, 0)$ and $(2, 1)$.

18. Find the value of k such that (k, k) is equidistant from $(-1, 0)$ and $(0, 2)$.

19. Describe the set of points (x, y) such that $x^2 + y^2 = 9$.

20. Describe the set of points (x, y) such that $x^2 + y^2 = 0$.

1

Functions and Applications

Website

www.WanerMath.com

At the Website you will find:

- Section-by-section tutorials, including game tutorials with randomized quizzes

- A detailed chapter summary

- A true/false quiz

- Additional review exercises

- Graphers, Excel tutorials, and other resources

- The following extra topic:

  New Functions from Old: Scaled and Shifted Functions

Case Study Modeling Spending on Internet Advertising

You are the new director of *Impact Advertising Inc.'s* Internet division, which has enjoyed a steady 0.25% of the Internet advertising market. You have drawn up an ambitious proposal to expand your division in light of your anticipation that Internet advertising will continue to skyrocket. The VP in charge of Financial Affairs feels that current projections (based on a linear model) do not warrant the level of expansion you propose. **How can you persuade the VP that those projections do not fit the data convincingly?**

Jeff Titcomb/Photographer's Choice / Getty Images

39

Introduction

To analyze recent trends in spending on Internet advertising and to make reasonable projections, we need a mathematical model of this spending. Where do we start? To apply mathematics to real-world situations like this, we need a good understanding of basic mathematical concepts. Perhaps the most fundamental of these concepts is that of a function: a relationship that shows how one quantity depends on another. Functions may be described numerically and, often, algebraically. They can also be described graphically—a viewpoint that is extremely useful.

The simplest functions—the ones with the simplest formulas and the simplest graphs—are linear functions. Because of their simplicity, they are also among the most useful functions and can often be used to model real-world situations, at least over short periods of time. In discussing linear functions, we will meet the concepts of slope and rate of change, which are the starting point of the mathematics of change.

algebra Review
For this chapter, you should be familiar with real numbers and intervals. To review this material, see **Chapter 0.**

In the last section of this chapter, we discuss *simple linear regression*: construction of linear functions that best fit given collections of data. Regression is used extensively in applied mathematics, statistics, and quantitative methods in business. The inclusion of regression utilities in computer spreadsheets like Excel® makes this powerful mathematical tool readily available for anyone to use.

1.1   Functions from the Numerical, Algebraic, and Graphical Viewpoints

The following table gives the approximate number of Facebook users at various times since its establishment early in 2004.[1]

Year t (Since start of 2004)	0	1	2	3	4	5	6
Facebook Members n (Millions)	0	1	5.5	12	58	150	450

Let's write $n(0)$ for the number of members (in millions) at time $t = 0$, $n(1)$ for the number at time $t = 1$, and so on (we read $n(0)$ as "n of 0"). Thus, $n(0) = 0$, $n(1) = 1, n(2) = 5.5, \ldots, n(6) = 450$. In general, we write $n(t)$ for the number of members (in millions) at time t. We call n a **function** of the variable t, meaning that for each value of t between 0 and 6, n gives us a single corresponding number $n(t)$ (the number of members at that time).

In general, we think of a function as a way of producing new objects from old ones. The functions we deal with in this text produce new numbers from old numbers. The numbers we have in mind are the *real* numbers, including not only positive and negative integers and fractions but also numbers like $\sqrt{2}$ or π. (See Chapter 0 for more on real numbers.) For this reason, the functions we use are called **real-valued functions of a real variable**. For example, the function n takes the year since the start of 2004 as input and returns the number of Facebook members as output (Figure 1).

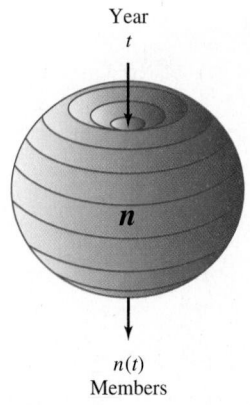

Year
t

n

$n(t)$
Members

Figure 1

[1]Sources: www.facebook.com, www.insidefacebook.com.

The variable t is called the **independent variable**, while n is called the **dependent variable** as its value depends on t. A function may be specified in several different ways. Here, we have specified the function n **numerically** by giving the values of the function for a number of values of the independent variable, as in the preceding table.

Q: *For which values of t does it make sense to ask for n(t)? In other words, for which years t is the function n defined?*

A: Because $n(t)$ refers to the number of members from the start of 2004 to the start of 2010, $n(t)$ is defined when t is any number between 0 and 6, that is, when $0 \leq t \leq 6$. Using interval notation (see Chapter 0), we can say that $n(t)$ is defined when t is in the interval [0, 6].

The set of values of the independent variable for which a function is defined is called its **domain** and is a necessary part of the definition of the function. Notice that the preceding table gives the value of $n(t)$ at only some of the infinitely many possible values in the domain [0, 6]. The domain of a function is not always specified explicitly; if no domain is specified for the function f, we take the domain to be the largest set of numbers x for which $f(x)$ makes sense. This "largest possible domain" is sometimes called the **natural domain**.

The previous Facebook data can also be represented on a graph by plotting the given pairs of numbers $(t, n(t))$ in the xy-plane. (See Figure 2. We have connected successive points by line segments.) In general, the **graph** of a function f consists of all points $(x, f(x))$ in the plane with x in the domain of f.

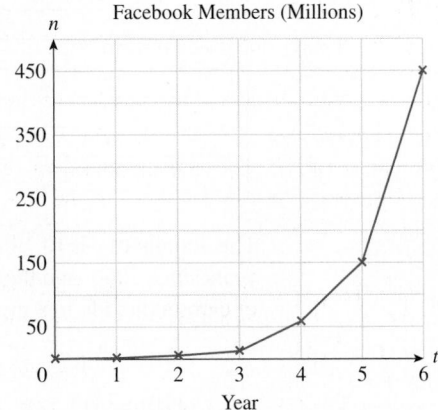

Facebook Members (Millions)

Figure 2

* In a graphically defined function, we can never know the y-coordinates of points exactly; no matter how accurately a graph is drawn, we can obtain only *approximate* values of the coordinates of points. That is why we have been using the word *estimate* rather than *calculate* and why we say $n(5) \approx 150$ rather than $n(5) = 150$.

In Figure 2 we specified the function n **graphically** by using a graph to display its values. Suppose now that we had only the graph without the table of data. We could use the graph to find approximate values of n. For instance, to find $n(5)$ from the graph, we do the following:

1. Find the desired value of t at the bottom of the graph ($t = 5$ in this case).

2. Estimate the height (n-coordinate) of the corresponding point on the graph (around 150 in this case).

Thus, $n(5) \approx 150$ million members.*

In some cases we may be able to use an algebraic formula to calculate the function, and we say that the function is specified **algebraically**. These are not the only ways in which a function can be specified; for instance, it could also be specified **verbally**, as in "Let $n(t)$ be the number of Facebook members, in millions, t years since the start of 2004."* Notice that any function can be represented graphically by plotting the points $(x, f(x))$ for a number of values of x in its domain.

Here is a summary of the terms we have just introduced.

* Specifying a function verbally in this way is useful for understanding what the function is doing, but it gives no numerical information.

Functions

A **real-valued function** f **of a real-valued variable** x assigns to each real number x in a specified set of numbers, called the **domain** of f, a unique real number $f(x)$, read "f of x." The variable x is called the **independent variable**, and f is called the **dependent variable**. A function is usually specified **numerically** using a table of values, **graphically** using a graph, or **algebraically** using a formula. The **graph of a function** consists of all points $(x, f(x))$ in the plane with x in the domain of f.

Quick Examples

1. **A function specified numerically:** Take $c(t)$ to be the world emission of carbon dioxide in year t since 2000, represented by the following table:[2]

t (Year Since 2000)	$c(t)$ (Billion Metric Tons of CO_2)
0	24
5	28
10	31
15	33
20	36
25	38
30	41

The domain of c is $[0, 30]$, the independent variable is t, the number of years since 2000, and the dependent variable is c, the world production of carbon dioxide in a given year. Some values of c are:

$c(0) = 24$ 24 billion metric tons of CO_2 were produced in 2000.

$c(10) = 31$ 31 billion metric tons of CO_2 were produced in 2010.

$c(30) = 41$ 41 billion metric tons of CO_2 were projected to be produced in 2030.

---

[2]Figures for 2015 and later are projections. Source: Energy Information Administration (EIA) (www.eia.doe.gov)

Graph of c: Plotting the pairs $(t, c(t))$ gives the following graph:

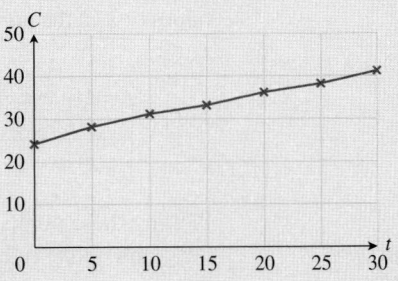

2. **A function specified graphically:** Take $m(t)$ to be the median U.S. home price in thousands of dollars, t years since 2000, as represented by the following graph:[3]

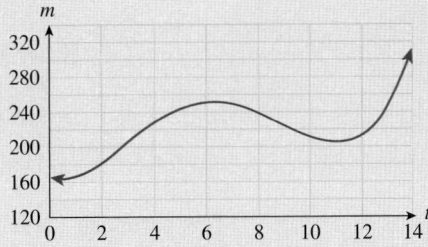

The domain of m is $[0, 14]$, the independent variable is t, the number of years since 2000, and the dependent variable is m, the median U.S. home price in thousands of dollars. Some values of m are:

$m(2) \approx 180$ The median home price in 2002 was about \$180,000.

$m(10) \approx 210$.   The median home price in 2010 was about \$210,000.

3. **A function specified algebraically:** Let $f(x) = \frac{1}{x}$. The function f is specified algebraically. The independent variable is x and the dependent variable is f. The natural domain of f consists of all real numbers except zero because $f(x)$ makes sense for all values of x other than $x = 0$. Some specific values of f are

$$f(2) = \frac{1}{2} \qquad f(3) = \frac{1}{3} \qquad f(-1) = \frac{1}{-1} = -1$$

$f(0)$ is not defined because 0 is not in the domain of f.

---

[3]Source for data through end of 2010: www.zillow.com/local-info.

4. The graph of a function: Let $f(x) = x^2$, with domain the set of all real numbers. To draw the graph of f, first choose some convenient values of x in the domain and compute the corresponding y-coordinates $f(x)$:

x	-3	-2	-1	0	1	2	3
$f(x) = x^2$	9	4	1	0	1	4	9

Plotting these points $(x, f(x))$ gives the picture on the left, suggesting the graph on the right.＊

＊ If you plot more points, you will find that they lie on a smooth curve as shown. That is why we did not use line segments to connect the points.

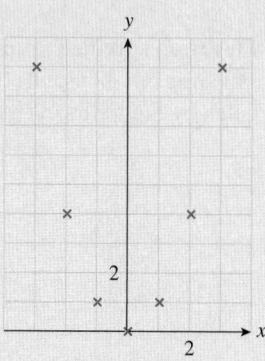
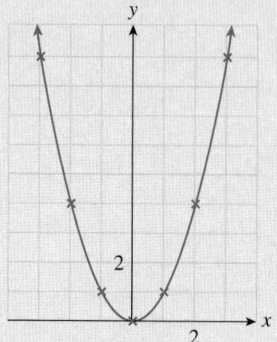

(This particular curve happens to be called a **parabola**, and its lowest point, at the origin, is called its **vertex**.)

EXAMPLE 1 iPod Sales

The total number of iPods sold by Apple up to the end of year x can be approximated by

$$f(x) = 4x^2 + 16x + 2 \text{ million iPods } (0 \le x \le 6),$$

where $x = 0$ represents 2003.[4]

a. What is the domain of f? Compute $f(0)$, $f(2)$, $f(4)$, and $f(6)$. What do these answers tell you about iPod sales? Is $f(-1)$ defined?

b. Compute $f(a)$, $f(-b)$, $f(a+h)$, and $f(a) + h$ assuming that the quantities a, $-b$, and $a + h$ are in the domain of f.

c. Sketch the graph of f. Does the shape of the curve suggest that iPod sales were accelerating or decelerating?

Solution

a. The domain of f is the set of numbers x with $0 \le x \le 6$—that is, the interval $[0, 6]$. If we substitute 0 for x in the formula for $f(x)$, we get

$$f(0) = 4(0)^2 + 16(0) + 2 = 2. \qquad \text{By the end of 2003 approximately 2 million iPods had been sold.}$$

---

[4]Source for data: Apple quarterly earnings reports at www.apple.com/investor/.

Similarly,

$$f(2) = 4(2)^2 + 16(2) + 2 = 50$$ By the end of 2005 approximately 50 million iPods had been sold.

$$f(4) = 4(4)^2 + 16(4) + 2 = 130$$ By the end of 2007 approximately 130 million iPods had been sold.

$$f(6) = 4(6)^2 + 16(6) + 2 = 242.$$ By the end of 2009 approximately 242 million iPods had been sold.

As -1 is not in the domain of f, $f(-1)$ is not defined.

b. To find $f(a)$ we substitute a for x in the formula for $f(x)$ to get

$$f(a) = 4a^2 + 16a + 2.$$ Substitute a for x.

Similarly,

$$f(-b) = 4(-b)^2 + 16(-b) + 2$$ Substitute $-b$ for x.
$$= 4b^2 - 16b + 2$$ $(-b)^2 = b^2$

$$f(a + h) = 4(a + h)^2 + 16(a + h) + 2$$ Substitute $(a + h)$ for x.
$$= 4(a^2 + 2ah + h^2) + 16a + 16h + 2$$ Expand.

$$= 4a^2 + 8ah + 4h^2 + 16a + 16h + 2$$

$$f(a) + h = 4a^2 + 16a + 2 + h.$$ Add h to $f(a)$.

Note how we placed parentheses around the quantities at which we evaluated the function. If we tried to do without any of these parentheses we would likely get an error:

Correct expression: $f(a + h) = 4(a + h)^2 + 16(a + h) + 2.$ ✓

NOT $4a + h^2 + 16a + h + 2x$

Also notice the distinction between $f(a + h)$ and $f(a) + h$: To find $f(a + h)$, we replace x by the quantity $(a + h)$; to find $f(a) + h$ we add h to $f(a)$.

c. To draw the graph of f we plot points of the form $(x, f(x))$ for several values of x in the domain of f. Let us use the values we computed in part (a):

x	0	2	4	6
$f(x) = 4x^2 + 16x + 2$	2	50	130	242

Graphing these points gives the graph shown in Figure 3, suggesting the curve shown on the right.

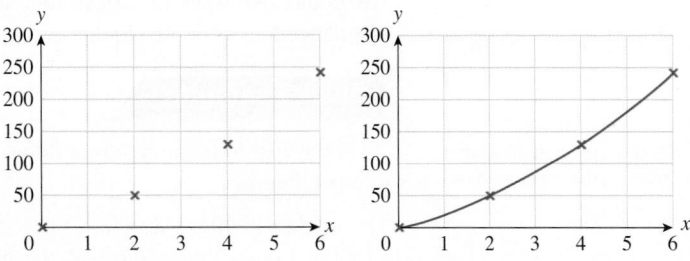

Figure 3

The graph becomes more steep as we move from left to right, suggesting that iPod sales were accelerating.

using Technology

See the Technology Guides at the end of the chapter for detailed instructions on how to obtain the table of values and graph in Example 1 using a TI-83/84 Plus or Excel. Here is an outline:

TI-83/84 Plus
Table of values:
Y₁=4X^2+16X+2
[2ND] [TABLE].
Graph: [WINDOW];
Xmin = 0, Xmax = 6
[ZOOM] [0].
[More details on page 114.]

Spreadsheet
Table of values: Headings x and $f(x)$ in A1–B1; x-values 0, 2, 4, 6 in A2–A5.
=4*A2^2+16*A2+2
in B2; copy down through B5.
Graph: Highlight A1 through B5 and insert a Scatter chart. [More details on page 119.]

Website
www.WanerMath.com
Go to the Function Evaluator and Grapher under Online Utilities, and enter
4x^2+16x+2
for y_1. To obtain a table of values, enter the x-values 0, 1, 2, 3 in the Evaluator box, and press "Evaluate" at the top of the box. Graph: Set Xmin = 0, Xmax = 6, and press "Plot Graphs".

➡ **Before we go on…** The following table compares the value of f in Example 1 with the actual sales figures:

x	0	2	4	6
$f(x) = 4x^2 + 16x + 2$	2	50	130	242
Actual iPod Sales (Millions)	2	32	141	240

The actual figures are only stated here for (some) integer values of x; for instance, $x = 4$ gives the total sales up to the end of 2007. But what were, for instance, the sales through June of 2008 ($x = 4.5$)? This is where our formula comes in handy: We can use the formula for f to **interpolate**—that is, to find sales at values of x other than those between values that are stated:

$$f(4.5) = 4(4.5)^2 + 16(4.5) + 2 = 155 \text{ million iPods.}$$

We can also use the formula to **extrapolate**—that is, to predict sales at values of x *outside* the domain—say, for $x = 6.5$ (that is, sales through June 2009):

$$f(6.5) = 4(6.5)^2 + 16(6.5) + 2 = 275 \text{ million iPods.}$$

As a general rule, extrapolation is far less reliable than interpolation: Predicting the future from current data is difficult, especially given the vagaries of the marketplace.

We call the algebraic function f an **algebraic model** of iPod sales because it uses an algebraic formula to model—or mathematically represent (approximately)—the annual sales. The particular kind of algebraic model we used is called a **quadratic model**. (See the end of this section for the names of some commonly used models.) ∎

Functions and Equations

Instead of using the usual "function notation" to specify a function, as in, say,

$$f(x) = 4x^2 + 16x + 2, \qquad \text{Function notation}$$

we could have specified it by an equation by replacing $f(x)$ by y:

$$y = 4x^2 + 16x + 2 \qquad \text{Equation notation}$$

(the choice of the letter y is a convention, but any letter will do).

Technically, $y = 4x^2 + 16x + 2$ is an equation and not a function. However, an equation of this type, $y = \textit{Expression in } x$, can be thought of as "specifying y as a function of x." When we specify a function in this way, the variable x is the independent variable and y is the dependent variable.

We could also write the above function as $f = 4x^2 + 16x + 2$, in which case the dependent variable would be f.

Quick Example

If the cost to manufacture x items is given by the "cost function"∗ C specified by

$$C(x) = 40x + 2{,}000, \qquad \text{Cost function}$$

we could instead write

$$C = 40x + 2{,}000 \qquad \text{Cost equation}$$

and think of C, the cost, as a function of x.

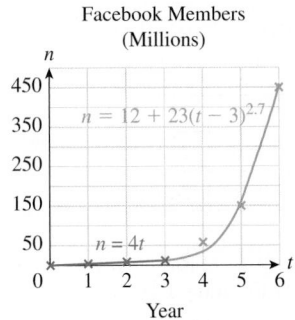

Figure 4

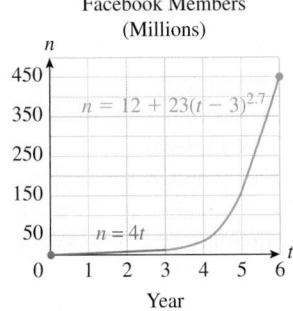

Figure 5

 **using** Technology

See the Technology Guides at the end of the chapter for detailed instructions on how to obtain the table of values and graph in Example 2 using a TI-83/84 Plus or Excel. Here is an outline:

TI-83/84 Plus
Table of values:
$Y_1=(X\le3)*(4X)+(X>3)*$
$(12+23*\text{abs}(X-3)^2.7)$
[2ND] [TABLE].
Graph: [WINDOW]; Xmin = 0,
Xmax = 6; [ZOOM] [0].
[More details on page 114.]

Spreadsheet
Table of values: Headings t and $n(t)$ in A1–B1; t-values 0, 1, . . . , 6 in A2–A8.
$=(A2<=3)*(4*A2)+(A2>3)*$
$(12+23*\text{abs}(A2-3)^2.7)$
in B2; copy down through B8.
Graph: Highlight A1 through B8 and insert a Scatter chart.
[More details on page 120.]

Function notation and equation notation, sometimes using the same letter for the function name and the dependent variable, are often used interchangeably. It is important to be able to switch back and forth between function notation and equation notation, and we shall do so when it is convenient.

Look again at the graph of the number of Facebook users in Figure 2. From year 0 through year 3 the membership appears to increase more-or-less linearly (that is, the graph is almost a straight line), but then curves upward quite sharply from year 3 to year 6. This behavior can be modeled by using two different functions: one for the interval [0, 3] and another for the interval [3, 6] (see Figure 4).

A function specified by two or more different formulas like this is called a **piecewise-defined function**.

EXAMPLE 2 A Piecewise-Defined Function: Facebook Membership

The number $n(t)$ of Facebook members can be approximated by the following function of time t in years ($t = 0$ represents January 2004):

$$n(t) = \begin{cases} 4t & \text{if } 0 \le t \le 3 \\ 12 + 23(t - 3)^{2.7} & \text{if } 3 < t \le 6 \end{cases} \quad \text{million members.}$$

What was the approximate membership of Facebook in January 2005, January 2007, and June 2009? Sketch the graph of n by plotting several points.

Solution We evaluate the given function at the corresponding values of t:

Jan. 2005 ($t = 1$): $n(1) = 4(1) = 4$ Use the first formula because $0 \le t \le 3$.

Jan. 2007 ($t = 3$): $n(3) = 4(3) = 12$ Use the first formula because $0 \le t \le 3$.

June 2009 ($t = 5.5$): $n(5.5) = 12 + 23(5.5 - 3)^{2.7} \approx 285$.   Use the second formula because $3 < t \le 6$.

Thus, the number of Facebook members was approximately 4 million in January 2005, 12 million in January 2007, and 285 million in June 2009.

To sketch the graph of n we use a table of rounded values of $n(t)$ (some of which we have already calculated above), plot the points, and connect them to sketch the graph:

t	0	1	2	3	4	5	6
$n(t)$	0	4	8	12	35	161	459

First Formula   Second Formula

The graph (Figure 5) has the following features:

1. The first formula (the line) is used for $0 \le t \le 3$.

2. The second formula (ascending curve) is used for $3 < t \le 6$.

3. The domain is [0, 6], so the graph is cut off at $t = 0$ and $t = 6$.

4. The heavy solid dots at the ends indicate the endpoints of the domain.

EXAMPLE 3 **More Complicated Piecewise-Defined Functions**

Let f be the function specified by

$$f(x) = \begin{cases} -1 & \text{if } -4 \le x < -1 \\ x & \text{if } -1 \le x \le 1 \\ x^2 - 1 & \text{if } 1 < x \le 2 \end{cases}.$$

a. What is the domain of f? Find $f(-2)$, $f(-1)$, $f(0)$, $f(1)$, and $f(2)$.
b. Sketch the graph of f.

Solution

a. The domain of f is $[-4, 2]$, because $f(x)$ is specified only when $-4 \le x \le 2$.

$$f(-2) = -1 \qquad \text{We used the first formula because } -4 \le x < -1.$$
$$f(-1) = -1 \qquad \text{We used the second formula because } -1 \le x \le 1.$$
$$f(0) = 0 \qquad \text{We used the second formula because } -1 \le x \le 1.$$
$$f(1) = 1 \qquad \text{We used the second formula because } -1 \le x \le 1.$$
$$f(2) = 2^2 - 1 = 3 \qquad \text{We used the third formula because } 1 < x \le 2.$$

b. To sketch the graph by hand, we first sketch the three graphs $y = -1$, $y = x$, and $y = x^2 - 1$, and then use the appropriate portion of each (Figure 6).

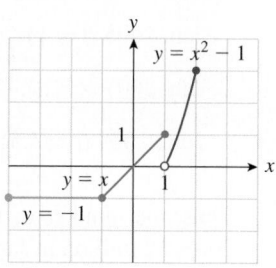

Figure 6

Note that solid dots indicate points on the graph, whereas the open dots indicate points not on the graph. For example, when $x = 1$, the inequalities in the formula tell us that we are to use the middle formula (x) rather than the bottom one ($x^2 - 1$). Thus, $f(1) = 1$, not 0, so we place a solid dot at $(1, 1)$ and an open dot at $(1, 0)$.

Vertical Line Test

Every point in the graph of a function has the form $(x, f(x))$ for some x in the domain of f. Because f assigns a *single* value $f(x)$ to each value of x in the domain, it follows that, in the graph of f, there should be only one y corresponding to any such value of x—namely, $y = f(x)$. In other words, *the graph of a function cannot contain two or more points with the same x-coordinate—that is, two or more points on the same vertical line.* On the other hand, a vertical line at a value of x not in the domain will not contain any points in the graph. This gives us the following rule.

Vertical-Line Test

For a graph to be the graph of a function, every vertical line must intersect the graph in *at most* one point.

Quick Examples

As illustrated below, only graph B passes the vertical line test, so only graph B is the graph of a function.

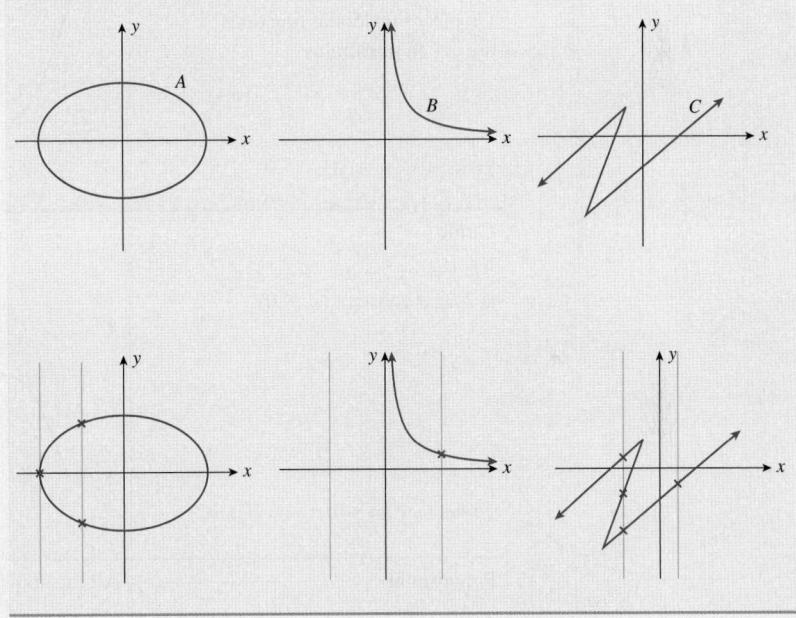

Table 1 lists some common types of functions that are often used to model real world situations.

Table 1 A Compendium of Functions and Their Graphs

Type of Function	*Examples*
Linear $f(x) = mx + b$ m, b constant Graphs of linear functions are straight lines. The quantity m is the **slope** of the line; the quantity b is the **y-intercept** of the line. [See Section 1.3.]	$y = x$ $\qquad$ $y = -2x + 2$
Technology formulas:	x $\qquad$ -2*x+2
Quadratic $f(x) = ax^2 + bx + c$ a, b, c constant ($a \neq 0$) Graphs of quadratic functions are called **parabolas**.	$y = x^2$ $\qquad$ $y = -2x^2 + 2x + 4$
Technology formulas:	x^2 $\qquad$ -2*x^2+2*x+4
Cubic $f(x) = ax^3 + bx^2 + cx + d$ a, b, c, d constant ($a \neq 0$)	$y = x^3$ $\qquad$ $y = -x^3 + 3x^2 + 1$
Technology formulas:	x^3 $\qquad$ -x^3+3*x^2+1
Polynomial $f(x) = ax^n + bx^{n-1} + \ldots + rx + s$ $a, b, \ldots, r, s$ constant (includes all of the above functions)	All the above, and $f(x) = x^6 - 2x^5 - 2x^4 + 4x^2$
Technology formula:	x^6-2x^5-2x^4+4x^2

Table 1 (*Continued*)

Type of Function	*Examples*	
Exponential $f(x) = Ab^x$ A, b constant $(b > 0$ and $b \neq 1)$ The y-coordinate is multiplied by b every time x increases by 1.	$y = 2^x$ y is doubled every time x increases by 1.	$y = 4(0.5)^x$ y is halved every time x increases by 1.
Technology formulas:	`2^x`	`4*0.5^x`
Rational $f(x) = \dfrac{P(x)}{Q(x)}$; $P(x)$ and $Q(x)$ polynomials The graph of $y = 1/x$ is a **hyperbola**. The domain excludes zero because $1/0$ is not defined.	$y = \dfrac{1}{x}$ 	$y = \dfrac{x}{x - 1}$
Technology formulas:	`1/x`	`x/(x-1)`
Absolute value For x positive or zero, the graph of $y = |x|$ is the same as that of $y = x$. For x negative or zero, it is the same as that of $y = -x$.	$y = |x|$ 	$y = |2x + 2|$
Technology formulas:	`abs(x)`	`abs(2*x+2)`
Square Root The domain of $y = \sqrt{x}$ must be restricted to the nonnegative numbers, because the square root of a negative number is not real. Its graph is the top half of a horizontally oriented parabola.	$y = \sqrt{x}$ 	$y = \sqrt{4x - 2}$
Technology formulas:	`x^0.5` or `√(x)`	`(4*x-2)^0.5` or `√(4*x-2)`

Go to the Website and follow the path

Online Text

→ New Functions from Old: Scaled and Shifted Functions

where you will find complete online interactive text, examples, and exercises on scaling and translating the graph of a function by changing the formula.

Functions and models other than linear ones are called **nonlinear**.

1.1 EXERCISES

▼ more advanced   ◆ challenging
▢ indicates exercises that should be solved using technology

In Exercises 1–4, evaluate or estimate each expression based on the following table. HINT *[See Quick Example 1 on page 42.]*

x	−3	−2	−1	0	1	2	3
f(x)	1	2	4	2	1	0.5	0.25

1. a. $f(0)$ **b.** $f(2)$ **2. a.** $f(-1)$ **b.** $f(1)$
3. a. $f(2) - f(-2)$ **b.** $f(-1)f(-2)$ **c.** $-2f(-1)$
4. a. $f(1) - f(-1)$ **b.** $f(1)f(-2)$ **c.** $3f(-2)$

In Exercises 5–8, use the graph of the function f to find approximations of the given values. HINT *[See Example 1.]*

5.
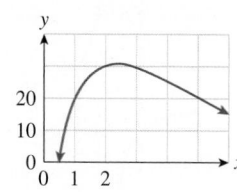
a. $f(1)$ **b.** $f(2)$
c. $f(3)$ **d.** $f(5)$
e. $f(3) - f(2)$

6.
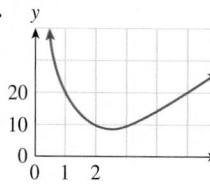
a. $f(1)$ **b.** $f(2)$
c. $f(3)$ **d.** $f(5)$
e. $f(3) - f(2)$

7.
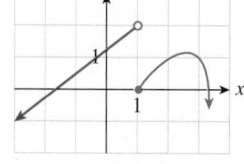
a. $f(-3)$ **b.** $f(0)$
c. $f(1)$ **d.** $f(2)$
e. $\dfrac{f(2) - f(1)}{2 - 1}$

8.
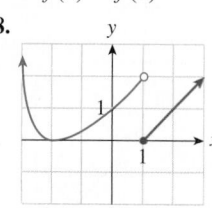
a. $f(-2)$ **b.** $f(0)$
c. $f(1)$ **d.** $f(3)$
e. $\dfrac{f(3) - f(1)}{3 - 1}$

In Exercises 9–12, say whether or not f(x) is defined for the given values of x. If it is defined, give its value. HINT *[See Quick Example 3 on page 43.]*

9. $f(x) = x - \dfrac{1}{x^2}$, with domain $(0, +\infty)$
  a. $x = 4$ **b.** $x = 0$ **c.** $x = -1$

10. $f(x) = \dfrac{2}{x} - x^2$, with domain $[2, +\infty)$
  a. $x = 4$ **b.** $x = 0$ **c.** $x = 1$

11. $f(x) = \sqrt{x + 10}$, with domain $[-10, 0)$
  a. $x = 0$ **b.** $x = 9$ **c.** $x = -10$

12. $f(x) = \sqrt{9 - x^2}$, with domain $(-3, 3)$
  a. $x = 0$ **b.** $x = 3$ **c.** $x = -3$

13. Given $f(x) = 4x - 3$, find **a.** $f(-1)$ **b.** $f(0)$
  c. $f(1)$ **d.** $f(y)$ **e.** $f(a + b)$ HINT *[See Example 1.]*

14. Given $f(x) = -3x + 4$, find
  a. $f(-1)$ **b.** $f(0)$ **c.** $f(1)$ **d.** $f(y)$ **e.** $f(a + b)$

15. Given $f(x) = x^2 + 2x + 3$, find
  a. $f(0)$ **b.** $f(1)$ **c.** $f(-1)$ **d.** $f(-3)$
  e. $f(a)$ **f.** $f(x + h)$ HINT *[See Example 1.]*

16. Given $g(x) = 2x^2 - x + 1$, find
  a. $g(0)$ **b.** $g(-1)$ **c.** $g(r)$ **d.** $g(x + h)$

17. Given $g(s) = s^2 + \dfrac{1}{s}$, find
  a. $g(1)$ **b.** $g(-1)$ **c.** $g(4)$ **d.** $g(x)$ **e.** $g(s + h)$
  f. $g(s + h) - g(s)$

18. Given $h(r) = \dfrac{1}{r + 4}$, find
  a. $h(0)$ **b.** $h(-3)$ **c.** $h(-5)$ **d.** $h(x^2)$
  e. $h(x^2 + 1)$ **f.** $h(x^2) + 1$

In Exercises 19–24, graph the given functions. Give the technology formula and use technology to check your graph. We suggest that you become familiar with these graphs in addition to those in Table 1. HINT *[See Quick Example 4 on page 44.]*

19. $f(x) = -x^3$ (domain $(-\infty, +\infty)$)
20. $f(x) = x^3$ (domain $[0, +\infty)$)
21. $f(x) = x^4$ (domain $(-\infty, +\infty)$)
22. $f(x) = \sqrt[3]{x}$ (domain $(-\infty, +\infty)$)
23. $f(x) = \dfrac{1}{x^2}$ $(x \neq 0)$ **24.** $f(x) = x + \dfrac{1}{x}$ $(x \neq 0)$

In Exercises 25 and 26, match the functions to the graphs. Using technology to draw the graphs is suggested, but not required.

25. ▢ **a.** $f(x) = x$ $(-1 \leq x \leq 1)$
   b. $f(x) = -x$ $(-1 \leq x \leq 1)$
   c. $f(x) = \sqrt{x}$ $(0 < x < 4)$
   d. $f(x) = x + \dfrac{1}{x} - 2$ $(0 < x < 4)$
   e. $f(x) = |x|$ $(-1 \leq x \leq 1)$
   f. $f(x) = x - 1$ $(-1 \leq x \leq 1)$

(I)
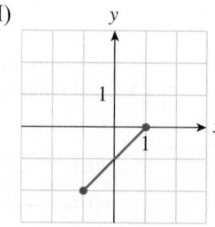

(II)
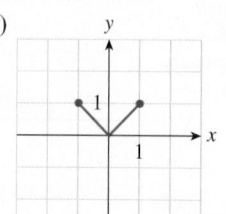

(III)

(IV)
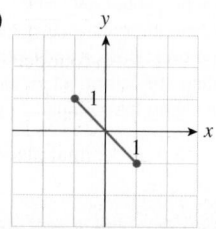

(V)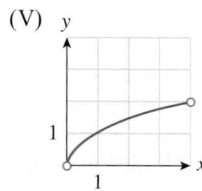

(VI)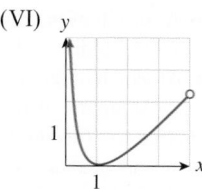

26. ⊤ **a.** $f(x) = -x + 4$ $(0 < x \le 4)$

b. $f(x) = 2 - |x|$ $(-2 < x \le 2)$

c. $f(x) = \sqrt{x+2}$ $(-2 < x \le 2)$

d. $f(x) = -x^2 + 2$ $(-2 < x \le 2)$

e. $f(x) = \dfrac{1}{x} - 1$ $(0 < x \le 4)$

f. $f(x) = x^2 - 1$ $(-2 < x \le 2)$

(I)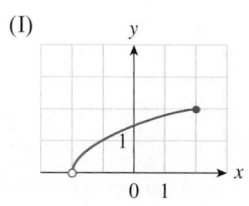

(II)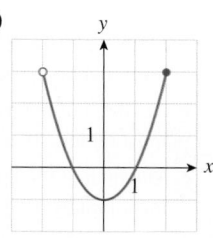

(III)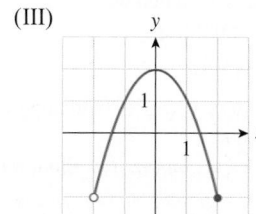

(IV)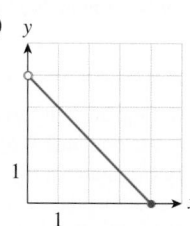

(V)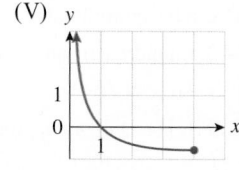

(VI)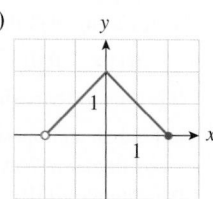

⊤ *In Exercises 27–30, first give the technology formula for the given function and then use technology to evaluate the function for the given values of x (when defined there).*

27. ⊤ $f(x) = 0.1x^2 - 4x + 5$; $x = 0, 1, \ldots, 10$

28. ⊤ $g(x) = 0.4x^2 - 6x - 0.1$; $x = -5, -4, \ldots, 4, 5$

29. ⊤ $h(x) = \dfrac{x^2 - 1}{x^2 + 1}$; $x = 0.5, 1.5, 2.5, \ldots, 10.5$ (Round all answers to four decimal places.)

30. ⊤ $r(x) = \dfrac{2x^2 + 1}{2x^2 - 1}$; $x = -1, 0, 1, \ldots, 9$ (Round all answers to four decimal places.)

In Exercises 31–36, sketch the graph of the given function, evaluate the given expressions, and then use technology to duplicate the graphs. Give the technology formula. HINT [See Example 2.]

31. $f(x) = \begin{cases} x & \text{if } -4 \le x < 0 \\ 2 & \text{if } 0 \le x \le 4 \end{cases}$

    a. $f(-1)$ **b.** $f(0)$ **c.** $f(1)$

32. $f(x) = \begin{cases} -1 & \text{if } -4 \le x \le 0 \\ x & \text{if } 0 < x \le 4 \end{cases}$

    a. $f(-1)$ **b.** $f(0)$ **c.** $f(1)$

33. $f(x) = \begin{cases} x^2 & \text{if } -2 < x \le 0 \\ 1/x & \text{if } 0 < x \le 4 \end{cases}$

    a. $f(-1)$ **b.** $f(0)$ **c.** $f(1)$

34. $f(x) = \begin{cases} -x^2 & \text{if } -2 < x \le 0 \\ \sqrt{x} & \text{if } 0 < x < 4 \end{cases}$

    a. $f(-1)$ **b.** $f(0)$ **c.** $f(1)$

35. $f(x) = \begin{cases} x & \text{if } -1 < x \le 0 \\ x + 1 & \text{if } 0 < x \le 2 \\ x & \text{if } 2 < x \le 4 \end{cases}$

    a. $f(0)$ **b.** $f(1)$ **c.** $f(2)$ **d.** $f(3)$ HINT [See Example 3.]

36. $f(x) = \begin{cases} -x & \text{if } -1 < x < 0 \\ x - 2 & \text{if } 0 \le x \le 2 \\ -x & \text{if } 2 < x \le 4 \end{cases}$

    a. $f(0)$ **b.** $f(1)$ **c.** $f(2)$ **d.** $f(3)$

In Exercises 37–40, find and simplify ***(a)*** $f(x + h) - f(x)$

(b) $\dfrac{f(x + h) - f(x)}{h}$

37. ▼ $f(x) = x^2$ **38.** ▼ $f(x) = 3x - 1$

39. ▼ $f(x) = 2 - x^2$ **40.** ▼ $f(x) = x^2 + x$

APPLICATIONS

41. *Petrochemical Sales: Mexico* The following table shows annual petrochemical sales in Mexico by Pemex, Mexico's national oil company, for 2005–2010 ($t = 0$ represents 2005):[5]

Year t (Year since 2005)	0	1	2	3	4	5
Petrochemical Sales s (Billion metric tons)	3.7	3.8	4.0	4.1	4.0	4.1

    a. Find $s(1)$, $s(4)$, and $s(5)$. Interpret your answers.

    b. What is the domain of s?

    c. Represent s graphically and use your graph to estimate $s(1.5)$. Interpret your answer. HINT [See Quick Example 1 on page 42.]

---

[5]2010 figure is a projection based on data through May 2010. Source: www.pemex.com (July 2010).

42. *Petrochemical Production: Mexico* The following table shows annual petrochemical production in Mexico by **Pemex**, Mexico's national oil company, for 2005–2010 ($t = 0$ represents 2005):[6]

Year t (Year since 2005)	0	1	2	3	4	5
Petrochemical Production p (Billion metric tons)	10.8	11.0	11.8	12.0	12.0	13.3

a. Find $p(0)$, $p(2)$, and $p(4)$. Interpret your answers.
b. What is the domain of p?
c. Represent p graphically and use your graph to estimate $p(2.5)$. Interpret your answer. HINT [See Quick Example 1 on page 42]

Housing Starts *Exercises 43–46 refer to the following graph, which shows the number $f(t)$ of housing starts in the U.S. each year from 2000 through 2010 ($t = 0$ represents 2000, and $f(t)$ is the number of housing starts in year t in thousands of units).*[7]

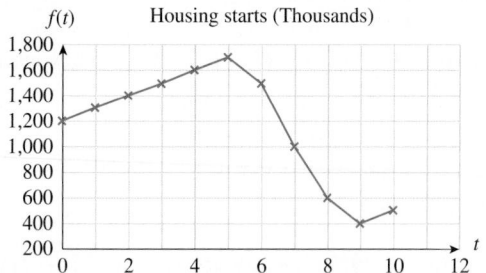
Housing starts (Thousands)

43. Estimate $f(4)$, $f(5)$, and $f(7.5)$. Interpret your answers.

44. Estimate $f(3)$, $f(6)$, and $f(8.5)$. Interpret your answers.

45. Which has the larger magnitude: $f(5) - f(0)$ or $f(8) - f(0)$? Interpret the answer.

46. Which has the larger magnitude: $f(5) - f(0)$ or $f(9) - f(7)$? Interpret the answer.

47. *Airline Net Income* The following graph shows the approximate annual after-tax net income $P(t)$, in millions of dollars, of **Continental Airlines** for 2005–2009 ($t = 0$ represents 2005):[8]

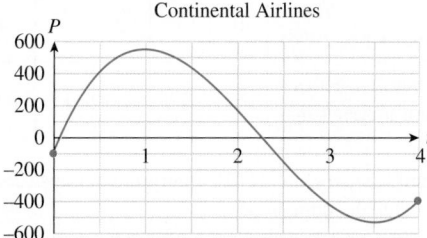
Continental Airlines

a. Estimate $P(0)$, $P(4)$, and $P(1.5)$ to the nearest 100. Interpret your answers.

b. At which of the following values of t is $P(t)$ *increasing* most rapidly: 0, 0.5, 1, 2.5, 3.5? Interpret the result.
c. At which of the following values of t is $P(t)$ *decreasing* most rapidly: 0, 0.5, 1, 2.5, 3.5? Interpret the result.

48. *Airline Net Income* The following graph shows the approximate annual after-tax net income $P(t)$, in millions of dollars, of **Delta Air Lines** for 2005–2009 ($t = 0$ represents 2005):[9]

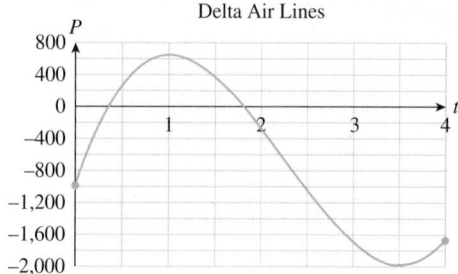
Delta Air Lines

a. Estimate $P(0)$, $P(1)$, and $P(3.5)$ to the nearest 100. Interpret your answers.

b. At which of the following values of t is $P(t)$ *increasing* most rapidly: 0, 0.5, 1, 2.5, 3.5? Interpret the result.

c. At which of the following values of t is $P(t)$ *decreasing* most rapidly: 0, 0.5, 1, 2.5, 3.5? Interpret the result.

49. *Trade with China* The value of U.S. trade with China from 1994 through 2004 can be approximated by

$$C(t) = 3t^2 - 7t + 50 \text{ billion dollars}$$

(t is time in years since 1994).[10]

a. Find an appropriate domain of C. Is $t \geq 0$ an appropriate domain? Why or why not?
b. Compute $C(10)$. What does the answer say about trade with China?

50. *Scientific Research* The number of research articles in *Physical Review* that were written by researchers in the United States from 1983 through 2003 can be approximated by

$$A(t) = -0.01\,t^2 + 0.24t + 3.4 \text{ hundred articles}$$

(t is time in years since 1983).[11]

a. Find an appropriate domain of A. Is $t \leq 20$ an appropriate domain? Why or why not?
b. Compute $A(10)$. What does the answer say about the number of research articles?

51. 🔢 ***Acquisition of Language*** The percentage $p(t)$ of children who can speak in at least single words by the age of t months can be approximated by the equation[12]

$$p(t) = 100\left(1 - \frac{12{,}200}{t^{4.48}}\right) \quad (t \geq 8.5).$$

[9]*Ibid.*

[10]Based on a regression by the authors. Source for data: U.S. Census Bureau/*New York Times*, September 23, 2004, p. C1.

[11]Based on a regression by the authors. Source for data: The American Physical Society/*New York Times*, May 3, 2003, p. A1.

[12]The model is the authors' and is based on data presented in the article *The Emergence of Intelligence* by William H. Calvin, *Scientific American*, October, 1994, pp. 101–107.

[6]*Ibid.*

[7]Sources for data: www.census.gov, www.forecast-chart.com

[8]"Net income" is an accounting term for profit (see Section 1.2). Model is the authors'. Source for data: Company reports.

a. Give a technology formula for p.

b. Graph p for $8.5 \le t \le 20$ and $0 \le p \le 100$.

c. Create a table of values of p for $t = 9, 10, \ldots, 20$ (rounding answers to one decimal place).

d. What percentage of children can speak in at least single words by the age of 12 months?

e. By what age are 90% or more children speaking in at least single words?

52. ▯ *Acquisition of Language* The percentage $p(t)$ of children who can speak in sentences of five or more words by the age of t months can be approximated by the equation[13]

$$p(t) = 100 \left(1 - \frac{5.27 \times 10^{17}}{t^{12}} \right) \quad (t \ge 30).$$

a. Give a technology formula for p.

b. Graph p for $30 \le t \le 45$ and $0 \le p \le 100$.

c. Create a table of values of p for $t = 30, 31, \ldots, 40$ (rounding answers to one decimal place).

d. What percentage of children can speak in sentences of five or more words by the age of 36 months?

e. By what age are 75% or more children speaking in sentences of five or more words?

53. ▼ *Processor Speeds* The processor speed, in megahertz (MHz), of Intel processors during the period 1980–2010 could be approximated by the following function of time t in years since the start of 1980:[14]

$$v(t) = \begin{cases} 8(1.22)^t & \text{if } 0 \le t < 16 \\ 400t - 6{,}200 & \text{if } 16 \le t < 25 \\ 3{,}800 & \text{if } 25 \le t \le 30 \end{cases}$$

a. Evaluate $v(10)$, $v(16)$, and $v(28)$ and interpret the results.

b. Write down a technology formula for v.

c. ▯ Use technology to sketch the graph of v and to generate a table of values for $v(t)$ with $t = 0, 2, \ldots, 30$. (Round values to two significant digits.)

d. When, to the nearest year, did processor speeds reach 3 gigahertz (1 gigahertz = 1,000 megahertz) according to the model?

54. ▼ *Processor Speeds* The processor speed, in megahertz (MHz), of Intel processors during the period 1970–2000 could be approximated by the following function of time t in years since the start of 1970:[15]

$$v(t) = \begin{cases} 0.12t^2 + 0.04t + 0.2 & \text{if } 0 \le t < 12 \\ 1.1(1.22)^t & \text{if } 12 \le t < 26 \\ 400t - 10{,}200 & \text{if } 26 \le t \le 30 \end{cases}$$

a. Evaluate $v(2)$, $v(12)$, and $v(28)$ and interpret the results.

b. Write down a technology formula for v.

c. ▯ Use technology to sketch the graph of v and to generate a table of values for $v(t)$ with $t = 0, 2, \ldots, 30$. (Round values to two significant digits.)

d. When, to the nearest year, did processor speeds reach 500 megahertz?

55. ▼ *Income Taxes* The U.S. Federal income tax is a function of taxable income. Write $T(x)$ for the tax owed on a taxable income of x dollars. For tax year 2010, the function T for a single taxpayer was specified as follows:

If your taxable income was			*of the amount*
Over...	*But not over...*	*Your tax is*	*over...*
$0	8,375	.... 10%	$0
8,375	34,000	$837.50 + 15%	$8,375
34,000	82,400	4,663.25 + 25%	$34,000
82,400	171,850	16,763.25 + 28%	$82,400
171,850	373,650	41,809.25 + 33%	$171,850
373,650	....	108,403.25 + 35%	$373,650

a. Represent T as a piecewise-defined function of income x. HINT [Each row of the table defines a formula with a condition.]

b. Use your function to compute the tax owed by a single taxpayer on a taxable income of $36,000.

56. ▼ *Income Taxes* Repeat Exercise 55 using the following information for tax year 2009.

If your taxable income was			*of the amount*
Over...	*But not over...*	*Your tax is*	*over...*
$0	8,350	.... 10%	$0
8,350	33,950	$835.00 + 15%	$8,350
33,950	82,250	4,675.00 + 25%	$33,950
82,250	171,550	16,750.00 + 28%	$82,250
171,550	372,950	41,754.00 + 33%	$171,550
372,950	....	108,216.00 + 35%	$372,950

COMMUNICATION AND REASONING EXERCISES

57. Complete the following sentence: If the market price m of gold varies with time t, then the independent variable is ___ and the dependent variable is ___.

58. Complete the following sentence: If weekly profit P is specified as a function of selling price s, then the independent variable is ___ and the dependent variable is ___.

59. Complete the following: The function notation for the equation $y = 4x^2 - 2$ is ___.

60. Complete the following: The equation notation for $C(t) = -0.34t^2 + 0.1t$ is ___.

---

[13] *Ibid.*

[14] Based on the fastest processors produced by Intel. Source for data: www.intel.com.

[15] *Ibid.*

61. True or false? Every graphically specified function can also be specified numerically. Explain.

62. True or false? Every algebraically specified function can also be specified graphically. Explain.

63. True or false? Every numerically specified function with domain [0, 10] can also be specified algebraically. Explain.

64. True or false? Every graphically specified function can also be specified algebraically. Explain.

65. ▼ True or false? Every function can be specified numerically. Explain.

66. ▼ Which supplies more information about a situation: a numerical model or an algebraic model?

67. ▼ Why is the following assertion false? "If $f(x) = x^2 - 1$, then $f(x + h) = x^2 + h - 1$."

68. ▼ Why is the following assertion false? "If $f(2) = 2$ and $f(4) = 4$, then $f(3) = 3$."

69. How do the graphs of two functions differ if they are specified by the same formula but have different domains?

70. How do the graphs of two functions f and g differ if $g(x) = f(x) + 10$? (Try an example.)

71. ▼ How do the graphs of two functions f and g differ if $g(x) = f(x - 5)$? (Try an example.)

72. ▼ How do the graphs of two functions f and g differ if $g(x) = f(-x)$? (Try an example.)

1.2 Functions and Models

The functions we used in Examples 1 and 2 in Section 1.1 are **mathematical models** of real-life situations, because they model, or represent, situations in mathematical terms.

Mathematical Modeling

To mathematically model a situation means to represent it in mathematical terms. The particular representation used is called a **mathematical model** of the situation. Mathematical models do not always represent a situation perfectly or completely. Some (like Example 1 of Section 1.1) represent a situation only approximately, whereas others represent only some aspects of the situation.

Quick Examples

1. The temperature is now 10°F and increasing by 20° per hour.

   Model: $T(t) = 10 + 20t$ (t = time in hours, T = temperature)

2. I invest $1,000 at 5% interest compounded quarterly. Find the value of the investment after t years.

   Model: $A(t) = 1,000 \left(1 + \dfrac{0.05}{4}\right)^{4t}$ (This is the compound interest formula we will study in Example 6.)

3. I am fencing a rectangular area whose perimeter is 100 ft. Find the area as a function of the width x.

   Model: Take y to be the length, so the perimeter is
   $$100 = x + y + x + y = 2(x + y).$$
   This gives
   $$x + y = 50.$$
   Thus the length is $y = 50 - x$, and the area is
   $$A = xy = x(50 - x).$$

4. You work 8 hours a day Monday through Friday, 5 hours on Saturday, and have Sunday off. Model the number of hours you work as a function of the day of the week n, with $n = 1$ being Sunday.
Model: Take $f(n)$ to be the number of hours you work on the nth day of the week, so

$$f(n) = \begin{cases} 0 & \text{if } n = 1 \\ 8 & \text{if } 2 \le n \le 6 \\ 5 & \text{if } n = 7 \end{cases}.$$

Note that the domain of f is $\{1, 2, 3, 4, 5, 6, 7\}$—a discrete set rather than a continuous interval of the real line.

5. The function

$$f(x) = 4x^2 + 16x + 2 \text{ million iPods sold } (x = \text{years since 2003})$$

in Example 1 of Section 1.1 is a model of iPod sales.

6. The function

$$n(t) = \begin{cases} 4t & \text{if } 0 \le t \le 3 \\ 12 + 23(t-3)^{2.7} & \text{if } 3 < t \le 6 \end{cases} \text{ million members}$$

($t = $ years since January 2004) in Example 2 of Section 1.1 is a model of Facebook membership.

Types of Models

Quick Examples 1–4 are **analytical models**, obtained by analyzing the situation being modeled, whereas Quick Examples 5 and 6 are **curve-fitting models**, obtained by finding mathematical formulas that approximate observed data. All the models except for Quick Example 4 are **continuous models**, defined by functions whose domains are intervals of the real line, whereas Quick Example 4 is a **discrete model** as its domain is a discrete set, as mentioned above. Discrete models are used extensively in probability and statistics.

Cost, Revenue, and Profit Models

EXAMPLE 1 Modeling Cost: Cost Function

As of August 2010, Yellow Cab Chicago's rates amounted to $2.05 on entering the cab plus $1.80 for each mile.[16]

a. Find the cost C of an x-mile trip.

b. Use your answer to calculate the cost of a 40-mile trip.

c. What is the cost of the second mile? What is the cost of the tenth mile?

d. Graph C as a function of x.

---

[16]According to their Web site at www.yellowcabchicago.com.

Solution

a. We are being asked to find how the cost C depends on the length x of the trip, or to find C as a function of x. Here is the cost in a few cases:

Cost of a 1-mile trip: $C = 1.80(1) + 2.05 = 3.85$ 1 mile at \$1.80 per mile plus \$2.05

Cost of a 2-mile trip: $C = 1.80(2) + 2.05 = 5.65$ 2 miles at \$1.80 per mile plus \$2.05

Cost of a 3-mile trip: $C = 1.80(3) + 2.05 = 7.45$ 3 miles at \$1.80 per mile plus \$2.05

Do you see the pattern? The cost of an x-mile trip is given by the linear function

$$C(x) = 1.80x + 2.05.$$

Notice that the cost function is a sum of two terms: The **variable cost** $1.80x$, which depends on x, and the **fixed cost** 2.05, which is independent of x:

Cost = Variable Cost + Fixed Cost.

The quantity 1.80 by itself is the incremental cost per mile; you might recognize it as the *slope* of the given linear function. In this context we call 1.80 the **marginal cost**. You might recognize the fixed cost 2.05 as the C-*intercept* of the given linear function.

b. We can use the formula for the cost function to calculate the cost of a 40-mile trip as

$$C(40) = 1.80(40) + 2.05 = \$74.05.$$

c. To calculate the cost of the second mile, we *could* proceed as follows:

Find the cost of a 1-mile trip: $C(1) = 1.80(1) + 2.05 = \$3.85$.

Find the cost of a 2-mile trip: $C(2) = 1.80(2) + 2.05 = \$5.65$.

Therefore, the cost of the second mile is $\$5.65 - \$3.85 = \$1.80$.

But notice that this is just the marginal cost. In fact, the marginal cost is the cost of each additional mile, so we could have done this more simply:

Cost of second mile = Cost of tenth mile = Marginal cost = \$1.80.

d. Figure 7 shows the graph of the cost function, which we can interpret as a *cost vs. miles* graph. The fixed cost is the starting height on the left, while the marginal cost is the slope of the line: It rises 1.80 units per unit of x. (See Section 1.3 for a discussion of properties of straight lines.)

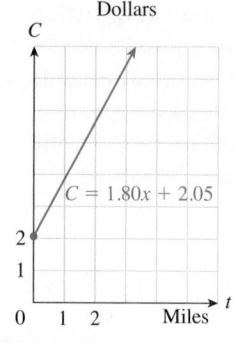

Figure 7

➡ **Before we go on...** The cost function in Example 1 is an example of an *analytical model:* We derived the form of the cost function from a knowledge of the cost per mile and the fixed cost.

As we discussed in Section 1.1, we can specify the cost function in Example 1 using equation notation:

$$C = 1.80x + 2.05.$$ Equation notation

Here, the independent variable is x, and the dependent variable is C. (This is the notation we have used in Figure 7. Remember that we will often switch between function and equation notation when it is convenient to do so.) ■

Here is a summary of some terms we used in Example 1, along with an introduction to some new terms:

Cost, Revenue, and Profit Functions

A **cost function** specifies the cost C as a function of the number of items x. Thus, $C(x)$ is the cost of x items, and has the form

$$\text{Cost} = \text{Variable cost} + \text{Fixed cost}$$

where the variable cost is a function of x and the fixed cost is a constant. A cost function of the form

$$C(x) = mx + b$$

is called a **linear cost function**; the variable cost is mx and the fixed cost is b. The slope m, the **marginal cost**, measures the incremental cost per item.

The **revenue**, or **net sales**, resulting from one or more business transactions is the total income received. If $R(x)$ is the revenue from selling x items at a price of m each, then R is the linear function $R(x) = mx$ and the selling price m can also be called the **marginal revenue**.

The **profit**, or **net income**, on the other hand, is what remains of the revenue when costs are subtracted. If the profit depends linearly on the number of items, the slope m is called the **marginal profit**. Profit, revenue, and cost are related by the following formula.

$$\text{Profit} = \text{Revenue} - \text{Cost}$$
$$P(x) = R(x) - C(x).\text{*}$$

* We say that the profit function P is the **difference** between the revenue and cost functions, and express this fact as a formula about functions: $P = R - C$. (We will discuss this further when we talk about the algebra of functions at the end of this section.)

If the profit is negative, say $-\$500$, we refer to a **loss** (of $\$500$ in this case). To **break even** means to make neither a profit nor a loss. Thus, breakeven occurs when $P = 0$, or

$$R(x) = C(x). \qquad \text{Breakeven}$$

The **break-even point** is the number of items x at which breakeven occurs.

Quick Example

If the daily cost (including operating costs) of manufacturing x T-shirts is $C(x) = 8x + 100$, and the revenue obtained by selling x T-shirts is $R(x) = 10x$, then the daily profit resulting from the manufacture and sale of x T-shirts is

$$P(x) = R(x) - C(x) = 10x - (8x + 100) = 2x - 100.$$

Breakeven occurs when $P(x) = 0$, or $x = 50$.

EXAMPLE 2   Cost, Revenue, and Profit

The annual operating cost of *YSport* Fitness gym is estimated to be

$$C(x) = 100,000 + 160x - 0.2x^2 \text{ dollars} \qquad (0 \le x \le 400),$$

where x is the number of members. Annual revenue from membership averages $800 per member. What is the variable cost? What is the fixed cost? What is the profit function? How many members must *YSport* have to make a profit? What will happen if it has fewer members? If it has more?

Solution The variable cost is the part of the cost function that depends on x:

$$\text{Variable cost} = 160x - 0.2x^2.$$

The fixed cost is the constant term:

$$\text{Fixed cost} = 100{,}000.$$

The annual revenue *YSport* obtains from a single member is $800. So, if it has x members, it earns an annual revenue of

$$R(x) = 800x.$$

For the profit, we use the formula

$$
\begin{aligned}
P(x) &= R(x) - C(x) &&\text{Formula for profit}\\
&= 800x - (100{,}000 + 160x - 0.2x^2) &&\text{Substitute } R(x) \text{ and } C(x).\\
&= -100{,}000 + 640x + 0.2x^2.
\end{aligned}
$$

To make a profit, *YSport* needs to do better than break even, so let us find the break-even point: the value of x such that $P(x) = 0$. All we have to do is set $P(x) = 0$ and solve for x:

$$-100{,}000 + 640x + 0.2x^2 = 0.$$

Notice that we have a quadratic equation $ax^2 + bx + c = 0$ with $a = 0.2$, $b = 640$, and $c = -100{,}000$. Its solution is given by the quadratic formula:

$$
\begin{aligned}
x &= \frac{-b \pm \sqrt{b^2 - 4ac}}{2a} = \frac{-640 \pm \sqrt{640^2 + 4(0.2)(100{,}000)}}{2(0.2)}\\
&\approx \frac{-640 \pm 699.71}{2(0.2)}\\
&\approx 149.3 \text{ or } -3{,}349.3.
\end{aligned}
$$

using Technology

Excel has a feature called "Goal Seek," which can be used to find the point of intersection of the cost and revenue graphs numerically rather than graphically. See the downloadable Excel tutorial for this section at the Website.

We reject the negative solution (as the domain is $[0, 400]$) and conclude that $x \approx 149.3$ members. To make a profit, should *YSport* have 149 members or 150 members? To decide, take a look at Figure 8, which shows two graphs: On the left we see the graph of revenue and cost, and on the right we see the graph of the profit function.

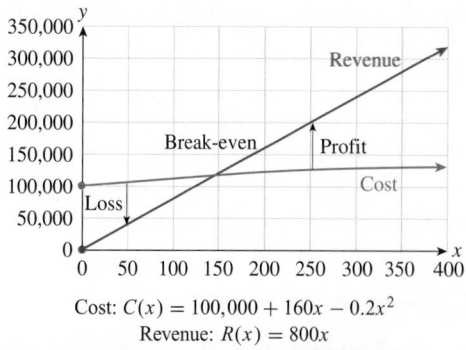

Cost: $C(x) = 100{,}000 + 160x - 0.2x^2$
Revenue: $R(x) = 800x$
Breakeven occurs at the point of intersection of the graphs of revenue and cost.

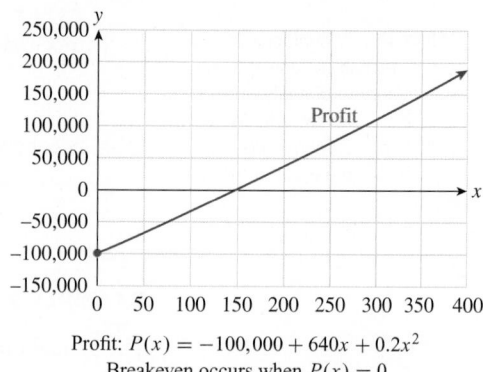

Profit: $P(x) = -100{,}000 + 640x + 0.2x^2$
Breakeven occurs when $P(x) = 0$

Figure 8

For values of x less than the break-even point of 149.3, $P(x)$ is negative, so the company will have a loss. For values of x greater than the break-even point, $P(x)$ is positive, so the company will make a profit. Thus, *YSport Fitness* needs at least 150 members to make a profit. (Note that we rounded 149.3 up to 150 in this case.)

Demand and Supply Models

The demand for a commodity usually goes down as its price goes up. It is traditional to use the letter q for the (quantity of) demand, as measured, for example, in sales. Consider the following example.

EXAMPLE 3 Demand: Private Schools

The demand for private schools in Michigan depends on the tuition cost and can be approximated by

$$q = 77.8p^{-0.11} \text{ thousand students} \qquad (200 \leq p \leq 2{,}200), \qquad \text{Demand curve}$$

where p is the net tuition cost in dollars.[17]

a. Use technology to plot the demand function.

b. What is the effect on demand if the tuition cost is increased from $1,000 to $1,500?

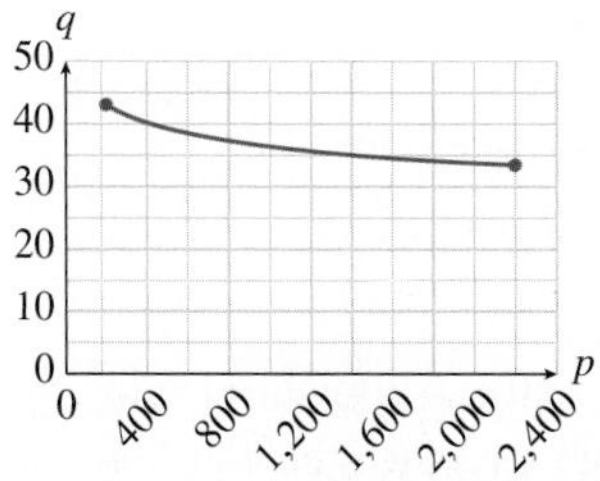

Solution

Technology Formula:
`y = 77.8x^(-0.11)`

Figure 9

a. The demand function is given by $q(p) = 77.8p^{-0.11}$. Its graph is known as a **demand curve** (Figure 9).

b. The demand at tuition costs of $1,000 and $1,500 is

$$q(1{,}000) = 77.8(1{,}000)^{-0.11} \approx 36.4 \text{ thousand students}$$
$$q(1{,}500) = 77.8(1{,}500)^{-0.11} \approx 34.8 \text{ thousand students}.$$

The change in demand is therefore
$$q(1{,}500) - q(1{,}000) \approx 34.8 - 36.4 = -1.6 \text{ thousand students}.$$

We have seen that a demand function gives the number of items consumers are willing to buy at a given price, and a higher price generally results in a lower demand. However, as the price rises, suppliers will be more inclined to produce these items (as opposed to spending their time and money on other products), so supply will generally rise. A **supply function** gives q, the number of items suppliers are willing to make available for sale*, as a function of p, the price per item.

✳ Although a bit confusing at first, it is traditional to use the same letter q for the quantity of supply and the quantity of demand, particularly when we want to compare them, as in the next example.

Demand, Supply, and Equilibrium Price

A **demand equation** or **demand function** expresses demand q (the number of items demanded) as a function of the unit price p (the price per item). A **supply equation** or **supply function** expresses supply q (the number of items a supplier is willing to make available) as a function of the unit price p (the price per item). It is usually the case that demand decreases and supply increases as the unit price increases.

[17]The tuition cost is net cost: tuition minus tax credit. The model is based on data in "The Universal Tuition Tax Credit: A Proposal to Advance Personal Choice in Education," Patrick L. Anderson, Richard McLellan, J.D., Joseph P. Overton, J.D., Gary Wolfram, Ph.D., Mackinac Center for Public Policy, www.mackinac.org/

Demand and supply are said to be in **equilibrium** when demand equals supply. The corresponding values of p and q are called the **equilibrium price** and **equilibrium demand**. To find the equilibrium price, determine the unit price p where the demand and supply curves cross (sometimes we can determine this value analytically by setting demand equal to supply and solving for p). To find the equilibrium demand, evaluate the demand (or supply) function at the equilibrium price.

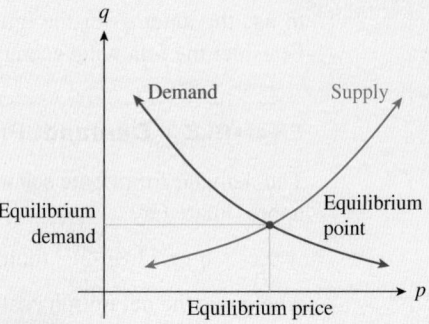

Quick Example

If the demand for your exclusive T-shirts is $q = -20p + 800$ shirts sold per day and the supply is $q = 10p - 100$ shirts per day, then the equilibrium point is obtained when demand = supply:

$$-20p + 800 = 10p - 100$$
$$30p = 900, \text{ giving } p = \$30.$$

The equilibrium price is therefore \$30 and the equilibrium demand is $q = -20(30) + 800 = 200$ shirts per day. What happens at prices other than the equilibrium price is discussed in Example 4.

Note In economics it is customary to plot the independent variable (price) on the vertical axis and the dependent variable (demand or supply) on the horizontal axis, but in this book we follow the usual mathematical convention for all graphs and plot the independent variable on the horizontal axis.

EXAMPLE 4 Demand, Supply, and Equilibrium Price

Continuing with Example 3, suppose that private school institutions are willing to create private schools to accommodate

$$q = 30.4 + 0.006p \text{ thousand students} \qquad (200 \le p \le 2{,}200) \qquad \text{Supply curve}$$

who pay a net tuition of p dollars.

a. Graph the demand curve of Example 3 and the supply curve given here on the same set of axes. Use your graph to estimate, to the nearest \$100, the tuition at which the demand equals the supply. Approximately how many students will be accommodated at that price, known as the **equilibrium price**?

b. What happens if the tuition is higher than the equilibrium price? What happens if it is lower?

c. Estimate the shortage or surplus of openings at private schools if tuition is set at \$1,200.

Solution

a. Figure 10 shows the graphs of demand $q = 77.8p^{-0.11}$ and supply $q = 30.4 + 0.006p$. (See the margin note for a brief description of how to plot them.)

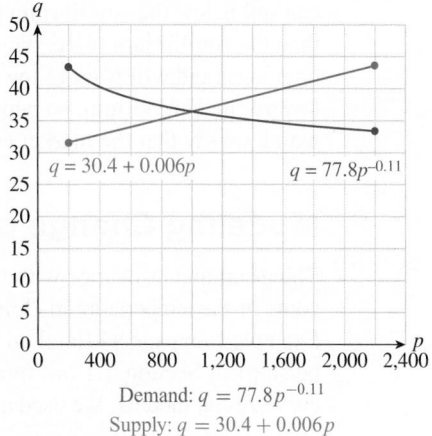

Demand: $q = 77.8p^{-0.11}$

Supply: $q = 30.4 + 0.006p$

Figure 10

The lines cross close to $p = \$1,000$, so we conclude that demand = supply when $p \approx \$1,000$ (to the nearest $100). This is the (approximate) equilibrium tuition price. At that price, we can estimate the demand or supply at around

Demand: $q = 77.8(1,000)^{-0.11} \approx 36.4$

Supply: $q = 30.4 + 0.006(1,000) = 36.4$ Demand = Supply at equilibrium

or 36,400 students.

b. Take a look at Figure 11, which shows what happens if schools charge more or less than the equilibrium price.

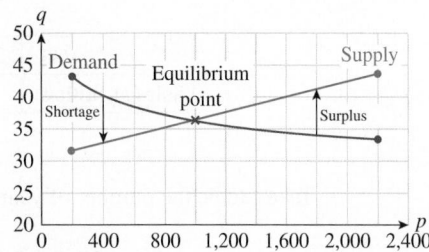

Figure 11

If tuition is, say, $1,800, then the supply will be larger than demand and there will be a surplus of available openings at private schools. Similarly, if tuition is less—say $400—then the supply will be less than the demand, and there will be a shortage of available openings.

c. The discussion in part (b) shows that if tuition is set at $1,200 there will be a surplus of available openings. To estimate that number, we calculate the projected demand and supply when $p = \$1,200$:

Demand: $q = 77.8(1,200)^{-0.11} \approx 35.7$ thousand seats

Supply: $q = 30.4 + 0.006(1,200) = 37.6$ thousand seats

Surplus = Supply − Demand $\approx 37.6 − 35.7 = 1.9$ thousand seats.

So, there would be a surplus of around 1,900 available seats.

 **using** Technology

See the Technology Guides at the end of the chapter for detailed instructions on how to obtain the table of values and graph in Example 4 using a TI-83/84 Plus or Excel. Here is an outline:

TI-83/84 Plus
Graphs:
$Y_1 = 77.8 \ast X^{\wedge}(-0.11)$
$Y_2 = 30.4 + 0.006 \ast X$
2ND TABLE Graph:
Xmin = 200, Xmax = 2200;
ZOOM 0 [More details on page 115.]

Spreadsheet
Headings p, Demand, Supply in A1–C1; p-values 200, 300, ..., 2200 in A2-A22.
$=77.8 \ast A2^{\wedge}(-0.11)$ in B2
$=30.4 + 0.006 \ast A2$ in C2
Copy down through C22.
Highlight A1–C22; insert Scatter chart. [More details on page 120.]

Website
www.WanerMath.com
Go to the Function Evaluator and Grapher under Online Utilities, and enter
$77.8 \ast x^{\wedge}(-0.11)$ for y_1 and $30.4 + 0.006 \ast x$ for y_2.
Graph: Set Xmin = 200, Xmax = 2200, and press "Plot Graphs".

➡ **Before we go on...** We just saw in Example 4 that if tuition is less than the equilibrium price there will be a shortage. If schools were to raise their tuition toward the equilibrium, they would create and fill more openings and increase revenue, because it is the supply equation—and not the demand equation—that determines what one can sell below the equilibrium price. On the other hand, if they were to charge more than the equilibrium price, they will be left with a possibly costly surplus of unused openings (and will want to lower tuition to reduce the surplus). Prices tend to move toward the equilibrium, so supply tends to equal demand. When supply equals demand, we say that the market **clears**. ∎

Modeling Change over Time

Things around us change with time. Thus, there are many quantities, such as your income or the temperature in Honolulu, that are natural to think of as functions of time. Example 1 on page 44 (on iPod sales) and Example 2 on page 47 (on Facebook membership) in Section 1.1 are models of change over time. Both of those models are curve-fitting models: We used algebraic functions to approximate observed data.

Note We usually use the independent variable t to denote time (in seconds, hours, days, years, etc.). If a quantity q changes with time, then we can regard q as a function of t. ∎

In the next example we are asked to select from among several curve-fitting models for given data.

ⓘ EXAMPLE 5 Model Selection: Sales

The following table shows annual sales, in billions of dollars, by Nike from 2005 through 2010:[18]

Year	2005	2006	2007	2008	2009	2010
Sales ($ billion)	13.5	15	16.5	18.5	19	19

Take t to be the number of years since 2005, and consider the following four models:

(1) $s(t) = 14 + 1.2t$ Linear model

(2) $s(t) = 13 + 2.2t - 0.2t^2$ Quadratic model

(3) $s(t) = 14(1.07^t)$ Exponential model

(4) $s(t) = \dfrac{19.5}{1 + 0.48(1.8^{-t})}$ Logistic model

a. Which models fit the data significantly better than the rest?

b. Of the models you selected in part (a), which gives the most reasonable prediction for 2013?

---

[18]Figures are rounded. Source: http://invest.nike.com.

Solution

a. The following table shows the original data together with the values, rounded to the nearest 0.5, for all four models:

	t	0	1	2	3	4	5
Sales ($billion)		13.5	15	16.5	18.5	19	19
Linear: $s(t) = 14 + 1.2t$ Technology: `14+1.2*x`		14	15	16.5	17.5	19	20
Quadratic: $s(t) = 13 + 2.2t - 0.2t^2$ Technology: `13+2.2*x-0.2*x^2`		13	15	16.5	18	18.5	19
Exponential: $s(t) = 14(1.07^t)$ Technology: `14*1.07^x`		14	15	16	17	18.5	19.5
Logistic: $s(t) = \dfrac{19.5}{1+0.48(1.8^{-t})}$ Technology: `19.5/(1+0.48*1.8^(-x))`		13	15.5	17	18	18.5	19

Notice that all the models give values that seem reasonably close to the actual sales values. However, the quadratic and logistic curves seem to model their behavior more accurately than the others (see Figure 12).

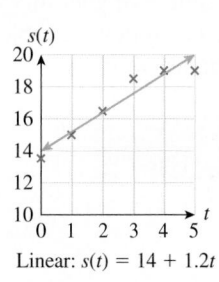
Linear: $s(t) = 14 + 1.2t$

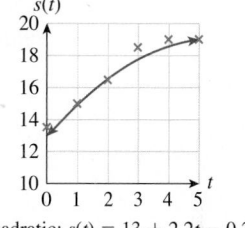
Quadratic: $s(t) = 13 + 2.2t - 0.2t^2$

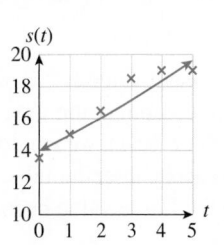
Exponential: $s(t) = 14(1.07^t)$

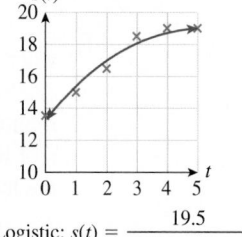
Logistic: $s(t) = \dfrac{19.5}{1 + 0.48(1.8^{-t})}$

Figure 12

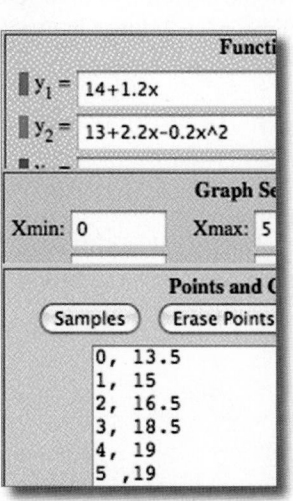

We therefore conclude that the quadratic and logistic models fit the data significantly better than the others.

b. Although the quadratic and logistic models both appear to fit the data well, they do not both extrapolate to give reasonable predictions for 2013:

Quadratic Model: $s(8) = 13 + 2.2(8) - 0.2(8)^2 = 17.8$

Logistic Model: $s(8) = \dfrac{19.5}{1 + 0.48(1.8^{-8})} \approx 19.4$.

Notice that the quadratic model predicts a significant *decline* in sales whereas the logistic model predicts a more reasonable modest increase. This discrepancy can be seen quite dramatically in Figure 13.

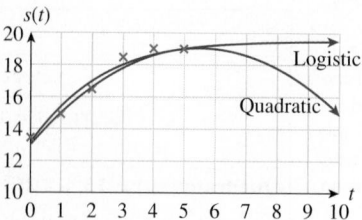

Figure 13

We now derive an analytical model of change over time based on the idea of **compound interest**. Suppose you invest \$500 (the **present value**) in an investment account with an annual yield of 15%, and the interest is reinvested at the end of every year (we say that the interest is **compounded** or **reinvested** once a year). Let t represent the number of years since you made the initial \$500 investment. Each year, the investment is worth 115% (or 1.15 times) of its value the previous year. The **future value** A of your investment changes over time t, so we think of A as a function of t. The following table illustrates how we can calculate the future value for several values of t:

t	0	1	2	3
FutureValue $A(t)$	500	575	661.25	760.44
A	$500(1.15)$	$500(1.15)^2$	$500(1.15)^3$

$$\times 1.15 \qquad \times 1.15 \qquad \times 1.15$$

Thus, $A(t) = 500(1.15)^t$. A traditional way to write this formula is

$$A(t) = P(1+r)^t,$$

where P is the present value ($P = 500$) and r is the annual interest rate ($r = 0.15$).

If, instead of compounding the interest once a year, we compound it every three months (four times a year), we would earn one quarter of the interest ($r/4$ of the current investment) every three months. Because this would happen $4t$ times in t years, the formula for the future value becomes

$$A(t) = P\left(1 + \frac{r}{4}\right)^{4t}.$$

Compound Interest

If an amount (**present value**) P is invested for t years at an annual rate of r, and if the interest is compounded (reinvested) m times per year, then the **future value** A is

$$A(t) = P\left(1 + \frac{r}{m}\right)^{mt}.$$

A special case is **interest compounded once a year**:

$$A(t) = P(1+r)^t.$$

> ### Quick Example
>
> If \$2,000 is invested for two and a half years in a mutual fund with an annual yield of 12.6% and the earnings are reinvested each month, then $P = 2,000, r = 0.126, m = 12$, and $t = 2.5$, which gives
>
> $$A(2.5) = 2,000\left(1 + \frac{0.126}{12}\right)^{12\times2.5}$$ `2000*(1+0.126/12)^(12*2.5)`
>
> $$= 2,000(1.0105)^{30} = \$2,736.02.$$

EXAMPLE 6 Compound Interest: Investments

Consider the scenario in the preceding Quick Example: You invest \$2,000 in a mutual fund with an annual yield of 12.6% and the interest is reinvested each month.

a. Find the associated exponential model.

b. ⊞ Use a table of values to estimate the year during which the value of your investment reaches \$5,000.

c. Use a graph to confirm your answer in part (b).

Solution

a. Apply the formula

$$A(t) = P\left(1 + \frac{r}{m}\right)^{mt}$$

with $P = 2,000, r = 0.126$, and $m = 12$. We get

$$A(t) = 2,000\left(1 + \frac{0.126}{12}\right)^{12t}$$

$$= 2,000(1.0105)^{12t}.$$ `2000*(1+0.126/12)^(12*t)`

This is the exponential model. (What would happen if we left out the last set of parentheses in the technology formula?)

b. We need to find the value of t for which $A(t) = \$5,000$, so we need to solve the equation

$$5,000 = 2,000(1.0105)^{12t}.$$

In Section 9.3 we will learn how to use logarithms to do this algebraically, but we can answer the question now using a graphing calculator, a spreadsheet, or the Function Evaluator and Grapher utility at the Website. Just enter the model and compute the balance at the end of several years. Here are examples of tables obtained using three forms of technology:

 using Technology

TI-83/84 Plus
Y₁=2000(1+0.126/12)^
(12X)
[2ND] [TABLE]

Spreadsheet
Headings t and A in A1–B1;
t-values 0–11 in A2–A13.
=2000*(1+0.126/12)^
(12*A2)
in B2; copy down through B13.

 Website
www.WanerMath.com
Go to the Function Evaluator
and Grapher under Online
Utilities, and enter
2000(1+0.126/12)^(12x)
for y_1. Scroll down to the
Evaluator, enter the values
0–11 under x-values and press
"Evaluate."

X	Y₁
5	3742.9
6	4242.7
7	4809.3
8	5451.5
9	6179.5
10	7004.7
11	7940.1

Y₁=5451.50618802

TI-83/84 Plus

	A	B
1	t	A
2	0	$ 2,000.00
3	1	$ 2,267.07
4	2	$ 2,569.81
5	3	$ 2,912.98
6	4	$ 3,301.97
7	5	$ 3,742.91
8	6	$ 4,242.72
9	7	$ 4,809.29
10	8	$ 5,451.51
11	9	$ 6,179.49

Excel

x-Values	y₁-Values
3	2912.98
4	3301.97
5	3742.91
6	4242.72
7	4809.29
8	5451.51

Website

Because the balance first exceeds $5,000 at $t = 8$ (the end of year 8), your investment has reached $5,000 during year 8.

c. Figure 14 shows the graph of $A(t) = 2,000(1.0105)^{12t}$ together with the horizontal line $y = 5,000$. The graphs cross between $t = 7$ and $t = 8$, confirming that year 8 is the first year during which the value of the investment reaches $5,000.

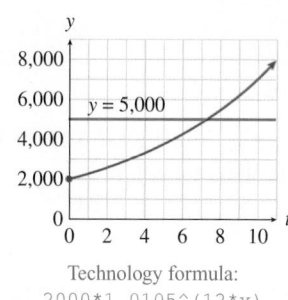

Technology formula:
`2000*1.0105^(12*x)`

Figure 14

The compound interest examples we saw above are instances of **exponential growth:** a quantity whose magnitude is an increasing exponential function of time. The decay of unstable radioactive isotopes provides instances of **exponential decay:** a quantity whose magnitude is a *decreasing* exponential function of time. For example, carbon 14, an unstable isotope of carbon, decays exponentially to nitrogen. Because carbon 14 decay is extremely slow, it has important applications in the dating of fossils.

EXAMPLE 7 Exponential Decay: Carbon Dating

The amount of carbon 14 remaining in a sample that originally contained A grams is approximately

$$C(t) = A(0.999879)^t,$$

where t is time in years.

a. What percentage of the original amount remains after one year? After two years?

b. Graph the function C for a sample originally containing 50 g of carbon 14, and use your graph to estimate how long, to the nearest 1,000 years, it takes for half the original carbon 14 to decay.

c. A fossilized plant unearthed in an archaeological dig contains 0.50 g of carbon 14 and is known to be 50,000 years old. How much carbon 14 did the plant originally contain?

Solution

Notice that the given model is exponential as it has the form $f(t) = Ab^t$. (See page 51.)

a. At the start of the first year, $t = 0$, so there are

$$C(0) = A(0.999879)^0 = A \text{ grams.}$$

At the end of the first year, $t = 1$, so there are

$$C(1) = A(0.999879)^1 = 0.999879A \text{ grams;}$$

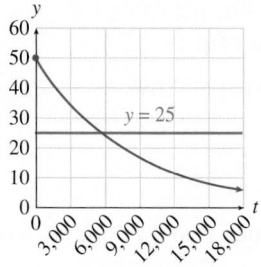

Figure 15

that is, 99.9879% of the original amount remains. After the second year, the amount remaining is

$$C(2) = A(0.999879)^2 \approx 0.999758A \text{ grams},$$

or about 99.9758% of the original sample.

b. For a sample originally containing 50 g of carbon 14, $A = 50$, so $C(t) = 50(0.999879)^t$. Its graph is shown in Figure 15. We have also plotted the line $y = 25$ on the same graph. The graphs intersect at the point where the original sample has decayed to 25 g: about $t = 6{,}000$ years.

c. We are given the following information: $C = 0.50$, $A =$ the unknown, and $t = 50{,}000$. Substituting gives

$$0.50 = A(0.999879)^{50{,}000}.$$

Solving for A gives

$$A = \frac{0.5}{0.999879^{50{,}000}} \approx 212 \text{ grams}.$$

Thus, the plant originally contained 212 g of carbon 14.

➡ **Before we go on...**

The formula we used for A in Example 7(c) has the form

$$A(t) = \frac{C}{0.999879^t},$$

which gives the original amount of carbon 14 t years ago in terms of the amount C that is left now. A similar formula can be used in finance to find the present value, given the future value. ∎

Algebra of Functions

If you look back at some of the functions considered in this section, you will notice that we frequently constructed them by combining simpler or previously constructed functions. For instance:

Quick Example 3 on page 56:   Area = Width × Length: $A(x) = x(50 - x)$

Example 1:   Cost = Variable Cost + Fixed Cost:
                                        $C(x) = 1.80x + 2.05$

Quick Example on page 59:   Profit = Revenue − Cost:
                                        $P(x) = 10x - (8x + 100)$.

Let us look a little more deeply at each of the above examples:

Area Example: $A(x) =$ Width × Length $= x(50 - x)$:
Think of the width and length as separate functions of x:

$$\text{Width: } W(x) = x; \quad \text{Length: } L(x) = 50 - x$$

so that

$$A(x) = W(x)L(x). \quad \text{\small Area = Width × Length}$$

We say that the area function A is the **product of the functions** W and L, and we write

$$A = WL. \quad \text{\small A is the product of the functions W and L.}$$

To calculate $A(x)$, we multiply $W(x)$ by $L(x)$.

Cost Example: $C(x) = $ Variable Cost $+$ Fixed Cost $= 1.80x + 2.05$:
Think of the variable and fixed costs as separate functions of x:

Variable Cost: $V(x) = 1.80x$;   Fixed Cost: $F(x) = 2.05$*

* F is called a constant function as its value, 2.05, is the same for every value of x.

so that

$$C(x) = V(x) + F(x).$$ Cost = Variable Cost + Fixed Cost

We say that the cost function C is the **sum of the functions V and F**, and we write

$$C = V + F.$$ C is the sum of the functions V and F.

To calculate $C(x)$, we add $V(x)$ to $F(x)$.

Profit Example: $P(x) = $ Revenue $-$ Cost $= 10x - (8x + 100)$:
Think of the revenue and cost as separate functions of x:

Revenue: $R(x) = 10x$;   Cost: $C(x) = 8x + 100$

so that

$$P(x) = R(x) - C(x).$$ Profit = Revenue − Cost

We say that the profit function P is the **difference between the functions R and C**, and we write

$$P = R - C.$$ P is the difference of the functions R and C.

To calculate $P(x)$, we subtract $C(x)$ from $R(x)$.

Algebra of Functions

If f and g are real-valued functions of the real variable x, then we define their **sum s, difference d, product p,** and **quotient q** as follows:

$s = f + g$ is the function specified by $s(x) = f(x) + g(x)$.

$d = f - g$ is the function specified by $d(x) = f(x) - g(x)$.

$p = fg$ is the function specified by $p(x) = f(x)g(x)$.

$q = \dfrac{f}{g}$ is the function specified by $q(x) = \dfrac{f(x)}{g(x)}$.

Also, if f is as above and c is a constant (real number), then we define the associated **constant multiple m of f** by

$m = cf$ is the function specified by $m(x) = cf(x)$.

Note on Domains
In order for any of the expressions $f(x) + g(x)$, $f(x) - g(x)$, $f(x)g(x)$, or $f(x)/g(x)$ to make sense, x must be simultaneously in the domains of both f and g. Further, for the quotient, the denominator $g(x)$ cannot be zero. Thus, we specify the domains of these functions as follows:

Domain of $f + g$, $f - g$, and fg: All real numbers x simultaneously in the domains of f and g

Domain of f/g: All real numbers x simultaneously in the domains of f and g such that $g(x) \neq 0$

Domain of cf: Same as the domain of f

Quick Examples

1. If $f(x) = x^2 - 1$ and $g(x) = \sqrt{x}$ with domain $[0, +\infty)$, then the sum s of f and g has domain $[0, +\infty)$ and is specified by
$$s(x) = f(x) + g(x) = x^2 - 1 + \sqrt{x}.$$

2. If $f(x) = x^2 - 1$ and $c = 3$, then the associated constant multiple m of f is specified by $m(x) = 3f(x) = 3(x^2 - 1)$.

3. If there are $N = 1{,}000t$ Mars shuttle passengers in year t who pay a total cost of $C = 40{,}000 + 800t$ million dollars, then the cost per passenger is given by the quotient of the two functions,

$$\text{Cost per passenger} = q(t) = \frac{C(t)}{N(t)}$$

$$= \frac{40{,}000 + 800t}{1{,}000t} \text{ million dollars per passenger.}$$

The largest possible domain of C/N is $(0, +\infty)$, as the quotient is not defined if $t = 0$.

1.2 EXERCISES

▼ more advanced   ◆ challenging

⊞ indicates exercises that should be solved using technology

Exercises 1–8 are based on the following functions:

$f(x) = x^2 + 1$ *with domain all real numbers*
$g(x) = x - 1$ *with domain all real numbers*
$h(x) = x + 4$ *with domain $x \geq 10$*
$u(x) = \sqrt{x + 10}$ *with domain $[-10, 0)$*
$v(x) = \sqrt{10 - x}$ *with domain $[0, 10]$*

*In each exercise, **a.** write a formula for the indicated function, **b.** give its domain, and **c.** specify its value at the given point a, if defined.*

1. $s = f + g$; $a = -3$
2. $d = g - f$; $a = -1$
3. $p = gu$; $a = -6$
4. $p = hv$; $a = 1$
5. $q = \dfrac{v}{g}$; $a = 1$
6. $q = \dfrac{g}{v}$; $a = 1$
7. $m = 5f$; $a = 1$
8. $m = 3u$; $a = -1$

APPLICATIONS

9. **Resources** You now have 200 music files on your hard drive, and this number is increasing by 10 music files each day. Find a mathematical model for this situation. HINT [See Quick Example 1 on page 56.]

10. **Resources** The amount of free space left on your hard drive is now 50 gigabytes (GB) and is decreasing by 5 GB/month. Find a mathematical model for this situation.

11. **Soccer** My rectangular soccer field site has a length equal to twice its width. Find its area in terms of its length x. HINT [See Quick Example 3 on page 56.]

12. **Cabbage** My rectangular cabbage patch has a total area of 100 sq. ft. Find its perimeter in terms of the width x.

13. **Vegetables** I want to fence in a square vegetable patch. The fencing for the east and west sides costs $4 per foot, and the fencing for the north and south sides costs only $2 per foot. Find the total cost of the fencing as a function of the length of a side x.

14. **Orchids** My square orchid garden abuts my house so that the house itself forms the northern boundary. The fencing for the southern boundary costs $4 per foot, and the fencing for the east and west sides costs $2 per foot. Find the total cost of the fencing as a function of the length of a side x.

15. **Study** You study math 4 hours a day Sunday through Thursday, and take the rest of the week off. Model the number of hours h you study math as a function of the day of the week n (with $n = 1$ being Sunday). HINT [See Quick Example 4 on page 57.]

16. **Recreation** You spend 5 hours per day on Saturdays and Sundays watching movies, but only two hours per day during the

week. Model the number of hours h you watch movies as a function of the day of the week n (with $n = 1$ being Sunday).

17. *Cost* A piano manufacturer has a daily fixed cost of $1,200 and a marginal cost of $1,500 per piano. Find the cost $C(x)$ of manufacturing x pianos in one day. Use your function to answer the following questions:

a. On a given day, what is the cost of manufacturing 3 pianos?
b. What is the cost of manufacturing the 3rd piano that day?
c. What is the cost of manufacturing the 11th piano that day?
d. What is the variable cost? What is the fixed cost? What is the marginal cost?
e. Graph C as a function of x. HINT [See Example 1.]

18. *Cost* The cost of renting tuxes for the Choral Society's formal is $20 down, plus $88 per tux. Express the cost C as a function of x, the number of tuxedos rented. Use your function to answer the following questions.

a. What is the cost of renting 2 tuxes?
b. What is the cost of the 2nd tux?
c. What is the cost of the 4,098th tux?
d. What is the variable cost? What is the fixed cost? What is the marginal cost?
e. Graph C as a function of x.

19. *Break-Even Analysis* Your college newspaper, *The Collegiate Investigator*, has fixed production costs of $70 per edition and marginal printing and distribution costs of 40¢ per copy. *The Collegiate Investigator* sells for 50¢ per copy.

a. Write down the associated cost, revenue, and profit functions. HINT [See Examples 1 and 2.]
b. What profit (or loss) results from the sale of 500 copies of *The Collegiate Investigator*?
c. How many copies should be sold in order to break even?

20. *Break-Even Analysis* The Audubon Society at Enormous State University (ESU) is planning its annual fund-raising "Eat-a-thon." The society will charge students 50¢ per serving of pasta. The only expenses the society will incur are the cost of the pasta, estimated at 15¢ per serving, and the $350 cost of renting the facility for the evening.

a. Write down the associated cost, revenue, and profit functions.
b. How many servings of pasta must the Audubon Society sell in order to break even?
c. What profit (or loss) results from the sale of 1,500 servings of pasta?

21. *Break-Even Analysis* Gymnast Clothing manufactures expensive hockey jerseys for sale to college bookstores in runs of up to 200. Its cost (in dollars) for a run of x hockey jerseys is

$$C(x) = 2,000 + 10x + 0.2x^2 \quad (0 \le x \le 200).$$

Gymnast Clothing sells the jerseys at $100 each. Find the revenue and profit functions. How many should Gymnast Clothing manufacture to make a profit? HINT [See Example 2.]

22. *Break-Even Analysis* Gymnast Clothing also manufactures expensive soccer cleats for sale to college bookstores in runs of up to 500. Its cost (in dollars) for a run of x pairs of cleats is

$$C(x) = 3,000 + 8x + 0.1x^2 \quad (0 \le x \le 500).$$

Gymnast Clothing sells the cleats at $120 per pair. Find the revenue and profit functions. How many should Gymnast Clothing manufacture to make a profit?

23. *Break-Even Analysis: School Construction Costs* The cost, in millions of dollars, of building a two-story high school in New York State was estimated to be

$$C(x) = 1.7 + 0.12x - 0.0001x^2 \quad (20 \le x \le 400),$$

where x is the number of thousands of square feet.[19] Suppose that you are contemplating building a for-profit two-story high school and estimate that your total revenue will be $0.1 million dollars per thousand square feet. What is the profit function? What size school should you build in order to break even?

24. *Break-Even Analysis: School Construction Costs* The cost, in millions of dollars, of building a three-story high school in New York State was estimated to be

$$C(x) = 1.7 + 0.14x - 0.0001x^2 \quad (20 \le x \le 400),$$

where x is the number of thousands of square feet.[20] Suppose that you are contemplating building a for-profit three-story high school and estimate that your total revenue will be $0.2 million dollars per thousand square feet. What is the profit function? What size school should you build in order to break even?

25. ▼ *Profit Analysis—Aviation* The hourly operating cost of a Boeing 747-100, which seats up to 405 passengers, is estimated to be $5,132.[21] If an airline charges each passenger a fare of $100 per hour of flight, find the hourly profit P it earns operating a 747-100 as a function of the number of passengers x. (Be sure to specify the domain.) What is the least number of passengers it must carry in order to make a profit? HINT [The cost function is constant (Variable cost = 0).]

26. ▼ *Profit Analysis—Aviation* The hourly operating cost of a McDonnell Douglas DC 10-10, which seats up to 295 passengers, is estimated to be $3,885.[22] If an airline charges each passenger a fare of $100 per hour of flight, find the hourly profit P it earns operating a DC10-10 as a function of the number of passengers x. (Be sure to specify the domain.) What is the least number of passengers it must carry in order to make a profit? HINT [The cost function is constant (Variable cost = 0).]

27. ▼ *Break-Even Analysis (based on a question from a CPA exam)* The Oliver Company plans to market a new product. Based on its market studies, Oliver estimates that it can sell up to 5,500 units in 2005. The selling price will be $2 per unit. Variable costs are estimated to be 40% of total revenue. Fixed costs are estimated to be $6,000 for 2005. How many units should the company sell to break even?

---

[19]The model is the authors'. Source for data: *Project Labor Agreements and Public Construction Cost in New York State,* Paul Bachman and David Tuerck, Beacon Hill Institute at Suffolk University, April 2006, www.beaconhill.org.
[20]*Ibid.*
[21]In 1992. Source: Air Transportation Association of America.
[22]*Ibid.*

28. ▼ *Break-Even Analysis (based on a question from a CPA exam)* The Metropolitan Company sells its latest product at a unit price of $5. Variable costs are estimated to be 30% of the total revenue, while fixed costs amount to $7,000 per month. How many units should the company sell per month in order to break even, assuming that it can sell up to 5,000 units per month at the planned price?

29. ◆ *Break-Even Analysis (from a CPA exam)* Given the following notations, write a formula for the break-even sales level.

SP = Selling price per unit
FC = Total fixed cost
VC = Variable cost per unit

30. ◆ *Break-Even Analysis (based on a question from a CPA exam)* Given the following notation, give a formula for the total fixed cost.

SP = Selling price per unit
VC = Variable cost per unit
BE = Break even sales level in units

31. ◆ *Break-Even Analysis—Organized Crime* The organized crime boss and perfume king Butch (Stinky) Rose has daily overheads (bribes to corrupt officials, motel photographers, wages for hit men, explosives, and so on) amounting to $20,000 per day. On the other hand, he has a substantial income from his counterfeit perfume racket: He buys imitation French perfume (Chanel № 22.5) at $20 per gram, pays an additional $30 per 100 grams for transportation, and sells it via his street thugs for $600 per gram. Specify Stinky's profit function, $P(x)$, where x is the quantity (in grams) of perfume he buys and sells, and use your answer to calculate how much perfume should pass through his hands per day in order that he break even.

32. ◆ *Break-Even Analysis—Disorganized Crime* Butch (Stinky) Rose's counterfeit Chanel № 22.5 racket has run into difficulties; it seems that the *authentic* Chanel № 22.5 perfume is selling for less than his counterfeit perfume. However, he has managed to reduce his fixed costs to zero, and his overall costs are now $400 per gram plus $30 per gram transportation costs and commission. (The perfume's smell is easily detected by specially trained Chanel Hounds, and this necessitates elaborate packaging measures.) He therefore decides to sell it for $420 per gram in order to undercut the competition. Specify Stinky's profit function, $P(x)$, where x is the quantity (in grams) of perfume he buys and sells, and use your answer to calculate how much perfume should pass through his hands per day in order that he break even. Interpret your answer.

33. *Demand for Monorail Service, Las Vegas* The demand for monorail service in Las Vegas can be approximated by

$$q(p) = 64p^{-0.76} \text{ thousand rides per day} \quad (3 \le p \le 5),$$

where p is the cost per ride in dollars.[23]

a. Graph the demand function.
b. What is the result on demand if the cost per ride is increased from $3.00 to $3.50? HINT [See Example 3.]

34. *Demand for Monorail Service, Mars* The demand for monorail service on the Utarek monorail, which links the three urbynes (or districts) of Utarek, Mars, can be approximated by

$$q(p) = 30p^{-0.49} \text{ million rides per day} \quad (3 \le p \le 5),$$

where p is the cost per ride in zonars ($\overline{\overline{Z}}$).[24]

a. Graph the demand function.
b. What is the result on demand if the cost per ride is decreased from $\overline{\overline{Z}}5.00$ to $\overline{\overline{Z}}3.50$?

35. ▼ *Demand: Smart Phones* The worldwide demand for smart phones may be modeled by

$$q(p) = 0.0009p^2 - 0.63p + 245 \text{ million units sold annually} \quad (50 \le p \le 150),$$

where p is the unit price in dollars.[25]

a. Graph the demand function.
b. Use the demand function to estimate, to the nearest million units, worldwide sales of smart phones if the price is $60.
c. Extrapolate the demand function to estimate, to the nearest million units, worldwide sales of smart phones if the price is $20.
d. Model the worldwide annual revenue from the sale of smart phones as a function of unit price, and use your model to estimate, to the nearest billion dollars, worldwide annual revenue when the price is set at $60. HINT [Revenue = Price × Quantity = p · q(p)]

36. ▼ *Demand: Smart Phones* (See Exercise 35.) Here is another model for worldwide demand for smart phones:

$$q(p) = 700p^{-0.284} \text{ million units sold annually} \quad (50 \le p \le 150),$$

where p is the unit price in dollars.[26]

a. Graph the demand function.
b. Use the demand function to estimate, to the nearest million units, worldwide sales of smart phones if the price is $70.
c. Extrapolate the demand function to estimate, to the nearest million units, worldwide sales of smart phones if the price is $180.
d. Model the worldwide annual revenue from the sale of smart phones as a function of unit price, and use your model to estimate, to the nearest billion dollars, worldwide annual revenue when the price is set at $70. HINT [Revenue = Price × Quantity = p · q(p)]

---

[23]Source: *New York Times,* February 10, 2007, p. A9.

[24]The zonar ($\overline{\overline{Z}}$) is the official currency in the city-state of Utarek, Mars (formerly www.Marsnext.com, a now extinct virtual society).

[25]The model is the authors' based on data available in 2010. Sources for Data: www.businessweek.com, http://techcrunch.com.

[26]*Ibid.*

37. Equilibrium Price The demand for your hand-made skateboards, in weekly sales, is $q = -3p + 700$ if the selling price is $\$p$. You are prepared to supply $q = 2p - 500$ per week at the price $\$p$. At what price should you sell your skateboards so that there is neither a shortage nor a surplus? HINT [See Quick Example on page 62.]

38. Equilibrium Price The demand for your factory-made skateboards, in weekly sales, is $q = -5p + 50$ if the selling price is $\$p$. If you are selling them at that price, you can obtain $q = 3p - 30$ per week from the factory. At what price should you sell your skateboards so that there is neither a shortage nor a surplus?

39. Equilibrium Price: Cell Phones Worldwide quarterly sales of Nokia® cell phones were approximately $q = -p + 156$ million phones when the wholesale price[27] was $\$p$.

a. If Nokia was prepared to supply $q = 4p - 394$ million phones per quarter at a wholesale price of $\$p$, what would have been the equilibrium price?

b. The actual wholesale price was $\$105$ in the fourth quarter of 2004. Estimate the projected shortage or surplus at that price. HINT [See Quick Example on page 62 and also Example 4.]

40. Equilibrium Price: Cell Phones Worldwide annual sales of all cell phones were approximately $-10p + 1,600$ million phones when the wholesale price[28] was $\$p$.

a. If manufacturers were prepared to supply $q = 14p - 800$ million phones per year at a wholesale price of $\$p$, what would have been the equilibrium price?

b. The actual wholesale price was projected to be $\$80$ in the fourth quarter of 2008. Estimate the projected shortage or surplus at that price.

41. 🔳 **Equilibrium Price: Las Vegas Monorail Service** The demand for monorail service in Las Vegas could be approximated by

$$q = 64p^{-0.76} \quad \text{thousand rides per day,}$$

where p was the fare the Las Vegas Monorail Company charges in dollars.[29] Assume the company was prepared to provide service for

$$q = 2.5p + 15.5 \quad \text{thousand rides per day}$$

at a fare of $\$p$.

a. Graph the demand and supply equations, and use your graph to estimate the equilibrium price (to the nearest 50¢).

b. Estimate, to the nearest 10 rides, the shortage or surplus of monorail service at the December 2005 fare of $\$5$ per ride.

42. 🔳 **Equilibrium Price: Mars Monorail Service** The demand for monorail service in the three urbynes (or districts) of Utarek, Mars can be approximated by

$$q = 31p^{-0.49} \text{ million rides per day,}$$

where p is the fare the Utarek Monorail Cooperative charges in zonars ($\mathbb{Z}$).[30] Assume the cooperative is prepared to provide service for

$$q = 2.5p + 17.5 \text{ million rides per day}$$

at a fare of $\mathbb{Z}p$.

a. Graph the demand and supply equations, and use your graph to estimate the equilibrium price (to the nearest 0.50 zonars).

b. Estimate the shortage or surplus of monorail service at the December 2085 fare of $\mathbb{Z}1$ per ride.

43. ▼ **Toxic Waste Treatment** The cost of treating waste by removing PCPs goes up rapidly as the quantity of PCPs removed goes up. Here is a possible model:

$$C(q) = 2,000 + 100q^2,$$

where q is the reduction in toxicity (in pounds of PCPs removed per day) and $C(q)$ is the daily cost (in dollars) of this reduction.

a. Find the cost of removing 10 pounds of PCPs per day.

b. Government subsidies for toxic waste cleanup amount to

$$S(q) = 500q,$$

where q is as above and $S(q)$ is the daily dollar subsidy. The *net cost* function is given by $N = C - S$. Give a formula for $N(q)$ and interpret your answer.

c. Find $N(20)$ and interpret your answer.

44. ▼ **Dental Plans** A company pays for its employees' dental coverage at an annual cost C given by

$$C(q) = 1,000 + 100\sqrt{q},$$

where q is the number of employees covered and $C(q)$ is the annual cost in dollars.

a. If the company has 100 employees, find its annual outlay for dental coverage.

b. Assume that the government subsidizes coverage by an annual dollar amount of

$$S(q) = 200q.$$

The *net cost* function is given by $N = C - S$. Give a formula for $N(q)$ and interpret your answer.

c. Find $N(100)$ and interpret your answer.

---

[27]Source: Embedded.com/Company reports December, 2004.

[28]Wholesale price projections are the authors'. Source for sales prediction: I-Stat/NDR December, 2004.

[29]The model is the authors'. Source for data: *New York Times,* February 10, 2007, p. A9.

[30]The official currency of Utarek, Mars. (See the footnote to Exercise 34.)

45. Spending on Corrections in the 1990s The following table shows the annual spending by all states in the United States on corrections ($t = 0$ represents the year 1990):[31]

Year (t)	0	2	4	6	7
Spending ($ billion)	16	18	22	28	30

a. Which of the following functions best fits the given data? (Warning: None of them fits exactly, but one fits more closely than the others.) HINT [See Example 5.]

(A) $S(t) = -0.2t^2 + t + 16$
(B) $S(t) = 0.2t^2 + t + 16$
(C) $S(t) = t + 16$

b. Use your answer to part (a) to "predict" spending on corrections in 1998, assuming that the trend continued.

46. Spending on Corrections in the 1990s Repeat Exercise 45, this time choosing from the following functions:

(A) $S(t) = 16 + 2t$
(B) $S(t) = 16 + t + 0.5t^2$
(C) $S(t) = 16 + t - 0.5t^2$

47. Soccer Gear The East Coast College soccer team is planning to buy new gear for its road trip to California. The cost per shirt depends on the number of shirts the team orders as shown in the following table:

x (Shirts ordered)	5	25	40	100	125
$A(x)$ (Cost/shirt, $)	22.91	21.81	21.25	21.25	22.31

a. Which of the following functions best models the data?

(A) $A(x) = 0.005x + 20.75$
(B) $A(x) = 0.01x + 20 + \dfrac{25}{x}$
(C) $A(x) = 0.0005x^2 - 0.07x + 23.25$
(D) $A(x) = 25.5(1.08)^{(x-5)}$

b. ⬛ Graph the model you chose in part (a) for $10 \le x \le 100$. Use your graph to estimate the lowest cost per shirt and the number of shirts the team should order to obtain the lowest price per shirt.

48. Hockey Gear The South Coast College hockey team wants to purchase wool hats for its road trip to Alaska. The cost per hat depends on the number of hats the team orders as shown in the following table:

x (Hats ordered)	5	25	40	100	125
$A(x)$ (Cost/hat, $)	25.50	23.50	24.63	30.25	32.70

a. Which of the following functions best models the data?

(A) $A(x) = 0.05x + 20.75$
(B) $A(x) = 0.1x + 20 + \dfrac{25}{x}$
(C) $A(x) = 0.0008x^2 - 0.07x + 23.25$
(D) $A(x) = 25.5(1.08)^{(x-5)}$

b. ⬛ Graph the model you chose in part (a) with $5 \le x \le 30$. Use your graph to estimate the lowest cost per hat and the number of hats the team should order to obtain the lowest price per hat.

Cost: Hard Drive Storage *Exercises 49 and 50 are based on the following data showing how the approximate retail cost of a gigabyte of hard drive storage has fallen since 2000:*[32]

t (Year since 2000)	$c(t)$ (Cost per Gigabyte ($))
0	7.5
2	2.5
4	1.2
6	0.6
8	0.2
10	0.1
12	0.06

49. ⬛ **a.** Graph each of the following models together with the data points above and use your graph to decide which two of the models best fit the data: HINT [See Example 5 and accompanying technology note.]

(A) $c(t) = 6.3(0.67)^t$
(B) $c(t) = 0.093t^2 - 1.6t + 6.7$
(C) $c(t) = 4.75 - 0.50t$
(D) $c(t) = \dfrac{12.8}{t^{1.7} + 1.7}$

b. Of the two models you chose in part (a), which predicts the lower price in 2020? What price does that model predict?

50. ⬛ **a.** Graph each of the following models together with the data points above and use your graph to decide which three of the models best fit the given data:

(A) $c(t) = \dfrac{15}{1 + 2^t}$
(B) $c(t) = (7.32)0.59^t + 0.10$
(C) $c(t) = 0.00085(t - 9.6)^4$
(D) $c(t) = 7.5 - 0.82t$

b. One of the three best-fit models in part (a) gives an unreasonable prediction for the price in 2020. Which is it, what price does it predict, and why is the prediction unreasonable?

[31]Data are rounded. Source: National Association of State Budget Officers/*New York Times*, February 28, 1999, p. A1.

[32]2012 price is estimated. Source for data: Historical Notes about the Cost of Hard Drive Storage Space http://ns1758.ca/winch/winchest.html.

51. *Value of Euro* The following table shows the approximate value V of one euro in U.S. dollars during three months in 2010. ($t = 1$ represents January, 2010.)[33]

t (Month)	1	5	6
V (Value in $)	1.42	1.26	1.22

Which of the following kinds of models would best fit the given data? Explain your choice of model. (A, a, b, c, and m are constants.)

(A) Linear: $V(t) = mt + b$
(B) Quadratic: $V(t) = at^2 + bt + c$
(C) Exponential: $V(t) = Ab^t$

52. *Value of Yen* The following table shows the approximate value V of one yen in U.S. dollars during three months in 2010. ($t = 1$ represents January, 2010.)[34]

t (Month)	1	4	10
V (Value in $)	0.0110	0.0107	0.0122

Which of the following kinds of models would best fit the given data? Explain your choice of model. (A, a, b, c, and m are constants.)

(A) Linear: $V(t) = mt + b$
(B) Quadratic: $V(t) = at^2 + bt + c$
(C) Exponential: $V(t) = Ab^t$

53. *Petrochemical Sales: Mexico* The following table shows annual petrochemial sales in Mexico by Pemex, Mexico's national oil company, for 2005–2010 ($t = 0$ represents 2005):[35]

Year t (Year since 2005)	0	2	3	4
Petrochemical Sales s (Billion metric tons)	3.7	4.0	4.1	4.0

Which of the following kinds of models would best fit the given data? Explain your choice of model. (A, a, b, and c are constants.)

(A) $s(t) = Ab^t$ ($b > 1$)
(B) $s(t) = Ab^t$ ($b < 1$)

(C) $s(t) = at^2 + bt + c$ ($a > 0$)
(D) $s(t) = at^2 + bt + c$ ($a < 0$)

54. *Petrochemical Production: Mexico* The following table shows annual petrochemical production in Mexico by Pemex, Mexico's national oil company, for 2005–2010 ($t = 0$ represents 2005):[36]

Year t (Year since 2005)	1	3	5
Petrochemical Production p (Billion metric tons)	11.0	12.0	13.3

Which of the following kinds of models would best fit the given data? Explain your choice of model. (A, a, b, c, and m are constants.)

(A) $s(t) = Ab^t$ ($b > 1$)
(B) $s(t) = Ab^t$ ($b < 1$)
(C) $s(t) = mt + b$ ($m > 0$)
(D) $s(t) = at^2 + bt + c$ ($a < 0$)

55. *Investments* In November 2010, E*Trade Financial was offering only 0.3% interest on its online savings accounts, with interest reinvested monthly.[37] Find the associated exponential model for the value of a $5,000 deposit after t years. Assuming this rate of return continued for seven years, how much would a deposit of $5,000 in November 2010 be worth in November 2017? (Answer to the nearest $1.) HINT [See Quick Example on page 67.]

56. *Investments* In November 2010, ING Direct was offering 2.4% interest on its Orange Savings Account, with interest reinvested quarterly.[38] Find the associated exponential model for the value of a $4,000 deposit after t years. Assuming this rate of return continued for eight years, how much would a deposit of $4,000 in November 2010 be worth in November 2018? (Answer to the nearest $1.)

57. **⊤** *Investments* Refer to Exercise 55. In November of which year will an investment of $5,000 made in November of 2010 first exceed $5,200? HINT [See Example 6.]

58. **⊤** *Investments* Refer to Exercise 56. In November of which year will an investment of $4,000 made in November of 2010 first exceed $5,200?

59. *Carbon Dating* A fossil originally contained 104 grams of carbon 14. Refer to the formula for $C(t)$ in Example 7 and estimate the amount of carbon 14 left in the sample after 10,000 years, 20,000 years, and 30,000 years. HINT [See Example 7.]

---

[33]Source: www.exchange-rates.org.

[34]*Ibid.*

[35]2010 figure is a projection based on data through May 2010. Source: www.pemex.com (July 2010)

[36]*Ibid.*

[37]Interest rate based on annual percentage yield. Source: https://us.etrade.com, November 2010.

[38]Interest rate based on annual percentage yield. Source: http://home.ingdirect.com, November 2010.

60. *Carbon Dating* A fossil contains 4.06 grams of carbon 14. Refer to the formula for $A(t)$ at the end of Example 7, and estimate the amount of carbon 14 in the sample 10,000 years, 20,000 years, and 30,000 years ago.

61. *Carbon Dating* A fossil contains 4.06 grams of carbon 14. It is estimated that the fossil originally contained 46 grams of carbon 14. By calculating the amount left after 5,000 years, 10,000 years, . . . , 35,000 years, estimate the age of the sample to the nearest 5,000 years. (Refer to the formula for $C(t)$ in Example 7.)

62. *Carbon Dating* A fossil contains 2.8 grams of carbon 14. It is estimated that the fossil originally contained 104 grams of carbon 14. By calculating the amount 5,000 years, 10,000 years, . . . , 35,000 years ago, estimate the age of the sample to the nearest 5,000 years. (Refer to the formula for $C(t)$ at the end of Example 7.)

63. *Radium Decay* The amount of radium 226 remaining in a sample that originally contained A grams is approximately

$$C(t) = A(0.999567)^t$$

where t is time in years.

a. Find, to the nearest whole number, the percentage of radium 226 left in an originally pure sample after 1,000 years, 2,000 years, and 3,000 years.

b. Use a graph to estimate, to the nearest 100 years, when one half of a sample of 100 grams will have decayed.

64. *Iodine Decay* The amount of iodine 131 remaining in a sample that originally contained A grams is approximately

$$C(t) = A(0.9175)^t$$

where t is time in days.

a. Find, to the nearest whole number, the percentage of iodine 131 left in an originally pure sample after 2 days, 4 days, and 6 days.

b. Use a graph to estimate, to the nearest day, when one half of a sample of 100 g will have decayed.

COMMUNICATION AND REASONING EXERCISES

65. If the population of the lunar station at Clavius has a population of $P = 200 + 30t$, where t is time in years since the station was established, then the population is increasing by _____ per year.

66. My bank balance can be modeled by $B(t) = 5,000 - 200t$ dollars, where t is time in days since I opened the account. The balance on my account is _____ by $200 per day.

67. Classify the following model as analytical or curve fitting, and give a reason for your choice: The price of gold was $700 on Monday, $710 on Tuesday, and $700 on Wednesday. Therefore, the price can be modeled by $p(t) = -10t^2 + 20t + 700$ where t is the day since Monday.

68. Classify the following model as analytical or curve fitting, and give a reason for your choice: The width of a small animated square on my computer screen is currently 10 mm and is growing by 2 mm per second. Therefore, its area can be modeled by $a(t) = (10 + 2t)^2$ square mm where t is time in seconds.

69. Fill in the missing information for the following *analytical model* (answers may vary): _____. Therefore, the cost of downloading a movie can be modeled by $c(t) = 4 - 0.2t$, where t is time in months since January.

70. Repeat Exercise 69, but this time regard the given model as a *curve-fitting model*.

71. Fill in the blanks: In a linear cost function, the _____ cost is x times the _____ cost.

72. Complete the following sentence: In a linear cost function, the marginal cost is the _____.

73. ▼ We said on page 61 that the demand for a commodity generally goes down as the price goes up. Assume that the demand for a certain commodity goes up as the price goes up. Is it still possible for there to be an equilibrium price? Explain with the aid of a demand and supply graph.

74. ▼ What would happen to the price of a certain commodity if the demand was always greater than the supply? Illustrate with a demand and supply graph.

75. You have a set of data points showing the sales of videos on your Web site versus time that are closely approximated by two different mathematical models. Give one criterion that would lead you to choose one over the other. (Answers may vary.)

76. Would it ever be reasonable to use a quadratic model $s(t) = at^2 + bt + c$ to predict long-term sales if a is negative? Explain.

77. If f and g are functions with $f(x) \geq g(x)$ for every x, what can you say about the values of the function $f - g$?

78. If f and g are functions with $f(x) > g(x) > 0$ for every x, what can you say about the values of the function $\frac{f}{g}$?

79. If f is measured in books, and g is measured in people, what are the units of measurement of the function $\frac{f}{g}$?

80. If f and g are linear functions, then what can you say about $f - g$?

1.3 Linear Functions and Models

Linear functions are among the simplest functions and are perhaps the most useful of all mathematical functions.

Linear Function

A **linear function** is one that can be written in the form

$$f(x) = mx + b \quad \text{Function form}$$

or

$$y = mx + b \quad \text{Equation form}$$

where m and b are fixed numbers. (The names m and b are traditional.*)

Quick Example

$$f(x) = 3x - 1$$

$$y = 3x - 1$$

✳ Actually, c is sometimes used instead of b. As for m, there has even been some research into the question of its origin, but no one knows exactly why the letter m is used.

Linear Functions from the Numerical and Graphical Point of View

The following table shows values of $y = 3x - 1 \,(m = 3, b = -1)$ for some values of x:

x	-4	-3	-2	-1	0	1	2	3	4
y	-13	-10	-7	-4	-1	2	5	8	11

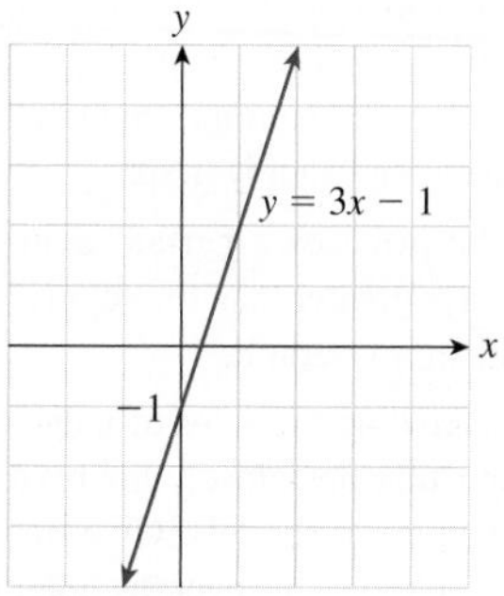

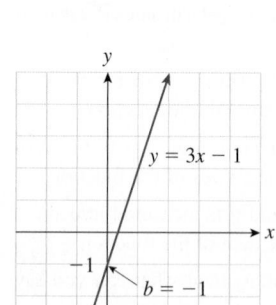

Figure 16

Its graph is shown in Figure 16.

Looking first at the table, notice that setting $x = 0$ gives $y = -1$, the value of b.

Numerically, b is the value of y when x = 0.

On the graph, the corresponding point $(0, -1)$ is the point where the graph crosses the y-axis, and we say that $b = -1$ is the **y-intercept** of the graph (Figure 17).

What about m? Looking once again at the table, notice that y increases by $m = 3$ units for every increase of 1 unit in x. This is caused by the term $3x$ in the formula: for every increase of 1 in x we get an increase of $3 \times 1 = 3$ in y.

Numerically, y increases by m units for every 1-unit increase of x.

Likewise, for every increase of 2 in x we get an increase of $3 \times 2 = 6$ in y. In general, if x increases by some amount, y will increase by three times that amount. We write:

Change in $y = 3 \times$ Change in x.

y-intercept $= b = -1$
Graphically, b is the y-intercept of the graph.

Figure 17

The Change in a Quantity: Delta Notation

If a quantity q changes from q_1 to q_2, the **change in q** is just the difference:

$$\text{Change in } q = \text{Second value} - \text{First value}$$
$$= q_2 - q_1.$$

Mathematicians traditionally use Δ (delta, the Greek equivalent of the Roman letter D) to stand for change, and write the change in q as Δq.

$$\Delta q = \text{Change in } q = q_2 - q_1$$

Quick Examples

1. If x is changed from 1 to 3, we write

$$\Delta x = \text{Second value} - \text{First value} = 3 - 1 = 2.$$

2. Looking at our linear function, we see that, when x changes from 1 to 3, y changes from 2 to 8. So,

$$\Delta y = \text{Second value} - \text{First value} = 8 - 2 = 6.$$

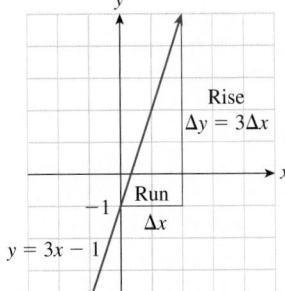

Using delta notation, we can now write, for our linear function,

$$\Delta y = 3\Delta x \qquad \text{Change in } y = 3 \times \text{Change in } x.$$

or

$$\frac{\Delta y}{\Delta x} = 3.$$

Because the value of y increases by exactly 3 units for every increase of 1 unit in x, the graph is a straight line rising by 3 units for every 1 unit we go to the right. We say that we have a **rise** of 3 units for each **run** of 1 unit. Because the value of y changes by $\Delta y = 3\Delta x$ units for every change of Δx units in x, in general we have a rise of $\Delta y = 3\Delta x$ units for each run of Δx units (Figure 18). Thus, we have a rise of 6 for a run of 2, a rise of 9 for a run of 3, and so on. So, $m = 3$ is a measure of the steepness of the line; we call m the **slope of the line**:

$$\text{Slope} = m = \frac{\Delta y}{\Delta x} = \frac{\text{Rise}}{\text{Run}}.$$

In general (replace the number 3 by a general number m), we can say the following.

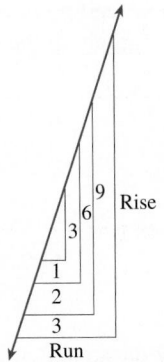

Slope $= m = 3$
Graphically, m is the slope of the graph.

Figure 18

The Roles of *m* and *b* in the Linear Function *f(x)* = *mx* + *b*

Role of *m*

Numerically If $y = mx + b$, then y changes by m units for every 1-unit change in x. A change of Δx units in x results in a change of $\Delta y = m\Delta x$ units in y. Thus,

$$m = \frac{\Delta y}{\Delta x} = \frac{\text{Change in } y}{\text{Change in } x}.$$

Graphically m is the slope of the line $y = mx + b$:

$$m = \frac{\Delta y}{\Delta x} = \frac{\text{Rise}}{\text{Run}} = \text{Slope}.$$

For positive m, the graph rises m units for every 1-unit move to the right, and rises $\Delta y = m\Delta x$ units for every Δx units moved to the right. For negative m, the graph drops $|m|$ units for every 1-unit move to the right, and drops $|m|\Delta x$ units for every Δx units moved to the right.

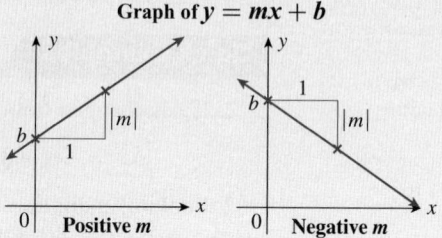

Graph of $y = mx + b$

Role of b

Numerically When $x = 0$, $y = b$.

Graphically b is the y-intercept of the line $y = mx + b$.

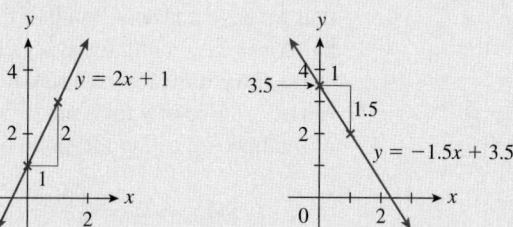

Figure 19 (left margin)

Slope $\frac{1}{2}$ Slope $-\frac{1}{2}$

Slope 1   Slope -1

Slope 2   Slope -2

Slope 3   Slope -3

Figure 19

> ### Quick Examples
>
> **1.** $f(x) = 2x + 1$ has slope $m = 2$ and y-intercept $b = 1$. To sketch the graph, we start at the y-intercept $b = 1$ on the y-axis, and then move 1 unit to the right and up $m = 2$ units to arrive at a second point on the graph. Now connect the two points to obtain the graph on the left.
>
> **2.** The line $y = -1.5x + 3.5$ has slope $m = -1.5$ and y-intercept $b = 3.5$. Because the slope is negative, the graph (above right) goes *down* 1.5 units for every 1 unit it moves to the right.

It helps to be able to picture what different slopes look like, as in Figure 19. Notice that the larger the absolute value of the slope, the steeper is the line.

using Technology

See the Technology Guides at the end of the chapter for detailed instructions on how to obtain a table with the successive quotients $m = \Delta y / \Delta x$ for the functions f and g in Example 1 using a TI-83/84 Plus or Excel. These tables show at a glance that f is not linear. Here is an outline:

TI-83/84 Plus
$\boxed{\text{STAT}}$ EDIT; Enter values of x and $f(x)$ in lists L_1 and L_2.
Highlight the heading L_3 then enter the following formula (including the quotes)
`"ΔList(L₂)/ΔList(L₁)"`
[More details on page 116.]

Spreadsheet
Enter headings x, $f(x)$, Df/Dx in cells A1–C1, and the corresponding values from one of the tables in cells A2–B8. Enter
`=(B3-B2)/(A3-A2)`
in cell C2, and copy down through C8.
[More details on page 122.]

EXAMPLE 1 Recognizing Linear Data Numerically and Graphically

Which of the following two tables gives the values of a linear function? What is the formula for that function?

x	0	2	4	6	8	10	12
$f(x)$	3	-1	-3	-6	-8	-13	-15

x	0	2	4	6	8	10	12
$g(x)$	3	-1	-5	-9	-13	-17	-21

Solution The function f cannot be linear: If it were, we would have $\Delta f = m \Delta x$ for some fixed number m. However, although the change in x between successive entries in the table is $\Delta x = 2$ each time, the change in f is not the same each time. Thus, the ratio $\Delta f / \Delta x$ is not the same for every successive pair of points.

On the other hand, the ratio $\Delta g / \Delta x$ is the same each time, namely,

$$\frac{\Delta g}{\Delta x} = \frac{-4}{2} = -2$$

as we see in the following table:

Δx		$2-0=2$	$4-2=2$	$6-4=2$	$8-6=2$	$10-8=2$	$12-10=2$
x	0	2	4	6	8	10	12
$g(x)$	3	-1	-5	-9	-13	-17	-21
Δg		$-1-3$ $=-4$	$-5-(-1)$ $=-4$	$-9-(-5)$ $=-4$	$-13-(-9)$ $=-4$	$-17-(-13)$ $=-4$	$-21-(-17)$ $=-4$

Thus, g is linear with slope $m = -2$. By the table, $g(0) = 3$, hence $b = 3$. Thus,

$$g(x) = -2x + 3. \qquad \text{Check that this formula gives the values in the table.}$$

If you graph the points in the tables defining f and g above, it becomes easy to see that g is linear and f is not; the points of g lie on a straight line (with slope -2), whereas the points of f do not lie on a straight line (Figure 20).

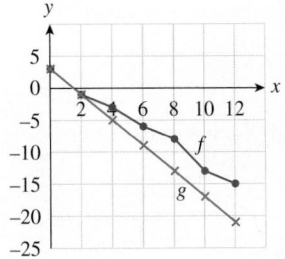

Figure 20

Finding a Linear Equation from Data

If we happen to know the slope and y-intercept of a line, writing down its equation is straightforward. For example, if we know that the slope is 3 and the y-intercept is -1, then the equation is $y = 3x - 1$. Sadly, the information we are given is seldom so

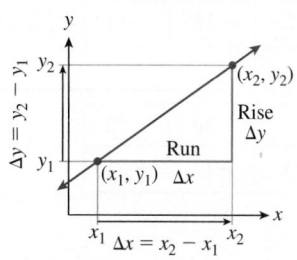

Figure 21

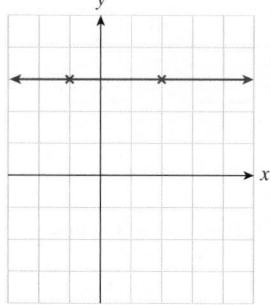

Figure 22

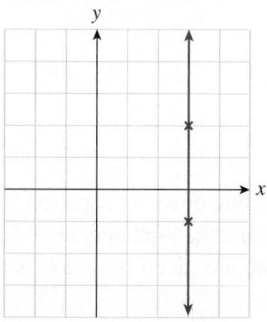

Vertical lines have undefined slope.

Figure 23

convenient. For instance, we may know the slope and a point other than the y-intercept, two points on the line, or other information. We therefore need to know how to use the information we are given to obtain the slope and the intercept.

Computing the Slope

We can always determine the slope of a line if we are given two (or more) points on the line, because any two points—say (x_1, y_1) and (x_2, y_2)—determine the line, and hence its slope. To compute the slope when given two points, recall the formula

$$\text{Slope} = m = \frac{\text{Rise}}{\text{Run}} = \frac{\Delta y}{\Delta x}.$$

To find its slope, we need a run Δx and corresponding rise Δy. In Figure 21, we see that we can use $\Delta x = x_2 - x_1$, the change in the x-coordinate from the first point to the second, as our run, and $\Delta y = y_2 - y_1$, the change in the y-coordinate, as our rise. The resulting formula for computing the slope is given in the box.

Computing the Slope of a Line

We can compute the slope m of the line through the points (x_1, y_1) and (x_2, y_2) using

$$m = \frac{\Delta y}{\Delta x} = \frac{y_2 - y_1}{x_2 - x_1}.$$

Quick Examples

1. The slope of the line through $(x_1, y_1) = (1, 3)$ and $(x_2, y_2) = (5, 11)$ is

$$m = \frac{\Delta y}{\Delta x} = \frac{y_2 - y_1}{x_2 - x_1} = \frac{11 - 3}{5 - 1} = \frac{8}{4} = 2.$$

Notice that we can use the points in the reverse order: If we take $(x_1, y_1) = (5, 11)$ and $(x_2, y_2) = (1, 3)$, we obtain the same answer:

$$m = \frac{\Delta y}{\Delta x} = \frac{y_2 - y_1}{x_2 - x_1} = \frac{3 - 11}{1 - 5} = \frac{-8}{-4} = 2.$$

2. The slope of the line through $(x_1, y_1) = (1, 2)$ and $(x_2, y_2) = (2, 1)$ is

$$m = \frac{\Delta y}{\Delta x} = \frac{y_2 - y_1}{x_2 - x_1} = \frac{1 - 2}{2 - 1} = \frac{-1}{1} = -1.$$

3. The slope of the line through $(2, 3)$ and $(-1, 3)$ is

$$m = \frac{\Delta y}{\Delta x} = \frac{y_2 - y_1}{x_2 - x_1} = \frac{3 - 3}{-1 - 2} = \frac{0}{-3} = 0.$$

A line of slope 0 has zero rise, so is a *horizontal* line, as shown in Figure 22.

4. The line through $(3, 2)$ and $(3, -1)$ has slope

$$m = \frac{\Delta y}{\Delta x} = \frac{y_2 - y_1}{x_2 - x_1} = \frac{-1 - 2}{3 - 3} = \frac{-3}{0},$$

which is undefined. The line passing through these points is *vertical*, as shown in Figure 23.

Computing the *y*-Intercept

Once we know the slope m of a line, and also the coordinates of a point (x_1, y_1), then we can calculate its *y*-intercept b as follows: The equation of the line must be

$$y = mx + b,$$

where b is as yet unknown. To determine b we use the fact that the line must pass through the point (x_1, y_1), so that (x_1, y_1) satisfies the equation $y = mx + b$. In other words,

$$y_1 = mx_1 + b.$$

Solving for b gives

$$b = y_1 - mx_1.$$

In summary:

Computing the *y*-Intercept of a Line

The *y*-intercept of the line passing through (x_1, y_1) with slope m is

$$b = y_1 - mx_1.$$

Quick Example

The line through $(2, 3)$ with slope 4 has

$$b = y_1 - mx_1 = 3 - (4)(2) = -5.$$

Its equation is therefore

$$y = mx + b = 4x - 5.$$

EXAMPLE 2 **Finding Linear Equations**

Find equations for the following straight lines.

a. Through the points $(1, 2)$ and $(3, -1)$
b. Through $(2, -2)$ and parallel to the line $3x + 4y = 5$
c. Horizontal and through $(-9, 5)$
d. Vertical and through $(-9, 5)$

Solution

a. To write down the equation of the line, we need the slope m and the *y*-intercept b.

• *Slope* Because we are given two points on the line, we can use the slope formula:

$$m = \frac{y_2 - y_1}{x_2 - x_1} = \frac{-1 - 2}{3 - 1} = -\frac{3}{2}.$$

using Technology

See the Technology Guides at the end of the chapter for detailed instructions on how to obtain the slope and intercept in Example 2(a) using a TI-83/84 Plus or a spreadsheet. Here is an outline:

TI-83/84 Plus
STAT EDIT; Enter values of x and y in lists L_1 and L_2.
Slope: Home screen
$(L_2(2)-L_2(1))/(L_1(2)-L_1(1))\to M$
Intercept: Home screen
$L_2(1)-M*L_1(1)$
[More details on page 117.]

Spreadsheet
Enter headings x, y, m, b, in cells A1–D1, and the values (x, y) in cells A2–B3. Enter
$=(B3-B2)/(A3-A2)$
in cell C2, and
$=B2-C2*A2$
in cell D2.
[More details on page 122.]

• *Intercept* We now have the slope of the line, $m = -3/2$, and also a point—we have two to choose from, so let us choose $(x_1, y_1) = (1, 2)$. We can now use the formula for the y-intercept:

$$b = y_1 - mx_1 = 2 - \left(-\frac{3}{2}\right)(1) = \frac{7}{2}.$$

Thus, the equation of the line is

$$y = -\frac{3}{2}x + \frac{7}{2}. \qquad y = mx + b$$

b. Proceeding as before,

• *Slope* We are not given two points on the line, but we are given a parallel line. We use the fact that *parallel lines have the same slope*. (Why?) We can find the slope of $3x + 4y = 5$ by solving for y and then looking at the coefficient of x:

$$y = -\frac{3}{4}x + \frac{5}{4} \qquad \text{To find the slope, solve for } y.$$

so the slope is $-3/4$.

• *Intercept* We now have the slope of the line, $m = -3/4$, and also a point $(x_1, y_1) = (2, -2)$. We can now use the formula for the y-intercept:

$$b = y_1 - mx_1 = -2 - \left(-\frac{3}{4}\right)(2) = -\frac{1}{2}.$$

Thus, the equation of the line is

$$y = -\frac{3}{4}x - \frac{1}{2}. \qquad y = mx + b$$

c. We are given a point: $(-9, 5)$. Furthermore, we are told that the line is horizontal, which tells us that the slope is $m = 0$. Therefore, all that remains is the calculation of the y-intercept:

$$b = y_1 - mx_1 = 5 - (0)(-9) = 5$$

so the equation of the line is

$$y = 5. \qquad y = mx + b$$

d. We are given a point: $(-9, 5)$. This time, we are told that the line is vertical, which means that the slope is undefined. Thus, we can't express the equation of the line in the form $y = mx + b$. (This formula makes sense only when the slope m of the line is defined.) What can we do? Well, here are some points on the desired line:

$$(-9, 1), (-9, 2), (-9, 3), \dots$$

so $x = -9$ and $y = $ *anything*. If we simply say that $x = -9$, then these points are all solutions, so the equation is $x = -9$.

Applications: Linear Models

Using linear functions to describe or approximate relationships in the real world is called **linear modeling**.

Recall from Section 1.2 that a **cost function** specifies the cost C as a function of the number of items x.

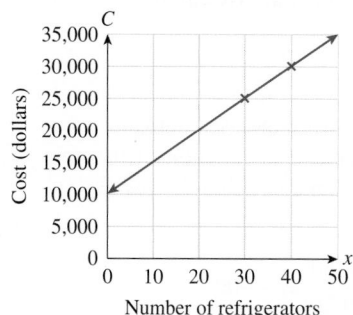

Figure 24

EXAMPLE 3 Linear Cost Function from Data

The manager of the FrozenAir Refrigerator factory notices that on Monday it cost the company a total of $25,000 to build 30 refrigerators and on Tuesday it cost $30,000 to build 40 refrigerators. Find a linear cost function based on this information. What is the daily fixed cost, and what is the marginal cost?

Solution We are seeking the cost C as a linear function of x, the number of refrigerators sold:

$$C = mx + b.$$

We are told that $C = 25,000$ when $x = 30$, and this amounts to being told that $(30, 25,000)$ is a point on the graph of the cost function. Similarly, $(40, 30,000)$ is another point on the line (Figure 24).

We can use the two points on the line to construct the linear cost equation:

• **Slope** $\quad m = \dfrac{C_2 - C_1}{x_2 - x_1} = \dfrac{30,000 - 25,000}{40 - 30} = 500 \qquad$ *C plays the role of y.*

• **Intercept** $b = C_1 - mx_1 = 25,000 - (500)(30) = 10,000$. $\quad$ We used the point $(x_1, C_1) = (30, 25,000)$.

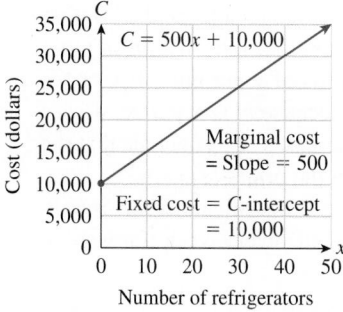

Figure 25

The linear cost function is therefore

$$C(x) = 500x + 10,000.$$

Because $m = 500$ and $b = 10,000$ the factory's fixed cost is $10,000 each day, and its marginal cost is $500 per refrigerator. (See page 59 in Section 1.2.) These are illustrated in Figure 25.

using Technology

To obtain the cost equation for Example 3 with technology, apply the Technology note for Example 2(a) to the given points (30, 25,000) and (40, 30,000) on the graph of the cost equation.

➡ **Before we go on...** Recall that, in general, the slope m measures the number of units of change in y per 1-unit change in x, so it is measured in units of y per unit of x:

Units of Slope = Units of y per unit of x.

In Example 3, y is the cost C, measured in dollars, and x is the number of items, measured in refrigerators. Hence,

Units of Slope = Units of y per Unit of x = Dollars per refrigerator.

The y-intercept b, being a value of y, is measured in the same units as y. In Example 3, b is measured in dollars. ∎

In Section 1.2 we saw that a **demand function** specifies the demand q as a function of the price p per item.

EXAMPLE 4 Linear Demand Function from Data

You run a small supermarket and must determine how much to charge for Hot'n'Spicy brand baked beans. The following chart shows weekly sales figures (the demand) for Hot'n'Spicy at two different prices.

Price/Can	$0.50	$0.75
Demand (cans sold/week)	400	350

a. Model these data with a linear demand function. (See Example 4 in Section 1.2.)

b. How do we interpret the slope and q-intercept of the demand function?

Solution

a. Recall that a demand equation—or demand function—expresses demand q (in this case, the number of cans of beans sold per week) as a function of the unit price p (in this case, price per can). We model the demand using the two points we are given: (0.50, 400) and (0.75, 350).

using Technology

To obtain the demand equation for Example 4 with technology, apply the Technology note for Example 2(a) to the given points (0.50, 400) and (0.75, 350) on the graph of the demand equation.

$$\textbf{\textit{Slope:}}\quad m = \frac{q_2 - q_1}{p_2 - p_1} = \frac{350 - 400}{0.75 - 0.50} = \frac{-50}{0.25} = -200$$

$$\textbf{\textit{Intercept:}}\ b = q_1 - mp_1 = 400 - (-200)(0.50) = 500$$

So, the demand equation is

$$q = -200p + 500. \qquad q = mp + b$$

b. The key to interpreting the slope in a demand equation is to recall (see the "Before we go on" note at the end of Example 3) that we measure the slope in *units of y per unit of x*. Here, $m = -200$, and the units of m are units of q per unit of p, or the number of cans sold per dollar change in the price. Because m is negative, we see that the number of cans sold decreases as the price increases. We conclude that the weekly sales will drop by 200 cans per \$1 increase in the price.

✳ Does this seem realistic? Demand is not always unlimited if items were given away. For instance, campus newspapers are sometimes given away, and yet piles of them are often left untaken. Also see the "Before we go on" discussion at the end of this example.

To interpret the q-intercept, recall that it gives the q-coordinate when $p = 0$. Hence it is the number of cans the supermarket can "sell" every week if it were to give them away.✳

➡ Before we go on...

Q: *Just how reliable is the linear model used in Example 4?*

A: The *actual* demand graph could in principle be obtained by tabulating demand figures for a large number of different prices. If the resulting points were plotted on the pq plane, they would probably suggest a curve and not a straight line. However, if you looked at a small enough portion of any curve, you could closely *approximate* it by a straight line. In other words, *over a small range of values of p, a linear model is accurate.* Linear models of real-world situations are generally reliable only for small ranges of the variables. (This point will come up again in some of the exercises.)

■

The next example illustrates modeling change over time t with a linear function of t.

EXAMPLE 5   Modeling Change Over Time: Growth of Sales

The worldwide market for portable navigation devices was expected to grow from 50 million units in 2007 to around 530 million units in 2015.[39]

---

[39]Sales were expected to grow to more than 500 million in 2015 according to a January 2008 press release by Telematics Research Group. Source: www.telematicsresearch.com.

a. Use this information to model annual worldwide sales of portable navigation devices as a linear function of time t in years since 2007. What is the significance of the slope?

b. Use the model to predict when annual sales of mobile navigation devices will reach 440 million units.

Solution

a. Since we are interested in worldwide sales s of portable navigation devices as a function of time, we take time t to be the independent coordinate (playing the role of x) and the annual sales s, in million of units, to be the dependent coordinate (in the role of y). Notice that 2007 corresponds to $t = 0$ and 2015 corresponds to $t = 8$, so we are given the coordinates of two points on the graph of sales s as a function of time t: $(0, 50)$ and $(8, 530)$. We model the sales using these two points:

$$m = \frac{s_2 - s_1}{t_2 - t_1} = \frac{530 - 50}{8 - 0} = \frac{480}{8} = 60$$

$$b = s_1 - mt_1 = 50 - (60)(0) = 50$$

So, $\qquad s = 60t + 50$ million units. $\qquad s = mt + b$

The slope m is measured in units of s per unit of t; that is, millions of devices per year, and is thus the *rate of change of annual sales*. To say that $m = 60$ is to say that annual sales are increasing at a rate of 60 million devices per year.

b. Our model of annual sales as a function of time is

$$s = 60t + 50 \text{ million units.}$$

Annual sales of mobile portable devices will reach 440 million when $s = 440$, or

$$440 = 60t + 50$$

Solving for t, $\qquad 60t = 440 - 50 = 390$

$$t = \frac{390}{60} = 6.5 \text{ years,}$$

which is midway through 2013. Thus annual sales are expected to reach 440 million midway through 2013.

using Technology

To use technology to obtain s as a function of t in Example 5, apply the Technology note for Example 2(a) to the points (0, 50) and (8, 530) on its graph.

EXAMPLE 6 **Velocity**

You are driving down the Ohio Turnpike, watching the mileage markers to stay awake. Measuring time in hours after you see the 20-mile marker, you see the following markers each half hour:

Time (h)	0	0.5	1	1.5	2
Marker (mi)	20	47	74	101	128

Find your location s as a function of t, the number of hours you have been driving. (The number s is also called your **position** or **displacement**.)

Solution

If we plot the location s versus the time t, the five markers listed give us the graph in Figure 26. These points appear to lie along a straight line. We can verify this by

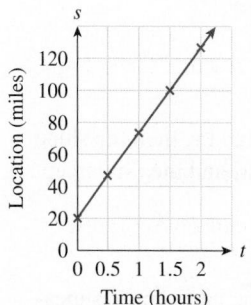

Figure 26

calculating how far you traveled in each half hour. In the first half hour, you traveled $47 - 20 = 27$ miles. In the second half hour you traveled $74 - 47 = 27$ miles also. In fact, you traveled exactly 27 miles each half hour. The points we plotted lie on a straight line that rises 27 units for every 0.5 unit we go to the right, for a slope of $27/0.5 = 54$.

To get the equation of that line, notice that we have the s-intercept, which is the starting marker of 20. Thus, the equation of s as a function of time t is

$$s(t) = 54t + 20.$$ We used s in place of y and t in place of x.

using Technology

To use technology to obtain s as a function of t in Example 6, apply the Technology note for Example 2(a) to the points (0, 20) and (1, 74) on its graph.

Notice the significance of the slope: For every hour you travel, you drive a distance of 54 miles. In other words, you are traveling at a constant velocity of 54 mph. We have uncovered a very important principle:

In the graph of displacement versus time, velocity is given by the slope.

Linear Change over Time

If a quantity q is a linear function of time t,

$$q = mt + b,$$

then the slope m measures the **rate of change** of q, and b is the quantity at time $t = 0$, the **initial quantity**. If q represents the position of a moving object, then the rate of change is also called the **velocity**.

Units of m and b

The units of measurement of m are units of q per unit of time; for instance, if q is income in dollars and t is time in years, then the rate of change m is measured in dollars per year.

The units of b are units of q; for instance, if q is income in dollars and t is time in years, then b is measured in dollars.

Quick Example

If the accumulated revenue from sales of your video game software is given by $R = 2{,}000t + 500$ dollars, where t is time in years from now, then you have earned $500 in revenue so far, and the accumulated revenue is increasing at a rate of $2,000 per year.

Examples 3–6 share the following common theme.

General Linear Models

If $y = mx + b$ is a linear model of changing quantities x and y, then the slope m is the rate at which y is increasing per unit increase in x, and the y-intercept b is the value of y that corresponds to $x = 0$.

Units of m and b

The slope m is measured in units of y per unit of x, and the intercept b is measured in units of y.

> ### Quick Example
>
> If the number n of spectators at a soccer game is related to the number g of goals your team has scored so far by the equation $n = 20g + 4$, then you can expect 4 spectators if no goals have been scored and 20 additional spectators per additional goal scored.

FAQs

What to Use as x and y, and How to Interpret a Linear Model

Q: *In a problem where I must find a linear relationship between two quantities, which quantity do I use as x and which do I use as y?*

A: The key is to decide which of the two quantities is the independent variable, and which is the dependent variable. Then use the independent variable as x and the dependent variable as y. In other words, *y depends on x.*

Here are examples of phrases that convey this information, usually of the form *Find y [dependent variable] in terms of x [independent variable]:*

- Find the cost in terms of the number of items. $y = \text{Cost}, x = \# \text{ Items}$
- How does color depend on wavelength? $y = \text{Color}, x = \text{Wavelength}$

If no information is conveyed about which variable is intended to be independent, then you can use whichever is convenient.

Q: *How do I interpret a general linear model $y = mx + b$?*

A: The key to interpreting a linear model is to remember the units we use to measure m and b:

> *The slope m is measured in units of y per unit of x; the intercept b is measured in units of y.*

For instance, if $y = 4.3x + 8.1$ and you know that x is measured in feet and y in kilograms, then you can already say, "y is 8.1 kilograms when $x = 0$ feet, and increases at a rate of 4.3 kilograms per foot" without knowing anything more about the situation!

1.3   EXERCISES

▼ more advanced   ◆ challenging
🔟 indicates exercises that should be solved using technology

In Exercises 1–6, a table of values for a linear function is given. Fill in the missing value and calculate m in each case.

1.

x	−1	0	1
y	5	8	

2.

x	−1	0	1
y	−1	−3	

3.

x	2	3	5
y	−1	−2	

4.

x	2	4	5
y	−1	−2	

5.

x	−2	0	2
y	4		10

6.

x	0	3	6
y	−1		−5

In Exercises 7–10, first find f(0), if not supplied, and then find the equation of the given linear function.

7.

x	−2	0	2	4
f(x)	−1	−2	−3	−4

8.

x	−6	−3	0	3
f(x)	1	2	3	4

9.

x	−4	−3	−2	−1
f(x)	−1	−2	−3	−4

10.

x	1	2	3	4
f(x)	4	6	8	10

In each of Exercises 11–14, decide which of the two given functions is linear and find its equation. HINT [See Example 1.]

11.

x	0	1	2	3	4
f(x)	6	10	14	18	22
g(x)	8	10	12	16	22

12.

x	−10	0	10	20	30
f(x)	−1.5	0	1.5	2.5	3.5
g(x)	−9	−4	1	6	11

13.

x	0	3	6	10	15
f(x)	0	3	5	7	9
g(x)	−1	5	11	19	29

14.

x	0	3	5	6	9
f(x)	2	6	9	12	15
g(x)	−1	8	14	17	26

In Exercises 15–24, find the slope of the given line, if it is defined.

15. $y = -\dfrac{3}{2}x - 4$ **16.** $y = \dfrac{2x}{3} + 4$

17. $y = \dfrac{x+1}{6}$

18. $y = -\dfrac{2x-1}{3}$

19. $3x + 1 = 0$ **20.** $8x - 2y = 1$

21. $3y + 1 = 0$ **22.** $2x + 3 = 0$

23. $4x + 3y = 7$ **24.** $2y + 3 = 0$

In Exercises 25–38, graph the given equation. HINT [See Quick Examples on page 80.]

25. $y = 2x - 1$ **26.** $y = x - 3$

27. $y = -\frac{2}{3}x + 2$ **28.** $y = -\frac{1}{2}x + 3$

29. $y + \frac{1}{4}x = -4$ **30.** $y - \frac{1}{4}x = -2$

31. $7x - 2y = 7$ **32.** $2x - 3y = 1$

33. $3x = 8$ **34.** $2x = -7$

35. $6y = 9$ **36.** $3y = 4$

37. $2x = 3y$ **38.** $3x = -2y$

In Exercises 39–58, calculate the slope, if defined, of the straight line through the given pair of points. Try to do as many as you can without writing anything down except the answer. HINT [See Quick Examples on page 82.]

39. $(0, 0)$ and $(1, 2)$ **40.** $(0, 0)$ and $(-1, 2)$

41. $(-1, -2)$ and $(0, 0)$ **42.** $(2, 1)$ and $(0, 0)$

43. $(4, 3)$ and $(5, 1)$ **44.** $(4, 3)$ and $(4, 1)$

45. $(1, -1)$ and $(1, -2)$ **46.** $(-2, 2)$ and $(-1, -1)$

47. $(2, 3.5)$ and $(4, 6.5)$ **48.** $(10, -3.5)$ and $(0, -1.5)$

49. $(300, 20.2)$ and $(400, 11.2)$

50. $(1, -20.2)$ and $(2, 3.2)$

51. $(0, 1)$ and $\left(-\frac{1}{2}, \frac{3}{4}\right)$

52. $\left(\frac{1}{2}, 1\right)$ and $\left(-\frac{1}{2}, \frac{3}{4}\right)$

53. (a, b) and (c, d) $(a \neq c)$

54. (a, b) and (c, b) $(a \neq c)$

55. (a, b) and (a, d) $(b \neq d)$

56. (a, b) and $(-a, -b)$ $(a \neq 0)$

57. $(-a, b)$ and $(a, -b)$ $(a \neq 0)$

58. (a, b) and (b, a) $(a \neq b)$

59. In the following figure, estimate the slopes of all line segments.

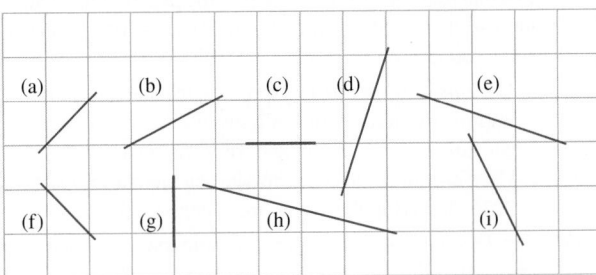

60. In the following figure, estimate the slopes of all line segments.

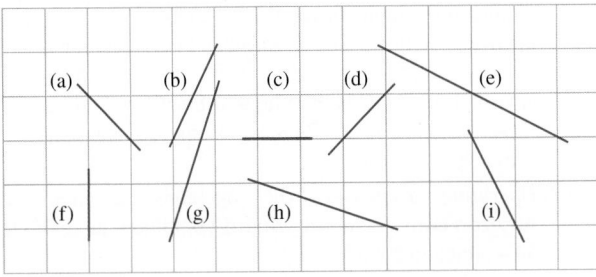

In Exercises 61–78, find a linear equation whose graph is the straight line with the given properties. HINT [See Example 2.]

61. Through $(1, 3)$ with slope 3

62. Through $(2, 1)$ with slope 2

63. Through $(1, -\frac{3}{4})$ with slope $\frac{1}{4}$

64. Through $(0, -\frac{1}{3})$ with slope $\frac{1}{3}$

65. Through $(20, -3.5)$ and increasing at a rate of 10 units of y per unit of x

66. Through $(3.5, -10)$ and increasing at a rate of 1 unit of y per 2 units of x

67. Through $(2, -4)$ and $(1, 1)$

68. Through $(1, -4)$ and $(-1, -1)$

69. Through $(1, -0.75)$ and $(0.5, 0.75)$

70. Through $(0.5, -0.75)$ and $(1, -3.75)$

71. Through $(6, 6)$ and parallel to the line $x + y = 4$

72. Through $(1/3, -1)$ and parallel to the line $3x - 4y = 8$

73. Through $(0.5, 5)$ and parallel to the line $4x - 2y = 11$

74. Through $(1/3, 0)$ and parallel to the line $6x - 2y = 11$

75. ▼ Through $(0, 0)$ and (p, q) $(p \neq 0)$

76. ▼ Through (p, q) parallel to $y = rx + s$

77. ▼ Through (p, q) and (r, q) $(p \neq r)$

78. ▼ Through (p, q) and (r, s) $(p \neq r)$

APPLICATIONS

79. *Cost* The RideEm Bicycles factory can produce 100 bicycles in a day at a total cost of $10,500 and it can produce 120 bicycles in a day at a total cost of $11,000. What are the company's daily fixed costs, and what is the marginal cost per bicycle? HINT [See Example 3.]

80. *Cost* A soft-drink manufacturer can produce 1,000 cases of soda in a week at a total cost of $6,000, and 1,500 cases of soda at a total cost of $8,500. Find the manufacturer's weekly fixed costs and marginal cost per case of soda.

81. *Cost: iPhones* If it costs Apple $1,900 to manufacture 10 iPhone 4s per hour and $3,780 to manufacture 20 per hour at a particular plant,[40] obtain the corresponding linear cost function. What was the cost to manufacture each additional iPhone? Use the cost function to estimate the cost of manufacturing 40 iPhones in an hour.

82. *Cost: Kinects* If it costs Microsoft $1,230 to manufacture 8 Kinects per hour for the Xbox 360 and $2,430 to manufacture 16 per hour at a particular plant,[41] obtain the corresponding linear cost function. What was the cost to manufacture each additional Kinect? Use the cost function to estimate the cost of manufacturing 30 Kinects in an hour.

83. *Demand* Sales figures show that your company sold 1,960 pen sets each week when they were priced at $1/pen set and 1,800 pen sets each week when they were priced at $5/pen set. What is the linear demand function for your pen sets? HINT [See Example 4.]

84. *Demand* A large department store is prepared to buy 3,950 of your tie-dye shower curtains per month for $5 each, but only 3,700 shower curtains per month for $10 each. What is the linear demand function for your tie-dye shower curtains?

85. *Demand for Cell Phones* The following table shows worldwide sales of cell phones by Nokia and their average wholesale prices in the first quarters of 2009 and 2010:[42]

Quarter	Q1 2009	Q1 2010
Wholesale Price ($)	85	81
Sales (millions)	103	111

a. Use the data to obtain a linear demand function for (Nokia) cell phones, and use your demand equation to predict sales if Nokia lowered the price further to $75.

b. Fill in the blanks: For every ____ increase in price, sales of cell phones decrease by ___ units.

---

[40]Based on marginal cost data at www.isuppli.com.

[41]Based on marginal cost data provided by a "highly-positioned, trusted source." www.develop-online.net.

[42]Data are approximate. Source: www.nokia.com.

86. *Demand for Smart Phones* The following table shows worldwide sales of smart phones and their average whole-sale prices in 2009 and 2010:[43]

Quarter	Q1 2009	Q1 2010
Wholesale Price ($)	162	152
Sales (millions)	174	213

a. Use the data to obtain a linear demand function for smart phones, and use your demand equation to predict sales to the nearest million phones if the price was set at $155.
b. Fill in the blanks: For every ____ increase in price, sales of smart phones decrease by ___ units.

87. *Demand for Monorail Service, Las Vegas* In 2005, the Las Vegas monorail charged $3 per ride and had an average ridership of about 28,000 per day. In December 2005 the Las Vegas Monorail Company raised the fare to $5 per ride, and average ridership in 2006 plunged to around 19,000 per day.[44]

a. Use the given information to find a linear demand equation.
b. Give the units of measurement and interpretation of the slope.
c. What would have been the effect on ridership of raising the fare to $6 per ride?

88. *Demand for Monorail Service, Mars* The Utarek monorail, which links the three urbynes (or districts) of Utarek, Mars, charged $\overline{\overline{Z}}5$ per ride[45] and sold about 14 million rides per day. When the Utarek City Council lowered the fare to $\overline{\overline{Z}}3$ per ride, the number of rides increased to 18 million per day.

a. Use the given information to find a linear demand equation.
b. Give the units of measurement and interpretation of the slope.
c. What would have been the effect on ridership of raising the fare to $\overline{\overline{Z}}10$ per ride?

89. *Pasta Imports in the 90s* During the period 1990–2001, U.S. imports of pasta increased from 290 million pounds in 1990 ($t = 0$) by an average of 40 million pounds/year.[46]

a. Use this information to express y, the annual U.S. imports of pasta (in millions of pounds), as a linear function of t, the number of years since 1990.
b. Use your model to estimate U.S. pasta imports in 2005, assuming the import trend continued.

90. *Mercury Imports in the 10s* During the period 2210–2220, Martian imports of mercury (from the planet of that name) increased from 550 million kg in 2210 ($t = 0$) by an average of 60 million kg/year.

a. Use this information to express y, the annual Martian imports of mercury (in millions of kilograms), as a linear function of t, the number of years since 2210.
b. Use your model to estimate Martian mercury imports in 2230, assuming the import trend continued.

91. *Net Income* The net income of Amazon.com increased from $10 billion in 2006 to $25 billion in 2009.[47]

a. Use this information to find a linear model for Amazon's net income N (in billions of dollars) as a function of time t in years since 2000.
b. Give the units of measurement and interpretation of the slope.
c. Use the model from part (a) to estimate the 2008 net income. (The actual 2008 net income was approximately $19 billion.)

92. *Operating Expenses* The operating expenses of Amazon.com increased from $2.0 billion in 2006 to $4.4 billion in 2009.[48]

a. Use this information to find a linear model for Amazon's operating expenses E (in billions of dollars) as a function of time t in years since 2000.
b. Give the units of measurement and interpretation of the slope.
c. Use the model from part (a) to estimate the 2007 operating expenses. (The actual 2007 operating expenses were approximately $2.5 billion.)

93. *Velocity* The position of a model train, in feet along a railroad track, is given by

$$s(t) = 2.5t + 10$$

after t seconds.

a. How fast is the train moving?
b. Where is the train after 4 seconds?
c. When will it be 25 feet along the track?

94. *Velocity* The height of a falling sheet of paper, in feet from the ground, is given by

$$s(t) = -1.8t + 9$$

after t seconds.

a. What is the velocity of the sheet of paper?
b. How high is it after 4 seconds?
c. When will it reach the ground?

95. ▼ *Fast Cars* A police car was traveling down Ocean Parkway in a high-speed chase from Jones Beach. It was at Jones Beach at exactly 10 pm ($t = 10$) and was at Oak Beach, 13 miles from Jones Beach, at exactly 10:06 pm.

a. How fast was the police car traveling? HINT [See Example 6.]
b. How far was the police car from Jones Beach at time t?

[43]Data are approximate. Sources: www.techcrunch.com, www.businessweek.com.

[44]Source: *New York Times,* February 10, 2007, p. A9.

[45]The zonar ($\overline{\overline{Z}}$) is the official currency in the city-state of Utarek, Mars (formerly www.Marsnext.com, a now extinct virtual society).

[46]Data are rounded. Sources: Department of Commerce/*New York Times*, September 5, 1995, p. D4; International Trade Administration, March 31, 2002, www.ita.doc.gov/.

[47]Recall that "net income" is another term for "profit." Source: www.wikinvest.com.
[48]*Ibid.*

96. ▼ *Fast Cars* The car that was being pursued by the police in Exercise 95 was at Jones Beach at exactly 9:54 pm ($t = 9.9$) and passed Oak Beach (13 miles from Jones Beach) at exactly 10:06 pm, where it was overtaken by the police.

 a. How fast was the car traveling? HINT [See Example 6.]
 b. How far was the car from Jones Beach at time t?

97. *Textbook Sizes* The second edition of *Applied Calculus* by Waner and Costenoble was 585 pages long. By the time we got to the fifth edition, the book had grown to 750 pages.

 a. Use this information to obtain the page length L as a linear function of the edition number n.
 b. What are the units of measurement of the slope? What does the slope tell you about the length of *Applied Calculus*?
 c. At this rate, by which edition will the book have grown to over 1,500 pages?

98. *Textbook Sizes* The second edition of *Finite Mathematics* by Waner and Costenoble was 603 pages long. By the time we got to the fifth edition, the book had grown to 690 pages.

 a. Use this information to obtain the page length L as a linear function of the edition number n.
 b. What are the units of measurement of the slope? What does the slope tell you about the length of *Finite Mathematics*?
 c. At this rate, by which edition will the book have grown to over 1,000 pages?

99. *Fahrenheit and Celsius* In the Fahrenheit temperature scale, water freezes at 32°F and boils at 212°F. In the Celsius scale, water freezes at 0°C and boils at 100°C. Further, the Fahrenheit temperature F and the Celsius temperature C are related by a linear equation. Find F in terms of C. Use your equation to find the Fahrenheit temperatures corresponding to 30°C, 22°C, -10°C, and -14°C, to the nearest degree.

100. *Fahrenheit and Celsius* Use the information about Celsius and Fahrenheit given in Exercise 99 to obtain a linear equation for C in terms of F, and use your equation to find the Celsius temperatures corresponding to 104°F, 77°F, 14°F, and -40°F, to the nearest degree.

Airline Net Income Exercises 101 and 102 are based on the following table, which compares the net incomes, in millions of dollars, of American Airlines, Continental Airlines, *and* Southwest Airlines:[49]

Year	2005	2006	2007	2008	2009	2010
American Airlines	−850	250	450	−2,100	−1,450	−300
Continental Airlines	−50	350	450	−600	−300	150
Southwest Airlines	550	500	650	200	100	300

101. a. Use the 2005 and 2009 data to obtain American's net income A as a linear function of Continental's net income C. (Use millions of dollars for all units.)

 b. How far off is your model in estimating American's net income based on Continental's 2007 income?
 c. What does the slope of the linear function from part (a) suggest about the net income of these two airlines?

102. a. Use the 2006 and 2010 data to obtain Continental's net income C as a linear function of Southwest's net income S. (Use millions of dollars for all units.)

 b. How far off is your model in estimating Continental's net income based on Southwest's 2008 income?
 c. What does the slope of the linear function from part (a) suggest about the net income of these two airlines?

103. ▼ *Income* The well-known romance novelist Celestine A. Lafleur (a.k.a. Bertha Snodgrass) has decided to sell the screen rights to her latest book, *Henrietta's Heaving Heart*, to Boxoffice Success Productions for $50,000. In addition, the contract ensures Ms. Lafleur royalties of 5% of the net profits.[50] Express her income I as a function of the net profit N, and determine the net profit necessary to bring her an income of $100,000. What is her marginal income (share of each dollar of net profit)?

104. ▼ *Income* Due to the enormous success of the movie *Henrietta's Heaving Heart* based on a novel by Celestine A. Lafleur (see Exercise 103), Boxoffice Success Productions decides to film the sequel, *Henrietta, Oh Henrietta*. At this point, Bertha Snodgrass (whose novels now top the best seller lists) feels she is in a position to demand $100,000 for the screen rights and royalties of 8% of the net profits. Express her income I as a function of the net profit N and determine the net profit necessary to bring her an income of $1,000,000. What is her marginal income (share of each dollar of net profit)?

105. *Processor Speeds* The processor speed, in megahertz (MHz), of Intel processors during the period 1996–2010 could be approximated by the following function of time t in years since the start of 1990:[51]

$$v(t) = \begin{cases} 400t - 2{,}200 & \text{if } 6 \le t < 15 \\ 3{,}800 & \text{if } 15 \le t \le 20. \end{cases}$$

How fast and in what direction was processor speed changing in 2000?

106. *Processor Speeds* The processor speed, in megahertz (MHz), of Intel processors during the period 1970–2000 could be approximated by the following function of time t in years since the start of 1970:[52]

$$v(t) = \begin{cases} 3t & \text{if } 0 \le t < 20 \\ 174t - 3{,}420 & \text{if } 20 \le t \le 30. \end{cases}$$

How fast and in what direction was processor speed changing in 1995?

---

[49]Data are rounded. 2010 net incomes projected based on first three quarters. Source: www.wikinvest.com.

[50]Percentages of net profit are commonly called "monkey points." Few movies ever make a net profit on paper, and anyone with any clout in the business gets a share of the *gross*, not the net.

[51]A rough model based on the fastest processors produced by Intel. Source: www.intel.com.

[52]*Ibid.*

Superbowl Advertising

Exercises 107 and 108 are based on the following graph and data from Wikipedia showing the increasing cost of a 30-second television ad during the Super Bowl.[53]

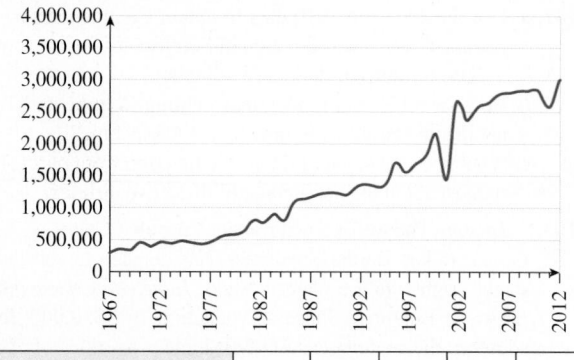

Year	1970	1980	1990	2000	2010
Cost (Thousands of $)	78	222	700	1,100	2,500

107. ▼ Take t to be the number of years since 1970 and y to be the cost, in thousands of dollars, of a Super Bowl ad.

  a. Model the 1970 and 1990 data with a linear equation.

  b. Model the 1990 and 2010 data with a linear equation.

  c. Use the results of parts (a) and (b) to obtain a piecewise linear model of the cost of a Super Bowl ad during 1970–2010.

  d. Use your model to estimate the cost in 2004. Is your answer in rough agreement with the graph? Explain any discrepancy.

108 ▼ Take t to be the number of years since 1980 and y to be the cost, in thousands of dollars, of a Super Bowl ad.

  a. Model the 1980 and 2000 data with a linear equation.

  b. Model the 2000 and 2010 data with a linear equation.

  c. Use the results of parts (a) and (b) to obtain a piecewise linear model of the cost of a Super Bowl ad during 1980–2010.

  d. Use your model to estimate the cost in 1992. Is your answer in rough agreement with the graph? Explain any discrepancy.

109. ▼ **Employment in Mexico** The number of workers employed in manufacturing jobs in Mexico was 3 million in 1995, rose to 4.1 million in 2000, and then dropped to 3.5 million in 2004.[54] Model this number N as a piecewise-linear function of the time t in years since 1995, and use your model to estimate the number of manufacturing jobs in Mexico in 2002. (Take the units of N to be millions.)

110. ▼ **Mortgage Delinquencies** The percentage of borrowers in the highest risk category who were delinquent on their payments decreased from 9.7% in 2001 to 4.3% in 2004 and

then shot up to 10.3% in 2007.[55] Model this percentage P as a piecewise-linear function of the time t in years since 2001, and use your model to estimate the percentage of delinquent borrowers in 2006.

COMMUNICATION AND REASONING EXERCISES

111. How would you test a table of values of x and y to see if it comes from a linear function?

112. You have ascertained that a table of values of x and y corresponds to a linear function. How do you find an equation for that linear function?

113. To what linear function of x does the linear equation $ax + by = c$ $(b \neq 0)$ correspond? Why did we specify $b \neq 0$?

114. Complete the following. The slope of the line with equation $y = mx + b$ is the number of units that ____ increases per unit increase in ____.

115. Complete the following. If, in a straight line, y is increasing three times as fast as x, then its ____ is ____.

116. Suppose that y is decreasing at a rate of 4 units per 3-unit increase of x. What can we say about the slope of the linear relationship between x and y? What can we say about the intercept?

117. If y and x are related by the linear expression $y = mx + b$, how will y change as x changes if m is positive? negative? zero?

118. Your friend April tells you that $y = f(x)$ has the property that, whenever x is changed by Δx, the corresponding change in y is $\Delta y = -\Delta x$. What can you tell her about f?

119. ▯ ▼ Consider the following worksheet:

◇	A	B	C	D
1	x	y	m	b
2		1	2	=(B3-B2)/(A3-A2)	=B2-C2*A2
3		3	-1	Slope	Intercept

What is the effect on the slope of increasing the y-coordinate of the second point (the point whose coordinates are in Row 3)? Explain.

120. ▯ ▼ Referring to the worksheet in Exercise 119, what is the effect on the slope of increasing the x-coordinate of the second point (the point whose coordinates are in row 3)? Explain.

121. If y is measured in bootlags[56] and x is measured in zonars[57], and $y = mx + b$, then m is measured in ____ and b is measured in ____.

---

[55]The 2007 figure was projected from data through October 2006. Source: *New York Times*, Februrary 18, 2007, p. BU9.

[56]An ancient Martian unit of length; one bootlag is the mean distance from a Martian's foreleg to its rearleg.

[57]The official currency of Utarek, Mars. (See the footnote to Exercise 88.)

---

[53]Source: http//en.wikipedia.org/wiki/Super_Bowl_advertising.

[54]Source: *New York Times*, Februrary 18, 2007, p. WK4.

122. If the slope in a linear relationship is measured in miles per dollar, then the independent variable is measured in _____ and the dependent variable is measured in _____.

123. If a quantity is changing linearly with time, and it increases by 10 units in the first day, what can you say about its behavior in the third day?

124. The quantities Q and T are related by a linear equation of the form

$$Q = mT + b.$$

When $T = 0$, Q is positive, but decreases to a negative quantity when T is 10. What are the signs of m and b? Explain your answers.

125. ▼ The velocity of an object is given by $v = 0.1t + 20$ m/sec, where t is time in seconds. The object is

(A) moving with fixed speed   **(B)** accelerating
(C) decelerating   **(D)** impossible to say from the given information

126. ▼ The position of an object is given by $x = 0.2t - 4$, where t is time in seconds. The object is

(A) moving with fixed speed   **(B)** accelerating
(C) decelerating   **(D)** impossible to say from the given information

127. If f and g are linear functions with slope m and n respectively, then what can you say about $f + g$?

128. If f and g are linear functions, then is $\frac{f}{g}$ linear? Explain.

129. Give examples of nonlinear functions f and g whose product is linear.

130. Give examples of nonlinear functions f and g whose quotient is linear (on a suitable domain).

131. ▼ Suppose the cost function is $C(x) = mx + b$ (with m and b positive), the revenue function is $R(x) = kx \ (k > m)$, and the number of items is increased from the break-even quantity. Does this result in a loss, a profit, or is it impossible to say? Explain your answer.

132. ▼ You have been constructing a demand equation, and you obtained a (correct) expression of the form $p = mq + b$, whereas you would have preferred one of the form $q = mp + b$. Should you simply switch p and q in the answer, should you start again from scratch, using p in the role of x and q in the role of y, or should you solve your demand equation for q? Give reasons for your decision.

1.4 Linear Regression

We have seen how to find a linear model given two data points: We find the equation of the line that passes through them. However, we often have more than two data points, and they will rarely all lie on a single straight line, but may often come close to doing so. The problem is to find the line coming *closest* to passing through all of the points.

Suppose, for example, that we are conducting research for a company interested in expanding into Mexico. Of interest to us would be current and projected growth in that country's economy. The following table shows past and projected per capita gross domestic product (GDP)[58] of Mexico for 2000–2014.[59]

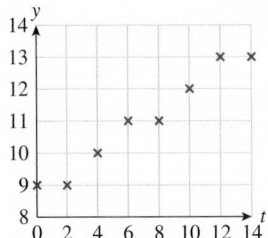

Figure 27(a)

Year t ($t = 0$ represents 2000)	0	2	4	6	8	10	12	14
Per Capita GDP y ($1,000)	9	9	10	11	11	12	13	13

A plot of these data suggests a roughly linear growth of the GDP (Figure 27(a)). These points suggest a roughly linear relationship between t and y, although they clearly do not all lie on a single straight line. Figure 27(b) shows the points together with several lines, some fitting better than others. Can we precisely measure which lines fit better than others? For instance, which of the two lines labeled as "good" fits in Figure 27(b) models the data more accurately? We begin by considering, for each value of t, the difference between the actual GDP (the **observed value**) and the GDP

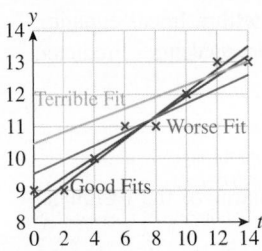

Figure 27(b)

---

[58]The GDP is a measure of the total market value of all goods and services produced within a country.

[59]Data are approximate and/or projected. Sources: CIA World Factbook/www.indexmundi.com, www.economist.com.

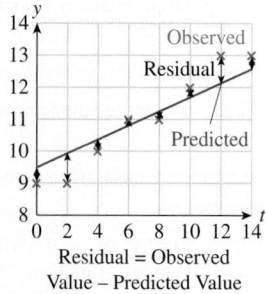

Residual = Observed
Value – Predicted Value

Figure 28

✱ Why not add the absolute values of the residuals instead? Mathematically, using the squares rather than the absolute values results in a simpler and more elegant solution. Further, using the squares always results in a *single* best-fit line in cases where the x-coordinates are all different, whereas this is not the case if we use absolute values.

predicted by a linear equation (the **predicted value**). The difference between the predicted value and the observed value is called the **residual**.

$$\text{Residual} = \text{Observed Value} - \text{Predicted Value}$$

On the graph, the residuals measure the vertical distances between the (observed) data points and the line (Figure 28) and they tell us how far the linear model is from predicting the actual GDP.

The more accurate our model, the smaller the residuals should be. We can combine all the residuals into a single measure of accuracy by adding their *squares*. (We square the residuals in part to make them all positive.✱) The sum of the squares of the residuals is called the **sum-of-squares error**, SSE. Smaller values of SSE indicate more accurate models.

Here are some definitions and formulas for what we have been discussing.

Observed and Predicted Values

Suppose we are given a collection of data points $(x_1, y_1), \ldots, (x_n, y_n)$. The n quantities $y_1, y_2, \ldots, y_n$ are called the **observed y-values**. If we model these data with a linear equation

$$\hat{y} = mx + b, \qquad \text{$\hat{y}$ stands for "estimated y" or "predicted y."}$$

then the y-values we get by substituting the given x-values into the equation are called the **predicted y-values**:

$$\hat{y}_1 = mx_1 + b \qquad \text{Substitute x_1 for x.}$$
$$\hat{y}_2 = mx_2 + b \qquad \text{Substitute x_2 for x.}$$
$$\ldots$$
$$\hat{y}_n = mx_n + b. \qquad \text{Substitute x_n for x.}$$

Quick Example

Consider the three data points $(0, 2)$, $(2, 5)$, and $(3, 6)$. The observed y-values are $y_1 = 2$, $y_2 = 5$, and $y_3 = 6$. If we model these data with the equation $\hat{y} = x + 2.5$, then the predicted values are:

$$\hat{y}_1 = x_1 + 2.5 = 0 + 2.5 = 2.5$$
$$\hat{y}_2 = x_2 + 2.5 = 2 + 2.5 = 4.5$$
$$\hat{y}_3 = x_3 + 2.5 = 3 + 2.5 = 5.5.$$

Residuals and Sum-of-Squares Error (SSE)

If we model a collection of data $(x_1, y_1), \ldots, (x_n, y_n)$ with a linear equation $\hat{y} = mx + b$, then the **residuals** are the n quantities (Observed Value – Predicted Value):

$$(y_1 - \hat{y}_1), (y_2 - \hat{y}_2), \ldots, (y_n - \hat{y}_n).$$

The **sum-of-squares error (SSE)** is the sum of the squares of the residuals:

$$\text{SSE} = (y_1 - \hat{y}_1)^2 + (y_2 - \hat{y}_2)^2 + \cdots + (y_n - \hat{y}_n)^2.$$

> **Quick Example**
>
> For the data and linear approximation given above, the residuals are:
>
> $$y_1 - \hat{y}_1 = 2 - 2.5 = -0.5$$
> $$y_2 - \hat{y}_2 = 5 - 4.5 = 0.5$$
> $$y_3 - \hat{y}_3 = 6 - 5.5 = 0.5$$
>
> and so SSE $= (-0.5)^2 + (0.5)^2 + (0.5)^2 = 0.75$.

EXAMPLE 1 Computing SSE

Using the data above on the GDP in Mexico, compute SSE for the linear models $y = 0.5t + 8$ and $y = 0.25t + 9$. Which model is the better fit?

Solution We begin by creating a table showing the values of t, the observed (given) values of y, and the values predicted by the first model.

Year t	Observed y	Predicted $\hat{y} = 0.5t + 8$
0	9	8
2	9	9
4	10	10
6	11	11
8	11	12
10	12	13
12	13	14
14	13	15

We now add two new columns for the residuals and their squares.

Year t	Observed y	Predicted $\hat{y} = 0.5t + 8$	Residual $y - \hat{y}$	Residual² $(y - \hat{y})^2$
0	9	8	$9 - 8 = 1$	$1^2 = 1$
2	9	9	$9 - 9 = 0$	$0^2 = 0$
4	10	10	$10 - 10 = 0$	$0^2 = 0$
6	11	11	$11 - 11 = 0$	$0^2 = 0$
8	11	12	$11 - 12 = -1$	$(-1)^2 = 1$
10	12	13	$12 - 13 = -1$	$(-1)^2 = 1$
12	13	14	$13 - 14 = -1$	$(-1)^2 = 1$
14	13	15	$13 - 15 = -2$	$(-2)^2 = 4$

using Technology

See the Technology Guides at the end of the chapter for detailed instructions on how to obtain the tables and graphs in Example 1 using a TI-83/84 Plus or a spreadsheet. Here is an outline:

TI-83/84 Plus

STAT EDIT

Values of t in L₁, and y in L₂.
Predicted y: Highlight L₃. Enter
`0.5*L₁+8`
Squares of residuals: Highlight L₄.
Enter
`(L₂-L₃)^2`
SSE: Home screen `sum(L₄)`
Graph: `Y₁=0.5X+8`
Y = screen: Turn on Plot 1 ZOOM
(STAT) [More details on page 117.]

Spreadsheet

Headings t, y, y-hat, Residual^2, m, b, and SSE in A1–F1.
t-values in A2–A9, y-values in B2–B9; 0.25 for m and 9 for b in E2–F2
Predicted y: `=$E$2*A2+$F$2` in C2 and copy down to C9.
Squares of residuals: `=(B2-C2)^2` in D2 and copy down to D9.
SSE: `=SUM(D2:D9)` in G2
Graph: Highlight A1–C9. Insert a Scatter chart.
[More details on page 122.]

SSE, the sum of the squares of the residuals, is then the sum of the entries in the last column,

$$SSE = 8.$$

Repeating the process using the second model, $0.25t + 9$, yields the following table:

Year t	Observed y	Predicted $\hat{y} = 0.25t + 9$	Residual $y - \hat{y}$	Residual^2 $(y - \hat{y})^2$
0	9	9	$9 - 9 = 0$	$0^2 = 0$
2	9	9.5	$9 - 9.5 = -0.5$	$(-0.5)^2 = 0.25$
4	10	10	$10 - 10 = 0$	$0^2 = 0$
6	11	10.5	$11 - 10.5 = 0.5$	$0.5^2 = 0.25$
8	11	11	$11 - 11 = 0$	$0^2 = 0$
10	12	11.5	$12 - 11.5 = 0.5$	$0.5^2 = 0.25$
12	13	12	$13 - 12 = 1$	$1^2 = 1$
14	13	12.5	$13 - 12.5 = 0.5$	$0.5^2 = 0.25$

This time, SSE = 2 and so the second model is a better fit.

Figure 29 shows the data points and the two linear models in question.

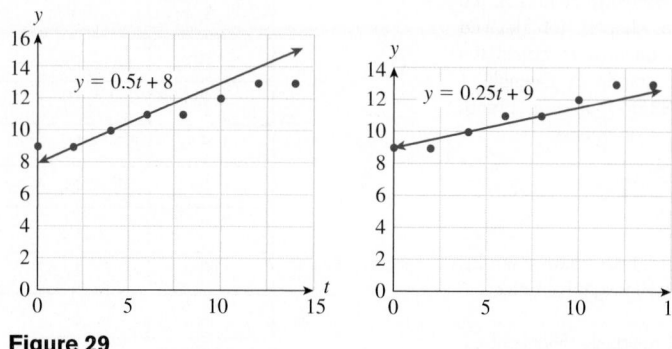

Figure 29

➡ **Before we go on...**

Q: *It seems clear from the figure that the second model in Example 1 gives a better fit. Why bother to compute SSE to tell me this?*

A: The difference between the two models we chose is so great that it is clear from the graphs which is the better fit. However, if we used a third model with $m = 0.25$ and $b = 9.1$, then its graph would be almost indistinguishable from that of the second, but a slightly better fit as measured by SSE = 1.68.

■

Among all possible lines, there ought to be one with the least possible value of SSE—that is, the greatest possible accuracy as a model. The line (and there is only one such line) that minimizes the sum of the squares of the residuals is called the **regression line**, the **least-squares line**, or the **best-fit line**.

To find the regression line, we need a way to find values of m and b that give the smallest possible value of SSE. As an example, let us take the second linear model in the example above. We said in the "Before we go on" discussion that increasing b from 9 to 9.1 had the desirable effect of decreasing SSE from 2 to 1.68. We could then increase m to 0.26, further reducing SSE to 1.328. Imagine this as a kind of game: Alternately alter the values of m and b by small amounts until SSE is as small as you can make it. This works, but is extremely tedious and time-consuming.

Fortunately, there is an algebraic way to find the regression line. Here is the calculation. To justify it rigorously requires calculus of several variables or linear algebra.

Regression Line

The **regression line** (**least squares line, best-fit line**) associated with the points (x_1, y_1), (x_2, y_2), . . . , (x_n, y_n) is the line that gives the minimum SSE. The regression line is

$$y = mx + b,$$

where m and b are computed as follows:

$$m = \frac{n\left(\sum xy\right) - \left(\sum x\right)\left(\sum y\right)}{n\left(\sum x^2\right) - \left(\sum x\right)^2}$$

$$b = \frac{\sum y - m\left(\sum x\right)}{n}$$

$n =$ number of data points.

The quantities m and b are called the **regression coefficients**.

Here, "$\sum$" means "the sum of." Thus, for example,

$$\sum x = \text{Sum of the } x\text{-values} = x_1 + x_2 + \cdots + x_n$$
$$\sum xy = \text{Sum of products} = x_1y_1 + x_2y_2 + \cdots + x_ny_n$$
$$\sum x^2 = \text{Sum of the squares of the } x\text{-values} = x_1^2 + x_2^2 + \cdots + x_n^2.$$

On the other hand,

$$\left(\sum x\right)^2 = \text{Square of } \sum x = \text{Square of the sum of the } x\text{-values}.$$

EXAMPLE 2 Per Capita Gross Domestic Product in Mexico

In Example 1 we considered the following data on the per capita gross domestic product (GDP) of Mexico:

Year x ($x = 0$ represents 2000)	0	2	4	6	8	10	12	14
Per Capita GDP y ($1,000)	9	9	10	11	11	12	13	13

Find the best-fit linear model for these data and use the model to predict the per capita GDP in Mexico in 2016.

using Technology

See the Technology Guides at the end of the chapter for detailed instructions on how to obtain the regression line and graph in Example 2 using a TI-83/84 Plus or a spreadsheet. Here is an outline:

TI-83/84 Plus

[STAT] EDIT

Values of x in L_1, and y in L_2.

Regression equation: [STAT] CALC option #4: `LinReg(ax+b)`

Graph: [Y=] [VARS] [5] EQ [1], then [ZOOM] [9]

[More details on page 118.]

Spreadsheet

x-values in A2–A9, y-values in B2–B9

Graph: Highlight A2–B9. Insert a Scatter Chart.

Regression line: Add a linear trendline. [More details on page 123.]

Website

www.WanerMath.com

The following two utilities will calculate and plot regression lines (link to either from Math Tools for Chapter 1):

Simple Regression

Function Evaluator and Grapher

Solution Let's organize our work in the form of a table, where the original data are entered in the first two columns and the bottom row contains the column sums.

	x	y	xy	x^2
	0	9	0	0
	2	9	18	4
	4	10	40	16
	6	11	66	36
	8	11	88	64
	10	12	120	100
	12	13	156	144
	14	13	182	196
$\sum$ (Sum)	56	88	670	560

Because there are $n = 8$ data points, we get

$$m = \frac{n(\sum xy) - (\sum x)(\sum y)}{n(\sum x^2) - (\sum x)^2} = \frac{8(670) - (56)(88)}{8(560) - (56)^2} \approx 0.321$$

and

$$b = \frac{\sum y - m(\sum x)}{n} \approx \frac{88 - (0.321)(56)}{8} \approx 8.75.$$

So, the regression line is

$$y = 0.321x + 8.75.$$

To predict the per capita GDP in Mexico in 2016 we substitute $x = 16$ and get $y \approx 14$, or \$14,000 per capita.

Figure 30 shows the data points and the regression line (which has SSE ≈ 0.643; a lot lower than in Example 1).

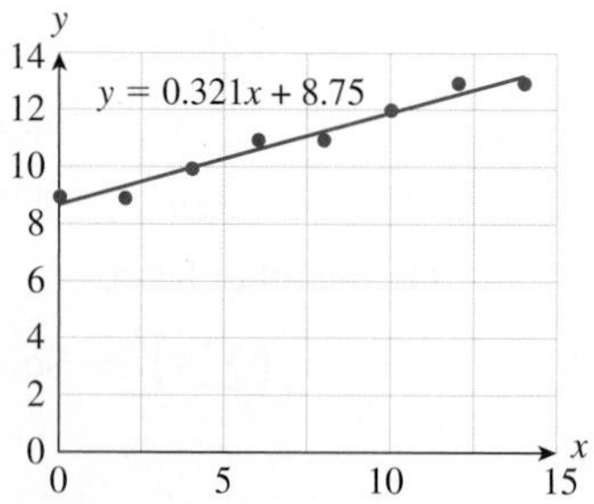

Figure 30

Coefficient of Correlation

If all the data points do not lie on one straight line, we would like to be able to measure how closely they can be approximated by a straight line. Recall that SSE measures the sum of the squares of the deviations from the regression line; therefore it constitutes a measurement of what is called "goodness of fit." (For instance, if SSE $= 0$, then all the points lie on a straight line.) However, SSE depends on the units we use to measure y, and also on the number of data points (the more data points we use, the larger SSE tends to be). Thus, while we can (and do) use SSE to compare the goodness of fit of two lines to the same data, we cannot use it to compare the goodness of fit of one line to one set of data with that of another to a different set of data.

To remove this dependency, statisticians have found a related quantity that can be used to compare the goodness of fit of lines to different sets of data. This quantity, called the **coefficient of correlation** or **correlation coefficient**, and usually denoted r, is between -1 and 1. The closer r is to -1 or 1, the better the fit. For an *exact* fit, we would have $r = -1$ (for a line with negative slope) or $r = 1$ (for a line with positive slope). For a bad fit, we would have r close to 0. Figure 31 shows several collections of data points with least-squares lines and the corresponding values of r.

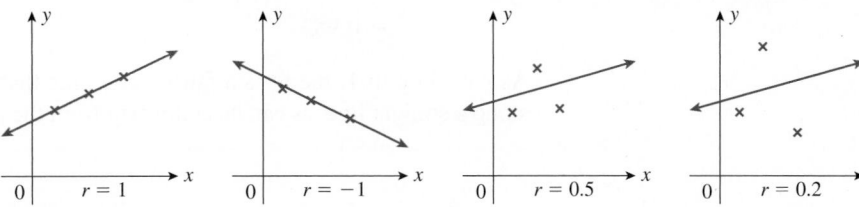

Figure 31

Correlation Coefficient

The coefficient of correlation of the n data points $(x_1, y_1), (x_2, y_2), \ldots, (x_n, y_n)$ is

$$r = \frac{n(\sum xy) - (\sum x)(\sum y)}{\sqrt{n(\sum x^2) - (\sum x)^2} \cdot \sqrt{n(\sum y^2) - (\sum y)^2}}.$$

It measures how closely the data points $(x_1, y_1), (x_2, y_2), \ldots, (x_n, y_n)$ fit the regression line. (The value r^2 is sometimes called the **coefficient of determination**.)

Interpretation

- If r is positive, the regression line has positive slope; if r is negative, the regression line has negative slope.

- If $r = 1$ or -1, then all the data points lie exactly on the regression line; if it is close to ± 1, then all the data points are close to the regression line.

- On the other hand, if r is not close to ± 1, then the data points are not close to the regression line, so the fit is not a good one. As a general rule of thumb, a value of $|r|$ less than around 0.8 indicates a poor fit of the data to the regression line.

using Technology

See the Technology Guides at the end of the chapter for detailed instructions on how to obtain the correlation coefficient in Example 3 using a TI-83/84 Plus or a spreadsheet. Here is an outline:

TI-83/84 Plus

2ND CATALOG DiagnosticOn
Then STAT CALC option #4:
LinReg(ax+b) [More details on page 119.]

Spreadsheet

Add a trendline and select the option to "Display R-squared value on chart."
[More details and other alternatives on page 124.]

 Website

www.WanerMath.com
The following two utilities will show regression lines and also r^2 (link to either from Math Tools for Chapter 1):

Simple Regression

Function Evaluator and Grapher

EXAMPLE 3 Computing the Coefficient of Correlation

Find the correlation coefficient for the data in Example 2. Is the regression line a good fit?

Solution The formula for r requires $\sum x, \sum x^2, \sum xy, \sum y$, and $\sum y^2$. We have all of these except for $\sum y^2$, which we find in a new column as shown.

x	y	xy	x^2	y^2
0	9	0	0	81
2	9	18	4	81
4	10	40	16	100
6	11	66	36	121
8	11	88	64	121
10	12	120	100	144
12	13	156	144	169
14	13	182	196	169
$\sum$ **(Sum)** 56	88	670	560	986

Substituting these values into the formula, we get

$$r = \frac{n\left(\sum xy\right) - \left(\sum x\right)\left(\sum y\right)}{\sqrt{n\left(\sum x^2\right) - \left(\sum x\right)^2} \cdot \sqrt{n\left(\sum y^2\right) - \left(\sum y\right)^2}}$$

$$= \frac{8(670) - (56)(88)}{\sqrt{8(560) - 56^2} \cdot \sqrt{8(986) - 88^2}}$$

$$\approx 0.982.$$

As r is close to 1, the fit is a fairly good one; that is, the original points lie nearly along a straight line, as can be confirmed from the graph in Example 2.

1.4 EXERCISES

▼ more advanced   ◆ challenging
Ⓣ indicates exercises that should be solved using technology

In Exercises 1–4, compute the sum-of-squares error (SSE) by hand for the given set of data and linear model. HINT [See Example 1.]

1. $(1, 1), (2, 2), (3, 4)$; $y = x - 1$

2. $(0, 1), (1, 1), (2, 2)$; $y = x + 1$

3. $(0, -1), (1, 3), (4, 6), (5, 0)$; $y = -x + 2$

4. $(2, 4), (6, 8), (8, 12), (10, 0)$; $y = 2x - 8$

Ⓣ *In Exercises 5–8, use technology to compute the sum-of-squares error (SSE) for the given set of data and linear models. Indicate which linear model gives the better fit.*

5. $(1, 1), (2, 2), (3, 4)$;   **a.** $y = 1.5x - 1$ **b.** $y = 2x - 1.5$

6. $(0, 1), (1, 1), (2, 2)$;   **a.** $y = 0.4x + 1.1$ **b.** $y = 0.5x + 0.9$

7. $(0, -1), (1, 3), (4, 6), (5, 0)$;   **a.** $y = 0.3x + 1.1$
    b. $y = 0.4x + 0.9$

8. $(2, 4), (6, 8), (8, 12), (10, 0)$;   **a.** $y = -0.1x + 7$
    b. $y = -0.2x + 6$

Find the regression line associated with each set of points in Exercises 9–12. Graph the data and the best-fit line. (Round all coefficients to 4 decimal places.) HINT [See Example 2.]

9. $(1, 1), (2, 2), (3, 4)$

10. $(0, 1), (1, 1), (2, 2)$

11. $(0, -1), (1, 3), (3, 6), (4, 1)$

12. $(2, 4), (4, 8), (8, 12), (10, 0)$

In the next two exercises, use correlation coefficients to determine which of the given sets of data is best fit by its associated regression line and which is fit worst. Is it a perfect fit for any of the data sets? HINT [See Example 3.]

13. a. $\{(1, 3), (2, 4), (5, 6)\}$
    b. $\{(0, -1), (2, 1), (3, 4)\}$
    c. $\{(4, -3), (5, 5), (0, 0)\}$

14. a. $\{(1, 3), (-2, 9), (2, 1)\}$
    b. $\{(0, 1), (1, 0), (2, 1)\}$
    c. $\{(0, 0), (5, -5), (2, -2.1)\}$

APPLICATIONS

15. *Mobile Broadband Subscriptions* The following table shows the number of mobile broadband subscribers worldwide (x is the number of years since 2000):[60]

Year x	2	6	10
Subscribers y (Millions)	0	70	900

Complete the following table and obtain the associated regression line. (Round coefficients to one decimal place.) HINT [See Example 3.]

x	y	xy	x^2
2	0		
6	70		
10	900		
$\sum$ (**Sum**)			

Use your regression equation to project the number in 2012.

16. *Investment in Gold* Following are approximate values of the Amex Gold BUGS Index ($x = 0$ represents 2000):[61]

Year x	0	5	10
Index y	50	200	500

[60]Data are rounded (2010 figure is estimated). Source: International Telecommunication Union www.itu.int/ITU-D/ict/statistics/at_glance/KeyTelecom.html.

[61]BUGS stands for "basket of unhedged gold stocks." Figures are approximate. Source: www.google.com/finance.

Complete the following table and obtain the associated regression line. (Round coefficients to the nearest whole number.)
HINT [See Example 3.]

x	y	xy	x^2
0	50		
5	200		
10	500		
$\sum$ (Sum)			

Use your regression equation to project the index in 2011.

17. E-Commerce The following chart shows second quarter total retail e-commerce sales in the U.S. in 2000, 2005, and 2010 ($t = 0$ represents 2000):[62]

Year t	0	5	10
Sales ($ billion) y	6	20	40

Find the regression line (round coefficients to one decimal place) and use it to estimate second quarter retail e-commerce sales in 2006. (The actual figure was approximately $25 billion.)

18. Net Sales The following table show the reported net sales by Nokia in the third quarters of 2008, 2009, and 2010 ($t = 0$ represents the third quarter of 2008):[63]

Year t	0	1	2
Net Sales (€ billion) y	12	10	10

Find the regression line (round coefficients to one decimal place) and use it to "predict" Nokia's net income in the third quarter of 2011. (The actual figure was approximately €9.0 billion.)

19. Oil Recovery The Texas Bureau of Economic Geology published a study on the economic impact of using carbon dioxide enhanced oil recovery (EOR) technology to extract additional oil from fields that have reached the end of their conventional economic life. The following table gives the approximate number of jobs for the citizens of Texas that would be created at various levels of recovery.[64]

Percent Recovery (%)	20	40	80	100
Jobs Created (Millions)	3	6	9	15

Find the regression line and use it to estimate the number of jobs that would be created at a recovery level of 50%.

20. Oil Recovery (Refer to Exercise 19.) The following table gives the approximate economic value associated with various levels of oil recovery in Texas.[65]

Percent Recovery (%)	10	40	50	80
Economic Value ($ billions)	200	900	1,000	2,000

Find the regression line and use it to estimate the economic value associated with a recovery level of 70%.

21. Profit: Amazon.com The following table shows Amazon.com's approximate net sales (revenue) and net income (profit) in the period 2006–2009:[66]

Net Sales ($ billions)	10	15	20	25
Net Income ($ millions)	200	500	600	900

a. Use this information to find a linear regression model for Amazon's net income I (in millions of dollars) as a function of net sales S (in billions of dollars).
b. Give the units of measurement and interpretation of the slope.
c. What, according to the model, would Amazon.com need to earn in net sales in order for its net income to be $1 billion? (Round answer to the nearest $ billion.)
d. Plot the data and regression line. Based on the graph, would you say that the linear model is reasonable? Why?

22. Operating Expenses: Amazon.com The following table shows Amazon.com's approximate net sales (revenue) and operating expenses in 2006–2009:[67]

Net Sales ($ billions)	10	15	20	25
Operating Expenses ($ billions)	2	2.5	3	4.5

a. Use this information to find a linear regression model for Amazon's operating expenses E (in billions of dollars) as a function of net sales S (in billions of dollars).
b. Give the units of measurement and interpretation of the slope.
c. What, according to the model, would Amazon.com need to earn in net sales in order for its operating expenses to be $5 billion? (Round answer to the nearest $ billion.)
d. Plot the data and regression line. Based on the graph, would you say that the linear model is reasonable? Why?

---

[62]Figures are rounded. Source: U.S. Census Bureau www.census.gov

[63]Data are approximate. Source: www.nokia.com.

[64]Source: "CO2–Enhanced Oil Recovery Resource Potential in Texas: Potential Positive Economic Impacts," Texas Bureau of Economic Geology, April 2004, www.rrc.state.tx.us/tepc/CO2-EOR_white_paper.pdf.

[65]Ibid.

[66]Figures are approximate. Source: www.wikinvest.com.

[67]Ibid.

23. 🖵 *Textbook Sizes* The following table shows the numbers of pages in previous editions of *Applied Calculus* by Waner and Costenoble:

Edition n	2	3	4	5
Number of Pages L	585	656	694	748

a. With the edition number as the independent variable, use technology to obtain a regression line and a plot of the points together with the regression line. (Round coefficients to two decimal places.)

b. Interpret the slope of the regression line.

24. 🖵 *Textbook Sizes* Repeat Exercise 23 using the following corresponding table for *Finite Mathematics* by Waner and Costenoble:

Edition n	2	3	4	5
Number of Pages L	603	608	676	692

25. 🖵 *Soybean Production: Cerrados* The following table shows soybean production, in millions of tons, in Brazil's *Cerrados* region, as a function of the cultivated area, in millions of acres.[68]

Area (millions of acres)	25	30	32	40	52
Production (millions of tons)	15	25	30	40	60

a. Use technology to obtain the regression line and to show a plot of the points together with the regression line. (Round coefficients to two decimal places.)

b. Interpret the slope of the regression line.

26. 🖵 *Soybean Production: U.S.* The following table shows soybean production, in millions of tons, in the U.S. as a function of the cultivated area, in millions of acres.[69]

Area (millions of acres)	30	42	69	59	74	74
Production (millions of tons)	20	33	55	57	83	88

a. Use technology to obtain the regression line and to show a plot of the points together with the regression line. (Round coefficients to two decimal places.)

b. Interpret the slope of the regression line.

27. 🖵 *Airline Profits and the Price of Oil* A common perception is that airline profits are strongly correlated with the price of oil. Following are annual net incomes of Continental Airlines together with the approximate price of oil in the period 2005–2010:[70]

Year	2005	2006	2007	2008	2009	2010
Price of Oil ($ per barrel)	56	63	67	92	54	71
Continental Net Income ($ million)	−70	370	430	−590	−280	150

a. Use technology to obtain a regression line showing Continental's net income as a function of the price of oil, and also the coefficient of correlation r.

b. What does the value of r suggest about the relationship of Continental's net income to the price of oil?

c. Support your answer to part (b) with a plot of the data and regression line.

28. 🖵 *Airline Profits and the Price of Oil* Repeat Exercise 27 using the following corresponding data for American Airlines:[71]

Year	2005	2006	2007	2008	2009	2010
Price of Oil ($ per barrel)	56	63	67	92	54	71
American Net Income ($ million)	−850	250	450	−2,100	−1,450	−700

🖵 *Doctorates in Mexico* Exercises 29–32 are based on the following table showing the annual number of PhD graduates in Mexico in various fields.[72]

	Natural Sciences	Engineering	Social Sciences
1990	84	8	98
1995	107	55	161
2000	174	247	222
2005	515	371	584
2010	979	578	1,230

29. a. With x = the number of natural science doctorates and y = the number of engineering doctorates, use technology to obtain the regression equation and graph the associated points and regression line. (Round coefficients to three significant digits.)

b. What does the slope tell you about the relationship between the number of natural science doctorates and the number of engineering doctorates?

c. Use technology to obtain the coefficient of correlation r. Does the value of r suggest a strong correlation between x and y?

d. Judging from the graph, would you say that a linear relationship between x and y is appropriate? Why?

30. a. With x = the number of natural science doctorates and y = the number of social science doctorates, use technology to obtain the regression equation and graph the associated points and regression line. (Round coefficients to three significant digits.)

---

[68]Source: Brazil Agriculture Ministry/*New York Times*, December 12, 2004, p. N32.

[69]Data are approximate. Source for data: L. David Roper, June 2010, *Crop Production in the World & the United States*, www.roperld.com/science/cropsworld&us.htm.

[70]Figures are rounded and oil prices are inflation adjusted. Sources: www.wikinvest.com, www.inflationdata.com.

[71]*Ibid.*

[72]2010 data is estimated. Source: Instituto Nacional de Estadística y Geografía www.inegi.org.mx.

b. What does the slope tell you about the relationship between the number of natural science doctorates and the number of social science doctorates?

c. Use technology to obtain the coefficient of correlation r. Does the value of r suggest a strong correlation between x and y?

d. Judging from the graph, would you say that a linear relationship between x and y is appropriate? Why?

31. ▼ **a.** Use technology to obtain the regression equation and the coefficient of correlation r for the number of natural science doctorates as a function of time t in years since 1990, and graph the associated points and regression line. (Round coefficients to three significant digits.)

b. What does the slope tell you about the number of natural science doctorates?

c. Judging from the graph, would you say that the number of natural science doctorates is increasing at a faster and faster rate, a slower and slower rate, or neither? Why?

d. If r had been equal to 1, could you have drawn the same conclusion as in part (c)? Explain.

32. ▼ **a.** Use technology to obtain the regression equation and the coefficient of correlation r for the number of engineering doctorates as a function of time t in years since 1990, and graph the associated points and regression line. (Round coefficients to three significant digits.)

b. What does the slope tell you about the number of engineering doctorates?

c. Judging from the graph, would you say that the number of engineering doctorates is increasing at a faster and faster rate, a slower and slower rate, or neither? Why?

d. If r had been close to 0, could you have drawn the same conclusion as in part (c)? Explain.

33. ▣ ▼ *NY City Housing Costs: Downtown* The following table shows the average price of a two-bedroom apartment in downtown New York City from 1994 to 2004 ($t = 0$ represents 1994).[73]

Year t	0	2	4	6	8	10
Price p ($ million)	0.38	0.40	0.60	0.95	1.20	1.60

a. Use technology to obtain the linear regression line and correlation coefficient r, with all coefficients rounded to two decimal places, and plot the regression line and the given points.

b. Does the graph suggest that a non-linear relationship between t and p would be more appropriate than a linear one? Why?

c. Use technology to obtain the residuals. What can you say about the residuals in support of the claim in part (b)?

34. ▣ ▼ *Fiber Optic Connections* The following table shows the number of fiber optic cable connections to homes in the U.S. from 2000 to 2004 ($t = 0$ represents 2000):[74]

Year t	0	1	2	3	4
Connections c (Thousands)	0	10	25	65	150

a. Use technology to obtain the linear regression line and correlation coefficient r, with all coefficients rounded to two decimal places, and plot the regression line and the given points.

b. Does the graph suggest that a non-linear relationship between t and c would be more appropriate than a linear one? Why?

c. Use technology to obtain the residuals. What can you say about the residuals in support of the claim in part (b)?

COMMUNICATION AND REASONING EXERCISES

35. Why is the regression line associated with the two points (a, b) and (c, d) the same as the line that passes through both? (Assume that $a \neq c$.)

36. What is the smallest possible sum-of-squares error if the given points happen to lie on a straight line? Why?

37. If the points (x_1, y_1), (x_2, y_2), . . . , (x_n, y_n) lie on a straight line, what can you say about the regression line associated with these points?

38. If all but one of the points (x_1, y_1), (x_2, y_2), . . . , (x_n, y_n) lie on a straight line, must the regression line pass through all but one of these points?

39. ▼ Verify that the regression line for the points $(0, 0)$, $(-a, a)$, and (a, a) has slope 0. What is the value of r? (Assume that $a \neq 0$.)

40. ▼ Verify that the regression line for the points $(0, a)$, $(0, -a)$, and $(a, 0)$ has slope 0. What is the value of r? (Assume that $a \neq 0$.)

41. ▼ Must the regression line pass through at least one of the data points? Illustrate your answer with an example.

42. ▼ Why must care be taken when using mathematical models to extrapolate?

43. ▼ Your friend Imogen tells you that if r for a collection of data points is more than 0.9, then the most appropriate relationship between the variables is a linear one. Explain why she is wrong by referring to one of the exercises.

44. ▼ Your other friend Mervyn tells you that if r for a collection of data points has an absolute value less than 0.8, then the most appropriate relationship between the variables is a quadratic and not a linear one. Explain why *he* is wrong.

---

[73]Data are rounded and 2004 figure is an estimate. Source: Miller Samuel/*New York Times*, March 28, 2004, p. RE 11.

[74]Source: Render, Vanderslice & Associates/*New York Times*, October 11, 2004, p. C1.

CHAPTER 1 REVIEW

KEY CONCEPTS

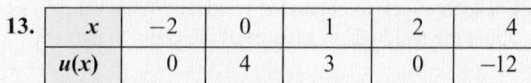

Website www.WanerMath.com

Go to the Website at www.WanerMath .com to find a comprehensive and interactive Web-based summary of Chapter 1.

1.1 Functions from the Numerical, Algebraic, and Graphical Viewpoints

Real-valued function f of a real-valued variable x, domain *p. 42*

Independent and dependent variables *p. 42*

Graph of the function f *p. 42*

Numerically specified function *p. 42*

Graphically specified function *p. 42*

Algebraically defined function *p. 42*

Piecewise-defined function *p. 47*

Vertical line test *p. 49*

Common types of algebraic functions and their graphs *p. 50*

1.2 Functions and Models

Mathematical model *p. 56*

Analytical model *p. 57*

Curve-fitting model *p. 57*

Cost, revenue, and profit; marginal cost, revenue, and profit; break-even point *p. 59*

Demand, supply, and equilibrium price *pp. 61–62*

Selecting a model *p. 64*

Compound interest *p. 66*

Exponential growth and decay *p. 68*

Algebra of functions (sum, difference, product, quotient) *p. 70*

1.3 Linear Functions and Models

Linear function $f(x) = mx + b$ *p. 78*

Change in q: $\Delta q = q_2 - q_1$ *p. 79*

Slope of a line:

$$m = \frac{\Delta y}{\Delta x} = \frac{\text{Change in } y}{\text{Change in } x} \quad p.\ 79$$

Interpretations of m *p. 80*

Interpretation of b: y-intercept *p. 80*

Recognizing linear data *p. 81*

Computing the slope of a line *p. 82*

Slopes of horizontal and vertical lines *p. 82*

Computing the y-intercept *p. 83*

Linear modeling *p. 84*

Linear cost *p. 85*

Linear demand *p. 85*

Linear change over time; rate of change; velocity *p. 88*

General linear models *p. 88*

1.4 Linear Regression

Observed and predicted values *p. 96*

Residuals and sum-of-squares error (SSE) *p. 96*

Regression line (least-squares line, best-fit line) *p. 99*

Correlation coefficient; coefficient of determination *p. 101*

REVIEW EXERCISES

In Exercises 1–4, use the graph of the function f to find approximations of the given values.

1.

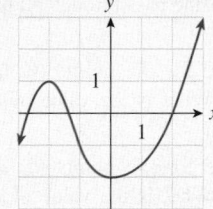

a. $f(-2)$ **b.** $f(0)$
c. $f(2)$ **d.** $f(2) - f(-2)$

2.

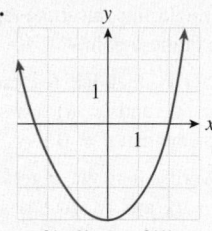

a. $f(-2)$ **b.** $f(0)$
c. $f(2)$ **d.** $f(2) - f(-2)$

3.

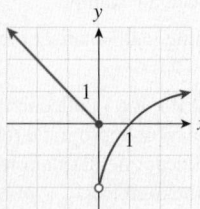

a. $f(-1)$ **b.** $f(0)$
c. $f(1)$ **d.** $f(1) - f(-1)$

4.

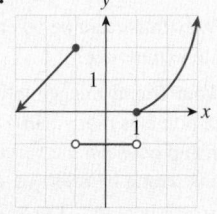

a. $f(-1)$ **b.** $f(0)$
c. $f(1)$ **d.** $f(1) - f(-1)$

In Exercises 5–8, graph the given function or equation.

5. $y = -2x + 5$ **6.** $2x - 3y = 12$

7. $y = \begin{cases} \frac{1}{2}x & \text{if } -1 \le x \le 1 \\ x - 1 & \text{if } 1 < x \le 3 \end{cases}$

8. $f(x) = 4x - x^2$ with domain $[0, 4]$

In Exercises 9–14, decide whether the specified values come from a linear, quadratic, exponential, or absolute value function.

9.

x	-2	0	1	2	4
$f(x)$	4	2	1	0	2

10.

x	-2	0	1	2	4
$g(x)$	-5	-3	-2	-1	1

11.

x	-2	0	1	2	4
$h(x)$	1.5	1	0.75	0.5	0

12.

x	-2	0	1	2	4
$k(x)$	0.25	1	2	4	16

13.

x	-2	0	1	2	4
$u(x)$	0	4	3	0	-12

14.

x	-2	0	1	2	4
$w(x)$	32	8	4	2	0.5

Year t	0	1	2	3	4	5
Cost $C(t)$	$5.42	$5.10	$5.00	$5.12	$5.40	$5.88

In Exercises 15–22, find the equation of the specified line.

15. Through $(3, 2)$ with slope -3

16. Through $(-2, 4)$ with slope -1

17. Through $(1, -3)$ and $(5, 2)$

18. Through $(-1, 2)$ and $(1, 0)$

19. Through $(1, 2)$ parallel to $x - 2y = 2$

20. Through $(-3, 1)$ parallel to $-2x - 4y = 5$

21. With slope 4 crossing $2x - 3y = 6$ at its x-intercept

22. With slope $1/2$ crossing $3x + y = 6$ at its x-intercept

In Exercises 23 and 24, determine which of the given lines better fits the given points.

23. $(-1, 1), (1, 2), (2, 0)$; $y = -x/2 + 1$ or $y = -x/4 + 1$

24. $(-2, -1), (-1, 1), (0, 1), (1, 2), (2, 4), (3, 3)$; $y = x + 1$ or $y = x/2 + 1$

In Exercises 25 and 26, find the line that best fits the given points and compute the correlation coefficient.

25 . $(-1, 1), (1, 2), (2, 0)$

26 . $(-2, -1), (-1, 1), (0, 1), (1, 2), (2, 4), (3, 3)$

APPLICATIONS: OHaganBooks.com

27. *Web Site Traffic* John Sean O'Hagan is CEO of the online bookstore OHaganBooks.com and notices that, since the establishment of the company Web site six years ago ($t = 0$), the number of visitors to the site has grown quite dramatically, as indicated by the following table:

Year t	0	1	2	3	4	5	6
Web site Traffic $V(t)$ (visits/day)	100	300	1,000	3,300	10,500	33,600	107,400

a. Graph the function V as a function of time t. Which of the following types of function seem to fit the curve best: linear, quadratic, or exponential?

b. Compute the ratios $\dfrac{V(1)}{V(0)}, \dfrac{V(2)}{V(1)}, \ldots,$ and $\dfrac{V(6)}{V(5)}$. What do you notice?

c. Use the result of part (b) to predict Web site traffic next year (to the nearest 100).

28. *Publishing Costs* Marjory Maureen Duffin is CEO of publisher Duffin House, a major supplier of paperback titles to OHaganBooks.com. She notices that publishing costs over the past five years have varied considerably as indicated by the following table, which shows the average cost to the company of publishing a paperback novel (t is time in years, and the current year is $t = 5$):

a. Graph the function C as a function of time t. Which of the following types of function seem to fit the curve best: linear, quadratic, or exponential?

b. Compute the differences $C(1) - C(0)$, $C(2) - C(1)$, ..., and $C(5) - C(4)$, rounded to one decimal place. What do you notice?

c. Use the result of part (b) to predict the cost of producing a paperback novel next year.

29. *Web Site Stability* John O'Hagan is considering upgrading the Web server equipment at OHaganBooks.com because of frequent crashes. The tech services manager has been monitoring the frequency of crashes as a function of Web site traffic (measured in thousands of visits per day) and has obtained the following model:

$$c(x) = \begin{cases} 0.03x + 2 & \text{if } 0 \le x \le 50 \\ 0.05x + 1 & \text{if } x > 50 \end{cases}$$

where $c(x)$ is the average number of crashes in a day in which there are x thousand visitors.

a. On average, how many times will the Web site crash on a day when there are 10,000 visits? 50,000 visits? 100,000 visits?

b. What does the coefficient 0.03 tell you about the Web site's stability?

c. Last Friday, the Web site went down 8 times. Estimate the number of visits that day.

30. *Book Sales* As OHaganBooks.com has grown in popularity, the sales manager has been monitoring book sales as a function of the Web site traffic (measured in thousands of visits per day) and has obtained the following model:

$$s(x) = \begin{cases} 1.55x & \text{if } 0 \le x \le 100 \\ 1.75x - 20 & \text{if } 100 < x \le 250 \end{cases}$$

where $s(x)$ is the average number of books sold in a day in which there are x thousand visitors.

a. On average, how many books per day does the model predict OHaganBooks.com will sell when it has 60,000 visits in a day? 100,000 visits in a day? 160,000 visits in a day?

b. What does the coefficient 1.75 tell you about book sales?

c. According to the model, approximately how many visitors per day will be needed in order to sell an average of 300 books per day?

31. *New Users* The number of registered users at OHaganBooks.com has increased substantially over the past few months. The following table shows the number of new users registering each month for the past six months:

Month t	1	2	3	4	5	6
New Users (thousands)	12.5	37.5	62.5	72.0	74.5	75.0

a. Which of the following models best approximates the data?

(A) $n(t) = \dfrac{300}{4 + 100(5^{-t})}$ **(B)** $n(t) = 13.3t + 8.0$

(C) $n(t) = -2.3t^2 + 30.0t - 3.3$

(D) $n(t) = 7(3^{0.5t})$

b. What do each of the above models predict for the number of new users in the next few months: rising, falling, or leveling off?

32. Purchases OHaganBooks.com has been promoting a number of books published at Duffin House. The following table shows the number of books purchased each month from Duffin House for the past five months:

Month t	1	2	3	4	5
Purchases (books)	1,330	520	520	1,340	2,980

a. Which of the following models best approximates the data?

(A) $n(t) = \dfrac{3,000}{1 + 12(2^{-t})}$ **(B)** $n(t) = \dfrac{2,000}{4.2 - 0.7t}$

(C) $n(t) = 300(1.6^t)$

(D) $n(t) = 100(4.1t^2 - 20.4t + 29.5)$

b. What do each of the above models predict for the number of new users in the next few months: rising, falling, leveling off, or something else?

33. Internet Advertising Several months ago. John O'Hagan investigated the effect on the popularity of OHaganBooks.com of placing banner ads at well-known Internet portals. The following model was obtained from available data:

$v(c) = -0.000005c^2 + 0.085c + 1,750$ new visits per day

where c is the monthly expenditure on banner ads.

a. John O'Hagan is considering increasing expenditure on banner ads from the current level of $5,000 to $6,000 per month. What will be the resulting effect on Web site popularity?

b. According to the model, would the Web site popularity continue to grow at the same rate if he continued to raise expenditure on advertising $1,000 each month? Explain.

c. Does this model give a reasonable prediction of traffic at expenditures larger than $8,500 per month? Why?

34. Production Costs Over at Duffin House, Marjory Duffin is trying to decide on the size of the print runs for the best-selling new fantasy novel *Larry Plotter and the Simplex Method*. The following model shows a calculation of the total cost to produce a million copies of the novel, based on an analysis of setup and storage costs:

$$c(n) = 0.0008n^2 - 72n + 2,000,000 \text{ dollars}$$

where n is the print run size (the number of books printed in each run).

a. What would be the effect on cost if the run size was increased from 20,000 to 30,000?

b. Would increasing the run size in further steps of 10,000 result in the same changes in the total cost? Explain.

c. What approximate run size would you recommend that Marjoy Duffin use for a minimum cost?

35. Internet Advertising When OHaganBooks.com actually went ahead and increased Internet advertising from $5,000 per month to $6,000 per month (see Exercise 33) it was noticed that the number of new visits increased from an estimated 2,050 per day to 2,100 per day. Use this information to construct a linear model giving the average number v of new visits per day as a function of the monthly advertising expenditure c.

a. What is the model?

b. Based on the model, how many new visits per day could be anticipated if OHaganBooks.com budgets $7,000 per month for Internet advertising?

c. The goal is to eventually increase the number of new visits to 2,500 per day. Based on the model, how much should be spent on Internet advertising in order to accomplish this?

36. Production Costs When Duffin House printed a million copies of *Larry Plotter and the Simplex Method* (see Exercise 34), it used print runs of 20,000, which cost the company $880,000. For the sequel, *Larry Plotter and the Simplex Method, Phase 2* it used print runs of 40,000 which cost the company $550,000. Use this information to construct a linear model giving the production cost c as a function of the run size n.

a. What is the model?

b. Based on the model, what would print runs of 25,000 have cost the company?

c. Marjory Duffin has decided to budget $418,000 for production of the next book in the *Simplex Method* series. Based on the model, how large should the print runs be to accomplish this?

37. Recreation John O'Hagan has just returned from a sales convention at Puerto Vallarta, Mexico where, in order to win a bet he made with Marjory Duffin (Duffin House was also at the convention), he went bungee jumping at a nearby mountain retreat. The bungee cord he used had the property that a person weighing 70 kg would drop a total distance of 74.5 meters, while a 90 kg person would drop 93.5 meters. Express the distance d a jumper drops as a linear function of the jumper's weight w. John OHagan dropped 90 m. What was his approximate weight?

38. Crickets The mountain retreat near Puerto Vallarta was so quiet at night that all one could hear was the chirping of the snowy tree crickets. These crickets behave in a rather interesting way: The rate at which they chirp depends linearly on the temperature. Early in the evening, John O'Hagan counted 140 chirps/minute and noticed that the temperature was 80°F. Later in the evening the temperature dropped to 75°F, and the chirping slowed down to 120 chirps/minute. Express the temperature T as a function of the rate of chirping r. The temperature that night dropped to a low of 65°F. At approximately what rate were the crickets chirping at that point?

39. *Break-Even Analysis* OHaganBooks.com has recently decided to start selling music albums online through a service it calls *o'Tunes*.[75] Users pay a fee to download an entire music album. Composer royalties and copyright fees cost an average of $5.50 per album, and the cost of operating and maintaining *o'Tunes* amounts to $500 per week. The company is currently charging customers $9.50 per album.

   a. What are the associated (weekly) cost, revenue, and profit functions?

   b. How many albums must be sold per week in order to make a profit?

   c. If the charge is lowered to $8.00 per album, how many albums must be sold per week in order to make a profit?

40. *Break-Even Analysis* OHaganBooks.com also generates revenue through its *o'Books* e-book service. Author royalties and copyright fees cost the company an average of $4 per novel, and the monthly cost of operating and maintaining the service amounts to $900 per month. The company is currently charging readers $5.50 per novel.

   a. What are the associated cost, revenue, and profit functions?

   b. How many novels must be sold per month in order to break even?

   c. If the charge is lowered to $5.00 per novel, how many books must be sold in order to break even?

41. *Demand and Profit* In order to generate a profit from its new *o'Tunes* service, OHaganBooks.com needs to know how the demand for music albums depends on the price it charges. During the first week of the service, it was charging $7 per album, and sold 500. Raising the price to $9.50 had the effect of lowering demand to 300 albums per week.

   a. Use the given data to construct a linear demand equation.

   b. Use the demand equation you constructed in part (a) to estimate the demand if the price was raised to $12 per album.

---

[75]The (highly original) name was suggested to John O'Hagan by Marjory Duffin over cocktails one evening.

   c. Using the information on cost given in Exercise 39, determine which of the three prices ($7, $9.50 and $12) would result in the largest weekly profit, and the size of that profit.

42. *Demand and Profit* In order to generate a profit from its *o'Books* e-book service, OHaganBooks.com needs to know how the demand for novels depends on the price it charges. During the first month of the service, it was charging $10 per novel, and sold 350. Lowering the price to $5.50 per novel had the effect of increasing demand to 620 novels per month.

   a. Use the given data to construct a linear demand equation.

   b. Use the demand equation you constructed in part (a) to estimate the demand if the price was raised to $15 per novel.

   c. Using the information on cost given in Exercise 40, determine which of the three prices ($5.50, $10 and $15) would result in the largest profit, and the size of that profit.

43. *Demand* OHaganBooks.com has tried selling music albums on *o'Tunes* at a variety of prices, with the following results:

Price	$8.00	$8.50	$10	$11.50
Demand (Weekly sales)	440	380	250	180

   a. Use the given data to obtain a linear regression model of demand.

   b. Use the demand model you constructed in part (a) to estimate the demand if the company charged $10.50 per album. (Round the answer to the nearest album.)

44. *Demand* OHaganBooks.com has tried selling novels through *o'Books* at a variety of prices, with the following results:

Price	$5.50	$10	$11.50	$12
Demand (Monthly sales)	620	350	350	300

   a. Use the given data to obtain a linear regression model of demand.

   b. Use the demand model you constructed in part (a) to estimate the demand if the company charged $8 per novel. (Round the answer to the nearest novel.)

Case Study   Modeling Spending on Internet Advertising

You are the new director of Impact Advertising Inc.'s Internet division, which has enjoyed a steady 0.25% of the Internet advertising market. You have drawn up an ambitious proposal to expand your division in light of your anticipation that Internet advertising will continue to skyrocket. However, upper management sees things differently and, based on the following email, does not seem likely to approve the budget for your proposal.

TO: JCheddar@impact.com (J. R. Cheddar)
CC: CVODoylePres@impact.com (C. V. O'Doyle, CEO)
FROM: SGLombardoVP@impact.com (S. G. Lombardo, VP Financial Affairs)
SUBJECT: Your Expansion Proposal
DATE: May 30, 2014

Hi John:

Your proposal reflects exactly the kind of ambitious planning and optimism we like
to see in our new upper management personnel. Your presentation last week was
most impressive, and obviously reflected a great deal of hard work and preparation.

I am in full agreement with you that Internet advertising is on the increase. Indeed,
our Market Research department informs me that, based on a regression of the
most recently available data, Internet advertising revenue in the United States will
continue to grow at a rate of approximately $2.7 billion per year. This translates
into approximately $6.75 million in increased revenues per year for Impact, given
our 0.25% market share. This rate of expansion is exactly what our planned 2015
budget anticipates. Your proposal, on the other hand, would require a budget of
approximately *twice* the 2015 budget allocation, even though your proposal
provides no hard evidence to justify this degree of financial backing.

At this stage, therefore, I am sorry to say that I am inclined not to approve the
funding for your project, although I would be happy to discuss this further with
you. I plan to present my final decision on the 2015 budget at next week's
divisional meeting.

Regards, Sylvia

Refusing to admit defeat, you contact the Market Research department and request
the details of their projections on Internet advertising. They fax you the following information:[76]

Year	2007	2008	2009	2010	2011	2012	2013	2014
Internet Advertising Revenue ($ Billion)	21.2	23.4	22.7	25.8	28.5	32.6	36	40.5

Regression Model: $y = 2.744x + 19.233$ (x = time in years since 2007)

Correlation Coefficient: $r = 0.970$

Now you see where the VP got that $2.7 billion figure: The slope of the regression
equation is close to 2.7, indicating a rate of increase of about $2.7 billion per year.
Also, the correlation coefficient is very high—an indication that the linear model fits
the data well. In view of this strong evidence, it seems difficult to argue that revenues
will increase by significantly more than the projected $2.7 billion per year. To get a

---

[76]The 2011–2014 figures are projections by eMarketer. Source: www.eMarketer.com.

Internet Advertising Revenue
($ billions)

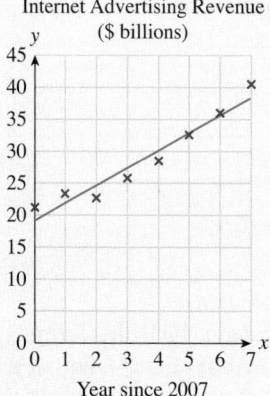

Year since 2007

Figure 32

✳ Note that this *r* is *not* the linear correlation coefficient we defined on page 101; what this *r* measures is how closely the *quadratic* regression model fits the data.

Internet Advertising Revenue
($ billions)

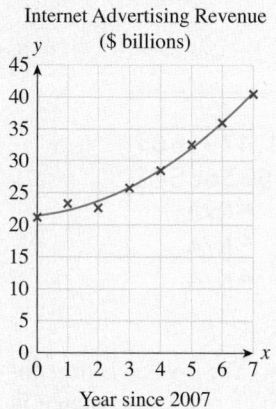

Year since 2007

Figure 33

† The number of degrees of freedom in a regression model is 1 less than the number of coefficients. For a linear model, it is 1 (there are two coefficients: the slope *m* and the intercept *b*), and for a quadratic model it is 2. For a detailed discussion, consult a text on regression analysis.

better picture of what's going on, you decide to graph the data together with the regression line in your spreadsheet. What you get is shown in Figure 32. You immediately notice that the data points seem to suggest a curve, and not a straight line. Then again, perhaps the suggestion of a curve is an illusion. Thus there are, you surmise, two possible interpretations of the data:

1. (Your first impression) As a function of time, Internet advertising revenue is nonlinear, and is in fact accelerating (the rate of change is increasing), so a linear model is inappropriate.

2. (Devil's advocate) Internet advertising revenue *is* a linear function of time; the fact that the points do not lie on the regression line is simply a consequence of random factors that do not reflect a long-term trend, such as world events, mergers and acquisitions, short-term fluctuations in economy or the stock market, etc.

You suspect that the VP will probably opt for the second interpretation and discount the graphical evidence of accelerating growth by claiming that it is an illusion: a "statistical fluctuation." That is, of course, a possibility, but you wonder how likely it really is.

For the sake of comparison, you decide to try a regression based on the simplest nonlinear model you can think of—a quadratic function.

$$y = ax^2 + bx + c$$

Your spreadsheet allows you to fit such a function with a click of the mouse. The result is the following.

$$y = 0.3208x^2 + 0.4982x + 21.479 \quad (x = \text{number of years since 2007})$$
$$r = 0.996 \qquad \text{See Note.}✳$$

Figure 33 shows the graph of the regression function together with the original data.

Aha! The fit is visually far better, and the correlation coefficient is even higher! Further, the quadratic model predicts 2015 revenue as

$$y = 0.3208(8)^2 + 0.4982(8) + 21.479 \approx \$46.0 \text{ billion,}$$

which is \$5.5 billion above the 2014 spending figure in the table above. Given Impact Advertising's 0.25% market share, this translates into an increase in revenues of \$13.75 million, which is about double the estimate predicted by the linear model!

You quickly draft an email to Lombardo, and are about to click "Send" when you decide, as a precaution, to check with a colleague who is knowledgeable in statistics. He tells you to be cautious: The value of *r* will always tend to increase if you pass from a linear model to a quadratic one because of the increase in "degrees of freedom."† A good way to test whether a quadratic model is more appropriate than a linear one is to compute a statistic called the "*p*-value" associated with the coefficient of x^2. A low value of *p* indicates a high degree of confidence that the coefficient of x^2 cannot be zero (see below). Notice that if the coefficient of x^2 *is* zero, then you have a linear model.

You can, your colleague explains, obtain the *p*-value using your spreadsheet as follows (the method we describe here works on all the popular spreadsheets, including *Excel, Google Docs,* and *Open Office Calc*).

First, set up the data in columns, with an extra column for the values of x^2:

◇	A	B	C
1	y	x	x^2
2	21.2	0	0
3	23.4	1	1
4	22.7	2	4
5	25.8	3	9
6	28.5	4	16
7	32.6	5	25
8	36	6	36
9	40.5	7	49

Then, highlight a vacant 5×3 block (the block E1:G5 say), type the formula =LINEST(A2:A9,B2:C9,,TRUE), and press Cntl+Shift+Enter (not just Enter!). You will see a table of statistics like the following:

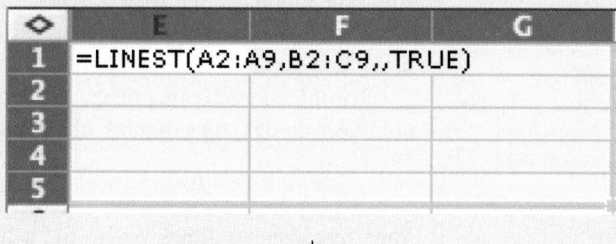

◇	E	F	G
1	=LINEST(A2:A9,B2:C9,,TRUE)		
2			
3			
4			
5			

↓
Cntl+Shift+Enter

◇	E	F	G
1	0.32083333	0.49821429	21.4791667
2	0.05808804	0.42288732	0.63366572
3	0.99157395	0.75290706	#N/A
4	294.198816	5	#N/A
5	333.544405	2.83434524	#N/A

(Notice the coefficients of the quadratic model in the first row.) The p-value is then obtained by the formula =TDIST(ABS(E1/E2),F4,2), which you can compute in any vacant cell. You should get $p \approx 0.00267$.

Q: *What does p actually measure?*

A: *Roughly speaking, $1 - p \approx 0.997733$ gives the degree of confidence you can have (99.7733%) in asserting that the coefficient of x^2 is not zero. (Technically, p is the probability—allowing for random fluctuation in the data—that, if the coefficient of x^2 were in fact zero, the ratio E1/E2 could be as large as it is.)*

In short, you can go ahead and send your email with almost 100% confidence!

EXERCISES

Suppose you are given the following data for the spending on Internet advertising in a hypothetical country in which Impact Advertising also has a 0.25% share of the market.

Year	2010	2011	2012	2013	2014	2015	2016
Spending on Advertising ($ Billion)	0	0.3	1.5	2.6	3.4	4.3	5.0

1. Obtain a linear regression model and the correlation coefficient r. (Take t to be time in years since 2010.) According to the model, at what rate is spending on Internet Advertising increasing in this country? How does this translate to annual revenues for Impact Advertising?

2. Use a spreadsheet or other technology to graph the data together with the best-fit line. Does the graph suggest a quadratic model (parabola)?

3. Test your impression in the preceding exercise by using technology to fit a quadratic function and graphing the resulting curve together with the data. Does the graph suggest that the quadratic model is appropriate?

4. Perform a regression analysis using the quadratic model and find the associated p-value. What does it tell you about the appropriateness of a quadratic model?

TI-83/84 Plus   Technology Guide

Section 1.1

Example 1(a) and (c) (page 44)   The total number of iPods sold by Apple up to the end of year x can be approximated by $f(x) = 4x^2 + 16x + 2$ million iPods $(0 \le x \le 6)$, where $x = 0$ represents 2003. Compute $f(0)$, $f(2)$, $f(4)$, and $f(6)$, and obtain the graph of f.

Solution with Technology

You can use the Y= screen to enter an algebraically defined function.

1. Enter the function in the Y= screen, as

   $Y_1 = 4X^2+16X+2$

or $Y_1 = 4X^2+16X+2$

(See Chapter 0 for a discussion of technology formulas.)

2. To evaluate $f(0)$, for example, enter $Y_1(0)$ in the Home screen to evaluate the function Y_1 at 0. Alternatively, you can use the table feature: After entering the function under Y_1, press [2ND] [TBLSET], and set Indpnt to Ask. (You do this once and for all; it will permit you to specify values for x in the table screen.) Then, press [2ND] [TABLE], and you will be able to evaluate the function at several values of x. Below (top) is a table showing the values requested:

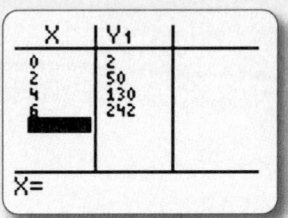

3. To obtain the graph above press [WINDOW], set Xmin = 0, Xmax = 6 (the range of x-values we are interested in), Ymin = 0, Ymax = 300 (we estimated Ymin and Ymax from the corresponding set of y-values in the table) and press [GRAPH] to obtain the curve. Alternatively, you can avoid having to estimate Ymin and Ymax by pressing ZoomFit ([ZOOM] [0]), which automatically sets Ymin and Ymax to the smallest and greatest values of y in the specified range for x.

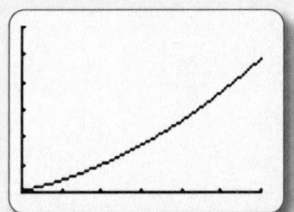

Example 2 (page 47)   The number $n(t)$ of Facebook members can be approximated by the following function of time t in years ($t = 0$ represents January 2004):

$$n(t) = \begin{cases} 4t & \text{if } 0 \le t \le 3 \\ 12 + 23(t-3)^{2.7} & \text{if } 3 < t \le 6 \end{cases}$$
million members.

Obtain a table showing the values $n(t)$ for $t = 0, \ldots, 6$ and also obtain the graph of n.

Solution with Technology

You can enter a piecewise-defined function using the logical inequality operators $<$, $>$, $\le$, and $\ge$, which are found by pressing [2ND] [TEST]:

1. Enter the function n in the Y= screen as:

   $Y_1=(X\le3)*(4X)+(X>3)*(12+23*abs(X-3)^2.7)$

When x is less than or equal to 3, the logical expression $(X\le3)$ evaluates to 1 because it is true, and the expression $(X>3)$ evaluates to 0 because it is false. The value of the function is therefore given by the expression $(4X)$. When x is greater than 3, the expression $(X\le3)$ evaluates to 0 while the expression $(X>3)$ evaluates to 1, so the value of the function is given by the expression $(12+23*abs(X-3)^2.7)$. (The reason we use the abs in the formula is to prevent an error in evaluating $(x-3)^{2.7}$ when $x < 3$; even though we don't use that formula when $x < 3$, we are in fact evaluating it and multiplying it by zero.)

2. As in Example 1, use the Table feature to compute several values of the function at once by pressing [2ND] [TABLE].

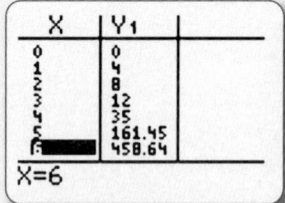

3. To obtain the graph, we proceed as in Example 1: Press [WINDOW], set Xmin = 0, Xmax = 6 (the range of x-values we are interested in), Ymin = 0, Ymax = 500 (see the y-values in the table) and press [GRAPH].

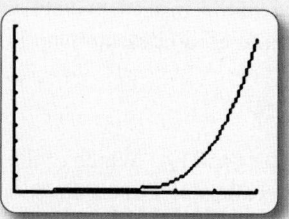

Section 1.2

Example 4(a) (page 62) The demand and supply curves for private schools in Michigan are $q = 77.8p^{-0.11}$ and $q = 30.4 + 0.006p$ thousand students, respectively $(200 \le p \le 2{,}200)$, where p is the net tuition cost in dollars. Graph the demand and supply curves on the same set of axes. Use your graph to estimate, to the nearest \$100, the tuition at which the demand equals the supply (equilibrium price). Approximately how many students will be accommodated at that price?

Solution with Technology

To obtain the graphs of demand and supply:

1. Enter $\texttt{Y}_1\texttt{=77.8*X\^{}(-0.11)}$ and $\texttt{Y}_2\texttt{=30.4+}$ $\texttt{0.006*X}$ in the "Y=" screen.
2. Press $\boxed{\text{WINDOW}}$, enter Xmin = 200, Xmax = 2200, Ymin = 0, Ymax = 50 and press $\boxed{\text{GRAPH}}$ for the graph shown below:

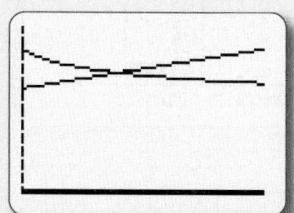

3. To estimate the equilibrium price, press $\boxed{\text{TRACE}}$ and use the arrow keys to follow the curve to the approximate point of intersection (around X = 1008) as shown below.

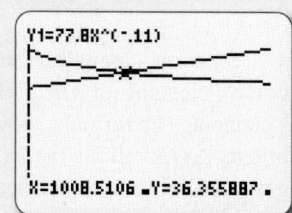

4. For a more accurate estimate, zoom in by pressing $\boxed{\text{ZOOM}}$ and selecting Option 1 ZBox.
5. Move the curser to a point slightly above and to the left of the intersection, press $\boxed{\text{ENTER}}$, and then move the curser to a point slightly below and to the right and press $\boxed{\text{ENTER}}$ again to obtain a box.

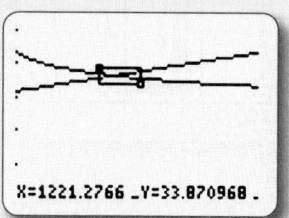

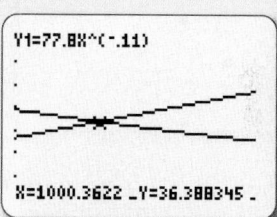

6. Now press $\boxed{\text{ENTER}}$ again for a zoomed-in view of the intersection.
7. You can now use $\boxed{\text{TRACE}}$ to obtain the intersection coordinates more accurately: X ≈ 1,000, representing a tuition cost of \$1,000. The associated demand is the Y-coordinate: around 36.4 thousand students.

Example 5(a) (page 64) The following table shows annual sales, in billions of dollars, by *Nike* from 2005 through 2010 ($t = 0$ represents 2005):

t	0	1	2	3	4	5
Sales (\$ billion)	13.5	15	16.5	18.5	19	19

Consider the following four models:

(1) $s(t) = 14 + 1.2t$ — Linear model

(2) $s(t) = 13 + 2.2t - 0.2t^2$ — Quadratic model

(3) $s(t) = 14(1.07^t)$ — Exponential model

(4) $s(t) = \dfrac{19.5}{1 + 0.48(1.8^{-t})}$. — Logistic model

a. Which models fit the data significantly better than the rest?

b. Of the models you selected in part (a), which gives the most reasonable prediction for 2013?

Solution with Technology

1. First enter the actual revenue data in the stat list editor ($\boxed{\text{STAT}}$ EDIT) with the values of t in L_1, and the values of $s(t)$ in L_2.

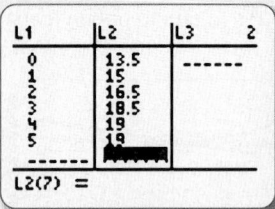

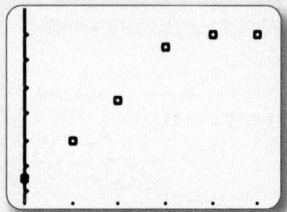

2. Now go to the Y= window and turn Plot1 on by selecting it and pressing $\boxed{\text{ENTER}}$. (You can also turn it on in the $\boxed{\text{2ND}}$ STAT PLOT screen.) Then press ZoomStat ($\boxed{\text{ZOOM}}$ $\boxed{9}$) to obtain a plot of the points (above).

3. To see any of the four curves plotted along with the points, enter its formula in the Y= screen (for instance, $Y_1=13+2.2x-0.2x^2$ for the second model) and press $\boxed{\text{GRAPH}}$ (figure on top below).

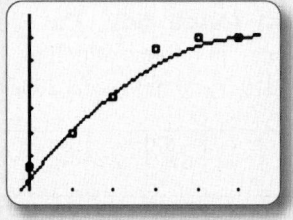

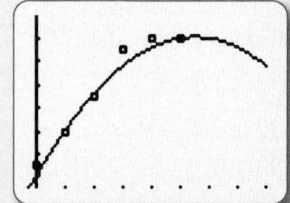

4. To see the extrapolation of the curve to 2013, just change Xmax to 8 (in the $\boxed{\text{WINDOW}}$ screen) and press $\boxed{\text{GRAPH}}$ again (lower figure above).

5. Now change Y_1 to see similar graphs for the remaining curves.

6. When you are done, turn Plot1 off again so that the points you entered do not show up in other graphs.

Section 1.3

Example 1 (page 81)   Which of the following two tables gives the values of a linear function? What is the formula for that function?

x	0	2	4	6	8	10	12
$f(x)$	3	-1	-3	-6	-8	-13	-15

x	0	2	4	6	8	10	12
$g(x)$	3	-1	-5	-9	-13	-17	-21

Solution with Technology

We can use the "List" feature in the TI-83/84 Plus to automatically compute the successive quotients $m = \Delta y/\Delta x$ for either f or g as follows:

1. Use the stat list editor ($\boxed{\text{STAT}}$ EDIT) to enter the values of x and $f(x)$ in the first two columns, called L_1 and L_2, as shown in the screenshot below. (If there is already data in a column you want to use, you can clear it by highlighting the column heading (e.g., L_1) using the arrow key, and pressing $\boxed{\text{CLEAR}}$ $\boxed{\text{ENTER}}$.)

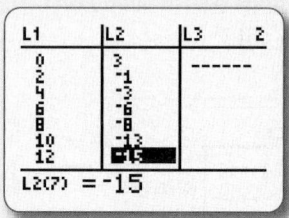

2. Highlight the heading L_3 by using the arrow keys, and enter the following formula (with the quotes, as explained below):

   "ΔList(L_2)/ΔList(L_1)"   ΔList is found under $\boxed{\text{2ND}}$ $\boxed{\text{LIST}}$ OPS. L_1 is $\boxed{\text{2ND}}$ $\boxed{1}$

   The "ΔList" function computes the differences between successive elements of a list, returning a list with one less element. The formula above then computes the quotients $\Delta y/\Delta x$ in the list L_3 as shown in the following screenshot. As you can see in the third column, $f(x)$ is not linear.

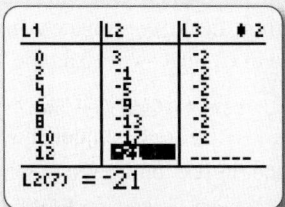

3. To redo the computation for $g(x)$, all you need to do is edit the values of L_2 in the stat list editor. By putting quotes around the formula we used for L_3, we told the calculator to remember the formula, so it automatically recalculates the values.

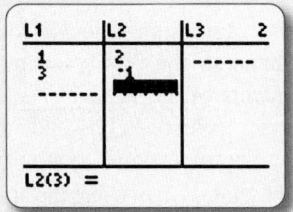

Example 2(a) (page 83)   Find the equation of the line through the points $(1, 2)$ and $(3, -1)$.

Solution with Technology

1. Enter the coordinates of the given points in the stat list editor ($\boxed{\text{STAT}}$ EDIT) with the values of x in L_1, and the values of y in L_2.

2. To compute the slope, enter the following formula in the Home screen:

$$(L_2(2)-L_2(1))/(L_1(2)-L_1(1)) \to M$$

L_1 and L_2 are under $\boxed{\text{2ND}}$ $\boxed{\text{LIST}}$ and the arrow is $\boxed{\text{STO}}$

3. Then, to compute the y-intercept, enter

$$L_2(1)-M*L_1(1)$$

Section 1.4

Example 1(a) (page 97)   Using the data on the per capita GDP in Mexico given at the beginning of Section 1.4, compute SSE, the sum-of-squares error, for the linear models $y = 0.5t + 8$ and $y = 0.25t + 9$, and graph the data with the given models.

Solution with Technology

We can use the "List" feature in the TI-83/84 Plus to automate the computation of SSE.

1. Use the stat list editor ($\boxed{\text{STAT}}$ EDIT) to enter the given data in the lists L_1 and L_2, as shown in the first screenshot below. (If there is already data in a column you want to use, you can clear it by highlighting the column heading (e.g., L_1) using the arrow key, and pressing $\boxed{\text{CLEAR}}$ $\boxed{\text{ENTER}}$.)

2. To compute the predicted values, highlight the heading L_3 using the arrow keys, and enter the following formula for the predicted values (figure on the top below):

$$0.5*L_1+8 \qquad L_1 \text{ is } \boxed{\text{2ND}} \boxed{1}$$

Pressing $\boxed{\text{ENTER}}$ again will fill column 3 with the predicted values (below bottom). Note that only seven of the eight data points can be seen on the screen at one time.

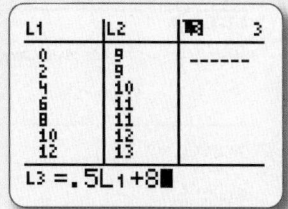

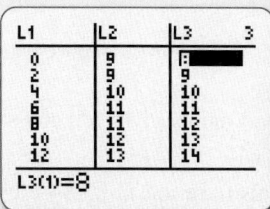

3. Highlight the heading L_4 and enter the following formula (including the quotes):

$$"(L_2-L_3)^2" \qquad \text{Squaring the residuals}$$

4. Pressing ENTER will fill L_4 with the squares of the residuals. (Putting quotes around the formula will allow us to easily check the second model, as we shall see.)

5. To compute SSE, the sum of the entries in L_4, go to the home screen and enter sum(L_4) (see below; "sum" is under 2ND LIST MATH.)

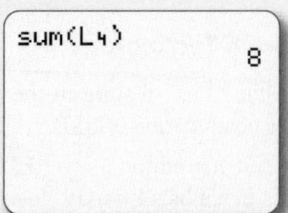

6. To check the second model, go back to the List screen, highlight the heading L_3, enter the formula for the second model, 0.25*L_1+9, and press ENTER. Because we put quotes around the formula for the residuals in L_4, the TI-83/84 Plus will remember the formula and automatically recalculate the values (below top). On the home screen we can again calculate sum(L_4) to get SSE for the second model (below bottom).

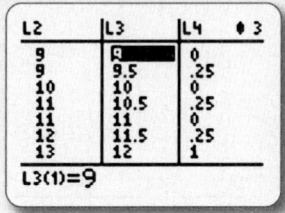

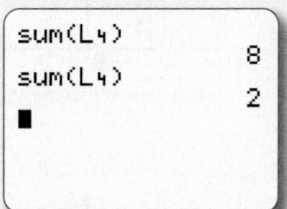

The second model gives a much smaller SSE, so is the better fit.

7. You can also use the TI-83/84 Plus to plot both the original data points and the two lines (see below). Turn Plot1 on in the STAT PLOT window, obtained by pressing 2ND STAT PLOT. To show the lines, enter them in the "Y=" screen as usual. To obtain a convenient window showing all the points and the lines, press ZOOM and choose 9: ZoomStat.

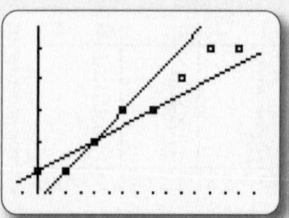

Example 2 (page 99) Use the data on the per capita GDP in Mexico to find the best-fit linear model.

Solution with Technology

1. Enter the data in the TI-83/84 Plus using the List feature, putting the x-coordinates in L_1 and the y-coordinates in L_2, just as in Example 1.

2. Press STAT, select CALC, and choose #4: LinReg(ax+b). Pressing ENTER will cause the equation of the regression line to be displayed in the home screen:

So, the regression line is $y \approx 0.321x + 8.75$.

3. To graph the regression line without having to enter it by hand in the "Y=" screen, press Y=, clear the contents of Y_1, press VARS, choose #5: Statistics, select EQ, and then choose #1:RegEQ. The regression equation will then be entered under Y_1.

4. To simultaneously show the data points, press 2ND STATPLOT and turn Plot1 on as in Example 1. To obtain a convenient window showing all the points and the line (see below), press ZOOM and choose #9: ZoomStat.

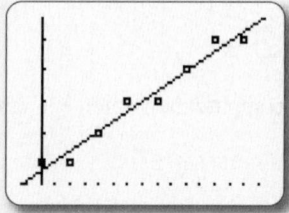

Example 3 (page 101) Find the correlation coefficient for the data in Example 2.

Solution with Technology

To find the correlation coefficient using a TI-83/84 Plus you need to tell the calculator to show you the coefficient at the same time that it shows you the regression line.

1. Press $\boxed{\text{2ND}}$ $\boxed{\text{CATALOG}}$ and select `DiagnosticOn` from the list. The command will be pasted to the home screen, and you should then press $\boxed{\text{ENTER}}$ to execute the command.

2. Once you have done this, the "`LinReg(ax+b)`" command (see the discussion for Example 2) will show you not only a and b, but r and r^2 as well:

```
LinReg
y=ax+b
a=.3214285714
b=8.75
r²=.9642857143
r=.9819805061
```

SPREADSHEET Technology Guide

Section 1.1

Example 1(a) and (c) (page 44) The total number of iPods sold by Apple up to the end of year x can be approximated by $f(x) = 4x^2 + 16x + 2$ million iPods ($0 \le x \le 6$), where $x = 0$ represents 2003. Compute $f(0)$, $f(2)$, $f(4)$, and $f(6)$, and obtain the graph of f.

Solution with Technology

To create a table of values of f using a spreadsheet:

1. Set up two columns: one for the values of x and one for the values of $f(x)$. Then enter the sequence of values 0, 2, 4, 6 in the x column as shown.

◇	A	B
1	x	f(x)
2	0	
3	2	
4	4	
5	6	

2. Now we enter a formula for $f(x)$ in cell B2 (below). The technology formula is `4*x^2+16*x+2`. To use this formula in a spreadsheet, we modify it slightly:

`=4*A2^2+16*A2+2` Spreadsheet version of tech formula

Notice that we have preceded the Excel formula by an equals sign (=) and replaced each occurrence of x by the name of the cell holding the value of x (cell A2 in this case).

◇	A	B	C
1	x	f(x)	
2		0	=4*A2^2+16*A2+2
3		2	
4		4	
5		6	

Note Instead of typing in the name of the cell "A2" each time, you can simply click on the cell A2, and "A2" will be automatically inserted. ∎

3. Now highlight cell B2 and drag the **fill handle** (the little square at the lower right-hand corner of the selection) down until you reach Row 5 as shown below on the top, to obtain the result shown on the bottom.

◇	A	B	C
1	x	f(x)	
2		0	=4*A2^2+16*A2+2
3		2	
4		4	
5		6	

◇	A	B
1	x	f(x)
2		0	2
3		2	50
4		4	130
5		6	242

4. To graph the data, highlight A1 through B5, and insert a "Scatter chart" (the exact method of doing this depends on the specific version of the spreadsheet program). When choosing the style of the chart,

choose a style that shows points connected by lines (if possible) to obtain a graph something like the following:

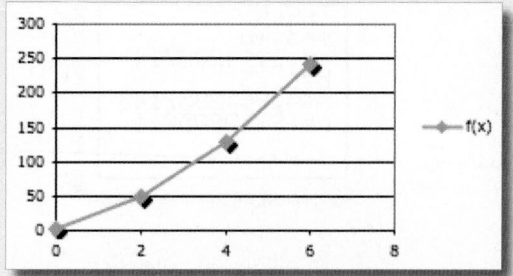

Example 2 (page 47)   The number $n(t)$ of Facebook members can be approximated by the following function of time t in years ($t = 0$ represents January 2004):

$$n(t) = \begin{cases} 4t & \text{if } 0 \le t \le 3 \\ 12 + 23(t-3)^{2.7} & \text{if } 3 < t \le 6 \end{cases} \quad \text{million members.}$$

Obtain a table showing the values $n(t)$ for $t = 0, \dots, 6$ and also obtain the graph of n.

Solution with Technology

We can generate a table of values of $n(t)$ for $t = 0, 1, \dots, 6$ as follows:

1. Set up two columns; one for the values of t and one for the values of $n(t)$, and enter the values $0, 1, \dots, 6$ in the t column as shown in the first screenshot below:

2. We must now enter the formula for n in cell B2. The following formula defines the function n in Excel:

```
=(x<=3)*(4*x)+(x>3)*(12+23*abs
(x-3)^2.7)
```

When x is less than or equal to 3, the logical expression $(x \le 3)$ evaluates to 1 because it is true, and the expression $(x>3)$ evaluates to 0 because it is false. The value of the function is therefore given by the expression $(4*x)$. When x is greater than 3, the expression $(x \le 3)$ evaluates to 0 while the expression $(x>3)$ evaluates to 1, so the value of the function is given by the expression $(12+23*abs$ $(x-3)^2.7)$. (The reason we use the abs in the formula is to prevent an error in evaluating $(x-3)^{2.7}$ when $x < 3$; even though we don't use that formula

when $x < 3$, we are in fact evaluating it and multiplying it by zero.) We therefore enter the formula

```
=(A2<=3)*(4*A2)+(A2>3)*(12+23
*ABS(A2-3)^2.7)
```

in cell B2 and then copy down to cell B8 (below top) to obtain the result shown on the bottom:

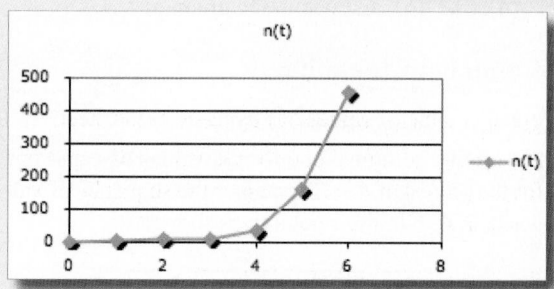

3. To graph the data, highlight A1 through B8, and insert a "Scatter chart" as in Example 1 to obtain the result shown below:

Section 1.2

Example 4(a) (page 62)   The demand and supply curves for private schools in Michigan are $q = 77.8p^{-0.11}$ and $q = 30.4 + 0.006p$ thousand students, respectively ($200 \le p \le 2{,}200$), where p is the net tuition cost in dollars. Graph the demand and supply curves on the same set of axes. Use your graph to estimate, to the nearest \$100, the tuition at which the demand equals the supply (equilibrium price). Approximately how many students will be accommodated at that price?

Solution with Technology

To obtain the graphs of demand and supply:

1. Enter the headings p, Demand, and Supply in cells A1–C1 and the p-values 200, 300, . . . , 2,200 in A2–A22.

2. Next, enter the formulas for the demand and supply functions in cells B2 and C2.

Demand: `=77.8*A2^(-0.11)`   in cell B2

Supply: `=30.4+0.006*A2`   in cell C2

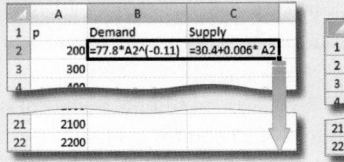 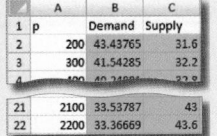

3. To graph the data, highlight A1 through C22, and insert a Scatter chart:

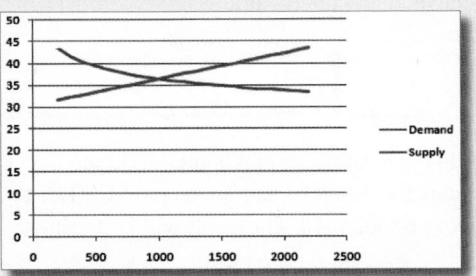

4. If you place the cursor as close as you can get to the intersection point (or just look at the table of values), you will see that the curves cross close to $p = \$1,000$ (to the nearest $100).

5. To more accurately determine where the curves cross, you can narrow down the range of values shown on the x-axis by changing the p-values to 990, 991, . . . , 1010.

Example 5(a) (page 64)   The following table shows annual sales, in billions of dollars, by *Nike* from 2005 through 2010 ($t = 0$ represents 2005):

t	0	1	2	3	4	5
Sales ($ billion)	13.5	15	16.5	18.5	19	19

Consider the following four models:

(1) $s(t) = 14t + 1.2t$ Linear model

(2) $s(t) = 13 + 2.2t - 0.2t^2$ Quadratic model

(3) $s(t) = 14(1.07^t)$ Exponential model

(4) $s(t) = \dfrac{19.5}{1 + 0.48(1.8^{-t})}$ Logistic model

a. Which models fit the data significantly better than the rest?

b. Of the models you selected in part (a), which gives the most reasonable prediction for 2013?

Solution with Technology

1. First create a scatter plot of the given data by tabulating the data as shown below, and selecting the Insert tab and choosing a "Scatter" chart:

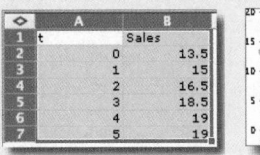 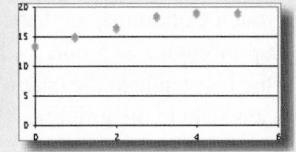

2. In column C use the formula for the model you are interested in seeing; for example, model (2):

`=13+2.2*A2-0.2*A2^2`

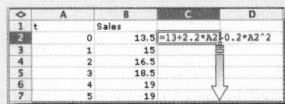 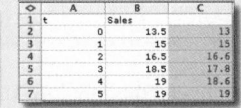

3. To adjust the graph to include the graph of the model you have added, you need to change the graph data from $A\$2:\$B\$7$ to $A\$2:\$C\$7$ so as to include column C. In Excel you can obtain this by right-clicking on the graph to select "Source Data". In OpenOffice, double-click on the graph and then right-click it to choose "Data Ranges". In Excel, you can also click once on the graph—the effect will be to outline the data you have graphed in columns A and B—and then use the fill handle at the bottom of Column B to extend the selection to Column C as shown:

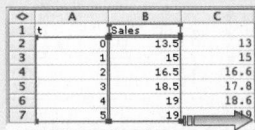 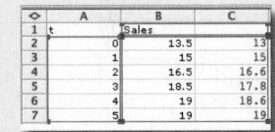

The graph will now include markers showing the values of both the actual sales and the model you inserted in Column C.

4. Right-click on any of the markers corresponding to column B in the graph (in OpenOffice you would first double-click on the graph), select "Format data series" to add lines connecting the points and remove the markers. The effect will be as shown below, with the model represented by a curve and the actual data points represented by dots:

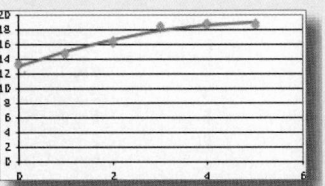

5. To see the extrapolation of the curve to 2013, add the values 6, 7, 8 to Column A. The values of $s(t)$ may automatically be computed in Column C as you type, depending on the spreadsheet. If not, you will need to copy the formula in column C down to C10. (Do not touch Column B, as that contains the observed data up through $t = 5$ only.) Click on the graph, and use the fill handle at the base of Column C to include the new data in the graph:

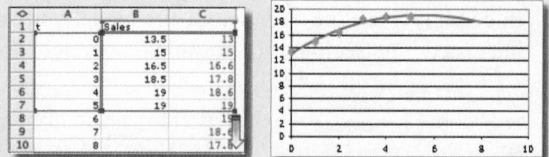

6. To see the plots for the remaining curves, change the formula in Column B (and don't forget to copy the new formula down to cell C10 when you do so).

Section 1.3

Example 1 (page 81)   Which of the following two tables gives the values of a linear function? What is the formula for that function?

x	0	2	4	6	8	10	12
$f(x)$	3	-1	-3	-6	-8	-13	-15

x	0	2	4	6	8	10	12
$g(x)$	3	-1	-5	-9	-13	-17	-21

Solution with Technology

1. The following worksheet shows how you can compute the successive quotients $m = \Delta y / \Delta x$, and hence check whether a given set of data shows a linear relationship, in which case all the quotients will be the same. (The shading indicates that the formula is to be copied down only as far as cell C7. Why not cell C8?)

	A	B	C
1	x	f(x)	m
2	0	3	=(B3-B2)/(A3-A2)
3	2	-1	
4	4	-3	
5	6	-6	
6	8	-8	
7	10	-13	
8	12	-15	

2. Here are the results for both f and g.

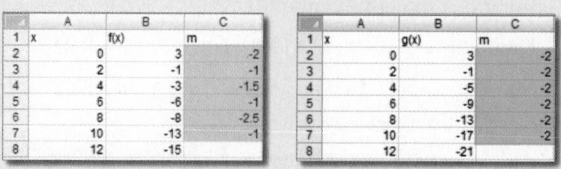

Example 2(a) (page 83)   Find the equation of the line through the points $(1, 2)$ and $(3, -1)$.

Solution with Technology

1. Enter the x- and y-coordinates in columns A and B, as shown below on the left.

	A	B			A	B	C	D
1	x	y		1	x	y	m	b
2		1	2		2		1	=(B3-B2)/(A3-A2)	=B2-C2*A2
3		3	-1		3		3	-1	Slope	Intercept

2. Add the headings m and b in C1-D1, and then the formulas for the slope and intercept in C2-D2, as shown above on the right. The result will be as shown below:

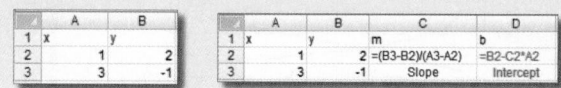

Section 1.4

Example 1(a) (page 97)   Using the data on the per capita GDP in Mexico given at the beginning of Section 1.4, compute SSE, the sum-of-squares error, for the linear models $y = 0.5t + 8$ and $y = 0.25t + 9$, and graph the data with the given models.

Solution with Technology

1. Begin by setting up your worksheet with the observed data in two columns, t and y, and the predicted data for the first model in the third.

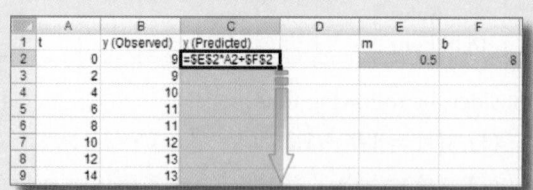

2. Notice that, instead of using the numerical equation for the first model in column C, we used absolute references to the cells containing the slope m and the intercept b. This way, we can switch from one linear model to the next by changing only m and b in cells E2 and F2. (We have deliberately left column D empty in anticipation of the next step.)

3. In column D we compute the squares of the residuals using the Excel formula $=(B2-C2)^2$.

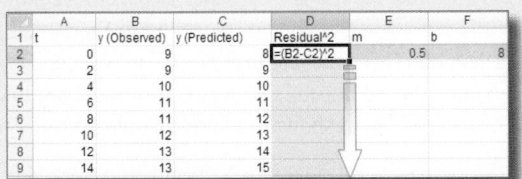

4. We now compute SSE in cell F4 by summing the entries in column D:

5. Here is the completed spreadsheet:

6. Changing m to 0.25 and b to 9 gives the sum of squares error for the second model, SSE $= 2$.

7. To plot both the original data points and each of the two lines, use a scatter plot to graph the data in columns A through C in each of the last two worksheets above.

$$y = 0.5t + 8 \qquad y = 0.25t + 9$$

Example 2 (page 99) Use the data on the per capita GDP in Mexico to find the best-fit linear model.

Solution with Technology

Here are two spreadsheet shortcuts for linear regression; one graphical and one based on a spreadsheet formula:

Using a Trendline

1. Start with the original data and insert a scatter plot (below left and right).

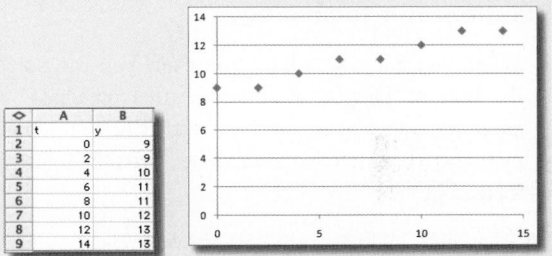

2. Insert a "linear trendline", choosing the option to display the equation on the chart. The method for doing so varies from spreadsheet to spreadsheet.[77] In Excel, you can right-click on one of the points in the graph and choose "Add Trendline" (in OpenOffice you would first double-click on the graph). Then, under "Trendline Options", select "Display Equation on chart". The procedure for OpenOffice is almost identical, but you first need to double-click on the graph. The result is shown below.

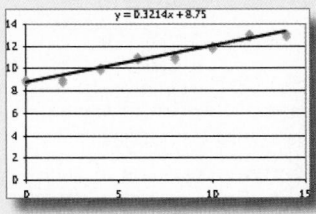

Using a Formula

1. Enter your data as above, and select a block of unused cells two wide and one tall; for example, C2:D2. Then enter the formula

```
=LINEST(B2:B9,A2:A9)
```

---

[77]At the time of this writing, Google Docs has no trendline feature for its spreadsheet, so you would need to use the formula method.

as shown on the left. Then press Control-Shift-Enter. The result should appear as on the right, with m and b appearing in cells C2 and D2 as shown:

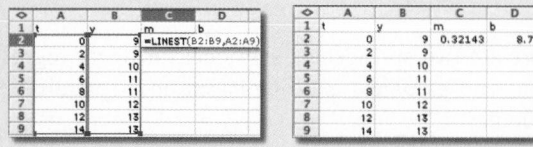

Example 3 (page 101)   Find the correlation coefficient for the data in Example 2.

Solution with Technology

1. When you add a trendline to a chart you can select the option "Display R-squared value on chart" to show the value of r^2 on the chart (it is common to examine r^2, which takes on values between 0 and 1, instead of r).

2. Alternatively, the LINEST function we used above in 2 can be used to display quite a few statistics about a best-fit line, including r^2. Instead of selecting a block of cells two wide and one tall as we did in Example 2, we select one two wide and *five* tall. We now enter the requisite LINEST formula with two additional arguments set to "TRUE" as shown, and press Control-Shift-Enter:

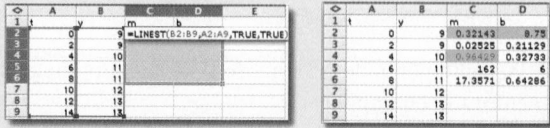

The values of m and b appear in cells C2 and D2 as before, and the value of r^2 in cell C4. (Among the other numbers shown is SSE in cell D6. For the meanings of the remaining numbers shown, do a Web search for "LINEST"; you will see numerous articles, including many that explain all the terms. A good course in statistics wouldn't hurt, either.)

2

The Mathematics of Finance

Case Study Adjustable Rate and Subprime Mortgages

Mr. and Mrs. Wong have an appointment tomorrow with you, their investment counselor, to discuss their plan to purchase a $400,000 house in Orlando, Florida. Their combined annual income is $80,000 per year, which they estimate will increase by 4% annually over the foreseeable future, and they are considering three different specialty 30-year mortgages:

Hybrid: The interest is fixed at a low introductory rate of 4% for 5 years.

Interest-Only: During the first 5 years, the rate is set at 4.2% and no principal is paid.

Negative Amortization: During the first 5 years, the rate is set at 4.7% based on a principal of 60% of the purchase price of the home.

How would you advise them?

Andy Dean Photography/Shutterstock

Introduction

A knowledge of the mathematics of investments and loans is important not only for business majors but also for everyone who deals with money, which is all of us. This chapter is largely about *interest*: interest paid by an investment, interest paid on a loan, and variations on these.

We focus on three forms of investment: investments that pay simple interest, investments in which interest is compounded, and annuities. An investment that pays *simple interest* periodically gives interest directly to the investor, perhaps in the form of a monthly check. If instead, the interest is reinvested, the interest is *compounded*, and the value of the account grows as the interest is added. An *annuity* is an investment earning compound interest into which periodic payments are made or from which periodic withdrawals are made; in the case of periodic payments, such an investment is more commonly called a *sinking fund*. From the point of view of the lender, a loan is a kind of annuity.

We also look at bonds, the primary financial instrument used by companies and governments to raise money. Although bonds nominally pay simple interest, determining their worth, particularly in the secondary market, requires an annuity calculation.

2.1   Simple Interest

You deposit $1,000, called the **principal** or **present value**, into a savings account. The bank pays you 5% interest, in the form of a check, each year. How much interest will you earn each year? Because the bank pays you 5% interest each year, your annual (or yearly) interest will be 5% of $1,000, or $1,000 \times 0.05 = \$50$.

Generalizing this calculation, call the present value PV and the interest rate (expressed as a decimal) r. Then INT, the annual interest paid to you, is given by[*]

$$INT = PVr.$$

If the investment is made for a period of t years, then the total interest accumulated is t times this amount, which gives us the following:

[*] Multiletter variables like PV and INT used here may be unusual in a math textbook but are almost universally used in finance textbooks, calculators (such as the TI-83/84 Plus), and such places as study guides for the finance portion of the Society of Actuaries exams. Just watch out for expressions like PVr, which is the product of two things, PV and r, not three.

Simple Interest

The **simple interest** on an investment (or loan) of PV dollars at an annual interest rate of r for a period of t years is

$$INT = PVrt.$$

Quick Example

The simple interest over a period of 4 years on a $5,000 investment earning 8% per year is

$$INT = PVrt$$
$$= (5,000)(0.08)(4) = \$1,600.$$

Given your $1,000 investment at 5% simple interest, how much money will you have after 2 years? To find the answer, we need to add the accumulated interest to the principal to get the **future value** (FV) of your deposit.

$$FV = PV + INT = \$1{,}000 + (1{,}000)(0.05)(2) = \$1{,}100$$

In general, we can compute the future value as follows:

$$FV = PV + INT = PV + PVrt = PV(1 + rt).$$

Future Value for Simple Interest

The **future value** of an investment of PV dollars at an annual simple interest rate of r for a period of t years is

$$FV = PV(1 + rt).$$

Quick Examples

1. The value, at the end of 4 years, of a $5,000 investment earning 8% simple interest per year is

$$FV = PV(1 + rt)$$
$$= 5{,}000[1 + (0.08)(4)] = \$6{,}600.$$

2. Writing the future value in Quick Example 1 as a function of time, we get

$$FV = 5{,}000(1 + 0.08t)$$
$$= 5{,}000 + 400t,$$

which is a linear function of time t. The intercept is $PV = \$5{,}000$, and the slope is the annual interest, $400 per year.

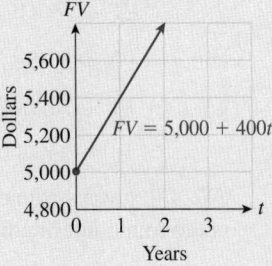

In general: *Simple interest growth is a linear function of time, with intercept given by the present value and slope given by annual interest.*

EXAMPLE 1 Savings Accounts

In November 2011, the Bank of Montreal was paying 1.30% interest on savings accounts.[1] If the interest is paid as simple interest, find the future value of a $2,000 deposit in 6 years. What is the total interest paid over the period?

---

[1]Source: Canoe Money (http://money.canoe.ca/rates/savings.html)

Solution We use the future value formula:

$$FV = PV(1 + rt)$$
$$= 2,000[1 + (0.013)(6)] = 2,000[1.078] = \$2,156.$$

The total interest paid is given by the simple interest formula:

$$INT = PVrt$$
$$= (2,000)(0.013)(6) = \$156.$$

Note To find the interest paid, we could also have computed

$$INT = FV - PV = 2,156 - 2,000 = \$156. \quad \blacksquare$$

➡ **Before we go on...** In the preceding example, we could look at the future value as a function of time:

$$FV = 2,000(1 + 0.013t) = 2,000 + 26t.$$

Thus, the future value is growing linearly at a rate of \$26 per year. ∎

EXAMPLE 2 **Bridge Loans**

When "trading up," homeowners sometimes have to buy a new house before they sell their old house. One way to cover the costs of the new house until they get the proceeds from selling the old house is to take out a short-term *bridge loan*. Suppose a bank charges 12% simple annual interest on such a loan. How much will be owed at the maturation (the end) of a 90-day bridge loan of \$90,000?

Solution We use the future value formula

$$FV = PV(1 + rt)$$

with $t = 90/365$, the fraction of a year represented by 90 days:

$$FV = 90,000[1 + (0.12)(90/365)]$$
$$= \$92,663.01.$$

(We will always round our answers to the nearest cent after calculation. Be careful not to round intermediate results.)

➡ **Before we go on...** Many banks use 360 days for this calculation rather than 365, which makes a "year" for the purposes of the loan slightly shorter than a calendar year. The effect is to increase the amount owed:

$$FV = 90,000[1 + (0.12)(90/360)] = \$92,700 \qquad \blacksquare$$

One of the primary ways companies and governments raise money is by selling **bonds**. At its most straightforward, a corporate bond promises to pay simple interest, usually twice a year, for a length of time until it **matures**, at which point it returns the original investment to the investor (U.S. Treasury notes and bonds are similar). Things get more complicated when the selling price is negotiable, as we will see later in this chapter.

EXAMPLE 3 **Corporate Bonds**

The Megabucks Corporation is issuing 10-year bonds paying an annual rate of 6.5%. If you buy $10,000 worth of bonds, how much interest will you earn every 6 months, and how much interest will you earn over the life of the bonds?

Solution Using the simple interest formula, every 6 months you will receive

$$INT = PVrt$$
$$= (10,000)(0.065)\left(\frac{1}{2}\right) = \$325.$$

Over the 10-year life of the bonds, you will earn

$$INT = PVrt$$
$$= (10,000)(0.065)(10) = \$6,500$$

in interest. So, at the end of 10 years, when your original investment is returned to you, your $10,000 will have turned into $16,500.

We often want to turn an interest calculation around: Rather than starting with the present value and finding the future value, there are times when we know the future value and need to determine the present value. Solving the future value formula for *PV* gives us the following.

Present Value for Simple Interest

The present value of an investment at an annual simple interest rate of *r* for a period of *t* years, with future value *FV*, is

$$PV = \frac{FV}{1+rt}.$$

Quick Example

If an investment earns 5% simple interest and will be worth $1,000 in 4 years, then its present value (its initial value) is

$$PV = \frac{FV}{1+rt}$$
$$= \frac{1,000}{1+(0.05)(4)} = \$833.33.$$

Here is a typical example. U.S. Treasury bills (T-bills) are short-term investments (up to 1 year) that pay you a set amount after a period of time; what you pay to buy a T-bill depends on the interest rate.

EXAMPLE 4 **Treasury Bills**

A U.S. Treasury bill paying $10,000 after 6 months earns 3.67% simple annual interest. How much did it cost to buy?

Solution The future value of the T-bill is $10,000; the price we paid is its present value. We know that

$$FV = \$10,000$$
$$r = 0.0367$$

and

$$t = 0.5.$$

Substituting into the present value formula, we have

$$PV = \frac{10,000}{1 + (0.0367)(0.5)} = 9,819.81$$

so we paid $9,819.81 for the T-bill.

➡ **Before we go on...** The simplest way to find the interest earned on the T-bill is by subtraction:

$$INT = FV - PV = 10,000 - 9,819.81 = \$180.19.$$

So, after 6 months we received back $10,000, which is our original investment plus $180.19 in interest. ∎

Here is some additional terminology on Treasury bills:

Treasury Bills (T-Bills): Maturity Value, Discount Rate, and Yield

The **maturity value** of a T-bill is the amount of money it will pay at the end of its life, that is, upon **maturity**.

Quick Example

A 1-year $10,000 T-bill has a maturity value of $10,000, and so will pay you $10,000 after one year.

The cost of a T-bill is generally less than its maturity value.* In other words, a T-bill will generally sell at a *discount*, and the **discount rate** is the *annualized* percentage of this discount; that is, the percentage is adjusted to give an annual percentage. (See Quick Examples 2 and 3.)

*An exception occurred during the financial meltdown of 2008, when T-bills were heavily in demand as "safe haven" investments and were sometimes selling at—or even above—their maturity values.

Quick Examples

1. A 1-year $10,000 T-bill with a discount rate of 5% will sell for 5% less than its maturity value of $10,000, that is, for $9,500.
2. A 6-month $10,000 T-bill with a discount rate of 5% will sell at an actual discount of half of that—2.5% less than its maturity value, or $9,750—because 6 months is half of a year.
3. A 3-month $10,000 T-bill with a discount rate of 5% will sell at an actual discount of a fourth of that: 1.25% less than its maturity value, or $9,875.

The annual **yield** of a T-bill is the simple annual interest rate an investor earns when the T-bill matures, as calculated in the next example.

EXAMPLE 5 **Treasury Bills**

A T-bill paying $10,000 after 6 months sells at a discount rate of 3.6%. What does it sell for? What is the annual yield?

Solution The (annualized) discount rate is 3.6%; so, for a 6-month bill, the actual discount will be half of that: $3.6\%/2 = 1.8\%$ below its maturity value. This makes the selling price

$$10,000 - (0.018)(10,000) = \$9,820. \qquad \text{Maturity value} - \text{Discount}$$

To find the annual yield, note that the present value of the investment is the price the investor pays, $9,820, and the future value is its maturity value, $10,000 six months later. So,

$$PV = \$9,820 \qquad FV = \$10,000 \qquad t = 0.5$$

and we wish to find the annual interest rate r. Substituting in the future value formula, we get

$$FV = PV(1 + rt)$$
$$10,000 = 9,820(1 + 0.5r)$$

so

$$1 + 0.5r = 10,000/9,820$$

and

$$r = (10,000/9,820 - 1)/0.5 \approx 0.0367.$$

Thus, the T-bill is paying 3.67% simple annual interest, so we say that its annual yield is 3.67%.

➡ **Before we go on...** The T-bill in Example 5 is the same one as in Example 4 (with a bit of rounding). The yield and the discount rate are two different ways of telling what the investment pays. One of the Communication and Reasoning Exercises asks you to find a formula for the yield in terms of the discount rate. ∎

Fees on loans can also be thought of as a form of interest.

EXAMPLE 6 **Tax Refunds**

You are expecting a tax refund of $800. Because it may take up to 6 weeks to get the refund, your tax preparation firm offers, for a fee of $40, to give you an "interest-free" loan of $800 to be paid back with the refund check. If we think of the fee as interest, what simple annual interest rate is the firm actually charging?

Solution If we view the $40 as interest, then the future value of the loan (the value of the loan to the firm, or the total you will pay the firm) is $840. Thus, we have

$$FV = 840$$
$$PV = 800$$
$$t = 6/52 \qquad \text{Using 52 weeks in a year}$$

and we wish to find r. Substituting, we get

$$FV = PV(1 + rt)$$

$$840 = 800(1 + 6r/52) = 800 + \frac{4{,}800r}{52}$$

so

$$\frac{4{,}800r}{52} = 840 - 800 = 40$$

$$r = \frac{40 \times 52}{4{,}800} \approx 0.43.$$

In other words, the firm is charging you 43% annual interest! Save your money and wait 6 weeks for your refund.

2.1 EXERCISES

▼ more advanced   ◆ challenging
T indicates exercises that should be solved using technology

In Exercises 1–6, compute the simple interest for the specified period and the future value at the end of the period. Round all answers to the nearest cent. HINT [See Example 1.]

1. $2,000 is invested for 1 year at 6% per year.

2. $1,000 is invested for 10 years at 4% per year.

3. $20,200 is invested for 6 months at 5% per year.

4. $10,100 is invested for 3 months at 11% per year.

5. You borrow $10,000 for 10 months at 3% per year.

6. You borrow $6,000 for 5 months at 9% per year.

In Exercises 7–12, find the present value of the given investment. HINT [See the Quick Example on page 129.]

7. An investment earns 2% per year and is worth $10,000 after 5 years.

8. An investment earns 5% per year and is worth $20,000 after 2 years.

9. An investment earns 7% per year and is worth $1,000 after 6 months.

10. An investment earns 10% per year and is worth $5,000 after 3 months.

11. An investment earns 3% per year and is worth $15,000 after 15 months.

12. An investment earns 6% per year and is worth $30,000 after 20 months.

APPLICATIONS

In Exercises 13–30, compute the specified quantity. Round all answers to the nearest month, the nearest cent, or the nearest 0.001%, as appropriate.

13. *Simple Loans* You take out a 6-month, $5,000 loan at 8% simple interest. How much would you owe at the end of the 6 months? HINT [See Example 2.]

14. *Simple Loans* You take out a 15-month, $10,000 loan at 11% simple interest. How much would you owe at the end of the 15 months? HINT [See Example 2.]

15. *Savings* How much would you have to deposit in an account earning 4.5% simple interest if you wanted to have $1,000 after 6 years? HINT [See Example 4.]

16. *Simple Loans* Your total payment on a 4-year loan, which charged 9.5% simple interest, amounted to $30,360. How much did you originally borrow? HINT [See Example 4.]

17. *Bonds* A 5-year bond costs $1,000 and will pay a total of $250 interest over its lifetime. What is its annual interest rate? HINT [See Example 3.]

18. *Bonds* A 4-year bond costs $10,000 and will pay a total of $2,800 in interest over its lifetime. What is its annual interest rate? HINT [See Example 3.]

19. *Treasury Bills* In December 2008 (in the midst of the financial crisis) a $5,000 6-month T-bill was selling at a discount rate of only 0.25%.[2] What was its simple annual yield? HINT [See Example 5.]

20. *Treasury Bills* In December 2008 (in the midst of the financial crisis) a $15,000 3-month T-bill was selling at a discount rate of only 0.06%.[3] What was its simple annual yield? HINT [See Example 5.]

21. ▼ *Simple Loans* A $4,000 loan, taken now, with a simple interest rate of 8% per year, will require a total repayment of $4,640. When will the loan mature?

22. ▼ *Simple Loans* The simple interest on a $1,000 loan at 8% per year amounted to $640. When did the loan mature?

---

[2]Discount rate on December 29, 2008. Source: U.S. Treasury (www.ustreas.gov/offices/domestic-finance/debt-management/interest-rate/daily_treas_bill_rates.shtml).
[3]*Ibid.*

23. ▼ *Treasury Bills* At auction on August 18, 2005, 6-month T-bills were sold at a discount of 3.705%.[4] What was the simple annual yield? HINT [See Example 5.]

24. ▼ *Treasury Bills* At auction on August 18, 2005, 3-month T-bills were sold at a discount of 3.470%.[5] What was the simple annual yield? HINT [See Example 5.]

25. ▼ *Fees* You are expecting a tax refund of $1,000 in 4 weeks. A tax preparer offers you a $1,000 loan for a fee of $50 to be repaid by your refund check when it arrives in 4 weeks. Thinking of the fee as interest, what annual simple interest rate would you be paying on this loan? HINT [See Example 6.]

26. ▼ *Fees* You are expecting a tax refund of $1,500 in 3 weeks. A tax preparer offers you a $1,500 loan for a fee of $60 to be repaid by your refund check when it arrives in 3 weeks. Thinking of the fee as interest, what annual simple interest rate would you be paying on this loan?

27. ▼ *Fees* You take out a 2-year, $5,000 loan at 9% simple annual interest. The lender charges you a $100 fee. Thinking of the fee as additional interest, what is the actual annual interest rate you will pay?

28. ▼ *Fees* You take out a 3-year, $7,000 loan at 8% simple annual interest. The lender charges you a $100 fee. Thinking of the fee as additional interest, what is the actual annual interest rate you will pay?

29. ▼ *Layaway Fees* Layaway plans allow you, for a fee, to pay for an item over a period of time and then receive the item when you finish paying for it. In November 2011, Senator Charles E. Schumer of New York warned that the holiday layaway programs recently reinstated by several popular retailers were, when you took the fees into account, charging interest at a rate significantly higher than the highest credit card rates.[6] Suppose that you bought a $69 item on November 15 on layaway, with the final payment due December 15, and that the retailer charged you a $5 service fee. Thinking of the fee as interest, what simple interest rate would you be paying for this layaway plan?

30. ▼ *Layaway Fees* Referring to Exercise 29, suppose that you bought a $99 item on October 15 on layaway, with the final payment due December 15, and that the retailer charged you a $10 service fee. Thinking of the fee as interest, what simple interest rate would you be paying for this layaway plan?

Stock Investments Exercises 31–36 are based on the following chart, which shows monthly figures for Apple Computer, Inc. stock in 2010.[7]

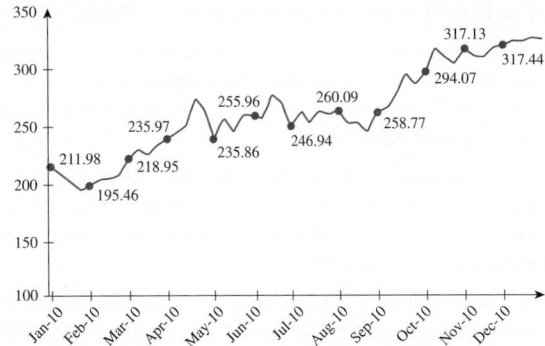

Marked are the following points on the chart:

Jan. 10	Feb. 10	Mar. 10	Apr. 10	May 10	June 10
211.98	195.46	218.95	235.97	235.86	255.96
July 10	Aug. 10	Sep. 10	Oct. 10	Nov. 10	Dec. 10
246.94	260.09	258.77	294.07	317.13	317.44

31. Calculate to the nearest 0.01% your annual percentage return (on a simple interest basis) if you had bought Apple stock in June and sold in December.

32. Calculate to the nearest 0.01% your annual percentage return (on a simple interest basis) if you had bought Apple stock in February and sold in June.

33. ▼ Suppose you bought Apple stock in April. If you later sold at one of the marked dates on the chart, which of those dates would have given you the largest annual return (on a simple interest basis), and what would that return have been?

34. ▼ Suppose you bought Apple stock in May. If you later sold at one of the marked dates on the chart, which of those dates would have given you the largest annual return (on a simple interest basis), and what would that return have been?

35. ▼ Did Apple's stock undergo simple interest change in the period January through May? (Give a reason for your answer.)

36. ▼ If Apple's stock had undergone simple interest change from April to June at the same simple rate as from March to April, what would the price have been in June?

Population Exercises 37–42 are based on the following graph, which shows the population of San Diego County from 1950 to 2000.[8]

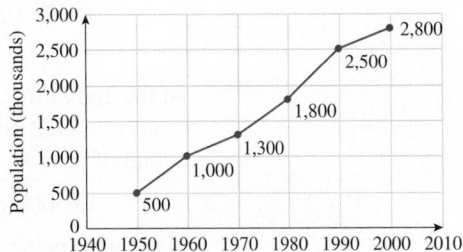

---

[4]Source: The Bureau of the Public Debt's Web site: www.publicdebt.treas.gov.

[5]*Ibid.*

[6]Source: November 14, 2011 press release from the office of Senator Charles E. Schumer (http://schumer.senate.gov/Newsroom/ releases.cfm)

[7]Source: Yahoo! Finance (http://finance.yahoo.com).

[8]Source: Census Bureau/*New York Times*, April 2, 2002, p. F3.

37. At what annual (simple interest) rate did the population of San Diego County increase from 1950 to 2000?

38. At what annual (simple interest) rate did the population of San Diego County increase from 1950 to 1990?

39. ▼ If you used your answer to Exercise 37 as the annual (simple interest) rate at which the population was growing since 1950, what would you predict the San Diego County population to be in 2010?

40. ▼ If you used your answer to Exercise 38 as the annual (simple interest) rate at which the population was growing since 1950, what would you predict the San Diego County population to be in 2010?

41. ▼ Use your answer to Exercise 37 to give a linear model for the population of San Diego County from 1950 to 2000. Draw the graph of your model over that period of time.

42. ▼ Use your answer to Exercise 38 to give a linear model for the population of San Diego County from 1950 to 2000. Draw the graph of your model over that period of time.

COMMUNICATION AND REASONING EXERCISES

43. One or more of the following three graphs represents the future value of an investment earning simple interest. Which one(s)? Give the reason for your choice(s).

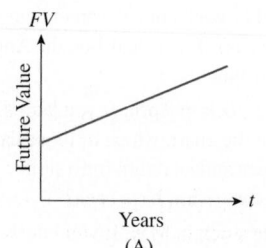

(A)

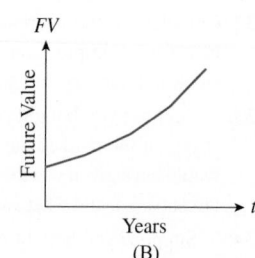

(B)

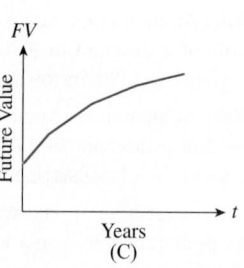

(C)

44. Given that $FV = 5t + 400$, for what interest rate is this the equation of future value (in dollars) as a function of time t (in years)?

45. ▼ *Interpreting the News* You hear the following on your local radio station's business news: "The economy last year grew by 1%. This was the second year in a row in which the economy showed a 1% growth." Because the rate of growth was the same two years in a row, this represents simple interest growth, right? Explain your answer.

46. ▼ *Interpreting the News* You hear the following on your local radio station's business news: "The economy last year grew by 1%. This was the second year in a row in which the economy showed a 1% growth." This means that, in dollar terms, the economy grew more last year than the year before. Why?

47. ▼ Explain why simple interest is not the appropriate way to measure interest on a savings account that pays interest directly into your account.

48. ▼ Suppose that a one-year T-bill sells at a discount rate d. Find a formula, in terms of d, for the simple annual interest rate r the bill will pay.

2.2 Compound Interest

You deposit $1,000 into a savings account. The bank pays you 5% interest, which it deposits into your account, or **reinvests**, at the end of each year. At the end of 5 years, how much money will you have accumulated? Let us compute the amount you have at the end of each year. At the end of the first year, the bank will pay you simple interest of 5% on your $1,000, which gives you

$$PV(1 + rt) = 1,000(1 + 0.05)$$
$$= \$1,050.$$

At the end of the second year, the bank will pay you another 5% interest, but this time computed on the total in your account, which is $1,050. Thus, you will have a total of

$$1,050(1 + 0.05) = \$1,102.50.$$

If you were being paid simple interest on your original $1,000, you would have only $1,100 at the end of the second year. The extra $2.50 is the interest earned on the $50 interest added to your account at the end of the first year. Having interest earn interest is called **compounding** the interest. We could continue like this until the end of the fifth year, but notice what we are doing: Each year we are multiplying by $1 + 0.05$. So, at the end of 5 years, you will have

$$1{,}000(1 + 0.05)^5 \approx \$1{,}276.28.$$

It is interesting to compare this to the amount you would have if the bank paid you simple interest:

$$1{,}000(1 + 0.05 \times 5) = \$1{,}250.00.$$

The extra \$26.28 is again the effect of compounding the interest.

Banks often pay interest more often than once a year. Paying interest quarterly (four times per year) or monthly is common. If your bank pays interest monthly, how much will your \$1,000 deposit be worth after 5 years? The bank will not pay you 5% interest every month, but will give you 1/12 of that,* or 5/12% interest each month. Thus, instead of multiplying by $1 + 0.05$ every year, we should multiply by $1 + 0.05/12$ each month. Because there are $5 \times 12 = 60$ months in 5 years, the total amount you will have at the end of 5 years is

$$1{,}000 \left(1 + \frac{0.05}{12}\right)^{60} \approx \$1{,}283.36.$$

Compare this to the \$1,276.28 you would get if the bank paid the interest every year. You earn an extra \$7.08 if the interest is paid monthly because interest gets into your account and starts earning interest earlier. The amount of time between interest payments is called the **compounding period**.

The following table summarizes the results above.

Time in Years	Amount with Simple Interest	Amount with Annual Compounding	Amount with Monthly Compounding
0	\$1,000	\$1,000	\$1,000
1	$1{,}000(1 + 0.05)$ $= \$1{,}050$	$1{,}000(1 + 0.05)$ $= \$1{,}050$	$1{,}000(1 + 0.05/12)^{12}$ $= \$1{,}051.16$
2	$1{,}000(1 + 0.05 \times 2)$ $= \$1{,}100$	$1{,}000(1 + 0.05)^2$ $= \$1{,}102.50$	$1{,}000(1 + 0.05/12)^{24}$ $= \$1{,}104.94$
5	$1{,}000(1 + 0.05 \times 5)$ $= \$1{,}250$	$1{,}000(1 + 0.05)^5$ $= \$1{,}276.28$	$1{,}000(1 + 0.05/12)^{60}$ $= \$1{,}283.36$

The preceding calculations generalize easily to give the general formula for future value when interest is compounded.

Future Value for Compound Interest

The future value of an investment of *PV* dollars earning interest at an annual rate of *r* compounded (reinvested) *m* times per year for a period of *t* years is

$$FV = PV\left(1 + \frac{r}{m}\right)^{mt}$$

or

$$FV = PV(1 + i)^n,$$

where $i = r/m$ is the interest paid each compounding period and $n = mt$ is the total number of compounding periods.

Quick Examples

1. To find the future value after 5 years of a $10,000 investment earning 6% interest, with interest reinvested every month, we set $PV = 10,000$, $r = 0.06$, $m = 12$, and $t = 5$. Thus,

$$FV = PV\left(1 + \frac{r}{m}\right)^{mt} = 10,000\left(1 + \frac{0.06}{12}\right)^{60} \approx \$13,488.50.$$

2. Writing the future value in Quick Example 1 as a function of time, we get

$$FV = 10,000\left(1 + \frac{0.06}{12}\right)^{12t}$$
$$= 10,000(1.005)^{12t},$$

which is an **exponential** function of time t.

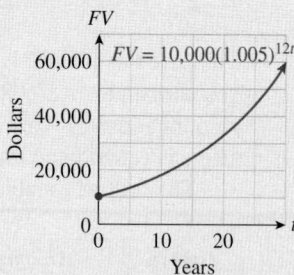

In general: *Compound interest growth is an exponential function of time.*

 using Technology

All three technologies discussed in this book have built-in mathematics of finance capabilities. See the Technology Guides at the end of the chapter for details on using a TI-83/84 Plus or a spreadsheet to do the calculations in Example 1. [Details: TI-83/84 Plus: page 166, Spreadsheet: page 170]

Website
www.WanerMath.com
→ On Line Utilities
→ Time Value of Money Utility

This utility is similar to the TVM Solver on the TI-83/84 Plus. To compute the future value, enter the values shown, and press "Compute" next to FV.

FV =	Compute
PV = -2000	Compute
PMT = 0	Compute
r = 2.4%	Compute
m = 4	Compute
t = 6	Compute
Clear All	

(For an explanation of the terms, see the Technology Guide for the TI-83/84 Plus.)

EXAMPLE 1 Savings Accounts

In November 2011, the Bank of Montreal was paying 1.30% interest on savings accounts.[9] If the interest is compounded quarterly, find the future value of a $2,000 deposit in 6 years. What is the total interest paid over the period?

Solution We use the future value formula with $m = 4$:

$$FV = PV\left(1 + \frac{r}{m}\right)^{mt}$$
$$= 2,000\left(1 + \frac{0.0130}{4}\right)^{4 \times 6} \approx \$2,161.97.$$

The total interest paid is

$$INT = FV - PV = 2,161.97 - 2,000 = \$161.97.$$

---

[9]Source: Canoe Money (http://money.canoe.ca/rates/savings.html)

Example 1 illustrates the concept of the **time value of money**: A given amount of money received now will usually be worth a different amount to us than the same amount received some time in the future. In the example above, we can say that $2,000 received now is worth the same as $2,161.97 received 6 years from now, because if we receive $2,000 now, we can turn it into $2,161.97 by the end of 6 years.

We often want to know, for some amount of money in the future, what is the equivalent value at present. As we did for simple interest, we can solve the future value formula for the present value and obtain the following formula.

Present Value for Compound Interest

The present value of an investment earning interest at an annual rate of r compounded m times per year for a period of t years, with future value FV, is

$$PV = \frac{FV}{\left(1 + \frac{r}{m}\right)^{mt}}$$

or

$$PV = \frac{FV}{(1 + i)^n} = FV(1 + i)^{-n},$$

where $i = r/m$ is the interest paid each compounding period and $n = mt$ is the total number of compounding periods.

Quick Example

To find the amount we need to invest in an investment earning 12% per year, compounded annually, so that we will have $1 million in 20 years, use $FV = \$1,000,000, r = 0.12, m = 1$, and $t = 20$:

$$PV = \frac{FV}{\left(1 + \frac{r}{m}\right)^{mt}} = \frac{1,000,000}{(1 + 0.12)^{20}} \approx \$103,666.77.$$

Put another way, $1,000,000 20 years from now is worth only $103,666.77 to us now, if we have a 12% investment available.

In the preceding section, we mentioned that a bond pays interest until it reaches maturity, at which point it pays you back an amount called its **maturity value** or **par value**. The two parts, the interest and the maturity value, can be separated and sold and traded by themselves. A **zero coupon bond** is a form of corporate bond that pays no interest during its life but, like U.S. Treasury bills, promises to pay you the maturity value when it reaches maturity. Zero coupon bonds are often created by removing or *stripping* the interest coupons from an ordinary bond, so are also known as **strips**. Zero coupon bonds sell for less than their maturity value, and the return on the investment is the difference between what the investor pays and the maturity value. Although no interest is actually paid, we measure the return on investment by thinking of the interest rate that would make the selling price (the present value) grow to become the maturity value (the future value).*

* The IRS refers to this kind of interest as **original issue discount (OID)** and taxes it as if it were interest actually paid to you each year.

＊ The return investors look for depends on a number of factors, including risk (the chance that the
company will go bankrupt and
you will lose your investment);
the higher the risk, the higher the
return. U.S. Treasuries are considered risk free because the
federal government has never
defaulted on its debts. On the
other hand, so-called junk bonds
are high-risk investments (below
investment grade) and have cor-
respondingly high yields.

using Technology

See the Technology Guides at the
end of the chapter for details on
using TVM Solver on the TI-83/84
Plus or the built-in finance functions
in spreadsheets to do the calculations in Example 2. [Details: TI-83/84
Plus: page 166, Spreadsheet:
page 171]

Website

www.WanerMath.com

→ On Line Utilities

→ Time Value of Money Utility

This utility is similar to the TVM
Solver on the TI-83/84 Plus. To
compute the present value, enter
the values shown, and press
"Compute" next to PV.

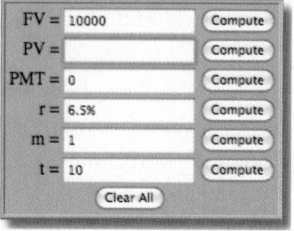

EXAMPLE 2 Zero Coupon Bonds

Megabucks Corporation is issuing 10-year zero coupon bonds. How much would you
pay for bonds with a maturity value of $10,000 if you wish to get a return of 6.5%
compounded annually?＊

Solution As we said earlier, we think of a zero coupon bond as if it were an account
earning (compound) interest. We are asked to calculate the amount you will pay for
the bond—the present value PV. We have

$$FV = \$10,000$$
$$r = 0.065$$
$$t = 10$$
$$m = 1.$$

We can now use the present value formula:

$$PV = \frac{FV}{\left(1 + \frac{r}{m}\right)^{mt}}$$

$$PV = \frac{10,000}{\left(1 + \frac{0.065}{1}\right)^{10 \times 1}} \approx \$5,327.26.$$

Thus, you should pay $5,327.26 to get a return of 6.5% annually.

➡ **Before we go on...** Particularly in financial applications, you will hear the word
"discounted" in place of "compounded" when discussing present value. Thus, the
result of Example 2 might be phrased, "The present value of $10,000 to be
received 10 years from now, with an interest rate of 6.5% discounted annually, is
$5,327.26." ∎

Time value of money calculations are often done to take into account inflation,
which behaves like compound interest. Suppose, for example, that inflation is running at 5% per year. Then prices will increase by 5% each year, so if PV represents
the price now, the price one year from now will be 5% higher, or $PV(1 + 0.05)$. The
price a year from then will be 5% higher still, or $PV(1 + 0.05)^2$. Thus, the effects of
inflation are compounded just as reinvested interest is.

EXAMPLE 3 Inflation

Inflation in East Avalon is 5% per year. TruVision television sets cost $200 today.
How much will a comparable set cost 2 years from now?

Solution To find the price of a television set 2 years from now, we compute the
future value of $200 at an inflation rate of 5% compounded yearly:

$$FV = 200(1 + 0.05)^2 = \$220.50.$$

EXAMPLE 4 **Constant Dollars**

Inflation in North Avalon is 6% per year. Which is really more expensive, a car costing $20,000 today or one costing $22,000 in 3 years?

Solution We cannot compare the two costs directly because inflation makes $1 today worth more (it buys more) than a dollar 3 years from now. We need the two prices expressed in comparable terms, so we convert to **constant dollars**. We take the car costing $22,000 three years from now and ask what it would cost in today's dollars. In other words, we convert the future value of $22,000 to its present value:

$$PV = FV(1 + i)^{-n}$$
$$= 22,000(1 + 0.06)^{-3}$$
$$\approx \$18,471.62.$$

Thus, the car costing $22,000 in 3 years actually costs less, after adjusting for inflation, than the one costing $20,000 now.

➡ **Before we go on...** In the presence of inflation, the only way to compare prices at different times is to convert all prices to constant dollars. We pick some fixed time and compute future or present values as appropriate to determine what things would have cost at that time. ■

    There are some other interesting calculations related to compound interest, besides present and future values.

EXAMPLE 5 **Effective Interest Rate**

You have just won $1 million in the lottery and are deciding what to do with it during the next year before you move to the South Pacific. Bank Ten offers 10% interest, compounded annually, while Bank Nine offers 9.8% compounded monthly. In which should you deposit your money?

Solution Let's calculate the future value of your $1 million after one year in each of the banks:

Bank Ten: $FV = 1(1 + 0.10)^1 = \$1.1$ million

Bank Nine: $FV = 1\left(1 + \dfrac{0.098}{12}\right)^{12} = \$1.1025$ million.

Bank Nine turns out to be better: It will pay you a total of $102,500 in interest over the year, whereas Bank Ten will pay only $100,000 in interest.

    Another way of looking at the calculation in Example 5 is that Bank Nine gave you a total of 10.25% interest on your investment over the year. We call 10.25% the **effective interest rate** of the investment (also referred to as the **annual percentage**

yield, or **APY** in the banking industry); the stated 9.8% is called the **nominal** interest rate. In general, to best compare two different investments, it is wisest to compare their *effective*—rather than nominal—interest rates.

Notice that we got 10.25% by computing

$$\left(1 + \frac{0.098}{12}\right)^{12} = 1.1025$$

and then subtracting 1 to get 0.1025, or 10.25%. Generalizing, we get the following formula.

Effective Interest Rate

The effective interest rate r_{eff} of an investment paying a nominal interest rate of r_{nom} compounded m times per year is

$$r_{\text{eff}} = \left(1 + \frac{r_{\text{nom}}}{m}\right)^m - 1.$$

To compare rates of investments with different compounding periods, always compare the effective interest rates rather than the nominal rates.

Quick Example

To calculate the effective interest rate of an investment that pays 8% per year, with interest reinvested monthly, set $r_{\text{nom}} = 0.08$ and $m = 12$, to obtain

$$r_{\text{eff}} = \left(1 + \frac{0.08}{12}\right)^{12} - 1 \approx 0.0830, \text{ or } 8.30\%.$$

EXAMPLE 6 How Long to Invest

You have $5,000 to invest at 6% interest compounded monthly. How long will it take for your investment to grow to $6,000?

Solution If we use the future value formula, we already have the values

$$FV = 6,000$$
$$PV = 5,000$$
$$r = 0.06$$
$$m = 12.$$

Substituting, we get

$$6,000 = 5,000\left(1 + \frac{0.06}{12}\right)^{12t}.$$

See the Technology Guides at the end of the chapter for details on using TVM Solver on the TI-83/84 Plus or the built-in finance functions in spreadsheets to do the calculations in Example 6. [Details: TI-83/84 Plus: page 167, Spreadsheet: page 171]

 Website
www.WanerMath.com
→ On Line Utilities
→ Time Value of Money Utility

This utility is similar to the TVM Solver on the TI-83/84 Plus. To compute the time needed, enter the values shown, and press "Compute" next to *t*.

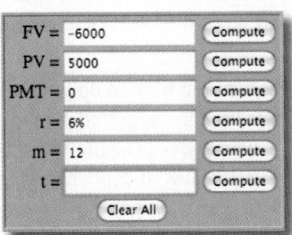

If you are familiar with logarithms, you can solve explicitly for *t* as follows:

$$\left(1 + \frac{0.06}{12}\right)^{12t} = \frac{6{,}000}{5{,}000} = 1.2$$

$$\log\left(1 + \frac{0.06}{12}\right)^{12t} = \log 1.2$$

$$12t \log\left(1 + \frac{0.06}{12}\right) = \log 1.2$$

$$t = \frac{\log 1.2}{12 \log\left(1 + \frac{0.06}{12}\right)} \approx 3.046 \approx 3 \text{ years.}$$

Another approach is to use a bit of trial and error to find the answer. Let's see what the future value is after 2, 3, and 4 years:

$$5{,}000\left(1 + \frac{0.06}{12}\right)^{12\times2} = 5{,}635.80 \quad \text{Future value after 2 years}$$

$$5{,}000\left(1 + \frac{0.06}{12}\right)^{12\times3} = 5{,}983.40 \quad \text{Future value after 3 years}$$

$$5{,}000\left(1 + \frac{0.06}{12}\right)^{12\times4} = 6{,}352.45. \quad \text{Future value after 4 years}$$

From these calculations, it looks as if the answer should be a bit more than 3 years, but is certainly between 3 and 4 years. We could try 3.5 years next and then narrow it down systematically until we have a pretty good approximation of the correct answer. Graphing calculators and spreadsheets give us alternative methods of solution.

FAQs

Recognizing When to Use Compound Interest and the Meaning of Present Value

Q: *How do I distinguish a problem that calls for compound interest from one that calls for simple interest?*

A: Study the scenario to ascertain whether the interest is being withdrawn as it is earned or reinvested (deposited back into the account). If the interest is being withdrawn, the problem is calling for simple interest because the interest is not itself earning interest. If it is being reinvested, the problem is calling for compound interest.

Q: *How do I distinguish present value from future value in a problem?*

A: The present value always refers to the value of an investment before any interest is included (or, in the case of a depreciating investment, before any depreciation takes place). As an example, the future value of a bond is its maturity value. The value of $1 today in constant 2010 dollars is its present value (even though 2010 is in the past).

2.2   EXERCISES

▼ more advanced   ♦ challenging
⊤ indicates exercises that should be solved using technology

In Exercises 1–8, calculate, to the nearest cent, the future value of an investment of $10,000 at the stated interest rate after the stated amount of time. HINT [See Example 1.]

1. 3% per year, compounded annually, after 10 years

2. 4% per year, compounded annually, after 8 years

3. 2.5% per year, compounded quarterly (4 times/year), after 5 years

4. 1.5% per year, compounded weekly (52 times/year), after 5 years

5. 6.5% per year, compounded daily (assume 365 days/year), after 10 years

6. 11.2% per year, compounded monthly, after 12 years

7. 0.2% per month, compounded monthly, after 10 years

8. 0.45% per month, compounded monthly, after 20 years

In Exercises 9–14, calculate, to the nearest cent, the present value of an investment that will be worth $1,000 at the stated interest rate after the stated amount of time. HINT [See Example 2.]

9. 10 years, at 5% per year, compounded annually

10. 5 years, at 6% per year, compounded annually

11. 5 years, at 4.2% per year, compounded weekly (assume 52 weeks per year)

12. 10 years, at 5.3% per year, compounded quarterly

13. 4 years, depreciating 5% each year

14. 5 years, depreciating 4% each year

In Exercises 15–20, find the effective annual interest rates of the given annual interest rates. Round your answers to the nearest 0.01%. HINT [See Quick Example on page 140.]

15. 5% compounded quarterly

16. 5% compounded monthly

17. 10% compounded monthly

18. 10% compounded daily (assume 365 days per year)

19. 10% compounded hourly (assume 365 days per year)

20. 10% compounded every minute (assume 365 days per year)

APPLICATIONS

21. *Savings* You deposit $1,000 in an account at the *Lifelong Trust Savings and Loan* that pays 6% interest compounded quarterly. By how much will your deposit have grown after 4 years?

22. *Investments* You invest $10,000 in *Rapid Growth Funds*, which appreciate by 2% per year, with yields reinvested quarterly. By how much will your investment have grown after 5 years?

23. *Depreciation* During 2008, the S&P 500 index depreciated by 37.6%.[10] Assuming that this trend had continued, how much would a $3,000 investment in an S&P index fund have been worth after 3 years?

24. *Depreciation* During 2008, the NASDAQ Composite Index depreciated by 42%.[11] Assuming that this trend had continued, how much would a $5,000 investment in a NASDAQ Composite Index fund have been worth after 4 years?

25. *Bonds* You want to buy a 10-year zero coupon bond with a maturity value of $5,000 and a yield of 5.5% annually. How much will you pay?

26. *Bonds* You want to buy a 15-year zero coupon bond with a maturity value of $10,000 and a yield of 6.25% annually. How much will you pay?

27. ▼ *Investments* When I was considering what to do with my $10,000 Lottery winnings, my broker suggested I invest half of it in gold, the value of which was growing by 10% per year, and the other half in certificates of deposit (CDs), which were yielding 5% per year, compounded every 6 months. Assuming that these rates are sustained, how much will my investment be worth in 10 years?

28. ▼ *Investments* When I was considering what to do with the $10,000 proceeds from my sale of technology stock, my broker suggested I invest half of it in municipal bonds, whose value was growing by 6% per year, and the other half in CDs, which were yielding 3% per year, compounded every 2 months. Assuming that these interest rates are sustained, how much will my investment be worth in 10 years?

29. ▼ *Depreciation* During a prolonged recession, property values on Long Island depreciated by 2% every 6 months. If my house cost $200,000 originally, how much was it worth 5 years later?

30. ▼ *Depreciation* Stocks in the health industry depreciated by 5.1% in the first 9 months of 1993.[12] Assuming that this trend continued, how much would a $40,000 investment be worth in 9 years? HINT [Nine years corresponds to 12 nine-month periods.]

31. ▼ *Retirement Planning* I want to be earning an annual salary of $100,000 when I retire in 15 years. I have been offered a job that guarantees an annual salary increase of 4% per year, and the starting salary is negotiable. What salary should I request in order to meet my goal?

---

[10]Source: http://finance.google.com.

[11]*Ibid.*

[12]Source: *New York Times*, October 9, 1993, p. 37.

32. ▼ Retirement Planning I want to be earning an annual salary of $80,000 when I retire in 10 years. I have been offered a job that guarantees an annual salary increase of 5% per year, and the starting salary is negotiable. What salary should I request in order to meet my goal?

33. ▼ Present Value Determine the amount of money, to the nearest dollar, you must invest at 6% per year, compounded annually, so that you will be a millionaire in 30 years.

34. ▼ Present Value Determine the amount of money, to the nearest dollar, you must invest now at 7% per year, compounded annually, so that you will be a millionaire in 40 years.

35. ▼ Stocks Six years ago, I invested some money in *Dracubunny Toy Co.* stock, acting on the advice of a "friend." As things turned out, the value of the stock decreased by 5% every 4 months, and I discovered yesterday (to my horror) that my investment was worth only $297.91. How much did I originally invest?

36. ▼ Sales My recent marketing idea, the *Miracle Algae Growing Kit*, has been remarkably successful, with monthly sales growing by 6% every 6 months over the past 5 years. Assuming that I sold 100 kits the first month, how many kits did I sell in the first month of this year?

37. ▼ Inflation Inflation has been running 2% per year. A car now costs $30,000. How much would it have cost 5 years ago? HINT [See Example 3.]

38. ▼ Inflation (Compare Exercise 37.) Inflation has been running 1% every 6 months. A car now costs $30,000. How much would it have cost 5 years ago?

39. ▼ Inflation Housing prices have been rising 6% per year. A house now costs $200,000. What would it have cost 10 years ago?

40. ▼ Inflation (Compare Exercise 39.) Housing prices have been rising 0.5% each month. A house now costs $200,000. What would it have cost 10 years ago? HINT [See Example 4.]

41. ▼ Constant Dollars Inflation is running 3% per year when you deposit $1,000 in an account earning interest of 5% per year compounded annually. In *constant dollars*, how much money will you have 2 years from now? HINT [First calculate the value of your account in 2 years' time, and then find its present value based on the inflation rate.]

42. ▼ Constant Dollars Inflation is running 1% per month when you deposit $10,000 in an account earning 8% compounded monthly. In *constant dollars*, how much money will you have 2 years from now? HINT [See Exercise 41.]

43. ▼ Investments You are offered two investments. One promises to earn 12% compounded annually. The other will earn 11.9% compounded monthly. Which is the better investment? HINT [See Example 5.]

44. ▼ Investments You are offered three investments. The first promises to earn 15% compounded annually, the second will earn 14.5% compounded quarterly, and the third will earn 14% compounded monthly. Which is the best investment? HINT [See Example 5.]

45. ▼ History Legend has it that a band of Lenape Indians known as the "Manhatta" sold Manhattan Island to the Dutch in 1626 for $24. In 2011, the total value of Manhattan real estate was estimated to be $314,119 million.[13] Suppose that the Lenape had instead taken that $24 and invested it at 6.3% compounded annually (a relatively conservative investment goal). Could they then have bought back the island in 2011?

46. ▼ History Repeat Exercise 45, assuming that the Lenape had invested the $24 at 6.2% compounded annually.

Inflation Exercises 47–54 are based on the following table, which shows the 2008 annual inflation rates in several Latin American countries.[14] Assume that the rates shown continue indefinitely.

Country	Argentina	Brazil	Bolivia	Nicaragua	Venezuela	Mexico	Uruguay
Currency	Peso	Real	Boliviano	Gold cordoba	Bolivar	Peso	Peso
Inflation Rate (%)	9.2	6.3	15.1	13.8	25.7	5.0	8.5

47. If an item in Brazil now costs 100 reals, what do you expect it to cost 5 years from now? (Answer to the nearest real.)

48. If an item in Argentina now costs 1,000 pesos, what do you expect it to cost 5 years from now? (Answer to the nearest peso.)

49. If an item in Bolivia will cost 1,000 bolivianos in 10 years, what does it cost now? (Answer to the nearest boliviano.)

50. If an item in Mexico will cost 20,000 pesos in 10 years, what does it cost now? (Answer to the nearest peso.)

51. ▼ You wish to invest 1,000 bolivars in Venezuela at 8% annually, compounded twice a year. Find the value of your investment in 10 years, expressing the answer in constant bolivars. (Answer to the nearest bolivar.)

52. ▼ You wish to invest 1,000 pesos in Uruguay at 8% annually, compounded twice a year. Find the value of your investment in 10 years, expressing the answer in constant pesos. (Answer to the nearest peso.)

53. ▼ Which is the better investment: an investment in Mexico yielding 5.3% per year, compounded annually, or an investment in Nicaragua yielding 14% per year, compounded every 6 months? Support your answer with figures that show the future value of an investment of 1 unit of currency in constant units.

---

[13]Source: FY12 Tentative Assessment Roll summary, New York City Department of Finance, January 14, 2011 (www.nyc.gov/finance)

[14]Sources: www.bloomberg.com, www.csmonitor.com, www.indexmundi.com.

54. ▼ Which is the better investment: an investment in Argentina yielding 10% per year, compounded annually, or an investment in Uruguay, yielding 9% per year, compounded every 6 months? Support your answer with figures that show the future value of an investment of 1 unit of currency in constant units.

Stock Investments *Exercises 55–60 are based on the following chart, which shows monthly figures for* **Apple Computer, Inc.** *stock in 2010.*[15]

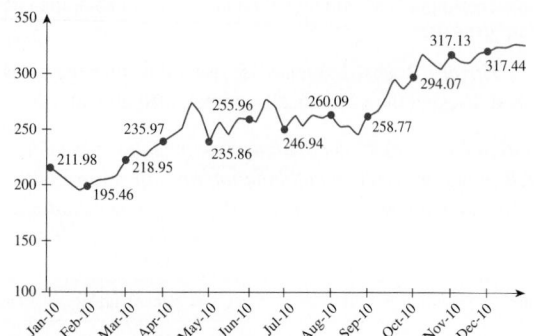

Marked are the following points on the chart:

Jan. 10	Feb. 10	Mar. 10	Apr. 10	May 10	June 10
211.98	195.46	218.95	235.97	235.86	255.96

July 10	Aug. 10	Sep. 10	Oct. 10	Nov. 10	Dec. 10
246.94	260.09	258.77	294.07	317.13	317.44

55. ▼ Calculate to the nearest 0.01% your annual percentage return (assuming annual compounding) if you had bought Apple stock in June and sold in December.

56. ▼ Calculate to the nearest 0.01% your annual percentage return (assuming annual compounding) if you had bought Apple stock in February and sold in June.

57. ▼ Suppose you bought Apple stock in April. If you later sold at one of the marked dates on the chart, which of those dates would have given you the largest annual return (assuming annual compounding), and what would that return have been?

58. ▼ Suppose you bought Apple stock in May. If you later sold at one of the marked dates on the chart, which of those dates would have given you the largest annual return (assuming annual compounding), and what would that return have been?

59. ▼ Did Apple's stock undergo compound interest change in the period January through May? (Give a reason for your answer.)

60. ▼ If Apple's stock had undergone compound interest change from April to June at the same monthly rate as from March to April, what would the price have been in June?

61. ▣ ▼ ***Competing Investments*** I just purchased $5,000 worth of municipal funds that are expected to yield 5.4% per year, compounded every 6 months. My friend has just purchased $6,000 worth of CDs that will earn 4.8% per year,

compounded every 6 months. Determine when, to the nearest year, the value of my investment will be the same as hers, and what this value will be. ▮▮ [You can either graph the values of both investments or make tables of the values of both investments.]

62. ▣ ▼ ***Investments*** Determine when, to the nearest year, $3,000 invested at 5% per year, compounded daily, will be worth $10,000.

63. ▣ ▼ ***Epidemics*** At the start of 1985, the incidence of AIDS was doubling every 6 months and 40,000 cases had been reported in the United States. Assuming this trend would have continued, determine when, to the nearest tenth of a year, the number of cases would have reached 1 million.

64. ▣ ▼ ***Depreciation*** My investment in *Genetic Splicing, Inc.*, is now worth $4,354 and is depreciating by 5% every 6 months. For some reason, I am reluctant to sell the stock and swallow my losses. Determine when, to the nearest year, my investment will drop below $50.

65. ◆ ***Bonds*** Once purchased, bonds can be sold in the secondary market. The value of a bond depends on the prevailing interest rates, which vary over time. Suppose that in January 2020, you buy a 30-year zero-coupon U.S. Treasury bond with a maturity value of $100,000 and a yield of 15% annually.
  a. How much do you pay for the bond?
  b. In January 2037, your bond has 13 years remaining until maturity. Rates on U.S. Treasury bonds of comparable length are now about 4.75%. If you sell your bond to an investor looking for a return of 4.75% annually, how much money do you receive?
  c. Using your answers to parts (a) and (b), what was the annual yield (assuming annual compounding) on your 17-year investment?

66. ◆ ***Bonds*** Suppose that in January 2040, you buy a 30-year zero-coupon U.S. Treasury bond with a maturity value of $100,000 and a yield of 5% annually.
  a. How much do you pay for the bond?
  b. Suppose that, 15 years later, interest rates have risen again, to 12%. If you sell your bond to an investor looking for a return of 12%, how much money will you receive?
  c. Using your answers to parts (a) and (b), what will be the annual yield (assuming annual compounding) on your 15-year investment?

COMMUNICATION AND REASONING EXERCISES

67. Why is the graph of the future value of a compound interest investment as a function of time not a straight line (assuming a nonzero rate of interest)?

68. An investment that earns 10% (compound interest) every year is the same as an investment that earns 5% (compound interest) every 6 months, right?

69. ▼ If a bacteria culture is currently 0.01 g and increases in size by 10% each day, then its growth is linear, right?

70. ▼ At what point is the future value of a compound interest investment the same as the future value of a simple interest investment at the same annual rate of interest?

[15]Source: Yahoo! Finance (http://finance.yahoo.com).

---

71. ▼ If two equal investments have the same effective interest rate and you graph the future value as a function of time for each of them, are the graphs necessarily the same? Explain your answer.

72. ▼ For what kind of compound interest investments is the effective rate the same as the nominal rate? Explain your answer.

73. ▼ For what kind of compound interest investments is the effective rate greater than the nominal rate? When is it smaller? Explain your answer.

74. ▼ If an investment appreciates by 10% per year for 5 years (compounded annually) and then depreciates by 10% per year (compounded annually) for 5 more years, will it have the same value as it had originally? Explain your answer.

75. ▼ You can choose between two investments that mature at different times in the future. If you knew the rate of inflation, how would you decide which is the better investment?

76. ▼ If you knew the various inflation rates for the years 2000 through 2011, how would you convert $100 in 2012 dollars to 2000 dollars?

77. ▮ ▼ On the same set of axes, graph the future value of a $100 investment earning 10% per year as a function of time over a 20-year period, compounded once a year, 10 times a year, 100 times a year, 1,000 times a year, and 10,000 times a year. What do you notice?

78. ▮ ▼ By graphing the future value of a $100 investment that is depreciating by 1% each year, convince yourself that, eventually, the future value will be less than $1.

2.3 Annuities, Loans, and Bonds

* Defined-contribution pension plans have largely replaced the defined-benefit pensions that were once the norm in private industry. In a defined-benefit plan, the size of your pension is guaranteed; it is typically a percentage of your final working salary. In a defined-contribution plan, the size of your pension depends on how well your investments do.

A typical defined-contribution pension fund works as follows*: Every month while you work, you and your employer deposit a certain amount of money in an account. This money earns (compound) interest from the time it is deposited. When you retire, the account continues to earn interest, but you may then start withdrawing money at a rate calculated to reduce the account to zero after some number of years. This account is an example of an **annuity**, an account earning interest into which you make periodic deposits or from which you make periodic withdrawals. In common usage, the term "annuity" is used for an account from which you make withdrawals. There are various terms used for accounts into which you make payments, based on their purpose. Examples include **savings account**, **pension fund**, and **sinking fund**. A sinking fund is generally used by businesses or governments to accumulate money to pay off an anticipated debt, but we'll use the term to refer to any account into which you make periodic payments.

Sinking Funds

Suppose you make a payment of $100 at the end of every month into an account earning 3.6% interest per year, compounded monthly. This means that your investment is earning $3.6\%/12 = 0.3\%$ per month. We write $i = 0.036/12 = 0.003$. What will be the value of the investment at the end of 2 years (24 months)?

Think of the deposits separately. Each earns interest from the time it is deposited, and the total accumulated after 2 years is the sum of these deposits and the interest they earn. In other words, the accumulated value is the sum of the future values of the deposits, taking into account how long each deposit sits in the account. Figure 1 shows a timeline with the deposits and the contribution of each to the final value. For example, the very last deposit (at the end of month 24) has no time to earn interest, so it contributes only $100. The very first deposit, which earns interest for 23 months, by the future value formula for compound interest contributes $100(1 + 0.003)^{23}$ to the total. Adding together all of the future values gives us the total future value:

$$FV = 100 + 100(1 + 0.003) + 100(1 + 0.003)^2 + \cdots + 100(1 + 0.003)^{23}$$
$$= 100[1 + (1 + 0.003) + (1 + 0.003)^2 + \cdots + (1 + 0.003)^{23}]$$

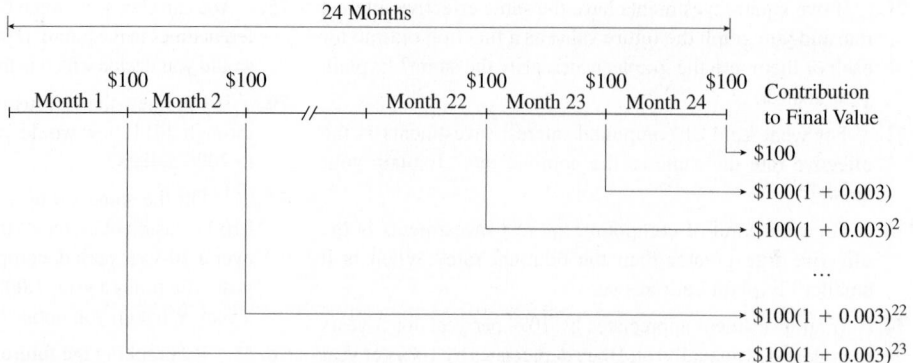

24 Months

| Month 1 | $100 Month 2 | // | $100 Month 22 | $100 Month 23 | $100 Month 24 | Contribution to Final Value |

Contribution to Final Value:
- 100
- $100(1 + 0.003)$
- $100(1 + 0.003)^2$
- $\ldots$
- $100(1 + 0.003)^{22}$
- $100(1 + 0.003)^{23}$

Figure 1

✱ It is called a **geometric series**.

† The quickest way to convince yourself that this formula is correct is to multiply out $(x - 1)(1 + x + x^2 + \cdots + x^{n-1})$ and see that you get $x^n - 1$. You should also try substituting some numbers. For example, $1 + 3 + 3^2 = 13 = (3^3 - 1)/(3 - 1)$.

Fortunately, this sort of sum is well-known (to mathematicians, anyway✱) and there is a convenient formula for its value†:

$$1 + x + x^2 + \cdots + x^{n-1} = \frac{x^n - 1}{x - 1}.$$

In our case, with $x = 1 + 0.003$, this formula allows us to calculate the future value:

$$FV = 100\frac{(1 + 0.003)^{24} - 1}{(1 + 0.003) - 1} = 100\frac{(1.003)^{24} - 1}{0.003} \approx \$2{,}484.65.$$

It is now easy to generalize this calculation.

Future Value of a Sinking Fund

A **sinking fund** is an account earning compound interest into which you make periodic deposits. Suppose that the account has an annual rate of r compounded m times per year, so that $i = r/m$ is the interest rate per compounding period. If you make a payment of PMT at the end of each period, then the future value after t years, or $n = mt$ periods, will be

$$FV = PMT\frac{(1 + i)^n - 1}{i}.$$

Quick Example

At the end of each month you deposit $50 into an account earning 2% annual interest compounded monthly. To find the future value after 5 years, we use $i = 0.02/12$ and $n = 12 \times 5 = 60$ compounding periods, so

$$FV = 50\frac{(1 + 0.02/12)^{60} - 1}{0.02/12} = \$3{,}152.37.$$

using Technology

To automate the computations in Example 1 using a graphing calculator or a spreadsheet, see the Technology Guides at the end of the chapter. Outline:

EXAMPLE 1 Retirement Account

Your retirement account has $5,000 in it and earns 5% interest per year compounded monthly. Every month for the next 10 years you will deposit $100 into the account. How much money will there be in the account at the end of those 10 years?

146   Chapter **2**   The Mathematics of Finance

Unless otherwise noted, all content on this page is © Cengage Learning.

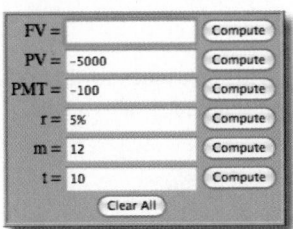

Solution This is a sinking fund with $PMT = \$100$, $r = 0.05$, $m = 12$, so $i = 0.05/12$, and $n = 12 \times 10 = 120$. Ignoring for the moment the $5,000 already in the account, your payments have the following future value:

$$FV = PMT \frac{(1+i)^n - 1}{i}$$

$$= 100 \frac{(1 + 0.05/12)^{120} - 1}{0.05/12}$$

$$\approx \$15,528.23.$$

What about the $5,000 that was already in the account? That sits there and earns interest, so we need to find its future value as well, using the compound interest formula:

$$FV = PV(1+i)^n$$

$$= 5,000(1 + 0.05/12)^{120}$$

$$= \$8,235.05.$$

Hence, the total amount in the account at the end of 10 years will be

$$\$15,528.23 + 8,235.05 = \$23,763.28.$$

Sometimes we know what we want the future value to be and need to determine the payments necessary to achieve that goal. We can simply solve the future value formula for the payment.

Payment Formula for a Sinking Fund

Suppose that an account has an annual rate of r compounded m times per year, so that $i = r/m$ is the interest rate per compounding period. If you want to accumulate a total of FV in the account after t years, or $n = mt$ periods, by making payments of PMT at the end of each period, then each payment must be

$$PMT = FV \frac{i}{(1+i)^n - 1}.$$

EXAMPLE 2 Education Fund

Tony and Maria have just had a son, José Phillipe. They establish an account to accumulate money for his college education, in which they would like to have $100,000 after 17 years. If the account pays 4% interest per year compounded quarterly, and they make deposits at the end of every quarter, how large must each deposit be for them to reach their goal?

Solution This is a sinking fund with $FV = \$100,000$, $m = 4$, $n = 4 \times 17 = 68$, and $r = 0.04$, so $i = 0.04/4 = 0.01$. From the payment formula, we get

$$PMT = 100,000 \frac{0.01}{(1 + 0.01)^{68} - 1} \approx \$1,033.89.$$

So, Tony and Maria must deposit $1,033.89 every quarter in order to meet their goal.

Spreadsheet

=PMT(4/4,17*4,0,100000)

[More details on page 172.]

 Website

www.WanerMath.com

→ On Line Utilities

→ Time Value of Money Utility

Enter the values shown, and press "Compute" next to PMT.

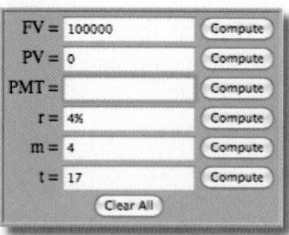

Annuities

Suppose we deposit an amount PV now in an account earning 3.6% interest per year, compounded monthly. Starting 1 month from now, the bank will send us monthly payments of $100. What must PV be so that the account will be drawn down to $0 in exactly 2 years?

As before, we write $i = r/m = 0.036/12 = 0.003$, and we have $PMT = 100$. The first payment of $100 will be made 1 month from now, so its present value is

$$\frac{PMT}{(1+i)^n} = \frac{100}{1+0.003} = 100(1+0.003)^{-1} \approx \$99.70.$$

In other words, that much of the original PV goes toward funding the first payment. The second payment, 2 months from now, has a present value of

$$\frac{PMT}{(1+i)^n} = \frac{100}{(1+0.003)^2} = 100(1+0.003)^{-2} \approx \$99.40.$$

That much of the original PV funds the second payment. This continues for 2 years, at which point we receive the last payment, which has a present value of

$$\frac{PMT}{(1+i)^n} = \frac{100}{(1+0.003)^{24}} = 100(1+0.003)^{-24} \approx \$93.06$$

and that exhausts the account. Figure 2 shows a timeline with the payments and the present value of each.

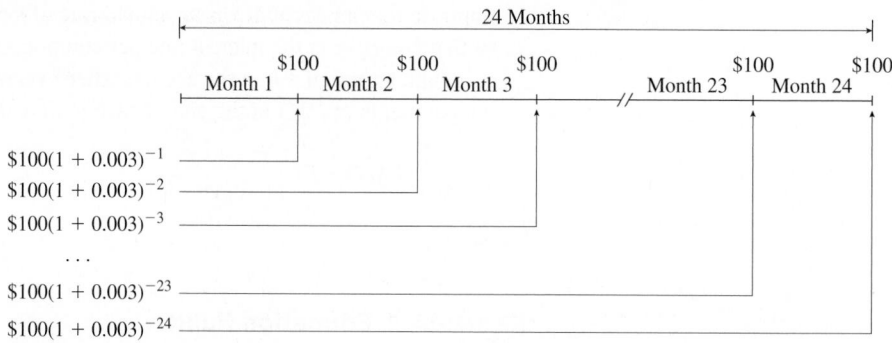

Figure 2

Because PV must be the sum of these present values, we get

$$PV = 100(1+0.003)^{-1} + 100(1+0.003)^{-2} + \cdots + 100(1+0.003)^{-24}$$
$$= 100[(1+0.003)^{-1} + (1+0.003)^{-2} + \cdots + (1+0.003)^{-24}].$$

We can again find a simpler formula for this sum:

$$x^{-1} + x^{-2} + \cdots + x^{-n} = \frac{1}{x^n}(x^{n-1} + x^{n-2} + \cdots + 1)$$
$$= \frac{1}{x^n} \cdot \frac{x^n - 1}{x - 1} = \frac{1 - x^{-n}}{x - 1}.$$

So, in our case,

$$PV = 100\frac{1 - (1 + 0.003)^{-24}}{(1 + 0.003) - 1}$$

or

$$PV = 100\frac{1 - (1.003)^{-24}}{0.003} \approx \$2,312.29.$$

If we deposit \$2,312.29 initially and the bank sends us \$100 per month for 2 years, our account will be exhausted at the end of that time.

Generalizing, we get the following formula:

Present Value of an Annuity

An **annuity** is an account earning compound interest from which periodic withdrawals are made. Suppose that the account has an annual rate of r compounded m times per year, so that $i = r/m$ is the interest rate per compounding period. Suppose also that the account starts with a balance of PV. If you receive a payment of PMT at the end of each compounding period, and the account is down to \$0 after t years, or $n = mt$ periods, then

$$PV = PMT\frac{1 - (1 + i)^{-n}}{i}.$$

Quick Example

At the end of each month you want to withdraw \$50 from an account earning 2% annual interest compounded monthly. If you want the account to last for 5 years (60 compounding periods), it must have the following amount to begin with:

$$PV = 50\frac{1 - (1 + 0.02/12)^{-60}}{0.02/12} = \$2,852.62.$$

Note If you make your withdrawals at the end of each compounding period, as we've discussed so far, you have an **ordinary annuity**. If, instead, you make withdrawals at the beginning of each compounding period, you have an **annuity due**. Because each payment occurs one period earlier, there is one less period in which to earn interest, hence the present value must be larger by a factor of $(1 + i)$ to fund each payment. So, the present value formula for an annuity due is

$$PV = PMT(1 + i)\frac{1 - (1 + i)^{-n}}{i}.$$

In this book, we will concentrate on ordinary annuities. ∎

EXAMPLE 3 Trust Fund

You wish to establish a trust fund from which your niece can withdraw \$2,000 every 6 months for 15 years, at the end of which time she will receive the remaining money in the trust, which you would like to be \$10,000. The trust will be invested at 7% per year compounded every 6 months. How large should the trust be?

---

using Technology

To automate the computations in Example 3 using a graphing calculator or a spreadsheet, see the Technology Guides at the end of the chapter. Outline:

TI-83/84 Plus
APPS 1:Finance, then
1:TVM Solver
N = 30, I% = 7, PMT = 2000,
FV = 10000, P/Y = 2, C/Y = 2
With cursor on PV line.
ALPHA SOLVE
[More details on page 168.]

Spreadsheet
=PV(7/2,15*2,2000,10000)
[More details on page 172.]

 Website
www.WanerMath.com
→ On Line Utilities
→ Time Value of Money Utility

Enter the values shown, and press "Compute" next to PV.

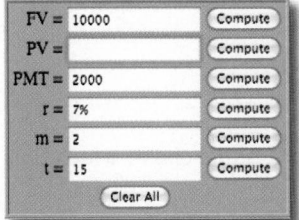

Solution We view this account as having two parts, one funding the semiannual payments and the other funding the $10,000 lump sum at the end. The amount of money necessary to fund the semiannual payments is the present value of an annuity, with $PMT = 2,000$, $r = 0.07$, and $m = 2$, so $i = 0.07/2 = 0.035$, and $n = 2 \times 15 = 30$. Substituting gives

$$PV = 2,000 \frac{1 - (1 + 0.035)^{-30}}{0.035}$$
$$= \$36,784.09.$$

To fund the lump sum of $10,000 after 15 years, we need the present value of $10,000 under compound interest:

$$PV = 10,000(1 + 0.035)^{-30}$$
$$= \$3,562.78.$$

Thus the trust should start with $36,784.09 + 3,562.78 = \$40,346.87$.

Sometimes we know how much money we begin with and for how long we want to make withdrawals. We then want to determine the amount of money we can withdraw each period. For this, we simply solve the present value formula for the payment.

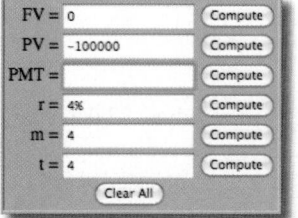

using Technology

To automate the computations in Example 4 using a graphing calculator or a spreadsheet, see the Technology Guides at the end of the chapter. Outline:

TI-83/84 Plus
APPS 1:Finance, then
1:TVM Solver
N = 16, I% = 4, PV = -100000,
FV = 0, P/Y = 4, C/Y = 4
With cursor on PMT line,
ALPHA SOLVE
[More details on page 168.]

Spreadsheet
=PMT(4/4,4*4,-100000,0)
[More details on page 172.]

Website
www.WanerMath.com
→ On Line Utilities
→ Time Value of Money Utility
Enter the values shown, and press "Compute" next to PMT.

FV =	0	Compute
PV =	-100000	Compute
PMT =		Compute
r =	4%	Compute
m =	4	Compute
t =	4	Compute
	Clear All	

Payment Formula for an Ordinary Annuity

Suppose that an account has an annual rate of r compounded m times per year, so that $i = r/m$ is the interest rate per compounding period. Suppose also that the account starts with a balance of PV. If you want to receive a payment of PMT at the end of each compounding period, and the account is down to $0 after t years, or $n = mt$ periods, then

$$PMT = PV \frac{i}{1 - (1 + i)^{-n}}.$$

EXAMPLE 4 Education Fund

Tony and Maria (see Example 2), having accumulated $100,000 for José Phillipe's college education, would now like to make quarterly withdrawals over the next 4 years. How much money can they withdraw each quarter in order to draw down the account to zero at the end of the 4 years? (Recall that the account pays 4% interest compounded quarterly.)

Solution Now Tony and Maria's account is acting as an annuity with a present value of $100,000. So, $PV = \$100,000$, $r = 0.04$, and $m = 4$, giving $i = 0.04/4 = 0.01$, and $n = 4 \times 4 = 16$. We use the payment formula to get

$$PMT = 100,000 \frac{0.01}{1 - (1 + 0.01)^{-16}} \approx \$6,794.46.$$

So, they can withdraw $6,794.46 each quarter for 4 years, at the end of which time their account balance will be 0.

EXAMPLE 5 Saving for Retirement

Jane Q. Employee has just started her new job with *Big Conglomerate, Inc.*, and is already looking forward to retirement. BCI offers her as a pension plan an annuity that is guaranteed to earn 6% annual interest compounded monthly. She plans to work for 40 years before retiring and would then like to be able to draw an income of $7,000 per month for 20 years. How much do she and BCI together have to deposit per month into the fund to accomplish this?

Solution Here we have the situation we described at the beginning of the section: a sinking fund accumulating money to be used later as an annuity. We know the desired payment out of the annuity, so we work backward. The first thing we need to do is calculate the present value of the annuity required to make the pension payments. We use the annuity present value formula with $PMT = 7,000$, $i = r/m = 0.06/12 = 0.005$ and $n = 12 \times 20 = 240$.

$$PV = PMT\frac{1 - (1+i)^{-n}}{i}$$

$$= 7,000\frac{1 - (1 + 0.005)^{-240}}{0.005} \approx \$977,065.40$$

This is the total that must be accumulated in the sinking fund during the 40 years she plans to work. In other words, this is the *future* value, *FV*, of the sinking fund. (Thus, the present value in the first step of our calculation is the future value in the second step.) To determine the payments necessary to accumulate this amount, we use the sinking fund payment formula with $FV = 977,065.40$, $i = 0.005$, and $n = 12 \times 40 = 480$.

$$PMT = FV\frac{i}{(1+i)^n - 1}$$

$$= 977,065.40\frac{0.005}{1.005^{480} - 1}$$

$$\approx \$490.62$$

So, if she and BCI collectively deposit $490.62 per month into her retirement fund, she can retire with the income she desires.

Installment Loans

In a typical installment loan, such as a car loan or a home mortgage, we borrow an amount of money and then pay it back with interest by making fixed payments (usually every month) over some number of years. From the point of view of the lender, this is an annuity. Thus, loan calculations are identical to annuity calculations.

EXAMPLE 6 Home Mortgages

Marc and Mira are buying a house, and have taken out a 30-year, $90,000 mortgage at 8% interest per year. What will their monthly payments be?

Solution From the bank's point of view, a mortgage is an annuity. In this case, the present value is $PV = \$90,000$, $r = 0.08$, $m = 12$, and $n = 12 \times 30 = 360$. To find the payments, we use the payment formula:

$$PMT = 90,000\frac{0.08/12}{1 - (1 + 0.08/12)^{-360}} \approx \$660.39.$$

The word "mortgage" comes from the French for "dead pledge." The process of paying off a loan is called **amortizing** the loan, meaning to kill the debt owed.

EXAMPLE 7 **Amortization Schedule**

Continuing Example 6: Mortgage interest is tax deductible, so it is important to know how much of a year's mortgage payments represents interest. How much interest will Marc and Mira pay in the first year of their mortgage?

Solution Let us calculate how much of each month's payment is interest and how much goes to reducing the outstanding principal. At the end of the first month Marc and Mira must pay 1 month's interest on $90,000, which is

$$\$90,000 \times \frac{0.08}{12} = \$600.$$

The remainder of their first monthly payment, $660.39 − 600 = \$60.39$, goes to reducing the principal. Thus, in the second month the outstanding principal is $\$90,000 − 60.39 = \$89,939.61$, and part of their second monthly payment will be for the interest on this amount, which is

$$\$89,939.61 \times \frac{0.08}{12} \approx \$599.60.$$

The remaining $660.39 − \$599.60 = \$60.79$ goes to further reduce the principal. If we continue this calculation for the 12 months of the first year, we get the beginning of the mortgage's **amortization schedule**.

using Technology

To automate the construction of the amortization schedule in Example 7 using a graphing calculator or a spreadsheet, see the Technology Guides at the end of the chapter. [Details: TI-83/84 Plus: page 169, Spreadsheet: page 173]

Month	Interest Payment	Payment on Principal	Outstanding Principal
0			$90,000.00
1	$600.00	$60.39	89,939.61
2	599.60	60.79	89,878.82
3	599.19	61.20	89,817.62
4	598.78	61.61	89,756.01
5	598.37	62.02	89,693.99
6	597.96	62.43	89,631.56
7	597.54	62.85	89,568.71
8	597.12	63.27	89,505.44
9	596.70	63.69	89,441.75
10	596.28	64.11	89,377.64
11	595.85	64.54	89,313.10
12	595.42	64.97	89,248.13
Total	$7,172.81	$751.87	

As we can see from the totals at the bottom of the columns, Marc and Mira will pay a total of $7,172.81 in interest in the first year.

Bonds

Suppose that a corporation offers a 10-year bond paying 6.5% with payments every 6 months. As we saw in Example 3 of Section 2.1, this means that if we pay $10,000 for bonds with a maturity value of $10,000, we will receive $6.5/2 = 3.25\%$ of $10,000, or $325, every 6 months for 10 years, at the end of which time the corporation will give us the original $10,000 back. But bonds are rarely sold at their maturity value. Rather, they are auctioned off and sold at a price the bond market determines they are worth.

For example, suppose that bond traders are looking for an investment that has a **rate of return** or **yield** of 7% rather than the stated 6.5% (sometimes called the **coupon interest rate** to distinguish it from the rate of return). How much would they be willing to pay for the bonds above with a maturity value of $10,000? Think of the bonds as an investment that will pay the owner $325 every 6 months for 10 years, and will pay an additional $10,000 on maturity at the end of the 10 years. We can treat the $325 payments as if they come from an annuity and determine how much an investor would pay for such an annuity if it earned 7% compounded semiannually. Separately, we determine the present value of an investment worth $10,000 ten years from now, if it earned 7% compounded semiannually. For the first calculation, we use the annuity present value formula, with $i = 0.07/2$ and $n = 2 \times 10 = 20$.

$$PV = PMT\frac{1 - (1 + i)^{-n}}{i}$$

$$= 325\frac{1 - (1 + 0.07/2)^{-20}}{0.07/2}$$

$$= \$4,619.03$$

For the second calculation, we use the present value formula for compound interest:

$$PV = 10,000(1 + 0.07/2)^{-20}$$

$$= \$5,025.66.$$

Thus, an investor looking for a 7% return will be willing to pay $4,619.03 for the semiannual payments of $325 and $5,025.66 for the $10,000 payment at the end of 10 years, for a total of $4,619.03 + 5,025.66 = \$9,644.69$ for the $10,000 bond.

EXAMPLE 8 **Bonds**

Suppose that bond traders are looking for only a 6% yield on their investment. How much would they pay per $10,000 for the 10-year bonds above, which have a coupon interest rate of 6.5% and pay interest every 6 months?

Solution We redo the calculation with $r = 0.06$. For the annuity calculation we now get

$$PV = 325\frac{1 - (1 + 0.06/2)^{-20}}{0.06/2} = \$4,835.18.$$

For the compound interest calculation we get

$$PV = 10,000(1 + 0.06/2)^{-20} = \$5,536.76.$$

Thus, traders would be willing to pay a total of $4,835.18 + 5,536.76 = \$10,371.94$ for bonds with a maturity value of $10,000.

➡ **Before we go on...** Notice how the selling price of the bonds behaves as the desired yield changes. As desired yield goes up, the price of the bonds goes down, and as desired yield goes down, the price of the bonds goes up. When the desired yield equals the coupon interest rate, the selling price will equal the maturity value. Therefore, when the yield is higher than the coupon interest rate, the price of the bond will be below its maturity value, and when the yield is lower than the coupon interest rate, the price will be above the maturity value.

As we've mentioned before, the desired yield depends on many factors, but it generally moves up and down with prevailing interest rates. And interest rates have historically gone up and down cyclically. The effect on the value of bonds can be quite dramatic (see Exercises 65 and 66 in the preceding section). Because bonds can be sold again once bought, someone who buys bonds while interest rates are high and then resells them when interest rates decline can make a healthy profit. ■

using Technology

To automate the computations in Example 9 using a graphing calculator or a spreadsheet, see the Technology Guides at the end of the chapter. Outline:

TI-83/84 Plus
APPS 1:Finance, then
1:TVM Solver
N = 40, PV = −9800, PMT = 250,
FV = 10000, P/Y = 2, C/Y = 2
With cursor on I% line, ALPHA
SOLVE
[More details on page 170.]

Spreadsheet
=RATE(20*2,250,-9800,
10000)*2
[More details on page 174.]

Website
www.WanerMath.com
→ On Line Utilities
→ Time Value of Money Utility
Enter the values shown, and press "Compute" next to r.

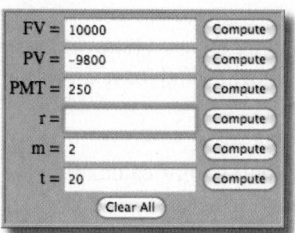

EXAMPLE 9 Rate of Return on a Bond

Suppose that a 5%, 20-year bond sells for $9,800 per $10,000 maturity value. What rate of return will investors get?

Solution Assuming the usual semiannual payments, we know the following about the annuity calculation:

$$PMT = 0.05 \times 10{,}000/2 = 250$$
$$n = 20 \times 2 = 40.$$

What we do not know is r or i, the annual or semiannual rate of return, respectively. So, we write

$$PV = 250\frac{1 - (1 + i)^{-40}}{i}.$$

For the compound interest calculation, we know $FV = 10{,}000$ and $n = 40$ again, so we write

$$PV = 10{,}000(1 + i)^{-40}.$$

Adding these together should give the selling price of $9,800:

$$250\frac{1 - (1 + i)^{-40}}{i} + 10{,}000(1 + i)^{-40} = 9{,}800.$$

This equation cannot be solved for i directly. The best we can do by hand is a sort of trial-and-error approach, substituting a few values for i in the left-hand side of the above equation to get an estimate; we use the fact that, because the selling price is below the maturity value, r must be larger than 0.05, so i must be more than 0.025:

i	0.025	0.03	0.035
$250 \dfrac{1 - (1 + i)^{-40}}{i} + 10{,}000(1 + i)^{-40}$	10,000	8,844	7,864

Since we want the value to be 9,800, we see that the correct answer is somewhere between $i = 0.025$ and $i = 0.03$. Let us try the value midway between 0.025 and 0.03; namely $i = 0.0275$.

i	0.025	0.0275	0.03
$250 \dfrac{1 - (1 + i)^{-40}}{i} + 10{,}000(1 + i)^{-40}$	10,000	9,398	8,844

Now we know that the correct value of i is somewhere between 0.025 and 0.0275, so we can choose for our next estimate of i the number midway between them: 0.02625. We could continue in this fashion to obtain i as accurately as we like. In fact, $i \approx 0.02581$, corresponding to an annual rate of return of approximately 5.162%.

FAQs

Which Formula to Use

Q : *We have retirement accounts, trust funds, loans, bonds, and so on. Some are sinking funds, others are annuities. How do we distinguish among them, so we can tell which formula to use?*

A : In general, remember that a sinking fund is an interest-bearing fund into which payments are made, while an annuity is an interest-bearing fund from which money is withdrawn. Here is a list of some of the accounts we have discussed in this section:

- *Retirement Accounts* A retirement account is a sinking fund while payments are being made into the account (prior to retirement) and an annuity while a pension is being withdrawn (after retirement).
- *Education Funds* These are similar to retirement accounts.
- *Trust Funds* A trust fund is an annuity if periodic withdrawals are made.
- *Installment Loans* We think of an installment loan as an investment a bank makes in the lender. In this way, the lender's payments can be viewed as the bank's withdrawals, and so a loan is an annuity.
- *Bonds* A bond pays regular fixed amounts until it matures, at which time it pays its maturity value. We think of the bond as an annuity coupled with a compound interest investment funding the payment of the maturity value. We can then determine its present value based on the current market interest rate.

From a mathematical point of view, sinking funds and annuities are really the same thing. See the Communication and Reasoning Exercises for more about this.

2.3 EXERCISES

▼ more advanced   ◆ challenging
Ⓣ indicates exercises that should be solved using technology

Find the amount accumulated in the sinking funds in Exercises 1–6. (Assume end-of-period deposits and compounding at the same intervals as deposits.) HINT [See Example 1.]

1. $100 deposited monthly for 10 years at 5% per year

2. $150 deposited monthly for 20 years at 3% per year

3. $1,000 deposited quarterly for 20 years at 7% per year

4. $2,000 deposited quarterly for 10 years at 7% per year

5. ▼ $100 deposited monthly for 10 years at 5% per year in an account containing $5,000 at the start

6. ▼ $150 deposited monthly for 20 years at 3% per year in an account containing $10,000 at the start

Find the periodic payments necessary to accumulate the amounts given in Exercises 7–12 in a sinking fund. (Assume end-of-period deposits and compounding at the same intervals as deposits.) HINT [See Example 2.]

7. $10,000 in a fund paying 5% per year, with monthly payments for 5 years

8. $20,000 in a fund paying 3% per year, with monthly payments for 10 years

9. $75,000 in a fund paying 6% per year, with quarterly payments for 20 years

10. $100,000 in a fund paying 7% per year, with quarterly payments for 20 years

11. ▼ $20,000 in a fund paying 5% per year, with monthly payments for 5 years, if the fund contains $10,000 at the start

12. ▼ $30,000 in a fund paying 3% per year, with monthly payments for 10 years, if the fund contains $10,000 at the start

Find the present value of the annuity necessary to fund the withdrawals given in Exercises 13–18. (Assume end-of-period withdrawals and compounding at the same intervals as withdrawals.) HINT [See Example 3.]

13. $500 per month for 20 years, if the annuity earns 3% per year

14. $1,000 per month for 15 years, if the annuity earns 5% per year

15. $1,500 per quarter for 20 years, if the annuity earns 6% per year

16. $2,000 per quarter for 20 years, if the annuity earns 4% per year

17. ▼ $500 per month for 20 years, if the annuity earns 3% per year and if there is to be $10,000 left in the annuity at the end of the 20 years

18. ▼ $1,000 per month for 15 years, if the annuity earns 5% per year and if there is to be $20,000 left in the annuity at the end of the 15 years

Find the periodic withdrawals for the annuities given in Exercises 19–24. (Assume end-of-period withdrawals and compounding at the same intervals as withdrawals.) HINT [See Example 4.]

19. $100,000 at 3%, paid out monthly for 20 years

20. $150,000 at 5%, paid out monthly for 15 years

21. $75,000 at 4%, paid out quarterly for 20 years

22. $200,000 at 6%, paid out quarterly for 15 years

23. ▼ $100,000 at 3%, paid out monthly for 20 years, leaving $10,000 in the account at the end of the 20 years

24. ▼ $150,000 at 5%, paid out monthly for 15 years, leaving $20,000 in the account at the end of the 15 years

Determine the periodic payments on the loans given in Exercises 25–28. HINT [See Example 6.]

25. $10,000 borrowed at 9% for 4 years, with monthly payments

26. $20,000 borrowed at 8% for 5 years, with monthly payments

27. $100,000 borrowed at 5% for 20 years, with quarterly payments

28. $1,000,000 borrowed at 4% for 10 years, with quarterly payments

Determine the selling price, per $1,000 maturity value, of the bonds[16] in Exercises 29–34. (Assume twice-yearly interest payments.) HINT [See Example 8.]

29. ▼ 10 year, 4.875% bond, with a yield of 4.880%

30. ▼ 30 year, 5.375% bond, with a yield of 5.460%

31. ▼ 2 year, 3.625% bond, with a yield of 3.705%

32. ▼ 5 year, 4.375% bond, with a yield of 4.475%

33. ▼ 10 year, 5.5% bond, with a yield of 6.643%

34. ▼ 10 year, 6.25% bond, with a yield of 33.409%

▼ *Determine the yield on the bonds[17] in Exercises 35–40. (Assume twice-yearly interest payments.)* HINT [See Example 9.]

35. ▼ 5 year, 3.5% bond, selling for $994.69 per $1,000 maturity value

36. ▼ 10 year, 3.375% bond, selling for $991.20 per $1,000 maturity value

37. ▼ 2 year, 3% bond, selling for $998.86 per $1,000 maturity value

38. ▼ 5 year, 4.625% bond, selling for $998.45 per $1,000 maturity value

39. ▼ 5 year, 4.25% bond, selling for $923.81 per $1,000 maturity value

40. ▼ 5 year 3.6% bond, selling for $216.96 per $1,000 maturity value

APPLICATIONS

41. *Car Loans* While shopping for a car loan, you get the following offers: Solid Savings & Loan is willing to loan you $10,000 at 9% interest for 4 years. Fifth Federal Bank & Trust will loan you the $10,000 at 7% interest for 3 years. Both require monthly payments. You can afford to pay $250 per month. Which loan, if either, can you take?

42. *Business Loans* You need to take out a loan of $20,000 to start up your T-shirt business. You have two possibilities: One bank is offering a 10% loan for 5 years, and another is offering a 9% loan for 4 years. Which will have the lower monthly payments? On which will you end up paying more interest total?

43. ▼ *Pensions* Your pension plan is an annuity with a guaranteed return of 3% per year (compounded monthly). You would like to retire with a pension of $5,000 per month for 20 years. If you work 40 years before retiring, how much must you and your employer deposit each month into the fund? HINT [See Example 5.]

44. ▼ *Pensions* Meg's pension plan is an annuity with a guaranteed return of 5% per year (compounded quarterly). She would like to retire with a pension of $12,000 per quarter for 25 years. If she works 45 years before retiring, how much money must she and her employer deposit each quarter? HINT [See Example 5.]

45. ▼ *Pensions* Your pension plan is an annuity with a guaranteed return of 4% per year (compounded quarterly). You can afford to put $1,200 per quarter into the fund, and you will work for 40 years before retiring. After you retire, you will

---

[16]The first four are actual U.S. Treasury notes and bonds auctioned in 2001 and 2002. The last two are, respectively, a Spanish and a Greek government bond auctioned November 17, 2011, during the Eurozone crisis. Sources: The Bureau of the Public Debt's Web site (www.publicdebt.treas.gov) and The Wall Street Journal's Web site (www.wsj.com).

[17]*Ibid.*

be paid a quarterly pension based on a 25-year payout. How much will you receive each quarter?

46. ▼ *Pensions* Jennifer's pension plan is an annuity with a guaranteed return of 5% per year (compounded monthly). She can afford to put $300 per month into the fund, and she will work for 45 years before retiring. If her pension is then paid out monthly based on a 20-year payout, how much will she receive per month?

Note: In Exercises 47–52 we suggest the use of the finance functions in the TI-83/84 Plus or amortization tables in a spreadsheet.

47. ▼ *Mortgages* You take out a 15-year mortgage for $50,000, at 8%, to be paid off monthly. Construct an amortization table showing how much you will pay in interest each year, and how much goes toward paying off the principal. HINT [See Example 7.]

48. ▼ *Mortgages* You take out a 30-year mortgage for $95,000 at 9.75%, to be paid off monthly. If you sell your house after 15 years, how much will you still owe on the mortgage? HINT [See Example 7.]

49. ▼ *Mortgages* This exercise describes a popular kind of mortgage. You take out a $75,000, 30-year mortgage. For the first 5 years the interest rate is held at 5%, but for the remaining 25 years it rises to 9.5%. The payments for the first 5 years are calculated as if the 5% rate were going to remain in effect for all 30 years, and then the payments for the last 25 years are calculated to amortize the debt remaining at the end of the fifth year. What are your monthly payments for the first 5 years, and what are they for the last 25 years?

50. ▼ *Adjustable Rate Mortgages* You take out an adjustable rate mortgage for $100,000 for 20 years. For the first 5 years, the rate is 4%. It then rises to 7% for the next 10 years, and then 9% for the last 5 years. What are your monthly payments in the first 5 years, the next 10 years, and the last 5 years? (Assume that each time the rate changes, the payments are recalculated to amortize the remaining debt if the interest rate were to remain constant for the remaining life of the mortgage.)

51. ▼ *Refinancing* Your original mortgage was a $96,000, 30-year 9.75% mortgage. After 4 years you refinance the remaining principal for 30 years at 6.875%. What was your original monthly payment? What is your new monthly payment? How much will you save in interest over the course of the loan by refinancing?

52. ▼ *Refinancing* Kara and Michael take out a $120,000, 30-year, 10% mortgage. After 3 years they refinance the remaining principal with a 15-year, 6.5% loan. What were their original monthly payments? What is their new monthly payment? How much did they save in interest over the course of the loan by refinancing?

53. ▼ *Fees* You take out a 2-year, $5,000 loan at 9% interest with monthly payments. The lender charges you a $100 fee that can be paid off, interest free, in equal monthly installments over the life of the loan. Thinking of the fee as additional interest, what is the actual annual interest rate you will pay?

54. ▼ *Fees* You take out a 3-year, $7,000 loan at 8% interest with monthly payments. The lender charges you a $100 fee that can be paid off, interest free, in equal monthly installments over the life of the loan. Thinking of the fee as additional interest, what is the actual annual interest rate you will pay?

55. ▼ *Savings* You wish to accumulate $100,000 through monthly payments of $500. If you can earn interest at an annual rate of 4% compounded monthly, how long (to the nearest year) will it take to accomplish your goal?

56. ▼ *Retirement* Alonzo plans to retire as soon as he has accumulated $250,000 through quarterly payments of $2,500. If Alonzo invests this money at 5.4% interest, compounded quarterly, when (to the nearest year) can he retire?

57. ▼ *Loans* You have a $2,000 credit card debt, and you plan to pay it off through monthly payments of $50. If you are being charged 15% interest per year, how long (to the nearest 0.5 years) will it take you to repay your debt?

58. ▼ *Loans* You owe $2,000 on your credit card, which charges you 15% interest. Determine, to the nearest 1¢, the minimum monthly payment that will allow you to eventually repay your debt.

59. ▼ *Savings* You are depositing $100 per month in an account that pays 4.5% interest per year (compounded monthly) while your friend Lucinda is depositing $75 per month in an account that earns 6.5% interest per year (compounded monthly). When, to the nearest year, will her balance exceed yours?

60. ▼ *Car Leasing* You can lease a $15,000 car for $300 per month. For how long (to the nearest year) should you lease the car so that your monthly payments are lower than if you were purchasing it with an 8%-per-year loan?

COMMUNICATION AND REASONING EXERCISES

61. Your cousin Simon claims that you have wasted your time studying annuities: If you wish to retire on an income of $1,000 per month for 20 years, you need to save $1,000 per month for 20 years, period. Explain why he is wrong.

62. ▼ Your other cousin Cecilia claims that you will earn more interest by depositing $10,000 through smaller more frequent payments than through larger less frequent payments. Is she correct? Give a reason for your answer.

63. ▼ A real estate broker tells you that doubling the period of a mortgage halves the monthly payments. Is he correct? Support your answer by means of an example.

64. ▼ Another real estate broker tells you that doubling the size of a mortgage doubles the monthly payments. Is he correct? Support your answer by means of an example.

65. ◆ Consider the formula for the future value of a sinking fund with given payments. Show algebraically that the present value of that future value is the same as the present value of the annuity required to fund the same payments.

66. ◆ Give a non-algebraic justification for the result from the preceding exercise.

CHAPTER 2 REVIEW

KEY CONCEPTS

WM Website www.WanerMath.com
Go to the Website at www.WanerMath
.com to find a comprehensive and
interactive Web-based summary
of Chapter 2.

2.1 Simple Interest
Simple interest *p. 126*
Future value *p. 127*
Bond, maturity *p. 128*
Present value *p. 129*
Treasury bills (T-bills); maturity value,
   discount rate, yield *p. 130*

2.2 Compound Interest
Compound interest *p. 134*
Future value for compound
   interest *p. 135*
Present value for compound
   interest *p. 137*
Zero coupon bond or strip *p. 137*
Inflation, constant dollars *pp. 138–139*
Effective interest rate, annual
   percentage yield (APY) *p. 139*

2.3 Annuities, Loans, and Bonds
Annuity, sinking fund *p. 145*
Future value of a sinking fund *p. 146*
Payment formula for a sinking
   fund *p. 147*
Present value of an annuity *p. 149*
Ordinary annuity, annuity due *p. 149*
Payment formula for an annuity *p. 150*
Installment loan *p. 151*
Amortization schedule *p. 152*
Bond *p. 153*

REVIEW EXERCISES

In each of Exercises 1–6, find the future value of the investment.

1. $6,000 for 5 years at 4.75% simple annual interest

2. $10,000 for 2.5 years at 5.25% simple annual interest

3. $6,000 for 5 years at 4.75% compounded monthly

4. $10,000 for 2.5 years at 5.25% compounded semiannually

5. $100 deposited at the end of each month for 5 years, at 4.75% interest compounded monthly

6. $2,000 deposited at the end of each half-year for 2.5 years, at 5.25% interest compounded semiannually

In each of Exercises 7–12, find the present value of the investment.

7. Worth $6,000 after 5 years at 4.75% simple annual interest

8. Worth $10,000 after 2.5 years at 5.25% simple annual interest

9. Worth $6,000 after 5 years at 4.75% compounded monthly

10. Worth $10,000 after 2.5 years at 5.25% compounded semiannually

11. Funding $100 withdrawals at the end of each month for 5 years, at 4.75% interest compounded monthly

12. Funding $2,000 withdrawals at the end of each half-year for 2.5 years, at 5.25% interest compounded semiannually

In each of Exercises 13–18, find the amounts indicated.

13. The monthly deposits necessary to accumulate $12,000 after 5 years in an account earning 4.75% compounded monthly

14. The semiannual deposits necessary to accumulate $20,000 after 2.5 years in an account earning 5.25% compounded semiannually

15. The monthly withdrawals possible over 5 years from an account earning 4.75% compounded monthly and starting with $6,000

16. The semiannual withdrawals possible over 2.5 years from an account earning 5.25% compounded semiannually and starting with $10,000

17. The monthly payments necessary on a 5-year loan of $10,000 at 4.75%

18. The semiannual payments necessary on a 2.5-year loan of $15,000 at 5.25%

19. How much would you pay for a $10,000, 5-year, 6% bond if you want a return of 7%? (Assume that the bond pays interest every 6 months.)

20. How much would you pay for a $10,000, 5-year, 6% bond if you want a return of 5%? (Assume that the bond pays interest every 6 months.)

21. ▪ A $10,000, 7-year, 5% bond sells for $9,800. What return does it give you? (Assume that the bond pays interest every 6 months.)

22. ▪ A $10,000, 7-year, 5% bond sells for $10,200. What return does it give you? (Assume that the bond pays interest every 6 months.)

In each of Exercises 23–28, find the time requested, to the nearest 0.1 year.

23. The time it would take $6,000 to grow to $10,000 at 4.75% simple annual interest

24. The time it would take $10,000 to grow to $15,000 at 5.25% simple annual interest

25. ▪ The time it would take $6,000 to grow to $10,000 at 4.75% interest compounded monthly

26. ▪ The time it would take $10,000 to grow to $15,000 at 5.25% interest compounded semiannually

27. ▪ The time it would take to accumulate $10,000 by depositing $100 at the end of each month in an account earning 4.75% interest compounded monthly

28. ▪ The time it would take to accumulate $15,000 by depositing $2,000 at the end of each half-year in an account earning 5.25% compounded semiannually

	A	B	C	D	E	F	G	H	I	J	K
		28% of Monthly Income		Interest Rate			Monthly Payment			Balance on Principal	
1	Year										
2			Scenario 1	Scenario 2	Scenario 3	Scenario 1	Scenario 2	Scenario 3	Scenario 1	Scenario 2	Scenario 3
3	1	$1,866.67	4.7	4.7	4.7	$1,182.49	$1,182.49	$1,182.49	$380,000.00	$380,000.00	$380,000.00
4	2	$1,941.33	4.7	4.7	4.7	$1,182.49	$1,182.49	$1,182.49	$383,750.17	$383,750.17	$383,750.17
5	3	$2,018.99	4.7	4.7	4.7	$1,182.49	$1,182.49	$1,182.49	$387,680.45	$387,680.45	$387,680.45
6	4	$2,099.75	4.7	4.7	4.7	$1,182.49	$1,182.49	$1,182.49	$391,799.48	$391,799.48	$391,799.48
7	5	$2,183.74	4.7	4.7	4.7	$1,182.49	$1,182.49	$1,182.49	$396,116.32	$396,116.32	$396,116.32
8	6	$2,271.09	9.25	9	15	$3,431.01	$3,362.16	$5,131.53	$400,640.49	$400,640.49	$400,640.49
9	7	$2,361.93	9.5	9	14.75	$3,498.87	$3,362.16	$5,054.87	$396,348.66	$396,170.82	$399,051.98
10	8	$2,456.41	9.75	9	14.5	$3,565.64	$3,362.16	$4,979.54	$391,821.64	$391,281.87	$397,127.00
11	9	$2,554.66	10	9	14.25	$3,631.26	$3,362.16	$4,905.69	$387,025.99	$385,934.29	$394,805.64
12	10	$2,656.85	10.25	9	14	$3,695.62	$3,362.16	$4,833.48	$381,923.88	$380,085.08	$392,019.86
13	11	$2,763.12	10.5	9	13.75	$3,758.62	$3,362.16	$4,763.06	$376,472.30	$373,687.17	$388,692.80
14	12	$2,873.65	10.75	9	13.5	$3,820.18	$3,362.16	$4,694.62	$370,622.18	$366,689.09	$384,738.24
15	13	$2,988.59	11	9	13.25	$3,880.16	$3,362.16	$4,628.32	$364,317.31	$359,034.55	$380,060.01
16	14	$3,108.14	11.25	9	13	$3,938.46	$3,362.16	$4,564.34	$357,493.03	$350,661.95	$374,551.54
17	15	$3,232.46	11.5	9	12.75	$3,994.93	$3,362.16	$4,502.85	$350,074.77	$341,503.95	$368,095.48
18	16	$3,361.76	11.75	9	12.5	$4,049.45	$3,362.16	$4,444.02	$341,976.15	$331,486.86	$360,563.39
19	17	$3,496.23	12	9	12.25	$4,101.85	$3,362.16	$4,388.03	$333,096.90	$320,530.10	$351,815.60
20	18	$3,636.08	12.25	9	12	$4,151.99	$3,362.16	$4,335.05	$323,320.15	$308,545.52	$341,701.22
21	19	$3,781.52	12.5	9	11.75	$4,199.68	$3,362.16	$4,285.23	$312,509.37	$295,436.71	$330,058.23
22	20	$3,932.79	12.75	9	11.5	$4,244.74	$3,362.16	$4,238.74	$300,504.55	$281,098.20	$316,713.83
23	21	$4,090.10	13	9	11.25	$4,286.97	$3,362.16	$4,195.73	$287,117.60	$265,414.64	$301,484.92
24	22	$4,253.70	13.25	9	11	$4,326.17	$3,362.16	$4,156.36	$272,126.84	$248,259.85	$284,178.86
25	23	$4,423.85	13.5	9	10.75	$4,362.10	$3,362.16	$4,120.77	$255,270.30	$229,495.82	$264,594.34
26	24	$4,600.80	13.75	9	10.5	$4,394.53	$3,362.16	$4,089.09	$236,237.47	$208,971.60	$242,522.55

Clearly the Wongs should steer clear of this type of loan in order to be able to continue to afford making payments!

In short, it seems unlikely that the Wongs will be able to afford payments on any of the three mortgages in question, and you decide to advise them to either seek a less expensive home or wait until their income has appreciated to enable them to afford a home of this price.

EXERCISES

1. ▣ In the case of a hybrid loan, what would the federal funds rate have to be in Scenario 2 to ensure that the Wongs can afford to make all payments?

2. ▣ Repeat the preceding exercise in the case of an interest-only loan.

3. ▣ Repeat the preceding exercise in the case of a negative-amortization loan.

4. ▣ What home price, to the nearest $5,000, could the Wongs afford if they took out a hybrid loan, regardless of scenario? HINT [Adjust the original value of the loan on your spreadsheet to obtain the desired result.]

5. ▣ What home price, to the nearest $5,000, could the Wongs afford if they took out a negative-amortization loan, regardless of scenario? HINT [Adjust the original value of the loan on your spreadsheet to obtain the desired result.]

6. ▣ How long would the Wongs need to wait before they could afford to purchase a $400,000 home, assuming that their income continues to increase as above, they still have $20,000 for a down payment, and the mortgage offers remain the same?

TECHNOLOGY GUIDE

TI-83/84 Plus   **Technology Guide**

Section 2.2

Example 1 (page 136)   In November 2011, the Bank of Montreal was paying 1.30% interest on savings accounts. If the interest is compounded quarterly, find the future value of a $2,000 deposit in 6 years. What is the total interest paid over the period?

Solution with Technology

We could calculate the future value using the TI-83/84 Plus by entering

$$2000(1+0.0130/4)^{(4*6)}$$

However, the TI-83/84 Plus has this and other useful calculations built into its TVM (Time Value of Money) Solver.

1. Press APPS then choose item 1:Finance... and then choose item 1:TVM Solver.... This brings up the TVM Solver window.

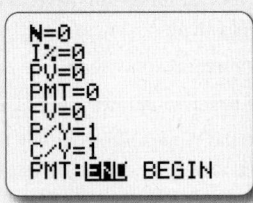

The second screen shows the values you should enter for this example. The various variables are:

N	Number of compounding periods
I%	Annual interest rate, as percent, not decimal
PV	Negative of present value
PMT	Payment per period (0 in this section)
FV	Future value
P/Y	Payments per year
C/Y	Compounding periods per year
PMT:	Not used in this section

Several things to notice:

- *I%* is the *annual* interest rate, corresponding to *r*, not *i*, in the compound interest formula.

- The present value, *PV*, is entered as a negative number. In general, when using the TVM Solver, any amount of money you give to someone else (such as the $2,000 you deposit in the bank) will be a negative number, whereas any amount of money someone gives to you (such as the future value of your deposit, which the bank will give back to you) will be a positive number.

- *PMT* is not used in this example (it will be used in the next section) and should be 0.

- *FV* is the future value, which we shall compute in a moment; it doesn't matter what you enter now.

- *P/Y* and *C/Y* stand for payments per year and compounding periods per year, respectively: They should both be set to the number of compounding periods per year for compound interest problems (setting *P/Y* automatically sets *C/Y* to the same value).

- *PMT*: *END* or *BEGIN* is not used in this example and it doesn't matter which you select.

2. To compute the future value, use the up or down arrow to put the cursor on the *FV* line, then press ALPHA SOLVE.

Example 2 (page 138)   Megabucks Corporation is issuing 10-year zero coupon bonds. How much would you pay for bonds with a maturity value of $10,000 if you wish to get a return of 6.5% compounded annually?

Solution with Technology

To compute the present value using a TI-83/84 Plus:

1. Enter the numbers shown below (top) in the TVM Solver window.
2. Put the cursor on the PV line, and press [ALPHA] [SOLVE].

```
N=10
I%=6.5
PV=■
PMT=0
FV=10000
P/Y=1
C/Y=1
PMT:END BEGIN
```

```
N=10
I%=6.5
•PV=-5327.260355
PMT=0
FV=10000
P/Y=1
C/Y=1
PMT:END BEGIN
```

Why is the present value given as negative?

Example 6 (page 141)
You have $5,000 to invest at 6% interest compounded monthly. How long will it take for your investment to grow to $6,000?

Solution with Technology

1. Enter the numbers shown below (top) in the TVM Solver window.
2. Put the cursor on the N line, and press [ALPHA] [SOLVE].

```
N=■
I%=6
PV=-5000
PMT=0
FV=6000
P/Y=12
C/Y=12
PMT:END BEGIN
```

```
•N=36.55539636
I%=6
PV=-5000
PMT=0
FV=6000
P/Y=12
C/Y=12
PMT:END BEGIN
```

Recall that I% is the annual interest rate, corresponding to r in the formula, but N is the number of compounding

periods, so number of months in this example. Thus, you will need to invest your money for about 36.5 months, or just over 3 years, before it grows to $6,000.

Section 2.3

Example 1 (page 146)
Your retirement account has $5,000 in it and earns 5% interest per year compounded monthly. Every month for the next 10 years, you will deposit $100 into the account. How much money will there be in the account at the end of those 10 years?

Solution with Technology

We can use the TVM Solver in the TI-83/84 Plus to calculate future values like these:

1. The TVM Solver allows you to put the $5,000 already in the account as the present value of the account. Following the TI-83/84 Plus's usual convention, set PV to the *negative* of the present value because this is money you paid into the account.
2. Likewise, set PMT to −100 since you are paying $100 each month.
3. Set the number of payment and compounding periods to 12 per year.
4. Set the payments to be made at the end of each period.
5. With the cursor on the FV line, press [ALPHA] [SOLVE] to find the future value.

```
N=120
I%=5
PV=-5000
PMT=-100
FV=■
P/Y=12
C/Y=12
PMT:END BEGIN
```

```
N=120
I%=5
PV=-5000
PMT=-100
•FV=23763.27543
P/Y=12
C/Y=12
PMT:END BEGIN
```

Example 2 (page 147)
Tony and Maria have just had a son, José Phillipe. They establish an account to accumulate money for his college education. They would like to have $100,000 in this account after 17 years. If the account pays 4% interest per year compounded quarterly, and they make deposits at the end of every quarter, how large must each deposit be for them to reach their goal?

Solution with Technology

1. In the TVM Solver in the TI-83/84 Plus, enter the values shown below.

2. Solve for *PMT*.

Why is *PMT* negative?

Example 3 (page 149)   You wish to establish a trust fund from which your niece can withdraw $2,000 every 6 months for 15 years, at which time she will receive the remaining money in the trust, which you would like to be $10,000. The trust will be invested at 7% per year compounded every 6 months. How large should the trust be?

Solution with Technology

1. In the TVM Solver in the TI-83/84 Plus, enter the values shown below.

2. Solve for *PV*.

The payment and future value are positive because you (or your niece) will be receiving these amounts from the investment.

Note We have assumed that your niece receives the withdrawals at the end of each compounding period, so that the trust fund is an ordinary annuity. If, instead, she receives the payments at the beginning of each compounding period, it is an annuity due. You switch between the two types of annuity by changing PMT: END at the bottom to PMT: BEGIN. ■

As mentioned in the text, the present value must be higher to fund payments at the beginning of each period, because the money in the account has less time to earn interest.

Example 4 (page 150)   Tony and Maria (see Example 2), having accumulated $100,000 for José Phillipe's college education, would now like to make quarterly withdrawals over the next 4 years. How much money can they withdraw each quarter in order to draw down the account to zero at the end of the 4 years? (Recall that the account pays 4% interest compounded quarterly.)

Solution with Technology

1. Enter the values shown below in the TI-83/84 Plus TVM Solver.

2. Solve for *PMT*.

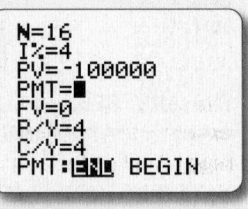

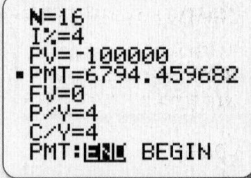

The present value is negative because Tony and Maria do not possess it; the bank does.

Example 7 (page 152)   Marc and Mira are buying a house and have taken out a 30-year, $90,000 mortgage at 8% interest per year. Mortgage interest is tax deductible, so it is important to know how much of a year's mortgage payments represents interest. How much interest will Marc and Mira pay in the first year of their mortgage?

Solution with Technology

The TI-83/84 Plus has built-in functions to compute the values in an amortization schedule.

1. First, use the TVM Solver to find the monthly payment.

```
N=360
I%=8
PV=90000
■PMT=-660.38811…
FV=0
P/Y=12
C/Y=12
PMT:END BEGIN
```

Three functions correspond to the last three columns of the amortization schedule given in the text: ΣInt, ΣPrn, and bal, (found in the Finance menu accessed through APPS). They all require that the values of I%, PV, and PMT be entered or calculated ahead of time; calculating the payment in the TVM Solver in Step 1 accomplishes this.

2. Use $\Sigma \text{Int}(m,n,2)$ to compute the sum of the interest payments from payment m through payment n. For example,

$$\Sigma \text{Int}(1,12,2)$$

will return −7,172.81, the total paid in interest in the first year, which answers the question asked in this example. (The last argument, 2, tells the calculator to round all intermediate calculations to two decimal places—that is, the nearest cent—as would the mortgage lender.)

3. Use $\Sigma \text{Prn}(m,n,2)$ to compute the sum of the payments on principal from payment m through payment n. For example,

$$\Sigma \text{Prn}(1,12,2)$$

will return −751.87, the total paid on the principal in the first year.

4. Finally, $\text{bal}(n,2)$ finds the balance of the principal outstanding after n payments. For example,

$$\text{bal}(12,2)$$

will return the value 89,248.13, the balance remaining at the end of one year.

5. To construct an amortization schedule as in the text, make sure that FUNC is selected in the MODE window; then enter the functions in the Y= window as shown below.

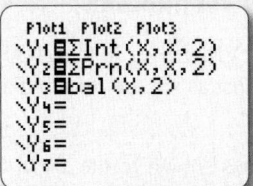

```
Plot1 Plot2 Plot3
\Y1■ΣInt(X,X,2)
\Y2■ΣPrn(X,X,2)
\Y3■bal(X,2)
\Y4=
\Y5=
\Y6=
\Y7=
```

6. Press 2ND TBLSET and enter the values shown here.

```
TABLE SETUP
 TblStart=0
 ΔTbl=1
 Indpnt: Auto Ask
 Depend: Auto Ask
```

7. Press 2ND TABLE, to get the table shown here.

X	Y1	Y2
0	ERROR	ERROR
1	-600	-60.39
2	-599.6	-60.79
3	-599.2	-61.2
4	-598.8	-61.61
5	-598.4	-62.02
6	-598	-62.43

Y1=-600

X	Y2	Y3
0	ERROR	90000
1	-60.39	89940
2	-60.79	89879
3	-61.2	89818
4	-61.61	89756
5	-62.02	89694
6	-62.43	89632

Y3=89939.61

The column labeled X gives the month, the column labeled Y1 gives the interest payment for each month, the column labeled Y2 gives the payment on principal for each month, and the column labeled Y3 (use the right arrow button to make it visible) gives the outstanding principal.

TECHNOLOGY GUIDE

8. To see later months, use the down arrow. As you can see, some of the values will be rounded in the table, but by selecting a value (as the outstanding principal at the end of the first month is selected in the second screen) you can see its exact value at the bottom of the screen.

Example 9 (page 154)   Suppose that a 5%, 20-year bond sells for $9,800 per $10,000 maturity value. What rate of return will investors get?

Solution with Technology

We can use the TVM Solver in the TI-83/84 Plus to find the interest rate just as we use it to find any other one of the variables.

1. Enter the values shown in the TVM Solver window.

2. Solve for *I%*. (Recall that *I%* is the annual interest rate, corresponding to *r* in the formula.)

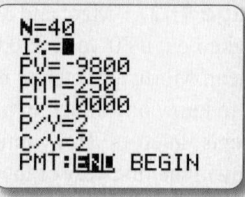

Thus, at $9,800 per $10,000 maturity value, these bonds yield 5.162% compounded semiannually.

SPREADSHEET **Technology Guide**

Section 2.2

Example 1 (page 136)   In November 2011, the Bank of Montreal was paying 1.30% interest on savings accounts. If the interest is compounded quarterly, find the future value of a $2,000 deposit in 6 years. What is the total interest paid over the period?

Solution with Technology

You can either compute compound interest directly or use financial functions built into your spreadsheet. The following worksheet has more than we need for this example, but will be useful for other examples in this and the next section.

	A	B	C	D
1		Entered	Calculated	
2	Rate	1.30%		
3	Years	6		
4	Payment	$0.00		
5	Present Value	-$2,000.00		
6	Future Value		=FV(B2/B7,B3*B7,B4,B5)	
7	Periods per year	4		

For this example the payment amount in B4 should be 0 (we shall use it in the next section).

1. Enter the other numbers as shown. As with other technologies, like the TVM Solver in the TI-83/84 Plus calculator, money that you pay to others (such as the $2,000 you deposit in the bank) should be entered as negative, whereas money that is paid to you is positive.

2. The formula entered in C6 uses the built-in FV function to calculate the future value based on the entries in column B. This formula has the following format:

$$=FV(i, n, PMT, PV)$$

$i =$ interest per period	We use B2/B7 for the interest.
$n =$ number of periods	We use B3*B7 for the number of periods.
$PMT =$ payment per period	The payment is 0 (cell B4).
$PV =$ present value	The present value is in cell B5.

Instead of using the built-in FV function, we could use

$$=-B5*(1+B2/B7)^{(B3*B7)}$$

based on the future value formula for compound interest. After calculation the result will appear in cell C6.

	A	B	C	D
1		Entered	Calculated	
2	Rate	1.30%		
3	Years	6		
4	Payment	$0.00		
5	Present Value	-$2,000.00		
6	Future Value		$2,161.97	
7	Periods per year	4		

Note that we have formatted the cells B4:C6 as currency with two decimal places. If you change the values in column B, the future value in column C will be automatically recalculated.

Example 2 (page 138)
Megabucks Corporation is issuing 10-year zero coupon bonds. How much would you pay for bonds with a maturity value of $10,000 if you wish to get a return of 6.5% compounded annually?

Solution with Technology

You can compute present value in your spreadsheet using the PV worksheet function. The following worksheet is similar to the one in the preceding example, except that we have entered a formula for computing the present value from the entered values.

	A	B	C	D
1		Entered	Calculated	
2	Rate	6.50%		
3	Years	10		
4	Payment	$0.00		
5	Present Value		=PV(B2/B7,B3*B7,B4,B6)	
6	Future Value	$10,000.00		
7	Periods per year	1		

The next worksheet shows the calculated value.

	A	B	C	D
1		Entered	Calculated	
2	Rate	6.50%		
3	Years	10		
4	Payment	$0.00		
5	Present Value		-$5,327.26	
6	Future Value	$10,000.00		
7	Periods per year	1		

Why is the present value negative?

Example 6 (page 141)
You have $5,000 to invest at 6% interest compounded monthly. How long will it take for your investment to grow to $6,000?

Solution with Technology

You can compute the requisite length of an investment in your spreadsheet using the NPER worksheet function. The following worksheets show the calculation.

	A	B	C	D
1		Entered	Calculated	
2	Rate	6.00%		
3	Years		=NPER(B2/B7,B4,B5,B6)/B7	
4	Payment	$0.00		
5	Present Value	-$5,000.00		
6	Future Value	$6,000.00		
7	Periods per year	12		

	A	B	C	D
1		Entered	Calculated	
2	Rate	6.00%		
3	Years		3.04628303	
4	Payment	$0.00		
5	Present Value	-$5,000.00		
6	Future Value	$6,000.00		
7	Periods per year	12		

The NPER function computes the number of compounding periods, months in this case, so we divide by B7, the number of periods per year, to calculate the number of years, which appears as 3.046. So, you need to invest your money for just over 3 years for it to grow to $6,000.

Section 2.3

Example 1 (page 146)
Your retirement account has $5,000 in it and earns 5% interest per year compounded monthly. Every month for the next 10 years you will deposit $100 into the account. How much money will there be in the account at the end of those 10 years?

Solution with Technology

We can use exactly the same worksheet that we used in Example 1 in the preceding section. In fact, we included the "Payment" row in that worksheet just for this purpose.

	A	B	C	D
1		Entered	Calculated	
2	Rate	5.00%		
3	Years	10		
4	Payment	-$100.00		
5	Present Value	-$5,000.00		
6	Future Value		=FV(B2/B7,B3*B7,B4,B5)	
7	Periods per year	12		

	A	B	C	D
1		Entered	Calculated	
2	Rate	5.00%		
3	Years	10		
4	Payment	-$100.00		
5	Present Value	-$5,000.00		
6	Future Value		$23,763.28	
7	Periods per year	12		

Note that the FV function allows us to enter, as the last argument, the amount of money already in the account. Following the usual convention, we enter the present value and the payment as *negative*, because these are amounts you pay into the account.

Example 2 (page 147)

Tony and Maria have just had a son, José Phillipe. They establish an account to accumulate money for his college education, in which they would like to have $100,000 after 17 years. If the account pays 4% interest per year compounded quarterly, and they make deposits at the end of every quarter, how large must each deposit be for them to reach their goal?

Solution with Technology

Use the following worksheet, in which the PMT worksheet function is used to calculate the required payments.

	A	B	C	D
1		Entered	Calculated	
2	Rate	4.00%		
3	Years	17		
4	Payment		=PMT(B2/B7,B3*B7,B5,B6)	
5	Present Value	$0.00		
6	Future Value	$100,000.00		
7	Periods per year	4		

	A	B	C	D
1		Entered	Calculated	
2	Rate	4.00%		
3	Years	17		
4	Payment		-$1,033.89	
5	Present Value	$0.00		
6	Future Value	$100,000.00		
7	Periods per year	4		

Why is the payment negative?

Example 3 (page 149)

You wish to establish a trust fund from which your niece can withdraw $2,000 every 6 months for 15 years, at which time she will receive the remaining money in the trust, which you would like to be $10,000. The trust will be invested at 7% per year compounded every 6 months. How large should the trust be?

Solution with Technology

You can use the same worksheet as in Example 2 in Section 2.2.

	A	B	C	D
1		Entered	Calculated	
2	Rate	7.00%		
3	Years	15		
4	Payment	$2,000.00		
5	Present Value		=PV(B2/B7,B3*B7,B4,B6)	
6	Future Value	$10,000.00		
7	Periods per year	2		

	A	B	C	D
1		Entered	Calculated	
2	Rate	7.00%		
3	Years	15		
4	Payment	$2,000.00		
5	Present Value		-$40,346.87	
6	Future Value	$10,000.00		
7	Periods per year	2		

The payment and future value are positive because you (or your niece) will be receiving these amounts from the investment.

Note We have assumed that your niece receives the withdrawals at the end of each compounding period, so that the trust fund is an ordinary annuity. If, instead, she receives the payments at the beginning of each compounding period, it is an annuity due. You switch to an annuity due by adding an optional last argument of 1 to the PV function (and similarly for the other finance functions in spreadsheets).

	A	B	C	D
1		Entered	Calculated	
2	Rate	7.00%		
3	Years	15		
4	Payment	$2,000.00		
5	Present Value		=PV(B2/B7,B3*B7,B4,B6,1)	
6	Future Value	$10,000.00		
7	Periods per year	2		

	A	B	C	D
1		Entered	Calculated	
2	Rate	7.00%		
3	Years	15		
4	Payment	$2,000.00		
5	Present Value		-$41,634.32	
6	Future Value	$10,000.00		
7	Periods per year	2		

As mentioned in the text, the present value must be higher to fund payments at the beginning of each period, because the money in the account has less time to earn interest. ■

Example 4 (page 150)

Tony and Maria (see Example 2), having accumulated $100,000 for José Phillipe's college education, would now like to make quarterly withdrawals over the next 4 years. How much money can they withdraw each quarter in order to draw down the account to zero at the end of the four years? (Recall that the account pays 4% interest compounded quarterly.)

Solution with Technology

You can use the same worksheet as in Example 2.

	A	B	C	D
1		Entered	Calculated	
2	Rate	4.00%		
3	Years	4		
4	Payment		=PMT(B2/B7,B3*B7,B5,B6)	
5	Present Value	-$100,000.00		
6	Future Value	$0.00		
7	Periods per year	4		

	A	B	C	D
1		Entered	Calculated	
2	Rate	4.00%		
3	Years	4		
4	Payment		$6,794.46	
5	Present Value	-$100,000.00		
6	Future Value	$0.00		
7	Periods per year	4		

The present value is negative since Tony and Maria do not possess it; the bank does.

Example 7 (page 152) Marc and Mira are buying a house and have taken out a 30-year, $90,000 mortgage at 8% interest per year. Mortgage interest is tax deductible, so it is important to know how much of a year's mortgage payments represents interest. How much interest will Marc and Mira pay in the first year of their mortgage?

Solution with Technology

We construct an amortization schedule with which we can answer the question.

1. Begin with the worksheet below.

	A	B	C	D	E	F	G	H
1	Month	Interest Payment	Payment on Principal	Outstanding Principal				
2	0			$90,000.00	Rate	8%		
3					Years	30		
4					Payment	=DOLLAR(-PMT(F2/12,F3*12,D2))		

	A	B	C	D	E	F
1	Month	Interest Payment	Payment on Principal	Outstanding Principal		
2	0			$90,000.00	Rate	8%
3					Years	30
4					Payment	$660.39

Note the formula for the monthly payment:

`=DOLLAR(-PMT(F2/12,F3*12,D2))`

The function DOLLAR rounds the payment to the nearest cent, as the bank would.

2. Calculate the interest owed at the end of the first month using the formula

`=DOLLAR(D2*F$2/12)`

in cell B3.

3. The payment on the principal is the remaining part of the payment, so enter

`=F$4-B3`

in cell C3.

4. Calculate the outstanding principal by subtracting the payment on the principal from the previous outstanding principal, by entering

`=D2-C3`

in cell D3.

5. Copy the formulas in cells B3, C3, and D3 into the cells below them to continue the table.

	A	B	C	D	E	F
1	Month	Interest Payment	Payment on Principal	Outstanding Principal		
2	0			$90,000.00	Rate	8%
3	1	$600.00	$60.39	$89,939.61	Years	30
4	2				Payment	$660.39
5	3					
6	4					
7	5					
8	6					
9	7					
10	8					
11	9					
12	10					
13	11					
14	12					

The result should be something like the following:

	A	B	C	D	E	F
1	Month	Interest Payment	Payment on Principal	Outstanding Principal		
2	0			$90,000.00	Rate	8%
3	1	$600.00	$60.39	$89,939.61	Years	30
4	2	$599.60	$60.79	$89,878.82	Payment	$660.39
5	3	$599.19	$61.20	$89,817.62		
6	4	$598.78	$61.61	$89,756.01		
7	5	$598.37	$62.02	$89,693.99		
8	6	$597.96	$62.43	$89,631.56		
9	7	$597.54	$62.85	$89,568.71		
10	8	$597.12	$63.27	$89,505.44		
11	9	$596.70	$63.69	$89,441.75		
12	10	$596.28	$64.11	$89,377.64		
13	11	$595.85	$64.54	$89,313.10		
14	12	$595.42	$64.97	$89,248.13		

6. Adding the calculated interest payments gives us the total interest paid in the first year: $7,172.81.

Note Spreadsheets have built-in functions that compute the interest payment (IPMT) or the payment on the principle (PPMT) in a given period. We could also have used the built-in future value function (FV) to calculate the outstanding principal each month. The main problem with using these functions is that, in a sense, they are too accurate. They do not take into account the fact that payments and interest are rounded to the nearest cent. Over time, this rounding causes the actual value of the outstanding principal to differ from what the FV function would tell us. In fact, because the actual payment is rounded slightly upward (to $660.39 from 660.38811…), the principal is reduced slightly faster than necessary and a last payment of $660.39 would be $2.95 larger than needed to clear out the debt. The lender would reduce the last payment by $2.95 for this reason; Marc and Mira will pay only $657.44 for their final payment. This is common: The last payment on an installment loan is usually slightly larger or smaller than the others, to compensate for the rounding of the monthly payment amount. ∎

Example 9 (page 154)   Suppose that a 5%, 20-year bond sells for $9,800 per $10,000 maturity value. What rate of return will investors get?

Solution with Technology

Use the following worksheet, in which the RATE worksheet function is used to calculate the interest rate.

	A	B	C	D
1		Entered	Calculated	
2	Rate		=RATE(B3*B7,B4,B5,B6)*B7	
3	Years	20		
4	Payment	$250.00		
5	Present Value	-$9,800.00		
6	Future Value	$10,000.00		
7	Periods per year	2		

	A	B	C	D
1		Entered	Calculated	
2	Rate		5.162%	
3	Years	20		
4	Payment	$250.00		
5	Present Value	-$9,800.00		
6	Future Value	$10,000.00		
7	Periods per year	2		

3

Systems of Linear Equations and Matrices

Website

www.WanerMath.com

At the Website you will find:

- Section-by-section tutorials, including game tutorials with randomized quizzes

- A detailed chapter summary

- A true/false quiz

- Additional review exercises

- Graphers, Excel tutorials, TI-83/84 Plus programs

- A Web page that pivots and does row operations

- An Excel worksheet that pivots and does row operations

Case Study Hybrid Cars—Optimizing the Degree of Hybridization

You are involved in new model development at a major automobile company. The company is planning to introduce two new plug-in hybrid electric vehicles: the subcompact "Green Town Hopper" and the midsize "Electra Supreme," and your department must decide on the degree of hybridization (DOH) for each of these models that will result in the largest reduction in gasoline consumption. The data you have available show the gasoline saving for only three values of the DOH. **How do you estimate the optimal value?**

Oleksiy Maksymenko/Alamy

175

Introduction

In Chapter 1 we studied single functions and equations. In this chapter we seek solutions to **systems** of two or more equations. For example, suppose we need to *find two numbers whose sum is* 3 *and whose difference is* 1. In other words, we need to find two numbers x and y such that $x + y = 3$ and $x - y = 1$. The only solution turns out to be $x = 2$ and $y = 1$, a solution you might easily guess. But, how do we know that this is the only solution, and how do we find solutions systematically? When we restrict ourselves to systems of *linear* equations, there is a very elegant method for determining the number of solutions and finding them all. Moreover, as we will see, many real-world applications give rise to just such systems of linear equations.

We begin in Section 3.1 with systems of two linear equations in two unknowns and some of their applications. In Section 3.2 we study a powerful matrix method, called *row reduction*, for solving systems of linear equations in any number of unknowns. In Section 3.3 we look at more applications.

Computers have been used for many years to solve the large systems of equations that arise in the real world. You probably already have access to devices that will do the row operations used in row reduction. Many graphing calculators can do them, as can spreadsheets and various special-purpose applications, including utilities available at the Website. Using such a device or program makes the calculations quicker and helps avoid arithmetic mistakes. Then there are programs (and calculators) into which you simply feed the system of equations and out pop the solutions. We can think of what we do in this chapter as looking inside the "black box" of such a program. More important, we talk about how, starting from a real-world problem, to get the system of equations to solve in the first place. No computer will do this conversion for us yet.

3.1 Systems of Two Equations in Two Unknowns

Suppose you have \$3 in your pocket to spend on snacks and a drink. If x represents the amount you'll spend on snacks and y represents the amount you'll spend on a drink, you can say that $x + y = 3$. On the other hand, if for some reason you want to spend \$1 more on snacks than on your drink, you can also say that $x - y = 1$. These are simple examples of **linear equations in two unknowns**.

Linear Equations in Two Unknowns

A **linear equation in two unknowns** is an equation that can be written in the form

$$ax + by = c$$

with a, b, and c being real numbers. The number a is called the **coefficient of x** and b is called the **coefficient of y**. A **solution** of an equation consists of a pair of numbers: a value for x and a value for y that satisfy the equation.

Quick Example

In the linear equation $3x - y = 15$, the coefficients are $a = 3$ and $b = -1$. The point $(x, y) = (5, 0)$ is a solution, because $3(5) - (0) = 15$.

In fact, a single linear equation such as $3x - y = 15$ has infinitely many solutions: We could solve for $y = 3x - 15$ and then, for every value of x we choose, we can get the corresponding value of y, giving a solution (x, y). As we saw in Chapter 1, these solutions are the points on a straight line, the *graph* of the equation.

In this section we are concerned with pairs (x, y) that are solutions of *two* linear equations at the same time. For example, $(2, 1)$ is a solution of both of the equations $x + y = 3$ and $x - y = 1$, because substituting $x = 2$ and $y = 1$ into these equations gives $2 + 1 = 3$ (true) and $2 - 1 = 1$ (also true), respectively. So, in the simple example we began with, you could spend \$2 on snacks and \$1 on a drink.

In the following set of examples, you will see how to graphically and algebraically solve a system of two linear equations in two unknowns. Then we'll return to some more interesting applications.

EXAMPLE 1 Two Ways of Solving a System: Graphically and Algebraically

Find all solutions (x, y) of the following system of two equations:

$$x + y = 3$$
$$x - y = 1.$$

Solution We will see how to find the solution(s) in two ways: graphically and algebraically. Remember that a solution is a pair (x, y) that simultaneously satisfies *both* equations.

Method 1: Graphical We already know that the solutions of a single linear equation are the points on its graph, which is a straight line. For a point to represent a solution of two linear equations, it must lie simultaneously on both of the corresponding lines. In other words, it must be a point where the two lines cross, or intersect. A look at Figure 1 should convince us that the lines cross only at the point $(2, 1)$, so this is the only possible solution.

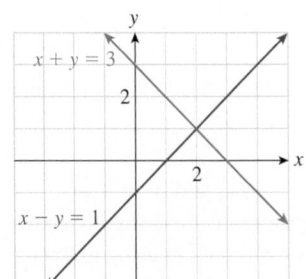

Figure 1

✱ We can add these equations because when we add equal amounts to both sides of an equation, the results are equal. That is, if $A = B$ and $C = D$, then $A + C = B + D$.

Method 2: Algebraic In the algebraic approach, we try to combine the equations in such a way as to eliminate one variable. In this case, notice that if we add the left-hand sides of the equations, the terms with y are eliminated. So, we add the first equation to the second (that is, add the left-hand sides and add the right-hand sides✱):

$$
\begin{array}{rl}
x + y &= 3 \\
x - y &= 1 \\
\hline
2x + 0 &= 4 \\
2x\ \ \ &= 4 \\
x\ \ \ &= 2.
\end{array}
$$

Now that we know that x has to be 2, we can substitute back into either equation to find y. Choosing the first equation (it doesn't matter which we choose), we have

$$2 + y = 3$$
$$y = 3 - 2$$
$$= 1.$$

We have found that the only possible solution is $x = 2$ and $y = 1$, or

$$(x, y) = (2, 1).$$

using Technology

See the Technology Guides at the end of the chapter for details on the graphical solution of Example 1 using a TI-83/84 Plus or a spreadsheet. Here is an outline:

TI-83/84 Plus

Y₁=-X+3 Y₂=X-1

Graph: WINDOW ; Xmin = −4,

Xmax = 4; ZOOM 0

Trace to estimate the point of intersection.
[More details on page 225.]

Website

www.WanerMath.com

→ On Line Utilities

→ Function Evaluator and Grapher

Enter -x+3 for y_1 and x-1 for y_2. Set Xmin $= -4$, Xmax $= 4$ and press "Plot Graphs."

Click on the graph and use the trace arrows below to estimate the point of intersection.

➡ **Before we go on...** There is another way we could find the solution in Example 1 algebraically: First we solve our two equations for y to obtain

$$y = -x + 3$$
$$y = x - 1$$

and then we equate the right-hand sides to get

$$-x + 3 = x - 1,$$

which we solve to find $x = 2$. The value for y is then found by substituting $x = 2$ in either equation.

Q: *So why don't we solve all systems of equations this way?*

A: The elimination method extends more easily to systems with more equations and unknowns. It is the basis for the matrix method of solving systems—a method we discuss in Section 3.2. So, we shall use it exclusively for the rest of this section. ■

Example 2 illustrates the drawbacks of the graphical method.

EXAMPLE 2 Solving a System: Algebraically vs. Graphically

Solve the system

$$3x + 5y = 0$$
$$2x + 7y = 1.$$

Solution

Method 1: Graphical First, solve for y, obtaining $y = -\frac{3}{5}x$ and $y = -\frac{2}{7}x + \frac{1}{7}$. Graphing these equations, we get Figure 2. The lines appear to intersect slightly above and to the left of the origin. Redrawing with a finer scale (or zooming in using graphing technology), we can get the graph in Figure 3.

If we look carefully at Figure 3, we see that the graphs intersect near $(-0.45, 0.27)$. Is the point of intersection *exactly* $(-0.45, 0.27)$? (Substitute these values into the equations to find out.) In fact, it is impossible to find the exact solution of this system graphically, but we now have a ballpark answer that we can use to help check the following algebraic solution.

Method 2: Algebraic We first see that adding the equations is not going to eliminate either x or y. Notice, however, that if we multiply (both sides of) the first equation by 2 and the second by -3, the coefficients of x will become 6 and -6. *Then* if we add them, x will be eliminated. So we proceed as follows:

$$2(3x + 5y) = 2(0)$$
$$-3(2x + 7y) = -3(1)$$

gives

$$6x + 10y = 0$$
$$-6x - 21y = -3.$$

Adding these equations, we get

$$-11y = -3$$

so that

$$y = \frac{3}{11} = 0.\overline{27}.$$

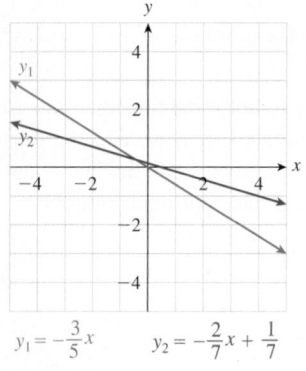

$$y_1 = -\frac{3}{5}x \qquad y_2 = -\frac{2}{7}x + \frac{1}{7}$$

Figure 2

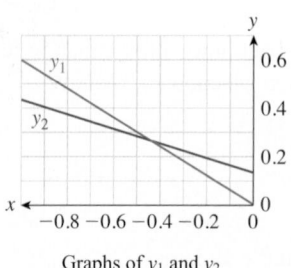

Graphs of y_1 and y_2 with $-1 \le x \le 0$

Figure 3

Substituting $y = \dfrac{3}{11}$ in the first equation gives

$$3x + 5\left(\frac{3}{11}\right) = 0$$

$$3x = -\frac{15}{11}$$

$$x = -\frac{5}{11} = -0.\overline{45}.$$

The solution is $(x, y) = \left(-\frac{5}{11}, \frac{3}{11}\right) = (-0.\overline{45}, 0.\overline{27})$.

Notice that the algebraic method gives us the exact solution that we could not find with the graphical method. Still, we can check that the graph and our algebraic solution agree to the accuracy with which we can read the graph. To be absolutely sure that our answer is correct, we should check it:

$$3\left(-\frac{5}{11}\right) + 5\left(\frac{3}{11}\right) = -\frac{15}{11} + \frac{15}{11} = 0 \quad \checkmark$$

$$2\left(-\frac{5}{11}\right) + 7\left(\frac{3}{11}\right) = -\frac{10}{11} + \frac{21}{11} = 1. \quad \checkmark$$

Get in the habit of checking your answers.

➡ **Before we go on...**

Q: *In solving the system in Example 2 algebraically, we multiplied (both sides of) the equations by numbers. How does that affect their graphs?*

A: Multiplying both sides of an equation by a nonzero number has no effect on its solutions, so the graph (which represents the set of all solutions) is unchanged. ∎

Before doing some more examples, let's summarize what we've said about solving systems of equations.

Graphical Method for Solving a System of Two Equations in Two Unknowns

Graph both equations on the same graph. (For example, solve each for y to find the slope and y-intercept.) A point of intersection gives the solution to the system. To find the point, you may need to adjust the range of x-values you use. To find the point accurately you may need to use a smaller range (or zoom in if using technology).

Algebraic Method for Solving a System of Two Equations in Two Unknowns

Multiply each equation by a nonzero number so that the coefficients of x are the same in absolute value but opposite in sign. Add the two equations to eliminate x; this gives an equation in y that we can solve to find its value. Substitute this value of y into one of the original equations to find the value of x. (Note that we could eliminate y first instead of x if it's more convenient.)

Sometimes, something appears to go wrong with these methods. The following examples show what can happen.

EXAMPLE 3 Inconsistent System

Solve the system

$$x - 3y = 5$$
$$-2x + 6y = 8.$$

Solution To eliminate x, we multiply the first equation by 2 and then add:

$$2x - 6y = 10$$
$$-2x + 6y = 8.$$

Adding gives

$$0 = 18.$$

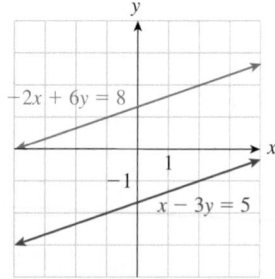

Figure 4

But this is absurd! This calculation shows that if we had two numbers x and y that satisfied both equations, it would be true that $0 = 18$. As 0 is *not* equal to 18, there can be no such numbers x and y. In other words, *the system has no solutions*, and is called an **inconsistent system**.

In slope-intercept form these lines are $y = \frac{1}{3}x - \frac{5}{3}$ and $y = \frac{1}{3}x + \frac{4}{3}$. Notice that they have the same slope but different y-intercepts. This means that they are parallel, but different lines. Plotting them confirms this fact (Figure 4).

Because they are parallel, they do not intersect. A solution must be a point of intersection, so we again conclude that there is no solution.

EXAMPLE 4 Redundant System

Solve the system

$$x + \ \ y = 2$$
$$2x + 2y = 4.$$

Solution Multiplying the first equation by -2 gives

$$-2x - 2y = -4$$
$$2x + 2y = 4.$$

Adding gives the not-very-enlightening result

$$0 = 0.$$

Now what has happened? Looking back at the original system, we note that the second equation is really the first equation in disguise. (It is the first equation multiplied by 2.) Put another way, if we solve both equations for y, we find that, in slope-intercept form, both equations become the same:

$$y = -x + 2.$$

The second equation gives us the same information as the first, so we say that this is a **redundant**, or **dependent system**. In other words, we really have only one equation in two unknowns. From Chapter 1, we know that a single linear equation in two unknowns has infinitely many solutions, one for each value of x. (Recall that to get the corresponding solution for y, we solve the equation for y and substitute the x-value.) The entire set of solutions can be summarized as follows:

x is arbitrary

$$y = 2 - x. \qquad \text{Solve the first equation for } y.$$

This set of solutions is called the **general solution** because it includes all possible solutions. When we write the general solution this way, we say that we have a **parameterized solution** and that x is the **parameter**.

We can also write the general solution as

$$(x, 2 - x) \quad x \text{ arbitrary.}$$

Different choices of the parameter x lead to different **particular solutions**. For instance, choosing $x = 3$ gives the particular solution

$$(x, y) = (3, -1).$$

Because there are infinitely many values of x from which to choose, there are infinitely many solutions.

What does this system of equations look like graphically? The two equations are really the same, so their graphs are identical, each being the line with x-intercept 2 and y-intercept 2. The "two" lines intersect at every point, so there is a solution for each point on the common line. In other words, we have a "whole line of solutions" (Figure 5).

We could also have solved the first equation for x instead and used y as a parameter, obtaining another form of the general solution:

$$(2 - y, y) \quad y \text{ arbitrary.} \qquad \text{Alternate form of the general solution}$$

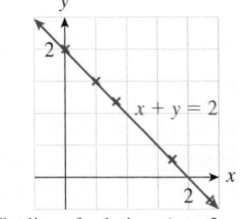

The line of solutions ($y = 2 - x$) and some particular solutions

Figure 5

We summarize the three possible outcomes we have encountered.

Possible Outcomes for a System of Two Linear Equations in Two Unknowns

1. **A single (or *unique*) solution** This happens when the lines corresponding to the two equations are distinct and not parallel so that they intersect at a single point. (See Example 1.)

2. **No solution** This happens when the two lines are parallel. We say that the system is **inconsistent**. (See Example 3.)

3. **An infinite number of solutions** This occurs when the two equations represent the same straight line, and we say that such a system is **redundant**, or **dependent**. In this case, we can represent the solutions by choosing one variable arbitrarily and solving for the other. (See Example 4.)

In cases 1 and 3, we say that the system of equations is **consistent** because it has at least one solution.

You should think about straight lines and convince yourself that these are the only three possibilities.

APPLICATIONS

EXAMPLE 5 Blending

Acme Baby Foods mixes two strengths of apple juice. One quart of Beginner's juice is made from 30 fluid ounces of water and 2 fluid ounces of apple juice concentrate. One quart of Advanced juice is made from 20 fluid ounces of water and 12 fluid ounces of concentrate. Every day Acme has available 30,000 fluid ounces of water and 3,600 fluid ounces of concentrate. If the company wants to use all the water and concentrate, how many quarts of each type of juice should it mix?

Solution In all applications we follow the same general strategy.

1. *Identify and label the unknowns.* What are we asked to find? To answer this question, it is common to respond by saying, "The unknowns are Beginner's juice and Advanced juice." Quite frankly, this is a baffling statement. Just what is unknown about juice? We need to be more precise:

The unknowns are (1) *the **number of quarts** of Beginner's juice and* (2) *the **number of quarts** of Advanced juice made each day.*

So, we label the unknowns as follows: Let

x = number of quarts of Beginner's juice made each day

y = number of quarts of Advanced juice made each day.

2. *Use the information given to set up equations in the unknowns.* This step is trickier, and the strategy varies from problem to problem. Here, the amount of juice the company can make is constrained by the fact that they have limited amounts of water and concentrate. This example shows a kind of application we will often see, and it is helpful in these problems to use a table to record the amounts of the resources used.

	Beginner's (x)	*Advanced (y)*	*Available*
Water (fl oz)	30	20	30,000
Concentrate (fl oz)	2	12	3,600

We can now set up an equation for each of the items listed in the left column of the table.

Water: We read across the first row. If Acme mixes x quarts of Beginner's juice, each quart using 30 fluid ounces of water, and y quarts of Advanced juice, each using 20 fluid ounces of water, it will use a total of $30x + 20y$ fluid ounces of water. But we are told that the total has to be 30,000 fluid ounces. Thus, $30x + 20y = 30,000$. This is our first equation.

Concentrate: We read across the second row. If Acme mixes x quarts of Beginner's juice, each using 2 fluid ounces of concentrate, and y quarts of Advanced juice, each using 12 fluid ounces of concentrate, it will use a total of $2x + 12y$ fluid ounces of concentrate. But we are told that the total has to be 3,600 fluid ounces. Thus, $2x + 12y = 3,600$.

Now we have two equations:

$$30x + 20y = 30,000$$
$$2x + 12y = 3,600.$$

To make the numbers easier to work with, let's divide (both sides of) the first equation by 10 and the second by 2:

$$3x + 2y = 3,000$$
$$x + 6y = 1,800.$$

using Technology
to Check Your
Answer

Website
www.WanerMath.com

→ Math Tools for Chapter 3

→ Pivot and Gauss-Jordan Tool

To check Example 5 enter the coefficients of x and y and the right-hand sides of the two equations as shown (no commas in numbers!):

Then press "Reduce Completely". The answer will appear in the last column. (In the next section we will see how this works.)

We can now eliminate x by multiplying the second equation by -3 and adding:

$$\begin{array}{r} 3x + 2y = 3{,}000 \\ -3x - 18y = -5{,}400 \\ \hline -16y = -2{,}400. \end{array}$$

So, $y = 2{,}400/16 = 150$. Substituting this into the equation $x + 6y = 1{,}800$ gives $x + 900 = 1{,}800$, and so $x = 900$. The solution is $(x, y) = (900, 150)$. In other words, the company should mix 900 quarts of Beginner's juice and 150 quarts of Advanced juice.

EXAMPLE 6 **Blending**

A medieval alchemist's love potion calls for a number of eyes of newt and toes of frog, the total being 20, but with twice as many newt eyes as frog toes. How many of each is required?

Solution As in the preceding example, the first step is to identify and label the unknowns. Let

$$x = \text{number of newt eyes}$$
$$y = \text{number of frog toes.}$$

As for the second step—setting up the equations—a table is less appropriate here than in the preceding example. Instead, we translate each phrase of the problem into an equation. The first sentence tells us that the total number of eyes and toes is 20. Thus,

$$x + y = 20.$$

The end of the first sentence gives us more information, but the phrase "twice as many newt eyes as frog toes" is a little tricky: does it mean that $2x = y$ or that $x = 2y$? We can decide which by rewording the statement using the phrases "the *number of* newt eyes," which is x, and "the *number of* frog toes," which is y. Rephrased, the statement reads:

The **number of** newt eyes is twice the **number of** frog toes.

(Notice how the word "twice" is forced into a different place.) With this rephrasing, we can translate directly into algebra:

$$x = 2y.$$

In standard form $(ax + by = c)$, this equation reads

$$x - 2y = 0.$$

Thus, we have the two equations:

$$\begin{array}{r} x + y = 20 \\ x - 2y = 0. \end{array}$$

To eliminate x, we multiply the second equation by -1 and then add:

$$\begin{array}{r} x + y = 20 \\ -x + 2y = 0. \end{array}$$

We'll leave it to you to finish solving the system and find that $x = 13\frac{1}{3}$ and $y = 6\frac{2}{3}$.

So, the recipe calls for exactly $13\frac{1}{3}$ eyes of newt and $6\frac{2}{3}$ toes of frog. The alchemist needs a very sharp scalpel and a very accurate balance (not to mention a very strong stomach).

We saw in Chapter 1 that the *equilibrium price* of an item (the price at which supply equals demand) and the *break-even point* (the number of items that must be sold to break even) can both be described as the intersection points of two graphs. If the graphs are straight lines, what we need to do to find the intersection is solve a system of two linear equations in two unknowns, as illustrated in the following problem.

EXAMPLE 7 **Equilibrium Price**

The demand for refrigerators in West Podunk is given by

$$q = -\frac{p}{10} + 100,$$

where q is the number of refrigerators that the citizens will buy each year if the refrigerators are priced at p dollars each. The supply is

$$q = \frac{p}{20} + 25,$$

where now q is the number of refrigerators the manufacturers will be willing to ship into town each year if they are priced at p dollars each. Find the equilibrium price and the number of refrigerators that will be sold at that price.

Solution   Figure 6 shows the demand and supply curves. The equilibrium price occurs at the point where these two lines cross, which is where demand equals supply. The graph suggests that the equilibrium price is $500, and zooming in confirms this.

To solve this system algebraically, first write both equations in standard form:

$$\frac{p}{10} + q = 100$$

$$-\frac{p}{20} + q = 25.$$

We can clear fractions and also prepare to eliminate p if we multiply the first equation by 10 and the second by 20:

$$p + 10q = 1{,}000$$

$$-p + 20q = 500,$$

and so:

$$30q = 1{,}500$$

$$q = 50.$$

Substituting this value of q into either equation gives us $p = 500$. Thus, the equilibrium price is $500, and 50 refrigerators will be sold at this price.

We could also have solved this system of equations by setting the two expressions for q (the supply and the demand) equal to each other:

$$-p/10 + 100 = p/20 + 25$$

and then solving for p. (See the **Before we go on** discussion at the end of Example 1.)

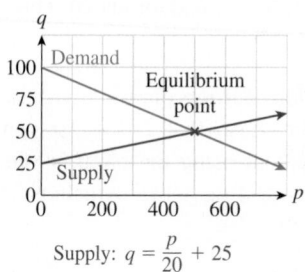

Supply: $q = \dfrac{p}{20} + 25$

Demand: $q = -\dfrac{p}{10} + 100$

Figure 6

FAQs

Setting Up the Equations

Q : *Looking through these examples, I notice that in some, we can tabulate the information given and read off the equations (as in Example 5), whereas in others (like Example 6), we have to reword each sentence to turn it into an equation. How do I know what approach to use?*

A : There is no hard-and-fast rule, and indeed some applications might call for a bit of each approach. However, it is generally not hard to see when it would be useful to tabulate values: Lists of the numbers of ingredients or components generally lend themselves to tabulation, whereas phrases like "twice as many of these as those" generally require direct translation into equations (after rewording if necessary).

3.1 EXERCISES

▼ more advanced   ◆ challenging
T indicates exercises that should be solved using technology

In Exercises 1–14, find all solutions of the given system of equations and check your answer graphically. HINT [See Examples 1–4.]

1. $x - y = 0$
$x + y = 4$

2. $x - y = 0$
$x + y = -6$

3. $x + y = 4$
$x - y = 2$

4. $2x + y = 2$
$-2x + y = 2$

5. $3x - 2y = 6$
$2x - 3y = -6$

6. $2x + 3y = 5$
$3x + 2y = 5$

7. $0.5x + 0.1y = 0.7$
$0.2x - 0.2y = 0.6$

8. $-0.3x + 0.5y = 0.1$
$0.1x - 0.1y = 0.4$

9. $\dfrac{x}{3} - \dfrac{y}{2} = 1$
$\dfrac{x}{4} + y = -2$

10. $-\dfrac{2x}{3} + \dfrac{y}{2} = -\dfrac{1}{6}$
$\dfrac{x}{4} - y = -\dfrac{3}{4}$

11. $2x + 3y = 1$
$-x - \dfrac{3y}{2} = -\dfrac{1}{2}$

12. $2x - 3y = 1$
$6x - 9y = 3$

13. $2x + 3y = 2$
$-x - \dfrac{3y}{2} = -\dfrac{1}{2}$

14. $2x - 3y = 2$
$6x - 9y = 3$

T *In Exercises 15–24, use technology to obtain approximate solutions graphically. All solutions should be accurate to one decimal place. (Zoom in for improved accuracy.)*

15. $2x + 8y = 10$
$x + y = 5$

16. $2x - y = 3$
$x + 3y = 5$

17. $3.1x - 4.5y = 6$
$4.5x + 1.1y = 0$

18. $0.2x + 4.5y = 1$
$1.5x + 1.1y = 2$

19. $10.2x + 14y = 213$
$4.5x + 1.1y = 448$

20. $100x + 4.5y = 540$
$1.05x + 1.1y = 0$

21. ▼ Find the intersection of the line through $(0, 1)$ and $(4.2, 2)$ and the line through $(2.1, 3)$ and $(5.2, 0)$.

22. ▼ Find the intersection of the line through $(2.1, 3)$ and $(4, 2)$ and the line through $(3.2, 2)$ and $(5.1, 3)$.

23. ▼ Find the intersection of the line through $(0, 0)$ and $(5.5, 3)$ and the line through $(5, 0)$ and $(0, 6)$.

24. ▼ Find the intersection of the line through $(4.3, 0)$ and $(0, 5)$ and the line through $(2.1, 2.2)$ and $(5.2, 1)$.

APPLICATIONS

25. *Resource Allocation* You manage an ice cream factory that makes two flavors: Creamy Vanilla and Continental Mocha. Into each quart of Creamy Vanilla go 2 eggs and 3 cups of cream. Into each quart of Continental Mocha go 1 egg and 3 cups of cream. You have in stock 500 eggs and 900 cups of cream. How many quarts of each flavor should you make in order to use up all the eggs and cream? HINT [See Example 5.]

26. *Class Scheduling* Enormous State University's Math Department offers two courses: Finite Math and Applied Calculus. Each section of Finite Math has 60 students, and each section of Applied Calculus has 50. The department will offer a total of 110 sections in a semester, and 6,000 students would like to take a math course. How many sections of each course should the department offer in order to fill all sections and accommodate all of the students? HINT [See Example 5.]

27. *Nutrition* Gerber Products' Gerber Mixed Cereal for Baby contains, in each serving, 60 calories and 11 grams of carbohydrates. Gerber Mango Tropical Fruit Dessert contains, in each serving, 80 calories and 21 grams of carbohydrates.[1] If you want to provide your child with 200 calories and 43 grams of carbohydrates, how many servings of each should you use?

28. *Nutrition* Anthony Altino is mixing food for his young daughter and would like the meal to supply 1 gram of protein and 5 milligrams of iron. He is mixing together cereal, with 0.5 grams of protein and 1 milligram of iron per ounce,

---

[1]Source: Nutrition information supplied with the products.

and fruit, with 0.2 grams of protein and 2 milligrams of iron per ounce. What mixture will provide the desired nutrition?

29. **Nutrition** One serving of Campbell Soup Company's Campbell's® Pork & Beans contains 5 grams of protein and 21 grams of carbohydrates.[2] A typical slice of white bread provides 2 grams of protein and 11 grams of carbohydrates per slice. The U.S. RDA (Recommended Daily Allowance) is 60 grams of protein each day.[3]

   a. I am planning a meal of "beans on toast" and wish to have it supply one-half of the RDA for protein and 139 grams of carbohydrates. How should I prepare my meal?

   b. Is it possible to have my meal supply the same amount of protein as in part (a) but only 100 g of carbohydrates?

30. **Nutrition** One serving of Campbell Soup Company's Campbell's® Pork & Beans contains 5 grams of protein and 21 grams of carbohydrates.[4] A typical slice of "lite" rye bread contains 4 grams of protein and 12 grams of carbohydrates.

   a. I am planning a meal of "beans on toast" and wish to have it supply one-third of the U.S. RDA for protein (see the preceding exercise) and 80 grams of carbohydrates. How should I prepare my meal?

   b. Is it possible to have my meal supply the same amount of protein as in part (a) but only 60 grams of carbohydrates?

Protein Supplements Exercises 31–34 are based on the following data on three popular protein supplements. (Figures shown correspond to a single serving.)[5]

	Protein (g)	Carbohydrates (g)	Sodium (mg)	Cost ($)
Designer Whey® (Next)	18	2	80	0.50
Muscle Milk® (Cytosport)	32	16	240	1.60
Pure Whey Protein Stack® (Champion)	24	3	100	0.60

31. You are thinking of combining Designer Whey and Muscle Milk to obtain a 7-day supply that provides exactly 280 grams of protein and 56 grams of carbohydrates. How many servings of each supplement should you combine in order to meet your requirements? What will it cost?

32. You are thinking of combining Muscle Milk and Pure Whey Protein Stack to obtain a supply that provides exactly 640 grams of protein and 3,200 mg of sodium. How many servings of each supplement should you combine in order to meet your requirements? What will it cost?

33. ▼ You have a mixture of Designer Whey and Pure Whey Protein Stack that costs a total of $14 and supplies exactly 540 g of protein. How many grams of carbohydrates does it supply?

34. ▼ You have a mixture of Designer Whey and Muscle Milk that costs a total of $14 and supplies exactly 104 g of carbohydrates. How many grams of protein does it supply?

35. **Investments: Tech Stocks** In July 2011, Apple (AAPL) stock increased from $350 to $400 per share, and Google (GOOG) stock increased from $500 to $600 per share.[6] If you invested a total of $22,000 in these stocks at the beginning of the month and sold them for $26,000 at the end of the month, how many shares of each stock did you buy?

36. **Investments: Energy Stocks** In the two-month period June 1–July 31, 2011, Hess Corp. (HES) stock decreased from $80 to $70 per share, and Exxon Mobil (XOM) stock decreased from $83 to $80 per share.[7] If you invested a total of $19,000 in these stocks at the beginning of June and sold them for $18,100 at the end of July, how many shares of each stock did you buy?

37. ▼ **Investments: Financial Stocks** In August 2011, Bank of Hawaii (BOH) stock cost $45 per share and yielded 4% per year in dividends, while JPMorgan Chase (JPM) stock cost $40 per share and yielded 2.5% per year in dividends.[8] If you invested a total of $25,000 in these stocks and earned $760 in dividends in a year, how many shares of each stock did you purchase? (Assume the dividend rate was unchanged for the year.)

38. ▼ **Investments: Utility Stocks** In April 2011, Consolidated Edison (ED) stock cost $50 per share and yielded 5% per year in dividends, while National Grid (NGG) stock cost $50 per share and yielded 6% per year in dividends.[9] If you invested a total of $45,000 in these stocks and earned $2,400 in dividends in a year, how many shares of each stock did you purchase? (Assume the dividend rate was unchanged for the year.)

39. **Voting** An appropriations bill passes the U.S. House of Representatives with 49 more members voting in favor than against. If all 435 members of the House vote for or against the bill, how many voted in favor and how many voted against?

40. **Voting** The U.S. Senate has 100 members. For a bill to pass with a supermajority, at least twice as many senators must vote in favor of the bill as vote against it. If all 100 senators vote in favor of or against a bill, how many must vote in favor for it to pass with a supermajority?

41. **Intramural Sports** The best sports dorm on campus, Lombardi House, has won a total of 12 games this semester. Some of these games were soccer games, and the others were football games. According to the rules of the university, each win in a soccer game earns the winning house 2 points, whereas each win in a football game earns them 4 points. If the total number of points Lombardi House earned was 38, how many of each type of game did they win?

---

[2]According to the label information on a 16 oz. can.

[3]Recommended Daily Allowance for a person weighing 75 kg (165 lb)

[4]Ibid.

[5]Source: Nutritional information supplied by the manufacturers (www.netrition.com). Cost per serving is approximate and varies.

[6]Approximate stock prices at or close to the dates cited. Source: http://finance.google.com.

[7]Ibid.

[8]Stock prices and yields are approximate. Source: http://finance.google.com.

[9]Ibid.

42. *Law* Enormous State University's campus publication, *The Campus Inquirer*, ran a total of 10 exposés five years ago, dealing with alleged recruiting violations by the football team and with theft by the student treasurer of the film society. Each exposé dealing with recruiting violations resulted in a $4 million libel suit, and the treasurer of the film society sued the paper for $3 million as a result of each exposé concerning his alleged theft. Unfortunately for *The Campus Inquirer*, all the lawsuits were successful, and the paper wound up being ordered to pay $37 million in damages. (It closed down shortly thereafter.) How many of each type of exposé did the paper run?

43. *Purchasing* *(from the GMAT)* Elena purchased brand *X* pens for $4.00 apiece and brand *Y* pens for $2.80 apiece. If Elena purchased a total of 12 of these pens for $42.00, how many brand *X* pens did she purchase?

44. *Purchasing* *(based on a question from the GMAT)* Earl is ordering supplies. Yellow paper costs $5.00 per ream while white paper costs $6.50 per ream. He would like to order 100 reams total, and has a budget of $560. How many reams of each color should he order?

45. *Equilibrium Price* The demand and supply functions for pet chias are, respectively, $q = -60p + 150$ and $q = 80p - 60$, where p is the price in dollars. At what price should the chias be marked so that there is neither a surplus nor a shortage of chias? HINT [See Example 7.]

46. *Equilibrium Price* The demand and supply functions for your college newspaper are, respectively, $q = -10,000p + 2,000$ and $q = 4,000p + 600$, where p is the price in dollars. At what price should the newspapers be sold so that there is neither a surplus nor a shortage of papers? HINT [See Example 7.]

47. *Supply and Demand* *(from the GRE Economics Test)* The demand curve for widgets is given by $D = 85 - 5P$, and the supply curve is given by $S = 25 + 5P$, where P is the price of widgets. When the widget market is in equilibrium, what is the quantity of widgets bought and sold?

48. *Supply and Demand* *(from the GRE Economics Test)* In the market for soybeans, the demand and supply functions are $Q_D = 100 - 10P$ and $Q_S = 20 + 5P$, where Q_D is quantity demanded, Q_S is quantity supplied, and P is price in dollars. If the government sets a price floor of $7, what will be the resulting surplus or shortage?

49. *Equilibrium Price* In June 2001, the retail price of a 25-kg bag of cornmeal was $8 in Zambia; by December, the price had risen to $11. The result was that one retailer reported a drop in sales from 15 bags/day to 3 bags/day.[10] Assume that the retailer is prepared to sell 3 bags/day at $8 and 15 bags/day at $11. Find linear demand and supply equations and then compute the retailer's equilibrium price.

50. *Equilibrium Price* At the start of December 2001, the retail price of a 25 kg bag of cornmeal was $10 in Zambia, while by the end of the month, the price had fallen to $6.[11] The

result was that one retailer reported an increase in sales from 3 bags/day to 5 bags/day. Assume that the retailer is prepared to sell 18 bags/day at $8 and 12 bags/day at $6. Obtain linear demand and supply equations, and hence compute the retailer's equilibrium price.

51. *Pollution* Joe Slo, a college sophomore, neglected to wash his dirty laundry for 6 weeks. By the end of that time, his roommate had had enough and tossed Joe's dirty socks and T-shirts into the trash, counting a total of 44 items. (A pair of dirty socks counts as one item.) The roommate noticed that there were three times as many pairs of dirty socks as T-shirts. How many of each item did he throw out?

52. *Diet* The local sushi bar serves 1-ounce pieces of raw salmon (consisting of 50% protein) and $1\frac{1}{4}$-ounce pieces of raw tuna (40% protein). A customer's total intake of protein amounts to $1\frac{1}{2}$ ounces after consuming a total of three pieces. How many of each did the customer consume? (Fractions of pieces are permitted.)

53. ▼ ***Management*** *(from the GMAT)* A manager has $6,000 budgeted for raises for four full-time and two part-time employees. Each of the full-time employees receives the same raise, which is twice the raise that each of the part-time employees receives. What is the amount of the raise that each full-time employee receives?

54. ▼ ***Publishing*** *(from the GMAT)* There were 36,000 hardback copies of a certain novel sold before the paperback version was issued. From the time the first paperback copy was sold until the last copy of the novel was sold, nine times as many paperback copies as hardback copies were sold. If a total of 441,000 copies of the novel were sold in all, how many paperback copies were sold?

COMMUNICATION AND REASONING EXERCISES

55. A system of three equations in two unknowns corresponds to three lines in the plane. Describe how these lines might be positioned if the system has a unique solution.

56. A system of three equations in two unknowns corresponds to three lines in the plane. Describe several ways that these lines might be positioned if the system has no solutions.

57. Both the supply and demand equations for a certain product have negative slope. Can there be an equilibrium price? Explain.

58. You are solving a system of equations with x representing the number of rocks and y representing the number of pebbles. The solution is $(200, -10)$. What do you conclude?

59. ▼ Referring to Exercise 25, suppose that the solution of the corresponding system of equations was 198.7 gallons of vanilla and 100.89 gallons of mocha. If your factory can produce only whole numbers of gallons, would you recommend rounding the answers to the nearest whole number? Explain.

60. ▼ Referring to Exercise 25 but using different data, suppose that the general solution of the corresponding system of equations was $(200 - y, y)$, where $x =$ number of gallons of vanilla, and $y =$ number of gallons of mocha. Your factory can

---

[10]The prices quoted are approximate. (Actual prices varied from retailer to retailer.) Source: *New York Times,* December 24, 2001, p. A4.
[11]*Ibid.*

produce only whole numbers of gallons. There are infinitely many ways of making the ice cream, mixes, right? Explain.

61. ▼ Select one: Multiplying both sides of a linear equation by a nonzero constant results in a linear equation whose graph is

(A) parallel to   (B) the same as
(C) not always parallel to   (D) not the same as

the graph of the original equation.

62. ▼ Select one: If the addition or subtraction of two linear equations results in the equation $3 = 3$, then the graphs of those equations are

(A) equal.   (B) parallel.
(C) perpendicular.   (D) none of the above.

63. ▼ Select one: If the addition or subtraction of two linear equations results in the equation $0 = 3$, then the graphs of those equations are

(A) equal.   (B) parallel.
(C) perpendicular.   (D) not parallel.

64. ▼ Select one: If adding two linear equations gives $x = 3$, and subtracting them gives $y = 3$, then the graphs of those equations are

(A) equal.   (B) parallel.
(C) perpendicular.   (D) not parallel.

65. ▼ Invent an interesting application that leads to a system of two equations in two unknowns with a unique solution.

66. ▼ Invent an interesting application that leads to a system of two equations in two unknowns with no solution.

67. ◆ How likely do you think it is that a "random" system of two equations in two unknowns has a unique solution? Give some justification for your answer.

68. ◆ How likely do you think it is that a "random" system of three equations in two unknowns has a unique solution? Give some justification for your answer.

3.2 Using Matrices to Solve Systems of Equations

In this section we describe a systematic method for solving systems of equations that makes solving large systems of equations in any number of unknowns straightforward. Although this method may seem a little cumbersome at first, it will prove *immensely* useful in this and the next several chapters. First, some terminology:

Linear Equation

A linear equation in the n variables $x_1, x_2, \ldots, x_n$ has the form

$$a_1x_1 + \cdots + a_nx_n = b. \qquad (a_1, a_2, \ldots, a_n, b \text{ constants})$$

The numbers $a_1, a_2, \ldots, a_n$ are called the **coefficients**, and the number b is called the **constant term**, or **right-hand side**.

Quick Examples

1. $3x - 5y = 0$
   Linear equation in x and y
   Coefficients: 3, −5 Constant term: 0

2. $x + 2y - z = 6$
   Linear equation in x, y, z
   Coefficients: 1, 2, −1 Constant term: 6

3. $30x_1 + 18x_2 + x_3 + x_4 = 19$
   Linear equation in x_1, x_2, x_3, x_4
   Coefficients: 30, 18, 1, 1 Constant term: 19

Note When the number of variables is small, we will almost always use $x, y, z, \ldots$ (as in Quick Examples 1 and 2) rather than $x_1, x_2, x_3, \ldots$ as the names of the variables. ■

Notice that a linear equation in any number of unknowns (for example, $2x - y = 3$) is entirely determined by its coefficients and its constant term. In other words, if we were simply given the row of numbers

$$[2 \quad -1 \quad 3]$$

we could easily reconstruct the original linear equation by multiplying the first number by x, the second by y, and inserting a plus sign and an equals sign, as follows:

$$2 \cdot x + (-1) \cdot y = 3$$

or $2x - y = 3.$

Similarly, the equation

$$-4x + 2y = 0$$

is represented by the row

$$[-4 \quad 2 \quad 0].$$

and the equation

$$-3y = \frac{1}{4}$$

is represented by

$$\left[0 \quad -3 \quad \frac{1}{4} \right].$$

As the last example shows, the first number is always the coefficient of x and the second is the coefficient of y. If an x or a y is missing, we write a zero for its coefficient. We shall call such a row the **coefficient row** of an equation.

If we have a system of equations, for example the system

$$2x - y = 3$$
$$-x + 2y = -4,$$

we can put the coefficient rows together like this:

$$\begin{bmatrix} 2 & -1 & 3 \\ -1 & 2 & -4 \end{bmatrix}.$$

We call this the **augmented matrix** of the system of equations. The term "augmented" means that we have included the right-hand sides 3 and -4. We will often drop the word "augmented" and simply refer to the matrix of the system. A **matrix** (plural: **matrices**) is nothing more than a rectangular array of numbers as above.

Matrix, Augmented Matrix

A **matrix** is a rectangular array of numbers. The **augmented matrix** of a system of linear equations is the matrix whose rows are the coefficient rows of the equations.

Quick Example

The augmented matrix of the system

$$x + y = 3$$
$$x - y = 1$$

is $\begin{bmatrix} 1 & 1 & 3 \\ 1 & -1 & 1 \end{bmatrix}.$

We'll be studying matrices in more detail in Chapter 4.

Q : *What good are coefficient rows and matrices?*

A : Think about what we do when we multiply both sides of an equation by a number. For example, consider multiplying both sides of the equation $2x - y = 3$ by -2 to get $-4x + 2y = -6$. All we are really doing is multiplying the coefficients and the right-hand side by -2. This corresponds to *multiplying the row* $[2 \ -1 \ 3]$ *by* -2, that is, multiplying every number in the row by -2. We shall see that any manipulation we want to do with equations can be done instead with rows, and this fact leads to a method of solving equations that is systematic and generalizes easily to larger systems.

Here is the same operation both in the language of equations and the language of rows. (We refer to the equation here as *Equation* 1, or simply E_1 for short, and to the row as *Row* 1, or R_1.)

	Equation	Row	
	E_1: $2x - \ y = 3$	$[\ 2 \quad -1 \quad 3]$	R_1
Multiply by -2:	$(-2)E_1$: $-4x + 2y = -6$	$[-4 \quad 2 \ -6]$	$(-2)R_1$

Multiplying both sides of an equation by the number a corresponds to multiplying the coefficient row by a.

Now look at what we do when we add two equations:

	Equation	Row	
	E_1: $2x - \ y = 3$	$[\ 2 \quad -1 \quad 3]$	R_1
	E_2: $-x + 2y = -4$	$[-1 \quad 2 \ -4]$	R_2
Add:	$E_1 + E_2$: $x + \ y = -1$	$[\ 1 \quad 1 \ -1]$	$R_1 + R_2$

All we are really doing is *adding the corresponding entries in the rows*, or *adding the rows*. In other words,

Adding two equations corresponds to adding their coefficient rows.

In short, the manipulations of equations that we saw in the preceding section can be done more easily with rows in a matrix because we don't have to carry x, y, and other unnecessary notation along with us; x and y can always be inserted at the end if desired.

The manipulations we are talking about are known as **row operations**. In particular, we use three **elementary row operations**.

✻ We are using the term elementary row operations a little more freely than most books. Some mathematicians insist that $a = 1$ in an operation of Type 2, but the less restrictive version is very useful.

† Multiplying an equation or row by zero gives us the not very surprising result $0 = 0$. In fact, we lose any information that the equation provided, which usually means that the resulting system has more solutions than the original system.

Elementary Row Operations✻

Type 1: Replacing R_i by aR_i (where $a \neq 0$)†
In words: multiplying or dividing a row by a nonzero number.

Type 2: Replacing R_i by $aR_i \pm bR_j$ (where $a \neq 0$)
Multiplying a row by a nonzero number and adding or subtracting a multiple of another row.

using Technology

See the Technology Guides at the end of the chapter to see how to do row operations using a TI-83/84 Plus or a spreadsheet.

 Website
www.WanerMath.com

→ Math Tools for Chapter 3

→ Pivot and Gauss-Jordan Tool

Enter the matrix in columns $x1, x2, x3, \ldots$. To do a row operation, type the instruction(s) next to the row(s) you are changing as shown, and press "Do Row Ops" once.

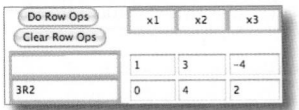

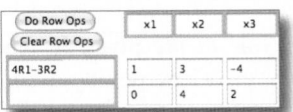

Type 3: Switching the order of the rows
This corresponds to switching the order in which we write the equations; occasionally this will be convenient.

For Types 1 and 2, we write the instruction for the row operation *next to the row we wish to replace.* (See the Quick Examples below.)

Quick Examples

Type 1: $\begin{bmatrix} 1 & 3 & -4 \\ 0 & 4 & 2 \end{bmatrix} 3R_2 \rightarrow \begin{bmatrix} 1 & 3 & -4 \\ 0 & 12 & 6 \end{bmatrix}$ Replace R_2 by $3R_2$.

Type 2: $\begin{bmatrix} 1 & 3 & -4 \\ 0 & 4 & 2 \end{bmatrix} 4R_1 - 3R_2 \rightarrow \begin{bmatrix} 4 & 0 & -22 \\ 0 & 4 & 2 \end{bmatrix}$ Replace R_1 by $4R_1 - 3R_2$.

Type 3: $\begin{bmatrix} 1 & 3 & -4 \\ 0 & 4 & 2 \\ 1 & 2 & 3 \end{bmatrix} R_1 \leftrightarrow R_2 \rightarrow \begin{bmatrix} 0 & 4 & 2 \\ 1 & 3 & -4 \\ 1 & 2 & 3 \end{bmatrix}$ Switch R_1 and R_2.

One very important fact about the elementary row operations is that they do not change the solutions of the corresponding system of equations. In other words, the new system of equations that we get by applying any one of these operations will have exactly the same solutions as the original system: It is easy to see that numbers that make the original equations true will also make the new equations true, because each of the elementary row operations corresponds to a valid operation on the original equations. That any solution of the new system is a solution of the old system follows from the fact that these row operations are *invertible:* The effects of a row operation can be reversed by applying another row operation, called its **inverse**. Here are some examples of this invertibility. (Try them out in the above Quick Examples.)

Operation	*Inverse Operation*
Replace R_2 by $3R_2$.	Replace R_2 by $\frac{1}{3}R_2$.
Replace R_1 by $4R_1 - 3R_2$.	Replace R_1 by $\frac{1}{4}R_1 + \frac{3}{4}R_2$.
Switch R_1 and R_2.	Switch R_1 and R_2.

Our objective, then, is to use row operations to change the system we are given into one with exactly the same set of solutions in which it is easy to see what the solutions are.

Solving Systems of Equations by Using Row Operations

Now we put rows to work for us in solving systems of equations. Let's start with a complicated-looking system of equations:

$$-\frac{2x}{3} + \frac{y}{2} = -3$$

$$\frac{x}{4} - y = \frac{11}{4}.$$

We begin by writing the matrix of the system:

$$\begin{bmatrix} -\frac{2}{3} & \frac{1}{2} & -3 \\ \frac{1}{4} & -1 & \frac{11}{4} \end{bmatrix}.$$

Now what do we do with this matrix?

Step 1 *Clear the fractions and/or decimals (if any) using operations of Type* **1.** To clear the fractions, we multiply the first row by 6 and the second row by 4. We record the operations by writing the symbolic form of an operation next to the row it will change, as follows.

$$\begin{bmatrix} -\frac{2}{3} & \frac{1}{2} & -3 \\ \frac{1}{4} & -1 & \frac{11}{4} \end{bmatrix} \begin{matrix} 6R_1 \\ 4R_2 \end{matrix}$$

By this we mean that we will replace the first row by $6R_1$ and the second by $4R_2$. Doing these operations gives

$$\begin{bmatrix} -4 & 3 & -18 \\ 1 & -4 & 11 \end{bmatrix}.$$

Step 2 *Designate the first nonzero entry in the first row as the* **pivot.** In this case we designate the entry -4 in the first row as the "pivot" by putting a box around it:

$$\begin{bmatrix} \boxed{-4} & 3 & -18 \\ 1 & -4 & 11 \end{bmatrix}. \quad \leftarrow \text{Pivot row}$$

$\uparrow$
Pivot column

Q: *What is a "pivot"?*

A: A **pivot** is an entry in a matrix that is used to "clear a column." (See Step 3.) In this procedure, we will always select the first nonzero entry of a row as our pivot. In Chapter 5, when we study the simplex method, we will select our pivots differently.

Step 3 *Use the pivot to clear its column using operations of Type* **2.** By **clearing a column**, we mean changing the matrix so that the pivot is the only nonzero number in its column. The procedure of clearing a column using a designated pivot is also called **pivoting**.

$$\begin{bmatrix} \boxed{-4} & 3 & -18 \\ 0 & \# & \# \end{bmatrix} \quad \leftarrow \text{Desired row 2 (the ``\#''s stand for as yet unknown numbers)}$$

$\uparrow$
Cleared pivot column

We want to replace R_2 by a row of the form $aR_2 \pm bR_1$ to get a zero in column 1. Moreover—and this will be important when we discuss the simplex method in Chapter 5—*we are going to choose positive values for both a and b.*[*] We need to choose a and b so that we get the desired cancellation. We can do this quite mechanically as follows:

a. Write the name of the row you need to change on the left and that of the pivot row on the right.

$$R_2 \qquad R_1$$

$\uparrow$ $\qquad$ $\uparrow$
Row to change   Pivot row

[*] Thus, the only place a negative sign may appear is between aR_2 and bR_1 as indicated in the formula $aR_2 \pm bR_1$.

b. Focus on the pivot column, $\begin{bmatrix} -4 \\ 1 \end{bmatrix}$. Multiply each row by the *absolute value* of the entry currently in the other. (We are not permitting a or b to be negative.)

$$4R_2 \qquad\qquad 1R_1$$
$$\uparrow \qquad\qquad\quad \uparrow$$

From Row 1   From Row 2

The effect is to make the two entries in the pivot column numerically the same. Sometimes, you can accomplish this by using smaller values of a and b.

c. If the entries in the pivot column have opposite signs, insert a plus ($+$). If they have the same sign, insert a minus ($-$). Here, we get the instruction

$$4R_2 + 1R_1,$$

or simply $4R_2 + R_1$.

d. Write the operation next to the row you want to change, and then replace that row using the operation:

$$\begin{bmatrix} \boxed{-4} & 3 & -18 \\ 1 & -4 & 11 \end{bmatrix}\, 4R_2 + 1R_1 \;\rightarrow\; \begin{bmatrix} -4 & 3 & -18 \\ 0 & -13 & 26 \end{bmatrix}.$$

We have cleared the pivot column and completed Step 3.

✳ We are deviating somewhat from the traditional procedure here. It is traditionally recommended first to divide the pivot row by the pivot, turning the pivot into a 1. This allows us to use $a = 1$, but usually results in fractions. The procedure we use here is easier for hand calculations and, we feel, mathematically more elegant, because it eliminates the need to introduce fractions. See the end of this section for an example done using the traditional procedure.

Note In general, the row operation you use should always have the following form[✳]:

$$a R_c \qquad \pm \qquad b R_p$$
$$\uparrow \qquad\qquad\quad \uparrow$$

Row to change   Pivot row

with a and b both positive. ∎

The next step is one that can be performed at any time.

Simplification Step (Optional) *If, at any stage of the process, all the numbers in a row are multiples of an integer, divide by that integer*—a Type 1 operation.

This is an optional but extremely helpful step: It makes the numbers smaller and easier to work with. In our case, the entries in R_2 are divisible by 13, so we divide that row by 13. (Alternatively, we could divide by -13. Try it.)

$$\begin{bmatrix} -4 & 3 & -18 \\ 0 & -13 & 26 \end{bmatrix}\, \tfrac{1}{13}R_2 \;\rightarrow\; \begin{bmatrix} -4 & 3 & -18 \\ 0 & -1 & 2 \end{bmatrix}$$

Step 4 *Select the first nonzero number in the second row as the pivot, and clear its column.* Here we have combined two steps in one: selecting the new pivot and clearing the column (pivoting). The pivot is shown below, as well as the desired result when the column has been cleared:

$$\begin{bmatrix} -4 & 3 & -18 \\ 0 & \boxed{-1} & 2 \end{bmatrix} \rightarrow \begin{bmatrix} \# & 0 & \# \\ 0 & -1 & 2 \end{bmatrix}. \quad \leftarrow \text{desired row}$$
$$\uparrow \qquad\qquad\qquad\qquad \uparrow$$

Pivot column   Cleared pivot column

We now wish to get a 0 in place of the 3 in the pivot column. Let's run once again through the mechanical steps to get the row operation that accomplishes this.

a. Write the name of the row you need to change on the left and that of the pivot row on the right:

$$R_1 \qquad\qquad\qquad R_2$$
$$\uparrow \qquad\qquad\qquad\quad \uparrow$$

Row to change   Pivot row

b. Focus on the pivot column, $\begin{bmatrix} 3 \\ -1 \end{bmatrix}$. Multiply each row by the absolute value of the entry currently in the other:

$$\underset{\underset{\text{From Row 2}}{\uparrow}}{1R_1} \qquad \underset{\underset{\text{From Row 1}}{\uparrow}}{3R_2}$$

c. If the entries in the pivot column have opposite signs, insert a plus ($+$). If they have the same sign, insert a minus ($-$). Here, we get the instruction

$$1R_1 + 3R_2.$$

d. Write the operation next to the row you want to change and then replace that row using the operation.

$$\begin{bmatrix} -4 & 3 & -18 \\ 0 & \boxed{-1} & 2 \end{bmatrix} \begin{matrix} R_1 + 3R_2 \\ \\ \end{matrix} \rightarrow \begin{bmatrix} -4 & 0 & -12 \\ 0 & -1 & 2 \end{bmatrix}$$

Now we are essentially done, except for one last step.

Final Step *Using operations of Type* **1,** *turn each pivot (the first nonzero entry in each row) into a* **1.** We can accomplish this by dividing the first row by -4 and multiplying the second row by -1:

$$\begin{bmatrix} -4 & 0 & -12 \\ 0 & -1 & 2 \end{bmatrix} \begin{matrix} -\frac{1}{4}R_1 \\ -R_2 \end{matrix} \rightarrow \begin{bmatrix} 1 & 0 & 3 \\ 0 & 1 & -2 \end{bmatrix}.$$

The matrix now has the following nice form:

$$\begin{bmatrix} \boxed{1} & 0 & \# \\ 0 & \boxed{1} & \# \end{bmatrix}.$$

(This is the form we will always obtain with two equations in two unknowns when there is a unique solution.) This form is nice because, when we translate back into equations, we get

$$1x + 0y = 3$$
$$0x + 1y = -2.$$

In other words,

$$x = 3 \text{ and } y = -2$$

and so we have found the solution, which we can also write as $(x, y) = (3, -2)$.

The procedure we've just demonstrated is called **Gauss-Jordan*** **reduction** or **row reduction.** It may seem too complicated a way to solve a system of two equations in two unknowns, and it is. However, for systems with more equations and more unknowns, it is very efficient.

In Example 1 below we use row reduction to solve a system of linear equations in *three* unknowns: $x, y,$ and z. Just as for a system in two unknowns, a **solution** of a system in any number of unknowns consists of values for each of the variables that, when substituted, satisfy all of the equations in the system. Again, just as for a system in two unknowns, any system of linear equations in any number of unknowns has either no solution, exactly one solution, or infinitely many solutions. There are no other possibilities.

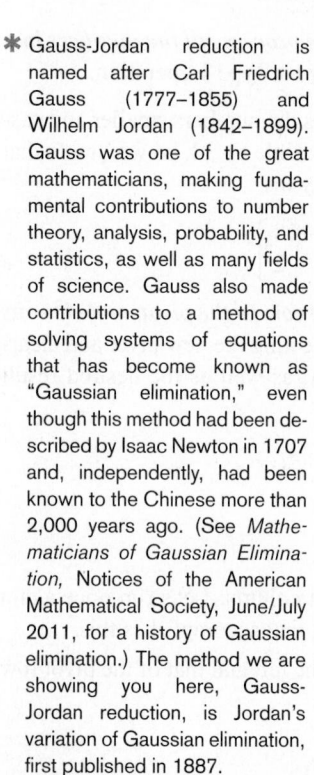

* Gauss-Jordan reduction is named after Carl Friedrich Gauss (1777–1855) and Wilhelm Jordan (1842–1899). Gauss was one of the great mathematicians, making fundamental contributions to number theory, analysis, probability, and statistics, as well as many fields of science. Gauss also made contributions to a method of solving systems of equations that has become known as "Gaussian elimination," even though this method had been described by Isaac Newton in 1707 and, independently, had been known to the Chinese more than 2,000 years ago. (See *Mathematicians of Gaussian Elimination,* Notices of the American Mathematical Society, June/July 2011, for a history of Gaussian elimination.) The method we are showing you here, Gauss-Jordan reduction, is Jordan's variation of Gaussian elimination, first published in 1887.

Solving a system in three unknowns graphically would require the graphing of planes (flat surfaces) in three dimensions. (The graph of a linear equation in three unknowns is a flat surface.) The use of row reduction makes three-dimensional graphing unnecessary.

EXAMPLE 1 Solving a System by Gauss-Jordan Reduction

Solve the system

$$x - y + 5z = -6$$
$$3x + 3y - z = 10$$
$$x + 3y + 2z = 5.$$

Solution The augmented matrix for this system is

$$\begin{bmatrix} 1 & -1 & 5 & -6 \\ 3 & 3 & -1 & 10 \\ 1 & 3 & 2 & 5 \end{bmatrix}.$$

Note that the columns correspond to x, y, z, and the right-hand side, respectively. We begin by selecting the pivot in the first row and clearing its column. Remember that clearing the column means that we turn *all* other numbers in the column into zeros. Thus, to clear the column of the first pivot, we need to change two rows, setting up the row operations in exactly the same way as above.

$$\begin{bmatrix} \boxed{1} & -1 & 5 & -6 \\ 3 & 3 & -1 & 10 \\ 1 & 3 & 2 & 5 \end{bmatrix} \begin{matrix} \\ R_2 - 3R_1 \rightarrow \\ R_3 - R_1 \end{matrix} \begin{bmatrix} 1 & -1 & 5 & -6 \\ 0 & 6 & -16 & 28 \\ 0 & 4 & -3 & 11 \end{bmatrix}$$

Notice that both row operations have the required form

$$a R_c \pm b R_1$$

    ↑   ↑

Row to change Pivot row

with a and b both positive.

Now we use the optional simplification step to simplify R_2:

$$\begin{bmatrix} 1 & -1 & 5 & -6 \\ 0 & 6 & -16 & 28 \\ 0 & 4 & -3 & 11 \end{bmatrix} \frac{1}{2}R_2 \rightarrow \begin{bmatrix} 1 & -1 & 5 & -6 \\ 0 & 3 & -8 & 14 \\ 0 & 4 & -3 & 11 \end{bmatrix}.$$

Next, we select the pivot in the second row and clear its column:

$$\begin{bmatrix} 1 & -1 & 5 & -6 \\ 0 & \boxed{3} & -8 & 14 \\ 0 & 4 & -3 & 11 \end{bmatrix} \begin{matrix} 3R_1 + R_2 \\ \rightarrow \\ 3R_3 - 4R_2 \end{matrix} \begin{bmatrix} 3 & 0 & 7 & -4 \\ 0 & 3 & -8 & 14 \\ 0 & 0 & 23 & -23 \end{bmatrix}.$$

            R_1 and R_3 are to be changed.

            R_2 is the pivot row.

We simplify R_3.

$$\begin{bmatrix} 3 & 0 & 7 & -4 \\ 0 & 3 & -8 & 14 \\ 0 & 0 & 23 & -23 \end{bmatrix} \frac{1}{23}R_3 \rightarrow \begin{bmatrix} 3 & 0 & 7 & -4 \\ 0 & 3 & -8 & 14 \\ 0 & 0 & 1 & -1 \end{bmatrix}$$

using Technology

See the Technology Guides at the end of the chapter to see how to use a TI-83/84 Plus or a spreadsheet to solve this system of equations.

 Website
www.WanerMath.com

  Everything

  → Math Tools for Chapter 3

  → Pivot and Gauss-Jordan Tool

Enter the augmented matrix in columns $x1$, $x2$, $x3$, ... At each step, type in the row operations exactly as written above next to the rows to which they apply. For example, for the first step type:
R2-3R1 next to Row 2, and
R3-R1 next to Row 3.
Press "Do Row Ops" once, and then "Clear Row Ops" to prepare for the next step.
For the last step, type
(1/3)R1 next to Row 1, and
(1/3)R2 next to Row 2.
Press "Do Row Ops" once.
The utility also does other things, such as automatic pivoting and complete reduction in one step. Use these features to check your work.

Now we select the pivot in the third row and clear its column:

$$\begin{bmatrix} 3 & 0 & 7 & -4 \\ 0 & 3 & -8 & 14 \\ 0 & 0 & \boxed{1} & -1 \end{bmatrix} \begin{matrix} R_1 - 7R_3 \\ R_2 + 8R_3 \\ \ \end{matrix} \rightarrow \begin{bmatrix} 3 & 0 & 0 & 3 \\ 0 & 3 & 0 & 6 \\ 0 & 0 & 1 & -1 \end{bmatrix}.$$

R_1 and R_2 are to be changed.
R_3 is the pivot row.

Finally, we turn all the pivots into 1s:

$$\begin{bmatrix} 3 & 0 & 0 & 3 \\ 0 & 3 & 0 & 6 \\ 0 & 0 & 1 & -1 \end{bmatrix} \begin{matrix} \frac{1}{3}R_1 \\ \frac{1}{3}R_2 \\ \ \end{matrix} \rightarrow \begin{bmatrix} 1 & 0 & 0 & 1 \\ 0 & 1 & 0 & 2 \\ 0 & 0 & 1 & -1 \end{bmatrix}.$$

The matrix is now reduced to a simple form, so we translate back into equations to obtain the solution:

$$x = 1, \ y = 2, \ z = -1, \text{ or } (x, y, z) = (1, 2, -1).$$

Notice the form of the very last matrix in the example:

$$\begin{bmatrix} 1 & 0 & 0 & \# \\ 0 & 1 & 0 & \# \\ 0 & 0 & 1 & \# \end{bmatrix}.$$

The 1s are on the **(main) diagonal** of the matrix; the goal in Gauss-Jordan reduction is to reduce our matrix to this form. If we can do so, then we can easily read off the solution, as we saw in Example 1. However, as we will see in several examples in this section, it is not always possible to achieve this ideal state. After Example 6, we will give a form that is always possible to achieve.

EXAMPLE 2 Solving a System by Gauss-Jordan Reduction

Solve the system:

$$\begin{aligned} 2x + \ y + 3z &= 1 \\ 4x + 2y + 4z &= 4 \\ x + 2y + \ z &= 4. \end{aligned}$$

Solution

$$\begin{bmatrix} \boxed{2} & 1 & 3 & 1 \\ 4 & 2 & 4 & 4 \\ 1 & 2 & 1 & 4 \end{bmatrix} \begin{matrix} \ \\ R_2 - 2R_1 \\ 2R_3 - R_1 \end{matrix} \rightarrow \begin{bmatrix} 2 & 1 & 3 & 1 \\ 0 & 0 & -2 & 2 \\ 0 & 3 & -1 & 7 \end{bmatrix}$$

Now we have a slight problem: The number in the position where we would like to have a pivot—the second column of the second row—is a zero and thus cannot be a pivot. There are two ways out of this problem. One is to move on to the third column and pivot on the -2. Another is to switch the order of the second and third rows so that we can use the 3 as a pivot. We will do the latter.

$$\begin{bmatrix} 2 & 1 & 3 & 1 \\ 0 & 0 & -2 & 2 \\ 0 & 3 & -1 & 7 \end{bmatrix} \begin{matrix} \\ \\ R_2 \leftrightarrow R_3 \end{matrix} \rightarrow \begin{bmatrix} 2 & 1 & 3 & 1 \\ 0 & \boxed{3} & -1 & 7 \\ 0 & 0 & -2 & 2 \end{bmatrix} \begin{matrix} 3R_1 - R_2 \\ \\ \end{matrix}$$

$$\rightarrow \begin{bmatrix} 6 & 0 & 10 & -4 \\ 0 & 3 & -1 & 7 \\ 0 & 0 & -2 & 2 \end{bmatrix} \begin{matrix} \\ \\ -\frac{1}{2}R_3 \end{matrix} \rightarrow \begin{bmatrix} 6 & 0 & 10 & -4 \\ 0 & 3 & -1 & 7 \\ 0 & 0 & \boxed{1} & -1 \end{bmatrix} \begin{matrix} R_1 - 10R_3 \\ R_2 + R_3 \\ \end{matrix}$$

$$\rightarrow \begin{bmatrix} 6 & 0 & 0 & 6 \\ 0 & 3 & 0 & 6 \\ 0 & 0 & 1 & -1 \end{bmatrix} \begin{matrix} \frac{1}{6}R_1 \\ \frac{1}{3}R_2 \\ \end{matrix} \rightarrow \begin{bmatrix} 1 & 0 & 0 & 1 \\ 0 & 1 & 0 & 2 \\ 0 & 0 & 1 & -1 \end{bmatrix}$$

Thus, the solution is $(x, y, z) = (1, 2, -1)$, as you can check in the original system.

EXAMPLE 3 Inconsistent System

Solve the system:

$$\begin{aligned} x + y + z &= 1 \\ 2x - y + z &= 0 \\ 4x + y + 3z &= 3. \end{aligned}$$

Solution

$$\begin{bmatrix} \boxed{1} & 1 & 1 & 1 \\ 2 & -1 & 1 & 0 \\ 4 & 1 & 3 & 3 \end{bmatrix} \begin{matrix} \\ R_2 - 2R_1 \\ R_3 - 4R_1 \end{matrix} \rightarrow \begin{bmatrix} 1 & 1 & 1 & 1 \\ 0 & \boxed{-3} & -1 & -2 \\ 0 & -3 & -1 & -1 \end{bmatrix} \begin{matrix} 3R_1 + R_2 \\ \\ R_3 - R_2 \end{matrix}$$

$$\rightarrow \begin{bmatrix} 3 & 0 & 2 & 1 \\ 0 & -3 & -1 & -2 \\ 0 & 0 & 0 & 1 \end{bmatrix}$$

Stop. That last row translates into $0 = 1$, which is nonsense, and so, as in Example 3 in Section 3.1, we can say that this system has no solution. We also say, as we did for systems with only two unknowns, that a system with no solution is **inconsistent**. A system with at least one solution is **consistent**.

⇒ **Before we go on...**

Q: *How, exactly, does the nonsensical equation $0 = 1$ tell us that there is no solution of the system in Example 3?*

A: Here is an argument similar to that in Example 3 in Section 3.1: If there *were* three numbers x, y, and z satisfying the original system of equations, then manipulating the equations according to the instructions in the row operations above leads us to conclude that $0 = 1$. Because 0 is *not* equal to 1, there can be no such numbers x, y, and z. ∎

EXAMPLE 4 Infinitely Many Solutions

Solve the system:

$$\begin{aligned} x + y + z &= 1 \\ \frac{1}{4}x - \frac{1}{2}y + \frac{3}{4}z &= 0 \\ x + 7y - 3z &= 3. \end{aligned}$$

Solution

$$
\begin{bmatrix} 1 & 1 & 1 & 1 \\ \frac{1}{4} & -\frac{1}{2} & \frac{3}{4} & 0 \\ 1 & 7 & -3 & 3 \end{bmatrix}
\begin{matrix} \\ 4R_2 \rightarrow \\ \\ \end{matrix}
\begin{bmatrix} \boxed{1} & 1 & 1 & 1 \\ 1 & -2 & 3 & 0 \\ 1 & 7 & -3 & 3 \end{bmatrix}
\begin{matrix} \\ R_2 - R_1 \\ R_3 - R_1 \end{matrix}
$$

$$
\rightarrow
\begin{bmatrix} 1 & 1 & 1 & 1 \\ 0 & -3 & 2 & -1 \\ 0 & 6 & -4 & 2 \end{bmatrix}
\begin{matrix} \\ \\ \frac{1}{2}R_3 \end{matrix}
\rightarrow
\begin{bmatrix} 1 & 1 & 1 & 1 \\ 0 & \boxed{-3} & 2 & -1 \\ 0 & 3 & -2 & 1 \end{bmatrix}
\begin{matrix} 3R_1 + R_2 \\ \\ R_3 + R_2 \end{matrix}
$$

$$
\rightarrow
\begin{bmatrix} 3 & 0 & 5 & 2 \\ 0 & -3 & 2 & -1 \\ 0 & 0 & 0 & 0 \end{bmatrix}
$$

There are no nonzero entries in the third row, so there can be no pivot in the third row. We skip to the final step and turn the pivots we did find into 1s.

$$
\begin{bmatrix} 3 & 0 & 5 & 2 \\ 0 & -3 & 2 & -1 \\ 0 & 0 & 0 & 0 \end{bmatrix}
\begin{matrix} \frac{1}{3}R_1 \\ -\frac{1}{3}R_2 \\ \\ \end{matrix}
\rightarrow
\begin{bmatrix} 1 & 0 & \frac{5}{3} & \frac{2}{3} \\ 0 & 1 & -\frac{2}{3} & \frac{1}{3} \\ 0 & 0 & 0 & 0 \end{bmatrix}
$$

Now we translate back into equations and obtain:

$$
\begin{aligned}
x \quad + \tfrac{5}{3}z &= \tfrac{2}{3} \\
y - \tfrac{2}{3}z &= \tfrac{1}{3} \\
0 &= 0.
\end{aligned}
$$

But how does this help us find a solution? The last equation doesn't tell us anything useful, so we ignore it. The thing to notice about the other equations is that we can easily solve the first equation for x and the second for y, obtaining

$$
\begin{aligned}
x &= \tfrac{2}{3} - \tfrac{5}{3}z \\
y &= \tfrac{1}{3} + \tfrac{2}{3}z.
\end{aligned}
$$

This is the solution! We can choose z to be any number and get corresponding values for x and y from the formulas above. This gives us infinitely many different solutions. Thus, the general solution (see Example 4 in Section 3.1) is

$$
\begin{aligned}
x &= \tfrac{2}{3} - \tfrac{5}{3}z \\
y &= \tfrac{1}{3} + \tfrac{2}{3}z \qquad\qquad \text{General solution} \\
z \;&\text{is arbitrary.}
\end{aligned}
$$

We can also write the general solution as

$$
\left(\tfrac{2}{3} - \tfrac{5}{3}z, \; \tfrac{1}{3} + \tfrac{2}{3}z, \; z \right) \; z \text{ arbitrary.} \qquad\qquad \text{General solution}
$$

This general solution has z as the parameter. Specific choices of values for the parameter z give particular solutions. For example, the choice $z = 6$ gives the particular solution

$$
\begin{aligned}
x &= \tfrac{2}{3} - \tfrac{5}{3}(6) = -\tfrac{28}{3} \\
y &= \tfrac{1}{3} + \tfrac{2}{3}(6) = \tfrac{13}{3} \qquad\qquad \text{Particular solution} \\
z &= 6,
\end{aligned}
$$

while the choice $z = 0$ gives the particular solution $(x, \; y, \; z) = \left(\tfrac{2}{3}, \tfrac{1}{3}, 0 \right)$.

Note that, unlike the system given in the preceding example, the system given in this example does have solutions, and is thus *consistent*.

➡ **Before we go on...** Why were there infinitely many solutions to Example 4? The reason is that the third equation was really a combination of the first and second equations to begin with, so we effectively had only two equations in three unknowns.[*] Choosing a specific value for z (say, $z = 6$) has the effect of supplying the "missing" equation. ∎

[*] In fact, you can check that the third equation, E_3, is equal to $3E_1 - 8E_2$. Thus, the third equation could have been left out because it conveys no more information than the first two. The process of row reduction always eliminates such a redundancy by creating a row of zeros.

Q: *How do we know when there are infinitely many solutions?*

A: When there are solutions (we have a consistent system, unlike the one in Example 3), and when the matrix we arrive at by row reduction has fewer pivots than there are unknowns. In Example 4 we had three unknowns but only two pivots.

Q: *How do we know which variables to use as parameters in a parameterized solution?*

A: The variables to use as parameters are those in the columns without pivots. In Example 4 there were pivots in the x and y columns, but no pivot in the z column, and it was z that we used as a parameter.

EXAMPLE 5 Four Unknowns

Solve the system:

$$
\begin{aligned}
x + 3y + 2z - \quad w &= 6 \\
2x + 6y + 6z + \ 3w &= 16 \\
x + 3y - 2z - 11w &= -2 \\
2x + 6y + 8z + \ 8w &= 20.
\end{aligned}
$$

Solution

$$
\begin{bmatrix}
\boxed{1} & 3 & 2 & -1 & 6 \\
2 & 6 & 6 & 3 & 16 \\
1 & 3 & -2 & -11 & -2 \\
2 & 6 & 8 & 8 & 20
\end{bmatrix}
\begin{matrix} \\ R_2 - 2R_1 \\ R_3 - R_1 \\ R_4 - 2R_1 \end{matrix}
\rightarrow
\begin{bmatrix}
1 & 3 & 2 & -1 & 6 \\
0 & 0 & 2 & 5 & 4 \\
0 & 0 & -4 & -10 & -8 \\
0 & 0 & 4 & 10 & 8
\end{bmatrix}
$$

There is no pivot available in the second column, so we move on to the third column.

$$
\begin{bmatrix}
1 & 3 & 2 & -1 & 6 \\
0 & 0 & \boxed{2} & 5 & 4 \\
0 & 0 & -4 & -10 & -8 \\
0 & 0 & 4 & 10 & 8
\end{bmatrix}
\begin{matrix} R_1 - R_2 \\ \\ R_3 + 2R_2 \\ R_4 - 2R_2 \end{matrix}
\rightarrow
\begin{bmatrix}
1 & 3 & 0 & -6 & 2 \\
0 & 0 & 2 & 5 & 4 \\
0 & 0 & 0 & 0 & 0 \\
0 & 0 & 0 & 0 & 0
\end{bmatrix}
\begin{matrix} \\ \frac{1}{2}R_2 \\ \\ \end{matrix}
$$

$$
\rightarrow
\begin{bmatrix}
1 & 3 & 0 & -6 & 2 \\
0 & 0 & 1 & \frac{5}{2} & 2 \\
0 & 0 & 0 & 0 & 0 \\
0 & 0 & 0 & 0 & 0
\end{bmatrix}
$$

Translating back into equations, we get:

$$
\begin{aligned}
x + 3y \ - 6w &= 2 \\
z + \tfrac{5}{2}w &= 2.
\end{aligned}
$$

(We have not written down the equations corresponding to the last two rows, each of which is $0 = 0$.) There are no pivots in the y or w columns, so we use these two variables as parameters. We bring them over to the right-hand sides of the equations above and write the general solution as

$$x = 2 - 3y + 6w$$

y is arbitrary

$$z = 2 - \tfrac{5}{2}w$$

w is arbitrary

or

$$(x, y, z, w) = (2 - 3y + 6w, y, 2 - 5w/2, w)\ y, w\ \text{arbitrary.}$$

➡ **Before we go on...** In Examples 4 and 5, you might have noticed an interesting phenomenon: If at any time in the process, two rows are equal or one is a multiple of the other, then one of those rows (eventually) becomes all zero. ∎

Up to this point, we have always been given as many equations as there are unknowns. However, we shall see in the next section that some applications lead to systems where the number of equations is not the same as the number of unknowns. As the following example illustrates, such systems can be handled the same way as any other.

EXAMPLE 6 Number of Equations ≠ Number of Unknowns

Solve the system:

$$\begin{aligned} x + y &= 1 \\ 13x - 26y &= -11 \\ 26x - 13y &= 2. \end{aligned}$$

Solution We proceed exactly as before and ignore the fact that there is one more equation than unknown.

$$\begin{bmatrix} \boxed{1} & 1 & 1 \\ 13 & -26 & -11 \\ 26 & -13 & 2 \end{bmatrix} \begin{matrix} \\ R_2 - 13R_1 \\ R_3 - 26R_1 \end{matrix} \rightarrow \begin{bmatrix} 1 & 1 & 1 \\ 0 & -39 & -24 \\ 0 & -39 & -24 \end{bmatrix} \begin{matrix} \\ \frac{1}{3}R_2 \\ \frac{1}{3}R_3 \end{matrix}$$

$$\rightarrow \begin{bmatrix} 1 & 1 & 1 \\ 0 & \boxed{-13} & -8 \\ 0 & -13 & -8 \end{bmatrix} \begin{matrix} 13R_1 + R_2 \\ \\ R_3 - R_2 \end{matrix} \rightarrow \begin{bmatrix} 13 & 0 & 5 \\ 0 & -13 & -8 \\ 0 & 0 & 0 \end{bmatrix} \begin{matrix} \frac{1}{13}R_1 \\ -\frac{1}{13}R_2 \\ \\ \end{matrix}$$

$$\rightarrow \begin{bmatrix} 1 & 0 & \frac{5}{13} \\ 0 & 1 & \frac{8}{13} \\ 0 & 0 & 0 \end{bmatrix}$$

Thus, the solution is $(x, y) = \left(\frac{5}{13}, \frac{8}{13}\right)$.

If, instead of a row of zeros, we had obtained, say, [0 0 6] in the last row, we would immediately have concluded that the system was inconsistent.

The fact that we wound up with a row of zeros indicates that one of the equations was actually a combination of the other two; you can check that the third equation can be obtained by multiplying the first equation by 13 and adding the result to the second. Because the third equation therefore tells us nothing that we don't already know from the first two, we call the system of equations **redundant**, or **dependent.** (Compare Example 4 in Section 3.1.)

➡ **Before we go on...** Example 5 above is another example of a redundant system; we could have started with the following smaller system of two equations in four unknowns

$$x + 3y + 2z - w = 6$$
$$2x + 6y + 6z + 3w = 16$$

and obtained the same general solution as we did with the larger system. Verify this by solving the smaller system. ■

The preceding examples illustrated that we cannot always reduce a matrix to the form shown before Example 2, with pivots going all the way down the diagonal. What we *can* always do is reduce a matrix to the following form:

Reduced Row Echelon Form

A matrix is said to be in **reduced row echelon form** or to be **row-reduced** if it satisfies the following properties.

P1. The first nonzero entry in each row (called the **leading entry** of that row) is a 1.

P2. The columns of the leading entries are **clear** (i.e., they contain zeros in all positions other than that of the leading entry).

P3. The leading entry in each row is to the right of the leading entry in the row above, and any rows of zeros are at the bottom.

Quick Examples

$$\begin{bmatrix} 1 & 0 & 0 & 2 \\ 0 & 1 & 0 & 4 \\ 0 & 0 & 1 & -3 \end{bmatrix}, \begin{bmatrix} 0 & 1 & -3 \\ 0 & 0 & 0 \end{bmatrix}, \text{ and } \begin{bmatrix} 1 & 0 & 0 & -2 \\ 0 & 0 & 1 & 4 \\ 0 & 0 & 0 & 0 \end{bmatrix} \text{ are row-reduced.}$$

$$\begin{bmatrix} 1 & 1 & 0 & 2 \\ 0 & 1 & 0 & 4 \\ 0 & 0 & 1 & -3 \end{bmatrix}, \begin{bmatrix} 0 & 1 & -3 \\ 0 & 0 & 1 \end{bmatrix}, \text{ and } \begin{bmatrix} 0 & 0 & 1 & 4 \\ 1 & 0 & 0 & -2 \\ 0 & 0 & 0 & 0 \end{bmatrix} \text{ are not row-reduced.}$$

You should check in the examples we did that the final matrices were all in reduced row echelon form.

It is an interesting and useful fact, though not easy to prove, that any two people who start with the same matrix and row-reduce it will reach exactly the same row-reduced matrix, even if they use different row operations.

The Traditional Gauss-Jordan Method (Optional)

In the version of the Gauss-Jordan method we have presented, we eliminated fractions and decimals in the first step and then worked with integer matrices, partly to make hand computation easier and partly for mathematical elegance. However, complicated fractions and decimals present no difficulty when we use technology. The following example illustrates the more traditional approach to Gauss-Jordan reduction used in many of the computer programs that solve the huge systems of equations that arise in practice.*

✱ Actually, for reasons of efficiency and accuracy, the methods used in commercial programs are closer to the method presented above. To learn more, consult a text on numerical methods.

EXAMPLE 7 Solving a System with the Traditional Gauss-Jordan Method

Solve the following system using the traditional Gauss-Jordan method:

$$2x + y + 3z = 5$$
$$3x + 2y + 4z = 7$$
$$2x + y + 5z = 10.$$

Solution We make two changes in our method. First, there is no need to get rid of decimals (because computers and calculators can handle decimals as easily as they can integers). Second, after selecting a pivot, *divide the pivot row by the pivot value, turning the pivot into a* 1. It is easier to determine the row operations that will clear the pivot column if the pivot is a 1.

If we use technology to solve this system of equations, the sequence of matrices might look like this:

using Technology

Website
www.WanerMath.com
Follow
→ Everything
→ Math Tools for Chapter 3
to find the following resources:
- An online Web page that pivots and does row operations automatically
- A TI-83/84 Plus program that pivots and does other row operations
- An Excel worksheet that pivots and does row operations automatically

$$\begin{bmatrix} \boxed{2} & 1 & 3 & 5 \\ 3 & 2 & 4 & 7 \\ 2 & 1 & 5 & 10 \end{bmatrix} \begin{matrix} \frac{1}{2}R_1 \\ \\ \\ \end{matrix} \rightarrow \begin{bmatrix} \boxed{1} & 0.5 & 1.5 & 2.5 \\ 3 & 2 & 4 & 7 \\ 2 & 1 & 5 & 10 \end{bmatrix} \begin{matrix} \\ R_2 - 3R_1 \\ R_3 - 2R_1 \end{matrix}$$

$$\rightarrow \begin{bmatrix} 1 & 0.5 & 1.5 & 2.5 \\ 0 & \boxed{0.5} & -0.5 & -0.5 \\ 0 & 0 & 2 & 5 \end{bmatrix} \begin{matrix} \\ 2R_2 \\ \\ \end{matrix} \rightarrow \begin{bmatrix} 1 & 0.5 & 1.5 & 2.5 \\ 0 & \boxed{1} & -1 & -1 \\ 0 & 0 & 2 & 5 \end{bmatrix} \begin{matrix} R_1 - 0.5R_2 \\ \\ \\ \end{matrix}$$

$$\rightarrow \begin{bmatrix} 1 & 0 & 2 & 3 \\ 0 & 1 & -1 & -1 \\ 0 & 0 & \boxed{2} & 5 \end{bmatrix} \begin{matrix} \\ \\ \frac{1}{2}R_3 \end{matrix} \rightarrow \begin{bmatrix} 1 & 0 & 2 & 3 \\ 0 & 1 & -1 & -1 \\ 0 & 0 & \boxed{1} & 2.5 \end{bmatrix} \begin{matrix} R_1 - 2R_3 \\ R_2 + R_3 \\ \\ \end{matrix}$$

$$\rightarrow \begin{bmatrix} 1 & 0 & 0 & -2 \\ 0 & 1 & 0 & 1.5 \\ 0 & 0 & 1 & 2.5 \end{bmatrix}.$$

The solution is $(x, y, z) = (-2, 1.5, 2.5)$.

Q: *The solution to Example 7 looked quite easy. Why didn't we use the traditional method from the start like the other textbooks?*

A: It looked easy because we deliberately chose an example that leads to simple decimals. In all but the most contrived examples, the decimals or fractions involved get very complicated very quickly.

FAQs

Getting Unstuck, Going Round in Circles, and Knowing When to Stop

Q: *Help! I have been doing row operations on this matrix for half an hour. I have filled two pages, and I am getting nowhere. What do I do?*

A: Here is a way of keeping track of where you are *at any stage of the process* and also deciding what to do next.

Starting at the top row of your current matrix:

1. Scan along the row until you get to the leading entry: the first nonzero entry. If there is none–that is, the row is all zero–go to the next row.
2. Having located the leading entry, scan up and down its *column*. If its column is not clear (that is, it contains other nonzero entries), use your leading entry as a pivot to clear its column as in the examples in this section.
3. Now go to the next row and start again at Step 1.

When you have scanned all the rows and find that all the columns of the leading entries are clear, it means you are done (except possibly for reordering the rows so that the leading entries go from left to right as you read down the matrix, and zero rows are at the bottom).

Q: *No good. I have been following these instructions, but every time I try to clear a column, I unclear a column I had already cleared. What is going on?*

A: Are you using *leading entries* as pivots? Also, are you *using the pivot* to clear its column? That is, are your row operations all of the following form?

$$aR_c \pm bR_p$$

      ↑   ↑

Row to change Pivot row

The instruction next to the row you are changing should involve only that row and the pivot row, even though you might be tempted to use some other row instead.

Q: *Must I continue until I get a matrix that has 1s down the leading diagonal and 0s above and below?*

A: Not necessarily. You are completely done when your matrix is row-reduced: Each leading entry is a 1, the column of each leading entry is clear, and the leading entries go from left to right. You are done *pivoting* when the column of each leading entry is clear. After that, all that remains is to turn each pivot into a 1 (the "Final Step") and, if necessary, rearrange the rows.

3.2 EXERCISES

▼ more advanced   ◆ challenging
T indicates exercises that should be solved using technology

In Exercises 1–42, use Gauss-Jordan row reduction to solve the given systems of equation. We suggest doing some by hand, and others using technology. HINT [See Examples 1–6.]

1. $x + y = 4$
    $x - y = 2$

2. $\quad 2x + y = 2$
    $-2x + y = 2$

3. $3x - 2y = 6$
    $2x - 3y = -6$

4. $2x + 3y = 5$
    $3x + 2y = 5$

5. $2x + 3y = 1$
    $-x - \dfrac{3y}{2} = -\dfrac{1}{2}$

6. $2x - 3y = 1$
    $6x - 9y = 3$

7. $2x + 3y = 2$

$-x - \dfrac{3y}{2} = -\dfrac{1}{2}$

8. $2x - 3y = 2$

$6x - 9y = 3$

9. $x + y = 1$

$3x - y = 0$

$x - 3y = -2$

10. $x + y = 1$

$3x - 2y = -1$

$5x - y = \dfrac{1}{5}$

11. $x + y = 0$

$3x - y = 1$

$x - y = -1$

12. $x + 2y = 1$

$3x - 2y = -2$

$5x - y = \dfrac{1}{5}$

13. $0.5x + 0.1y = 1.7$

$0.1x - 0.1y = 0.3$

$x + y = \dfrac{11}{3}$

14. $-0.3x + 0.5y = 0.1$

$x - y = 4$

$\dfrac{x}{17} + \dfrac{y}{17} = 1$

15. $-x + 2y - z = 0$

$-x - y + 2z = 0$

$2x \quad - z = 4$

16. $x + 2y \quad = 4$

$y - z = 0$

$x + 3y - 2z = 5$

17. $x + y + 6z = -1$

$\dfrac{1}{3}x - \dfrac{1}{3}y + \dfrac{2}{3}z = 1$

$\dfrac{1}{2}x \quad + z = 0$

18. $x - \dfrac{1}{2}y \quad = 0$

$\dfrac{1}{3}x + \dfrac{1}{3}y + \dfrac{1}{3}z = 2$

$\dfrac{1}{2}x \quad - \dfrac{1}{2}z = -1$

19. $-\dfrac{1}{2}x + y - \dfrac{1}{2}z = 0$

$-\dfrac{1}{2}x - \dfrac{1}{2}y + z = 0$

$x - \dfrac{1}{2}y - \dfrac{1}{2}z = 0$

20. $x - \dfrac{1}{2}y \quad = 0$

$\dfrac{1}{2}x - \dfrac{1}{2}z = -1$

$3x - y - z = -2$

21. $x + y + 2z = -1$

$2x + 2y + 2z = 2$

$\dfrac{3}{5}x + \dfrac{3}{5}y + \dfrac{3}{5}z = \dfrac{2}{5}$

22. $x + y - z = -2$

$x - y - 7z = 0$

$\dfrac{2}{7}x \quad - \dfrac{8}{7}z = 14$

23. $-0.5x + 0.5y + 0.5z = 1.5$

$4.2x + 2.1y + 2.1z = 0$

$0.2x \quad + 0.2z = 0$

24. $0.25x - 0.5y \quad = 0$

$0.2x + 0.2y - 0.2z = -0.6$

$0.5x - 1.5y + \quad z = 0.5$

25. $2x - y + z = 4$

$3x - y + z = 5$

26. $3x - y - z = 0$

$x + y + z = 4$

27. $0.75x - 0.75y - z = 4$

$x - y + 4z = 0$

28. $2x - y + z = 4$

$-x + 0.5y - 0.5z = 1.5$

29. ▼ $3x + y - z = 12$

30. ▼ $x + y - 3z = 21$
(Yes: One equation in three unknowns!)

31. ▼ $\quad x + \quad y + 2z = -1$

$2x + \quad 2y + 2z = 2$

$0.75x + 0.75y + \quad z = 0.25$

$-x \quad\quad - 2z = 21$

32. ▼ $\quad x + \quad y - \quad z = -2$

$x - \quad y - \quad 7z = 0$

$0.75x - 0.5y + 0.25z = 14$

$x + \quad y + \quad z = 4$

33. ▼ $x + y + 5z \quad = 1$

$y + 2z + w = 1$

$x + 3y + 7z + 2w = 2$

$x + y + 5z + w = 1$

34. ▼ $x + y \quad + 4w = 1$

$2x - 2y - 3z + 2w = -1$

$4y + 6z + w = 4$

$2x + 4y + 9z \quad = 6$

35. ▼ $x + y + 5z \quad = 1$

$y + 2z + w = 1$

$x + y + 5z + w = 1$

$x + 2y + 7z + 2w = 2$

36. ▼ $x + y + \quad 4w = 1$

$2x - 2y - 3z + 2w = -1$

$4y + 6z + w = 4$

$3x + 3y + 3z + 7w = 4$

37. ▼ $x - 2y + z - 4w = 1$

$x + 3y + 7z + 2w = 2$

$2x + y + 8z - 2w = 3$

38. ▼ $x - 3y - 2z - w = 1$

$x + 3y + z + 2w = 2$

$2x \quad - z + w = 3$

39. ▼ $x + y + z + u + v = 15$

$y - z + u - v = -2$

$z + u + v = 12$

$u - v = -1$

$v = 5$

40. ▼ $x - y + z - u + v = 1$

$y + z + u + v = 2$

$z - u + v = 1$

$u + v = 1$

$v = 1$

41. ▼ $x - y + z - u + v = 0$

$y - z + u - v = -2$

$x \quad - 2v = -2$

$2x - y + z - u - 3v = -2$

$4x - y + z - u - 7v = -6$

42. ▼ $x + y + z + u + v = 15$

$y + z + u + v = 3$

$x + 2y + 2z + 2u + 2v = 18$

$x - y - z - u - v = 9$

$x - 2y - 2z - 2u - 2v = 6$

T *In Exercises 43–46, use technology to solve the systems of equations. Express all solutions as fractions.*

43.
$$x + 2y - z + w = 30$$
$$2x \quad\ - z + 2w = 30$$
$$x + 3y + 3z - 4w = 2$$
$$2x - 9y \quad\ + w = 4$$

44.
$$4x - 2y + z + w = 20$$
$$3y + 3z - 4w = 2$$
$$2x + 4y \quad\ - w = 4$$
$$x + 3y + 3z \quad\ = 2$$

45.
$$x + 2y + 3z + 4w + 5t = 6$$
$$2x + 3y + 4z + 5w + t = 5$$
$$3x + 4y + 5z + w + 2t = 4$$
$$4x + 5y + z + 2w + 3t = 3$$
$$5x + y + 2z + 3w + 4t = 2$$

46.
$$x - 2y + 3z - 4w \quad\ = 0$$
$$-2x + 3y - 4z + \quad\ t = 0$$
$$3x - 4y \quad\ + w - 2t = 0$$
$$-4x \quad\ + z - 2w + 3t = 0$$
$$y - 2z + 3w - 4t = 1$$

T *In Exercises 47–50, use technology to solve the systems of equations. Express all solutions as decimals, rounded to one decimal place.*

47.
$$1.6x + 2.4y - 3.2z = 4.4$$
$$5.1x - 6.3y + 0.6z = -3.2$$
$$4.2x + 3.5y + 4.9z = 10.1$$

48.
$$2.1x + 0.7y - 1.4z = -2.3$$
$$3.5x - 4.2y - 4.9z = 3.3$$
$$1.1x + 2.2y - 3.3z = -10.2$$

49.
$$-0.2x + 0.3y + 0.4z - t = 4.5$$
$$2.2x + 1.1y - 4.7z + 2t = 8.3$$
$$9.2y \quad\ - 1.3t = 0$$
$$3.4x \quad\ + 0.5z - 3.4t = 0.1$$

50.
$$1.2x - 0.3y + 0.4z - 2t = 4.5$$
$$1.9x \quad\ - 0.5z - 3.4t = 0.2$$
$$12.1y \quad\ - 1.3t = 0$$
$$3x + 2y - 1.1z \quad\ = 9$$

COMMUNICATION AND REASONING EXERCISES

51. What is meant by a pivot? What does pivoting do?

52. Give instructions to check whether or not a matrix is row-reduced.

53. You are row-reducing a matrix and have chosen a -6 as a pivot in Row 4. Directly above the pivot, in Row 1, is a 15. What row operation can you use to clear the 15?

54. You are row-reducing a matrix and have chosen a -4 as a pivot in Row 2. Directly below the pivot, in Row 4, is a -6. What row operation can you use to clear the -6?

55. In the matrix of a system of linear equations, suppose that two of the rows are equal. What can you say about the row-reduced form of the matrix?

56. In the matrix of a system of linear equations, suppose that one of the rows is a multiple of another. What can you say about the row-reduced form of the matrix?

57. ▼ Your friend Frans tells you that the system of linear equations you are solving cannot have a unique solution because the reduced matrix has a row of zeros. Comment on his claim.

58. ▼ Your other friend Hans tells you that because he is solving a consistent system of five linear equations in six unknowns, he will get infinitely many solutions. Comment on his claim.

59. ▼ If the reduced matrix of a consistent system of linear equations has five rows, three of which are zero, and five columns, how many parameters does the general solution contain?

60. ▼ If the reduced matrix of a consistent system of linear equations has five rows, two of which are zero, and seven columns, how many parameters does the general solution contain?

61. ▼ Suppose a system of equations has a unique solution. What must be true of the number of pivots in the reduced matrix of the system? Why?

62. ▼ Suppose a system has infinitely many solutions. What must be true of the number of pivots in the reduced matrix of the system? Why?

63. ▼ Give an example of a system of three linear equations with the general solution $x = 1$, $y = 1 + z$, z arbitrary. (Check your system by solving it.)

64. ▼ Give an example of a system of three linear equations with the general solution $x = y - 1$, y arbitrary, $z = y$. (Check your system by solving it.)

65. ◆ A system of linear equations is called **homogeneous** if the right-hand side of each equation in the system is 0. If a homogeneous system has a unique solution, what can you say about that solution? Why?

66. ◆ A certain homogeneous system of linear equations in three unknowns (see the preceding exercise) has a solution $(1, 2, 0)$. Is this solution unique? Explain.

67. ◆ Can a homogeneous system (see Exercise 65) of linear equations be inconsistent? Explain.

68. ◆ Can a non-homogeneous system (see Exercise 65) of linear equations have the zero solution? Explain.

Applications of Systems of Linear Equations

In the examples and the exercises of this section, we consider scenarios that lead to systems of linear equations in three or more unknowns. Some of these applications will strike you as a little idealized or even contrived compared with the kinds of problems you might encounter in the real world.* One reason is that we will not have tools to handle more realistic versions of these applications until we have studied linear programming in Chapter 5.

✱ See the discussion at the end of the first example below.

In each example that follows, we set up the problem as a linear system and then give the solution; the emphasis in this section is on modeling a scenario by a system of linear equations and then interpreting the solution that results, rather than on obtaining the solution. For practice, you should do the row reduction necessary to get the solution.

EXAMPLE 1 Resource Allocation

The Arctic Juice Company makes three juice blends: PineOrange, using 2 quarts of pineapple juice and 2 quarts of orange juice per gallon; PineKiwi, using 3 quarts of pineapple juice and 1 quart of kiwi juice per gallon; and OrangeKiwi, using 3 quarts of orange juice and 1 quart of kiwi juice per gallon. Each day the company has 800 quarts of pineapple juice, 650 quarts of orange juice, and 350 quarts of kiwi juice available. How many gallons of each blend should it make each day if it wants to use up all of the supplies?

Solution We take the same steps to understand the problem that we took in Section 3.1. The first step is to identify and label the unknowns. Looking at the question asked in the last sentence, we see that we should label the unknowns like this:

x = number of gallons of PineOrange made each day

y = number of gallons of PineKiwi made each day

z = number of gallons of OrangeKiwi made each day.

Next, we can organize the information we are given in a table:

	PineOrange (x)	*PineKiwi (y)*	*OrangeKiwi (z)*	*Total Available*
Pineapple Juice (qt)	2	3	0	800
Orange Juice (qt)	2	0	3	650
Kiwi Juice (qt)	0	1	1	350

Notice how we have arranged the table; we have placed headings corresponding to the unknowns along the top, rather than down the side, and we have added a heading for the available totals. This gives us a table that is essentially the matrix of the system of linear equations we are looking for. (However, read the caution in the **Before we go on** section.)

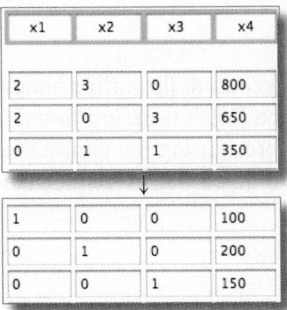
Now we read across each row of the table. The fact that we want to use exactly the amount of each juice that is available leads to the following three equations:

$$\begin{aligned} 2x + 3y &= 800 \\ 2x + 3z &= 650 \\ y + z &= 350. \end{aligned}$$

The solution of this system is $(x, y, z) = (100, 200, 150)$, so Arctic Juice should make 100 gallons of PineOrange, 200 gallons of PineKiwi, and 150 gallons of OrangeKiwi each day.

⇒ **Before we go on...**
Caution

We do not recommend relying on the coincidence that the table we created to organize the information in Example 1 happened to be the matrix of the system; it is too easy to set up the table "sideways" and get the wrong matrix. You should always write down the system of equations *and be sure you understand each equation.* For example, the equation $2x + 3y = 800$ in Example 1 indicates that the number of quarts of pineapple juice that will be used $(2x + 3y)$ is equal to the amount available (800 quarts). By thinking of the reason for each equation, you can check that you have the correct system. If you have the wrong system of equations to begin with, solving it won't help you.

Q : *Just how realistic is the scenario in Example 1?*

A : This is a very unrealistic scenario, for several reasons:

1. Isn't it odd that we happened to end up with exactly the same number of equations as unknowns? Real scenarios are rarely so considerate. If there had been four equations, there would in all likelihood have been no solution at all. However, we need to understand these idealized problems before we can tackle the real world.

2. Even if a real-world scenario does give the same number of equations as unknowns, there is still no guarantee that there will be a unique solution consisting of positive values. What, for instance, would we have done in this example if x had turned out to be negative?

3. The requirement that we use exactly all the ingredients would be an unreasonable constraint in real life. When we discuss linear programming, we will be able to substitute the more reasonable constraint that you use no more than is available, and we will add the more reasonable objective that you maximize profit. ∎

EXAMPLE 2 Aircraft Purchases: Airbus **and** Boeing

A new airline has recently purchased a fleet of Airbus A330-300s, Boeing 767-200ERs, and Boeing Dreamliner 787-9s to meet an estimated demand for 4,800 seats. The A330-300s seat 320 passengers and cost $200 million each, the 767-200ERs each seat 250 passengers and cost $125 million each, while the Dreamliner 787-9s seat 275 passengers and cost $200 million each.[12] The total cost of the fleet, which had twice as many Dreamliners as 767s, was $3,100 million. How many of each type of aircraft did the company purchase?

---

[12]The prices are approximate 2008 prices. Prices and seating capacities found at the companies' Web sites and in Wikipedia.

Solution We label the unknowns as follows:

$$x = \text{number of Airbus A330-300s}$$
$$y = \text{number of Boeing 767-200ERs}$$
$$z = \text{number of Boeing Dreamliner 787-9s.}$$

We must now set up the equations. We can organize some (but not all) of the given information in a table:

	A330-300	767-200ER	787-9 Dreamliner	Total
Capacity	320	250	275	4,800
Cost ($ million)	200	125	200	3,100

Reading across, we get the equations expressing the facts that the airline needed to seat 4,800 passengers and that it spent $3,100 million:

$$320x + 250y + 275z = 4,800$$
$$200x + 125y + 200z = 3,100.$$

There is an additional piece of information we have not yet used: the airline bought twice as many Dreamliners as 767s. As we said in Section 3.1, it is easiest to translate a statement like this into an equation if we first reword it using the phrase "the number of." Thus, we say: "The number of Dreamliners ordered was twice the number of 767s ordered," or

$$z = 2y$$
$$2y - z = 0.$$

We now have a system of three equations in three unknowns:

$$320x + 250y + 275z = 4,800$$
$$200x + 125y + 200z = 3,100$$
$$2y - z = 0.$$

Solving the system, we get the solution $(x, y, z) = (5, 4, 8)$. Thus, the airline ordered five A330-300s, four 767-200ERs, and eight Dreamliner 787-9s.

EXAMPLE 3 Traffic Flow

Traffic through downtown Urbanville flows through the one-way system shown in Figure 7.

Traffic counting devices installed in the road (shown as boxes) count 200 cars entering town from the west each hour, 150 leaving town on the north each hour, and 50 leaving town on the south each hour.

a. From this information, is it possible to determine how many cars drive along Allen, Baker, and Coal streets every hour?

b. What is the maximum possible traffic flow along Baker Street?

c. What is the minimum possible traffic along Allen Street?

d. What is the maximum possible traffic flow along Coal Street?

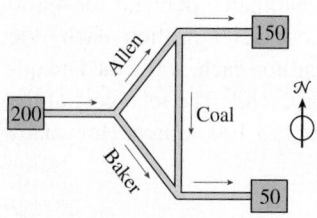

Figure 7

Solution

a. Our unknowns are:

$$x = \text{number of cars per hour on Allen Street}$$
$$y = \text{number of cars per hour on Baker Street}$$
$$z = \text{number of cars per hour on Coal Street.}$$

Assuming that, at each intersection, cars do not fall into a pit or materialize out of thin air, the number of cars entering each intersection has to equal the number exiting. For example, at the intersection of Allen and Baker Streets there are 200 cars entering and $x + y$ cars exiting:

$$\text{Traffic in} = \text{Traffic out}$$
$$200 = x + y.$$

At the intersection of Allen and Coal Streets, we get:

$$\text{Traffic in} = \text{Traffic out}$$
$$x = z + 150$$

and at the intersection of Baker and Coal Streets, we get:

$$\text{Traffic in} = \text{Traffic out}$$
$$y + z = 50.$$

We now have the following system of equations:

$$
\begin{aligned}
x + y \quad\;\;\; &= 200 \\
x \quad\;\;\; - z &= 150 \\
y + z &= 50.
\end{aligned}
$$

If we solve this system using the methods of the preceding section, we find that it has infinitely many solutions. The general solution is:

$$
\begin{aligned}
x &= z + 150 \\
y &= -z + 50 \\
z &\text{ is arbitrary.}
\end{aligned}
$$

Because we do not have a unique solution, it is *not* possible to determine how many cars drive along Allen, Baker and Coal Streets every hour.

b. The traffic flow along Baker Street is measured by y. From the general solution,

$$y = -z + 50$$

where z is arbitrary. How arbitrary is z? It makes no sense for any of the variables x, y, or z to be negative in this scenario, so $z \geq 0$. Therefore, the largest possible value y can have is

$$y = -0 + 50 = 50 \text{ cars per hour.}$$

c. The traffic flow along Allen Street is measured by x. From the general solution,

$$x = z + 150,$$

where $z \geq 0$, as we saw in part (b). Therefore, the smallest possible value x can have is

$$x = 0 + 150 = 150 \text{ cars per hour.}$$

d. The traffic flow along Coal Street is measured by z. Referring to the general solution, we see that z shows up in the expressions for both x and y:

$$x = z + 150$$
$$y = -z + 50.$$

In the first of these equations, there is nothing preventing z from being as big as we like; the larger we make z, the larger x becomes. However, the second equation places a limit on how large z can be: If $z > 50$, then y is negative, which is impossible. Therefore, the largest value z can take is 50 cars per hour.

From the discussion above, we see that z is not completely arbitrary: We must have $z \geq 0$ and $z \leq 50$. Thus, z has to satisfy $0 \leq z \leq 50$ for us to get a realistic answer.

➡ **Before we go on...** Here are some questions to think about in Example 3: If you wanted to nail down x, y, and z to see where the cars are really going, how would you do it with only one more traffic counter? Would it make sense for z to be fractional? What if you interpreted x, y, and z as *average* numbers of cars per hour over a long period of time?

Traffic flow is only one kind of flow in which we might be interested. Water and electricity flows are others. In each case, to analyze the flow, we use the fact that the amount entering an intersection must equal the amount leaving it. ■

EXAMPLE 4 **Transportation**

A car rental company has four locations in the city: Southwest, Northeast, Southeast, and Northwest. The Northwest location has 20 more cars than it needs, and the Northeast location has 15 more cars than it needs. The Southwest location needs 10 more cars than it has, and the Southeast location needs 25 more cars than it has. It costs $10 (in salary and gas) to have an employee drive a car from Northwest to Southwest. It costs $20 to drive a car from Northwest to Southeast. It costs $5 to drive a car from Northeast to Southwest, and it costs $10 to drive a car from Northeast to Southeast. If the company will spend a total of $475 rearranging its cars, how many cars will it drive from each of Northwest and Northeast to each of Southwest and Southeast?

Solution

Figure 8 shows a diagram of this situation. Each arrow represents a route along which the rental company can drive cars. At each location is written the number of extra cars the location has or the number it needs. Along each route is written the cost of driving a car along that route.

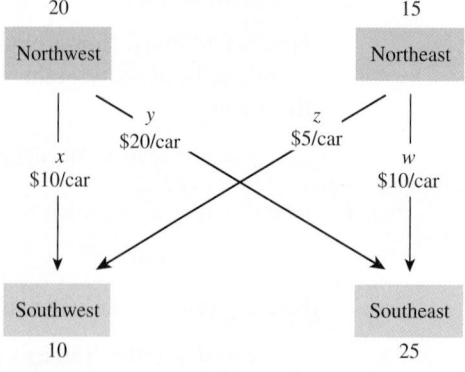

Figure 8

The unknowns are the number of cars the company will drive along each route, so we have the following four unknowns, as indicated in the figure:

$$x = \text{number of cars driven from Northwest to Southwest}$$
$$y = \text{number of cars driven from Northwest to Southeast}$$
$$z = \text{number of cars driven from Northeast to Southwest}$$
$$w = \text{number of cars driven from Northeast to Southeast.}$$

Consider the Northwest location. It has 20 more cars than it needs, so the total number of cars being driven out of Northwest should be 20. This gives us the equation

$$x + y = 20.$$

Similarly, the total number of cars being driven out of Northeast should be 15, so

$$z + w = 15.$$

Considering the number of cars needed at the Southwest and Southeast locations, we get the following two equations as well:

$$x + z = 10$$
$$y + w = 25.$$

There is one more equation that we should write down, the equation that says that the company will spend $475:

$$10x + 20y + 5z + 10w = 475.$$

Thus, we have the following system of five equations in four unknowns:

$$
\begin{aligned}
x + y & = 20 \\
z + w &= 15 \\
x + z & = 10 \\
y + w &= 25 \\
10x + 20y + 5z + 10w &= 475
\end{aligned}
$$

Solving this system, we find that $(x, y, z, w) = (5, 15, 5, 10)$. In words, the company will drive 5 cars from Northwest to Southwest, 15 from Northwest to Southeast, 5 from Northeast to Southwest, and 10 from Northeast to Southeast.

➡ **Before we go on...** A very reasonable question to ask in Example 4 is, Can the company rearrange its cars for less than $475? Even better, what is the least possible cost? In general, a question asking for the optimal cost may require the techniques of linear programming, which we will discuss in Chapter 5. However, in this case we can approach the problem directly. If we remove the equation that says that the total cost is $475 and solve the system consisting of the other four equations, we find that there are infinitely many solutions and that the general solution may be written as

$$x = w - 5$$
$$y = 25 - w$$
$$z = 15 - w$$
$$w \text{ is arbitrary.}$$

This allows us to write the total cost as a function of w:

$$\text{Cost} = 10x + 20y + 5z + 10w = 10(w - 5) + 20(25 - w) + 5(15 - w) + 10w$$
$$= 525 - 5w.$$

So, the larger we make w, the smaller the total cost will be. The largest we can make w is 15 (why?), and if we do so we get $(x, y, z, w) = (10, 10, 0, 15)$ and a total cost of $450. ■

3.3   EXERCISES

▼ more advanced   ◆ challenging
 T indicates exercises that should be solved using technology

APPLICATIONS

1. *Resource Allocation* You manage an ice cream factory that makes three flavors: Creamy Vanilla, Continental Mocha, and Succulent Strawberry. Into each batch of Creamy Vanilla go 2 eggs, 1 cup of milk, and 2 cups of cream. Into each batch of Continental Mocha go 1 egg, 1 cup of milk, and 2 cups of cream, while into each batch of Succulent Strawberry go 1 egg, 2 cups of milk, and 1 cup of cream. You have in stock 350 eggs, 350 cups of milk, and 400 cups of cream. How many batches of each flavor should you make in order to use up all of your ingredients? HINT [See Example 1.]

2. *Resource Allocation* You own a hamburger franchise and are planning to shut down operations for the day, but you are left with 13 bread rolls, 19 defrosted beef patties, and 15 opened cheese slices. Rather than throw them out, you decide to use them to make burgers that you will sell at a discount. Plain burgers each require 1 beef patty and 1 bread roll, double cheeseburgers each require 2 beef patties, 1 bread roll, and 2 slices of cheese, while regular cheeseburgers each require 1 beef patty, 1 bread roll, and 1 slice of cheese. How many of each should you make? HINT [See Example 1.]

3. *Resource Allocation* Urban Community College is planning to offer courses in Finite Math, Applied Calculus, and Computer Methods. Each section of Finite Math has 40 students and earns the college $40,000 in revenue. Each section of Applied Calculus has 40 students and earns the college $60,000, while each section of Computer Methods has 10 students and earns the college $20,000. Assuming the college wishes to offer a total of six sections, to accommodate 210 students, and to bring in $260,000 in revenues, how many sections of each course should it offer? HINT [See Example 2.]

4. *Resource Allocation* The Enormous State University History Department offers three courses, Ancient, Medieval, and Modern History, and the chairperson is trying to decide how many sections of each to offer this semester. The department is allowed to offer 45 sections total, there are 5,000 students who would like to take a course, and there are 60 professors

to teach them. Sections of Ancient History have 100 students each, sections of Medieval History hold 50 students each, and sections of Modern History have 200 students each. Modern History sections are taught by a team of 2 professors, while Ancient and Medieval History need only 1 professor per section. How many sections of each course should the chair schedule in order to offer all the sections that they are allowed to, accommodate all of the students, and give one teaching assignment to each professor? HINT [See Example 2.]

5. *Latin Music Sales (Digital)* In 2010, total revenues from digital sales of regional (Mexican/Tejano), pop/rock, and tropical (salsa/merengue/cumbia/bachata) Latin music in the U.S. amounted to $46 million. Regional music brought in fives times as much as tropical music and $9 million more than pop/rock music.[13] How much revenue was earned from digital sales in each of the three categories?

6. *Latin Music Sales (Digital)* In 2009, total revenues from digital sales of pop/rock, tropical (salsa/merengue/cumbia/bachata), and urban (reggaeton) Latin music in the U.S. amounted to $28 million. Tropical music brought in twice as much as urban music, and pop/rock music brought in $10 million more than the others combined.[14] How much revenue was earned from digital sales in each of the three categories?

7. *Purchasing Aircraft* In Example 2 we saw that Airbus A330-300s seat 320 passengers and cost $200 million each, Boeing 767-200ERs seat 250 passengers and cost $125 million each, while Boeing Dreamliner 787-9s seat 275 passengers and cost $200 million each. You are the purchasing manager of an airline company and have a spending goal of $2,900 million for the purchase of new aircraft to seat a total of 4,480 passengers. Your company has a policy of supporting U.S. industries, and you have been instructed to buy twice as many Dreamliners as Airbus A330-300s. Given the selection of three aircraft, how many of each should you order?

8. *Purchasing Aircraft* Refer to Exercise 7. René DuFleur has just been appointed the new CEO of your airline, and you

---

[13]Revenues are based on suggested list prices, and rounded to the nearest $1 million. Source: Recording Industry Association of America http://riaa.org
[14]*Ibid.*

have received instructions that the company policy is now to purchase 25% more Airbus A330-300s than Boeing 767-200ERs (that is, 1.25 times as many Airbus A330-300s as 767-200ERs). Further, the desired seating capacity has been revised downward to 2,600 passengers, and the cost target to $1,500 million. Given the selection of three aircraft, how many of each should you order?

9. *Supply* A bagel store orders cream cheese from three suppliers, Cheesy Cream Corp. (CCC), Super Smooth & Sons (SSS), and Bagel's Best Friend Co. (BBF). One month, the total order of cheese came to 100 tons (they do a booming trade). The costs were $80, $50, and $65 per ton from the three suppliers, respectively, with total cost amounting to $5,990. Given that the store ordered the same amount from CCC and BBF, how many tons of cream cheese were ordered from each supplier?

10. *Supply* Refer to Exercise 9. The bagel store's outlay for cream cheese the following month was $2,310, when it purchased a total of 36 tons. Two more tons of cream cheese came from Bagel's Best Friend Co. than from Super Smooth & Sons. How many tons of cream cheese came from each supplier?

11. *Pest Control* Aragorn the Great has boasted to his hordes of followers that many a notorious villain has fallen to his awesome sword: His total of 560 victims consists of evil sorcerers, trolls, and orcs. These he has slain with a total of 620 mighty thrusts of his sword; evil sorcerers and trolls each requiring two thrusts (to the chest) and orcs each requiring one thrust (to the neck). When asked about the number of trolls he has slain, he replies, "I, the mighty Aragorn, despise trolls five times as much as I despise evil sorcerers. Accordingly, five times as many trolls as evil sorcerers have fallen to my sword!" How many of each type of villain has he slain?

12. *Manufacturing Perfume* The Fancy French Perfume Company recently had its secret formula divulged. It turned out that it was using, as the three ingredients, rose oil, oil of fermented prunes, and alcohol. Moreover, each 22-ounce econo-size bottle contained 4 more ounces of alcohol than oil of fermented prunes, while the amount of alcohol was equal to the combined volume of the other two ingredients. How much of each ingredient did it use in an econo-size bottle?[15]
HINT [The answer is the brand-name of a famous eau de cologne.]

13. ▼ *Donations* The Enormous State University Good Works Society recently raised funds for three worthwhile causes: the Math Professors' Benevolent Fund (MPBF), the Society of Computer Nerds (SCN), and the NY Jets. Because the society's members are closet jocks, the society donated twice as much to the NY Jets as to the MPBF, and equal amounts to the first two funds (it is unable to distinguish between mathematicians and nerds). Further, for every $1 it gave to the MPBF, it decided to keep $1 for itself; for every $1 it gave to the SCN, it kept $2, and for every $1 to the Jets, it also kept $2. The treasurer of the Society, Johnny Treasure, was required to itemize all donations for the Dean of Students, but discovered to his consternation that he had lost

the receipts! The only information available to him was that the society's bank account had swelled by $4,200. How much did the society donate to each cause?

14. ▼ *Tenure* Professor Walt is up for tenure, and wishes to submit a portfolio of written student evaluations as evidence of his good teaching. He begins by grouping all the evaluations into four categories: good reviews, bad reviews (a typical one being "GET RID OF WALT! THE MAN CAN'T TEACH!"), mediocre reviews (such as "I suppose he's OK, given the general quality of teaching at this college"), and reviews left blank. When he tallies up the piles, Walt gets a little worried: There are 280 more bad reviews than good ones and only half as many blank reviews as bad ones. The good reviews and blank reviews together total 170. On an impulse, he decides to even up the piles a little by removing 280 of the bad reviews, and this leaves him with a total of 400 reviews of all types. How many of each category of reviews were there originally?

■ *Airline Costs* Exercises 15 and 16 are based on the following table, which shows the amount spent by four U.S. airlines to fly one available seat one mile in the second quarter of 2011.[16] Set up each system and then solve using technology.
HINT [See the technology note accompanying Example 1.]

Airline	American	Continental	Delta	Southwest
Cost	13.1¢	13.0¢	14.4¢	12.5¢

15. ▼ Suppose that, on a 3,000-mile New York–Los Angeles flight, Continental, American, and Southwest flew a total of 210 empty seats, costing them a total of $81,270. If Continental had three times as many empty seats as American, how many empty seats did each of these three airlines carry on its flight?

16. ▼ Suppose that, on a 2,000-mile Miami–Memphis flight, Continental, Delta, and Southwest flew a total of 200 empty seats, costing them a total of $54,300. If Delta had twice as many empty seats as Southwest, how many empty seats did each of these three airlines carry on its flight?

Investing: Inverse Mutual Funds Inverse mutual funds, sometimes referred to as "bear market" or "short" funds, seek to deliver the opposite of the performance of the index or category they track, and can thus be used by traders to bet against the stock market. Exercises 17 and 18 are based on the following table, which shows the performance of three such funds as of August 12, 2011.[17]

	Year-to-date Loss
SHPIX (Short Smallcap Profund)	6%
RYURX (Rydex Inverse S&P 500)	5%
RYIHX (Rydex Inverse High Yield)	7%

---

[15]Most perfumes consist of 10 to 20% perfume oils dissolved in alcohol. This may or may not be reflected in this company's formula.

[16]Costs are rounded to the nearest 0.1¢. Source: Company financial statements, obtained from their respective Web sites. The cost per available seat-mile (CASM) is a widely used operating statistic in the airline industry.

[17]Based on prices at the close of the stock market on August 12, 2011. YTD losses rounded to the nearest percentage point. Source: www.fidelity.com.

17. You invested a total of $9,000 in the three funds at the beginning of 2011, including an equal amount in RYURX and RYIHX. Your year-to-date loss from the first two funds amounted to $400. How much did you invest in each of the three funds?

18. You invested a total of $6,000 in the three funds at the beginning of 2011, including an equal amount in SHPIX and RYURX. Your total year-to-date loss amounted to $360. How much did you invest in each of the three funds?

Investing: Lesser-Known Stocks *Exercises 19 and 20 are based on the following information about the stocks of* KT Corporation, International Flavors & Fragrances, *and* Quality Systems.[18]

	Price	*Dividend Yield*
KT (KT Corporation)	$16	7%
IFF (International Flavors & Fragrances)	56	2
QSII (Quality Systems)	80	2

19. ▼ You invested a total of $8,400 in shares of the three stocks at the given prices, and expected to earn $248 in annual dividends. If you purchased a total of 200 shares, how many shares of each stock did you purchase?

20. ▼ You invested a total of $11,200 in shares of the three stocks at the given prices, and expected to earn $304 in annual dividends. If you purchased a total of 250 shares, how many shares of each stock did you purchase?

21. ⊤ ***Internet Audience*** At the end of 2003, the four companies with the largest number of home Internet users in the United States were Microsoft, Time Warner, Yahoo, and Google, with a combined audience of 284 million users.[19] Taking x to be the Microsoft audience in millions, y the Time Warner audience in millions, z the Yahoo audience in millions, and u the Google audience in millions, it was observed that

$$z - u = 3(x - y) + 6$$
$$x + y = 50 + z + u$$
and $\quad x - y + z - u = 42.$

How large was the audience of each of the four companies in November 2003?

22. ⊤ ***Internet Audience*** At the end of 2003, the four organizations ranking 5 through 8 in home Internet users in the United States were eBay, the U.S. government, Amazon, and Lycos, with a combined audience of 112 million users.[20] Taking x to be the eBay audience in millions, y the U.S. Government audience in millions, z the Amazon audience in millions, and u the Lycos audience in millions, it was observed that

$$y - z = z - u$$
$$x - y = 3(y - u) + 5$$
and $\quad x - y + z - u = 12.$

How large was the audience of each of the four companies in November 2003?

23. ▼ ***Market Share: Homeowners Insurance*** The three market leaders in homeowners insurance in Missouri are State Farm, American Family Insurance Group, and Allstate. Based on data from 2007, two relationships between the Missouri homeowners insurance percentage market shares are found to be

$$x = 1 + y + z$$
$$z = 16 - 0.2w,$$

where x, y, z, and w are, respectively, the percentages of the market held by State Farm, American Family, Allstate, and other companies.[21] Given that the four groups account for the entire market, obtain a third equation relating x, y, z, and w, and solve the associated system of three linear equations to show how the market shares of State Farm, American Family, and Allstate depend on the share held by other companies. Which of the three companies' market share is most impacted by the share held by other companies?

24. ▼ ***Market Share: Auto Insurance*** Repeat the preceding exercise using the following relationships among the auto insurance percentage market shares:

$$x = -40 + 5y + z$$
$$z = 3 - 2y + w.$$

25. ***Inventory Control*** Red Bookstore wants to ship books from its warehouses in Brooklyn and Queens to its stores, one on Long Island and one in Manhattan. Its warehouse in Brooklyn has 1,000 books and its warehouse in Queens has 2,000. Each store orders 1,500 books. It costs $5 to ship each book from Brooklyn to Long Island and $1 to ship each book from Brooklyn to Manhattan. It costs $4 to ship each book from Queens to Long Island and $2 to ship each book from Queens to Manhattan.

a. If Red has a transportation budget of $9,000 and is willing to spend all of it, how many books should Red ship from each warehouse to each store in order to fill all the orders?

b. Is there a way of doing this for less money?
HINT [See Example 4.]

26. ***Inventory Control*** The Tubular Ride Boogie Board Company has manufacturing plants in Tucson, AZ and Toronto, Ontario. You have been given the job of coordinating distribution of the latest model, the Gladiator, to outlets in Honolulu and Venice Beach. The Tucson plant, when operating at full capacity, can manufacture 620 Gladiator boards per week, while the

---

[18]Yields rounded to the nearest percentage point and stock prices at the close of the stock market on August 12, 2011 rounded to the nearest $1. Source: www.google.com/finance.

[19]Source: Nielsen/NetRatings www.nielsen-netratings.com/ January 1, 2004.

[20]*Ibid.*

[21]Source: Missouri Dept. of Insurance (www.insurance.mo.gov).

Toronto plant, beset by labor disputes, can produce only 410 boards per week. The outlet in Honolulu orders 500 Gladiator boards per week, while Venice Beach orders 530 boards per week. Transportation costs are as follows:

Tucson to Honolulu: $10 per board; Tucson to Venice Beach: $5 per board.

Toronto to Honolulu: $20 per board; Toronto to Venice Beach: $10 per board.

a. Assuming that you wish to fill all orders and assure full capacity production at both plants, is it possible to meet a total transportation budget of $10,200? If so, how many Gladiator boards are shipped from each manufacturing plant to each distribution outlet?

b. Is there a way of doing this for less money? HINT [See Example 4.]

27. ▼ *Tourism in the 1990s* In the 1990s, significant numbers of tourists traveled from North America and Europe to Australia and South Africa. In 1998, a total of 1,390,000 of these tourists visited Australia, while 1,140,000 of them visited South Africa. Further, 630,000 of them came from North America and 1,900,000 of them came from Europe.[22] (Assume no single tourist visited both destinations or traveled from both North America and Europe.)

a. The given information is not sufficient to determine the number of tourists from each region to each destination. Why?

b. If you were given the additional information that a total of 2,530,000 tourists traveled from these two regions to these two destinations, would you now be able to determine the number of tourists from each region to each destination? If so, what are these numbers?

c. If you were given the additional information that the same number of people from Europe visited South Africa as visited Australia, would you now be able to determine the number of tourists from each region to each destination? If so, what are these numbers?

28. ▼ *Tourism in the 1990s* In the 1990s, significant numbers of tourists traveled from North America and Asia to Australia and South Africa. In 1998, a total of 2,230,000 of these tourists visited Australia, while 390,000 of them visited South Africa. Also, 630,000 of these tourists came from North America, and a total of 2,620,000 tourists traveled from these two regions to these two destinations.[23] (Assume no single tourist visited both destinations or traveled from both North America and Asia.)

a. The given information is not sufficient to determine the number of tourists from each region to each destination. Why?

b. If you were given the additional information that a total of 1,990,000 tourists came from Asia, would you now be

able to determine the number of tourists from each region to each destination? If so, what are these numbers?

c. If you were given the additional information that 200,000 tourists visited South Africa from Asia, would you now be able to determine the number of tourists from each region to each destination? If so, what are these numbers?

29. ▼ *Alcohol* The following table shows some data from a 2000 study on substance use among 10th graders in the United States and Europe.[24]

	Used Alcohol	*Alcohol-Free*	*Totals*
U.S.	x	y	14,000
Europe	z	w	95,000
Totals	63,550	45,450	

a. The table leads to a linear system of four equations in four unknowns. What is the system? Does it have a unique solution? What does this indicate about the given and the missing data?

b. 🖥 Given that the number of U.S. 10th graders who were alcohol-free was 50% more than the number who had used alcohol, find the missing data.

30. ▼ *Tobacco* The following table shows some data from the same study cited in Exercise 30.[25]

	Smoked Cigarettes	*Cigarette-Free*	*Totals*
U.S.	x	y	14,000
Europe	z	w	95,000
Totals		70,210	109,000

a. The table leads to a linear system of four equations in four unknowns. What is the system? Does it have a unique solution? What does this indicate about the missing data?

b. 🖥 Given that 31,510 more European 10th graders smoked cigarettes than U.S. 10th graders, find the missing data.

31. *Traffic Flow* One-way traffic through Enormous State University is shown in the figure, where the numbers indicate daily counts of vehicles.

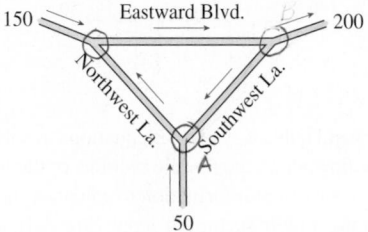

[22]Figures are rounded to the nearest 10,000. Sources: South African Dept. of Environmental Affairs and Tourism; Australia Tourist Commission/*New York Times*, January 15, 2000, p. C1.

[23]*Ibid.*

[24]"Used Alcohol" indicates consumption of alcohol at least once in the past 30 days. Source: Council of Europe/University of Michigan. "Monitoring the Future"/*New York Times*, February 21, 2001, p. A10.

[25]"Smoked Cigarettes" indicates that at least one cigarette was smoked in the past 30 days. Source: *Ibid.*

a. Is it possible to determine the daily flow of traffic along each of the three streets from the information given? If your answer is *yes*, what is the traffic flow along each street? If your answer is *no*, what additional information would suffice?

b. Is a flow of 60 vehicles per day along Southwest Lane consistent with the information given?

c. What is the minimum traffic flow possible along Northwest Lane consistent with the information given? HINT [See Example 3.]

32. *Traffic Flow* The traffic through downtown East Podunk flows through the one-way system shown below.

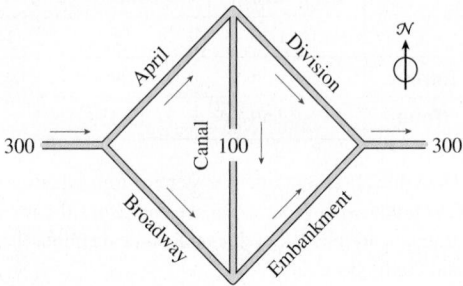

Traffic counters find that 300 vehicles enter town from the west each hour, and 300 leave town toward the east each hour. Also, 100 cars drive down Canal Street each hour.

a. Write down the general solution of the associated system of linear equations. Is it possible to determine the number of vehicles on each street per hour?

b. On which street could you put another traffic counter in order to determine the flow completely?

c. What is the minimum traffic flow along April Street consistent with the information given? HINT [See Example 3.]

33. *Traffic Flow* The traffic through downtown Johannesburg follows the one-way system shown below, with traffic movement recorded at incoming and outgoing streets (in cars per minute) as shown.

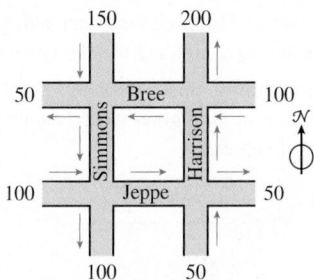

a. Set up and solve a system of equations to solve for the traffic flow along the middle sections of the four streets.

b. Is there sufficient information to calculate the traffic along the middle section of Jeppe Street? If so, what is it; if not, why not?

c. Given that 400 cars per minute flow down the middle section of Bree Street, how many cars per minute flow down the middle section of Simmons?

d. What is the minimum traffic flow along the middle section of Harrison Street?

e. Is there an upper limit to the possible traffic down Simmons Street consistent with the information given? Explain.

34. *Traffic Flow* Town of Hempstead officials were planning to make Hempstead Turnpike and the surrounding streets into one-way streets to prepare for the opening of Hofstra USA. In an experiment, they restricted traffic flow along the streets near Hofstra as shown in the diagram. The numbers show traffic flow per minute and the arrows indicate the direction of traffic.

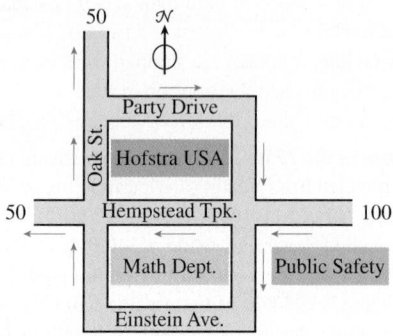

a. Set up and solve the associated traffic flow problem with the following unknowns:
 $x =$ traffic per minute along the middle stretch of Oak St.,
 $y =$ traffic per minute along the middle stretch of Hempstead Tpk.,
 $z =$ traffic per minute along Einstein Ave., and
 $u =$ traffic per minute along Party Drive.

b. If 20 cars per minute drive along Party Drive, what is the traffic like along the middle stretch of Oak Street?

c. If 20 vehicles per minute drive along Einstein Ave., and 90 vehicles per minute drive down the middle stretch of Hempstead Tpk., how many cars per minute drive along the middle stretch of Oak Street?

d. If Einstein Avenue is deserted, what is the minimum traffic along the middle stretch of Hempstead Tpk.?

35. ▼ *Traffic Management* The Outer Village Town Council has decided to convert its (rather quiet) main street, Broadway, to a one-way street, but is not sure of the direction of most of the traffic. The accompanying diagram illustrates the downtown area of Outer Village, as well as the *net* traffic flow along the intersecting streets (in vehicles per day). (There are no one-way streets; a net traffic flow in a certain direction is defined as the traffic flow in that direction minus the flow in the opposite direction.)

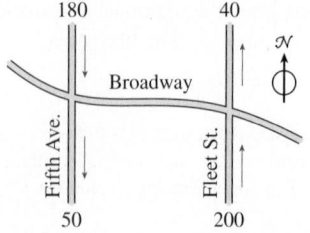

a. Is the given information sufficient to determine the net traffic flow along the three portions of Broadway shown? If your answer is *yes,* give the traffic flow along each stretch. If your answer is *no,* what additional information would suffice? HINT [For the direction of net traffic flow, choose either east or west. If a corresponding value is negative, it indicates net flow in the opposite direction.]

b. Assuming that there is little traffic (less than 160 vehicles per day) east of Fleet Street, in what direction is the net flow of traffic along the remaining stretches of Broadway?

36. ▼ *Electric Current* Electric current measures (in **amperes** or **amps**) the flow of electrons through wires. Like traffic flow, the current entering an intersection of wires must equal the current leaving it.[26] Here is an electrical circuit known as a **Wheatstone bridge.**

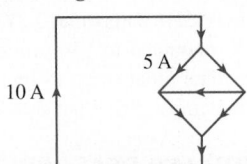

a. If the currents in two of the wires are 10 amps and 5 amps as shown, determine the currents in the unlabeled wires in terms of suitable parameters.

b. In which wire should you measure the current in order to know all of the currents exactly?

37. ▼ *Econometrics* (*from the GRE Economics Test*) This and the next exercise are based on the following simplified model of the determination of the money stock.

$$M = C + D$$
$$C = 0.2D$$
$$R = 0.1D$$
$$H = R + C$$

where

$$M = \text{money stock}$$
$$C = \text{currency in circulation}$$
$$R = \text{bank reserves}$$
$$D = \text{deposits of the public}$$
$$H = \text{high-powered money}$$

If the money stock were $120 billion, what would bank reserves have to be?

38. ▼ *Econometrics* (*from the GRE Economics Test*) With the model in the previous exercise, if H were equal to $42 billion, what would M equal?

CAT Scans CAT (*Computerized Axial Tomographic*) scans are used to map the exact location of interior features of the human body. CAT scan technology is based on the following principles: (1) different components of the human body (water, gray matter, bone, etc.) absorb X-rays to different extents; and (2) to measure the X-ray absorption by a specific region of, say, the brain, it suffices to pass a *number of line-shaped pencil beams of X-rays through the brain at different angles and measure the total absorption for each beam, which is the sum of the absorptions of the regions through which it passes. The accompanying diagram illustrates a simple example. (The number in each region shows its absorption, and the number on each X-ray beam shows the total absorption for that beam.)[27]*

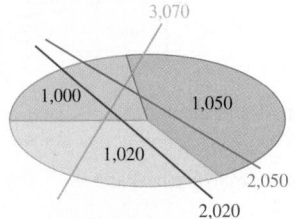

In Exercises 39–44, use the table and the given X-ray absorption diagrams to identify the composition of each of the regions marked by a letter.

Type	Air	Water	Gray Matter	Tumor	Blood	Bone
Absorption	0	1,000	1,020	1,030	1,050	2,000

39.

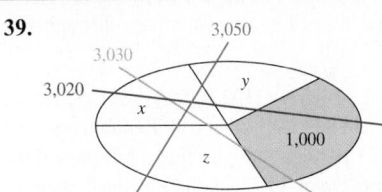

40.

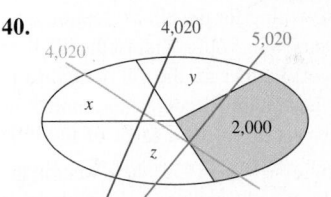

41.

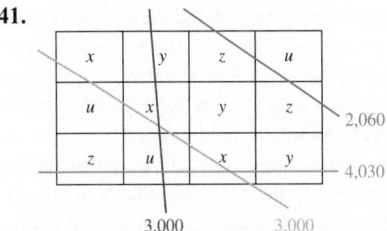

42.

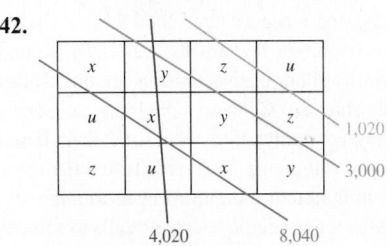

---

[26]This is known as **Kirchhoff's current law,** named after Gustav Robert Kirchhoff (1824–1887). Kirchhoff made important contributions to the fields of geometric optics, electromagnetic radiation, and electrical network theory.

[27]Based on a COMAP article *Geometry, New Tools for New Technologies,* by J. Malkevitch, Video Applications Library, COMAP, 1992. The absorptions are actually calibrated on a logarithmic scale. In real applications, the size of the regions is very small, and very large numbers of beams must be used.

43. ▼ Identify the composition of site *x*.

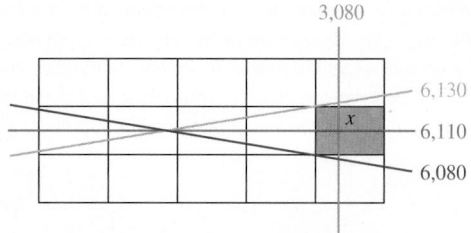

(The horizontal and slanted beams each pass through five regions.)

44. ▼ Identify the composition of site *x*.

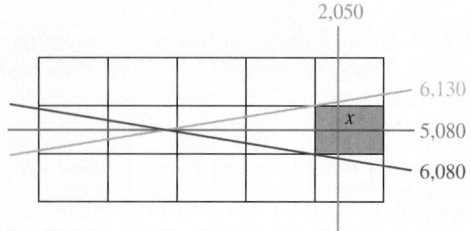

(The horizontal and slanted beams each pass through five regions.)

45. ◆ *Voting* In the 75th Congress (1937–1939) the U.S. House of Representatives had 333 Democrats, 89 Republicans, and 13 members of other parties. Suppose that a bill passed the House with 31 more votes in favor than against, with 10 times as many Democrats voting for the bill as Republicans, and with 36 more non-Democrats voting against the bill than for it. If every member voted either for the bill or against it, how many Democrats, how many Republicans, and how many members of other parties voted in favor of the bill?

46. ◆ *Voting* In the 75th Congress (1937–1939) there were in the Senate 75 Democrats, 17 Republicans, and 4 members of other parties. Suppose that a bill passed the Senate with 16 more votes in favor than against, with three times as many Democrats voting in favor as non-Democrats voting in favor, and 32 more Democrats voting in favor than Republicans voting in favor. If every member voted either for the bill or against it, how many Democrats, how many Republicans, and how many members of other parties voted in favor of the bill?

47. ◆ *Investments* Things have not been going too well here at Accurate Accounting, Inc. since we hired Todd Smiley. He has a tendency to lose important documents, especially around April, when tax returns of our business clients are due. Today Smiley accidentally shredded Colossal Conglomerate Corp.'s investment records. We must therefore reconstruct them based on the information he can gather. Todd recalls that the company earned an $8 million return on investments totaling $65 million last year. After a few frantic telephone calls to sources in Colossal, he learned that Colossal had made investments in four companies last year: X, Y, Z, and W. (For reasons of confidentiality, we are withholding their names.) Investments in company X earned 15% last year, investments in Y depreciated by 20% last year, investments in Z neither appreciated nor

depreciated last year, while investments in W earned 20% last year. Smiley was also told that Colossal invested twice as much in company X as in company Z, and three times as much in company W as in company Z. Does Smiley have sufficient information to piece together Colossal's investment portfolio before its tax return is due next week? If so, what does the investment portfolio look like?

48. ◆ *Investments* Things are going from bad to worse here at Accurate Accounting, Inc.! Colossal Conglomerate Corp.'s tax return is due tomorrow and the accountant Todd Smiley seems to have no idea how Colossal earned a return of $8 million on a $65 million investment last year. It appears that, although the returns from companies X, Y, Z, and W were as listed in Exercise 47, the rest of the information there was wrong. What Smiley is now being told is that Colossal only invested in companies X, Y, and Z and that the investment in X amounted to $30 million. His sources in Colossal still maintain that twice as much was invested in company X as in company Z. What should Smiley do?

COMMUNICATION AND REASONING EXERCISES

49. Are Exercises 1 and 2 realistic in their expectation of using up all the ingredients? What does your answer have to do with the solution(s) of the associated system of equations?

50. Suppose that you obtained a solution for Exercise 3 or 4 consisting of positive values that were not all whole numbers. What would such a solution signify about the situation in the exercise? Should you round these values to the nearest whole numbers?

In Exercises 51–56, x, y and z represent the weights of the three ingredients X, Y, and Z in a gasoline blend. Say which of the following is represented by a linear equation in x, y, and z, and give a form of the equation when it is:

51. The blend consists of 100 pounds of ingredient *X*.

52. The blend is free of ingredient *X*.

53. The blend contains 30% ingredient *Y* by weight.

54. The weight of ingredient *X* is the product of the weights of ingredients *Y* and *Z*.

55. There is at least 30% ingredient *Y* by weight.

56. There is twice as much ingredient *X* by weight as *Y* and *Z* combined.

57. Make up an entertaining word problem leading to the following system of equations.

$$10x + 20y + 10z = 100$$
$$5x + 15y \quad\quad = 50$$
$$x + \quad y + \quad z = 10$$

58. Make up an entertaining word problem leading to the following system of equations.

$$10x + 20y \quad\quad\quad = 300$$
$$10z + 20w = 400$$
$$20x \quad\quad + 10z \quad\quad = 400$$
$$10y \quad\quad + 20w = 300$$

CHAPTER 3 REVIEW

KEY CONCEPTS

Website www.WanerMath.com
Go to the Website at www.WanerMath .com to find a comprehensive and interactive Web-based summary of Chapter 3.

3.1 Systems of Two Equations in Two Unknowns
Linear equation in two unknowns *p. 176*
Coefficient *p. 176*
Solution of an equation in two unknowns *p. 176*
Graphical method for solving a system of two linear equations *p. 179*

Algebraic method for solving a system of two linear equations *p. 179*
Redundant or dependent system *p. 180*
Possible outcomes for a system of two linear equations *p. 181*
Consistent system *p. 181*

3.2 Using Matrices to Solve Systems of Equations
Linear equation (in any number of unknowns) *p. 188*
Matrix *p. 189*

Augmented matrix of a system of linear equations *p. 189*
Elementary row operations *p. 190*
Pivot *p. 192*
Clearing a column; pivoting *p. 192*
Gauss-Jordan or row reduction *p. 194*
Reduced row echelon form *p. 201*

3.3 Applications of Systems of Linear Equations
Resource allocation *p. 206*
(Traffic) flow *p. 208*
Transportation *p. 210*

REVIEW EXERCISES

In each of Exercises 1–6, graph the equations and determine how many solutions the system has, if any.

1.
$$x + 2y = 4$$
$$2x - y = 1$$

2. $0.2x - 0.1y = 0.3$
$$0.2x + 0.2y = 0.4$$

3.
$$\frac{1}{2}x - \frac{3}{4}y = 0$$
$$6x - 9y = 0$$

4. $2x + 3y = 2$
$$-x - 3y/2 = 1/2$$

5.
$$x + y = 1$$
$$2x + y = 0.3$$
$$3x + 2y = \frac{13}{10}$$

6. $3x + 0.5y = 0.1$
$$6x + y = 0.2$$
$$\frac{3x}{10} - 0.05y = 0.01$$

Solve each of the systems of linear equations in Exercises 7–18.

7.
$$x + 2y = 4$$
$$2x - y = 1$$

8. $0.2x - 0.1y = 0.3$
$$0.2x + 0.2y = 0.4$$

9.
$$\frac{1}{2}x - \frac{3}{4}y = 0$$
$$6x - 9y = 0$$

10. $2x + 3y = 2$
$$-x - 3y/2 = 1/2$$

11.
$$x + y = 1$$
$$2x + y = 0.3$$
$$3x + 2y = \frac{13}{10}$$

12. $3x + 0.5y = 0.1$
$$6x + y = 0.2$$
$$\frac{3x}{10} - 0.05y = 0.01$$

13. $x + 2y = -3$
$$x - z = 0$$
$$x + 3y - 2z = -2$$

14. $x - y + z = 2$
$$7x + y - z = 6$$
$$x - \frac{1}{2}y + \frac{1}{3}z = 1$$
$$x + y + z = 6$$

15.
$$x - \frac{1}{2}y + z = 0$$
$$\frac{1}{2}x - \frac{1}{2}z = -1$$
$$\frac{3}{2}x - \frac{1}{2}y + \frac{1}{2}z = -1$$

16. $x + y - 2z = -1$
$$-2x - 2y + 4z = 2$$
$$0.75x + 0.75y - 1.5z = -0.75$$

17.
$$x = \frac{1}{2}y$$
$$\frac{1}{2}x = -\frac{1}{2}z + 2$$
$$z = -3x + y$$

18. $x - y + z = 1$
$$y - z + w = 1$$
$$x + z - w = 1$$
$$2x + z = 3$$

Exercises 19–22 are based on the following equation relating the Fahrenheit and Celsius (or centigrade) temperature scales:

$$5F - 9C = 160,$$

where F is the Fahrenheit temperature of an object and C is its Celsius temperature.

19. What temperature should an object be if its Fahrenheit and Celsius temperatures are the same?

20. What temperature should an object be if its Celsius temperature is half its Fahrenheit temperature?

21. Is it possible for the Fahrenheit temperature of an object to be 1.8 times its Celsius temperature? Explain.

22. Is it possible for the Fahrenheit temperature of an object to be 30° more than 1.8 times its Celsius temperature? Explain.

In Exercises 23–28, let x, y, z, and w represent the population in millions of four cities A, B, C, and D, respectively. Express the given statement as an equation in x, y, z, and w. If the equation is linear, say so and express it in the standard form $ax + by + cz + dw = k$.

23. The total population of the four cities is 10 million people.

24. City *A* has three times as many people as cities *B* and *C* combined.

25. City *D* is actually a ghost town; there are no people living in it.

26. The population of City *A* is the sum of the squares of the populations of the other three cities.

27. City *C* has 30% more people than City *B*.

28. City *C* has 30% fewer people than City *B*.

APPLICATIONS: OHaganBooks.com

Purchasing You are the buyer for OHaganBooks.com and are considering increasing stocks of romance and horror novels at the new OHaganBooks.com warehouse in Texas. You have offers from two publishers: Duffin House and Higgins Press. Duffin offers a package of 5 horror novels and 5 romance novels for $50, and Higgins offers a package of 5 horror and 11 romance novels for $15. Exercises 29–32 deal with your purchasing options given these offers.

29. How many packages should you purchase from each publisher to get exactly 4,500 horror novels and 6,600 romance novels?

30. You want to spend a total of $50,000 on books and have promised to buy twice as many packages from Duffin as from Higgins. How many packages should you purchase from each publisher?

31. The accountant tells you that the company can actually afford to spend a total of $90,000 on romance and horror books. She also reminds you that you had signed an agreement to spend twice as much money for books from Duffin as from Higgins. How many packages should you purchase from each publisher?

32. Upon revising her records, the accountant now tells you that the company can afford to spend a total of only $60,000 on romance and horror books, and that it is company policy to spend the same amount of money at both publishers. How many packages should you purchase from each publisher?

33. *Equilibrium* The demand per year for *Finite Math the OHagan Way* is given by $q = -1,000p + 140,000$, where *p* is the price per book in dollars. The supply is given by $q = 2,000p + 20,000$. Find the price at which supply and demand balance.

34. *Equilibrium* OHaganbooks.com CEO John O'Hagan announces to a stunned audience at the annual board meeting that he is considering expanding into the jumbo jet airline manufacturing business. The demand per year for jumbo jets is given by $q = -2p + 18$ where *p* is the price per jet in millions of dollars. The supply is given by $q = 3p + 3$. Find the price the envisioned O'Hagan jumbo jet division should charge to balance supply and demand.

35. *Feeding Schedules* Billy-Sean O'Hagan is John O'Hagan's son and a freshman in college. Billy's 36-gallon tropical fish tank contains three types of carnivorous creatures: baby sharks, piranhas and squids, and he feeds them three types of delicacies: goldfish, angelfish and butterfly fish. Each baby shark can consume 1 goldfish, 2 angelfish, and 2 butterfly fish per day; each piranha can consume 1 goldfish and 3 butterfly fish per day (the piranhas are rather large as a result of their diet); while each squid can consume 1 goldfish and 1 angelfish per day. After a trip to the local pet store, he was able to feed his creatures to capacity, and noticed that 21 goldfish, 21 angelfish, and 35 butterfly fish were eaten. How many of each type of creature does he have?

36. *Resource Allocation* Duffin House is planning its annual Song Festival, when it will serve three kinds of delicacies: granola treats, nutty granola treats, and nuttiest granola treats. The following table shows the ingredients required (in ounces) for a single serving of each delicacy, as well as the total amount of each ingredient available.

	Granola	Nutty Granola	Nuttiest Granola	Total Available
Toasted Oats	1	1	5	1,500
Almonds	4	8	8	10,000
Raisins	2	4	8	4,000

The Song Festival planners at Duffin House would like to use up all the ingredients. Is this possible? If so, how many servings of each kind of delicacy can they make?

37. *Web Site Traffic* OHaganBooks.com has two principal competitors: JungleBooks.com and FarmerBooks.com. Combined Web site traffic at the three sites is estimated at 10,000 hits per day. Only 10% of the hits at OHaganBooks.com result in orders, whereas JungleBooks.com and FarmerBooks.com report that 20% of the hits at their sites result in book orders. Together, the three sites process 1,500 book orders per day. FarmerBooks.com appears to be the most successful of the three, and gets as many book orders as the other two combined. What is the traffic (in hits per day) at each of the sites?

38. *Sales* As the buyer at OHaganBooks.com you are planning to increase stocks of books about music, and have been monitoring worldwide sales. Last year, worldwide sales of books about rock, rap, and classical music amounted to $5.8 billion. Books on rock music brought in twice as much revenue as books on rap music and they brought in 900% the revenue of books on classical music. How much revenue was earned in each of the three categories of books?

39. *Investing in Stocks* Billy-Sean O'Hagan is the treasurer at his college fraternity, which recently earned $12,400 in its annual carwash fundraiser. Billy-Sean decided to invest all the proceeds in the purchase of three computer stocks: HAL, POM, and WELL.

	Price per Share	*Dividend Yield*
HAL	$100	0.5%
POM	$20	1.50%
WELL	$25	0%

If the investment was expected to earn $56 in annual dividends, and he purchased a total of 200 shares, how many shares of each stock did he purchase?

40. *Initial Public Offerings (IPOs)* Duffin House, Higgins Press, and Sickle Publications all went public on the same day recently. John O'Hagan had the opportunity to participate in all three initial public offerings (partly because he and Marjory Duffin are good friends). He made a considerable profit when he sold all of the stock two days later on the open market. The following table shows the purchase price and percentage yield on the investment in each company.

	Purchase Price per Share	Yield
Duffin House (DHS)	$8	20%
Higgins Press (HPR)	$10	15%
Sickle Publications (SPUB)	$15	15%

He invested $20,000 in a total of 2,000 shares, and made a $3,400 profit from the transactions. How many shares in each company did he purchase?

41. *Degree Requirements* During his lunch break, John O'Hagan decides to devote some time to assisting his son Billy-Sean, who is having a terrible time coming up with a college course schedule. One reason for this is the very complicated Bulletin of Suburban State University. It reads as follows:

> *All candidates for the degree of Bachelor of Science at SSU must take a total of 124 credits from the Sciences, Fine Arts, Liberal Arts, and Mathematics,[28] including an equal number of Science and Fine Arts credits, and twice as many Mathematics credits as Science credits and Fine Arts credits combined, but with Liberal Arts credits exceeding Mathematics credits by exactly one-third of the number of Fine Arts credits.*

What are all the possible degree programs for Billy-Sean?

42. *Degree Requirements* Having finally decided on his degree program, Billy-Sean learns that the Suburban State University Senate (under pressure from the English Department) has revised the Bulletin to include a "Verbal Expression" component in place of the Fine Arts requirement in all programs (including the sciences):

> *All candidates for the degree of Bachelor of Science at SSU must take a total of 120 credits from the Liberal Arts, Sciences, Verbal Expression, and Mathematics, including an equal number of Science and Liberal Arts credits, and twice as many Verbal Expression credits as Science credits and Liberal Arts credits combined, but with Liberal Arts credits exceeding Mathematics credits by one quarter of the number of Verbal Expression Credits.*

What are now the possible degree programs for Billy-Sean?

---

[28]Strictly speaking, mathematics is not a science; it is the Queen of the Sciences, although we like to think of it as the Mother of all Sciences.

43. *Network Traffic* All book orders received at the Order Department at OHaganBooks.com are transmitted through a small computer network to the Shipping Department. The following diagram shows the network (which uses two intermediate computers as routers), together with some of the average daily traffic measured in book orders.

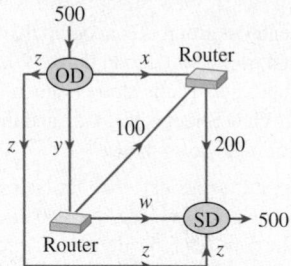

OD = Order department
SD = Shipping department

a. Set up a system of linear equations in which the unknowns give the average traffic along the paths labeled x, y, z, w, and find the general solution.
b. What is the minimum volume of traffic along y?
c. What is the maximum volume of traffic along w?
d. If there is no traffic along z, find the volume of traffic along all the paths.
e. If there is the same volume of traffic along y and z, what is the volume of traffic along w?

44. *Business Retreats* Marjory Duffin is planning a joint business retreat for Duffin House and OHaganBooks.com at Laguna Surf City, but is concerned about traffic conditions (she feels that too many cars tend to spoil the ambiance of a seaside retreat). She managed to obtain the following map from the Laguna Surf City Engineering Department (all the streets are one-way as indicated). The counters show traffic every five minutes.

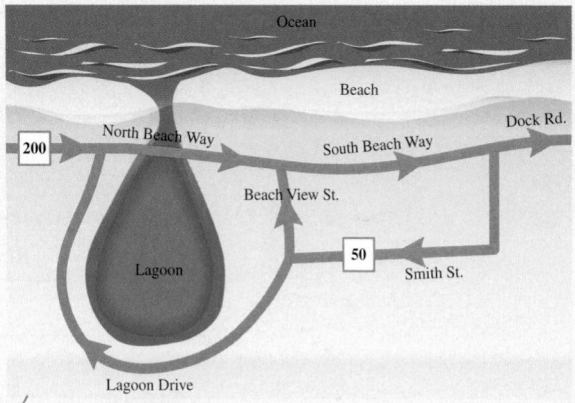

a. Set up and solve the associated system of linear equations. *Be sure to give the general solution.* (Take x = traffic along North Beach Way, y = traffic along South Beach Way, z = traffic along Beach View

St., u = traffic along Lagoon Drive and v = traffic along Dock Road.)

b. Assuming that all roads are one-way in the directions shown, what, if any, is the maximum possible traffic along Lagoon Drive?

c. The Laguna Surf City Traffic Department is considering opening up Beach View Street to two-way traffic, but an environmentalist group is concerned that this will result in increased traffic on Lagoon Drive. What, if any, is the maximum possible traffic along Lagoon Drive assuming that Beach View Street is two-way and the traffic counter readings are as shown?

45. *Shipping* On the same day that the sales department at Duffin House received an order for 600 packages from the OHaganBooks.com Texas headquarters, it received an additional order for 200 packages from FantasyBooks.com, based in California. Duffin House has warehouses in New York and Illinois. The Illinois warehouse is closing down

and must clear all 300 packages it has in stock. Shipping costs per package of books are as follows:

New York to Texas: $20   New York to California: $50
Illinois to Texas: $30   Illinois to California: $40

Is it possible to fill both orders and clear the Illinois warehouse at a cost of $22,000? If so, how many packages should be sent from each warehouse to each online bookstore?

46. *Transportation Scheduling* Duffin House is about to start a promotional blitz for its new book, *Physics for the Liberal Arts.* The company has 20 salespeople stationed in Chicago and 10 in Denver, and would like to fly 15 to sales fairs at each of Los Angeles and New York. A round-trip plane flight from Chicago to LA costs $200; from Chicago to NY costs $150; from Denver to LA costs $400; and from Denver to NY costs $200. For tax reasons, Duffin House needs to budget exactly $6,500 for the total cost of the plane flights. How many salespeople should the company fly from each of Chicago and Denver to each of LA and NY?

Case Study   Hybrid Cars—Optimizing the Degree of Hybridization

You are involved in new model development at a major automobile company. The company is planning to introduce two new plug-in hybrid electric vehicles: the subcompact "Green Town Hopper" and the midsize "Electra Supreme," and your department must decide on the degree of hybridization (DOH) for each of these models that will result in the largest reduction in gasoline consumption. (The DOH of a vehicle is defined as the ratio of electric motor power to the total power, and typically ranges from 10% to 50%. For example, a model with a 20% DOH has an electric motor that delivers 20% of the total power of the vehicle.)

The tables below show the benefit for each of the two models, measured as the estimated reduction in annual gasoline consumption, as well as an estimate of retail cost increment, for various DOH percentages[29]: (The retail cost increment estimate is given by the formula $5,000 + 50(DOH - 10)$ for the Green Town Hopper and $7,000 + 50(DOH - 10)$ for the Electra Supreme.)

Green Town Hopper

DOH (%)	10	20	50
Reduction in Annual Consumption (gals)	180	230	200
Retail Cost Increment ($)	5,000	5,500	7,000

Electra Supreme

DOH (%)	10	20	50
Reduction in Annual Consumption (gals)	220	270	260
Retail Cost Increment ($)	7,000	7,500	9,000

---

[29]The figures are approximate and based on data for two actual vehicles as presented in a 2006 paper entitled *Cost-Benefit Analysis of Plug-In Hybrid Electric Vehicle Technology* by A. Simpson. Source: National Renewable Energy Laboratory, U.S. Department of Energy (www.nrel.gov).

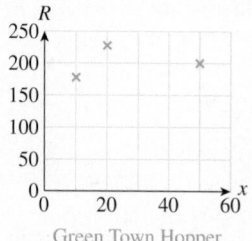

Green Town Hopper

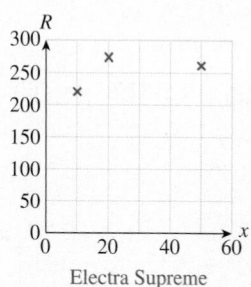

Electra Supreme

Figure 9

Notice that increasing the DOH toward 50% results in a decreased benefit. This is due in part to the need to increase the weight of the batteries while keeping the vehicle performance at a desirable level, thus necessitating a more powerful gasoline engine. The *optimum* DOH is the percentage that gives the largest reduction in gasoline consumption, and this is what you need to determine. Since the optimum DOH may not be 20%, you would like to create a mathematical model to compute the reduction R in gas consumption as a function of the DOH x. Your first inclination is to try linear equations—that is, an equation of the form

$$R = ax + b \quad (a \text{ and } b \text{ constants}),$$

but you quickly discover that the data simply won't fit, no matter what the choice of the constants. The reason for this can be seen graphically by plotting R versus x (Figure 9). In neither case do the three points lie on a straight line. In fact, the data are not even *close* to being linear. Thus, you will need curves to model these data. After giving the matter further thought, you remember something your mathematics instructor once said: The simplest curve passing through any three points not all on the same line is a parabola. Since you are looking for a simple model of the data, you decide to try a parabola. A general parabola has the equation

$$R = ax^2 + bx + c,$$

where a, b, and c are constants. The problem now is: What are a, b, and c? You decide to try substituting the values of R and x for the Green Town Hopper into the general equation, and you get the following:

$$x = 10, R = 180 \quad \text{gives} \quad 180 = 100a + 10b + c$$
$$x = 20, R = 230 \quad \text{gives} \quad 230 = 400a + 20b + c$$
$$x = 50, R = 200 \quad \text{gives} \quad 200 = 2{,}500a + 50b + c.$$

Now you notice that you have three linear equations in three unknowns! You solve the system:

$$a = -0.15, b = 9.5, c = 100.$$

Thus your reduction equation for the Green Town Hopper becomes

$$R = -0.15x^2 + 9.5x + 100.$$

For the Electra Supreme, you get

$$x = 10, R = 220: \quad 220 = 100a + 10b + c$$
$$x = 20, R = 270: \quad 270 = 400a + 20b + c$$
$$x = 50, R = 260: \quad 260 = 2{,}500a + 50b + c.$$

$$a = -0.1\bar{3}, b = 9, c = 143.\bar{3}$$

and so

$$R = -0.1\bar{3}x^2 + 9x + 143.\bar{3}.$$

Figure 10 shows the parabolas superimposed on the data points. You can now estimate a value for the optimal DOH as the value of x that gives the largest benefit R.

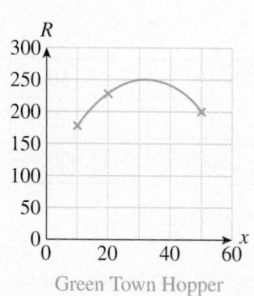

Green Town Hopper

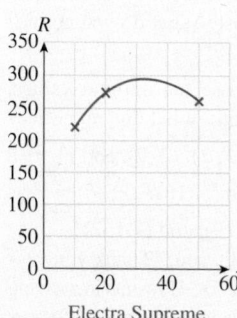

Electra Supreme

Figure 10

Recalling that the x-coordinate of the vertex of the parabola $y = ax^2 + bx + c$ is $x = -\frac{b}{2a}$, you obtain the following estimates:

$$\text{Green Town Hopper: Optimal DOH} = -\frac{9.5}{2(-0.15)} \approx 31.67\%$$

$$\text{Electra Supreme: Optimal DOH} = -\frac{9}{2(-0.1\bar{3})} \approx 33.75\%.$$

You can now use the formulas given earlier to estimate the resulting reductions in gasoline consumption and increases in cost:

Green Town Hopper:

Reduction in gasoline consumption $= R \approx -0.15(31.67)^2 + 9.5(31.67) + 100$

$$\approx 250.4 \text{ gals/year}$$

Retail Cost Increment $\approx 5{,}000 + 50(31.67 - 10) \approx \$6{,}080$

Electra Supreme:

Reduction in gasoline consumption $= R = -0.1\bar{3}(33.75)^2 + 9(33.75) + 143.\bar{3}$

$$\approx 295.2 \text{ gals/year}$$

Retail Cost Increment $\approx 7{,}000 + 50(33.75 - 10) \approx \$8{,}190.$

You thus submit the following estimates: The optimal degree of hybridization for the Green Town Hopper is about 31.67% and will result in a reduction in gasoline consumption of 250.4 gals/year and a retail cost increment of around \$6,080. The optimal degree of hybridization for the Electra Supreme is about 33.75% and will result in a reduction in gasoline consumption of 295.2 gals/year and a retail cost increment of around \$8,190.

EXERCISES

1. Repeat the computations above for the "Earth Suburban" using the following data:

Earth Suburban

DOH (%)	10	20	50
Reduction in Annual Consumption (gals)	240	330	300
Retail Cost Increment ($)	9,000	9,500	11,000

Retail cost increment $= 9{,}000 + 50(DOH - 10)$

2. ⊤ Repeat the analysis for the Green Town Hopper, but this time take x to be the cost increment, in thousands of dollars. (The curve of benefit versus cost is referred to as a *cost-benefit* curve.) What value of DOH corresponds to the optimal cost? What do you notice? Comment on the answer.

3. ⊤ Repeat the analysis for the Electra Supreme, but this time use the optimal values, DOH $= 33.8$, Reduction $= 295.2$ gals/year in place of the 20% data. What do you notice?

4. Find the equation of the parabola that passes through the points $(1, 2)$, $(2, 9)$, and $(3, 19)$.

5. Is there a parabola that passes through the points $(1, 2)$, $(2, 9)$, and $(3, 16)$?

6. Is there a parabola that passes though the points $(1, 2)$, $(2, 9)$, $(3, 19)$, and $(-1, 2)$?

7. ⊤ You submit your recommendations to your manager, and she tells you, "Thank you very much, but we have additional data for the Green Town Hopper: A 30% DOH results in a saving of 255 gals/year. Please resubmit your recommendations taking this into account by tomorrow." HINT [You now have four data points on each graph, so try a general cubic instead: $R = ax^3 + bx^2 + cx + d$. Use a graph to estimate the optimal DOH.]

TI-83/84 Plus   Technology Guide

Section 3.1

Example 1 (page 177)   Find all solutions (x, y) of the following system of two equations:

$$x + y = 3$$
$$x - y = 1.$$

Solution with Technology

You can use a graphing calculator to draw the graphs of the two equations on the same set of axes and to check the solution. First, solve the equations for y, obtaining $y = -x + 3$ and $y = x - 1$. On the TI-83/84 Plus:

1. Set

    Y₁=-X+3
    Y₂=X-1

2. Decide on the range of x-values you want to use. As in Figure 1, let us choose the range $[-4, 4]$.[30]

3. In the WINDOW menu set Xmin $= -4$ and Xmax $= 4$.

4. Press ZOOM and select Zoomfit to set the y range.

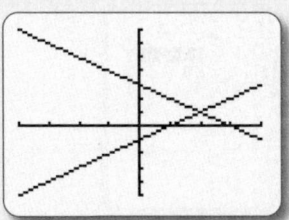

You can now zoom in for a more accurate view by choosing a smaller x-range that includes the point of intersection, like $[1.5, 2.5]$, and using Zoomfit again. You can also use the trace feature to see the coordinates of points near the point of intersection.

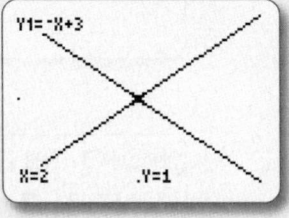

To check that $(2, 1)$ is the correct solution, use the table feature to compare the two values of y corresponding to $x = 2$:

1. Press 2ND TABLE.

2. Set X $= 2$, and compare the corresponding values of Y_1 and Y_2; they should each be 1.

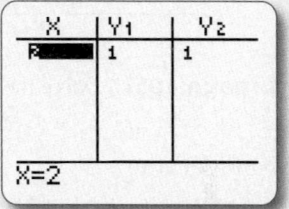

Q: *How accurate is the answer shown using the trace feature?*

A: That depends. We can increase the accuracy up to a point by zooming in on the point of intersection of the two graphs. But there is a limit to this: Most graphing calculators are capable of giving an answer correct to about 13 decimal places. This means, for instance, that, in the eyes of the TI-83/84 Plus, 2.000 000 000 000 1 is exactly the same as 2 (subtracting them yields 0). It follows that if you attempt to use a window so narrow that you need approximately 13 significant digits to distinguish the left and right edges, you will run into accuracy problems.

Section 3.2

Row Operations with a TI-83/84 Plus   Start by entering the matrix into [A] using MATRIX EDIT. You can then do row operations on [A] using the following instructions (*row, *row+, and rowSwap are found in the MATRIX MATH menu).

---

[30]How did we come up with this interval? Trial and error. You might need to try several intervals before finding one that gives a graph showing the point of intersection clearly.

TECHNOLOGY GUIDE

Row Operation	TI-83/84 Plus Instruction (Matrix name is [A])
$R_i \to kR_i$	`*row(k,[A],i)→[A]`
Example: $R_2 \to 3R_2$	`*row(3,[A],2)→[A]`
$R_i \to R_i + kR_j$	`*row+(k,[A],j,i)→[A]`
Examples: $R_1 \to R_1 - 3R_2$	`*row+(-3,[A],2,1)→[A]`
$R_1 \to 4R_1 - 3R_2$	`*row(4,[A],1)→[A]` `*row+(-3,[A],2,1)→[A]`
Swap R_i and R_j	`rowSwap([A],i,j)→[A]`
Example: Swap R_1 and R_2	`rowSwap([A],1,2)→[A]`

Example 1 (page 195)   Solve the system:

$$x - y + 5z = -6$$
$$3x + 3y - z = 10$$
$$x + 3y + 2z = 5.$$

Solution with Technology

1. Begin by entering the matrix into `[A]` using MATRX EDIT. Only three columns can be seen at a time; you can see the rest of the matrix by scrolling left or right.

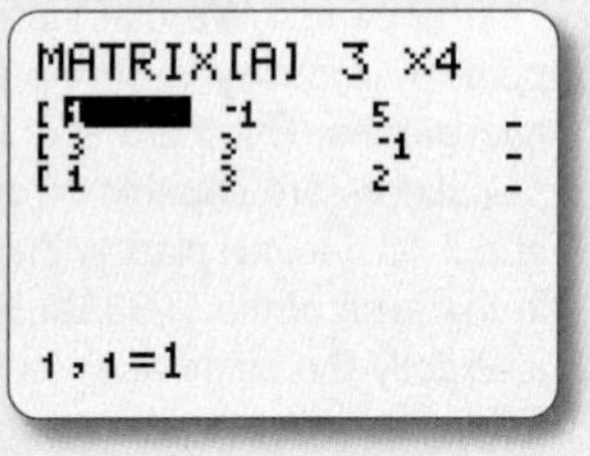

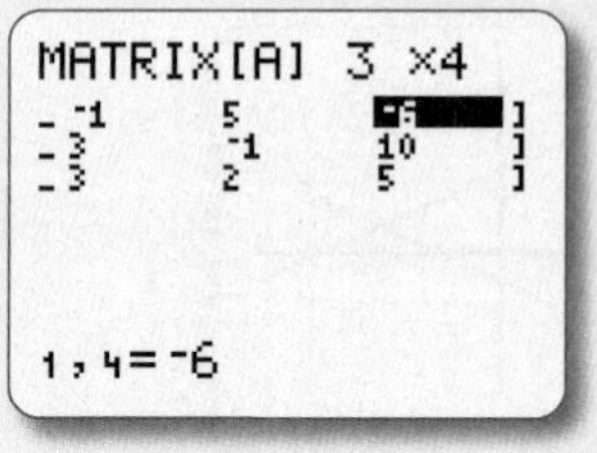

```
MATRIX[A]  3 ×4
[ 1      -1      5     -
[ 3       3     -1     -
[ 1       3      2     -


1,1=1
```

```
MATRIX[A]  3 ×4
 -1      5     -6   ]
 -3     -1     10   ]
 -3      2      5   ]


1,4=-6
```

2. Now perform the operations given in Example 1:

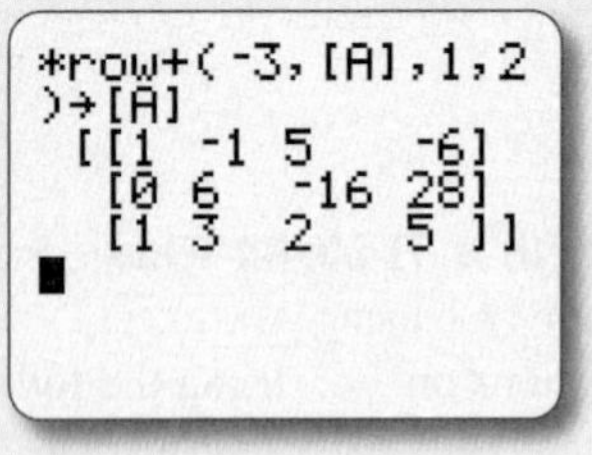

```
*row+(-3,[A],1,2
)→[A]
[[1  -1  5   -6]
 [0  6  -16 28]
 [1  3   2   5 ]]
■
```

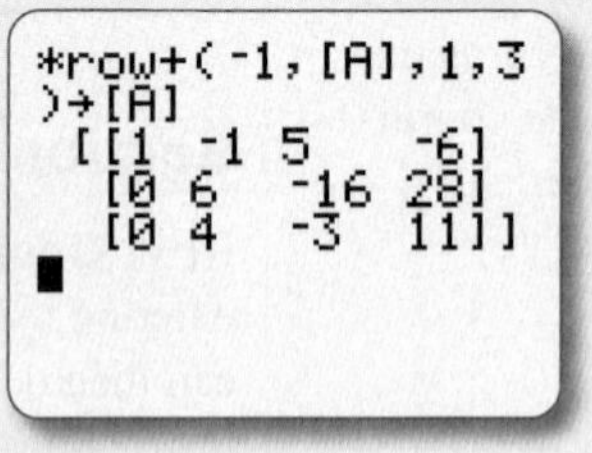

```
*row+(-1,[A],1,3
)→[A]
[[1  -1  5   -6]
 [0  6  -16 28]
 [0  4  -3  11]]
■
```

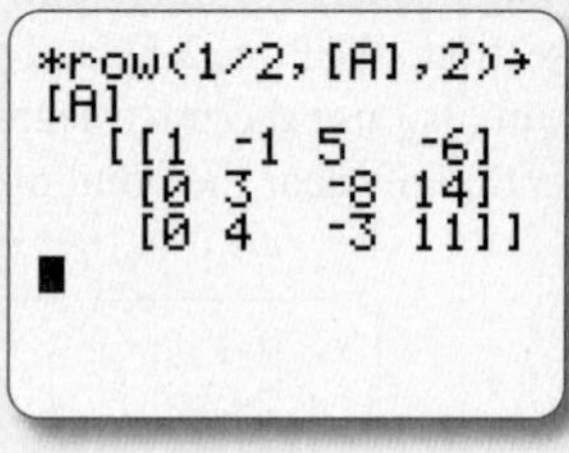

```
*row(1/2,[A],2)→
[A]
[[1  -1  5   -6]
 [0  3  -8  14]
 [0  4  -3  11]]
■
```

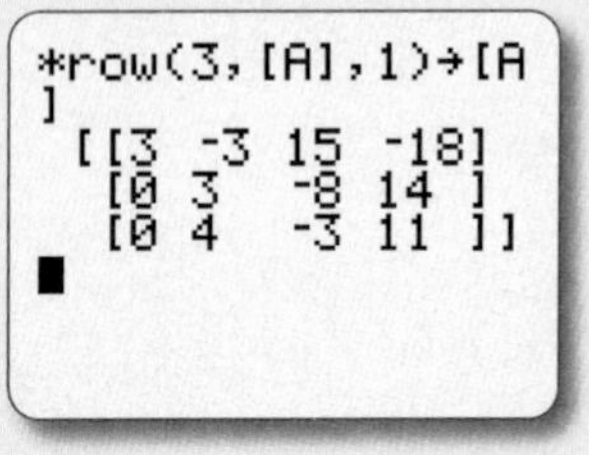

```
*row(3,[A],1)→[A
]
[[3  -3  15 -18]
 [0  3  -8  14]
 [0  4  -3  11 ]]
■
```

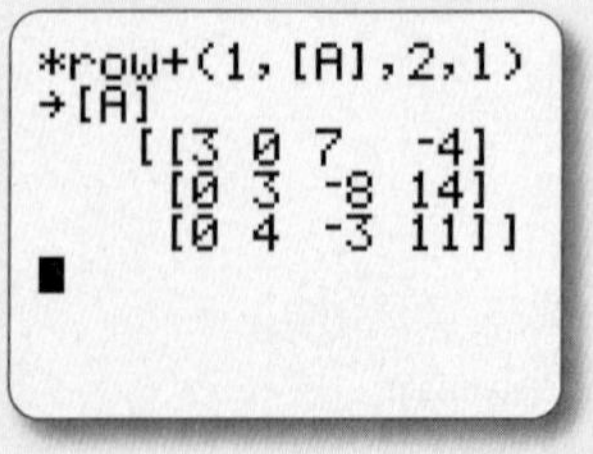

```
*row+(1,[A],2,1)
→[A]
[[3  0  7   -4]
 [0  3  -8  14]
 [0  4  -3  11]]
■
```

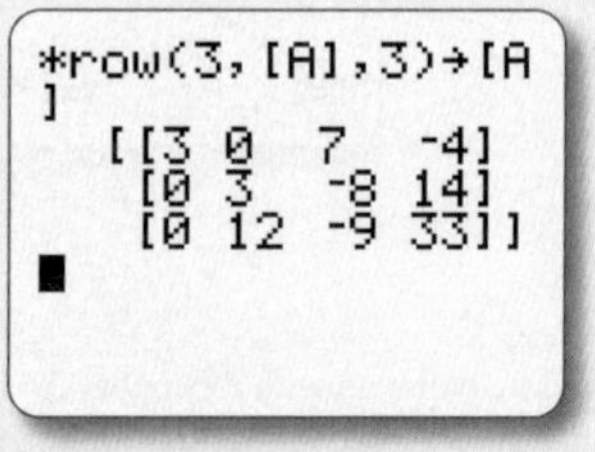

```
*row(3,[A],3)→[A
]
[[3  0  7   -4]
 [0  3  -8  14]
 [0  12 -9  33]]
■
```

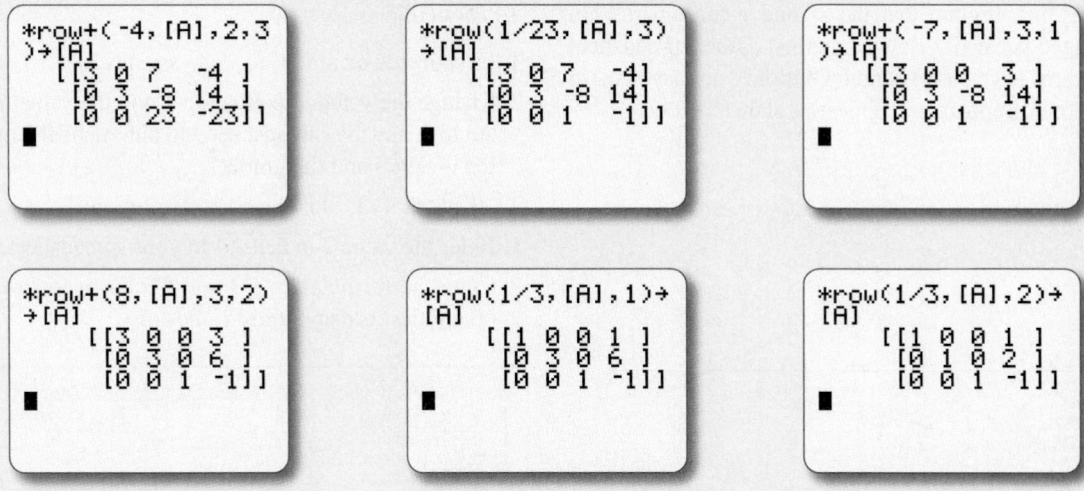

As in Example 1, we can now read the solution from the right-hand column: $x = 1, y = 2, z = -1$.

Note The TI-83/84 Plus has a function, `rref`, that gives the reduced row echelon form of a matrix in one step. (See the text for the definition of reduced row echelon form.) Internally it uses a variation of Gauss-Jordan reduction to do this. ■

SPREADSHEET Technology Guide

Section 3.1

Example 1 (page 177) Find all solutions (x, y) of the following system of two equations:

$$x + y = 3$$
$$x - y = 1.$$

Solution with Technology

You can use a spreadsheet to draw the graphs of the two equations on the same set of axes, and to check the solution.

1. Solve the equations for y, obtaining $y = -x + 3$ and $y = x - 1$.

2. To graph these lines we can use the following simple worksheet:

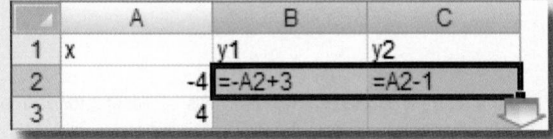

The two values of x give the x-coordinates of the two points we will use as endpoints of the lines. (We have—somewhat arbitrarily—chosen the range $[-4, 4]$ for x.) The formula for the first line, $y = -x + 3$, is in cell B2, and the formula for the second line, $y = x - 1$, is in C2.

3. Copy these two cells as shown to yield the following result:

	A	B	C
1	x	y1	y2
2	-4	7	-5
3	4	-1	3

4. For the graph, select all nine cells and create a scatter graph with line segments joining the data points. Instruct the spreadsheet to insert a chart and select the "scatter" option. In the same dialogue box, select the option that shows points connected by lines. If you are using Excel, press "Next" to bring up a new dialogue box called "Data Type," where you should make sure that the "Series in Columns" option is selected,

telling the program that the x- and y-coordinates are arranged vertically, down columns. (Other spreadsheet programs have corresponding options you may need to set.) Your graph should appear as shown below.

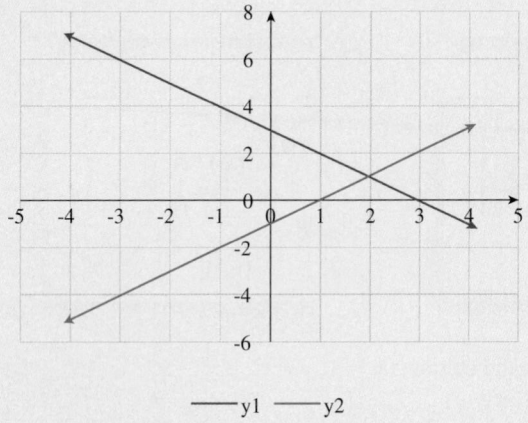

—— y1 —— y2

To zoom in:

1. First decide on a new x-range, say, [1, 3].
2. Change the value in cell A2 to 1 and the value in cell A3 to 3, and the spreadsheet will automatically update the y-values and the graph.*

To check that (2, 1) is the correct solution:

1. Enter the value 2 in cell A4 in your spreadsheet.
2. Copy the formulas in B3 and C3 down to row 4 to obtain the corresponding values of y.

	A	B	C
1	x	y1	y2
2	-4	7	-5
3	4	-1	3
4	2	1	1

Because the values of y agree, we have verified that (2, 1) is a solution of both equations.

Section 3.2

Row Operations with a Spreadsheet   To a spreadsheet, a block of data with one or more rows or columns is an **array**, and a spreadsheet has built in the capability to handle arrays in much the same way as it handles single cells. Consider the following example:

$$\begin{bmatrix} 1 & 3 & -4 \\ 0 & 4 & 2 \end{bmatrix} 3R_2 \rightarrow \begin{bmatrix} 1 & 3 & -4 \\ 0 & 12 & 6 \end{bmatrix}.$$ Replace R_2 by $3R_2$.

1. Enter the original matrix in a convenient location, say the cells A1 through C2.
2. To get the first row of the new matrix, which is simply a copy of the first row of the old matrix, decide where you want to place the new matrix, and highlight *the whole row of the new matrix,* say A4:C4.
3. Enter the formula =A1:C1 (the easiest way to do this is to type "=" and then use the mouse to select cells A1 through C1, i.e., the first row of the old matrix).

	A	B	C	D
1	1	3	-4	
2	0	4	2	3R2
3				
4	=A1:C1			

Enter formula for new first row.

$\rightarrow$

	A	B	C	D
1	1	3	-4	
2	0	4	2	3R2
3				
4	1	3	-4	

Press Control-Shift-Enter.

---

* NOTE Your spreadsheet will set the y-range of the graph automatically, and the range it chooses may not always be satisfactory. It may sometimes be necessary to "narrow down" the range of one or both axes (in Excel, double-click on the axis in question).

this follows the "RC" convention: The row number is specified first and the column number second.

Quick Example

With $A = \begin{bmatrix} 2 & 0 & 1 \\ 33 & -22 & 0 \end{bmatrix}$,

$a_{13} = 1$ First row, third column

$a_{21} = 33.$ Second row, first column

According to the labeling convention, the entries of the matrix A above are

$$A = \begin{bmatrix} a_{11} & a_{12} & a_{13} \\ a_{21} & a_{22} & a_{23} \end{bmatrix}.$$

In general, the $m \times n$ matrix A has its entries labeled as follows:

$$A = \begin{bmatrix} a_{11} & a_{12} & a_{13} & \cdots & a_{1n} \\ a_{21} & a_{22} & a_{23} & \cdots & a_{2n} \\ \vdots & \vdots & \vdots & \ddots & \vdots \\ a_{m1} & a_{m2} & a_{m3} & \cdots & a_{mn} \end{bmatrix}.$$

We say that two matrices A and B are **equal** if they have the same dimensions and the corresponding entries are equal. Note that a 3×4 matrix can never equal a 3×5 matrix because they do not have the same dimensions.

EXAMPLE 1 Matrix Equality

Let $A = \begin{bmatrix} 7 & 9 & x \\ 0 & -1 & y+1 \end{bmatrix}$ and $B = \begin{bmatrix} 7 & 9 & 0 \\ 0 & -1 & 11 \end{bmatrix}$. Find the values of x and y such that $A = B$.

Solution For the two matrices to be equal, we must have corresponding entries equal, so

$\begin{aligned} x &= 0 \\ y + 1 &= 11 \quad \text{or} \quad y = 10. \end{aligned}$ $\begin{aligned} a_{13} &= b_{13} \\ a_{23} &= b_{23} \end{aligned}$

➡ **Before we go on...** Note in Example 1 that the matrix equation

$$\begin{bmatrix} 7 & 9 & x \\ 0 & -1 & y+1 \end{bmatrix} = \begin{bmatrix} 7 & 9 & 0 \\ 0 & -1 & 11 \end{bmatrix}$$

is really six equations in one: $7 = 7, 9 = 9, x = 0, 0 = 0, -1 = -1,$ and $y + 1 = 11$. We used only the two that were interesting. ∎

Row Matrix, Column Matrix, and Square Matrix

A matrix with a single row is called a **row matrix**, or **row vector**. A matrix with a single column is called a **column matrix** or **column vector**. A matrix with the same number of rows as columns is called a **square matrix**.

Quick Examples

The 1×5 matrix $C = [3 \quad -4 \quad 0 \quad 1 \quad -11]$ is a row matrix.

The 4×1 matrix $D = \begin{bmatrix} 2 \\ 10 \\ -1 \\ 8 \end{bmatrix}$ is a column matrix.

The 3×3 matrix $E = \begin{bmatrix} 1 & -2 & 0 \\ 0 & 1 & 4 \\ -4 & 32 & 1 \end{bmatrix}$ is a square matrix.

Matrix Addition and Subtraction

The first matrix operations we discuss are matrix addition and subtraction. The rules for these operations are simple.

Matrix Addition and Subtraction

Two matrices can be added (or subtracted) if and only if they have the same dimensions. To add (or subtract) two matrices of the same dimensions, we add (or subtract) the corresponding entries. More formally, if A and B are $m \times n$ matrices, then $A + B$ and $A - B$ are the $m \times n$ matrices whose entries are given by:

$(A + B)_{ij} = A_{ij} + B_{ij}$ ijth entry of the sum = sum of the ijth entries
$(A - B)_{ij} = A_{ij} - B_{ij}.$ ijth entry of the difference = difference of the ijth entries

Visualizing Matrix Addition

$$\begin{bmatrix} 2 & -3 \\ 1 & 0 \end{bmatrix} + \begin{bmatrix} 1 & 1 \\ -2 & 1 \end{bmatrix} = \begin{bmatrix} 3 & -2 \\ -1 & 1 \end{bmatrix}$$

Quick Examples

1. $\begin{bmatrix} 2 & -3 \\ 1 & 0 \\ -1 & 3 \end{bmatrix} + \begin{bmatrix} 9 & -5 \\ 0 & 13 \\ -1 & 3 \end{bmatrix} = \begin{bmatrix} 11 & -8 \\ 1 & 13 \\ -2 & 6 \end{bmatrix}$ Corresponding entries added

2. $\begin{bmatrix} 2 & -3 \\ 1 & 0 \\ -1 & 3 \end{bmatrix} - \begin{bmatrix} 9 & -5 \\ 0 & 13 \\ -1 & 3 \end{bmatrix} = \begin{bmatrix} -7 & 2 \\ 1 & -13 \\ 0 & 0 \end{bmatrix}$ Corresponding entries subtracted

using Technology

Technology can be used to enter, add, and subtract matrices. Here is an outline (see the Technology Guides at the end of the chapter for additional details on using a TI-83/84 Plus or a spreadsheet):

TI-83/84 Plus
Entering a matrix: MATRX ; EDIT
Select a name, ENTER ; type in the entries.
Adding two matrices:
Home Screen: [A]+[B]
(Use MATRX ; NAMES to enter them on the home screen.)
[More details on page 304.]

Spreadsheet
Entering a matrix: Type entries in a convenient block of cells.
Adding two matrices: Highlight block where you want the answer to appear.
Type "="; highlight first matrix; type "+"; highlight second matrix; press Control+Shift+Enter
[More details on page 307.]

Website
www.WanerMath.com
    Student Home
        → On Line Utilities
        → Matrix Algebra Tool

Enter matrices as shown (use a single letter for the name).

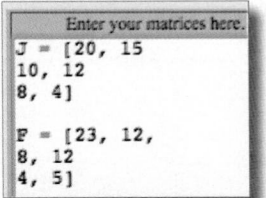

To compute their difference, type F-J in the formula box and press "Compute". (You can enter multiple formulas separated by commas in the formula box. For instance, F+J, F-J will compute both the sum and the difference. *Note:* The utility is case sensitive, so be consistent.)

EXAMPLE 2 Sales

The A-Plus auto parts store chain has two outlets, one in Vancouver and one in Quebec. Among other things, it sells wiper blades, windshield cleaning fluid, and floor mats. The monthly sales of these items at the two stores for two months are given in the following tables:

January Sales

	Vancouver	Quebec
Wiper Blades	20	15
Cleaning Fluid (bottles)	10	12
Floor Mats	8	4

February Sales

	Vancouver	Quebec
Wiper Blades	23	12
Cleaning Fluid (bottles)	8	12
Floor Mats	4	5

Use matrix arithmetic to calculate the change in sales of each product in each store from January to February.

Solution The tables suggest two matrices:

$$J = \begin{bmatrix} 20 & 15 \\ 10 & 12 \\ 8 & 4 \end{bmatrix} \quad \text{and} \quad F = \begin{bmatrix} 23 & 12 \\ 8 & 12 \\ 4 & 5 \end{bmatrix}.$$

To compute the change in sales of each product for both stores, we want to subtract corresponding entries in these two matrices. In other words, we want to compute the difference of the two matrices:

$$F - J = \begin{bmatrix} 23 & 12 \\ 8 & 12 \\ 4 & 5 \end{bmatrix} - \begin{bmatrix} 20 & 15 \\ 10 & 12 \\ 8 & 4 \end{bmatrix} = \begin{bmatrix} 3 & -3 \\ -2 & 0 \\ -4 & 1 \end{bmatrix}.$$

Thus, the change in sales of each product is the following:

	Vancouver	Quebec
Wiper Blades	3	−3
Cleaning Fluid (bottles)	−2	0
Floor Mats	−4	1

Scalar Multiplication

A matrix A can be added to itself because the expression $A + A$ is the sum of two matrices that have the same dimensions. When we compute $A + A$, we end up doubling every entry in A. So we can think of the expression $2A$ as telling us to *multiply every element in A by* 2.

In general, to multiply a matrix by a number, multiply every entry in the matrix by that number. For example,

$$6 \begin{bmatrix} \frac{5}{2} & -3 \\ 1 & 0 \\ -1 & \frac{5}{6} \end{bmatrix} = \begin{bmatrix} 15 & -18 \\ 6 & 0 \\ -6 & 5 \end{bmatrix}.$$

It is traditional when talking about matrices to call individual numbers **scalars**. For this reason, we call the operation of multiplying a matrix by a number **scalar multiplication**.

using Technology

Technology can be used to compute scalar multiples:

TI-83/84 Plus
Enter the matrix [A] using MATRX; EDIT
Home Screen: 0.65 [A]

Spreadsheets
Enter the matrix A in a convenient 3×2 block of cells.
Highlight block where you want the answer to appear.
Type =0.65*; highlight matrix A; press Control+Shift+Enter.

 Website
www.WanerMath.com
    Student Home
        → On Line Utilities
        → Matrix Algebra Tool
Enter the matrix A as shown.

```
Enter your matrices here.
A = [140, 105
30, 36
96, 48]
```

Type 0.65*A in the formula box and press "Compute".

EXAMPLE 3 Sales

The revenue generated by sales in the Vancouver and Quebec branches of the A-Plus auto parts store (see Example 2) was as follows:

January Sales in Canadian Dollars

	Vancouver	Quebec
Wiper Blades	140.00	105.00
Cleaning Fluid	30.00	36.00
Floor Mats	96.00	48.00

If the Canadian dollar was worth $0.65 U.S. at the time, compute the revenue in U.S. dollars.

Solution We need to multiply each revenue figure by 0.65. Let A be the matrix of revenue figures in Canadian dollars:

$$A = \begin{bmatrix} 140.00 & 105.00 \\ 30.00 & 36.00 \\ 96.00 & 48.00 \end{bmatrix}.$$

The revenue figures in U.S. dollars are then given by the scalar multiple

$$0.65A = 0.65 \begin{bmatrix} 140.00 & 105.00 \\ 30.00 & 36.00 \\ 96.00 & 48.00 \end{bmatrix} = \begin{bmatrix} 91.00 & 68.25 \\ 19.50 & 23.40 \\ 62.40 & 31.20 \end{bmatrix}.$$

In other words, in U.S. dollars, $91 worth of wiper blades was sold in Vancouver, $68.25 worth of wiper blades was sold in Quebec, and so on.

Formally, scalar multiplication is defined as follows:

Scalar Multiplication

If A is an $m \times n$ matrix and c is a real number, then cA is the $m \times n$ matrix obtained by multiplying all the entries of A by c. (We usually use lowercase letters $c, d, e, \ldots$ to denote scalars.) Thus, the ijth entry of cA is given by

$$(cA)_{ij} = c(A_{ij}).$$

In words, this rule is: To get the ijth entry of cA, multiply the ijth entry of A by c.

EXAMPLE 4 Combining Operations

Let $A = \begin{bmatrix} 2 & -1 & 0 \\ 3 & 5 & -3 \end{bmatrix}$, $B = \begin{bmatrix} 1 & 3 & -1 \\ 5 & -6 & 0 \end{bmatrix}$, and $C = \begin{bmatrix} x & y & w \\ z & t+1 & 3 \end{bmatrix}$.

Evaluate the following: $4A$, xB, and $A + 3C$.

Solution First, we find $4A$ by multiplying each entry of A by 4:

$$4A = 4 \begin{bmatrix} 2 & -1 & 0 \\ 3 & 5 & -3 \end{bmatrix} = \begin{bmatrix} 8 & -4 & 0 \\ 12 & 20 & -12 \end{bmatrix}.$$

Similarly, we find xB by multiplying each entry of B by x:

$$xB = x \begin{bmatrix} 1 & 3 & -1 \\ 5 & -6 & 0 \end{bmatrix} = \begin{bmatrix} x & 3x & -x \\ 5x & -6x & 0 \end{bmatrix}.$$

We get $A + 3C$ in two steps as follows:

$$A + 3C = \begin{bmatrix} 2 & -1 & 0 \\ 3 & 5 & -3 \end{bmatrix} + 3 \begin{bmatrix} x & y & w \\ z & t+1 & 3 \end{bmatrix}$$

$$= \begin{bmatrix} 2 & -1 & 0 \\ 3 & 5 & -3 \end{bmatrix} + \begin{bmatrix} 3x & 3y & 3w \\ 3z & 3t+3 & 9 \end{bmatrix}$$

$$= \begin{bmatrix} 2+3x & -1+3y & 3w \\ 3+3z & 3t+8 & 6 \end{bmatrix}.$$

Addition and scalar multiplication of matrices have nice properties, reminiscent of the properties of addition and multiplication of real numbers. Before we state them, we need to introduce some more notation.

If A is any matrix, then $-A$ is the matrix $(-1)A$. In other words, $-A$ is A multiplied by the scalar -1. This amounts to changing the signs of all the entries in A. For example,

$$-\begin{bmatrix} 4 & -2 & 0 \\ 6 & 10 & -6 \end{bmatrix} = \begin{bmatrix} -4 & 2 & 0 \\ -6 & -10 & 6 \end{bmatrix}.$$

For any two matrices A and B, $A - B$ is the same as $A + (-B)$. (Why?)

Also, a **zero matrix** is a matrix all of whose entries are zero. Thus, for example, the 2×3 zero matrix is

$$O = \begin{bmatrix} 0 & 0 & 0 \\ 0 & 0 & 0 \end{bmatrix}.$$

Now we state the most important properties of the operations that we have been talking about:

Properties of Matrix Addition and Scalar Multiplication

If A, B, and C are any $m \times n$ matrices and if O is the zero $m \times n$ matrix, then the following hold:

$A + (B + C) = (A + B) + C$	*Associative law*
$A + B = B + A$	*Commutative law*
$A + O = O + A = A$	*Additive identity law*
$A + (-A) = O = (-A) + A$	*Additive inverse law*
$c(A + B) = cA + cB$	*Distributive law*
$(c + d)A = cA + dA$	*Distributive law*
$1A = A$	*Scalar unit*
$0A = O$	*Scalar zero*

These properties would be obvious if we were talking about addition and multiplication of *numbers*, but here we are talking about addition and multiplication of *matrices*. We are using "+" to mean something new: matrix addition. There is no reason why matrix addition has to obey *all* the properties of addition of numbers. It happens that it does obey many of them, which is why it is convenient to call it *addition* in the first place. This means that we can manipulate equations involving matrices in much the same way that we manipulate equations involving numbers. One word of caution: We haven't yet discussed how to multiply matrices, and it probably isn't what you think. It will turn out that multiplication of matrices does *not* obey all the same properties as multiplication of numbers.

Transposition

We mention one more operation on matrices:

Transposition

 using Technology

Technology can be used to compute transposes like Quick Example 1 on the right:

TI-83/84 Plus
Enter the matrix [A] using
MATRX ; EDIT
Home Screen: [A] ^{T}
(The symbol ^{T} is found in
MATRX ; MATH.)

Spreadsheet
Enter the matrix A in a convenient 3×4 block of cells (e.g., A1–D3). Highlight a 4×3 block for the answer (e.g., A5–C8).
Type =TRANSPOSE(A1:D3)
Press Control+Shift+Enter

Website
www.WanerMath.com
  Student Home
    → On Line Utilities
    → Matrix Algebra Tool

Enter the matrix A as shown.

```
Enter your matrices here.
A = [2, 0, 1, 0
33, -22, 0, 5
1, -1, 2, -2]
```

Type A^T in the formula box and press "Compute".

If A is an $m \times n$ matrix, then its **transpose** is the $n \times m$ matrix obtained by writing its rows as columns, so that the ith row of the original matrix becomes the ith column of the transpose. We denote the transpose of the matrix A by A^T.

Visualizing Transposition

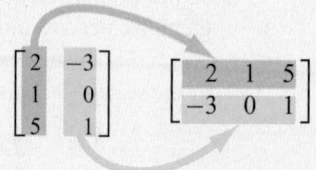

Quick Examples

1. Let $A = \begin{bmatrix} 2 & 0 & 1 & 0 \\ 33 & -22 & 0 & 5 \\ 1 & -1 & 2 & -2 \end{bmatrix}$. Then $A^T = \begin{bmatrix} 2 & 33 & 1 \\ 0 & -22 & -1 \\ 1 & 0 & 2 \\ 0 & 5 & -2 \end{bmatrix}$.

3×4 matrix 4×3 matrix

2. $\begin{bmatrix} -1 & 1 & 2 \end{bmatrix}^T = \begin{bmatrix} -1 \\ 1 \\ 2 \end{bmatrix}$.

1×3 matrix 3×1 matrix

Properties of Transposition

If A and B are $m \times n$ matrices, then the following hold:

$$(A + B)^T = A^T + B^T$$
$$(cA)^T = c(A^T)$$
$$(A^T)^T = A.$$

To see why the laws of transposition are true, let us consider the first one: $(A + B)^T = A^T + B^T$. The left-hand side is the transpose of $A + B$, and so is obtained by first adding A and B, and then writing the rows as columns. This is the same as first writing the rows of A and B individually as columns before adding, which gives the right-hand side. Similar arguments can be used to establish the other laws of transposition.

4.1 EXERCISES

▼ more advanced   ◆ challenging
🔲 indicates exercises that should be solved using technology

In each of Exercises 1–10, find the dimensions of the given matrix and identify the given entry.

1. $A = [1 \quad 5 \quad 0 \quad \frac{1}{4}]; a_{13}$ **2.** $B = [44 \quad 55]; b_{12}$

3. $C = \begin{bmatrix} \frac{5}{2} \\ 1 \\ -2 \\ 8 \end{bmatrix}; C_{11}$ **4.** $D = \begin{bmatrix} 15 & -18 \\ 6 & 0 \\ -6 & 5 \\ 48 & 18 \end{bmatrix}; d_{31}$

5. $E = \begin{bmatrix} e_{11} & e_{12} & e_{13} & \cdots & e_{1q} \\ e_{21} & e_{22} & e_{23} & \cdots & e_{2q} \\ \vdots & \vdots & \vdots & \ddots & \vdots \\ e_{p1} & e_{p2} & e_{p3} & \cdots & e_{pq} \end{bmatrix}; E_{22}$

6. $A = \begin{bmatrix} 2 & -1 & 0 \\ 3 & 5 & -3 \end{bmatrix}; A_{21}$ **7.** $B = \begin{bmatrix} 1 & 3 \\ 5 & -6 \end{bmatrix}; b_{12}$

8. $C = \begin{bmatrix} x & y & w & e \\ z & t+1 & 3 & 0 \end{bmatrix}; C_{23}$

9. $D = [d_1 \quad d_2 \quad \cdots \quad d_n]; D_{1r}$ (any r)

10. $E = [d \quad d \quad d \quad d]; E_{1r}$ (any r)

11. Solve for $x, y, z,$ and w. HINT [See Example 1.]
$$\begin{bmatrix} x+y & y+z \\ z+w & w \end{bmatrix} = \begin{bmatrix} 3 & 5 \\ 7 & 4 \end{bmatrix}$$

12. Solve for $x, y, z,$ and w. HINT [See Example 1.]
$$\begin{bmatrix} x-y & x-z \\ y-w & w \end{bmatrix} = \begin{bmatrix} 0 & 0 \\ 0 & 6 \end{bmatrix}$$

In Exercises 13–20, evaluate the given expression. Take
$$A = \begin{bmatrix} 0 & -1 \\ 1 & 0 \\ -1 & 2 \end{bmatrix}, B = \begin{bmatrix} 0.25 & -1 \\ 0 & 0.5 \\ -1 & 3 \end{bmatrix}, \text{ and }$$
$$C = \begin{bmatrix} 1 & -1 \\ 1 & 1 \\ -1 & -1 \end{bmatrix}.$$

HINT [See Example 4 and Quick Examples page 234.]

13. $A + B$ **14.** $A - C$

15. $A + B - C$ **16.** $12B$

17. $2A - C$ **18.** $2A + 0.5C$

19. $2A^T$ **20.** $A^T + 3C^T$

In Exercises 21–28, evaluate the given expression. Take
$$A = \begin{bmatrix} 1 & -1 & 0 \\ 0 & 2 & -1 \end{bmatrix}, B = \begin{bmatrix} 3 & 0 & -1 \\ 5 & -1 & 1 \end{bmatrix}, \text{ and }$$
$$C = \begin{bmatrix} x & 1 & w \\ z & r & 4 \end{bmatrix}.$$

HINT [See Example 4 and Quick Examples page 234.]

21. $A + B$ **22.** $B - C$

23. $A - B + C$ **24.** $\frac{1}{2}B$

25. $2A - B$ **26.** $2A - 4C$

27. $3B^T$ **28.** $2A^T - C^T$

🔲 *Use technology in Exercises 29–36.*
$$\text{Let } A = \begin{bmatrix} 1.5 & -2.35 & 5.6 \\ 44.2 & 0 & 12.2 \end{bmatrix}, B = \begin{bmatrix} 1.4 & 7.8 \\ 5.4 & 0 \\ 5.6 & 6.6 \end{bmatrix},$$
$$C = \begin{bmatrix} 10 & 20 & 30 \\ -10 & -20 & -30 \end{bmatrix}.$$

29. $A - C$ **30.** $C - A$

31. $1.1B$ **32.** $-0.2B$

33. $A^T + 4.2B$ **34.** $(A + 2.3C)^T$

35. $(2.1A - 2.3C)^T$ **36.** $(A - C)^T - B$

APPLICATIONS

37. *Sales* The following table shows the number of Mac computers, iPods, and iPhones sold, in millions of units, in the first quarter (Q1) of 2009, and the changes over the previous year in each of 2010 Q1 and 2011 Q1.[1]

	Macs	**iPods**	**iPhones**
Q1 2009	2.5	22.7	4.4
Change in 2010	0.8	−1.7	4.3
Change in 2011	0.8	−1.6	7.5

Use matrix algebra to find the sales of Macs, iPods, and iPhones in Q1 2010 and Q1 2011. HINT [See Example 2.]

---

[1] Apple's first quarter ends at the end of December on the preceding year. Source for data: Apple company reports (www.apple.com/investor).

38. Home Prices The following table shows median home prices, in thousands of dollars, in four regions of the U.S. in December 2010, as well as the month-over-month changes for the subsequent two months.[2]

	Northeast	Midwest	South	West
Dec. 2010	238	140	149	205
Change in Jan. 2011	-2	-13	-13	-14
Change in Feb. 2011	-6	-5	-1	-1

Use matrix algebra to find the median home price in each region in January and February 2011. HINT [See Example 2.]

39. Inventory The *Left Coast Bookstore* chain has two stores, one in San Francisco and one in Los Angeles. It stocks three kinds of book: hardcover, softcover, and plastic (for infants). At the beginning of January, the central computer showed the following books in stock:

	Hard	Soft	Plastic
San Francisco	1,000	2,000	5,000
Los Angeles	1,000	5,000	2,000

Suppose its sales in January were as follows: 700 hardcover books, 1,300 softcover books, and 2,000 plastic books sold in San Francisco, and 400 hardcover, 300 softcover, and 500 plastic books sold in Los Angeles. Write these sales figures in the form of a matrix, and then show how matrix algebra can be used to compute the inventory remaining in each store at the end of January.

40. Inventory The *Left Coast Bookstore* chain discussed in Exercise 39 actually maintained the same sales figures for the first 6 months of the year. Each month, the chain restocked the stores from its warehouse by shipping 600 hardcover, 1,500 softcover, and 1,500 plastic books to San Francisco and 500 hardcover, 500 softcover, and 500 plastic books to Los Angeles.

a. Use matrix operations to determine the total sales over the 6 months, broken down by store and type of book.

b. Use matrix operations to determine the inventory in each store at the end of June.

41. Profit Annual revenues and production costs at *Luddington's Wellington Boots & Co.* are shown in the following spreadsheet.

▲	A	B	C	D
1	Revenue			
2		2004	2005	2006
3	Full Boots	$10,000	$9,000	$11,000
4	Half Boots	$8,000	$7,200	$8,800
5	Sandals	$4,000	$5,000	$6,000
6				
7	Production Costs			
8		2004	2005	2006
9	Full Boots	$2,000	$1,800	$2,200
10	Half Boots	$2,400	$1,440	$1,760
11	Sandals	$1,200	$1,500	$2,000

Use matrix algebra to compute the profits from each sector each year.

42. Revenue The following spreadsheet gives annual production costs and profits at *Gauss-Jordan Sneakers, Inc.*

▲	A	B	C	D
1	Production Costs			
2		2004	2005	2006
3	Gauss Grip	$1,800	$2,200	$2,400
4	Air Gauss	$1,400	$1,700	$1,200
5	Gauss Gel	$1,500	$2,000	$1,300
6				
7	Profit			
8		2004	2005	2006
9	Gauss Grip	$10,000	$14,000	$16,000
10	Air Gauss	$8,000	$12,000	$14,000
11	Gauss Gel	$9,000	$14,000	$12,000

Use matrix algebra to compute the revenues from each sector each year.

43. Population Movement In 2000 the U.S. population, broken down by regions, was 53.6 million in the Northeast, 64.4 million in the Midwest, 100.2 million in the South, and 63.2 million in the West.[3] In 2010 the population was 55.3 million in the Northeast, 66.9 million in the Midwest, 114.6 million in the South, and 71.9 million in the West. Set up the population figures for each year as a row vector, and then show how to use matrix operations to find the net increase or decrease of population in each region from 2000 to 2010. Assuming the same population growth from 2010 to 2020 as from 2000 to 2010, use matrix operations to predict the population in each region in 2020.

44. Population Movement In 1990 the U.S. population, broken down by regions, was 50.8 million in the Northeast, 59.7 million in the Midwest, 85.4 million in the South, and 52.8 million in the West.[4] Between 1990 and 2000, the population in the Northeast grew by 2.8 million, the population in the Midwest grew by 4.7 million, the population in the South grew by 14.8 million, and the population in the West grew by 10.4 million. Set up the population figures for 1990 and the growth figures for the decade as row vectors. Assuming the same population growth from 2000 to 2010 as from 1990 to 2000, use matrix operations to estimate the population in each region in 2010. Compare the predicted 2010 population in the Northeast with the actual population given in the preceding exercise.

Foreclosures *Exercises 45–48 are based on the following table, which shows the numbers of foreclosures in three states in April through August of 2011.[5]*

	April	May	June	July	Aug.
California	55,900	51,900	54,100	56,200	59,400
Florida	19,600	19,200	23,800	22,400	23,600
Texas	8,800	9,100	9,300	10,600	10,100

---

[2]Source for data: Census Bureau (www.census.gov).

[3]Source: U.S. Census Bureau http://2010.census.gov/2010census/data/apportionment-pop-text.php.

[4]*Ibid.*

[5]Figures are rounded. Source: www.realtytrac.com.

45. Use matrix algebra to determine the total number of foreclosures in each of the given months.

46. Use matrix algebra to determine the total number of foreclosures in each of the given states during the entire period shown.

47. Use matrix algebra to determine in which month the difference between the number of foreclosures in California and in Florida was greatest.

48. Use matrix algebra to determine in which region the difference between the number of foreclosures in April and August was greatest.

49. ▼ *Inventory Microbucks Computer Company* makes two computers, the Pomegranate II and the Pomegranate Classic, at two different factories. The Pom II requires 2 processor chips, 16 memory chips, and 20 vacuum tubes, while the Pom Classic requires 1 processor chip, 4 memory chips, and 40 vacuum tubes. Microbucks has in stock at the beginning of the year 500 processor chips, 5,000 memory chips, and 10,000 vacuum tubes at the Pom II factory, and 200 processor chips, 2,000 memory chips, and 20,000 vacuum tubes at the Pom Classic factory. It manufactures 50 Pom IIs and 50 Pom Classics each month.

    a. Find the company's inventory of parts after 2 months, using matrix operations.

    b. When (if ever) will the company run out of one of the parts?

50. ▼ *Inventory Microbucks Computer Company*, besides having the stock mentioned in Exercise 49, gets shipments of parts every month in the amounts of 100 processor chips, 1,000 memory chips, and 3,000 vacuum tubes at the Pom II factory, and 50 processor chips, 1,000 memory chips, and 2,000 vacuum tubes at the Pom Classic factory.

    a. What will the company's inventory of parts be after 6 months?

    b. When (if ever) will the company run out of one of the parts?

51. ▼ *Tourism in the 1990s* The following table gives the number of people (in thousands) who visited Australia and South Africa in 1998[6]:

	To	Australia	South Africa
From	**North America**	440	190
	Europe	950	950
	Asia	1,790	200

    It was predicted that in 2008, 20,000 fewer people from North America would visit Australia and 40,000 more would visit South Africa, 50,000 more people from Europe would visit each of Australia and South Africa, and 100,000 more people from Asia would visit South Africa, but there would be no change in the number visiting Australia.

    a. Represent the changes predicted in 2008 in the form of a matrix, and use matrix algebra to predict the number

of visitors from the three regions to Australia and South Africa in 2008.

    b. Take A to be the 3×2 matrix whose entries are the 1998 tourism figures and take B to be the 3×2 matrix whose entries are the 2008 tourism figures. Give a formula (in terms of A and B) that predicts the average of the numbers of visitors from the three regions to Australia and South Africa in 1998 and 2008. Compute its value.

52. ▼ *Tourism in the 1990s* Referring to the 1998 tourism figures given in the preceding exercise, assume that the following (fictitious) figures represent the corresponding numbers from 1988.

	To	Australia	South Africa
From	**North America**	500	100
	Europe	900	800
	Asia	1,400	50

Take A to be the 3×2 matrix whose entries are the 1998 tourism figures and take B to be the 3×2 matrix whose entries are the 1988 tourism figures.

    a. Compute the matrix $A - B$. What does this matrix represent?

    b. Assuming that the changes in tourism over 1988–1998 are repeated in 1998–2008, give a formula (in terms of A and B) that predicts the number of visitors from the three regions to Australia and South Africa in 2008.

COMMUNICATION AND REASONING EXERCISES

53. Is it possible for a 2×3 matrix to equal a 3×2 matrix? Explain.

54. If A and B are 2×3 matrices and $A = B$, what can you say about $A - B$? Explain.

55. What does it mean when we say that $(A + B)_{ij} = A_{ij} + B_{ij}$?

56. What does it mean when we say that $(cA)_{ij} = c(A_{ij})$?

57. What would a 5×5 matrix A look like if $A_{ii} = 0$ for every i?

58. What would a matrix A look like if $A_{ij} = 0$ whenever $i \neq j$?

59. ▼ Give a formula for the ijth entry of the transpose of a matrix A.

60. ▼ A matrix is **symmetric** if it is equal to its transpose. Give an example of **a.** a nonzero 2×2 symmetric matrix and **b.** a nonzero 3×3 symmetric matrix.

61. ▼ A matrix is **skew-symmetric** or **antisymmetric** if it is equal to the negative of its transpose. Give an example of **a.** a nonzero 2×2 skew-symmetric matrix and **b.** a nonzero 3×3 skew-symmetric matrix.

62. ▼ Referring to Exercises 60 and 61, what can be said about a matrix that is both symmetric and skew-symmetric?

63. ▼ Why is matrix addition associative?

64. Describe a scenario (possibly based on one of the preceding examples or exercises) in which you might wish to compute $A - 2B$ for certain matrices A and B.

65. Describe a scenario (possibly based on one of the preceding examples or exercises) in which you might wish to compute $A + B - C$ for certain matrices A, B, and C.

---

[6]Figures are rounded to the nearest 10,000. Sources: South African Dept. of Environmental Affairs and Tourism; Australia Tourist Commission/*New York Times*, January 15, 2000, p. C1.

4.2 Matrix Multiplication

Suppose we download 3 movies at $10 each and 5 Chopin albums at $8 each. We calculate our total cost by computing the products' price × quantity and adding:

$$\text{Cost} = 10 \times 3 + 8 \times 5 = \$70.$$

Let us instead put the prices in a row vector

$$P = [10 \quad 8] \qquad \text{The price matrix}$$

and the quantities purchased in a column vector,

$$Q = \begin{bmatrix} 3 \\ 5 \end{bmatrix}. \qquad \text{The quantity matrix}$$

Q: *Why a row and a column instead of, say, two rows?*

A: It's rather a long story, but mathematicians found that it works best this way . . .

Because P represents the prices of the items we are purchasing and Q represents the quantities, it would be useful if the product PQ represented the total cost, a *single number* (which we can think of as a 1×1 matrix). For this to work, PQ should be calculated the same way we calculated the total cost:

$$PQ = [10 \quad 8] \begin{bmatrix} 3 \\ 5 \end{bmatrix} = [10 \times 3 + 8 \times 5] = [70].$$

Notice that we obtain the answer by multiplying each entry in P (going from left to right) by the corresponding entry in Q (going from top to bottom) and then adding the results.

The Product *Row × Column*

The **product** AB of a row matrix A and a column matrix B is a 1×1 matrix. The length of the row in A must match the length of the column in B for the product to be defined. To find the product, multiply each entry in A (going from left to right) by the corresponding entry in B (going from top to bottom) and then add the results.

Visualizing Matrix Multiplication

$$[2 \quad 4 \quad 1] \begin{bmatrix} 2 \\ 10 \\ -1 \end{bmatrix}$$

2×2	$= 4$	Product of first entries $= 4$
4×10	$= 40$	Product of second entries $= 40$
$1 \times (-1)$	$= \underline{-1}$	Product of third entries $= -1$
	43	Sum of products $= 43$

Quick Examples

1. $[2 \quad 1] \begin{bmatrix} -3 \\ 1 \end{bmatrix} = [2 \times (-3) + 1 \times 1] = [-6 + 1] = [-5]$

2. $[2 \quad 4 \quad 1] \begin{bmatrix} 2 \\ 10 \\ -1 \end{bmatrix} = [2 \times 2 + 4 \times 10 + 1 \times (-1)] = [4 + 40 + (-1)]$

$$= [43]$$

Notes

1. In the discussion so far, *the row is on the left and the column is on the right* (RC again). (Later we will consider products where the column matrix is on the left and the row matrix is on the right.)

2. The row size has to match the column size. This means that, if we have a 1×3 row on the left, then the column on the right must be 3×1 in order for the product to make sense. For example, the product

$$[a \quad b \quad c]\begin{bmatrix} x \\ y \end{bmatrix}$$

is not defined. ■

EXAMPLE 1 Revenue

The A-Plus auto parts store mentioned in examples in the previous section had the following sales in its Vancouver store:

	Vancouver
Wiper Blades	20
Cleaning Fluid (bottles)	10
Floor Mats	8

The store sells wiper blades for $7.00 each, cleaning fluid for $3.00 per bottle, and floor mats for $12.00 each. Use matrix multiplication to find the total revenue generated by sales of these items.

Solution We need to multiply each sales figure by the corresponding price and then add the resulting revenue figures. We represent the sales by a column vector, as suggested by the table:

$$Q = \begin{bmatrix} 20 \\ 10 \\ 8 \end{bmatrix}.$$

We put the selling prices in a row vector:

$$P = [\,7.00 \quad 3.00 \quad 12.00\,].$$

We can now compute the total revenue as the product

$$R = PQ = [7.00 \quad 3.00 \quad 12.00]\begin{bmatrix} 20 \\ 10 \\ 8 \end{bmatrix}$$

$$= [140.00 + 30.00 + 96.00] = [266.00].$$

So, the sale of these items generated a total revenue of $266.00.

Note We could also have written the quantity sold as a row vector (which would be Q^T) and the prices as a column vector (which would be P^T) and then multiplied them in the opposite order ($Q^T P^T$). Try this. ■

EXAMPLE 2 Relationship with Linear Equations

a. Represent the matrix equation

$$[2 \quad -4 \quad 1]\begin{bmatrix} x \\ y \\ z \end{bmatrix} = [5]$$

as an ordinary equation.

b. Represent the linear equation $3x + y - z + 2w = 8$ as a matrix equation.

Solution

a. If we perform the multiplication on the left, we get the 1×1 matrix $[2x - 4y + z]$. Thus the equation may be rewritten as

$$[2x - 4y + z] = [5]. \quad \text{1 × 1 matrix on the left = 1 × 1 matrix on the right}$$

Saying that these two 1×1 matrices are equal means that their entries are equal, so we get the equation

$$2x - 4y + z = 5.$$

b. This is the reverse of part (a):

$$[3 \quad 1 \quad -1 \quad 2]\begin{bmatrix} x \\ y \\ z \\ w \end{bmatrix} = [8].$$

➡ **Before we go on...** The row matrix $[3 \quad 1 \quad -1 \quad 2]$ in Example 2 is the row of **coefficients** of the original equation. (See Section 3.1.) ∎

Now to the general case of matrix multiplication:

The Product of Two Matrices: General Case

In general, for matrices A and B, we can take the product AB only if the number of columns of A equals the number of rows of B (so that we can multiply the rows of A by the columns of B as above). The product AB is then obtained by taking its ijth entry to be

$$ij\text{th entry of } AB = \text{Row } i \text{ of } A \times \text{Column } j \text{ of } B. \quad \text{As defined above}$$

Quick Examples

(R stands for row; C stands for column.)

$$\begin{matrix} & C_1 & C_2 & C_3 \\ & \downarrow & \downarrow & \downarrow \end{matrix}$$

1. $R_1 \to [2 \quad 0 \quad -1 \quad 3]\begin{bmatrix} 1 & 1 & -8 \\ 1 & -6 & 0 \\ 0 & 5 & 2 \\ -3 & 8 & 1 \end{bmatrix} = [R_1 \times C_1 \quad R_1 \times C_2 \quad R_1 \times C_3]$

$$= [-7 \quad 21 \quad -15]$$

$$\begin{matrix} & C_1 & C_2 \\ & \downarrow & \downarrow \end{matrix}$$

2. $\begin{matrix} R_1 \to \\ R_2 \to \end{matrix} \begin{bmatrix} 1 & -1 \\ 0 & 2 \end{bmatrix}\begin{bmatrix} 3 & 0 \\ 5 & -1 \end{bmatrix} = \begin{bmatrix} R_1 \times C_1 & R_1 \times C_2 \\ R_2 \times C_1 & R_2 \times C_2 \end{bmatrix} = \begin{bmatrix} -2 & 1 \\ 10 & -2 \end{bmatrix}$

In matrix multiplication we always take

Rows on the left × Columns on the right.

Look at the dimensions in the two Quick Examples on the previous page.

$$
\begin{array}{cc}
\text{Match} & \text{Match} \\
\downarrow\ \downarrow & \downarrow\ \downarrow \\
(1 \times 4)(4 \times 3) \to 1 \times 3 & (2 \times 2)(2 \times 2) \to 2 \times 2
\end{array}
$$

The fact that the number of columns in the left-hand matrix equals the number of rows in the right-hand matrix amounts to saying that the middle two numbers must match as above. If we "cancel" the middle matching numbers, we are left with the dimensions of the product.

Before continuing with examples, we state the rule for matrix multiplication formally.

Multiplication of Matrices: Formal Definition

If A is an $m \times n$ matrix and B is an $n \times k$ matrix, then the product AB is the $m \times k$ matrix whose ijth entry is the product

$$
\text{Row } i \text{ of } A \times \text{Column } j \text{ of } B
$$

$$
(AB)_{ij} = [a_{i1}\ a_{i2}\ a_{i3}\ldots a_{in}]\begin{bmatrix} b_{1j} \\ b_{2j} \\ b_{3j} \\ \vdots \\ b_{nj} \end{bmatrix} = a_{i1}b_{1j} + a_{i2}b_{2j} + a_{i3}b_{3j} + \cdots + a_{in}b_{nj}.
$$

EXAMPLE 3 Matrix Product

Calculate:

a. $\begin{bmatrix} 2 & 0 & -1 & 3 \\ 1 & -1 & 2 & -2 \end{bmatrix}\begin{bmatrix} 1 & 1 & -8 \\ 1 & 0 & 0 \\ 0 & 5 & 2 \\ -2 & 8 & -1 \end{bmatrix}$ **b.** $\begin{bmatrix} -3 \\ 1 \end{bmatrix}[2\ \ 1]$

Solution

a. Before we start the calculation, we check that the dimensions of the matrices match up.

$$
\begin{array}{cc}
 & \text{Match} \\
 & \swarrow\ \searrow \\
2 \times 4 & 4 \times 3
\end{array}
$$

$$
\begin{bmatrix} 2 & 0 & -1 & 3 \\ 1 & -1 & 2 & -2 \end{bmatrix}\begin{bmatrix} 1 & 1 & -8 \\ 1 & 0 & 0 \\ 0 & 5 & 2 \\ -2 & 8 & -1 \end{bmatrix}
$$

using Technology

Technology can be used to multiply the matrices in Example 3. Here is an outline for part (a) (see the Technology Guides at the end of the chapter for additional details on using a TI-83/84 Plus or a spreadsheet):

TI-83/84 Plus
Enter the matrices A and B using MATRX ; EDIT
Home Screen: [A]*[B]
[More details on page 304.]

Spreadsheet
Enter the matrices in convenient blocks of cells (e.g., A1–D2 and F1–H4).
Highlight a 2×3 block for the answer.
Type =MMULT(A1:D2,F1:H4)
Press Control+Shift+Enter
[More details on page 308.]

 Website
www.WanerMath.com
Student Home
→ On Line Utilities
→ Matrix Algebra Tool
Enter matrices as shown:

Enter your matrices here.
A = [2, 0, −1, 3
1, −1, 2, −2]
B = [1, 1, −8
1, 0, 0
0, 5, 2
−2, 8, −1]

Type A*B in the formula box and press "Compute". If you try to multiply two matrices whose product is not defined, you will get an error alert box telling you that.

The product of the two matrices is defined, and the product will be a 2×3 matrix (we remove the matching 4s: $(2 \times 4)(4 \times 3) \to 2 \times 3$). To calculate the product, we follow the previous prescription:

$$
\begin{array}{c} \\ R_1 \to \\ R_2 \to \end{array}
\begin{bmatrix} 2 & 0 & -1 & 3 \\ 1 & -1 & 2 & -2 \end{bmatrix}
\begin{array}{c} C_1\ \ C_2\ \ C_3 \\ \downarrow\ \ \downarrow\ \ \downarrow \\ \begin{bmatrix} 1 & 1 & -8 \\ 1 & 0 & 0 \\ 0 & 5 & 2 \\ -2 & 8 & -1 \end{bmatrix} \end{array}
= \begin{bmatrix} R_1 \times C_1 & R_1 \times C_2 & R_1 \times C_3 \\ R_2 \times C_1 & R_2 \times C_2 & R_2 \times C_3 \end{bmatrix}
$$

$$
= \begin{bmatrix} -4 & 21 & -21 \\ 4 & -5 & -2 \end{bmatrix}.
$$

b. The dimensions of the two matrices given are 2×1 and 1×2. Because the 1s match, the product is defined, and the result will be a 2×2 matrix.

$$
\begin{array}{c} \\ R_1 \to \\ R_2 \to \end{array}
\begin{bmatrix} -3 \\ 1 \end{bmatrix}
\begin{array}{c} C_1\ C_2 \\ \downarrow\ \ \downarrow \\ [2\ \ 1] \end{array}
= \begin{bmatrix} R_1 \times C_1 & R_1 \times C_2 \\ R_2 \times C_1 & R_2 \times C_2 \end{bmatrix}
= \begin{bmatrix} -6 & -3 \\ 2 & 1 \end{bmatrix}.
$$

Note In part (a) we *cannot* multiply the matrices in the opposite order—the dimensions do not match. We say simply that the product in the opposite order is **not defined**. In part (b) we *can* multiply the matrices in the opposite order, but we would get a 1×1 matrix if we did so. Thus, order is important when multiplying matrices. In general, if AB is defined, then BA need not even be defined. If BA is also defined, it may not have the same dimensions as AB. And even if AB and BA have the same dimensions, they may have different entries. (See the next example.) ∎

EXAMPLE 4 *AB* **versus** *BA*

Let $A = \begin{bmatrix} 1 & -1 \\ 0 & 2 \end{bmatrix}$ and $B = \begin{bmatrix} 3 & 0 \\ 5 & -1 \end{bmatrix}$. Find AB and BA.

Solution Note first that A and B are both 2×2 matrices, so the products AB and BA are both defined and are both 2×2 matrices—unlike the case in Example 3(b). We first calculate AB:

$$
AB = \begin{bmatrix} 1 & -1 \\ 0 & 2 \end{bmatrix} \begin{bmatrix} 3 & 0 \\ 5 & -1 \end{bmatrix} = \begin{bmatrix} -2 & 1 \\ 10 & -2 \end{bmatrix}.
$$

Now let's calculate BA:

$$
BA = \begin{bmatrix} 3 & 0 \\ 5 & -1 \end{bmatrix} \begin{bmatrix} 1 & -1 \\ 0 & 2 \end{bmatrix} = \begin{bmatrix} 3 & -3 \\ 5 & -7 \end{bmatrix}.
$$

Notice that BA has no resemblance to AB! Thus, we have discovered that, even for square matrices:

Matrix multiplication is not commutative.

In other words, $AB \neq BA$ in general, even when AB and BA both exist and have the same dimensions. (There are instances when $AB = BA$ for particular matrices A and B, but this is an exception, not the rule.)

EXAMPLE 5 Revenue

January sales at the A-Plus auto parts stores in Vancouver and Quebec are given in the following table.

	Vancouver	Quebec
Wiper Blades	20	15
Cleaning Fluid (bottles)	10	12
Floor Mats	8	4

The usual selling prices for these items are $7.00 each for wiper blades, $3.00 per bottle for cleaning fluid, and $12.00 each for floor mats. The discount prices for A-Plus Club members are $6.00 each for wiper blades, $2.00 per bottle for cleaning fluid, and $10.00 each for floor mats. Use matrix multiplication to compute the total revenue at each store, assuming first that all items were sold at the usual prices, and then that they were all sold at the discount prices.

Solution We can do all of the requested calculations at once with a single matrix multiplication. Consider the following two labeled matrices.

$$Q = \begin{matrix} & \begin{matrix} \mathbf{V} & \mathbf{Q} \end{matrix} \\ \begin{matrix} \mathbf{Wb} \\ \mathbf{Cf} \\ \mathbf{Fm} \end{matrix} & \begin{bmatrix} 20 & 15 \\ 10 & 12 \\ 8 & 4 \end{bmatrix} \end{matrix}$$

$$P = \begin{matrix} & \begin{matrix} \mathbf{Wb} & \mathbf{Cf} & \mathbf{Fm} \end{matrix} \\ \begin{matrix} \mathbf{Usual} \\ \mathbf{Discount} \end{matrix} & \begin{bmatrix} 7.00 & 3.00 & 12.00 \\ 6.00 & 2.00 & 10.00 \end{bmatrix} \end{matrix}$$

The first matrix records the quantities sold, while the second records the sales prices under the two assumptions. To compute the revenue at both stores under the two different assumptions, we calculate $R = PQ$.

$$R = PQ = \begin{bmatrix} 7.00 & 3.00 & 12.00 \\ 6.00 & 2.00 & 10.00 \end{bmatrix} \begin{bmatrix} 20 & 15 \\ 10 & 12 \\ 8 & 4 \end{bmatrix}$$

$$= \begin{bmatrix} 266.00 & 189.00 \\ 220.00 & 154.00 \end{bmatrix}$$

We can label this matrix as follows:

$$R = \begin{matrix} & \begin{matrix} \mathbf{V} & \mathbf{Q} \end{matrix} \\ \begin{matrix} \mathbf{Usual} \\ \mathbf{Discount} \end{matrix} & \begin{bmatrix} 266.00 & 189.00 \\ 220.00 & 154.00 \end{bmatrix} \end{matrix}.$$

In other words, if the items were sold at the usual price, then Vancouver had a revenue of $266 while Quebec had a revenue of $189, and so on.

➡ **Before we go on...** In Example 5 we were able to calculate PQ because the dimensions matched correctly: $(2 \times 3)(3 \times 2) \rightarrow 2 \times 2$. We could also have multiplied them in the opposite order and gotten a 3×3 matrix. Would the product QP be meaningful? In an application like this, not only do the dimensions have to match, but also the *labels* have to match for the result to be meaningful. The labels on the three columns of P are the parts that were sold, and these are also the labels on the three rows of Q. Therefore, we can "cancel labels" at the same time that we cancel the dimensions in the product. However, the labels on the two columns of Q do not match the labels on the two rows of P, and there is no useful interpretation of the product QP in this situation. ■

There are very special square matrices of every size: $1 \times 1, 2 \times 2, 3 \times 3$, and so on, called the **identity** matrices.

Identity Matrix

The $n \times n$ identity matrix I is the matrix with 1s down the **main diagonal** (the diagonal starting at the top left) and 0s everywhere else. In symbols,

$$I_{ii} = 1, \quad \text{and}$$

$$I_{ij} = 0 \quad \text{if } i \neq j.$$

Quick Examples

1. 1×1 identity matrix $\qquad I = [1]$

2. 2×2 identity matrix $\qquad I = \begin{bmatrix} 1 & 0 \\ 0 & 1 \end{bmatrix}$

3. 3×3 identity matrix $\qquad I = \begin{bmatrix} 1 & 0 & 0 \\ 0 & 1 & 0 \\ 0 & 0 & 1 \end{bmatrix}$

4. 4×4 identity matrix $\qquad I = \begin{bmatrix} 1 & 0 & 0 & 0 \\ 0 & 1 & 0 & 0 \\ 0 & 0 & 1 & 0 \\ 0 & 0 & 0 & 1 \end{bmatrix}$

Note Identity matrices are always square matrices, meaning that they have the same number of rows as columns. There is no such thing, for example, as the "2×4 identity matrix."

The next example shows why I is interesting. ■

EXAMPLE 6 Identity Matrix

Evaluate the products AI and IA, where $A = \begin{bmatrix} a & b & c \\ d & e & f \\ g & h & i \end{bmatrix}$ and I is the 3×3 identity matrix.

Solution

First notice that A is arbitrary; it could be any 3×3 matrix.

$$AI = \begin{bmatrix} a & b & c \\ d & e & f \\ g & h & i \end{bmatrix} \begin{bmatrix} 1 & 0 & 0 \\ 0 & 1 & 0 \\ 0 & 0 & 1 \end{bmatrix} = \begin{bmatrix} a & b & c \\ d & e & f \\ g & h & i \end{bmatrix}$$

using Technology

Technology can be used to obtain an identity matrix as in Example 6. Here is an outline (see the Technology Guides at the end of the chapter for additional details on using a TI-83/84 Plus or a spreadsheet):

TI-83/84 Plus
3×3 identity matrix: identity(3)
Obtained from MATRX ; MATH
[More details on page 305.]

Spreadsheet
Click on a cell for the top left corner (e.g., B1).
Type =IF(ROW(B1)-ROW($B$1)=COLUMN(B1)-COLUMN($B$1),1,0)
Copy across and down.
[More details on page 308.]

Website
www.WanerMath.com
   Student Home
      → On Line Utilities
      → Matrix Algebra Tool

Just use I in the formula box to refer to the identity matrix of any dimension. The program will choose the correct dimension in the context of the formula. For example, if A is a 3×3 matrix, then the expression I-A uses the 3×3 identity matrix for I.

and

$$IA = \begin{bmatrix} 1 & 0 & 0 \\ 0 & 1 & 0 \\ 0 & 0 & 1 \end{bmatrix} \begin{bmatrix} a & b & c \\ d & e & f \\ g & h & i \end{bmatrix} = \begin{bmatrix} a & b & c \\ d & e & f \\ g & h & i \end{bmatrix}.$$

In both cases, the answer is the matrix A we started with. In symbols,

$$AI = A$$

and

$$IA = A$$

no matter which 3×3 matrix A you start with. Now this should remind you of a familiar fact from arithmetic:

$$a \cdot 1 = a$$

and

$$1 \cdot a = a.$$

That is why we call the matrix I the 3×3 *identity* matrix, because it appears to play the same role for 3×3 matrices that the identity 1 does for numbers.

➡ **Before we go on...** Try a similar calculation using 2×2 matrices: Let $A = \begin{bmatrix} a & b \\ c & d \end{bmatrix}$, let I be the 2×2 identity matrix, and check that $AI = IA = A$. In fact, the equation

$$AI = IA = A$$

works for square matrices of every dimension. It is also interesting to notice that $AI = A$ if I is the 2×2 identity matrix and A is any 3×2 matrix (try one). In fact, if I is any identity matrix, then $AI = A$ whenever the product is defined, and $IA = A$ whenever this product is defined. ■

We can now add to the list of properties we gave for matrix arithmetic at the end of Section 4.1 by writing down properties of matrix multiplication. In stating these properties, we shall assume that all matrix products we write are defined—that is, that the matrices have correctly matching dimensions. The first eight properties are the ones we've already seen; the rest are new.

Properties of Matrix Addition and Multiplication

If A, B, and C are matrices, if O is a zero matrix, and if I is an identity matrix, then the following hold:

$A + (B + C) = (A + B) + C$	*Additive associative law*
$A + B = B + A$	*Additive commutative law*
$A + O = O + A = A$	*Additive identity law*
$A + (-A) = O = (-A) + A$	*Additive inverse law*
$c(A + B) = cA + cB$	*Distributive law*
$(c + d)A = cA + dA$	*Distributive law*
$1A = A$	*Scalar unit*

$$0A = O \qquad \textit{Scalar zero}$$
$$A(BC) = (AB)C \qquad \textit{Multiplicative associative law}$$
$$c(AB) = (cA)B \qquad \textit{Multiplicative associative law}$$
$$c(dA) = (cd)A \qquad \textit{Multiplicative associative law}$$
$$AI = IA = A \qquad \textit{Multiplicative identity law}$$
$$A(B + C) = AB + AC \qquad \textit{Distributive law}$$
$$(A + B)C = AC + BC \qquad \textit{Distributive law}$$
$$OA = AO = O \qquad \textit{Multiplication by zero matrix}$$

Note that we have not included a multiplicative commutative law for matrices, because the equation $AB = BA$ does not hold in general. In other words, matrix multiplication is *not* exactly like multiplication of numbers. (You have to be a little careful because it is easy to apply the commutative law without realizing it.)

We should also say a bit more about transposition. Transposition and multiplication have an interesting relationship. We write down the properties of transposition again, adding one new one.

Properties of Transposition

$$(A + B)^T = A^T + B^T$$
$$(cA)^T = c(A^T)$$
$$(AB)^T = B^T A^T$$

Notice the change in order in the last one. The order is crucial.

Quick Examples

1. $\left(\begin{bmatrix} 1 & -1 \\ 0 & 2 \end{bmatrix} \begin{bmatrix} 3 & 0 \\ 5 & -1 \end{bmatrix} \right)^T = \begin{bmatrix} -2 & 1 \\ 10 & -2 \end{bmatrix}^T = \begin{bmatrix} -2 & 10 \\ 1 & -2 \end{bmatrix}$ $(AB)^T$

2. $\begin{bmatrix} 3 & 0 \\ 5 & -1 \end{bmatrix}^T \begin{bmatrix} 1 & -1 \\ 0 & 2 \end{bmatrix}^T = \begin{bmatrix} 3 & 5 \\ 0 & -1 \end{bmatrix} \begin{bmatrix} 1 & 0 \\ -1 & 2 \end{bmatrix} = \begin{bmatrix} -2 & 10 \\ 1 & -2 \end{bmatrix}$ $B^T A^T$

3. $\begin{bmatrix} 1 & -1 \\ 0 & 2 \end{bmatrix}^T \begin{bmatrix} 3 & 0 \\ 5 & -1 \end{bmatrix}^T = \begin{bmatrix} 1 & 0 \\ -1 & 2 \end{bmatrix} \begin{bmatrix} 3 & 5 \\ 0 & -1 \end{bmatrix} = \begin{bmatrix} 3 & 5 \\ -3 & -7 \end{bmatrix}$ $A^T B^T$

These properties give you a glimpse of the field of mathematics known as **abstract algebra**. Algebraists study operations like these that resemble the operations on numbers but differ in some way, such as the lack of commutativity for multiplication seen here.

We end this section with more on the relationship between linear equations and matrix equations, which is one of the important applications of matrix multiplication.

EXAMPLE 7 Matrix Form of a System of Linear Equations

a. If

$$A = \begin{bmatrix} 1 & -2 & 3 \\ 2 & 0 & -1 \\ -3 & 1 & 1 \end{bmatrix}, X = \begin{bmatrix} x \\ y \\ z \end{bmatrix}, \text{ and } B = \begin{bmatrix} 3 \\ -1 \\ 0 \end{bmatrix},$$

rewrite the matrix equation $AX = B$ as a system of linear equations.

b. Express the following system of equations as a matrix equation of the form $AX = B$:

$$2x + y = 3$$
$$4x - y = -1.$$

Solution

a. The matrix equation $AX = B$ is

$$\begin{bmatrix} 1 & -2 & 3 \\ 2 & 0 & -1 \\ -3 & 1 & 1 \end{bmatrix} \begin{bmatrix} x \\ y \\ z \end{bmatrix} = \begin{bmatrix} 3 \\ -1 \\ 0 \end{bmatrix}.$$

As in Example 2(a), we first evaluate the left-hand side and then set it equal to the right-hand side.

$$\begin{bmatrix} 1 & -2 & 3 \\ 2 & 0 & -1 \\ -3 & 1 & 1 \end{bmatrix} \begin{bmatrix} x \\ y \\ z \end{bmatrix} = \begin{bmatrix} x - 2y + 3z \\ 2x - z \\ -3x + y + z \end{bmatrix}$$

$$\begin{bmatrix} x - 2y + 3z \\ 2x - z \\ -3x + y + z \end{bmatrix} = \begin{bmatrix} 3 \\ -1 \\ 0 \end{bmatrix}$$

Because these two matrices are equal, their corresponding entries must be equal.

$$x - 2y + 3z = 3$$
$$2x \quad\quad - z = -1$$
$$-3x + y + z = 0$$

In other words, the matrix equation $AX = B$ is equivalent to this system of linear equations. Notice that the coefficients of the left-hand sides of these equations are the entries of the matrix A. We call A the **coefficient matrix** of the system of equations. The entries of X are the unknowns and the entries of B are the right-hand sides $3, -1$, and 0.

b. As we saw in part (a), the coefficient matrix A has entries equal to the coefficients of the left-hand sides of the equations. Thus,

$$A = \begin{bmatrix} 2 & 1 \\ 4 & -1 \end{bmatrix}.$$

X is the column matrix consisting of the unknowns, while B is the column matrix consisting of the right-hand sides of the equations, so

$$X = \begin{bmatrix} x \\ y \end{bmatrix} \quad \text{and} \quad B = \begin{bmatrix} 3 \\ -1 \end{bmatrix}.$$

The system of equations can be rewritten as the matrix equation $AX = B$ with this A, X, and B.

This translation of systems of linear equations into matrix equations is really the first step in the method of solving linear equations discussed in Chapter 3. There we worked with the **augmented matrix** of the system, which is simply A with B adjoined as an extra column.

Q: *When we write a system of equations as $AX = B$, couldn't we solve for the unknown X by dividing both sides by A?*

A: If we interpret division as multiplication by the inverse (for example, $2 \div 3 = 2 \times 3^{-1}$), we shall see in the next section that *certain* systems of the form $AX = B$ can be solved in this way, by multiplying both sides by A^{-1}. We first need to discuss what we mean by A^{-1} and how to calculate it.

4.2 EXERCISES

▼ more advanced   ◆ challenging
■ indicates exercises that should be solved using technology

In Exercises 1–28, compute the products. Some of these may be undefined. Exercises marked ■ *should be done using technology. The others should be done two ways: by hand and by using technology where possible.* HINT *[See Example 3.]*

1. $[1 \quad 3 \quad -1]\begin{bmatrix} 9 \\ 1 \\ -1 \end{bmatrix}$

2. $[4 \quad 0 \quad -1]\begin{bmatrix} -4 \\ 1 \\ 8 \end{bmatrix}$

3. $\left[-1 \quad \frac{1}{2}\right]\begin{bmatrix} -\frac{1}{3} \\ 1 \end{bmatrix}$

4. $[-1 \quad 1]\begin{bmatrix} \frac{3}{4} \\ \frac{1}{4} \end{bmatrix}$

5. $[0 \quad -2 \quad 1]\begin{bmatrix} x \\ y \\ z \end{bmatrix}$

6. $[4 \quad -1 \quad 1]\begin{bmatrix} -x \\ x \\ y \end{bmatrix}$

7. $[1 \quad 3 \quad 2]\begin{bmatrix} 1 \\ -1 \end{bmatrix}$

8. $[3 \quad 2][1 \quad -2]$

9. $[-1 \quad 1]\begin{bmatrix} -3 & 1 & 4 & 3 \\ 0 & 1 & -2 & 1 \end{bmatrix}$

10. $[2 \quad -1]\begin{bmatrix} -3 & 1 & 4 & 3 \\ 4 & 0 & 1 & 3 \end{bmatrix}$

11. $[1 \quad -1 \quad 2 \quad 3]\begin{bmatrix} -1 & 2 & 0 \\ 2 & -1 & 0 \\ 0 & 5 & 2 \\ -1 & 8 & 1 \end{bmatrix}$

12. $[0 \quad 1 \quad -1 \quad 2]\begin{bmatrix} 1 & -2 & 1 \\ 0 & 1 & 3 \\ 6 & 0 & 2 \\ -1 & -2 & 11 \end{bmatrix}$

13. $\begin{bmatrix} 1 & 0 & -1 \\ 1 & 1 & 2 \end{bmatrix}\begin{bmatrix} 0 & 1 & -1 \\ 1 & 0 & 1 \\ 4 & 8 & 0 \end{bmatrix}$

14. $\begin{bmatrix} 0 & 1 & -1 \\ 3 & 1 & -1 \end{bmatrix}\begin{bmatrix} 1 & 1 \\ 4 & 2 \\ 0 & 1 \end{bmatrix}$

15. $\begin{bmatrix} 1 & 0 \\ 1 & -1 \end{bmatrix}\begin{bmatrix} 0 & 1 \\ 0 & 1 \end{bmatrix}$

16. $\begin{bmatrix} 1 & -1 \\ 1 & -1 \end{bmatrix}\begin{bmatrix} 3 & -3 \\ 5 & -7 \end{bmatrix}$

17. $\begin{bmatrix} 0 & 1 \\ 0 & 1 \end{bmatrix}\begin{bmatrix} 1 & 0 \\ 1 & -1 \end{bmatrix}$

18. $\begin{bmatrix} 3 & -3 \\ 5 & -7 \end{bmatrix}\begin{bmatrix} 1 & -1 \\ 1 & -1 \end{bmatrix}$

19. $\begin{bmatrix} 1 & -1 \\ 1 & -1 \end{bmatrix}\begin{bmatrix} 2 & 3 \\ 2 & 3 \end{bmatrix}$

20. $\begin{bmatrix} 0 & 1 \\ 1 & 0 \end{bmatrix}\begin{bmatrix} 3 & -3 \\ 2 & -1 \end{bmatrix}$

21. $\begin{bmatrix} 1 & -1 \\ -1 & 1 \end{bmatrix}\begin{bmatrix} 2 & 3 \\ 2 & 3 \\ 1 & 1 \end{bmatrix}$

22. $\begin{bmatrix} 0 & 1 & -1 \\ 0 & -1 & 1 \end{bmatrix}\begin{bmatrix} 3 & -3 \\ 2 & -1 \end{bmatrix}$

23. $\begin{bmatrix} 1 & 0 & -1 \\ 2 & -2 & 1 \\ 0 & 0 & 1 \end{bmatrix}\begin{bmatrix} 1 & -1 & 4 \\ 1 & 1 & 0 \\ 0 & 4 & 1 \end{bmatrix}$

24. $\begin{bmatrix} 1 & 2 & 0 \\ 4 & -1 & 1 \\ 1 & 0 & 1 \end{bmatrix}\begin{bmatrix} 1 & 2 & -4 \\ 4 & 1 & 0 \\ 0 & -2 & 1 \end{bmatrix}$

25. $\begin{bmatrix} 1 & 0 & 1 & 0 \\ -1 & 1 & 0 & 1 \\ -2 & 0 & 1 & 4 \\ 0 & -1 & 0 & 1 \end{bmatrix}\begin{bmatrix} 1 \\ -3 \\ 2 \\ 0 \end{bmatrix}$

26. $\begin{bmatrix} 1 & 1 & -7 & 0 \\ -1 & 0 & 2 & 4 \\ -1 & 0 & -2 & 1 \\ 1 & -1 & 1 & 1 \end{bmatrix}\begin{bmatrix} 1 \\ -3 \\ 2 \\ 1 \end{bmatrix}$

27. ■ $\begin{bmatrix} 1.1 & 2.3 & 3.4 & -1.2 \\ 3.4 & 4.4 & 2.3 & 1.1 \\ 2.3 & 0 & -2.2 & 1.1 \\ 1.2 & 1.3 & 1.1 & 1.1 \end{bmatrix}\begin{bmatrix} -2.1 & 0 & -3.3 \\ -3.4 & -4.8 & -4.2 \\ 3.4 & 5.6 & 1 \\ 1 & 2.2 & 9.8 \end{bmatrix}$

28. ■ $\begin{bmatrix} 1.2 & 2.3 & 3.4 & 4.5 \\ 3.3 & 4.4 & 5.5 & 6.6 \\ 2.3 & -4.3 & -2.2 & 1.1 \\ 2.2 & -1.2 & -1 & 1.1 \end{bmatrix}\begin{bmatrix} 9.8 & 1 & -1.1 \\ 8.8 & 2 & -2.2 \\ 7.7 & 3 & -3.3 \\ 6.6 & 4 & -4.4 \end{bmatrix}$

29. Find[7] $A^2 = A \cdot A$, $A^3 = A \cdot A \cdot A$, A^4, and A^{100}, given that

$$A = \begin{bmatrix} 0 & 1 & 1 & 1 \\ 0 & 0 & 1 & 1 \\ 0 & 0 & 0 & 1 \\ 0 & 0 & 0 & 0 \end{bmatrix}.$$

30. Repeat the preceding exercise with $A = \begin{bmatrix} 0 & 2 & 0 & -1 \\ 0 & 0 & 2 & 0 \\ 0 & 0 & 0 & 2 \\ 0 & 0 & 0 & 0 \end{bmatrix}$.

Exercises 31–38 should be done two ways: by hand and by using technology where possible.

Let

$$A = \begin{bmatrix} 0 & -1 & 0 & 1 \\ 10 & 0 & 1 & 0 \end{bmatrix}, B = \begin{bmatrix} 0 & -1 \\ 1 & 1 \\ -1 & 3 \\ 5 & 0 \end{bmatrix}, C = \begin{bmatrix} 1 & -1 \\ 1 & 1 \\ 1 & 1 \\ 1 & 1 \end{bmatrix}.$$

Evaluate:

31. AB **32.** AC **33.** $A(B - C)$ **34.** $(B - C)A$

Let $A = \begin{bmatrix} 1 & -1 \\ 0 & 2 \\ 0 & -2 \end{bmatrix}$, $B = \begin{bmatrix} 3 & 0 & -1 \\ 5 & -1 & 1 \end{bmatrix}$, $C = \begin{bmatrix} x & 1 & w \\ z & r & 4 \end{bmatrix}$.

Evaluate:

35. AB **36.** AC **37.** $A(B + C)$ **38.** $(B + C)A$

In Exercises 39–44, calculate (a) $P^2 = P \cdot P$ (b) $P^4 = P^2 \cdot P^2$ and (c) P^8. (Round all entries to four decimal places.) (d) Without computing it explicitly, find $P^{1,000}$.

39. ▼ $P = \begin{bmatrix} 0.2 & 0.8 \\ 0.2 & 0.8 \end{bmatrix}$ **40.** ▼ $P = \begin{bmatrix} 0.1 & 0.1 \\ 0.9 & 0.9 \end{bmatrix}$

41. ▼ $P = \begin{bmatrix} 0.1 & 0.9 \\ 0 & 1 \end{bmatrix}$ **42.** ▼ $P = \begin{bmatrix} 1 & 0 \\ 0.8 & 0.2 \end{bmatrix}$

43. ▼ $P = \begin{bmatrix} 0.3 & 0.3 & 0.4 \\ 0.3 & 0.3 & 0.4 \\ 0.3 & 0.3 & 0.4 \end{bmatrix}$

What do you notice about the rows of P? Compare with Exercise 39 and state a general result about square matrices that these two exercises seem to suggest.

44. ▼ $P = \begin{bmatrix} -0.3 & -0.3 & -0.3 \\ 0.9 & 0.9 & 0.9 \\ 0.4 & 0.4 & 0.4 \end{bmatrix}$

What do you notice about the columns of P? Compare with Exercise 40 and state a general result about square matrices that these two exercises seem to suggest.

In Exercises 45–48, translate the given matrix equations into systems of linear equations. HINT [See Example 7.]

45. $\begin{bmatrix} 2 & -1 & 4 \\ -4 & \frac{3}{4} & \frac{1}{3} \\ -3 & 0 & 0 \end{bmatrix} \begin{bmatrix} x \\ y \\ z \end{bmatrix} = \begin{bmatrix} 3 \\ -1 \\ 0 \end{bmatrix}$

46. $\begin{bmatrix} 1 & -1 & 4 \\ -\frac{1}{3} & -3 & \frac{1}{3} \\ 3 & 0 & 1 \end{bmatrix} \begin{bmatrix} x \\ y \\ z \end{bmatrix} = \begin{bmatrix} -3 \\ -1 \\ 2 \end{bmatrix}$

47. $\begin{bmatrix} 1 & -1 & 0 & 1 \\ 1 & 1 & 2 & 4 \end{bmatrix} \begin{bmatrix} x \\ y \\ z \\ w \end{bmatrix} = \begin{bmatrix} -1 \\ 2 \end{bmatrix}$

48. $\begin{bmatrix} 0 & 1 & 6 & 1 \\ 1 & -5 & 0 & 0 \end{bmatrix} \begin{bmatrix} x \\ y \\ z \\ w \end{bmatrix} = \begin{bmatrix} -2 \\ 9 \end{bmatrix}$

In Exercises 49–52, translate the given systems of equations into matrix form. HINT [See Example 7.]

49. $\begin{aligned} x - y &= 4 \\ 2x - y &= 0 \end{aligned}$ **50.** $\begin{aligned} 2x + y &= 7 \\ -x \quad &= 9 \end{aligned}$

51. $\begin{aligned} x + y - z &= 8 \\ 2x + y + z &= 4 \\ \frac{3x}{4} \quad + \frac{z}{2} &= 1 \end{aligned}$ **52.** $\begin{aligned} x + y + 2z &= -2 \\ 4x + 2y - z &= -8 \\ \frac{x}{2} - \frac{y}{3} \quad &= 4 \end{aligned}$

APPLICATIONS

53. *Revenue* Your T-shirt operation is doing a booming trade. Last week you sold 50 tie-dye shirts for $15 each, 40 Suburban State University Crew shirts for $10 each, and 30 Lacrosse T-shirts for $12 each. Use matrix operations to calculate your total revenue for the week. HINT [See Example 1.]

54. *Revenue* Karen Sandberg, your competitor in Suburban State U's T-shirt market, has apparently been undercutting your prices and outperforming you in sales. Last week she sold 100 tie-dye shirts for $10 each, 50 (low quality) Crew shirts at $5 apiece, and 70 Lacrosse T-shirts for $8 each. Use matrix operations to calculate her total revenue for the week. HINT [See Example 1.]

55. *Real Estate* The following table shows the cost of 1,000 square feet of "business executive"-grade residential real estate in three cities at the end of 2011[8] together with the quantity your development company intends to purchase in each city.

	London	New York	Mumbai
Cost ($ millions)	1.15	0.80	0.50
Quantity Purchased (thousands of square feet)	10	20	10

Use matrix multiplication to estimate the total cost of the purchase.

---

[7] $A \cdot A \cdot A$ is $A(A \cdot A)$, or the equivalent $(A \cdot A)A$ by the associative law. Similarly, $A \cdot A \cdot A \cdot A = A(A \cdot A \cdot A) = (A \cdot A \cdot A)A = (A \cdot A)(A \cdot A)$; it doesn't matter where we place parentheses.

[8] Source: *Insights, World Cities Review* World Press, Savallis, Autumn 2011.

56. *Real Estate* Repeat Exercise 55 using the following table for Hong Kong, Paris, and Tokyo:[9]

	Hong Kong	**Paris**	**Tokyo**
Cost ($ millions)	2.10	1.15	1.30
Quantity Purchased (thousands of square feet)	10	30	5

Use matrix multiplication to estimate the total cost of the purchase.

57. *Revenue* Recall the *Left Coast Bookstore* chain from the preceding section. In January, it sold 700 hardcover books, 1,300 softcover books, and 2,000 plastic books in San Francisco; it sold 400 hardcover, 300 softcover, and 500 plastic books in Los Angeles. Now, hardcover books sell for $30 each, softcover books sell for $10 each, and plastic books sell for $15 each. Write a column matrix with the price data and show how matrix multiplication (using the sales and price data matrices) may be used to compute the total revenue at the two stores. HINT [See Example 5.]

58. *Profit* Refer back to Exercise 57, and now suppose that each hardcover book costs the stores $10, each softcover book costs $5, and each plastic book costs $10. Use matrix operations to compute the total *profit* at each store in January. HINT [See Example 5.]

T **Income** *Exercises 59–62 are based on the following spreadsheet, which shows the projected 2020 and 2030 U.S. male and female population in various age groups, as well as per capita incomes.*[10]

	A	B	C	D	E	F
1			2020 Population		2030 population	
2	Age	Mean Income ($1000)	Female (Millions)	Male (Millions)	Female (Millions)	Male (Millions)
3	15 to 24	14	23	24	25	26
4	25 to 44	42	43	43	45	46
5	45 to 64	48	43	41	42	41
6	65 to 84	29	31	20	40	31

59. Use matrix algebra to estimate the total income for females in 2020. (Round the answer to two significant digits.)

60. Use matrix algebra to estimate the total income for males in 2030. (Round the answer to two significant digits.)

61. Give a single matrix formula that expresses the difference in total income between males and females in 2020, and compute its value, rounded to two significant digits.

62. Give a single matrix formula that expresses the total income in 2030, and compute its value, rounded to two significant digits.

63. *Cheese Production* The total amount of cheese, in billions of pounds, produced in the 13 western and 12 north central U.S. states in 1999 and 2000 was as follows:[11]

	1999	**2000**
Western States	2.7	3.0
North Central States	3.9	4.0

Thinking of this table as a (labeled) 2×2 matrix P, compute the matrix product $[-1 \quad 1]P$. What does this product represent?

64. *Milk Production* The total amount of milk, in billions of pounds, produced in the 13 western and 12 north central U.S. states in 1999 and 2000 was as follows.[12]

	1999	**2000**
Western States	56	60
North Central States	57	59

Thinking of this table as a (labeled) 2×2 matrix P, compute the matrix product $P\begin{bmatrix} -1 \\ 1 \end{bmatrix}$. What does this product represent?

Foreclosures *Exercises 65–70 are based on the following table, which shows the numbers of foreclosures in three states during three months of 2011.*[13]

	June	**July**	**Aug.**
California	54,100	56,200	59,400
Florida	23,800	22,400	23,600
Texas	9,300	10,600	10,100

65. Each month, your law firm handles 10% of all foreclosures in California, 5% of all foreclosures in Florida, and 20% of all foreclosures in Texas. Use matrix multiplication to compute the total number of foreclosures handled by your firm in each of the months shown.

66. Your law firm handled 10% of all foreclosures in each state in June, 30% of all foreclosures in July, and 20% of all foreclosures in August 2011. Use matrix multiplication to compute the total number of foreclosures handled by your firm in each of the states shown.

67. Let A be the 3×3 matrix whose entries are the figures in the table, and let $B = [1 \quad 1 \quad 0]$. What does the matrix BA represent?

68. Let A be the 3×3 matrix whose entries are the figures in the table, and let $B = [1 \quad 1 \quad 0]^T$. What does the matrix AB represent?

---

[9]*Ibid.*

[10]The population figures are Census Bureau estimates, and the income figures are 2007 mean per capita incomes. All figures are approximate. Source: U.S. Census Bureau (www.census.gov).

[11]Figures are approximate. Source: Department of Agriculture/*New York Times*, June 28, 2001, p. C1.

[12]*Ibid.*

[13]Figures are rounded. Source: www.realtytrac.com.

69. ▼ Write a matrix product whose computation gives the total number by which the combined foreclosures for all three months in California and Texas exceeded the foreclosures in Florida. Calculate the product.

70. ▼ Write a matrix product whose computation gives the total number by which combined foreclosures in August exceeded foreclosures in June. Calculate the product.

71. ▼ *Costs* *Microbucks Computer Co.* makes two computers, the Pomegranate II and the Pomegranate Classic. The Pom II requires 2 processor chips, 16 memory chips, and 20 vacuum tubes, while the Pom Classic requires 1 processor chip, 4 memory chips, and 40 vacuum tubes. There are two companies that can supply these parts: Motorel can supply them at $100 per processor chip, $50 per memory chip, and $10 per vacuum tube, while Intola can supply them at $150 per processor chip, $40 per memory chip, and $15 per vacuum tube. Write down all of this data in two matrices, one showing the parts required for each model computer, and the other showing the prices for each part from each supplier. Then show how matrix multiplication allows you to compute the total cost for parts for each model when parts are bought from either supplier.

72. ▼ *Profits* Refer back to Exercise 71. It actually costs Motorel only $25 to make each processor chip, $10 for each memory chip, and $5 for each vacuum tube. It costs Intola $50 per processor chip, $10 per memory chip, and $7 per vacuum tube. Use matrix operations to find the total profit Motorel and Intola would make on each model.

73. ▼ *Tourism in the 1990s* The following table gives the number of people (in thousands) who visited Australia and South Africa in 1998.[14]

To	Australia	South Africa
From **North America**	440	190
Europe	950	950
Asia	1,790	200

You estimate that 5% of all visitors to Australia and 4% of all visitors to South Africa decide to settle there permanently. Take A to be the 3×2 matrix whose entries are the 1998 tourism figures in the above table and take

$$B = \begin{bmatrix} 0.05 \\ 0.04 \end{bmatrix} \quad \text{and} \quad C = \begin{bmatrix} 0.05 & 0 \\ 0 & 0.04 \end{bmatrix}.$$

Compute the products AB and AC. What do the entries in these matrices represent?

74. ▼ *Tourism in the 1990s* Referring to the tourism figures in the preceding exercise, you estimate that from 1998 to 2018, tourism from North America to each of Australia and South Africa will have increased by 20%, tourism from Europe by 30%, and tourism from Asia by 10%. Take A to be the 3×2 matrix whose entries are the 1998 tourism figures and take

$$B = \begin{bmatrix} 1.2 & 1.3 & 1.1 \end{bmatrix} \quad \text{and} \quad C = \begin{bmatrix} 1.2 & 0 & 0 \\ 0 & 1.3 & 0 \\ 0 & 0 & 1.1 \end{bmatrix}.$$

Compute the products BA and CA. What do the entries in these matrices represent?

75. 🔲▼ *Population Movement* In 2008, the population of the U.S., broken down by regions, was 54.1 million in the Northeast, 65.7 million in the Midwest, 112.9 million in the South, and 70.3 million in the West. The table below shows the population movement during the period 2008–2009. (Thus, 99.23% of the population in the Northeast stayed there, while 0.16% of the population in the Northeast moved to the Midwest, and so on.)[15]

To	Northeast	Midwest	South	West
From **Northeast**	0.9923	0.0016	0.0042	0.0019
Midwest	0.0018	0.9896	0.0047	0.0039
South	0.0056	0.0059	0.9827	0.0058
West	0.0024	0.0033	0.0044	0.9899

Set up the 2008 population figures as a row vector. Then use matrix multiplication to estimate the population in each region in 2009. (Round all answers to the nearest 0.1 million.)

76. 🔲▼ *Population Movement* Assume that the percentages given in the preceding exercise also describe the population movements from 2009 to 2010. Use two matrix multiplications to estimate from the data in the preceding exercise the population in each region in 2010.

COMMUNICATION AND REASONING EXERCISES

77. Give an example of two matrices A and B such that AB is defined but BA is not defined.

78. Give an example of two matrices A and B of different dimensions such that both AB and BA are defined.

79. Compare addition and multiplication of 1×1 matrices to the arithmetic of numbers.

80. In comparing the algebra of 1×1 matrices, as discussed so far, to the algebra of real numbers (see Exercise 79), what important difference do you find?

81. Comment on the following claim: Every matrix equation represents a system of equations.

---

[14]Figures are rounded to the nearest 10,000. Sources: South African Dept. of Environmental Affairs and Tourism; Australia Tourist Commission/*New York Times*, January 15, 2000, p. C1.

[15]Note that this exercise ignores migration into or out of the country. Source: U.S. Census Bureau, Current Population Survey, 2009 Annual Social and Economic Supplement.

82. When is it true that both AB and BA are defined, even though neither A nor B is a square matrix?

83. ▼ Find a scenario in which it would be useful to "multiply" two row vectors according to the rule

$$[a \quad b \quad c][d \quad e \quad f] = [ad \quad be \quad cf].$$

84. ▼ Make up an application whose solution reads as follows.

$$\text{"Total revenue} = [10 \quad 100 \quad 30] \begin{bmatrix} 10 & 0 & 3 \\ 1 & 2 & 0 \\ 0 & 1 & 40 \end{bmatrix}."$$

85. ▼ What happens in a spreadsheet if, instead of using the function MMULT, you use "ordinary multiplication" as shown here?

	A	B	C	D	E	F	G
1	2	0	7		1	1	-8
2	1	-1	0		1	0	0
3	-2	1	1		0	5	2
4							
5	=A1:C3*E1:G3						
6							
7							

86. ▼ Define the *naïve product* $A \square B$ of two $m \times n$ matrices A and B by

$$(A \square B)_{ij} = A_{ij} B_{ij}.$$

(This is how someone who has never seen matrix multiplication before might think to multiply matrices.) Referring to Example 1 in this section, compute and comment on the meaning of $P \square (Q^T)$.

4.3 Matrix Inversion

Now that we've discussed matrix addition, subtraction, and multiplication, you may well be wondering about matrix *division*. In the realm of real numbers, division can be thought of as a form of multiplication: Dividing 3 by 7 is the same as multiplying 3 by 1/7, the inverse of 7. In symbols, $3 \div 7 = 3 \times (1/7)$, or 3×7^{-1}. In order to imitate division of real numbers in the realm of matrices, we need to discuss the multiplicative **inverse**, A^{-1}, of a matrix A.

Note Because multiplication of real numbers is commutative, we can write, for example, $\frac{3}{7}$ as either 3×7^{-1} or $7^{-1} \times 3$. In the realm of matrices, multiplication is not commutative, so from now on we shall *never* talk about "division" of matrices (by "$\frac{B}{A}$" should we mean $A^{-1}B$ or BA^{-1}?). ■

Before we try to find the inverse of a matrix, we must first know exactly what we *mean* by the inverse. Recall that the inverse of a number a is the number, often written a^{-1}, with the property that $a^{-1} \cdot a = a \cdot a^{-1} = 1$. For example, the inverse of 76 is the number $76^{-1} = 1/76$, because $(1/76) \cdot 76 = 76 \cdot (1/76) = 1$. This is the number calculated by the x^{-1} button found on most calculators. Not all numbers have an inverse. For example—and this is the only example—the number 0 has no inverse, because you cannot get 1 by multiplying 0 by anything.

The inverse of a matrix is defined similarly. To make life easier, we shall restrict attention to **square** matrices, matrices that have the same number of rows as columns.*

✱ Nonsquare matrices *cannot* have inverses in the sense that we are talking about here. This is not a trivial fact to prove.

Inverse of a Matrix

The **inverse** of an $n \times n$ matrix A is that $n \times n$ matrix A^{-1} which, when multiplied by A on either side, yields the $n \times n$ identity matrix I. Thus,

$$AA^{-1} = A^{-1}A = I.$$

If A has an inverse, it is said to be **invertible**. Otherwise, it is said to be **singular**.

> **Quick Examples**
>
> **1.** The inverse of the 1×1 matrix $[3]$ is $[1/3]$, because $[3][1/3] = [1] = [1/3][3]$.
>
> **2.** The inverse of the $n \times n$ identity matrix I is I itself, because $I \times I = I$. Thus, $I^{-1} = I$.
>
> **3.** The inverse of the 2×2 matrix $A = \begin{bmatrix} 1 & -1 \\ -1 & -1 \end{bmatrix}$ is $A^{-1} = \begin{bmatrix} \frac{1}{2} & -\frac{1}{2} \\ -\frac{1}{2} & -\frac{1}{2} \end{bmatrix}$,
>
> because $\begin{bmatrix} 1 & -1 \\ -1 & -1 \end{bmatrix} \begin{bmatrix} \frac{1}{2} & -\frac{1}{2} \\ -\frac{1}{2} & -\frac{1}{2} \end{bmatrix} = \begin{bmatrix} 1 & 0 \\ 0 & 1 \end{bmatrix}$ $AA^{-1} = I$
>
> and $\begin{bmatrix} \frac{1}{2} & -\frac{1}{2} \\ -\frac{1}{2} & -\frac{1}{2} \end{bmatrix} \begin{bmatrix} 1 & -1 \\ -1 & -1 \end{bmatrix} = \begin{bmatrix} 1 & 0 \\ 0 & 1 \end{bmatrix}$. $A^{-1}A = I$

Notes

1. It is possible to show that if A and B are square matrices with $AB = I$, then it must also be true that $BA = I$. In other words, once we have checked that $AB = I$, we know that B is the inverse of A. The second check, that $BA = I$, is unnecessary.

2. If B is the inverse of A, then we can also say that A is the inverse of B (why?). Thus, we sometimes refer to such a pair of matrices as an **inverse pair** of matrices. ∎

EXAMPLE 1 Singular Matrix

Can $A = \begin{bmatrix} 1 & 1 \\ 0 & 0 \end{bmatrix}$ have an inverse?

Solution No. To see why not, notice that both entries in the second row of AB will be 0, no matter what B is. So AB cannot equal I, no matter what B is. Hence, A is singular.

➡ **Before we go on...** If you think about it, you can write down many similar examples of singular matrices. There is only one number with no multiplicative inverse (0), but there are many matrices having no inverses. ∎

Finding the Inverse of a Square Matrix

Q: *In the box, it was stated that the inverse of* $\begin{bmatrix} 1 & -1 \\ -1 & -1 \end{bmatrix}$ *is* $\begin{bmatrix} \frac{1}{2} & -\frac{1}{2} \\ -\frac{1}{2} & -\frac{1}{2} \end{bmatrix}$. *How was that obtained?*

A: We can think of the problem of finding A^{-1} as a problem of finding four unknowns, the four unknown entries of A^{-1}:

$$A^{-1} = \begin{bmatrix} x & y \\ z & w \end{bmatrix}.$$

These unknowns must satisfy the equation $AA^{-1} = I$, or

$$\begin{bmatrix} 1 & -1 \\ -1 & -1 \end{bmatrix} \begin{bmatrix} x & y \\ z & w \end{bmatrix} = \begin{bmatrix} 1 & 0 \\ 0 & 1 \end{bmatrix}.$$

If we were to try to find the first column of A^{-1}, consisting of x and z, we would have to solve

$$\begin{bmatrix} 1 & -1 \\ -1 & -1 \end{bmatrix}\begin{bmatrix} x \\ z \end{bmatrix} = \begin{bmatrix} 1 \\ 0 \end{bmatrix}$$

or

$$x - z = 1$$
$$-x - z = 0.$$

To solve this system by Gauss-Jordan reduction, we would row-reduce the augmented matrix, which is A with the column $\begin{bmatrix} 1 \\ 0 \end{bmatrix}$ adjoined.

$$\begin{bmatrix} 1 & -1 & | & 1 \\ -1 & -1 & | & 0 \end{bmatrix} \rightarrow \begin{bmatrix} 1 & 0 & | & x \\ 0 & 1 & | & z \end{bmatrix}$$

To find the second column of A^{-1} we would similarly row-reduce the augmented matrix obtained by tacking on to A the second column of the identity matrix.

$$\begin{bmatrix} 1 & -1 & | & 0 \\ -1 & -1 & | & 1 \end{bmatrix} \rightarrow \begin{bmatrix} 1 & 0 & | & y \\ 0 & 1 & | & w \end{bmatrix}$$

The row operations used in doing these two reductions would be exactly the same. We could do both reductions simultaneously by "doubly augmenting" A, putting both columns of the identity matrix to the right of A.

$$\begin{bmatrix} 1 & -1 & | & 1 & 0 \\ -1 & -1 & | & 0 & 1 \end{bmatrix} \rightarrow \begin{bmatrix} 1 & 0 & | & x & y \\ 0 & 1 & | & z & w \end{bmatrix}$$

We carry out this reduction in the following example.

EXAMPLE 2 **Computing Matrix Inverse**

Find the inverse of each matrix.

a. $P = \begin{bmatrix} 1 & -1 \\ -1 & -1 \end{bmatrix}$ **b.** $Q = \begin{bmatrix} 1 & 0 & 1 \\ 2 & -2 & -1 \\ 3 & 0 & 0 \end{bmatrix}$

Solution

a. As described above, we put the matrix P on the left and the identity matrix I on the right to get a 2×4 matrix.

$$\begin{bmatrix} 1 & -1 & | & 1 & 0 \\ -1 & -1 & | & 0 & 1 \end{bmatrix}$$
$$\quad\;\; P \qquad\quad I$$

We now row-reduce the whole matrix:

$$\begin{bmatrix} 1 & -1 & 1 & 0 \\ -1 & -1 & 0 & 1 \end{bmatrix} \begin{matrix} \\ R_2 + R_1 \end{matrix} \rightarrow \begin{bmatrix} 1 & -1 & 1 & 0 \\ 0 & -2 & 1 & 1 \end{bmatrix} \begin{matrix} 2R_1 - R_2 \\ \\ \end{matrix} \rightarrow$$

$$\begin{bmatrix} 2 & 0 & 1 & -1 \\ 0 & -2 & 1 & 1 \end{bmatrix} \begin{matrix} \frac{1}{2}R_1 \\ -\frac{1}{2}R_2 \end{matrix} \rightarrow \begin{bmatrix} 1 & 0 & | & \frac{1}{2} & -\frac{1}{2} \\ 0 & 1 & | & -\frac{1}{2} & -\frac{1}{2} \end{bmatrix}.$$
$$\qquad\qquad\qquad\qquad\qquad\quad\;\; I \qquad\quad P^{-1}$$

We have now solved the systems of linear equations that define the entries of P^{-1}. Thus,

$$P^{-1} = \begin{bmatrix} \frac{1}{2} & -\frac{1}{2} \\ -\frac{1}{2} & -\frac{1}{2} \end{bmatrix}.$$

```
Enter your matrices here.
Q = [1, 0, 1
2, -2,-1
3, 0, 0]
```

Type Q^-1 or Q^(-1) in the formula box and press "Compute".

b. The procedure to find the inverse of a 3 × 3 matrix (or larger) is just the same as for a 2 × 2 matrix. We place Q on the left and the identity matrix (now 3 × 3) on the right, and reduce.

$$
\begin{array}{c} Q \\[-2pt] \left[\begin{array}{ccc|ccc} 1 & 0 & 1 & 1 & 0 & 0 \\ 2 & -2 & -1 & 0 & 1 & 0 \\ 3 & 0 & 0 & 0 & 0 & 1 \end{array}\right] \begin{array}{l} \\ R_2 - 2R_1 \\ R_3 - 3R_1 \end{array} \to \end{array}
\left[\begin{array}{ccc|ccc} 1 & 0 & 1 & 1 & 0 & 0 \\ 0 & -2 & -3 & -2 & 1 & 0 \\ 0 & 0 & -3 & -3 & 0 & 1 \end{array}\right] \begin{array}{l} 3R_1 + R_3 \\ R_2 - R_3 \end{array} \to
$$

$$
\left[\begin{array}{ccc|ccc} 3 & 0 & 0 & 0 & 0 & 1 \\ 0 & -2 & 0 & 1 & 1 & -1 \\ 0 & 0 & -3 & -3 & 0 & 1 \end{array}\right] \begin{array}{l} \frac{1}{3}R_1 \\ -\frac{1}{2}R_2 \\ -\frac{1}{3}R_3 \end{array} \to
\left[\begin{array}{ccc|ccc} 1 & 0 & 0 & 0 & 0 & \frac{1}{3} \\ 0 & 1 & 0 & -\frac{1}{2} & -\frac{1}{2} & \frac{1}{2} \\ 0 & 0 & 1 & 1 & 0 & -\frac{1}{3} \end{array}\right].
$$

$$ I Q^{-1}$$

Thus,

$$
Q^{-1} = \begin{bmatrix} 0 & 0 & \frac{1}{3} \\ -\frac{1}{2} & -\frac{1}{2} & \frac{1}{2} \\ 1 & 0 & -\frac{1}{3} \end{bmatrix}.
$$

We have already checked that P^{-1} is the inverse of P. You should also check that Q^{-1} is the inverse of Q.

The method we used in Example 2 can be summarized as follows:

Inverting an *n* × *n* Matrix

In order to determine whether an $n \times n$ matrix A is invertible or not, and to find A^{-1} if it does exist, follow this procedure:

1. Write down the $n \times 2n$ matrix $[A\,|\,I]$ (this is A with the $n \times n$ identity matrix set next to it).

2. Row-reduce $[A\,|\,I]$.

3. If the reduced form is $[I\,|\,B]$ (i.e., has the identity matrix in the left part), then A is invertible and $B = A^{-1}$. If you cannot obtain I in the left part, then A is singular. (See Example 3.)

Although there is a general formula for the inverse of a matrix, it is not a simple one. In fact, using the formula for anything larger than a 3 × 3 matrix is so inefficient that the row-reduction procedure is the method of choice even for computers. However, the general formula is very simple for the special case of 2 × 2 matrices:

Formula for the Inverse of a 2 × 2 Matrix

The inverse of a 2 × 2 matrix is

$$
\begin{bmatrix} a & b \\ c & d \end{bmatrix}^{-1} = \frac{1}{ad - bc}\begin{bmatrix} d & -b \\ -c & a \end{bmatrix}, \quad \text{provided } ad - bc \neq 0.
$$

If the quantity $ad - bc$ is zero, then the matrix is singular (noninvertible). The quantity $ad - bc$ is called the **determinant** of the matrix $\begin{bmatrix} a & b \\ c & d \end{bmatrix}$.

Quick Examples

1. $\begin{bmatrix} 1 & 2 \\ 3 & 4 \end{bmatrix}^{-1} = \dfrac{1}{(1)(4) - (2)(3)} \begin{bmatrix} 4 & -2 \\ -3 & 1 \end{bmatrix} = -\dfrac{1}{2} \begin{bmatrix} 4 & -2 \\ -3 & 1 \end{bmatrix}$

$= \begin{bmatrix} -2 & 1 \\ \frac{3}{2} & -\frac{1}{2} \end{bmatrix}$

2. $\begin{bmatrix} 1 & -1 \\ 2 & -2 \end{bmatrix}$ has determinant $ad - bc = (1)(-2) - (-1)(2) = 0$ and so is singular.

The formula for the inverse of a 2×2 matrix can be obtained using the technique of row reduction. (See the Communication and Reasoning Exercises at the end of the section.)

As we have mentioned earlier, not every square matrix has an inverse, as we see in the next example.

EXAMPLE 3 Singular 3 × 3 Matrix

Find the inverse of the matrix $S = \begin{bmatrix} 1 & 1 & 2 \\ -2 & 0 & 4 \\ 3 & 1 & -2 \end{bmatrix}$, if it exists.

Solution We proceed as before.

$$
\overset{\displaystyle S \qquad\qquad\qquad I}{\left[\begin{array}{ccc|ccc} 1 & 1 & 2 & 1 & 0 & 0 \\ -2 & 0 & 4 & 0 & 1 & 0 \\ 3 & 1 & -2 & 0 & 0 & 1 \end{array}\right]} \begin{array}{l} \\ R_2 + 2R_1 \\ R_3 - 3R_1 \end{array} \rightarrow \left[\begin{array}{ccc|ccc} 1 & 1 & 2 & 1 & 0 & 0 \\ 0 & 2 & 8 & 2 & 1 & 0 \\ 0 & -2 & -8 & -3 & 0 & 1 \end{array}\right] \begin{array}{l} 2R_1 - R_2 \\ \\ R_3 + R_2 \end{array}
$$

$$
\rightarrow \left[\begin{array}{ccc|ccc} 2 & 0 & -4 & 0 & -1 & 0 \\ 0 & 2 & 8 & 2 & 1 & 0 \\ 0 & 0 & 0 & -1 & 1 & 1 \end{array}\right]
$$

We stopped here, even though the reduction is incomplete, because there is *no hope* of getting the identity on the left-hand side. Completing the row reduction will not change the three zeros in the bottom row. So what is wrong? Nothing. As in Example 1, we have here a singular matrix. Any square matrix that, after row reduction, winds up with a row of zeros is singular. (See Exercise 77.)

➡ **Before we go on...** In practice, deciding whether a given matrix is invertible or singular is easy: Simply try to find its inverse. If the process works, then the matrix is invertible, and we get its inverse. If the process fails, then the matrix is singular. If you try to invert a singular matrix using a spreadsheet, calculator, or computer program, you should get an error. Sometimes, instead of an error, you will get a spurious answer due to round-off errors in the device. ■

Using the Inverse to Solve a System of *n* Linear Equations in *n* Unknowns

Having used systems of equations and row reduction to find matrix inverses, we will now use matrix inverses to solve systems of equations. Recall that, at the end of the previous section, we saw that a system of linear equations could be written in the form

$AX = B,$

Summing to get A's total winnings and then dividing by the number of times the game is played gives the average value of

$$(45 - 60 - 10 + 60)/100 = 0.35$$

so that A can expect to win an average of 0.35 points per play of the game. We call 0.35 the **expected payoff** of the game resulting from these particular strategies for A and B.

This calculation was somewhat tedious and it would only get worse if A and B had many moves to choose from. There is a far more convenient way of doing exactly the same calculation, using matrix multiplication: We start by representing the player's strategies as matrices. For reasons to become clear in a moment, we record A's strategy as a row matrix:

$$R = [0.75 \quad 0.25].$$

We record B's strategy as a column matrix:

$$C = \begin{bmatrix} 0.20 \\ 0.80 \end{bmatrix}.$$

using Technology

The use of technology becomes indispensable when we need to do several calculations or when the matrices involved are big. See the technology note accompanying Example 3 in Section 4.2 on page 246 for instructions on multiplying matrices using a TI-83/84 Plus, a spreadsheet, and the Matrix Algebra Tool at the Website.

(We will sometimes write column vectors using transpose notation, writing, for example, $[0.20 \quad 0.80]^T$ for the column above, to save space.) Now: *The expected payoff is the matrix product RPC, where P is the payoff matrix!*

$$\text{Expected payoff} = RPC = [0.75 \quad 0.25] \begin{bmatrix} 3 & -1 \\ -2 & 3 \end{bmatrix} \begin{bmatrix} 0.20 \\ 0.80 \end{bmatrix}$$

$$= [1.75 \quad 0] \begin{bmatrix} 0.20 \\ 0.80 \end{bmatrix} = [0.35].$$

Why does this work? Write out the arithmetic involved in the matrix product RPC to see what we calculated:

$$[0.75 \times 3 + 0.25 \times (-2)] \times 0.20 + [0.75 \times (-1) + 0.25 \times 3] \times 0.80$$
$$= 0.75 \times 3 \times 0.20 + 0.25 \times (-2) \times 0.20 + 0.75 \times (-1) \times 0.80 + 0.25 \times 3 \times 0.80$$
$$= \quad \text{Case 1} \quad + \quad \text{Case 3} \quad + \quad \text{Case 2} \quad + \quad \text{Case 4.}$$

So, the matrix product does all at once the various cases we considered above.

To summarize what we just saw:

The Expected Payoff resulting from Mixed Strategies *R* and *C*

The **expected payoff of a game resulting from given mixed strategies** is the average payoff that occurs if the game is played a large number of times with the row and column players using the given strategies.

To compute the expected payoff resulting from mixed strategies R and C:

1. Write the row player's mixed strategy as a row matrix R.

2. Write the column player's mixed strategy as a column matrix C.

3. Calculate the product RPC, where P is the payoff matrix. This product is a 1×1 matrix whose entry is the expected payoff e.

Quick Example

Consider "Rock, Paper, Scissors."

$$\mathbf{A} \; \begin{array}{c} \\ r \\ p \\ s \end{array} \overset{\displaystyle\mathbf{B}}{\begin{array}{ccc} r & p & s \\ \begin{bmatrix} 0 & -1 & 1 \\ 1 & 0 & -1 \\ -1 & 1 & 0 \end{bmatrix} \end{array}}$$

Suppose that the row player plays *rock* half the time and each of the other two strategies a quarter of the time, and the column player always plays *paper*. We write

$$R = \begin{bmatrix} \dfrac{1}{2} & \dfrac{1}{4} & \dfrac{1}{4} \end{bmatrix}, \quad C = \begin{bmatrix} 0 \\ 1 \\ 0 \end{bmatrix}.$$

So,

$$e = RPC = \begin{bmatrix} \dfrac{1}{2} & \dfrac{1}{4} & \dfrac{1}{4} \end{bmatrix} \begin{bmatrix} 0 & -1 & 1 \\ 1 & 0 & -1 \\ -1 & 1 & 0 \end{bmatrix} \begin{bmatrix} 0 \\ 1 \\ 0 \end{bmatrix}$$

$$= \begin{bmatrix} \dfrac{1}{2} & \dfrac{1}{4} & \dfrac{1}{4} \end{bmatrix} \begin{bmatrix} -1 \\ 0 \\ 1 \end{bmatrix} = -\dfrac{1}{4}.$$

Thus, player A can expect to lose an average of once every four plays.

Solving a Game

Now that we know how to evaluate particular strategies, we want to find the *best* strategy. The next example takes us another step toward that goal.

EXAMPLE 2 Television Ratings Wars

Commercial TV station RTV and cultural station CTV are competing for viewers in the Tuesday prime-time 9–10 PM time slot. RTV is trying to decide whether to show a sitcom, a docudrama, a reality show, or a movie, while CTV is thinking about either a nature documentary, a symphony concert, a ballet, or an opera. A television rating company estimates the payoffs for the various alternatives as follows. (Each point indicates a shift of 1,000 viewers from one channel to the other; thus, for instance, -2 indicates a shift of 2,000 viewers from RTV to CTV.)

		CTV			
		Nature Doc.	**Symphony**	**Ballet**	**Opera**
	Sitcom	2	1	-2	2
RTV	**Docudrama**	-1	1	-1	2
	Reality Show	-2	0	0	1
	Movie	3	1	-1	1

a. If RTV notices that CTV is showing nature documentaries half the time and symphonies the other half, what would RTV's best strategy be, and how many viewers would it gain if it followed this strategy?

b. If, on the other hand, CTV notices that RTV is showing docudramas half the time and reality shows the other half, what would CTV's best strategy be, and how many viewers would it gain or lose if it followed this strategy?

Solution

a. We are given the matrix of the game, P, in the table above, and we are given CTV's strategy $C = [0.50 \ \ 0.50 \ \ 0 \ \ 0]^T$. We are not given RTV's strategy R. To say that RTV is looking for its best strategy is to say that it wants the resulting expected payoff $e = RPC$ to be as high as possible. So, we take $R = [x \ \ y \ \ z \ \ t]$ and look for values for x, y, z, and t that make RPC as high as possible. First, we calculate e in terms of these unknowns:

$$e = RPC = [x \quad y \quad z \quad t] \begin{bmatrix} 2 & 1 & -2 & 2 \\ -1 & 1 & -1 & 2 \\ -2 & 0 & 0 & 1 \\ 3 & 1 & -1 & 1 \end{bmatrix} \begin{bmatrix} 0.50 \\ 0.50 \\ 0 \\ 0 \end{bmatrix}$$

$$= [x \quad y \quad z \quad t] \begin{bmatrix} 1.5 \\ 0 \\ -1 \\ 2 \end{bmatrix} = 1.5x - z + 2t.$$

Now, the unknowns x, y, z, and t must be nonnegative and add up to 1 (why?). Because t has the largest coefficient, 2, we'll get the best result by making it as large as possible, namely, $t = 1$, leaving $x = y = z = 0$. Thus, RTV's best strategy is $R = [0 \ \ 0 \ \ 0 \ \ 1]$. In other words, RTV should use the pure strategy of showing a movie every Tuesday evening. If it does so, the expected payoff will be

$$e = 1.5(0) - 0 + 2(1) = 2$$

so RTV can expect to gain 2,000 viewers.

b. Here, we are given $R = [0 \ \ 0.50 \ \ 0.50 \ \ 0]$ and are not given CTV's strategy C, so this time we take $C = [x \ \ y \ \ z \ \ t]^T$ and calculate the resulting expected payoff e:

$$e = RPC = [0 \quad 0.50 \quad 0.50 \quad 0] \begin{bmatrix} 2 & 1 & -2 & 2 \\ -1 & 1 & -1 & 2 \\ -2 & 0 & 0 & 1 \\ 3 & 1 & -1 & 1 \end{bmatrix} \begin{bmatrix} x \\ y \\ z \\ t \end{bmatrix}$$

$$= [-1.5 \quad 0.5 \quad -0.5 \quad 1.5] \begin{bmatrix} x \\ y \\ z \\ t \end{bmatrix} = -1.5x + 0.5y - 0.5z + 1.5t.$$

Now, CTV wants e to be as *low* as possible (why?). Because x has the largest negative coefficient, CTV would like it to be as large as possible: $x = 1$, so the rest of the unknowns must be zero. Thus, CTV's best strategy is $C = [1 \ \ 0 \ \ 0 \ \ 0]^T$; that is, show a nature documentary every night. If it does so, the expected payoff will be

$$e = -1.5(1) + 0.5(0) - 0.5(0) + 1.5(0) = -1.5.$$

So CTV can expect to gain 1,500 viewers.

This example illustrates the fact that, no matter what mixed strategy one player selects, the other player can choose an appropriate *pure* counterstrategy in order to maximize its gain. How does this affect what decisions you should make as one of the players? If you were on the board of directors of RTV, you might reason as follows: Since for every mixed strategy you try, CTV can find a best counterstrategy (as in part (b)), it is in your company's best interest to select a mixed strategy that *minimizes* the effect of CTV's best counterstrategy. This is called the **minimax criterion**.

Minimax Criterion

A player using the **minimax criterion** chooses a strategy that, among all possible strategies, minimizes the effect of the other player's best counterstrategy. That is, an optimal (best) strategy according to the minimax criterion is one that minimizes the maximum damage the opponent can cause.

This criterion assumes that your opponent is determined to win. More precisely, it assumes the following.

Fundamental Principle of Game Theory

Each player tries to use its best possible strategy, and assumes that the other player is doing the same.

This principle is not always followed by every player. For example, one of the players may be nature and may choose its move at random, with no particular purpose in mind. In such a case, criteria other than the minimax criterion may be more appropriate. For example, there is the "maximax" criterion, which maximizes the maximum possible payoff (also known as the "reckless" strategy), or the criterion that seeks to minimize "regret" (the difference between the payoff you get and the payoff you *would have gotten* if you had known beforehand what was going to happen).[*] But, we shall assume here the fundamental principle and try to find optimal strategies under the minimax criterion.

Finding the optimal strategy is called **solving the game**. In general, solving a game can be done using linear programming, as we shall see in the next chapter. However, we can solve 2×2 games "by hand," as we shall see in the next example. First, we notice that some large games can be reduced to smaller games.

Consider the game in the preceding example, which had the following matrix:

$$P = \begin{bmatrix} 2 & 1 & -2 & 2 \\ -1 & 1 & -1 & 2 \\ -2 & 0 & 0 & 1 \\ 3 & 1 & -1 & 1 \end{bmatrix}.$$

* See *Location in Space: Theoretical Perspectives in Economic Geography,* 3rd Edition, by Peter Dicken and Peter E. Lloyd, HarperCollins Publishers, 1990, p. 276.

Compare the second and third columns through the eyes of the column player, CTV. Every payoff in the third column is as good as or better, from CTV's point of view, than the corresponding entry in the second column. Thus, no matter what RTV does, CTV will do better showing a ballet (third column) than a symphony (second column). We say that the third column **dominates** the second column. As far as CTV is concerned, we might as well forget about symphonies entirely, so we remove the second column. Similarly, the third column dominates the fourth, so we can remove the fourth column, too. This gives us a smaller game to work with:

$$P = \begin{bmatrix} 2 & -2 \\ -1 & -1 \\ -2 & 0 \\ 3 & -1 \end{bmatrix}.$$

Now compare the first and last rows. Every payoff in the last row is larger than the corresponding payoff in the first row, so the last row is always better to RTV. Again, we say that the last row dominates the first row, and we can discard the first row. Similarly, the last row dominates the second row, so we discard the second row as well. This reduces us to the following game:

$$P = \begin{bmatrix} -2 & 0 \\ 3 & -1 \end{bmatrix}.$$

In this matrix, neither row dominates the other and neither column dominates the other. So, this is as far as we can go with this line of argument. We call this **reduction by dominance**.

Reduction by Dominance

One *row* **dominates** another if every entry in the former is greater than or equal to the corresponding entry in the latter. Put another way, one row dominates another if it is always at least as good for the row player.

One *column* dominates another if every entry in the former is less than or equal to the corresponding entry in the latter. Put another way, one column dominates another if it is always at least as good for the column player.

Procedure for Reducing by Dominance:
1. Check whether there is any row in the (remaining) matrix that is dominated by another row. Remove all dominated rows.
2. Check whether there is any column in the (remaining) matrix that is dominated by another column. Remove all dominated columns.
3. Repeat steps 1 and 2 until there are no dominated rows or columns.

Let us now go back to the "television ratings wars" example and see how we can solve a game using the minimax criterion once we are down to a 2×2 payoff matrix.

EXAMPLE 3 Solving a 2 × 2 Game

Continuing the preceding example:

a. Find the optimal strategy for RTV.

b. Find the optimal strategy for CTV.

c. Find the expected payoff of the game if RTV and CTV use their optimal strategies.

Solution As in the text, we begin by reducing the game by dominance, which brings us down to the following 2×2 game:

		CTV	
		Nature Doc.	**Ballet**
RTV	**Reality Show**	-2	0
	Movie	3	-1

a. Now let's find RTV's optimal strategy. Because we don't yet know what it is, we write down a general strategy:

$$R = [x \quad y].$$

Because $x + y = 1$, we can replace y by $1 - x$:

$$R = [x \quad 1 - x].$$

We know that CTV's best counterstrategy to R will be a pure strategy (see the discussion after Example 2), so let's compute the expected payoff that results from each of CTV's possible pure strategies:

$$e = [x \quad 1 - x]\begin{bmatrix} -2 & 0 \\ 3 & -1 \end{bmatrix}\begin{bmatrix} 1 \\ 0 \end{bmatrix}$$

$$= (-2)x + 3(1 - x) = -5x + 3$$

$$f = [x \quad 1 - x]\begin{bmatrix} -2 & 0 \\ 3 & -1 \end{bmatrix}\begin{bmatrix} 0 \\ 1 \end{bmatrix}$$

$$= 0x - (1 - x) = x - 1.$$

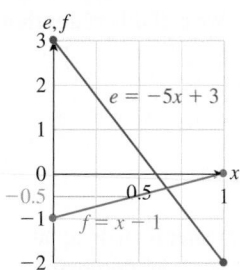

Figure 1

Because both e and f depend on x, we can graph them as in Figure 1.

If, for instance, RTV happened to choose $x = 0.5$, then the expected payoffs resulting from CTV's two pure strategies are $e = -5(1/2) + 3 = 1/2$ and $f = 1/2 - 1 = -1/2$. The worst outcome for RTV is the lower of the two, f, and this will be true wherever the graph of f is below the graph of e. On the other hand, if RTV chose $x = 1$, the graph of e would be lower and the worst possible expected value would be $e = -5(1) + 3 = -2$. Since RTV can choose x to be any value between 0 and 1, the worst possible outcomes are those shown by the colored portion of the graph in Figure 2.

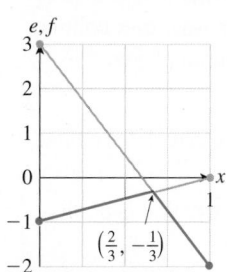

Figure 2

Because RTV is trying to make the worst possible outcome as large as possible (that is, to minimize damages), it is seeking the point on the colored portion of the graph that is highest. This is the intersection point of the two lines. To calculate its coordinates, it's easiest to equate the two functions of x:

$$-5x + 3 = x - 1,$$

$$-6x = -4,$$

or $\qquad x = \dfrac{2}{3}.$

The e (or f) coordinate is then obtained by substituting $x = 2/3$ into the expression for e (or f), giving:

$$e = -5\left(\frac{2}{3}\right) + 3$$

$$= -\frac{1}{3}.$$

We conclude that RTV's best strategy is to take $x = 2/3$, giving an expected value of $-1/3$. In other words, RTV's optimal mixed strategy is:

$$R = \begin{bmatrix} \dfrac{2}{3} & \dfrac{1}{3} \end{bmatrix}.$$

Going back to the original game, RTV should show reality shows $2/3$ of the time and movies $1/3$ of the time. It should not bother showing any sitcoms or docudramas. It expects to lose, on average, 333 viewers to CTV, but all of its other options are worse.

b. To find CTV's optimal strategy, we must reverse roles and start by writing its unknown strategy as follows:

$$C = \begin{bmatrix} x \\ 1 - x \end{bmatrix}.$$

We calculate the expected payoffs for the two pure row strategies:

$$e = \begin{bmatrix} 1 & 0 \end{bmatrix} \begin{bmatrix} -2 & 0 \\ 3 & -1 \end{bmatrix} \begin{bmatrix} x \\ 1 - x \end{bmatrix}$$

$$= -2x$$

and

$$f = \begin{bmatrix} 0 & 1 \end{bmatrix} \begin{bmatrix} -2 & 0 \\ 3 & -1 \end{bmatrix} \begin{bmatrix} x \\ 1 - x \end{bmatrix}$$

$$= 3x - (1 - x)$$

$$= 4x - 1.$$

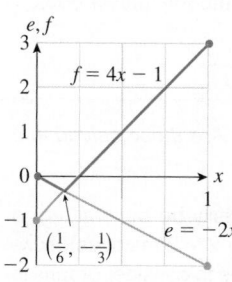

Figure 3

As with the row player, we know that the column player's best strategy will correspond to the intersection of the graphs of e and f (Figure 3). (Why is the upper edge colored, rather than the lower edge?) The graphs intersect when

$$-2x = 4x - 1$$

or

$$x = \frac{1}{6}.$$

The corresponding value of e (or f) is

$$e = -2\left(\frac{1}{6}\right) = -\frac{1}{3}.$$

Thus, CTV's optimal mixed strategy is $\begin{bmatrix} \frac{1}{6} & \frac{5}{6} \end{bmatrix}^T$ and the expected payoff is $-1/3$. So, CTV should show nature documentaries $1/6$ of the time and ballets $5/6$ of the time. It should not bother to show symphonies or operas. It expects to gain, on average, 333 viewers from RTV.

c. We can now calculate the expected payoff as usual, using the optimal strategies we found in parts (a) and (b).

$$e = RPC$$

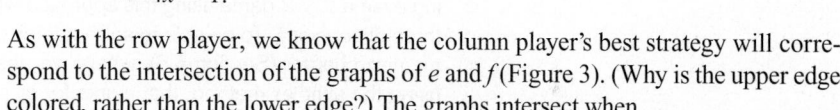

$$= \begin{bmatrix} \dfrac{2}{3} & \dfrac{1}{3} \end{bmatrix} \begin{bmatrix} -2 & 0 \\ 3 & -1 \end{bmatrix} \begin{bmatrix} \dfrac{1}{6} \\ \dfrac{5}{6} \end{bmatrix}$$

$$= -\frac{1}{3}$$

➡ **Before we go on...** In Example 3, it is no accident that the expected payoff resulting from the optimal strategies equals the expected payoff we found in (a) and (b). If we call the expected payoff resulting from the optimal strategies the **expected value of the game**, the row player's optimal strategy guarantees an expected payoff no smaller than the expected value, while the column player's optimal strategy guarantees an expected payoff no larger. Together they force the average payoff to be the expected value of the game. ■

Expected Value of a Game

The **expected value of a game** is its expected payoff that results when the row and column players use their optimal (minimax) strategies. By using its optimal strategy, the row player guarantees an expected payoff no lower than the expected value of the game, no matter what the column player does. Similarly, by using its optimal strategy, the column player guarantees an expected payoff no higher than the expected value of the game, no matter what the row player does.

Q: *What about games that don't reduce to 2 × 2 matrices? Are these solved in a similar way?*

A: The method illustrated in the previous example cannot easily be generalized to solve bigger games (i.e., games that cannot be reduced to 2 × 2 matrices); solving even a 2 × 3 game using this approach would require us to consider graphs in three dimensions. To be able to solve games of arbitrary size, we need to wait until the next chapter (Section 5.5), where we describe a method for solving a game, using the simplex method, that works for all payoff matrices.

Strictly Determined Games

Although we haven't yet discussed how to solve general $m \times n$ games, there are certain kinds of games that can be solved quite simply regardless of size, as illustrated by the following example.

EXAMPLE 4 Strictly Determined Game

Solve the following game:

$$
\mathbf{A} \begin{array}{c} \\ s \\ t \\ u \end{array} \overset{\displaystyle \overset{\mathbf{B}}{\begin{array}{ccc} p & q & r \end{array}}}{\begin{bmatrix} -4 & -3 & 3 \\ 2 & -1 & -2 \\ 1 & 0 & 2 \end{bmatrix}}.
$$

Solution If we look carefully at this matrix, we see that no row dominates another and no column dominates another, so we can't reduce it. Nor do we know how to solve a 3 × 3 game, so it looks as if we're stuck. However, there is a way to understand this particular game. With the minimax criterion in mind, let's begin by considering the worst possible outcomes for the row player for each possible move. We do this by circling the smallest payoff in each row, the **row minima**:

$$
\begin{array}{cc}
 & \mathbf{B} \\
\begin{array}{ccc} p & q & r \end{array} & \text{Row minima} \\
\mathbf{A}\begin{array}{c} s \\ t \\ u \end{array}
\left[\begin{array}{ccc}
\boxed{-4} & -3 & 3 \\
2 & -1 & \boxed{-2} \\
1 & \boxed{0} & 2
\end{array}\right]
\begin{array}{c} -4 \\ -2 \\ 0 \leftarrow \text{(largest)} \end{array}
\end{array}
$$

So, for example, if A plays move s, the worst possible outcome is to lose 4. Player A takes the least risk by using move u, which has the largest row minimum.

We do the same thing for the column player, remembering that smaller payoffs are better for B and larger payoffs worse. We draw a box around the largest payoff in each column, the **column maxima**:

$$
\mathbf{A}\begin{array}{c} s \\ t \\ u \end{array}
\left[\begin{array}{ccc}
-4 & -3 & \boxed{3} \\
\boxed{2} & -1 & -2 \\
1 & \boxed{0} & 2
\end{array}\right].
$$

$$
\begin{array}{ccc}
\text{Column maxima} & 2 & 0 & 3 \\
 & & \uparrow & \\
 & & \text{(smallest)} &
\end{array}
$$

Player B takes the least risk by using move q, which has the smallest column maximum.

Now put the circles and boxes together:

$$
\mathbf{A}\begin{array}{c} s \\ t \\ u \end{array}
\left[\begin{array}{ccc}
\boxed{-4} & -3 & \boxed{3} \\
\boxed{2} & -1 & \boxed{-2} \\
1 & \boxed{0} & 2
\end{array}\right].
$$

Notice that the uq entry is both circled and boxed: It is both a row minimum and a column maximum. We call such an entry a **saddle point**.

Now we claim that the optimal strategy for A is to always play u while the optimal strategy for B is to always play q. By playing u, A guarantees that the payoff will be 0 or higher, no matter what B does, so the expected value of the game has to be *at least* 0. On the other hand, by playing q, B guarantees that the payoff will be 0 or less, so the expected value of the game has to be *no more than* 0. Combining these facts, we conclude that the expected value of the game must be exactly 0, and A and B have no strategies that could do any better for them than the pure strategies u and q.

➡ **Before we go on...** You should consider what happens in an example like the television rating wars game of Example 3. In that game, the largest row minimum is -1 while the smallest column maximum is 0; there is no saddle point. The row player can force a payoff of at least -1 by playing a pure strategy (always showing movies, for example), but can do better, forcing an expected payoff of $-1/3$, by playing a mixed strategy as we saw in Example 3. Similarly, the column player can force the payoff to be 0 or less with a pure strategy, but can do better, forcing an expected payoff of $-1/3$, with a mixed strategy. Only when there is a saddle point will pure strategies be optimal. ■

Strictly Determined Game

A **saddle point** is a payoff v that is simultaneously a row minimum and a column maximum (both boxed and circled in our approach). If a game has a saddle point, the corresponding row and column strategies are the optimal ones, the expected value of the game is the payoff v, and we say that the game is **strictly determined**.

using Technology

Website
www.WanerMath.com
To play a strictly determined game against your computer, and also to automatically reduce by dominance and solve arbitrary games up to 5×5, follow

Student Website

→ On Line Utilities

→ Game Theory Tool

FAQs

Solving a Game

Q: *We've seen several ways of trying to solve a game. What should I do and in what order?*

A: Here are the steps you should take when trying to solve a game:

1. Reduce by dominance. This should always be your first step.
2. If you were able to reduce to a 1×1 game, you're done. The optimal strategies are the corresponding pure strategies, as they dominate all the others.
3. Look for a saddle point in the reduced game. If it has one, the game is strictly determined, and the corresponding pure strategies are optimal.
4. If your reduced game is 2×2 and has no saddle point, use the method of Example 3 to find the optimal mixed strategies.
5. If your reduced game is larger than 2×2 and has no saddle point, you have to use linear programming to solve it, but that will have to wait until the following chapter.

4.4 EXERCISES

▼ more advanced   ◆ challenging

T indicates exercises that should be solved using technology

In Exercises 1–4, calculate the expected payoff of the game with payoff matrix

$$P = \begin{bmatrix} 2 & 0 & -1 & 2 \\ -1 & 0 & 0 & -2 \\ -2 & 0 & 0 & 1 \\ 3 & 1 & -1 & 1 \end{bmatrix}$$

using the mixed strategies supplied. HINT [See Example 1.]

1. $R = [0 \quad 1 \quad 0 \quad 0], C = [1 \quad 0 \quad 0 \quad 0]^T$

2. $R = [0 \quad 0 \quad 0 \quad 1], C = [0 \quad 1 \quad 0 \quad 0]^T$

3. $R = [0.5 \quad 0.5 \quad 0 \quad 0], C = [0 \quad 0 \quad 0.5 \quad 0.5]^T$

4. $R = [0 \quad 0.5 \quad 0 \quad 0.5], C = [0.5 \quad 0.5 \quad 0 \quad 0]^T$

In Exercises 5–8, either a mixed column or row strategy is given. In each case, use

$$P = \begin{bmatrix} 0 & -1 & 5 \\ 2 & -2 & 4 \\ 0 & 3 & 0 \\ 1 & 0 & -5 \end{bmatrix}$$

and find the optimal pure strategy (or strategies) the other player should use. Express the answer as a row or column matrix. Also determine the resulting expected payoff. HINT [See Example 2.]

5. $C = [0.25 \quad 0.75 \quad 0]^T$

6. $C = \left[\frac{1}{3} \quad \frac{1}{3} \quad \frac{1}{3}\right]^T$

7. $R = \left[\frac{1}{2} \quad 0 \quad \frac{1}{4} \quad \frac{1}{4}\right]$

8. $R = [0.8 \quad 0.2 \quad 0 \quad 0]$

Reduce the payoff matrices in Exercises 9–14 by dominance.

9.
$$\begin{array}{c c} & \begin{array}{c c c} p & q & r \end{array} \\ \begin{array}{c} a \\ b \end{array} & \begin{bmatrix} 1 & 1 & 10 \\ 2 & 3 & -4 \end{bmatrix} \end{array}$$

10.
$$\begin{array}{c c} & \begin{array}{c c c} p & q & r \end{array} \\ \begin{array}{c} a \\ b \end{array} & \begin{bmatrix} 2 & 0 & 10 \\ 15 & -4 & -5 \end{bmatrix} \end{array}$$

11.
$$\begin{array}{c c} & \begin{array}{c c c} a & b & c \end{array} \\ \begin{array}{c} 1 \\ 2 \\ 3 \end{array} & \begin{bmatrix} 2 & -4 & -9 \\ -1 & -2 & -3 \\ 5 & 0 & -1 \end{bmatrix} \end{array}$$

12.
$$\begin{array}{c c} & \begin{array}{c c c} a & b & c \end{array} \\ \begin{array}{c} 1 \\ 2 \\ 3 \end{array} & \begin{bmatrix} 0 & -1 & -5 \\ -3 & -10 & 10 \\ 2 & 3 & -4 \end{bmatrix} \end{array}$$

13.

$$\begin{array}{c} & \begin{array}{ccc} a & b & c \end{array} \\ \begin{array}{c} p \\ q \\ r \\ s \end{array} & \left[\begin{array}{ccc} 1 & -1 & -5 \\ 4 & 0 & 2 \\ 3 & -3 & 10 \\ 3 & -5 & -4 \end{array} \right] \end{array}$$

14.

$$\begin{array}{c} & \begin{array}{ccc} a & b & c \end{array} \\ \begin{array}{c} p \\ q \\ r \\ s \end{array} & \left[\begin{array}{ccc} 2 & -4 & 9 \\ 1 & 1 & 0 \\ -1 & -2 & -3 \\ 1 & 1 & -1 \end{array} \right] \end{array}$$

In Exercises 15–20, decide whether the game is strictly determined. If it is, give the players' optimal pure strategies and the value of the game. HINT [See Example 4.]

15.

$$\begin{array}{c} & \begin{array}{cc} p & q \end{array} \\ \begin{array}{c} a \\ b \end{array} & \left[\begin{array}{cc} 1 & 1 \\ 2 & -4 \end{array} \right] \end{array}$$

16.

$$\begin{array}{c} & \begin{array}{cc} p & q \end{array} \\ \begin{array}{c} a \\ b \end{array} & \left[\begin{array}{cc} -1 & 2 \\ 10 & -1 \end{array} \right] \end{array}$$

17.

$$\begin{array}{c} & \begin{array}{ccc} p & q & r \end{array} \\ \begin{array}{c} a \\ b \end{array} & \left[\begin{array}{ccc} 2 & 0 & -2 \\ -1 & 3 & 0 \end{array} \right] \end{array}$$

18.

$$\begin{array}{c} & \begin{array}{ccc} p & q & r \end{array} \\ \begin{array}{c} a \\ b \end{array} & \left[\begin{array}{ccc} -2 & 1 & -3 \\ -2 & 3 & -2 \end{array} \right] \end{array}$$

19.

$$\begin{array}{c} & \begin{array}{ccc} a & b & c \end{array} \\ \begin{array}{c} P \\ Q \\ R \\ S \end{array} & \left[\begin{array}{ccc} 1 & -1 & -5 \\ 4 & -4 & 2 \\ 3 & -3 & -10 \\ 5 & -5 & -4 \end{array} \right] \end{array}$$

20.

$$\begin{array}{c} & \begin{array}{ccc} a & b & c \end{array} \\ \begin{array}{c} P \\ Q \\ R \\ S \end{array} & \left[\begin{array}{ccc} -2 & -4 & 9 \\ 1 & 1 & 0 \\ -1 & -2 & -3 \\ 1 & 1 & -1 \end{array} \right] \end{array}$$

*In Exercises 21–24, find: **(a)** the optimal mixed row strategy; **(b)** the optimal mixed column strategy, and **(c)** the expected value of the game in the event that each player uses his or her optimal mixed strategy.* HINT [See Example 3.]

21. $P = \left[\begin{array}{cc} -1 & 2 \\ 0 & -1 \end{array} \right]$

22. $P = \left[\begin{array}{cc} -1 & 0 \\ 1 & -1 \end{array} \right]$

23. $P = \left[\begin{array}{cc} -1 & -2 \\ -2 & 1 \end{array} \right]$

24. $P = \left[\begin{array}{cc} -2 & -1 \\ -1 & -3 \end{array} \right]$

APPLICATIONS

Set up the payoff matrix in each of Exercises 25–32.

25. Games to Pass the Time You and your friend have come up with the following simple game to pass the time: at each round, you simultaneously call "heads" or "tails." If you have both called the same thing, your friend wins 1 point; if your calls differ, you win 1 point.

26. Games to Pass the Time Bored with the game in Exercise 25, you decide to use the following variation instead: If you both call "heads" your friend wins 2 points; if you both call "tails" your friend wins 1 point; if your calls differ, then you win 2 points if you called "heads," and 1 point if you called "tails."

27. War Games You are deciding whether to invade France, Sweden, or Norway, and your opponent is simultaneously deciding which of these three countries to defend. If you invade a country that your opponent is defending, you will be defeated (payoff: -1), but if you invade a country your opponent is not defending, you will be successful (payoff: $+1$).

28. War Games You must decide whether to attack your opponent by sea or air, and your opponent must simultaneously decide whether to mount an all-out air defense, an all-out coastal defense (against an attack from the sea), or a combined air and coastal defense. If there is no defense for your mode of attack, you win 100 points. If your attack is met by a shared air and coastal defense, you win 50 points. If your attack is met by an all-out defense, you lose 200 points.

29. ▼ Marketing Your fast-food outlet, *Burger Queen*, has obtained a license to open branches in three closely situated South African cities: Brakpan, Nigel, and Springs. Your market surveys show that Brakpan and Nigel each provide a potential market of 2,000 burgers a day, while Springs provides a potential market of 1,000 burgers per day. Your company can finance an outlet in only one of those cities. Your main competitor, Burger Princess, has also obtained licenses for these cities, and is similarly planning to open only one outlet. If you both happen to locate at the same city, you will share the total business from all three cities equally, but if you locate in different cities, you will each get all the business in the cities in which you have located, plus half the business in the third city. The payoff is the number of burgers you will sell per day minus the number of burgers your competitor will sell per day.

30. ▼ Marketing Repeat Exercise 29, given that the potential sales markets in the three cities are: Brakpan, 2,500 per day; Nigel, 1,500 per day; Springs, 1,200 per day.

31. ▼ Betting When you bet on a racehorse with odds of *m–n*, you stand to win *m* dollars for every bet of *n* dollars if your horse wins; for instance, if the horse you bet is running at 5–2 and wins, you will win $5 for every $2 you bet. (Thus a $2 bet will return $7.) Here are some actual odds from a 1992 race at Belmont Park, NY.[21] The favorite at 5–2 was Pleasant Tap. The second choice was Thunder Rumble at 7–2, while the third choice was Strike the Gold at 4–1. Assume you are making a $10 bet on one of these horses. The payoffs are your winnings. (If your horse does not win, you lose your entire bet. Of course, it is possible for none of your horses to win.)

32. ▼ Betting Referring to Exercise 31, suppose that just before the race there has been frantic betting on Thunder Rumble with the result that the odds have dropped to 2–5. The odds on the other two horses remain unchanged.

33. Retail Discount Wars Just one week after your *Abercrom B* men's fashion outlet has opened at a new location opposite *Burger Prince* in the Mall, your rival, *Abercrom A*, opens up

---

[21]Source: *New York Times*, September 18, 1992, p. B14.

directly across from you. You have been informed that Abercrom A is about to launch either a 30% off everything sale or a 50% off everything sale. You, on the other hand, have decided to either *increase* prices (to make your store seem more exclusive) or do absolutely nothing. You construct the following payoff matrix, where the payoffs represent the number of customers your outlet can expect to gain from Abercrom A:

		Abercrom A	
		30% Off	50% Off
Abercrom B	Do Nothing	−60	−40
	Increase Prices	30	−50

There is a 20% chance that Abercrom A will opt for the "30% off" sale and an 80% chance that they will opt for the "50% Off" sale. Your sense from upper management at Abercrom B is that there is a 50% chance you will be given the go-ahead to raise prices. What is the expected resulting effect on your customer base?

34. ***More Retail Discount Wars*** Your *Abercrom B* men's fashion outlet has a 30% chance of launching an expensive new line of used auto-mechanic dungarees (complete with grease stains) and a 70% chance of staying instead with its traditional torn military-style dungarees. Your rival across from you in the mall, *Abercrom A*, appears to be deciding between a line of torn gym shirts and a more daring line of "empty shirts" (that is, empty shirt boxes). Your corporate spies reveal that there is a 20% chance that Abercrom A will opt for the empty shirt option. The following payoff matrix gives the number of customers your outlet can expect to gain from Abercrom A in each situation:

		Abercrom A	
		Torn Shirts	Empty Shirts
Abercrom B	Mechanics	10	−40
	Military	−30	50

What is the expected resulting effect on your customer base?

35. ***Factory Location***[22] A manufacturer of electrical machinery is located in a cramped, though low-rent, factory close to the center of a large city. The firm needs to expand, and it could do so in one of three ways: (1) Remain where it is and install new equipment, (2) move to a suburban site in the same city, or (3) relocate in a different part of the country where labor is cheaper. Its decision will be influenced by the fact that one of the following will happen: (I) The government may introduce a program of equipment grants, (II) a new suburban highway may be built, or (III) the government may institute a policy of financial help to companies who move into regions of high unemployment. The value to the company of each combination is given in the following payoff matrix.

		Government's Options		
		I	II	III
Manufacturer's Options	1	200	150	140
	2	130	220	130
	3	110	110	220

If the manufacturer judges that there is a 20% probability that the government will go with option I, a 50% probability that they will go with option II, and a 30% probability that they will go with option III, what is the manufacturer's best option?

36. ***Crop Choice***[23] A farmer has a choice of growing wheat, barley, or rice. Her success will depend on the weather, which could be dry, average, or wet. Her payoff matrix is as follows.

		Weather		
		Dry	Average	Wet
Crop Choices	Wheat	20	20	10
	Barley	10	15	20
	Rice	10	20	20

If the probability that the weather will be dry is 10%, the probability that it will be average is 60%, and the probability that it will be wet is 30%, what is the farmer's best choice of crop?

37. ***Study Techniques*** Your mathematics test is tomorrow and will cover the following topics: game theory, linear programming, and matrix algebra. You have decided to do an "all-nighter" and must determine how to allocate your eight hours of study time among the three topics. If you were to spend the entire eight hours on any one of these topics (thus using a pure strategy), you feel confident that you would earn a 90% score on that portion of the test but would not do so well on the other topics. You have come up with the following table, where the entries are your expected scores. (The fact that linear programming and matrix algebra are used in game theory is reflected in these numbers.)

	Test		
Your Strategies	**Game Theory**	**Linear Programming**	**Matrix Algebra**
Game Theory	90	70	70
Linear Programming	40	90	40
Matrix Algebra	60	40	90

You have been told that the test will be weighted as follows: game theory: 25%; linear programming: 50%; matrix algebra: 25%.

[22] Adapted from an example in *Location in Space: Theoretical Perspectives in Economic Geography* by P. Dicken and P.E. Lloyd (Harper & Row, 1990).

[23] *Ibid.*

a. If you spend 25% of the night on game theory, 50% on linear programming, and 25% on matrix algebra, what score do you expect to get on the test?

b. Is it possible to improve on this by altering your study schedule? If so, what is the highest score you can expect on the test?

c. If your study schedule is according to part (a) and your teacher decides to forget her promises about how the test will be weighted and instead base it all on a single topic, which topic would be worst for you, and what score could you expect on the test?

38. *Study Techniques* Your friend Joe has been spending all of his time on fraternity activities and thus knows absolutely nothing about any of the three topics on tomorrow's math test. (See Exercise 37.) Because you are recognized as an expert on the use of game theory to solve study problems, he has turned to you for advice as to how to spend his "all-nighter." As the following table shows, his situation is not so rosy. (Since he knows no linear programming or matrix algebra, the table shows, for instance, that studying game theory all night will not be much use in preparing him for this topic.)

Test

Joe's Strategies	Game Theory	Linear Programming	Matrix Algebra
Game Theory	30	0	20
Linear Programming	0	70	0
Matrix Algebra	0	0	70

Assuming that the test will be weighted as described in Exercise 37, what are the answers to parts (a), (b), and (c) as they apply to Joe?

39. ▼ *Staff Cutbacks* Frank Tempest manages a large snowplow service in Manhattan, Kansas, and is alarmed by the recent weather trends; there have been no significant snowfalls in recent years. He is therefore contemplating laying off some of his workers, but is unsure about whether to lay off 5, 10, or 15 of his 50 workers. Being very methodical, he estimates his annual net profits based on four possible annual snowfall figures: 0 inches, 20 inches, 40 inches, and 60 inches. (He takes into account the fact that, if he is running a small operation in the face of a large annual snowfall, he will lose business to his competitors because he will be unable to discount on volume.)

	0 Inches	20 Inches	40 Inches	60 Inches
5 Laid Off	−$500,000	−$200,000	$10,000	$200,000
10 Laid Off	−$200,000	$0	$0	$0
15 Laid Off	−$100,000	$10,000	−$200,000	−$300,000

a. During the past 10 years, the region has had 0 inches twice, 20 inches twice, 40 inches three times, and 60 inches three times. Based on this information, how

many workers should Tempest lay off, and how much would it cost him?

b. There is a 50% chance that Tempest will lay off 5 workers and a 50% chance that he will lay off 15 workers. What is the worst thing Nature can do to him in terms of snowfall? How much would it cost him?

c. The Gods of Chaos (who control the weather) know that Tempest is planning to use the strategy in part (a), and are determined to hurt Tempest as much as possible. Tempest, being somewhat paranoid, suspects it too. What should he do?

40. ▼ *Textbook Writing* You are writing a college-level textbook on finite mathematics and are trying to come up with the best combination of word problems. Over the years, you have accumulated a collection of amusing problems, serious applications, long complicated problems, and "generic" problems.[24] Before your book is published, it must be scrutinized by several reviewers, who, it seems, are never satisfied with the mix you use. You estimate that there are three kinds of reviewers: the "no-nonsense" types, who prefer applications and generic problems; the "dead serious" types, who feel that a college-level text should be contain little or no humor and lots of long complicated problems; and the "laid-back" types, who believe that learning best takes place in a light-hearted atmosphere bordering on anarchy. You have drawn up the following chart, where the payoffs represent the reactions of reviewers on a scale of −10 (ballistic) to +10 (ecstatic):

Reviewers

You	No-Nonsense	Dead Serious	Laid-Back
Amusing	−5	−10	10
Serious	5	3	0
Long	−5	5	3
Generic	5	3	−10

a. Your first draft of the book contained no generic problems, and equal numbers of the other categories. If half the reviewers of your book were "dead serious" and the rest were equally divided between the "no-nonsense" and "laid-back" types, what score would you expect?

b. In your second draft of the book, you tried to balance the content by including some generic problems and eliminating several amusing ones, and wound up with a mix of which one eighth were amusing, one quarter were serious, three eighths were long, and a quarter were generic. What kind of reviewer would be *least* impressed by this mix?

c. What kind of reviewer would be *most* impressed by the mix in your second draft?

---

[24]Of the following type: "A certain company has three processing plants: A, B, and C, each of which uses three processes: P_1, P_2, and P_3. Process P_1 uses 100 units of chemical C_1, 50 units of C_2, . . ." and so on.

41. *Price Wars* Computer Electronics, Inc. (CE) and the *Gigantic Computer Store* (GCS) are planning to discount the price they charge for the HAL Laptop Computer, of which they are the only distributors. Because Computer Electronics provides a free warranty service, they can generally afford to charge more. A market survey provides the following data on the gains to CE's market share that will result from different pricing decisions:

GCS

		$900	$1,000	$1,200
	$1,000	15%	60%	80%
CE	$1,200	15%	60%	60%
	$1,300	10%	20%	40%

a. Use reduction by dominance to determine how much each company should charge. What is the effect on CE's market share?

b. CE is aware that GCS is planning to use reduction by dominance to determine its pricing policy, and wants its market share to be as large as possible. What effect, if any, would the information about GCS have on CE's best strategy?

42. *More Price Wars* (Refer to Exercise 41.) A new market survey results in the following revised data:

GCS

		$900	$1,000	$1,200
	$1,000	20%	60%	60%
CE	$1,200	15%	60%	60%
	$1,300	10%	20%	40%

a. Use reduction by dominance to determine how much each company should charge. What is the effect on CE's market share?

b. In general, why do price wars tend to force prices down?

43. *Wrestling Tournaments* City Community College (CCC) plans to host Midtown Military Academy (MMA) for a wrestling tournament. Each school has three wrestlers in the 190 lb. weight class: CCC has Pablo, Sal, and Edison, while MMA has Carlos, Marcus, and Noto. Pablo can beat Carlos and Marcus, Marcus can beat Edison and Sal, Noto can beat Edison, while the other combinations will result in an even match. Set up a payoff matrix, and use reduction by dominance to decide which wrestler each team should choose as their champion. Does one school have an advantage over the other?

44. *Wrestling Tournaments* (Refer to Exercise 43.) One day before the wrestling tournament discussed in Exercise 43, Pablo sustains a hamstring injury and is replaced by Hans, who (unfortunately for CCC) can be beaten by both Carlos and Marcus. Set up the payoff matrix, and use reduction by dominance to decide which wrestler each team should

choose as their champion. Does one school have an advantage over the other?

45. *The Battle of Rabaul-Lae*[25] In the Second World War, during the struggle for New Guinea, intelligence reports revealed that the Japanese were planning to move a troop and supply convoy from the port of Rabaul at the Eastern tip of New Britain to Lae, which lies just west of New Britain on New Guinea. It could either travel via a northern route, which was plagued by poor visibility, or by a southern route, where the visibility was clear. General Kenney, who was the commander of the Allied Air Forces in the area, had the choice of concentrating reconnaissance aircraft on one route or the other and bombing the Japanese convoy once it was sighted. Kenney's staff drafted the following outcomes for his choices, where the payoffs are estimated days of bombing time:

		Japanese Commander's Strategies	
		Northern Route	Southern Route
Kenney's Strategies	Northern Route	2	2
	Southern Route	1	3

What would you have recommended to General Kenney? What would you have recommended to the Japanese Commander?[26] How much bombing time results if these recommendations are followed?

46. *The Battle of Rabaul-Lae* Referring to Exercise 45, suppose that General Kenney had a third alternative: splitting his reconnaissance aircraft between the two routes, which would result in the following estimates:

		Japanese Commander's Strategies	
		Northern Route	Southern Route
Kenney's Strategies	Northern Route	2	2
	Split Reconnaissance	1.5	2.5
	Southern Route	1	3

What would you have recommended to General Kenney? What would you have recommended to the Japanese Commander? How much bombing time results if these recommendations are followed?

---

[25]As discussed in *Games and Decisions* by R. D. Luce and H. Raiffa Section 11.3 (New York: Wiley, 1957). This is based on an article in the *Journal of the Operations Research Society of America* 2 (1954) 365–385.

[26]The correct answers to these two questions correspond to the actual decisions both commanders made.

47. ▼ *The Prisoner's Dilemma* Slim Lefty and Joe Rap have been arrested for grand theft auto, having been caught red-handed driving away in a stolen 2012 Porsche. Although the police have more than enough evidence to convict them both, they feel that a confession would simplify the work of the prosecution. They decide to interrogate the prisoners separately. Slim and Joe are both told of the following plea-bargaining arrangement: If both confess, they will each receive a 2-year sentence; if neither confesses, they will both receive 5-year sentences, and if only one confesses (and thus squeals on the other), he will receive a suspended sentence, while the other will receive a 10-year sentence. What should Slim do?

48. ▼ *More Prisoners' Dilemmas* Jane Good and Prudence Brown have been arrested for robbery, but the police lack sufficient evidence for a conviction, and so decide to interrogate them separately in the hope of extracting a confession. Both Jane and Prudence are told the following: If they both confess, they will each receive a 5-year sentence; if neither confesses, they will be released; if one confesses, she will receive a suspended sentence, while the other will receive a 10-year sentence. What should Jane do?

49. ▼ *Campaign Strategies*[27] Florida and Ohio are "swing states" that have a large bounty of electoral votes and are therefore highly valued by presidential campaign strategists. Suppose it is now the weekend before Election Day 2012, and each candidate (Romney and Obama) can visit only one more state. Further, to win the election, Romney needs to win both of these states. Currently Romney has a 40% chance of winning Ohio and a 60% chance of winning Florida. Therefore, he has a $0.40 \times 0.60 = 0.24$, or 24% chance of winning the election. Assume that each candidate can increase his probability of winning a state by 10% if he, and not his opponent, visits that state. If both candidates visit the same state, there is no effect.

   a. Set up a payoff matrix with Romney as the row player and Obama as the column player, where the payoff for a specific set of circumstances is the probability (expressed as a percentage) that Romney will win both states.

   b. Where should each candidate visit under the circumstances?

50. ▼ *Campaign Strategies* Repeat Exercise 49, this time assuming that Romney has an 80% chance of winning Ohio and a 90% chance of winning Florida.

51. *Advertising* The *Softex Shampoo Company* is considering how to split its advertising budget between ads on two radio stations: WISH and WASH. Its main competitor, Splish Shampoo, Inc. has found out about this and is considering countering Softex's ads with its own, on the same radio stations. (Proposed jingle: *Softex, Shmoftex; Splash with Splish*) Softex has calculated that, were it to devote its entire advertising budget to ads on WISH, it would increase revenues in the coming month by $100,000 in the event that Splish was running all its ads on the less popular WASH, but would lose $20,000 in revenues if Splish ran its ads on WISH. If, on the other hand, it devoted its entire budget to WASH ads, it would neither increase nor decrease revenues in the event that Splish was running all its ads on the more popular WISH, and would in fact lose $20,000 in revenues if Splish ran its ads on WASH. What should Softex do, and what effect will this have on revenues?

52. *Labor Negotiations* The management team of the *Abstract Concrete Company* is negotiating a 3-year contract with the labor unions at one of its plants and is trying to decide on its offer for a salary increase. If it offers a 5% increase and the unions accept the offer, Abstract Concrete will gain $20 million in projected profits in the coming year, but if labor rejects the offer, the management team predicts that it will be forced to increase the offer to the union demand of 15%, thus halving the projected profits. If Abstract Concrete offers a 15% increase, the company will earn $10 million in profits over the coming year if the unions accept. If the unions reject, they will probably go out on strike (because management has set 15% as its upper limit) and management has decided that it can then in fact gain $12 million in profits by selling out the defunct plant in retaliation. What intermediate percentage should the company offer, and what profit should it project?

COMMUNICATION AND REASONING EXERCISES

53. Why is a saddle point called a "saddle point"?

54. Can the payoff in a saddle point ever be larger than all other payoffs in a game? Explain.

55. ◆One day, while browsing through an old *Statistical Abstract of the United States,* you came across the following data, which show the number of females employed (in thousands) in various categories according to their educational attainment.[28]

	Managerial/ Professional	Technical/Sales/ Administrative	Service	Precision Production	Operators/ Fabricators
Less Than 4 Years High School	260	1,080	2,020	260	1,400
4 Years of High School Only	2,430	9,510	3,600	570	2,130
1 to 3 Years of College	2,690	5,080	1,080	160	350
At Least 4 Years of College	7,210	2,760	380	70	110

Because you had been studying game theory that day, the first thing you did was to search for a saddle point. Having

---

[27]Based on *Game Theory for Swingers: What states should the candidates visit before Election Day?* by Jordan Ellenberg. Source: www.slate.com.

[28]Source: *Statistical Abstract of the United States 1991* (111th Ed.) U.S. Department of Commerce, Economics and Statistics Administration, and Bureau of the Census.

found one, you conclude that, as a female, your best strategy in the job market is to forget about a college career. Find the flaw in this reasoning.

56. ◆Exercises 37 and 38 seem to suggest that studying a single topic prior to an exam is better than studying all the topics in that exam. Comment on this discrepancy between the game theory result and common sense.

57. ◆Explain what is wrong with a decision to play the mixed strategy [0.5 0.5] by alternating the two strategies: Play the first strategy on the odd-numbered moves and the second strategy on the even-numbered moves. Illustrate your argument by devising a game in which your best strategy is [0.5 0.5].

58. ◆Describe a situation in which a both a mixed strategy and a pure strategy are equally effective.

4.5   Input-Output Models

In this section we look at an application of matrix algebra developed by Wassily Leontief (1906–1999) in the middle of the twentieth century. In 1973, he won the Nobel Prize in Economics for this work. The application involves analyzing national and regional economies by looking at how various parts of the economy interrelate. We'll work out some of the details by looking at a simple scenario.

First, we can think of the economy of a country or a region as being composed of various **sectors**, or groups of one or more industries. Typical sectors are the manufacturing sector, the utilities sector, and the agricultural sector. To introduce the basic concepts, we shall consider two specific sectors: the coal-mining sector (Sector 1) and the electric utilities sector (Sector 2). Both produce a commodity: The coal-mining sector produces coal, and the electric utilities sector produces electricity. We measure these products by their dollar value. By **one unit** of a product, we mean $1 worth of that product.

Here is the scenario.

1. To produce one unit ($1 worth) of coal, assume that the coal-mining sector uses 50¢ worth of coal (to power mining machinery, say) and 10¢ worth of electricity.

2. To produce one unit ($1 worth) of electricity, assume that the electric utilities sector uses 25¢ worth of coal and 25¢ worth of electricity.

These are *internal* usage figures. In addition to this, assume that there is an *external* demand (from the rest of the economy) of 7,000 units ($7,000 worth) of coal and 14,000 units ($14,000 worth) of electricity over a specific time period (one year, say). Our basic question is: How much should each of the two sectors supply in order to meet both internal and external demand?

The key to answering this question is to set up equations of the form:

Total supply = Total demand.

The unknowns, the values we are seeking, are

x_1 = the total supply (in units) from Sector 1 (coal) and
x_2 = the total supply (in units) from Sector 2 (electricity).

Our equations then take the following form:

Total supply from Sector 1 = Total demand for Sector 1 products

$$x_1 = 0.50x_1   +   0.25x_2   +   7,000$$

↑   ↑   ↑
Coal required by Sector 1   Coal required by Sector 2   External demand for coal

Total supply from Sector 2 = Total demand for Sector 2 products

$$x_2 = 0.10x_1 \qquad + \qquad 0.25x_2 \qquad + \qquad 14{,}000.$$

$\qquad\qquad\quad \uparrow \qquad\qquad\qquad\quad \uparrow \qquad\qquad\qquad\quad \uparrow$

Electricity required by Sector 1   Electricity required by Sector 2   External demand for electricity

This is a system of two linear equations in two unknowns:

$$x_1 = 0.50x_1 + 0.25x_2 + 7{,}000$$
$$x_2 = 0.10x_1 + 0.25x_2 + 14{,}000.$$

We can rewrite this system of equations in matrix form as follows:

$$\underbrace{\begin{bmatrix} x_1 \\ x_2 \end{bmatrix}}_{\text{Production}} = \underbrace{\begin{bmatrix} 0.50 & 0.25 \\ 0.10 & 0.25 \end{bmatrix} \begin{bmatrix} x_1 \\ x_2 \end{bmatrix}}_{\text{Internal demand}} + \underbrace{\begin{bmatrix} 7{,}000 \\ 14{,}000 \end{bmatrix}}_{\text{External demand}}.$$

In symbols,

$$X = AX + D.$$

Here,

$$X = \begin{bmatrix} x_1 \\ x_2 \end{bmatrix}$$

is called the **production vector**. Its entries are the amounts produced by the two sectors. The matrix

$$D = \begin{bmatrix} 7{,}000 \\ 14{,}000 \end{bmatrix}$$

is called the **external demand** vector, and

$$A = \begin{bmatrix} 0.50 & 0.25 \\ 0.10 & 0.25 \end{bmatrix} \qquad \text{Organization: } \begin{bmatrix} 1 \to 1 & 1 \to 2 \\ 2 \to 1 & 2 \to 2 \end{bmatrix}$$

is called the **technology matrix**. The entries of the technology matrix have the following meanings:

a_{11} = units of Sector 1 needed to produce one unit of Sector 1
a_{12} = units of Sector 1 needed to produce one unit of Sector 2
a_{21} = units of Sector 2 needed to produce one unit of Sector 1
a_{22} = units of Sector 2 needed to produce one unit of Sector 2.

Now that we have the matrix equation

$$X = AX + D$$

we can solve it as follows. First, subtract AX from both sides:

$$X - AX = D.$$

Because $X = IX$, where I is the 2×2 identity matrix, we can rewrite this as

$$IX - AX = D.$$

Now factor out X:

$$(I - A)X = D.$$

If we multiply both sides by the inverse of $(I - A)$, we get the solution

$$X = (I - A)^{-1}D.$$

Input-Output Model

In an input-output model, an economy (or part of one) is divided into n **sectors**. We then record the $n \times n$ **technology matrix** A, whose ijth entry is the number of units from Sector i used in producing one unit from Sector j (in symbols, "$i \to j$"). To meet an **external demand** of D, the economy must produce X, where X is the **production vector**. These are related by the equations

$$X = AX + D$$

or

$$X = (I - A)^{-1}D. \qquad \text{Provided } (I - A) \text{ is invertible}$$

Quick Example

In the previous scenario, $A = \begin{bmatrix} 0.50 & 0.25 \\ 0.10 & 0.25 \end{bmatrix}$, $X = \begin{bmatrix} x_1 \\ x_2 \end{bmatrix}$, and $D = \begin{bmatrix} 7,000 \\ 14,000 \end{bmatrix}$.

The solution is

$$X = (I - A)^{-1}D$$

$$\begin{bmatrix} x_1 \\ x_2 \end{bmatrix} = \left(\begin{bmatrix} 1 & 0 \\ 0 & 1 \end{bmatrix} - \begin{bmatrix} 0.50 & 0.25 \\ 0.10 & 0.25 \end{bmatrix} \right)^{-1} \begin{bmatrix} 7,000 \\ 14,000 \end{bmatrix} \quad \text{Calculate } I - A.$$

$$= \begin{bmatrix} 0.50 & -0.25 \\ -0.10 & 0.75 \end{bmatrix}^{-1} \begin{bmatrix} 7,000 \\ 14,000 \end{bmatrix} \qquad \text{Calculate } (I - A)^{-1}.$$

$$= \begin{bmatrix} \frac{15}{7} & \frac{5}{7} \\ \frac{2}{7} & \frac{10}{7} \end{bmatrix} \begin{bmatrix} 7,000 \\ 14,000 \end{bmatrix}$$

$$= \begin{bmatrix} 25,000 \\ 22,000 \end{bmatrix}.$$

In other words, to meet the demand, the economy must produce \$25,000 worth of coal and \$22,000 worth of electricity.

The next example uses actual data from the U.S. economy (we have rounded the figures to make the computations less complicated). It is rare to find input-output data already packaged for you as a technology matrix. Instead, the data commonly found in statistical sources come in the form of "input-output tables," from which we will have to construct the technology matrix.

EXAMPLE 1   Petroleum and Natural Gas

Consider two sectors of the U.S. economy: crude petroleum and natural gas (*crude*) and petroleum refining and related industries (*refining*). According to government figures,[29] in 1998 the crude sector used \$27,000 million worth of its own products and \$750 million worth of the products of the refining sector to produce \$87,000 million worth of goods (crude oil and natural gas). The refining sector in the same

---

[29]The data have been rounded to two significant digits. Source: *Survey of Current Business*, December, 2001, U.S. Department of Commerce. The *Survey of Current Business* and the input-output tables themselves are available at the Web site of the Department of Commerce's Bureau of Economic Analysis (www.bea.gov).

year used $59,000 million worth of the products of the crude sector and $15,000 million worth of its own products to produce $140,000 million worth of goods (refined oil and the like). What was the technology matrix for these two sectors? What was left over from each of these sectors for use by other parts of the economy or for export?

Solution First, for convenience, we record the given data in the form of a table, called the **input-output table**. (All figures are in millions of dollars.)

To	**Crude**	**Refining**
From **Crude**	27,000	59,000
Refining	750	15,000
Total Output	87,000	140,000

The entries in the top portion are arranged in the same way as those of the technology matrix: The ijth entry represents the number of units of Sector i that went to Sector j. Thus, for instance, the 59,000 million entry in the 1, 2 position represents the number of units of Sector 1, crude, that were used by Sector 2, refining. ("From the side, to the top.")

We now construct the technology matrix. The technology matrix has entries a_{ij} = units of Sector i used to produce *one* unit of Sector j. Thus,

a_{11} = units of crude to produce one unit of crude. We are told that 27,000 million units of crude were used to produce 87,000 million units of crude. Thus, to produce *one* unit of crude, $27{,}000/87{,}000 \approx 0.31$ units of crude were used, and so $a_{11} \approx 0.31$. (We have rounded this value to two significant digits; further digits are not reliable due to rounding of the original data.)

a_{12} = units of crude to produce one unit of refined:
$a_{12} = 59{,}000/140{,}000 \approx 0.42$

a_{21} = units of refined to produce one unit of crude:
$a_{21} = 750/87{,}000 \approx 0.0086$

a_{22} = units of refined to produce one unit of refined:
$a_{22} = 15{,}000/140{,}000 \approx 0.11$.

This gives the technology matrix

$$A = \begin{bmatrix} 0.31 & 0.42 \\ 0.0086 & 0.11 \end{bmatrix}. \quad \text{Technology matrix}$$

In short *we obtained the technology matrix from the input-output table by dividing the Sector 1 column by the Sector 1 total, and the Sector 2 column by the Sector 2 total.*

Now we also know the total output from each sector, so *we have already been given the production vector:*

$$X = \begin{bmatrix} 87{,}000 \\ 140{,}000 \end{bmatrix}. \quad \text{Production vector}$$

What we are asked for is the external demand vector D, the amount available for the outside economy. To find D, we use the equation

$$X = AX + D, \quad \text{Relationship of } X, A, \text{ and } D$$

where, this time, we are given A and X, and must solve for D. Solving for D gives

$$D = X - AX$$

$$= \begin{bmatrix} 87{,}000 \\ 140{,}000 \end{bmatrix} - \begin{bmatrix} 0.31 & 0.42 \\ 0.0086 & 0.11 \end{bmatrix} \begin{bmatrix} 87{,}000 \\ 140{,}000 \end{bmatrix}$$

✱ Why?

$$\approx \begin{bmatrix} 87{,}000 \\ 140{,}000 \end{bmatrix} - \begin{bmatrix} 86{,}000 \\ 16{,}000 \end{bmatrix} = \begin{bmatrix} 1{,}000 \\ 124{,}000 \end{bmatrix}.$$ We rounded to 2 digits.✱

The first number, $1,000 million, is the amount produced by the crude sector that is available to be used by other parts of the economy or to be exported. (In fact, because something has to happen to all that crude petroleum and natural gas, this is the amount actually used or exported, where use can include stockpiling.) The second number, $124,000 million, represents the amount produced by the refining sector that is available to be used by other parts of the economy or to be exported.

Note that we could have calculated D more simply from the input-output table. The internal use of units from the crude sector was the sum of the outputs from that sector:

$$27{,}000 + 59{,}000 = 86{,}000.$$

using Technology

See the Technology Guides at the end of the chapter to see how to compute the technology matrix and the external demand vector in Example 1 using a TI-83/84 Plus or a spreadsheet.

Because 87,000 units were actually produced by the sector, that left a surplus of $87{,}000 - 86{,}000 = 1{,}000$ units for export. We could compute the surplus from the refining sector similarly. (The two calculations actually come out slightly different, because we rounded the intermediate results.) The calculation in Example 2 below cannot be done as trivially, however.

Input-Output Table

National economic data are often given in the form of an **input-output table**. The ijth entry in the top portion of the table is the number of units that go from Sector i to Sector j. The "Total outputs" are the total numbers of units produced by each sector. We obtain the technology matrix from the input-output table by dividing the Sector 1 column by the Sector 1 total, the Sector 2 column by the Sector 2 total, and so on.

Quick Example

Input-Output Table:

† The production of skateboards required skateboards due to the fact that skateboard workers tend to commute to work on (what else?) skateboards!

	To	**Skateboards**	**Wood**
From	**Skateboards**	20,000†	0
	Wood	100,000	500,000
	Total Output	200,000	5,000,000

Technology Matrix:

$$A = \begin{bmatrix} \frac{20{,}000}{200{,}000} & \frac{0}{5{,}000{,}000} \\ \frac{100{,}000}{200{,}000} & \frac{500{,}000}{5{,}000{,}000} \end{bmatrix} = \begin{bmatrix} 0.1 & 0 \\ 0.5 & 0.1 \end{bmatrix}.$$

EXAMPLE 2 Rising Demand

Suppose that external demand for refined petroleum rises to $200,000 million, but the demand for crude remains $1,000 million (as in Example 1). How do the production levels of the two sectors considered in Example 1 have to change?

Solution We are being told that now

$$D = \begin{bmatrix} 1,000 \\ 200,000 \end{bmatrix}$$

and we are asked to find X. Remember that we can calculate X from the formula

$$X = (I - A)^{-1}D.$$

Now

$$I - A = \begin{bmatrix} 1 & 0 \\ 0 & 1 \end{bmatrix} - \begin{bmatrix} 0.31 & 0.42 \\ 0.0086 & 0.11 \end{bmatrix} = \begin{bmatrix} 0.69 & -0.42 \\ -0.0086 & 0.89 \end{bmatrix}.$$

✳ Because A is accurate to two digits, we should use more than two significant digits in intermediate calculations so as not to lose additional accuracy. We must, of course, round the final answer to two digits.

We take the inverse using our favorite technique and find that, to four significant digits,✳

$$(I - A)^{-1} \approx \begin{bmatrix} 1.458 & 0.6880 \\ 0.01409 & 1.130 \end{bmatrix}.$$

Now we can compute X:

$$X = (I - A)^{-1}D = \begin{bmatrix} 1.458 & 0.6880 \\ 0.01409 & 1.130 \end{bmatrix} \begin{bmatrix} 1,000 \\ 200,000 \end{bmatrix} \approx \begin{bmatrix} 140,000 \\ 230,000 \end{bmatrix}.$$

(As in Example 1, we have rounded all the entries in the answer to two significant digits.) Comparing this vector to the production vector used in Example 1, we see that production in the crude sector has to increase from $87,000 million to $140,000 million, while production in the refining sector has to increase from $140,000 million to $230,000 million.

Note Using the matrix $(I - A)^{-1}$, we have a slightly different way of solving Example 2. We are asking for the effect on production of a *change* in the final demand of 0 for crude and $200,000 - 124,000 = \$76,000$ million for refined products. If we multiply $(I - A)^{-1}$ by the matrix representing this *change,* we obtain

$$\begin{bmatrix} 1.458 & 0.6880 \\ 0.01409 & 1.130 \end{bmatrix} \begin{bmatrix} 0 \\ 76,000 \end{bmatrix} \approx \begin{bmatrix} 53,000 \\ 90,000 \end{bmatrix}.$$

$(I - A)^{-1} \times$ Change in Demand = Change in Production

We see the changes required in production: an increase of $53,000 million in the crude sector and an increase of $90,000 million in the refining sector.

Notice that the increase in external demand for the products of the refining sector requires the crude sector to increase production as well, even though there is no increase in the *external* demand for its products. The reason is that, in order to increase production, the refining sector needs to use more crude oil, so that the *internal* demand for crude oil goes up. The inverse matrix $(I - A)^{-1}$ takes these **indirect effects** into account in a nice way.

By replacing the $76,000 by $1 in the computation we just did, we see that a $1 increase in external demand for refined products will require an increase in production of $0.6880 in the crude sector, as well as an increase in production of $1.130 in the refining sector. This is how we interpret the entries in $(I - A)^{-1}$, and this is why it is useful to look at this matrix inverse rather than just solve $(I - A)X = D$ for X using, say, Gauss-Jordan reduction. Looking at $(I - A)^{-1}$, we can also find the effects of an increase of $1 in external demand for crude: an increase in production of $1.458 in the crude sector and an increase of $0.01409 in the refining sector.

Here are some questions to think about: Why are the diagonal entries of $(I - A)^{-1}$ (slightly) larger than 1? Why is the entry in the lower left so small compared to the others? ∎

Interpreting $(I - A)^{-1}$: Indirect Effects

If A is the technology matrix, then the ijth entry of $(I - A)^{-1}$ is the change in the number of units Sector i must produce in order to meet a one-unit increase in external demand for Sector j products. To meet a rising external demand, the necessary change in production for each sector is given by

$$\text{Change in production} = (I - A)^{-1}D^+,$$

where D^+ is the change in external demand.

Quick Example

Take Sector 1 to be skateboards, and Sector 2 to be wood, and assume that

$$(I - A)^{-1} = \begin{bmatrix} 1.1 & 0 \\ 0.6 & 1.1 \end{bmatrix}.$$

Then

$a_{11} = 1.1 =$ number of additional units of skateboards that must be produced to meet a one-unit increase in the demand for skateboards (Why is this number larger than 1?)

$a_{12} = 0 =$ number of additional units of skateboards that must be produced to meet a one-unit increase in the demand for wood (Why is this number 0?)

$a_{21} = 0.6 =$ number of additional units of wood that must be produced to meet a one-unit increase in the demand for skateboards

$a_{22} = 1.1 =$ number of additional units of wood that must be produced to meet a one-unit increase in the demand for wood.

To meet an increase in external demand of 100 skateboards and 400 units of wood, the necessary change in production is

$$(I - A)^{-1}D^+ = \begin{bmatrix} 1.1 & 0 \\ 0.6 & 1.1 \end{bmatrix}\begin{bmatrix} 100 \\ 400 \end{bmatrix} = \begin{bmatrix} 110 \\ 500 \end{bmatrix}$$

so 110 additional skateboards and 500 additional units of wood will need to be produced.

In the preceding examples, we used only two sectors of the economy. The data used in Examples 1 and 2 were taken from an input-output table published by the U.S. Department of Commerce, in which the whole U.S. economy was broken down into 85 sectors. This in turn was a simplified version of a model in which the economy was broken into about 500 sectors. Obviously, computers are required to make a realistic input-output analysis possible. Many governments collect and publish input-output data as part of their national planning. The United Nations collects these data and publishes collections of national statistics. The United Nations also has a useful set of links to government statistics at the following URL:

www.un.org/Depts/unsd/sd_natstat.htm

EXAMPLE 3 **Kenya Economy**

Consider four sectors of the economy of Kenya[30]: (1) the traditional economy, (2) agriculture, (3) manufacture of metal products and machinery, and (4) wholesale and retail trade. The input-output table for these four sectors for 1976 looks like this (all numbers are thousands of K£):

		To	**1**	**2**	**3**	**4**
From		**1**	8,600	0	0	0
		2	0	20,000	24	0
		3	1,500	530	15,000	660
		4	810	8,500	5,800	2,900
Total Output			87,000	530,000	110,000	180,000

 using Technology

Technology can be used to do the computations in Example 3. Here is an outline for part (a) (see the Technology Guides at the end of the chapter for additional details on using a TI-83/84 Plus or a spreadsheet):

TI-83/84 Plus
Enter the matrices A and D using MATRX ; EDIT.
(For A type the entries as quotients: Column entry/Column total.) Home Screen:
`(identity(4)-[A])`$^{-1}$`[D]`
[More details on page 306.]

Spreadsheet
Obtain the technology matrix as in Example 1.
Insert the 4×4 identity matrix and then use MINVERSE and MMULT to compute $(I - A)^{-1}D$.
[More details on page 310.]

Website
www.WanerMath.com
Student Website
  → On Line Utilities
  → Matrix Algebra Tool
Enter the matrices A and D. (For A type the entries as quotients: Column entry/Column total.)
Type `(I-A)^(-1)*D` in the formula box and press "Compute".

Suppose that external demand for agriculture increased by K£50,000,000 and that external demand for metal products and machinery increased by K£10,000,000. How would production in these four sectors have to change to meet this rising demand?

Solution To find the change in production necessary to meet the rising demand, we need to use the formula

$$\text{Change in production} = (I - A)^{-1}D^+,$$

where A is the technology matrix and D^+ is the change in demand:

$$D^+ = \begin{bmatrix} 0 \\ 50,000 \\ 10,000 \\ 0 \end{bmatrix}.$$

With entries shown rounded to two significant digits, the matrix A is

$$A = \begin{bmatrix} 0.099 & 0 & 0 & 0 \\ 0 & 0.038 & 0.00022 & 0 \\ 0.017 & 0.001 & 0.14 & 0.0037 \\ 0.0093 & 0.016 & 0.053 & 0.016 \end{bmatrix}.$$

Entries shown are rounded to two significant digits.

---

[30]Figures are rounded. Source: *Input-Output Tables for Kenya 1976*, Central Bureau of Statistics of the Ministry of Economic Planning and Community Affairs, Kenya.

The next calculation is best done using technology:

$$\text{Change in production} = (I - A)^{-1}D^+ = \begin{bmatrix} 0 \\ 52{,}000 \\ 12{,}000 \\ 1{,}500 \end{bmatrix}.$$

Entries shown are rounded to two significant digits.

Looking at this result, we see that the changes in external demand will leave the traditional economy unaffected, production in agriculture will rise by K£52 million, production in the manufacture of metal products and machinery will rise by K£12 million, and activity in wholesale and retail trade will rise by K£1.5 million.

➡ **Before we go on...** Can you see why the traditional economy was unaffected in Example 3? Although it takes inputs from other parts of the economy, it is not itself an input to any other part. In other words, there is no intermediate demand for the products of the traditional economy coming from any other part of the economy, and so an increase in production in any other sector of the economy will require no increase from the traditional economy. On the other hand, the wholesale and retail trade sector does provide input to the agriculture and manufacturing sectors, so increases in those sectors do require an increase in the trade sector.

One more point: If you calculate $(I - A)^{-1}$ you will notice how small the off-diagonal entries are. This says that increases in each sector have relatively small effects on the other sectors. We say that these sectors are **loosely coupled**. Regional economies, where many products are destined to be shipped out to the rest of the country, tend to show this phenomenon even more strongly. Notice in Example 2 that those two sectors are **strongly coupled**, because a rise in demand for refined products requires a comparable rise in the production of crude. ■

4.5 EXERCISES

▼ more advanced   ◆ challenging
[T] indicates exercises that should be solved using technology

1. Let A be the technology matrix $A = \begin{bmatrix} 0.2 & 0.05 \\ 0.8 & 0.01 \end{bmatrix}$, where Sector 1 is paper and Sector 2 is wood. Fill in the missing quantities.

   a. ___ units of wood are needed to produce one unit of paper.

   b. ___ units of paper are used in the production of one unit of paper.

   c. The production of each unit of wood requires the use of ___ units of paper.

2. Let A be the technology matrix $A = \begin{bmatrix} 0.01 & 0.001 \\ 0.2 & 0.004 \end{bmatrix}$, where Sector 1 is computer chips and Sector 2 is silicon. Fill in the missing quantities.

   a. ___ units of silicon are required in the production of one unit of silicon.

   b. ___ units of computer chips are used in the production of one unit of silicon.

   c. The production of each unit of computer chips requires the use of ___ units of silicon.

3. Each unit of television news requires 0.2 units of television news and 0.5 units of radio news. Each unit of radio news requires 0.1 units of television news and no radio news. With Sector 1 as television news and Sector 2 as radio news, set up the technology matrix A.

4. Production of one unit of cologne requires no cologne and 0.5 units of perfume. Into one unit of perfume go 0.1 units of cologne and 0.3 units of perfume. With Sector 1 as cologne and Sector 2 as perfume, set up the technology matrix A.

In each of Exercises 5–12, you are given a technology matrix A and an external demand vector D. Find the corresponding production vector X. HINT [See Quick Example page 286.]

5. $A = \begin{bmatrix} 0.5 & 0.4 \\ 0 & 0.5 \end{bmatrix}$, $D = \begin{bmatrix} 10{,}000 \\ 20{,}000 \end{bmatrix}$

6. $A = \begin{bmatrix} 0.5 & 0.4 \\ 0 & 0.5 \end{bmatrix}$, $D = \begin{bmatrix} 20{,}000 \\ 10{,}000 \end{bmatrix}$

7. $A = \begin{bmatrix} 0.1 & 0.4 \\ 0.2 & 0.5 \end{bmatrix}$, $D = \begin{bmatrix} 25{,}000 \\ 15{,}000 \end{bmatrix}$

8. $A = \begin{bmatrix} 0.1 & 0.2 \\ 0.4 & 0.5 \end{bmatrix}$, $D = \begin{bmatrix} 24{,}000 \\ 14{,}000 \end{bmatrix}$

9. $A = \begin{bmatrix} 0.5 & 0.1 & 0 \\ 0 & 0.5 & 0.1 \\ 0 & 0 & 0.5 \end{bmatrix}, D = \begin{bmatrix} 1{,}000 \\ 1{,}000 \\ 2{,}000 \end{bmatrix}$

10. $A = \begin{bmatrix} 0.5 & 0.1 & 0 \\ 0 & 0.5 & 0.1 \\ 0 & 0 & 0.5 \end{bmatrix}, D = \begin{bmatrix} 3{,}000 \\ 3{,}800 \\ 2{,}000 \end{bmatrix}$

11. $A = \begin{bmatrix} 0.2 & 0.2 & 0 \\ 0.2 & 0.4 & 0.2 \\ 0 & 0.2 & 0.2 \end{bmatrix}, D = \begin{bmatrix} 16{,}000 \\ 8{,}000 \\ 8{,}000 \end{bmatrix}$

12. $A = \begin{bmatrix} 0.2 & 0.2 & 0.2 \\ 0.2 & 0.4 & 0.2 \\ 0.2 & 0.2 & 0.2 \end{bmatrix}, D = \begin{bmatrix} 7{,}000 \\ 14{,}000 \\ 7{,}000 \end{bmatrix}$

13. Given $A = \begin{bmatrix} 0.1 & 0.4 \\ 0.2 & 0.5 \end{bmatrix}$, find the changes in production required to meet an increase in demand of 50 units of Sector 1 products and 30 units of Sector 2 products.

14. Given $A = \begin{bmatrix} 0.5 & 0.4 \\ 0 & 0.5 \end{bmatrix}$, find the changes in production required to meet an increase in demand of 20 units of Sector 1 products and 10 units of Sector 2 products.

15. Let $(I - A)^{-1} = \begin{bmatrix} 1.5 & 0.1 & 0 \\ 0.2 & 1.2 & 0.1 \\ 0.1 & 0.7 & 1.6 \end{bmatrix}$ and assume that the external demand for the products in Sector 1 increases by 1 unit. By how many units should each sector increase production? What do the columns of the matrix $(I - A)^{-1}$ tell you? HINT [See Quick Example on page 290.]

16. Let $(I - A)^{-1} = \begin{bmatrix} 1.5 & 0.1 & 0 \\ 0.1 & 1.1 & 0.1 \\ 0 & 0 & 1.3 \end{bmatrix}$, and assume that the external demand for the products in each of the sectors increases by 1 unit. By how many units should each sector increase production? HINT [See Quick Example on page 290.]

In Exercises 17 and 18, obtain the technology matrix from the given input-output table. HINT [See Example 1.]

17.

	To	A	B	C
From	A	1,000	2,000	3,000
	B	0	4,000	0
	C	0	1,000	3,000
Total Output		5,000	5,000	6,000

18.

	To	A	B	C
From	A	0	100	300
	B	500	400	300
	C	0	0	600
Total Output		1,000	2,000	3,000

APPLICATIONS

19. *Campus Food* The two campus cafeterias, the Main Dining Room and Bits & Bytes, typically use each other's food in doing business on campus. One weekend, the input-output table was as follows.[31]

	To	Main DR	Bits & Bytes
From	Main DR	$10,000	$20,000
	Bits & Bytes	5,000	0
Total Output		50,000	40,000

Given that the demand for food on campus last weekend was $45,000 from the Main Dining Room and $30,000 from Bits & Bytes, how much did the two cafeterias have to produce to meet the demand last weekend? HINT [See Example 1.]

20. *Plagiarism* Two student groups at Enormous State University, the Choral Society and the Football Club, maintain files of term papers that they write and offer to students for research purposes. Some of these papers they use themselves in generating more papers. In order to avoid suspicion of plagiarism by faculty members (who seem to have astute memories), each paper is given to students or used by the clubs only once (no copies are kept). The number of papers that were used in the production of new papers last year is shown in the following input-output table:

	To	Choral Soc.	Football Club
From	Choral Soc.	20	10
	Football Club	10	30
Total Output		100	200

Given that 270 Choral Society papers and 810 Football Club papers will be used by students outside of these two clubs next year, how many new papers do the two clubs need to write?

21. 🖳 *Communication Equipment* Two sectors of the U.S. economy are (1) audio, video, and communication equipment and (2) electronic components and accessories. In 1998, the input-output table involving these two sectors was as follows (all figures are in millions of dollars):[32]

	To	Equipment	Components
From	Equipment	6,000	500
	Components	24,000	30,000
Total Output		90,000	140,000

Determine the production levels necessary in these two sectors to meet an external demand for $80,000 million of communication equipment and $90,000 million of electronic components. Round answers to two significant digits.

---

[31]For some reason, the Main Dining Room consumes a lot of its own food!

[32]The data have been rounded. Source: *Survey of Current Business*, December, 2001, U.S. Department of Commerce.

22. ▣ *Wood and Paper* Two sectors of the U.S. economy are (1) lumber and wood products and (2) paper and allied products. In 1998 the input-output table involving these two sectors was as follows (all figures are in millions of dollars).[33]

	To	**Wood**	**Paper**
From	**Wood**	36,000	7,000
	Paper	100	17,000
	Total Output	120,000	120,000

If external demand for lumber and wood products rises by $10,000 million and external demand for paper and allied products rises by $20,000 million, what increase in output of these two sectors is necessary? Round answers to two significant digits.

23. *Australia Economy* Two sectors of the Australian economy are (1) textiles and (2) clothing and footwear. The 1977 input-output table[34] involving these two sectors results in the following value for $(I - A)^{-1}$:

$$(I - A)^{-1} = \begin{bmatrix} 1.228 & 0.182 \\ 0.006 & 1.1676 \end{bmatrix}.$$

Complete the following sentences.

a. ____ additional dollars worth of clothing and footwear must be produced to meet a $1 increase in the demand for textiles.

b. 0.182 additional dollars worth of ____ must be produced to meet a one-dollar increase in the demand for ____.

24. *Australia Economy* Two sectors of the Australian economy are (1) community services and (2) recreation services. The 1978–79 input-output table[35] involving these two sectors results in the following value for $(I - A)^{-1}$:

$$(I - A)^{-1} = \begin{bmatrix} 1.0066 & 0.00576 \\ 0.00496 & 1.04206 \end{bmatrix}.$$

Complete the following sentences.

a. 0.00496 additional dollars worth of ____ must be produced to meet a $1 increase in the demand for ____.

b. ____ additional dollars worth of community services must be produced to meet a one-dollar increase in the demand for community services.

▣ *Exercises 25–28 require the use of technology.*

25. ▣ *United States Input-Output Table* Four sectors of the U.S. economy are (1) livestock and livestock products, (2) other agricultural products, (3) forestry and fishery products, and (4) agricultural, forestry, and fishery services. In 1977 the input-output table involving these four sectors was as follows (all figures are in millions of dollars):[36]

	To	**1**	**2**	**3**	**4**
From	**1**	11,937	9	109	855
	2	26,649	4,285	0	4,744
	3	0	0	439	61
	4	5,423	10,952	3,002	216
Total Output		97,795	120,594	14,642	47,473

Determine how these four sectors would react to an increase in demand for livestock (Sector 1) of $1,000 million, how they would react to an increase in demand for other agricultural products (Sector 2) of $1,000 million, and so on.

26. ▣ *United States Input-Output Table* Four sectors of the U.S. economy are (1) motor vehicles, (2) truck and bus bodies, trailers, and motor vehicle parts, (3) aircraft and parts, and (4) other transportation equipment. In 1998 the input-output table involving these four sectors was (all figures in millions of dollars):[37]

	To	**1**	**2**	**3**	**4**
From	**1**	75	1,092	0	1,207
	2	64,858	13,081	7	1,070
	3	0	0	21,782	0
	4	0	0	0	1,375
Total Output		230,676	135,108	129,376	44,133

Determine how these four sectors would react to an increase in demand for motor vehicles (Sector 1) of $1,000 million, how they would react to an increase in demand for truck and bus bodies (Sector 2) of $1,000 million, and so on.

27. ▼ *Australia Input-Output Table* Four sectors of the Australian economy are (1) agriculture, (2) forestry, fishing, and hunting, (3) meat and milk products, and (4) other food products. In 1978–79 the input-output table involving these four sectors was as follows (all figures are in millions of Australian dollars).[38]

[33]The data have been rounded. Source: *Survey of Current Business,* December, 2001, U.S. Department of Commerce.

[34]Source: *Australian National Accounts and Input-Output Tables 1978–1979,* Australian Bureau of Statistics.

[35]*Ibid.*

[36]Source: *Survey of Current Business,* December 2001, U.S. Department of Commerce.

[37]*Ibid.*

[38]Source: *Australian National Accounts and Input-Output Tables 1978–1979,* Australian Bureau of Statistics.

	To	**1**	**2**	**3**	**4**
From	**1**	678.4	3.7	3,341.5	1,023.5
	2	15.5	6.9	17.1	124.5
	3	47.3	4.3	893.1	145.8
	4	312.5	22.1	83.2	693.5
Total Output		9,401.3	685.8	6,997.3	4,818.3

a. How much additional production by the meat and milk sector is necessary to accommodate a $100 increase in the demand for agriculture?

b. Which sector requires the most of its own product in order to meet a $1 increase in external demand for that product?

28. ▼ *Australia Input-Output Table* Four sectors of the Australian economy are (1) petroleum and coal products, (2) non-metallic mineral products, (3) basic metals and products, and (4) fabricated metal products. In 1978–79 the input-output table involving these four sectors was as follows (all figures are in millions of Australian dollars).[39]

	To	**1**	**2**	**3**	**4**
From	**1**	174.1	30.5	120.3	14.2
	2	0	190.1	55.8	12.6
	3	2.1	40.2	1,418.7	1,242.0
	4	0.1	7.3	40.4	326.0
Total Output		3,278.0	2,188.8	6,541.7	4,065.8

---

[39]Source: *Australian National Accounts and Input-Output Tables 1978–1979,* Australian Bureau of Statistics.

a. How much additional production by the petroleum and coal products sector is necessary to accommodate a $1,000 increase in the demand for fabricated metal products?

b. Which sector requires the most of the product of some other sector in order to meet a $1 increase in external demand for that product?

COMMUNICATION AND REASONING EXERCISES

29. What would it mean if the technology matrix A were the zero matrix?

30. Can an external demand be met by an economy whose technology matrix A is the identity matrix? Explain.

31. ▼ What would it mean if the total output figure for a particular sector of an input-output table were equal to the sum of the figures in the row for that sector?

32. ▼ What would it mean if the total output figure for a particular sector of an input-output table were less than the sum of the figures in the row for that sector?

33. ▼ What does it mean if an entry in the matrix $(I - A)^{-1}$ is zero?

34. ▼ Why do we expect the diagonal entries in the matrix $(I - A)^{-1}$ to be slightly larger than 1?

35. ▼ Why do we expect the off-diagonal entries of $(I - A)^{-1}$ to be less than 1?

36. ▼ Why do we expect all the entries of $(I - A)^{-1}$ to be nonnegative?

CHAPTER 4 REVIEW

KEY CONCEPTS

 Website www.WanerMath.com
Go to the Website at www.WanerMath
.com to find a comprehensive and
interactive Web-based summary
of Chapter 4.

4.1 Matrix Addition and Scalar Multiplication

$m \times n$ matrix, dimensions, entries
p. 232
Referring to the entries of a matrix
p. 232
Matrix equality p. 233
Row, column, and square matrices
p. 234
Addition and subtraction of matrices
p. 234
Scalar multiplication p. 236
Properties of matrix addition and scalar
multiplication p. 237
The transpose of a matrix p. 238
Properties of transposition p. 238

4.2 Matrix Multiplication

Multiplying a row by a column
p. 242
Linear equation as a matrix
equation p. 244

The product of two matrices: general
case p. 244
Identity matrix p. 248
Properties of matrix addition and
multiplication p. 249
Properties of transposition and
multiplication p. 250
A system of linear equations can be
written as a single matrix equation
p. 250

4.3 Matrix Inversion

The inverse of a matrix, singular matrix
p. 256
Procedure for finding the inverse of a
matrix p. 256
Formula for the inverse of a 2×2
matrix; determinant of a 2×2
matrix p. 257
Using an inverse matrix to solve a
system of equations p. 261

4.4 Game Theory

Two-person zero-sum game, payoff
matrix p. 268
A strategy specifies how a player
chooses a move p. 268
The expected payoff of a game for given
mixed strategies R and C p. 269

An optimal strategy, according to the
minimax criterion, is one that minimizes the maximum damage your
opponent can cause you. p. 272
The Fundamental Principle of Game
Theory p. 272
Procedure for reducing by dominance
p. 273
Procedure for solving a 2×2 game
p. 273
The expected value of a game is its
expected payoff when the players use
their optimal strategies p. 276
A strictly determined game is one with
a saddle point p. 278
Steps to follow in solving a game
p. 278

4.5 Input-Output Models

An input-output model divides an
economy into sectors. The technology matrix records the interactions of
these sectors and allows us to relate
external demand to the production
vector. p. 286
Procedure for finding a technology
matrix from an input-output table
p. 288
The entries of $(I - A)^{-1}$ p. 290

REVIEW EXERCISES

For Exercises 1–10, let

$$A = \begin{bmatrix} 1 & 2 & 3 \\ 4 & 5 & 6 \end{bmatrix}, \ B = \begin{bmatrix} 1 & -1 \\ 0 & 1 \end{bmatrix},$$

$$C = \begin{bmatrix} -1 & 0 \\ 1 & 1 \\ 0 & 1 \end{bmatrix}, \text{ and } D = \begin{bmatrix} -3 & -2 & -1 \\ 1 & 2 & 3 \end{bmatrix}.$$

Determine whether each expression is defined, and if it is,
evaluate it.

1. $A + B$ **2.** $A - D$

3. $2A^T + C$ **4.** AB

5. $A^T B$ **6.** A^2

7. B^2 **8.** B^3

9. $AC + B$ **10.** $CD + B$

In Exercises 11–16, find the inverse of the given matrix or
determine that the matrix is singular.

11. $\begin{bmatrix} 1 & -1 \\ 0 & 1 \end{bmatrix}$ **12.** $\begin{bmatrix} 1 & 2 \\ 0 & 0 \end{bmatrix}$

13. $\begin{bmatrix} 1 & 2 & 3 \\ 0 & 4 & 1 \\ 0 & 0 & 1 \end{bmatrix}$ **14.** $\begin{bmatrix} 1 & 2 & 3 & 4 \\ 1 & 3 & 4 & 2 \\ 0 & 1 & 2 & 3 \\ 0 & 0 & 1 & 2 \end{bmatrix}$

15. $\begin{bmatrix} 1 & 2 & 3 & 4 \\ 2 & 3 & 3 & 3 \\ 0 & 1 & 2 & 3 \\ 0 & 0 & 1 & 2 \end{bmatrix}$ **16.** $\begin{bmatrix} 0 & 1 & 0 & 0 \\ 1 & 0 & 0 & 0 \\ 0 & 0 & 0 & 1 \\ 0 & 0 & 1 & 0 \end{bmatrix}$

In Exercises 17–20, write the given system of linear equations as
a matrix equation, and solve by inverting the coefficient matrix.

17. $\begin{aligned} x + 2y &= 0 \\ 3x + 4y &= 2 \end{aligned}$ **18.** $\begin{aligned} x + y + z &= 3 \\ y + 2z &= 4 \\ y - z &= 1 \end{aligned}$

19.
$$\begin{aligned} x + y + z &= 2 \\ x + 2y + z &= 3 \\ x + y + 2z &= 1 \end{aligned}$$

20.
$$\begin{aligned} x + y &= 0 \\ y + z &= 1 \\ z + w &= 0 \\ x - w &= 3 \end{aligned}$$

In each of Exercises 21–24, solve the game with the given payoff matrix and give the expected value of the game.

21. $P = \begin{bmatrix} 2 & 1 & 3 & 2 \\ -1 & 0 & -2 & 1 \\ 2 & 0 & 1 & 3 \end{bmatrix}$ **22.** $P = \begin{bmatrix} 3 & -3 & -2 \\ -1 & 3 & 0 \\ 2 & 2 & 1 \end{bmatrix}$

23. $P = \begin{bmatrix} -1 & -3 & -2 \\ -1 & 3 & 0 \\ 3 & 3 & -1 \end{bmatrix}$ **24.** $P = \begin{bmatrix} 1 & 4 & 3 & 3 \\ 0 & -1 & 2 & 3 \\ 2 & 0 & -1 & 2 \end{bmatrix}$

In each of Exercises 25–28, find the production vector X corresponding to the given technology matrix A and external demand vector D.

25. $A = \begin{bmatrix} 0.3 & 0.1 \\ 0 & 0.3 \end{bmatrix}$, $D = \begin{bmatrix} 700 \\ 490 \end{bmatrix}$

26. $A = \begin{bmatrix} 0.7 & 0.1 \\ 0.1 & 0.7 \end{bmatrix}$, $D = \begin{bmatrix} 1,000 \\ 2,000 \end{bmatrix}$

27. $A = \begin{bmatrix} 0.2 & 0.2 & 0.2 \\ 0 & 0.2 & 0.2 \\ 0 & 0 & 0.2 \end{bmatrix}$, $D = \begin{bmatrix} 32,000 \\ 16,000 \\ 8,000 \end{bmatrix}$

28. $A = \begin{bmatrix} 0.5 & 0.1 & 0 \\ 0.1 & 0.5 & 0.1 \\ 0 & 0.1 & 0.5 \end{bmatrix}$, $D = \begin{bmatrix} 23,000 \\ 46,000 \\ 23,000 \end{bmatrix}$

APPLICATIONS: OHaganBooks.com

It is now July 1 and online sales of romance, science fiction, and horror novels at OHaganBooks.com were disappointingly slow over the past month. Exercises 29–34 are based on the following tables:

Inventory of books in stock on June 1 at the OHaganBooks.com warehouses in Texas and Nevada:

Books in Stock (June 1)

	Romance	Sci Fi	Horror
Texas	2,500	4,000	3,000
Nevada	1,500	3,000	1,000

Online sales during June:

June Sales

	Romance	Sci Fi	Horror
Texas	300	500	100
Nevada	100	600	200

New books purchased each month:

Monthly Purchases

	Romance	Sci Fi	Horror
Texas	400	400	300
Nevada	200	400	300

July Sales (Projected)

	Romance	Sci Fi	Horror
Texas	280	550	100
Nevada	50	500	120

29. *Inventory* Use matrix algebra to compute the inventory at each warehouse at the end of June.

30. *Inventory* Use matrix algebra to compute the change in inventory at each warehouse during June.

31. *Inventory* Assuming that sales continue at the level projected for July for the next few months, write down a matrix equation showing the inventory N at each warehouse x months after July 1. How many months from now will OHaganBooks.com run out of Sci Fi novels at the Nevada warehouse?

32. *Inventory* Assuming that sales continue at the level projected for July for the next few months, write down a matrix equation showing the change in inventory N at each warehouse x months after July 1. How many months from now will OHaganBooks.com have 1,000 more horror novels in stock in Texas than currently?

33. *Revenue* It is now the end of July and OHaganBooks.com's e-commerce manager bursts into the CEO's office. "I thought you might want to know, John, that our sales figures are exactly what I projected a month ago. Is that good market analysis or what?" OHaganBooks.com has charged an average of $5 for romance novels, $6 for science fiction novels, and $5.50 for horror novels. Use the projected July sales figures from above and matrix arithmetic to compute the total revenue OHaganBooks.com earned at each warehouse in July.

34. *Cost* OHaganBooks.com pays an average of $2 for romance novels, $3.50 for science fiction novels, and $1.50 for horror novels. Use this information together with the monthly purchasing information to compute the monthly purchasing cost.

Acting on a "tip" from Marjory Duffin, John O'Hagan decided that his company should invest a significant sum in shares of Duffin House Publishers (DHP) and Duffin Subprime Ventures (DSV). Exercises 35–38 are based on the following table, which shows what information John was able to piece together later, after some of the records had been deleted by an angry student intern.

Date	Number of Shares: DHP	Price per Share: DHP	Number of Shares: DSV	Price per Share: DSV
July 1	?	$20	?	$10
August 1	?	$10	?	$20
September 1	?	$5	?	$40
Total	5,000		7,000	

35. *Investments* Over the three months shown, the company invested a total of $50,000 in DHP stock, and, on August 15, was paid dividends of 10¢ per share held on that date, for a total of $300. Use matrix inversion to determine how many shares of DHP OHaganBooks.com purchased on each of the three dates shown.

36. *Investments* Over the three months shown, the company invested a total of $150,000 in DSV stock, and, on July 15, was paid dividends of 20¢ per share held on that date, for a total of $600. Use matrix inversion to determine how many shares of DSV OHaganBooks.com purchased on each of the three dates shown.

37. *Investments* (Refer to Exercise 35.) On October 1, the shares of DHP purchased on July 1 were sold at $3 per share. The remaining shares were sold one month later at $1 per share. Use matrix algebra to determine the total loss (taking into account the dividends paid on August 15) incurred as a result of the Duffin stock debacle.

38. *Investments* (Refer to Exercise 36.) On September 15, DSV announced a two-for-one stock split, so that each share originally purchased was converted into two shares. On October 1, the company paid an additional special dividend of 10¢ per share. On October 15, OHaganBooks.com sold 3,000 shares at $20 per share. One week later the subprime market crashed, Duffin Subprime Ventures declared bankruptcy (after awarding its fund manager a $10 million bonus), and DSV stock became worthless. Use matrix algebra to determine the total loss (taking into account the dividends paid on July 15) incurred as a result of the Duffin stock debacle.

OHaganBooks.com has two main competitors: JungleBooks.com and FarmerBooks.com, and no other competitors of any significance on the horizon. Exercises 39–42 are based on the following table, which shows the movement of customers during July.[40] (Thus, for instance, the first row tells us that 80% of OHaganBooks.com's customers remained loyal, 10% of them went to JungleBooks.com, and the remaining 10% went to FarmerBooks.com.)

	To OHagan	To Jungle	To Farmer
From OHagan	0.8	0.1	0.1
From Jungle	0.4	0.6	0
From Farmer	0.2	0	0.8

At the beginning of July, OHaganBooks.com had an estimated 2,000 customers, while its two competitors had 4,000 each.

39. *Competition* Set up the July 1 customer numbers in a row matrix, and use matrix arithmetic to estimate the number of customers each company has at the end of July.

40. *Competition* Assuming the July trends continue in August, predict the number of customers each company will have at the end of August.

41. *Competition* Assuming the July trends continue in August, why is it not possible for a customer of FarmerBooks.com on July 1 to have ended up as a JungleBooks.com customer two months later without having ever been an OHaganBooks.com customer?

42. *Competition* Name one or more important factors that the model we have used does not take into account.

Publisher Marjory Duffin reveals that JungleBooks may be launching a promotional scheme in which it will offer either two books for the price of one, or three books for the price of two (Marjory can't quite seem to remember which, and is not certain whether they will go with the scheme at all). John O'Hagan's marketing advisers Flood and O'Lara seem to have different ideas as to how to respond. Flood suggests that the company counter by offering three books for the price of one, while O'Lara suggests that it offer instead a free copy of the Finite Mathematics Student Solutions Manual *with every purchase. After a careful analysis, O'Hagan comes up with the following payoff matrix, where the payoffs represent the number of customers, in thousands, he expects to gain from JungleBooks.*

	JungleBooks		
	No Promo	2 for Price of 1	3 for Price of 2
O'Hagan No Promo	0	−60	−40
3 for Price of 1	30	20	10
Finite Math	20	0	15

Use the above information in Exercises 43–48.

43. *Competition* After a very expensive dinner at an exclusive restaurant, Marjory suddenly "remembers" that the JungleBooks CEO mentioned to her (at a less expensive restaurant) that there is only a 20% chance JungleBooks will launch a "2 for the price of 1" promotion, and a 40% chance that it will launch a "3 for the price of 2" promotion. What should OHaganBooks.com do in view of this information, and what will the expected effect be on its customer base?

44. *Competition* JungleBooks CEO François Dubois has been told by someone with personal ties to OHaganBooks.com staff that OHaganBooks is in fact 80% certain to opt for the "3 for the price of 1" option and will certainly go with one of the two possible promos. What should JungleBooks.com do in view of this information, and what will the expected effect be on its customer base?

45. *Corporate Spies* One of John O'Hagan's trusted marketing advisers has, without knowing it, accidentally sent him a copy of the following e-mail:

> To: René, JungleBooks Marketing Department
> From: O'Lara
>
> Hey René somehow the CEO here says he has learned that there is a 20% chance that you will opt for the "2 for the price of 1" promotion, and a 40% chance that you will opt for the "3 for the price of 2" promotion. Thought you might want to know. So when do I get my "commission"? —Jim

---

[40]By a "customer" of one of the three e-commerce sites, we mean someone who purchases more at that site than at either of the two competitors.

What will each company do in view of this new information, and what will the expected effect be on its customer base?

46. *More Corporate Spies* The next day, everything changes: OHaganBooks.com's mole at JungleBooks, Davíde DuPont, is found unconscious in the coffee room at JungleBooks.com headquarters, clutching in his hand the following correspondence he had apparently received moments earlier:

> To: Davíde
> From: John O'Hagan
> Subject: Re: Urgent Information
> Davíde: This information is much appreciated and definitely changes my plans—J
>
> >
> >To: John O
> >From: Davíde
> >Subject: Urgent Information
> >Thought you might want to know that JungleBooks
> >thinks you are 80% likely to opt for 3 for 1 and 20%
> >likely to opt for Finite Math.
> >—D

What will each company do in view of this information, and what will the expected effect be on its customer base?

47. *Things Unravel* John O'Hagan is about to go with the option chosen in Exercise 45 when he hears word about an exposé in the *Publisher Enquirer* on corporate spying in the two companies. Each company now knows that no information about the other's intentions can be trusted. Now what should OHaganBooks.com do, and how many customers should it expect to gain or lose?

48. *Competition* It is now apparent as a result of the *Publisher Enquirer* exposé that, not only can each company make no assumptions about the strategies the other might be using, but the payoff matrix they have been using is wrong: A crack

investigative reporter at the *Enquirer* publishes the following revised matrix:

$$P = \begin{bmatrix} 0 & -60 & -40 \\ 30 & 20 & 10 \\ 20 & 0 & 20 \end{bmatrix}.$$

Now what should JungleBooks.com do, and how many customers should it expect to gain or lose?

Some of the books sold by OHaganBooks.com are printed at Bruno Mills, Inc., a combined paper mill and printing company. Exercises 49–52 are based on the following typical monthly input-output table for Bruno Mills' paper and book printing sectors.

	To	**Paper**	**Books**
From	**Paper**	$20,000	$50,000
	Books	2,000	5,000
Total Output		200,000	100,000

49. *Production* Find the technology matrix for Bruno Mills' paper and book printing sectors.

50. *Production* Compute $(I - A)^{-1}$. What is the significance of the $(1, 2)$-entry?

51. *Production* Approximately $1,700 worth of the books sold each month by OHaganBooks.com are printed at Bruno Mills, Inc., and OHaganBooks.com uses approximately $170 worth of Bruno Mills' paper products each month. What is the total value of paper and books that must be produced by Bruno Mills, Inc. in order to meet demand from OHaganBooks.com?

52. *Production* Currently, Bruno Mills, Inc. has a monthly capacity of $500,000 of paper products and $200,000 of books. What level of external demand would cause Bruno to meet the capacity for both products?

Case Study   **Projecting Market Share**

You are the sales director at *Selular*, a cellphone provider, and things are not looking good for your company: The recently launched *iClone* competitor is beginning to chip away at Selular's market share. Particularly disturbing are rumors of fierce brand loyalty by iClone customers, with several bloggers suggesting that iClone retains close to 100% of their customers. Worse, you will shortly be presenting a sales report to the board of directors, and the CEO has "suggested" that your report include 2-, 5-, and 10-year projections of Selular's market share given the recent impact on the market by iClone, and also a "worst-case" scenario projecting what would happen if iClone customers are so loyal that none of them ever switch services.

The sales department has conducted two market surveys, taken one quarter apart, of the major cellphone providers, which are *iClone, Selular, AB&C*, and some smaller companies lumped together as "Other," and has given you the data shown in Figure 4, which shows the percentages of subscribers who switched from one service to another during the quarter.*

* If you go on to study probability theory in Chapter 7, you will see how this scenario can be interpreted as a Markov system, and you will revisit the analysis below in that context.

The percentages in the figure and market shares below reflect actual data for several cellphone services in the United States during a single quarter of 2003. (See the exercises for Section 7.7.)

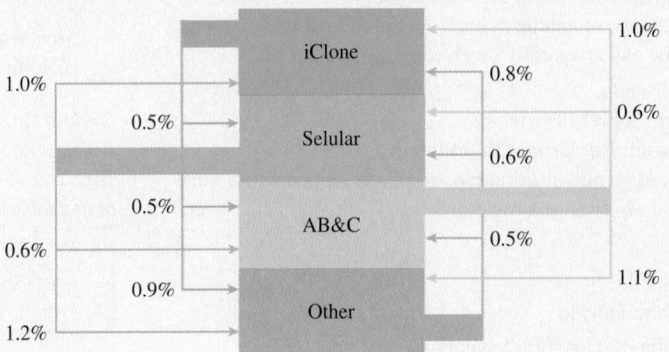

Figure 4

For example, by the end of the quarter 0.5% of iClone's customers had switched to Selular, 0.5% to AB&C, and 0.9% to Other. The current market shares are as follows: iClone: 29.7%, Selular: 19.3%, AB&C: 18.1%, Other: 32.9%.

"This is simple," you tell yourself. "Since I know the current market shares and percentage movements over one quarter, I can easily calculate the market shares next quarter (assuming the percentages that switch services remain the same), then repeat the calculation for the following quarter, and so on, until I get the long-term prediction I am seeking." So you begin your calculations by computing the market shares next quarter. First you note that, since a total of $0.5 + 0.5 + 0.9 = 1.9\%$ of iClone users switched to other brands, the rest, 98.1%, stayed with iClone. Similarly, 97.2% of Selular users, 97.3% of AB&C, and 98.1% of Other stayed with their respective brands. Then you compute:

iClone share after one quarter

$= 98.1\%$ of iClone $+ 1.0\%$ of Selular $+ 1.0\%$ of AB&C $+ 0.8\%$ of Other

$= (0.981)(0.297) + (0.010)(0.193) + (0.010)(0.181) + (0.008)(0.329)$

$= 0.297729$, or 29.7729%.

Similarly,

Selular share:

$= (0.005)(0.297) + (0.972)(0.193) + (0.006)(0.181) + (0.006)(0.329)$

$= 0.192141$

AB&C share:

$= (0.005)(0.297) + (0.006)(0.193) + (0.973)(0.181) + (0.005)(0.329)$

$= 0.180401$

Other share:

$= (0.009)(0.297) + (0.012)(0.193) + (0.011)(0.181) + (0.981)(0.329)$

$= 0.329729$.

So, you project the market shares after one quarter to be: iClone: 29.7729%, Selular: 19.2141%, AB&C: 18.0401%, Other: 32.9729%. You now begin to realize that repeating this kind of calculation for the large number of quarters required for long-term projections will be tedious. You call in your student intern (who happens to be a mathematics major) to see if she can help. After taking one look at the calculations, she makes the observation that all you have really done is compute the product of two matrices:

$$[0.297729 \quad 0.192141 \quad 0.180401 \quad 0.329729]$$

$$= [0.297 \quad 0.193 \quad 0.181 \quad 0.329] \begin{bmatrix} .981 & .005 & .005 & .009 \\ .010 & .972 & .006 & .012 \\ .010 & .006 & .973 & .011 \\ .008 & .006 & .005 & .981 \end{bmatrix}$$

Market shares after one quarter = Market shares at start of quarter $\times A$

The 4×4 matrix A is organized as follows:

$$\begin{array}{cc} & \textit{To} \\ & \begin{array}{cccc} \textbf{iClone} & \textbf{Selular} & \textbf{AB\&C} & \textbf{Other} \end{array} \\ \textit{From} \begin{array}{r} \textbf{iClone} \\ \textbf{Selular} \\ \textbf{AB\&C} \\ \textbf{Other} \end{array} & \begin{bmatrix} .981 & .005 & .005 & .009 \\ .010 & .972 & .006 & .012 \\ .010 & .006 & .973 & .011 \\ .008 & .006 & .005 & .981 \end{bmatrix} \end{array}.$$

Since the market shares one quarter later can be obtained from the shares at the start of the quarter, you realize that you can now obtain the shares *two* quarters later by multiplying the result by A—and at this point you start using technology (such as the Matrix Algebra Tool at www.WanerMath.com) to continue the calculation:

Market shares after two quarters = Market shares after one quarter $\times A$

$$= [0.297729 \quad 0.192141 \quad 0.180401 \quad 0.329729] \begin{bmatrix} .981 & .005 & .005 & .009 \\ .010 & .972 & .006 & .012 \\ .010 & .006 & .973 & .011 \\ .008 & .006 & .005 & .981 \end{bmatrix}$$

$$\approx [0.298435 \quad 0.191310 \quad 0.179820 \quad 0.330434].$$

Although the use of matrices has simplified your work, continually multiplying the result by A over and over again to get the market shares for successive months is still tedious. Would it not be possible to get, say, the market share after 10 years (40 quarters) with a single calculation? To explore this, you decide to use symbols for the various market shares:

m_0 = Starting market shares = $[0.297 \quad 0.193 \quad 0.181 \quad 0.329]$

m_1 = Market shares after 1 quarter
$\quad = [0.297729 \quad 0.192141 \quad 0.180401 \quad 0.329729]$

m_2 = Market shares after 2 quarters

$\cdots$

m_n = Market share after n quarters.

You then rewrite the relationships above as

$m_1 = m_0 A \quad$ Shares after one quarter = Shares at start of quarter $\times A$

$m_2 = m_1 A \quad$ Shares after two quarters = Shares after one quarter $\times A$

On an impulse, you substitute the first equation in the second:

$$m_2 = m_1 A = (m_0 A)A = m_0 A^2.$$

Continuing,

$$m_3 = m_2 A = (m_0 A^2)A = m_0 A^3$$

$$\cdots$$

$$m_n = m_0 A^n,$$

which is exactly the formula you need! You can now obtain the 2-, 5-, and 10-year projections each in a single step (with the aid of technology):

2-year projection: $m_8 = m_0 A^8$

$$= [0.297 \quad 0.193 \quad 0.181 \quad 0.329] \begin{bmatrix} .981 & .005 & .005 & .009 \\ .010 & .972 & .006 & .012 \\ .010 & .006 & .973 & .011 \\ .008 & .006 & .005 & .981 \end{bmatrix}^8$$

$$\approx [0.302237 \quad 0.186875 \quad 0.176691 \quad 0.334198]$$

5-year projection: $m_{20} = m_0 A^{20} \approx [0.307992 \quad 0.180290 \quad 0.171938 \quad 0.33978]$

10-year projection: $m_{40} = m_0 A^{40} \approx [0.313857 \quad 0.173809 \quad 0.167074 \quad 0.34526].$

In particular, Selular's market shares are projected to decline slightly: 2-year projection: 18.7%, 5-year projection: 18.0%, 10-year projection: 17.4%.

You now move on to the "worst-case" scenario, which you represent by assuming 100% loyalty by iClone users with the remaining percentages staying the same (Figure 5).

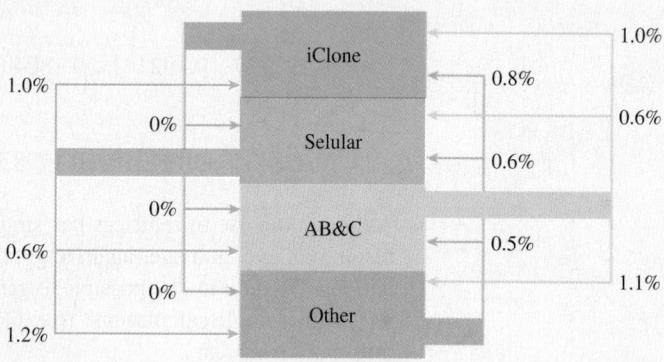

Figure 5

The matrix A corresponding to this diagram is $\begin{bmatrix} 1 & 0 & 0 & 0 \\ .010 & .972 & .006 & .012 \\ .010 & .006 & .973 & .011 \\ .008 & .006 & .005 & .981 \end{bmatrix}$, and

you find:

2-year projection: $m_8 = m_0 A^8 \approx [0.346335 \quad 0.175352 \quad 0.165200 \quad 0.313113]$

5-year projection: $m_{20} = m_0 A^{20} \approx [0.413747 \quad 0.152940 \quad 0.144776 \quad 0.288536]$

10-year projection: $m_{40} = m_0 A^{40} \approx [0.510719 \quad 0.123553 \quad 0.117449 \quad 0.248279].$

Thus, in the worst-case scenario, Selular's market shares are projected to decline more rapidly: 2-year projection: 17.5%, 5-year projection: 15.3%, 10-year projection: 12.4%.

EXERCISES

1. Project Selular's market share 20 and 30 years from now based on the original data shown in Figure 4. (Round all figures to the nearest 0.1%.)

2. Using the original data, what can you say about each company's share in 60 and 80 years (assuming current trends continue)? (Round all figures to the nearest 0.1%.)

3. Obtain a sequence of projections several hundreds of years into the future using the scenarios in both Figure 4 and Figure 5. What do you notice?

4. Compute a sequence of larger and larger powers of the matrix A for both scenarios with all figures rounded to four decimal places. What do you notice?

5. Suppose the trends noted in the market survey were in place for several quarters *before* the current quarter. Using the scenario in Figure 4, determine what the companies' market shares were one quarter before the present and one year before the present (to the nearest 0.1%). Do the same for the scenario in Figure 5.

6. If A is the matrix from the scenario in Figure 4, compute A^{-1}. In light of the preceding exercise, what do the entries in A^{-1} mean?

TECHNOLOGY GUIDE

TI-83/84 Plus   Technology Guide

Section 4.1

Example 2 (page 235)   The A-Plus auto parts store chain has two outlets, one in Vancouver and one in Quebec. Among other things, it sells wiper blades, windshield cleaning fluid, and floor mats. The monthly sales of these items at the two stores for two months are given in the following tables:

January Sales

	Vancouver	Quebec
Wiper Blades	20	15
Cleaning Fluid (bottles)	10	12
Floor Mats	8	4

February Sales

	Vancouver	Quebec
Wiper Blades	23	12
Cleaning Fluid (bottles)	8	12
Floor Mats	4	5

Use matrix arithmetic to calculate the change in sales of each product in each store from January to February.

Solution with Technology

On the TI-83/84 Plus, matrices are referred to as [A], [B], and so on through [J]. To enter a matrix, press MATRX to bring up the matrix menu, select EDIT, select a matrix, and press ENTER. Then enter the dimensions of the matrix followed by its entries. When you want to use a matrix, press MATRX, select the matrix and press ENTER.

On the TI-83/84 Plus, adding matrices is similar to adding numbers. The sum of the matrices [A] and [B] is [A] + [B]; their difference, of course, is [A] - [B]. As in the text, for this example,

1. Create two matrices, [J] and [F].
2. Compute their difference, [F] - [J] using [F] - [J] → [D].

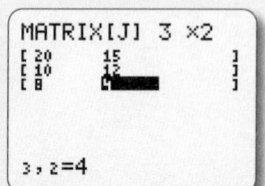
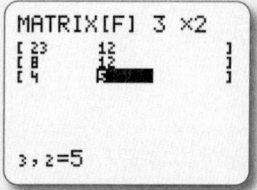

Note that we have stored the difference in the matrix [D] in case we need it for later use.

Section 4.2

Example 3(a) (page 245)   Calculate

$$\begin{bmatrix} 2 & 0 & -1 & 3 \\ 1 & -1 & 2 & -2 \end{bmatrix} \begin{bmatrix} 1 & 1 & -8 \\ 1 & 0 & 0 \\ 0 & 5 & 2 \\ -2 & 8 & -1 \end{bmatrix}.$$

Solution with Technology

On the TI-83/84 Plus, the format for multiplying matrices is the same as for multiplying numbers: [A][B] or [A]*[B] will give the product. We enter the matrices and then multiply them. (Note that, while editing, you can see only three columns of [A] at a time.)

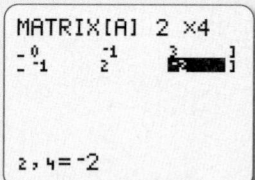
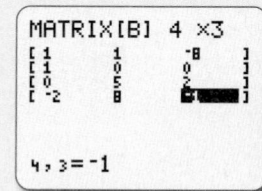

Note that if you try to multiply two matrices whose product is not defined, you will get the error "DIM MISMATCH" (dimension mismatch).

Example 6 (page 248)—Identity Matrix   On the TI-83/84 Plus, the function `identity(n)` (in the MATRX MATH menu) returns the $n \times n$ identity matrix.

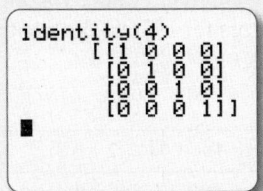

Section 4.3

Example 2(b) (page 259)

Find the inverse of

$$Q = \begin{bmatrix} 1 & 0 & 1 \\ 2 & -2 & -1 \\ 3 & 0 & 0 \end{bmatrix}.$$

Solution with Technology

On a TI-83/84 Plus, you can invert the square matrix [A] by entering [A] X^{-1} ENTER.

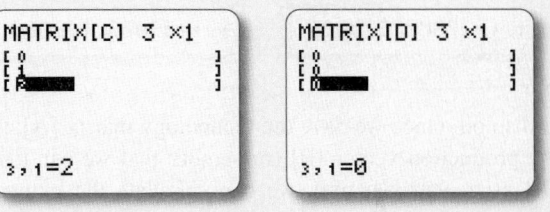

(Note that you can use the right and left arrow keys to scroll the answer, which cannot be shown on the screen all at once.) You could also use the calculator to help you go through the row reduction, as described in Chapter 2.

Example 4 (page 262)

Solve the following three systems of equations.

a. $2x \quad + z = 1$
$2x + y - z = 1$
$3x + y - z = 1$

b. $2x \quad + z = 0$
$2x + y - z = 1$
$3x + y - z = 2$

c. $2x \quad + z = 0$
$2x + y - z = 0$
$3x + y - z = 0$

Solution with Technology

1. Enter the four matrices A, B, C, and D:

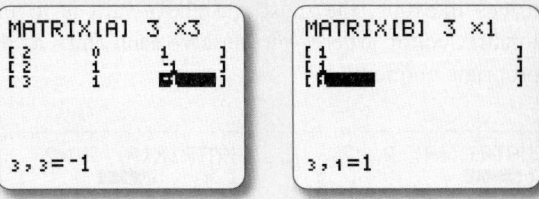

2. Compute the solutions $A^{-1}B$, $A^{-1}C$, and $A^{-1}D$:

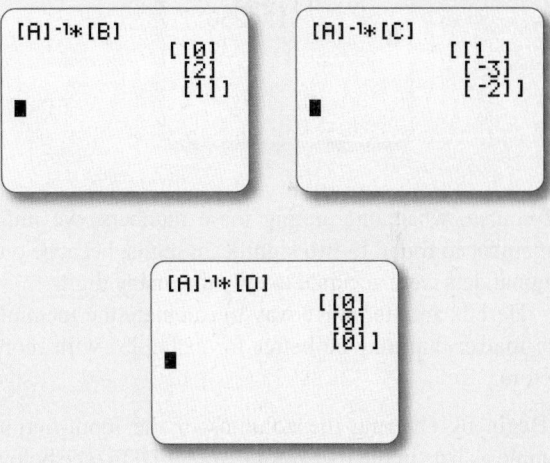

Section 4.5

Example 1 (page 286)   Recall that the input-output table in Example 1 looks like this:

	To	Crude	Refining
From	Crude	27,000	59,000
	Refining	750	15,000
	Total Output	87,000	140,000

What was the technology matrix for these two sectors? What was left over from each of these sectors for use by other parts of the economy or for export?

Solution with Technology

There are several ways to use these data to create the technology matrix in your TI-83/84 Plus. For small matrices like this, the most straightforward is to use the matrix editor, where you can give each entry as the appropriate quotient:

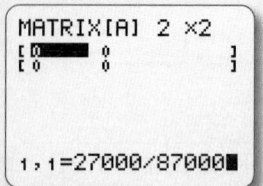

and so on. Once we have the technology matrix [A] and the production vector [B] (remember that we can't use [X] as a matrix name), we can calculate the external demand vector:

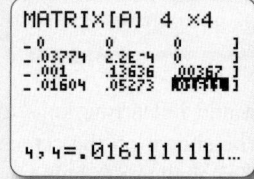

Of course, when interpreting these numbers, we must remember to round to two significant digits, because our original data were accurate to only that many digits.

Here is an alternative way to calculate the technology matrix that may be better for examples with more sectors.

1. Begin by entering the columns of the input-output table as lists in the list editor ([STAT] EDIT) (see below left).

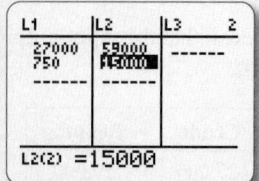 

2. We now want to divide each column by the total output of its sector and assemble the results into a matrix. We can do this using the List ▶ matr function (under the [MATRX] MATH menu) shown on the right above.

Example 3 (page 291)   Consider four sectors of the economy of Kenya[41]: (1) the traditional economy, (2) agriculture, (3) manufacture of metal products and machinery, and (4) wholesale and retail trade. The input-output table for these four sectors for 1976 looks like this (all numbers are thousands of K£):

	To	1	2	3	4
From	1	8,600	0	0	0
	2	0	20,000	24	0
	3	1,500	530	15,000	660
	4	810	8,500	5,800	2,900
Total Output		87,000	530,000	110,000	180,000

Suppose that external demand for agriculture increased by K£50,000,000 and that external demand for metal products and machinery increased by K£10,000,000. How would production in these four sectors have to change to meet this rising demand?

Solution with Technology

1. Enter the technology matrices A as [A] and D^+ as [D] using one of the techniques above:

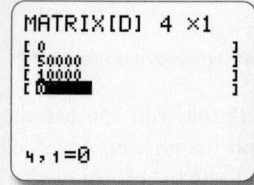 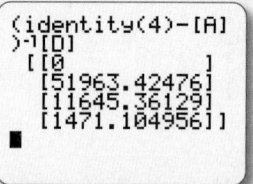

2. You can then compute the change in production with the formula: (identity(4)-[A])$^{-1}$ [D].

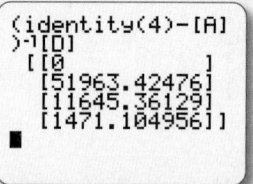

---

[41]Figures are rounded. Source: *Input-Output Tables for Kenya 1976*, Central Bureau of Statistics of the Ministry of Economic Planning and Community Affairs, Kenya.

SPREADSHEET Technology Guide

Section 4.1

Example 2 (page 235) The A-Plus auto parts store chain has two outlets, one in Vancouver and one in Quebec. Among other things, it sells wiper blades, windshield cleaning fluid, and floor mats. The monthly sales of these items at the two stores for two months are given in the following tables:

January Sales

	Vancouver	Quebec
Wiper Blades	20	15
Cleaning Fluid (bottles)	10	12
Floor Mats	8	4

February Sales

	Vancouver	Quebec
Wiper Blades	23	12
Cleaning Fluid (bottles)	8	12
Floor Mats	4	5

Use matrix arithmetic to calculate the change in sales of each product in each store from January to February.

Solution with Technology

To enter a matrix in a spreadsheet, we put its entries in any convenient block of cells. For example, the matrix A in the first Quick Example of this section might look like this:

	A	B	C
1	2	0	1
2	33	-22	0

Spreadsheets refer to such blocks of data as **arrays**, which it can handle in much the same way as it handles single cells of data. For instance, when typing a formula, just as clicking on a cell creates a reference to that cell, selecting a whole array of cells will create a reference to that array. An array is referred to using an **array range** consisting of the top-left and bottom-right cell coordinates, separated by a colon. For example, the array range A1:C2 refers to the 2×3 matrix above, with top-left corner A1 and bottom-right corner C2.

1. To add or subtract two matrices in a spreadsheet, first input their entries in two separate arrays in the spreadsheet (we've also added labels as in the previous tables, which you might do if you wanted to save the spreadsheet for later use):

	A	B	C
1	January Sales		
2		Vancouver	Quebec
3	wiper blades	20	15
4	cleaning fluid (bottles)	10	12
5	floor mats	8	4
6			
7	February Sales		
8		Vancouver	Quebec
9	wiper blades	23	12
10	cleaning fluid (bottles)	8	12
11	floor mats	4	5

2. Select (highlight) a block of the same size (3×2 in this case) where you would like the answer, $F - J$, to appear, enter the formula =B9:C11-B3:C5, and then type Control+Shift+Enter. The easiest way to do this is as follows:

- Highlight cells B15:C17. *Where you want the answer to appear*

- Type "=".
- Highlight the matrix F. *Cells B9 through C11*
- Type "-".
- Highlight the matrix J. *Cells B3 through C5*
- Press Control+Shift+Enter. *Not just Enter*

	A	B	C
1	January Sales		
2		Vancouver	Quebec
3	wiper blades	20	15
4	cleaning fluid (bottles)	10	12
5	floor mats	8	4
6			
7	February Sales		
8		Vancouver	Quebec
9	wiper blades	23	12
10	cleaning fluid (bottles)	8	12
11	floor mats	4	5
12			
13	Change in Sales		
14		Vancouver	Quebec
15	wiper blades	=B9:C11-B3:C5	
16	cleaning fluid (bottles)		
17	floor mats		

Typing Control+Shift+Enter (instead of Enter) tells the spreadsheet that your formula is an *array formula,* one that returns a matrix rather than a single number.[42] Once entered, the formula bar will show the formula you entered enclosed in "curly braces," indicating that it is an array formula. Note that you must use Control+Shift+Enter to delete any array you create: Select the block you wish to delete and press Delete followed by Control+Shift+Enter.

Section 4.2

Example 3(a) (page 245)   Calculate the product

$$\begin{bmatrix} 2 & 0 & -1 & 3 \\ 1 & -1 & 2 & -2 \end{bmatrix} \begin{bmatrix} 1 & 1 & -8 \\ 1 & 0 & 0 \\ 0 & 5 & 2 \\ -2 & 8 & -1 \end{bmatrix}.$$

Solution with Technology

In a spreadsheet, the function we use for matrix multiplication is MMULT. (Ordinary multiplication, *, will *not* work.)

1. Enter the two matrices as shown in the spreadsheet and highlight a block where you want the answer to appear. (Note that it should have the correct dimensions for the product: 2×3.)

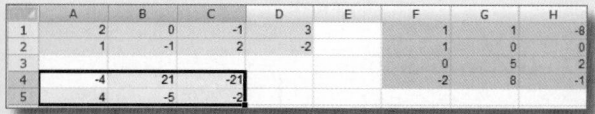

2. Enter the formula =MMULT(A1:D2,F1:H4) (using the mouse to avoid typing the array ranges if you like) and press Control+Shift+Enter. The product will appear in the region you highlighted. If you try to multiply two matrices whose product is not defined, you will get the error "#VALUE!".

Example 6—Identity Matrix (page 248)   There is no spreadsheet function that returns an identity matrix. If you need a small identity matrix, it's simplest to just

enter the 1s and 0s by hand. If you need a large identity matrix, here is one way to get it quickly.

1. Say we want a 4×4 identity matrix in the cells B1:E4. Enter the following formula in cell B1:

```
=IF(ROW(B1)-ROW($B$1)
=COLUMN(B1)-COLUMN($B$1),1,0)
```

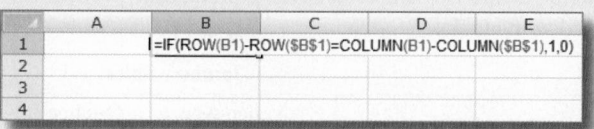

2. Press Enter, then copy cell B1 to cells B1:E4. The formula will return 1s along the diagonal of the matrix and 0s elsewhere, giving you the identity matrix. Why does this formula work?

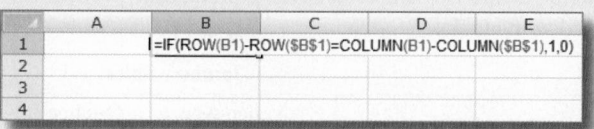

Section 4.3

Example 2(b) (page 259)   Find the inverse of

$$Q = \begin{bmatrix} 1 & 0 & 1 \\ 2 & -2 & -1 \\ 3 & 0 & 0 \end{bmatrix}.$$

Solution with Technology

In a spreadsheet, the function MINVERSE computes the inverse of a matrix.

1. Enter Q somewhere convenient, for example, in cells A1:C3.

2. Choose the block where you would like the inverse to appear, highlight the whole block.

3. Enter the formula =MINVERSE(A1:C3) and press Control+Shift+Enter.

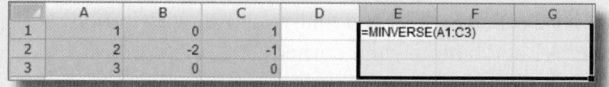

---

[42]Note that on a Mac, Command-Enter has the same effect as Control+Shift+Enter.

The inverse will appear in the region you highlighted. (To convert the answer to fractions, format the cells as fractions.)

If a matrix is singular, a spreadsheet will register an error by showing #NUM! in each cell.

	A	B	C	D	E	F	G
1	1	0	1		0	0	0.3333333
2	2	-2	-1		-0.5	-0.5	0.5
3	3	0	0		1	0	-0.3333333

Although the spreadsheet appears to invert the matrix in one step, it is going through the procedure in the text or some variation of it to find the inverse. Of course, you could also use the spreadsheet to help you go through the row reduction, just as in Chapter 3.

Example 4 (page 262) Solve the following three systems of equations.

a. $2x \quad + z = 1$
$2x + y - z = 1$
$3x + y - z = 1$

b. $2x \quad + z = 0$
$2x + y - z = 1$
$3x + y - z = 2$

c. $2x \quad + z = 0$
$2x + y - z = 0$
$3x + y - z = 0$

Solution with Technology

Spreadsheets instantly update calculated results every time the contents of a cell are changed. We can take advantage of this to solve the three systems of equations given above using the same worksheet as follows.

1. Enter the matrices A and B from the matrix equation $AX = B$.

2. Select a 3×1 block of cells for the matrix X.

3. The Excel formula we can use to calculate X is:

`=MMULT(MINVERSE(A1:C3),E1:E3)` $A^{-1}B$

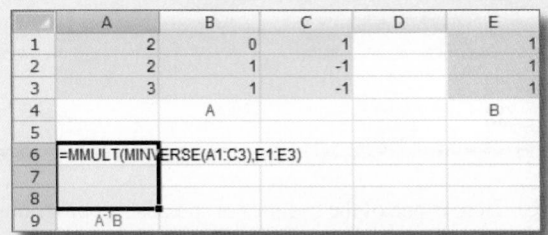

(As usual, use the mouse to select the ranges for A and B while typing the formula, and don't forget to press Control+Shift+Enter.) Having obtained the solution to

part (a), you can now simply modify the entries in Column E to see the solutions for parts (b) and (c).

Note Your spreadsheet for part (a) may look like this:

	A	B	C	D	E
1	2	0	1		1
2	2	1	-1		1
3	3	1	-1		1
4		A			B
5					
6	1.11022E-16				
7	2				
8	1				
9	$A^{-1}B$				

What is that strange number doing in cell A6? "E-16" represents "$\times 10^{-16}$", so the entry is really

$1.11022 \times 10^{-16} = 0.000\,000\,000\,000\,000\,111022 \approx 0.$

Mathematically, it is supposed to be *exactly* zero (see the solution to part (a) in the text) but Excel made a small error in computing the inverse of A, resulting in this spurious value. Note, however, that it is accurate (agrees with zero) to 15 decimal places! In practice, when we see numbers arise in matrix calculations that are far smaller than all the other entries, we can usually assume they are supposed to be zero. ■

Section 4.5

Example 1 (page 286) Recall that the input-output table in Example 1 looks like this:

	To	Crude	Refining
From	Crude	27,000	59,000
	Refining	750	15,000
	Total Output	87,000	140,000

What was the technology matrix for these two sectors? What was left over from each of these sectors for use by other parts of the economy or for export?

Solution with Technology

1. Enter the input-output table in a spreadsheet:

	A	B	C	D
1			To	
2			Crude	Refining
3	From	Crude	27000	59000
4		Refining	750	15000
5		Total Output	87000	140000

TECHNOLOGY GUIDE

2. To obtain the technology matrix, we divide each column by the total output of its sector:

	A	B	C	D
1			To	
2			Crude	Refining
3	From	Crude	27000	59000
4		Refining	750	15000
5		Total Output	87000	140000
6				
7			=C3/C$5	
8				

	A	B	C	D
1			To	
2			Crude	Refining
3	From	Crude	27000	59000
4		Refining	750	15000
5		Total Output	87000	140000
6				
7			0.3103448	0.4214286
8			0.0086207	0.1071429

The formula =C3/C$5 is copied into the shaded 2 × 2 block shown above. (The $ sign in front of the 5 forces the program to always divide by the total in Row 5 even when the formula is copied from Row 7 to Row 8.) The result is the technology matrix shown in the bottom screenshot above.

3. Using the techniques discussed in the second section, we can now compute $D = X - AX$ to find the demand vector.

Example 3 (page 291)   Consider four sectors of the economy of Kenya[43]: (1) the traditional economy, (2) agriculture, (3) manufacture of metal products and machinery, and (4) wholesale and retail trade. The input-output table for these four sectors for 1976 looks like this (all numbers are thousands of K£):

	To	1	2	3	4
From	1	8,600	0	0	0
	2	0	20,000	24	0
	3	1,500	530	15,000	660
	4	810	8,500	5,800	2,900
Total Output		87,000	530,000	110,000	180,000

[43]Figures are rounded. Source: *Input-Output Tables for Kenya 1976*, Central Bureau of Statistics of the Ministry of Economic Planning and Community Affairs, Kenya.

Suppose that external demand for agriculture increased by K£50,000,000 and that external demand for metal products and machinery increased by K£10,000,000. How would production in these four sectors have to change to meet this rising demand?

Solution with Technology

1. Enter the input-output table in the spreadsheet.

2. Compute the technology matrix by dividing each column by the column total.

3. Insert the identity matrix I in preparation for the next step.

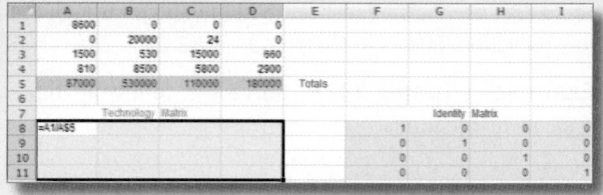

4. To see how each sector reacts to rising external demand, you must calculate the inverse matrix $(I - A)^{-1}$, as shown below. (Remember to use Control+Shift+Enter each time.)

5. To compute $(I - A)^{-1}D^+$, enter D^+ as a column and use the MMULT operation. (See Example 3 in Section 4.2.)

Note Here is one of the beauties of spreadsheet programs: Once you are done with the calculation, you can use the spreadsheet as a template for any 4 × 4 input-output table by just changing the entries of the input-output matrix and/or D^+. The rest of the computation will then be done automatically as the spreadsheet is updated. In other words, you can use it to do your homework! ∎

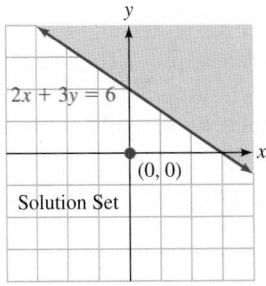

Figure 4

The same kind of argument can be used to show that the solution set of every inequality of the form $ax + by \leq c$ or $ax + by \geq c$ consists of the half plane above or below the line $ax + by = c$. The "test-point" procedure we describe below gives us an easy method for deciding whether the solution set includes the region above or below the corresponding line.

Now we are going to do something that will appear backward at first (but makes it simpler to sketch sets of solutions of *systems* of linear inequalities). For our standard drawing of the region of solutions of $2x + 3y \leq 6$, we are going to *shade only the part that we do not want and leave the solution region blank*. Think of covering over or "blocking out" the unwanted points, leaving those that we do want in full view (but remember that the points on the boundary line are also points that we want). The result is Figure 4. The reason we do this should become clear in Example 2.

Sketching the Region Represented by a Linear Inequality in Two Variables

1. Sketch the straight line obtained by replacing the given inequality with an equality.

2. Choose a test point not on the line; $(0, 0)$ is a good choice if the line does not pass through the origin.

3. If the test point satisfies the inequality, then the set of solutions is the entire region on the same side of the line as the test point. Otherwise, it is the region on the other side of the line. In either case, shade (block out) the side that does *not* contain the solutions, leaving the solution set unshaded.

Quick Example

Here are the three steps used to graph the inequality $x + 2y \geq 5$:

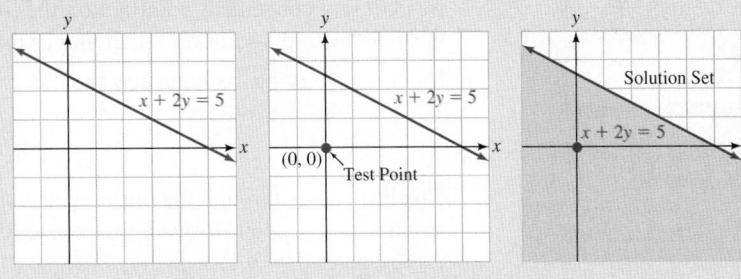

1. Sketch the line $x + 2y = 5$.

2. Test the point $(0, 0)$. $0 + 2(0) \ngeq 5$. Inequality is not satisfied.

3. Because the inequality is not satisfied, shade the region containing the test point.

EXAMPLE 1 Graphing Single Inequalities

Sketch the regions determined by each of the following inequalities:

   a. $3x - 2y \leq 6$ **b.** $6x \leq 12 + 4y$ **c.** $x \leq -1$ **d.** $y \geq 0$ **e.** $x \geq 3y$

Solution

a. The boundary line $3x - 2y = 6$ has *x*-intercept 2 and *y*-intercept -3 (Figure 5). We use $(0, 0)$ as a test point (because it is not on the line). Because $3(0) - 2(0) \leq 6$, the inequality is satisfied by the test point $(0, 0)$, and so it lies inside the solution set. The solution set is shown in Figure 5.

b. The given inequality, $6x \leq 12 + 4y$, can be rewritten in the form $ax + by \leq c$ by subtracting $4y$ from both sides:

$$6x - 4y \leq 12.$$

Dividing both sides by 2 gives the inequality $3x - 2y \leq 6$, which we considered in part (a). Now, *applying the rules for manipulating inequalities does not affect the set of solutions.* Thus, the inequality $6x \leq 12 + 4y$ has the same set of solutions as $3x - 2y \leq 6$. (See Figure 5.)

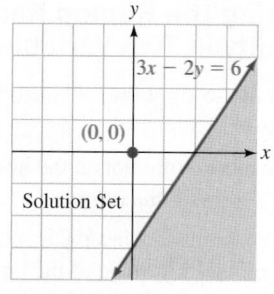

Figure 5

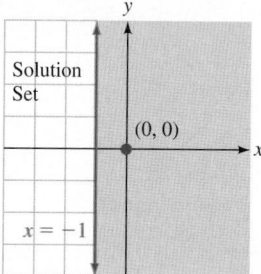

Figure 6

c. The region $x \leq -1$ has as boundary the vertical line $x = -1$. The test point $(0, 0)$ is not in the solution set, as shown in Figure 6.

d. The region $y \geq 0$ has as boundary the horizontal line $y = 0$ (that is, the *x*-axis). We cannot use $(0, 0)$ for the test point because it lies on the boundary line. Instead, we choose a convenient point not on the line $y = 0$—say, $(0, 1)$. Because $1 \geq 0$, this point is in the solution set, giving us the region shown in Figure 7.

e. The line $x \geq 3y$ has as boundary the line $x = 3y$ or, solving for *y*,

$$y = \frac{1}{3}x.$$

This line passes through the origin with slope 1/3, so again we cannot choose the origin as a test point. Instead, we choose $(0, 1)$. Substituting these coordinates in $x \geq 3y$ gives $0 \geq 3(1)$, which is false, so $(0, 1)$ is not in the solution set, as shown in Figure 8.

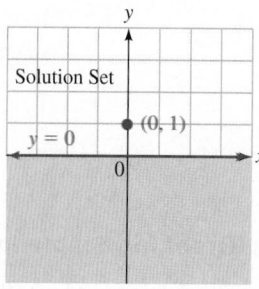

Figure 7

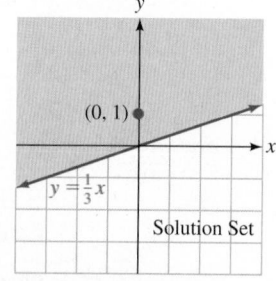

Figure 8

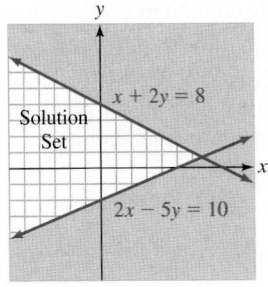

Figure 9

✳ Although these graphs are quite easy to do by hand, the more lines we have to graph the more difficult it becomes to get everything in the right place, and this is where graphing technology can become important. This is especially true when, for instance, three or more lines intersect in points that are very close together and hard to distinguish in hand-drawn graphs.

EXAMPLE 2 Graphing Simultaneous Inequalities

Sketch the region of points that satisfy both inequalities:

$$2x - 5y \leq 10$$
$$x + 2y \leq 8.$$

Solution Each inequality has a solution set that is a half plane. If a point is to satisfy *both* inequalities, it must lie in both sets of solutions. Put another way, if we cover the points that are not solutions to $2x - 5y \leq 10$ and then also cover the points that are not solutions to $x + 2y \leq 8$, the points that remain uncovered must be the points we want, those that are solutions to both inequalities. The result is shown in Figure 9, where the unshaded region is the set of solutions.✳

As a check, we can look at points in various regions in Figure 9. For example, our graph shows that (0, 0) should satisfy both inequalities, and it does:

$$2(0) - 5(0) = 0 \leq 10 \qquad ✔$$
$$0 + 2(0) = 0 \leq 8. \qquad ✔$$

On the other hand, (0, 5) should fail to satisfy one of the inequalities:

$$2(0) - 5(5) = -25 \leq 10 \qquad ✔$$
$$0 + 2(5) = 10 > 8 \qquad ✗$$

One more: (5, −1) should fail one of the inequalities:

$$2(5) - 5(-1) = 15 > 10 \qquad ✗$$
$$5 + 2(-1) = 3 \leq 8. \qquad ✔$$

EXAMPLE 3 Corner Points

Sketch the region of solutions of the following system of inequalities and list the coordinates of all the corner points.

$$3x - 2y \leq 6$$
$$x + y \geq -5$$
$$y \leq 4$$

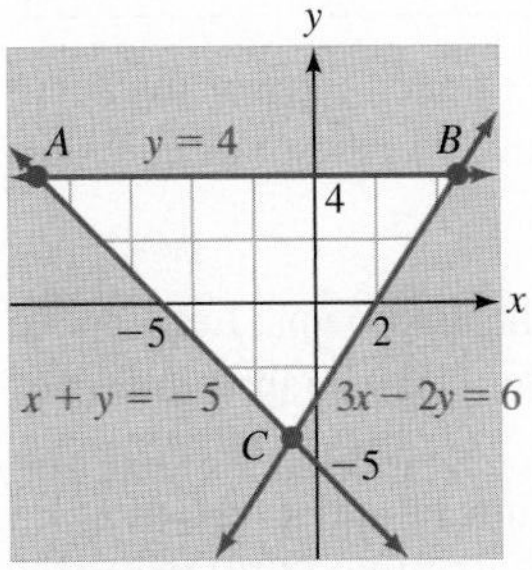

Solution Shading the regions that we do not want leaves us with the triangle shown in Figure 10. We label the corner points A, B, and C as shown.

Each of these corner points lies at the intersection of two of the bounding lines. So, to find the coordinates of each corner point, we need to solve the system of equations given by the two lines. To do this systematically, we make the following table:

Figure 10

Point	Lines through Point	Coordinates
A	$y = 4$ $x + y = -5$	$(-9, 4)$
B	$y = 4$ $3x - 2y = 6$	$\left(\dfrac{14}{3}, 4\right)$
C	$x + y = -5$ $3x - 2y = 6$	$\left(-\dfrac{4}{5}, -\dfrac{21}{5}\right)$

✳ **Technology Note** Using the trace feature makes it easy to locate corner points graphically. Remember to zoom in for additional accuracy when appropriate. Of course, you can also use technology to help solve the systems of equations, as we discussed in Chapter 3.

Here, we have solved each system of equations in the middle column to get the point on the right, using the techniques of Chapter 2. You should do this for practice.✳

As a partial check that we have drawn the correct region, let us choose any point in its interior—say, (0, 0). We can easily check that (0, 0) satisfies all three given inequalities. It follows that all of the points in the triangular region containing (0, 0) are also solutions.

Take another look at the regions of solutions in Examples 2 and 3 (Figure 11).

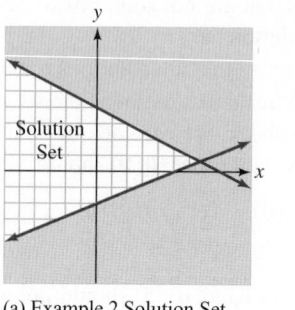

(a) Example 2 Solution Set

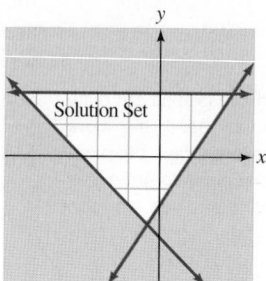

(b) Example 3 Solution Set

Figure 11

Notice that the solution set in Figure 11(a) extends infinitely far to the left, whereas the one in Figure 11(b) is completely enclosed by a boundary. Sets that are completely enclosed are called **bounded**, and sets that extend infinitely in one or more directions are **unbounded**. For example, all the solution sets in Example 1 are unbounded.

EXAMPLE 4 **Resource Allocation**

Socaccio Pistachio Inc. makes two types of pistachio nuts: Dazzling Red and Organic. Pistachio nuts require food color and salt, and the following table shows the amount of food color and salt required for a 1-kilogram batch of pistachios, as well as the total amount of these ingredients available each day.

	Dazzling Red	*Organic*	*Total Available*
Food Color (g)	2	1	20
Salt (g)	10	20	220

Use a graph to show the possible numbers of batches of each type of pistachio Socaccio can produce each day. This region (the solution set of a system of inequalities) is called the **feasible region**.

Solution As we did in Chapter 3, we start by identifying the unknowns: Let x be the number of batches of Dazzling Red manufactured per day and let y be the number of batches of Organic manufactured each day.

Now, because of our experience with systems of linear equations, we are tempted to say: For food color $2x + y = 20$ and for salt, $10x + 20y = 220$. However, no one is saying that Socaccio has to use all available ingredients; the company might choose to use fewer than the total available amounts if this proves more profitable. Thus, $2x + y$ can be anything *up to a total of* 20. In other words,

$$2x + y \leq 20.$$

Similarly,

$$10x + 20y \leq 220.$$

Maximize $p = x + y$ Objective function
subject to $x + 2y \leq 12$
$\left.\begin{array}{l} 2x + y \leq 12 \\ x \geq 0, y \geq 0. \end{array}\right\}$ Constraints

See Example 1 for a method of solving this LP problem (that is, finding an optimal solution and value).

The set of points (x, y) satisfying all the constraints is the **feasible region** for the problem. Our methods of solving LP problems rely on the following facts:

Fundamental Theorem of Linear Programming

- If an LP problem has optimal solutions, then at least one of these solutions occurs at a corner point of the feasible region.
- Linear programming problems with bounded, nonempty feasible regions always have optimal solutions.

Let's see how we can use this to solve an LP problem, and then we'll discuss why it's true.

EXAMPLE 1 Solving an LP Problem

Maximize $p = x + y$
subject to $x + 2y \leq 12$
$2x + y \leq 12$
$x \geq 0, y \geq 0.$

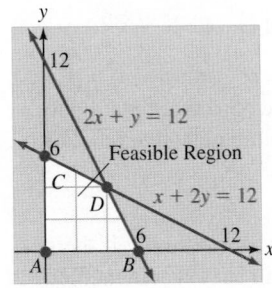

Figure 13

Solution We begin by drawing the feasible region for the problem. We do this using the techniques of Section 5.1, and we get Figure 13.

Each **feasible point** (point in the feasible region) gives an x and a y satisfying the constraints. The question now is, which of these points gives the largest value of the objective function $p = x + y$? The Fundamental Theorem of Linear Programming tells us that the largest value must occur at one (or more) of the corners of the feasible region. In the following table, we list the coordinates of each corner point and we compute the value of the objective function at each corner.

Corner Point	Lines through Point	Coordinates	$p = x + y$
A		$(0, 0)$	0
B		$(6, 0)$	6
C		$(0, 6)$	6
D	$x + 2y = 12$ $2x + y = 12$	$(4, 4)$	8

Now we simply pick the one that gives the largest value for p, which is D. Therefore, the optimal value of p is 8, and an optimal solution is $(4, 4)$.

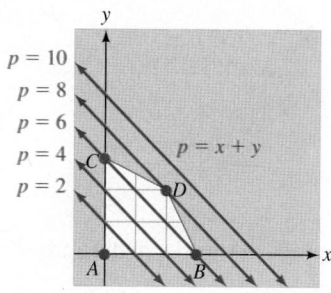

Figure 14

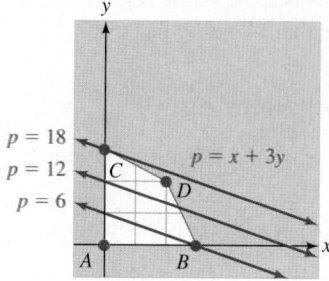

Figure 15

Now we owe you an explanation of why one of the corner points should be an optimal solution. The question is, which point in the feasible region gives the largest possible value of $p = x + y$?

Consider first an easier question: Which points result in a *particular value* of p? For example, which points result in $p = 2$? These would be the points on the line $x + y = 2$, which is the line labeled $p = 2$ in Figure 14.

Now suppose we want to know which points make $p = 4$: These would be the points on the line $x + y = 4$, which is the line labeled $p = 4$ in Figure 14. Notice that this line is parallel to but higher than the line $p = 2$. (If p represented profit in an application, we would call these **isoprofit lines**, or **constant-profit lines**.) Imagine moving this line up or down in the picture. As we move the line down, we see smaller values of p, and as we move it up, we see larger values. Several more of these lines are drawn in Figure 14. Look, in particular, at the line labeled $p = 10$. This line does not meet the feasible region, meaning that no feasible point makes p as large as 10. Starting with the line $p = 2$, as we move the line up, increasing p, there will be a last line that meets the feasible region. In the figure it is clear that this is the line $p = 8$, and this meets the feasible region in only one point, which is the corner point D. Therefore, D gives the greatest value of p of all feasible points.

If we had been asked to maximize some other objective function, such as $p = x + 3y$, then the optimal solution might be different. Figure 15 shows some of the isoprofit lines for this objective function. This time, the last point that is hit as p increases is C, not D. This tells us that the optimal solution is $(0, 6)$, giving the optimal value $p = 18$.

This discussion should convince you that the optimal value in an LP problem will always occur at one of the corner points. By the way, it is possible for the optimal value to occur at *two* corner points and at all points along an edge connecting them. (Do you see why?) We will see this in Example 3(b).

Here is a summary of the method we have just been using.

> ## Graphical Method for Solving Linear Programming Problems in Two Unknowns (Bounded Feasible Regions)
>
> **1.** Graph the feasible region and check that it is bounded.
>
> **2.** Compute the coordinates of the corner points.
>
> **3.** Substitute the coordinates of the corner points into the objective function to see which gives the maximum (or minimum) value of the objective function.
>
> **4.** Any such corner point is an optimal solution.
>
> **Note** If the feasible region is unbounded, this method will work only if there are optimal solutions; otherwise, it will not work. We will show you a method for deciding this on page 328. ∎

APPLICATIONS

EXAMPLE 2 Resource Allocation

Acme Baby Foods mixes two strengths of apple juice. One quart of Beginner's juice is made from 30 fluid ounces of water and 2 fluid ounces of apple juice concentrate. One quart of Advanced juice is made from 20 fluid ounces of water and 12 fluid ounces of concentrate. Every day Acme has available 30,000 fluid ounces of water and 3,600 fluid ounces of concentrate. Acme makes a profit of 20¢ on each quart of Beginner's juice and 30¢ on each quart of Advanced juice. How many quarts of each should Acme make each day to get the largest profit? How would this change if Acme made a profit of 40¢ on Beginner's juice and 20¢ on Advanced juice?

Solution Looking at the question that we are asked, we see that our unknown quantities are

$$x = \text{number of quarts of Beginner's juice made each day}$$
$$y = \text{number of quarts of Advanced juice made each day.}$$

(In this context, x and y are often called the **decision variables**, because we must decide what their values should be in order to get the largest profit.) We can write down the data given in the form of a table (the numbers in the first two columns are amounts per quart of juice):

	Beginner's, x	Advanced, y	Available
Water (ounces)	30	20	30,000
Concentrate (ounces)	2	12	3,600
Profit (¢)	20	30	

Because nothing in the problem says that Acme must use up all the water or concentrate, just that it can use no more than what is available, the first two rows of the table give us two inequalities:

$$30x + 20y \leq 30{,}000$$
$$2x + 12y \leq 3{,}600.$$

Dividing the first inequality by 10 and the second by 2 gives

$$3x + 2y \leq 3{,}000$$
$$x + 6y \leq 1{,}800.$$

We also have that $x \geq 0$ and $y \geq 0$ because Acme can't make a negative amount of juice. To finish setting up the problem, we are asked to maximize the profit, which is

$$p = 20x + 30y. \qquad \text{Expressed in ¢}$$

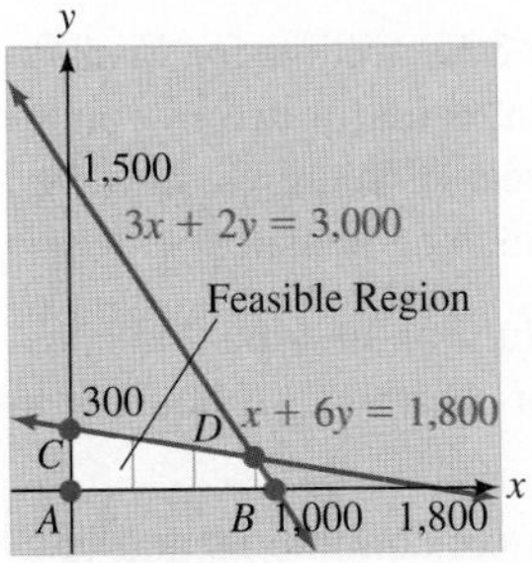

This gives us our LP problem:

$$\begin{aligned} \text{Maximize} \quad & p = 20x + 30y \\ \text{subject to} \quad & 3x + 2y \leq 3{,}000 \\ & x + 6y \leq 1{,}800 \\ & x \geq 0, y \geq 0. \end{aligned}$$

The (bounded) feasible region is shown in Figure 16.

The corners and the values of the objective function are listed in the following table:

Figure 16

Point	Lines through Point	Coordinates	p = 20x + 30y
A		(0, 0)	0
B		(1,000, 0)	20,000
C		(0, 300)	9,000
D	$3x + 2y = 3{,}000$ $x + 6y = 1{,}800$	(900, 150)	22,500

We are seeking to maximize the objective function p, so we look for corner points that give the maximum value for p. Because the maximum occurs at the point D, we conclude that the (only) optimal solution occurs at D. Thus, the company should make 900 quarts of Beginner's juice and 150 quarts of Advanced juice, for a largest possible profit of 22,500¢, or $225.

If, instead, the company made a profit of 40¢ on each quart of Beginner's juice and 20¢ on each quart of Advanced juice, then we would have $p = 40x + 20y$. This gives the following table:

Point	Lines through Point	Coordinates	$p = 40x + 20y$
A		$(0, 0)$	0
B		$(1,000, 0)$	40,000
C		$(0, 300)$	6,000
D	$3x + 2y = 3,000$ $x + 6y = 1,800$	$(900, 150)$	39,000

We can see that, in this case, Acme should make 1,000 quarts of Beginner's juice and no Advanced juice, for a largest possible profit of 40,000¢, or $400.

➡ **Before we go on...** Notice that, in the first version of the problem in Example 2, the company used all the water and juice concentrate:

Water: $30(900) + 20(150) = 30,000$
Concentrate: $2(900) + 12(150) = 3,600.$

In the second version, it used all the water but not all the concentrate:

Water: $30(100) + 20(0) = 30,000$
Concentrate: $2(100) + 12(0) = 200 < 3,600.$ ∎

EXAMPLE 3 Investments

The Solid Trust Savings & Loan Company has set aside $25 million for loans to home buyers. Its policy is to allocate at least $10 million annually for luxury condominiums. A government housing development grant it receives requires, however, that at least one third of its total loans be allocated to low-income housing.

a. Solid Trust's return on condominiums is 12% and its return on low-income housing is 10%. How much should the company allocate for each type of housing to maximize its total return?

b. Redo part (a), assuming that the return is 12% on both condominiums and low-income housing.

Solution

a. We first identify the unknowns: Let x be the annual amount (in millions of dollars) allocated to luxury condominiums and let y be the annual amount allocated to low-income housing.

We now look at the constraints. The first constraint is mentioned in the first sentence: The total the company can invest is $25 million. Thus,

$$x + y \le 25.$$

(The company is not required to invest all of the $25 million; rather, it can invest *up to* $25 million.) Next, the company has allocated at least $10 million to condos. Rephrasing this in terms of the unknowns, we get

The amount allocated to condos is at least $10 *million.*

The phrase "is at least" means $\geq$. Thus, we obtain a second constraint:

$$x \geq 10.$$

The third constraint is that at least one third of the total financing must be for low-income housing. Rephrasing this, we say:

The amount allocated to low-income housing is at least one third of the total.

Because the total investment will be $x + y$, we get

$$y \geq \frac{1}{3}(x + y).$$

We put this in the standard form of a linear inequality as follows:

$$3y \geq x + y \qquad \text{Multiply both sides by 3.}$$
$$-x + 2y \geq 0. \qquad \text{Subtract } x + y \text{ from both sides.}$$

There are no further constraints.

Now, what about the return on these investments? According to the data, the annual return is given by

$$p = 0.12x + 0.10y.$$

We want to make this quantity p as large as possible. In other words, we want to

$$\begin{aligned} \text{Maximize} \quad & p = 0.12x + 0.10y \\ \text{subject to} \quad & x + y \leq 25 \\ & x \geq 10 \\ & -x + 2y \geq 0 \\ & x \geq 0, y \geq 0. \end{aligned}$$

(Do you see why the inequalities $x \geq 0$ and $y \geq 0$ are slipped in here?) The feasible region is shown in Figure 17.

We now make a table that gives the return on investment at each corner point:

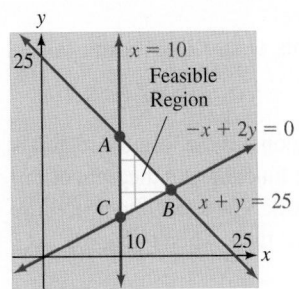

Figure 17

Point	Lines through Point	Coordinates	$p = 0.12x + 0.10y$
A	$x = 10$ $x + y = 25$	(10, 15)	2.7
B	$x + y = 25$ $-x + 2y = 0$	(50/3, 25/3)	2.833
C	$x = 10$ $-x + 2y = 0$	(10, 5)	1.7

From the table, we see that the values of x and y that maximize the return are $x = 50/3$ and $y = 25/3$, which give a total return of $2.833 million. In other words, the most profitable course of action is to invest $16.667 million in loans for condominiums and $8.333 million in loans for low-income housing, giving a maximum annual return of $2.833 million.

b. The LP problem is the same as for part (a) except for the objective function:

$$\text{Maximize} \quad p = 0.12x + 0.12y$$
subject to
$$x + y \leq 25$$
$$x \geq 10$$
$$-x + 2y \geq 0$$
$$x \geq 0, y \geq 0.$$

Here are the values of p at the three corners:

Point	Coordinates	$p = 0.12x + 0.12y$
A	(10, 15)	3
B	(50/3, 25/3)	3
C	(10, 5)	1.8

Looking at the table, we see that a curious thing has happened: We get the same maximum annual return at both A and B. Thus, we could choose either option to maximize the annual return. In fact, any point along the line segment AB will yield an annual return of $3 million. For example, the point (12, 13) lies on the line segment AB and also yields an annual revenue of $3 million. This happens because the "isoreturn" lines are parallel to that edge.

➡ **Before we go on...** What breakdowns of investments would lead to the *lowest* return for parts (a) and (b)? ■

The preceding examples all had bounded feasible regions. If the feasible region is unbounded, then, *provided there are optimal solutions,* the fundamental theorem of linear programming guarantees that the above method will work. The following procedure determines whether or not optimal solutions exist and finds them when they do.

Solving Linear Programming Problems in Two Unknowns (Unbounded Feasible Regions)

If the feasible region of an LP problem is unbounded, proceed as follows:

1. Draw a rectangle large enough so that all the corner points are inside the rectangle (and not on its boundary):

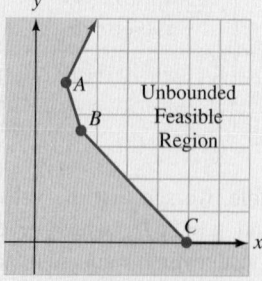

Corner points: A, B, C

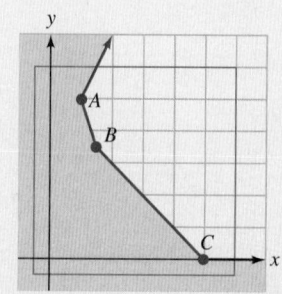

Corner points inside the rectangle

2. Shade the outside of the rectangle so as to define a new bounded feasible region, and locate the new corner points:

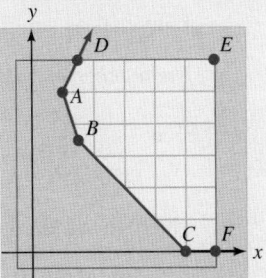

New corner points: D, E, and F

3. Obtain the optimal solutions using this bounded feasible region.
4. If any optimal solutions occur at one of the original corner points (A, B, and C in the figure), then the LP problem has that corner point as an optimal solution. Otherwise, the LP problem has no optimal solutions. When the latter occurs, we say that the **objective function is unbounded**, because it can assume arbitrarily large (positive or negative) values.

In the next two examples, we work with unbounded feasible regions.

EXAMPLE 4 **Cost**

You are the manager of a small store that specializes in hats, sunglasses, and other accessories. You are considering a sales promotion of a new line of hats and sunglasses. You will offer the sunglasses only to those who purchase two or more hats, so you will sell at least twice as many hats as pairs of sunglasses. Moreover, your supplier tells you that, due to seasonal demand, your order of sunglasses cannot exceed 100 pairs. To ensure that the sale items fill out the large display you have set aside, you estimate that you should order at least 210 items in all.

a. Assume that you will lose $3 on every hat and $2 on every pair of sunglasses sold. Given the constraints above, how many hats and pairs of sunglasses should you order to lose the least amount of money in the sales promotion?

b. Suppose instead that you lose $1 on every hat sold but make a profit of $5 on every pair of sunglasses sold. How many hats and pairs of sunglasses should you order to make the largest profit in the sales promotion?

c. Now suppose that you make a profit of $1 on every hat sold but lose $5 on every pair of sunglasses sold. How many hats and pairs of sunglasses should you order to make the largest profit in the sales promotion?

Solution

a. The unknowns are:

x = number of hats you order
y = number of pairs of sunglasses you order.

The objective is to minimize the total loss:

$$c = 3x + 2y.$$

Now for the constraints. The requirement that you will sell at least twice as many hats as sunglasses can be rephrased as:

The number of hats is at least twice the number of pairs of sunglasses,

or

$$x \geq 2y$$

which, in standard form, is

$$x - 2y \geq 0.$$

Next, your order of sunglasses cannot exceed 100 pairs, so

$$y \leq 100.$$

Finally, you would like to sell at least 210 items in all, giving

$$x + y \geq 210.$$

Thus, the LP problem is the following:

Minimize $c = 3x + 2y$
subject to $x - 2y \geq 0$
$y \leq 100$
$x + y \geq 210$
$x \geq 0, y \geq 0.$

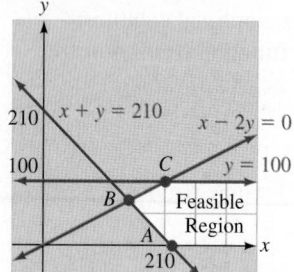

Figure 18

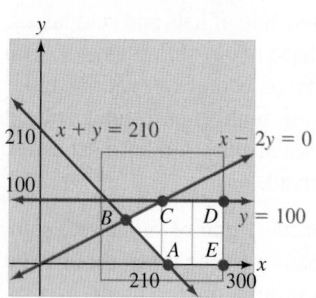

Figure 19

The feasible region is shown in Figure 18. This region is unbounded, so there is no guarantee that there are any optimal solutions. Following the procedure described above, we enclose the corner points in a rectangle as shown in Figure 19. (There are many infinitely many possible rectangles we could have used. We chose one that gives convenient coordinates for the new corners.)

We now list all the corners of this bounded region along with the corresponding values of the objective function c:

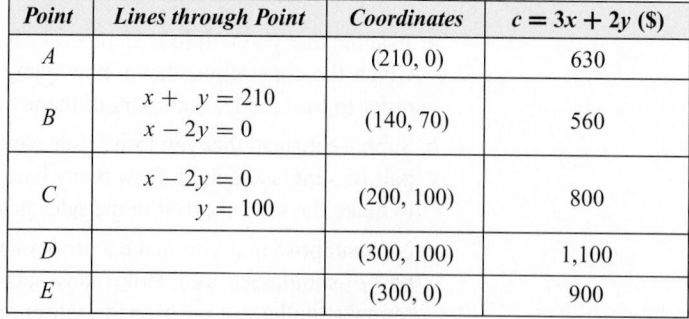

Point	Lines through Point	Coordinates	$c = 3x + 2y$ ($)
A		(210, 0)	630
B	$x + y = 210$ $x - 2y = 0$	(140, 70)	560
C	$x - 2y = 0$ $y = 100$	(200, 100)	800
D		(300, 100)	1,100
E		(300, 0)	900

The corner point that gives the minimum value of the objective function c is B. Because B is one of the corner points of the original feasible region, we conclude that our linear programming problem has an optimal solution at B. Thus, the combination that gives the smallest loss is 140 hats and 70 pairs of sunglasses.

b. The LP problem is the following:

$$\text{Maximize} \quad p = -x + 5y$$
$$\text{subject to} \quad x - 2y \geq 0$$
$$y \leq 100$$
$$x + y \geq 210$$
$$x \geq 0, y \geq 0.$$

Because most of the work is already done for us in part (a), all we need to do is change the objective function in the table that lists the corner points:

Point	Lines through Point	Coordinates	$p = -x + 5y$ ($)
A		(210, 0)	−210
B	$x + y = 210$ $x - 2y = 0$	(140, 70)	210
C	$x - 2y = 0$ $y = 100$	(200, 100)	300
D		(300, 100)	200
E		(300, 0)	−300

The corner point that gives the maximum value of the objective function p is C. Because C is one of the corner points of the original feasible region, we conclude that our LP problem has an optimal solution at C. Thus, the combination that gives the largest profit ($300) is 200 hats and 100 pairs of sunglasses.

c. The objective function is now $p = x - 5y$, which is the negative of the objective function used in part (b). Thus, the table of values of p is the same as in part (b), except that it has opposite signs in the p column. This time we find that the maximum value of p occurs at E. However, E is not a corner point of the original feasible region, so the LP problem has no optimal solution. Referring to Figure 18, we can make the objective p as large as we like by choosing a point far to the right in the unbounded feasible region. Thus, the objective function is unbounded; that is, it is possible to make an arbitrarily large profit.

EXAMPLE 5 Resource Allocation

You are composing a very avant-garde ballade for violins and bassoons. In your ballade, each violinist plays a total of two notes and each bassoonist only one note. To make your ballade long enough, you decide that it should contain at least 200 instrumental notes. Furthermore, after playing the requisite two notes, each violinist will sing one soprano note, while each bassoonist will sing three soprano notes.[*] To make the ballade sufficiently interesting, you have decided on a minimum of 300 soprano notes. To give your composition a sense of balance, you wish to have no more than three times as many bassoonists as violinists. Violinists charge $200 per performance and bassoonists $400 per performance. How many of each should your ballade call for in order to minimize personnel costs?

[*] Whether or not these musicians are capable of singing decent soprano notes will be left to chance. You reason that a few bad notes will add character to the ballade.

Solution   First, the unknowns are $x =$ number of violinists and $y =$ number of bassoonists. The constraint on the number of instrumental notes implies that

$$2x + y \geq 200$$

because the total number is to be *at least* 200. Similarly, the constraint on the number of soprano notes is

$$x + 3y \geq 300.$$

The next one is a little tricky. As usual, we reword it in terms of the quantities x and y.

The number of bassoonists should be no more than three times the number of violinists.

Thus, $y \leq 3x$

or $3x - y \geq 0.$

Finally, the total cost per performance will be

$$c = 200x + 400y.$$

We wish to minimize total cost. So, our linear programming problem is as follows:

$$\begin{aligned}
\text{Minimize} \quad & c = 200x + 400y \\
\text{subject to} \quad & 2x + y \geq 200 \\
& x + 3y \geq 300 \\
& 3x - y \geq 0 \\
& x \geq 0, y \geq 0.
\end{aligned}$$

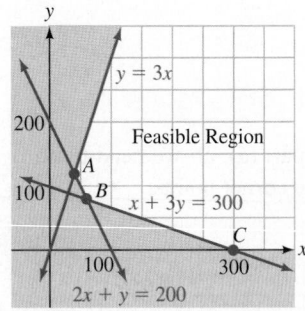

Figure 20

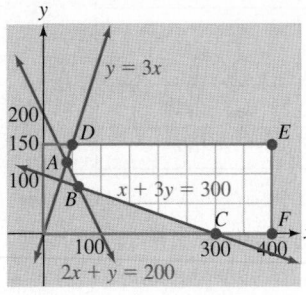

Figure 21

We get the feasible region shown in Figure 20.✻ The feasible region is unbounded, and so we add a convenient rectangle as before (Figure 21).

Point	Lines through Point	Coordinates	$c = 200x + 400y$
A	$2x + y = 200$ $3x - y = 0$	(40, 120)	56,000
B	$2x + y = 200$ $x + 3y = 300$	(60, 80)	44,000
C		(300, 0)	60,000
D	$3x - y = 0$ $y = 150$	(50, 150)	70,000
E		(400, 150)	140,000
F		(400, 0)	80,000

From the table we see that the minimum cost occurs at B, a corner point of the original feasible region. The linear programming problem thus has an optimal solution, and the minimum cost is \$44,000 per performance, employing 60 violinists and 80 bassoonists. (Quite a wasteful ballade, one might say.)

these stocks, and would like to earn at least $2,400 in dividends over the course of a year. (Assume the dividend to be unchanged for the year.) How many shares of each stock should you purchase to meet your requirements and minimize the total risk index for your portfolio? What is the minimum total risk index?

39. ▼ *Planning* My friends: I, the mighty Brutus, have decided to prepare for retirement by instructing young warriors in the arts of battle and diplomacy. For each hour spent in battle instruction, I have decided to charge 50 ducats. For each hour in diplomacy instruction I shall charge 40 ducats. Due to my advancing years, I can spend no more than 50 hours per week instructing the youths, although the great Jove knows that they are sorely in need of instruction! Due to my fondness for physical pursuits, I have decided to spend no more than one third of the total time in diplomatic instruction. However, the present border crisis with the Gauls is a sore indication of our poor abilities as diplomats. As a result, I have decided to spend at least 10 hours per week instructing in diplomacy. Finally, to complicate things further, there is the matter of Scarlet Brew: I have estimated that each hour of battle instruction will require 10 gallons of Scarlet Brew to quench my students' thirst, and that each hour of diplomacy instruction, being less physically demanding, requires half that amount. Because my harvest of red berries has far exceeded my expectations, I estimate that I'll have to use at least 400 gallons per week in order to avoid storing the fine brew at great expense. Given all these restrictions, how many hours per week should I spend in each type of instruction to maximize my income?

40. ▼ *Planning* Repeat the preceding exercise with the following changes: I would like to spend no more than half the total time in diplomatic instruction, and I must use at least 600 gallons of Scarlet Brew.

41. ▼ *Resource Allocation* One day, Gillian the magician summoned the wisest of her women. "Devoted followers," she began, "I have a quandary: As you well know, I possess great expertise in sleep spells and shock spells, but unfortunately, these are proving to be a drain on my aural energy resources; each sleep spell costs me 500 pico-shirleys of aural energy, while each shock spell requires 750 pico-shirleys. Clearly, I would like to hold my overall expenditure of aural energy to a minimum, and still meet my commitments in protecting the Sisterhood from the ever-present threat of trolls. Specifically, I have estimated that each sleep spell keeps us safe for an average of two minutes, while every shock spell protects us for about three minutes. We certainly require enough protection to last 24 hours of each day, and possibly more, just to be safe. At the same time, I have noticed that each of my sleep spells can immobilize three trolls at once, while one of my typical shock spells (having a narrower range) can immobilize only two trolls at once. We are faced, my sisters, with an onslaught of 1,200 trolls per day! Finally, as you are no doubt aware, the bylaws dictate that for a Magician of the Order to remain in good standing, the number of shock spells must be between one quarter and one third the number of shock and sleep spells combined. What do I do, oh Wise Ones?"

42. ▼ *Risk Management* The Grand Vizier of the Kingdom of Um is being blackmailed by numerous individuals and is having a very difficult time keeping his blackmailers from going public. He has been keeping them at bay with two kinds of payoff: gold bars from the Royal Treasury and political favors. Through bitter experience, he has learned that each payoff in gold gives him peace for an average of about 1 month, while each political favor seems to earn him about a month and a half of reprieve. To maintain his flawless reputation in the Court, he feels he cannot afford any revelations about his tainted past to come to light within the next year. Thus it is imperative that his blackmailers be kept at bay for 12 months. Furthermore, he would like to keep the number of gold payoffs at no more than one quarter of the combined number of payoffs because the outward flow of gold bars might arouse suspicion on the part of the Royal Treasurer. The Grand Vizier feels that he can do no more than seven political favors per year without arousing undue suspicion in the Court. The gold payoffs tend to deplete his travel budget. (The treasury has been subsidizing his numerous trips to the Himalayas.) He estimates that each gold bar removed from the treasury will cost him four trips. On the other hand, because the administering of political favors tends to cost him valuable travel time, he suspects that each political favor will cost him about two trips. Now, he would obviously like to keep his blackmailers silenced and lose as few trips as possible. What is he to do? How many trips will he lose in the next year?

43. ◆ *Management*[21] You are the service manager for a supplier of closed-circuit television systems. Your company can provide up to 160 hours per week of technical service for your customers, although the demand for technical service far exceeds this amount. As a result, you have been asked to develop a model to allocate service technicians' time between new customers (those still covered by service contracts) and old customers (whose service contracts have expired). To ensure that new customers are satisfied with your company's service, the sales department has instituted a policy that at least 100 hours per week be allocated to servicing new customers. At the same time, your superiors have informed you that the company expects your department to generate at least $1,200 per week in revenues. Technical service time for new customers generates an average of $10 per hour (because much of the service is still under warranty) and for old customers generates $30 per hour. How many hours per week should you allocate to each type of customer to generate the most revenue?

---

[21]Loosely based on a similiar problem in *An Introduction to Management Science* (6th Ed.) by D. R. Anderson, D. J. Sweeney, and T. A. Williams (West, 1991).

44. ◆ *Scheduling*[22] The *Scottsville Textile Mill* produces several different fabrics on eight dobby looms which operate 24 hours per day and are scheduled for 30 days in the coming month. The Scottsville Textile Mill will produce only Fabric 1 and Fabric 2 during the coming month. Each dobby loom can turn out 4.63 yards of either fabric per hour. Assume that there is a monthly demand of 16,000 yards of Fabric 1 and 12,000 yards of Fabric 2. Profits are calculated as 33¢ per yard for each fabric produced on the dobby looms.

a. Will it be possible to satisfy total demand?

b. In the event that total demand is not satisfied, the Scottsville Textile Mill will need to purchase the fabrics from another mill to make up the shortfall. Its profits on resold fabrics ordered from another mill amount to 20¢ per yard for Fabric 1 and 16¢ per yard for Fabric 2. How many yards of each fabric should it produce to maximize profits?

COMMUNICATION AND REASONING EXERCISES

45. If a linear programming problem has a bounded, nonempty feasible region, then optimal solutions

(A) must exist   (B) may or may not exist

(C) cannot exist

46. If a linear programming problem has an unbounded, nonempty feasible region, then optimal solutions

(A) must exist   (B) may or may not exist

(C) cannot exist

47. What can you say if the optimal value occurs at two adjacent corner points?

48. Describe at least one drawback to using the graphical method to solve a linear programming problem arising from a real-life situation.

49. Create a linear programming problem in two variables that has no optimal solution.

[22]Adapted from *The Calhoun Textile Mill Case* by J. D. Camm, P. M. Dearing, and S. K. Tadisina as presented for case study in *An Introduction to Management Science* (6th Ed.) by D. R. Anderson, D. J. Sweeney, and T. A. Williams (West, 1991). Our exercise uses a subset of the data given in the cited study.

50. Create a linear programming problem in two variables that has more than one optimal solution.

51. Create an interesting scenario leading to the following linear programming problem:

Maximize $p = 10x + 10y$

subject to $20x + 40y \leq 1{,}000$

           $30x + 20y \leq 1{,}200$

           $x \geq 0, y \geq 0.$

52. Create an interesting scenario leading to the following linear programming problem:

Minimize $c = 10x + 10y$

subject to $20x + 40y \geq 1{,}000$

           $30x + 20y \geq 1{,}200$

           $x \geq 0, y \geq 0.$

53. ▼ Use an example to show why there may be no optimal solution to a linear programming problem if the feasible region is unbounded.

54. ▼ Use an example to illustrate why, in the event that an optimal solution does occur despite an unbounded feasible region, that solution corresponds to a corner point of the feasible region.

55. ▼ You are setting up an LP problem for Fly-by-Night Airlines with the unknowns x and y, where x represents the number of first-class tickets it should issue for a specific flight and y represents the number of business-class tickets it should issue for that flight, and the problem is to maximize profit. You find that there are two different corner points that maximize the profit. How do you interpret this?

56. ▼ In the situation described in the preceding exercise, you find that there are no optimal solutions. How do you interpret this?

57. ◆ Consider the following example of a *nonlinear* programming problem: Maximize $p = xy$ subject to $x \geq 0$, $y \geq 0$, $x + y \leq 2$. Show that p is zero on every corner point, but is greater than zero at many non-corner points.

58. ◆ Solve the nonlinear programming problem in Exercise 57.

5.3 The Simplex Method: Solving Standard Maximization Problems

The method discussed in Section 5.2 works quite well for LP problems in two unknowns, but what about three or more unknowns? Because we need an axis for each unknown, we would need to draw graphs in three dimensions (where we have x-, y-, and z-coordinates) to deal with problems in three unknowns, and we would have to draw in hyperspace to answer questions involving four or more unknowns. Given the state of technology as this book is being written, we can't easily do this. So we need another method for solving LP problems that will work for any number of unknowns. One such method, called the **simplex method**, has been the method of choice since it was invented by George Dantzig in 1947. (See the Introduction to this chapter for more about Dantzig.) To illustrate it best, we first use it to solve only so-called standard maximization problems.

General Linear Programming Problem

A **linear programming problem in n unknowns** $x_1, x_2, \ldots, x_n$ is one in which we are to find the maximum or minimum value of a linear **objective function**

$$a_1x_1 + a_2x_2 + \cdots + a_nx_n,$$

where $a_1, a_2, \ldots, a_n$ are numbers, subject to a number of linear **constraints** of the form

$$b_1x_1 + b_2x_2 + \cdots + b_nx_n \leq c \quad \text{or} \quad b_1x_1 + b_2x_2 + \cdots + b_nx_n \geq c,$$

where $b_1, b_2, \ldots, b_n, c$ are numbers.

Standard Maximization Problem

A **standard maximization problem** is an LP problem in which we are required to *maximize* (not minimize) an objective function of the form

$$p = a_1x_1 + a_2x_2 + \cdots + a_nx_n$$

subject to the constraints

$$x_1 \geq 0, x_2 \geq 0, \ldots, x_n \geq 0$$

and further constraints of the form

$$b_1x_1 + b_2x_2 + \cdots + b_nx_n \leq c$$

with c *nonnegative*. It is important that the inequality here be $\leq$, *not* $=$ or $\geq$.

Note As in the chapter on linear equations, we will almost always use $x, y, z, \ldots$ for the unknowns. Subscripted variables $x_1, x_2, \ldots$ are very useful names when you start running out of letters of the alphabet, but we should not find ourselves in that predicament. ■

Quick Examples

1. Maximize $p = 2x - 3y + 3z$
    subject to $2x \quad\quad + z \leq 7$

$$-x + 3y - 6z \leq 6$$
$$x \geq 0, y \geq 0, z \geq 0.$$

This is a standard maximization problem.

2. Maximize $p = 2x_1 + x_2 - x_3 + x_4$
    subject to $x_1 - 2x_2 \quad\quad + x_4 \leq 0$

$$3x_1 \quad\quad\quad\quad \leq 1$$
$$x_2 + x_3 \quad \leq 2$$
$$x_1 \geq 0, x_2 \geq 0, x_3 \geq 0, x_4 \geq 0.$$

This is a standard maximization problem.

3. Maximize $p = 2x - 3y + 3z$
  subject to $2x \quad + \; z \geq 7$
      $-x + 3y - 6z \leq 6$ This is *not* a standard maximization problem.
      $x \geq 0, y \geq 0, z \geq 0.$

The inequality $2x + z \geq 7$ cannot be written in the required form. If we reverse the inequality by multiplying both sides by -1, we get $-2x - z \leq -7$, but a negative value on the right side is not allowed.

The idea behind the simplex method is this: In any linear programming problem, there is a feasible region. If there are only two unknowns, we can draw the region; if there are three unknowns, it is a solid region in space; and if there are four or more unknowns, it is an abstract higher-dimensional region. But it is a faceted region with corners (think of a diamond), and it is at one of these corners that we will find the optimal solution. Geometrically, what the simplex method does is to start at the corner where all the unknowns are 0 (possible because we are talking of standard maximization problems) and then walk around the region, from corner to adjacent corner, always increasing the value of the objective function, until the best corner is found. In practice, we will visit only a small number of the corners before finding the right one. Algebraically, as we are about to see, this walking around is accomplished by matrix manipulations of the same sort as those used in the chapter on systems of linear equations.

We describe the method while working through an example.

EXAMPLE 1   **Meet the Simplex Method**

Maximize $p = 3x + 2y + z$
subject to $2x + 2y + \; z \leq 10$
      $x + 2y + 3z \leq 15$
      $x \geq 0, \; y \geq 0, \; z \geq 0.$

Solution

Step 1   *Convert to a system of linear equations.* The inequalities $2x + 2y + z \leq 10$ and $x + 2y + 3z \leq 15$ are less convenient than equations. Look at the first inequality. It says that the left-hand side, $2x + 2y + z$, must have some positive number (or zero) *added to it* if it is to equal 10. Because we don't yet know what x, y, and z are, we are not yet sure what number to add to the left-hand side. So we invent a new unknown, $s \geq 0$, called a **slack variable**, to "take up the slack," so that

$$2x + 2y + z + s = 10.$$

Turning to the next inequality, $x + 2y + 3z \leq 15$, we now add a slack variable to its left-hand side, to get it up to the value of the right-hand side. We might have to add a different number than we did the last time, so we use a new slack variable, $t \geq 0$, and obtain

$$x + 2y + 3z + t = 15.$$ Use a different slack variable for each constraint.

Now we write the system of equations we have (including the one that defines the objective function) in standard form.

$$
\begin{aligned}
2x + 2y + \ z + s\ \ \ \ \ \ \ \ \ \ \ &= 10 \\
x + 2y + 3z\ \ \ \ \ + t\ \ \ \ &= 15 \\
-3x - 2y - \ z\ \ \ \ \ \ \ \ \ + p &= 0
\end{aligned}
$$

Note three things: First, all the variables are neatly aligned in columns, as they were in Chapter 3. Second, in rewriting the objective function $p = 3x + 2y + z$, we have left the coefficient of p as $+1$ and brought the other variables over to the same side of the equation as p. This will be our standard procedure from now on. *Don't* write $3x + 2y + z - p = 0$ (even though it means the same thing) because the negative coefficients will be important in later steps. Third, the above system of equations has fewer equations than unknowns, and hence cannot have a unique solution.

Step 2 *Set up the initial tableau.* We represent our system of equations by the following table (which is simply the augmented matrix in disguise), called **the initial tableau**:

	x	y	z	s	t	p	
	2	2	1	1	0	0	10
	1	2	3	0	1	0	15
	-3	-2	-1	0	0	1	0

The labels along the top keep track of which columns belong to which variables.

Now notice a peculiar thing. If we rewrite the matrix using the variables s, t, and p first, we get the matrix

$$
\begin{array}{cccccc}
s & t & p & x & y & z \\
\end{array}
$$
$$
\left[\begin{array}{cccccc|c}
1 & 0 & 0 & 2 & 2 & 1 & 10 \\
0 & 1 & 0 & 1 & 2 & 3 & 15 \\
0 & 0 & 1 & -3 & -2 & -1 & 0
\end{array}\right],
\qquad \text{Matrix with } s, t, \text{ and } p \text{ columns first}
$$

which is already in reduced form. We can therefore read off the general solution (see Section 3.2) to our system of equations as

$$
\begin{aligned}
s &= 10 - 2x - 2y - z \\
t &= 15 - \ x - 2y - 3z \\
p &= 0 + \ 3x + 2y + z \\
\end{aligned}
$$
$$
x, y, z \text{ arbitrary.}
$$

Thus, we get a whole family of solutions, one for each choice of x, y, and z. One possible choice is to set x, y, and z all equal to 0. This gives the particular solution

$$
s = 10, t = 15, p = 0, x = 0, y = 0, z = 0. \qquad \text{Set } x = y = z = 0 \text{ above.}
$$

This solution is called the **basic solution** associated with the tableau. The variables s and t are called the **active** variables, and x, y, and z are the **inactive** variables. (Other terms used are **basic** and **nonbasic** variables.)

We can obtain the basic solution directly from the tableau as follows.

- The active variables correspond to the cleared columns (columns with only one nonzero entry).

- The values of the active variables are calculated as shown below.
- All other variables are inactive, and set equal to zero.

	Inactive $x=0$	Inactive $y=0$	Inactive $z=0$	Active $s=\frac{10}{1}$	Active $t=\frac{15}{1}$	Active $p=\frac{0}{1}$	
	x	y	z	s	t	p	
	2	2	1	1	0	0	10
	1	2	3	0	1	0	15
	-3	-2	-1	0	0	1	0

As an additional aid to recognizing which variables are active and which are inactive, we label each row with the name of the corresponding active variable. Thus, the complete initial tableau looks like this.

	x	y	z	s	t	p	
s	2	2	1	1	0	0	10
t	1	2	3	0	1	0	15
p	-3	-2	-1	0	0	1	0

This basic solution represents our starting position $x = y = z = 0$ in the feasible region in xyz space.

We now need to move to another corner point. To do so, we choose a pivot* in one of the first three columns of the tableau and clear its column. Then we will get a different basic solution, which corresponds to another corner point. Thus, in order to move from corner point to corner point, all we have to do is choose suitable pivots and clear columns in the usual manner.

The next two steps give the procedure for choosing the pivot.

Step 3 *Select the pivot column* (the column that contains the pivot we are seeking).

✳ Also see Section 3.2 for a discussion of pivots and pivoting.

Selecting the Pivot Column

Choose the negative number with the largest magnitude on the left-hand side of the bottom row (that is, don't consider the last number in the bottom row). Its column is the pivot column. (If there are two or more candidates, choose any one.) If all the numbers on the left-hand side of the bottom row are zero or positive, then we are done, and the basic solution is the optimal solution.

Simple enough. The most negative number in the bottom row is -3, so we choose the x column as the pivot column:

	x	y	z	s	t	p	
s	**2**	2	1	1	0	0	10
t	**1**	2	3	0	1	0	15
p	**−3**	-2	-1	0	0	1	0

↑
Pivot column

Q: *Why choose the pivot column this way?*

A: The variable labeling the pivot column is going to be increased from 0 to something positive. In the equation $p = 3x + 2y + z$, the fastest way to increase p is to increase x because p would increase by 3 units for every 1-unit increase in x.

(If we chose to increase y, then p would increase by only 2 units for every 1-unit increase in y, and if we increased z instead, p would grow even more slowly.) In short, choosing the pivot column this way makes it likely that we'll increase p as much as possible.

Step 4 *Select the pivot in the pivot column.*

Selecting the Pivot

1. The pivot must always be a positive number. (This rules out zeros and negative numbers, such as the -3 in the bottom row.)

2. For each positive entry b in the pivot column, compute the ratio a/b, where a is the number in the rightmost column in that row. We call this a **test ratio**.

3. Of these ratios, choose the smallest one. (If there are two or more candidates, choose any one.) The corresponding number b is the pivot.

In our example, the test ratio in the first row is $10/2 = 5$, and the test ratio in the second row is $15/1 = 15$. Here, 5 is the smallest, so the 2 in the upper left is our pivot.

	x	y	z	s	t	p		Test Ratios
s	$\boxed{2}$	2	1	1	0	0	10	$10/2 = 5$
t	1	2	3	0	1	0	15	$15/1 = 15$
p	-3	-2	-1	0	0	1	0	

Q: *Why select the pivot this way?*

A: The rule given above guarantees that, after pivoting, all variables will be nonnegative in the basic solution. In other words, it guarantees that we will remain in the feasible region. We will explain further after finishing this example.

Step 5 *Use the pivot to clear the column in the normal manner and then relabel the pivot row with the label from the pivot column.* It is important to follow the exact prescription described in Section 3.2 for formulating the row operations:

$$a R_c \pm b R_p. \qquad \text{\textit{a} and \textit{b} both positive}$$

$$\underset{\text{Row to change}}{\uparrow} \qquad \underset{\text{Pivot row}}{\uparrow}$$

All entries in the last column should remain nonnegative after pivoting. Furthermore, because the x column (and no longer the s column) will be cleared, x will become an active variable. In other words, the s on the left of the pivot will be replaced by x. We call s the **departing**, or **exiting variable** and x the **entering variable** for this step.

Entering variable
$\downarrow$

		x	y	z	s	t	p		
Departing variable $\rightarrow$	s	$\boxed{2}$	2	1	1	0	0	10	
	t	1	2	3	0	1	0	15	$2R_2 - R_1$
	p	-3	-2	-1	0	0	1	0	$2R_3 + 3R_1$

This gives

	x	y	z	s	t	p	
x	2	2	1	1	0	0	10
t	0	2	5	−1	2	0	20
p	0	2	1	3	0	2	30

This is the second tableau.

Step 6 *Go to Step 3.* But wait! According to Step 3, we are finished because there are no negative numbers in the bottom row. Thus, we can read off the answer. Remember, though, that the solution for x, the first active variable, is not just $x = 10$, but is $x = 10/2 = 5$ because the pivot has not been reduced to a 1. Similarly, $t = 20/2 = 10$ and $p = 30/2 = 15$. All the other variables are zero because they are inactive. Thus, the solution is as follows: p has a maximum value of 15, and this occurs when $x = 5$, $y = 0$, and $z = 0$. (The slack variables then have the values $s = 0$ and $t = 10$.)

Q: *Why can we stop when there are no negative numbers in the bottom row? Why does this tableau give an optimal solution?*

A: The bottom row corresponds to the equation $2y + z + 3s + 2p = 30$, or

$$p = 15 - y - \frac{1}{2}z - \frac{3}{2}s.$$

Think of this as part of the general solution to our original system of equations, with y, z, and s as the parameters. Because these variables must be nonnegative, *the largest possible value of p in any feasible solution of the system comes when all three of the parameters are 0.* Thus, the current basic solution must be an optimal solution.*

---

* Calculators or spreadsheets could obviously be a big help in the calculations here, just as in Chapter 3. We'll say more about that after the next couple of examples.

We owe some further explanation for Step 4 of the simplex method. After Step 3, we knew that x would be the entering variable, and we needed to choose the departing variable. In the next basic solution, x was to have some positive value and we wanted this value to be as large as possible (to make p as large as possible) without making any other variables negative. Look again at the equations written in Step 2:

$$s = 10 - 2x - 2y - z$$
$$t = 15 - x - 2y - 3z.$$

We needed to make either s or t into an inactive variable and hence zero. Also, y and z were to remain inactive. If we had made s inactive, then we would have had $0 = 10 - 2x$, so $x = 10/2 = 5$. This would have made $t = 15 - 5 = 10$, which would be fine. On the other hand, if we had made t inactive, then we would have had $0 = 15 - x$, so $x = 15$, and this would have made $s = 10 - 2 \cdot 15 = -20$, which would *not* be fine, because slack variables must be nonnegative. In other words, we had a choice of making $x = 10/2 = 5$ or $x = 15/1 = 15$, but making x larger than 5 would have made another variable negative. We were thus compelled to choose the smaller ratio, 5, and make s the departing variable. Of course, we do not have to think it through this way every time. We just use the rule stated in Step 4. (For a graphical explanation, see Example 3.)

EXAMPLE 2 Simplex Method

Find the maximum value of $p = 12x + 15y + 5z$, subject to the constraints:

$$2x + 2y + z \le 8$$
$$x + 4y - 3z \le 12$$
$$x \ge 0, \ y \ge 0, \ z \ge 0.$$

Solution Following Step 1, we introduce slack variables and rewrite the constraints and objective function in standard form:

$$2x + 2y + z + s = 8$$
$$x + 4y - 3z + t = 12$$
$$-12x - 15y - 5z + p = 0.$$

We now follow with Step 2, setting up the initial tableau:

	x	y	z	s	t	p	
s	2	2	1	1	0	0	8
t	1	4	-3	0	1	0	12
p	-12	-15	-5	0	0	1	0

For Step 3, we select the column over the negative number with the largest magnitude in the bottom row, which is the y column. For Step 4, finding the pivot, we see that the test ratios are 8/2 and 12/4, the smallest being $12/4 = 3$. So we select the pivot in the t row and clear its column:

	x	y	z	s	t	p		
s	2	2	1	1	0	0	8	$2R_1 - R_2$
t	1	[4]	-3	0	1	0	12	
p	-12	-15	-5	0	0	1	0	$4R_3 + 15R_2$

The departing variable is t and the entering variable is y. This gives the second tableau.

	x	y	z	s	t	p	
s	3	0	5	2	-1	0	4
y	1	4	-3	0	1	0	12
p	-33	0	-65	0	15	4	180

We now go back to Step 3. Because we still have negative numbers in the bottom row, we choose the one with the largest magnitude (which is –65), and thus our pivot column is the z column. Because negative numbers can't be pivots, the only possible choice for the pivot is the 5. (We need not compute the test ratios because there would only be one from which to choose.) We now clear this column, remembering to take care of the departing and entering variables.

	x	y	z	s	t	p		
s	3	0	[5]	2	-1	0	4	
y	1	4	-3	0	1	0	12	$5R_2 + 3R_1$
p	-33	0	-65	0	15	4	180	$R_3 + 13R_1$

This gives

	x	y	z	s	t	p	
z	3	0	5	2	−1	0	4
y	14	20	0	6	2	0	72
p	6	0	0	26	2	4	232

Notice how the value of p keeps climbing: It started at 0 in the first tableau, went up to $180/4 = 45$ in the second, and is currently at $232/4 = 58$. Because there are no more negative numbers in the bottom row, we are done and can write down the solution: p has a maximum value of $232/4 = 58$, and this occurs when

$$x = 0$$

$$y = \frac{72}{20} = \frac{18}{5} \quad \text{and}$$

$$z = \frac{4}{5}.$$

The slack variables are both zero.

As a partial check on our answer, we can substitute these values into the objective function and the constraints:

$$58 = 12(0) + 15(18/5) + 5(4/5) \qquad ✔$$

$$2(0) + 2(18/5) + (4/5) = 8 \le 8 \qquad ✔$$

$$0 + 4(18/5) - 3(4/5) = 12 \le 12. \qquad ✔$$

We say that this is only a partial check, because it shows only that our solution is feasible and that we have correctly calculated p. It does not show that we have the optimal solution. This check will *usually* catch any arithmetic mistakes we make, but it is not foolproof.

APPLICATIONS

In the next example (further exploits of Acme Baby Foods—compare Example 2 in Section 2) we show how the simplex method relates to the graphical method.

EXAMPLE 3 Resource Allocation

Acme Baby Foods makes two puddings, vanilla and chocolate. Each serving of vanilla pudding requires 2 teaspoons of sugar and 25 fluid ounces of water, and each serving of chocolate pudding requires 3 teaspoons of sugar and 15 fluid ounces of water. Acme has available each day 3,600 teaspoons of sugar and 22,500 fluid ounces of water. Acme makes no more than 600 servings of vanilla pudding because that is all that it can sell each day. If Acme makes a profit of 10¢ on each serving of vanilla pudding and 7¢ on each serving of chocolate, how many servings of each should it make to maximize its profit?

Solution We first identify the unknowns. Let

$x =$ the number of servings of vanilla pudding

$y =$ the number of servings of chocolate pudding.

The objective function is the profit $p = 10x + 7y$, which we need to maximize. For the constraints, we start with the fact that Acme will make no more than 600 servings of vanilla: $x \leq 600$. We can put the remaining data in a table as follows:

	Vanilla	*Chocolate*	*Total Available*
Sugar (teaspoons)	2	3	3,600
Water (ounces)	25	15	22,500

Because Acme can use no more sugar and water than is available, we get the two constraints:

$$2x + 3y \leq 3,600$$
$$25x + 15y \leq 22,500. \qquad \text{Note that all the terms are divisible by 5.}$$

Thus our linear programming problem is this:

Maximize $\quad p = 10x + 7y$

subject to $\quad x \leq 600$
$\qquad\qquad 2x + 3y \leq 3,600$
$\qquad\qquad 5x + 3y \leq 4,500 \qquad \text{We divided } 25x + 15y \leq 22,500 \text{ by 5.}$
$\qquad\qquad x \geq 0, \ y \geq 0.$

Next, we introduce the slack variables and set up the initial tableau.

$$
\begin{aligned}
x \qquad\quad + s \qquad\qquad\quad &= 600 \\
2x + 3y \quad\ + t \qquad\quad\ &= 3,600 \\
5x + 3y \qquad\quad + u \quad\ &= 4,500 \\
-10x - 7y \qquad\qquad + p &= 0
\end{aligned}
$$

Note that we have had to introduce a third slack variable, u. There need to be as many slack variables as there are constraints (other than those of the $x \geq 0$ variety).

Q: *What do the slack variables say about Acme puddings?*

A: The first slack variable, s, represents the number you must add to the number of servings of vanilla pudding actually made to obtain the maximum of 600 servings. The second slack variable, t, represents the amount of sugar that is left over once the puddings are made, and u represents the amount of water left over.

We now use the simplex method to solve the problem:

	x	y	s	t	u	p		
s	$\boxed{1}$	0	1	0	0	0	600	
t	2	3	0	1	0	0	3,600	$R_2 - 2R_1$
u	5	3	0	0	1	0	4,500	$R_3 - 5R_1$
p	-10	-7	0	0	0	1	0	$R_4 + 10R_1$

	x	y	s	t	u	p		
x	1	0	1	0	0	0	600	
t	0	3	-2	1	0	0	2,400	$R_2 - R_3$
u	0	$\boxed{3}$	-5	0	1	0	1,500	
p	0	-7	10	0	0	1	6,000	$3R_4 + 7R_3$

using Technology

See the Technology Guides at the end of the chapter for a discussion of using a TI-83/84 Plus to help with the simplex method in Example 3. Or, go to the Website at www.WanerMath.com and follow the path

→ On Line Utilities

→ Pivot and Gauss-Jordan Tool

to find a utility that allows you to avoid doing the calculations in each pivot step: Just highlight the entry you wish to use as a pivot, and press "Pivot on Selection."

	x	y	s	t	u	p		
x	1	0	1	0	0	0	600	$3R_1 - R_2$
t	0	0	3	1	-1	0	900	
y	0	3	-5	0	1	0	1,500	$3R_3 + 5R_2$
p	0	0	-5	0	7	3	28,500	$3R_4 + 5R_2$

	x	y	s	t	u	p	
x	3	0	0	-1	1	0	900
s	0	0	3	1	-1	0	900
y	0	9	0	5	-2	0	9,000
p	0	0	0	5	16	9	90,000

Thus, the solution is as follows: The maximum value of p is $90,000/9 = 10,000¢ = \$100$, which occurs when $x = 900/3 = 300$, and $y = 9,000/9 = 1,000$. (The slack variables are $s = 900/3 = 300$ and $t = u = 0$.)

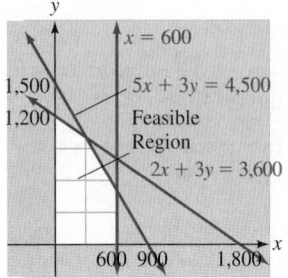

Figure 22

➡ **Before we go on...** Because the problem in Example 3 had only two variables, we could have solved it graphically. It is interesting to think about the relationship between the two methods. Figure 22 shows the feasible region. Each tableau in the simplex method corresponds to a corner of the feasible region, given by the corresponding basic solution. In this example, the sequence of basic solutions is

$$(x, y) = (0, 0), (600, 0), (600, 500), (300, 1,000).$$

This is the sequence of corners shown in Figure 23. In general, we can think of the simplex method as walking from corner to corner of the feasible region, until it locates the optimal solution. In problems with many variables and many constraints, the simplex method usually visits only a small fraction of the total number of corners.

We can also explain again, in a different way, the reason we use the test ratios when choosing the pivot. For example, when choosing the first pivot we had to choose among the test ratios 600, 1,800, and 900 (look at the first tableau). In Figure 22, you can see that those are the three x-intercepts of the lines that bound the feasible region. If we had chosen 1,800 or 900, we would have jumped along the x-axis to a point outside of the feasible region, which we do not want to do. In general, the test ratios measure the distance from the current corner to the constraint lines, and we must choose the smallest such distance to avoid crossing any of them into the unfeasible region.

It is also interesting in an application like this to think about the values of the slack variables. We said above that s is the difference between the maximum 600 servings of vanilla that might be made and the number that is actually made. In the optimal solution, $s = 300$, which says that 300 fewer servings of vanilla were made than the maximum possible. Similarly, t was the amount of sugar left over. In the optimal solution, $t = 0$, which tells us that all of the available sugar is used. Finally, $u = 0$, so all of the available water is used as well. ∎

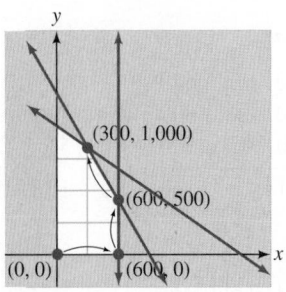

Figure 23

using Technology

Website
www.WanerMath.com
On the
→ On Line Utilities
page you will find utilities that automate the simplex method to varying
extents:

• Pivot and Gauss-Jordan Tool
  (Pivots and does row operations)
• Simplex Method Tool (Solves
  entire LP problems; shows all
  tableaux)

Summary: The Simplex Method for Standard Maximization Problems

To solve a standard maximization problem using the simplex method, we take the following steps:

1. Convert to a system of equations by introducing **slack variables** to turn the constraints into equations and by rewriting the objective function in standard form.

2. Write down the initial **tableau**.

3. Select the pivot column: Choose the negative number with the largest magnitude in the left-hand side of the bottom row. Its column is the pivot column. (If there are two or more candidates, choose any one.) If all the numbers in the left-hand side of the bottom row are zero or positive, then we are finished, and the basic solution maximizes the objective function. (See below for the basic solution.)

4. Select the pivot in the pivot column: The pivot must always be a positive number. For each positive entry b in the pivot column, compute the ratio a/b, where a is the number in the last column in that row. Of these **test ratios**, choose the smallest one. (If there are two or more candidates, choose any one.) The corresponding number b is the pivot.

5. Use the pivot to clear the column in the normal manner (taking care to follow the exact prescription for formulating the row operations described in Chapter 3) and then relabel the pivot row with the label from the pivot column. The variable originally labeling the pivot row is the **departing**, or **exiting**, **variable**, and the variable labeling the column is the **entering variable**.

6. Go to Step 3.

To get the **basic solution** corresponding to any tableau in the simplex method, set to zero all variables that do not appear as row labels. The value of a variable that does appear as a row label (an **active variable**) is the number in the rightmost column in that row divided by the number in that row in the column labeled by the same variable.

FAQs

Troubleshooting the Simplex Method

Q: *What if there is no candidate for the pivot in the pivot column? For example, what do we do with a tableau like the following?*

	x	y	z	s	t	p	
z	0	0	5	2	0	0	4
y	−8	20	0	6	5	0	72
p	−20	0	0	26	15	4	232

A: Here, the pivot column is the *x* column, but there is no suitable entry for a pivot (because zeros and negative numbers can't be pivots). This happens when the feasible region is unbounded and there is also no optimal solution. In other words, *p* can be made as large as we like without violating the constraints.

Q: *What should we do if there is a negative number in the rightmost column?*

A: A negative number will not appear above the bottom row in the rightmost column if we follow the procedure correctly. (The bottom right entry is allowed to be negative if the objective takes on negative values as in a negative profit, or loss.) Following are the most likely errors leading to this situation:

- The pivot was chosen incorrectly. (Don't forget to choose the *smallest* test ratio.) When this mistake is made, one or more of the variables will be negative in the corresponding basic solution.
- The row operation instruction was written backward or performed backward (for example, instead of $R_2 - R_1$, it was $R_1 - R_2$). This mistake can be corrected by multiplying the row by -1.
- An arithmetic error occurred. (We all make those annoying errors from time to time.)

Q: *What about zeros in the rightmost column?*

A: Zeros are permissible in the rightmost column. For example, the constraint $x - y \leq 0$ will lead to a zero in the rightmost column.*

Q: *What happens if we choose a pivot column other than the one with the most negative number in the bottom row?*

A: There is no harm in doing this as long as we choose the pivot in that column using the smallest test ratio. All it might do is slow the whole calculation by adding extra steps.

* When there are zeros in the rightmost column there is a potential problem of *cycling*, where a sequence of pivots brings you back to a tableau you already considered, with no change in the objective function. You can usually break out of a cycle by choosing a different pivot. This problem should not arise in the exercises.

One last suggestion: If it is possible to do a simplification step (dividing a row by a positive number) *at any stage*, we should do so. As we saw in Chapter 3, this can help prevent the numbers from getting out of hand.

5.3 EXERCISES

▼ more advanced   ◆ challenging

T indicates exercises that should be solved using technology

1. Maximize $p = 2x + y$
subject to $x + 2y \leq 6$
$-x + y \leq 4$
$x + y \leq 4$
$x \geq 0, y \geq 0.$

HINT [See Examples 1 and 2.]

2. Maximize $p = x$
subject to $x - y \leq 4$
$-x + 3y \leq 4$
$x \geq 0, y \geq 0.$

HINT [See Examples 1 and 2.]

3. Maximize $p = x - y$
subject to $5x - 5y \leq 20$
$2x - 10y \leq 40$
$x \geq 0, y \geq 0.$

4. Maximize $p = 2x + 3y$
subject to $3x + 8y \leq 24$
$6x + 4y \leq 30$
$x \geq 0, y \geq 0.$

5. Maximize $p = 5x - 4y + 3z$
subject to $5x + 5z \leq 100$
$5y - 5z \leq 50$
$5x - 5y \leq 50$
$x \geq 0, y \geq 0, z \geq 0.$

see which rows should now be starred. Repeat until no starred rows remain, and then go on to Phase II.

Phase II: Use the Simplex Method for Standard Maximization Problems

If there are any negative entries on the left side of the bottom row after Phase I, use the method described in the preceding section.

Because there is a starred row, we need to use Phase I. The largest positive number in the starred row is 1, which occurs three times. Arbitrarily select the first, which is in the first column. In that column, the smallest test ratio happens to be given by the 1 in the u row, so this is our first pivot.

Pivot column
↓

	x	y	z	s	t	u	p		
s	1	1	1	1	0	0	0	100	$R_1 - R_3$
t	4	10	7	0	1	0	0	480	$R_2 - 4R_3$
*u	[1]	1	1	0	0	-1	0	60	
p	-4	-12	-6	0	0	0	1	0	$R_4 + 4R_3$

This gives

	x	y	z	s	t	u	p	
s	0	0	0	1	0	1	0	40
t	0	6	3	0	1	4	0	240
x	1	1	1	0	0	-1	0	60
p	0	-8	-2	0	0	-4	1	240

Notice that we removed the star from row 3. To see why, look at the basic solution given by this tableau:

$$x = 60, y = 0, z = 0, s = 40, t = 240, u = 0.$$

None of the variables is negative anymore, so there are no rows to star. The basic solution is therefore feasible—it satisfies all the constraints.

Now that there are no more stars, we have completed Phase I, so we proceed to Phase II, which is just the method of the preceding section.

	x	y	z	s	t	u	p		
s	0	0	0	1	0	1	0	40	
t	0	[6]	3	0	1	4	0	240	
x	1	1	1	0	0	-1	0	60	$6R_3 - R_2$
p	0	-8	-2	0	0	-4	1	240	$3R_4 + 4R_2$

	x	y	z	s	t	u	p	
s	0	0	0	1	0	1	0	40
y	0	6	3	0	1	4	0	240
x	6	0	3	0	-1	-10	0	120
p	0	0	6	0	4	4	3	1,680

And we are finished. Thus the solution is

$$p = 1{,}680/3 = 560, x = 120/6 = 20, y = 240/6 = 40, z = 0.$$

The slack and surplus variables are

$$s = 40, t = 0, u = 0.$$

➡ **Before we go on...** We owe you an explanation of why this method works. When we perform a pivot in Phase I, one of two things will happen. As in Example 1, we may pivot in a starred row. In that case, the negative active variable in that row will become inactive (hence zero) and some other variable will be made active with a positive value because we are pivoting on a positive entry. Thus, at least one star will be eliminated. (We will not introduce any new stars because pivoting on the entry with the smallest test ratio will keep all nonnegative variables nonnegative.)

The second possibility is that we may pivot on some row other than a starred row. Choosing the pivot via test ratios again guarantees that no new starred rows are created. A little bit of algebra shows that the value of the negative variable in the first starred row must increase toward zero. (Choosing the *largest* positive entry in the starred row will make it a little more likely that we will increase the value of that variable as much as possible; the rationale for choosing the largest entry is the same as that for choosing the most negative entry in the bottom row during Phase II.) Repeating this procedure as necessary, the value of the variable must eventually become zero or positive, assuming that there are feasible solutions to begin with.

So, one way or the other, we can eventually get rid of all of the stars. ■

Here is an example that begins with two starred rows.

EXAMPLE 2 **More Mixed Constraints**

Maximize $p = 2x + y$
subject to $x + y \geq 35$
$x + 2y \leq 60$
$2x + y \geq 60$
$x \leq 25$
$x \geq 0, y \geq 0.$

Solution We introduce slack and surplus variables, and write down the initial tableau:

$$x + y - s = 35$$
$$x + 2y + t = 60$$
$$2x + y - u = 60$$
$$x + v = 25$$
$$-2x - y + p = 0.$$

	x	y	s	t	u	v	p	
*s	1	1	−1	0	0	0	0	35
t	1	2	0	1	0	0	0	60
*u	2	1	0	0	−1	0	0	60
v	1	0	0	0	0	1	0	25
p	−2	−1	0	0	0	0	1	0

We locate the largest positive entry in the first starred row (row 1). There are two to choose from (both 1s); let's choose the one in the x column. The entry with the smallest test ratio in that column is the 1 in the v row, so that is the entry we use as the pivot:

Pivot column
↓

	x	y	s	t	u	v	p		
*s	1	1	−1	0	0	0	0	35	$R_1 - R_4$
t	1	2	0	1	0	0	0	60	$R_2 - R_4$
*u	2	1	0	0	−1	0	0	60	$R_3 - 2R_4$
v	①	0	0	0	0	1	0	25	
p	−2	−1	0	0	0	0	1	0	$R_5 + 2R_4$

	x	y	s	t	u	v	p	
*s	0	1	−1	0	0	−1	0	10
t	0	2	0	1	0	−1	0	35
*u	0	1	0	0	−1	−2	0	10
x	1	0	0	0	0	1	0	25
p	0	−1	0	0	0	2	1	50

Notice that both stars are still there because the basic solutions for s and u remain negative (but less so). The only positive entry in the first starred row is the 1 in the y column, and that entry also has the smallest test ratio in its column. (Actually, it is tied with the 1 in the u column, so we could choose either one.)

	x	y	s	t	u	v	p		
*s	0	①	−1	0	0	−1	0	10	
t	0	2	0	1	0	−1	0	35	$R_2 - 2R_1$
*u	0	1	0	0	−1	−2	0	10	$R_3 - R_1$
x	1	0	0	0	0	1	0	25	
p	0	−1	0	0	0	2	1	50	$R_5 + R_1$

	x	y	s	t	u	v	p	
y	0	1	−1	0	0	−1	0	10
t	0	0	2	1	0	1	0	15
u	0	0	1	0	−1	−1	0	0
x	1	0	0	0	0	1	0	25
p	0	0	−1	0	0	1	1	60

The basic solution is $x = 25$, $y = 10$, $s = 0$, $t = 15$, $u = 0/(-1) = 0$, and $v = 0$. Because there are no negative variables left (even u has become 0), we are in the feasible region, so we can go on to Phase II, shown next. (Filling in the instructions for the row operations is an exercise.)

	x	y	s	t	u	v	p	
y	0	1	−1	0	0	−1	0	10
t	0	0	2	1	0	1	0	15
u	0	0	[1]	0	−1	−1	0	0
x	1	0	0	0	0	1	0	25
p	0	0	−1	0	0	1	1	60

	x	y	s	t	u	v	p	
y	0	1	0	0	−1	−2	0	10
t	0	0	0	1	[2]	3	0	15
s	0	0	1	0	−1	−1	0	0
x	1	0	0	0	0	1	0	25
p	0	0	0	0	−1	0	1	60

	x	y	s	t	u	v	p	
y	0	2	0	1	0	−1	0	35
u	0	0	0	1	2	3	0	15
s	0	0	2	1	0	1	0	15
x	1	0	0	0	0	1	0	25
p	0	0	0	1	0	3	2	135

The optimal solution is

$$x = 25, \ y = 35/2 = 17.5, \ p = 135/2 = 67.5 \quad (s = 7.5, t = 0, u = 7.5).$$

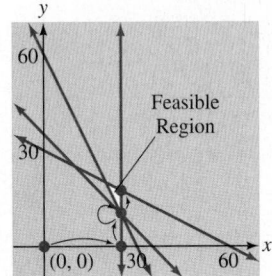

Figure 24

➡ **Before we go on...** Because Example 2 had only two unknowns, we can picture the sequence of basic solutions on the graph of the feasible region. This is shown in Figure 24.

You can see that there was no way to jump from (0, 0) in the initial tableau directly into the feasible region because the first jump must be along an axis. (Why?) Also notice that the third jump did not move at all. To which step of the simplex method does this correspond? ∎

Minimization Problems

Now that we know how to deal with nonstandard constraints, we consider **minimization** problems, problems in which we have to minimize, rather than maximize, the objective function. The idea is to *convert a minimization problem into a maximization problem*, which we can then solve as usual.

Suppose, for instance, we want to minimize $c = 10x - 30y$ subject to some constraints. The technique is as follows: Define a new variable p by taking p to be the negative of c, so that $p = -c$. Then, the larger we make p, the smaller c becomes. For example, if we can make p increase from -10 to -5, then c will decrease from 10 to 5. So, if we are looking for the smallest value of c, we might as well look for the largest value of p instead. More concisely,

Minimizing c is the same as maximizing p = −c.

Now because $c = 10x - 30y$, we have $p = -10x + 30y$, and the requirement that we "minimize $c = 10x - 30y$" is now replaced by "maximize $p = -10x + 30y$."

Minimization Problems

We convert a minimization problem into a maximization problem by taking the negative of the objective function. All the constraints remain unchanged.

Quick Example

Minimization Problem	→	Maximization Problem
Minimize $\quad c = 10x - 30y$		Maximize $\quad p = -10x + 30y$
subject to $\quad 2x + \ y \le 160$		subject to $\quad 2x + \ y \le 160$
$x + 3y \ge 120$		$x + 3y \ge 120$
$x \ge 0, y \ge 0.$		$x \ge 0, y \ge 0.$

EXAMPLE 3 Purchasing

You are in charge of ordering furniture for your company's new headquarters. You need to buy at least 200 tables, 500 chairs, and 300 computer desks. Wall-to-Wall Furniture (WWF) is offering a package of 20 tables, 25 chairs, and 18 computer desks for $2,000, whereas rival Acme Furniture (AF) is offering a package of 10 tables, 50 chairs, and 24 computer desks for $3,000. How many packages should you order from each company to minimize your total cost?

Solution The unknowns here are

$x =$ number of packages ordered from WWF
$y =$ number of packages ordered from AF.

We can put the information about the various kinds of furniture in a table:

	WWF	*AF*	*Needed*
Tables	20	10	200
Chairs	25	50	500
Computer Desks	18	24	300
Cost ($)	2,000	3,000	

From this table we get the following LP problem:

$$\begin{aligned}\text{Minimize} \quad & c = 2{,}000x + 3{,}000y \\ \text{subject to} \quad & 20x + 10y \ge 200 \\ & 25x + 50y \ge 500 \\ & 18x + 24y \ge 300 \\ & x \ge 0, y \ge 0.\end{aligned}$$

Before we start solving this problem, notice that all the inequalities may be simplified. The first is divisible by 10, the second by 25, and the third by 6. (However, this affects the meaning of the surplus variables; see Before we go on below.) Dividing gives the following simpler problem:

$$\begin{aligned}\text{Minimize} \quad & c = 2{,}000x + 3{,}000y \\ \text{subject to} \quad & 2x + \ y \ge 20 \\ & x + 2y \ge 20 \\ & 3x + 4y \ge 50 \\ & x \ge 0, y \ge 0.\end{aligned}$$

Following the discussion that preceded this example, we convert to a maximization problem:

$$\text{Maximize } p = -2{,}000x - 3{,}000y$$
$$\text{subject to } \quad 2x + y \geq 20$$
$$x + 2y \geq 20$$
$$3x + 4y \geq 50$$
$$x \geq 0, y \geq 0.$$

We introduce surplus variables.

$$2x + \quad y - s \qquad\qquad = 20$$
$$x + \quad 2y \quad - t \qquad\quad = 20$$
$$3x + \quad 4y \qquad\quad - u \quad = 50$$
$$2{,}000x + 3{,}000y \qquad\qquad + p = 0.$$

The initial tableau is then

	x	y	s	t	u	p	
*s	2	1	-1	0	0	0	20
*t	1	2	0	-1	0	0	20
*u	3	4	0	0	-1	0	50
p	2,000	3,000	0	0	0	1	0

The largest entry in the first starred row is the 2 in the upper left, which happens to give the smallest test ratio in its column.

	x	y	s	t	u	p		
*s	$\boxed{2}$	1	-1	0	0	0	20	
*t	1	2	0	-1	0	0	20	$2R_2 - R_1$
*u	3	4	0	0	-1	0	50	$2R_3 - 3R_1$
p	2,000	3,000	0	0	0	1	0	$R_4 - 1{,}000R_1$

	x	y	s	t	u	p		
x	2	1	-1	0	0	0	20	$3R_1 - R_2$
*t	0	$\boxed{3}$	1	-2	0	0	20	
*u	0	5	3	0	-2	0	40	$3R_3 - 5R_2$
p	0	2,000	1,000	0	0	1	$-20{,}000$	$3R_4 - 2{,}000R_2$

	x	y	s	t	u	p		
x	6	0	-4	2	0	0	40	$5R_1 - R_3$
y	0	3	1	-2	0	0	20	$5R_2 + R_3$
*u	0	0	4	$\boxed{10}$	-6	0	20	
p	0	0	1,000	4,000	0	3	$-100{,}000$	$R_4 - 400R_3$

	x	y	s	t	u	p		
x	30	0	-24	0	6	0	180	$R_1/6$
y	0	15	9	0	-6	0	120	$R_2/3$
t	0	0	4	10	-6	0	20	$R_3/2$
p	0	0	-600	0	2,400	3	$-108{,}000$	$R_4/3$

This completes Phase I. We are not yet at the optimal solution, so after performing the simplifications indicated we proceed with Phase II.

	x	y	s	t	u	p		
x	5	0	−4	0	1	0	30	$R_1 + 2R_3$
y	0	5	3	0	−2	0	40	$2R_2 - 3R_3$
t	0	0	②	5	−3	0	10	
p	0	0	−200	0	800	1	−36,000	$R_4 + 100R_3$

	x	y	s	t	u	p	
x	5	0	0	10	−5	0	50
y	0	10	0	−15	5	0	50
s	0	0	2	5	−3	0	10
p	0	0	0	500	500	1	−35,000

The optimal solution is

$x = 50/5 = 10$, $y = 50/10 = 5$, $p = -35,000$, so $c = 35,000$
($s = 5, t = 0, u = 0$).

You should buy 10 packages from Wall-to-Wall Furniture and 5 from Acme Furniture, for a minimum cost of $35,000.

➡ **Before we go on...** The surplus variables in the preceding example represent pieces of furniture over and above the minimum requirements. The order you place will give you 50 extra tables ($s = 5$, but s was introduced after we divided the first inequality by 10, so the actual surplus is $10 \times 5 = 50$), the correct number of chairs ($t = 0$), and the correct number of computer desks ($u = 0$). ∎

The preceding LP problem is an example of a **standard minimization problem**—in a sense the opposite of a standard maximization problem: We are *minimizing* an objective function, where all the constraints have the form $Ax + By + Cz + \cdots \geq N$. We will discuss standard minimization problems more fully in Section 5.5, as well as another method of solving them.

FAQs

When to Switch to Phase II, Equality Constraints, and Troubleshooting

Q: *How do I know when to switch to Phase II?*

A: After each step, check the basic solution for starred rows. You are not ready to proceed with Phase II until all the stars are gone.

Q: *How do I deal with an equality constraint, such as $2x + 7y - z = 90$?*

A: Although we haven't given examples of equality constraints, they can be treated by the following trick: Replace an equality by two inequalities. For example, replace the equality $2x + 7y - z = 90$ by the two inequalities $2x + 7y - z \leq 90$ and $2x + 7y - z \geq 90$. A little thought will convince you that these two inequalities amount to the same thing as the original equality!

> **Q** : *What happens if it is impossible to choose a pivot using the instructions in Phase I?*
>
> **A** : In that case, the LP problem has no solution. In fact, the feasible region is empty. If it is impossible to choose a pivot in Phase II, then the feasible region is unbounded and there is no optimal solution.

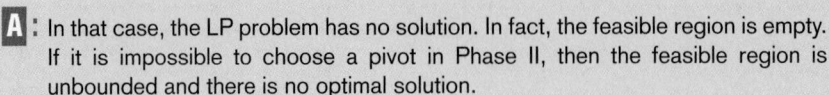

5.4 EXERCISES

▼ more advanced   ◆ challenging
T indicates exercises that should be solved using technology

1. Maximize $p = x + y$
subject to
$$x + 2y \geq 6$$
$$-x + y \leq 4$$
$$2x + y \leq 8$$
$$x \geq 0, y \geq 0.$$ HINT [See Examples 1 and 2.]

2. Maximize $p = 3x + 2y$
subject to
$$x + 3y \geq 6$$
$$-x + y \leq 4$$
$$2x + y \leq 8$$
$$x \geq 0, y \geq 0.$$ HINT [See Examples 1 and 2.]

3. Maximize $p = 12x + 10y$
subject to
$$x + y \leq 25$$
$$x \geq 10$$
$$-x + 2y \geq 0$$
$$x \geq 0, y \geq 0.$$

4. Maximize $p = x + 2y$
subject to
$$x + y \leq 25$$
$$y \geq 10$$
$$2x - y \geq 0$$
$$x \geq 0, y \geq 0.$$

5. Maximize $p = 2x + 5y + 3z$
subject to
$$x + y + z \leq 150$$
$$x + y + z \geq 100$$
$$x \geq 0, y \geq 0, z \geq 0.$$

6. Maximize $p = 3x + 2y + 2z$
subject to
$$x + y + 2z \leq 38$$
$$2x + y + z \geq 24$$
$$x \geq 0, y \geq 0, z \geq 0.$$

7. Maximize $p = 10x + 20y + 15z$
subject to
$$x + 2y + z \leq 40$$
$$2y - z \geq 10$$
$$2x - y + z \geq 20$$
$$x \geq 0, y \geq 0, z \geq 0.$$

8. Maximize $p = 10x + 10y + 15z$
subject to
$$x - y + z \leq 12$$
$$2x - 2y + z \geq 15$$
$$-y + z \geq 3$$
$$x \geq 0, y \geq 0, z \geq 0.$$

9. Maximize $p = x + y + 3z + w$
subject to
$$x + y + z + w \leq 40$$
$$2x + y - z - w \geq 10$$
$$x + y + z + w \geq 10$$
$$x \geq 0, y \geq 0, z \geq 0, w \geq 0.$$

10. Maximize $p = x + y + 4z + 2w$
subject to
$$x + y + z + w \leq 50$$
$$2x + y - z - w \geq 10$$
$$x + y + z + w \geq 20$$
$$x \geq 0, y \geq 0, z \geq 0, w \geq 0.$$

11. Minimize $c = 6x + 6y$
subject to
$$x + 2y \geq 20$$
$$2x + y \geq 20$$
$$x \geq 0, y \geq 0.$$ HINT [See Example 3.]

12. Minimize $c = 3x + 2y$
subject to
$$x + 2y \geq 20$$
$$2x + y \geq 10$$
$$x \geq 0, y \geq 0.$$ HINT [See Example 3.]

13. Minimize $c = 2x + y + 3z$
subject to
$$x + y + z \geq 100$$
$$2x + y \geq 50$$
$$y + z \geq 50$$
$$x \geq 0, y \geq 0, z \geq 0.$$

14. Minimize $c = 2x + 2y + 3z$
subject to
$$x + z \geq 100$$
$$2x + y \geq 50$$
$$y + z \geq 50$$
$$x \geq 0, y \geq 0, z \geq 0.$$

15. Minimize $c = 50x + 50y + 11z$
subject to $2x \quad + z \geq 3$
$2x + y - z \geq 2$
$3x + y - z \leq 3$
$x \geq 0, y \geq 0, z \geq 0.$

16. Minimize $c = 50x + 11y + 50z$
subject to $3x \quad + z \geq 8$
$3x + y - z \geq 6$
$4x + y - z \leq 8$
$x \geq 0, y \geq 0, z \geq 0.$

17. Minimize $c = x + y + z + w$
subject to $5x - y \quad + w \geq 1,000$
$z + w \leq 2,000$
$x + y \quad \leq 500$
$x \geq 0, y \geq 0, z \geq 0, w \geq 0.$

18. Minimize $c = 5x + y + z + w$
subject to $5x - y \quad + w \geq 1,000$
$z + w \leq 2,000$
$x + y \quad \leq 500$
$x \geq 0, y \geq 0, z \geq 0, w \geq 0.$

▓ *In Exercises 19–24, we suggest the use of technology. Round all answers to two decimal places.*

19. Maximize $p = 2x + 3y + 1.1z + 4w$
subject to $1.2x + y + z + \quad w \leq 40.5$
$2.2x + y - z - \quad w \geq 10$
$1.2x + y + z + 1.2w \geq 10.5$
$x \geq 0, y \geq 0, z \geq 0, w \geq 0.$

20. Maximize $p = 2.2x + 2y + 1.1z + 2w$
subject to $x + 1.5y + 1.5z + \quad w \leq 50.5$
$2x + 1.5y - \quad z - \quad w \geq 10$
$x + 1.5y + \quad z + 1.5w \geq 21$
$x \geq 0, y \geq 0, z \geq 0, w \geq 0.$

21. Minimize $c = 2.2x + y + 3.3z$
subject to $x + 1.5y + 1.2z \geq 100$
$2x + 1.5y \quad \geq 50$
$1.5y + 1.1z \geq 50$
$x \geq 0, y \geq 0, z \geq 0.$

22. Minimize $c = 50.3x + 10.5y + 50.3z$
subject to $3.1x \quad + 1.1z \geq 28$
$3.1x + y - 1.1z \geq 23$
$4.2x + y - 1.1z \geq 28$
$x \geq 0, y \geq 0, z \geq 0.$

23. Minimize $c = 1.1x + y + 1.5z - w$
subject to $5.12x - y \quad + w \leq 1,000$
$z + w \geq 2,000$
$1.22x + y \quad \leq 500$
$x \geq 0, y \geq 0, z \geq 0, w \geq 0.$

24. Minimize $c = 5.45x + y + 1.5z + w$
subject to $5.12x - y \quad + w \geq 1,000$
$z + w \geq 2,000$
$1.12x + y \quad \leq 500$
$x \geq 0, y \geq 0, z \geq 0, w \geq 0.$

APPLICATIONS

25. *Agriculture* (Compare Exercise 27 in Section 5.3.) Your small farm encompasses 100 acres, and you are planning to grow tomatoes, lettuce, and carrots in the coming planting season. Fertilizer costs per acre are: $5 for tomatoes, $4 for lettuce, and $2 for carrots. Based on past experience, you estimate that each acre of tomatoes will require an average of 4 hours of labor per week, while tending to lettuce and carrots will each require an average of 2 hours per week. You estimate a profit of $2,000 for each acre of tomatoes, $1,500 for each acre of lettuce, and $500 for each acre of carrots. You would like to spend at least $400 on fertilizer (your niece owns the company that manufactures it) and your farm laborers can supply up to 500 hours per week. How many acres of each crop should you plant to maximize total profits? In this event, will you be using all 100 acres of your farm? HINT [See Example 3.]

26. *Agriculture* (Compare Exercise 28 in Section 5.3.) Your farm encompasses 900 acres, and you are planning to grow soybeans, corn, and wheat in the coming planting season. Fertilizer costs per acre are: $5 for soybeans, $2 for corn, and $1 for wheat. You estimate that each acre of soybeans will require an average of 5 hours of labor per week, while tending to corn and wheat will each require an average of 2 hours per week. Based on past yields and current market prices, you estimate a profit of $3,000 for each acre of soybeans, $2,000 for each acre of corn, and $1,000 for each acre of wheat. You can afford to spend no more than $3,000 on fertilizer, but your labor union contract stipulates at least 2,000 hours per week of labor. How many acres of each crop should you plant to maximize total profits? In this event, will you be using more than 2,000 hours of labor? HINT [See Example 3.]

27. *Politics* The political pollster Canter is preparing for a national election. It would like to poll at least 1,500 Democrats and 1,500 Republicans. Each mailing to the East Coast gets responses from 100 Democrats and 50 Republicans. Each mailing to the Midwest gets responses from 100 Democrats and 100 Republicans. And each mailing to the West Coast gets responses from 50 Democrats and 100 Republicans. Mailings to the East Coast cost $40 each to produce and mail, mailings to the Midwest cost $60 each, and mailings to the West Coast cost $50 each. How many mailings should Canter send to each area of the country to get the responses it needs at the least possible cost? What will it cost?

28. *Purchasing* Bingo's Copy Center needs to buy white paper and yellow paper. Bingo's can buy from three suppliers. Harvard Paper sells a package of 20 reams of white and 10 reams of yellow for $60, Yale Paper sells a package of

10 reams of white and 10 reams of yellow for $40, and Dartmouth Paper sells a package of 10 reams of white and 20 reams of yellow for $50. If Bingo's needs 350 reams of white and 400 reams of yellow, how many packages should it buy from each supplier to minimize the cost? What is the least possible cost?

29. *Resource Allocation Succulent Citrus* produces orange juice and orange concentrate. This year the company anticipates a demand of at least 10,000 quarts of orange juice and 1,000 quarts of orange concentrate. Each quart of orange juice requires 10 oranges, and each quart of concentrate requires 50 oranges. The company also anticipates using at least 200,000 oranges for these products. Each quart of orange juice costs the company 50¢ to produce, and each quart of concentrate costs $2.00 to produce. How many quarts of each product should Succulent Citrus produce to meet the demand and minimize total costs?

30. *Resource Allocation Fancy Pineapple* produces pineapple juice and canned pineapple rings. This year the company anticipates a demand of at least 10,000 pints of pineapple juice and 1,000 cans of pineapple rings. Each pint of pineapple juice requires 2 pineapples, and each can of pineapple rings requires 1 pineapple. The company anticipates using at least 20,000 pineapples for these products. Each pint of pineapple juice costs the company 20¢ to produce, and each can of pineapple rings costs 50¢ to produce. How many pints of pineapple juice and cans of pineapple rings should Fancy Pineapple produce to meet the demand and minimize total costs?

31. *Latin Music Sales (Digital)* You are about to go live with a Latin music download service called *iYayay* that will compete head-to-head with **Apple's** *iTunes*® (good luck!). You will be selling digital albums of regional (Mexican/Tejano) music for $5 each, pop/rock albums for $4 each, and tropical (salsa/merengue/cumbia/bachata) at $6 per album. Your servers can handle up to 40,000 downloaded albums per day, and you anticipate that, based on national sales,[30] revenues from regional music will be at least fives times those from tropical music. You also anticipate that that you will sell at least 10,000 pop/rock albums per day due to the very attractive $4 price. Based on these assumptions, how many of each type of album should you sell for a maximum daily revenue, and what will your daily revenue be?

32. *Latin Music Sales (Digital)* It seems that your biggest rival, Lupita Pelogrande, has learned about your music store and has decided to launch *iRico*, a competitor Latin Music download service that will also offer regional (Mexican/Tejano) music albums for $5 each and pop/rock albums for $4 each. However, instead of tropical music, *iRico* will offer reggaeton albums at only $3 per album. Lupita's servers are more robust than yours, and can handle up to 60,000 downloaded albums per day. She anticipates that revenues from regional music can total to up to double those from reggaeton, and that *iRico* can sell at least 50,000 albums per day, not even counting reggaeton. Based on these assumptions, how many of each type of album should *iRico* sell for a maximum daily revenue, and what will the daily revenue be?

33. ⊓ *Nutrition* Gerber Products' Gerber Mixed Cereal for Baby contains, in each serving, 60 calories and no Vitamin C. Gerber Mango Tropical Fruit Dessert contains, in each serving, 80 calories and 45% of the U.S. Recommended Daily Allowance (RDA) of Vitamin C for infants. Gerber Apple Banana Juice contains, in each serving, 60 calories and 120% of the RDA of Vitamin C for infants.[31] The cereal costs 10¢/serving, the dessert costs 53¢/serving, and the juice costs 27¢/serving. If you want to provide your child with at least 120 calories and at least 120% of the RDA of Vitamin C, how can you do so at the least cost?

34. ⊓ *Nutrition* Gerber Products' Gerber Mixed Cereal for Baby contains, in each serving, 60 calories, no Vitamin C, and 11 grams of carbohydrates. Gerber Mango Tropical Fruit Dessert contains, in each serving, 80 calories, 45% of the RDA of Vitamin C for infants, and 21 grams of carbohydrates. Gerber Apple Banana Juice contains, in each serving, 60 calories, 120% of the RDA of Vitamin C for infants, and 15 grams of carbohydrates.[32] Assume that the cereal costs 11¢/serving, the dessert costs 50¢/serving, and the juice costs 30¢/serving. If you want to provide your child with at least 180 calories, at least 120% of the RDA of Vitamin C, and at least 37 grams of carbohydrates, how can you do so at the least cost?

35. ⊓ *Purchasing Cheapskate Electronics Store* needs to update its inventory of stereos, TVs, and DVD players. There are three suppliers it can buy from: Nadir offers a bundle consisting of 5 stereos, 10 TVs, and 15 DVD players for $3,000. Blunt offers a bundle consisting of 10 stereos, 10 TVs, and 10 DVD players for $4,000. Sonny offers a bundle consisting of 15 stereos, 10 TVs, and 10 DVD players for $5,000. Cheapskate Electronics needs at least 150 stereos, 200 TVs, and 150 DVD players. How can it update its inventory at the least possible cost? What is the least possible cost?

36. ⊓ *Purchasing Federal Rent-a-Car* is putting together a new fleet. It is considering package offers from three car manufacturers. Fred Motors is offering 5 small cars, 5 medium cars, and 10 large cars for $500,000. Admiral Motors is offering 5 small, 10 medium, and 5 large cars for $400,000. Chrysalis is offering 10 small, 5 medium, and 5 large cars for $300,000. Federal would like to buy at least 550 small cars, at least 500 medium cars, and at least 550 large cars. How many packages should it buy from each car maker to keep the total cost as small as possible? What will be the total cost?

---

[30]In 2010, total U.S. revenues from regional music were about five times those from tropical music. Source: Recording Industry Association of America http://riaa.org

[31]Source: Nutrition information supplied with the products.
[32]*Ibid.*

⊡ **Bodybuilding Supplements** *Exercises 37 and 38 are based on the following data on four bodybuilding supplements. (Figures shown correspond to a single serving.)*[33]

	Creatine (g)	Carbohydrates (g)	Taurine (g)	Alpha Lipoic Acid (mg)	Cost ($)
Cell-Tech® (MuscleTech)	10	75	2	200	2.20
RiboForce HP® (EAS)	5	15	1	0	1.60
Creatine Transport® (Kaizen)	5	35	1	100	0.60
Pre-Load Creatine® (Optimum)	6	35	1	25	0.50

37. (Compare Exercise 29 in Section 5.2.) You are thinking of combining Cell-Tech, RiboForce HP, and Creatine Transport to obtain a 10-day supply that provides at least 80 grams of creatine and at least 10 grams of taurine, but no more than 750 grams of carbohydrates and 1,000 milligrams of alpha lipoic acid. How many servings of each supplement should you combine to meet your specifications for the least cost?

38. (Compare Exercise 30 in Section 5.2.) You are thinking of combining RiboForce HP, Creatine Transport, and Pre-Load Creatine to obtain a 10-day supply that provides at least 80 grams of creatine and at least 10 grams of taurine, but no more than 600 grams of carbohydrates and 2,000 milligrams of alpha lipoic acid. How many servings of each supplement should you combine to meet your specifications for the least cost?

39. ▼ **Subsidies** The Miami Beach City Council has offered to subsidize hotel development in Miami Beach, and it is hoping for at least two hotels with a total capacity of at least 1,400. Suppose that you are a developer interested in taking advantage of this offer by building a small group of hotels in Miami Beach. You are thinking of three prototypes: a convention-style hotel with 500 rooms costing $100 million, a vacation-style hotel with 200 rooms costing $20 million, and a small motel with 50 rooms costing $4 million. The City Council will approve your plans, provided you build at least one convention-style hotel and no more than two small motels.

a. How many of each type of hotel should you build to satisfy the city council's wishes and stipulations while minimizing your total cost?

b. Now assume that the city council will give developers 20 percent of the cost of building new hotels in Miami Beach, up to $50 million.[34] Will the city's $50 million subsidy be sufficient to cover 20 percent of your total costs?

40. ▼ **Subsidies** Refer back to the preceding exercise. You are about to begin the financial arrangements for your new hotels when the city council informs you that it has changed its mind and now requires at least two vacation-style hotels and no more than four small motels.

a. How many of each type of hotel should you build to satisfy the city council's wishes and stipulations while minimizing your total costs?

b. Will the city's $50 million subsidy limit still be sufficient to cover 20 percent of your total costs?

41. ▼ **Transportation Scheduling** We return to your exploits coordinating distribution for the Tubular Ride Boogie Board Company.[35] You will recall that the company has manufacturing plants in Tucson, Arizona, and Toronto, Ontario, and you have been given the job of coordinating distribution of their latest model, the Gladiator, to their outlets in Honolulu and Venice Beach. The Tucson plant can manufacture up to 620 boards per week, while the Toronto plant, beset by labor disputes, can produce no more than 410 Gladiator boards per week. The outlet in Honolulu orders 500 Gladiator boards per week, while Venice Beach orders 530 boards per week. Transportation costs are as follows: Tucson to Honolulu: $10/board; Tucson to Venice Beach: $5/board; Toronto to Honolulu: $20/board; Toronto to Venice Beach: $10/board. Your manager has said that you are to be sure to fill all orders and ship the boogie boards at a minimum total transportation cost. How will you do it?

42. ▼ **Transportation Scheduling** In the situation described in the preceding exercise, you have just been notified that workers at the Toronto boogie board plant have gone on strike, resulting in a total work stoppage. You are to come up with a revised delivery schedule by tomorrow with the understanding that the Tucson plant can push production to a maximum of 1,000 boards per week. What should you do?

43. ▼ **Finance** Senator Porkbarrel habitually overdraws his three bank accounts, at the *Congressional Integrity Bank, Citizens' Trust,* and *Checks R Us.* There are no penalties because the overdrafts are subsidized by the taxpayer. The Senate Ethics Committee tends to let slide irregular banking activities as long as they are not flagrant. At the moment (due to Congress's preoccupation with a Supreme Court nominee), a total overdraft of up to $10,000 will be overlooked. Porkbarrel's conscience makes him hesitate to overdraw accounts at banks whose names include expressions like "integrity" and "citizens' trust." The effect is that his overdrafts at the first two banks combined amount to no more than one quarter of the total. On the other hand, the financial officers at Integrity Bank, aware that Senator Porkbarrel is a member of the Senate Banking Committee, "suggest" that he overdraw at least $2,500 from their bank. Find the amount he should overdraw from each bank in order to avoid investigation by the Ethics Committee and overdraw his account at Integrity by as much as his sense of guilt will allow.

---

[33]Source: Nutritional information supplied by the manufacturers (www.netrition.com). Cost per serving is approximate.

[34]The Miami Beach City Council made such an offer in 1993. (*Chicago Tribune,* June 20, 1993, Section 7, p. 8).

[35]See Exercise 26 in Section 3.3 and also Exercise 41 in Section 5.3. This time, we will use the simplex method to solve the version of this problem we first considered in the chapter on systems of equations.

44. ▼ *Scheduling* Because Joe Slim's brother was recently elected to the State Senate, Joe's financial advisement concern, *Inside Information Inc.*, has been doing a booming trade, even though the financial counseling he offers is quite worthless. (None of his seasoned clients pays the slightest attention to his advice.) Slim charges different hourly rates to different categories of individuals: $5,000/hour for private citizens, $50,000/hour for corporate executives, and $10,000/hour for presidents of universities. Due to his taste for leisure, he feels that he can spend no more than 40 hours/week in consultation. On the other hand, Slim feels that it would be best for his intellect were he to devote at least 10 hours of consultation each week to university presidents. However, Slim always feels somewhat uncomfortable dealing with academics, so he would prefer to spend no more than half his consultation time with university presidents. Furthermore, he likes to think of himself as representing the interests of the common citizen, so he wishes to offer at least 2 more hours of his time each week to private citizens than to corporate executives and university presidents combined. Given all these restrictions, how many hours each week should he spend with each type of client in order to maximize his income?

45. ▼ *Transportation Scheduling* Your publishing company is about to start a promotional blitz for its new book, *Advanced Quantum Mechanics for the Liberal Arts*. You have 20 salespeople stationed in Chicago and 10 in Denver. You would like to fly at least 10 into Los Angeles and at least 15 into New York. A round-trip plane flight from Chicago to LA costs $200; from Chicago to NY costs $125; from Denver to LA costs $225; and from Denver to NY costs $280.[36] How many salespeople should you fly from each of Chicago and Denver to each of LA and NY to spend the least amount on plane flights?

46. ▼ *Transportation Scheduling* Repeat Exercise 45, but now suppose that you would like at least 15 salespeople in Los Angeles.

47. ▼ *Hospital Staffing* As the staff director of a new hospital, you are planning to hire cardiologists, rehabilitation specialists, and infectious disease specialists. According to recent data, each cardiology case averages $12,000 in revenue, each physical rehabilitation case $19,000, and each infectious disease case, $14,000.[37] You judge that each specialist you employ will expand the hospital caseload by about 10 patients per week. You already have 3 cardiologists on staff, and the hospital is equipped to admit up to 200 patients per week. Based on past experience, each cardiologist and rehabilitation specialist brings in one government research grant per year, while each infectious disease specialist brings in three. Your board of directors would like to see

a total of at least 30 grants per year and would like your weekly revenue to be as large as possible. How many of each kind of specialist should you hire?

48. ⊤ ▼ *Hospital Staffing* Referring to Exercise 47, you completely misjudged the number of patients each type of specialist would bring to the hospital per week. It turned out that each cardiologist brought in 120 new patients per *year*, each rehabilitation specialist brought in 90 per year, and each infectious disease specialist brought in 70 per year.[38] It also turned out that your hospital could deal with no more than 1,960 new patients per year. Repeat the preceding exercise in light of this corrected data.

COMMUNICATION AND REASONING EXERCISES

49. Explain the need for Phase I in a nonstandard LP problem.

50. Explain the need for Phase II in a nonstandard LP problem.

51. Explain briefly why we would need to use Phase I in solving a linear programming problem with the constraint $x + 2y - z \geq 3$.

52. Which rows do we star, and why?

53. Consider the following linear programming problem:

Maximize $p = x + y$
subject to $x - 2y \geq 0$
                    $2x + y \leq 10$
                    $x \geq 0, y \geq 0.$

This problem

(A) must be solved using the techniques of Section 5.4.
(B) must be solved using the techniques of Section 5.3.
(C) can be solved using the techniques of either section.

54. Consider the following linear programming problem:

Maximize $p = x + y$
subject to $x - 2y \geq 1$
                    $2x + y \leq 10$
                    $x \geq 0, y \geq 0.$

This problem

(A) must be solved using the techniques of Section 5.4.
(B) must be solved using the techniques of Section 5.3.
(C) can be solved using the techniques of either section.

55. ▼ Find a linear programming problem in three variables that requires one pivot in Phase I.

56. ▼ Find a linear programming problem in three variables that requires two pivots in Phase I.

57. ▼ Find a linear programming problem in two or three variables with no optimal solution, and show what happens when you try to solve it using the simplex method.

58. ▼ Find a linear programming problem in two or three variables with more than one optimal solution, and investigate which solution is found by the simplex method.

---

[36]Approximate prices advertised on various Web sites in September 2011.

[37]These (rounded) figures are based on an Illinois survey of 1.3 million hospital admissions (*Chicago Tribune*, March 29, 1993, Section 4, p. 1). Source: Lutheran General Health System, Argus Associates, Inc.

[38]These (rounded) figures were obtained from the survey referenced in the preceding exercise by dividing the average hospital revenue per physician by the revenue per case.

5.5 The Simplex Method and Duality

We mentioned **standard minimization problems** in the last section. These problems have the following form.

Standard Minimization Problem

A **standard minimization problem** is an LP problem in which we are required to *minimize* (not maximize) a linear objective function

$$c = as + bt + cu + \cdots$$

of the variables $s, t, u, \ldots$ (in this section, we will always use the letters $s, t, u, \ldots$ for the unknowns in a standard minimization problem) subject to the constraints

$$s \geq 0, t \geq 0, u \geq 0, \ldots$$

and further constraints of the form

$$As + Bt + Cu + \cdots \geq N$$

where $A, B, C, \ldots$ and N are numbers with N nonnegative.

A **standard linear programming problem** is an LP problem that is either a standard maximization problem or a standard minimization problem. An LP problem satisfies the **nonnegative objective condition** if all the coefficients in the objective function are nonnegative.

Standard Minimization and Maximization Problems

1. Minimize $c = 2s + 3t + 3u$
   subject to $2s \quad\;\; + \;\; u \geq 10$ This is a standard minimization problem
   $\qquad\qquad s + 3t - 6u \geq 5$ satisfying the nonnegative objective
   $\qquad\qquad s \geq 0, t \geq 0, u \geq 0.$ condition.

2. Maximize $p = 2x + 3y + 3z$
   subject to $2x \quad\;\; + \;\; z \leq 7$ This is a standard maximization problem
   $\qquad\qquad x + 3y - 6z \leq 6$ satisfying the nonnegative objective
   $\qquad\qquad x \geq 0, y \geq 0, z \geq 0.$ condition.

3. Minimize $c = 2s - 3t + 3u$
   subject to $2s \quad\;\; + \;\; u \geq 10$ This is a standard minimization problem
   $\qquad\qquad s + 3t - 6u \geq 5$ that does *not* satisfy the nonnegative
   $\qquad\qquad s \geq 0, t \geq 0, u \geq 0.$ objective condition.

We saw a way of solving minimization problems in Section 5.4, but a mathematically elegant relationship between maximization and minimization problems gives us another way of solving minimization problems that satisfy the nonnegative objective condition. This relationship is called **duality**.

To describe duality, we must first represent an LP problem by a matrix. This matrix is *not* the first tableau but something simpler: Pretend you forgot all about slack

* Forgetting these things is exactly what happens to many students under test conditions!

† Although duality does not require the problems to be standard, it does require them to be written in so-called *standard form*: In the case of a maximization problem all constraints need to be (re)written using $\leq$, while for a minimization problem all constraints need to be (re)written using $\geq$. It is least confusing to stick with standard problems, which is what we will do in this section.

variables and also forgot to change the signs of the objective function.* As an example, consider the following two standard† problems:

Problem 1

$$\text{Maximize} \quad p = 20x + 20y + 50z$$
$$\text{subject to} \quad 2x + y + 3z \leq 2{,}000$$
$$x + 2y + 4z \leq 3{,}000$$
$$x \geq 0, y \geq 0, z \geq 0.$$

We represent this problem by the matrix

$$\begin{bmatrix} 2 & 1 & 3 & 2{,}000 \\ 1 & 2 & 4 & 3{,}000 \\ 20 & 20 & 50 & 0 \end{bmatrix}$$

Constraint 1
Constraint 2
Objective

Notice that the coefficients of the objective function go in the bottom row, and we place a zero in the bottom right corner.

Problem 2 (from Example 3 in Section 5.4)

$$\text{Minimize} \quad c = 2{,}000s + 3{,}000t$$
$$\text{subject to} \quad 2s + t \geq 20$$
$$s + 2t \geq 20$$
$$3s + 4t \geq 50$$
$$s \geq 0, t \geq 0.$$

Problem 2 is represented by

$$\begin{bmatrix} 2 & 1 & 20 \\ 1 & 2 & 20 \\ 3 & 4 & 50 \\ 2{,}000 & 3{,}000 & 0 \end{bmatrix}$$

Constraint 1
Constraint 2
Constraint 3
Objective

These two problems are related: The matrix for Problem 1 is the transpose of the matrix for Problem 2. (Recall that the transpose of a matrix is obtained by writing its rows as columns; see Section 4.1.) When we have a pair of LP problems related in this way, we say that the two are *dual* LP problems.

Dual Linear Programming Problems

Two LP problems, one a maximization and one a minimization problem, are **dual** if the matrix that represents one is the transpose of the matrix that represents the other.

Finding the Dual of a Given Problem

Given an LP problem, we find its dual as follows:

1. Represent the problem as a matrix (see above).

2. Take the transpose of the matrix.

3. Write down the dual, which is the LP problem corresponding to the new matrix. If the original problem was a maximization problem, its dual will be a minimization problem, and vice versa.

The original problem is called the **primal problem**, and its dual is referred to as the **dual problem**.

> ### Quick Example
>
> Primal problem
> Minimize $c = s + 2t$
> subject to $5s + 2t \geq 60$
> $3s + 4t \geq 80$
> $s + t \geq 20$
> $s \geq 0, t \geq 0.$
>
> $\xrightarrow{1}$
>
> $$\begin{bmatrix} 5 & 2 & 60 \\ 3 & 4 & 80 \\ 1 & 1 & 20 \\ 1 & 2 & 0 \end{bmatrix}$$
>
> Dual problem
> Maximize $p = 60x + 80y + 20z$
> subject to $5x + 3y + z \leq 1$
> $2x + 4y + z \leq 2$
> $x \geq 0, y \geq 0, z \geq 0.$
>
> $\xrightarrow{2}$
>
> $$\begin{bmatrix} 5 & 3 & 1 & 1 \\ 2 & 4 & 1 & 2 \\ 60 & 80 & 20 & 0 \end{bmatrix}$$
>
> $\xrightarrow{3}$

The following theorem justifies what we have been doing, and says that solving the dual problem of an LP problem is equivalent to solving the original problem.

Fundamental Theorem of Duality

a. If an LP problem has an optimal solution, then so does its dual. Moreover, the primal problem and the dual problem have the same optimal value for their objective functions.

b. Contained in the final tableau of the simplex method applied to an LP problem is the solution to its dual problem: It is given by the bottom entries in the columns associated with the slack variables, divided by the entry under the objective variable.

*The proof of the theorem is beyond the scope of this book but can be found in a textbook devoted to linear programming, like *Linear Programming* by Vašek Chvátal (San Francisco: W. H. Freeman and Co., 1983), which has a particularly well-motivated discussion.

The theorem* gives us an alternative way of solving minimization problems that satisfy the nonnegative objective condition. Let's illustrate by solving Problem 2 above.

EXAMPLE 1 Solving by Duality

Minimize $c = 2{,}000s + 3{,}000t$
subject to $2s + t \geq 20$
$s + 2t \geq 20$
$3s + 4t \geq 50$
$s \geq 0, t \geq 0.$

Solution

Step 1 *Find the dual problem.* Write the primal problem in matrix form and take the transpose:

$$\begin{bmatrix} 2 & 1 & 20 \\ 1 & 2 & 20 \\ 3 & 4 & 50 \\ 2{,}000 & 3{,}000 & 0 \end{bmatrix} \rightarrow \begin{bmatrix} 2 & 1 & 3 & 2{,}000 \\ 1 & 2 & 4 & 3{,}000 \\ 20 & 20 & 50 & 0 \end{bmatrix}.$$

The dual problem is:

Maximize $p = 20x + 20y + 50z$
subject to $2x + y + 3z \leq 2{,}000$
$x + 2y + 4z \leq 3{,}000$
$x \geq 0, y \geq 0, z \geq 0.$

Step 2 *Use the simplex method to solve the dual problem.* Because we have a standard maximization problem, we do not have to worry about Phase I but go straight to Phase II.

	x	y	z	s	t	p	
s	2	1	③	1	0	0	2,000
t	1	2	4	0	1	0	3,000
p	−20	−20	−50	0	0	1	0

	x	y	z	s	t	p	
z	2	1	3	1	0	0	2,000
t	−5	②	0	−4	3	0	1,000
p	40	−10	0	50	0	3	100,000

	x	y	z	s	t	p	
z	9	0	6	6	−3	0	3,000
y	−5	2	0	−4	3	0	1,000
p	15	0	0	30	15	3	105,000

Note that the maximum value of the objective function is $p = 105,000/3 = 35,000$. By the theorem, this is also the optimal value of c in the primal problem!

Step 3 *Read off the solution to the primal problem by dividing the bottom entries in the columns associated with the slack variables by the entry in the p column.* Here is the final tableau again with the entries in question highlighted.

	x	y	z	s	t	p	
z	9	0	6	6	−3	0	3,000
y	−5	2	0	−4	3	0	1,000
p	15	0	0	**30**	**15**	**3**	**105,000**

The solution to the primal problem is

$$s = 30/3 = 10, t = 15/3 = 5, c = 105,000/3 = 35,000.$$

(Compare this with the method we used to solve Example 3 in the preceding section. Which method seems more efficient?)

➡ **Before we go on...**   Can you now see the reason for using the variable names $s, t, u, \ldots$ in standard minimization problems? ■

Q: *Is the theorem also useful for solving problems that do not satisfy the nonnegative objective condition?*

A: Consider a standard minimization problem that does not satisfy the nonnegative objective condition, such as

Minimize $c = 2s - t$
subject to $2s + 3t \geq 2$
$s + 2t \geq 2$
$s \geq 0, t \geq 0$.

Its dual would be

Maximize $\quad p = 2x + 2y$
subject to $\quad 2x + y \leq 2$
$\qquad\qquad 3x + 2y \leq -1$
$\qquad\qquad x \geq 0,\, y \geq 0.$

This is not a standard maximization problem because the right-hand side of the second constraint is negative. In general, if a problem does not satisfy the nonnegative objective condition, its dual is not standard. Therefore, to solve the dual by the simplex method will require using Phase I as well as Phase II, and we may as well just solve the primal problem that way to begin with. Thus, duality helps us solve problems only when the primal problem satisfies the nonnegative objective condition.

In many economic applications, the solution to the dual problem also gives us useful information about the primal problem, as we will see in the following example.

EXAMPLE 2 Shadow Costs

You are trying to decide how many vitamin pills to take. SuperV brand vitamin pills each contain 2 milligrams of vitamin X, 1 milligram of vitamin Y, and 1 milligram of vitamin Z. Topper brand vitamin pills each contain 1 milligram of vitamin X, 1 milligram of vitamin Y, and 2 milligrams of vitamin Z. You want to take enough pills daily to get at least 12 milligrams of vitamin X, 10 milligrams of vitamin Y, and 12 milligrams of vitamin Z. However, SuperV pills cost 4¢ each and Toppers cost 3¢ each, and you would like to minimize the total cost of your daily dosage. How many of each brand of pill should you take? How would changing your daily vitamin requirements affect your minimum cost?

Solution This is a straightforward minimization problem. The unknowns are

$s = $ number of SuperV brand pills
$t = $ number of Topper brand pills.

The linear programming problem is

Minimize $\quad c = 4s + 3t$
subject to $\quad 2s + t \geq 12$
$\qquad\qquads + t \geq 10$
$\qquad\qquads + 2t \geq 12$
$\qquad\qquad s \geq 0,\, t \geq 0.$

We solve this problem by using the simplex method on its dual, which is

Maximize $\quad p = 12x + 10y + 12z$
subject to $\quad 2x + y + z \leq 4$
$\qquad\qquadx + y + 2z \leq 3$
$\qquad\qquad x \geq 0,\, y \geq 0,\, z \geq 0.$

After pivoting three times, we arrive at the final tableau:

	x	y	z	s	t	p	
x	6	0	-6	6	-6	0	6
y	0	1	3	-1	2	0	2
p	0	0	6	2	8	1	32

Therefore, the answer to the original problem is that you should take two SuperV vitamin pills and eight Toppers at a cost of 32¢ per day.

Now, the key to answering the last question, which asks you to determine how changing your daily vitamin requirements would affect your minimum cost, is to look at the solution to the dual problem. From the tableau we see that $x = 1$, $y = 2$, and $z = 0$. To see what x, y, and z might tell us about the original problem, let's look at their units. In the inequality $2x + y + z \leq 4$, the coefficient 2 of x has units "mg of vitamin X/SuperV pill," and the 4 on the right-hand side has units "¢/SuperV pill." For $2x$ to have the same units as the 4 on the right-hand side, x must have units "¢/mg of vitamin X." Similarly, y must have units "¢/mg of vitamin Y" and z must have units "¢/mg of vitamin Z." One can show (although we will not do it here) that x gives the amount that would be added to the minimum cost for each increase* of 1 milligram of vitamin X in our daily requirement. For example, if we were to increase our requirement from 12 milligrams to 14 milligrams, an increase of 2 milligrams, the minimum cost would change by $2x = 2$¢, from 32¢ to 34¢. (Try it; you'll end up taking four SuperV pills and six Toppers.) Similarly, each increase of 1 milligram of vitamin Y in the requirements would increase the cost by $y = 2$¢. These costs are called the **marginal costs** or the **shadow costs** of the vitamins.

> * To be scrupulously correct, this works only for relatively small changes in the requirements, not necessarily for very large ones.

What about $z = 0$? The shadow cost of vitamin Z is 0¢/mg, meaning that you can increase your requirement of vitamin Z without changing your cost. In fact, the solution $s = 2$ and $t = 8$ provides you with 18 milligrams of vitamin Z, so you can increase the required amount of vitamin Z up to 18 milligrams without changing the solution at all.

We can also interpret the shadow costs as the effective cost to you of each milligram of each vitamin in the optimal solution. You are paying 1¢/milligram of vitamin X, 2¢/milligram of vitamin Y, and getting the vitamin Z for free. This gives a total cost of $1 \times 12 + 2 \times 10 + 0 \times 12 = 32$¢, as we know. Again, if you change your requirements slightly, these are the amounts you will pay per milligram of each vitamin.

Game Theory

We return to a topic we discussed in Section 4.4: solving two-person zero-sum games. In that section, we described how to solve games that could be reduced to 2×2 games or smaller. It turns out that we can solve larger games using linear programming and duality. We summarize the procedure, work through an example, and then discuss why it works.

Solving a Matrix Game

Step 1 Reduce the payoff matrix by dominance.

Step 2 Add a fixed number k to each of the entries so that they all become nonnegative and no column is all zero.

Step 3 Write 1s to the right of and below the matrix, and then write down the associated standard maximization problem. Solve this primal problem using the simplex method.

Step 4 Find the optimal strategies and the expected value as follows:

Column Strategy

1. Express the solution to the primal problem as a column vector.
2. Normalize by dividing each entry of the solution vector by p (which is also the sum of all the entries).
3. Insert zeros in positions corresponding to the columns deleted during reduction.

Row Strategy

1. Express the solution to the dual problem as a row vector.
2. Normalize by dividing each entry by p, which will once again be the sum of all the entries.
3. Insert zeros in positions corresponding to the rows deleted during reduction.

Value of the Game

$$e = \frac{1}{p} - k$$

EXAMPLE 3 **Restaurant Inspector**

You manage two restaurants, Tender Steaks Inn (TSI) and Break for a Steak (BFS). Even though you run the establishments impeccably, the Department of Health has been sending inspectors to your restaurants on a daily basis and fining you for minor infractions. You've found that you can head off a fine if you're present, but you can cover only one restaurant at a time. The Department of Health, on the other hand, has two inspectors, who sometimes visit the same restaurant and sometimes split up, one to each restaurant. The average fines you have been getting are shown in the following matrix.

		Health Inspectors		
		Both at BFS	*Both at TSI*	*One at Each*
You go to	**TSI**	$8,000	0	$2,000
	BFS	0	$10,000	$4,000

How should you choose which restaurant to visit to minimize your expected fine?

Solution This matrix is not quite the payoff matrix because fines, being penalties, should be negative payoffs. Thus, the payoff matrix is the following:

$$P = \begin{bmatrix} -8,000 & 0 & -2,000 \\ 0 & -10,000 & -4,000 \end{bmatrix}.$$

We follow the steps above to solve the game using the simplex method.

Step 1 There are no dominated rows or columns, so this game does not reduce.

Step 2 We add $k = 10,000$ to each entry so that none are negative, getting the following new matrix (with no zero column):

$$\begin{bmatrix} 2,000 & 10,000 & 8,000 \\ 10,000 & 0 & 6,000 \end{bmatrix}.$$

Step 3 We write 1s to the right and below this matrix:

$$\begin{bmatrix} 2,000 & 10,000 & 8,000 & 1 \\ 10,000 & 0 & 6,000 & 1 \\ 1 & 1 & 1 & 0 \end{bmatrix}.$$

The corresponding standard maximization problem is the following:

$$\begin{aligned} \text{Maximize} \quad & p = x + y + z \\ \text{subject to} \quad & 2,000x + 10,000y + 8,000z \le 1 \\ & 10,000x + 6,000z \le 1 \\ & x \ge 0, y \ge 0, z \ge 0. \end{aligned}$$

Step 4 We use the simplex method to solve this problem. After pivoting twice, we arrive at the final tableau:

	x	y	z	s	t	p	
y	0	50,000	34,000	5	−1	0	4
x	10,000	0	6,000	0	1	0	1
p	0	0	14,000	5	4	50,000	9

Column Strategy The solution to the primal problem is

$$\begin{bmatrix} x \\ y \\ z \end{bmatrix} = \begin{bmatrix} \frac{1}{10,000} \\ \frac{4}{50,000} \\ 0 \end{bmatrix}.$$

We divide each entry by $p = 9/50,000$, which is also the sum of the entries. This gives the optimal column strategy:

$$C = \begin{bmatrix} \frac{5}{9} \\ \frac{4}{9} \\ 0 \end{bmatrix}.$$

Thus, the inspectors' optimal strategy is to stick together, visiting BFS with probability 5/9 and TSI with probability 4/9.

Row Strategy The solution to the dual problem is

$$[s \quad t] = \begin{bmatrix} \dfrac{5}{50,000} & \dfrac{4}{50,000} \end{bmatrix}.$$

Once again, we divide by $p = 9/50,000$ to find the optimal row strategy:

$$R = \begin{bmatrix} \dfrac{5}{9} & \dfrac{4}{9} \end{bmatrix}.$$

Thus, you should visit TSI with probability 5/9 and BFS with probability 4/9.

Value of the Game Your expected average fine is

$$e = \frac{1}{p} - k = \frac{50,000}{9} - 10,000 = -\frac{40,000}{9} \approx -\$4,444.$$

➡ **Before we go on...** We owe you an explanation of why the procedure we used in Example 3 works. The main point is to understand how we turn a game into a linear programming problem. It's not hard to see that adding a fixed number k to all the payoffs will change only the payoff, increasing it by k, and not change the optimal strategies. So let's pick up Example 3 from the point where we were considering the following game:

$$P = \begin{bmatrix} 2{,}000 & 10{,}000 & 8{,}000 \\ 10{,}000 & 0 & 6{,}000 \end{bmatrix}.$$

We are looking for the optimal strategies R and C for the row and column players, respectively; if e is the value of the game, we will have $e = RPC$. Let's concentrate first on the column player's strategy $C = [u \ v \ w]^T$, where u, v, and w are the unknowns we want to find. Because e is the value of the game, if the column player uses the optimal strategy C and the row player uses any old strategy S, the expected value with these strategies has to be e or better for the column player, so $SPC \leq e$. Let's write that out for two particular choices of S. First, consider $S = [1 \ 0]$:

$$[1 \ 0] \begin{bmatrix} 2{,}000 & 10{,}000 & 8{,}000 \\ 10{,}000 & 0 & 6{,}000 \end{bmatrix} \begin{bmatrix} u \\ v \\ w \end{bmatrix} \leq e.$$

Multiplied out, this gives

$$2{,}000u + 10{,}000v + 8{,}000w \leq e.$$

Next, do the same thing for $S = [0 \ 1]$:

$$[0 \ 1] \begin{bmatrix} 2{,}000 & 10{,}000 & 8{,}000 \\ 10{,}000 & 0 & 6{,}000 \end{bmatrix} \begin{bmatrix} u \\ v \\ w \end{bmatrix} \leq e$$

$$10{,}000u + 6{,}000w \leq e.$$

It turns out that if these two inequalities are true, then $SPC \leq e$ for any S at all, which is what the column player wants. These are starting to look like constraints in a linear programming problem, but the e appearing on the right is in the way. We get around this by dividing by e, which we know to be positive because all of the payoffs are nonnegative and no column is all zero (so the column player can't force the value of the game to be 0; here is where we need these assumptions). We get the following inequalities:

$$2{,}000 \left(\frac{u}{e}\right) + 10{,}000 \left(\frac{v}{e}\right) + 8{,}000 \left(\frac{w}{e}\right) \leq 1$$

$$10{,}000 \left(\frac{u}{e}\right) + 6{,}000 \left(\frac{w}{e}\right) \leq 1.$$

Now we're getting somewhere. To make these look even more like linear constraints, we replace our unknowns u, v, and w with new unknowns, $x = u/e$, $y = v/e$, and $z = w/e$. Our inequalities then become.

$$2{,}000x + 10{,}000y + 8{,}000z \leq 1$$

$$10{,}000x + 6{,}000z \leq 1.$$

What about an objective function? From the point of view of the column player, the objective is to find a strategy that will minimize the expected value e. In order to write e in terms of our new variables x, y, and z, we use the fact that our original variables, being the entries in the column strategy, have to add up to 1: $u + v + w = 1$. Dividing by e gives

$$\frac{u}{e} + \frac{v}{e} + \frac{w}{e} = \frac{1}{e}$$

or

$$x + y + z = \frac{1}{e}.$$

Now we notice that, if we *maximize* $p = x + y + z = 1/e$, it will have the effect of minimizing e, which is what we want. So, we get the following linear programming problem:

Maximize $p = x + y + z$
subject to $2{,}000x + 10{,}000y + 8{,}000z \leq 1$
$10{,}000x \qquad\qquad\;\; + 6{,}000z \leq 1$
$x \geq 0, y \geq 0, z \geq 0.$

Why can we say that x, y, and z should all be nonnegative? Because the unknowns u, v, w, and e must all be nonnegative.

So now, if we solve this linear programming problem to find x, y, z, and p, we can find the column player's optimal strategy by computing $u = xe = x/p$, $v = y/p$, and $w = z/p$. Moreover, the value of the game is $e = 1/p$. (If we added k to all the payoffs, we should now adjust by subtracting k again to find the correct value of the game.)

Turning now to the row player's strategy, if we repeat the above type of argument from the row player's viewpoint, we'll end up with the following linear programming problem to solve:

Minimize $c = s + t$
subject to $2{,}000s + 10{,}000t \geq 1$
$10{,}000s \qquad\qquad\; \geq 1$
$8{,}000s + \;\; 6{,}000t \geq 1$
$s \geq 0, t \geq 0.$

This is, of course, the dual to the problem we solved to find the column player's strategy, so we know that we can read its solution off of the same final tableau. The optimal value of c will be the same as the value of p, so $c = 1/e$ also. The entries in the optimal row strategy will be s/c and t/c. ∎

FAQs

When to Use Duality

Q: *Given a minimization problem, when should I use duality, and when should I use the two-phase method in Section 5.4?*

A: If the original problem satisfies the nonnegative objective condition (none of the coefficients in the objective function are negative), then you can use duality to convert the problem to a standard maximization one, which can be solved with the one-phase method. If the original problem does not satisfy the nonnegative objective condition, then dualizing results in a nonstandard LP problem, so dualizing may not be worthwhile.

Q: *When is it absolutely necessary to use duality?*

A: Never. Duality gives us an efficient but not necessary alternative for solving standard minimization problems.

5.5 EXERCISES

▼ more advanced   ◆ challenging
Ⓣ indicates exercises that should be solved using technology

In Exercises 1–8, write down (without solving) the dual LP problem. HINT [See the Quick Example on page 371.]

1. Maximize $p = 2x + y$
subject to $x + 2y \le 6$
         $-x + y \le 2$
         $x \ge 0, y \ge 0.$

2. Maximize $p = x + 5y$
subject to $x + y \le 6$
         $-x + 3y \le 4$
         $x \ge 0, y \ge 0.$

3. Minimize $c = 2s + t + 3u$
subject to $s + t + u \ge 100$
        $2s + t \ge 50$
        $s \ge 0, t \ge 0, u \ge 0.$

4. Minimize $c = 2s + 2t + 3u$
subject to $s + u \ge 100$
        $2s + t \ge 50$
        $s \ge 0, t \ge 0, u \ge 0.$

5. Maximize $p = x + y + z + w$
subject to $x + y + z \le 3$
        $y + z + w \le 4$
        $x + z + w \le 5$
        $x + y + w \le 6$
        $x \ge 0, y \ge 0, z \ge 0, w \ge 0.$

6. Maximize $p = x + y + z + w$
subject to $x + y + z \le 3$
        $y + z + w \le 3$
        $x + z + w \le 4$
        $x + y + w \le 4$
        $x \ge 0, y \ge 0, z \ge 0, w \ge 0.$

7. Minimize $c = s + 3t + u$
subject to $5s - t + v \ge 1,000$
          $u - v \ge 2,000$
       $s + t \ge 500$
       $s \ge 0, t \ge 0, u \ge 0, v \ge 0.$

8. Minimize $c = 5s + 2u + v$
subject to $s - t + 2v \ge 2,000$
          $u + v \ge 3,000$
       $s + t \ge 500$
       $s \ge 0, t \ge 0, u \ge 0, v \ge 0.$

In Exercises 9–22, solve the standard minimization problems using duality. (You may already have seen some of them in earlier sections, but now you will be solving them using a different method.) HINT [See Example 1.]

9. Minimize $c = s + t$
subject to $s + 2t \ge 6$
        $2s + t \ge 6$
        $s \ge 0, t \ge 0.$

10. Minimize $c = s + 2t$
subject to $s + 3t \ge 30$
        $2s + t \ge 30$
        $s \ge 0, t \ge 0.$

11. Minimize $c = 6s + 6t$
subject to $s + 2t \ge 20$
        $2s + t \ge 20$
        $s \ge 0, t \ge 0.$

12. Minimize $c = 3s + 2t$
subject to $s + 2t \ge 20$
        $2s + t \ge 10$
        $s \ge 0, t \ge 0.$

13. Minimize $c = 0.2s + 0.3t$
subject to $2s + t \ge 10$
        $s + 2t \ge 10$
        $s + t \ge 8$
        $s \ge 0, t \ge 0.$

14. Minimize $c = 0.4s + 0.1t$
subject to $3s + 2t \ge 60$
        $s + 2t \ge 40$
        $2s + 3t \ge 45$
        $s \ge 0, t \ge 0.$

15. Minimize $c = 2s + t$
subject to $3s + t \ge 30$
        $s + t \ge 20$
        $s + 3t \ge 30$
        $s \ge 0, t \ge 0.$

16. Minimize $c = s + 2t$
subject to $4s + t \ge 100$
        $2s + t \ge 80$
        $s + 3t \ge 150$
        $s \ge 0, t \ge 0.$

17. Minimize $c = s + 2t + 3u$
subject to $3s + 2t + u \ge 60$
        $2s + t + 3u \ge 60$
        $s \ge 0, t \ge 0, u \ge 0.$

18. Minimize $c = s + t + 2u$
subject to $s + 2t + 2u \geq 60$
$2s + t + 3u \geq 60$
$s \geq 0, t \geq 0, u \geq 0.$

19. Minimize $c = 2s + t + 3u$
subject to $s + t + u \geq 100$
$2s + t \geq 50$
$t + u \geq 50$
$s \geq 0, t \geq 0, u \geq 0.$

20. Minimize $c = 2s + 2t + 3u$
subject to $s + u \geq 100$
$2s + t \geq 50$
$t + u \geq 50$
$s \geq 0, t \geq 0, u \geq 0.$

21. Minimize $c = s + t + u$
subject to $3s + 2t + u \geq 60$
$2s + t + 3u \geq 60$
$s + 3t + 2u \geq 60$
$s \geq 0, t \geq 0, u \geq 0.$

22. Minimize $c = s + t + 2u$
subject to $s + 2t + 2u \geq 60$
$2s + t + 3u \geq 60$
$s + 3t + 6u \geq 60$
$s \geq 0, t \geq 0, u \geq 0.$

In Exercises 23–28, solve the games with the given payoff matrices. HINT [See Example 3.]

23. $P = \begin{bmatrix} -1 & 1 & 2 \\ 2 & -1 & -2 \end{bmatrix}$

24. $P = \begin{bmatrix} 1 & -1 & 2 \\ 1 & 2 & 0 \end{bmatrix}$

25. $P = \begin{bmatrix} -1 & 1 & 2 \\ 2 & -1 & -2 \\ 1 & 2 & 0 \end{bmatrix}$

26. $P = \begin{bmatrix} 1 & -1 & 2 \\ 1 & 2 & 0 \\ 0 & 1 & 1 \end{bmatrix}$

27. ⊤ $P = \begin{bmatrix} -1 & 1 & 2 & -1 \\ 2 & -1 & -2 & -3 \\ 1 & 2 & 0 & 1 \\ 0 & 2 & 3 & 3 \end{bmatrix}$

28. ⊤ $P = \begin{bmatrix} 1 & -1 & 2 & 0 \\ 1 & 2 & 0 & 1 \\ 0 & 1 & 1 & 0 \\ 2 & 0 & -2 & 2 \end{bmatrix}$

APPLICATIONS

Exercises 29–38 are similar to ones in preceding exercise sets. Use duality to answer them.

29. *Nutrition* Meow makes cat food out of fish and cornmeal. Fish has 8 grams of protein and 4 grams of fat per ounce, and cornmeal has 4 grams of protein and 8 grams of fat. A jumbo can of cat food must contain at least 48 grams of protein and 48 grams of fat. If fish and cornmeal both cost 5¢/ounce, how many ounces of each should Meow use in each can of cat food to minimize costs? What are the shadow costs of protein and of fat? HINT [See Example 2.]

30. *Nutrition* Oz makes lion food out of giraffe and gazelle meat. Giraffe meat has 18 grams of protein and 36 grams of fat per pound, while gazelle meat has 36 grams of protein and 18 grams of fat per pound. A batch of lion food must contain at least 36,000 grams of protein and 54,000 grams of fat. Giraffe meat costs $2/pound and gazelle meat costs $4/pound. How many pounds of each should go into each batch of lion food in order to minimize costs? What are the shadow costs of protein and fat? HINT [See Example 2.]

31. *Nutrition* Ruff makes dog food out of chicken and grain. Chicken has 10 grams of protein and 5 grams of fat/ounce, and grain has 2 grams of protein and 2 grams of fat/ounce. A bag of dog food must contain at least 200 grams of protein and at least 150 grams of fat. If chicken costs 10¢/ounce and grain costs 1¢/ounce, how many ounces of each should Ruff use in each bag of dog food in order to minimize cost? What are the shadow costs of protein and fat?

32. *Purchasing* The Enormous State University's Business School is buying computers. The school has two models to choose from, the Pomegranate and the iZac. Each Pomegranate comes with 400 GB of memory and 80 TB of disk space, while each iZac has 300 GB of memory and 100 TB of disk space. For reasons related to its accreditation, the school would like to be able to say that it has a total of at least 48,000 GB of memory and at least 12,800 TB of disk space. If both the Pomegranate and the iZac cost $2,000 each, how many of each should the school buy to keep the cost as low as possible? What are the shadow costs of memory and disk space?

33. *Nutrition* Each serving of **Gerber** Mixed Cereal for Baby contains 60 calories and no vitamin C. Each serving of Gerber Mango Tropical Fruit Dessert contains 80 calories and 45 percent of the U.S. Recommended Daily Allowance (RDA) of vitamin C for infants. Each serving of Gerber Apple Banana Juice contains 60 calories and 120 percent of the U.S. RDA of vitamin C for infants.[39] The cereal costs 10¢/serving, the dessert costs 53¢/serving, and the juice costs 27¢/serving. If you want to provide your child with at least 120 calories and at least 120 percent of the U.S. RDA of vitamin C, how can you do so at the least cost? What are your shadow costs for calories and vitamin C?

34. *Nutrition* Each serving of Gerber Mixed Cereal for Baby contains 60 calories, no vitamin C, and 11 grams of carbohydrates. Each serving of Gerber Mango Tropical Fruit Dessert contains 80 calories, 45 percent of the U.S. Recommended Daily Allowance (RDA) of vitamin C for infants, and 21 grams of carbohydrates. Each serving of Gerber Apple Banana Juice contains 60 calories, 120 percent of the U.S. RDA of vitamin C for infants, and 15 grams of carbohydrates.[40]

---

[39]Source: Nutrition information supplied with the products.
[40]*Ibid.*

Assume that the cereal costs 11¢/serving, the dessert costs 50¢/serving, and the juice costs 30¢/serving. If you want to provide your child with at least 180 calories, at least 120 percent of the U.S. RDA of vitamin C, and at least 37 grams of carbohydrates, how can you do so at the least cost? What are your shadow costs for calories, vitamin C, and carbohydrates?

35. *Politics* The political pollster Canter is preparing for a national election. It would like to poll at least 1,500 Democrats and 1,500 Republicans. Each mailing to the East Coast gets responses from 100 Democrats and 50 Republicans. Each mailing to the Midwest gets responses from 100 Democrats and 100 Republicans. And each mailing to the West Coast gets responses from 50 Democrats and 100 Republicans. Mailings to the East Coast cost $40 each to produce and mail, mailings to the Midwest cost $60 each, and mailings to the West Coast cost $50 each. How many mailings should Canter send to each area of the country to get the responses it needs at the least possible cost? What will it cost? What are the shadow costs of a Democratic response and a Republican response?

36. *Purchasing* *Bingo's Copy Center* needs to buy white paper and yellow paper. Bingo's can buy from three suppliers. Harvard Paper sells a package of 20 reams of white and 10 reams of yellow for $60; Yale Paper sells a package of 10 reams of white and 10 reams of yellow for $40, and Dartmouth Paper sells a package of 10 reams of white and 20 reams of yellow for $50. If Bingo's needs 350 reams of white and 400 reams of yellow, how many packages should it buy from each supplier so as to minimize the cost? What is the lowest possible cost? What are the shadow costs of white paper and yellow paper?

37. ▼ *Resource Allocation* One day Gillian the Magician summoned the wisest of her women. "Devoted followers," she began, "I have a quandary: As you well know, I possess great expertise in sleep spells and shock spells, but unfortunately, these are proving to be a drain on my aural energy resources; each sleep spell costs me 500 pico-shirleys of aural energy, while each shock spell requires 750 pico-shirleys. Clearly, I would like to hold my overall expenditure of aural energy to a minimum, and still meet my commitments in protecting the Sisterhood from the ever-present threat of trolls. Specifically, I have estimated that each sleep spell keeps us safe for an average of 2 minutes, while every shock spell protects us for about 3 minutes. We certainly require enough protection to last 24 hours of each day, and possibly more, just to be safe. At the same time, I have noticed that each of my sleep spells can immobilize 3 trolls at once, while one of my typical shock spells (having a narrower range) can immobilize only 2 trolls at once. We are faced, my sisters, with an onslaught of 1,200 trolls per day! Finally, as you are no doubt aware, the bylaws dictate that for a Magician of the Order to remain in good standing, the number of shock spells must be between one quarter and one third the number of shock and sleep spells combined. What do I do, oh Wise Ones?"

38. ▼ *Risk Management* The Grand Vizier of the Kingdom of Um is being blackmailed by numerous individuals and is having a very difficult time keeping his blackmailers from going public. He has been keeping them at bay with two kinds of payoff: gold bars from the Royal Treasury and political favors. Through bitter experience, he has learned that each payoff in gold gives him peace for an average of about one month, and each political favor seems to earn him about a month and a half of reprieve. To maintain his flawless reputation in the court, he feels he cannot afford any revelations about his tainted past to come to light within the next year. Thus, it is imperative that his blackmailers be kept at bay for 12 months. Furthermore, he would like to keep the number of gold payoffs at no more than one quarter of the combined number of payoffs because the outward flow of gold bars might arouse suspicion on the part of the Royal Treasurer. The gold payoffs tend to deplete the Grand Vizier's travel budget. (The treasury has been subsidizing his numerous trips to the Himalayas.) He estimates that each gold bar removed from the treasury will cost him four trips. On the other hand, because the administering of political favors tends to cost him valuable travel time, he suspects that each political favor will cost him about two trips. Now, he would obviously like to keep his blackmailers silenced and lose as few trips as possible. What is he to do? How many trips will he lose in the next year?

39. ▼ *Game Theory—Politics* Incumbent Tax N. Spend and challenger Trick L. Down are running for county executive, and polls show them to be in a dead heat. The election hinges on three cities: Littleville, Metropolis, and Urbantown. The candidates have decided to spend the last weeks before the election campaigning in those three cities; each day each candidate will decide in which city to spend the day. Pollsters have determined the following payoff matrix, where the payoff represents the number of votes gained or lost for each one-day campaign trip.

T. N. Spend

		Littleville	Metropolis	Urbantown
	Littleville	−200	−300	300
T. L. Down	Metropolis	−500	500	−100
	Urbantown	−500	0	0

What percentage of time should each candidate spend in each city in order to maximize votes gained? If both candidates use their optimal strategies, what is the expected vote?

40. ▼ *Game Theory—Marketing* Your company's new portable phone/music player/PDA/bottle washer, the RunMan, will compete against the established market leader, the iNod, in a saturated market. (Thus, for each device you sell, one fewer iNod is sold.) You are planning to launch the RunMan with a traveling road show, concentrating on two cities,

New York and Boston. The makers of the iNod will do the same to try to maintain their sales. If, on a given day, you both go to New York, you will lose 1,000 units in sales to the iNod. If you both go to Boston, you will lose 750 units in sales. On the other hand, if you go to New York and your competitor to Boston, you will gain 1,500 units in sales from them. If you go to Boston and they to New York, you will gain 500 units in sales. What percentage of time should you spend in New York and what percentage in Boston, and how do you expect your sales to be affected?

41. ▼ *Game Theory—Morra Games* A three-finger *Morra game* is a game in which two players simultaneously show one, two, or three fingers at each round. The outcome depends on a predetermined set of rules. Here is an interesting example: If the numbers of fingers shown by A and B differ by 1, then A loses one point. If they differ by more than 1, the round is a draw. If they show the same number of fingers, A wins an amount equal to the sum of the fingers shown. Determine the optimal strategy for each player and the expected value of the game.

42. ⊞ ▼ *Game Theory—Morra Games* Referring to the preceding exercise, consider the following rules for a three-finger Morra game: If the sum of the fingers shown is odd, then A wins an amount equal to that sum. If the sum is even, B wins the sum. Determine the optimal strategy for each player and the expected value of the game. HINT [Use technology to do the pivoting in the associated linear programming problem.]

43. ⊞ ◆ *Game Theory—Military Strategy* Colonel Blotto is a well-known game in military strategy.[41] Here is a version of this game: Colonel Blotto has four regiments under his command, while his opponent, Captain Kije, has three. The armies are to try to occupy two locations, and each commander must decide how many regiments to send to each location. The army that sends more regiments to a location captures that location as well as the other army's regiments. If both armies send the same number of regiments to a location, then there is a draw. The payoffs are one point for each

location captured and one point for each regiment captured. Find the optimum strategy for each commander and also the value of the game.

44. ⊞ ◆ *Game Theory—Military Strategy* Referring to the preceding exercise, consider the version of Colonel Blotto with the same payoffs given there except that Captain Kije earns two points for each location captured, while Colonel Blotto continues to earn only one point. Find the optimum strategy for each commander and also the value of the game. Round all figures to two decimal places.

COMMUNICATION AND REASONING EXERCISES

45. Give one possible advantage of using duality to solve a standard minimization problem.

46. To ensure that the dual of a minimization problem will result in a standard maximization problem,
   (A) the primal problem should satisfy the nonnegative objective condition.
   (B) the primal problem should be a standard minimization problem.
   (C) the primal problem should not satisfy the nonnegative objective condition.

47. Give an example of a standard minimization problem whose dual is *not* a standard maximization problem. How would you go about solving your problem?

48. Give an example of a nonstandard minimization problem whose dual is a standard maximization problem.

49. ▼ Given a minimization problem, when would you solve it by applying the simplex method to its dual, and when would you apply the simplex method to the minimization problem itself?

50. ▼ Create an interesting application that leads to a standard maximization problem. Solve it using the simplex method and note the solution to its dual problem. What does the solution to the dual tell you about your application?

[41]See Samuel Karlin, *Mathematical Methods and Theory in Games, Programming and Economics* (Addison-Wesley, 1959).

KEY CONCEPTS

Website www.WanerMath.com
Go to the Website at www.WanerMath
.com to find a comprehensive and
interactive Web-based summary
of Chapter 5.

5.1 Graphing Linear Inequalities

Inequalities, strict and nonstrict *p. 313*
Linear inequalities *p. 313*
Solution of an inequality *p. 313*
Sketching the region represented by a
linear inequality in two variables
p. 315
Bounded and unbounded regions *p. 318*
Feasible region *p. 318*

5.2 Solving Linear Programming Problems Graphically

Linear programming (LP) problem in
two unknowns; objective function;
constraints; optimal value; optimal
solution *p. 322*
Feasible region *p. 323*
Fundamental Theorem of Linear
Programming *p. 323*

Graphical method for solving an
LP problem *p. 324*
Decision variables *p. 325*
Procedure for solving an LP problem
with an unbounded feasible
region *p. 328*

5.3 The Simplex Method: Solving Standard Maximization Problems

General linear programming problem in
n unknowns *p. 339*
Standard maximization problem
p. 339
Slack variable *p. 340*
Tableau *p. 341*
Active (or basic) variables; inactive
(or nonbasic) variables; basic
solution *p. 341*
Rules for selecting the pivot
column *p. 342*
Rules for selecting the pivot; test
ratios *p. 343*
Departing or exiting variable, entering
variable *p. 343*

5.4 The Simplex Method: Solving General Linear Programming Problems

Surplus variable *p. 356*
Phase I and Phase II for solving
general LP problems *p. 356*
Using the simplex method to solve
a minimization problem *p. 361*

5.5 The Simplex Method and Duality

Standard minimization problem
p. 369
Standard LP problem *p. 369*
Nonnegative objective condition
p. 369
Dual LP problems; primal problem;
dual problem *p. 370*
Fundamental Theorem of Duality
p. 371
Shadow costs *p. 374*
Game theory: The LP problem
associated with a two-person
zero-sum game *p. 374*

REVIEW EXERCISES

In each of Exercises 1–4, sketch the region corresponding to the given inequalities, say whether it is bounded, and give the coordinates of all corner points.

1. $2x - 3y \leq 12$

2. $x \leq 2y$

3. $x + 2y \leq 20$
$3x + 2y \leq 30$
$x \geq 0, y \geq 0$

4. $3x + 2y \geq 6$
$2x - 3y \leq 6$
$3x - 2y \geq 0$
$x \geq 0, y \geq 0$

In each of Exercises 5–8, solve the given linear programming problem graphically.

5. Maximize $p = 2x + y$
subject to $3x + y \leq 30$
$x + y \leq 12$
$x + 3y \leq 30$
$x \geq 0, y \geq 0.$

6. Maximize $p = 2x + 3y$
subject to $x + y \geq 10$
$2x + y \geq 12$
$x + y \leq 20$
$x \geq 0, y \geq 0.$

7. Minimize $c = 2x + y$
subject to $3x + y \geq 30$
$x + 2y \geq 20$
$2x - y \geq 0$
$x \geq 0, y \geq 0.$

8. Minimize $c = 3x + y$
subject to $3x + 2y \geq 6$
$2x - 3y \leq 0$
$3x - 2y \geq 0$
$x \geq 0, y \geq 0.$

In each of Exercises 9–18, solve the given linear programming problem using the simplex method. If no optimal solution exists, indicate whether the feasible region is empty or the objective function is unbounded.

9. Maximize $p = x + y + 2z$
subject to $x + 2y + 2z \leq 60$
$2x + y + 3z \leq 60$
$x \geq 0, y \geq 0, z \geq 0.$

10. Maximize $p = x + y + 2z$
subject to $x + 2y + 2z \leq 60$
$2x + y + 3z \leq 60$
$x + 3y + 6z \leq 60$
$x \geq 0, y \geq 0, z \geq 0.$

11. Maximize $p = x + y + 3z$
subject to $x + y + z \geq 100$
$y + z \leq 80$
$x \quad + z \leq 80$
$x \geq 0, y \geq 0, z \geq 0.$

12. Maximize $p = 2x + y$
subject to $x + 2y \geq 12$
$2x + y \leq 12$
$x + y \leq 5$
$x \geq 0, y \geq 0.$

13. Minimize $c = x + 2y + 3z$
subject to $3x + 2y + z \geq 60$
$2x + y + 3z \geq 60$
$x \geq 0, y \geq 0, z \geq 0.$

14. Minimize $c = 5x + 4y + 3z$
subject to $x + y + 4z \geq 30$
$2x + y + 3z \geq 60$
$x \geq 0, y \geq 0, z \geq 0.$

15. ⬛ Minimize $c = x - 2y + 4z$
subject to $3x + 2y - z \geq 10$
$2x + y + 3z \geq 20$
$x + 3y - 2z \geq 30$
$x \geq 0, y \geq 0, z \geq 0.$

16. ⬛ Minimize $c = x + y - z$
subject to $3x + 2y + z \geq 60$
$2x + y + 3z \geq 60$
$x + 3y + 2z \geq 60$
$x \geq 0, y \geq 0, z \geq 0.$

17. Minimize $c = x + y + z + w$
subject to $x + y \qquad \geq 30$
$x \quad + z \quad \geq 20$
$x + y \quad - w \leq 10$
$y + z - w \leq 10$
$x \geq 0, y \geq 0, z \geq 0, w \geq 0.$

18. Minimize $c = 4x + y + z + w$
subject to $x + y \qquad \geq 30$
$y - z \qquad \leq 20$
$z - w \leq 10$
$x \geq 0, y \geq 0, z \geq 0, w \geq 0.$

In each of Exercises 19–22, solve the given linear programming problem using duality.

19. Minimize $c = 2x + y$
subject to $3x + 2y \geq 60$
$2x + y \geq 60$
$x + 3y \geq 60$
$x \geq 0, y \geq 0.$

20. Minimize $c = 2x + y + 2z$
subject to $3x + 2y + z \geq 100$
$2x + y + 3z \geq 200$
$x \geq 0, y \geq 0, z \geq 0.$

21. Minimize $c = 2x + y$
subject to $3x + 2y \geq 10$
$2x - y \leq 30$
$x + 3y \geq 60$
$x \geq 0, y \geq 0.$

22. Minimize $c = 2x + y + 2z$
subject to $3x - 2y + z \geq 100$
$2x + y - 3z \leq 200$
$x \geq 0, y \geq 0, z \geq 0.$

In each of Exercises 23–26, solve the game with the given payoff matrix.

23. $P = \begin{bmatrix} -1 & 2 & -1 \\ 1 & -2 & 1 \\ 3 & -1 & 0 \end{bmatrix}$ **24.** $P = \begin{bmatrix} -3 & 0 & 1 \\ -4 & 0 & 0 \\ 0 & -1 & -2 \end{bmatrix}$

25. $P = \begin{bmatrix} -3 & -2 & 3 \\ 1 & 0 & 0 \\ -2 & 2 & 1 \end{bmatrix}$ **26.** $P = \begin{bmatrix} -4 & -2 & -3 \\ 1 & -3 & -2 \\ -3 & 1 & -4 \end{bmatrix}$

Exercises 27–30 are adapted from the Actuarial Exam on Operations Research.

27. You are given the following linear programming problem:

Minimize $c = x + 2y$
subject to $-2x + y \geq 1$
$x - 2y \geq 1$
$x \geq 0, y \geq 0.$

Which of the following is true?

(A) The problem has no feasible solutions.
(B) The objective function is unbounded.
(C) The problem has optimal solutions.

28. Repeat the preceding exercise with the following linear programming problem:

Maximize $p = x + y$
subject to $-2x + y \leq 1$
$x - 2y \leq 2$
$x \geq 0, y \geq 0.$

29. Determine the optimal value of the objective function. You are given the following linear programming problem.

Maximize $Z = x_1 + 4x_2 + 2x_3 - 10$
subject to $4x_1 + x_2 + x_3 \leq 45$
$-x_1 + x_2 + 2x_3 \leq 0$
$x_1, x_2, x_3 \geq 0.$

30. Determine the optimal value of the objective function. You are given the following linear programming problem.

Minimize $Z = x_1 + 4x_2 + 2x_3 + x_4 + 40$
subject to $4x_1 + x_2 + x_3 \leq 45$
$-x_1 + 2x_2 + x_4 \geq 40$
$x_1, x_2, x_3 \geq 0.$

APPLICATIONS: OHaganBooks.com

In Exercises 31–34, you are the buyer for OHaganBooks.com and are considering increasing stocks of romance and horror novels at the new OHaganBooks.com warehouse in Texas. You have offers from several publishers: Duffin House, Higgins Press, McPhearson Imprints, and O'Conell Books. Duffin offers a package of 5 horror novels and 5 romance novels for $50, Higgins offers a package of 5 horror and 10 romance novels for $80, McPhearson offers a package of 10 horror novels and 5 romance novels for $80, and

O'Conell offers a package of 10 horror novels and 10 romance novels for $90.

31. How many packages should you purchase from Duffin House and Higgins Press to obtain at least 4,000 horror novels and 6,000 romance novels at minimum cost? What is the minimum cost?

32. How many packages should you purchase from McPhearson Imprints and O'Conell Books to obtain at least 5,000 horror novels and 4,000 romance novels at minimum cost? What is the minimum cost?

33. Refer to the scenario in Exercise 31. As it turns out, John O'Hagan promised Marjory Duffin that OHaganBooks.com would buy at least 20 percent more packages from Duffin as from Higgins, but you still want to obtain at least 4,000 horror novels and 6,000 romance novels at minimum cost.

   a. *Without solving the problem,* say which of the following statements are possible:

   (A) The cost will stay the same.
   (B) The cost will increase.
   (C) The cost will decrease.
   (D) It will be impossible to meet all the conditions.
   (E) The cost will become unbounded.

   b. If you wish to meet all the requirements at minimum cost, how many packages should you purchase from each publisher? What is the minimum cost?

34. Refer to Exercise 32. You are about to place the order meeting the requirements of Exercise 32 when you are told that you can order no more than a total of 500 packages, and that at least half of the packages should be from McPhearson. Explain why this is impossible by referring to the feasible region for Exercise 32.

35. *Investments* Marjory Duffin's portfolio manager has suggested two high-yielding stocks: European Emerald Emporium (EEE) and Royal Ruby Retailers (RRR).[42] EEE shares cost $50, yield 4.5% in dividends, and have a risk index of 2.0 per share. RRR shares cost $55, yield 5% in dividends, and have a risk index of 3.0 per share. Marjory has up to $12,100 to invest and would like to earn at least $550 in dividends. How many shares of each stock should she purchase to meet her requirements and minimize the total risk index for her portfolio? What is the minimum total risk index?

36. *Investments* Marjory Duffin's other portfolio manager has suggested another two high-yielding stocks: Countrynarrow Mortgages (CNM) and Scotland Subprime (SS).[43] CNM shares cost $40, yield 5.5% in dividends, and have a risk index of 1.0 per share. SS shares cost $25, yield 7.5% in

dividends, and have a risk index of 1.5 per share. Marjory can invest up to $30,000 in these stocks and would like to earn at least $1,650 in dividends. How many shares of each stock should she purchase in order to meet her requirements and minimize the total risk index for her portfolio?

37. *Resource Allocation* Billy-Sean O'Hagan has joined the Physics Society at Suburban State University, and the group is planning to raise money to support the dying space program by making and selling umbrellas. The society intends to make three models: the Sprinkle, the Storm, and the Hurricane. The amounts of cloth, metal, and wood used in making each model are given in this table:

	Sprinkle	Storm	Hurricane	Total Available
Cloth (sq. yd)	1	2	2	600
Metal (lbs)	2	1	3	600
Wood (lbs)	1	3	6	600
Profit ($)	1	1	2	

The table also shows the amounts of each material available in a given day and the profits to be made from each model. How many of each model should the society make in order to maximize its profit?

38. *Profit* Duffin House, which is now the largest publisher of books sold at the OHaganBooks.com site, prints three kinds of books: paperback, quality paperback, and hardcover. The amounts of paper, ink, and time on the presses required for each kind of book are given in this table:

	Paperback	Quality Paperback	Hardcover	Total Available
Paper (pounds)	3	2	1	6,000
Ink (gallons)	2	1	3	6,000
Time (minutes)	10	10	10	22,000
Profit ($)	1	2	3	

The table also lists the total amounts of paper, ink, and time available in a given day and the profits made on each kind of book. How many of each kind of book should Duffin print to maximize profit?

39. *Purchases* You are just about to place book orders from Duffin and Higgins (see Exercise 31) when everything changes: Duffin House informs you that, due to a global romance crisis, its packages now each will contain 5 horror novels but only 2 romance novels and still cost $50 per package. Packages from Higgins will now contain 10 of each type of novel, but now cost $150 per package. Ewing Books enters the fray and offers its own package of 5 horror and 5 romance novels for $100. The sales manager now tells you that at least 50% of the packages must come from Higgins Press

---

[42]RRR and EEE happen to be, respectively, the ticker symbols of RSC Holdings (an equipment rental provider) and Evergreen Energy Inc. (a environmentally friendly energy technology company) and thus have nothing to do with rubies and emeralds.

[43]CNM is actually the ticker symbol of Carnegie Wave, whereas SS is not the ticker symbol of any U.S.-based company we are aware of.

and, as before, you want to obtain at least 4,000 horror novels and 6,000 romance novels at minimum cost. Taking all of this into account, how many packages should you purchase from each publisher? What is the minimum cost?

40. *Purchases* You are about to place book orders from McPhearson and O'Conell (see Exercise 32) when you get an e-mail from McPhearson Imprints saying that, sorry, but they have stopped publishing romance novels due to the global romance crisis and can now offer only packages of 10 horror novels for $50. O'Conell is still offering packages of 10 horror novels and 10 romance novels for $90, and now the United States Treasury, in an attempt to bolster the floundering romance industry, is offering its own package of 20 romance novels for $120. Furthermore, Congress, in approving this measure, has passed legislation dictating that at least two thirds of the packages in every order must come from the U.S. Treasury. As before, you wish to obtain at least 5,000 horror novels and 4,000 romance novels at minimum cost. Taking all of this into account, how many packages should you purchase from each supplier? What is the minimum cost?

41. *Degree Requirements* During his lunch break, John O'Hagan decides to devote some time to assisting his son Billy-Sean, who continues to have a terrible time planning his college course schedule. The latest Bulletin of Suburban State University claims to have added new flexibility to its course requirements, but it remains as complicated as ever. It reads as follows:

> *All candidates for the degree of Bachelor of Arts at SSU must take at least 120 credits from the Sciences, Fine Arts, Liberal Arts, and Mathematics combined, including at least as many Science credits as Fine Arts credits, and at most twice as many Mathematics credits as Science credits, but with Liberal Arts credits exceeding Mathematics credits by no more than one third of the number of Fine Arts credits.*

Science and fine arts credits cost $300 each, and liberal arts and mathematics credits cost $200 each. John would like to have Billy-Sean meet all the requirements at a minimum total cost.

a. Set up (without solving) the associated linear programming problem.
b. ⊞ Use technology to determine how many of each type of credit Billy-Sean should take. What will the total cost be?

42. *Degree Requirements* No sooner had the "new and flexible" course requirement been released than the English Department again pressured the University Senate to include their vaunted "Verbal Expression" component in place of the fine arts requirement in all programs (including the sciences):

> *All candidates for the degree of Bachelor of Science at SSU must take at least 120 credits from the Liberal Arts, Sciences, Verbal Expression, and Mathematics, including at most as many Science credits as Liberal Arts credits, and at least twice as many Verbal Expression credits as Science credits and Liberal Arts*

credits combined, with Liberal Arts credits exceeding Mathematics credits by at least a quarter of the number of Verbal Expression credits.

Science credits cost $300 each, while each credit in the remaining subjects now costs $400. John would like to have Billy-Sean meet all the requirements at a minimum total cost.

a. Set up (without solving) the associated linear programming problem.
b. ⊞ Use technology to determine how many of each type of credit Billy-Sean should take. What will the total cost be?

43. *Shipping* On the same day that the sales department at Duffin House received an order for 600 packages from the OHaganBooks.com Texas headquarters, it received an additional order for 200 packages from FantasyBooks.com, based in California. Duffin House has warehouses in New York and Illinois. The New York warehouse has 600 packages in stock, but the Illinois warehouse is closing down and has only 300 packages in stock. Shipping costs per package of books are as follows: New York to Texas: $20; New York to California: $50; Illinois to Texas: $30; Illinois to California: $40. What is the lowest total shipping cost for which Duffin House can fill the orders? How many packages should be sent from each warehouse to each online bookstore at a minimum shipping cost?

44. *Transportation Scheduling* Duffin House is about to start a promotional blitz for its new book, *Advanced String Theory for the Liberal Arts*. The company has 25 salespeople stationed in Austin and 10 in San Diego, and would like to fly at least 15 to sales fairs in each of Houston and Cleveland. A round-trip plane flight from Austin to Houston costs $200; from Austin to Cleveland costs $150; from San Diego to Houston costs $400; and from San Diego to Cleveland costs $200. How many salespeople should the company fly from each of Austin and San Diego to each of Houston and Cleveland for the lowest total cost in air fare?

45. *Marketing* Marjory Duffin, head of Duffin House, reveals to John O'Hagan that FantasyBooks.com is considering several promotional schemes: It may offer two books for the price of one, three books for the price of two, or possibly a free copy of *Brain Surgery for Klutzes* with each order. OHaganBooks.com's marketing advisers Floody and O'Lara seem to have different ideas as to how to respond. Floody suggests offering *three* books for the price of one, while O'Lara suggests instead offering a free copy of the *Finite Mathematics Student Solutions Manual* with every purchase. After a careful analysis, O'Hagan comes up with the following payoff matrix, where the payoffs represent the number of customers, in thousands, O'Hagan expects to gain from FantasyBooks.com.

	FantasyBooks.com			
	No Promo	2 for Price of 1	3 for Price of 2	Brain Surgery
No Promo	0	−60	−40	10
OHaganBooks.com 3 for Price of 1	30	20	10	15
Finite Math	20	0	15	10

Find the optimal strategies for both companies and the expected shift in customers.

46. *Study Techniques* Billy-Sean's friend Pat from college has been spending all of his time in fraternity activities, and thus knows absolutely nothing about any of the three topics on tomorrow's math test. He has turned to Billy-Sean for advice as to how to spend his "all-nighter." The table below shows the scores Pat could expect to earn if the entire test were to be in a specific subject. (Because he knows no linear programming or matrix algebra, the table shows, for instance, that studying game theory all night will not be much use in preparing him for this topic.)

	Test		
Pat's Strategies ↓	**Game Theory**	**Linear Programming**	**Matrix Algebra**
Study Game Theory	30	0	20
Study Linear Programming	0	70	0
Study Matrix Algebra	0	0	70

What percentage of the night should Pat spend on each topic, assuming the principles of game theory, and what score can he expect to get?

Case Study

The Diet Problem

Image Studios/UpperCut Images/Getty Images

The Galaxy Nutrition health-food mega-store chain provides free online nutritional advice and support to its customers. As Web site technical consultant, you are planning to construct an interactive Web page to assist customers in preparing a diet tailored to their nutritional and budgetary requirements. Ideally, the customer would select foods to consider and specify nutritional and/or budgetary constraints, and the tool should return the optimal diet meeting those requirements. You would also like the Web page to allow the customer to decide whether, for instance, to find the cheapest possible diet meeting the requirements, the diet with the lowest number of calories, or the diet with the least total carbohydrates.

After doing a little research, you notice that the kind of problem you are trying is solve is quite well known and referred to as the *diet problem*, and that solving the diet problem is a famous example of linear programming. Indeed, there are already some online pages that solve versions of the problem that minimize total cost, so you have adequate information to assist you as you plan the page.*

You decide to start on a relatively small scale, starting with a program that uses a list of 10 foods, and minimizes either total caloric intake or total cost and satisfies a small list of requirements. Following is a small part of a table of nutritional information from the demo at the NEOS Wiki (all the values shown are for a single serving) as well as approximate minimum daily requirements:

✱ See, for instance, the Diet Problem Demo at the NEOS Wiki: www.neos-guide.org/NEOS/index.php/Diet_Problem_Demo

	Price per Serving	Calories	Total Fat g	Carbs g	Dietary Fiber g	Protein g	Vit C IU
Tofu	$0.31	88.2	5.5	2.2	1.4	9.4	0.1
Roast Chicken	$0.84	277.4	10.8	0	0	42.2	0
Spaghetti w/Sauce	$0.78	358.2	12.3	58.3	11.6	8.2	27.9
Tomato	$0.27	25.8	0.4	5.7	1.4	1.0	23.5
Oranges	$0.15	61.6	0.2	15.4	3.1	1.2	69.7
Wheat Bread	$0.05	65.0	1.0	12.4	1.3	2.2	0
Cheddar Cheese	$0.25	112.7	9.3	0.4	0	7.0	0
Oatmeal	$0.82	145.1	2.3	25.3	4.0	6.1	0
Peanut Butter	$0.07	188.5	16.0	6.9	2.1	7.7	0
White Tuna in Water	$0.69	115.6	2.1	0	0	22.7	0
Minimum Requirements		2,200	20	80	25	60	90

Source: www.neos-guide.org/NEOS/index.php/Diet_Problem_Demo

Now you get to work. As always, you start by identifying the unknowns. Since the output of the Web page will consist of a recommended diet, the unknowns should logically be the number of servings of each item of food selected by the user. In your first trial run, you decide to include all the 10 food items listed, so you take

x_1 = Number of servings of tofu

x_2 = Number of servings of roast chicken

$\vdots$

x_{10} = Number of servings of white tuna in water.

You now set up a linear programming problem for two sample scenarios:

Scenario 1 (Minimum Cost): Satisfy all minimum nutritional requirements at a minimum cost. Here the linear programming problem is:

Minimize

$$c = 0.31x_1 + 0.84x_2 + 0.78x_3 + 0.27x_4 + 0.15x_5 + 0.05x_6 + 0.25x_7$$
$$+ 0.82x_8 + 0.07x_9 + 0.69x_{10}$$

subject to

$$88.2x_1 + 277.4x_2 + 358.2x_3 + 25.8x_4 + 61.6x_5 + 65x_6 + 112.7x_7$$
$$+ 145.1x_8 + 188.5x_9 + 115.6x_{10} \geq 2{,}200$$

$$5.5x_1 + 10.8x_2 + 12.3x_3 + 0.4x_4 + 0.2x_5 + 1x_6 + 9.3x_7 + 2.3x_8$$
$$+ 16x_9 + 2.1x_{10} \geq 20$$

$$2.2x_1 + 58.3x_3 + 5.7x_4 + 15.4x_5 + 12.4x_6 + 0.4x_7 + 25.3x_8 + 6.9x_9 \geq 80$$

$$1.4x_1 + 11.6x_3 + 1.4x_4 + 3.1x_5 + 1.3x_6 + 4x_8 + 2.1x_9 \geq 25$$

$$9.4x_1 + 42.2x_2 + 8.2x_3 + 1x_4 + 1.2x_5 + 2.2x_6 + 7x_7 + 6.1x_8 + 7.7x_9$$
$$+ 22.7x_{10} \geq 60$$

$$0.1x_1 + 27.9x_3 + 23.5x_4 + 69.7x_5 \geq 90.$$

This is clearly the kind of linear programming problem no one in their right mind would like to do by hand (solving it requires 16 tableaus!) so you decide to use the online simplex method tool at the Website (www.WanerMath.com → Student Web Site → On Line Utilities → Simplex Method Tool).

Here is a picture of the input, entered almost exactly as written above (you need to enter each constraint on a new line, and "Minimize c = 0.31x1 + ··· Subject to" must be typed on a single line):

```
        Type your linear programming problem below. (Press "Example" to see how to set it up.)
Minimize   c =
0.31x1+0.84x2+0.78x3+0.27x4+0.15x5+0.05x6+0.25x7+0.82x8+0.07x9+0.69x10 Subject to
88.2x1+277.4x2+358.2x3+25.8x4+61.6x5+65x6+112.7x7+145.1x8+188.5x9+115.6x10 >= 2200
5.5x1+10.8x2+12.3x3+0.4x4+0.2x5+1x6+9.3x7+2.3x8+16x9+2.1x10 >= 20
2.2x1+58.3x3+5.7x4+15.4x5+12.4x6+0.4x7+25.3x8+6.9x9 >= 80
1.4x1+11.6x3+1.4x4+3.1x5+1.3x6+4x8+2.1x9 >= 25
9.4x1+42.2x2+8.2x3+1x4+1.2x5+2.2x6+7x7+6.1x8+7.7x9+22.7x10 >= 60
0.1x1+0x2+27.9x3+23.5x4+69.7x5 >= 90
```

Clicking "Solve" results in the following solution:

$$c = 0.981126;\ x_1 = 0,\ x_2 = 0,\ x_3 = 0,\ x_4 = 0,\ x_5 = 1.29125,\ x_6 = 0,\ x_7 = 0,$$
$$x_8 = 0,\ x_9 = 11.2491,\ x_{10} = 0.$$

6

Sets and Counting

Website

www.WanerMath.com

At the Website you will find:

- Section-by-section tutorials, including game tutorials with randomized quizzes

- A detailed chapter summary

- A true/false quiz

- A utility to compute factorials, permutations, and combinations

- Additional review exercises

Case Study Designing a Puzzle

As Product Design Manager for Cerebral Toys, Inc., you are constantly on the lookout for ideas for intellectually stimulating yet inexpensive toys. Your design team recently came up with an idea for a puzzle consisting of a number of plastic cubes. Each cube will have two faces colored red, two white, and two blue, and there will be exactly two cubes with each possible configuration of colors. The goal of the puzzle is to seek out the matching pairs, thereby enhancing a child's geometric intuition and three-dimensional manipulation skills. **If the kit is to include every possible configuration of colors, how many cubes will the kit contain?**

Image Source/Getty Images

395

Introduction

The theory of sets is the foundation for most of mathematics. It also has direct applications—for example, in searching computer databases. We will use set theory extensively in the chapter on probability, and thus much of this chapter revolves around the idea of a **set of outcomes** of a procedure such as rolling a pair of dice or choosing names from a database. Also important in probability is the theory of **counting** the number of elements in a set, which is called **combinatorics**.

Counting elements is not a trivial proposition; for example, the betting game Lotto (used in many state lotteries) has you pick six numbers from some range—say, 1–55. If your six numbers match the six numbers chosen in the "official drawing," you win the top prize. How many Lotto tickets would you need to buy to guarantee that you will win? That is, how many Lotto tickets are possible? By the end of this chapter, we will be able to answer these questions.

6.1   Sets and Set Operations

In this section we introduce some of the basic ideas of set theory. Some of the examples and applications we see here are derived from the theory of probability and will recur throughout the rest of this chapter and the next.

Sets

Sets and Elements

A **set** is a collection of items, referred to as the **elements** of the set.

Visualizing a Set

Set

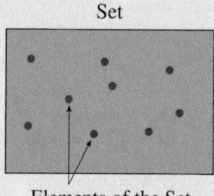

Elements of the Set

Quick Examples

We usually use a capital letter to name a set and braces to enclose the elements of a set.

$W = \{\text{Amazon, eBay, Apple}\}$
$N = \{1, 2, 3, \ldots\}$

$x \in A$ means that x **is an element of** the set A. If x is not an element of A, we write $x \notin A$.

Amazon $\in W$ (W as above)
Microsoft $\notin W$ $2 \in N$

$B = A$ means that A and B have the same elements. The order in which the elements are listed does not matter.

$\{5, -9, 1, 3\} = \{-9, 1, 3, 5\}$
$\{1, 2, 3, 4\} \neq \{1, 2, 3, 6\}$

$B \subseteq A$ means that B is a **subset** of A; every element of B is also an element of A.

$\{\text{eBay, Apple}\} \subseteq W$
$\{1, 2, 3, 4\} \subseteq \{1, 2, 3, 4\}$

$B \subset A$ means that B is a **proper subset** of A: $B \subseteq A$, but $B \neq A$.

$\{\text{eBay, Apple}\} \subset W$
$\{1, 2, 3\} \subset \{1, 2, 3, 4\}$
$\{1, 2, 3\} \subset N$ (N as above)

Ø is the **empty set**, the set containing no elements. It is a subset of every set.	$\emptyset \subseteq W$ $\emptyset \subset W$
A **finite** set has finitely many elements. An **infinite** set does not have finitely many elements.	$W = \{$Amazon, eBay, Apple$\}$ is a finite set. $N = \{1, 2, 3, \ldots\}$ is an infinite set.

One type of set we'll use often is the **set of outcomes** of some activity or experiment. For example, if we toss a coin and observe which side faces up, there are two possible outcomes, heads (H) and tails (T). The set of outcomes of tossing a coin once can be written

$$S = \{H, T\}.$$

As another example, suppose we roll a die that has faces numbered 1 through 6, as usual, and observe which number faces up. The set of outcomes *could* be represented as

$$S = \left\{ \boxed{\,\cdot\,}, \boxed{\because}, \boxed{\therefore}, \boxed{::}, \boxed{\vdots\cdot}, \boxed{:::} \right\}.$$

However, we can much more easily write

$$S = \{1, 2, 3, 4, 5, 6\}.$$

EXAMPLE 1 Two Dice: Distinguishable vs. Indistinguishable

a. Suppose we have two dice that we can distinguish in some way—say, one is green and one is red. If we roll both dice, what is the set of outcomes?

b. Describe the set of outcomes if the dice are indistinguishable.

Solution

a. A systematic way of laying out the set of outcomes for a distinguishable pair of dice is shown in Figure 1.

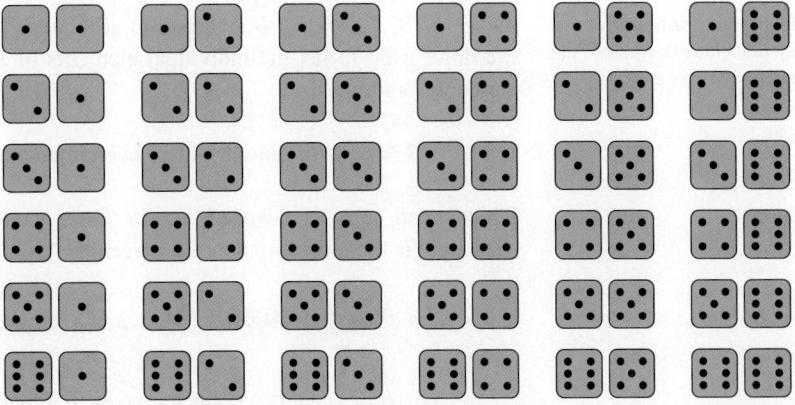

Figure 1

In the first row all the green dice show a 1, in the second row a 2, in the third row a 3, and so on. Similarly, in the first column all the red dice show a 1, in the second column a 2, and so on. The diagonal pairs (top left to bottom right) show all the "doubles." Using the picture as a guide, we can write the set of 36 outcomes as follows.

$$S = \begin{Bmatrix} (1,1),\ (1,2),\ (1,3),\ (1,4),\ (1,5),\ (1,6), \\ (2,1),\ (2,2),\ (2,3),\ (2,4),\ (2,5),\ (2,6), \\ (3,1),\ (3,2),\ (3,3),\ (3,4),\ (3,5),\ (3,6), \\ (4,1),\ (4,2),\ (4,3),\ (4,4),\ (4,5),\ (4,6), \\ (5,1),\ (5,2),\ (5,3),\ (5,4),\ (5,5),\ (5,6), \\ (6,1),\ (6,2),\ (6,3),\ (6,4),\ (6,5),\ (6,6) \end{Bmatrix}$$

Distinguishable dice

Notice that S is also the set of outcomes if we roll a single die twice, if we take the first number in each pair to be the outcome of the first roll and the second number the outcome of the second roll.

b. If the dice are truly indistinguishable, we will have no way of knowing which die is which once they are rolled. Think of placing two identical dice in a closed box and then shaking the box. When we look inside afterward, there is no way to tell which die is which. (If we make a small marking on one of the dice or somehow keep track of it as it bounces around, we are *distinguishing* the dice.) We regard two dice as **indistinguishable** if we make no attempt to distinguish them. Thus, for example, the two different outcomes $(1,3)$ and $(3,1)$ from part (a) would represent the same outcome in part (b) (one die shows a 3 and the other a 1). Because the set of outcomes should contain each outcome only once, we can remove $(3,1)$. Following this approach gives the following smaller set of outcomes:

$$S = \begin{Bmatrix} (1,1),\ (1,2),\ (1,3),\ (1,4),\ (1,5),\ (1,6), \\ (2,2),\ (2,3),\ (2,4),\ (2,5),\ (2,6), \\ (3,3),\ (3,4),\ (3,5),\ (3,6), \\ (4,4),\ (4,5),\ (4,6), \\ (5,5),\ (5,6), \\ (6,6) \end{Bmatrix}.$$

Indistinguishable dice

EXAMPLE 2 Set-Builder Notation

✳ Note that the **nonnegative** integers *include* 0, whereas the **positive** integers *exclude* 0.

Let $B = \{0, 2, 4, 6, 8\}$. B is the set of all nonnegative✳ even integers less than 10. If we don't want to list the individual elements of B, we can instead use "set-builder notation," and write

$$B = \{n \mid n \text{ is a nonnegative even integer less than } 10\}.$$

This is read "*B is the set of all n such that n is a nonnegative even integer less than 10.*" Here is the correspondence between the words and the symbols:

B is the set of all n such that n is a nonnegative even integer less than 10

$$B = \{n \mid n \text{ is a nonnegative even integer less than } 10\}.$$

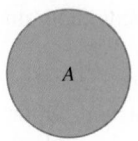

Figure 2

Venn Diagrams

We can visualize sets and relations between sets using **Venn diagrams**. In a Venn diagram, we represent a set as a region, often a disk (Figure 2).

   The elements of A are the points inside the region. The following Venn diagrams illustrate the relations we've discussed so far.

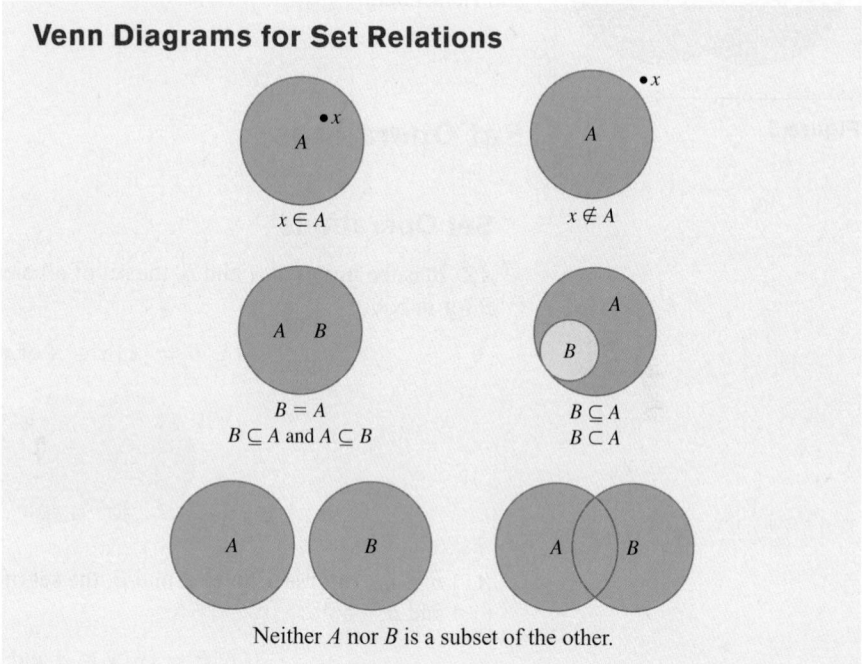

Venn Diagrams for Set Relations

Note Although the diagram for $B \subseteq A$ suggests a proper subset, it is customary to use the same diagram for both subsets and proper subsets. ■

EXAMPLE 3 **Customer Interests**

NobelBooks.com (a fierce competitor of OHaganBooks.com) maintains a database of customers and the types of books they have purchased. In the company's database is the set of customers

   $S = \{$Einstein, Bohr, Millikan, Heisenberg, Schrödinger, Dirac$\}$.

A search of the database for customers who have purchased cookbooks yields the subset

   $A = \{$Einstein, Bohr, Heisenberg, Dirac$\}$.

Another search, this time for customers who have purchased mysteries, yields the subset

   $B = \{$Bohr, Heisenberg, Schrödinger$\}$.

NobelBooks.com wants to promote a new combination mystery/cookbook, and wants to target two subsets of customers: those who have purchased either cookbooks or mysteries (or both) and, for additional promotions, those who have purchased both cookbooks and mysteries. Name the customers in each of these subsets.

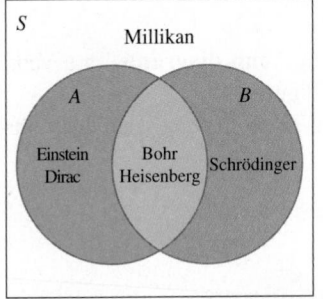

Figure 3

Solution We can picture the database and the two subsets using the Venn diagram in Figure 3.

The set of customers who have purchased either cookbooks or mysteries (or both) consists of the customers who are in A or B or both: Einstein, Bohr, Heisenberg, Schrödinger, and Dirac. The set of customers who have purchased both cookbooks and mysteries consists of the customers in the overlap of A and B, Bohr and Heisenberg.

Set Operations

Set Operations

$A \cup B$ is the **union** of A and B, the set of all elements that are either in A or in B (or in both).

$$A \cup B = \{x \mid x \in A \text{ or } x \in B\}$$

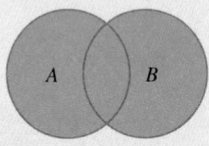

$A \cap B$ is the **intersection** of A and B, the set of all elements that are common to A and B.

$$A \cap B = \{x \mid x \in A \text{ and } x \in B\}$$

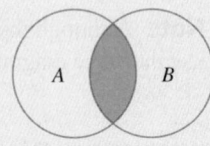

Logical Equivalents
Union: For an element to be in $A \cup B$, it must be in A **or** in B.
Intersection: For an element to be in $A \cap B$, it must be in A **and** in B.

Quick Examples

If $A = \{a, b, c, d\}$ and $B = \{c, d, e, f\}$, then

$$A \cup B = \{a, b, c, d, e, f\}$$
$$A \cap B = \{c, d\}.$$

Note Mathematicians always use "or" in its *inclusive* sense: one thing or another *or both.* ∎

There is one other operation we use, called the **complement** of a set A, which, roughly speaking, is the set of things *not* in A.

Q: *Why only "roughly"? Why not just form the set of things not in A?*

A: This would amount to assuming that there is a set of *all things*. (It would be the complement of the empty set.) Although tempting, talking about entities such as the "set of all things" leads to paradoxes.* Instead, we first need to fix a set S of *all objects under consideration,* or the *universe of discourse,* which we generally call the **universal set** for the discussion. For example, when we search the Web, we take S to be the set of all Web pages. When talking about integers, we take S to be the set of all integers. In other words, our choice of universal set depends on the context. The complement of a set $A \subseteq S$ is then the set of *things in S* that are not in A.

Complement

If S is the universal set and $A \subseteq S$, then A' is the **complement** of A (in S), the set of all elements of S not in A.

$$A' = \{x \in S \mid x \notin A\} = \text{Green Region Below}$$

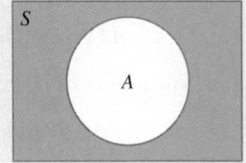

Logical Equivalent

For an element to be in A', it must be in S but **not** in A.

Quick Example

If $S = \{a, b, c, d, e, f, g\}$ and $A = \{a, b, c, d\}$, then

$$A' = \{e, f, g\}.$$

In the following example we use set operations to describe the sets we found in Example 3, as well as some others.

EXAMPLE 4 **Customer Interests**

NobelBooks.com maintains a database of customers and the types of books they have purchased. In the company's database is the set of customers

$S = \{$Einstein, Bohr, Millikan, Heisenberg, Schrödinger, Dirac$\}$.

A search of the database for customers who have purchased cookbooks yields the subset

$A = \{$Einstein, Bohr, Heisenberg, Dirac$\}$.

Another search, this time for customers who have purchased mysteries, yields the subset

$B = \{$Bohr, Heisenberg, Schrödinger$\}$.

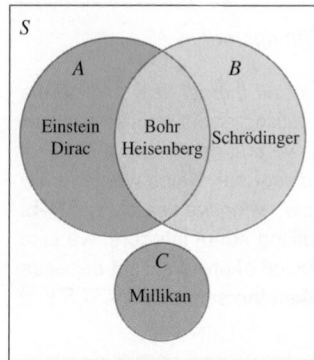

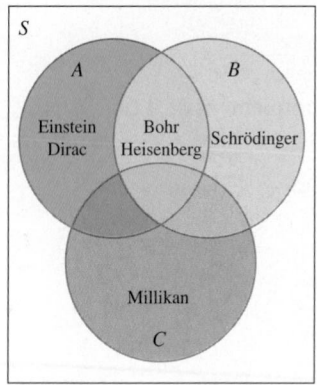

Figure 4

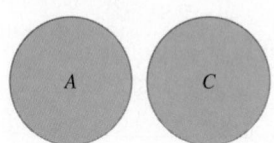

Figure 5

A third search, for customers who had registered with the site but not used their first-time customer discount, yields the subset

$$C = \{\text{Millikan}\}.$$

Use set operations to describe the following subsets:

a. The subset of customers who have purchased either cookbooks or mysteries

b. The subset of customers who have purchased both cookbooks and mysteries

c. The subset of customers who have not purchased cookbooks

d. The subset of customers who have purchased cookbooks but have not used their first-time customer discount

Solution Figure 4 shows two alternative Venn diagram representations of the database. Although the second version shows C overlapping A and B, the placement of the names inside shows that there are no customers in those overlaps.

a. The subset of customers who have bought either cookbooks *or* mysteries is

$$A \cup B = \{\text{Einstein, Bohr, Heisenberg, Schrödinger, Dirac}\}.$$

b. The subset of customers who have bought both cookbooks *and* mysteries is

$$A \cap B = \{\text{Bohr, Heisenberg}\}.$$

c. The subset of customers who have *not* bought cookbooks is

$$A' = \{\text{Millikan, Schrödinger}\}.$$

Note that, for the universal set, we are using the set S of all customers in the database.

d. The subset of customers who have bought cookbooks but have not used their first-time purchase discount is the empty set

$$A \cap C = \emptyset.$$

When the intersection of two sets is empty, we say that the two sets are **disjoint**. In a Venn diagram, disjoint sets are drawn as regions that don't overlap, as in Figure 5.*

---

* People new to set theory sometimes find it strange to consider the empty set a valid set. Here is one of the times where it is very useful to do so. If we did not, we would have to say that $A \cap C$ was defined only when A and C had something in common. Having to deal with the fact that this set operation was not always defined would quickly get tiresome.

➡ **Before we go on...** Computer databases and the Web can be searched using so-called "Boolean searches." These are search requests using "and," "or," and "not." Using "and" gives the intersection of separate searches, using "or" gives the union, and using "not" gives the complement. In the next section we'll see how Web search engines allow such searches. ■

Cartesian Product

There is one more set operation we need to discuss.

Cartesian Product

The **Cartesian product** of two sets, A and B, is the set of all ordered pairs (a, b) with $a \in A$ and $b \in B$.

$$A \times B = \{(a, b) \mid a \in A \text{ and } b \in B\}$$

In words, $A \times B$ is the set of all ordered pairs whose first component is in A and whose second component is in B.

Quick Examples

1. If $A = \{a, b\}$ and $B = \{1, 2, 3\}$, then

$$A \times B = \{(a, 1), (a, 2), (a, 3), (b, 1), (b, 2), (b, 3)\}.$$

Visualizing A ✕ B

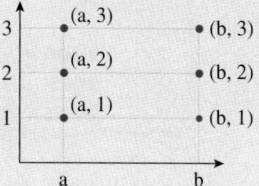

2. If $S = \{H, T\}$, then

$$S \times S = \{(H, H), (H, T), (T, H), (T, T)\}.$$

In other words, if S is the set of outcomes of tossing a coin once, then $S \times S$ is the set of outcomes of tossing a coin twice.

3. If $S = \{1, 2, 3, 4, 5, 6\}$, then

$$S \times S = \begin{cases} (1,1), & (1,2), & (1,3), & (1,4), & (1,5), & (1,6), \\ (2,1), & (2,2), & (2,3), & (2,4), & (2,5), & (2,6), \\ (3,1), & (3,2), & (3,3), & (3,4), & (3,5), & (3,6), \\ (4,1), & (4,2), & (4,3), & (4,4), & (4,5), & (4,6), \\ (5,1), & (5,2), & (5,3), & (5,4), & (5,5), & (5,6), \\ (6,1), & (6,2), & (6,3), & (6,4), & (6,5), & (6,6) \end{cases}.$$

In other words, if S is the set of outcomes of rolling a die once, then $S \times S$ is the set of outcomes of rolling a die twice (or rolling two distinguishable dice).

4. If $A = \{\text{red, yellow}\}$ and $B = \{\text{Mustang, Firebird}\}$, then

$A \times B = \{(\text{red, Mustang}), (\text{red, Firebird}), (\text{yellow, Mustang}), (\text{yellow, Firebird})\}$ which we might also write as

$A \times B = \{\text{red Mustang, red Firebird, yellow Mustang, yellow Firebird}\}.$

EXAMPLE 5 Representing Cartesian Products

The manager of an automobile dealership has collected data on the number of pre-owned Acura, Infiniti, Lexus, and Mercedes cars the dealership has from the 2009, 2010, and 2011 model years. In entering this information on a spreadsheet, the manager would like to have each spreadsheet cell represent a particular year and make. Describe this set of cells.

Solution Because each cell represents a year and a make, we can think of the cell as a pair (year, make), as in (2009, Acura). Thus, the set of cells can be thought of as a Cartesian product:

$$Y = \{2009, 2010, 2011\}$$ Year of car

$$M = \{\text{Acura, Infiniti, Lexus, Mercedes}\}$$ Make of car

$$Y \times M = \left\{ \begin{array}{l} \text{(2009, Acura) (2009, Infiniti) (2009, Lexus) (2009, Mercedes)} \\ \text{(2010, Acura) (2010, Infiniti) (2010, Lexus) (2010, Mercedes)} \\ \text{(2011, Acura) (2011, Infiniti) (2011, Lexus) (2011, Mercedes)} \end{array} \right\}.$$ Cells

Thus, the manager might arrange the spreadsheet as follows:

	A	B	C	D	E
1		**Acura**	**Infiniti**	**Lexus**	**Mercedes**
2	**2009**	(2009 Acura)	(2009 Infiniti)	(2009 Lexus)	(2009 Mercedes)
3	**2010**	(2010 Acura)	(2010 Infiniti)	(2010 Lexus)	(2010 Mercedes)
4	**2011**	(2011 Acura)	(2011 Infiniti)	(2011 Lexus)	(2011 Mercedes)

The highlighting shows the 12 cells to be filled in, representing the numbers of cars of each year and make. For example, in cell B2 should go the number of 2009 Acuras the dealership has.

➡ **Before we go on...** The arrangement in the spreadsheet in Example 5 is consistent with the matrix notation in Chapter 2. We could also have used the elements of Y as column labels along the top and the elements of M as row labels down the side. Along those lines, we can also visualize the Cartesian product $Y \times M$ as a set of points in the xy-plane ("Cartesian plane") as shown in Figure 6.

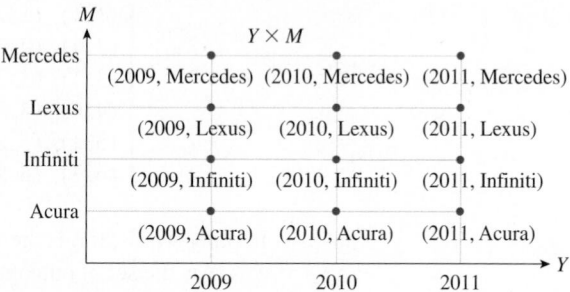

Figure 6

FAQs

The Many Meanings of "And"

Q: *Suppose A is the set of actors and B is the set of all baseball players. Then the set of all actors and baseball players is A ∩ B, right?*

A: Wrong. The fact that the word "and" appears in the description of a set does not always mean that the set is an intersection; the word "and" can mean different things in different contexts. $A \cap B$ refers to the set of elements that are in both A and B, hence to actors who are also baseball players. On the other hand, the set of all actors and baseball players is the set of people who are either actors or baseball players (or both), which is $A \cup B$. We can use the word "and" to describe both sets:

$A \cap B = \{$people who are both actors *and* baseball players$\}$

$A \cup B = \{$people who are actors *or* baseball players$\}$

     $= \{$all actors *and* baseball players$\}$.

6.1 EXERCISES

▼ more advanced   ◆ challenging

T indicates exercises that should be solved using technology

List the elements in each of the sets in Exercises 1–16.

1. The set F consisting of the four seasons

2. The set A consisting of the authors of this book

3. The set I of all positive integers no greater than 6

4. The set N of all negative integers greater than -3

5. $A = \{n \mid n$ is a positive integer and $0 \le n \le 3\}$ HINT [See Example 2.]

6. $A = \{n \mid n$ is a positive integer and $0 < n < 8\}$ HINT [See Example 2.]

7. $B = \{n \mid n$ is an even positive integer and $0 \le n \le 8\}$

8. $B = \{n \mid n$ is an odd positive integer and $0 \le n \le 8\}$

9. The set of all outcomes of tossing a pair of **(a)** distinguishable coins **(b)** indistinguishable coins HINT [See Example 1.]

10. The set of outcomes of tossing three **(a)** distinguishable coins **(b)** indistinguishable coins HINT [See Example 1.]

11. The set of all outcomes of rolling two distinguishable dice such that the numbers add to 6

12. The set of all outcomes of rolling two distinguishable dice such that the numbers add to 8

13. The set of all outcomes of rolling two indistinguishable dice such that the numbers add to 6

14. The set of all outcomes of rolling two indistinguishable dice such that the numbers add to 8

15. The set of all outcomes of rolling two distinguishable dice such that the numbers add to 13

16. The set of all outcomes of rolling two distinguishable dice such that the numbers add to 1

In each of Exercises 17–20, draw a Venn diagram that illustrates the relationships among the given sets. HINT [See Example 3.]

17. $S = \{$eBay, Google™, Amazon, OHaganBooks, Hotmail$\}$, $A = \{$Amazon, OHaganBooks$\}$, $B = \{$eBay, Amazon$\}$, $C = \{$Amazon, Hotmail$\}$

18. $S = \{$Apple, Dell, Gateway, Pomegranate, Compaq$\}$, $A = \{$Gateway, Pomegranate, Compaq$\}$, $B = \{$Dell, Gateway, Pomegranate, Compaq$\}$, $C = \{$Apple, Dell, Compaq$\}$

19. $S = \{$eBay, Google™, Amazon, OHaganBooks, Hotmail$\}$, $A = \{$Amazon, Hotmail$\}$, $B = \{$eBay, Google™, Amazon, Hotmail$\}$, $C = \{$Amazon, Hotmail$\}$

20. $S = \{$Apple, Dell, Gateway, Pomegranate, Compaq$\}$, $A = \{$Apple, Dell, Pomegranate, Compaq$\}$, $B = \{$Pomegranate$\}$, $C = \{$Pomegranate$\}$

Let $A = \{$June, Janet, Jill, Justin, Jeffrey, Jello$\}$, $B = \{$Janet, Jello, Justin$\}$, and $C = \{$Sally, Solly, Molly, Jolly, Jello$\}$. Find each set in Exercises 21–34. HINT [See Quick Examples on page 400.]

21. $A \cup B$ **22.** $A \cup C$

23. $A \cup \varnothing$ **24.** $B \cup \varnothing$

25. $A \cup (B \cup C)$ **26.** $(A \cup B) \cup C$

27. $C \cap B$ **28.** $C \cap A$

29. $A \cap \varnothing$ **30.** $\varnothing \cap B$

31. $(A \cap B) \cap C$ **32.** $A \cap (B \cap C)$

33. $(A \cap B) \cup C$ **34.** $A \cup (B \cap C)$

In Exercises 35–42, $A = \{$small, medium, large$\}$, $B = \{$blue, green$\}$, and $C = \{$triangle, square$\}$. HINT [See Quick Examples on page 403.]

35. List the elements of $A \times C$.

36. List the elements of $B \times C$.

37. List the elements of $A \times B$.

38. The elements of $A \times B \times C$ are the ordered triples (a, b, c) with $a \in A$, $b \in B$, and $c \in C$. List all the elements of $A \times B \times C$.

39. Represent $B \times C$ as cells in a spreadsheet. HINT [See Example 5.]

40. Represent $A \times C$ as cells in a spreadsheet. HINT [See Example 5.]

41. Represent $A \times B$ as cells in a spreadsheet.

42. Represent $A \times A$ as cells in a spreadsheet.

Let $A = \{$H, T$\}$ be the set of outcomes when a coin is tossed, and let $B = \{1, 2, 3, 4, 5, 6\}$ be the set of outcomes when a die is rolled. Write each set in Exercises 43–46 in terms of A and/or B and list its elements.

43. The set of outcomes when a die is rolled and then a coin tossed

44. The set of outcomes when a coin is tossed twice

45. The set of outcomes when a coin is tossed three times

46. The set of outcomes when a coin is tossed twice and then a die is rolled

Let S be the set of outcomes when two distinguishable dice are rolled, let E be the subset of outcomes in which at least one die shows an even number, and let F be the subset of outcomes in which at least one die shows an odd number. List the elements in each subset given in Exercises 47–52.

47. E' **48.** F'

49. $(E \cup F)$ **50.** $(E \cap F)'$

51. $E' \cup F'$ **52.** $E' \cap F'$

Use Venn diagrams to illustrate the following identities for subsets A, B, and C of S.

53. ▼ $(A \cup B)' = A' \cap B'$ DeMorgan's law
54. ▼ $(A \cap B)' = A' \cup B'$ DeMorgan's law
55. ▼ $(A \cap B) \cap C = A \cap (B \cap C)$ Associative law
56. ▼ $(A \cup B) \cup C = A \cup (B \cup C)$ Associative law
57. ▼ $A \cup (B \cap C) = (A \cup B) \cap (A \cup C)$ Distributive law
58. ▼ $A \cap (B \cup C) = (A \cap B) \cup (A \cap C)$ Distributive law
59. ▼ $S' = \emptyset$ **60.** ▼ $\emptyset' = S$

APPLICATIONS

Databases A freelance computer consultant keeps a database of her clients, which contains the names

$S = \{$Acme, Brothers, Crafts, Dion, Effigy, Floyd, Global, Hilbert$\}$.

The following clients owe her money:

$A = \{$Acme, Crafts, Effigy, Global$\}$.

The following clients have done at least $10,000 worth of business with her:

$B = \{$Acme, Brothers, Crafts, Dion$\}$.

The following clients have employed her in the last year:

$C = \{$Acme, Crafts, Dion, Effigy, Global, Hilbert$\}$.

In Exercises 61–68, a subset of clients is described that the consultant could find using her database. Write each subset in terms of A, B, and C and list the clients in that subset.
HINT [See Example 4.]

61. The clients who owe her money and have done at least $10,000 worth of business with her

62. The clients who owe her money or have done at least $10,000 worth of business with her

63. The clients who have done at least $10,000 worth of business with her or have employed her in the last year

64. The clients who have done at least $10,000 worth of business with her and have employed her in the last year

65. The clients who do not owe her money and have employed her in the last year

66. The clients who do not owe her money or have employed her in the last year

67. ▼ The clients who owe her money, have not done at least $10,000 worth of business with her, and have not employed her in the last year

68. ▼ The clients who either do not owe her money, have done at least $10,000 worth of business with her, or have employed her in the last year

69. 🔲 *Boat Sales* You are given data on revenues from sales of sail boats, motor boats, and yachts for each of the years 2003 through 2006. How would you represent these data in a spreadsheet? The cells in your spreadsheet represent elements of which set?

70. 🔲 *Health-Care Spending* Spending in most categories of health care in the United States increased dramatically in the last 30 years of the 1900s.[1] You are given data showing total spending on prescription drugs, nursing homes, hospital care, and professional services for each of the last three decades of the 1900s. How would you represent these data in a spreadsheet? The cells in your spreadsheet represent elements of which set?

COMMUNICATION AND REASONING EXERCISES

71. You sell iPads® and jPads. Let I be the set of all iPads you sold last year, and let J be the set of all jPads you sold last year. What set represents the collection of all iPads and jPads you sold combined?

72. You sell two models of music players: the *yoVaina Grandote* and the *yoVaina Minúsculito,* and each comes in three colors: Infraroja, Ultravioleta, and Radiografía. Let M be the set of models and let C be the set of colors. What set represents the different choices a customer can make?

73. You are searching online for techno music that is neither European nor Dutch. In set notation, which set of music files are you searching for?

(A) Techno $\cap$ (European $\cap$ Dutch)$'$
(B) Techno $\cap$ (European $\cup$ Dutch)$'$
(C) Techno $\cup$ (European $\cap$ Dutch)$'$
(D) Techno $\cup$ (European $\cup$ Dutch)$'$

74. You would like to see either a World War II movie, or one that that is based on a comic book character but does not feature aliens. Which set of movies are you interested in seeing?

(A) WWII $\cap$ (Comix $\cap$ Aliens$'$)
(B) WWII $\cap$ (Comix $\cup$ Aliens$'$)
(C) WWII $\cup$ (Comix $\cap$ Aliens$'$)
(D) WWII $\cup$ (Comix $\cup$ Aliens$'$)

75. ▼ Explain, illustrating by means of an example, why $(A \cap B) \cup C \neq A \cap (B \cup C)$.

76. ▼ Explain, making reference to operations on sets, why the statement "He plays soccer or rugby and cricket" is ambiguous.

77. ▼ Explain the meaning of a universal set, and give two different universal sets that could be used in a discussion about sets of positive integers.

78. ▼ Is the set of outcomes when two indistinguishable dice are rolled (Example 1) a Cartesian product of two sets? If so, which two sets; if not, why not?

79. ▼ Design a database scenario that leads to the following statement: To keep the factory operating at maximum capacity, the plant manager should select the suppliers in $A \cap (B \cup C')$.

80. ▼ Design a database scenario that leads to the following statement: To keep her customers happy, the bookstore owner should stock periodicals in $A \cup (B \cap C')$.

[1]Source: Department of Health and Human Services/*New York Times,* January 8, 2002, p. A14.

81. ▼ Rewrite in set notation: She prefers movies that are not violent, are shorter than two hours, and have neither a tragic ending nor an unexpected ending.

82. ▼ Rewrite in set notation: He will cater for any event as long as there are no more than 1,000 people, it lasts for at least three hours, and it is within a 50-mile radius of Toronto.

83. ▼ When this book was being written, the copy editor wanted to delete the comma in the following sentence (see Exercise 74): "You would like to see either a World War II movie, or one that that is based on a comic book character but does not feature aliens." Explain why this would have resulted in an ambiguity.

84. ▼ When an older version of this book was being written, the copy editor wanted to delete the comma in the following sentence: "You would like to see a World War II movie, based on a comic book character but not featuring aliens." Explain why removing the comma would have had no effect on the meaning of the sentence.

6.2 Cardinality

In this section, we begin to look at a deceptively simple idea: the size of a set, which we call its **cardinality**.

Cardinality

If A is a finite set, then its **cardinality** is

$n(A)$ = number of elements in A.

Visualizing Cardinality

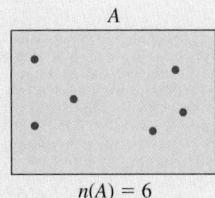

A

$n(A) = 6$

Quick Examples

1. Let $S = \{a, b, c\}$. Then $n(S) = 3$.

2. Let S be the set of outcomes when two distinguishable dice are rolled. Then $n(S) = 36$ (see Example 1 in Section 6.1).

3. $n(\emptyset) = 0$ because the empty set has no elements.

Counting the elements in a small, simple set is straightforward. To count the elements in a large, complicated set, we try to describe the set as built of simpler sets using the set operations. We then need to know how to calculate the number of elements in, for example, a union, based on the number of elements in the simpler sets whose union we are taking.

The Cardinality of a Union

How can we calculate $n(A \cup B)$ if we know $n(A)$ and $n(B)$? Our first guess might be that $n(A \cup B)$ is $n(A) + n(B)$. But consider a simple example. Let

$$A = \{a, b, c\}$$

and

$$B = \{b, c, d\}.$$

Then $A \cup B = \{a, b, c, d\}$ so $n(A \cup B) = 4$, but $n(A) + n(B) = 3 + 3 = 6$. The calculation $n(A) + n(B)$ gives the wrong answer because the elements b and c are counted twice, once for being in A and again for being in B. To correct for this overcounting, we need to subtract the number of elements that get counted twice, which is the number of elements that A and B have in common, or $n(A \cap B) = 2$ in this case. So, we get the right number for $n(A \cup B)$ from the following calculation:

$$n(A) + n(B) - n(A \cap B) = 3 + 3 - 2 = 4.$$

This argument leads to the following general formula.

Cardinality of a Union

If A and B are finite sets, then

$$n(A \cup B) = n(A) + n(B) - n(A \cap B).$$

In particular, if A and B are disjoint (meaning that $A \cap B = \emptyset$), then

$$n(A \cup B) = n(A) + n(B).$$

(When A and B are disjoint we say that $A \cup B$ is a **disjoint union**.)

Visualizing Cardinality of a Union

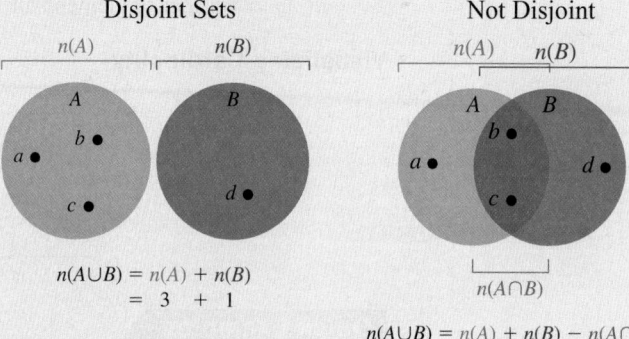

Disjoint Sets

$$n(A \cup B) = n(A) + n(B)$$
$$= 3 + 1$$

Not Disjoint

$$n(A \cap B)$$

$$n(A \cup B) = n(A) + n(B) - n(A \cap B)$$
$$= 3 + 3 - 2$$

Quick Examples

1. If $A = \{a, b, c, d\}$ and $B = \{b, c, d, e, f\}$, then

$$n(A \cup B) = n(A) + n(B) - n(A \cap B) = 4 + 5 - 3 = 6.$$

In fact, $A \cup B = \{a, b, c, d, e, f\}$.

2. If $A = \{a, b, c\}$ and $B = \{d, e, f\}$, then $A \cap B = \emptyset$, so

$$n(A \cup B) = n(A) + n(B) = 3 + 3 = 6.$$

digitallife/Alamy

EXAMPLE 1 Web Searches

In October 2011, a search on Bing™ for "Lt. William Burrows" yielded 29 Web sites containing that phrase, and a search for "Romulan probe" yielded 17 sites. A search for sites containing both phrases yielded 4 Web sites. How many Web sites contained either "Lt. William Burrows," "Romulan probe," or both?

Solution Let A be the set of sites containing "Lt. William Burrows" and let B be the set of sites containing "Romulan probe." We are told that

$$n(A) = 29$$
$$n(B) = 17$$
$$n(A \cap B) = 4. \quad \text{"Lt. William Burrows" AND "Romulan probe"}$$

The formula for the cardinality of the union tells us that

$$n(A \cup B) = n(A) + n(B) - n(A \cap B) = 29 + 17 - 4 = 42.$$

So, 42 sites in the Bing database contained one or both of the phrases "Lt. William Burrows" or "Romulan probe."

➡ **Before we go on...** Each search engine has a different way of specifying a search for a union or an intersection. At Bing or Google™, you can use "OR" for the union and "AND" for the intersection.

Although the formula $n(A \cup B) = n(A) + n(B) - n(A \cap B)$ always holds mathematically, you may sometimes find that, in an actual search, the numbers don't add up. Google, for example, appears not to adhere strictly to the search rule you enter, but instead presents results that it thinks you want.* ■

✳ For example, in October 2011, a search on Google gave the following results:
"Romulan probe": 471 results
"Abraham Lincoln is born on Earth": 8 results
"Romulan probe" OR "Abraham Lincoln is born on Earth": 1,600 results!

Q: *Is there a similar formula for $n(A \cap B)$?*

A: The formula for the cardinality of a union can also be thought of as a formula for the cardinality of an intersection. We can solve for $n(A \cap B)$ to get

$$n(A \cap B) = n(A) + n(B) - n(A \cup B).$$

In fact, we can think of this formula as an equation relating four quantities. If we know any three of them, we can use the equation to find the fourth. (See Example 2 on next page).

Q: *Is there a similar formula for $n(A')$?*

A: We can get a formula for the cardinality of a complement as follows: If S is our universal set and $A \subseteq S$, then S is the disjoint union of A and its complement. That is,

$$S = A \cup A' \quad \text{and} \quad A \cap A' = \emptyset.$$

Applying the cardinality formula for a disjoint union, we get

$$n(S) = n(A) + n(A').$$

We can then solve for $n(A')$ or for $n(A)$ to get the formulas shown below:

Cardinality of a Complement

If S is a finite universal set and A is a subset of S, then

$$n(A') = n(S) - n(A)$$

and

$$n(A) = n(S) - n(A').$$

Visualizing Cardinality of a Complement

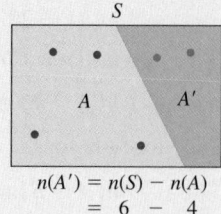

$$n(A') = n(S) - n(A)$$
$$= 6 - 4$$

Quick Example

If $S = \{a, b, c, d, e, f\}$ and $A = \{a, b, c, d\}$, then

$$n(A') = n(S) - n(A) = 6 - 4 = 2.$$

In fact, $A' = \{e, f\}$.

EXAMPLE 2 **Cookbooks**

In November 2011, a search at Amazon.com found 132,000 books on cooking.[2] Of these, 20,000 were on regional cooking, 5,000 were on vegetarian cooking, and 24,000 were on either regional or vegetarian cooking (or both). How many of these books were not on both regional and vegetarian cooking?

Solution Let S be the set of all 132,000 books on cooking, let A be the set of books on regional cooking, and let B be the set of books on vegetarian cooking. We wish to find the size of the complement of the set of books on both regional and vegetarian cooking—that is, $n((A \cap B)')$. Using the formula for the cardinality of a complement, we have

$$n((A \cap B)') = n(S) - n(A \cap B) = 132,000 - n(A \cap B).$$

To find $n(A \cap B)$, we use the formula for the cardinality of a union:

$$n(A \cup B) = n(A) + n(B) - n(A \cap B).$$

Substituting the values we were given, we find

$$24,000 = 20,000 + 5,000 - n(A \cap B),$$

which we can solve to get

$$n(A \cap B) = 1,000.$$

Therefore,

$$n((A \cap B)') = 132,000 - n(A \cap B) = 132,000 - 1,000 = 131,000.$$

So, 131,000 of the books on cooking were not on both regional and vegetarian cooking.

---

[2]Precisely, it found that many books under the subject "Cooking, Food & Wine." Regional cooking falls under "Regional & International." Figures are rounded to the nearest 1,000.

EXAMPLE 3 iPods, iPhones, and iPads

The following table shows sales, in millions of units, of iPods®, iPhones®, and iPads® in the last three quarters of 2011.[3]

	iPods (A)	iPhones (B)	iPads (C)	Total
2011 Q2 (U)	9.0	18.7	4.7	32.4
2011 Q3 (V)	7.5	20.3	9.3	37.1
2011 Q4 (W)	6.6	17.1	11.1	34.8
Total	23.1	56.1	25.1	104.3

Let S be the set of all these iPods, iPhones, and iPads, and label the sets representing the sales in each row and column as shown (so that, for example, A is the set of all iPods sold during the three quarters). Describe the following sets and compute their cardinality:

a. U' **b.** $A \cap U'$ **c.** $(A \cap U)'$ **d.** $C \cup U$

Solution Because all the figures are stated in millions of units, we'll give our calculations and results in millions of units as well.

a. U' is the set of all items not sold in the second quarter of 2011. To compute its cardinality, we could add the totals for all the other quarters listed in the rightmost column:

$$n(U') = n(V) + n(W) = 37.1 + 34.8 = 71.9 \text{ million items.}$$

Alternatively, we can use the formula for the cardinality of a complement (referring again to the totals in the table):

$$n(U') = n(S) - n(U) = 104.3 - 32.4 = 71.9 \text{ million items.}$$

b. $A \cap U'$ is the intersection of the set of all iPods and the set of all items not sold in 2011 Q2. In other words, it is the set of all iPods not sold in the second quarter of 2011. Here is the table with the corresponding sets A and U' shaded ($A \cap U'$ is the overlap):

	iPods (A)	iPhones (B)	iPads (C)	Total
2011 Q2 (U)	9.0	18.7	4.7	32.4
2011 Q3 (V)	7.5	20.3	9.3	37.1
2011 Q4 (W)	6.6	17.1	11.1	34.8
Total	23.1	56.1	25.1	104.3

From the table:

$$n(A \cap U') = 7.5 + 6.6 = 14.1 \text{ million items.}$$

---

[3]Figures are rounded to one decimal place. Source: Apple quarterly press releases (www.investor.apple.com).

c. $A \cap U$ is the set of all iPods sold in 2011 Q2, and so $(A \cap U)'$ is the set of all items remaining if we exclude iPods sold in 2011 Q2:

	iPods (*A*)	iPhones (*B*)	iPads (*C*)	Total
2011 Q2 (*U*)	9.0	18.7	4.7	32.4
2011 Q3 (*V*)	7.5	20.3	9.3	37.1
2011 Q4 (*W*)	6.6	17.1	11.1	34.8
Total	23.1	56.1	25.1	104.3

From the formula for the cardinality of a complement:

$$n((A \cap U)') = n(S) - n(A \cap U)$$
$$= 104.3 - 9.0 = 95.3 \text{ million items.}$$

d. $C \cup U$ is the set of items that were either iPads or sold in 2011 Q2:

	iPods (*A*)	iPhones (*B*)	iPads (*C*)	Total
2011 Q2 (*U*)	9.0	18.7	4.7	32.4
2011 Q3 (*V*)	7.5	20.3	9.3	37.1
2011 Q4 (*W*)	6.6	17.1	11.1	34.8
Total	23.1	56.1	25.1	104.3

To compute it, we can use the formula for the cardinality of a union:

$$n(C \cup U) = n(C) + n(U) - n(C \cap U)$$
$$= 25.1 + 32.4 - 4.7 = 52.8 \text{ million items.}$$

To determine the cardinality of a union of three or more sets, like $n(A \cup B \cup C)$, we can think of $A \cup B \cup C$ as a union of two sets, $(A \cup B)$ and C, and then analyze each piece using the techniques we already have. Alternatively, there are formulas for the cardinalities of unions of any number of sets, but these formulas get more and more complicated as the number of sets grows. In many applications, like the following example, we can use Venn diagrams instead.

EXAMPLE 4 **Reading Lists**

A survey of 300 college students found that 100 had read *War and Peace,* 120 had read *Crime and Punishment,* and 100 had read *The Brothers Karamazov.* It also found that 40 had read only *War and Peace,* 70 had read *War and Peace* but not *The Brothers Karamazov,* and 80 had read *The Brothers Karamazov* but not *Crime and Punishment.* Only 10 had read all three novels. How many had read none of these three novels?

Solution There are four sets mentioned in the problem: the universe S consisting of the 300 students surveyed, the set W of students who had read *War and Peace,* the set C of students who had read *Crime and Punishment,* and the set K of students who had read *The Brothers Karamazov.* Figure 7 shows a Venn diagram representing these sets.

We have put labels in the various regions of the diagram to represent the number of students in each region. For example, x represents the number of students in

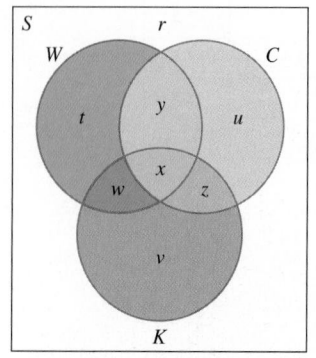

Figure 7

$W \cap C \cap K$, which is the number of students who have read all three novels. We are told that this number is 10, so

$$x = 10.$$

(You should draw the diagram for yourself and fill in the numbers as we go along.) We are also told that 40 students had read only *War and Peace*, so

$$t = 40.$$

We are given none of the remaining regions directly. However, because 70 had read *War and Peace* but not *The Brothers Karamazov*, we see that t and y must add up to 70. Because we already know that $t = 40$, it follows that $y = 30$. Further, because a total of 100 students had read *War and Peace*, we have

$$x + y + t + w = 100.$$

Substituting the known values of x, y, and t gives

$$10 + 30 + 40 + w = 100$$

so $w = 20$. Because 80 students had read *The Brothers Karamazov* but not *Crime and Punishment*, we see that $v + w = 80$, so $v = 60$ (because we know $w = 20$). We can now calculate z using the fact that a total of 100 students had read *The Brothers Karamazov*:

$$60 + 20 + 10 + z = 100$$

giving $z = 10$. Similarly, we can now get u using the fact that 120 students had read *Crime and Punishment*:

$$10 + 30 + 10 + u = 120,$$

giving $u = 70$. Of the 300 students surveyed, we've now found $x + y + z + w + t + u + v = 240$. This leaves

$$r = 60$$

who had read none of the three novels.

The Cardinality of a Cartesian Product

We've covered all the operations except Cartesian product. To find a formula for $n(A \times B)$, consider the following simple example:

$$A = \{\text{H}, \text{T}\}$$
$$B = \{1, 2, 3, 4, 5, 6\}$$

so that

$$A \times B = \{\text{H1}, \text{H2}, \text{H3}, \text{H4}, \text{H5}, \text{H6}, \text{T1}, \text{T2}, \text{T3}, \text{T4}, \text{T5}, \text{T6}\}.$$

As we saw in Example 5 in Section 6.1, the elements of $A \times B$ can be arranged in a table or spreadsheet with $n(A) = 2$ rows and $n(B) = 6$ elements in each row.

	A	B	C	D	E	F	G
1		1	2	3	4	5	6
2	**H**	H1	H2	H3	H4	H5	H6
3	**T**	T1	T2	T3	T4	T5	T6

In a region with 2 rows and 6 columns, there are $2 \times 6 = 12$ cells. So,

$$n(A \times B) = n(A)n(B)$$

in this case. There is nothing particularly special about this example, however, and that formula holds true in general.

Cardinality of a Cartesian Product

If A and B are finite sets, then
$$n(A \times B) = n(A)n(B).$$

Quick Example

If $A = \{a, b, c\}$ and $B = \{x, y, z, w\}$, then
$$n(A \times B) = n(A)n(B) = 3 \times 4 = 12.$$

EXAMPLE 5 Coin Tosses

a. If we toss a coin twice and observe the sequence of heads and tails, how many possible outcomes are there?

b. If we toss a coin three times, how many possible outcomes are there?

c. If we toss a coin ten times, how many possible outcomes are there?

Solution

a. Let $A = \{H, T\}$ be the set of possible outcomes when a coin is tossed once. The set of outcomes when a coin is tossed twice is $A \times A$, which has

$$n(A \times A) = n(A)n(A) = 2 \times 2 = 4$$

possible outcomes.

b. When a coin is tossed three times, we can think of the set of outcomes as the product of the set of outcomes for the first two tosses, which is $A \times A$, and the set of outcomes for the third toss, which is just A. The set of outcomes for the three tosses is then $(A \times A) \times A$, which we usually write as $A \times A \times A$ or A^3. The number of outcomes is

$$n((A \times A) \times A) = n(A \times A)n(A) = (2 \times 2) \times 2 = 8.$$

c. Considering the result of part (b), we can easily see that the set of outcomes here is $A^{10} = A \times A \times \cdots \times A$ (10 copies of A), or the set of ordered sequences of ten Hs and Ts. It's also easy to see that

$$n(A^{10}) = [n(A)]^{10} = 2^{10} = 1{,}024.$$

➡ **Before we go on...** We can start to see the power of these formulas for cardinality. In Example 5, we were able to calculate that there are 1,024 possible outcomes when we toss a coin 10 times without writing out all 1,024 possibilities and counting them. ∎

6.2 EXERCISES

▼ more advanced   ◆ challenging

T indicates exercises that should be solved using technology

Let $A = \{$Dirk, Johan, Frans, Sarie$\}$, $B = \{$Frans, Sarie, Tina, Klaas, Henrika$\}$, $C = \{$Hans, Frans$\}$. *Find the numbers indicated in Exercises 1–6.* HINT [See Quick Examples on page 408.]

1. $n(A) + n(B)$ **2.** $n(A) + n(C)$ **3.** $n(A \cup B)$

4. $n(A \cup C)$ **5.** $n(A \cup (B \cap C))$ **6.** $n(A \cap (B \cup C))$

7. Verify that $n(A \cup B) = n(A) + n(B) - n(A \cap B)$ with A and B as above.

8. Verify that $n(A \cup C) = n(A) + n(C) - n(A \cap C)$ with A and C as above.

Let $A = \{$H, T$\}$, $B = \{1, 2, 3, 4, 5, 6\}$, *and* $C = \{$red, green, blue$\}$. *Find the numbers indicated in Exercises 9–14.* HINT [See Example 5.]

9. $n(A \times A)$ **10.** $n(B \times B)$ **11.** $n(B \times C)$

12. $n(A \times C)$ **13.** $n(A \times B \times B)$ **14.** $n(A \times B \times C)$

15. If $n(A) = 43$, $n(B) = 20$, and $n(A \cap B) = 3$, find $n(A \cup B)$.

16. If $n(A) = 60$, $n(B) = 20$, and $n(A \cap B) = 1$, find $n(A \cap B)$.

17. If $n(A \cup B) = 100$ and $n(A) = n(B) = 60$, find $n(A \cap B)$.

18. If $n(A) = 100$, $n(A \cup B) = 150$, and $n(A \cap B) = 40$, find $n(B)$.

Let $S = \{$Barnsley, Manchester United, Southend, Sheffield United, Liverpool, Maroka Swallows, Witbank Aces, Royal Tigers, Dundee United, Lyon$\}$ *be a universal set,* $A = \{$Southend, Liverpool, Maroka Swallows, Royal Tigers$\}$, *and* $B = \{$Barnsley, Manchester United, Southend$\}$. *Find the numbers indicated in Exercises 19–24.* HINT [See Quick Example on page 410.]

19. $n(A')$ **20.** $n(B')$ **21.** $n((A \cap B)')$

22. $n((A \cup B)')$ **23.** $n(A' \cap B')$ **24.** $n(A' \cup B')$

25. With S, A, and B as above, verify that $n((A \cap B)') = n(A') + n(B') - n((A \cup B)')$.

26. With S, A, and B as above, verify that $n(A' \cap B') + n(A \cup B) = n(S)$.

In Exercises 27–30, use the given information to complete the solution of each partially solved Venn diagram. HINT [See Example 4.]

27.

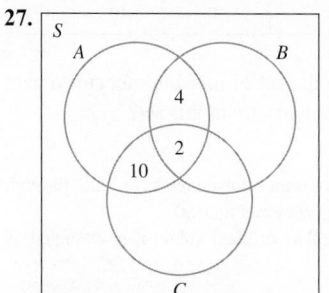

$n(A) = 20$, $n(B) = 20$,
$n(C) = 28$, $n(B \cap C) = 8$,
$n(S) = 50$

28.

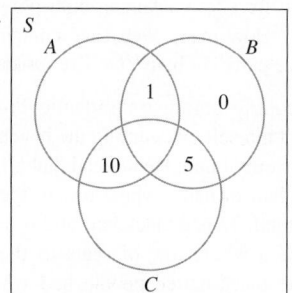

$n(A) = 16$, $n(B) = 11$, $n(C) = 30$, $n(S) = 40$

29. ▼

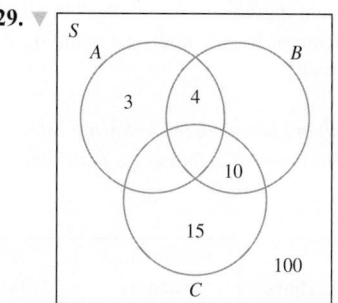

$n(A) = 10$, $n(B) = 19$, $n(S) = 140$

30. ▼

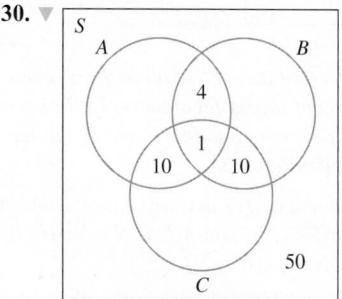

$n(A \cup B) = 30$, $n(B \cup C) = 30$, $n(A \cup C) = 35$

APPLICATIONS

31. *Web Searches* In November 2011, a search using the Web search engine Bing™ for "asteroid" yielded 25.0 million Web sites containing that word. A search for "comet" yielded 93.5 million sites. A search for sites containing both words yielded 3.1 million sites.[4] How many Web sites contained either "asteroid" or "comet" or both? HINT [See Example 1.]

---

[4]Rounded to the nearest 0.1 million.

32. *Web Searches* In November 2011, a search using the Web search engine Bing™ for "tea party" yielded 58.0 million Web sites containing that phrase. A search for "coffee party" yielded 0.6 million sites. A search for sites containing both phrases yielded 0.1 million sites.[5] How many Web sites contained either "tea party" or "coffee party" or both? HINT [See Example 1.]

33. *Amusement* On a particularly boring transatlantic flight, one of the authors amused himself by counting the heads of the people in the seats in front of him. He noticed that all 37 of them either had black hair or had a whole row to themselves (or both). Of this total, 33 had black hair and 6 were fortunate enough to have a whole row of seats to themselves. How many of the black-haired people had whole rows to themselves?

34. *Restaurant Menus* Your favorite restaurant offers a total of 14 desserts, of which 8 have ice cream as a main ingredient and 9 have fruit as a main ingredient. Assuming that all of them have either ice cream or fruit or both as a main ingredient, how many have both?

Publishing Exercises 35–40 are based on the following table, which shows the results of a survey of authors by a fictitious publishing company.

	New Authors	Established Authors	Total
Successful	5	25	30
Unsuccessful	15	55	70
Total	20	80	100

Consider the following subsets of the set S of all authors represented in the table: C, the set of successful authors; U, the set of unsuccessful authors; N, the set of new authors; and E, the set of established authors. HINT [See Example 3.]

35. Describe the sets $C \cap N$ and $C \cup N$ in words. Use the table to compute $n(C), n(N), n(C \cap N)$, and $n(C \cup N)$. Verify that $n(C \cup N) = n(C) + n(N) - n(C \cap N)$.

36. Describe the sets $N \cap U$ and $N \cup U$ in words. Use the table to compute $n(N), n(U), n(N \cap U)$, and $n(N \cup U)$. Verify that $n(N \cup U) = n(N) + n(U) - n(N \cap U)$.

37. Describe the set $C \cap N'$ in words, and find the number of elements it contains.

38. Describe the set $U \cup E'$ in words, and find the number of elements it contains.

39. ▼ What percentage of established authors are successful? What percentage of successful authors are established?

40. ▼ What percentage of new authors are unsuccessful? What percentage of unsuccessful authors are new?

Home Sales Exercises 41–46 are based on the following table, which shows the number of sales of existing homes, in millions, in the different regions of the United States during 2009–2011.[6] Take S to be the set of all home sales represented in the table and label the sets representing the home sales in each row and column as shown (so that, for example, N is the set of all sales of existing homes in the Northeast during 2009–2011)

	Northeast (N)	Midwest (M)	South (T)	West (W)	Total
2009 (A)	0.9	1.2	1.9	1.2	5.2
2010 (B)	0.8	1.1	1.9	1.2	5.0
2011 (C)	0.8	1.1	1.9	1.1	4.9
Total	2.5	3.4	5.7	3.5	15.1

In each exercise, use symbols to describe the given set, and compute its cardinality.

41. The set of sales of existing homes in the Midwest in 2011

42. The set of sales of existing homes either in the West or in 2009

43. The set of sales of existing homes in 2010 excluding sales in the South

44. The set of sales of existing homes in the Northeast after 2009

45. ▼ The set of sales of existing homes in 2010 in the West and Midwest

46. ▼ The set of sales of existing homes in the South and West in years other than 2009

Stock Prices Exercises 47–52 are based on the following table, which shows the performance of a selection of 100 fictitious stocks after one year. (Take S to be the set of all stocks represented in the table.)

	Companies			
	Pharmaceutical (P)	Electronic (E)	Internet (I)	Total
Increased (V)	10	5	15	30
Unchanged[7] (N)	30	0	10	40
Decreased (D)	10	5	15	30
Total	50	10	40	100

47. Use symbols to describe the set of non-Internet stocks that increased. How many elements are in this set?

48. Use symbols to describe the set of Internet stocks that did not increase. How many elements are in this set?

49. Compute $n(P' \cup N)$. What does this number represent?

50. Compute $n(P \cup N')$. What does this number represent?

51. Calculate $\dfrac{n(V \cap I)}{n(I)}$. What does the answer represent?

52. Calculate $\dfrac{n(D \cap I)}{n(D)}$. What does the answer represent?

53. ▼ *Medicine* In a study of Tibetan children,[8] a total of 1,556 children were examined. Of these, 1,024 had rickets. Of the 243 urban children in the study, 93 had rickets.

a. How many children living in nonurban areas had rickets?

b. How many children living in nonurban areas did not have rickets?

54. ▼ *Medicine* In a study of Tibetan children,[9] a total of 1,556 children were examined. Of these, 615 had caries (cavities). Of the 1,313 children living in nonurban areas, 504 had caries.

a. How many children living in urban areas had caries?

b. How many children living in urban areas did not have caries?

55. *Entertainment* According to a survey of 100 people regarding their movie attendance in the last year, 40 had seen a science fiction movie, 55 had seen an adventure movie, and 35 had seen a horror movie. Moreover, 25 had seen a science fiction movie and an adventure movie, 5 had seen an adventure movie and a horror movie, and 15 had seen a science fiction movie and a horror movie. Only 5 people had seen a movie from all three categories.

a. Use the given information to set up a Venn diagram and solve it. HINT [See Example 4.]

b. Complete the following sentence: The survey suggests that __ % of science fiction movie fans are also horror movie fans.

56. *Athletics* Of the 4,700 students at Medium Suburban College (MSC), 50 play collegiate soccer, 60 play collegiate lacrosse, and 96 play collegiate football. Only 4 students play both collegiate soccer and lacrosse, 6 play collegiate soccer and football, and 16 play collegiate lacrosse and football. No students play all three sports.

a. Use the given information to set up a Venn diagram and solve it. HINT [See Example 4.]

b. Complete the following sentence: __ % of the college soccer players also play one of the other two sports at the collegiate level.

57. *Entertainment* In a survey of 100 Enormous State University students, 21 enjoyed classical music, 22 enjoyed rock music, and 27 enjoyed house music. Five of the students enjoyed both classical and rock. How many of those that enjoyed rock did not enjoy classical music?

58. *Entertainment* Refer back to the preceding exercise. You are also told that 5 students enjoyed all three kinds of music while 53 enjoyed music in none of these categories. How many students enjoyed both classical and rock but disliked house music?

COMMUNICATION AND REASONING EXERCISES

59. If A and B are finite sets with $A \subset B$, how are $n(A)$ and $n(B)$ related?

60. If A and B are subsets of the finite set S with $A \subset B$, how are $n(A')$ and $n(B')$ related?

61. Why is the Cartesian product referred to as a "product"? HINT [Think about cardinality.]

62. Refer back to your answer to Exercise 61. What set operation could you use to represent the *sum* of two disjoint sets A and B? Why?

63. Formulate an interesting application whose answer is $n(A \cap B) = 20$.

64. Formulate an interesting application whose answer is $n(A \times B) = 120$.

65. ▼ When is $n(A \cup B) \neq n(A) + n(B)$?

66. ▼ When is $n(A \times B) = n(A)$?

67. ▼ When is $n(A \cup B) = n(A)$?

68. ▼ When is $n(A \cap B) = n(A)$?

69. ◆ Use a Venn diagram or some other method to obtain a formula for $n(A \cup B \cup C)$ in terms of $n(A)$, $n(B)$, $n(C)$, $n(A \cap B)$, $n(A \cap C)$, $n(B \cap C)$, and $n(A \cap B \cap C)$.

70. ◆ Suppose that A and B are sets with $A \subset B$ and $n(A)$ at least 2. Arrange the following numbers from smallest to largest (if two numbers are equal, say so): $n(A)$, $n(A \times B)$, $n(A \cap B)$, $n(A \cup B)$, $n(B \times A)$, $n(B)$, $n(B \times B)$.

---

[8]Source: N. S. Harris et al., Nutritional and Health Status of Tibetan Children Living at High Altitudes, *New England Journal of Medicine,* 344(5), February 1, 2001, pp. 341–347.
[9]*Ibid.*

6.3   Decision Algorithms: The Addition and Multiplication Principles

Let's start with a really simple example. You walk into an ice cream parlor and find that you can choose between ice cream, of which there are 15 flavors, and frozen yogurt, of which there are 5 flavors. How many different selections can you make? Clearly, you have $15 + 5 = 20$ different desserts from which to choose. Mathematically, this is an example of the formula for the cardinality of a disjoint union: If we let A be the set of ice creams you can choose from, and B the set of frozen yogurts, then $A \cap B = \emptyset$ and we want $n(A \cup B)$. But the formula for the cardinality of a disjoint union is $n(A \cup B) = n(A) + n(B)$, which gives $15 + 5 = 20$ in this case.

This example illustrates a very useful general principle.

Addition Principle

When choosing among r disjoint alternatives, suppose that

alternative 1 has n_1 possible outcomes,

alternative 2 has n_2 possible outcomes,

. . .

alternative r has n_r possible outcomes,

with no two of these outcomes the same. Then there are a total of $n_1 + n_2 + \cdots + n_r$ possible outcomes.

Quick Example

At a restaurant you can choose among 8 chicken dishes, 10 beef dishes, 4 seafood dishes, and 12 vegetarian dishes. This gives a total of $8 + 10 + 4 + 12 = 34$ different dishes to choose from.

Here is another simple example. In that ice cream parlor, not only can you choose from 15 flavors of ice cream, but you can also choose from 3 different sizes of cone. How many different ice cream cones can you select from? This time, we want to choose both a flavor and a size, or, in other words, a pair (flavor, size). Therefore, if we let A again be the set of ice cream flavors and now let C be the set of cone sizes, the pair we want to choose is an element of $A \times C$, the Cartesian product. To find the number of choices we have, we use the formula for the cardinality of a Cartesian product: $n(A \times C) = n(A)n(C)$. In this case, we get $15 \times 3 = 45$ different ice cream cones we can select.

This example illustrates another general principle.

Multiplication Principle

When making a sequence of choices with r steps, suppose that

step 1 has n_1 possible outcomes

step 2 has n_2 possible outcomes

. . .

step r has n_r possible outcomes

✳ See Example 3 for a case in which different sequences of choices can lead to the same outcome, with the result that the multiplication principle does not apply.

and that each sequence of choices results in a distinct outcome.✳ Then there are a total of $n_1 \times n_2 \times \cdots \times n_r$ possible outcomes.

Quick Example

At a restaurant you can choose among 5 appetizers, 34 main dishes, and 10 desserts. This gives a total of $5 \times 34 \times 10 = 1,700$ different meals (each including one appetizer, one main dish, and one dessert) from which you can choose.

Things get more interesting when we have to use the addition and multiplication principles in tandem.

EXAMPLE 1 Desserts

You walk into an ice cream parlor and find that you can choose between ice cream, of which there are 15 flavors, and frozen yogurt, of which there are 5 flavors. In addition, you can choose among 3 different sizes of cones for your ice cream or 2 different sizes of cups for your yogurt. How many different desserts can you choose from?

Solution It helps to think about a definite procedure for deciding which dessert you will choose. Here is one we can use:

Alternative 1: An ice cream cone
    Step 1 Choose a flavor.
    Step 2 Choose a size.

Alternative 2: A cup of frozen yogurt
    Step 1 Choose a flavor.
    Step 2 Choose a size.

That is, we can choose between alternative 1 and alternative 2. If we choose alternative 1, we have a sequence of two choices to make: flavor and size. The same is true of alternative 2. We shall call a procedure in which we make a sequence of decisions a **decision algorithm**.✳ Once we have a decision algorithm, we can use the addition and multiplication principles to count the number of possible outcomes.

✳ An algorithm is a procedure with definite rules for what to do at every step.

Alternative 1: An ice cream cone
    Step 1 Choose a flavor: 15 choices.
    Step 2 Choose a size: 3 choices.
    There are $15 \times 3 = 45$ possible choices in alternative 1.   Multiplication Principle

Alternative 2: A cup of frozen yogurt
    Step 1 Choose a flavor: 5 choices.
    Step 2 Choose a size: 2 choices.
    There are $5 \times 2 = 10$ possible choices in alternative 2.   Multiplication Principle

So, there are $45 + 10 = 55$ possible choices of desserts.   Addition Principle

➡ **Before we go on...** Decision algorithms can be illustrated by **decision trees**. To simplify the picture, suppose we had fewer choices in Example 1—say, only two choices of ice cream flavor: vanilla and chocolate, and two choices of yogurt flavor: banana and raspberry. This gives us a total of $2 \times 3 + 2 \times 2 = 10$ possible desserts.

We can illustrate the decisions we need to make when choosing what to buy in the diagram in Figure 8, called a decision tree.

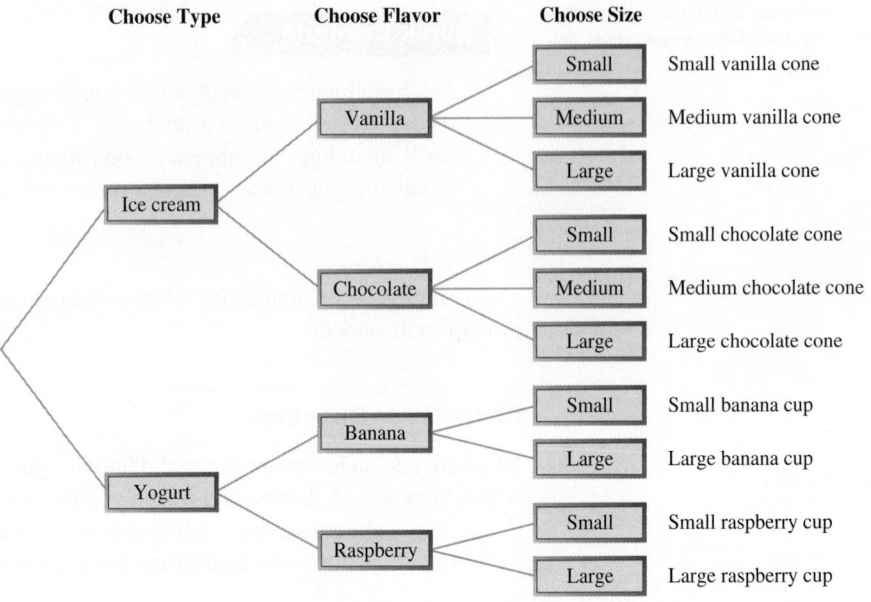

Figure 8

We do not use decision trees much in this chapter because, while they provide a good way of thinking about decision algorithms, they're not really practical for counting large sets. Similar diagrams will be very useful, however, in the chapter on probability. ■

To count the number of possible gadgets, pretend you are *designing* a gadget, and list the decisions to be made at each stage.

Decision Algorithm

A **decision algorithm** is a procedure in which we make a sequence of decisions. We can use decision algorithms to determine the number of possible items by pretending we are *designing* such an item (for example, an ice-cream cone) and listing the decisions or choices we should make at each stage of the process.

Quick Example

An iPod is available in two sizes. The larger size comes in two colors and the smaller size (the Mini) comes in four colors. A decision algorithm for "designing" an iPod is:

Alternative 1: Select Large:
   Step 1 Choose a color: Two choices.
   (So, there are two choices for alternative 1.)

Alternative 2: Select a Mini:
   Step 1 Choose a color: Four choices.
   (So, there are four choices for alternative 2.)

Thus, there are $2 + 4 = 6$ possible choices of iPods.

> **Caution**
> For a decision algorithm to give the correct number of possible items, there must be a one-to-one correspondence between sequences of choices and resulting items. So, it is necessary that each sequence of choices results in a distinct item. In other words, *changing one or more choices must result in a different item.* (See Example 3.)

EXAMPLE 2 **Exams**

An exam is broken into two parts, Part A and Part B, both of which you are required to do. In Part A you can choose between answering 10 true-false questions or answering 4 multiple-choice questions, each of which has 5 answers to choose from. In Part B you can choose between answering 8 true-false questions and answering 5 multiple-choice questions, each of which has 4 answers to choose from. How many different collections of answers are possible?

Solution While deciding what answers to write down, we use the following decision algorithm:

Step 1 Do Part A.
   Alternative 1: Answer the 10 true-false questions.
      Steps 1–10 Choose true or false for each question: 2 choices each.
      There are $2 \times 2 \times \cdots \times 2 = 2^{10} = 1{,}024$ choices in alternative 1.

   Alternative 2: Answer the 4 multiple-choice questions.
      Steps 1–4 Choose one answer for each question: 5 choices each.
      There are $5 \times 5 \times 5 \times 5 = 5^4 = 625$ choices in alternative 2.
   $1{,}024 + 625 = 1{,}649$ choices in step 1

Step 2 Do Part B.
   Alternative 1: Answer the 8 true-false questions: 2 choices each.
      $2^8 = 256$ choices in alternative 1

   Alternative 2: Answer the 5 multiple-choice questions, 4 choices each.
      $4^5 = 1{,}024$ choices in alternative 2
   $256 + 1{,}024 = 1{,}280$ choices in step 2

There are $1{,}649 \times 1{,}280 = 2{,}110{,}720$ different collections of answers possible.

The next example illustrates the need to select your decision algorithm with care.

EXAMPLE 3 **Scrabble®**

You are playing Scrabble and have the following letters to work with: k, e, r, e. Because you are losing the game, you would like to use all your letters to make a single word, but you can't think of any four-letter words using all these letters. In desperation, you decide to list *all* the four-letter sequences possible to see if there are any valid words among them. How large is your list?

Solution It may first occur to you to try the following decision algorithm.

Step 1 Select the first letter: 4 choices.
Step 2 Select the second letter: 3 choices.

Step 3 Select the third letter: 2 choices.
Step 4 Select the last letter: 1 choice.

This gives $4 \times 3 \times 2 \times 1 = 24$ choices. However, something is wrong with the algorithm.

Q : *What is wrong with this decision algorithm?*

A : We didn't take into account the fact that there are two "e"s;* different decisions in Steps 1–4 can produce the same sequence. Suppose, for example, that we selected the first "e" in Step 1, the second "e" in Step 2, and then the "k" and the "r." This would produce the sequence "eekr." If we selected the *second* "e" in Step 1, the *first* "e" in Step 2, and then the "k" and "r," we would obtain the *same* sequence: "eekr." In other words, the decision algorithm produces two copies of the sequence "eekr" in the associated decision tree. (In fact, it produces two copies of each possible sequence of the letters.) In short, different sequences of choices produce the same result, violating the requirement in the note of caution that precedes Example 2:

> *For a decision algorithm to be valid, each sequence of choices must produce a different result.*

---

* Consider the following extreme case: If all four letters were "e", then there would be only a single sequence: "eeee" and not the 24 predicted by the decision algorithm.

Because our original algorithm is not valid, we need a new one. Here is a strategy that works nicely for this example. Imagine, as before, that we are going to construct a sequence of four letters. This time we are going to imagine that we have a sequence of four empty slots: ☐☐☐☐, and instead of selecting letters to fill the slots from left to right, we are going to select *slots* in which to place each of the letters. Remember that we have to use these letters: k, e, r, e. We proceed as follows, leaving the "e"s until last.

Step 1 Select an empty slot for the k: 4 choices. (e.g., ☐☐k☐)
Step 2 Select an empty slot for the r: 3 choices. (e.g., r☐k☐)
Step 3 Place the "e"s in the remaining two slots: 1 choice!

Thus the multiplication principle yields $4 \times 3 \times 1 = 12$ choices.

➡ **Before we go on...** You should try constructing a decision tree for Example 3, and you will see that each sequence of four letters is produced exactly once when we use the correct (second) decision algorithm. ∎

FAQs

Creating and Testing a Decision Algorithm

Q : *How do I set up a decision algorithm to count how many items there are in a given scenario?*

A : Pretend that you are *designing* such an item (for example, pretend that you are designing an ice-cream cone) and come up with a step-by-step procedure for doing so, listing the decisions you should make at each stage.

Q : *Once I have my decision algorithm, how do I check if it is valid?*

A : Ask yourself the following question: "Is it possible to get the same item (the exact same ice-cream cone, say) by making different decisions when applying the algorithm?" If the answer is "yes," then your decision algorithm is invalid. Otherwise, it is valid.

6.3 EXERCISES

▼ more advanced   ◆ challenging
🔳 indicates exercises that should be solved using technology

1. An experiment requires a choice among 3 initial setups. The first setup can result in 2 possible outcomes, the second in 3 possible outcomes, and the third in 5 possible outcomes. What is the total number of outcomes possible? HINT [See Quick Example on page 418.]

2. A surgical procedure requires choosing among 4 alternative methodologies. The first can result in 4 possible outcomes, the second in 3 possible outcomes, and the remaining methodologies can each result in 2 possible outcomes. What is the total number of outcomes possible? HINT [See Quick Example on page 418.]

3. An experiment requires a sequence of 3 steps. The first step can result in 2 possible outcomes, the second in 3 possible outcomes, and the third in 5 possible outcomes. What is the total number of outcomes possible? HINT [See Quick Example on page 419.]

4. A surgical procedure requires 4 steps. The first can result in 4 possible outcomes, the second in 3 possible outcomes, and the remaining 2 can each result in 2 possible outcomes. What is the total number of outcomes possible? HINT [See Quick Example on page 419.]

For the decision algorithms in Exercises 5–12, find how many outcomes are possible. HINT [See Example 1.]

5. Alternative 1:   Alternative 2:
   Step 1: 1 outcome   Step 1: 2 outcomes
   Step 2: 2 outcomes   Step 2: 2 outcomes
                     Step 3: 1 outcome

6. Alternative 1:   Alternative 2:
   Step 1: 1 outcome   Step 1: 2 outcomes
   Step 2: 2 outcomes   Step 2: 2 outcomes
   Step 3: 2 outcomes

7. Step 1:   Step 2:
   Alternative 1: 1 outcome   Alternative 1: 2 outcomes
   Alternative 2: 2 outcomes   Alternative 2: 2 outcomes
                     Alternative 3: 1 outcome

8. Step 1:   Step 2:
   Alternative 1: 1 outcome   Alternative 1: 2 outcomes
   Alternative 2: 2 outcomes   Alternative 2: 2 outcomes
   Alternative 3: 2 outcomes

9. Alternative 1:   Alternative 2: 5 outcomes
   Step 1:
     Alternative 1: 3 outcomes
     Alternative 2: 1 outcome
   Step 2: 2 outcomes

10. Alternative 1: 2 outcomes   Alternative 2:
                            Step 1:
                              Alternative 1: 4 outcomes
                              Alternative 2: 1 outcome
                            Step 2: 2 outcomes

11. Step 1:   Step 2: 5 outcomes
   Alternative 1:
     Step 1: 3 outcomes
     Step 2: 1 outcome
   Alternative 2: 2 outcomes

12. Step 1: 2 outcomes   Step 2:
                           Alternative 1:
                             Step 1: 4 outcomes
                             Step 2: 1 outcome
                           Alternative 2: 2 outcomes

13. How many different four-letter sequences can be formed from the letters a, a, a, b? HINT [See Example 3.]

14. How many different five-letter sequences can be formed from the letters a, a, a, b, c? HINT [See Example 3.]

APPLICATIONS

15. *Ice Cream* When Baskin-Robbins was founded in 1945, it made 31 different flavors of ice cream.[10] If you had a choice of having your ice cream in a cone, a cup, or a sundae, how many different desserts could you have?

16. *Ice Cream* At the beginning of 2002, Baskin-Robbins claimed to have "nearly 1,000 different ice cream flavors."[11] Assuming that you could choose from 1,000 different flavors, that you could have your ice cream in a cone, a cup, or a sundae, and that you could choose from a dozen different toppings, how many different desserts could you have?

17. *Binary Codes* A binary digit, or "bit," is either 0 or 1. A nybble is a four-bit sequence. How many different nybbles are possible?

18. *Ternary Codes* A ternary digit is either 0, 1, or 2. How many sequences of six ternary digits are possible?

19. *Ternary Codes* A ternary digit is either 0, 1, or 2. How many sequences of six ternary digits containing a single 1 and a single 2 are possible?

20. *Binary Codes* A binary digit, or "bit," is either 0 or 1. A nybble is a four-bit sequence. How many different nybbles containing a single 1 are possible?

21. *Reward* While selecting candy for students in his class, Professor Murphy must choose between gummy candy and licorice nibs. Gummy candy packets come in three sizes, while packets of licorice nibs come in two. If he chooses gummy candy, he must select gummy bears, gummy worms, or gummy dinos. If he chooses licorice nibs, he must choose between red and black. How many choices does he have? HINT [See Example 2.]

---

[10]Source: Company Web site (www.baskinrobbins.com).
[11]*Ibid.*

22. *Productivity* Professor Oger must choose between an extra writing assignment and an extra reading assignment for the upcoming spring break. For the writing assignment, there are two essay topics to choose from and three different mandatory lengths (30 pages, 35 pages, or 40 pages). The reading topic would consist of one scholarly biography combined with one volume of essays. There are five biographies and two volumes of essays to choose from. How many choices does she have? **HINT** [See Example 2.]

23. *CD Discs* CD discs at your local computer store are available in two types (CD-R and CD-RW), packaged singly, in spindles of 50, or in spindles of 100. When purchasing singly, you can choose from five colors; when purchasing in spindles of 50 or 100 you have two choices, silver or an assortment of colors. If you are purchasing CD discs, how many possibilities do you have to choose from?

24. *Radar Detectors* Radar detectors are either powered by their own battery or plug into the cigarette lighter socket. All radar detectors come in two models: no-frills and fancy. In addition, detectors powered by their own batteries detect either radar or laser, or both, whereas the plug-in types come in models that detect either radar or laser, but not both. How many different radar detectors can you buy?

25. *Multiple-Choice Tests* Professor Easy's final examination has 10 true-false questions followed by 2 multiple-choice questions. In each of the multiple-choice questions, you must select the correct answer from a list of five. How many answer sheets are possible?

26. *Multiple-choice Tests* Professor Tough's final examination has 20 true-false questions followed by 3 multiple-choice questions. In each of the multiple-choice questions, you must select the correct answer from a list of six. How many answer sheets are possible?

27. *Tests* A test requires that you answer either Part A or Part B. Part A consists of eight true-false questions, and Part B consists of five multiple-choice questions with one correct answer out of five. How many different completed answer sheets are possible?

28. *Tests* A test requires that you answer first Part A and then either Part B or Part C. Part A consists of four true-false questions, Part B consists of four multiple-choice questions with one correct answer out of five, and Part C consists of three questions with one correct answer out of six. How many different completed answer sheets are possible?

29. *Stock Portfolios* Your broker has suggested that you diversify your investments by splitting your portfolio among mutual funds, municipal bond funds, stocks, and precious metals. She suggests four good mutual funds, three municipal bond funds, eight stocks, and three precious metals (gold, silver, and platinum).

   a. Assuming your portfolio is to contain one of each type of investment, how many different portfolios are possible?

   b. Assuming your portfolio is to contain three mutual funds, two municipal bond funds, one stock, and two precious metals, how many different portfolios are possible?

30. *Menus* The local diner offers a meal combination consisting of an appetizer, a soup, a main course, and a dessert. There are five appetizers, two soups, four main courses, and five desserts. Your diet restricts you to choosing between a dessert and an appetizer. (You cannot have both.) Given this restriction, how many three-course meals are possible?

31. *Computer Codes* A computer "byte" consists of eight "bits," each bit being either a 0 or a 1. If characters are represented using a code that uses a byte for each character, how many different characters can be represented?

32. *Computer Codes* Some written languages, like Chinese and Japanese, use tens of thousands of different characters. If a language uses roughly 50,000 characters, a computer code for this language would have to use how many bytes per character? (See Exercise 31.)

33. *Symmetries of a Five-Pointed Star* A five-pointed star will appear unchanged if it is rotated through any one of the angles 0°, 72°, 144°, 216°, or 288°. It will also appear unchanged if it is flipped about the axis shown in the figure. A *symmetry* of the five-pointed star consists of either a rotation, or a rotation followed by a flip. How many different symmetries are there altogether?

34. *Symmetries of a Six-Pointed Star* A six-pointed star will appear unchanged if it is rotated through any one of the angles 0°, 60°, 120°, 180°, 240°, or 300°. It will also appear unchanged if it is flipped about the axis shown in the figure. A *symmetry* of the six-pointed star consists of either a rotation, or a rotation followed by a flip. How many different symmetries are there altogether?

35. *Variables in BASIC*[12] A variable name in the programming language BASIC can be either a letter or a letter followed by a decimal digit—that is, one of the numbers 0, 1, . . . , 9. How many different variables are possible?

36. *Employee IDs* A company assigns to each of its employees an ID code that consists of one, two, or three letters followed by a digit from 0 through 9. How many employee codes does the company have available?

37. *Tournaments* How many ways are there of filling in the blanks for the following (fictitious) soccer tournament?

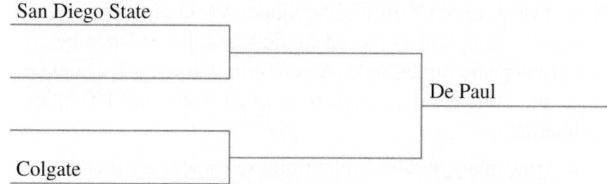

North Carolina

Central Connecticut

Virginia

Syracuse

38. *Tournaments* How many ways are there of filling in the blanks for a (fictitious) soccer tournament involving the four teams San Diego State, De Paul, Colgate, and Hofstra?

San Diego State

De Paul

Colgate

39. ▼ *Telephone Numbers* A telephone number consists of a sequence of seven digits not starting with 0 or 1.

a. How many telephone numbers are possible?
b. How many of them begin with either 463, 460, or 400?
c. How many telephone numbers are possible if no two adjacent digits are the same? (For example, 235-9350 is permitted, but not 223-6789.)

40. ▼ *Social Security Numbers* A Social Security Number is a sequence of nine digits.

a. How many Social Security Numbers are possible?
b. How many of them begin with either 023 or 003?
c. How many Social Security Numbers are possible if no two adjacent digits are the same? (For example, 235-93-2345 is permitted, but not 126-67-8189.)

41. ▼ *DNA Chains* DNA (deoxyribonucleic acid) is the basic building block of reproduction in living things. A DNA chain is a sequence of chemicals called *bases*. There are four possible bases: thymine (T), cytosine (C), adenine (A), and guanine (G).

a. How many three-element DNA chains are possible?
b. How many n-element DNA chains are possible?

c. A human DNA chain has 2.1×10^{10} elements. How many human DNA chains are possible?

42. ▼ *Credit Card Numbers*[13] Each customer of Mobil Credit Corporation is given a 9-digit number for computer identification purposes.

a. If each digit can be any number from 0 to 9, are there enough different account numbers for 10 million credit-card holders?
b. Would there be if the digits were only 0 or 1?

43. ▼ *HTML* Colors in HTML (the language in which many Web pages are written) can be represented by 6-digit hexadecimal codes: sequences of six integers ranging from 0 to 15 (represented as 0, . . . , 9, A, B, . . . , F).

a. How many different colors can be represented?
b. Some monitors can only display colors encoded with pairs of repeating digits (such as 44DD88). How many colors can these monitors display?
c. Grayscale shades are represented by sequences $xyxyxy$ consisting of a repeated pair of digits. How many grayscale shades are possible?
d. The pure colors are pure red: $xy0000$; pure green: $00xy00$; and pure blue: $0000xy$. ($xy = FF$ gives the brightest pure color, while $xy = 00$ gives the darkest: black.) How many pure colors are possible?

44. ▼ *Telephone Numbers* In the past, a local telephone number in the United States consisted of a sequence of two letters followed by five digits. Three letters were associated with each number from 2 to 9 (just as in the standard telephone layout shown in the figure) so that each telephone number corresponded to a sequence of seven digits. How many different sequences of seven digits were possible?

45. ▼ *Romeo and Juliet* Here is a list of the main characters in Shakespeare's *Romeo and Juliet*. The first 7 characters are male and the last 4 are female.

Escalus, *prince of Verona*
Paris, *kinsman to the prince*
Romeo, *of Montague Household*
Mercutio, *friend of Romeo*
Benvolio, *friend of Romeo*

[12]From *Applied Combinatorics* by F. S. Roberts (Prentice-Hall, New Jersey, 1984).

[13]Taken from a exercise in *Applied Combinatorics* by F. S. Roberts (Prentice-Hall, New Jersey, 1984).

Tybalt, *nephew to Lady Capulet*
Friar Lawrence, *a Franciscan*
Lady Montague, *of Montague Household*
Lady Capulet, *of Capulet Household*
Juliet, *of Capulet Household*
Juliet's Nurse

A total of 10 male and 8 female actors are available to play these roles. How many possible casts are there? (All roles are to be played by actors of the same gender as the character.)

46. ▼ *Swan Lake* The Enormous State University's Accounting Society has decided to produce a version of the ballet *Swan Lake,* in which all the female roles (including all of the swans) will be danced by men, and vice versa. Here are the main characters:

Prince Siegfried
Prince Siegfried's Mother
Princess Odette, *the White Swan*
The Evil Duke Rotbart
Odile, *the Black Swan*
Cygnet #1, *young swan*
Cygnet #2, *young swan*
Cygnet #3, *young swan*

The ESU Accounting Society has on hand a total of 4 female dancers and 12 male dancers who are to be considered for the main roles. How many possible casts are there?

47. ▼ *License Plates* Many U.S. license plates display a sequence of three letters followed by three digits.

   a. How many such license plates are possible?
   b. In order to avoid confusion of letters with digits, some states do not issue standard plates with the last letter an I, O, or Q. How many license plates are still possible?
   c. Assuming that the letter combinations VET, MDZ, and DPZ are reserved for disabled veterans, medical practitioners, and disabled persons, respectively, how many license plates are possible, also taking the restriction in part (b) into account?

48. ▼ *License Plates*[14] License plates in Montana have a sequence consisting of: (1) a digit from 1 to 9, (2) a letter, (3) a dot, (4) a letter, and (5) a four-digit number.

   a. How many different license plates are possible?
   b. How many different license plates are available for citizens if numbers that end with 0 are reserved for official state vehicles?

49. ▼ *Mazes*

   a. How many four-letter sequences are possible that contain only the letters R and D, with D occurring only once?
   b. Use part (a) to calculate the number of possible routes from Start to Finish in the maze shown in the figure, where each move is either to the right or down.

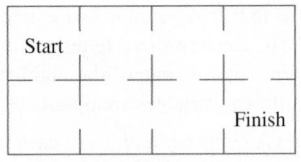

   c. Comment on what would happen if we also allowed left and up moves.

50. ▼ *Mazes*

   a. How many six-letter sequences are possible that contain only the letters R and D, with D occurring only once?
   b. Use part (a) to calculate the number of possible routes from Start to Finish in the maze shown in the figure, where each move is either to the right or down.

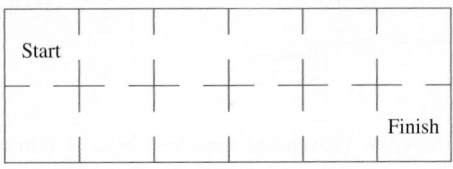

   c. Comment on what would happen if we also allowed left and up moves.

51. ▼ *Car Engines*[15] In a six-cylinder V6 engine, the even-numbered cylinders are on the left and the odd-numbered cylinders are on the right. A good firing order is a sequence of the numbers 1 through 6 in which right and left sides alternate.

   a. How many possible good firing sequences are there?
   b. How many good firing sequences are there that start with a cylinder on the left?

52. ▼ *Car Engines* Repeat the preceding exercise for an eight-cylinder V8 engine.

53. ▼ *Minimalist Art* You are exhibiting your collection of minimalist paintings. Art critics have raved about your paintings, each of which consists of 10 vertical colored lines set against a white background. You have used the following rule to produce your paintings: Every second line, starting with the first, is to be either blue or gray, while the remaining five lines are to be either all light blue, all red, or all purple. Your collection is complete: Every possible combination that satisfies the rules occurs. How many paintings are you exhibiting?

54. ▼ *Combination Locks* Dripping wet after your shower, you have clean forgotten the combination of your lock. It is one

---

[14]Source: The License Plates of the World Web site (http://servo.oit .gatech.edu/~mk5/).

[15]Adapted from an exercise in *Basic Techniques of Combinatorial Theory* by D. I. A. Cohen (New York: John Wiley, 1978).

of those "standard" combination locks, which uses a three-number combination with each number in the range 0 through 39. All you remember is that the second number is either 27 or 37, while the third number ends in a 5. In desperation, you decide to go through all possible combinations using the information you remember. Assuming that it takes about 10 seconds to try each combination, what is the longest possible time you may have to stand dripping in front of your locker?

55. ▼ *Product Design* Your company has patented an electronic digital padlock that a user can program with his or her own four-digit code. (Each digit can be 0 through 9.) The padlock is designed to open if either the correct code is keyed in or—and this is helpful for forgetful people—if exactly one of the digits is incorrect.

   a. How many incorrect codes will open a programmed padlock?

   b. How many codes will open a programmed padlock?

56. ▼ *Product Design* Your company has patented an electronic digital padlock that has a telephone-style keypad. Each digit from 2 through 9 corresponds to three letters of the alphabet (see the figure for Exercise 44). How many different four-letter sequences correspond to a single four-digit sequence using digits in the range 2 through 9?

57. ▼ *Calendars* The World Almanac[16] features a "perpetual calendar," a collection of 14 possible calendars. Why does this suffice to be sure there is a calendar for every conceivable year?

58. ▼ *Calendars* How many possible calendars are there that have February 12 falling on a Sunday, Monday, or Tuesday?

59. ▼ *Programming in BASIC* (Some programming knowledge assumed for this exercise.) How many iterations will be carried out in the following routine?

```
For i = 1 to 10
    For j = 2 to 20
        For k = 1 to 10
            Print i, j, k
        Next k
    Next j
Next i
```

60. ▼ *Programming in JAVASCRIPT* (Some programming knowledge assumed for this exercise.) How many iterations will be carried out in the following routine?

```
for (i = 1; i <= 2; i++) {
    for (j = 1; j <= 2; j++) {
        for (k = 1; k <= 2; k++)
            sum += i+j+k;
    }
}
```

61. ◆ *Building Blocks* Use a decision algorithm to show that a rectangular solid with dimensions $m \times n \times r$ can be constructed with $m \cdot n \cdot r$ cubical $1 \times 1 \times 1$ blocks. (See the figure.)

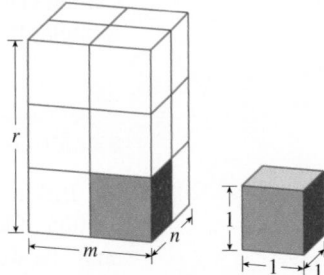

Rectangular Solid Made Up of $1 \times 1 \times 1$ Cubes

62. ◆ *Matrices* (Some knowledge of matrices is assumed for this exercise.) Use a decision algorithm to show that an $m \times n$ matrix must have $m \cdot n$ entries.

63. ◆ *Morse Code* In Morse Code, each letter of the alphabet is encoded by a different sequence of dots and dashes. Different letters may have sequences of different lengths. How long should the longest sequence be in order to allow for every possible letter of the alphabet?

64. ◆ *Numbers* How many odd numbers between 10 and 99 have distinct digits?

COMMUNICATION AND REASONING EXERCISES

65. Complete the following sentences: The multiplication principle is based on the cardinality of the _____ of two sets.

66. Complete the following sentences: The addition principle is based on the cardinality of the _____ of two disjoint sets.

67. You are packing for a short trip and want to take 2 of the 10 shirts you have hanging in your closet. Critique the following decision algorithm and calculation of how many different ways you can choose two shirts to pack: Step 1, choose one shirt, 10 choices. Step 2, choose another shirt, 9 choices. Hence there are 90 possible choices of two shirts.

68. You are designing an advertising logo that consists of a tower of five squares. Three are yellow, one is blue, and one is green. Critique the following decision algorithm and calculation of the number of different five-square sequences: Step 1: Choose the first square, 5 choices. Step 2: Choose the second square, 4 choices. Step 3: Choose the third square: 3 choices. Step 4: Choose the fourth square, 2 choices. Step 5: Choose the last square: 1 choice. Hence there are 120 possible five-square sequences.

69. ▼ Construct a decision algorithm that gives the correct number of five-square sequences in Exercise 68.

70. ▼ Find an interesting application that requires a decision algorithm with two steps in which each step has two alternatives.

---

[16]Source: *The World Almanac and Book of Facts* 1992 (New York: Pharos Books, 1992).

6.4 Permutations and Combinations

Certain classes of counting problems come up frequently, and it is useful to develop formulas to deal with them.

EXAMPLE 1 Casting

Ms. Birkitt, the English teacher at Brakpan Girls High School, wanted to stage a production of R. B. Sheridan's play, *The School for Scandal*. The casting was going well until she was left with five unfilled characters and five seniors who were yet to be assigned roles. The characters were Lady Sneerwell, Lady Teazle, Mrs. Candour, Maria, and Snake; while the unassigned seniors were April, May, June, Julia, and Augusta. How many possible assignments are there?

Solution To decide on a specific assignment, we use the following algorithm:

> **Step 1** Choose a senior to play Lady Sneerwell: 5 choices.
>
> **Step 2** Choose one of the remaining seniors to play Lady Teazle: 4 choices.
>
> **Step 3** Choose one of the now remaining seniors to play Mrs. Candour: 3 choices.
>
> **Step 4** Choose one of the now remaining seniors to play Maria: 2 choices.
>
> **Step 5** Choose the remaining senior to play Snake: 1 choice.

Thus, there are $5 \times 4 \times 3 \times 2 \times 1 = 120$ possible assignments of seniors to roles.

What the situation in Example 1 has in common with many others is that we start with a set—here the set of seniors—and we want to know how many ways we can put the elements of that set in order in a list. In this example, an ordered list of the five seniors, for instance

1. May

2. Augusta

3. June

4. Julia

5. April

corresponds to a particular casting:

Cast	
Lady Sneerwell	May
Lady Teazle	Augusta
Mrs. Candour	June
Maria	Julia
Snake	April

We call an ordered list of items a **permutation** of those items.

If we have *n* items, how many permutations of those items are possible? We can use a decision algorithm similar to the one we used in the preceding example to select a permutation.

Solution

a. We simply need to find the number of ways of choosing 4 marbles out of 11, which is

$$C(11, 4) = 330 \text{ possible sets of 4 marbles}$$

b. We use a decision algorithm for choosing such a set of marbles.

Step 1 Choose one red one from the three red ones: $C(3, 1) = 3$ choices.
Step 2 Choose one blue one from the three blue ones: $C(3, 1) = 3$ choices.
Step 3 Choose one green one from the three green ones: $C(3, 1) = 3$ choices.
Step 4 Choose one yellow one from the two yellow ones: $C(2, 1) = 2$ choices.

This gives a total of $3 \times 3 \times 3 \times 2 = 54$ possible sets.

c. We need another decision algorithm. To say that at least two must be red means that either two are red or three are red (with a total of three red ones). In other words, we have two *alternatives*.

Alternative 1: Exactly two red marbles
Step 1 Choose two red ones: $C(3, 2) = 3$ choices.
Step 2 Choose two nonred ones. There are eight of these, so we get $C(8, 2) = 28$ possible choices.
Thus, the total number of choices for this alternative is $3 \times 28 = 84$.

Alternative 2: Exactly three red marbles.
Step 1 Choose the three red ones: $C(3, 3) = 1$ choice.
Step 2 Choose one nonred one: $C(8, 1) = 8$ choices.
Thus, the total number of choices for this alternative is $1 \times 8 = 8$.

By the addition principle, we get a total of $84 + 8 = 92$ sets.

d. The phrase "at least one green" tells us that we again have some alternatives.

Alternative 1: One green marble
Step 1 Choose one green marble from the three: $C(3, 1) = 3$ choices.
Step 2 Choose three nongreen, nonred marbles: $C(5, 3) = 10$ choices.
Thus, the total number of choices for alternative 1 is $3 \times 10 = 30$.

Alternative 2: Two green marbles
Step 1 Choose two green marbles from the three: $C(3, 2) = 3$ choices.
Step 2 Choose two nongreen, nonred marbles: $C(5, 2) = 10$ choices.
Thus, the total number of choices for alternative 2 is $3 \times 10 = 30$.

Alternative 3: Three green marbles
Step 1 Choose three green marbles from the three: $C(3, 3) = 1$ choice.
Step 2 Choose one nongreen, nonred marble: $C(5, 1) = 5$ choices.
Thus, the total number of choices for alternative 3 is $1 \times 5 = 5$.

The addition principle now tells us that the number of sets of four marbles with none red, but at least one green, is $30 + 30 + 5 = 65$.

➡ **Before we go on...** Here is an easier way to answer part (d) of Example 7. First, the total number of sets having *no* red marbles is $C(8, 4) = 70$. Next, of those, the number containing no green marbles is $C(5, 4) = 5$. This leaves $70 - 5 = 65$ sets that contain no red marbles but have at least one green marble. (We have really used here the formula for the cardinality of the complement of a set.) ∎

The last example concerns poker hands. For those unfamiliar with playing cards, here is a short description. A standard deck consists of 52 playing cards. Each card is in one of 13 denominations: ace, 2, 3, 4, 5, 6, 7, 8, 9, 10, jack (J), queen (Q), and king (K), and in one of four suits: hearts (♥), diamonds (♦), clubs (♣), and spades (♠). Thus, for instance, the jack of spades, J♠, refers to the denomination of jack in the suit of spades. The entire deck of cards is thus

A♥	2♥	3♥	4♥	5♥	6♥	7♥	8♥	9♥	10♥	J♥	Q♥	K♥
A♦	2♦	3♦	4♦	5♦	6♦	7♦	8♦	9♦	10♦	J♦	Q♦	K♦
A♣	2♣	3♣	4♣	5♣	6♣	7♣	8♣	9♣	10♣	J♣	Q♣	K♣
A♠	2♠	3♠	4♠	5♠	6♠	7♠	8♠	9♠	10♠	J♠	Q♠	K♠

EXAMPLE 8 Poker Hands

In the card game poker, a hand consists of a set of 5 cards from a standard deck of 52. A **full house** is a hand consisting of three cards of one denomination ("three of a kind"—e.g., three 10s) and two of another ("two of a kind"—e.g., two queens). Here is an example of a full house: 10♣, 10♦, 10♠, Q♥, Q♣.

a. How many different poker hands are there?

b. How many different full houses are there that contain three 10s and two queens?

c. How many different full houses are there altogether?

Solution

a. Because the order of the cards doesn't matter, we simply need to know the number of ways of choosing a set of 5 cards out of 52, which is

$$C(52, 5) = 2{,}598{,}960 \text{ hands.}$$

b. Here is a decision algorithm for choosing a full house with three 10s and two queens.

Step 1 Choose three 10s. Because there are four 10s to choose from, we have $C(4, 3) = 4$ choices.

Step 2 Choose two queens: $C(4, 2) = 6$ choices.

Thus, there are $4 \times 6 = 24$ possible full houses with three 10s and two queens.

c. Here is a decision algorithm for choosing a full house.

Step 1 Choose a denomination for the three of a kind; 13 choices.

Step 2 Choose three cards of that denomination. Because there are four cards of each denomination (one for each suit), we get $C(4, 3) = 4$ choices.

Step 3 Choose a different denomination for the two of a kind. There are only 12 denominations left, so we have 12 choices.

Step 4 Choose two of that denomination: $C(4, 2) = 6$ choices.

Thus, by the multiplication principle, there are a total of $13 \times 4 \times 12 \times 6 = 3{,}744$ possible full houses.

FAQs

Recognizing When to Use Permutations or Combinations

Q: *How can I tell whether a given application calls for permutations or combinations?*

A: Decide whether the application calls for ordered lists (as in situations where order is implied) or for unordered sets (as in situations where order is not relevant). Ordered lists are permutations, whereas unordered sets are combinations.

6.4 EXERCISES

▼ more advanced   ◆ challenging
⊤ indicates exercises that should be solved using technology

Evaluate each number in Exercises 1–16. HINT [See Quick Examples on pp. 430 and 432.]

1. 6!   **2.** 7!

3. 8!/6!   **4.** 10!/8!

5. $P(6, 4)$ **6.** $P(8, 3)$

7. $P(6, 4)/4!$ **8.** $P(8, 3)/3!$

9. $C(3, 2)$ **10.** $C(4, 3)$

11. $C(10, 8)$ **12.** $C(11, 9)$

13. $C(20, 1)$ **14.** $C(30, 1)$

15. $C(100, 98)$ **16.** $C(100, 97)$

17. How many ordered lists are there of four items chosen from six?

18. How many ordered sequences are possible that contain three objects chosen from seven?

19. How many unordered sets are possible that contain three objects chosen from seven?

20. How many unordered sets are there of four items chosen from six?

21. How many five-letter sequences are possible that use the letters b, o, g, e, y once each?

22. How many six-letter sequences are possible that use the letters q, u, a, k, e, s once each?

23. How many three-letter sequences are possible that use the letters q, u, a, k, e, s at most once each?

24. How many three-letter sequences are possible that use the letters b, o, g, e, y at most once each?

25. How many three-letter (unordered) sets are possible that use the letters q, u, a, k, e, s at most once each?

26. How many three-letter (unordered) sets are possible that use the letters b, o, g, e, y at most once each?

27. ▼ How many six-letter sequences are possible that use the letters a, u, a, a, u, k? HINT [Use the decision algorithm discussed in Example 3 of Section 6.3.]

28. ▼ How many six-letter sequences are possible that use the letters f, f, a, a, f, f? HINT [See the hint for the preceding exercise.]

Marbles *For Exercises 29–42, a bag contains 3 red marbles, 2 green ones, 1 lavender one, 2 yellows, and 2 orange marbles.* HINT [See Example 7.]

29. How many possible sets of four marbles are there?

30. How many possible sets of three marbles are there?

31. How many sets of four marbles include all the red ones?

32. How many sets of three marbles include all the yellow ones?

33. How many sets of four marbles include none of the red ones?

34. How many sets of three marbles include none of the yellow ones?

35. How many sets of four marbles include one of each color other than lavender?

36. How many sets of five marbles include one of each color?

37. How many sets of five marbles include at least two red ones?

38. How many sets of five marbles include at least one yellow one?

39. How many sets of five marbles include at most one of the yellow ones?

40. How many sets of five marbles include at most one of the red ones?

41. ▼ How many sets of five marbles include either the lavender one or exactly one yellow one but not both colors?

42. ▼ How many sets of five marbles include at least one yellow one but no green ones?

Dice If a die is rolled 30 times, there are 6^{30} different sequences possible. Exercises 43–46 ask how many of these sequences satisfy certain conditions. HINT [Use the decision algorithm discussed in Example 3 of Section 6.3.]

43. ▼ What fraction of these sequences have exactly five 1s?

44. ▼ What fraction of these sequences have exactly five ones and five 2s?

45. ▼ What fraction of these sequences have exactly 15 even numbers?

46. ▼ What fraction of these sequences have exactly 10 numbers less than or equal to 2?

In Exercises 47–52, calculate how many different sequences can be formed that use the letters of each given word. [Decide where, for example, all the s's will go, rather than what will go in each position. Leave your answer as a product of terms of the form $C(n, r)$.]

47. ▼ mississippi

48. ▼ mesopotamia

49. ▼ megalomania

50. ▼ schizophrenia

51. ▼ casablanca

52. ▼ desmorelda

APPLICATIONS

53. *Itineraries* Your international diplomacy trip requires stops in Thailand, Singapore, Hong Kong, and Bali. How many possible itineraries are there? HINT [See Examples 1 and 2.]

54. *Itineraries* Refer back to the preceding exercise. How many possible itineraries are there in which the last stop is Thailand?

Poker Hands *A poker hand consists of 5 cards from a standard deck of 52. (See the chart preceding Example 8.) In Exercises 55–60, find the number of different poker hands of the specified type.* HINT [See Example 8.]

55. Two pairs (two of one denomination, two of another denomination, and one of a third)

56. Three of a kind (three of one denomination, one of another denomination, and one of a third)

57. Two of a kind (two of one denomination and three of different denominations)

58. Four of a kind (all four of one denomination and one of another)

59. ▼ Straight (five cards of consecutive denominations: A, 2, 3, 4, 5 up through 10, J, Q, K, A, not all of the same suit) (Note that the ace counts either as a 1 or as the denomination above king.)

60. ▼ Flush (five cards all of the same suit, but not consecutive denominations)

Dogs of the Dow *The "Dogs of the Dow" refer to the stocks listed on the Dow with the highest dividend yield. Exercises 61 and 62 are based on the following table, which shows the top 10 stocks of the "Dogs of the Dow" list in November 2011:*[18]

Symbol	Company	Price	Yield
T	AT&T	29.46	5.84%
VZ	Verizon	37.52	5.20%
MRK	Merck	34.47	4.41%
PFE	Pfizer	20.08	3.98%
JNJ	Johnson & Johnson	64.86	3.52%
INTC	Intel	24.75	3.39%
KFT	Kraft	35.48	3.27%
DD	DuPont	49.81	3.29%
CVX	Chevron	108.86	2.87%
MCD	McDonald's	94.60	2.58%

61. You decide to make a small portfolio consisting of a collection of 5 of the top 10 Dogs of the Dow.
   a. How many portfolios are possible?
   b. How many of these portfolios contain VZ and KFT but neither JNJ nor DD?
   c. How many of these portfolios contain at least four stocks with yields above 3.5%?

62. You decide to make a small portfolio consisting of a collection of 6 of the top 10 Dogs of the Dow.
   a. How many portfolios are possible?
   b. How many of these portfolios contain MRK but not PFE?
   c. How many of these portfolios contain at most one stock priced above $60?

Day Trading *Day traders typically buy and sell stocks (or other investment instruments) during the trading day and sell all investments by the end of the day. Exercises 63 and 64 are based on the following table, which shows the closing prices on November 20, 2011, of 12 stocks selected by your broker, Prudence Swift, as well as the change that day.*

Tech Stocks	Close	Change
AAPL (Apple)	$385.22	−10.06
MSFT (Microsoft)	$26.28	0.08
GOOG (Google)	$595.08	−5.87
INTC (Intel)	$24.06	0.22
SNE (Sony Corp.)	$17.11	−0.09
NOK (Nokia)	$6.45	0.07
Non-Tech Stocks		
F (Ford Motor Co.)	$10.99	−0.05
LCC (US Airways)	$4.86	−0.20
ED (Con Ed)	$58.88	0.66
HD (Home Depot)	$37.20	0.04
MO (Altria Group)	$27.63	0.46
ANF (Abercrombie & Fitch)	$55.46	−0.82

63. On the morning of November 20, 2011, Swift advised you to purchase a collection of three tech stocks and two non-tech stocks, all chosen at random from those listed in the table. You were to sell all the stocks at the end of the trading day.
   a. How many possible collections are possible?
   b. You tend to have bad luck with stocks—they usually start going down the moment you buy them. How many of the collections in part (a) consist entirely of stocks that declined in value by the end of the day?
   c. Using the answers to parts (a) and (b), what would you say your chances were of choosing a collection consisting entirely of stocks that declined in value by the end of the day?

64. On the morning of November 20, 2011, Swift advised your friend to purchase a collection of three stocks chosen at random from those listed in the table. Your friend was to sell all the stocks at the end of the trading day.
   a. How many possible collections are possible?
   b. How many of the collections in part (a) included exactly two tech stocks that increased in value by the end of the day?

[18]Source: www.dogsofthedow.com.

c. Using the answers to parts (a) and (b), what would you say the chances were that your friend chose a collection that included exactly two tech stocks that increased in value by the end of the day?

Movies Exercises 65 and 66 are based on the following list of top DVD rentals (based on revenue) for the weekend ending November 6, 2011:[19]

Name	Rank
Captain America: The First Avenger	1
Bad Teacher	2
Fast Five	3
Cars 2	4
Trespass	5
Bridesmaids	6
Zookeeper	7
Transformers: Dark of the Moon	8
Scream	9
Thor	10

65. ▼ Rather than study for math, you and your buddies decide to get together for a marathon movie-watching, popcorn-guzzling event on Saturday night. You decide to watch four movies selected at random from the above list.

a. How many sets of four movies are possible?

b. Your best friends, the Lara twins, refuse to see *Bridesmaids* on the grounds that it is "for girlie men" and also insist that at least one of *Captain America* or *Thor* be among the movies selected. How many of the possible groups of four will satisfy the twins?

c. Comparing the answers in parts (a) and (b), would you say the Lara twins are more likely than not to be satisfied with your random selection?

66. ▼ Rather than study for astrophysics, you and your friends decide to get together for a marathon movie-watching, gummy-bear-munching event on Saturday night. You decide to watch three movies selected at random from the above list.

a. How many sets of three movies are possible?

b. Your best friends, the Pelogrande twins, refuse to see either *Cars 2* or *Zookeeper* on the grounds that they are "for idiots" and also insist that no more than one of *Bad Teacher* and *Fast Five* should be among the movies selected. How many of the possible groups of three will satisfy the twins?

c. Comparing the answers in parts (a) and (b), would you say the Pelogrande twins are more likely than not to be satisfied with your random selection?

67. ▼ *Traveling Salesperson* Suppose you are a salesperson who must visit the following 23 cities: Dallas, Tampa, Orlando, Fairbanks, Seattle, Detroit, Chicago, Houston, Arlington, Grand Rapids, Urbana, San Diego, Aspen, Little Rock, Tuscaloosa, Honolulu, New York, Ithaca, Charlottesville, Lynchville, Raleigh, Anchorage, and Los Angeles. Leave all your answers in factorial form.

a. How many possible itineraries are there that visit each city exactly once?

b. Repeat part (a) in the event that the first five stops have already been determined.

c. Repeat part (a) in the event that your itinerary must include the sequence Anchorage, Fairbanks, Seattle, Chicago, and Detroit, in that order.

68. ▼ *Traveling Salesperson* Refer back to the preceding exercise (and leave all your answers in factorial form).

a. How many possible itineraries are there that start and end at Detroit, and visit every other city exactly once?

b. How many possible itineraries are there that start and end at Detroit, and visit Chicago twice and every other city once?

c. Repeat part (a) in the event that your itinerary must include the sequence Anchorage, Fairbanks, Seattle, Chicago, and New York, in that order.

69. ▼ (*From the GMAT*) Ben and Ann are among seven contestants from which four semifinalists are to be selected. Of the different possible selections, how many contain neither Ben nor Ann?

(A) 5   **(B)** 6   **(C)** 7   **(D)** 14   **(E)** 21

70. ▼ (*Based on a Question from the GMAT*) Ben and Ann are among seven contestants from which four semifinalists are to be selected. Of the different possible selections, how many contain Ben but not Ann?

(A) 5   **(B)** 8   **(C)** 9   **(D)** 10   **(E)** 20

71. ◆ (*From the GMAT exam*) If 10 persons meet at a reunion and each person shakes hands exactly once with each of the others, what is the total number of handshakes?

(A) $10 \cdot 9 \cdot 8 \cdot 7 \cdot 6 \cdot 5 \cdot 4 \cdot 3 \cdot 2 \cdot 1$ **(B)** $10 \cdot 10$
(C) $10 \cdot 9$ **(D)** 45   **(E)** 36

72. ◆ (*Based on a question from the GMAT exam*) If 12 businesspeople have a meeting and each pair exchanges business cards, how many business cards, total, get exchanged?

(A) $12 \cdot 11 \cdot 10 \cdot 9 \cdot 8 \cdot 7 \cdot 6 \cdot 5 \cdot 4 \cdot 3 \cdot 2 \cdot 1$ **(B)** $12 \cdot 12$
(C) $12 \cdot 11$ **(D)** 66   **(E)** 72

73. ◆ *Product Design* The Honest Lock Company plans to introduce what it refers to as the "true combination lock." The lock will open if the correct set of three numbers from 0 through 39 is entered in any order.

a. How many different combinations of three different numbers are possible?

b. If it is allowed that a number appear twice (but not three times), how many more possibilities are created?

---

[19]Source: Home Media Magazine (www.imdb.com/Charts/videolast).

c. If it is allowed that any or all of the numbers may be the same, what is the total number of combinations that will open the lock?

74. ◆ *Product Design* Repeat Exercise 73 for a lock based on selecting from the numbers 0 through 19.

75. ◆ *Theory of Linear Programming* (Some familiarity with linear programming is assumed for this exercise.) Suppose you have a linear programming problem with two unknowns and 20 constraints. You decide that graphing the feasible region would take a lot of work, but then you recall that corner points are obtained by solving a system of two equations in two unknowns obtained from two of the constraints. Thus, you decide that it might pay instead to locate all the possible corner points by solving all possible combinations of two equations and then checking whether each solution is a feasible point.

a. How many systems of two equations in two unknowns will you be required to solve?

b. Generalize this to n constraints.

76. ◆ *More Theory of Linear Programming* (Some familiarity with linear programming is assumed for this exercise.) Before the advent of the simplex method for solving linear programming problems, the following method was used: Suppose you have a linear programming problem with three unknowns and 20 constraints. You locate corner points as follows: Selecting three of the constraints, you turn them into equations (by replacing the inequalities with equalities), solve the resulting system of three equations in three unknowns, and then check to see whether the solution is feasible.

a. How many systems of three equations in three unknowns will you be required to solve?

b. Generalize this to n constraints.

COMMUNICATION AND REASONING EXERCISES

77. If you were hard pressed to study for an exam on counting and had only enough time to study one topic, would you choose the formula for the number of permutations or the multiplication principle? Give reasons for your choice.

78. The formula for $C(n, r)$ is written as a ratio of two whole numbers. Can $C(n, r)$ ever be a fraction and not a whole number? Explain (without actually discussing the formula itself).

79. Which of the following represent permutations?

(A) An arrangement of books on a shelf

(B) A group of 10 people in a bus

(C) A committee of 5 senators chosen from 100

(D) A presidential cabinet of 5 portfolios chosen from 20

80. Which of the following represent combinations?

(A) A portfolio of five stocks chosen from the S&P Top Ten

(B) A group of 5 tenors for a choir chosen from 12 singers

(C) A new company CEO and a new CFO chosen from five

(D) The *New York Times* Top Ten Bestseller list

81. When you click on "Get Driving Directions" on Mapquest .com do you get a permutation or a combination of driving instructions? Explain, and give a simple example to illustrate why the answer is important.

82. If you want to know how many possible lists, arranged in alphabetical order, there are of five students selected from your class, you would use the formula for permutations, right? (Explain your answer.)

83. You are tutoring your friend for a test on sets and counting and she asks the question: "How do I know what formula to use for a given problem?" What is a good way to respond?

84. Complete the following. If a counting procedure has five alternatives, each of which has four steps of two choices each, then there are ___ outcomes. On the other hand, if there are five steps, each of which has four alternatives of two choices each, then there are ___ outcomes.

85. ▼ A textbook has the following exercise. "Three students from a class of 50 are selected to take part in a play. How many casts are possible?" Comment on this exercise.

86. ▼ Explain why the coefficient of a^2b^4 in $(a + b)^6$ is $C(6, 2)$ (this is a consequence of the **binomial theorem**). HINT [In the product $(a + b)(a + b) \cdots (a + b)$ (six times), in how many different ways can you pick two a's and four b's to multiply together?]

CHAPTER 6 REVIEW

KEY CONCEPTS

Website www.WanerMath.com
Go to the Website at www.WanerMath .com to find a comprehensive and interactive Web-based summary of Chapter 6, along with section-by-section tutorials.

6.1 Sets and Set Operations

Sets, elements, subsets, proper subsets, empty set, finite and infinite sets *pp. 396 & 397*

Visualizing sets and relations between sets using Venn diagrams *p. 399*

Union:
$$A \cup B = \{x \mid x \in A \text{ or } x \in B\} \text{ } p. 400$$

Intersection: $A \cap B =$
$\{x \mid x \in A \text{ and } x \in B\}$ *p. 400*

Universal sets, complements *p. 401*

Disjoint sets: $A \cap B = \emptyset$ *p. 402*

Cartesian product: $A \times B =$
$\{(a, b) \mid a \in A \text{ and } b \in B\}$ *p. 402*

6.2 Cardinality

Cardinality: $n(A) =$
number of elements in A. *p. 407*

If A and B are finite sets, then $n(A \cup B) = n(A) + n(B) - n(A \cap B)$. *p. 408*

If A and B are disjoint finite sets, then $n(A \cup B) = n(A) + n(B)$. In this case, we say that $A \cup B$ is a **disjoint union**. *p. 408*

If S is a finite universal set and A is a subset of S, then $n(A') = n(S) - n(A)$ and $n(A) = n(S) - n(A')$. *p. 409*

If A and B are finite sets, then $n(A \times B) = n(A)n(B)$. *p. 414*

6.3 Decision Algorithms: The Addition and Multiplication Principles

Addition principle *p. 418*

Multiplication principle *p. 418*

Decision algorithm: a procedure for making a sequence of decisions

to choose an element of a set *p. 420*

6.4 Permutations and Combinations

n factorial:
$$n! = n \times (n - 1) \times (n - 2) \times \cdots \times 2 \times 1$$
p. 429

Permutation of n items taken r at a time:
$$P(n, r) = \frac{n!}{(n - r)!} \quad p. 430$$

Combination of n items taken r at a time:
$$C(n, r) = P(n, r)/r! = \frac{n!}{r!(n - r)!}$$
p. 432

Using the equality
$C(n, r) = C(n, n - r)$ to simplify calculations of combinations *p. 433*

REVIEW EXERCISES

In each of Exercises 1–5, list the elements of the given set.

1. The set N of all negative integers greater than or equal to -3

2. The set of all outcomes of tossing a coin five times

3. The set of all outcomes of tossing two distinguishable dice such that the numbers are different

4. The sets $(A \cap B) \cup C$ and $A \cap (B \cup C)$, where $A = \{1, 2, 3, 4, 5\}$, $B = \{3, 4, 5\}$, and $C = \{1, 2, 5, 6, 7\}$

5. The sets $A \cup B'$ and $A \times B'$, where $A = \{a, b\}$, $B = \{b, c\}$, and $S = \{a, b, c, d\}$

In each of Exercises 6–10, write the indicated set in terms of the given sets.

6. S: the set of all customers; A: the set of all customers who owe money; B: the set of all customers who owe at least \$1,000. The set of all customers who owe money but owe less than \$1,000

7. A: the set of outcomes when a day in August is selected; B: the set of outcomes when a time of day is selected. The set of outcomes when a day in August and a time of that day are selected

8. S: the set of outcomes when two dice are rolled; E: those outcomes in which at most one die shows an even number, F: those outcomes in which the sum of the numbers is 7. The set of outcomes in which both dice show an even number or sum to seven

9. S: the set of all integers; P: the set of all positive integers; E: the set of all even integers; Q: the set of all integers that are perfect squares ($Q = \{0, 1, 4, 9, 16, 25, \ldots\}$). The set of all integers that are not positive odd perfect squares

10. S: the set of all integers; N: the set of all negative integers; E: the set of all even integers; T: the set of all integers that are multiples of 3 ($T = \{0, 3, -3, 6, -6, 9, -9, \ldots\}$). The set of all even integers that are neither negative nor multiples of three

In each of Exercises 11–14, give a formula for the cardinality rule or rules needed to answer the question, and then give the solution.

11. You have read 150 of the 400 novels in your home, but your sister Roslyn has read 200, of which only 50 are novels you have read as well. How many have neither of you read?

12. There are 32 students in categories A and B combined; 24 are in A and 24 are in B. How many are in both A and B?

13. You roll two dice, one red and one green. Losing combinations are doubles (both dice show the same number) and outcomes in which the green die shows an odd number and the red die shows an even number. The other combinations are winning ones. How many winning combinations are there?

14. The *Apple iMac* used to come in three models, each with five colors to choose from. How many combinations were possible?

Recall that a poker hand consists of 5 cards from a standard deck of 52. In each of Exercises 15–18, find the number of different poker hands of the specified type. Leave your answer in terms of combinations.

15. Two of a kind with no aces

16. A full house with either two kings and three queens or two queens and three kings

17. Straight Flush (five cards of the same suit with consecutive denominations: A, 2, 3, 4, 5 up through 10, J, Q, K, A)

18. Three of a kind with no aces

In each of Exercises 19–24, consider a bag containing four red marbles, two green ones, one transparent one, three yellow ones, and two orange ones.

19. How many possible sets of five marbles are there in which all of them are red or green?

20. How many possible sets of five marbles are there in which none of them are red or green?

21. How many sets of five marbles include all the red ones?

22. How many sets of five marbles do not include all the red ones?

23. How many sets of five marbles include at least two yellow ones?

24. How many sets of five marbles include at most one of the red ones but no yellow ones?

APPLICATIONS: OHaganBooks.com

Inventories OHaganBooks.com currently operates three warehouses: one in Washington, one in California, and the new one in Texas. Exercises 25–30 are based on the following table, which shows the book inventories at each warehouse:

	Sci Fi	Horror	Romance	Other	Total
Washington	10,000	12,000	12,000	30,000	64,000
California	8,000	12,000	6,000	16,000	42,000
Texas	15,000	15,000	20,000	44,000	94,000
Total	33,000	39,000	38,000	90,000	200,000

Take the first letter of each category to represent the corresponding set of books; for instance, S is the set of sci fi books in stock, W is the set of books in the Washington warehouse, and so on. In each exercise describe the given set in words and compute its cardinality.

25. $S \cup T$

26. $H \cap C$

27. $C \cup S'$

28. $(R \cap T) \cup H$

29. $R \cap (T \cup H)$

30. $(S \cap W) \cup (H \cap C')$

Customers OHaganBooks.com has two main competitors: JungleBooks.com and FarmerBooks.com. At the beginning of

August, OHaganBooks.com had 3,500 customers. Of these, a total of 2,000 customers were shared with JungleBooks.com and 1,500 with FarmerBooks.com. Furthermore, 1,000 customers were shared with both. JungleBooks.com has a total of 3,600 customers, FarmerBooks.com has 3,400, and they share 1,100 customers between them. Use these data for Exercises 31–36.

31. How many of all these customers are exclusive OHaganBooks.com customers?

32. How many customers of the other two companies are not customers of OHaganBooks.com?

33. Which of the three companies has the largest number of exclusive customers?

34. Which of the three companies has the smallest number of exclusive customers?

35. OHaganBooks.com is interested in merging with one of its two competitors. Which merger would give it the largest combined customer base, and how large would that be?

36. Referring to the preceding exercise, which merger would give OHaganBooks.com the largest *exclusive* customer base, and how large would that be?

Online IDs *As the customer base at OHaganBooks.com grows, software manager Ruth Nabarro is thinking of introducing identity codes for all the online customers. Help her with the questions in Exercises 37–40.*

37. If she uses 3-letter codes, how many different customers can be identified?

38. If she uses codes with three different letters, how many different customers can be identified?

39. It appears that Nabarro has finally settled on codes consisting of two letters followed by two digits. For technical reasons, the letters must be different and the first digit cannot be a zero. How many different customers can be identified?

40. O'Hagan sends Nabarro the following memo:

```
To: Ruth Nabarro, Software Manager
From: John O'Hagan, CEO
Subject: Customer Identity Codes

I have read your proposal for the cus-
tomer ID codes. However, due to our
ambitious expansion plans, I would
like our system software to allow for
at least 500,000 customers. Please
adjust your proposal accordingly.
```

Nabarro is determined to have a sequence of letters followed by some digits, and, for reasons too complicated to explain, there cannot be more than two letters, the letters must be different, the digits must all be different, and the first digit cannot be a zero. What is the form of the shortest code she can use to satisfy the CEO, and how many different customers can be identified?

Degree Requirements After an exhausting day at the office, John O'Hagan returns home and finds himself having to assist his son Billy-Sean, who continues to have a terrible time planning his first-year college course schedule. The latest Bulletin of Suburban State University reads as follows:

All candidates for the degree of Bachelor of Arts at SSU must take, in their first year, at least 10 courses in the Sciences, Fine Arts, Liberal Arts, and Mathematics combined, of which at least 2 must be in each of the Sciences and Fine Arts, and exactly 3 must be in each of the Liberal Arts and Mathematics.

Help him with the answers to Exercises 41–43.

41. If the Bulletin lists exactly five first-year-level science courses and six first-year-level courses in each of the other categories, how many course combinations are possible that meet the minimum requirements?

42. Reading through the course descriptions in the bulletin a second time, John O'Hagan notices that Calculus I (listed as one of the mathematics courses) is a required course for many of the other courses, and so decides that it would be best if Billy-Sean included Calculus I. Further, two of the Fine Arts courses cannot both be taken in the first year. How many course combinations are possible that meet the minimum requirements and include Calculus I?

43. To complicate things further, in addition to the requirement in Exercise 42, Physics II has Physics I as a prerequisite (both are listed as first-year science courses, but it is not necessary to take both). How many course combinations are possible that include Calculus I and meet the minimum requirements?

Case Study   Designing a Puzzle

As Product Design Manager for Cerebral Toys Inc., you are constantly on the lookout for ideas for intellectually stimulating yet inexpensive toys. You recently received the following memo from Felix Frost, the developmental psychologist on your design team.

> To: Felicia
> From: Felix
> Subject: Crazy Cubes
>
> We've hit on an excellent idea for a new educational puzzle (which we are calling "Crazy Cubes" until Marketing comes up with a better name). Basically, Crazy Cubes will consist of a set of plastic cubes. Two faces of each cube will be colored red, two will be colored blue, and two white, and there will be exactly two cubes with each possible configuration of colors. The goal of the puzzle is to seek out the matching pairs, thereby enhancing a child's geometric intuition and three-dimensional manipulation skills. The kit will include every possible configuration of colors. We are, however, a little stumped on the following question: How many cubes will the kit contain? In other words, how many possible ways can one color the faces of a cube so that two faces are red, two are blue, and two are white?

Looking at the problem, you reason that the following three-step decision algorithm ought to suffice:

Step 1 Choose a pair of faces to color red; $C(6, 2) = 15$ choices.

Step 2 Choose a pair of faces to color blue; $C(4, 2) = 6$ choices.

Step 3 Choose a pair of faces to color white; $C(2, 2) = 1$ choice.

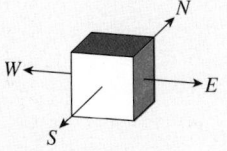

Figure 9

This algorithm appears to give a total of $15 \times 6 \times 1 = 90$ possible cubes. However, before sending your reply to Felix, you realize that something is wrong, because there are different choices that result in the same cube. To describe some of these choices, imagine a cube oriented so that four of its faces are facing the four compass directions (Figure 9). Consider choice 1, with the top and bottom faces blue, north and south faces white, and east and west faces red; and choice 2, with the top and bottom

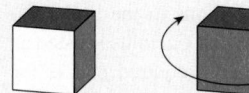

Figure 10

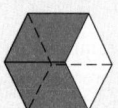

Figure 11

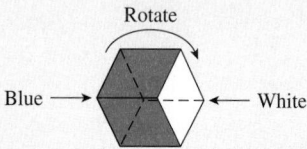

Figure 12

faces blue, north and south faces red, and east and west faces white. These cubes are actually the same, as you see by rotating the second cube 90 degrees (Figure 10).

You therefore decide that you need a more sophisticated decision algorithm. Here is one that works.

Alternative 1: Faces with the same color opposite each other. Place one of the blue faces down. Then the top face is also blue. The cube must look like the one drawn in Figure 10. Thus there is only one choice here.

Alternative 2: Red faces opposite each other and the other colors on adjacent pairs of faces. Again there is only one choice, as you can see by putting the red faces on the top and bottom and then rotating.

Alternative 3: White faces opposite each other and the other colors on adjacent pairs of faces; one possibility.

Alternative 4: Blue faces opposite each other and the other colors on adjacent pairs of faces; one possibility.

Alternative 5: Faces with the same color adjacent to each other. Look at the cube so that the edge common to the two red faces is facing you and horizontal (Figure 11). Then the faces on the left and right must be of different colors because they are opposite each other. Assume that the face on the right is white. (If it's blue, then rotate the die with the red edge still facing you to move it there, as in Figure 12.) This leaves two choices for the other white face, on the upper or the lower of the two back faces. This alternative gives two choices.

It follows that there are $1 + 1 + 1 + 1 + 2 = 6$ choices. Because the Crazy Cubes kit will feature two of each cube, the kit will require 12 different cubes.*

✳ There is a beautiful way of calculating this and similar numbers, called Pólya enumeration, but it requires a discussion of topics well outside the scope of this book. Take this as a hint that counting techniques can use some of the most sophisticated mathematics.

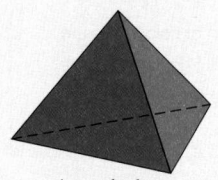

A tetrahedron

EXERCISES

In all of the following exercises, there are three colors to choose from: red, white, and blue.

1. In order to enlarge the kit, Felix suggests including two each of two-colored cubes (using two of the colors red, white, and blue) with three faces one color and three another. How many additional cubes will be required?

2. If Felix now suggests adding two each of cubes with two faces one color, one face another color, and three faces the third color, how many additional cubes will be required?

3. Felix changes his mind and suggests the kit use tetrahedral blocks with two colors instead (see the figure). How many of these would be required?

4. Once Felix finds the answer to the preceding exercise, he decides to go back to the cube idea, but this time insists that all possible combinations of up to three colors should be included. (For instance, some cubes will be all one color, others will be two colors.) How many cubes should the kit contain?

7

Probability

WV Website

www.WanerMath.com

At the Website you will find:

- Section-by-section tutorials, including game tutorials with randomized quizzes

- A detailed chapter summary

- A true/false quiz

- A Markov system simulation and matrix algebra tool

- Additional review exercises

Case Study The Monty Hall Problem

On the game show *Let's Make a Deal*, you are shown three doors, A, B, and C, and behind one of them is the Big Prize. After you select one of them—say, door A—to make things more interesting the host (Monty Hall) opens one of the other doors—say, door B—revealing that the Big Prize is not there. He then offers you the opportunity to change your selection to the remaining door, door C. Should you switch or stick with your original guess? **Does it make any difference?**

Everett Collection

445

Introduction

What is the probability of winning the lottery twice? What are the chances that a college athlete whose drug test is positive for steroid use is actually using steroids? You are playing poker and have been dealt two jacks. What is the likelihood that one of the next three cards you are dealt will also be a jack? These are all questions about probability.

Understanding probability is important in many fields, ranging from risk management in business through hypothesis testing in psychology to quantum mechanics in physics. Historically, the theory of probability arose in the sixteenth and seventeenth centuries from attempts by mathematicians such as Gerolamo Cardano, Pierre de Fermat, Blaise Pascal, and Christiaan Huygens to understand games of chance. Andrey Nikolaevich Kolmogorov set forth the foundations of modern probability theory in his 1933 book *Foundations of the Theory of Probability.*

The goal of this chapter is to familiarize you with the basic concepts of modern probability theory and to give you a working knowledge that you can apply in a variety of situations. In the first two sections, the emphasis is on translating real-life situations into the language of sample spaces, events, and probability. Once we have mastered the language of probability, we spend the rest of the chapter studying some of its theory and applications. The last section gives an interesting application of both probability and matrix arithmetic.

7.1 Sample Spaces and Events

Sample Spaces

At the beginning of a football game, to ensure fairness, the referee tosses a coin to decide who will get the ball first. When the ref tosses the coin and observes which side faces up, there are two possible results: heads (H) and tails (T). These are the *only* possible results, ignoring the (remote) possibility that the coin lands on its edge. The act of tossing the coin is an example of an **experiment**. The two possible results, H and T, are possible **outcomes** of the experiment, and the set $S = \{H, T\}$ of all possible outcomes is the **sample space** for the experiment.

Experiments, Outcomes, and Sample Spaces

An **experiment** is an occurrence with a result, or **outcome**, that is uncertain before the experiment takes place. The set of all possible outcomes is called the **sample space** for the experiment.

Quick Examples

1. *Experiment:* Flip a coin and observe the side facing up.
   Outcomes: H, T
   Sample Space: $S = \{H, \ T\}$
2. *Experiment:* Select a student in your class.
   Outcomes: The students in your class
   Sample Space: The set of students in your class

3. *Experiment:* Select a student in your class and observe the color of his or her hair.
Outcomes: red, black, brown, blond, green, ...
Sample Space: {red, black, brown, blond, green, ...}

4. *Experiment:* Cast a die and observe the number facing up.
Outcomes: 1, 2, 3, 4, 5, 6
Sample Space: $S = \{1, 2, 3, 4, 5, 6\}$

5. *Experiment:* Cast two distinguishable dice (see Example 1(a) of Section 6.1) and observe the numbers facing up.
Outcomes: (1, 1), (1, 2), ... , (6, 6)   (36 outcomes)

$$
\text{Sample Space: } S = \begin{Bmatrix}
(1, 1), & (1, 2), & (1, 3), & (1, 4), & (1, 5), & (1, 6), \\
(2, 1), & (2, 2), & (2, 3), & (2, 4), & (2, 5), & (2, 6), \\
(3, 1), & (3, 2), & (3, 3), & (3, 4), & (3, 5), & (3, 6), \\
(4, 1), & (4, 2), & (4, 3), & (4, 4), & (4, 5), & (4, 6), \\
(5, 1), & (5, 2), & (5, 3), & (5, 4), & (5, 5), & (5, 6), \\
(6, 1), & (6, 2), & (6, 3), & (6, 4), & (6, 5), & (6, 6)
\end{Bmatrix}
$$

$n(S) =$ the number of outcomes in $S = 36$

6. *Experiment:* Cast two indistinguishable dice (see Example 1(b) of Section 6.1) and observe the numbers facing up.
Outcomes: (1, 1), (1, 2), ... , (6, 6)   (21 outcomes)

$$
\text{Sample Space: } S = \begin{Bmatrix}
(1, 1), & (1, 2), & (1, 3), & (1, 4), & (1, 5), & (1, 6), \\
 & (2, 2), & (2, 3), & (2, 4), & (2, 5), & (2, 6), \\
 & & (3, 3), & (3, 4), & (3, 5), & (3, 6), \\
 & & & (4, 4), & (4, 5), & (4, 6), \\
 & & & & (5, 5), & (5, 6), \\
 & & & & & (6, 6)
\end{Bmatrix}
$$

$n(S) = 21$

7. *Experiment:* Cast two dice and observe the *sum* of the numbers facing up.
Outcomes: 2, 3, 4, 5, 6, 7, 8, 9, 10, 11, 12
Sample Space: $S = \{2, 3, 4, 5, 6, 7, 8, 9, 10, 11, 12\}$

8. *Experiment:* Choose 2 cars (without regard to order) at random from a fleet of 10.
Outcomes: Collections of 2 cars chosen from 10
Sample Space: The set of all collections of 2 cars chosen from 10

$n(S) = C(10, 2) = 45$

The following example introduces a sample space that we'll use in several other examples.

EXAMPLE 1 **School and Work**

In a survey conducted by the Bureau of Labor Statistics,[1] the high school graduating class of 2010 was divided into those who went on to college and those who did not. Those who went on to college were further divided into those who went to 2-year colleges and those who went to 4-year colleges. All graduates were also asked whether they were working or not. Find the sample space for the experiment "Select a member of the high school graduating class of 2010 and classify his or her subsequent school and work activity."

Solution The tree in Figure 1 shows the various possibilities.

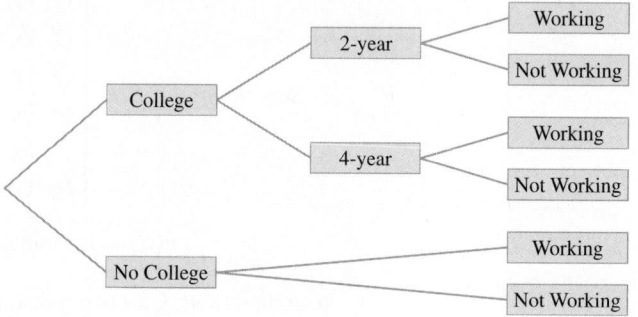

Figure 1

The sample space is

$$S = \{\text{2-year college \& working, 2-year college \& not working,}$$
$$\text{4-year college \& working, 4-year college \& not working,}$$
$$\text{no college \& working, no college \& not working}\}.$$

Events

In Example 1, suppose we are interested in the event that a 2010 high school graduate was working. In mathematical language, we are interested in the *subset* of the sample space consisting of all outcomes in which the graduate was working.

Events

Given a sample space S, an **event** E is a subset of S. The outcomes in E are called the **favorable** outcomes. We say that E **occurs** in a particular experiment if the outcome of that experiment is one of the elements of E—that is, if the outcome of the experiment is favorable.

---

[1]"College Enrollment and Work Activity of High School Graduates," U.S. Bureau of Labor Statistics, www.bls.gov/news.release/hsgec.htm.

Visualizing an Event

In the following figure, the favorable outcomes (events in E) are shown in green.

Sample Space S

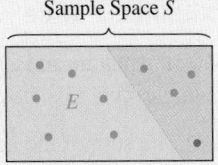

Quick Examples

1. *Experiment:* Roll a die and observe the number facing up.

   $S = \{1, 2, 3, 4, 5, 6\}$

   Event: E: The number observed is odd.

   $E = \{1, 3, 5\}$

2. *Experiment:* Roll two distinguishable dice and observe the numbers facing up.

   $S = \{(1, 1), (1, 2), \ldots, (6, 6)\}$

   Event: F: The dice show the same number.

   $F = \{(1, 1), (2, 2), (3, 3), (4, 4), (5, 5), (6, 6)\}$

3. *Experiment:* Roll two distinguishable dice and observe the numbers facing up.

   $S = \{(1, 1), (1, 2), \ldots, (6, 6)\}$

   Event: G: The sum of the numbers is 1.

   $G = \emptyset$ There are no favorable outcomes.

4. *Experiment:* Select a city beginning with "J."
   Event: E: The city is Johannesburg.

   $E = \{\text{Johannesburg}\}$ An event can consist of a single outcome.

5. *Experiment:* Roll a die and observe the number facing up.
   Event: E: The number observed is either even or odd.

   $E = S = \{1, 2, 3, 4, 5, 6\}$ An event can consist of all possible outcomes.

6. *Experiment:* Select a student in your class.
   Event: E: The student has red hair.

   $E = \{\text{red-haired students in your class}\}$

7. *Experiment:* Draw a hand of 2 cards from a deck of 52.
   Event: H: Both cards are diamonds.

   H is the set of all hands of 2 cards chosen from 52 such that both cards are diamonds.

Here are some more examples of events.

EXAMPLE 2 Dice

We roll a red die and a green die and observe the numbers facing up. Describe the following events as subsets of the sample space.

a. E: The sum of the numbers showing is 6.

b. F: The sum of the numbers showing is 2.

Solution Here (again) is the sample space for the experiment of throwing two dice.

$$S = \begin{Bmatrix} (1,1), & (1,2), & (1,3), & (1,4), & (1,5), & (1,6), \\ (2,1), & (2,2), & (2,3), & (2,4), & (2,5), & (2,6), \\ (3,1), & (3,2), & (3,3), & (3,4), & (3,5), & (3,6), \\ (4,1), & (4,2), & (4,3), & (4,4), & (4,5), & (4,6), \\ (5,1), & (5,2), & (5,3), & (5,4), & (5,5), & (5,6), \\ (6,1), & (6,2), & (6,3), & (6,4), & (6,5), & (6,6) \end{Bmatrix}$$

a. In mathematical language, E is the subset of S that consists of all those outcomes in which the sum of the numbers showing is 6. Here is the sample space once again, with the outcomes in question shown in color:

$$S = \begin{Bmatrix} (1,1), & (1,2), & (1,3), & (1,4), & (1,5), & (1,6), \\ (2,1), & (2,2), & (2,3), & (2,4), & (2,5), & (2,6), \\ (3,1), & (3,2), & (3,3), & (3,4), & (3,5), & (3,6), \\ (4,1), & (4,2), & (4,3), & (4,4), & (4,5), & (4,6), \\ (5,1), & (5,2), & (5,3), & (5,4), & (5,5), & (5,6), \\ (6,1), & (6,2), & (6,3), & (6,4), & (6,5), & (6,6) \end{Bmatrix}$$

Thus, $E = \{(1,5), (2,4), (3,3), (4,2), (5,1)\}$.

b. The only outcome in which the numbers showing add to 2 is $(1,1)$. Thus,

$$F = \{(1,1)\}.$$

EXAMPLE 3 School and Work

Let S be the sample space of Example 1. List the elements in the following events:

a. The event E that a 2010 high school graduate was working.

b. The event F that a 2010 high school graduate was not going to a 2-year college.

Solution

a. We had this sample space:

$S = \{$2-year college & working, two-year college & not working,
4-year college & working, four-year college & not working,
no college & working, no college & not working$\}$.

We are asked for the event that a graduate was working. Whenever we encounter a phrase involving "the event that . . . ," we mentally translate this into mathematical language by changing the wording.

Replace the phrase "the event that . . ." by the phrase "the subset of the sample space consisting of all outcomes in which . . . ."

Thus we are interested in the subset of the sample space consisting of all outcomes in which the graduate was working. This gives

$$E = \{2\text{-year college \& working, 4-year college \& working, no college \& working}\}.$$

The outcomes in E are illustrated by the shaded cells in Figure 2.

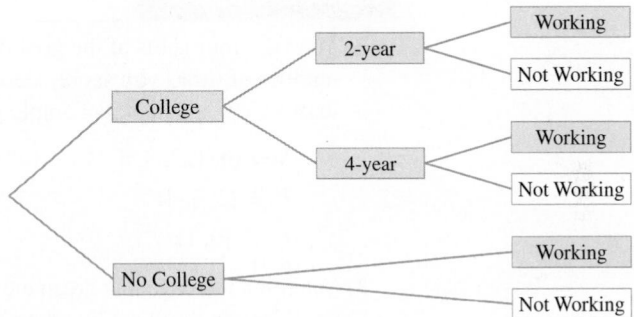

Figure 2

b. We are looking for the event that a graduate was not going to a 2-year college; that is, the subset of the sample space consisting of all outcomes in which the graduate was not going to a 2-year college. Thus,

$$F = \{4\text{-year college \& working, 4-year college \& not working, no college \& working, no college \& not working}\}.$$

The outcomes in F are illustrated by the shaded cells in Figure 3.

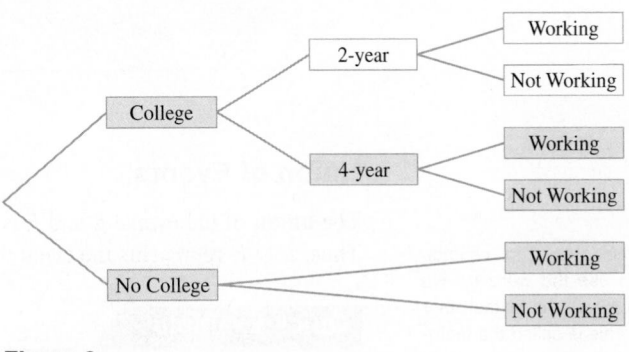

Figure 3

Complement, Union, and Intersection of Events

Events may often be described in terms of other events, using set operations such as complement, union, and intersection.

Complement of an Event

The **complement** of an event E is the set of outcomes not in E. Thus, the complement of E represents the event that E *does not occur*.

Visualizing the Complement

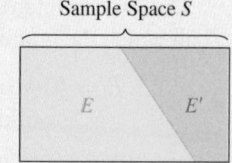

Sample Space S

Quick Examples

1. You take four shots at the goal during a soccer game and record the number of times you score. Describe the event that you score at least twice, and also its complement.

   $S = \{0, 1, 2, 3, 4\}$ Set of outcomes

   $E = \{2, 3, 4\}$ Event that you score at least twice

   $E' = \{0, 1\}$ Event that you do not score at least twice

2. You roll a red die and a green die and observe the two numbers facing up. Describe the event that the sum of the numbers is not 6.

   $S = \{(1, 1), (1, 2), \ldots, (6, 6)\}$

   $F = \{(1, 5), (2, 4), (3, 3), (4, 2), (5, 1)\}$ Sum of numbers is 6.

$$
F' = \begin{cases}
(1, 1), & (1, 2), & (1, 3), & (1, 4), & & (1, 6), \\
(2, 1), & (2, 2), & (2, 3), & & (2, 5), & (2, 6), \\
(3, 1), & (3, 2), & & (3, 4), & (3, 5), & (3, 6), \\
(4, 1), & & & (4, 3), & (4, 4), & (4, 5), & (4, 6), \\
& & (5, 2), & (5, 3), & (5, 4), & (5, 5), & (5, 6), \\
(6, 1), & (6, 2), & (6, 3), & (6, 4), & (6, 5), & (6, 6)
\end{cases}
$$

Sum of numbers is not 6.

Union of Events

The **union** of the events E and F is the set of all outcomes in E or F (or both). Thus, $E \cup F$ represents the event that E occurs *or* F occurs (or both).*

* As in the preceding chapter, when we use the word *or*, we agree to mean one or the other *or both*. This is called the **inclusive or** and mathematicians have agreed to take this as the meaning of *or* to avoid confusion.

Quick Example

Roll a die.

   E: The outcome is a 5; $E = \{5\}$.

   F: The outcome is an even number; $F = \{2, 4, 6\}$.

   $E \cup F$: The outcome is either a 5 *or* an even number;
   $E \cup F = \{2, 4, 5, 6\}$.

Intersection of Events

The **intersection** of the events E and F is the set of all outcomes common to E and F. Thus, $E \cap F$ represents the event that both E *and* F occur.

> ### Quick Example
>
> Roll two dice; one red and one green.
>
> E: The red die is 2.
>
> F: The green die is odd.
>
> $E \cap F$: The red die is 2 and the green die is odd;
> $$E \cap F = \{(2, 1), (2, 3), (2, 5)\}.$$

EXAMPLE 4 Weather

Let R be the event that it will rain tomorrow, let P be the event that it will be pleasant, let C be the event that it will be cold, and let H be the event that it will be hot.

a. Express in words: $R \cap P'$, $R \cup (P \cap C)$.

b. Express in symbols: Tomorrow will be either a pleasant day or a cold and rainy day; it will not, however, be hot.

Solution The key here is to remember that intersection corresponds to *and* and union to *or*.

a. $R \cap P'$ is the event that it will rain *and* it will not be pleasant.

$R \cup (P \cap C)$ is the event that either it will rain, or it will be pleasant and cold.

b. If we rephrase the given statement using *and* and *or* we get "Tomorrow will be either a pleasant day or a cold and rainy day, and it will not be hot."

$$[P \cup (C \cap R)] \cap H' \qquad \text{Pleasant, or cold and rainy, and not hot.}$$

The nuances of the English language play an important role in this formulation. For instance, the effect of the pause (comma) after "rainy day" suggests placing the preceding clause $P \cup (C \cap R)$ in parentheses. In addition, the phrase "cold and rainy" suggests that C and R should be grouped together in their own parentheses.

EXAMPLE 5 iPods, iPhones, and iPads

(Compare Example 3 in Section 6.2.) The following table shows sales, in millions of units, of iPods®, iPhones®, and iPads® in the last three quarters of 2011.[2]

	iPods (A)	iPhones (B)	iPads (C)	**Total**
2011 Q2 (U)	9.0	18.7	4.7	32.4
2011 Q3 (V)	7.5	20.3	9.3	37.1
2011 Q4 (W)	6.6	17.1	11.1	34.8
Total	23.1	56.1	25.1	104.3

---

[2]Figures are rounded to one decimal place. Source: Apple quarterly press releases (www.investor.apple.com).

Consider the experiment in which a device is selected at random from those represented in the table. Let A be the event that it was an iPod, let C be the event that it was an iPad, and let U be the event that it was sold in the second quarter of 2011. Describe the following events and compute their cardinality:

a. U' **b.** $A \cap U'$ **c.** $C \cup U$

Solution Before we answer the questions, note that S is the set of all items represented in the table, so S has a total of 104.3 million outcomes.

a. U' is the event that the item was not sold in the second quarter of 2011. Its cardinality is

$$n(U') = n(S) - n(U) = 104.3 - 32.4 = 71.9 \text{ million items.}$$

b. $A \cap U'$ is the event that it was an iPod not sold in the second quarter of 2011.

	iPods (A)	iPhones (B)	iPads (C)	Total
2011 Q2 (U)	9.0	18.7	4.7	32.4
2011 Q3 (V)	7.5	20.3	9.3	37.1
2011 Q4 (W)	6.6	17.1	11.1	34.8
Total	23.1	56.1	25.1	104.3

Referring to the table, we find

$$n(A \cap U') = 7.5 + 6.6 = 14.1 \text{ million items.}$$

c. $C \cup U$ is the event that either it was an iPad or it was sold in the second quarter of 2011:

	iPods (A)	iPhones (B)	iPads (C)	Total
2011 Q2 (U)	9.0	18.7	4.7	32.4
2011 Q3 (V)	7.5	20.3	9.3	37.1
2011 Q4 (W)	6.6	17.1	11.1	34.8
Total	23.1	56.1	25.1	104.3

To compute its cardinality, we can use the formula for the cardinality of a union:

$$n(C \cup U) = n(C) + n(U) - n(C \cap U)$$
$$= 25.1 + 32.4 - 4.7 = 52.8 \text{ million items.}$$

The case where $E \cap F$ is empty is interesting, and we give it a name.

Mutually Exclusive Events

If E and F are events, then E and F are said to be **disjoint** or **mutually exclusive** if $E \cap F$ is empty. (Hence, they have no outcomes in common.)

Visualizing Mutually Exclusive Events

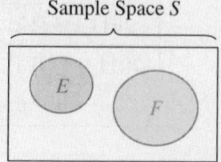

Interpretation
It is impossible for mutually exclusive events to occur simultaneously.

Quick Examples

In each of the following examples, E and F are mutually exclusive events.

1. Roll a die and observe the number facing up. E: The outcome is even; F: The outcome is odd.

$$E = \{2, 4, 6\}, F = \{1, 3, 5\}$$

2. Toss a coin three times and record the sequence of heads and tails. E: All three tosses land the same way up, F: One toss shows heads and the other two show tails.

$$E = \{\text{HHH, TTT}\}, F = \{\text{HTT, THT, TTH}\}$$

3. Observe tomorrow's weather. E: It is raining; F: There is not a cloud in the sky.

FAQs

Specifying the Sample Space

Q: *How do I determine the sample space in a given application?*

A: Strictly speaking, an experiment should include a description of what kinds of objects are in the sample space, as in:

Cast a die and observe the number facing up.
Sample space: the possible numbers facing up, {1, 2, 3, 4, 5, 6}.

Choose a person at random and record her Social Security number and whether she is blonde.
Sample space: pairs (9-digit number, Y/N).

However, in many of the scenarios discussed in this chapter and the next, an experiment is specified more vaguely, as in "Select a student in your class." In cases like this, the nature of the sample space should be determined from the context. For example, if the discussion is about grade-point averages and gender, the sample space can be taken to consist of pairs (grade-point average, M/F).

7.1 EXERCISES

▼ more advanced   ◆ challenging
Ⓣ indicates exercises that should be solved using technology

In Exercises 1–18, describe the sample space S of the experiment and list the elements of the given event. (Assume that the coins are distinguishable and that what is observed are the faces or numbers that face up.) HINT [See Examples 1–3.]

1. Two coins are tossed; the result is at most one tail.

2. Two coins are tossed; the result is one or more heads.

3. Three coins are tossed; the result is at most one head.

4. Three coins are tossed; the result is more tails than heads.

5. Two distinguishable dice are rolled; the numbers add to 5.

6. Two distinguishable dice are rolled; the numbers add to 9.

7. Two indistinguishable dice are rolled; the numbers add to 4.

8. Two indistinguishable dice are rolled; one of the numbers is even and the other is odd.

9. Two indistinguishable dice are rolled; both numbers are prime.[3]

10. Two indistinguishable dice are rolled; neither number is prime.

11. A letter is chosen at random from those in the word *Mozart*; the letter is a vowel.

12. A letter is chosen at random from those in the word *Mozart*; the letter is neither *a* nor *m*.

13. A sequence of two different letters is randomly chosen from those of the word *sore*; the first letter is a vowel.

14. A sequence of two different letters is randomly chosen from those of the word *hear*; the second letter is not a vowel.

15. A sequence of two different digits is randomly chosen from the digits 0–4; the first digit is larger than the second.

16. A sequence of two different digits is randomly chosen from the digits 0–4; the first digit is twice the second.

17. You are considering purchasing either a domestic car, an imported car, a van, an antique car, or an antique truck; you do not buy a car.

18. You are deciding whether to enroll for Psychology 1, Psychology 2, Economics 1, General Economics, or Math for Poets; you decide to avoid economics.

19. A packet of gummy candy contains four strawberry gums, four lime gums, two black currant gums, and two orange gums. April May sticks her hand in and selects four at random. Complete the following sentences:
    a. The sample space is the set of . . . .
    b. April is particularly fond of combinations of two strawberry and two black currant gums. The event that April will get the combination she desires is the set of . . . .

20. A bag contains three red marbles, two blue ones, and four yellow ones, and Alexandra Great pulls out three of them at random. Complete the following sentences:
    a. The sample space is the set of . . . .
    b. The event that Alexandra gets one of each color is the set of . . . .

21. ▼ President Barack H. Obama's cabinet consisted of the Secretaries of Agriculture, Commerce, Defense, Education, Energy, Health and Human Services, Homeland Security, Housing and Urban Development, Interior, Labor, State, Transportation, Treasury, Veterans Affairs, and the Attorney General.[4] Assuming that President Obama had 20 candidates, including Hillary Clinton, to fill these posts (and wished to

assign no one to more than one post), complete the following sentences:
    a. The sample space is the set of . . . .
    b. The event that Hillary Clinton is the Secretary of State is the set of . . . .

22. ▼ A poker hand consists of a set of 5 cards chosen from a standard deck of 52 playing cards. You are dealt a poker hand. Complete the following sentences:
    a. The sample space is the set of . . . .
    b. The event "a full house" is the set of . . . . (Recall that a full house is three cards of one denomination and two of another.)

Suppose two dice (one red, one green) are rolled. Consider the following events. A: the red die shows 1; B: the numbers add to 4; C: at least one of the numbers is 1; and D: the numbers do not add to 11. In Exercises 23–30, express the given event in symbols and say how many elements it contains. HINT [See Example 5.]

23. The red die shows 1 and the numbers add to 4.

24. The red die shows 1 but the numbers do not add to 11.

25. The numbers do not add to 4.

26. The numbers add to 11.

27. The numbers do not add to 4, but they do add to 11.

28. Either the numbers add to 11 or the red die shows a 1.

29. At least one of the numbers is 1 or the numbers add to 4.

30. Either the numbers add to 4, or they add to 11, or at least one of them is 1.

Let W be the event that you will use the book's Website tonight, let I be the event that your math grade will improve, and let E be the event that you will use the Website every night. In Exercises 31–38, express the given event in symbols.

31. You will use the Website tonight and your math grade will improve.

32. You will use the Website tonight or your math grade will not improve.

33. Either you will use the Website every night, or your math grade will not improve.

34. Your math grade will not improve even though you use the Website every night.

35. ▼ Either your math grade will improve, or you will use the Website tonight but not every night.

36. ▼ You will use the Website either tonight or every night, and your grade will improve.

37. ▼ (Compare Exercise 35.) Either your math grade will improve or you will use the Website tonight, but you will not use it every night.

38. ▼ (Compare Exercise 36.) Either you will use the Website tonight, or you will use it every night and your grade will improve.

---

[3]A positive integer is **prime** if it is neither 1 nor a product of smaller integers.

[4]Source: The White House Web site (www.whitehouse.gov).

In Exercises 39–42, Pablo randomly picks three marbles from a bag of eight marbles (four red ones, two green ones, and two yellow ones).

39. How many outcomes are there in the sample space? How many outcomes in the event that none of the marbles he picks are red?

40. How many outcomes are there in the sample space? How many outcomes in the event that all of the marbles he picks are red?

41. ▼ How many outcomes are there in the event that Pablo picks one marble of each color?

42. ▼ How many outcomes are there in the event that the marbles Pablo picks are not all the same color?

APPLICATIONS

Housing Prices Exercises 43–48 are based on the map below, which shows the percentage change in housing prices from June 2010 to June 2011 in each of nine regions (U.S. Census divisions).[5]

43. You are choosing a region of the country to move to. Describe the event E that the region you choose saw a decrease in housing prices of 7% or more.

44. You are choosing a region of the country to move to. Describe the event E that the region you choose saw a decrease in housing prices of less than 3%.

45. You are choosing a region of the country to move to. Let E be the event that the region you choose saw a decrease in housing prices of 7% or more, and let F be the event that the region you choose is on the east coast. Describe the events $E \cup F$ and $E \cap F$ both in words and by listing the outcomes of each.

46. You are choosing a region of the country to move to. Let E be the event that the region you choose saw a decrease in housing prices of less than 3%, and let F be the event that the region you choose is not on the east coast. Describe the events $E \cup F$ and $E \cap F$ both in words and by listing the outcomes of each.

47. ▼ You are choosing a region of the country to move to. Which of the following pairs of events are mutually exclusive?

a. E: You choose a region from among the two with the highest percentage decrease in housing prices.
F: You choose a region that is not on the east or west coast.

b. E: You choose a region from among the two with the highest percentage decrease in housing prices.
F: You choose a region that is on the east coast.

48. ▼ You are choosing a region of the country to move to. Which of the following pairs of events are mutually exclusive?

a. E: You choose a region from among the three with the least decline in housing prices.
F: You choose a region from among the central divisions.

b. E: You choose a region from among the three with the least decline in housing prices.
F: You choose a region on the east or west coast.

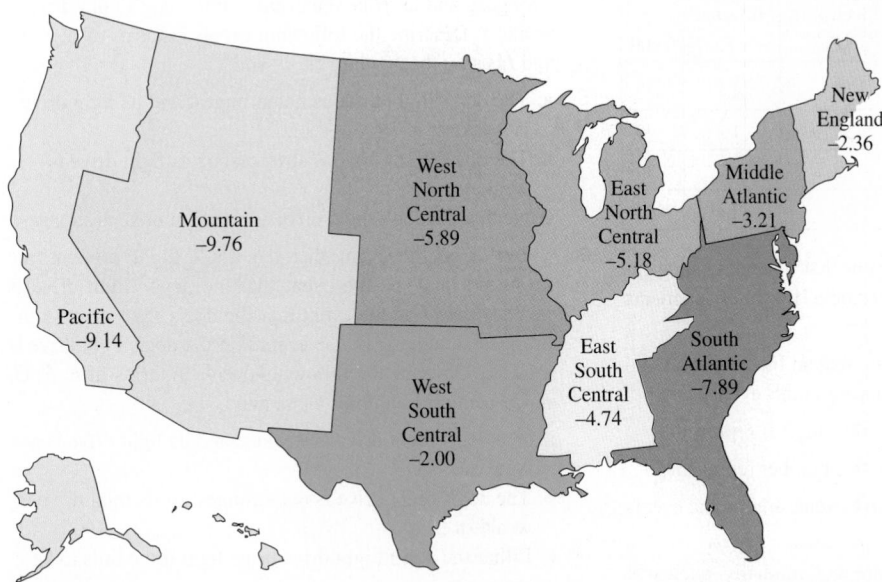

Publishing Exercises 49–56 are based on the accompanying table, which shows the results of a survey of authors by a (fictitious) publishing company. HINT [See Example 5.]

	New Authors	Established Authors	Total
Successful	5	25	30
Unsuccessful	15	55	70
Total	20	80	100

[5]Source: Office of Federal Housing Enterprise Oversight (www.ofheo.gov).

Consider the following events: S: an author is successful; U: an author is unsuccessful; N: an author is new; and E: an author is established.

49. Describe the events $S \cap N$ and $S \cup N$ in words. Use the table to compute $n(S \cap N)$ and $n(S \cup N)$.

50. Describe the events $N \cap U$ and $N \cup U$ in words. Use the table to compute $n(N \cap U)$ and $n(N \cup U)$.

51. Which of the following pairs of events are mutually exclusive: N and E; N and S; S and E?

52. Which of the following pairs of events are mutually exclusive: U and E; U and S; S and N?

53. Describe the event $S \cap N'$ in words and find the number of elements it contains.

54. Describe the event $U \cup E'$ in words and find the number of elements it contains.

55. ▼ What percentage of established authors are successful? What percentage of successful authors are established?

56. ▼ What percentage of new authors are unsuccessful? What percentage of unsuccessful authors are new?

Stock Prices Exercises 57–64 are based on the following table, which shows the performance of a selection of 100 stocks after one year. (Take S to be the set of all stocks represented in the table.)

	Companies			
	Pharmaceutical P	Electronic E	Internet I	**Total**
Increased V	10	5	15	30
Unchanged[6] N	30	0	10	40
Decreased D	10	5	15	30
Total	50	10	40	100

57. Use symbols to describe the event that a stock's value increased but it was not an Internet stock. How many elements are in this event?

58. Use symbols to describe the event that an Internet stock did not increase. How many elements are in this event?

59. Compute $n(P' \cup N)$. What does this number represent?

60. Compute $n(P \cup N')$. What does this number represent?

61. Find all pairs of mutually exclusive events among the events $P, E, I, V, N,$ and D.

62. Find all pairs of events that are not mutually exclusive among the events $P, E, I, V, N,$ and D.

63. ▼ Calculate $\dfrac{n(V \cap I)}{n(I)}$. What does the answer represent?

64. ▼ Calculate $\dfrac{n(D \cap I)}{n(D)}$. What does the answer represent?

Animal Psychology Exercises 65–70 concern the following chart, which shows the way in which a dog moves its facial muscles when torn between the drives of fight and flight.[7] The "fight" drive increases from left to right; the "flight" drive increases from top to bottom. (Notice that an increase in the "fight" drive causes its upper lip to lift, while an increase in the "flight" drive draws its ears downward.)

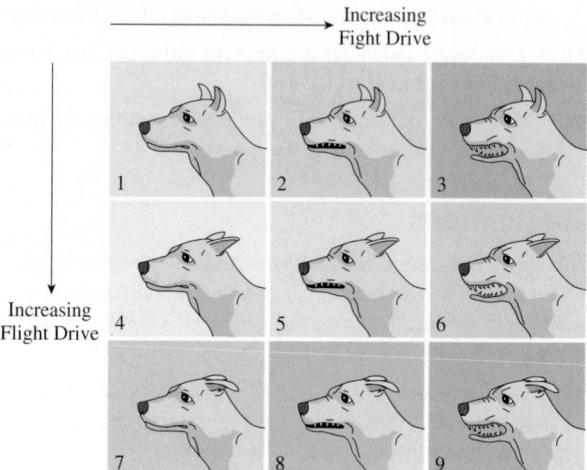

65. ▼ Let E be the event that the dog's flight drive is the strongest, let F be the event that the dog's flight drive is weakest, let G be the event that the dog's fight drive is the strongest, and let H be the event that the dog's fight drive is weakest. Describe the following events in terms of $E, F, G,$ and H using the symbols $\cap$, $\cup$, and $'$.

  a. The dog's flight drive is not strongest and its fight drive is weakest.

  b. The dog's flight drive is strongest or its fight drive is weakest.

  c. Neither the dog's flight drive nor its fight drive is strongest.

66. ▼ Let E be the event that the dog's flight drive is the strongest, let F be the event that the dog's flight drive is weakest, let G be the event that the dog's fight drive is the strongest, and let H be the event that the dog's fight drive is weakest. Describe the following events in terms of $E, F, G,$ and H using the symbols $\cap$, $\cup$, and $'$.

  a. The dog's flight drive is weakest and its fight drive is not weakest.

  b. The dog's flight drive is not strongest or its fight drive is weakest.

  c. Either the dog's flight drive or its fight drive fails to be strongest.

67. ▼ Describe the following events explicitly (as subsets of the sample space):

  a. The dog's fight and flight drives are both strongest.

  b. The dog's fight drive is strongest, but its flight drive is neither weakest nor strongest.

---

[6]If a stock stayed within 20% of its original value, it is classified as "unchanged."

[7]Source: *On Aggression* by Konrad Lorenz (Fakenham, Norfolk: University Paperback Edition, Cox & Wyman Limited, 1967).

68. ▼ Describe the following events explicitly (as subsets of the sample space):

  a. Neither the dog's fight drive nor its flight drive is strongest.

  b. The dog's fight drive is weakest, but its flight drive is neither weakest nor strongest.

69. ▼ Describe the following events in words:

  a. $\{1, 4, 7\}$
  b. $\{1, 9\}$
  c. $\{3, 6, 7, 8, 9\}$

70. ▼ Describe the following events in words:

  a. $\{7, 8, 9\}$
  b. $\{3, 7\}$
  c. $\{1, 2, 3, 4, 7\}$

Exercises 71–74 use counting arguments from the preceding chapter.

71. ▼ *Gummy Bears* A bag contains six gummy bears. Noel picks four at random. How many possible outcomes are there? If one of the gummy bears is raspberry, how many of these outcomes include the raspberry gummy bear?

72. ▼ *Chocolates* My couch potato friend enjoys sitting in front of the TV and grabbing handfuls of 5 chocolates at random from his snack jar. Unbeknownst to him, I have replaced one of the 20 chocolates in his jar with a cashew. (He hates cashews with a passion.) How many possible outcomes are there the first time he grabs 5 chocolates? How many of these include the cashew?

73. ▼ *Horse Races* The seven contenders in the fifth horse race at Aqueduct on February 18, 2002, were: Pipe Bomb, Expect a Ship, All That Magic, Electoral College, Celera, Cliff Glider, and Inca Halo.[8] You are interested in the first three places (winner, second place, and third place) for the race.

  a. Find the cardinality $n(S)$ of the sample space S of all possible finishes of the race. (A finish for the race consists of a first, second, and third place winner.)

  b. Let E be the event that Electoral College is in second or third place, and let F be the event that Celera is the winner. Express the event $E \cap F$ in words, and find its cardinality.

74. ▼ *Intramurals* The following five teams will be participating in Urban University's hockey intramural tournament: the Independent Wildcats, the Phi Chi Bulldogs, the Gate Crashers, the Slide Rule Nerds, and the City Slickers. Prizes will be awarded for the winner and runner-up.

  a. Find the cardinality $n(S)$ of the sample space S of all possible outcomes of the tournament. (An outcome of the tournament consists of a winner and a runner-up.)

  b. Let E be the event that the City Slickers are runners-up, and let F be the event that the Independent Wildcats are neither the winners nor runners-up. Express the event $E \cup F$ in words, and find its cardinality.

COMMUNICATION AND REASONING EXERCISES

75. Complete the following sentence. An event is a ____.

76. Complete the following sentence. Two events E and F are mutually exclusive if their intersection is ____.

77. If E and F are events, then $(E \cap F)'$ is the event that ____.

78. If E and F are events, then $(E' \cap F')$ is the event that ____.

79. Let E be the event that you meet a tall dark stranger. Which of the following could reasonably represent the experiment and sample space in question?

  (A) You go on vacation and lie in the sun; S is the set of cloudy days.

  (B) You go on vacation and spend an evening at the local dance club; S is the set of people you meet.

  (C) You go on vacation and spend an evening at the local dance club; S is the set of people you do not meet.

80. Let E be the event that you buy a Porsche. Which of the following could reasonably represent the experiment and sample space in question?

  (A) You go to an auto dealership and select a Mustang; S is the set of colors available.

  (B) You go to an auto dealership and select a red car; S is the set of cars you decide not to buy.

  (C) You go to an auto dealership and select a red car; S is the set of car models available.

81. ▼ True or false? Every set S is the sample space for some experiment. Explain.

82. ▼ True or false? Every sample space S is a finite set. Explain.

83. ▼ Describe an experiment in which a die is cast and the set of outcomes is $\{0, 1\}$.

84. ▼ Describe an experiment in which two coins are flipped and the set of outcomes is $\{0, 1, 2\}$.

85. ▼ Two distinguishable dice are rolled. Could there be two mutually exclusive events that both contain outcomes in which the numbers facing up add to 7?

86. ▼ Describe an experiment in which two dice are rolled and describe two mutually exclusive events that both contain outcomes in which both dice show a 1.

[8]Source: *Newsday*, Feb. 18, 2002, p. A36.

7.2 Relative Frequency

Suppose you have a coin that you think is not fair and you would like to determine the likelihood that heads will come up when it is tossed. You could estimate this likelihood by tossing the coin a large number of times and counting the number of times heads comes up. Suppose, for instance, that in 100 tosses of the coin, heads comes up 58 times. The fraction of times that heads comes up, $58/100 = .58$, is the **relative frequency**, or **estimated probability** of heads coming up when the coin is tossed. In other words, saying that the relative frequency of heads coming up is .58 is the same as saying that heads came up 58% of the time in your series of experiments.

Now let's think about this example in terms of sample spaces and events. First of all, there is an experiment that has been repeated $N = 100$ times: Toss the coin and observe the side facing up. The sample space for this experiment is $S = \{H, T\}$. Also, there is an event E in which we are interested: the event that heads comes up, which is $E = \{H\}$. The number of times E has occurred, or the **frequency** of E, is $fr(E) = 58$. The relative frequency of the event E is then

$$P(E) = \frac{fr(E)}{N} \qquad \frac{\text{Frequency of event } E}{\text{Number of repetitions } N}$$

$$= \frac{58}{100} = .58.$$

Notes

1. The relative frequency gives us an *estimate* of the likelihood that heads will come up when that particular coin is tossed. This is why statisticians often use the alternative term *estimated probability* to describe it.

2. The larger the number of times the experiment is performed, the more accurate an estimate we expect this estimated probability to be. ■

Relative Frequency

When an experiment is performed a number of times, the **relative frequency** or **estimated probability** of an event E is the fraction of times that the event E occurs. If the experiment is performed N times and the event E occurs $fr(E)$ times, then the relative frequency is given by

$$P(E) = \frac{fr(E)}{N}. \qquad \text{Fraction of times } E \text{ occurs}$$

The number $fr(E)$ is called the **frequency** of E. N, the number of times that the experiment is performed, is called the number of **trials** or the **sample size**. If E consists of a single outcome s, then we refer to $P(E)$ as the relative frequency or estimated probability of the outcome s, and we write $P(s)$.

Visualizing Relative Frequency

$$P(E) = \frac{fr(E)}{N} = \frac{4}{10} = .4$$

The collection of the estimated probabilities of *all* the outcomes is the **relative frequency distribution** or **estimated probability distribution**.

Publishing *Exercises 37–46 are based on the following table, which shows the results of a survey of 100 authors by a (fictitious) publishing company.*

	New Authors	Established Authors	Total
Successful	5	25	30
Unsuccessful	15	55	70
Total	20	80	100

Compute the relative frequencies of the given events if an author as specified is chosen at random.

37. ▼ An author is established and successful.

38. ▼ An author is unsuccessful and new.

39. ▼ An author is a new author.

40. ▼ An author is successful.

41. ▼ An author is unsuccessful.

42. ▼ An author is established.

43. ▼ A successful author is established.

44. ▼ An unsuccessful author is established.

45. ▼ An established author is successful.

46. ▼ A new author is unsuccessful.

47. ▼ ***Public Health*** A random sampling of chicken in supermarkets revealed that approximately 80% was contaminated with the organism *Campylobacter*.[23] Of the contaminated chicken, 20% had the strain resistant to antibiotics. Construct a relative frequency distribution showing the following outcomes when chicken is purchased at a supermarket: *U:* the chicken is not infected with *Campylobacter*; *C:* the chicken is infected with nonresistant *Campylobacter*; *R:* the chicken is infected with resistant *Campylobacter*.

48. ▼ ***Public Health*** A random sampling of turkey in supermarkets found 58% to be contaminated with *Campylobacter*, and 84% of those to be resistant to antibiotics.[24] Construct a relative frequency distribution showing the following outcomes when turkey is purchased at a supermarket: *U:* the turkey is not infected with *Campylobacter*; *C:* the turkey is infected with nonresistant *Campylobacter*; and *R:* the turkey is infected with resistant *Campylobacter*.

49. ▼ ***Organic Produce*** A 2001 Agriculture Department study of more than 94,000 samples from more than 20 crops showed that 73% of conventionally grown foods had residues from at least one pesticide. Moreover, conventionally grown foods were six times as likely to contain multiple pesticides as organic foods. Of the organic foods tested, 23% had pesticide residues, which includes 10% with multiple pesticide residues.[25] Compute two estimated probability distributions: one for conventional produce and one for organic produce, showing the relative frequencies that a randomly selected product has no pesticide residues, has residues from a single pesticide, and has residues from multiple pesticides.

50. ▼ ***Organic Produce*** Repeat Exercise 49 using the following information for produce from California: 31% of conventional food and 6.5% of organic food had residues from at least one pesticide. Assume that, as in Exercise 49, conventionally grown foods were six times as likely to contain multiple pesticides as organic foods. Also assume that 3% of the organic food has residues from multiple pesticides.

51. ▼ ***Steroids Testing*** A pharmaceutical company is running trials on a new test for anabolic steroids. The company uses the test on 400 athletes known to be using steroids and 200 athletes known not to be using steroids. Of those using steroids, the new test is positive for 390 and negative for 10. Of those not using steroids, the test is positive for 10 and negative for 190. What is the relative frequency of a **false negative** result (the probability that an athlete using steroids will test negative)? What is the relative frequency of a **false positive** result (the probability that an athlete not using steroids will test positive)?

52. ▼ ***Lie Detectors*** A manufacturer of lie detectors is testing its newest design. It asks 300 subjects to lie deliberately and another 500 to tell the truth. Of those who lied, the lie detector caught 200. Of those who told the truth, the lie detector accused 200 of lying. What is the relative frequency of the machine wrongly letting a liar go, and what is the probability that it will falsely accuse someone who is telling the truth?

53. ▮ ◆ ***Public Health*** Refer back to Exercise 47. Simulate the experiment of selecting chicken at a supermarket and determining the following outcomes: *U:* The chicken is not infected with *campylobacter*; *C:* The chicken is infected with nonresistant *campylobacter*; *R:* The chicken is infected with resistant *campylobacter*. HINT [Generate integers in the range 1–100. The outcome is determined by the range. For instance if the number is in the range 1–20, regard the outcome as *U*, etc.]

54. ▮ ◆ ***Public Health*** Repeat the preceding exercise, but use turkeys and the data given in Exercise 48.

---

[23]*Campylobacter* is one of the leading causes of food poisoning in humans. Thoroughly cooking the meat kills the bacteria. Source: *New York Times*, October 20, 1997, p. A1. Publication of this article first brought *Campylobacter* to the attention of a wide audience.
[24]*Ibid.*

---

[25]The 10% figure is an estimate. Source: *New York Times*, May 8, 2002, p. A29.

COMMUNICATION AND REASONING EXERCISES

55. Complete the following. The relative frequency of an event *E* is defined to be _____.

56. If two people each flip a coin 100 times and compute the relative frequency that heads comes up, they will both obtain the same result, right?

57. How many different answers are possible if you flip a coin 100 times and compute the relative frequency that heads comes up? What are the possible answers?

58. Interpret the popularity rating of the student council president as a relative frequency by specifying an appropriate experiment and also what is observed.

59. ▼ Ruth tells you that when you roll a pair of fair dice, the probability of obtaining a pair of matching numbers is 1/6. To test this claim, you roll a pair of fair dice 20 times, and never once get a pair of matching numbers.

This proves that either Ruth is wrong or the dice are not fair, right?

60. ▼ Juan tells you that when you roll a pair of fair dice, the probability that the numbers add up to 7 is 1/6. To test this claim, you roll a pair of fair dice 24 times, and the numbers add up to 7 exactly four times. This proves that Juan is right, right?

61. ▼ How would you measure the relative frequency that the weather service accurately predicts the next day's high temperature?

62. ▼ Suppose that you toss a coin 100 times and get 70 heads. If you continue tossing the coin, the estimated probability of heads overall should approach 50% if the coin is fair. Will you have to get more tails than heads in subsequent tosses to "correct" for the 70 heads you got in the first 100 tosses?

7.3 Probability and Probability Models

It is understandable if you are a little uncomfortable with using relative frequency as the estimated probability because it does not always agree with what you intuitively feel to be true. For instance, if you toss a fair coin (one as likely to come up heads as tails) 100 times and heads happen to come up 62 times, the experiment seems to suggest that the probability of heads is .62, even though you *know* that the "actual" probability is .50 (because the coin is fair).

Q: *So what do we mean by "actual" probability?*

A: There are various philosophical views as to exactly what we should mean by "actual" probability. For example, (finite) *frequentists* say that there is no such thing as "actual probability"—all we should really talk about is what we can actually measure, the relative frequency. *Propensitists* say that the actual probability *p* of an event is a (often physical) property of the event that makes its relative frequency tend to *p* in the long run; that is, *p* will be the limiting value of the relative frequency as the number of trials in a repeated experiment gets larger and larger. (See Figure 4 in the preceding section.) *Bayesians*, on the other hand, argue that the actual probability of an event is the degree to which we *expect* it to occur, given our knowledge about the nature of the experiment. These and other viewpoints have been debated in considerable depth in the literature.✱

✱ The interested reader should consult references in the philosophy of probability. For an online summary, see, for example, the Stanford Encyclopedia of Philosophy (http://plato.stanford.edu/contents.html).

Mathematicians tend to avoid the whole debate, and talk instead about *abstract* probability, or **probability distributions**, based purely on the properties of relative frequency listed in Section 7.2. Specific probability distributions can then be used as *models* in real-life situations, such as flipping a coin or tossing a die, to predict (or model) relative frequency.

Probability Distribution; Probability

(Compare with the properties of relative frequency distributions in Section 7.2.) A (finite) **probability distribution** is an assignment of a number $P(s_i)$, the **probability of** s_i, to each outcome of a finite sample space $S = \{s_1, s_2, \ldots, s_n\}$. The probabilities must satisfy

1. $0 \le P(s_i) \le 1$

and

2. $P(s_1) + P(s_2) + \cdots + P(s_n) = 1.$

We find the **probability of an event** E, written $P(E)$, by adding up the probabilities of the outcomes in E.

If $P(E) = 0$, we call E an **impossible event**. The empty event Ø is always impossible, since *something* must happen.

Quick Examples

1. All the examples of estimated probability distributions in Section 7.2 are examples of probability distributions. (See page 463.)

2. Let us take $S = \{H, T\}$ and make the assignments $P(H) = .5$, $P(T) = .5$. Because these numbers are between 0 and 1 and add to 1, they specify a probability distribution.

3. In Quick Example 2, we can instead make the assignments $P(H) = .2$, $P(T) = .8$. Because these numbers are between 0 and 1 and add to 1, they, too, specify a probability distribution.

4. With $S = \{H, T\}$ again, we could also take $P(H) = 1$, $P(T) = 0$, so that $\{T\}$ is an impossible event.

5. The following table gives a probability distribution for the sample space $S = \{1, 2, 3, 4, 5, 6\}$.

Outcome	1	2	3	4	5	6
Probability	.3	.3	0	.1	.2	.1

It follows that

$$P(\{1, 6\}) = .3 + .1 = .4$$
$$P(\{2, 3\}) = .3 + 0 = .3$$
$$P(3) = 0. \qquad \text{\{3\} is an impossible event.}$$

* Just how large is a "large number of times"? That depends on the nature of the experiment. For example, if you toss a fair coin 100 times, then the relative frequency of heads will be between .45 and .55 about 73% of the time. If an outcome is extremely unlikely (such as winning the lotto), you might need to repeat the experiment billions or trillions of times before the relative frequency approaches any specific number.

The above Quick Examples included models for the experiments of flipping fair and unfair coins. In general:

Probability Models

A **probability model** for a particular experiment is a probability distribution that predicts the relative frequency of each outcome if the experiment is performed a large number of times (see Figure 4 at the end of the preceding section).* Just as we think of relative frequency as *estimated probability*, we can think of modeled probability as *theoretical probability*.

Quick Examples

1. **Fair Coin Model:** (See Quick Example 2 on the previous page.) Flip a fair coin and observe the side that faces up. Because we expect that heads is as likely to come up as tails, we model this experiment with the probability distribution specified by $S = \{H, T\}$, $P(H) = .5$, $P(T) = .5$. Figure 4 on p. 463 suggests that the relative frequency of heads approaches .5 as the number of coin tosses gets large, so the fair coin model predicts the relative frequency for a large number of coin tosses quite well.

2. **Unfair Coin Model:** (See Quick Example 3 on the previous page.) Take $S = \{H, T\}$ and $P(H) = .2$, $P(T) = .8$. We can think of this distribution as a model for the experiment of flipping an unfair coin that is four times as likely to land with tails uppermost than heads.

3. **Fair Die Model:** Roll a fair die and observe the uppermost number. Because we expect to roll each specific number one sixth of the time, we model the experiment with the probability distribution specified by $S = \{1, 2, 3, 4, 5, 6\}$, $P(1) = 1/6$, $P(2) = 1/6, \ldots, P(6) = 1/6$. This model predicts, for example, that the relative frequency of throwing a 5 approaches $1/6$ as the number of times you roll the die gets large.

4. Roll a pair of fair dice (recall that there are a total of 36 outcomes if the dice are distinguishable). Then an appropriate model of the experiment has

$$S = \left\{ \begin{array}{llllll} (1,1), & (1,2), & (1,3), & (1,4), & (1,5), & (1,6), \\ (2,1), & (2,2), & (2,3), & (2,4), & (2,5), & (2,6), \\ (3,1), & (3,2), & (3,3), & (3,4), & (3,5), & (3,6), \\ (4,1), & (4,2), & (4,3), & (4,4), & (4,5), & (4,6), \\ (5,1), & (5,2), & (5,3), & (5,4), & (5,5), & (5,6), \\ (6,1), & (6,2), & (6,3), & (6,4), & (6,5), & (6,6) \end{array} \right\}$$

with each outcome being assigned a probability of $1/36$.

5. In the experiment in Quick Example 4, take E to be the event that the sum of the numbers that face up is 5, so

$$E = \{(1,4), (2,3), (3,2), (4,1)\}.$$

By the properties of probability distributions,

$$P(E) = \frac{1}{36} + \frac{1}{36} + \frac{1}{36} + \frac{1}{36} = \frac{4}{36} = \frac{1}{9}.$$

Notice that, in all of the Quick Examples above except for the unfair coin, all the outcomes are equally likely, and each outcome s has a probability of

$$P(s) = \frac{1}{\text{Total number of outcomes}} = \frac{1}{n(S)}.$$

More generally, in the last Quick Example we saw that adding the probabilities of the individual outcomes in an event E amounted to computing the ratio (Number of favorable outcomes)/(Total number of outcomes):

$$P(E) = \frac{\text{Number of favorable outcomes}}{\text{Total number of outcomes}} = \frac{n(E)}{n(S)}.$$

Probability Model for Equally Likely Outcomes

In an experiment in which all outcomes are equally likely, we model the experiment by taking the probability of an event E to be

$$P(E) = \frac{\text{Number of favorable outcomes}}{\text{Total number of outcomes}} = \frac{n(E)}{n(S)}.$$

Visualizing Probability for Equally Likely Outcomes

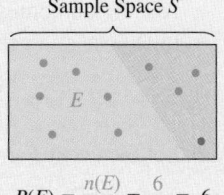

Sample Space S

$$P(E) = \frac{n(E)}{n(S)} = \frac{6}{10} = .6$$

Note

Remember that this formula will work *only* when the outcomes are equally likely. If, for example, a die is *weighted*, then the outcomes may not be equally likely, and the formula above will not give an appropriate probability model.

Quick Examples

1. Toss a fair coin three times, so $S = \{$HHH, HHT, HTH, HTT, THH, THT, TTH, TTT$\}$. The probability that we throw exactly two heads is

$$P(E) = \frac{n(E)}{n(S)} = \frac{3}{8}.$$ There are eight equally likely outcomes and $E = \{$HHT, HTH, THH$\}$.

2. Roll a pair of fair dice. The probability that we roll a double (both dice show the same number) is

$$P(E) = \frac{n(E)}{n(S)} = \frac{6}{36} = \frac{1}{6}.$$ $E = \{(1, 1), (2, 2), (3, 3), (4, 4), (5, 5), (6, 6)\}$

3. Randomly choose a person from a class of 40, in which 6 have red hair. If E is the event that a randomly selected person in the class has red hair, then

$$P(E) = \frac{n(E)}{n(S)} = \frac{6}{40} = .15.$$

EXAMPLE 1 Sales of Hybrid Vehicles

(Compare Example 1 in Section 7.2.) A total of 1.9 million hybrid vehicles had been sold in the United States through October of 2011. Of these, 955,000 were Toyota Prii, 205,000 were Honda Civics, 170,000 were Toyota Camrys, 105,000 were Ford Escapes, and the rest were other makes.[26]

a. What is the probability that a randomly selected hybrid vehicle sold in the United States was either a Toyota Prius or a Honda Civic?

b. What is the probability that a randomly selected hybrid vehicle sold in the United States was not a Toyota Camry?

[26]Source for sales data: www.wikipedia.com.

Solution

a. The experiment suggested by the question consists of randomly choosing a hybrid vehicle sold in the United States and determining its make. We are interested in the event E that the hybrid vehicle was either a Toyota Prius or a Honda Civic. So,

$$S = \text{the set of hybrid vehicles sold; } n(S) = 1{,}900{,}000$$

$$E = \text{the set of Toyota Prii and Honda Civics sold;}$$
$$n(E) = 955{,}000 + 205{,}000 = 1{,}160{,}000.$$

Are the outcomes equally likely in this experiment? Yes, because we are as likely to choose one vehicle as another. Thus,

$$P(E) = \frac{n(E)}{n(S)} = \frac{1{,}160{,}000}{1{,}900{,}000} \approx .61.$$

b. Let the event F consist of those hybrid vehicles sold that were not Toyota Camrys.

$$n(F) = 1{,}900{,}000 - 170{,}000 = 1{,}730{,}000$$

Hence,

$$P(F) = \frac{n(F)}{n(S)} = \frac{1{,}730{,}000}{1{,}900{,}000} \approx .91.$$

Q: *In Example 1 of Section 7.2 we had a similar example about hybrid vehicles, but we called the probabilities calculated there relative frequencies. Here they are probabilities. What is the difference?*

A: In Example 1 of Section 7.2, the data were based on the results of a survey, or sample, of only 250 hybrid vehicles (out of a total of about 1.9 million sold in the United States), and were therefore incomplete. (A statistician would say that we were given *sample data*.) It follows that any inference we draw from the 250 surveyed, such as the probability that a hybrid vehicle sold in the United States is not a Toyota Camry, is uncertain, and this is the cue that tells us that we are working with relative frequency, or estimated probability. Think of the survey as an experiment (choosing a hybrid vehicle) repeated 250 times—exactly the setting for estimated probability.

In Example 1 above, on the other hand, the data do not describe how *some* hybrid vehicle sales are broken down into the categories described, but they describe how *all 1.9 million* hybrid vehicle sales in the United States are broken down. (The statistician would say that we were given *population data* in this case, because the data describe the entire "population" of hybrid vehicles sold in the United States.)

EXAMPLE 2 Indistinguishable Dice

We recall from Section 7.1 that the sample space when we roll a pair of indistinguishable dice is

$$S = \left\{ \begin{array}{l} (1,1),\ (1,2),\ (1,3),\ (1,4),\ (1,5),\ (1,6), \\ (2,2),\ (2,3),\ (2,4),\ (2,5),\ (2,6), \\ (3,3),\ (3,4),\ (3,5),\ (3,6), \\ (4,4),\ (4,5),\ (4,6), \\ (5,5),\ (5,6), \\ (6,6) \end{array} \right\}.$$

Construct a probability model for this experiment.

Solution Because there are 21 outcomes, it is tempting to say that the probability of each outcome should be taken to be 1/21. However, the outcomes are not all equally likely. For instance, the outcome (2, 3) is twice as likely as (2, 2), because (2, 3) can occur in two ways (it corresponds to the event {(2, 3), (3, 2)} for distinguishable dice). For purposes of calculating probability, it is easiest to use calculations for distinguishable dice.* Here are some examples.

* Note that any pair of real dice can be distinguished in principle because they possess slight differences, although we may regard them as indistinguishable by not attempting to distinguish them. Thus, the probabilities of events must be the same as for the corresponding events for distinguishable dice.

Outcome (indistinguishable dice)	(1, 1)	(1, 2)	(2, 2)	(1, 3)	(2, 3)	(3, 3)
Corresponding Event (distinguishable dice)	{(1, 1)}	{(1, 2), (2, 1)}	{(2, 2)}	{(1, 3), (3, 1)}	{(2, 3), (3, 2)}	{(3, 3)}
Probability	$\frac{1}{36}$	$\frac{2}{36} = \frac{1}{18}$	$\frac{1}{36}$	$\frac{2}{36} = \frac{1}{18}$	$\frac{2}{36} = \frac{1}{18}$	$\frac{1}{36}$

If we continue this process for all 21 outcomes, we will find that they add to 1. Figure 5 illustrates the complete probability distribution:

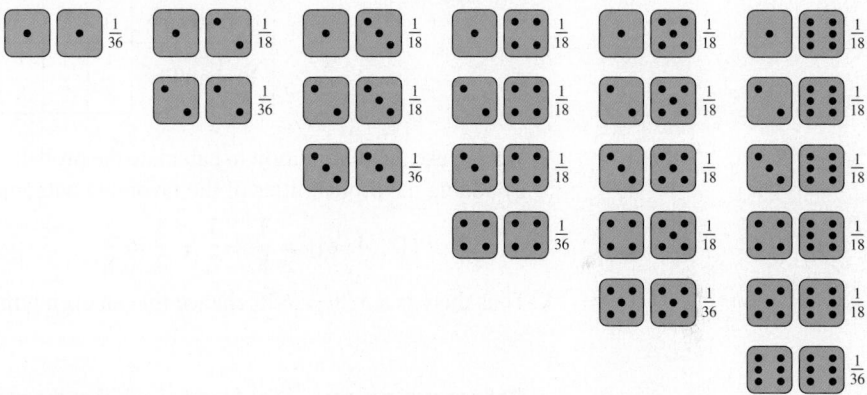

Figure 5

EXAMPLE 3 **Weighted Dice**

In order to impress your friends with your die-rolling skills, you have surreptitiously weighted your die in such a way that 6 is three times as likely to come up as any one of the other numbers. (All the other outcomes are equally likely.) Obtain a probability distribution for a roll of the die and use it to calculate the probability of an even number coming up.

Solution Let us label our unknowns (there appear to be two of them):

x = probability of rolling a 6

y = probability of rolling any one of the other numbers

We are first told that "6 is three times as likely to come up as any one of the other numbers." If we rephrase this in terms of our unknown probabilities we get, "the probability of rolling a 6 is three times the probability of rolling any one of the other numbers." In symbols,

$x = 3y.$

We must also use a piece of information not given to us, but one we know must be true: The sum of the probabilities of all the outcomes is 1:

$$x + y + y + y + y + y = 1$$

or

$$x + 5y = 1.$$

We now have two linear equations in two unknowns, and we solve for x and y. Substituting the first equation ($x = 3y$) in the second ($x + 5y = 1$) gives

$$8y = 1$$

or

$$y = \frac{1}{8}.$$

To get x, we substitute the value of y back into either equation and find

$$x = \frac{3}{8}.$$

Thus, the probability model we seek is the one shown in the following table.

Outcome	1	2	3	4	5	6
Probability	$\frac{1}{8}$	$\frac{1}{8}$	$\frac{1}{8}$	$\frac{1}{8}$	$\frac{1}{8}$	$\frac{3}{8}$

We can use the distribution to calculate the probability of an even number coming up by adding the probabilities of the favorable outcomes.

$$P(\{2, 4, 6\}) = \frac{1}{8} + \frac{1}{8} + \frac{3}{8} = \frac{5}{8}.$$

Thus there is a $5/8 = .625$ chance that an even number will come up.

➡ **Before we go on...** We should check that the probability distribution in Example 3 satisfies the requirements: 6 is indeed three times as likely to come up as any other number. Also, the probabilities we calculated do add up to 1:

$$\frac{1}{8} + \frac{1}{8} + \frac{1}{8} + \frac{1}{8} + \frac{1}{8} + \frac{3}{8} = 1. \quad \blacksquare$$

Probability of Unions, Intersections, and Complements

So far, all we know about computing the probability of an event E is that $P(E)$ is the sum of the probabilities of the individual outcomes in E. Suppose, though, that we do not know the probabilities of the individual outcomes in E but we do know that $E = A \cup B$, where we happen to know $P(A)$ and $P(B)$. How do we compute the probability of $A \cup B$? We might be tempted to say that $P(A \cup B)$ is $P(A) + P(B)$, but let us look at an example using the probability distribution in Quick Example 5 at the beginning of this section:

Outcome	1	2	3	4	5	6
Probability	.3	.3	0	.1	.2	.1

For A let us take the event $\{2, 4, 5\}$, and for B let us take $\{2, 4, 6\}$. $A \cup B$ is then the event $\{2, 4, 5, 6\}$. We know that we can find the probabilities $P(A)$, $P(B)$, and $P(A \cup B)$ by adding the probabilities of all the outcomes in these events, so

$$P(A) = P(\{2, 4, 5\}) = .3 + .1 + .2 = .6$$
$$P(B) = P(\{2, 4, 6\}) = .3 + .1 + .1 = .5, \text{ and}$$
$$P(A \cup B) = P(\{2, 4, 5, 6\}) = .3 + .1 + .2 + .1 = .7.$$

Our first guess was wrong: $P(A \cup B) \neq P(A) + P(B)$. Notice, however, that the outcomes in $A \cap B$ are counted twice in computing $P(A) + P(B)$, but only once in computing $P(A \cup B)$:

$$P(A) + P(B) = P(\{2, 4, 5\}) + P(\{2, 4, 6\}) \qquad A \cap B = \{2, 4\}$$
$$= (.3 + .1 + .2) + (.3 + .1 + .1) \qquad P(A \cap B) \text{ counted twice}$$
$$= 1.1$$

whereas

$$P(A \cup B) = P(\{2, 4, 5, 6\}) = .3 + .1 + .2 + .1 \qquad P(A \cap B) \text{ counted once}$$
$$= .7.$$

Thus, if we take $P(A) + P(B)$ and then subtract the surplus $P(A \cap B)$, we get $P(A \cup B)$. In symbols,

$$P(A \cup B) = P(A) + P(B) - P(A \cap B)$$
$$.7 = .6 + .5 - .4$$

(see Figure 6). We call this formula the **addition principle**. One more thing: Notice that our original guess $P(A \cup B) = P(A) + P(B)$ would have worked if we had chosen A and B with no outcomes in common; that is, if $A \cap B = \emptyset$. When $A \cap B = \emptyset$, recall that we say that A and B are mutually exclusive.

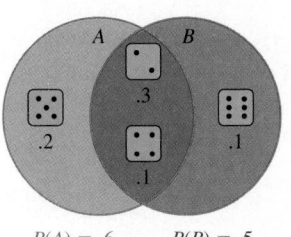

$P(A) = .6 \qquad P(B) = .5$
$P(A \cap B) = .4$
$P(A \cup B) = .6 + .5 - .4$

Figure 6

Addition Principle

If A and B are any two events, then

$$P(A \cup B) = P(A) + P(B) - P(A \cap B).$$

Visualizing the Addition Principle
In the figure, the area of the union is obtained by adding the areas of A and B and then subtracting the overlap (because it is counted twice when we add the areas).

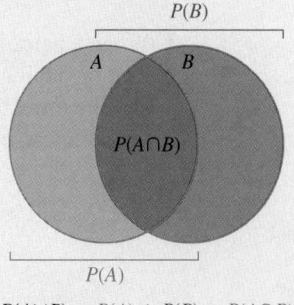

$P(A \cup B) = P(A) + P(B) - P(A \cap B)$

Addition Principle for Mutually Exclusive Events

If $A \cap B = \emptyset$, we say that A and B are **mutually exclusive**, and we have

$$P(A \cup B) = P(A) + P(B).$$ Because $P(A \cap B) = 0$

Visualizing the Addition Principle for Mutually Exclusive Events

If A and B do not overlap, then the area of the union is obtained by adding the areas of A and B.

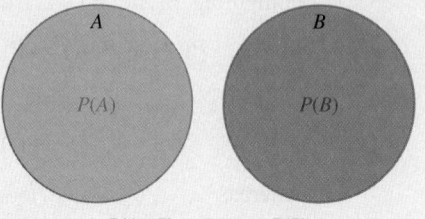

$P(A \cup B) = P(A) + P(B)$

This holds true also for more than two events: If $A_1, A_2, \ldots, A_n$ are mutually exclusive events (that is, the intersection of every pair of them is empty), then

$$P(A_1 \cup A_2 \cup \cdots \cup A_n) = P(A_1) + P(A_2) + \cdots + P(A_n).$$ Addition principle for many mutually exclusive events

Quick Examples

1. There is a 10% chance of rain (R) tomorrow, a 20% chance of high winds (W), and a 5% chance of both. The probability of either rain or high winds (or both) is

$$P(R \cup W) = P(R) + P(W) - P(R \cap W)$$
$$= .10 + .20 - .05 = .25.$$

2. The probability that you will be in Cairo at 6:00 am tomorrow (C) is .3, while the probability that you will be in Alexandria at 6:00 am tomorrow (A) is .2. Thus, the probability that you will be either in Cairo or Alexandria at 6:00 am tomorrow is

$$P(C \cup A) = P(C) + P(A)$$ A and C are mutually exclusive.
$$= .3 + .2 = .5.$$

3. When a pair of fair dice is rolled, the probability of the numbers that face up adding to 7 is 6/36, the probability of their adding to 8 is 5/36, and the probability of their adding to 9 is 4/36. Thus, the probability of the numbers adding to 7, 8, or 9 is

$$P(\{7\} \cup \{8\} \cup \{9\}) = P(7) + P(8) + P(9)$$ The events are mutually exclusive.*
$$= \frac{6}{36} + \frac{5}{36} + \frac{4}{36} = \frac{15}{36} = \frac{5}{12}.$$

* The sum of the numbers that face up cannot equal two different numbers at the same time.

You can use the formula
$P(A \cup B) = P(A) + P(B) - P(A \cap B)$
to calculate any of the four quantities in the formula if you know the other three.

EXAMPLE 4 **School and Work**

A survey[27] conducted by the Bureau of Labor Statistics found that 68% of the high school graduating class of 2010 went on to college the following year, while 42% of the class was working. Furthermore, 92% were either in college or working, or both.

a. What percentage went on to college and work at the same time?

b. What percentage went on to college but not work?

Solution We can think of the experiment of choosing a member of the high school graduating class of 2010 at random. The sample space is the set of all these graduates.

a. We are given information about two events:

A: A graduate went on to college; $P(A) = .68$.

B: A graduate went on to work; $P(B) = .42$.

We are also told that $P(A \cup B) = .92$. We are asked for the probability that a graduate went on to both college and work, $P(A \cap B)$. To find $P(A \cap B)$, we take advantage of the fact that the formula

$$P(A \cup B) = P(A) + P(B) - P(A \cap B)$$

can be used to calculate any one of the four quantities that appear in it as long as we know the other three. Substituting the quantities we know, we get

$$.92 = .68 + .42 - P(A \cap B)$$

so

$$P(A \cap B) = .68 + .42 - .92 = .18.$$

Thus, 18% of the graduates went on to college and work at the same time.

b. We are asked for the probability of a new event:

C: A graduate went on to college but not work.

C is the part of A outside of $A \cap B$, so $C \cup (A \cap B) = A$, and C and $A \cap B$ are mutually exclusive. (See Figure 7.)
Thus, applying the addition principle, we have

$$P(C) + P(A \cap B) = P(A).$$

From part (a), we know that $P(A \cap B) = .18$, so

$$P(C) + .18 = .68$$

giving

$$P(C) = .50.$$

In other words, 50% of the graduates went on to college but not work.

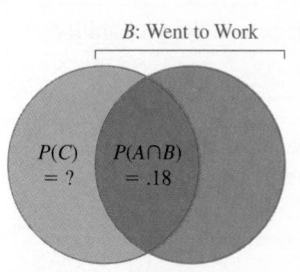

B: Went to Work

$P(C)$ = ? $P(A \cap B)$ = .18

A: Went to College
$P(A) = .68$

Figure 7

We can use the addition principle to deduce other useful properties of probability distributions.

---

[27]Source: "College Enrollment and Work Activity of High School Graduates," U.S. Bureau of Labor Statistics (www.bls.gov/news.release/hsgec.htm).

More Principles of Probability Distributions

The following rules hold for any sample space S and any event A:

$$P(S) = 1 \qquad \text{The probability of } \textit{something} \text{ happening is 1.}$$

$$P(\emptyset) = 0 \qquad \text{The probability of } \textit{nothing} \text{ happening is 0.}$$

$$P(A') = 1 - P(A). \qquad \text{The probability of } A \textit{ not} \text{ happening is 1 minus the probability of } A.$$

Note

We can also write the third equation as

$$P(A) = 1 - P(A')$$

or

$$P(A) + P(A') = 1. \qquad \blacksquare$$

Visualizing the Rule for Complements

Think of A' as the portion of S outside of A. Adding the two areas gives the area of all of S, equal to 1.

Sample Space S

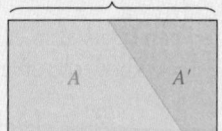

$$P(A) + P(A') = 1$$

Quick Examples

1. There is a 10% chance of rain (R) tomorrow. Therefore, the probability that it will *not* rain is

   $$P(R') = 1 - P(R) = 1 - .10 = .90.$$

2. The probability that Eric Ewing will score at least two goals is .6. Therefore, the probability that he will score at most one goal is

   $$1 - .6 = .4.$$

Q: *Can you persuade me that all of these principles are true?*

A: Let us take them one at a time.

We know that $S = \{s_1, s_2, \ldots, s_n\}$ is the set of all outcomes, and so

$$P(S) = P(\{s_1, s_2, \ldots, s_n\})$$

$$= P(s_1) + P(s_2) + \cdots + P(s_n) \qquad \text{We add the probabilities of the outcomes to obtain the probability of an event.}$$

$$= 1. \qquad \text{By the definition of a probability distribution}$$

Now, note that $S \cap \emptyset = \emptyset$, so that S and $\emptyset$ are mutually exclusive. Applying the addition principle gives

$$P(S) = P(S \cup \emptyset) = P(S) + P(\emptyset).$$

Subtracting $P(S)$ from both sides gives $0 = P(\emptyset)$.

If A is any event in S, then we can write

$$S = A \cup A'$$

where A and A' are mutually exclusive. (Why?) Thus, by the addition principle,

$$P(S) = P(A) + P(A').$$

Because $P(S) = 1$, we get

$$1 = P(A) + P(A')$$

or $P(A') = 1 - P(A).$

EXAMPLE 5 Subprime Mortgages during the Housing Bubble

A home loan is either current, 30–59 days past due, 60–89 days past due, 90 or more days past due, in foreclosure, or repossessed by the lender. In November 2008, the probability that a randomly selected subprime home mortgage in California was not current was .51. The probability that a mortgage was not current, but neither in foreclosure nor repossessed, was .28.[28] Calculate the probabilities of the following events.

a. A California home mortgage was current.

b. A California home mortgage was in foreclosure or repossessed.

Solution

a. Let us write C for the event that a randomly selected subprime home mortgage in California was current. The event that the home mortgage was *not* current is its complement C', and we are given that $P(C') = .51$. We have

$$P(C) + P(C') = 1$$
$$P(C) + .51 = 1,$$

so $P(C) = 1 - .51 = .49.$

b. Take

F: A mortgage was in foreclosure or repossessed.

N: A mortgage was neither current, in foreclosure, nor repossessed.

We are given $P(N) = .28$. Further, the events F and N are mutually exclusive with union C', the set of all non-current mortgages. Hence,

$$P(C') = P(F) + P(N)$$
$$.51 = P(F) + .28$$

giving

$$P(F) = .51 - .28 = .23.$$

Thus, there was a 23% chance that a subprime home mortgage was either in foreclosure or repossessed.

---

[28]Source: Federal Reserve Bank of New York (www.newyorkfed.org/regional/subprime.html).

EXAMPLE 6 iPods, iPhones, and iPads

The following table shows sales, in millions of units, of iPods®, iPhones®, and iPads® in the last three quarters of 2011.[29]

	iPods (A)	iPhones (B)	iPads (C)	Total
2011 Q2 (U)	9.0	18.7	4.7	32.4
2011 Q3 (V)	7.5	20.3	9.3	37.1
2011 Q4 (W)	6.6	17.1	11.1	34.8
Total	23.1	56.1	25.1	104.3

If one of the items sold is selected at random, find the probabilities of the following events:

a. It is an iPod.

b. It was sold in the third quarter of 2011.

c. It is an iPod sold in the third quarter of 2011.

d. It either is an iPod or was sold in the third quarter of 2011.

e. It is not an iPod.

Solution Before we answer the questions, note that S is the set of all items represented in the table, so S has a total of 104.3 million outcomes.

a. Because an item is being selected at random, all the outcomes are equally likely. If A is the event that the selected item is an iPod, then

$$P(A) = \frac{n(A)}{n(S)} = \frac{23.1}{104.3} \approx .221.$$

The event A is represented by the pink shaded region in the table:

	iPods (A)	iPhones (B)	iPads (C)	Total
2011 Q2 (U)	9.0	18.7	4.7	32.4
2011 Q3 (V)	7.5	20.3	9.3	37.1
2011 Q4 (W)	6.6	17.1	11.1	34.8
Total	23.1	56.1	25.1	104.3

b. If V is the event that the selected item was sold in the third quarter of 2011, then

$$P(V) = \frac{n(V)}{n(S)} = \frac{37.1}{104.3} \approx .356.$$

---

[29] Figures are rounded to one decimal place. Source: Apple quarterly press releases (www.investor.apple.com).

In the table, V is represented as shown:

	iPods (A)	iPhones (B)	iPads (C)	Total
2011 Q2 (U)	9.0	18.7	4.7	32.4
2011 Q3 (V)	7.5	20.3	9.3	37.1
2011 Q4 (W)	6.6	17.1	11.1	34.8
Total	23.1	56.1	25.1	104.3

c. The event that the selected item is an iPod sold in the third quarter of 2011 is the event $A \cap V$.

$$P(A \cap V) = \frac{n(A \cap V)}{n(S)} = \frac{7.5}{104.3} \approx .072$$

In the table, $A \cap V$ is represented by the overlap of the regions representing A and V:

	iPods (A)	iPhones (B)	iPads (C)	Total
2011 Q2 (U)	9.0	18.7	4.7	32.4
2011 Q3 (V)	7.5	20.3	9.3	37.1
2011 Q4 (W)	6.6	17.1	11.1	34.8
Total	23.1	56.1	25.1	104.3

d. The event that the selected item either is an iPod or was sold in the third quarter of 2011 is the event $A \cup V$, and is represented by the pink shaded area in the table:

	iPods (A)	iPhones (B)	iPads (C)	Total
2011 Q2 (U)	9.0	18.7	4.7	32.4
2011 Q3 (V)	7.5	20.3	9.3	37.1
2011 Q4 (W)	6.6	17.1	11.1	34.8
Total	23.1	56.1	25.1	104.3

We can compute its probability in two ways:

1. Directly from the table:

$$P(A \cup V) = \frac{n(A \cup V)}{n(S)} = \frac{23.1 + 37.1 - 7.5}{104.3} \approx .505$$

2. Using the addition principle:

$$P(A \cup V) = P(A) + P(V) - P(A \cap V)$$
$$\approx .221 + .356 - .072 = .505$$

e. The event that the selected item is not an iPod is the event A'. Its probability may be computed using the formula for the probability of the complement:

$$P(A') = 1 - P(A) \approx 1 - .221 = .779.$$

FAQs

Distinguishing Probability from Relative Frequency

Q: *Relative frequency and modeled probability using equally likely outcomes have essentially the same formula: (Number of favorable outcomes)/(Total number of outcomes). How do I know whether a given probability is one or the other?*

A: Ask yourself this: Has the probability been arrived at experimentally, by performing a number of trials and counting the number of times the event occurred? If so, the probability is estimated; that is, relative frequency. If, on the other hand, the probability was computed by analyzing the experiment under consideration rather than by performing actual trials of the experiment, it is a probability model.

Q: *Out of every 100 homes, 22 have broadband Internet service. Thus, the probability that a house has broadband service is .22. Is this probability estimated (relative frequency) or theoretical (a probability model)?*

A: That depends on how the ratio 22 out of 100 was arrived at. If it is based on a poll of *all* homes, then the probability is theoretical. If it is based on a survey of only a *sample* of homes, it is estimated (see the Q/A following Example 1).

7.3 EXERCISES

▼ more advanced   ◆ challenging

T indicates exercises that should be solved using technology

1. Complete the following probability distribution table and then calculate the stated probabilities. HINT [See Quick Example 5, page 469.]

Outcome	a	b	c	d	e
Probability	.1	.05	.6	.05	

   a. $P(\{a, c, e\})$
   b. $P(E \cup F)$, where $E = \{a, c, e\}$ and $F = \{b, c, e\}$
   c. $P(E')$, where E is as in part (b)
   d. $P(E \cap F)$, where E and F are as in part (b)

2. Repeat the preceding exercise using the following table. HINT [See Quick Example 5, page 469.]

Outcome	a	b	c	d	e
Probability	.1		.65	.1	.05

In Exercises 3–8, calculate the (modeled) probability $P(E)$ using the given information, assuming that all outcomes are equally likely. HINT [See Quick Examples on page 471.]

3. $n(S) = 20$, $n(E) = 5$ 4. $n(S) = 8$, $n(E) = 4$

5. $n(S) = 10$, $n(E) = 10$ 6. $n(S) = 10$, $n(E) = 0$

7. $S = \{a, b, c, d\}$, $E = \{a, b, d\}$

8. $S = \{1, 3, 5, 7, 9\}$, $E = \{3, 7\}$

In Exercises 9–18, an experiment is given together with an event. Find the (modeled) probability of each event, assuming that the coins and dice are distinguishable and fair, and that what is observed are the faces or numbers uppermost. (Compare with Exercises 1–10 in Section 7.1.)

9. Two coins are tossed; the result is at most one tail.

10. Two coins are tossed; the result is one or more heads.

11. Three coins are tossed; the result is at most one head.

12. Three coins are tossed; the result is more tails than heads.

13. Two dice are rolled; the numbers add to 5.

14. Two dice are rolled; the numbers add to 9.

15. Two dice are rolled; the numbers add to 1.

16. Two dice are rolled; one of the numbers is even, the other is odd.

17. Two dice are rolled; both numbers are prime.[30]

18. Two dice are rolled; neither number is prime.

19. If two indistinguishable dice are rolled, what is the probability of the event $\{(4, 4), (2, 3)\}$? What is the corresponding event for a pair of distinguishable dice? HINT [See Example 2.]

20. If two indistinguishable dice are rolled, what is the probability of the event $\{(5, 5), (2, 5), (3, 5)\}$? What is the corresponding event for a pair of distinguishable dice? HINT [See Example 2.]

---

[30]A positive integer is prime if it is neither 1 nor a product of smaller integers.

21. A die is weighted in such a way that each of 2, 4, and 6 is twice as likely to come up as each of 1, 3, and 5. Find the probability distribution. What is the probability of rolling less than 4? HINT [See Example 3.]

22. Another die is weighted in such a way that each of 1 and 2 is three times as likely to come up as each of the other numbers. Find the probability distribution. What is the probability of rolling an even number?

23. A tetrahedral die has four faces, numbered 1–4. If the die is weighted in such a way that each number is twice as likely to land facing down as the next number (1 twice as likely as 2, 2 twice as likely as 3, and 3 twice as likely as 4), what is the probability distribution for the face landing down?

24. A dodecahedral die has 12 faces, numbered 1–12. If the die is weighted in such a way that 2 is twice as likely to land facing up as 1, 3 is three times as likely to land facing up as 1, and so on, what is the probability distribution for the face landing up?

In Exercises 25–40, use the given information to find the indicated probability. HINT [See Quick Examples, page 476.]

25. $P(A) = .1$, $P(B) = .6$, $P(A \cap B) = .05$. Find $P(A \cup B)$.

26. $P(A) = .3$, $P(B) = .4$, $P(A \cap B) = .02$. Find $P(A \cup B)$.

27. $A \cap B = \emptyset$, $P(A) = .3$, $P(A \cup B) = .4$. Find $P(B)$.

28. $A \cap B = \emptyset$, $P(B) = .8$, $P(A \cup B) = .8$. Find $P(A)$.

29. $A \cap B = \emptyset$, $P(A) = .3$, $P(B) = .4$. Find $P(A \cup B)$.

30. $A \cap B = \emptyset$, $P(A) = .2$, $P(B) = .3$. Find $P(A \cup B)$.

31. $P(A \cup B) = .9$, $P(B) = .6$, $P(A \cap B) = .1$. Find $P(A)$.

32. $P(A \cup B) = 1.0$, $P(A) = .6$, $P(A \cap B) = .1$. Find $P(B)$.

33. $P(A) = .75$. Find $P(A')$.

34. $P(A) = .22$. Find $P(A')$.

35. A, B, and C are mutually exclusive. $P(A) = .3$, $P(B) = .4$, $P(C) = .3$. Find $P(A \cup B \cup C)$.

36. A, B, and C are mutually exclusive. $P(A) = .2$, $P(B) = .6$, $P(C) = .1$. Find $P(A \cup B \cup C)$.

37. A and B are mutually exclusive. $P(A) = .3$, $P(B) = .4$. Find $P((A \cup B)')$.

38. A and B are mutually exclusive. $P(A) = .4$, $P(B) = .4$. Find $P((A \cup B)')$.

39. $A \cup B = S$ and $A \cap B = \emptyset$. Find $P(A) + P(B)$.

40. $P(A \cup B) = .3$ and $P(A \cap B) = .1$. Find $P(A) + P(B)$.

In Exercises 41–46, determine whether the information shown is consistent with a probability distribution. If not, say why.

41. $P(A) = .2$; $P(B) = .1$; $P(A \cup B) = .4$

42. $P(A) = .2$; $P(B) = .4$; $P(A \cup B) = .2$

43. $P(A) = .2$; $P(B) = .4$; $P(A \cap B) = .2$

44. $P(A) = .2$; $P(B) = .4$; $P(A \cap B) = .3$

45. $P(A) = 0.1$; $P(B) = 0$; $P(A \cup B) = 0$

46. $P(A) = .1$; $P(B) = 0$; $P(A \cap B) = 0$

APPLICATIONS

47. *Subprime Mortgages during the 2000–2008 Housing Bubble* (Compare Exercise 27 in Section 7.2.) The following chart shows the (approximate) total number of subprime home mortgages in Texas in November 2008, broken down into four categories:[31]

Mortgage Status	Current	Past Due	In Foreclosure	Repossessed	Total
Number	136,330	53,310	8,750	5,090	203,480

(The four categories are mutually exclusive; for instance, "Past Due" refers to a mortgage whose payment status is past due but is not in foreclosure, and "In Foreclosure" refers to a mortgage that is in the process of being foreclosed but not yet repossessed.)

a. Find the probability that a randomly selected subprime mortgage in Texas during November 2008 was neither in foreclosure nor repossessed. HINT [See Example 1.]

b. What is the probability that a randomly selected subprime mortgage in Texas during November 2008 was not current?

48. *Subprime Mortgages during the 2000–2008 Housing Bubble* (Compare Exercise 28 in Section 7.2.) The following chart shows the (approximate) total number of subprime home mortgages in Florida in November 2008, broken down into four categories:[32]

Mortgage Status	Current	Past Due	In Foreclosure	Repossessed	Total
Number	130,400	73,260	72,380	17,000	293,040

(The four categories are mutually exclusive; for instance, "Past Due" refers to a mortgage whose payment status is past due but is not in foreclosure, and "In Foreclosure" refers to a mortgage that is in the process of being foreclosed but not yet repossessed.)

a. Find the probability that a randomly selected subprime mortgage in Florida during November 2008 was either in foreclosure or repossessed. HINT [See Example 1.]

b. What is the probability that a randomly selected subprime mortgage in Florida during November 2008 was not repossessed?

49. *Ethnic Diversity* The following pie chart shows the ethnic makeup of California schools in the 2006–2007 academic year.[33]

---

[31]Data are rounded to the nearest 10 units. Source: Federal Reserve Bank of New York (www.newyorkfed.org/regional/subprime.html).
[32]*Ibid.*
[33]Source: CBEDS data collection, Educational Demographics, October 2006 (www.cde.ca.gov/resrc/factbook/).

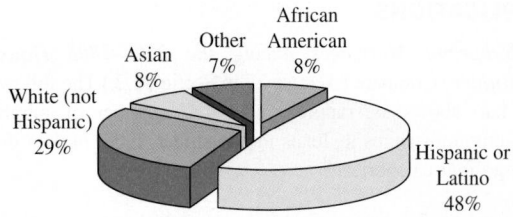

Write down the probability distribution showing the probability that a randomly selected California student in 2006–2007 belonged to one of the ethnic groups named. What is the probability that a student is neither white nor Asian?

50. ***Ethnic Diversity*** (Compare Exercise 49.) The following pie chart shows the ethnic makeup of California schools in the 1981–1982 academic year.[34]

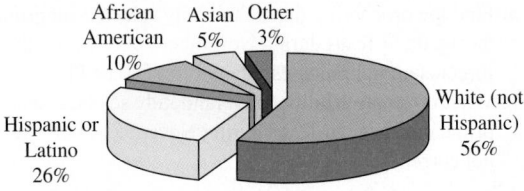

Write down the probability distribution showing the probability that a randomly selected California student in 1981–1982 belonged to one of the ethnic groups named. What is the probability that a student is neither Hispanic, Latino, nor African American?

51. ▼ ***Internet Investments in the 1990s*** The following excerpt is from an article in *The New York Times* in July 1999.[35]

While statistics are not available for Web entrepreneurs who fail, the venture capitalists that finance such Internet start-up companies have a rule of thumb. For every 10 ventures that receive financing—and there are plenty who do not—2 will be stock market successes, which means spectacular profits for early investors; 3 will be sold to other concerns, which translates into more modest profits; and the rest will fail.

a. What is a sample space for the scenario?
b. Write down the associated probability distribution.
c. What is the probability that a start-up venture that receives financing will realize profits for early investors?

52. ▼ ***Internet Investments in the 1990s*** The following excerpt is from an article in *The New York Times* in July 1999.[36]

Right now, the market for Web stocks is sizzling. Of the 126 initial public offerings of Internet stocks priced this year, 73 are trading above the price they closed on their first day of trading. . . . Still, 53 of the offerings have failed to live up to their fabulous first-day billings, and 17 [of these] are below the initial offering price.

Assume that, on the first day of trading, all stocks closed higher than their initial offering price.

a. What is a sample space for the scenario?
b. Write down the associated probability distribution. (Round your answers to two decimal places.)
c. What is the probability that an Internet stock purchased during the period reported ended either below its initial offering price or above the price it closed on its first day of trading? HINT [See Example 3.]

53. ▼ ***Market Share: Light Vehicles*** In 2003, 25% of all light vehicles sold (SUVs, pickups, passenger cars, and minivans) in the United States were SUVs and 15% were pickups. Moreover, a randomly chosen vehicle sold that year was five times as likely to be a passenger car as a minivan.[37] Find the associated probability distribution.

54. ▼ ***Market Share: Light Vehicles*** In 2000, 15% of all light vehicles (SUVs, pickups, passenger cars, and minivans) sold in the United States were pickups and 55% were passenger cars. Moreover, a randomly chosen vehicle sold that year was twice as likely to be an SUV as a minivan.[38] Find the associated probability distribution.

Gambling In Exercises 55–62 are detailed some of the nefarious dicing practices of the Win Some/Lose Some Casino. In each case, find the probabilities of all the possible outcomes and also the probability that an odd number or an odd sum faces up. HINT [See Example 3.]

55. Some of the dice are specially designed so that 1 and 6 never come up, and all the other outcomes are equally likely.

56. Other dice are specially designed so that 1 comes up half the time, 6 never comes up, and all the other outcomes are equally likely.

57. Some of the dice are cleverly weighted so that each of 2, 3, 4, and 5 is twice as likely to come up as 1 is, and 1 and 6 are equally likely.

58. Other dice are weighted so that each of 2, 3, 4, and 5 is half as likely to come up as 1 is, and 1 and 6 are equally likely.

59. ▼ Some pairs of dice are magnetized so that each pair of mismatching numbers is twice as likely to come up as each pair of matching numbers.

[34]Source: CBEDS data collection, Educational Demographics, October 2006 (www.cde.ca.gov/resrc/factbook/).

[35]Article: "Not All Hit It Rich in the Internet Gold Rush," *New York Times*, July 20, 1999, p. A1.

[36]Article: Ibid. Source for data: Comm-Scan/*New York Times*, July 20, 1999, p. A1.

[37]Source: Environmental Protection Agency/*New York Times*, June 28, 2003.

[38]*Ibid.*

60. ▼ Other pairs of dice are so strongly magnetized that mismatching numbers never come up.

61. ▼ Some dice are constructed in such a way that deuce (2) is five times as likely to come up as 4 and three times as likely to come up as each of 1, 3, 5, and 6.

62. ▼ Other dice are constructed in such a way that deuce is six times as likely to come up as 4 and four times as likely to come up as each of 1, 3, 5, and 6.

63. *Astrology* The astrology software package, Turbo Kismet,[39] works by first generating random number sequences and then interpreting them numerologically. When I ran it yesterday, it informed me that there was a 1/3 probability that I would meet a tall dark stranger this month, a 2/3 probability that I would travel this month, and a 1/6 probability that I would meet a tall dark stranger and also travel this month. What is the probability that I will either meet a tall dark stranger or that I will travel this month? HINT [See Quick Example 1, page 476 and Example 4.]

64. *Astrology* Another astrology software package, Java Kismet, is designed to help day traders choose stocks based on the position of the planets and constellations. When I ran it yesterday, it informed me that there was a .5 probability that Amazon.com will go up this afternoon, a .2 probability that Yahoo.com will go up this afternoon, and a .2 chance that both will go up this afternoon. What is the probability that either Amazon.com or Yahoo.com will go up this afternoon? HINT [See Quick Example 1 on page 476 and Example 4.]

65. *Polls* According to a *New York Times*/CBS poll released in March 2005, 61% of those polled ranked jobs or health care as the top domestic priority.[40] What is the probability that a randomly selected person polled did not rank either as the top domestic priority? HINT [See Example 5.]

66. *Polls* According to *The New York Times*/CBS poll of March 2005 referred to in Exercise 65, 72% of those polled ranked neither Iraq nor North Korea as the top foreign policy issue.[41] What is the probability that a randomly selected person polled ranked either Iraq or North Korea as the top foreign policy issue? HINT [See Example 5.]

67. *Hybrid Auto Sales* In 2010, the probability that a randomly chosen hybrid vehicle sold in the United States was manufactured by Toyota was .70, while the probability that it was manufactured by Honda was .05.[42] What is the probability that a randomly chosen hybrid vehicle was manufactured by neither company?

68. *Hybrid Auto Sales* In 2010, the probability that a randomly chosen hybrid vehicle sold in the United States was manufactured by Ford was .12, while the probability that it was manufactured by Nissan was .02.[43] What is the probability that a randomly chosen hybrid vehicle was manufactured by neither company?

Student Admissions *Exercises 69–84 are based on the following table, which shows the profile, by the math section of the SAT Reasoning Test, of admitted students at UCLA for the Fall 2011 semester.*[44]

SAT Reasoning Test—Math Section

	700–800	600–699	500–599	400–499	200–399	Total
Admitted	6,611	3,981	1,388	385	8	12,373
Not admitted	6,622	13,876	9,890	4,974	1,363	36,725
Total applicants	13,233	17,857	11,278	5,359	1,371	49,098

Determine the probabilities of the following events. (Round your answers to the nearest .01.) HINT [Example 6.]

69. An applicant was admitted.

70. An applicant had a Math SAT below 400.

71. An applicant had a Math SAT below 400 and was admitted.

72. An applicant had a Math SAT of 700 or above and was admitted.

73. An applicant was not admitted.

74. An applicant did not have a Math SAT below 400.

75. An applicant had a Math SAT in the range 500–599 or was admitted.

76. An applicant had a Math SAT of 700 or above or was admitted.

77. An applicant neither was admitted nor had a Math SAT in the range 500–599.

78. An applicant neither had a Math SAT of 700 or above nor was admitted.

79. ▼ An applicant who had a Math SAT below 400 was admitted.

80. ▼ An applicant who had a Math SAT of 700 or above was admitted.

81. ▼ An admitted student had a Math SAT of 700 or above.

82. ▼ An admitted student had a Math SAT below 400.

83. ▼ A rejected applicant had a Math SAT below 600.

84. ▼ A rejected applicant had a Math SAT of at least 600.

---

[39]The name and concept were borrowed from a hilarious (as yet unpublished) novel by the science-fiction writer William Orr, who also happens to be a faculty member at Hofstra University.

[40]Source: *New York Times*, March 3, 2005, p. A20.

[41]*Ibid.*

[42]Source for sales data: www.wikipedia.com.

---

[43]Source for sales data: www.wikipedia.com.

[44]Source: University of California (www.admissions.ucla.edu/Prospect/Adm_fr/Frosh_Prof11.htm).

85. ▼ *Social Security* According to *The New York Times*/CBS poll of March 2005 referred to in Exercise 65, 79% agreed that it should be the government's responsibility to provide a decent standard of living for the elderly, and 43% agreed that it would be a good idea to invest part of their Social Security taxes on their own. What is the smallest percentage of people who could have agreed with both statements? What is the largest percentage of people who could have agreed with both statements?

86. ▼ *Social Security* According to *The New York Times*/CBS poll of March 2005 referred to in Exercise 65, 49% agreed that Social Security taxes should be raised if necessary to keep the system afloat, and 43% agreed that it would be a good idea to invest part of their Social Security taxes on their own. What is the largest percentage of people who could have agreed with at least one of these statements? What is the smallest percentage of people who could have agreed with at least one of these statements?

87. ▼ *Greek Life* The ΤΦΦ Sorority has a tough pledging program—it requires its pledges to master the Greek alphabet forward, backward, and "sideways." During the last pledge period, two-thirds of the pledges failed to learn it backward and three quarters of them failed to learn it sideways; 5 of the 12 pledges failed to master it either backward or sideways. Because admission into the sisterhood requires both backward and sideways mastery, what fraction of the pledges were disqualified on this basis?

88. ▼ *Swords and Sorcery* Lance the Wizard has been informed that tomorrow there will be a 50% chance of encountering the evil Myrmidons and a 20% chance of meeting up with the dreadful Balrog. Moreover, Hugo the Elf has predicted that there is a 10% chance of encountering both tomorrow. What is the probability that Lance will be lucky tomorrow and encounter neither the Myrmidons nor the Balrog?

89. ▼ *Public Health* A study shows that 80% of the population has been vaccinated against the Venusian flu, but 2% of the vaccinated population gets the flu anyway. If 10% of the total population gets this flu, what percent of the population either gets the vaccine or gets the disease?

90. ▼ *Public Health* A study shows that 75% of the population has been vaccinated against the Martian ague, but 4% of this group gets this disease anyway. If 10% of the total population gets this disease, what is the probability that a randomly selected person has been neither vaccinated nor has contracted Martian ague?

COMMUNICATION AND REASONING EXERCISES

91. Design an experiment based on rolling a fair die for which there are exactly three outcomes with the same probabilities.

92. Design an experiment based on rolling a fair die for which there are at least three outcomes with different probabilities.

93. ▼ Tony has had a "losing streak" at the casino—the chances of winning the game he is playing are 40%, but he has lost five times in a row. Tony argues that, because he should have won two times, the game must obviously be "rigged." Comment on his reasoning.

94. ▼ Maria is on a winning streak at the casino. She has already won four times in a row and concludes that her chances of winning a fifth time are good. Comment on her reasoning.

95. Complete the following sentence. The probability of the union of two events is the sum of the probabilities of the two events if _____.

96. A friend of yours asserted at lunch today that, according to the weather forecast for tomorrow, there is a 52% chance of rain and a 60% chance of snow. "But that's impossible!" you blurted out, "the percentages add up to more than 100%." Explain why you were wrong.

97. ▼ A certain experiment is performed a large number of times, and the event E has relative frequency equal to zero. This means that it should have modeled probability zero, right? HINT [See the definition of a probability model on page 469.]

98. ▼ (Refer to the preceding exercise.) How can the modeled probability of winning the lotto be nonzero if you have never won it, despite playing 600 times? HINT [See the definition of a probability model on page 469.]

99. ▼ Explain how the addition principle for mutually exclusive events follows from the general addition principle.

100. ▼ Explain how the property $P(A') = 1 - P(A)$ follows directly from the properties of a probability distribution.

101. ◆ It is said that lightning never strikes twice in the same spot. Assuming this to be the case, what should be the modeled probability that lightning will strike a given spot during a thunderstorm? Explain. HINT [See the definition of a probability model on page 469.]

102. ◆ A certain event has modeled probability equal to zero. This means it will never occur, right? HINT [See the definition of a probability model on page 469.]

103. ◆ Find a formula for the probability of the union of three (not necessarily mutually exclusive) events A, B, and C.

104. ◆ Four events A, B, C, and D have the following property: If any two events have an outcome in common, that outcome is common to all four events. Find a formula for the probability of their union.

<table>
<tr><td>**7.4**</td></tr>
</table>

7.4 Probability and Counting Techniques

We saw in the preceding section that, when all outcomes in a sample space are equally likely, we can use the following formula to model the probability of each event:

Modeling Probability: Equally Likely Outcomes

In an experiment in which all outcomes are equally likely, the probability of an event E is given by

$$P(E) = \frac{\text{Number of favorable outcomes}}{\text{Total number of outcomes}} = \frac{n(E)}{n(S)}.$$

This formula is simple, but calculating $n(E)$ and $n(S)$ may not be. In this section, we look at some examples in which we need to use the counting techniques discussed in Chapter 6.

EXAMPLE 1 Marbles

A bag contains four red marbles and two green ones. Upon seeing the bag, Suzan (who has compulsive marble-grabbing tendencies) sticks her hand in and grabs three at random. Find the probability that she will get both green marbles.

S is the set of all outcomes that can occur, and has nothing to do with having green marbles.

Solution According to the formula, we need to know these numbers:

- The number of elements in the sample space S.
- The number of elements in the event E.

First of all, what is the sample space? The sample space is the set of all possible outcomes, and each outcome consists of a set of three marbles (in Suzan's hand). So, the set of outcomes is the set of all sets of three marbles chosen from a total of six marbles (four red and two green). Thus,

$$n(S) = C(6, 3) = 20.$$

Now what about E? This is the event that Suzan gets both green marbles. We must *rephrase this as a subset of S* in order to deal with it: "E is the collection of sets of three marbles such that one is red and two are green." Thus, $n(E)$ is the *number* of such sets, which we determine using a decision algorithm.

    Step 1 Choose a red marble: $C(4, 1) = 4$ possible outcomes.

    Step 2 Choose the two green marbles: $C(2, 2) = 1$ possible outcome.

We get $n(E) = 4 \times 1 = 4$. Now,

$$P(E) = \frac{n(E)}{n(S)} = \frac{4}{20} = \frac{1}{5}.$$

Thus, there is a one in five chance of Suzan's getting both the green marbles.

EXAMPLE 2 Investment Lottery

After a down day on the stock market, you decide to ignore your broker's cautious advice and purchase three stocks at random from the six most active stocks listed on the New York Stock Exchange at the end of the day's trading.[45]

Company	Symbol	Price	Change
Bank of America	**BAC**	$5.80	−$1.00
PowerShares QQQ	**QQQ**	$55.83	−$1.34
Och-Ziff Capital Mgmt	**OZM**	$7.98	−$0.43
UBS AG Common Stock	**UBS**	$11.21	−$0.30
iShares Silver Trust	**SLV**	$30.64	−$2.18
Direxion Small Cap Bull	**TNA**	$42.80	−$1.92

Find the probabilities of the following events:

a. You purchase BAC and QQQ.

b. At most two of the stocks you purchase declined in value by more than $1.

Solution First, the sample space is the set of all collections of 3 stocks chosen from the 6. Thus,

$$n(S) = C(6, 3) = 20.$$

a. The event E of interest is the event that you purchase BAC and QQQ. Thus, E is the set of all groups of 3 stocks that include BAC and QQQ. Because there is only one more stock left to choose,

$$n(E) = C(4, 1) = 4.$$

We now have

$$P(E) = \frac{n(E)}{n(S)} = \frac{4}{20} = \frac{1}{5} = .2.$$

b. Let F be the event that at most two of the stocks you purchase declined in value by more than $1. Thus, F is the set of all groups of three stocks of which at most two declined in value by more than $1. To calculate $n(F)$, we use the following decision algorithm.

Alternative 1: None of the stocks declined in value by more than $1.
  Step 1 Choose three stocks that did not decline in value by more than $1:
    $C(3, 3) = 1$ possibility.

Alternative 2: One of the stocks declined in value by more than $1.
  Step 1 Choose one stock that declined in value by more than $1: $C(3, 1) = 3$ possibilities.
  Step 2 Choose two stocks that did not decline in value by more than $1:
    $C(3, 2) = 3$ possibilities.
  This gives $3 \times 3 = 9$ possibilities for this alternative.

Alternative 3: Two of the stocks declined in value by more than $1.
  Step 1 Choose two stocks that declined in value by more than $1:
    $C(3, 2) = 3$ possibilities.
  Step 2 Choose one stock that did not decline in value by more than $1:
    $C(3, 1) = 3$ possibilities.
  This gives $3 \times 3 = 9$ possibilities for this alternative.

---

[45]Most active stocks on November 17, 2011. Source: Yahoo! Finance (http://finance.yahoo.com).

So, we have a total of $1 + 9 + 9 = 19$ possible outcomes. Thus,

$$n(F) = 19$$

and

$$P(F) = \frac{n(F)}{n(S)} = \frac{19}{20} = .95.$$

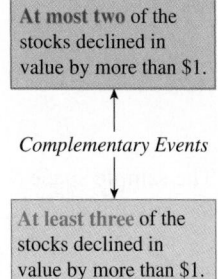

Complementary Events

➡ **Before we go on...** When counting the number of outcomes in an event, the calculation is sometimes easier if we look at the *complement* of that event. In the case of part (b) of Example 2, the complement of the event F is

> F': At least three of the stocks you purchase declined in value by more than $1.

Because there are only three stocks in your portfolio, this is the same as the event that all three you purchase declined in value by more than $1. The decision algorithm for $n(F')$ is far simpler:

> **Step 1**   Choose three stocks that declined in value by more than $1: $C(3, 3) = $ 1 possibility.

So, $n(F') = 1$, giving

$$n(F) = n(S) - n(F') = 20 - 1 = 19$$

as we calculated above. ∎

EXAMPLE 3 **Poker Hands**

You are dealt 5 cards from a well-shuffled standard deck of 52. Find the probability that you have a full house. (Recall that a full house consists of 3 cards of one denomination and 2 of another.)

Solution   The sample space S is the set of all possible 5-card hands dealt from a deck of 52. Thus,

$$n(S) = C(52, 5) = 2{,}598{,}960.$$

If the deck is thoroughly shuffled, then each of these 5-card hands is equally likely. Now consider the event E, the set of all possible 5-card hands that constitute a full house. To calculate $n(E)$, we use a decision algorithm, which we show in the following compact form.

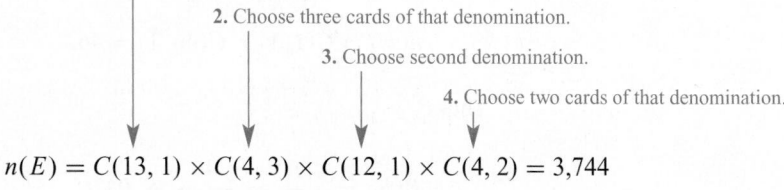

$$n(E) = C(13, 1) \times C(4, 3) \times C(12, 1) \times C(4, 2) = 3{,}744$$

Thus,

$$P(E) = \frac{n(E)}{n(S)} = \frac{3{,}744}{2{,}598{,}960} \approx .00144.$$

In other words, there is an approximately 0.144% chance that you will be dealt a full house.

EXAMPLE 4 More Poker Hands

You are playing poker, and you have been dealt the following hand:

$$J\spadesuit, J\diamondsuit, J\heartsuit, 2\clubsuit, 10\spadesuit.$$

You decide to exchange the last two cards. The exchange works as follows: The two cards are discarded (not replaced in the deck), and you are dealt two new cards.

a. Find the probability that you end up with a full house.

b. Find the probability that you end up with four jacks.

c. What is the probability that you end up with either a full house or four jacks?

Solution

a. In order to get a full house, you must be dealt two of a kind. The sample space S is the set of all pairs of cards selected from what remains of the original deck of 52. You were dealt 5 cards originally, so there are $52 - 5 = 47$ cards left in the deck. Thus, $n(S) = C(47, 2) = 1{,}081$. The event E is the set of all pairs of cards that constitute two of a kind. Note that you cannot get two jacks because only one is left in the deck. Also, only three 2s and three 10s are left in the deck. We have

1. Choose a denomination other than Jacks, 2s, and 10s.

1. Choose either 2s or 10s.

$$n(E) = C(10, 1) \times C(4, 2) \quad + \quad C(2, 1) \times C(3, 2) = 66.$$

OR

2. Choose two cards of that denomination.

2. Choose two cards of that denomination.

Thus,

$$P(E) = \frac{n(E)}{n(S)} = \frac{66}{1{,}081} \approx .0611.$$

b. We have the same sample space as in part (a). Let F be the set of all pairs of cards that include the missing jack of clubs. So,

1. Choose the jack of clubs.

2. Choose 1 card from the remaining 46.

$$n(F) = C(1, 1) \times C(46, 1) = 46.$$

Thus,

$$P(F) = \frac{n(F)}{n(S)} = \frac{46}{1{,}081} \approx .0426.$$

c. We are asked to calculate the probability of the event $E \cup F$. From the addition principle, we have

$$P(E \cup F) = P(E) + P(F) - P(E \cap F).$$

Because $E \cap F$ means "E and F," $E \cap F$ is the event that the pair of cards you are dealt are two of a kind and include the jack of clubs. But this is impossible because only one jack is left. Thus $E \cap F = \emptyset$, and so $P(E \cap F) = 0$. This gives us

$$P(E \cup F) = P(E) + P(F) \approx .0611 + .0426 = .1037.$$

In other words, there is slightly better than a one in ten chance that you will wind up with either a full house or four of a kind, given the original hand.

➡ **Before we go on...** A more accurate answer to part (c) of Example 4 is $(66 + 46)/1,081 \approx .1036$; we lost some accuracy in rounding the answers to parts (a) and (b). ■

EXAMPLE 5 **Committees**

The University Senate bylaws at Hofstra University state the following:[46]

> The Student Affairs Committee shall consist of one elected faculty senator, one faculty senator-at-large, one elected student senator, five student senators-at-large (including one from the graduate school), two delegates from the Student Government Association, the President of the Student Government Association or his/her designate, and the President of the Graduate Student Organization. It shall be chaired by the elected student senator on the Committee and it shall be advised by the Dean of Students or his/her designate.

You are an undergraduate student and, even though you are not an elected student senator, you would very much like to serve on the Student Affairs Committee. The senators-at-large as well as the Student Government delegates are chosen by means of a random drawing from a list of candidates. There are already 13 undergraduate candidates for the position of senator-at-large, and 6 candidates for Student Government delegates, and you have been offered a position on the Student Government Association by the President (who happens to be a good friend of yours), should you wish to join it. (This would make you ineligible for a senator-at-large position.) What should you do?

Solution You have two options. Option 1 is to include your name on the list of candidates for the senator-at-large position. Option 2 is to join the Student Government Association (SGA) and add your name to its list of candidates. Let us look at the two options separately.

Option 1: Add your name to the senator-at-large list.
This will result in a list of 14 undergraduates for 4 undergraduate positions. The sample space is the set of all possible outcomes of the random drawing. Each outcome consists of a set of 4 lucky students chosen from 14. Thus,

$$n(S) = C(14, 4) = 1,001.$$

We are interested in the probability that you are among the chosen four. Thus, E is the set of sets of four that include you.

1. Choose yourself.

2. Choose 3 from the remaining 13.

$$n(E) = C(1, 1) \times C(13, 3) = 286.$$

---

[46]As of 2011. Source: Hofstra University Senate Bylaws.

So,

$$P(E) = \frac{n(E)}{n(S)} = \frac{286}{1,001} = \frac{2}{7} \approx .2857.$$

Option 2: Join the SGA and add your name to its list.
This results in a list of seven candidates from which two are selected. For this case, the sample space consists of all sets of two chosen from seven, so

$$n(S) = C(7, 2) = 21$$

and

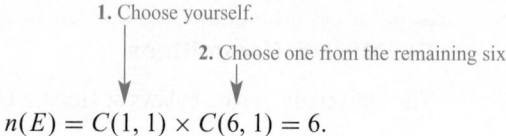

1. Choose yourself.

2. Choose one from the remaining six.

$$n(E) = C(1, 1) \times C(6, 1) = 6.$$

Thus,

$$P(E) = \frac{n(E)}{n(S)} = \frac{6}{21} = \frac{2}{7} \approx .2857.$$

In other words, the probability of being selected is exactly the same for Option 1 as it is for Option 2! Thus, you can choose either option, and you will have slightly less than a 29% chance of being selected.

7.4   EXERCISES

▼ more advanced   ◆ challenging
🔲 indicates exercises that should be solved using technology

Recall from Example 1 that whenever Suzan sees a bag of marbles, she grabs a handful at random. In Exercises 1–10, she has seen a bag containing four red marbles, three green ones, two white ones, and one purple one. She grabs five of them. Find the probabilities of the following events, expressing each as a fraction in lowest terms. HINT [See Example 1.]

1. She has all the red ones.

2. She has none of the red ones.

3. She has at least one white one.

4. She has at least one green one.

5. She has two red ones and one of each of the other colors.

6. She has two green ones and one of each of the other colors.

7. She has at most one green one.

8. She has no more than one white one.

9. She does not have all the red ones.

10. She does not have all the green ones.

Dogs of the Dow *The phrase "Dogs of the Dow" refers to the stocks listed on the Dow with the highest dividend yield.*

Exercises 11–16 are based on the following table, which shows the top ten stocks of the "Dogs of the Dow" list in November 2011:[47] HINT [See Example 2.]

Symbol	Company	Price	Yield
T	AT&T	29.46	5.84%
VZ	Verizon	37.52	5.20%
MRK	Merck	34.47	4.41%
PFE	Pfizer	20.08	3.98%
JNJ	Johnson & Johnson	64.86	3.52%
INTC	Intel	24.75	3.39%
KFT	Kraft	35.48	3.27%
DD	DuPont	49.81	3.29%
CVX	Chevron	108.86	2.87%
MCD	McDonald's	94.60	2.58%

[47]Source: www.dogsofthedow.com.

11. If you selected two of these stocks at random, what is the probability that both the stocks in your selection had yields of 4% or more?

12. If you selected three of these stocks at random, what is the probability that all three of the stocks in your selection had yields of 4% or more?

13. If you selected four of these stocks at random, what is the probability that your selection included the company with the highest yield and excluded the company with the lowest yield?

14. If you selected four of these stocks at random, what is the probability that your selection included T and VZ but excluded PFE and DD?

15. ▼ If your portfolio included 100 shares of PFE and you then purchased 100 shares each of any two companies on the list at random, find the probability that you ended up with a total of 200 shares of PFE.

16. ▼ If your portfolio included 100 shares of PFE and you then purchased 100 shares each of any three companies on the list at random, find the probability that you ended up with a total of 200 shares of PFE.

17. *Tests* A test has three parts. Part A consists of eight true-false questions, Part B consists of five multiple choice questions with five choices each, and Part C requires you to match five questions with five different answers one-to-one. Assuming that you make random guesses in filling out your answer sheet, what is the probability that you will earn 100% on the test? (Leave your answer as a formula.)

18. *Tests* A test has three parts. Part A consists of four true-false questions, Part B consists of four multiple choice questions with five choices each, and Part C requires you to match six questions with six different answers one-to-one. Assuming that you make random choices in filling out your answer sheet, what is the probability that you will earn 100% on the test? (Leave your answer as a formula.)

Poker In Exercises 19–24, you are asked to calculate the probability of being dealt various poker hands. (Recall that a poker player is dealt 5 cards at random from a standard deck of 52.) Express each of your answers as a decimal rounded to four decimal places, unless otherwise stated. HINT [See Example 3.]

19. Two of a kind: Two cards with the same denomination and three cards with other denominations (different from each other and that of the pair). Example: K♣, K♥, 2♠, 4♦, J♠

20. Three of a kind: Three cards with the same denomination and two cards with other denominations (different from each other and that of the three). Example: Q♣, Q♥, Q♠, 4♦, J♠

21. Two pairs: Two cards with one denomination, two with another, and one with a third. Example: 3♣, 3♥, Q♦, Q♥, 10♠

22. Straight Flush: Five cards of the same suit with consecutive denominations but not a royal flush (a royal flush consists of the 10, J, Q, K, and A of one suit). Round the answer to one significant digit. Examples: A♣, 2♣, 3♣, 4♣, 5♣, or

9♦, 10♦, J♦, Q♦, K♦, or A♥, 2♥, 3♥, 4♥, 5♥, but *not* 10♦, J♦, Q♦, K♦, A♦

23. Flush: Five cards of the same suit, but not a straight flush or royal flush. Example: A♣, 5♣, 7♣, 8♣, K♣

24. Straight: Five cards with consecutive denominations, but not all of the same suit. Examples: 9♦, 10♦, J♣, Q♥, K♦, and 10♥, J♦, Q♦, K♦, A♠.

25. *The Monkey at the Typewriter* Suppose that a monkey is seated at a computer keyboard and randomly strikes the 26 letter keys and the space bar. Find the probability that its first 39 characters (including spaces) will be "to be or not to be that is the question". (Leave your answer as a formula.)

26. *The Cat on the Piano* A standard piano keyboard has 88 different keys. Find the probability that a cat, jumping on 4 keys in sequence and at random (possibly with repetition), will strike the first four notes of Beethoven's Fifth Symphony. (Leave your answer as a formula.)

27. *(Based on a question from the GMAT)* Tyler and Gebriella are among seven contestants from which four semifinalists are to be selected at random. Find the probability that neither Tyler nor Gebriella is selected.

28. *(Based on a question from the GMAT)* Tyler and Gebriella are among seven contestants from which four semifinalists are to be selected at random. Find the probability that Tyler but not Gebriella is selected.

29. ▼ *Lotteries* The Sorry State Lottery requires you to select five different numbers from 0 through 49. (Order is not important.) You are a Big Winner if the five numbers you select agree with those in the drawing, and you are a Small-Fry Winner if four of your five numbers agree with those in the drawing. What is the probability of being a Big Winner? What is the probability of being a Small-Fry Winner? What is the probability that you are either a Big Winner or a Small-Fry winner?

30. ▼ *Lotto* The Sad State Lottery requires you to select a sequence of three different numbers from 0 through 49. (Order is important.) You are a Winner if your sequence agrees with that in the drawing, and you are a Booby Prize Winner if your selection of numbers is correct, but in the wrong order. What is the probability of being a Winner? What is the probability of being a Booby Prize Winner? What is the probability that you are either a Winner or a Booby Prize Winner?

31. ▼ *Transfers* Your company is considering offering 400 employees the opportunity to transfer to its new headquarters in Ottawa and, as personnel manager, you decide that it would be fairest if the transfer offers are decided by means of a lottery. Assuming that your company currently employs 100 managers, 100 factory workers, and 500 miscellaneous staff, find the following probabilities, leaving the answers as formulas:

a. All the managers will be offered the opportunity.
b. You will be offered the opportunity.

32. ▼ *Transfers* (Refer back to the preceding exercise.) After thinking about your proposed method of selecting employees for the opportunity to move to Ottawa, you decide it might be a better idea to select 50 managers, 50 factory workers, and 300 miscellaneous staff, all chosen at random. Find the probability that you will be offered the opportunity. (Leave your answer as a formula.)

33. ▼ *Lotteries* In a New York State daily lottery game, a sequence of three digits (not necessarily different) in the range 0–9 are selected at random. Find the probability that all three are different.

34. ▼ *Lotteries* Refer back to the preceding exercise. Find the probability that two of the three digits are the same.

35. ▼ *Sports* The following table shows the results of the Big Eight Conference for the 1988 college football season.[48]

Team	Won	Lost
Nebraska (NU)	7	0
Oklahoma (OU)	6	1
Oklahoma State (OSU)	5	2
Colorado (CU)	4	3
Iowa State (ISU)	3	4
Missouri (MU)	2	5
Kansas (KU)	1	6
Kansas State (KSU)	0	7

This is referred to as a "perfect progression." Assuming that the "Won" score for each team is chosen at random in the range 0–7, find the probability that the results form a perfect progression.[49] (Leave your answer as a formula.)

36. ▼ *Sports* Refer back to Exercise 35. Find the probability of a perfect progression with Nebraska scoring seven wins and zero losses. (Leave your answer as a formula.)

37. ▼ *Graph Searching* A graph consists of a collection of **nodes** (the dots in the figure) connected by **edges** (line segments from one node to another). A **move on a graph** is a move from one node to another along a single edge. Find the probability of going from Start to Finish in a sequence of

[48]Source: On the probability of a perfect progression, *The American Statistician*, August 1991, vol. 45, no. 3, p. 214.

[49]Even if all the teams are equally likely to win each game, the chances of a perfect progression actually coming up are a little more difficult to estimate, because the number of wins by one team impacts directly on the number of wins by the others. For instance, it is impossible for all eight teams to show a score of seven wins and zero losses at the end of the season—someone must lose! It is, however, not too hard to come up with a counting argument to estimate the total number of win-loss scores actually possible.

two random moves in the graph shown. (All directions are equally likely.)

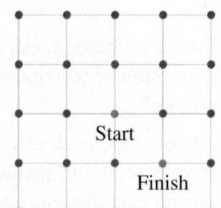

38. ▼ *Graph Searching* Refer back to Exercise 37. Find the probability of going from Start to one of the Finish nodes in a sequence of two random moves in the following figure. (All directions are equally likely.)

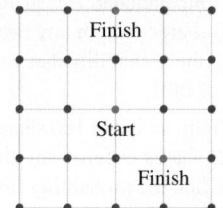

39. ▼ *Tournaments* What is the probability that North Carolina will beat Central Connecticut but lose to Virginia in the following (fictitious) soccer tournament? (Assume that all outcomes are equally likely.)

North Carolina

Central Connecticut

Virginia

Syracuse

40. ▼ *Tournaments* In a (fictitious) soccer tournament involving the four teams San Diego State, De Paul, Colgate, and Hofstra, find the probability that Hofstra will play Colgate in the finals and win. (Assume that all outcomes are equally likely and that the teams not listed in the first round slots are placed at random.)

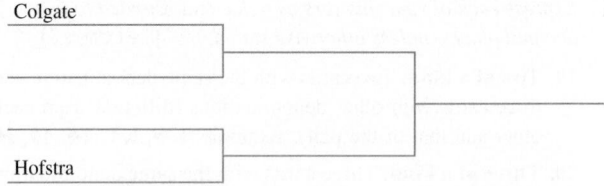

Colgate

Hofstra

41. ◆ *Product Design* Your company has patented an electronic digital padlock that a user can program with his or her own four-digit code. (Each digit can be 0 through 9, and repetitions are allowed.) The padlock is designed to open either if the correct code is keyed in or—and this is helpful for forgetful people—if exactly one of the digits is incorrect. What is the probability that a randomly chosen sequence of four digits will open a programmed padlock?

42. ◆ *Product Design* Assume that you already know the first digit of the combination for the lock described in Exercise 41. Find the probability that a random guess of the remaining three digits will open the lock. HINT [See Example 5.]

43. ◆ *Committees* An investigatory committee in the Kingdom of Utopia consists of a chief investigator (a Royal Party member), an assistant investigator (a Birthday Party member), two at-large investigators (either party), and five ordinary members (either party). Royal Party member Larry Sifford is hoping to avoid serving on the committee, unless he is the Chief Investigator and Otis Taylor, a Birthday Party member, is the Assistant Investigator. The committee is to be selected at random from a pool of 12 candidates (including Larry Sifford and Otis Taylor), half of whom are Royal Party and half of whom are Birthday Party.

   a. How many different committees are possible? HINT [See Example 5.]
   b. How many committees are possible in which Larry's hopes are fulfilled? (This includes the possibility that he's not on the committee at all.)
   c. What is the probability that he'll be happy with a randomly selected committee?

44. ◆ *Committees* A committee is to consist of a chair, three hagglers, and four do-nothings. The committee is formed by choosing randomly from a pool of 10 people and assigning them to the various "jobs."

   a. How many different committees are possible? HINT [See Example 5.]
   b. Norman is eager to be the chair of the committee. What is the probability that he will get his wish?
   c. Norman's girlfriend Norma is less ambitious and would be happy to hold any position on the committee provided Norman is also selected as a committee member. What is the probability that she will get her wish and serve on the committee?

   d. Norma does not get along with Oona (who is also in the pool of prospective members) and would be most unhappy if Oona were to chair the committee. Find the probability that all her wishes will be fulfilled: She and Norman are on the committee and it is not chaired by Oona.

COMMUNICATION AND REASONING EXERCISES

45. What is wrong with the following argument? A bag contains two blue marbles and two red ones; two are drawn at random. Because there are four possibilities—(red, red), (blue, blue), (red, blue) and (blue, red)—the probability that both are red is 1/4.

46. What is wrong with the following argument? When we roll two indistinguishable dice, the number of possible outcomes (unordered groups of two not necessarily distinct numbers) is 21 and the number of outcomes in which both numbers are the same is 6. Hence, the probability of throwing a double is $6/21 = 2/7$.

47. ▼ Suzan grabs two marbles out of a bag of five red marbles and four green ones. She could do so in two ways: She could take them out one at a time, so that there is a first and a second marble, or she could grab two at once so that there is no order. Does the method she uses to grab the marbles affect the probability that she gets two red marbles?

48. ▼ If Suzan grabs two marbles, one at a time, out of a bag of five red marbles and four green ones, find an event with a probability that depends on the order in which the two marbles are drawn.

49. Create an interesting application whose solution requires finding a probability using combinations.

50. Create an interesting application whose solution requires finding a probability using permutations.

7.5 Conditional Probability and Independence

Cyber Video Games, Inc., ran a television ad in advance of the release of its latest game, "Ultimate Hockey." As Cyber Video's director of marketing, you would like to assess the ad's effectiveness, so you ask your market research team to survey video game players. The results of its survey of 2,000 video game players are summarized in the following table:

	Saw Ad	Did Not See Ad	Total
Purchased Game	100	200	300
Did Not Purchase Game	200	1,500	1,700
Total	300	1,700	2,000

The market research team concludes in its report that the ad is highly persuasive, and recommends using the company that produced the ad for future projects.

But wait, how could the ad possibly have been persuasive? Only 100 people who saw the ad purchased the game, while 200 people purchased the game without seeing the ad at all! At first glance, it looks as though potential customers are being *put off* by the ad. But let us analyze the figures a little more carefully.

First, let us restrict attention to those players who saw the ad (first column of data: "Saw Ad") and compute the estimated probability that a player *who saw the ad* purchased Ultimate Hockey.

	Saw Ad
Purchased Game	100
Did Not Purchase Game	200
Total	300

To compute this probability, we calculate

Probability that someone who saw the ad purchased the game

$$= \frac{\text{Number of people who saw the ad and bought the game}}{\text{Total number of people who saw the ad}} = \frac{100}{300} \approx .33.$$

In other words, 33% of game players who saw the ad went ahead and purchased the game. Let us compare this with the corresponding probability for those players who did *not* see the ad (second column of data "Did Not See Ad"):

	Did Not See Ad
Purchased Game	200
Did Not Purchase Game	1,500
Total	1,700

Probability that someone who did not see the ad purchased the game

$$= \frac{\text{Number of people who did not see the ad and bought the game}}{\text{Total number of people who did not see the ad}} = \frac{200}{1,700} \approx .12.$$

In other words, only 12% of game players who did not see the ad purchased the game, whereas 33% of those who *did* see the ad purchased the game. Thus, it appears that the ad *was* highly persuasive.

Here's some terminology. In this example there were two related events of importance:

A: A video game player purchased Ultimate Hockey.

B: A video game player saw the ad.

The first probability we computed was the estimated probability that a video game player purchased Ultimate Hockey *given that* he or she saw the ad. We call the latter probability the (estimated) **probability of *A*, given *B***, and we write it as $P(A \mid B)$. We call $P(A \mid B)$ a **conditional probability**—it is the probability of *A* under the condition that B occurred. Put another way, it is the probability of A occurring if the sample space is reduced to just those outcomes in B.

$$P(\text{Purchased game } given \text{ } that \text{ saw the ad}) = P(A \mid B) \approx .33$$

The second probability we computed was the estimated probability that a video game player purchased Ultimate Hockey *given that* he or she did not see the ad, or the **probability of A, given B'**.

$$P(\text{Purchased game } given \text{ } that \text{ did not see the ad}) = P(A \mid B') \approx .12$$

Calculating Conditional Probabilities

How do we calculate conditional probabilities? In the example above we used the ratio

$$P(A \mid B) = \frac{\text{Number of people who saw the ad and bought the game}}{\text{Total number of people who saw the ad}}.$$

The numerator is the frequency of $A \cap B$, and the denominator is the frequency of B:

$$P(A \mid B) = \frac{fr(A \cap B)}{fr(B)}.$$

Now, we can write this formula in another way:

$$P(A \mid B) = \frac{fr(A \cap B)}{fr(B)} = \frac{fr(A \cap B)/N}{fr(B)/N} = \frac{P(A \cap B)}{P(B)}.$$

We therefore have the following definition, which applies to general probability distributions.

Conditional Probability

If A and B are events with $P(B) \neq 0$, then the probability of A given B is

$$P(A \mid B) = \frac{P(A \cap B)}{P(B)}.$$

Visualizing Conditional Probability

In the figure, $P(A \mid B)$ is represented by the fraction of B that is covered by A.

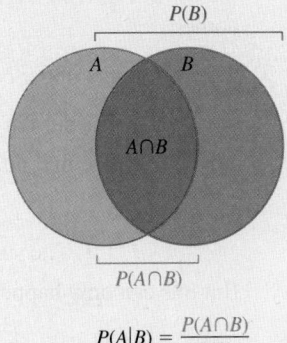

$$P(A|B) = \frac{P(A \cap B)}{P(B)}$$

Quick Examples

1. If there is a 50% chance of rain (R) and a 10% chance of both rain and lightning (L), then the probability of lightning, given that it rains, is

$$P(L \mid R) = \frac{P(L \cap R)}{P(R)} = \frac{.10}{.50} = .20.$$

Here are two more ways to express the result:

- If it rains, the probability of lightning is .20.
- Assuming that it rains, there is a 20% chance of lightning.

2. Referring to the Cyber Video data at the beginning of this section, the probability that a video game player did not purchase the game (A'), given that she did not see the ad (B'), is

$$P(A' \mid B') = \frac{P(A' \cap B')}{P(B')} = \frac{1{,}500/2{,}000}{1{,}700/2{,}000} = \frac{15}{17} \approx .88.$$

Q: *Returning to the video game sales survey, how do we compute the* ordinary *probability of A, not "given" anything?*

A: We look at the event A that a randomly chosen game player purchased Ultimate Hockey *regardless of whether or not he or she saw the ad*. In the "Purchased Game" row we see that a total of 300 people purchased the game out of a total of 2,000 surveyed. Thus, the (estimated) probability of A is

$$P(A) = \frac{fr(A)}{N} = \frac{300}{2{,}000} = .15.$$

We sometimes refer to $P(A)$ as the **unconditional** probability of A to distinguish it from conditional probabilities like $P(A|B)$ and $P(A|B')$.

Now, let's see some more examples involving conditional probabilities.

EXAMPLE 1 Dice

If you roll a fair die twice and observe the numbers that face up, find the probability that the sum of the numbers is 8, given that the first number is 3.

Solution We begin by recalling that the sample space when we roll a fair die twice is the set $S = \{(1, 1), (1, 2), \ldots, (6, 6)\}$ containing the 36 different equally likely outcomes.

The two events under consideration are

A: The sum of the numbers is 8.

B: The first number is 3.

We also need

$A \cap B$: The sum of the numbers is 8 and the first number is 3.

But this can only happen in one way: $A \cap B = \{(3, 5)\}$. From the formula, then,

$$P(A \mid B) = \frac{P(A \cap B)}{P(B)} = \frac{1/36}{6/36} = \frac{1}{6}.$$

➡ **Before we go on...** There is another way to think about Example 1. When we say that the first number is 3, we are restricting the sample space to the six outcomes (3, 1), (3, 2), . . . , (3, 6), all still equally likely. Of these six, only one has a sum of 8, so the probability of the sum being 8, given that the first number is 3, is 1/6. ■

Notes

1. Remember that, in the expression $P(A \mid B)$, A is the event whose probability you want, given that you know the event B has occurred.
2. From the formula, notice that $P(A \mid B)$ is not defined if $P(B) = 0$. Could $P(A \mid B)$ make any sense if the event B were impossible? ■

EXAMPLE 2 School and Work

A survey[50] of the high school graduating class of 2010, conducted by the Bureau of Labor Statistics, found that, if a graduate went on to college, there was a 40% chance that he or she would work at the same time. On the other hand, there was a 68% chance that a randomly selected graduate would go on to college. What is the probability that a graduate went to college and work at the same time?

Solution To understand what the question asks and what information is given, it is helpful to rephrase everything using the standard wording "*the probability that ___* " and "*the probability that ___ given that ___.*" Now we have, "The probability that a graduate worked, given that the graduate went on to college, equals .40. (See Figure 8.) The probability that a graduate went on to college is .68." The events in question are as follows:

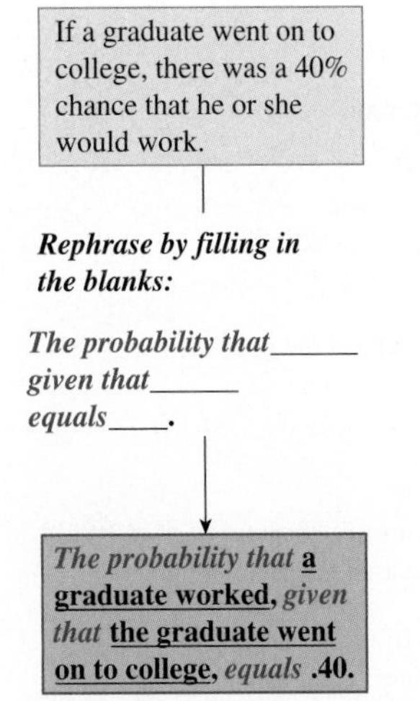

> W: A high school graduate went on to work.
>
> C: A high school graduate went on to college.

From our rephrasing of the question we can write:

$$P(W \mid C) = .40. \qquad P(C) = .68. \qquad \text{Find } P(W \cap C).$$

The definition

$$P(W \mid C) = \frac{P(W \cap C)}{P(C)}$$

can be used to find $P(W \cap C)$:

$$P(W \cap C) = P(W \mid C)P(C)$$
$$= (.40)(.68) \approx .27.$$

Thus there is a 27% chance that a member of the high school graduating class of 2010 went on to college and work at the same time.

If a graduate went on to college, there was a 40% chance that he or she would work.

|

Rephrase by filling in the blanks:

The probability that_____ given that_____ equals____.

↓

The probability that **a graduate worked,** *given that* **the graduate went on to college,** *equals* **.40.**

P(Worked | Went to college) = .40

Figure 8

The Multiplication Principle and Trees

In Example 2, we saw that the formula

$$P(A \mid B) = \frac{P(A \cap B)}{P(B)}$$

---

[50]Source: "College Enrollment and Work Activity of High School Graduates," U.S. Bureau of Labor Statistics (www.bls.gov/news.release/hsgec.htm).

can be used to calculate $P(A \cap B)$ if we rewrite the formula in the following form, known as the **multiplication principle for conditional probability**:

Multiplication Principle for Conditional Probability

If A and B are events, then

$$P(A \cap B) = P(A \mid B)P(B).$$

Quick Example

If there is a 50% chance of rain (R) and a 20% chance of a lightning (L) if it rains, then the probability of both rain and lightning is

$$P(R \cap L) = P(L \mid R)P(R) = (.20)(.50) = .10.$$

The multiplication principle is often used in conjunction with **tree diagrams**. Let's return to Cyber Video Games, Inc., and its television ad campaign. Its marketing survey was concerned with the following events:

A: A video game player purchased Ultimate Hockey.

B: A video game player saw the ad.

We can illustrate the various possibilities by means of the two-stage "tree" shown in Figure 9.

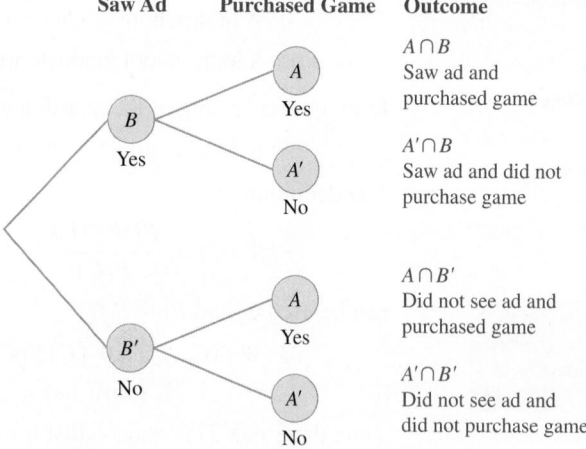

Figure 9

Consider the outcome $A \cap B$. To get there from the starting position on the left, we must first travel up to the B node. (In other words, B must occur.) Then we must travel up the branch from the B node to the A node. We are now going to associate a probability with each branch of the tree: the probability of traveling along that branch *given that we have gotten to its beginning node*. For instance, the probability of traveling up the branch from the starting position to the B node is $P(B) = 300/2,000 = .15$ (see the data in the survey). The probability of going up the branch from the B node to the A node is the probability that A occurs, given that B has occurred. In other words, it is the *conditional* probability $P(A \mid B) \approx .33$.

(We calculated this probability at the beginning of the section.) The probability of the outcome $A \cap B$ can then be computed using the multiplication principle:

$$P(A \cap B) = P(B)P(A \mid B) \approx (.15)(.33) \approx .05.$$

In other words, *to obtain the probability of the outcome $A \cap B$, we multiply the probabilities on the branches leading to that outcome* (Figure 10).

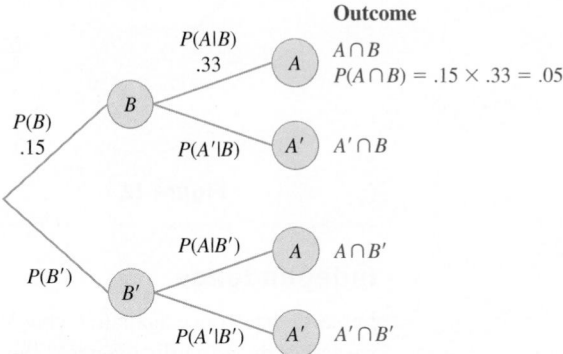

Figure 10

The same argument holds for the remaining three outcomes, and we can use the table given at the beginning of this section to calculate all the conditional probabilities shown in Figure 11.

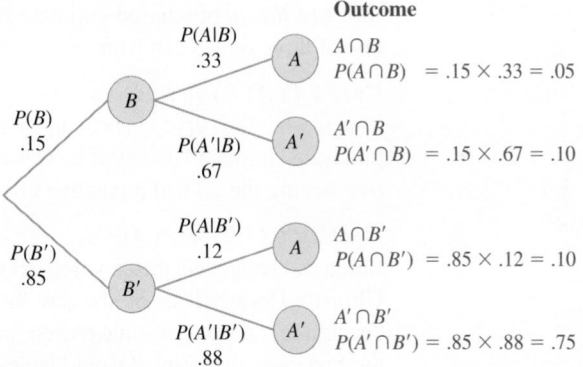

Figure 11

Note The sum of the probabilities on the branches leaving any node is always 1 (why?). This observation often speeds things up because after we have labeled one branch (or, all but one, if a node has more than two branches leaving it), we can easily label the remaining one. ∎

EXAMPLE 3 **Unfair Coins**

An experiment consists of tossing two coins. The first coin is fair, while the second coin is twice as likely to land with heads facing up as it is with tails facing up. Draw a tree diagram to illustrate all the possible outcomes, and use the multiplication principle to compute the probabilities of all the outcomes.

Solution A quick calculation shows that the probability distribution for the second coin is $P(\text{H}) = 2/3$ and $P(\text{T}) = 1/3$. (How did we get that?) Figure 12 shows the tree diagram and the calculations of the probabilities of the outcomes.

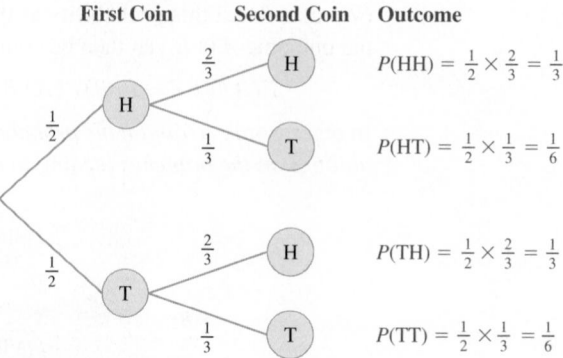

Figure 12

Independence

Let us go back once again to Cyber Video Games, Inc., and its ad campaign. How did we assess the ad's effectiveness? We considered the following events.

> *A*: A video game player purchased Ultimate Hockey.
>
> *B*: A video game player saw the ad.

We used the survey data to calculate $P(A)$, the probability that a video game player purchased Ultimate Hockey, and $P(A \mid B)$, the probability that a video game player *who saw the ad* purchased Ultimate Hockey. When these probabilities are compared, one of three things can happen.

Case 1 $P(A \mid B) > P(A)$
This is what the survey data actually showed: A video game player was more likely to purchase Ultimate Hockey if he or she saw the ad. This indicates that the ad is effective; seeing the ad had a positive effect on a player's decision to purchase the game.

Case 2 $P(A \mid B) < P(A)$
If this had happened, then a video game player would have been *less* likely to purchase Ultimate Hockey if he or she saw the ad. This would have indicated that the ad had "backfired"; it had, for some reason, put potential customers off. In this case, just as in the first case, the event *B* would have had an effect—a negative one—on the event *A*.

Case 3 $P(A \mid B) = P(A)$
In this case seeing the ad would have had absolutely no effect on a potential customer's buying Ultimate Hockey. Put another way, the probability of *A* occurring *does not depend* on whether *B* occurred or not. We say in a case like this that the events *A* and *B* are **independent**.

In general, we say that two events *A* and *B* are independent if $P(A \mid B) = P(A)$. When this happens, we have

$$P(A) = P(A \mid B) = \frac{P(A \cap B)}{P(B)}$$

so

$$P(A \cap B) = P(A)P(B).$$

＊ We shall only discuss the independence of two events in cases where their probabilities are both nonzero.

Conversely, if $P(A \cap B) = P(A)P(B)$, then, assuming $P(B) \neq 0,$＊ $P(A) = P(A \cap B)/P(B) = P(A \mid B)$. Thus, saying that $P(A) = P(A \mid B)$ is the same as saying that $P(A \cap B) = P(A)P(B)$. Also, we can switch *A* and *B* in this last formula and conclude that saying that $P(A \cap B) = P(A)P(B)$ is the same as saying that $P(B \mid A) = P(B)$.

Independent Events

The events A and B are **independent** if

$$P(A \cap B) = P(A)P(B).$$

Equivalent formulas (assuming neither A nor B is impossible) are

$$P(A \mid B) = P(A)$$

and $\quad P(B \mid A) = P(B).$

If two events A and B are not independent, then they are **dependent**.

The property $P(A \cap B) = P(A)P(B)$ can be extended to three or more independent events. If, for example, A, B, and C are three mutually independent events (that is, each one of them is independent of each of the other two and of their intersection), then, among other things,

$$P(A \cap B \cap C) = P(A)P(B)P(C).$$

To test for independence, calculate the three quantities $P(A)$, $P(B)$, and $P(A \cap B)$ separately, and then see if $P(A \cap B) = P(A) \cdot P(B)$.

Quick Examples

1. If A and B are independent, and if A has a probability of .2 and B has a probability of .3, then $A \cap B$ has a probability of $(.2)(.3) = .06$.

2. Let us assume that the phase of the moon has no effect on whether or not my newspaper is delivered. The probability of a full moon (M) on a randomly selected day is about .034, and the probability that my newspaper will be delivered (D) on the random day is .20. Therefore, the probability that it is a full moon and my paper is delivered is

$$P(M \cap D) = P(M)P(D) = (.034)(.20) = .0068.$$

Testing for Independence

To check whether two events A and B are independent, we compute $P(A)$, $P(B)$, and $P(A \cap B)$. If $P(A \cap B) = P(A)P(B)$, the events are independent; otherwise, they are dependent. Sometimes it is obvious that two events, by their nature, are independent, so a test is not necessary. For example, the event that a die you roll comes up 1 is clearly independent of whether or not a coin you toss comes up heads.

* Two numbers have the **same parity** if both are even or both are odd. Otherwise, they have **opposite parity**.

Quick Examples

1. Roll two distinguishable dice (one red, one green) and observe the numbers that face up.

   A: The red die is even; $P(A) = \dfrac{18}{36} = \dfrac{1}{2}.$

   B: The dice have the same parity*; $P(B) = \dfrac{18}{36} = \dfrac{1}{2}.$

   $A \cap B$: Both dice are even; $P(A \cap B) = \dfrac{9}{36} = \dfrac{1}{4}.$

   $P(A \cap B) = P(A)P(B)$, and so A and B are independent.

2. Roll two distinguishable dice and observe the numbers that face up.

A: The sum of the numbers is 6; $P(A) = \dfrac{5}{36}$.

B: Both numbers are odd; $P(B) = \dfrac{9}{36} = \dfrac{1}{4}$.

$A \cap B$: The sum is 6, and both are odd; $P(A \cap B) = \dfrac{3}{36} = \dfrac{1}{12}$.

$P(A \cap B) \neq P(A)P(B)$, and so A and B are dependent.

EXAMPLE 4 Weather Prediction

According to the weather service, there is a 50% chance of rain in New York and a 30% chance of rain in Honolulu. Assuming that New York's weather is independent of Honolulu's, find the probability that it will rain in at least one of these cities.

Solution We take A to be the event that it will rain in New York and B to be the event that it will rain in Honolulu. We are asked to find the probability of $A \cup B$, the event that it will rain in at least one of the two cities. We use the addition principle:

$$P(A \cup B) = P(A) + P(B) - P(A \cap B).$$

We know that $P(A) = .5$ and $P(B) = .3$. But what about $P(A \cap B)$? Because the events A and B are independent, we can compute

$$P(A \cap B) = P(A)P(B)$$
$$= (.5)(.3) = .15.$$

Thus,

$$P(A \cup B) = P(A) + P(B) - P(A \cap B)$$
$$= .5 + .3 - .15$$
$$= .65.$$

So, there is a 65% chance that it will rain either in New York or in Honolulu (or in both).

EXAMPLE 5 Roulette

You are playing roulette and have decided to leave all 10 of your $1 chips on black for five consecutive rounds, hoping for a sequence of five blacks which, according to the rules, will leave you with $320. There is a 50% chance of black coming up on each spin, ignoring the complicating factor of zero or double zero. What is the probability that you will be successful?

Solution Because the roulette wheel has no memory, each spin is independent of the others. Thus, if A_1 is the event that black comes up the first time, A_2 the event that it comes up the second time, and so on, then

$$P(A_1 \cap A_2 \cap A_3 \cap A_4 \cap A_5) = P(A_1)P(A_2)P(A_3)P(A_4)P(A_5) = \left(\frac{1}{2}\right)^5 = \frac{1}{32}.$$

The next example is a version of a well known "brain teaser" that forces one to think carefully about conditional probability.

EXAMPLE 6 **Legal Argument**

A man was arrested for attempting to smuggle a bomb on board an airplane. During the subsequent trial, his lawyer claimed that, by means of a simple argument, she would prove beyond a shadow of a doubt that her client was not only innocent of any crime, but was in fact contributing to the safety of the other passengers on the flight. This was her eloquent argument: "Your Honor, first of all, my client had absolutely no intention of setting off the bomb. As the record clearly shows, the detonator was unarmed when he was apprehended. In addition—and your Honor is certainly aware of this—there is a small but definite possibility that there will be a bomb on any given flight. On the other hand, the chances of there being *two* bombs on a flight are so remote as to be negligible. There is in fact no record of this having *ever* occurred. Thus, because my client had already brought one bomb on board (with no intention of setting it off) and because we have seen that the chances of there being a second bomb on board were vanishingly remote, it follows that the flight was far safer as a result of his action! I rest my case." This argument was so elegant in its simplicity that the judge acquitted the defendant. Where is the flaw in the argument? (Think about this for a while before reading the solution.)

Solution The lawyer has cleverly confused the phrases "two bombs on board" and "a second bomb on board." To pinpoint the flaw, let us take B to be the event that there is one bomb on board a given flight, and let A be the event that there are two independent bombs on board. Let us assume for argument's sake that $P(B) = 1/1,000,000 = .000\,001$. Then the probability of the event A is

$$(.000\,001)(.000\,001) = .000\,000\,000\,001.$$

This *is* vanishingly small, as the lawyer contended. It was at this point that the lawyer used a clever maneuver: She assumed in concluding her argument that the probability of having two bombs on board was the same as the probability of having a *second* bomb on board. But to say that there is a *second* bomb on board is to imply that there already is one bomb on board. This is therefore a *conditional* event: the event that there are two bombs on board, *given that there is already one bomb on board*. Thus, the probability that there is a second bomb on board is the probability that there are two bombs on board, given that there is already one bomb on board, which is

$$P(A \mid B) = \frac{P(A \cap B)}{P(B)} = \frac{.000\,000\,000\,001}{.000\,001} = .000\,001.$$

> ✳ If we want to be picky, there was a *slight* decrease in the probability of a second bomb because there was one less seat for a potential second bomb bearer to occupy. In terms of our analysis, this is saying that the event of one passenger with a bomb and the event of a second passenger with a bomb are not completely independent.

In other words, it is the same as the probability of there being a single bomb on board to begin with! Thus the man's carrying the bomb onto the plane did not improve the flight's safety at all.✳

7.5   EXERCISES

▼ more advanced   ◆ challenging
T indicates exercises that should be solved using technology

In Exercises 1–10, compute the indicated quantity.

1. $P(B) = .5$, $P(A \cap B) = .2$. Find $P(A \mid B)$.

2. $P(B) = .6$, $P(A \cap B) = .3$. Find $P(A \mid B)$.

3. $P(A \mid B) = .2$, $P(B) = .4$. Find $P(A \cap B)$.

4. $P(A \mid B) = .1$, $P(B) = .5$. Find $P(A \cap B)$.

5. $P(A \mid B) = .4$, $P(A \cap B) = .3$. Find $P(B)$.

6. $P(A \mid B) = .4$, $P(A \cap B) = .1$. Find $P(B)$.

7. $P(A) = .5$, $P(B) = .4$. A and B are independent. Find $P(A \cap B)$.

8. $P(A) = .2$, $P(B) = .2$. A and B are independent. Find $P(A \cap B)$.

9. $P(A) = .5$, $P(B) = .4$. A and B are independent. Find $P(A \mid B)$.

10. $P(A) = .3$, $P(B) = .6$. A and B are independent. Find $P(B \mid A)$.

In Exercises 11–14, supply the missing quantities.

11.

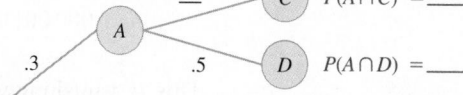

12.

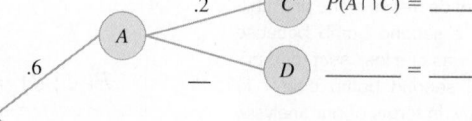
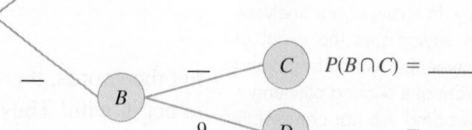

13.

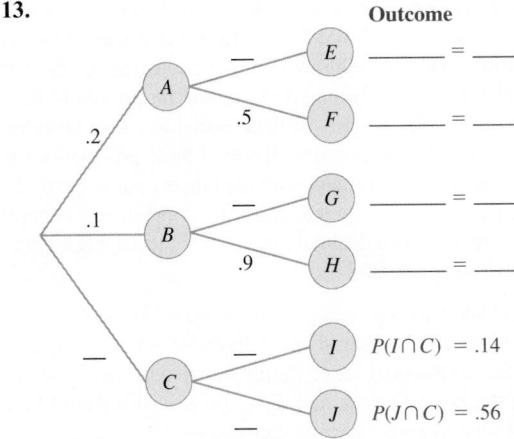

Outcome

$P(I \cap C) = .14$

$P(J \cap C) = .56$

14.

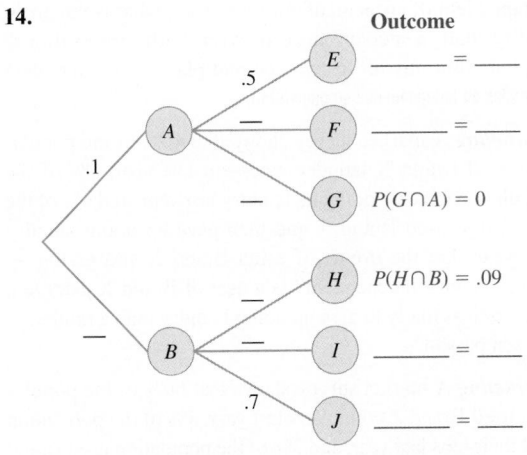

Outcome

$P(G \cap A) = 0$

$P(H \cap B) = .09$

In Exercises 15–20, find the conditional probabilities of the indicated events when two fair dice (one red and one green) are rolled. HINT [See Example 1.]

15. The sum is 5, given that the green one is not a 1.

16. The sum is 6, given that the green one is either 4 or 3.

17. The red one is 5, given that the sum is 6.

18. The red one is 4, given that the green one is 4.

19. The sum is 5, given that the dice have opposite parity.

20. The sum is 6, given that the dice have opposite parity.

Exercises 21–26 require the use of counting techniques from the last chapter. A bag contains three red marbles, two green ones, one fluorescent pink one, two yellow ones, and two orange ones. Suzan grabs four at random. Find the probabilities of the indicated events.

21. She gets all the red ones, given that she gets the fluorescent pink one.

22. She gets all the red ones, given that she does not get the fluorescent pink one.

23. She gets none of the red ones, given that she gets the fluorescent pink one.

24. She gets one of each color other than fluorescent pink, given that she gets the fluorescent pink one.

25. She gets one of each color other than fluorescent pink, given that she gets at least one red one.

26. She gets at least two red ones, given that she gets at least one green one.

In Exercises 27–30, say whether the given pairs of events are independent, mutually exclusive, or neither.

27. *A*: Your new skateboard design is a success.

   B: Your new skateboard design is a failure.

28. *A*: Your new skateboard design is a success.

   B: There is life in the Andromeda galaxy.

29. *A*: Your new skateboard design is a success.

   B: Your competitor's new skateboard design is a failure.

30. *A*: Your first coin flip results in heads.

   B: Your second coin flip results in heads.

In Exercises 31–36, two dice (one red and one green) are rolled, and the numbers that face up are observed. Test the given pairs of events for independence. HINT [See Quick Example on Testing for Independence, page 503.]

31. *A*: The red die is 1, 2, or 3; *B*: The green die is even.

32. *A*: The red die is 1; *B*: The sum is even.

33. *A*: Exactly one die is 1; *B*: The sum is even.

34. *A*: Neither die is 1 or 6; *B*: The sum is even.

35. *A*: Neither die is 1; *B*: Exactly one die is 2.

36. *A*: Both dice are 1; *B*: Neither die is 2.

37. If a coin is tossed 11 times, find the probability of the sequence H, T, T, H, H, H, T, H, H, T, T. HINT [See Example 5.]

38. If a die is rolled four times, find the probability of the sequence 4, 3, 2, 1. HINT [See Example 5.]

In Exercises 39–44, fill in the blanks using the named events. HINT [See Example 2 and also the FAQ on page 506.]

39. 10% of all Anchovians detest anchovies (*D*), whereas 30% of all married Anchovians (*M*) detest them. $P(\_\_) = \_\_$; $P(\_\_ | \_\_) = \_\_$

40. 95% of all music composers can read music (*M*), whereas 99% of all classical music composers (*C*) can read music. $P(\_\_) = \_\_$; $P(\_\_ | \_\_) = \_\_$

41. 30% of all lawyers who lost clients (*L*) were antitrust lawyers (*A*), whereas 10% of all antitrust lawyers lost clients. $P(\_\_ | \_\_) = \_\_$; $P(\_\_ | \_\_) = \_\_$

42. 2% of all items bought on my auction site (*B*) were works of art (*A*), whereas only 1% of all works of art on the site were bought. $P(\_\_ | \_\_) = \_\_$; $P(\_\_ | \_\_) = \_\_$

43. 55% of those who go out in the midday sun (M) are Englishmen (E) whereas only 5% of those who do not go out in the midday sun are Englishmen. $P(\_\_\_ \mid \_\_\_) = \_\_\_$; $P(\_\_\_ \mid \_\_\_) = \_\_\_$

44. 80% of those who have a Mac now (M) will purchase a Mac next time (X) whereas 20% of those who do not have a Mac now will purchase a Mac next time. $P(\_\_\_ \mid \_\_\_) = \_\_\_$; $P(\_\_\_ \mid \_\_\_) = \_\_\_$

APPLICATIONS

45. *Personal Bankruptcy* In 2004, the probability that a person in the United States would declare personal bankruptcy was .006. The probability that a person in the United States would declare personal bankruptcy and had recently experienced a "big three" event (loss of job, medical problem, or divorce or separation) was .005.[51] What was the probability that a person had recently experienced one of the "big three" events, given that she had declared personal bankruptcy? (Round your answer to one decimal place.)

46. *Personal Bankruptcy* In 2004, the probability that a person in the United States would declare personal bankruptcy was .006. The probability that a person in the United States would declare personal bankruptcy and had recently overspent credit cards was .0004.[52] What was the probability that a person had recently overspent credit cards given that he had declared personal bankruptcy?

47. *Existing Home Sales* During the year ending February 28, 2011, there were approximately 4.9 million sales of existing homes in the United States, of which 1.2 million were sold in the West. In February 2011 there were a total of 293,000 existing homes sold in the United States, of which 71,000 were sold in the West.[53]

a. Find the probability that a home sale in the year ending February 28, 2011, took place in the West, given that the home was sold in February 2011.

b. Find the probability that a home sale in the year ending February 28, 2011, took place in February 2011, given that it took place in the West.

48. *Existing Home Sales* Refer to the data given in the preceding exercise.

a. Find the probability that a home sale in the year ending February 28, 2011, took place outside the West, given that the home was sold in February 2011.

b. Find the probability that a home sale in the year ending February 28, 2011, took place in February 2011, given that it took place outside the West.

49. *Social Security* According to a *New York Times*/CBS poll released in March 2005, 79% agreed that it should be the government's responsibility to provide a decent standard of living for the elderly, and 43% agreed that it would be a good idea to invest part of their Social Security taxes on their own.[54] If agreement with one of these propositions is independent of agreement with the other, what is the probability that a person agreed with both propositions? (Round your answer to two decimal places.) HINT [See Quick Examples on Independence on page 503.]

50. *Social Security* According to *The New York Times*/CBS poll of March 2005 referred to in Exercise 43, 49% agreed that Social Security taxes should be raised if necessary to keep the system afloat, and 43% agreed that it would be a good idea to invest part of their Social Security taxes on their own.[55] If agreement with one of these propositions is independent of agreement with the other, what is the probability that a person agreed with both propositions? (Round your answer to two decimal places.) HINT [See Quick Examples on Independence on page 503.]

51. *Marketing* A market survey shows that 40% of the population used Brand X laundry detergent last year, 5% of the population gave up doing its laundry last year, and 4% of the population used Brand X and then gave up doing laundry last year. Are the events of using Brand X and giving up doing laundry independent? Is a user of Brand X detergent more or less likely to give up doing laundry than a randomly chosen person?

52. *Marketing* A market survey shows that 60% of the population used Brand Z computers last year, 5% of the population quit their jobs last year, and 3% of the population used Brand Z computers and then quit their jobs. Are the events of using Brand Z computers and quitting your job independent? Is a user of Brand Z computers more or less likely to quit a job than a randomly chosen person?

53. *Road Safety* In 1999, the probability that a randomly selected vehicle would be involved in a deadly tire-related accident was approximately 3×10^{-6}, whereas the probability that a tire-related accident would prove deadly was .02.[56] What was the probability that a vehicle would be involved in a tire-related accident?

54. *Road Safety* In 1998, the probability that a randomly selected vehicle would be involved in a deadly tire-related accident was approximately 2.8×10^{-6}, while the probability that a tire-related accident would prove deadly

---

[51] Probabilities are approximate. Source: *New York Times*, March 13, 2005, p, WK3.

[52] Ibid. The .0004 figure is an estimate by the authors.

[53] Source: National Association of Realtors (www.realtor.org).

[54] Source: *New York Times*, March 3, 2005, p. A20.

[55] *Ibid.*

[56] The original data reported three tire-related deaths per million vehicles. Source: *New York Times* analysis of National Traffic Safety Administration crash data/Polk Company vehicle registration data/ *New York Times*, Nov. 22, 2000, p. C5.

was .016.[57] What was the probability that a vehicle would be involved in a tire-related accident?

Publishing *Exercises 55–62 are based on the following table, which shows the results of a survey of 100 authors by a publishing company.*

	New Authors	Established Authors	Total
Successful	5	25	30
Unsuccessful	15	55	70
Total	20	80	100

Compute the following conditional probabilities:

55. An author is established, given that she is successful.

56. An author is successful, given that he is established.

57. An author is unsuccessful, given that he is a new author.

58. An author is a new author, given that she is unsuccessful.

59. An author is unsuccessful, given that she is established.

60. An author is established, given that he is unsuccessful.

61. An unsuccessful author is established.

62. An established author is successful.

In Exercises 63–68, draw an appropriate tree diagram and use the multiplication principle to calculate the probabilities of all the outcomes. **HINT** [See Example 3.]

63. *Sales* Each day, there is a 40% chance that you will sell an automobile. You know that 30% of all the automobiles you sell are two-door models, and the rest are four-door models.

64. *Product Reliability* You purchase Brand X memory chips one quarter of the time and Brand Y memory chips the rest of the time. Brand X memory chips have a 1% failure rate, while Brand Y memory chips have a 3% failure rate.

65. *Car Rentals* Your auto rental company rents out 30 small cars, 24 luxury sedans, and 46 slightly damaged "budget" vehicles. The small cars break down 14% of the time, the luxury sedans break down 8% of the time, and the "budget" cars break down 40% of the time.

66. *Travel* It appears that there is only a one in five chance that you will be able to take your spring vacation to the Greek Islands. If you are lucky enough to go, you will visit either Corfu (20% chance) or Rhodes. On Rhodes, there is a 20% chance of meeting a tall dark stranger, while on Corfu, there is no such chance.

67. *Weather Prediction* There is a 50% chance of rain today and a 50% chance of rain tomorrow. Assuming that the event that it rains today is independent of the event that it rains tomorrow, draw a tree diagram showing the probabilities of all outcomes. What is the probability that there will be no rain today or tomorrow?

68. *Weather Prediction* There is a 20% chance of snow today and a 20% chance of snow tomorrow. Assuming that the event that it snows today is independent of the event that it snows tomorrow, draw a tree diagram showing the probabilities of all outcomes. What is the probability that it will snow by the end of tomorrow?

Education and Employment *Exercises 69–78 are based on the following table, which shows U.S. employment figures for August 2011, broken down by educational attainment.[58] All numbers are in millions, and represent civilians aged 25 years and over. Those classed as "not in labor force" were not employed nor actively seeking employment. Round all answers to two decimal places.*

	Employed	Unemployed	Not in Labor Force	Total
Less Than High School Diploma	9.9	1.6	17.7	29.2
High School Diploma Only	34.1	3.5	24.4	62.0
Some College or Associate's Degree	34.4	3.0	15.6	53.0
Bachelor's Degree or Higher	46.4	2.0	12.2	60.6
Total	124.8	10.1	69.9	204.8

69. Find the probability that a person was employed, given that the person had a bachelor's degree or higher.

70. Find the probability that a person was employed, given that the person had attained less than a high school diploma.

71. Find the probability that a person had a bachelor's degree or higher, given that the person was employed.

72. Find the probability that a person had attained less than a high school diploma, given that the person was employed.

73. ▼ Find the probability that a person who had not completed a bachelor's degree or higher was not in the labor force.

74. ▼ Find the probability that a person who had completed at least a high school diploma was not in the labor force.

75. ▼ Find the probability that a person who had completed a bachelor's degree or higher and was in the labor force was employed.

[57] The original data reported 2.8 tire-related deaths per million vehicles. Source: *Ibid.*

[58] Source: Bureau of Labor Statistics (www.bls.gov).

76. ▼ Find the probability that a person who had completed less than a high school diploma and was in the labor force was employed.

77. ▼ Your friend claims that an unemployed person is more likely to have a high school diploma only than an employed person. Respond to this claim by citing actual probabilities.

78. ▼ Your friend claims that a person not in the labor force is more likely to have less than a high school diploma than an employed person. Respond to this claim by citing actual probabilities.

79. ▼ *Airbag Safety* According to a study conducted by the Harvard School of Public Health, a child seated in the front seat who was wearing a seatbelt was 31% more likely to be killed in an accident if the car had an air bag that deployed than if it did not.[59] Let the sample space S be the set of all accidents involving a child seated in the front seat wearing a seatbelt. Let K be the event that the child was killed and let D be the event that the airbag deployed. Fill in the missing terms and quantities: $P(\_\_ \mid \_\_) = \_\_ \times P(\_\_ \mid \_\_)$.
HINT [When we say "A is 31% more likely than B" we mean that the probability of A is 1.31 times the probability of B.]

80. ▼ *Airbag Safety* According to the study cited in Exercise 73, a child seated in the front seat not wearing a seatbelt was 84% more likely to be killed in an accident if the car had an air bag that deployed than if it did not.[60] Let the sample space S be the set of all accidents involving a child seated in the front seat not wearing a seatbelt. Fill in the missing terms and quantities: $P(\_\_ \mid \_\_) = \_\_ \times P(\_\_ \mid \_\_)$.
HINT [When we say "A is 84% more likely than B" we mean that the probability of A is 1.84 times the probability of B.]

81. ▼ *Productivity* A company wishes to enhance productivity by running a one-week training course for its employees. Let T be the event that an employee participated in the course, and let I be the event that an employee's productivity improved the week after the course was run.

   a. Assuming that the course has a positive effect on productivity, how are $P(I|T)$ and $P(I)$ related?
   b. If T and I are independent, what can one conclude about the training course?

82. ▼ *Productivity* Consider the events T and I in the preceding exercise.

   a. Assuming that everyone who improved took the course but that not everyone took the course, how are $P(T \mid I)$ and $P(T)$ related?
   b. If half the employees who improved took the course and half the employees took the course, are T and I independent?

83. ▼ *Internet Use in 2000* The following pie chart shows the percentage of the population that used the Internet in 2000, broken down further by family income, and based on a survey taken in August 2000.[61]

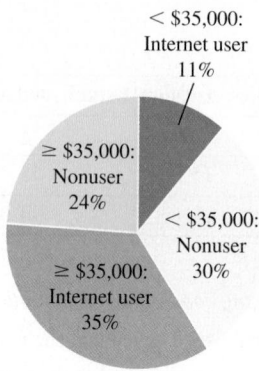

   a. Determine the probability that a randomly chosen person was an Internet user, given that his or her family income was at least $35,000.
   b. Based on the data, was a person more likely to be an Internet user if his or her family income was less than $35,000 or $35,000 or more? (Support your answer by citing the relevant conditional probabilities.)

84. ▼ *Internet Use in 2001* Repeat Exercise 83 using the following pie chart, which shows the results of a similar survey taken in September 2001.[62]

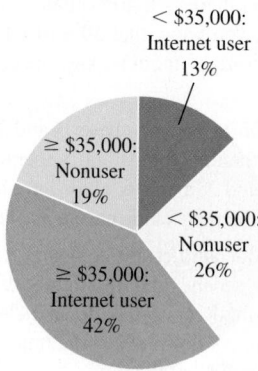

[59] The study was conducted by Dr. Segul-Gomez at the Harvard School of Public Health. Source: *New York Times*, December 1, 2000, p. F1.
[60] *Ibid.*

[61] Source: *Falling Through the Net: Toward Digital Inclusion, A Report on Americans' Access to Technology Tools*, U.S. Department of Commerce, October, 2000. Available at www.ntia.doc.gov/ntiahome/fttn00/contents00.html.

[62] Source: *A Nation Online: How Americans Are Expanding Their Use of the Internet*, U.S. Department of Commerce, February, 2002. Available at www.ntia.doc.gov/ntiahome/dn/index.html.

Auto Theft *Exercises 85–90 are based on the following table, which shows the probability that an owner of the given model would report his or her vehicle stolen in a one-year period.*[63]

Brand	Jeep Wrangler	Suzuki Sidekick (two-door)	Toyota Land Cruiser	Geo Tracker (two-door)	Acura Integra (two-door)
Probability	.0170	.0154	.0143	.0142	.0123
Brand	**Mitsubishi Montero**	**Acura Integra (four-door)**	**BMW 3-series (two-door)**	**Lexus GS300**	**Honda Accord (two-door)**
Probability	.0108	.0103	.0077	.0074	.0070

In an experiment in which a vehicle is selected, consider the following events:

R: The vehicle was reported stolen.
J: The vehicle was a Jeep Wrangler.
A2: The vehicle was an Acura Integra (two-door).
A4: The vehicle was an Acura Integra (four-door).
A: The vehicle was an Acura Integra (either two-door or four-door).

85. ▼ Fill in the blanks: $P(\_\_\_ \mid \_\_\_) = .0170$.

86. ▼ Fill in the blanks: $P(\_\_\_ \mid A4) = \_\_\_$.

87. ▼ Which of the following is true?

   (A) There is a 1.43% chance that a vehicle reported stolen was a Toyota Land Cruiser.
   (B) Of all the vehicles reported stolen, 1.43% of them were Toyota Land Cruisers.
   (C) Given that a vehicle was reported stolen, there is a .0143 probability that it was a Toyota Land Cruiser.
   (D) Given that a vehicle was a Toyota Land Cruiser, there was a 1.43% chance that it was reported stolen.

88. ▼ Which of the following is true?

   (A) $P(R \mid A) = .0123 + .0103 = .0226$
   (B) $P(R' \mid A2) = 1 - .0123 = .9877$
   (C) $P(A2 \mid A) = .0123/(.0123 + .0103) \approx .544$
   (D) $P(R \mid A2') = 1 - .0123 = .9877$

89. ▼ It is now January, and I own a BMW 3-series and a Lexus GS300. Because I house my vehicles in different places, the event that one of my vehicles gets stolen does not depend on the event that the other gets stolen. Compute each probability to six decimal places.

   a. Both my vehicles will get stolen this year.
   b. At least one of my vehicles will get stolen this year.

90. ▼ It is now December, and I own a Mitsubishi Montero and a Jeep Wrangler. Because I house my vehicles in different places, the event that one of my vehicles gets stolen does not depend on the event that the other gets stolen. I have just returned from a one-year trip to the Swiss Alps.

   a. What is the probability that my Montero, but not my Wrangler, has been stolen?

   b. Which is more likely: the event that my Montero was stolen or the event that *only* my Montero was stolen?

91. ▼ ***Drug Tests*** If 90% of the athletes who test positive for steroids in fact use them, and 10% of all athletes use steroids and test positive, what percentage of athletes test positive?

92. ▼ ***Fitness Tests*** If 80% of candidates for the soccer team pass the fitness test, and only 20% of all athletes are soccer team candidates who pass the test, what percentage of the athletes are candidates for the soccer team?

93. ▼ ***Food Safety*** According to a University of Maryland study of 200 samples of ground meats,[64] the probability that a sample was contaminated by *salmonella* was .20. The probability that a *salmonella*-contaminated sample was contaminated by a strain resistant to at least three antibiotics was .53. What was the probability that a ground meat sample was contaminated by a strain of *salmonella* resistant to at least three antibiotics?

94. ▼ ***Food Safety*** According to the study mentioned in Exercise 87,[65] the probability that a ground meat sample was contaminated by *salmonella* was .20. The probability that a *salmonella*-contaminated sample was contaminated by a strain resistant to at least one antibiotic was .84. What was the probability that a ground meat sample was contaminated by a strain of *salmonella* resistant to at least one antibiotic?

95. ◆ ***Food Safety*** According to a University of Maryland study of 200 samples of ground meats,[66] the probability that one of the samples was contaminated by *salmonella* was .20. The probability that a *salmonella*-contaminated sample was contaminated by a strain resistant to at least one antibiotic was .84, and the probability that a *salmonella*-contaminated sample was contaminated by a strain resistant to at least three antibiotics was .53. Find the probability that a ground meat sample that was contaminated by an antibiotic-resistant strain was contaminated by a strain resistant to at least three antibiotics.

96. ◆ ***Food Safety*** According to a University of Maryland study of 200 samples of ground meats,[67] the probability that

---

[63] Data are for insured vehicles, for 1995 to 1997 models except Wrangler, which is for 1997 models only. Source: Highway Loss Data Institute/*New York Times,* March 28, 1999, p. WK3.

[64] As cited in the *New York Times*, October 16, 2001, p. A12.
[65] *Ibid.*
[66] *Ibid.*
[67] *Ibid.*

a ground meat sample was contaminated by a strain of *salmonella* resistant to at least three antibiotics was .11. The probability that someone infected with any strain of *salmonella* will become seriously ill is .10. What is the probability that someone eating a randomly chosen ground meat sample will not become seriously ill with a strain of *salmonella* resistant to at least three antibiotics?

COMMUNICATION AND REASONING EXERCISES

97. Name three events, each independent of the others, when a fair coin is tossed four times.

98. Name three pairs of independent events when a pair of distinguishable and fair dice is rolled and the numbers that face up are observed.

99. You wish to ascertain the probability of an event E, but you happen to know that the event F has occurred. Is the probability you are seeking $P(E)$ or $P(E|F)$? Give the reason for your answer.

100. Your television advertising campaign seems to have been very persuasive: 10,000 people who saw the ad purchased your product, while only 2,000 people purchased the product without seeing the ad. Explain how additional data could show that your ad campaign was, in fact, unpersuasive.

101. You are having trouble persuading your friend Iliana that conditional probability is different from unconditional prob-

ability. She just said: "Look here, Saul, the probability of throwing a double-six is 1/36, and that's that! That probability is not affected by anything, including the 'given' that the sum is larger than 7." How do you persuade her otherwise?

102. Your other friend Giuseppe is spreading rumors that the conditional probability $P(E|F)$ is always bigger than $P(E)$. Is he right? (If he is, explain why; if not, give an example to prove him wrong.)

103. ▼ If $A \subseteq B$ and $P(B) \neq 0$, why is $P(A|B) = \dfrac{P(A)}{P(B)}$?

104. ▼ If $B \subseteq A$ and $P(B) \neq 0$, why is $P(A|B) = 1$?

105. ▼ Your best friend thinks that it is impossible for two mutually exclusive events with nonzero probabilities to be independent. Establish whether or not he is correct.

106. ▼ Another of your friends thinks that two mutually exclusive events with nonzero probabilities can never be dependent. Establish whether or not she is correct.

107. ◆ Show that if A and B are independent, then so are A' and B' (assuming none of these events has zero probability). HINT [$A' \cap B'$ is the complement of $A \cup B$.]

108. ◆ Show that if A and B are independent, then so are A and B' (assuming none of these events has zero probability). HINT [$P(B'|A) + P(B|A) = 1$.]

7.6 Bayes' Theorem and Applications

Should schools test their athletes for drug use? A problem with drug testing is that there are always false positive results, so one can never be certain that an athlete who tests positive is in fact using drugs. Here is a typical scenario.

EXAMPLE 1 Steroids Testing

Gamma Chemicals advertises its anabolic steroid detection test as being 95% effective at detecting steroid use, meaning that it will show a positive result on 95% of all anabolic steroid users. It also states that its test has a false positive rate of 6%. This means that the probability of a nonuser testing positive is .06. Estimating that about 10% of its athletes are using anabolic steroids, Enormous State University (ESU) begins testing its football players. The quarterback, Hugo V. Huge, tests positive and is promptly dropped from the team. Hugo claims that he is not using anabolic steroids. How confident can we be that he is not telling the truth?

Solution There are two events of interest here: the event T that a person tests positive, and the event A that the person tested uses anabolic steroids. Here are the probabilities we are given:

$$P(T\,|\,A) = .95$$
$$P(T\,|\,A') = .06$$
$$P(A) = .10$$

We are asked to find $P(A \mid T)$, the probability that someone who tests positive is using anabolic steroids. We can use a tree diagram to calculate $P(A \mid T)$. The trick to setting up the tree diagram is to use as the first branching the events with *unconditional* probabilities we know. Because the only unconditional probability we are given is $P(A)$, we use A and A' as our first branching (Figure 13).

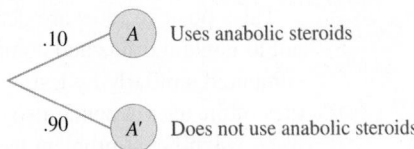

Figure 13

For the second branching, we use the outcomes of the drug test: positive (T) or negative (T'). The probabilities on these branches are conditional probabilities because they depend on whether or not an athlete uses steroids. (See Figure 14.) (We fill in the probabilities that are not supplied by remembering that the sum of the probabilities on the branches leaving any node must be 1.)

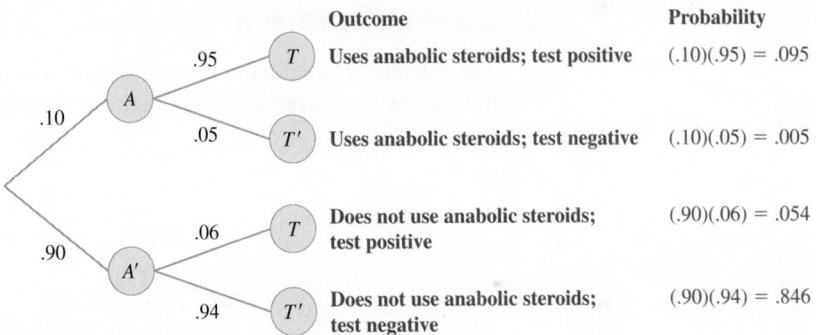

Figure 14

We can now calculate the probability we are asked to find:

$$P(A \mid T) = \frac{P(A \cap T)}{P(T)} = \frac{P(\text{Uses anabolic steroids and tests positive})}{P(\text{Tests positive})}$$

$$= \frac{P(\text{Using } A \text{ and } T \text{ branches})}{\text{Sum of } P(\text{Using branches ending in } T)}.$$

From the tree diagram, we see that $P(A \cap T) = .095$. To calculate $P(T)$, the probability of testing positive, notice that there are two outcomes on the tree diagram that reflect a positive test result. The probabilities of these events are .095 and .054. Because these two events are mutually exclusive (an athlete either uses steroids or does not, but not both), the probability of a test being positive (ignoring whether or not steroids are used) is the sum of these probabilities, .149. Thus,

$$P(A \mid T) = \frac{.095}{.095 + .054} = \frac{.095}{.149} \approx .64.$$

Thus there is a 64% chance that a randomly selected athlete who tests positive, like Hugo, is using steroids. In other words, we can be 64% confident that Hugo is lying.

➡ **Before we go on...** Note that the correct answer in Example 1 is 64%, *not* the 94% we might suspect from the test's false positive rating. In fact, we can't answer the question asked without knowing the percentage of athletes who actually use steroids. For instance, if *no* athletes at all use steroids, then Hugo must be telling the truth, and so the test result has no significance whatsoever. On the other hand, if *all* athletes use steroids, then Hugo is definitely lying, regardless of the outcome of the test.

False positive rates are determined by testing a large number of samples known not to contain drugs and computing estimated probabilities. False negative rates are computed similarly by testing samples known to contain drugs. However, the accuracy of the tests depends also on the skill of those administering them. False positives were a significant problem when drug testing started to become common, with estimates of false positive rates for common immunoassay tests ranging from 10% to 30% on the high end,[*] but the accuracy has improved since then. Because of the possibility of false positive results, positive immunoassay tests need to be confirmed by the more expensive and much more reliable gas chromatograph/mass spectrometry (GC/MS) test. See also the National Collegiate Athletic Association's (NCAA) Drug-Testing Program Handbook, available at www.ncaa.org. (The section on Institutional Drug Testing addresses the problem of false positives.) ∎

[*] *Drug Testing in the Workplace,* ACLU Briefing Paper, 1996.

Bayes' Theorem

The calculation we used to answer the question in Example 1 can be recast as a formula known as **Bayes' theorem**. Figure 15 shows a general form of the tree we used in Example 1.

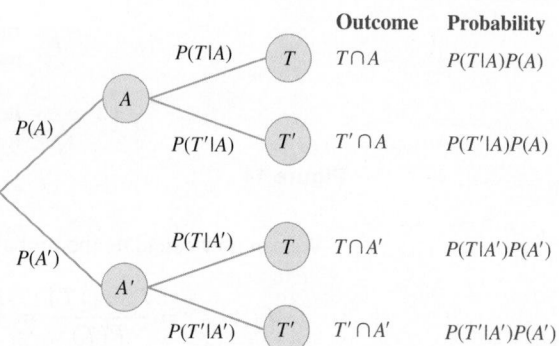

Figure 15

We first calculated

$$P(A \mid T) = \frac{P(A \cap T)}{P(T)}$$

as follows. We first calculated the numerator $P(A \cap T)$ using the multiplication principle:

$$P(A \cap T) = P(T \mid A)P(A).$$

We then calculated the denominator $P(T)$ by using the addition principle for mutually exclusive events together with the multiplication principle:

$$P(T) = P(A \cap T) + P(A' \cap T)$$
$$= P(T \mid A)P(A) + P(T \mid A')P(A').$$

Substituting gives

$$P(A \mid T) = \frac{P(T \mid A)P(A)}{P(T \mid A)P(A) + P(T \mid A')P(A')}.$$

This is the short form of Bayes' theorem.

Bayes' Theorem (Short Form)

If A and T are events, then

Bayes' Formula

$$P(A \mid T) = \frac{P(T \mid A)P(A)}{P(T \mid A)P(A) + P(T \mid A')P(A')}.$$

Using a Tree

$$P(A \mid T) = \frac{P(\text{Using } A \text{ and } T \text{ branches})}{\text{Sum of } P(\text{Using branches ending in } T)}$$

Quick Example

Let us calculate the probability that an athlete from Example 1 who tests positive is actually using steroids if only 5% of ESU athletes are using steroids. Thus,

$$P(T \mid A) = .95$$
$$P(T \mid A') = .06$$
$$P(A) = .05$$
$$P(A') = .95$$

and so

$$P(A \mid T) = \frac{P(T \mid A)P(A)}{P(T \mid A)P(A) + P(T \mid A')P(A')}$$
$$= \frac{(.95)(.05)}{(.95)(.05) + (.06)(.95)} \approx .45.$$

In other words, it is actually more likely that such an athlete does *not* use steroids than he does.[*]

---

[*] Without knowing the results of the test, we would have said that there was a probability of $P(A) = 0.05$ that the athlete is using steroids. The positive test result raises the probability to $P(A \mid T) = 0.45$, but the test gives too many false positives for us to be any more than 45% certain that the athlete is actually using steroids.

Remembering the Formula

Although the formula looks complicated at first sight, it is not hard to remember if you notice the pattern. Or, you could re-derive it yourself by thinking of the tree diagram.

The next example illustrates that we can use either a tree diagram or the Bayes' theorem formula.

EXAMPLE 2 Lie Detectors

[†] The reason for this is that many people show physical signs of distress when asked accusatory questions. Many people are nervous around police officers even if they have done nothing wrong.

The Sherlock Lie Detector Company manufactures the latest in lie detectors, and the Count-Your-Pennies (CYP) store chain is eager to use them to screen its employees for theft. Sherlock's advertising claims that the test misses a lie only once in every 100 instances. On the other hand, an analysis by a consumer group reveals 20% of people who are telling the truth fail the test anyway.[†] The local police department estimates that 1 out of every 200 employees has engaged in theft. When the CYP store first screened its employees, the test indicated Mrs. Prudence V. Good was lying when she claimed that she had never stolen from CYP. What is the probability that she was lying and had in fact stolen from the store?

Solution We are asked for the probability that Mrs. Good was lying, and in the preceding sentence we are told that the lie detector test showed her to be lying. So, we are looking for a conditional probability: the probability that she is lying, given that the lie detector test is positive. Now we can start to give names to the events:

   L: A subject is lying.

   T: The test is positive (indicated that the subject was lying).

We are looking for $P(L \mid T)$. We know that 1 out of every 200 employees engages in theft; let us assume that no employee admits to theft while taking a lie detector test, so the probability $P(L)$ that a test subject is lying is 1/200. We also know the false negative and false positive rates $P(T' \mid L)$ and $P(T \mid L')$.

Using a Tree Diagram
Figure 16 shows the tree diagram.

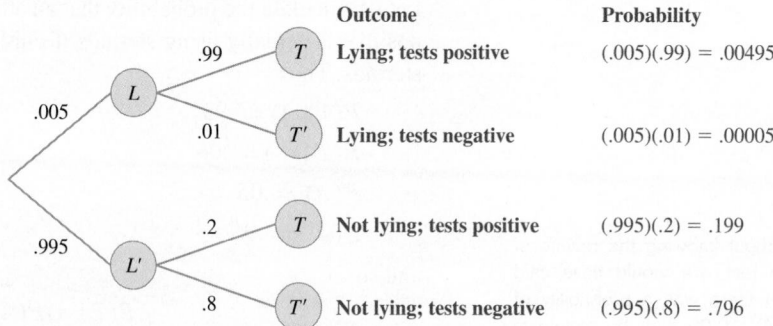

Figure 16

We see that

$$P(L \mid T) = \frac{P(\text{Using } L \text{ and } T \text{ branches})}{\text{Sum of } P(\text{Using branches ending in } T)}$$

$$= \frac{.00495}{.00495 + .199} \approx .024.$$

This means that there was only a 2.4% chance that poor Mrs. Good was lying and had stolen from the store!

Using Bayes' Theorem
We have

   $P(L) = .005$

   $P(T' \mid L) = .01$, from which we obtain

   $P(T \mid L) = .99$

   $P(T \mid L') = .2$

and so

$$P(L \mid T) = \frac{P(T \mid L)P(L)}{P(T \mid L)P(L) + P(T \mid L')P(L')} = \frac{(.99)(.005)}{(.99)(.005) + (.2)(.995)} \approx .024.$$

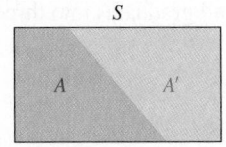

A and A′ form a partition of S.

Figure 17

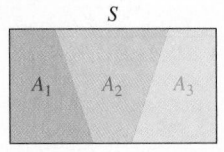

A_1, A_2, and A_3 form a partition of S.

Figure 18

Expanded Form of Bayes' Theorem

We have seen the "short form" of Bayes' theorem. What is the "long form?" To motivate an expanded form of Bayes' theorem, look again at the formula we've been using:

$$P(A \mid T) = \frac{P(T \mid A)P(A)}{P(T \mid A)P(A) + P(T \mid A')P(A')}.$$

The events A and A' form a **partition** of the sample space S; that is, their union is the whole of S and their intersection is empty (Figure 17).

The expanded form of Bayes' theorem applies to a partition of S into three or more events, as shown in Figure 18.

By saying that the events A_1, A_2, and A_3 form a partition of S, we mean that their union is the whole of S and the intersection of any two of them is empty, as in the figure. When we have a partition into three events as shown, the formula gives us $P(A_1 \mid T)$ in terms of $P(T \mid A_1)$, $P(T \mid A_2)$, $P(T \mid A_3)$, $P(A_1)$, $P(A_2)$, and $P(A_3)$.

Bayes' Theorem (Expanded Form)

If the events A_1, A_2, and A_3 form a partition of the sample space S, then

$$P(A_1 \mid T) = \frac{P(T \mid A_1)P(A_1)}{P(T \mid A_1)P(A_1) + P(T \mid A_2)P(A_2) + P(T \mid A_3)P(A_3)}$$

As for why this is true, and what happens when we have a partition into *four or more* events, we will wait for the exercises. In practice, as was the case with a partition into two events, we can often compute $P(A_1 \mid T)$ by constructing a tree diagram.

EXAMPLE 3 School and Work

A survey[68] conducted by the Bureau of Labor Statistics found that approximately 27% of the high school graduating class of 2010 went on to a 2-year college, 41% went on to a 4-year college, and the remaining 32% did not go on to college. Of those who went on to a 2-year college, 52% worked at the same time, 32% of those going on to a 4-year college worked, and 78% of those who did not go on to college worked. What percentage of those working had not gone on to college?

Solution We can interpret these percentages as probabilities if we consider the experiment of choosing a member of the high school graduating class of 2010 at random. The events we are interested in are these:

R_1: A graduate went on to a 2-year college

R_2: A graduate went on to a 4-year college

R_3: A graduate did not go to college

A: A graduate went on to work.

[68] Source: "College Enrollment and Work Activity of High School Graduates," U.S. Bureau of Labor Statistics (www.bls.gov/news.release/hsgec.htm).

The three events R_1, R_2, and R_3 partition the sample space of all graduates into three events. We are given the following probabilities:

$$P(R_1) = .27 \qquad P(R_2) = .41 \qquad P(R_3) = .32$$
$$P(A \mid R_1) = .52 \quad P(A \mid R_2) = .32 \quad P(A \mid R_3) = .78.$$

We are asked to find the probability that a graduate who went on to work did not go to college, so we are looking for $P(R_3 \mid A)$. Bayes' formula for these events is

$$P(R_3 \mid A) = \frac{P(A \mid R_3)P(R_3)}{P(A \mid R_1)P(R_1) + P(A \mid R_2)P(R_2) + P(A \mid R_3)P(R_3)}$$

$$= \frac{(.78)(.32)}{(.52)(.27) + (.32)(.41) + (.78)(.32)} \approx .48.$$

Thus we conclude that 48% of all those working had not gone on to college.

➡ **Before we go on…** We could also solve Example 3 using a tree diagram. As before, the first branching corresponds to the events with unconditional probabilities that we know: R_1, R_2, and R_3. You should complete the tree and check that you obtain the same result as above. ■

7.6 EXERCISES

▼ more advanced   ◆ challenging

T indicates exercises that should be solved using technology

In Exercises 1–8, use Bayes' theorem or a tree diagram to calculate the indicated probability. Round all answers to four decimal places. HINT [See Quick Example on page 515 and Example 3.]

1. $P(A \mid B) = .8$, $P(B) = .2$, $P(A \mid B') = .3$. Find $P(B \mid A)$.

2. $P(A \mid B) = .6$, $P(B) = .3$, $P(A \mid B') = .5$. Find $P(B \mid A)$.

3. $P(X \mid Y) = .8$, $P(Y') = .3$, $P(X \mid Y') = .5$. Find $P(Y \mid X)$.

4. $P(X \mid Y) = .6$, $P(Y') = .4$, $P(X \mid Y') = .3$. Find $P(Y \mid X)$.

5. Y_1, Y_2, Y_3 form a partition of S. $P(X \mid Y_1) = .4$, $P(X \mid Y_2) = .5$, $P(X \mid Y_3) = .6$, $P(Y_1) = .8$, $P(Y_2) = .1$. Find $P(Y_1 \mid X)$.

6. Y_1, Y_2, Y_3 form a partition of S. $P(X \mid Y_1) = .2$, $P(X \mid Y_2) = .3$, $P(X \mid Y_3) = .6$, $P(Y_1) = .3$, $P(Y_2) = .4$. Find $P(Y_1 \mid X)$.

7. Y_1, Y_2, Y_3 form a partition of S. $P(X \mid Y_1) = .4$, $P(X \mid Y_2) = .5$, $P(X \mid Y_3) = .6$, $P(Y_1) = .8$, $P(Y_2) = .1$. Find $P(Y_2 \mid X)$.

8. Y_1, Y_2, Y_3 form a partition of S. $P(X \mid Y_1) = .2$, $P(X \mid Y_2) = .3$, $P(X \mid Y_3) = .6$, $P(Y_1) = .3$, $P(Y_2) = .4$. Find $P(Y_2 \mid X)$.

APPLICATIONS

9. *Music Downloading* According to a study on the effect of music downloading on spending on music, 11% of all Internet users had decreased their spending on music.[69] We estimate that 40% of all music fans used the Internet at the time of the study.[70] If 20% of non–Internet users had decreased their spending on music, what percentage of those who had decreased their spending on music were Internet users? HINT [See Examples 1 and 2.]

10. *Music Downloading* According to the study cited in the preceding exercise, 36% of experienced file-sharers with broadband access had decreased their spending on music. Let us estimate that 3% of all music fans were experienced file-sharers with broadband access at the time of the study.[71] If 20% of the other music fans had decreased their spending on music, what percentage of those who had decreased their spending on music were experienced file-sharers with broadband access? HINT [See Examples 1 and 2.]

---

[69] Regardless of whether they used the Internet to download music. Source: *New York Times*, May 6, 2002, p. C6.

[70] According to the U.S. Department of Commerce, 51% of all U.S. households had computers in 2001.

[71] Around 15% of all online households had broadband access in 2001 according to a *New York Times* article (Dec. 24, 2001, p. C1).

Notice that, because the system must go somewhere at each time step, the transition probabilities originating at a particular state always add up to 1. For example, in the transition diagram above, when we add the probabilities originating at state 1, we get $.8 + .2 = 1$.

The transition probabilities may be conveniently arranged in a matrix.

Transition Matrix

The **transition matrix** associated with a given Markov system is the matrix P whose ijth entry is the transition probability p_{ij}, the transition probability of going *from* state i *to* state j. In other words, the entry in position ij is the *label on the arrow going from state i to state j* in a state transition diagram.

Thus, the transition matrix for a system with two states would be set up as follows:

$$\text{From:} \begin{array}{c} \\ 1 \\ 2 \end{array} \overset{\begin{array}{cc} \quad \text{To:} \\ 1 \qquad 2 \end{array}}{\begin{bmatrix} p_{11} & p_{12} \\ p_{21} & p_{22} \end{bmatrix}}. \quad \begin{array}{l} \text{Arrows Originating in State 1} \\ \text{Arrows Originating in State 2} \end{array}$$

Quick Example

In the system pictured in Figure 19, the transition matrix is

$$P = \begin{bmatrix} .8 & .2 \\ .1 & .9 \end{bmatrix}.$$

Note Notice that because the sum of the transition probabilities that originate at any state is 1, *the sum of the entries in any row of a transition matrix is 1.* ■

Now, let's start doing some calculations.

EXAMPLE 1 Laundry Detergent Switching

Consider the Markov system found by Gamble Detergents at the beginning of this section. Suppose that 70% of consumers are now using powdered detergent, while the other 30% are using liquid.

a. What will be the distribution one year from now? (That is, what percentage will be using powdered and what percentage liquid detergent?)

b. Assuming that the probabilities remain the same, what will be the distribution two years from now? Three years from now?

Solution

a. First, let us think of the statement that 70% of consumers are using powdered detergent as telling us a probability: The probability that a randomly chosen consumer uses powdered detergent is .7. Similarly, the probability that a randomly chosen consumer uses liquid detergent is .3. We want to find the corresponding probabilities one year from now. To do this, consider the tree diagram in Figure 20.

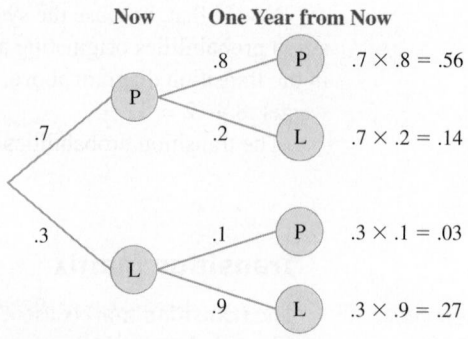

Figure 20

The first branching shows the probabilities now, while the second branching shows the (conditional) transition probabilities. So, if we want to know the probability that a consumer is using powdered detergent one year from now, it will be

$$\text{Probability of using powder after one year} = .7 \times .8 + .3 \times .1 = .59.$$

On the other hand, we have:

$$\text{Probability of using liquid after one year} = .7 \times .2 + .3 \times .9 = .41.$$

Now, here's the crucial point: *These are exactly the same calculations as in the matrix product*

$$[.7 \quad .3]\begin{bmatrix} .8 & .2 \\ .1 & .9 \end{bmatrix} = [.59 \quad .41].$$

↑   ↑   ↑

Initial distribution   Transition matrix   Distribution after 1 step

Thus, to get the distribution of detergent users after one year, all we have to do is multiply the **initial distribution vector** [.7 .3] by the transition matrix *P*. The result is [.59 .41], the **distribution vector after one step**.

b. Now what about the distribution after *two* years? If we assume that the same fraction of consumers switch or stay put in the second year as in the first, we can simply repeat the calculation we did above, using the new distribution vector:

$$[.59 \quad .41]\begin{bmatrix} .8 & .2 \\ .1 & .9 \end{bmatrix} = [.513 \quad .487].$$

↑   ↑   ↑

Distribution after 1 step   Transition matrix   Distribution after 2 steps

Thus, after two years we can expect 51.3% of consumers to be using powdered detergent and 48.7% to be using liquid detergent. Similarly, after three years we have

$$[.513 \quad .487]\begin{bmatrix} .8 & .2 \\ .1 & .9 \end{bmatrix} = [.4591 \quad .5409].$$

So, after three years, 45.91% of consumers will be using powdered detergent and 54.09% will be using liquid. Slowly but surely, liquid detergent seems to be winning.

➡ **Before we go on...** Note that the sum of the entries is 1 in each of the distribution vectors in Example 1. In fact, these vectors are giving the probability distributions for each year of finding a randomly chosen consumer using either powdered or liquid detergent. A vector having nonnegative entries adding up to 1 is called a **probability vector**. ∎

 using Technology

Technology can be used to compute the distribution vectors in Example 1:

TI-83/84 Plus
Define [A] as the initial distribution and [B] as the transition matrix. Then compute [A]*[B] for the distribution after 1 step. Ans*[B] gives the distribution after successive steps. [More details on page 542.]

Spreadsheet
Use the MMULT command to multiply the associated matrices. [More details on page 544.]

 Website
www.WanerMath.com
Student Home
   → On Line Utilities
   → Matrix Algebra Tool
You can enter the transition matrix *P* and the initial distribution vector *v* in the input area, and compute the various distribution vectors using the following formulas:
v*P   1 step
v*P^2   2 steps
v*P^3   3 steps.

Distribution Vector after m Steps

A **distribution vector** is a probability vector giving the probability distribution for finding a Markov system in its various possible states. If v is a distribution vector, then the distribution vector one step later will be vP. The distribution m steps later will be

$$\text{Distribution after } m \text{ steps} = v \cdot P \cdot P \cdot \ldots \cdot P \ (m \text{ times}) = vP^m.$$

Quick Example

If $P = \begin{bmatrix} 0 & 1 \\ .5 & .5 \end{bmatrix}$ and $v = [.2 \quad .8]$, then we can calculate the following distribution vectors:

$$vP = [.2 \quad .8]\begin{bmatrix} 0 & 1 \\ .5 & .5 \end{bmatrix} = [.4 \quad .6] \qquad \text{Distribution after one step}$$

$$vP^2 = (vP)P = [.4 \quad .6]\begin{bmatrix} 0 & 1 \\ .5 & .5 \end{bmatrix} = [.3 \quad .7] \qquad \text{Distribution after two steps}$$

$$vP^3 = (vP^2)P = [.3 \quad .7]\begin{bmatrix} 0 & 1 \\ .5 & .5 \end{bmatrix} = [.35 \quad .65]. \qquad \text{Distribution after three steps}$$

What about the matrix P^m that appears above? Multiplying a distribution vector v times P^m gives us the distribution m steps later, so we can think of P^m as the m-step transition matrix. More explicitly, consider the following example.

EXAMPLE 2 Powers of the Transition Matrix

Continuing the example of detergent switching, suppose that a consumer is now using powdered detergent. What are the probabilities that the consumer will be using powdered or liquid detergent two years from now? What if the consumer is now using liquid detergent?

Solution To record the fact that we know that the consumer is using powdered detergent, we can take as our initial distribution vector $v = [1 \quad 0]$. To find the distribution two years from now, we compute vP^2. To make a point, we do the calculation slightly differently:

$$vP^2 = [1 \quad 0]\begin{bmatrix} .8 & .2 \\ .1 & .9 \end{bmatrix}\begin{bmatrix} .8 & .2 \\ .1 & .9 \end{bmatrix}$$

$$= [1 \quad 0]\begin{bmatrix} .66 & .34 \\ .17 & .83 \end{bmatrix}$$

$$= [.66 \quad .34].$$

So, the probability that our consumer is using powdered detergent two years from now is .66, while the probability of using liquid detergent is .34. The point to notice is that these are the entries in the first row of P^2. Similarly, if we consider a consumer now using liquid detergent, we should take the initial distribution vector to be $v = [0 \quad 1]$ and compute

$$vP^2 = [0 \quad 1]\begin{bmatrix} .66 & .34 \\ .17 & .83 \end{bmatrix} = [.17 \quad .83].$$

Thus, the bottom row gives the probabilities that a consumer, now using liquid detergent, will be using either powdered or liquid detergent two years from now.

In other words, the *ij*th entry of P^2 gives the probability that a consumer, starting in state *i*, will be in state *j* after two time steps.

What is true in Example 2 for two time steps is true for any number of time steps:

Powers of the Transition Matrix

P^m ($m = 1, 2, 3, \ldots$) is the ***m*-step transition matrix**. The *ij*th entry in P^m is the probability of a transition from state *i* to state *j* in *m* steps.

Quick Example

If $P = \begin{bmatrix} 0 & 1 \\ .5 & .5 \end{bmatrix}$ then

$$P^2 = P \cdot P = \begin{bmatrix} .5 & .5 \\ .25 & .75 \end{bmatrix} \qquad \text{2-step transition matrix}$$

$$P^3 = P \cdot P^2 = \begin{bmatrix} .25 & .75 \\ .375 & .625 \end{bmatrix}. \qquad \text{3-step transition matrix}$$

The probability of going from state 1 to state 2 in two steps = (1, 2)-entry of $P^2 = .5$.

The probability of going from state 1 to state 2 in three steps = (1, 2)-entry of $P^3 = .75$.

What happens if we follow our laundry detergent–using consumers for many years?

EXAMPLE 3 Long-Term Behavior

Suppose that 70% of consumers are now using powdered detergent while the other 30% are using liquid. Assuming that the transition matrix remains valid the whole time, what will be the distribution 1, 2, 3, . . . , and 50 years later?

Solution Of course, to do this many matrix multiplications, we're best off using technology. We already did the first three calculations in an earlier example.

Distribution after 1 year: $\quad [.7 \quad .3] \begin{bmatrix} .8 & .2 \\ .1 & .9 \end{bmatrix} = [.59 \quad .41].$

Distribution after 2 years: $\quad [.59 \quad .41] \begin{bmatrix} .8 & .2 \\ .1 & .9 \end{bmatrix} = [.513 \quad .487].$

Distribution after 3 years: $\quad [.513 \quad .487] \begin{bmatrix} .8 & .2 \\ .1 & .9 \end{bmatrix} = [.4591 \quad .5409].$

. . .

Distribution after 48 years: $\quad [.33333335 \quad .66666665].$

Distribution after 49 years: $\quad [.33333334 \quad .66666666].$

Distribution after 50 years: $\quad [.33333334 \quad .66666666].$

Thus, the distribution after 50 years is approximately $[.33333334 \quad .66666666]$.

Something interesting seems to be happening in Example 3. The distribution seems to be getting closer and closer to

$$[.333333\ldots \quad .666666\ldots] = \begin{bmatrix} \dfrac{1}{3} & \dfrac{2}{3} \end{bmatrix}.$$

Let's call this distribution vector v_∞. Notice two things about v_∞:

- v_∞ is a probability vector.
- If we calculate $v_\infty P$, we find

$$v_\infty P = \begin{bmatrix} \dfrac{1}{3} & \dfrac{2}{3} \end{bmatrix} \begin{bmatrix} .8 & .2 \\ .1 & .9 \end{bmatrix} = \begin{bmatrix} \dfrac{1}{3} & \dfrac{2}{3} \end{bmatrix} = v_\infty.$$

In other words,

$$v_\infty P = v_\infty.$$

We call a probability vector v with the property that $vP = v$ a **steady-state (probability) vector**.

Q: *Where does the name steady-state vector come from?*

A: If $vP = v$, then v is a distribution that will not change from time step to time step. In the example above, because [1/3 2/3] is a steady-state vector, if 1/3 of consumers use powdered detergent and 2/3 use liquid detergent one year, then the proportions will be the same the next year. Individual consumers may still switch from year to year, but as many will switch from powder to liquid as switch from liquid to powder, so the number using each will remain constant.

But how do we find a steady-state vector?

---

using Technology

Technology can be used to compute the steady-state vector in Example 4.

TI-83/84 Plus
Define [A] as the coefficient matrix of the system of equations being solved, and [B] as the column matrix of the right-hand sides. Then compute $[A]^{-1}[B]$
[More details on page 543.]

Spreadsheet
Use the MMULT and MINVERSE commands to solve the necessary system of equations.
[More details on page 545.]

 Website
www.WanerMath.com
  Student Home
   → On Line Utilities
   → Pivot and Gauss-Jordan Tool
You can use the Pivot and Gauss-Jordan Tool to solve the system of equations that gives you the steady-state vector.

EXAMPLE 4   Calculating the Steady-State Vector

Calculate the steady-state probability vector for the transition matrix in the preceding examples:

$$P = \begin{bmatrix} .8 & .2 \\ .1 & .9 \end{bmatrix}.$$

Solution   We are asked to find

$$v_\infty = [x \quad y].$$

This vector must satisfy the equation

$$v_\infty P = v_\infty$$

or

$$[x \quad y] \begin{bmatrix} .8 & .2 \\ .1 & .9 \end{bmatrix} = [x \quad y].$$

Doing the matrix multiplication gives

$$[.8x + .1y \quad .2x + .9y] = [x \quad y].$$

Equating corresponding entries gives

$$.8x + .1y = x$$
$$.2x + .9y = y$$

or

$$-.2x + .1y = 0$$
$$.2x - .1y = 0.$$

Now these equations are really the same equation. (Do you see that?) There is one more thing we know, though: Because $[x \; y]$ is a probability vector, its entries must add up to 1. This gives one more equation:

$$x + y = 1.$$

Taking this equation together with one of the two equations above gives us the following system:

$$x + \quad y = 1$$
$$-.2x + .1y = 0.$$

We now solve this system using any of the techniques we learned for solving systems of linear equations. We find that the solution is $x = 1/3$, and $y = 2/3$, so the steady-state vector is

$$v_\infty = [x \; y] = \begin{bmatrix} \dfrac{1}{3} & \dfrac{2}{3} \end{bmatrix}$$

as suggested in Example 3.

The method we just used works for any size transition matrix and can be summarized as follows.

Calculating the Steady-State Distribution Vector

To calculate the steady-state probability vector for a Markov system with transition matrix P, we solve the system of equations given by

$$x + y + z + \cdots = 1$$
$$[x \; y \; z \ldots]P = [x \; y \; z \ldots],$$

where we use as many unknowns as there are states in the Markov system. The steady-state probability vector is then

$$v_\infty = [x \; y \; z \ldots].$$

Q: *Is there always a steady-state distribution vector?*

A: Yes, although the explanation why is more involved than we can give here.

Q: *In Example 3, we started with a distribution vector v and found that vP^m got closer and closer to v_∞ as m got larger. Does that always happen?*

A: It does if the Markov system is **regular**, as we define below, but may not for other kinds of systems. Again, we shall not prove this fact here.

Regular Markov Systems

A **regular** Markov system is one for which some power of its transition matrix P has no zero entries. If a Markov system is regular, then

1. It has a unique steady-state probability vector v_∞, and

2. If v is any probability vector whatsoever, then vP^m approaches v_∞ as m gets large. We say that the **long-term behavior** of the system is to have distribution (close to) v_∞.

Interpreting the Steady-State Vector

In a regular Markov system, the entries in the steady-state probability vector give the long-term probabilities that the system will be in the corresponding states, or the fractions of time one can expect to find the Markov system in the corresponding states.

Quick Examples

1. The system with transition matrix $P = \begin{bmatrix} .8 & .2 \\ .1 & .9 \end{bmatrix}$ is regular because $P(= P^1)$ has no zero entries.

2. The system with transition matrix $P = \begin{bmatrix} 0 & 1 \\ .5 & .5 \end{bmatrix}$ is regular because
$P^2 = \begin{bmatrix} .5 & .5 \\ .25 & .75 \end{bmatrix}$ has no zero entries.

3. The system with transition matrix $P = \begin{bmatrix} 0 & 1 \\ 1 & 0 \end{bmatrix}$ is *not* regular:
$P^2 = \begin{bmatrix} 1 & 0 \\ 0 & 1 \end{bmatrix}$ and $P^3 = P$ again, so the powers of P alternate between these two matrices. Thus, every power of P has zero entries. Although this system has a steady-state vector, namely $[.5 \ .5]$, if we take $v = [1 \ 0]$, then $vP = [0 \ 1]$ and $vP^2 = v$, so the distribution vectors vP^m just alternate between these two vectors, not approaching v_∞.

We finish with one more example.

EXAMPLE 5 **Gambler's Ruin**

A timid gambler, armed with her annual bonus of $20, decides to play roulette using the following scheme. At each spin of the wheel, she places $10 on red. If red comes up, she wins an additional $10; if black comes up, she loses her $10. For the sake of simplicity, assume that she has a probability of 1/2 of winning. (In the real game, the probability is slightly lower—a fact that many gamblers forget.) She keeps playing until she has either gotten up to $30 or lost it all. In either case, she then packs up and leaves. Model this situation as a Markov system and find the associated transition matrix. What can we say about the long-term behavior of this system?

Solution We must first decide on the states of the system. A good choice is the gambler's financial state, the amount of money she has at any stage of the game. According to her rules, she can be broke, have $10, $20, or $30. Thus, there are four states: $1 = \$0, 2 = \$10, 3 = \$20,$ and $4 = \$30$. Because she bets $10 each time, she moves down $10 if she loses (with probability 1/2) and up $10 if she wins (with probability also 1/2), until she reaches one of the extremes. The transition diagram is shown in Figure 21.

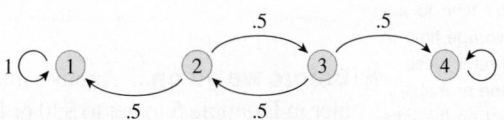

Figure 21

Note that once the system enters state 1 or state 4, it does not leave; with probability 1, it stays in the same state.* We call such states **absorbing states**. We can now write down the transition matrix:

$$P = \begin{bmatrix} 1 & 0 & 0 & 0 \\ .5 & 0 & .5 & 0 \\ 0 & .5 & 0 & .5 \\ 0 & 0 & 0 & 1 \end{bmatrix}.$$

(Notice all the 0 entries, corresponding to possible transitions that we did not draw in the transition diagram because they have 0 probability of occurring. We usually leave out such arrows.) Is this system regular? Take a look at P^2:

$$P^2 = \begin{bmatrix} 1 & 0 & 0 & 0 \\ .5 & .25 & 0 & .25 \\ .25 & 0 & .25 & .5 \\ 0 & 0 & 0 & 1 \end{bmatrix}.$$

Notice that the first and last rows haven't changed. After two steps, there is still no chance of leaving states 1 or 4. In fact, no matter how many powers we take, no matter how many steps we look at, there will still be no way to leave either of those states, and the first and last rows will still have plenty of zeros. This system is not regular.

Nonetheless, we can try to find a steady-state probability vector. If we do this (and you should set up the system of linear equations and solve it), we find that there are infinitely many steady-state probability vectors, namely all vectors of the form $[x \ 0 \ 0 \ 1 - x]$ for $0 \le x \le 1$. (You can check directly that these are all steady-state vectors.) As with a regular system, if we start with any distribution, the system will tend toward one of these steady-state vectors. In other words, eventually the gambler will either lose all her money or leave the table with $30.

But which outcome is more likely, and with what probability? One way to approach this question is to try computing the distribution after many steps. The distribution that represents the gambler starting with $20 is $v = [0 \ 0 \ 1 \ 0]$. Using technology, it's easy to compute vP^n for some large values of n:

$$vP^{10} \approx [.333008 \ 0 \ .000976 \ .666016]$$
$$vP^{50} \approx [.333333 \ 0 \ 0 \ .666667].$$

So, it looks like the probability that she will leave the table with $30 is approximately 2/3, while the probability that she loses it all is 1/3.

What if she started with only $10? Then our initial distribution would be $v = [0 \ 1 \ 0 \ 0]$ and

$$vP^{10} \approx [.666016 \ .000976 \ 0 \ .333008]$$
$$vP^{50} \approx [.666667 \ 0 \ 0 \ .333333].$$

So, this time, the probability of her losing everything is about 2/3 while the probability of her leaving with $30 is 1/3. There is a way of calculating these probabilities exactly using matrix arithmetic; however, it would take us too far afield to describe it here.

Before we go on... Another interesting question is, how long will it take the gambler in Example 5 to get to $30 or lose it all? This is called the *time to absorption* and can be calculated using matrix arithmetic. ∎

7.7 EXERCISES

▼ more advanced   ◆ challenging

T indicates exercises that should be solved using technology

In Exercises 1–10, write down the transition matrix associated with each state transition diagram.

1.

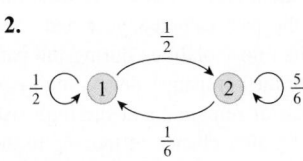

2.

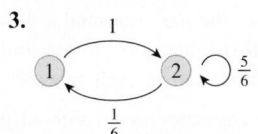

3. **4.**

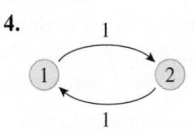

5. **6.**

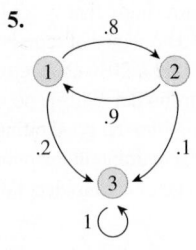

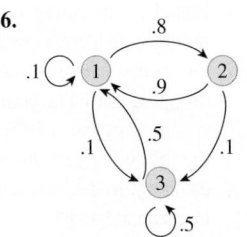

7. **8.**

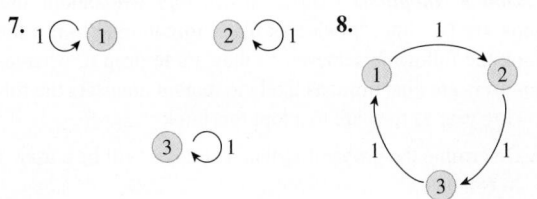

9.

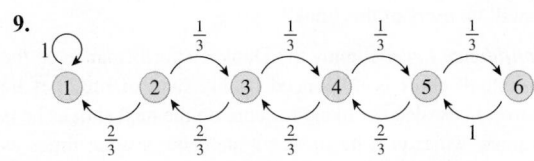

10.

In each of Exercises 11–24, you are given a transition matrix P and initial distribution vector v. Find

(a) *the two-step transition matrix and*

(b) *the distribution vectors after one, two, and three steps.*
HINT [See Quick Examples on pages 525 and 526.]

11. $P = \begin{bmatrix} .5 & .5 \\ 0 & 1 \end{bmatrix}, v = [1 \ \ 0]$

12. $P = \begin{bmatrix} 1 & 0 \\ .5 & .5 \end{bmatrix}, v = [0 \ \ 1]$

13. $P = \begin{bmatrix} .2 & .8 \\ .4 & .6 \end{bmatrix}, v = [.5 \ \ .5]$

14. $P = \begin{bmatrix} 1/3 & 2/3 \\ 1/2 & 1/2 \end{bmatrix}, v = [1/4 \ \ 3/4]$

15. $P = \begin{bmatrix} 1/2 & 1/2 \\ 1 & 0 \end{bmatrix}, v = [2/3 \ \ 1/3]$

16. $P = \begin{bmatrix} 0 & 1 \\ 1/4 & 3/4 \end{bmatrix}, v = [1/5 \ \ 4/5]$

17. $P = \begin{bmatrix} 3/4 & 1/4 \\ 3/4 & 1/4 \end{bmatrix}, v = [1/2 \ \ 1/2]$

18. $P = \begin{bmatrix} 2/3 & 1/3 \\ 2/3 & 1/3 \end{bmatrix}, v = [1/7 \ \ 6/7]$

19. $P = \begin{bmatrix} .5 & .5 & 0 \\ 0 & 1 & 0 \\ 0 & .5 & .5 \end{bmatrix}, v = [1 \ \ 0 \ \ 0]$

20. $P = \begin{bmatrix} .5 & 0 & .5 \\ 1 & 0 & 0 \\ 0 & .5 & .5 \end{bmatrix}, v = [0 \ \ 1 \ \ 0]$

21. $P = \begin{bmatrix} 0 & 1 & 0 \\ 1/3 & 1/3 & 1/3 \\ 1 & 0 & 0 \end{bmatrix}, v = [1/2 \ \ 0 \ \ 1/2]$

22. $P = \begin{bmatrix} 1/2 & 1/2 & 0 \\ 1/2 & 1/2 & 0 \\ 1/2 & 0 & 1/2 \end{bmatrix}, v = [0 \ \ 0 \ \ 1]$

23. $P = \begin{bmatrix} .1 & .9 & 0 \\ 0 & 1 & 0 \\ 0 & .2 & .8 \end{bmatrix}, v = [.5 \ \ 0 \ \ .5]$

24. $P = \begin{bmatrix} .1 & .1 & .8 \\ .5 & 0 & .5 \\ .5 & 0 & .5 \end{bmatrix}, v = [0 \ \ 1 \ \ 0]$

In each of Exercises 25–36, you are given a transition matrix P. Find the steady-state distribution vector. HINT [See Example 4.]

25. $P = \begin{bmatrix} 1/2 & 1/2 \\ 1 & 0 \end{bmatrix}$ **26.** $P = \begin{bmatrix} 0 & 1 \\ 1/4 & 3/4 \end{bmatrix}$

27. $P = \begin{bmatrix} 1/3 & 2/3 \\ 1/2 & 1/2 \end{bmatrix}$ **28.** $P = \begin{bmatrix} .2 & .8 \\ .4 & .6 \end{bmatrix}$

29. $P = \begin{bmatrix} .1 & .9 \\ .6 & .4 \end{bmatrix}$ **30.** $P = \begin{bmatrix} .2 & .8 \\ .7 & .3 \end{bmatrix}$

31. $P = \begin{bmatrix} .5 & 0 & .5 \\ 1 & 0 & 0 \\ 0 & .5 & .5 \end{bmatrix}$ **32.** $P = \begin{bmatrix} 0 & .5 & .5 \\ .5 & .5 & 0 \\ 1 & 0 & 0 \end{bmatrix}$

33. $P = \begin{bmatrix} 0 & 1 & 0 \\ 1/3 & 1/3 & 1/3 \\ 1 & 0 & 0 \end{bmatrix}$ **34.** $P = \begin{bmatrix} 1/2 & 1/2 & 0 \\ 1/2 & 1/2 & 0 \\ 1/2 & 0 & 1/2 \end{bmatrix}$

35. $P = \begin{bmatrix} .1 & .9 & 0 \\ 0 & 1 & 0 \\ 0 & .2 & .8 \end{bmatrix}$ **36.** $P = \begin{bmatrix} .1 & .1 & .8 \\ .5 & 0 & .5 \\ .5 & 0 & .5 \end{bmatrix}$

APPLICATIONS

37. *Marketing* A market survey shows that half the owners of Sorey State Boogie Boards became disenchanted with the product and switched to C&T Super Professional Boards the next surf season, while the other half remained loyal to Sorey State. On the other hand, three quarters of the C&T Boogie Board users remained loyal to C&T, while the rest switched to Sorey State. Set these data up as a Markov transition matrix, and calculate the probability that a Sorey State Board user will be using the same brand two seasons later. HINT [See Example 1.]

38. *Major Switching* At Suburban Community College, 10% of all business majors switched to another major the next semester, while the remaining 90% continued as business majors. Of all non–business majors, 20% switched to a business major the following semester, while the rest did not. Set up these data as a Markov transition matrix, and calculate the probability that a business major will no longer be a business major in two semesters' time. HINT [See Example 1.]

39. *Pest Control* In an experiment to test the effectiveness of the latest roach trap, the "Roach Resort," 50 roaches were placed in the vicinity of the trap and left there for an hour. At the end of the hour, it was observed that 30 of them had "checked in," while the rest were still scurrying around. (Remember that "once a roach checks in, it never checks out.")

  a. Set up the transition matrix P for the system with decimal entries, and calculate P^2 and P^3.

  b. If a roach begins outside the "Resort," what is the probability of it "checking in" by the end of 1 hour? 2 hours? 3 hours?

  c. What do you expect to be the long-term impact on the number of roaches? HINT [See Example 5.]

40. *Employment* You have worked for the Department of Administrative Affairs (DAA) for 27 years, and you still have little or no idea exactly what your job entails. To make your life a little more interesting, you have decided on the following course of action. Every Friday afternoon, you will use your desktop computer to generate a random digit from 0 to 9 (inclusive). If the digit is a zero, you will immediately quit your job, never to return. Otherwise, you will return to work the following Monday.

  a. Use the states (1) employed by the DAA and (2) not employed by the DAA to set up a transition probability matrix P with decimal entries, and calculate P^2 and P^3.

  b. What is the probability that you will still be employed by the DAA after each of the next three weeks?

  c. What are your long-term prospects for employment at the DAA? HINT [See Example 5.]

41. *Risk Analysis* An auto insurance company classifies each motorist as "high risk" if the motorist has had at least one moving violation during the past calendar year and "low risk" if the motorist has had no violations during the past calendar year. According to the company's data, a high-risk motorist has a 50% chance of remaining in the high-risk category the next year and a 50% chance of moving to the low-risk category. A low-risk motorist has a 10% chance of moving to the high-risk category the next year and a 90% chance of remaining in the low-risk category. In the long term, what percentage of motorists fall in each category?

42. *Debt Analysis* A credit card company classifies its cardholders as falling into one of two credit ratings: "good" and "poor." Based on its rating criteria, the company finds that a cardholder with a good credit rating has an 80% chance of remaining in that category the following year and a 20% chance of dropping into the poor category. A cardholder with a poor credit rating has a 40% chance of moving into the good rating the following year and a 60% chance of remaining in the poor category. In the long term, what percentage of cardholders fall in each category?

43. *Textbook Adoptions* College instructors who adopt this book are (we hope!) twice as likely to continue to use the book the following semester as they are to drop it, whereas nonusers are nine times as likely to remain nonusers the following year as they are to adopt this book.

  a. Determine the probability that a nonuser will be a user in two years.

  b. In the long term, what proportion of college instructors will be users of this book?

44. *Confidence Level* Tommy the Dunker's performance on the basketball court is influenced by his state of mind: If he scores, he is twice as likely to score on the next shot as he is to miss, whereas if he misses a shot, he is three times as likely to miss the next shot as he is to score.

  a. If Tommy has missed a shot, what is the probability that he will score two shots later?

  b. In the long term, what percentage of shots are successful?

45. *Debt Analysis* As the manager of a large retailing outlet, you have classified all credit customers as falling into one of the following categories: Paid Up, Outstanding 0–90 Days, Bad

Debts. Based on an audit of your company's records, you have come up with the following table, which gives the probabilities that a single credit customer will move from one category to the next in the period of 1 month.

		To		
		Paid Up	**0–90 Days**	**Bad Debts**
From	**Paid Up**	.5	.5	0
	0–90 Days	.5	.3	.2
	Bad Debts	0	.5	.5

How do you expect the company's credit customers to be distributed in the long term?

46. *Debt Analysis* Repeat the preceding exercise using the following table:

		To		
		Paid Up	**0–90 Days**	**Bad Debts**
From	**Paid Up**	.8	.2	0
	0–90 Days	.5	.3	.2
	Bad Debts	0	.5	.5

47. ▼ *Income Brackets* The following diagram shows the movement of U.S. households among three income groups—affluent, middle class, and poor—over the 11-year period 1980–1991.[90]

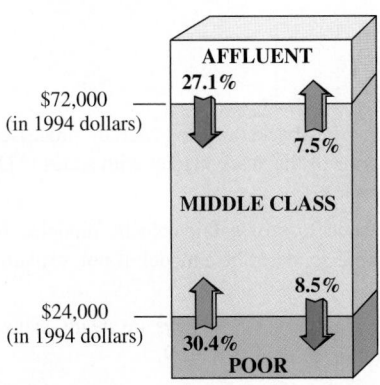

a. Use the transitions shown in the diagram to construct a transition matrix (assuming zero probabilities for the transitions between affluent and poor).

b. Assuming that the trend shown was to continue, what percent of households classified as affluent in 1980 were predicted to become poor in 2002? (Give your answer to the nearest 0.1%.)

c. ▆ According to the model, what percentage of all U.S. households will be in each income bracket in the long term? (Give your answer to the nearest 0.1%.)

48. ▼ *Income Brackets* The following diagram shows the movement of U.S. households among three income groups—affluent, middle class, and poor—over the 12-year period 1967–1979.[91]

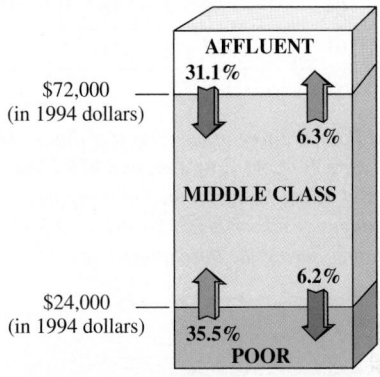

a. Use the transitions shown in the diagram to construct a transition matrix (assuming zero probabilities for the transitions between affluent and poor).

b. Assuming that the trend shown had continued, what percent of households classified as affluent in 1967 would have been poor in 1991? (Give your answer to the nearest 0.1%.)

c. ▆ According to the model, what percentage of all U.S. households will be in each income bracket in the long term? (Give your answer to the nearest 0.1%.)

49. ▆ ▼ *Income Distribution* A University of Michigan study shows the following one-generation transition probabilities among four major income groups.[92]

Father's Income	Eldest Son's Income			
	Bottom 10%	**10–50%**	**50–90%**	**Top 10%**
Bottom 10%	.30	.52	.17	.01
10–50%	.10	.48	.38	.04
50–90%	.04	.38	.48	.10
Top 10%	.01	.17	.52	.30

In the long term, what percentage of male earners would you expect to find in each category? Why are the long-range figures not necessarily 10% in the lowest 10% income bracket, 40% in the 10–50% range, 40% in the 50–90% range, and 10% in the top 10% range? (Use technology to compute an approximation of the steady-state transition matrix.)

---

[90] The figures were based on household after-tax income. The study was conducted by G. J. Duncan of Northwestern University and T. Smeeding of Syracuse University and based on annual surveys of the personal finances of 5,000 households since the late 1960s. (The surveys were conducted by the University of Michigan.) Source: *New York Times*, June 4, 1995, p. E4.

[91] *Ibid.*

[92] Source: Gary Solon, University of Michigan/*New York Times*, May 18, 1992, p. D5. We have adjusted some of the figures so that the probabilities add to 1; they did not do so in the original table due to rounding.

50. ⊤ ▼ *Income Distribution* Repeat Exercise 49, using the following data:

Father's Income	Son's income			
	Bottom 10%	10–50%	50–90%	Top 10%
Bottom 10%	.50	.32	.17	.01
10–50%	.10	.48	.38	.04
50–90%	.04	.38	.48	.10
Top 10%	.01	.17	.32	.50

Market Share: Cell Phones Three of the largest cellular phone companies in 2004 were **Verizon**, **Cingular**, *and* **AT&T Wireless**. *Exercises 51 and 52 are based on the following figure, which shows percentages of subscribers who switched from one company to another during the third quarter of 2003:*[93]

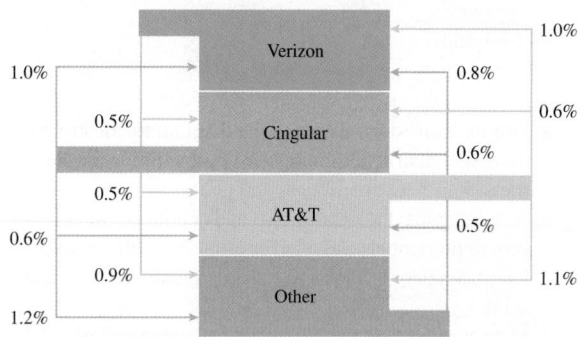

51. a. ⊤ ▼ Use the diagram to set up an associated transition matrix.
  b. At the end of the third quarter of 2003, the market shares were: **Verizon**: 29.7%, **Cingular**: 19.3%, **AT&T**: 18.1%, and Other: 32.9%. Use your Markov system to estimate the percentage shares at the *beginning* of the third quarter of 2003.
  c. Using the information from part (b), estimate the market shares at the end of 2005. Which company is predicted to gain the most in market share?

52. a. ⊤ ▼ Use the diagram to set up an associated transition matrix.
  b. At the end of the third quarter of 2003, the market shares were: **Verizon**: 29.7%, **Cingular**: 19.3%, **AT&T**: 18.1%, Other: 32.9%. Use your Markov system to estimate the percentage shares one year earlier.
  c. Using the information from part (b), estimate the market shares at the end of 2010. Which company is predicted to lose the most in market share?

---

[93] Published market shares and "churn rate" (percentage drops for each company) were used to estimate the individual transition percentages. "Other" consists of Sprint, Nextel, and T Mobile. No other cellular companies are included in this analysis. Source: The Yankee Group/ *New York Times*, January 21, 2004, p. C1.

53. ▼ *Dissipation* Consider the following five-state model of one-dimensional dissipation without drift:

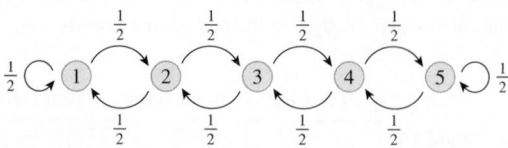

Find the steady-state distribution vector.

54. ▼ *Dissipation with Drift* Consider the following five-state model of one-dimensional dissipation with drift:

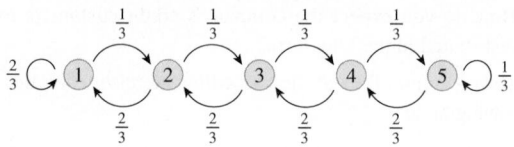

Find the steady-state distribution vector.

COMMUNICATION AND REASONING EXERCISES

55. Describe an interesting situation that can be modeled by the transition matrix

$$P = \begin{bmatrix} .2 & .8 & 0 \\ 0 & 1 & 0 \\ .4 & .6 & 0 \end{bmatrix}.$$

56. Describe an interesting situation that can be modeled by the transition matrix

$$P = \begin{bmatrix} .8 & .1 & .1 \\ 1 & 0 & 0 \\ .3 & .3 & .4 \end{bmatrix}.$$

57. ▼ Describe some drawbacks to using Markov processes to model the behavior of the stock market with states (1) bull market (2) bear market.

58. ▼ Can the repeated toss of a fair coin be modeled by a Markov process? If so, describe a model; if not, explain the reason.

59. ▼ Explain: If Q is a matrix whose rows are steady-state distribution vectors for P, then $QP = Q$.

60. ▼ Construct a four-state Markov system so that both [.5 .5 0 0] and [0 0 .5 .5] are steady-state vectors. HINT [Try one in which no arrows link the first two states to the last two.]

61. ▼ Refer to the following state transition diagram and explain in words (without doing any calculation) why the steady-state vector has a zero in position 1.

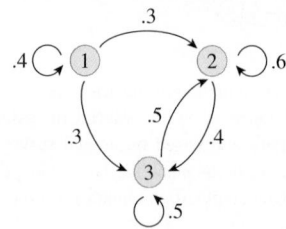

TI-83/84 Plus   Technology Guide

Section 7.2

The TI-83/84 Plus has a random number generator that we can use to simulate experiments. For the following example, recall that a fair coin has probability 1/2 of coming up heads and 1/2 of coming up tails.

Example   Use a simulated experiment to check the following.

a. The estimated probability of heads coming up in a toss of a fair coin approaches 1/2 as the number of trials gets large.

b. The estimated probability of heads coming up in two consecutive tosses of a fair coin approaches 1/4 as the number of trials gets large.[95]

Solution with Technology

a. Let us use 1 to represent heads and 0 to represent tails. We need to generate a list of **random binary digits** (0 or 1). One way to do this—and a method that works for most forms of technology—is to generate a random number between 0 and 1 and then round it to the nearest whole number, which will be either 0 or 1.

We generate random numbers on the TI-83/84 Plus using the "rand" function. To round the number X to the nearest whole number on the TI-83/84 Plus, follow MATH →NUM, select "round," and enter round(X,0). This instruction rounds X to zero decimal places—that is, to the nearest whole number. Since we wish to round a random number we need to enter

round(rand,0)   To obtain rand,
follow MATH → PRB.

The result will be either 0 or 1. Each time you press ENTER you will now get another 0 or 1. The TI-83/84 Plus can also generate a random integer directly (without the need for rounding) through the instruction

randInt(0,1)   To obtain randInt,
follow MATH → PRB.

In general, the command randInt (m, n) generates a random integer in the range $[m, n]$. The following sequence of 100 random binary digits was produced using technology.[96]

0	1	0	0	1	1	0	1	0	0
0	1	0	0	0	0	0	0	1	0
1	1	0	0	0	1	0	0	1	1
1	1	1	0	1	0	0	0	1	0
1	1	1	1	1	1	1	0	0	1
1	0	1	1	1	0	0	1	1	0
0	1	0	1	1	1	0	1	1	1
1	0	0	0	0	0	0	1	1	1
1	1	1	1	0	0	1	1	1	0
1	1	1	0	1	1	0	1	0	0

If we use only the first row of data (corresponding to the first ten tosses), we find

$$P(\text{H}) = \frac{fr(1)}{N} = \frac{4}{10} = .4.$$

Using the first two rows ($N = 20$) gives

$$P(\text{H}) = \frac{fr(1)}{N} = \frac{6}{20} = .3.$$

Using all ten rows ($N = 100$) gives

$$P(\text{H}) = \frac{fr(1)}{N} = \frac{54}{100} = .54.$$

This is somewhat closer to the theoretical probability of 1/2 and supports our intuitive notion that the larger the number of trials, the more closely the estimated probability should approximate the theoretical value.[97]

b. We need to generate pairs of random binary digits and then check whether they are both 1s. Although the TI-83/84 Plus will generate a pair of random digits if you enter round(rand(2),0), it would be a lot more convenient if the calculator could tell you right away whether both digits are 1s (corresponding to two consecutive heads in a coin toss). Here is a simple way of accomplishing this. Notice that if we *add* the two random binary digits, we obtain either 0, 1, or 2, telling us the number of heads that result from the two consecutive throws. Therefore, all we need to do is add the pairs of random digits and then count the number of times 2 comes up. A formula we can use is

randInt(0,1)+randInt(0,1)

---

[95] Since the set of outcomes of a pair of coin tosses is {HH, HT, TH, TT}, we expect HH to come up once in every four trials, on average.

[96] The instruction randInt(0,1,100)→L₁ will generate a list of 100 random 0s and 1s and store it in L₁, where it can be summed with Sum(L₁) (under 2ND LIST →MATH).

[97] Do not expect this to happen every time. Compare, for example, $P(\text{H})$ for the first five rows and for all ten rows.

What would be even *more* convenient is if the result of the calculation would be either 0 or 1, with 1 signifying success (two consecutive heads) and 0 signifying failure. Then, we could simply add up all the results to obtain the number of times two heads occurred. To do this, we first divide the result of the previous calculation above by 2 (obtaining 0, .5, or 1, where now 1 signifies success) and then round *down* to an integer using a function called "int":

```
int(0.5*(randInt(0,1)+randInt(0,1)))
```

Following is the result of 100 such pairs of coin tosses, with 1 signifying success (two heads) and 0 signifying failure (all other outcomes). The last column records the number of successes in each row and the total number at the end.

1	1	0	0	0	0	0	0	0	0	2
0	1	0	0	0	0	0	1	0	1	3
0	1	0	0	1	1	0	0	0	1	4
0	0	0	0	0	0	0	0	1	0	1
0	1	0	0	1	0	0	1	0	0	3
1	0	1	0	0	0	0	0	0	0	2
0	0	0	0	0	0	0	0	0	1	1
0	1	1	1	1	0	0	0	0	1	5
1	1	0	1	0	0	1	1	0	0	5
0	0	0	0	0	0	0	0	1	0	1
										27

Now, as in part (a), we can compute estimated probabilities, with D standing for the outcome "two heads":

First 10 trials: $P(D) = \dfrac{fr(1)}{N} = \dfrac{2}{10} = .2$

First 20 trials: $P(D) = \dfrac{fr(1)}{N} = \dfrac{5}{20} = .25$

First 50 trials: $P(D) = \dfrac{fr(1)}{N} = \dfrac{13}{50} = .26$

100 trials: $P(D) = \dfrac{fr(1)}{N} = \dfrac{27}{100} = .27.$

Q: *What is happening with the data? The probabilities seem to be getting less accurate as N increases!*

A: Quite by chance, exactly 5 of the first 20 trials resulted in success, which matches the theoretical probability. Figure 23 shows an Excel plot of estimated probability versus N (for N a multiple of 10). Notice that, as N increases, the graph seems to meander within smaller distances of .25.

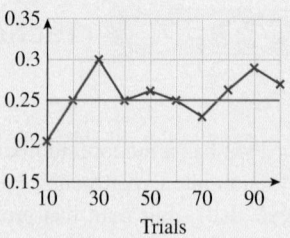

Figure 23

Q: *The previous techniques work fine for simulating coin tosses. What about rolls of a fair die, where we want outcomes between 1 and 6?*

A: We can simulate a roll of a die by generating a random integer in the range 1 through 6. The following formula accomplishes this:

$$1 + int(5.99999*rand).$$

(We used **5.99999** instead of **6** to avoid the outcome **7**.)

Section 7.7

Example 1 (page 523)   Consider the Markov system found by Gamble Detergents at the beginning of this section. Suppose that 70% of consumers are now using powdered detergent while the other 30% are using liquid. What will be the distribution one year from now? Two years from now? Three years from now?

Solution with Technology

In Chapter 4 we saw how to set up and multiply matrices. For this example, we can use the matrix editor to define [A] as the initial distribution and [B] as the transition matrix (remember that the only names we can use are [A] through [J]).

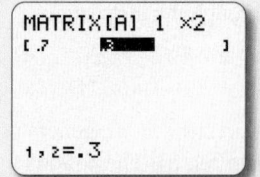
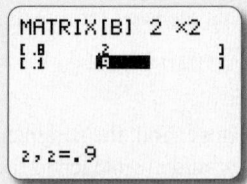

Entering [A] (obtained by pressing MATRX 1 ENTER) will show you the initial distribution. To obtain the distribution after 1 step, press X MATRX 2 ENTER, which has the effect of multiplying the previous answer by the transition matrix [B]. Now, just press ENTER repeatedly to continue multiplying by the transition matrix and obtain the distribution after any number of steps. The screenshot shows the initial distribution [A] and the distributions after 1, 2, and 3 steps.

```
[A]
        [[.7 .3]]
Ans*[B]
        [[.59 .41]]
      [[.513 .487]]
    [[.4591 .5409]]
```

Example 4 (page 527)   Calculate the steady-state probability vector for the transition matrix in the preceding examples.

Solution with Technology

Finding the steady-state probability vector comes down to solving a system of equations. As discussed in Chapters 3 and 4, there are several ways to use a calculator to help. The most straightforward is to use matrix inversion to solve the matrix form of the system. In this case, as in the text, the system of equations we need to solve is

$$x + y = 1$$
$$-.2x + .1y = 0.$$

We write this as the matrix equation $AX = B$ with

$$A = \begin{bmatrix} 1 & 1 \\ -.2 & .1 \end{bmatrix} \qquad B = \begin{bmatrix} 1 \\ 0 \end{bmatrix}.$$

To find $X = A^{-1}B$ using the TI-83/84 Plus, we first use the matrix editor to enter these matrices as [A] and [B], then compute $[A]^{-1}[B]$ on the home screen.

```
[A]⁻¹[B]
   [[.3333333333]
    [.6666666667]]
```

To convert the entries to fractions, we can follow this by the command.

➤ Frac   MATH ENTER ENTER

```
[A]⁻¹[B]
   [[.3333333333]
    [.6666666667]]
Ans▸Frac
         [[1/3]
          [2/3]]
```

SPREADSHEET Technology Guide

Section 7.2

Spreadsheets have random number generators that we can use to simulate experiments. For the following example, recall that a fair coin has probability 1/2 of coming up heads and 1/2 of coming up tails.

Example   Use a simulated experiment to check the following.

a. The estimated probability of heads coming up in a toss of a fair coin approaches 1/2 as the number of trials gets large.
b. The estimated probability of heads coming up in two consecutive tosses of a fair coin approaches 1/4 as the number of trials gets large.[98]

Solution with Technology

a. Let us use 1 to represent heads and 0 to represent tails. We need to generate a list of **random binary digits** (0 or 1). One way to do this—and a method that works for most forms of technology—is to generate a random number between 0 and 1 and then round it to the nearest whole number, which will be either 0 or 1.

In a spreadsheet, the formula RAND() gives a random number between 0 and 1.[99] The function ROUND(X,0) rounds X to zero decimal places—that is, to the nearest integer. Therefore, to obtain a random binary digit in any cell, just enter the following formula:

    =ROUND(RAND(),0)

[98] Because the set of outcomes of a pair of coin tosses is {HH, HT, TH, TT}, we expect HH to come up once in every four trials, on average.

[99] The parentheses after RAND are necessary even though the function takes no arguments.

Spreadsheets can also generate a random integer directly (without the need for rounding) through the formula

```
=RANDBETWEEN(0,1)
```

To obtain a whole array of random numbers, just drag this formula into the cells you wish to use.

b. We need to generate pairs of random binary digits and then check whether they are both 1s. It would be convenient if the spreadsheet could tell you right away whether both digits are 1s (corresponding to two consecutive heads in a coin toss). Here is a simple way of accomplishing this. Notice that if we *add* two random binary digits, we obtain either 0, 1, or 2, telling us the number of heads that result from the two consecutive throws. Therefore, all we need to do is add pairs of random digits and then count the number of times 2 comes up. Formulas we can use are

```
=RANDBETWEEN(0,1)+RANDBETWEEN(0,1)
```

What would be even *more* convenient is if the result of the calculation would be either 0 or 1, with 1 signifying success (two consecutive heads) and 0 signifying failure. Then, we could simply add up all the results to obtain the number of times two heads occurred. To do this, we first divide the result of the calculation above by 2 (obtaining 0, .5, or 1, where now 1 signifies success) and then round *down* to an integer using a function called "int":

```
=INT(0.5*(RANDBETWEEN(0,1)+
RANDBETWEEN(0,1)))
```

Following is the result of 100 such pairs of coin tosses, with 1 signifying success (two heads) and 0 signifying failure (all other outcomes). The last column records the number of successes in each row and the total number at the end.

1	1	0	0	0	0	0	0	0	0	2
0	1	0	0	0	0	0	1	0	1	3
0	1	0	0	1	1	0	0	0	1	4
0	0	0	0	0	0	0	0	1	0	1
0	1	0	0	1	0	0	1	0	0	3
1	0	1	0	0	0	0	0	0	0	2
0	0	0	0	0	0	0	0	0	1	1
0	1	1	1	1	0	0	0	0	1	5
1	1	0	1	0	0	1	1	0	0	5
0	0	0	0	0	0	0	0	1	0	1

27

Now, as in part (a), we can compute estimated probabilities, with D standing for the outcome "two heads":

First 10 trials: $P(D) = \dfrac{fr(1)}{N} = \dfrac{2}{10} = .2$

First 20 trials: $P(D) = \dfrac{fr(1)}{N} = \dfrac{5}{20} = .25$

First 50 trials: $P(D) = \dfrac{fr(1)}{N} = \dfrac{13}{50} = .26$

100 trials: $P(D) = \dfrac{fr(1)}{N} = \dfrac{27}{100} = .27.$

Q: *What is happening with the data? The probabilities seem to be getting less accurate as N increases!*

A: Quite by chance, exactly 5 of the first 20 trials resulted in success, which matches the theoretical probability. Figure 24 shows an Excel plot of estimated probability versus N (for N a multiple of 10). Notice that, as N increases, the graph seems to meander within smaller distances of .25.

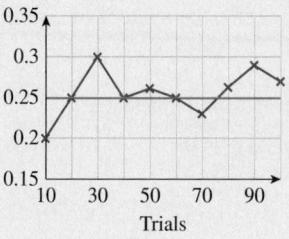

Figure 24

Q: *The previous techniques work fine for simulating coin tosses. What about rolls of a fair die, where we want outcomes between 1 and 6?*

A: We can simulate a roll of a die by generating a random integer in the range 1 through 6. The following formula accomplishes this:

```
=1 + INT(5.99999*RAND())
```

(We used 5.99999 instead of 6 to avoid the outcome 7.)

Section 7.7

Example 1 (page 523)   Consider the Markov system found by Gamble Detergents at the beginning of this section. Suppose that 70% of consumers are now using powdered detergent while the other 30% are using liquid. What will be the distribution one year from now? Two years from now? Three years from now?

Solution with Technology

In your spreadsheet, enter the initial distribution vector in cells A1 and B1 and the transition matrix to the right of that, as shown.

	A	B	C	D	E
1	0.7	0.3		0.8	0.2
2				0.1	0.9

To calculate the distribution after one step, use the array formula

$$=\text{MMULT(A1:B1,\$D\$1:\$E\$2)}$$

The absolute cell references (dollar signs) ensure that the formula always refers to the same transition matrix, even if we copy it into other cells. To use the array formula, select cells A2 and B2, where the distribution vector will go, enter this formula, and then press Control+Shift+Enter.[100]

	A	B	C	D	E
1	0.7	0.3		0.8	0.2
2	=MMULT(A1:B1,$D$1:$E$2)			0.1	0.9

The result is the following, with the distribution after one step highlighted.

	A	B	C	D	E
1	0.7	0.3		0.8	0.2
2	0.59	0.41		0.1	0.9

To calculate the distribution after two steps, select cells A2 and B2 and drag the fill handle down to copy the formula to cells A3 and B3. Note that the formula now takes the vector in A2:B2 and multiplies it by the transition matrix to get the vector in A3:B3. To calculate several more steps, drag down as far as desired.

	A	B	C	D	E
1	0.7	0.3		0.8	0.2
2	0.59	0.41		0.1	0.9
3					
4					

	A	B	C	D	E
1	0.7	0.3		0.8	0.2
2	0.59	0.41		0.1	0.9
3	0.513	0.487			
4	0.4591	0.5409			

Example 4 (page 527) Calculate the steady-state probability vector for the transition matrix in the preceding examples.

Solution with Technology

Finding the steady-state probability vector comes down to solving a system of equations. As discussed in Chapters 3 and 4, there are several ways to use Excel to help. The most straightforward is to use matrix inversion to solve the matrix form of the system. In this case, as in the text, the system of equations we need to solve is

$$x + \quad y = 1$$
$$-.2x + .1y = 0.$$

We write this as the matrix equation $AX = B$ with

$$A = \begin{bmatrix} 1 & 1 \\ -.2 & .1 \end{bmatrix} \qquad B = \begin{bmatrix} 1 \\ 0 \end{bmatrix}.$$

We enter A in cells A1:B2, B in cells D1:D2, and the formula for $X = A^{-1}B$ in a convenient location, say B4:B5.

	A	B	C	D
1	1	1		1
2	-0.2	0.1		0
3				
4		=MMULT(MINVERSE(A1:B2),D1:D2)		
5				

When we press Control+Shift+Enter we see the result:

	A	B	C	D
1	1	1		1
2	-0.2	0.1		0
3				
4		0.3333333		
5		0.6666667		

If we want to see the answer in fraction rather than decimal form, we format the cells as fractions.

	A	B	C	D
1	1	1		1
2	-0.2	0.1		0
3				
4		1/3		
5		2/3		

---

[100] On a Macintosh, you can also use Command+Enter.

8

Random Variables and Statistics

 Website

www.WanerMath.com

At the Website you will find:

- Section-by-section tutorials, including game tutorials with randomized quizzes

- A detailed chapter summary

- A true/false quiz

- Additional review exercises

- Histogram, Bernoulli trials, and normal distribution utilities

- The following optional extra sections:

    Sampling Distributions and the Central Limit Theorem

    Confidence Intervals

    Calculus and Statistics

Case Study Spotting Tax Fraud with Benford's Law

You are a tax fraud specialist working for the Internal Revenue Service (IRS), and you have just been handed a portion of the tax return from Colossal Conglomerate. The IRS suspects that the portion you were handed may be fraudulent and would like your opinion. **Is there any mathematical test, you wonder, that can point to a suspicious tax return based on nothing more than the numbers entered?**

age fotostock/SuperStock

547

Introduction

Statistics is the branch of mathematics concerned with organizing, analyzing, and interpreting numerical data. For example, given the current annual incomes of 1,000 lawyers selected at random, you might wish to answer some questions: If I become a lawyer, what income am I likely to earn? Do lawyers' salaries vary widely? How widely?

To answer questions like these, it helps to begin by organizing the data in the form of tables or graphs. This is the topic of the first section of the chapter. The second section describes an important class of examples that are applicable to a wide range of situations, from tossing a coin to product testing.

Once the data are organized, the next step is to apply mathematical tools for analyzing the data and answering questions like those posed above. Numbers such as the **mean** and the **standard deviation** can be computed to reveal interesting facts about the data. These numbers can then be used to make predictions about future events.

The chapter ends with a section on one of the most important distributions in statistics, the **normal distribution**. This distribution describes many sets of data and also plays an important role in the underlying mathematical theory.

8.1   Random Variables and Distributions

Random Variables

In many experiments we can assign numerical values to the outcomes. For instance, if we roll a die, each outcome has a value from 1 through 6. If you select a lawyer and ascertain his or her annual income, the outcome is again a number. We call a rule that assigns a number to each outcome of an experiment a **random variable**.

✻ In the language of functions (Chapter 1), a random variable is a *real-valued function* whose domain is the sample space.

Random Variable

A **random variable** X is a rule that assigns a number, or **value**, to each outcome in the sample space of an experiment.✻

Visualizing a Random Variable

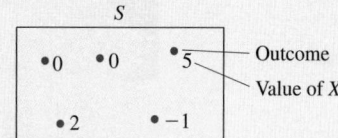

Quick Examples

1. Roll a die; X = the number facing up.
2. Select a mutual fund; X = the number of companies in the fund portfolio.
3. Select a computer; X = the number of gigabytes of memory it has.
4. Survey a group of 20 college students; X = the mean SAT.

Discrete and Continuous Random Variables

A **discrete** random variable can take on only specific, isolated numerical values, like the outcome of a roll of a die, or the number of dollars in a randomly chosen bank account. A **continuous** random variable, on the other hand, can take on any values within a continuum or an interval, like the temperature in Central Park, or the height of an athlete in centimeters. Discrete random variables that can take on only finitely many values (like the outcome of a roll of a die) are called **finite** random variables.

Quick Examples

Random Variable	Values	Type
1. Select a mutual fund; $X =$ the number of companies in the fund portfolio.	$\{1, 2, 3, \ldots\}$	Discrete Infinite
2. Take five shots at the goal during a soccer match; $X =$ the number of times you score.	$\{0, 1, 2, 3, 4, 5\}$	Finite
3. Measure the length of an object; $X =$ its length in centimeters.	Any positive real number	Continuous
4. Roll a die until you get a 6; $X =$ the number of times you roll the die.	$\{1, 2, \ldots\}$	Discrete Infinite
5. Bet a whole number of dollars in a race where the betting limit is \$100; $X =$ the amount you bet.	$\{0, 1, \ldots, 100\}$	Finite
6. Bet a whole number of dollars in a race where there is no betting limit; $X =$ the amount you bet.	$\{0, 1, \ldots, 100, 101, \ldots\}$	Discrete Infinite

Notes

1. In Chapter 7, the only sample spaces we considered in detail were finite sample spaces. However, in general, sample spaces can be infinite, as in many of the experiments mentioned above.

2. There are some "borderline" situations. For instance, if X is the salary of a factory worker, then X is, strictly speaking, discrete. However, the values of X are so numerous and close together that in some applications it makes sense to model X as a continuous random variable. ■

For the moment, we shall consider only finite random variables.

EXAMPLE 1 Finite Random Variable

Let X be the number of heads that come up when a coin is tossed three times. List the value of X for each possible outcome. What are the possible values of X?

Solution First, we describe X as a random variable.

X is the rule that assigns to each outcome the number of heads that come up.

2 Heads ($X = 2$)

We take as the outcomes of this experiment all possible sequences of three heads and tails. Then, for instance, if the outcome is HTH, the value of X is 2. An easy way to list the values of X for all the outcomes is by means of a table.

Outcome	HHH	HHT	HTH	HTT	THH	THT	TTH	TTT
Value of X	3	2	2	1	2	1	1	0

Website
www.WanerMath.com
Go to the Chapter 8 Topic Summary
to find an interactive simulation based
on Example 1.

From the table, we also see that the possible values of X are 0, 1, 2, and 3.

➡ **Before we go on...** Remember that X is just a rule we decide on. In Example 1, we could have taken X to be a different rule, such as the number of tails or perhaps the number of heads minus the number of tails. These different rules are examples of different random variables associated with the same experiment. ∎

EXAMPLE 2 Stock Prices

You have purchased $10,000 worth of stock in a biotech company whose newest arthritis drug is awaiting approval by the Food and Drug Administration (FDA). If the drug is approved this month, the value of the stock will double by the end of the month. If the drug is rejected this month, the stock's value will decline by 80%. If no decision is reached this month, its value will decline by 10%. Let X be the value of your investment at the end of this month. List the value of X for each possible outcome.

Solution There are three possible outcomes: the drug is approved this month, it is rejected this month, and no decision is reached. Once again, we express the random variable as a rule.

The random variable X is the rule that assigns to each outcome the value of your investment at the end of this month.

We can now tabulate the values of X as follows:

Outcome	Approved This Month	Rejected This Month	No Decision
Value of X	$20,000	$2,000	$9,000

Probability Distribution of a Finite Random Variable

Given a random variable X, it is natural to look at certain *events*—for instance, the event that $X = 2$. By this, we mean the event consisting of all outcomes that have an assigned X-value of 2. Looking once again at the chart in Example 1, with X being the number of heads that face up when a coin is tossed three times, we find the following events:

The event that $X = 0$ is {TTT}.
The event that $X = 1$ is {HTT, THT, TTH}.
The event that $X = 2$ is {HHT, HTH, THH}.
The event that $X = 3$ is {HHH}.
The event that $X = 4$ is ∅.   There are no outcomes with four heads.

Each of these events has a certain probability. For instance, the probability of the event that $X = 2$ is 3/8 because the event in question consists of three of the eight possible (equally likely) outcomes. We shall abbreviate this by writing

$$P(X = 2) = \frac{3}{8}.$$ The probability that $X = 2$ is 3/8.

Similarly,

$$P(X = 4) = 0.$$ The probability that $X = 4$ is 0.

When X is a finite random variable, the collection of the probabilities of X equaling each of its possible values is called the **probability distribution** of X. Because the probabilities in a probability distribution can be estimated or theoretical, we shall discuss both *estimated probability distributions* (or *relative frequency distributions*) and *theoretical (modeled) probability distributions* of random variables. (See the next two examples.)

Probability Distribution of a Finite Random Variable

If X is a finite random variable, with values $n_1, n_2, \ldots$ then its **probability distribution** lists the probabilities that $X = n_1$, $X = n_2$, $\ldots$ The sum of these probabilities is always 1.

Visualizing the Probability Distribution of a Random Variable

If each outcome in S is equally likely, we get the probability distribution shown for the random variable X.

S

•0	•0	•5
•2	•−1	

Probability Distribution of X

x	−1	0	2	5
$P(X = x)$	$\frac{1}{5} = .2$	$\frac{2}{5} = .4$	$\frac{1}{5} = .2$	$\frac{1}{5} = .2$

Here $P(X = x)$ means "the probability that the random variable X has the specific value x."

Quick Example

Roll a fair die; $X =$ the number facing up. Then, the probability that any specific value of X occurs is $\frac{1}{6}$. So, the probability distribution of X is the following (notice that the probabilities add up to 1):

x	1	2	3	4	5	6
$P(X = x)$	$\frac{1}{6}$	$\frac{1}{6}$	$\frac{1}{6}$	$\frac{1}{6}$	$\frac{1}{6}$	$\frac{1}{6}$

Using this probability distribution, we can calculate the probabilities of certain events; for instance:

$$P(X < 3) = \frac{1}{3}$$ The event that $X < 3$ is the event $\{1, 2\}$.

$$P(1 < X < 5) = \frac{1}{2}$$ The event that $1 < X < 5$ is the event $\{2, 3, 4\}$.

Note The distinction between X (uppercase) and x (lowercase) in the tables above is important; X stands for the random variable in question, whereas x stands for a specific *value* of X (so that x is always a number). Thus, if, say $x = 2$, then $P(X = x)$ means $P(X = 2)$, the probability that X is 2. Similarly, if Y is a random variable, then $P(Y = y)$ is the probability that Y has the specific value y. ■

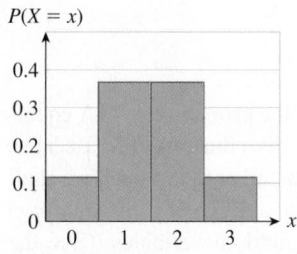

$P(X = x)$

Figure 1

EXAMPLE 3 Probability Distribution

Let X be the number of heads that face up in three tosses of a coin. Give the probability distribution of X. What is the probability of throwing at least two heads?

Solution X is the random variable of Example 1, so its values are 0, 1, 2, and 3. The probability distribution of X is given in the following table:

x	0	1	2	3
$P(X = x)$	$\dfrac{1}{8}$	$\dfrac{3}{8}$	$\dfrac{3}{8}$	$\dfrac{1}{8}$

Notice that the probabilities add to 1, as we might expect. From the distribution, the probability of throwing at least two heads is

$$P(X \ge 2) = P(\{2, 3\}) = \frac{3}{8} + \frac{1}{8} = \frac{1}{2}.$$

We can use a bar graph to visualize a probability distribution. Figure 1 shows the bar graph for the probability distribution we obtained. Such a graph is sometimes called a **histogram**.

➡ **Before we go on...** The probabilities in the table in Example 3 are *modeled* probabilities. To obtain a similar table of relative frequencies, we would have to repeatedly toss a coin three times and calculate the fraction of times we got 0, 1, 2, and 3 heads. ∎

Note The table of probabilities in Example 3 looks like the probability distribution associated with an experiment, as we studied in Section 7.3. In fact, the probability distribution of a random variable is not really new. Consider the following experiment: toss three coins and count the number of heads. The associated probability distribution (as per Section 7.3) would be this:

Outcome	0	1	2	3
Probability	$\dfrac{1}{8}$	$\dfrac{3}{8}$	$\dfrac{3}{8}$	$\dfrac{1}{8}$

The difference is that in this chapter we are thinking of 0, 1, 2, and 3 not as the outcomes of the experiment, but as values of the random variable X. ∎

EXAMPLE 4 Relative Frequency Distribution

The following table shows the (fictitious) income brackets of a sample of 1,000 lawyers in their first year out of law school.

Income Bracket	$20,000–$29,999	$30,000–$39,999	$40,000–$49,999	$50,000–$59,999	$60,000–$69,999	$70,000–$79,999	$80,000–$89,999
Number	20	80	230	400	170	70	30

Think of the experiment of choosing a first-year lawyer at random (all being equally likely) and assign to each lawyer the number X that is the midpoint of his or her income bracket. Find the relative frequency distribution of X.

Solution Statisticians refer to the income brackets as **measurement classes**. Because the first measurement class contains incomes that are at least $20,000, but less than $30,000, its midpoint is $25,000.* Similarly the second measurement class has midpoint $35,000, and so on. We can rewrite the table with the midpoints, as follows:

x	25,000	35,000	45,000	55,000	65,000	75,000	85,000
Frequency	20	80	230	400	170	70	30

We have used the term *frequency* rather than *number*, although it means the same thing. This table is called a **frequency table**. It is *almost* the relative frequency distribution for X, except that we must replace frequencies by relative frequencies. (We did this in calculating relative frequencies in the preceding chapter.) We start with the lowest measurement class. Because 20 of the 1,000 lawyers fall in this group, we have

$$P(X = 25,000) = \frac{20}{1,000} = .02.$$

We can calculate the remaining relative frequencies similarly to obtain the following distribution:

x	25,000	35,000	45,000	55,000	65,000	75,000	85,000
$P(X = x)$	.02	.08	.23	.40	.17	.07	.03

Note again the distinction between X and x: X stands for the random variable in question, whereas x stands for a specific value (25,000, 35,000, . . . , or 85,000) of X.

EXAMPLE 5 Probability Distribution: Greenhouse Gases

The following table shows per capita emissions of greenhouse gases for the 30 countries with the highest per capita carbon dioxide emissions. (Emissions are rounded to the nearest 5 metric tons.)[1]

Country	Per Capita Emissions (metric tons)	Country	Per Capita Emissions (metric tons)
Qatar	70	Estonia	15
Kuwait	40	Faroe Islands	15
United Arab Emirates	40	Saudi Arabia	15
Luxembourg	25	Kazakhstan	15
Trinidad and Tobago	25	Gibraltar	15
Brunei	25	Finland	15
Bahrain	25	Oman	15
Netherlands Antilles	20	Singapore	10
Aruba	20	Palau	10
United States	20	Montserrat	10
Canada	20	Czech Republic	10
Norway	20	Equatorial Guinea	10
Australia	15	New Caledonia	10
Falkland Islands	15	Israel	10
Nauru	15	Russia	10

[1]Figures are based on 2004 data. Source: United Nations Millennium Development Goals Indicators (http://mdgs.un.org/unsd/mdg/Data.aspx).

---

* One might argue that the midpoint should be (20,000 + 29,999)/2 = 24,999.50, but we round this to 25,000. So, technically we are using "rounded" midpoints of the measurement classes.

using Technology

Technology can be used to automate the calculations in Example 4. Here is an outline.

TI-83/84 Plus

STAT EDIT values of x in L_1 and frequencies in L_2.
Home screen: $L_2/\text{sum}(L_2) \to L_3$
[More details on page 615.]

Spreadsheet

Headings x, Fr, and $P(X = x)$ in A1–C1
x-values and frequencies in Columns A2–B8
=B2/SUM(B:B) in C2
Copy down column C.
[More details on page 616.]

 Website
www.WanerMath.com
Student Home
  → On Line Utilities
  → Histogram Utility
Enter the x-values and frequencies as shown:

```
25000, 20
35000, 80
45000, 230
55000, 400
65000, 170
75000, 70
85000, 30
```

Make sure "Show probability distribution" is checked, and press "Results". The relative frequency distribution will appear at the bottom of the page.

Consider the experiment in which a country is selected at random from this list, and let X be the per capita carbon dioxide emissions for that country. Find the probability distribution of X and graph it with a histogram. Use the probability distribution to compute $P(X \geq 20)$ (the probability that X is 20 or more) and interpret the result.

Solution The values of X are the possible emissions figures, which we can take to be 0, 5, 10, 15, ..., 70. In the table below, we first compute the frequency of each value of X by counting the number of countries that produce that per capita level of greenhouse gases. For instance, there are seven countries that have $X = 15$. Then, we divide each frequency by the sample size $N = 30$ to obtain the probabilities.*

* Even though we are using the term "frequency," we are really calculating *modeled* probability based on the assumption of equally likely outcomes. In this context, the frequencies are the number of favorable outcomes for each value of X. (See the Q&A discussion at the end of Section 7.3.)

x	0	5	10	15	20	25	30	35	40	45	50	55	60	65	70
Frequency	0	0	8	10	5	4	0	0	2	0	0	0	0	0	1
$P(X=x)$	0	0	$\frac{8}{30}$	$\frac{10}{30}$	$\frac{5}{30}$	$\frac{4}{30}$	0	0	$\frac{2}{30}$	0	0	0	0	0	$\frac{1}{30}$

Figure 2 shows the resulting histogram.

Finally, we compute $P(X \geq 20)$, the probability of the event that X has a value of 20 or more, which is the sum of the probabilities $P(X = 20)$, $P(X = 25)$, and so on. From the table, we obtain

$$P(X \geq 20) = \frac{5}{30} + \frac{4}{30} + \frac{2}{30} + \frac{1}{30} = \frac{12}{30} = .4.$$

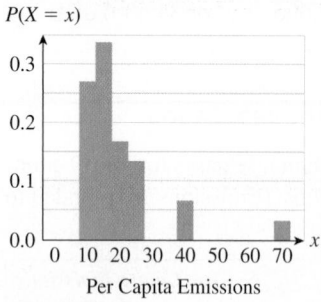
Figure 2

Thus, there is a 40% chance that a country randomly selected from the given list produces 20 or more metric tons per capita of carbon dioxide.

FAQs

Recognizing What to Use as a Random Variable and Deciding on Its Values

Q: *In an application, how, exactly, do I decide what to use as a random variable X?*

A: Be as systematic as possible: First, decide what the experiment is and what its sample space is. Then, based on what is asked for in the application, complete the following sentence: "X assigns ___ to each outcome." For instance, "X assigns the number of flavors to each packet of gummy bears selected," or "X assigns the average faculty salary to each college selected."

Q: *Once I have decided what X should be, how do I decide what values to assign it?*

A: Ask yourself: What are the conceivable values I could get for X? Then choose a collection of values that includes all of these. For instance, if X is the number of heads obtained when a coin is tossed five times, then the possible values of X are 0, 1, 2, 3, 4, and 5. If X is the average faculty salary in dollars, rounded to the nearest $5,000, then possible values of X could be 20,000, 25,000, 30,000, and so on, up to the highest salary in your data.

8.1 EXERCISES

▼ more advanced   ◆ challenging
▪ indicates exercises that should be solved using technology

In Exercises 1–10, classify each random variable X as finite, discrete infinite, or continuous, and indicate the values that X can take. HINT [See Quick Examples on page 549.]

1. Roll two dice; X = the sum of the numbers facing up.

2. Open a 500-page book on a random page; X = the page number.

3. Select a stock at random; X = your profit, to the nearest dollar, if you purchase one share and sell it one year later.

4. Select an electric utility company at random; X = the exact amount of electricity, in gigawatt hours, it supplies in a year.

5. Look at the second hand of your watch; X is the time it reads in seconds.

6. Watch a soccer game; X = the total number of goals scored.

7. Watch a soccer game; X = the total number of goals scored, up to a maximum of 10.

8. Your class is given a mathematics exam worth 100 points; X is the average score, rounded to the nearest whole number.

9. According to quantum mechanics, the energy of an electron in a hydrogen atom can assume only the values $k/1$, $k/4$, $k/9$, $k/16$, ... for a certain constant value k. X = the energy of an electron in a hydrogen atom.

10. According to classical mechanics, the energy of an electron in a hydrogen atom can assume any positive value. X = the energy of an electron in a hydrogen atom.

In Exercises 11–18, (a) say what an appropriate sample space is; (b) complete the following sentence: "X is the rule that assigns to each . . . "; (c) list the values of X for all the outcomes. HINT [See Example 1.]

11. X is the number of tails that come up when a coin is tossed twice.

12. X is the largest number of consecutive times heads comes up in a row when a coin is tossed three times.

13. X is the sum of the numbers that face up when two dice are rolled.

14. X is the value of the larger number when two dice are rolled.

15. X is the number of red marbles that Tonya has in her hand after she selects four marbles from a bag containing four red marbles and two green ones and then notes how many there are of each color.

16. X is the number of green marbles that Stej has in his hand after he selects four marbles from a bag containing three red

marbles and two green ones and then notes how many there are of each color.

17. The mathematics final exam scores for the students in your study group are 89%, 85%, 95%, 63%, 92%, and 80%.

18. The capacities of the hard drives of your dormitory suite mates' computers are 1,000 GB, 1,500 GB, 2,000 GB, 2,500 GB, 3,000 GB, and 3,500 GB.

19. The random variable X has this probability distribution table:

x	2	4	6	8	10
$P(X = x)$	.1	.2	–	–	.1

a. Assuming $P(X = 8) = P(X = 6)$, find each of the missing values. HINT [See Quick Example on page 551.]

b. Calculate $P(X \geq 6)$ and $P(2 < X < 8)$.

20. The random variable X has the probability distribution table shown below:

x	−2	−1	0	1	2
$P(X = x)$	–	–	.4	.1	.1

a. Calculate $P(X \geq 0)$ and $P(X < 0)$. HINT [See Quick Example on page 551.]

b. Assuming $P(X = -2) = P(X = -1)$, find each of the missing values.

In Exercises 21–28, give the probability distribution for the indicated random variable and draw the corresponding histogram, and calculate the indicated probability. HINT [See Example 3.]

21. A fair die is rolled, and X is the number facing up. Calculate $P(X < 5)$.

22. A fair die is rolled, and X is the square of the number facing up. Calculate $P(X > 9)$.

23. Three fair coins are tossed, and X is the square of the number of heads showing. Calculate $P(1 \leq X \leq 9)$.

24. Three fair coins are tossed, and X is the number of heads minus the number of tails. Calculate $P(-3 \leq X \leq -1)$.

25. A red die and a green die are rolled, and X is the sum of the numbers facing up. Calculate $P(X \neq 7)$.

26. A red die and a green die are rolled, and
$$X = \begin{cases} 0 & \text{if the numbers are the same} \\ 1 & \text{if the numbers are different.} \end{cases}$$
Calculate $P(X > 1)$.

27. ▼ A red die and a green die are rolled, and X is the larger of the two numbers facing up. Calculate $P(X \leq 3)$.

28. ▼ A red die and a green die are rolled, and X is the smaller of the two numbers facing up. Calculate $P(X \geq 4)$.

APPLICATIONS

29. *2010 Income Distribution up to $100,000* The following table shows the distribution of household incomes in 2010 for a sample of 1,000 households in the U.S. with incomes up to $100,000.[2]

Income Bracket ($)	0– 19,999	20,000– 39,999	40,000– 59,000	60,000– 79,999	80,000– 99,999
Households	240	290	180	170	120

a. Let X be the (rounded) midpoint of a bracket in which a household falls. Find the relative frequency distribution of X and graph its histogram. HINT [See Example 4.]

b. Shade the area of your histogram corresponding to the probability that a randomly selected U.S. household in the sample has a value of X above 50,000. What is this probability?

30. *2003 Income Distribution up to $100,000* Repeat Exercise 29, using the following data from a sample of 1,000 households in the U.S. in 2003.[3]

Income Bracket ($)	0– 19,999	20,000– 39,999	40,000– 59,000	60,000– 79,999	80,000– 99,999
Households	270	280	200	150	100

31. *Population Age in Mexico* The following chart shows the ages of 250 randomly selected residents of Mexico:[4]

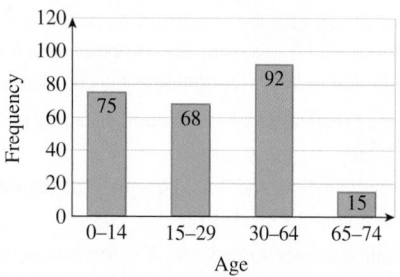

What is the associated random variable? Represent the data as a relative frequency distribution using the (rounded) midpoints of the given measurement classes. HINT [See Example 4.]

32. *Population Age in the U.S.* Repeat the preceding exercise, using the following chart, which shows the ages of 250 randomly selected residents of the U.S.:[5]

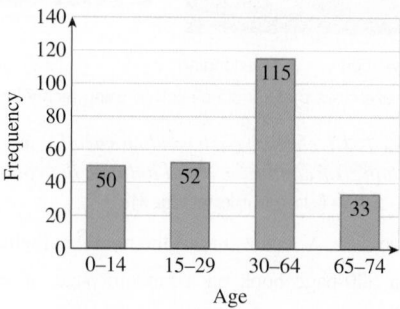

33. *Sport Utility Vehicles—Tow Ratings* The following table shows tow ratings (in pounds) for some popular sports utility vehicles in 2000:[6]

Mercedes Grand Marquis V8	2,000
Jeep Wrangler I6	2,000
Ford Explorer V6	3,000
Dodge Dakota V6	4,000
Mitsubishi Montero V6	5,000
Ford Explorer V8	6,000
Dodge Durango V8	6,000
Dodge Ram 1500 V8	8,000
Ford Expedition V8	8,000
Hummer 2-door Hardtop	8,000

Let X be the tow rating of a randomly chosen popular SUV from the list above.

a. What are the values of X?

b. Compute the frequency and probability distributions of X. HINT [See Example 5.]

c. What is the probability that an SUV (from the list above) is rated to tow no more than 5,000 pounds?

34. *Housing Prices Going into the Real Estate Bubble* The following table shows the average percentage increase in the price of a house from 1980 to 2001 in nine regions of the United States.[7]

---

[2]Based on actual income distribution in 2010. Source: U.S. Census Bureau, Current Population Survey, 2010 American Community Survey (www.census.gov).

[3]Based on actual income distribution in 2003 (not adjusted for inflation). Source: U.S. Census Bureau, Current Population Survey, 2004 Annual Social and Economic Supplement (www.census.gov).

[4]Based on population distribution in 2010. Source: Instituto Nacional de Estadística y Geografía (www.inegi.org.mx).

[5]*Ibid.* (The data for the U.S. was also provided by the Instituto Nacional de Estadística y Geografía.)

[6]Tow ratings are for 2000 models and vary considerably within each model. Figures cited are rounded. For more detailed information, consult www.rvsafety.com/towrate2k.htm.

[7]Percentages are rounded to the nearest 25%. Source: Third Quarter 2001 House Price Index, released November 30, 2001 by the Office of Federal Housing Enterprise Oversight; available online at www.ofheo.gov/house/3q01hpi.pdf.

c. Compute $P(X \leq 2)$.

d. Compute $P(X \geq 2)$.

Solution

a. The experiment is a sequence of Bernoulli trials; in each trial we select a person and ascertain his or her age. If we take "success" to mean selection of a person aged 65 or older, then the probability distribution is

$$P(X = x) = C(n, x)p^x q^{n-x}$$

where n = number of trials = 6
p = probability of success = .2
q = probability of failure = .8

So,
$$P(X = 4) = C(6, 4)(.2)^4(.8)^2$$
$$= 15 \times .0016 \times .64 = .01536$$

b. We have already computed $P(X = 4)$. Here are all the calculations:

$$P(X = 0) = C(6, 0)(.2)^0(.8)^6$$
$$= 1 \times 1 \times .262144 = .262144$$

$$P(X = 1) = C(6, 1)(.2)^1(.8)^5$$
$$= 6 \times .2 \times .32768 = .393216$$

$$P(X = 2) = C(6, 2)(.2)^2(.8)^4$$
$$= 15 \times .04 \times .4096 = .24576$$

$$P(X = 3) = C(6, 3)(.2)^3(.8)^3$$
$$= 20 \times .008 \times .512 = .08192$$

$$P(X = 4) = C(6, 4)(.2)^4(.8)^2$$
$$= 15 \times .0016 \times .64 = .01536$$

$$P(X = 5) = C(6, 5)(.2)^5(.8)^1$$
$$= 6 \times .00032 \times .8 = .001536$$

$$P(X = 6) = C(6, 6)(.2)^6(.8)^0$$
$$= 1 \times .000064 \times 1 = .000064.$$

The probability distribution is therefore as follows:

x	0	1	2	3	4	5	6
$P(X = x)$	.262144	.393216	.24576	.08192	.01536	.001536	.000064

Figure 3 shows its histogram.

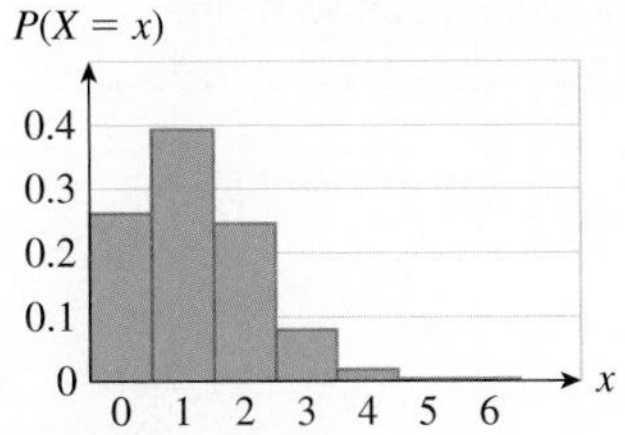

$P(X = x)$

Figure 3

 using Technology

Technology can be used to replicate the histogram in Example 2.

TI-83/84 Plus

Y= screen: `Y₁=6 nCr`
`X*0.2^X*0.8^(6-X)`
`2ND` `TBLSET` Set Indpt to Ask.
`2ND` `TABLE` Enter x-values
0, 1, . . . , 6.

[More details on page 613.]

c. $P(X \leq 2)$—the probability that the number of people selected who are at least 65 years old is either 0, 1, or 2—is the probability of the union of these events and is thus the sum of the three probabilities:

$$P(X \leq 2) = P(X = 0) + P(X = 1) + P(X = 2)$$
$$= .262144 + .393216 + .24576$$
$$= .90112.$$

Spreadsheet

Headings x and $P(X = x)$ in A1, B1

x-values 0, 1, . . . , 6 in A2–A8

`=BINOMDIST(A2,6,0,.2,0)` in B2

Copy down to B6.

[More details on page 616.]

Website

www.WanerMath.com

  Student Home

    → On Line Utilities

    → Binomial Distribution Utility

Enter $n = 6$ and $p = .2$ and press "Generate Distribution."

d. To compute $P(X \geq 2)$, we *could* compute the sum

$$P(X \geq 2) = P(X = 2) + P(X = 3) + P(X = 4) + P(X = 5) + P(X = 6)$$

but it is far easier to compute the probability of the complement of the event:

$$P(X < 2) = P(X = 0) + P(X = 1)$$
$$= .262144 + .393216 = .65536$$

and then subtract the answer from 1:

$$P(X \geq 2) = 1 - P(X < 2)$$
$$= 1 - .65536 = .34464.$$

FAQs

Terminology and Recognizing when to Use the Binomial Distribution

Q :*What is the difference between Bernoulli trials and a binomial random variable?*

A :A Bernoulli trial is a type of experiment, whereas a binomial random variable is the resulting kind of random variable. More precisely, if your experiment consists of performing a sequence of n Bernoulli trials (think of throwing a dart n times at random points on a dartboard hoping to hit the bull's eye), then the random variable X that counts the number of successes (the number of times you actually hit the bull's eye) is a binomial random variable.

Q :*How do I recognize when a situation gives a binomial random variable?*

A :Make sure that the experiment consists of a sequence of independent Bernoulli trials; that is, a sequence of a fixed number of trials of an experiment that has two outcomes, where the outcome of each trial does not depend on the outcomes in previous trials, and where the probability of success is the same for all the trials. For instance, repeatedly throwing a dart at a dartboard hoping to hit the bull's eye does not constitute a sequence of Bernoulli trials if you adjust your aim each time depending on the outcome of your previous attempt. This dart-throwing experiment can be modeled by a sequence of Bernoulli trials if you make no adjustments after each attempt and your aim does not improve (or deteriorate) with time.

8.2   EXERCISES

▼ more advanced   ◆ challenging

Ⓣ indicates exercises that should be solved using technology

In Exercises 1–10, you are performing 5 independent Bernoulli trials with $p = .1$ and $q = .9$. Calculate the probability of each of the stated outcomes. Check your answer using technology. HINT [See Quick Example on page 562.]

1. Two successes   **2.** Three successes

3. No successes   **4.** No failures

5. All successes   **6.** All failures

7. At most two successes   **8.** At least four successes

9. At least three successes   **10.** At most three successes

In Exercises 11–18, X is a binomial variable with $n = 6$ and $p = .4$. Compute the given probabilities. Check your answer using technology. HINT [See Example 2.]

11. $P(X = 3)$ **12.** $P(X = 4)$

13. $P(X \leq 2)$ **14.** $P(X \leq 1)$

15. $P(X \geq 5)$ **16.** $P(X \geq 4)$

17. $P(1 \leq X \leq 3)$ **18.** $P(3 \leq X \leq 5)$

In Exercises 19 and 20, graph the histogram of the given binomial distribution. Check your answer using technology.

19. $n = 5, p = \dfrac{1}{4}, q = \dfrac{3}{4}$ **20.** $n = 5, p = \dfrac{1}{3}, q = \dfrac{2}{3}$

In Exercises 21 and 22, graph the histogram of the given binomial distribution and compute the given quantity, indicating the corresponding region on the graph.

21. $n = 4$, $p = \dfrac{1}{3}$, $q = \dfrac{2}{3}$; $P(X \le 2)$

22. $n = 4$, $p = \dfrac{1}{4}$, $q = \dfrac{3}{4}$; $P(X \le 1)$

APPLICATIONS

23. *Internet Addiction* The probability that a randomly chosen person in the Netherlands connects to the Internet immediately upon waking is approximately .25.[12] What is the probability that, in a randomly selected sample of five people, two connect to the Internet immediately upon waking? HINT [See Example 2.]

24. *Alien Retirement* The probability that a randomly chosen citizen-entity of Cygnus is of pension age[13] is approximately .8. What is the probability that, in a randomly selected sample of four citizen-entities, all of them are of pension age? HINT [See Example 2.]

25. *1990s Internet Stock Boom* According to a July 1999 article in *The New York Times*,[14] venture capitalists had this "rule of thumb": The probability that an Internet start-up company will be a "stock market success" resulting in "spectacular profits for early investors" is .2. If you were a venture capitalist who invested in 10 Internet start-up companies, what was the probability that at least 1 of them would be a stock market success? (Round your answer to four decimal places.)

26. *1990s Internet Stock Boom* According to the article cited in Exercise 25, 13.5% of Internet stocks that entered the market in 1999 ended up trading below their initial offering prices. If you were an investor who purchased five Internet stocks at their initial offering prices, what was the probability that at least four of them would end up trading at or above their initial offering price? (Round your answer to four decimal places.)

27. *Job Training (from the GRE Exam in Economics)* In a large on-the-job training program, half of the participants are female and half are male. In a random sample of three participants, what is the probability that an investigator will draw at least one male?

28. *Job Training (based on a question from the GRE Exam in Economics)* In a large on-the-job training program, half of the participants are female and half are male. In a random sample of five participants, what is the probability that an investigator will draw at least two males?

29. *Manufacturing* Your manufacturing plant produces air bags, and it is known that 10% of them are defective. Five air bags are tested.
 a. Find the probability that three of them are defective.
 b. Find the probability that at least two of them are defective.

30. *Manufacturing* Compute the probability distribution of the binomial variable described in the preceding exercise, and use it to compute the probability that if five air bags are tested, at least one will be defective and at least one will not.

31. *Teenage Pastimes* According to a study,[15] the probability that a randomly selected teenager watched a rented video at least once during a week was .71. What is the probability that at least 8 teenagers in a group of 10 watched a rented movie at least once last week?

32. *Other Teenage Pastimes* According to the study cited in the preceding exercise, the probability that a randomly selected teenager studied at least once during a week was only .52. What is the probability that less than half of the students in your study group of 10 have studied in the last week?

33. *Subprime Mortgages during the Housing Bubble* In November 2008 the probability that a randomly selected subprime home mortgage in Florida was in foreclosure was .24.[16] Choose 10 subprime home mortgages at random.
 a. What is the probability that exactly 5 of them were in foreclosure?
 b. ⊤ Use technology to generate the probability distribution for the associated binomial random variable.
 c. Fill in the blank: If 10 subprime home mortgages were chosen at random, the number of them most likely to have been in foreclosure was _____.

34. *Subprime Mortgages during the Housing Bubble* In November 2008 the probability that a randomly selected subprime home mortgage in Texas was current in its payments was .67.[17] Choose 10 subprime home mortgages at random.
 a. What is the probability that exactly 4 of them were current?
 b. ⊤ Use technology to generate the probability distribution for the associated binomial random variable.
 c. Fill in the blank: If 10 subprime home mortgages were chosen at random, the number of them most likely to have been current was _____.

35. ▼ *Triple Redundancy* In order to ensure reliable performance of vital computer systems, aerospace engineers sometimes employ the technique of "triple redundancy," in which three identical computers are installed in a space vehicle. If one of the three computers gives results different

[12]Source: *Webwereld* November 17, 2008 (http://webwereld.nl/article/view/id/53599).

[13]The retirement age in Cygnus is 12,000 bootlags, which is equivalent to approximately 20 minutes Earth time.

[14]"Not All Hit It Rich in the Internet Gold Rush," *New York Times*, July 20, 1999, p. A1.

[15]Sources: Rand Youth Poll/Teen-Age Research Unlimited/*New York Times*, March 14, 1998, p. D1.

[16]Source: Federal Reserve Bank of New York (www.newyorkfed.org/regional/subprime.html).

[17]*Ibid.*

from the other two, it is assumed to be malfunctioning and it is ignored. This technique will work as long as no more than one computer malfunctions. Assuming that an onboard computer is 99% reliable (that is, the probability of its failing is .01), what is the probability that at least two of the three computers will malfunction?

36. ▼ *IQ Scores* Mensa is a club for people who have high IQ scores. To qualify, your IQ must be at least 132, putting you in the top 2% of the general population. If a group of 10 people are chosen at random, what is the probability that at least 2 of them qualify for Mensa?

37. ⊡ ▼ *Standardized Tests* Assume that on a standardized test of 100 questions, a person has a probability of 80% of answering any particular question correctly. Find the probability of answering between 75 and 85 questions, inclusive, correctly. (Assume independence, and round your answer to 4 decimal places.)

38. ⊡ ▼ *Standardized Tests* Assume that on a standardized test of 100 questions, a person has a probability of 80% of answering any particular question correctly. Find the probability of answering at least 90 questions correctly. (Assume independence, and round your answer to 4 decimal places.)

39. ⊡ ▼ *Product Testing* It is known that 43% of all the ZeroFat hamburger patties produced by your factory actually contain more than 10 grams of fat. Compute the probability distribution for $n = 50$ Bernoulli trials.

   a. What is the most likely value for the number of burgers in a sample of 50 that contain more than 10 grams of fat?

   b. Complete the following sentence: There is an approximately 71% chance that a batch of 50 ZeroFat patties contains _____ or more patties with more than 10 grams of fat.

   c. Compare the graphs of the distributions for $n = 50$ trials and $n = 20$ trials. What do you notice?

40. ⊡ ▼ *Product Testing* It is known that 65% of all the ZeroCal hamburger patties produced by your factory actually contain more than 1,000 calories. Compute the probability distribution for $n = 50$ Bernoulli trials.

   a. What is the most likely value for the number of burgers in a sample of 50 that contain more than 1,000 calories?

   b. Complete the following sentence: There is an approximately 73% chance that a batch of 50 ZeroCal patties contains _____ or more patties with more than 1,000 calories.

   c. Compare the graphs of the distributions for $n = 50$ trials and $n = 20$ trials. What do you notice?

41. ⊡ ▼ *Quality Control* A manufacturer of light bulbs chooses bulbs at random from its assembly line for testing. If the probability of a bulb's being bad is .01, how many bulbs does the manufacturer need to test before the probability of finding at least one bad one rises to more than .5? (You may have to use trial and error to solve this.)

42. ⊡ ▼ *Quality Control* A manufacturer of light bulbs chooses bulbs at random from its assembly line for testing. If the probability of a bulb's being bad is .01, how many bulbs does the manufacturer need to test before the probability of finding at least two bad ones rises to more than .5? (You may have to use trial and error to solve this.)

43. ▼ *Highway Safety* According to a study,[18] a male driver in the United States will average 562 accidents per 100 million miles. Regard an n-mile trip as a sequence of n Bernoulli trials with "success" corresponding to having an accident during a particular mile. What is the probability that a male driver will have an accident in a 1-mile trip?

44. ▼ *Highway Safety*: According to the study cited in the preceding exercise, a female driver in the United States will average 611 accidents per 100 million miles. Regard an n-mile trip as a sequence of n Bernoulli trials with "success" corresponding to having an accident during a particular mile. What is the probability that a female driver will have an accident in a 1-mile trip?

45. ◆ *Mad Cow Disease* In March 2004, the U.S. Department of Agriculture announced plans to test approximately 243,000 slaughtered cows per year for mad cow disease (bovine spongiform encephalopathy).[19] When announcing the plan, the Agriculture Department stated that "by the laws of probability, that many tests should detect mad cow disease even if it is present in only 5 cows out of the 45 million in the nation."[20] Test the Department's claim by computing the probability that, if only 5 out of 45 million cows had mad cow disease, at least 1 cow would test positive in a year (assuming the testing was done randomly).

46. ◆ *Mad Cow Disease* According to the article cited in Exercise 45, only 223,000 of the cows being tested for bovine spongiform encephalopathy were to be "downer cows"; cows unable to walk to their slaughter. Assuming that just one downer cow in 500,000 is infected on average, use a binomial distribution to find the probability that 2 or more cows would test positive in a year. Your associate claims that "by the laws of probability, that many tests should detect at least 2 cases of mad cow disease even if it is present in only 2 cows out of a million downers." Comment on that claim.

COMMUNICATION AND REASONING EXERCISES

47. A soccer player is more likely to score on his second shot if he was successful on his first. Can we model a succession of shots a player takes as a sequence of Bernoulli trials? Explain.

---

[18]Data are based on a report by National Highway Traffic Safety Administration released in January, 1996. Source for data: U.S. Department of Transportation/*New York Times*, April 9, 1999, p. F1.

[19]Source: *New York Times*, March 17, 2004, p. A19.

[20]As stated in the *New York Times* article.

48. A soccer player takes repeated shots on goal. What assumption must we make if we want to model a succession of shots by a player as a sequence of Bernoulli trials?

49. Your friend just told you that "misfortunes always occur in threes." If life is just a sequence of Bernoulli trials, is this possible? Explain.

50. Suppose an experiment consists of repeatedly (every week) checking whether your graphing calculator battery has died. Is this a sequence of Bernoulli trials? Explain.

51. In an experiment with sample space S, a certain event E has a probability p of occurring. What has this scenario to do with Bernoulli trials?

52. An experiment consists of removing a gummy bear from a bag originally containing 10, and then eating it. Regard eating a lime-flavored bear as success. Repeating the experiment five times is a sequence of Bernoulli trials, right?

53. ▼ Why is the following not a binomial random variable? Select, without replacement, five marbles from a bag containing six red marbles and two blue ones, and let X be the number of red marbles you have selected.

54. ▼ By contrast with Exercise 53, why can the following be modeled by a binomial random variable? Select, without replacement, 5 electronic components from a batch of 10,000 in which 1,000 are defective, and let X be the number of defective components you select.

8.3   Measures of Central Tendency

Mean, Median, and Mode of a Set of Data

One day you decide to measure the popularity rating of your statistics instructor, Mr. Pelogrande. Ideally, you should poll all of Mr. Pelogrande's students, which is what statisticians would refer to as the **population**. However, it would be difficult to poll all the members of the population in question (Mr. Pelogrande teaches more than 400 students). Instead, you decide to survey 10 of his students, chosen at random, and ask them to rate Mr. Pelogrande on a scale of 0–100. The survey results in the following set of data:

$$60, 50, 55, 0, 100, 90, 40, 20, 40, 70.$$

Such a collection of data is called a **sample**, because the 10 people polled represent only a (small) sample of Mr. Pelogrande's students. We should think of the individual scores 60, 50, 55, . . . as values of a random variable: Choose one of Mr. Pelogrande's students at random and let X be the rating the student gives to Mr. Pelogrande.

How do we distill a single measurement, or **statistic**, from this sample that would describe Mr. Pelogrande's popularity? Perhaps the most commonly used statistic is the **average**, or **mean**, which is computed by adding the scores and dividing the sum by the number of scores in the sample:

$$\text{Sample Mean} = \frac{60 + 50 + 55 + 0 + 100 + 90 + 40 + 20 + 40 + 70}{10}$$

$$= \frac{525}{10} = 52.5.$$

We might then conclude, based on the sample, that Mr. Pelogrande's average popularity rating is about 52.5. The usual notation for the sample mean is $\bar{x}$, and the formula we use to compute it is

$$\bar{x} = \frac{x_1 + x_2 + \cdots + x_n}{n},$$

where $x_1, x_2, \ldots, x_n$ are the values of X in the sample.

A convenient way of writing the sum that appears in the numerator is to use **summation** or **sigma notation**. We write the sum $x_1 + x_2 + \cdots + x_n$ as

$$\sum_{i=1}^{n} x_i.$$

$\sum_{i=1}^{n}$ by itself stands for "the sum, from $i = 1$ to n."

$\sum_{i=1}^{n} x_i$ stands for "the sum of the x_i, from $i = 1$ to n."

We think of i as taking on the values $1, 2, \ldots, n$ in turn, making x_i equal $x_1, x_2, \ldots, x_n$ in turn, and we then add up these values.

Sample and Mean

A **sample** is a sequence of values (or scores) of a random variable X. (The process of collecting such a sequence is sometimes called **sampling** X.) The **sample mean** is the average of the values, or **scores**, in the sample. To compute the sample mean, we use the following formula:

$$\bar{x} = \frac{x_1 + x_2 + \cdots + x_n}{n} = \frac{\sum_{i=1}^{n} x_i}{n}$$

or simply

$$\bar{x} = \frac{\sum_i x_i}{n}. \qquad \sum_i \text{ stands for "sum over all } i." *$$

* In Section 1.4 we simply wrote $\sum x$ for the sum of all the x_i, but here we will use the subscripts to make it easier to interpret formulas in this and the next section.

† When we talk about *populations*, the understanding is that the underlying experiment consists of selecting a member of a given population and ascertaining the value of X.

Here, n is the **sample size** (number of scores), and $x_1, x_2, \ldots, x_n$ are the individual values.

If the sample $x_1, x_2, \ldots, x_n$ consists of all the values of X from the entire population† (for instance, the ratings given Mr. Pelogrande by *all* of his students), we refer to the mean as the **population mean**, and write it as μ (Greek "mu") instead of $\bar{x}$.

Visualizing the Mean

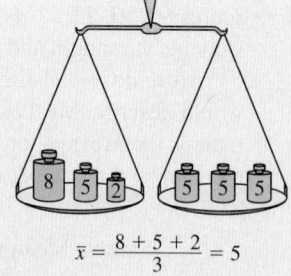

$$\bar{x} = \frac{8 + 5 + 2}{3} = 5$$

Quick Examples

1. The mean of the sample $1, 2, 3, 4, 5$ is $\bar{x} = 3$.

2. The mean of the sample $-1, 0, 2$ is $\bar{x} = \dfrac{-1 + 0 + 2}{3} = \dfrac{1}{3}$.

3. The mean of the population $-3, -3, 0, 0, 1$ is
$$\mu = \frac{-3 - 3 + 0 + 0 + 1}{5} = -1.$$

Note: Sample Mean versus Population Mean

Determining a population mean can be difficult or even impossible. For instance, computing the mean household income for the United States would entail recording the income of every single household in the United States. Instead of attempting to do this, we usually use sample means instead. The larger the sample used, the more accurately we expect the sample mean to approximate the population mean. Estimating how accurately a sample mean based on a given sample size approximates the population mean is possible, but we will not go into that in this book. ■

The mean $\bar{x}$ is an attempt to describe where the "center" of the sample is. It is therefore called a **measure of central tendency**. There are two other common measures of central tendency: the "middle score," or **median**, and the "most frequent score," or **mode**. These are defined as follows.

Median and Mode

The **sample median** m is the middle score (in the case of an odd-size sample), or average of the two middle scores (in the case of an even-size sample) when the scores in a sample are arranged in ascending order.

A **sample mode** is a score that appears most often in the collection. (There may be more than one mode in a sample.)

Visualizing the Median and Mode

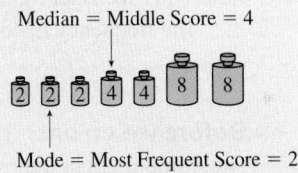

As before, we refer to the **population median** and **population mode** if the sample consists of the data from the entire population.

Quick Examples

1. The sample median of 2, −3, −1, 4, 2 is found by first arranging the scores in ascending order: −3, −1, 2, 2, 4 and then selecting the middle (third) score: $m = 2$. The sample mode is also 2 because this is the score that appears most often.

2. The sample 2, 5, 6, −1, 0, 6 has median $m = (2 + 5)/2 = 3.5$ and mode 6.

The mean tends to give more weight to scores that are further away from the center than does the median. For example, if you take the largest score in a collection of more than two numbers and make it larger, the mean will increase but the median will remain the same. For this reason the median is often preferred for collections that contain a wide range of scores. The mode can sometimes lie far from the center and is thus used less often as an indication of where the "center" of a sample lies.

EXAMPLE 1 **Teenage Spending in the 1990s**

A 10-year survey of spending patterns of U.S. teenagers in the 1990s yielded the following figures (in billions of dollars spent in a year):[21] 90, 90, 85, 80, 80, 80, 80, 85, 90, 100. Compute and interpret the mean, median, and mode, and illustrate the data on a graph.

Solution The *mean* is given by

$$\bar{x} = \frac{\sum_i x_i}{n}$$

$$= \frac{90 + 90 + 85 + 80 + 80 + 80 + 80 + 85 + 90 + 100}{10} = \frac{860}{10} = 86.$$

Thus, spending by teenagers averaged $86 billion per year.

For the *median*, we arrange the sample data in ascending order:

$$80, 80, 80, 80, 85, 85, 90, 90, 90, 100.$$

We then take the average of the two middle scores:

$$m = \frac{85 + 85}{2} = 85.$$

This means that in half the years in question, teenagers spent $85 billion or less, and in half they spent $85 billion or more.

For the *mode* we choose the score (or scores) that occurs most frequently: $80 billion. Thus, teenagers spent $80 billion per year more often than any other amount.

The frequency histogram in Figure 4 illustrates these three measures.

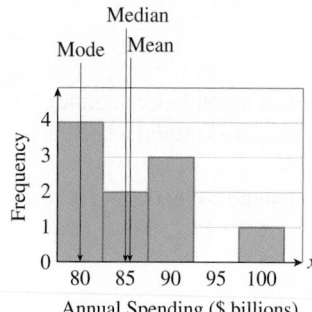

Figure 4

➡ **Before we go on...** There is a nice geometric interpretation of the difference between the median and mode: The median line shown in Figure 4 divides the total area of the histogram into two equal pieces, whereas the mean line passes through its "center of gravity"; if you placed the histogram on a knife-edge along the mean line, it would balance. ∎

Expected Value of a Finite Random Variable

Now, instead of looking at a sample of values of a given random variable, let us look at the probability distribution of the random variable itself and see if we can predict the sample mean without actually taking a sample. This prediction is what we call the *expected value* of the random variable.

EXAMPLE 2 **Expected Value of a Random Variable**

Suppose you roll a fair die a large number of times. What do you expect to be the average of the numbers that face up?

Solution Suppose we take a sample of n rolls of the die (where n is large). Because the probability of rolling a 1 is $1/6$, we would expect that we would roll a 1 one sixth

---

[21]Spending figures are rounded, and cover the years 1988 through 1997. Source: Rand Youth Poll/Teen-Age Research Unlimited/*New York Times*, March 14, 1998, p. D1.

of the time, or $n/6$ times. Similarly, each other number should also appear $n/6$ times. The frequency table should then look like this:

x	1	2	3	4	5	6
Number of Times x Is Rolled (frequency)	$\dfrac{n}{6}$	$\dfrac{n}{6}$	$\dfrac{n}{6}$	$\dfrac{n}{6}$	$\dfrac{n}{6}$	$\dfrac{n}{6}$

Note that we would not really expect the scores to be evenly distributed in practice, although for very large values of n we would expect the frequencies to vary only by a small percentage. To calculate the sample mean, we would add up all the scores and divide by the sample size. Now, the table tells us that there are $n/6$ ones, $n/6$ twos, $n/6$ threes, and so on, up to $n/6$ sixes. Adding these all up gives

$$\sum_i x_i = \frac{n}{6} \cdot 1 + \frac{n}{6} \cdot 2 + \frac{n}{6} \cdot 3 + \frac{n}{6} \cdot 4 + \frac{n}{6} \cdot 5 + \frac{n}{6} \cdot 6.$$

(Notice that we can obtain this number by multiplying the frequencies by the values of X and then adding.) Thus, the mean is

$$\begin{aligned}
\bar{x} &= \frac{\sum_i x_i}{n} \\
&= \frac{\frac{n}{6} \cdot 1 + \frac{n}{6} \cdot 2 + \frac{n}{6} \cdot 3 + \frac{n}{6} \cdot 4 + \frac{n}{6} \cdot 5 + \frac{n}{6} \cdot 6}{n} \\
&= \frac{1}{6} \cdot 1 + \frac{1}{6} \cdot 2 + \frac{1}{6} \cdot 3 + \frac{1}{6} \cdot 4 + \frac{1}{6} \cdot 5 + \frac{1}{6} \cdot 6 \qquad \text{Divide top and bottom by } n. \\
&= 3.5.
\end{aligned}$$

This is the average value we expect to get after a large number of rolls or, in short, the **expected value** of a roll of the die. More precisely, we say that this is the expected value of the random variable X whose value is the number we get by rolling a die. Notice that n, the number of rolls, does not appear in the expected value. In fact, we could redo the calculation more simply by dividing the frequencies in the table by n *before* adding. Doing this replaces the frequencies with the *probabilities*, $1/6$. That is, it *replaces the frequency distribution with the probability distribution.*

x	1	2	3	4	5	6
$P(X = x)$	$\dfrac{1}{6}$	$\dfrac{1}{6}$	$\dfrac{1}{6}$	$\dfrac{1}{6}$	$\dfrac{1}{6}$	$\dfrac{1}{6}$

The expected value of X is then the sum of the products $x \cdot P(X = x)$. This is how we shall compute it from now on.

Expected Value of a Finite Random Variable

If X is a finite random variable that takes on the values $x_1, x_2, \ldots, x_n$, then the **expected value** of X, written $E(X)$ or μ, is

$$\begin{aligned}
\mu &= E(X) = x_1 \cdot P(X = x_1) + x_2 \cdot P(X = x_2) + \cdots + x_n \cdot P(X = x_n) \\
&= \sum_i x_i \cdot P(X = x_i).
\end{aligned}$$

To obtain the expected value, multiply the values of X by their probabilities, and then add the results.

In Words

To compute the expected value from the probability distribution of X, we multiply the values of X by their probabilities and add up the results.

Interpretation

We interpret the expected value of X as a *prediction* of the mean of a large random sample of measurements of X; in other words, it is what we "expect" the mean of a large number of scores to be. (The larger the sample, the more accurate this prediction will tend to be.)

Quick Example

If X has the distribution shown,

x	-1	0	4	5
$P(X = x)$	.3	.5	.1	.1

then $\mu = E(X) = -1(.3) + 0(.5) + 4(.1) + 5(.1) = -.3 + 0 + .4 + .5$
$= .6$.

EXAMPLE 3 Sports Injuries

According to historical data, the number of injuries that a member of the Enormous State University women's soccer team will sustain during a typical season is given by the following probability distribution table:

Injuries	0	1	2	3	4	5	6
Probability	.20	.20	.22	.20	.15	.01	.02

If X denotes the number of injuries sustained by a player during one season, compute $E(X)$ and interpret the result.

Solution We can compute the expected value using the following tabular approach: Take the probability distribution table, add another row in which we compute the product $x P(X = x)$, and then add these products together.

x	0	1	2	3	4	5	6	
$P(X = x)$	.20	.20	.22	.20	.15	.01	.02	**Total:**
$xP(X = x)$	0	.20	.44	.60	.60	.05	.12	2.01

The total of the entries in the bottom row is the expected value. Thus,

$$E(X) = 2.01.$$

We interpret the result as follows: If many soccer players are observed for a season, we predict that the average number of injuries each will sustain is about two.

EXAMPLE 4 **Roulette**

A roulette wheel (of the kind used in the United States) has the numbers 1 through 36, 0 and 00. A bet on a single number pays 35 to 1. This means that if you place a $1 bet on a single number and win (your number comes up), you get your $1 back plus $35 (that is, you gain $35). If your number does not come up, you lose the $1 you bet. What is the expected gain from a $1 bet on a single number?

Solution The probability of winning is 1/38, so the probability of losing is 37/38. Let X be the gain from a $1 bet. X has two possible values: $X = -1$ if you lose and $X = 35$ if you win. $P(X = -1) = 37/38$ and $P(X = 35) = 1/38$. This probability distribution and the calculation of the expected value are given in the following table:

x	-1	35	
$P(X = x)$	$\frac{37}{38}$	$\frac{1}{38}$	**Total:**
$xP(X = x)$	$-\frac{37}{38}$	$\frac{35}{38}$	$-\frac{2}{38}$

So, we expect to average a small loss of $2/38 \approx \$0.0526$ on each spin of the wheel.

➡ **Before we go on...** Of course, you cannot actually lose the expected $0.0526 on one spin of the roulette wheel in Example 4. However, if you play many times, this is what you expect your *average* loss per bet to be. For example, if you played 100 times, you could expect to lose about $100 \times 0.0526 = \$5.26$. ∎

    A betting game in which the expected value is zero is called a **fair game**. For example, if you and I flip a coin, and I give you $1 each time it comes up heads but you give me $1 each time it comes up tails, then the game is fair. Over the long run, we expect to come out even. On the other hand, a game like roulette, in which the expected value is not zero, is **biased**. Most casino games are slightly biased in favor of the house.* Thus, most gamblers will lose only a small amount and many gamblers will actually win something (and return to play some more). However, when the earnings are averaged over the huge numbers of people playing, the house is guaranteed to come out ahead. This is how casinos make (lots of) money.

＊ Only rarely are games not biased in favor of the house. However, blackjack played without continuous shuffle machines can be beaten by card counting.

Expected Value of a Binomial Random Variable

Suppose you guess all the answers to the questions on a multiple-choice test. What score can you expect to get? This scenario is an example of a sequence of Bernoulli trials (see the preceding section), and the number of correct guesses is therefore a binomial random variable whose expected value we wish to know. There is a simple formula for the expected value of a binomial random variable.

Expected Value of Binomial Random Variable

If X is the binomial random variable associated with n independent Bernoulli trials, each with probability p of success, then the expected value of X is

$$\mu = E(X) = np.$$

Quick Examples

1. If X is the number of successes in 20 Bernoulli trials with $p = .7$, then the expected number of successes is $\mu = E(X) = (20)(.7) = 14$.

2. If an event F in some experiment has $P(F) = .25$, the experiment is repeated 100 times, and X is the number of times F occurs, then $E(X) = (100)(.25) = 25$ is the number of times we expect F to occur.

Where does this formula come from? We *could* use the formula for expected value and compute the sum

$$E(X) = 0C(n, 0)p^0 q^n + 1C(n, 1)p^1 q^{n-1} + 2C(n, 2)p^2 q^{n-2} + \cdots + nC(n, n)p^n q^0$$

directly (using the binomial theorem), but this is one of the many places in mathematics where a less direct approach is much easier. X is the number of successes in a sequence of n Bernoulli trials, each with probability p of success. Thus, p is the fraction of time we expect a success, so out of n trials we expect np successes. Because X counts successes, we expect the value of X to be np. (With a little more effort, this can be made into a formal proof that the sum above equals np.)

EXAMPLE 5 Guessing on an Exam

An exam has 50 multiple-choice questions, each having four choices. If a student randomly guesses on each question, how many correct answers can he or she expect to get?

Solution Each guess is a Bernoulli trial with probability of success 1 in 4, so $p = .25$. Thus, for a sequence of $n = 50$ trials,

$$\mu = E(X) = np = (50)(.25) = 12.5.$$

Thus, the student can expect to get about 12.5 correct answers.

Q: *Wait a minute. How can a student get a fraction of a correct answer?*

A: Remember that the expected value is the average number of correct answers a student will get if he or she guesses on a large number of such tests. Or, we can say that if many students use this strategy of guessing, they will average about 12.5 correct answers each.

Estimating the Expected Value from a Sample

It is not always possible to know the probability distribution of a random variable. For instance, if we take X to be the income of a randomly selected lawyer, we could not be expected to know the probability distribution of X. However, we can still obtain a good *estimate* of the expected value of X (the average income of all lawyers) by using the relative frequency distribution based on a large random sample.

EXAMPLE 6 **Estimating an Expected Value**

The following table shows the (fictitious) incomes of a random sample of 1,000 lawyers in the United States in their first year out of law school.

Income Bracket	$20,000–$29,999	$30,000–$39,999	$40,000–$49,999	$50,000–$59,999	$60,000–$69,999	$70,000–$79,999	$80,000–$89,999
Number	20	80	230	400	170	70	30

Estimate the average of the incomes of all lawyers in their first year out of law school.

Solution We first interpret the question in terms of a random variable. Let X be the income of a lawyer selected at random from among all currently practicing first-year lawyers in the United States. We are given a sample of 1,000 values of X, and we are asked to find the expected value of X. First, we use the midpoints of the income brackets to set up a relative frequency distribution for X:

x	25,000	35,000	45,000	55,000	65,000	75,000	85,000
$P(X = x)$	.02	.08	.23	.40	.17	.07	.03

Our estimate for $E(X)$ is then

$$E(X) = \sum_i x_i \cdot P(X = x_i)$$

$$= (25,000)(.02) + (35,000)(.08) + (45,000)(.23) + (55,000)(.40)$$
$$+ (65,000)(.17) + (75,000)(.07) + (85,000)(.03) = \$54,500.$$

Thus, $E(X)$ is approximately \$54,500. That is, the average income of all currently practicing first-year lawyers in the United States is approximately \$54,500.

FAQ

Recognizing when to Compute the Mean and when to Compute the Expected Value

Q: *When am I supposed to compute the mean (add the values of X and divide by n) and when am I supposed to use the expected value formula?*

A: The formula for the mean (adding and dividing by the number of observations) is used to compute the mean of a sequence of random scores, or sampled values of X. If, on the other hand, you are given the probability distribution for X (even if it is only an estimated probability distribution), then you need to use the expected value formula.

8.3 EXERCISES

Compute the mean, median, and mode of the data samples in Exercises 1–8. HINT [See Quick Examples on pages 568 and 569.]

1. −1, 5, 5, 7, 14

2. 2, 6, 6, 7, −1

3. 2, 5, 6, 7, −1, −1

4. 3, 1, 6, −3, 0, 5

5. $\frac{1}{2}, \frac{3}{2}, -4, \frac{5}{4}$

6. $-\frac{3}{2}, \frac{3}{8}, -1, \frac{5}{2}$

7. 2.5, −5.4, 4.1, −0.1, −0.1

8. 4.2, −3.2, 0, 1.7, 0

9. ▼ Give a sample of six scores with mean 1 and with median ≠ mean. (Arrange the scores in ascending order.)

10. ▼ Give a sample of five scores with mean 100 and median 1. (Arrange the scores in ascending order.)

In Exercises 11–16, calculate the expected value of X for the given probability distribution. HINT [See Quick Example on page 572.]

11.

x	0	1	2	3
P(X = x)	.5	.2	.2	.1

12.

x	1	2	3	4
P(X = x)	.1	.2	.5	.2

13.

x	10	20	30	40
P(X = x)	$\frac{15}{50}$	$\frac{20}{50}$	$\frac{10}{50}$	$\frac{5}{50}$

14.

x	2	4	6	8
P(X = x)	$\frac{1}{20}$	$\frac{15}{20}$	$\frac{2}{20}$	$\frac{2}{20}$

15.

x	−5	−1	0	2	5	10
P(X = x)	.2	.3	.2	.1	.2	0

16.

x	−20	−10	0	10	20	30
P(X = x)	.2	.4	.2	.1	0	.1

In Exercises 17–28, calculate the expected value of the given random variable X. [Exercises 23, 24, 27, and 28 assume familiarity with counting arguments and probability (Section 7.4).] HINT [See Quick Example page 572.]

17. *X* is the number that faces up when a fair die is rolled.

18. *X* is a number selected at random from the set {1, 2, 3, 4}.

19. *X* is the number of tails that come up when a coin is tossed twice.

20. *X* is the number of tails that come up when a coin is tossed three times.

21. ▼ *X* is the higher number when two dice are rolled.

22. ▼ *X* is the lower number when two dice are rolled.

23. ▼ *X* is the number of red marbles that Suzan has in her hand after she selects four marbles from a bag containing four red marbles and two green ones.

24. ▼ *X* is the number of green marbles that Suzan has in her hand after she selects four marbles from a bag containing three red marbles and two green ones.

25. ▼ Twenty darts are thrown at a dartboard. The probability of hitting a bull's-eye is .1. Let *X* be the number of bull's-eyes hit.

26. ▼ Thirty darts are thrown at a dartboard. The probability of hitting a bull's-eye is $\frac{1}{5}$. Let *X* be the number of bull's-eyes hit.

27. **T** ▼ Select 5 cards without replacement from a standard deck of 52, and let *X* be the number of queens you draw.

28. **T** ▼ Select 5 cards without replacement from a standard deck of 52, and let *X* be the number of red cards you draw.

APPLICATIONS

29. *Stock Market Gyrations* Following is a sample of the day-by-day change, rounded to the nearest 100 points, in the Dow Jones Industrial Average during 10 successive business days around the start of the financial crisis in October 2008:[22]

−400, −500, −200, −700, −100, 900, −100, −700, 400, −100

Compute the mean and median of the given sample. Fill in the blank: There were as many days with a change in the Dow above _____ points as there were with changes below that. HINT [See Quick Examples on pages 568 and 569.]

30. *Stock Market Gyrations* Following is a sample of the day-by-day change, rounded to the nearest 100 points, in the Dow Jones Industrial Average during 10 successive business days around the start of the financial crisis in October 2008:[23]

−100, 400, −200, −500, 200, −300, −200, 900, −100, 200

Compute the mean and median of the given sample. Fill in the blank: There were as many days with a change in the Dow above _____ points as there were with changes below that. HINT [See Quick Examples on pages 568 and 569.]

31. *Gold* The following figures show the price of gold per ounce, in dollars, for the 10-business-day period Nov. 17–Nov. 30, 2011:[24]

1,756, 1,730, 1,704, 1,698, 1,686, 1,699, 1,676, 1,714, 1,717, 1,704

Find the sample mean, median, and mode(s). What do your answers tell you about the price of gold?

---

[22]Source: http://finance.google.com.

[23]*Ibid.*

[24]Prices rounded to the nearest $1. Source: www.kitco.com/gold.londonfix.html.

32. *Silver* The following figures show the price of silver per ounce, in dollars, for the 10-business-day period Nov. 17–Nov. 30, 2011:[25]

$$33, 33, 31, 32, 32, 32, 31, 32, 32, 32$$

Find the sample mean, median, and mode(s). What do your answers tell you about the price of silver?

33. *Supermarkets* A survey of 52 U.S. supermarkets yielded the following relative frequency table, where X is the number of checkout lanes at a randomly chosen supermarket.[26]

x	1	2	3	4	5	6	7	8	9	10
$P(X = x)$	.01	.04	.04	.08	.10	.15	.25	.20	.08	.05

a. Compute $\mu = E(X)$ and interpret the result.
HINT [See Example 3.]
b. Which is larger, $P(X < \mu)$ or $P(X > \mu)$? Interpret the result.

34. *Video Arcades* Your company, *Sonic Video, Inc.*, has conducted research that shows the following probability distribution, where X is the number of video arcades in a randomly chosen city with more than 500,000 inhabitants.

x	0	1	2	3	4	5	6	7	8	9
$P(X = x)$	.07	.09	.35	.25	.15	.03	.02	.02	.01	.01

a. Compute $\mu = E(X)$ and interpret the result.
HINT [See Example 3.]
b. Which is larger, $P(X < \mu)$ or $P(X > \mu)$? Interpret the result.

35. *School Enrollment* The following table shows the approximate numbers of school goers in the United States (residents who attended some educational institution) in 1998, broken down by age group.[27]

Age	3–6.9	7–12.9	13–16.9	17–22.9	23–26.9	27–42.9
Population (millions)	12	24	15	14	2	5

Use the rounded midpoints of the given measurement classes to compute the probability distribution of the age X of a school goer. (Round probabilities to 2 decimal places.) Hence compute the expected value of X. What information does the expected value give about residents enrolled in schools?

36. *School Enrollment* Repeat Exercise 35, using the following data from 1980.[28]

Age	3–6.9	7–12.9	13–16.9	17–22.9	23–26.9	27–42.9
Population (millions)	8	20	11	13	1	3

37. *Population Age in Mexico* The following chart shows the ages of 250 randomly selected residents of Mexico:[29]

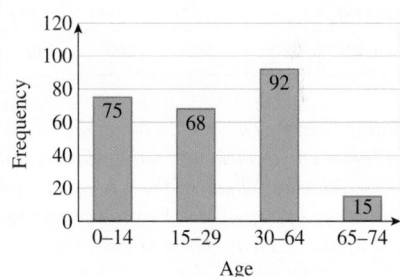

Use the relative frequency distribution based on the (rounded) midpoints of the given measurement classes to obtain an estimate of the average age of a resident in Mexico. (Round the answer to 1 decimal place). HINT [See Example 6.]

38. *Population Age in the U.S.* Repeat the preceding exercise, using the following chart, which shows the ages of 250 randomly selected residents of the U.S.:[30]

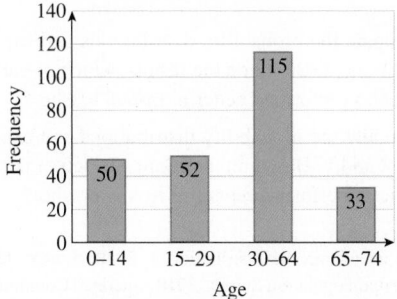

---

[25]Prices rounded to the nearest $1. Source: www.kitco.com/gold.londonfix.html.

[26]Sources: J.T. McClave, P.G. Benson, T. Sincich, *Statistics for Business and Economics,* 7th Ed. (Prentice Hall, 1998), p. 177; W. Chow *et al*. "A model for predicting a supermarket's annual sales per square foot," Graduate School of Management, Rutgers University.

[27]Data are approximate. Source: Statistical Abstract of the United States: 2000.

[28]*Ibid.*

[29]Based on population distribution in 2010. Source: Instituto Nacional de Estadística y Geografía (www.inegi.org.mx).

[30]*Ibid.*

39. *2010 Income Distribution up to $100,000* The following table shows the distribution of household incomes in 2010 for a sample of 1,000 households in the U.S. with incomes up to $100,000.[31]

Income Bracket ($)	0–19,999	20,000–39,999	40,000–59,000	60,000–79,999	80,000–99,999
Households	240	290	180	170	120

Use this information to estimate, to the nearest $1,000, the average household income for such households. HINT [See Example 6.]

40. *2003 Income Distribution up to $100,000* Repeat Exercise 39, using the following data from a sample of 1,000 households in the U.S. in 2003.[32]

Income Bracket ($)	0–19,999	20,000–39,999	40,000–59,000	60,000–79,999	80,000–99,999
Households	270	280	200	150	100

Highway Safety Exercises 41–44 are based on the following table, which shows crashworthiness ratings for several categories of motor vehicles.[33] In all of these exercises, take X as the crash-test rating of a small car, Y as the crash-test rating for a small SUV, and so on as shown in the table.

	Number Tested	**Overall Frontal Crash-Test Rating**			
		3 (Good)	2 (Acceptable)	1 (Marginal)	0 (Poor)
Small Cars X	16	1	11	2	2
Small SUVs Y	10	1	4	4	1
Medium SUVs Z	15	3	5	3	4
Passenger Vans U	13	3	0	3	7
Midsize Cars V	15	3	5	0	7
Large Cars W	19	9	5	3	2

41. ▼ Compute the probability distributions and expected values of X and Y. Based on the results, which of the two types of vehicles performed better in frontal crashes?

42. ▼ Compute the probability distributions and expected values of Z and V. Based on the results, which of the two types of vehicles performed better in frontal crashes?

43. ⬛ ▼ Based on expected values, which of the following categories performed best in crash tests: small cars, midsize cars, or large cars?

44. ⬛ ▼ Based on expected values, which of the following categories performed best in crash tests: small SUVs, medium SUVs, or passenger vans?

45. ▼ *Roulette* A roulette wheel has the numbers 1 through 36, 0, and 00. Half of the numbers from 1 through 36 are red, and a bet on red pays even money (that is, if you win, you will get back your $1 plus another $1). How much do you expect to win with a $1 bet on red? HINT [See Example 4.]

46. ▼ *Roulette* A roulette wheel has the numbers 1 through 36, 0, and 00. A bet on two numbers pays 17 to 1 (that is, if one of the two numbers you bet comes up, you get back your $1 plus another $17). How much do you expect to win with a $1 bet on two numbers? HINT [See Example 4.]

47. *Teenage Pastimes* According to a study,[34] the probability that a randomly selected teenager shopped at a mall at least once during a week was .63. How many teenagers in a randomly selected group of 40 would you expect to shop at a mall during the next week? HINT [See Example 5.]

48. *Other Teenage Pastimes* According to the study referred to in the preceding exercise, the probability that a randomly selected teenager played a computer game at least once during a week was .48. How many teenagers in a randomly selected group of 30 would you expect to play a computer game during the next 7 days? HINT [See Example 5.]

49. ▼ *Manufacturing* Your manufacturing plant produces air bags, and it is known that 10% of them are defective. A random collection of 20 air bags is tested.

 a. How many of them would you expect to be defective?
 b. In how large a sample would you expect to find 12 defective airbags?

50. ▼ *Spiders* Your pet tarantula, Spider, has a .12 probability of biting an acquaintance who comes into contact with him. Next week, you will be entertaining 20 friends (all of whom will come into contact with Spider).

 a. How many guests should you expect Spider to bite?
 b. At your last party, Spider bit 6 of your guests. Assuming that Spider bit the expected number of guests, how many guests did you have?

Exercises 51 and 52 assume familiarity with counting arguments and probability (Section 7.4).

51. ▼ *Camping* Kent's Tents has four red tents and three green tents in stock. Karin selects four of them at random. Let X be the number of red tents she selects. Give the probability distribution of X and find the expected number of red tents selected.

52. ▼ *Camping* Kent's Tents has five green knapsacks and four yellow ones in stock. Curt selects four of them at random.

---

[31]Based on actual income distribution in 2010. Source: U.S. Census Bureau, Current Population Survey, 2010 American Community Survey (www.census.gov).

[32]Based on actual income distribution in 2003 (not adjusted for inflation). Source: U.S. Census Bureau, Current Population Survey, 2004 Annual Social and Economic Supplement (www.census.gov).

[33]Ratings are by the Insurance Institute for Highway Safety. Sources: Oak Ridge National Laboratory: "An Analysis of the Impact of Sport Utility Vehicles in the United States," Stacy C. Davis, Lorena F. Truett, August 2000, Insurance Institute for Highway Safety (www-cta.ornl.gov/Publications/Final SUV report.pdf, www.highwaysafety.org/vehicle_ratings).

[34]Source: Rand Youth Poll/Teen-Age Research Unlimited/*New York Times*, March 14, 1998, p. D1.

Let X be the number of green knapsacks he selects. Give the probability distribution of X, and find the expected number of green knapsacks selected.

53. ![1] ▼ *Stock Portfolios* You are required to choose between two stock portfolios, *FastForward Funds* and *SolidState Securities*. Stock analysts have constructed the following probability distributions for next year's rate of return for the two funds.

FastForward Funds

Rate of Return	−.4	−.3	−.2	−.1	0	.1	.2	.3	.4
Probability	.015	.025	.043	.132	.289	.323	.111	.043	.019

SolidState Securities

Rate of Return	−.4	−.3	−.2	−.1	0	.1	.2	.3	.4
Probability	.012	.023	.050	.131	.207	.330	.188	.043	.016

Which of the two funds gives the higher expected rate of return?

54. ![1] ▼ *Risk Management* Before making your final decision whether to invest in *FastForward Funds* or *SolidState Securities* (see Exercise 53), you consult your colleague in the risk management department of your company. She informs you that, in the event of a stock market crash, the following probability distributions for next year's rate of return would apply:

FastForward Funds

Rate of Return	−.8	−.7	−.6	−.5	−.4	−.2	−.1	0	.1
Probability	.028	.033	.043	.233	.176	.230	.111	.044	.102

SolidState Securities

Rate of Return	−.8	−.7	−.6	−.5	−.4	−.2	−.1	0	.1
Probability	.033	.036	.038	.167	.176	.230	.211	.074	.035

Which of the two funds offers the lowest risk in case of a market crash?

55. ◆ *Insurance Schemes* The *Acme Insurance Company* is launching a drive to generate greater profits, and it decides to insure racetrack drivers against wrecking their cars. The company's research shows that, on average, a racetrack driver races 4 times a year and has a 1 in 10 chance of wrecking a vehicle, worth an average of $100,000, in every race. The annual premium is $5,000, and Acme automatically drops any driver who is involved in an accident (after paying for a new car), but does not refund the premium. How much profit (or loss) can the company expect to earn from a typical driver in a year? HINT [Use a tree diagram to compute the probabilities of the various outcomes.]

56. ◆ *Insurance* The *Blue Sky Flight Insurance Company* insures passengers against air disasters, charging a prospective passenger $20 for coverage on a single plane ride. In the event of a fatal air disaster, it pays out $100,000 to the named beneficiary. In the event of a nonfatal disaster, it pays out an average of $25,000 for hospital expenses. Given that the probability of a plane's crashing on a single trip is .00000087,[35]

and that a passenger involved in a plane crash has a .9 chance of being killed, determine the profit (or loss) per passenger that the insurance company expects to make on each trip. HINT [Use a tree to compute the probabilities of the various outcomes.]

COMMUNICATION AND REASONING EXERCISES

57. In a certain set of scores, there are as many values above the mean as below it. It follows that

(A) The median and mean are equal.
(B) The mean and mode are equal.
(C) The mode and median are equal.
(D) The mean, mode, and median are all equal.

58. In a certain set of scores, the median occurs more often than any other score. It follows that

(A) The median and mean are equal.
(B) The mean and mode are equal.
(C) The mode and median are equal.
(D) The mean, mode, and median are all equal.

59. Your friend Charlesworth claims that the median of a collection of data is always close to the mean. Is he correct? If so, say why; if not, give an example to prove him wrong.

60. Your other friend Imogen asserts that Charlesworth is wrong and that it is the mode and the median that are always close to each other. Is she correct? If so, say why; if not, give an example to prove her wrong.

61. Must the expected number of times you hit a bull's-eye after 50 attempts always be a whole number? Explain.

62. Your statistics instructor tells you that the expected score of the upcoming midterm test is 75%. That means that 75% is the most likely score to occur, right?

63. ▼ Your grade in a recent midterm was 80%, but the class average was 83%. Most people in the class scored better than you, right?

64. ▼ Your grade in a recent midterm was 80%, but the class median was 100%. Your score was lower than the average score, right?

65. ▼ Slim tells you that the population mean is just the mean of a suitably large sample. Is he correct? Explain.

66. ▼ Explain how you can use a sample to estimate an expected value.

67. ▼ Following is an excerpt from a full-page ad by MoveOn.org in *The New York Times* criticizing President G.W. Bush:[36]

On Tax Cuts:

George Bush: "... Americans will keep, this year, an average of almost $1,000 more of their own money."

The Truth: Nearly half of all taxpayers get less than $100. And 31% of all taxpayers get nothing at all.

The statements referred to as "The Truth" contradict the statement attributed to President Bush, right? Explain.

---

[35]This was the probability of a passenger plane crashing per departure in 1990. (Source: National Transportation Safety Board)

[36]Source: Full-page ad in the *New York Times*, September 17, 2003, p. A25.

68. ▼ Following is an excerpt from a five-page ad by WeissneggerForGov.org in *The Martian Enquirer* criticizing Supreme Martian Administrator, Gov. Red Davis:

On Worker Accommodation:

Gov. Red Davis: "The median size of Government worker habitats in Valles Marineris is at least 400 square feet."

Weissnegger: "The average size of a Government worker habitat in Valles Marineris is a mere 150 square feet."

The statements attributed to Weissnegger do not contradict the statement attributed to Gov. Davis, right? Explain.

69. ▼ Sonia has just told you that the expected household income in the United States is the same as the population mean of all U.S. household incomes. Clarify her statement by describing an experiment and an associated random variable X so that the expected household income is the expected value of X.

70. ▼ If X is a random variable, what is the difference between a sample mean of measurements of X and the expected value of X? Illustrate by means of an example.

8.4 Measures of Dispersion

Variance and Standard Deviation of a Set of Scores

Your grade on a recent midterm was 68%; the class average was 72%. How do you stand in comparison with the rest of the class? If the grades were widely scattered, then your grade may be close to the mean and a fair number of people may have done a lot worse than you (Figure 5a). If, on the other hand, almost all the grades were within a few points of the average, then your grade may not be much higher than the lowest grade in the class (Figure 5b).

This scenario suggests that it would be useful to have a way of measuring not only the central tendency of a set of scores (mean, median, or mode) but also the amount of "scatter" or "dispersion" of the data.

If the scores in our set are $x_1, x_2, \ldots, x_n$ and their mean is $\bar{x}$ (or μ in the case of a population mean), we are really interested in the distribution of the differences $x_i - \bar{x}$. We could compute the *average* of these differences, but this average will always be 0. (Why?) It is really the *sizes* of these differences that interest us, so we might try computing the average of the absolute values of the differences. This idea is reasonable, but it leads to technical difficulties that are avoided by a slightly different approach: The statistic we use is based on the average of the *squares* of the differences, as explained in the following definitions.

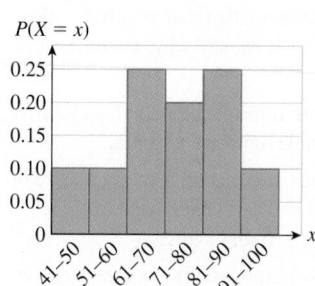

$P(X = x)$

Figure 5(a)

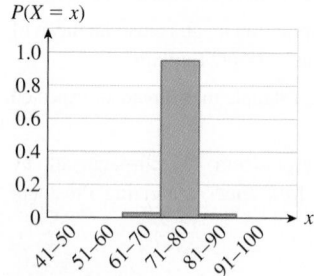

$P(X = x)$

Figure 5(b)

Population Variance and Standard Deviation

If the values $x_1, x_2, \ldots, x_n$ are all the measurements of X in the entire population, then the **population variance** is given by

$$\sigma^2 = \frac{(x_1 - \mu)^2 + (x_2 - \mu)^2 + \cdots + (x_n - \mu)^2}{n} = \frac{\sum_{i=1}^{n} (x_i - \mu)^2}{n}.$$

(Remember that μ is the symbol we use for the *population* mean.) The **population standard deviation** is the square root of the population variance:

$$\sigma = \sqrt{\sigma^2}.$$

8.4 EXERCISES

▼ more advanced   ◆ challenging
⧉ indicates exercises that should be solved using technology

Compute the (sample) variance and standard deviation of the data samples given in Exercises 1–8. (You calculated the means in the last exercise set. Round all answers to two decimal places.) HINT [See Quick Examples on page 581.]

1. −1, 5, 5, 7, 14

2. 2, 6, 6, 7, −1

3. 2, 5, 6, 7, −1, −1

4. 3, 1, 6, −3, 0, 5

5. $\frac{1}{2}, \frac{3}{2}, -4, \frac{5}{4}$

6. $-\frac{3}{2}, \frac{3}{8}, -1, \frac{5}{2}$

7. 2.5, −5.4, 4.1, −0.1, −0.1

8. 4.2, −3.2, 0, 1.7, 0

In Exercises 9–14, calculate the standard deviation of X for each probability distribution. (You calculated the expected values in the last exercise set. Round all answers to two decimal places.) HINT [See Quick Example on page 585.]

9.

x	0	1	2	3
$P(X = x)$	.5	.2	.2	.1

10.

x	1	2	3	4
$P(X = x)$	.1	.2	.5	.2

11.

x	10	20	30	40
$P(X = x)$	$\frac{3}{10}$	$\frac{2}{5}$	$\frac{1}{5}$	$\frac{1}{10}$

12.

x	2	4	6	8
$P(X = x)$	$\frac{1}{20}$	$\frac{15}{20}$	$\frac{2}{20}$	$\frac{2}{20}$

13.

x	−5	−1	0	2	5	10
$P(X = x)$	.2	.3	.2	.1	.2	0

14.

x	−20	−10	0	10	20	30
$P(X = x)$	.2	.4	.2	.1	0	.1

In Exercises 15–24, calculate the expected value, the variance, and the standard deviation of the given random variable X. (You calculated the expected values in the last exercise set. Round all answers to two decimal places.)

15. *X* is the number that faces up when a fair die is rolled.

16. *X* is the number selected at random from the set {1, 2, 3, 4}.

17. *X* is the number of tails that come up when a coin is tossed twice.

18. *X* is the number of tails that come up when a coin is tossed three times.

19. ▼ *X* is the higher number when two dice are rolled.

20. ▼ *X* is the lower number when two dice are rolled.

21. ▼ *X* is the number of red marbles that Suzan has in her hand after she selects four marbles from a bag containing four red marbles and two green ones.

22. ▼ *X* is the number of green marbles that Suzan has in her hand after she selects four marbles from a bag containing three red marbles and two green ones.

23. ▼ Twenty darts are thrown at a dartboard. The probability of hitting a bull's-eye is .1. Let *X* be the number of bull's-eyes hit.

24. ▼ Thirty darts are thrown at a dartboard. The probability of hitting a bull's-eye is $\frac{1}{5}$. Let *X* be the number of bull's-eyes hit.

APPLICATIONS

25. *Popularity Ratings* In your bid to be elected class representative, you have your election committee survey five randomly chosen students in your class and ask them to rank you on a scale of 0–10. Your rankings are 3, 2, 0, 9, 1.

   a. Find the sample mean and standard deviation. (Round your answers to two decimal places.) HINT [See Example 1 and Quick Examples on page 581.]

   b. Assuming the sample mean and standard deviation are indicative of the class as a whole, in what range does the empirical rule predict that approximately 68% of the class will rank you? What other assumptions must we make to use the rule?

26. *Popularity Ratings* Your candidacy for elected class representative is being opposed by Slick Sally. Your election committee has surveyed six of the students in your class and had them rank Sally on a scale of 0–10. The rankings were 2, 8, 7, 10, 5, 8.

   a. Find the sample mean and standard deviation. (Round your answers to two decimal places.) HINT [See Example 1 and Quick Examples on page 581.]

   b. Assuming the sample mean and standard deviation are indicative of the class as a whole, in what range does the empirical rule predict that approximately 95% of the class will rank Sally? What other assumptions must we make to use the rule?

27. *Unemployment* Following is a sample of unemployment rates (in percentage points) in the United States sampled from the period 1990–2004:[38]

   4.2, 4.7, 5.4, 5.8, 4.9.

   a. Compute the mean and standard deviation of the given sample. (Round your answers to one decimal place.)

---

[38] Sources for data: Bureau of Labor Statistics (BLS) (www.bls.gov).

b. Assuming the distribution of unemployment rates in the population is symmetric and bell shaped, 95% of the time, the unemployment rate is between ____ and ____ percent.

28. Unemployment Following is a sample of unemployment rates among Hispanics (in percentage points) in the US sampled from the period 1990–2004:[39]

$$7.7, 7.5, 9.3, 6.9, 8.6$$

a. Compute the mean and standard deviation of the given sample. (Round your answers to one decimal place.)

b. Assuming the distribution of unemployment rates in the population of interest is symmetric and bell shaped, 68% of the time, the unemployment rate is between ____ and ____ percent.

29. Stock Market Gyrations Following is a sample of the day-by-day change, rounded to the nearest 100 points, in the Dow Jones Industrial Average during 10 successive business days around the start of the financial crisis in October 2008:[40]

$$-400, -500, -200, -700, -100, 900, -100, -700, 400, -100$$

a. Compute the mean and standard deviation of the given sample. (Round your answers to the nearest whole number.)

b. Assuming the distribution of day-by-day changes of the Dow during financial crises is symmetric and bell shaped, then the Dow falls by more than ____ points 16% of the time. What is the percentage of times in the sample that the Dow actually fell by more than that amount? HINT [See Example 2(b).]

30. Stock Market Gyrations Following is a sample of the day-by-day change, rounded to the nearest 100 points, in the Dow Jones Industrial Average during 10 successive business days around the start of the financial crisis in October 2008:[41]

$$-100, 400, -200, -500, 200, -300, -200, 900, -100, 200.$$

a. Compute the mean and standard deviation of the given sample. (Round your answers to the nearest whole number.)

b. Assuming the distribution of day-by-day changes of the Dow during financial crises is symmetric and bell shaped, then the Dow rises by more than ____ points 2.5% of the time. What is the percentage of times in the sample that the Dow actually rose by more than that amount? HINT [See Example 2(b).]

31. ⊤ **Sport Utility Vehicles** Following are highway driving gas mileages of a selection of medium-sized sport utility vehicles (SUVs):[42]

17, 18, 17, 18, 21, 16, 21, 18, 16, 14, 15, 22, 17, 19, 17, 18.

a. Find the sample standard deviation (rounded to two decimal places).

b. In what gas mileage range does Chebyshev's inequality predict that at least 8/9 (approximately 89%) of the selection will fall?

c. What is the actual percentage of SUV models of the sample that fall in the range predicted in part (b)? Which gives the more accurate prediction of this percentage: Chebyshev's rule or the empirical rule?

32. ⊤ **Sport Utility Vehicles** Following are the city driving gas mileages of a selection of sport utility vehicles (SUVs):[43]

14, 15, 14, 15, 13, 16, 12, 14, 19, 18, 16, 16, 12, 15, 15, 13.

a. Find the sample standard deviation (rounded to two decimal places).

b. In what gas mileage range does Chebyshev's inequality predict that at least 75% of the selection will fall?

c. What is the actual percentage of SUV models of the sample that fall in the range predicted in part (b)? Which gives the more accurate prediction of this percentage: Chebyshev's rule or the empirical rule?

33. Shopping Malls A survey of all the shopping malls in your region yields the following probability distribution, where X is the number of movie theater screens in a selected mall:

Number of Movie Screens	0	1	2	3	4
Probability	.4	.1	.2	.2	.1

Compute the expected value μ and the standard deviation σ of X. (Round answers to two decimal places.) What percentage of malls have a number of movie theater screens within two standard deviations of μ?

34. Pastimes A survey of all the students in your school yields the following probability distribution, where X is the number of movies that a selected student has seen in the past week:

Number of Movies	0	1	2	3	4
Probability	.5	.1	.2	.1	.1

Compute the expected value μ and the standard deviation σ of X. (Round answers to two decimal places.) For what percentage of students is X within two standard deviations of μ?

35. ⊤ **2010 Income Distribution up to $100,000** The following table shows the distribution of household incomes in 2010 for a sample of 1,000 households in the U.S. with incomes up to $100,000.[44]

---

[39]Sources for data: Bureau of Labor Statistics (BLS) (www.bls.gov).

[40]Source: http://finance.google.com.

[41]Ibid.

[42]Figures are the low-end of ranges for 1999 models tested. Source: Oak Ridge National Laboratory: "An Analysis of the Impact of Sport Utility Vehicles in the United States" Stacy C. Davis, Lorena F. Truett (August 2000)/Insurance Institute for Highway Safety (http://cta.ornl.gov/cta/Publications/pdf/ORNL_TM_2000_147.pdf).

---

[43]Ibid.

[44]Based on actual income distribution in 2010. Source: U.S. Census Bureau, Current Population Survey, 2010 American Community Survey (www.census.gov).

Income ($ Thousands)	10	30	50	70	90
Households	240	290	180	170	120

Compute the expected value μ and the standard deviation σ of the associated random variable X. If we define a "lower income" family as one whose income is more than one standard deviation below the mean, and a "higher income" family as one whose income is at least one standard deviation above the mean, what is the income gap between higher- and lower-income families in the U.S.? (Round your answers to the nearest $1,000.)

36. ▥ *2003 Income Distribution up to $100,000* Repeat Exercise 35, using the following data from a sample of 1,000 households in the U.S. in 2003.[45]

Income ($ thousands)	10	30	50	70	90
Households	270	280	200	150	100

37. *Hispanic Employment: Male* The following table shows the approximate number of males of Hispanic origin employed in the United States in 2005, broken down by age group.[46]

Age	15–24.9	25–54.9	55–64.9
Employment (thousands)	16,000	13,000	1,600

  a. Use the rounded midpoints of the given measurement classes to compute the expected value and the standard deviation of the age X of a male Hispanic worker in the United States. (Round all probabilities and intermediate calculations to two decimal places.)
  b. In what age interval does the empirical rule predict that 68% of all male Hispanic workers will fall? (Round answers to the nearest year.)

38. *Hispanic Employment: Female* Repeat Exercise 37, using the corresponding data for females of Hispanic origin.[47]

Age	15–24.9	25–54.9	55–64.9
Employment (thousands)	1,200	5,000	600

39. *Commerce* You have been told that the average life span of an Internet-based company is 2 years, with a standard deviation of 0.15 years. Further, the associated distribution is highly skewed (not symmetric). Your Internet company is now 2.6 years old. What percentage of all Internet-based companies have enjoyed a life span at least as long as yours? Your answer should contain one of the following phrases: *At least; At most; Approximately*. HINT [See Example 2.]

40. *Commerce* You have been told that the average life span of a car-compounding service is 3 years, with a standard deviation of 0.2 years. Further, the associated distribution is symmetric but not bell shaped. Your car-compounding service is exactly 2.6 years old. What fraction of car-compounding services last at most as long as yours? Your answer should contain one of the following phrases: *At least; At most; Approximately*. HINT [See Example 2.]

41. *Batmobiles* The average life span of a Batmobile is 9 years, with a standard deviation of 2 years.[48] Further, the probability distribution of the life spans of Batmobiles is symmetric, but not known to be bell shaped.

Because my old Batmobile has been sold as bat-scrap, I have decided to purchase a new one. According to the above information, there is

(A) at least   (B) at most   (C) approximately

a ____ percent chance that my new Batmobile will last 13 years or more.

42. *Spiderman Coupés* The average life span of a Spiderman Coupé is 8 years, with a standard deviation of 2 years. Further, the probability distribution of the life spans of Spiderman Coupés is not known to be bell shaped or symmetric. I have just purchased a brand-new Spiderman Coupé. According to the above information, there is

(A) at least   (B) at most   (C) approximately

a ____ percent chance that my new Spiderman Coupé will last for less than 4 years.

43. *Teenage Pastimes* According to a study,[49] the probability that a randomly selected teenager shopped at a mall at least once during a week was .63. Let X be the number of students in a randomly selected group of 40 that will shop at a mall during the next week.

---

[45]Based on actual income distribution in 2003 (not adjusted for inflation). Source: U.S. Census Bureau, Current Population Survey, 2004 Annual Social and Economic Supplement (www.census.gov).

[46]Figures are rounded. Bounds for the age groups for the first and third categories were adjusted for computational convenience. Source: Bureau of Labor Statistics (ftp://ftp.bls.gov/pub/suppl/empsit.cpseed15.txt).

[47]*Ibid.*

[48]See Example 2.

[49]Source: Rand Youth Poll/Teen-Age Research Unlimited/*New York Times*, March 14, 1998, p. D1.

a. Compute the expected value and standard deviation of X. (Round answers to two decimal places.) HINT [See Example 5.]

b. Fill in the missing quantity: There is an approximately 2.5% chance that ___ or more teenagers in the group will shop at a mall during the next week.

44. *Other Teenage Pastimes* According to the study referred to in the preceding exercise, the probability that a randomly selected teenager played a computer game at least once during a week was .48. Let X be the number of teenagers in a randomly selected group of 30 who will play a computer game during the next 7 days.

a. Compute the expected value and standard deviation of X. (Round answers to two decimal places.) HINT [See Example 5.]

b. Fill in the missing quantity: There is an approximately 16% chance that ___ or more teenagers in the group will play a computer game during the next 7 days.

45. ▼ *Teenage Marketing* In 2000, 22% of all teenagers in the United States had checking accounts.[50] Your bank, *TeenChex Inc.,* is interested in targeting teenagers who do not already have a checking account.

a. If TeenChex selects a random sample of 1,000 teenagers, what number of teenagers *without* checking accounts can it expect to find? What is the standard deviation of this number? (Round the standard deviation to one decimal place.)

b. Fill in the missing quantities: There is an approximately 95% chance that between ___ and ___ teenagers in the sample will not have checking accounts. (Round answers to the nearest whole number.)

46. ▼ *Teenage Marketing* In 2000, 18% of all teenagers in the United States owned stocks or bonds.[51] Your brokerage company, *TeenStox Inc.,* is interested in targeting teenagers who do not already own stocks or bonds.

a. If TeenStox selects a random sample of 2,000 teenagers, what number of teenagers who *do not* own stocks or bonds can it expect to find? What is the standard deviation of this number? (Round the standard deviation to one decimal place.)

b. Fill in the missing quantities: There is an approximately 99.7% chance that between ___ and ___ teenagers in the sample will not own stocks or bonds. (Round answers to the nearest whole number.)

47. ⊡ ▼ *Supermarkets* A survey of supermarkets in the United States yielded the following relative frequency table, where X is the number of checkout lanes at a randomly chosen supermarket:[52]

x	1	2	3	4	5	6	7	8	9	10
$P(X=x)$	.01	.04	.04	.08	.10	.15	.25	.20	.08	.05

a. Compute the mean, variance, and standard deviation (accurate to one decimal place).

b. As financial planning manager at *Express Lane Mart,* you wish to install a number of checkout lanes that is in the range of at least 75% of all supermarkets. What is this range according to Chebyshev's inequality? What is the *least* number of checkout lanes you should install so as to fall within this range?

48. ⊡ ▼ *Video Arcades* Your company, *Sonic Video, Inc.,* has conducted research that shows the following probability distribution, where X is the number of video arcades in a randomly chosen city with more than 500,000 inhabitants:

x	0	1	2	3	4	5	6	7	8	9
$P(X=x)$	.07	.09	.35	.25	.15	.03	.02	.02	.01	.01

a. Compute the mean, variance, and standard deviation (accurate to one decimal place).

b. As CEO of Sonic Video, you wish to install a chain of video arcades in Sleepy City, U.S.A. The city council regulations require that the number of arcades be within the range shared by at least 75% of all cities. What is this range? What is the *largest* number of video arcades you should install so as to comply with this regulation?

Distribution of Wealth *If we model after-tax household income by a normal distribution, then the figures of a 1995 study imply the information in the following table, which should be used for Exercises 49–60.[53] Assume that the distribution of incomes in each country is bell shaped and symmetric.*

Country	United States	Canada	Switzerland	Germany	Sweden
Mean Household Income	$38,000	$35,000	$39,000	$34,000	$32,000
Standard Deviation	$21,000	$17,000	$16,000	$14,000	$11,000

[50]Source: Teen-Age Research Unlimited, January 25, 2001 (www.teenresearch.com).

[51]*Ibid.*

[52]Source: J.T. McClave, P.G. Benson, T. Sincich, *Statistics for Business and Economics,* 7th Ed. (Prentice Hall, 1998) p. 177, W. Chow *et al.* "A model for predicting a supermarket's annual sales per square foot," Graduate School of Management, Rutgers University.

[53]The data are rounded to the nearest $1,000 and based on a report published by the Luxembourg Income Study. The report shows after-tax income, including government benefits (such as food stamps) of households with children. Our figures were obtained from the published data by assuming a normal distribution of incomes. All data were based on constant 1991 U.S. dollars and converted foreign currencies (adjusted for differences in buying power). Source: Luxembourg Income Study/*New York Times,* August 14, 1995, p. A9.

49. If we define a "poor" household as one whose after-tax income is at least 1.3 standard deviations below the mean, what is the household income of a poor family in the United States?

50. If we define a "poor" household as one whose after-tax income is at least 1.3 standard deviations below the mean, what is the household income of a poor family in Switzerland?

51. If we define a "rich" household as one whose after-tax income is at least 1.3 standard deviations above the mean, what is the household income of a rich family in the United States?

52. If we define a "rich" household as one whose after-tax income is at least 1.3 standard deviations above the mean, what is the household income of a rich family in Sweden?

53. ▼ Refer to Exercise 49. Which of the five countries listed has the poorest households (i.e., the lowest cutoff for considering a household poor)?

54. ▼ Refer to Exercise 52. Which of the five countries listed has the wealthiest households (i.e., the highest cutoff for considering a household rich)?

55. ▼ Which of the five countries listed has the largest gap between rich and poor?

56. ▼ Which of the five countries listed has the smallest gap between rich and poor?

57. What percentage of U.S. families earned an after-tax income of $17,000 or less?

58. What percentage of U.S. families earned an after-tax income of $80,000 or more?

59. What was the after-tax income range of approximately 99.7% of all Germans?

60. What was the after-tax income range of approximately 99.7% of all Swedes?

▣ *Aging Exercises 61–68 are based on the following list, which shows the percentage of aging population (residents of age 65 and older) in each of the 50 states in 1990, 2000, and 2010:*[54]

1990

> 4, 9, 10, 10, 10, 10, 10, 11, 11, 11, 11, 11, 11, 11, 11, 12, 12, 12, 12, 12, 12, 13, 13, 13, 13, 13, 13, 13, 13, 13, 13, 13, 13, 13, 13, 14, 14, 14, 14, 14, 14, 14, 14, 15, 15, 15, 15, 15, 15, 18

2000

> 6, 9, 10, 10, 10, 11, 11, 11, 11, 11, 11, 11, 12, 12, 12, 12, 12, 12, 12, 12, 12, 12, 12, 13, 13, 13, 13, 13, 13, 13, 13, 13, 13, 13, 14, 14, 14, 14, 14, 14, 14, 15, 15, 15, 15, 16, 18

2010

> 8, 9, 10, 11, 11, 11, 12, 12, 12, 12, 12, 12, 12, 13, 13, 13, 13, 13, 13, 13, 13, 13, 14, 14, 14, 14, 14, 14, 14, 14, 14, 14, 14, 14, 14, 14, 14, 15, 15, 15, 15, 15, 16, 16, 17

61. ▣ Compute the population mean and standard deviation for the 2000 and 2010 data.

62. ▣ Compute the population mean and standard deviation for the 1990 and 2000 data.

63. The answer to Exercise 61 suggests that the 2010 population was:

(A) older on average and more diverse with respect to age

(B) older on average and less diverse with respect to age

(C) younger on average and more diverse with respect to age

(D) younger on average and less diverse with respect to age than the 2000 population.

64. The answer to Exercise 62 suggests that the 1990 population was:

(A) older on average and more diverse with respect to age

(B) older on average and less diverse with respect to age

(C) younger on average and more diverse with respect to age

(D) younger on average and less diverse with respect to age

than the 2000 population.

65. ▣ Compare the actual percentage of states whose aging population in 2010 was within one standard deviation of the mean to the percentage predicted by the empirical rule. Comment on your answer.

66. ▣ Compare the actual percentage of states whose aging population in 1990 was within one standard deviation of the mean to the percentage predicted by the empirical rule. Comment on your answer.

67. ▣ What was the actual percentage of states whose aging population in 2010 was within two standard deviations of the mean? Does Chebyshev's rule apply to the data? Explain.

68. ▣ What was the actual percentage of states whose aging population in 2000 was within two standard deviations of the mean? Does Chebyshev's rule apply to the data? Explain.

Electric Grid Stress In early 2000, a Federal Energy Regulatory Commission order (FERC Order 888, first issued in 1996 and advocated by various energy companies, including Enron*) went into effect, mandating that owners of power transmission lines make them available on the open market. The subsequent levels of stress on the electric grid are believed by many to have led to the Northeast blackout of August 14, 2003. Exercises 69–72 deal with the stress on the national electric grid before and after Order 888 went into effect.*

69. ▼ The following chart shows the approximate standard deviation of the power grid frequency, in $1/1,000$ cycles per second, taken over 6-month periods. (0.9 is the average standard deviation.)[55]

---

[54] Percentages are rounded and listed in ascending order. Source: U.S. Census Bureau, Census 2000 Summary File 1 (www.census.gov/prod/2001pubs/c2kbr01-10.pdf), Age and Sex Composition: 2010 (www.census.gov/prod/cen2010/briefs/c2010br-03.pdf).

[55] Source Robert Blohm, energy consultant and adviser to the North American Electric Reliability Council/*New York Times*, August 20, 2003, p. A16.

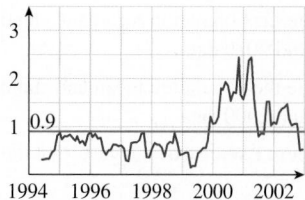

Which of the following statements are true? (More than one may be true.)

(A) The power grid frequency was at or below the mean until late 1999.

(B) The power grid frequency was more stable in mid-1999 than in 1995.

(C) The power grid frequency was more stable in mid-2002 than in mid-1999.

(D) The greatest fluctuations in the power grid frequency occurred in 2000–2001.

(E) The power grid frequency was more stable around January 1995 than around January 1999.

70. ▼ The following chart shows the approximate monthly means of the power grid frequency, in 1/1,000 cycles per second. 0.0 represents the desired frequency of exactly 60 cycles per second.[56]

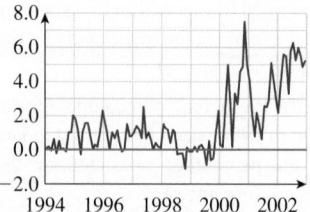

Which of the following statements are true? (More than one may be true.)

(A) Both the mean and the standard deviation show an upward trend from 2000 on.

(B) The mean, but not the standard deviation, shows an upward trend from 2000 on.

(C) The demand for electric power peaked in the second half of 2001.

(D) The standard deviation was larger in the second half of 2002 than in the second half of 1999.

(E) The mean of the monthly means in 2000 was lower than that for 2002, but the standard deviation of the monthly means was higher.

71. ▼ The following chart shows the approximate monthly means of the power grid frequency, in 1/1,000 cycles per second. 0.0 represents the desired frequency of exactly 60 cycles per second. (Note that this is the same data as in Exercise 70 but over a different period of time.)[57]

[56]Source Robert Blohm, energy consultant and adviser to the North American Electric Reliability Council/*New York Times*, August 20, 2003, p. A16.

[57]Source: Eric J. Lerner, *What's wrong with the electric grid? The Industrial Physicist,* Oct./Nov. 2003, American Institute of Physics (www.aip.org/tip/INPHFA/vol-9/iss-5/p8.pdf).

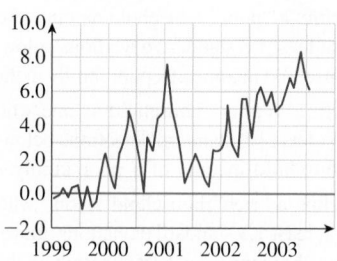

Which of the following statements are true? (More than one may be true.)

(A) Both the mean and the standard deviation show an upward trend from 2002 on.

(B) The mean, but not the standard deviation, shows an upward trend from 2002 on.

(C) The standard deviation, but not the mean, shows an upward trend from 2002 on.

(D) The standard deviation was greater in 2003 than in 2001.

(E) The mean of the monthly means in 2002 was higher than that for 2000, but the standard deviation of the monthly means was lower.

72. ▼ The following chart shows the number of transmission loading relief procedures (procedures undertaken to relieve excessive transmission loads by shifting power to other lines) per month.[58]

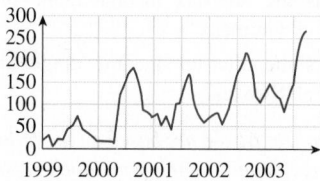

Which of the following statements are true? (More than one may be true.)

(A) Both the annual mean and the standard deviation show a significant upward trend from 2000 on.

(B) The annual mean, but not the standard deviation, shows a significant upward trend from 2000 on.

(C) The annual standard deviation, but not the mean, shows a significant upward trend from 2000 on.

(D) The standard deviation in the second half of 2000 is significantly greater than that in the first half.

(E) The standard deviation in 2000 is significantly greater than that in 2001.

COMMUNICATION AND REASONING EXERCISES

73. Which is greater for a given set of data: the sample standard deviation or the population standard deviation? Explain.

[58] *Ibid.*

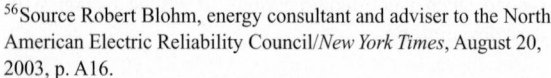

74. Suppose you take larger and larger samples of a given population. Would you expect the sample and population standard deviations to get closer or further apart? Explain.

75. In one Finite Math class, the average grade was 75 and the standard deviation of the grades was 5. In another Finite Math class, the average grade was 65 and the standard deviation of the grades was 20. What conclusions can you draw about the distributions of the grades in each class?

76. You are a manager in a precision manufacturing firm and you must evaluate the performance of two employees. You do so by examining the quality of the parts they produce. One particular item should be 50.0 ± 0.3 mm long to be usable. The first employee produces parts that are an average of 50.1 mm long with a standard deviation of 0.15 mm.

The second employee produces parts that are an average of 50.0 mm long with a standard deviation of 0.4 mm. Which employee do you rate higher? Why? (Assume that the empirical rule applies.)

77. ▼ If a finite random variable has an expected value of 10 and a standard deviation of 0, what must its probability distribution be?

78. ▼ If the values of X in a population consist of an equal number of 1s and –1s, what is its standard deviation?

79. ◆ Find an algebraic formula for the population standard deviation of a sample $\{x, y\}$ of two scores ($x \le y$).

80. ◆ Find an algebraic formula for the sample standard deviation of a sample $\{x, y\}$ of two scores ($x \le y$).

8.5 Normal Distributions

Continuous Random Variables

Figure 9 shows the probability distributions for the number of successes in sequences of 10 and 15 independent Bernoulli trials, each with probability of success $p = .5$.

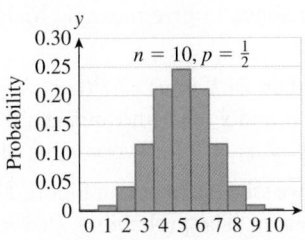

Figure 9(a)   **Figure 9(b)**

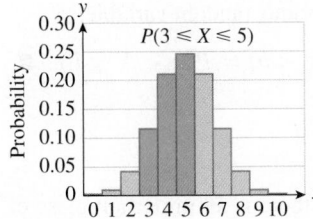

Figure 10

Because each column is 1 unit wide, its area is numerically equal to its height. Thus, the area of each rectangle can be interpreted as a probability. For example, in Figure 9(a) the area of the rectangle over $X = 3$ represents $P(X = 3)$. If we want to find $P(3 \le X \le 5)$, we can add up the areas of the three rectangles over 3, 4, and 5, shown shaded in Figure 10. Notice that if we add up the areas of *all* the rectangles in Figure 9(a), the total is 1 because $P(0 \le X \le 10) = 1$. We can summarize these observations.

Properties of the Probability Distribution Histogram

In a probability distribution histogram where each column is 1 unit wide:

• The total area enclosed by the histogram is 1 square unit.

• $P(a \le X \le b)$ is the area enclosed by the rectangles lying between and including $X = a$ and $X = b$.

This discussion is motivation for considering another kind of random variable, one whose probability distribution is specified not by a bar graph, as above, but by the graph of a function.

Continuous Random Variable; Probability Density Function

A **continuous random variable** X may take on any real value whatsoever. The probabilities $P(a \leq X \leq b)$ are defined by means of a **probability density function**, a function whose graph lies above the x-axis with the total area between the graph and the x-axis being 1. The probability $P(a \leq X \leq b)$ is defined to be the area enclosed by the curve, the x-axis, and the lines $x = a$ and $x = b$ (see Figure 11).

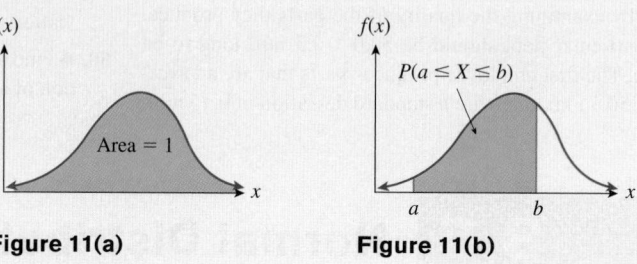

Figure 11(a)   **Figure 11(b)**

Notes

1. In Chapter 7, we defined probability distributions only for *finite* sample spaces. Because continuous random variables have infinite sample spaces, we need the definition above to give meaning to $P(a \leq X \leq b)$ if X is a continuous random variable.

2. If $a = b$, then $P(X = a) = P(a \leq X \leq a)$ is the area under the curve between the lines $x = a$ and $x = a$—no area at all! Thus, when X is a continuous random variable, $P(X = a) = 0$ for every value of a.

3. Whether we take the region in Figure 11(b) to include the boundary or not does not affect the area. The probability $P(a < X < b)$ is defined as the area strictly between the vertical lines $x = a$ and $x = b$, but is, of course, the same as $P(a \leq X \leq b)$, because the boundary contributes nothing to the area. When we are calculating probabilities associated with a continuous random variable,

$$P(a \leq X \leq b) = P(a < X \leq b) = P(a \leq X < b) = P(a < X < b). \quad \blacksquare$$

Normal Density Functions

Among all the possible probability density functions, there is an important class of functions called **normal density functions**, or **normal distributions**. The graph of a normal density function is bell shaped and symmetric, as the following figure shows. The formula for a normal density function is rather complicated looking:

$$f(x) = \frac{1}{\sigma \sqrt{2\pi}} e^{-\frac{(x-\mu)^2}{2\sigma^2}}.$$

The quantity μ is called the **mean** and can be any real number. The quantity σ is called the **standard deviation** and can be any positive real number. The number $e = 2.7182\ldots$ is a useful constant that shows up many places in mathematics, much as the constant π does. Finally, the constant $1/(\sigma \sqrt{2\pi})$ that appears in front is there to make the total area come out to be 1. We rarely use the actual formula in computations; instead, we use tables or technology.

Normal Density Function; Normal Distribution

A **normal density function**, or **normal distribution**, is a function of the form

$$f(x) = \frac{1}{\sigma\sqrt{2\pi}} e^{-\frac{(x-\mu)^2}{2\sigma^2}},$$

where μ is the mean and σ is the standard deviation. Its graph is bell shaped and symmetric, and has the following form:

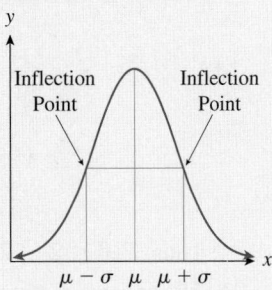

The "inflection points" are the points where the curve changes from bending in one direction to bending in another.*

* Pretend you were driving along the curve in a car. Then the points of inflection are the points where you would change the direction in which you are steering (from left to right or right to left).

Figure 12 shows the graphs of several normal density functions. The third of these has mean 0 and standard deviation 1, and is called the **standard normal distribution**. We use Z rather than X to refer to the standard normal variable.

 using Technology

The graphs in Figure 12 can be drawn on a TI-83/84 Plus or the Website grapher.

TI-83/84 Plus
Figure 12(a):
`Y₁=normalpdf(x,2,1)`
Figure 12(b):
`Y₁=normalpdf(x,0,2)`
Figure 12(c):
`Y₁=normalpdf(x)`

 Website
www.WanerMath.com
   Student Home
     → On Line Utilities
     → Function Evaluator and
       Grapher
Figure 12(a):
`normalpdf(x,2,1)`
Figure 12(b):
`normalpdf(x,0,2)`
Figure 12(c):
`normalpdf(x)`

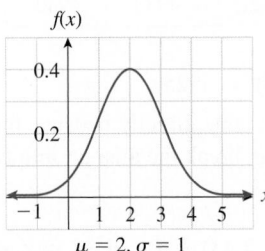

$\mu = 2, \sigma = 1$

Figure 12(a)

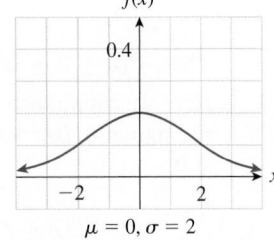

$\mu = 0, \sigma = 2$

Figure 12(b)

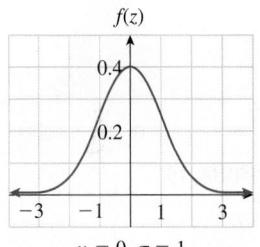

$\mu = 0, \sigma = 1$
Standard Normal Distribution

Figure 12(c)

Calculating Probabilities for the Standard Normal Distribution

The standard normal distribution has $\mu = 0$ and $\sigma = 1$. The corresponding variable is called the **standard normal variable**, which we always denote by Z. Recall that to calculate the probability $P(a \leq Z \leq b)$, we need to find the area under the distribution curve between the vertical lines $z = a$ and $z = b$. We can use the table in the Appendix to look up these areas, or we can use technology. Here is an example.

EXAMPLE 1 Standard Normal Distribution

Let Z be the standard normal variable. Calculate the following probabilities:

a. $P(0 \leq Z \leq 2.4)$ **b.** $P(0 \leq Z \leq 2.43)$
c. $P(-1.37 \leq Z \leq 2.43)$ **d.** $P(1.37 \leq Z \leq 2.43)$

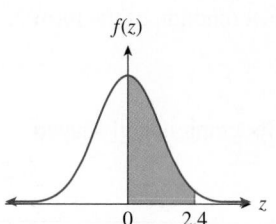

Figure 13

Solution

a. We are asking for the shaded area under the standard normal curve shown in Figure 13. We can find this area, correct to four decimal places, by looking at the table in the Appendix, which lists the area under the standard normal curve from $Z = 0$ to $Z = b$ for any value of b between 0 and 3.09. To use the table, write 2.4 as 2.40, and read the entry in the row labeled 2.4 and the column labeled 0.00 $(2.4 + 0.00 = 2.40)$. Here is the relevant portion of the table:

Z	0.00	0.01	0.02	0.03
2.3	.4893	.4896	.4898	.4901
2.4	.4918	.4920	.4922	.4925
2.5	.4938	.4940	.4941	.4943

Thus, $P(0 \le Z \le 2.40) = .4918$.

b. The area we require can be read from the same portion of the table shown above. Write 2.43 as $2.4 + 0.03$, and read the entry in the row labeled 2.4 and the column labeled 0.03:

Z	0.00	0.01	0.02	0.03
2.3	.4893	.4896	.4898	.4901
2.4	.4918	.4920	.4922	.4925
2.5	.4938	.4940	.4941	.4943

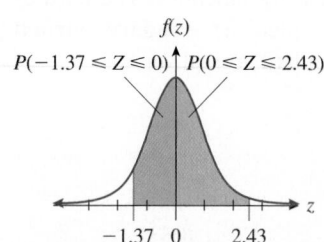

Figure 14

Thus, $P(0 \le Z \le 2.43) = .4925$.

c. Here we cannot use the table directly because the range $-1.37 \le Z \le 2.43$ does not start at 0. But we can break the area up into two smaller areas that start or end at 0:

$$P(-1.37 \le Z \le 2.43) = P(-1.37 \le Z \le 0) + P(0 \le Z \le 2.43).$$

In terms of the graph, we are splitting the desired area into two smaller areas (Figure 14).

We already calculated the area of the right-hand piece in part (b):

$$P(0 \le Z \le 2.43) = .4925.$$

For the left-hand piece, the symmetry of the normal curve tells us that

$$P(-1.37 \le Z \le 0) = P(0 \le Z \le 1.37).$$

This we can find on the table. Look at the row labeled 1.3 and the column labeled .07, and read

$$P(-1.37 \le Z \le 0) = P(0 \le Z \le 1.37) = .4147.$$

Thus,

$$P(-1.37 \le Z \le 2.43) = P(-1.37 \le Z \le 0) + P(0 \le Z \le 2.43)$$
$$= .4147 + .4925$$
$$= .9072.$$

using Technology

Technology can be used to calculate the probabilities in Example 1. For instance, the calculation for part (c) is as follows:

TI-83/84 Plus

Home Screen: normal-cdf(-1.37, 2.43)

(normalcdf is in 2ND VARS.)

[More details on page 614.]

Spreadsheet

=NORMSDIST(2.43)
-NORMSDIST(-1.37)

[More details on page 617.]

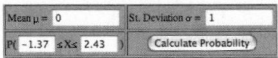

d. The range $1.37 \leq Z \leq 2.43$ does not contain 0, so we cannot use the technique of part (c). Instead, the corresponding area can be computed as the *difference* of two areas:

$$P(1.37 \leq Z \leq 2.43) = P(0 \leq Z \leq 2.43) - P(0 \leq Z \leq 1.37)$$
$$= .4925 - .4147$$
$$= .0778.$$

Calculating Probabilities for Any Normal Distribution

Although we have tables to compute the area under the *standard* normal curve, there are no readily available tables for nonstandard distributions. For example, if $\mu = 2$ and $\sigma = 3$, then how would we calculate $P(0.5 \leq X \leq 3.2)$? The following conversion formula provides a method for doing so:

Standardizing a Normal Distribution

If X has a normal distribution with mean μ and standard deviation σ, and if Z is the standard normal variable, then

$$P(a \leq X \leq b) = P\left(\frac{a - \mu}{\sigma} \leq Z \leq \frac{b - \mu}{\sigma}\right).$$

Quick Example

If $\mu = 2$ and $\sigma = 3$, then

$$P(0.5 \leq X \leq 3.2) = P\left(\frac{0.5 - 2}{3} \leq Z \leq \frac{3.2 - 2}{3}\right)$$
$$= P(-0.5 \leq Z \leq 0.4) = .1915 + .1554 = .3469.$$

To completely justify the above formula requires more mathematics than we shall discuss here. However, here is the main idea: If X is normal with mean μ and standard deviation σ, then $X - \mu$ is normal with mean 0 and standard deviation still σ, while $(X - \mu)/\sigma$ is normal with mean 0 and standard deviation 1. In other words, $(X - \mu)/\sigma = Z$. Therefore,

$$P(a \leq X \leq b) = P\left(\frac{a - \mu}{\sigma} \leq \frac{X - \mu}{\sigma} \leq \frac{b - \mu}{\sigma}\right) = P\left(\frac{a - \mu}{\sigma} \leq Z \leq \frac{b - \mu}{\sigma}\right).$$

EXAMPLE 2 **Quality Control**

Pressure gauges manufactured by Precision Corp. must be checked for accuracy before being placed on the market. To test a pressure gauge, a worker uses it to measure the pressure of a sample of compressed air known to be at a pressure of exactly 50 pounds per square inch. If the gauge reading is off by more than 1% (0.5 pounds), it is rejected. Assuming that the reading of a pressure gauge under these circumstances is a normal random variable with mean 50 and standard deviation 0.4, find the percentage of gauges rejected.

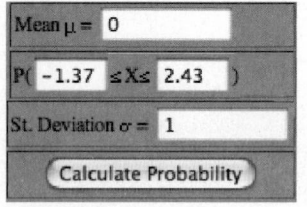
Solution If X is the reading of the gauge, then X has a normal distribution with $\mu = 50$ and $\sigma = 0.4$. We are asking for $P(X < 49.5 \text{ or } X > 50.5) = 1 - P(49.5 \leq X \leq 50.5)$. We calculate

$$P(49.5 \leq X \leq 50.5) = P\left(\frac{49.5 - 50}{0.4} \leq Z \leq \frac{50.5 - 50}{0.4}\right) \quad \text{Standardize}$$
$$= P(-1.25 \leq Z \leq 1.25)$$
$$= 2 \cdot P(0 \leq Z \leq 1.25)$$
$$= 2(.3944) = .7888.$$

So, $P(X < 49.5 \text{ or } X > 50.5) = 1 - P(49.5 \leq X \leq 50.5)$
$$= 1 - .7888 = .2112.$$

In other words, about 21% of the gauges will be rejected.

In many applications, we need to know the probability that a value of a normal random variable will lie within one standard deviation of the mean, or within two standard deviations, or within some number of standard deviations. To compute these probabilities, we first notice that, if X has a normal distribution with mean μ and standard deviation σ, then

$$P(\mu - k\sigma \leq X \leq \mu + k\sigma) = P(-k \leq Z \leq k)$$

by the standardizing formula. We can compute these probabilities for various values of k using the table in the Appendix, and we obtain the following results.

Probability of a Normal Distribution Being within k Standard Deviations of Its Mean

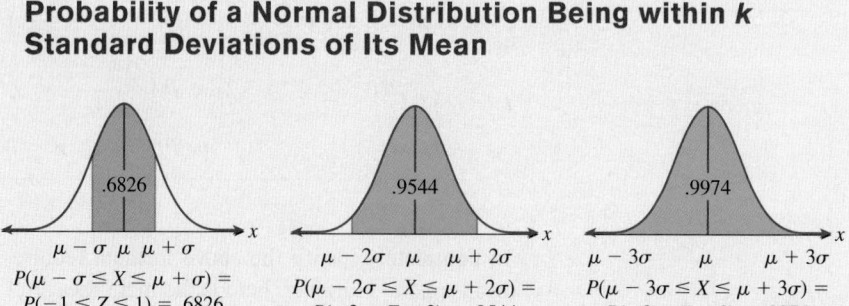

Now you can see where the empirical rule in Section 8.4 comes from! Notice also that the probabilities above are a good deal larger than the lower bounds given by Chebyshev's rule. Chebyshev's rule must work for distributions that are skew or any shape whatsoever.

EXAMPLE 3 Loans

The values of mortgage loans made by a certain bank one year were normally distributed with a mean of $120,000 and a standard deviation of $40,000.

a. What is the probability that a randomly selected mortgage loan was in the range of $40,000–$200,000?

b. You would like to state in your annual report that 50% of all mortgage loans were in a certain range with the mean in the center. What is that range?

Solution

a. We are asking for the probability that a loan was within two standard deviations ($80,000) of the mean. By the calculation done previously, this probability is .9544.

b. We look for the k such that

$$P(120{,}000 - k \cdot 40{,}000 \le X \le 120{,}000 + k \cdot 40{,}000) = .5.$$

Because

$$P(120{,}000 - k \cdot 40{,}000 \le X \le 120{,}000 + k \cdot 40{,}000) = P(-k \le Z \le k)$$

we look in the Appendix to see for which k we have

$$P(0 \le Z \le k) = .25$$

so that $P(-k \le Z \le k) = .5$. That is, we look *inside* the table to see where 0.25 is, and find the corresponding k. We find

$$P(0 \le Z \le 0.67) = .2486$$
and $P(0 \le Z \le 0.68) = .2517.$

Therefore, the k we want is about halfway between 0.67 and 0.68, call it 0.675. This tells us that 50% of all mortgage loans were in the range

$$120{,}000 - 0.675 \cdot 40{,}000 = \$93{,}000$$
to $120{,}000 + 0.675 \cdot 40{,}000 = \$147{,}000.$

Normal Approximation to a Binomial Distribution

You might have noticed that the histograms of some of the binomial distributions we have drawn (for example, those in Figure 9) have a very rough bell shape. In fact, in many cases it is possible to draw a normal curve that closely approximates a given binomial distribution.

Normal Approximation to a Binomial Distribution

If X is the number of successes in a sequence of n independent Bernoulli trials, with probability p of success in each trial, and if the range of values of X within three standard deviations of the mean lies entirely within the range 0 to n (the possible values of X), then

$$P(a \le X \le b) \approx P(a - 0.5 \le Y \le b + 0.5)$$

where Y has a normal distribution with the same mean and standard deviation as X; that is, $\mu = np$ and $\sigma = \sqrt{npq}$, where $q = 1 - p$.

Notes

1. The condition that $0 \le \mu - 3\sigma < \mu + 3\sigma \le n$ is satisfied if n is sufficiently large and p is not too close to 0 or 1; it ensures that most of the normal curve lies in the range 0 to n.

2. In the formula $P(a \leq X \leq b) \approx P(a - 0.5 \leq Y \leq b + 0.5)$, we assume that a and b are integers. The use of $a - 0.5$ and $b + 0.5$ is called the **continuity correction**. To see that it is necessary, think about what would happen if you wanted to approximate, say, $P(X = 2) = P(2 \leq X \leq 2)$. Should the answer be 0? ∎

Figures 15 and 16 show two binomial distributions with their normal approximations superimposed, and illustrate how closely the normal approximation fits the binomial distribution.

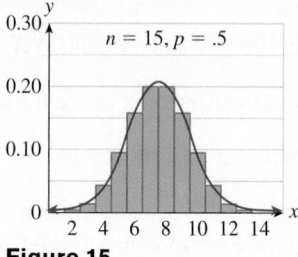

Figure 15   **Figure 16**

EXAMPLE 4 Coin Tosses

a. If you flip a fair coin 100 times, what is the probability of getting more than 55 heads or fewer than 45 heads?

b. What number of heads (out of 100) would make you suspect that the coin is not fair?

Solution

a. We are asking for

$$P(X < 45 \text{ or } X > 55) = 1 - P(45 \leq X \leq 55).$$

We *could* compute this by calculating

$$1 - [C(100, 45)(.5)^{45}(.5)^{55} + C(100, 46)(.5)^{46}(.5)^{54} + \cdots + C(100, 55)(.5)^{55}(.5)^{45}]$$

but we can much more easily *approximate* it by looking at a normal distribution with mean $\mu = 50$ and standard deviation $\sigma = \sqrt{(100)(.5)(.5)} = 5$. (Notice that three standard deviations above and below the mean is the range 35 to 65, which is well within the range of possible values for X, which is 0 to 100, so the approximation should be a good one.) Let Y have this normal distribution. Then

$$P(45 \leq X \leq 55) \approx P(44.5 \leq Y \leq 55.5)$$
$$= P(-1.1 \leq Z \leq 1.1)$$
$$= .7286.$$

Therefore,

$$P(X < 45 \text{ or } X > 55) \approx 1 - .7286 = .2714.$$

b. This is a deep question that touches on the concept of **statistical significance**: What evidence is strong enough to overturn a reasonable assumption (the assumption that the coin is fair)? Statisticians have developed sophisticated ways of answering this question, but we can look at one simple test now. Suppose we tossed a coin 100 times and got 66 heads. If the coin were fair, then $P(X > 65) \approx P(Y > 65.5) = P(Z > 3.1) \approx .001$. This is small enough to raise a reasonable doubt that the coin is fair. However, we should not be too surprised if we threw 56 heads because we can calculate $P(X > 55) \approx .1357$, which is not such a small probability. As we said, the actual tests of statistical significance are more sophisticated than this, but we shall not go into them.

FAQ

When to Subtract from .5 and when Not to

Q : *When computing probabilities like, say $P(Z \leq 1.2)$, $P(Z \geq 1.2)$, or $P(1.2 \leq Z \leq 2.1)$ using a table, just looking up the given values (1.2, 2.1, or whatever) is not enough. Sometimes you have to subtract from .5, sometimes not. Is there a simple rule telling me what to do when?*

A : The simplest—and also most instructive—way of knowing what to do is to draw a picture of the standard normal curve, and shade in the area you are looking for. Drawing pictures also helps you come up with the following mechanical rules:

1. To compute $P(a \leq Z \leq b)$, look up the areas corresponding to $|a|$ and $|b|$ in the table. If a and b have opposite signs, add these areas. Otherwise, subtract the smaller area from the larger.

2. To compute $P(Z \leq a)$, look up the area corresponding to $|a|$. If a is positive, add .5. Otherwise, subtract from .5.

3. To compute $P(Z \geq a)$, look up the area corresponding to $|a|$. If a is positive, subtract from .5. Otherwise, add .5.

8.5 EXERCISES

▼ more advanced   ◆ challenging

T indicates exercises that should be solved using technology

Note: Answers for Section 8.5 were computed using the 4-digit table in the Appendix, and may differ slightly from the more accurate answers generated using technology.

In Exercises 1–8, Z is the standard normal distribution. Find the indicated probabilities. HINT *[See Example 1.]*

1. $P(0 \leq Z \leq 0.5)$ **2.** $P(0 \leq Z \leq 1.5)$

3. $P(-0.71 \leq Z \leq 0.71)$ **4.** $P(-1.71 \leq Z \leq 1.71)$

5. $P(-0.71 \leq Z \leq 1.34)$ **6.** $P(-1.71 \leq Z \leq 0.23)$

7. $P(0.5 \leq Z \leq 1.5)$ **8.** $P(0.71 \leq Z \leq 1.82)$

In Exercises 9–14, X has a normal distribution with the given mean and standard deviation. Find the indicated probabilities. HINT *[See Quick Example on page 599.]*

9. $\mu = 50$, $\sigma = 10$, find $P(35 \leq X \leq 65)$

10. $\mu = 40$, $\sigma = 20$, find $P(35 \leq X \leq 45)$

11. $\mu = 50$, $\sigma = 10$, find $P(30 \leq X \leq 62)$

12. $\mu = 40$, $\sigma = 20$, find $P(30 \leq X \leq 53)$

13. $\mu = 100$, $\sigma = 15$, find $P(110 \leq X \leq 130)$

14. $\mu = 100$, $\sigma = 15$, find $P(70 \leq X \leq 80)$

15. ▼ Find the probability that a normal variable takes on values within 0.5 standard deviations of its mean.

16. ▼ Find the probability that a normal variable takes on values within 1.5 standard deviations of its mean.

17. ▼ Find the probability that a normal variable takes on values more than $\frac{2}{3}$ standard deviations away from its mean.

18. ▼ Find the probability that a normal variable takes on values more than $\frac{5}{3}$ standard deviations away from its mean.

19. ▼ Suppose X is a normal random variable with mean $\mu = 100$ and standard deviation $\sigma = 10$. Find b such that $P(100 \leq X \leq b) = .3$. HINT *[See Example 3.]*

20. ▼ Suppose X is a normal random variable with mean $\mu = 10$ and standard deviation $\sigma = 5$. Find b such that $P(10 \leq X \leq b) = .4$. HINT *[See Example 3.]*

21. ▼ Suppose X is a normal random variable with mean $\mu = 100$ and standard deviation $\sigma = 10$. Find a such that $P(X \geq a) = .04$.

22. ▼ Suppose X is a normal random variable with mean $\mu = 10$ and standard deviation $\sigma = 5$. Find a such that $P(X \geq a) = .03$.

23. If you roll a die 100 times, what is the approximate probability that you will roll between 10 and 15 ones, inclusive? (Round your answer to two decimal places.) HINT [See Example 4.]

24. If you roll a die 100 times, what is the approximate probability that you will roll between 15 and 20 ones, inclusive? (Round your answer to two decimal places.) HINT [See Example 4.]

25. If you roll a die 200 times, what is the approximate probability that you will roll fewer than 25 ones, inclusive? (Round your answer to two decimal places.)

26. If you roll a die 200 times, what is the approximate probability that you will roll more than 40 ones? (Round your answer to two decimal places.)

APPLICATIONS

27. **SAT Scores** SAT test scores are normally distributed with a mean of 500 and a standard deviation of 100. Find the probability that a randomly chosen test-taker will score between 450 and 550. HINT [See Example 3.]

28. **SAT Scores** SAT test scores are normally distributed with a mean of 500 and a standard deviation of 100. Find the probability that a randomly chosen test-taker will score 650 or higher. HINT [See Example 3.]

29. **LSAT Scores** LSAT test scores are normally distributed with a mean of 151 and a standard deviation of 7. Find the probability that a randomly chosen test-taker will score between 137 and 158.

30. **LSAT Scores** LSAT test scores are normally distributed with a mean of 151 and a standard deviation of 7. Find the probability that a randomly chosen test-taker will score 144 or lower.

31. **IQ Scores** IQ scores (as measured by the Stanford-Binet intelligence test) are normally distributed with a mean of 100 and a standard deviation of 16. What percentage of the population has an IQ score between 110 and 140? (Round your answer to the nearest percentage point.)

32. **IQ Scores** Refer to Exercise 31. What percentage of the population has an IQ score between 80 and 90? (Round your answer to the nearest percentage point.)

33. **IQ Scores** Refer to Exercise 31. Find the approximate number of people in the United States (assuming a total population of 313,000,000) with an IQ of 120 or higher.

34. **IQ Scores** Refer to Exercise 31. Find the approximate number of people in the United States (assuming a total population of 313,000,000) with an IQ of 140 or higher.

35. **SAT Scores** SAT test scores are normally distributed with a mean of 500 and a standard deviation of 100. What score would place you in the top 5% of test-takers? HINT [See Example 3.]

36. **LSAT Scores** LSAT test scores are normally distributed with a mean of 151 and a standard deviation of 7. What score would place you in the top 2% of test-takers? HINT [See Example 3.]

37. **Baseball** The mean batting average in major league baseball is about 0.250. Supposing that batting averages are normally distributed, that the standard deviation in the averages is 0.03, and that there are 250 batters, what is the expected number of batters with an average of at least 0.400?

38. **Baseball** The mean batting average in major league baseball is about 0.250. Supposing that batting averages are normally distributed, that the standard deviation in the averages is 0.05, and that there are 250 batters, what is the expected number of batters with an average of at least 0.400?[59]

39. **Marketing** Your pickle company rates its pickles on a scale of spiciness from 1 to 10. Market research shows that customer preferences for spiciness are normally distributed, with a mean of 7.5 and a standard deviation of 1. Assuming that you sell 100,000 jars of pickles, how many jars with a spiciness of 9 or above do you expect to sell?

40. **Marketing** Your hot sauce company rates its sauce on a scale of spiciness of 1 to 20. Market research shows that customer preferences for spiciness are normally distributed, with a mean of 12 and a standard deviation of 2.5. Assuming that you sell 300,000 bottles of sauce, how many bottles with a spiciness below 9 do you expect to sell?

Distribution of Income If we model after-tax household income with a normal distribution, then the figures of a 1995 study imply the information in the following table, which should be used for Exercises 41–46.[60] *Assume that the distribution of incomes in each country was normal, and round all percentages to the nearest whole number.*

Country	United States	Canada	Switzerland	Germany	Sweden
Mean Household Income	$38,000	$35,000	$39,000	$34,000	$32,000
Standard Deviation	$21,000	$17,000	$16,000	$14,000	$11,000

---

[59] The last time that a batter ended the year with an average above 0.400 was in 1941. The batter was Ted Williams of the Boston Red Sox, and his average was 0.406. Over the years, as pitching and batting have improved, the standard deviation in batting averages has declined from around 0.05 when professional baseball began to around 0.03 by the end of the twentieth century. For a very interesting discussion of statistics in baseball and in evolution, see Stephen Jay Gould, *Full House: The Spread of Excellence from Plato to Darwin*, Random House, 1997.

[60] The data are rounded to the nearest $1,000 and are based on a report published by the Luxembourg Income Study. The report shows after-tax income, including government benefits (such as food stamps) of households with children. Our figures were obtained from the published data by assuming a normal distribution of incomes. All data were based on constant 1991 U.S. dollars and converted foreign currencies (adjusted for differences in buying power). Source: Luxembourg Income Study/ *New York Times*, August 14, 1995, p. A9.

41. What percentage of U.S. households had an income of $50,000 or more?

42. What percentage of German households had an income of $50,000 or more?

43. What percentage of Swiss households were either very wealthy (income at least $100,000) or very poor (income at most $12,000)?

44. What percentage of Swedish households were either very wealthy (income at least $100,000) or very poor (income at most $12,000)?

45. Which country had a higher proportion of very poor families (income $12,000 or less): the United States or Canada?

46. Which country had a higher proportion of very poor families (income $12,000 or less): Canada or Switzerland?

47. ▼ *Comparing IQ Tests* IQ scores as measured by both the Stanford-Binet intelligence test and the Wechsler intelligence test have a mean of 100. The standard deviation for the Stanford-Binet test is 16, while that for the Wechsler test is 15. For which test do a smaller percentage of test-takers score less than 80? Why?

48. ▼ *Comparing IQ Tests* Referring to Exercise 47, for which test do a larger percentage of test-takers score more than 120?

49. ▼ *Product Repairs* The new copier your business bought lists a mean time between failures of 6 months, with a standard deviation of 1 month. One month after a repair, it breaks down again. Is this surprising? (Assume that the times between failures are normally distributed.)

50. ▼ *Product Repairs* The new computer your business bought lists a mean time between failures of 1 year, with a standard deviation of 2 months. Ten months after a repair, it breaks down again. Is this surprising? (Assume that the times between failures are normally distributed.)

Software Testing Exercises 51–56 are based on the following information, gathered from student testing of a statistical software package called MODSTAT.[61] *Students were asked to complete certain tasks using the software, without any instructions. The results were as follows. (Assume that the time for each task is normally distributed.)*

Task	Mean Time (minutes)	Standard Deviation
Task 1: Descriptive Analysis of Data	11.4	5.0
Task 2: Standardizing Scores	11.9	9.0
Task 3: Poisson Probability Table	7.3	3.9
Task 4: Areas under Normal Curve	9.1	5.5

51. Find the probability that a student will take at least 10 minutes to complete Task 1.

52. Find the probability that a student will take at least 10 minutes to complete Task 3.

53. ▼ Assuming that the time it takes a student to complete each task is independent of the others, find the probability that a student will take at least 10 minutes to complete each of Tasks 1 and 2.

54. ▼ Assuming that the time it takes a student to complete each task is independent of the others, find the probability that a student will take at least 10 minutes to complete each of Tasks 3 and 4.

55. ♦ It can be shown that if X and Y are independent normal random variables with means μ_X and μ_Y, and standard deviations σ_X and σ_Y respectively, then their sum $X + Y$ is also normally distributed and has mean $\mu = \mu_X + \mu_Y$ and standard deviation $\sigma = \sqrt{\sigma_X^2 + \sigma_Y^2}$. Assuming that the time it takes a student to complete each task is independent of the others, find the probability that a student will take at least 20 minutes to complete both Tasks 1 and 2.

56. ♦ Referring to Exercise 55, compute the probability that a student will take at least 20 minutes to complete both Tasks 3 and 4.

57. *Internet Access* In 2010, 71% of all households in the United States had Internet access.[62] Find the probability that, in a small town with 1,200 households, at least 840 had Internet access in 2010. HINT [See Example 4.]

58. *Television Ratings* Based on data from the Nielsen Company, there is a 1.8% chance that any television that is turned on during the time of the evening newscasts will be tuned to ABC's evening news show.[63] Your company wishes to advertise on a local station carrying ABC that serves a community with 5,000 households that regularly tune in during this time slot. Find the approximate probability that at least 100 households will be tuned in to the show. HINT [See Example 4.]

59. *Aviation* The probability of a plane crashing on a single trip in 2010 was .00000276.[64] Find the approximate probability that in 10,000,000 flights, there will be fewer than 35 crashes.

60. *Aviation* The probability of a plane crashing on a single trip in 1990 was .00000087. Find the approximate probability that in 100,000,000 flights, there will be more than 110 crashes.

---

[61] Data are rounded to one decimal place. Source: *Student Evaluations of MODSTAT,* by Joseph M. Nowakowski, Muskingum College, New Concord, OH, 1997.

[62] Source: *Digital Nation: Expanding Internet Usage,* National Telecommunications and Information Administration, February 2011 (www.ntia.doc.gov/data).

[63] As of November 2011. Source: The Nielsen Company via the TVbytheNumbers Web site (http://tvbythenumbers.zap2it.com/category/ratings/evening-news).

[64] Figures are for scheduled commercial flights. Source for this exercise and the following three: National Transportation Safety Board (www.ntsb.gov).

61. ▼ *Insurance* Your company issues flight insurance. You charge $3 and, in the event of a plane crash, you will pay out $1 million to the victim or his or her family. In 2010, the probability of a plane crashing on a single trip was .00000276. If 10 people per flight buy insurance from you, what was your approximate probability of losing money over the course of 10 million flights in 2010? HINT [First determine how many crashes there must be for you to lose money.]

62. ▼ *Insurance* Refer back to the preceding exercise. What is your approximate probability of losing money over the course of 100 million flights?

63. ◆ *Polls* In a certain political poll, each person polled has a 90% probability of telling his or her real preference. Suppose that 55% of the population really prefer candidate Goode, and 45% prefer candidate Slick. First find the probability that a person polled will say that he or she prefers Goode. Then find the approximate probability that, if 1,000 people are polled, more than 52% will say they prefer Goode.

64. ◆ *Polls* In a certain political poll, each person polled has a 90% probability of telling his or her real preference. Suppose that 1,000 people are polled and 51% say that they prefer candidate Goode, while 49% say that they prefer candidate Slick. Find the approximate probability that Goode could do at least this well if, in fact, only 49% prefer Goode.

65. ◆ *IQ Scores* Mensa is a club for people with high IQs. To qualify, you must be in the top 2% of the population. One way of qualifying is by having an IQ of at least 148, as measured by the Cattell intelligence test. Assuming that scores on this test are normally distributed with a mean of 100, what is the standard deviation? HINT [Use the table in the Appendix "backward."]

66. ◆ *SAT Scores* Another way to qualify for Mensa (see the previous exercise) is to score at least 1,250 on the SAT [combined Critical Reading (Verbal, before March 2005) and Math

scores], which puts you in the top 2%. Assuming that SAT scores are normally distributed with a mean of 1,000, what is the standard deviation? (See the hint for the previous exercise.)

COMMUNICATION AND REASONING EXERCISES

67. Under what assumptions are the estimates in the empirical rule exact?

68. If X is a continuous random variable, what values can the quantity $P(X = a)$ have?

69. Which is larger for a continuous random variable, $P(X \leq a)$ or $P(X < a)$?

70. Which of the following is greater: $P(X \leq b)$ or $P(a \leq X \leq b)$?

71. ▼ A uniform continuous distribution is one with a probability density curve that is a horizontal line. If X takes on values between the numbers a and b with a uniform distribution, find the height of its probability density curve.

72. ▼ Which would you expect to have the greater variance: the standard normal distribution or the uniform distribution taking values between -1 and 1? Explain.

73. ◆ Which would you expect to have a density curve that is higher at the mean: the standard normal distribution, or a normal distribution with standard deviation 0.5? Explain.

74. ◆ Suppose students must perform two tasks: Task 1 and Task 2. Which of the following would you expect to have a smaller standard deviation?

(A) The time it takes a student to perform both tasks if the time it takes to complete Task 2 is independent of the time it takes to complete Task 1.

(B) The time it takes a student to perform both tasks if students will perform similarly in both tasks.

Explain.

CHAPTER 8 REVIEW

KEY CONCEPTS

 Website www.WanerMath.com

Go to the Website at www.WanerMath .com to find a comprehensive and interactive Web-based summary of Chapter 8.

8.1 Random Variables and Distributions

Random variable; discrete vs continuous random variable *pp. 548–549*

Probability distribution of a finite random variable *p. 551*

Using measurement classes *p. 552*

8.2 Bernoulli Trials and Binomial Random Variables

Bernoulli trial; binomial random variable *pp. 559, 560*

Probability distribution of binomial random variable:

$P(X = x) = C(n, x)p^x q^{n-x}$
p. 562

8.3 Measures of Central Tendency

Sample, sample mean; population, population mean *p. 568*

Sample median, sample mode *p. 569*

Expected value of a random variable:

$\mu = E(X) = \sum_i x_i \cdot P(X = x_i)$
p. 571

Expected value of a binomial random variable: $\mu = E(X) = np$ *p. 573*

8.4 Measures of Dispersion

Population variance:

$$\sigma^2 = \frac{\sum_{i=1}^{n}(x_i - \mu)^2}{n}$$

Population standard deviation:
$\sigma = \sqrt{\sigma^2}$ *p. 580*

Sample variance:

$$s^2 = \frac{\sum_{i=1}^{n}(x_i - \bar{x})^2}{n - 1}$$

Sample standard deviation:
$s = \sqrt{s^2}$ *p. 581*

Chebyshev's rule *p. 583*

Empirical rule *p. 583*

Variance of a random variable:

$\sigma^2 = \sum_i (x_i - \mu)^2 P(X = x_i)$ *p. 585*

Standard deviation of X: $\sigma = \sqrt{\sigma^2}$ *p. 585*

Variance and standard deviation of a binomial random variable:

$\sigma^2 = npq, \sigma = \sqrt{npq}$ *p. 588*

8.5 Normal Distributions

Probability density function *p. 596*

Normal density function; normal distribution; standard normal distribution *p. 597*

Calculating probabilities based on the standard normal distribution *p. 597*

Standardizing a normal distribution *p. 599*

Calculating probabilities based on nonstandard normal distributions *p. 599*

Normal approximation to a binomial distribution *p. 601*

REVIEW EXERCISES

In Exercises 1–6, find the probability distribution for the given random variable and draw a histogram.

1. A couple has two children; $X =$ the number of boys. (Assume an equal likelihood of a child being a boy or a girl.)

2. A couple has three children; $X =$ the number of girls. (Assume an equal likelihood of a child being a boy or a girl.)

3. A four-sided die (with sides numbered 1 through 4) is rolled twice in succession; $X =$ the sum of the two numbers.

4. 48.2% of *Xbox* players are in their teens, 38.6% are in their twenties, 11.6% are in their thirties, and the rest are in their forties; $X =$ age of an *Xbox* player. (Use the midpoints of the measurement classes.)

5. From a bin that contains 20 defective joysticks and 30 good ones, 3 are chosen at random; $X =$ the number of defective joysticks chosen. (Round all probabilities to four decimal places.)

6. Two dice are weighted so that each number 2, 3, 4, and 5 is half as likely to face up as each 1 and 6; $X =$ the number of 1s that face up when both are thrown.

7. Use any method to calculate the sample mean, median, and standard deviation of the following sample of scores: $-1, 2,$ $0, 3, 6.$

8. Use any method to calculate the sample mean, median, and standard deviation of the following sample of scores: 4, 4, 5, 6, 6.

9. Give an example of a sample of four scores with mean 1 and median 0. (Arrange them in ascending order.)

10. Give an example of a sample of six scores with sample standard deviation 0 and mean 2.

11. Give an example of a population of six scores with mean 0 and population standard deviation 1.

12. Give an example of a sample of five scores with mean 0 and sample standard deviation 1.

A die is constructed in such a way that rolling a 6 is twice as likely as rolling each other number. That die is rolled four times. Let X be the number of times a 6 is rolled. Evaluate the probabilities in Exercises 13–20.

13. $P(X = 1)$

14. $P(X = 3)$

15. The probability that 6 comes up at most twice

16. The probability that 6 comes up at most once

17. The probability that X is more than 3

18. The probability that X is at least 2

19. $P(1 \leq X \leq 3)$

20. $P(X \leq 3)$

21. A couple has three children; $X =$ the number of girls. (Assume an equal likelihood of a child being a boy or a girl.) Find the expected value and standard deviation of X, and

complete the following sentence with the smallest possible whole number: All values of X lie within ___ standard deviations of the expected value.

22. A couple has four children; X = the number of boys. (Assume only a 25% chance of a child being a boy.) Find the expected value and standard deviation of X, and complete the following sentence with the smallest possible whole number: All values of X lie within ___ standard deviations of the expected value.

23. A random variable X has the following frequency distribution.

x	-3	-2	-1	0	1	2	3
$fr(X = x)$	1	2	3	4	3	2	1

Find the probability distribution, expected value, and standard deviation of X, and complete the following sentence: 87.5% (or 14/16) of the time, X is within ___ (round to one decimal place) standard deviations of the expected value.

24. A random variable X has the following frequency distribution.

x	-4	-2	0	2	4	6
$fr(X = x)$	3	3	4	5	3	2

Find the probability distribution, expected value, and standard deviation of X, and complete the following sentence: ___ percent of the values of X lie within one standard deviation of the expected value.

25. A random variable X has expected value $\mu = 100$ and standard deviation $\sigma = 16$. Use Chebyshev's rule to find an interval in which X is guaranteed to lie with a probability of at least 90%.

26. A random variable X has a symmetric distribution and an expected value $\mu = 200$ and standard deviation $\sigma = 5$. Use Chebyshev's rule to find a value that X is guaranteed to exceed with a probability of at most 10%.

27. A random variable X has a bell shaped, symmetric distribution, with expected value $\mu = 200$ and standard deviation $\sigma = 20$. The empirical rule tells us that X has a value greater than ___ approximately 0.15% of the time.

28. A random variable X has a bell shaped, symmetric distribution, with expected value $\mu = 100$ and standard deviation $\sigma = 30$. Use the empirical rule to give an interval in which X lies approximately 95% of the time.

In Exercises 29–34 the mean and standard deviation of a normal variable X are given. Find the indicated probability.

29. X is the standard normal variable Z; $P(0 \leq X \leq 1.5)$.

30. X is the standard normal variable Z; $P(X \leq -1.5)$.

31. X is the standard normal variable Z; $P(|X| \geq 2.1)$.

32. $\mu = 100, \sigma = 16$; $P(80 \leq X \leq 120)$

33. $\mu = 0, \sigma = 2$; $P(X \leq -1)$

34. $\mu = -1, \sigma = 0.5$; $P(X \geq 1)$

APPLICATIONS: OHaganBooks.com

Marketing As a promotional gimmick, OHaganBooks.com has been selling copies of the Encyclopædia Galactica at an extremely low price that is changed each week at random in a nationally televised drawing. Exercises 35–40 are based on the following table, which summarizes the anticipated sales.

Price	$5.50	$10	$12	$15
Frequency (weeks)	1	2	3	4
Weekly Sales	6,200	3,500	3,000	1,000

35. What is the expected value of the price of *Encyclopædia Galactica*?

36. What are the expected weekly sales of *Encyclopædia Galactica*?

37. What is the expected weekly revenue from sales of *Encyclopædia Galactica*? (Revenue = Price per copy sold × Number of copies sold.)

38. OHaganBooks.com originally paid Duffin House $20 per copy for the *Encyclopædia Galactica*. What is the expected weekly loss from sales of the encyclopædia? (Loss = Loss per copy sold × Number of copies sold.)

39. True or false? If X and Y are two random variables, then $E(XY) = E(X)E(Y)$ (the expected value of the product of two random variables is the product of the expected values). Support your claim by referring to the answers of Exercises 35, 36, and 37.

40. True or false? If X and Y are two random variables, then $E(X/Y) = E(X)/E(Y)$ (the expected value of the ratio of two random variables is the ratio of the expected values). Support your claim by referring to the answers of Exercises 36 and 38.

41. *Online Sales* The following table shows the number of online orders at OHaganBooks.com per million residents in 100 U.S. cities during one month:

Orders (per million residents)	1–2.9	3–4.9	5–6.9	7–8.9	9–10.9
Number of Cities	25	35	15	15	10

a. Let X be the number of orders per million residents in a randomly chosen U.S. city (use rounded midpoints of the given measurement classes). Construct the probability distribution for X and hence compute the expected value μ of X and standard deviation σ. (Round answers to four decimal places.)

b. What range of orders per million residents does the empirical rule predict from approximately 68% of all cities? Would you judge that the empirical rule applies? Why?

c. The actual percentage of cities from which you obtain between 3 and 8 orders per million residents is (choose the correct answer that gives the most specific information):
 (A) Between 50% and 65%   (B) At least 65%
 (C) At least 50%   (D) 57.5%

42. *Pollen* Marjory Duffin is planning a joint sales meeting with OHaganBooks.com in Atlanta at the end of March, but is extremely allergic to pollen, so she went online to find pollen counts for the period. The following table shows the results of her search:

Pollen Count	0–1.9	2–3.9	4–5.9	6–7.9	8–9.9	10–11.9
Number of Days	3	5	7	2	1	2

a. Let X be the pollen count on a given day (use rounded midpoints of the given measurement classes). Construct the probability distribution for X and hence compute the expected value μ of X and standard deviation σ. (Round answers to four decimal places.)

b. What range of pollen counts does the empirical rule predict on approximately 95% of the days? Would you judge that the empirical rule applies? Why?

c. The actual percentage of days on which the pollen count is between 2 and 7 is (choose the correct answer that gives the most specific information):

(A) Between 50% and 60%   **(B)** At least 60%

(C) At most 70%   **(D)** Between 60% and 70%

Mac vs. Windows *On average, 5% of all hits by Mac OS users and 10% of all hits by Windows users result in orders for books at OHaganBooks.com. Due to online promotional efforts, the site traffic is approximately 10 hits per hour by Mac OS users, and 20 hits per hour by Windows users. Compute the probabilities in Exercises 43–48. (Round all answers to three decimal places.)*

43. What is the probability that exactly three Windows users will order books in the next hour?

44. What is the probability that at most three Windows users will order books in the next hour?

45. What is the probability that exactly one Mac OS user and three Windows users will order books in the next hour?

46. What assumption must you make to justify your calculation in Exercise 45?

47. How many orders for books can OHaganBooks.com expect in the next hour from Mac OS users?

48. How many orders for books can OHaganBooks.com expect in the next hour from Windows users?

Online Cosmetics *OHaganBooks.com has launched a subsidiary, GnuYou.com, which sells beauty products online. Most products sold by GnuYou.com are skin creams and hair products. Exercises 49–52 are based on the following table, which shows monthly revenues earned through sales of these products. (Assume a normal distribution. Round all answers to three decimal places.)*

Product	Skin Creams	Hair Products
Mean Monthly Revenue	$38,000	$34,000
Standard Deviation	$21,000	$14,000

49. What is the probability that GnuYou.com will sell *at least* $50,000 worth of skin cream next month?

50. What is the probability that GnuYou.com will sell *at most* $50,000 worth of hair products next month?

51. What is the probability that GnuYou.com will sell less than $12,000 of skin creams next month?

52. What is the probability that GnuYou.com will sell less than $12,000 of hair products next month?

53. *Intelligence* Billy-Sean O'Hagan, now a senior at Suburban State University, has done exceptionally well and has just joined Mensa, a club for people with high IQs. Within Mensa is a group called the Three Sigma Club because their IQ scores are at least 3 standard deviations higher than the U.S. mean. Assuming a U.S. population of 313,000,000, how many people in the United States are qualified for the Three Sigma Club? (Round your answer to the nearest 1,000 people.)

54. *Intelligence* To join Mensa (not necessarily the Three Sigma Club), one needs an IQ of at least 132, corresponding to the top 2% of the population. Assuming that scores on this test are normally distributed with a mean of 100, what is the standard deviation? (Round your answer to the nearest whole number.)

55. *Intelligence* Based on the information given in Exercises 53 and 54, what score must Billy-Sean have to get into the Three Sigma Club? (Assume that IQ scores are normally distributed with a mean of 100, and use the rounded standard deviation.)

56. *Intelligence* Mensa allows the results of various standardized tests to be used to gain membership. Suppose that there was such a test with a mean of 500 on which one needed to score at least 600 to be in the top 2% of the population, hence eligible to join Mensa. What would Billy-Sean need to score on this test to get into the Three Sigma Club?

Case Study

Spotting Tax Fraud with Benford's Law[65]

You are a tax fraud specialist working for the Internal Revenue Service (IRS), and you have just been handed a portion of the tax return from Colossal Conglomerate. The IRS suspects that the portion you were handed may be fraudulent, and would like your opinion. Is there any mathematical test, you wonder, that can point to a suspicious tax return based on nothing more than the numbers entered?

[65] The discussion is based on the article "Following Benford's Law, or Looking Out for No. 1" by Malcolm W. Browne, *New York Times*, August 4, 1998, p. F4. The use of Benford's Law in detecting tax evasion is discussed in a Ph.D. dissertation by Dr. Mark J. Nigrini (Southern Methodist University, Dallas).

You decide, on an impulse, to make a list of the first digits of all the numbers entered in the portion of the Colossal Conglomerate tax return (there are 625 of them). You reason that, if the tax return is an honest one, the first digits of the numbers should be uniformly distributed. More precisely, if the experiment consists of selecting a number at random from the tax return, and the random variable X is defined to be the first digit of the selected number, then X should have the following probability distribution:

x	1	2	3	4	5	6	7	8	9
$P(X=x)$	$\frac{1}{9}$	$\frac{1}{9}$	$\frac{1}{9}$	$\frac{1}{9}$	$\frac{1}{9}$	$\frac{1}{9}$	$\frac{1}{9}$	$\frac{1}{9}$	$\frac{1}{9}$

You then do a quick calculation based on this probability distribution and find an expected value of $E(X) = 5$. Next, you turn to the Colossal Conglomerate tax return data and calculate the relative frequency (estimated probability) of the actual numbers in the tax return. You find the following results.

Colossal Conglomerate Return

y	1	2	3	4	5	6	7	8	9
$P(Y=y)$	.29	.1	.04	.15	.31	.08	.01	.01	.01

It certainly does look suspicious! For one thing, the digits 1 and 5 seem to occur a lot more often than any of the other digits, and roughly three times what you predicted. Moreover, when you compute the expected value, you obtain $E(Y) = 3.48$, considerably lower than the value of 5 you predicted. Gotcha! you exclaim.

You are about to file a report recommending a detailed audit of Colossal Conglomerate when you recall an article you once read about first digits in lists of numbers. The article dealt with a remarkable discovery in 1938 by Dr. Frank Benford, a physicist at General Electric. What Dr. Benford noticed was that the pages of logarithm tables that listed numbers starting with the digits 1 and 2 tended to be more soiled and dog-eared than the pages that listed numbers starting with higher digits—say, 8. For some reason, numbers that start with low digits seemed more prevalent than numbers that start with high digits. He subsequently analyzed more than 20,000 sets of numbers, such as tables of baseball statistics, listings of widths of rivers, half-lives of radioactive elements, street addresses, and numbers in magazine articles. The result was always the same: Inexplicably, numbers that start with low digits tended to appear more frequently than those that start with high ones, with numbers beginning with the digit 1 most prevalent of all.[66] Moreover, the expected value of the first digit was not the expected 5, but 3.44.

Because the first digits in Colossal Conglomerate's return have an expected value of 3.48, very close to Benford's value, it might appear that your suspicion was groundless after all. (Back to the drawing board . . .)

Out of curiosity, you decide to investigate Benford's discovery more carefully. What you find is that Benford did more than simply observe a strange phenomenon in lists of numbers. He went further and derived the following formula for the probability distribution of first digits in lists of numbers:

$$P(X=x) = \log(1 + 1/x) \qquad (x = 1, 2, \ldots, 9).$$

---

[66] This does not apply to all lists of numbers. For instance, a list of randomly chosen numbers between 100 and 999 will have first digits uniformly distributed between 1 and 9.

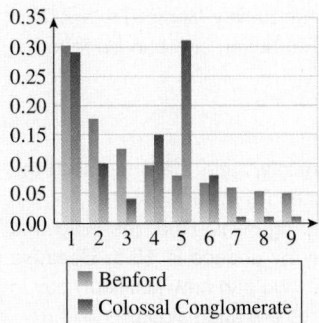

Figure 17

You compute these probabilities, and find the following distribution (the probabilities are all rounded, and thus do not add to exactly 1).

x	1	2	3	4	5	6	7	8	9
$P(X = x)$	.30	.18	.12	.10	.08	.07	.06	.05	.05

You then enter these data along with the Colossal Conglomerate tax return data in your spreadsheet program and obtain the graph shown in Figure 17.

The graph shows something awfully suspicious happening with the digit 5. The percentage of numbers in the Colossal Conglomerate return that begin with 5 far exceeds Benford's prediction that approximately 8% of all numbers should begin with 5.

Now it seems fairly clear that you are justified in recommending Colossal Conglomerate for an audit, after all.

Q: *Because no given set of data can reasonably be expected to satisfy Benford's law exactly, how can I be certain that the Colossal Conglomerate data is not simply due to chance?*

A: You can never be 100% certain. It is certainly conceivable that the tax figures just happen to result in the "abnormal" distribution in the Colossal Conglomerate tax return. However—and this is the subject of "inferential statistics"—there is a method for deciding whether you can be, say "95% certain" that the anomaly reflected in the data is not due to chance. To check, you must first compute a statistic that determines how far a given set of data deviates from satisfying a theoretical prediction (Benford's law, in this case). This statistic is called a **sum-of-squares error** and is given by the following formula (reminiscent of the variance):

$$\text{SSE} = n\left[\frac{[P(y_1) - P(x_1)]^2}{P(x_1)} + \frac{[P(y_2) - P(x_2)]^2}{P(x_2)} + \cdots + \frac{[P(y_9) - P(x_9)]^2}{P(x_9)}\right].$$

Here, n is the sample size: 625 in the case of Colossal Conglomerate. The quantities $P(x_i)$ are the theoretically predicted probabilities according to Benford's Law, and the $P(y_i)$ are the probabilities in the Colossal Conglomerate return. Notice that if the Colossal Conglomerate return probabilities had exactly matched the theoretically predicted probabilities, then SSE would have been zero. Notice also the effect of multiplying by the sample size n: The larger the sample, the more likely that the discrepancy between the $P(x_i)$ and the $P(y_i)$ is not due to chance. Substituting the numbers gives

$$\text{SSE} \approx 625\left[\frac{[.29 - .30]^2}{.30} + \frac{[.1 - .18]^2}{.18} + \cdots + \frac{[.01 - .05]^2}{.05}\right]$$
$$\approx 552.^{67}$$

Q: *The value of SSE does seem quite large. But how can I use this figure in my report? I would like to say something impressive, such as "Based on the portion of the Colossal Conglomerate tax return analyzed, one can be 95% certain that the figures are anomalous."*

A: The error SSE is used by statisticians to answer exactly such a question. What they would do is compare this figure to the largest SSE we would have expected to get

---

[67] If you use more accurate values for the probabilities in Benford's distribution, the value is approximately 560.

by chance in 95 out of 100 selections of data that *do* satisfy Benford's law. This "biggest error" is computed using a "chi-squared" distribution and can be found in Excel by entering

```
=CHIINV(0.05,8)
```

Here, the 0.05 is $1 - 0.95$, encoding the "95% certainty," and the 8 is called the "number of degrees of freedom" = number of outcomes (9) minus 1.

You now find, using Excel, that the chi-squared figure is 15.5, meaning that the largest SSE that you could have expected purely by chance is 15.5. Because Colossal Conglomerate's error is much larger at 552, you can now justifiably say in your report that there is a 95% certainty that the figures are anomalous.[68]

EXERCISES

Which of the following lists of data would you expect to follow Benford's law? If the answer is "no," give a reason.

1. Distances between cities in France, measured in kilometers

2. Distances between cities in France, measured in miles

3. The grades (0–100) in your math instructor's grade book

4. The Dow Jones averages for the past 100 years

5. Verbal SAT scores of college-bound high school seniors

6. Life spans of companies

[T] *Use technology to determine whether the given distribution of first digits fails, with 95% certainty, to follow Benford's law.*

7. Good Neighbor Inc.'s tax return ($n = 1,000$)

y	1	2	3	4	5	6	7	8	9
$P(Y = y)$	.31	.16	.13	.11	.07	.07	.05	.06	.04

8. Honest Growth Funds Stockholder Report ($n = 400$)

y	1	2	3	4	5	6	7	8	9
$P(Y = y)$	.28	.16	.1	.11	.07	.09	.05	.07	.07

---

[68] What this actually means is that, if you were to do a similar analysis on a large number of tax returns, and you designated as "not conforming to Benford's law" all of those whose value of SSE was larger than 15.5, you would be justified in 95% of the cases.

TI-83/84 Plus   Technology Guide

Section 8.1

Example 3 (page 552) Let X be the number of heads that face up in three tosses of a coin. We obtained the following probability distribution of X in the text:

x	0	1	2	3
$P(X = x)$	$\frac{1}{8}$	$\frac{3}{8}$	$\frac{3}{8}$	$\frac{1}{8}$

Use technology to obtain the corresponding histogram.

Solution with Technology

1. In the TI-83/84 Plus, you can enter a list of probabilities as follows: press [STAT], choose EDIT, and then press [ENTER]. Clear columns L_1 and L_2 if they are not already cleared. (Select the heading of a column and press [CLEAR] [ENTER] to clear it.) Enter the values of X in the column under L_1 (pressing [ENTER] after each entry) and enter the frequencies in the column under L_2.

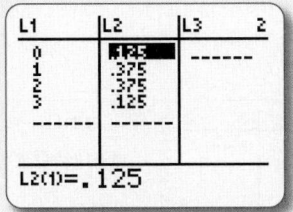

2. To graph the data as in Figure 1, first set the [WINDOW] to $0 \leq X \leq 4$, $0 \leq Y \leq 0.5$, and Xscl = 1 (the width of the bars). Then turn STAT PLOT on ([2nd] [Y=]), and configure it by selecting the histogram icon, setting Xlist = L_1 and Freq = L_2. Then hit [GRAPH].

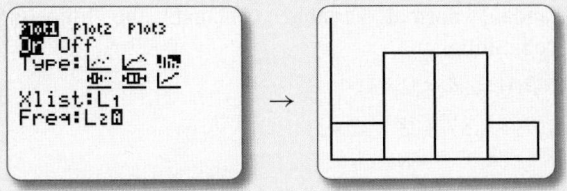

Example 4 (page 552) We obtained the following frequency table in the text:

x	25,000	35,000	45,000	55,000	65,000	75,000	85,000
Frequency	20	80	230	400	170	70	30

Find the probability distribution of X.

Solution with Technology

We need to divide each frequency by the sum. Although the computations in this example (dividing the seven frequencies by 1,000) are simple to do by hand, they could become tedious in general, so technology is helpful.

1. On the TI-83/84 Plus, press [STAT], select EDIT, enter the values of X in the L_1 list, and enter the frequencies in the L_2 list as in Example 3 (below left).

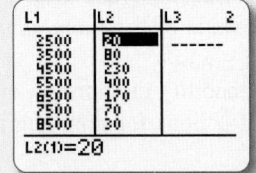

2. Then, on the home screen, enter

   $L_2/1{,}000 \rightarrow L_3$ L_2 is [2nd] [2], L_3 is [2nd] [3].

   or, better yet,

   $L_2/\text{sum}(L_2) \rightarrow L_3$ Sum is found in [2nd] [STAT], under MATH.

3. After pressing [ENTER] you can now go back to the [STAT] EDIT screen, and you will find the probabilities displayed in L_3 as shown above on the right.

Section 8.2

Example 2(b) (page 562) By 2030, the probability that a randomly chosen resident in the United States will be 65 years old or older is projected to be .2. If X is the number of people aged 65 or older in a sample of 6, construct the probability distribution of X.

Solution with Technology

In the "Y=" screen, you can enter the binomial distribution formula

$$Y_1 = 6 \text{ nCr } X*0.2^X*0.8^{(6-X)}$$

directly (to get nCr, press [MATH] and select PRB), and hit [TABLE]. You can then replicate the table in the text by choosing $X = 0, 1, \ldots, 6$ (use the TBLSET screen to set "Indpnt" to "Ask" if you have not already done so).

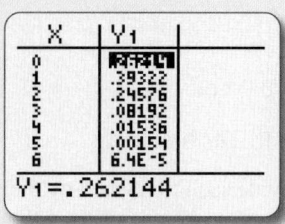

$Y_1 = .262144$

The TI-83/84 Plus also has a built-in binomial distribution function that you can use in place of the explicit formula:

$Y_1 = $ binompdf(6, 0.2, X)   Press [2nd] [VARS] [0].

The TI-83/84 Plus function binompcf (directly following binompdf) gives the value of the *cumulative* distribution function, $P(0 \le X \le x)$.

To graph the resulting probability distribution on your calculator, follow the instructions for graphing a histogram in Section 8.1.

Section 8.3

Example 3 (page 572) According to historical data, the number of injuries that a member of the Enormous State University women's soccer team will sustain during a typical season is given by the following probability distribution table:

Injuries	0	1	2	3	4	5	6
Probability	.20	.20	.22	.20	.15	.01	.02

If X denotes the number of injuries sustained by a player during one season, compute $E(X)$.

Solution with Technology

To obtain the expected value of a probability distribution on the TI-83/84 Plus, press [STAT], select EDIT, and then press [ENTER], and enter the values of X in the L_1 list and the probabilities in the column in the L_2 list. Then, on the home screen, you can obtain the expected value as

sum ($L_1 * L_2$) L_1 is [2nd] [1] L_2 is [2nd] [2]

Sum is found in [2nd] [STAT], under MATH

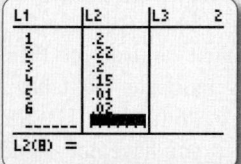
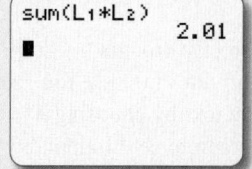

Section 8.4

Example 3 (page 586) Compute the variance and standard deviation for the following probability distribution.

x	10	20	30	40	50	60
$P(X = x)$	.2	.2	.3	.1	.1	.1

Solution with Technology

1. As in Example 3 in the preceding section, begin by entering the probability distribution of X into columns L_1 and L_2 in the LIST screen (press [STAT] and select EDIT). (See below left.)

2. Then, on the home screen, enter

sum($L_1 * L_2$)→M   Stores the value of μ as M
                      Sum is found in [2nd] [STAT],
                      under MATH.

3. To obtain the variance, enter the following.

sum((L_1-M)^2*L_2)   Computation of
                          $\sum(x - \mu)^2 P(X = x)$

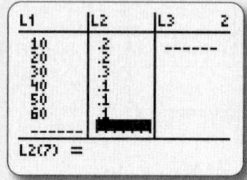
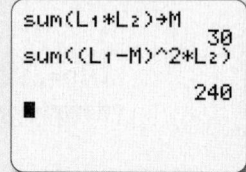

Section 8.5

Example 1(b), (c) (page 598) Let Z be the standard normal variable. Calculate the following probabilities.

b. $P(0 \le Z \le 2.43)$

c. $P(-1.37 \le Z \le 2.43)$

Solution with Technology

On the TI-83/84 Plus, press [2nd] [VARS] to obtain the selection of distribution functions. The first function, normalpdf, gives the values of the normal density function (whose graph is the normal curve). The second, normalcdf, gives $P(a \le Z \le b)$. For example, to compute $P(0 \le Z \le 2.43)$, enter

normalcdf(0, 2.43)

To compute $P(-1.37 \le Z \le 2.43)$, enter

```
normalcdf(-1.37, 2.43)
```

```
normalcdf(0,2.43
)
        .4924505885
normalcdf(-1.37,
2.43)
        .9071070809
■
```

Example 2 (page 599)   Pressure gauges manufactured by Precision Corp. must be checked for accuracy before being placed on the market. To test a pressure gauge, a worker uses it to measure the pressure of a sample of compressed air known to be at a pressure of exactly 50 pounds per square inch. If the gauge reading is off by more than 1% (0.5 pounds), it is rejected. Assuming that the reading of a pressure gauge under these circumstances is a normal random variable with mean 50 and standard deviation 0.4, find the percentage of gauges rejected.

Solution with Technology

As seen in the text, we need to compute $1 - P(49.5 \le X \le 50.5)$ with $\mu = 50$ and $\sigma = 0.4$. On the TI-83/84 Plus, the built-in `normalcdf` function permits us to compute $P(a \le X \le b)$ for nonstandard normal distributions as well. The format is

$$\texttt{normalcdf(a, b, } \mu \texttt{, } \sigma \texttt{)} \quad P(a \le X \le b)$$

For example, we can compute $P(49.5 \le X \le 50.5)$ by entering

```
normalcdf(49.5, 50.5, 50, 0.4)
```

Then subtract it from 1 to obtain the answer:

```
normalcdf(49.5,5
0.5,50,.4)
        .7887003221
1-Ans
        .2112996779
■
```

SPREADSHEET **Technology Guide**

Section 8.1

Example 3 (page 552)   Let X be the number of heads that face up in three tosses of a coin. We obtained the following probability distribution of X in the text:

x	0	1	2	3
$P(X = x)$	$\frac{1}{8}$	$\frac{3}{8}$	$\frac{3}{8}$	$\frac{1}{8}$

Use technology to obtain the corresponding histogram.

Solution with Technology

1. In your spreadsheet, enter the values of X in one column and the probabilities in another.

	A	B
1	x	P(X=x)
2	0	0.125
3	1	0.375
4	2	0.375
5	3	0.125

2. Next, select *only* the column of probabilities (B2–B5) and then choose Insert → Column Chart. The procedure for doing this depends heavily on the spreadsheet and platform you are using.

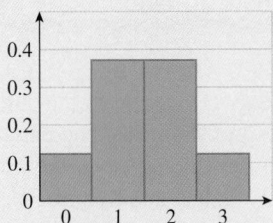

Example 4 (page 552)   We obtained the following frequency table in the text:

x	25,000	35,000	45,000	55,000	65,000	75,000	85,000
Frequency	20	80	230	400	170	70	30

Find the probability distribution of X.

Solution with Technology

We need to divide each frequency by the sum. Although the computations in this example (dividing the seven frequencies by 1,000) are simple to do by hand, they could become tedious in general, so technology is helpful. Spreadsheets manipulate lists with ease. Set up your spreadsheet as shown.

	A	B	C
1	x	Fr	P(X=x)
2	25000	20	=B2/SUM(B:B)
3	35000	80	
4	45000	230	
5	55000	400	
6	65000	170	
7	75000	70	
8	85000	30	

↓

	A	B	C
1	x	Fr	P(X=x)
2	25000	20	0.02
3	35000	80	0.08
4	45000	230	0.23
5	55000	400	0.4
6	65000	170	0.17
7	75000	70	0.07
8	85000	30	0.03

The formula SUM(B:B) gives the sum of all the numerical entries in Column B. You can now change the frequencies to see the effect on the probabilities. You can also add new values and frequencies to the list if you copy the formula in column C further down the column.

Section 8.2

Example 2(b) (page 562)   By 2030, the probability that a randomly chosen resident in the United States will be 65 years old or older is projected to be .2. If X is the number of people aged 65 or older in a sample of 6, construct the probability distribution of X.

Solution with Technology

You can generate the binomial distribution as follows in your spreadsheet:

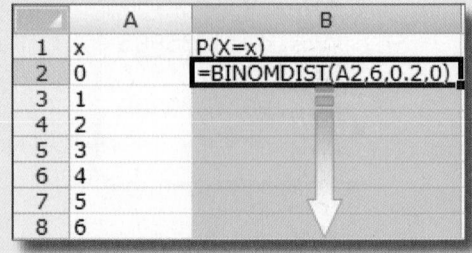

	A	B
1	x	P(X=x)
2	0	=BINOMDIST(A2,6,0.2,0)
3	1	
4	2	
5	3	
6	4	
7	5	
8	6	

↓

	A	B
1	x	P(X=x)
2	0	0.262144
3	1	0.393216
4	2	0.24576
5	3	0.08192
6	4	0.01536
7	5	0.001536
8	6	6.4E-05

The values of X are shown in column A, and the probabilities are computed in column B. The arguments of the BINOMDIST function are as follows:

BINOMDIST(x, n, p, Cumulative (0 = no, 1 = yes)).

Setting the last argument to 0 (as shown) gives $P(X = x)$. Setting it to 1 gives $P(X \le x)$.

To graph the resulting probability distribution using your spreadsheet, insert a bar chart as in Section 8.1.

Section 8.3

Example 3 (page 572)   According to historical data, the number of injuries that a member of the Enormous State University women's soccer team will sustain during a typical season is given by the following probability distribution table:

Injuries	0	1	2	3	4	5	6
Probability	.20	.20	.22	.20	.15	.01	.02

If X denotes the number of injuries sustained by a player during one season, compute $E(X)$.

Solution with Technology

As the method we used suggests, the calculation of the expected value from the probability distribution is particularly easy to do using a spreadsheet program such as Excel.

The following worksheet shows one way to do it. (The first two columns contain the probability distribution of X; the quantities $xP(X=x)$ are summed in cell C9.)

	A	B	C
1	x	P(X=x)	x*P(X=x)
2	0	0.2	=A2*B2
3	1	0.2	
4	2	0.22	
5	3	0.2	
6	4	0.15	
7	5	0.01	
8	6	0.02	
9			=SUM(C2:C8)

↓

	A	B	C
1	x	P(X=x)	x*P(X=x)
2	0	0.2	0
3	1	0.2	0.2
4	2	0.22	0.44
5	3	0.2	0.6
6	4	0.15	0.6
7	5	0.01	0.05
8	6	0.02	0.12
9			2.01

An alternative is to use the SUMPRODUCT function: Once we enter the first two columns above, the formula

$$=\text{SUMPRODUCT}(A2:A8,B2:B8)$$

computes the sum of the products of corresponding entries in the columns, giving us the expected value.

Section 8.4

Example 3 (page 586)   Compute the variance and standard deviation for the following probability distribution.

x	10	20	30	40	50	60
$P(X = x)$	.2	.2	.3	.1	.1	.1

Solution with Technology

As in Example 3 in the preceding section, begin by entering the probability distribution into columns A and B, and then proceed as shown:

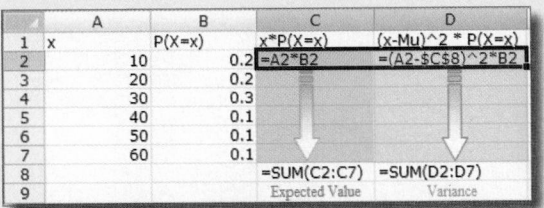

The variance then appears in cell D8:

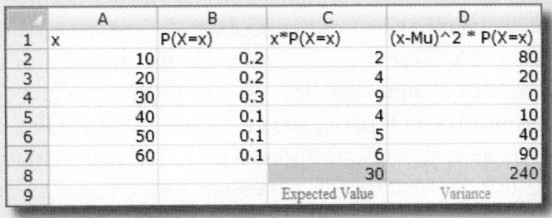

Section 8.5

Example 1(b), (c) (page 598)   Let Z be the standard normal variable. Calculate the following probabilities.
b. $P(0 \leq Z \leq 2.43)$
c. $P(-1.37 \leq Z \leq 2.43)$

Solution with Technology

In spreadsheets, the function NORMSDIST (Normal Standard Distribution) gives the area shown on the left in Figure 18. (Tables such as the one in the Appendix give the area shown on the right.)

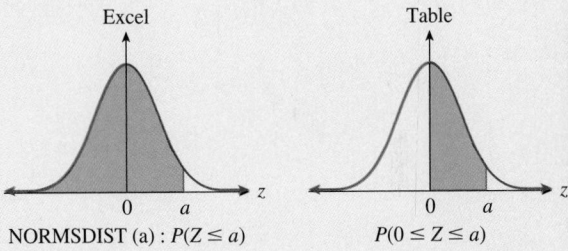

Figure 18

To compute a general area, $P(a \leq Z \leq b)$ in your spreadsheet, subtract the cumulative area to a from that to b:

$$=\text{NORMSDIST}(b)-\text{NORMSDIST}(a) \quad P(a \leq Z \leq b)$$

In particular, to compute $P(0 \leq Z \leq 2.43)$, use

$$=\text{NORMSDIST}(2.43)-\text{NORMSDIST}(0)$$

and to compute $P(-1.37 \leq Z \leq 2.43)$, use

$$=\text{NORMSDIST}(2.43)-\text{NORMSDIST}(-1.37)$$

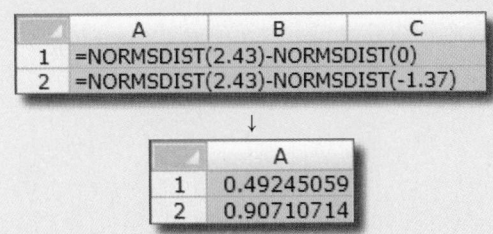

Example 2 (page 599) Pressure gauges manufactured by Precision Corp. must be checked for accuracy before being placed on the market. To test a pressure gauge, a worker uses it to measure the pressure of a sample of compressed air known to be at a pressure of exactly 50 pounds per square inch. If the gauge reading is off by more than 1% (0.5 pounds), it is rejected. Assuming that the reading of a pressure gauge under these circumstances is a normal random variable with mean 50 and standard deviation 0.4, find the percentage of gauges rejected.

Solution with Technology

In spreadsheets, we use the function NORMDIST instead of NORMSDIST. Its format is similar to NORMSDIST, but includes extra arguments as shown.

$$=\text{NORMDIST}(a, \mu, \sigma, 1) \quad P(X \le a)$$

(The last argument, set to 1, tells the spreadsheet that we want the cumulative distribution.) To compute $P(a \le X \le b)$ we enter the following in any vacant cell:

$$=\text{NORMDIST}(b, \mu, \sigma, 1)$$
$$-\text{NORMDIST}(a, \mu, \sigma, 1) \quad P(a \le X \le b)$$

For example, we can compute $P(49.5 \le X \le 50.5)$ by entering

$$=\text{NORMDIST}(50.5, 50, 0.4, 1)$$
$$-\text{NORMDIST}(49.5, 50, 0.4, 1)$$

We then subtract it from 1 to obtain the answer:

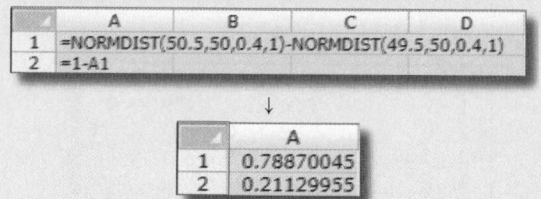

	A	B	C	D
1	=NORMDIST(50.5,50,0.4,1)-NORMDIST(49.5,50,0.4,1)			
2	=1-A1			

↓

	A
1	0.78870045
2	0.21129955

9

Nonlinear Functions and Models

Website

www.WanerMath.com

At the Website you will find:

- Section-by-section tutorials, including game tutorials with randomized quizzes

- A detailed chapter summary

- A true/false quiz

- Additional review exercises

- Graphers, Excel tutorials, and other resources

- The following extra topics:

   Inverse functions

   Using and deriving algebraic properties of logarithms

Case Study Checking up on Malthus

In 1798 Thomas R. Malthus (1766–1834) published an influential pamphlet, later expanded into a book, titled *An Essay on the Principle of Population as It Affects the Future Improvement of Society*. One of his main contentions was that population grows geometrically (exponentially), while the supply of resources such as food grows only arithmetically (linearly). Some 200+ years later, you have been asked to check the validity of Malthus's contention. **How do you go about doing so?**

Robert Nickelsberg/Getty Images

Introduction

To see if Malthus was right, we need to see if the data fit the models (linear and exponential) that he suggested or if other models would be better. We saw in Chapter 1 how to fit a linear model. In this chapter we discuss how to construct models that use various *nonlinear* functions.

The nonlinear functions we consider in this chapter are the *quadratic* functions, the simplest nonlinear functions; the *exponential* functions, essential for discussing many kinds of growth and decay, including the growth (and decay) of money in finance and the initial growth of an epidemic; the *logarithmic* functions, needed to fully understand the exponential functions; and the *logistic* functions, used to model growth with an upper limit, such as the spread of an epidemic.

algebra Review
For this chapter, you should be familiar with the algebra reviewed in **Chapter 0, Section 2.**

9.1   Quadratic Functions and Models

In Chapter 1 we studied linear functions. Linear functions are useful, but in real-life applications, they are often accurate for only a limited range of values of the variables. The relationship between two quantities is often best modeled by a curved line rather than a straight line. The simplest function with a graph that is not a straight line is a *quadratic* function.

> ### Quadratic Function
>
> A **quadratic function** of the variable x is a function that can be written in the form
>
> $$f(x) = ax^2 + bx + c \qquad \text{Function form}$$
>
> or
>
> $$y = ax^2 + bx + c \qquad \text{Equation form}$$
>
> where a, b, and c are fixed numbers (with $a \neq 0$).
>
> #### Quick Examples
>
> 1. $f(x) = 3x^2 - 2x + 1$ $a = 3, b = -2, c = 1$
> 2. $g(x) = -x^2$ $a = -1, b = 0, c = 0$
> 3. $R(p) = -5{,}600p^2 + 14{,}000p$ $a = -5{,}600, b = 14{,}000, c = 0$

✱ We shall not fully justify the formula for the vertex and the axis of symmetry until we have studied some calculus, although it is possible to do so with just algebra.

Every quadratic function $f(x) = ax^2 + bx + c$ $(a \neq 0)$ has a **parabola** as its graph. Following is a summary of some features of parabolas that we can use to sketch the graph of any quadratic function.✱

Features of a Parabola

The graph of $f(x) = ax^2 + bx + c$ $(a \neq 0)$ is a **parabola**. If $a > 0$ the parabola opens upward (concave up) and if $a < 0$ it opens downward (concave down):

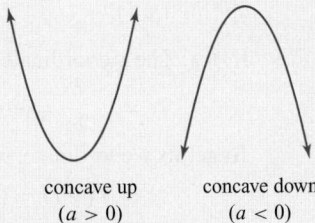

concave up   concave down
$(a > 0)$ $(a < 0)$

Vertex, Intercepts, and Symmetry

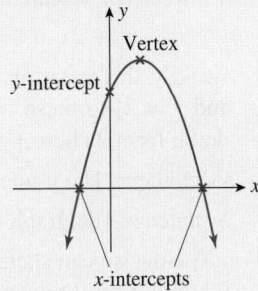

Vertex   The vertex is the highest or lowest point of the parabola (see the figure above). Its x-coordinate is $-\dfrac{b}{2a}$. Its y-coordinate is $f\left(-\dfrac{b}{2a}\right)$.

x**-Intercepts** (if any)   These occur when $f(x) = 0$; that is, when

$$ax^2 + bx + c = 0.$$

Solve this equation for x by either factoring or using the quadratic formula. The x-intercepts are

$$x = \frac{-b \pm \sqrt{b^2 - 4ac}}{2a}.$$

If the **discriminant** $b^2 - 4ac$ is positive, there are two x-intercepts. If it is zero, there is a single x-intercept (at the vertex). If it is negative, there are no x-intercepts (so the parabola doesn't touch the x-axis at all).

y**-Intercept**   This occurs when $x = 0$, so

$$y = a(0)^2 + b(0) + c = c.$$

Symmetry   The parabola is symmetric with respect to the vertical line through the vertex, which is the line $x = -\dfrac{b}{2a}$.

Note that the x-intercepts can also be written as

$$x = -\frac{b}{2a} \pm \frac{\sqrt{b^2 - 4ac}}{2a},$$

making it clear that they are located symmetrically on either side of the line $x = -b/(2a)$. This partially justifies the claim that the whole parabola is symmetric with respect to this line.

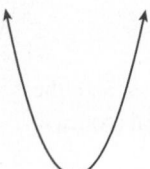

Figure 1

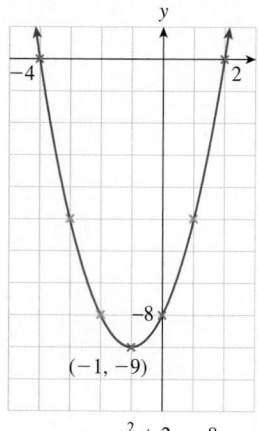

$y = x^2 + 2x - 8$

Figure 2

EXAMPLE 1 **Sketching the Graph of a Quadratic Function**

Sketch the graph of $f(x) = x^2 + 2x - 8$ by hand.

Solution   Here, $a = 1$, $b = 2$, and $c = -8$. Because $a > 0$, the parabola is concave up (Figure 1).

Vertex: The x coordinate of the vertex is

$$x = -\frac{b}{2a} = -\frac{2}{2} = -1.$$

To get its y coordinate, we substitute the value of x back into $f(x)$ to get

$$y = f(-1) = (-1)^2 + 2(-1) - 8 = 1 - 2 - 8 = -9.$$

Thus, the coordinates of the vertex are $(-1, -9)$.

x-Intercepts: To calculate the x-intercepts (if any), we solve the equation

$$x^2 + 2x - 8 = 0.$$

Luckily, this equation factors as $(x + 4)(x - 2) = 0$. Thus, the solutions are $x = -4$ and $x = 2$, so these values are the x-intercepts. (We could also have used the quadratic formula here.)

y-Intercept: The y-intercept is given by $c = -8$.

Symmetry: The graph is symmetric around the vertical line $x = -1$.

Now we can sketch the curve as in Figure 2. (As we see in the figure, it is helpful to plot additional points by using the equation $y = x^2 + 2x - 8$, and to use symmetry to obtain others.)

EXAMPLE 2 **One x-Intercept and No x-Intercepts**

Sketch the graph of each quadratic function, showing the location of the vertex and intercepts.

a. $f(x) = 4x^2 - 12x + 9$

b. $g(x) = -\dfrac{1}{2}x^2 + 4x - 12$

Solution

a. We have $a = 4$, $b = -12$, and $c = 9$. Because $a > 0$, this parabola is concave up.

Vertex: $x = -\dfrac{b}{2a} = \dfrac{12}{8} = \dfrac{3}{2}$ *x coordinate of vertex*

$$y = f\left(\frac{3}{2}\right) = 4\left(\frac{3}{2}\right)^2 - 12\left(\frac{3}{2}\right) + 9 = 0$$ *y coordinate of vertex*

Thus, the vertex is at the point $(3/2, 0)$.

x-Intercepts: $4x^2 - 12x + 9 = 0$

$$(2x - 3)^2 = 0$$

The only solution is $2x - 3 = 0$, or $x = 3/2$. Note that this coincides with the vertex, which lies on the x-axis.

y-Intercept: $c = 9$

Symmetry: The graph is symmetric around the vertical line $x = 3/2$.

The graph is the narrow parabola shown in Figure 3. (As we remarked in Example 1, plotting additional points and using symmetry helps us obtain an accurate sketch.)

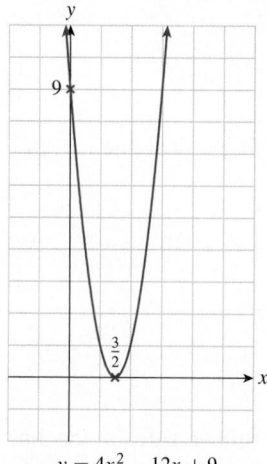

$y = 4x^2 - 12x + 9$

Figure 3

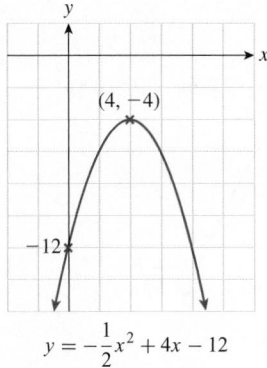

$$y = -\frac{1}{2}x^2 + 4x - 12$$

Figure 4

 using Technology

To automate the computations in Example 2 using a graphing calculator or a spreadsheet, see the Technology Guides at the end of the chapter. Outline:

TI-83/84 Plus
Y$_1$=AX^2+BX+C
4→A:-12→B:9→C
`WINDOW` Xmin=0, Xmax=3
`ZOOM` `0`
[More details on page 680.]

Spreadsheet
Enter x values in column A.
Compute the corresponding y values in column B.
Graph the data in columns A and B.
[More details on page 682.]

 Website
www.WanerMath.com
In the Function Evaluator and Grapher, enter 4x^2-12x+9 for y_1.
For a table of values, enter the various x-values in the Evaluator box, and press "Evaluate".

b. Here, $a = -1/2$, $b = 4$, and $c = -12$. Because $a < 0$, the parabola is concave down. The vertex has x coordinate $-b/(2a) = 4$, with corresponding y coordinate $f(4) = -\frac{1}{2}(4)^2 + 4(4) - 12 = -4$. Thus, the vertex is at $(4, -4)$.

For the x-intercepts, we must solve $-\frac{1}{2}x^2 + 4x - 12 = 0$. If we try to use the quadratic formula, we discover that the discriminant is $b^2 - 4ac = 16 - 24 = -8$. Because the discriminant is negative, there are no solutions of the equation, so there are no x-intercepts.

The y-intercept is given by $c = -12$, and the graph is symmetric around the vertical line $x = 4$.

Because there are no x-intercepts, the graph lies entirely below the x-axis, as shown in Figure 4. (Again, you should plot additional points and use symmetry to ensure that your sketch is accurate.)

APPLICATIONS

Recall that the **revenue** resulting from one or more business transactions is the total payment received. Thus, if q units of some item are sold at p dollars per unit, the revenue resulting from the sale is

$$\text{revenue} = \text{price} \times \text{quantity}$$
$$R = pq.$$

EXAMPLE 3 Demand and Revenue

Alien Publications, Inc. predicts that the demand equation for the sale of its latest illustrated sci-fi novel *Episode 93: Yoda vs. Alien* is

$$q = -2{,}000p + 150{,}000$$

where q is the number of books it can sell each year at a price of $\$p$ per book. What price should Alien Publications, Inc., charge to obtain the maximum annual revenue?

Solution The total revenue depends on the price, as follows:

$$R = pq \qquad\qquad \text{Formula for revenue}$$
$$= p(-2{,}000p + 150{,}000) \qquad \text{Substitute for } q \text{ from demand equation.}$$
$$= -2{,}000p^2 + 150{,}000p. \qquad \text{Simplify.}$$

We are after the price p that gives the maximum possible revenue. Notice that what we have is a quadratic function of the form $R(p) = ap^2 + bp + c$, where $a = -2{,}000$, $b = 150{,}000$, and $c = 0$. Because a is negative, the graph of the function is a parabola, concave down, so its vertex is its highest point (Figure 5). The p coordinate of the vertex is

$$p = -\frac{b}{2a} = -\frac{150{,}000}{-4{,}000} = 37.5.$$

This value of p gives the highest point on the graph and thus gives the largest value of $R(p)$. We may conclude that Alien Publications, Inc., should charge $\$37.50$ per book to maximize its annual revenue.

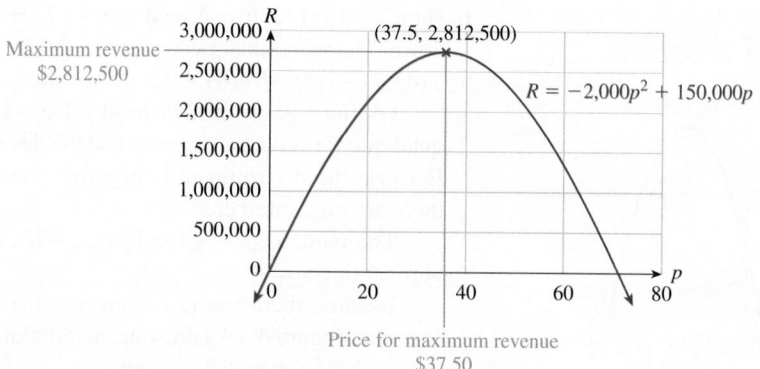

Figure 5

⟹ **Before we go on...** You might ask what the maximum annual revenue is for the publisher in Example 3. Because $R(p)$ gives us the revenue at a price of $\$p$, the answer is $R(37.5) = -2,000\,(37.5)^2 + 150,000(37.5) = 2,812,500$. In other words, the company will earn total annual revenues from this book amounting to $\$2,812,500$. ◼

EXAMPLE 4 Demand, Revenue, and Profit

As the operator of *YSport Fitness* gym, you calculate your demand equation to be

$$q = -0.06p + 84,$$

where q is the number of members in the club and p is the annual membership fee you charge.

a. Your annual operating costs are a fixed cost of $\$20,000$ per year plus a variable cost of $\$20$ per member. Find the annual revenue and profit as functions of the membership price p.

b. At what price should you set the annual membership fee to obtain the maximum revenue? What is the maximum possible revenue?

c. At what price should you set the annual membership fee to obtain the maximum profit? What is the maximum possible profit? What is the corresponding revenue?

Solution

a. The annual revenue is given by

$$R = pq \qquad\qquad \text{Formula for revenue}$$
$$= p(-0.06p + 84) \qquad \text{Substitute for } q \text{ from demand equation.}$$
$$= -0.06p^2 + 84p. \qquad \text{Simplify.}$$

The annual cost C is given by

$$C = 20,000 + 20q. \qquad \text{\$20,000 plus \$20 per member}$$

However, this is a function of q, and not p. To express C as a function of p we substitute for q using the demand equation $q = -0.06p + 84$:

$$C = 20,000 + 20(-0.06p + 84)$$
$$= 20,000 - 1.2p + 1,680$$
$$= -1.2p + 21,680.$$

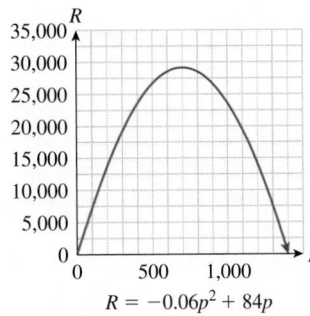

$$R = -0.06p^2 + 84p$$

Figure 6

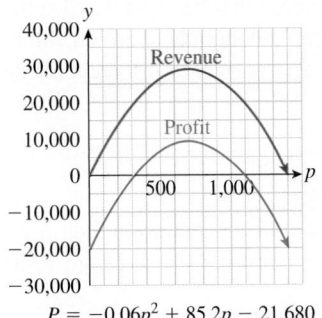

$$P = -0.06p^2 + 85.2p - 21,680$$

Figure 7

Thus, the profit function is:

$$P = R - C \qquad \text{Formula for profit}$$
$$= -0.06p^2 + 84p - (-1.2p + 21,680) \qquad \text{Substitute for revenue and cost.}$$
$$= -0.06p^2 + 85.2p - 21,680.$$

b. From part (a) the revenue function is given by

$$R = -0.06p^2 + 84p.$$

This is a quadratic function ($a = -0.06, b = 84, c = 0$) whose graph is a concave-down parabola (Figure 6). The maximum revenue corresponds to the highest point of the graph: the vertex, of which the p coordinate is

$$p = -\frac{b}{2a} = -\frac{84}{2(-0.06)} \approx \$700.$$

This is the membership fee you should charge for the maximum revenue. The corresponding maximum revenue is given by the y coordinate of the vertex in Figure 6:

$$R(700) = -0.06(700)^2 + 84(700) = \$29,400.$$

c. From part (a), the profit function is given by

$$P = -0.06p^2 + 85.2p - 21,680.$$

Like the revenue function, the profit function is quadratic ($a = -0.06$, $b = 85.2$, $c = -21,680$). Figure 7 shows both the revenue and profit functions. The maximum profit corresponds to the vertex, whose p coordinate is

$$p = -\frac{b}{2a} = -\frac{85.2}{2(-0.06)} \approx \$710.$$

This is the membership fee you should charge for the maximum profit. The corresponding maximum profit is given by the y coordinate of the vertex of the profit curve in Figure 7:

$$P(710) = -0.06(710)^2 + 85.2(710) - 21,680 = \$8,566.$$

The corresponding revenue is

$$R(710) = -0.06(710)^2 + 84(710) = \$29,394,$$

slightly less than the maximum possible revenue of \$29,400.

➡ **Before we go on...** The result of part (c) of Example 4 tells us that the vertex of the profit curve in Figure 7 is slightly to the right of the vertex in the revenue curve. However, the difference is tiny compared to the scale of the graphs, so the graphs appear to be parallel. ∎

> **Q:** *Charging $710 membership brings in less revenue than charging $700. So why charge $710?*
>
> **A:** A membership fee of $700 does bring in slightly larger revenue than a fee of $710, but it also brings in a slightly larger membership, which in turn raises the operating expense and has the effect of *lowering* the profit slightly (to $8,560). In other words, the slightly higher fee, while bringing in less revenue, also lowers the cost, and the net result is a larger profit.

Fitting a Quadratic Function to Data: Quadratic Regression

In Section 1.4 we saw how to fit a regression line to a collection of data points. Here, we see how to use technology to obtain the **quadratic regression curve** associated with a set of points. The quadratic regression curve is the quadratic curve $y = ax^2 + bx + c$ that best fits the data points in the sense that the associated sum-of-squares error (SSE—see Section 1.4) is a minimum. Although there are algebraic methods for obtaining the quadratic regression curve, it is normal to use technology to do this.

EXAMPLE 5 Carbon Dioxide Concentration

The following table shows the annual mean carbon dioxide concentration measured at Mauna Loa Observatory in Hawaii, in parts per million, every 10 years from 1960 through 2010 ($t = 0$ represents 1960).[1]

Year t	0	10	20	30	40	50
PPM CO$_2$ C	317	326	339	354	369	390

a. Is a linear model appropriate for these data?

b. Find the quadratic model

$$C(t) = at^2 + bt + c$$

that best fits the data.

Solution

a. To see whether a linear model is appropriate, we plot the data points and the regression line using one of the methods of Example 2 in Section 1.4 (Figure 8).

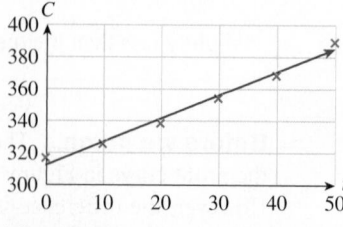

Figure 8

---

[1]Figures are approximate. Source: U.S. Department of Commerce/National Oceanic and Atmospheric Administration (NOAA) Earth System Research Laboratory, data downloaded from www.esrl.noaa.gov/gmd/ccgg/trends/ on March 13, 2011.

using Technology

For detailed instructions on how to find and graph the regression curve in Example 5 using a graphing calculator or a spreadsheet, see the Technology Guides at the end of the chapter. Outline:

TI-83/84 Plus

STAT EDIT values of t in L_1 and values of C in L_2
Regression curve: STAT
CALC option 5
QuadReg ENTER
Graph: Y= VARS 5
EQ 1 , then ZOOM 9
[More details on page 680.]

Spreadsheet

t- and C-values in Columns A and B. Graph: Highlight data and insert a Scatter chart.
Regression curve: Right-click a datapoint and add polynomial order 2 trendline with option to show equation on chart.
[More details on page 683.]

Website
www.WanerMath.com
In the Simple Regression utility, enter the data in the x- and y-columns and press
`"y=ax^2+bx+c"`.

From the graph, we can see that the given data suggest a curve and not a straight line: The observed points are above the regression line at the ends but below in the middle. (We would expect the data points from a linear relation to fall randomly above and below the regression line.)

b. The quadratic model that best fits the data is the quadratic regression model. As with linear regression, there are algebraic formulas to compute a, b, and c, but they are rather involved. However, we exploit the fact that these formulas are built into graphing calculators, spreadsheets, and other technology and obtain the regression curve using technology (see Figure 9):

$$C(t) = 0.012t^2 + 0.85t + 320. \qquad \text{Coefficients rounded to two significant digits}$$

Notice from the graphs that the quadratic regression model appears to give a better fit than the linear regression model. This impression is supported by the values of SSE: For the linear regression model, SSE $\approx$ 58, while for the quadratic regression model, SSE is much smaller, approximately 2.6, indicating a much better fit.

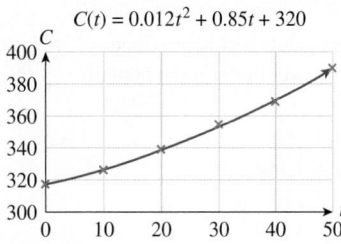

Figure 9

9.1 EXERCISES

▼ more advanced   ◆ challenging
⊤ indicates exercises that should be solved using technology

In Exercises 1–10, sketch the graphs of the quadratic functions, indicating the coordinates of the vertex, the y-intercept, and the x-intercepts (if any). HINT [See Example 1.]

1. $f(x) = x^2 + 3x + 2$

2. $f(x) = -x^2 - x$

3. $f(x) = -x^2 + 4x - 4$

4. $f(x) = x^2 + 2x + 1$

5. $f(x) = -x^2 - 40x + 500$

6. $f(x) = x^2 - 10x - 600$

7. $f(x) = x^2 + x - 1$

8. $f(x) = x^2 + \sqrt{2}x + 1$

9. $f(x) = x^2 + 1$

10. $f(x) = -x^2 + 5$

In Exercises 11–14, for each demand equation, express the total revenue R as a function of the price p per item, sketch the graph of the resulting function, and determine the price p that maximizes total revenue in each case. HINT [See Example 3.]

11. $q = -4p + 100$

12. $q = -3p + 300$

13. $q = -2p + 400$

14. $q = -5p + 1,200$

⊤ *In Exercises 15–18, use technology to find the quadratic regression curve through the given points. (Round all coefficients to four decimal places.)* HINT [See Example 5.]

15. $\{(1, 2), (3, 5), (4, 3), (5, 1)\}$

16. $\{(-1, 2), (-3, 5), (-4, 3), (-5, 1)\}$

17. $\{(-1, 2), (-3, 5), (-4, 3)\}$

18. $\{(2, 5), (3, 5), (5, 3)\}$

APPLICATIONS

19. *World Military Expenditure* The following chart shows total military and arms trade expenditure from 1990 to 2008 ($t = 0$ represents 1990).[2]

---

[2]Approximate figures in constant 2005 dollars. The 2008 figure is an estimate, based on the increase in U.S. military expenditure.

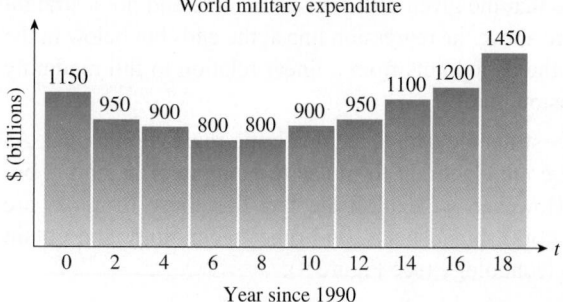

World military expenditure

Source: www.globalissues.org/Geopolitics/ArmsTrade/Spending.asp

a. If you want to model the expenditure figures with a function of the form

$$f(t) = at^2 + bt + c,$$

would you expect the coefficient a to be positive or negative? Why? HINT [See "Features of a Parabola," page 621.]

b. Which of the following models best approximates the data given? (Try to answer this without actually computing values.)

(A) $f(t) = 5t^2 - 80t - 1{,}150$
(B) $f(t) = -5t^2 - 80t + 1{,}150$
(C) $f(t) = 5t^2 - 80t + 1{,}150$
(D) $f(t) = -5t^2 - 80t - 1{,}150$

c. What is the nearest year that would correspond to the vertex of the graph of the correct model from part (b)? What is the danger of extrapolating the data in either direction?

20. Education Expenditure The following chart shows the percentage of the U.S. Discretionary Budget allocated to education from 2003 to 2009 ($t = 3$ represents the start of 2003).

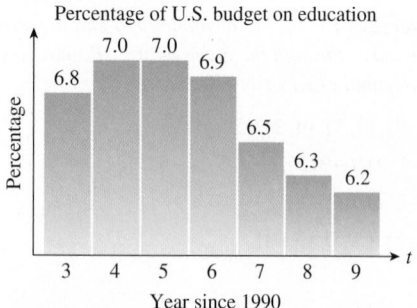

Percentage of U.S. budget on education

Source: www.globalissues.org/Geopolitics/ArmsTrade/Spending.asp

a. If you want to model the percentage figures with a function of the form

$$f(t) = at^2 + bt + c,$$

would you expect the coefficient a to be positive or negative? Why? HINT [See "Features of a Parabola," page 621.]

b. Which of the following models best approximates the data given? (Try to answer this without actually computing values)

(A) $f(t) = 0.04t^2 + 0.3t - 6$
(B) $f(t) = -0.04t^2 + 0.3t + 6$
(C) $f(t) = 0.04t^2 + 0.3t + 6$
(D) $f(t) = -0.04t^2 + 0.3t - 6$

c. What is the nearest year that would correspond to the vertex of the graph of the correct model from part (b)? What is the danger of extrapolating the data in either direction?

21. Oil Imports from Mexico Daily oil imports to the United States from Mexico can be approximated by

$$I(t) = -0.015t^2 + 0.1t + 1.4$$
$$\text{million barrels/day} \ (0 \le t \le 8)$$

where t is time in years since the start of 2000.[3] According to the model, in what year were oil imports to the United States greatest? How many barrels per day were imported that year? HINT [See Example 1.]

22. Oil Production in Mexico Daily oil production by Pemex, Mexico's national oil company, for 2001–2009 can be approximated by

$$P(t) = -0.022t^2 + 0.2t + 2.9$$
$$\text{million barrels/day} \ (1 \le t \le 9)$$

where t is time in years since the start of 2000.[4] According to the model, in what year was oil production by Pemex greatest? How many barrels per day were produced that year?

23. Net Income The annual net income of General Electric for the period 2005–2010 could be approximated by

$$P(t) = -2.0t^2 + 6.6t + 16 \text{ billion dollars } (0 \le t \le 5)$$

where t is time in years since 2005.[5] According to the model, in what year in this period was General Electric's net income highest? What was its net income that year? Would you trust this model to continue to be valid long past this period? Why?

24. Net Income The annual net income of General Electric for the period 2007–2012 could be approximated by

$$P(t) = 3.0t^2 - 24t + 59 \text{ billion dollars } (2 \le t \le 7)$$

where t is time in years since 2005.[6] According to the model, in what year in this period was General Electric's net income lowest? What was its net income that year? Would you trust this model to continue to be valid long past this period? Why?

---

[3] Source: Energy Information Administration/Pemex/www.eia.gov
[4] Figures are approximate, and 2008–2009 figures are projections by the Department of Energy. Source: Energy Information Administration: Pemex.
[5] Source for data: www.wikinvest.com/
[6] *Ibid.* 2011 net income is projected.

25. Revenue The market research department of the *Better Baby Buggy Co.* predicts that the demand equation for its buggies is given by $q = -0.5p + 140$, where q is the number of buggies it can sell in a month if the price is $\$p$ per buggy. At what price should it sell the buggies to get the largest revenue? What is the largest monthly revenue? HINT [See Example 3.]

26. Revenue The *Better Baby Buggy Co.* has just come out with a new model, the Turbo. The market research department predicts that the demand equation for Turbos is given by $q = -2p + 320$, where q is the number of buggies it can sell in a month if the price is $\$p$ per buggy. At what price should it sell the buggies to get the largest revenue? What is the largest monthly revenue?

27. Revenue *Pack-Em-In Real Estate* is building a new housing development. The more houses it builds, the less people will be willing to pay, due to the crowding and smaller lot sizes. In fact, if it builds 40 houses in this particular development, it can sell them for $200,000 each, but if it builds 60 houses, it will only be able to get $160,000 each. Obtain a linear demand equation and hence determine how many houses Pack-Em-In should build to get the largest revenue. What is the largest possible revenue? HINT [See Example 3.]

28. Revenue *Pack-Em-In* has another development in the works. If it builds 50 houses in this development, it will be able to sell them at $190,000 each, but if it builds 70 houses, it will get only $170,000 each. Obtain a linear demand equation and hence determine how many houses it should build to get the largest revenue. What is the largest possible revenue?

29. ▼Revenue from Monorail Service, Las Vegas In 2005, the Las Vegas monorail charged $3 per ride and had an average ridership of about 28,000 per day. In December 2005 the Las Vegas Monorail Company raised the fare to $5 per ride, and average ridership in 2006 plunged to around 19,000 per day.[7]

a. Use the given information to find a linear demand equation.
b. Find the price the company should have charged to maximize revenue from ridership. What is the corresponding daily revenue?
c. The Las Vegas Monorail Company would have needed $44.9 million in revenues from ridership to break even in 2006. Would it have been possible to break even in 2006 by charging a suitable price?

30. ▼Revenue from Monorail Service, Mars The Utarek monorail, which links the three urbynes (or districts) of Utarek, Mars, charged $\bar{\bar{Z}}5$ per ride[8] and sold about 14 million rides per day. When the Utarek City Council lowered the fare to $\bar{\bar{Z}}3$ per ride, the number of rides increased to 18 million per day.

a. Use the given information to find a linear demand equation.

b. Find the price the City Council should have charged to maximize revenue from ridership. What is the corresponding daily revenue?
c. The City Council would have needed to raise $\bar{\bar{Z}}48$ billion in revenues from ridership each Martian year (670 days[9]) to finance the new Mars organism research lab. Would this have been possible by charging a suitable price?

31. Web Site Profit You operate a gaming Web site, www.mudbeast.net, where users must pay a small fee to log on. When you charged $2 the demand was 280 log-ons per month. When you lowered the price to $1.50, the demand increased to 560 log-ons per month.

a. Construct a linear demand function for your Web site and hence obtain the monthly revenue R as a function of the log-on fee x.
b. Your Internet provider charges you a monthly fee of $30 to maintain your site. Express your monthly profit P as a function of the log-on fee x, and hence determine the log-on fee you should charge to obtain the largest possible monthly profit. What is the largest possible monthly profit? HINT [See Example 4.]

32. T-Shirt Profit Two fraternities, Sig Ep and Ep Sig, plan to raise money jointly to benefit homeless people on Long Island. They will sell *Yoda vs. Alien* T-shirts in the student center, but are not sure how much to charge. Sig Ep treasurer Augustus recalls that they once sold 400 shirts in a week at $8 per shirt, but Ep Sig treasurer Julius has solid research indicating that it is possible to sell 600 per week at $4 per shirt.

a. Based on this information, construct a linear demand equation for Yoda vs. Alien T-shirts, and hence obtain the weekly revenue R as a function of the unit price x.
b. The university administration charges the fraternities a weekly fee of $500 for use of the Student Center. Write down the monthly profit P as a function of the unit price x, and hence determine how much the fraternities should charge to obtain the largest possible weekly profit. What is the largest possible weekly profit? HINT [See Example 4.]

33. Web Site Profit The latest demand equation for your gaming Web site, www.mudbeast.net, is given by

$$q = -400x + 1,200$$

where q is the number of users who log on per month and x is the log-on fee you charge. Your Internet provider bills you as follows:

| Site maintenance fee: | $20 per month |
| High-volume access fee: | 50¢ per log-on |

Find the monthly cost as a function of the log-on fee x. Hence, find the monthly profit as a function of x and determine the log-on fee you should charge to obtain the largest possible monthly profit. What is the largest possible monthly profit?

[7]Source: *New York Times*, Februrary 10, 2007, p. A9.

[8]The zonar ($\bar{\bar{Z}}$) is the official currency in the city-state of Utarek, Mars (formerly www.Marsnext.com, a now extinct virtual society).

[9]As measured in Mars days. The actual length of a Mars year is about 670.55 Mars days, so frequent leap years are designated by the Mars Planetary Authority to adjust.

34. *T-Shirt Profit* The latest demand equation for your *Yoda vs. Alien* T-shirts is given by

$$q = -40x + 600$$

where q is the number of shirts you can sell in one week if you charge $\$x$ per shirt. The Student Council charges you $\$400$ per week for use of their facilities, and the T-shirts cost you $\$5$ each. Find the weekly cost as a function of the unit price x. Hence, find the weekly profit as a function of x and determine the unit price you should charge to obtain the largest possible weekly profit. What is the largest possible weekly profit?

35. ▼ *Nightclub Management* You have just opened a new nightclub, *Russ' Techno Pitstop,* but are unsure of how high to set the cover charge (entrance fee). One week you charged $\$10$ per guest and averaged 300 guests per night. The next week you charged $\$15$ per guest and averaged 250 guests per night.

a. Find a linear demand equation showing the number of guests q per night as a function of the cover charge p.

b. Find the nightly revenue R as a function of the cover charge p.

c. The club will provide two free nonalcoholic drinks for each guest, costing the club $\$3$ per head. In addition, the nightly overheads (rent, salaries, dancers, DJ, etc.) amount to $\$3,000$. Find the cost C as a function of the cover charge p.

d. Now find the profit in terms of the cover charge p, and hence determine the entrance fee you should charge for a maximum profit.

36. ▼ *Television Advertising* As sales manager for *Montevideo Productions, Inc.,* you are planning to review the prices you charge clients for television advertisement development. You currently charge each client an hourly development fee of $\$2,500$. With this pricing structure, the demand, measured by the number of contracts Montevideo signs per month, is 15 contracts. This is down 5 contracts from the figure last year, when your company charged only $\$2,000$.

a. Construct a linear demand equation giving the number of contracts q as a function of the hourly fee p Montevideo charges for development.

b. On average, Montevideo bills for 50 hours of production time on each contract. Give a formula for the total revenue obtained by charging $\$p$ per hour.

c. The costs to Montevideo Productions are estimated as follows:

Fixed costs:	$\$120,000$ per month
Variable costs:	$\$80,000$ per contract

Express Montevideo Productions' monthly cost **(i)** as a function of the number q of contracts and **(ii)** as a function of the hourly production charge p.

d. Express Montevideo Productions' monthly profit as a function of the hourly development fee p and hence find the price it should charge to maximize the profit.

37. ▮ ***World Military Expenditure*** The following table shows total military and arms trade expenditure in 1994, 1998, and 2006. (See Exercise 19; $t = 4$ represents 1994.)[10]

Year t	4	8	16
Military Expenditure ($ billion)	900	800	1,200

Find a quadratic model for these data, and use your model to estimate world military expenditure in 2008. Compare your answer with the actual figure shown in Exercise 19. HINT [See Example 5.]

38. ▮ ***Education Expenditure*** The following table shows the percentage of the U.S. Discretionary Budget allocated to education in 2003, 2005, and 2009. (See Exercise 20; $t = 3$ represents the start of 2003.)[11]

Year t	3	5	9
Percentage	6.8	7	6.2

Find a quadratic model for these data, and use your model to estimate the percentage of the U.S. Discretionary Budget allocated to education in 2008. Compare your answer with the actual figure shown in Exercise 20.

39. ▮ ***iPod Sales*** The following table shows Apple iPod sales from the last quarter of 2009 through the last quarter of 2010.[12]

Quarter t	4	5	6	7	8
iPod Sales (millions)	21.0	10.9	9.4	9.1	19.5

a. Find a quadratic regression model for these data. (Round coefficients to two significant digits.) Graph the model together with the data.

b. What does the model predict for iPod sales in the first quarter of 2011 ($t = 9$), to the nearest million? Comment on the answer.

40. ▮ ***iPod Sales*** The following table shows Apple iPod sales from the second quarter of 2009 through the second quarter of 2010.[13]

Quarter t	2	3	4	5	6
iPod Sales (millions)	10.2	10.2	21.0	10.9	9.4

---

[10] Approximate figures in constant 2005 dollars. The 2008 figure is an estimate, based on the increase in U.S. military expenditure.
Source: www.globalissues.org/Geopolitics/ArmsTrade/Spending.asp.

[11] Source: www.globalissues.org/Geopolitics/ArmsTrade/Spending.asp.

[12] Source: Apple quarterly press releases (www.apple.com/investor/).

[13] *Ibid.*

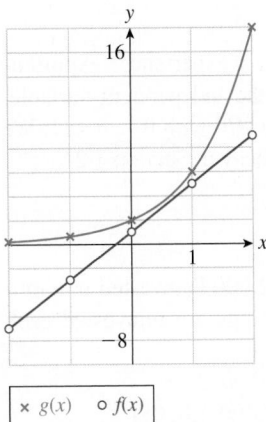

× g(x)   ○ f(x)

Figure 11

Solution Remember that a linear function increases (or decreases) by the same amount every time x increases by 1. The values of f behave this way: Every time x increases by 1, the value of $f(x)$ increases by 4. Therefore, f is a linear function with a *slope* of 4. Because $f(0) = 1$, we see that

$$f(x) = 4x + 1$$

is a linear formula that fits the data.

On the other hand, every time x increases by 1, the value of $g(x)$ is *multiplied* by 3. Because $g(0) = 2$, we find that

$$g(x) = 2(3^x)$$

is an exponential function fitting the data.

We can visualize the two functions f and g by plotting the data points (Figure 11). The data points for $f(x)$ clearly lie along a straight line, whereas the points for $g(x)$ lie along a curve. The y coordinate of each point for $g(x)$ is 3 times the y coordinate of the preceding point, demonstrating that the curve is an exponential one.

In Section 1.3, we discussed a method for calculating the equation of the line that passes through two given points. In the following example, we show a method for calculating the equation of the exponential curve through two given points.

EXAMPLE 2 Finding the Exponential Curve through Two Points

Find an equation of the exponential curve through $(1, 6.3)$ and $(4, 170.1)$.

Solution We want an equation of the form

$$y = Ab^x \quad (b > 0).$$

Substituting the coordinates of the given points, we get

$$6.3 = Ab^1 \qquad \text{Substitute } (1, 6.3).$$
$$170.1 = Ab^4. \qquad \text{Substitute } (4, 170.1).$$

If we now divide the second equation by the first, we get

$$\frac{170.1}{6.3} = \frac{Ab^4}{Ab} = b^3$$
$$b^3 = 27$$
$$b = 27^{1/3} \qquad \text{Take reciprocal power of both sides.}$$
$$b = 3.$$

Now that we have b, we can substitute its value into the first equation to obtain

$$6.3 = 3A \qquad \text{Substitute } b = 3 \text{ into the equation } 6.3 = Ab^1.$$
$$A = \frac{6.3}{3} = 2.1.$$

We have both constants, $A = 2.1$ and $b = 3$, so the model is

$$y = 2.1(3^x).$$

Example 6 will show how to use technology to fit an exponential function to two or more data points.

APPLICATIONS

Recall some terminology we mentioned earlier: A quantity y experiences **exponential growth** if $y = Ab^t$ with $b > 1$. (Here we use t for the independent variable, thinking of time.) It experiences **exponential decay** if $y = Ab^t$ with $0 < b < 1$. We already saw several examples of exponential growth and decay in Section 1.2.

EXAMPLE 3   Exponential Growth and Decay

a. Compound Interest (See Section 1.2 Example 6.) If \$2,000 is invested in a mutual fund with an annual yield of 12.6% and the earnings are reinvested each month, then the future value after t years is

$$A(t) = P\left(1 + \frac{r}{m}\right)^{mt} = 2{,}000\left(1 + \frac{0.126}{12}\right)^{12t} = 2{,}000(1.0105)^{12t},$$

which can be written as $2{,}000(1.0105^{12})^t$, so $A = 2{,}000$ and $b = 1.0105^{12}$. This is an example of exponential growth, because $b > 1$.

b. Carbon Decay (See Section 1.2 Example 7.) The amount of carbon 14 remaining in a sample that originally contained A grams is approximately

$$C(t) = A(0.999879)^t.$$

This is an instance of exponential decay, because $b < 1$.

➡ **Before we go on...** Refer again to part (a). In Example 6(b) of Section 1.2 we showed how to use technology to answer questions such as the following: "When, to the nearest year, will the value of your investment reach \$5,000?" ■

The next example shows an application to public health.

EXAMPLE 4   Exponential Growth: Epidemics

In the early stages of the AIDS epidemic during the 1980s, the number of cases in the United States was increasing by about 50% every 6 months. By the start of 1983, there were approximately 1,600 AIDS cases in the United States.[14]

a. Assuming an exponential growth model, find a function that predicts the number of people infected t years after the start of 1983.

b. Use the model to estimate the number of people infected by October 1, 1986, and also by the end of that year.

Solution

a. One way of finding the desired exponential function is to reason as follows: At time $t = 0$ (January 1, 1983), the number of people infected was 1,600, so $A = 1{,}600$. Every 6 months, the number of cases increased to 150% of the number 6 months earlier—that is, to 1.50 times that number. Each year, it therefore increased to $(1.50)^2 = 2.25$ times the number one year earlier. Hence, after t years, we need to multiply the original 1,600 by 2.25^t, so the model is

$$y = 1{,}600(2.25^t) \text{ cases.}$$

[14]Data based on regression of the 1982–1986 figures. Source for data: Centers for Disease Control and Prevention. HIV/AIDS Surveillance Report, 2000;12 (No. 2).

Alternatively, if we wish to use the method of Example 2, we need two data points. We are given one point: (0, 1,600). Because y increased by 50% every 6 months, 6 months later it reached $1,600 + 800 = 2,400$ ($t = 0.5$). This information gives a second point: (0.5, 2,400). We can now apply the method in Example 2 to find the model above.

b. October 1, 1986, corresponds to $t = 3.75$ (because October 1 is 9 months, or $9/12 = 0.75$ of a year after January 1). Substituting this value of t in the model gives

$$y = 1,600(2.25^{3.75}) \approx 33,481 \text{ cases} \qquad \texttt{1600*2.25\^{}3.75}$$

By the end of 1986, the model predicts that

$$y = 1,600(2.25^4) = 41,006 \text{ cases.}$$

(The actual number of cases was around 41,700.)

➡ **Before we go on...** Increasing the number of cases by 50% every 6 months couldn't continue for very long and this is borne out by observations. If increasing by 50% every 6 months did continue, then by January 2003 ($t = 20$), the number of infected people would have been

$$1,600(2.25^{20}) \approx 17,700,000,000$$

a number that is more than 50 times the size of the U.S. population! Thus, although the exponential model is fairly reliable in the early stages of the epidemic, it is unreliable for predicting long-term trends. ■

Epidemiologists use more sophisticated models to measure the spread of epidemics, and these models predict a leveling-off phenomenon as the number of cases becomes a significant part of the total population. We discuss such a model, the **logistic function**, in Section 9.4.

The Number e and More Applications

In nature we find examples of growth that occurs *continuously*, as though "interest" is being added more often than every second or fraction of a second. To model this, we need to see what happens to the compound interest formula of Section 1.2 as we let m (the number of times interest is added per year) become extremely large. Something very interesting does happen: We end up with a more compact and elegant formula than we began with. To see why, let's look at a very simple situation.

Suppose we invest \$1 in the bank for 1 year at 100% interest, compounded m times per year. If $m = 1$, then 100% interest is added every year, and so our money doubles at the end of the year. In general, the accumulated capital at the end of the year is

	A	B
1	m	(1+1/m)^m
2	1	2
3	10	2.59374246
4	100	2.704813829
5	1000	2.716923932
6	10000	2.718145927
7	100000	2.718268237
8	1000000	2.718280469
9	10000000	2.718281694
10	100000000	2.718281786
11	1000000000	2.718282031

$$A = 1\left(1 + \frac{1}{m}\right)^m = \left(1 + \frac{1}{m}\right)^m. \qquad \texttt{(1+1/m)\^{}m}$$

Now, we are interested in what A becomes for large values of m. On the left is a spreadsheet showing the quantity $\left(1 + \frac{1}{m}\right)^m$ for larger and larger values of m.

Something interesting *does* seem to be happening! The numbers appear to be getting closer and closer to a specific value. In mathematical terminology, we say that the numbers **converge** to a fixed number, 2.71828..., called the **limiting value**[*] of the quantities $\left(1 + \frac{1}{m}\right)^m$. This number, called e, is one of the most important in mathematics. The number e is irrational, just as the more familiar number π is, so we cannot write down its exact numerical value. To 20 decimal places,

$$e = 2.71828182845904523536....$$

[*] See Chapter 10 for more on limits.

We now say that, if $1 is invested for 1 year at 100% interest **compounded continuously**, the accumulated money at the end of that year will amount to $e = \$2.72$ (to the nearest cent). But what about the following more general question?

Q : *What about a more general scenario: If we invest an amount $P for t years at an interest rate of r, compounded continuously, what will be the accumulated amount A at the end of that period?*

A : In the special case above (P, t, and r all equal 1), we took the compound interest formula and let m get larger and larger. We do the same more generally, after a little preliminary work with the algebra of exponentials.

$$A = P\left(1 + \frac{r}{m}\right)^{mt}$$

$$= P\left(1 + \frac{1}{(m/r)}\right)^{mt} \qquad \text{Substituting } \frac{r}{m} = \frac{1}{(m/r)}$$

$$= P\left(1 + \frac{1}{(m/r)}\right)^{(m/r)rt} \qquad \text{Substituting } m = \left(\frac{m}{r}\right)r$$

$$= P\left[\left(1 + \frac{1}{(m/r)}\right)^{(m/r)}\right]^{rt} \qquad \text{Using the rule } a^{bc} = (a^b)^c$$

For continuous compounding of interest, we let m, and hence m/r, get very large. This affects only the term in brackets, which converges to e, and we get the formula

$$A = Pe^{rt}.$$

Q : *How do I obtain powers of e or e itself on a TI-83/84 Plus or in a spreadsheet?*

A : On the TI-83/84 Plus, enter e^x as `e^(x)`, where `e^(` can be obtained by pressing [2ND] [LN] . To obtain the number e on the TI-83/84 Plus, enter `e^(1)`. Spreadsheets have a built-in function called `EXP`; `EXP(x)` gives the value of e^x. To obtain the number e in a spreadsheet, enter `=EXP(1)`.

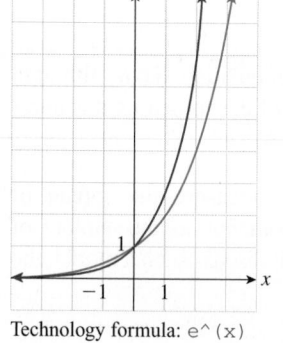

y $y = e^x$ $y = 2^x$

Technology formula: `e^(x)`
or `EXP(x)`

Figure 12

Figure 12 shows the graph of $y = e^x$ with that of $y = 2^x$ for comparison.

The Number e and Continuous Compounding

The number e is the limiting value of the quantities $\left(1 + \frac{1}{m}\right)^m$ as m gets larger and larger, and has the value $2.71828182845904523536\ldots$

If P is invested at an annual interest rate r compounded continuously, the accumulated amount after t years is

$$A(t) = Pe^{rt}. \qquad \text{P*e^(r*t) or P*EXP(r*t)}$$

Quick Examples

1. If $100 is invested in an account that bears 15% interest compounded continuously, at the end of 10 years the investment will be worth

$$A(10) = 100e^{(0.15)(10)} = \$448.17. \qquad \begin{array}{l}\texttt{100*e^(0.15*10) or}\\ \texttt{100*EXP(0.15*10)}\end{array}$$

2. If $1 is invested in an account that bears 100% interest compounded continuously, at the end of x years the investment will be worth

$$A(x) = e^x \text{dollars.}$$

EXAMPLE 5 Continuous Compounding

a. You invest $10,000 at *Fastrack Savings & Loan,* which pays 6% compounded continuously. Express the balance in your account as a function of the number of years t and calculate the amount of money you will have after 5 years.

b. Your friend has just invested $20,000 in *Constant Growth Funds,* whose stocks are continuously *declining* at a rate of 6% per year. How much will her investment be worth in 5 years?

c. ☐ During which year will the value of your investment first exceed that of your friend?

Solution

a. We use the continuous growth formula with $P = 10,000$, $r = 0.06$, and t variable, getting

$$A(t) = Pe^{rt} = 10,000e^{0.06t}.$$

In 5 years,

$$A(5) = 10,000e^{0.06(5)}$$
$$= 10,000e^{0.3}$$
$$\approx \$13,498.59.$$

b. Because the investment is depreciating, we use a negative value for r and take $P = 20,000$, $r = -0.06$, and $t = 5$, getting

$$A(t) = Pe^{rt} = 20,000e^{-0.06t}$$
$$A(5) = 20,000e^{-0.06(5)}$$
$$= 20,000e^{-0.3}$$
$$\approx \$14,816.36.$$

c. We can answer the question now using a graphing calculator, a spreadsheet, or the Function Evaluator and Grapher tool at the Website. Just enter the exponential models of parts (a) and (b) and create tables to compute the values at the end of several years:

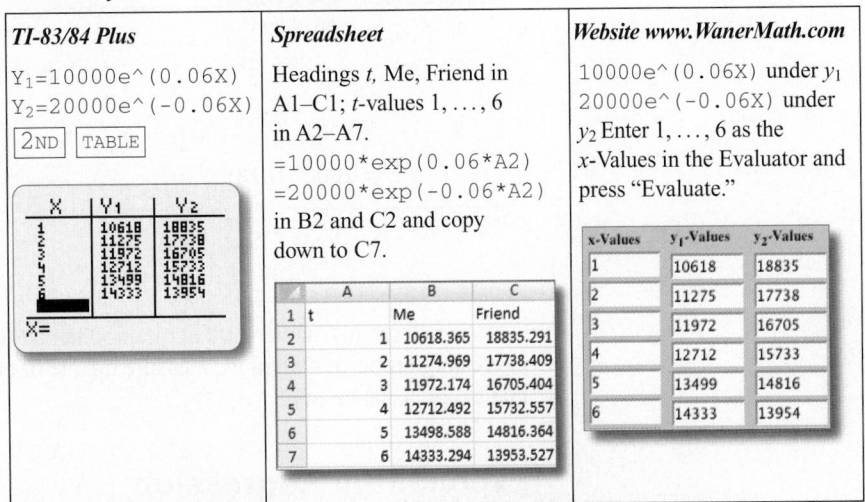

TI-83/84 Plus	Spreadsheet	Website www.WanerMath.com
Y₁=10000e^(0.06X)	Headings *t*, Me, Friend in A1–C1; *t*-values 1, …, 6 in A2–A7.	10000e^(0.06X) under y_1
Y₂=20000e^(-0.06X)		20000e^(-0.06X) under y_2 Enter 1, …, 6 as the
[2ND] [TABLE]	=10000*exp(0.06*A2) =20000*exp(-0.06*A2) in B2 and C2 and copy down to C7.	*x*-Values in the Evaluator and press "Evaluate."

TI-83/84 Plus table:

X	Y₁	Y₂
1	10618	18835
2	11275	17738
3	11972	16705
4	12712	15733
5	13499	14816
6	14333	13954

X=

Spreadsheet:

	A	B	C
1	t	Me	Friend
2	1	10618.365	18835.291
3	2	11274.969	17738.409
4	3	11972.174	16705.404
5	4	12712.492	15732.557
6	5	13498.588	14816.364
7	6	14333.294	13953.527

Website:

x-Values	y₁-Values	y₂-Values
1	10618	18835
2	11275	17738
3	11972	16705
4	12712	15733
5	13499	14816
6	14333	13954

From the table, we see that the value of your investment overtakes that of your friend after $t = 5$ (the end of year 5) and before $t = 6$ (the end of year 6). Thus your investment first exceeds that of your friend sometime during year 6.

➡ **Before we go on...**

Q: *How does continuous compounding compare with monthly compounding?*

A: To repeat the calculation in part (a) of Example 5 using monthly compounding instead of continuous compounding, we use the compound interest formula with $P = 10,000$, $r = 0.06$, $m = 12$, and $t = 5$ and find

$$A(5) = 10,000(1 + 0.06/12)^{60} \approx \$13,488.50.$$

Thus, continuous compounding earns you approximately \$10 more than monthly compounding on a 5-year, \$10,000 investment. This is little to get excited about.

■

If we write the continuous compounding formula $A(t) = Pe^{rt}$ as $A(t) = P(e^r)^t$, we see that $A(t)$ is an exponential function of t, where the base is $b = e^r$, so we have really not introduced a new kind of function. In fact, exponential functions are often written in this way:

Exponential Functions: Alternative Form

We can write any exponential function in the following alternative form:

$$f(x) = Ae^{rx}$$

where A and r are constants. If r is positive, f models exponential growth; if r is negative, f models exponential decay.

Quick Examples

1. $f(x) = 100e^{0.15x}$ Exponential growth $A = 100, r = 0.15$

2. $f(t) = Ae^{-0.000\,121\,01t}$ Exponential decay of carbon 14; $r = -0.000\,121\,01$

3. $f(t) = 100e^{0.15t} = 100\left(e^{0.15}\right)^t$
   $= 100(1.1618)^t$ Converting Ae^{rt} to the form Ab^t

We will see in Chapter 11 that the exponential function with base e exhibits some interesting properties when we measure its rate of change, and this is the real mathematical importance of e.

Exponential Regression

Starting with a set of data that suggests an exponential curve, we can use technology to compute the exponential regression curve in much the same way as we did for the quadratic regression curve in Example 5 of Section 9.1.

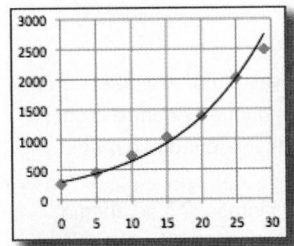

Figure 13

EXAMPLE 6 ▣ Exponential Regression: Health Expenditures

The following table shows annual expenditure on health in the U.S. from 1980 through 2009 ($t = 0$ represents 1980).[15]

Year t	0	5	10	15	20	25	29
Expenditure ($ billion)	256	444	724	1,030	1,380	2,020	2,490

a. Find the exponential regression model

$$C(t) = Ab^t$$

for the annual expenditure.

b. Use the regression model to estimate the expenditure in 2002 ($t = 22$; the actual expenditure was approximately $1,640 billion).

Solution

a. We use technology to obtain the exponential regression curve (See Figure 13):

$$C(t) \approx 296(1.08)^t. \qquad \text{Coefficients rounded}$$

b. Using the model $C(t) \approx 296(1.08)^t$ we find that

$$C(22) \approx 296(1.08)^{22} \approx \$1{,}609 \text{ billion,}$$

which is close to the actual number of about $1,640 billion.

➡ **Before we go on...** We said in the preceding section that the regression curve gives the smallest value of the sum-of-squares error, SSE (the sum of the squares of the residuals). However, exponential regression as computed via technology generally minimizes the sum of the squares of the residuals of the *logarithms* (logarithms are discussed in the next section). Using logarithms allows one easily to convert an exponential function into a linear one and then use linear regression formulas. However, in Section 9.4, we will discuss a way of using Excel's Solver to minimize SSE directly, which allows us to find the best-fit exponential curve directly without the need for devices to simplify the mathematics. If we do this, we obtain a very different equation:

$$C(t) \approx 353(1.07)^t.$$

If you plot this function, you will notice that it seems to fit the data more closely than the regression curve. ∎

FAQs

When to Use an Exponential Model for Data Points, and when to Use e in Your Model

Q: *Given a set of data points that appear to be curving upward, how can I tell whether to use a quadratic model or an exponential model?*

---

[15]Data are rounded. Source: U.S. Department of Health & Human Services/Centers for Medicare & Medicaid Services, National Health Expenditure Data, downloaded April 2011 from www.cms.gov.

A : Here are some things to look for:

- Do the data values appear to double at regular intervals? (For example, do the values approximately double every 5 units?) If so, then an exponential model is appropriate. If it takes longer and longer to double, then a quadratic model may be more appropriate.
- Do the values first decrease to a low point and then increase? If so, then a quadratic model is more appropriate.

It is also helpful to use technology to graph both the regression quadratic and exponential curves and to visually inspect the graphs to determine which gives the closest fit to the data.

Q : *We have two ways of writing exponential functions: $f(x) = Ab^x$ and $f(x) = Ae^{rx}$. How do we know which one to use?*

A : The two forms are equivalent, and it is always possible to convert from one form to the other.* So, use whichever form seems to be convenient for a particular situation. For instance, $f(t) = A(3^t)$ conveniently models exponential growth that is tripling every unit of time, whereas $f(t) = Ae^{0.06t}$ conveniently models an investment with continuous compounding at 6%.

* Quick Example 3 on page 640 shows how to convert Ae^{rx} to Ab^x. Conversion from Ab^x to Ae^{rx} involves logarithms: $r = \ln b$.

9.2   EXERCISES

▼ more advanced   ◆ challenging

T indicates exercises that should be solved using technology

For each function in Exercises 1–12, compute the missing values in the following table and supply a valid technology formula for the given function: HINT *[See Quick Examples on page 632.]*

x	-3	-2	-1	0	1	2	3
$f(x)$							

1. $f(x) = 4^x$ **2.** $f(x) = 3^x$

3. $f(x) = 3^{-x}$ **4.** $f(x) = 4^{-x}$

5. $g(x) = 2(2^x)$ **6.** $g(x) = 2(3^x)$

7. $h(x) = -3(2^{-x})$ **8.** $h(x) = -2(3^{-x})$

9. $r(x) = 2^x - 1$ **10.** $r(x) = 2^{-x} + 1$

11. $s(x) = 2^{x-1}$ **12.** $s(x) = 2^{1-x}$

Using a chart of values, graph each of the functions in Exercises 13–18. (Use $-3 \le x \le 3$.)

13. $f(x) = 3^{-x}$ **14.** $f(x) = 4^{-x}$

15. $g(x) = 2(2^x)$ **16.** $g(x) = 2(3^x)$

17. $h(x) = -3(2^{-x})$ **18.** $h(x) = -2(3^{-x})$

In Exercises 19–24, the values of two functions, f and g, are given in a table. One, both, or neither of them may be exponential. Decide which, if any, are exponential, and give the exponential models for those that are. HINT *[See Example 1.]*

19.

x	-2	-1	0	1	2
$f(x)$	0.5	1.5	4.5	13.5	40.5
$g(x)$	8	4	2	1	$\frac{1}{2}$

20.

x	-2	-1	0	1	2
$f(x)$	$\frac{1}{2}$	1	2	4	8
$g(x)$	3	0	-1	0	3

21.

x	-2	-1	0	1	2
$f(x)$	22.5	7.5	2.5	7.5	22.5
$g(x)$	0.3	0.9	2.7	8.1	16.2

22.

x	-2	-1	0	1	2
$f(x)$	0.3	0.9	2.7	8.1	24.3
$g(x)$	3	1.5	0.75	0.375	0.1875

23.

x	-2	-1	0	1	2
$f(x)$	100	200	400	600	800
$g(x)$	100	20	4	0.8	0.16

24.

x	-2	-1	0	1	2
$f(x)$	0.8	0.2	0.1	0.05	0.025
$g(x)$	80	40	20	10	2

For each function in Exercises 25–30, supply a valid technology formula and then use technology to compute the missing values in the following table: HINT [See Quick Examples on page 632.]

x	-3	-2	-1	0	1	2	3
$f(x)$							

25. $f(x) = e^{-2x}$ **26.** $g(x) = e^{x/5}$

27. $h(x) = 1.01(2.02^{-4x})$ **28.** $h(x) = 3.42(3^{-x/5})$

29. $r(x) = 50\left(1 + \dfrac{1}{3.2}\right)^{2x}$

30. $r(x) = 0.043\left(4.5 - \dfrac{5}{1.2}\right)^{-x}$

In Exercises 31–38, supply a valid technology formula for the given function.

31. 2^{x-1} **32.** 2^{-4x} **33.** $\dfrac{2}{1 - 2^{-4x}}$ **34.** $\dfrac{2^{3-x}}{1 - 2^x}$

35. $\dfrac{(3+x)^{3x}}{x+1}$ **36.** $\dfrac{20.3^{3x}}{1 + 20.3^{2x}}$ **37.** $2e^{(1+x)/x}$ **38.** $\dfrac{2e^{2/x}}{x}$

On the same set of axes, use technology to graph the pairs of functions in Exercises 39–46 with $-3 \le x \le 3$. Identify which graph corresponds to which function. HINT [See Quick Examples on page 633.]

39. $f_1(x) = 1.6^x$, $f_2(x) = 1.8^x$

40. $f_1(x) = 2.2^x$, $f_2(x) = 2.5^x$

41. $f_1(x) = 300(1.1^x)$, $f_2(x) = 300(1.1^{2x})$

42. $f_1(x) = 100(1.01^{2x})$, $f_2(x) = 100(1.01^{3x})$

43. $f_1(x) = 2.5^{1.02x}$, $f_2(x) = e^{1.02x}$

44. $f_1(x) = 2.5^{-1.02x}$, $f_2(x) = e^{-1.02x}$

45. $f_1(x) = 1,000(1.045^{-3x})$, $f_2(x) = 1,000(1.045^{3x})$

46. $f_1(x) = 1,202(1.034^{-3x})$, $f_2(x) = 1,202(1.034^{3x})$

For Exercises 47–54, model the data using an exponential function $f(x) = Ab^x$. HINT [See Example 1.]

47.

x	0	1	2
$f(x)$	500	250	125

48.

x	0	1	2
$f(x)$	500	1,000	2,000

49.

x	0	1	2
$f(x)$	10	30	90

50.

x	0	1	2
$f(x)$	90	30	10

51.

x	0	1	2
$f(x)$	500	225	101.25

52.

x	0	1	2
$f(x)$	5	3	1.8

53.

x	1	2
$f(x)$	-110	-121

54.

x	1	2
$f(x)$	-41	-42.025

Find equations for exponential functions that pass through the pairs of points given in Exercises 55–62. (Round all coefficients to 4 decimal places when necessary.) HINT [See Example 2.]

55. Through (2, 36) and (4, 324)

56. Through (2, −4) and (4, −16)

57. Through (−2, −25) and (1, −0.2)

58. Through (1, 1.2) and (3, 0.108)

59. Through (1, 3) and (3, 6)   **60.** Through (1, 2) and (4, 6)

61. Through (2, 3) and (6, 2)   **62.** Through (−1, 2) and (3, 1)

Obtain exponential functions in the form $f(t) = Ae^{rt}$ in Exercises 63–66. HINT [See Example 5.]

63. $f(t)$ is the value after t years of a \$5,000 investment earning 10% interest compounded continuously.

64. $f(t)$ is the value after t years of a \$2,000 investment earning 5.3% interest compounded continuously.

65. $f(t)$ is the value after t years of a \$1,000 investment depreciating continuously at an annual rate of 6.3%.

66. $f(t)$ is the value after t years of a \$10,000 investment depreciating continuously at an annual rate of 60%.

In Exercises 67–70, use technology to find the exponential regression function through the given points. (Round all coefficients to 4 decimal places.) HINT [See Example 6.]

67. {(1, 2), (3, 5), (4, 9), (5, 20)}

68. {(−1, 2), (−3, 5), (−4, 9), (−5, 20)}

69. {(−1, 10), (−3, 5), (−4, 3)}

70. {(3, 3), (4, 5), (5, 10)}

APPLICATIONS

71. *Aspirin* Soon after taking an aspirin, a patient has absorbed 300 mg of the drug. After 2 hours, only 75 mg remain. Find an exponential model for the amount of aspirin in the bloodstream after t hours, and use your model to find the amount of aspirin in the bloodstream after 5 hours. HINT [See Example 2.]

72. *Alcohol* After a large number of drinks, a person has a blood alcohol level of 200 mg/dL (milligrams per deciliter). If the amount of alcohol in the blood decays exponentially, and after 2 hours, 112.5 mg/dL remain, find an exponential model for the person's blood alcohol level, and use your model to estimate the person's blood alcohol level after 4 hours. HINT [See Example 2.]

73. *Freon Production* The production of ozone-layer-damaging Freon 22 (chlorodifluoromethane) in developing countries rose from 200 tons in 2004 to a projected 590 tons in 2010.[16]

---

[16]Figures are approximate. Source: Lampert Kuijpers (Panel of the Montreal Protocol), National Bureau of Statistics in China, via CEIC Data/*New York Times*, February 23, 2007, p. C1.

a. Use this information to find both a linear model and an exponential model for the amount F of Freon 22 (in tons) as a function of time t in years since 2000. (Round all coefficients to three significant digits.) HINT [See Example 2.] Which of these models would you judge to be more appropriate to the data shown below?

t (year since 2000)	0	2	4	6	8	10
F (tons of Freon 22)	100	140	200	270	400	590

b. Use the better of the two models from part (a) to predict the 2008 figure and compare it with the projected figure above.

74. *Revenue* The annual revenue of Amazon.com rose from approximately \$10.7 billion in 2006 to \$34.2 billion in 2010.[17]

a. Use this information to find both a linear model and an exponential model for Amazon.com's annual revenue I (in billions of dollars) as a function of time t in years since 2000. (Round all coefficients to three significant digits.) HINT [See Example 2.] Which of these models would you judge to be more appropriate to the data shown below?

t (Year since 2000)	6	7	8	9	10
I (\$ billions)	10.7	14.8	19.2	24.5	34.2

b. Use the better of the two models from part (a) to predict the 2008 figure and compare it with the actual figure above.

75. ▼ *U.S. Population* The U.S. population was 180 million in 1960 and 309 million in 2010.[18]

a. Use these data to give an exponential growth model showing the U.S. population P as a function of time t in years since 1960. Round coefficients to 6 significant digits. HINT [See Example 2.]

b. By experimenting, determine the smallest number of significant digits to which you should round the coefficients in part (a) in order to obtain the correct 2010 population figure accurate to 3 significant digits.

c. Using the model in part (a), predict the population in 2020.

76. ▼ *World Population* World population was estimated at 2.56 billion people in 1950 and 6.91 billion people in 2011.[19]

a. Use these data to give an exponential growth model showing the world population P as a function of time t in years since 1950. Round coefficients to 6 significant digits. HINT [See Example 2.]

b. By experimenting, determine the smallest number of significant digits to which you should round the coefficients in part (a) in order to obtain the correct 2011 population figure to 3 significant digits.

c. Assuming the exponential growth model from part (a), estimate the world population in the year 1000. Comment on your answer.

77. ▼ *Frogs* Frogs have been breeding like flies at the Enormous State University (ESU) campus! Each year, the pledge class of the Epsilon Delta fraternity is instructed to tag all the frogs residing on the ESU campus. Two years ago they managed to tag all 50,000 of them (with little Epsilon Delta Fraternity tags). This year's pledge class discovered that last year's tags had all fallen off, and they wound up tagging a total of 75,000 frogs.

a. Find an exponential model for the frog population.

b. Assuming exponential population growth, and that all this year's tags have fallen off, how many tags should Epsilon Delta order for next year's pledge class?

78. ▼ *Flies* Flies in Suffolk County have been breeding like frogs! Three years ago the Health Commission caught 4,000 flies in a trap in 1 hour. This year it caught 7,000 flies in 1 hour.

a. Find an exponential model for the fly population.

b. Assuming exponential population growth, how many flies should the commission expect to catch next year?

79. *Bacteria* A bacteria culture starts with 1,000 bacteria and doubles in size every 3 hours. Find an exponential model for the size of the culture as a function of time t in hours and use the model to predict how many bacteria there will be after 2 days. HINT [See Example 4.]

80. *Bacteria* A bacteria culture starts with 1,000 bacteria. Two hours later there are 1,500 bacteria. Find an exponential model for the size of the culture as a function of time t in hours, and use the model to predict how many bacteria there will be after 2 days. HINT [See Example 4.]

81. *SARS* In the early stages of the deadly SARS (Severe Acute Respiratory Syndrome) epidemic in 2003, the number of cases was increasing by about 18% each day.[20] On March 17, 2003 (the first day for which statistics were reported by the World Health Organization), there were 167 cases. Find an exponential model that predicts the number of cases t days after March 17, 2003, and use it to estimate the number of cases on March 31, 2003. (The actual reported number of cases was 1,662.)

82. *SARS* A few weeks into the deadly SARS (Severe Acute Respiratory Syndrome) epidemic in 2003, the number of cases was increasing by about 4% each day.[21] On April 1,

[17]Source for data: www.wikinvest.com/

[18]Figures are rounded to 3 significant digits. Source: U.S. Census Bureau, www.census.gov.

[19]*Ibid.*

[20]Source: World Health Organization, www.who.int.

[21]*Ibid.*

2003 there were 1,804 cases. Find an exponential model that predicts the number of cases t days after April 1, 2003, and use it to estimate the number of cases on April 30, 2003. (The actual reported number of cases was 5,663.)

83. **Investments** In November 2010, E*TRADE Financial was offering only 0.3% interest on its online savings accounts, with interest reinvested monthly.[22] Find the associated exponential model for the value of a $5,000 deposit after t years. Assuming this rate of return continued for seven years, how much would a deposit of $5,000 in November 2010 be worth in November 2017? (Answer to the nearest $1.) HINT [See Example 3; you saw this exercise before in Section 1.2.]

84. **Investments** In November 2010, ING Direct was offering 2.4% interest on its Orange Savings Account, with interest reinvested quarterly.[23] Find the associated exponential model for the value of a $4,000 deposit after t years. Assuming this rate of return continued for 8 years, how much would a deposit of $4,000 in November 2010 be worth in November 2018? (Answer to the nearest $1.) HINT [See Example 3; you saw this exercise before in Section 1.2.]

85. ▊ **Investments** Refer to Exercise 83. In November of which year will an investment of $5,000 made in November of 2010 first exceed $5,200? HINT [See Example 5; you saw this exercise before in Section 1.2.]

86. ▊ **Investments** Refer to Exercise 84. In November of which year will an investment of $4,000 made in November of 2010 first exceed $5,200? HINT [See Example 5; you saw this exercise before in Section 1.2.]

87. **Investments** *Rock Solid Bank & Trust* is offering a CD (certificate of deposit) that pays 4% compounded continuously. How much interest would a $1,000 deposit earn over 10 years? HINT [See Example 5.]

88. **Savings** *FlybynightSavings.com* is offering a savings account that pays 31% interest compounded continuously. How much interest would a deposit of $2,000 earn over 10 years?

89. **Home Sales** Sales of existing homes in the U.S. rose continuously over the period 2008–2011 at the rate of 3.2% per year from 4.9 million in 2008.[24] Write down a formula that predicts sales of existing homes t years after 2008. Use your model to estimate, to the nearest 0.1 million, sales of existing homes in 2010 and 2012.

90. **Home Prices** The median selling price of an existing home in the U.S. declined continuously over the period 2008–2011 at the rate of 7.6% per year from approximately $198 thousand in 2008.[25] Write down a formula that predicts the median selling price of an existing home t years after 2008. Use your model to estimate, to the nearest $1,000, the median selling price of an existing home in 2011 and 2013.

91. **Global Warming** The most abundant greenhouse gas is carbon dioxide. According to figures from the Intergovernmental Panel on Climate Change (IPCC), the amount of carbon dioxide in the atmosphere (in parts of volume per million) can be approximated by

$$C(t) \approx 280e^{0.00119t} \text{ parts per million}$$

where t is time in years since 1750.[26]
a. Use the model to estimate the amount of carbon dioxide in the atmosphere in 1950, 2000, 2050, and 2100.
b. According to the model, when, to the nearest decade, will the level surpass 390 parts per million?

92. **Global Warming** Another greenhouse gas is methane. According to figures from the Intergovernmental Panel on Climate Change (IPCC), the amount of methane in the atmosphere (in parts of volume per billion) can be approximated by

$$C(t) \approx 715e^{0.00356t} \text{ parts per billion}$$

where t is time in years since 1750.[27]
a. Use the model to estimate the amount of methane in the atmosphere in 1950, 2000, 2050, and 2100. (Round your answers to the nearest 10 parts per billion.)
b. According to the model, when, to the nearest decade, will the level surpass 2,000 parts per billion?

93. ▊ **New York City Housing Costs: Downtown** The following table shows the average price of a two-bedroom apartment in downtown New York City during the real estate boom from 1994 to 2004.[28]

t	0 (1994)	2	4	6	8	10 (2004)
Price ($ million)	0.38	0.40	0.60	0.95	1.20	1.60

a. Use exponential regression to model the price $P(t)$ as a function of time t since 1994. Include a sketch of the points and the regression curve. (Round the coefficients to 3 decimal places.) HINT [See Example 6.]
b. Extrapolate your model to estimate the cost of a two-bedroom downtown apartment in 2005.

94. ▊ **New York City Housing Costs: Uptown** The following table shows the average price of a two-bedroom apartment in uptown New York City during the real estate boom from 1994 to 2004.[29]

t	0 (1994)	2	4	6	8	10 (2004)
Price ($ million)	0.18	0.18	0.19	0.2	0.35	0.4

[22]Interest rate based on annual percentage yield. Source: us.etrade.com, November 2010.

[23]Interest rate based on annual percentage yield. Source: home.ingdirect.com, November 2010.

[24]Source: National Association of Realtors, www.realtor.org.

[25]*Ibid.*

[26]Authors' exponential model based on the 1750 and 2005 figures. Source for data: IPCC Fourth Assessment Report: Climate Change 2007, www.ipcc.ch.

[27]*Ibid.*

[28]Data are rounded and 2004 figure is an estimate. Source: Miller Samuel/*New York Times*, March 28, 2004, p. RE 11.

[29]*Ibid.*

a. Use exponential regression to model the price $P(t)$ as a function of time t since 1994. Include a sketch of the points and the regression curve. (Round the coefficients to 3 decimal places.)

b. Extrapolate your model to estimate the cost of a two-bedroom uptown apartment in 2005.

95. 🔲 *Facebook* The following table gives the approximate numbers of Facebook members at various times early in its history.[30]

Year t (Since start of 2005)	0	0.5	1	1.5	2	2.5	3	3.5
Facebook Members n (millions)	1	2	5.5	7	12	30	58	80

a. Use exponential regression to model Facebook membership as a function of time in years since the start of 2005, and graph the data points and regression curve. (Round coefficients to 3 decimal places.)

b. Fill in the missing quantity: According to the model, Facebook membership each year was _____ times that of the year before.

c. Use your model to estimate Facebook membership in early 2009 to the nearest million.

96. 🔲 *Freon Production* The following table shows Freon 22 production in developing countries in various years since 2000.[31]

t (Year since 2000)	0	2	4	6	8	10
F (Tons of Freon 22)	100	140	200	270	400	590

a. Use exponential regression to model Freon production as a function of time in years since 2000, and graph the data points and regression curve. (Round coefficients to 3 decimal places.)

b. Fill in the missing quantity: According to the model, Freon production each year was _____ times that of the year before.

c. Use your model to estimate freon production in 2009 to the nearest ton.

COMMUNICATION AND REASONING EXERCISES

97. Which of the following three functions will be largest for large values of x?

(A) $f(x) = x^2$ (B) $r(x) = 2^x$ (C) $h(x) = x^{10}$

98. Which of the following three functions will be smallest for large values of x?

(A) $f(x) = x^{-2}$ (B) $r(x) = 2^{-x}$ (C) $h(x) = x^{-10}$

99. What limitations apply to using an exponential function to model growth in real-life situations? Illustrate your answer with an example.

100. Explain in words why 5% per year compounded continuously yields more interest than 5% per year compounded monthly.

101. ▼ The following commentary and graph appeared in politicalcalculations.blogspot.com on August 30, 2005:[32]

> One of the neater blogs I've recently encountered is The Real Returns, which offers a wealth of investing, market and economic data. Earlier this month, The Real Returns posted data related to the recent history of U.S. median house prices over the period from 1963 to 2004. The original source of the housing data is the U.S. Census Bureau.
>
> Well, that kind of data deserves some curve-fitting and a calculator to estimate what the future U.S. median house price might be, so Political Calculations has extracted the data from 1973 onward to create the following chart:

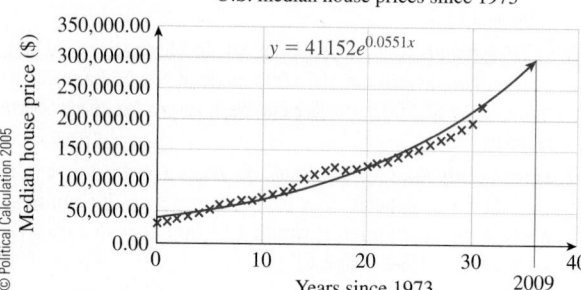

U.S. median house prices since 1973

$y = 41152e^{0.0551x}$

Comment on the article and graph. HINT [See Exercise 90.]

102. ▼ Refer to Exercise 101. Of what possible predictive use, then, is the kind of exponential model given by the blogger in the article referred to?

103. ▼ Describe two real-life situations in which a linear model would be more appropriate than an exponential model, and two situations in which an exponential model would be more appropriate than a linear model.

104. ▼ Describe a real-life situation in which a quadratic model would be more appropriate than an exponential model and one in which an exponential model would be more appropriate than a quadratic model.

105. How would you check whether data points of the form $(1, y_1), (2, y_2), (3, y_3)$ lie on an exponential curve?

---

[30]Sources: www.facebook.com, www.insidehighered.com. (Some data are interpolated.)

[31]Figures are approximate. Source: Lampert Kuijpers (Panel of the Montreal Protocol), National Bureau of Statistics in China, via CEIC Data/*New York Times*, February 23, 2007, p. C1.

[32]The graph was re-created by the authors using the blog author's data source.

Source for article: politicalcalculations.blogspot.com/2005/08/projecting-us-median-housing-prices.html.

Source for data: therealreturns.blogspot.com/2005_08_01_archive.html.

106. ▼ You are told that the points $(1, y_1)$, $(2, y_2)$, $(3, y_3)$ lie on an exponential curve. Express y_3 in terms of y_1 and y_2.

107. ▼ Your local banker tells you that the reason his bank doesn't compound interest continuously is that it would be too demanding of computer resources because the computer would need to spend a great deal of time keeping all accounts updated. Comment on his reasoning.

108. ▼ Your other local banker tells you that the reason *her* bank doesn't offer continuously compounded interest is that it is equivalent to offering a fractionally higher interest rate compounded daily. Comment on her reasoning.

9.3 Logarithmic Functions and Models

Logarithms were invented by John Napier (1550–1617) in the late sixteenth century as a means of aiding calculation. His invention made possible the prodigious hand calculations of astronomer Johannes Kepler (1571–1630), who was the first to describe accurately the orbits and the motions of the planets. Today, computers and calculators have done away with that use of logarithms, but many other uses remain. In particular, the logarithm is used to model real-world phenomena in numerous fields, including physics, finance, and economics.

From the equation

$$2^3 = 8$$

we can see that the power to which we need to raise 2 in order to get 8 is 3. We abbreviate the phrase "the power to which we need to raise 2 in order to get 8" as "$\log_2 8$." Thus, another way of writing the equation $2^3 = 8$ is

$$\log_2 8 = 3. \qquad \text{The power to which we need to raise 2 in order to get 8 is 3.}$$

This is read "the base 2 logarithm of 8 is 3" or "the log, base 2, of 8 is 3."

Here is the general definition.

Base b Logarithm

The **base b logarithm of** x, $\log_b x$, is the power to which we need to raise b in order to get x. Symbolically,

$$\log_b x = y \qquad\qquad \text{means} \qquad\qquad b^y = x.$$

Logarithmic form   *Exponential form*

Quick Examples

1. The following table lists some exponential equations and their equivalent logarithmic forms:

Exponential Form	$10^3 = 1000$	$4^2 = 16$	$3^3 = 27$	$5^1 = 5$	$7^0 = 1$	$4^{-2} = \dfrac{1}{16}$	$25^{1/2} = 5$
Logarithmic Form	$\log_{10} 1000 = 3$	$\log_4 16 = 2$	$\log_3 27 = 3$	$\log_5 5 = 1$	$\log_7 1 = 0$	$\log_4 \dfrac{1}{16} = -2$	$\log_{25} 5 = \dfrac{1}{2}$

2. $\log_3 9 = $ the power to which we need to raise 3 in order to get 9. Because $3^2 = 9$, this power is 2, so $\log_3 9 = 2$.

3. $\log_{10} 10{,}000 =$ the power to which we need to raise 10 in order to get 10,000. Because $10^4 = 10{,}000$, this power is 4, so $\log_{10} 10{,}000 = 4$.

4. $\log_3 \frac{1}{27}$ is the power to which we need to raise 3 in order to get $\frac{1}{27}$. Because $3^{-3} = \frac{1}{27}$ this power is –3, so $\log_3 \frac{1}{27} = -3$.

5. $\log_b 1 = 0$ for every positive number b other than 1 because $b^0 = 1$.

Note The number $\log_b x$ is defined only if b and x are both positive and $b \neq 1$. Thus, it is impossible to compute, say, $\log_3(-9)$ (because there is no power of 3 that equals –9), or $\log_1(2)$ (because there is no power of 1 that equals 2). ∎

Logarithms with base 10 and base e are frequently used, so they have special names and notations.

Common Logarithm, Natural Logarithm

The following are standard abbreviations.

		TI-83/84 Plus & Spreadsheet Formula
Base 10: $\log_{10} x = \log x$	*Common Logarithm*	`log(x)`
Base e: $\log_e x = \ln x$	*Natural Logarithm*	`ln(x)`

Quick Examples

Logarithmic Form	**Exponential Form**
1. $\log 10{,}000 = 4$	$10^4 = 10{,}000$
2. $\log 10 = 1$	$10^1 = 10$
3. $\log \frac{1}{10{,}000} = -4$	$10^{-4} = \frac{1}{10{,}000}$
4. $\ln e = 1$	$e^1 = e$
5. $\ln 1 = 0$	$e^0 = 1$
6. $\ln 2 = 0.69314718\ldots$	$e^{0.69314718\ldots} = 2$

Some technologies (such as calculators) do not permit direct calculation of logarithms other than common and natural logarithms. To compute logarithms with other bases with these technologies, we can use the following formula:

Change-of-Base Formula

$$\log_b a = \frac{\log a}{\log b} = \frac{\ln a}{\ln b} \qquad \text{Change-of-base formula}^*$$

* Here is a quick explanation of why this formula works: To calculate $\log_b a$, we ask, "to what power must we raise b to get a?" To check the formula, we try using $\log a/\log b$ as the exponent.

$$b^{\frac{\log a}{\log b}} = (10^{\log b})^{\frac{\log a}{\log b}}$$
$$\text{(because } b = 10^{\log b})$$
$$= 10^{\log a} = a$$

so this exponent works!

Quick Examples

1. $\log_{11} 9 = \dfrac{\log 9}{\log 11} \approx 0.91631$ `log(9)/log(11)`

2. $\log_{11} 9 = \dfrac{\ln 9}{\ln 11} \approx 0.91631$ `ln(9)/ln(11)`

3. $\log_{3.2}\left(\dfrac{1.42}{3.4}\right) \approx -0.75065$ `log(1.42/3.4)/log(3.2)`

Using Technology to Compute Logarithms

To compute $\log_b x$ using technology, use the following formulas:

TI-83/84 Plus	`log(x)/log(b)`	Example: $\log_2(16)$ is `log(16)/log(2)`
Spreadsheet:	`=LOG(x,b)`	Example: $\log_2(16)$ is `= LOG(16,2)`

One important use of logarithms is to solve equations in which the unknown is in the exponent.

EXAMPLE 1 Solving Equations with Unknowns in the Exponent

Solve the following equations.

a. $5^{-x} = 125$ **b.** $3^{2x-1} = 6$ **c.** $100(1.005)^{3x} = 200$

Solution

a. Write the given equation $5^{-x} = 125$ in logarithmic form:

$$-x = \log_5 125$$

This gives $x = -\log_5 125 = -3$.

b. In logarithmic form, $3^{2x-1} = 6$ becomes

$$2x - 1 = \log_3 6$$
$$2x = 1 + \log_3 6$$

giving $\qquad x = \dfrac{1 + \log_3 6}{2} \approx \dfrac{1 + 1.6309}{2} \approx 1.3155.$

c. We cannot write the given equation, $100(1.005)^{3x} = 200$, directly in logarithmic form. We must first divide both sides by 100:

$$1.005^{3x} = \frac{200}{100} = 2$$
$$3x = \log_{1.005} 2$$
$$x = \frac{\log_{1.005} 2}{3} \approx \frac{138.9757}{3} \approx 46.3252.$$

Now that we know what logarithms are, we can talk about functions based on logarithms:

Logarithmic Function

A **logarithmic function** has the form

$$f(x) = \log_b x + C \qquad \text{(b and C are constants with $b > 0$, $b \neq 1$)}$$

or, alternatively,

$$f(x) = A \ln x + C. \qquad \text{(A, C constants with $A \neq 0$)}$$

Quick Examples

1. $f(x) = \log x$
2. $g(x) = \ln x - 5$
3. $h(x) = \log_2 x + 1$
4. $k(x) = 3.2 \ln x + 7.2$

Q: *What is the difference between the two forms of the logarithmic function?*

A: None, really—they're equivalent: We can start with an equation in the first form and use the change-of-base formula to rewrite it:

$$f(x) = \log_b x + C$$

$$= \frac{\ln x}{\ln b} + C \qquad \text{Change-of-base formula}$$

$$= \left(\frac{1}{\ln b}\right) \ln x + C.$$

Our function now has the form $f(x) = A \ln x + C$, where $A = 1/\ln b$. We can go the other way as well, to rewrite $A \ln x + C$ in the form $\log_b x + C$.

EXAMPLE 2 **Graphs of Logarithmic Functions**

a. Sketch the graph of $f(x) = \log_2 x$ by hand.

b. Use technology to compare the graph in part (a) with the graphs of $\log_b x$ for $b = 1/4, 1/2,$ and 4.

Solution

a. To sketch the graph of $f(x) = \log_2 x$ by hand, we begin with a table of values. Because $\log_2 x$ is not defined when $x = 0$, we choose several values of x close to zero and also some larger values, all chosen so that their logarithms are easy to compute:

x	$\frac{1}{8}$	$\frac{1}{4}$	$\frac{1}{2}$	1	2	4	8
$f(x) = \log_2 x$	-3	-2	-1	0	1	2	3

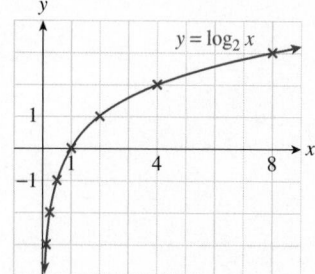

Figure 14

Graphing these points and joining them by a smooth curve gives us Figure 14.

b. We enter the logarithmic functions in graphing utilities as follows (note the use of the change-of-base formula in the TI-83/84 Plus version):

TI-83/84 Plus	Spreadsheet
Y₁=log(X)/log(0.25)	=LOG(x,0.25)
Y₂=log(X)/log(0.5)	=LOG(x,0.5)
Y₃=log(X)/log(2)	=LOG(x,2)
Y₄=log(X)/log(4)	=LOG(x,4)

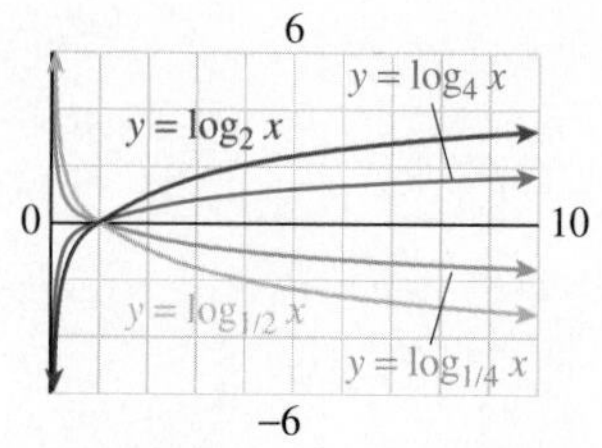

Figure 15

Figure 15 shows the resulting graphs.

➡ **Before we go on...** Notice that the graphs of the logarithmic functions in Example 2 all pass through the point (1, 0). (Why?) Notice further that the graphs of the logarithmic functions with bases less than 1 are upside-down versions of the others. Finally, how are these graphs related to the graphs of exponential functions? ∎

Below are some important algebraic properties of logarithms we shall use throughout the rest of this section.

Website
www.WanerMath.com
Follow the path
  Chapter 9
  → Using and Deriving
     Algebraic Properties of
     Logarithms
to find a list of logarithmic identities and a discussion on where they come from.

Follow the path
  Chapter 9
  → Inverse Functions
for a general discussion of inverse functions, including further discussion of the relationship between logarithmic and exponential functions.

Logarithm Identities

The following identities hold for all positive bases $a \neq 1$ and $b \neq 1$, all positive numbers x and y, and every real number r. These identities follow from the laws of exponents.

Identity

1. $\log_b(xy) = \log_b x + \log_b y$

2. $\log_b\left(\dfrac{x}{y}\right) = \log_b x - \log_b y$

3. $\log_b(x^r) = r \log_b x$

4. $\log_b b = 1; \log_b 1 = 0$

5. $\log_b\left(\dfrac{1}{x}\right) = -\log_b x$

6. $\log_b x = \dfrac{\log_a x}{\log_a b}$

Quick Examples

$\log_2 16 = \log_2 8 + \log_2 2$

$\log_2\left(\dfrac{5}{3}\right) = \log_2 5 - \log_2 3$

$\log_2(6^5) = 5\log_2 6$

$\log_2 2 = 1; \ln e = 1; \log_{11} 1 = 0$

$\log_2\left(\dfrac{1}{3}\right) = -\log_2 3$

$\log_2 5 = \dfrac{\log_{10} 5}{\log_{10} 2} = \dfrac{\log 5}{\log 2}$

Relationship with Exponential Functions

The following two identities demonstrate that the operations of taking the base b logarithm and raising b to a power are *inverse* to each other.

Identity

1. $\log_b(b^x) = x$

In words: The power to which you raise b in order to get b^x is x. (!)

2. $b^{\log_b x} = x$

In words: Raising b to the power to which it must be raised to get x yields x. (!)

Quick Examples

$\log_2(2^7) = 7$

$5^{\log_5 8} = 8$

APPLICATIONS

EXAMPLE 3 Investments: How Long?

Global bonds sold by Mexico are yielding an average of 2.51% per year.[33] At that interest rate, how long will it take a \$1,000 investment to be worth \$1,200 if the interest is compounded monthly?

---

[33] In 2011 (Bonds maturing 03/03/2015). Source: www.bloomberg.com.

Solution Substituting $A = 1,200$, $P = 1,000$, $r = 0.0251$, and $m = 12$ in the compound interest equation gives

$$A(t) = P \left(1 + \frac{r}{m}\right)^{mt}$$

$$1,200 = 1,000 \left(1 + \frac{0.0251}{12}\right)^{12t}$$

$$\approx 1,000(1.002092)^{12t}$$

and we must solve for t. We first divide both sides by 1,000, getting an equation in exponential form:

$$1.2 = 1.002092^{12t}.$$

In logarithmic form, this becomes

$$12t = \log_{1.002092}(1.2).$$

We can now solve for t:

$$t = \frac{\log_{1.002092}(1.2)}{12} \approx 7.3 \text{ years.} \qquad \texttt{log(1.2)/(log(1.002092)*12)}$$

Thus, it will take approximately 7.3 years for a \$1,000 investment to be worth \$1,200.

➡ **Before we go on...** We can use the logarithm identities to solve the equation

$$1.2 = 1.002092^{12t}$$

that arose in Example 3 (and also more general equations with unknowns in the exponent) by taking the natural logarithm of both sides:

$$\ln 1.2 = \ln(1.002092^{12t})$$

$$= 12t \ln 1.002092. \qquad \text{By Identity 3}$$

We can now solve this for t to get

$$t = \frac{\ln 1.2}{12 \ln 1.002092},$$

which, by the change-of-base formula, is equivalent to the answer we got in Example 3.

■

EXAMPLE 4 Half-Life

a. The weight of carbon 14 that remains in a sample that originally contained A grams is given by

$$C(t) = A(0.999879)^t$$

where t is time in years. Find the **half-life**, the time it takes half of the carbon 14 in a sample to decay.

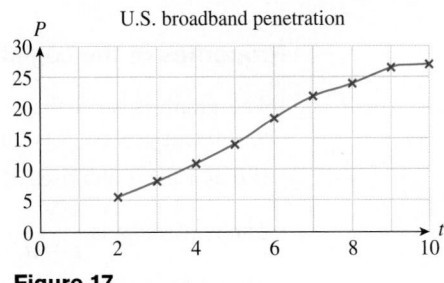

U.S. broadband penetration

Figure 17

kind of behavior, growing exponentially at first and then leveling off. In addition to modeling the demand for a new technology or product, logistic functions are often used in epidemic and population modeling. In an epidemic, the number of infected people often grows exponentially at first and then slows when a significant proportion of the entire susceptible population is infected and the epidemic has "run its course." Similarly, populations may grow exponentially at first and then slow as they approach the capacity of the available resources.

Logistic Function

A **logistic function** has the form

$$f(x) = \frac{N}{1 + Ab^{-x}}$$

for nonzero constants N, A, and b (A and b positive and $b \neq 1$).

Quick Example

$N = 6$, $A = 2$, $b = 1.1$ gives $f(x) = \dfrac{6}{1 + 2(1.1^{-x})}$ 6/(1+2*1.1^-x)

$f(0) = \dfrac{6}{1 + 2} = 2$ The y-intercept is $N/(1 + A)$.

$f(1,000) = \dfrac{6}{1 + 2(1.1^{-1,000})} \approx \dfrac{6}{1 + 0} = 6 = N$ When x is large, $f(x) \approx N$.

Graph of a Logistic Function

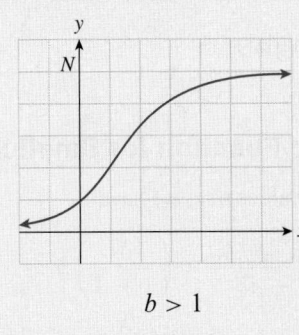

$b > 1$

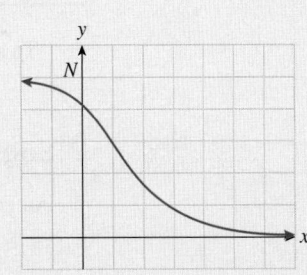

$0 < b < 1$

$$y = \frac{N}{1 + Ab^{-x}}$$

Properties of the Logistic Curve $y = \dfrac{N}{1 + Ab^{-x}}$

- The graph is an S-shaped curve sandwiched between the horizontal lines $y = 0$ and $y = N$. N is called the **limiting value** of the logistic curve.
- If $b > 1$ the graph rises; if $b < 1$, the graph falls.
- The y-intercept is $\dfrac{N}{1 + A}$.
- The curve is steepest when $t = \dfrac{\ln A}{\ln b}$.   We will see why in Chapter 12.

Note If we write b^{-x} as e^{-kx} (so that $k = \ln b$), we get the following alternative form of the logistic function:

$$f(x) = \frac{N}{1 + Ae^{-kx}}.$$ ■

Q: *How does the constant b affect the graph?*

A: To understand the role of b, we first rewrite the logistic function by multiplying top and bottom by b^x:

$$f(x) = \frac{N}{1 + Ab^{-x}}$$

$$= \frac{Nb^x}{(1 + Ab^{-x})b^x}$$

$$= \frac{Nb^x}{b^x + A}$$ Because $b^{-x}b^x = 1$

For values of x close to 0, the quantity b^x is close to 1, so the denominator is approximately $1 + A$, giving

$$f(x) \approx \frac{Nb^x}{1 + A} = \left(\frac{N}{1 + A}\right)b^x.$$

In other words, $f(x)$ is approximately exponential with base b for values of x close to 0. Put another way, if x represents time, then initially the logistic function behaves like an exponential function.

To summarize:

Logistic Function for Small x and the Role of b

For small values of x, we have

$$\frac{N}{1 + Ab^{-x}} \approx \left(\frac{N}{1 + A}\right)b^x.$$

Thus, for small x, the logistic function grows approximately exponentially with base b.

Quick Example

Let

$$f(x) = \frac{50}{1 + 24(3^{-x})}. \qquad N = 50, A = 24, b = 3$$

Then

$$f(x) \approx \left(\frac{50}{1 + 24}\right)(3^x) = 2(3^x)$$

for small values of x. The following figure compares their graphs:

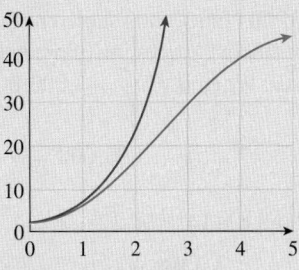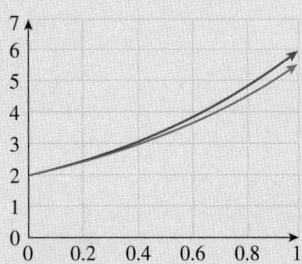

The upper curve is the exponential curve.

Modeling with the Logistic Function

EXAMPLE 1 **Epidemics**

A flu epidemic is spreading through the U.S. population. An estimated 150 million people are susceptible to this particular strain, and it is predicted that all susceptible people will eventually become infected. There are 10,000 people already infected, and the number is doubling every 2 weeks. Use a logistic function to model the number of people infected. Hence predict when, to the nearest week, 1 million people will be infected.

Solution   Let t be time in weeks, and let $P(t)$ be the total number of people infected at time t. We want to express P as a logistic function of t, so that

$$P(t) = \frac{N}{1 + Ab^{-t}}.$$

We are told that, in the long run, 150 million people will be infected, so that

$$N = 150,000,000. \qquad \text{Limiting value of } P$$

At the current time $(t = 0)$, 10,000 people are infected, so

$$10,000 = \frac{N}{1 + A} = \frac{150,000,000}{1 + A}. \qquad \text{Value of } P \text{ when } t = 0$$

Solving for A gives

$$10,000(1 + A) = 150,000,000$$
$$1 + A = 15,000$$
$$A = 14,999.$$

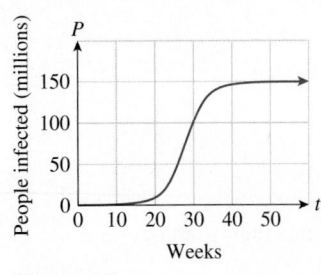

Figure 18

What about b? At the beginning of the epidemic (t near 0), P is growing approximately exponentially, doubling every 2 weeks. Using the technique of Section 9.2, we find that the exponential curve passing through the points (0, 10,000) and (2, 20,000) is

$$y = 10,000(\sqrt{2})^t$$

giving us $b = \sqrt{2}$. Now that we have the constants N, A, and b, we can write down the logistic model:

$$P(t) = \frac{150,000,000}{1 + 14,999(\sqrt{2})^{-t}}.$$

The graph of this function is shown in Figure 18.

Now we tackle the question of prediction: When will 1 million people be infected? In other words: When is $P(t) = 1,000,000$?

$$1,000,000 = \frac{150,000,000}{1 + 14,999(\sqrt{2})^{-t}}$$

$$1,000,000[1 + 14,999(\sqrt{2})^{-t}] = 150,000,000$$

$$1 + 14,999(\sqrt{2})^{-t} = 150$$

$$14,999(\sqrt{2})^{-t} = 149$$

$$(\sqrt{2})^{-t} = \frac{149}{14,999}$$

$$-t = \log_{\sqrt{2}}\left(\frac{149}{14,999}\right) \approx -13.31 \quad \text{Logarithmic form}$$

Thus, 1 million people will be infected by about the thirteenth week.

➡ **Before we go on...** We said earlier that the logistic curve is steepest when $t = \dfrac{\ln A}{\ln b}$. In Example 1, this occurs when $t = \dfrac{\ln 14,999}{\ln \sqrt{2}} \approx 28$ weeks into the epidemic. At that time, the number of cases is growing most rapidly (look at the apparent slope of the graph at the corresponding point). ∎

Logistic Regression

Let's go back to the data on broadband penetration in the United States with which we began this section and try to determine the long-term percentage of broadband penetration. In order to be able to make predictions such as this, we require a model for the data, so we will need to do some form of regression.

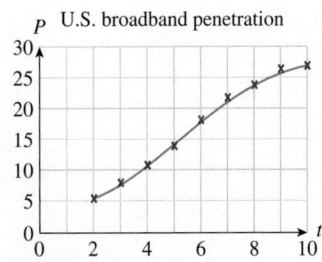

Figure 19

EXAMPLE 2 🔢 **Broadband Penetration**

Here are the data graphed in Figure 17:

Year (t)	2	3	4	5	6	7	8	9	10
Penetration (%) (P)	5.5	7.9	10.9	14.2	18.2	21.9	23.9	26.5	27.1

using Technology

See the Technology Guide at the end of the chapter for detailed instructions on how to obtain the regression curve and graph for Example 2 using a graphing calculator or spreadsheet. Outline:

TI-83/84 Plus
STAT EDIT values of t in L_1 and values of P in L_2
Regression curve: STAT
CALC option #B Logistic ENTER
Graph: Y= VARS 5 EQ 1,
then ZOOM 9 [More details on page 682.]

Spreadsheet
Use the Solver Add-in to obtain the best-fit logistic curve. [More details on page 684.]

Website
www.WanerMath.com
In the Function Evaluator and Grapher, enter the data as shown, press "Examples" until the logistic model $1/(1+\$2*\$3^(-x))$ shows in the first box, and press "Fit Curve".

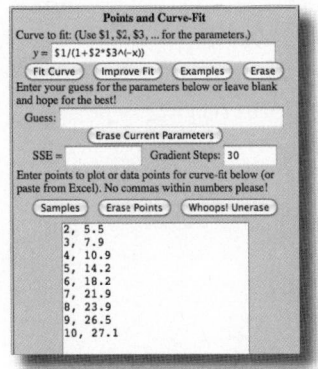

Find a logistic regression curve of the form

$$P(t) = \frac{N}{1 + Ab^{-t}}.$$

In the long term, what percentage of broadband penetration in the United States does the model predict?

Solution We can use technology to obtain the following regression model:

$$P(t) \approx \frac{29.842}{1 + 12.502(1.642)^{-t}}. \quad \text{Coefficients rounded to 3 decimal places}$$

Its graph and the original data are shown in Figure 19. Because $N = 29.842$, this model predicts that, in the long term, the percentage of broadband penetration in the United States will be 29.842%, or about 30%.

➡ **Before we go on...** Logistic regression programs generally estimate all three constants N, A, and b for a model $y = \frac{N}{1 + Ab^{-x}}$. However, there are times, as in Example 1, when we already know the limiting value N and require estimates of only A and b. In such cases, we can use technology like Excel Solver to find A and b for the best-fit curve with N fixed. Alternatively, we can use exponential regression to compute estimates of A and b as follows: First rewrite the logistic equation as

$$\frac{N}{y} = 1 + Ab^{-x},$$

so that

$$\frac{N}{y} - 1 = Ab^{-x} = A(b^{-1})^x.$$

This equation gives $N/y - 1$ as an exponential function of x. Thus, if we do exponential regression using the data points $(x, N/y - 1)$, we can obtain estimates for A and b^{-1} (and hence b). This is done in Exercises 35 and 36.

It is important to note that the resulting curve is not the best-fit curve (in the sense of minimizing SSE; see the "Before we go on" discussion on page 641 after Example 6 in Section 9.2) and will be thus be different from that obtained using the method in Example 2. ∎

9.4 EXERCISES

▼ more advanced ◆ challenging
🔲 indicates exercises that should be solved using technology

In Exercises 1–6, find N, A, and b, give a technology formula for the given function, and use technology to sketch its graph for the given range of values of x. HINT [See Quick Examples on page 661.]

1. $f(x) = \dfrac{7}{1 + 6(2^{-x})}$; [0, 10]

2. $g(x) = \dfrac{4}{1 + 0.333(4^{-x})}$; [0, 2]

3. $f(x) = \dfrac{10}{1 + 4(0.3^{-x})}$; [−5, 5]

4. $g(x) = \dfrac{100}{1 + 5(0.5^{-x})}$; [−5, 5]

5. $h(x) = \dfrac{2}{0.5 + 3.5(1.5^{-x})}$; [0, 15]
(First divide top and bottom by 0.5.)

6. $k(x) = \dfrac{17}{2 + 6.5(1.05^{-x})}$; $[0, 100]$

(First divide top and bottom by 2.)

In Exercises 7–10, find the logistic function f with the given properties. HINT [See Example 1.]

7. $f(0) = 10$, f has limiting value 200, and for small values of x, f is approximately exponential and doubles with every increase of 1 in x.

8. $f(0) = 1$, f has limiting value 10, and for small values of x, f is approximately exponential and grows by 50% with every increase of 1 in x.

9. f has limiting value 6 and passes through $(0, 3)$ and $(1, 4)$. HINT [First find **A**, then substitute.]

10. f has limiting value 4 and passes through $(0, 1)$ and $(1, 2)$. HINT [First find **A**, then substitute.]

In Exercises 11–16, choose the logistic function that best approximates the given curve.

11.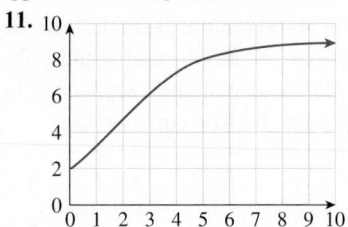

(A) $f(x) = \dfrac{6}{1 + 0.5(3^{-x})}$

(B) $f(x) = \dfrac{9}{1 + 3.5(2^{-x})}$

(C) $f(x) = \dfrac{9}{1 + 0.5(1.01)^{-x}}$

12.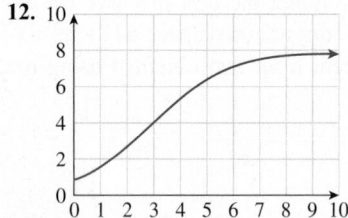

(A) $f(x) = \dfrac{8}{1 + 7(2)^{-x}}$ **(B)** $f(x) = \dfrac{8}{1 + 3(2)^{-x}}$

(C) $f(x) = \dfrac{6}{1 + 11(5)^{-x}}$

13.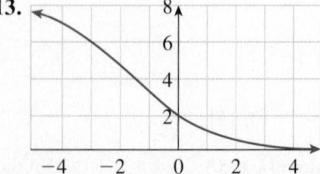

(A) $f(x) = \dfrac{8}{1 + 7(0.5)^{-x}}$ **(B)** $f(x) = \dfrac{8}{1 + 3(0.5)^{-x}}$

(C) $f(x) = \dfrac{8}{1 + 3(2)^{-x}}$

14.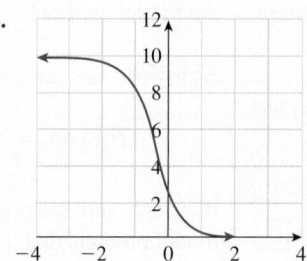

(A) $f(x) = \dfrac{10}{1 + 3(1.01)^{-x}}$ **(B)** $f(x) = \dfrac{8}{1 + 7(0.1)^{-x}}$

(C) $f(x) = \dfrac{10}{1 + 3(0.1)^{-x}}$

15.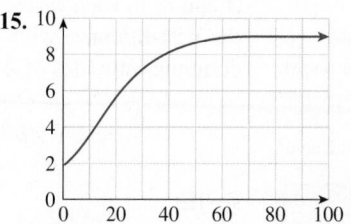

(A) $f(x) = \dfrac{18}{2 + 7(5)^{-x}}$ **(B)** $f(x) = \dfrac{18}{2 + 3(1.1)^{-x}}$

(C) $f(x) = \dfrac{18}{2 + 7(1.1)^{-x}}$

16.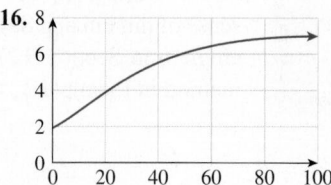

(A) $f(x) = \dfrac{14}{2 + 5(15)^{-x}}$ **(B)** $f(x) = \dfrac{14}{1 + 13(1.05)^{-x}}$

(C) $f(x) = \dfrac{14}{2 + 5(1.05)^{-x}}$

T *In Exercises 17–20, use technology to find a logistic regression curve* $y = \dfrac{N}{1 + Ab^{-x}}$ *approximating the given data. Draw a graph showing the data points and regression curve. (Round b to 3 significant digits and A and N to 2 significant digits.)* HINT [See Example 2.]

17.

x	0	20	40	60	80	100
y	2.1	3.6	5.0	6.1	6.8	6.9

18.

x	0	30	60	90	120	150
y	2.8	5.8	7.9	9.4	9.7	9.9

19.

x	0	20	40	60	80	100
y	30.1	11.6	3.8	1.2	0.4	0.1

20.

x	0	30	60	90	120	150
y	30.1	20	12	7.2	3.8	2.4

APPLICATIONS

21. *Subprime Mortgages during the Housing Bubble* The following graph shows the approximate percentage of mortgages issued in the United States during the real-estate run-up in 2000–2008 that were subprime (normally classified as risky) as well as the logistic regression curve:[43]

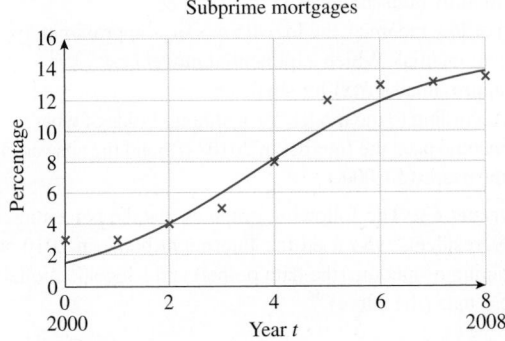

Subprime mortgages

a. Which of the following logistic functions best approximates the curve? (*t* is the number of years since the start of 2000.) Try to determine the correct model without actually computing data points. HINT [See Properties of the Logistic Curve on page 662.]

(A) $A(t) = \dfrac{15.0}{1 + 8.6(1.8)^{-t}}$

(B) $A(t) = \dfrac{2.0}{1 + 6.8(0.8)^{-t}}$

(C) $A(t) = \dfrac{2.0}{1 + 6.8(1.8)^{-t}}$

(D) $A(t) = \dfrac{15.0}{1 + 8.6(0.8)^{-t}}$

b. According to the model you selected, during which year was the percentage growing fastest? HINT [See the "Before we go on" discussion after Example 1.]

22. *Subprime Mortgage Debt during the Housing Bubble* The following graph shows the approximate value of subprime (normally classified as risky) mortgage debt outstanding in the United States during the real-estate run-up in 2000–2008 as well as the logistic regression curve:[44]

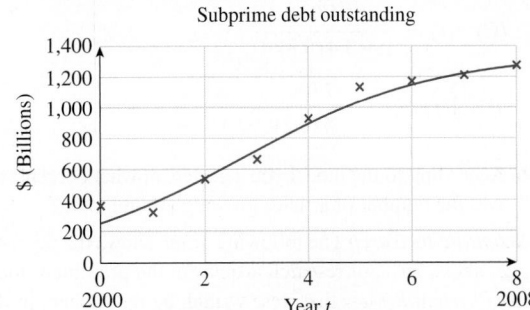

Subprime debt outstanding

a. Which of the following logistic functions best approximates the curve? (*t* is the number of years since the start of 2000.) Try to determine the correct model without actually computing data points. HINT [See Properties of the Logistic Curve on page 662.]

(A) $A(t) = \dfrac{1,850}{1 + 5.36(1.8)^{-t}}$

(B) $A(t) = \dfrac{1,350}{1 + 4.2(1.7)^{-t}}$

(C) $A(t) = \dfrac{1,020}{1 + 5.3(1.8)^{-t}}$

(D) $A(t) = \dfrac{1,300}{1 + 4.2(0.9)^{-t}}$

b. According to the model you selected, during which year was outstanding debt growing fastest? HINT [See the "Before we go on" discussion after Example 1.]

23. *Scientific Research* The following graph shows the number of research articles in the prominent journal *Physical Review* that were written by researchers in Europe during 1983–2003 (*t* = 0 represents 1983).[45]

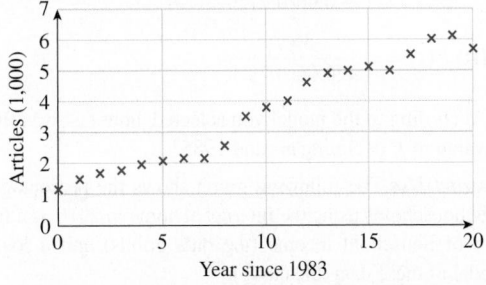

a. Which of the following logistic functions best models the data? (*t* is the number of years since 1983.) Try to determine the correct model without actually computing data points.

---

[43]2008 figure is an estimate. Sources: Mortgage Bankers Association, UBS.

[44]2008–2009 figures are estimates. Source: www.data360.org/dataset.aspx?Data_Set_Id=9549.

[45]Source: The American Physical Society/*New York Times* May 3, 2003, p. A1.

(A) $A(t) = \dfrac{7.0}{1+5.4(1.2)^{-t}}$

(B) $A(t) = \dfrac{4.0}{1+3.4(1.2)^{-t}}$

(C) $A(t) = \dfrac{4.0}{1+3.4(0.8)^{-t}}$

(D) $A(t) = \dfrac{7.0}{1+5.4(6.2)^{-t}}$

b. According to the model you selected, at what percentage was the number of articles growing around 1985?

24. *Scientific Research* The following graph shows the percentage, above 25%, of research articles in the prominent journal *Physical Review* that were written by researchers in the United States during 1983–2003 ($t = 0$ represents 1983).[46]

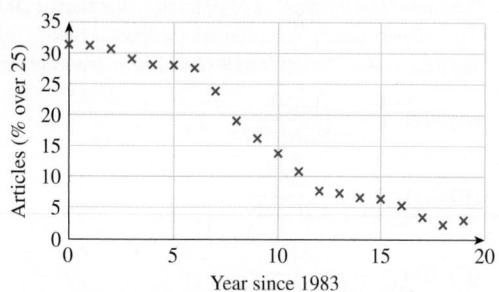

a. Which of the following logistic functions best models the data? (t is the number of years since 1983 and P is the actual percentage.) Try to determine the correct model without actually computing data points.

(A) $P(t) = \dfrac{36}{1+0.06(0.02)^{-t}}$

(B) $P(t) = \dfrac{12}{1+0.06(1.7)^{-t}}$

(C) $P(t) = \dfrac{12}{1+0.06(0.7)^{-t}}$

(D) $P(t) = \dfrac{36}{1+0.06(0.7)^{-t}}$

b. According to the model you selected, how fast was the value of P declining around 1985?

25. *Internet Use* The following graph shows the percentage of U.S. households using the Internet at home in 2010 as a function of household income (the data points) and a logistic model of these data (the curve).[47]

[46]Source: The American Physical Society/*New York Times* May 3, 2003, p. A1.

[47]Income levels are midpoints of income brackets. Source: *Current Population Survey (CPS) Internet Use 2010,* National Telecommunications and Information Administration, www.ntia.doc.gov, January 2011.

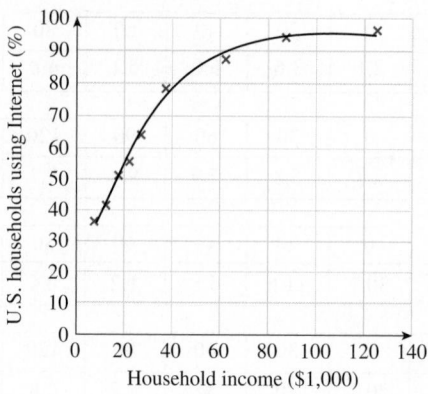

The logistic model is

$$P(x) = \dfrac{95.0}{1+2.78(1.064)^{-x}} \text{ percent}$$

where x is the household income in thousands of dollars.

a. According to the model, what percentage of extremely wealthy households used the Internet?

b. For low incomes, the logistic model is approximately exponential. Which exponential model best approximates $P(x)$ for small x?

c. According to the model, 50% of households of what income used the Internet in 2010? (Round the answer to the nearest $1,000.)

26. *Internet Use* The following graph shows the percentage of U.S. residents who used the Internet at home in 2010 as a function of income (the data points) and a logistic model of these data (the curve).[48]

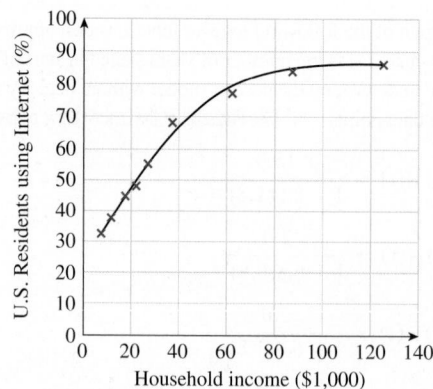

The logistic model is given by

$$P(x) = \dfrac{86.2}{1+2.49(1.054)^{-x}} \text{ percent}$$

where x is the household income in thousands of dollars.

[48]*Ibid.*

a. According to the model, what percentage of extremely wealthy people used the Internet at home?

b. For low incomes, the logistic model is approximately exponential. Which exponential model best approximates $P(x)$ for small x?

c. According to the model, 50% of individuals with what household income used the Internet at home in 2010? (Round the answer to the nearest $1,000.)

27. *Epidemics* There are currently 1,000 cases of Venusian flu in a total susceptible population of 10,000 and the number of cases is increasing by 25% each day. Find a logistic model for the number of cases of Venusian flu and use your model to predict the number of flu cases a week from now. HINT [See Example 1.]

28. *Epidemics* Last year's epidemic of Martian flu began with a single case in a total susceptible population of 10,000. The number of cases was increasing initially by 40% per day. Find a logistic model for the number of cases of Martian flu and use your model to predict the number of flu cases 3 weeks into the epidemic. HINT [See Example 1.]

29. *Sales* You have sold 100 "I ♥ Calculus" T-shirts and sales appear to be doubling every 5 days. You estimate the total market for "I ♥ Calculus" T-shirts to be 3,000. Give a logistic model for your sales and use it to predict, to the nearest day, when you will have sold 700 T-shirts.

30. *Sales* In Russia the average consumer drank two servings of Coca-Cola® in 1993. This amount appeared to be increasing exponentially with a doubling time of 2 years.[49] Given a long-range market saturation estimate of 100 servings per year, find a logistic model for the consumption of Coca-Cola in Russia and use your model to predict when, to the nearest year, the average consumption reached 50 servings per year.

31. ▇ ***Scientific Research*** The following chart shows some the data shown in the graph in Exercise 23 ($t = 0$ represents 1983).[50]

Year t	0	5	10	15	20
Research Articles A (1,000)	1.2	2.1	3.8	5.1	5.7

a. What is the logistic regression model for the data? (Round all coefficients to 2 significant digits.) At what value does the model predict that the number of research articles will level off? HINT [See Example 2.]

b. According to the model, how many *Physical Review* articles were published by U.S. researchers in 2000 ($t = 17$)? (The actual number was about 5,500 articles.)

32. ▇ ***Scientific Research*** The following chart shows some the data shown in the graph in Exercise 24 ($t = 0$ represents 1983).[51]

Year t	0	5	10	15	20
Percentage P (over 25)	36	28	16	7	3

a. What is the logistic regression model for the data? (Round all coefficients to 2 significant digits.) HINT [See Example 2.]

b. According to the model, what percentage of *Physical Review* articles were published by researchers in the United States in 2000 ($t = 17$)? (The actual figure was about 30.1%.)

33. ▇ ***College Basketball: Men*** The following table shows the number of NCAA men's college basketball teams in the U.S. for various years since 1990.[52]

t (Year since 1990)	0	5	10	11	12	13	14
Teams	767	868	932	937	936	967	981
t (Year since 1990)	15	16	17	18	19	20	
Teams	983	984	982	1,017	1,017	1,011	

a. What is the logistic regression model for the data? (Round all coefficients to 3 significant digits.) At what value does the model predict that the number of basketball teams will level off?

b. According to the model, for what value of t is the regression curve steepest? Interpret the answer.

c. Interpret the coefficient b in the context of the number of men's basketball teams.

34. ▇ ***College Basketball: Women*** The following table shows the number of NCAA women's college basketball teams in the U.S. for various years since 1990.[53]

t (Year since 1990)	0	5	10	11	12	13	14
Teams	782	864	956	958	975	1,009	1,008
t (Year since 1990)	15	16	17	18	19	20	
Teams	1,036	1,018	1,003	1,013	1,032	1,036	

a. What is the logistic regression model for the data? (Round all coefficients to 3 significant digits.) At what value does the model predict that the number of basketball teams will level off?

---

[49]The doubling time is based on retail sales of Coca-Cola products in Russia. Sales in 1993 were double those in 1991, and were expected to double again by 1995. Source: *New York Times*, September 26, 1994, p. D2.

[50]Source: The American Physical Society/*New York Times* May 3, 2003, p. A1.

[51]*Ibid.*

[52]2010 figure is an estimate. Source: www.census.gov.

[53]*Ibid.*

b. According to the model, for what value of t is the regression curve steepest? Interpret the answer.

c. Interpret the coefficient b in the context of the number of women's basketball teams.

⊤ *Exercises 35 and 36 are based on the discussion following Example 2. If the limiting value N is known, then*

$$\frac{N}{y} - 1 = A(b^{-1})^x$$

and so $N/y - 1$ is an exponential function of x. In Exercises 35 and 36, use the given value of N and the data points $(x, N/y - 1)$ to obtain A and b, and hence a logistic model.

35. ⊤ ◆ *Population: Puerto Rico* The following table and graph show the population of Puerto Rico in thousands from 1950 to 2025.[54]

t (year since 1950)	0	10	20	30	40	50
Population (thousands)	2,220	2,360	2,720	3,210	3,540	3,820
t (year since 1950)	55	60	65	70	75	
Population (thousands)	3,910	3,990	4,050	4,080	4,100	

Puerto Rico population

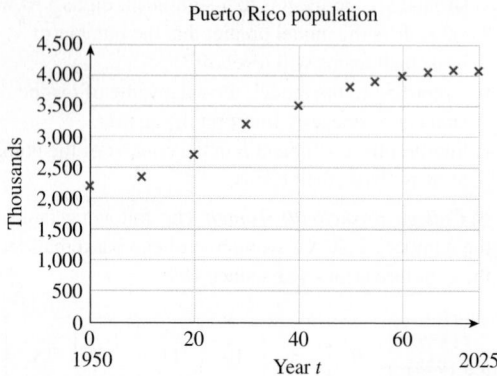

Take t to be the number of years since 1950, and find a logistic model based on the assumption that, eventually, the population of Puerto Rico will grow to 4.5 million. (Round coefficients to 4 decimal places.) In what year does your model predict the population of Puerto Rico will first reach 4.0 million?

36. ⊤ ◆ *Population: Virgin Islands* The following table and graph show the population of the Virgin Islands in thousands from 1950 to 2025.[55]

t (year since 1950)	0	10	20	30	40	50
Population (thousands)	27	33	63	98	104	106
t (year since 1950)	55	60	65	70	75	
Population (thousands)	108	108	107	107	108	

Virgin Islands population

Take t to be the number of years since 1950, and find a logistic model based on the assumption that, eventually, the population of the Virgin Islands will grow to 110,000. (Round coefficients to 4 decimal places.) In what year does your model predict the population of the Virgin Islands first reached 80,000?

COMMUNICATION AND REASONING EXERCISES

37. Logistic functions are commonly used to model the spread of epidemics. Given this fact, explain why a logistic function is also useful to model the spread of a new technology.

38. Why is a logistic function more appropriate than an exponential function for modeling the spread of an epidemic?

39. Give one practical use for logistic regression.

40. Refer to an exercise or example in this section to find a scenario in which a logistic model may not be a good predictor of long-term behavior.

41. What happens to the function $P(t) = \dfrac{N}{1 + Ab^{-t}}$ if we replace b^{-t} by b^t when $b > 1$? When $b < 1$?

42. ▼ What happens to the function $P(t) = \dfrac{N}{1 + Ab^{-t}}$ if $A = 0$? If $A < 0$?

43. ▼ We said that the logistic curve $y = \dfrac{N}{1 + Ab^{-t}}$ is steepest when $t = \dfrac{\ln A}{\ln b}$. Show that the corresponding value of y is $N/2$. HINT [Use the fact that $\dfrac{\ln A}{\ln b} = \log_b A$.]

44. ▼ We said that the logistic curve $y = \dfrac{N}{1 + Ab^{-t}}$ is steepest when $t = \dfrac{\ln A}{\ln b}$. For which values of A and b is this value of t positive, zero, and negative?

---
[54]Figures from 2010 on are U.S. census projections. Source: The 2008 Statistical Abstract, www.census.gov.
[55]*Ibid.*

2. Press $\boxed{\text{STAT}}$, select CALC, and choose option #5 QuadReg. Pressing $\boxed{\text{ENTER}}$ gives the quadratic regression curve in the home screen:

$$y \approx 0.01214x^2 + 0.8471x + 316.9. \quad \text{Coefficients rounded to 4 decimal places}$$

3. Now go to the Y= window and turn Plot1 on by selecting it and pressing $\boxed{\text{ENTER}}$. (You can also turn it on in the $\boxed{\text{2ND}}$ STAT PLOT screen.)

4. Next, enter the regression equation in the $\boxed{\text{Y=}}$ screen by pressing $\boxed{\text{Y=}}$, clearing out whatever function is there, and pressing $\boxed{\text{VARS}}$ $\boxed{5}$ and selecting EQ option #1: RegEq as shown below left.

 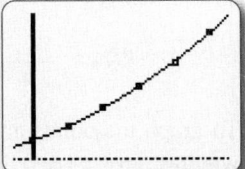

5. To obtain a convenient window showing all the points and the lines, press $\boxed{\text{ZOOM}}$ and choose option #9: ZoomStat as shown above on the right.

Note When you are done viewing the graph, it is a good idea to turn Plot1 off again to avoid errors in graphing or data points showing up in your other graphs. ■

Section 9.2

Example 6(a) (page 641) The following table shows annual expenditure on health in the U.S. from 1980 through 2009 ($t = 0$ represents 1980).

Year t	0	5	10	15	20	25	29
Expenditure ($ billion)	256	444	724	1,030	1,380	2,020	2,490

Find the exponential regression model

$$C(t) = Ab^t$$

for the annual expenditure.

Solution with Technology

This is very similar to Example 5 in Section 9.1 (see the Technology Guide for Section 9.1):

1. Use $\boxed{\text{STAT}}$ EDIT to enter the table of values.

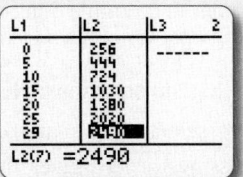

2. Press $\boxed{\text{STAT}}$, select CALC, and choose option #0 ExpReg. Pressing $\boxed{\text{ENTER}}$ gives the exponential regression curve in the home screen:

$$C(t) \approx 296.25(1.0798)^t. \quad \text{Coefficients rounded}$$

3. To graph the points and regression line in the same window, turn Plot1 on (see the Technology Guide for Example 5 in Section 9.1) and enter the regression equation in the Y= screen by pressing $\boxed{\text{Y=}}$, clearing out whatever function is there, and pressing $\boxed{\text{VARS}}$ $\boxed{5}$ and selecting EQ option #1: RegEq. Then press $\boxed{\text{ZOOM}}$ and choose option #9: ZoomStat to see the graph.

 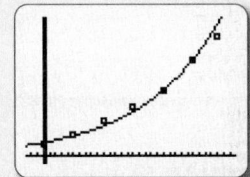

Note When you are done viewing the graph, it is a good idea to turn Plot1 off again to avoid errors in graphing or data points showing up in your other graphs. ■

Section 9.3

Example 5 (page 655) The following table shows the total spent on research and development by universities and colleges in the U.S., in billions of dollars, for the period 1998–2008 (t is the year since 1990).

Year t	8	9	10	11	12	13
Spending ($ billions)	27	29	31	33	36	38
Year t	14	15	16	17	18	
Spending ($ billions)	39	40	40	41	42	

Find the best-fit logarithmic model of the form

$$S(t) = A \ln t + C$$

and use the model to project total spending on research by universities and colleges in 2012, assuming the trend continues.

Solution with Technology

This is very similar to Example 5 in Section 9.1 (see the Technology Guide for Section 9.1):

1. Use $\boxed{\text{STAT}}$ EDIT to enter the table of values.
2. Press $\boxed{\text{STAT}}$, select CALC, and choose option #9 LnReg. Pressing $\boxed{\text{ENTER}}$ gives the logarithmic regression curve in the home screen:

$$S(t) \approx 19.25 \ln t - 12.78. \qquad \text{Coefficients rounded}$$

3. To graph the points and regression line in the same window, turn Plot1 on (see the Technology Guide for Example 5 in Section 9.1) and enter the regression equation in the Y= screen by pressing $\boxed{\text{Y=}}$, clearing out whatever function is there, and pressing $\boxed{\text{VARS}}$ $\boxed{5}$ and selecting EQ option #1: RegEq. Then press $\boxed{\text{ZOOM}}$ and choose option #9: ZoomStat to see the graph.

 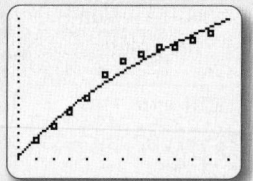

Section 9.4

Example 2 (page 664) The following table shows wired broadband penetration in the United States as a function of time t in years ($t = 0$ represents 2000).

Year (t)	2	3	4	5	6	7	8	9	10
Penetration (%) (P)	5.5	7.9	10.9	14.2	18.2	21.9	23.9	26.5	27.1

Find a logistic regression curve of the form

$$P(t) = \frac{N}{1 + Ab^{-t}}.$$

Solution with Technology

This is very similar to Example 5 in Section 9.1 (see the Technology Guide for Section 9.1):

1. Use $\boxed{\text{STAT}}$ EDIT to enter the table of values.
2. Press $\boxed{\text{STAT}}$, select CALC, and choose option #B Logistic. Pressing $\boxed{\text{ENTER}}$ gives the logistic regression curve in the home screen:

$$P(t) \approx \frac{29.842}{1 + 12.502e^{-0.49592t}}. \qquad \text{Coefficients rounded}$$

This is not exactly the form we are seeking, but we can convert it to that form by writing

$$e^{-0.49592t} = (e^{0.49592})^{-t} \approx 1.642^{-t}$$

so

$$P(t) \approx \frac{29.842}{1 + 12.502(1.642)^{-t}}.$$

3. To graph the points and regression line in the same window, turn Plot1 on (see the Technology Guide for Example 5 in Section 9.1) and enter the regression equation in the Y= screen by pressing $\boxed{\text{Y=}}$, clearing out whatever function is there, and pressing $\boxed{\text{VARS}}$ $\boxed{5}$ and selecting EQ option #1: RegEq. Then press $\boxed{\text{ZOOM}}$ and choose option #9: ZoomStat to see the graph.

 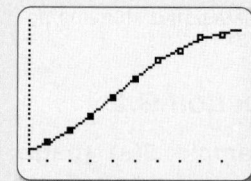

SPREADSHEET Technology Guide

Section 9.1

Example 2 (page 622) Sketch the graph of each quadratic function, showing the location of the vertex and intercepts.

a. $f(x) = 4x^2 - 12x + 9$

b. $g(x) = -\dfrac{1}{2}x^2 + 4x - 12$

Solution with Technology

We can set up a worksheet so that all we have to enter are the coefficients a, b, and c, and a range of x-values for the graph. Here is a possible layout that will plot 101 points using the coefficients for part (a).

1. First, we compute the *x* coordinates:

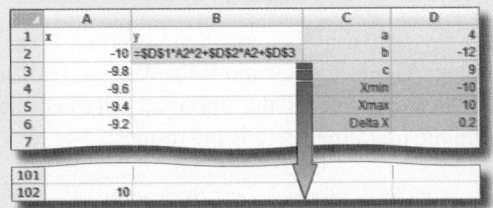

2. To add the *y* coordinates, we use the technology formula

$$a*x^2+b*x+c$$

replacing *a*, *b*, and *c* with (absolute) references to the cells containing their values.

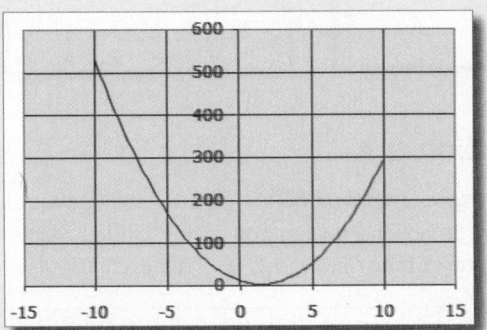

3. Graphing the data in columns A and B gives the graph shown here:

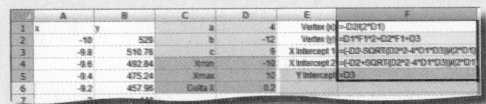

$$y = 4x^2 - 12x + 9$$

4. We can go further and compute the exact coordinates of the vertex and intercepts:

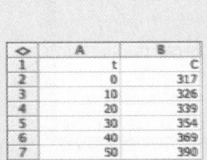

The completed sheet should look like this:

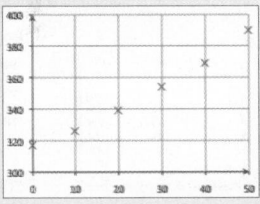

We can now save this sheet as a template to handle all quadratic functions. For instance, to do part (b), we just change the values of *a*, *b*, and *c* in column D to $a = -1/2$, $b = 4$, and $c = -12$.

Example 5 (page 626) The following table shows the annual mean carbon dioxide concentration measured at Mauna Loa Observatory in Hawaii, in parts per million, every 10 years from 1960 through 2010 ($t = 0$ represents 1960).

Year t	0	10	20	30	40	50
PPM CO_2 C	317	326	339	354	369	390

Find the quadratic model

$$C(t) = at^2 + bt + c$$

that best fits the data.

Solution with Technology

As in Section 1.4, Example 2, we start with a scatter plot of the original data, and add a trendline:

1. Start with the original data and a "Scatter plot" (see Section 1.2 Example 5).

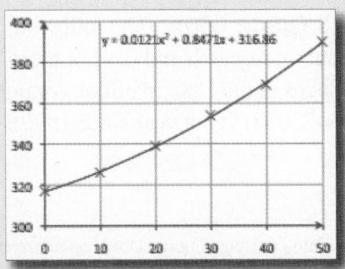

2. Add a quadratic trend line. (As of the time of this writing, among the common spreadsheets only Excel has the ability to add a polynomial trendline.) Right-click on any data point in the chart and select "Add Trendline," then select a "Polynomial" type of order 2 and check the option to "Display Equation on chart."

Section 9.2

Example 6(a) (page 641) The following table shows annual expenditure on health in the U.S. from 1980 through 2009 ($t = 0$ represents 1980).

Year t	0	5	10	15	20	25	29
Expenditure ($ billion)	256	444	724	1,030	1,380	2,020	2,490

Find the exponential regression model

$$C(t) = Ab^t$$

for the annual expenditure.

Solution with Technology

This is very similar to Example 5 in Section 9.1 (see the Technology Guide for Section 9.1):

1. Start with a "Scatter plot" of the observed data.

2. Add an exponential trendline:[60] The details vary from spreadsheet to spreadsheet—in OpenOffice, first double-click on the graph. Right-click on any data point in the chart and select "Add Trendline," then select an "Exponential" type and check the option to "Display Equation on chart."

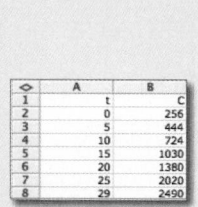
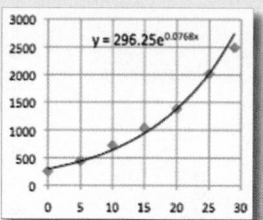

Notice that the regression curve is given in the form Ae^{kt} rather than Ab^t. To transform it, write

$$296.25e^{0.0768t} = 296.25(e^{0.0768})^t$$
$$\approx 296.25(1.0798)^t. \quad e^{0.0768} \approx 1.0798$$

Section 9.3

Example 5 (page 655) The following table shows the total spent on research and development by universities and colleges in the U.S., in billions of dollars, for the period 1998–2008 (t is the year since 1990).

Year t	8	9	10	11	12	13
Spending ($ billions)	27	29	31	33	36	38
Year t	14	15	16	17	18	
Spending ($ billions)	39	40	40	41	42	

Find the best-fit logarithmic model of the form

$$S(t) = A \ln t + C$$

and use the model to project total spending on research by universities and colleges in 2012, assuming the trend continues.

Solution with Technology

This is very similar to Example 5 in Section 9.1 (see the Technology Guide for Section 9.1): We start, as usual, with a "Scatter plot" of the observed data and add a Logarithmic trendline. Here is the result:

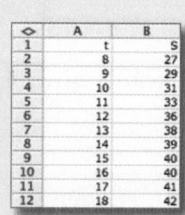
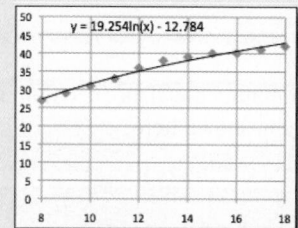

Section 9.4

Example 2 (page 664) The following table shows wired broadband penetration in the United States as a function of time t in years ($t = 0$ represents 2000).

Year (t)	2	3	4	5	6	7	8	9	10
Penetration (%) (P)	5.5	7.9	10.9	14.2	18.2	21.9	23.9	26.5	27.1

Find a logistic regression curve of the form

$$P(t) = \frac{N}{1 + Ab^{-t}}.$$

Solution with Technology

At the time of this writing, available spreadsheets did not have a built-in logistic regression calculation, so we use an alternative method that works for any type of regression curve. The Solver included with Windows versions of Excel and some Mac versions can find logistic regression curves while the Solver included with some other spreadsheets is not yet capable of this, so the instructions here are specific to Excel.

---

[60] At the time of this writing, Google Docs has no trendline feature for its spreadsheet.

1. First use rough estimates for N, A, and b, and compute the sum-of-squares error (SSE; see Section 1.4) directly:

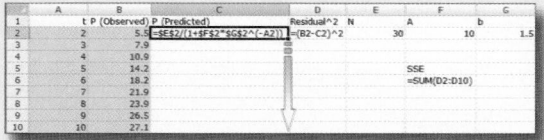

Cells E2:G2 contain our initial rough estimates of N, A, and b. For N, we used 30 (notice that the y-coordinates do appear to level off around 30). For A, we used the fact that the y-intercept is $N/(1 + A)$ and the y-intercept appears to be approximately 3. In other words,

$$3 = \frac{30}{1 + A}.$$

Because a very rough estimate is all we are after, using $A = 10$ will do just fine. For b, we chose 1.5 as the values of P appear to be increasing by around 50% per year initially (again, this is rough).

2. Cell C2 contains the formula for $P(t)$, and the square of the resulting residual is computed in D2.

3. Cell F6 will contain SSE. The completed spreadsheet should look like this:

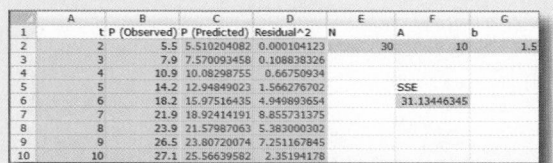

The best-fit curve will result from values of N, A, and b that give a minimum value for SSE. We shall use Excel's "Solver," found in the "Analysis" group on the "Data" tab. (If "Solver" does not appear in the Analysis group, you will have to install the Solver

Add-in using the Excel Options dialogue.) The Solver dialogue box with the necessary fields completed to solve the problem looks like this:

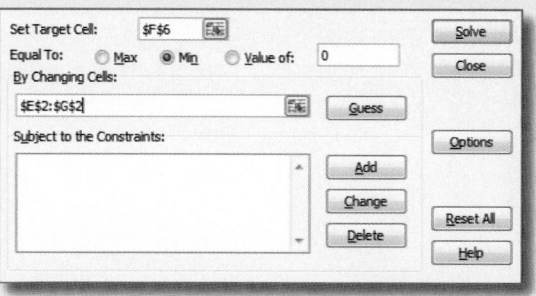

- The Target Cell refers to the cell that contains SSE.
- "Min" is selected because we are minimizing SSE.
- "Changing Cells" are obtained by selecting the cells that contain the current values of N, A, and b.

4. When you have filled in the values for the three items above, press "Solve" and tell Solver to Keep Solver Solution when done. You will find $N \approx 29.842$, $A \approx 12.501$, and $b \approx 1.642$ so that

$$P(t) \approx \frac{29.842}{1 + 12.501(1.642)^{-t}}.$$

If you use a scatter plot to graph the data in columns A, B and C, you will obtain the following graph:

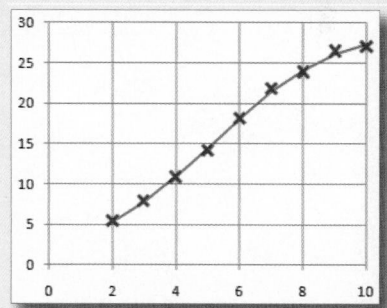

10

Introduction to the Derivative

 Website
www.WanerMath.com
At the Website you will find:

- Section-by-section tutorials, including game tutorials with randomized quizzes

- A detailed chapter summary

- A true/false quiz

- Additional review exercises

- Graphers, Excel tutorials, and other resources

- The following extra topics:

  Sketching the graph of the derivative

  Continuity and differentiability

Case Study Reducing Sulfur Emissions

The Environmental Protection Agency (EPA) wants to formulate a policy that will encourage utilities to reduce sulfur emissions. Its goal is to reduce annual emissions of sulfur dioxide by a total of 10 million tons from the current level of 25 million tons by imposing a fixed charge for every ton of sulfur released into the environment per year. The EPA has some data showing the marginal cost to utilities of reducing sulfur emissions. As a consultant to the EPA, you must determine the amount to be charged per ton of sulfur emissions in light of these data.

Norbert Schaefer/CORBIS

687

Introduction

In the world around us, everything is changing. The mathematics of change is largely about the rate of change: how fast and in which direction the change is occurring. Is the Dow Jones average going up, and if so, how fast? If I raise my prices, how many customers will I lose? If I launch this missile, how fast will it be traveling after two seconds, how high will it go, and where will it come down?

We have already discussed the concept of rate of change for linear functions (straight lines), where the slope measures the rate of change. But this works only because a straight line maintains a constant rate of change along its whole length. Other functions rise faster here than there—or rise in one place and fall in another—so that the rate of change varies along the graph. The first achievement of calculus is to provide a systematic and straightforward way of calculating (hence the name) these rates of change. To describe a changing world, we need a language of change, and that is what calculus is.

The history of calculus is an interesting story of personalities, intellectual movements, and controversy. Credit for its invention is given to two mathematicians: Isaac Newton (1642–1727) and Gottfried Leibniz (1646–1716). Newton, an English mathematician and scientist, developed calculus first, probably in the 1660s. We say "probably" because, for various reasons, he did not publish his ideas until much later. This allowed Leibniz, a German mathematician and philosopher, to publish his own version of calculus first, in 1684. Fifteen years later, stirred up by nationalist fervor in England and on the continent, controversy erupted over who should get the credit for the invention of calculus. The debate got so heated that the Royal Society (of which Newton and Leibniz were both members) set up a commission to investigate the question. The commission decided in favor of Newton, who happened to be president of the society at the time. The consensus today is that both mathematicians deserve credit because they came to the same conclusions working independently. This is not really surprising: Both built on well-known work of other people, and it was almost inevitable that someone would put it all together at about that time.

algebra Review
For this chapter, you should be familiar with the algebra reviewed in **Chapter 0, Section 2.**

10.1 Limits: Numerical and Graphical Viewpoints

Rates of change are calculated by derivatives, but an important part of the definition of the derivative is something called a **limit**. Arguably, much of mathematics since the eighteenth century has revolved around understanding, refining, and exploiting the idea of the limit. The basic idea is easy, but getting the technicalities right is not.

Evaluating Limits Numerically

Start with a very simple example: Look at the function $f(x) = 2 + x$ and ask: What happens to $f(x)$ as x approaches 3? The following table shows the value of $f(x)$ for values of x close to and on either side of 3:

	x approaching 3 from the left →					← x approaching 3 from the right			
x	2.9	2.99	2.999	2.9999	3	3.0001	3.001	3.01	3.1
$f(x) = 2 + x$	4.9	4.99	4.999	4.9999		5.0001	5.001	5.01	5.1

We have left the entry under 3 blank to emphasize that when calculating the limit of $f(x)$ as x *approaches* 3, we are not interested in its value when x *equals* 3.

Notice from the table that the closer x gets to 3 from either side, the closer $f(x)$ gets to 5. We write this as

$$\lim_{x \to 3} f(x) = 5.$$ The limit of $f(x)$, as x approaches 3, equals 5.

Q : *Why all the fuss? Can't we simply substitute $x = 3$ and avoid having to use a table?*

A : This happens to work for *some* functions, but not for *all* functions. The following example illustrates this point.

EXAMPLE 1 Estimating a Limit Numerically

Use a table to estimate the following limits:

a. $\lim\limits_{x \to 2} \dfrac{x^3 - 8}{x - 2}$ **b.** $\lim\limits_{x \to 0} \dfrac{e^{2x} - 1}{x}$

Solution

a. We cannot simply substitute $x = 2$, because the function $f(x) = \dfrac{x^3 - 8}{x - 2}$ is not defined at $x = 2$. (Why?)* Instead, we use a table of values as we did above, with x approaching 2 from both sides.

* However, if you factor $x^3 - 8$, you will find that $f(x)$ can be simplified to a function that *is* defined at $x = 2$. This point will be discussed (and this example redone) in Section 10.3. The function in part (b) cannot be simplified by factoring.

	x approaching 2 from the left →					← x approaching 2 from the right			
x	1.9	1.99	1.999	1.9999	2	2.0001	2.001	2.01	2.1
$f(x) = \dfrac{x^3 - 8}{x - 2}$	11.41	11.9401	11.9940	11.9994		12.0006	12.0060	12.0601	12.61

We notice that as x approaches 2 from either side, $f(x)$ appears to be approaching 12. This suggests that the limit is 12, and we write

$$\lim_{x \to 2} \frac{x^3 - 8}{x - 2} = 12.$$

b. The function $g(x) = \dfrac{e^{2x} - 1}{x}$ is not defined at $x = 0$ (nor can it even be simplified to one which *is* defined at $x = 0$). In the following table, we allow x to approach 0 from both sides:

	x approaching 0 from the left →					← x approaching 0 from the right			
x	−0.1	−0.01	−0.001	−0.0001	0	0.0001	0.001	0.01	0.1
$g(x) = \dfrac{e^{2x} - 1}{x}$	1.8127	1.9801	1.9980	1.9998		2.0002	2.0020	2.0201	2.2140

The table suggests that $\lim\limits_{x \to 0} \dfrac{e^{2x} - 1}{x} = 2$.

using Technology

To automate the computations in Example 1 using a graphing calculator or a spreadsheet, see the Technology Guides at the end of the chapter. Outline for part (a):

TI-83/84 Plus
Home screen: $Y_1 = (X^3 - 8) / (X - 2)$
[2ND] [TBLSET] Indpnt
set to Ask
[2ND] [TABLE] Enter some values of x from the example:
1.9, 1.99, 1.999 . . .
[More details on page 778.]

Spreadsheet

Headings x, $f(x)$ in A1–B1 and again in C1–D1.
In A2–A5 enter 1.9, 1.99, 1.999, 1.9999.
In C1–C5 enter 2.1, 2.01, 2.001, 2.0001. Enter
`=(A2^3-8)/(A2-2)`
in B2 and copy down to B5.
Copy and paste the same formula in D2–D5.
[More details on page 780.]

 Website

www.WanerMath.com
In the Function Evaluator and Grapher, enter `(x^3-8)/(x-2)` for y_1. For a table of values, enter the various x-values in the Evaluator box, and press "Evaluate".

➡ **Before we go on...** Although the table *suggests* that the limit in Example 1 part (b) is 2, it by no means establishes that fact conclusively. It is *conceivable* (though not in fact the case here) that putting $x = 0.000000087$ could result in, say, $g(x) = 426$. Using a table can only *suggest* a value for the limit. In the next two sections we shall discuss algebraic techniques to allow us to actually *calculate* limits. ■

Before we continue, let us make a more formal definition.

Definition of a Limit

If $f(x)$ approaches the number L as x approaches (but is not equal to) a from both sides, then we say that $f(x)$ **approaches L as $x \to a$** ("x approaches a") or that the **limit** of $f(x)$ as $x \to a$ is L. More precisely, *we can make $f(x)$ be as close to L as we like by choosing any x sufficiently close to (but not equal to) a on either side.* We write

$$\lim_{x \to a} f(x) = L$$

or

$$f(x) \to L \text{ as } x \to a.$$

If $f(x)$ *fails* to approach *a single fixed number* as x approaches a from both sides, then we say that $f(x)$ **has no limit** as $x \to a$, or

$$\lim_{x \to a} f(x) \text{ does not exist}.$$

Quick Examples

1. $\lim\limits_{x \to 3}(2 + x) = 5$ See discussion before Example 1.

2. $\lim\limits_{x \to -2}(3x) = -6$ As x approaches -2, $3x$ approaches -6.

3. $\lim\limits_{x \to 0}(x^2 - 2x + 1)$ exists.   In fact, the limit is 1.

4. $\lim\limits_{x \to 5} \dfrac{1}{x} = \dfrac{1}{5}$ As x approaches 5, $\dfrac{1}{x}$ approaches $\dfrac{1}{5}$.

5. $\lim\limits_{x \to 2} \dfrac{x^3 - 8}{x - 2} = 12$ See Example 1. (We cannot just put $x = 2$ here.)

(For examples where the limit does not exist, see Example 2.)

Notes

1. It is important that $f(x)$ approach the same number as x approaches a from *both sides*. For instance, if $f(x)$ approaches 5 for $x = 1.9$, 1.99, 1.999, ... , but approaches 4 for $x = 2.1$, 2.01, 2.001, ... , then the limit as $x \to 2$ does not exist. (See Example 2 for such a situation.)

2. It may happen that $f(x)$ does not approach any fixed number at all as $x \to a$ from either side. In this case, we also say that the limit does not exist.

3. If a happens to be an endpoint of the domain of f, then the function is only defined on one side of a, and so the limit as $x \to a$ does not exist. For example, the natural domain of $f(x) = \sqrt{x}$ is $[0, +\infty)$, so the limit of $f(x)$ as $x \to 0$ does not exist. The appropriate kind of limit to consider in such a case is a **one-sided limit** (see Example 2). ■

The next example gives instances in which a stated limit does not exist.

EXAMPLE 2 Limits Do Not Always Exist

Do the following limits exist?

a. $\lim\limits_{x \to 0} \dfrac{1}{x^2}$ **b.** $\lim\limits_{x \to 0} \dfrac{|x|}{x}$ **c.** $\lim\limits_{x \to 2} \dfrac{1}{x - 2}$ **d.** $\lim\limits_{x \to 1} \sqrt{x - 1}$

Solution

a. Here is a table of values for $f(x) = \dfrac{1}{x^2}$, with x approaching 0 from both sides.

x approaching 0 from the left $\to$ $\leftarrow$ x approaching 0 from the right

x	-0.1	-0.01	-0.001	-0.0001	0	0.0001	0.001	0.01	0.1
$f(x) = \dfrac{1}{x^2}$	100	10,000	1,000,000	100,000,000		100,000,000	1,000,000	10,000	100

The table shows that as x gets closer to zero on either side, $f(x)$ gets larger and larger **without bound**—that is, if you name any number, no matter how large, $f(x)$ will be even larger than that if x is sufficiently close to 0. Because $f(x)$ is not approaching any real number, we conclude that $\lim\limits_{x \to 0} \dfrac{1}{x^2}$ does not exist. Because $f(x)$ is becoming arbitrarily large, we also say that $\lim\limits_{x \to 0} \dfrac{1}{x^2}$ **diverges to** $+\infty$, or just

$$\lim\limits_{x \to 0} \dfrac{1}{x^2} = +\infty.$$

Note This is not meant to imply that the limit exists; the symbol $+\infty$ does not represent any real number. We write $\lim_{x \to a} f(x) = +\infty$ to indicate two things: (1) The limit does not exist and (2) the function gets large without bound as x approaches a. ∎

b. Here is a table of values for $f(x) = \dfrac{|x|}{x}$, with x approaching 0 from both sides.

x approaching 0 from the left $\to$ $\leftarrow$ x approaching 0 from the right

x	-0.1	-0.01	-0.001	-0.0001	0	0.0001	0.001	0.01	0.1
$f(x) = \dfrac{|x|}{x}$	-1	-1	-1	-1		1	1	1	1

The table shows that $f(x)$ does not approach the same limit as x approaches 0 from both sides. There appear to be two *different* limits: the limit as we approach 0 from the left and the limit as we approach from the right. We write

$$\lim\limits_{x \to 0^-} f(x) = -1$$

read as "the limit as x approaches 0 from the left (or from below) is -1" and

$$\lim\limits_{x \to 0^+} f(x) = 1$$

read as "the limit as x approaches 0 from the right (or from above) is 1." These are called the **one-sided limits** of $f(x)$. In order for f to have a **two-sided limit**, the two one-sided limits must be equal. Because they are not, we conclude that $\lim_{x \to 0} f(x)$ does not exist.

c. Near $x = 2$, we have the following table of values for $f(x) = \dfrac{1}{x-2}$:

	x approaching 2 from the left $\rightarrow$					$\leftarrow$ x approaching 2 from the right			
x	1.9	1.99	1.999	1.9999	2	2.0001	2.001	2.01	2.1
$f(x) = \dfrac{1}{x-2}$	-10	-100	$-1{,}000$	$-10{,}000$		$10{,}000$	$1{,}000$	100	10

Because $f(x)$ is approaching no (single) real number as $x \to 2$, we see that $\lim\limits_{x\to2} \dfrac{1}{x-2}$ does not exist. Notice also that $\dfrac{1}{x-2}$ diverges to $+\infty$ as $x \to 2$ from the positive side (right half of the table) and to $-\infty$ as $x \to 2$ from the left (left half of the table). In other words,

$$\lim_{x\to2^-} \frac{1}{x-2} = -\infty$$

$$\lim_{x\to2^+} \frac{1}{x-2} = +\infty$$

$$\lim_{x\to2} \frac{1}{x-2} \text{ does not exist.}$$

d. The natural domain of $f(x) = \sqrt{x-1}$ is $[1, +\infty)$, as $f(x)$ is defined only when $x \geq 1$. Thus we cannot evaluate $f(x)$ if x is to the left of 1. This means that

$$\lim_{x\to1^-} \sqrt{x-1} \text{ does not exist,}$$

and our table looks like this:

		$\leftarrow$ x approaching 1 from the right				
x	1	1.00001	1.0001	1.001	1.01	1.1
$f(x) = \sqrt{x-1}$		0.0032	0.0100	0.0316	0.1000	0.3162

The values suggest that

$$\lim_{x\to1^+} \sqrt{x-1} = 0.$$

In fact, we can obtain this limit by substituting $x = 1$ in the formula for $f(x)$ (see the comments after the example).

Q: *In Example 2(d) (and also in some of the Quick Examples before that) we could find a limit of an algebraically specified function by simply substituting the value of x in the formula for f(x). Does this always work?*

A: Short answer: Yes, when it makes sense. If the function is specified by a *single* algebraic formula and if $x = a$ is in the domain of f, then the limit can be obtained by substituting. We will say more about this when we discuss the algebraic approach to limits in Section 10.3. Remember, however, that, by definition, the limit of a function as $x \to a$ has nothing to do with its value at $x = a$, but rather its values for x close to a.

Q: If f(x) is undefined when x = a, then the limit does not exist, right?

A: Wrong. If f(a) is not defined, then the limit may or may not exist. Example 1 shows instances where the limit *does* exist, and Example 2 shows instances where it does not. Again, the limit of a function as x → a has nothing to do with its value at x = a, but rather its values for x *close to a*.

In another useful kind of limit, we let x approach either $+\infty$ or $-\infty$, by which we mean that we let x get arbitrarily large or let x become an arbitrarily large negative number. The next example illustrates this.

EXAMPLE 3 Limits at Infinity

Use a table to estimate: **a.** $\lim_{x \to +\infty} \dfrac{2x^2 - 4x}{x^2 - 1}$ and **b.** $\lim_{x \to -\infty} \dfrac{2x^2 - 4x}{x^2 - 1}$.

Solution

a. By saying that x is "approaching $+\infty$," we mean that x is getting larger and larger without bound, so we make the following table:

x approaching $+\infty \to$

x	10	100	1,000	10,000	100,000
$f(x) = \dfrac{2x^2 - 4x}{x^2 - 1}$	1.6162	1.9602	1.9960	1.9996	2.0000

(Note that we are only approaching $+\infty$ from the left because we can hardly approach it from the right!) What seems to be happening is that $f(x)$ is approaching 2. Thus we write

$$\lim_{x \to +\infty} f(x) = 2.$$

b. Here, x is approaching $-\infty$, so we make a similar table, this time with x assuming negative values of greater and greater magnitude (read this table from right to left):

$\leftarrow x$ approaching $-\infty$

x	$-100,000$	$-10,000$	$-1,000$	-100	-10
$f(x) = \dfrac{2x^2 - 4x}{x^2 - 1}$	2.0000	2.0004	2.0040	2.0402	2.4242

Once again, $f(x)$ is approaching 2. Thus, $\lim_{x \to -\infty} f(x) = 2$.

Estimating Limits Graphically

We can often estimate a limit from a graph, as the next example shows.

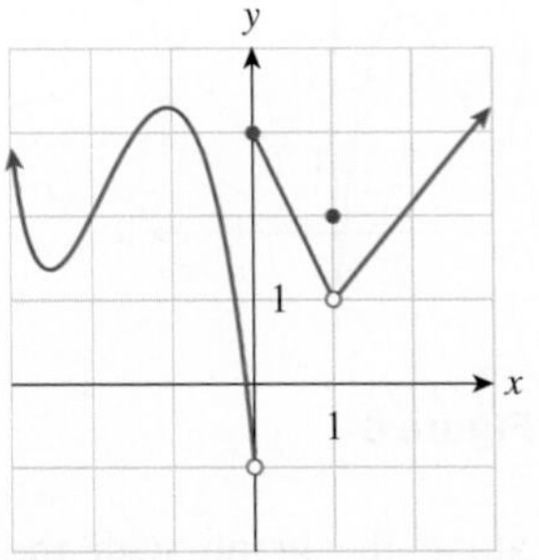

EXAMPLE 4 Estimating Limits Graphically

The graph of a function f is shown in Figure 1. (Recall that the solid dots indicate points on the graph, and the hollow dots indicate points not on the graph.) From the graph, analyze the following limits.

Figure 1

a. $\lim_{x \to -2} f(x)$ **b.** $\lim_{x \to 0} f(x)$ **c.** $\lim_{x \to 1} f(x)$ **d.** $\lim_{x \to +\infty} f(x)$

Solution Since we are given only a graph of f, we must analyze these limits graphically.

a. Imagine that Figure 1 was drawn on a graphing calculator equipped with a trace feature that allows us to move a cursor along the graph and see the coordinates as we go. To simulate this, place a pencil point on the graph to the left of $x = -2$, and move it along the curve so that the x-coordinate approaches -2. (See Figure 2.) We evaluate the limit numerically by noting the behavior of the y-coordinates.*

* For a visual animation of this process, look at the online tutorial for this section at the Website.

However, we can see directly from the graph that the y-coordinate approaches 2. Similarly, if we place our pencil point to the right of $x = -2$ and move it to the left, the y coordinate will approach 2 from that side as well (Figure 3). Therefore, as x approaches -2 from either side, $f(x)$ approaches 2, so

$$\lim_{x \to -2} f(x) = 2.$$

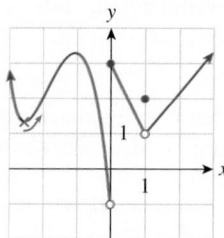

Figure 2

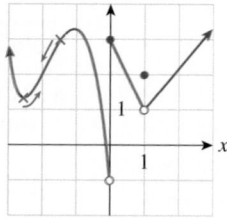

Figure 3

b. This time we move our pencil point toward $x = 0$. Referring to Figure 4, if we start from the left of $x = 0$ and approach 0 (by moving right), the y-coordinate approaches -1. However, if we start from the right of $x = 0$ and approach 0 (by moving left), the y-coordinate approaches 3. Thus (see Example 2),

$$\lim_{x \to 0^-} f(x) = -1$$

and

$$\lim_{x \to 0^+} f(x) = 3.$$

Because these limits are not equal, we conclude that

$$\lim_{x \to 0} f(x) \text{ does not exist.}$$

In this case there is a "break" in the graph at $x = 0$, and we say that the function is **discontinuous** at $x = 0$ (see Section 10.2).

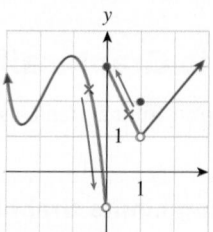

Figure 4

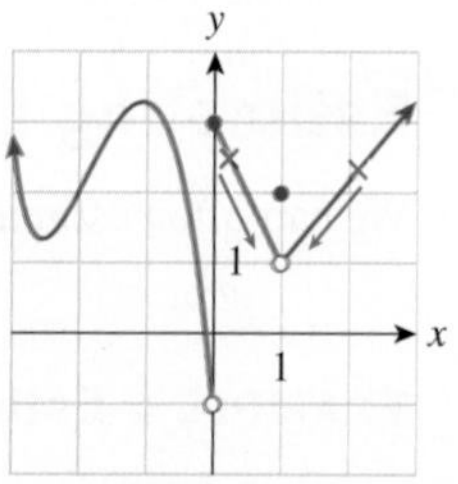

Figure 5

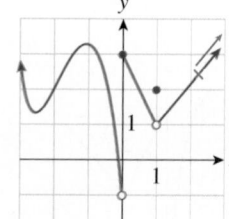

Figure 6

c. Once more we think about a pencil point moving along the graph with the x-coordinate this time approaching $x = 1$ from the left and from the right (Figure 5).

As the x-coordinate of the point approaches 1 from either side, the y-coordinate approaches 1 also. Therefore,

$$\lim_{x \to 1} f(x) = 1.$$

d. For this limit, x is supposed to approach infinity. We think about a pencil point moving along the graph further and further to the right as shown in Figure 6.

As the x-coordinate gets larger, the y-coordinate also gets larger and larger without bound. Thus, $f(x)$ diverges to $+\infty$:

$$\lim_{x \to +\infty} f(x) = +\infty.$$

Similarly,

$$\lim_{x \to -\infty} f(x) = +\infty.$$

➡ **Before we go on...** In Example 4(c) $\lim_{x \to 1} f(x) = 1$ but $f(1) = 2$ (why?). Thus, $\lim_{x \to 1} f(x) \neq f(1)$. In other words, the limit of $f(x)$ as x *approaches* 1 is not the same as the value of f *at* $x = 1$. Always keep in mind that when we evaluate a limit as $x \to a$, *we do not care about the value of the function at $x = a$*. We only care about the value of $f(x)$ as x *approaches a*. In other words, $f(a)$ may or may not equal $\lim_{x \to a} f(x)$. ∎

Here is a summary of the graphical method we used in Example 4, together with some additional information:

Evaluating Limits Graphically

To decide whether $\lim_{x \to a} f(x)$ exists and to find its value if it does:

1. Draw the graph of $f(x)$ by hand or with graphing technology.
2. Position your pencil point (or the Trace cursor) on a point of the graph to the right of $x = a$.
3. Move the point *along the graph* toward $x = a$ from the right and read the y-coordinate as you go. The value the y-coordinate approaches (if any) is the limit $\lim_{x \to a^+} f(x)$.
4. Repeat Steps 2 and 3, this time starting from a point on the graph to the left of $x = a$, and approaching $x = a$ along the graph from the left. The value the y-coordinate approaches (if any) is $\lim_{x \to a^-} f(x)$.
5. If the left and right limits both exist and have the same value L, then $\lim_{x \to a} f(x) = L$. Otherwise, the limit does not exist. The value $f(a)$ has no relevance whatsoever.
6. To evaluate $\lim_{x \to +\infty} f(x)$, move the pencil point toward the far right of the graph and estimate the value the y-coordinate approaches (if any). For $\lim_{x \to -\infty} f(x)$, move the pencil point toward the far left.
7. If $x = a$ happens to be an endpoint of the domain of f, then only a one-sided limit is possible at $x = a$. For instance, if the domain is $(-\infty, 4]$, then $\lim_{x \to 4^-} f(x)$ may exist, but neither $\lim_{x \to 4^+} f(x)$ nor $\lim_{x \to 4} f(x)$ exists.

In the next example we use both the numerical and graphical approaches.

EXAMPLE 5 **Infinite Limit**

Does $\lim\limits_{x\to 0^+} \dfrac{1}{x}$ exist?

Solution

Numerical Method Because we are asked for only the right-hand limit, we need only list values of x approaching 0 from the right.

$\leftarrow$ x approaching 0 from the right

x	0	0.0001	0.001	0.01	0.1
$f(x) = \dfrac{1}{x}$		10,000	1,000	100	10

What seems to be happening as x approaches 0 from the right is that $f(x)$ is increasing without bound, as in Example 4(d). That is, if you name any number, no matter how large, $f(x)$ will be even larger than that if x is sufficiently close to zero. Thus, the limit diverges to $+\infty$, so

$$\lim_{x\to 0^+} \frac{1}{x} = +\infty$$

Graphical Method Recall that the graph of $f(x) = \dfrac{1}{x}$ is the standard hyperbola shown in Figure 7. The figure also shows the pencil point moving so that its x-coordinate approaches 0 from the right. Because the point moves along the graph, it is forced to go higher and higher. In other words, its y-coordinate becomes larger and larger, approaching $+\infty$. Thus, we conclude that

$$\lim_{x\to 0^+} \frac{1}{x} = +\infty.$$

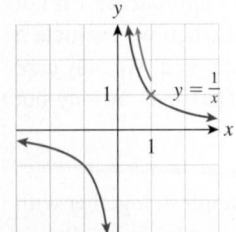

Figure 7

➡ **Before we go on...** In Example 5 you should check that

$$\lim_{x\to 0^-} \frac{1}{x} = -\infty. \qquad\qquad \frac{1}{x} \text{ diverges to } -\infty \text{ as } x \to 0^-.$$

Also, check that

$$\lim_{x\to +\infty} \frac{1}{x} = \lim_{x\to -\infty} \frac{1}{x} = 0. \ \blacksquare$$

APPLICATION

EXAMPLE 6 **Broadband Penetration**

Wired broadband penetration in the United States can be modeled by

$$P(t) = \frac{29.842}{1 + 12.502(1.642)^{-t}} \quad (t \geq 0),$$

where t is time in years since 2000.[1]

---

[1] See Example 2 in Section 9.4. Broadband penetration is the number of broadband installations divided by the total population. Source for data: Organization for Economic Cooperation and Development (OECD) Directorate for Science, Technology, and Industry, table of Historical Penetration Rates, June 2010, downloaded April 2011 from www.oecd.org/sti /ict/broadband.

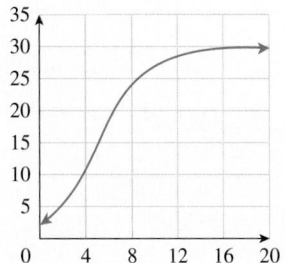

Figure 8

a. Estimate $\lim_{t \to +\infty} P(t)$ and interpret the answer.

b. Estimate $\lim_{t \to 0^+} P(t)$ and interpret the answer.

Solution

a. Figure 8 shows a plot of $P(t)$ for $0 \le t \le 20$. Using either the numerical or the graphical approach, we find

$$\lim_{t \to +\infty} P(t) = \lim_{t \to +\infty} \frac{29.842}{1 + 12.502(1.642)^{-t}} \approx 30.$$

(The actual limit is 29.842. Why?) Thus, in the long term (as t gets larger and larger), broadband penetration in the United States is expected to approach 30%; that is, the number of installations is expected to approach 30% of the total population.

b. The limit here is

$$\lim_{t \to 0^+} P(t) = \lim_{t \to 0^+} \frac{29.842}{1 + 12.502(1.642)^{-t}} \approx 2.21.$$

(Notice that in this case, we can simply put $t = 0$ to evaluate this limit.) Thus, the closer t gets to 0 (representing 2000) from the right, the closer $P(t)$ gets to 2.21%, meaning that, in 2000, broadband penetration was about 2.2% of the population.

FAQs

Determining when a Limit Does or Does Not Exist

Q: *If I substitute $x = a$ in the formula for a function and find that the function is defined there, it means that $\lim_{x \to a} f(x)$ exists and equals f(a), right?*

A: Correct, provided the function is specified by *a single algebraic formula* and is not, say, piecewise-defined. We shall say more about this in the next two sections.

Q: *If I substitute $x = a$ in the formula for a function and find that the function is not defined there, it means that $\lim_{x \to a} f(x)$ does not exist, right?*

A: Wrong. The limit may still exist, as in Example 1, or may not exist, as in Example 2. In general, whether or not $\lim_{x \to a} f(x)$ exists has nothing to do with f(a), but rather the values of f when x is *very close to, but not equal to a.*

Q: *Is there a quick and easy way of telling from a graph whether $\lim_{x \to a} f(x)$ exists?*

A: Yes. If you cover up the portion of the graph corresponding to $x = a$ and it appears as though the visible part of the graph could be made into a continuous curve by filling in a suitable point at $x = a$, then the limit exists. (The "suitable point" need not be (a, f(a)).) Otherwise, it does not. Try this method with the curves in Example 4.

10.1   EXERCISES

▼ more advanced   ◆ challenging
🚹 indicates exercises that should be solved using technology

Estimate the limits in Exercises 1–20 numerically.
HINT [See Examples 1–3.]

1. $\lim\limits_{x\to 0} \dfrac{x^2}{x+1}$

2. $\lim\limits_{x\to 0} \dfrac{x-3}{x-1}$

3. $\lim\limits_{x\to 2} \dfrac{x^2-4}{x-2}$

4. $\lim\limits_{x\to 2} \dfrac{x^2-1}{x-2}$

5. $\lim\limits_{x\to -1} \dfrac{x^2+1}{x+1}$

6. $\lim\limits_{x\to -1} \dfrac{x^2+2x+1}{x+1}$

7. $\lim\limits_{x\to +\infty} \dfrac{3x^2+10x-1}{2x^2-5x}$

8. $\lim\limits_{x\to +\infty} \dfrac{6x^2+5x+100}{3x^2-9}$

9. $\lim\limits_{x\to -\infty} \dfrac{x^5-1,000x^4}{2x^5+10,000}$

10. $\lim\limits_{x\to -\infty} \dfrac{x^6+3,000x^3+1,000,000}{2x^6+1,000x^3}$

11. $\lim\limits_{x\to +\infty} \dfrac{10x^2+300x+1}{5x+2}$

12. $\lim\limits_{x\to +\infty} \dfrac{2x^4+20x^3}{1,000x^6+6}$

13. $\lim\limits_{x\to +\infty} \dfrac{10x^2+300x+1}{5x^3+2}$

14. $\lim\limits_{x\to +\infty} \dfrac{2x^4+20x^3}{1,000x^3+6}$

15. $\lim\limits_{x\to 2} e^{x-2}$

16. $\lim\limits_{x\to +\infty} e^{-x}$

17. $\lim\limits_{x\to +\infty} xe^{-x}$

18. $\lim\limits_{x\to -\infty} xe^{x}$

19. $\lim\limits_{x\to -\infty} (x^{10}+2x^5+1)e^{x}$

20. $\lim\limits_{x\to +\infty} (x^{50}+x^{30}+1)e^{-x}$

In each of Exercises 21–34, the graph of f is given. Use the graph to compute the quantities asked for. HINT [See Examples 4–5.]

21. a. $\lim\limits_{x\to 1} f(x)$ **b.** $\lim\limits_{x\to -1} f(x)$

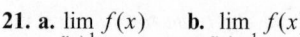

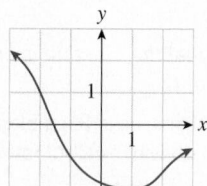

22. a. $\lim\limits_{x\to -1} f(x)$ **b.** $\lim\limits_{x\to 1} f(x)$

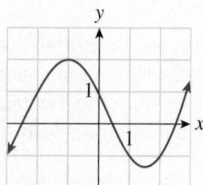

23. a. $\lim\limits_{x\to 0} f(x)$ **b.** $\lim\limits_{x\to 2} f(x)$ **c.** $\lim\limits_{x\to -\infty} f(x)$ **d.** $\lim\limits_{x\to +\infty} f(x)$

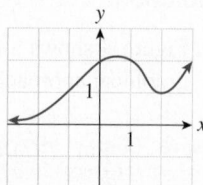

24. a. $\lim\limits_{x\to -1} f(x)$ **b.** $\lim\limits_{x\to 1} f(x)$ **c.** $\lim\limits_{x\to +\infty} f(x)$ **d.** $\lim\limits_{x\to -\infty} f(x)$

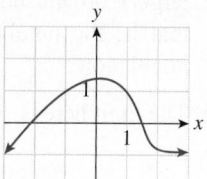

25. a. $\lim\limits_{x\to 2} f(x)$ **b.** $\lim\limits_{x\to 0^+} f(x)$ **c.** $\lim\limits_{x\to 0^-} f(x)$
d. $\lim\limits_{x\to 0} f(x)$ **e.** $f(0)$ **f.** $\lim\limits_{x\to -\infty} f(x)$

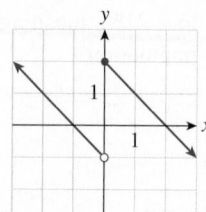

26. a. $\lim\limits_{x\to 3} f(x)$ **b.** $\lim\limits_{x\to 1^+} f(x)$ **c.** $\lim\limits_{x\to 1^-} f(x)$
d. $\lim\limits_{x\to 1} f(x)$ **e.** $f(1)$ **f.** $\lim\limits_{x\to +\infty} f(x)$

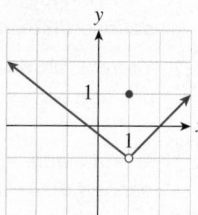

27. a. $\lim\limits_{x\to -2} f(x)$ **b.** $\lim\limits_{x\to -1^+} f(x)$ **c.** $\lim\limits_{x\to -1^-} f(x)$
d. $\lim\limits_{x\to -1} f(x)$ **e.** $f(-1)$ **f.** $\lim\limits_{x\to +\infty} f(x)$

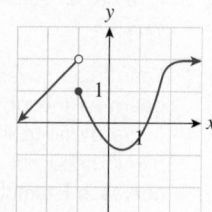

28. a. $\lim\limits_{x \to -1} f(x)$ **b.** $\lim\limits_{x \to 0^+} f(x)$ **c.** $\lim\limits_{x \to 0^-} f(x)$
d. $\lim\limits_{x \to 0} f(x)$ **e.** $f(0)$ **f.** $\lim\limits_{x \to -\infty} f(x)$

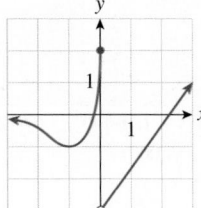

29. a. $\lim\limits_{x \to -1^+} f(x)$ **b.** $\lim\limits_{x \to -1^-} f(x)$ **c.** $\lim\limits_{x \to -1} f(x)$ **d.** $f(-1)$

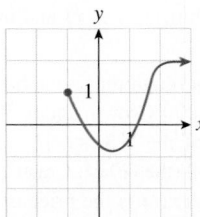

30. a. $\lim\limits_{x \to 0^+} f(x)$ **b.** $\lim\limits_{x \to 0^-} f(x)$ **c.** $\lim\limits_{x \to 0} f(x)$ **d.** $f(0)$

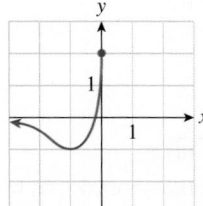

31. a. $\lim\limits_{x \to -1} f(x)$ **b.** $\lim\limits_{x \to 0^+} f(x)$ **c.** $\lim\limits_{x \to 0^-} f(x)$
d. $\lim\limits_{x \to 0} f(x)$ **e.** $f(0)$ **f.** $\lim\limits_{x \to +\infty} f(x)$

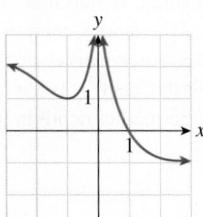

32. a. $\lim\limits_{x \to 1} f(x)$ **b.** $\lim\limits_{x \to 0^+} f(x)$ **c.** $\lim\limits_{x \to 0^-} f(x)$
d. $\lim\limits_{x \to 0} f(x)$ **e.** $f(0)$ **f.** $\lim\limits_{x \to -\infty} f(x)$

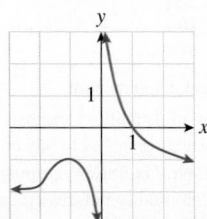

33. a. $\lim\limits_{x \to -1} f(x)$ **b.** $\lim\limits_{x \to 0^+} f(x)$ **c.** $\lim\limits_{x \to 0^-} f(x)$
d. $\lim\limits_{x \to 0} f(x)$ **e.** $f(0)$ **f.** $f(-1)$

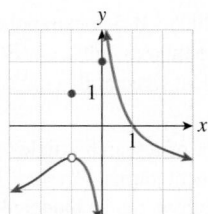

34. a. $\lim\limits_{x \to 0^-} f(x)$ **b.** $\lim\limits_{x \to 1^+} f(x)$ **c.** $\lim\limits_{x \to 0} f(x)$
d. $\lim\limits_{x \to 1} f(x)$ **e.** $f(0)$ **f.** $f(1)$

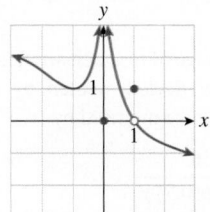

APPLICATIONS

35. *Economic Growth* The value of sold goods in Mexico can be approximated by

$$v(t) = 210 - 62e^{-0.05t} \text{ trillion pesos per month} \quad (t \geq 0),$$

where t is time in months since January 2005.[2] Numerically estimate $\lim\limits_{t \to +\infty} v(t)$ and interpret the answer. HINT [See Example 6.]

36. *Economic Growth* The number of housing starts for single family homes in the U.S. can be approximated by

$$n(t) = 400 + 1{,}420e^{-0.48t} \text{ thousand housing starts per year} \quad (t \geq 0),$$

where t is time in years since January 2005.[3] Numerically estimate $\lim\limits_{t \to +\infty} n(t)$ and interpret the answer.

37. *Revenue* The annual revenue of Amazon.com for the period 2006–2010 could be approximated by

$$P(t) = 870t^2 + 2{,}200t + 11{,}000 \text{ million dollars } (0 \leq t \leq 4),$$

where t is time in years since 2006.[4] If one extrapolates the function and numerically estimates $\lim\limits_{t \to +\infty} P(t)$, what does the answer suggest about Amazon's revenue?

---

[2]Source: Instituto Nacional de Estadística y Geografía (INEGI), www.inegi.org.mx.
[3]Source for data: www.census.gov.
[4]Source for data: www.wikinvest.com.

38. *Net Income* The annual net income of General Electric for the period 2005–2010 could be approximated by

$$P(t) = -2.0t^2 + 6.6t + 16 \text{ billion dollars} \quad (0 \le t \le 5),$$

where t is time in years since 2005.[5] If one extrapolates the function and numerically estimates $\lim_{t\to+\infty} P(t)$, what does the answer suggest about General Electric's net income?

39. *Scientific Research* The number of research articles per year, in thousands, in the prominent journal *Physical Review* written by researchers in Europe can be modeled by

$$A(t) = \frac{7.0}{1 + 5.4(1.2)^{-t}},$$

where t is time in years ($t = 0$ represents 1983).[6] Numerically estimate $\lim_{t\to+\infty} A(t)$ and interpret the answer.

40. *Scientific Research* The percentage of research articles in the prominent journal *Physical Review* written by researchers in the United States can be modeled by

$$A(t) = 25 + \frac{36}{1 + 0.6(0.7)^{-t}},$$

where t is time in years ($t = 0$ represents 1983).[7] Numerically estimate $\lim_{t\to+\infty} A(t)$ and interpret the answer.

41. *SAT Scores by Income* The following bar graph shows U.S. math SAT scores as a function of household income:[8]

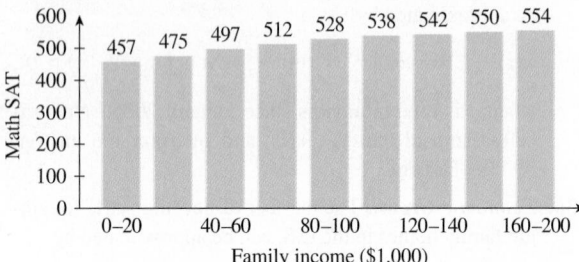

Family income ($1,000)

These data can be modeled by

$$S(x) = 573 - 133(0.987)^x,$$

where $S(x)$ is the average math SAT score of students whose household income is x thousand dollars per year. Numerically estimate $\lim_{x\to+\infty} S(x)$ and interpret the answer.

42. *SAT Scores by Income* The following bar graph shows U.S. critical reading SAT scores as a function of household income:[9]

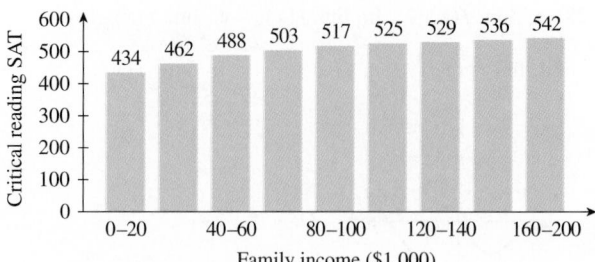

Family income ($1,000)

These data can be modeled by

$$S(x) = 550 - 136(0.985)^x,$$

where $S(x)$ is the average critical reading SAT score of students whose household income is x thousand dollars per year. Numerically estimate $\lim_{x\to+\infty} S(x)$ and interpret the answer.

43. *Flash Crash* The graph shows a rough representation of what happened to the Russel 1000 Growth Index Fund (IWF) stock price on the day of the U.S. stock market crash at 2:45 pm on May 6, 2010 (the "Flash Crash"; t is the time of the day in hours, and $r(t)$ is the price of the stock in dollars).[10]

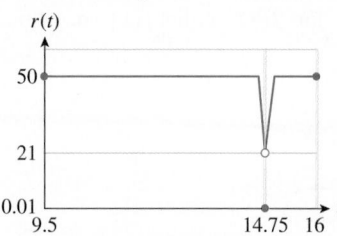

a. Compute the following (if a limit does not exist, say why):

$$\lim_{t\to14.75^-} r(t), \quad \lim_{t\to14.75^+} r(t), \quad \lim_{t\to14.75} r(t), \quad r(14.75).$$

b. What do the answers to part (a) tell you about the IWF stock price?

44. *Flash Crash* The graph shows a rough representation of the (aggregate) market depth[11] of the stocks comprising the S&P 500 on the day of the U.S. stock market crash at 2:45 pm on May 6, 2010 (the "Flash Crash"; t is the time of the day in hours, and $m(t)$ is the market depth in millions of shares).

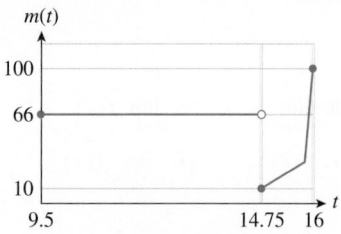

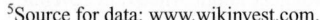

[5]Source for data: www.wikinvest.com.

[6]Based on data from 1983 to 2003. Source: The American Physical Society/*New York Times*, May 3, 2003, p. A1.

[7]*Ibid.*

[8]2009 data. Source: College Board/*New York Times* http://economix.blogs.nytimes.com.

[9]*Ibid.*

[10]The actual graph can be seen at http://seekingalpha.com.

[11]The market depth of a stock is a measure of its ability to withstand relatively large market orders, and is measured in orders to buy or sell the given stock. Source for data on graph: *Findings Regarding the Market Events of May 6, 2010,* U.S. Commodity Futures Trading Commission, U.S. Securities & Exchange Commission.

(E)

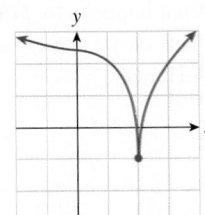

(F)

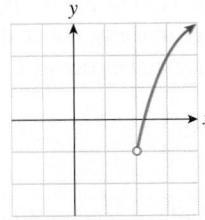

In Exercises 17–24, use a graph of f or some other method to determine what, if any, value to assign to f(a) to make f continuous at x = a. HINT [See Example 2.]

17. $f(x) = \dfrac{x^2 - 2x + 1}{x - 1}; a = 1$

18. $f(x) = \dfrac{x^2 + 3x + 2}{x + 1}; a = -1$

19. $f(x) = \dfrac{x}{3x^2 - x}; a = 0$

20. $f(x) = \dfrac{x^2 - 3x}{x + 4}; a = -4$

21. $f(x) = \dfrac{3}{3x^2 - x}; a = 0$

22. $f(x) = \dfrac{x - 1}{x^3 - 1}; a = 1$

23. $f(x) = \dfrac{1 - e^x}{x}; a = 0$

24. $f(x) = \dfrac{1 + e^x}{1 - e^x}; a = 0$

In Exercises 25–34, use a graph to determine whether the given function is continuous on its domain. If it is not continuous on its domain, list the points of discontinuity. HINT [See Example 1.]

25. $f(x) = |x|$

26. $f(x) = \dfrac{|x|}{x}$

27. $g(x) = \dfrac{1}{x^2 - 1}$

28. $g(x) = \dfrac{x - 1}{x + 2}$

29. $f(x) = \begin{cases} x + 2 & \text{if } x < 0 \\ 2x - 1 & \text{if } x \geq 0 \end{cases}$

30. $f(x) = \begin{cases} 1 - x & \text{if } x \leq 1 \\ x - 1 & \text{if } x > 1 \end{cases}$

31. $h(x) = \begin{cases} \dfrac{|x|}{x} & \text{if } x \neq 0 \\ 0 & \text{if } x = 0 \end{cases}$

32. $h(x) = \begin{cases} \dfrac{1}{x^2} & \text{if } x \neq 0 \\ 2 & \text{if } x = 0 \end{cases}$

33. $g(x) = \begin{cases} x + 2 & \text{if } x < 0 \\ 2x + 2 & \text{if } x \geq 0 \end{cases}$

34. $g(x) = \begin{cases} 1 - x & \text{if } x \leq 1 \\ x + 1 & \text{if } x > 1 \end{cases}$

COMMUNICATION AND REASONING EXERCISES

35. If a function is continuous on its domain, is it continuous at every real number? Explain.

36. True or false? The graph of a function that is continuous on its domain is a continuous curve with no breaks in it. Explain your answer.

37. True or false? The graph of a function that is continuous at every real number is a continuous curve with no breaks in it. Explain your answer.

38. True or false? If the graph of a function is a continuous curve with no breaks in it, then the function is continuous on its domain. Explain your answer.

39. ▼ Give a formula for a function that is continuous on its domain but whose graph consists of three distinct curves.

40. ▼ Give a formula for a function that is not continuous at $x = -1$ but is not discontinuous there either.

41. ▼ Draw the graph of a function that is discontinuous at every integer.

42. ▼ Draw the graph of a function that is continuous on its domain but whose graph has a break at every integer.

43. ▼ Describe a real-life scenario in the stock market that can be modeled by a discontinuous function.

44. ▼ Describe a real-life scenario in your room that can be modeled by a discontinuous function.

10.3 Limits and Continuity: Algebraic Viewpoint

Although numerical and graphical estimation of limits is effective, the estimates these methods yield may not be perfectly accurate. The algebraic method, when it can be used, will always yield an exact answer. Moreover, algebraic analysis of a function often enables us to take a function apart and see "what makes it tick."

Let's start with the function $f(x) = 2 + x$ and ask: What happens to $f(x)$ as x approaches 3? To answer this algebraically, notice that as x gets closer and closer to 3, the quantity $2 + x$ must get closer and closer to $2 + 3 = 5$. Hence,

$$\lim_{x \to 3} f(x) = \lim_{x \to 3}(2 + x) = 2 + 3 = 5$$

Q: *Is that all there is to the algebraic method? Just substitute $x = a$?*

A: Under certain circumstances: Notice that by substituting $x = 3$ we *evaluated the function at $x = 3$.* In other words, we relied on the fact that

$$\lim_{x \to 3} f(x) = f(3).$$

In Section 10.2 we said that a function satisfying this equation is *continuous* at $x = 3$.

Thus,

If we know that the function f is continuous at a point a, we can compute $\lim_{x \to a} f(x)$ by simply substituting $x = a$ into $f(x)$.

To use this fact, we need to know how to recognize continuous functions when we see them. Geometrically, they are easy to spot: A function is continuous at $x = a$ if its graph has no break at $x = a$. Algebraically, a large class of functions are known to be continuous on their domains—those, roughly speaking, that are *specified by a single formula.*

We can be more precise: A **closed-form function** is any function that can be obtained by combining constants, powers of x, exponential functions, radicals, logarithms, absolute values, trigonometric functions (and some other functions we do not encounter in this text) into a *single* mathematical formula by means of the usual arithmetic operations and composition of functions. (They can be as complicated as we like.)

Closed-Form Functions

A function is **written in closed form** if it is specified by combining constants, powers of x, exponential functions, radicals, logarithms, absolute values, trigonometric functions (and some other functions we do not encounter in this text) into a *single* mathematical formula by means of the usual arithmetic operations and composition of functions. A **closed-form function** is any function that can be written in closed form.

Quick Examples

1. $3x^2 - |x| + 1$, $\dfrac{\sqrt{x^2 - 1}}{6x - 1}$, $e^{-\frac{4x^2-1}{x}}$, and $\sqrt{\log_3(x^2 - 1)}$ are written in closed form, so they are all closed-form functions.

2. $f(x) = \begin{cases} -1 & \text{if } x \leq -1 \\ x^2 + x & \text{if } -1 < x \leq 1 \\ 2 - x & \text{if } 1 < x \leq 2 \end{cases}$ is not written in closed-form because $f(x)$ is not expressed by a *single* mathematical formula.*

* It is possible to rewrite some piecewise-defined functions in closed form (using a single formula), but not this particular function, so $f(x)$ is not a closed-form function.

What is so special about closed-form functions is the following theorem.

> ## Theorem 10.1 Continuity of Closed-Form Functions
>
> *Every closed-form function is continuous on its domain.* Thus, if f is a closed-form function and $f(a)$ is defined, then $\lim_{x \to a} f(x)$ exists, and equals $f(a)$. (When a is an endpoint of the domain of f, we understand the limit to be the appropriate one-sided limit.)
>
> ### Quick Example
>
> > $f(x) = 1/x$ is a closed-form function, and its natural domain consists of all real numbers except 0. Thus, f is continuous at every nonzero real number. That is,
> >
> > $$\lim_{x \to a} \frac{1}{x} = \frac{1}{a}$$
> >
> > provided $a \neq 0$.

Mathematics majors spend a great deal of time studying the proof of this theorem. We ask you to accept it without proof.

EXAMPLE 1 Limit of a Closed-Form Function

Evaluate the following limits algebraically:

a. $\displaystyle\lim_{x \to 1} \frac{x^3 - 8}{x - 2}$ **b.** $\displaystyle\lim_{x \to 2} \frac{x^3 - 8}{x - 2}$.

Solution

a. First, notice that $(x^3 - 8)/(x - 2)$ is a closed-form function because it is specified by a single algebraic formula. Also, $x = 1$ is in the domain of this function. Therefore the theorem applies, and

$$\lim_{x \to 1} \frac{x^3 - 8}{x - 2} = \frac{1^3 - 8}{1 - 2} = 7.$$

b. Although $(x^3 - 8)/(x - 2)$ is a closed-form function, $x = 2$ is not in its domain. Thus, the theorem does not apply and we cannot obtain the limit by substitution. However—and this is the key to finding such limits—*some preliminary algebraic simplification will allow us to obtain a closed-form function with $x = 2$ in its domain*. To do this, notice first that the numerator can be factored as

$$x^3 - 8 = (x - 2)(x^2 + 2x + 4).$$

Thus,

$$\frac{x^3 - 8}{x - 2} = \frac{(x - 2)(x^2 + 2x + 4)}{x - 2} = x^2 + 2x + 4.$$

Once we have canceled the offending $(x - 2)$ in the denominator, we are left with a closed-form function *with 2 in its domain.* Thus,

$$\lim_{x \to 2} \frac{x^3 - 8}{x - 2} = \lim_{x \to 2}(x^2 + 2x + 4)$$

$$= 2^2 + 2(2) + 4 = 12. \quad \text{Substitute } x = 2.$$

This confirms the answer we found numerically in Example 1 in Section 10.1.

➡ **Before we go on...** Notice that in Example 1(b) before simplification, the substitution $x = 2$ yields

$$\frac{x^3 - 8}{x - 2} = \frac{8 - 8}{2 - 2} = \frac{0}{0}.$$

Worse than the fact that 0/0 is undefined, it also conveys absolutely no information as to what the limit might be. (The limit turned out to be 12!) We therefore call the expression 0/0 an **indeterminate form.** Once simplified, the function became $x^2 + 2x + 4$, which, upon the substitution $x = 2$, yielded 12—no longer an indeterminate form. In general, we have the following rule of thumb:

If the substitution $x = a$ yields the indeterminate form 0/0, try simplifying by the method in Example 1.

We will say more about indeterminate forms in Example 2. ■

Q : *There is something suspicious about Example 1(b). If 2 was not in the domain before simplifying but was in the domain after simplifying, we must have changed the function, right?*

A : Correct. In fact, when we said that

$$\frac{x^3 - 8}{x - 2} = x^2 + 2x + 4$$

Domain excludes 2   Domain includes 2

we were lying a little bit. What we really meant is that these two expressions are equal *where both are defined.* The functions $(x^3 - 8)/(x - 2)$ and $x^2 + 2x + 4$ are different functions. The difference is that $x = 2$ is not in the domain of $(x^3 - 8)/(x - 2)$ and is in the domain of $x^2 + 2x + 4$. Since $\lim_{x \to 2} f(x)$ explicitly *ignores* any value that f may have at 2, this does not affect the limit. From the point of view of the limit at 2, these functions *are* equal. In general we have the following rule.

Functions with Equal Limits

If $f(x) = g(x)$ for all x except possibly $x = a$, then

$$\lim_{x \to a} f(x) = \lim_{x \to a} g(x).$$

Quick Example

$$\frac{x^2 - 1}{x - 1} = x + 1 \text{ for all } x \text{ except } x = 1. \quad \text{Write } \frac{x^2 - 1}{x - 1} \text{ as } \frac{(x + 1)(x - 1)}{x - 1}$$

$$\text{and cancel the } (x - 1).$$

Therefore,

$$\lim_{x \to 1} \frac{x^2 - 1}{x - 1} = \lim_{x \to 1}(x + 1) = 1 + 1 = 2.$$

Q : *How do we find* $\lim_{x \to a} f(x)$ *when* $x = a$ *is not in the domain of the function f and we cannot simplify the given function to make a a point of the domain?*

A : In such a case, it might be necessary to analyze the function by some other method, such as numerically or graphically. However, if we do not obtain the indeterminate form 0/0 upon substitution, we can often say what the limit is, as the following example shows.

EXAMPLE 2 Limit of a Closed-Form Function at a Point Not in Its Domain: The Determinate Form *k*/0

Evaluate the following limits, if they exist:

a. $\displaystyle\lim_{x \to 1^+} \frac{x^2 - 4x + 1}{x - 1}$ **b.** $\displaystyle\lim_{x \to 1} \frac{x^2 - 4x + 1}{x - 1}$ **c.** $\displaystyle\lim_{x \to 1} \frac{x^2 - 4x + 1}{x^2 - 2x + 1}$

Solution

a. Although the function $f(x) = \dfrac{x^2 - 4x + 1}{x - 1}$ is a closed-form function, $x = 1$ is not in its domain. Notice that substituting $x = 1$ gives

$$\frac{x^2 - 4x + 1}{x - 1} = \frac{1^2 - 4 + 1}{1 - 1} = \frac{-2}{0} \qquad \text{The \textbf{determinate} form } \frac{k}{0}$$

which, although not defined, conveys important information to us: As x gets closer and closer to 1, the numerator approaches -2 and the denominator gets closer and closer to 0. Now, if we divide a number close to -2 by a number close to zero, we get a number of large absolute value; for instance,

$$\frac{-2.1}{0.0001} = -21,000 \qquad \text{and} \qquad \frac{-2.1}{-0.0001} = 21,000$$

$$\frac{-2.01}{0.00001} = -201,000 \qquad \text{and} \qquad \frac{-2.01}{-0.00001} = 201,000.$$

(Compare Example 5 in Section 10.1.) In our limit for part (a), x is approaching 1 from the right, so the denominator $x - 1$ is positive (as x is to the right of 1). Thus we have the scenario illustrated previously on the left, and we can conclude that

$$\lim_{x \to 1^+} \frac{x^2 - 4x + 1}{x - 1} = -\infty. \qquad \text{Think of this as } \frac{-2}{0^+} = -\infty.$$

b. This time, x could be approaching 1 from either side. We already have, from part (a)

$$\lim_{x \to 1^+} \frac{x^2 - 4x + 1}{x - 1} = -\infty.$$

The same reasoning we used in part (a) gives

$$\lim_{x \to 1^-} \frac{x^2 - 4x + 1}{x - 1} = +\infty \qquad \text{Think of this as } \frac{-2}{0^-} = +\infty.$$

because now the denominator is negative and still approaching zero while the numerator still approaches -2 and therefore is also negative. (See the numerical calculations above on the right.) Because the left and right limits do not agree, we conclude that

$$\lim_{x \to 1} \frac{x^2 - 4x + 1}{x - 1} \quad \text{does not exist.}$$

c. First notice that the denominator factors:

$$\lim_{x \to 1} \frac{x^2 - 4x + 1}{x^2 - 2x + 1} = \lim_{x \to 1} \frac{x^2 - 4x + 1}{(x - 1)^2}.$$

As x approaches 1, the numerator approaches -2 as before, and the denominator approaches 0. However, this time, the denominator $(x - 1)^2$, being a square, is ≥ 0, regardless of from which side x is approaching 1. Thus, the entire function is negative as x approaches 1, and

$$\lim_{x \to 1} \frac{x^2 - 4x + 1}{(x - 1)^2} = -\infty. \qquad \frac{-2}{0^+} = -\infty$$

➡ **Before we go on...** In general, the determinate forms $\dfrac{k}{0^+}$ and $\dfrac{k}{0^-}$ will always yield $\pm\infty$, with the sign depending on the sign of the overall expression as $x \to a$. (When we write the form $\dfrac{k}{0}$ we always mean $k \neq 0$.) This and other determinate forms are discussed further after Example 4.

Figure 17 shows the graphs of $\dfrac{x^2 - 4x + 1}{x - 1}$ and $\dfrac{x^2 - 4x + 1}{(x - 1)^2}$ from Example 2. You should check that results we obtained above agree with a geometric analysis of these graphs near $x = 1$.

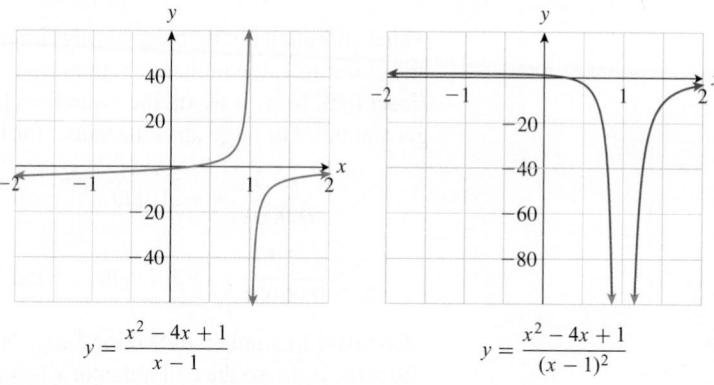

$$y = \frac{x^2 - 4x + 1}{x - 1} \qquad\qquad y = \frac{x^2 - 4x + 1}{(x - 1)^2}$$

Figure 17

We can also use algebraic techniques to analyze functions that are not given in closed form.

EXAMPLE 3 Functions Not Written in Closed Form

For which values of x are the following piecewise defined functions continuous?

a. $f(x) = \begin{cases} x^2 + 2 & \text{if } x < 1 \\ 2x - 1 & \text{if } x \geq 1 \end{cases}$ **b.** $g(x) = \begin{cases} x^2 - x + 1 & \text{if } x \leq 0 \\ 1 - x & \text{if } 0 < x \leq 1 \\ x - 3 & \text{if } x > 1 \end{cases}$

Solution

a. The function $f(x)$ is given in closed form over the intervals $(-\infty, 1)$ and $[1, +\infty)$. At $x = 1$, $f(x)$ suddenly switches from one closed-form formula to another, so

$x = 1$ is the only place where there is a potential problem with continuity. To investigate the continuity of $f(x)$ at $x = 1$, let's calculate the limit there:

$$\lim_{x \to 1^-} f(x) = \lim_{x \to 1^-} (x^2 + 2) \quad f(x) = x^2 + 2 \text{ for } x < 1.$$
$$= (1)^2 + 2 = 3 \quad x^2 + 2 \text{ is closed-form.}$$

$$\lim_{x \to 1^+} f(x) = \lim_{x \to 1^+} (2x - 1) \quad f(x) = 2x - 1 \text{ for } x > 1.$$
$$= 2(1) - 1 = 1. \quad 2x - 1 \text{ is closed-form.}$$

Because the left and right limits are different, $\lim_{x \to 1} f(x)$ does not exist, and so $f(x)$ is discontinuous at $x = 1$.

b. The only potential points of discontinuity for $g(x)$ occur at $x = 0$ and $x = 1$:

$$\lim_{x \to 0^-} g(x) = \lim_{x \to 0^-} (x^2 - x + 1) = 1$$
$$\lim_{x \to 0^+} g(x) = \lim_{x \to 0^+} (1 - x) = 1.$$

Thus, $\lim_{x \to 0} g(x) = 1$. Further, $g(0) = 0^2 - 0 + 1 = 1$ from the formula, and so

$$\lim_{x \to 0} g(x) = g(0),$$

which shows that $g(x)$ is continuous at $x = 0$. At $x = 1$ we have

$$\lim_{x \to 1^-} g(x) = \lim_{x \to 1^-} (1 - x) = 0$$
$$\lim_{x \to 1^+} g(x) = \lim_{x \to 1^+} (x - 3) = -2$$

so that $\lim_{x \to 1} g(x)$ does not exist. Thus, $g(x)$ is discontinuous at $x = 1$. We conclude that $g(x)$ is continuous at every real number x except $x = 1$.

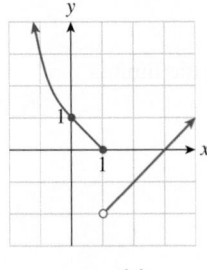

$y = g(x)$

Figure 18

➡ **Before we go on...** Figure 18 shows the graph of g from Example 3(b). Notice how the discontinuity at $x = 1$ shows up as a break in the graph, whereas at $x = 0$ the two pieces "fit together" at the point $(0, 1)$. ∎

Limits at Infinity

Let's look once again at some limits similar to those in Examples 3 and 6 in Section 10.1.

EXAMPLE 4 **Limits at Infinity**

Compute the following limits, if they exist:

a. $\lim\limits_{x \to +\infty} \dfrac{2x^2 - 4x}{x^2 - 1}$ **b.** $\lim\limits_{x \to -\infty} \dfrac{2x^2 - 4x}{x^2 - 1}$

c. $\lim\limits_{x \to +\infty} \dfrac{-x^3 - 4x}{2x^2 - 1}$ **d.** $\lim\limits_{x \to +\infty} \dfrac{2x^2 - 4x}{5x^3 - 3x + 5}$

e. $\lim\limits_{t \to +\infty} (e^{0.1t} - 20)$ **f.** $\lim\limits_{t \to +\infty} \dfrac{80}{1 + 2.2(3.68)^{-t}}$

Solution a and **b.** While calculating the values for the tables used in Example 3 in Section 10.1, you might have noticed that the highest power of x in both the numerator and denominator dominated the calculations. For instance, when $x = 100,000$, the term $2x^2$ in the numerator has the value of 20,000,000,000, whereas the term $4x$ has the comparatively insignificant value of 400,000. Similarly, the term x^2 in the denominator overwhelms the term -1. In other words, for large values of x (or negative values with large magnitude),

$$\frac{2x^2 - 4x}{x^2 - 1} \approx \frac{2x^2}{x^2} \qquad \text{Use only the highest powers top and bottom.}$$

$$= 2.$$

Therefore,

$$\lim_{x \to \pm\infty} \frac{2x^2 - 4x}{x^2 - 1} = \lim_{x \to \pm\infty} \frac{2x^2}{x^2}$$

$$= \lim_{x \to \pm\infty} 2 = 2.$$

The procedure of using only the highest powers of x to compute the limit is stated formally and justified after this example.

c. Applying the previous technique of looking only at highest powers gives

$$\lim_{x \to +\infty} \frac{-x^3 - 4x}{2x^2 - 1} = \lim_{x \to +\infty} \frac{-x^3}{2x^2} \qquad \text{Use only the highest powers top and bottom.}$$

$$= \lim_{x \to +\infty} \frac{-x}{2}. \qquad \text{Simplify.}$$

As x gets large, $-x/2$ gets large in magnitude but negative, so the limit is

$$\lim_{x \to +\infty} \frac{-x}{2} = -\infty. \qquad \frac{-\infty}{2} = -\infty \text{ (See below.)}$$

d. $\lim_{x \to +\infty} \dfrac{2x^2 - 4x}{5x^3 - 3x + 5} = \lim_{x \to +\infty} \dfrac{2x^2}{5x^3}$ \qquad Use only the highest powers top and bottom.

$$= \lim_{x \to +\infty} \frac{2}{5x}.$$

As x gets large, $2/(5x)$ gets close to zero, so the limit is

$$\lim_{x \to +\infty} \frac{2}{5x} = 0. \qquad \frac{2}{\infty} = 0 \text{ (See below.)}$$

e. Here we do not have a ratio of polynomials. However, we know that, as t becomes large and positive, so does $e^{0.1t}$, and hence also $e^{0.1t} - 20$. Thus,

$$\lim_{t \to +\infty} (e^{0.1t} - 20) = +\infty \qquad e^{+\infty} = +\infty \text{ (See below.)}$$

f. As $t \to +\infty$, the term $(3.68)^{-t} = \dfrac{1}{3.68^t}$ in the denominator, being 1 divided by a very large number, approaches zero. Hence the denominator $1 + 2.2(3.68)^{-t}$ approaches $1 + 2.2(0) = 1$ as $t \to +\infty$. Thus,

$$\lim_{t \to +\infty} \frac{80}{1 + 2.2(3.68)^{-t}} = \frac{80}{1 + 2.2(0)} = 80 \qquad (3.68)^{-\infty} = 0 \text{ (See below.)}$$

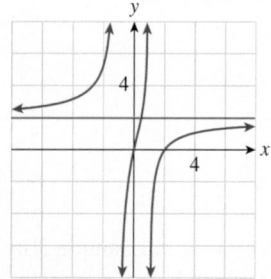

Figure 19

➡ **Before we go on...** Let's now look at the graph of the function $\dfrac{2x^2 - 4x}{x^2 - 1}$ in parts (a) and (b) of Example 4. We say that the graph of f has a **horizontal asymptote** at $y = 2$ because of the limits we have just calculated. This means that the graph approaches the horizontal line $y = 2$ far to the right or left (in this case, to both the right and left). Figure 19 shows the graph of f together with the line $y = 2$.

The graph reveals some additional interesting information: as $x \to 1^+$, $f(x) \to -\infty$, and as $x \to 1^-$, $f(x) \to +\infty$. Thus,

$$\lim_{x \to 1} f(x) \text{ does not exist.}$$

See if you can determine what happens as $x \to -1$.

If you graph the functions in parts (d) and (f) of Example 4, you will again see a horizontal asymptote. Do the limits in parts (c) and (e) show horizontal asymptotes? ∎

It is worthwhile looking again at what we did in each of the limits in Example 4:

a and **b.** We saw that $\dfrac{2x^2 - 4x}{x^2 - 1} \approx \dfrac{2x^2}{x^2}$, and then we canceled the x^2. Notice that, before we cancel, letting x approach $\pm\infty$ in the numerator and denominator yields the ratio $\dfrac{\infty}{\infty}$, which, like $\dfrac{0}{0}$, is another *indeterminate form*, and indicates to us that further work is needed—in this case cancellation—before we can write down the limit.

c. We obtained $\dfrac{-x^3 - 4x}{2x^2 - 1} \approx \dfrac{-x^3}{2x^2}$, which results in another indeterminate form, $\dfrac{-\infty}{\infty}$, as $x \to +\infty$. Cancellation of the x^2 gave us $\dfrac{-x}{2}$, resulting in the *determinate* form $\dfrac{-\infty}{2} = -\infty$ (a very large number divided by 2 is again a very large number).

d. Here, $\dfrac{2x^2 - 4x}{5x^3 - 3x + 5} \approx \dfrac{2x^2}{5x^3} = \dfrac{2}{5x}$, and the cancellation step turns the indeterminate form $\dfrac{\infty}{\infty}$ into the determinate form $\dfrac{2}{\infty} = 0$ (dividing 2 by a very large number yields a very small number).

e. We reasoned that e raised to a large positive number is large and positive. Putting $t = +\infty$ gives us the determinate form $e^{+\infty} = +\infty$.

f. Here we reasoned that 3.68 raised to a large *negative* number is close to zero. Putting $t = -\infty$ gives us the determinate form $3.68^{-\infty} = 1/3.68^{+\infty} = 1/\infty = 0$ (see (d)).

In parts (a)–(d) of Example 4, $f(x)$ was a **rational function**: a quotient of polynomial functions. We calculated the limit of $f(x)$ at $\pm\infty$ by ignoring all powers of x in both the numerator and denominator except for the largest. Following is a theorem that justifies this procedure.

Theorem 10.2 Evaluating the Limit of a Rational Function at $\pm\infty$

If $f(x)$ has the form

$$f(x) = \frac{c_n x^n + c_{n-1} x^{n-1} + \cdots + c_1 x + c_0}{d_m x^m + d_{m-1} x^{m-1} + \cdots + d_1 x + d_0}$$

with the c_i and d_i constants ($c_n \neq 0$ and $d_m \neq 0$), then we can calculate the limit of $f(x)$ as $x \to \pm\infty$ by ignoring all powers of x except the highest in both the numerator and denominator. Thus,

$$\lim_{x \to \pm\infty} f(x) = \lim_{x \to \pm\infty} \frac{c_n x^n}{d_m x^m}.$$

Quick Examples

(See Example 4.)

1. $\lim\limits_{x \to +\infty} \dfrac{2x^2 - 4x}{x^2 - 1} = \lim\limits_{x \to +\infty} \dfrac{2x^2}{x^2} = \lim\limits_{x \to +\infty} 2 = 2$

2. $\lim\limits_{x \to +\infty} \dfrac{-x^3 - 4x}{2x^2 - 1} = \lim\limits_{x \to +\infty} \dfrac{-x^3}{2x^2} = \lim\limits_{x \to +\infty} \dfrac{-x}{2} = -\infty$

3. $\lim\limits_{x \to +\infty} \dfrac{2x^2 - 4x}{5x^3 - 3x + 5} = \lim\limits_{x \to +\infty} \dfrac{2x^2}{5x^3} = \lim\limits_{x \to +\infty} \dfrac{2}{5x} = 0$

Proof Our function $f(x)$ is a polynomial of degree n divided by a polynomial of degree m. If n happens to be larger than m, then dividing the top and bottom by the largest power x^n of x gives

$$f(x) = \frac{c_n x^n + c_{n-1} x^{n-1} + \cdots + c_1 x + c_0}{d_m x^m + d_{m-1} x^{m-1} + \cdots + d_1 x + d_0}$$

$$= \frac{c_n x^n / x^n + c_{n-1} x^{n-1} / x^n + \cdots + c_1 x / x^n + c_0 / x^n}{d_m x^m / x^n + d_{m-1} x^{m-1} / x^n + \cdots + d_1 x / x^n + d_0 / x^n}.$$

Canceling powers of x in each term and remembering that $n > m$ leaves us with

$$f(x) = \frac{c_n + c_{n-1}/x + \cdots + c_1/x^{n-1} + c_0/x^n}{d_m/x^{n-m} + d_{m-1}/x^{n-m+1} + \cdots + d_1/x^{n-1} + d_0/x^n}.$$

As $x \to \pm\infty$, all the terms shown in red approach 0, so we can ignore them in taking the limit. (The first term in the denominator happens to approach 0 as well, but we retain it for convenience.) Thus,

$$\lim_{x \to \pm\infty} f(x) = \lim_{x \to \pm\infty} \frac{c_n}{d_m/x^{n-m}} = \lim_{x \to \pm\infty} \frac{c_n x^n}{d_m x^m},$$

as required. The cases when n is smaller than m and $m = n$ are proved similarly by dividing top and bottom by the largest power of x in each case.

Some Determinate and Indeterminate Forms

The following table brings these ideas together with our observations in Example 2.

Some Determinate and Indeterminate Forms

$\dfrac{0}{0}$ and $\pm\dfrac{\infty}{\infty}$ are **indeterminate**; evaluating limits in which these arise requires simplification or further analysis.[†]

The following are **determinate** forms for any nonzero number k:

$\dfrac{k}{0^{\pm}} = \pm\infty$ $\qquad\qquad$ $\dfrac{k}{\text{Small}} = \text{Big*}$ (See Example 2.)

$k(\pm\infty) = \pm\infty$ $\qquad\qquad$ $k \times \text{Big} = \text{Big*}$

$k \pm \infty = \pm\infty$ $\qquad\qquad$ $k \pm \text{Big} = \pm\text{Big}$

[†] Some other indeterminate forms are: $\pm\infty \cdot 0$, $\infty - \infty$, and 1^{∞}. (These are not discussed in this text, but see the Communication and Reasoning exercises for this section.)

$$\pm\frac{\infty}{k} = \pm\infty \qquad\qquad \frac{\text{Big}}{k} = \text{Big*}$$

$$\pm\frac{k}{\infty} = 0 \qquad\qquad \frac{k}{\text{Big}} = \text{Small}$$

and, if $k > 1$, then

$$k^{+\infty} = +\infty \qquad\qquad k^{\text{Big positive}} = \text{Big}$$

$$k^{-\infty} = 0. \qquad\qquad k^{\text{Big negative}} = \text{Small}$$

---

*The sign gets switched in these forms if k is negative.

Quick Examples

1. $\lim\limits_{x\to 0} \dfrac{60}{2x^2} = +\infty$ $\qquad\qquad \dfrac{k}{0^+} = +\infty$

2. $\lim\limits_{x\to -1^-} \dfrac{2x-6}{x+1} = +\infty$ $\qquad\qquad \dfrac{-8}{0^-} = +\infty$

3. $\lim\limits_{x\to -\infty} 3x - 5 = -\infty$ $\qquad\qquad 3(-\infty) - 5 = -\infty - 5 = -\infty$

4. $\lim\limits_{x\to +\infty} \dfrac{2x}{60} = +\infty$ $\qquad\qquad \dfrac{2(\infty)}{60} = \infty$

5. $\lim\limits_{x\to -\infty} \dfrac{60}{2x} = 0$ $\qquad\qquad \dfrac{60}{2(-\infty)} = 0$

6. $\lim\limits_{x\to +\infty} \dfrac{60x}{2x} = 30$ $\qquad\qquad \dfrac{\infty}{\infty}$ is indeterminate but we can cancel.

7. $\lim\limits_{x\to -\infty} \dfrac{60}{e^x - 1} = \dfrac{60}{0-1} = -60$ $\quad e^{-\infty} = 0$

FAQs

Strategy for Evaluating Limits Algebraically

Q: *Is there a systematic way to evaluate a limit $\lim_{x\to a} f(x)$ algebraically?*

A: The following approach is often successful:

Case 1: a Is a Finite Number (Not $\pm\infty$)

1. Decide whether f is a closed-form function. If it is not, then find the left and right limits at the values of x where the function changes from one formula to another.

2. If f is a closed-form function, try substituting $x = a$ in the formula for $f(x)$. Then one of the following three things may happen:

   $f(a)$ is defined. Then $\lim_{x\to a} f(x) = f(a)$.

   $f(a)$ is not defined and has the indeterminate form 0/0. Try to simplify the expression for f to cancel one of the terms that gives 0.

   $f(a)$ is not defined and has one of the determinate forms listed above in the above table. Use the table to determine the limit as in the Quick Examples.

> **Case 2:** $a = \pm\infty$
>
> Remember that we can use the determinate forms $k^{+\infty} = \infty$ and $k^{-\infty} = 0$ if $k > 1$. Further, if the given function is a polynomial or ratio of polynomials, use the technique of Example 4: Focus only on the highest powers of x and then simplify to obtain either a number L, in which case the limit exists and equals L, or one of the determinate forms $\pm\infty/k = \pm\infty$ or $\pm k/\infty = 0$.

There is another technique for evaluating certain difficult limits, called *l'Hospital's rule,* but this uses derivatives, so we'll have to wait to discuss it until Section 11.1.

10.3   EXERCISES

▼ more advanced   ◆ challenging

Ⓣ indicates exercises that should be solved using technology

In Exercises 1–4, complete the given sentence.

1. The closed-form function $f(x) = \dfrac{1}{x-1}$ is continuous for all x except _____. HINT [See Quick Example on page 709.]

2. The closed-form function $f(x) = \dfrac{1}{x^2-1}$ is continuous for all x except _____. HINT [See Quick Example on page 709.]

3. The closed-form function $f(x) = \sqrt{x+1}$ has $x = 3$ in its domain. Therefore, $\lim_{x\to 3} \sqrt{x+1} = $ ___. HINT [See Example 1.]

4. The closed-form function $f(x) = \sqrt{x-1}$ has $x = 10$ in its domain. Therefore, $\lim_{x\to 10} \sqrt{x-1} = $ ___. HINT [See Example 1.]

In Exercises 5–20, determine whether the given limit leads to a determinate or indeterminate form. Evaluate the limit if it exists, or say why if not. HINT [See Example 2 and Quick Examples on page 717.]

5. $\lim_{x\to 0} \dfrac{60}{x^4}$

6. $\lim_{x\to 0} \dfrac{2x^2}{x^2}$

7. $\lim_{x\to 0} \dfrac{x^3-1}{x^3}$

8. $\lim_{x\to 0} \dfrac{-2}{x^2}$

9. $\lim_{x\to -\infty} (-x^2 + 5)$

10. $\lim_{x\to 0} \dfrac{2x^2+4}{x}$

11. $\lim_{x\to +\infty} 4^{-x}$

12. $\lim_{x\to +\infty} \dfrac{60+e^{-x}}{2-e^{-x}}$

13. $\lim_{x\to 0} \dfrac{-x^3}{3x^3}$

14. $\lim_{x\to -\infty} 3x^2 + 6$

15. $\lim_{x\to -\infty} \dfrac{-x^3}{3x^6}$

16. $\lim_{x\to +\infty} \dfrac{-x^6}{3x^3}$

17. $\lim_{x\to -\infty} \dfrac{4}{-x+2}$

18. $\lim_{x\to -\infty} e^x$

19. $\lim_{x\to -\infty} \dfrac{60}{e^x - 1}$

20. $\lim_{x\to -\infty} \dfrac{2}{2x^2+3}$

Calculate the limits in Exercises 21–74 algebraically. If a limit does not exist, say why.

21. $\lim_{x\to 0} (x+1)$

HINT [See Example 1(a).]

22. $\lim_{x\to 0} (2x-4)$

HINT [See Example 1(a).]

23. $\lim_{x\to 2} \dfrac{2+x}{x}$

24. $\lim_{x\to -1} \dfrac{4x^2+1}{x}$

25. $\lim_{x\to -1} \dfrac{x+1}{x}$

26. $\lim_{x\to 4} (x + \sqrt{x})$

27. $\lim_{x\to 8} (x - \sqrt[3]{x})$

28. $\lim_{x\to 1} \dfrac{x-2}{x+1}$

29. $\lim_{h\to 1} (h^2 + 2h + 1)$

30. $\lim_{h\to 0} (h^3 - 4)$

31. $\lim_{h\to 3} 2$

32. $\lim_{h\to 0} -5$

33. $\lim_{h\to 0} \dfrac{h^2}{h+h^2}$

HINT [See Example 1(b).]

34. $\lim_{h\to 0} \dfrac{h^2+h}{h^2+2h}$

HINT [See Example 1(b).]

35. $\lim_{x\to 1} \dfrac{x^2-2x+1}{x^2-x}$

36. $\lim_{x\to -1} \dfrac{x^2+3x+2}{x^2+x}$

37. $\lim_{x\to 2} \dfrac{x^3-8}{x-2}$

38. $\lim_{x\to -2} \dfrac{x^3+8}{x^2+3x+2}$

39. $\lim_{x\to 0+} \dfrac{1}{x^2}$ HINT [See Example 2.]

40. $\lim_{x\to 0+} \dfrac{1}{x^2-x}$ HINT [See Example 2.]

41. $\lim_{x\to -1} \dfrac{x^2+1}{x+1}$

42. $\lim_{x\to -1^-} \dfrac{x^2+1}{x+1}$

43. $\lim_{x\to -2^+} \dfrac{x^2+8}{x^2+3x+2}$

44. $\lim_{x\to -1} \dfrac{x^2+3x}{x^2+x}$

45. $\lim\limits_{x \to -2} \dfrac{x^2 + 8}{x^2 + 3x + 2}$

46. $\lim\limits_{x \to -1} \dfrac{x^2 + 3x}{x^2 + 2x + 1}$

47. $\lim\limits_{x \to 2} \dfrac{x^2 + 8}{x^2 - 4x + 4}$

48. $\lim\limits_{x \to -1} \dfrac{x^2 + 3x}{x^2 + 3x + 2}$

49. ▼ $\lim\limits_{x \to 2^+} \dfrac{x - 2}{\sqrt{x - 2}}$

50. ▼ $\lim\limits_{x \to 3^-} \dfrac{\sqrt{3 - x}}{3 - x}$

51. ▼ $\lim\limits_{x \to 9} \dfrac{\sqrt{x} - 3}{x - 9}$

52. ▼ $\lim\limits_{x \to 4} \dfrac{x - 4}{\sqrt{x} - 2}$

53. $\lim\limits_{x \to +\infty} \dfrac{3x^2 + 10x - 1}{2x^2 - 5x}$ HINT [See Example 4.]

54. $\lim\limits_{x \to +\infty} \dfrac{6x^2 + 5x + 100}{3x^2 - 9}$ HINT [See Example 4.]

55. $\lim\limits_{x \to +\infty} \dfrac{x^5 - 1,000x^4}{2x^5 + 10,000}$

56. $\lim\limits_{x \to +\infty} \dfrac{x^6 + 3,000x^3 + 1,000,000}{2x^6 + 1,000x^3}$

57. $\lim\limits_{x \to +\infty} \dfrac{10x^2 + 300x + 1}{5x + 2}$

58. $\lim\limits_{x \to +\infty} \dfrac{2x^4 + 20x^3}{1,000x^3 + 6}$

59. $\lim\limits_{x \to -\infty} \dfrac{3x^2 + 10x - 1}{2x^2 - 5x}$

60. $\lim\limits_{x \to -\infty} \dfrac{6x^2 + 5x + 100}{3x^2 - 9}$

61. $\lim\limits_{x \to -\infty} \dfrac{x^5 - 1,000x^4}{2x^5 + 10,000}$

62. $\lim\limits_{x \to -\infty} \dfrac{x^6 + 3000x^3 + 1,000,000}{2x^6 + 1,000x^3}$

63. $\lim\limits_{x \to -\infty} \dfrac{10x^2 + 300x + 1}{5x + 2}$

64. $\lim\limits_{x \to -\infty} \dfrac{2x^4 + 20x^3}{1,000x^3 + 6}$

65. $\lim\limits_{x \to -\infty} \dfrac{10x^2 + 300x + 1}{5x^3 + 2}$

66. $\lim\limits_{x \to -\infty} \dfrac{2x^4 + 20x^3}{1,000x^6 + 6}$

67. $\lim\limits_{x \to +\infty} (4e^{-3x} + 12)$

68. $\lim\limits_{x \to +\infty} \dfrac{2}{5 - 5.3e^{-3x}}$

69. $\lim\limits_{t \to +\infty} \dfrac{2}{5 - 5.3(3^{3t})}$

70. $\lim\limits_{t \to +\infty} (4.1 - 2e^{3t})$

71. $\lim\limits_{t \to +\infty} \dfrac{2^{3t}}{1 + 5.3e^{-t}}$

72. $\lim\limits_{x \to -\infty} \dfrac{4.2}{2 - 3^{2x}}$

73. $\lim\limits_{x \to -\infty} \dfrac{-3^{2x}}{2 + e^x}$

74. $\lim\limits_{x \to +\infty} \dfrac{2^{-3x}}{1 + 5.3e^{-x}}$

In each of Exercises 75–82, find all points of discontinuity of the given function. HINT [See Example 3.]

75. $f(x) = \begin{cases} x + 2 & \text{if } x < 0 \\ 2x - 1 & \text{if } x \geq 0 \end{cases}$

76. $g(x) = \begin{cases} 1 - x & \text{if } x \leq 1 \\ x - 1 & \text{if } x > 1 \end{cases}$

77. $g(x) = \begin{cases} x + 2 & \text{if } x < 0 \\ 2x + 2 & \text{if } 0 \leq x < 2 \\ x^2 + 2 & \text{if } x \geq 2 \end{cases}$

78. $f(x) = \begin{cases} 1 - x & \text{if } x \leq 1 \\ x + 2 & \text{if } 1 < x < 3 \\ x^2 - 4 & \text{if } x \geq 3 \end{cases}$

79. ▼ $h(x) = \begin{cases} x + 2 & \text{if } x < 0 \\ 0 & \text{if } x = 0 \\ 2x + 2 & \text{if } x > 0 \end{cases}$

80. ▼ $h(x) = \begin{cases} 1 - x & \text{if } x < 1 \\ 1 & \text{if } x = 1 \\ x + 2 & \text{if } x > 1 \end{cases}$

81. ▼ $f(x) = \begin{cases} 1/x & \text{if } x < 0 \\ x & \text{if } 0 \leq x \leq 2 \\ 2^{x-1} & \text{if } x > 2 \end{cases}$

82. ▼ $f(x) = \begin{cases} x^3 + 2 & \text{if } x \leq -1 \\ x^2 & \text{if } -1 < x < 0 \\ x & \text{if } x \geq 0 \end{cases}$

APPLICATIONS

83. *Processor Speeds* The processor speeds, in megahertz (MHz), of Intel processors during the period 1996–2010 can be approximated by the following function of time t in years since the start of 1990:[17]

$$v(t) = \begin{cases} 400t - 2{,}200 & \text{if } 6 \leq t < 15 \\ 3{,}800 & \text{if } 15 \leq t \leq 20. \end{cases}$$

a. Compute $\lim_{t \to 15^-} v(t)$ and $\lim_{t \to 15^+} v(t)$ and interpret each answer. HINT [See Example 3.]

b. Is the function v continuous at $t = 15$? According to the model, was there any abrupt change in processor speeds during the period 1996–2010?

84. *Processor Speeds* The processor speeds, in megahertz (MHz), of Intel processors during the period 1970–2000 can be approximated by the following function of time t in years since the start of 1970:[18]

$$v(t) = \begin{cases} 3t & \text{if } 0 \leq t < 20 \\ 174t - 3{,}420 & \text{if } 20 \leq t \leq 30. \end{cases}$$

---

[17]A rough model based on the fastest processors produced by Intel. Source for data: www.intel.com.

[18]*Ibid.*

a. Compute $\lim_{t\to 20^-} v(t)$ and $\lim_{t\to 20^+} v(t)$ and interpret each answer.

b. Is the function v continuous at $t = 20$? According to the model, was there any abrupt change in processor speeds during the period 1970–2000?

85. *Movie Advertising* Movie expenditures, in billions of dollars, on advertising in newspapers from 1995 to 2004 can be approximated by

$$f(t) = \begin{cases} 0.04t + 0.33 & \text{if } t \le 4 \\ -0.01t + 1.2 & \text{if } t > 4 \end{cases}$$

where t is time in years since 1995.[19]

a. Compute $\lim_{t\to 4^-} f(t)$ and $\lim_{t\to 4^+} f(t)$, and interpret each answer. HINT [See Example 3.]

b. Is the function f continuous at $t = 4$? What does the answer tell you about movie advertising expenditures?

86. *Movie Advertising* The percentage of movie advertising as a share of newspapers' total advertising revenue from 1995 to 2004 can be approximated by

$$p(t) = \begin{cases} -0.07t + 6.0 & \text{if } t \le 4 \\ 0.3t + 17.0 & \text{if } t > 4 \end{cases}$$

where t is time in years since 1995.[20]

a. Compute $\lim_{t\to 4^-} p(t)$ and $\lim_{t\to 4^+} p(t)$, and interpret each answer. HINT [See Example 3.]

b. Is the function p continuous at $t = 4$? What does the answer tell you about newspaper revenues?

87. *Law Enforcement in the 1980s and 1990s* The cost of fighting crime in the United States increased significantly during the period 1982–1999. Total spending on police and courts can be approximated, respectively, by[21]

$$P(t) = 1.745t + 29.84 \text{ billion dollars} \quad (2 \le t \le 19)$$
$$C(t) = 1.097t + 10.65 \text{ billion dollars} \quad (2 \le t \le 19),$$

where t is time in years since 1980. Compute $\lim_{t\to +\infty} \dfrac{P(t)}{C(t)}$ to two decimal places and interpret the result. HINT [See Example 4.]

88. *Law Enforcement in the 1980s and 1990s* Refer to Exercise 87. Total spending on police, courts, and prisons in

the period 1982–1999 could be approximated, respectively, by[22]

$$P(t) = 1.745t + 29.84 \text{ billion dollars} \quad (2 \le t \le 19)$$
$$C(t) = 1.097t + 10.65 \text{ billion dollars} \quad (2 \le t \le 19)$$
$$J(t) = 1.919t + 12.36 \text{ billion dollars} \quad (2 \le t \le 19),$$

where t is time in years since 1980. Compute $\lim_{t\to +\infty} \dfrac{P(t)}{P(t) + C(t) + J(t)}$ to two decimal places and interpret the result. HINT [See Example 4.]

89. *SAT Scores by Income* The following bar graph shows U.S. math SAT scores as a function of household income:[23]

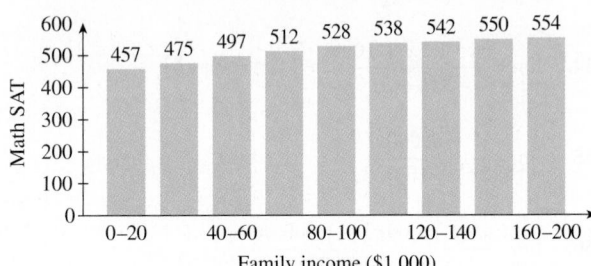

These data can be modeled by

$$S(x) = 573 - 33e^{-0.0131x},$$

where $S(x)$ is the average math SAT score of students whose household income is x thousand dollars per year. Calculate $\lim_{x\to +\infty} S(x)$ and interpret the answer.

90. *SAT Scores by Income* The following bar graph shows U.S. critical reading SAT scores as a function of household income:[24]

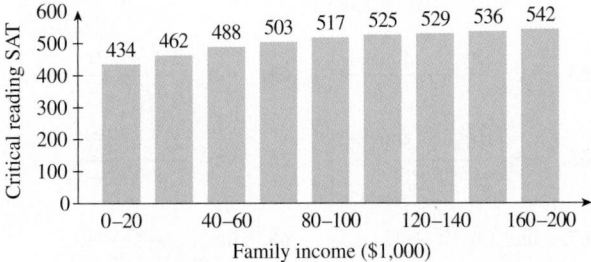

These data can be modeled by

$$S(x) = 550 - 136e^{-0.0151x},$$

where $S(x)$ is the average critical reading SAT score of students whose household income is x thousand dollars per year. Calculate $\lim_{x\to +\infty} S(x)$ and interpret the answer.

---

[19]Model by the authors. Source for data: Newspaper Association of America Business Analysis and Research/*New York Times*, May 16, 2005.

[20]*Ibid.*

[21]Spending is adjusted for inflation and shown in 1999 dollars. Models are based on a linear regression. Source for data: Bureau of Justice Statistics/*New York Times*, February 11, 2002, p. A14.

[22]*Ibid.*

[23]2009 data. Source: College Board/*New York Times* http://economix .blogs.nytimes.com.

[24]*Ibid.*

Foreign Trade *Annual U.S. imports from China in the years 1996 through 2003 could be approximated by*

$$I(t) = t^2 + 3.5t + 50 \quad (1 \leq t \leq 9)$$

billion dollars, where t represents time in years since 1995. Annual U.S. exports to China in the same years could be approximated by

$$E(t) = 0.4t^2 - 1.6t + 14 \quad (0 \leq t \leq 10)$$

billion dollars.[25] *Exercises 91 and 92 are based on these models.*

91. Assuming that the trends shown in the above models continue indefinitely, calculate the limits

$$\lim_{t \to +\infty} I(t) \text{ and } \lim_{t \to +\infty} \frac{I(t)}{E(t)}$$

algebraically, interpret your answers, and comment on the results. HINT [See Example 4.]

92. Repeat Exercise 91, this time calculating

$$\lim_{t \to +\infty} E(t) \text{ and } \lim_{t \to +\infty} \frac{E(t)}{I(t)} \text{ HINT [See Example 4.]}$$

93. *Acquisition of Language* The percentage $p(t)$ of children who can speak in at least single words by the age of t months can be approximated by the equation[26]

$$p(t) = 100 \left(\frac{1 - 12{,}200}{t^{4.48}} \right) \quad (t \geq 8.5).$$

Calculate $\lim_{t \to +\infty} p(t)$ and interpret the result. HINT [See Example 4.]

94. *Acquisition of Language* The percentage $q(t)$ of children who can speak in sentences of five or more words by the age of t months can be approximated by the equation[27]

$$q(t) = 100 \left(1 - \frac{5.27 \times 10^{17}}{t^{12}} \right) \quad (t \geq 30).$$

If p is the function referred to in the preceding exercise, calculate $\lim_{t \to +\infty} [p(t) - q(t)]$ and interpret the result. HINT [See Example 4.]

COMMUNICATION AND REASONING EXERCISES

95. Describe the algebraic method of evaluating limits as discussed in this section and give at least one disadvantage of this method.

96. What is a closed-form function? What can we say about such functions?

---

[25]Based on quadratic regression using data from the U.S. Census Bureau Foreign Trade Division Web site www.census.gov/foreign-trade/sitc1/ as of December 2004.

[26]The model is the authors' and is based on data presented in the article *The Emergence of Intelligence* by William H. Calvin, *Scientific American,* October, 1994, pp. 101–107.

[27]*Ibid.*

97. Why was the following marked wrong? What is the correct answer?

$$\lim_{x \to 3} \frac{x^3 - 27}{x - 3} = \frac{0}{0} \text{ undefined} \qquad \textit{✗ WRONG!}$$

98. Why was the following marked wrong? What is the correct answer?

$$\lim_{x \to 1^-} \frac{x - 1}{x^2 - 2x + 1} = \frac{0}{0} = 0 \qquad \textit{✗ WRONG!}$$

99. ▼ Your friend Karin tells you that $f(x) = 1/(x - 2)^2$ cannot be a closed-form function because it is not continuous at $x = 2$. Comment on her assertion.

100. ▼ Give an example of a function f specified by means of algebraic formulas such that the domain of f consists of all real numbers and f is not continuous at $x = 2$. Is f a closed-form function?

101. Give examples of two limits that lead to two different indeterminate forms, but where both limits exist.

102. Give examples of two limits: one that leads to a determinate form and another that leads to an indeterminate form, but where neither limit exists.

103. ▼ (Compare Exercise 59 in Section 10.1.) Which indeterminate form results from $\lim_{x \to +\infty} \dfrac{p(x)}{e^x}$ if $p(x)$ is a polynomial? Numerically or graphically estimate these limits for various polynomials $p(x)$. What does this suggest about limits that result in $\dfrac{p(\infty)}{e^{\infty}}$?

104. ▼ (Compare Exercise 60 in Section 10.1.) Which indeterminate form results from $\lim_{x \to -\infty} p(x)e^x$ if $p(x)$ is a polynomial? What does this suggest about the limits that result in $p(\infty)e^{-\infty}$?

105. ▼ What is wrong with the following statement? If $f(x)$ is specified algebraically and $f(a)$ is defined, then $\lim_{x \to a} f(x)$ exists and equals $f(a)$. How can it be corrected?

106. ▼ What is wrong with the following statement? If $f(x)$ is specified algebraically and $f(a)$ is not defined, then $\lim_{x \to a} f(x)$ does not exist.

107. ▼ Give the formula for a function that is continuous everywhere except at two points.

108. ▼ Give the formula for a function that is continuous everywhere except at three points.

109. ◆ *The Indeterminate Form* $\infty - \infty$ An indeterminate form not mentioned in Section 10.3 is $\infty - \infty$. Give examples of three limits that lead to this indeterminate form, and where the first limit exists and equals 5, where the second limit diverges to $+\infty$, and where the third exists and equals -5.

110. ◆ *The Indeterminate Form* 1^{∞} An indeterminate form not mentioned in Section 10.3 is 1^{∞}. Give examples of three limits that lead to this indeterminate form, and where the first limit exists and equals 1, where the second limit exists and equals e, and where the third diverges to $+\infty$. HINT [For the third, consider modifying the second.]

10.4 Average Rate of Change

Calculus is the mathematics of change, inspired largely by observation of continuously changing quantities around us in the real world. As an example, the Consumer Price Index (CPI) C increased from 211 points in January 2009 to 220 points in January 2011.[28] As we saw in Chapter 1, the **change** in this index can be measured as the difference:

$$\Delta C = \text{Second value} - \text{First value} = 220 - 211 = 9 \text{ points.}$$

(The fact that the CPI increased is reflected in the positive sign of the change.) The kind of question we will concentrate on is *how fast* the CPI was changing. Because C increased by 9 points in 2 years, we say it averaged a $9/2 = 4.5$ point rise each year. (It actually rose 5 points the first year and 4 the second, giving an average rise of 4.5 points each year.)

Alternatively, we might want to measure this rate in points per month rather than points per year. Because C increased by 9 points in 24 months, it increased at an average rate of $9/24 = 0.375$ points per month.

In both cases, we obtained the average rate of change by dividing the change by the corresponding length of time:

$$\text{Average rate of change} = \frac{\text{Change in } C}{\text{Change in time}} = \frac{9}{2} = 4.50 \text{ points per year}$$

$$\text{Average rate of change} = \frac{\text{Change in } C}{\text{Change in time}} = \frac{9}{24} = 0.375 \text{ points per month.}$$

Average Rate of Change of a Function Numerically and Graphically

EXAMPLE 1 Standard and Poor's 500

The following table lists the approximate value of Standard and Poor's 500 stock market index (S&P) during the period 2005–2011[29] ($t = 5$ represents 2005):

t (year)	5	6	7	8	9	10	11
$S(t)$ (points)	1,200	1,300	1,400	1,400	900	1,150	1,300

a. What was the average rate of change in the S&P over the 2-year period 2005–2007 (the period $5 \le t \le 7$ or $[5, 7]$ in interval notation); over the 4-year period 2005–2009 (the period $5 \le t \le 9$ or $[5, 9]$); and over the period $[6, 11]$?

b. Graph the values shown in the table. How are the rates of change reflected in the graph?

---

[28]Figures are approximate. Source: Bureau of Labor Statistics www.bls.gov.

[29]The values are approximate values at the start of the given year. Source: http://finance.google.com.

Solution

a. During the 2-year period [5, 7], the S&P changed as follows:

Start of the period ($t = 5$):	$S(5) = 1{,}200$
End of the period ($t = 7$):	$S(7) = 1{,}400$

Change during the period [5, 7]: $S(7) - S(5) = 200$

Thus, the S&P increased by 200 points in 2 years, giving an average rate of change of $200/2 = 100$ points per year. We can write the calculation this way:

$$\text{Average rate of change of } S = \frac{\text{Change in } S}{\text{Change in } t}$$

$$= \frac{\Delta S}{\Delta t}$$

$$= \frac{S(7) - S(5)}{7 - 5}$$

$$= \frac{1{,}400 - 1{,}200}{7 - 5} = \frac{200}{2} = 100 \text{ points per year.}$$

Interpreting the result: During the period [5, 7] (that is, 2005–2007), the S&P increased at an average rate of 100 points per year.

Similarly, the average rate of change during the period [5, 9] was

$$\text{Average rate of change of } S = \frac{\Delta S}{\Delta t} = \frac{S(9) - S(5)}{9 - 5} = \frac{900 - 1{,}200}{9 - 5}$$

$$= \frac{-300}{4} = -75 \text{ points per year.}$$

Interpreting the result: During the period [5, 9] (that is, 2005–2009), the S&P *decreased* at an average rate of 75 points per year.

Finally, during the period [6, 11], the average change was

$$\text{Average rate of change of } S = \frac{\Delta S}{\Delta t} = \frac{S(11) - S(6)}{11 - 6} = \frac{1{,}300 - 1{,}300}{11 - 6}$$

$$= \frac{0}{5} = 0 \text{ points per year.}$$

Interpreting the result: During the period [6, 11] the average rate of change of the S&P was zero points per year (even though its value did fluctuate during that period).

b. In Chapter 1, we saw that the rate of change of a quantity that changes linearly with time is measured by the slope of its graph. However, the S&P index does not change linearly with time. Figure 20 shows the data plotted two different ways: (a) as a bar chart and (b) as a piecewise linear graph. Bar charts are more commonly used in the media, but Figure 20(b) on the right illustrates the changing index more clearly.

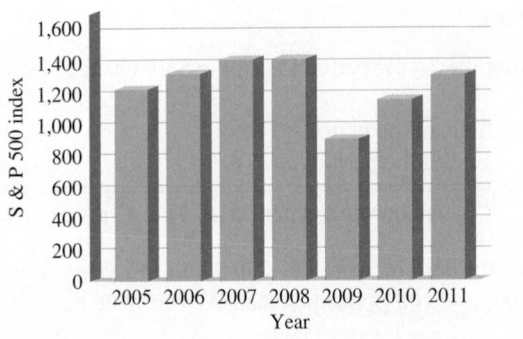

Figure 20(a)

Figure 20(b)

We saw in part (a) that the average rate of change of S over the interval $[5, 9]$ is the ratio

$$\text{Average rate of change of } S = \frac{\Delta S}{\Delta t} = \frac{S(9) - S(5)}{9 - 5} = -75 \text{ points per year.}$$

Notice that this rate of change is also the slope of the line through P and Q shown in Figure 21, and we can estimate this slope directly from the graph as shown.

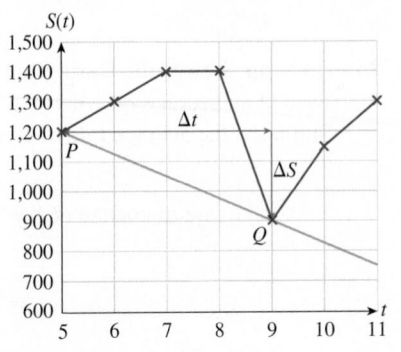
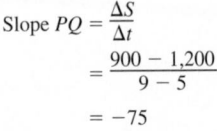

$$\text{Slope } PQ = \frac{\Delta S}{\Delta t}$$
$$= \frac{900 - 1{,}200}{9 - 5}$$
$$= -75$$

Figure 21

Average Rate of Change as Slope: The average rate of change of the S&P over the interval $[5, 9]$ is the slope of the line passing through the points on the graph where $t = 5$ and $t = 9$.

Similarly, the average rates of change of the S&P over the intervals $[5, 7]$ and $[6, 11]$ are the slopes of the lines through pairs of corresponding points.

Here is the formal definition of the average rate of change of a function over an interval.

Change and Average Rate of Change of *f* over [*a*, *b*]: Difference Quotient

The **change** in $f(x)$ over the interval $[a, b]$ is

$$\text{Change in } f = \Delta f$$
$$= \text{Second value} - \text{First value}$$
$$= f(b) - f(a).$$

The **average rate of change** of $f(x)$ over the interval $[a, b]$ is

$$\text{Average rate of change of } f = \frac{\text{Change in } f}{\text{Change in } x}$$

$$= \frac{\Delta f}{\Delta x} = \frac{f(b) - f(a)}{b - a}$$

$$= \text{Slope of line through points } P \text{ and } Q$$
$$\text{(see figure).}$$

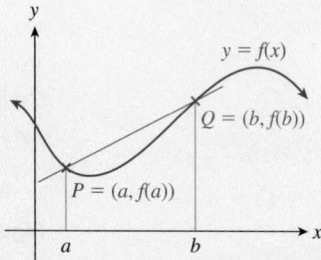

Average rate of change = Slope of PQ

We also call this average rate of change the **difference quotient** of f over the interval $[a, b]$. (It is the *quotient* of the *differences* $f(b) - f(a)$ and $b - a$.) A line through two points of a graph like P and Q is called a **secant line** of the graph.

Units

The units of the change Δf in f are the units of $f(x)$.
The units of the average rate of change of f are units of $f(x)$ per unit of x.*

* The average rate of change is a slope, and so it is measured in the same units as the slope: units of *y* (or *f(x)*) per unit of *x*.

Quick Example

If $f(3) = -1$ billion dollars, $f(5) = 0.5$ billion dollars, and x is measured in years, then the change and average rate of change of f over the interval $[3, 5]$ are given by

$$\text{Change in } f = f(5) - f(3) = 0.5 - (-1) = 1.5 \text{ billion dollars}$$

$$\text{Average rate of change} = \frac{f(5) - f(3)}{5 - 3} = \frac{0.5 - (-1)}{2}$$

$$= 0.75 \text{ billion dollars/year.}$$

Alternative Formula: Average Rate of Change of *f* over [*a, a + h*]

(Replace b in the formula for the average rate of change by $a + h$.) The average rate of change of f over the interval $[a, a + h]$ is

$$\text{Average rate of change of } f = \frac{f(a + h) - f(a)}{h}. \qquad \text{\small Replace } b \text{ by } a + h.$$

In Example 1 we saw that the average rate of change of a quantity can be estimated directly from a graph. Here is an example that further illustrates the graphical approach.

EXAMPLE 2 **Freon 22 Production**

Figure 22 shows the number of tons $f(t)$ of ozone-layer-damaging Freon 22 (chlorodi-fluoromethane) produced annually in developing countries for the period 2000–2010 (t is time in years, and $t = 0$ represents 2000).[30]

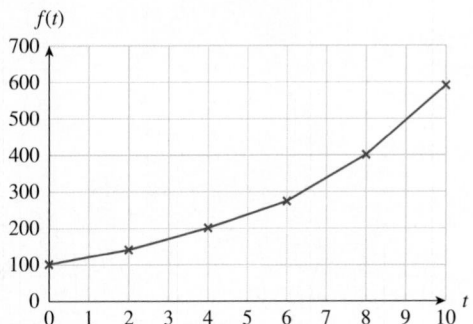

Figure 22

a. Use the graph to estimate the average rate of change of $f(t)$ with respect to t over the interval [4, 8] and interpret the result.

b. Over which 2-year period(s) was the average rate of change of Freon 22 production the greatest?

c. Multiple choice: For the period of time under consideration, Freon 22 production was

 (A) increasing at a faster and faster rate.
 (B) increasing at a slower and slower rate.
 (C) decreasing at a faster and faster rate.
 (D) decreasing at a slower and slower rate.

Solution

a. The average rate of change of f over the interval [4, 8] is given by the slope of the line through the points P and Q shown in Figure 23.

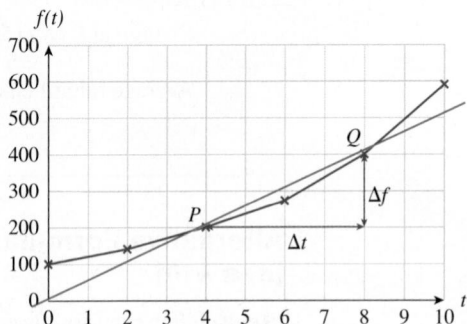

Figure 23

From the figure,

$$\text{Average rate of change of } f = \frac{\Delta f}{\Delta t} = \text{slope } PQ \approx \frac{400 - 200}{8 - 4} = \frac{200}{4} = 50.$$

Thus, the rate of change of f over the interval [4, 8] is approximately 50.

---

[30]Figures from 2007 on were projected. Source: *New York Times*, February 23, 2007, p. C1.

Q: *How do we interpret the result?*

A: A clue is given by the units of the average rate of change: units of *f* per unit of *t*. The units of *f* are tons of Freon 22 and the units of *t* are years. Thus, the average rate of change of *f* is measured in tons of Freon 22 per year, and we can now interpret the result as follows:

Interpreting the average rate of change: Annual production of Freon 22 was increasing at an average rate of 50 tons of Freon 22 per year from 2004 to 2008.

---

b. The rates of change of Freon 22 production over successive 2-year periods are given by the slopes of the individual line segments that make up the graph in Figure 22. Thus, the greatest average rate of change in a single 2-year period corresponds to the segment(s) with the largest slope. If you look at the figure, you will notice that the segment corresponding to [8, 10] is the steepest. Thus, the average rate of change of Freon 22 production was largest over the 2-year period from 2008 to 2010.

c. Looking again at the figure, notice that the graph rises as we go from left to right; that is, the value of the function (Freon 22 production) is increasing with increasing *t*. At the same time, the fact that the curve bends up (is concave up) with increasing *t* tells us that the successive slopes get steeper, and so the average rates of change increase as well (Choice (A)).

➡ **Before we go on...** Notice in Example 2 that we do not get exact answers from a graph; the best we can do is *estimate* the rates of change: Was the exact answer to part (a) closer to 49 or 51? Two people can reasonably disagree about results read from a graph, and you should bear this in mind when you check the answers to the exercises. ∎

Perhaps the most sophisticated way to compute the average rate of change of a quantity is through the use of a mathematical formula or model for the quantity in question.

Average Rate of Change of a Function Using Algebraic Data

EXAMPLE 3 Average Rate of Change from a Formula

You are a commodities trader and you monitor the price of gold on the spot market very closely during an active morning. Suppose you find that the price of an ounce of gold can be approximated by the function

$$G(t) = 5t^2 - 85t + 1{,}762 \qquad (7.5 \le t \le 10.5),$$

where *t* is time in hours. (See Figure 24. $t = 8$ represents 8:00 am.)

Looking at the graph on the right, we can see that the price of gold was falling at the beginning of the time period, but by $t = 8.5$ the fall had slowed to a stop, whereupon the market turned around and the price began to rise more and more rapidly toward the end of the period. What was the average rate of change of the price of gold over the $1\frac{1}{2}$-hour period starting at 8:00 am (the interval [8, 9.5] on the *t*-axis)?

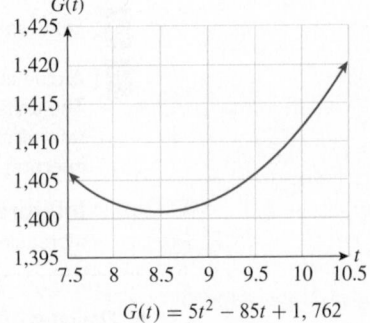

$$G(t) = 5t^2 - 85t + 1,762$$

Source: www.kitco.com

Figure 24

Solution We have

$$\text{Average rate of change of } G \text{ over } [8, 9.5] = \frac{\Delta G}{\Delta t} = \frac{G(9.5) - G(8)}{9.5 - 8}.$$

From the formula for $G(t)$, we find

$$G(9.5) = 5(9.5)^2 - 85(9.5) + 1,762 = 1,405.75$$
$$G(8) = 5(8)^2 - 85(8) + 1,762 = 1,402.$$

Thus, the average rate of change of G is given by

$$\frac{G(9.5) - G(8)}{9.5 - 8} = \frac{1,405.75 - 1,402}{1.5} = \frac{3.75}{1.5} = \$2.50 \text{ per hour.}$$

In other words, the price of gold increased at an average rate of $2.50 per hour over
the $1\frac{1}{2}$-hour period.

EXAMPLE 4   **Rates of Change over Shorter Intervals**

Continuing with Example 3, use technology to compute the average rate of change of

$$G(t) = 5t^2 - 85t + 1,762 \qquad (7.5 \le t \le 10.5)$$

over the intervals $[8, 8 + h]$, where $h = 1, 0.1, 0.01, 0.001,$ and 0.0001. What do the
answers tell you about the price of gold?

Solution

We use the "alternative" formula

$$\text{Average rate of change of } G \text{ over } [a, a + h] = \frac{G(a + h) - G(a)}{h}$$

so

$$\text{Average rate of change of } G \text{ over } [8, 8 + h] = \frac{G(8 + h) - G(8)}{h}.$$

using Technology

Example 4 is the kind of example where the use of technology can make a huge difference. See the Technology Guides at the end of the chapter to find out how to do the above computations almost effortlessly using a TI-83/84 Plus or a spreadsheet. Here is an outline:

TI-83/84 Plus
Y₁=5X^2-85X+1762
Home screen:
(Y₁(8+1)-Y₁(8))/1
(Y₁(8+0.1)-Y₁(8))/0.1
(Y₁(8+0.01)-Y₁(8))/0.01
etc.
[More details on page 778.]

Spreadsheet
Headings a, h, t, G(t), Rate of Change in A1–E1
8 in A2, 1 in B2,
=A2 in C2, =A2+B2 in C3
=5*C2^2-85*C2+1762 in D2; copy down to D3
=(D3-D2)/(C3-C2) in E2
[More details on page 780.]

Let us calculate this average rate of change for some of the values of h listed:

$h = 1$: $G(8+h) = G(8+1) = G(9) = 5(9)^2 - 85(9) + 1{,}762 = 1{,}402$

$G(8) = 5(8)^2 - 85(8) + 1{,}762 = 1{,}402$

Average rate of change of $G = \dfrac{G(9) - G(8)}{1} = \dfrac{1{,}402 - 1{,}402}{1} = 0$

$h = 0.1$: $G(8+h) = G(8+0.1) = G(8.1) = 5(8.1)^2 - 85(8.1) + 1{,}762$
$= 1{,}401.55$

$G(8) = 5(8)^2 - 85(8) + 1{,}762 = 1{,}402$

Average rate of change of $G = \dfrac{G(8.1) - G(8)}{0.1} = \dfrac{1{,}401.55 - 1{,}402}{0.1} = \dfrac{-0.45}{0.1}$
$= -4.5$

$h = 0.01$: $G(8+h) = G(8+0.01) = G(8.01) = 5(8.01)^2 - 85(8.01) + 1{,}762$
$= 1{,}401.9505$

$G(8) = 5(8)^2 - 85(8) + 1{,}762 = 1{,}402$

Average rate of change of $G = \dfrac{G(8.01) - G(8)}{0.01} = \dfrac{1{,}401.9505 - 1{,}402}{0.01} = \dfrac{-0.0495}{0.01}$
$= -4.95$

Continuing in this way, we get the values in the following table:

h	1	0.1	0.01	0.001	0.0001
Ave. Rate of Change $\dfrac{G(8+h) - G(8)}{h}$	0	−4.5	−4.95	−4.995	−4.9995

Each value is an average rate of change of G. For example, the value corresponding to $h = 0.01$ is -4.95, which tells us:

Over the interval $[8, 8.01]$ *the price of gold was decreasing at an average rate of* $4.95 *per hour.*

In other words, during the first one hundredth of an hour (or 36 seconds) starting at $t = 8{:}00$ am, the price of gold was decreasing at an average rate of $4.95 per hour. Put another way, in those 36 seconds, the price of gold decreased at a rate that, if continued, would have produced a decrease of $4.95 in the price of gold during the next hour. We will return to this example at the beginning of Section 10.5.

FAQs

Recognizing When and How to Compute the Average Rate of Change and How to Interpret the Answer

Q : *How do I know, by looking at the wording of a problem, that it is asking for an average rate of change?*

A : If a problem does not ask for an average rate of change directly, it might do so indirectly, as in "On average, how fast is quantity q increasing?"

Q : *If I know that a problem calls for computing an average rate of change, how should I compute it? By hand or using technology?*

A : All the computations can be done by hand, but when hand calculations are not called for, using technology might save time.

Q : *Lots of problems ask us to "interpret" the answer. How do I do that for questions involving average rates of change?*

A : The *units* of the average rate of change are often the key to interpreting the results:

The units of the average rate of change of f(x) are units of f(x) per unit of x.

Thus, for instance, if $f(x)$ is the cost, in dollars, of a trip of x miles in length, and the average rate of change of f is calculated to be 10, then the units of the average rate of change are dollars per mile, and so we can interpret the answer by saying that the cost of a trip rises an average of $10 for each additional mile.

10.4 EXERCISES

▼ more advanced   ◆ challenging
T indicates exercises that should be solved using technology

In Exercises 1–18, calculate the average rate of change of the given function over the given interval. Where appropriate, specify the units of measurement. HINT [See Example 1.]

1.

x	0	1	2	3
$f(x)$	3	5	2	−1

Interval: [1, 3]

2.

x	0	1	2	3
$f(x)$	−1	3	2	1

Interval: [0, 2]

3.

x	−3	−2	−1	0
$f(x)$	−2.1	0	−1.5	0

Interval: [−3, −1]

4.

x	−2	−1	0	1
$f(x)$	−1.5	−0.5	4	6.5

Interval: [−1, 1]

5.

t (months)	2	4	6
$R(t)$ ($ millions)	20.2	24.3	20.1

Interval: [2, 6]

6.

x (kilos)	1	2	3
$C(x)$ (£)	2.20	3.30	4.00

Interval: [1, 3]

7.

p ($)	5.00	5.50	6.00
$q(p)$ (items)	400	300	150

Interval: [5, 5.5]

8.

t (hours)	0	0.1	0.2
$D(t)$ (miles)	0	3	6

Interval: [0.1, 0.2]

9. Apple Computer Stock Price ($)

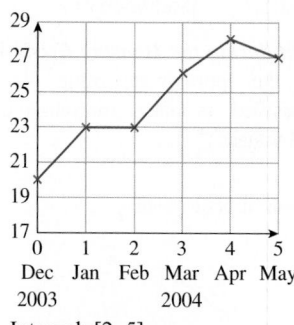

Interval: [2, 5]
HINT [See Example 2.]

10. Cisco Systems Stock Price ($)

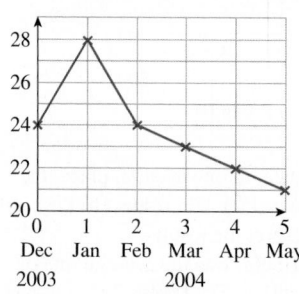

Interval: [1, 5]
HINT [See Example 2.]

11. Unemployment (%)

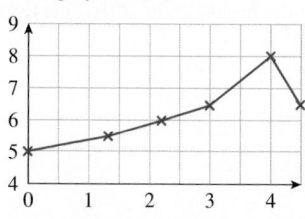

Budget deficit (% of GNP)

Interval: [0, 4]

12. Inflation (%)

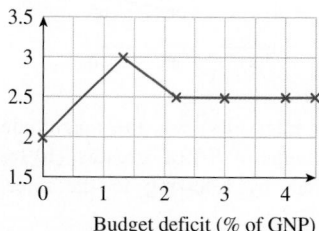

Budget deficit (% of GNP)

Interval: [0, 4]

13. $f(x) = x^2 - 3$; [1, 3] HINT [See Example 3.]

14. $f(x) = 2x^2 + 4$; [−1, 2] HINT [See Example 3.]

15. $f(x) = 2x + 4$; [−2, 0]

16. $f(x) = \dfrac{1}{x}$; [1, 4]

17. $f(x) = \dfrac{x^2}{2} + \dfrac{1}{x}$; [2, 3]   **18.** $f(x) = 3x^2 - \dfrac{x}{2}$; [3, 4]

In Exercises 19–24, calculate the average rate of change of the given function f over the intervals [a, a + h] where h = 1, 0.1, 0.01, 0.001, and 0.0001. (Technology is recommended for the cases h = 0.01, 0.001, and 0.0001.) HINT [See Example 4.]

19. $f(x) = 2x^2$; $a = 0$ **20.** $f(x) = \dfrac{x^2}{2}$; $a = 1$

21. $f(x) = \dfrac{1}{x}$; $a = 2$ **22.** $f(x) = \dfrac{2}{x}$; $a = 1$

23. $f(x) = x^2 + 2x$; $a = 3$ **24.** $f(x) = 3x^2 - 2x$; $a = 0$

APPLICATIONS

25. *World Military Expenditure* The following table shows total military and arms trade expenditure in 2000, 2005, and 2010 ($t = 0$ represents 2000):[31]

Year t	0	5	10
Military Expenditure $C(t)$ ($ billion)	1,100	1,300	1,600

Compute and interpret the average rate of change of $C(t)$ **(a)** over the period 2005–2010 (that is, [5, 10]), and **(b)** over the period [0, 10]. Be sure to state the units of measurement. HINT [See Example 1.]

26. *Education Expenditure* The following table shows the percentage of the U.S. Discretionary Budget allocated to education in 2003, 2005, and 2009 ($t = 0$ represents 2000):[32]

Year t	3	5	9
Percentage $P(t)$	6.8	7	6.2

Compute and interpret the average rate of change of $P(t)$ **(a)** over the period 2003–2009 (that is, [3, 9]), and **(b)** over the period [5, 9]. Be sure to state the units of measurement. HINT [See Example 1.]

27. *Oil Production in Mexico* The following table shows approximate daily oil production by Pemex, Mexico's national oil company, for 2001–2009 ($t = 1$ represents the start of 2001):[33]

t (year since 2000)	1	2	3	4	5	6	7	8	9
$P(t)$ (million barrels)	3.1	3.3	3.4	3.4	3.4	3.3	3.2	3.1	3.0

---

[31] Source: www.globalissues.org/Geopolitics/ArmsTrade/Spending.asp.
[32] *Ibid.*
[33] Figures are approximate, and 2008–2009 figures are projections by the Department of Energy. Source: Energy Information Administration/Pemex (www.eia.doe.gov).

a. Compute the average rate of change of $P(t)$ over the period 2002–2007. Interpret the result. HINT [See Example 1.]

b. Which of the following is true? From 2001 to 2008, the one-year average rate of change of oil production by Pemex

(**A**) increased in value.

(**B**) decreased in value.

(**C**) never increased in value.

(**D**) never decreased in value.

HINT [See Example 2.]

28. *Oil Imports from Mexico* The following table shows U.S. daily oil imports from Mexico, for 2001–2009 ($t = 1$ represents the start of 2001):[34]

t (year since 2000)	1	2	3	4	5	6	7	8	9
$I(t)$ (million barrels)	1.4	1.35	1.5	1.55	1.6	1.5	1.5	1.5	1.2

a. Use the data in the table to compute the average rate of change of $I(t)$ over the period 2001–2006. Interpret the result.

b. Which of the following is true? From 2002 to 2006, the one-year average rate of change of oil imports from Mexico

(**A**) increased in value.

(**B**) decreased in value.

(**C**) never increased in value.

(**D**) never decreased in value.

29. *Subprime Mortgages during the Housing Crisis* The following graph shows the approximate percentage $P(t)$ of mortgages issued in the U.S. that were subprime (normally classified as risky):[35]

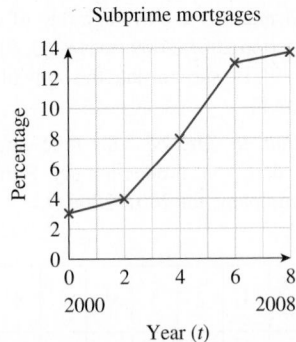

Subprime mortgages

a. Use the graph to estimate, to one decimal place, the average rate of change of $P(t)$ with respect to t over the interval [0, 6] and interpret the result.

b. Over which 2-year period(s) was the average rate of change of $P(t)$ the greatest? HINT [See Example 2.]

30. *Subprime Mortgage Debt during the Housing Crisis* The following graph shows the approximate value $V(t)$ of subprime (normally classified as risky) mortgage debt outstanding in the United States:[36]

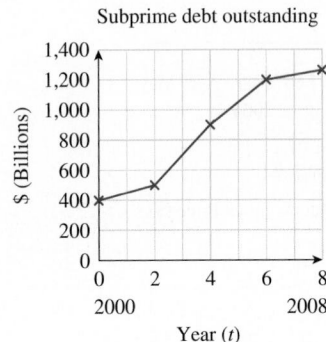

Subprime debt outstanding

a. Use the graph to estimate, to one decimal place, the average rate of change of $V(t)$ with respect to t over the interval [2, 6] and interpret the result.

b. Over which 2-year period(s) was the average rate of change of $V(t)$ the least? HINT [See Example 2.]

31. *Immigration to Ireland* The following graph shows the approximate number (in thousands) of people who immigrated to Ireland during the period 2006–2010 (t is time in years since 2000):[37]

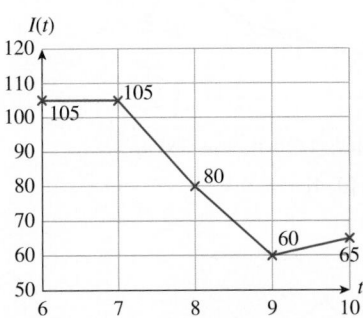

During which 2-year interval(s) was the magnitude of the average rate of change of $I(t)$ (**a**) greatest (**b**) least? Interpret your answers by referring to the rates of change.

32. *Emigration from Ireland* The following graph shows the approximate number (in thousands) of people who emigrated from Ireland during the period 2006–2010.[38]

---

[34] Figures are approximate, and 2008–2009 figures are projections by the Department of Energy. Source: Energy Information Administration/Pemex (www.eia.doe.gov).

[35] Sources: Mortgage Bankers Association, UBS.

[36] Source: Data 360 www.data360.org.

[37] Sources: Ireland Central Statistic Office/Data 360 www.data360.org.

[38] *Ibid.*

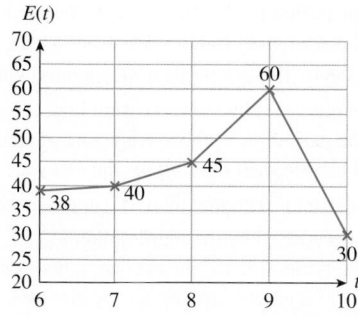

During which 2-year interval(s) was the magnitude of the average rate of change of $E(t)$ (**a**) greatest (**b**) least? Interpret your answers by referring to the rates of change.

33. ▼ *Physics Research in the U.S.* The following table shows the number of research articles in the journal *Physical Review* authored by U.S researchers during the period 1993–2003 ($t = 3$ represents 1993):[39]

t (year since 1990)	3	5	7	9	11	13
$N(t)$ (articles, thousands)	5.1	4.6	4.3	4.3	4.5	4.2

a. Find the interval(s) over which the average rate of change of N was the most negative. What was that rate of change? Interpret your answer.

b. The **percentage change of N over the interval $[a, b]$** is defined to be

$$\text{Percentage change of } N = \frac{\text{Change in } N}{\text{First value}} = \frac{N(b) - N(a)}{N(a)}.$$

Compute the percentage change of N over the interval [3, 13] and also the average rate of change. Interpret the answers.

34. ▼ *Physics Research in Europe* The following table shows the number of research articles in the journal *Physical Review* authored by researchers in Europe during the period 1993–2003 ($t = 3$ represents 1993):[40]

t (year since 1990)	3	5	7	9	11	13
$N(t)$ (articles, thousands)	3.8	4.6	5.0	5.0	6.0	5.7

a. Find the interval(s) over which the average rate of change of N was the most positive. What was that rate of change? Interpret your answer.

b. The **percentage change of N over the interval $[a, b]$** is defined to be

$$\text{Percentage change of } N = \frac{\text{Change in } N}{\text{First value}} = \frac{N(b) - N(a)}{N(a)}.$$

Compute the percentage change of N over the interval
[7, 13] and also the average rate of change. Interpret the answers.

35. *College Basketball: Men* The following chart shows the number of NCAA men's college basketball teams in the United States during the period 2000–2010:[41]

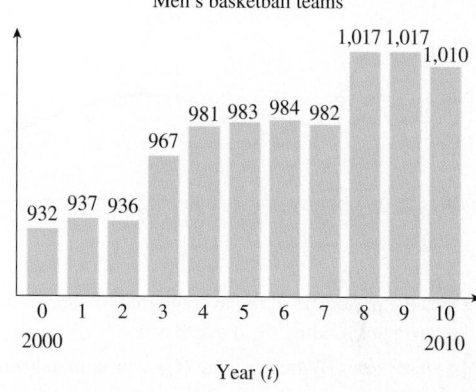

Men's basketball teams

a. On average, how fast was the number of men's college basketball teams growing over the 4-year period beginning in 2002?

b. By inspecting the chart, determine whether the 3-year average rates of change increased or decreased beginning in 2005. HINT [See Example 2.]

36. *College Basketball: Women* The following chart shows the number of NCAA women's college basketball teams in the United States during the period 2000–2010:[42]

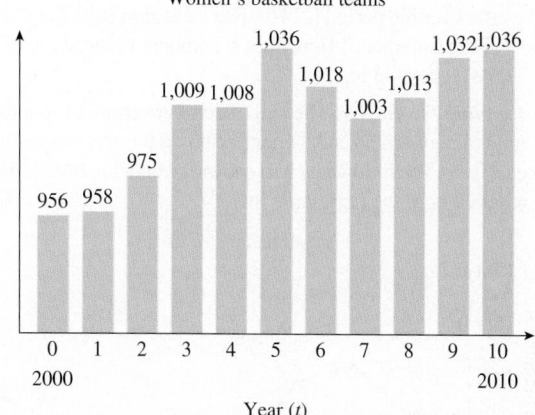

Women's basketball teams

a. On average, how fast was the number of women's college basketball teams growing over the 4-year period beginning in 2004?

b. By inspecting the graph, find the 3-year period over which the average rate of change was largest.

[39] Source: The Americal Physical Society/*New York Times*, May 3, 2003, p. A1.
[40] *Ibid.*

[41] 2010 figure is an estimate. Source: www.census.gov.
[42] *Ibid.*

37. *Funding for the Arts* State governments in the United States spend a total of between $1 and $2 per person on the arts and culture each year. The following chart shows the data for 2002–2010, together with the regression line:[43]

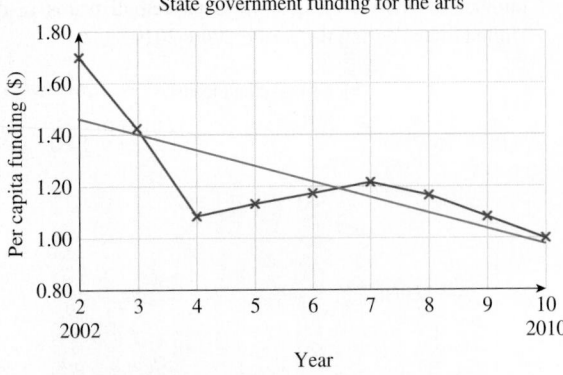

State government funding for the arts

a. Over the period [2, 6] the average rate of change of state government funding for the arts was

(A) less than (B) greater than (C) approximately equal to

the rate predicted by the regression line.

b. Over the period [3, 10] the average rate of change of state government funding for the arts was

(A) less than (B) greater than (C) approximately equal to

the rate predicted by the regression line.

c. Over the period [4, 8] the average rate of change of state government funding for the arts was

(A) less than (B) greater than (C) approximately equal to

the rate predicted by the regression line.

d. Estimate, to two significant digits, the average rate of change of per capita state government funding for the arts over the period [2, 10]. (Be careful to state the units of measurement.) How does it compare to the slope of the regression line?

38. *Funding for the Arts* The U.S. federal government spends a total of between $6 and $7 per person on the arts and culture each year. The following chart shows the data for 2002–2010, together with the regression line:[44]

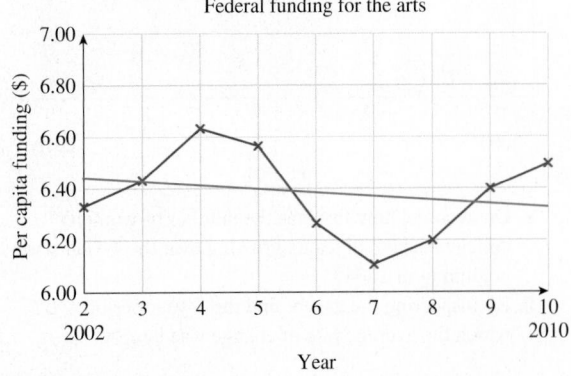

Federal funding for the arts

a. Over the period [4, 10] the average rate of change of federal government funding for the arts was

(A) less than (B) greater than (C) approximately equal to

the rate predicted by the regression line.

b. Over the period [2, 7] the average rate of change of federal government funding for the arts was

(A) less than (B) greater than (C) approximately equal to

the rate predicted by the regression line.

c. Over the period [3, 10] the average rate of change of federal government funding for the arts was

(A) less than (B) greater than (C) approximately equal to

the rate predicted by the regression line.

d. Estimate, to one significant digit, the average rate of change of per capita federal government funding for the arts over the period [2, 10]. (Be careful to state the units of measurement.) How does it compare to the slope of the regression line?

39. ▼ ***Market Volatility during the Dot-com Boom*** A volatility index generally measures the extent to which a market undergoes sudden changes in value. The volatility of the S&P 500 (as measured by one such index) was decreasing at an average rate of 0.2 points per year during 1991–1995, and was increasing at an average rate of about 0.3 points per year during 1995–1999. In 1995, the volatility of the S&P was 1.1.[45] Use this information to give a rough sketch of the volatility of the S&P 500 as a function of time, showing its values in 1991 and 1999.

40. ▼***Market Volatility during the Dot-com Boom*** The volatility (see the preceding exercise) of the NASDAQ had an average rate of change of 0 points per year during 1992–1995, and increased at an average rate of 0.2 points per year during 1995–1998. In 1995, the volatility of the NASDAQ was 1.1.[46] Use this information to give a rough sketch of the volatility of the NASDAQ as a function of time.

41. *Market Index* Joe Downs runs a small investment company from his basement. Every week he publishes a report on the success of his investments, including the progress of the "Joe Downs Index." At the end of one particularly memorable week, he reported that the index for that week had the value $I(t) = 1{,}000 + 1{,}500t - 800t^2 + 100t^3$ points, where t represents the number of business days into the week; t ranges from 0 at the beginning of the week to 5 at the end of the week. The graph of I is shown below.

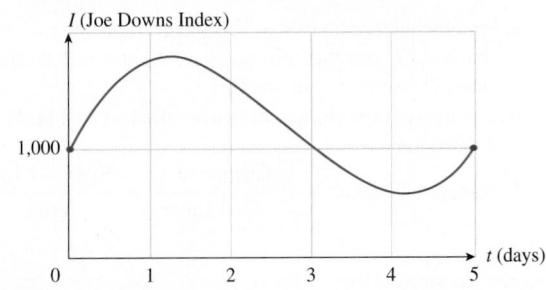

I (Joe Downs Index)

t (days)

[43]Figures are in constant 2008 dollars, and the 2010 figure is the authors' estimate. Source: *Americans for the Arts* www.artsusa.org.
[44]*Ibid.*

[45]Source for data: Sanford C. Bernstein Company/*New York Times*, March 24, 2000, p. C1.
[46]*Ibid.*

IMPORTANT NOTES

1. Sections 10.1–10.3 discuss limits in some detail. If you have not (yet) covered those sections, you can trust to your intuition.

2. The formula for the derivative tells us that the instantaneous rate of change is the limit of the average rates of change $[f(a + h) - f(a)]/h$ over smaller and smaller intervals. Thus, the value of $f'(a)$ can be approximated by computing the average rate of change for smaller and smaller values of h, both positive and negative.

3. If a happens to be an endpoint of the domain of f, then $f'(a)$ does not exist, as then $[f(a + h) - f(a)]/h$ only has a one-sided limit as $h \to 0$, and so

$$\lim_{h \to 0} \frac{f(a + h) - f(a)}{h} \text{ does not exist.}^*$$

4. In this section we will only *approximate* derivatives. In Section 10.6 we will begin to see how we find the *exact* values of derivatives.

5. $f'(a)$ is a number we can calculate, or at least approximate, for various values of a, as we have done in the earlier example. Since $f'(a)$ depends on the value of a, we can think of f' as *a function of a*. (We return to this idea at the end of this section.) An old name for f' is "the function *derived from f*," which has been shortened to the *derivative* of f.

6. It is because f' is a function that we sometimes refer to $f'(a)$ as "the derivative of f evaluated at a," or the "derivative of $f(x)$ evaluated at $x = a$."

    It may happen that the average rates of change $[f(a + h) - f(a)]/h$ do not approach any fixed number at all as h approaches zero, or that they approach one number on the intervals using positive h, and another on those using negative h. If this happens, $\lim_{h \to 0}[f(a + h) - f(a)]/h$ does not exist, and we say that f is **not differentiable** at $x = a$, or $f'(a)$ **does not exist**. When the limit *does* exist, we say that f is **differentiable** at the point $x = a$, or $f'(a)$ **exists**. It is comforting to know that all polynomials and exponential functions are differentiable at *every* point. On the other hand, certain functions are not differentiable. Examples are $f(x) = |x|$ and $f(x) = x^{1/3}$, neither of which is differentiable at $x = 0$. (See Section 11.1.)

* One could define the derivative at an endpoint by instead using the associated one-sided limit; for instance, if a is a left endpoint of the domain of f, then one could define

$$f'(a) = \lim_{h \to 0^+} \frac{f(a + h) - f(a)}{h}$$

as we did in previous editions of this book. However, in this edition we have decided to follow the usual convention and say that the derivative at an endpoint does not exist.

EXAMPLE 1 **Instantaneous Rate of Change: Numerically and Graphically**

The air temperature one spring morning, t hours after 7:00 am, was given by the function $f(t) = 50 + 0.1t^4$ degrees Fahrenheit $(0 \le t \le 4)$.

a. How fast was the temperature rising at 9:00 am?

b. How is the instantaneous rate of change of temperature at 9:00 am reflected in the graph of temperature vs. time?

Solution

a. We are being asked to find the instantaneous rate of change of the temperature at $t = 2$, so we need to find $f'(2)$. To do this we examine the average rates of change

$$\frac{f(2 + h) - f(2)}{h} \quad \text{Average rate of change = difference quotient}$$

✱ We can quickly compute these values using technology as in Example 4 in Section 10.4. (See the Technology Guides at the end of the chapter.)

for values of h approaching 0. Calculating the average rate of change over $[2, 2 + h]$ for $h = 1$, 0.1, 0.01, 0.001, and 0.0001 we get the following values (rounded to four decimal places):✱

h	1	0.1	0.01	0.001	0.0001
Average Rate of Change Over $[2, 2 + h]$	6.5	3.4481	3.2241	3.2024	3.2002

Here are the values we get using negative values of h:

h	-1	-0.1	-0.01	-0.001	-0.0001
Average Rate of Change Over $[2 + h, 2]$	1.5	2.9679	3.1761	3.1976	3.1998

The average rates of change are clearly approaching the number 3.2, so we can say that $f'(2) = 3.2$. Thus, at 9:00 in the morning, the temperature was rising at the rate of 3.2 degrees per hour.

b. We saw in Section 10.4 that the average rate of change of f over an interval is the slope of the secant line through the corresponding points on the graph of f. Figure 25 illustrates this for the intervals $[2, 2 + h]$ with $h = 1, 0.5$, and 0.1.

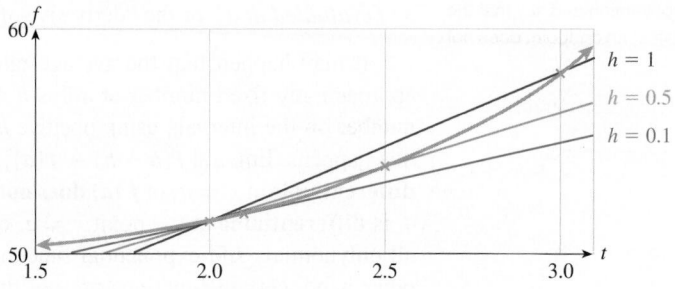

Figure 25

All three secant lines pass though the point $(2, f(2)) = (2, 51.6)$ on the graph of f. Each of them passes through a second point on the curve (the second point is different for each secant line) and this second point gets closer and closer to $(2, 51.6)$ as h gets closer to 0. What seems to be happening is that the secant lines are getting closer and closer to a line that just touches the curve at $(2, 51.6)$: the **tangent line** at $(2, 51.6)$, shown in Figure 26.

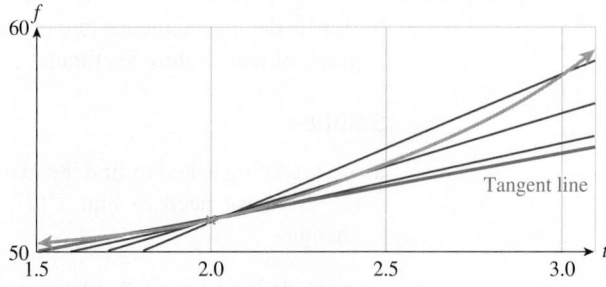

Figure 26

Q: *What is the slope of this tangent line?*

A: Because the slopes of the secant lines are getting closer and closer to 3.2, and because the secant lines are approaching the tangent line, the tangent line must have slope 3.2. In other words,

At the point on the graph where x = 2, the slope of the tangent line is f′(2).

Q: *What is the difference between f(2) and f′(2)?*

A: An important question. Briefly, $f(2)$ is the *value of f* when $t = 2$, while $f′(2)$ is the *rate at which f is changing* when $t = 2$. Here,

$$f(2) = 50 + 0.1(2)^4 = 51.6 \text{ degrees.}$$

Thus, at 9:00 am ($t = 2$), the temperature was 51.6 degrees. On the other hand,

$$f′(2) = 3.2 \text{ degrees per hour.} \qquad \text{Units of slope are units of } f \text{ per unit of } t.$$

This means that, at 9:00 am ($t = 2$), the temperature was increasing at a rate of 3.2 degrees per hour.

Because we have been talking about tangent lines, we should say more about what they *are*. A tangent line to a *circle* is a line that touches the circle in just one point. A tangent line gives the circle "a glancing blow," as shown in Figure 27.

For a smooth curve other than a circle, a tangent line may touch the curve at more than one point, or pass through it (Figure 28).

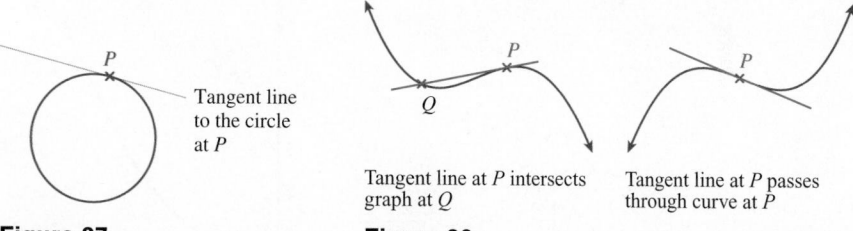

Tangent line to the circle at *P*

Figure 27

Tangent line at *P* intersects graph at *Q*

Tangent line at *P* passes through curve at *P*

Figure 28

However, all tangent lines have the following interesting property in common: If we focus on a small portion of the curve very close to the point *P*—in other words, if we "zoom in" to the graph near the point *P*—the curve will appear almost straight, and almost indistinguishable from the tangent line (Figure 29).

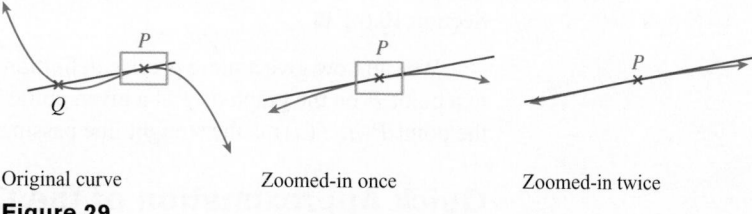

Original curve

Zoomed-in once

Zoomed-in twice

Figure 29

You can check this property by zooming in on the curve shown in Figures 25 and 26 in the previous example near the point where $x = 2$.

Secant and Tangent Lines

The *slope of the secant line* through the points on the graph of f where $x = a$ and $x = a + h$ is given by the average rate of change, or difference quotient,

$$m_{sec} = \text{slope of secant} = \text{average rate of change} = \frac{f(a+h) - f(a)}{h}.$$

The *slope of the tangent line* through the point on the graph of f where $x = a$ is given by the instantaneous rate of change, or derivative

$$m_{tan} = \text{slope of tangent} = \text{instantaneous rate of change} = \text{derivative}$$

$$= f'(a) = \lim_{h \to 0} \frac{f(a+h) - f(a)}{h},$$

assuming the limit exists.

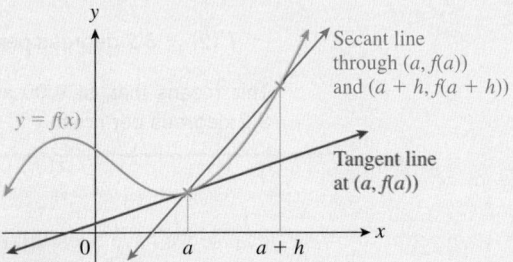

Quick Example

In the following graph, the tangent line at the point where $x = 2$ has slope 3. Therefore, the derivative at $x = 2$ is 3. That is, $f'(2) = 3$.

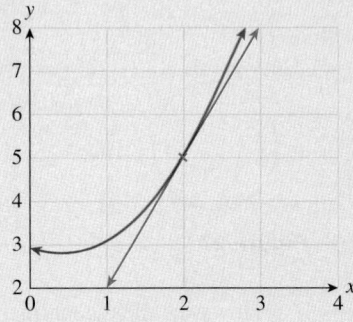

Note It might happen that the tangent line is vertical at some point or does not exist at all. These are the cases in which f is not differentiable at the given point. (See Section 10.6.) ∎

We can now give a more precise definition of what we mean by the tangent line to a point P on the graph of f at a given point: The **tangent line** to the graph of f at the point $P(a, f(a))$ is the straight line passing through P with slope $f'(a)$.

Quick Approximation of the Derivative

Q: *Do we always need to make tables of difference quotients as above in order to calculate an approximate value for the derivative? That seems like a large amount of work just to get an approximation.*

A : We can usually *approximate* the value of the derivative by using a single, small value of h. In the example above, the value $h = 0.0001$ would have given a pretty good approximation. The problems with using a fixed value of h are that (1) we do not get an *exact* answer, only an *approximation* of the derivative, and (2) how good an approximation it is depends on the function we're differentiating.✻ However, with most of the functions we'll be considering, setting $h = 0.0001$ does give us a good approximation.

✻ In fact, no matter how small the value we decide to use for h, it is possible to craft a function f for which the difference quotient at a is not even close to $f'(a)$.

Calculating a Quick Approximation of the Derivative

We can calculate an approximate value of $f'(a)$ by using the formula

$$f'(a) \approx \frac{f(a+h) - f(a)}{h} \qquad \text{Rate of change over } [a,\, a+h]$$

with a small value of h. The value $h = 0.0001$ works for most examples we encounter (students of numerical methods study the question of exactly how accurate this approximation is).

Alternative Formula: The Balanced Difference Quotient

The following alternative formula, which measures the rate of change of f over the interval $[a - h, a + h]$, often gives a more accurate result, and is the one used in many calculators:

$$f'(a) \approx \frac{f(a+h) - f(a-h)}{2h}. \qquad \text{Rate of change over } [a - h,\, a + h]$$

Note For the quick approximations to be valid, the function f must be differentiable; that is, $f'(a)$ must exist. ■

EXAMPLE 2 Quick Approximation of the Derivative

a. Calculate an approximate value of $f'(1.5)$ if $f(x) = x^2 - 4x$.

b. Find the equation of the tangent line at the point on the graph where $x = 1.5$.

Solution

a. We shall compute both the ordinary difference quotient and the balanced difference quotient.

Ordinary Difference Quotient: Using $h = 0.0001$, the ordinary difference quotient is:

$$f'(1.5) \approx \frac{f(1.5 + 0.0001) - f(1.5)}{0.0001} \qquad \text{Ordinary difference quotient}$$

$$= \frac{f(1.5001) - f(1.5)}{0.0001}$$

$$= \frac{(1.5001^2 - 4 \times 1.5001) - (1.5^2 - 4 \times 1.5)}{0.0001} = -0.9999.$$

This answer is accurate to 0.0001; in fact, $f'(1.5) = -1$.

Graphically, we can picture this approximation as follows: Zoom in on the curve using the window $1.5 \leq x \leq 1.5001$ and measure the slope of the secant line joining both ends of the curve segment. Figure 30 shows close-up views of the curve and tangent line near the point P in which we are interested, the third view being the zoomed-in view used for this approximation.

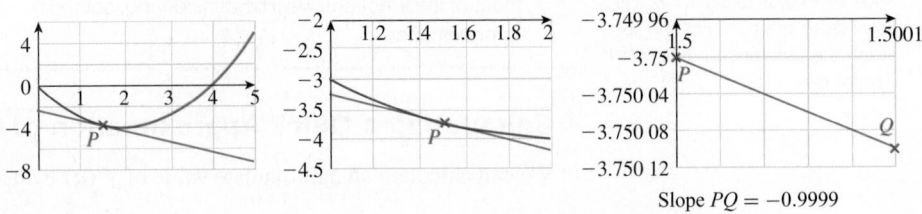

Slope $PQ = -0.9999$

Figure 30

Notice that in the third window the tangent line and curve are indistinguishable. Also, the point P in which we are interested is on the left edge of the window.

Balanced Difference Quotient: For the balanced difference quotient, we get

$$f'(1.5) \approx \frac{f(1.5 + 0.0001) - f(1.5 - 0.0001)}{2(0.0001)} \quad \text{Balanced difference quotient}$$

$$= \frac{f(1.5001) - f(1.4999)}{0.0002}$$

$$= \frac{(1.5001^2 - 4 \times 1.5001) - (1.4999^2 - 4 \times 1.4999)}{0.0002} = -1.$$

* The balanced difference quotient always gives the exact derivative for a quadratic function.

This balanced difference quotient gives the exact answer in this case!* Graphically, it is as though we have zoomed in using a window that puts the point P in the *center* of the screen (Figure 31) rather than at the left edge.

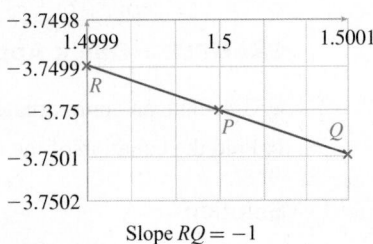

Slope $RQ = -1$

Figure 31

using Technology

See the Technology Guides at the end of the chapter to find out how to calculate the quick approximations to the derivative in Example 2 using a TI-83/84 Plus or a spreadsheet. Here is an outline:

TI-83/84 Plus
Y₁=X^2-4*X
Home screen:
  (Y₁(1.5001)-Y₁(1.5))/
  0.0001
  (Y₁(1.5001)-
  Y₁(1.4999))/0.0002
[More details on page 779.]

Spreadsheet
Headings *a*, *h*, *x*, *f(x)*, Diff Quotient, Balanced Diff Quotient in A1–F1
  1.5 in A2, 0.0001 in B2,
  =A2-B2 in C2, =A2 in C3,
  =A2+B2 in C4
  =C2^2-4*C2 in D2; copy down to D4
  =(D3-D2)/(C3-C2) in E2
  =(D4-D2)/(C4-C2) in E3
[More details on page 781.]

b. We find the equation of the tangent line from a point on the line and its slope, as we did in Chapter 1:

• **Point** $(1.5, f(1.5)) = (1.5, -3.75)$.
• **Slope** $m = f'(1.5) = -1$.   Slope of the tangent line = derivative.

The equation is

$$y = mx + b,$$

where $m = -1$ and $b = y_1 - mx_1 = -3.75 - (-1)(1.5) = -2.25$. Thus, the equation of the tangent line is

$$y = -x - 2.25.$$

Q: *Why can't we simply put* $h = 0.000\,000\,000\,000\,000\,000\,01$ *for an incredibly accurate approximation to the instantaneous rate of change and be done with it?*

A: This approach would certainly work if you were patient enough to do the (thankless) calculation by hand! However, doing it with the help of technology—even an ordinary calculator—will cause problems: The issue is that calculators and spreadsheets represent numbers with a maximum number of significant digits (15 in the case of Excel). As the value of h gets smaller, the value of $f(a + h)$ gets closer and closer to the value of $f(a)$. For example, if $f(x) = 50 + 0.1x^4$, Excel might compute

$$f(2 + 0.000\,000\,000\,000\,1) - f(2)$$
$$= 51.600\,000\,000\,000\,3 - 51.6 \qquad \text{\small Rounded to 15 digits}$$
$$= 0.000\,000\,000\,000\,3$$

and the corresponding difference quotient would be 3, not 3.2 as it should be. If h gets even smaller, Excel will not be able to distinguish between $f(a + h)$ and $f(a)$ at all, in which case it will compute 0 for the rate of change. This loss in accuracy when subtracting two very close numbers is called **subtractive error**.

Thus, there is a trade-off in lowering the value of h: Smaller values of h yield *mathematically* more accurate approximations of the derivative, but if h gets too small, subtractive error becomes a problem and decreases the accuracy of computations that use technology.

Leibniz *d* Notation

We introduced the notation $f'(x)$ for the derivative of f at x, but there is another interesting notation. We have written the average rate of change as

$$\text{Average rate of change} = \frac{\Delta f}{\Delta x}. \qquad \frac{\text{\small Change in } f}{\text{\small Change in } x}$$

As we use smaller and smaller values for Δx, we approach the instantaneous rate of change, or derivative, for which we also have the notation df/dx, due to Leibniz:

$$\text{Instantaneous rate of change} = \lim_{\Delta x \to 0} \frac{\Delta f}{\Delta x} = \frac{df}{dx}.$$

That is, df/dx is just another notation for $f'(x)$. Do not think of df/dx as an actual quotient of two numbers: Remember that we only use an actual quotient $\Delta f/\Delta x$ to *approximate* the value of df/dx.

In Example 3, we apply the quick approximation method of estimating the derivative.

EXAMPLE 3   Velocity

* Eric's claim is difficult to believe; 100 ft/s corresponds to around 68 mph, and professional pitchers can throw *forward* at about 100 mph.

My friend Eric, an enthusiastic baseball player, claims he can "probably" throw a ball upward at a speed of 100 feet per second (ft/s).* Our physicist friends tell us that its height s (in feet) t seconds later would be $s = 100t - 16t^2$. Find its average velocity over the interval $[2, 3]$ and its instantaneous velocity exactly 2 seconds after Eric throws it.

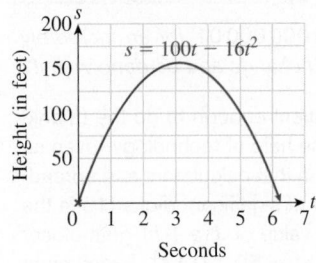

Figure 32

Solution The graph of the ball's height as a function of time is shown in Figure 32. Asking for the velocity is really asking for the rate of change of height with respect to time. (Why?) Consider average velocity first. To compute the **average velocity** of the ball from time 2 to time 3, we first compute the change in height:

$$\Delta s = s(3) - s(2) = 156 - 136 = 20 \text{ ft}.$$

Since it rises 20 feet in $\Delta t = 1$ second, we use the defining formula *speed = distance/time* to get the average velocity:

$$\text{Average velocity} = \frac{\Delta s}{\Delta t} = \frac{20}{1} = 20 \text{ ft/sec}.$$

from time $t = 2$ to $t = 3$. This is just the difference quotient, so

The average velocity is the average rate of change of height.

To get the **instantaneous velocity** at $t = 2$, we find the instantaneous rate of change of height. In other words, we need to calculate the derivative ds/dt at $t = 2$. Using the balanced quick approximation described earlier, we get

$$\frac{ds}{dt} \approx \frac{s(2 + 0.0001) - s(2 - 0.0001)}{2(0.0001)}$$

$$= \frac{s(2.0001) - s(1.9999)}{0.0002}$$

$$= \frac{100(2.0001) - 16(2.0001)^2 - (100(1.9999) - 16(1.9999)^2)}{0.0002}$$

$$= 36 \text{ ft/sec}.$$

In fact, this happens to be the exact answer; the instantaneous velocity at $t = 2$ is exactly 36 ft/sec. (Try an even smaller value of h to persuade yourself.)

➡ **Before we go on...** If we repeat the calculation in Example 3 at time $t = 5$, we get

$$\frac{ds}{dt} = -60 \text{ ft/sec}.$$

The negative sign tells us that the ball is *falling* at a rate of 60 feet per second at time $t = 5$. (How does the fact that it is falling at $t = 5$ show up on the graph?) ■

The preceding example gives another interpretation of the derivative.

Average and Instantaneous Velocity

For an object moving in a straight line with position $s(t)$ at time t, the **average velocity** from time t to time $t + h$ is the average rate of change of position with respect to time:

$$v_{ave} = \frac{s(t + h) - s(t)}{h} = \frac{\Delta s}{\Delta t}.$$

Average velocity = Average rate of change of position

The **instantaneous velocity** at time t is

$$v = \lim_{h \to 0} \frac{s(t + h) - s(t)}{h} = \frac{ds}{dt}.$$

Instantaneous velocity = Instantaneous rate of change of position

In other words, *instantaneous velocity is the derivative of position with respect to time.*

Here is one last comment on Leibniz notation. In Example 3, we could have written the velocity either as s' or as ds/dt, as we chose to do. To write the answer to the question, that the velocity at $t = 2$ sec was 36 ft/sec, we can write either

$$s'(2) = 36$$

or

$$\left.\frac{ds}{dt}\right|_{t=2} = 36.$$

The notation "$|_{t=2}$" is read "evaluated at $t = 2$." Similarly, if $y = f(x)$, we can write the instantaneous rate of change of f at $x = 5$ in either functional notation as

$$f'(5) \qquad \text{The derivative of } f, \text{ evaluated at } x = 5$$

or in Leibniz notation as

$$\left.\frac{dy}{dx}\right|_{x=5}. \qquad \text{The derivative of } y, \text{ evaluated at } x = 5$$

The latter notation is obviously more cumbersome than the functional notation $f'(5)$, but the notation dy/dx has compensating advantages. You should practice using both notations.

The Derivative Function

The derivative $f'(x)$ is a number we can calculate, or at least approximate, for various values of x. Because $f'(x)$ depends on the value of x, we may think of f' as a function of x. This function is the **derivative function**.

Derivative Function

If f is a function, its **derivative function** f' is the function whose value $f'(x)$ is the derivative of f at x. Its domain is the set of all x at which f is differentiable. Equivalently, f' associates to each x the slope of the tangent to the graph of the function f at x, or the rate of change of f at x. The formula for the derivative function is

$$f'(x) = \lim_{h \to 0} \frac{f(x+h) - f(x)}{h}. \qquad \text{Derivative function}$$

Quick Examples

1. Let $f(x) = 3x - 1$. The graph of f is a straight line that has slope 3 everywhere. In other words, $f'(x) = 3$ for every choice of x; that is, f' is a constant function.

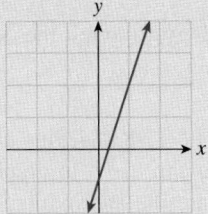

Original Function f
$f(x) = 3x - 1$

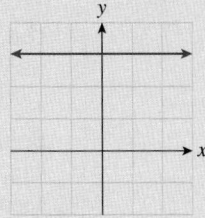

Derivative Function f'
$f'(x) = 3$

Chapter **10**   Introduction to the Derivative

* This method is discussed in
detail on the Website at

Online Text → Sketching the
Graph of the Derivative.

2. Given the graph of a function f, we can get a rough sketch of the graph of f' by estimating the slope of the tangent to the graph of f at several points, as illustrated below.*

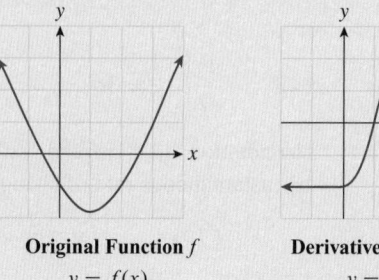

Original Function f
$y = f(x)$

Derivative Function f'
$y = f'(x)$

For x between -2 and 0, the graph of f is linear with slope -2. As x increases from 0 to 2, the slope increases from -2 to 2. For x larger than 2, the graph of f is linear with slope 2. (Notice that, when $x = 1$, the graph of f has a horizontal tangent, so $f'(1) = 0$.)

3. Look again at the graph on the left in Quick Example 2. When $x < 1$ the derivative $f'(x)$ is negative, so the graph has negative slope and f is **decreasing**; its values are going down as x increases. When $x > 1$ the derivative $f'(x)$ is positive, so the graph has positive slope and f is **increasing**; its values are going up as x increases.

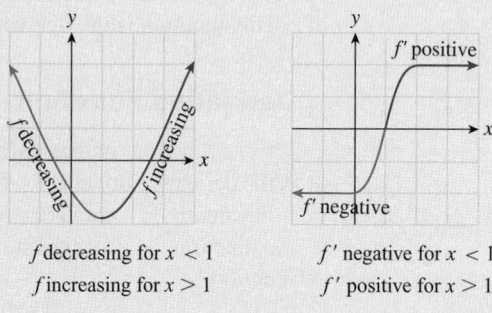

f decreasing for $x < 1$
f increasing for $x > 1$

f' negative for $x < 1$
f' positive for $x > 1$

The following example shows how we can use technology to graph the (approximate) derivative of a function, where it exists.

EXAMPLE 4 🎛 Graphing the Derivative with Technology

Use technology to graph the derivative of $f(x) = -2x^2 + 6x + 5$ for values of x starting at -5.

Solution The TI-83/84 Plus has a built-in function that approximates the derivative, and we can use it to graph the derivative of a function. In a spreadsheet, we need to create the approximation using one of the quick approximation formulas and we can then graph a table of its values. See the technology note in the margin on the next page to find out how to graph the derivative (Figure 33) using the Website graphing utility, the TI-83/84 Plus, and a spreadsheet.

using Technology

See the Technology Guides at the end of the chapter to find out how to obtain a table of values of and graph the derivative in Example 4 using a TI-83/84 Plus or a spreadsheet. Here is an outline:

TI-83/84 Plus
```
Y₁=-2X^2+6X+5
Y₂=nDeriv(Y₁,X,X)
```
[More details on page 779.]

Spreadsheet
Value of h in E2
Values of x from A2 down increasing by h
-2*A2^2+6*A2+5 from B2 down
=(B3-B2)/$E$2 from C2 down
Insert scatter chart using columns A and C [More details on page 781.]

 Website
www.WanerMath.com
Web grapher:

Online Utilities→ Function Evaluator and Grapher

Enter
`deriv(-2*x^2+6*x+5)` for
y₁. Alternatively, enter
`-2*x^2+6*x+5` for y₁ and
`deriv(y1)` for y₂.

Excel grapher:
Student Home→ Online
Utilities→ Excel First and
Second Derivative Graphing Utility
Function: `-2*x^2+6*x+5`

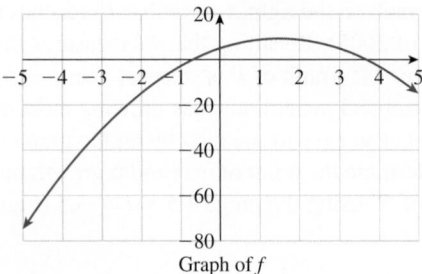

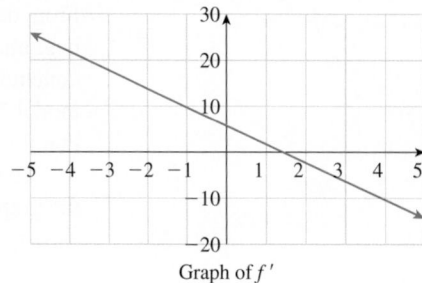

Figure 33

We said that f' records the slope of (the tangent line to) the function f at each point. Notice that the graph of f' confirms that the slope of the graph of f is decreasing as x increases from -5 to 5. Note also that the graph of f reaches a high point at $x = 1.5$ (the vertex of the parabola). At that point, the slope of the tangent is zero; that is, $f'(1.5) = 0$, as we see in the graph of f'.

EXAMPLE 5 ⓣ An Application: Broadband Penetration

Wired broadband penetration in the United States can be modeled by

$$P(t) = \frac{29.842}{1 + 12.502(1.642)^{-t}} \qquad (0 \le t \le 12),$$

where t is time in years since 2000.[52] Graph both P and its derivative, and determine when broadband penetration was growing most rapidly.

Solution Using one of the methods in Example 4, we obtain the graphs shown in Figure 34.

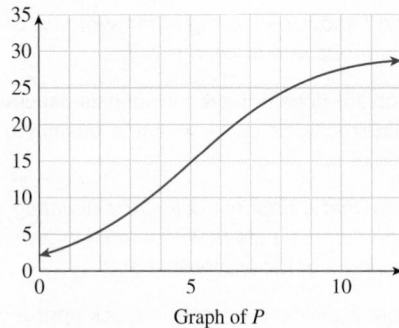

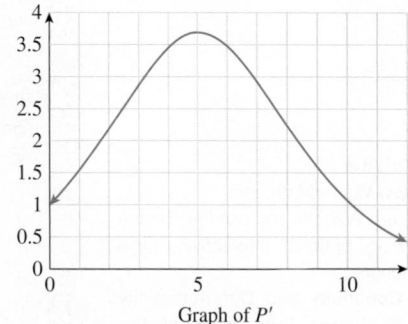

Figure 34

---

[52]Broadband penetration is the number of broadband installations divided by the total population. Source for data: Organisation for Economic Co-operation and Development (OECD) Directorate for Science, Technology, and Industry, table of Historical Penetration Rates, June 2010, downloaded April 2011 from www.oecd.org/sti/ict/broadband.

From the graph on the right, we see that P' reaches a peak somewhere near $t = 5$ (the beginning of 2005). Recalling that P' measures the *slope* of the graph of P, we can conclude that the graph of P is steepest near $t = 5$, indicating that, according to the model, broadband penetration was growing most rapidly at the start of 2005. Notice that this is not so easy to see directly on the graph of P.

To determine the point of maximum growth more accurately, we can zoom in on the graph of P' using the range $4.5 \le t \le 5.5$ (Figure 35).

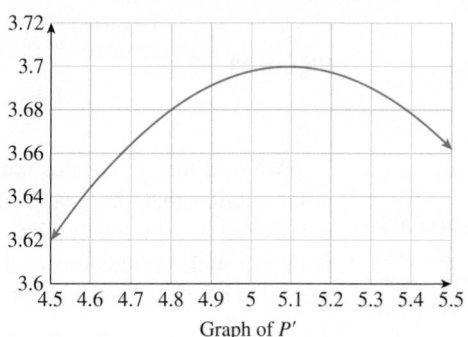

Graph of P'

Figure 35

We can now see that P' reaches its highest point around $t = 5.1$, so we conclude that broadband penetration was growing most rapidly in early 2005.

➡ **Before we go on...** Besides helping us to determine the point of maximum growth, the graph of P' in Example 5 gives us a great deal of additional information. As just one example, in Figure 35 we can see that the maximum value of P' is about 3.7, indicating that broadband penetration grew at a fastest rate of about 3.7 percentage points per year. ■

FAQs

Recognizing When and How to Compute the Instantaneous Rate of Change

Q: *How do I know, by looking at the wording of a problem, that it is asking for an instantaneous rate of change?*

A: If a problem does not ask for an instantaneous rate of change directly, it might do so indirectly, as in "How fast is quantity q increasing?" or "Find the rate of increase of q."

Q: *If I know that a problem calls for estimating an instantaneous rate of change, how should I estimate it: with a table showing smaller and smaller values of h, or by using a quick approximation?*

A: For most practical purposes, a quick approximation is accurate enough. Use a table showing smaller and smaller values of h when you would like to check the accuracy.

Q: *Which should I use in computing a quick approximation: the balanced difference quotient or the ordinary difference quotient?*

A: In general, the balanced difference quotient gives a more accurate answer.

 Website
www.WanerMath.com
At the Website you can find the following optional interactive online sections:
- **Continuity and Differentiability**
- **Sketching the Graph of the Derivative**
You can find these sections by following
Everything → Chapter 10 (Online Sections)

10.5 EXERCISES

In Exercises 1–4, estimate the derivative from the table of average rates of change. HINT [See discussion at the beginning of the section.]

1.

h	1	0.1	0.01	0.001	0.0001
Average Rate of Change of f Over [5, 5 + h]	12	6.4	6.04	6.004	6.0004
h	−1	−0.1	−0.01	−0.001	−0.0001
Average Rate of Change of f Over [5 + h, 5]	3	5.6	5.96	5.996	5.9996

Estimate $f'(5)$.

2.

h	1	0.1	0.01	0.001	0.0001
Average Rate of Change of g Over [7, 7 + h]	4	4.8	4.98	4.998	4.9998
h	−1	−0.1	−0.01	−0.001	−0.0001
Average Rate of Change of g Over [7 + h, 7]	5	5.3	5.03	5.003	5.0003

Estimate $g'(7)$.

3.

h	1	0.1	0.01	0.001	0.0001
Average Rate of Change of r Over [−6, −6 + h]	−5.4	−5.498	−5.4998	−5.499982	−5.49999822
h	−1	−0.1	−0.01	−0.001	−0.0001
Average Rate of Change of r Over [−6 + h, −6]	−7.52	−6.13	−5.5014	−5.5000144	−5.500001444

Estimate $r'(-6)$.

4.

h	1	0.1	0.01	0.001	0.0001
Average Rate of Change of s Over [0, h]	−2.52	−1.13	−0.6014	−0.6000144	−0.600001444
h	−1	−0.1	−0.01	−0.001	−0.0001
Average Rate of Change of s Over [h, 0]	−0.4	−0.598	−0.5998	−0.599982	−0.59999822

Estimate $s'(0)$.

Consider the functions in Exercises 5–8 as representing the value of an ounce of palladium in U.S. dollars as a function of the time t in days.[53] Find the average rates of change of R(t) over the time intervals $[t, t + h]$, *where t is as indicated and* $h = 1, 0.1,$ *and* 0.01 *days. Hence, estimate the instantaneous rate of change of R at time t, specifying the units of measurement. (Use smaller values of h to check your estimates.)* HINT [See Example 1.]

5. $R(t) = 60 + 50t - t^2; t = 5$

6. $R(t) = 60t - 2t^2; t = 3$

7. $R(t) = 270 + 20t^3; t = 1$

8. $R(t) = 200 + 50t - t^3; t = 2$

Each of the functions in Exercises 9–12 gives the cost to manufacture x items. Find the average cost per unit of manufacturing h more items (i.e., the average rate of change of the total cost) at a production level of x, where x is as indicated and $h = 10$ *and* 1. *Hence, estimate the instantaneous rate of change of the total cost at the given production level x, specifying the units of measurement. (Use smaller values of h to check your estimates.)* HINT [See Example 1.]

9. $C(x) = 10{,}000 + 5x - \dfrac{x^2}{10{,}000}; x = 1{,}000$

10. $C(x) = 20{,}000 + 7x - \dfrac{x^2}{20{,}000}; x = 10{,}000$

11. $C(x) = 15{,}000 + 100x + \dfrac{1{,}000}{x}; x = 100$

12. $C(x) = 20{,}000 + 50x + \dfrac{10{,}000}{x}; x = 100$

In Exercises 13–16, the graph of a function is shown together with the tangent line at a point P. Estimate the derivative of f at the corresponding x value. HINT [See Quick Example page 742.]

13.
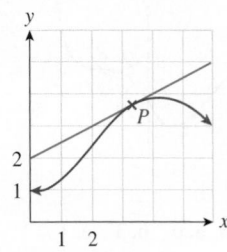

14.
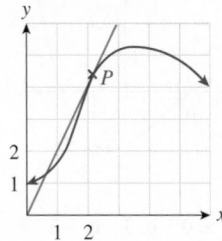

15.
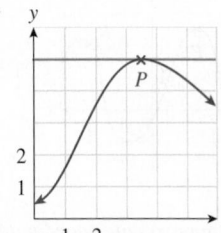

16.
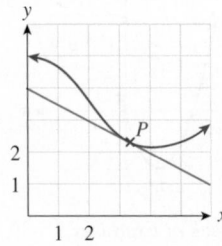

---

[53]Palladium was trading at around $290 in August 2008.

*In each of the graphs given in Exercises 17–22, say at which labeled point the slope of the tangent is (**a**) greatest and (**b**) least (in the sense that −7 is less than 1).* HINT [See Quick Example page 742.]

17.
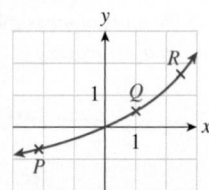

18.
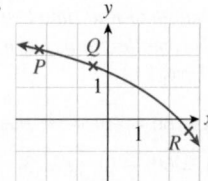

19.
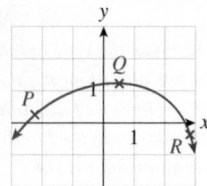

20.
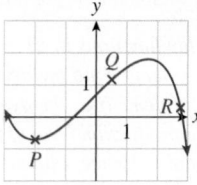

21.
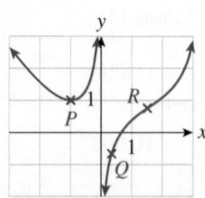

22.
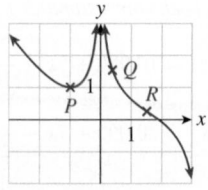

In each of Exercises 23–26, three slopes are given. For each slope, determine at which of the labeled points on the graph the tangent line has that slope.

23. a. 0   **b.** 4   **c.** −1
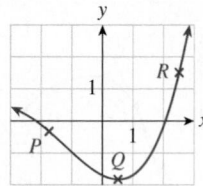

24. a. 0   **b.** 1   **c.** −1
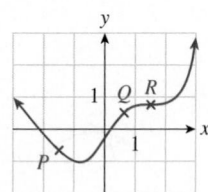

25. a. 0   **b.** 3   **c.** −3
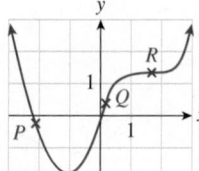

26. a. 0   **b.** 3   **c.** 1
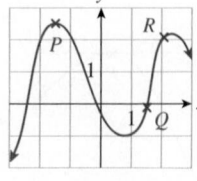

*In each of Exercises 27–30, find the approximate coordinates of all points (if any) where the slope of the tangent is: (**a**) 0, (**b**) 1, (**c**) −1.* HINT [See Quick Example page 742.]

27.
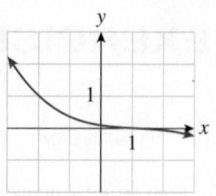

28.
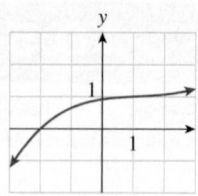

29.
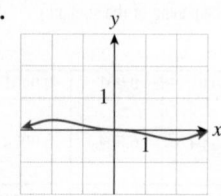

30.
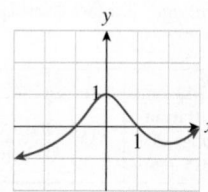

31. Complete the following: The tangent to the graph of the function f at the point where $x = a$ is the line passing through the point ____ with slope ____ .

32. Complete the following: The difference quotient for f at the point where $x = a$ gives the slope of the ____ line that passes through ____ .

33. Which is correct? The derivative function assigns to each value x

(**A**) the average rate of change of f at x.
(**B**) the slope of the tangent to the graph of f at $(x, f(x))$.
(**C**) the rate at which f is changing over the interval $[x, x + h]$ for $h = 0.0001$.
(**D**) the balanced difference quotient $[f(x + h) - f(x - h)]/(2h)$ for $h \approx 0.0001$.

34. Which is correct? The derivative function $f'(x)$ tells us
(**A**) the slope of the tangent line at each of the points $(x, f(x))$.
(**B**) the approximate slope of the tangent line at each of the points $(x, f(x))$.
(**C**) the slope of the secant line through $(x, f(x))$ and $(x + h, f(x + h))$ for $h = 0.0001$.
(**D**) the slope of a certain secant line through each of the points $(x, f(x))$.

35. ▼ Let f have the graph shown.

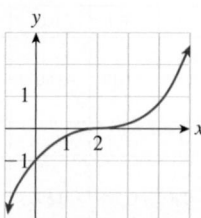

a. The average rate of change of f over the interval $[2, 4]$ is
(**A**) greater than $f'(2)$.   (**B**) less than $f'(2)$.
(**C**) approximately equal to $f'(2)$.

b. The average rate of change of f over the interval $[-1, 1]$ is
(**A**) greater than $f'(0)$.   (**B**) less than $f'(0)$.
(**C**) approximately equal to $f'(0)$.

c. Over the interval $[0, 2]$, the instantaneous rate of change of f is
(A) increasing.   (B) decreasing.   (C) neither.

d. Over the interval $[0, 4]$, the instantaneous rate of change of f is
(A) increasing, then decreasing.
(B) decreasing, then increasing.
(C) always increasing.
(D) always decreasing.

e. When $x = 4$, $f(x)$ is
(A) approximately 0, and increasing at a rate of about 0.7 units per unit of x.
(B) approximately 0, and decreasing at a rate of about 0.7 units per unit of x.
(C) approximately 0.7, and increasing at a rate of about 1 unit per unit of x.
(D) approximately 0.7, and increasing at a rate of about 3 units per unit of x.

36. ▼ A function f has the following graph.

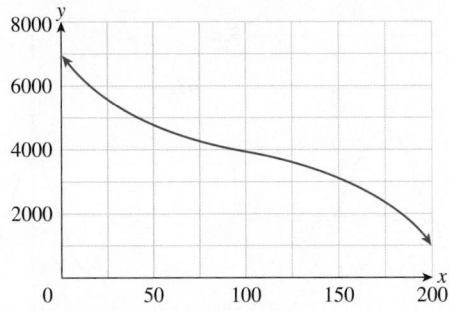

a. The average rate of change of f over $[0, 200]$ is
(A) greater than
(B) less than
(C) approximately equal to
the instantaneous rate of change at $x = 100$.

b. The average rate of change of f over $[0, 200]$ is
(A) greater than
(B) less than
(C) approximately equal to
the instantaneous rate of change at $x = 150$.

c. Over the interval $[0, 50]$ the instantaneous rate of change of f is
(A) increasing, then decreasing.
(B) decreasing, then increasing.
(C) always increasing.
(D) always decreasing.

d. On the interval $[0, 200]$, the instantaneous rate of change of f is
(A) always positive.   (B) always negative.
(C) negative, positive, and then negative.

e. $f'(100)$ is
(A) greater than $f'(25)$.   (B) less than $f'(25)$.
(C) approximately equal to $f'(25)$.

In Exercises 37–40, use a quick approximation to estimate the derivative of the given function at the indicated point. HINT [See Example 2(a).]

37. $f(x) = 1 - 2x$; $x = 2$ **38.** $f(x) = \dfrac{x}{3} - 1$; $x = -3$

39. $f(x) = \dfrac{x^2}{4} - \dfrac{x^3}{3}$; $x = -1$ **40.** $f(x) = \dfrac{x^2}{2} + \dfrac{x}{4}$; $x = 2$

In Exercises 41–48, estimate the indicated derivative by any method. HINT [See Example 2.]

41. $g(t) = \dfrac{1}{t^5}$; estimate $g'(1)$

42. $s(t) = \dfrac{1}{t^3}$; estimate $s'(-2)$

43. $y = 4x^2$; estimate $\left.\dfrac{dy}{dx}\right|_{x=2}$

44. $y = 1 - x^2$; estimate $\left.\dfrac{dy}{dx}\right|_{x=-1}$

45. $s = 4t + t^2$; estimate $\left.\dfrac{ds}{dt}\right|_{t=-2}$

46. $s = t - t^2$; estimate $\left.\dfrac{ds}{dt}\right|_{t=2}$

47. $R = \dfrac{1}{p}$; estimate $\left.\dfrac{dR}{dp}\right|_{p=20}$

48. $R = \sqrt{p}$; estimate $\left.\dfrac{dR}{dp}\right|_{p=400}$

In Exercises 49–54, (a) use any method to estimate the slope of the tangent to the graph of the given function at the point with the given x-coordinate, and (b) find an equation of the tangent line in part (a). In each case, sketch the curve together with the appropriate tangent line. HINT [See Example 2(b).]

49. $f(x) = x^3$; $x = -1$

50. $f(x) = x^2$; $x = 0$

51. $f(x) = x + \dfrac{1}{x}$; $x = 2$

52. $f(x) = \dfrac{1}{x^2}$; $x = 1$

53. $f(x) = \sqrt{x}$; $x = 4$

54. $f(x) = 2x + 4$; $x = -1$

In each of Exercises 55–58, estimate the given quantity.

55. $f(x) = e^x$; estimate $f'(0)$

56. $f(x) = 2e^x$; estimate $f'(1)$

57. $f(x) = \ln x$; estimate $f'(1)$

58. $f(x) = \ln x$; estimate $f'(2)$

In Exercises 59–64, match the graph of f to the graph of f′ (the graphs of f′ are shown after Exercise 64).

59. ▼

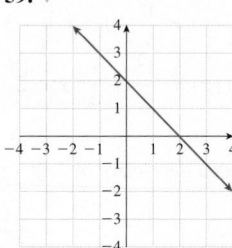

60. ▼

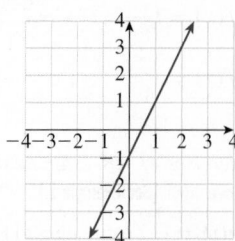

61. ▼

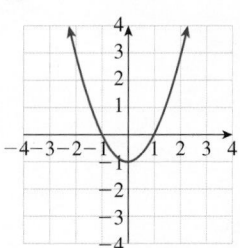

62. ▼

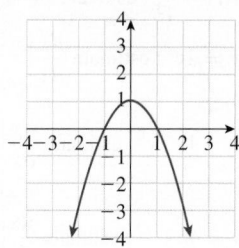

63. ▼

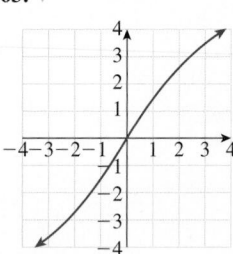

64. ▼

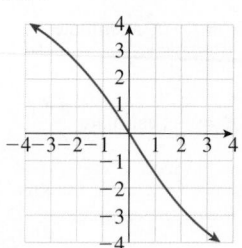

Graphs of derivatives for Exercises 59–64:

(A)

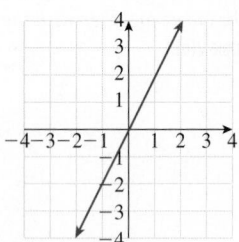

(B)

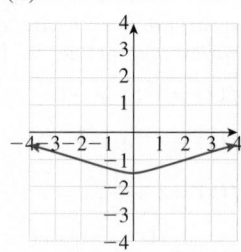

(C)

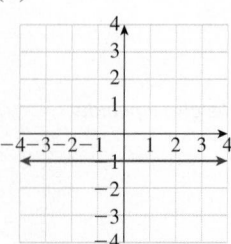

(D)

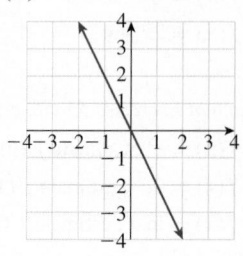

(E)

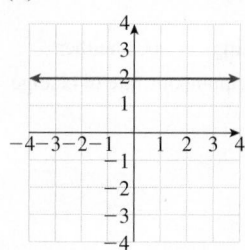

(F)

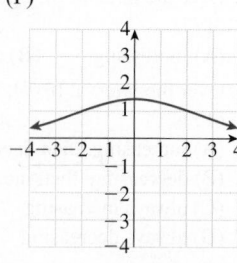

In Exercises 65–68, the graph of a function is given. For which x in the range shown is the function increasing? For which x is the function decreasing? HINT [See Quick Example 3 page 748.]

65.

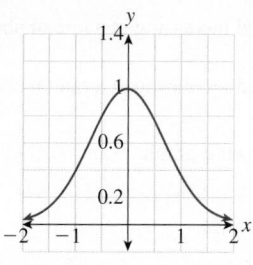

66.

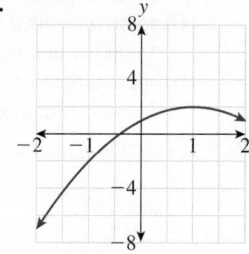

67.

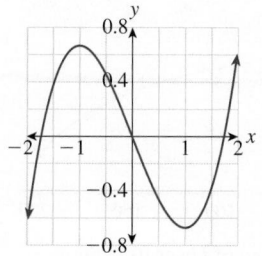

68.

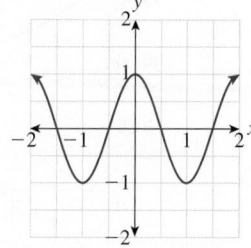

In Exercises 69–72, the graph of the derivative of a function is given. For which x is the (original) function increasing? For which x is the (original) function decreasing? HINT [See Quick Example 3 page 748.]

69.

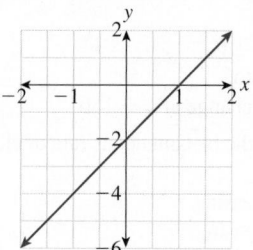

70.

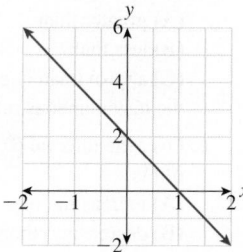

71.

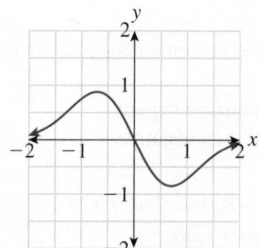

72.

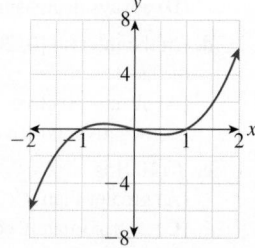

T *In Exercises 73 and 74, use technology to graph the derivative of the given function for the given range of values of x. Then use your graph to estimate all values of x (if any) where the tangent line to the graph of the given function is horizontal. Round answers to one decimal place.* HINT [See Example 4.]

73. $f(x) = x^4 + 2x^3 - 1; \quad -2 \leq x \leq 1$

74. $f(x) = -x^3 - 3x^2 - 1; \quad -3 \leq x \leq 1$

T *In Exercises 75 and 76, use the method of Example 4 to list approximate values of f'(x) for x in the given range. Graph f(x) together with f'(x) for x in the given range.*

75. $f(x) = \dfrac{x+2}{x-3}; \quad 4 \leq x \leq 5$

76. $f(x) = \dfrac{10x}{x-2}; \quad 2.5 \leq x \leq 3$

APPLICATIONS

77. *Demand* Suppose the demand for a new brand of sneakers is given by

$$q = \frac{5{,}000{,}000}{p}$$

where p is the price per pair of sneakers, in dollars, and q is the number of pairs of sneakers that can be sold at price p. Find $q(100)$ and estimate $q'(100)$. Interpret your answers. HINT [See Example 1.]

78. *Demand* Suppose the demand for an old brand of TV is given by

$$q = \frac{100{,}000}{p+10}$$

where p is the price per TV set, in dollars, and q is the number of TV sets that can be sold at price p. Find $q(190)$ and estimate $q'(190)$. Interpret your answers. HINT [See Example 1.]

79. *Oil Imports from Mexico* The following graph shows approximate daily oil imports to the U.S. from Mexico.[54] Also shown is the tangent line at the point corresponding to year 2005.

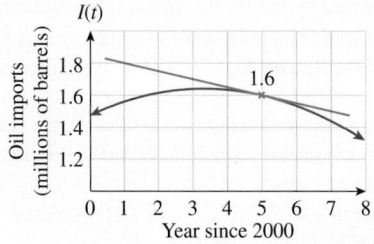

a. Estimate the slope of the tangent line shown on the graph. What does the graph tell you about oil imports from Mexico in 2005? HINT [Identify two points on the tangent line. Then see Quick Example page 742.]

b. According to the graph, is the rate of change of oil imports from Mexico increasing, decreasing, or increasing then decreasing? Why?

80. *Oil Production in Mexico* The following graph shows approximate daily oil production by **Pemex**, Mexico's national oil company.[55] Also shown is the tangent line at the point corresponding to year 2003.

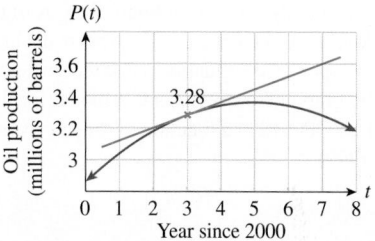

a. Estimate the slope of the tangent line shown on the graph. What does the graph tell you about oil production by Pemex in 2003? HINT [Identify two points on the tangent line. Then see Quick Example page 742.]

b. According to the graph, is the rate of change of oil production by Pemex increasing or decreasing over the range [0, 4]? Why?

81. ▼ *Prison Population* The following curve is a model of the total U.S. prison population as a function of time in years ($t = 0$ represents 2000).[56]

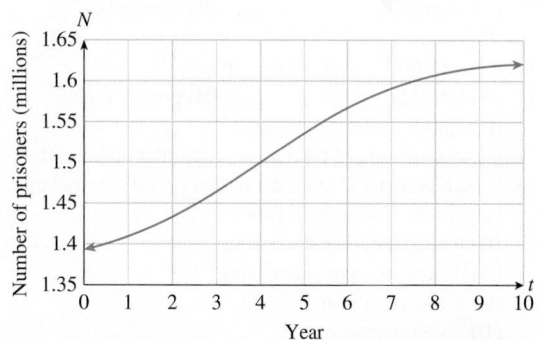

a. Which is correct? Over the period [5, 10] the instantaneous rate of change of N is
(A) increasing. (B) decreasing.

b. Which is correct? The instantaneous rate of change of prison population at $t = 4$ was
(A) less than (B) greater than
(C) approximately equal to
the average rate of change over the interval [0, 10].

[54]Figures are approximate, and 2008–2009 figures are projections by the Department of Energy. Source: Energy Information Administration/Pemex (www.eia.doe.gov).

[55]Ibid.

[56]Source: Bureau of Justice Statistics http://bjs.ojp.usdoj.gov.

c. Which is correct? Over the period [0, 10] the instantaneous rate of change of N is
(A) increasing, then decreasing.
(B) decreasing, then increasing.
(C) always increasing.
(D) always decreasing.

d. According to the model, the U.S. prison population was increasing fastest around what year?

e. Roughly estimate the instantaneous rate of change of N at $t = 4$ by using a balanced difference quotient with $h = 1.5$. Interpret the result.

82. ▼ *Demand for Freon 12* The demand for chlorofluorocarbon-12 (CFC-12)—the ozone-depleting refrigerant commonly known as Freon 12[57]—has been declining significantly in response to regulation and concern about the ozone layer. The graph below represents a model for the projected demand for CFC-12 as a function of time in years ($t = 0$ represents 1990).[58]

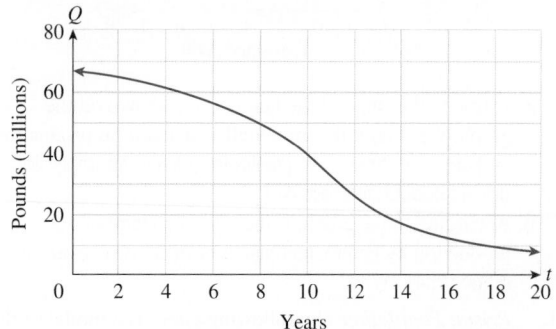

a. Which is correct? Over the period [12, 20] the instantaneous rate of change of Q is
(A) increasing.   (B) decreasing.

b. Which is correct? The instantaneous rate of change of demand for Freon 12 at $t = 10$ was
(A) less than   (B) greater than
(C) approximately equal to
the average rate of change over the interval [0, 20].

c. Which is correct? Over the period [0, 20] the instantaneous rate of change of Q is
(A) increasing, then decreasing.
(B) decreasing, then increasing.
(C) always increasing.
(D) always decreasing.

d. According to the model, the demand for Freon 12 was decreasing most rapidly around what year?

e. Roughly estimate the instantaneous rate of change of Q at $t = 13$ by using a balanced difference quotient with $h = 5$. Interpret the result.

83. *Velocity* If a stone is dropped from a height of 400 feet, its height after t seconds is given by $s = 400 - 16t^2$.
a. Find its average velocity over the period [2, 4].
b. Estimate its instantaneous velocity at time $t = 4$. HINT [See Example 3.]

84. *Velocity* If a stone is thrown down at 120 ft/s from a height of 1,000 feet, its height after t seconds is given by $s = 1,000 - 120t - 16t^2$.
a. Find its average velocity over the period [1, 3].
b. Estimate its instantaneous velocity at time $t = 3$. HINT [See Example 3.]

85. *Crude Oil Prices* The price per barrel of crude oil in constant 2008 dollars can be approximated by
$$P(t) = 0.45t^2 - 12t + 105 \text{ dollars} \quad (0 \le t \le 28),$$
where t is time in years since the start of 1980.[59]
a. Compute the average rate of change of $P(t)$ over the interval [0, 28], and interpret your answer. HINT [See Section 10.4 Example 3.]
b. Estimate the instantaneous rate of change of $P(t)$ at $t = 0$, and interpret your answer. HINT [See Example 2(a).]
c. The answers to part (a) and part (b) have opposite signs. What does this indicate about the price of oil?

86. *Median Home Prices* The median home price in the U.S. over the period 2003–2011 can be approximated by
$$P(t) = -5t^2 + 75t - 30 \text{ thousand dollars} \quad (3 \le t \le 11),$$
where t is time in years since the start of 2000.[60]
a. Compute the average rate of change of $P(t)$ over the interval [5, 9], and interpret your answer. HINT [See Section 10.4 Example 3.]
b. Estimate the instantaneous rate of change of $P(t)$ at $t = 5$, and interpret your answer. HINT [See Example 2(a).]
c. The answer to part (b) has larger absolute value than the answer to part (a). What does this indicate about the median home price?

87. *SARS* In the early stages of the deadly SARS (Severe Acute Respiratory Syndrome) epidemic in 2003, the number of reported cases could be approximated by
$$A(t) = 167(1.18)^t \quad (0 \le t \le 20)$$
t days after March 17, 2003 (the first day in which statistics were reported by the World Health Organization).
a. What, approximately, was the instantaneous rate of change of $A(t)$ on March 27 ($t = 10$)? Interpret the result.
b. Which of the following is true? For the first 20 days of the epidemic, the instantaneous rate of change of the number of cases
(A) increased.   (B) decreased.
(C) increased and then decreased.
(D) decreased and then increased.

---

[57]The name given to it by DuPont. Freon 12 (dichlorodifluoromethane) is distinct from Freon 22 (chlorodifluoromethane, also registered by DuPont; see p. 232).

[58]Source for data: The Automobile Consulting Group (*New York Times*, December 26, 1993, p. F23). The exact figures were not given, and the chart is a reasonable facsimile of the chart that appeared in *New York Times*.

---

[59]Source for data: www.inflationdata.com.
[60]Source for data: www.zillow.com.

Solution

a. $f'(x) = \lim_{h\to 0} \dfrac{f(x+h) - f(x)}{h}$ Derivative formula

$$= \lim_{h\to 0} \dfrac{\overbrace{(x+h)^3}^{f(x+h)} - \overbrace{x^3}^{f(x)}}{h}$$ Substitute for $f(x)$ and $f(x+h)$.

$$= \lim_{h\to 0} \dfrac{(x^3 + 3x^2h + 3xh^2 + h^3) - x^3}{h}$$ Expand $(x+h)^3$.

$$= \lim_{h\to 0} \dfrac{3x^2h + 3xh^2 + h^3}{h}$$ Cancel the x^3.

$$= \lim_{h\to 0} \dfrac{h(3x^2 + 3xh + h^2)}{h}$$ Factor out h.

$$= \lim_{h\to 0} (3x^2 + 3xh + h^2)$$ Cancel the h.

$$= 3x^2.$$ Let h approach 0.

b. $f'(x) = \lim_{h\to 0} \dfrac{f(x+h) - f(x)}{h}$ Derivative formula

$$= \lim_{h\to 0} \dfrac{\overbrace{(2(x+h)^2 - (x+h))}^{f(x+h)} - \overbrace{(2x^2 - x)}^{f(x)}}{h}$$ Substitute for $f(x)$ and $f(x+h)$.

$$= \lim_{h\to 0} \dfrac{(2x^2 + 4xh + 2h^2 - x - h) - (2x^2 - x)}{h}$$ Expand.

$$= \lim_{h\to 0} \dfrac{4xh + 2h^2 - h}{h}$$ Cancel the $2x^2$ and x.

$$= \lim_{h\to 0} \dfrac{h(4x + 2h - 1)}{h}$$ Factor out h.

$$= \lim_{h\to 0} (4x + 2h - 1)$$ Cancel the h.

$$= 4x - 1.$$ Let h approach 0.

c. $f'(x) = \lim_{h\to 0} \dfrac{f(x+h) - f(x)}{h}$ Derivative formula

$$= \lim_{h\to 0} \dfrac{\left[\overbrace{\dfrac{1}{x+h}}^{f(x+h)} - \overbrace{\dfrac{1}{x}}^{f(x)}\right]}{h}$$ Substitute for $f(x)$ and $f(x+h)$.

$$= \lim_{h\to 0} \dfrac{\left[\dfrac{x - (x+h)}{(x+h)x}\right]}{h}$$ Subtract the fractions.

$$= \lim_{h\to 0} \dfrac{1}{h}\left[\dfrac{x - (x+h)}{(x+h)x}\right]$$ Dividing by h = Multiplying by $1/h$.

$$= \lim_{h \to 0} \left[\frac{-h}{h(x+h)x} \right] \qquad \text{Simplify.}$$

$$= \lim_{h \to 0} \left[\frac{-1}{(x+h)x} \right] \qquad \text{Cancel the } h.$$

$$= \frac{-1}{x^2} \qquad \text{Let } h \text{ approach } 0.$$

In Example 4, we redo Example 3 of Section 10.5, this time getting an exact, rather than approximate, answer.

EXAMPLE 4   Velocity

My friend Eric, an enthusiastic baseball player, claims he can "probably" throw a ball upward at a speed of 100 feet per second (ft/sec). Our physicist friends tell us that its height s (in feet) t seconds later would be $s(t) = 100t - 16t^2$. Find the ball's instantaneous velocity function and its velocity exactly 2 seconds after Eric throws it.

Solution The instantaneous velocity function is the derivative ds/dt, which we calculate as follows:

$$\frac{ds}{dt} = \lim_{h \to 0} \frac{s(t+h) - s(t)}{h}.$$

Let us compute $s(t+h)$ and $s(t)$ separately:

$$s(t) = 100t - 16t^2$$
$$s(t+h) = 100(t+h) - 16(t+h)^2$$
$$= 100t + 100h - 16(t^2 + 2th + h^2)$$
$$= 100t + 100h - 16t^2 - 32th - 16h^2.$$

Therefore,

$$\frac{ds}{dt} = \lim_{h \to 0} \frac{s(t+h) - s(t)}{h}$$

$$= \lim_{h \to 0} \frac{100t + 100h - 16t^2 - 32th - 16h^2 - (100t - 16t^2)}{h}$$

$$= \lim_{h \to 0} \frac{100h - 32th - 16h^2}{h}$$

$$= \lim_{h \to 0} \frac{h(100 - 32t - 16h)}{h}$$

$$= \lim_{h \to 0} (100 - 32t - 16h)$$

$$= 100 - 32t \text{ ft/sec.}$$

Thus, the velocity exactly 2 seconds after Eric throws it is

$$\left. \frac{ds}{dt} \right|_{t=2} = 100 - 32(2) = 36 \text{ ft/sec.}$$

This verifies the accuracy of the approximation we made in Section 10.5.

➡ **Before we go on...** From the derivative function in Example 4, we can now describe the behavior of the velocity of the ball: Immediately on release ($t = 0$) the ball is traveling at 100 feet per second upward. The ball then slows down; precisely, it loses 32 feet per second of speed every second. When, exactly, does the velocity become zero and what happens after that? ∎

Q : *Do we always have to calculate the limit of the difference quotient to find a formula for the derivative function?*

A : As it turns out, no. In Section 11.1 we will start to look at shortcuts for finding derivatives that allow us to bypass the definition of the derivative in many cases.

A Function Not Differentiable at a Point

Recall from Section 10.5 that a function is **differentiable** at a point a if $f'(a)$ exists; that is, if the difference quotient $[f(a + h) - f(a)]/h$ approaches a fixed value as h approaches 0. In Section 10.5, we mentioned that the function $f(x) = |x|$ is not differentiable at $x = 0$. In Example 5, we find out why.

EXAMPLE 5 A Function Not Differentiable at 0

Numerically, graphically, and algebraically investigate the differentiability of the function $f(x) = |x|$ at the points **(a)** $x = 1$ and **(b)** $x = 0$.

Solution

a. We compute

$$f'(1) = \lim_{h \to 0} \frac{f(1 + h) - f(1)}{h}$$

$$= \lim_{h \to 0} \frac{|1 + h| - 1}{h}.$$

Numerically, we can make tables of the values of the average rate of change $(|1 + h| - 1)/h$ for h positive or negative and approaching 0:

h	1	0.1	0.01	0.001	0.0001
Average Rate of Change Over [1, 1 + h]	1	1	1	1	1

h	−1	−0.1	−0.01	−0.001	−0.0001
Average Rate of Change Over [1 + h, 1]	1	1	1	1	1

From these tables it appears that $f'(1)$ is equal to 1. We can verify that algebraically: For h that is sufficiently small, $1 + h$ is positive (even if h is negative) and so

$$f'(1) = \lim_{h \to 0} \frac{1 + h - 1}{h}$$

$$= \lim_{h \to 0} \frac{h}{h} \qquad \text{Cancel the 1s.}$$

$$= \lim_{h \to 0} 1 \qquad \text{Cancel the } h.$$

$$= 1.$$

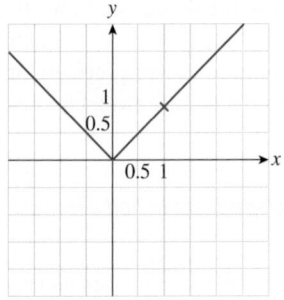

Figure 38

Graphically, we are seeing the fact that the tangent line at the point $(1, 1)$ has slope 1 because the graph is a straight line with slope 1 near that point (Figure 38).

b. $f'(0) = \lim\limits_{h \to 0} \dfrac{f(0 + h) - f(0)}{h}$

$\qquad = \lim\limits_{h \to 0} \dfrac{|0 + h| - 0}{h}$

$\qquad = \lim\limits_{h \to 0} \dfrac{|h|}{h}$

If we make tables of values in this case we get the following:

h	1	0.1	0.01	0.001	0.0001
Average Rate of Change over $[0, 0 + h]$	1	1	1	1	1

h	-1	-0.1	-0.01	-0.001	-0.0001
Average Rate of Change over $[0 + h, 0]$	-1	-1	-1	-1	-1

For the limit and hence the derivative $f'(0)$ to exist, the average rates of change should approach the same number for both positive and negative h. Because they do not, f is not differentiable at $x = 0$. We can verify this conclusion algebraically: If h is positive, then $|h| = h$, and so the ratio $|h|/h$ is 1, regardless of how small h is. Thus, according to the values of the difference quotients with $h > 0$, the limit should be 1. On the other hand if h is negative, then $|h| = -h$ (positive) and so $|h|/h = -1$, meaning that the limit should be -1. Because the limit cannot be both -1 and 1 (it must be a single number for the derivative to exist), we conclude that $f'(0)$ does not exist.

To see what is happening graphically, take a look at Figure 39, which shows zoomed-in views of the graph of f near $x = 0$.

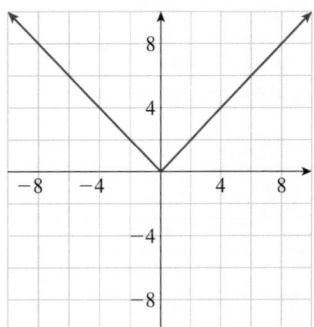 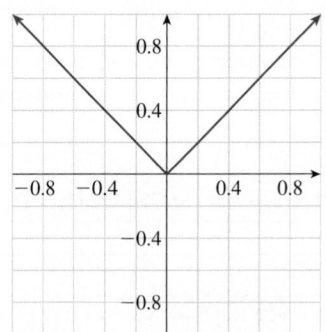 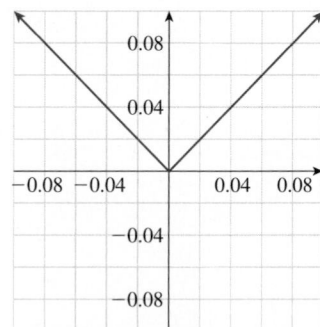

Figure 39

No matter what scale we use to view the graph, it has a sharp corner at $x = 0$ and hence has no tangent line there. Since there is no tangent line at $x = 0$, the function is not differentiable there.

➡ **Before we go on...** Notice that $|x| = \begin{cases} -x & \text{if } x < 0 \\ x & \text{if } x \geq 0 \end{cases}$ is an example of a piecewise-linear function whose graph comes to a point at $x = 0$. In general, if $f(x)$ is any piecewise linear function whose graph comes to a point at $x = a$, it will be nondifferentiable at $x = a$ for the same reason that $|x|$ fails to be differentiable at $x = 0$.

If we repeat the computation in Example 5(a) using any nonzero value for a in place of 1, we see that f is differentiable there as well. If a is positive, we find that $f'(a) = 1$ and, if a is negative, $f'(a) = -1$. In other words, the derivative function is

$$f'(x) = \begin{cases} -1 & \text{if } x < 0 \\ 1 & \text{if } x > 0 \end{cases}.$$

Immediately to the left of $x = 0$, we see that $f'(x) = -1$, immediately to the right, $f'(x) = 1$; and when $x = 0$, $f'(x)$ is not defined. ∎

Q : *So does that mean there is no single formula for the derivative of $|x|$?*

A : Actually, there is a convenient formula. Consider the ratio $\dfrac{|x|}{x}$. If x is positive, then $|x| = x$, so $\dfrac{|x|}{x} = \dfrac{x}{x} = 1$. On the other hand, if x is negative, then $|x| = -x$, so $\dfrac{|x|}{x} = \dfrac{-x}{x} = -1$. In other words,

$$\frac{|x|}{x} = \begin{cases} -1 & \text{if } x < 0 \\ 1 & \text{if } x > 0 \end{cases},$$

which is exactly the formula we obtained for $f'(x)$. We have therefore obtained a convenient closed-form formula for the derivative of $|x|$!

Derivative of |x|

If $f(x) = |x|$, then $f'(x) = \dfrac{|x|}{x}$.

Note that $|x|/x$ is not defined if $x = 0$, reflecting the fact that $f'(x)$ does not exist when $x = 0$.

We will use the above formula extensively in the next chapter.

FAQ

Computing Derivatives Algebraically

Q : *The algebraic computation of $f'(x)$ seems to require a number of steps. How do I remember what to do, and when?*

A : If you examine the computations in the examples above, you will find the following pattern:

1. Write out the formula for $f'(x)$, as the limit of the difference quotient, then substitute $f(x + h)$ and $f(x)$.
2. Expand and simplify the *numerator* of the expression, but not the denominator.
3. After simplifying the numerator, factor out an h to cancel with the h in the denominator. If h does not factor out of the numerator, you might have made an error. (A frequent error is a wrong sign.)
4. After canceling the h, you should be able to see what the limit is by letting $h \to 0$.

10.6 EXERCISES

▼ more advanced   ◆ challenging
T indicates exercises that should be solved using technology

In Exercises 1–14, compute $f'(a)$ algebraically for the given value of a. HINT [See Example 1.]

1. $f(x) = x^2 + 1; a = 2$

2. $f(x) = x^2 - 3; a = 1$

3. $f(x) = 3x - 4; a = -1$

4. $f(x) = -2x + 4; a = -1$

5. $f(x) = 3x^2 + x; a = 1$

6. $f(x) = 2x^2 + x; a = -2$

7. $f(x) = 2x - x^2; a = -1$

8. $f(x) = -x - x^2; a = 0$

9. $f(x) = x^3 + 2x; a = 2$

10. $f(x) = x - 2x^3; a = 1$

11. $f(x) = \dfrac{-1}{x}; a = 1$ HINT [See Example 3.]

12. $f(x) = \dfrac{2}{x}; a = 5$ HINT [See Example 3.]

13. ▼ $f(x) = mx + b; a = 43$

14. ▼ $f(x) = \dfrac{x}{k} - b \ (k \neq 0); a = 12$

In Exercises 15–28, compute the derivative function $f'(x)$ algebraically. (Notice that the functions are the same as those in Exercises 1–14.) HINT [See Examples 2 and 3.]

15. $f(x) = x^2 + 1$

16. $f(x) = x^2 - 3$

17. $f(x) = 3x - 4$

18. $f(x) = -2x + 4$

19. $f(x) = 3x^2 + x$

20. $f(x) = 2x^2 + x$

21. $f(x) = 2x - x^2$

22. $f(x) = -x - x^2$

23. $f(x) = x^3 + 2x$

24. $f(x) = x - 2x^3$

25. ▼ $f(x) = \dfrac{-1}{x}$

26. ▼ $f(x) = \dfrac{2}{x}$

27. ▼ $f(x) = mx + b$

28. ▼ $f(x) = \dfrac{x}{k} - b \ (k \neq 0)$

In Exercises 29–38, compute the indicated derivative.

29. $R(t) = -0.3t^2; R'(2)$

30. $S(t) = 1.4t^2; S'(-1)$

31. $U(t) = 5.1t^2 + 5.1; U'(3)$

32. $U(t) = -1.3t^2 + 1.1; U'(4)$

33. $U(t) = -1.3t^2 - 4.5t; U'(1)$

34. $U(t) = 5.1t^2 - 1.1t; U'(1)$

35. $L(r) = 4.25r - 5.01; L'(1.2)$

36. $L(r) = -1.02r + 5.7; L'(3.1)$

37. ▼ $q(p) = \dfrac{2.4}{p} + 3.1; q'(2)$

38. ▼ $q(p) = \dfrac{1}{0.5p} - 3.1; q'(2)$

In Exercises 39–44, find the equation of the tangent to the graph at the indicated point. HINT [Compute the derivative algebraically; then see Example 2(b) in Section 10.5.]

39. ▼ $f(x) = x^2 - 3; a = 2$ 40. ▼ $f(x) = x^2 + 1; a = 2$
41. ▼ $f(x) = -2x - 4; a = 3$ 42. ▼ $f(x) = 3x + 1; a = 1$
43. ▼ $f(x) = x^2 - x; a = -1$ 44. ▼ $f(x) = x^2 + x; a = -1$

APPLICATIONS

45. **Velocity** If a stone is dropped from a height of 400 feet, its height after t seconds is given by $s = 400 - 16t^2$. Find its instantaneous velocity function and its velocity at time $t = 4$. HINT [See Example 4.]

46. **Velocity** If a stone is thrown down at 120 feet per second from a height of 1,000 feet, its height after t seconds is given by $s = 1{,}000 - 120t - 16t^2$. Find its instantaneous velocity function and its velocity at time $t = 3$. HINT [See Example 4.]

47. **Oil Imports from Mexico** Daily oil imports to the United States from Mexico can be approximated by
$$I(t) = -0.015t^2 + 0.1t + 1.4 \text{ million barrels} \quad (0 \le t \le 8),$$
where t is time in years since the start of 2000.[68] Find the derivative function $\dfrac{dI}{dt}$. At what rate were oil imports changing at the start of 2007 ($t = 7$)? HINT [See Example 4.]

48. **Oil Production in Mexico** Daily oil production by Pemex, Mexico's national oil company, can be approximated by
$$P(t) = -0.022t^2 + 0.2t + 2.9 \text{ million barrels} \quad (1 \le t \le 9),$$
where t is time in years since the start of 2000.[69] Find the derivative function $\dfrac{dP}{dt}$. At what rate was oil production changing at the start of 2004 ($t = 4$)? HINT [See Example 4.]

49. **Bottled Water Sales** The following chart shows the amount of bottled water sold in the United States for the period 2000–2010:[70]

---

[68] Source for data: Energy Information Administration/Pemex (www.eia.doe.gov).

[69] *Ibid.*

[70] The 2010 figure is an estimate. Source: Beverage Marketing Corporation/www.bottledwater.org.

Bottled water sales in the U.S.

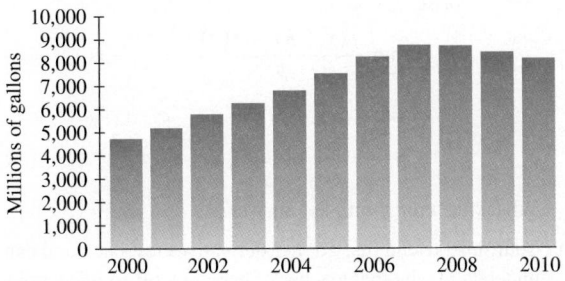

The function

$$R(t) = -45t^2 + 900t + 4,200 \text{ million gallons} \quad (0 \le t \le 10)$$

gives a good approximation, where t is time in years since 2000. Find the derivative function $R'(t)$. According to the model, how fast were annual sales of bottled water changing in 2005?

50. *Bottled Water Sales* The following chart shows annual per capita sales of bottled water in the United States for the period 2000–2010:[71]

Per capita bottled water sales in the U.S.

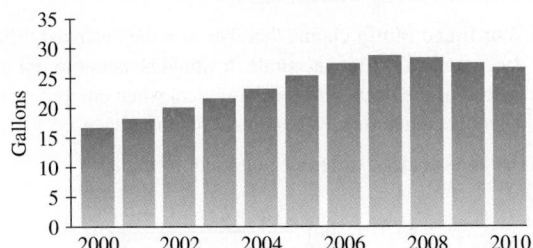

The function

$$R(t) = -0.18t^2 + 3t + 15 \text{ gallons} \quad (0 \le t \le 10)$$

gives a good approximation, where t is time in years since 2000. Find the derivative function $R'(t)$. According to the model, how fast were per capita sales of bottled water changing in 2009?

51. ▼ *Ecology* Increasing numbers of manatees ("sea sirens") have been killed by boats off the Florida coast. The following graph shows the relationship between the number of boats registered in Florida and the number of manatees killed each year.

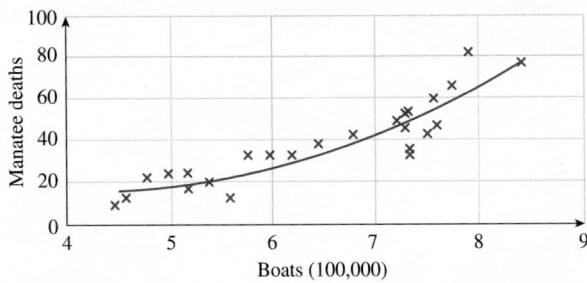

Boats (100,000)

The regression curve shown is given by

$$f(x) = 3.55x^2 - 30.2x + 81 \text{ manatee deaths} \\ (4.5 \le x \le 8.5),$$

where x is the number of boats (hundreds of thousands) registered in Florida in a particular year and $f(x)$ is the number of manatees killed by boats in Florida that year.[72] Compute and interpret $f'(8)$.

52. ▼ *SAT Scores by Income* The following graph shows U.S. math SAT scores as a function of parents' income level.[73]

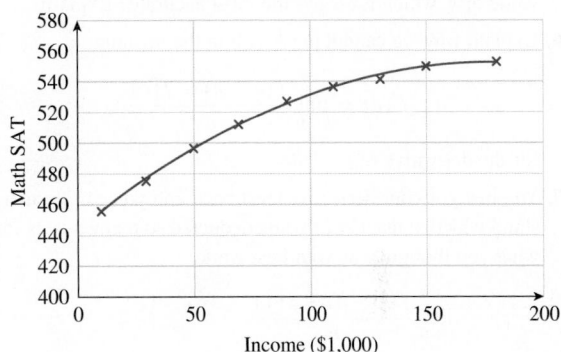

Income ($1,000)

The regression curve shown is given by

$$f(x) = -0.0034x^2 + 1.2x + 444 \quad (10 \le x \le 180),$$

where $f(x)$ is the average math SAT score of a student whose parents earn x thousand dollars per year. Compute and interpret $f'(30)$.

53. ▼ *Television Advertising* The cost, in thousands of dollars, of a 30-second television ad during the Super Bowl in the years 1970 to 2010 can be approximated by the following piecewise linear function ($t = 0$ represents 1970):[74]

$$C(t) = \begin{cases} 31.1t + 78 & \text{if } 0 \le t < 20 \\ 90t - 1{,}100 & \text{if } 20 \le t \le 40 \end{cases}.$$

a. Is C a continuous function of t? Why? HINT [See Example 4 of Section 10.3.]

b. Is C a differentiable function of t? Compute $\lim_{t \to 20^-} C'(t)$ and $\lim_{t \to 20^+} C'(t)$ and interpret the results. HINT [See *Before we go on...* after Example 5.]

54. ▼ *Television Advertising* (Compare Exercise 53.) The cost, in thousands of dollars, of a 30-second television ad during the Super Bowl in the years 1980 to 2010 can be approximated by the following piecewise linear function ($t = 0$ represents 1980):[75]

$$C(t) = \begin{cases} 43.9t + 222 & \text{if } 0 \le t \le 20 \\ 140t - 1{,}700 & \text{if } 20 < t \le 30 \end{cases}.$$

---

[71] The 2010 figure is an estimate. Source: Beverage Marketing Corporation/www.bottledwater.org.

[72] Regression model is based on data from 1976 to 2000. Sources for data: Florida Department of Highway Safety & Motor Vehicles, Florida Marine Institute/*New York Times*, February 12, 2002, p. F4.

[73] Regression model is based on 2009 data. Source: College Board/*New York Times* http://economix.blogs.nytimes.com.

[74] Source: http://en.wikipedia.org/wiki/Super_Bowl_advertising.

[75] *Ibid.*

a. Is C a continuous function of t? Why? HINT [See Example 4 of Section 10.3.]

b. Is C a differentiable function of t? Compute $\lim_{t \to 20^-} C'(t)$ and $\lim_{t \to 20^+} C'(t)$ and interpret the results. HINT [See *Before we go on...* after Example 5.]

COMMUNICATION AND REASONING EXERCISES

55. Of the three methods (numerical, graphical, algebraic) we can use to estimate the derivative of a function at a given value of x, which is always the most accurate? Explain.

56. Explain why we cannot put $h = 0$ in the formula

$$f'(a) = \lim_{h \to 0} \frac{f(a+h) - f(a)}{h}$$

for the derivative of f.

57. You just got your derivatives test back and you can't understand why that teacher of yours deducted so many points for what you thought was your best work:

$$\lim_{h \to 0} \frac{f(x+h) - f(x)}{h}$$

$$= \lim_{h \to 0} \frac{f(x) + h - f(x)}{h}$$

$$= \lim_{h \to 0} \frac{h}{h}$$ Canceled the $f(x)$

$$= 1.$$ ✗ *WRONG* -10

What was wrong with your answer?

58. Your friend just got his derivatives test back and can't understand why that teacher of his deducted so many points for the following:

$$\lim_{h \to 0} \frac{f(x+h) - f(x)}{h}$$

$$= \lim_{h \to 0} \frac{f(x) + f(h) - f(x)}{h}$$

$$= \lim_{h \to 0} \frac{f(h)}{h}$$ Canceled the $f(x)$

$$= \lim_{h \to 0} \frac{f(\cancel{h})}{\cancel{h}}$$ Now cancel the h.

$$= f.$$ ✗ *WRONG* -50

What was wrong with his answer?

59. Your other friend just got her derivatives test back and can't understand why that teacher of hers took off so many points for the following:

$$\lim_{h \to 0} \frac{f(x+h) - f(x)}{h}$$

$$= \lim_{h \to 0} \frac{f(x+h) - f(x)}{\cancel{h}}$$ Now cancel the h.

$$= \lim_{h \to 0} f(x) - f(x)$$ Cancel the $f(x)$

$$= 0.$$ ✗ *WRONG* -15

What was wrong with her answer?

60. Your third friend just got her derivatives test back and can't understand why that teacher of hers took off so many points for the following:

$$\lim_{h \to 0} \frac{f(x+h) - f(x)}{h}$$

$$= \lim_{h \to 0} \frac{f(x) + h - f(x)}{h}$$

$$= \lim_{h \to 0} \frac{f(x) + \cancel{h} - f(x)}{\cancel{h}}$$ Now cancel the h.

$$= \lim_{h \to 0} f(x) - f(x)$$ Cancel the $f(x)$

$$= 0.$$ ✗ *WRONG* -25

What was wrong with her answer?

61. Your friend Muffy claims that, because the balanced difference quotient is more accurate, it would be better to use that instead of the usual difference quotient when computing the derivative algebraically. Comment on this advice.

62. Use the balanced difference quotient formula,

$$f'(a) = \lim_{h \to 0} \frac{f(a+h) - f(a-h)}{2h},$$

to compute $f'(3)$ when $f(x) = x^2$. What do you find?

63. ▼ A certain function f has the property that $f'(a)$ does not exist. How is that reflected in the attempt to compute $f'(a)$ algebraically?

64. ▼ One cannot put $h = 0$ in the formula

$$f'(a) = \lim_{h \to 0} \frac{f(a+h) - f(a)}{h}$$

for the derivative of f. (See Exercise 56.) However, in the last step of each of the computations in the text, we are effectively setting $h = 0$ when taking the limit. What is going on here?

KEY CONCEPTS

 Website www.WanerMath.com

Go to the Website at www.WanerMath .com to find a comprehensive and interactive Web-based summary of Chapter 10.

10.1 Limits: Numerical and Graphical Viewpoints

$\lim_{x \to a} f(x) = L$ means that $f(x)$ approaches L as x approaches a p. 690

What it means for a limit to exist p. 691

Limits at infinity p. 693

Estimating limits graphically p. 693

Interpreting limits in real-world situations p. 696

10.2 Limits and Continuity

f is continuous at a if $\lim_{x \to a} f(x)$ exists and $\lim_{x \to a} f(x) = f(a)$ p. 702

Discontinuous, continuous on domain p. 702

Determining whether a given function is continuous p. 703

10.3 Limits and Continuity: Algebraic Viewpoint

Closed-form function p. 708

Limits of closed form functions p. 709

Simplifying to obtain limits p. 709

The indeterminate form 0/0 p. 710

The determinate form k/0 p. 711

Limits of piecewise-defined functions p. 712

Limits at infinity p. 713

Determinate and indeterminate forms p. 716

10.4 Average Rate of Change

Average rate of change of $f(x)$ over
$[a, b]$: $\dfrac{\Delta f}{\Delta x} = \dfrac{f(b) - f(a)}{b - a}$ p. 724

Average rate of change as slope of the secant line p. 725

Computing the average rate of change from a graph p. 726

Computing the average rate of change from a formula p. 727

Computing the average rate of change over short intervals $[a, a + h]$ p. 728

10.5 Derivatives: Numerical and Graphical Viewpoints

Instantaneous rate of change of $f(x)$ (derivative of f at a);
$f'(a) = \lim_{h \to 0} \dfrac{f(a + h) - f(a)}{h}$ p. 738

The derivative as slope of the tangent line p. 742

Quick approximation of the derivative p. 743

$\dfrac{d}{dx}$ Notation p. 745

The derivative as velocity p. 745

Average and instantaneous velocity p. 746

The derivative function p. 747

Graphing the derivative function with technology p. 748

10.6 Derivatives: Algebraic Viewpoint

Derivative at the point $x = a$:
$f'(a) = \lim_{h \to 0} \dfrac{f(a + h) - f(a)}{h}$ p. 761

Derivative function:
$f'(x) = \lim_{h \to 0} \dfrac{f(x + h) - f(x)}{h}$ p. 762

Examples of the computation of $f'(x)$ p. 763

$f(x) = |x|$ is not differentiable at $x = 0$ p. 765

REVIEW EXERCISES

⊤ indicates exercises that must be solved using technology

Numerically *estimate whether the limits in Exercises 1–4 exist. If a limit does exist, give its approximate value.*

1. $\lim_{x \to 3} \dfrac{x^2 - x - 6}{x - 3}$

2. $\lim_{x \to 3} \dfrac{x^2 - 2x - 6}{x - 3}$

3. $\lim_{x \to -1} \dfrac{|x + 1|}{x^2 - x - 2}$

4. $\lim_{x \to -1} \dfrac{|x + 1|}{x^2 + x - 2}$

In Exercises 5 and 6, the graph of a function f is shown. Graphically determine whether the given limits exist. If a limit does exist, give its approximate value.

5.

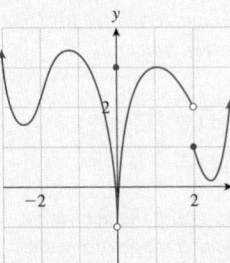

a. $\lim_{x \to 0} f(x)$ b. $\lim_{x \to 1} f(x)$

c. $\lim_{x \to 2} f(x)$

6.

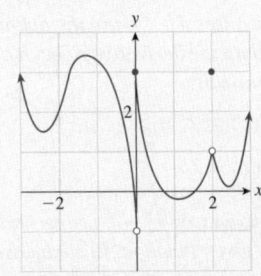

a. $\lim_{x \to 0} f(x)$ b. $\lim_{x \to -2} f(x)$

c. $\lim_{x \to 2} f(x)$

Calculate the limits in Exercises 7–30 algebraically. If a limit does not exist, say why.

7. $\lim_{x \to -2} \dfrac{x^2}{x - 3}$

8. $\lim_{x \to 3} \dfrac{x^2 - 9}{2x - 6}$

9. $\lim_{x \to -2} \dfrac{x^2 - 4}{x^3 + 2x^2}$

10. $\lim_{x \to -1} \dfrac{x^2 - 9}{2x - 6}$

11. $\lim_{x \to 0} \dfrac{x}{2x^2 - x}$

12. $\lim_{x \to 1} \dfrac{x^2 - 9}{x - 1}$

13. $\lim_{x \to -1} \dfrac{x^2 + 3x}{x^2 - x - 2}$

14. $\lim_{x \to -1^+} \dfrac{x^2 + 1}{x^2 + 3x + 2}$

15. $\lim_{x \to 8} \dfrac{x^2 - 6x - 16}{x^2 - 9x + 8}$

16. $\lim_{x \to 4} \dfrac{x^2 + 3x}{x^2 - 8x + 16}$

17. $\lim_{x \to 4} \dfrac{x^2 + 8}{x^2 - 2x - 8}$

18. $\lim_{x \to 6} \dfrac{x^2 - 5x - 6}{x^2 - 36}$

19. $\lim_{x \to 1/2} \dfrac{x^2 + 8}{4x^2 - 4x + 1}$

20. $\lim_{x \to 1/2} \dfrac{x^2 + 3x}{2x^2 + 3x - 1}$

21. $\lim_{x \to +\infty} \dfrac{10x^2 + 300x + 1}{5x^3 + 2}$

22. $\lim_{x \to +\infty} \dfrac{2x^4 + 20x^3}{1{,}000x^6 + 6}$

23. $\lim_{x \to -\infty} \dfrac{x^2 - x - 6}{x - 3}$

24. $\lim_{x \to +\infty} \dfrac{x^2 - x - 6}{4x^2 - 3}$

25. $\lim\limits_{t\to+\infty} \dfrac{-5}{5+5.3(3^{2t})}$

26. $\lim\limits_{t\to+\infty} \left(3+\dfrac{2}{e^{4t}}\right)$

27. $\lim\limits_{x\to+\infty} \dfrac{2}{5+4e^{-3x}}$

28. $\lim\limits_{x\to+\infty} (4e^{3x}+12)$

29. $\lim\limits_{t\to+\infty} \dfrac{1+2^{-3t}}{1+5.3e^{-t}}$

30. $\lim\limits_{x\to-\infty} \dfrac{8+0.5^x}{2-3^{2x}}$

In Exercises 31–34, find the average rate of change of the given function over the interval $[a, a+h]$ for $h = 1, 0.01,$ and 0.001. (Round answers to four decimal places.) Then estimate the slope of the tangent line to the graph of the function at a.

31. $f(x) = \frac{1}{x+1}$; $a = 0$

32. $f(x) = x^x$; $a = 2$

33. $f(x) = e^{2x}$; $a = 0$

34. $f(x) = \ln(2x)$; $a = 1$

In Exercises 35–38, you are given the graph of a function with four points marked. Determine at which (if any) of these points the derivative of the function is: (a) -1, (b) 0, (c) 1, and (d) 2.

35.

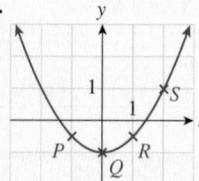

36.

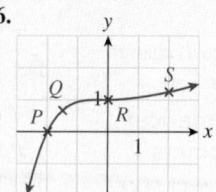

37.

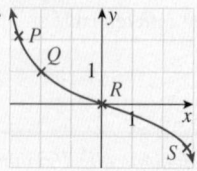

38.

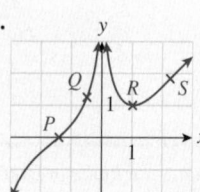

39. Let f have the graph shown.

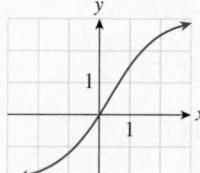

Select the correct answer.

a. The average rate of change of f over the interval $[0, 2]$ is
   (A) greater than $f'(0)$.   (B) less than $f'(0)$.
   (C) approximately equal to $f'(0)$.
b. The average rate of change of f over the interval $[-1, 1]$ is
   (A) greater than $f'(0)$.   (B) less than $f'(0)$.
   (C) approximately equal to $f'(0)$.
c. Over the interval $[0, 2]$, the instantaneous rate of change of f is
   (A) increasing.   (B) decreasing.
   (C) neither increasing nor decreasing.
d. Over the interval $[-2, 2]$, the instantaneous rate of change of f is
   (A) increasing, then decreasing.

(B) decreasing, then increasing.
(C) approximately constant.
e. When $x = 2$, $f(x)$ is
   (A) approximately 1 and increasing at a rate of about 2.5 units per unit of x.
   (B) approximately 1.2 and increasing at a rate of about 1 unit per unit of x.
   (C) approximately 2.5 and increasing at a rate of about 0.5 units per unit of x.
   (D) approximately 2.5 and increasing at a rate of about 2.5 units per unit of x.

40. Let f have the graph shown.

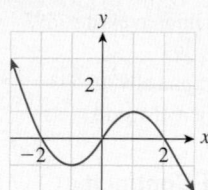

Select the correct answer.

a. The average rate of change of f over the interval $[0, 1]$ is
   (A) greater than $f'(0)$.   (B) less than $f'(0)$.
   (C) approximately equal to $f'(0)$.
b. The average rate of change of f over the interval $[0, 2]$ is
   (A) greater than $f'(1)$.   (B) less than $f'(1)$.
   (C) approximately equal to $f'(1)$.
c. Over the interval $[-2, 0]$, the instantaneous rate of change of f is
   (A) increasing.   (B) decreasing.
   (C) neither increasing nor decreasing.
d. Over the interval $[-2, 2]$, the instantaneous rate of change of f is
   (A) increasing, then decreasing.
   (B) decreasing, then increasing.
   (C) approximately constant.
e. When $x = 0$, $f(x)$ is
   (A) approximately 0 and increasing at a rate of about 1.5 units per unit of x.
   (B) approximately 0 and decreasing at a rate of about 1.5 units per unit of x.
   (C) approximately 1.5 and neither increasing nor decreasing.
   (D) approximately 0 and neither increasing nor decreasing.

In Exercises 41–44, use the definition of the derivative to calculate the derivative of each of the given functions algebraically.

41. $f(x) = x^2 + x$

42. $f(x) = 3x^2 - x + 1$

43. $f(x) = 1 - \dfrac{2}{x}$

44. $f(x) = \dfrac{1}{x} + 1$

 In Exercises 45–48, use technology to graph the derivative of the given function. In each case, choose a range of x-values and y-values that shows the interesting features of the graph.

45. $f(x) = 10x^5 + \dfrac{1}{2}x^4 - x + 2$

Section 10.5

Example 2 (page 743) Calculate an approximate value of $f'(1.5)$ if $f(x) = x^2 - 4x$, and then find the equation of the tangent line at the point on the graph where $x = 1.5$.

Solution with Technology

You can compute both the difference quotient and the balanced difference quotient approximations in a spreadsheet using the following extension of the worksheet in Example 4 in Section 10.4:

	A	B	C	D	E	F
1	a	h	x	f(x)	Diff Quotients	Balanced Diff Quotient
2	1.5	0.0001	=A2-B2	=C2^2-4*C2	=(D3-D2)/(C3-C2)	=(D4-D2)/(C4-C2)
3			=A2			
4			=A2+B2			

Notice that we get two difference quotients in column E. The first uses $h = -0.0001$ while the second uses $h = 0.0001$ and is the one we use for our quick approximation. The balanced quotient is their average (column F). The results are as follows.

	A	B	C	D	E	F
1	a	h	x	f(x)	Diff Quotients	Balanced Diff Quotient
2	1.5	0.0001	1.4999	-3.7499	-1.0001	-1
3			1.5	-3.75	-0.9999	
4			1.5001	-3.7501		

From the results shown above, we find that the difference quotient quick approximation is -0.9999 and that the balanced difference quotient quick approximation is -1, which is in fact the exact value of $f'(1.5)$. See the discussion in the text for the calculation of the equation of the tangent line.

Example 4 (page 748) Use technology to graph the derivative of $f(x) = -2x^2 + 6x + 5$ for values of x starting at -5.

Solution with Technology

1. Start with a table of values for the function f:

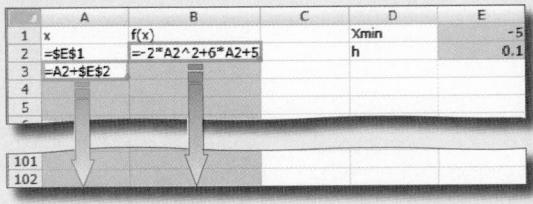

2. Next, compute approximate derivatives in Column C:

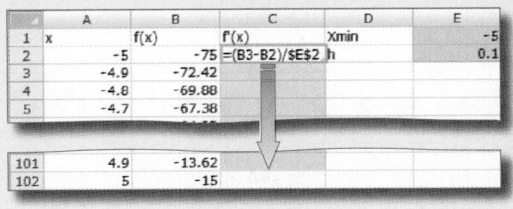

	A	B	C	D	E
1	x	f(x)	f'(x)	Xmin	-5
2	-5	-75	=(B3-B2)/$E$2	h	0.1
3	-4.9	-72.42			
4	-4.8	-69.88			
5	-4.7	-67.38			
101	4.9	-13.62			
102	5	-15			

	A	B	C	D	E
1	x	f(x)	f'(x)	Xmin	-5
2	-5	-75	25.8	h	0.1
3	-4.9	-72.42	25.4		
4	-4.8	-69.88	25		
5	-4.7	-67.38	24.6		
101	4.9	-13.62	-13.8		
102	5	-15			

You cannot paste the difference quotient formula into cell C102. (Why?) Notice that this worksheet uses the ordinary difference quotients, $[f(x + h) - f(x)]/h$. If you prefer, you can use balanced difference quotients $[f(x + h) - f(x - h)]/(2h)$, in which case cells C2 and C102 would both have to be left blank.

We now graph the function and the derivative on different graphs as follows:

1. First, graph the function f in the usual way, using Columns A and B.

2. Make a copy of this graph and click on it once. Columns A and B should be outlined, indicating that these are the columns used in the graph.

3. By dragging from the center of the bottom edge of the box, move the Column B box over to Column C as shown:

	A	B	C
96	4.4	-7.32	-11.8
97	4.5	-8.5	-12.2
98	4.6	-9.72	-12.6
99	4.7	-10.98	-13
100	4.8	-12.28	-13.4
101	4.9	-13.62	-13.8
102	5	-15	

↓

	A	B	C
96	4.4	-7.32	-11.8
97	4.5	-8.5	-12.2
98	4.6	-9.72	-12.6
99	4.7	-10.98	-13
100	4.8	-12.28	-13.4
101	4.9	-13.62	-13.8
102	5	-15	

The graph will then show the derivative (Columns A and C):

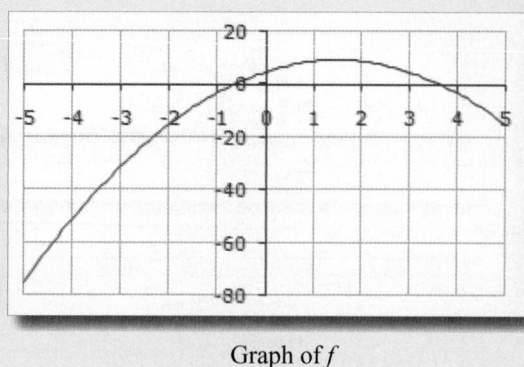

Graph of f

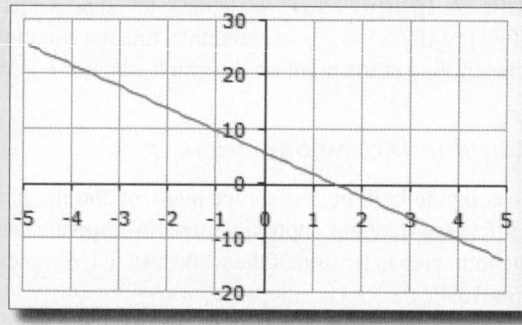

Graph of f'

Theorem 11.2 Derivatives of Sums, Differences, and Constant Multiples

If f and g are any two differentiable functions, and if c is any constant, then the sum, $f + g$, the difference, $f - g$, and the constant multiple, cf, are differentiable, and

$$[f \pm g]'(x) = f'(x) \pm g'(x) \qquad \text{Sum Rule}$$

$$[cf]'(x) = cf'(x). \qquad \text{Constant Multiple Rule}$$

In Words:

- The derivative of a sum is the sum of the derivatives, and the derivative of a difference is the difference of the derivatives.
- The derivative of c times a function is c times the derivative of the function.

Differential Notation:

$$\frac{d}{dx}[f(x) \pm g(x)] = \frac{d}{dx}f(x) \pm \frac{d}{dx}g(x)$$

$$\frac{d}{dx}[cf(x)] = c\frac{d}{dx}f(x)$$

Quick Examples

1. $\dfrac{d}{dx}(x^2 - x^4) = \dfrac{d}{dx}(x^2) - \dfrac{d}{dx}(x^4) = 2x - 4x^3$

2. $\dfrac{d}{dx}(7x^3) = 7\dfrac{d}{dx}(x^3) = 7(3x^2) = 21x^2$

   In other words, we multiply the coefficient (7) by the exponent (3), and then decrease the exponent by 1.

3. $\dfrac{d}{dx}(12x) = 12\dfrac{d}{dx}(x) = 12(1) = 12$

   In other words, the derivative of a constant times x is that constant.

4. $\dfrac{d}{dx}(-x^{0.5}) = \dfrac{d}{dx}[(-1)x^{0.5}] = (-1)\dfrac{d}{dx}(x^{0.5}) = (-1)(0.5)x^{-0.5}$
   $= -0.5x^{-0.5}$

5. $\dfrac{d}{dx}(12) = \dfrac{d}{dx}[12(1)] = 12\dfrac{d}{dx}(1) = 12(0) = 0.$

   In other words, the derivative of a constant is zero.

6. If my company earns twice as much (annual) revenue as yours and the derivative of your revenue function is the curve on the left, then the derivative of my revenue function is the curve on the right.

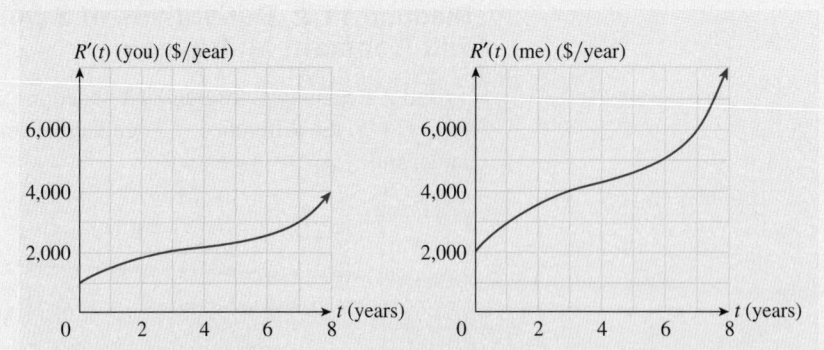

7. Suppose that a company's revenue R and cost C are changing with time. Then so is the profit, $P(t) = R(t) - C(t)$, and the rate of change of the profit is

$$P'(t) = R'(t) - C'(t).$$

In words: *The derivative of the profit is the derivative of revenue minus the derivative of cost.*

Proof of the Sum Rule

By the definition of the derivative of a function,

$$\frac{d}{dx}[f(x) + g(x)] = \lim_{h \to 0} \frac{[f(x+h) + g(x+h)] - [f(x) + g(x)]}{h}$$

$$= \lim_{h \to 0} \frac{[f(x+h) - f(x)] + [g(x+h) - g(x)]}{h}$$

$$= \lim_{h \to 0} \left[\frac{f(x+h) - f(x)}{h} + \frac{g(x+h) - g(x)}{h} \right]$$

$$= \lim_{h \to 0} \frac{f(x+h) - f(x)}{h} + \lim_{h \to 0} \frac{g(x+h) - g(x)}{h}$$

$$= \frac{d}{dx}[f(x)] + \frac{d}{dx}[g(x)].$$

The next-to-last step uses a property of limits: The limit of a sum is the sum of the limits. Think about why this should be true. The last step uses the definition of the derivative again (and the fact that the functions are differentiable).

The proof of the rule for constant multiples is similar.

EXAMPLE 2 Combining the Sum and Constant Multiple Rules, and Dealing with x in the Denominator

Find the derivatives of the following:

a. $f(x) = 3x^2 + 2x - 4$

b. $f(x) = \dfrac{2x}{3} - \dfrac{6}{x} + \dfrac{2}{3x^{0.2}} - \dfrac{x^4}{2}$

c. $f(x) = \dfrac{|x|}{4} + \dfrac{1}{2\sqrt{x}}$

a. Compute $P'(t)$. How fast was GE's annual net income changing in 2010? (Be careful to give correct units of measurement.)

b. According to the model, GE's annual net income

   (A) increased at a faster and faster rate
   (B) increased at a slower and slower rate
   (C) decreased at a faster and faster rate
   (D) decreased at a slower and slower rate

during the first two years shown (the interval [2, 4]). Justify your answer in two ways: geometrically, reasoning entirely from the graph; and algebraically, reasoning from the derivative of P. HINT [See Example 4.]

97. *Ecology* Increasing numbers of manatees ("sea sirens") have been killed by boats off the Florida coast. The following graph shows the relationship between the number of boats registered in Florida and the number of manatees killed each year.

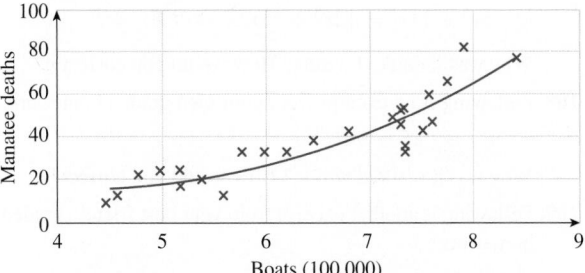

The regression curve shown is given by

$$f(x) = 3.55x^2 - 30.2x + 81 \quad (4.5 \le x \le 8.5),$$

where x is the number of boats (hundreds of thousands) registered in Florida in a particular year and $f(x)$ is the number of manatees killed by boats in Florida that year.[7]

a. Find $f'(x)$, and use your formula to compute $f'(8)$, stating its units of measurement. What does the answer say about manatee deaths?

b. Is $f'(x)$ increasing or decreasing with increasing x? Interpret the answer. HINT [See Example 4.]

98. *SAT Scores by Income* The following graph shows U.S. math SAT scores as a function of parents' income level.[8]

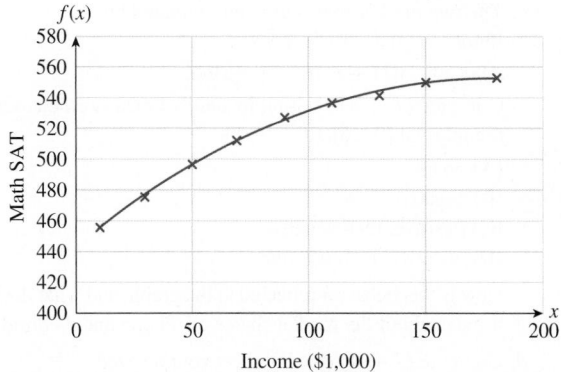

The regression curve shown is given by

$$f(x) = -0.0034x^2 + 1.2x + 444 \quad (10 \le x \le 180),$$

where $f(x)$ is the average math SAT score of a student whose parents earn x thousand dollars per year.

a. Find $f'(x)$, and use your formula to compute $f'(100)$, stating its units of measurement. What does the answer say about math SAT scores?

b. Does $f'(x)$ increase or decrease with increasing x? What does your answer say about math SAT scores? HINT [See Example 4.]

99. ▼ *Market Share: Smart Phones* The following graph shows the approximate market shares, in percentage points, of **Apple's** *iPhone* and **Google's** *Android*-based smart phones from January 2009 to May 2010 (t is time in months, and $t = 0$ represents January, 2009):[9]

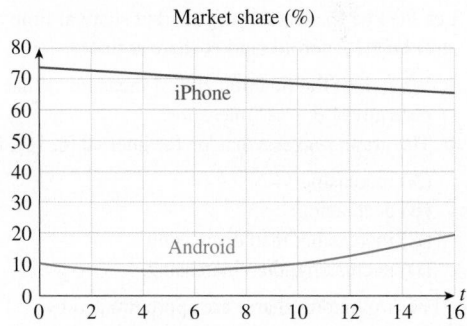

Let $I(t)$ be the iPhone market share at time t, and let $A(t)$ be the Android market share at time t.

a. What does the function $I - A$ measure? What does its derivative $(I - A)'$ measure?

b. The graph suggests that, on the interval [6, 16], $I - A$ is

   (A) increasing.
   (B) decreasing.
   (C) increasing, then decreasing.
   (D) decreasing, then increasing.

---

[7]Regression model is based on data from 1976 to 2000. Sources for data: Florida Department of Highway Safety & Motor Vehicles, Florida Marine Institute/*New York Times*, February 12, 2002, p. F4.

[8]Regression model is based on 2009 data. Source: College Board/ *New York Times* http://economix.blogs.nytimes.com.

[9]Source for data: Quantcast www.quantcast.com.

c. The two market shares are approximated by
iPhone: $I(t) = -0.5t + 73$
Android: $A(t) = 0.1t^2 - t + 10$.

Compute $(I - A)'$, stating its units of measurement. On the interval [0, 16], $(I - A)'$ is

(A) positive.
(B) negative.
(C) positive, then negative.
(D) negative, then positive.

How is this behavior reflected in the graph, and what does it mean about the market shares of iPhone and Android?

d. Compute $(I - A)'(4)$. Interpret your answer.

100. ▼ **Market Share: Smart Phones** The following graph shows the approximate market shares, in percentage points, of Research In Motion's *BlackBerry* smartphones and Google's *Android*-based smartphones from July 2009 to May 2010 (t is time in months, and $t = 0$ represents January 2009):[10]

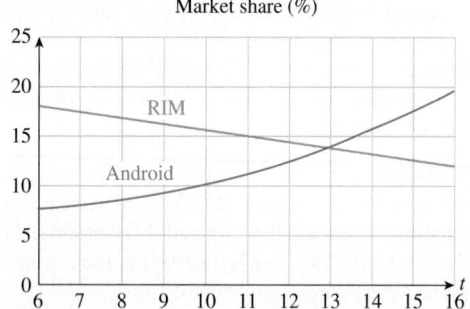

Market share (%)

Let $B(t)$ be the BlackBerry market share at time t, and let $A(t)$ be the Android market share at time t.

a. What does the function $B - A$ measure? What does its derivative $(B - A)'$ measure?

b. The graph suggests that, on the interval [6, 16], $B - A$ is

(A) increasing.
(B) decreasing.
(C) increasing, then decreasing.
(D) decreasing, then increasing.

c. The two market shares are approximated by
BlackBerry: $B(t) = -0.6t + 21.6$
Android: $A(t) = 0.1t^2 - t + 10$.

Compute $(B - A)'$, stating its units of measurement. On the interval [6, 16], $(B - A)'$ is

(A) positive.
(B) negative.
(C) positive, then negative.
(D) negative, then positive.

How is this reflected in the graph, and what does it mean about the market shares of BlackBerry and Android?

d. Compute $(B - A)'(10)$. Interpret your answer.

COMMUNICATION AND REASONING EXERCISES

101. What instructions would you give to a fellow student who wanted to accurately graph the tangent line to the curve $y = 3x^2$ at the point $(-1, 3)$?

102. What instructions would you give to a fellow student who wanted to accurately graph a line at right angles to the curve $y = 4/x$ at the point where $x = 0.5$?

103. Consider $f(x) = x^2$ and $g(x) = 2x^2$. How do the slopes of the tangent lines of f and g at the same x compare?

104. Consider $f(x) = x^3$ and $g(x) = x^3 + 3$. How do the slopes of the tangent lines of f and g compare?

105. Suppose $g(x) = -f(x)$. How do the derivatives of f and g compare?

106. Suppose $g(x) = f(x) - 50$. How do the derivatives of f and g compare?

107. Following is an excerpt from your best friend's graded homework:

$$3x^4 + 11x^5 = 12x^3 + 55x^4. \quad ✗ \; WRONG \quad -8$$

Why was it marked wrong? How would you correct it?

108. Following is an excerpt from your own graded homework:

$$x^n = nx^{n-1}. \quad ✗ \; WRONG \quad -10$$

Why was it marked wrong? How would you correct it?

109. Following is another excerpt from your best friend's graded homework:

$$y = \frac{1}{2x} = 2x^{-1}, \text{ so } \frac{dy}{dx} = -2x^{-2}. \quad ✗ \; WRONG \quad -5$$

Why was it marked wrong? How would you correct it?

110. Following is an excerpt from your second best friend's graded homework:

$$f(x) = \frac{3}{4x^2}; f'(x) = \frac{3}{8x}. \quad ✗ \; WRONG \quad -10$$

Why was it marked wrong? How would you correct it?

111. Following is an excerpt from your worst enemy's graded homework:

$$f(x) = 4x^2; f'(x) = (0)(2x) = 0. \quad ✗ \; WRONG \quad -6$$

Why was it marked wrong? How would you correct it?

112. Following is an excerpt from your second worst enemy's graded homework:

$$f(x) = \frac{3}{4x}; f'(x) = \frac{0}{4} = 0. \quad ✗ \; WRONG \quad -10$$

Why was it marked wrong? How would you correct it?

113. One of the questions in your last calculus test was "**Question 1(a)** Give the definition of the derivative of a function f." Following is your answer and the grade you received:

$$nx^{n-1}. \quad ✗ \; WRONG \quad -10$$

Why was it marked wrong? What is the correct answer?

114. ▼ How would you respond to an acquaintance who says, "I finally understand what the derivative is: It is nx^{n-1}! Why weren't we taught that in the first place instead of the difficult way using limits?"

115. ▼ Sketch the graph of a function whose derivative is undefined at exactly two points but that has a tangent line at all but one point.

116. ▼ Sketch the graph of a function that has a tangent line at each of its points, but whose derivative is undefined at exactly two points.

11.2 A First Application: Marginal Analysis

In Chapter 1, we considered linear *cost functions* of the form $C(x) = mx + b$, where C is the total cost, x is the number of items, and m and b are constants. The slope m is the *marginal cost.* It measures the *cost of one more item.* Notice that the derivative of $C(x) = mx + b$ is $C'(x) = m$. In other words, for a linear cost function, *the marginal cost is the derivative of the cost function.*

In general, we make the following definition.

Marginal Cost

Recall from Section 1.2 that a **cost function** C specifies the total cost as a function of the number of items x, so that $C(x)$ is the total cost of x items. The **marginal cost function** is the derivative C' of the cost function C. Thus, $C'(x)$ measures the rate of change of cost with respect to x.

Units
The units of marginal cost are units of cost (dollars, say) per item.

Interpretation
We interpret $C'(x)$ as the approximate cost of one more item.*

＊ See Example 1.

> #### Quick Example
>
> If $C(x) = 400x + 1{,}000$ dollars, then the marginal cost function is $C'(x) = \$400$ per item (a constant).

EXAMPLE 1 Marginal Cost

Suppose that the cost in dollars to manufacture portable music players is given by

$$C(x) = 150{,}000 + 20x - 0.0001x^2$$

where x is the number of music players manufactured.† Find the marginal cost function C' and use it to estimate the cost of manufacturing the 50,001st music player.

† The term $0.0001x^2$ may reflect a cost saving for high levels of production, such as a bulk discount in the cost of electronic components.

Solution Since

$$C(x) = 150{,}000 + 20x - 0.0001x^2$$

the marginal cost function is

$$C'(x) = 20 - 0.0002x.$$

The units of $C'(x)$ are units of C (dollars) per unit of x (music players). Thus, $C'(x)$ is measured in dollars per music player.

The cost of the 50,001st music player is the amount by which the total cost would rise if we increased production from 50,000 music players to 50,001. Thus, we need to know the rate at which the total cost rises as we increase production. This rate of change is measured by the derivative, or marginal cost, which we just computed. At $x = 50,000$, we get

$$C'(50,000) = 20 - 0.0002(50,000) = \$10 \text{ per music player.}$$

In other words, we estimate that the 50,001st music player will cost approximately \$10.

➡ **Before we go on...** In Example 1, the marginal cost is really only an *approximation* to the cost of the 50,001st music player:

$$C'(50,000) \approx \frac{C(50,001) - C(50,000)}{1} \qquad \text{Set } h = 1 \text{ in the definition of the derivative.}$$

$$= C(50,001) - C(50,000)$$

$$= \text{cost of the 50,001st music player}$$

The exact cost of the 50,001st music player is

$$C(50,001) - C(50,000) = [150,000 + 20(50,001) - 0.0001(50,001)^2]$$

$$- [150,000 + 20(50,000) - 0.0001(50,000)^2]$$

$$= \$9.9999$$

So, the marginal cost is a good approximation to the actual cost.

Graphically, we are using the tangent line to approximate the cost function near a production level of 50,000. Figure 3 shows the graph of the cost function together with the tangent line at $x = 50,000$. Notice that the tangent line is essentially indistinguishable from the graph of the function for some distance on either side of 50,000.

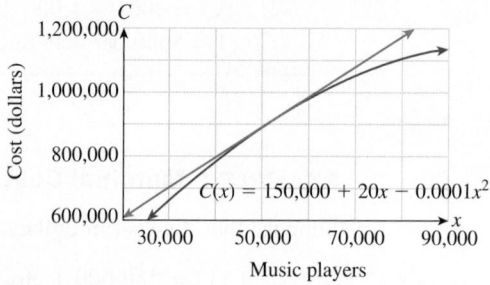

Figure 3

Notes

1. In general, the difference quotient $[C(x + h) - C(x)]/h$ gives the **average cost per item** to produce h more items at a current production level of x items. (Why?)

2. Notice that $C'(x)$ is much easier to calculate than $[C(x + h) - C(x)]/h$. (Try it.) ■

We can extend the idea of marginal cost to include other functions we discussed in Section 1.2, like revenue and profit:

Marginal Revenue and Profit

Recall that a **revenue** or **profit function** specifies the total revenue R or profit P as a function of the number of items x. The derivatives, R' and P', of these functions are called the **marginal revenue** and **marginal profit** functions. They measure the rate of change of revenue and profit with respect to the number of items.

Units

The units of marginal revenue and profit are the same as those of marginal cost: dollars (or euros, pesos, etc.) per item.

Interpretation

We interpret $R'(x)$ and $P'(x)$ as the approximate revenue and profit from the sale of one more item.

EXAMPLE 2 Marginal Revenue and Profit

You operate an *iPod* refurbishing service (a typical refurbished iPod might have a custom color case with blinking lights and a personalized logo). The cost to refurbish x iPods in a month is calculated to be

$$C(x) = 0.25x^2 + 40x + 1{,}000 \text{ dollars.}$$

You charge customers $80 per iPod for the work.

a. Calculate the marginal revenue and profit functions. Interpret the results.

b. Compute the revenue and profit, and also the marginal revenue and profit, if you have refurbished 20 units this month. Interpret the results.

c. For which value of x is the marginal profit zero? Interpret your answer.

Solution

a. We first calculate the revenue and profit functions:

$$R(x) = 80x \qquad \text{Revenue} = \text{Price} \times \text{Quantity}$$
$$P(x) = R(x) - C(x) \qquad \text{Profit} = \text{Revenue} - \text{Cost}$$
$$= 80x - (0.25x^2 + 40x + 1{,}000)$$
$$P(x) = -0.25x^2 + 40x - 1{,}000.$$

The marginal revenue and profit functions are then the derivatives:

$$\text{Marginal revenue} = R'(x) = 80$$
$$\text{Marginal profit} = P'(x) = -0.5x + 40.$$

Interpretation: $R'(x)$ gives the approximate revenue from the refurbishing of one more item, and $P'(x)$ gives the approximate profit from the refurbishing of one more item. Thus, if x iPods have been refurbished in a month, you will earn a revenue of $80 and make a profit of approximately $(-0.5x + 40)$ if you refurbish one more that month.

Notice that the marginal revenue is a constant, so you earn the same revenue ($80) for each iPod you refurbish. However, the marginal profit, $(−0.5x + 40)$, decreases as x increases, so your additional profit is about 50¢ less for each additional iPod you refurbish.

b. From part (a), the revenue, profit, marginal revenue, and marginal profit functions are

$$R(x) = 80x$$
$$P(x) = -0.25x^2 + 40x - 1,000$$
$$R'(x) = 80$$
$$P'(x) = -0.5x + 40$$

Because you have refurbished $x = 20$ iPods this month, $x = 20$, so

$R(20) = 80(20) = \$1,600$	Total revenue from 20 iPods
$P(20) = -0.25(20)^2 + 40(20) - 1,000 = -\$300$	Total profit from 20 iPods
$R'(20) = \$80$ per unit	Approximate revenue from the 21st iPod
$P'(20) = -0.5(20) + 40 = \$30$ per unit	Approximate profit from the 21st iPod

Interpretation: If you refurbish 20 iPods in a month, you will earn a total revenue of $160 and a profit of –$300 (indicating a loss of $300). Refurbishing one more iPod that month will earn you an additional revenue of $80 and an additional profit of about $30.

c. The marginal profit is zero when $P'(x) = 0$:

$$-0.5x + 40 = 0$$
$$x = \frac{40}{0.5} = 80 \text{ iPods}$$

Thus, if you refurbish 80 iPods in a month, refurbishing one more will get you (approximately) zero additional profit. To understand this further, let us take a look at the graph of the profit function, shown in Figure 4. Notice that the graph is a parabola (the profit function is quadratic) with vertex at the point $x = 80$, where $P'(x) = 0$, so the profit is a maximum at this value of x.

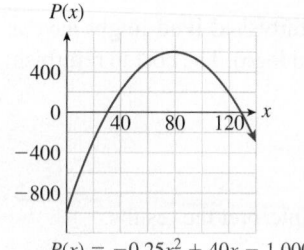

$P(x)$

$P(x) = -0.25x^2 + 40x - 1,000$

Figure 4

➡ **Before we go on...** In general, setting $P'(x) = 0$ and solving for x will always give the exact values of x for which the profit peaks as in Figure 4, assuming there is such a value. We recommend that you graph the profit function to check whether the profit is indeed a maximum at such a point. ∎

EXAMPLE 3 **Marginal Product**

A consultant determines that *Precision Manufacturers'* annual profit (in dollars) is given by

$$P(n) = -200,000 + 400,000n - 4,600n^2 - 10n^3 \qquad (10 \le n \le 50),$$

where n is the number of assembly-line workers it employs.

a. Compute $P'(n)$. $P'(n)$ is called the **marginal product** at the employment level of n assembly-line workers. What are its units?

b. Calculate $P(20)$ and $P'(20)$, and interpret the results.

c. Precision Manufacturers currently employs 20 assembly-line workers and is considering laying off some of them. What advice would you give the company's management?

Solution

a. Taking the derivative gives

$$P'(n) = 400{,}000 - 9{,}200n - 30n^2.$$

The units of $P'(n)$ are profit (in dollars) per worker.

b. Substituting into the formula for $P(n)$, we get

$$P(20) = -200{,}000 + 400{,}000(20) - 4{,}600(20)^2 - 10(20)^3 = \$5{,}880{,}000.$$

Thus, Precision Manufacturers will make an annual profit of \$5,880,000 if it employs 20 assembly-line workers. On the other hand,

$$P'(20) = 400{,}000 - 9{,}200(20) - 30(20)^2 = \$204{,}000/\text{worker}.$$

Thus, at an employment level of 20 assembly-line workers, annual profit is increasing at a rate of \$204,000 per additional worker. In other words, if the company were to employ one more assembly-line worker, its annual profit would increase by approximately \$204,000.

c. Because the marginal product is positive, profits will increase if the company increases the number of workers and will decrease if it decreases the number of workers, so your advice would be to hire additional assembly-line workers. Downsizing the assembly-line workforce would reduce its annual profits.

➡ **Before we go on...** In Example 3, it would be interesting for Precision Manufacturers to ascertain how many additional assembly-line workers it should hire to obtain the *maximum* annual profit. Taking our cue from Example 2, we suspect that such a value of n would correspond to a point where $P'(n) = 0$. Figure 5 shows the graph of P, and on it we see that the highest point of the graph is indeed a point where the tangent line is horizontal; that is, $P'(n) = 0$, and occurs somewhere between $n = 35$ and 40. To compute this value of n more accurately, set $P'(n) = 0$ and solve for n:

$$P'(n) = 400{,}000 - 9{,}200n - 30n^2 = 0 \quad \text{or} \quad 40{,}000 - 920n - 3n^2 = 0.$$

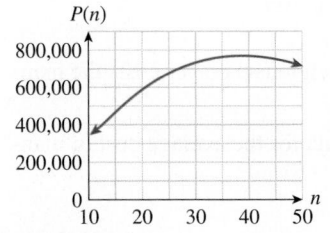

Figure 5

We can now obtain n using the quadratic formula:

$$n = \frac{-b \pm \sqrt{b^2 - 4ac}}{2a} = \frac{920 \pm \sqrt{920^2 - 4(-3)(40{,}000)}}{2(-3)}$$

$$= \frac{920 \pm \sqrt{1{,}326{,}400}}{-6} \approx -345.3 \text{ or } 38.6.$$

The only meaningful solution is the positive one, $n \approx 38.6$ workers, and we conclude that the company should employ between 38 and 39 assembly-line workers for a maximum profit. To see which gives the larger profit, 38 or 39, we check:

$$P(38) = \$7{,}808{,}880$$

while

$$P(39) = \$7{,}810{,}210.$$

This tells us that the company should employ 39 assembly-line workers for a maximum profit. Thus, instead of laying off any of its 20 assembly-line workers, the company should hire 19 additional assembly-line workers for a total of 39. ∎

Average Cost

EXAMPLE 4 Average Cost

Suppose the cost in dollars to manufacture portable music players is given by

$$C(x) = 150{,}000 + 20x - 0.0001x^2$$

where x is the number of music players manufactured. (This is the cost equation we saw in Example 1.)

a. Find the average cost per music player if 50,000 music players are manufactured.

b. Find a formula for the average cost per music player if x music players are manufactured. This function of x is called the **average cost function, $\bar{C}(x)$.**

Solution

a. The total cost of manufacturing 50,000 music players is given by

$$C(50{,}000) = 150{,}000 + 20(50{,}000) - 0.0001(50{,}000)^2$$
$$= \$900{,}000.$$

Because 50,000 music players cost a total of \$900,000 to manufacture, the average cost of manufacturing one music player is this total cost divided by 50,000:

$$\bar{C}(50{,}000) = \frac{900{,}000}{50{,}000} = \$18.00 \text{ per music player.}$$

Thus, if 50,000 music players are manufactured, each music player costs the manufacturer an average of \$18.00 to manufacture.

b. If we replace 50,000 by x, we get the general formula for the average cost of manufacturing x music players:

$$\bar{C}(x) = \frac{C(x)}{x}$$

$$= \frac{1}{x}(150{,}000 + 20x - 0.0001x^2)$$

$$= \frac{150{,}000}{x} + 20 - 0.0001x. \qquad \text{Average cost function}$$

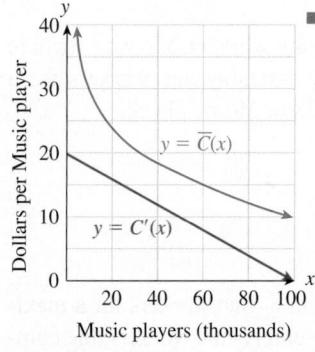

Figure 6

➡ **Before we go on...** Average cost and marginal cost convey different but related information. The average cost $\bar{C}(50{,}000) = \$18$ that we calculated in Example 4 is the cost per item of manufacturing the first 50,000 music players, whereas the marginal cost $C'(50{,}000) = \$10$ that we calculated in Example 1 gives the (approximate) cost of manufacturing the *next* music player. Thus, according to our calculations, the first 50,000 music players cost an average of \$18 to manufacture, but it costs only about \$10 to manufacture the next one. Note that the marginal cost at a production level of 50,000 music players is lower than the average cost. This means that the average cost to manufacture CDs is going down with increasing volume. (Think about why.)

Figure 6 shows the graphs of average and marginal cost. Notice how the decreasing marginal cost seems to pull the average cost down with it. ∎

To summarize:

Average Cost

Given a cost function C, the **average cost** of the first x items is given by

$$\bar{C}(x) = \frac{C(x)}{x}.$$

The average cost is distinct from the **marginal cost** $C'(x)$, which tells us the approximate cost of the *next* item.

Quick Example

For the cost function $C(x) = 20x + 100$ dollars

Marginal Cost $= C'(x) = \$20$ per additional item.

Average Cost $= \bar{C}(x) = \dfrac{C(x)}{x} = \dfrac{20x + 100}{x} = \$(20 + 100/x)$ per item.

11.2 EXERCISES

In Exercises 1–4, for each cost function, find the marginal cost at the given production level x, and state the units of measurement. (All costs are in dollars.) HINT [See Example 1.]

1. $C(x) = 10,000 + 5x - 0.0001x^2$; $\; x = 1,000$

2. $C(x) = 20,000 + 7x - 0.00005x^2$; $\; x = 10,000$

3. $C(x) = 15,000 + 100x + \dfrac{1,000}{x}$; $\; x = 100$

4. $C(x) = 20,000 + 50x + \dfrac{10,000}{x}$; $\; x = 100$

In Exercises 5 and 6, find the marginal cost, marginal revenue, and marginal profit functions, and find all values of x for which the marginal profit is zero. Interpret your answer. HINT [See Example 2.]

5. $C(x) = 4x$; $R(x) = 8x - 0.001x^2$

6. $C(x) = 5x^2$; $R(x) = x^3 + 7x + 10$

7. ▼ A certain cost function has the following graph:

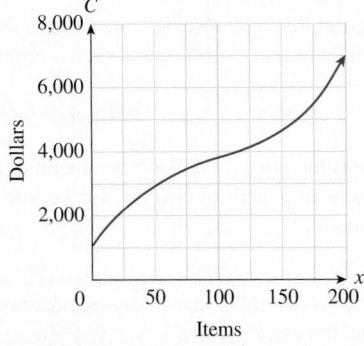

a. The associated marginal cost is

   (A) increasing, then decreasing.
   (B) decreasing, then increasing.
   (C) always increasing.
   (D) always decreasing.

b. The marginal cost is least at approximately

   (A) $x = 0$.   **(B)** $x = 50$.   **(C)** $x = 100$.   **(D)** $x = 150$.

c. The cost of 50 items is

   (A) approximately $20, and increasing at a rate of about $3,000 per item.
   (B) approximately $0.50, and increasing at a rate of about $3,000 per item.
   (C) approximately $3,000, and increasing at a rate of about $20 per item.
   (D) approximately $3,000, and increasing at a rate of about $0.50 per item.

8. ▼ A certain cost function has the following graph:

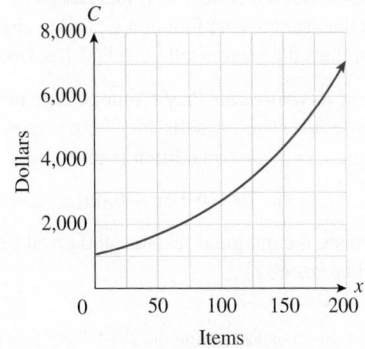

a. The associated marginal cost is

  (A) increasing, then decreasing.
  (B) decreasing, then increasing.
  (C) always increasing.
  (D) always decreasing.

b. When $x = 100$, the marginal cost is

  (A) greater than the average cost.
  (B) less than the average cost.
  (C) approximately equal to the average cost.

c. The cost of 150 items is

  (A) approximately $4,400, and increasing at a rate of about $40 per item.
  (B) approximately $40, and increasing at a rate of about $4,400 per item.
  (C) approximately $4,400, and increasing at a rate of about $1 per item.
  (D) approximately $1, and increasing at a rate of about $4,400 per item.

APPLICATIONS

9. *Advertising Costs* The cost, in thousands of dollars, of airing x television commercials during a Super Bowl game is given by[11]

$$C(x) = 150 + 2,500x - 0.02x^2.$$

a. Find the marginal cost function and use it to estimate how fast the cost is increasing when $x = 4$. Compare this with the exact cost of airing the fifth commercial. HINT [See Example 1.]

b. Find the average cost function $\bar{C}$, and evaluate $\bar{C}(4)$. What does the answer tell you? HINT [See Example 4.]

10. *Marginal Cost and Average Cost* The cost of producing x teddy bears per day at the *Cuddly Companion Co.* is calculated by the company's marketing staff to be given by the formula

$$C(x) = 100 + 40x - 0.001x^2.$$

a. Find the marginal cost function and use it to estimate how fast the cost is going up at a production level of 100 teddy bears. Compare this with the exact cost of producing the 101st teddy bear. HINT [See Example 1.]

b. Find the average cost function $\bar{C}$, and evaluate $\bar{C}(100)$. What does the answer tell you? HINT [See Example 4.]

11. *Marginal Revenue and Profit* Your college newspaper, *The Collegiate Investigator*, sells for 90¢ per copy. The cost of producing x copies of an edition is given by

$$C(x) = 70 + 0.10x + 0.001x^2 \text{ dollars.}$$

a. Calculate the marginal revenue and profit functions. HINT [See Example 2.]

b. Compute the revenue and profit, and also the marginal revenue and profit, if you have produced and sold 500 copies of the latest edition. Interpret the results.

c. For which value of x is the marginal profit zero? Interpret your answer.

12. *Marginal Revenue and Profit* The Audubon Society at Enormous State University (ESU) is planning its annual fund-raising "Eatathon." The society will charge students $1.10 per serving of pasta. The society estimates that the total cost of producing x servings of pasta at the event will be

$$C(x) = 350 + 0.10x + 0.002x^2 \text{ dollars.}$$

a. Calculate the marginal revenue and profit functions. HINT [See Example 2.]

b. Compute the revenue and profit, and also the marginal revenue and profit, if you have produced and sold 200 servings of pasta. Interpret the results.

c. For which value of x is the marginal profit zero? Interpret your answer.

13. *Marginal Profit* Suppose $P(x)$ represents the profit in dollars on the sale of x DVDs. If $P(1,000) = 3,000$ and $P'(1,000) = -3$, what do these values tell you about the profit?

14. *Marginal Loss* An automobile retailer calculates that its loss in dollars on the sale of *Type M* cars is given by $L(50) = 5,000$ and $L'(50) = -200$, where $L(x)$ represents the loss on the sale of x Type M cars. What do these values tell you about losses?

15. *Marginal Profit* Your monthly profit (in dollars) from selling magazines is given by

$$P = 5x + \sqrt{x}$$

where x is the number of magazines you sell in a month. If you are currently selling 50 magazines per month, find your profit and your marginal profit. Interpret your answers.

16. *Marginal Profit* Your monthly profit (in dollars) from your newspaper route is given by

$$P = 2n - \sqrt{n}$$

where n is the number of subscribers on your route. If you currently have 100 subscribers, find your profit and your marginal profit. Interpret your answers.

17. ▼ *Marginal Revenue: Pricing Tuna* Assume that the demand equation for tuna in a small coastal town is given by

$$p = \frac{20,000}{q^{1.5}} \qquad (200 \le q \le 800),$$

where p is the price (in dollars) per pound of tuna, and q is the number of pounds of tuna that can be sold at the price p in one month.[12]

---

[11]The cost of a 30-second ad during the 2010 Super Bowl game was about $2.5 million. This explains the coefficient of x in the cost function. Source: http://en.wikipedia.org/wiki/Super_Bowl_advertising.

[12]Notice that here we have specified p as a function of q, and not the other way around as we did in Section 1.2. Economists frequently specify demand curves this way.

a. Calculate the price that the town's fishery should charge for tuna in order to produce a demand of 400 pounds of tuna per month.

b. Calculate the monthly revenue R as a function of the number of pounds of tuna q.

c. Calculate the revenue and marginal revenue (derivative of the revenue with respect to q) at a demand level of 400 pounds per month, and interpret the results.

d. If the town fishery's monthly tuna catch amounted to 400 pounds of tuna, and the price is at the level in part (a), would you recommend that the fishery raise or lower the price of tuna in order to increase its revenue?

18. ▼ ***Marginal Revenue: Pricing Tuna*** Repeat Exercise 17, assuming a demand equation of

$$p = \frac{60}{q^{0.5}} \quad (200 \le q \le 800).$$

19. ***Marginal Product*** A car wash firm calculates that its daily profit (in dollars) depends on the number n of workers it employs according to the formula

$$P = 400n - 0.5n^2.$$

Calculate the marginal product at an employment level of 50 workers, and interpret the result. HINT [See Example 3.]

20. ***Marginal Product*** Repeat the preceding exercise using the formula

$$P = -100n + 25n^2 - 0.005n^4.$$

HINT [See Example 3.]

21. ***Average and Marginal Cost*** The daily cost to manufacture generic trinkets for gullible tourists is given by the cost function

$$C(x) = -0.001x^2 + 0.3x + 500 \text{ dollars}$$

where x is the number of trinkets.

a. As x increases, the marginal cost
   (A) increases.   **(B)** decreases.   **(C)** increases, then decreases.   **(D)** decreases, then increases.

b. As x increases, the average cost
   (A) increases.   **(B)** decreases.   **(C)** increases, then decreases.   **(D)** decreases, then increases.

c. The marginal cost is
   (A) greater than   **(B)** equal to   **(C)** less than
   the average cost when $x = 100$. HINT [See Example 4.]

22. ***Average and Marginal Cost*** Repeat Exercise 21, using the following cost function for imitation oil paintings (x is the number of "oil paintings" manufactured):

$$C(x) = 0.1x^2 - 3.5x + 500 \text{ dollars}.$$

HINT [See Example 4.]

23. ***Advertising Cost*** Your company is planning to air a number of television commercials during the ABC Television Network's presentation of the Academy Awards. ABC is charg-ing your company $1.6 million per 30-second spot.[13] Addi-tional fixed costs (development and personnel costs) amount to $500,000, and the network has agreed to provide a discount of $10,000\sqrt{x}$ for x television spots.

a. Write down the cost function C, marginal cost function C', and average cost function $\bar{C}$.

b. Compute $C'(3)$ and $\bar{C}(3)$. (Round all answers to three significant digits.) Use these two answers to say whether the average cost is increasing or decreasing as x increases.

24. ***Housing Costs*** The cost C of building a house is related to the number k of carpenters used and the number x of electri-cians used by the formula[14]

$$C = 15,000 + 50k^2 + 60x^2.$$

a. Assuming that 10 carpenters are currently being used, find the cost function C, marginal cost function C', and average cost function $\bar{C}$, all as functions of x.

b. Use the functions you obtained in part (a) to compute $C'(15)$ and $\bar{C}(15)$. Use these two answers to say whether the average cost is increasing or decreasing as the number of electricians increases.

25. ▼ ***Emission Control*** The cost of controlling emissions at a firm rises rapidly as the amount of emissions reduced in-creases. Here is a possible model:

$$C(q) = 4,000 + 100q^2$$

where q is the reduction in emissions (in pounds of pollutant per day) and C is the daily cost (in dollars) of this reduction.

a. If a firm is currently reducing its emissions by 10 pounds each day, what is the marginal cost of reducing emissions further?

b. Government clean-air subsidies to the firm are based on the formula

$$S(q) = 500q$$

where q is again the reduction in emissions (in pounds per day) and S is the subsidy (in dollars). At what reduc-tion level does the marginal cost surpass the marginal subsidy?

c. Calculate the net cost function, $N(q) = C(q) - S(q)$, given the cost function and subsidy above, and find the value of q that gives the lowest net cost. What is this lowest net cost? Compare your answer to that for part (b) and comment on what you find.

26. ▼ ***Taxation Schemes*** In order to raise revenues during the recent recession, the governor of your state proposed the following taxation formula:

$$T(i) = 0.001i^{0.5},$$

---

[13]ABC charged an average of $1.6 million for a 30-second spot during the 2005 Academy Awards presentation. Source: CNN/Reuters, www.cnn.com, February 9, 2005.

[14]Based on an exercise in *Introduction to Mathematical Economics* by A. L. Ostrosky, Jr., and J. V. Koch (Waveland Press, Prospect Heights, Illinois, 1979).

where i represents total annual income earned by an individual in dollars and $T(i)$ is the income tax rate as a percentage of total annual income. (Thus, for example, an income of $50,000 per year would be taxed at about 22%, while an income of double that amount would be taxed at about 32%.)[15]

a. Calculate the after-tax (net) income $N(i)$ an individual can expect to earn as a function of income i.

b. Calculate an individual's marginal after-tax income at income levels of $100,000 and $500,000.

c. At what income does an individual's marginal after-tax income become negative? What is the after-tax income at that level, and what happens at higher income levels?

d. What do you suspect is the most anyone can earn after taxes? (See NOTE at the bottom of this page.)

27. ▼ *Fuel Economy* Your Porsche's gas mileage (in miles per gallon) is given as a function $M(x)$ of speed x in miles per hour. It is found that

$$M'(x) = \frac{3,600x^{-2} - 1}{(3,600x^{-1} + x)^2}.$$

Estimate $M'(10)$, $M'(60)$, and $M'(70)$. What do the answers tell you about your car?

28. ▼ *Marginal Revenue* The estimated marginal revenue for sales of ESU soccer team T-shirts is given by

$$R'(p) = \frac{(8 - 2p)e^{-p^2 + 8p}}{10,000,000}$$

where p is the price (in dollars) that the soccer players charge for each shirt. Estimate $R'(3)$, $R'(4)$, and $R'(5)$. What do the answers tell you?

29. ◆ *Marginal Cost (from the GRE Economics Test)* In a multiplant firm in which the different plants have different and continuous cost schedules, if costs of production for a given output level are to be minimized, which of the following is essential?

(A) Marginal costs must equal marginal revenue.
(B) Average variable costs must be the same in all plants.
(C) Marginal costs must be the same in all plants.
(D) Total costs must be the same in all plants.
(E) Output per worker per hour must be the same in all plants.

30. ◆ *Study Time (from the GRE economics test)* A student has a fixed number of hours to devote to study and is certain of the relationship between hours of study and the final grade for each course. Grades are given on a numerical scale (0 to 100), and each course is counted equally in computing the grade average. In order to maximize his or her grade average, the student should allocate these hours to different courses so that

(A) the grade in each course is the same.
(B) the marginal product of an hour's study (in terms of final grade) in each course is zero.

(C) the marginal product of an hour's study (in terms of final grade) in each course is equal, although not necessarily equal to zero.
(D) the average product of an hour's study (in terms of final grade) in each course is equal.
(E) the number of hours spent in study for each course is equal.

31. ◆ *Marginal Product (from the GRE Economics Test)* Assume that the marginal product of an additional senior professor is 50% higher than the marginal product of an additional junior professor and that junior professors are paid one half the amount that senior professors receive. With a fixed overall budget, a university that wishes to maximize its quantity of output from professors should do which of the following?

(A) Hire equal numbers of senior professors and junior professors.
(B) Hire more senior professors and junior professors.
(C) Hire more senior professors and discharge junior professors.
(D) Discharge senior professors and hire more junior professors.
(E) Discharge all senior professors and half of the junior professors.

32. ◆ *Marginal Product (based on a question from the GRE Economics Test)* Assume that the marginal product of an additional senior professor is twice the marginal product of an additional junior professor and that junior professors are paid two thirds the amount that senior professors receive. With a fixed overall budget, a university that wishes to maximize its quantity of output from professors should do which of the following?

(A) Hire equal numbers of senior professors and junior professors.
(B) Hire more senior professors and junior professors.
(C) Hire more senior professors and discharge junior professors.
(D) Discharge senior professors and hire more junior professors.
(E) Discharge all senior professors and half of the junior professors.

COMMUNICATION AND REASONING EXERCISES

33. The marginal cost of producing the 1,001st item is

(A) equal to
(B) approximately equal to
(C) always slightly greater than
(D) always slightly less than

the actual cost of producing the 1,001st item.

34. For the cost function $C(x) = mx + b$, the marginal cost of producing the 1,001st item is,

(A) equal to
(B) approximately equal to
(C) always slightly greater than
(D) always slightly less than

the actual cost of producing the 1,001st item.

---

[15]This model has the following interesting feature: An income of $1 million per year would be taxed at 100%, leaving the individual penniless!

35. What is a cost function? Carefully explain the difference between *average cost* and *marginal cost* in terms of **(a)** their mathematical definition, **(b)** graphs, and **(c)** interpretation.

36. The cost function for your grand piano manufacturing plant has the property that $\bar{C}(1,000) = \$3,000$ per unit and $C'(1,000) = \$2,500$ per unit. Will the average cost increase or decrease if your company manufactures a slightly larger number of pianos? Explain your reasoning.

37. Give an example of a cost function for which the marginal cost function is the same as the average cost function.

38. Give an example of a cost function for which the marginal cost function is always less than the average cost function.

39. If the average cost to manufacture one grand piano increases as the production level increases, which is greater, the marginal cost or the average cost?

40. If your analysis of a manufacturing company yielded positive marginal profit but negative profit at the company's current production levels, what would you advise the company to do?

41. ▼ If the marginal cost is decreasing, is the average cost necessarily decreasing? Explain.

42. ▼ If the average cost is decreasing, is the marginal cost necessarily decreasing? Explain.

43. ◆ If a company's marginal average cost is zero at the current production level, positive for a slightly higher production level, and negative for a slightly lower production level, what should you advise the company to do?

44. ◆ The **acceleration** of cost is defined as the derivative of the marginal cost function: that is, the derivative of the derivative—or *second derivative*—of the cost function. What are the units of acceleration of cost, and how does one interpret this measure?

11.3 The Product and Quotient Rules

We know how to find the derivatives of functions that are sums of powers, such as polynomials. In general, if a function is a sum or difference of functions whose derivatives we know, then we know how to find its derivative. But what about *products and quotients* of functions whose derivatives we know? For instance, how do we calculate the derivative of something like $x^2/(x + 1)$? The derivative of $x^2/(x + 1)$ is not, as one might suspect, $2x/1 = 2x$. That calculation is based on an assumption that the derivative of a quotient is the quotient of the derivatives. But it is easy to see that this assumption is false: For instance, the derivative of $1/x$ is not $0/1 = 0$, but $-1/x^2$. Similarly, the derivative of a product is not the product of the derivatives: For instance, the derivative of $x = 1 \cdot x$ is not $0 \cdot 1 = 0$, but 1.

To identify the correct method of computing the derivatives of products and quotients, let's look at a simple example. We know that the daily revenue resulting from the sale of q items per day at a price of p dollars per item is given by the product, $R = pq$ dollars. Suppose you are currently selling wall posters on campus. At this time your daily sales are 50 posters, and sales are increasing at a rate of 4 per day. Furthermore, you are currently charging $10 per poster, and you are also raising the price at a rate of $2 per day. Let's use this information to estimate how fast your daily revenue is increasing. In other words, let us estimate the rate of change, dR/dt, of the revenue R.

There are two contributions to the rate of change of daily revenue: the increase in daily sales and the increase in the unit price. We have

$$\frac{dR}{dt} \text{ due to increasing price:} \quad \$2 \text{ per day} \times 50 \text{ posters} = \$100 \text{ per day}$$

$$\frac{dR}{dt} \text{ due to increasing sales:} \quad \$10 \text{ per poster} \times 4 \text{ posters per day} = \$40 \text{ per day.}$$

Thus, we estimate the daily revenue to be increasing at a rate of $100 + $40 = $140 per day. Let us translate what we have said into symbols:

$$\frac{dR}{dt} \text{ due to increasing price:} \qquad \frac{dp}{dt} \times q$$

$$\frac{dR}{dt} \text{ due to increasing sales:} \qquad p \times \frac{dq}{dt}.$$

Thus, the rate of change of revenue is given by

$$\frac{dR}{dt} = \frac{dp}{dt}q + p\frac{dq}{dt}.$$

Because $R = pq$, we have discovered the following rule for differentiating a product:

$$\frac{d}{dt}(pq) = \frac{dp}{dt}q + p\frac{dq}{dt}.$$

The derivative of a product is the derivative of the first times the second, plus the first times the derivative of the second.

This rule and a similar rule for differentiating quotients are given next, and also a discussion of how these results are proved rigorously.

Product Rule

If f and g are differentiable functions of x, then so is their product fg, and

$$\frac{d}{dx}[f(x)g(x)] = f'(x)g(x) + f(x)g'(x).$$

Product Rule in Words
The derivative of a product is the derivative of the first times the second, plus the first times the derivative of the second.

Quick Example

Let $f(x) = x^2$ and $g(x) = 3x - 1$. Because f and g are both differentiable functions of x, so is their product fg, and its derivative is

$$\frac{d}{dx}[x^2(3x - 1)] = 2x \cdot (3x - 1) + x^2 \cdot (3).$$

Derivative of first   Second   First   Derivative of second

Quotient Rule

If f and g are differentiable functions of x, then so is their quotient f/g, and

$$\frac{d}{dx}\left(\frac{f(x)}{g(x)}\right) = \frac{f'(x)g(x) - f(x)g'(x)}{[g(x)]^2}.$$

✳ If $g(x)$ is zero, then the quotient $f(x)/g(x)$ is not defined in the first place.

provided $g(x) \neq 0$.✳

Quotient Rule in Words
The derivative of a quotient is the derivative of the top times the bottom, minus the top times the derivative of the bottom, all over the bottom squared.

> ## Quick Example
>
> Let $f(x) = x^3$ and $g(x) = x^2 - 1$. Because f and g are both differentiable functions of x, so is their quotient f/g, and its derivative is
>
> Derivative of top   Bottom   Top   Derivative of bottom
> $$\frac{d}{dx}\left(\frac{x^3}{x^2-1}\right) = \frac{3x^2(x^2-1) - x^3 \cdot 2x}{(x^2-1)^2},$$
>
> Bottom squared
>
> provided $x \neq 1$ or -1.

Notes

1. Don't try to remember the rules by the symbols we have used, but remember them in words. (The slogans are easy to remember, even if the terms are not precise.)

2. One more time: *The derivative of a product is* NOT *the product of the derivatives, and the derivative of a quotient is* NOT *the quotient of the derivatives.* To find the derivative of a product, you must use the product rule, and to find the derivative of a quotient, you must use the quotient rule.* ∎

✱ Leibniz made this mistake at first, too, so you would be in good company if you forgot to use the product or quotient rule.

Q : *Wait a minute! The expression $2x^3$ is a product, and we already know that its derivative is $6x^2$. Where did we use the product rule?*

A : To differentiate functions such as $2x^3$, we have used the rule from Section 11.1:

The derivative of c times a function is c times the derivative of the function.

However, the product rule gives us the same result:

Derivative of first Second   First Derivative of second
$$\frac{d}{dx}(2x^3) = (0)(x^3) \quad + \quad (2)(3x^2) = 6x^2 \qquad \text{Product rule}$$

$$\frac{d}{dx}(2x^3) = (2)(3x^2) = 6x^2 \qquad \text{Derivative of a constant times a function}$$

We do not recommend that you use the product rule to differentiate functions such as $2x^3$; continue to use the simpler rule when one of the factors is a constant.

Derivation of the Product Rule

Before we look at more examples of using the product and quotient rules, let's see why the product rule is true. To calculate the derivative of the product $f(x)g(x)$ of two differentiable functions, we go back to the definition of the derivative:

$$\frac{d}{dx}[f(x)g(x)] = \lim_{h \to 0} \frac{f(x+h)g(x+h) - f(x)g(x)}{h}.$$

We now rewrite this expression so that we can evaluate the limit: Notice that the numerator reflects a simultaneous change in f [from $f(x)$ to $f(x+h)$] and g [from

$g(x)$ to $g(x + h)$]. To separate the two effects, we add and subtract a quantity in the numerator that reflects a change in only one of the functions:

$$\frac{d}{dx}[f(x)g(x)] = \lim_{h\to 0}\frac{f(x+h)g(x+h) - f(x)g(x)}{h}$$

$$= \lim_{h\to 0}\frac{f(x+h)g(x+h) - f(x)g(x+h) + f(x)g(x+h) - f(x)g(x)}{h}$$

We subtracted and added the quantity* $f(x)g(x+h)$.

$$= \lim_{h\to 0}\frac{[f(x+h) - f(x)]\,g(x+h) + f(x)[g(x+h) - g(x)]}{h}$$

Common factors

$$= \lim_{h\to 0}\left(\frac{f(x+h) - f(x)}{h}\right)g(x+h) + \lim_{h\to 0}f(x)\left(\frac{g(x+h) - g(x)}{h}\right)$$

Limit of sum

$$= \lim_{h\to 0}\left(\frac{f(x+h) - f(x)}{h}\right)\lim_{h\to 0}g(x+h) + \lim_{h\to 0}f(x)\lim_{h\to 0}\left(\frac{g(x+h) - g(x)}{h}\right)$$

Limit of product

* Adding an appropriate form of zero is an age-old mathematical ploy.

Now we already know the following four limits:

$$\lim_{h\to 0}\frac{f(x+h) - f(x)}{h} = f'(x)$$ Definition of derivative of f; f is differentiable.

$$\lim_{h\to 0}\frac{g(x+h) - g(x)}{h} = g'(x)$$ Definition of derivative of g; g is differentiable.

$$\lim_{h\to 0}g(x+h) = g(x)$$ If g is differentiable, it must be continuous.†

$$\lim_{h\to 0}f(x) = f(x)$$ Limit of a constant

† For a proof of the fact that, if g is differentiable, it must be continuous, go to the Website and follow the path
  Everything
  → Chapter 11
  → Continuity and Differentiability

Putting these limits into the one we're calculating, we get

$$\frac{d}{dx}[f(x)g(x)] = f'(x)g(x) + f(x)g'(x)$$

which is the product rule.

EXAMPLE 1 Using the Product Rule

Compute the following derivatives.

a. $\dfrac{d}{dx}[(x^{3.2} + 1)(1 - x)]$ Simplify the answer.

b. $\dfrac{d}{dx}[(x + 1)(x^2 + 1)(x^3 + 1)]$ Do not expand the answer.

c. $\dfrac{d}{dx}\left[\dfrac{x|x|}{2}\right]$

Website
www.WanerMath.com
The quotient rule can be proved in a very similar way. Go to the Website and follow the path
  Everything
  → Chapter 11
  → Proof of Quotient Rule

Solution

a. We can do the calculation in two ways.

Using the Product Rule:

$$\frac{d}{dx}[(x^{3.2} + 1)(1 - x)] = (3.2x^{2.2})(1 - x) + (x^{3.2} + 1)(-1)$$

Derivative of first, Second, First, Derivative of second

$$= 3.2x^{2.2} - 3.2x^{3.2} - x^{3.2} - 1$$ Expand the answer.

$$= -4.2x^{3.2} + 3.2x^{2.2} - 1$$

Not Using the Product Rule: First, expand the given expression.

$$(x^{3.2} + 1)(1 - x) = -x^{4.2} + x^{3.2} - x + 1$$

Thus,

$$\frac{d}{dx}[(x^{3.2} + 1)(1 - x)] = \frac{d}{dx}(-x^{4.2} + x^{3.2} - x + 1)$$
$$= -4.2x^{3.2} + 3.2x^{2.2} - 1$$

In this example the product rule saves us little or no work, but in later sections we shall see examples that can be done in no other way. Learn how to use the product rule now!

b. Here we have a product of *three* functions, not just two. We can find the derivative by using the product rule twice:

$$\frac{d}{dx}[(x + 1)(x^2 + 1)(x^3 + 1)]$$
$$= \frac{d}{dx}(x + 1) \cdot [(x^2 + 1)(x^3 + 1)] + (x + 1) \cdot \frac{d}{dx}[(x^2 + 1)(x^3 + 1)]$$
$$= (1)(x^2 + 1)(x^3 + 1) + (x + 1)[(2x)(x^3 + 1) + (x^2 + 1)(3x^2)]$$
$$= (1)(x^2 + 1)(x^3 + 1) + (x + 1)(2x)(x^3 + 1) + (x + 1)(x^2 + 1)(3x^2)$$

We can see here a more general product rule:

$$(fgh)' = f'gh + fg'h + fgh'$$

Notice that every factor has a chance to contribute to the rate of change of the product. There are similar formulas for products of four or more functions.

c. First write $\dfrac{x|x|}{2}$ as $\dfrac{1}{2}x|x|$.

$$\frac{d}{dx}\left[\frac{1}{2}x|x|\right] = \frac{1}{2}\frac{d}{dx}[x|x|] \qquad \text{Constant multiple rule}$$
$$= \frac{1}{2}\left((1) \cdot |x| + x \cdot \frac{|x|}{x}\right) \qquad \text{Recall that } \frac{d}{dx}|x| = \frac{|x|}{x}.$$
$$= \frac{1}{2}(|x| + |x|) \qquad \text{Cancel the } x.$$
$$= \frac{1}{2}(2|x|) = |x| \qquad \text{See the note.}^*$$

✱ Notice that we have found a function whose derivative is $|x|$; namely $x|x|/2$. Notice also that the derivation we gave assumes that $x \neq 0$ because we divided by x in the third step. However, one can verify, using the definition of the derivative as a limit, that $x|x|/2$ is differentiable at $x = 0$ as well, and that its derivative at $x = 0$ is 0, implying that the formula $\dfrac{d}{dx}(x|x|/2) = |x|$ is valid for all values of x, including 0.

EXAMPLE 2 Using the Quotient Rule

Compute the derivatives **a.** $\dfrac{d}{dx}\left[\dfrac{1 - 3.2x^{-0.1}}{x + 1}\right]$ **b.** $\dfrac{d}{dx}\left[\dfrac{(x + 1)(x + 2)}{x - 1}\right]$

Solution

$$\overset{\text{Derivative of top}}{\downarrow} \quad \overset{\text{Bottom}}{\downarrow} \qquad\qquad \overset{\text{Top}}{\downarrow} \quad \overset{\text{Derivative of bottom}}{\downarrow}$$

a. $\dfrac{d}{dx}\left[\dfrac{1 - 3.2x^{-0.1}}{x + 1}\right] = \dfrac{(0.32x^{-1.1})(x + 1) - (1 - 3.2x^{-0.1})(1)}{(x + 1)^2}$

$$\underset{\text{Bottom squared}}{\uparrow}$$

$$= \frac{0.32x^{-0.1} + 0.32x^{-1.1} - 1 + 3.2x^{-0.1}}{(x+1)^2} \qquad \text{Expand the numerator.}$$

$$= \frac{3.52x^{-0.1} + 0.32x^{-1.1} - 1}{(x+1)^2}$$

b. Here we have both a product and a quotient. Which rule do we use, the product or the quotient rule? Here is a way to decide. Think about how we would calculate, step by step, the value of $(x+1)(x+2)/(x-1)$ for a specific value of x—say $x = 11$. Here is how we would probably do it:

1. Calculate $(x+1)(x+2) = (11+1)(11+2) = 156$.

2. Calculate $x - 1 = 11 - 1 = 10$.

3. Divide 156 by 10 to get 15.6.

Now ask: *What was the last operation we performed?* The last operation we performed was division, so we can regard the whole expression as a *quotient*—that is, as $(x+1)(x+2)$ *divided by* $(x-1)$. Therefore, we should use the quotient rule.

    The first thing the quotient rule tells us to do is to take the derivative of the numerator. Now, the numerator is a product, so we must use the product rule to take its derivative. Here is the calculation:

$$\frac{d}{dx}\left[\frac{(x+1)(x+2)}{x-1}\right] = \frac{\overbrace{[(1)(x+2)+(x+1)(1)]}^{\text{Derivative of top}}\overbrace{(x-1)}^{\text{Bottom}} - \overbrace{[(x+1)(x+2)]}^{\text{Top}}\overbrace{(1)}^{\text{Derivative of bottom}}}{\underset{\uparrow}{(x-1)^2}}$$

$$\text{Bottom squared}$$

$$= \frac{(2x+3)(x-1) - (x+1)(x+2)}{(x-1)^2}$$

$$= \frac{x^2 - 2x - 5}{(x-1)^2}$$

What is important is to determine the *order of operations* and, in particular, to determine the last operation to be performed. Pretending to do an actual calculation reminds us of the order of operations; we call this technique the **calculation thought experiment**.

⇒ **Before we go on...**   We used the quotient rule in Example 2 because the function was a quotient; we used the product rule to calculate the derivative of the numerator because the numerator was a product. Get used to this: Differentiation rules usually must be used in combination.

    Here is another way we could have done this problem: Our calculation thought experiment could have taken the following form.

1. Calculate $(x+1)/(x-1) = (11+1)/(11-1) = 1.2$.

2. Calculate $x + 2 = 11 + 2 = 13$.

3. Multiply 1.2 by 13 to get 15.6.

We would have then regarded the expression as a *product*—the product of the factors $(x + 1)/(x - 1)$ and $(x + 2)$—and used the product rule instead. We can't escape the quotient rule, however: We need to use it to take the derivative of the first factor, $(x + 1)/(x - 1)$. Try this approach for practice and check that you get the same answer. ∎

Calculation Thought Experiment

The **calculation thought experiment** is a technique to determine whether to treat an algebraic expression as a product, quotient, sum, or difference. Given an expression, consider the steps you would use in computing its value. If the last operation is multiplication, treat the expression as a product; if the last operation is division, treat the expression as a quotient; and so on.

Quick Examples

1. $(3x^2 - 4)(2x + 1)$ can be computed by first calculating the expressions in parentheses and then multiplying. Because the last step is multiplication, we can treat the expression as a product.

2. $\dfrac{2x - 1}{x}$ can be computed by first calculating the numerator and denominator and then dividing one by the other. Because the last step is division, we can treat the expression as a quotient.

3. $x^2 + (4x - 1)(x + 2)$ can be computed by first calculating x^2, then calculating the product $(4x - 1)(x + 2)$, and finally adding the two answers. Thus, we can treat the expression as a sum.

4. $(3x^2 - 1)^5$ can be computed by first calculating the expression in parentheses and then raising the answer to the fifth power. Thus, we can treat the expression as a power. (We shall see how to differentiate powers of expressions in Section 11.4.)

5. The expression $(x + 1)(x + 2)/(x - 1)$ can be treated as either a quotient or a product: We can write it as a quotient: $\dfrac{(x + 1)(x + 2)}{x - 1}$ or as a product: $(x + 1)\left(\dfrac{x + 2}{x - 1}\right)$. (See Example 2(b).)

EXAMPLE 3 Using the Calculation Thought Experiment

Find $\dfrac{d}{dx}\left[6x^2 + 5\left(\dfrac{x}{x - 1}\right)\right]$.

Solution The calculation thought experiment tells us that the expression we are asked to differentiate can be treated as a *sum*. Because the derivative of a sum is the sum of the derivatives, we get

$$\frac{d}{dx}\left[6x^2 + 5\left(\frac{x}{x - 1}\right)\right] = \frac{d}{dx}(6x^2) + \frac{d}{dx}\left[5\left(\frac{x}{x - 1}\right)\right].$$

In other words, we must take the derivatives of $6x^2$ and $5\left(\dfrac{x}{x-1}\right)$ separately and then add the answers. The derivative of $6x^2$ is $12x$. There are two ways of taking the derivative of $5\left(\dfrac{x}{x-1}\right)$: We could either first multiply the expression $\left(\dfrac{x}{x-1}\right)$ by 5 to get $\left(\dfrac{5x}{x-1}\right)$ and then take its derivative using the quotient rule, or we could pull the 5 out, as we do next.

$$\frac{d}{dx}\left[6x^2 + 5\left(\frac{x}{x-1}\right)\right] = \frac{d}{dx}(6x^2) + \frac{d}{dx}\left[5\left(\frac{x}{x-1}\right)\right] \qquad \text{Derivative of sum}$$

$$= 12x + 5\frac{d}{dx}\left(\frac{x}{x-1}\right) \qquad \text{Constant} \times \text{Function}$$

$$= 12x + 5\left(\frac{(1)(x-1) - (x)(1)}{(x-1)^2}\right) \qquad \text{Quotient rule}$$

$$= 12x + 5\left(\frac{-1}{(x-1)^2}\right)$$

$$= 12x - \frac{5}{(x-1)^2}$$

APPLICATIONS

In the next example, we return to a scenario similar to the one discussed at the start of this section.

EXAMPLE 4 Applying the Product and Quotient Rules: Revenue and Average Cost

Sales of your newly launched miniature wall posters for college dorms, *iMiniPosters*, are really taking off. (Those old-fashioned large wall posters no longer fit in today's "downsized" college dorm rooms.) Monthly sales to students at the start of this year were 1,500 iMiniPosters, and since that time, sales have been increasing by 300 posters each month, even though the price you charge has also been going up.

a. The price you charge for iMiniPosters is given by

$$p(t) = 10 + 0.05t^2 \text{ dollars per poster},$$

where t is time in months since the start of January of this year. Find a formula for the monthly revenue, and then compute its rate of change at the beginning of March.

b. The number of students who purchase iMiniPosters in a month is given by

$$n(t) = 800 + 0.2t,$$

where t is as in part (a). Find a formula for the average number of posters each student buys, and hence estimate the rate at which this number was growing at the beginning of March.

Solution

a. To compute monthly revenue as a function of time t, we use

$$R(t) = p(t)q(t). \qquad \text{Revenue} = \text{Price} \times \text{Quantity}$$

We already have a formula for $p(t)$. The function $q(t)$ measures sales, which were 1,500 posters/month at time $t = 0$, and were rising by 300 per month:

$$q(t) = 1{,}500 + 300t.$$

Therefore, the formula for revenue is

$$R(t) = p(t)q(t)$$
$$R(t) = (10 + 0.05t^2)(1{,}500 + 300t).$$

Rather than expand this expression, we shall leave it as a product so that we can use the product rule in computing its rate of change:

$$R'(t) = p'(t)q(t) + p(t)q'(t)$$
$$= [0.10t][1{,}500 + 300t] + [10 + 0.05t^2][300].$$

Because the beginning of March corresponds to $t = 2$, we have

$$R'(2) = [0.10(2)][1{,}500 + 300(2)] + [10 + 0.05(2)^2][300]$$
$$= (0.2)(2{,}100) + (10.2)(300) = \$3{,}480 \text{ per month.}$$

Therefore, your monthly revenue was increasing at a rate of $\$3{,}480$ per month at the beginning of March.

b. The average number of posters sold to each student is

$$k(t) = \frac{\text{Number of posters}}{\text{Number of students}} = \frac{q(t)}{n(t)} = \frac{1{,}500 + 300t}{800 + 0.2t}.$$

The rate of change of $k(t)$ is computed with the quotient rule:

$$k'(t) = \frac{q'(t)n(t) - q(t)n'(t)}{n(t)^2}$$
$$= \frac{(300)(800 + 0.2t) - (1{,}500 + 300t)(0.2)}{(800 + 0.2t)^2}$$

so that

$$k'(2) = \frac{(300)[800 + 0.2(2)] - [1{,}500 + 300(2)](0.2)}{[800 + 0.2(2)]^2}$$
$$= \frac{(300)(800.4) - (2{,}100)(0.2)}{800.4^2} \approx 0.37 \text{ posters/student per month.}$$

Therefore, the average number of posters sold to each student was increasing at a rate of about 0.37 posters/student per month.

11.3 EXERCISES

▼ more advanced   ◆ challenging
🔲 indicates exercises that should be solved using technology

In Exercises 1–12:

a. *Calculate the derivative of the given function without using either the product or quotient rule.*

b. *Use the product or quotient rule to find the derivative. Check that you obtain the same answer.* HINT *[See Quick Examples on pages 810–811.]*

1. $f(x) = 3x$ **2.** $f(x) = 2x^2$

3. $g(x) = x \cdot x^2$ **4.** $g(x) = x \cdot x$

5. $h(x) = x(x + 3)$ **6.** $h(x) = x(1 + 2x)$

7. $r(x) = 100x^{2.1}$ **8.** $r(x) = 0.2x^{-1}$ **9.** $s(x) = \dfrac{2}{x}$

10. $t(x) = \dfrac{x}{3}$ **11.** $u(x) = \dfrac{x^2}{3}$ **12.** $s(x) = \dfrac{3}{x^2}$

Calculate $\dfrac{dy}{dx}$ *in Exercises 13–28. Simplify your answer.*
HINT [See Examples 1 and 2.]

13. $y = 3x(4x^2 - 1)$ **14.** $y = 3x^2(2x + 1)$

15. $y = x^3(1 - x^2)$ **16.** $y = x^5(1 - x)$

17. $y = (2x + 3)^2$ **18.** $y = (4x - 1)^2$

19. $y = \dfrac{4x}{5x - 2}$ **20.** $y = \dfrac{3x}{-3x + 2}$

21. $y = \dfrac{2x + 4}{3x - 1}$ **22.** $y = \dfrac{3x - 9}{2x + 4}$

23. $y = \dfrac{|x|}{x}$ **24.** $y = \dfrac{x}{|x|}$

25. $y = \dfrac{|x|}{x^2}$ **26.** $y = \dfrac{x^2}{|x|}$

27. $y = x\sqrt{x}$ **28.** $y = x^2\sqrt{x}$

Calculate $\dfrac{dy}{dx}$ *in Exercises 29–56. You need not expand your answers.* HINT [See Examples 1 and 2.]

29. $y = (x + 1)(x^2 - 1)$

30. $y = (4x^2 + x)(x - x^2)$

31. $y = (2x^{0.5} + 4x - 5)(x - x^{-1})$

32. $y = (x^{0.7} - 4x - 5)(x^{-1} + x^{-2})$

33. $y = (2x^2 - 4x + 1)^2$

34. $y = (2x^{0.5} - x^2)^2$

35. $y = \left(\dfrac{x}{3.2} + \dfrac{3.2}{x}\right)(x^2 + 1)$

36. $y = \left(\dfrac{x^{2.1}}{7} + \dfrac{2}{x^{2.1}}\right)(7x - 1)$

37. $y = x^2(2x + 3)(7x + 2)$ HINT [See Example 1(b).]

38. $y = x(x^2 - 3)(2x^2 + 1)$ HINT [See Example 1(b).]

39. $y = (5.3x - 1)(1 - x^{2.1})(x^{-2.3} - 3.4)$

40. $y = (1.1x + 4)(x^{2.1} - x)(3.4 - x^{-2.1})$

41. ▼$y = \left(\sqrt{x} + 1\right)\left(\sqrt{x} + \dfrac{1}{x^2}\right)$

42. ▼$y = \left(4x^2 - \sqrt{x}\right)\left(\sqrt{x} - \dfrac{2}{x^2}\right)$

43. $y = \dfrac{2x^2 + 4x + 1}{3x - 1}$ **44.** $y = \dfrac{3x^2 - 9x + 11}{2x + 4}$

45. $y = \dfrac{x^2 - 4x + 1}{x^2 + x + 1}$ **46.** $y = \dfrac{x^2 + 9x - 1}{x^2 + 2x - 1}$

47. $y = \dfrac{x^{0.23} - 5.7x}{1 - x^{-2.9}}$ **48.** $y = \dfrac{8.43x^{-0.1} - 0.5x^{-1}}{3.2 + x^{2.9}}$

49. ▼$y = \dfrac{\sqrt{x} + 1}{\sqrt{x} - 1}$ **50.** ▼$y = \dfrac{\sqrt{x} - 1}{\sqrt{x} + 1}$

51. ▼$y = \dfrac{\left(\dfrac{1}{x} + \dfrac{1}{x^2}\right)}{x + x^2}$ **52.** ▼$y = \dfrac{\left(1 - \dfrac{1}{x^2}\right)}{x^2 - 1}$

53. $y = \dfrac{(x + 3)(x + 1)}{3x - 1}$ HINT [See Example 2(b).]

54. $y = \dfrac{x}{(x - 5)(x - 4)}$ HINT [See Example 2(b).]

55. $y = \dfrac{(x + 3)(x + 1)(x + 2)}{3x - 1}$

56. $y = \dfrac{3x - 1}{(x - 5)(x - 4)(x - 1)}$

In Exercises 57–62, compute the indicated derivatives.

57. $\dfrac{d}{dx}[(x^2 + x)(x^2 - x)]$

58. $\dfrac{d}{dx}[(x^2 + x^3)(x + 1)]$

59. $\dfrac{d}{dx}[(x^3 + 2x)(x^2 - x)]\Big|_{x=2}$

60. $\dfrac{d}{dx}[(x^2 + x)(x^2 - x)]\Big|_{x=1}$

61. $\dfrac{d}{dt}[(t^2 - t^{0.5})(t^{0.5} + t^{-0.5})]\Big|_{t=1}$

62. $\dfrac{d}{dt}[(t^2 + t^{0.5})(t^{0.5} - t^{-0.5})]\Big|_{t=1}$

In Exercises 63–70, use the calculation thought experiment to say whether the expression is written as a sum, difference, scalar multiple, product, or quotient. Then use the appropriate rules to find its derivative. HINT [See Quick Examples on page 815 and Example 3.]

63. $y = x^4 - (x^2 + 120)(4x - 1)$

64. $y = x^4 - \dfrac{x^2 + 120}{4x - 1}$

65. $y = x + 1 + 2\left(\dfrac{x}{x + 1}\right)$

66. $y = (x + 2) - 4(x^2 - x)\left(x + \dfrac{1}{x}\right)$

   (Do not simplify the answer.)

67. $y = (x + 2)\left(\dfrac{x}{x + 1}\right)$ (Do not simplify the answer.)

68. $y = \dfrac{(x + 2)x}{x + 1}$ (Do not simplify the answer.)

69. $y = (x + 1)(x - 2) - 2\left(\dfrac{x}{x + 1}\right)$

70. $y = \dfrac{x + 2}{x + 1} + (x + 1)(x - 2)$

In Exercises 71–76, find the equation of the line tangent to the graph of the given function at the point with the indicated x-coordinate.

71. $f(x) = (x^2 + 1)(x^3 + x);\ x = 1$

72. $f(x) = (x^{0.5} + 1)(x^2 + x);\ x = 1$

73. $f(x) = \dfrac{x+1}{x+2}; x = 0$ **74.** $f(x) = \dfrac{\sqrt{x}+1}{\sqrt{x}+2}; x = 4$

75. $f(x) = \dfrac{x^2+1}{x}; x = -1$ **76.** $f(x) = \dfrac{x}{x^2+1}; x = 1$

APPLICATIONS

77. *Revenue* The monthly sales of *Sunny Electronics'* new sound system are given by $q(t) = 2{,}000t - 100t^2$ units per month, t months after its introduction. The price Sunny charges is $p(t) = 1{,}000 - t^2$ dollars per sound system, t months after introduction. Find the rate of change of monthly sales, the rate of change of the price, and the rate of change of monthly revenue 5 months after the introduction of the sound system. Interpret your answers. HINT [See Example 4(a).]

78. *Revenue* The monthly sales of *Sunny Electronics'* new *iSun* walkman is given by $q(t) = 2{,}000t - 100t^2$ units per month, t months after its introduction. The price Sunny charges is $p(t) = 100 - t^2$ dollars per iSun, t months after introduction. Find the rate of change of monthly sales, the rate of change of the price, and the rate of change of monthly revenue 6 months after the introduction of the iSun. Interpret your answers. HINT [See Example 4(a).]

79. *Saudi Oil Revenues* The price of crude oil during the period 2000–2010 can be approximated by

$$P(t) = 6t + 18 \text{ dollars per barrel} \quad (0 \le t \le 10)$$

in year t, where $t = 0$ represents 2000. Saudi Arabia's crude oil production over the same period can be approximated by[16]

$$Q(t) = -0.036t^2 + 0.62t + 8 \text{ million barrels per day.}$$
$$(0 \le t \le 10).$$

Use these models to estimate Saudi Arabia's daily oil revenue and also its rate of change in 2008. (Round your answers to the nearest $1 million.)

80. *Russian Oil Revenues* The price of crude oil during the period 2000–2010 can be approximated by

$$P(t) = 6t + 18 \text{ dollars per barrels} \quad (0 \le t \le 10)$$

in year t, where $t = 0$ represents 2000. Russia's crude oil production over the same period can be approximated by[17]

$$Q(t) = -0.08t^2 + 1.2t + 5.5 \text{ million barrels per day}$$
$$(0 \le t \le 10).$$

Use these models to estimate Russia's daily oil revenue and also its rate of change in 2005. (Round your answers to the nearest $1 million.)

81. *Revenue* Dorothy Wagner is currently selling 20 "I ♥ Calculus" T-shirts per day, but sales are dropping at a rate of 3 per day. She is currently charging $7 per T-shirt, but to compensate for dwindling sales, she is increasing the unit price by $1 per day. How fast, and in what direction, is her daily revenue currently changing?

82. *Pricing Policy* Let us turn Exercise 81 around a little: Dorothy Wagner is currently selling 20 "I ♥ Calculus" T-shirts per day, but sales are dropping at a rate of 3 per day. She is currently charging $7 per T-shirt, and she wishes to increase her daily revenue by $10 per day. At what rate should she increase the unit price to accomplish this (assuming that the price increase does not affect sales)?

83. *Bus Travel* *Thoroughbred Bus Company* finds that its monthly costs for one particular year were given by $C(t) = 10{,}000 + t^2$ dollars after t months. After t months the company had $P(t) = 1{,}000 + t^2$ passengers per month. How fast is its cost per passenger changing after 6 months? HINT [See Example 4(b).]

84. *Bus Travel* *Thoroughbred Bus Company* finds that its monthly costs for one particular year were given by $C(t) = 100 + t^2$ dollars after t months. After t months, the company had $P(t) = 1{,}000 + t^2$ passengers per month. How fast is its cost per passenger changing after 6 months? HINT [See Example 4(b).]

85. *Fuel Economy* Your muscle car's gas mileage (in miles per gallon) is given as a function $M(x)$ of speed x in mph, where

$$M(x) = \frac{3{,}000}{x + 3{,}600x^{-1}}.$$

Calculate $M'(x)$, and then $M'(10)$, $M'(60)$, and $M'(70)$. What do the answers tell you about your car?

86. *Fuel Economy* Your used Chevy's gas mileage (in miles per gallon) is given as a function $M(x)$ of speed x in mph, where

$$M(x) = \frac{4{,}000}{x + 3{,}025x^{-1}}.$$

Calculate $M'(x)$ and hence determine *the sign* of each of the following: $M'(40)$, $M'(55)$, and $M'(60)$. Interpret your results.

87. ▼ *Oil Imports from Mexico* Daily oil production in Mexico and daily U.S. oil imports from Mexico during 2005–2009 could be approximated by

$$P(t) = 3.9 - 0.10t \text{ million barrels} \quad (5 \le t \le 9)$$
$$I(t) = 2.1 - 0.11t \text{ million barrels} \quad (5 \le t \le 9),$$

where t is time in years since the start of 2000.[18]

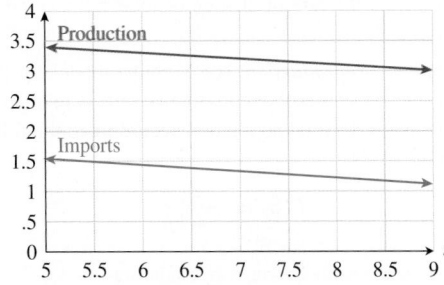

a. What are represented by the functions $P(t) - I(t)$ and $I(t)/P(t)$?

[16]Sources for data: Oil price: InflationData.com www.inflationdata.com, Production: Energy Bulletin www.energybulletin.net.
[17]Ibid.
[18]Source for data: Energy Information Administration (www.eia.doe.gov)/Pemex.

b. Compute $\dfrac{d}{dt}\left[\dfrac{I(t)}{P(t)}\right]\Big|_{t=8}$ to two significant digits. What does the answer tell you about oil imports from Mexico?

88. ▼ *Oil Imports from Mexico* Daily oil production in Mexico and daily U.S. oil imports from Mexico during 2000–2004 could be approximated by

$$P(t) = 3.0 + 0.13t \text{ million barrels} \quad (0 \le t \le 4)$$

$$I(t) = 1.4 + 0.06t \text{ million barrels} \quad (0 \le t \le 4),$$

where t is time in years since the start of 2000.[19]

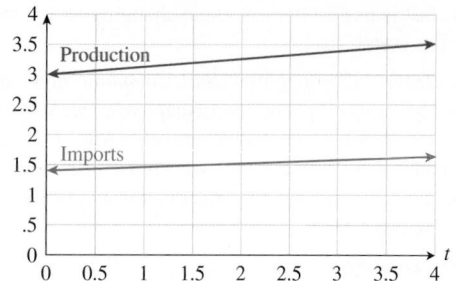

a. What are represented by the functions $P(t) - I(t)$ and $I(t)/P(t)$?

b. Compute $\dfrac{d}{dt}\left[\dfrac{I(t)}{P(t)}\right]\Big|_{t=3}$ to two significant digits. What does the answer tell you about oil imports from Mexico?

89. ▼ *Military Spending* The annual cost per active-duty armed service member in the United States increased from $80,000 in 1995 to a projected $120,000 in 2007. In 1995, there were 1.5 million armed service personnel, and this number was projected to decrease to 1.4 million in 2003.[20] Use linear models for annual cost and personnel to estimate, to the nearest $10 million, the rate of change of total military personnel costs in 2002.

90. ▼ *Military Spending in the 1990s* The annual cost per active-duty armed service member in the United States increased from $80,000 in 1995 to $90,000 in 2000. In 1990, there were 2 million armed service personnel and this number decreased to 1.5 million in 2000.[21] Use linear models for annual cost and personnel to estimate, to the nearest $10 million, the rate of change of total military personnel costs in 1995.

91. ▼ *Biology—Reproduction* The Verhulst model for population growth specifies the reproductive rate of an organism as a function of the total population according to the following formula:

$$R(p) = \frac{r}{1 + kp}$$

where p is the total population in thousands of organisms, r and k are constants that depend on the particular circumstances and the organism being studied, and $R(p)$ is the reproduction rate in thousands of organisms per hour.[22] If $k = 0.125$ and $r = 45$, find $R'(p)$ and then $R'(4)$. Interpret the result.

92. ▼ *Biology—Reproduction* Another model, the predator satiation model for population growth, specifies that the reproductive rate of an organism as a function of the total population varies according to the following formula:

$$R(p) = \frac{rp}{1 + kp}$$

where p is the total population in thousands of organisms, r and k are constants that depend on the particular circumstances and the organism being studied, and $R(p)$ is the reproduction rate in new organisms per hour.[23] Given that $k = 0.2$ and $r = 0.08$, find $R'(p)$ and $R'(2)$. Interpret the result.

93. ▼ *Embryo Development* Bird embryos consume oxygen from the time the egg is laid through the time the chick hatches. For a typical galliform bird egg, the oxygen consumption (in milliliters) t days after the egg was laid can be approximated by[24]

$$C(t) = -0.016t^4 + 1.1t^3 - 11t^2 + 3.6t \quad (15 \le t \le 30).$$

(An egg will usually hatch at around $t = 28$.) Suppose that at time $t = 0$ you have a collection of 30 newly laid eggs and that the number of eggs decreases linearly to zero at time $t = 30$ days. How fast is the total oxygen consumption of your collection of embryos changing after 25 days? (Round your answers to two significant digits.) Comment on the result. HINT [Total oxygen consumption = Oxygen consumption per egg × Number of eggs.]

94. ▼ *Embryo Development* Turkey embryos consume oxygen from the time the egg is laid through the time the chick hatches. For a brush turkey, the oxygen consumption (in milliliters) t days after the egg was laid can be approximated by[25]

$$C(t) = -0.0071t^4 + 0.95t^3 - 22t^2 + 95t \quad (25 \le t \le 50).$$

(An egg will typically hatch at around $t = 50$.) Suppose that at time $t = 0$ you have a collection of 100 newly laid eggs and that the number of eggs decreases linearly to zero at time $t = 50$ days. How fast is the total oxygen consumption of your collection of embryos changing after 40 days? (Round your answer to two significant digits.) Interpret the result. HINT [Total oxygen consumption = Oxygen consumption per egg × Number of eggs.]

[19]Source for data: Energy Information Administration (www.eia.doe.gov)/Pemex.

[20]Annual costs are adjusted for inflation. Sources: Department of Defense, Stephen Daggett, military analyst, Congressional Research Service/*New York Times*, April 19, 2002, p. A21.

[21]*Ibid.*

[22]Source: *Mathematics in Medicine and the Life Sciences* by F. C. Hoppensteadt and C. S. Peskin (Springer-Verlag, New York, 1992) pp. 20–22.

[23]*Ibid.*

[24]The model is derived from graphical data published in the article "The Brush Turkey" by Roger S. Seymour, *Scientific American*, December, 1991, pp. 108–114.

[25]*Ibid.*

COMMUNICATION AND REASONING EXERCISES

95. If f and g are functions of time, and at time $t = 3$, f equals 5 and is rising at a rate of 2 units per second, and g equals 4 and is rising at a rate of 5 units per second, then the product fg equals _____ and is rising at a rate of _____ units per second.

96. If f and g are functions of time, and at time $t = 2$, f equals 3 and is rising at a rate of 4 units per second, and g equals 5 and is rising at a rate of 6 units per second, then fg equals _____ and is rising at a rate of _____ units per second.

97. If f and g are functions of time, and at time $t = 3$, f equals 5 and is rising at a rate of 2 units per second, and g equals 4 and is rising at a rate of 5 units per second, then f/g equals _____ and is changing at a rate of _____ units per second.

98. If f and g are functions of time, and at time $t = 2$, f equals 3 and is rising at a rate of 4 units per second, and g equals 5 and is rising at a rate of 6 units per second, then f/g equals _____ and is changing at a rate of _____ units per second.

99. You have come across the following in a newspaper article: "Revenues of HAL Home Heating Oil Inc. are rising by $4.2 million per year. This is due to an annual increase of 70¢ per gallon in the price HAL charges for heating oil and an increase in sales of 6 million gallons of oil per year." Comment on this analysis.

100. Your friend says that because average cost is obtained by dividing the cost function by the number of units x, it follows that the derivative of average cost is the same as marginal cost because the derivative of x is 1. Comment on this analysis.

101. ▼Find a demand function $q(p)$ such that, at a price per item of $p = \$100$, revenue will rise if the price per item is increased.

102. ▼What must be true about a demand function $q(p)$ so that, at a price per item of $p = \$100$, revenue will decrease if the price per item is increased?

103. ▼You and I are both selling a steady 20 T-shirts per day. The price I am getting for my T-shirts is increasing twice as fast as yours, but your T-shirts are currently selling for twice the price of mine. Whose revenue is increasing faster: yours, mine, or neither? Explain.

104. ▼You and I are both selling T-shirts for a steady $20 per shirt. Sales of my T-shirts are increasing at twice the rate of yours, but you are currently selling twice as many as I am. Whose revenue is increasing faster: yours, mine, or neither? Explain.

105. ◆*Marginal Product (from the GRE Economics Test)* Which of the following statements about average product and marginal product is correct?

(A) If average product is decreasing, marginal product must be less than average product.
(B) If average product is increasing, marginal product must be increasing.
(C) If marginal product is decreasing, average product must be less than marginal product.
(D) If marginal product is increasing, average product must be decreasing.
(E) If marginal product is constant over some range, average product must be constant over that range.

106. ◆*Marginal Cost (based on a question from the GRE Economics Test)* Which of the following statements about average cost and marginal cost is correct?

(A) If average cost is increasing, marginal cost must be increasing.
(B) If average cost is increasing, marginal cost must be decreasing.
(C) If average cost is increasing, marginal cost must be more than average cost.
(D) If marginal cost is increasing, average cost must be increasing.
(E) If marginal cost is increasing, average cost must be larger than marginal cost.

11.4 The Chain Rule

We can now find the derivatives of expressions involving powers of x combined using addition, subtraction, multiplication, and division, but we still cannot take the derivative of an expression like $(3x + 1)^{0.5}$. For this we need one more rule. The function $h(x) = (3x + 1)^{0.5}$ is not a sum, difference, product, or quotient. To find out what it is, we can use the calculation thought experiment and think about the last operation we would perform in calculating $h(x)$.

1. Calculate $3x + 1$.

2. Take the 0.5 power (square root) of the answer.

The last operation is "take the 0.5 power." We do not yet have a rule for finding the derivative of the 0.5 power of a quantity other than x.

There is a way to build $h(x) = (3x + 1)^{0.5}$ out of two simpler functions: $u(x) = 3x + 1$ (the function that corresponds to the first step in the calculation above) and $f(x) = x^{0.5}$ (the function that corresponds to the second step):

$$
\begin{aligned}
h(x) &= (3x + 1)^{0.5} \\
&= [u(x)]^{0.5} \qquad u(x) = 3x + 1 \\
&= f(u(x)). \qquad f(x) = x^{0.5}
\end{aligned}
$$

We say that h is the **composite** of f and u. We read $f(u(x))$ as "f of u of x."

To compute $h(1)$, say, we first compute $3 \cdot 1 + 1 = 4$ and then take the square root of 4, giving $h(1) = 2$. To compute $f(u(1))$ we follow exactly the same steps: First compute $u(1) = 4$ and then $f(u(1)) = f(4) = 2$. We always compute $f(u(x))$ from the inside out: Given x, first compute $u(x)$ and then $f(u(x))$.

Now, f and u are functions *whose derivatives we know*. The *chain rule* allows us to use our knowledge of the derivatives of f and u to find the derivative of $f(u(x))$. For the purposes of stating the rule, let us avoid some of the nested parentheses by abbreviating $u(x)$ as u. Thus, we write $f(u)$ instead of $f(u(x))$ and remember that u is a function of x.

Chain Rule

If f is a differentiable function of u and u is a differentiable function of x, then the composite $f(u)$ is a differentiable function of x, and

$$
\frac{d}{dx}[f(u)] = f'(u)\frac{du}{dx}. \qquad \text{Chain rule}
$$

In words *The derivative of f(quantity) is the derivative of f, evaluated at that quantity, times the derivative of the quantity.*

Quick Examples

In the Quick Examples that follow, u, "the quantity," is some (unspecified) differentiable function of x.

1. Take $f(u) = u^2$. Then

$$
\frac{d}{dx}(u^2) = 2u\frac{du}{dx}. \qquad \text{Because } f'(u) = 2u
$$

The derivative of a quantity squared is two times the quantity, times the derivative of the quantity.

2. Take $f(u) = u^{0.5}$. Then

$$
\frac{d}{dx}(u^{0.5}) = 0.5u^{-0.5}\frac{du}{dx}. \qquad \text{Because } f'(u) = 0.5u^{-0.5}
$$

The derivative of a quantity raised to the 0.5 is 0.5 times the quantity raised to the −0.5, times the derivative of the quantity.

As the quick examples illustrate, for every power of a function u whose derivative we know, we now get a "generalized" differentiation rule. The following table gives more examples.

Original Rule	Generalized Rule	In Words
$\dfrac{d}{dx}(x^2) = 2x$	$\dfrac{d}{dx}(u^2) = 2u\dfrac{du}{dx}$	The derivative of a quantity squared is twice the quantity, times the derivative of the quantity.
$\dfrac{d}{dx}(x^3) = 3x^2$	$\dfrac{d}{dx}(u^3) = 3u^2\dfrac{du}{dx}$	The derivative of a quantity cubed is 3 times the quantity squared, times the derivative of the quantity.
$\dfrac{d}{dx}\left(\dfrac{1}{x}\right) = -\dfrac{1}{x^2}$	$\dfrac{d}{dx}\left(\dfrac{1}{u}\right) = -\dfrac{1}{u^2}\dfrac{du}{dx}$	The derivative of 1 over a quantity is negative 1 over the quantity squared, times the derivative of the quantity.
Power Rule	**Generalized Power Rule**	**In Words**
$\dfrac{d}{dx}(x^n) = nx^{n-1}$	$\dfrac{d}{dx}(u^n) = nu^{n-1}\dfrac{du}{dx}$	The derivative of a quantity raised to the n is n times the quantity raised to the $n-1$, times the derivative of the quantity.
$\dfrac{d}{dx}|x| = \dfrac{|x|}{x}$	$\dfrac{d}{dx}|u| = \dfrac{|u|}{u}\dfrac{du}{dx}$	The derivative of the absolute value of a quantity is the absolute value of the quantity divided by the quantity, times the derivative of the quantity.

To motivate the chain rule, let us see why it is true in the special case when $f(u) = u^3$, where the chain rule tells us that

$$\frac{d}{dx}(u^3) = 3u^2\frac{du}{dx}. \qquad \text{Generalized power rule with } n = 3$$

But we could have done this using the product rule instead:

$$\frac{d}{dx}(u^3) = \frac{d}{dx}(u \cdot u \cdot u) = \frac{du}{dx}u \cdot u + u\frac{du}{dx}u + u \cdot u\frac{du}{dx} = 3u^2\frac{du}{dx},$$

which gives us the same result. A similar argument works for $f(u) = u^n$, where $n = 2, 3, 4, \ldots$. We can then use the quotient rule and the chain rule for positive powers to verify the generalized power rule for *negative* powers as well. For the case of a general differentiable function f, the proof of the chain rule is beyond the scope of this book, but you can find one on the Website by following the path

Website → Everything → Chapter 11 → Proof of Chain Rule.

EXAMPLE 1   Using the Chain Rule

Compute the following derivatives.

a. $\dfrac{d}{dx}[(2x^2 + x)^3]$ **b.** $\dfrac{d}{dx}[(x^3 + x)^{100}]$ **c.** $\dfrac{d}{dx}\sqrt{3x + 1}$ **d.** $\dfrac{d}{dx}|4x^2 - x|$

Solution

a. Using the calculation thought experiment, we see that the last operation we would perform in calculating $(2x^2 + x)^3$ is that of *cubing*. Thus we think of $(2x^2 + x)^3$ as *a quantity cubed*. There are two similar methods we can use to calculate its derivative.

Method 1: Using the formula We think of $(2x^2 + x)^3$ as u^3, where $u = 2x^2 + x$. By the formula,

$$\frac{d}{dx}(u^3) = 3u^2\frac{du}{dx}. \qquad \text{Generalized power rule}$$

Now substitute for u:

$$\frac{d}{dx}[(2x^2 + x)^3] = 3(2x^2 + x)^2\frac{d}{dx}(2x^2 + x)$$
$$= 3(2x^2 + x)^2(4x + 1)$$

Method 2: Using the verbal form If we prefer to use the verbal form, we get:

The derivative of $(2x^2 + x)$ cubed is three times $(2x^2 + x)$ squared, times the derivative of $(2x^2 + x)$.

In symbols,

$$\frac{d}{dx}[(2x^2 + x)^3] = 3(2x^2 + x)^2(4x + 1),$$

as we obtained above.

b. First, the calculation thought experiment: If we were computing $(x^3 + x)^{100}$, the last operation we would perform is *raising a quantity to the power* 100. Thus we are dealing with *a quantity raised to the power* 100, and so we must again use the generalized power rule. According to the verbal form of the generalized power rule, the derivative of a quantity raised to the power 100 is 100 times that quantity to the power 99, times the derivative of that quantity. In symbols,

$$\frac{d}{dx}[(x^3 + x)^{100}] = 100(x^3 + x)^{99}(3x^2 + 1).$$

c. We first rewrite the expression $\sqrt{3x + 1}$ as $(3x + 1)^{0.5}$ and then use the generalized power rule as in parts (a) and (b):

The derivative of a quantity raised to the 0.5 is 0.5 times the quantity raised to the −0.5, times the derivative of the quantity.

Thus,

$$\frac{d}{dx}[(3x + 1)^{0.5}] = 0.5(3x + 1)^{-0.5} \cdot 3 = 1.5(3x + 1)^{-0.5}.$$

d. The calculation thought experiment tells us that $|4x^2 - x|$ is the absolute value of a quantity, so we use the generalized rule for absolute values (above):

$$\frac{d}{dx}|u| = \frac{|u|}{u}\frac{du}{dx}, \text{ or, in words,}$$

The derivative of the absolute value of a quantity is the absolute value of the quantity divided by the quantity, times the derivative of the quantity.

Thus,

$$\frac{d}{dx}|4x^2 - x| = \frac{|4x^2 - x|}{4x^2 - x} \cdot (8x - 1). \qquad \frac{d}{dx}|u| = \frac{|u|}{u}\frac{du}{dx}$$

➡ **Before we go on...** The following are examples of common errors in solving Example 1(b):

$$\text{``}\frac{d}{dx}[(x^3 + x)^{100}] = 100(3x^2 + 1)^{99}\text{''} \qquad \text{✗ } \textit{WRONG!}$$

$$\text{``}\frac{d}{dx}[(x^3 + x)^{100}] = 100(x^3 + x)^{99}.\text{''} \qquad \text{✗ } \textit{WRONG!}$$

Remember that the generalized power rule says that the derivative of a quantity to the power 100 is 100 times *that same quantity* raised to the power 99, *times the derivative of that quantity.* ∎

Q: *It seems that there are now two formulas for the derivative of an nth power:*

1. $\dfrac{d}{dx}[x^n] = nx^{n-1}$

2. $\dfrac{d}{dx}[u^n] = nu^{n-1}\dfrac{du}{dx}.$

Which one do I use?

A: Formula 1 is actually a special case of Formula 2: Formula 1 is the original power rule, which applies only to a power of x. For instance, it applies to x^{10}, but it does not apply to $(2x + 1)^{10}$ because the quantity that is being raised to a power is not x. Formula 2 applies to a power of any *function of x*, such as $(2x + 1)^{10}$. It can even be used in place of the original power rule. For example, if we take $u = x$ in Formula 2, we obtain

$$\frac{d}{dx}[x^n] = nx^{n-1}\frac{dx}{dx}$$

$$= nx^{n-1}. \qquad \text{The derivative of } x \text{ with respect to } x \text{ is 1.}$$

Thus, the generalized power rule really *is* a generalization of the original power rule, as its name suggests.

EXAMPLE 2 **More Examples Using the Chain Rule**

Find: **a.** $\dfrac{d}{dx}[(2x^5 + x^2 - 20)^{-2/3}]$ **b.** $\dfrac{d}{dx}\left[\dfrac{1}{\sqrt{x+2}}\right]$ **c.** $\dfrac{d}{dx}\left[\dfrac{1}{x^2 + x}\right]$

Solution Each of the given functions is, or can be rewritten as, a power of a function whose derivative we know. Thus, we can use the method of Example 1.

a. $\dfrac{d}{dx}[(2x^5 + x^2 - 20)^{-2/3}] = -\dfrac{2}{3}(2x^5 + x^2 - 20)^{-5/3}(10x^4 + 2x)$

b. $\dfrac{d}{dx}\left[\dfrac{1}{\sqrt{x+2}}\right] = \dfrac{d}{dx}(x + 2)^{-1/2} = -\dfrac{1}{2}(x + 2)^{-3/2} \cdot 1 = -\dfrac{1}{2(x + 2)^{3/2}}$

c. $\dfrac{d}{dx}\left[\dfrac{1}{x^2 + x}\right] = \dfrac{d}{dx}(x^2 + x)^{-1} = -(x^2 + x)^{-2}(2x + 1) = -\dfrac{2x + 1}{(x^2 + x)^2}$

➡ **Before we go on...** In Example 2(c), we could have used the quotient rule instead of the generalized power rule. We can think of the quantity $1/(x^2 + x)$ in two different ways using the calculation thought experiment:

1. As 1 divided by something—in other words, as a quotient

2. As something raised to the -1 power

Of course, we get the same derivative using either approach. ∎

We now look at some more complicated examples.

EXAMPLE 3 **Harder Examples Using the Chain Rule**

Find $\dfrac{dy}{dx}$ in each case.   **a.** $y = [(x+1)^{-2.5} + 3x]^{-3}$ **b.** $y = (x+10)^3 \sqrt{1-x^2}$

Solution

a. The calculation thought experiment tells us that the last operation we would perform in calculating y is raising the quantity $[(x+1)^{-2.5} + 3x]$ to the power -3. Thus, we use the generalized power rule.

$$\frac{dy}{dx} = -3[(x+1)^{-2.5} + 3x]^{-4} \frac{d}{dx}[(x+1)^{-2.5} + 3x]$$

We are not yet done; we must still find the derivative of $(x+1)^{-2.5} + 3x$. Finding the derivative of a complicated function in several steps helps to keep the problem manageable. Continuing, we have

$$\frac{dy}{dx} = -3[(x+1)^{-2.5} + 3x]^{-4} \frac{d}{dx}[(x+1)^{-2.5} + 3x]$$

$$= -3[(x+1)^{-2.5} + 3x]^{-4} \left(\frac{d}{dx}[(x+1)^{-2.5}] + \frac{d}{dx}(3x) \right). \quad \text{Derivative of a sum}$$

Now we have two derivatives left to calculate. The second of these we know to be 3, and the first is the derivative of a quantity raised to the -2.5 power. Thus

$$\frac{dy}{dx} = -3[(x+1)^{-2.5} + 3x]^{-4}[-2.5(x+1)^{-3.5} \cdot 1 + 3].$$

b. The expression $(x+10)^3 \sqrt{1-x^2}$ is a product, so we use the product rule:

$$\frac{d}{dx}[(x+10)^3 \sqrt{1-x^2}] = \left(\frac{d}{dx}[(x+10)^3] \right) \sqrt{1-x^2} + (x+10)^3 \left(\frac{d}{dx}\sqrt{1-x^2} \right)$$

$$= 3(x+10)^2 \sqrt{1-x^2} + (x+10)^3 \frac{1}{2\sqrt{1-x^2}}(-2x)$$

$$= 3(x+10)^2 \sqrt{1-x^2} - \frac{x(x+10)^3}{\sqrt{1-x^2}}.$$

APPLICATIONS

The next example is a new way of looking at Example 3 from Section 11.2.

EXAMPLE 4 **Marginal Product**

A consultant determines that *Precision Manufacturers'* annual profit (in dollars) is given by

$$P = -200{,}000 + 4{,}000q - 0.46q^2 - 0.00001q^3,$$

where q is the number of surgical lasers it sells each year. The consultant also informs Precision that the number of surgical lasers it can manufacture each year depends on the number n of assembly-line workers it employs according to the equation

$$q = 100n. \qquad \text{Each worker contributes 100 lasers per year.}$$

Use the chain rule to find the marginal product $\dfrac{dP}{dn}$.

Solution We could calculate the marginal product by substituting the expression for q in the expression for P to obtain P as a function of n (as given in Example 3 from Section 11.2) and then finding dP/dn. Alternatively—and this will simplify the calculation—we can use the chain rule. To see how the chain rule applies, notice that P is a function of q, where q in turn is given as a function of n. By the chain rule,

$$\frac{dP}{dn} = P'(q)\frac{dq}{dn} \qquad \text{Chain rule}$$

$$= \frac{dP}{dq}\frac{dq}{dn}. \qquad \text{Notice how the "quantities" } dq \text{ appear to cancel.}$$

Now we compute

$$\frac{dP}{dq} = 4{,}000 - 0.92q - 0.00003q^2$$

and $\quad \dfrac{dq}{dn} = 100.$

Substituting into the equation for $\dfrac{dP}{dn}$ gives

$$\frac{dP}{dn} = (4{,}000 - 0.92q - 0.00003q^2)(100)$$

$$= 400{,}000 - 92q - 0.003q^2.$$

Notice that the answer has q as a variable. We can express dP/dn as a function of n by substituting $100n$ for q:

$$\frac{dP}{dn} = 400{,}000 - 92(100n) - 0.003(100n)^2$$

$$= 400{,}000 - 9{,}200n - 30n^2.$$

The equation

$$\frac{dP}{dn} = \frac{dP}{dq}\frac{dq}{dn}$$

in the example above is an appealing way of writing the chain rule because it suggests that the "quantities" dq cancel. In general, we can write the chain rule as follows.

Chain Rule in Differential Notation

If y is a differentiable function of u, and u is a differentiable function of x, then

$$\frac{dy}{dx} = \frac{dy}{du}\frac{du}{dx}. \qquad \text{The terms } du \text{ cancel.}$$

Notice how the units of measurement also cancel:

$$\frac{\text{Units of } y}{\text{Units of } x} = \frac{\text{Units of } y}{\cancel{\text{Units of } u}}\frac{\cancel{\text{Units of } u}}{\text{Units of } x}.$$

Quick Examples

1. If $y = u^3$, where $u = 4x + 1$, then

$$\frac{dy}{dx} = \frac{dy}{du}\frac{du}{dx} = 3u^2 \cdot 4 = 12u^2 = 12(4x + 1)^2.$$

2. If $q = 43p^2$, where p (and hence q also) is a differentiable function of t, then

$$\frac{dq}{dt} = \frac{dq}{dp}\frac{dp}{dt}$$

$$= 86p\frac{dp}{dt}. \qquad p \text{ is not specified, so we leave } dp/dt \text{ as is.}$$

EXAMPLE 5 Marginal Revenue

Suppose a company's weekly revenue R is given as a function of the unit price p, and p in turn is given as a function of weekly sales q (by means of a demand equation). If

$$\left.\frac{dR}{dp}\right|_{q=1,000} = \$40 \text{ per } \$1 \text{ increase in price}$$

and

$$\left.\frac{dp}{dq}\right|_{q=1,000} = -\$20 \text{ per additional item sold per week}$$

find the marginal revenue when sales are 1,000 items per week.

Solution The marginal revenue is $\dfrac{dR}{dq}$. By the chain rule, we have

$$\frac{dR}{dq} = \frac{dR}{dp}\frac{dp}{dq}. \qquad \text{Units: Revenue per item}$$
$$= \text{Revenue per } \$1 \text{ price increase} \times \text{price increase per additional item.}$$

Because we are interested in the marginal revenue at a demand level of 1,000 items per week, we have

$$\left.\frac{dR}{dq}\right|_{q=1,000} = (40)(-20) = -\$800 \text{ per additional item sold.}$$

Thus, if the price is lowered to increase the demand from 1,000 to 1,001 items per week, the weekly revenue will drop by approximately $800.

Look again at the way the terms "du" appeared to cancel in the differential formula $\dfrac{dy}{dx} = \dfrac{dy}{du}\dfrac{du}{dx}$. In fact, the chain rule tells us more:

✱ The notion of "thinking of *x* as a function of *y*" will be made more precise in Section 11.6.

Manipulating Derivatives in Differential Notation

1. Suppose y is a function of x. Then, thinking of x as a function of y (as, for instance, when we can solve for x)✱ we have

$$\frac{dx}{dy} = \frac{1}{\left(\dfrac{dy}{dx}\right)}, \text{ provided } \frac{dy}{dx} \neq 0.$$ Notice again how $\frac{dy}{dx}$ behaves like a fraction.

Quick Example

In the demand equation $q = -0.2p - 8$, we have $\dfrac{dq}{dp} = -0.2$. Therefore,

$$\frac{dp}{dq} = \frac{1}{\left(\dfrac{dq}{dp}\right)} = \frac{1}{-0.2} = -5.$$

2. Suppose x and y are functions of t. Then, thinking of y as a function of x (as, for instance, when we can solve for t as a function of x, and hence obtain y as a function of x), we have

$$\frac{dy}{dx} = \frac{dy/dt}{dx/dt}.$$ The terms dt appear to cancel.

Quick Example

If $x = 3 - 0.2t$ and $y = 6 + 6t$, then

$$\frac{dy}{dx} = \frac{dy/dt}{dx/dt} = \frac{6}{-0.2} = -30.$$

To see why the above formulas work, notice that the second formula,

$$\frac{dy}{dx} = \frac{\left(\dfrac{dy}{dt}\right)}{\left(\dfrac{dx}{dt}\right)}$$

can be written as

$$\frac{dy}{dx}\frac{dx}{dt} = \frac{dy}{dt},$$ Multiply both sides by $\frac{dx}{dt}$.

which is just the differential form of the chain rule. For the first formula, use the second formula with y playing the role of t:

$$\frac{dy}{dx} = \frac{dy/dy}{dx/dy}$$

$$= \frac{1}{dx/dy}.$$ $\frac{dy}{dy} = \frac{d}{dy}[y] = 1$

FAQs

Using the Chain Rule

Q: *How do I decide whether or not to use the chain rule when taking a derivative?*

A: Use the calculation thought experiment (Section 11.3): Given an expression, consider the steps you would use in computing its value.

- If the last step is *raising a quantity to a power,* as in $\left(\dfrac{x^2-1}{x+4}\right)^4$, then the first step to use is the chain rule (in the form of the generalized power rule):

$$\frac{d}{dx}\left(\frac{x^2-1}{x+4}\right)^4 = 4\left(\frac{x^2-1}{x+4}\right)^3 \frac{d}{dx}\left(\frac{x^2-1}{x+4}\right).$$

Then use the appropriate rules to finish the computation. You may need to again use the calculation thought experiment to decide on the next step (here the quotient rule):

$$= 4\left(\frac{x^2-1}{x+4}\right)^3 \frac{(2x)(x+4)-(x^2-1)(1)}{(x+4)^2}.$$

- If the last step is *division,* as in $\dfrac{(x^2-1)}{(3x+4)^4}$, then the first step to use is the quotient rule:

$$\frac{d}{dx}\frac{(x^2-1)}{(3x+4)^4} = \frac{(2x)(3x+4)^4-(x^2-1)\dfrac{d}{dx}(3x+4)^4}{(3x+4)^8}.$$

Then use the appropriate rules to finish the computation (here the chain rule):

$$= \frac{(2x)(3x+4)^4-(x^2-1)[4(3x+4)^3(3)]}{(3x+4)^8}.$$

- If the last step is *multiplication, addition, subtraction, or multiplication by a constant,* then the first rule to use is the product rule, or the rule for sums, differences, or constant multiples as appropriate.

Q: *Every time I compute a derivative, I leave something out. How do I make sure I am really done when taking the derivative of a complicated-looking expression?*

A: Until you are an expert at taking derivatives, the key is to use one rule at a time and write out each step, rather than trying to compute the derivative in a single step.

To illustrate this, try computing the derivative of $(x+10)^3\sqrt{1-x^2}$ in Example 3(b) in two ways: First try to compute it in a single step, and then compute it by writing out each step as shown in the example. How do your results compare? For more practice, try Exercises 87 and 88 following.

11.4   **EXERCISES**

▼ more advanced   ◆ challenging
T indicates exercises that should be solved using technology

Calculate the derivatives of the functions in Exercises 1–50.
HINT [See Example 1.]

1. $f(x) = (2x + 1)^2$ **2.** $f(x) = (3x - 1)^2$

3. $f(x) = (x - 1)^{-1}$ **4.** $f(x) = (2x - 1)^{-2}$

5. $f(x) = (2 - x)^{-2}$ **6.** $f(x) = (1 - x)^{-1}$

7. $f(x) = (2x + 1)^{0.5}$ **8.** $f(x) = (-x + 2)^{1.5}$

9. $f(x) = \dfrac{1}{3x - 1}$ **10.** $f(x) = \dfrac{1}{(x + 1)^2}$

11. $f(x) = (x^2 + 2x)^4$ **12.** $f(x) = (x^3 - x)^3$

13. $f(x) = (2x^2 - 2)^{-1}$ **14.** $f(x) = (2x^3 + x)^{-2}$

15. $g(x) = (x^2 - 3x - 1)^{-5}$ **16.** $g(x) = (2x^2 + x + 1)^{-3}$

17. $h(x) = \dfrac{1}{(x^2 + 1)^3}$ **18.** $h(x) = \dfrac{1}{(x^2 + x + 1)^2}$

HINT [See Example 2.]   HINT [See Example 2.]

19. $r(x) = (0.1x^2 - 4.2x + 9.5)^{1.5}$

20. $r(x) = (0.1x - 4.2x^{-1})^{0.5}$

21. $r(s) = (s^2 - s^{0.5})^4$ **22.** $r(s) = (2s + s^{0.5})^{-1}$

23. $f(x) = \sqrt{1 - x^2}$ **24.** $f(x) = \sqrt{x + x^2}$

25. $f(x) = |3x - 6|$ **26.** $f(x) = |-5x + 1|$

HINT [See Example 1(d).]   HINT [See Example 1(d).]

27. $f(x) = |-x^3 + 5x|$ **28.** $f(x) = |x - x^4|$

29. $h(x) = 2[(x + 1)(x^2 - 1)]^{-1/2}$ HINT [See Example 3.]
30. $h(x) = 3[(2x - 1)(x - 1)]^{-1/3}$ HINT [See Example 3.]

31. $h(x) = (3.1x - 2)^2 - \dfrac{1}{(3.1x - 2)^2}$

32. $h(x) = \left[3.1x^2 - 2 - \dfrac{1}{3.1x - 2}\right]^2$

33. $f(x) = [(6.4x - 1)^2 + (5.4x - 2)^3]^2$

34. $f(x) = (6.4x - 3)^{-2} + (4.3x - 1)^{-2}$

35. $f(x) = (x^2 - 3x)^{-2}(1 - x^2)^{0.5}$

36. $f(x) = (3x^2 + x)(1 - x^2)^{0.5}$

37. $s(x) = \left(\dfrac{2x + 4}{3x - 1}\right)^2$ **38.** $s(x) = \left(\dfrac{3x - 9}{2x + 4}\right)^3$

39. $g(z) = \left(\dfrac{z}{1 + z^2}\right)^3$ **40.** $g(z) = \left(\dfrac{z^2}{1 + z}\right)^2$

41. $f(x) = [(1 + 2x)^4 - (1 - x)^2]^3$

42. $f(x) = [(3x - 1)^2 + (1 - x)^5]^2$

43. $f(x) = (3x - 1)|3x - 1|$

44. $f(x) = |(x - 3)^{1/3}|$

45. $f(x) = |x - (2x - 3)^{1/2}|$

46. $f(x) = (3 - |3x - 1|)^{-2}$

47. ▼ $r(x) = (\sqrt{2x + 1} - x^2)^{-1}$

48. ▼ $r(x) = (\sqrt{x + 1} + \sqrt{x})^3$

49. ▼ $f(x) = (1 + (1 + (1 + 2x)^3)^3)^3$

50. ▼ $f(x) = 2x + (2x + (2x + 1)^3)^3$

Find the indicated derivatives in Exercises 51–58. In each case, the independent variable is a (unspecified) differentiable function of t. HINT [See Quick Example 2 on page 828.]

51. $y = x^{100} + 99x^{-1}$. Find $\dfrac{dy}{dt}$.

52. $y = x^{0.5}(1 + x)$. Find $\dfrac{dy}{dt}$.

53. $s = \dfrac{1}{r^3} + r^{0.5}$. Find $\dfrac{ds}{dt}$.

54. $s = r + r^{-1}$. Find $\dfrac{ds}{dt}$.

55. $V = \dfrac{4}{3}\pi r^3$. Find $\dfrac{dV}{dt}$.

56. $A = 4\pi r^2$. Find $\dfrac{dA}{dt}$.

57. ▼ $y = x^3 + \dfrac{1}{x}$, $x = 2$ when $t = 1$, $\left.\dfrac{dx}{dt}\right|_{t=1} = -1$

Find $\left.\dfrac{dy}{dt}\right|_{t=1}$.

58. ▼ $y = \sqrt{x} + \dfrac{1}{\sqrt{x}}$, $x = 9$ when $t = 1$, $\left.\dfrac{dx}{dt}\right|_{t=1} = -1$

Find $\left.\dfrac{dy}{dt}\right|_{t=1}$.

In Exercises 59–64, compute the indicated derivative using the chain rule. HINT [See Quick Examples on page 829.]

59. $y = 3x - 2$; $\dfrac{dx}{dy}$ **60.** $y = 8x + 4$; $\dfrac{dx}{dy}$

61. $x = 2 + 3t$, $y = -5t$; $\dfrac{dy}{dx}$

62. $x = 1 - t/2$, $y = 4t - 1$; $\dfrac{dy}{dx}$

63. $y = 3x^2 - 2x$; $\left.\dfrac{dx}{dy}\right|_{x=1}$ **64.** $y = 3x - \dfrac{2}{x}$; $\left.\dfrac{dx}{dy}\right|_{x=2}$

APPLICATIONS

65. *Marginal Product* Paramount Electronics has an annual profit given by

$$P = -100,000 + 5,000q - 0.25q^2 \text{ dollars,}$$

where q is the number of laptop computers it sells each year. The number of laptop computers it can make and sell each year depends on the number n of electrical engineers Paramount employs, according to the equation

$$q = 30n + 0.01n^2.$$

Use the chain rule to find $\dfrac{dP}{dn}\bigg|_{n=10}$ and interpret the result. HINT [See Example 3.]

66. **Marginal Product** Refer back to Exercise 65. The average profit $\bar{P}$ per computer is given by dividing the total profit P by q:

$$\bar{P} = -\frac{100{,}000}{q} + 5{,}000 - 0.25q \text{ dollars.}$$

Determine the **marginal average product**, $d\bar{P}/dn$ at an employee level of 10 engineers. Interpret the result. HINT [See Example 3.]

67. **Food versus Education** The percentage y (of total personal consumption) an individual spends on food is approximately

$$y = 35x^{-0.25} \text{ percentage points} \quad (6.5 \le x \le 17.5),$$

where x is the percentage the individual spends on education.[26] An individual finds that she is spending

$$x = 7 + 0.2t$$

percent of her personal consumption on education, where t is time in months since January 1. Use direct substitution to express the percentage y as a function of time t (do not simplify the expression) and then use the chain rule to estimate how fast the percentage she spends on food is changing on November 1. Be sure to specify the units.

68. **Food versus Recreation** The percentage y (of total personal consumption) an individual spends on food is approximately

$$y = 33x^{-0.63} \text{ percentage points} \quad (2.5 \le x \le 4.5),$$

where x is the percentage the individual spends on recreation.[27] A college student finds that he is spending

$$x = 3.5 + 0.1t$$

percent of his personal consumption on recreation, where t is time in months since January 1. Use direct substitution to express the percentage y as a function of time t (do not simplify the expression) and then use the chain rule to estimate how fast the percentage he spends on food is changing on November 1. Be sure to specify the units.

69. **Marginal Revenue** The weekly revenue from the sale of rubies at *Royal Ruby Retailers* (RRR) is increasing at a rate of $40 per $1 increase in price, and the price is decreasing at a rate of $0.75 per additional ruby sold. What is the marginal revenue? (Be sure to state the units of measurement.) Interpret the result. HINT [See Example 5.]

70. **Marginal Revenue** The weekly revenue from the sale of emeralds at *Eduardo's Emerald Emporium* (EEE) is decreasing at a rate of €500 per €1 increase in price, and the price is decreasing at a rate of €0.45 per additional emerald sold. What is the marginal revenue? (Be sure to state the units of measurement.) Interpret the result. HINT [See Example 5.]

71. **Crime Statistics** The murder rate in large cities (over 1 million residents) can be related to that in smaller cities (500,000–1,000,000 residents) by the following linear model:[28]

$$y = 1.5x - 1.9 \quad (15 \le x \le 25),$$

where y is the murder rate (in murders per 100,000 residents each year) in large cities and x is the murder rate in smaller cities. During the period 1991–1998, the murder rate in small cities was decreasing at an average rate of 2 murders per 100,000 residents each year. Use the chain rule to estimate how fast the murder rate was changing in larger cities during that period. (Show how you used the chain rule in your answer.)

72. **Crime Statistics** Following is a quadratic model relating the murder rates described in the preceding exercise:

$$y = 0.1x^2 - 3x + 39 \quad (15 \le x \le 25).$$

In 1996, the murder rate in smaller cities was approximately 22 murders per 100,000 residents each year and was decreasing at a rate of approximately 2.5 murders per 100,000 residents each year. Use the chain rule to estimate how fast the murder rate was changing for large cities. (Show how you used the chain rule in your answer.)

73. **Existing Home Sales** The following graph shows the approximate value of home prices and existing home sales in 2006–2010 as a percentage change from 2003, together with quadratic approximations.[29]

Home prices and sales of existing homes

The quadratic approximations are given by

$$\text{Home Prices: } P(t) = t^2 - 10t + 41 \quad (0 \le t \le 4)$$

$$\text{Existing Home Sales: } S(t) = 1.5t^2 - 11t \quad (0 \le t \le 4),$$

[26]Model based on historical and projected data from 1908–2010. Sources: Historical data, Bureau of Economic Analysis; projected data, Bureau of Labor Statistics/*New York Times*, December 1, 2003, p. C2.
[27]*Ibid.*

[28]The model is a linear regression model. Source for data: Federal Bureau of Investigation, Supplementary Homicide Reports/*New York Times*, May 29, 2000, p. A12.
[29]Sources: Standard & Poors/Bloomberg Financial Markets/*New York Times*, September 29, 2007, p. C3. Projection is the authors'.

where t is time in years since the start of 2006. Use the chain rule to estimate $\dfrac{dS}{dP}\bigg|_{t=2}$. What does the answer tell you about home sales and prices? HINT [See the second Quick Example on page 829.]

74. *Existing Home Sales Leading to the Financial Crisis* The following graph shows the approximate value of home prices and existing home sales in 2004–2007 (the 3 years prior to the 2008 economic crisis) as a percentage change from 2003, together with quadratic approximations.[30]

Home prices and sales of existing homes

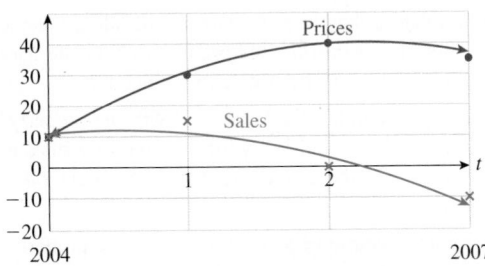

The quadratic approximations are given by

Home Prices: $P(t) = -6t^2 + 27t + 10 \quad (0 \le t \le 3)$

Existing Home Sales: $S(t) = -4t^2 + 4t + 11 \quad (0 \le t \le 3),$

where t is time in years since the start of 2004. Use the chain rule to estimate $\dfrac{dS}{dP}\bigg|_{t=2}$. What does the answer tell you about home sales and prices? HINT [See the second Quick Example on page 829.]

75. ▼ *Pollution* An offshore oil well is leaking oil and creating a circular oil slick. If the radius of the slick is growing at a rate of 2 miles/hour, find the rate at which the area is increasing when the radius is 3 miles. (The area of a disc of radius r is $A = \pi r^2$.) HINT [See Quick Example 2 on page 828.]

76. ▼ *Mold* A mold culture in a dorm refrigerator is circular and growing. The radius is growing at a rate of 0.3 cm/day. How fast is the area growing when the culture is 4 centimeters in radius? (The area of a disc of radius r is $A = \pi r^2$.) HINT [See Quick Example 2 on page 828.]

77. ▼ *Budget Overruns* The Pentagon is planning to build a new, spherical satellite. As is typical in these cases, the specifications keep changing, so that the size of the satellite keeps growing. In fact, the radius of the planned satellite is growing 0.5 feet per week. Its cost will be $1,000 per cubic foot. At the point when the plans call for a satellite 10 feet in radius, how fast is the cost growing? (The volume of a solid sphere of radius r is $V = \frac{4}{3}\pi r^3$.)

78. ▼ *Soap Bubbles* The soap bubble I am blowing has a radius that is growing at a rate of 4 cm/sec. How fast is the surface area growing when the radius is 10 cm? (The surface area of a sphere of radius r is $S = 4\pi r^2$.)

79. ▣ ▼ *Revenue Growth* The demand for the Cyberpunk II arcade video game is modeled by the logistic curve

$$q(t) = \frac{10,000}{1 + 0.5e^{-0.4t}}$$

where $q(t)$ is the total number of units sold t months after its introduction.

a. Use technology to estimate $q'(4)$.
b. Assume that the manufacturers of Cyberpunk II sell each unit for $800. What is the company's marginal revenue dR/dq?
c. Use the chain rule to estimate the rate at which revenue is growing 4 months after the introduction of the video game.

80. ▣ ▼ *Information Highway* The amount of information transmitted each month in the early years of the Internet (1988 to 1994) can be modeled by the equation

$$q(t) = \frac{2e^{0.69t}}{3 + 1.5e^{-0.4t}} \qquad (0 \le t \le 6),$$

where q is the amount of information transmitted each month in billions of data packets and t is the number of years since the start of 1988.[31]

a. Use technology to estimate $q'(2)$.
b. Assume that it costs $5 to transmit a million packets of data. What is the marginal cost $C'(q)$?
c. How fast was the cost increasing at the start of 1990?

Money Stock Exercises 81–84 are based on the following demand function for money (taken from a question on the GRE Economics Test):

$$M_d = 2 \times y^{0.6} \times r^{-0.3} \times p,$$

where
   M_d = *demand for nominal money balances (money stock)*
   y = *real income*
   r = *an index of interest rates*
   p = *an index of prices*

*These exercises also use the idea of **percentage rate of growth**:*

Percentage Rate of Growth of $M = \dfrac{\text{Rate of Growth of } M}{M}$

$$= \frac{dM/dt}{M}.$$

81. ◆ (from the GRE Economics Test) If the interest rate and price level are to remain constant while real income grows at 5 percent per year, the money stock must grow at what percent per year?

82. ◆ (from the GRE Economics Test) If real income and price level are to remain constant while the interest rate grows at 5 percent per year, the money stock must change by what percent per year?

---

[30]Sources: Standard & Poors /Bloomberg Financial Markets/*New York Times*, September 29, 2007, p. C3. Projection is the authors'.

[31]This is the authors' model, based on figures published in *New York Times*, Nov. 3, 1993.

83. ◆ (from the GRE Economics Test) If the interest rate is to remain constant while real income grows at 5 percent per year and the price level rises at 5 percent per year, the money stock must grow at what percent per year?

84. ◆ (from the GRE Economics Test) If real income grows by 5 percent per year, the interest rate grows by 2 percent per year, and the price level drops by 3 percent per year, the money stock must change by what percent per year?

COMMUNICATION AND REASONING EXERCISES

85. Complete the following: The derivative of 1 over a glob is -1 over . . . .

86. Complete the following: The derivative of the square root of a glob is 1 over . . . .

87. Say why the following was marked wrong and give the correct answer.

$$\frac{d}{dx}[(3x^3 - x)^3] = 3(9x^2 - 1)^2 \qquad ✗ \quad WRONG!$$

88. Say why the following was marked wrong and give the correct answer.

$$\frac{d}{dx}\left[\left(\frac{3x^2 - 1}{2x - 2}\right)^3\right] = 3\left(\frac{3x^2 - 1}{2x - 2}\right)^2\left(\frac{6x}{2}\right) \qquad ✗ \quad WRONG!$$

89. Name two major errors in the following graded test question and give the correct answer.

$$\frac{d}{dx}\left[\left(\frac{3x^2 - 1}{2x - 2}\right)^3\right] = 3\left(\frac{6x}{2}\right)^2 \qquad ✗ \quad WRONG! \; SEE \; ME!$$

90. Name two major errors in the following graded test question and give the correct answer.

$$\frac{d}{dx}[(3x^3 - x)(2x + 1)]^4 = 4[(9x^2 - 1)(2)]^3 \qquad ✗ \quad WRONG! \; SEE \; ME!$$

91. ▼ Formulate a simple procedure for deciding whether to apply first the chain rule, the product rule, or the quotient rule when finding the derivative of a function.

92. ▼ Give an example of a function f with the property that calculating $f'(x)$ requires use of the following rules in the given order: (1) the chain rule, (2) the quotient rule, and (3) the chain rule.

93. ◆ Give an example of a function f with the property that calculating $f'(x)$ requires use of the chain rule five times in succession.

94. ◆ What can you say about the composite of two linear functions and what can you say about its derivative?

11.5 # Derivatives of Logarithmic and Exponential Functions

At this point, we know how to take the derivative of any algebraic expression in x (involving powers, radicals, and so on). We now turn to the derivatives of logarithmic and exponential functions.

Derivative of the Natural Logarithm

$$\frac{d}{dx}[\ln x] = \frac{1}{x} \qquad \text{Recall that } \ln x = \log_e x.$$

Quick Examples

1. $\dfrac{d}{dx}[3 \ln x] = 3 \cdot \dfrac{1}{x} = \dfrac{3}{x}$ Derivative of a constant times a function

2. $\dfrac{d}{dx}[x \ln x] = 1 \cdot \ln x + x \cdot \dfrac{1}{x}$ Product rule, because $x \ln x$ is a product

      $= \ln x + 1.$

The above simple formula works only for the natural logarithm (the logarithm with base e). For logarithms with bases other than e, we have the following:

Derivative of the Logarithm with Base b

$$\frac{d}{dx}[\log_b x] = \frac{1}{x \ln b}$$

Notice that, if $b = e$, we get the same formula as previously.

Quick Examples

1. $\dfrac{d}{dx}[\log_3 x] = \dfrac{1}{x \ln 3} \approx \dfrac{1}{1.0986x}$

2. $\dfrac{d}{dx}[\log_2(x^4)] = \dfrac{d}{dx}(4 \log_2 x)$ We used the logarithm identity $\log_b(x^r) = r \log_b x$.

$$= 4 \cdot \frac{1}{x \ln 2} \approx \frac{4}{0.6931x}$$

Derivation of the formulas $\dfrac{d}{dx}[\ln x] = \dfrac{1}{x}$ and $\dfrac{d}{dx}[\log_b x] = \dfrac{1}{x \ln b}$

To compute $\dfrac{d}{dx}[\ln x]$, we need to use the definition of the derivative. We also use properties of the logarithm to help evaluate the limit.

$$\frac{d}{dx}[\ln x] = \lim_{h \to 0} \frac{\ln(x+h) - \ln x}{h}$$ Definition of the derivative

$$= \lim_{h \to 0} \frac{1}{h}[\ln(x+h) - \ln x]$$ Algebra

$$= \lim_{h \to 0} \frac{1}{h} \ln\left(\frac{x+h}{x}\right)$$ Properties of the logarithm

$$= \lim_{h \to 0} \frac{1}{h} \ln\left(1 + \frac{h}{x}\right)$$ Algebra

$$= \lim_{h \to 0} \ln\left(1 + \frac{h}{x}\right)^{1/h}$$ Properties of the logarithm

which we rewrite as

$$\lim_{h \to 0} \ln\left[\left(1 + \frac{1}{(x/h)}\right)^{x/h}\right]^{1/x}.$$

As $h \to 0^+$, the quantity x/h gets large and positive, and so the quantity in brackets approaches e (see the definition of e in Section 9.2), which leaves us with

$$\ln[e]^{1/x} = \frac{1}{x} \ln e = \frac{1}{x}$$

* We actually used the fact that the logarithm function is continuous when we took the limit.

† Here is an outline of the argument for negative *h*. Because *x* must be positive for ln *x* to be defined, we find that $x/h \to -\infty$ as $h \to 0^-$, and so we must consider the quantity $(1 + 1/m)^m$ for large *negative m*. It turns out the limit is still *e* (check it numerically!) and so the computation above still works.

which is the derivative we are after.* What about the limit as $h \to 0^-$? We will glide over that case and leave it for the interested reader to pursue.†

The rule for the derivative of $\log_b x$ follows from the fact that $\log_b x = \ln x/\ln b$.

If we were to take the derivative of the natural logarithm of a *quantity* (a function of *x*), rather than just *x*, we would need to use the chain rule:

Derivatives of Logarithms of Functions

Original Rule	Generalized Rule	In Words
$\dfrac{d}{dx}[\ln x] = \dfrac{1}{x}$	$\dfrac{d}{dx}[\ln u] = \dfrac{1}{u}\dfrac{du}{dx}$	The derivative of the natural logarithm of a quantity is 1 over that quantity, times the derivative of that quantity.
$\dfrac{d}{dx}[\log_b x] = \dfrac{1}{x \ln b}$	$\dfrac{d}{dx}[\log_b u] = \dfrac{1}{u \ln b}\dfrac{du}{dx}$	The derivative of the log to base b of a quantity is 1 over the product of ln b and that quantity, times the derivative of that quantity.

Quick Examples

§ If we were to evaluate $\ln(x^2+1)$, the last operation we would perform would be to take the natural logarithm of a quantity. Thus, the calculation thought experiment tells us that we are dealing with ln of a quantity, and so we need the generalized logarithm rule as stated above.

1. $\dfrac{d}{dx}\ln[x^2+1] = \dfrac{1}{x^2+1}\dfrac{d}{dx}(x^2+1)$ $u = x^2+1$ (See the margin note.§)

$= \dfrac{1}{x^2+1}(2x) = \dfrac{2x}{x^2+1}$

2. $\dfrac{d}{dx}\log_2[x^3+x] = \dfrac{1}{(x^3+x)\ln 2}\dfrac{d}{dx}(x^3+x)$ $u = x^3+x$

$= \dfrac{1}{(x^3+x)\ln 2}(3x^2+1) = \dfrac{3x^2+1}{(x^3+x)\ln 2}$

EXAMPLE 1 Derivative of Logarithmic Function

Compute the following derivatives:

a. $\dfrac{d}{dx}[\ln\sqrt{x+1}]$ **b.** $\dfrac{d}{dx}[\ln[(1+x)(2-x)]]$ **c.** $\dfrac{d}{dx}[\ln|x|]$

Solution

a. The calculation thought experiment tells us that we have the natural logarithm of a quantity, so

$\dfrac{d}{dx}[\ln\sqrt{x+1}] = \dfrac{1}{\sqrt{x+1}}\dfrac{d}{dx}\sqrt{x+1}$ $\dfrac{d}{dx}\ln u = \dfrac{1}{u}\dfrac{du}{dx}$

$= \dfrac{1}{\sqrt{x+1}}\cdot\dfrac{1}{2\sqrt{x+1}}$ $\dfrac{d}{dx}\sqrt{u} = \dfrac{1}{2\sqrt{u}}\dfrac{du}{dx}$

$= \dfrac{1}{2(x+1)}.$

Q: *What happened to the square root?*

A: As with many problems involving logarithms, we could have done this one differently and much more easily if we had simplified the expression $\ln \sqrt{x+1}$ using the properties of logarithms *before* differentiating. Doing this, we get the following:

Part (a) redone by simplifying first:

$$\ln \sqrt{x+1} = \ln(x+1)^{1/2} = \frac{1}{2}\ln(x+1). \quad \text{Simplify the logarithm first.}$$

Thus,

$$\frac{d}{dx}[\ln \sqrt{x+1}] = \frac{d}{dx}\left[\frac{1}{2}\ln(x+1)\right]$$

$$= \frac{1}{2}\left[\frac{1}{x+1}\right] \cdot 1 = \frac{1}{2(x+1)}.$$

A *lot* easier!

b. This time, we simplify the expression $\ln[(1+x)(2-x)]$ before taking the derivative:

$$\ln[(1+x)(2-x)] = \ln(1+x) + \ln(2-x). \quad \text{Simplify the logarithm first.}$$

Thus,

$$\frac{d}{dx}[\ln[(1+x)(2-x)]] = \frac{d}{dx}[\ln[(1+x)]] + \frac{d}{dx}[\ln[(2-x)]]$$

$$= \frac{1}{1+x} - \frac{1}{2-x}. \qquad \frac{d}{dx}\ln u = \frac{1}{u}\frac{du}{dx}$$

For practice, try doing this calculation without simplifying first. What other differentiation rule do you need to use?

c. Before we start, we note that $\ln x$ is defined only for positive values of x, so its domain is the set of positive real numbers. The domain of $\ln |x|$, on the other hand, is the set of *all* nonzero real numbers. For example, $\ln|-2| = \ln 2 \approx 0.6931$. For this reason, $\ln |x|$ often turns out to be more useful than the ordinary logarithm function.

$$\frac{d}{dx}[\ln |x|] = \frac{1}{|x|}\frac{d}{dx}|x| \qquad \frac{d}{dx}\ln u = \frac{1}{u}\frac{du}{dx}$$

$$= \frac{1}{|x|}\frac{|x|}{x} \qquad \text{Recall that } \frac{d}{dx}|x| = \frac{|x|}{x}.$$

$$= \frac{1}{x}$$

⟹ **Before we go on...** Figure 7(a) shows the graphs of $y = \ln |x|$ and $y = 1/x$. Figure 7(b) shows the graphs of $y = \ln |x|$ and $y = 1/|x|$. You should be able to see from these graphs why the derivative of $\ln |x|$ is $1/x$ and not $1/|x|$.

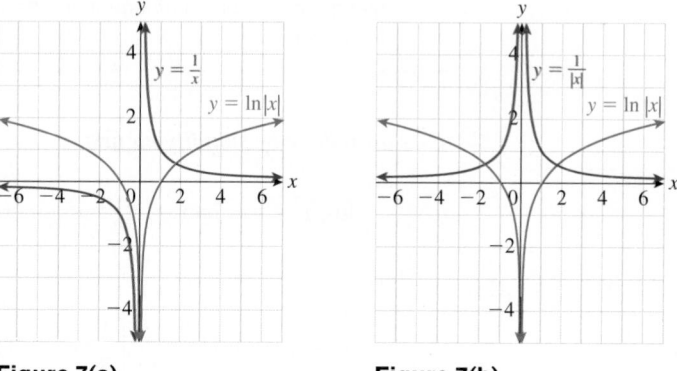

Figure 7(a)   **Figure 7(b)**

This last example, in conjunction with the chain rule, gives us the following formulas.

Derivative of Logarithms of Absolute Values

Original Rule	Generalized Rule	In Words
$\dfrac{d}{dx}[\ln |x|] = \dfrac{1}{x}$	$\dfrac{d}{dx}[\ln |u|] = \dfrac{1}{u}\dfrac{du}{dx}$	The derivative of the natural logarithm of the absolute value of a quantity is 1 over that quantity, times the derivative of that quantity.
$\dfrac{d}{dx}[\log_b |x|] = \dfrac{1}{x \ln b}$	$\dfrac{d}{dx}[\log_b |u|] = \dfrac{1}{u \ln b}\dfrac{du}{dx}$	The derivative of the log to base b of the absolute value of a quantity is 1 over the product of $\ln b$ and that quantity, times the derivative of that quantity.

Note: Compare the above formulas with those on page 836. They tell us that we can simply ignore the absolute values in $\ln |u|$ or $\log_b |u|$ when taking the derivative.

Quick Examples

1. $\dfrac{d}{dx}[\ln |x^2 - 1|] = \dfrac{1}{x^2 - 1}\dfrac{d}{dx}(x^2 - 1)$ $u = x^2 - 1$

$= \dfrac{1}{x^2 - 1}(2x) = \dfrac{2x}{x^2 - 1}$

$$2. \ \frac{d}{dx}[\log_2|x^3 + x|] = \frac{1}{(x^3 + x)\ln 2}\frac{d}{dx}(x^3 + x) \qquad u = x^3 + x$$

$$= \frac{1}{(x^3 + x)\ln 2}(3x^2 + 1) = \frac{3x^2 + 1}{(x^3 + x)\ln 2}$$

We now turn to the derivatives of *exponential* functions—that is, functions of the form $f(x) = b^x$. We begin by showing how *not* to differentiate them.

Caution The derivative of b^x is *not* xb^{x-1}. The power rule applies only to *constant* exponents. In this case the exponent is decidedly *not* constant, and so the power rule does not apply.

The following shows the correct way of differentiating b^x, beginning with a special case.

Derivative of e^x

$$\frac{d}{dx}[e^x] = e^x$$

Quick Examples

$$1. \ \frac{d}{dx}[3e^x] = 3\frac{d}{dx}[e^x] = 3e^x \qquad \text{Constant multiple rule}$$

$$2. \ \frac{d}{dx}\left[\frac{e^x}{x}\right] = \frac{e^x x - e^x(1)}{x^2} \qquad \text{Quotient rule}$$

$$= \frac{e^x(x - 1)}{x^2}$$

* There is another—very simple—function that is its own derivative. What is it?

Thus, e^x has the amazing property that its derivative is itself!* For bases other than e, we have the following generalization:

Derivative of b^x

If b is any positive number, then

$$\frac{d}{dx}[b^x] = b^x \ln b.$$

Note that if $b = e$, we obtain the previous formula.

Quick Example

$$\frac{d}{dx}[3^x] = 3^x \ln 3$$

Derivation of the Formula $\frac{d}{dx}[e^x] = e^x$

* This shortcut is an example of a technique called *logarithmic differentiation*, which is occasionally useful. We will see it again in the next section.

To find the derivative of e^x, we use a shortcut.* Write $g(x) = e^x$. Then

$$\ln g(x) = x.$$

Take the derivative of both sides of this equation to get

$$\frac{g'(x)}{g(x)} = 1$$

or

$$g'(x) = g(x) = e^x.$$

In other words, the exponential function with base e is its own derivative. The rule for exponential functions with other bases follows from the equality $b^x = e^{x\ln b}$ (why?) and the chain rule. (Try it.)

If we were to take the derivative of e raised to a *quantity*, not just x, we would need to use the chain rule, as follows.

Derivatives of Exponentials of Functions

Original Rule	Generalized Rule	In Words
$\dfrac{d}{dx}[e^x] = e^x$	$\dfrac{d}{dx}[e^u] = e^u\dfrac{du}{dx}$	The derivative of e raised to a quantity is e raised to that quantity, times the derivative of that quantity.
$\dfrac{d}{dx}[b^x] = b^x \ln b$	$\dfrac{d}{dx}[b^u] = b^u \ln b\dfrac{du}{dx}$	The derivative of b raised to a quantity is b raised to that quantity, times $\ln b$, times the derivative of that quantity.

Quick Examples

† The calculation thought experiment tells us that we have e raised to a quantity.

1. $\dfrac{d}{dx}\left[e^{x^2+1}\right] = e^{x^2+1}\dfrac{d}{dx}[x^2+1]$ $u = x^2 + 1$ (See margin note.†)

   $= e^{x^2+1}(2x) = 2x\,e^{x^2+1}$

2. $\dfrac{d}{dx}[2^{3x}] = 2^{3x}\ln 2\dfrac{d}{dx}[3x]$ $u = 3x$

   $= 2^{3x}(\ln 2)(3) = (3\ln 2)2^{3x}$

3. $\dfrac{d}{dt}[30e^{1.02t}] = 30e^{1.02t}(1.02) = 30.6e^{1.02t}$ $u = 1.02t$

4. If \$1,000 is invested in an account earning 5% per year compounded continuously, then the rate of change of the account balance after t years is

   $$\frac{d}{dt}[1,000e^{0.05t}] = 1,000(0.05)e^{0.05t} = 50e^{0.05t} \text{ dollars/year.}$$

APPLICATIONS

EXAMPLE 2 Epidemics

In the early stages of the AIDS epidemic during the 1980s, the number of cases in the United States was increasing by about 50% every 6 months. By the start of 1983, there were approximately 1,600 AIDS cases in the United States.[32] Had this trend continued, how many new cases per year would have been occurring by the start of 1993?

Solution To find the answer, we must first model this exponential growth using the methods of Chapter 9. Referring to Example 4 in Section 9.2, we find that t years after the start of 1983 the number of cases is

$$A = 1,600(2.25^t).$$

We are asking for the number of new cases each year. In other words, we want the rate of change, dA/dt:

$$\frac{dA}{dt} = 1,600(2.25)^t \ln 2.25 \text{ cases per year.}$$

At the start of 1993, $t = 10$, so the number of new cases per year is

$$\left.\frac{dA}{dt}\right|_{r=10} = 1,600(2.25)^{10} \ln 2.25 \approx 4,300,000 \text{ cases per year.}$$

➡ **Before we go on...** In Example 2, the figure for the number of new cases per year is so large because we assumed that exponential growth—the 50% increase every 6 months—would continue. A more realistic model for the spread of a disease is the logistic model. (See Section 9.4, as well as the next example.) ∎

EXAMPLE 3 Sales Growth

The sales of the *Cyberpunk II* video game can be modeled by the logistic curve

$$q(t) = \frac{10,000}{1 + 0.5e^{-0.4t}}$$

where $q(t)$ is the total number of units sold t months after its introduction. How fast is the game selling 2 years after its introduction?

Solution We are asked for $q'(24)$. We can find the derivative of $q(t)$ using the quotient rule, or we can first write

$$q(t) = 10,000(1 + 0.5e^{-0.4t})^{-1}$$

and then use the generalized power rule:

$$q'(t) = -10,000(1 + 0.5e^{-0.4t})^{-2}(0.5e^{-0.4t})(-0.4)$$
$$= \frac{2,000e^{-0.4t}}{(1 + 0.5e^{-0.4t})^2}.$$

---

[32]Data based on regression of 1982–1986 figures. Source for data: Centers for Disease Control and Prevention. HIV/AIDS Surveillance Report, 2000;12 (No. 2).

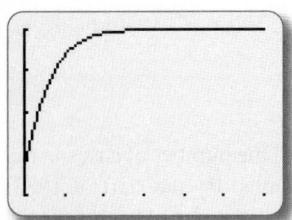

Figure 8

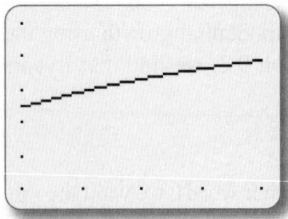

Figure 9

✱ We can also say this using limits:
$$\lim_{t \to +\infty} q(t) = 10,000.$$

Thus,

$$q'(24) = \frac{2,000e^{-0.4(24)}}{(1 + 0.5e^{-0.4(24)})^2} \approx 0.135 \text{ units per month.}$$

So, after 2 years, sales are quite slow.

➡ **Before we go on...** We can check the answer in Example 3 graphically. If we plot the total sales curve for $0 \le t \le 30$ and $6,000 \le q \le 10,000$, on a TI-83/84 Plus, for example, we get the graph shown in Figure 8. Notice that total sales level off at about 10,000 units.✱ We computed $q'(24)$, which is the slope of the curve at the point with t-coordinate 24. If we zoom in to the portion of the curve near $t = 24$, we obtain the graph shown in Figure 9, with $23 \le t \le 25$ and $9,999 \le q \le 10,000$. The curve is almost linear in this range. If we use the two endpoints of this segment of the curve, $(23, 9,999.4948)$ and $(25, 9,999.7730)$, we can approximate the derivative as

$$\frac{9,999.7730 - 9,999.4948}{25 - 23} = 0.1391$$

which is accurate to two decimal places. ∎

11.5   EXERCISES

▼ more advanced   ◆ challenging
Ⓣ indicates exercises that should be solved using technology

Find the derivatives of the functions in Exercises 1–76.
HINT [See Quick Examples on page 836.]

1. $f(x) = \ln(x - 1)$ **2.** $f(x) = \ln(x + 3)$

3. $f(x) = \log_2 x$ **4.** $f(x) = \log_3 x$

5. $g(x) = \ln|x^2 + 3|$ **6.** $g(x) = \ln|2x - 4|$

7. $h(x) = e^{x+3}$ **8.** $h(x) = e^{x^2}$
   HINT [See Quick Examples on   HINT [See Quick Examples on
   page 840.]   page 840.]

9. $f(x) = e^{-x}$ **10.** $f(x) = e^{1-x}$

11. $g(x) = 4^x$ **12.** $g(x) = 5^x$

13. $h(x) = 2^{x^2-1}$ **14.** $h(x) = 3^{x^2-x}$

15. $f(x) = x \ln x$ **16.** $f(x) = 3 \ln x$

17. $f(x) = (x^2 + 1) \ln x$ **18.** $f(x) = (4x^2 - x) \ln x$

19. $f(x) = (x^2 + 1)^5 \ln x$ **20.** $f(x) = (x + 1)^{0.5} \ln x$

21. $g(x) = \ln|3x - 1|$ **22.** $g(x) = \ln|5 - 9x|$

23. $g(x) = \ln|2x^2 + 1|$ **24.** $g(x) = \ln|x^2 - x|$

25. $g(x) = \ln(x^2 - 2.1x^{0.3})$ **26.** $g(x) = \ln(x - 3.1x^{-1})$

27. $h(x) = \ln[(-2x + 1)(x + 1)]$ HINT [See Example 1b.]

28. $h(x) = \ln[(3x + 1)(-x + 1)]$ HINT [See Example 1b.]

29. $h(x) = \ln\left(\dfrac{3x + 1}{4x - 2}\right)$ **30.** $h(x) = \ln\left(\dfrac{9x}{4x - 2}\right)$

31. $r(x) = \ln\left|\dfrac{(x + 1)(x - 3)}{-2x - 9}\right|$ **32.** $r(x) = \ln\left|\dfrac{-x + 1}{(3x - 4)(x - 9)}\right|$

33. $s(x) = \ln(4x - 2)^{1.3}$ **34.** $s(x) = \ln(x - 8)^{-2}$
   HINT [See Example 1a.]   HINT [See Example 1a.]

35. $s(x) = \ln\left|\dfrac{(x + 1)^2}{(3x - 4)^3(x - 9)}\right|$

36. $s(x) = \ln\left|\dfrac{(x + 1)^2(x - 3)^4}{2x + 9}\right|$

37. $h(x) = \log_2(x + 1)$ **38.** $h(x) = \log_3(x^2 + x)$

39. $r(t) = \log_3(t + 1/t)$ **40.** $r(t) = \log_3(t + \sqrt{t})$

41. $f(x) = (\ln|x|)^2$ **42.** $f(x) = \dfrac{1}{\ln|x|}$

43. $r(x) = \ln(x^2) - [\ln(x - 1)]^2$ **44.** $r(x) = (\ln(x^2))^2$

45. $f(x) = xe^x$ **46.** $f(x) = 2e^x - x^2e^x$

47. $r(x) = \ln(x + 1) + 3x^3e^x$ **48.** $r(x) = \ln|x + e^x|$

49. $f(x) = e^x \ln|x|$ **50.** $f(x) = e^x \log_2|x|$

51. $f(x) = e^{2x+1}$ **52.** $f(x) = e^{4x-5}$

53. $h(x) = e^{x^2-x+1}$ **54.** $h(x) = e^{2x^2-x+1/x}$

55. $s(x) = x^2e^{2x-1}$ **56.** $s(x) = \dfrac{e^{4x-1}}{x^3 - 1}$

57. $r(x) = (e^{2x-1})^2$ **58.** $r(x) = (e^{2x^2})^3$

59. $t(x) = 3^{2x-4}$ **60.** $t(x) = 4^{-x+5}$

61. $v(x) = 3^{2x+1} + e^{3x+1}$ **62.** $v(x) = e^{2x}4^{2x}$

63. $u(x) = \dfrac{3^{x^2}}{x^2 + 1}$

64. $u(x) = (x^2 + 1)4^{x^2 - 1}$

65. $g(x) = \dfrac{e^x + e^{-x}}{e^x - e^{-x}}$

66. $g(x) = \dfrac{1}{e^x + e^{-x}}$

67. ▼$g(x) = e^{3x - 1}e^{x - 2}e^x$

68. ▼$g(x) = e^{-x + 3}e^{2x - 1}e^{-x + 11}$

69. ▼$f(x) = \dfrac{1}{x \ln x}$

70. ▼$f(x) = \dfrac{e^{-x}}{xe^x}$

71. ▼$f(x) = [\ln(e^x)]^2 - \ln[(e^x)^2]$

72. ▼$f(x) = e^{\ln x} - e^{2\ln(x^2)}$

73. ▼$f(x) = \ln|\ln x|$

74. ▼$f(x) = \ln|\ln|\ln x||$

75. ▼$s(x) = \ln\sqrt{\ln x}$

76. ▼$s(x) = \sqrt{\ln(\ln x)}$

Find the equations of the straight lines described in Exercises 77–82. Use graphing technology to check your answers by plotting the given curve together with the tangent line.

77. Tangent to $y = e^x \log_2 x$ at the point $(1, 0)$

78. Tangent to $y = e^x + e^{-x}$ at the point $(0, 2)$

79. Tangent to $y = \ln\sqrt{2x + 1}$ at the point where $x = 0$

80. Tangent to $y = \ln\sqrt{2x^2 + 1}$ at the point where $x = 1$

81. At right angles to $y = e^{x^2}$ at the point where $x = 1$

82. At right angles to $y = \log_2(3x + 1)$ at the point where $x = 1$

APPLICATIONS

83. *Research and Development: Industry* The total spent on research and development by industry in the United States during 1995–2007 can be approximated by

$$S(t) = 57.5 \ln t + 31 \text{ billion dollars} \quad (5 \le t \le 17),$$

where t is the year since 1990.[33] What was the total spent in 2000 ($t = 10$) and how fast was it increasing? HINT [See Quick Examples on page 834.]

84. *Research and Development: Federal* The total spent on research and development by the federal government in the United States during 1995–2007 can be approximated by

$$S(t) = 7.4 \ln t + 3 \text{ billion dollars} \quad (5 \le t \le 17),$$

where t is the year since 1990.[34] What was the total spent in 2005 ($t = 15$) and how fast was it increasing? HINT [See Quick Examples on page 834.]

---

[33]Spending is in constant 2000 dollars. Source for data through 2006: National Science Foundation, Division of Science Resources Statistics, National Patterns of R&D Resources (www.nsf.gov/statistics) August 2008.

[34]Federal funding excluding grants to industry and nonprofit organizations. Spending is in constant 2000 dollars. Source for data through 2006: National Science Foundation, Division of Science Resources Statistics, National Patterns of R&D Resources (www.nsf.gov/statistics) August 2008.

85. *Research and Development: Industry* The function $S(t)$ in Exercise 83 can also be written (approximately) as

$$S(t) = 57.5 \ln (1.71t + 17.1) \text{ billion dollars} \quad (-5 \le t \le 7),$$

where this time t is the year since 2000. Use this alternative formula to estimate the amount spent in 2000 and its rate of change, and check your answers by comparing them with those in Exercise 83.

86. *Research and Development: Federal* The function $S(t)$ in Exercise 84 can also be written (approximately) as

$$S(t) = 7.4 \ln (1.5t + 15) \text{ billion dollars} \quad (-5 \le t \le 7),$$

where this time t is the year since 2000. Use this alternative formula to estimate the amount spent in 2005 and its rate of change, and check your answers by comparing them with those in Exercise 84.

87. ▼ ***Carbon Dating*** The age in years of a specimen that originally contained 10g of carbon 14 is given by

$$y = \log_{0.999879}(0.1x),$$

where x is the amount of carbon 14 it currently contains. Compute $\left.\dfrac{dy}{dx}\right|_{x=5}$ and interpret your answer. HINT [For the calculation, see Quick Examples on page 836.]

88. ▼ ***Iodine Dating*** The age in years of a specimen that originally contained 10g of iodine 131 is given by

$$y = \log_{0.999567}(0.1x),$$

where x is the amount of iodine 131 it currently contains. Compute $\left.\dfrac{dy}{dx}\right|_{x=8}$ and interpret your answer. HINT [For the calculation, see Quick Examples on page 836.]

89. *New York City Housing Costs: Downtown* The average price of a two-bedroom apartment in downtown New York City during the real estate boom from 1994 to 2004 can be approximated by

$$p(t) = 0.33e^{0.16t} \text{ million dollars} \quad (0 \le t \le 10),$$

where t is time in years ($t = 0$ represents 1994).[35] What was the average price of a two-bedroom apartment in downtown New York City in 2003, and how fast was it increasing? (Round your answers to two significant digits.) HINT [See Quick Example 3 on page 840.]

90. *New York City Housing Costs: Uptown* The average price of a two-bedroom apartment in uptown New York City during the real estate boom from 1994 to 2004 can be approximated by

$$p(t) = 0.14e^{0.10t} \text{ million dollars} \quad (0 \le t \le 10),$$

where t is time in years ($t = 0$ represents 1994).[36] What was the average price of a two-bedroom apartment in uptown

---

[35]Model is based on a exponential regression. Source for data: Miller Samuel/*New York Times*, March 28, 2004, p. RE 11.

[36]*Ibid.*

New York City in 2002, and how fast was it increasing? (Round your answers to two significant digits.) HINT [See Quick Example 3 on page 840.]

91. Big Brother The following chart shows the total number of wiretaps authorized each year by U.S. state and federal courts from 1990 to 2009 ($t = 0$ represents 1990):[37]

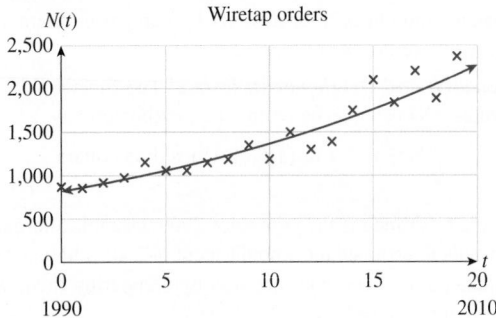

These data can be approximated with the model
$$N(t) = 820e^{0.051t} \quad (0 \le t \le 19).$$

a. Find $N(15)$ and $N'(15)$. Be sure to state the units of measurement. To how many significant digits should we round the answers? Why?

b. The number of people whose communications are intercepted averages around 100 per wiretap order. What does the answer to part (a) tell you about the number of people whose communications were intercepted?[38]

c. According to the model, the number of wiretaps orders each year (choose one)

(A) increased at a linear rate
(B) decreased at a quadratic rate
(C) increased at an exponential rate
(D) increased at a logarithmic rate

over the period shown.

92. Big Brother The following chart shows the total number of wiretaps authorized each year by U.S. state courts from 1990 to 2009 ($t = 0$ represents 1990):[39]

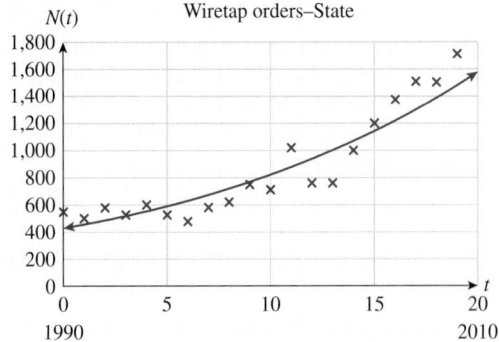

These data can be approximated with the model
$$N(t) = 430e^{0.065t} \quad (0 \le t \le 19).$$

a. Find $N(10)$ and $N'(10)$. Be sure to state the units of measurement. To how many significant digits should we round the answers? Why?

b. The number of people whose communications are intercepted averages around 100 per wiretap order. What does the answer to part (a) tell you about the number of people whose communications were intercepted?[40]

c. According to the model, the number of wiretaps orders each year (choose one)

(A) increased at a linear rate
(B) decreased at a quadratic rate
(C) increased at an exponential rate
(D) increased at a logarithmic rate

over the period shown.

93. Investments If $10,000 is invested in a savings account offering 4% per year, compounded continuously, how fast is the balance growing after 3 years?

94. Investments If $20,000 is invested in a savings account offering 3.5% per year, compounded continuously, how fast is the balance growing after 3 years?

95. Investments If $10,000 is invested in a savings account offering 4% per year, compounded semiannually, how fast is the balance growing after 3 years?

96. Investments If $20,000 is invested in a savings account offering 3.5% per year, compounded semiannually, how fast is the balance growing after 3 years?

97. SARS In the early stages of the deadly SARS (Severe Acute Respiratory Syndrome) epidemic in 2003, the number of cases was increasing by about 18% each day.[41] On March 17, 2003 (the first day for which statistics were reported by the World Health Organization) there were 167 cases. Find an exponential model that predicts the number of people infected t days after March 17, 2003, and use it to estimate how fast the epidemic was spreading on March 31, 2003. (Round your answer to the nearest whole number of new cases per day.) HINT [See Example 2.]

98. SARS A few weeks into the deadly SARS (Severe Acute Respiratory Syndrome) epidemic in 2003, the number of cases was increasing by about 4% each day.[42] On April 1, 2003 there were 1,804 cases. Find an exponential model that predicts the number $A(t)$ of people infected t days after April 1, 2003, and use it to estimate how fast the epidemic was spreading on April 30, 2003. (Round your answer to the nearest whole number of new cases per day.) HINT [See Example 2.]

[37]Source for data: Wiretap Reports, Administrative Office of the United States Courts www.uscourts.gov/Statistics/WiretapReports.

[38]Assume there is no significant overlap between the people whose communications are intercepted in different wiretap orders.

[39]See Footnote 37.

[40]See Footnote 38.

[41]World Health Organization (www.who.int).

[42]*Ibid.*

99. ▼ *SAT Scores by Income* The following bar graph shows U.S. math SAT scores as a function of household income:[43]

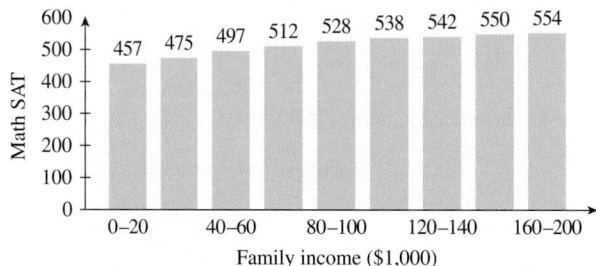

a. Which of the following best models the data (C is a constant)?

(A) $S(x) = C - 133e^{-0.0131x}$
(B) $S(x) = C + 133e^{-0.0131x}$
(C) $S(x) = C + 133e^{0.0131x}$
(D) $S(x) = C - 133e^{0.0131x}$

($S(x)$ is the average math SAT score of students whose household income is x thousand dollars per year.)

b. Use $S'(x)$ to predict how a student's math SAT score is affected by a $1,000 increase in parents' income for a student whose parents earn $45,000.

c. Does $S'(x)$ increase or decrease as x increases? Interpret your answer.

100. *SAT Scores by Income* The following bar graph shows U.S. critical reading SAT scores as a function of household income:[44]

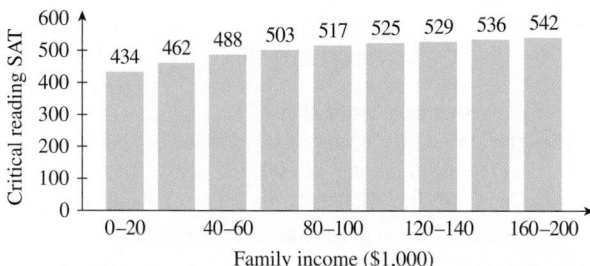

a. Which of the following best models the data (C is a constant)?

(A) $S(x) = C + \dfrac{1}{136e^{0.015x}}$

(B) $S(x) = C - 136e^{0.015x}$

(C) $S(x) = C - \dfrac{136}{e^{0.015x}}$

(D) $S(x) = C - \dfrac{e^{0.015x}}{136}$

($S(x)$ is the average critical reading SAT score of students whose household income is x thousand dollars per year.)

b. Use $S'(x)$ to predict how a student's critical reading SAT score is affected by a $1,000 increase in parents' income for a student whose parents earn $45,000.

c. Does $S'(x)$ increase or decrease as x increases? Interpret your answer.

101. ▼ *Demographics: Average Age and Fertility* The following graph shows a plot of average age of a population versus fertility rate (the average number of children each woman has in her lifetime) in the United States and Europe over the period 1950–2005.[45]

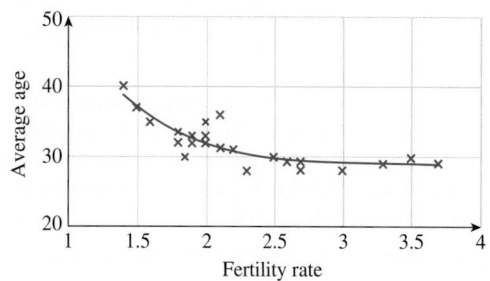

The equation of the accompanying curve is

$$a = 28.5 + 120(0.172)^x \quad (1.4 \le x \le 3.7),$$

where a is the average age (in years) of the population and x is the fertility rate.

a. Compute $a'(2)$. What does the answer tell you about average age and fertility rates?

b. Use the answer to part (a) to estimate how much the fertility rate would need to increase from a level of 2 children per woman to lower the average age of a population by about 1 year.

102. ▼ *Demographics: Average Age and Fertility* The following graph shows a plot of average age of a population versus fertility rate (the average number of children each woman has in her lifetime) in Europe over the period 1950–2005.[46]

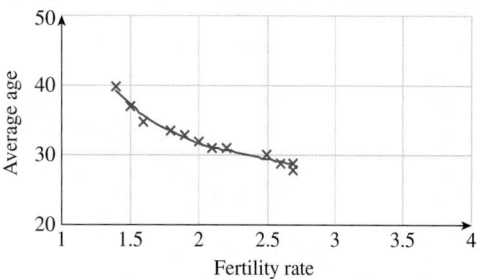

The equation of the accompanying curve is

$$g = 27.6 + 128(0.181)^x \quad (1.4 \le x \le 3.7),$$

where g is the average age (in years) of the population and x is the fertility rate.

a. Compute $g'(2.5)$. What does the answer tell you about average age and fertility rates?

---

[43]2009 data. Source: College Board/*New York Times* http://economix.blogs.nytimes.com.

[44]*Ibid.*

[45]The separate data for Europe and the United States are collected in the same graph. 2005 figures are estimates. Source: United Nations World Population Division/*New York Times*, June 29, 2003, p. 3.

[46]All European countries including the Russian Federation. 2005 figures are estimates. Source: United Nations World Population Division/*New York Times*, June 29, 2003, p. 3.

b. Referring to the model that combines the data for Europe and the United States in Exercise 101, which population's average age is affected more by a changing fertility rate at the level of 2.5 children per woman?

103. *Epidemics* A flu epidemic described in Example 1 in Section 9.4 approximately followed the curve

$$P = \frac{150}{1 + 15,000e^{-0.35t}} \text{ million people,}$$

where P is the number of people infected and t is the number of weeks after the start of the epidemic. How fast is the epidemic growing (that is, how many new cases are there each week) after 20 weeks? After 30 weeks? After 40 weeks? (Round your answers to two significant digits.) HINT [See Example 3.]

104. *Epidemics* Another epidemic follows the curve

$$P = \frac{200}{1 + 20,000e^{-0.549t}} \text{ million people,}$$

where t is in years. How fast is the epidemic growing after 10 years? After 20 years? After 30 years? (Round your answers to two significant digits.) HINT [See Example 3.]

105. *Subprime Mortgages during the Housing Bubble* During the real estate run-up in 2000–2008, the percentage of mortgages issued in the U.S. that were subprime (normally classified as risky) could be approximated by

$$A(t) = \frac{15.0}{1 + 8.6e^{-0.59t}} \text{ percent} \quad (0 \le t \le 8)$$

t years after the start of 2000.[47]

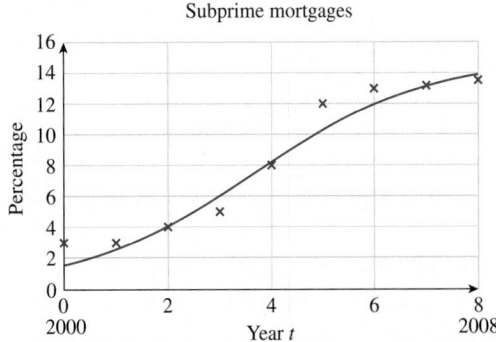

Subprime mortgages

How fast, to the nearest 0.1%, was the percentage increasing at the start of 2003? How would you check that the answer is approximately correct by looking at the graph? HINT [See Example 3.]

106. *Subprime Mortgage Debt during the Housing Bubble* During the real estate run-up in 2000–2008, the value of

subprime (normally classified as risky) mortgage debt outstanding in the U.S. was approximately

$$A(t) = \frac{1,350}{1 + 4.2e^{-0.53t}} \text{ billion dollars} \quad (0 \le t \le 8)$$

t years after the start of 2000.[48]

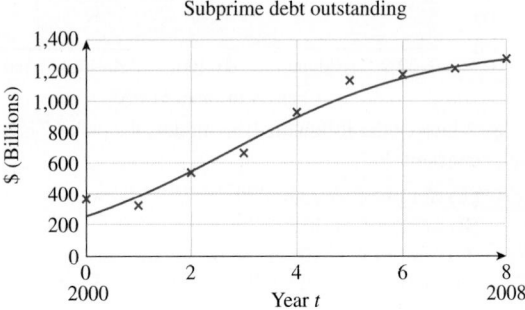

Subprime debt outstanding

How fast, to the nearest $1 billion, was subprime mortgage debt increasing at the start of 2005? How would you check that the answer is approximately correct by looking at the graph? HINT [See Example 3.]

107. *Subprime Mortgages during the Housing Bubble* (Compare Exercise 105.) During the real estate run-up in 2000–2008, the percentage of mortgages issued in the U.S. that were subprime (normally classified as risky) could be approximated by

$$A(t) = \frac{15.0}{1 + 8.6(1.8)^{-t}} \text{ percent} \quad (0 \le t \le 8)$$

t years after the start of 2000.[49]
  a. How fast, to the nearest 0.1%, was the percentage increasing at the start of 2003?
  b. Compute $\lim_{t \to +\infty} A(t)$ and $\lim_{t \to +\infty} A'(t)$. What do the answers tell you about subprime mortgages?

108. *Subprime Mortgage Debt during the Housing Bubble* (Compare Exercise 106.) During the real estate run-up in 2000–2008, the value of subprime (normally classified as risky) mortgage debt outstanding in the U.S. could be approximated by

$$A(t) = \frac{1,350}{1 + 4.2(1.7)^{-t}} \text{ billion dollars} \quad (0 \le t \le 8)$$

t years after the start of 2000.[50]
  a. How fast, to the nearest $1 billion, was subprime mortgage debt increasing at the start of 2005?
  b. Compute $\lim_{t \to +\infty} A(t)$ and $\lim_{t \to +\infty} A'(t)$. What do the answers tell you about subprime mortgages?

---

[47]2009 figure is an estimate. Sources: Mortgage Bankers Association, UBS.

[48]2008–2009 figures are estimates. Source: www.data360.org/dataset.aspx? Data_Set_Id=9549.

[49]*Ibid.*

[50]*Ibid.*

109. ▼ Population Growth The population of Lower Anchovia was 4,000,000 at the start of 2010 and was doubling every 10 years. How fast was it growing per year at the start of 2010? (Round your answer to three significant digits.) HINT [Use the method of Example 2 in Section 9.2 to obtain an exponential model for the population.]

110. ▼ Population Growth The population of Upper Anchovia was 3,000,000 at the start of 2011 and doubling every 7 years. How fast was it growing per year at the start of 2011? (Round your answer to three significant digits.) HINT [Use the method of Example 2 in Section 9.2 to obtain an exponential model for the population.]

111. ▼ Radioactive Decay Plutonium 239 has a half-life of 24,400 years. How fast is a lump of 10 grams decaying after 100 years?

112. ▼ Radioactive Decay Carbon 14 has a half-life of 5,730 years. How fast is a lump of 20 grams decaying after 100 years?

113. ◆ Cellphone Revenues The number of cellphone subscribers in China for the period 2000–2005 was projected to follow the equation[51]

$$N(t) = 39t + 68 \text{ million subscribers}$$

in year t ($t = 0$ represents 2000). The average annual revenue per cellphone user was \$350 in 2000. Assuming that, due to competition, the revenue per cellphone user decreases continuously at an annual rate of 10%, give a formula for the annual revenue in year t. Hence, project the annual revenue and its rate of change in 2002. Round all answers to the nearest billion dollars or billion dollars per year.

114. ◆ Cellphone Revenues The annual revenue for cellphone use in China for the period 2000–2005 was projected to follow the equation[52]

$$R(t) = 14t + 24 \text{ billion dollars}$$

in year t ($t = 0$ represents 2000). At the same time, there were approximately 68 million subscribers in 2000. Assuming that the number of subscribers increases continuously at an annual rate of 10%, give a formula for the annual revenue per subscriber in year t. Hence, project to the nearest dollar the annual revenue per subscriber and its rate of change in 2002. (Be careful with units!)

---

[51]Based on a regression of projected figures (coefficients are rounded). Source: Intrinsic Technology/*New York Times*, Nov. 24, 2000, p. C1.

[52]Not allowing for discounting due to increased competition. Source: Ibid.

COMMUNICATION AND REASONING EXERCISES

115. Complete the following: The derivative of e raised to a glob is . . . .

116. Complete the following: The derivative of the natural logarithm of a glob is . . . .

117. Complete the following: The derivative of 2 raised to a glob is . . . .

118. Complete the following: The derivative of the base 2 logarithm of a glob is . . . .

119. What is wrong with the following?

$$\frac{d}{dx} \ln |3x + 1| = \frac{3}{|3x + 1|} \qquad \text{✗ WRONG!}$$

120. What is wrong with the following?

$$\frac{d}{dx} 2^{2x} = (2)2^{2x} \qquad \text{✗ WRONG!}$$

121. What is wrong with the following?

$$\frac{d}{dx} 3^{2x} = (2x)3^{2x-1} \qquad \text{✗ WRONG!}$$

122. What is wrong with the following?

$$\frac{d}{dx} \ln(3x^2 - 1) = \frac{1}{6x} \qquad \text{✗ WRONG!}$$

123. ▼ The number N of music downloads on campus is growing exponentially with time. Can $N'(t)$ grow linearly with time? Explain.

124. ▼ The number N of graphing calculators sold on campus is decaying exponentially with time. Can $N'(t)$ grow with time? Explain.

*The **percentage rate of change** or **fractional rate of change** of a function is defined to be the ratio $f'(x)/f(x)$. (It is customary to express this as a percentage when speaking about percentage rate of change.)*

125. ◆ Show that the fractional rate of change of the exponential function e^{kx} is equal to k, which is often called its **fractional growth rate**.

126. ◆ Show that the fractional rate of change of $f(x)$ is the rate of change of $\ln(f(x))$.

127. ◆ Let $A(t)$ represent a quantity growing exponentially. Show that the percentage rate of change, $A'(t)/A(t)$, is constant.

128. ◆ Let $A(t)$ be the amount of money in an account that pays interest that is compounded some number of times per year. Show that the percentage rate of growth, $A'(t)/A(t)$, is constant. What might this constant represent?

11.6 Implicit Differentiation

Consider the equation $y^5 + y + x = 0$, whose graph is shown in Figure 10.

How did we obtain this graph? We did not solve for y as a function of x; that is impossible. In fact, we solved for x in terms of y to find points to plot. Nonetheless, the graph in Figure 10 is the graph of a function because it passes the vertical line test: Every vertical line crosses the graph no more than once, so for each value of x there is no more than one corresponding value of y. Because we cannot solve for y explicitly in terms of x, we say that the equation $y^5 + y + x = 0$ determines y as an **implicit function** of x.

Now, suppose we want to find the slope of the tangent line to this curve at, say, the point $(2, -1)$ (which, you should check, is a point on the curve). In the following example we find, surprisingly, that it is possible to obtain a formula for dy/dx without having to first solve the equation for y.

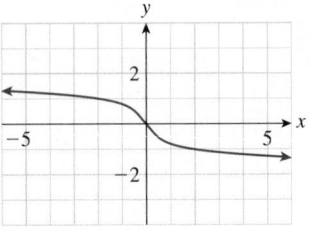

Figure 10

EXAMPLE 1 Implicit Differentiation

Find $\dfrac{dy}{dx}$, given that $y^5 + y + x = 0$.

Solution We use the chain rule and a little cleverness. Think of y as a function of x and take the derivative with respect to x of both sides of the equation:

$$y^5 + y + x = 0 \qquad \text{Original equation}$$

$$\frac{d}{dx}[y^5 + y + x] = \frac{d}{dx}[0] \qquad \text{Derivative with respect to } x \text{ of both sides}$$

$$\frac{d}{dx}[y^5] + \frac{d}{dx}[y] + \frac{d}{dx}[x] = 0. \qquad \text{Derivative rules}$$

Now we must be careful. The derivative *with respect to* x of y^5 is *not* $5y^4$. Rather, because y is a function of x, we must use the chain rule, which tells us that

$$\frac{d}{dx}[y^5] = 5y^4 \frac{dy}{dx}.$$

Thus, we get

$$5y^4 \frac{dy}{dx} + \frac{dy}{dx} + 1 = 0.$$

We want to find dy/dx, so we *solve for it*:

$$(5y^4 + 1)\frac{dy}{dx} = -1 \qquad \text{Isolate } dy/dx \text{ on one side.}$$

$$\frac{dy}{dx} = -\frac{1}{5y^4 + 1}. \qquad \text{Divide both sides by } 5y^4 + 1.$$

➡ **Before we go on...** Note that we should not expect to obtain dy/dx as an explicit function of x if y was not an explicit function of x to begin with. For example, the formula we found for dy/dx in Example 1 is not a function of x because there is a y in

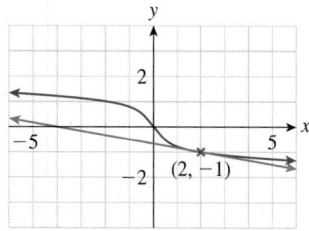

Figure 11

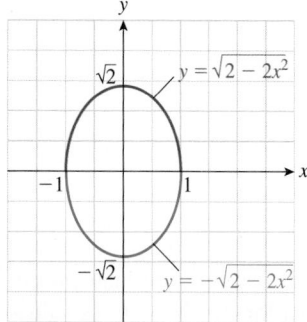

Figure 12

it. However, the result is still useful because we can evaluate the derivative at any point on the graph. For instance, at the point $(2, -1)$ on the graph, we get

$$\frac{dy}{dx} = -\frac{1}{5y^4 + 1} = -\frac{1}{5(-1)^4 + 1} = -\frac{1}{6}.$$

Thus, the slope of the tangent line to the curve $y^5 + y + x = 0$ at the point $(2, -1)$ is $-1/6$. Figure 11 shows the graph and this tangent line. ■

This procedure we just used—differentiating an equation to find dy/dx without first solving the equation for y—is called **implicit differentiation**.

In Example 1 we were given an equation in x and y that determined y as an (implicit) function of x, even though we could not solve for y. But an equation in x and y need not always determine y as a function of x. Consider, for example, the equation

$$2x^2 + y^2 = 2.$$

Solving for y yields $y = \pm\sqrt{2 - 2x^2}$. The $\pm$ sign reminds us that for some values of x there are two corresponding values for y. We can graph this equation by superimposing the graphs of

$$y = \sqrt{2 - 2x^2} \quad \text{and} \quad y = -\sqrt{2 - 2x^2}.$$

The graph, an *ellipse*, is shown in Figure 12.

The graph of $y = \sqrt{2 - 2x^2}$ constitutes the top half of the ellipse, and the graph of $y = -\sqrt{2 - 2x^2}$ constitutes the bottom half.

EXAMPLE 2 Slope of Tangent Line

Refer to Figure 12. Find the slope of the tangent line to the ellipse $2x^2 + y^2 = 2$ at the point $(1/\sqrt{2}, 1)$.

Solution Because $(1/\sqrt{2}, 1)$ is on the top half of the ellipse in Figure 12, we *could* differentiate the function $y = \sqrt{2 - 2x^2}$ to obtain the result, but it is actually easier to apply implicit differentiation to the original equation.

$$2x^2 + y^2 = 2 \qquad \text{Original equation}$$

$$\frac{d}{dx}[2x^2 + y^2] = \frac{d}{dx}[2] \qquad \text{Derivative with respect to } x \text{ of both sides}$$

$$4x + 2y\frac{dy}{dx} = 0$$

$$2y\frac{dy}{dx} = -4x \qquad \text{Solve for } dy/dx.$$

$$\frac{dy}{dx} = -\frac{4x}{2y} = -\frac{2x}{y}$$

To find the slope at $(1/\sqrt{2}, 1)$ we now substitute for x and y:

$$\left.\frac{dy}{dx}\right|_{(1/\sqrt{2},1)} = -\frac{2/\sqrt{2}}{1} = -\sqrt{2}.$$

Thus, the slope of the tangent to the ellipse at the point $(1/\sqrt{2}, 1)$ is $-\sqrt{2} \approx -1.414$.

EXAMPLE 3 Tangent Line for an Implicit Function

Find the equation of the tangent line to the curve $\ln y = xy$ at the point where $y = 1$.

Solution   First, we use implicit differentiation to find dy/dx:

$$\frac{d}{dx}[\ln y] = \frac{d}{dx}[xy] \qquad \text{Take } d/dx \text{ of both sides.}$$

$$\frac{1}{y}\frac{dy}{dx} = (1)y + x\frac{dy}{dx}. \qquad \text{Chain rule on left, product rule on right}$$

To solve for dy/dx, we bring all the terms containing dy/dx to the left-hand side and all terms not containing it to the right-hand side:

$$\frac{1}{y}\frac{dy}{dx} - x\frac{dy}{dx} = y \qquad \text{Bring the terms with } dy/dx \text{ to the left.}$$

$$\frac{dy}{dx}\left(\frac{1}{y} - x\right) = y \qquad \text{Factor out } dy/dx.$$

$$\frac{dy}{dx}\left(\frac{1 - xy}{y}\right) = y$$

$$\frac{dy}{dx} = y\left(\frac{y}{1 - xy}\right) = \frac{y^2}{1 - xy}. \qquad \text{Solve for } dy/dx.$$

The derivative gives the slope of the tangent line, so we want to evaluate the derivative at the point where $y = 1$. However, the formula for dy/dx requires values for both x and y. We get the value of x by substituting $y = 1$ in the original equation:

$$\ln y = xy$$

$$\ln 1 = x \cdot 1$$

But $\ln 1 = 0$, and so $x = 0$ for this point. Thus,

$$\left.\frac{dy}{dx}\right|_{(0,1)} = \frac{1^2}{1 - (0)(1)} = 1.$$

Therefore, the tangent line is the line through $(x, y) = (0, 1)$ with slope 1, which is

$$y = x + 1.$$

➡ **Before we go on...** Example 3 presents an instance of an implicit function in which it is simply not possible to solve for y. Try it. ∎

Sometimes, it is easiest to differentiate a complicated function of x by first taking the logarithm and then using implicit differentiation—a technique called **logarithmic differentiation**.

EXAMPLE 4   Logarithmic Differentiation

Find $\dfrac{d}{dx}\left[\dfrac{(x+1)^{10}(x^2+1)^{11}}{(x^3+1)^{12}}\right]$ without using the product or quotient rules.

Solution   Write

$$y = \frac{(x+1)^{10}(x^2+1)^{11}}{(x^3+1)^{12}}$$

and then take the natural logarithm of both sides:

$$\ln y = \ln\left[\frac{(x+1)^{10}(x^2+1)^{11}}{(x^3+1)^{12}}\right].$$

We can use properties of the logarithm to simplify the right-hand side:

$$\ln y = \ln(x+1)^{10} + \ln(x^2+1)^{11} - \ln(x^3+1)^{12}$$

$$= 10\ln(x+1) + 11\ln(x^2+1) - 12\ln(x^3+1).$$

Now we can find $\dfrac{dy}{dx}$ using implicit differentiation:

$$\frac{1}{y}\frac{dy}{dx} = \frac{10}{x+1} + \frac{22x}{x^2+1} - \frac{36x^2}{x^3+1} \qquad \text{Take } d/dx \text{ of both sides.}$$

$$\frac{dy}{dx} = y\left(\frac{10}{x+1} + \frac{22x}{x^2+1} - \frac{36x^2}{x^3+1}\right) \qquad \text{Solve for } dy/dx.$$

$$= \frac{(x+1)^{10}(x^2+1)^{11}}{(x^3+1)^{12}}\left(\frac{10}{x+1} + \frac{22x}{x^2+1} - \frac{36x^2}{x^3+1}\right). \qquad \text{Substitute for } y.$$

➡ **Before we go on...** Redo Example 4 using the product and quotient rules (and the chain rule) instead of logarithmic differentiation and compare the answers. Compare also the amount of work involved in both methods. ■

APPLICATION

Productivity usually depends on both labor and capital. Suppose, for example, you are managing a surfboard manufacturing company. You can measure its productivity by counting the number of surfboards the company makes each year. As a measure of labor, you can use the number of employees, and as a measure of capital you can use its operating budget. The so-called *Cobb-Douglas* model uses a function of the form:

$$P = Kx^a y^{1-a}, \qquad \text{Cobb-Douglas model for productivity}$$

where P stands for the number of surfboards made each year, x is the number of employees, and y is the operating budget. The numbers K and a are constants that depend on the particular situation studied, with a between 0 and 1.

EXAMPLE 5 **Cobb-Douglas Production Function**

The surfboard company you own has the Cobb-Douglas production function

$$P = x^{0.3}y^{0.7}$$

where P is the number of surfboards it produces per year, x is the number of employees, and y is the daily operating budget (in dollars). Assume that the production level P is constant.

a. Find $\dfrac{dy}{dx}$.

b. Evaluate this derivative at $x = 30$ and $y = 10,000$, and interpret the answer.

Solution

a. We are given the equation $P = x^{0.3}y^{0.7}$, in which P is constant. We find $\dfrac{dy}{dx}$ by implicit differentiation.

$$0 = \frac{d}{dx}[x^{0.3}y^{0.7}] \qquad d/dx \text{ of both sides}$$

$$0 = 0.3x^{-0.7}y^{0.7} + x^{0.3}(0.7)y^{-0.3}\frac{dy}{dx} \qquad \text{Product and chain rules}$$

$$-0.7x^{0.3}y^{-0.3}\frac{dy}{dx} = 0.3x^{-0.7}y^{0.7} \qquad \text{Bring term with } dy/dx \text{ to left.}$$

$$\frac{dy}{dx} = -\frac{0.3x^{-0.7}y^{0.7}}{0.7x^{0.3}y^{-0.3}} \qquad \text{Solve for } dy/dx.$$

$$= -\frac{3y}{7x}. \qquad \text{Simplify.}$$

b. Evaluating this derivative at $x = 30$ and $y = 10{,}000$ gives

$$\left.\frac{dy}{dx}\right|_{x=30,\ y=10{,}000} = -\frac{3(10{,}000)}{7(30)} \approx -143.$$

To interpret this result, first look at the units of the derivative: We recall that the units of dy/dx are units of y per unit of x. Because y is the daily budget, its units are dollars; because x is the number of employees, its units are employees. Thus,

$$\left.\frac{dy}{dx}\right|_{x=30,\ y=10{,}000} \approx -\$143 \text{ per employee.}$$

Next, recall that dy/dx measures the rate of change of y as x changes. Because the answer is negative, the daily budget to maintain production at the fixed level is decreasing by approximately $143 per additional employee at an employment level of 30 employees and a daily operating budget of $10,000. In other words, increasing the workforce by one worker will result in a savings of approximately $143 per day. Roughly speaking, *a new employee is worth $143 per day* at the current levels of employment and production.

11.6   EXERCISES

▼ more advanced   ◆ challenging
🔲 indicates exercises that should be solved using technology

In Exercises 1–10, find dy/dx, using implicit differentiation. In each case, compare your answer with the result obtained by first solving for y as a function of x and then taking the derivative. HINT [See Example 1.]

1. $2x + 3y = 7$

2. $4x - 5y = 9$

3. $x^2 - 2y = 6$

4. $3y + x^2 = 5$

5. $2x + 3y = xy$

6. $x - y = xy$

7. $e^x y = 1$

8. $e^x y - y = 2$

9. $y \ln x + y = 2$

10. $\dfrac{\ln x}{y} = 2 - x$

In Exercises 11–30, find the indicated derivative using implicit differentiation. HINT [See Example 1.]

11. $x^2 + y^2 = 5; \dfrac{dy}{dx}$

12. $2x^2 - y^2 = 4; \dfrac{dy}{dx}$

13. $x^2 y - y^2 = 4; \dfrac{dy}{dx}$

14. $xy^2 - y = x; \dfrac{dy}{dx}$

15. $3xy - \dfrac{y}{3} = \dfrac{2}{x}; \dfrac{dy}{dx}$

16. $\dfrac{xy}{2} - y^2 = 3; \dfrac{dy}{dx}$

17. $x^2 - 3y^2 = 8; \dfrac{dx}{dy}$

18. $(xy)^2 + y^2 = 8; \dfrac{dx}{dy}$

19. $p^2 - pq = 5p^2q^2; \dfrac{dp}{dq}$

20. $q^2 - pq = 5p^2q^2; \dfrac{dp}{dq}$

21. $xe^y - ye^x = 1; \dfrac{dy}{dx}$

22. $x^2e^y - y^2 = e^x; \dfrac{dy}{dx}$

23. ▼ $e^{st} = s^2; \dfrac{ds}{dt}$

24. ▼ $e^{s^2t} - st = 1; \dfrac{ds}{dt}$

25. ▼ $\dfrac{e^x}{y^2} = 1 + e^y; \dfrac{dy}{dx}$

26. ▼ $\dfrac{x}{e^y} + xy = 9y; \dfrac{dy}{dx}$

27. ▼ $\ln(y^2 - y) + x = y; \dfrac{dy}{dx}$

28. ▼ $\ln(xy) - x \ln y = y; \dfrac{dy}{dx}$

29. ▼ $\ln(xy + y^2) = e^y; \dfrac{dy}{dx}$

30. ▼ $\ln(1 + e^{xy}) = y; \dfrac{dy}{dx}$

In Exercises 31–42, use implicit differentiation to find (a) the slope of the tangent line, and (b) the equation of the tangent line at the indicated point on the graph. (Round answers to four decimal places as needed.) If only the x-coordinate is given, you must also find the y-coordinate. HINT [See Examples 2, 3.]

31. $4x^2 + 2y^2 = 12, (1, -2)$

32. $3x^2 - y^2 = 11, (-2, 1)$

33. $2x^2 - y^2 = xy, (-1, 2)$

34. $2x^2 + xy = 3y^2, (-1, -1)$

35. $x^2y - y^2 + x = 1, (1, 0)$

36. $(xy)^2 + xy - x = 8, (-8, 0)$

37. $xy - 2000 = y, x = 2$

38. $x^2 - 10xy = 200, x = 10$

39. ▼ $\ln(x + y) - x = 3x^2, x = 0$

40. ▼ $\ln(x - y) + 1 = 3x^2, x = 0$

41. ▼ $e^{xy} - x = 4x, x = 3$

42. ▼ $e^{-xy} + 2x = 1, x = -1$

In Exercises 43–52, use logarithmic differentiation to find dy/dx. Do not simplify the result. HINT [See Example 4.]

43. $y = \dfrac{2x + 1}{4x - 2}$

44. $y = (3x + 2)(8x - 5)$

45. $y = \dfrac{(3x + 1)^2}{4x(2x - 1)^3}$

46. $y = \dfrac{x^2(3x + 1)^2}{(2x - 1)^3}$

47. $y = (8x - 1)^{1/3}(x - 1)$

48. $y = \dfrac{(3x + 2)^{2/3}}{3x - 1}$

49. $y = (x^3 + x)\sqrt{x^3 + 2}$

50. $y = \sqrt{\dfrac{x - 1}{x^2 + 2}}$

51. ▼ $y = x^x$

52. ▼ $y = x^{-x}$

APPLICATIONS

53. Productivity The number of CDs per hour that *Snappy Hardware* can manufacture at its plant is given by
$$P = x^{0.6}y^{0.4},$$
where x is the number of workers at the plant and y is the monthly budget (in dollars). Assume P is constant, and compute $\dfrac{dy}{dx}$ when $x = 100$ and $y = 200,000$. Interpret the result. HINT [See Example 5.]

54. Productivity The number of cellphone accessory kits (neon lights, matching covers, and earpods) per day that *USA Cellular Makeover Inc,* can manufacture at its plant in Cambodia is given by
$$P = x^{0.5}y^{0.5},$$
where x is the number of workers at the plant and y is the monthly budget (in dollars). Assume P is constant, and compute $\dfrac{dy}{dx}$ when $x = 200$ and $y = 100,000$. Interpret the result. HINT [See Example 5.]

55. Demand The demand equation for soccer tournament T-shirts is
$$xy - 2,000 = y,$$
where y is the number of T-shirts the Enormous State University soccer team can sell at a price of $\$x$ per shirt. Find $\dfrac{dy}{dx}\bigg|_{x=5}$, and interpret the result.

56. Cost Equations The cost y (in cents) of producing x gallons of *Ectoplasm* hair gel is given by the cost equation
$$y^2 - 10xy = 200.$$
Evaluate $\dfrac{dy}{dx}$ at $x = 1$ and interpret the result.

57. Housing Costs[53] The cost C (in dollars) of building a house is related to the number k of carpenters used and the number e of electricians used by the formula
$$C = 15,000 + 50k^2 + 60e^2.$$
If the cost of the house is fixed at $\$200,000$, find $\dfrac{dk}{de}\bigg|_{e=15}$ and interpret your result.

58. Employment An employment research company estimates that the value of a recent MBA graduate to an accounting company is
$$V = 3e^2 + 5g^3,$$
where V is the value of the graduate, e is the number of years of prior business experience, and g is the graduate school grade-point average. If V is fixed at 200, find $\dfrac{de}{dg}$ when $g = 3.0$ and interpret the result.

59. ▼ **Grades**[54] A productivity formula for a student's performance on a difficult English examination is
$$g = 4tx - 0.2t^2 - 10x^2 \quad (t < 30),$$
where g is the score the student can expect to obtain, t is the number of hours of study for the examination, and x is the student's grade-point average.

a. For how long should a student with a 3.0 grade-point average study in order to score 80 on the examination?

b. Find $\dfrac{dt}{dx}$ for a student who earns a score of 80, evaluate it when $x = 3.0$, and interpret the result.

60. ▼ **Grades** Repeat the preceding exercise using the following productivity formula for a basket-weaving examination:
$$g = 10tx - 0.2t^2 - 10x^2 \quad (t < 10).$$
Comment on the result.

[53]Based on an exercise in *Introduction to Mathematical Economics* by A. L. Ostrosky Jr., and J. V. Koch (Waveland Press, Springfield, Illinois, 1979).

[54]*Ibid.*

Exercises 61 and 62 are based on the following demand function for money (taken from a question on the GRE Economics Test):

$$M_d = (2) \times (y)^{0.6} \times (r)^{-0.3} \times (p)$$

where

M_d = *demand for nominal money balances (money stock)*

y = *real income*

r = *an index of interest rates*

p = *an index of prices.*

61. ◆ ***Money Stock*** If real income grows while the money stock and the price level remain constant, the interest rate must change at what rate? (First find dr/dy, then dr/dt; your answers will be expressed in terms of r, y, and $\dfrac{dy}{dt}$.)

62. ◆ ***Money Stock*** If real income grows while the money stock and the interest rate remain constant, the price level must change at what rate?

COMMUNICATION AND REASONING EXERCISES

63. Fill in the missing terms: The equation $x = y^3 + y - 3$ specifies ___ as a function of ___, and ___ as an implicit function of ___.

64. Fill in the missing terms: When $x \neq 0$ in the equation $xy = x^3 + 4$, it is possible to specify ___ as a function of ___. However, ___ is only an implicit function of ___.

65. ▼ Use logarithmic differentiation to give another proof of the product rule.

66. ▼ Use logarithmic differentiation to give a proof of the quotient rule.

67. ▼ If y is given explicitly as a function of x by an equation $y = f(x)$, compare finding dy/dx by implicit differentiation to finding it explicitly in the usual way.

68. ▼ Explain why one should not expect dy/dx to be a function of x if y is not a function of x.

69. ◆ If y is a function of x and $dy/dx \neq 0$ at some point, regard x as an implicit function of y and use implicit differentiation to obtain the equation

$$\frac{dx}{dy} = \frac{1}{dy/dx}.$$

70. ◆ If you are given an equation in x and y such that dy/dx is a function of x only, what can you say about the graph of the equation?

Q : *Then what is the point of using any model to project in the first place?*

A : Projections are always tricky as we cannot foresee the future. But a *good* model is not merely one that seems to fit the data well, but rather a model whose structure is based on the situation being modeled. For instance, a *good* model of student graduation rates should take into account such factors as the birth rate, current school populations at all levels, and the relative popularity of private schools as opposed to public schools. It is by using models of this kind that the National Center for Educational Statistics is able to make the projections shown in the data above.

EXERCISES

1. In 1994 there were 246,000 private high school graduates. What do the two logistic models (unshifted and shifted) "predict" for 1994? (Round answer to the nearest 1,000.) Which gives the better prediction?

2. What is the long-term prediction of each of the two models? (Round answer to the nearest 1,000.)

3. Find $\lim\limits_{t \to +\infty} N'(t)$ for both models, and interpret the results.

4. 🖳 You receive a last-minute memo from TJM to the effect that, sorry, the 2011 and 2012 figures are not accurate. Use technology to re-estimate M, A, b, and C for the shifted logistic model in the absence of this data and obtain new estimates for the 2011 and 2012 data. What does the new model predict the rate of change in the number of high school seniors will be in 2015?

5. 🖳 *Another Model* Using the original data, find the best-fit shifted logistic curve of the form

$$N(t) = c + b \frac{a(t - m)}{1 + a|t - m|}. \qquad (a, b, c, m \text{ constant})$$

Its graph is shown below:

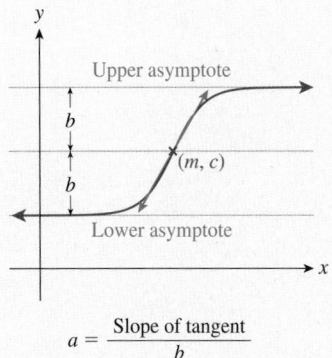

$$a = \frac{\text{Slope of tangent}}{b}$$

(Use the model `$1+$2*$3*(x-$4)/(1+$3*abs(x-$4))` and start with the following values: $a = 0.05$, $b = 160$, $c = 250$, $m = 5$; that is, input `250, 160, 0.05, 5` in the "Guess" field.) Graph the data together with the model. What is SSE? Is the model as accurate a fit as the model used in the text? What does this model predict will be the growth rate of the number of high school graduates in 2015? Comment on the answer. (Round the coefficients in the model and all answers to four decimal places.)

6. 🖳 *Demand for Freon* The demand for chlorofluorocarbon-12 (CFC-12)—the ozone-depleting refrigerant commonly known as Freon 12 (the name given to it by Du Pont)—has

been declining significantly in response to regulation and concern about the ozone layer. The chart below shows the projected demand for CFC-12 for the period 1994–2005.[56]

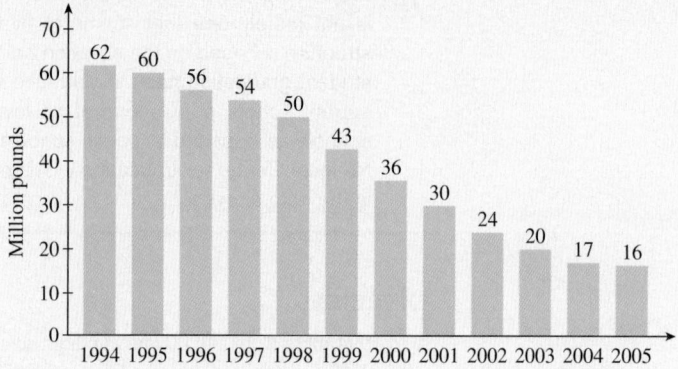

a. Use technology to obtain the best-fit equation of the form

$$N(t) = c + b\frac{a(t - m)}{1 + a|t - m|}, \qquad (a, b, c, m \text{ constant})$$

where t is the number of years since 1990. Use your function to estimate the total demand for CFC-12 from the start of the year 2000 to the start of 2010. [Start with the following values: $a = 1$, $b = -25$, $c = 35$, and $m = 10$, and round your answers to four decimal places.]

b. According to your model, how fast is the demand for Freon declining in 2000?

---

[56]Source: The Automobile Consulting Group (*New York Times,* December 26, 1993, p. F23). The exact figures were not given, and the chart is a reasonable facsimile of the chart that appeared in *New York Times.*

Q: *Are there any other types of relative extrema?*

A: No; relative extrema of a function always occur at critical points or endpoints. (A rigorous proof is beyond the scope of this book.)*

* Here is an outline of the argument. Suppose f has a maximum, say, at $x = a$, at some interior point of its domain. Then either f is differentiable there, or it is not. If it is not, then we have a singular point. If f is differentiable at $x = a$, then consider the slope of the secant line through the points where $x = a$ and $x = a + h$ for small positive h. Because f has a maximum at $x = a$, it is falling (or level) to the right of $x = a$, and so the slope of this secant line must be ≤ 0. Thus, we must have $f'(a) \leq 0$ in the limit as $h \to 0$. On the other hand, if h is small and *negative*, then the corresponding secant line must have slope ≥ 0 because f is also falling (or level) as we move left from $x = a$, and so $f'(a) \geq 0$. Because $f'(a)$ is both ≥ 0 and ≤ 0, it must be zero, and so we have a stationary point at $x = a$.

Locating Candidates for Extrema

If f is a real-valued function, then its extrema occur among the following types of points:

1. **Stationary Points:** f has a stationary point at x if x is in the interior of the domain and $f'(x) = 0$. To locate stationary points, set $f'(x) = 0$ and solve for x.

2. **Singular Points:** f has a singular point at x if x is in the interior of the domain and $f'(x)$ is not defined. To locate singular points, find values of x where $f'(x)$ is *not* defined, but $f(x)$ *is* defined.

3. **Endpoints:** These are the endpoints, if any, of the domain. Recall that closed intervals contain endpoints, but open intervals do not. If the domain of f is an open interval or the whole real line, then there are no endpoints.

Once we have a candidate for an extremum of f, we find the corresponding point (x, y) on the graph of f using $y = f(x)$.

Quick Examples

1. **Stationary Points:** Let $f(x) = x^3 - 12x$. Then to locate the stationary points, set $f'(x) = 0$ and solve for x. This gives $3x^2 - 12 = 0$, so f has stationary points at $x = \pm 2$. The corresponding points on the graph are $(-2, f(-2)) = (-2, 16)$ and $(2, f(2)) = (2, -16)$.

2. **Singular Points:** Let $f(x) = 3(x-1)^{1/3}$. Then $f'(x) = (x-1)^{-2/3} = 1/(x-1)^{2/3}$. $f'(1)$ is not defined, although $f(1)$ *is* defined. Thus, the (only) singular point occurs at $x = 1$. The corresponding point on the graph is $(1, f(1)) = (1, 0)$.

3. **Endpoints:** Let $f(x) = 1/x$, with domain $(-\infty, 0) \cup [1, +\infty)$. Then the only endpoint in the domain of f occurs at $x = 1$. The corresponding point on the graph is $(1, 1)$. The natural domain of $1/x$, on the other hand, has no endpoints.

Remember, though, that the three types of points we identify above are only *candidates* for extrema. It is quite possible, as we shall see, to have a stationary point or a singular point that is neither a maximum nor a minimum. (It is also possible for an endpoint to be neither a maximum nor a minimum, but only in functions whose graphs are rather bizarre—see Exercise 65.)

Now let's look at some examples of finding maxima and minima. In all of these examples, we will use the following procedure: First, we find the derivative, which we examine to find the stationary points and singular points. Next, we make a table listing the x-coordinates of the critical points and endpoints, together with their y-coordinates. We use this table to make a rough sketch of the graph. From the table and rough sketch, we usually have enough data to be able to say where the extreme points are and what kind they are.

EXAMPLE 1 **Maxima and Minima**

Find the relative and absolute maxima and minima of

$$f(x) = x^2 - 2x$$

on the interval $[0, 4]$.

Solution We first calculate $f'(x) = 2x - 2$. We use this derivative to locate the critical points (stationary and singular points).

Stationary Points To locate the stationary points, we solve the equation $f'(x) = 0$, or

$$2x - 2 = 0,$$

getting $x = 1$. The domain of the function is $[0, 4]$, so $x = 1$ is in the interior of the domain. Thus, the only candidate for a stationary relative extremum occurs when $x = 1$.

Singular Points We look for interior points where the derivative is not defined. However, the derivative is $2x - 2$, which is defined for every x. Thus, there are no singular points and hence no candidates for singular relative extrema.

Endpoints The domain is $[0, 4]$, so the endpoints occur when $x = 0$ and $x = 4$.

We record these values of x in a table, together with the corresponding y-coordinates (values of f):

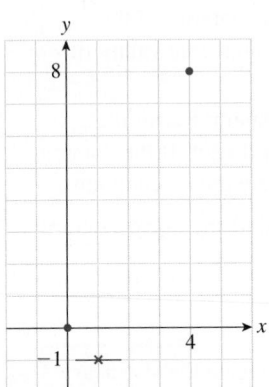

y

8

4

−1

Figure 9

x	0	1	4
$f(x) = x^2 - 2x$	0	−1	8

This gives us three points on the graph, $(0, 0)$, $(1, -1)$, and $(4, 8)$, which we plot in Figure 9. We remind ourselves that the point $(1, -1)$ is a stationary point of the graph by drawing in a part of the horizontal tangent line. Connecting these points must give us a graph something like that in Figure 10.

From Figure 10 we can see that f has the following extrema:

x	$y = x^2 - 2x$	*Classification*
0	0	Relative maximum (endpoint)
1	−1	Absolute minimum (stationary point)
4	8	Absolute maximum (endpoint)

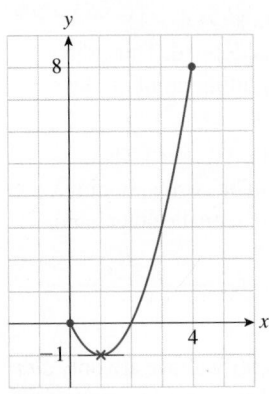

y

8

4

−1

Figure 10

➡ **Before we go on...** A little terminology: If the point (a, b) on the graph of f represents a maximum (or minimum) of f, we will sometimes say that **f has a maximum (or minimum) value of b at $x = a$**, or simply that **f has a maximum (or minimum) at (a, b)**. Thus, in the above example, we could have said the following:

- "f has a relative maximum value of 0 at $x = 0$," or "f has a relative maximum at $(0, 0)$."
- "f has an absolute minimum value of -1 at $x = 1$," or "f has an absolute minimum at $(1, -1)$."
- "f has an absolute maximum value of 8 at $x = 4$," or "f has an absolute maximum at $(4, 8)$."

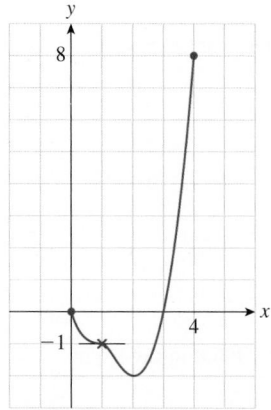

Figure 11

* Why "first" derivative test? To distinguish it from a test based on the **second derivative** of a function, which we shall discuss in Section 12.3.

Q: *How can we be sure that the graph in Example 1 doesn't look like Figure 11?*

A: If it did, there would be another critical point somewhere between $x = 1$ and $x = 4$. But we already know that there aren't any other critical points. The table we made listed all of the possible extrema; there can be no more.

First Derivative Test

The **first derivative test*** gives another, very systematic, way of checking whether a critical point is a maximum or minimum. To motivate the first derivative test, consider again the critical point $x = 1$ in Example 1. If we look at some values of $f'(x)$ to the left and right of the critical point, we obtain the information shown in the following table:

	Point to the Left	Critical Point	Point to the Right
x	0.5	1	2
$f'(x) = 2x - 2$	-1	0	2
Direction of Graph	↘	→	↗

At $x = 0.5$ (to the left of the critical point) we see that $f'(0.5) = -1 < 0$, so the graph has negative slope and f is decreasing. We note this with the downward pointing arrow. At $x = 2$ (to the right of the critical point), we find $f'(2) = 2 > 0$, so the graph has positive slope and f is increasing. In fact, because $f'(x) = 0$ only at $x = 1$, we know that $f'(x) < 0$ for all x in $(0, 1)$, and we can say that f is decreasing on the interval $(0, 1)$. Similarly, f is increasing on $(1, 4)$.

So, starting at $x = 0$, the graph of f goes down until we reach $x = 1$ and then it goes back up, telling us that $x = 1$ must be a minimum. Notice how the minimum is suggested by the arrows to the left and right.

First Derivative Test for Extrema

Suppose that c is a critical point of the continuous function f, and that its derivative is defined for x close to, and on both sides of, $x = c$. Then, determine the sign of the derivative to the left and right of $x = c$.

1. If $f'(x)$ is positive to the left of $x = c$ and negative to the right, then f has a maximum at $x = c$.

2. If $f'(x)$ is negative to the left of $x = c$ and positive to the right, then f has a minimum at $x = c$.

3. If $f'(x)$ has the same sign on both sides of $x = c$, then f has neither a maximum nor a minimum at $x = c$.

Quick Examples

1. In Example 1 above, we saw that $f(x) = x^2 - 2x$ has a critical point at $x = 1$ with $f'(x)$ negative to the left of $x = 1$ and positive to the right (see the table). Therefore, f has a minimum at $x = 1$.

2. Here is a graph showing a function f with a singular point at $x = 1$:

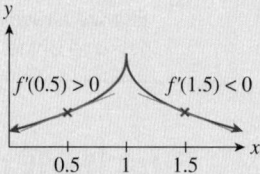

The graph gives us the information shown in the table:

	Point to the Left	Critical Point	Point to the Right
x	0.5	1	1.5
$f'(x)$	$+$	Undefined	$-$
Direction of Graph	↗		↘

Since $f'(x)$ is positive to the left of $x = 1$ and negative to the right, we see that f has a maximum at $x = 1$. (Notice again how this is suggested by the direction of the arrows.)

EXAMPLE 2 **Unbounded Interval**

Find all extrema of $f(x) = 3x^4 - 4x^3$ on $[-1, \infty)$.

Solution We first calculate $f'(x) = 12x^3 - 12x^2$.

   Stationary points We solve the equation $f'(x) = 0$, which is

$$12x^3 - 12x^2 = 0 \text{ or}$$
$$12x^2(x - 1) = 0.$$

There are two solutions, $x = 0$ and $x = 1$, and both are in the domain. These are our candidates for the x-coordinates of stationary extrema.

   Singular points There are no points where $f'(x)$ is not defined, so there are no singular points.

   Endpoints The domain is $[-1, \infty)$, so there is one endpoint, at $x = -1$.

We record these points in a table with the corresponding y-coordinates:

x	-1	0	1
$f(x) = 3x^4 - 4x^3$	7	0	-1

We will illustrate three methods we can use to determine which are minima, which are maxima, and which are neither:

1. Plot these points and sketch the graph by hand.

2. Use the first derivative test.

3. Use technology to help us.

Use the method you find most convenient.

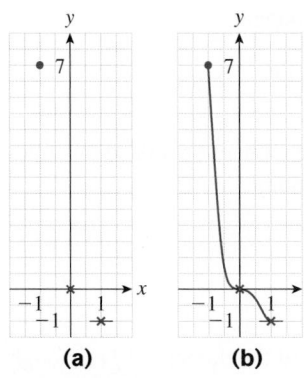

(a)   **(b)**

Figure 12

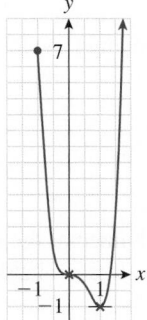

Figure 13

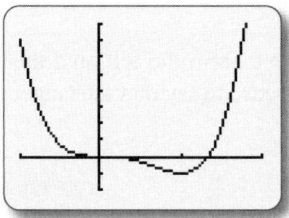

Figure 14

Using a Hand Plot: If we plot these points by hand, we obtain Figure 12(a), which suggests Figure 12(b).

We can't be sure what happens to the right of $x = 1$. Does the curve go up, or does it go down? To find out, let's plot a "test point" to the right of $x = 1$. Choosing $x = 2$, we obtain $y = 3(2)^4 - 4(2)^3 = 16$, so $(2, 16)$ is another point on the graph. Thus, it must turn upward to the right of $x = 1$, as shown in Figure 13.

From the graph, we find that f has the following extrema:

A relative (endpoint) maximum at $(-1, 7)$

An absolute (stationary) minimum at $(1, -1)$

Using the First Derivative Test: List the critical and endpoints in a table, and add additional points as necessary so that each critical point has a noncritical point on either side. Then compute the derivative at each of these points, and draw an arrow to indicate the direction of the graph.

	Endpoint		Critical Point		Critical Point	
x	-1	-0.5	0	0.5	1	2
$f'(x) = 12x^3 - 12x^2$	-24	-1.5	0	-1.5	0	48
Direction of Graph		↘	→	↘	→	↗

Notice that the arrows now suggest the shape of the curve in Figure 13. The first derivative test tells us that the function has a relative maximum at $x = -1$, neither a maximum nor a minimum at $x = 0$, and a relative minimum at $x = 1$. Deciding which of these extrema are absolute and which are relative requires us to compute y-coordinates and plot the corresponding points on the graph by hand, as we did in the first method.

---

using Technology

If we use technology to show the graph, we should choose the viewing window so that it contains the three interesting points we found: $x = -1$, $x = 0$, and $x = 1$. Again, we can't be sure yet what happens to the right of $x = 1$; does the graph go up or down from that point? If we set the viewing window to an interval of $[-1, 2]$ for x and $[-2, 8]$ for y, we will leave enough room to the right of $x = 1$ and below $y = -1$ to see what the graph will do. The result will be something like Figure 14.

Now we can tell what happens to the right of $x = 1$: the function increases. We know that it cannot later decrease again because if it did, there would have to be another critical point where it turns around, and we found that there are no other critical points. ■

➡ **Before we go on...** Notice that the stationary point at $x = 0$ in Example 2 is neither a relative maximum nor a relative minimum. It is simply a place where the graph of f flattens out for a moment before it continues to fall. Notice also that f has no absolute maximum because $f(x)$ increases without bound as x gets large. ■

EXAMPLE 3 Singular Point

Find all extrema of $f(t) = t^{2/3}$ on $[-1, 1]$.

Solution First, $f'(t) = \dfrac{2}{3}t^{-1/3}$.

Stationary points We need to solve

$$\frac{2}{3}t^{-1/3} = 0.$$

We can rewrite this equation without the negative exponent:

$$\frac{2}{3t^{1/3}} = 0.$$

Now, the only way that a fraction can equal 0 is if the numerator is 0, so this fraction can never equal 0. Thus, there are no stationary points.

Singular Points The derivative

$$f'(t) = \frac{2}{3t^{1/3}}$$

is not defined for $t = 0$. However, 0 is in the interior of the domain of f (although f' is not defined at $t = 0$, f itself is). Thus, f has a singular point at $t = 0$.

Endpoints There are two endpoints, -1 and 1.

We now put these three points in a table with the corresponding y-coordinates:

t	-1	0	1
$f(t)$	1	0	1

Using a Hand Plot: The derivative, $f'(t) = 2/(3t^{1/3})$, is not defined at the singular point $t = 0$. To help us sketch the graph, let's use limits to investigate what happens to the derivative as we approach 0 from either side:

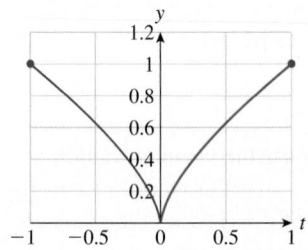

Figure 15

$$\lim_{t \to 0^-} f'(t) = \lim_{t \to 0^-} \frac{2}{3t^{1/3}} = -\infty$$

$$\lim_{t \to 0^+} f'(t) = \lim_{t \to 0^+} \frac{2}{3t^{1/3}} = +\infty.$$

Thus, the graph decreases very steeply, approaching $t = 0$ from the left, and then rises very steeply as it leaves to the right. It would make sense to say that the tangent line at $x = 0$ is vertical, as seen in Figure 15.

From this graph, we find the following extrema for f:

An absolute (endpoint) maximum at $(-1, 1)$

An absolute (singular) minimum at $(0, 0)$

An absolute (endpoint) maximum at $(1, 1)$.

Notice that the absolute maximum value of f is achieved at two values of t: $t = -1$ and $t = 1$.

First Derivative Test: Here is the corresponding table for the first derivative test.

	t	-0.5	0	0.5
$f'(t) = \dfrac{2}{3t^{1/3}}$		$-\dfrac{2}{3(0.5)^{1/3}}$	Undefined	$\dfrac{2}{3(0.5)^{1/3}}$
Direction of Graph		↘	↕	↗

(We drew a vertical arrow at $t = 0$ to indicate a vertical tangent.) Again, notice how the arrows suggest the shape of the curve in Figure 15, and the first derivative test confirms that we have a minimum at $x = 0$.

✱ Many graphing calculators will give you only the right-hand half of the graph shown in Figure 15 because fractional powers of negative numbers are not, in general, real numbers. To obtain the whole curve, enter the formula as `Y=(x^2)^(1/3)`, a fractional power of the non-negative function x^2.

using Technology

Because there is only one critical point, at $t = 0$, it is clear from this table that f must decrease from $t = -1$ to $t = 0$ and then increase from $t = 0$ to $t = 1$. To graph f using technology, choose a viewing window with an interval of $[-1, 1]$ for t and $[0, 1]$ for y. The result will be something like Figure 15.✱ ∎

In Examples 1 and 3, we could have found the absolute maxima and minima without doing any graphing. In Example 1, after finding the critical points and endpoints, we created the following table:

x	0	1	4
$f(x)$	0	-1	8

From this table we can see that f must decrease from its value of 0 at $x = 0$ to -1 at $x = 1$, and then increase to 8 at $x = 4$. The value of 8 must be the largest value it takes on, and the value of -1 must be the smallest, on the interval $[0, 4]$. Similarly, in Example 3 we created the following table:

t	-1	0	1
$f(t)$	1	0	1

From this table we can see that the largest value of f on the interval $[-1, 1]$ is 1 and the smallest value is 0. We are taking advantage of the following fact, the proof of which uses some deep and beautiful mathematics (alas, beyond the scope of this book):

Extreme Value Theorem

If f is *continuous* on a *closed interval* $[a, b]$, then it will have an absolute maximum and an absolute minimum value on that interval. Each absolute extremum must occur at either an endpoint or a critical point. Therefore, the absolute maximum is the largest value in a table of the values of f at the endpoints and critical points, and the absolute minimum is the smallest value.

Quick Example

The function $f(x) = 3x - x^3$ on the interval $[0, 2]$ has one critical point at $x = 1$. The values of f at the critical point and the endpoints of the interval are given in the following table:

	Endpoint	Critical point	Endpoint
x	0	1	2
$f(x)$	0	2	-2

From this table we can say that the absolute maximum value of f on $[0, 2]$ is 2, which occurs at $x = 1$, and the absolute minimum value of f is -2, which occurs at $x = 2$.

As we can see in Example 2 and the following examples, if the domain is not a closed interval, then f may not have an absolute maximum and minimum, and a table of values as above is of little help in determining whether it does.

EXAMPLE 4 **Domain Not a Closed Interval**

Find all extrema of $f(x) = x + \dfrac{1}{x}$.

Solution Because no domain is specified, we take the domain to be as large as possible. The function is not defined at $x = 0$ but is at all other points, so we take its domain to be $(-\infty, 0) \cup (0, +\infty)$. We calculate

$$f'(x) = 1 - \frac{1}{x^2}.$$

Stationary Points Setting $f'(x) = 0$, we solve

$$1 - \frac{1}{x^2} = 0$$

to find $x = \pm 1$. Calculating the corresponding values of f, we get the two stationary points $(1, 2)$ and $(-1, -2)$.

Singular Points The only value of x for which $f'(x)$ is not defined is $x = 0$, but then f is not defined there either, so there are no singular points in the domain.

Endpoints The domain, $(-\infty, 0) \cup (0, +\infty)$, has no endpoints.

From this scant information, it is hard to tell what f does. If we are sketching the graph by hand, or using the first derivative test, we will need to plot additional "test points" to the left and right of the stationary points $x = \pm 1$.

---

 using Technology

For the technology approach, let's choose a viewing window with an interval of $[-3, 3]$ for x and $[-4, 4]$ for y, which should leave plenty of room to see how f behaves near the stationary points. The result is something like Figure 16.
    From this graph we can see that f has:

A relative (stationary) maximum at $(-1, -2)$

A relative (stationary) minimum at $(1, 2)$

Curiously, the relative maximum is lower than the relative minimum! Notice also that, because of the break in the graph at $x = 0$, the graph did not need to rise to get from $(-1, -2)$ to $(1, 2)$. ∎

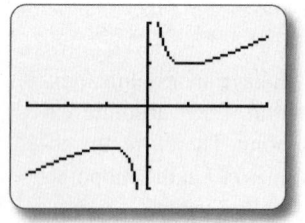

Figure 16

So far we have been solving the equation $f'(x) = 0$ to obtain our candidates for stationary extrema. However, it is often not easy—or even possible—to solve equations analytically. In the next example, we show a way around this problem by using graphing technology.

EXAMPLE 5 ■ **Finding Approximate Extrema Using Technology**

Graph the function $f(x) = (x - 1)^{2/3} - \dfrac{x^2}{2}$ with domain $[-2, +\infty)$. Also graph its derivative and hence locate and classify all extrema of f, with coordinates accurate to two decimal places.

Solution In Example 4 of Section 10.5, we saw how to draw the graphs of *f* and *f'* using technology. Note that the technology formula to use for the graph of *f* is

```
((x-1)^2)^(1/3)-0.5*x^2
```

instead of

```
(x-1)^(2/3)-0.5*x^2
```

(Why?)

Figure 17 shows the resulting graphs of *f* and *f'*.

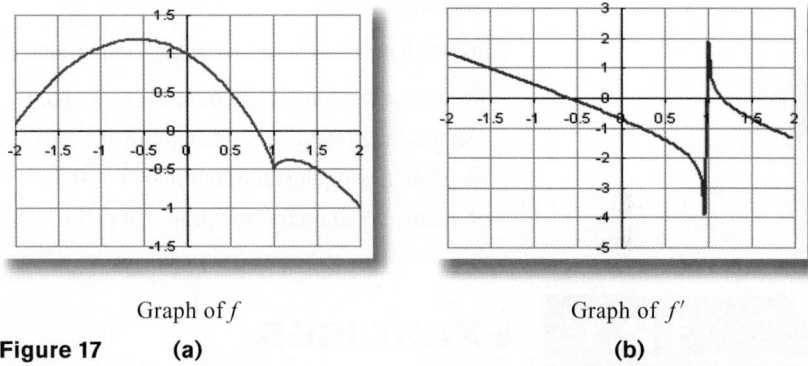

Graph of *f*   Graph of *f'*

Figure 17   **(a)**   **(b)**

If we extend Xmax beyond $x = 2$, we find that the graph continues downward, apparently without any further interesting behavior.

Stationary Points The graph of *f* shows two stationary points, both maxima, at around $x = -0.6$ and $x = 1.2$. Notice that the graph of *f'* is zero at these points. Moreover, it is easier to locate these values accurately on the graph of *f'* because it is easier to pinpoint where a graph crosses the *x*-axis than to locate a stationary point. Zooming in to the stationary point at $x \approx -0.6$ results in Figure 18.

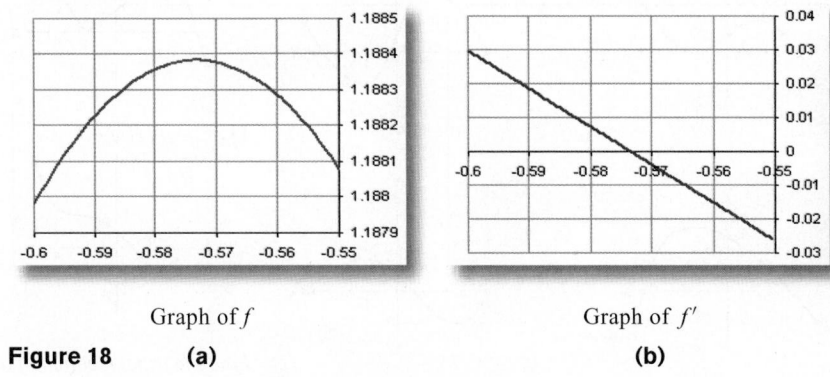

Graph of *f*   Graph of *f'*

Figure 18   **(a)**   **(b)**

From the graph of *f*, we can see that the stationary point is somewhere between -0.58 and -0.57. The graph of *f'* shows more clearly that the zero of *f'*, hence the stationary point of *f* lies somewhat closer to -0.57 than to -0.58. Thus, the stationary point occurs at $x \approx -0.57$, rounded to two decimal places.

In a similar way, we find the second stationary point at $x \approx 1.18$.

Singular Points Going back to Figure 17, we notice what appears to be a cusp (singular point) at the relative minimum around $x = 1$, and this is confirmed by a glance at the graph of f', which seems to take a sudden jump at that value. Zooming in closer suggests that the singular point occurs at exactly $x = 1$. In fact, we can calculate

$$f'(x) = \frac{2}{3(x-1)^{1/3}} - x.$$

From this formula we see clearly that $f'(x)$ is defined everywhere except at $x = 1$.

Endpoints The only endpoint in the domain is $x = -2$, which gives a relative minimum.

Thus, we have found the following approximate extrema for f:

A relative (endpoint) minimum at $(-2, 0.08)$

An absolute (stationary) maximum at $(-0.57, 1.19)$

A relative (singular) minimum at $(1, -0.5)$

A relative (stationary) maximum at $(1.18, -0.38)$.

12.1   **EXERCISES**

In Exercises 1–12, locate and classify all extrema in each graph. (By classifying the extrema, we mean listing whether each extremum is a relative or absolute maximum or minimum.) Also, locate any stationary points or singular points that are not relative extrema. HINT *[See Figure 8.]*

1.
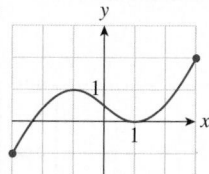

2.
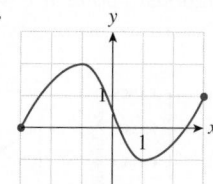

3.
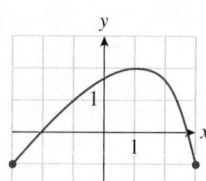

4.
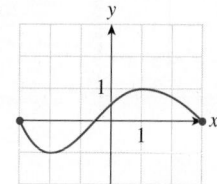

5.
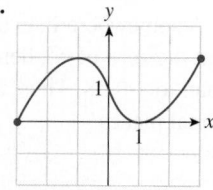

6.
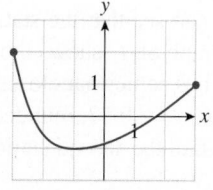

7.
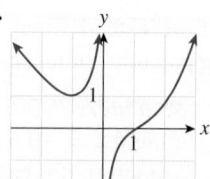

8.
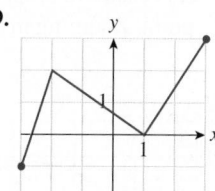

9.
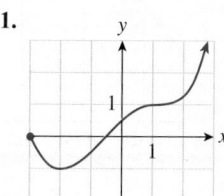

10.
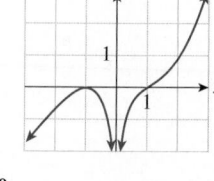

11.
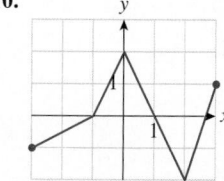

12.
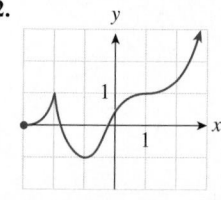

Find the exact location of all the relative and absolute extrema of each function in Exercises 13–44. HINT *[See Example 1.]*

13. $f(x) = x^2 - 4x + 1$ with domain $[0, 3]$

14. $f(x) = 2x^2 - 2x + 3$ with domain $[0, 3]$

15. $g(x) = x^3 - 12x$ with domain $[-4, 4]$

16. $g(x) = 2x^3 - 6x + 3$ with domain $[-2, 2]$

17. $f(t) = t^3 + t$ with domain $[-2, 2]$

18. $f(t) = -2t^3 - 3t$ with domain $[-1, 1]$

19. $h(t) = 2t^3 + 3t^2$ with domain $[-2, +\infty)$ HINT [See Example 2.]

20. $h(t) = t^3 - 3t^2$ with domain $[-1, +\infty)$ HINT [See Example 2.]

21. $f(x) = x^4 - 4x^3$ with domain $[-1, +\infty)$

22. $f(x) = 3x^4 - 2x^3$ with domain $[-1, +\infty)$

23. $g(t) = \frac{1}{4}t^4 - \frac{2}{3}t^3 + \frac{1}{2}t^2$ with domain $(-\infty, +\infty)$

24. $g(t) = 3t^4 - 16t^3 + 24t^2 + 1$ with domain $(-\infty, +\infty)$

25. $h(x) = (x - 1)^{2/3}$ with domain $[0, 2]$ HINT [See Example 3.]

26. $h(x) = (x + 1)^{2/5}$ with domain $[-2, 0]$ HINT [See Example 3.]

27. $k(x) = \frac{2x}{3} + (x + 1)^{2/3}$ with domain $(-\infty, 0]$

28. $k(x) = \frac{2x}{5} - (x - 1)^{2/5}$ with domain $[0, +\infty)$

29. ▼ $f(t) = \frac{t^2 + 1}{t^2 - 1}; -2 \le t \le 2, t \ne \pm 1$

30. ▼ $f(t) = \frac{t^2 - 1}{t^2 + 1}$ with domain $[-2, 2]$

31. ▼ $f(x) = \sqrt{x}(x - 1); x \ge 0$

32. ▼ $f(x) = \sqrt{x}(x + 1); x \ge 0$

33. ▼ $g(x) = x^2 - 4\sqrt{x}$

34. ▼ $g(x) = \frac{1}{x} - \frac{1}{x^2}$

35. ▼ $g(x) = \frac{x^3}{x^2 + 3}$

36. ▼ $g(x) = \frac{x^3}{x^2 - 3}$

37. ▼ $f(x) = x - \ln x$ with domain $(0, +\infty)$

38. ▼ $f(x) = x - \ln x^2$ with domain $(0, +\infty)$

39. ▼ $g(t) = e^t - t$ with domain $[-1, 1]$

40. ▼ $g(t) = e^{-t^2}$ with domain $(-\infty, +\infty)$

41. ▼ $f(x) = \frac{2x^2 - 24}{x + 4}$

42. ▼ $f(x) = \frac{x - 4}{x^2 + 20}$

43. ▼ $f(x) = xe^{1-x^2}$

44. ▼ $f(x) = x \ln x$ with domain $(0, +\infty)$

In Exercises 45–48, use graphing technology and the method in Example 5 to find the x-coordinates of the critical points, accurate to two decimal places. Find all relative and absolute maxima and minima. HINT [See Example 5.]

45. ▯ $y = x^2 + \frac{1}{x - 2}$ with domain $(-3, 2) \cup (2, 6)$

46. ▯ $y = x^2 - 10(x - 1)^{2/3}$ with domain $(-4, 4)$

47. ▯ $f(x) = (x - 5)^2(x + 4)(x - 2)$ with domain $[-5, 6]$

48. ▯ $f(x) = (x + 3)^2(x - 2)^2$ with domain $[-5, 5]$

In Exercises 49–56, the graph of the derivative *of a function f is shown. Determine the x-coordinates of all stationary and singular points of f, and classify each as a relative maximum, relative minimum, or neither. (Assume that f(x) is defined and continuous everywhere in* $[-3, 3]$.) HINT [See Example 5.]

49. ▼
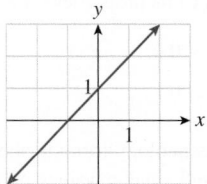

50. ▼
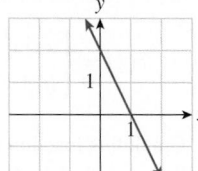

51. ▼
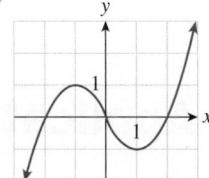

52. ▼
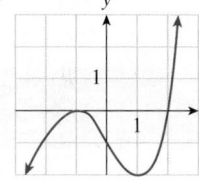

53. ▼
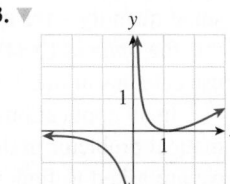

54. ▼
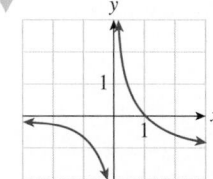

55. ▼
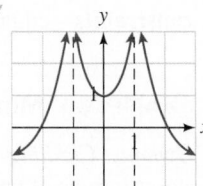

56. ▼
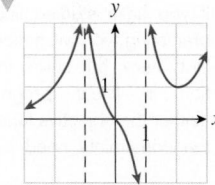

COMMUNICATION AND REASONING EXERCISES

57. Draw the graph of a function f with domain the set of all real numbers, such that f is not linear and has no relative extrema.

58. Draw the graph of a function g with domain the set of all real numbers, such that g has a relative maximum and minimum but no absolute extrema.

59. Draw the graph of a function that has stationary and singular points but no relative extrema.

60. Draw the graph of a function that has relative, not absolute, maxima and minima, but has no stationary or singular points.

61. If a stationary point is not a relative maximum, then must it be a relative minimum? Explain your answer.

62. If one endpoint is a relative maximum, must the other be a relative minimum? Explain your answer.

63. ▼ We said that if f is continuous on a closed interval $[a, b]$, then it will have an absolute maximum and an absolute minimum. Draw the graph of a function with domain $[0, 1]$ having an absolute maximum but no absolute minimum.

64. ▼ Refer to Exercise 63. Draw the graph of a function with domain $[0, 1]$ having no absolute extrema.

65. ⓘ ▼ Must endpoints always be extrema? Consider the following function (based on the trigonometric sine function—see Chapter 16 for a discussion of its properties):

$$f(x) = \begin{cases} x \sin\left(\dfrac{1}{x}\right) & \text{if } x > 0 \\ 0 & \text{if } x = 0 \end{cases}.$$
Technology formula: x*sin(1/x)

Graph this function using the technology formula above for $0 \le x \le h$, choosing smaller and smaller values of h, and decide whether f has a either a relative maximum or relative

minimum at the endpoint $x = 0$. Explain your answer. (Note: Very few graphers can draw this curve accurately; the grapher on the Website does a good job (you can increase the number of points to plot for more beautiful results), the grapher that comes with Mac computers is probably among the best, while the TI-83/84 Plus is probably among the worst.)

66. ⓘ ▼ Refer to the preceding exercise, and consider the function

$$f(x) = \begin{cases} x^2 \sin\left(\dfrac{1}{x}\right) & \text{if } x \ne 0 \\ 0 & \text{if } x = 0 \end{cases}.$$
Technology formula: x^2*sin(1/x)

Graph this function using the technology formula above for $-h \le x \le h$, choosing smaller and smaller values of h, and decide **(a)** whether $x = 0$ is a stationary point, and **(b)** whether f has either a relative maximum or a relative minimum at $x = 0$. Explain your answers. HINT [For part (a), use technology to estimate the derivative at $x = 0$.]

12.2 Applications of Maxima and Minima

In many applications we would like to find the largest or smallest possible value of some quantity—for instance, the greatest possible profit or the lowest cost. We call this the *optimal* (best) value. In this section we consider several such examples and use calculus to find the optimal value in each.

In all applications the first step is to translate a written description into a mathematical problem. In the problems we look at in this section, there are *unknowns* that we are asked to find, there is an expression involving those unknowns that must be made as large or as small as possible—the **objective function**—and there may be **constraints**—equations or inequalities relating the variables.*

* If you have studied linear programming, you will notice a similarity here, but unlike the situation in linear programming, neither the objective function nor the constraints need be linear.

EXAMPLE 1 Minimizing Average Cost

Gymnast Clothing manufactures expensive hockey jerseys for sale to college bookstores in runs of up to 500. Its cost (in dollars) for a run of x hockey jerseys is

$$C(x) = 2,000 + 10x + 0.2x^2.$$

How many jerseys should Gymnast produce per run in order to minimize average cost?[†]

† Why don't we seek to minimize total cost? The answer would be uninteresting; to minimize total cost, we would make *no* jerseys at all. Minimizing the average cost is a more practical objective.

Solution Here is the procedure we will follow to solve problems like this.

1. Identify the unknown(s). There is one unknown: x, the number of hockey jerseys Gymnast should produce per run. (We know this because the question is, How many jerseys . . . ?)

2. Identify the objective function. The objective function is the quantity that must be made as small (in this case) as possible. In this example it is the average cost, which is given by

$$\bar{C}(x) = \frac{C(x)}{x} = \frac{2,000 + 10x + 0.2x^2}{x}$$
$$= \frac{2,000}{x} + 10 + 0.2x \text{ dollars/jersey.}$$

3. ***Identify the constraints (if any).*** At most 500 jerseys can be manufactured in a run. Also, $\bar{C}(0)$ is not defined. Thus, x is constrained by

$$0 < x \le 500.$$

Put another way, the domain of the objective function $\bar{C}(x)$ is (0, 500].

4. ***State and solve the resulting optimization problem.*** Our optimization problem is:

$$\text{Minimize } \bar{C}(x) = \frac{2{,}000}{x} + 10 + 0.2x \qquad \text{Objective function}$$

$$\text{subject to } 0 < x \le 500. \qquad \text{Constraint}$$

We now solve this problem as in Section 12.1. We first calculate

$$\bar{C}'(x) = -\frac{2{,}000}{x^2} + 0.2.$$

We solve $\bar{C}'(x) = 0$ to find $x = \pm 100$. We reject $x = -100$ because -100 is not in the domain of $\bar{C}$ (and makes no sense), so we have one stationary point, at $x = 100$. There, the average cost is $\bar{C}(100) = \$50$ per jersey.

The only point at which the formula for $\bar{C}'$ is not defined is $x = 0$, but that is not in the domain of $\bar{C}$, so we have no singular points. We have one endpoint in the domain, at $x = 500$. There, the average cost is $\bar{C}(500) = \$114$.

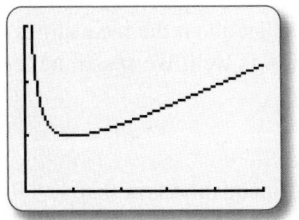

Figure 19

using Technology

Let's plot $\bar{C}$ in a viewing window with the intervals [0, 500] for x and [0, 150] for y, which will show the whole domain and the two interesting points we've found so far. The result is Figure 19.

From the graph of $\bar{C}$, we can see that the stationary point at $x = 100$ gives the absolute minimum. We can therefore say that Gymnast Clothing should produce 100 jerseys per run, for a lowest possible average cost of $\$50$ per jersey. ■

EXAMPLE 2 Maximizing Area

Slim wants to build a rectangular enclosure for his pet rabbit, Killer, against the side of his house, as shown in Figure 20. He has bought 100 feet of fencing. What are the dimensions of the largest area that he can enclose?

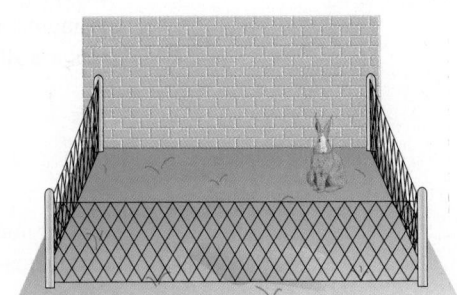

Figure 20

Solution

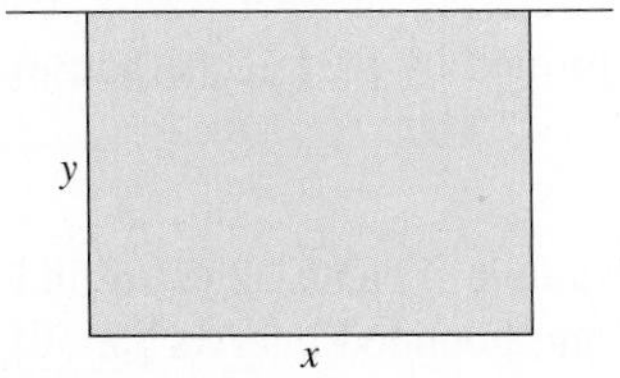

1. ***Identify the unknown(s).*** To identify the unknown(s), we look at the question: What are the *dimensions* of the largest area he can enclose? Thus, the unknowns are the dimensions of the fence. We call these x and y, as shown in Figure 21.

2. ***Identify the objective function.*** We look for what it is that we are trying to maximize (or minimize). The phrase "largest area" tells us that our object is to *maximize the area*, which is the product of length and width, so our objective function is

$$A = xy, \text{ where } A \text{ is the area of the enclosure.}$$

Figure 21

3. *Identify the constraints (if any).* What stops Slim from making the area as large as he wants? He has only 100 feet of fencing to work with. Looking again at Figure 21, we see that the sum of the lengths of the three sides must equal 100, so

$$x + 2y = 100.$$

One more point: Because x and y represent the lengths of the sides of the enclosure, neither can be a negative number.

4. *State and solve the resulting optimization problem.* Our mathematical problem is:

Maximize $A = xy$ Objective function

subject to $x + 2y = 100$, $x \geq 0$, and $y \geq 0$.   Constraints

We know how to find maxima and minima of a function of one variable, but A appears to depend on two variables. We can remedy this by using a constraint to express one variable in terms of the other. Let's take the constraint $x + 2y = 100$ and solve for x in terms of y:

$$x = 100 - 2y.$$

Substituting into the objective function gives

$$A = xy = (100 - 2y)y = 100y - 2y^2$$

and we have eliminated x from the objective function. What about the inequalities? One says that $x \geq 0$, but we want to eliminate x from this as well. We substitute for x again, getting

$$100 - 2y \geq 0.$$

Solving this inequality for y gives $y \leq 50$. The second inequality says that $y \geq 0$. Now, we can restate our problem with x eliminated:

Maximize $A(y) = 100y - 2y^2$ subject to $0 \leq y \leq 50$.

We now proceed with our usual method of solving such problems. We calculate $A'(y) = 100 - 4y$. Solving $100 - 4y = 0$, we get one stationary point at $y = 25$. There, $A(25) = 1,250$. There are no points at which $A'(y)$ is not defined, so there are no singular points. We have two endpoints, at $y = 0$ and $y = 50$. The corresponding areas are $A(0) = 0$ and $A(50) = 0$. We record the three points we found in a table:

y	0	25	50
$A(y)$	0	1,250	0

It's clear now how A must behave: It increases from 0 at $y = 0$ to 1,250 at $y = 25$ and then decreases back to 0 at $y = 50$. Thus, the largest possible value of A is 1,250 square feet, which occurs when $y = 25$. To completely answer the question that was asked, we need to know the corresponding value of x. We have $x = 100 - 2y$, so $x = 50$ when $y = 25$. Thus, Slim should build his enclosure 50 feet across and 25 feet deep (with the "missing" 50-foot side being formed by part of the house).

➡ **Before we go on...** Notice that the problem in Example 2 came down to finding the absolute maximum value of A on the closed and bounded interval $[0, 50]$. As we noted in the preceding section, the table of values of A at its critical points and the endpoints of the interval gives us enough information to find the absolute maximum. ∎

Let's stop for a moment and summarize the steps we've taken in these two examples.

Solving an Optimization Problem

1. **Identify the unknown(s), possibly with the aid of a diagram.** These are usually the quantities asked for in the problem.

2. **Identify the objective function.** This is the quantity you are asked to maximize or minimize. You should name it explicitly, as in "Let S = surface area."

3. **Identify the constraint(s).** These can be equations relating variables or inequalities expressing limitations on the values of variables.

4. **State the optimization problem.** This will have the form "Maximize [minimize] the objective function subject to the constraint(s)."

5. **Eliminate extra variables.** If the objective function depends on several variables, solve the constraint equations to express all variables in terms of one particular variable. Substitute these expressions into the objective function to rewrite it as a function of a single variable. In short, if there is only one constraint equation:

   Solve the constraint for one of the unknowns and substitute into the objective.

   Also substitute the expressions into any inequality constraints to help determine the domain of the objective function.

6. **Find the absolute maximum (or minimum) of the objective function.** Use the techniques of the preceding section.

Now for some further examples.

EXAMPLE 3 Maximizing Revenue

Cozy Carriage Company builds baby strollers. Using market research, the company estimates that if it sets the price of a stroller at p dollars, then it can sell $q = 300,000 - 10p^2$ strollers per year.* What price will bring in the greatest annual revenue?

*This equation is, of course, the demand equation for the baby strollers. However, coming up with a suitable demand equation in real life is hard, to say the least. In this regard, the very entertaining and also insightful article, *Camels and Rubber Duckies* by Joel Spolsky at www.joelonsoftware.com/articles/CamelsandRubberDuckies.html is a must-read.

Solution The question we are asked identifies our main unknown, the price p. However, there is another quantity that we do not know, q, the number of strollers the company will sell per year. The question also identifies the objective function, revenue, which is

$$R = pq.$$

Including the equality constraint given to us, that $q = 300,000 - 10p^2$, and the "reality" inequality constraints $p \geq 0$ and $q \geq 0$, we can write our problem as

Maximize $R = pq$ subject to $q = 300,000 - 10p^2$, $p \geq 0$, and $q \geq 0$.

We are given q in terms of p, so let's substitute to eliminate q:

$$R = pq = p(300,000 - 10p^2) = 300,000p - 10p^3.$$

Substituting in the inequality $q \geq 0$, we get

$$300,000 - 10p^2 \geq 0.$$

Thus, $p^2 \leq 30,000$, which gives $-100\sqrt{3} \leq p \leq 100\sqrt{3}$. When we combine this with $p \geq 0$, we get the following restatement of our problem:

Maximize $R(p) = 300,000p - 10p^3$ such that $0 \leq p \leq 100\sqrt{3}$.

We solve this problem in much the same way we did the preceding one. We calculate $R'(p) = 300{,}000 - 30p^2$. Setting $300{,}000 - 30p^2 = 0$, we find one stationary point at $p = 100$. There are no singular points and we have the endpoints $p = 0$ and $p = 100\sqrt{3}$. Putting these points in a table and computing the corresponding values of R, we get the following:

p	0	100	$100\sqrt{3}$
$R(p)$	0	20,000,000	0

Thus, Cozy Carriage should price its strollers at \$100 each, which will bring in the largest possible revenue of \$20,000,000.

Figure 22

EXAMPLE 4 **Optimizing Resources**

The Metal Can Company has an order to make cylindrical cans with a volume of 250 cubic centimeters. What should be the dimensions of the cans in order to use the least amount of metal in their production?

Solution We are asked to find the dimensions of the cans. It is traditional to take as the dimensions of a cylinder the height h and the radius of the base r, as in Figure 22.

We are also asked to minimize the amount of metal used in the can, which is the area of the surface of the cylinder. We can look up the formula or figure it out ourselves: Imagine removing the circular top and bottom and then cutting vertically and flattening out the hollow cylinder to get a rectangle, as shown in Figure 23.

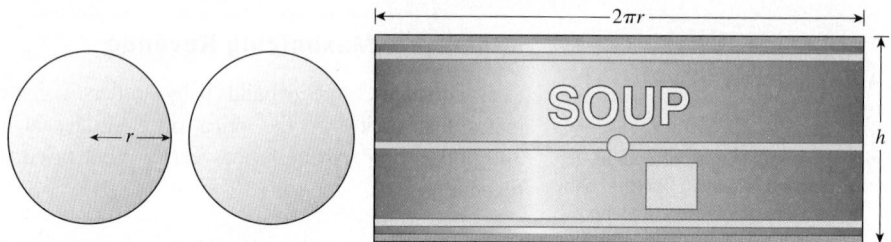

Figure 23

Our objective function is the (total) surface area S of the can. The area of each disc is πr^2, while the area of the rectangular piece is $2\pi rh$. Thus, our objective function is

$$S = 2\pi r^2 + 2\pi rh.$$

As usual, there is a constraint: The volume must be exactly 250 cubic centimeters. The formula for the volume of a cylinder is $V = \pi r^2 h$, so

$$\pi r^2 h = 250.$$

It is easiest to solve this constraint for h in terms of r:

$$h = \frac{250}{\pi r^2}.$$

Substituting in the objective function, we get

$$S = 2\pi r^2 + 2\pi r \frac{250}{\pi r^2} = 2\pi r^2 + \frac{500}{r}.$$

Now r cannot be negative or 0, but it can become very large (a very wide but very short can could have the right volume). We therefore take the domain of $S(r)$ to be $(0, +\infty)$, so our mathematical problem is as follows:

$$\text{Minimize } S(r) = 2\pi r^2 + \frac{500}{r} \text{ subject to } r > 0.$$

Now we calculate

$$S'(r) = 4\pi r - \frac{500}{r^2}.$$

To find stationary points, we set this equal to 0 and solve:

$$4\pi r - \frac{500}{r^2} = 0$$

$$4\pi r = \frac{500}{r^2}$$

$$4\pi r^3 = 500$$

$$r^3 = \frac{125}{\pi}.$$

So

$$r = \sqrt[3]{\frac{125}{\pi}} = \frac{5}{\sqrt[3]{\pi}} \approx 3.41.$$

The corresponding surface area is approximately $S(3.41) \approx 220$. There are no singular points or endpoints in the domain.

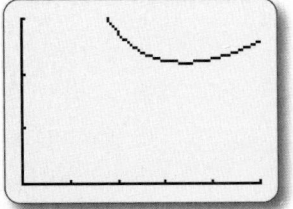

Figure 24

using Technology

To see how S behaves near the one stationary point, let's graph it in a viewing window with interval $[0, 5]$ for r and $[0, 300]$ for S. The result is Figure 24.

From the graph we can clearly see that the smallest surface area occurs at the stationary point at $r \approx 3.41$. The height of the can will be

$$h = \frac{250}{\pi r^2} \approx 6.83.$$

■

Thus, the can that uses the least amount of metal has a height of approximately 6.83 centimeters and a radius of approximately 3.41 centimeters. Such a can will use approximately 220 square centimeters of metal.

➡ **Before we go on...** We obtained the value of r in Example 4 by solving the equation

$$4\pi r = \frac{500}{r^2}.$$

This time, let us do things differently: Divide both sides by 4π to obtain

$$r = \frac{500}{4\pi r^2} = \frac{125}{\pi r^2}$$

and compare what we got with the expression for h:

$$h = \frac{250}{\pi r^2},$$

which we see is exactly twice the expression for r. Put another way, the height is exactly equal to the diameter so that the can looks square when viewed from the side. Have you ever seen cans with that shape? Why do you think most cans do not have this shape? ■

EXAMPLE 5 Allocation of Labor

The Gym Sock Company manufactures cotton athletic socks. Production is partially automated through the use of robots. Daily operating costs amount to $50 per laborer and $30 per robot. The number of pairs of socks the company can manufacture in a day is given by a Cobb-Douglas* production formula

$$q = 50n^{0.6}r^{0.4},$$

where q is the number of pairs of socks that can be manufactured by n laborers and r robots. Assuming that the company wishes to produce 1,000 pairs of socks per day at a minimum cost, how many laborers and how many robots should it use?

* Cobb-Douglas production formulas were discussed in Section 11.6.

Solution The unknowns are the number of laborers n and the number of robots r. The objective is to minimize the daily cost:

$$C = 50n + 30r.$$

The constraints are given by the daily quota

$$1,000 = 50n^{0.6}r^{0.4}$$

and the fact that n and r are nonnegative. We solve the constraint equation for one of the variables; let's solve for n:

$$n^{0.6} = \frac{1,000}{50r^{0.4}} = \frac{20}{r^{0.4}}.$$

Taking the $1/0.6$ power of both sides gives

$$n = \left(\frac{20}{r^{0.4}}\right)^{1/0.6} = \frac{20^{1/0.6}}{r^{0.4/0.6}} = \frac{20^{5/3}}{r^{2/3}} \approx \frac{147.36}{r^{2/3}}.$$

Substituting in the objective equation gives us the cost as a function of r:

$$C(r) \approx 50\left(\frac{147.36}{r^{2/3}}\right) + 30r$$

$$= 7,368r^{-2/3} + 30r.$$

The only remaining constraint on r is that $r > 0$. To find the minimum value of $C(r)$, we first take the derivative:

$$C'(r) \approx -4,912r^{-5/3} + 30.$$

Setting this equal to zero, we solve for r:

$$r^{-5/3} \approx 0.006107$$

$$r \approx (0.006107)^{-3/5} \approx 21.3.$$

The corresponding cost is $C(21.3) \approx \$1,600$. There are no singular points or endpoints in the domain of C.

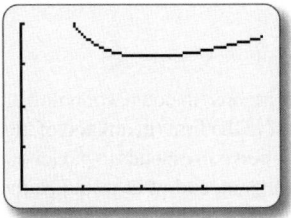

Figure 25

using Technology

To see how C behaves near its stationary point, let's draw its graph in a viewing window with an interval of $[0, 40]$ for r and $[0, 2{,}000]$ for C. The result is Figure 25.

From the graph we can see that C does have its minimum at the stationary point. The corresponding value of n is

$$n \approx \frac{147.36}{r^{2/3}} \approx 19.2.$$

At this point, our solution appears to be this: Use (approximately) 19.2 laborers and (approximately) 21.3 robots to meet the manufacturing quota at a minimum cost. However, we are not interested in fractions of robots or people, so we need to find integer solutions for n and r. If we round these numbers, we get the solution $(n, r) = (19, 21)$. However, a quick calculation shows that

$$q = 50(19)^{0.6}(21)^{0.4} \approx 989 \text{ pairs of socks},$$

which fails to meet the quota of 1,000. Thus, we need to round at least one of the quantities n and r *upward* in order to meet the quota. The three possibilities, with corresponding values of q and C, are as follows:

$$(n, r) = (20, 21), \text{ with } q \approx 1{,}020 \text{ and } C = \$1{,}630$$
$$(n, r) = (19, 22), \text{ with } q \approx 1{,}007 \text{ and } C = \$1{,}610$$
$$(n, r) = (20, 22), \text{ with } q \approx 1{,}039 \text{ and } C = \$1{,}660.$$

Of these, the solution that meets the quota at a minimum cost is $(n, r) = (19, 22)$. Thus, the Gym Sock Co. should use 19 laborers and 22 robots, at a cost of $50 \times 19 + 30 \times 22 = \$1{,}610$, to manufacture $50 \times 19^{0.6} \times 22^{0.4} \approx 1{,}007$ pairs of socks.

FAQs

Constraints and Objectives

Q: *How do I know whether or not there are constraints in an applied optimization problem?*

A: There are usually at least *inequality* constraints; the variables usually represent real quantities, such as length or number of items, and so cannot be negative, leading to constraints like $x \geq 0$ (or $0 \leq x \leq 100$ in the event that there is an upper limit). *Equation* constraints usually arise when there is more than one unknown in the objective, and dictate how one unknown is related to others; as in, say "the length is twice the width," or "the demand is 8 divided by the price" (a demand equation).

Q: *How do I know what to use as the objective, and what to use as the constraint(s)?*

A: To identify the objective, look for a phrase such as "find the maximum (or minimum) value of." The amount you are trying to maximize or minimize is the objective. For example,

- . . . at the least cost. . . The objective function is the equation for cost, $C = \dots$.
- . . . the greatest area. . . The objective function is the equation for area, $A = \dots$.

To determine the constraint *inequalities*, ask yourself what limitations are placed on the unknown variables as above—are they nonnegative? are there upper limits? To identify the constraint *equations*, look for sentences that dictate restrictions in the form of relationships between the variables, as in the answer to the first question above.

12.2   EXERCISES

▼ more advanced   ◆ challenging
🔲 indicates exercises that should be solved using technology

Solve the optimization problems in Exercises 1–8. HINT [See Example 2.]

1. Maximize $P = xy$ subject to $x + y = 10$.

2. Maximize $P = xy$ subject to $x + 2y = 40$.

3. Minimize $S = x + y$ subject to $xy = 9$ and both x and $y > 0$.

4. Minimize $S = x + 2y$ subject to $xy = 2$ and both x and $y > 0$.

5. Minimize $F = x^2 + y^2$ subject to $x + 2y = 10$.

6. Minimize $F = x^2 + y^2$ subject to $xy^2 = 16$.

7. Maximize $P = xyz$ subject to $x + y = 30$, $y + z = 30$, and $x, y, z \geq 0$.

8. Maximize $P = xyz$ subject to $x + z = 12$, $y + z = 12$, and $x, y, z \geq 0$.

9. For a rectangle with perimeter 20 to have the largest area, what dimensions should it have?

10. For a rectangle with area 100 to have the smallest perimeter, what dimensions should it have?

APPLICATIONS

11. *Average Cost: iPods* Assume that it costs Apple approximately
$$C(x) = 22,500 + 100x + 0.01x^2$$
dollars to manufacture x 32GB iPods in a day.[1] How many iPods should be manufactured in order to minimize average cost? What is the resulting average cost of an iPod? (Give your answer to the nearest dollar.) HINT [See Example 1.]

12. *Average Cost: Xboxes* Assume that it costs Microsoft approximately
$$C(x) = 14,400 + 550x + 0.01x^2$$
dollars to manufacture x Xbox 360s in a day.[2] How many Xboxes should be manufactured in order to minimize average cost? What is the resulting average cost of an Xbox? (Give your answer to the nearest dollar.) HINT [See Example 1.]

13. *Pollution Control* The cost of controlling emissions at a firm rises rapidly as the amount of emissions reduced increases. Here is a possible model:
$$C(q) = 4,000 + 100q^2$$

where q is the reduction in emissions (in pounds of pollutant per day) and C is the daily cost to the firm (in dollars) of this reduction. What level of reduction corresponds to the lowest average cost per pound of pollutant, and what would be the resulting average cost to the nearest dollar?

14. *Pollution Control* Repeat the preceding exercise using the following cost function:
$$C(q) = 2,000 + 200q^2.$$

15. *Pollution Control* (Compare Exercise 13.) The cost of controlling emissions at a firm is given by
$$C(q) = 4,000 + 100q^2,$$
where q is the reduction in emissions (in pounds of pollutant per day) and C is the daily cost to the firm (in dollars) of this reduction. Government clean-air subsidies amount to $500 per pound of pollutant removed. How many pounds of pollutant should the firm remove each day in order to minimize *net* cost (cost minus subsidy)?

16. *Pollution Control* (Compare Exercise 14.) Repeat the preceding exercise, using the following cost function:
$$C(q) = 2,000 + 200q^2$$
with government subsidies amounting to $100 per pound of pollutant removed per day.

17. *Fences* I would like to create a rectangular vegetable patch. The fencing for the east and west sides costs $4 per foot, and the fencing for the north and south sides costs only $2 per foot. I have a budget of $80 for the project. What are the dimensions of the vegetable patch with the largest area I can enclose? HINT [See Example 2.]

18. *Fences* I would like to create a rectangular orchid garden that abuts my house so that the house itself forms the northern boundary. The fencing for the southern boundary costs $4 per foot, and the fencing for the east and west sides costs $2 per foot. If I have a budget of $80 for the project, what are the dimensions of the garden with the largest area I can enclose? HINT [See Example 2.]

19. *Fences* You are building a right-angled triangular flower garden along a stream as shown in the figure. (The borders can be in any directions as long as they are at right angles as shown.)

[1] Not the actual cost equation; the authors do not know Apple's actual cost equation. The marginal cost in the model given is in rough agreement with the actual marginal cost for reasonable values of x for one of the 2007 models. Source for cost data: *Manufacturing & Technology News*, July 31, 2007 Volume 14, No. 14 (www.manufacturingnews.com).

[2] Not the actual cost equation; the authors do not know Microsoft's actual cost equation. The marginal cost in the model given is in rough agreement with the actual marginal cost for reasonable values of x. Source for estimate of marginal cost: iSuppli (www.isuppli.com).

The fencing of the left border costs $5 per foot, while the fencing of the lower border costs $1 per foot. (No fencing is required along the river.) You want to spend $100 and enclose as much area as possible. What are the dimensions of your garden, and what area does it enclose? HINT [The area of a right-triangle is given by $A = xy/2$.]

20. *Fences* Repeat Exercise 19, this time assuming that the fencing of the left border costs $8 per foot, while the fencing of the lower border costs $2 per foot, and that you can spend $400.

21. ▼ *Fences* (Compare Exercise 17.) For tax reasons, I need to create a rectangular vegetable patch with an area of exactly 242 sq. ft. The fencing for the east and west sides costs $4 per foot, and the fencing for the north and south sides costs only $2 per foot. What are the dimensions of the vegetable patch with the least expensive fence? HINT [Compare Exercise 3.]

22. ▼ *Fences* (Compare Exercise 18.) For reasons too complicated to explain, I need to create a rectangular orchid garden with an area of exactly 324 sq. ft. abutting my house so that the house itself forms the northern boundary. The fencing for the southern boundary costs $4 per foot, and the fencing for the east and west sides costs $2 per foot. What are the dimensions of the orchid garden with the least expensive fence? HINT [Compare Exercise 4.]

23. *Revenue* Hercules Films is deciding on the price of the video release of its film *Son of Frankenstein*. Its marketing people estimate that at a price of p dollars, it can sell a total of $q = 200,000 - 10,000p$ copies. What price will bring in the greatest revenue? HINT [See Example 3.]

24. *Profit* Hercules Films is also deciding on the price of the video release of its film *Bride of the Son of Frankenstein*. Again, marketing estimates that at a price of p dollars, it can sell $q = 200,000 - 10,000p$ copies, but each copy costs $4 to make. What price will give the greatest *profit*?

25. *Revenue: Cellphones* Worldwide quarterly sales of Nokia cellphones were approximately $q = -p + 156$ million phones when the wholesale price was p. At what wholesale price should Nokia have sold its phones to maximize its quarterly revenue? What would have been the resulting revenue?[3]

26. *Revenue: Cellphones* Worldwide annual sales of all cellphones were approximately $-10p + 1,600$ million phones when the wholesale price was p. At what wholesale price should cellphones have been sold to maximize annual revenue? What would have been the resulting revenue?[4]

27. *Revenue: Monorail Service* The demand, in rides per day, for monorail service in Las Vegas in 2005 can be approximated by $q = -4,500p + 41,500$ when the fare was p. What price should have been charged to maximize total revenue?[5]

28. *Revenue: Mars Monorail* The demand, in rides per day, for monorail service in the three urbynes (or districts) of Utarek, Mars, can be approximated by $q = -2p + 24$ million riders when the fare is $\overline{\overline{Z}}p$. What price should be charged to maximize total revenue?[6]

29. *Revenue* Assume that the demand for tuna in a small coastal town is given by

$$p = \frac{500,000}{q^{1.5}},$$

where q is the number of pounds of tuna that can be sold in a month at p dollars per pound. Assume that the town's fishery wishes to sell at least 5,000 pounds of tuna per month.

a. How much should the town's fishery charge for tuna in order to maximize monthly revenue? HINT [See Example 3, and don't neglect endpoints.]

b. How much tuna will it sell per month at that price?

c. What will be its resulting revenue?

30. *Revenue* In the 1930s the economist Henry Schultz devised the following demand function for corn:

$$p = \frac{6,570,000}{q^{1.3}},$$

where q is the number of bushels of corn that could be sold at p dollars per bushel in one year.[7] Assume that at least 10,000 bushels of corn per year must be sold.

a. How much should farmers charge per bushel of corn to maximize annual revenue? HINT [See Example 3, and don't neglect endpoints.]

b. How much corn can farmers sell per year at that price?

c. What will be the farmers' resulting revenue?

31. *Revenue* During the 1950s the wholesale price for chicken in the United States fell from 25¢ per pound to 14¢ per pound, while per capita chicken consumption rose from 22 pounds per year to 27.5 pounds per year.[8] Assuming that the demand for chicken depended linearly on the price, what wholesale price for chicken would have maximized revenues for poultry farmers, and what would that revenue have amounted to?

32. *Revenue* Your underground used-book business is booming. Your policy is to sell all used versions of *Calculus and You* at the same price (regardless of condition). When you set the price at $10, sales amounted to 120 volumes during the first week of classes. The following semester, you set the price at $30 and sales dropped to zero. Assuming that the demand for

---

[3] Demand equation based on second- and fourth-quarter sales. Source: Embedded.com/Company reports December, 2004.

[4] Demand equation based on estimated 2004 sales and projected 2008 sales. Source: I-Stat/NDR, December 2004.

[5] Source for ridership data: *New York Times*, February 10, 2007, p. A9.

[6] The zonar ($\overline{\overline{Z}}$) is the official currency in the city-state of Utarek, Mars (formerly www.Marsnext.com, a now extinct virtual society).

[7] Based on data for the period 1915–1929. Source: Henry Schultz (1938), *The Theory and Measurement of Demand,* University of Chicago Press, Chicago.

[8] Data are provided for the years 1951–1958. Source: U.S. Department of Agriculture, *Agricultural Statistics.*

books depends linearly on the price, what price gives you the maximum revenue, and what does that revenue amount to?

33. *Profit: Cellphones* (Compare Exercise 25.) Worldwide quarterly sales of Nokia cellphones were approximately $q = -p + 156$ million phones when the wholesale price was \$$p$. Assuming that it cost Nokia \$40 to manufacture each cellphone, at what wholesale price should Nokia have sold its phones to maximize its quarterly profit? What would have been the resulting profit?[9] (The actual wholesale price was \$105 in the fourth quarter of 2004.) HINT [See Example 3, and recall that Profit = Revenue − Cost.]

34. *Profit: Cellphones* (Compare Exercise 26.) Worldwide annual sales of all cellphones were approximately $-10p + 1,600$ million phones when the wholesale price was \$$p$. Assuming that it costs \$30 to manufacture each cellphone, at what wholesale price should cellphones have been sold to maximize annual profit? What would have been the resulting profit?[10] HINT [See Example 3, and recall that Profit = Revenue − Cost.]

35. ▼ *Profit* The demand equation for your company's virtual reality video headsets is

$$p = \frac{1,000}{q^{0.3}},$$

where q is the total number of headsets that your company can sell in a week at a price of p dollars. The total manufacturing and shipping cost amounts to \$100 per headset.

a. What is the greatest profit your company can make in a week, and how many headsets will your company sell at this level of profit? (Give answers to the nearest whole number.)

b. How much, to the nearest \$1, should your company charge per headset for the maximum profit?

36. ▼ *Profit* Due to sales by a competing company, your company's sales of virtual reality video headsets have dropped, and your financial consultant revises the demand equation to

$$p = \frac{800}{q^{0.35}},$$

where q is the total number of headsets that your company can sell in a week at a price of p dollars. The total manufacturing and shipping cost still amounts to \$100 per headset.

a. What is the greatest profit your company can make in a week, and how many headsets will your company sell at this level of profit? (Give answers to the nearest whole number.)

b. How much, to the nearest \$1, should your company charge per headset for the maximum profit?

37. *Paint Cans* A company manufactures cylindrical paint cans with open tops with a volume of 27,000 cubic centimeters. What should be the dimensions of the cans in order to use the least amount of metal in their production? HINT [See Example 4.]

38. *Metal Drums* A company manufactures cylindrical metal drums with open tops with a volume of 1 cubic meter. What should be the dimensions of the drum in order to use the least amount of metal in their production? HINT [See Example 4.]

39. *Tin Cans* A company manufactures cylindrical tin cans with closed tops with a volume of 250 cubic centimeters. The metal used to manufacture the cans costs \$0.01 per square cm for the sides and \$0.02 per square cm for the (thicker) top and bottom. What should be the dimensions of the cans in order to minimize the cost of metal in their production? What is the ratio height/radius? HINT [See Example 4.]

40. *Metal Drums* A company manufactures cylindrical metal drums with open tops with a volume of 2 cubic meters. The metal used to manufacture the cans costs \$2 per square meter for the sides and \$3 per square meter for the (thicker) bottom. What should be the dimensions of the drums in order to minimize the cost of metal in their production? What is the ratio height/radius? HINT [See Example 4.]

41. ▼ *Box Design* Chocolate Box Company is going to make open-topped boxes out of 6 × 16-inch rectangles of cardboard by cutting squares out of the corners and folding up the sides. What is the largest volume box it can make this way?

42. ▼ *Box Design* Vanilla Box Company is going to make open-topped boxes out of 12 × 12-inch rectangles of cardboard by cutting squares out of the corners and folding up the sides. What is the largest volume box it can make this way?

43. ▼ *Box Design* A packaging company is going to make closed boxes, with square bases, that hold 125 cubic centimeters. What are the dimensions of the box that can be built with the least material?

44. ▼ *Box Design* A packaging company is going to make open-topped boxes, with square bases, that hold 108 cubic centimeters. What are the dimensions of the box that can be built with the least material?

45. ▼ *Luggage Dimensions* American Airlines requires that the total outside dimensions (length + width + height) of a checked bag not exceed 62 inches.[11] Suppose you want to check a bag whose height equals its width. What is the largest volume bag of this shape that you can check on an American flight?

46. ▼ *Carry-on Dimensions* American Airlines requires that the total outside dimensions (length + width + height) of a carry-on bag not exceed 45 inches.[12] Suppose you want to carry on a bag whose length is twice its height. What is the largest volume bag of this shape that you can carry on an American flight?

47. ▼ *Luggage Dimensions* Fly-by-Night Airlines has a peculiar rule about luggage: The length and width of a bag must add up to at most 45 inches, and the width and height must also add up to at most 45 inches. What are the dimensions of the bag with the largest volume that Fly-by-Night will accept?

---

[9] See Exercise 25.

[10] See Exercise 26.

[11] According to information on its Web site (www.aa.com/).

[12] *Ibid.*

48. ▼ *Luggage Dimensions Fair Weather Airlines* has a similar rule. It will accept only bags for which the sum of the length and width is at most 36 inches, while the sum of length, height, and twice the width is at most 72 inches. What are the dimensions of the bag with the largest volume that Fair Weather will accept?

49. ▼ *Package Dimensions* The U.S. Postal Service (USPS) will accept packages only if the length plus girth is no more than 108 inches.[13] (See the figure.)

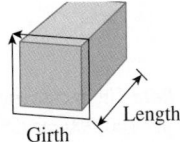

Length

Girth

Assuming that the front face of the package (as shown in the figure) is square, what is the largest volume package that the USPS will accept?

50. ▼ *Package Dimensions* United Parcel Service (UPS) will accept only packages with a length of no more than 108 inches and length plus girth of no more than 165 inches.[14] (See figure for the preceding exercise.) Assuming that the front face of the package (as shown in the figure) is square, what is the largest volume package that UPS will accept?

51. ▼ *Cellphone Revenues* The number of cellphone subscribers in China in the years 2000–2005 was projected to follow the equation $N(t) = 39t + 68$ million subscribers in year t ($t = 0$ represents January 2000). The average annual revenue per cellphone user was $350 in 2000.[15] If we assume that due to competition the revenue per cellphone user decreases continuously at an annual rate of 30%, we can model the annual revenue as

$$R(t) = 350(39t + 68)e^{-0.3t} \text{ million dollars.}$$

Determine **a.** when to the nearest 0.1 year the revenue was projected to peak and **b.** the revenue, to the nearest $1 million, at that time.

52. ▼ *Cellphone Revenues* (Refer to Exercise 51.) If we assume instead that the revenue per cellphone user decreases continuously at an annual rate of 20%, we obtain the revenue model

$$R(t) = 350(39t + 68)e^{-0.2t} \text{ million dollars.}$$

Determine **a.** when to the nearest 0.1 year the revenue was projected to peak and **b.** the revenue, to the nearest $1 million, at that time.

53. ▼ *Research and Development* Spending on research and development by drug companies in the United States t years after 1970 can be modeled by

$$S(t) = 2.5e^{0.08t} \text{ billion dollars} \quad (0 \le t \le 31).$$

The number of new drugs approved by the Federal Drug Administration (FDA) over the same period can be modeled by

$$D(t) = 10 + t \text{ drugs per year}[16] \quad (0 \le t \le 31).$$

When was the function $D(t)/S(t)$ at a maximum? What is the maximum value of $D(t)/S(t)$? What does the answer tell you about the cost of developing new drugs?

54. ▼ *Research and Development* (Refer to Exercise 53.) If the number of new drugs approved by the FDA had been $10 + 2t$ new drugs each year, when would the function $D(t)/S(t)$ have reached a maximum? What does the answer tell you about the cost of developing new drugs?

55. ▼ *Asset Appreciation* As the financial consultant to a classic auto dealership, you estimate that the total value (in dollars) of its collection of 1959 Chevrolets and Fords is given by the formula

$$v = 300{,}000 + 1{,}000t^2 \quad (t \ge 5),$$

where t is the number of years from now. You anticipate a continuous inflation rate of 5% per year, so that the discounted (present) value of an item that will be worth $\$v$ in t years' time is

$$p = ve^{-0.05t}.$$

When would you advise the dealership to sell the vehicles to maximize their discounted value?

56. ▼ *Plantation Management* The value of a fir tree in your plantation increases with the age of the tree according to the formula

$$v = \frac{20t}{1 + 0.05t},$$

where t is the age of the tree in years. Given a continuous inflation rate of 5% per year, the discounted (present) value of a newly planted seedling is

$$p = ve^{-0.05t}.$$

At what age (to the nearest year) should you harvest your trees in order to ensure the greatest possible discounted value?

57. ▼ *Marketing Strategy FeatureRich Software Company* has a dilemma. Its new program, Doors-X 10.27, is almost ready to go on the market. However, the longer the company works on it, the better it can make the program and the more it can charge for it. The company's marketing analysts estimate that if it delays t days, it can set the price at $100 + 2t$ dollars. On the other hand, the longer it delays, the more market share it will lose to its main competitor (see the next exercise) so that if it delays t days it will be able to sell $400{,}000 - 2{,}500t$ copies of

---

[13] The requirement for packages sent other than Parcel Post, as of August 2011 (www.usps.com/).

[14] The requirement as of August, 2011 (www.ups.com/).

[15] Based on a regression of projected figures (coefficients are rounded). Source: Intrinsic Technology/*New York Times*, Nov. 24, 2000, p. C1.

[16] The exponential model for R&D is based on the 1970 and 2001 spending in constant 2001 dollars, while the linear model for new drugs approved is based on the 6-year moving average from data from 1970 to 2000. Source for data: Pharmaceutical Research and Manufacturers of America, FDA/*New York Times*, April 19, 2002, p. C1.

the program. How many days should FeatureRich delay the release in order to get the greatest revenue?

58. ▼ *Marketing Strategy FeatureRich Software's* main competitor (see previous exercise) is Moon Systems, and Moon is in a similar predicament. Its product, Walls-Y 11.4, could be sold now for $200, but for each day Moon delays, it could increase the price by $4. On the other hand, it could sell 300,000 copies now, but each day it waits will cut sales by 1,500. How many days should Moon delay the release in order to get the greatest revenue?

59. ▼ *Average Profit* The *FeatureRich Software Company* sells its graphing program, Dogwood, with a volume discount. If a customer buys x copies, then he or she pays[17] $500\sqrt{x}$. It cost the company $10,000 to develop the program and $2 to manufacture each copy. If a single customer were to buy all the copies of Dogwood, how many copies would the customer have to buy for FeatureRich Software's average profit per copy to be maximized? How are average profit and marginal profit related at this number of copies?

60. ▼ *Average Profit* Repeat the preceding exercise with the charge to the customer $600\sqrt{x}$ and the cost to develop the program $9,000.

61. *Resource Allocation* Your company manufactures automobile alternators, and production is partially automated through the use of robots. Daily operating costs amount to $100 per laborer and $16 per robot. In order to meet production deadlines, the company calculates that the numbers of laborers and robots must satisfy the constraint

$$xy = 10,000,$$

where x is the number of laborers and y is the number of robots. Assuming that the company wishes to meet production deadlines at a minimum cost, how many laborers and how many robots should it use? HINT [See Example 5.]

62. *Resource Allocation* Your company is the largest sock manufacturer in the Solar System, and production is automated through the use of androids and robots. Daily operating costs amount to ₩200 per android and ₩8 per robot.[18] In order to meet production deadlines, the company calculates that the numbers of androids and robots must satisfy the constraint

$$xy = 1,000,000,$$

where x is the number of androids and y is the number of robots. Assuming that the company wishes to meet production deadlines at a minimum cost, how many androids and how many robots should it use? HINT [See Example 5.]

63. ▼ *Resource Allocation* Your automobile assembly plant has a Cobb-Douglas production function given by

$$q = x^{0.4}y^{0.6},$$

where q is the number of automobiles it produces per year, x is the number of employees, and y is the daily operating budget (in dollars). Annual operating costs amount to an average of $20,000 per employee plus the operating budget of $365y$. Assume that you wish to produce 1,000 automobiles per year at a minimum cost. How many employees should you hire? HINT [See Example 5.]

64. ▼ *Resource Allocation* Repeat the preceding exercise using the production formula

$$q = x^{0.5}y^{0.5}.$$

HINT [See Example 5.]

65. ▼ *Incarceration Rate* The incarceration rate (the number of persons in prison per 100,000 residents) in the United States can be approximated by

$$N(t) = 0.04t^3 - 2t^2 + 40t + 460 \quad (0 \le t \le 18)$$

(t is the year since 1990).[19] When, to the nearest year, was the incarceration rate increasing most rapidly? When was it increasing least rapidly? HINT [You are being asked to find the extreme values of the rate of change of the incarceration rate.]

66. ▼ *Prison Population* The prison population in the United States can be approximated by

$$N(t) = 0.02t^3 - 2t^2 + 100t + 1,100 \text{ thousand people}$$
$$(0 \le t \le 18)$$

(t is the year since 1990).[20] When, to the nearest year, was the prison population increasing most rapidly? When was it increasing least rapidly? HINT [You are being asked to find the extreme values of the rate of change of the prison population.]

67. ▼ *Embryo Development* The oxygen consumption of a bird embryo increases from the time the egg is laid through the time the chick hatches. In a typical galliform bird, the oxygen consumption can be approximated by

$$c(t) = -0.065t^3 + 3.4t^2 - 22t + 3.6 \text{ milliliters per day}$$
$$(8 \le t \le 30),$$

where t is the time (in days) since the egg was laid.[21] (An egg will typically hatch at around $t = 28$.) When, to the nearest day, is $c'(t)$ a maximum? What does the answer tell you?

68. ▼ *Embryo Development* The oxygen consumption of a turkey embryo increases from the time the egg is laid through the time the chick hatches. In a brush turkey, the oxygen consumption can be approximated by

$$c(t) = -0.028t^3 + 2.9t^2 - 44t + 95 \text{ milliliters per day}$$
$$(20 \le t \le 50),$$

---

[17] This is similar to the way site licenses have been structured for the program Maple®.

[18] ₩ are Neptunian Standard Solar Units of currency.

[19] Source for data: Sourcebook of Criminal Justice Statistics Online (www.albany.edu/sourcebook).

[20] *Ibid.*

[21] The model approximates graphical data published in the article "The Brush Turkey" by Roger S. Seymour, *Scientific American,* December, 1991, pp. 108–114.

where t is the time (in days) since the egg was laid.[22] (An egg will typically hatch at around $t = 50$.) When, to the nearest day, is $c'(t)$ a maximum? What does the answer tell you?

69. 🔲 ▼ *Subprime Mortgages during the Housing Bubble* During the real estate run-up in 2000–2008, the percentage of mortgages issued in the United States that were subprime (normally classified as risky) could be approximated by

$$A(t) = \frac{15.0}{1 + 8.6(1.8)^{-t}} \text{ percent} \quad (0 \le t \le 8)$$

t years after the start of 2000.[23] Graph the *derivative* of $A(t)$ and determine the year during which this derivative had an absolute maximum and also its value at that point. What does the answer tell you?

70. 🔲 ▼ *Subprime Mortgage Debt during the Housing Bubble* During the real estate run-up in 2000–2008, the value of subprime (normally classified as risky) mortgage debt outstanding in the United States was approximately

$$A(t) = \frac{1,350}{1 + 4.2(1.7)^{-t}} \text{ billion dollars} \quad (0 \le t \le 8)$$

t years after the start of 2000.[24] Graph the *derivative* of $A(t)$ and determine the year during which this derivative had an absolute maximum and also its value at that point. What does the answer tell you?

71. 🔲 ▼ *Asset Appreciation* You manage a small antique company that owns a collection of Louis XVI jewelry boxes. Their value v is increasing according to the formula

$$v = \frac{10,000}{1 + 500e^{-0.5t}},$$

where t is the number of years from now. You anticipate an inflation rate of 5% per year, so that the present value of an item that will be worth $\$v$ in t years' time is given by

$$p = v \cdot (1.05)^{-t}.$$

When (to the nearest year) should you sell the jewelry boxes to maximize their present value? How much (to the nearest constant dollar) will they be worth at that time?

72. 🔲 ▼ *Harvesting Forests* The following equation models the approximate volume in cubic feet of a typical Douglas fir tree of age t years.[25]

$$V = \frac{22,514}{1 + 22,514t^{-2.55}}$$

The lumber will be sold at $10 per cubic foot, and you do not expect the price of lumber to appreciate in the foreseeable future. On the other hand, you anticipate a general inflation

rate of 5% per year, so that the present value of an item that will be worth $\$v$ in t years' time is given by

$$p = v \cdot (1.05)^{-t}.$$

At what age (to the nearest year) should you harvest a Douglas fir tree in order to maximize its present value? How much (to the nearest constant dollar) will a Douglas fir tree be worth at that time?

73. ◆ *Agriculture* The fruit yield per tree in an orchard containing 50 trees is 100 pounds per tree each year. Due to crowding, the yield decreases by 1 pound per season for every additional tree planted. How many additional trees should be planted for a maximum total annual yield?

74. ◆ *Agriculture* Two years ago your orange orchard contained 50 trees and the yield per tree was 75 bags of oranges. Last year you removed 10 of the trees and noticed that the yield per tree increased to 80 bags. Assuming that the yield per tree depends linearly on the number of trees in the orchard, what should you do this year to maximize your total yield?

75. ◆ *Revenue* (based on a question on the GRE Economics Test[26]) If total revenue (TR) is specified by $TR = a + bQ - cQ^2$, where Q is quantity of output and a, b, and c are positive parameters, then TR is maximized for this firm when it produces Q equal to:

(A) $b/2ac$. **(B)** $b/4c$. **(C)** $(a + b)/c$. **(D)** $b/2c$. **(E)** $c/2b$.

76. ◆ *Revenue* (based on a question on the GRE Economics Test) If total demand (Q) is specified by $Q = -aP + b$, where P is unit price and a and b are positive parameters, then total revenue is maximized for this firm when it charges P equal to:

(A) $b/2a$. **(B)** $b/4a$. **(C)** a/b. **(D)** $a/2b$. **(E)** $-b/2a$.

COMMUNICATION AND REASONING EXERCISES

77. You are interested in knowing the height of the tallest condominium complex that meets the city zoning requirements that the height H should not exceed eight times the distance D from the road and that it must provide parking for at least 50 cars. The objective function of the associated optimization problem is then:

(A) H. **(B)** $H - 8D$. **(C)** D. **(D)** $D - 8H$.

One of the constraints is:

(A) $8H = D$. **(B)** $8D = H$.
(C) $H'(D) = 0$. **(D)** $D'(H) = 0$.

78. You are interested in building a condominium complex with a height H of at least 8 times the distance D from the road and parking area of at least 1,000 sq. ft. at the cheapest cost C. The objective function of the associated optimization problem is then:

(A) H. **(B)** D. **(C)** C. **(D)** $H + D - C$.

---

[22] The model approximates graphical data published in the article "The Brush Turkey" by Roger S. Seymour, *Scientific American,* December, 1991, pp. 108–114.

[23] Sources: Mortgage Bankers Association, UBS.

[24] Source: www.data360.org/dataset.aspx?Data_Set_Id=9549.

[25] The model is the authors' and is based on data in *Environmental and Natural Resource Economics* by Tom Tietenberg, third edition (New York: HarperCollins, 1992), p. 282.

[26] Source: GRE Economics Test, by G. Gallagher, G. E. Pollock, W. J. Simeone, G. Yohe (Piscataway, NJ: Research and Education Association, 1989).

One of the constraints is:

(A) $H - 8D = 0.$ (B) $H + D - C = 0.$
(C) $C'(D) = 0.$ (D) $8H = D.$

79. Explain why the following problem is uninteresting: A packaging company wishes to make cardboard boxes with open tops by cutting square pieces from the corners of a square sheet of cardboard and folding up the sides. What is the box with the least surface area it can make this way?

80. Explain why finding the production level that minimizes a cost function is frequently uninteresting. What would a more interesting objective be?

81. Your friend Margo claims that all you have to do to find the absolute maxima and minima in applications is set the derivative equal to zero and solve. "All that other stuff about endpoints and so on is a waste of time just to make life hard for us," according to Margo. Explain why she is wrong, and find at least one exercise in this exercise set to illustrate your point.

82. You are having a hard time persuading your friend Marco that maximizing revenue is not the same as maximizing profit. "How on earth can you expect to obtain the largest profit if you are not taking in the largest revenue?" Explain why he is wrong, and find at least one exercise in this exercise set to illustrate your point.

83. ▼ If demand q decreases as price p increases, what does the minimum value of dq/dp measure?

84. ▼ Explain how you would solve an optimization problem of the following form. Maximize $P = f(x, y, z)$ subject to $z = g(x, y)$ and $y = h(x)$.

12.3 Higher Order Derivatives: Acceleration and Concavity

The **second derivative** is simply the derivative of the derivative function. To explain why we would be interested in such a thing, we start by discussing one of its interpretations.

Acceleration

Recall that if $s(t)$ represents the position of a car at time t, then its velocity is given by the derivative: $v(t) = s'(t)$. But one rarely drives a car at a constant speed; the velocity itself may be changing. The rate at which the velocity is changing is the **acceleration**. Because the derivative measures the rate of change, acceleration is the derivative of velocity: $a(t) = v'(t)$. Because v is the derivative of s, we can express the acceleration in terms of s:

$$a(t) = v'(t) = (s')'(t) = s''(t).$$

That is, a is the derivative of the derivative of s; in other words, the second derivative of s, which we write as s''. (In this context you will often hear the derivative s' referred to as the **first derivative**.)

Second Derivative, Acceleration

If a function f has a derivative that is in turn differentiable, then its **second derivative** is the derivative of the derivative of f, written as f''. If $f''(a)$ exists, we say that f is **twice differentiable at** $x = a$.

Quick Examples

1. If $f(x) = x^3 - x$, then $f'(x) = 3x^2 - 1$, so $f''(x) = 6x$ and $f''(-2) = -12$.
2. If $f(x) = 3x + 1$, then $f'(x) = 3$, so $f''(x) = 0$.
3. If $f(x) = e^x$, then $f'(x) = e^x$, so $f''(x) = e^x$ as well.

The **acceleration** of a moving object is the derivative of its velocity—that is, the second derivative of the position function.

Quick Example

If t is time in hours and the position of a car at time t is $s(t) = t^3 + 2t^2$ miles, then the car's velocity is $v(t) = s'(t) = 3t^2 + 4t$ miles per hour and its acceleration is $a(t) = s''(t) = v'(t) = 6t + 4$ miles per hour per hour.

Differential Notation for the Second Derivative

We have written the second derivative of $f(x)$ as $f''(x)$. We could also use differential notation:

$$f''(x) = \frac{d^2 f}{dx^2}.$$

This notation comes from writing the second derivative as the derivative of the derivative in differential notation:

$$f''(x) = \frac{d}{dx}\left[\frac{df}{dx}\right] = \frac{d^2 f}{dx^2}.$$

Similarly, if $y = f(x)$, we write $f''(x)$ as $\frac{d}{dx}\left[\frac{dy}{dx}\right] = \frac{d^2 y}{dx^2}$. For example, if $y = x^3$, then $\frac{d^2 y}{dx^2} = 6x$.

An important example of acceleration is the acceleration due to gravity.

EXAMPLE 1 Acceleration Due to Gravity

According to the laws of physics, the height of an object near the surface of the earth falling in a vacuum from an initial rest position 100 feet above the ground under the influence of gravity is approximately

$$s(t) = 100 - 16t^2 \text{ feet}$$

in t seconds. Find its acceleration.

Solution The velocity of the object is

$$v(t) = s'(t) = -32t \text{ ft/sec.} \qquad \text{Differential notation: } v = \frac{ds}{dt} = -32t \text{ ft/sec}$$

The reason for the negative sign is that the height of the object is decreasing with time, so its velocity is negative. Hence, the acceleration is

$$a(t) = s''(t) = -32 \text{ ft/sec}^2. \qquad \text{Differential notation: } a = \frac{d^2 s}{dt^2} = -32 \text{ ft/sec}^2$$

(We write ft/sec^2 as an abbreviation for feet/second/second—that is, feet per second per second. It is often read "feet per second squared.") Thus, the *downward* velocity is increasing by 32 ft/sec every second. We say that 32 ft/sec^2 is the **acceleration due to gravity**. In the absence of air resistance, all falling bodies near the surface of the earth, no matter what their weight, will fall with this acceleration.*

⇒ **Before we go on...** In very careful experiments using balls rolling down inclined planes, Galileo made one of his most important discoveries—that the acceleration due to gravity is constant and does not depend on the weight or composition of the object falling.† A famous, though probably apocryphal, story has him dropping cannonballs of different weights off the Leaning Tower of Pisa to prove his point.§ ■

* On other planets the acceleration due to gravity is different. For example, on Jupiter, it is about three times as large as on Earth.

† An interesting aside: Galileo's experiments depended on getting extremely accurate timings. Because the timepieces of his day were very inaccurate, he used the most accurate time measurement he could: He sang and used the beat as his stopwatch.

§ Here is a true story: The point was made again during the Apollo 15 mission to the moon (July 1971) when astronaut David R. Scott dropped a feather and a hammer from the same height. The moon has no atmosphere, so the two hit the surface of the moon simultaneously.

EXAMPLE 2 **Acceleration of Sales**

For the first 15 months after the introduction of a new video game, the total sales can be modeled by the curve

$$S(t) = 20e^{0.4t} \text{ units sold,}$$

where t is the time in months since the game was introduced. After about 25 months total sales follow more closely the curve

$$S(t) = 100,000 - 20e^{17-0.4t}.$$

How fast are total sales accelerating after 10 months? How fast are they accelerating after 30 months? What do these numbers mean?

Solution By acceleration we mean the rate of change of the rate of change, which is the second derivative. During the first 15 months, the first derivative of sales is

$$\frac{dS}{dt} = 8e^{0.4t}$$

and so the second derivative is

$$\frac{d^2S}{dt^2} = 3.2e^{0.4t}.$$

Thus, after 10 months the acceleration of sales is

$$\frac{d^2S}{dt^2}\bigg|_{t=10} = 3.2e^4 \approx 175 \text{ units/month/month, or units/month}^2.$$

We can also compute total sales

$$S(10) = 20e^4 \approx 1,092 \text{ units}$$

and the rate of change of sales

$$\frac{dS}{dt}\bigg|_{t=10} = 8e^4 \approx 437 \text{ units/month.}$$

What do these numbers mean? By the end of the tenth month, a total of 1,092 video games have been sold. At that time the game is selling at the rate of 437 units per month. This rate of sales is increasing by 175 units per month per month. More games will be sold each month than the month before.

Analysis of the sales after 30 months is done similarly, using the formula

$$S(t) = 100,000 - 20e^{17-0.4t}.$$

The derivative is

$$\frac{dS}{dt} = 8e^{17-0.4t}$$

and the second derivative is

$$\frac{d^2S}{dt^2} = -3.2e^{17-0.4t}.$$

After 30 months,

$$S(30) = 100{,}000 - 20e^{17-12} \approx 97{,}032 \text{ units}$$

$$\left.\frac{dS}{dt}\right|_{t=30} = 8e^{17-12} \approx 1{,}187 \text{ units/month}$$

$$\left.\frac{d^2S}{dt^2}\right|_{t=30} = -3.2e^{17-12} \approx -475 \text{ units/month}^2.$$

By the end of the thirtieth month, 97,032 video games have been sold, the game is selling at a rate of 1,187 units per month, and the rate of sales is *decreasing* by 475 units per month. Fewer games are sold each month than the month before.

Geometric Interpretation of Second Derivative: Concavity

The first derivative of f tells us where the graph of f is rising [where $f'(x) > 0$] and where it is falling [where $f'(x) < 0$]. The second derivative tells in what direction the graph of f *curves* or *bends*. Consider the graphs in Figures 26 and 27.

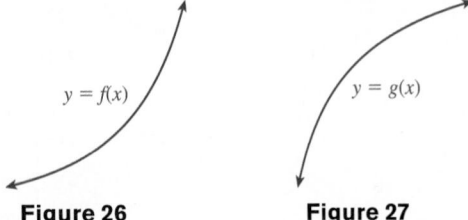

$y = f(x)$ $y = g(x)$

Figure 26   **Figure 27**

Think of a car driving from left to right along each of the roads shown in the two figures. A car driving along the graph of f in Figure 26 will turn to the left (upward); a car driving along the graph of g in Figure 27 will turn to the right (downward). We say that the graph of f is **concave up** and the graph of g is **concave down**. Now think about the derivatives of f and g. The derivative $f'(x)$ starts small but *increases* as the graph gets steeper. Because $f'(x)$ is increasing, its derivative $f''(x)$ must be positive. On the other hand, $g'(x)$ *decreases* as we go to the right. Because $g'(x)$ is decreasing, its derivative $g''(x)$ must be negative. Summarizing, we have the following.

Concavity and the Second Derivative

A curve is **concave up** if its slope is increasing, in which case the second derivative is positive. A curve is **concave down** if its slope is decreasing, in which case the second derivative is negative. A point in the domain of f where the graph of f changes concavity, from concave up to concave down or vice versa, is called a **point of inflection**. At a point of inflection, the second derivative is either zero or undefined.

Locating Points of Inflection

To locate possible points of inflection, list points where $f''(x) = 0$ and also interior points where $f''(x)$ is not defined.

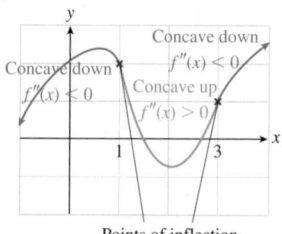

Figure 28

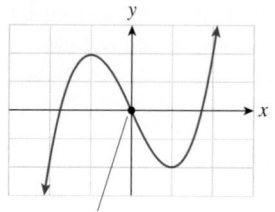

Point of inflection

Figure 29

Quick Examples

1. The graph of the function f shown in Figure 28 is concave up when $1 < x < 3$, so $f''(x) > 0$ for $1 < x < 3$. It is concave down when $x < 1$ and $x > 3$, so $f''(x) < 0$ when $x < 1$ and $x > 3$. It has points of inflection at $x = 1$ and $x = 3$.

2. Consider $f(x) = x^3 - 3x$, whose graph is shown in Figure 29. $f''(x) = 6x$ is negative when $x < 0$ and positive when $x > 0$. The graph of f is concave down when $x < 0$ and concave up when $x > 0$. f has a point of inflection at $x = 0$, where the second derivative is 0.

The following example shows one of the reasons it's useful to look at concavity.

EXAMPLE 3 Inflation

Figure 30 shows the value of the U.S. Consumer Price Index (CPI) from January 2010 through April 2011.[27]

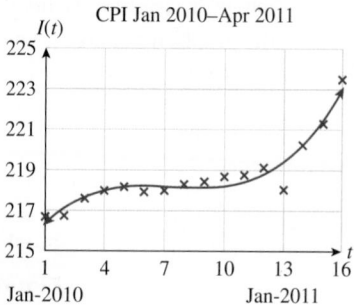

CPI Jan 2010–Apr 2011

Figure 30

The approximating curve shown on the figure is given by

$$I(t) = 0.0081t^3 - 0.18t^2 + 1.3t + 215 \qquad (1 \le t \le 16),$$

where t is time in months ($t = 1$ represents January 2010). When the CPI is increasing, the U.S. economy is **experiencing inflation**. In terms of the model, this means that the derivative is positive: $I'(t) > 0$. Notice that $I'(t) > 0$ for most of the period shown (the graph is sloping upward), so the U.S. economy experienced inflation for most of $1 \le t \le 16$. We could measure **inflation** by the first derivative $I'(t)$ of the CPI, but we traditionally measure it as a ratio:

$$\text{Inflation rate} = \frac{I'(t)}{I(t)}, \qquad \text{Relative rate of change of the CPI}$$

expressed as a percentage per unit time (per month in this case).

a. Use the model to estimate the inflation rate in March 2010.

b. Was inflation slowing or speeding up in March 2010?

c. When was inflation slowing? When was inflation speeding up? When was inflation slowest?

---

[27]The CPI is compiled by the Bureau of Labor Statistics and is based upon a 1982 value of 100. For instance, a CPI of 200 means the CPI has doubled since 1982. Source: InflationData.com (www.inflationdata.com).

Solution

a. We need to compute $I'(t)$:

$$I'(t) = 0.0243t^2 - 0.36t + 1.3.$$

Thus, the inflation rate in March 2010 was given by

$$\text{Inflation rate} = \frac{I'(3)}{I(3)} = \frac{0.0243(3)^2 - 0.36(3) + 1.3}{0.0081(3)^3 - 0.18(3)^2 + 1.3(3) + 215}$$

$$= \frac{0.4387}{217.4987} \approx 0.0020,$$

or 0.20% per month.*

* The 0.20% monthly inflation rate corresponds to a $12 \times 0.20 = 2.40\%$ annual inflation rate. This result could be obtained directly by changing the units of the t-axis from months to years and then redoing the calculation.

† When the CPI is falling, the inflation rate is negative and we experience *deflation*.

b. We say that inflation is "slowing" when the CPI is decelerating ($I''(t) < 0$; the index rises at a slower rate or falls at a faster rate†). Similarly, inflation is "speeding up" when the CPI is accelerating ($I''(t) > 0$; the index rises at a faster rate or falls at a slower rate). From the formula for $I'(t)$, the second derivative is

$$I''(t) = 0.0486t - 0.36$$
$$I''(3) = 0.0486(3) - 0.36 = -0.2142.$$

Because this quantity is negative, we conclude that inflation was slowing down in March 2010.

c. When inflation is slowing, $I''(t)$ is negative, so the graph of the CPI is concave down. When inflation is speeding up, it is concave up. At the point at which it switches, there is a point of inflection (Figure 31).

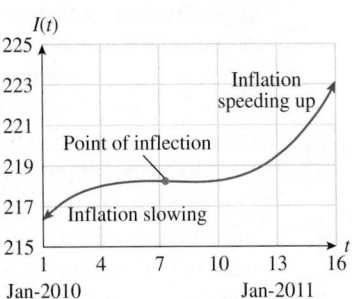

Figure 31

The point of inflection occurs when $I''(t) = 0$; that is,

$$0.0486t - 0.36 = 0$$

$$t = \frac{0.36}{0.0486} \approx 7.4.$$

Thus, inflation was slowing when $t < 7.4$ (that is, until around the middle of July), and speeding up when $t > 7.4$ (after that time). Inflation was slowest at the point when it stopped slowing down and began to speed up, $t \approx 7.4$ (in fact there was slight deflation at that particular point as $I'(7.4)$ is negative); notice that the graph has the least slope at that point.

EXAMPLE 4 **The Point of Diminishing Returns**

After the introduction of a new video game, the total worldwide sales are modeled by the curve

$$S(t) = \frac{1}{1 + 50e^{-0.2t}} \text{ million units sold,}$$

where t is the time in months since the game was introduced (compare Example 2). The graphs of $S(t)$, $S'(t)$, and $S''(t)$ are shown in Figure 32. Where is the graph of S concave up, and where is it concave down? Where are any points of inflection? What does this all mean?

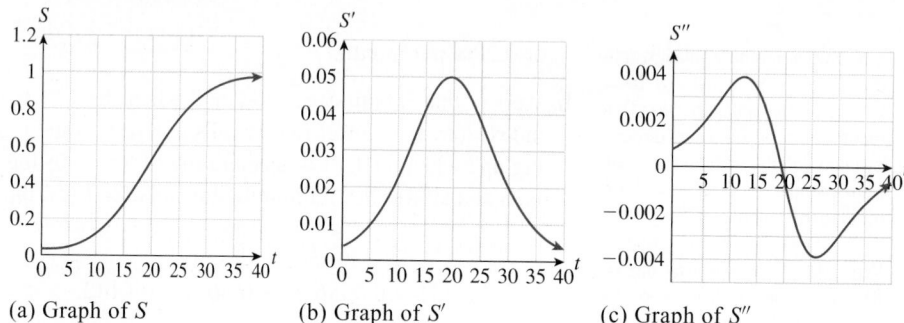

(a) Graph of S (b) Graph of S' (c) Graph of S''

Figure 32

 using Technology

You can use a TI-83/84 Plus, a downloadable Excel sheet at the Website, or the Function Evaluator and Grapher at the Website to graph the second derivative of the function in Example 4:

TI-83/84 Plus

Y₁=1/(1+50*e^(-0.2X))
Y₂=nDeriv(Y₁,X,X)
Y₃=nDeriv(Y₂,X,X)

Website

www.WanerMath.com
In the Function Evaluator and Grapher, enter the functions as shown.

Functions
y₁ = 1/(1+50e^(-0.2x))
y₂ = deriv(y1)
y₃ = deriv(y2)
y₄ =
y₅ =

(Use Ymin = 0, yMax = 0.1 for a nice view of S' and Ymin = −0.01 and yMax = 0.01 for S''.)

For an Excel utility, try:
On Line Utilities
↓
Excel First and Second
Derivative Graphing Utility

(This utility needs macros, so ensure they are enabled.)

Solution Look at the graph of S. We see that the graph of S is concave up in the early months and then becomes concave down later. The point of inflection, where the concavity changes, is somewhere between 15 and 25 months.

Now look at the graph of S''. This graph crosses the t-axis very close to $t = 20$, is positive before that point, and negative after that point. Because positive values of S'' indicate S is concave up and negative values concave down, we conclude that the graph of S is concave up for about the first 20 months; that is, for $0 < t < 20$ and concave down for $20 < t < 40$. The concavity switches at the point of inflection, which occurs at about $t = 20$ (when $S''(t) = 0$; a more accurate answer is $t \approx 19.56$).

What does this all mean? Look at the graph of S', which shows sales per unit time, or monthly sales. From this graph we see that monthly sales are increasing for $t < 20$: more units are being sold each month than the month before. Monthly sales reach a peak of 0.05 million = 50,000 games per month at the point of inflection $t = 20$ and then begin to drop off. Thus, the point of inflection occurs at the time when monthly sales stop increasing and start to fall off; that is, the time when monthly sales peak. The point of inflection is sometimes called the **point of diminishing returns**. Although the total sales figure continues to rise (see the graph of S: game units continue to be sold), the *rate* at which units are sold starts to drop. (See Figure 33.)

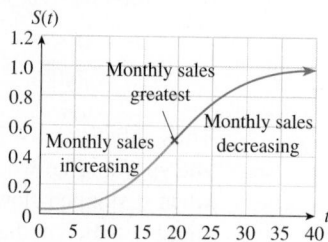

Figure 33

The Second Derivative Test for Relative Extrema

The second derivative often gives us a way of knowing whether or not a stationary point is a relative extremum. Figure 34 shows a graph with two stationary points: a relative maximum at $x = a$ and a relative minimum at $x = b$.

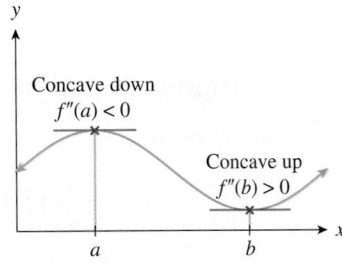

Figure 34

Notice that the curve is *concave down* at the relative maximum ($x = a$), so that $f''(a) < 0$, and *concave up* at the relative minimum ($x = b$), so that $f''(b) > 0$. This suggests the following (compare the First Derivative Test in Section 12.1).

Second Derivative Test for Relative Extrema

Suppose that the function f has a stationary point at $x = c$, and that $f''(c)$ exists. Determine the sign of $f''(c)$.

1. If $f''(c) > 0$, then f has a relative minimum at $x = c$.

2. If $f''(c) < 0$, then f has a relative maximum at $x = c$.

If $f''(c) = 0$, then the test is inconclusive and you need to use one of the methods of Section 12.1 (such as the first derivative test) to determine whether or not f has a relative extremum at $x = c$.

Quick Examples

1. $f(x) = x^2 - 2x$ has $f'(x) = 2x - 2$ and hence a stationary point at $x = 1$. $f''(x) = 2$, and so $f''(1) = 2$, which is positive, so f has a relative minimum at $x = 1$.

2. Let $f(x) = x^3 - 3x^2 - 9x$. Then
$f'(x) = 3x^2 - 6x - 9 = 3(x + 1)(x - 3)$
Stationary points at $x = -1, x = 3$
$f''(x) = 6x - 6$
$f''(-1) = -12$, so there is a relative maximum at $x = -1$
$f''(3) = 12$, so there is a relative minimum at $x = 3$.

3. $f(x) = x^4$ has $f'(x) = 4x^3$ and hence a stationary point at $x = 0$. $f''(x) = 12x^2$ and so $f''(0) = 0$, telling us that the second derivative test is inconclusive. However, we can see from the graph of f or the first derivative test that f has a minimum at $x = 0$.

Higher Order Derivatives

There is no reason to stop at the second derivative; we could once again take the *derivative* of the second derivative to obtain the **third derivative**, f''', and we could take the derivative once again to obtain the **fourth derivative**, written $f^{(4)}$, and then continue to obtain $f^{(5)}$, $f^{(6)}$, and so on (assuming we get a differentiable function at each stage).

Higher Order Derivatives

We define

$$f'''(x) = \frac{d}{dx}[f''(x)]$$

$$f^{(4)}(x) = \frac{d}{dx}[f'''(x)]$$

$$f^{(5)}(x) = \frac{d}{dx}[f^{(4)}(x)],$$

and so on, assuming all these derivatives exist.

Different Notations

$$f'(x), f''(x), f'''(x), f^{(4)}(x), \ldots, f^{(n)}(x), \ldots$$

$$\frac{df}{dx}, \frac{d^2f}{dx^2}, \frac{d^3f}{dx^3}, \frac{d^4f}{dx^4}, \ldots, \frac{d^nf}{dx^n}, \ldots$$

$$\frac{dy}{dx}, \frac{d^2y}{dx^2}, \frac{d^3y}{dx^3}, \frac{d^4y}{dx^4}, \ldots, \frac{d^ny}{dx^n}, \ldots \qquad \text{When } y = f(x)$$

$$y, y', y'', y''', y^{(4)}, \ldots, y^{(n)}, \ldots \qquad \text{When } y = f(x)$$

Quick Examples

1. If $f(x) = x^3 - x$, then $f'(x) = 3x^2 - 1$, $f''(x) = 6x$, $f'''(x) = 6$,
   $f^{(4)}(x) = f^{(5)}(x) = \cdots = 0$.
2. If $f(x) = e^x$, then $f'(x) = e^x$, $f''(x) = e^x$, $f'''(x) = e^x$, $f^{(4)}(x) = f^{(5)}(x) = \cdots = e^x$.

Q: *We know that the second derivative can be interpreted as acceleration. How do we interpret the third derivative; and the fourth, fifth, and so on?*

A: Think of a car traveling down the road (with position $s(t)$ at time t) in such a way that its acceleration $\dfrac{d^2s}{dt^2}$ is changing with time (for instance, the driver may be slowly increasing pressure on the accelerator, causing the car to accelerate at a greater and greater rate). Then $\dfrac{d^3s}{dt^3}$ is the rate of change of acceleration. $\dfrac{d^4s}{dt^4}$ would then be the *acceleration* of the acceleration, and so on.

Q: *How are these higher order derivatives reflected in the graph of a function f?*

A: Because the concavity is measured by f'', its derivative f''' tells us the rate of change of concavity. Similarly, $f^{(4)}$ would tell us the *acceleration* of concavity, and so on. These properties are very subtle and hard to discern by simply looking at the curve; the higher the order, the more subtle the property. There is a remarkable theorem by Taylor* that tells us that, for a large class of functions (including polynomial, exponential, logarithmic, and trigonometric functions) the values of all orders of derivative $f(a)$, $f'(a)$, $f''(a)$, $f'''(a)$, and so on at the single point $x = a$ are enough to describe the entire graph (even at points very far from $x = a$)! In other words, the smallest piece of a graph near any point a contains sufficient information to "clone" the entire graph!

✳ Brook Taylor (1685–1731) was an English mathematician.

FAQs

Interpreting Points of Inflection and Using the Second Derivative Test

Q: *It says in Example 4 that monthly sales reach a maximum at the point of inflection (second derivative is zero), but the second derivative test says that, for a maximum, the second derivative must be negative. What is going on here?*

A: What is a maximum in Example 4 is the *rate of change of* sales: which is measured in sales per unit time (monthly sales in the example). In other words, it is the *derivative* of the total sales function that is a maximum, so we located the maximum by setting its derivative (which is the *second* derivative of total sales) equal to zero. In general: To find relative (stationary) extrema of the *original* function, set $f'(x)$ equal to zero and solve for x as usual. The second derivative test can then be used to test the stationary point obtained. To find relative (stationary) extrema of the *rate of change of f*, set $f''(x) = 0$ and solve for x.

Q: *I used the second derivative test and it was inconclusive. That means that there is neither a relative maximum nor a relative minimum at $x = a$, right?*

A: Wrong. If (as is often the case) the second derivative is zero at a stationary point, all it means is that the second derivative test itself cannot determine whether the given point is a relative maximum, minimum, or neither. For instance, $f(x) = x^4$ has a stationary minimum at $x = 0$, but the second derivative test is inconclusive. In such cases, one should use another test (such as the first derivative test) to decide if the point is a relative maximum, minimum, or neither.

12.3 EXERCISES

▼ more advanced   ◆ challenging
🔲 indicates exercises that should be solved using technology

In Exercises 1–10, calculate $\dfrac{d^2y}{dx^2}$. HINT [See Quick Examples on page 892.]

1. $y = 3x^2 - 6$ **2.** $y = -x^2 + x$

3. $y = \dfrac{2}{x}$ **4.** $y = -\dfrac{2}{x^2}$

5. $y = 4x^{0.4} - x$ **6.** $y = 0.2x^{-0.1}$

7. $y = e^{-(x-1)} - x$ **8.** $y = e^{-x} + e^x$

9. $y = \dfrac{1}{x} - \ln x$ **10.** $y = x^{-2} + \ln x$

In Exercises 11–16, the position s of a point (in feet) is given as a function of time t (in seconds). Find (a) its acceleration as a function of t and (b) its acceleration at the specified time. HINT [See Example 1.]

11. $s = 12 + 3t - 16t^2; t = 2$

12. $s = -12 + t - 16t^2; t = 2$

13. $s = \dfrac{1}{t} + \dfrac{1}{t^2}; t = 1$ **14.** $s = \dfrac{1}{t} - \dfrac{1}{t^2}; t = 2$

15. $s = \sqrt{t} + t^2; t = 4$ **16.** $s = 2\sqrt{t} + t^3; t = 1$

In Exercises 17–24, the graph of a function is given. Find the approximate coordinates of all points of inflection of each function (if any). HINT [See Quick Examples on page 896.]

17.
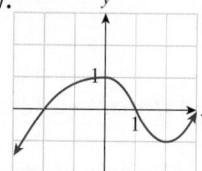

18.
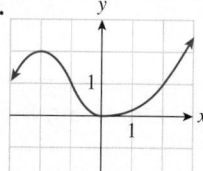

19.
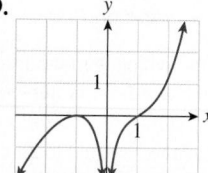

20.
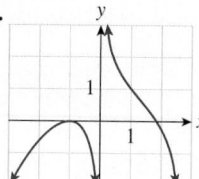

21.
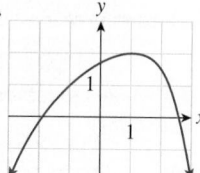

22.
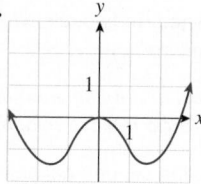

23.
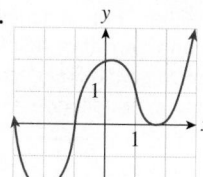

24.
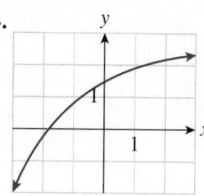

In Exercises 25–28, the graph of the derivative, $f'(x)$, *is given. Determine the x-coordinates of all points of inflection of $f(x)$, if any. (Assume that $f(x)$ is defined and continuous everywhere in $[-3, 3]$.)* HINT [See the **Before we go on** discussion in Example 4.]

25.
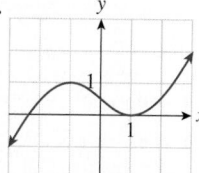

26.
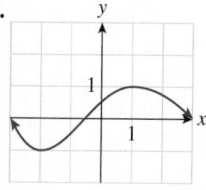

27.
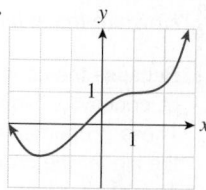

28.
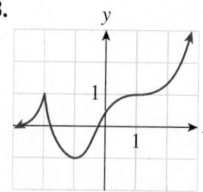

In Exercises 29–32, the graph of the second derivative, $f''(x)$, *is given. Determine the x-coordinates of all points of inflection of $f(x)$, if any. (Assume that $f(x)$ is defined and continuous everywhere in $[-3, 3]$.)* HINT [Remember that a point of inflection of f corresponds to a point at which f'' changes sign, from positive to negative or vice versa. This could be a point where its graph crosses the x-axis, or a point where its graph is broken: positive on one side of the break and negative on the other.]

29.
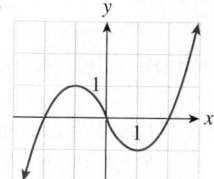

30.
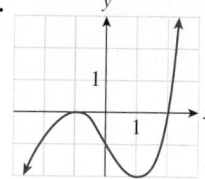

31.
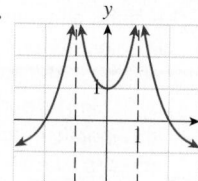

32.
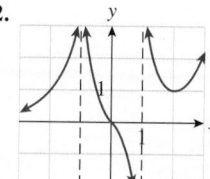

In Exercises 33–44, find the x-coordinates of all critical points of the given function. Determine whether each critical point is a relative maximum, minimum, or neither by first applying the second derivative test, and, if the test fails, by some other method. HINT [See Quick Examples on page 899.]

33. $f(x) = x^2 - 4x + 1$ **34.** $f(x) = 2x^2 - 2x + 3$

35. $g(x) = x^3 - 12x$ **36.** $g(x) = 2x^3 - 6x + 3$

37. $f(t) = t^3 - t$ **38.** $f(t) = -2t^3 + 3t$

39. $f(x) = x^4 - 4x^3$ **40.** $f(x) = 3x^4 - 2x^3$

41. $f(x) = e^{-x^2}$ **42.** $f(x) = e^{2-x^2}$

43. $f(x) = xe^{1-x^2}$ **44.** $f(x) = xe^{-x^2}$

In Exercises 45–54, calculate the derivatives of all orders: $f'(x), f''(x), f'''(x), f^{(4)}(x), \ldots, f^{(n)}(x), \ldots$ HINT [See Quick Examples on page 900.]

45. $f(x) = 4x^2 - x + 1$ **46.** $f(x) = -3x^3 + 4x$

47. $f(x) = -x^4 + 3x^2$ **48.** $f(x) = x^4 + x^3$

49. $f(x) = (2x + 1)^4$ **50.** $f(x) = (-2x + 1)^3$

51. $f(x) = e^{-x}$ **52.** $f(x) = e^{2x}$

53. $f(x) = e^{3x-1}$ **54.** $f(x) = 2e^{-x+3}$

APPLICATIONS

55. Acceleration on Mars If a stone is dropped from a height of 40 meters above the Martian surface, its height in meters after t seconds is given by $s = 40 - 1.9t^2$. What is its acceleration? HINT [See Example 1.]

56. Acceleration on the Moon If a stone is thrown up at 10 m per second from a height of 100 meters above the surface of the Moon, its height in meters after t seconds is given by $s = 100 + 10t - 0.8t^2$. What is its acceleration? HINT [See Example 1.]

57. Motion in a Straight Line The position of a particle moving in a straight line is given by $s = t^3 - t^2$ ft after t seconds. Find an expression for its acceleration after a time t. Is its velocity increasing or decreasing when $t = 1$?

58. Motion in a Straight Line The position of a particle moving in a straight line is given by $s = 3e^t - 8t^2$ ft after t seconds. Find an expression for its acceleration after a time t. Is its velocity increasing or decreasing when $t = 1$?

59. Bottled Water Sales Annual sales of bottled water in the United States in the period 2000–2010 could be approximated by

$$R(t) = -45t^2 + 900t + 4{,}200 \text{ million gallons} \quad (0 \le t \le 10),$$

where t is time in years since 2000.[28] According to the model, were annual sales of bottled water accelerating or decelerating in 2009? How fast? HINT [See Example 2.]

60. Bottled Water Sales Annual U.S. per capita sales of bottled water in the period 2000–2010 could be approximated by

$$Q(t) = -0.18t^2 + 3t + 15 \text{ gallons} \quad (0 \le t \le 10),$$

where t is time in years since 2000.[29] According to the model, were annual U.S. per capita sales of bottled water accelerating or decelerating in 2009? How fast?

61. Embryo Development The daily oxygen consumption of a bird embryo increases from the time the egg is laid through the time the chick hatches. In a typical galliform bird, the oxygen consumption can be approximated by

$$c(t) = -0.065t^3 + 3.4t^2 - 22t + 3.6 \text{ ml} \quad (8 \le t \le 30),$$

where t is the time (in days) since the egg was laid.[30] (An egg will typically hatch at around $t = 28$.) Use the model to estimate the following (give the units of measurement for each answer and round all answers to two significant digits):

a. The daily oxygen consumption 20 days after the egg was laid

b. The rate at which the oxygen consumption is changing 20 days after the egg was laid

c. The rate at which the oxygen consumption is accelerating 20 days after the egg was laid

62. Embryo Development The daily oxygen consumption of a turkey embryo increases from the time the egg is laid through the time the chick hatches. In a brush turkey, the oxygen consumption can be approximated by

$$c(t) = -0.028t^3 + 2.9t^2 - 44t + 95 \text{ ml} \quad (20 \le t \le 50),$$

where t is the time (in days) since the egg was laid.[31] (An egg will typically hatch at around $t = 50$.) Use the model to estimate the following (give the units of measurement for each answer and round all answers to two significant digits):

a. The daily oxygen consumption 40 days after the egg was laid

b. The rate at which the oxygen consumption is changing 40 days after the egg was laid

c. The rate at which the oxygen consumption is accelerating 40 days after the egg was laid

63. Inflation The following graph shows the approximate value of the United States Consumer Price Index (CPI) from December 2006 through July 2007.[32]

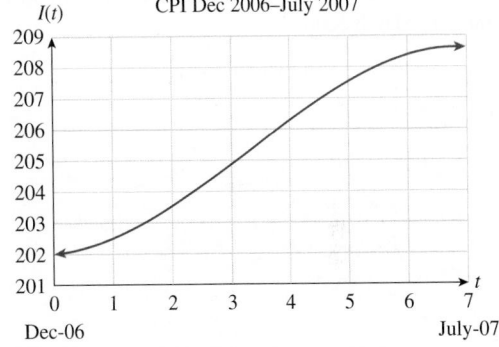

The approximating curve shown on the figure is given by

$$I(t) = -0.04t^3 + 0.4t^2 + 0.1t + 202 \quad (0 \le t \le 7),$$

where t is time in months ($t = 0$ represents December 2006).

a. Use the model to estimate the monthly inflation rate in February 2007 ($t = 2$). [Recall that the inflation *rate* is $I'(t)/I(t)$.]

b. Was inflation slowing or speeding up in February 2007?

c. When was inflation speeding up? When was inflation slowing? HINT [See Example 3.]

64. Inflation The following graph shows the approximate value of the U.S. Consumer Price Index (CPI) from September 2004 through November 2005.[33]

---

[28]Source for data: Beverage Marketing Corporation (www.bottledwater.org).

[29]Ibid.

[30]The model approximates graphical data published in the article "The Brush Turkey" by Roger S. Seymour, *Scientific American*, December, 1991, pp. 108–114.

[31]Ibid.

[32]The CPI is compiled by the Bureau of Labor Statistics and is based upon a 1982 value of 100. For instance, a CPI of 200 means the CPI has doubled since 1982. Source: InflationData.com (www.inflationdata.com).

[33]Ibid.

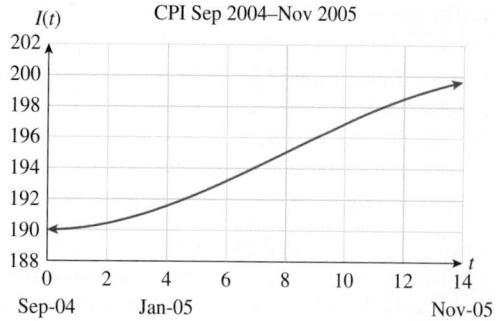

CPI Sep 2004–Nov 2005

The approximating curve shown on the figure is given by

$$I(t) = -0.005t^3 + 0.12t^2 - 0.01t + 190 \quad (0 \le t \le 14),$$

where t is time in months ($t = 0$ represents September 2004).

a. Use the model to estimate the monthly inflation rate in July 2005 ($t = 10$). [Recall that the inflation *rate* is $I'(t)/I(t)$.]

b. Was inflation slowing or speeding up in July 2005?

c. When was inflation speeding up? When was inflation slowing? HINT [See Example 3.]

65. *Inflation* The following graph shows the approximate value of the U.S. Consumer Price Index (CPI) from July 2005 through March 2006.[34]

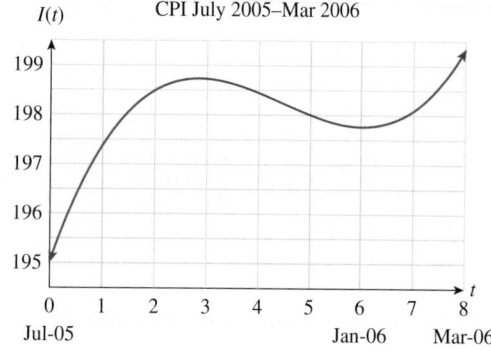

CPI July 2005–Mar 2006

The approximating curve shown on the figure is given by

$$I(t) = 0.06t^3 - 0.8t^2 + 3.1t + 195 \quad (0 \le t \le 8),$$

where t is time in months ($t = 0$ represents July 2005).

a. Use the model to estimate the monthly inflation rates in December 2005 and February 2006 ($t = 5$ and $t = 7$).

b. Was inflation slowing or speeding up in February 2006?

c. When was inflation speeding up? When was inflation slowing?

66. *Inflation* The following graph shows the approximate value of the U.S. Consumer Price Index (CPI) from March 2006 through May 2007.[35]

---

[34]The CPI is compiled by the Bureau of Labor Statistics and is based upon a 1982 value of 100. For instance, a CPI of 200 means the CPI has doubled since 1982. Source: InflationData.com (www.inflationdata.com).

[35]*Ibid.*

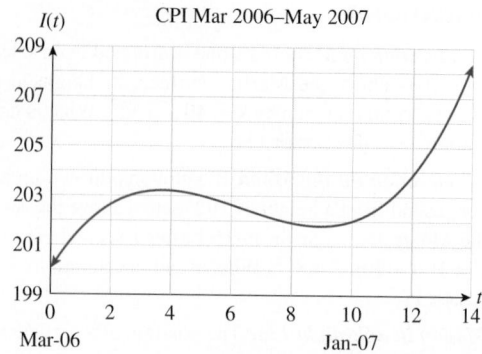

CPI Mar 2006–May 2007

The approximating curve shown on the figure is given by

$$I(t) = 0.02t^3 - 0.38t^2 + 2t + 200 \quad (0 \le t \le 14),$$

where t is time in months ($t = 0$ represents March, 2006).

a. Use the model to estimate the monthly inflation rates in September 2006 and January 2007 ($t = 6$ and $t = 10$).

b. Was inflation slowing or speeding up in January 2007?

c. When was inflation speeding up? When was inflation slowing?

67. *Scientific Research* The percentage of research articles in the prominent journal *Physical Review* that were written by researchers in the United States during the years 1983–2003 can be modeled by

$$P(t) = 25 + \frac{36}{1 + 0.06(0.7)^{-t}},$$

where t is time in years since 1983.[36] The graphs of P, P', and P'' are shown here:

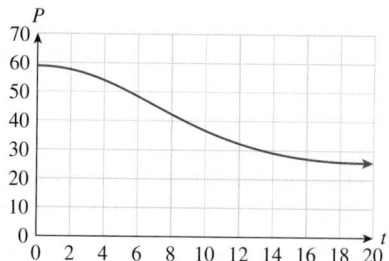

Graph of P

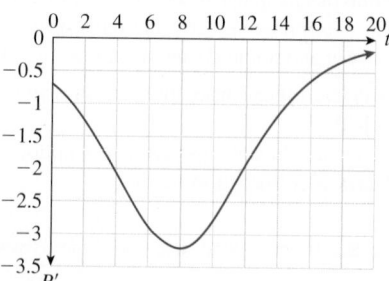

Graph of P'

---

[36]Source: The American Physical Society/*New York Times*, May 3, 2003, p. A1.

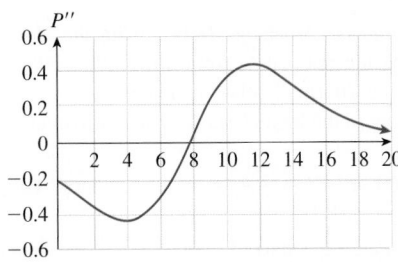

Graph of P''

Determine, to the nearest whole number, the values of t for which the graph of P is concave up and where it is concave down, and locate any points of inflection. What does the point of inflection tell you about science articles? HINT [See Example 4.]

68. *Scientific Research* The number of research articles in the prominent journal *Physical Review* that were written by researchers in Europe during the years 1983–2003 can be modeled by

$$P(t) = \frac{7.0}{1 + 5.4(1.2)^{-t}},$$

where t is time in years since 1983.[37] The graphs of P, P', and P'' are shown here:

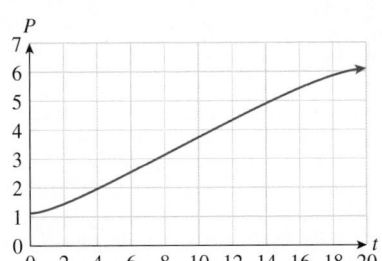

Graph of P

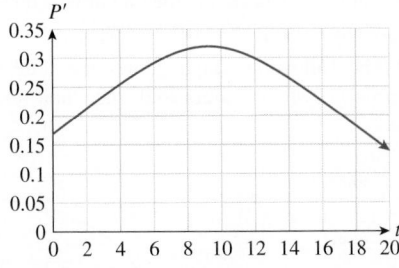

Graph of P'

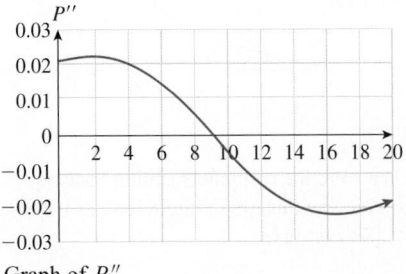

Graph of P''

[37] Source: The American Physical Society/*New York Times*, May 3, 2003, p. A1.

Determine, to the nearest whole number, the values of t for which the graph of P is concave up and where it is concave down, and locate any points of inflection. What does the point of inflection tell you about science articles?

69. *Embryo Development* Here are sketches of the graphs of c, c', and c'' from Exercise 61:

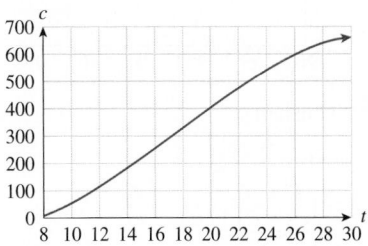

Graph of c

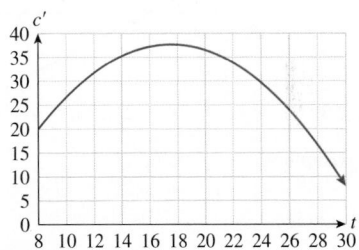

Graph of c'

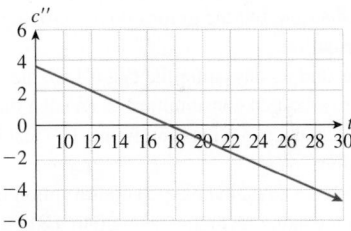

Graph of c''

Multiple choice:

a. The graph of c' **(A)** has a point of inflection. **(B)** has no points of inflection.

b. At around 18 days after the egg is laid, daily oxygen consumption is: **(A)** at a maximum. **(B)** increasing at a maximum rate. **(C)** just beginning to decrease.

c. For $t > 18$ days, the oxygen consumption is **(A)** increasing at a decreasing rate. **(B)** decreasing at an increasing rate. **(C)** increasing at an increasing rate.

70. *Embryo Development* Here are sketches of the graphs of c, c', and c'' from Exercise 62:

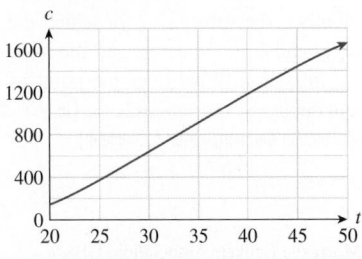

Graph of c

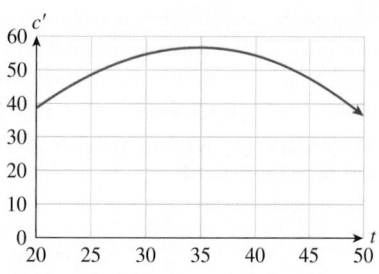

Graph of c'

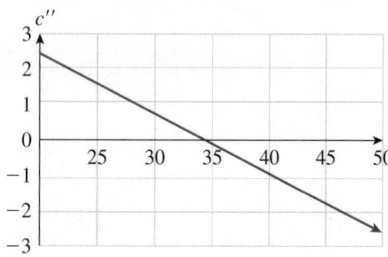

Graph of c''

Multiple choice:

a. The graph of c: **(A)** has points of inflection. **(B)** has no points of inflection. **(C)** may or may not have a point of inflection, but the graphs do not provide enough information.

b. At around 35 days after the egg is laid, the rate of change of daily oxygen consumption is: **(A)** at a maximum. **(B)** increasing at a maximum rate. **(C)** just becoming negative.

c. For $t < 35$ days, the oxygen consumption is: **(A)** increasing at an increasing rate. **(B)** increasing at a decreasing rate. **(C)** decreasing at an increasing rate.

71. ◻ *Subprime Mortgages during the Housing Bubble* During the real estate run-up in 2000–2008, the percentage of mortgages issued in the United States that were subprime (normally classified as risky) could be approximated by

$$A(t) = \frac{15.0}{1 + 8.6(1.8)^{-t}} \text{ percent } \quad (0 \le t \le 8)$$

t years after the start of 2000.[38] Graph the function as well as its first and second derivatives. Determine, to the nearest whole number, the values of t for which the graph of A is concave up and concave down, and the t-coordinate of any points of inflection. What does the point of inflection tell you about subprime mortgages? HINT [To graph the second derivative, see the note in the margin after Example 4.]

72. ◻ *Subprime Mortgage Debt during the Housing Bubble* During the real estate run-up in 2000–2008, the value of subprime (normally classified as risky) mortgage debt outstanding in the United States was approximately

$$A(t) = \frac{1,350}{1 + 4.2(1.7)^{-t}} \text{ billion dollars } \quad (0 \le t \le 8)$$

t years after the start of 2000.[39] Graph the function as well as its first and second derivatives. Determine, to the nearest whole number, the values of t for which the graph of A is concave up and concave down, and the t-coordinate of any points of inflection. What does the point of inflection tell you about subprime mortgages? HINT [To graph the second derivative, see the note in the margin after Example 4.]

73. *Epidemics* The following graph shows the total number n of people (in millions) infected in an epidemic as a function of time t (in years):

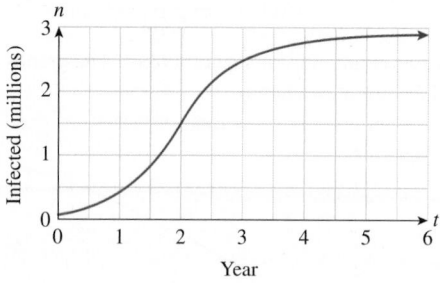

a. When to the nearest year was the rate of new infection largest?

b. When could the Centers for Disease Control and Prevention announce that the rate of new infection was beginning to drop? HINT [See Example 4.]

74. *Sales* The following graph shows the total number of *Pomegranate Q4* computers sold since their release (t is in years):

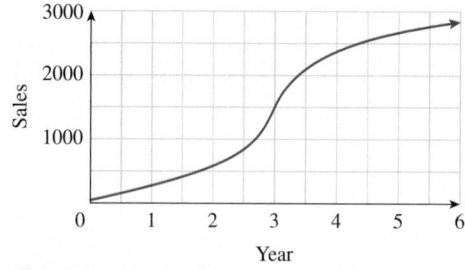

a. When were the computers selling fastest?

b. Explain why this graph might look as it does. HINT [See Example 4.]

---

[38] Sources: Mortgage Bankers Association, UBS.

[39] 2008 figure is an estimate.
Source: www.data360.org/dataset.aspx?Data_Set_Id=9549.

75. *Industrial Output* The following graph shows the yearly industrial output (measured in billions of zonars) of the city-state of Utarek, Mars over a seven-year period:

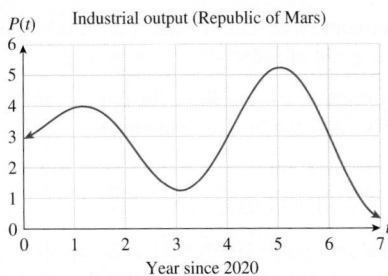

Industrial output (Republic of Mars)

Year since 2020

a. When to the nearest year did the rate of change of yearly industrial output reach a maximum?

b. When to the nearest year did the rate of change of yearly industrial output reach a minimum?

c. When to the nearest year does the graph first change from concave down to concave up? The result tells you that:

   (A) In that year the rate of change of industrial output reached a minimum compared with nearby years.

   (B) In that year the rate of change of industrial output reached a maximum compared with nearby years.

76. *Profits* The following graph shows the yearly profits of Gigantic Conglomerate, Inc. (GCI) from 2020 to 2035:

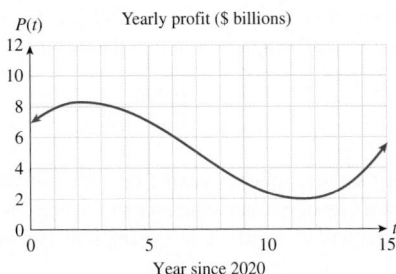

Yearly profit ($ billions)

Year since 2020

a. Approximately when were the profits rising most rapidly?

b. Approximately when were the profits falling most rapidly?

c. Approximately when could GCI's board of directors legitimately tell stockholders that they had "turned the company around"?

77. ▼ *Education and Crime* The following graph compares the total U.S. prison population and the average combined SAT score in the United States during the 1970s and 1980s:

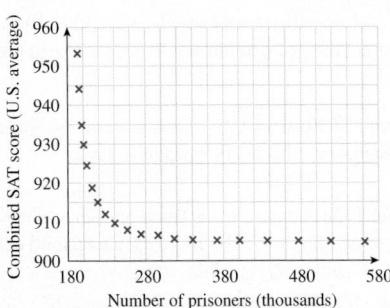

Number of prisoners (thousands)

These data can be accurately modeled by

$$S(n) = 904 + \frac{1,326}{(n - 180)^{1.325}} \quad (192 \leq n \leq 563).$$

Here, $S(n)$ is the combined U.S. average SAT score at a time when the total U.S. prison population was n thousand.[40]

a. Are there any points of inflection on the graph of S?

b. What does the concavity of the graph of S tell you about prison populations and SAT scores?

78. ▼ *Education and Crime* Refer back to the model in the preceding exercise.

a. Are there any points of inflection on the graph of S'?

b. What does the concavity of the graph of S' tell you about prison populations and SAT scores?

79. ▼ *Patents* In 1965, the economist F. M. Scherer modeled the number, n, of patents produced by a firm as a function of the size, s, of the firm (measured in annual sales in millions of dollars). He came up with the following equation based on a study of 448 large firms:[41]

$$n = -3.79 + 144.42s - 23.86s^2 + 1.457s^3.$$

a. Find $\left.\dfrac{d^2n}{ds^2}\right|_{s=3}$. Is the rate at which patents are produced as the size of a firm goes up increasing or decreasing with size when $s = 3$? Comment on Scherer's words, "... we find diminishing returns dominating."

b. Find $\left.\dfrac{d^2n}{ds^2}\right|_{s=7}$ and interpret the answer.

c. Find the s-coordinate of any points of inflection and interpret the result.

80. ▼ *Returns on Investments* A company finds that the number of new products it develops per year depends on the size of its annual R&D budget, x (in thousands of dollars), according to the formula

$$n(x) = -1 + 8x + 2x^2 - 0.4x^3.$$

a. Find $n''(1)$ and $n''(3)$, and interpret the results.

b. Find the size of the budget that gives the largest rate of return as measured in new products per dollar (again, called the point of diminishing returns).

81. ▣ ▼ *Oil Imports from Mexico* Daily oil production in Mexico and daily U.S. oil imports from Mexico during 2005–2009 can be approximated by

$$P(t) = 3.9 - 0.10t \text{ million barrels} \quad (5 \leq t \leq 9)$$
$$I(t) = 2.1 - 0.11t \text{ million barrels} \quad (5 \leq t \leq 9),$$

[40] Based on data for the years 1967–1989. Sources: *Sourcebook of Criminal Justice Statistics*, 1990, p. 604/Educational Testing Service.

[41] Source: F. M. Scherer, "Firm Size, Market Structure, Opportunity, and the Output of Patented Inventions," *American Economic Review* 55 (December 1965): pp. 1097–1125.

where t is time in years since the start of 2000.[42]

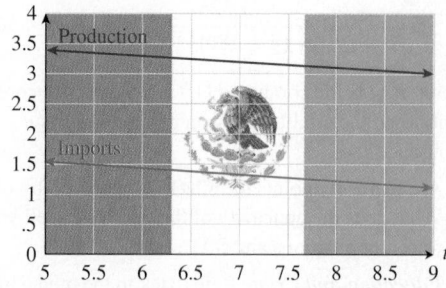

Graph the function $I(t)/P(t)$ and its derivative. Is the graph of $I(t)/P(t)$ concave up or concave down? The concavity of $I(t)/P(t)$ tells you that

(A) the percentage of oil produced in Mexico that was exported to the United States was decreasing.

(B) the percentage of oil produced in Mexico that was not exported to the United States was increasing.

(C) the percentage of oil produced in Mexico that was exported to the United States was decreasing at a slower rate.

(D) the percentage of oil produced in Mexico that was exported to the United States was decreasing at a faster rate.

82. ◆ ▼ *Oil Imports from Mexico* Repeat Exercise 81 using instead the models for 2000–2004 shown below:

$$P(t) = 3.0 + 0.13t \text{ million barrels} \quad (0 \le t \le 4)$$
$$I(t) = 1.4 + 0.06t \text{ million barrels} \quad (0 \le t \le 4)$$

(t is time in years since the start of 2000).[43]

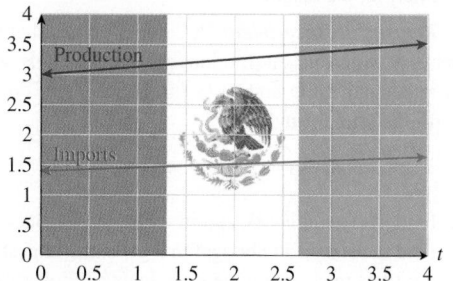

83. ◆ *Logistic Models* Let

$$f(x) = \frac{N}{1 + Ab^{-x}}$$

for constants N, A, and b (A and b positive and $b \ne 1$). Show that f has a single point of inflection at $x = \ln A / \ln b$.

84. ◆ *Logistic Models* Let

$$f(x) = \frac{N}{1 + Ae^{-kx}}$$

for constants N, A, and k (A and k positive). Show that f has a single point of inflection at $x = \ln A / k$.

85. ⊞ *Population: Puerto Rico* The population of Puerto Rico in 1950–2025 can be approximated by

$$P(t) = \frac{4{,}500}{1 + 1.1466\,(1.0357)^{-t}} \text{ thousand people } (0 \le t \le 75)$$

(t is the year since 1950).[44] Use the result of Exercise 83 to find the location of the point of inflection in the graph of P. What does the result tell you about the population of Puerto Rico?

86. ⊞ *Population: Virgin Islands* The population of the Virgin Islands in 1950–2025 can be approximated by

$$P(t) = \frac{110}{1 + 2.3596\,(1.0767)^{-t}} \text{ thousand people } (0 \le t \le 75)$$

(t is the year since 1950).[45] Use the result of Exercise 83 to find the location of the point of inflection in the graph of P. What does the result tell you about the population of the Virgin Islands?

87. ⊞ ▼ *Asset Appreciation* You manage a small antique store that owns a collection of Louis XVI jewelry boxes. Their value v is increasing according to the formula

$$v = \frac{10{,}000}{1 + 500e^{-0.5t}},$$

where t is the number of years from now. You anticipate an inflation rate of 5% per year, so that the present value of an item that will be worth $\$v$ in t years' time is given by

$$p = v \cdot (1.05)^{-t}.$$

What is the greatest rate of increase of the value of your antiques, and when is this rate attained?

88. ⊞ ▼ *Harvesting Forests* The following equation models the approximate volume in cubic feet of a typical Douglas fir tree of age t years[46]:

$$V = \frac{22{,}514}{1 + 22{,}514t^{-2.55}}.$$

The lumber will be sold at $10 per cubic foot, and you do not expect the price of lumber to appreciate in the foreseeable

---

[42] Source for data: Energy Information Administration/Pemex (www.eia.doe.gov).

[43] *Ibid.*

[44] Figures from 2010 on are U.S. census projections. Source for data: The 2008 Statistical Abstract (www.census.gov/).

[45] *Ibid.*

[46] The model is the authors', and is based on data in *Environmental and Natural Resource Economics* by Tom Tietenberg, third edition (New York: HarperCollins, 1992), p. 282.

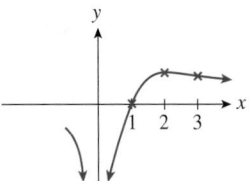

Figure 39

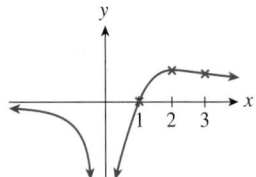

Figure 40

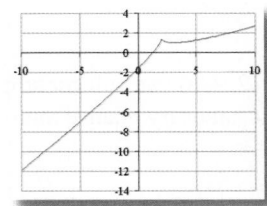

Technology:
2*x/3-((x-2)^2)^(1/3)

Figure 41

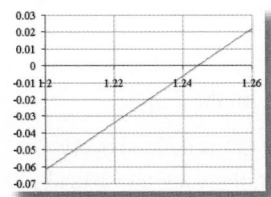

Figure 42

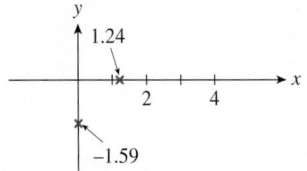

Figure 43

From these limits, we see the following:

(1) Immediately to the *left* of $x = 0$, the graph plunges down toward $-\infty$.

(2) Immediately to the *right* of $x = 0$, the graph also plunges down toward $-\infty$.

Figure 39 shows our graph with these features added. We say that f has a **vertical asymptote** at $x = 0$, meaning that the points on the graph of f get closer and closer to points on a vertical line (the y-axis in this case) further and further from the origin.

5. *Behavior at infinity:* Both $1/x$ and $1/x^2$ go to 0 as x goes to $-\infty$ or $+\infty$; that is,

$$\lim_{x \to -\infty} f(x) = 0$$

and

$$\lim_{x \to +\infty} f(x) = 0.$$

Thus, on the extreme left and right of our picture, the height of the curve levels off toward zero. Figure 40 shows the completed freehand sketch of the graph.

We say that f has a **horizontal asymptote** at $y = 0$. (Notice another thing: We haven't plotted a single point to the left of the y-axis, and yet we have a pretty good idea of what the curve looks like there! Compare the technology-drawn curve in Figure 35.)

In summary, there is one x-intercept at $x = 1$; there is one relative maximum (which, we can now see, is also an absolute maximum) at $x = 2$; there is one point of inflection at $x = 3$, where the graph changes from being concave down to concave up. There is a vertical asymptote at $x = 0$, on both sides of which the graph goes down toward $-\infty$, and a horizontal asymptote at $y = 0$.

EXAMPLE 2 Analyzing a Graph

Analyze the graph of $f(x) = \dfrac{2x}{3} - (x - 2)^{2/3}$.

Solution Figure 41 shows a technology-generated version of the graph. Note that in the technology formulation $(x - 2)^{2/3}$ is written as $[(x - 2)^2]^{1/3}$ to avoid problems with some graphing calculators and Excel.

Let us now re-create this graph by hand, and in the process identify the features we see in Figure 41.

1. *The x- and y-intercepts:* We consider $y = \dfrac{2x}{3} - (x - 2)^{2/3}$. For the y-intercept, we set $x = 0$ and solve for y:

$$y = \frac{2(0)}{3} - (0 - 2)^{2/3} = -2^{2/3} \approx -1.59.$$

To find the x-intercept(s), we set $y = 0$ and solve for x. However, if we attempt this, we will find ourselves with a cubic equation that is hard to solve. (Try it!) Following the advice in the note on page 911, we use graphing technology to locate the x-intercept we see in Figure 41 by zooming in (Figure 42). From Figure 42, we find $x \approx 1.24$. We shall see in the discussion to follow that there can be no other x-intercepts.

Figure 43 shows our freehand sketch so far.

2. *Relative extrema:* We calculate

$$f'(x) = \frac{2}{3} - \frac{2}{3}(x-2)^{-1/3}$$

$$= \frac{2}{3} - \frac{2}{3(x-2)^{1/3}}.$$

To find any stationary points, we set the derivative equal to 0 and solve for x:

$$\frac{2}{3} - \frac{2}{3(x-2)^{1/3}} = 0$$

$$(x-2)^{1/3} = 1$$

$$x - 2 = 1^3 = 1$$

$$x = 3.$$

To check for singular points, look for points where $f(x)$ is defined and $f'(x)$ is not defined. The only such point is $x = 2$: $f'(x)$ is not defined at $x = 2$, whereas $f(x)$ is defined there, so we have a singular point at $x = 2$.

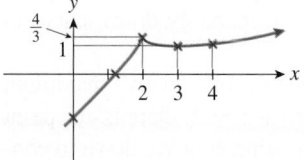

x	2 (Singular point)	3 (Stationary point)	4 (Test point)
$y = \dfrac{2x}{3} - (x-2)^{2/3}$	$\dfrac{4}{3}$	1	1.079

Figure 44

Figure 44 shows our graph so far.

We see that there is a singular relative maximum at $(2, 4/3)$ (we will confirm that the graph eventually gets higher on the right) and a stationary relative minimum at $x = 3$.

3. *Points of inflection:* We calculate

$$f''(x) = \frac{2}{9(x-2)^{4/3}}.$$

To find points of inflection, we set the second derivative equal to 0 and solve for x. But the equation

$$0 = \frac{2}{9(x-2)^{4/3}}$$

has no solution for x, so there are no points of inflection on the graph.

4. *Behavior near points where f is not defined:* Because $f(x)$ is defined everywhere, there are no such points to consider. In particular, there are no vertical asymptotes.

5. *Behavior at infinity:* We estimate the following limits numerically:

$$\lim_{x \to -\infty} \left[\frac{2x}{3} - (x-2)^{2/3} \right] = -\infty$$

and

$$\lim_{x \to +\infty} \left[\frac{2x}{3} - (x-2)^{2/3} \right] = +\infty.$$

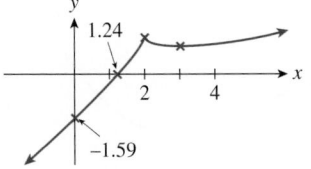

y

1.24

2   4

x

−1.59

Figure 45

Thus, on the extreme left the curve goes down toward $-\infty$, and on the extreme right the curve rises toward $+\infty$. In particular, there are no horizontal asymptotes. (There can also be no other x-intercepts.)

Figure 45 shows the completed graph.

<hr/>

12.4 **EXERCISES**

▼ more advanced   ◆ challenging
🔲 indicates exercises that should be solved using technology

In Exercises 1–26, sketch the graph of the given function, indicating (a) x- and y-intercepts, (b) extrema, (c) points of inflection, (d) behavior near points where the function is not defined, and (e) behavior at infinity. Where indicated, technology should be used to approximate the intercepts, coordinates of extrema, and/or points of inflection to one decimal place. Check your sketch using technology. HINT [See Example 1.]

1. $f(x) = x^2 + 2x + 1$

2. $f(x) = -x^2 - 2x - 1$

3. $g(x) = x^3 - 12x$, domain $[-4, 4]$

4. $g(x) = 2x^3 - 6x$, domain $[-4, 4]$

5. $h(x) = 2x^3 - 3x^2 - 36x$ [Use technology for x-intercepts.]

6. $h(x) = -2x^3 - 3x^2 + 36x$ [Use technology for x-intercepts.]

7. $f(x) = 2x^3 + 3x^2 - 12x + 1$ [Use technology for x-intercepts.]

8. $f(x) = 4x^3 + 3x^2 + 2$ [Use technology for x-intercepts.]

9. $k(x) = -3x^4 + 4x^3 + 36x^2 + 10$ [Use technology for x-intercepts.]

10. $k(x) = 3x^4 + 4x^3 - 36x^2 - 10$ [Use technology for x-intercepts.]

11. $g(t) = \frac{1}{4}t^4 - \frac{2}{3}t^3 + \frac{1}{2}t^2$

12. $g(t) = 3t^4 - 16t^3 + 24t^2 + 1$

13. $f(x) = x + \frac{1}{x}$

14. $f(x) = x^2 + \frac{1}{x^2}$

15. $g(x) = x^3/(x^2 + 3)$

16. $g(x) = x^3/(x^2 - 3)$

17. $f(t) = \frac{t^2 + 1}{t^2 - 1}$, domain $[-2, 2]$, $t \neq \pm 1$

18. $f(t) = \frac{t^2 - 1}{t^2 + 1}$, domain $[-2, 2]$

19. $k(x) = \frac{2x}{3} + (x + 1)^{2/3}$ [Use technology for x-intercepts.]
    HINT [See Example 2.]

20. $k(x) = \frac{2x}{5} - (x - 1)^{2/5}$ [Use technology for x-intercepts.]
    HINT [See Example 2.]

21. $f(x) = x - \ln x$, domain $(0, +\infty)$

22. $f(x) = x - \ln x^2$, domain $(0, +\infty)$

23. $f(x) = x^2 + \ln x^2$ [Use technology for x-intercepts.]

24. $f(x) = 2x^2 + \ln x$ [Use technology for x-intercepts.]

25. $g(t) = e^t - t$, domain $[-1, 1]$

26. $g(t) = e^{-t^2}$

🔲 *In Exercises 27–30, use technology to sketch the graph of the given function, labeling all relative and absolute extrema and points of inflection, and vertical and horizontal asymptotes. The coordinates of the extrema and points of inflection should be accurate to two decimal places.* HINT [To locate extrema accurately, plot the first derivative; to locate points of inflection accurately, plot the second derivative.]

27. ▼ $f(x) = x^4 - 2x^3 + x^2 - 2x + 1$

28. ▼ $f(x) = x^4 + x^3 + x^2 + x + 1$

29. ▼ $f(x) = e^x - x^3$

30. ▼ $f(x) = e^x - \frac{x^4}{4}$

APPLICATIONS

31. *Home Prices* The following graph approximates historical and projected median home prices in the United States for the period 2000–2020:[47]

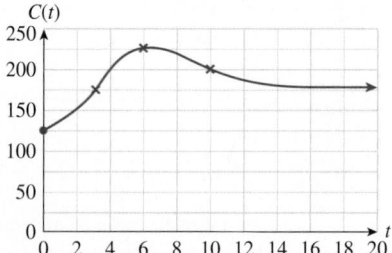

Here, t is time in years since the start of 2000 and $C(t)$ is the median home price in thousands of dollars. The locations of extrema and points of inflection are indicated on the graph. Analyze the graph's important features and interpret each feature in terms of the median home price.

32. *Housing Starts* The following graph approximates historical and projected numbers of housing starts of single-family homes each year in the United States for the period 2000–2020:[48]

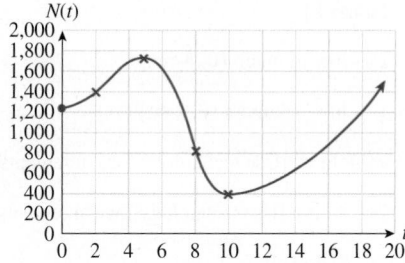

Here, t is time in years since 2000 and $N(t)$ is the number, in thousands, of housing starts per year. The locations of extrema and points of inflection are indicated on the graph. Analyze the graph's important features and interpret each feature in terms of the number of housing starts.

33. *Consumer Price Index* The following graph shows the approximate value of the U.S. Consumer Price Index (CPI) from July 2005 through March 2006.[49]

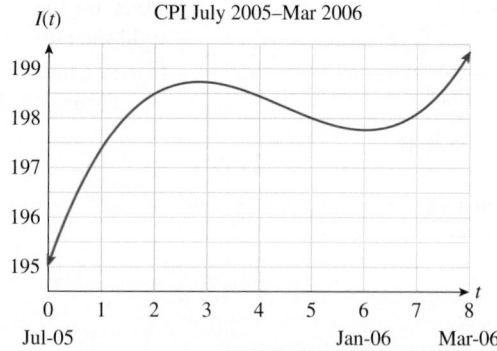

The approximating curve shown on the figure is given by

$$I(t) = 0.06t^3 - 0.8t^2 + 3.1t + 195 \quad (0 \le t \le 8),$$

where t is time in months ($t = 0$ represents July 2005).

a. Locate the intercepts, extrema, and points of inflection of the curve and interpret each feature in terms of the CPI. (Approximate all coordinates to one decimal place.) HINT [See Example 1.]

b. Recall from Section 12.2 that the inflation rate is defined to be $\dfrac{I'(t)}{I(t)}$. What do the stationary extrema of the curve shown above tell you about the inflation rate?

34. *Consumer Price Index* The following graph shows the approximate value of the U.S. Consumer Price Index (CPI) from March 2006 through May 2007.[50]

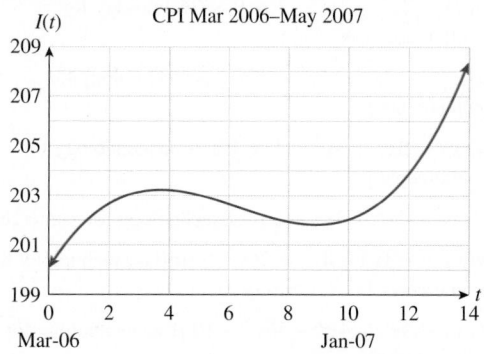

The approximating curve shown on the figure is given by

$$I(t) = 0.02t^3 - 0.38t^2 + 2t + 200 \quad (0 \le t \le 14),$$

where t is time in months ($t = 0$ represents March, 2006).

a. Locate the intercepts, extrema, and points of inflection of the curve and interpret each feature in terms of the CPI. (Approximate all coordinates to one decimal place.) HINT [See Example 1.]

[47] Values from 2012 on are authors' projections. Source for data through 2011: Zillow (www.zillow.com).

[48] Values from 2011 on are authors' projections. Source for data through 2010: U.S. Census Bureau (www.census.gov).

[49] The CPI is compiled by the Bureau of Labor Statistics and is based upon a 1982 value of 100. For instance, a CPI of 200 means the CPI has doubled since 1982. Source: InflationData.com. (www.inflationdata.com).

[50] *Ibid.*

b. Recall from Section 12.2 that the inflation rate is defined to be $\dfrac{I'(t)}{I(t)}$. What do the stationary extrema of the curve shown above tell you about the inflation rate?

35. *Motion in a Straight Line* The distance of a UFO from an observer is given by $s = 2t^3 - 3t^2 + 100$ feet after t seconds ($t \geq 0$). Obtain the extrema, points of inflection, and behavior at infinity. Sketch the curve and interpret these features in terms of the movement of the UFO.

36. *Motion in a Straight Line* The distance of the Mars orbiter from your location in Utarek, Mars is given by $s = 2(t-1)^3 -3(t-1)^2 + 100$ km after t seconds ($t \geq 0$). Obtain the extrema, points of inflection, and behavior at infinity. Sketch the curve and interpret these features in terms of the movement of the Mars orbiter.

37. *Average Cost: iPods* Assume that it costs Apple approximately

$$C(x) = 22,500 + 100x + 0.01x^2$$

dollars to manufacture x 32 GB iPods in a day.[51] Obtain the average cost function, sketch its graph, and analyze the graph's important features. Interpret each feature in terms of iPods. HINT [Recall that the average cost function is $\bar{C}(x) = C(x)/x$.]

38. *Average Cost: Xboxes* Assume that it costs Microsoft approximately

$$C(x) = 14,400 + 550x + 0.01x^2$$

dollars to manufacture x Xbox 360s in a day.[52] Obtain the average cost function, sketch its graph, and analyze the graph's important features. Interpret each feature in terms of Xboxes. HINT [Recall that the average cost function is $\bar{C}(x) = C(x)/x$.]

39. ▣ ▾ *Subprime Mortgages during the Housing Bubble* During the real estate run-up in 2000–2008, the percentage of mortgages issued in the United States that were subprime (normally classified as risky) could be approximated by

$$A(t) = \frac{15.0}{1 + 8.6(1.8)^{-t}} \text{ percent} \qquad (0 \leq t \leq 8)$$

t years after the start of 2000.[53] Graph the *derivative* $A'(t)$ of $A(t)$ using an extended domain of $0 \leq t \leq 15$. Determine

the approximate coordinates of the maximum and determine the behavior of $A'(t)$ at infinity. What do the answers tell you?

40. ▣ ▾ *Subprime Mortgage Debt during the Housing Bubble* During the real estate run-up in 2000–2008, the value of subprime (normally classified as risky) mortgage debt outstanding in the United States was approximately

$$A(t) = \frac{1,350}{1 + 4.2(1.7)^{-t}} \text{ billion dollars} \qquad (0 \leq t \leq 8)$$

t years after the start of 2000.[54] Graph the *derivative* $A'(t)$ of $A(t)$ using an extended domain of $0 \leq t \leq 15$. Determine the approximate coordinates of the maximum and determine the behavior of $A'(t)$ at infinity. What do the answers tell you?

COMMUNICATION AND REASONING EXERCISES

41. A function is *bounded* if its entire graph lies between two horizontal lines. Can a bounded function have vertical asymptotes? Can a bounded function have horizontal asymptotes? Explain.

42. A function is *bounded above* if its entire graph lies below some horizontal line. Can a bounded above function have vertical asymptotes? Can a bounded above function have horizontal asymptotes? Explain.

43. If the graph of a function has a vertical asymptote at $x = a$ in such a way that y increases to $+\infty$ as $x \to a$, what can you say about the graph of its derivative? Explain.

44. If the graph of a function has a horizontal asymptote at $y = a$ in such a way that y decreases to a as $x \to +\infty$, what can you say about the graph of its derivative? Explain.

45. Your friend tells you that he has found a continuous function defined on $(-\infty, +\infty)$ with exactly two critical points, each of which is a relative maximum. Can he be right?

46. Your other friend tells you that she has found a continuous function with two critical points, one a relative minimum and one a relative maximum, and no point of inflection between them. Can she be right?

47. ▾ By thinking about extrema, show that, if $f(x)$ is a polynomial, then between every pair of zeros (x-intercepts) of $f(x)$ there is a zero of $f'(x)$.

48. ▾ If $f(x)$ is a polynomial of degree 2 or higher, show that between every pair of relative extrema of $f(x)$ there is a point of inflection of $f(x)$.

---

[51] Not the actual cost equation; the authors do not know Apple's actual cost equation. The marginal cost in the model given is in rough agreement with the actual marginal cost for reasonable values of x for one of the 2007 models. Source for cost data: *Manufacturing & Technology News,* July 31, 2007 Volume 14, No. 14 (www.manufacturingnews.com).

[52] Not the actual cost equation; the authors do not know Microsoft's actual cost equation. The marginal cost in the model given is in rough agreement with the actual marginal cost for reasonable values of x. Source for estimate of marginal cost: iSuppli: (www.isuppli.com/news/xbox/).

[53] 2009 figure is an estimate. Sources: Mortgage Bankers Association, UBS.

[54] 2008–2009 figure are estimates. Source: www.data360.org/dataset.aspx?Data_Set_Id=9549.

12.5 | Related Rates

We start by recalling some basic facts about the rate of change of a quantity:

Rate of Change of Q

If Q is a quantity changing over time t, then the derivative dQ/dt is the rate at which Q changes over time.

Quick Examples

1. If A is the area of an expanding circle, then dA/dt is the rate at which the area is increasing.
2. *Words:* The radius r of a sphere is currently 3 cm and increasing at a rate of 2 cm/sec.
   Symbols: $r = 3$ cm and $dr/dt = 2$ cm/sec.

In this section we are concerned with what are called **related rates** problems. In such a problem we have two (sometimes more) related quantities, we know the rate at which one is changing, and we wish to find the rate at which another is changing. A typical example is the following.

EXAMPLE 1 The Expanding Circle

The radius of a circle is increasing at a rate of 10 cm/sec. How fast is the area increasing at the instant when the radius has reached 5 cm?

Solution We have two related quantities: the radius of the circle, r, and its area, A. The first sentence of the problem tells us that r is increasing at a certain rate. When we see a sentence referring to speed or change, it is very helpful to rephrase the sentence using the phrase "the rate of change of." Here, we can say

> *The rate of change of r is* 10 cm/sec.

Because the rate of change is the derivative, we can rewrite this sentence as the equation

$$\frac{dr}{dt} = 10.$$

Similarly, the second sentence of the problem asks how A is changing. We can rewrite that question:

> *What is the rate of change of A when the radius is 5 cm?*

Using mathematical notation, the question is:

> *What is* $\dfrac{dA}{dt}$ *when r = 5?*

Thus, knowing one rate of change, dr/dt, we wish to find a related rate of change, dA/dt. To find exactly how these derivatives are related, we need the equation relating the variables, which is

$$A = \pi r^2.$$

To find the relationship between the derivatives, we take the derivative of both sides of this equation *with respect to t*. On the left we get dA/dt. On the right we need to remember that r is a function of t and use the chain rule. We get

$$\frac{dA}{dt} = 2\pi r \frac{dr}{dt}.$$

Now we substitute the given values $r = 5$ and $dr/dt = 10$. This gives

$$\left.\frac{dA}{dt}\right|_{r=5} = 2\pi(5)(10) = 100\pi \approx 314 \text{ cm}^2/\text{sec}.$$

Thus, the area is increasing at the rate of 314 cm^2/sec when the radius is 5 cm.

We can organize our work as follows:

Solving a Related Rates Problem

A. The Problem

1. List the related, changing quantities.
2. Restate the problem in terms of rates of change. Rewrite the problem using mathematical notation for the changing quantities and their derivatives.

B. The Relationship

1. Draw a diagram, if appropriate, showing the changing quantities.
2. Find an equation or equations relating the changing quantities.
3. Take the derivative with respect to time of the equation(s) relating the quantities to get the **derived equation(s)**, which relate the rates of change of the quantities.

C. The Solution

1. Substitute into the derived equation(s) the given values of the quantities and their derivatives.
2. Solve for the derivative required.

We can illustrate the procedure with the "ladder problem" found in almost every calculus textbook.

EXAMPLE 2 The Falling Ladder

Jane is at the top of a 5-foot ladder when it starts to slide down the wall at a rate of 3 feet per minute. Jack is standing on the ground behind her. How fast is the base of the ladder moving when it hits him if Jane is 4 feet from the ground at that instant?

Solution The first sentence talks about (the top of) the ladder sliding down the wall. Thus, one of the changing quantities is the height of the top of the ladder. The question asked refers to the motion of the base of the ladder, so another changing quantity is the distance of the base of the ladder from the wall. Let's record these variables and follow the outline above to obtain the solution.

A. The Problem

1. The changing quantities are

h = height of the top of the ladder
b = distance of the base of the ladder from the wall

2. We rephrase the problem in words, using the phrase "rate of change":

The rate of change of the height of the top of the ladder is −3 feet per minute. What is the rate of change of the distance of the base from the wall when the top of the ladder is 4 feet from the ground?

We can now rewrite the problem mathematically:

$$\frac{dh}{dt} = -3. \text{ Find } \frac{db}{dt} \text{ when } h = 4.$$

B. The Relationship

Figure 46

1. Figure 46 shows the ladder and the variables h and b. Notice that we put in the figure the fixed length, 5, of the ladder, but any changing quantities, like h and b, we leave as variables. We shall not use any specific values for h or b until the very end.

2. From the figure, we can see that h and b are related by the Pythagorean theorem:

$$h^2 + b^2 = 25.$$

3. Taking the derivative with respect to time of the equation above gives us the derived equation:

$$2h\frac{dh}{dt} + 2b\frac{db}{dt} = 0.$$

C. The Solution

1. We substitute the known values $dh/dt = -3$ and $h = 4$ into the derived equation:

$$2(4)(-3) + 2b\frac{db}{dt} = 0.$$

We would like to solve for db/dt, but first we need the value of b, which we can determine from the equation $h^2 + b^2 = 25$, using the value $h = 4$:

$$16 + b^2 = 25$$
$$b^2 = 9$$
$$b = 3.$$

Substituting into the derived equation, we get

$$-24 + 2(3)\frac{db}{dt} = 0.$$

2. Solving for db/dt gives

$$\frac{db}{dt} = \frac{24}{6} = 4.$$

Thus, the base of the ladder is sliding away from the wall at 4 ft/min when it hits Jack.

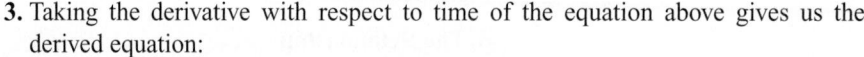

EXAMPLE 3 **Average Cost**

The cost to manufacture x cellphones in a day is

$$C(x) = 10,000 + 20x + \frac{x^2}{10,000} \text{ dollars.}$$

The daily production level is currently $x = 5,000$ cellphones and is increasing at a rate of 100 units per day. How fast is the average cost changing?

Solution

A. The Problem

1. The changing quantities are the production level x and the average cost, $\bar{C}$.

2. We rephrase the problem as follows:

> *The daily production level is $x = 5,000$ units and the rate of change of x is 100 units/day. What is the rate of change of the average cost, $\bar{C}$?*

In mathematical notation,

$$x = 5,000 \text{ and } \frac{dx}{dt} = 100. \text{ Find } \frac{d\bar{C}}{dt}.$$

B. The Relationship

1. In this example the changing quantities cannot easily be depicted geometrically.

2. We are given a formula for the *total* cost. We get the *average* cost by dividing the total cost by x:

$$\bar{C} = \frac{C}{x}.$$

So,

$$\bar{C} = \frac{10,000}{x} + 20 + \frac{x}{10,000}.$$

3. Taking derivatives with respect to t of both sides, we get the derived equation:

$$\frac{d\bar{C}}{dt} = \left(-\frac{10,000}{x^2} + \frac{1}{10,000} \right) \frac{dx}{dt}.$$

C. The Solution

Substituting the values from part A into the derived equation, we get

$$\frac{d\bar{C}}{dt} = \left(-\frac{10,000}{5,000^2} + \frac{1}{10,000} \right) 100$$

$$= -0.03 \text{ dollars/day.}$$

Thus, the average cost is decreasing by 3¢ per day.

The scenario in the following example is similar to Example 5 in Section 12.2.

EXAMPLE 4 **Allocation of Labor**

The Gym Sock Company manufactures cotton athletic socks. Production is partially automated through the use of robots. The number of pairs of socks the company can manufacture in a day is given by a Cobb-Douglas production formula:

$$q = 50n^{0.6}r^{0.4},$$

where q is the number of pairs of socks that can be manufactured by n laborers and r robots. The company currently produces 1,000 pairs of socks each day and employs 20 laborers. It is bringing one new robot on line every month. At what rate are laborers being laid off, assuming that the number of socks produced remains constant?

Solution

A. The Problem

1. The changing quantities are the number of laborers n and the number of robots r.

2. $\dfrac{dr}{dt} = 1$. Find $\dfrac{dn}{dt}$ when $n = 20$.

B. The Relationship

1. No diagram is appropriate here.

2. The equation relating the changing quantities:

$$1,000 = 50n^{0.6}r^{0.4} \qquad \text{Productivity is constant at 1,000 pairs of socks each day.}$$

or

$$20 = n^{0.6}r^{0.4}.$$

3. The derived equation is

$$0 = 0.6n^{-0.4}\left(\frac{dn}{dt}\right)r^{0.4} + 0.4n^{0.6}r^{-0.6}\left(\frac{dr}{dt}\right)$$

$$= 0.6\left(\frac{r}{n}\right)^{0.4}\left(\frac{dn}{dt}\right) + 0.4\left(\frac{n}{r}\right)^{0.6}\left(\frac{dr}{dt}\right).$$

We solve this equation for dn/dt because we shall want to find dn/dt below and because the equation becomes simpler when we do this:

$$0.6\left(\frac{r}{n}\right)^{0.4}\left(\frac{dn}{dt}\right) = -0.4\left(\frac{n}{r}\right)^{0.6}\left(\frac{dr}{dt}\right)$$

$$\frac{dn}{dt} = -\frac{0.4}{0.6}\left(\frac{n}{r}\right)^{0.6}\left(\frac{n}{r}\right)^{0.4}\left(\frac{dr}{dt}\right)$$

$$= -\frac{2}{3}\left(\frac{n}{r}\right)\left(\frac{dr}{dt}\right).$$

C. The Solution

Substituting the numbers in A into the last equation in B, we get

$$\frac{dn}{dt} = -\frac{2}{3}\left(\frac{20}{r}\right) \quad (1).$$

We need to compute r by substituting the known value of n in the original formula:

$$20 = n^{0.6}r^{0.4}$$
$$20 = 20^{0.6}r^{0.4}$$
$$r^{0.4} = \frac{20}{20^{0.6}} = 20^{0.4}$$
$$r = 20.$$

Thus,

$$\frac{dn}{dt} = -\frac{2}{3}\left(\frac{20}{20}\right)(1) = -\frac{2}{3} \text{ laborers per month.}$$

The company is laying off laborers at a rate of 2/3 per month, or two every three months.

We can interpret this result as saying that, at the current level of production and number of laborers, one robot is as productive as 2/3 of a laborer, or 3 robots are as productive as 2 laborers.

12.5 EXERCISES

▼ more advanced   ◆ challenging
T indicates exercises that should be solved using technology

Rewrite the statements and questions in Exercises 1–8 in mathematical notation. HINT [See Quick Examples on page 918.]

1. The population P is currently 10,000 and growing at a rate of 1,000 per year.

2. There are presently 400 cases of Bangkok flu, and the number is growing by 30 new cases every month.

3. The annual revenue of your tie-dye T-shirt operation is currently $7,000 but is decreasing by $700 each year. How fast are annual sales changing?

4. A ladder is sliding down a wall so that the distance between the top of the ladder and the floor is decreasing at a rate of 3 feet per second. How fast is the base of the ladder receding from the wall?

5. The price of shoes is rising $5 per year. How fast is the demand changing?

6. Stock prices are rising $1,000 per year. How fast is the value of your portfolio increasing?

7. The average global temperature is 60°F and rising by 0.1°F per decade. How fast are annual sales of Bermuda shorts increasing?

8. The country's population is now 260,000,000 and is increasing by 1,000,000 people per year. How fast is the annual demand for diapers increasing?

APPLICATIONS

9. *Sun Spots* The area of a circular sun spot is growing at a rate of 1,200 km²/sec.
   a. How fast is the radius growing at the instant when it equals 10,000 km? HINT [See Example 1.]
   b. How fast is the radius growing at the instant when the sun spot has an area of 640,000 km²? HINT [Use the area formula to determine the radius at that instant.]

10. *Puddles* The radius of a circular puddle is growing at a rate of 5 cm/sec.
   a. How fast is its area growing at the instant when the radius is 10 cm? HINT [See Example 1.]
   b. How fast is the area growing at the instant when it equals 36 cm²? HINT [Use the area formula to determine the radius at that instant.]

11. *Balloons* A spherical party balloon is being inflated with helium pumped in at a rate of 3 cubic feet per minute. How fast is the radius growing at the instant when the radius has reached 1 foot? (The volume of a sphere of radius r is $V = \frac{4}{3}\pi r^3$.) HINT [See Example 1.]

12. *More Balloons* A rather flimsy spherical balloon is designed to pop at the instant its radius has reached 10 centimeters. Assuming the balloon is filled with helium at a rate of 10 cubic centimeters per second, calculate how fast the radius is growing at the instant it pops. (The volume of a sphere of radius r is $V = \frac{4}{3}\pi r^3$.) HINT [See Example 1.]

13. **Sliding Ladders** The base of a 50-foot ladder is being pulled away from a wall at a rate of 10 feet per second. How fast is the top of the ladder sliding down the wall at the instant when the base of the ladder is 30 feet from the wall? HINT [See Example 2.]

14. **Sliding Ladders** The top of a 5-foot ladder is sliding down a wall at a rate of 10 feet per second. How fast is the base of the ladder sliding away from the wall at the instant when the top of the ladder is 3 feet from the ground? HINT [See Example 2.]

15. **Average Cost** The average cost function for the weekly manufacture of portable CD players is given by

$$\bar{C}(x) = 150{,}000x^{-1} + 20 + 0.0001x \text{ dollars per player,}$$

where x is the number of CD players manufactured that week. Weekly production is currently 3,000 players and is increasing at a rate of 100 players per week. What is happening to the average cost? HINT [See Example 3.]

16. **Average Cost** Repeat the preceding exercise, using the revised average cost function

$$\bar{C}(x) = 150{,}000x^{-1} + 20 + 0.01x \text{ dollars per player.}$$

HINT [See Example 3.]

17. **Demand** Demand for your tie-dyed T-shirts is given by the formula

$$q = 500 - 100p^{0.5},$$

where q is the number of T-shirts you can sell each month at a price of p dollars. If you currently sell T-shirts for $15 each and you raise your price by $2 per month, how fast will the demand drop? (Round your answer to the nearest whole number.)

18. **Supply** The number of portable CD players you are prepared to supply to a retail outlet every week is given by the formula

$$q = 0.1p^2 + 3p,$$

where p is the price it offers you. The retail outlet is currently offering you $40 per CD player. If the price it offers decreases at a rate of $2 per week, how will this affect the number you supply?

19. **Revenue** You can now sell 50 cups of lemonade per week at 30¢ per cup, but demand is dropping at a rate of 5 cups per week each week. Assuming that raising the price does not affect demand, how fast do you have to raise your price if you want to keep your weekly revenue constant? HINT [Revenue = Price × Quantity.]

20. **Revenue** You can now sell 40 cars per month at $20,000 per car, and demand is increasing at a rate of 3 cars per month each month. What is the fastest you could drop your price before your monthly revenue starts to drop? HINT [Revenue = Price × Quantity.]

21. ▼ **Oil Revenues** Daily oil production by **Pemex**, Mexico's national oil company, can be approximated by

$$q(t) = -0.022t^2 + 0.2t + 2.9 \text{ million barrels} \quad (1 \le t \le 9),$$

where t is time in years since the start of 2000.[55]

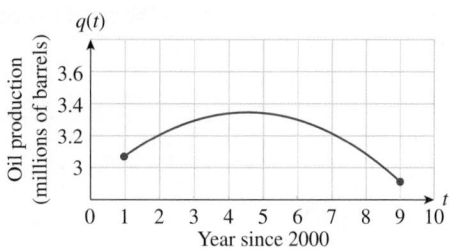

At the start of 2008 the price of oil was $90 per barrel and increasing at a rate of $80 per year.[56] How fast was Pemex's oil (daily) revenue changing at that time?

22. ▼ **Oil Expenditures** Daily oil imports to the United States from Mexico can be approximated by

$$q(t) = -0.015t^2 + 0.1t + 1.4 \text{ million barrels} \quad (0 \le t \le 8),$$

where t is time in years since the start of 2000.[57]

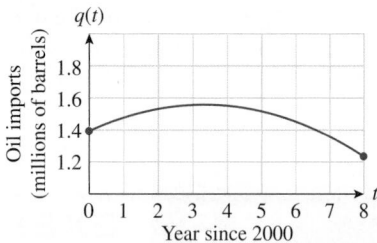

At the start of 2004 the price of oil was $30 per barrel and increasing at a rate of $40 per year.[58] How fast was (daily) oil expenditure for imports from Mexico changing at that time?

23. **Resource Allocation** Your company manufactures automobile alternators, and production is partially automated through the use of robots. In order to meet production deadlines, your company calculates that the numbers of laborers and robots must satisfy the constraint

$$xy = 10{,}000,$$

where x is the number of laborers and y is the number of robots. Your company currently uses 400 robots and is increasing robot deployment at a rate of 16 per month. How fast is it laying off laborers? HINT [See Example 4.]

24. **Resource Allocation** Your company is the largest sock manufacturer in the Solar System, and production is automated through the use of androids and robots. In order to meet production deadlines, your company calculates that the numbers of androids and robots must satisfy the constraint

$$xy = 1{,}000{,}000,$$

[55] Source for data: Energy Information Administration/Pemex (www.eia.doe.gov).

[56] Based on NYMEX crude oil futures; average rate of change during January–June, 2008.

[57] Source for data: Energy Information Administration/Pemex (www.eia.doe.gov).

[58] Based on NYMEX crude oil futures; average rate of change during 2004–2005.

where x is the number of androids and y is the number of robots. Your company currently uses 5000 androids and is increasing android deployment at a rate of 200 per month. How fast is it scrapping robots? HINT [See Example 4.]

25. Production The automobile assembly plant you manage has a Cobb-Douglas production function given by

$$P = 10x^{0.3}y^{0.7},$$

where P is the number of automobiles it produces per year, x is the number of employees, and y is the daily operating budget (in dollars). You maintain a production level of 1,000 automobiles per year. If you currently employ 150 workers and are hiring new workers at a rate of 10 per year, how fast is your daily operating budget changing? HINT [See Example 4.]

26. Production Refer back to the Cobb-Douglas production formula in the preceding exercise. Assume that you maintain a constant workforce of 200 workers and wish to increase production in order to meet a demand that is increasing by 100 automobiles per year. The current demand is 1000 automobiles per year. How fast should your daily operating budget be increasing? HINT [See Example 4.]

27. Demand Assume that the demand equation for tuna in a small coastal town is

$$pq^{1.5} = 50,000,$$

where q is the number of pounds of tuna that can be sold in one month at the price of p dollars per pound. The town's fishery finds that the demand for tuna is currently 900 pounds per month and is increasing at a rate of 100 pounds per month each month. How fast is the price changing?

28. Demand The demand equation for rubies at *Royal Ruby Retailers* is

$$q + \frac{4}{3}p = 80,$$

where q is the number of rubies RRR can sell per week at p dollars per ruby. RRR finds that the demand for its rubies is currently 20 rubies per week and is dropping at a rate of one ruby per week. How fast is the price changing?

29. ▼ Ships Sailing Apart The H.M.S. *Dreadnaught* is 40 miles south of Montauk and steaming due south at 20 miles/hour, while the U.S.S. *Mona Lisa* is 50 miles east of Montauk and steaming due east at an even 30 miles/hour. How fast is their distance apart increasing?

30. ▼ Near Miss My aunt and I were approaching the same intersection, she from the south and I from the west. She was traveling at a steady speed of 10 miles/hour, while I was approaching the intersection at 60 miles/hour. At a certain instant in time, I was one tenth of a mile from the intersection, while she was one twentieth of a mile from it. How fast were we approaching each other at that instant?

31. ▼ Baseball A baseball diamond is a square with side 90 ft.

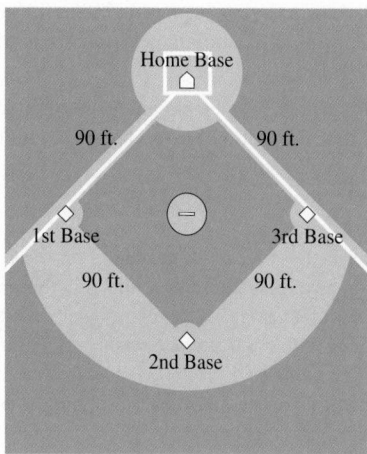

A batter at home base hits the ball and runs toward first base with a speed of 24 ft/sec. At what rate is his distance from third base increasing when he is halfway to first base?

32. ▼ Baseball Refer to Exercise 31. Another player is running from third base to home at 30 ft/sec. How fast is her distance from second base increasing when she is 60 feet from third base?

33. ▼ Movement along a Graph A point on the graph of $y = 1/x$ is moving along the curve in such a way that its x-coordinate is increasing at a rate of 4 units per second. What is happening to the y-coordinate at the instant the y-coordinate is equal to 2?

34. ▼ Motion around a Circle A point is moving along the circle $x^2 + (y - 1)^2 = 8$ in such a way that its x-coordinate is decreasing at a rate of 1 unit per second. What is happening to the y-coordinate at the instant when the point has reached $(-2, 3)$?

35. ▼ Education In 1991, the expected income of an individual depended on his or her educational level according to the following formula:

$$I(n) = 2.929n^3 - 115.9n^2 + 1,530n -$$
$$6,760 \text{ thousand dollars } (12 \le n \le 15).$$

Here, n is the number of school years completed and $I(n)$ is the individual's expected income in thousands of dollars.[59] It is 1991, and you have completed 13 years of school and are currently a part-time student. Your schedule is such that you will complete the equivalent of one year of college every three years. Assuming that your salary is linked to the above model, how fast is your income going up? (Round your answer to the nearest $1.)

---

[59]The model is a based on Table 358, U.S. Department of Education, *Digest of Education Statistics, 1991*, Washington, DC: Government Printing Office, 1991.

36. ▼ *Education* Refer back to the model in the preceding exercise. Assume that you have completed 14 years of school and that your income is increasing by $5,000 per year. How much schooling per year is this rate of increase equivalent to?

37. ▼ *Employment* An employment research company estimates that the value of a recent MBA graduate to an accounting company is

$$V = 3e^2 + 5g^3,$$

where V is the value of the graduate, e is the number of years of prior business experience, and g is the graduate school grade-point average. A company that currently employs graduates with a 3.0 average wishes to maintain a constant employee value of $V = 200$, but finds that the grade-point average of its new employees is dropping at a rate of 0.2 per year. How fast must the experience of its new employees be growing in order to compensate for the decline in grade point average?

38. ▼ *Grades*[60] A production formula for a student's performance on a difficult English examination is given by

$$g = 4hx - 0.2h^2 - 10x^2,$$

where g is the grade the student can expect to obtain, h is the number of hours of study for the examination, and x is the student's grade point average. The instructor finds that students' grade point averages have remained constant at 3.0 over the years, and that students currently spend an average of 15 hours studying for the examination. However, scores on the examination are dropping at a rate of 10 points per year. At what rate is the average study time decreasing?

39. ▼ *Cones* A right circular conical vessel is being filled with green industrial waste at a rate of 100 cubic meters per second. How fast is the level rising after 200π cubic meters have been poured in? The cone has a height of 50 m and a radius of 30 m at its brim. (The volume of a cone of height h and cross-sectional radius r at its brim is given by $V = \frac{1}{3}\pi r^2 h$.)

40. ▼ *More Cones* A circular conical vessel is being filled with ink at a rate of 10 cm³/sec. How fast is the level rising after 20 cm³ have been poured in? The cone has height 50 cm and radius 20 cm at its brim. (The volume of a cone of height h and cross-sectional radius r at its brim is given by $V = \frac{1}{3}\pi r^2 h$.)

41. ▼ *Cylinders* The volume of paint in a right cylindrical can is given by $V = 4t^2 - t$ where t is time in seconds and V is the volume in cm³. How fast is the level rising when the height is 2 cm? The can has a height of 4 cm and a radius of 2 cm. HINT [To get h as a function of t, first solve the volume $V = \pi r^2 h$ for h.]

42. ▼ *Cylinders* A cylindrical bucket is being filled with paint at a rate of 6 cm³ per minute. How fast is the level rising when the bucket starts to overflow? The bucket has a radius of 30 cm and a height of 60 cm.

43. ▼ *Computers vs. Income* The demand for personal computers in the home goes up with household income. For a given community, we can approximate the average number of computers in a home as

$$q = 0.3454 \ln x - 3.047 \quad (10{,}000 \le x \le 125{,}000),$$

where x is mean household income.[61] Your community has a mean income of $30,000, increasing at a rate of $2,000 per year. How many computers per household are there, and how fast is the number of computers in a home increasing? (Round your answer to four decimal places.)

44. ▼ *Computers vs. Income* Refer back to the model in the preceding exercise. The average number of computers per household in your town is 0.5 and is increasing at a rate of 0.02 computers per household per year. What is the average household income in your town, and how fast is it increasing? (Round your answers to the nearest $10).

Education and Crime *The following graph compares the total U.S. prison population and the average combined SAT score in the United States during the 1970s and 1980s:*

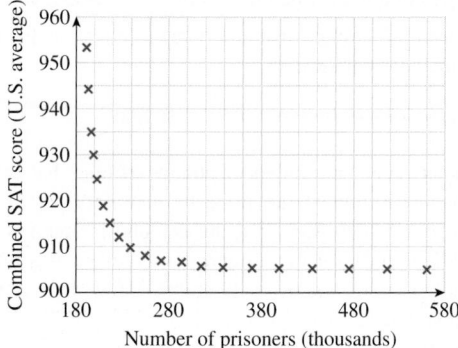

Exercises 45 and 46 are based on the following model for these data:

$$S(n) = 904 + \frac{1{,}326}{(n - 180)^{1.325}} \quad (192 \le n \le 563).$$

Here, $S(n)$ is the combined average SAT score at a time when the total prison population is n thousand.[62]

45. ▼ In 1985, the U.S. prison population was 475,000 and increasing at a rate of 35,000 per year. What was the average SAT score, and how fast, and in what direction, was it changing? (Round your answers to two decimal places.)

---

[60] Based on an Exercise in *Introduction to Mathematical Economics* by A.L. Ostrosky Jr. and J.V. Koch (Waveland Press, Illinois, 1979).

[61] The model is a regression model. Source for data: Income distribution: Computer data: Forrester Research/*New York Times*, August 8, 1999, p. BU4.

[62] Based on data for the years 1967–1989. Sources: Sourcebook of Criminal Justice Statistics, 1990, p. 604/Educational Testing Service.

46. ▼ In 1970, the U.S. combined SAT average was 940 and dropping by 10 points per year. What was the U.S. prison population, and how fast, and in what direction, was it changing? (Round your answers to the nearest 100.)

Divorce Rates *A study found that the divorce rate d (given as a percentage) appears to depend on the ratio r of available men to available women.*[63] *This function can be approximated by*

$$d(r) = \begin{cases} -40r + 74 & \text{if } r \le 1.3 \\ \dfrac{130r}{3} - \dfrac{103}{3} & \text{if } r > 1.3 \end{cases}.$$

Exercises 47 and 48 are based on this model.

47. ◆ There are currently 1.1 available men per available woman in Littleville, and this ratio is increasing by 0.05 per year. What is happening to the divorce rate?

48. ◆ There are currently 1.5 available men per available woman in Largeville, and this ratio is decreasing by 0.03 per year. What is happening to the divorce rate?

COMMUNICATION AND REASONING EXERCISES

49. Why is this section titled "related rates"?

50. If you know how fast one quantity is changing and need to compute how fast a second quantity is changing, what kind of information do you need?

51. In a related rates problem, there is no limit to the number of changing quantities we can consider. Illustrate this by creating a related rates problem with four changing quantities.

---

[63]The cited study, by Scott J. South and associates, appeared in the *American Sociological Review* (February, 1995). Figures are rounded. Source: *New York Times*, February 19, 1995, p. 40.

52. If three quantities are related by a single equation, how would you go about computing how fast one of them is changing based on a knowledge of the other two?

53. ▼ The demand and unit price for your store's checkered T-shirts are changing with time. Show that the percentage rate of change of revenue equals the sum of the percentage rates of change of price and demand. (The percentage rate of change of a quantity Q is $Q'(t)/Q(t)$.)

54. ▼ The number N of employees and the total floor space S of your company are both changing with time. Show that the percentage rate of change of square footage per employee equals the percentage rate of change of S minus the percentage rate of change of N. (The percentage rate of change of a quantity Q is $Q'(t)/Q(t)$.)

55. ▼ In solving a related rates problem, a key step is solving the derived equation for the unknown rate of change (once we have substituted the other values into the equation). Call the unknown rate of change X. The derived equation is what kind of equation in X?

56. ▼ On a recent exam, you were given a related rates problem based on an algebraic equation relating two variables x and y. Your friend told you that the correct relationship between dx/dt and dy/dt was given by

$$\left(\frac{dx}{dt}\right) = \left(\frac{dy}{dt}\right)^2.$$

Could he be correct?

57. ▼ Transform the following into a mathematical statement about derivatives: If my grades are improving at twice the speed of yours, then your grades are improving at half the speed of mine.

58. ▼ If two quantities x and y are related by a linear equation, how are their rates of change related?

12.6 Elasticity

You manufacture an extremely popular brand of sneakers and want to know what will happen if you increase the selling price. Common sense tells you that demand will drop as you raise the price. But will the drop in demand be enough to cause your revenue to fall? Or will it be small enough that your revenue will rise because of the higher selling price? For example, if you raise the price by 1%, you might suffer only a 0.5% loss in sales. In this case, the loss in sales will be more than offset by the increase in price and your revenue will rise. In such a case, we say that the demand is **inelastic**, because it is not very sensitive to the increase in price. On the other hand, if your 1% price increase results in a 2% drop in demand, then raising the price will cause a drop in revenues. We then say that the demand is **elastic** because it reacts strongly to a price change.

✳ Coming up with a good demand equation is not always easy. We saw in Chapter 1 that it is possible to find a linear demand equation if we know the sales figures at two different prices. However, such an equation is only a first approximation. To come up with a more accurate demand equation, we might need to gather data corresponding to sales at several different prices and use curve-fitting techniques like regression. Another approach would be an analytic one, based on mathematical modeling techniques that an economist might use.

That said, we refer you again to *Camels and Rubber Duckies* by Joel Spolsky at www.joelon software.com/articles/Camelsand RubberDuckies.html just in case you think there is nothing more to demand curves.

We can use calculus to measure the response of demand to price changes if we have a demand equation for the item we are selling.✳ We need to know the *percentage drop in demand per percentage increase in price*. This ratio is called the **elasticity of demand**, or **price elasticity of demand**, and is usually denoted by E. Let's derive a formula for E in terms of the demand equation.

Assume that we have a demand equation

$$q = f(p),$$

where q stands for the number of items we would sell (per week, per month, or what have you) if we set the price per item at p. Now suppose we increase the price p by a very small amount, Δp. Then our percentage increase in price is $(\Delta p / p) \times 100\%$. This increase in p will presumably result in a decrease in the demand q. Let's denote this corresponding decrease in q by $-\Delta q$ (we use the minus sign because, by convention, Δq stands for the *increase* in demand). Thus, the percentage decrease in demand is $(-\Delta q / q) \times 100\%$.

Now E is the ratio

$$E = \frac{\text{Percentage decrease in demand}}{\text{Percentage increase in price}}$$

so

$$E = \frac{-\dfrac{\Delta q}{q} \times 100\%}{\dfrac{\Delta p}{p} \times 100\%}.$$

Canceling the 100%s and reorganizing, we get

$$E = -\frac{\Delta q}{\Delta p} \cdot \frac{p}{q}.$$

Q : *What small change in price will we use for* Δp?

A : It should probably be pretty small. If, say, we increased the price of sneakers to $1 million per pair, the sales would likely drop to zero. But knowing this tells us nothing about how the market would respond to a modest increase in price. In fact, we'll do the usual thing we do in calculus and let Δp approach 0.

In the expression for E, if we let Δp go to 0, then the ratio $\Delta q / \Delta p$ goes to the derivative dq/dp. This gives us our final and most useful definition of the elasticity.

Price Elasticity of Demand

The **price elasticity of demand** E is the percentage rate of decrease of demand per percentage increase in price. E is given by the formula

$$E = -\frac{dq}{dp} \cdot \frac{p}{q}.$$

We say that the demand is **elastic** if $E > 1$, is **inelastic** if $E < 1$, and has **unit elasticity** if $E = 1$.

> **Quick Example**
>
> Suppose that the demand equation is $q = 20,000 - 2p$, where p is the price in dollars. Then
>
> $$E = -(-2)\frac{p}{20,000 - 2p} = \frac{p}{10,000 - p}.$$
>
> If $p = \$2,000$, then $E = 1/4$, and demand is inelastic at this price.
> If $p = \$8,000$, then $E = 4$, and demand is elastic at this price.
> If $p = \$5,000$, then $E = 1$, and the demand has unit elasticity at this price.

✱ For another—more rigorous—argument, see Exercise 29.

† See, for example, Exercise 53 in Section 12.5.

We are generally interested in the price that maximizes revenue and, in ordinary cases, the price that maximizes revenue must give unit elasticity. One way of seeing this is as follows:✱ If the demand is inelastic (which ordinarily occurs at a low unit price), then raising the price by a small percentage—1% say—results in a smaller percentage drop in demand. For example, in the Quick Example above, if $p = \$2,000$, then the demand would drop by only $\frac{1}{4}$% for every 1% increase in price. To see the effect on revenue, we use the fact† that, for small changes in price,

Percentage change in revenue ≈ Percentage change in price
$$+ \text{ Percentage change in demand}$$
$$= 1 + \left(-\frac{1}{4}\right) = \frac{3}{4}\%.$$

Thus, the revenue will increase by about 3/4%. Put another way:

If the demand is inelastic, raising the price increases revenue.

On the other hand, if the price is elastic (which ordinarily occurs at a high unit price), then increasing the price slightly will lower the revenue, so:

If the demand is elastic, lowering the price increases revenue.

The price that results in the largest revenue must therefore be at unit elasticity.

EXAMPLE 1 Price Elasticity of Demand: Dolls

Suppose that the demand equation for *Bobby Dolls* is given by $q = 216 - p^2$, where p is the price per doll in dollars and q is the number of dolls sold per week.

a. Compute the price elasticity of demand when $p = \$5$ and $p = \$10$, and interpret the results.

b. Find the range of prices for which the demand is elastic and the range for which the demand is inelastic.

c. Find the price at which the weekly revenue is maximized. What is the maximum weekly revenue?

Solution

a. The price elasticity of demand is

$$E = -\frac{dq}{dp} \cdot \frac{p}{q}.$$

Taking the derivative and substituting for q gives

$$E = 2p \cdot \frac{p}{216 - p^2} = \frac{2p^2}{216 - p^2}.$$

using Technology

See the Technology Guides at the end of the chapter to find out how to automate computations like those in part (a) of Example 1 using a graphing calculator or Excel. Here is an outline:

TI-83/84 Plus
$Y_1 = 216 - X^2$
$Y_2 = -\text{nDeriv}(Y_1, X, X) * X / Y_1$
2ND   TABLE   Enter $x = 5$
[More details on page 943.]

Spreadsheet
Enter values of p: 4.9, 4.91, ..., 5.0, 5.01, ..., 5.1 in A5–A25. In B5 enter `216-A5^2` and copy down to B25. In C5 enter `=(A6-A5)/A5` and paste the formula in C5–D24. In E5 enter `=-D5/C5` and copy down to E24. This column contains the values of E for the values of p in column A. [More details on page 943.]

When $p = \$5$,

$$E = \frac{2(5)^2}{216 - 5^2} = \frac{50}{191} \approx 0.26.$$

Thus, when the price is set at \$5, the demand is dropping at a rate of 0.26% per 1% increase in the price. Because $E < 1$, the demand is inelastic at this price, so raising the price will increase revenue.

When $p = \$10$,

$$E = \frac{2(10)^2}{216 - 10^2} = \frac{200}{116} \approx 1.72.$$

Thus, when the price is set at \$10, the demand is dropping at a rate of 1.72% per 1% increase in the price. Because $E > 1$, demand is elastic at this price, so raising the price will decrease revenue; lowering the price will increase revenue.

b. and **c.** We answer part (c) first. Setting $E = 1$, we get

$$\frac{2p^2}{216 - p^2} = 1$$

$$p^2 = 72.$$

Thus, we conclude that the maximum revenue occurs when $p = \sqrt{72} \approx \$8.49$. We can now answer part (b): The demand is elastic when $p > \$8.49$ (the price is too high), and the demand is inelastic when $p < \$8.49$ (the price is too low). Finally, we calculate the maximum weekly revenue, which equals the revenue corresponding to the price of \$8.49:

$$R = qp = (216 - p^2)p = (216 - 72)\sqrt{72} = 144\sqrt{72} \approx \$1,222.$$

The concept of elasticity can be applied in other situations. In the following example we consider *income* elasticity of demand—the percentage increase in demand for a particular item per percentage increase in personal income.

EXAMPLE 2 Income Elasticity of Demand: Porsches

You are the sales director at *Suburban Porsche* and have noticed that demand for Porsches depends on income according to

$$q = 0.005e^{-0.05x^2 + x} \qquad (1 \le x \le 10).$$

Here, x is the income of a potential customer in hundreds of thousands of dollars and q is the probability that the person will actually purchase a Porsche.* The **income elasticity of demand** is

$$E = \frac{dq}{dx} \frac{x}{q}.$$

Compute and interpret E for $x = 2$ and 9.

* In other words, q is the fraction of visitors to your showroom having income x who actually purchase a Porsche.

Solution

Q: Why is there no negative sign in the formula?

A: Because we anticipate that the demand will increase as income increases, the ratio

$$\frac{\text{Percentage increase in demand}}{\text{Percentage increase in income}}$$

will be positive, so there is no need to introduce a negative sign.

Turning to the calculation, since $q = 0.005e^{-0.05x^2+x}$,

$$\frac{dq}{dx} = 0.005e^{-0.05x^2+x}(-0.1x + 1)$$

and so

$$E = \frac{dq}{dx}\frac{x}{q}$$

$$= 0.005e^{-0.05x^2+x}(-0.1x + 1)\frac{x}{0.005e^{-0.05x^2+x}}$$

$$= x(-0.1x + 1).$$

When $x = 2$, $E = 2[-0.1(2) + 1)] = 1.6$. Thus, at an income level of $200,000, the probability that a customer will purchase a Porsche increases at a rate of 1.6% per 1% increase in income.

When $x = 9$, $E = 9[-0.1(9) + 1)] = 0.9$. Thus, at an income level of $900,000, the probability that a customer will purchase a Porsche increases at a rate of 0.9% per 1% increase in income.

12.6 EXERCISES

▼ more advanced   ◆ challenging
🔧 indicates exercises that should be solved using technology

APPLICATIONS

1. Demand for Oranges The weekly sales of *Honolulu Red Oranges* is given by $q = 1,000 - 20p$. Calculate the price elasticity of demand when the price is $30 per orange (yes, $30 per orange[64]). Interpret your answer. Also, calculate the price that gives a maximum weekly revenue, and find this maximum weekly revenue. HINT [See Example 1.]

2. Demand for Oranges Repeat the preceding exercise for weekly sales of $1,000 - 10p$. HINT [See Example 1.]

3. Tissues The consumer demand equation for tissues is given by $q = (100 - p)^2$, where p is the price per case of tissues and q is the demand in weekly sales.

a. Determine the price elasticity of demand E when the price is set at $30, and interpret your answer.

---

[64]They are very hard to find, and their possession confers considerable social status.

b. At what price should tissues be sold in order to maximize the revenue?

c. Approximately how many cases of tissues would be demanded at that price?

4. Bodybuilding The consumer demand curve for *Professor Stefan Schwarzenegger* dumbbells is given by $q = (100 - 2p)^2$, where p is the price per dumbbell, and q is the demand in weekly sales. Find the price Professor Schwarzenegger should charge for his dumbbells in order to maximize revenue.

5. T-Shirts The Physics Club sells $E = mc^2$ T-shirts at the local flea market. Unfortunately, the club's previous administration has been losing money for years, so you decide to do an analysis of the sales. A quadratic regression based on old sales data reveals the following demand equation for the T-shirts:

$$q = -2p^2 + 33p \quad (9 \le p \le 15).$$

Here, p is the price the club charges per T-shirt, and q is the number it can sell each day at the flea market.

a. Obtain a formula for the price elasticity of demand for $E = mc^2$ T-shirts.

b. Compute the elasticity of demand if the price is set at $10 per shirt. *Interpret the result.*

c. How much should the Physics Club charge for the T-shirts in order to obtain the maximum daily revenue? What will this revenue be?

6. Comics The demand curve for original *Iguanawoman* comics is given by

$$q = \frac{(400 - p)^2}{100} \quad (0 \le p \le 400),$$

where q is the number of copies the publisher can sell per week if it sets the price at $\$p$.

a. Find the price elasticity of demand when the price is set at $40 per copy.

b. Find the price at which the publisher should sell the books in order to maximize weekly revenue.

c. What, to the nearest $1, is the maximum weekly revenue the publisher can realize from sales of *Iguanawoman* comics?

7. College Tuition A study of about 1,800 U.S. colleges and universities resulted in the demand equation $q = 9,900 - 2.2p$, where q is the enrollment at a college or university, and p is the average annual tuition (plus fees) it charges.[65]

a. The study also found that the average tuition charged by universities and colleges was $2,900. What is the corresponding price elasticity of demand? Is the price elastic or inelastic? Should colleges charge more or less on average to maximize revenue?

b. Based on the study, what would you advise a college to charge its students in order to maximize total revenue, and what would the revenue be?

8. Demand for Fried Chicken A fried chicken franchise finds that the demand equation for its new roast chicken product, "Roasted Rooster," is given by

$$p = \frac{40}{q^{1.5}},$$

where p is the price (in dollars) per quarter-chicken serving and q is the number of quarter-chicken servings that can be sold per hour at this price. Express q as a function of p and find the price elasticity of demand when the price is set at $4 per serving. Interpret the result.

9. Paint-By-Number The estimated monthly sales of *Mona Lisa* paint-by-number sets is given by the formula $q = 100e^{-3p^2 + p}$, where q is the demand in monthly sales and p is the retail price in hundreds of yen.

a. Determine the price elasticity of demand E when the retail price is set at ¥300 and interpret your answer.

b. At what price will revenue be a maximum?

c. Approximately how many paint-by-number sets will be sold per month at the price in part (b)?

10. Paint-By-Number Repeat the preceding exercise using the demand equation $q = 100e^{p - 3p^2/2}$.

11. ▼ Linear Demand Functions A general linear demand function has the form $q = mp + b$ (m and b constants, $m \neq 0$).

a. Obtain a formula for the price elasticity of demand at a unit price of p.

b. Obtain a formula for the price that maximizes revenue.

12. ▼ Exponential Demand Functions A general exponential demand function has the form $q = Ae^{-bp}$ (A and b nonzero constants).

a. Obtain a formula for the price elasticity of demand at a unit price of p.

b. Obtain a formula for the price that maximizes revenue.

13. ▼ Hyperbolic Demand Functions A general hyperbolic demand function has the form $q = \dfrac{k}{p^r}$ (r and k nonzero constants).

a. Obtain a formula for the price elasticity of demand at unit price p.

b. How does E vary with p?

c. What does the answer to part (b) say about the model?

14. ▼ Quadratic Demand Functions A general quadratic demand function has the form $q = ap^2 + bp + c$ (a, b, and c constants with $a \neq 0$).

a. Obtain a formula for the price elasticity of demand at a unit price p.

b. Obtain a formula for the price or prices that could maximize revenue.

15. ▼ Modeling Linear Demand You have been hired as a marketing consultant to *Johannesburg Burger Supply, Inc.*, and you wish to come up with a unit price for its hamburgers in order to maximize its weekly revenue. To make life as simple as possible, you assume that the demand equation for Johannesburg hamburgers has the linear form $q = mp + b$, where p is the price per hamburger, q is the demand in weekly sales, and m and b are certain constants you must determine.

a. Your market studies reveal the following sales figures: When the price is set at $2.00 per hamburger, the sales amount to 3,000 per week, but when the price is set at $4.00 per hamburger, the sales drop to zero. Use these data to calculate the demand equation.

b. Now estimate the unit price that maximizes weekly revenue and predict what the weekly revenue will be at that price.

16. ▼ Modeling Linear Demand You have been hired as a marketing consultant to *Big Book Publishing, Inc.*, and you have been approached to determine the best selling price for the hit calculus text by Whiner and Istanbul entitled *Fun with Derivatives*. You decide to make life easy and assume that the demand equation for *Fun with Derivatives* has the linear form $q = mp + b$, where p is the price per book, q is the demand in annual sales, and m and b are certain constants you must determine.

---

[65]Based on a study by A.L. Ostrosky Jr. and J.V. Koch, as cited in their book *Introduction to Mathematical Economics* (Waveland Press, Illinois, 1979) p. 133.

a. Your market studies reveal the following sales figures: When the price is set at $50.00 per book, the sales amount to 10,000 per year; when the price is set at $80.00 per book, the sales drop to 1000 per year. Use these data to calculate the demand equation.

b. Now estimate the unit price that maximizes annual revenue and predict what Big Book Publishing, Inc.'s annual revenue will be at that price.

17. *Income Elasticity of Demand: Live Drama* The likelihood that a child will attend a live theatrical performance can be modeled by

$$q = 0.01(-0.0078x^2 + 1.5x + 4.1) \qquad (15 \le x \le 100).$$

Here, q is the fraction of children with annual household income x thousand dollars who will attend a live dramatic performance at a theater during the year.[66] Compute the income elasticity of demand at an income level of $20,000 and interpret the result. (Round your answer to two significant digits.) HINT [See Example 2.]

18. *Income Elasticity of Demand: Live Concerts* The likelihood that a child will attend a live musical performance can be modeled by

$$q = 0.01(0.0006x^2 + 0.38x + 35) \quad (15 \le x \le 100).$$

Here, q is the fraction of children with annual household income x who will attend a live musical performance during the year.[67] Compute the income elasticity of demand at an income level of $30,000 and interpret the result. (Round your answer to two significant digits.) HINT [See Example 2.]

19. *Income Elasticity of Demand: Broadband in 2010* The following graph shows the percentage q of people in households with annual income x thousand dollars using broadband Internet access in 2010, together with the exponential curve $q = -74e^{-0.021x} + 92$.[68]

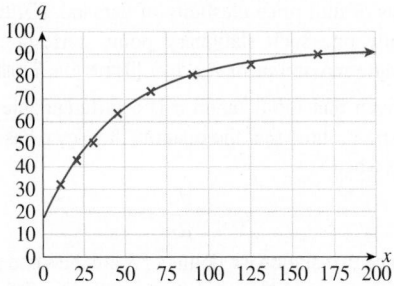

a. Find an equation for the income elasticity of demand for broadband usage, and use it to compute the

elasticity for a household with annual income $100,000 to two decimal places. Interpret the result.

b. What does the model predict as the elasticity of demand for households with very large incomes?

20. *Income Elasticity of Demand: Broadband in 2007* The following graph shows the percentage q of people in households with annual income x thousand dollars using broadband Internet access in 2007, together with the exponential curve $q = -86e^{-0.013x} + 92$.[69]

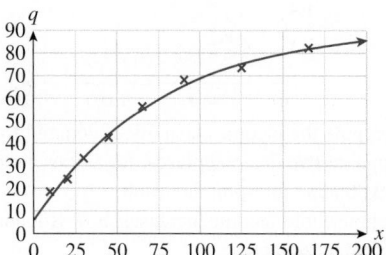

a. Find an equation for the income elasticity of demand for broadband usage, and use it to compute the elasticity for a household with annual income $60,000 to two decimal places. Interpret the result.

b. What does the model predict as the elasticity of demand for households with very large incomes?

21. *Income Elasticity of Demand: Computer Usage in the 1990s* The following graph shows the probability q that a household in the 1990s with annual income x dollars had a computer, together with the logarithmic curve $q = 0.3454 \ln x - 3.047$.[70]

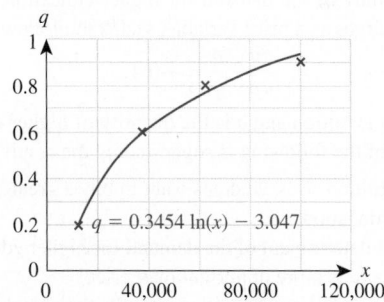

a. Compute the income elasticity of demand for computers, to two decimal places, for a household income of $60,000 and interpret the result.

b. As household income increases, how is income elasticity of demand affected?

c. How reliable is the given model of demand for incomes well above $120,000? Explain.

d. What can you say about E for incomes much larger than those shown?

[66] Based on a quadratic regression of data from a 2001 survey. Source for data: New York Foundation of the Arts (www.nyfa.org/culturalblueprint).

[67] *Ibid.*

[68] Source for data: *Digital Nation: Expanding Internet Usage,* National Telecommunications and Information Administration (U.S. Department of Commerce) (http://search.ntia.doc.gov).

[69] *Ibid.*

[70] Source for data: Income distribution computer data: Forrester Research/*New York Times*, August 8, 1999, p. BU4.

22. ***Income Elasticity of Demand: Internet Usage in the 1990s*** The following graph shows the probability q that a person in the 1990s with household annual income x dollars used the Internet, together with the logarithmic curve $q = 0.2802 \ln x - 2.505$.[71]

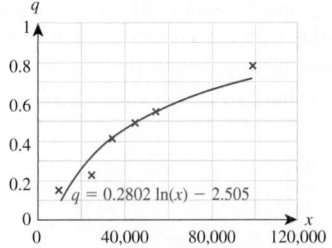

a. Compute the income elasticity of demand for Internet usage, to two decimal places, for a household income of $60,000 and interpret the result.

b. As household income increases, how is income elasticity of demand affected?

c. The logarithmic model shown above is not appropriate for incomes well above $100,000. Suggest a model that might be more appropriate.

d. In the model you propose, how would E behave for very large incomes?

23. ▼ ***Income Elasticity of Demand*** *(based on a question on the GRE Economics Test)* If $Q = a P^{\alpha} Y^{\beta}$ is the individual's demand function for a commodity, where P is the (fixed) price of the commodity, Y is the individual's income, and a, α, and β are parameters, explain why β can be interpreted as the income elasticity of demand.

24. ▼ ***College Tuition*** *(from the GRE Economics Test)* A time-series study of the demand for higher education, using tuition charges as a price variable, yields the following result:
$$\frac{dq}{dp} \cdot \frac{p}{q} = -0.4,$$
where p is tuition and q is the quantity of higher education. Which of the following is suggested by the result?

(A) As tuition rises, students want to buy a greater quantity of education.

(B) As a determinant of the demand for higher education, income is more important than price.

(C) If colleges lowered tuition slightly, their total tuition receipts would increase.

(D) If colleges raised tuition slightly, their total tuition receipts would increase.

(E) Colleges cannot increase enrollments by offering larger scholarships.

25. ▼ ***Modeling Exponential Demand*** As the new owner of a supermarket, you have inherited a large inventory of unsold imported Limburger cheese, and you would like to set the price so that your revenue from selling it is as large as

possible. Previous sales figures of the cheese are shown in the following table:

Price per Pound, p	$3.00	$4.00	$5.00
Monthly Sales in Pounds, q	407	287	223

a. Use the sales figures for the prices $3 and $5 per pound to construct a demand function of the form $q = Ae^{-bp}$, where A and b are constants you must determine. (Round A and b to two significant digits.)

b. Use your demand function to find the price elasticity of demand at each of the prices listed.

c. At what price should you sell the cheese in order to maximize monthly revenue?

d. If your total inventory of cheese amounts to only 200 pounds, and it will spoil one month from now, how should you price it in order to receive the greatest revenue? Is this the same answer you got in part (c)? If not, give a brief explanation.

26. ▼ ***Modeling Exponential Demand*** Repeat the preceding exercise, but this time use the sales figures for $4 and $5 per pound to construct the demand function.

COMMUNICATION AND REASONING EXERCISES

27. Complete the following: When demand is inelastic, revenue will decrease if _____ .

28. Complete the following: When demand has unit elasticity, revenue will decrease if _____ .

29. ▼ Given that the demand q is a differentiable function of the unit price p, show that the revenue $R = pq$ has a stationary point when
$$q + p \frac{dq}{dp} = 0.$$
Deduce that the stationary points of R are the same as the points of unit price elasticity of demand. (Ordinarily, there is only one such stationary point, corresponding to the absolute maximum of R.) HINT [Differentiate R with respect to p.]

30. ▼ Given that the demand q is a differentiable function of income x, show that the quantity $R = q/x$ has a stationary point when
$$q - x \frac{dq}{dx} = 0.$$
Deduce that stationary points of R are the same as the points of unit income elasticity of demand. HINT [Differentiate R with respect to x.]

31. ◆ Your calculus study group is discussing price elasticity of demand, and a member of the group asks the following question: "Since elasticity of demand measures the response of demand to change in unit price, what is the difference between elasticity of demand and the quantity $-dq/dp$?" How would you respond?

32. ◆ Another member of your study group claims that unit price elasticity of demand need not always correspond to maximum revenue. Is he correct? Explain your answer.

[71] Sources: Luxembourg Income Study/*New York Times*, August 14, 1995, p. A9, Commerce Department, Deloitte & Touche Survey/ *New York Times*, November 24, 1999, p. C1.

KEY CONCEPTS

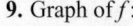 Website www.WanerMath.com
Go to the Website at www.WanerMath
.com to find a comprehensive and
interactive Web-based summary
of Chapter 12.

12.1 Maxima and Minima

Relative maximum, relative minimum
p. 865

Absolute maximum, absolute
minimum *p. 865*

Stationary points, singular points,
endpoints *p. 867*

Finding and classifying maxima
and minima *p. 868*

First derivative test for relative
extrema *p. 869*

Extreme value theorem *p. 873*

Using technology to locate approximate
extrema *p. 874*

12.2 Applications of Maxima and Minima

Minimizing average cost *p. 878*

Maximizing area *p. 879*

Steps in solving optimization
problems *p. 881*

Maximizing revenue *p. 881*

Optimizing resources *p. 882*

Allocation of labor *p. 884*

12.3 Higher Order Derivatives: Acceleration and Concavity

The second derivative of a function f is
the derivative of the derivative of f,
written as f'' *p. 892*

The acceleration of a moving object is
the second derivative of the position
function *p. 892*

Acceleration due to gravity *p. 893*

Acceleration of sales *p. 894*

Concave up, concave down, point of
inflection *p. 895*

Locating points of inflection *p. 895*

Application to inflation *p. 896*

Second derivative test for relative
extrema *p. 899*

Higher order derivatives *p. 900*

12.4 Analyzing Graphs

Features of a graph: x- and y-intercepts,
relative extrema, points of inflection;
behavior near points where the
function is not defined, behavior at
infinity *pp. 910–911*

Analyzing a graph *p. 911*

12.5 Related Rates

If Q is a quantity changing over time t,
then the derivative dQ/dt is the rate at
which Q changes over time *p. 918*

The expanding circle *p. 918*

Steps in solving related rates
problems *p. 919*

The falling ladder *p. 919*

Average cost *p. 921*

Allocation of labor *p. 922*

12.6 Elasticity

Price elasticity of demand

$$E = -\frac{dq}{dp} \cdot \frac{p}{q}; \text{ demand is elastic}$$

if $E > 1$, inelastic if $E < 1$, has unit
elasticity if $E = 1$ *p. 928*

Computing and interpreting elasticity,
and maximizing revenue *p. 929*

Using technology to compute
elasticity *p. 930*

Income elasticity of demand *p. 930*

REVIEW EXERCISES

In Exercises 1–8, find all the relative and absolute extrema of
the given functions on the given domain (if supplied) or on the
largest possible domain (if no domain is supplied).

1. $f(x) = 2x^3 - 6x + 1$ on $[-2, +\infty)$

2. $f(x) = x^3 - x^2 - x - 1$ on $(-\infty, \infty)$

3. $g(x) = x^4 - 4x$ on $[-1, 1]$

4. $f(x) = \dfrac{x+1}{(x-1)^2}$ for $-2 \le x \le 2, x \ne 1$

5. $g(x) = (x-1)^{2/3}$ **6.** $g(x) = x^2 + \ln x$ on $(0, +\infty)$

7. $h(x) = \dfrac{1}{x} + \dfrac{1}{x^2}$ **8.** $h(x) = e^{x^2} + 1$

In Exercises 9–12, the graph of the function f or its derivative is
given. Find the approximate x-coordinates of all relative extrema
and points of inflection of the original function f (if any).

9. Graph of f:

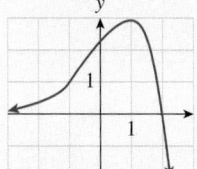

10. Graph of f:

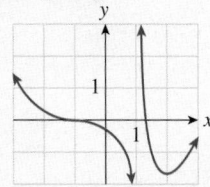

11. Graph of f':

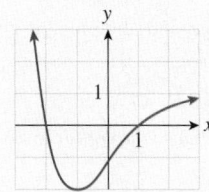

12. Graph of f':

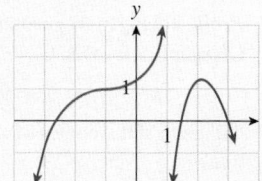

In Exercises 13 and 14, the graph of the second derivative of a
function f is given. Find the approximate x-coordinates of all
points of inflection of the original function f (if any).

13. Graph of f''

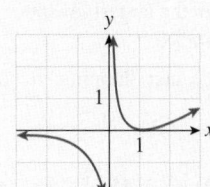

14. Graph of f''

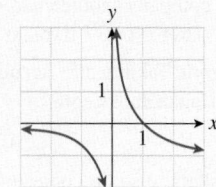

*In Exercises 15 and 16, the position s of a point (in meters) is given as a function of time t (in seconds). Find **(a)** its acceleration as a function of t and **(b)** its acceleration at the specified time.*

15. $s = \dfrac{2}{3t^2} - \dfrac{1}{t}; t = 1$ **16.** $s = \dfrac{4}{t^2} - \dfrac{3t}{4}; t = 2$

In Exercises 17–22, sketch the graph of the given function, indicating all relative and absolute extrema and points of inflection. Find the coordinates of these points exactly, where possible. Also indicate any horizontal and vertical asymptotes.

17. $f(x) = x^3 - 12x$ on $[-2, +\infty)$

18. $g(x) = x^4 - 4x$ on $[-1, 1]$

19. $f(x) = \dfrac{x^2 - 3}{x^3}$

20. $f(x) = (x - 1)^{2/3} + \dfrac{2x}{3}$

21. $g(x) = (x - 3)\sqrt{x}$

22. $g(x) = (x + 3)\sqrt{x}$

APPLICATIONS: OHaganBooks.com

23. *Revenue* Demand for the latest best-seller at OHaganBooks.com, *A River Burns through It*, is given by

$$q = -p^2 + 33p + 9 \qquad (18 \le p \le 28)$$

copies sold per week when the price is p dollars. What price should the company charge to obtain the largest revenue?

24. *Revenue* Demand for *The Secret Loves of John O*, a romance novel by Margó Dufón that flopped after two weeks on the market, is given by

$$q = -2p^2 + 5p + 6 \qquad (0 \le p \le 3.3)$$

copies sold per week when the price is p dollars. What price should OHaganBooks charge to obtain the largest revenue?

25. *Profit* Taking into account storage and shipping, it costs OHaganBooks.com

$$C = 9q + 100$$

dollars to sell q copies of *A River Burns through It* in a week (see Exercise 23).

a. If demand is as in Exercise 23, express the weekly profit earned by OHaganBooks.com from the sale of *A River Burns through It* as a function of unit price p.

b. What price should the company charge to get the largest weekly profit? What is the maximum possible weekly profit?

c. Compare your answer in part (b) with the price the company should charge to obtain the largest revenue (Exercise 23). Explain any difference.

26. *Profit* Taking into account storage and shipping, it costs OHaganBooks.com

$$C = 3q$$

dollars to sell q copies of Margó Dufón's *The Secret Loves of John O* in a week (see Exercise 24).

a. If demand is as in Exercise 24, express the weekly profit earned by OHaganBooks.com from the sale of *The Secret Loves of John O* as a function of unit price p.

b. What price should the company charge to get the largest weekly profit? What is the maximum possible weekly profit?

c. Compare your answer in part (b) with the price the company should charge to obtain the largest revenue (Exercise 24). Explain any difference.

27. *Box Design* The sales department at OHaganBooks.com, which has decided to send chocolate lobsters to each of its customers, is trying to design a shipping box with a square base. It has a roll of cardboard 36 inches wide from which to make the boxes. Each box will be obtained by cutting out corners from a rectangle of cardboard as shown in the following diagram:

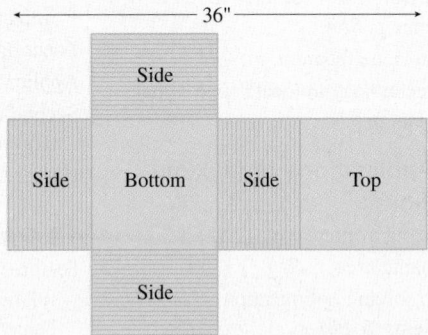

(Notice that the top and bottom of each box will be square, but the sides will not necessarily be square.) What are the dimensions of the boxes with the largest volume that can be made in this way? What is the maximum volume?

28. *Box Redesign* The sales department at OHaganBooks.com was not pleased with the result of the box design in the preceding exercise; the resulting box was too large for the chocolate lobsters, so, following a suggestion by a math major student intern, the department decided to redesign the boxes to meet the following specifications: As in Exercise 27, each box would be obtained by cutting out corners from a rectangle of cardboard as shown in the following diagram:

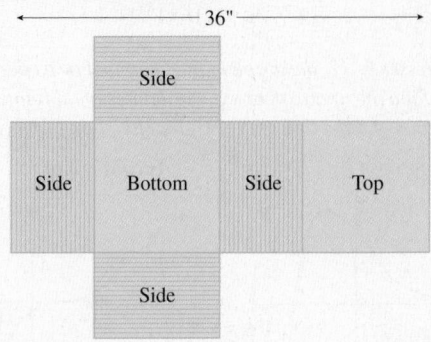

TI-83/84 Plus Technology Guide

Section 12.6

Example 1(a) (page 929) Suppose that the demand equation for *Bobby Dolls* is given by $q = 216 - p^2$, where p is the price per doll in dollars and q is the number of dolls sold per week. Compute the price elasticity of demand when $p = \$5$ and $p = \$10$, and interpret the results.

Solution with Technology

The TI-83/84 Plus function nDeriv can be used to compute approximations of the elasticity E at various prices.

1. Set

$$Y_1 = 216-X^2 \qquad \text{Demand equation}$$
$$Y_2 = -\text{nDeriv}(Y_1,X,X) * X/Y_1 \quad \text{Formula for } E$$

2. Use the table feature to list the values of elasticity for a range of prices. For part (a) we chose values of X close to 5:

X	Y₁	Y₂
4.7	193.91	.22784
4.8	192.96	.23881
4.9	191.99	.25012
5	191	.26178
5.1	189.99	.2738
5.2	188.96	.2862
5.3	187.91	.29897

Y₂=.261780104712

SPREADSHEET Technology Guide

Section 12.6

Example 1(a) (page 929) Suppose that the demand equation for *Bobby Dolls* is given by $q = 216 - p^2$, where p is the price per doll in dollars and q is the number of dolls sold per week. Compute the price elasticity of demand when $p = \$5$ and $p = \$10$, and interpret the results.

Solution with Technology

To approximate E in a spreadsheet, we can use the following approximation of E.

$$E \approx \frac{\text{Percentage decrease in demand}}{\text{Percentage increase in price}} \approx -\frac{\left(\dfrac{\Delta q}{q}\right)}{\left(\dfrac{\Delta p}{p}\right)}$$

The smaller Δp is, the better the approximation. Let's use $\Delta p = 1¢$, or 0.01 (which is small compared with the typical prices we consider—around \$5 to \$10).

1. We start by setting up our worksheet to list a range of prices, in increments of Δp, on either side of a price in which we are interested, such as $p_0 = \$5$:

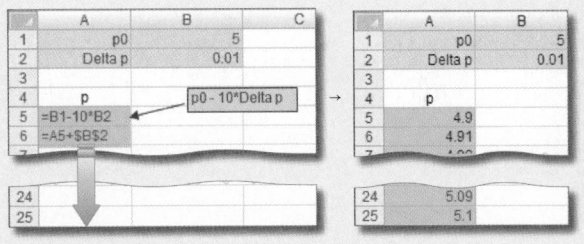

We start in cell A5 with the formula for $p_0 - 10\Delta p$ and then successively add Δp going down column A. You will find that the value $p_0 = 5$ appears midway down the list.

2. Next, we compute the corresponding values for the demand q in column B.

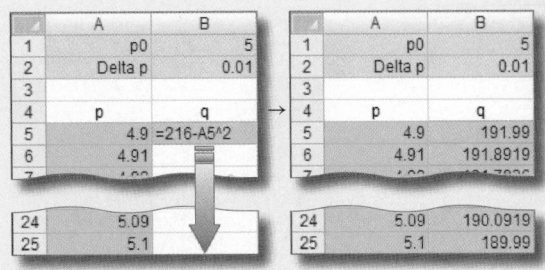

3. We add two new columns for the percentage changes in p and q. The formula shown in cell C5 is copied down columns C and D, to Row 24. (Why not row 25?)

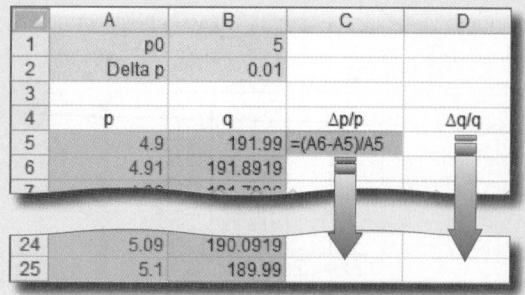

	A	B	C	D
1	p0	5		
2	Delta p	0.01		
3				
4	p	q	Δp/p	Δq/q
5	4.9	191.99	=(A6-A5)/A5	
6	4.91	191.8919		
7				
24	5.09	190.0919		
25	5.1	189.99		

4. The elasticity can now be computed in column E as shown:

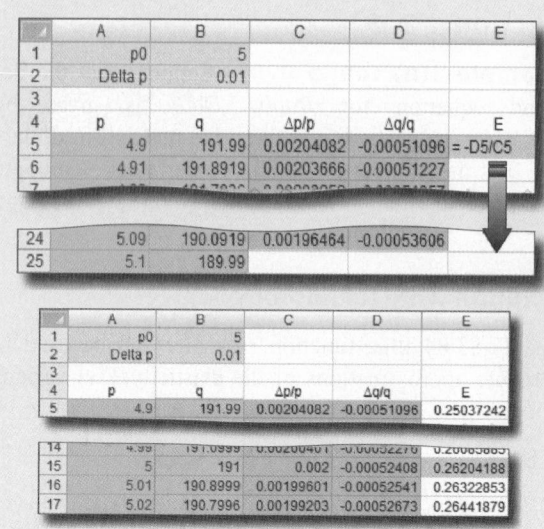

	A	B	C	D	E
1	p0	5			
2	Delta p	0.01			
3					
4	p	q	Δp/p	Δq/q	E
5	4.9	191.99	0.00204082	-0.00051096	= -D5/C5
6	4.91	191.8919	0.00203666	-0.00051227	
7					
24	5.09	190.0919	0.00196464	-0.00053606	
25	5.1	189.99			

	A	B	C	D	E
1	p0	5			
2	Delta p	0.01			
3					
4	p	q	Δp/p	Δq/q	E
5	4.9	191.99	0.00204082	-0.00051096	0.25037242
14	4.99	191.0999	0.00200401	-0.00052270	0.26065865
15	5	191	0.002	-0.00052408	0.26204188
16	5.01	190.8999	0.00199601	-0.00052541	0.26322853
17	5.02	190.7996	0.00199203	-0.00052673	0.26441879

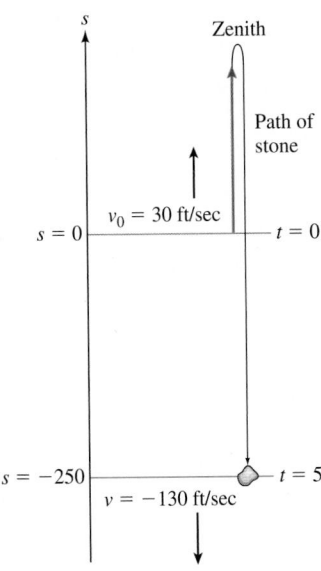

Figure 2

b. We wish to know the position, but position is an antiderivative of velocity. Thus,

$$s(t) = \int v(t)\, dt = \int (-32t + 30)\, dt = -16t^2 + 30t + C.$$

Now to find C, we need to know the initial position $s(0)$. We are not told this, so let's measure heights so that the initial position is zero. Then

$$0 = s(0) = C$$

and

$$s(t) = -16t^2 + 30t \text{ ft.} \qquad \begin{array}{l} s(t) = -16t^2 + v_0 t + s_0 \\ s_0 = \text{initial position} \end{array}$$

In particular, after 5 seconds the stone has a height of

$$s(5) = -16(5)^2 + 30(5) = -250 \text{ ft.}$$

In other words, the stone is now 250 ft *below* where it was when you first threw it, as shown in Figure 2.

c. The stone reaches its zenith when its height $s(t)$ is at its maximum value, which occurs when $v(t) = s'(t)$ is zero. So we solve

$$v(t) = -32t + 30 = 0$$

getting $t = 30/32 = 15/16 = 0.9375$ sec. This is the time when the stone reaches its zenith. The height of the stone at that time is

$$s(15/16) = -16(15/16)^2 + 30(15/16) = 14.0625 \text{ ft.}$$

➡ **Before we go on...** Here again are the formulas we obtained in Example 8, together with their metric equivalents:

Vertical Motion under Gravity: Velocity and Position

If we ignore air resistance, the vertical velocity and position of an object moving under gravity are given by

British Units	Metric Units
Velocity: $v(t) = -32t + v_0$ ft/sec	$v(t) = -9.8t + v_0$ m/sec
Position: $s(t) = -16t^2 + v_0 t + s_0$ ft	$s(t) = -4.9t^2 + v_0 t + s_0$ m

$v_0 = $ initial velocity $=$ velocity at time 0
$s_0 = $ initial position $=$ position at time 0

Quick Example

If a ball is thrown down at 2 ft/sec from a height of 200 ft, then its velocity and position after t seconds are $v(t) = -32t - 2$ ft/sec and $s(t) = -16t^2 - 2t + 200$ ft. ∎

13.1　EXERCISES

▼ more advanced　◆ challenging
T indicates exercises that should be solved using technology

Evaluate the integrals in Exercises 1–42.
HINT [for 1–6: See Quick Examples on page 947.]

1. $\int x^5 \, dx$　　　**2.** $\int x^7 \, dx$　　　**3.** $\int 6 \, dx$

4. $\int (-5) \, dx$　　　**5.** $\int x \, dx$　　　**6.** $\int (-x) \, dx$

HINT [for 7–18: See Example 2.]

7. $\int (x^2 - x) \, dx$　　　　　**8.** $\int (x + x^3) \, dx$

9. $\int (1 + x) \, dx$　　　　　**10.** $\int (4 - x) \, dx$

11. $\int x^{-5} \, dx$　　　　　**12.** $\int x^{-7} \, dx$

13. $\int (x^{2.3} + x^{-1.3}) \, dx$　　　**14.** $\int (x^{-0.2} - x^{0.2}) \, dx$

15. $\int (u^2 - u^{-1}) \, du$ HINT [See Example 4.]

16. $\int (v^{-2} + 2v^{-1}) \, dv$ HINT [See Example 4.]

17. $\int \sqrt[4]{x} \, dx$　　　　　**18.** $\int \sqrt[3]{x} \, dx$

HINT [for 19–42: See Example 3.]

19. $\int (3x^4 - 2x^{-2} + x^{-5} + 4) \, dx$　**20.** $\int (4x^7 - x^{-3} + 1) \, dx$

21. $\int \left(\frac{2}{u} + \frac{u}{4} \right) du$　　　**22.** $\int \left(\frac{2}{u^2} + \frac{u^2}{4} \right) du$

23. $\int \left(\frac{1}{x} + \frac{2}{x^2} - \frac{1}{x^3} \right) dx$

24. $\int \left(\frac{3}{x} - \frac{1}{x^5} + \frac{1}{x^7} \right) dx$

25. $\int (3x^{0.1} - x^{4.3} - 4.1) \, dx$　**26.** $\int \left(\frac{x^{2.1}}{2} - 2.3 \right) dx$

27. $\int \left(\frac{3}{x^{0.1}} - \frac{4}{x^{1.1}} \right) dx$　**28.** $\int \left(\frac{1}{x^{1.1}} - \frac{1}{x} \right) dx$

29. $\int \left(5.1t - \frac{1.2}{t} + \frac{3}{t^{1.2}} \right) dt$

30. $\int \left(3.2 + \frac{1}{t^{0.9}} + \frac{t^{1.2}}{3} \right) dt$

31. $\int (2e^x + 5|x| + 1/4) \, dx$

32. $\int (-4e^x + |x|/3 - 1/8) \, dx$

33. $\int \left(\frac{6.1}{x^{0.5}} + \frac{x^{0.5}}{6} - e^x \right) dx$

34. $\int \left(\frac{4.2}{x^{0.4}} + \frac{x^{0.4}}{3} - 2e^x \right) dx$

35. $\int (2^x - 3^x) \, dx$

36. $\int (1.1^x + 2^x) \, dx$

37. $\int \left(100(1.1^x) - \frac{2|x|}{3} \right) dx$

38. $\int \left(1{,}000(0.9^x) + \frac{4|x|}{5} \right) dx$

39. ▼ $\int x^{-2} \left(x^4 - \frac{3}{2x^4} \right) dx$

40. ▼ $\int 3x^4 \left(\frac{2}{x^4} + \frac{3}{5x^6} \right) dx$

41. ▼ $\int \frac{x + 2}{x^3} \, dx$　　　　**42.** ▼ $\int \frac{x^2 - 2}{x} \, dx$

43. Find $f(x)$ if $f(0) = 1$ and the tangent line at $(x, f(x))$ has slope x. HINT [See Example 5.]

44. Find $f(x)$ if $f(1) = 1$ and the tangent line at $(x, f(x))$ has slope $\frac{1}{x}$. HINT [See Example 5.]

45. Find $f(x)$ if $f(0) = 0$ and the tangent line at $(x, f(x))$ has slope $e^x - 1$.

46. Find $f(x)$ if $f(1) = -1$ and the tangent line at $(x, f(x))$ has slope $2e^x + 1$.

APPLICATIONS

47. *Marginal Cost* The marginal cost of producing the xth box of light bulbs is $5 - \frac{x}{10{,}000}$ and the fixed cost is \$20,000. Find the cost function $C(x)$. HINT [See Example 5.]

48. *Marginal Cost* The marginal cost of producing the xth box of DVDs is $10 + \frac{x^2}{100{,}000}$ and the fixed cost is \$100,000. Find the cost function $C(x)$. HINT [See Example 5.]

49. *Marginal Cost* The marginal cost of producing the xth roll of film is $5 + 2x + \frac{1}{x}$. The total cost to produce one roll is \$1,000. Find the cost function $C(x)$. HINT [See Example 5.]

50. *Marginal Cost* The marginal cost of producing the xth box of CDs is $10 + x + \frac{1}{x^2}$. The total cost to produce 100 boxes is \$10,000. Find the cost function $C(x)$. HINT [See Example 5.]

51. *Facebook Membership* At the start of 2005, Facebook had 1 million members. Since that time, new members joined at a rate of roughly

$$m(t) = 3.2t^3 - 12t + 10 \text{ million members per year}$$
$$(0 \le t \le 5),$$

where t is time in years since the start of 2005.[2]

a. Find an expression for total Facebook membership $M(t)$ at time t. HINT [See Example 6.]

b. Use the answer to part (a) to estimate Facebook membership midway through 2008. (Round your answer to the nearest 1 million members. The actual figure was 80 million.)

52. *Uploads to YouTube* Since YouTube first became available to the public in mid-2005, the rate at which video has been uploaded to the site can be approximated by

$$v(t) = 525,600(0.42t^2 + 2.7t - 1) \text{ hours of video per year}$$
$$(0.5 \le t \le 7),$$

where t is time in years since the start of 2005.[3]

a. Find an expression for the total number of hours $V(t)$ of video at time t (starting from zero hours of video at $t = 0.5$). HINT [See Example 6.]

b. Use the answer to part (a) to estimate the total number of hours of video uploaded by the start of 2012. (Round your answer to the nearest million hours of video.)

53. *Median Household Income* From 2000 to 2007, median household income in the United States rose by an average of approximately $1,200 per year.[4] Given that the median household income in 2000 was approximately $42,000, use an indefinite integral to find a formula for the median household income I as a function of the year t ($t = 0$ represents 2000), and use your formula to estimate the median household income in 2005. (You can do this exercise without integration using the methods of Section 1.3, but here you should use an indefinite integral.)

54. *Mean Household Income* From 2000 to 2007, the mean household income in the United States rose by an average of approximately $1,500 per year.[5] Given that the mean household income in 2000 was approximately $57,000, use an indefinite integral to find a formula for the mean household income I as a function of the year t ($t = 0$ represents 2000), and use your formula to estimate the mean household income in 2006. (You can do this exercise without integration using the methods of Section 1.3, but here you should use an indefinite integral.)

55. *Bottled Water Sales* The rate of U.S. sales of bottled water for the period 2000–2010 can be approximated by

$$s(t) = -45t^2 + 900t + 4,200 \text{ million gallons per year}$$
$$(0 \le t \le 10),$$

where t is time in years since the start of 2000.[6] Use an indefinite integral to approximate the total sales $S(t)$ of bottled water since the start of 2000. Approximately how much bottled water was sold from the start of 2000 to the start of 2008? HINT [At the start of 2000, sales since that time are zero.]

56. *Bottled Water Sales* The rate of U.S. per capita sales of bottled water for the period 2000–2010 can be approximated by

$$s(t) = -0.18t^2 + 3t + 15 \text{ gallons per year} \quad (0 \le t \le 10),$$

where t is time in years since the start of 2000.[7] Use an indefinite integral to approximate the total per capita sales $S(t)$ of bottled water since the start of 2000. Approximately how much bottled water was sold, per capita, from the start of 2000 to the end of 2009? HINT [At the start of 2000, sales since that time are zero.]

57. ▼ *Health-Care Spending in the 1990s* Write $H(t)$ for the amount spent in the United States on health care in year t, where t is measured in years since 1990. The rate of increase of $H(t)$ was approximately $65 billion per year in 1990 and rose to $100 billion per year in 2000.[8]

a. Find a linear model for the rate of change $H'(t)$.

b. Given that $700 billion was spent on health care in the United States in 1990, find the function $H(t)$.

58. ▼ *Health-Care Spending in the 2000s* Write $H(t)$ for the amount spent in the United States on health care in year t, where t is measured in years since 2000. The rate of increase of $H(t)$ was projected to rise from $100 billion per year in 2000 to approximately $190 billion per year in 2010.[9]

a. Find a linear model for the rate of change $H'(t)$.

b. Given that $1,300 billion was spent on health care in the United States in 2000, find the function $H(t)$.

59. ▼ *Subprime Mortgage during the Housing Bubble* At the start of 2007, the percentage of U.S. mortgages that were subprime was about 13%, was increasing at a rate of 1 percentage point per year, but was decelerating at 0.4 percentage points per year per year.[10]

a. Find an expression for the rate of change (velocity) of this percentage at time t in years since the start of 2007.

b. Use the result of part (a) to find an expression for the percentage of mortgages that were subprime at time t, and use it to estimate the rate at the start of 2008. HINT [See Quick Examples 2 and 3 on page 955.]

———
[2]Sources for data: www.facebook.com, www.insidefacebook.com.

[3]Source for data: www.YouTube.com.

[4]In current dollars, unadjusted for inflation. Source for data: U.S. Census Bureau (www.census.gov).

[5]*Ibid.*

———
[6]Source for data: Beverage Marketing Corporation (www.bottledwater.org).

[7]*Ibid.*

[8]Source: Centers for Medicare and Medicaid Services, "National Health Expenditures," 2002 version, released January 2004 (www.cms.hhs.gov/statistics/nhe/).

[9]Source: Centers for Medicare and Medicaid Services, "National Health Expenditures 1965–2013, History and Projections" (www.cms.hhs.gov/statistics/nhe/).

[10]Sources: Mortgage Bankers Association, UBS.

60. ▼ *Subprime Mortgage Debt during the Housing Bubble* At the start of 2008, the value of subprime mortgage debt outstanding in the U.S. was about $1,300 billion, was increasing at a rate of 40 billion dollars per year, but was decelerating at 20 billion dollars per year per year.[11]

  a. Find an expression for the rate of change (velocity) of the value of subprime mortgage debt at time t in years since the start of 2008.

  b. Use the result of part (a) to find an expression for the value of subprime mortgage debt at time t, and use it to estimate the value at the start of 2009. HINT [See Quick Examples 2 and 3 on page 955.]

61. *Motion in a Straight Line* The velocity of a particle moving in a straight line is given by $v = t^2 + 1$.

  a. Find an expression for the position s after a time t.

  b. Given that $s = 1$ at time $t = 0$, find the constant of integration C, and hence find an expression for s in terms of t without any unknown constants. HINT [See Example 7.]

62. *Motion in a Straight Line* The velocity of a particle moving in a straight line is given by $v = 3e^t + t$.

  a. Find an expression for the position s after a time t.

  b. Given that $s = 3$ at time $t = 0$, find the constant of integration C, and hence find an expression for s in terms of t without any unknown constants. HINT [See Example 7.]

63. *Vertical Motion under Gravity* If a stone is dropped from a rest position above the ground, how fast (in feet per second) and in what direction will it be traveling after 10 seconds? (Neglect the effects of air resistance.) HINT [See Example 8.]

64. *Vertical Motion under Gravity* If a stone is thrown upward at 10 feet per second, how fast (in feet per second) and in what direction will it be traveling after 10 seconds? (Neglect the effects of air resistance.) HINT [See Example 8.]

65. *Vertical Motion under Gravity* Your name is Galileo Galilei and you toss a weight upward at 16 feet per second from the top of the Leaning Tower of Pisa (height 185 ft).

  a. Neglecting air resistance, find the weight's velocity as a function of time t in seconds.

  b. Find the height of the weight above the ground as a function of time. Where and when will it reach its zenith? HINT [See Example 8 and the formulas that follow.]

66. *Vertical Motion under Gravity* Your name is Spaghettini Bologna (an assistant of Galileo Galilei) and, to impress your boss, you toss a weight upward at 24 feet per second from the top of the Leaning Tower of Pisa (height 185 ft).

  a. Neglecting air resistance, find the weight's velocity as a function of time t in seconds.

  b. Find the height of the weight above the ground as a function of time. Where and when will it reach its zenith? HINT [See Example 8 and the formulas that follow.]

67. ▼ *Tail Winds* The ground speed of an airliner is obtained by adding its air speed and the tail-wind speed. On your recent trip from Mexico to the United States your plane was traveling at an air speed of 500 miles per hour and experienced tail winds of $25 + 50t$ miles per hour, where t is the time in hours since takeoff.

  a. Obtain an expression for the distance traveled in terms of the time since takeoff. HINT [Ground speed = Air speed + Tail-wind speed.]

  b. Use the result of part (a) to estimate the time of your 1,800-mile trip.

  c. The equation solved in part (b) leads mathematically to two solutions. Explain the meaning of the solution you rejected.

68. ▼ *Head Winds* The ground speed of an airliner is obtained by subtracting its head-wind speed from its air speed. On your recent trip to Mexico from the United States your plane was traveling at an air speed of 500 miles per hour and experienced head winds of $25 + 50t$ miles per hour, where t is the time in hours since takeoff. HINT [Ground speed = Air speed − Head-wind speed.]

  a. Obtain an expression for the distance traveled in terms of the time since takeoff.

  b. Use the result of part (a) to estimate the time of your 1,500-mile trip.

  c. The equation solved in part (b) leads mathematically to two solutions. Explain the meaning of the solution you rejected.

69. ▼ *Vertical Motion* Show that if a projectile is thrown upward with a velocity of v_0 ft/sec, then (neglecting air resistance) it will reach its highest point after $v_0/32$ seconds. HINT [See the formulas after Example 8.]

70. ▼ *Vertical Motion* Use the result of the preceding exercise to show that if a projectile is thrown upward with a velocity of v_0 ft/sec, its highest point will be $v_0^2/64$ feet above the starting point (if we neglect the effects of air resistance).

Exercises 71–76 use the results in the preceding two exercises.

71. ▼ I threw a ball up in the air to a height of 20 feet. How fast was the ball traveling when it left my hand?

72. ▼ I threw a ball up in the air to a height of 40 feet. How fast was the ball traveling when it left my hand?

73. ▼ A piece of chalk is tossed vertically upward by Prof. Schwarzenegger and hits the ceiling 100 feet above with a *BANG*.

  a. What is the minimum speed the piece of chalk must have been traveling to enable it to hit the ceiling?

  b. Assuming that Prof. Schwarzenegger in fact tossed the piece of chalk up at 100 ft/sec, how fast was it moving when it struck the ceiling?

  c. Assuming that Prof. Schwarzenegger tossed the chalk up at 100 ft/sec, and that it recoils from the ceiling with the same speed it had at the instant it hit, how long will it take the chalk to make the return journey and hit the ground?

---

[11]Source: www.data360.org/dataset.aspx?Data_Set_Id=9549.

74. ▼ A projectile is fired vertically upward from ground level at 16,000 feet per second.

   a. How high does the projectile go?
   b. How long does it take to reach its zenith (highest point)?
   c. How fast is it traveling when it hits the ground?

75. ▼ *Strength* Prof. Strong can throw a 10-pound dumbbell twice as high as Prof. Weak can. How much faster can Prof. Strong throw it?

76. ▼ *Weakness* Prof. Weak can throw a book three times as high as Prof. Strong can. How much faster can Prof. Weak throw it?

COMMUNICATION AND REASONING EXERCISES

77. Why is this section called "The *Indefinite* Integral?"

78. If the derivative of Julius is Augustus, then Augustus is _____ of Julius.

79. Linear functions are antiderivatives of what kind of function? Explain.

80. Constant functions are antiderivatives of what kind of function? Explain.

81. If we know the *derivative* of a function, do we know the function? Explain. If not, what further information will suffice?

82. If we know an *antiderivative* of a function, do we know the function? Explain. If not, what further information will suffice?

83. If $F(x)$ and $G(x)$ are both antiderivatives of $f(x)$, how are $F(x)$ and $G(x)$ related?

84. Your friend Marco claims that once you have one antiderivative of $f(x)$, you have all of them. Explain what he means.

85. Complete the following: The total cost function is a(n) _____ of the _____ cost function.

86. Complete the following: The distance covered is an antiderivative of the _____ function, and the velocity is an antiderivative of the _____ function.

87. If x represents the number of items manufactured and $f(x)$ represents dollars per item, what does $\int f(x)\,dx$ represent? In general, how are the units of $f(x)$ and the units of $\int f(x)\,dx$ related?

88. If t represents time in seconds since liftoff and $g(t)$ represents the volume of rocket fuel burned per second, what does $\int g(t)\,dt$ represent?

89. Why was the following marked wrong? What is the correct answer?

$$\int (3x + 1)\,dx = \frac{3x^2}{2} + 0 + C = \frac{3x^2}{2} + C \quad \text{✗ WRONG!}$$

90. Why was the following marked wrong? What is the correct answer?

$$\int (3x^2 - 11x)\,dx = x^3 - 11 + C \quad \text{✗ WRONG!}$$

91. Why was the following marked wrong? What is the correct answer?

$$\int (12x^5 - 4x)\,dx = \int 2x^6 - 2x^2 + C \quad \text{✗ WRONG!}$$

92. Why was the following marked wrong? What is the correct answer?

$$\int 5\,dt = 5x + C \quad \text{✗ WRONG!}$$

93. Why was the following marked wrong? What is the correct answer?

$$\int 4(e^x - 2x)\,dx = (4x)(e^x - x^2) + C \quad \text{✗ WRONG!}$$

94. Why was the following marked wrong? What is the correct answer?

$$\int (2^x - 1)\,dx = \frac{2^{x+1}}{x+1} - x + C \quad \text{✗ WRONG!}$$

95. Why was the following marked wrong? How should it be corrected?

$$\frac{1}{x} = \ln|x| + C \quad \text{✗ WRONG!}$$

96. Why was the following marked wrong? What is the correct answer?

$$\int \frac{1}{x^3}\,dx = \ln|x^3| + C \quad \text{✗ WRONG!}$$

97. ▼ Give an argument for the rule that the integral of a sum is the sum of the integrals.

98. ▼ Give an argument for the rule that the integral of a constant multiple is the constant multiple of the integrals.

99. ▼ Give an example to show that the integral of a product is not the product of the integrals.

100. ▼ Give an example to show that the integral of a quotient is not the quotient of the integrals.

101. ▼ Complete the following: If you take the _____ of the _____ of $f(x)$, you obtain $f(x)$ back. On the other hand, if you take the _____ of the _____ of $f(x)$, you obtain $f(x) + C$.

102. ▼ If a Martian told you that the *Institute of Alien Mathematics,* after a long and difficult search, has announced the discovery of a new antiderivative of $x - 1$ called $M(x)$ [the formula for $M(x)$ is classified information and cannot be revealed here], how would you respond?

13.2   Substitution

The chain rule for derivatives gives us an extremely useful technique for finding antiderivatives. This technique is called **change of variables** or **substitution**.

Recall that to differentiate a function like $(x^2 + 1)^6$, we first think of the function as $g(u)$, where $u = x^2 + 1$ and $g(u) = u^6$. We then compute the derivative, using the chain rule, as

$$\frac{d}{dx} g(u) = g'(u) \frac{du}{dx}.$$

Any rule for derivatives can be turned into a technique for finding antiderivatives by writing it in integral form. The integral form of the above formula is

$$\int g'(u) \frac{du}{dx} \, dx = g(u) + C.$$

But, if we write $g(u) + C = \int g'(u) \, du$, we get the following interesting equation:

$$\int g'(u) \frac{du}{dx} \, dx = \int g'(u) \, du.$$

This equation is the one usually called the *change of variables formula.* We can turn it into a more useful integration technique as follows. Let $f = g'(u)(du/dx)$. We can rewrite the above change of variables formula using f:

$$\int f \, dx = \int \left(\frac{f}{du/dx} \right) du.$$

In essence, we are making the formal substitution

$$dx = \frac{1}{du/dx} \, du.$$

Here's the technique:

Substitution Rule

If u is a function of x, then we can use the following formula to evaluate an integral:

$$\int f \, dx = \int \left(\frac{f}{du/dx} \right) du.$$

Rather than use the formula directly, we use the following step-by-step procedure:

1. Write u as a function of x.

2. Take the derivative du/dx and solve for the quantity dx in terms of du.

3. Use the expression you obtain in step 2 to substitute for dx in the given integral and substitute u for its defining expression.

Now let's see how this procedure works in practice.

EXAMPLE 1 Substitution

Find $\int 4x(x^2 + 1)^6\, dx$.

Solution To use substitution we need to choose an expression to be u. There is no hard and fast rule, but here is one hint that often works:

Take u to be an expression that is being raised to a power.

In this case, let's set $u = x^2 + 1$. Continuing the procedure above, we place the calculations for Step 2 in a box.

$u = x^2 + 1$	Write u as a function of x.
$\dfrac{du}{dx} = 2x$	Take the derivative of u with respect to x.
$dx = \dfrac{1}{2x}\, du$	Solve for dx: $dx = \dfrac{1}{du/dx}\, du$.

Now we *substitute u for its defining expression and substitute for dx* in the original integral:

* This step is equivalent to using the formula stated in the Substitution Rule box. If it should bother you that the integral contains both x and u, note that x is now a function of u.

$$\int 4x(x^2 + 1)^6\, dx = \int 4xu^6 \frac{1}{2x}\, du \qquad \text{Substitute* for } u \text{ and } dx.$$

$$= \int 2u^6\, du. \qquad \text{Cancel the } xs \text{ and simplify.}$$

We have boiled the given integral down to the much simpler integral $\int 2u^6\, du$, and we can now write down the solution:

$$2\frac{u^7}{7} + C = \frac{2(x^2 + 1)^7}{7} + C. \qquad \text{Substitute } (x^2 + 1) \text{ for } u \text{ in the answer.}$$

➡ **Before we go on...** There are two points to notice in Example 1. First, before we can actually integrate with respect to u, *we must eliminate all x's from the integrand*. If we cannot, we may have chosen the wrong expression for u. Second, after integrating, we must substitute back to obtain an expression involving x.

It is easy to check our answer. We differentiate:

$$\frac{d}{dx}\left[\frac{2(x^2 + 1)^7}{7}\right] = \frac{2(7)(x^2 + 1)^6(2x)}{7} = 4x(x^2 + 1)^6. \qquad ✔$$

Notice how we used the chain rule to check the result obtained by substitution. ∎

When we use substitution, the first step is always to decide what to take as u. Again, there are no set rules, but we see some common cases in the examples.

EXAMPLE 2 **More Substitution**

Evaluate the following:

a. $\int x^2(x^3+1)^2\,dx$ **b.** $\int 3xe^{x^2}\,dx$ **c.** $\int \dfrac{1}{2x+5}\,dx$ **d.** $\int \left(\dfrac{1}{2x+5}+4x^2+1\right)dx$

Solution

a. As we said in Example 1, it often works to take u to be an expression that is being raised to a power. We usually also want to see the derivative of u as a factor in the integrand so that we can cancel terms involving x. In this case, x^3+1 is being raised to a power, so let's set $u = x^3+1$. Its derivative is $3x^2$; in the integrand we see x^2, which is missing the factor 3, but missing or incorrect constant factors are not a problem.

$$u = x^3 + 1 \qquad \text{Write } u \text{ as a function of } x.$$

$$\frac{du}{dx} = 3x^2 \qquad \text{Take the derivative of } u \text{ with respect to } x.$$

$$dx = \frac{1}{3x^2}\,du \qquad \text{Solve for } dx\colon dx = \frac{1}{du/dx}\,du.$$

$$\int x^2(x^3+1)^2\,dx = \int x^2 u^2 \frac{1}{3x^2}\,du \qquad \text{Substitute for } u \text{ and } dx.$$

$$= \int \frac{1}{3}u^2\,du \qquad \text{Cancel the terms with } x.$$

$$= \frac{1}{9}u^3 + C \qquad \text{Take the antiderivative.}$$

$$= \frac{1}{9}(x^3+1)^3 + C \qquad \text{Substitute for } u \text{ in the answer.}$$

b. When we have an exponential with an expression in the exponent, it often works to substitute u for that expression. In this case, let's set $u = x^2$. (Notice again that we see a constant multiple of its derivative $2x$ as a factor in the integrand—a good sign.)

$$u = x^2$$

$$\frac{du}{dx} = 2x$$

$$dx = \frac{1}{2x}\,du$$

Substituting into the integral, we have

$$\int 3xe^{x^2}\,dx = \int 3xe^u \frac{1}{2x}\,du = \int \frac{3}{2}e^u\,du$$

$$= \frac{3}{2}e^u + C = \frac{3}{2}e^{x^2} + C.$$

c. We begin by rewriting the integrand as a power:

$$\int \frac{1}{2x+5}\,dx = \int (2x+5)^{-1}\,dx.$$

Now we take our earlier advice and set u equal to the expression that is being raised to a power:

$u = 2x + 5$
$\dfrac{du}{dx} = 2$
$dx = \dfrac{1}{2}\,du$

Substituting into the integral, we have

$$\int \frac{1}{2x+5}\,dx = \int \frac{1}{2}u^{-1}\,du = \frac{1}{2}\ln|u| + C$$

$$= \frac{1}{2}\ln|2x+5| + C.$$

d. Here, the substitution $u = 2x + 5$ works for the first part of the integrand, $1/(2x+5)$, but not for the rest of it, so we break up the integral:

$$\int \left(\frac{1}{2x+5} + 4x^2 + 1\right) dx = \int \frac{1}{2x+5}\,dx + \int (4x^2 + 1)\,dx.$$

For the first integral we can use the substitution $u = 2x + 5$ [which we did in part (a)], and for the second, no substitution is necessary:

$$\int \frac{1}{2x+5}\,dx + \int (4x^2 + 1)\,dx = \frac{1}{2}\ln|2x+5| + \frac{4x^3}{3} + x + C.$$

EXAMPLE 3 Choosing u

Evaluate $\displaystyle\int (x+3)\sqrt{x^2 + 6x}\,dx$.

Solution There are two parenthetical expressions. Notice, however, that the derivative of the expression $(x^2 + 6x)$ is $2x + 6$, which is twice the term $(x + 3)$ in front of the radical. Recall that we would like the derivative of u to appear as a factor. Thus, let's take $u = x^2 + 6x$.

$u = x^2 + 6x$
$\dfrac{du}{dx} = 2x + 6 = 2(x+3)$
$dx = \dfrac{1}{2(x+3)}\,du$

Substituting into the integral, we have

$$\int (x+3)\sqrt{x^2 + 6x}\,dx$$

$$= \int (x+3)\sqrt{u}\left(\frac{1}{2(x+3)}\right) du$$

$$= \int \frac{1}{2}\sqrt{u}\,du = \frac{1}{2}\int u^{1/2}\,du$$

$$= \frac{1}{2}\frac{2}{3}u^{3/2} + C = \frac{1}{3}(x^2 + 6x)^{3/2} + C.$$

Some cases require a little more work.

EXAMPLE 4 When the *x* Terms Do Not Cancel

Evaluate $\int \dfrac{2x}{(x-5)^2}\,dx$.

Solution We first rewrite

$$\int \dfrac{2x}{(x-5)^2}\,dx = \int 2x(x-5)^{-2}\,dx.$$

This suggests that we should set $u = x - 5$.

$$
\begin{array}{l}
u = x - 5 \\[4pt]
\dfrac{du}{dx} = 1 \\[4pt]
dx = du
\end{array}
$$

Substituting, we have

$$\int \dfrac{2x}{(x-5)^2}\,dx = \int 2xu^{-2}\,du.$$

Now, there is nothing in the integrand to cancel the *x* that appears. If, as here, there is still an *x* in the integrand after substituting, we go back to the expression for *u*, solve for *x*, and substitute the expression we obtain for *x* in the integrand. So, we take $u = x - 5$ and solve for $x = u + 5$. Substituting, we get

$$\int 2xu^{-2}\,du = \int 2(u+5)u^{-2}\,du$$

$$= 2\int (u^{-1} + 5u^{-2})\,du$$

$$= 2\ln|u| - \dfrac{10}{u} + C$$

$$= 2\ln|x-5| - \dfrac{10}{x-5} + C.$$

EXAMPLE 5 Application: Bottled Water for Pets

Annual sales of bottled spring water for pets can be modeled by the logistic function

$$s(t) = \dfrac{3{,}000e^{0.5t}}{3 + e^{0.5t}} \text{ million gallons per year} \qquad (0 \le t \le 12),$$

where *t* is time in years since the start of 2000.[12]

a. Find an expression for the total amount of bottled spring water for pets sold since the start of 2000.

b. How much bottled spring water for pets was sold from the start of 2005 to the start of 2008?

Solution

a. If we write the total amount of pet spring water sold since the start of 2000 as $S(t)$, then the information we are given says that

$$S'(t) = s(t) = \dfrac{3{,}000e^{0.5t}}{3 + e^{0.5t}}.$$

---

[12]Based on data through 2008 and the recovery in 2010 of the general bottled water market to pre-recession levels. Sources: "*Liquid Assets: America's Expensive Love Affair with Bottled Water*" Daniel Gross, April 26, 2011 (finance.yahoo.com), Beverage Marketing Corporation (www.beveragemarketing.com).

92. Scientific Research The number of research articles in the prominent journal *Physical Review* written by researchers in the United States can be approximated by

$$U(t) = \frac{4.6e^{0.6t}}{0.4 + e^{0.6t}} \text{ thousand articles per year} \quad (t \geq 0),$$

where t is time in years ($t = 0$ represents 1983).[18]

a. Find an (approximate) expression for the total number of articles written by researchers in the United States since 1983 ($t = 0$). HINT [See Example 5.]
b. Roughly how many articles were written by researchers in the United States from 1983 to 2003?

93. Sales The rate of sales of your company's *Jackson Pollock Advanced Paint-by-Number* sets can be modeled by

$$s(t) = \frac{900e^{0.25t}}{3 + e^{0.25t}} \text{ sets per month}$$

t months after their introduction. Find an expression for the total number of paint-by-number sets $S(t)$ sold t months after their introduction, and use it to estimate the total sold in the first 12 months. HINT [See Example 5.]

94. Sales The rate of sales of your company's *Jackson Pollock Beginners Paint-by-Number* sets can be modeled by

$$s(t) = \frac{1,800e^{0.75t}}{10 + e^{0.75t}} \text{ sets per month}$$

t months after their introduction. Find an expression for the total number of paint-by-number sets $S(t)$ sold t months after their introduction, and use it to estimate the total sold in the first 12 months. HINT [See Example 5.]

95. Motion in a Straight Line The velocity of a particle moving in a straight line is given by $v = t(t^2 + 1)^4 + t$.

a. Find an expression for the position s after a time t. HINT [See Example 2(d).]
b. Given that $s = 1$ at time $t = 0$, find the constant of integration C and hence an expression for s in terms of t without any unknown constants.

96. Motion in a Straight Line The velocity of a particle moving in a straight line is given by $v = 3te^{t^2} + t$.

a. Find an expression for the position s after a time t. HINT [See Example 2(d).]
b. Given that $s = 3$ at time $t = 0$, find the constant of integration C and hence an expression for s in terms of t without any unknown constants.

97. Bottled Water Sales The rate of U.S. sales of bottled water for the period 2000–2010 could be approximated by

$$s(t) = -45(t - 2,000)^2 + 900(t - 2,000) + 4,200 \text{ million}$$
$$\text{gallons per year} \quad (2,000 \leq t \leq 2,010),$$

where t is the year.[19] Use an indefinite integral to approximate the total sales $S(t)$ of bottled water since 2003 ($t = 2003$). Approximately how much bottled water was sold from $t = 2003$ to $t = 2008$?

98. Bottled Water Sales The rate of U.S. per capita sales of bottled water for the period 2000–2010 could be approximated by

$$s(t) = -0.18(t - 2,000)^2 + 3(t - 2,000) + 15 \text{ gallons per year}$$
$$(2,000 \leq t \leq 2,010),$$

where t is the year.[20] Use an indefinite integral to approximate the total per capita sales $S(t)$ of bottled water since the start of 2006. Approximately how much bottled water was sold, per capita, from $t = 2006$ to $t = 2008$?

COMMUNICATION AND REASONING EXERCISES

99. Are there any circumstances in which you should use the substitution $u = x$? Illustrate your answer by giving an example that shows the effect of this substitution.

100. You are asked to calculate $\int \frac{u}{u^2 + 1} \, du$. What is wrong with the substitution $u = u^2 + 1$?

101. If the x's do not cancel in a substitution, that means you chose the wrong expression for u, right?

102. At what stage of a calculation using a u substitution should you substitute back for u in terms of x: before or after taking the antiderivative?

103. Consider $\int \left(\frac{x}{x^2 - 1} + \frac{3x}{x^2 + 1} \right) dx$. To compute it, you should use which of the following?

(A) $u = x^2 - 1$ **(B)** $u = x^2 + 1$ **(C)** Neither **(D)** Both

Explain your answer.

104. If the substitution $u = x^2 - 1$ works in $\int \frac{x}{x^2 - 1} \, dx$, why does it not work nearly so easily in $\int \frac{x^2 - 1}{x} \, dx$? How would you do the second integral most simply?

105. Why was the following calculation marked wrong? What is the correct answer?

$u = x^2 - 1$
$\dfrac{du}{dx} = 2x$
$dx = \dfrac{1}{2x} du$

$$\int 3x(x^2 - 1) = \int 3xu$$
$$= 3x\frac{u^2}{2} + C = 3x\frac{(x^2 - 1)^2}{2} + C$$
$$\text{✗ WRONG!}$$

---

[18]Based on data from 1983 to 2003. Source: The American Physical Society/*New York Times*, May 3, 2003, p. A1.

[19]Source for data: Beverage Marketing Corporation (www.bottledwater.org).
[20]*Ibid.*

106. Why was the following calculation marked wrong? What is the correct answer?

$$u = x^3 - 1$$
$$\frac{du}{dx} = 3x^2$$
$$dx = \frac{1}{3x^2}\,du$$

$$\int x^2(x^3 - 1)^2 dx = \int x^2 u^2 \frac{1}{3x^2}\,du$$
$$= \frac{1}{3}u^2 + C = \frac{1}{3}(x^3 - 1)^2 + C$$
$$\textbf{✗ WRONG!}$$

107. Why was the following calculation marked wrong? What is the correct answer?

$$u = x^3 - 1$$
$$\frac{du}{dx} = 3x^2$$
$$dx = \frac{1}{3x^2}\,du$$

$$\int x^2(x^3 - 1)\,dx = \int x^2 u \frac{1}{3x^2}\,du$$
$$= \int \frac{1}{3}u\,du = \int \frac{1}{3}(x^3 - 1)\,dx$$
$$= \frac{1}{3}\left(\frac{x^4}{4} - x\right) + C \quad \textbf{✗ WRONG!}$$

108. Why was the following calculation marked wrong? What is the correct answer?

$$u = x^2 - 1$$
$$\frac{du}{dx} = 2x$$
$$dx = \frac{1}{2x}\,du$$

$$\int \frac{x}{x^2 - 1}\,dx = \int \frac{x}{u}\,du$$
$$= x\int \frac{1}{u}\,du = x \ln|u| + C$$
$$= x \ln|x^2 - 1| + C \quad \textbf{✗ WRONG!}$$

109. ▼ Show that *none* of the following substitutions work for $\int e^{-x^2}\,dx$: $u = -x$, $u = x^2$, $u = -x^2$. (The antiderivative of e^{-x^2} involves the *error function* erf(x).)

110. ▼ Show that *none* of the following substitutions work for $\int \sqrt{1 - x^2}\,dx$: $u = 1 - x^2$, $u = x^2$, and $u = -x^2$. (The antiderivative of $\sqrt{1 - x^2}$ involves inverse trigonometric functions, discussion of which is beyond the scope of this book.)

13.3   The Definite Integral: Numerical and Graphical Viewpoints

In Sections 13.1 and 13.2, we discussed the indefinite integral. There is an older, related concept called the **definite integral**. Let's introduce this new idea with an example. (We'll drop hints now and then about how the two types of integral are related. In Section 13.4 we discuss the exact relationship, which is one of the most important results in calculus.)

In Section 13.1, we used antiderivatives to answer questions of the form "Given the marginal cost, compute the total cost." (See Example 5 in Section 13.1.) In this section we approach such questions more directly, and we will forget about antiderivatives for now.

EXAMPLE 1 Oil Spill

Your deep ocean oil rig has suffered a catastrophic failure, and oil is leaking from the ocean floor wellhead at a rate of

$$v(t) = 0.08t^2 - 4t + 60 \text{ thousand barrels per day } (0 \le t \le 20),$$

where t is time in days since the failure.[21] Use a numerical calculation to estimate the total volume of oil released during the first 20 days.

---

[21]The model is consistent with the order of magnitude of the BP Deepwater Horizon oil spill of April 20–June 15, 2010, when the rate of flow of oil was estimated by the Federal Emergency Management Agency's Flow Rate Technical Group to be between 35,000 and 60,000 barrels per day (one barrel of oil is equivalent to about 0.16 cubic meters). Source: www.doi.gov/deepwaterhorizon.

Solution The graph of $v(t)$ is shown in Figure 3.

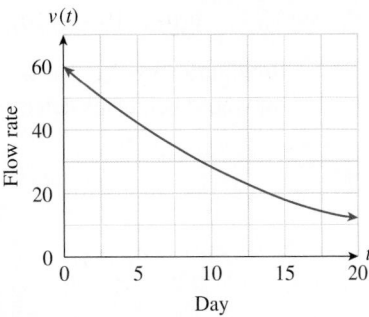

Figure 3

Let's start with a very crude estimate of the total volume of oil released, using the graph as a guide. The rate of change of this total volume at the beginning of the time period is $v(0) = 60$ thousand barrels per day. If this rate were to remain constant for the entire 20-day period, the total volume of oil released would be

$$\text{Total volume} = \text{Volume per day} \times \text{Number of days} = 60 \times 20$$
$$= 1{,}200 \text{ thousand barrels.}$$

Figure 4 shows how we can represent this calculation on the graph of $v(t)$.

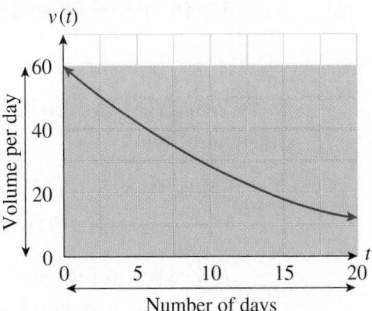

Figure 4

The volume per day based on $v(0) = 60$ is represented by the y-coordinate of the graph at its left edge, while the number of days is represented by the width of the interval $[0, 20]$ on the x-axis. Therefore, computing the area of the shaded rectangle in the figure gives the same calculation:

$$\text{Area of rectangle} = \text{Volume per day} \times \text{Number of days}$$
$$= 60 \times 20 = 1{,}200 = \text{Total volume.}$$

But, as we see in the graph, the flow rate does not remain constant, but goes down quite significantly over the course of the 20-day interval. We can obtain a somewhat more accurate estimate of the total volume by re-estimating the volume using 10-day periods—that is, by dividing the interval $[0, 20]$ into two equal intervals, or subdivisions. We estimate the volume over each 10-day period using the flow rate at the beginning of that period.

$$\text{Volume in first period} = \text{Volume per day} \times \text{Number of days}$$
$$= v(0) \times 10 = 60 \times 10 = 600 \text{ thousand barrels}$$

$$\text{Volume in second period} = \text{Volume per day} \times \text{Number of days}$$
$$= v(10) \times 10 = 28 \times 10 = 280 \text{ thousand barrels}$$

Adding these volumes gives us the more accurate estimate

$$v(0) \times 10 + v(10) \times 10 = 880 \text{ thousand barrels.}$$ Calculation using 2 subdivisions

In Figure 5 we see that we are computing the combined area of two rectangles, each of whose heights is determined by the height of the graph at its left edge:

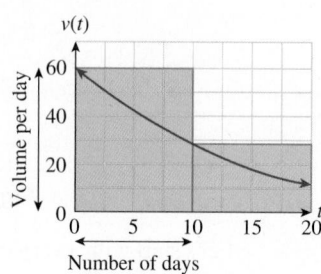

The areas of the rectangles are estimates of the volumes for successive 10-day periods.

Figure 5

Area of first rectangle = Volume per day × Number of days

$$= v(0) \times 10 = 60 \times 10 = 600 = \text{Volume for first 10 days}$$

Area of second rectangle = Volume per day × Number of days = $v(10) \times 10$

$$= 28 \times 10 = 280 = \text{Volume for second 10 days.}$$

We can get an even better estimate of the volume by using four divisions of [0, 20] instead of two:

$$v(0) \times 5 + v(5) \times 5 + v(10) \times 5 + v(15) \times 5$$ Calculation using 4 subdivisions

$$= 300 + 210 + 140 + 90 = 740 \text{ thousand barrels.}$$

As we see in Figure 6, we have now computed the combined area of *four* rectangles, each of whose heights is again determined by the height of the graph at its left edge.

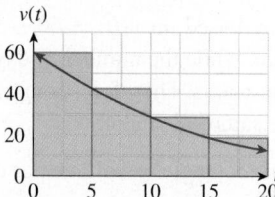

Estimated Volume Using 4 Subdivisions
The areas of the rectangles are estimates of the volumes for successive 5-day periods.

Figure 6

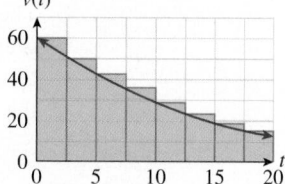

Estimated Volume Using 8 Subdivisions
The areas of the rectangles are estimates of the volumes for successive 2.5-day periods.

Figure 7

Notice how the volume seems to be decreasing as we use more subdivisions. More importantly, total volume seems to be getting closer to the area under the graph. Figure 7 illustrates the calculation for 8 equal subdivisions. The approximate total volume using 8 subdivisions is the total area of the shaded region in Figure 7:

$$v(0) \times 2.5 + v(2.5) \times 2.5 + v(5) \times 2.5 + \cdots + v(17.5) \times 2.5$$

$$= 675 \text{ thousand barrels.}$$ Calculation using 8 subdivisions

Looking at Figure 7, we still get the impression that we are overestimating the volume. If we want to be *really* accurate in our estimation of the volume, we should really be calculating the volume *continuously* every few hours or, better yet, minute by minute, as illustrated in Figure 8.

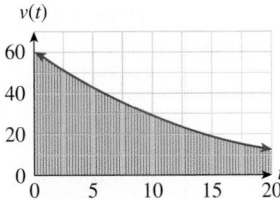
Every Six Hours (80 subdivisions)
Volume ≈ 619.35 thousand barrels

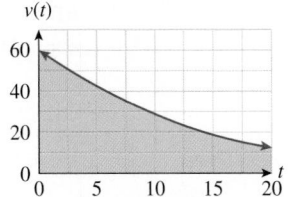
Every Minute (28,800 subdivisions)
Volume ≈ 613.35 thousand barrels

Figure 8

Figure 8 strongly suggests that the more accurately we estimate the total volume, the closer the answer gets to the exact area under the portion of the graph of $v(t)$ with $0 \leq t \leq 20$, and leads us to the conclusion that the *exact* total volume is the exact area under the rate of change of volume curve for $0 \leq t \leq 20$. In other words, we have made the following remarkable discovery:

Total volume is the area under the rate of change of volume curve!

➡ **Before we go on...** The 80-subdivision calculation in Example 1 is tedious to do by hand, and no one in his or her right mind would even *attempt* to do the minute-by-minute calculation by hand! Below we discuss ways of doing these calculations with the aid of technology. ∎

The type of calculation done in Example 1 is useful in many applications. Let's look at the general case and give the result a name.

In general, we have a function f (such as the function v in the example), and we consider an interval $[a, b]$ of possible values of the independent variable x. We subdivide the interval $[a, b]$ into some number of segments of equal length. Write n for the number of segments, or **subdivisions**.

Next, we label the endpoints of these subdivisions x_0 for a, x_1 for the end of the first subdivision, x_2 for the end of the second subdivision, and so on until we get to x_n, the end of the nth subdivision, so that $x_n = b$. Thus,

$$a = x_0 < x_1 < \cdots < x_n = b.$$

The first subdivision is the interval $[x_0, x_1]$, the second subdivision is $[x_1, x_2]$, and so on until we get to the last subdivision, which is $[x_{n-1}, x_n]$. We are dividing the interval $[a, b]$ into n subdivisions of equal length, so each segment has length $(b - a)/n$. We write Δx for $(b - a)/n$ (Figure 9).

Figure 9

Having established this notation, we can write the calculation that we want to do as follows: For each subdivision $[x_{k-1}, x_k]$, compute $f(x_{k-1})$, the value of the function

✶ After Georg Friedrich Bernhard Riemann (1826–1866).

f at the left endpoint. Multiply this value by the length of the interval, which is Δx. Then add together all n of these products to get the number

$$f(x_0)\Delta x + f(x_1)\Delta x + \cdots + f(x_{n-1})\Delta x.$$

This sum is called a **(left) Riemann*** **sum** for f. In Example 1 we computed several different Riemann sums. Here is the computation for $n = 4$ we used in the oil spill example (see Figure 10):

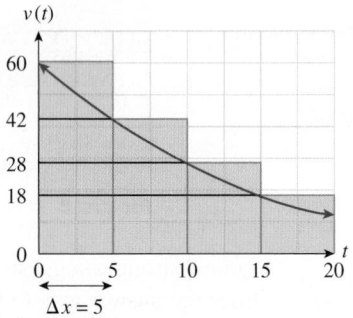

Figure 10

$$
\begin{aligned}
\text{Left Riemann sum} &= f(x_0)\Delta x + f(x_1)\Delta x + \cdots + f(x_{n-1})\Delta x \\
&= f(0)(5) + f(5)(5) + f(10)(5) + f(15)(5) \\
&= 60(5) + 42(5) + 28(5) + 18(5) = 740.
\end{aligned}
$$

Because sums are often used in mathematics, mathematicians have developed a shorthand notation for them. We write

$$f(x_0)\Delta x + f(x_1)\Delta x + \cdots + f(x_{n-1})\Delta x \quad \text{as} \quad \sum_{k=0}^{n-1} f(x_k)\Delta x.$$

The symbol $\sum$ is the Greek letter sigma and stands for **summation**. The letter k here is called the index of summation, and we can think of it as counting off the segments. We read the notation as "the sum from $k = 0$ to $n - 1$ of the quantities $f(x_k)\Delta x$." Think of it as a set of instructions:

Set $k = 0$, and calculate $f(x_0)\Delta x$. $f(0)(5)$ in the above calculation

Set $k = 1$, and calculate $f(x_1)\Delta x$. $f(5)(5)$ in the above calculation

. . .

Set $k = n - 1$, and calculate $f(x_{n-1})\Delta x$. $f(15)(5)$ in the above calculation

Then sum all the quantities so calculated.

Riemann Sum

If f is a continuous function, the **left Riemann sum** with n equal subdivisions for f over the interval $[a, b]$ is defined to be

$$
\begin{aligned}
\text{Left Riemann sum} &= \sum_{k=0}^{n-1} f(x_k)\Delta x \\
&= f(x_0)\Delta x + f(x_1)\Delta x + \cdots + f(x_{n-1})\Delta x \\
&= [f(x_0) + f(x_1) + \cdots + f(x_{n-1})]\Delta x,
\end{aligned}
$$

where $a = x_0 < x_1 < \cdots < x_n = b$ are the endpoints of the subdivisions, and $\Delta x = (b - a)/n$.

Interpretation of the Riemann Sum

If f is the rate of change of a quantity F (that is, $f = F'$), then the Riemann sum of f approximates the total change of F from $x = a$ to $x = b$. The approximation improves as the number of subdivisions increases toward infinity.

Quick Examples

1. If $f(t)$ is the rate of change in the number of bats in a belfry and $[a, b] = [2, 3]$, then the Riemann sum approximates the total change in the number of bats in the belfry from time $t = 2$ to time $t = 3$.

2. If $c(x)$ is the marginal cost of producing the xth item and $[a, b] = [10, 20]$, then the Riemann sum approximates the cost of producing items 11 through 20.

Visualizing a Left Riemann Sum (Non-negative Function)

Graphically, we can represent a left Riemann sum of a non-negative function as an approximation of the area under a curve:

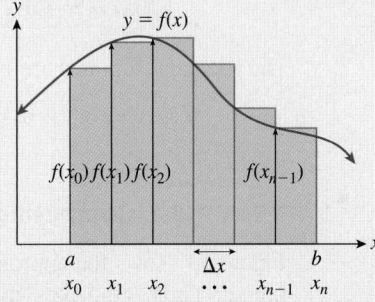

Riemann sum = Shaded area = Area of first rectangle + Area of second rectangle + $\cdots$ + Area of nth rectangle = $f(x_0)\Delta x + f(x_1)\Delta x + f(x_2)\Delta x + \cdots + f(x_{n-1})\Delta x$.

Quick Example

In Example 1 we computed several Riemann sums, including these:

$n = 1$:   Riemann sum $= v(0)\Delta t = 60 \times 20 = 1{,}200$

$n = 2$:   Riemann sum $= [v(t_0) + v(t_1)]\Delta t$
$$= [v(0) + v(10)](10) = 880$$

$n = 4$:   Riemann sum $= [v(t_0) + v(t_1) + v(t_2) + v(t_3)]\Delta t$
$$= [v(0) + v(5) + v(10) + v(15)](5) = 740$$

$n = 8$:   Riemann sum $= [v(t_0) + v(t_1) + \cdots + v(t_7)]\Delta t$
$$= [v(0) + v(2.5) + \cdots + v(17.5)](2.5) = 675$$

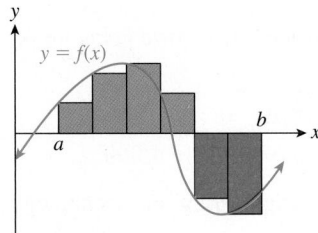

Riemann sum = Area above x-axis
– Area below x-axis

Figure 11

Note To visualize the Riemann sum of a function that is negative, look again at the formula $f(x_0)\Delta x + f(x_1)\Delta x + f(x_2)\Delta x + \cdots + f(x_{n-1})\Delta x$ for the Riemann sum. Each term $f(x_k)\Delta x_k$ represents the area of one rectangle in the figure above. So, the areas of the rectangles with negative values of $f(x_k)$ are automatically counted as negative. They appear as red rectangles in Figure 11. ∎

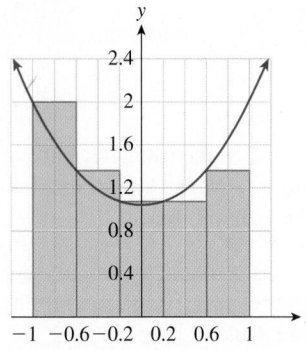

Figure 12

EXAMPLE 2 **Computing a Riemann Sum from a Formula**

Compute the left Riemann sum for $f(x) = x^2 + 1$ over the interval $[-1, 1]$, using $n = 5$ subdivisions.

Solution Because the interval is $[a, b] = [-1, 1]$ and $n = 5$, we have

$$\Delta x = \frac{b - a}{n} = \frac{1 - (-1)}{5} = 0.4.$$ Width of subdivisions

Thus, the subdivisions of $[-1, 1]$ are determined by

$$-1 < -0.6 < -0.2 < 0.2 < 0.6 < 1.$$ Start with -1 and keep adding $\Delta x = 0.4$.

Figure 12 shows the graph with a representation of the Riemann sum.

The Riemann sum we want is

$$[f(x_0) + f(x_1) + \cdots + f(x_4)]\Delta x$$
$$= [f(-1) + f(-0.6) + f(-0.2) + f(0.2) + f(0.6)]0.4.$$

We can conveniently organize this calculation in a table as follows:

x	-1	-0.6	-0.2	0.2	0.6	**Total**
$f(x) = x^2 + 1$	2	1.36	1.04	1.04	1.36	6.8

The Riemann sum is therefore

$$6.8\Delta x = 6.8 \times 0.4 = 2.72.$$

EXAMPLE 3 **Computing a Riemann Sum from a Graph**

Figure 13 shows the approximate annual production $n(t)$ of engineering and technology Ph.D. graduates in Mexico during the period 2000–2010.[22] Use a left Riemann sum with four subdivisions to estimate the total number of Ph.D. graduates from 2002 to 2010.

Solution Let us represent the total number of (engineering and technology) Ph.D. graduates up to time t (measured in years since 2000) by $N(t)$. The total number of Ph.D. graduates from 2002 to 2010 is then the total change in $N(t)$ over the interval $[2, 10]$. In view of the above discussion, we can approximate the total change in $N(t)$ using a Riemann sum of its rate of change $n(t)$. Because $n = 4$ subdivisions are specified, the width of each subdivision is

$$\Delta t = \frac{b - a}{n} = \frac{10 - 2}{4} = 2.$$

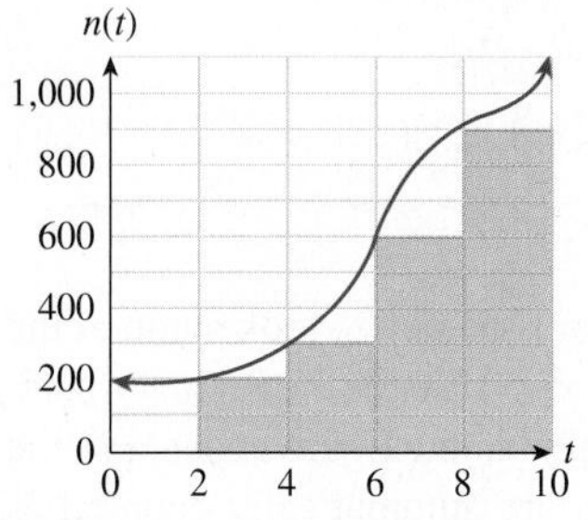

We can therefore represent the left Riemann sum by the shaded area shown in Figure 14.

From the graph,

$$\text{Left sum} = n(2)\Delta t + n(4)\Delta t + n(6)\Delta t + n(8)\Delta t$$
$$= (200)(2) + (300)(2) + (600)(2) + (900)(2) = 4,000.$$

So, we estimate that there was a total of about 4,000 engineering and technology Ph.D. graduates during the given period.

Figure 13

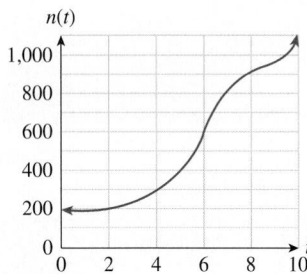

Figure 14

[22] Source for data: Instituto Nacional de Estadística y Geografía (www.inegi.org.mx).

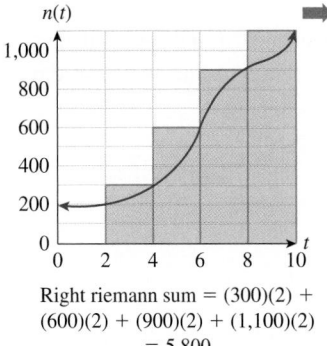

$n(t)$

Right riemann sum = $(300)(2) +$ $(600)(2) + (900)(2) + (1,100)(2)$ $= 5,800$

Height of each ractangle is determined by height of graph at right edge.

Figure 15

*This applies to some other functions as well, including "piecewise continuous" functions discussed in the next section.

Before we go on... A glance at Figure 14 tells us that we have significantly underestimated the actual number of graduates, as the actual area under the curve is considerably larger. Figure 15 shows a **right Riemann sum** and gives us a much larger estimate. For continuous functions, the difference between these two types of Riemann sums approaches zero as the number of subdivisions approaches infinity (see below) so we will focus primarily on only one type: left Riemann sums. ■

As in Example 1, we're most interested in what happens to the Riemann sum when we let n get very large. When f is continuous,* its Riemann sums will always approach a limit as n goes to infinity. (This is not meant to be obvious. Proofs may be found in advanced calculus texts.) We give the limit a name.

The Definite Integral

If f is a continuous function, the **definite integral of f from a to b** is defined to be the limit of the Riemann sums as the number of subdivisions approaches infinity:

$$\int_a^b f(x)\,dx = \lim_{n\to\infty} \sum_{k=0}^{n-1} f(x_k)\,\Delta x.$$

In Words: The integral, from a to b, of $f(x)\,dx$ equals the limit, as $n \to \infty$, of the Riemann Sum with a partition of n subdivisions.

The function f is called the **integrand**, the numbers a and b are the **limits of integration**, and the variable x is the **variable of integration**. A Riemann sum with a large number of subdivisions may be used to approximate the definite integral.

Interpretation of the Definite Integral

If f is the rate of change of a quantity F (that is, $f = F'$), then $\int_a^b f(x)\,dx$ is the (exact) total change of F from $x = a$ to $x = b$.

Quick Examples

1. If $f(t)$ is the rate of change in the number of bats in a belfry and $[a, b] = [2, 3]$, then $\int_2^3 f(t)\,dt$ is the total change in the number of bats in the belfry from time $t = 2$ to time $t = 3$.

2. If, at time t hours, you are selling wall posters at a rate of $s(t)$ posters per hour, then

   Total number of posters sold from hour 3 to hour $5 = \int_3^5 s(t)\,dt.$

Visualizing the Definite Integral

Non-negative Functions: If $f(x) \geq 0$ for all x in $[a, b]$, then $\int_a^b f(x)\,dx$ is the area under the graph of f over the interval $[a, b]$, as shaded in the figure.

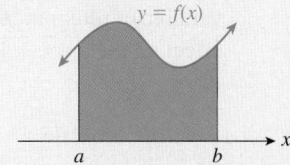

$y = f(x)$

a b x

General Functions: $\int_a^b f(x)\,dx$ is the area between $x = a$ and $x = b$ that is above the x-axis and below the graph of f, minus the area that is below the x-axis and above the graph of f:

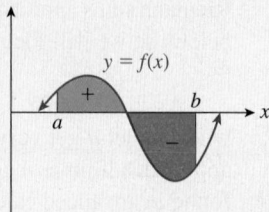

$$\int_a^b f(x)\,dx = \text{Area above } x\text{-axis} - \text{Area below } x\text{-axis}$$

Quick Examples

1.
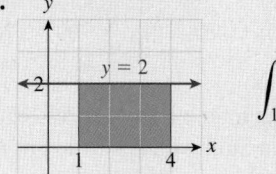

$$\int_1^4 2\,dx = \text{Area of rectangle} = 6$$

2.
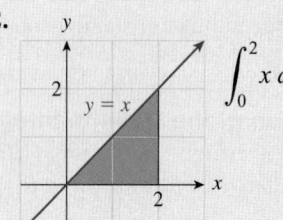

$$\int_0^2 x\,dx = \text{Area of triangle} = \frac{1}{2}\,\text{base} \times \text{height} = 2$$

3.
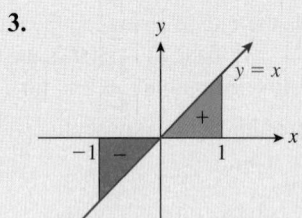

$$\int_{-1}^1 x\,dx = 0 \qquad \text{The areas above and below the } x\text{-axis are equal.}$$

Notes

1. Remember that $\int_a^b f(x)\,dx$ stands for a number that depends on f, a, and b. The variable of integration x that appears is called a **dummy variable** because it has no effect on the answer. In other words,

$$\int_a^b f(x)\,dx = \int_a^b f(t)\,dt. \qquad x \text{ or } t \text{ is just a name we give the variable.}$$

2. The notation for the definite integral (due to Leibniz) comes from the notation for the Riemann sum. The integral sign $\int$ is an elongated S, the Roman equivalent of the Greek $\sum$. The d in dx is the lowercase Roman equivalent of the Greek Δ.

3. The definition above is adequate for continuous functions, but more complicated definitions are needed to handle other functions. For example, we broke the interval $[a, b]$ into n subdivisions of equal length, but other definitions allow a **partition** of the interval into subdivisions of possibly unequal lengths. We have evaluated f at the left endpoint of each subdivision, but we could equally well have used the right endpoint or any other point in the subdivision. All of these variations lead to the same answer when f is continuous.

4. The similarity between the notations for the definite integral and the indefinite integral is no mistake. We will discuss the exact connection in the next section. ■

Computing Definite Integrals

In some cases, we can compute the definite integral directly from the graph (see the quick examples above and the next example below). In general, the only method of computing definite integrals we have discussed so far is numerical estimation: Compute the Riemann sums for larger and larger values of n and then estimate the number it seems to be approaching as we did in Example 1. (In the next section we will discuss an algebraic method for computing them.)

EXAMPLE 4 Estimating a Definite Integral from a Graph

Figure 16 shows the graph of the (approximate) rate $f'(t)$, at which the United States consumed gasoline from 2000 through 2008. (t is time in years since 2000.)[23]

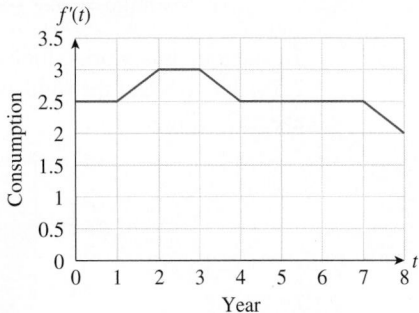

$f'(t)$ = Gasoline consumption (100 million gals per year)

Figure 16

Use the graph to estimate the total U.S. consumption of gasoline over the period shown.

Solution The derivative $f'(t)$ represents the rate of change of the total U.S. consumption of gasoline, and so the total U.S. consumption of gasoline over the given period $[0, 8]$ is given by the definite integral:

$$\text{Total U.S. consumption of gasoline} = \text{Total change in } f(t) = \int_0^8 f'(t)\, dt$$

and is given by the area under the graph (Figure 17).

---

[23] Source: Energy Information Administration (Department of Energy). (www.eia.doe.gov).

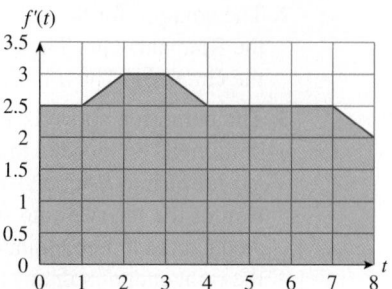

Figure 17

One way to determine the area is to count the number of filled rectangles as defined by the grid. Each rectangle has an area of $1 \times 0.5 = 0.5$ units (and the half-rectangles determined by diagonal portions of the graph have half that area). Counting rectangles, we find a total of 41.5 complete rectangles, so

$$\text{Total area} = 20.75.$$

Because $f'(t)$ is in 100 million gallons per year, we conclude that the total U.S. consumption of gasoline over the given period was about 2,075 million gallons, or 2.075 billion gallons.

While counting rectangles might seem easy, it becomes awkward in cases involving large numbers of rectangles or partial rectangles whose area is not easy to determine. In a case like this, in which the graph consists of straight lines, rather than counting rectangles, we can get the area by averaging the left and right Riemann sums whose subdivisions are determined by the grid:

$$\text{Left sum} = (2.5 + 2.5 + 3 + 3 + 2.5 + 2.5 + 2.5 + 2.5)(1) = 21$$
$$\text{Right sum} = (2.5 + 3 + 3 + 2.5 + 2.5 + 2.5 + 2.5 + 2)(1) = 20.5$$
$$\text{Average} = \frac{21 + 20.5}{2} = 20.75.$$

To see why this works, look at the single interval $[1, 2]$. The left sum contributes $2.5 \times 1 = 2.5$ and the right sum contributes $3 \times 1 = 3$. The exact area is their average, 2.75 (Figure 18).

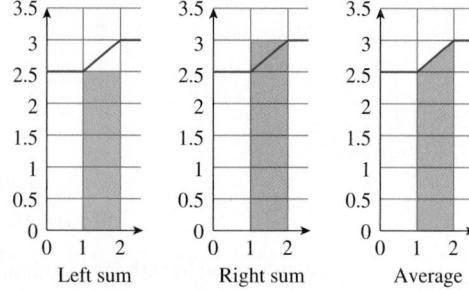

Figure 18

The average of the left and right Riemann sums is frequently a better estimate of the definite integral than either is alone.

➡ **Before we go on...** It is important to check that the units we are using in Example 4 match up correctly: t is given in *years* and $f'(t)$ is given in 100 million gallons per *year*. The integral is then given in

$$\text{Years} \times \frac{100 \text{ million gallons}}{\text{Year}} = 100 \text{ million gallons}.$$

If we had specified $f'(t)$ in, say, 100 million gallons per *day* but t in years, then we would have needed to convert either t or $f'(t)$ so that the units of time match. ■

The next example illustrates the use of technology in estimating definite integrals using Riemann sums.

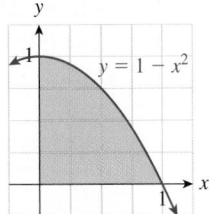

Figure 19

▦ **using** Technology

See the Technology Guides at the end of the chapter to find out how to compute these sums on a TI-83/84 Plus and a spreadsheet.

WW Website
www.WanerMath.com
At the Website, select the Online Utilities tab, and choose the Numerical Integration Utility and Grapher. This utility computes left and right Riemann sums, as well as a good numerical approximation of the integral. There is also a downloadable Excel spreadsheet that computes and graphs Riemann sums (Riemann Sum Grapher).

EXAMPLE 5 ■ **Using Technology to Approximate the Definite Integral**

Use technology to estimate the area under the graph of $f(x) = 1 - x^2$ over the interval $[0, 1]$ using $n = 100$, $n = 200$, and $n = 500$ subdivisions.

Solution We need to estimate the area under the parabola shown in Figure 19.

From the discussion above,

$$\text{Area} = \int_0^1 (1 - x^2)\, dx.$$

The Riemann sum with $n = 100$ has $\Delta x = (b - a)/n = (1 - 0)/100 = 0.01$ and is given by

$$\sum_{k=0}^{99} f(x_k)\Delta x = [f(0) + f(0.01) + \cdots + f(0.99)](0.01).$$

Similarly, the Riemann sum with $n = 200$ has $\Delta x = (b - a)/n = (1 - 0)/200 = 0.005$ and is given by

$$\sum_{k=0}^{199} f(x_k)\Delta x = [f(0) + f(0.005) + \cdots + f(0.995)](0.005).$$

For $n = 500$, $x = (b - a)/n = (1 - 0)/500 = 0.002$ and the Riemann sum is

$$\sum_{k=0}^{499} f(x_k)\Delta x = [f(0) + f(0.002) + \cdots + f(0.998)](0.002).$$

Using technology to evaluate these Riemann sums, we find:

$$n = 100: \sum_{k=0}^{99} f(x_k)\Delta x = 0.67165$$

$$n = 200: \sum_{k=0}^{199} f(x_k)\Delta x = 0.6691625$$

$$n = 500: \sum_{k=0}^{499} f(x_k)\Delta x = 0.667666$$

so we estimate that the area under the curve is about 0.667. (The exact answer is $2/3$, as we will be able to verify using the techniques in the next section.)

EXAMPLE 6 **Motion**

A fast car has velocity $v(t) = 6t^2 + 10t$ ft/sec after t seconds (as measured by a radar gun). Use several values of n to find the distance covered by the car from time $t = 3$ seconds to time $t = 4$ seconds.

Solution Because the velocity $v(t)$ is rate of change of position, the total change in position over the interval $[3, 4]$ is

$$\text{Distance covered} = \text{Total change in position} = \int_3^4 v(t)\,dt = \int_3^4 (6t^2 + 10t)\,dt.$$

As in Examples 1 and 5, we can subdivide the one-second interval $[3, 4]$ into smaller and smaller pieces to get more and more accurate approximations of the integral. By computing Riemann sums for various values of n, we get the following results.

$$n = 10: \sum_{k=0}^{9} v(t_k)\Delta t = 106.41 \qquad n = 100: \sum_{k=0}^{99} v(t_k)\Delta t \approx 108.740$$

$$n = 1{,}000: \sum_{k=0}^{999} v(t_k)\Delta t \approx 108.974 \qquad n = 10{,}000: \sum_{k=0}^{9999} v(t_k)\Delta t \approx 108.997$$

These calculations suggest that the total distance covered by the car, the value of the definite integral, is approximately 109 feet.

➡ **Before we go on...** Do Example 6 using antiderivatives instead of Riemann sums, as in Section 13.1. Do you notice a relationship between antiderivatives and definite integrals? This will be explored in the next section. ∎

Website
www.WanerMath.com
At the Website you will find the following optional online interactive section: Internet Topic: Numerical Integration.

13.3 EXERCISES

▼ more advanced   ◆ challenging
T indicates exercises that should be solved using technology

Calculate the left Riemann sums for the given functions over the given intervals in Exercises 1–10, using the given values of n. (When rounding, round answers to four decimal places.)
HINT [See Example 2.]

1. $f(x) = 4x - 1$ over $[0, 2]$, $n = 4$
2. $f(x) = 1 - 3x$ over $[-1, 1]$, $n = 4$
3. $f(x) = x^2$ over $[-2, 2]$, $n = 4$
4. $f(x) = x^2$ over $[1, 5]$, $n = 4$
5. $f(x) = \dfrac{1}{1+x}$ over $[0, 1]$, $n = 5$
6. $f(x) = \dfrac{x}{1+x^2}$ over $[0, 1]$, $n = 5$
7. $f(x) = e^{-x}$ over $[0, 10]$, $n = 5$
8. $f(x) = e^{-x}$ over $[-5, 5]$, $n = 5$
9. $f(x) = e^{-x^2}$ over $[0, 10]$, $n = 4$
10. $f(x) = e^{-x^2}$ over $[0, 100]$, $n = 4$

In Exercises 11–18, use the given graph to estimate the left Riemann sum for the given interval with the stated number of subdivisions. HINT [See Example 3.]

11. $[0, 5]$, $n = 5$
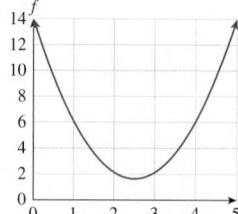

12. $[0, 8]$, $n = 4$
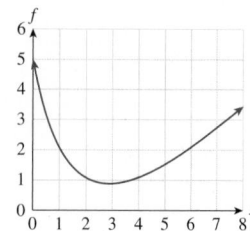

13. $[1, 9]$, $n = 4$
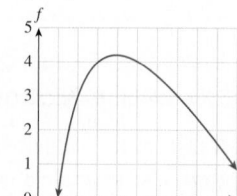

14. $[0.5, 2.5]$, $n = 4$
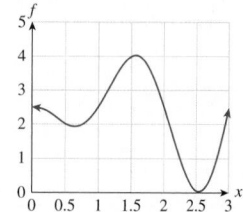

15. $[1, 3.5]$, $n = 5$

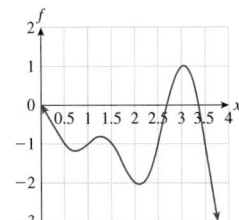

16. $[0.5, 3.5]$, $n = 3$

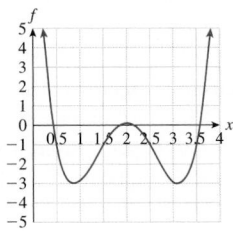

31. $[2, 6]$

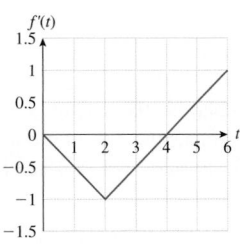

32. $[0, 5]$

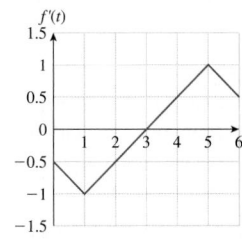

17. $[0, 3]$; $n = 3$

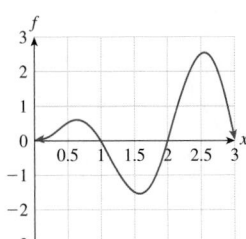

18. $[0.5, 3]$; $n = 5$

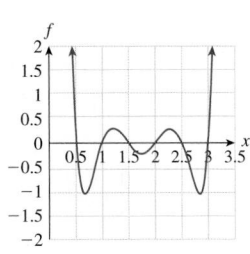

33. $[-1, 2]$

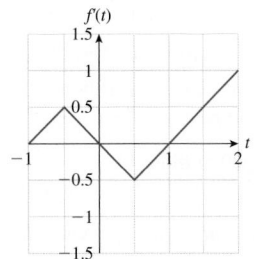

34. $[-1, 2]$

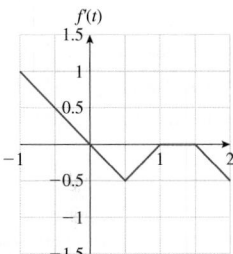

Use geometry (not Riemann sums) to compute the integrals in Exercises 19–28. HINT [See Quick Examples page 982.]

19. $\int_0^1 1\,dx$

20. $\int_0^2 5\,dx$

21. $\int_0^1 x\,dx$

22. $\int_1^2 x\,dx$

23. $\int_0^1 \frac{x}{2}\,dx$

24. $\int_1^2 \frac{x}{2}\,dx$

25. $\int_2^4 (x - 2)\,dx$

26. $\int_3^6 (x - 3)\,dx$

27. $\int_{-1}^1 x^3\,dx$

28. $\int_{-2}^2 \frac{x}{2}\,dx$

In Exercises 29–34, the graph of the derivative $f'(t)$ of $f(t)$ is shown. Compute the total change of $f(t)$ over the given interval. HINT [See Example 4.]

29. $[1, 5]$

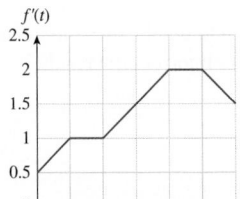

30. $[2, 6]$

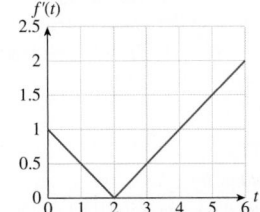

⊤ *In Exercises 35–38, use technology to approximate the given integrals with Riemann sums, using **(a)** $n = 10$, **(b)** $n = 100$, and **(c)** $n = 1,000$. Round all answers to four decimal places.* HINT [See Example 5.]

35. $\int_0^1 4\sqrt{1 - x^2}\,dx$

36. $\int_0^1 \frac{4}{1 + x^2}\,dx$

37. $\int_2^3 \frac{2x^{1.2}}{1 + 3.5x^{4.7}}\,dx$

38. $\int_3^4 3xe^{1.3x}\,dx$

APPLICATIONS

39. *Pumps* A pump is delivering water into a tank at a rate of

$$r(t) = 3t^2 + 5 \text{ liters/minute,}$$

where t is time in minutes since the pump is turned on. Use a Riemann sum with $n = 5$ subdivisions to estimate the total volume of water pumped in during the first two minutes. HINT [See Examples 1 and 2. This exercise is also discussed in the tutorial on the Website.]

40. *Pumps* A pump is delivering water into a tank at a rate of

$$r(t) = 6t^2 + 40 \text{ liters/minute,}$$

where t is time in minutes since the pump is turned on. Use a Riemann sum with $n = 6$ subdivisions to estimate the total volume of water pumped in during the first three minutes. HINT [See Examples 1 and 2. A similar exercise is also discussed in the tutorial on the Website.]

41. *Cost* The marginal cost function for the manufacture of portable MP3 players is given by

$$C'(x) = 20 - \frac{x}{200},$$

where x is the number of MP3 players manufactured. Use a Riemann sum with $n = 5$ to estimate the cost of producing the first 5 MP3 players. HINT [See Examples 1 and 2 and Quick Example 2 on page 979.]

42. *Cost* Repeat the preceding exercise using the marginal cost function

$$C'(x) = 25 - \frac{x}{50}.$$

HINT [See Examples 1 and 2 and Quick Example 2 on page 979.]

43. *Bottled Water Sales* The rate of U.S. sales of bottled water for the period 2000–2010 could be approximated by

$$s(t) = -45t^2 + 900t + 4{,}200 \text{ million gallons per year}$$
$$(0 \le t \le 10),$$

where t is time in years since the start of 2000.[24] Use a Riemann sum with $n = 5$ to estimate the total U.S. sales of bottled water from the start of 2005 to the start of 2010. (Round your answer to the nearest billion gallons.) HINT [See Example 2.]

44. *Bottled Water Sales* The rate of U.S. per capita sales of bottled water for the period 2000–2010 could be approximated by

$$s(t) = -0.18t^2 + 3t + 15 \text{ gallons per year} \quad (0 \le t \le 10),$$

where t is time in years since the start of 2000.[25] Use a Riemann sum with $n = 5$ to estimate the total U.S. per capita sales of bottled water from the start of 2003 to the start of 2008. (Round your answer to the nearest gallon.) HINT [See Example 2.]

45. *Online Vehicle Sales* The following graph shows the rate of change $v(t)$ of the value of vehicle transactions on **eBay**, in billions of dollars per quarter from the first quarter of 2008 through the first quarter of 2011 (t is time in quarters; $t = 1$ represents the first quarter of 2008).[26]

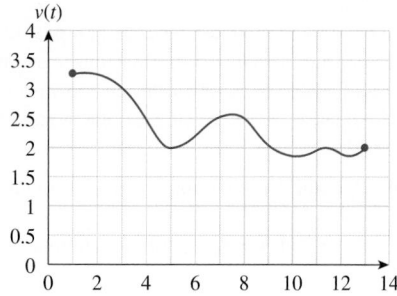

Use a left Riemann sum with four subdivisions to estimate the total value of vehicle transactions on eBay from the third quarter of 2008 to the third quarter of 2010 (the interval [3, 11]). HINT [See Example 3.]

46. *Online Auctions* The following graph shows the rate of change $n(t)$ of the number of active **eBay** users, in millions of users per quarter, from the first quarter of 2008 through the first quarter of 2011 (t is time in quarters; $t = 1$ represents the first quarter of 2008).[27]

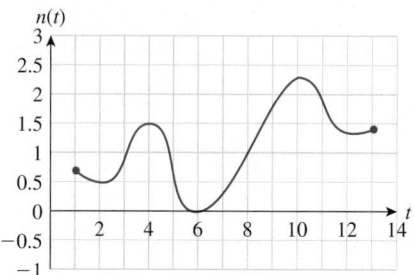

Use a left Riemann sum with four subdivisions to estimate the total change in the number of active users from the second quarter of 2008 to the second quarter of 2010 (the interval [2, 10]). HINT [See Example 3.]

47. *Scientific Research* The rate of change $r(t)$ of the total number of research articles in the prominent journal *Physical Review* written by researchers in Europe is shown in the following graph:

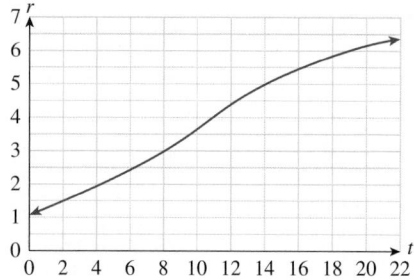

Here, t is time in years ($t = 0$ represents the start of 1983).[28]

a. Use both left and right Riemann sums with eight subdivisions to estimate the total number of articles in *Physical Review* written by researchers in Europe during the 16-year period beginning at the start of 1983. (Estimate each value of $r(t)$ to the nearest 0.5.) HINT [See Example 3.]

b. Use the answers from part (a) to obtain an estimate of $\int_0^{16} r(t)\,dt$. HINT [See Example 4.] Interpret the result.

---

[24] Source for data: Beverage Marketing Corporation (www.bottledwater.org).

[25] *Ibid.*

[26] Source for data: eBay company reports (http://investor.ebay.com).

[27] *Ibid.*

[28] Based on data from 1983 to 2003. Source: The American Physical Society/*New York Times*, May 3, 2003, p. A1.

48. *Scientific Research* The rate of change $r(t)$ of the total number of research articles in the prominent journal *Physical Review* written by researchers in the United States is shown in the following graph:

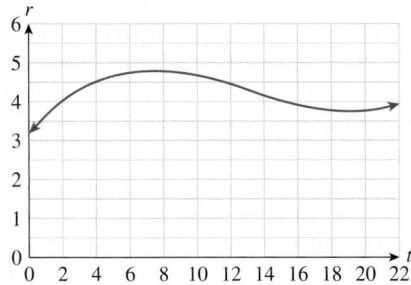

Here, t is time in years ($t = 0$ represents the start of 1983).[29]

a. Use both left and right Riemann sums with six subdivisions to estimate the total number of articles in *Physical Review* written by researchers in the United States during the 12-year period beginning at the start of 1993. (Estimate each value of $r(t)$ to the nearest 0.25.) HINT [See Example 3.]

b. Use the answers from part (a) to obtain an estimate of $\int_{10}^{22} r(t)\, dt$. HINT [See Example 4.] Interpret the result.

49. *Graduate Degrees: Women* The following graph shows the approximate number $f'(t)$ of doctoral degrees per year awarded to women in the United States during 2005–2014 ($t = 0$ represents 2005 and each unit of the y-axis represents 10,000 degrees).[30]

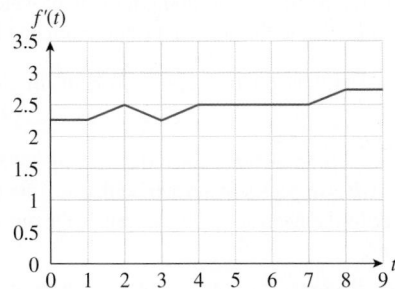

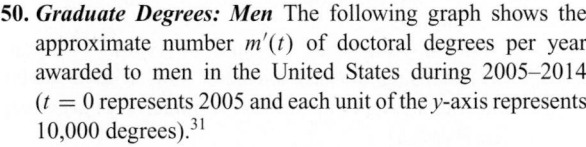

Use the graph to estimate the total number of doctoral degrees awarded to women from 2005 to 2010. HINT [See Example 4.]

50. *Graduate Degrees: Men* The following graph shows the approximate number $m'(t)$ of doctoral degrees per year awarded to men in the United States during 2005–2014 ($t = 0$ represents 2005 and each unit of the y-axis represents 10,000 degrees).[31]

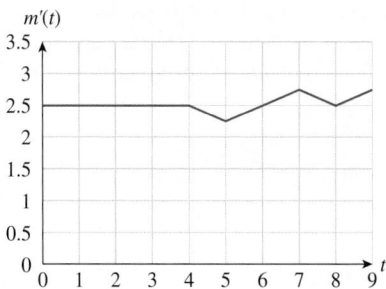

Use the graph to estimate the total number of doctoral degrees awarded to men from 2009 to 2014. HINT [See Example 4.]

51. *Visiting Students in the Early 2000s* The aftermath of the September 11 attacks saw a decrease in the number of students visiting the United States. The following graph shows the approximate rate of change $c'(t)$ in the number of students from China who had taken the GRE exam required for admission to U.S. universities (t is time in years since 2000):[32]

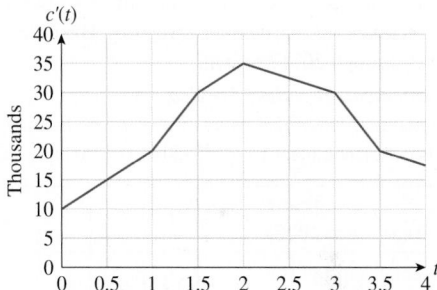

Use the graph to estimate, to the nearest 1,000, the total number of students from China who took the GRE exams during the period 2002–2004.

52. *Visiting Students in the Early 2000s* Repeat Exercise 51, using the following graph for students from India:[33]

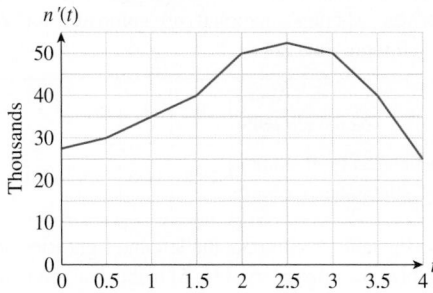

---

[29]Based on data from 1983 to 2003. Source: The American Physical Society/*New York Times*, May 3, 2003, p. A1.

[30]Source for data and projections: National Center for Education Statistics (www.nces.edu.gov).

[31]*Ibid.*

[32]Source: Educational Testing Services/Shanghai and Jiao Tong University/*New York Times*, December 21, 2004, p. A25.

[33]*Ibid.*

Net Income: Exercises 53 and 54 are based on the following graph, which shows **General Electric's** approximate net income in billions of dollars each year from 2005 to 2011.[34]

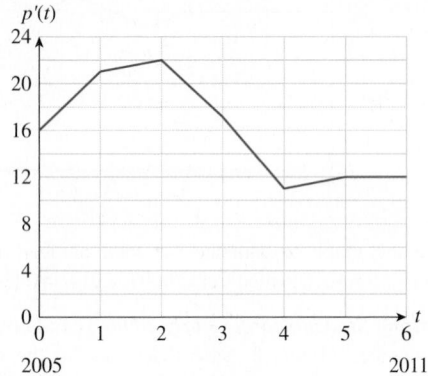

53. Compute the left and right Riemann sum estimates of $\int_0^4 p'(t)\,dt$ using $\Delta t = 1$. Which of these two sums gives the actual total net income earned by **GE** during the years 2005 through 2008? Explain.

54. Compute the left and right Riemann sum estimates of $\int_2^6 p'(t)\,dt$ using $\Delta t = 1$. Which of these two sums gives the actual total net income earned by **GE** during the years 2008 through 2011? Explain.

55. *Petrochemical Sales: Mexico* The following table shows annual petrochemical sales in Mexico by **Pemex**, Mexico's national oil company, for 2005–2010 ($t = 0$ represents 2005).[35]

Year t (Year since 2005)	0	1	2	3	4	5
Petrochemical Sales s (Billion metric tons)	3.7	3.8	4.0	4.1	4.0	4.1

a. Use the table to compute the left and right Riemann sums for $s(t)$ over the interval [0, 4] using 4 subdivisions.

b. What does the *right* Riemann sum in part (a) tell you about petrochemical sales by Pemex?

56. *Petrochemical Production: Mexico* The following table shows annual petrochemical production in Mexico by **Pemex**, Mexico's national oil company, for 2005–2010 ($t = 0$ represents 2005):[36]

Year t (Year since 2005)	0	1	2	3	4	5
Petrochemical Production p (Billion metric tons)	10.8	11.0	11.8	12.0	12.0	13.3

a. Use the table to compute the left and right Riemann sums for $p(t)$ over the interval [1, 5] using 4 subdivisions.

b. What does the *left* Riemann sum in part (a) tell you about production of petrochemicals by Pemex?

57. *Motion under Gravity* The velocity of a stone moving under gravity t seconds after being thrown up at 30 ft/s is given by $v(t) = -32t + 30$ ft/sec. Use a Riemann sum with 5 subdivisions to estimate $\int_0^4 v(t)\,dt$. What does the answer represent? HINT [See Example 6.]

58. *Motion under Gravity* The velocity of a stone moving under gravity t seconds after being thrown up at 4 m/s is given by $v(t) = -9.8t + 4$ m/sec. Use a Riemann sum with 5 subdivisions to estimate $\int_0^1 v(t)\,dt$. What does the answer represent? HINT [See Example 6.]

59. ⊤ *Motion* A model rocket has upward velocity $v(t) = 40t^2$ ft/sec, t seconds after launch. Use a Riemann sum with $n = 10$ to estimate how high the rocket is 2 seconds after launch. (Use technology to compute the Riemann sum.)

60. ⊤ *Motion* A race car has a velocity of $v(t) = 600(1 - e^{-0.5t})$ ft/sec, t seconds after starting. Use a Riemann sum with $n = 10$ to estimate how far the car has traveled in the first 4 seconds. (Round your answer to the nearest whole number.) (Use technology to compute the Riemann sum.)

61. ⊤ *Facebook Membership* Since the start of 2005, new members joined **Facebook** at a rate of roughly

$$m(t) = 3.2t^3 - 12t + 10 \text{ million members per year}$$
$$(0 \le t \le 5),$$

where t is time in years since the start of 2005.[37] Estimate $\int_0^5 m(t)\,dt$ using a Riemann sum with $n = 150$. (Round your answer to the nearest whole number.) Interpret the answer. HINT [See Example 5.]

62. ⊤ *Uploads to YouTube* Since **YouTube** first became available to the public in mid-2005, the rate at which video has been uploaded to the site can be approximated by

$$v(t) = 525{,}600(0.42t^2 + 2.7t - 1) \text{ hours of video per year}$$
$$(0.5 \le t \le 7),$$

where t is time in years since the start of 2005.[38] Estimate $\int_1^7 v(t)\,dt$ using a Riemann sum with $n = 150$. (Round your answer to the nearest 1,000.) Interpret the answer. HINT [See Example 5.]

63. ⊤ *Big Brother* The total number of wiretaps authorized each year by U.S. state and federal courts from 1990 to 2010 can be approximated by

$$w(t) = 820e^{0.051t} \quad (0 \le t \le 20)$$

(t is time in years since the start of 1990).[39] Estimate $\int_0^{15} w(t)\,dt$ using a (left) Riemann sum with $n = 100$. (Round your answer to the nearest 10.) Interpret the answer. HINT [See Example 5.]

---

[34]Source: Company reports (www.ge.com/investors).

[35]2010 figure is a projection based on data through May 2010. Source: www.pemex.com (July 2010).

[36]*Ibid.*

[37]Sources for data: www.facebook.com, www.insidefacebook.com.

[38]Source for data: www.YouTube.com.

[39]Source for data: Wiretap Reports, Administrative Office of the United States Courts (www.uscourts.gov/Statistics/WiretapReports.aspx).

64. ▯ *Big Brother* The number of wiretaps authorized each year by U.S. state courts from 1990 to 2010 can be approximated by

$$w(t) = 430e^{0.065t} \quad (0 \le t \le 20)$$

(t is time in years since the start of 1990).[40] Estimate $\int_{10}^{15} w(t)\,dt$ using a (left) Riemann sum with $n = 100$. (Round your answer to the nearest 10.) Interpret the answer. HINT [See Example 5.]

65. ▼ *Surveying* My uncle intends to build a kidney-shaped swimming pool in his small yard, and the town zoning board will approve the project only if the total area of the pool does not exceed 500 square feet. The accompanying figure shows a diagram of the planned swimming pool, with measurements of its width at the indicated points. Will my uncle's plans be approved? Use a (left) Riemann sum to approximate the area.

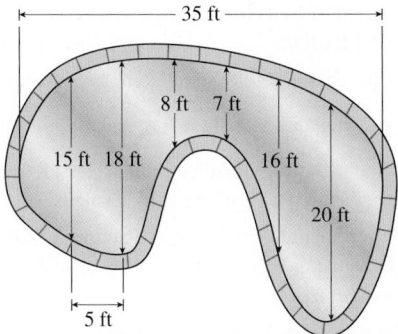

66. ▼ *Pollution* An aerial photograph of an ocean oil spill shows the pattern in the accompanying diagram. Assuming that the oil slick has a uniform depth of 0.01 m, how many cubic meters of oil would you estimate to be in the spill? (Volume = Area × Thickness. Use a (left) Riemann sum to approximate the area.)

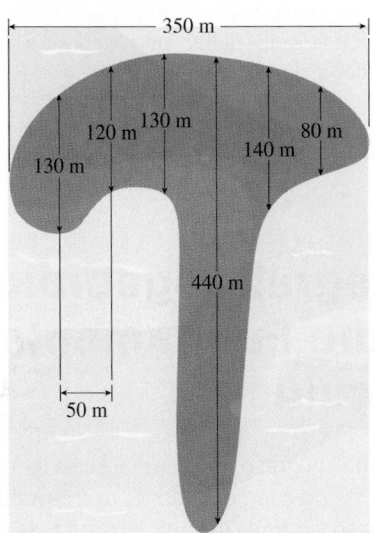

67. ▯ ▼ *Oil Consumption: United States* During the period 1980–2008 the United States was consuming oil at a rate of about

$$q(t) = 76t + 5{,}540 \text{ million barrels per year} \quad (0 \le t \le 28),$$

where t is time in years since the start of 1980.[41] During the same period, the price per barrel of crude oil in constant 2008 dollars was about

$$p(t) = 0.45t^2 - 12t + 105 \text{ dollars}[42] \quad (0 \le t \le 28).$$

a. Graph the function $r(t) = p(t)q(t)$ for $0 \le t \le 28$, indicating the area that represents $\int_{10}^{20} r(t)\,dt$. What does this area signify?

b. Estimate the area in part (a) using a Riemann sum with $n = 200$. (Round the answer to 3 significant digits.) Interpret the answer.

68. ▯ ▼ *Oil Consumption: China* Repeat Exercise 67 using instead the rate of consumption of oil in China:

$$q(t) = 82t + 221 \text{ million barrels per year}[43] \quad (0 \le t \le 28).$$

The Normal Curve *The normal distribution curve, which models the distributions of data in a wide range of applications, is given by the function*

$$p(x) = \frac{1}{\sqrt{2\pi}\,\sigma} e^{-(x-\mu)^2/2\sigma^2},$$

where $\pi = 3.14159265\ldots$ *and* σ *and* μ *are constants called the* ***standard deviation*** *and the* ***mean***, *respectively. The graph of the normal distribution (when* $\sigma = 1$ *and* $\mu = 2$*) is shown in the figure. Exercises 69 and 70 illustrate its use.*

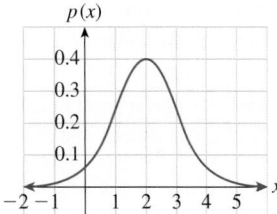

69. ▯ ▼ *Test Scores* Enormous State University's Calculus I test scores are modeled by a normal distribution with $\mu = 72.6$ and $\sigma = 5.2$. The percentage of students who obtained scores between a and b on the test is given by

$$\int_a^b p(x)\,dx.$$

a. Use a Riemann sum with $n = 40$ to estimate the percentage of students who obtained between 60 and 100 on the test.

b. What percentage of students scored less than 30?

---

[40] Source for data: Wiretap Reports, Administrative Office of the United States Courts (www.uscourts.gov/Statistics/WiretapReports.aspx).

[41] Source for data: BP Statistical Review of World Energy (www.bp.com/statisticalreview).

[42] Source for data: www.inflationdata.com.

[43] Source for data: BP Statistical Review of World Energy (www.bp.com/statisticalreview).

70. ⊓ ▼ *Consumer Satisfaction* In a survey, consumers were asked to rate a new toothpaste on a scale of 1–10. The resulting data are modeled by a normal distribution with $\mu = 4.5$ and $\sigma = 1.0$. The percentage of consumers who rated the toothpaste with a score between a and b on the test is given by

$$\int_a^b p(x)\,dx.$$

a. Use a Riemann sum with $n = 10$ to estimate the percentage of customers who rated the toothpaste 5 or above. (Use the range 4.5 to 10.5.)

b. What percentage of customers rated the toothpaste 0 or 1? (Use the range −0.5 to 1.5.)

COMMUNICATION AND REASONING EXERCISES

71. If $f(x) = 6$, then the left Riemann sum _____ (increases/decreases/stays the same) as n increases.

72. If $f(x) = -1$, then the left Riemann sum _____ (increases/decreases/stays the same) as n increases.

73. If f is an increasing function of x, then the left Riemann sum _____ (increases/decreases/stays the same) as n increases.

74. If f is a decreasing function of x, then the left Riemann sum _____ (increases/decreases/stays the same) as n increases.

75. If $\int_a^b f(x)\,dx = 0$, what can you say about the graph of f?

76. Sketch the graphs of two (different) functions f and g such that $\int_a^b f(x)\,dx = \int_a^b g(x)\,dx$.

77. ▼ The definite integral counts the area under the x-axis as negative. Give an example that shows how this can be useful in applications.

78. ▼ Sketch the graph of a nonconstant function whose Riemann sum with $n = 1$ gives the exact value of the definite integral.

79. ▼ Sketch the graph of a nonconstant function whose Riemann sums with $n = 1, 5$, and 10 are all zero.

80. ▼ Besides left and right Riemann sums, another approximation of the integral is the **midpoint** approximation, in which we compute the sum

$$\sum_{k=1}^n f(\bar{x}_k)\,\Delta x$$

where $\bar{x}_k = (x_{k-1} + x_k)/2$ is the point midway between the left and right endpoints of the interval $[x_{k-1}, x_k]$. Why is it true that the midpoint approximation is exact if f is linear? (Draw a picture.)

81. ▼ Your cellphone company charges you $c(t) = \dfrac{20}{t + 100}$ dollars for the tth minute. You make a 60-minute phone call. What kind of (left) Riemann sum represents the total cost of the call? Explain.

82. ▼ Your friend's cellphone company charges her $c(t) = \dfrac{20}{t + 100}$ dollars for the $(t + 1)$st minute. Your friend makes a 60-minute phone call. What kind of (left) Riemann sum represents the total cost of the call? Explain.

83. ▼ Give a formula for the **right Riemann Sum** with n equal subdivisions $a = x_0 < x_1 < \cdots < x_n = b$ for f over the interval $[a, b]$.

84. ▼ Refer to Exercise 83. If f is continuous, what happens to the difference between the left and right Riemann sums as $n \to \infty$? Explain.

85. ▼ Sketch the graph of a nonzero function whose left Riemann sum with n subdivisions is zero for every *even* number n.

86. ▼ When approximating a definite integral by computing Riemann sums, how might you judge whether you have chosen n large enough to get your answer accurate to, say, three decimal places?

13.4 The Definite Integral: Algebraic Viewpoint and the Fundamental Theorem of Calculus

In Section 13.3 we saw that the definite integral of the marginal cost function gives the total cost. However, in Section 13.1 we used antiderivatives to recover the cost function from the marginal cost function, so we *could* use antiderivatives to compute total cost. The following example, based on Example 5 in Section 13.1, compares these two approaches.

EXAMPLE 1 Finding Cost from Marginal Cost

The marginal cost of producing baseball caps at a production level of x caps is $4 - 0.001x$ dollars per cap. Find the total change of cost if production is increased from 100 to 200 caps.

Solution

Method 1: Using an Antiderivative (based on Example 5 in Section 13.1): Let $C(x)$ be the cost function. Because the marginal cost function is the derivative of the cost function, we have $C'(x) = 4 - 0.001x$ and so

$$C(x) = \int (4 - 0.001x)\, dx$$

$$= 4x - 0.001\frac{x^2}{2} + K \qquad K \text{ is the constant of integration.}$$

$$= 4x - 0.0005x^2 + K.$$

Although we do not know what to use for the value of the constant K, we can say:

$$\text{Cost at production level of 100 caps} = C(100)$$

$$= 4(100) - 0.0005(100)^2 + K$$

$$= \$395 + K$$

$$\text{Cost at production level of 200 caps} = C(200)$$

$$= 4(200) - 0.0005(200)^2 + K$$

$$= \$780 + K.$$

Therefore,

$$\text{Total change in cost} = C(200) - C(100)$$

$$= (\$780 + K) - (\$395 + K) = \$385.$$

Notice how the constant of integration simply canceled out! So, we could choose any value for K that we wanted (such as $K = 0$) and still come out with the correct total change. Put another way, we could use *any antiderivative* of $C'(x)$, such as

$$F(x) = 4x - 0.0005x^2 \qquad \begin{array}{l}F(x) \text{ is } \textit{any} \text{ antiderivative of } C'(x) \\ \text{whereas } C(x) \text{ is the actual cost function.}\end{array}$$

or

$$F(x) = 4x - 0.0005x^2 + 4,$$

compute $F(200) - F(100)$, and obtain the total change, \$385.

Summarizing this method: To compute the total change of $C(x)$ over the interval $[100, 200]$, use any antiderivative $F(x)$ of $C'(x)$, and compute $F(200) - F(100)$.

Method 2: Using a Definite Integral (based on the interpretation of the definite integral as total change discussed in Section 13.3): Because the marginal cost $C'(x)$ is the rate of change of the total cost function $C(x)$, the total change in $C(x)$ over the interval $[100, 200]$ is given by

$$\text{Total change in cost} = \text{Area under the marginal cost function curve}$$

$$= \int_{100}^{200} C'(x)\, dx$$

$$= \int_{100}^{200} (4 - 0.001x)\, dx \qquad \text{See Figure 20.}$$

$$= \$385. \qquad \text{Using geometry or Riemann sums}$$

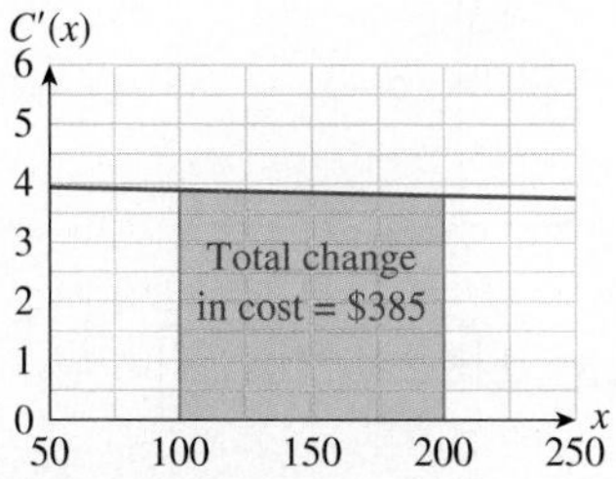

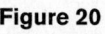

$C'(x)$

Total change in cost = \$385

Figure 20

Putting these two methods together gives us the following surprising result:

$$\int_{100}^{200} C'(x)\,dx = F(200) - F(100),$$

where $F(x)$ is any antiderivative of $C'(x)$.

Now, there is nothing special in Example 1 about the specific function $C'(x)$ or the choice of endpoints of integration. So if we replace $C'(x)$ by a general continuous function $f(x)$, we can write

$$\int_a^b f(x)\,dx = F(b) - F(a),$$

where $F(x)$ is any antiderivative of $f(x)$. This result is known as the **Fundamental Theorem of Calculus**.

The Fundamental Theorem of Calculus (FTC)

Let f be a continuous function defined on the interval $[a, b]$ and let F be *any* antiderivative of f defined on $[a, b]$. Then

$$\int_a^b f(x)\,dx = F(b) - F(a).$$

Moreover, an antiderivative of f is guaranteed to exist.

In Words: Every continuous function has an antiderivative. To compute the definite integral of $f(x)$ over $[a, b]$, first find an antiderivative $F(x)$, then evaluate it at $x = b$, evaluate it at $x = a$, and subtract the two answers.

Quick Example

Because $F(x) = x^2$ is an antiderivative of $f(x) = 2x$,

$$\int_0^1 2x\,dx = F(1) - F(0) = 1^2 - 0^2 = 1.$$

Note The Fundamental Theorem of Calculus actually applies to some other functions besides the continuous ones. The function f is **piecewise continuous** on $[a, b]$ if it is defined and continuous at all but finitely many points in the interval, and at each point where the function is not defined or is discontinuous, the left and right limits of f exist and are finite. (See Figure 21.)

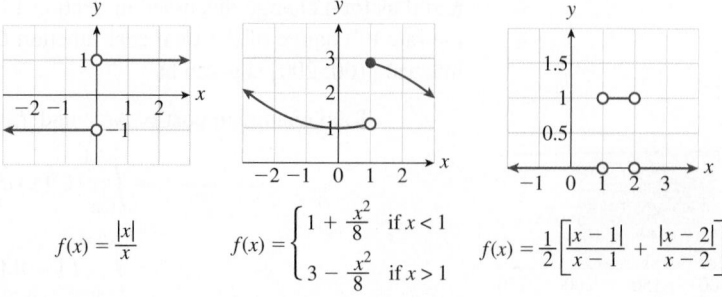

$$f(x) = \frac{|x|}{x}$$

$$f(x) = \begin{cases} 1 + \dfrac{x^2}{8} & \text{if } x < 1 \\[2mm] 3 - \dfrac{x^2}{8} & \text{if } x > 1 \end{cases}$$

$$f(x) = \frac{1}{2}\left[\frac{|x - 1|}{x - 1} + \frac{|x - 2|}{x - 2}\right]$$

Figure 21

The FTC also applies to any piecewise continuous function f, as long as we specify that the antiderivative F that we choose be continuous. To be precise, to say that F is an antiderivative of f means here that $F'(x) = f(x)$ except at the points at which f is discontinuous or not defined, where $F'(x)$ may not exist. For example, if f is the step function $f(x) = |x|/x$, shown on the left in Figure 21, then we can use $F(x) = |x|$. (Note that F is continuous, and $F'(x) = |x|/x$ except when $x = 0$.) ∎

EXAMPLE 2 **Using the FTC to Calculate a Definite Integral**

Calculate $\displaystyle\int_0^1 (1 - x^2)\, dx$.

Solution To use the FTC, we need to find an antiderivative of $1 - x^2$. But we know that

$$\int (1 - x^2)\, dx = x - \frac{x^3}{3} + C.$$

We need only one antiderivative, so let's take $F(x) = x - x^3/3$. The FTC tells us that

$$\int_0^1 (1 - x^2)\, dx = F(1) - F(0) = \left(1 - \frac{1}{3}\right) - (0) = \frac{2}{3}$$

which is the value we estimated in Section 13.3.

➡ **Before we go on...** A useful piece of notation is often used here. We write*

$$\left[F(x)\right]_a^b = F(b) - F(a).$$

Thus, we can rewrite the computation in Example 2 more simply as follows:

$$\int_0^1 (1 - x^2)\, dx = \left[x - \frac{x^3}{3}\right]_0^1$$

Substitute $x = 1$.　　Substitute $x = 0$.

$$= \left(1 - \frac{1}{3}\right) - \left(0 - \frac{0}{3}\right)$$

$$= \left(1 - \frac{1}{3}\right) - (0) = \frac{2}{3}.$$ ∎

* There seem to be several notations in use, actually. Another common notation is $\left.F(x)\right|_a^b$.

EXAMPLE 3 **More Use of the FTC**

Compute the following definite integrals.

a. $\displaystyle\int_0^1 (2x^3 + 10x + 1)\, dx$　　　　**b.** $\displaystyle\int_1^5 \left(\frac{1}{x^2} + \frac{1}{x}\right) dx$

Solution

a. $\displaystyle\int_0^1 (2x^3 + 10x + 1)\, dx = \left[\frac{1}{2}x^4 + 5x^2 + x\right]_0^1$

Substitute $x = 1$.　　Substitute $x = 0$.

$$= \left(\frac{1}{2} + 5 + 1\right) - \left(\frac{1}{2}(0) + 5(0) + 0\right)$$

$$= \left(\frac{1}{2} + 5 + 1\right) - (0) = \frac{13}{2}$$

b. $\displaystyle\int_1^5 \left(\frac{1}{x^2} + \frac{1}{x}\right) dx = \int_1^5 (x^{-2} + x^{-1}) \, dx$

$$= \left[-x^{-1} + \ln|x| \right]_1^5$$

Substitute $x = 5$.   Substitute $x = 1$.

$$= \left(-\frac{1}{5} + \ln 5 \right) - (-1 + \ln 1)$$

$$= \frac{4}{5} + \ln 5$$

When calculating a definite integral, we may have to use substitution to find the necessary antiderivative. We could substitute, evaluate the indefinite integral with respect to u, express the answer in terms of x, and then evaluate at the limits of integration. However, there is a shortcut, as we shall see in the next example.

EXAMPLE 4   Using the FTC with Substitution

Evaluate $\displaystyle\int_1^2 (2x - 1)e^{2x^2 - 2x} \, dx$.

Solution The shortcut we promised is to put *everything* in terms of u, including the limits of integration.

$$u = 2x^2 - 2x$$

$$\frac{du}{dx} = 4x - 2$$

$$dx = \frac{1}{4x - 2}\, du$$

When $x = 1$, $u = 0$.   Substitute $x = 1$ in the formula for u.
When $x = 2$, $u = 4$.   Substitute $x = 2$ in the formula for u.

We get the value $u = 0$, for example, by substituting $x = 1$ in the equation $u = 2x^2 - 2x$. We can now rewrite the integral.

$$\int_1^2 (2x - 1)e^{2x^2 - 2x} \, dx = \int_0^4 (2x - 1)e^u \frac{1}{4x - 2} \, du$$

$$= \int_0^4 \frac{1}{2} e^u \, du$$

$$= \left[\frac{1}{2} e^u \right]_0^4 = \frac{1}{2} e^4 - \frac{1}{2}$$

➡ **Before we go on...** The alternative, longer calculation in Example 4 is first to calculate the indefinite integral:

$$\int (2x - 1)e^{2x^2 - 2x} \, dx = \int \frac{1}{2} e^u \, du$$

$$= \frac{1}{2} e^u + C = \frac{1}{2} e^{2x^2 - 2x} + C.$$

Then we can say that

$$\int_1^2 (2x - 1)e^{2x^2 - 2x}\, dx = \left[\frac{1}{2}e^{2x^2 - 2x}\right]_1^2 = \frac{1}{2}e^4 - \frac{1}{2}.$$

■

APPLICATIONS

EXAMPLE 5 Oil Spill

In Section 13.3 we considered the following example: Your deep ocean oil rig has suffered a catastrophic failure, and oil is leaking from the ocean floor wellhead at a rate of

$$v(t) = 0.08t^2 - 4t + 60 \text{ thousand barrels per day } (0 \le t \le 20),$$

where t is time in days since the failure. Compute the total volume of oil released during the first 20 days.

Solution We calculate

$$\text{Total volume} = \int_0^{20} (0.08t^2 - 4t + 60)\, dt = \left[0.08\frac{t^3}{3} - 2t^2 + 60t\right]_0^{20}$$

$$= \left[0.08\frac{20^3}{3} - 2(20)^2 + 60(20)\right] - [0.08(0) - 2(0) + 60(0)]$$

$$= \frac{640}{3} - 800 + 1{,}200 \approx 613.3 \text{ thousand barrels.}$$

using Technology

TI-83/84 Plus
Home Screen:
`fnInt(0.08x^2-4x+60,x,0,20)`
(`fnInt` is MATH → 9)

Website
www.WanerMath.com
At the Website, select the Online Utilities tab, and choose the Numerical Integration Utility and Grapher. There, enter the formula

`0.08x^2-4x+60`

for $f(x)$, enter 0 and 20 for the lower and upper limits, and press "Integral".

EXAMPLE 6 Computing Area

Find the total area of the region enclosed by the graph of $y = xe^{x^2}$, the x-axis, and the vertical lines $x = -1$ and $x = 1$.

Solution The region whose area we want is shown in Figure 22. Notice the symmetry of the graph. Also, half the region we are interested in is above the x-axis, while the other half is below. If we calculated the integral $\int_{-1}^1 xe^{x^2}\, dx$, the result would be

$$\text{Area above } x\text{-axis} - \text{Area below } x\text{-axis} = 0,$$

which does not give us the total area. To prevent the area below the x-axis from being combined with the area above the axis, we do the calculation in two parts, as illustrated in Figure 23.

(In Figure 23 we broke the integral at $x = 0$ because that is where the graph crosses the x-axis.) These integrals can be calculated using the substitution $u = x^2$:

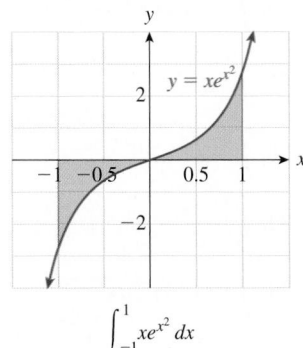

Figure 22

$$\int_{-1}^0 xe^{x^2}\, dx = \frac{1}{2}\left[e^{x^2}\right]_{-1}^0 = \frac{1}{2}(1 - e) \approx -0.85914 \qquad \text{Why is it negative?}$$

$$\int_0^1 xe^{x^2}\, dx = \frac{1}{2}\left[e^{x^2}\right]_0^1 = \frac{1}{2}(e - 1) \approx 0.85914$$

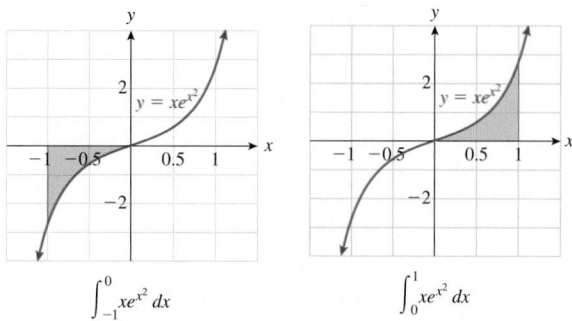

$$\int_{-1}^{0} xe^{x^2}\, dx \qquad\qquad \int_{0}^{1} xe^{x^2}\, dx$$

Figure 23

To obtain the total area, we should add the *absolute* values of these answers because we don't wish to count any area as negative. Thus,

Total area $\approx 0.85914 + 0.85914 = 1.71828.$

13.4 EXERCISES

▼ more advanced   ◆ challenging
Ⓣ indicates exercises that should be solved using technology

Evaluate the integrals in Exercises 1–44. HINT [See Example 2.]

1. $\displaystyle\int_{-1}^{1} (x^2 + 2)\, dx$ **2.** $\displaystyle\int_{-2}^{1} (x - 2)\, dx$

3. $\displaystyle\int_{0}^{1} (12x^5 + 5x^4 - 6x^2 + 4)\, dx$

4. $\displaystyle\int_{0}^{1} (4x^3 - 3x^2 + 4x - 1)\, dx$

5. $\displaystyle\int_{-2}^{2} (x^3 - 2x)\, dx$ **6.** $\displaystyle\int_{-1}^{1} (2x^3 + x)\, dx$

7. $\displaystyle\int_{1}^{3} \left(\frac{2}{x^2} + 3x\right) dx$ **8.** $\displaystyle\int_{2}^{3} \left(x + \frac{1}{x}\right) dx$

9. $\displaystyle\int_{0}^{1} (2.1x - 4.3x^{1.2})\, dx$ **10.** $\displaystyle\int_{-1}^{0} (4.3x^2 - 1)\, dx$

11. $\displaystyle\int_{0}^{1} 2e^x\, dx$ **12.** $\displaystyle\int_{-1}^{0} 3e^x\, dx$

13. $\displaystyle\int_{0}^{1} \sqrt{x}\, dx$ **14.** $\displaystyle\int_{-1}^{1} \sqrt[3]{x}\, dx$

15. $\displaystyle\int_{0}^{1} 2^x\, dx$ **16.** $\displaystyle\int_{0}^{1} 3^x\, dx$

HINT [In Exercises 17–44 use a shortcut or see Example 4.]

17. $\displaystyle\int_{0}^{1} 18(3x + 1)^5\, dx$ **18.** $\displaystyle\int_{0}^{1} 8(-x + 1)^7\, dx$

19. $\displaystyle\int_{-1}^{1} e^{2x-1}\, dx$ **20.** $\displaystyle\int_{0}^{2} e^{-x+1}\, dx$

21. $\displaystyle\int_{0}^{2} 2^{-x+1}\, dx$ **22.** $\displaystyle\int_{-1}^{1} 3^{2x-1}\, dx$

23. $\displaystyle\int_{0}^{4} |-3x + 4|\, dx$ **24.** $\displaystyle\int_{-4}^{4} |-x - 2|\, dx$

25. $\displaystyle\int_{0}^{1} 5x(8x^2 + 1)^{-1/2}\, dx$ **26.** $\displaystyle\int_{0}^{\sqrt{2}} x\sqrt{2x^2 + 1}\, dx$

27. $\displaystyle\int_{-\sqrt{2}}^{\sqrt{2}} 3x\sqrt{2x^2 + 1}\, dx$ **28.** $\displaystyle\int_{-2}^{2} xe^{-x^2+1}\, dx$

29. $\displaystyle\int_{0}^{1} 5xe^{x^2+2}\, dx$ **30.** $\displaystyle\int_{0}^{2} \frac{3x}{x^2 + 2}\, dx$

31. $\displaystyle\int_{2}^{3} \frac{x^2}{x^3 - 1}\, dx$ **32.** $\displaystyle\int_{2}^{3} \frac{x}{2x^2 - 5}\, dx$

33. $\displaystyle\int_{0}^{1} x(1.1)^{-x^2}\, dx$ **34.** $\displaystyle\int_{0}^{1} x^2(2.1)^{x^3}\, dx$

35. $\displaystyle\int_{1}^{2} \frac{e^{1/x}}{x^2}\, dx$ **36.** $\displaystyle\int_{1}^{2} \frac{\sqrt{\ln x}}{x}\, dx$

37. $\displaystyle\int_{0}^{2} \frac{e^{-2x}}{1 + 3e^{-2x}}\, dx$ (Round the answer to four decimal places.)

38. $\displaystyle\int_{0}^{1} \frac{e^{2x}}{1 - 3e^{2x}}\, dx$ (Round the answer to four decimal places.)

39. ▼ $\displaystyle\int_0^2 \frac{x}{x+1}\, dx$

40. ▼ $\displaystyle\int_{-1}^1 \frac{2x}{x+2}\, dx$

41. ▼ $\displaystyle\int_1^2 x(x-2)^5\, dx$

42. ▼ $\displaystyle\int_1^2 x(x-2)^{1/3}\, dx$

43. ▼ $\displaystyle\int_0^1 x\sqrt{2x+1}\, dx$

44. ▼ $\displaystyle\int_{-1}^0 2x\sqrt{x+1}\, dx$

Calculate the total area of the regions described in Exercises 45–54. Do not count area beneath the x-axis as negative. HINT [See Example 6.]

45. Bounded by the line $y = x$, the x-axis, and the lines $x = 0$ and $x = 1$.

46. Bounded by the line $y = 2x$, the x-axis, and the lines $x = 1$ and $x = 2$.

47. Bounded by the curve $y = \sqrt{x}$, the x-axis, and the lines $x = 0$ and $x = 4$.

48. Bounded by the curve $y = 2\sqrt{x}$, the x-axis, and the lines $x = 0$ and $x = 16$.

49. Bounded by the graph of $y = |2x - 3|$, the x-axis, and the lines $x = 0$ and $x = 3$.

50. Bounded by the graph of $y = |3x - 2|$, the x-axis, and the lines $x = 0$ and $x = 3$.

51. ▼ Bounded by the curve $y = x^2 - 1$, the x-axis, and the lines $x = 0$ and $x = 4$.

52. ▼ Bounded by the curve $y = 1 - x^2$, the x-axis, and the lines $x = -1$ and $x = 2$.

53. ▼ Bounded by the x-axis, the curve $y = xe^{x^2}$, and the lines $x = 0$ and $x = (\ln 2)^{1/2}$.

54. ▼ Bounded by the x-axis, the curve $y = xe^{x^2-1}$, and the lines $x = 0$ and $x = 1$.

APPLICATIONS

55. *Cost* The marginal cost of producing the xth box of light bulbs is $5 + x^2/1{,}000$ dollars. Determine how much is added to the total cost by a change in production from $x = 10$ to $x = 100$ boxes. HINT [See Example 5.]

56. *Revenue* The marginal revenue of the xth box of flash cards sold is $100e^{-0.001x}$ dollars. Find the revenue generated by selling items 101 through 1,000. HINT [See Example 5.]

57. *Motion* A car traveling down a road has a velocity of $v(t) = 60 - e^{-t/10}$ mph at time t hours. Find the distance it has traveled from time $t = 1$ hour to time $t = 6$ hours. (Round your answer to the nearest mile.)

58. *Motion* A ball thrown in the air has a velocity of $v(t) = 100 - 32t$ ft/sec at time t seconds. Find the total displacement of the ball between times $t = 1$ second and $t = 7$ seconds, and interpret your answer.

59. *Motion* A car slows to a stop at a stop sign, then starts up again, in such a way that its speed at time t seconds after it starts to slow is $v(t) = |-10t + 40|$ ft/sec. How far does the car travel from time $t = 0$ to time $t = 10$ sec?

60. *Motion* A truck slows, doesn't quite stop at a stop sign, and then speeds up again in such a way that its speed at time t seconds is $v(t) = 10 + |-5t + 30|$ ft/sec. How far does the truck travel from time $t = 0$ to time $t = 10$?

61. *Bottled Water Sales* (Compare Exercise 55 in Section 13.1.) The rate of U.S. sales of bottled water for the period 2000–2010 could be approximated by

$$s(t) = -45t^2 + 900t + 4{,}200 \text{ million gallons per year}$$
$$(0 \le t \le 10),$$

where t is time in years since the start of 2000.[44] Use a definite integral to estimate the total sales $S(t)$ of bottled water from the start of 2000 to the start of 2008.

62. *Bottled Water Sales* (Compare Exercise 56 in Section 13.1.) The rate of U.S. per capita sales of bottled water for the period 2000–2010 could be approximated by

$$s(t) = -0.18t^2 + 3t + 15 \text{ gallons per year} \quad (0 \le t \le 10),$$

where t is time in years since the start of 2000.[45] Use a definite integral to estimate the total U.S. per capita sales of bottled water from the start of 2000 to the end of 2009.

63. *Facebook Membership* At the start of 2005, Facebook had 1 million members. Since that time, new members joined at a rate of roughly

$$m(t) = 3.2t^3 - 12t + 10 \text{ million members per year}$$
$$(0 \le t \le 5),$$

where t is time in years since the start of 2005.[46] Use a definite integral to estimate the total number of new Facebook members from the start of 2007 to the start of 2010.

64. *Uploads to YouTube* Since YouTube first became available to the public in mid-2005, the rate at which video has been uploaded to the site can be approximated by

$$v(t) = 525{,}600(0.42t^2 + 2.7t - 1) \text{ hours of video per year}$$
$$(0.5 \le t \le 7),$$

where t is time in years since the start of 2005.[47] Use a definite integral to estimate the total number of hours of video uploaded from the start of 2006 to the start of 2012. (Round your answer to the nearest million hours of video.)

65. *Big Brother* (Compare Exercise 69 in Section 13.3.) The total number of wiretaps authorized each year by U.S. state and federal courts from 1990 to 2010 can be approximated by

$$w(t) = 820e^{0.051t} \quad (0 \le t \le 20)$$

(t is time in years since the start of 1990).[48] Compute $\int_0^{15} w(t)\, dt$. (Round your answer to the nearest 10.) Interpret the answer.

————————
[44] Source for data: Beverage Marketing Corporation (www.bottledwater.org).
[45] *Ibid.*
[46] Sources for data: www.facebook.com, www.insidefacebook.com.
[47] Source for data: www.YouTube.com.
[48] Source for data: 2007 Wiretap Report, Administrative Office of the United States Courts (www.uscourts.gov/wiretap07/2007WTText.pdf).

66. *Big Brother* (Compare Exercise 64 in Section 13.3.) The number of wiretaps authorized each year by U.S. state courts from 1990 to 2010 can be approximated by

$$w(t) = 430e^{0.065t} \quad (0 \le t \le 20)$$

(t is time in years since the start of 1990).[49] Compute $\int_{10}^{15} w(t)\, dt$. (Round your answer to the nearest 10.) Interpret the answer.

67. *Economic Growth* (Compare Exercise 85 in Section 13.2.) The value of sold goods in Mexico can be approximated by

$$v(x) = 210 - 62e^{-0.05x} \text{ trillion pesos per month } (x \ge 0),$$

where x is time in months since January 2005.[50] Find an expression for the total value $V(t)$ of sold goods in Mexico from January 2005 to time t.

68. *Economic Contraction* (Compare Exercise 86 in Section 13.2.) Housing starts in the United States can be approximated by

$$n(x) = 1.1 + 1.2e^{-0.08x} \text{ million homes per year}$$
$$(x \ge 0),$$

where x is time in years since January 2006.[51] Find an expression for the total number $N(t)$ of housing starts in the United States from January 2006 to time t.

69. *Fuel Consumption* The way Professor Waner drives, he burns gas at the rate of $1 - e^{-t}$ gallons each hour, t hours after a fill-up. Find the number of gallons of gas he burns in the first 10 hours after a fill-up.

70. *Fuel Consumption* The way Professor Costenoble drives, he burns gas at the rate of $1/(t+1)$ gallons each hour, t hours after a fill-up. Find the number of gallons of gas he burns in the first 10 hours after a fill-up.

71. ▼ *Sales* Weekly sales of your *Lord of the Rings* T-shirts have been falling by 5% per week. Assuming you are now selling 50 T-shirts per week, how many shirts will you sell during the coming year? (Round your answer to the nearest shirt.)

72. ▼ *Sales* Annual sales of fountain pens in Littleville are 4,000 per year and are increasing by 10% per year. How many fountain pens will be sold over the next five years?

73. ▣ *Embryo Development* The oxygen consumption of a bird embryo increases from the time the egg is laid through the time the chick hatches. In a typical galliform bird, the oxygen consumption can be approximated by

$$c(t) = -0.065t^3 + 3.4t^2 - 22t + 3.6 \text{ milliliters per day}$$
$$(8 \le t \le 30),$$

where t is the time (in days) since the egg was laid.[52] (An egg will typically hatch at around $t = 28$.) Use technology to estimate the total amount of oxygen consumed during the ninth and tenth days ($t = 8$ to $t = 10$). Round your answer to the nearest milliliter. HINT [See the technology note in the margin on page 997.]

74. ▣ *Embryo Development* The oxygen consumption of a turkey embryo increases from the time the egg is laid through the time the chick hatches. In a brush turkey, the oxygen consumption can be approximated by

$$c(t) = -0.028t^3 + 2.9t^2 - 44t + 95 \text{ milliliters per day}$$
$$(20 \le t \le 50),$$

where t is the time (in days) since the egg was laid.[53] (An egg will typically hatch at around $t = 50$.) Use technology to estimate the total amount of oxygen consumed during the 21st and 22nd days ($t = 20$ to $t = 22$). Round your answer to the nearest 10 milliliters. HINT [See the technology note in the margin on page 997.]

75. ▣ *Online Vehicle Sales* The rate of change $v(t)$ of the value of vehicle transactions on **eBay** could be approximated by

$$v(t) = -0.002t^3 + 0.06t^2 - 0.55t + 3.9 \text{ billion dollars}$$
$$\text{per quarter } (1 \le t \le 13)$$

(t is time in quarters; $t = 1$ represents the first quarter of 2008).[54] Use technology to estimate $\int_1^9 v(t)\, dt$. Interpret your answer.

76. ▣ *Online Auctions* The rate of change $n(t)$ of the number of active **eBay** users could be approximated by

$$n(t) = -0.002t^4 + 0.06t^3 - 0.55t^2 + 1.9t - 1 \text{ million}$$
$$\text{users per quarter } (1 \le t \le 13)$$

(t is time in quarters; $t = 1$ represents the first quarter of 2008).[55] Use technology to compute $\int_5^{13} n(t)\, dt$ correct to the nearest whole number. Interpret your answer.

77. ▼ *Cost* Use the Fundamental Theorem of Calculus to show that if $m(x)$ is the marginal cost at a production level of x items, then the cost function $C(x)$ is given by

$$C(x) = C(0) + \int_0^x m(t)\, dt.$$

What do we call $C(0)$?

78. ▼ *Cost* The total cost of producing x items is given by

$$C(x) = 246.76 + \int_0^x 5t\, dt.$$

Find the fixed cost and the marginal cost of producing the 10th item.

---

[49] Source for data: Wiretap Reports, Administrative Office of the United States Courts (www.uscourts.gov/Statistics/WiretapReports.aspx).

[50] Source: Instituto Nacional de Estadística y Geografía (INEGI) (www.inegi.org.mx).

[51] Source for data: *New York Times*, February 17, 2007, p. C3.

[52] The model approximates graphical data published in the article "The Brush Turkey" by Roger S. Seymour, *Scientific American*, December 1991, pp. 108–114.

[53] *Ibid.*

[54] Source for data: eBay company reports (http://investor.ebay.com).

[55] *Ibid.*

79. *Scientific Research* (Compare Exercise 91 in Section 13.2.) The number of research articles in the prominent journal *Physical Review* written by researchers in Europe can be approximated by

$$E(t) = \frac{7e^{0.2t}}{5 + e^{0.2t}} \text{ thousand articles per year} \quad (t \geq 0),$$

where t is time in years ($t = 0$ represents 1983).[56] Use a definite integral to estimate the number of articles written by researchers in Europe from 1983 to 2003. (Round your answer to the nearest 1,000 articles.) HINT [See Example 5 in Section 13.2.]

80. *Scientific Research* (Compare Exercise 92 in Section 13.2.) The number of research articles in the prominent journal *Physical Review* written by researchers in the United States can be approximated by

$$U(t) = \frac{4.6e^{0.6t}}{0.4 + e^{0.6t}} \text{ thousand articles per year} \quad (t \geq 0),$$

where t is time in years ($t = 0$ represents 1983).[57] Use a definite integral to estimate the total number of articles written by researchers in the United States from 1983 to 2003. (Round your answer to the nearest 1,000 articles.) HINT [See Example 5 in Section 13.2.]

81. ▼ *The Logistic Function and High School Graduates*
  a. Show that the logistic function $f(x) = \dfrac{N}{1 + Ab^{-x}}$ can be written in the form
  $$f(x) = \frac{Nb^x}{A + b^x}.$$
  HINT [See the note after Example 5 in Section 13.2.]
  b. Use the result of part (a) and a suitable substitution to show that
  $$\int \frac{N}{1 + Ab^{-x}} \, dx = \frac{N \, \ln(A + b^x)}{\ln b} + C.$$
  c. The rate of graduation of private high school students in the United States for the period 1994–2008 was approximately
  $$r(t) = 220 + \frac{110}{1 + 3.8(1.27)^{-t}} \text{ thousand students per year}$$
  $$(0 \leq t \leq 14)$$
  t years since 1994.[58] Use the result of part (b) to estimate the total number of private high school graduates over the period 2000–2008.

82. ▼ *The Logistic Function and Grant Spending*
  a. Show that the logistic function $f(x) = \dfrac{N}{1 + Ae^{-kx}}$ can be written in the form
  $$f(x) = \frac{Ne^{kx}}{A + e^{kx}}.$$
  HINT [See the note after Example 5 in Section 13.2.]

  b. Use the result of part (a) and a suitable substitution to show that
  $$\int \frac{N}{1 + Ae^{-kx}} \, dx = \frac{N \ln(A + e^{kx})}{k} + C.$$
  c. The rate of spending on grants by U.S. foundations in the period 1993–2003 was approximately
  $$s(t) = 11 + \frac{20}{1 + 1{,}800e^{-0.9t}} \text{ billion dollars per year}$$
  $$(3 \leq t \leq 13),$$
  where t is the number of years since 1990.[59] Use the result of part (b) to estimate, to the nearest \$10 billion, the total spending on grants from 1998 to 2003.

83. ♦ *Kinetic Energy* The work done in accelerating an object from velocity v_0 to velocity v_1 is given by

$$W = \int_{v_0}^{v_1} v \frac{dp}{dv} \, dv,$$

where p is its momentum, given by $p = mv$ (m = mass). Assuming that m is a constant, show that

$$W = \frac{1}{2}mv_1^2 - \frac{1}{2}mv_0^2.$$

The quantity $\frac{1}{2}mv^2$ is referred to as the **kinetic energy** of the object, so the work required to accelerate an object is given by its change in kinetic energy.

84. ♦ *Einstein's Energy Equation* According to the special theory of relativity, the apparent mass of an object depends on its velocity according to the formula

$$m = \frac{m_0}{\left(1 - \dfrac{v^2}{c^2}\right)^{1/2}},$$

where v is its velocity, m_0 is the "rest mass" of the object (that is, its mass when $v = 0$), and c is the velocity of light: approximately 3×10^8 meters per second.
  a. Show that, if $p = mv$ is the momentum,
  $$\frac{dp}{dv} = \frac{m_0}{\left(1 - \dfrac{v^2}{c^2}\right)^{3/2}}.$$

  b. Use the integral formula for W in the preceding exercise, together with the result in part (a) to show that the work required to accelerate an object from a velocity of v_0 to v_1 is given by
  $$W = \frac{m_0 c^2}{\sqrt{1 - \dfrac{v_1^2}{c^2}}} - \frac{m_0 c^2}{\sqrt{1 - \dfrac{v_0^2}{c^2}}}.$$

---

[56] Based on data from 1983 to 2003. Source: The American Physical Society/*New York Times*, May 3, 2003, p. A1.

[57] *Ibid.*

[58] Based on a logistic regression. Source for data: National Center for Educational Statistics (www.nces.ed.gov).

---

[59] Based on a logistic regression. Source for data: The Foundation Center, *Foundation Growth and Giving Estimates*, 2004, downloaded from the Center's Web site (www.fdncenter.org).

We call the quantity $\dfrac{m_0 c^2}{\sqrt{1 - \frac{v^2}{c^2}}}$ the **total relativistic energy** of an object moving at velocity v. Thus, the work to accelerate an object from one velocity to another is given by the change in its total relativistic energy.

c. Deduce (as Albert Einstein did) that the total relativistic energy E of a body at rest with rest mass m is given by the famous equation

$$E = mc^2.$$

COMMUNICATION AND REASONING EXERCISES

85. Explain how the indefinite integral and the definite integral are related.

86. What is "definite" about the definite integral?

87. The total change of a quantity from time a to time b can be obtained from its rate of change by doing what?

88. Complete the following: The total sales from time a to time b are obtained from the marginal sales by taking its _____ _____ from _____ to _____ .

89. What does the Fundamental Theorem of Calculus permit one to do?

90. If Felice and Philipe have different antiderivatives of f and each uses his or her own antiderivative to compute $\int_a^b f(x)\,dx$, they might get different answers, right?

91. ▼ Give an example of a nonzero velocity function that will produce a displacement of 0 from time $t = 0$ to time $t = 10$.

92. ▼ Give an example of a nonzero function whose definite integral over the interval $[4, 6]$ is zero.

93. ▼ Give an example of a decreasing function $f(x)$ with the property that $\int_a^b f(x)\,dx$ is positive for every choice of a and $b > a$.

94. ▼ Explain why, in computing the total change of a quantity from its rate of change, it is useful to have the definite integral subtract area below the x-axis.

95. ◆ If $f(x)$ is a continuous function defined for $x \geq a$, define a new function $F(x)$ by the formula

$$F(x) = \int_a^x f(t)\,dt.$$

Use the Fundamental Theorem of Calculus to deduce that $F'(x) = f(x)$. What, if anything, is interesting about this result?

96. ▮ ◆ Use the result of Exercise 95 and technology to compute a table of values for $x = 1, 2, 3$ for an antiderivative $A(x)$ of e^{-x^2} with the property that $A(0) = 0$. (Round answers to two decimal places.)

KEY CONCEPTS

Website www.WanerMath.com
Go to the Website at www.WanerMath
.com to find a comprehensive and
interactive Web-based summary
of Chapter 13.

13.1 The Indefinite Integral

An antiderivative of a function f is a
function F such that $F' = f$. *p. 946*
Indefinite integral $\int f(x)\, dx$ *p. 946*
Power rule for the indefinite integral:

$$\int x^n\, dx = \frac{x^{n+1}}{n+1} + C$$

$$(\text{if } n \neq -1) \ \ p. \ 948$$

$$\int x^{-1}\, dx = \ln|x| + C \ \ p. \ 948$$

Indefinite integral of e^x and b^x:

$$\int e^x\, dx = e^x + C$$

$$\int b^x\, dx = \frac{b^x}{\ln b} + C \ \ p. \ 949$$

Indefinite integral of $|x|$:

$$\int |x|\, dx = \frac{x|x|}{2} + C \ \ p. \ 949$$

Sums, differences, and constant
multiples:

$$\int [f(x) \pm g(x)]\, dx$$

$$= \int f(x)\, dx \pm \int g(x)\, dx$$

$$\int k f(x)\, dx = k \int f(x)\, dx$$

$$(k \text{ constant}) \ \ p. \ 949$$

Combining the rules *p. 951*
Position, velocity, and acceleration:

$$v = \frac{ds}{dt} \qquad s(t) = \int v(t)\, dt$$

$$a = \frac{dv}{dt} \qquad v(t) = \int a(t)\, dt$$

pp. 954–955

Motion in a straight line *p. 955*
Vertical motion under gravity *p. 957*

13.2 Substitution

Substitution rule:

$$\int f\, dx = \int \left(\frac{f}{du/dx}\right) du \ \ p. \ 962$$

Using the substitution rule *p. 963*
Shortcuts: integrals of expressions
involving $(ax + b)$:

$$\int (ax+b)^n\, dx = \frac{(ax+b)^{n+1}}{a(n+1)} + C$$

$$(\text{if } n \neq -1)$$

$$\int (ax+b)^{-1}\, dx = \frac{1}{a}\ln|ax+b| + C$$

$$\int e^{ax+b}\, dx = \frac{1}{a}e^{ax+b} + C$$

$$\int c^{ax+b}\, dx = \frac{1}{a\ln c}c^{ax+b} + C$$

$$\int |ax+b|\, dx$$

$$= \frac{1}{2a}(ax+b)|ax+b| + C \ \ p. \ 968$$

Mike's shortcut rule:

If $\int f(x)\, dx = F(x) + C$ and g and u

are differentiable functions of x, then

$$\int g \cdot f(u)\, dx = \frac{g}{u'} \cdot F(u) + C$$

provided that $\dfrac{g}{u'}$ is constant. *p. 969*

13.3 The Definite Integral: Numerical and Graphical Viewpoints

Left Riemann sum:

$$\sum_{k=0}^{n-1} f(x_k)\Delta x$$

$$= [f(x_0) + f(x_1) + \cdots + f(x_{n-1})]\Delta x$$

p. 978

Computing the Riemann sum from a
formula *p. 980*
Computing the Riemann sum from a
graph *p. 980*
Definite integral of f from a to b:

$$\int_a^b f(x)\, dx = \lim_{n \to \infty} \sum_{k=0}^{n-1} f(x_k)\Delta x.$$

p. 981

Estimating the definite integral from a
graph *p. 983*
Estimating the definite integral using
technology *p. 985*
Application to motion in a straight line
p. 985

13.4 The Definite Integral: Algebraic Viewpoint and the Fundamental Theorem of Calculus

Computing total cost from marginal
cost *p. 993*
The Fundamental Theorem of Calculus
(FTC) *p. 994*
Using the FTC to compute definite
integrals *p. 995*
Using the FTC with substitution *p. 996*
Computing area *p. 997*

REVIEW EXERCISES

Evaluate the indefinite integrals in Exercises 1–18.

1. $\displaystyle\int (x^2 - 10x + 2)\, dx$

2. $\displaystyle\int (e^x + \sqrt{x})\, dx$

3. $\displaystyle\int \left(\frac{4x^2}{5} - \frac{4}{5x^2}\right) dx$

4. $\displaystyle\int \left(\frac{3x}{5} - \frac{3}{5x}\right) dx$

5. $\displaystyle\int (2x)^{-1}\, dx$

6. $\displaystyle\int (-2x + 2)^{-2}\, dx$

7. $\displaystyle\int e^{-2x+11}\, dx$

8. $\displaystyle\int \frac{dx}{(4x - 3)^2}$

9. $\displaystyle\int x(x^2 + 1)^{1.3}\, dx$

10. $\displaystyle\int x(x^2 + 4)^{10}\, dx$

11. $\displaystyle\int \frac{4x}{(x^2 - 7)}\, dx$

12. $\displaystyle\int \frac{x}{(3x^2 - 1)^{0.4}}\, dx$

13. $\displaystyle\int (x^3 - 1)\sqrt{x^4 - 4x + 1}\, dx$

14. $\displaystyle\int \frac{x^2+1}{(x^3+3x+2)^2}\,dx$ **15.** $\displaystyle\int (-xe^{x^2/2})\,dx$

16. $\displaystyle\int xe^{-x^2/2}\,dx$ **17.** $\displaystyle\int \frac{x+1}{x+2}\,dx$

18. $\displaystyle\int x\sqrt{x-1}\,dx$

In Exercises 19 and 20, use the given graph to estimate the left Riemann sum for the given interval with the stated number of subdivisions.

19. $[0, 3], n = 6$ **20.** $[1, 3], n = 4$

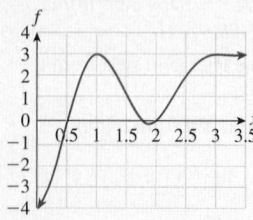   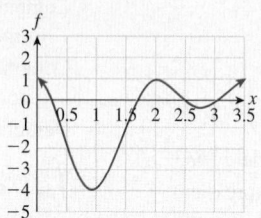

Calculate the left Riemann sums for the given functions over the given interval in Exercises 21–24, using the given values of n. (When rounding, round answers to four decimal places.)

21. $f(x) = x^2 + 1$ over $[-1, 1]$, $n = 4$

22. $f(x) = (x-1)(x-2) - 2$ over $[0, 4]$, $n = 4$

23. $f(x) = x(x^2 - 1)$ over $[0, 1]$, $n = 5$

24. $f(x) = \dfrac{x-1}{x-2}$ over $[0, 1.5]$, $n = 3$

⊤ *In Exercises 25 and 26, use technology to approximate the given definite integrals using left Riemann sums with $n = 10$, 100, and 1,000. (Round answers to four decimal places.)*

25. $\displaystyle\int_0^1 e^{-x^2}\,dx$ **26.** $\displaystyle\int_1^3 x^{-x}\,dx$

In Exercises 27 and 28, the graph of the derivative $f'(x)$ of $f(x)$ is shown. Compute the total change of $f(x)$ over the given interval.

27. $[-1, 2]$ **28.** $[0, 2]$

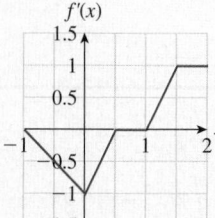   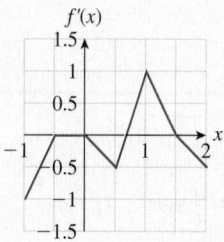

Evaluate the definite integrals in Exercises 29–38, using the Fundamental Theorem of Calculus.

29. $\displaystyle\int_{-1}^1 (x - x^3 + |x|)\,dx$ **30.** $\displaystyle\int_0^9 (x + \sqrt{x})\,dx$

31. $\displaystyle\int_{-1}^1 \frac{3}{(2x-5)^2}\,dx$ **32.** $\displaystyle\int_0^9 \frac{1}{x+1}\,dx$

33. $\displaystyle\int_0^{50} e^{-0.02x-1}\,dx$ **34.** $\displaystyle\int_{-20}^0 3e^{2.2x}\,dx$

35. $\displaystyle\int_0^2 x^2\sqrt{x^3+1}\,dx$ **36.** $\displaystyle\int_0^2 \frac{x^2}{\sqrt{x^3+1}}\,dx$

37. $\displaystyle\int_0^{\ln 2} \frac{e^{-2x}}{1+4e^{-2x}}\,dx$ **38.** $\displaystyle\int_0^{\ln 3} e^{2x}(1-3e^{2x})^2\,dx$

In Exercises 39–42, find the areas of the specified regions. (Do not count area below the x-axis as negative.)

39. The area bounded by $y = 4 - x^2$, the x-axis, and the lines $x = -2$ and $x = 2$

40. The area bounded by $y = 4 - x^2$, the x-axis, and the lines $x = 0$ and $x = 5$

41. The area bounded by $y = xe^{-x^2}$, the x-axis, and the lines $x = 0$ and $x = 5$

42. The area bounded by $y = |2x|$, the x-axis, and the lines $x = -1$ and $x = 1$

APPLICATIONS: OHaganBooks.com

43. *Sales* The rate of net sales (sales minus returns) of *The Secret Loves of John O*, a romance novel by Margó Dufón, can be approximated by

$$n(t) = 196 + t^2 - 0.16t^5 \text{ copies per week}$$

t weeks since its release.

a. Find the total net sales N as a function of time t.

b. How many books are still held by customers after 6 weeks? (Round your answer to the nearest book.)

44. *Demand* If OHaganBooks.com were to give away its latest bestseller, *A River Burns through It*, the demand q would be 100,000 books. The marginal demand (dq/dp) for the book is $-20p$ at a price of p dollars.

a. What is the demand function for this book?

b. At what price does demand drop to zero?

45. *Motion under Gravity* Billy-Sean O'Hagan's friend Juan (Billy-Sean is John O'Hagan's son, currently a senior in college) says he can throw a baseball vertically upward at 100 feet per second. Assuming Juan's claim is true,

a. Where would the baseball be at time t seconds?

b. How high would the ball go?

c. When would it return to Juan's hand?

46. *Motion under Gravity* An overworked employee at OHaganBooks.com goes to the top of the company's 100-foot-tall headquarters building and flings a book up into the air at a speed of 60 feet per second.

a. When will the book hit the ground 100 feet below? (Neglect air resistance.)

b. How fast will it be traveling when it hits the ground?

c. How high will the book go?

(See the graph.)

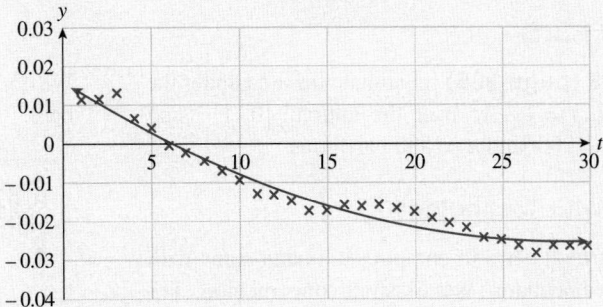

Use this formula and the given January 2006 spending figure to obtain corresponding models for $p(t)$ and $\bar{P}$.

6. Compare the model in the text with the quadratic model in Exercise 5 in terms of both short- and long-term predictions; in particular, when does the quadratic model predict construction spending will have reached its biggest monthly decrease? Are either of these models realistic in the near term? in the long term?

TECHNOLOGY GUIDE

TI-83/84 Plus   Technology Guide

Section 13.3

Example 5 (page 985)   Estimate the area under the graph of $f(x) = 1 - x^2$ over the interval [0, 1] using $n = 100$, $n = 200$, and $n = 500$ partitions.

Solution with Technology

There are several ways to compute Riemann sums with a graphing calculator. We illustrate one method. For $n = 100$, we need to compute the sum

$$\sum_{k=0}^{99} f(x_k)\Delta x = [f(0) + f(0.01) + \cdots + f(0.99)](0.01).$$

See discussion in Example 5.

Thus, we first need to calculate the numbers $f(0)$, $f(0.01)$, and so on, and add them up. The TI-83/84 Plus has a built-in sum function (available in the LIST MATH menu), which, like the SUM function in a spreadsheet, sums the entries in a list.

1. To generate a list that contains the numbers we want to add together, use the seq function (available in the LIST OPS menu). If we enter

$$\text{seq}(1-X^2,X,0,0.99,0.01)$$

seq: 2ND LIST OPS 5

the calculator will calculate a list by evaluating $1-X^2$ for values of X from 0 to 0.99 in steps of 0.01.

2. To take the sum of all these numbers, we wrap the seq function in a call to sum:

$$\text{sum}(\text{seq}(1-X^2,X,0,0.99,0.01))$$

sum: 2ND LIST MATH 5

This gives the sum

$$f(0) + f(0.01) + \cdots + f(0.99) = 67.165.$$

3. To obtain the Riemann sum, we need to multiply this sum by $\Delta x = 0.01$, and we obtain the estimate of $67.165 \times 0.01 = 0.67165$ for the Riemann sum:

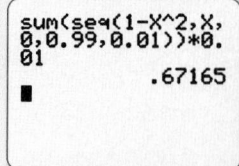

We obtain the other Riemann sums similarly, as shown here:

$n = 200$ $n = 500$

One disadvantage of this method is that the TI-83/84 Plus can generate and sum a list of at most 999 entries. The LEFTSUM program below calculates left Riemann sums for any n. The TI-83/84 Plus also has a built-in function fnInt, which finds a very accurate approximation of a definite integral, using a more sophisticated technique than the one we are discussing here.

The LEFTSUM program for the TI-83/84 Plus

The following program calculates (left) Riemann sums for any n. The latest version of this program (and others) is available at the Website.

```
PROGRAM: LEFTSUM
:Input "LEFT ENDPOINT? ",A
                    Prompts for the left endpoint a
:Input "RIGHT ENDPOINT? ",B
                    Prompts for the right endpoint b
:Input "N? ",N      Prompts for the number
                      of rectangles
:(B-A)/N→D          D is Δx = (b − a)/n.
:∅→L                L will eventually be the left sum.
:A→X                X is the current x-coordinate.
:For(I,1,N)         Start of a loop—recall the sigma
                      notation.
:L+Y₁→L             Add f(xᵢ₋₁) to L.
:A+I*D→X            Uses formula xᵢ = a + iΔx
:End                End of loop
:L*D→L              Multiply by Δx.
:Disp "LEFT SUM IS ",L
:Stop
```

SPREADSHEET **Technology Guide**

Section 13.3

Example 5 (page 985) Estimate the area under the graph of $f(x) = 1 - x^2$ over the interval $[0, 1]$ using $n = 100$, $n = 200$, and $n = 500$ partitions.

Solution with Technology

We need to compute various sums.

$$\sum_{k=0}^{99} f(x_k)\Delta x = [f(0) + f(0.01) + \cdots + f(0.99)](0.01)$$

See discussion in Example 5.

$$\sum_{k=0}^{199} f(x_k)\Delta x = [f(0) + f(0.005) + \cdots + f(0.995)](0.005)$$

$$\sum_{k=0}^{499} f(x_k)\Delta x = [f(0) + f(0.002) + \cdots + f(0.998)](0.002).$$

Here is how you can compute them all on the same spreadsheet.

1. Enter the values for the endpoints a and b, the number of subdivisions n, and the formula $\Delta x = (b - a)/n$:

	A	B	C	D
1	x	f(x)	a	0
2			b	1
3			n	100
4			Delta x	=(D2-D1)/D3

2. Next, we compute all the x-values we might need in column A. Because the largest value of n that we will be using is 500, we will need a total of 501 values of x. Note that the value in each cell below A3 is obtained from the one above by adding Δx.

	A	B	C	D
1	x	f(x)	a	0
2	=D1		b	1
3	=A2+$D$4		n	100
4			Delta x	0.01
5				
501				
502				

	A	B	C	D
1	x	f(x)	a	0
2	0		b	1
3	0.01		n	100
4	0.02		Delta x	0.01
5				
501	4.99			
502	5			

(The fact that the values of x currently go too far will be corrected in the next step.)

3. We need to calculate the numbers $f(0)$, $f(0.01)$, and so on, but only those for which the corresponding x-value is less than b. To do this, we use a logical formula like we did with piecewise-defined functions in Chapter 1:

	A	B	C	D
1	x	f(x)	a	0
2	0	=(1-A2^2)*(A2<$D$2)	b	1
3	0.01		n	100
4	0.02		Delta x	0.01
5	0.03			
501	4.99			
502	5			

When the value of x is b or above, the function will evaluate to zero, because we do not want to count it.

4. Finally, we compute the Riemann sum by adding up everything in column B and multiplying by Δx:

	A	B	C	D
1	x	f(x)	a	0
2	0	1	b	1
3	0.01	0.9999	n	100
4	0.02	0.9996	Delta x	0.01
5	0.03	0.9991	Left Sum	=SUM(B:B)*D4
6	0.04	0.9984		

↓

	A	B	C	D
1	x	f(x)	a	0
2	0	1	b	1
3	0.01	0.9999	n	100
4	0.02	0.9996	Delta x	0.01
5	0.03	0.9991	Left Sum	0.67165
6	0.04	0.9984		

Now it is easy to obtain the sums for $n = 200$ and $n = 500$: Simply change the value of n in cell D3:

	A	B	C	D
1	x	f(x)	a	0
2	0	1	b	1
3	0.005	0.999975	n	200
4	0.01	0.9999	Delta x	0.005
5	0.015	0.999775	Left Sum	0.6691625
6	0.02	0.9996		

	A	B	C	D
1	x	f(x)	a	0
2	0	1	b	1
3	0.002	0.999996	n	500
4	0.004	0.999984	Delta x	0.002
5	0.006	0.999964	Left Sum	0.667666
6	0.008	0.999936		

14

Further Integration Techniques and Applications of the Integral

 Website

www.WanerMath.com

At the Website you will find:

- A detailed chapter summary

- A true/false quiz

- Additional review exercises

- A numerical integration utility

- Graphing calculator programs for numerical integration

- Graphers, Excel tutorials, and other resources

Case Study Estimating Tax Revenues

You have just been hired by the incoming administration to coordinate national tax policy, and the so-called experts on your staff can't seem to agree on which of two tax proposals will result in more revenue for the government. The data you have are the two income tax proposals (graphs of tax vs. income) and the distribution of incomes in the country. **How do you use this information to decide which tax policy will result in more revenue?**

Dmitriy Shironosov/Shutterstock.com

Introduction

In the preceding chapter, we learned how to compute many integrals and saw some of the applications of the integral. In this chapter, we look at some further techniques for computing integrals and then at more applications of the integral. We also see how to extend the definition of the definite integral to include integrals over infinite intervals, and show how such integrals can be used for long-term forecasting. Finally, we introduce the beautiful theory of differential equations and some of its numerous applications.

14.1   Integration by Parts

Integration by parts is an integration technique that comes from the product rule for derivatives. The tabular method we present here has been around for some time and makes integration by parts quite simple, particularly in problems where it has to be used several times.*

* The version of the tabular method we use was developed and taught to us by Dan Rosen at Hofstra University.

We start with a little notation to simplify things while we introduce integration by parts. (We use this notation only in the next few pages.) If u is a function, denote its derivative by $D(u)$ and an antiderivative by $I(u)$. Thus, for example, if $u = 2x^2$, then

$$D(u) = 4x$$

and

$$I(u) = \frac{2x^3}{3}.$$

[If we wished, we could instead take $I(u) = \dfrac{2x^3}{3} + 46$, but we usually opt to take the simplest antiderivative.]

Integration by Parts

If u and v are continuous functions of x, and u has a continuous derivative, then

$$\int u \cdot v \, dx = u \cdot I(v) - \int D(u) I(v) \, dx.$$

Quick Example

(Discussed more fully in Example 1 below)

$$\int x \cdot e^x \, dx = x I(e^x) - \int D(x) I(e^x) \, dx$$

$$= xe^x - \int 1 \cdot e^x \, dx \qquad I(e^x) = e^x;\; D(x) = 1$$

$$= xe^x - e^x + C. \qquad \int e^x \, dx = e^x + C$$

As the Quick Example shows, although we could not immediately integrate $u \cdot v = x \cdot e^x$, we could easily integrate $D(u)I(v) = 1 \cdot e^x = e^x$.

Derivation of Integration-by-Parts Formula

As we mentioned, the integration-by-parts formula comes from the product rule for derivatives. We apply the product rule to the function $uI(v)$

$$D[u \cdot I(v)] = D(u)I(v) + uD(I(v))$$
$$= D(u)I(v) + uv$$

because $D(I(v))$ is the derivative of an antiderivative of v, which is v. Integrating both sides gives

$$u \cdot I(v) = \int D(u)I(v)\,dx + \int uv\,dx.$$

A simple rearrangement of the terms now gives us the integration-by-parts formula.

The integration-by-parts formula is easiest to use via the tabular method illustrated in the following example, where we repeat the calculation we did in the Quick Example above.

EXAMPLE 1 Integration by Parts: Tabular Method

Calculate $\int xe^x\,dx$.

Solution First, the reason we *need* to use integration by parts to evaluate this integral is that none of the other techniques of integration that we've talked about up to now will help us. Furthermore, we cannot simply find antiderivatives of x and e^x and multiply them together. [You should check that $(x^2/2)e^x$ is *not* an antiderivative of xe^x.] However, as we saw above, this integral can be found by integration by parts. We want to find the integral of the *product* of x and e^x. We must make a decision: Which function will play the role of u and which will play the role of v in the integration-by-parts formula? Because the derivative of x is just 1, differentiating makes it simpler, so we try letting x be u and letting e^x be v. We need to calculate $D(u)$ and $I(v)$, which we record in the following table.

	D	**I**
$+$	x	e^x
$-\int$	1	e^x

The table is read as
$+x \cdot e^x - \int 1 \cdot e^x\,dx.$

Below x in the D column, we put $D(x) = 1$; below e^x in the I column, we put $I(e^x) = e^x$. The arrow at an angle connecting x and $I(e^x)$ reminds us that the product $xI(e^x)$ will appear in the answer; the plus sign on the left of the table reminds us that it is $+xI(e^x)$ that appears. The integral sign and the horizontal arrow connecting $D(x)$ and $I(e^x)$ remind us that the *integral* of the product $D(x)I(e^x)$ also appears in the answer; the minus sign on the left reminds us that we need to subtract this integral. Combining these two contributions, we get

$$\int xe^x\,dx = xe^x - \int e^x\,dx.$$

The integral that appears on the right is much easier than the one we began with, so we can complete the problem:

$$\int xe^x\,dx = xe^x - \int e^x\,dx = xe^x - e^x + C.$$

➡ **Before we go on...** In Example 1, what if we had made the opposite decision and put e^x in the D column and x in the I column? Then we would have had the following table:

	D	I
$+$	e^x	x
$-\int$	$e^x \longrightarrow$	$x^2/2$

This gives

$$\int xe^x \, dx = \frac{x^2}{2}e^x - \int \frac{x^2}{2}e^x \, dx.$$

The integral on the right is harder than the one we started with, not easier! How do we know beforehand which way to go? We don't. We have to be willing to do a little trial and error: We try it one way, and if it doesn't make things simpler, we try it another way. *Remember, though, that the function we put in the I column must be one that we can integrate.* ■

EXAMPLE 2 Repeated Integration by Parts

Calculate $\displaystyle\int x^2 e^{-x} \, dx$.

Solution Again, we have a product—the integrand is the product of x^2 and e^{-x}. Because differentiating x^2 makes it simpler, we put it in the D column and get the following table:

	D	I
$+$	x^2	e^{-x}
$-\int$	$2x \longrightarrow$	$-e^{-x}$

This table gives us

$$\int x^2 e^{-x} \, dx = x^2(-e^{-x}) - \int 2x(-e^{-x}) \, dx.$$

The last integral is simpler than the one we started with, but it still involves a product. It's a good candidate for another integration by parts. The table we would use would start with $2x$ in the D column and $-e^{-x}$ in the I column, which is exactly what we see in the last row of the table we've already made. Therefore, we *continue the process*, elongating the table above:

	D	I
$+$	x^2	e^{-x}
$-$	$2x$	$-e^{-x}$
$+\int$	$2 \longrightarrow$	e^{-x}

(Notice how the signs on the left alternate. Here's why: To compute $-\int 2x(-e^{-x})\,dx$ we use the negative of the following table:

	D	*I*
$+$	$2x$	$-e^{-x}$
$-\int$	2	e^{-x}

so we reverse all the signs.)

Now, we still have to compute an integral (the integral of the product of the functions in the bottom row) to complete the computation. But why stop here? Let's continue the process one more step:

	D	*I*
$+$	x^2	e^{-x}
$-$	$2x$	$-e^{-x}$
$+$	2	e^{-x}
$-\int$	0	$-e^{-x}$

In the bottom line we see that all that is left to integrate is $0(-e^{-x}) = 0$. Because the indefinite integral of 0 is C, we can read the answer from the table as

$$\int x^2 e^{-x}\,dx = x^2(-e^{-x}) - 2x(e^{-x}) + 2(-e^{-x}) + C$$
$$= -x^2 e^{-x} - 2x e^{-x} - 2e^{-x} + C$$
$$= -e^{-x}(x^2 + 2x + 2) + C.$$

In Example 2 we saw a technique that we can summarize as follows:

Integrating a Polynomial Times a Function

If one of the factors in the integrand is a polynomial and the other factor is a function that can be integrated repeatedly, put the polynomial in the *D* column and keep differentiating until you get zero. Then complete the *I* column to the same depth, and read off the answer.

For practice, redo Example 1 using this technique.

It is not always the case that the integrand is a polynomial times something easy to integrate, so we can't always expect to end up with a zero in the *D* column. In that case we hope that at some point we will be able to integrate the product of the functions in the last row. Here are some examples.

EXAMPLE 3 Polynomial Times a Logarithm

Calculate: **a.** $\int x \ln x \, dx$ **b.** $\int (x^2 - x) \ln x \, dx$ **c.** $\int \ln x \, dx$

Solution

a. This is a product and therefore a good candidate for integration by parts. Our first impulse is to differentiate x, but that would mean integrating $\ln x$, and we do not (yet) know how to do that. So we try it the other way around and hope for the best.

	D	I
$+$	$\ln x$	x
$-\int$	$\dfrac{1}{x}$	$\dfrac{x^2}{2}$

Why did we stop? If we continued the table, both columns would get more complicated. However, if we stop here we get

$$\int x \ln x \, dx = (\ln x)\left(\frac{x^2}{2}\right) - \int \left(\frac{1}{x}\right)\left(\frac{x^2}{2}\right) dx$$

$$= \frac{x^2}{2} \ln x - \frac{1}{2} \int x \, dx$$

$$= \frac{x^2}{2} \ln x - \frac{x^2}{4} + C.$$

b. We can use the same technique we used in part (a) to integrate any polynomial times the logarithm of x:

	D	I
$+$	$\ln x$	$x^2 - x$
$-\int$	$\dfrac{1}{x}$	$\dfrac{x^3}{3} - \dfrac{x^2}{2}$

$$\int (x^2 - x) \ln x \, dx = (\ln x)\left(\frac{x^3}{3} - \frac{x^2}{2}\right) - \int \left(\frac{1}{x}\right)\left(\frac{x^3}{3} - \frac{x^2}{2}\right) dx$$

$$= \left(\frac{x^3}{3} - \frac{x^2}{2}\right) \ln x - \int \left(\frac{x^2}{3} - \frac{x}{2}\right) dx$$

$$= \left(\frac{x^3}{3} - \frac{x^2}{2}\right) \ln x - \frac{x^3}{9} + \frac{x^2}{4} + C.$$

c. The integrand $\ln x$ is not a product. We can, however, *make* it into a product by thinking of it as $1 \cdot \ln x$. Because this is a polynomial times $\ln x$, we proceed as in parts (a) and (b):

	D	I
$+$	$\ln x$	1
$-\int$	$1/x$	x

The closing stock prices and moving averages are plotted in Figure 11.

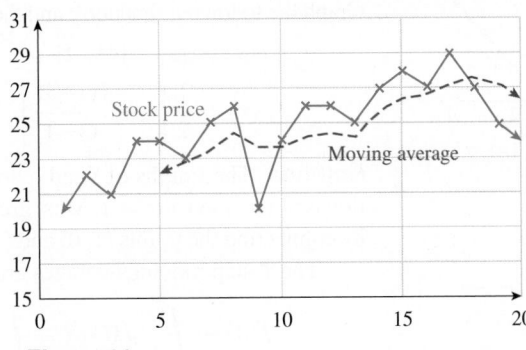

Figure 11

As you can see, the moving average is less volatile than the closing price. Because the moving average incorporates the stock's performance over 5 days at a time, a single day's fluctuation is smoothed out. Look at day 9 in particular. The moving average also tends to lag behind the actual performance because it takes past performance into account. Look at the downturns at days 6 and 18 in particular.

The period of 5 days for a moving average, as used in Example 3, is arbitrary. Using a longer period of time would smooth the data more but increase the lag. For data used as economic indicators, such as housing prices or retail sales, it is common to compute the four-quarter moving average to smooth out seasonal variations.

It is also sometimes useful to compute moving averages of continuous functions. We may want to do this if we use a mathematical model of a large collection of data. Also, some physical systems have the effect of converting an input function (an electrical signal, for example) into its moving average. By an ***n*-unit moving average** of a function $f(x)$ we mean the function $\bar{f}$ for which $\bar{f}(x)$ is the average of the value of $f(x)$ on $[x - n, x]$. Using the formula for the average of a function, we get the following formula.

---

n-Unit Moving Average of a Function

The *n*-unit moving average of a function f is

$$\bar{f}(x) = \frac{1}{n} \int_{x-n}^{x} f(t)\, dt.$$

Quick Example

The 2-unit moving average of $f(x) = x^2$ is

$$\bar{f}(x) = \frac{1}{2} \int_{x-2}^{x} t^2\, dt = \frac{1}{6} \left[t^3 \right]_{x-2}^{x} = x^2 - 2x + \frac{4}{3}.$$

The graphs of $f(x)$ and $\bar{f}(x)$ are shown in Figure 12.

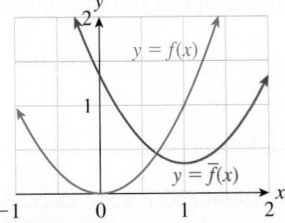

Figure 12

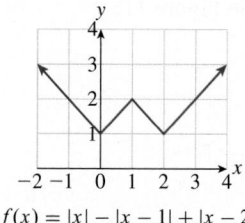

$$f(x) = |x| - |x - 1| + |x - 2|$$

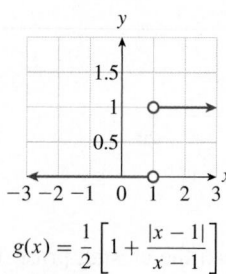

$$g(x) = \frac{1}{2}\left[1 + \frac{|x-1|}{x-1}\right]$$

Figure 13

EXAMPLE 4 Moving Averages: Sawtooth and Step Functions

Graph the following functions, and then compute and graph their 1-unit moving averages.

$$f(x) = |x| - |x - 1| + |x - 2| \qquad \text{Sawtooth}$$

$$g(x) = \frac{1}{2}\left[1 + \frac{|x-1|}{x-1}\right] \qquad \text{Unit step at } x = 1$$

Solution The graphs of f and g are shown in Figure 13. (Notice that the step function is not defined at $x = 1$. Most graphers will show the step function as an actual step by connecting the points $(1, 0)$ and $(1, 1)$ with a vertical line.)

The 1-step moving averages are:

$$\bar{f}(x) = \int_{x-1}^{x} f(t)\, dt = \int_{x-1}^{x} [|t| - |t - 1| + |t - 2|]\, dt$$

$$= \frac{1}{2}[t|t| - (t - 1)|t - 1| + (t - 2)|t - 2|]_{x-1}^{x}$$

$$= \frac{1}{2}([x|x| - (x - 1)|x - 1| + (x - 2)|x - 2|]$$
$$\qquad - [(x - 1)|x - 1| - (x - 2)|x - 2| + (x - 3)|x - 3|])$$

$$= \frac{1}{2}[x|x| - 2(x - 1)|x - 1| + 2(x - 2)|x - 2| - (x - 3)|x - 3|].$$

$$\bar{g}(x) = \int_{x-1}^{x} g(t)\, dt = \int_{x-1}^{x} \frac{1}{2}\left[1 + \frac{|t-1|}{t-1}\right] dt$$

$$= \frac{1}{2}[t + |t - 1|]_{x-1}^{x}$$

$$= \frac{1}{2}[(x + |x - 1|) - (x - 1 + |x - 2|)]$$

$$= \frac{1}{2}[1 + |x - 1| - |x - 2|]$$

The graphs of $\bar{f}$ and $\bar{g}$ are shown in Figure 14.

using Technology

TI-83/84 Plus
We can graph the moving average of f in Example 4 on a TI-83/84 Plus as follows:

```
Y₁=abs(X)-abs(X-1)
+abs(X-2)
Y₂=fnInt(Y1(T),T,
X-1,X)
ZOOM  0
```

For g, change Y_1 to
```
Y₁=.5*(1+abs(X-1)/
(X-1))
```

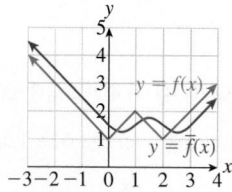

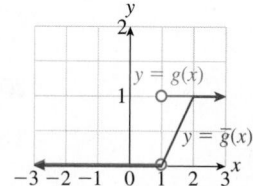

Figure 14

Notice how the graph of $\bar{f}$ smooths out the zig-zags of the sawtooth function.

➡️ **Before we go on...** Figure 15 shows the 2-unit moving average of f in Example 4,

$$\bar{f}(x) = \frac{1}{2}\int_{x-2}^{x} f(t)\, dt$$

$$= \frac{1}{4}(x|x| - (x - 1)|x - 1| + (x - 3)|x - 3| - (x - 4)|x - 4|).$$

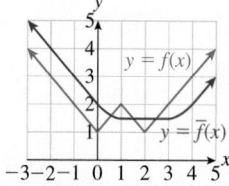

Figure 15

Notice how the 2-point moving average has completely eliminated the zig-zags, illustrating how moving averages can be used to remove seasonal fluctuations in real-life situations. ∎

14.3 EXERCISES

▼ more advanced   ◆ challenging

Ⓣ indicates exercises that should be solved using technology

Find the averages of the functions in Exercises 1–8 over the given intervals. Plot each function and its average on the same graph (as in Figure 8). HINT [See Quick Example page 1035.]

1. $f(x) = x^3$ over $[0, 2]$ **2.** $f(x) = x^3$ over $[-1, 1]$

3. $f(x) = x^3 - x$ over $[0, 2]$ **4.** $f(x) = x^3 - x$ over $[0, 1]$

5. $f(x) = e^{-x}$ over $[0, 2]$ **6.** $f(x) = e^x$ over $[-1, 1]$

7. $f(x) = |2x - 5|$ over $[0, 4]$

8. $f(x) = |-x + 2|$ over $[-1, 3]$

In Exercises 9 and 10, complete the given table with the values of the 3-unit moving average of the given function. HINT [See Example 3.]

9.

x	0	1	2	3	4	5	6	7
$r(x)$	3	5	10	3	2	5	6	7
$\bar{r}(x)$								

10.

x	0	1	2	3	4	5	6	7
$s(x)$	2	9	7	3	2	5	7	1
$\bar{s}(x)$								

In Exercises 11 and 12, some values of a function and its 3-unit moving average are given. Supply the missing information.

11.

x	0	1	2	3	4	5	6	7
$r(x)$	1	2			11		10	2
$\bar{r}(x)$			3	5		11		

12.

x	0	1	2	3	4	5	6	7
$s(x)$	1	5		1				
$\bar{s}(x)$			5		5	2	3	2

Calculate the 5-unit moving average of each function in Exercises 13–24. Plot each function and its moving average on the same graph, as in Example 4. (You may use graphing technology for these plots, but you should compute the moving averages analytically.) HINT [See Quick Example page 1037, and Example 4.]

13. $f(x) = x^3$ **14.** $f(x) = x^3 - x$

15. $f(x) = x^{2/3}$ **16.** $f(x) = x^{2/3} + x$

17. $f(x) = e^{0.5x}$ **18.** $f(x) = e^{-0.02x}$

19. $f(x) = \sqrt{x}$ **20.** $f(x) = x^{1/3}$

21. $f(x) = 1 - \dfrac{|2x - 1|}{2x - 1}$ **22.** $f(x) = 2 + \dfrac{|3x + 1|}{3x + 1}$

23. ▼ $f(x) = 2 - |x + 1| + |x|$ [Do not simplify the answer.]

24. ▼ $f(x) = |2x + 1| - |2x| - 2$ [Do not simplify the answer.]

Ⓣ *In Exercises 25–34, use graphing technology to plot the given functions together with their 3-unit moving averages.* HINT [See Technology Note for Example 4.]

25. $f(x) = \dfrac{10x}{1 + 5|x|}$ **26.** $f(x) = \dfrac{1}{1 + e^x}$

27. $f(x) = \ln(1 + x^2)$ **28.** $f(x) = e^{1-x^2}$

29. $f(x) = |x| - |x - 1| + |x - 2| - |x - 3| + |x - 4|$

30. $f(x) = |x| - 2|x - 1| + 2|x - 2| - 2|x - 3| + |x - 4|$

31. $f(x) = \dfrac{|x|}{x} - \dfrac{|x - 1|}{x - 1} + \dfrac{|x - 2|}{x - 2} - \dfrac{|x - 3|}{x - 3}$

32. $f(x) = \dfrac{|x|}{x} - 2\dfrac{|x - 1|}{x - 1} + 2\dfrac{|x - 2|}{x - 2} - \dfrac{|x - 3|}{x - 3}$

33. $f(x) = \dfrac{|x|}{x} + \dfrac{|x - 1|}{x - 1} + \dfrac{|x - 2|}{x - 2} + \dfrac{|x - 3|}{x - 3}$

34. $f(x) = 4 - \dfrac{|x|}{x} - \dfrac{|x - 1|}{x - 1} - \dfrac{|x - 2|}{x - 2} - \dfrac{|x - 3|}{x - 3}$

APPLICATIONS

35. *Television Advertising* The cost, in millions of dollars, of a 30-second television ad during the Super Bowl in the years 2000 to 2010 can be approximated by

$$C(t) = 0.14t + 1.1 \text{ million dollars} \quad (0 \le t \le 10)$$

($t = 0$ represents 2000).[19] What was the average cost of a Super Bowl ad during the given period? HINT [See Example 1.]

36. *Television Advertising* The cost, in millions of dollars, of a 30-second television ad during the Super Bowl in the years 1980 to 2000 can be approximated by

$$C(t) = 0.044t + 0.222 \text{ million dollars} \quad (0 \le t \le 20)$$

($t = 0$ represents 1980).[20] What was the average cost of a Super Bowl ad during the given period? HINT [See Example 1.]

37. *Membership: Facebook* The number of new members joining Facebook each year in the period from 2005 to the middle of 2008 can be modeled by

$$m(t) = 12t^2 - 20t + 10 \text{ million members per year}$$
$$(0 \le t \le 3.5),$$

---

[19] Source for data: en.wikipedia.org/wiki/Super_Bowl_advertising.

[20] *Ibid.*

where t is time in years since the start of 2005.[21] What was the average number of new members joining Facebook each year from the start of 2005 to the start of 2008?

38. *Membership:* MySpace The number of new members joining Myspace each year in the period from 2004 to the middle of 2007 can be modeled by

$$m(t) = 10.5t^2 + 14t - 6 \text{ million members per year}$$
$$(0 \leq t \leq 3.5),$$

where t is time in years since the start of 2004.[22] What was the average number of new members joining MySpace each year from the start of 2004 to the start of 2007?

39. *Freon Production* Annual production of ozone-layer-damaging Freon 22 (chlorodifluoromethane) in developing countries from 2000 to 2010 can be modeled by

$$F(t) = 97.2(1.20)^t \text{ million tons} \quad (0 \leq t \leq 10)$$

(t is the number of years since 2000).[23] What was the average annual production over the period shown? (Round your answer to the nearest million tons.) HINT [See Example 2.]

40. *Health Expenditures* Annual expenditures on health in the United States from 1980 to 2010 could be modeled by

$$F(t) = 296(1.08)^t \text{ billion dollars} \quad (0 \leq t \leq 30)$$

($t = 0$ represents 1980).[24] What was the average annual expenditure over the period shown? (Round your answer to the nearest billion dollars.) HINT [See Example 2.]

41. *Investments* If you invest \$10,000 at 8% interest compounded continuously, what is the average amount in your account over one year?

42. *Investments* If you invest \$10,000 at 12% interest compounded continuously, what is the average amount in your account over one year?

43. ▼ *Average Balance* Suppose you have an account (paying no interest) into which you deposit \$3,000 at the beginning of each month. You withdraw money continuously so that the amount in the account decreases linearly to 0 by the end of the month. Find the average amount in the account over a period of several months. (Assume that the account starts at \$0 at $t = 0$ months.)

44. ▼ *Average Balance* Suppose you have an account (paying no interest) into which you deposit \$4,000 at the beginning of each month. You withdraw \$3,000 during the course of each month, in such a way that the amount decreases linearly. Find the average amount in the account in the first two months. (Assume that the account starts at \$0 at $t = 0$ months.)

45. ◫ *Online Vehicle Sales* The value of vehicle transactions on eBay each quarter from the first quarter of 2008 through the last quarter of 2010 could be approximated by

$$v(t) = -0.002x^3 + 0.06x^2 - 0.55x + 3.9 \text{ billion dollars}$$
$$\text{per quarter} \quad (1 \leq t \leq 13)$$

(t is time in quarters; $t = 1$ represents the first quarter of 2008).[25] Use technology to estimate the average quarterly value of vehicle transactions on eBay during the given period. (Round your answer to the nearest hundred million dollars.)

46. ◫ *Online Auctions* The net number of new active eBay users each quarter from the first quarter of 2008 through the last quarter of 2010 could be approximated by

$$n(t) = -0.002x^4 + 0.06x^3 - 0.55x^2 + 1.9x - 1$$
$$\text{million users per quarter} \quad (1 \leq t \leq 13)$$

(t is time in quarters; $t = 1$ represents the first quarter of 2008).[26] Use technology to estimate the average number of new active eBay users each quarter during the given period. (Round your answer to the nearest hundred thousand users.)

47. *Stock Prices:* Exxon Mobil The following table shows the approximate price of Exxon Mobil stock in December of each year from 2001 through 2010.[27] Complete the table by computing the 4-year moving averages. (Note the peak in 2007 and the drop in subsequent years.) Round each average to the nearest dollar.

Year t	2001	2002	2003	2004	2005	2006	2007	2008	2009	2010
Stock Price	39	35	41	51	56	77	94	80	68	73
Moving Average (rounded)										

The stock price spiked in 2007 and then dropped steeply in 2008. What happened to the corresponding moving average? HINT [See Example 3.]

48. *Stock Prices:* Nokia The following table shows the approximate price of Nokia stock in December of each year from 2001 through 2010.[28] Complete the table by computing the 4-year moving averages. (Note the peak in 2007 and the drop in subsequent years.) Round each average to the nearest dollar.

Year t	2001	2002	2003	2004	2005	2006	2007	2008	2009	2010
Stock Price	25	16	17	16	18	20	38	16	13	10
Moving Average (rounded)										

---

[21] Sources for data: www.facebook.com, insidehighered.com. (Some data are interpolated.)

[22] Source for data: www.swivel.com/data_sets.

[23] Figures are approximate. Source: Lampert Kuijpers (Panel of the Montreal Protocol), National Bureau of Statistics in China, via CEIC Data/*New York Times*, February 23, 2007, p. C1.

[24] Source for data: U.S. Department of Health & Human Services/Centers for Medicare & Medicaid Services, National Health Expenditure Data, downloaded April 2011 from www.cms.gov.

[25] Source for data: eBay company reports http://investor.ebay.com/.

[26] *Ibid.*

[27] Source: finance.yahoo.com.

[28] *Ibid.*

How does the average year-by-year change in the moving average compare with the average year-by-year change in the stock price? HINT [See Example 3.]

49. ▼ *Cancun* The *Playa Loca Hotel* in Cancun has an advertising brochure with the following chart, showing the year-round temperature.[29]

a. Estimate and plot the year-round 6-month moving average. (Use graphing technology, if available, to check your graph.)

b. What can you say about the 12-month moving average?

50. ▼ *Reykjavik* Repeat the preceding exercise, using the following data from the brochure of the *Tough Traveler Lodge* in Reykjavik.[30]

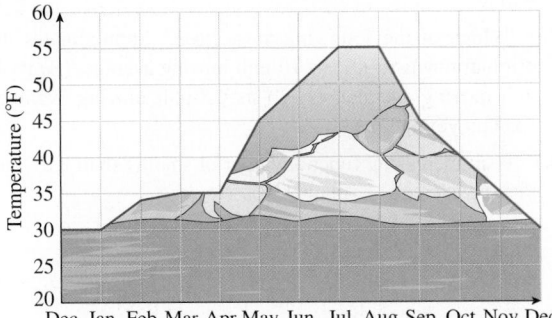

51. ⊡ ▼ *Sales: Apple* The following table shows approximate quarterly sales of iPods in millions of units, starting in the first quarter of 2006.[31]

Quarter	2006 Q1	2006 Q2	2006 Q3	2006 Q4	2007 Q1	2007 Q2	2007 Q3	2007 Q4	2008 Q1	2008 Q2
Sales (millions)	8.5	8.1	8.7	21.1	10.5	9.8	10.2	22.1	10.6	11.0

Quarter	2008 Q3	2008 Q4	2009 Q1	2009 Q2	2009 Q3	2009 Q4	2010 Q1	2010 Q2	2010 Q3	2010 Q4
Sales (millions)	11.1	22.7	11.0	10.2	10.2	21.0	10.9	9.4	9.1	19.5

a. Use technology to compute and plot the four-quarter moving average of these data.

b. The graph of the moving average for the last eight quarters will appear almost linear during 2009 and 2010. Use the 2009 Q1 and 2010 Q4 figures of the moving average to give an estimate (to the nearest 0.1 million units) of the rate of change of iPod sales during 2009–2010.

52. ⊡ ▼ *Housing Starts* The following table shows the number of housing starts for one-family units, in thousands of units, starting in the first quarter of 2006.[32]

Quarter	2006 Q1	2006 Q2	2006 Q3	2006 Q4	2007 Q1	2007 Q2	2007 Q3	2007 Q4	2008 Q1	2008 Q2
Housing Starts (1,000)	382	433	372	278	260	333	265	188	162	194

Quarter	2008 Q3	2008 Q4	2009 Q1	2009 Q2	2009 Q3	2009 Q4	2010 Q1	2010 Q2	2010 Q3	2010 Q4
Housing Starts (1,000)	163	103	78	124	138	105	114	142	119	96

a. Use technology to compute and plot the four-quarter moving average of these data.

b. The graph of the moving average for the eight quarters in 2007 and 2008 will appear almost linear. Use the 2007 Q1 and 2008 Q4 figures of the moving average to give an estimate (to the nearest thousand units) of the rate of change of housing starts during 2007–2008.

53. *Bottled Water Sales* The rate of U.S. sales of bottled water for the period 2000–2010 could be approximated by

$$s(t) = -45t^2 + 900t + 4{,}200 \text{ million gallons per year}$$
$$(0 \le t \le 10),$$

where t is time in years since the start of 2000.[33]

a. Estimate the average annual sales of bottled water over the period 2000–2010, to the nearest 100 million gallons per year. HINT [See Quick Example page 1035.]

b. Compute the two-year moving average of s. (You need not simplify the answer.) HINT [See Quick Example page 1037.]

c. Without simplifying the answer in part (b), say what kind of function the moving average is.

54. *Bottled Water Sales* The rate of U.S. per capita sales of bottled water for the period 2000–2010 coud be approximated by

$$s(t) = -0.18t^2 + 3t + 15 \text{ gallons per year} \quad (0 \le t \le 10),$$

where t is the time in years since the start of 2000.[34] Repeat the preceding exercise as applied to per capita sales. (Give your answer to (a) to the nearest gallon per year.)

---

[29] Source: www.holiday-weather.com. (Temperatures are rounded.)

[30] *Ibid.*

[31] Source: Apple quarterly press releases, www.apple.com/investor.

[32] Source: U.S. Census Bureau, www.census.gov/const/www/ newresconstindex.html.

[33] Source for data: Beverage Marketing Corporation/ www.bottledwater.org.

[34] *Ibid.*

55. *Medicare Spending* Annual spending on Medicare was projected to increase from $526 billion in 2010 to around $977
billion in 2021.[35]

   a. Use this information to express s, the annual spending
   on Medicare (in billions of dollars), as a linear function
   of t, the number of years since 2010.
   b. Find the 4-year moving average of your model.
   c. What can you say about the slope of the moving
   average?

56. *Pasta Imports in the 1990s* In 1990, the United States imported 290 million pounds of pasta. From 1990 to 2000 imports increased by an average of 40 million pounds per
year.[36]

   a. Use these data to express q, the annual U.S. imports of
   pasta (in millions of pounds), as a linear function of t,
   the number of years since 1990.
   b. Find the 4-year moving average of your model.
   c. What can you say about the slope of the moving
   average?

57. ▼ *Moving Average of a Linear Function* Find a formula
for the a-unit moving average of a general linear function
$f(x) = mx + b$.

58. ▼ *Moving Average of an Exponential Function* Find a formula for the a-unit moving average of a general exponential
function $f(x) = Ae^{kx}$.

---

[35]Source: Congressional Budget Office, *March 2011 Medicare Baseline*
(www.cbo.gov).

[36]Data are rounded. Sources: Department of Commerce/*New York Times*,
September 5, 1995, p. D4, International Trade Administration
(www.ita.doc.gov) March 31, 2002.

COMMUNICATION AND REASONING EXERCISES

59. Explain why it is sometimes more useful to consider the moving average of a stock price rather than the stock price itself.

60. Sales this month were sharply lower than they were last
month, but the 12-unit moving average this month was
higher than it was last month. How can that be?

61. Your company's 6-month moving average of sales is constant. What does that say about the sales figures?

62. Your monthly salary has been increasing steadily for the past
year, and your average monthly salary over the past year was
x dollars. Would you have earned more money if you had
been paid x dollars per month? Explain your answer.

63. ▼ What property does the graph of a (nonconstant) function
have if its average value over an interval is zero? Give an example of such a function.

64. ▼ Can the average value of a function f on an interval
be greater than its value at every point in that interval?
Explain.

65. ▼ Criticize the following claim: The average value of a function on an interval is midway between its highest and lowest
value.

66. ▼ Your manager tells you that 12-month moving averages
give at least as much information as shorter-term moving
averages and very often more. How would you argue that he is
wrong?

67. ▼ Which of the following most closely approximates the
original function: (A) its 10-unit moving average, (B) its 1-
unit moving average, or (C) its 0.8-unit moving average?
Explain your answer.

68. ▼ Is an increasing function larger or smaller than its 1-unit
moving average? Explain.

14.4 # Applications to Business and Economics: Consumers' and Producers' Surplus and Continuous Income Streams

Consumers' Surplus

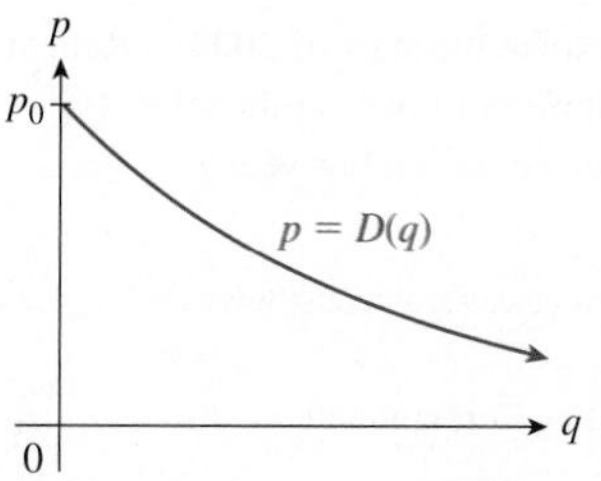

Figure 16

Consider a general demand curve presented, as is traditional in economics, as
$p = D(q)$, where p is unit price and q is demand measured, say, in annual sales
(Figure 16). Thus, $D(q)$ is the price at which the demand will be q units per year.
The price p_0 shown on the graph is the highest price that customers are willing
to pay.

Suppose, for example, that the graph in Figure 16 is the demand curve for a particular new model of computer. When the computer first comes out and supplies are low (q is small), "early adopters" will be willing to pay a high price. This is the part of the graph on the left, near the p-axis. As supplies increase and the price drops, more consumers will be willing to pay and more computers will be sold. We can ask the following question: How much are consumers willing to spend for the first $\bar{q}$ units?

Consumers' Willingness to Spend

We can approximate consumers' willingness to spend on the first $\bar{q}$ units as follows. We partition the interval $[0, \bar{q}]$ into n subintervals of equal length, as we did when discussing Riemann sums. Figure 17 shows a typical subinterval, $[q_{k-1}, q_k]$.

The price consumers are willing to pay for each of units q_{k-1} through q_k is approximately $D(q_{k-1})$, so the total that consumers are willing to spend for these units is approximately $D(q_{k-1})(q_k - q_{k-1}) = D(q_{k-1})\Delta q$, the area of the shaded region in Figure 17. Thus, the total amount that consumers are willing to spend for items 0 through $\bar{q}$ is

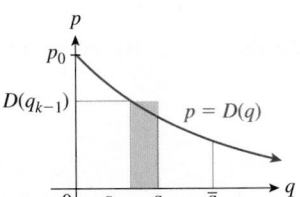

Figure 17

$$W \approx D(q_0)\Delta q + D(q_1)\Delta q + \cdots + D(q_{n-1})\Delta q = \sum_{k=0}^{n-1} D(q_k)\Delta q,$$

which is a Riemann sum. The approximation becomes better the larger n becomes, and in the limit the Riemann sums converge to the integral

$$W = \int_0^{\bar{q}} D(q)\, dq.$$

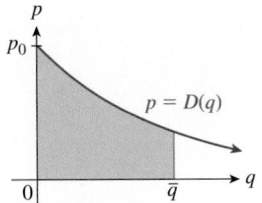

Figure 18

This quantity, the area shaded in Figure 18, is the total consumers' willingness to spend to buy the first $\bar{q}$ units.

Consumers' Expenditure

Now suppose that the manufacturer simply sets the price at $\bar{p}$, with a corresponding demand of $\bar{q}$, so $D(\bar{q}) = \bar{p}$. Then the amount that consumers will actually spend to buy these $\bar{q}$ is $\bar{p}\bar{q}$, the product of the unit price and the quantity sold. This is the area of the rectangle shown in Figure 19. Notice that we can write $\bar{p}\bar{q} = \int_0^{\bar{q}} \bar{p}\, dq$, as suggested by the figure.

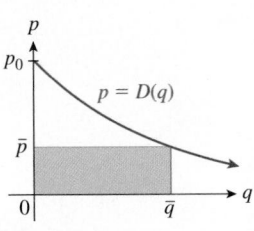

Figure 19

The difference between what consumers are willing to pay and what they actually pay is money in their pockets and is called the **consumers' surplus**.

✱ Multiletter variables like *CS* used here may be unusual in a math textbook but are traditional in the math of finance. In particular, the notations *PV* and *FV* used later in this section are almost universally used in finance textbooks, calculators (such as the TI-83/84 Plus), and such places as study guides for the finance portion of the Society of Actuaries exams.

Consumers' Surplus

If demand for an item is given by $p = D(q)$, the selling price is $\bar{p}$, and $\bar{q}$ is the corresponding demand [so that $D(\bar{q}) = \bar{p}$], then the **consumers' surplus** is the difference between willingness to spend and actual expenditure:✱

$$CS = \int_0^{\bar{q}} D(q)\, dq - \bar{p}\bar{q} = \int_0^{\bar{q}} (D(q) - \bar{p})\, dq.$$

Graphically, it is the area between the graphs of $p = D(q)$ and $p = \bar{p}$, as shown in the figure.

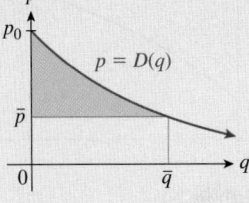

EXAMPLE 1 Consumers' Surplus

Your used-CD store has an exponential demand equation of the form

$$p = 15e^{-0.01q}$$

where q represents daily sales of used CDs and p is the price you charge per CD. Calculate the daily consumers' surplus if you sell your used CDs at \$5 each.

Solution We are given $D(q) = 15e^{-0.01q}$ and $\bar{p} = 5$. We also need $\bar{q}$. By definition,

$$D(\bar{q}) = \bar{p}$$

or $15e^{-0.01\bar{q}} = 5,$

which we must solve for $\bar{q}$:

$$e^{-0.01\bar{q}} = \frac{1}{3}$$

$$-0.01\bar{q} = \ln\left(\frac{1}{3}\right) = -\ln 3$$

$$\bar{q} = \frac{\ln 3}{0.01} \approx 109.8612.$$

We now have

$$CS = \int_{0}^{\bar{q}} (D(q) - \bar{p})\, dq$$

$$= \int_{0}^{109.8612} (15e^{-0.01q} - 5)\, dq$$

$$= \left[\frac{15}{-0.01} e^{-0.01q} - 5q \right]_{0}^{109.8612}$$

$$\approx (-500 - 549.31) - (-1,500 - 0)$$

$$= \$450.69 \text{ per day.}$$

Producers' Surplus

We can also calculate extra income earned by producers. Consider a supply equation of the form $p = S(q)$, where $S(q)$ is the price at which a supplier is willing to supply q items (per time period). Because a producer is generally willing to supply more units at a higher price per unit, a supply curve usually has a positive slope, as shown in Figure 20. The price p_0 is the lowest price that a producer is willing to charge.

Arguing as before, we see that the minimum amount of money producers are willing to receive in exchange for $\bar{q}$ items is $\int_{0}^{\bar{q}} S(q)\, dq$. On the other hand, if the producers charge $\bar{p}$ per item for $\bar{q}$ items, their actual revenue is $\bar{p}\bar{q} = \int_{0}^{\bar{q}} \bar{p}\, dq$.

The difference between the producers' actual revenue and the minimum they would have been willing to receive is the **producers' surplus**.

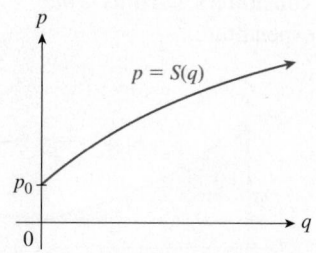

Figure 20

⟹ **Before we go on...** The interest earned in the account in Example 5 was fairly small. (Compare this answer to that in Example 4.) Not only was the money in the account for only three months, but much of it was put in the account toward the end of that period, so had very little time to earn interest. ∎

Generalizing again, we have the following:

Future Value of a Continuous Income Stream

If the rate of receipt of income from time $t = a$ to $t = b$ is $R(t)$ dollars per unit of time and the income is deposited as it is received in an account paying interest at rate r per unit of time, compounded continuously, then the amount of money in the account at time $t = b$ is

$$\text{Future value} = FV = \int_a^b R(t)e^{r(b-t)}dt.$$

EXAMPLE 6 Present Value

You are thinking of buying the ice cream store discussed in the preceding two examples. What is its income stream worth to you on June 1? Assume that you have access to the same account paying 5% per year compounded continuously.

Solution The value of the income stream on June 1 is the amount of money that, if deposited June 1, would give you the same future value as the income stream will. If we let PV denote this "present value," its value after 92 days will be

$$PVe^{0.05 \times 92/365}.$$

We equate this with the future value of the income stream to get

$$PVe^{0.05 \times 92/365} = \int_0^{92} R(t)e^{0.05(92-t)/365}\,dt$$

so

$$PV = \int_0^{92} R(t)e^{-0.05t/365}\,dt.$$

Substituting the formula for $R(t)$ and integrating using technology or integration by parts, we get

$$PV \approx \$33,455.$$

The general formula is the following:

Present Value of a Continuous Income Stream

If the rate of receipt of income from time $t = a$ to $t = b$ is $R(t)$ dollars per unit of time and the income is deposited as it is received in an account paying interest at rate r per unit of time, compounded continuously, then the value of the income stream at time $t = a$ is

$$\text{Present value} = PV = \int_a^b R(t)e^{r(a-t)}dt.$$

We can derive this formula from the relation

$$FV = PVe^{r(b-a)}$$

because the present value is the amount that would have to be deposited at time $t = a$ to give a future value of FV at time $t = b$.

Note These formulas are more general than we've said. They still work when $R(t) < 0$ if we interpret negative values as money flowing *out* rather than in. That is, we can use these formulas for income we receive or for payments that we make, or for situations where we sometimes receive money and sometimes pay it out. These formulas can also be used for flows of quantities other than money. For example, if we use an exponential model for population growth and we let $R(t)$ represent the rate of immigration [$R(t) > 0$] or emigration [$R(t) < 0$], then the future value formula gives the future population. ∎

14.4 EXERCISES

▼ more advanced   ◆ challenging
T indicates exercises that should be solved using technology

Calculate the consumers' surplus at the indicated unit price $\bar{p}$ for each of the demand equations in Exercises 1–12. HINT [See Example 1.]

1. $p = 10 - 2q$; $\bar{p} = 5$
2. $p = 100 - q$; $\bar{p} = 20$
3. $p = 100 - 3\sqrt{q}$; $\bar{p} = 76$
4. $p = 10 - 2q^{1/3}$; $\bar{p} = 6$
5. $p = 500e^{-2q}$; $\bar{p} = 100$
6. $p = 100 - e^{0.1q}$; $\bar{p} = 50$
7. $q = 100 - 2p$; $\bar{p} = 20$
8. $q = 50 - 3p$; $\bar{p} = 10$
9. $q = 100 - 0.25p^2$; $\bar{p} = 10$
10. $q = 20 - 0.05p^2$; $\bar{p} = 5$
11. $q = 500e^{-0.5p} - 50$; $\bar{p} = 1$
12. $q = 100 - e^{0.1p}$; $\bar{p} = 20$

Calculate the producers' surplus for each of the supply equations in Exercises 13–24 at the indicated unit price $\bar{p}$. HINT [See Example 2.]

13. $p = 10 + 2q$; $\bar{p} = 20$
14. $p = 100 + q$; $\bar{p} = 200$
15. $p = 10 + 2q^{1/3}$; $\bar{p} = 12$
16. $p = 100 + 3\sqrt{q}$; $\bar{p} = 124$

17. $p = 500e^{0.5q}$; $\bar{p} = 1,000$
18. $p = 100 + e^{0.01q}$; $\bar{p} = 120$
19. $q = 2p - 50$; $\bar{p} = 40$
20. $q = 4p - 1,000$; $\bar{p} = 1,000$
21. $q = 0.25p^2 - 10$; $\bar{p} = 10$
22. $q = 0.05p^2 - 20$; $\bar{p} = 50$
23. $q = 500e^{0.05p} - 50$; $\bar{p} = 10$
24. $q = 10(e^{0.1p} - 1)$; $\bar{p} = 5$

In Exercises 25–30, find the total value of the given income stream and also find its future value (at the end of the given interval) using the given interest rate. HINT [See Examples 4 and 5.]

25. $R(t) = 30,000$, $0 \le t \le 10$, at 7%
26. $R(t) = 40,000$, $0 \le t \le 5$, at 10%
27. $R(t) = 30,000 + 1,000t$, $0 \le t \le 10$, at 7%
28. $R(t) = 40,000 + 2,000t$, $0 \le t \le 5$, at 10%
29. $R(t) = 30,000e^{0.05t}$, $0 \le t \le 10$, at 7%
30. $R(t) = 40,000e^{0.04t}$, $0 \le t \le 5$, at 10%

In Exercises 31–36, find the total value of the given income stream and also find its present value (at the beginning of the given interval) using the given interest rate. HINT [See Examples 4 and 6.]

31. $R(t) = 20,000$, $0 \le t \le 5$, at 8%
32. $R(t) = 50,000$, $0 \le t \le 10$, at 5%

33. $R(t) = 20,000 + 1,000t$, $0 \le t \le 5$, at 8%

34. $R(t) = 50,000 + 2,000t$, $0 \le t \le 10$, at 5%

35. $R(t) = 20,000e^{0.03t}$, $0 \le t \le 5$, at 8%

36. $R(t) = 50,000e^{0.06t}$, $0 \le t \le 10$, at 5%

APPLICATIONS

37. *College Tuition* A study of U.S. colleges and universities resulted in the demand equation $q = 20,000 - 2p$, where q is the enrollment at a public college or university and p is the average annual tuition (plus fees) it charges.[37] Officials at Enormous State University have developed a policy whereby the number of students it will accept per year at a tuition level of p dollars is given by $q = 7,500 + 0.5p$. Find the equilibrium tuition price $\bar{p}$ and the consumers' and producers' surpluses at this tuition level. What is the total social gain at the equilibrium price? HINT [See Example 3.]

38. *Fast Food* A fast-food outlet finds that the demand equation for its new side dish, "Sweetdough Tidbit," is given by

$$p = \frac{128}{(q+1)^2},$$

where p is the price (in cents) per serving and q is the number of servings that can be sold per hour at this price. At the same time, the franchise is prepared to sell $q = 0.5p - 1$ servings per hour at a price of p cents. Find the equilibrium price $\bar{p}$ and the consumers' and producers' surpluses at this price level. What is the total social gain at the equilibrium price? HINT [See Example 3.]

39. *Revenue:* Nokia The annual net sales (revenue) earned by Nokia in the years January 2004 to January 2010 can be approximated by

$$R(t) = -1.75t^2 + 12.5t + 30 \text{ billion euros per year}$$
$$(0 \le t \le 6),$$

where t is time in years ($t = 0$ represents January 2004).[38] Estimate, to the nearest €10 billion, Nokia's total revenue from January 2006 to January 2010. HINT [See Example 4.]

40. *Revenue:* Nintendo The annual net sales (revenue) earned by Nintendo Co., Ltd., in the fiscal years from April 1, 2004, to April 1, 2010, can be approximated by

$$R(t) = -14.5t^3 + 38t^2 + 380t + 510 \text{ billion yen per year}$$
$$(0 \le t \le 6),$$

where t is time in years ($t = 0$ represents April 1, 2004).[39] Estimate, to the nearest ¥100 billion, Nintendo's total revenue from April 1, 2006, to April 1, 2010. HINT [See Example 4.]

41. *Revenue:* Walmart The annual revenue earned by Walmart in the years from January 2004 to January 2010 can be approximated by

$$R(t) = 242e^{0.098t} \text{ billion dollars per year} \quad (0 \le t \le 6),$$

where t is time in years ($t = 0$ represents January 2004).[40] Estimate, to the nearest $10 billion, Wal-Mart's total revenue from January 2004 to January 2008.

42. *Revenue:* Target The annual revenue earned by Target for fiscal years 2004 through 2010 can be approximated by

$$R(t) = 41e^{0.094t} \text{ billion dollars per year} \quad (0 \le t \le 6),$$

where t is time in years ($t = 0$ represents the beginning of fiscal year 2004).[41] Estimate, to the nearest $10 billion, Target's total revenue from the beginning of fiscal year 2006 to the beginning of fiscal year 2010.

43. ▼ *Revenue* Refer back to Exercise 39. Suppose that, from January 2004 on, Nokia invested its revenue in an investment yielding 4% compounded continuously. What, to the nearest €10 billion, would the total value of Nokia's revenue from January 2006 to January 2010 have been in January 2010? HINT [See Example 5.]

44. ▼ *Revenue* Refer back to Exercise 40. Suppose that, from April 2004 on, Nintendo invested its revenue in an investment yielding 5% compounded continuously. What, to the nearest ¥100 billion, would the total value of Nintendo's revenue from April 1, 2006, to April 1, 2010, have been in April 2010? HINT [See Example 5.]

45. ▼ *Revenue* Refer back to Exercise 41. Suppose that, from January 2004 on, Walmart invested its revenue in an investment that depreciated continuously at a rate of 5% per year. What, to the nearest $10 billion, would the total value of Walmart's revenues from January 2004 to January 2008 have been in January 2008?

46. ▼ *Revenue* Refer back to Exercise 42. Suppose that, from the start of fiscal year 2004 on, Target invested its revenue in an investment that depreciated continuously at a rate of 3% per year. What, to the nearest $10 billion, would the total value of Target's revenue from the beginning of fiscal year 2006 to the beginning of fiscal year 2010 have been at the beginning of fiscal year 2010?

---

[37]Idea based on a study by A. L. Ostrosky, Jr. and J. V. Koch, as cited in their book *Introduction to Mathematical Economics* (Waveland Press, Illinois, 1979, p. 133). The data used here are fictitious, however.

[38]Source for data: Nokia financial statements (www.investors.nokia.com).

[39]Source for data: Nintendo annual reports (www.nintendo.com/corp).

[40]Source for data: Wal-Mart annual reports (www.Walmartstores.com/Investors).

[41]Source for data: Target annual reports (investors.target.com).

47. ▼ *Saving for Retirement* You are saving for your retirement by investing $700 per month in an annuity with a guaranteed interest rate of 6% per year. With a continuous stream of investment and continuous compounding, how much will you have accumulated in the annuity by the time you retire in 45 years?

48. ▼ *Saving for College* When your first child is born, you begin to save for college by depositing $400 per month in an account paying 12% interest per year. With a continuous stream of investment and continuous compounding, how much will you have accumulated in the account by the time your child enters college 18 years later?

49. ▼ *Saving for Retirement* You begin saving for your retirement by investing $700 per month in an annuity with a guaranteed interest rate of 6% per year. You increase the amount you invest at the rate of 3% per year. With continuous investment and compounding, how much will you have accumulated in the annuity by the time you retire in 45 years?

50. ▼ *Saving for College* When your first child is born, you begin to save for college by depositing $400 per month in an account paying 12% interest per year. You increase the amount you save by 2% per year. With continuous investment and compounding, how much will have accumulated in the account by the time your child enters college 18 years later?

51. ▼ *Bonds* The U.S. Treasury issued a 30-year bond on February 10, 2011, paying 4.750% interest.[42] Thus, if you bought $100,000 worth of these bonds you would receive $4,750 per year in interest for 30 years. An investor wishes to buy the rights to receive the interest on $100,000 worth of these bonds. The amount the investor is willing to pay is the present value of the interest payments, assuming a 5% rate of return. Assuming (incorrectly, but approximately) that the interest payments are made continuously, what will the investor pay? HINT [See Example 6.]

52. ▼ *Bonds* The Megabucks Corporation is issuing a 20-year bond paying 7% interest. (See the preceding exercise.) An investor wishes to buy the rights to receive the interest on $50,000 worth of these bonds, and seeks a 6% rate of return. Assuming that the interest payments are made continuously, what will the investor pay? HINT [See Example 6.]

53. ▼ *Valuing Future Income* Inga was injured and can no longer work. As a result of a lawsuit, she is to be awarded the present value of the income she would have received over the next 20 years. Her income at the time she was injured was $100,000 per year, increasing by $5,000 per year. What will be the amount of her award, assuming continuous income and a 5% interest rate?

54. ▼ *Valuing Future Income* Max was injured and can no longer work. As a result of a lawsuit, he is to be awarded the present value of the income he would have received over the next 30 years. His income at the time he was injured was $30,000 per year, increasing by $1,500 per year. What will be the amount of his award, assuming continuous income and a 6% interest rate?

COMMUNICATION AND REASONING EXERCISES

55. Complete the following: The future value of a continuous income stream earning 0% interest is the same as the _____ value.

56. Complete the following: The present value of a continuous income stream earning 0% interest is the same as the _____ value.

57. ▼ *Linear Demand* Given a linear demand equation $q = -mp + b$ $(m > 0)$, find a formula for the consumers' surplus at a price level of $\bar{p}$ per unit.

58. ▼ *Linear Supply* Given a linear supply equation of the form $q = mp + b$ $(m > 0)$, find a formula for the producers' surplus at a price level of $\bar{p}$ per unit.

59. ▼ Your study group friend says that the future value of a continuous stream of income is always greater than the total value, assuming a positive rate of return. Is she correct? Why?

60. ▼ Your other study group friend says that the present value of a continuous stream of income can sometimes be greater than the total value, depending on the (positive) interest rate. Is he correct? Explain.

61. ▼ Arrange from smallest to largest: Total Value, Future Value, Present Value of a continuous stream of income (assuming a positive income and positive rate of return).

62. ▼ **a.** Arrange the following functions from smallest to largest $R(t)$, $R(t)e^{r(b-t)}$, $R(t)e^{r(a-t)}$, where $a \le t \le b$, and r and $R(t)$ are positive.
    b. Use the result from part (a) to justify your answers in Exercises 59–61.

---

[42]Source: The Bureau of the Public Debt (www.publicdebt.treas.gov).

14.5 Improper Integrals and Applications

All the definite integrals we have seen so far have had the form $\int_a^b f(x)\,dx$, with a and b finite and $f(x)$ piecewise continuous on the closed interval $[a, b]$. If we relax one or both of these requirements somewhat, we obtain what are called **improper integrals**. There are various types of improper integrals.

Integrals in Which a Limit of Integration Is Infinite

Integrals in which one or more limits of integration are infinite can be written as

$$\int_a^{+\infty} f(x)\,dx, \quad \int_{-\infty}^b f(x)\,dx, \quad \text{or} \quad \int_{-\infty}^{+\infty} f(x)\,dx.$$

Let's concentrate for a moment on the first form, $\int_a^{+\infty} f(x)\,dx$. What does the $+\infty$ mean here? As it often does, it means that we are to take a limit as something gets large. Specifically, it means the limit as the upper bound of integration gets large.

Improper Integral with an Infinite Limit of Integration

We define

$$\int_a^{+\infty} f(x)\,dx = \lim_{M \to +\infty} \int_a^M f(x)\,dx,$$

provided the limit exists. If the limit exists, we say that $\int_a^{+\infty} f(x)\,dx$ **converges**. Otherwise, we say that $\int_a^{+\infty} f(x)\,dx$ **diverges**. Similarly, we define

$$\int_{-\infty}^b f(x)\,dx = \lim_{M \to -\infty} \int_M^b f(x)\,dx,$$

provided the limit exists. Finally, we define

$$\int_{-\infty}^{+\infty} f(x)\,dx = \int_{-\infty}^a f(x)\,dx + \int_a^{+\infty} f(x)\,dx$$

for some convenient a, provided *both* integrals on the right converge.

Quick Examples

1. $\displaystyle \int_1^{+\infty} \frac{dx}{x^2} = \lim_{M \to +\infty} \int_1^M \frac{dx}{x^2} = \lim_{M \to +\infty} \left[-\frac{1}{x}\right]_1^M = \lim_{M \to +\infty} \left(-\frac{1}{M} + 1\right) = 1$
   Converges

2. $\displaystyle \int_1^{+\infty} \frac{dx}{x} = \lim_{M \to +\infty} \int_1^M \frac{dx}{x} = \lim_{M \to +\infty} \left[\ln|x|\right]_1^M = \lim_{M \to +\infty} (\ln M - \ln 1) = +\infty$
   Diverges

3. $\displaystyle \int_{-\infty}^{-1} \frac{dx}{x^2} = \lim_{M \to -\infty} \int_M^{-1} \frac{dx}{x^2} = \lim_{M \to -\infty} \left[-\frac{1}{x}\right]_M^{-1} = \lim_{M \to -\infty} \left(1 + \frac{1}{M}\right) = 1$
   Converges

4. $\displaystyle\int_{-\infty}^{+\infty} e^{-x}\,dx = \int_{-\infty}^{0} e^{-x}\,dx + \int_{0}^{+\infty} e^{-x}\,dx$

$$= \lim_{M\to-\infty}\int_{M}^{0} e^{-x}\,dx + \lim_{M\to+\infty}\int_{0}^{M} e^{-x}\,dx$$

$$= \lim_{M\to-\infty}\;-\,[e^{-x}]_{M}^{0} + \lim_{M\to+\infty}\;-\,[e^{-x}]_{0}^{M}$$

$$= \lim_{M\to-\infty}\,(e^{-M}-1) + \lim_{M\to+\infty}\,(1-e^{-M})$$

$$= +\infty + 1 \qquad\qquad\qquad \text{Diverges}$$

5. $\displaystyle\int_{-\infty}^{+\infty} xe^{-x^2}\,dx = \int_{-\infty}^{0} xe^{-x^2}\,dx + \int_{0}^{+\infty} xe^{-x^2}\,dx$

$$= \lim_{M\to-\infty}\int_{M}^{0} xe^{-x^2}\,dx + \lim_{M\to+\infty}\int_{0}^{M} xe^{-x^2}\,dx$$

$$= \lim_{M\to-\infty}\left[-\frac{1}{2}e^{-x^2}\right]_{M}^{0} + \lim_{M\to+\infty}\left[-\frac{1}{2}e^{-x^2}\right]_{0}^{M}$$

$$= \lim_{M\to-\infty}\left(-\frac{1}{2}+\frac{1}{2}e^{-M^2}\right) + \lim_{M\to+\infty}\left(-\frac{1}{2}e^{-M^2}+\frac{1}{2}\right)$$

$$= -\frac{1}{2}+\frac{1}{2} = 0 \qquad\qquad\qquad \text{Converges}$$

Q : *We learned that the integral can be interpreted as the area under the curve. Is this still true for improper integrals?*

A : Yes. Figure 22 illustrates how we can represent an improper integral as the area of an infinite region.

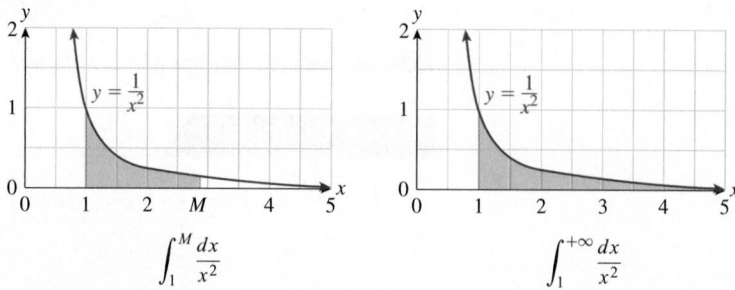

$$\int_{1}^{M}\frac{dx}{x^2} \qquad\qquad\qquad \int_{1}^{+\infty}\frac{dx}{x^2}$$

Figure 22

On the left we see the area represented by $\int_{1}^{M} dx/x^2$. As M gets larger, the integral approaches $\int_{1}^{+\infty} dx/x^2$. In the picture, think of M being moved farther and farther along the x-axis in the direction of increasing x, resulting in the region shown on the right.

Ⓠ : *Wait! We calculated $\int_1^{+\infty} dx/x^2 = 1$. Does this mean that the infinitely long area in Figure 22 has an area of only 1 square unit?*

Ⓐ : That is exactly what it means. If you had enough paint to cover 1 square unit, you would never run out of paint while painting the region in Figure 22. This is one of the places where mathematics seems to contradict common sense. But common sense is notoriously unreliable when dealing with infinities.

Andy Crawford/Dorling Kindersley/ Getty Images

EXAMPLE 1 Future Sales of CDs

By 2009, music downloads were making serious inroads into the sales of physical CDs. Approximately 290 million CD albums were sold in 2009 and sales declined by about 23% per year the following year.[43] Suppose that this rate of decrease were to continue indefinitely and continuously. How many CD albums, total, would be sold from 2009 on?

Solution Recall that the total sales between two dates can be computed as the definite integral of the rate of sales. So, if we wanted the sales between 2009 and a time far in the future, we would compute $\int_0^M s(t)\,dt$ with a large M, where $s(t)$ is the annual sales t years after 2009. Because we want to know the *total* number of CD albums sold from 2009 on, we let $M \to +\infty$; that is, we compute $\int_0^{+\infty} s(t)\,dt$.

Because sales of CD albums are decreasing by 23% per year, we can model $s(t)$ by

$$s(t) = 290(0.77)^t \text{ million CD albums per year,}$$

where t is the number of years since 2009.

$$
\begin{aligned}
\text{Total sales from 2009 on} &= \int_0^{+\infty} 290(0.77)^t\,dt \\
&= \lim_{M \to +\infty} \int_0^M 290(0.77)^t\,dt \\
&= \frac{290}{\ln 0.77} \lim_{M \to +\infty} [(0.77)^t]_0^M \\
&= \frac{290}{\ln 0.77} \lim_{M \to +\infty} (0.77^M - 0.77^0) \\
&= \frac{290}{\ln 0.77}(-1) \\
&\approx 1{,}110 \text{ million CD albums}
\end{aligned}
$$

using Technology

You can estimate the integral in Example 1 with technology by computing $\int_0^M 290(0.77)^t\,dt$ for $M =$ 10, 100, 1000, ... You will find that the resulting values appear to converge to about 1,110. (Stop when the effect of further increases of M has no effect at this level of accuracy.)

TI-83/84 Plus
Y₁=290*0.77^X
Home screen:
fnInt(Y₁,X,0,10)
fnInt(Y₁,X,0,100)
fnInt(Y₁,X,0,1000)

 Website
www.WanerMath.org:
  On-Line Utilities
  → Numerical Integration
    Utility and Grapher
Enter
290*0.77^X
for $f(x)$. Enter 0 and 10 for the lower and upper limits and press "Integral" for the most accurate estimate of the integral.
Repeat with the upper limit set to 100, 1000, and higher.

Integrals in Which the Integrand Becomes Infinite

We can sometimes compute integrals $\int_a^b f(x)\,dx$ in which $f(x)$ becomes infinite. As we'll see in Example 4, the Fundamental Theorem of Calculus does not work for such integrals. The first case to consider is when $f(x)$ approaches $\pm\infty$ at either a or b.

_____
[43]Source: *2010 Year-End Shipment Statistics*, Recording Industry Association of America (www.riaa.com).

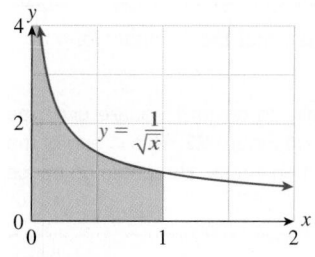

Figure 23

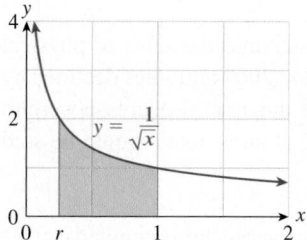

Figure 24

EXAMPLE 2 Integrand Infinite at One Endpoint

Calculate $\displaystyle\int_0^1 \frac{1}{\sqrt{x}}\,dx$.

Solution Notice that the integrand approaches $+\infty$ as x approaches 0 from the right and is not defined at 0. This makes the integral an improper integral. Figure 23 shows the region whose area we are trying to calculate; it extends infinitely vertically rather than horizontally.

Now, if $0 < r < 1$, the integral $\int_r^1 (1/\sqrt{x})\,dx$ is a proper integral because we avoid the bad behavior at 0. This integral gives the area shown in Figure 24. If we let r approach 0 from the right, the area in Figure 24 will approach the area in Figure 23. So, we calculate

$$\int_0^1 \frac{1}{\sqrt{x}}\,dx = \lim_{r\to 0^+}\int_r^1 \frac{1}{\sqrt{x}}\,dx$$

$$= \lim_{r\to 0^+}[2\sqrt{x}]_r^1$$

$$= \lim_{r\to 0^+}(2 - 2\sqrt{r})$$

$$= 2.$$

Thus, we again have an infinitely long region with finite area.

Generalizing, we make the following definition.

Improper Integral in Which the Integrand Becomes Infinite

If $f(x)$ is defined for all x with $a < x \le b$ but approaches $\pm\infty$ as x approaches a, we define

$$\int_a^b f(x)\,dx = \lim_{r\to a^+}\int_r^b f(x)\,dx$$

provided the limit exists. Similarly, if $f(x)$ is defined for all x with $a \le x < b$ but approaches $\pm\infty$ as x approaches b, we define

$$\int_a^b f(x)\,dx = \lim_{r\to b^-}\int_a^r f(x)\,dx$$

provided the limit exists. In either case, if the limit exists, we say that $\int_a^b f(x)\,dx$ **converges**. Otherwise, we say that $\int_a^b f(x)\,dx$ **diverges**.

Note We saw in Chapter 13 that the Fundamental Theorem of Calculus applies to piecewise continuous functions as well as continuous ones. Examples are $f(x) = |x|/x$ and $(x^2 - 1)/(x - 1)$. The integrals of such functions are not improper, and we can use the Fundamental Theorem of Calculus to evaluate such integrals in the usual way. ∎

EXAMPLE 3 Testing for Convergence

Does $\displaystyle\int_{-1}^{3} \frac{x}{x^2 - 9}\, dx$ converge? If so, to what?

Solution We first check to see where, if anywhere, the integrand approaches $\pm\infty$. That will happen where the denominator becomes 0, so we solve $x^2 - 9 = 0$.

$$x^2 - 9 = 0$$
$$x^2 = 9$$
$$x = \pm 3$$

The solution $x = -3$ is outside of the range of integration, so we ignore it. The solution $x = 3$ is, however, the right endpoint of the range of integration, so the integral is improper. We need to investigate the following limit:

$$\int_{-1}^{3} \frac{x}{x^2 - 9}\, dx = \lim_{r \to 3^-} \int_{-1}^{r} \frac{x}{x^2 - 9}\, dx.$$

Now, to calculate the integral we use a substitution:

$$u = x^2 - 9$$
$$\frac{du}{dx} = 2x$$
$$dx = \frac{1}{2x}\, du$$
when $x = r,\, u = r^2 - 9$
when $x = -1,\, u = (-1)^2 - 9 = -8$

Thus,

$$\int_{-1}^{r} \frac{x}{x^2 - 9}\, dx = \int_{-8}^{r^2 - 9} \frac{1}{2u}\, du$$

$$= \frac{1}{2}[\ln|u|]_{-8}^{r^2 - 9}$$

$$= \frac{1}{2}(\ln|r^2 - 9| - \ln 8).$$

Now we take the limit:

$$\int_{-1}^{3} \frac{x}{x^2 - 9}\, dx = \lim_{r \to 3^-} \int_{-1}^{r} \frac{x}{x^2 - 9}\, dx$$

$$= \lim_{r \to 3^-} \frac{1}{2}(\ln|r^2 - 9| - \ln 8)$$

$$= -\infty$$

because, as $r \to 3,\, r^2 - 9 \to 0$, and so $\ln|r^2 - 9| \to -\infty$. Thus, this integral diverges.

EXAMPLE 4 **Integrand Infinite between the Endpoints**

Does $\int_{-2}^{3} \frac{1}{x^2}\,dx$ converge? If so, to what?

Solution Again we check to see if there are any points at which the integrand approaches $\pm\infty$. There is such a point, at $x = 0$. This is between the endpoints of the range of integration. To deal with this we break the integral into two integrals:

$$\int_{-2}^{3} \frac{1}{x^2}\,dx = \int_{-2}^{0} \frac{1}{x^2}\,dx + \int_{0}^{3} \frac{1}{x^2}\,dx.$$

Each integral on the right is an improper integral with the integrand approaching $\pm\infty$ at an endpoint. If both of the integrals on the right converge, we take the sum as the value of the integral on the left. So now we compute

$$\int_{-2}^{0} \frac{1}{x^2}\,dx = \lim_{r \to 0^-} \int_{-2}^{r} \frac{1}{x^2}\,dx$$

$$= \lim_{r \to 0^-} \left[-\frac{1}{x} \right]_{-2}^{r}$$

$$= \lim_{r \to 0^-} \left(-\frac{1}{r} - \frac{1}{2} \right),$$

which diverges to $+\infty$. There is no need now to check $\int_{0}^{3}(1/x^2)\,dx$; because one of the two pieces of the integral diverges, we simply say that $\int_{-2}^{3}(1/x^2)\,dx$ diverges.

➡ **Before we go on...** What if we had been sloppy in Example 4 and had not checked first whether the integrand approached $\pm\infty$ somewhere? Then we probably would have applied the Fundamental Theorem of Calculus and done the following:

$$\int_{-2}^{3} \frac{1}{x^2}\,dx = \left(-\frac{1}{x} \right)_{-2}^{3} = \left(-\frac{1}{3} - \frac{1}{2} \right) = -\frac{5}{6}. \qquad ✗ \quad WRONG!$$

Notice that the answer this "calculation" gives is patently ridiculous. Because $1/x^2 > 0$ for all x for which it is defined, any definite integral of $1/x^2$ must give a positive answer. *Moral:* Always check to see whether the integrand blows up anywhere in the range of integration. If it does, the FTC does not apply, and we must use the methods of this example. ∎

We end with an example of what to do if an integral is improper for more than one reason.

EXAMPLE 5 **An Integral Improper in Two Ways**

Does $\int_{0}^{+\infty} \frac{1}{\sqrt{x}}\,dx$ converge? If so, to what?

Solution This integral is improper for two reasons. First, the range of integration is infinite. Second, the integrand blows up at the endpoint 0. In order to separate these two problems, we break up the integral at some convenient point:

$$\int_0^{+\infty} \frac{1}{\sqrt{x}}\, dx = \int_0^1 \frac{1}{\sqrt{x}}\, dx + \int_1^{+\infty} \frac{1}{\sqrt{x}}\, dx$$

We chose to break the integral at 1. Any positive number would have sufficed, but 1 is generally easier to use in calculations.

The first piece, $\int_0^1 (1/\sqrt{x})\, dx$, we discussed in Example 2; it converges to 2. For the second piece we have

$$\int_1^{+\infty} \frac{1}{\sqrt{x}}\, dx = \lim_{M \to +\infty} \int_1^M \frac{1}{\sqrt{x}}\, dx$$

$$= \lim_{M \to +\infty} [2\sqrt{x}]_1^M$$

$$= \lim_{M \to +\infty} (2\sqrt{M} - 2),$$

which diverges to $+\infty$. Because the second piece of the integral diverges, we conclude that $\int_0^{+\infty} (1/\sqrt{x})\, dx$ diverges.

14.5 EXERCISES

▼ more advanced   ◆ challenging
Ⓣ indicates exercises that should be solved using technology

Note: For some of the exercises in this section you need to assume the fact that $\lim_{M \to +\infty} M^n e^{-M} = 0$ for all $n \geq 0$. (See Exercises 59 and 60 in Section 10.1 and Exercise 103 in Section 10.3.)

In Exercises 1–26, decide whether or not the given integral converges. If the integral converges, compute its value.
HINT [See Quick Examples page 1053.]

1. $\int_1^{+\infty} x\, dx$

2. $\int_0^{+\infty} e^{-x}\, dx$

3. $\int_{-2}^{+\infty} e^{-0.5x}\, dx$

4. $\int_1^{+\infty} \frac{1}{x^{1.5}}\, dx$

5. $\int_{-\infty}^2 e^x\, dx$

6. $\int_{-\infty}^{-1} \frac{1}{x^{1/3}}\, dx$

7. $\int_{-\infty}^{-2} \frac{1}{x^2}\, dx$

8. $\int_{-\infty}^0 e^{-x}\, dx$

9. $\int_0^{+\infty} x^2 e^{-6x}\, dx$

10. $\int_0^{+\infty} (2x - 4)e^{-x}\, dx$

11. $\int_0^5 \frac{2}{x^{1/3}}\, dx$ HINT [See Example 2.]

12. $\int_0^2 \frac{1}{x^2}\, dx$

13. $\int_{-1}^2 \frac{3}{(x+1)^2}\, dx$ HINT [See Example 3.]

14. $\int_{-1}^2 \frac{3}{(x+1)^{1/2}}\, dx$

15. $\int_{-1}^2 \frac{3x}{x^2 - 1}\, dx$ HINT [See Example 4.]

16. $\int_{-1}^2 \frac{3}{x^{1/3}}\, dx$

17. $\int_{-2}^2 \frac{1}{(x+1)^{1/5}}\, dx$

18. $\int_{-2}^2 \frac{2x}{\sqrt{4-x^2}}\, dx$

19. $\int_{-1}^1 \frac{2x}{x^2 - 1}\, dx$

20. $\int_{-1}^2 \frac{2x}{x^2 - 1}\, dx$

21. $\int_{-\infty}^{+\infty} xe^{-x^2}\, dx$

22. $\int_{-\infty}^{\infty} xe^{1-x^2}\, dx$

23. $\int_0^{+\infty} \frac{1}{x \ln x}\, dx$ HINT [See Example 5.]

24. $\int_0^{+\infty} \ln x\, dx$

25. ▼ $\int_0^{+\infty} \frac{2x}{x^2 - 1}\, dx$

26. ▼ $\int_{-\infty}^0 \frac{2x}{x^2 - 1}\, dx$

⊞ *In Exercises 27–34, use technology to approximate the given integrals with M = 10, 100, 1,000, . . . and hence decide whether the associated improper integral converges and estimate its value to four significant digits if it does.* HINT [See the Technology Note for Example 1.]

27. $\int_1^M \frac{1}{x^2}\, dx$

28. $\int_0^M e^{-x^2}\, dx$

29. $\int_0^M \frac{x}{1+x}\, dx$

30. $\int_{1/M}^1 \frac{1}{\sqrt{x}}\, dx$

31. $\int_{1+1/M}^2 \frac{1}{\sqrt{x-1}}\, dx$

32. $\int_1^M \frac{1}{x}\, dx$

33. $\int_0^{1-1/M} \frac{1}{(1-x)^2}\, dx$

34. $\int_0^{2-1/M} \frac{1}{(2-x)^3}\, dx$

APPLICATIONS

35. *New Home Sales* Sales of new homes in the United States decreased dramatically from 2005 to 2010 as shown in the model

$$n(t) = 1.33e^{-0.299t} \text{ million homes per year} \quad (0 \le t \le 5),$$

where *t* is the year since 2005.[44] If this trend were to have continued into the indefinite future, estimate the total number of new homes that would have been sold in the United States from 2005 on. HINT [See Example 1.]

36. *Revenue from New Home Sales* Revenue from the sale of new homes in the United States decreased dramatically from 2005 to 2010 as shown in the model

$$r(t) = 412e^{-0.323t} \text{ billion dollars per year} \quad (0 \le t \le 5),$$

where *t* is the year since 2005.[45] If this trend were to have continued into the indefinite future, estimate the total revenue from the sale of new homes in the United States from 2005 on. HINT [See Example 1.]

37. *Cigarette Sales* According to data published by the Federal Trade Commission, the number of cigarettes sold domestically has been decreasing by about 3% per year from the 2000 total of about 415 billion.[46] Use an exponential model to forecast the total number of cigarettes sold from 2000 on. (Round your answer to the nearest 100 billion cigarettes.) HINT [Use a model of the form Ab^t.]

38. *Sales* Sales of the text *Calculus and You* have been declining continuously at a rate of 5% per year. Assuming that

Calculus and You currently sells 5,000 copies per year and that sales will continue this pattern of decline, calculate total future sales of the text. HINT [Use a model of the form Ae^{rt}.]

39. ▼ *Sales* My financial adviser has predicted that annual sales of Frodo T-shirts will continue to decline by 10% each year. At the moment, I have 3,200 of the shirts in stock and am selling them at a rate of 200 per year. Will I ever sell them all?

40. ▼ *Revenue* Alarmed about the sales prospects for my Frodo T-shirts (see the preceding exercise), I will try to make up lost revenues by increasing the price by $1 each year. I now charge $10 per shirt. What is the total amount of revenue I can expect to earn from sales of my T-shirts, assuming the sales levels described in the previous exercise? (Give your answer to the nearest $1,000.)

41. ▼ *Education* Let $N(t)$ be the number of high school students graduated in the United States in year *t*. This number has been changing at a rate of about

$$N'(t) = 0.22t^{-0.91} \text{ million graduates per year} \quad (2 \le t \le 9),$$

where *t* is time in years since 2000.[47] In 2002, there were about 2.7 million high school students graduated. By extrapolating the model, what can you say about the number of high school students graduated in a year far in the future?

42. ▼ *Education, Martian* Let $M(t)$ be the number of high school students graduated in the Republic of Mars in year *t*. This number is projected to change at a rate of about

$$M'(t) = 0.321t^{-1.10} \text{ thousand graduates per year} \quad (1 \le t \le 50),$$

where *t* is time in years since 2020. In 2021, there were about 1,300 high school students graduated. By extrapolating the model, what can you say about the number of high school students graduated in a year far in the future?

43. ▼ *Cellphone Revenues* The number of cellphone subscribers in China in the early 2000s was projected to follow the equation,[48]

$$N(t) = 39t + 68 \text{ million subscribers}$$

in year $t(t = 0$ represents 2000). The average annual revenue per cellphone user was $350 in 2000.

a. Assuming that, due to competition, the revenue per cellphone user decreases continuously at an annual rate of 10%, give a formula for the annual revenue in year *t*.

b. Using the model you obtained in part (a) as an estimate of the rate of change of total revenue, estimate the total revenue from 2000 into the indefinite future.

[44]Based on new home sales data from the U.S. Census Bureau (www.census.gov/const/www/newressalesindex.html).

[45]*Ibid.*

[46]Source for data: Federal Trade Commission Cigarette Report for 2007 and 2008, issued July 2011 (www.ftc.gov).

[47]Based on a regression model. Source for Data: U.S. Department of Education, National Center for Education Statistics, *Digest of Education Statistics: 2010,* April 2011 (nces.ed.gov).

[48]Based on a regression of projected figures (coefficients are rounded). Source: Intrinsic Technology/*New York Times*, Nov. 24, 2000, p. C1.

44. ▼ *Vidphone Revenues* The number of vidphone subscribers in the Republic of Mars for the period 2200–2300 was projected to follow the equation

$$N(t) = 18t - 10 \text{ thousand subscribers}$$

in year t ($t = 0$ represents 2200). The average annual revenue per vidphone user was $\bar{\bar{Z}}40$ in 2200.[49]

a. Assuming that, due to competition, the revenue per vidphone user decreases continuously at an annual rate of 20%, give a formula for the annual revenue in year t.

b. Using the model you obtained in part (a) as an estimate of the rate of change of total revenue, estimate the total revenue from 2200 into the indefinite future.

45. ▣ *Development Assistance* According to data published by the World Bank, development assistance to low-income countries from 2000 through 2008 was approximately

$$q(t) = 0.2t^2 + 3.5t + 60 \text{ billion dollars per year}$$

where t is time in years since 2000.[50] Assuming a worldwide inflation rate of 3% per year, and that the above model remains accurate into the indefinite future, find the value of all development assistance to low-income countries from 2000 on in constant dollars. (The constant dollar value of $q(t)$ dollars t years from now is given by $q(t)e^{-rt}$, where r is the fractional rate of inflation. Give your answer to the nearest $100 billion.) HINT [See Technology Note for Example 1.]

46. ▣ *Humanitarian Aid* Repeat the preceding exercise, using the following model for humanitarian aid.[51]

$$q(t) = 0.07t^2 + 0.3t + 5 \text{ billion dollars per year}$$

47. ▼ *Hair Mousse Sales* The amount of extremely popular hair mousse sold online at your Web site can be approximated by

$$N(t) = \frac{80(7)^t}{20 + 7^t} \text{ million gallons per year.}$$

($t = 0$ represents the current year.) Investigate the integrals $\int_0^{+\infty} N(t)\, dt$ and $\int_{-\infty}^0 N(t)\, dt$ and interpret your answers.

48. ▼ *Chocolate Mousse Sales* The weekly demand for your company's *Lo-Cal Chocolate Mousse* is modeled by the equation

$$q(t) = \frac{50e^{2t-1}}{1 + e^{2t-1}} \text{ gallons per week,}$$

where t is time from now in weeks. Investigate the integrals $\int_0^{+\infty} q(t)\, dt$ and $\int_{-\infty}^0 q(t)\, dt$ and interpret your answers.

▣ *The Normal Curve* Exercises 49–52 require the use of a graphing calculator or computer programmed to do numerical integration. The normal distribution curve, which models the distributions of data in a wide range of applications, is given by the function

$$p(x) = \frac{1}{\sqrt{2\pi}\,\sigma} e^{-(x-\mu)^2 / 2\sigma^2},$$

where $\pi = 3.14159265\ldots$ and σ and μ are constants called the standard deviation and the mean, respectively. Its graph (for $\sigma = 1$ and $\mu = 2$) is shown in the figure.

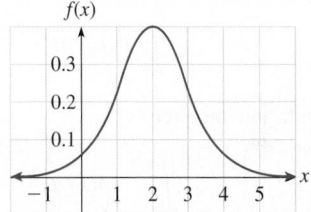

49. ▼ With $\sigma = 4$ and $\mu = 1$, approximate $\int_{-\infty}^{+\infty} p(x)\, dx$. HINT [See Example 5 and Technology Note for Example 1.]

50. ▼ With $\sigma = 1$ and $\mu = 0$, approximate $\int_0^{+\infty} p(x)\, dx$.

51. ▼ With $\sigma = 1$ and $\mu = 0$, approximate $\int_1^{+\infty} p(x)\, dx$.

52. ▼ With $\sigma = 1$ and $\mu = 0$, approximate $\int_{-\infty}^1 p(x)\, dx$.

53. ◆ *Variable Sales* The value of your Chateau Petit Mont Blanc 1963 vintage burgundy is increasing continuously at an annual rate of 40%, and you have a supply of 1,000 bottles worth $85 each at today's prices. In order to ensure a steady income, you have decided to sell your wine at a diminishing rate—starting at 500 bottles per year, and then decreasing this figure continuously at a fractional rate of 100% per year. How much income (to the nearest dollar) can you expect to generate by this scheme? HINT [Use the formula for continuously compounded interest.]

54. ◆ *Panic Sales* Unfortunately, your large supply of Chateau Petit Mont Blanc is continuously turning to vinegar at a fractional rate of 60% per year! You have thus decided to sell off your Petit Mont Blanc at $50 per bottle, but the market is a little thin, and you can only sell 400 bottles per year. Because you have no way of knowing which bottles now contain vinegar until they are opened, you shall have to give refunds for all the bottles of vinegar. What will your net income be before all the wine turns to vinegar?

55. ◆ *Meteor Impacts* The frequency of meteor impacts on Earth can be modeled by

$$n(k) = \frac{1}{5.6997k^{1.081}}$$

where $n(k) = N'(k)$, and $N(k)$ is the average number of meteors of energy less than or equal to k megatons that will

---

[49]The zonar ($\bar{\bar{Z}}$) is the official currency in the city-state of Utarek, Mars. Source: www.Marsnext.com, a now extinct virtual society.

[50]The authors' approximation, based on data from the World Bank, obtained from www.worldbank.org.

[51]*Ibid.*

hit the Earth in one year.[52] (A small nuclear bomb releases on the order of one megaton of energy.)

a. How many meteors of energy at least $k = 0.2$ hit the Earth each year?

b. Investigate and interpret the integral $\int_0^1 n(k)\, dk$.

56. ◆ **Meteor Impacts** (continuing the previous exercise)

a. Explain why the integral

$$\int_a^b kn(k)\, dk$$

computes the total energy released each year by meteors with energies between a and b megatons.

b. Compute and interpret

$$\int_0^1 kn(k)\, dk.$$

c. Compute and interpret

$$\int_1^{+\infty} kn(k)\, dk.$$

57. ◆ **The Gamma Function** The gamma function is defined by the formula

$$\Gamma(x) = \int_0^{+\infty} t^{x-1} e^{-t}\, dt.$$

a. Find $\Gamma(1)$ and $\Gamma(2)$.

b. Use integration by parts to show that for every positive integer n, $\Gamma(n+1) = n\Gamma(n)$.

c. Deduce that $\Gamma(n) = (n-1)![= (n-1)(n-2)\cdots 2 \cdot 1]$ for every positive integer n.

58. ◆ **Laplace Transforms** The Laplace transform $F(x)$ of a function $f(t)$ is given by the formula

$$F(x) = \int_0^{+\infty} f(t)e^{-xt}\, dt \quad (x > 0).$$

a. Find $F(x)$ for $f(t) = 1$ and for $f(t) = t$.

b. Find a formula for $F(x)$ if $f(t) = t^n$ ($n = 1, 2, 3, \ldots$).

c. Find a formula for $F(x)$ if $f(t) = e^{at}$ (a constant).

---

[52]The authors' model, based on data published by NASA International Near-Earth-Object Detection Workshop (*The New York Times*, Jan. 25, 1994, p. C1).

COMMUNICATION AND REASONING EXERCISES

59. Why can't the Fundamental Theorem of Calculus be used to evaluate $\int_{-1}^{1} \dfrac{1}{x}\, dx$?

60. Why can't the Fundamental Theorem of Calculus be used to evaluate $\int_{1}^{+\infty} \dfrac{1}{x^2}\, dx$?

61. It sometimes happens that the Fundamental Theorem of Calculus gives the correct answer for an improper integral. Does the FTC give the correct answer for improper integrals of the form

$$\int_{-a}^{a} \frac{1}{x^{1/r}}\, dx$$

if $r = 3, 5, 7, \ldots$?

62. Does the FTC give the correct answer for improper integrals of the form

$$\int_{-a}^{a} \frac{1}{x^r}\, dx$$

if $r = 3, 5, 7, \ldots$?

63. Which of the following integrals are improper, and why? (Do not evaluate any of them.)

a. $\displaystyle\int_{-1}^{1} \frac{|x|}{x}\, dx$ **b.** $\displaystyle\int_{-1}^{1} x^{-1/3}\, dx$

c. $\displaystyle\int_{0}^{2} \frac{x-2}{x^2 - 4x + 4}\, dx$

64. Which of the following integrals are improper, and why? (Do not evaluate any of them.)

a. $\displaystyle\int_{-1}^{1} \frac{|x-1|}{x-1}\, dx$ **b.** $\displaystyle\int_{0}^{1} \frac{1}{x^{2/3}}\, dx$

c. $\displaystyle\int_{0}^{2} \frac{x^2 - 4x + 4}{x-2}\, dx$

65. 🔲 ▼ How could you use technology to approximate improper integrals? (Your discussion should refer to each type of improper integral.)

66. 🔲 ▼ Use technology to approximate the integrals $\int_0^M e^{-(x-10)^2}\, dx$ for larger and larger values of M, using Riemann sums with 500 subdivisions. What do you find? Comment on the answer.

67. ▼ Make up an interesting application whose solution is $\int_{10}^{+\infty} 100te^{-0.2t}\, dt = \$1,015.01$.

68. ▼ Make up an interesting application whose solution is $\int_{100}^{+\infty} \frac{1}{r^2}\, dr = 0.01$.

14.6 Differential Equations and Applications

A **differential equation** is an equation that involves a derivative of an unknown function. A **first-order differential equation** involves only the first derivative of the unknown function. A **second-order differential equation** involves the second derivative of the unknown function (and possibly the first derivative). Higher order differential equations are defined similarly. In this book, we will deal only with first-order differential equations.

To **solve** a differential equation means to find the unknown function. Many of the laws of science and other fields describe how things change. When expressed mathematically, these laws take the form of equations involving derivatives—that is, differential equations. The field of differential equations is a large and very active area of study in mathematics, and we shall see only a small part of it in this section.

EXAMPLE 1 **Motion**

A dragster accelerates from a stop so that its speed t seconds after starting is $40t$ ft/sec. How far will the car go in 8 seconds?

Solution We wish to find the car's position function $s(t)$. We are told about its speed, which is ds/dt. Precisely, we are told that

$$\frac{ds}{dt} = 40t.$$

This is the differential equation we have to solve to find $s(t)$. But we already know how to solve this kind of differential equation; we integrate:

$$s(t) = \int 40t \, dt = 20t^2 + C.$$

We now have the **general solution** to the differential equation. By letting C take on different values, we get all the possible solutions. We can specify the one **particular solution** that gives the answer to our problem by imposing the **initial condition** that $s(0) = 0$. Substituting into $s(t) = 20t^2 + C$, we get

$$0 = s(0) = 20(0)^2 + C = C$$

so $C = 0$ and $s(t) = 20t^2$. To answer the question, the car travels $20(8)^2 = 1,280$ feet in 8 seconds.

We did not have to work hard to solve the differential equation in Example 1. In fact, any differential equation of the form $dy/dx = f(x)$ can (in theory) be solved by integrating. (Whether we can actually carry out the integration is another matter!)

Simple Differential Equations

A **simple** differential equation has the form

$$\frac{dy}{dx} = f(x).$$

Its general solution is

$$y = \int f(x)\, dx.$$

Quick Example

The differential equation

$$\frac{dy}{dx} = 2x^2 - 4x^3$$

is simple and has general solution

$$y = \int f(x)\, dx = \frac{2x^3}{3} - x^4 + C.$$

Not all differential equations are simple, as the next example shows.

EXAMPLE 2 Separable Differential Equation

Consider the differential equation $\dfrac{dy}{dx} = \dfrac{x}{y^2}$.

a. Find the general solution.

b. Find the particular solution that satisfies the initial condition $y(0) = 2$.

Solution

a. This is not a simple differential equation because the right-hand side is a function of both x and y. We cannot solve this equation by just integrating; the solution to this problem is to "separate" the variables.

Step 1: *Separate the variables algebraically.* We rewrite the equation as

$$y^2\, dy = x\, dx.$$

Step 2: *Integrate both sides.*

$$\int y^2\, dy = \int x\, dx$$

giving

$$\frac{y^3}{3} = \frac{x^2}{2} + C$$

Step 3: *Solve for the dependent variable.* We solve for y:

$$y^3 = \frac{3}{2}x^2 + 3C = \frac{3}{2}x^2 + D$$

(Rewriting $3C$ as D, an equally arbitrary constant), so

$$y = \left(\frac{3}{2}x^2 + D\right)^{1/3}.$$

This is the general solution of the differential equation.

b. We now need to find the value for D that will give us the solution satisfying the condition $y(0) = 2$. Substituting 0 for x and 2 for y in the general solution, we get

$$2 = \left(\frac{3}{2}(0)^2 + D\right)^{1/3} = D^{1/3}$$

so

$$D = 2^3 = 8.$$

Thus, the particular solution we are looking for is

$$y = \left(\frac{3}{2}x^2 + 8\right)^{1/3}.$$

➡ **Before we go on...** We can check the general solution in Example 2 by calculating both sides of the differential equation and comparing.

$$\frac{dy}{dx} = \frac{d}{dx}\left(\frac{3}{2}x^2 + D\right)^{1/3} = x\left(\frac{3}{2}x^2 + D\right)^{-2/3}$$

$$\frac{x}{y^2} = \frac{x}{\left(\frac{3}{2}x^2 + 8\right)^{2/3}} = x\left(\frac{3}{2}x^2 + D\right)^{-2/3} \qquad ✔ \qquad ■$$

Q : *In Example 2, we wrote $y^2\,dy$ and $x\,dx$. What do they mean?*

A : Although it is possible to give meaning to these symbols, for us they are just a notational convenience. We could have done the following instead:

$$y^2\frac{dy}{dx} = x.$$

Now we integrate both sides with respect to x.

$$\int y^2\frac{dy}{dx}\,dx = \int x\,dx$$

We can use substitution to rewrite the left-hand side:

$$\int y^2\frac{dy}{dx}\,dx = \int y^2\,dy,$$

which brings us back to the equation

$$\int y^2\,dy = \int x\,dx.$$

We were able to separate the variables in the preceding example because the right-hand side, x/y^2, was a *product* of a function of x and a function of y—namely,

$$\frac{x}{y^2} = x\left(\frac{1}{y^2}\right).$$

In general, we can say the following:

Separable Differential Equation

A **separable** differential equation has the form

$$\frac{dy}{dx} = f(x)g(y).$$

We solve a separable differential equation by separating the xs and the ys algebraically, writing

$$\frac{1}{g(y)}\,dy = f(x)\,dx$$

and then integrating:

$$\int \frac{1}{g(y)}\,dy = \int f(x)\,dx.$$

EXAMPLE 3 **Rising Medical Costs**

Spending on Medicare from 2010 to 2021 was projected to rise continuously at an instantaneous rate of 5.6% per year.[53] Find a formula for Medicare spending y as a function of time t in years since 2010.

Solution When we say that Medicare spending y was going up continuously at an instantaneous rate of 5.6% per year, we mean that

the instantaneous rate of increase of y was 5.6% of its value

or $\dfrac{dy}{dt} = 0.056y.$

This is a separable differential equation. Separating the variables gives

$$\frac{1}{y}dy = 0.056\,dt.$$

Integrating both sides, we get

$$\int \frac{1}{y}\,dy = \int 0.056\,dt$$

so $\ln y = 0.056t + C.$

(We should write $\ln|y|$, but we know that the medical costs are positive.) We now solve for y.

$$y = e^{0.056t+C} = e^C e^{0.056t} = Ae^{0.056t},$$

where A is a positive constant. This is the formula we used before for continuous percentage growth.

---

[53]Spending is in constant 2010 dollars. Source for projected data: Congressional Budget Office, *March 2011 Medicare Baseline* (www.cbo.gov).

➡ **Before we go on...** To determine A in Example 3 we need to know, for example, Medicare spending at time $t = 0$ (the initial condition). The source cited gives Medicare spending as \$525.6 billion in 2010. Substituting $t = 0$ in the equation above gives

$$525.6 = Ae^0 = A.$$

Thus, projected Medicare spending is

$$y = 525.6e^{0.056t} \text{ billion dollars}$$

t years after 2010. ∎

EXAMPLE 4 Newton's Law of Cooling

Newton's Law of Cooling states that a hot object cools at a rate proportional to the difference between its temperature and the temperature of the surrounding environment (the **ambient temperature**). If a hot cup of coffee, at 170°F, is left to sit in a room at 70°F, how will the temperature of the coffee change over time?

Solution We let $H(t)$ denote the temperature of the coffee at time t. Newton's Law of Cooling tells us that $H(t)$ *decreases* at a rate proportional to the difference between $H(t)$ and 70°F, the ambient temperature. In other words,

$$\frac{dH}{dt} = -k(H - 70),$$

* When we say that a quantity Q is *proportional* to a quantity R, we mean that $Q = kR$ for some constant k. The constant k is referred to as the **constant of proportionality**.

where k is some positive constant.* Note that $H \geq 70$: The coffee will never cool to less than the ambient temperature.

The variables here are H and t, which we can separate as follows:

$$\frac{dH}{H - 70} = -k\,dt.$$

Integrating, we get

$$\int \frac{dH}{H - 70} = \int (-k)\,dt$$

so $\ln(H - 70) = -kt + C.$

(Note that $H - 70$ is positive, so we don't need absolute values.) We now solve for H:

$$\begin{aligned} H - 70 &= e^{-kt+C} \\ &= e^C e^{-kt} \\ &= Ae^{-kt} \end{aligned}$$

so

$$H(t) = 70 + Ae^{-kt},$$

where A is some positive constant. We can determine the constant A using the initial condition $H(0) = 170$:

$$170 = 70 + Ae^0 = 70 + A$$

so $A = 100.$

Therefore,

$$H(t) = 70 + 100e^{-kt}.$$

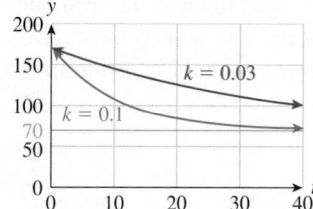

Figure 25

Q : *But what is k?*

A : The constant k determines the rate of cooling. Its value depends on the units of time we are using, on the substance cooling—in this case the coffee—and on its container. Because k depends so heavily on the particular circumstances, it's usually easiest to determine it experimentally. Figure 25 shows two possible graphs, one with $k = 0.1$ and the other with $k = 0.03$ ($k \approx 0.03$ would be reasonable for a cup of coffee in a polystyrene container with t measured in minutes).

In any case, we can see from the graph or the formula for $H(t)$ that the temperature of the coffee will approach the ambient temperature exponentially.

➡ **Before we go on...** Notice that the calculation in Example 4 shows that the temperature of an object cooling according to Newton's Law is given in general by

$$H(t) = T_a + (T_0 - T_a)e^{-kt},$$

where T_a is the ambient temperature (70° in the example) and T_0 is the initial temperature (170° in the example). The formula also holds if the ambient temperature is higher than the initial temperature ("Newton's Law of Heating"). ∎

14.6 EXERCISES

▼ more advanced   ◆ challenging
T indicates exercises that should be solved using technology

Find the general solution of each differential equation in Exercises 1–10. Where possible, solve for y as a function of x.

1. $\dfrac{dy}{dx} = x^2 + \sqrt{x}$

HINT [See Quick Example page 1064.]

2. $\dfrac{dy}{dx} = \dfrac{1}{x} + 3$

HINT [See Quick Example page 1064.]

3. $\dfrac{dy}{dx} = \dfrac{x}{y}$

HINT [See Example 2(a).]

4. $\dfrac{dy}{dx} = \dfrac{y}{x}$

HINT [See Example 2(a).]

5. $\dfrac{dy}{dx} = xy$

6. $\dfrac{dy}{dx} = x^2 y$

7. $\dfrac{dy}{dx} = (x+1)y^2$

8. $\dfrac{dy}{dx} = \dfrac{1}{(x+1)y^2}$

9. $x\dfrac{dy}{dx} = \dfrac{1}{y} \ln x$

10. $\dfrac{1}{x}\dfrac{dy}{dx} = \dfrac{1}{y} \ln x$

For each differential equation in Exercises 11–20, find the particular solution indicated. HINT [See Example 2(b).]

11. $\dfrac{dy}{dx} = x^3 - 2x$; $y = 1$ when $x = 0$

12. $\dfrac{dy}{dx} = 2 - e^{-x}$; $y = 0$ when $x = 0$

13. $\dfrac{dy}{dx} = \dfrac{x^2}{y^2}$; $y = 2$ when $x = 0$

14. $\dfrac{dy}{dx} = \dfrac{y^2}{x^2}$; $y = \dfrac{1}{2}$ when $x = 1$

15. $x\dfrac{dy}{dx} = y$; $y(1) = 2$

16. $x^2\dfrac{dy}{dx} = y$; $y(1) = 1$

17. $\dfrac{dy}{dx} = x(y+1)$; $y(0) = 0$ **18.** $\dfrac{dy}{dx} = \dfrac{y+1}{x}$; $y(1) = 2$

19. $\dfrac{dy}{dx} = \dfrac{xy^2}{x^2+1}$; $y(0) = -1$ **20.** $\dfrac{dy}{dx} = \dfrac{xy}{(x^2+1)^2}$; $y(0) = 1$

APPLICATIONS

21. *Sales* Your monthly sales of Green Tea Ice Cream are falling at an instantaneous rate of 5% per month. If you currently sell 1,000 quarts per month, find the differential equation describing your change in sales, and then solve it to predict your monthly sales. HINT [See Example 3.]

22. *Profit* Your monthly profit on sales of Avocado Ice Cream is rising at an instantaneous rate of 10% per month. If you currently make a profit of $15,000 per month, find the differential equation describing your change in profit, and solve it to predict your monthly profits. HINT [See Example 3.]

23. *Newton's Law of Cooling* For coffee in a ceramic cup, suppose $k \approx 0.05$ with time measured in minutes. **(a)** Use Newton's Law of Cooling to predict the temperature of the coffee, initially at a temperature of 200°F, that is left to sit in a room at 75°F. **(b)** When will the coffee have cooled to 80°F? HINT [See Example 4.]

24. *Newton's Law of Cooling* For coffee in a paper cup, suppose $k \approx 0.08$ with time measured in minutes. **(a)** Use Newton's Law of Cooling to predict the temperature of the coffee, initially at a temperature of 210°F, that is left to sit in a room at 60°F. **(b)** When will the coffee have cooled to 70°F? HINT [See Example 4.]

25. *Cooling* A bowl of clam chowder at 190°F is placed in a room whose air temperature is 75°F. After 10 minutes, the soup has cooled to 150°F. Find the value of k in Newton's Law of Cooling, and hence find the temperature of the chowder as a function of time.

26. *Heating* Suppose that a pie, at 20°F, is put in an oven at 350°F. After 15 minutes, its temperature has risen to 80°F. Find the value of k in Newton's Law of Heating (see the note after Example 4), and hence find the temperature of the pie as a function of time.

27. *Market Saturation* You have just introduced a new 3D monitor to the market. You predict that you will eventually sell 100,000 monitors and that your monthly rate of sales will be 10% of the difference between the saturation value of 100,000 and the total number you have sold up to that point. Find a differential equation for your total sales (as a function of the month) and solve. (What are your total sales at the moment when you first introduce the monitor?)

28. *Market Saturation* Repeat the preceding exercise, assuming that monthly sales will be 5% of the difference between the saturation value (of 100,000 monitors) and the total sales to that point, and assuming that you sell 5,000 monitors to corporate customers before placing the monitor on the open market.

29. *Determining Demand* *Nancy's Chocolates* estimates that the elasticity of demand for its dark chocolate truffles is $E = 0.05p - 1.5$ where p is the price per pound. Nancy's sells 20 pounds of truffles per week when the price is $20 per pound. Find the formula expressing the demand q as a function of p. Recall that the elasticity of demand is given by

$$E = -\frac{dq}{dp} \times \frac{p}{q}.$$

30. *Determining Demand* *Nancy's Chocolates* estimates that the elasticity of demand for its chocolate strawberries is $E = 0.02p - 0.5$ where p is the price per pound. It sells 30 pounds of chocolate strawberries per week when the price is $30 per pound. Find the formula expressing the demand q as a function of p. Recall that the elasticity of demand is given by

$$E = -\frac{dq}{dp} \times \frac{p}{q}.$$

Linear Differential Equations *Exercises 31–36 are based on first-order linear differential equations with constant coefficients. These have the form*

$$\frac{dy}{dt} + py = f(t) \quad (p \text{ constant})$$

and the general solution is

$$y = e^{-pt} \int f(t)e^{pt}\, dt. \quad (\textit{Check this by substituting!})$$

31. Solve the linear differential equation

$$\frac{dy}{dt} + y = e^{-t}; y = 1 \text{ when } t = 0.$$

32. Solve the linear differential equation

$$\frac{dy}{dt} - y = e^{2t}; y = 2 \text{ when } t = 0.$$

33. Solve the linear differential equation

$$2\frac{dy}{dt} - y = 2t; y = 1 \text{ when } t = 0.$$

HINT [First rewrite the differential equation in the form $\frac{dy}{dt} + py = f(t)$.]

34. Solve the linear differential equation

$$2\frac{dy}{dt} + y = -t. \quad y = 1 \text{ when } t = 0$$

HINT [First rewrite the differential equation in the form $\frac{dy}{dt} + py = f(t)$.]

35. ▼ ***Electric Circuits*** The flow of current $i(t)$ in an electric circuit without capacitance satisfies the linear differential equation

$$L\frac{di}{dt} + Ri = V(t),$$

where L and R are constants (the *inductance* and *resistance*, respectively) and $V(t)$ is the applied voltage. (See figure.)

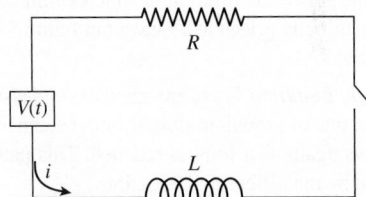

If the voltage is supplied by a 10-volt battery and the switch is turned on at time $t = 1$, then the voltage V is a step function that jumps from 0 to 10 at $t = 1$: $V(t) = 5\left[1 + \frac{|t - 1|}{t - 1}\right]$. Find the current as a function of time for $L = R = 1$. Use a grapher to plot the resulting current as a function of time. (Assume there is no current flowing at time $t = 0$.) [Use the following integral formula: $\int\left[1 + \frac{|t - 1|}{t - 1}\right]e^t\, dt = \left[1 + \frac{|t - 1|}{t - 1}\right](e^t - e) + C.$]

36. ▼ ***Electric Circuits*** Repeat Exercise 35 for $L = 1$, $R = 5$, and $V(t) = 5\left[1 + \frac{|t - 2|}{t - 2}\right]$. (The switch flipped on at time $t = 2$.) [Use the following integral formula: $\int\left[1 + \frac{|t - 2|}{t - 2}\right]e^{5t}\, dt = \left(1 + \frac{|t - 2|}{t - 2}\right)\left(\frac{e^{5t} - e^{10}}{5}\right) + C.$]

1070 Chapter 14 Further Integration Techniques and Applications of the Integral

37. ▼ Approach to Equilibrium The *Extrasoft Toy Co.* has just released its latest creation, a plush platypus named "Eggbert." The demand function for Eggbert dolls is $D(p) = 50,000 - 500p$ dolls per month when the price is p dollars. The supply function is $S(p) = 30,000 + 500p$ dolls per month when the price is p dollars. This makes the equilibrium price $20. The **Evans price adjustment model** assumes that if the price is set at a value other than the equilibrium price, it will change over time in such a way that its rate of change is proportional to the shortage $D(p) - S(p)$.

a. Write the differential equation given by the Evans price adjustment model for the price p as a function of time.
b. Find the general solution of the differential equation you wrote in (a). (You will have two unknown constants, one being the constant of proportionality.)
c. Find the particular solution in which Eggbert dolls are initially priced at $10 and the price rises to $12 after one month.

38. ▼ Approach to Equilibrium *Spacely Sprockets* has just released its latest model, the Dominator. The demand function is $D(p) = 10,000 - 1,000p$ sprockets per year when the price is p dollars. The supply function is $S(p) = 8,000 + 1,000p$ sprockets per year when the price is p dollars.

a. Using the Evans price adjustment model described in the preceding exercise, write the differential equation for the price $p(t)$ as a function of time.
b. Find the general solution of the differential equation you wrote in (a).
c. Find the particular solution in which Dominator sprockets are initially priced at $5 each but fall to $3 each after one year.

39. ▼ Logistic Equation There are many examples of growth in which the rate of growth is slow at first, becomes faster, and then slows again as a limit is reached. This pattern can be described by the differential equation

$$\frac{dy}{dt} = ay(L - y),$$

where a is a constant and L is the limit of y. Show by substitution that

$$y = \frac{CL}{e^{-aLt} + C}$$

is a solution of this equation, where C is an arbitrary constant.

40. ▼ Logistic Equation Using separation of variables and integration with a table of integrals or a symbolic algebra program, solve the differential equation in the preceding exercise to derive the solution given there.

⊤ *Exercises 41–44 require the use of technology.*

41. ▼ Market Saturation You have just introduced a new model of Blu-ray disc player. You predict that the market will saturate at 2,000,000 Blu-ray disc players and that your total sales will be governed by the equation

$$\frac{dS}{dt} = \frac{1}{4}S(2 - S),$$

where S is the total sales in millions of Blu-ray disc players and t is measured in months. If you give away 1,000 Blu-ray disc players when you first introduce them, what will S be? Sketch the graph of S as a function of t. About how long will it take to saturate the market? (See Exercise 39.)

42. ▼ Epidemics A certain epidemic of influenza is predicted to follow the function defined by

$$\frac{dA}{dt} = \frac{1}{10}A(20 - A),$$

where A is the number of people infected in millions and t is the number of months after the epidemic starts. If 20,000 cases are reported initially, find $A(t)$ and sketch its graph. When is A growing fastest? How many people will eventually be affected? (See Exercise 39.)

43. ▼ Growth of Tumors The growth of tumors in animals can be modeled by the Gompertz equation:

$$\frac{dy}{dt} = -ay \ln\left(\frac{y}{b}\right),$$

where y is the size of a tumor, t is time, and a and b are constants that depend on the type of tumor and the units of measurement.

a. Solve for y as a function of t.
b. If $a = 1$, $b = 10$, and $y(0) = 5$ cm³ (with t measured in days), find the specific solution and graph it.

44. ▼ Growth of Tumors Refer back to the preceding exercise. Suppose that $a = 1$, $b = 10$, and $y(0) = 15$ cm³. Find the specific solution and graph it. Comparing its graph to the one obtained in the preceding exercise, what can you say about tumor growth in these instances?

COMMUNICATION AND REASONING EXERCISES

45. What is the difference between a particular solution and the general solution of a differential equation? How do we get a particular solution from the general solution?

46. Why is there always an arbitrary constant in the general solution of a differential equation? Why are there not two or more arbitrary constants in a first-order differential equation?

47. ▼ Show by example that a **second-order** differential equation, one involving the second derivative y'', usually has two arbitrary constants in its general solution.

48. ▼ Find a differential equation that is not separable.

49. ▼ Find a differential equation whose general solution is $y = 4e^{-x} + 3x + C$.

50. ▼ Explain how, knowing the elasticity of demand as a function of either price or demand, you may find the demand equation. (See Exercise 29.)

CHAPTER 14 REVIEW

KEY CONCEPTS

 Website www.WanerMath.com
Go to the Website at www.WanerMath
.com to find a comprehensive and
interactive Web-based summary
of Chapter 14.

14.1 Integration by Parts

Integration-by-parts formula:

$$\int u \cdot v\, dx = u \cdot I(v) - \int D(u)I(v)\, dx$$

p. 1016

Tabular method for integration by parts
p. 1017

Integrating a polynomial times a
logarithm *p. 1020*

14.2 Area between Two Curves and Applications

If $f(x) \geq g(x)$ for all x in $[a, b]$, then
the area of the region between the
graphs of f and g and between $x = a$
and $x = b$ is given by

$$A = \int_a^b [f(x) - g(x)]\, dx \quad p.\ 1025$$

Regions enclosed by crossing curves
p. 1027

Area enclosed by two curves *p. 1029*

General instructions for finding the area
between the graphs of $f(x)$ and $g(x)$
p. 1029

Approximating the area between two
curves using technology:

$$A = \int_a^b |f(x) - g(x)|\, dx \quad p.\ 1030$$

14.3 Averages and Moving Averages

Average, or mean, of a collection of
values

$$\bar{y} = \frac{y_1 + y_2 + \cdots + y_n}{n} \quad p.\ 1034$$

The *average*, or *mean*, of a function
$f(x)$ on an interval $[a, b]$ is

$$\bar{f} = \frac{1}{b - a} \int_a^b f(x)\, dx. \quad p.\ 1035$$

Average balance *p. 1035*

Computing the moving average of a set
of data *p. 1036*

n-Unit moving average of a function:

$$\bar{f}(x) = \frac{1}{n} \int_{x-n}^x f(t)\, dt \quad p.\ 1037$$

Computing moving averages of
sawtooth and step functions *p. 1038*

14.4 Applications to Business and Economics: Consumers' and Producers' Surplus and Continuous Income Streams

Consumers' surplus:

$$CS = \int_0^{\bar{q}} [D(q) - \bar{p}]\, dq \quad p.\ 1043$$

Producers' surplus:

$$PS = \int_0^{\bar{q}} [\bar{p} - S(q)]\, dq \quad p.\ 1045$$

Equilibrium price *p. 1046*

Social gain $= CS + PS$ *p. 1047*

Total value of a continuous income

$$\text{stream: } TV = \int_a^b R(t)\, dt \quad p.\ 1048$$

Future value of a continuous income

$$\text{stream: } FV = \int_a^b R(t)e^{r(b-t)}\, dt$$

p. 1049

Present value of a continuous income

$$\text{stream: } PV = \int_a^b R(t)e^{r(a-t)}\, dt$$

p. 1049

14.5 Improper Integrals and Applications

Improper integral with an infinite limit
of integration:

$$\int_a^{+\infty} f(x)\, dx, \quad \int_{-\infty}^b f(x)\, dx,$$

$$\int_{-\infty}^{+\infty} f(x)\, dx \quad p.\ 1053$$

Improper integral in which the
integrand becomes infinite *p. 1056*

Testing for convergence *p. 1057*

Integrand infinite between the endpoints
p. 1058

Integral improper in two ways *p. 1058*

14.6 Differential Equations and Applications

Simple differential equations:

$$\frac{dy}{dx} = f(x) \quad p.\ 1063$$

Separable differential equations:

$$\frac{dy}{dx} = f(x)g(y) \quad p.\ 1066$$

Newton's Law of Cooling *p. 1067*

REVIEW EXERCISES

Evaluate the integrals in Exercises 1–10.

1. $\int (x^2 + 2)e^x\, dx$

2. $\int (x^2 - x)e^{-3x+1}\, dx$

3. $\int x^2 \ln(2x)\, dx$

4. $\int \log_5 x\, dx$

5. $\int 2x|2x + 1|\, dx$

6. $\int 3x|-x + 5|\, dx$

7. $\int 5x \frac{|-x + 3|}{-x + 3}\, dx$

8. $\int 2x \frac{|3x + 1|}{3x + 1}\, dx$

9. $\int_{-2}^2 (x^3 + 1)e^{-x}\, dx$

10. $\int_1^e x^2 \ln x\, dx$

In Exercises 11–14, find the areas of the given regions.

11. Between $y = x^3$ and $y = 1 - x^3$ for x in $[0, 1]$

12. Between $y = e^x$ and $y = e^{-x}$ for x in $[0, 2]$

13. Enclosed by $y = 1 - x^2$ and $y = x^2$

14. Between $y = x$ and $y = xe^{-x}$ for x in $[0, 2]$

In Exercises 15–18, find the average value of the given function over the indicated interval.

15. $f(x) = x^3 - 1$ over $[-2, 2]$

16. $f(x) = \dfrac{x}{x^2 + 1}$ over $[0, 1]$

17. $f(x) = x^2 e^x$ over $[0, 1]$

18. $f(x) = (x + 1) \ln x$ over $[1, 2e]$

1071

In Exercises 19–22, find the 2-unit moving averages of the given function.

19. $f(x) = 3x + 1$ **20.** $f(x) = 6x^2 + 12$

21. $f(x) = x^{4/3}$ **22.** $f(x) = \ln x$

In Exercises 23 and 24, calculate the consumers' surplus at the indicated unit price $\bar{p}$ for the given demand equation.

23. $p = 50 - \dfrac{1}{2}q; \bar{p} = 10$ **24.** $p = 10 - q^{1/2}; \bar{p} = 4$

In Exercises 25 and 26, calculate the producers' surplus at the indicated unit price $\bar{p}$ for the given supply equation.

25. $p = 50 + \dfrac{1}{2}q; \bar{p} = 100$ **26.** $p = 10 + q^{1/2}; \bar{p} = 40$

In Exercises 27–32, decide whether the given integral converges. If the integral converges, compute its value.

27. $\displaystyle\int_1^\infty \frac{1}{x^5}\, dx$ **28.** $\displaystyle\int_0^1 \frac{1}{x^5}\, dx$

29. $\displaystyle\int_{-1}^1 \frac{x}{(x^2-1)^{5/3}}\, dx$ **30.** $\displaystyle\int_0^2 \frac{x}{(x^2-1)^{1/3}}\, dx$

31. $\displaystyle\int_0^{+\infty} 2xe^{-x^2}\, dx$ **32.** $\displaystyle\int_0^{+\infty} x^2 e^{-6x^3}\, dx$

Solve the differential equations in Exercises 33–36.

33. $\dfrac{dy}{dx} = x^2 y^2$ **34.** $\dfrac{dy}{dx} = xy + 2x$

35. $xy\dfrac{dy}{dx} = 1; y(1) = 1$

36. $y(x^2+1)\dfrac{dy}{dx} = xy^2; y(0) = 2$

APPLICATIONS: OHaganBooks.com

37. *Spending on Stationery* Alarmed by the volume of pointless memos and reports being copied and circulated by management at OHaganBooks.com, John O'Hagan ordered a 5-month audit of paper usage at the company. He found that management consumed paper at a rate of

$q(t) = 45t + 200$ thousand sheets per month $(0 \le t \le 5)$

(t is the time in months since the audit began). During the same period, the price of paper was escalating; the company was charged approximately

$$p(t) = 9e^{0.09t} \text{ dollars per thousand sheets.}$$

Use an integral to estimate, to the nearest hundred dollars, the total spent on paper for management during the given period.

38. *Spending on Shipping* During the past 10 months, OHaganBooks.com shipped orders at a rate of about

$q(t) = 25t + 3,200$ packages per month $(0 \le t \le 10)$

(t is the time in months since the beginning of the year). During the same period, the cost of shipping a package averaged approximately

$$p(t) = 4e^{0.04t} \text{ dollars per package.}$$

Use an integral to estimate, to the nearest thousand dollars, the total spent on shipping orders during the given period.

39. *Education Costs* Billy-Sean O'Hagan, having graduated *summa cum laude* from college, has been accepted by the doctoral program in biophysics at Oxford. John O'Hagan estimates that the total cost (minus scholarships) he will need to pay is $2,000 per month, but that this cost will escalate at a continuous compounding rate of 1% per month.

a. What, to the nearest dollar, is the average monthly cost over the course of two years?

b. Find the four-month moving average of the monthly cost.

40. *Investments* OHaganBooks.com keeps its cash reserves in a hedge fund paying 6% compounded continuously. It starts a year with $1 million in reserves and does not withdraw or deposit any money.

a. What is the average amount it will have in the fund over the course of two years?

b. Find the one-month moving average of the amount it has in the fund.

41. *Consumers' and Producers' Surplus* Currently, the hottest selling item at OHaganBooks.com is *Mensa for Dummies*[54] with a demand curve of $q = 20,000(28 - p)^{1/3}$ books per week, and a supply curve of $q = 40,000(p - 19)^{1/3}$ books per week.

a. Find the equilibrium price and demand.

b. Find the consumers' and producers' surpluses at the equilibrium price.

42. *Consumers' and Producers' Surplus* OHaganBooks.com is about to start selling a new coffee table book, *Computer Designs of the Late Twentieth Century*. It estimates the demand curve to be $q = 1,000\sqrt{200 - 2p}$, and its willingness to order books from the publisher is given by the supply curve $q = 1,000\sqrt{10p - 400}$.

a. Find the equilibrium price and demand.

b. Find the consumers' and producers' surpluses at the equilibrium price.

43. *Revenue* Sales of the bestseller *A River Burns through It* are dropping at OHaganBooks.com. To try to bolster sales, the company is dropping the price of the book, now $40, at a rate of $2 per week. As a result, this week OHaganBooks.com will sell 5,000 copies, and it estimates that sales will fall continuously at a rate of 10% per week. How much revenue will it earn on sales of this book over the next 8 weeks?

44. *Foreign Investments* Panicked by the performance of the U.S. stock market, Marjory Duffin is investing her 401(k) money in a Russian hedge fund at a rate of approximately

$$q(t) = 1.7t^2 - 0.5t + 8 \text{ thousand shares per month,}$$

---

[54]The actual title is: *Let Us Just Have A Ball! Mensa for Dummies,* by Wendu Mekbib, Silhouette Publishing Corporation.

where t is time in months since the stock market began to plummet. At the time she started making the investments, the hedge fund was selling for $1 per share, but subsequently declined in value at a continuous rate of 5% per month. What was the total amount of money Marjory Duffin invested after one year? (Answer to the nearest $1,000.)

45. Investments OHaganBooks.com CEO John O'Hagan has started a gift account for the Marjory Duffin Foundation. The account pays 6% compounded continuously and is initially empty. OHaganBooks.com deposits money continuously into it, starting at the rate of $100,000 per month and increasing continuously by $10,000 per month.

   a. How much money will the company have in the account at the end of two years?

   b. How much of the amount you found in part (a) was principal deposited and how much was interest earned? (Round answers to the nearest $1,000.)

46. Savings John O'Hagan had been saving money for Billy-Sean's education since he had been a wee lad. O'Hagan began depositing money at the rate of $1,000 per month and increased his deposits continuously by $50 per month. If the account earned 5% compounded continuously and O'Hagan continued these deposits for 15 years,

   a. How much money did he accumulate?

   b. How much was money deposited and how much was interest?

47. Acquisitions The Megabucks Corporation is considering buying OHaganBooks.com. It estimates OHaganBooks.com's revenue stream at $50 million per year, growing continuously at a 10% rate. Assuming interest rates of 6%, how much is OHaganBooks.com's revenue for the next year worth now?

48. More Acquisitions OHaganBooks.com is thinking of buying JungleBooks and would like to recoup its investment after three years. The estimated net profit for JungleBooks is $40 million per year, growing linearly by $5 million per year. Assuming interest rates of 4%, how much should OHaganBooks.com pay for JungleBooks?

49. Incompetence OHaganBooks.com is shopping around for a new bank. A junior executive at one bank offers it the following interesting deal: The bank will pay OHaganBooks.com interest continuously at a rate numerically equal to 0.01% of the square of the amount of money it has in the account at any time. By considering what would happen if $10,000 was deposited in such an account, explain why the junior executive was fired shortly after this offer was made.

50. Shrewd Bankers The new junior officer at the bank (who replaced the one fired in the preceding exercise) offers OHaganBook.com the following deal for the $800,000 they plan to deposit: While the amount in the account is less than $1 million, the bank will pay interest continuously at a rate equal to 10% of the difference between $1 million and the amount of money in the account. When it rises over $1 million, the bank will pay interest of 20%. Why should OHaganBooks.com not take this offer?

Case Study   Estimating Tax Revenues

✱ To simplify our discussion, we are assuming that (1) all tax revenues are based on earned income and that (2) everyone in the population we consider earns some income.

You have just been hired by the incoming administration of your country as chief consultant for national tax policy, and you have been getting conflicting advice from the finance experts on your staff. Several of them have come up with plausible suggestions for new tax structures, and your job is to choose the plan that results in more revenue for the government.

Before you can evaluate their plans, you realize that it is essential to know your country's income distribution—that is, how many people earn how much money.✱ You might think that the most useful way of specifying income distribution would be to use a function that gives the exact number $f(x)$ of people who earn a given salary x. This would necessarily be a discrete function—it makes sense only if x happens to be a whole number of cents. There is, after all, no one earning a salary of exactly $22,000.142567! Furthermore, this function would behave rather erratically, because there are, for example, probably many more people making a salary of exactly $30,000 than exactly $30,000.01. Given these problems, it is far more convenient to start with the function defined by

$$N(x) = \textit{the total number of people earning between } 0 \textit{ and } x \textit{ dollars.}$$

Actually, you would want a "smoothed" version of this function. The graph of $N(x)$ might look like the one shown in Figure 26.

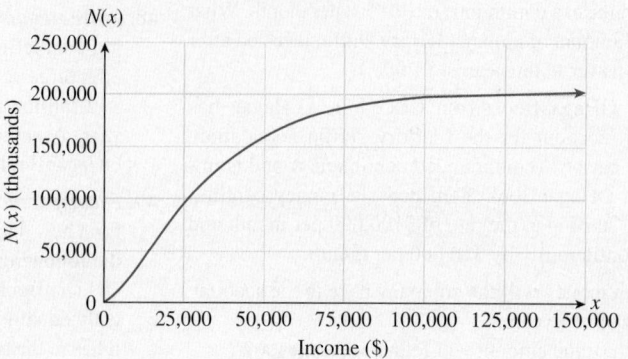

Figure 26

If we take the *derivative* of $N(x)$, we get an income distribution function. Its graph might look like the one shown in Figure 27.

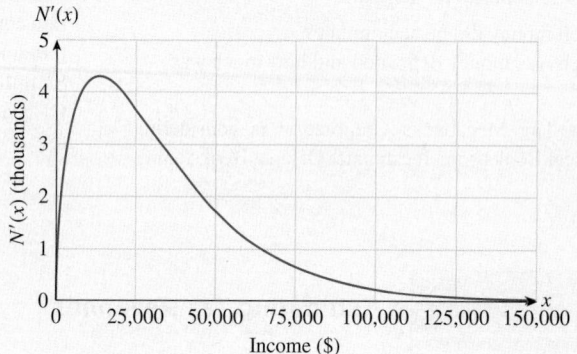

Figure 27

* A very similar idea is used in probability. See the optional chapter "Calculus Applied to Probability and Statistics" on the Website.

† Gamma distributions are often good models for income distributions. The one used in the text is the authors' approximation of the income distribution in the United States in 2009. Source for data: U.S. Census Bureau, Current Population Survey, 2010 Annual Social and Economic Supplement. (www.census.gov).

Because the derivative measures the rate of change, its value at x is the additional number of taxpayers per \$1 increase in salary. Thus, the fact that $N'(25,000) \approx 3,700$ tells us that approximately 3,700 people are earning a salary of between \$25,000 and \$25,001. In other words, N' shows the distribution of incomes among the population—hence, the name "distribution function."*

You thus send a memo to your experts requesting the income distribution function for the nation. After much collection of data, they tell you that the income distribution function is

$$N'(x) = 14x^{0.672}e^{-x/20,400}.$$

This is in fact the function whose graph is shown in Figure 27 and is an example of a **gamma distribution.**† (You might find it odd that you weren't given the original function N, but it will turn out that you don't need it. How would you compute it?)

Given this income distribution, your financial experts have come up with the two possible tax policies illustrated in Figures 28 and 29.

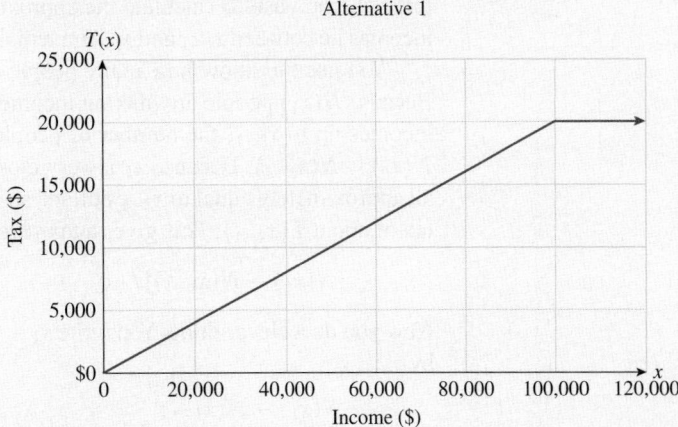

Figure 28

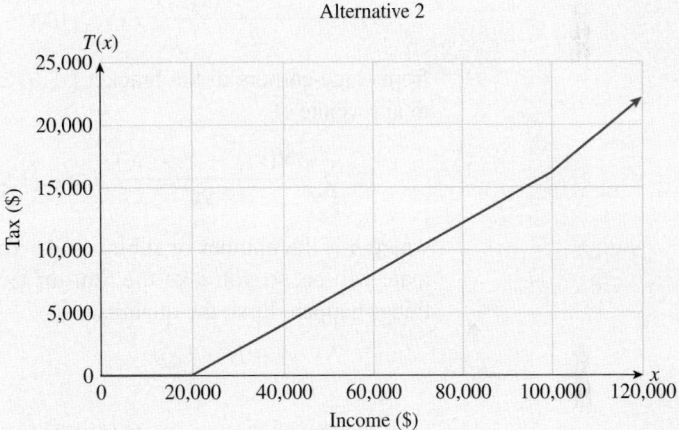

Figure 29

In the first alternative, all taxpayers pay 20% of their income in taxes, except that no one pays more than $20,000 in taxes. In the second alternative, there are three tax brackets, described by the following table:

Income	Marginal tax rate
$0–20,000	0%
$20,000–100,000	20%
Above $100,000	30%

Now you must determine which alternative will generate more annual tax revenue.

Each of Figures 28 and 29 is the graph of a function, T. Rather than using the formulas for these particular functions, you begin by working with the general situation. You have an income distribution function N' and a tax function T, both functions of annual income. You need to find a formula for total tax revenues. First you decide to use a cutoff so that you need to work only with incomes in some finite interval $[0, M]$; you might use, for example, $M = \$10$ million. (Later you will let M approach $+\infty$.) Next, you subdivide the interval $[0, M]$ into a large number of intervals of small width, Δx. If $[x_{k-1}, x_k]$ is a typical such interval, you wish to calculate the approximate tax revenue from people whose total incomes lie between x_{k-1} and x_k. You will then sum over k to get the total revenue.

You need to know how many people are making incomes between x_{k-1} and x_k. Because $N(x_k)$ people are making incomes *up to* x_k and $N(x_{k-1})$ people are making incomes up to x_{k-1}, the number of people making incomes between x_{k-1} and x_k is $N(x_k) - N(x_{k-1})$. Because x_k is very close to x_{k-1}, the incomes of these people are all approximately equal to x_{k-1} dollars, so each of these taxpayers is paying an annual tax of about $T(x_{k-1})$. This gives a tax revenue of

$$[N(x_k) - N(x_{k-1})]T(x_{k-1}).$$

Now you do a clever thing. You write $x_k - x_{k-1} = \Delta x$ and replace $N(x_k) - N(x_{k-1})$ by

$$\frac{N(x_k) - N(x_{k-1})}{\Delta x}\Delta x.$$

This gives you a tax revenue of about

$$\frac{N(x_k) - N(x_{k-1})}{\Delta x}T(x_{k-1})\Delta x$$

from wage-earners in the bracket $[x_{k-1}, x_k]$. Summing over k gives an approximate total revenue of

$$\sum_{k=1}^{n} \frac{N(x_k) - N(x_{k-1})}{\Delta x}T(x_{k-1})\Delta x,$$

where n is the number of subintervals. The larger n is, the more accurate your estimate will be, so you take the limit of the sum as $n \to \infty$. When you do this, two things happen. First, the quantity

$$\frac{N(x_k) - N(x_{k-1})}{\Delta x}$$

approaches the derivative, $N'(x_{k-1})$. Second, the sum, which you recognize as a Riemann sum, approaches the integral

$$\int_0^M N'(x)T(x)\,dx.$$

You now take the limit as $M \to +\infty$ to get

$$\text{Total tax revenue} = \int_0^{+\infty} N'(x)T(x)\,dx.$$

This improper integral is fine in theory, but the actual calculation will have to be done numerically, so you stick with the upper limit of $10 million for now. You will have to check that it is reasonable at the end. (Notice that, by the graph of N', it appears that extremely few, if any, people earn that much.) Now you already have a formula for $N'(x)$, but you still need to write formulas for the tax functions $T(x)$ for both alternatives.

Alternative 1 The graph in Figure 28 rises linearly from 0 to 20,000 as x ranges from 0 to 100,000, and then stays constant at 20,000. The slope of the first part is $20,000/100,000 = 0.2$. The taxation function is therefore

$$T_1(x) = \begin{cases} 0.2x & \text{if } 0 \le x < 100,000 \\ 20,000 & \text{if } x \ge 100,000 \end{cases}.$$

> **Quick Examples**
>
> 1. $f(x, y) = 3x - 5y$ Linear function of x and y
> 2. $C(x, y) = 10,000 + 20x + 40y$ Example 1
> 3. $S(x_1, x_2, x_3) = 13,005 + 230x_1 + 18x_2 + 102x_3$ Example 2

Interaction Function

If we add to a linear function one or more terms of the form $bx_i x_j$ (b a nonzero constant and $i \neq j$), we get a **second-order interaction function**.

> **Quick Examples**
>
> 1. $C(x, y) = 10,000 + 20x + 40y + 0.1xy$
> 2. $R(x_1, x_2, x_3) = 10,000 - 0.01x_1 - 0.02x_2 - 0.01x_3 + 0.00001x_2 x_3$

So far, we have been specifying functions of several variables **algebraically**—by using algebraic formulas. If you have ever studied statistics, you are probably familiar with statistical tables. These tables may also be viewed as representing functions **numerically**, as the next example shows.

EXAMPLE 3 Function Represented Numerically: Body Mass Index

The following table lists some values of the "body mass index," which gives a measure of the massiveness of your body, taking height into account.* The variable w represents your weight in pounds, and h represents your height in inches. An individual with a body mass index of 25 or above is generally considered overweight.

* It is interesting that weight-lifting competitions are usually based on weight, rather than body mass index. As a consequence, taller people are at a significant disadvantage in these competitions because they must compete with shorter, stockier people of the same weight. (An extremely thin, very tall person can weigh as much as a muscular short person, although his body mass index would be significantly lower.)

$w \rightarrow$

h $\downarrow$	130	140	150	160	170	180	190	200	210
60	25.2	27.1	29.1	31.0	32.9	34.9	36.8	38.8	40.7
61	24.4	26.2	28.1	30.0	31.9	33.7	35.6	37.5	39.4
62	23.6	25.4	27.2	29.0	30.8	32.7	34.5	36.3	38.1
63	22.8	24.6	26.4	28.1	29.9	31.6	33.4	35.1	36.9
64	22.1	23.8	25.5	27.2	28.9	30.7	32.4	34.1	35.8
65	21.5	23.1	24.8	26.4	28.1	29.7	31.4	33.0	34.7
66	20.8	22.4	24.0	25.6	27.2	28.8	30.4	32.0	33.6
67	20.2	21.8	23.3	24.9	26.4	28.0	29.5	31.1	32.6
68	19.6	21.1	22.6	24.1	25.6	27.2	28.7	30.2	31.7
69	19.0	20.5	22.0	23.4	24.9	26.4	27.8	29.3	30.8
70	18.5	19.9	21.4	22.8	24.2	25.6	27.0	28.5	29.9
71	18.0	19.4	20.8	22.1	23.5	24.9	26.3	27.7	29.1
72	17.5	18.8	20.2	21.5	22.9	24.2	25.6	26.9	28.3
73	17.0	18.3	19.6	20.9	22.3	23.6	24.9	26.2	27.5
74	16.6	17.8	19.1	20.4	21.7	22.9	24.2	25.5	26.7
75	16.1	17.4	18.6	19.8	21.1	22.3	23.6	24.8	26.0
76	15.7	16.9	18.1	19.3	20.5	21.7	22.9	24.2	25.4

As the table shows, the value of the body mass index depends on two quantities: w and h. Let us write $M(w, h)$ for the body mass index function. What are $M(140, 62)$ and $M(210, 63)$?

Solution We can read the answers from the table:

$$M(140, 62) = 25.4 \qquad {\scriptstyle w = 140 \text{ lb}, \, h = 62 \text{ in}}$$

and $\quad M(210, 63) = 36.9. \qquad {\scriptstyle w = 210 \text{ lb}, \, h = 63 \text{ in}}$

The function $M(w, h)$ is actually given by the formula

$$M(w, h) = \frac{0.45w}{(0.0254h)^2}.$$

[The factor 0.45 converts the weight to kilograms, and 0.0254 converts the height to meters. If w is in kilograms and h is in meters, the formula is simpler: $M(w, h) = w/h^2$.]

Geometric Viewpoint: Three-Dimensional Space and the Graph of a Function of Two Variables

Just as functions of a single variable have graphs, so do functions of two or more variables. Recall that the graph of $f(x)$ consists of all points $(x, f(x))$ in the xy-plane. By analogy, we would like to say that the graph of a function of *two* variables, $f(x, y)$, consists of all points of the form $(x, y, f(x, y))$. Thus, we need three axes: the x-, y-, and z-axes. In other words, our graph will live in **three-dimensional space**, or **3-space**.*

> * If we were dealing instead with a function of *three* variables, then we would need to go to *four-dimensional* space. Here we run into visualization problems (to say the least!) so we won't discuss the graphs of functions of three or more variables in this text.

Just as we had two mutually perpendicular axes in two-dimensional space (the xy-plane; see Figure 4(a)), so we have three mutually perpendicular axes in three-dimensional space (Figure 4(b)).

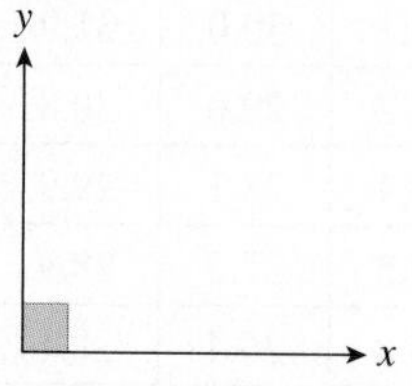

Two-dimensional space
Figure 4(a)

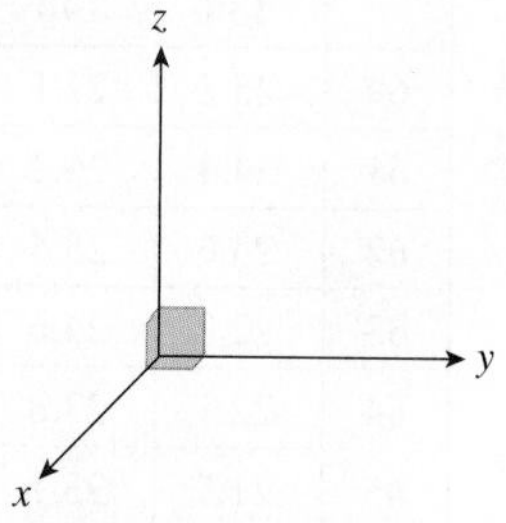

Three-dimensional space
Figure 4(b)

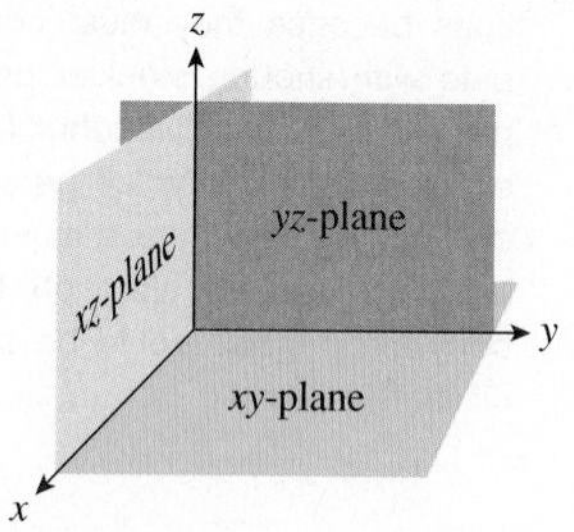

Figure 5

In both 2-space and 3-space, the axis labeled with the last letter goes up. Thus, the z-direction is the "up" direction in 3-space, rather than the y-direction.

Three important planes are associated with these axes: the xy-plane, the yz-plane, and the xz-plane. These planes are shown in Figure 5. Any two of these planes intersect in one of the axes (for example, the xy- and xz-planes intersect in the x-axis) and all three meet at the origin. Notice that the xy-plane consists of all points with z-coordinate zero, the xz-plane consists of all points with $y = 0$, and the yz-plane consists of all points with $x = 0$.

In 3-space, each point has *three* coordinates, as you might expect: the x-coordinate, the y-coordinate, and the z-coordinate. To see how this works, look at the following examples.

The z-coordinate of a point is its height above the xy-plane.

Figure 7

EXAMPLE 4 Plotting Points in Three Dimensions

Locate the points $P(1, 2, 3)$, $Q(-1, 2, 3)$, $R(1, -1, 0)$, and $S(1, 2, -2)$ in 3-space.

Solution To locate P, the procedure is similar to the one we used in 2-space: Start at the origin, proceed 1 unit in the x direction, then proceed 2 units in the y direction, and finally, proceed 3 units in the z direction. We wind up at the point P shown in Figures 6(a) and 6(b).

Here is another, extremely useful way of thinking about the location of P. First, look at the x- and y-coordinates, obtaining the point $(1, 2)$ in the xy-plane. The point we want is then 3 units vertically above the point $(1, 2)$ because the z-coordinate of a point is just its height above the xy-plane. This strategy is shown in Figure 6(c).

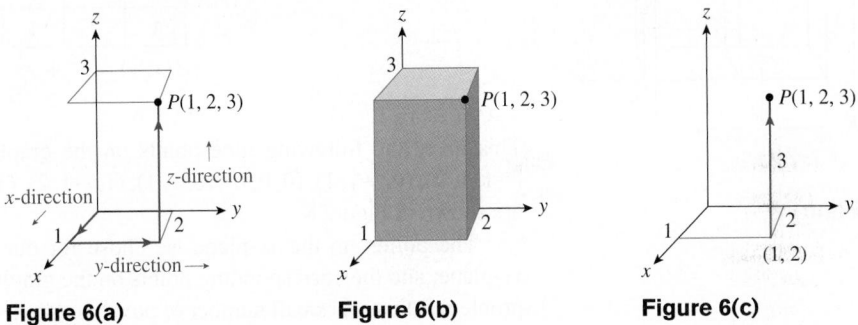

Figure 6(a)   **Figure 6(b)**   **Figure 6(c)**

Plotting the points Q, R, and S is similar, using the convention that negative coordinates correspond to moves back, left, or down. (See Figure 7.)

Our next task is to describe the graph of a function $f(x, y)$ of two variables.

Graph of a Function of Two Variables

The **graph of the function f of two variables** is the set of all points $(x, y, f(x, y))$ in three-dimensional space, where we restrict the values of (x, y) to lie in the domain of f. In other words, the graph is the set of all the points (x, y, z) with $z = f(x, y)$.

Note For *every* point (x, y) in the domain of f, the z-coordinate of the corresponding point on the graph is given by evaluating the function at (x, y). Thus, there will be a point of the graph on the vertical line through *every* point in the domain of f, so that the graph is usually a *surface* of some sort (see the figure).

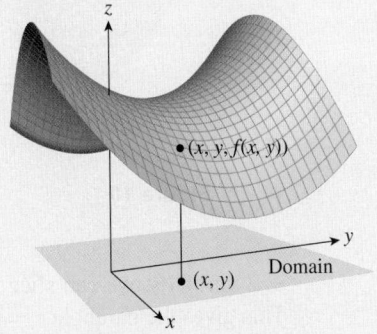

EXAMPLE 5 **Graph of a Function of Two Variables**

Describe the graph of $f(x, y) = x^2 + y^2$.

Solution Your first thought might be to make a table of values. You could choose some values for x and y and then, for each such pair, calculate $z = x^2 + y^2$. For example, you might get the following table:

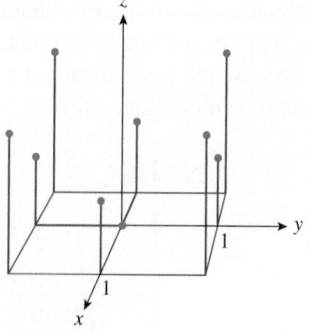

Figure 8

		$x \rightarrow$		
		−1	**0**	**1**
y ↓	**−1**	2	1	2
	0	1	0	1
	1	2	1	2

$$f(x, y) = x^2 + y^2$$

This gives the following nine points on the graph of f: $(-1, -1, 2)$, $(-1, 0, 1)$, $(-1, 1, 2)$, $(0, -1, 1)$, $(0, 0, 0)$, $(0, 1, 1)$, $(1, -1, 2)$, $(1, 0, 1)$, and $(1, 1, 2)$. These points are shown in Figure 8.

The points on the xy-plane we chose for our table are the grid points in the xy-plane, and the corresponding points on the graph are marked with solid dots. The problem is that this small number of points hardly tells us what the surface looks like, and even if we plotted more points, it is not clear that we would get anything more than a mass of dots on the page.

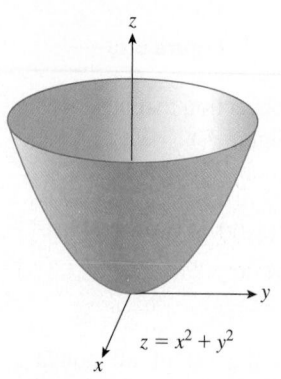

Figure 9

What can we do? There are several alternatives. One place to start is to use technology to draw the graph. (See the technology note on the next page.) We then obtain something like Figure 9. This particular surface is called a **paraboloid**.

If we slice vertically through this surface along the yz-plane, we get the picture in Figure 10. The shape of the front edge, where we cut, is a parabola. To see why, note that the yz-plane is the set of points where $x = 0$. To get the intersection of $x = 0$ and $z = x^2 + y^2$, we substitute $x = 0$ in the second equation, getting $z = y^2$. This is the equation of a parabola in the yz-plane.

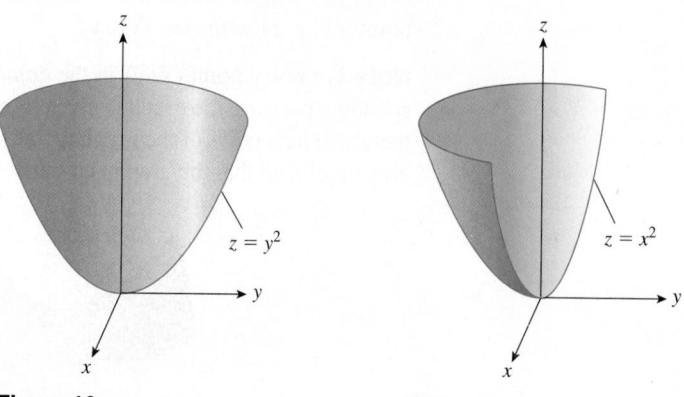

Figure 10   **Figure 11**

Similarly, we can slice through the surface with the xz-plane by setting $y = 0$. This gives the parabola $z = x^2$ in the xz-plane (Figure 11).

We can also look at horizontal slices through the surface, that is, slices by planes parallel to the xy-plane. These are given by setting $z = c$ for various numbers c. For example, if we set $z = 1$, we will see only the points with height 1. Substituting in the equation $z = x^2 + y^2$ gives the equation

$$1 = x^2 + y^2,$$

* See Section 0.7 for a discussion of equations of circles.

which is the equation of a circle of radius 1.* If we set $z = 4$, we get the equation of a circle of radius 2:

$$4 = x^2 + y^2.$$

using Technology

We can use technology to obtain the graph of the function in Example 5:

Spreadsheet
Table of values:
x-values −3 to 3 in B1–H1
y-values −3 to 3 in A2–A8
=B1^2+A2^2
in B2; copy down and across through H8.
Graph: Highlight A1 through H8 and insert a Surface chart. [More details on page 1148.]

Website
www.WanerMath.com
  Online Utilities
  → Surface Graphing Utility

Enter x^2+y^2 for $f(x, y)$
Set xMin = −3, xMax = 3, yMin = −3, yMax = 3
Press "Graph".

In general, if we slice through the surface at height $z = c$, we get a circle (of radius $\sqrt{c}$). Figure 12 shows several of these circles.

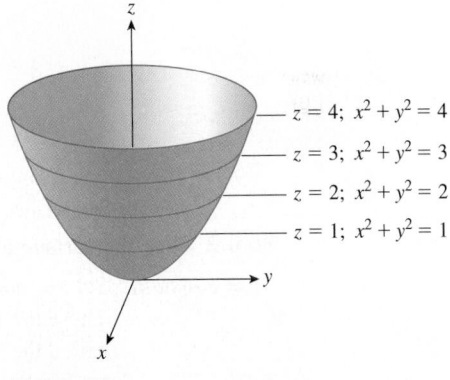

Figure 12

Looking at these circular slices, we see that this surface is the one we get by taking the parabola $z = x^2$ and spinning it around the z-axis. This is an example of what is known as a **surface of revolution**.

➡ **Before we go on...** The graph of any function of the form $f(x, y) = Ax^2 + By^2 + Cxy + Dx + Ey + F$ ($A, B, \ldots, F$ constants), with $4AB - C^2$ positive, can be shown to be a paraboloid of the same general shape as that in Example 5 if A and B are positive, or upside-down if A and B are negative. If $A \neq B$, the horizontal slices will be ellipses rather than circles.

Notice that each horizontal slice through the surface in Example 5 was obtained by putting $z = constant$. This gave us an equation in x and y that described a curve. These curves are called the **level curves** of the surface $z = f(x, y)$ (see the discussion on the next page). In Example 5, the equations are of the form $x^2 + y^2 = c$ (c constant), and so the level curves are circles. Figure 13 shows the level curves for $c = 0$, 1, 2, 3, and 4.

The level curves give a contour map or topographical map of the surface. Each curve shows all of the points on the surface at a particular height c. You can use this contour map to visualize the shape of the surface. Imagine moving the contour at $c = 1$ to a height of 1 unit above the xy-plane, the contour at $c = 2$ to a height of 2 units above the xy-plane, and so on. You will end up with something like Figure 12. ∎

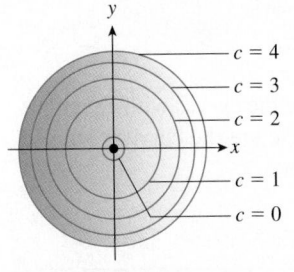

Level curves of the paraboloid $z = x^2 + y^2$

Figure 13

The following summary includes the techniques we have just used plus some additional ones:

Analyzing the Graph of a Function of Two Variables

If possible, use technology to render the graph $z = f(x, y)$ of a given function f of two variables. You can analyze its graph as follows:

Step 1 Obtain the **x-, y-, and z-intercepts** (the places where the surface crosses the coordinate axes).

x-Intercept(s): Set $y = 0$ and $z = 0$ and solve for x.

y-Intercept(s): Set $x = 0$ and $z = 0$ and solve for y.

z-Intercept: Set $x = 0$ and $y = 0$ and compute z.

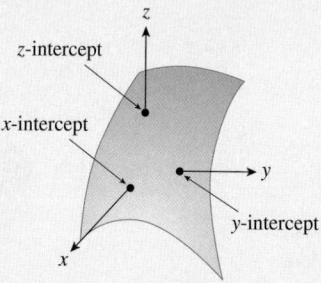

Step 2 Slice the surface along planes parallel to the xy-, yz-, and xz-planes.

$z = constant$ Set $z = constant$ and analyze the resulting curves.
These are the curves resulting from horizontal slices, and are called the **level curves** (see below).

$x = constant$ Set $x = constant$ and analyze the resulting curves.
These are the curves resulting from slices parallel to the yz-plane.

$y = constant$ Set $y = constant$ and analyze the resulting curves.
These are the curves resulting from slices parallel to the xz-plane.

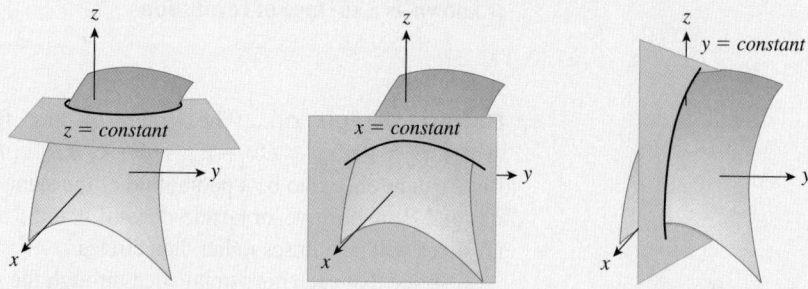

Level Curves

The **level curves** of a function f of two variables are the curves with equations of the form $f(x, y) = c$, where c is constant. These are the curves obtained from the graph of f by slicing it horizontally as above.

Quick Examples

1. Figure 13 above shows some level curves of $f(x, y) = x^2 + y^2$. The ones shown have equations $f(x, y) = 0, 1, 2, 3$, and 4.

2. Let $f(x, y) = y - x^2 + 4$. Its level curves have the form $y - x^2 + 4 = c$ (c constant). If we solve this equation for y, we see that $y = x^2 + c - 4$, the equation of a parabola with its vertex on the y-axis at the point $c - 4$. The following figure shows a portion of the graph of f and some of its level curves.

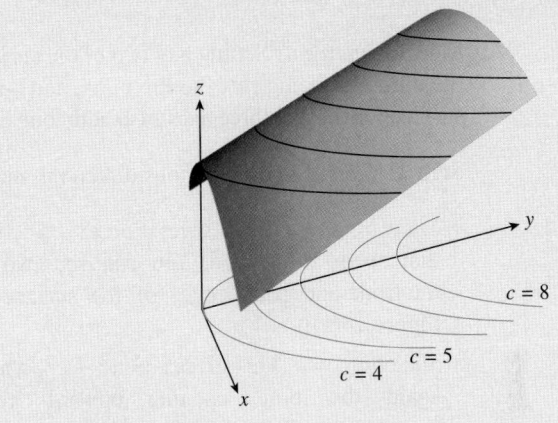

Spreadsheets often have built-in features to render surfaces such as the paraboloid in Example 5. In the following example, we use Excel to graph another surface and then analyze it as above.

EXAMPLE 6 ⊓ Analyzing a Surface

Describe the graph of $f(x, y) = x^2 - y^2$.

Solution First we obtain a picture of the graph using technology. Figure 14 shows two graphs obtained using resources at the Website.

Chapter 15 → Math Tools for Chapter 15
→ Surface Graphing Utility

Chapter 15 → Math Tools for Chapter 15
→ Excel Surface Graphing Utility

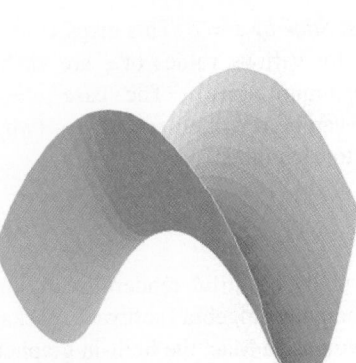

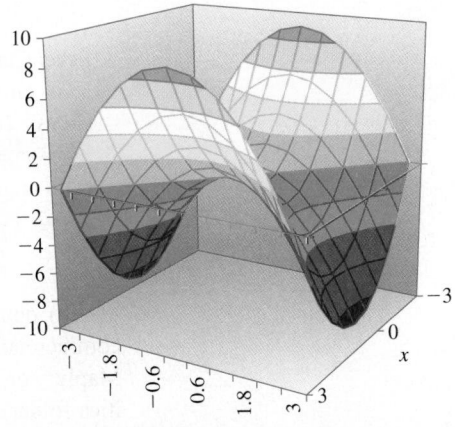

Figure 14

See the Technology Guides at the end of the chapter to find out how to obtain a similar graph from scratch using a spreadsheet.

The graph shows an example of a "saddle point" at the origin. (We return to this idea in Section 15.3.) To analyze the graph for the features shown in the box above, replace $f(x, y)$ by z to obtain

$$z = x^2 - y^2.$$

Step 1: *Intercepts* Setting any two of the variables x, y, and z equal to zero results in the third also being zero, so the x-, y-, and z-intercepts are all 0. In other words, the surface touches all three axes in exactly one point, the origin.

Step 2: *Slices* Slices in various directions show more interesting features.

Slice by $x = c$ This gives $z = c^2 - y^2$, which is the equation of a parabola that opens downward. You can see two of these slices ($c = -3, c = 3$) as the front and back edges of the surface in Figure 14. (More are shown in Figure 15(a).)

Slice by $y = c$ This gives $z = x^2 - c^2$, which is the equation of a parabola once again—this time, opening upward. You can see two of these slices ($c = -3, c = 3$) as the left and right edges of the surface in Figure 14. (More are shown in Figure 15(b).)

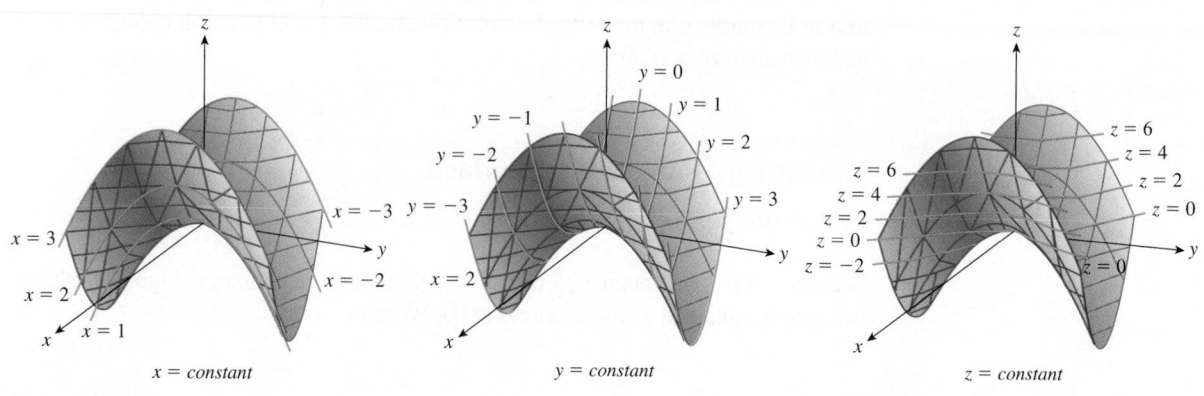

Figure 15(a)   **Figure 15(b)**   **Figure 15(c)**

Level Curves: Slice by $z = c$ This gives $x^2 - y^2 = c$, which is a hyperbola. The level curves for various values of c are visible in Figure 14 as the horizontal slices. (See Figure 15(c).) The case $c = 0$ is interesting: The equation $x^2 - y^2 = 0$ can be rewritten as $x = \pm y$ (why?), which represents two lines at right angles to each other.

To obtain really beautiful renderings of surfaces, you could use one of the commercial computer algebra software packages, such as Mathematica® or Maple®, or, if you use a Mac, the built-in grapher (grapher.app located in the Utilities folder).

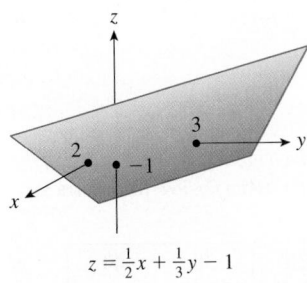

$$z = \tfrac{1}{2}x + \tfrac{1}{3}y - 1$$

Figure 16

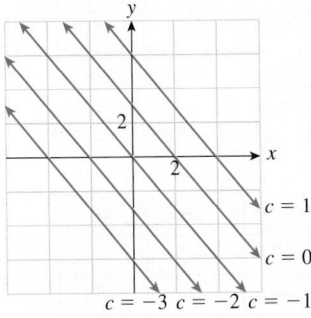

Level curves: $3x + 2y = 12$

Figure 17

✷ Think about what happens when the function is constant.

EXAMPLE 7 Graph and Level Curves of a Linear Function

Describe the graph of $g(x, y) = \dfrac{1}{2}x + \dfrac{1}{3}y - 1$.

Solution Notice first that g is a linear function of x and y. Figure 16 shows a portion of the graph, which is a plane.

We can get a good idea of what plane this is by looking at the x-, y-, and z-intercepts.

x-intercept Set $y = z = 0$, which gives $x = 2$.

y-intercept Set $x = z = 0$, which gives $y = 3$.

z-intercept Set $x = y = 0$, which gives $z = -1$.

Three points are enough to define a plane, so we can say that the plane is the one passing through the three points $(2, 0, 0)$, $(0, 3, 0)$, and $(0, 0, -1)$. It can be shown that the graph of every linear function of two variables is a plane.

Level curves: Set $g(x, y) = c$ to obtain $\tfrac{1}{2}x + \tfrac{1}{3}y - 1 = c$, or $\tfrac{1}{2}x + \tfrac{1}{3}y = c + 1$. We can rewrite this equation as $3x + 2y = 6(c + 1)$, which is the equation of a straight line. Choosing different values of c gives us a family of parallel lines as shown in Figure 17. (For example, the line corresponding to $c = 1$ has equation $3x + 2y = 6(1 + 1) = 12$.) In general, the set of level curves of every non-constant linear function is a set of parallel straight lines.✷

EXAMPLE 8 Using Level Curves

A certain function f of two variables has level curves $f(x, y) = c$ for $c = -2, -1, 0, 1$, and 2, as shown in Figure 18. (Each grid square is 1×1.)

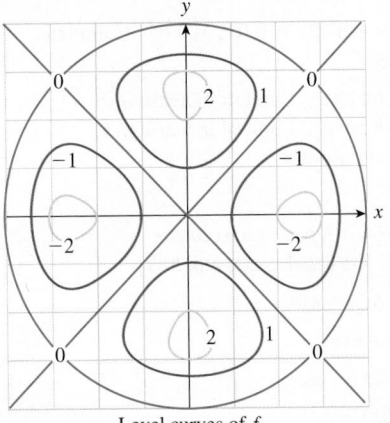

Level curves of f

Figure 18

Estimate the following: $f(1, 1)$, $f(1.5, -2)$, $f(1.5, 0)$, and $f(1, 2)$.

Solution The point $(1, 1)$ appears to lie exactly on the red level curve $c = 0$, so $f(1, 1) \approx 0$. Similarly, the point $(1.5, -2)$ appears to lie exactly on the blue level curve $c = 1$, so $f(1.5, -2) \approx 1$. The point $(1.5, 0)$ appears to lie midway between the level curves $c = -1$ and $c = -2$, so we estimate $f(1.5, 2) \approx -1.5$. Finally, the point $(1, 2)$ lies between the level curves $c = 1$ and $c = 2$, but closer to $c = 1$, so we can estimate $f(1, 2)$ at around 1.3.

15.1   EXERCISES

▼ more advanced   ◆ challenging

⊤ indicates exercises that should be solved using technology

*For each function in Exercises 1–4, evaluate **(a)** $f(0, 0)$;*
***(b)** $f(1, 0)$; **(c)** $f(0, -1)$; **(d)** $f(a, 2)$; **(e)** $f(y, x)$;*
***(f)** $f(x + h, y + k)$* HINT [See Quick Examples page 1082.]

1. $f(x, y) = x^2 + y^2 - x + 1$

2. $f(x, y) = x^2 - y - xy + 1$

3. $f(x, y) = 0.2x + 0.1y - 0.01xy$

4. $f(x, y) = 0.4x - 0.5y - 0.05xy$

*For each function in Exercises 5–8, evaluate **(a)** $g(0, 0, 0)$;*
***(b)** $g(1, 0, 0)$; **(c)** $g(0, 1, 0)$; **(d)** $g(z, x, y)$;*
***(e)** $g(x + h, y + k, z + l)$, provided such a value exists.*

5. $g(x, y, z) = e^{x+y+z}$

6. $g(x, y, z) = \ln(x + y + z)$

7. $g(x, y, z) = \dfrac{xyz}{x^2 + y^2 + z^2}$

8. $g(x, y, z) = \dfrac{e^{xyz}}{x + y + z}$

9. Let $f(x, y, z) = 1.5 + 2.3x - 1.4y - 2.5z$. Complete the following sentences. HINT [See Example 1.]

   a. f ___ by ___ units for every 1 unit of increase in x.

   b. f ___ by ___ units for every 1 unit of increase in y.

   c. _____ by 2.5 units for every _____.

10. Let $g(x, y, z) = 0.01x + 0.02y - 0.03z - 0.05$. Complete the following sentences.

   a. g ___ by ___ units for every 1 unit of increase in z.

   b. g ___ by ___ units for every 1 unit of increase in x.

   c. _____ by 0.02 units for every _____.

In Exercises 11–18, classify each function as linear, interaction, or neither. HINT [See Quick Examples page 1085.]

11. $L(x, y) = 3x - 2y + 6xy - 4y^2$

12. $L(x, y, z) = 3x - 2y + 6xz$

13. $P(x_1, x_2, x_3) = 0.4 + 2x_1 - x_3$

14. $Q(x_1, x_2) = 4x_2 - 0.5x_1 - x_1^2$

15. $f(x, y, z) = \dfrac{x + y - z}{3}$

16. $g(x, y, z) = \dfrac{xz - 3yz + z^2}{4z}$ $(z \neq 0)$

17. $g(x, y, z) = \dfrac{xz - 3yz + z^2y}{4z}$ $(z \neq 0)$

18. $f(x, y) = x + y + xy + x^2y$

In Exercises 19 and 20, use the given tabular representation of the function f to compute the quantities asked for. HINT [See Example 3.]

19.

	$x \to$			
	10	**20**	**30**	**40**
y ↓ **10**	-1	107	162	-3
20	-6	194	294	-14
30	-11	281	426	-25
40	-16	368	558	-36

   a. $f(20, 10)$ **b.** $f(40, 20)$ **c.** $f(10, 20) - f(20, 10)$

20.

	$x \to$			
	10	**20**	**30**	**40**
y ↓ **10**	162	107	-5	-7
20	294	194	-22	-30
30	426	281	-39	-53
40	558	368	-56	-76

   a. $f(10, 30)$ **b.** $f(20, 10)$ **c.** $f(10, 40) + f(10, 20)$

⊤ *In Exercises 21 and 22, use a spreadsheet or some other method to complete the given tables.*

21. $P(x, y) = x - 0.3y + 0.45xy$

	$x \to$			
	10	**20**	**30**	**40**
y ↓ **10**				
20				
30				
40				

22. $Q(x, y) = 0.4x + 0.1y - 0.06xy$

	$x \to$			
	10	**20**	**30**	**40**
y ↓ **10**				
20				
30				
40				

118. ▼ ***Hiking*** The following diagram shows some level curves of the altitude of a mountain valley, as well as the location, on the 2,000-ft curve, of a hiker. The hiker is currently moving at the greatest possible rate of descent. In which of the three directions shown is he moving? Explain your answer.

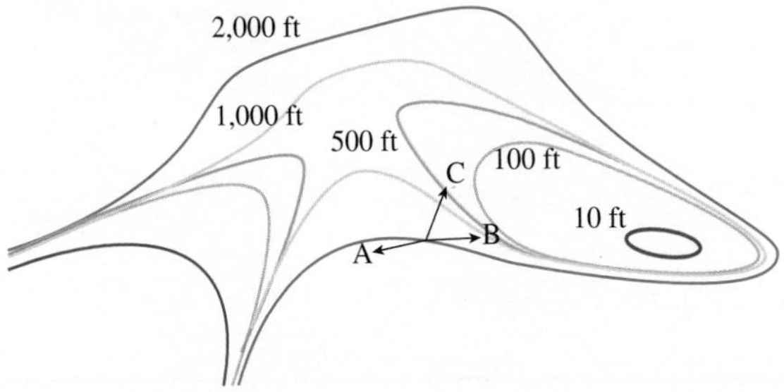

119. Your study partner Slim claims that because the surface $z = f(x, y)$ you have been studying is a plane, it follows that all the slices $x = constant$ and $y = constant$ are straight lines. Do you agree or disagree? Explain.

120. Your other study partner Shady just told you that the surface $z = xy$ you have been trying to graph must be a plane because you've already found that the slices $x = constant$ and $y = constant$ are all straight lines. Do you agree or disagree? Explain.

121. Why do we not sketch the graphs of functions of three or more variables?

122. The surface of a mountain can be thought of as the graph of what function?

123. Why is three-dimensional space used to represent the graph of a function of two variables?

124. Why is it that we can sketch the graphs of functions of two variables on the two-dimensional flat surfaces of these pages?

15.2 Partial Derivatives

Recall that if f is a function of x, then the derivative df/dx measures how fast f changes as x increases. If f is a function of two or more variables, we can ask how fast f changes as each variable increases while the others remain fixed. These rates of change are called the "partial derivatives of f," and they measure how each variable contributes to the change in f. Here is a more precise definition.

Partial Derivatives

The **partial derivative of f with respect to x** is the derivative of f with respect to x, when all other variables are treated as constant. Similarly, the **partial derivative of f with respect to y** is the derivative of f with respect to y, with all other variables treated as constant, and so on for other variables. The partial derivatives are written as $\frac{\partial f}{\partial x}, \frac{\partial f}{\partial y}$, and so on. The symbol ∂ is used (instead of d) to remind us that there is more than one variable and that we are holding the other variables fixed.

Quick Examples

1. Let $f(x, y) = x^2 + y^2$.

$$\frac{\partial f}{\partial x} = 2x + 0 = 2x$$ Because y^2 is treated as a constant

$$\frac{\partial f}{\partial y} = 0 + 2y = 2y$$ Because x^2 is treated as a constant

2. Let $z = x^2 + xy$.

$$\frac{\partial z}{\partial x} = 2x + y \qquad\qquad \frac{\partial}{\partial x}(xy) = \frac{\partial}{\partial x}(x \cdot \text{constant}) = \text{constant} = y$$

$$\frac{\partial z}{\partial y} = 0 + x \qquad\qquad \frac{\partial}{\partial y}(xy) = \frac{\partial}{\partial y}(\text{constant} \cdot y) = \text{constant} = x$$

3. Let $f(x, y) = x^2 y + y^2 x - xy + y$.

$$\frac{\partial f}{\partial x} = 2xy + y^2 - y \qquad\qquad \text{\textit{y} is treated as a constant.}$$

$$\frac{\partial f}{\partial y} = x^2 + 2xy - x + 1 \qquad\qquad \text{\textit{x} is treated as a constant.}$$

Interpretation

$\dfrac{\partial f}{\partial x}$ is the rate at which f changes as x changes, for a fixed (constant) y.

$\dfrac{\partial f}{\partial y}$ is the rate at which f changes as y changes, for a fixed (constant) x.

EXAMPLE 1 Marginal Cost: Linear Model

We return to Example 1 from Section 15.1. Suppose that you own a company that makes two models of speakers, the Ultra Mini and the Big Stack. Your total monthly cost (in dollars) to make x Ultra Minis and y Big Stacks is given by

$$C(x, y) = 10{,}000 + 20x + 40y.$$

What is the significance of $\dfrac{\partial C}{\partial x}$ and of $\dfrac{\partial C}{\partial y}$?

Solution First we compute these partial derivatives:

$$\frac{\partial C}{\partial x} = 20$$

$$\frac{\partial C}{\partial y} = 40.$$

We interpret the results as follows: $\dfrac{\partial C}{\partial x} = 20$ means that the cost is increasing at a rate of \$20 per additional Ultra Mini (if production of Big Stacks is held constant); $\dfrac{\partial C}{\partial y} = 40$ means that the cost is increasing at a rate of \$40 per additional Big Stack (if production of Ultra Minis is held constant). In other words, these are the **marginal costs** of each model of speaker.

➡ **Before we go on...** How much does the cost rise if you increase x by Δx and y by Δy? In Example 1, the change in cost is given by

$$\Delta C = 20\Delta x + 40\Delta y = \frac{\partial C}{\partial x}\Delta x + \frac{\partial C}{\partial y}\Delta y.$$

This suggests the **chain rule for several variables**. Part of this rule says that if x and y are both functions of t, then C is a function of t through them, and the rate of change of C with respect to t can be calculated as

$$\frac{dC}{dt} = \frac{\partial C}{\partial x} \cdot \frac{dx}{dt} + \frac{\partial C}{\partial y} \cdot \frac{dy}{dt}.$$

See the optional section on the chain rule for several variables for further discussion and applications of this interesting result. ∎

EXAMPLE 2 Marginal Cost: Interaction Model

Another possibility for the cost function in the preceding example is an interaction model

$$C(x, y) = 10,000 + 20x + 40y + 0.1xy.$$

a. *Now* what are the marginal costs of the two models of speakers?

b. What is the marginal cost of manufacturing Big Stacks at a production level of 100 Ultra Minis and 50 Big Stacks per month?

Solution

a. We compute the partial derivatives:

$$\frac{\partial C}{\partial x} = 20 + 0.1y$$

$$\frac{\partial C}{\partial y} = 40 + 0.1x.$$

Thus, the marginal cost of manufacturing Ultra Minis increases by $0.1 or 10¢ for each Big Stack that is manufactured. Similarly, the marginal cost of manufacturing Big Stacks increases by 10¢ for each Ultra Mini that is manufactured.

b. From part (a), the marginal cost of manufacturing Big Stacks is

$$\frac{\partial C}{\partial y} = 40 + 0.1x.$$

At a production level of 100 Ultra Minis and 50 Big Stacks per month, we have $x = 100$ and $y = 50$. Thus, the marginal cost of manufacturing Big Stacks at these production levels is

$$\left.\frac{\partial C}{\partial y}\right|_{(100,50)} = 40 + 0.1(100) = \$50 \text{ per Big Stack}.$$

Partial derivatives of functions of three variables are obtained in the same way as those for functions of two variables, as the following example shows.

EXAMPLE 3 **Function of Three Variables**

Calculate $\dfrac{\partial f}{\partial x}, \dfrac{\partial f}{\partial y}$, and $\dfrac{\partial f}{\partial z}$ if $f(x, y, z) = xy^2z^3 - xy$.

Solution Although we now have three variables, the calculation remains the same: $\partial f/\partial x$ is the derivative of f with respect to x, with *both* other variables, y and z, held constant:

$$\frac{\partial f}{\partial x} = y^2z^3 - y.$$

Similarly, $\partial f/\partial y$ is the derivative of f with respect to y, with both x and z held constant:

$$\frac{\partial f}{\partial y} = 2xyz^3 - x.$$

Finally, to find $\partial f/\partial z$, we hold both x and y constant and take the derivative with respect to z.

$$\frac{\partial f}{\partial z} = 3xy^2z^2.$$

Note The procedure for finding a partial derivative is the same for any number of variables: To get the partial derivative with respect to any one variable, we treat all the others as constants. ∎

Geometric Interpretation of Partial Derivatives

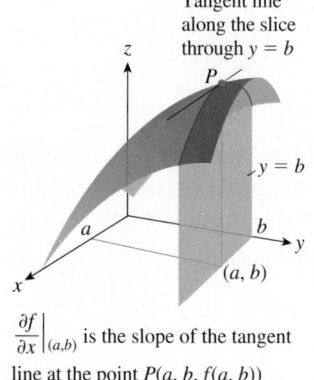

Tangent line along the slice through $y = b$

$\left.\dfrac{\partial f}{\partial x}\right|_{(a,b)}$ is the slope of the tangent line at the point $P(a, b, f(a, b))$ along the slice through $y = b$.

Figure 19

Recall that if f is a function of one variable x, then the derivative df/dx gives the slopes of the tangent lines to its graph. Now, suppose that f is a function of x and y. By definition, $\partial f/\partial x$ is the derivative of the function of x we get by holding y fixed. If we evaluate this derivative at the point (a, b), we are holding y fixed at the value b, taking the ordinary derivative of the resulting function of x, and evaluating this at $x = a$. Now, holding y fixed at b amounts to slicing through the graph of f along the plane $y = b$, resulting in a curve. Thus, the partial derivative is the slope of the tangent line to this curve at the point where $x = a$ and $y = b$, along the plane $y = b$ (Figure 19). This fits with our interpretation of $\partial f/\partial x$ as the rate of increase of f with increasing x when y is held fixed at b.

The other partial derivative, $\partial f/\partial y|_{(a,b)}$, is, similarly, the slope of the tangent line at the same point $P(a, b, f(a, b))$ but along the slice by the plane $x = a$. You should draw the corresponding picture for this on your own.

EXAMPLE 4 **Marginal Cost**

Referring to the interactive cost function $C(x, y) = 10{,}000 + 20x + 40y + 0.1xy$ in Example 2, we can identify the marginal costs $\partial C/\partial x$ and $\partial C/\partial y$ of manufacturing Ultra Minis and Big Stacks at a production level of 100 Ultra Minis and 50 Big Stacks per month as the slopes of the tangent lines to the two slices by $y = 50$ and $x = 100$ at the point on the graph where $(x, y) = (100, 50)$ as seen in Figure 20.

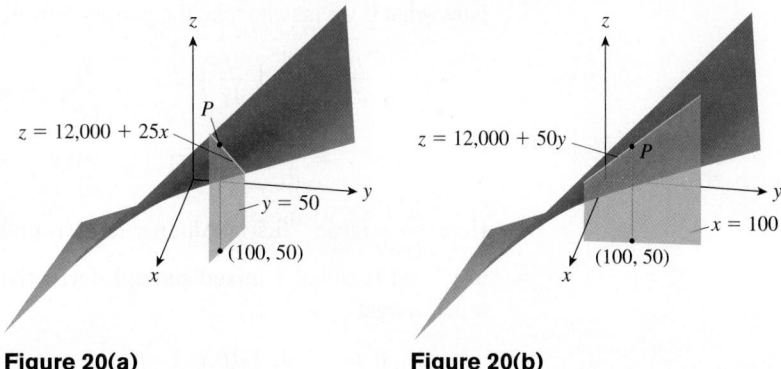

Figure 20(a)   **Figure 20(b)**

Figure 20(a) shows the slice at $y = 50$ through the point $P = (100, 50, C(100, 50)) = (100, 50, 14,500)$. The equation of that slice is given by substituting $y = 50$ in the cost equation:

$$C(x, 50) = 10,000 + 20x + 40(50) + 0.1x(50) = 12,000 + 25x.$$ A line of slope 25

Because the slice is already a line, it coincides with the tangent line through P as depicted in Figure 19. This slope is equal to $\partial C / \partial x|_{(100,50)}$:

$$\frac{\partial C}{\partial x} = 20 + 0.1y$$ See Example 2.

so $$\left.\frac{\partial C}{\partial x}\right|_{(100,50)} = 20 + 0.1(50) = 25.$$

Similarly, Figure 20(b) shows the slice at $x = 100$ through the same point P. The equation of that slice is given by substituting $x = 100$ in the cost equation:

$$C(100, y) = 10,000 + 20(100) + 40y + 0.1(100)y = 12,000 + 50y.$$ A line of slope 50

This slope is equal to $\partial C / \partial y|_{(100,50)} = 50$ as we calculated in Example 2.

Second-Order Partial Derivatives

Just as for functions of a single variable, we can calculate second derivatives. Suppose, for example, that we have a function of x and y, say, $f(x, y) = x^2 - x^2y^2$. We know that

$$\frac{\partial f}{\partial x} = 2x - 2xy^2.$$

If we take the partial derivative with respect to x once again, we obtain

$$\frac{\partial}{\partial x}\left(\frac{\partial f}{\partial x}\right) = 2 - 2y^2.$$ Take $\frac{\partial}{\partial x}$ of $\frac{\partial f}{\partial x}$.

(The symbol $\partial / \partial x$ means "the partial derivative with respect to x," just as d/dx stands for "the derivative with respect to x.") This is called the **second-order partial derivative** and is written $\dfrac{\partial^2 f}{\partial x^2}$. We get the following derivatives similarly:

$$\frac{\partial f}{\partial y} = -2x^2y$$

$$\frac{\partial^2 f}{\partial y^2} = -2x^2.$$ Take $\frac{\partial}{\partial y}$ of $\frac{\partial f}{\partial y}$.

Now what if we instead take the partial derivative with respect to y of $\partial f/\partial x$?

$$\frac{\partial^2 f}{\partial y \partial x} = \frac{\partial}{\partial y}\left(\frac{\partial f}{\partial x}\right) \qquad \text{Take } \frac{\partial}{\partial y} \text{ of } \frac{\partial f}{\partial x}.$$

$$= \frac{\partial}{\partial y}[2x - 2xy^2] = -4xy$$

Here, $\dfrac{\partial^2 f}{\partial y \partial x}$ means "first take the partial derivative with respect to x and then with respect to y" and is called a **mixed partial derivative**. If we differentiate in the opposite order, we get

$$\frac{\partial^2 f}{\partial x \partial y} = \frac{\partial}{\partial x}\left(\frac{\partial f}{\partial y}\right) = \frac{\partial}{\partial x}[-2x^2 y] = -4xy,$$

the same expression as $\dfrac{\partial^2 f}{\partial y \partial x}$. This is no coincidence: The mixed partial derivatives $\dfrac{\partial^2 f}{\partial x \partial y}$ and $\dfrac{\partial^2 f}{\partial y \partial x}$ are always the same as long as the first partial derivatives are both differentiable functions of x and y and the mixed partial derivatives are continuous. Because all the functions we shall use are of this type, we can take the derivatives in any order we like when calculating mixed derivatives.

Here is another notation for partial derivatives that is especially convenient for second-order partial derivatives:

$$f_x \text{ means } \frac{\partial f}{\partial x}$$

$$f_y \text{ means } \frac{\partial f}{\partial y}$$

$$f_{xy} \text{ means } (f_x)_y = \frac{\partial^2 f}{\partial y \partial x} \quad \text{(Note the order in which the derivatives are taken.)}$$

$$f_{yx} \text{ means } (f_y)_x = \frac{\partial^2 f}{\partial x \partial y}.$$

15.2   EXERCISES

▼ more advanced   ◆ challenging
🔢 indicates exercises that should be solved using technology

In Exercises 1–18, calculate $\dfrac{\partial f}{\partial x}, \dfrac{\partial f}{\partial y}, \dfrac{\partial f}{\partial x}\Big|_{(1,-1)}$, *and* $\dfrac{\partial f}{\partial y}\Big|_{(1,-1)}$

when defined. HINT [See Quick Examples pages 1101–1102.]

1. $f(x, y) = 10,000 - 40x + 20y$

2. $f(x, y) = 1,000 + 5x - 4y$

3. $f(x, y) = 3x^2 - y^3 + x - 1$

4. $f(x, y) = x^{1/2} - 2y^4 + y + 6$

5. $f(x, y) = 10,000 - 40x + 20y + 10xy$

6. $f(x, y) = 1,000 + 5x - 4y - 3xy$

7. $f(x, y) = 3x^2 y$

8. $f(x, y) = x^4 y^2 - x$

9. $f(x, y) = x^2 y^3 - x^3 y^2 - xy$

10. $f(x, y) = x^{-1} y^2 + xy^2 + xy$

11. $f(x, y) = (2xy + 1)^3$

12. $f(x, y) = \dfrac{1}{(xy + 1)^2}$

13. ▼ $f(x, y) = e^{x+y}$

14. ▼ $f(x, y) = e^{2x+y}$

15. ▼ $f(x, y) = 5x^{0.6} y^{0.4}$

16. ▼ $f(x, y) = -2x^{0.1} y^{0.9}$

17. ▼ $f(x, y) = e^{0.2xy}$

18. ▼ $f(x, y) = xe^{xy}$

In Exercises 19–28, find $\dfrac{\partial^2 f}{\partial x^2}, \dfrac{\partial^2 f}{\partial y^2}, \dfrac{\partial^2 f}{\partial x \partial y}$, *and* $\dfrac{\partial^2 f}{\partial y \partial x}$, *and evaluate them all at* $(1, -1)$ *if possible.* HINT [See discussion on pages 1105–1106.]

19. $f(x, y) = 10,000 - 40x + 20y$

20. $f(x, y) = 1,000 + 5x - 4y$

21. $f(x, y) = 10,000 - 40x + 20y + 10xy$

22. $f(x, y) = 1,000 + 5x - 4y - 3xy$

23. $f(x, y) = 3x^2 y$　　**24.** $f(x, y) = x^4 y^2 - x$

25. ▼ $f(x, y) = e^{x+y}$　　**26.** ▼ $f(x, y) = e^{2x+y}$

27. ▼ $f(x, y) = 5x^{0.6} y^{0.4}$　　**28.** ▼ $f(x, y) = -2x^{0.1} y^{0.9}$

In Exercises 29–40, find $\dfrac{\partial f}{\partial x}, \dfrac{\partial f}{\partial y}, \dfrac{\partial f}{\partial z}$, *and their values at* $(0, -1, 1)$ *if possible.* HINT [See Example 3.]

29. $f(x, y, z) = xyz$

30. $f(x, y, z) = xy + xz - yz$

31. ▼ $f(x, y, z) = -\dfrac{4}{x + y + z^2}$

32. ▼ $f(x, y, z) = \dfrac{6}{x^2 + y^2 + z^2}$

33. ▼ $f(x, y, z) = xe^{yz} + ye^{xz}$

34. ▼ $f(x, y, z) = xye^z + xe^{yz} + e^{xyz}$

35. ▼ $f(x, y, z) = x^{0.1} y^{0.4} z^{0.5}$

36. ▼ $f(x, y, z) = 2x^{0.2} y^{0.8} + z^2$

37. ▼ $f(x, y, z) = e^{xyz}$

38. ▼ $f(x, y, z) = \ln(x + y + z)$

39. ▼ $f(x, y, z) = \dfrac{2,000z}{1 + y^{0.3}}$

40. ▼ $f(x, y, z) = \dfrac{e^{0.2x}}{1 + e^{-0.1y}}$

APPLICATIONS

41. *Marginal Cost (Linear Model)* Your weekly cost (in dollars) to manufacture x cars and y trucks is

$$C(x, y) = 240,000 + 6,000x + 4,000y.$$

Calculate and interpret $\dfrac{\partial C}{\partial x}$ and $\dfrac{\partial C}{\partial y}$. HINT [See Example 1.]

42. *Marginal Cost (Linear Model)* Your weekly cost (in dollars) to manufacture x bicycles and y tricycles is

$$C(x, y) = 24,000 + 60x + 20y.$$

Calculate and interpret $\dfrac{\partial C}{\partial x}$ and $\dfrac{\partial C}{\partial y}$.

43. *Scientific Research* In each year from 1983 to 2003, the percentage y of research articles in *Physical Review* written by researchers in the United States can be approximated by

$$y = 82 - 0.78t - 1.02x \text{ percentage points} \quad (0 \le t \le 20),$$

where t is the year since 1983 and x is the percentage of articles written by researchers in Europe.[14] Calculate and interpret $\dfrac{\partial y}{\partial t}$ and $\dfrac{\partial y}{\partial x}$.

44. *Scientific Research* The number z of research articles in *Physical Review* that were written by researchers in the United States from 1993 through 2003 can be approximated by

$$z = 5,960 - 0.71x + 0.50y \quad (3,000 \le x, y \le 6,000)$$

articles each year, where x is the number of articles written by researchers in Europe and y is the number written by researchers in other countries (excluding Europe and the United States).[15] Calculate and interpret $\dfrac{\partial z}{\partial x}$ and $\dfrac{\partial z}{\partial y}$.

45. *Marginal Cost (Interaction Model)* Your weekly cost (in dollars) to manufacture x cars and y trucks is

$$C(x, y) = 240,000 + 6,000x + 4,000y - 20xy.$$

(Compare with Exercise 41.) Compute the marginal cost of manufacturing cars at a production level of 10 cars and 20 trucks. HINT [See Example 2.]

46. *Marginal Cost (Interaction Model)* Your weekly cost (in dollars) to manufacture x bicycles and y tricycles is

$$C(x, y) = 24,000 + 60x + 20y + 0.3xy.$$

(Compare with Exercise 42.) Compute the marginal cost of manufacturing tricycles at a production level of 10 bicycles and 20 tricycles. HINT [See Example 2.]

47. *Brand Loyalty* The fraction of Mazda car owners who chose another new Mazda can be modeled by the following function:[16]

$$M(c, f, g, h, t) = 1.1 - 3.8c + 2.2f + 1.9g - 1.7h - 1.3t.$$

Here, c is the fraction of Chrysler car owners who remained loyal to Chrysler, f is the fraction of Ford car owners remaining loyal to Ford, g the corresponding figure for General Motors, h the corresponding figure for Honda, and t for Toyota.

a. Calculate $\dfrac{\partial M}{\partial c}$ and $\dfrac{\partial M}{\partial f}$ and interpret the answers.

b. One year it was observed that $c = 0.56, f = 0.56$, $g = 0.72, h = 0.50$, and $t = 0.43$. According to the model, what percentage of Mazda owners remained loyal to Mazda? (Round your answer to the nearest percentage point.)

48. *Brand Loyalty* The fraction of Mazda car owners who chose another new Mazda can be modeled by the following function:[17]

$$M(c, f) = 9.4 + 7.8c + 3.6c^2 - 38f - 22cf + 43f^2,$$

---

[15]*Ibid.*

[16]The model is an approximation of a linear regression based on data from the period 1988–1995. Source for data: Chrysler, Maritz Market Research, Consumer Attitude Research, and Strategic Vision/*New York Times*, November 3, 1995, p. D2.

[17]The model is an approximation of a second-order regression based on data from the period 1988–1995. Source for data: Chrysler, Maritz Market Research, Consumer Attitude Research, and Strategic Vision/ *New York Times*, November 3, 1995, p. D2.

[14]Based on a linear regression. Source for data: The American Physical Society/*New York Times*, May 3, 2003, p. A1.

where c is the fraction of **Chrysler** car owners who remained loyal to Chrysler and f is the fraction of **Ford** car owners remaining loyal to Ford.

a. Calculate $\dfrac{\partial M}{\partial c}$ and $\dfrac{\partial M}{\partial f}$ evaluated at the point $(0.7, 0.7)$, and interpret the answers.

b. One year it was observed that $c = 0.56$, and $f = 0.56$. According to the model, what percentage of Mazda owners remained loyal to Mazda? (Round your answer to the nearest percentage point.)

49. *Marginal Cost* Your weekly cost (in dollars) to manufacture x cars and y trucks is

$$C(x, y) = 200{,}000 + 6{,}000x + 4{,}000y - 100{,}000e^{-0.01(x+y)}.$$

What is the marginal cost of a car? Of a truck? How do these marginal costs behave as total production increases?

50. *Marginal Cost* Your weekly cost (in dollars) to manufacture x bicycles and y tricycles is

$$C(x, y) = 20{,}000 + 60x + 20y + 50\sqrt{xy}.$$

What is the marginal cost of a bicycle? Of a tricycle? How do these marginal costs behave as x and y increase?

51. ▼ *Income Gap* The following model is based on data on the median family incomes of Hispanic and white families in the United States for the period 1980–2008:[18]

$$z(t, x) = 31{,}200 + 270t + 13{,}500x + 140xt,$$

where

$z(t, x) = $ median family income
$t = $ year ($t = 0$ represents 1980)
$x = \begin{cases} 0 & \text{if the income was for a Hispanic family} \\ 1 & \text{if the income was for a white family} \end{cases}$.

a. Use the model to estimate the median income of a Hispanic family and of a white family in 2000.

b. According to the model, how fast was the median income for a Hispanic family increasing in 2000? How fast was the median income for a white family increasing in 2000?

c. Do the answers in part (b) suggest that the income gap between white and Hispanic families was widening or narrowing during the given period?

d. What does the coefficient of xt in the formula for $z(t, x)$ represent in terms of the income gap?

52. ▼ *Income Gap* The following model is based on data on the median family incomes of black and white families in the United States for the period 1980–2008:[19]

$$z(t, x) = 24{,}500 + 390t + 20{,}200x + 20xt,$$

where

$z(t, x) = $ median family income
$t = $ year ($t = 0$ represents 1980)

---
[18] Incomes are in 2007 dollars. Source for data: U.S. Census Bureau (www.census.gov).
[19] *Ibid.*

$x = \begin{cases} 0 & \text{if the income was for a black family} \\ 1 & \text{if the income was for a white family} \end{cases}$.

a. Use the model to estimate the median income of a black family and of a white family in 2000.

b. According to the model, how fast was the median income for a black family increasing in 2000? How fast was the median income for a white family increasing in 2000?

c. Do the answers in part (b) suggest that the income gap between white and black families was widening or narrowing during the given period?

d. What does the coefficient of xt in the formula for $z(t, x)$ represent in terms of the income gap?

53. ▼ *Average Cost* If you average your costs over your total production, you get the **average cost**, written $\bar{C}$:

$$\bar{C}(x, y) = \frac{C(x, y)}{x + y}.$$

Find the average cost for the cost function in Exercise 49. Then find the marginal average cost of a car and the marginal average cost of a truck at a production level of 50 cars and 50 trucks. Interpret your answers.

54. ▼ *Average Cost* Find the average cost for the cost function in Exercise 50. (See the preceding exercise.) Then find the marginal average cost of a bicycle and the marginal average cost of a tricycle at a production level of five bicycles and five tricycles. Interpret your answers.

55. ▼ *Marginal Revenue* As manager of an auto dealership, you offer a car rental company the following deal: You will charge $15,000 per car and $10,000 per truck, but you will then give the company a discount of $5,000 times the square root of the total number of vehicles it buys from you. Looking at your marginal revenue, is this a good deal for the rental company?

56. ▼ *Marginal Revenue* As marketing director for a bicycle manufacturer, you come up with the following scheme: You will offer to sell a dealer x bicycles and y tricycles for

$$R(x, y) = 3{,}500 - 3{,}500e^{-0.02x - 0.01y} \text{ dollars.}$$

Find your marginal revenue for bicycles and for tricycles. Are you likely to be fired for your suggestion?

57. ▼ *Research Productivity* Here we apply a variant of the Cobb-Douglas function to the modeling of research productivity. A mathematical model of research productivity at a particular physics laboratory is

$$P = 0.04x^{0.4}y^{0.2}z^{0.4},$$

where P is the annual number of groundbreaking research papers produced by the staff, x is the number of physicists on the research team, y is the laboratory's annual research budget, and z is the annual National Science Foundation subsidy to the laboratory. Find the rate of increase of research papers per government-subsidy dollar at a subsidy level of $1,000,000 per year and a staff level of 10 physicists if the annual budget is $100,000.

58. ▼ *Research Productivity* A major drug company estimates that the annual number P of patents for new drugs developed by its research team is best modeled by the formula

$$P = 0.3x^{0.3}y^{0.4}z^{0.3},$$

where x is the number of research biochemists on the payroll, y is the annual research budget, and z is the size of the bonus awarded to discoverers of new drugs. Assuming that the company has 12 biochemists on the staff, has an annual research budget of $500,000, and pays $40,000 bonuses to developers of new drugs, calculate the rate of growth in the annual number of patents per new research staff member.

59. ▼ *Utility* Your newspaper is trying to decide between two competing desktop publishing software packages, Macro Publish and Turbo Publish. You estimate that if you purchase x copies of Macro Publish and y copies of Turbo Publish, your company's daily productivity will be

$$U(x, y) = 6x^{0.8}y^{0.2} + x \text{ pages per day.}$$

a. Calculate $U_x(10, 5)$ and $U_y(10, 5)$ to two decimal places, and interpret the results.

b. What does the ratio $\dfrac{U_x(10, 5)}{U_y(10, 5)}$ tell about the usefulness of these products?

60. ▼ *Grades*[20] A production formula for a student's performance on a difficult English examination is given by

$$g(t, x) = 4tx - 0.2t^2 - x^2,$$

where g is the grade the student can expect to get, t is the number of hours of study for the examination, and x is the student's grade-point average.

a. Calculate $g_t(10, 3)$ and $g_x(10, 3)$ and interpret the results.

b. What does the ratio $\dfrac{g_t(10, 3)}{g_x(10, 3)}$ tell about the relative merits of study and grade-point average?

61. ▼ *Electrostatic Repulsion* If positive electric charges of Q and q coulombs are situated at positions (a, b, c) and (x, y, z), respectively, then the force of repulsion they experience is given by

$$F = K\dfrac{Qq}{(x - a)^2 + (y - b)^2 + (z - c)^2},$$

where $K \approx 9 \times 10^9$, F is given in newtons, and all positions are measured in meters. Assume that a charge of 10 coulombs is situated at the origin, and that a second charge of 5 coulombs is situated at $(2, 3, 3)$ and moving in the y-direction at one meter per second. How fast is the electrostatic force it experiences decreasing? (Round the answer to one significant digit.)

62. ▼ *Electrostatic Repulsion* Repeat the preceding exercise, assuming that a charge of 10 coulombs is situated at the origin and that a second charge of 5 coulombs is situated at $(2, 3, 3)$ and moving in the negative z direction at one meter per second. (Round the answer to one significant digit.)

63. ▼ *Investments* Recall that the compound interest formula for annual compounding is

$$A(P, r, t) = P(1 + r)^t,$$

where A is the future value of an investment of P dollars after t years at an interest rate of r.

a. Calculate $\dfrac{\partial A}{\partial P}, \dfrac{\partial A}{\partial r}$, and $\dfrac{\partial A}{\partial t}$, all evaluated at $(100, 0.10, 10)$. (Round your answers to two decimal places.) Interpret your answers.

b. What does the function $\dfrac{\partial A}{\partial P}\Big|_{(100, 0.10, t)}$ of t tell about your investment?

64. ▼ *Investments* Repeat the preceding exercise, using the formula for continuous compounding:

$$A(P, r, t) = Pe^{rt}.$$

65. ▼ *Modeling with the Cobb-Douglas Production Formula* Assume you are given a production formula of the form

$$P(x, y) = Kx^ay^b \quad (a + b = 1).$$

a. Obtain formulas for $\dfrac{\partial P}{\partial x}$ and $\dfrac{\partial P}{\partial y}$, and show that $\dfrac{\partial P}{\partial x} = \dfrac{\partial P}{\partial y}$ precisely when $x/y = a/b$.

b. Let x be the number of workers a firm employs and let y be its monthly operating budget in thousands of dollars. Assume that the firm currently employs 100 workers and has a monthly operating budget of $200,000. If each additional worker contributes as much to productivity as each additional $1,000 per month, find values of a and b that model the firm's productivity.

66. ▼ *Housing Costs*[21] The cost C of building a house is related to the number k of carpenters used and the number e of electricians used by

$$C(k, e) = 15,000 + 50k^2 + 60e^2.$$

If three electricians are currently employed in building your new house and the marginal cost per additional electrician is the same as the marginal cost per additional carpenter, how many carpenters are being used? (Round your answer to the nearest carpenter.)

67. ▼ *Nutrient Diffusion* Suppose that one cubic centimeter of nutrient is placed at the center of a circular petri dish filled with water. We might wonder how the nutrient is distributed after a time of t seconds. According to the classical theory of diffusion, the concentration of nutrient (in parts of nutrient per part of water) after a time t is given by

$$u(r, t) = \dfrac{1}{4\pi Dt}e^{-\frac{r^2}{4Dt}}.$$

Here D is the *diffusivity*, which we will take to be 1, and r is the distance from the center in centimeters. How fast is the concentration increasing at a distance of 1 cm from the center 3 seconds after the nutrient is introduced?

---

[20]Based on an Exercise in *Introduction to Mathematical Economics* by A. L. Ostrosky Jr. and J. V. Koch (Waveland Press, Illinois, 1979).

[21]*Ibid.*

68. ▼ *Nutrient Diffusion* Refer back to the preceding exercise. How fast is the concentration increasing at a distance of 4 cm from the center 4 seconds after the nutrient is introduced?

COMMUNICATION AND REASONING EXERCISES

69. Given that $f(a, b) = r$, $f_x(a, b) = s$, and $f_y(a, b) = t$, complete the following: _____ is increasing at a rate of _____ units per unit of x, _____ is increasing at a rate of _____ units per unit of y, and the value of _____ is _____ when $x =$ _____ and $y =$ _____.

70. A firm's productivity depends on two variables, x and y. Currently, $x = a$ and $y = b$, and the firm's productivity is 4,000 units. Productivity is increasing at a rate of 400 units per unit *decrease* in x, and is decreasing at a rate of 300 units per unit increase in y. What does all of this information tell you about the firm's productivity function $g(x, y)$?

71. Complete the following: Let $f(x, y, z)$ be the cost to build a development of x cypods (one-bedroom units) in the city-state of Utarek, Mars, y argaats (two-bedroom units), and z orbici (singular: orbicus; three-bedroom units) in $\overline{\overline{Z}}$ (zonars, the designated currency in Utarek). Then $\dfrac{\partial f}{\partial z}$ measures _____ and has units _____ .

72. Complete the following: Let $f(t, x, y)$ be the projected number of citizens of the Principality State of Voodice, Luna in year t since its founding, assuming the presence of x lunar vehicle factories and y domed settlements. Then $\dfrac{\partial f}{\partial x}$ measures _____ and has units _____ .

73. Give an example of a function $f(x, y)$ with $f_x(1, 1) = -2$ and $f_y(1, 1) = 3$.

74. Give an example of a function $f(x, y, z)$ that has all of its partial derivatives equal to nonzero constants.

75. ▼ The graph of $z = b + mx + ny$ (b, m, and n constants) is a plane.

   a. Explain the geometric significance of the numbers b, m, and n.

   b. Show that the equation of the plane passing through (h, k, l) with slope m in the x direction (in the sense of $\partial/\partial x$) and slope n in the y direction is

$$z = l + m(x - h) + n(y - k).$$

76. ▼ The **tangent plane** to the graph of $f(x, y)$ at $P(a, b, f(a, b))$ is the plane containing the lines tangent to the slice through the graph by $y = b$ (as in Figure 19) and the slice through the graph by $x = a$. Use the result of the preceding exercise to show that the equation of the tangent plane is

$$z = f(a, b) + f_x(a, b)(x - a) + f_y(a, b)(y - b).$$

15.3   Maxima and Minima

In Chapter 12, on applications of the derivative, we saw how to locate relative extrema of a function of a single variable. In this section we extend our methods to functions of two variables. Similar techniques work for functions of three or more variables.

Figure 21 shows a portion of the graph of the function

$$f(x, y) = 2(x^2 + y^2) - (x^4 + y^4) + 1.$$

The graph in Figure 21 resembles a "flying carpet," and several interesting points, marked a, b, c, and d, are shown.

1. The point a has coordinates $(0, 0, f(0, 0))$, is directly above the origin $(0, 0)$, and is the lowest point in its vicinity; water would puddle there. We say that f has a **relative minimum** at $(0, 0)$ because $f(0, 0)$ is smaller than $f(x, y)$ for any (x, y) near $(0, 0)$.

2. Similarly, the point b is higher than any point in its vicinity. Thus, we say that f has a **relative maximum** at $(1, 1)$.

3. The points c and d represent a new phenomenon and are called **saddle points**. They are neither relative maxima nor relative minima but seem to be a little of both.

To see more clearly what features a saddle point has, look at Figure 22, which shows a portion of the graph near the point d.

If we slice through the graph along $y = 1$, we get a curve on which d is the *lowest* point. Thus, d looks like a relative minimum along this slice. On the other hand, if we slice through the graph along $x = 0$, we get another curve, on which d is the *highest*

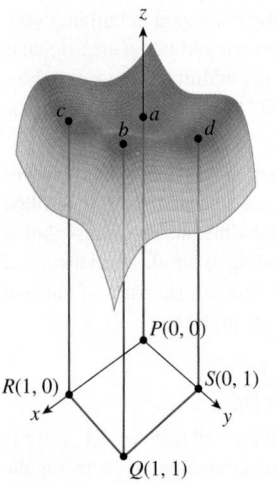

Figure 21

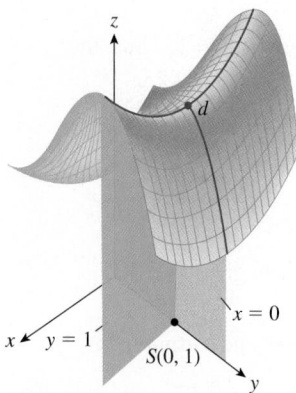

Figure 22

✳ For $(x_1, x_2, \ldots, x_n)$ to be near $(r_1, r_2, \ldots, r_n)$ we mean that x_1 is in some open interval centered at r_1, x_2 is in some open interval centered at r_2, and so on.

† One can use the techniques of the next section to find extrema on the *boundary* of the domain of a function; for a complete discussion, see the optional extra section: *Maxima and Minima: Boundaries and the Extreme Value Theorem*. (We shall not consider the analogs of the singular points.)

point, so d looks like a relative maximum along this slice. This kind of behavior characterizes a saddle point: f has a **saddle point** at (r, s) if f has a relative minimum at (r, s) along some slice through that point and a relative maximum along another slice through that point. If you look at the other saddle point, c, in Figure 21, you see the same characteristics.

While numerical information can help us locate the approximate positions of relative extrema and saddle points, calculus permits us to locate these points accurately, as we did for functions of a single variable. Look once again at Figure 21, and notice the following:

- The points P, Q, R, and S are all in the **interior** of the domain of f; that is, none lie on the boundary of the domain. Said another way, we can move some distance in any direction from any of these points without leaving the domain of f.

- The tangent lines along the slices through these points parallel to the x- and y-axes are *horizontal*. Thus, the partial derivatives $\partial f/\partial x$ and $\partial f/\partial y$ are zero when evaluated at any of the points P, Q, R, and S. This gives us a way of locating candidates for relative extrema and saddle points.

The following summary generalizes and also expands on some of what we have just said:

Relative and Absolute Maxima and Minima

The function f of n variables has a **relative maximum** at $(x_1, x_2, \ldots, x_n) = (r_1, r_2, \ldots, r_n)$ if $f(r_1, r_2, \ldots, r_n) \geq f(x_1, x_2, \ldots, x_n)$ for every point $(x_1, x_2, \ldots, x_n)$ near✳ $(r_1, r_2, \ldots, r_n)$ in the domain of f. We say that f has an **absolute maximum** at $(r_1, r_2, \ldots, r_n)$ if $f(r_1, r_2, \ldots, r_n) \geq f(x_1, x_2, \ldots, x_n)$ for every point $(x_1, x_2, \ldots, x_n)$ in the domain of f. The terms **relative minimum** and **absolute minimum** are defined in a similar way. Note that, as with functions of a single variable, absolute extrema are special kinds of relative extrema.

Locating Candidates for Extrema and Saddle Points in the Interior of the Domain of f

- Set $\dfrac{\partial f}{\partial x_1} = 0, \dfrac{\partial f}{\partial x_2} = 0, \ldots, \dfrac{\partial f}{\partial x_n} = 0$, simultaneously, and solve for $x_1, x_2, \ldots, x_n$.

- Check that the resulting points $(x_1, x_2, \ldots, x_n)$ are in the interior of the domain of f.

Points at which all the partial derivatives of f are zero are called **critical points**. The critical points are the only candidates for extrema and saddle points in the interior of the domain of f, assuming that its partial derivatives are defined at every point.†

Quick Examples

In each of the following Quick Examples, the domain is the whole Cartesian plane, and the partial derivatives are defined at every point, so the critical points give us the only candidates for extrema and saddle points:

1. Let $f(x, y) = x^3 + (y - 1)^2$. Then $\dfrac{\partial f}{\partial x} = 3x^2$ and $\dfrac{\partial f}{\partial y} = 2(y - 1)$.

Thus, we solve the system

$$3x^2 = 0 \quad \text{and} \quad 2(y - 1) = 0.$$

The first equation gives $x = 0$, and the second gives $y = 1$. Thus, the only critical point is $(0, 1)$.

2. Let $f(x, y) = x^2 - 4xy + 8y$. Then $\dfrac{\partial f}{\partial x} = 2x - 4y$ and $\dfrac{\partial f}{\partial y} = -4x + 8$. Thus, we solve

$$2x - 4y = 0 \quad \text{and} \quad -4x + 8 = 0.$$

The second equation gives $x = 2$, and the first then gives $y = 1$. Thus, the only critical point is $(2, 1)$.

3. Let $f(x, y) = e^{-(x^2 + y^2)}$. Taking partial derivatives and setting them equal to zero gives

$$-2xe^{-(x^2 + y^2)} = 0 \qquad \text{We set } \dfrac{\partial f}{\partial x} = 0.$$

$$-2ye^{-(x^2 + y^2)} = 0. \qquad \text{We set } \dfrac{\partial f}{\partial y} = 0.$$

The first equation implies that $x = 0$,* and the second implies that $y = 0$. Thus, the only critical point is $(0, 0)$.

✱ Recall that if a product of two numbers is zero, then one or the other must be zero. In this case the number $e^{-(x^2 + y^2)}$ can't be zero (because e^u is never zero), which gives the result claimed.

In the remainder of this section we will be interested in locating all critical points of a given function and then classifying each one as a relative maximum, minimum, saddle point, or none of these. Whether or not any relative extrema we find are in fact absolute is a subject we discuss in the next section.†

† In some of the applications in the exercises you will, however, need to consider whether the extrema you find are absolute.

EXAMPLE 1 **Locating and Classifying Critical Points**

Locate all critical points of $f(x, y) = x^2 y - x^2 - 2y^2$. Graph the function to classify the critical points as relative maxima, minima, saddle points, or none of these.

Solution The partial derivatives are

$$f_x = 2xy - 2x = 2x(y - 1)$$
$$f_y = x^2 - 4y.$$

Setting these equal to zero gives

$$x = 0 \text{ or } y = 1$$
$$x^2 = 4y.$$

We get a solution by choosing either $x = 0$ or $y = 1$ and substituting into $x^2 = 4y$.

Case 1: $x = 0$ Substituting into $x^2 = 4y$ gives $0 = 4y$ and hence $y = 0$. Thus, the critical point for this case is $(x, y) = (0, 0)$.

Case 2: $y = 1$ Substituting into $x^2 = 4y$ gives $x^2 = 4$ and hence $x = \pm 2$. Thus, we get two critical points for this case: $(2, 1)$ and $(-2, 1)$.

We now have three critical points altogether: $(0, 0)$, $(2, 1)$, and $(-2, 1)$. Because the domain of f is the whole Cartesian plane and the partial derivatives are defined at every point, these critical points are the only candidates for relative extrema and saddle points. We get the corresponding points on the graph by substituting for x and y in the equation for f to get the z-coordinates. The points are $(0, 0, 0)$, $(2, 1, -2)$, and $(-2, 1, -2)$.

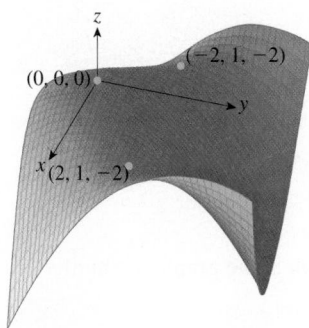

Figure 23

T *Classifying the Critical Points Graphically*   To classify the critical points graphically, we look at the graph of f shown in Figure 23.

Examining the graph carefully, we see that the point $(0, 0, 0)$ is a relative maximum. As for the other two critical points, are they saddle points or are they relative maxima? They are relative maxima along the y-direction, but they are relative minima along the lines $y = \pm x$ (see the top edge of the picture, which shows a dip at $(-2, 1, 2)$) and so they are saddle points. If you don't believe this, we will get more evidence following and in a later example.

T *Classifying the Critical Points Numerically*   We can use a tabular representation of the function to classify the critical points numerically. The following tabular representation of the function can be obtained using a spreadsheet. (See the Spreadsheet Technology Guide discussion of Section 15.1 Example 3 at the end of the chapter for information on using a spreadsheet to generate such a table.)

		$x \rightarrow$						
		-3	**-2**	**-1**	**0**	**1**	**2**	**3**
y	**-3**	-54	-34	-22	-18	-22	-34	-54
$\downarrow$	**-2**	-35	-20	-11	-8	-11	-20	-35
	-1	-20	-10	-4	-2	-4	-10	-20
	0	-9	-4	-1	0	-1	-4	-9
	1	-2	-2	-2	-2	-2	-2	-2
	2	1	-4	-7	-8	-7	-4	1
	3	0	-10	-16	-18	-16	-10	0

The shaded and colored cells show rectangular neighborhoods of the three critical points $(0, 0)$, $(2, 1)$, and $(-2, 1)$. (Notice that they overlap.) The values of f at the critical points are at the centers of these rectangles. Looking at the gray neighborhood of $(x, y) = (0, 0)$, we see that $f(0, 0) = 0$ is the largest value of f in the shaded cells, suggesting that f has a maximum at $(0, 0)$. The shaded neighborhood of $(2, 1)$ on the right shows $f(2, 1) = -2$ as the maximum along some slices (e.g., the vertical slice), and a minimum along the diagonal slice from top left to bottom right. This is what results in a saddle point on the graph. The point $(-2, 1)$ is similar, and thus f also has a saddle point at $(-2, 1)$.

Q: *Is there an algebraic way of deciding whether a given point is a relative maximum, relative minimum, or saddle point?*

A: There is a "second derivative test" for functions of two variables, stated as follows.

Second Derivative Test for Functions of Two Variables

Suppose (a, b) is a critical point in the interior of the domain of the function f of two variables. Let H be the quantity

$$H = f_{xx}(a, b)f_{yy}(a, b) - [f_{xy}(a, b)]^2.$$ *H* is called the *Hessian*.

Then, if H is *positive*,

- f has a relative minimum at (a, b) if $f_{xx}(a, b) > 0$.
- f has a relative maximum at (a, b) if $f_{xx}(a, b) < 0$.

If H is *negative*,

- f has a saddle point at (a, b).

If $H = 0$, the test tells us nothing, so we need to look at the graph or a numerical table to see what is going on.

Quick Examples

1. Let $f(x, y) = x^2 - y^2$. Then
$$f_x = 2x \quad \text{and} \quad f_y = -2y,$$
which gives $(0, 0)$ as the only critical point. Also,
$$f_{xx} = 2, f_{xy} = 0, \quad \text{and} \quad f_{yy} = -2, \qquad \text{Note that these are constant.}$$
which gives $H = (2)(-2) - 0^2 = -4$. Because H is negative, we have a saddle point at $(0, 0)$.

2. Let $f(x, y) = x^2 + 2y^2 + 2xy + 4x$. Then
$$f_x = 2x + 2y + 4 \quad \text{and} \quad f_y = 2x + 4y.$$
Setting these equal to zero gives a system of two linear equations in two unknowns:
$$x + y = -2$$
$$x + 2y = 0.$$
This system has solution $(-4, 2)$, so this is our only critical point. The second partial derivatives are $f_{xx} = 2$, $f_{xy} = 2$, and $f_{yy} = 4$, so $H = (2)(4) - 2^2 = 4$. Because $H > 0$ and $f_{xx} > 0$, we have a relative minimum at $(-4, 2)$.

Note There is a second derivative test for functions of three or more variables, but it is considerably more complicated. We stick with functions of two variables for the most part in this book. The justification of the second derivative test is beyond the scope of this book. ∎

EXAMPLE 2 Using the Second Derivative Test

Use the second derivative test to analyze the function $f(x, y) = x^2y - x^2 - 2y^2$ discussed in Example 1, and confirm the results we got there.

Solution We saw in Example 1 that the first-order derivatives are
$$f_x = 2xy - 2x = 2x(y - 1)$$
$$f_y = x^2 - 4y$$
and the critical points are $(0, 0), (2, 1)$, and $(-2, 1)$. We also need the second derivatives:
$$f_{xx} = 2y - 2$$
$$f_{xy} = 2x$$
$$f_{yy} = -4.$$

The point (0, 0): $f_{xx}(0, 0) = -2$, $f_{xy}(0, 0) = 0$, $f_{yy}(0, 0) = -4$, so $H = 8$. Because $H > 0$ and $f_{xx}(0, 0) < 0$, the second derivative test tells us that f has a relative maximum at $(0, 0)$.

The point (2, 1): $f_{xx}(2, 1) = 0$, $f_{xy}(2, 1) = 4$ and $f_{yy}(2, 1) = -4$, so $H = -16$. Because $H < 0$, we know that f has a saddle point at $(2, 1)$.

The point (−2, 1): $f_{xx}(-2, 1) = 0$, $f_{xy}(-2, 1) = -4$ and $f_{yy}(-2, 1) = -4$, so once again $H = -16$, and f has a saddle point at $(-2, 1)$.

Deriving the Formulas for Linear Regression

Back in Section 1.4, we presented the following set of formulas for the **regression** or **best-fit** line associated with a given set of data points $(x_1, y_1), (x_2, y_2), \ldots, (x_n, y_n)$.

Regression Line

The line that best fits the n data points $(x_1, y_1), (x_2, y_2), \ldots, (x_n, y_n)$ has the form

$$y = mx + b,$$

where

$$m = \frac{n\left(\sum xy\right) - \left(\sum x\right)\left(\sum y\right)}{n\left(\sum x^2\right) - \left(\sum x\right)^2}$$

$$b = \frac{\sum y - m\left(\sum x\right)}{n}$$

n = number of data points.

Derivation of the Regression Line Formulas

Recall that the regression line is defined to be the line that minimizes the sum of the squares of the **residuals**, measured by the vertical distances shown in Figure 24, which shows a regression line associated with $n = 5$ data points. In the figure, the points $P_1, \ldots, P_n$ on the regression line have coordinates $(x_1, mx_1 + b), (x_2, mx_2 + b), \ldots, (x_n, mx_n + b)$. The residuals are the quantities $y_{\text{Observed}} - y_{\text{Predicted}}$:

$$y_1 - (mx_1 + b), \, y_2 - (mx_2 + b), \ldots, y_n - (mx_n + b).$$

The sum of the squares of the residuals is therefore

$$S(m, b) = [y_1 - (mx_1 + b)]^2 + [y_2 - (mx_2 + b)]^2 + \cdots + [y_n - (mx_n + b)]^2$$

and this is the quantity we must minimize by choosing m and b. Because we reason that there is a line that minimizes this quantity, there must be a relative minimum at that point. We shall see in a moment that the function S has at most one critical point, which must therefore be the desired absolute minimum. To obtain the critical points of S, we set the partial derivatives equal to zero and solve:

$$S_m = 0: \quad -2x_1[y_1 - (mx_1 + b)] - \cdots - 2x_n[y_n - (mx_n + b)] = 0$$

$$S_b = 0: \quad -2[y_1 - (mx_1 + b)] - \cdots - 2[y_n - (mx_n + b)] = 0.$$

Dividing by -2 and gathering terms allows us to rewrite the equations as

$$m\left(x_1^2 + \cdots + x_n^2\right) + b(x_1 + \cdots + x_n) = x_1 y_1 + \cdots + x_n y_n$$

$$m(x_1 + \cdots + x_n) + nb \qquad\qquad = y_1 + \cdots + y_n.$$

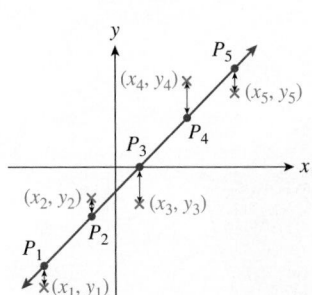

Figure 24

We can rewrite these equations more neatly using $\sum$-notation:

$$m\left(\sum x^2\right) + b\left(\sum x\right) = \sum xy$$
$$m\left(\sum x\right) + nb \qquad = \sum y.$$

This is a system of two linear equations in the two unknowns m and b. It may or may not have a unique solution. When there is a unique solution, we can conclude that the best-fit line is given by solving these two equations for m and b. Alternatively, there is a general formula for the solution of any system of two equations in two unknowns, and if we apply this formula to our two equations, we get the regression formulas above.

15.3   EXERCISES

▼ more advanced   ◆ challenging
⊤ indicates exercises that should be solved using technology

In Exercises 1–4, classify each labeled point on the graph as one of the following:

> *Relative maximum*
> *Relative minimum*
> *Saddle point*
> *Critical point but neither a relative extremum nor a saddle point*
> *None of the above* HINT *[See Example 1.]*

1.

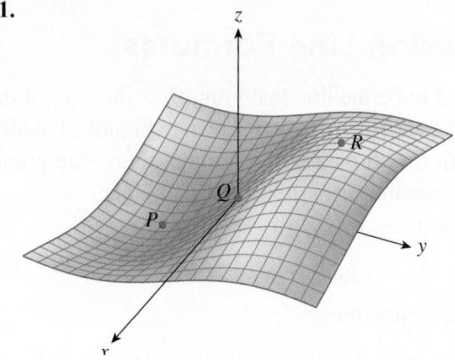

2.

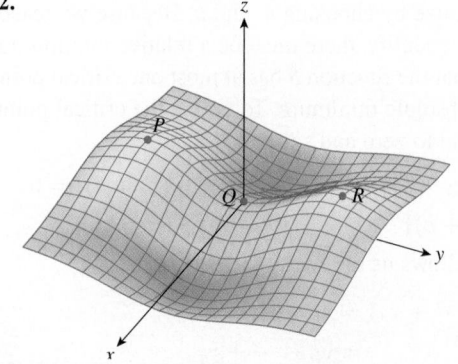

3.

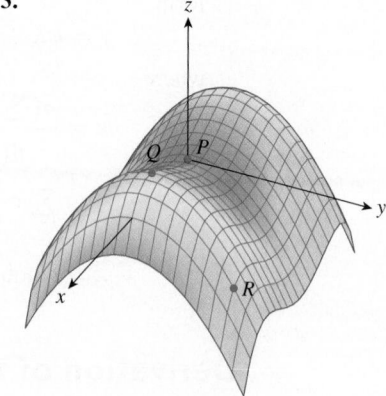

4.

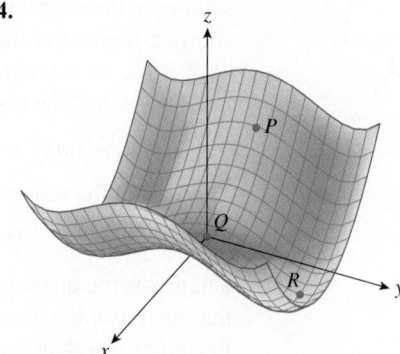

In Exercises 5–10, classify the shaded value in each table as one of the following:

> *Relative maximum*
> *Relative minimum*
> *Saddle point*
> *Neither a relative extremum nor a saddle point*

Assume that the shaded value represents a critical point.

5.

$x \rightarrow$

$y \downarrow$	-3	-2	-1	0	1	2
-3	10	5	2	1	2	5
-2	9	4	1	0	1	4
-1	10	5	2	1	2	5
0	13	8	5	4	5	8
1	18	13	10	9	10	13
2	25	20	17	16	17	20
3	34	29	26	25	26	29

6.

$x \rightarrow$

$y \downarrow$	-3	-2	-1	0	1	2
-3	5	0	-3	-4	-3	0
-2	8	3	0	-1	0	3
-1	9	4	1	0	1	4
0	8	3	0	-1	0	3
1	5	0	-3	-4	-3	0
2	0	-5	-8	-9	-8	-5
3	-7	-12	-15	-16	-15	-12

7.

$x \rightarrow$

$y \downarrow$	-3	-2	-1	0	1	2
-3	5	0	-3	-4	-3	0
-2	8	3	0	-1	0	3
-1	9	4	1	0	1	4
0	8	3	0	-1	0	3
1	5	0	-3	-4	-3	0
2	0	-5	-8	-9	-8	-5
3	-7	-12	-15	-16	-15	-12

8.

$x \rightarrow$

$y \downarrow$	-3	-2	-1	0	1	2
-3	2	3	2	-1	-6	-13
-2	3	4	3	0	-5	-12
-1	2	3	2	-1	-6	-13
0	-1	0	-1	-4	-9	-16
1	-6	-5	-6	-9	-14	-21
2	-13	-12	-13	-16	-21	-28
3	-22	-21	-22	-25	-30	-37

9.

$x \rightarrow$

$y \downarrow$	-3	-2	-1	0	1	2
-3	4	5	4	1	-4	-11
-2	3	4	3	0	-5	-12
-1	4	5	4	1	-4	-11
0	7	8	7	4	-1	-8
1	12	13	12	9	4	-3
2	19	20	19	16	11	4
3	28	29	28	25	20	13

10.

$x \rightarrow$

$y \downarrow$	-3	-2	-1	0	1	2
-3	100	101	100	97	92	85
-2	99	100	99	96	91	84
-1	98	99	98	95	90	83
0	91	92	91	88	83	76
1	72	73	72	69	64	57
2	35	36	35	32	27	20
3	-26	-25	-26	-29	-34	-41

Locate and classify all the critical points of the functions in Exercises 11–36. HINT [See Example 2.]

11. $f(x, y) = x^2 + y^2 + 1$

12. $f(x, y) = 4 - (x^2 + y^2)$

13. $g(x, y) = 1 - x^2 - x - y^2 + y$

14. $g(x, y) = x^2 + x + y^2 - y - 1$

15. $k(x, y) = x^2 - 3xy + y^2$

16. $k(x, y) = x^2 - xy + 2y^2$

17. $f(x, y) = x^2 + 2xy + 2y^2 - 2x + 4y$

18. $f(x, y) = x^2 + xy - y^2 + 3x - y$

19. $g(x, y) = -x^2 - 2xy - 3y^2 - 3x - 2y$

20. $g(x, y) = -x^2 - 2xy + y^2 + x - 4y$

21. $h(x, y) = x^2 y - 2x^2 - 4y^2$

22. $h(x, y) = x^2 + y^2 - y^2 x - 4$

23. $f(x, y) = x^2 + 2xy^2 + 2y^2$

24. $f(x, y) = x^2 + x^2 y + y^2$

25. $s(x, y) = e^{x^2 + y^2}$ **26.** $s(x, y) = e^{-(x^2 + y^2)}$

27. $t(x, y) = x^4 + 8xy^2 + 2y^4$ **28.** $t(x, y) = x^3 - 3xy + y^3$

29. $f(x, y) = x^2 + y - e^y$ **30.** $f(x, y) = xe^y$

31. $f(x, y) = e^{-(x^2 + y^2 + 2x)}$ **32.** $f(x, y) = e^{-(x^2 + y^2 - 2x)}$

33. ▼ $f(x, y) = xy + \dfrac{2}{x} + \dfrac{2}{y}$ **34.** ▼ $f(x, y) = xy + \dfrac{4}{x} + \dfrac{2}{y}$

35. ▼ $g(x, y) = x^2 + y^2 + \dfrac{2}{xy}$

36. ▼ $g(x, y) = x^3 + y^3 + \dfrac{3}{xy}$

37. ▼ Refer back to Exercise 11. Which (if any) of the critical points of $f(x, y) = x^2 + y^2 + 1$ are absolute extrema?

38. ▼ Refer back to Exercise 12. Which (if any) of the critical points of $f(x, y) = 4 - (x^2 + y^2)$ are absolute extrema?

39. ⬛ ▼ Refer back to Exercise 21. Which (if any) of the critical points of $h(x, y) = x^2y - 2x^2 - 4y^2$ are absolute extrema?

40. ⬛ ▼ Refer back to Exercise 22. Which (if any) of the critical points of $h(x, y) = x^2 + y^2 - y^2x - 4$ are absolute extrema?

APPLICATIONS

41. *Brand Loyalty* Suppose the fraction of Mazda car owners who chose another new Mazda can be modeled by the following function:[22]

$$M(c, f) = 11 + 8c + 4c^2 - 40f - 20cf + 40f^2,$$

where c is the fraction of Chrysler car owners who remained loyal to Chrysler and f is the fraction of Ford car owners remaining loyal to Ford. Locate and classify all the critical points and interpret your answer. HINT [See Example 2.]

42. *Brand Loyalty* Repeat the preceding exercise using the function:

$$M(c, f) = -10 - 8f - 4f^2 + 40c + 20fc - 40c^2.$$

HINT [See Example 2.]

43. ▼ *Pollution Control* The cost of controlling emissions at a firm goes up rapidly as the amount of emissions reduced goes up. Here is a possible model:

$$C(x, y) = 4,000 + 100x^2 + 50y^2,$$

where x is the reduction in sulfur emissions, y is the reduction in lead emissions (in pounds of pollutant per day), and C is the daily cost to the firm (in dollars) of this reduction. Government clean-air subsidies amount to $500 per pound of sulfur and $100 per pound of lead removed. How many pounds of pollutant should the firm remove each day in order to minimize *net* cost (cost minus subsidy)?

44. ▼ *Pollution Control* Repeat the preceding exercise using the following information:

$$C(x, y) = 2,000 + 200x^2 + 100y^2$$

with government subsidies amounting to $100 per pound of sulfur and $500 per pound of lead removed per day.

45. ▼ *Revenue* Your company manufactures two models of speakers, the Ultra Mini and the Big Stack. Demand for each depends partly on the price of the other. If one is expensive, then more people will buy the other. If p_1 is the price of the Ultra Mini, and p_2 is the price of the Big Stack, demand for the Ultra Mini is given by

$$q_1(p_1, p_2) = 100,000 - 100p_1 + 10p_2,$$

where q_1 represents the number of Ultra Minis that will be sold in a year. The demand for the Big Stack is given by

$$q_2(p_1, p_2) = 150,000 + 10p_1 - 100p_2.$$

Find the prices for the Ultra Mini and the Big Stack that will maximize your total revenue.

46. ▼ *Revenue* Repeat the preceding exercise, using the following demand functions:

$$q_1(p_1, p_2) = 100,000 - 100p_1 + p_2$$
$$q_2(p_1, p_2) = 150,000 + p_1 - 100p_2.$$

47. ▼ *Luggage Dimensions: American Airlines* American Airlines requires that the total outside dimensions (length + width + height) of a checked bag not exceed 62 inches.[23] What are the dimensions of the largest volume bag that you can check on an American flight?

48. ▼ *Carry-on Bag Dimensions: American Airlines* American Airlines requires that the total outside dimensions (length + width + height) of a carry-on bag not exceed 45 inches.[24] What are the dimensions of the largest volume bag that you can carry on an American flight?

49. ▼ *Package Dimensions: USPS* The U.S. Postal Service (USPS) will accept only packages with a length plus girth no more than 108 inches.[25] (See the figure.)

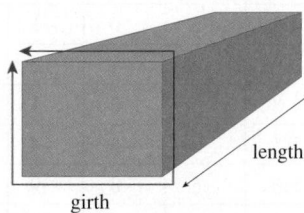

What are the dimensions of the largest volume package that the USPS will accept? What is its volume?

50. ▼ *Package Dimensions: UPS* United Parcel Service (UPS) will accept only packages with length no more than 108 inches and length plus girth no more than 165 inches.[26] (See figure for the preceding exercise.) What are the dimensions of the largest volume package that UPS will accept? What is its volume?

---

[22]This model is not accurate, although it was inspired by an approximation of a second-order regression based on data from the period 1988–1995. Source for original data: Chrysler, Maritz Market Research, Consumer Attitude Research, and Strategic Vision/*New York Times*, November 3, 1995, p. D2.

[23]According to information on its Web site (www.aa.com).

[24]*Ibid.*

[25]The requirement for packages sent other than Parcel Post, as of August 2011 (www.usps.com).

[26]The requirement as of August 2011 (www.ups.com).

COMMUNICATION AND REASONING EXERCISES

51. Sketch the graph of a function that has one extremum and no saddle points.

52. Sketch the graph of a function that has one saddle point and one extremum.

53. ▼ Sketch the graph of a function that has one relative extremum, no absolute extrema, and no saddle points.

54. ▼ Sketch the graph of a function that has infinitely many absolute maxima.

55. Let $H = f_{xx}(a, b) f_{yy}(a, b) - f_{xy}(a, b)^2$. What condition on H guarantees that f has a relative extremum at the point (a, b)?

56. Let H be as in the preceding exercise. Give an example to show that it is possible to have $H = 0$ and a relative minimum at (a, b).

57. ▼ Suppose that when the graph of $f(x, y)$ is sliced by a vertical plane through (a, b) parallel to either the xz-plane or the yz-plane, the resulting curve has a relative maximum at (a, b). Does this mean that f has a relative maximum at (a, b)? Explain your answer.

58. ▼ Suppose that f has a relative maximum at (a, b). Does it follow that, if the graph of f is sliced by a vertical plane parallel to either the xz-plane or the yz-plane, the resulting curve has a relative maximum at (a, b)? Explain your answer.

59. ▼ *Average Cost* Let $C(x, y)$ be any cost function. Show that when the average cost is minimized, the marginal costs C_x and C_y both equal the average cost. Explain why this is reasonable.

60. ▼ *Average Profit* Let $P(x, y)$ be any profit function. Show that when the average profit is maximized, the marginal profits P_x and P_y both equal the average profit. Explain why this is reasonable.

61. ◆ The tangent plane to a graph was introduced in Exercise 76 in the preceding section. Use the equation of the tangent plane given there to explain why the tangent plane is parallel to the xy-plane at a relative maximum or minimum of $f(x, y)$.

62. ◆ Use the equation of the tangent plane given in Exercise 76 in the preceding section to explain why the tangent plane is parallel to the xy-plane at a saddle point of $f(x, y)$.

15.4 Constrained Maxima and Minima and Applications

So far we have looked only at the relative extrema of functions with no constraints. However, in Section 12.2 we saw examples in which we needed to find the maximum or minimum of an objective function subject to one or more constraints on the independent variables. For instance, consider the following problem:

$$\text{Minimize } S = xy + 2xz + 2yz \quad \text{subject to } xyz = 4 \text{ with } x > 0, y > 0, z > 0.$$

One strategy for solving such problems is essentially the same as the strategy we used earlier: Solve the constraint equation for one of the variables, substitute into the objective function, and then optimize the resulting function using the methods of the preceding section. We will call this the *substitution method*.* An alternative method, called the *method of Lagrange multipliers*, is useful when it is difficult or impossible to solve the constraint equation for one of the variables, and even when it is possible to do so.

✱ Although often the method of choice, the substitution method is not infallible (see Exercises 19 and 20).

Substitution Method

EXAMPLE 1 Using Substitution

Minimize $S = xy + 2xz + 2yz$ subject to $xyz = 4$ with $x > 0, y > 0, z > 0$.

Solution As suggested in the above discussion, we proceed as follows:

Solve the constraint equation for one of the variables and then substitute in the objective function. The constraint equation is $xyz = 4$. Solving for z gives

$$z = \frac{4}{xy}.$$

The objective function is $S = xy + 2xz + 2yz$, so substituting $z = 4/xy$ gives

$$S = xy + 2x\frac{4}{xy} + 2y\frac{4}{xy}$$

$$= xy + \frac{8}{y} + \frac{8}{x}.$$

Minimize the resulting function of two variables. We use the method in Section 15.4 to find the minimum of $S = xy + \frac{8}{y} + \frac{8}{x}$ for $x > 0$ and $y > 0$. We look for critical points:

$$S_x = y - \frac{8}{x^2} \qquad S_y = x - \frac{8}{y^2}$$

$$S_{xx} = \frac{16}{x^3} \qquad S_{xy} = 1 \qquad S_{yy} = \frac{16}{y^3}.$$

We now equate the first partial derivatives to zero:

$$y = \frac{8}{x^2} \qquad \text{and} \qquad x = \frac{8}{y^2}.$$

To solve for x and y, we substitute the first of these equations in the second, getting

$$x = \frac{x^4}{8}$$

$$x^4 - 8x = 0$$

$$x(x^3 - 8) = 0.$$

The two solutions are $x = 0$, which we reject because x cannot be zero, and $x = 2$. Substituting $x = 2$ in $y = 8/x^2$ gives $y = 2$ also. Thus, the only critical point is $(2, 2)$. To apply the second derivative test, we compute

$$S_{xx}(2, 2) = 2 \qquad S_{xy}(2, 2) = 1 \qquad S_{yy}(2, 2) = 2$$

and find that $H = 3 > 0$ and $S_{xx}(2, 2) > 0$, so we have a relative minimum at $(2, 2)$. The corresponding value of z is given by the constraint equation:

$$z = \frac{4}{xy} = \frac{4}{4} = 1.$$

The corresponding value of the objective function is

$$S = xy + \frac{8}{y} + \frac{8}{x} = 4 + \frac{8}{2} + \frac{8}{2} = 12.$$

Figure 25 shows a portion of the graph of $S = xy + \frac{8}{y} + \frac{8}{x}$ for positive x and y (drawn using the Excel Surface Grapher in the Chapter 15 utilities at the Website) and suggests that there is a single absolute minimum, which must be at our only candidate point $(2, 2)$.

We conclude that the minimum of S is 12 and occurs at $(2, 2, 1)$.

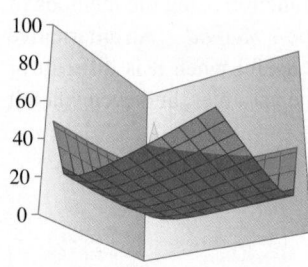

Graph of $S = xy + \dfrac{8}{y} + \dfrac{8}{x}$

$(0.2 \le x \le 5, 0.2 \le y \le 5)$

Figure 25

The Method of Lagrange Multipliers

As we mentioned above, the method of Lagrange multipliers has the advantage that it can be used in constrained optimization problems when it is difficult or impossible to solve a constraint equation for one of the variables. We restrict attention to the case of a single constraint equation, although the method also generalizes to any number of constraint equations.

Locating Relative Extrema Using the Method of Lagrange Multipliers

To locate the candidates for relative extrema of a function $f(x, y, \ldots)$ subject to the constraint $g(x, y, \ldots) = 0$:

1. Construct the **Lagrangian function**

$$L(x, y, \ldots) = f(x, y, \ldots) - \lambda g(x, y, \ldots)$$

where λ is a new unknown called a **Lagrange multiplier.**

2. The candidates for the relative extrema occur at the critical points of $L(x, y, \ldots)$. To find them, set all the partial derivatives of $L(x, y, \ldots)$ equal to zero and solve the resulting system, together with the constraint equation $g(x, y, \ldots) = 0$, for the unknowns $x, y, \ldots$ and λ.

The points $(x, y, \ldots)$ that occur in solutions are then the candidates for the relative extrema of f subject to $g = 0$.

Although the justification for the method of Lagrange multipliers is beyond the scope of this text (a derivation can be found in many vector calculus textbooks), we will demonstrate by example how it is used.

EXAMPLE 2 Using Lagrange Multipliers

Use the method of Lagrange multipliers to find the maximum value of $f(x, y) = 2xy$ subject to $x^2 + 4y^2 = 32$.

Solution We start by rewriting the problem with the constraint in the form $g(x, y) = 0$:

Maximize $f(x, y) = 2xy$ subject to $x^2 + 4y^2 - 32 = 0$.

Here, $g(x, y) = x^2 + 4y^2 - 32$, and the Lagrangian function is

$$L(x, y) = f(x, y) - \lambda g(x, y)$$
$$= 2xy - \lambda(x^2 + 4y^2 - 32).$$

The system of equations we need to solve is thus

$$L_x = 0: \qquad 2y - 2\lambda x = 0$$
$$L_y = 0: \qquad 2x - 8\lambda y = 0$$
$$g = 0: \qquad x^2 + 4y^2 - 32 = 0.$$

It is often convenient to solve such a system by first solving one of the equations for λ and then substituting in the remaining equations. Thus, we start by solving the first equation to obtain

$$\lambda = \frac{y}{x}.$$

(A word of caution: Because we divided by x, we made the implicit assumption that $x \neq 0$, so before continuing we should check what happens if $x = 0$. But if $x = 0$, then the first equation, $2y = 2\lambda x$, tells us that $y = 0$ as well, and this contradicts the third equation: $x^2 + 4y^2 - 32 = 0$. Thus, we can rule out the possibility that $x = 0$.) Substituting the value of λ in the second equation gives

$$2x - 8\left(\frac{y}{x}\right)y = 0 \quad \text{or} \quad x^2 = 4y^2.$$

We can now substitute $x^2 = 4y^2$ in the constraint equation, obtaining

$$4y^2 + 4y^2 - 32 = 0$$
$$8y^2 = 32$$
$$y = \pm 2.$$

We now substitute back to obtain

$$x^2 = 4y^2 = 16,$$

or $x = \pm 4.$

We don't need the value of λ, so we won't solve for it. Thus, the candidates for relative extrema are given by $x = \pm 4$ and $y = \pm 2$; that is, the four points $(-4, -2)$, $(-4, 2)$, $(4, -2)$, and $(4, 2)$. Recall that we are seeking the values of x and y that give the maximum value for $f(x, y) = 2xy$. Because we now have only four points to choose from, we compare the values of f at these four points and conclude that the maximum value of f occurs when $(x, y) = (-4, -2)$ or $(4, 2)$.

Something is suspicious in Example 2. We didn't check to see whether these candidates were relative extrema to begin with, let alone absolute extrema! How do we justify this omission? One of the difficulties with using the method of Lagrange multipliers is that it does not provide us with a test analogous to the second derivative test for functions of several variables. However, if you grant that the function in question does have an absolute maximum, then we require no test, because one of the candidates must give this maximum.

Q: *But how do we know that the given function has an absolute maximum?*

A: The best way to see this is by giving a geometric interpretation. The constraint $x^2 + 4y^2 = 32$ tells us that the point (x, y) must lie on the ellipse shown in Figure 26. The function $f(x, y) = 2xy$ gives the area of the rectangle shaded in the figure. There must be a *largest* such rectangle, because the area varies continuously from 0 when (x, y) is on the x-axis, to positive when (x, y) is in the first quadrant, to 0 again when (x, y) is on the y-axis, so f must have an absolute maximum for at least one pair of coordinates (x, y).

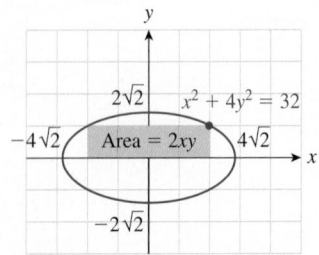

Figure 26

We now show how to use Lagrange multipliers to solve the minimization problem in Example 1:

EXAMPLE 3 Using Lagrange Multipliers: Function of Three Variables

Use the method of Lagrange multipliers to find the minimum value of $S = xy + 2xz + 2yz$ subject to $xyz = 4$ with $x > 0$, $y > 0$, $z > 0$.

Solution We start by rewriting the problem in standard form:

$$\text{Maximize } f(x, y, z) = xy + 2xz + 2yz$$
$$\text{subject to } xyz - 4 = 0 \text{ (with } x > 0, y > 0, z > 0).$$

Here, $g(x, y, z) = xyz - 4$, and the Lagrangian function is

$$L(x, y, z) = f(x, y, z) - \lambda g(x, y, z)$$
$$= xy + 2xz + 2yz - \lambda(xyz - 4).$$

The system of equations we need to solve is thus

$$L_x = 0: \quad y + 2z - \lambda yz = 0$$
$$L_y = 0: \quad x + 2z - \lambda xz = 0$$
$$L_z = 0: \quad 2x + 2y - \lambda xy = 0$$
$$g = 0: \quad xyz - 4 = 0.$$

As in the last example, we solve one of the equations for λ and substitute in the others. The first equation gives

$$\lambda = \frac{1}{z} + \frac{2}{y}.$$

Substituting this into the second equation gives

$$x + 2z = x + \frac{2xz}{y}$$

or $\qquad 2 = \dfrac{2x}{y},$ $\qquad\qquad$ Subtract x from both sides and then divide by z.

giving $\qquad y = x.$

Substituting the expression for λ into the third equation gives

$$2x + 2y = \frac{xy}{z} + 2x$$

or $\qquad 2 = \dfrac{x}{z},$ $\qquad\qquad$ Subtract $2x$ from both sides and then divide by y.

giving $\qquad z = \dfrac{x}{2}.$

Now we have both y and z in terms of x. We substitute these values in the last (constraint) equation:

$$x(x)\left(\frac{x}{2}\right) - 4 = 0$$
$$x^3 = 8$$
$$x = 2.$$

Thus, $y = x = 2$, and $z = \dfrac{x}{2} = 1$. Therefore, the only critical point occurs at $(2, 2, 1)$, as we found in Example 1, and the corresponding value of S is

$$S = xy + 2xz + 2yz = (2)(2) + 2(2)(1) + 2(2)(1) = 12.$$

➡ **Before we go on...** Again, the method of Lagrange multipliers does not tell us whether the critical point in Example 3 is a maximum, minimum, or neither. However, if you grant that the function in question does have an absolute minimum, then the values we found must give this minimum value. ■

APPLICATIONS

EXAMPLE 4 **Minimizing Area**

Find the dimensions of an open-top rectangular box that has a volume of 4 cubic feet and the smallest possible surface area.

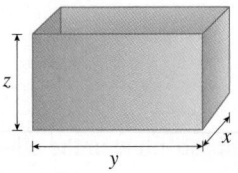

Figure 27

Solution Our first task is to rephrase this request as a mathematical optimization problem. Figure 27 shows a picture of the box with dimensions x, y, and z. We want to minimize the total surface area, which is given by

$$S = xy + 2xz + 2yz. \qquad \text{Base + Sides + Front and Back}$$

This is our objective function. We can't simply choose x, y, and z to all be zero, however, because the enclosed volume must be 4 cubic feet. So,

$$xyz = 4. \qquad \text{Constraint}$$

This is our constraint equation. Other unstated constraints are $x > 0$, $y > 0$, and $z > 0$, because the dimensions of the box must be positive. We now restate the problem as follows:

Minimize $S = xy + 2xz + 2yz$ subject to $xyz = 4$, $x > 0$, $y > 0$, $z > 0$.

But this is exactly the problem in Examples 1 and 3, and has a solution $x = 2$, $y = 2$, $z = 1$, $S = 12$. Thus, the required dimensions of the box are

$$x = 2 \text{ ft}, y = 2 \text{ ft}, z = 1 \text{ ft},$$

requiring a total surface area of 12 ft².

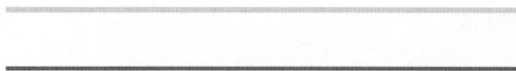

Q: *In Example 1 we checked that we had a relative minimum at* $(x, y) = (2, 2)$ *and we were persuaded graphically that this was probably an absolute minimum. Can we be* sure *that this relative minimum is an absolute minimum?*

A: Yes. There must be a least surface area among all boxes that hold 4 cubic feet. (Why?) Because this would give a relative minimum of S and because the only possible relative minimum of S occurs at $(2, 2)$, this is the absolute minimum.

Here, A is the area of R. We can compute the area A geometrically, or by using the techniques from the chapter on applications of the integral, or by computing

$$A = \iint_R 1 \, dx \, dy.$$

Quick Example

The average value of $f(x, y) = xy$ on the rectangle given by $0 \le x \le 1$ and $0 \le y \le 2$ is

$$\bar{f} = \frac{1}{2} \iint_R xy \, dx \, dy \qquad \text{The area of the rectangle is 2.}$$

$$= \frac{1}{2} \int_0^2 \int_0^1 xy \, dx \, dy$$

$$= \frac{1}{2} \cdot 1 = \frac{1}{2}. \qquad \text{We calculated the integral in Example 1.}$$

EXAMPLE 4 Average Revenue

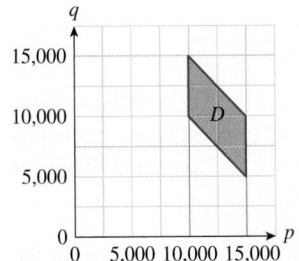

Figure 37

Your company is planning to price its new line of subcompact cars at between $10,000 and $15,000. The marketing department reports that if the company prices the cars at p dollars per car, the demand will be between $q = 20,000 - p$ and $q = 25,000 - p$ cars sold in the first year. What is the average of all the possible revenues your company could expect in the first year?

Solution Revenue is given by $R = pq$ as usual, and we are told that

$$10,000 \le p \le 15,000$$

and $$20,000 - p \le q \le 25,000 - p.$$

This domain D of prices and demands is shown in Figure 37.

To average the revenue R over the domain D, we need to compute the area A of D. Using either calculus or geometry, we get $A = 25,000,000$. We then need to integrate R over D:

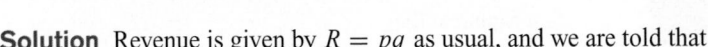

$$\iint_D pq \, dp \, dq = \int_{10,000}^{15,000} \int_{20,000-p}^{25,000-p} pq \, dq \, dp$$

$$= \int_{10,000}^{15,000} \left[\frac{pq^2}{2} \right]_{q=20,000-p}^{25,000-p} dp$$

$$= \frac{1}{2} \int_{10,000}^{15,000} [p(25,000-p)^2 - p(20,000-p)^2] \, dp$$

$$= \frac{1}{2} \int_{10,000}^{15,000} [225,000,000p - 10,000p^2] \, dp$$

$$\approx 3,072,900,000,000,000.$$

The average of all the possible revenues your company could expect in the first year is therefore

$$\bar{R} = \frac{3{,}072{,}900{,}000{,}000{,}000}{25{,}000{,}000} \approx \$122{,}900{,}000.$$

➡ **Before we go on...** To check that the answer obtained in Example 4 is reasonable, notice that the revenues at the corners of the domain are $100,000,000 per year, $150,000,000 per year (at two corners), and $75,000,000 per year. Some of these are smaller than the average and some larger, as we would expect. ■

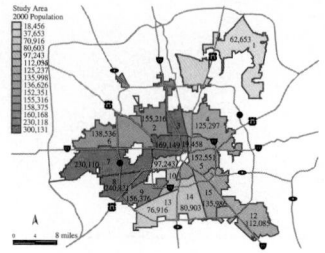

Darker regions have higher population density

Figure 38

Another useful application of the double integral comes about when we consider density. For example, suppose that $P(x, y)$ represents the population density (in people per square mile, say) in the city of Houston, shown in Figure 38.

If we break the city up into small rectangles (for example, city blocks), then the population in the small rectangle $x_{i-1} \le x \le x_i$ and $y_{j-1} \le y \le y_j$ is approximately $P(x_i, y_j)\Delta x \Delta y$. Adding up all of these population estimates, we get

$$\text{Total population} \approx \sum_{j=1}^{n} \sum_{i=1}^{m} P(x_i, y_j)\, \Delta x\, \Delta y.$$

Because this is a double Riemann sum, when we take the limit as m and n go to infinity, we get the following calculation of the population of the city:

$$\text{Total population} = \iint_{\text{City}} P(x, y)\, dx\, dy.$$

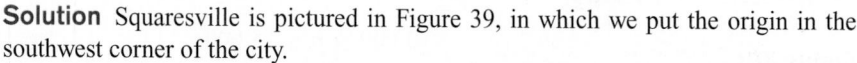

EXAMPLE 5 **Population**

Squaresville is a city in the shape of a square 5 miles on a side. The population density at a distance of x miles east and y miles north of the southwest corner is $P(x, y) = x^2 + y^2$ thousand people per square mile. Find the total population of Squaresville.

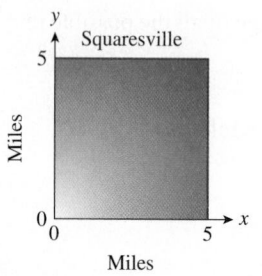

Figure 39

Solution Squaresville is pictured in Figure 39, in which we put the origin in the southwest corner of the city.

To compute the total population, we integrate the population density over the city S.

$$\text{Total population} = \iint_{\text{Squaresville}} P(x, y)\, dx\, dy$$

$$= \int_0^5 \int_0^5 (x^2 + y^2)\, dx\, dy$$

$$= \int_0^5 \left[\frac{x^3}{3} + xy^2 \right]_{x=0}^5 dy$$

$$= \int_0^5 \left[\frac{125}{3} + 5y^2 \right] dy$$

$$= \frac{1{,}250}{3} \approx 417 \text{ thousand people}$$

➡ **Before we go on...** Note that the average population density is the total population divided by the area of the city, which is about 17,000 people per square mile. Compare this calculation with the calculations of averages in the previous two examples. ■

15.5 EXERCISES

▼ more advanced   ◆ challenging
T indicates exercises that should be solved using technology

Compute the integrals in Exercises 1–16. HINT [See Example 1.]

1. $\int_0^1 \int_0^1 (x - 2y)\, dx\, dy$ **2.** $\int_{-1}^1 \int_0^2 (2x + 3y)\, dx\, dy$

3. $\int_0^1 \int_0^2 (ye^x - x - y)\, dx\, dy$ **4.** $\int_1^2 \int_2^3 \left(\frac{1}{x} + \frac{1}{y}\right) dx\, dy$

5. $\int_0^2 \int_0^3 e^{x+y}\, dx\, dy$ **6.** $\int_0^1 \int_0^1 e^{x-y}\, dx\, dy$

7. $\int_0^1 \int_0^{2-y} x\, dx\, dy$ **8.** $\int_0^1 \int_0^{2-y} y\, dx\, dy$

9. $\int_{-1}^1 \int_{y-1}^{y+1} e^{x+y}\, dx\, dy$ **10.** $\int_0^1 \int_y^{y+2} \frac{1}{\sqrt{x+y}}\, dx\, dy$

HINT [See Example 2.]   HINT [See Example 2.]

11. $\int_0^1 \int_{-x^2}^{x^2} x\, dy\, dx$ **12.** $\int_1^4 \int_{-\sqrt{x}}^{\sqrt{x}} \frac{1}{x}\, dy\, dx$

13. $\int_0^1 \int_0^x e^{x^2}\, dy\, dx$ **14.** $\int_0^1 \int_0^{x^2} e^{x^3+1}\, dy\, dx$

15. $\int_0^2 \int_{1-x}^{8-x} (x+y)^{1/3}\, dy\, dx$ **16.** $\int_1^2 \int_{1-2x}^{x^2} \frac{x+1}{(2x+y)^3}\, dy\, dx$

In Exercises 17–24, find $\iint_R f(x, y)\, dx\, dy$, where R is the indicated domain. (Remember that you often have a choice as to the order of integration.) HINT [See Example 2.]

17. $f(x, y) = 2$ **18.** $f(x, y) = x$

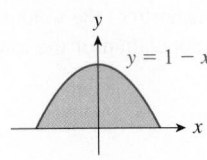   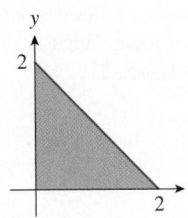

19. $f(x, y) = 1 + y$ **20.** $f(x, y) = e^{x+y}$
    HINT [See Example 3.]   HINT [See Example 3.]

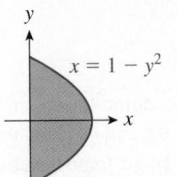   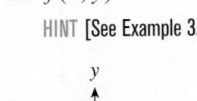
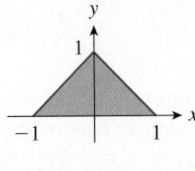

21. $f(x, y) = xy^2$ **22.** $f(x, y) = xy^2$

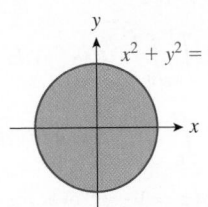   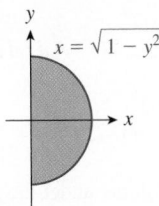

23. $f(x, y) = x^2 + y^2$ **24.** $f(x, y) = x^2$

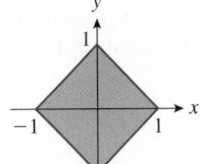   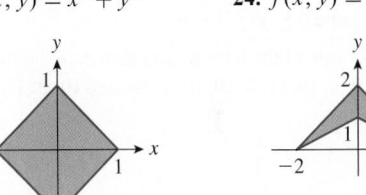

In Exercises 25–30, find the average value of the given function over the indicated domain. HINT [See Quick Example page 1133.]

25. $f(x, y) = y$ **26.** $f(x, y) = 2 + x$

   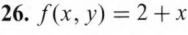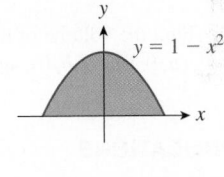

27. $f(x, y) = e^y$ **28.** $f(x, y) = y$

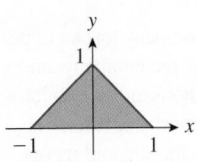   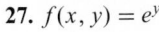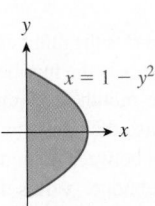

29. $f(x, y) = x^2 + y^2$ **30.** $f(x, y) = x^2$

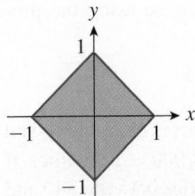   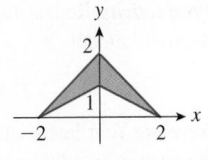

In Exercises 31–36, sketch the region over which you are integrating, and then write down the integral with the order of integration reversed (changing the limits of integration as necessary).

31. ▼ $\int_0^1 \int_0^{1-y} f(x, y)\, dx\, dy$ **32.** ▼ $\int_{-1}^1 \int_0^{1+y} f(x, y)\, dx\, dy$

33. ▼ $\int_{-1}^1 \int_0^{\sqrt{1+y}} f(x, y)\, dx\, dy$ **34.** ▼ $\int_{-1}^1 \int_0^{\sqrt{1-y}} f(x, y)\, dx\, dy$

35. ▼ $\int_1^2 \int_1^{4/x^2} f(x, y)\, dy\, dx$ **36.** ▼ $\int_1^{e^2} \int_0^{\ln x} f(x, y)\, dy\, dx$

37. Find the volume under the graph of $z = 1 - x^2$ over the region $0 \le x \le 1$ and $0 \le y \le 2$.

38. Find the volume under the graph of $z = 1 - x^2$ over the triangle $0 \le x \le 1$ and $0 \le y \le 1 - x$.

39. ▼ Find the volume of the tetrahedron shown in the figure. Its corners are $(0, 0, 0)$, $(1, 0, 0)$, $(0, 1, 0)$, and $(0, 0, 1)$.

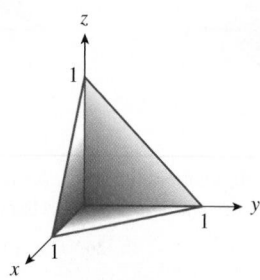

40. ▼ Find the volume of the tetrahedron with corners at $(0, 0, 0)$, $(a, 0, 0)$, $(0, b, 0)$, and $(0, 0, c)$.

APPLICATIONS

41. *Productivity* A productivity model at the *Handy Gadget Company* is

$$P = 10,000x^{0.3}y^{0.7},$$

where P is the number of gadgets the company turns out per month, x is the number of employees at the company, and y is the monthly operating budget in thousands of dollars. Because the company hires part-time workers, it uses anywhere between 45 and 55 workers each month, and its operating budget varies from \$8,000 to \$12,000 per month. What is the average of the possible numbers of gadgets it can turn out per month? (Round the answer to the nearest 1,000 gadgets.) HINT [See Quick Example page 1133.]

42. *Productivity* Repeat the preceding exercise using the productivity model

$$P = 10,000x^{0.7}y^{0.3}.$$

43. *Revenue* Your latest CD-ROM of clip art is expected to sell between $q = 8,000 - p^2$ and $q = 10,000 - p^2$ copies if priced at p dollars. You plan to set the price between \$40 and

\$50. What is the average of all the possible revenues you can make? HINT [See Example 4.]

44. *Revenue* Your latest DVD drive is expected to sell between $q = 180,000 - p^2$ and $q = 200,000 - p^2$ units if priced at p dollars. You plan to set the price between \$300 and \$400. What is the average of all the possible revenues you can make? HINT [See Example 4.]

45. *Revenue* Your self-published novel has demand curves between $p = 15,000/q$ and $p = 20,000/q$. You expect to sell between 500 and 1,000 copies. What is the average of all the possible revenues you can make?

46. *Revenue* Your self-published book of poetry has demand curves between $p = 80,000/q^2$ and $p = 100,000/q^2$. You expect to sell between 50 and 100 copies. What is the average of all the possible revenues you can make?

47. *Population Density* The town of West Podunk is shaped like a rectangle 20 miles from west to east and 30 miles from north to south. (See the figure.) It has a population density of $P(x, y) = e^{-0.1(x+y)}$ hundred people per square mile x miles east and y miles north of the southwest corner of town. What is the total population of the town? HINT [See Example 5.]

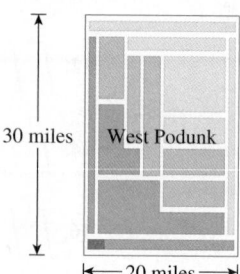

30 miles   West Podunk

← 20 miles →

48. *Population Density* The town of East Podunk is shaped like a triangle with an east-west base of 20 miles and a north-south height of 30 miles. (See the figure.) It has a population density of $P(x, y) = e^{-0.1(x+y)}$ hundred people per square mile x miles east and y miles north of the southwest corner of town. What is the total population of the town? HINT [See Example 5.]

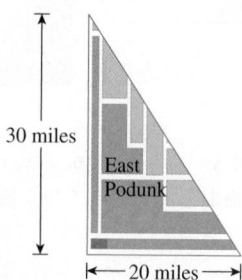

30 miles   East Podunk

← 20 miles →

49. *Temperature* The temperature at the point (x, y) on the square with vertices $(0, 0)$, $(0, 1)$, $(1, 0)$, and $(1, 1)$ is given by $T(x, y) = x^2 + 2y^2$. Find the average temperature on the square.

50. *Temperature* The temperature at the point (x, y) on the square with vertices $(0, 0)$, $(0, 1)$, $(1, 0)$, and $(1, 1)$ is given by $T(x, y) = x^2 + 2y^2 - x$. Find the average temperature on the square.

COMMUNICATION AND REASONING EXERCISES

51. Explain how double integrals can be used to compute the area between two curves in the xy plane.

52. Explain how double integrals can be used to compute the volume of solids in 3-space.

53. Complete the following: The first step in calculating an integral of the form $\int_a^b \int_{r(x)}^{s(x)} f(x, y)\, dy\, dx$ is to evaluate the integral _____, obtained by holding ____ constant and integrating with respect to ____ .

54. If the units of $f(x, y)$ are zonars per square meter, and x and y are given in meters, what are the units of $\int_a^b \int_{r(x)}^{s(x)} f(x, y)\, dy\, dx$?

55. If the units of $\int_a^b \int_{r(x)}^{s(x)} f(x, y)\, dy\, dx$ are paintings, the units of x are picassos, and the units of y are dalis, what are the units of $f(x, y)$?

56. Complete the following: If the region R is bounded on the left and right by vertical lines and on the top and bottom by the graphs of functions of x, then we integrate over R by first integrating with respect to ____ and then with respect to ____.

57. ▼ Show that if a, b, c, and d are constant, then $\int_a^b \int_c^d f(x)\, g(y)\, dx\, dy = \int_c^d f(x)\, dx \int_a^b g(y)\, dy$. Test this result on the integral $\int_0^1 \int_1^2 y e^x\, dx\, dy$.

58. ▼ Refer to Exercise 57. If a, b, c, and d are constants, can $\int_a^b \int_c^d \dfrac{f(x)}{g(y)}\, dx\, dy$ be expressed as a product of two integrals? Explain.

KEY CONCEPTS

 Website www.WanerMath.com
Go to the Website at www.WanerMath
.com to find a comprehensive and
interactive Web-based summary
of Chapter 15.

15.1 Functions of Several Variables from the Numerical, Algebraic, and Graphical Viewpoints

A real-valued function, f, of $x, y, z, \ldots$
p. 1082
Cost functions p. 1083
A linear function of the variables
$x_1, x_2, \ldots, x_n$ is a function of the form
$f(x_1, x_2, \ldots, x_n) = a_0 + a_1 x_1 + \cdots + a_n x_n$ ($a_0, a_1, \ldots, a_n$ constants)
p. 1084
Representing functions of two variables
numerically p. 1085
Using a spreadsheet to represent a function of two variables p. 1086
Plotting points in three dimensions p. 1086
Graph of a function of two variables p. 1087
Analyzing the graph of a function of two
variables p. 1090
Graph of a linear function p. 1093

15.2 Partial Derivatives
Definition of partial derivatives p. 1101

Application to marginal cost: linear cost
function p. 1102
Application to marginal cost: interaction
cost function p. 1103
Geometric interpretation of partial
derivatives p. 1104
Second-order partial derivatives
p. 1105

15.3 Maxima and Minima

Definition of relative maximum and
minimum p. 1111
Locating candidates for relative maxima
and minima p. 1112
Classifying critical points graphically
p. 1113
Classifying critical points numerically
p. 1113
Second derivative test for a function of
two variables p. 1113
Using the second derivative test
p. 1114
Formulas for Linear Regression
$$m = \frac{n\left(\sum xy\right) - \left(\sum x\right)\left(\sum y\right)}{n\left(\sum x^2\right) - \left(\sum x\right)^2}$$
$$b = \frac{\sum y - m\left(\sum x\right)}{n}$$
$n =$ number of data points p. 1115

15.4 Constrained Maxima and Minima and Applications

Constrained maximum and minimum
problem p. 1119
Solving constrained maxima and minima
problems using substitution p. 1120
The method of Lagrange multipliers
p. 1121
Using Lagrange multipliers p. 1121

15.5 Double Integrals and Applications

Geometric definition of the double
integral p. 1128
Algebraic definition of the double
integral
$$\iint_R f(x, y)\, dx\, dy =$$
$$\lim_{n \to \infty} \lim_{m \to \infty} \sum_{j=1}^{n} \sum_{i=1}^{m} f(x_i, y_j) \Delta x \Delta y$$
p. 1129
Computing the double integral over a
rectangle p. 1130
Computing the double integral over
nonrectangular regions p. 1131
Average of $f(x, y)$ on the region R:
$$\bar{f} = \frac{1}{A} \iint_R f(x, y)\, dx\, dy \quad \text{p. 1132}$$

REVIEW EXERCISES

1. Let $f(x, y, z) = \dfrac{x}{y + xz} + x^2 y$. Evaluate $f(0, 1, 1), f(2, 1, 1)$,
$f(-1, 1, -1)$, $f(z, z, z)$, and $f(x + h, y + k, z + l)$.

2. Let $g(x, y, z) = xy(x + y - z) + x^2$. Evaluate $g(0, 0, 0)$,
$g(1, 0, 0), g(0, 1, 0), g(x, x, x)$, and $g(x, y + k, z)$.

3. Let $f(x, y, z) = 2.72 - 0.32x - 3.21y + 12.5z$. Complete
the following: f____ by ____ units for every 1 unit of increase
in x, and ____ by ____ units for every unit of increase in z.

4. Let $g(x, y, z) = 2.16x + 11y - 1.53z + 31.4$. Complete the
following: g ____ by ____ units for every 1 unit of increase in
y, and ____ by ____ units for every unit of increase in z.

In Exercises 5–6 complete the given table for values for
$h(x, y) = 2x^2 + xy - x$.

5.

$x \to$			
$y \downarrow$	-1	0	1
-1			
0			
1			

6.

$x \to$			
$y \downarrow$	-2	2	3
-2			
2			
3			

7. Give a formula for a (single) function f with the property
that $f(x, y) = -f(y, x)$ and $f(1, -1) = 3$.

8. Let $f(x, y) = x^2 + (y + 1)^2$. Show that $f(y, x) = f(x + 1, y - 1)$.

Sketch the graphs of the functions in Exercises 9–14.

9. $r(x, y) = x + y$

10. $r(x, y) = x - y$

11. $t(x, y) = x^2 + 2y^2$. Show cross sections at $x = 0$ and
$z = 1$.

12. $t(x, y) = \dfrac{1}{2}x^2 + y^2$. Show cross sections at $x = 0$ and
$z = 1$.

13. $f(x, y) = -2\sqrt{x^2 + y^2}$. Show cross sections at $z = -4$ and $y = 1$.

14. $f(x, y) = 2 + 2\sqrt{x^2 + y^2}$. Show cross sections at $z = 4$ and $y = 1$.

In Exercises 15–20, compute the partial derivatives shown for the given function.

15. $f(x, y) = x^2 + xy$; find f_x, f_y, and f_{yy}

16. $f(x, y) = \dfrac{6}{xy} + \dfrac{xy}{6}$; find f_x, f_y, and f_{yy}

17. $f(x, y) = 4x + 5y - 6xy$; find $f_{xx}(1, 0) - f_{xx}(3, 2)$

18. $f(x, y) = e^{xy} + e^{3x^2 - y^2}$; find $\dfrac{\partial f}{\partial x}$ and $\dfrac{\partial^2 f}{\partial x \partial y}$

19. $f(x, y, z) = \dfrac{x}{x^2 + y^2 + z^2}$; find $\dfrac{\partial f}{\partial x}, \dfrac{\partial f}{\partial y}, \dfrac{\partial f}{\partial z}$, and $\dfrac{\partial f}{\partial x}\Big|_{(0,1,0)}$.

20. $f(x, y, z) = x^2 + y^2 + z^2 + xyz$; find $f_{xx} + f_{yy} + f_{zz}$

In Exercises 21–26, locate and classify all critical points.

21. $f(x, y) = (x - 1)^2 + (2y - 3)^2$

22. $g(x, y) = (x - 1)^2 - 3y^2 + 9$

23. $k(x, y) = x^2 y - x^2 - y^2$

24. $j(x, y) = xy + x^2$

25. $h(x, y) = e^{xy}$

26. $f(x, y) = \ln(x^2 + y^2) - (x^2 + y^2)$

In Exercises 27–30, solve the given constrained optimization problem by using substitution to eliminate a variable. (Do not use Lagrange multipliers.)

27. Find the largest value of xyz subject to $x + y + z = 1$ with $x > 0, y > 0, z > 0$. Also find the corresponding point(s) (x, y, z).

28. Find the minimum value of $f(x, y, z) = x^2 + y^2 + z^2 - 1$ subject to $x = y + z$. Also find the corresponding point(s) (x, y, z).

29. Find the point on the surface $z = \sqrt{x^2 + 2(y - 3)^2}$ closest to the origin.

30. Minimize $S = xy + x^2 z^2 + 4yz$ subject to $xyz = 1$ with $x > 0, y > 0, z > 0$.

In Exercises 31–34, use Lagrange multipliers to solve the given optimization problem.

31. Find the minimum value of $f(x, y) = x^2 + y^2$ subject to $xy = 2$. Also find the corresponding point(s) (x, y).

32. The problem in Exercise 28.

33. The problem in Exercise 29.

34. The problem in Exercise 30.

In Exercises 35–40, compute the given quantities.

35. $\displaystyle\int_0^1 \int_0^2 (2xy)\, dx\, dy$

36. $\displaystyle\int_1^2 \int_0^1 xye^{x+y}\, dx\, dy$

37. $\displaystyle\int_0^2 \int_0^{2x} \dfrac{1}{x^2 + 1}\, dy\, dx$

38. The average value of xye^{x+y} over the rectangle $0 \le x \le 1$, $1 \le y \le 2$

39. $\iint_R (x^2 - y^2)\, dx\, dy$, where R is the region shown in the figure

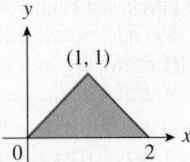

40. The volume under the graph of $z = 1 - y$ over the region in the xy plane between the parabola $y = 1 - x^2$ and the x-axis

APPLICATIONS: OHaganBooks.com

41. *Web Site Traffic* OHaganBooks.com has two principal competitors: JungleBooks.com and FarmerBooks.com. Current Web site traffic at OHaganBooks.com is estimated at 5,000 hits per day. This number is predicted to decrease by 0.8 for every new customer of JungleBooks.com and by 0.6 for every new customer of FarmerBooks.com.

   a. Use this information to model the daily Website traffic at OHaganBooks.com as a linear function of the new customers of its two competitors.

   b. According to the model, if Junglebooks.com gets 100 new customers and OHaganBooks.com traffic drops to 4,770 hits per day, how many new customers has FarmerBooks.com obtained?

   c. The model in part (a) did not take into account the growth of the total online consumer base. OHaganBooks.com expects to get approximately one additional hit per day for every 10,000 new Internet shoppers. Modify your model in part (a) so as to include this information using a new independent variable.

   d. How many new Internet shoppers would it take to offset the effects on traffic at OHaganBooks.com of 100 new customers at each of its competitor sites?

42. *Productivity* Billy-Sean O'Hagan is writing up his Ph.D. thesis in biophysics but finds that his productivity is affected by the temperature and the number of text messages he receives per hour. On a brisk winter's day when the temperature is 0°C and there are no text messages, Billy-Sean can produce 15 pages of his thesis. His productivity goes down by 0.3 pages per degree Celsius increase in the temperature and by 1.2 pages for each additional text message per hour.

   a. Use this information to model Billy-Sean's productivity p as a function of the temperature and the hourly rate of text messages.

   b. The other day the temperature was 20°C and Billy-Sean managed to produce only three pages of his thesis. What was the hourly rate of incoming text messages?

c. Billy-Sean reasons that each cup of coffee he drinks per hour can counter the effect on his productivity of two text messages per hour. Modify the model in part (a) to take consumption of coffee into account.

d. What would the domain of your function look like to ensure that p is never negative?

43. *Internet Advertising* To increase business at OHaganBooks .com, you have purchased banner ads at well-known Internet portals and have advertised on television. The following interaction model shows the average number h of hits per day as a function of monthly expenditures x on banner ads and y on television advertising (x and y are in dollars).

$$h(x, y) = 1,800 + 0.05x + 0.08y + 0.00003xy$$

a. Based on your model, how much traffic can you anticipate if you spend $2,000 per month for banner ads and $3,000 per month on television advertising?

b. Evaluate $\dfrac{\partial h}{\partial y}$, specify its units of measurement, and indicate whether it increases or decreases with increasing x.

c. How much should the company spend on banner ads to obtain 1 hit per day for each $5 spent per month on television advertising?

44. *Company Retreats* Their companies having recently been bailed out by the government at taxpayer expense, Marjory Duffin and John O'Hagan are planning a joint winter business retreat in Cancun, but they are not sure how many sales reps to take along. The following interaction model shows the estimated cost C to their companies (in dollars) as a function of the number of sales reps x and the length of time t in days.

$$C(x, t) = 20,000 - 100x + 600t + 300xt$$

a. Based on the model, how much would it cost to take five sales reps along for a 10-day retreat?

b. Evaluate $\dfrac{\partial C}{\partial t}$, specify its units of measurement, and indicate whether it increases or decreases with increasing x.

c. How many reps should they take along if they wish to limit the marginal daily cost to $1,200?

45. *Internet Advertising* Refer to the model in Exercise 43. One or more of the following statements is correct. Identify which one(s).

(A) If nothing is spent on television advertising, one more dollar spent per month in banner ads will buy approximately 0.05 hits per day at OHaganBooks.com.

(B) If nothing is spent on television advertising, one more hit per day at OHaganBooks.com will cost the company about 5¢ per month in banner ads.

(C) If nothing is spent on banner ads, one more hit per day at OHaganBooks.com will cost the company about 5¢ per month in banner ads.

(D) If nothing is spent on banner ads, one more dollar spent per month in banner ads will buy approximately 0.05 hits per day at OHaganBooks.com.

(E) Hits at OHaganBooks.com cost approximately 5¢ per month spent on banner ads, and this cost increases at a rate of 0.003¢ per month, per hit.

46. *Company Retreats* Refer to the model in Exercise 44. One or more of the following statements is correct. Identify which one(s).

(A) If the retreat lasts for 10 days, the daily cost per sales rep is $400.

(B) If the retreat lasts for 10 days, each additional day will cost the company $2,900.

(C) If the retreat lasts for 10 days, each additional sales rep will cost the company $800.

(D) If the retreat lasts for 10 days, the daily cost per sales rep is $2,900.

(E) If the retreat lasts for 10 days, each additional sales rep will cost the company $2,900.

47. *Productivity* The holiday season is now at its peak and OHaganBooks.com has been understaffed and swamped with orders. The current backlog (orders unshipped for two or more days) has grown to a staggering 50,000, and new orders are coming in at a rate of 5,000 per day. Research based on productivity data at OHaganBooks.com results in the following model:

$$P(x, y) = 1,000x^{0.9}y^{0.1} \text{ additional orders filled per day,}$$

where x is the number of additional personnel hired and y is the daily budget (excluding salaries) allocated to eliminating the backlog.

a. How many additional orders will be filled per day if the company hires 10 additional employees and budgets an additional $1,000 per day? (Round the answer to the nearest 100.)

b. In addition to the daily budget, extra staffing costs the company $150 per day for every new staff member hired. In order to fill at least 15,000 additional orders per day at a minimum total daily cost, how many new staff members should the company hire? (Use the method of Lagrange multipliers.)

48. *Productivity* The holiday season has now ended, and orders at OHaganBooks.com have plummeted, leaving staff members in the shipping department with little to do besides spend their time on their Facebook pages, so the company is considering laying off a number of personnel and slashing the shipping budget. Research based on productivity data at OHaganBooks.com results in the following model:

$$C(x, y) = 1,000x^{0.8}y^{0.2} \text{ fewer orders filled per day,}$$

where x is the number of personnel laid off and y is the cut in the shipping budget (excluding salaries).

a. How many fewer orders will be filled per day if the company lays off 15 additional employees and cuts the budget by an additional $2,000 per day? (Round the answer to the nearest 100.)

b. In addition to the cut in the shipping budget, the layoffs will save the company $200 per day for every new staff member laid off. The company needs to meet a target of 20,000 fewer orders per day but, for tax reasons, it must minimize the total resulting savings. How many new staff members should the company lay off? (Use the method of Lagrange multipliers.)

49. *Profit* If OHaganBooks.com sells x paperback books and y hardcover books per week, it will make an average weekly profit of

$$P(x, y) = 3x + 10y \text{ dollars.}$$

If it sells between 1,200 and 1,500 paperback books and between 1,800 and 2,000 hardcover books per week, what is the average of all its possible weekly profits?

50. *Cost* It costs Duffin House

$$C(x, y) = x^2 + 2y \text{ dollars}$$

to produce x coffee table art books and y paperback books per week. If it produces between 100 and 120 art books and between 800 and 1,000 paperbacks per week, what is the average of all its possible weekly costs?

Case Study   Modeling College Population

College Malls, Inc. is planning to build a national chain of shopping malls in college neighborhoods. However, malls in general have been experiencing large numbers of store closings due to, among other things, misjudgments of the shopper demographics. As a result, the company is planning to lease only to stores that target the specific age demographics of the national college student population.

As a marketing consultant to College Malls, you will be providing the company with a report that addresses the following specific issues:

- A quick way of estimating the number of students of any specified age and in any particular year, and the effect of increasing age on the college population
- The ages that correspond to relatively high and low college populations
- How fast the 20-year old and 25-year old student populations are increasing
- Some near-term projections of the student population trend

You decide that a good place to start would be with a visit to the Census Bureau's Web site at www.census.gov. After some time battling with search engines, all you can find is some data on college enrollment for three age brackets for the period 1980–2009, as shown in the following table:[29]

College Enrollment (Thousands)

Year	1980	1985	1990	1995	2000	2001	2002	2003	2004	2005	2006	2007	2008	2009
18–24	7,229	7,537	7,964	8,541	9,451	9,629	10,033	10,365	10,611	10,834	10,587	11,161	11,466	12,072
25–34	2,703	3,063	3,161	3,349	3,207	3,422	3,401	3,494	3,690	3,600	3,658	3,838	4,013	6,141
35–44	700	963	1,344	1,548	1,454	1,557	1,678	1,526	1,615	1,657	1,548	1,520	1,672	1,848

The data are inadequate for several reasons: The data are given only for certain years, and in age brackets rather than year-by-year; nor is it obvious as to how you would project the figures. However, you notice that the table is actually a numerical representation of a function of two variables: year and age. Since the age brackets are of different sizes, you "normalize" the data by dividing each figure by the number of years represented in the corresponding age bracket; for instance, you divide the 1980 figure for the first age group by 7 in order to obtain the average enrollment for each year of age in that group. You then rewrite the resulting table representing the years by values of t and each age bracket by the (rounded) age x at its center (enrollment values are rounded):

david pearson/Alamy

---

[29]Source: Census Bureau (www.census.gov/population/www/socdemo/school.html).

x $\downarrow$	0	5	10	15	20	21	22	23	24	25	26	27	28	29
21	1,033	1,077	1,138	1,220	1,350	1,376	1,433	1,481	1,516	1,548	1,512	1,594	1,638	1,725
30	270	306	316	335	321	342	340	349	369	360	366	384	401	614
40	70	96	134	155	145	156	168	153	162	166	155	152	167	185

$t \rightarrow$

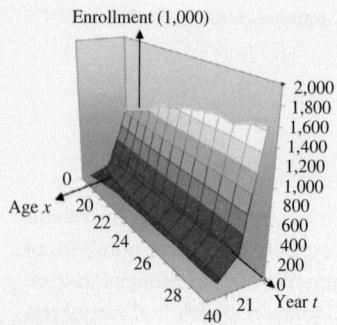

Figure 40

In order to see a visual representation of what the data are saying, you use Excel to graph the data as a surface (Figure 40). It is important to notice that Excel does not scale the t-axis as you would expect: It uses one subdivision for each year shown in the chart, and the result is an uneven scaling of the t-axis. Despite this drawback, you do see two trends after looking at views of the graph from various angles. First, enrollment of 21-year olds (the back edge of the graph) seems to be increasing faster than enrollment of other age groups. Second, the enrollment for all ages seem to be increasing approximately linearly with time, although at different rates for different age groups; for instance, the front and rear edges rise more-or-less linearly, but do not seem to be parallel.

At this point you realize that a mathematical model of these data would be useful; not only would it "smooth out the bumps" but it would give you a way to estimate enrollment N at each specific age, project the enrollments, and thereby complete the project for College Malls. Although technology can give you a regression model for data such as this, it is up to you to decide on the form of the model. It is in choosing an appropriate model that your analysis of the graph comes in handy. Because N should vary linearly with time t for each value of x, you would like

$$N = mt + k$$

for each value of x. Also, because there are three values of x for every value of time, you try a quadratic model for N as a function of x:

$$N = a + bx + cx^2.$$

Putting these together, you get the following candidate model:

$$N(t, x) = a_1 + a_2t + a_3x + a_4x^2,$$

where a_1, a_2, a_3, and a_4 are constants. However, for each specific age $x = k$, you get

$$N(t, k) = a_1 + a_2t + a_3k + a_4k^2 = \text{Constant} + a_2t$$

with the same slope a_2 for every choice of the age k, contrary to your observation that enrollment for different age groups is rising at different rates, so you will need a more elaborate model. You recall from your applied calculus course that interaction functions give a way to model the effect of one variable on the rate of change of another, so, as an experiment, you try adding interaction terms to your model:

Model 1: $N(t, x) = a_1 + a_2t + a_3x + a_4x^2 + a_5xt$ Second-order model

Model 2: $N(t, x) = a_1 + a_2t + a_3x + a_4x^2 + a_5xt + a_6x^2t$.   Third-order model

(Model 1 is referred to as a second-order model because it contains no products of more than two independent variables, whereas Model 2 contains the third-order term $x^2t = x \cdot x \cdot t$.) If you study these two models for specific values k of x you get:

Model 1: $N = \text{Constant} + (a_2 + a_5k)t$ Slope depends linearly on age.

Model 2: $N = \text{Constant} + (a_2 + a_5k + a_6k^2)t$.   Slope depends quadratically on age.

◇	A	B	C
1	N	t	x
2	1033	0	21
3	1077	5	21
4	1138	10	21
5	1220	15	21
6	1350	20	21
7	1376	21	21
8	1433	22	21
9	1481	23	21
10	1516	24	21
11	1548	25	21
12	1512	26	21
13	1594	27	21
14	1638	28	21
15	1725	29	21
16	270	0	30
17	306	5	30
18	316	10	30
19	335	15	30
20	321	20	30
21	342	21	30
22	340	22	30
23	349	23	30
24	369	24	30
25	360	25	30
26	366	26	30
27	384	27	30
28	401	28	30
29	614	29	30
30	70	0	40
31	96	5	40
32	134	10	40
33	155	15	40
34	145	20	40
35	156	21	40
36	168	22	40
37	153	23	40
38	162	24	40
39	166	25	40
40	155	26	40
41	152	27	40
42	167	28	40
43	185	29	40

Figure 41

This is encouraging: Both models show different slopes for different ages. Model 1 would predict that the slope either increases with increasing age (a_5 positive) or decreases with increasing age (a_5 negative). However, the graph suggests that the slope is larger for both younger and older students, but smaller for students of intermediate age, contrary to what Model 1 predicts, so you decide to go with the more flexible Model 2, which permits the slope to decrease and then increase with increasing age, which is exactly what you observe on the graph, and so you decide to go ahead with Model 2.

You decide to use Excel to generate your model. However, the data as shown in the table are not in a form Excel can use for regression; the data need to be organized into columns; Column A for the dependent variable N and Columns B–C for the independent variables, as shown in Figure 41.

You then add columns for the higher order terms x^2, xt, and x^2t as shown below:

◇	A	B	C	D	E	F
1	N	t	x	x^2	x*t	x^2*t
2	1033	0	21	=C2^2	=C2*B2	=C2^2*B2
3	1077	5	21			
4	1138	10	21			
5	1220	15	21			
6	1350	20	21			
7	1376	21	21			
8	1433	22	21			
9	1481	23	21			

Figure 42

◇	A	B	C	D	E	F
1	N	t	x	x^2	x*t	x^2*t
2	1033	0	21	441	0	0
3	1077	5	21	441	105	2205
4	1138	10	21	441	210	4410
5	1220	15	21	441	315	6615
6	1350	20	21	441	420	8820
7	1376	21	21	441	441	9261
8	1433	22	21	441	462	9702
9	1481	23	21	441	483	10143

Figure 43

Next, highlight a vacant 5×6 block (the block A46:F50 say), type the formula =LINEST(A2:A43,B2:F43,,TRUE), and press Ctrl+Shift+Enter (not just Enter!). You will see a table of statistics like the following:

42						
43	185	29	40	1600	1160	46400
44						
45						
46	=LINEST(A2:A43,B2:F43,,TRUE)					
47						
48						
49						
50						

Figure 44

42						
43	185	29	40	1600	1160	46400
44						
45						
46	0.088570354	-6.46546922	3.21423521	-241.273271	120.009278	4594.62978
47	0.02096866	1.28767845	0.44955302	27.6069027	18.6958148	400.824866
48	0.99337478	49.506512	#N/A	#N/A	#N/A	#N/A
49	1079.5571	36	#N/A	#N/A	#N/A	#N/A
50	13229404.1	88232.2104	#N/A	#N/A	#N/A	#N/A

Figure 45

The desired constants $a_1, a_2, a_3, a_4, a_5, a_6$ appear in the first row of the data, but in *reverse order*. Thus, if we round to 5 significant digits, we have

$$a_1 = 4{,}594.6 \quad a_2 = 120.01 \quad a_3 = -241.27$$
$$a_4 = 3.2142 \quad a_5 = -6.4655 \quad a_6 = 0.088570,$$

which gives our regression model:

$$N(t, x) = 4{,}594.6 + 120.01t - 241.27x + 3.2142x^2 - 6.4655xt + 0.088570x^2t.$$

Fine, you say to yourself, now you have the model, but how good a fit is it to the data? That is where rest of the data shown in the output comes in: In the second row are the "standard errors" corresponding to the corresponding coefficients. Notice that each of the standard errors is small compared with the magnitude of the coefficient above it; for instance, 0.021 is only around 1/4 of the magnitude of $a_6 \approx 0.088$ and indicates that the dependence of N on x^2t is statistically significant. (What we do not want to see are standard errors of magnitudes comparable to the coefficients, as those could indicate the wrong choice of independent variables.) The third figure in the left column, 0.99337478, is R^2, where R generalizes the coefficient of correlation discussed in the section on regression in Chapter 1: The closer R is to 1, the better the fit. We can interpret R^2 as indicating that approximately 99.3% of the variation in college enrollment is explained by the regression model, indicating an excellent fit. The figure 1,079.5571 beneath R^2 is called the "F-statistic." The higher the F-statistic (typically, anything above 4 or so would be considered "high"), the more confident we can be that N does depend on the independent variables we are using.*

* We are being deliberately vague about the exact meaning of these statistics, which are discussed fully in many applied statistics texts.

As comforting as these statistics are, nothing can be quite as persuasive as a graph. You turn to the graphing software of your choice and notice that the graph of the model appears to be a faithful representation of the data. (See Figure 46.)

Now you get to work, using the model to address the questions posed by College Malls.

1. *A quick way of estimating the number of students of any specified age and in any particular year, and the effect of increasing age on the college population.* You already have a quantitative relationship in the form of the regression model. As for the second part of the question, the rate of change of college enrollment with respect to age is given by the partial derivative:

$$\frac{\partial N}{\partial x} = -241.27 + 6.4284x - 6.4655t + 0.17714xt \text{ thousand students per additional year of age.}$$

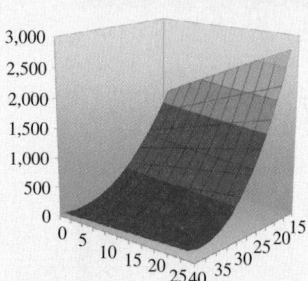

Figure 46

Thus, for example, with $x = 20$ in 2004 ($t = 24$), we have

$$\frac{\partial N}{\partial x} = -241.27 + 6.4284(20) - 6.4655(24) + 0.17714(20)(24)$$

$$\approx -183 \text{ thousand students per additional year of age,}$$

so there were about 183,000 fewer students of age 21 than age 20 in 2004.

On the other hand, when $x = 38$ in the same year, we have

$$\frac{\partial N}{\partial x} = -241.27 + 6.4284(38) - 6.4655(24) + 0.17714(38)(24)$$

$$\approx 9.4 \text{ thousand students per additional year of age,}$$

so there were about 9,400 more students of age 39 than age 38 that year.

2. *The ages that correspond to relatively high and low college populations.* Although a glance at the graph shows you that there are no relative maxima, holding t constant (that is, on any given year) gives a parabola along the corresponding slice and hence a minimum somewhere along the slice.

$$\frac{\partial N}{\partial x} = 0$$

when $\qquad -241.27 + 6.4284x - 6.4655t + 0.17714xt = 0,$

which gives $x = \dfrac{241.27 + 6.4655t}{6.4284 + 0.17714t}$ years of age

For instance, in 2010 ($t = 30$; we are extrapolating the model slightly), the age at which there were fewest students (in the given range) is 37 years of age. The relative maxima for each slice occur at the front and back edges of the surface, meaning that there are relatively more students of the lowest and highest ages represented. The absolute maximum for each slice occurs, as expected, at the lowest age. In short, a mall catering to college students in 2010 should have focused mostly on freshman age students, least on 37-year-olds, and somewhat more on people around age 40.

3. *How fast the 20-year-old and 25-year-old student populations are increasing.* The rate of change of student population with respect to time is

$$\frac{\partial N}{\partial t} = 120.01 - 6.4655x + 0.088570x^2 \text{ thousand students per year.}$$

For the two age groups in question, we obtain

$$x = 20: 120.01 - 6.4655(20) + 0.088570(20)^2 \approx 26.1 \text{ thousand students/year}$$

$$x = 25: 120.01 - 6.4655(25) + 0.088570(25)^2 \approx 13.7 \text{ thousand students/year.}$$

(Note that these rates of change are independent of time as we chose a model that is linear in time.)

4. *Some near-term projections of the student population trend.* As we have seen throughout the book, extrapolation can be a risky venture; however, near-term extrapolation from a good model can be reasonable. You enter the model in an Excel spreadsheet to obtain the following predicted college enrollments (in thousands) for the years 2010–2015:

$t \rightarrow$

x ↓	30	31	32	33	34	35
21	1,644	1,668	1,691	1,714	1,737	1,761
30	422	428	434	439	445	451
40	180	183	186	189	192	195

EXERCISES

1. Use a spreadsheet to obtain Model 1:

$$N(t, x) = a_1 + a_2t + a_3x + a_4x^2 + a_5xt.$$

Compare the fit of this model with that of the quadratic model above. Comment on the result.

2. Obtain Model 2 using only the data through 2005, and also obtain the projections for 2010–2015 using the resulting model. Compare the projections with those based on the more complete set of data in the text.

3. Compute and interpret $\left.\dfrac{\partial N}{\partial t}\right|_{(10,18)}$ and $\left.\dfrac{\partial^2 N}{\partial t \partial x}\right|_{(10,18)}$ for the model in the text. What are their units of measurement?

4. Notice that the derivatives in the preceding exercise do not depend on time. What additional polynomial term(s) would make both $\partial N/\partial t$ and $\partial^2 N/\partial t \partial x$ depend on time? (Write down the entire model.) Of what order is your model?

5. Test the model you constructed in the preceding question by inspecting the standard errors associated with the additional coefficients.

TI-83/84 Plus Technology Guide

Section 15.1

Example 1 (page 1083) You own a company that makes two models of speakers: the Ultra Mini and the Big Stack. Your total monthly cost (in dollars) to make x Ultra Minis and y Big Stacks is given by

$$C(x, y) = 10,000 + 20x + 40y.$$

Compute several values of this function.

Solution with Technology

You can have a TI-83/84 Plus compute $C(x, y)$ numerically as follows:

1. In the "Y=" screen, enter

$$Y_1=10000+20X+40Y$$

2. To evaluate, say, $C(10, 30)$ (the cost to make 10 Ultra Minis and 30 Big Stacks), enter

$$10 \rightarrow X$$
$$30 \rightarrow Y$$
$$Y_1$$

```
10→X
                    10
30→Y
                    30
Y₁
                 11400
■
```

and the calculator will evaluate the function and give the answer $C(10, 30) = 11,400$.

This procedure is too laborious if you want to calculate $f(x, y)$ for a large number of different values of x and y.

SPREADSHEET Technology Guide

Section 15.1

Example 1 (page 1083) You own a company that makes two models of speakers: the Ultra Mini and the Big Stack. Your total monthly cost (in dollars) to make x Ultra Minis and y Big Stacks is given by
$$C(x, y) = 10,000 + 20x + 40y.$$
Compute several values of this function.

Solution with Technology

Spreadsheets handle functions of several variables easily. The following setup shows how a table of values of C can be created, using values of x and y you enter:

	A	B	C
1	x	y	C(x, y)
2	10	30	=10000+20*A2+40*B2
3	20	30	
4	15	0	
5	0	30	
6	30	30	

↓

	A	B	C
1	x	y	C(x, y)
2	10	30	11400
3	20	30	11600
4	15	0	10300
5	0	30	11200
6	30	30	11800

A disadvantage of this layout is that it's not easy to enter values of x and y systematically in two columns. Can you find a way to remedy this? (See Example 3 for one method.)

Example 3 (page 1085) Use technology to create a table of values of the body mass index
$$M(w, h) = \frac{0.45w}{(0.0254h)^2}.$$

Solution with Technology

We can use this formula to recreate a table in a spreadsheet, as follows:

	A	B	C	D
1		130	140	150
2		60	=0.45*B$1/(0.0254*$A2)^2	
3		61		
4		62		
5		63		
6		64		
7		65		
8		66		
9		67		

In the formula in cell B2 we have used B$1 instead of B1 for the w-coordinate because we want all references to w to use the same row (1). Similarly, we want all references to h to refer to the same column (A), so we used $A2 instead of A2.

We copy the formula in cell B2 to all of the red shaded area to obtain the desired table:

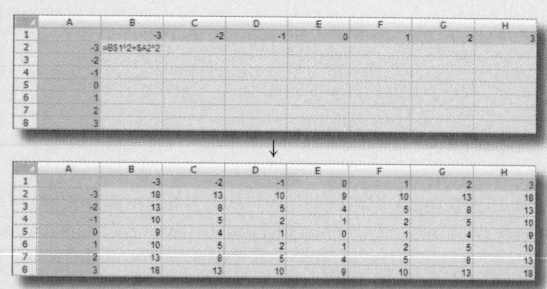

	A	B	C	D
1		130	140	150
2	60	25.18755038	27.12505425	29.06255813	3
3	61	24.36849808	26.24299793	28.11749778	29.9
4	62	23.58875685	25.40327661	27.21779637	29.0
5	63	22.84585068	24.60322381	26.36059694	28.1
6	64	22.13749545	23.84037971	25.54326398	27.2
7	65	21.46158138	23.11247226	24.76336314	26.4
8	66	20.81615733	22.41740021	24.01864308	25.6
9	67	20.19941665	21.75321793	23.30701921	24.8

Example 5 (page 1088) Obtain the graph of $f(x, y) = x^2 + y^2$.

Solution with Technology

1. Set up a table showing a range of values of x and y and the corresponding values of the function (see Example 3):

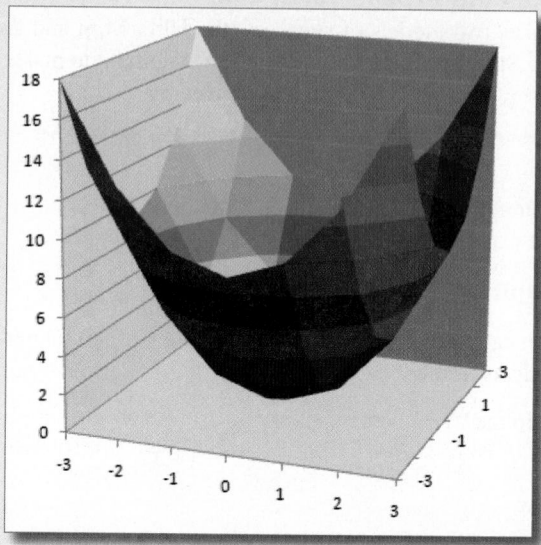

2. Select the cells with the values (B2: H8) and insert a chart, with the "Surface" option selected and "Series in Columns" selected as the data option, to obtain a graph like the following:

16

Trigonometric Models

Website
www.WanerMath.com
At the Website you will find:

- A detailed chapter summary
- A true/false quiz
- Graphers, Excel tutorials, and other resources

Case Study Predicting Airline Empty Seat Volume

You are a consultant to the Department of Transportation's Special Task Force on Air Traffic Congestion and have been asked to model the volume of empty seats on U.S. airline flights, to make short-term projections of this volume, and to give a formula that estimates the accumulated volume over a specified period of time. You have data from the Bureau of Transportation Statistics on the number of seats and passengers each month starting with January 2002. **How will you analyze these data to prepare your report?**

Lawrence Manning/Flirt/Corbis

Introduction

Cyclical behavior is common in the business world: There are seasonal fluctuations in the demand for surfing equipment, swim wear, snow shovels, and many other items. The nonlinear functions we have studied up to now cannot model this kind of behavior. To model cyclical behavior, we need the **trigonometric** functions.

In the first section, we study the basic trigonometric functions—especially the **sine** and **cosine** functions from which all the trigonometric functions are built—and see how to model various kinds of periodic behavior using these functions. The rest of the chapter is devoted to the calculus of the trigonometric functions—their derivatives and integrals—and to its numerous applications.

16.1 Trigonometric Functions, Models, and Regression

The Sine Function

Figure 1 shows the approximate average daily high temperatures in New York's Central Park.[1] If we draw the graph for several years, we get the repeating pattern shown in Figure 2, where the x-coordinate represents time in years, with $x = 0$ corresponding to August 1, and where the y-coordinate represents the temperature in degrees F. This is an example of **cyclical** or **periodic** behavior.

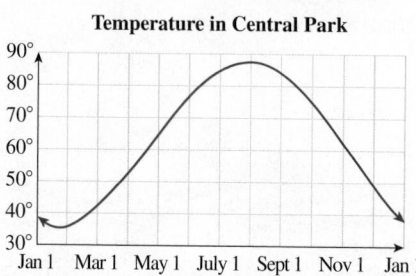

Figure 1

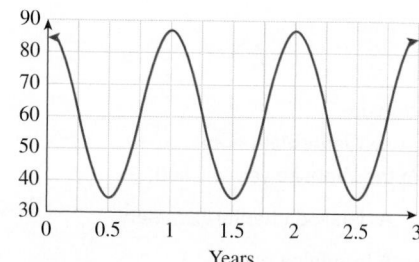

Figure 2

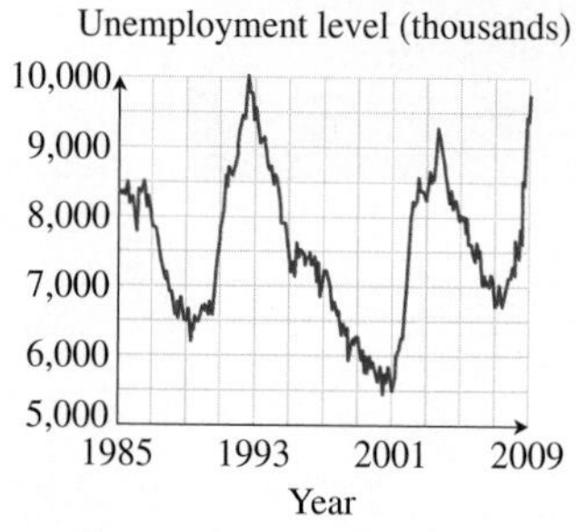

Cyclical behavior is also common in the business world. The graph in Figure 3 suggests cyclical behavior in the U.S. unemployment level.

From a mathematical point of view, the simplest models of cyclical behavior are the **sine** and **cosine** functions. An easy way to describe these functions is as follows. Imagine a bicycle wheel whose radius is one unit, with a marker attached to the rim of the rear wheel, as shown in Figure 4.

Unemployment level (thousands)

Source: Bureau of Labor Statistics, December 2008 (www.data.bls.gov).

Figure 3

[1] Source: National Weather Service/*New York Times*, January 7, 1996, p. 36.

Note that the basepoint of the cosine curve is at the highest point of the curve. All the constants have the same meaning as for the general sine curve:

- A is the **amplitude** (the height of each peak above the baseline).
- C is the **vertical offset** (height of the baseline).
- P is the **period** or **wavelength** (the length of each cycle) and is related to ω by

$$P = 2\pi/\omega \quad \text{or} \quad \omega = 2\pi/P.$$

- ω is the **angular frequency** (the number of cycles in every interval of length 2π).
- β is the phase shift.

Notes

1. We can also describe the above curve as a generalized sine function: Observe by comparing the picture on the preceding page to the one on page 1153 that $\beta = \alpha + P/4$. Thus, $\alpha = \beta - P/4$, and the above curve is also

$$f(x) = A \sin [\omega(x - \beta + P/4)] + C.$$

2. As is the case with the generalized sine function, the cosine function above can be written in the form

$$f(x) = A \cos (\omega t + d) + C. \qquad d = -\omega\beta \qquad \blacksquare$$

EXAMPLE 4 Cash Flows into Stock Funds

The annual cash flow into stock funds (measured as a percentage of total assets) has fluctuated in cycles of approximately 40 years since 1955, when it was at a high point. The highs were roughly +15% of total assets, whereas the lows were roughly −10% of total assets.[2]

a. Model this cash flow with a cosine function of the time t in years, with $t = 0$ representing 1955.

b. Convert the answer in part (a) to a sine function model.

Solution

a. Cosine modeling is similar to sine modeling; we are seeking a function of the form

$$P(t) = A \cos [\omega(t - \beta)] + C.$$

Amplitude A and Vertical Offset C: The cash flow fluctuates between −10% and +15%. We can express this as a fluctuation of $A = 12.5$ about the average $C = 2.5$.

Period P: This is given as $P = 40$.

Angular Frequency ω: We find ω from the formula

$$\omega = \frac{2\pi}{P} = \frac{2\pi}{40} = \frac{\pi}{20} \approx 0.157.$$

---

[2]Source: Investment Company Institute/*New York Times*, February 2, 1997, p. F8.

Phase Shift β: The basepoint is at the high point of the curve, and we are told that cash flow was at its high point at $t = 0$. Therefore, the basepoint occurs at $t = 0$, and so $\beta = 0$.

Putting the model together gives

$$P(t) = A \cos[\omega(t - \beta)] + C$$
$$\approx 12.5 \cos(0.157t) + 2.5,$$

where t is time in years since 1955.

b. To convert between a sine and cosine model, we can use one of the relationships given earlier. Let us use the formula

$$\cos x = \sin(x + \pi/2).$$

Therefore,

$$P(t) \approx 12.5 \cos(0.157t) + 2.5$$
$$= 12.5 \sin(0.157t + \pi/2) + 2.5.$$

The Other Trigonometric Functions

The ratios and reciprocals of sine and cosine are given their own names.

Tangent, Cotangent, Secant, Cosecant

Tangent: $\tan x = \dfrac{\sin x}{\cos x}$

Cotangent: $\cot x = \cotan x = \dfrac{\cos x}{\sin x} = \dfrac{1}{\tan x}$

Secant: $\sec x = \dfrac{1}{\cos x}$

Cosecant: $\csc x = \cosec x = \dfrac{1}{\sin x}$

Trigonometric Regression

In the examples so far, we were given enough information to obtain a sine (or cosine) model directly. Often, however, we are given data that only *suggest* a sine curve. In such cases we can use regression to find the best-fit generalized sine (or cosine) curve.

EXAMPLE 5 ⓘ Spam

The authors of this book tend to get inundated with spam email. One of us systematically documented the number of spam emails arriving at his email account, and noticed a curious cyclical pattern in the average number of emails arriving each week.[3] Figure 9 shows the daily spam for a 16-week period[4] (each point is a one-week average):

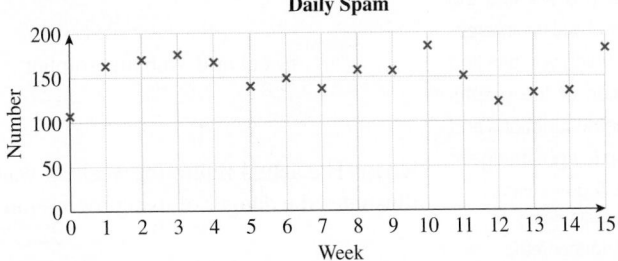

Figure 9

Week	0	1	2	3	4	5	6	7	8	9	10	11	12	13	14	15
Number	107	163	170	176	167	140	149	137	158	157	185	151	122	132	134	182

a. Use technology to find the best-fit sine curve of the form $S(t) = A \sin[\omega(t - \alpha)] + C$.

b. Use your model to estimate the period of the cyclical pattern in spam mail, and also to predict the daily spam average for week 23.

Solution

a. Following are the models obtained by using the TI-83/84 Plus, Excel with Solver, and the Function Evaluator and Grapher on the Website. (See the Technology Guides at the end of the chapter to find out how to obtain these models. For the Website utility, we used the initial guess $A = 50$, $\omega = 1$, $\alpha = 0$, and $C = 150$.)

TI-83/84 Plus: $S(t) \approx 11.6 \sin[0.910(t - 1.63)] + 155$

Excel and Website grapher: $S(t) \approx 25.8 \sin[0.96(t - 1.22)] + 153$

Q : *Why do the models from the TI-83/84 Plus differ so drastically from the Solver and Website model?*

A : Not all regression algorithms are identical, and it seems that the TI-83/84 Plus's algorithm is not very efficient at finding the best-fit sine curve. Indeed, the value for the sum-of-squares error (SSE) for the TI-83/84 Plus regression curve is around 5,030, whereas it is around 2,148 for the Excel curve, indicating a far better fit.* Notice another thing: The sine curve does not appear to fit the data well in either graph. In general, we can expect better agreement between the different forms of technology for data that follow a sine curve more closely.

---

* This comparison is actually unfair: The method using Excel's Solver or the Website grapher starts with an initial guess of the coefficients, so the TI-83/84 Plus algorithm, which does not require an initial guess, is starting at a significant disadvantage. An initial guess that is way off can result in Solver or the Website grapher coming up with a very different result! On the other hand, the TI-83/84 Plus algorithm seems problematic and tends to fail (giving an error message) on many sets of data.

b. This model gives a period of approximately

TI-83/84 Plus: $P = \dfrac{2\pi}{\omega} \approx \dfrac{2\pi}{0.910} \approx 6.9 \text{ weeks}$

Excel and Website grapher: $P = \dfrac{2\pi}{\omega} \approx \dfrac{2\pi}{0.96} \approx 6.5 \text{ weeks}.$

So, both models predict a very similar period.

---

[3]Confirming the notion that academics have little else to do but fritter away their time in pointless pursuits.
[4]Beginning June 6, 2005.

Website

www.WanerMath.com

At the Website you will find the following optional online interactive section, in which you can find further discussion of the graphs of the trigonometric functions and their relationship to right triangles:

Internet Topic: Trigonometric Functions and Calculus

→ The Six Trigonometric Functions.

In week 23, we obtain the following predictions:

TI-83/84 Plus:
$$S(23) \approx 11.6 \sin [0.910(23 - 1.63)] + 155$$
$$\approx 162 \text{ spam emails per day}$$

Excel and Website grapher:
$$S(23) \approx 25.8 \sin [0.96(23 - 1.22)] + 153$$
$$\approx 176 \text{ spam emails per day.}$$

Note The actual figure for week 23 was 213 spam emails per day. The discrepancy illustrates the danger of using regression models to extrapolate. ∎

16.1   EXERCISES

▼ more advanced   ◆ challenging

 indicates exercises that should be solved using technology

In Exercises 1–12, graph the given functions or pairs of functions on the same set of axes.

***a.** Sketch the curves without any technological help by consulting the discussion in Example 1.*

b. *Use technology to check your sketches.* HINT [See Example 1.]

1. $f(t) = \sin(t)$; $g(t) = 3 \sin(t)$

2. $f(t) = \sin(t)$; $g(t) = 2.2 \sin(t)$

3. $f(t) = \sin(t)$; $g(t) = \sin(t - \pi/4)$

4. $f(t) = \sin(t)$; $g(t) = \sin(t + \pi)$

5. $f(t) = \sin(t)$; $g(t) = \sin(2t)$

6. $f(t) = \sin(t)$; $g(t) = \sin(-t)$

7. $f(t) = 2 \sin[3\pi(t - 0.5)] - 3$

8. $f(t) = 2 \sin[3\pi(t + 1.5)] + 1.5$

9. $f(t) = \cos(t)$; $g(t) = 5 \cos[3(t - 1.5\pi)]$

10. $f(t) = \cos(t)$; $g(t) = 3.1 \cos(3t)$

11. $f(t) = \cos(t)$; $g(t) = -2.5 \cos(t)$

12. $f(t) = \cos(t)$; $g(t) = 2 \cos(t - \pi)$

In Exercises 13–18, model each curve with a sine function. (Note that not all are drawn with the same scale on the two axes.) HINT [See Example 2.]

13.

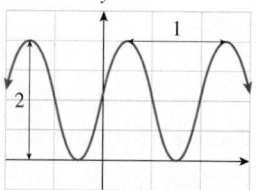

14.

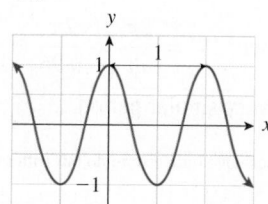

15.

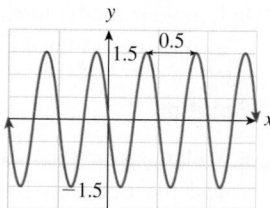

16.

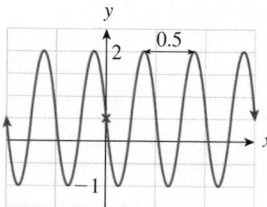

17.

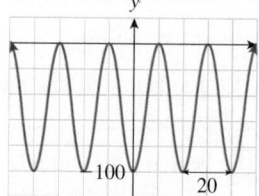

18.

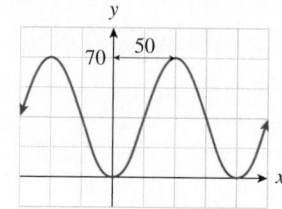

In Exercises 19–24, model each curve with a cosine function. (Note that not all are drawn with the same scale on the two axes.) HINT [See Example 2.]

19.

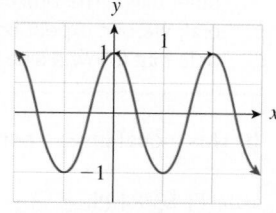

20.

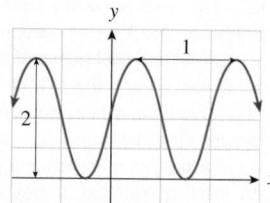

21.

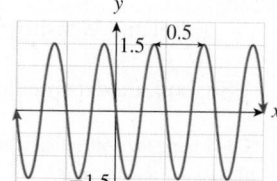

22.

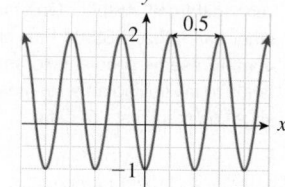

23.

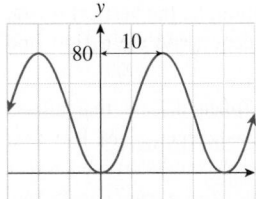

24.

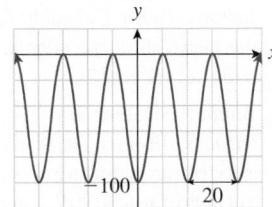

In Exercises 25–28, use the conversion formula $\cos x = \sin(\pi/2 - x)$ *to replace each expression by a sine function.*

25. ▼ $f(t) = 4.2\cos(2\pi t) + 3$

26. ▼ $f(t) = 3 - \cos(t - 4)$

27. ▼ $g(x) = 4 - 1.3\cos[2.3(x - 4)]$

28. ▼ $g(x) = 4.5\cos[2\pi(3x - 1)] + 7$

Some Identities *Starting with the identity* $\sin^2 x + \cos^2 x = 1$ *and then dividing both sides of the equation by a suitable trigonometric function, derive the trigonometric identities in Exercises 29 and 30.*

29. ▼ $\sec^2 x = 1 + \tan^2 x$ **30.** ▼ $\csc^2 x = 1 + \cot^2 x$

Exercises 31–38 are based on the ***addition formulas:***

$$\sin(x + y) = \sin x \cos y + \cos x \sin y$$
$$\sin(x - y) = \sin x \cos y - \cos x \sin y$$
$$\cos(x + y) = \cos x \cos y - \sin x \sin y$$
$$\cos(x - y) = \cos x \cos y + \sin x \sin y$$

31. ▼ Calculate $\sin(\pi/3)$, given that $\sin(\pi/6) = 1/2$ and $\cos(\pi/6) = \sqrt{3}/2$.

32. ▼ Calculate $\cos(\pi/3)$, given that $\sin(\pi/6) = 1/2$ and $\cos(\pi/6) = \sqrt{3}/2$.

33. ▼ Use the formula for $\sin(x + y)$ to obtain the identity $\sin(t + \pi/2) = \cos t$.

34. ▼ Use the formula for $\cos(x + y)$ to obtain the identity $\cos(t - \pi/2) = \sin t$.

35. ▼ Show that $\sin(\pi - x) = \sin x$.

36. ▼ Show that $\cos(\pi - x) = -\cos x$.

37. ▼ Use the addition formulas to express $\tan(x + \pi)$ in terms of $\tan x$.

38. ▼ Use the addition formulas to express $\cot(x + \pi)$ in terms of $\cot x$.

APPLICATIONS

39. ***Sunspot Activity*** The activity of the sun (sunspots, solar flares, and coronal mass ejection) fluctuates in cycles of around 10–11 years. Sunspot activity can be modeled by the following function:[5]

$$N(t) = 57.7\sin[0.602(t - 1.43)] + 58.8,$$

where t is the number of years since January 1, 1997, and $N(t)$ is the number of sunspots observed at time t. HINT [See Example 3.]

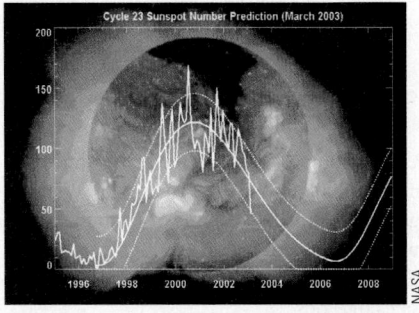

Cycle 23 Sunspot Number Prediction (March 2003)

a. What is the period of sunspot activity according to this model? (Round your answer to the nearest 0.1 year.)

b. What is the maximum number of sunspots observed? What is the minimum number? (Round your answers to the nearest sunspot.)

c. When, to the nearest year, is sunspot activity expected to reach the first high point beyond 2012?

40. ***Solar Emissions*** The following model gives the flux of radio emission from the sun:

$$F(t) = 49.6\sin[0.602(t - 1.48)] + 111,$$

where t is the number of years since January 1, 1997, and $F(t)$ is the flux of solar emissions of a specified wavelength at time t.[6] HINT [See Example 3.]

a. What is the period of radio activity according to this model? (Round your answer to the nearest 0.1 year.)

b. What is the maximum flux of radio emissions? What is the minimum flux? (Round your answers to the nearest whole number.)

c. When, to the nearest year, is radio activity expected to reach the first low point beyond 2012?

---

[5]The model is based on a regression obtained from predicted data for 1997–2006 and the mean historical period of sunspot activity from 1755 to 1995. Source: NASA Science Directorate; Marshall Space Flight Center, August 2002 (http://science.nasa.gov/ssl/pad/solar/predict.htm).

[6]*Ibid.* Flux is measured at a wavelength of 10.7 cm.

41. ▮ *iPod Sales* Sales of personal electronic devices like Apple's iPods are subject to seasonal fluctuations. Apple's sales of iPods in 2008 through 2010 can be approximated by the function

$$s(t) = 6.00 \sin(1.51t + 1.85) + 15 \text{ million iPods}$$
$$\text{per quarter} \quad (0 \le t \le 12),$$

where t is time in quarters ($t = 0$ represents the start of the first quarter of 2008).[7]

a. Use technology to plot sales versus time from the beginning of 2008 through the end of 2010. Then use your graph to estimate the value of t and the quarters during which sales were lowest and highest.

b. Estimate Apple's maximum and minimum quarterly sales of iPods.

c. Indicate how the answers to part (b) can be obtained directly from the equation for $s(t)$.

42. ▮ *Housing Starts* Housing construction is subject to seasonal fluctuations. The number of housing starts (number of new privately owned housing units started) from 2009 through mid-2011 can be approximated by the function

$$s(t) = 10.2 \sin(0.537t - 1.46) + 47.6 \text{ thousand housing}$$
$$\text{starts per month} \quad (0 \le t \le 30),$$

where t is time in months ($t = 0$ represents the start of January 2009).[8]

a. Use technology to plot housing starts versus time from the beginning of 2009 through the end of June 2011. Then use your graph to estimate the value of t and the months during which housing starts were lowest and highest.

b. Estimate the maximum and minimum number of housing starts.

c. Indicate how the answers to part (b) can be obtained directly from the equation for $s(t)$.

43. *iPod Sales* (Based on Exercise 41, but no technology required) Apple's sales of iPods in 2008 through 2010 can be approximated by the function

$$s(t) = 6.00 \sin(1.51t + 1.85) + 15 \text{ million iPods}$$
$$\text{per quarter} \quad (0 \le t \le 12),$$

where t is time in quarters ($t = 0$ represents the start of the first quarter of 2008). Calculate the amplitude, the vertical offset, the phase shift, the angular frequency, and the period, and interpret the results.

44. *Housing Starts* (Based on Exercise 42, but no technology required) The number of housing starts (number of new privately owned housing units started) from 2009 through mid-2011 can be approximated by the function

$$s(t) = 10.2 \sin(0.537t - 1.46) + 47.6 \text{ thousand housing}$$
$$\text{starts per month} \quad (0 \le t \le 30),$$

where t is time in months ($t = 0$ represents the start of January 2009). Calculate the amplitude, the vertical offset, the phase shift, the angular frequency, and the period, and interpret the results.

45. *Biology* Sigatoka leaf spot is a plant disease that affects bananas. In an infected plant, the percentage of leaf area affected varies from a low of around 5% at the start of each year to a high of around 20% at the middle of each year.[9] Use the sine function to model the percentage of leaf area affected by Sigatoka leaf spot t weeks since the start of a year. HINT [See Example 2.]

46. *Biology* Apple powdery mildew is an epidemic that affects apple shoots. In a new infection, the percentage of apple shoots infected varies from a low of around 10% at the start of May to a high of around 60% 6 months later.[10] Use the sine function to model the percentage of apple shoots affected by apple powdery mildew t months since the start of a year.

47. *Cancun* The *Playa Loca Hotel* in Cancun has an advertising brochure with a chart showing the year-round temperature.[11] The added curve is an approximate 5-month moving average.

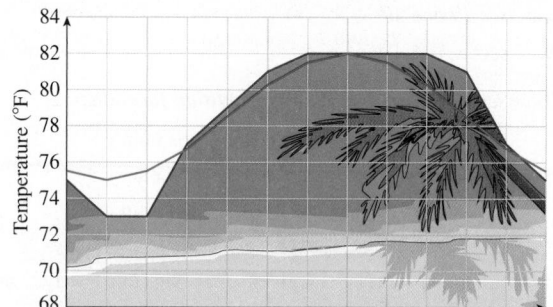

Use a cosine function to model the temperature (moving average) in Cancun as a function of time t in months since December.

48. *Reykjavik* Repeat the preceding exercise, using the following data from the brochure of the *Tough Traveler Lodge* in Reykjavik.[12]

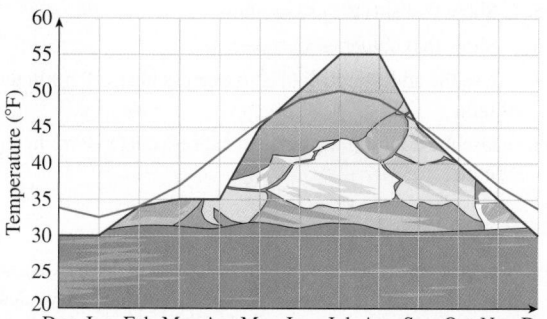

[7]Authors' model. Source for data: Apple quarterly press releases (www.apple.com/investor).

[8]Authors' regression model. Source for data: U.S. Census Bureau (www.census.gov/const/www/newresconstindex.html).

[9]Based on graphical data. Source: American Phytopathological Society, July 2002 (www.apsnet.org/education/AdvancedPlantPath/Topics/Epidemiology/CyclicalNature.htm).

[10]*Ibid.*

[11]Source: www.holiday-weather.com.

[12]*Ibid.* Temperatures are rounded.

49. Net Income Fluctuations: General Electric General Electric's quarterly net income $n(t)$ fluctuated from a low of \$4.5 billion in Quarter 1 of 2007 ($t = 0$) to a high of \$7 billion, and then back down to \$4.5 billion in Quarter 1 of 2008 ($t = 4$).[13] Use a sine function to model General Electric's quarterly net income $n(t)$, where t is time in quarters.

50. Sales Fluctuations Sales of cypods (one-bedroom units) in the city-state of Utarek, Mars[14] fluctuate from a low of 5 units per week each February 1 ($t = 1$) to a high of 35 units per week each August 1 ($t = 7$). Use a sine function to model the weekly sales $s(t)$ of cypods, where t is time in months.

51. Net Income Fluctuations Repeat Exercise 49, but this time use a cosine function for your model. HINT [See Example 4.]

52. Sales Fluctuations Repeat Exercise 50, but this time use a cosine function for your model.

53. Tides The depth of water at my favorite surfing spot varies from 5 to 15 feet, depending on the time. Last Sunday, high tide occurred at 5:00 am and the next high tide occurred at 6:30 pm. Use a sine function model to describe the depth of water as a function of time t in hours since midnight on Sunday morning.

54. Tides Repeat Exercise 53 using data from the depth of water at my second favorite surfing spot, where the tide last Sunday varied from a low of 6 feet at 4:00 am to a high of 10 feet at noon.

55. ▼ Inflation The uninflated cost of *Dugout* brand snow shovels currently varies from a high of \$10 on January 1 ($t = 0$) to a low of \$5 on July 1 ($t = 0.5$).

 a. Assuming this trend continues indefinitely, calculate the uninflated cost $u(t)$ of Dugout snow shovels as a function of time t in years. (Use a sine function.)

 b. Assuming a 4% annual rate of inflation in the cost of snow shovels, the cost of a snow shovel t years from now, adjusted for inflation, will be 1.04^t times the uninflated cost. Find the cost $c(t)$ of Dugout snow shovels as a function of time t.

56. ▼ Deflation Sales of my exclusive 2010 vintage *Chateau Petit Mont Blanc* vary from a high of 10 bottles per day on April 1 ($t = 0.25$) to a low of 4 bottles per day on October 1.

 a. Assuming this trend continues indefinitely, find the undeflated sales $u(t)$ of Chateau Petit Mont Blanc as a function of time t in years. (Use a sine function.)

 b. Regrettably, ever since that undercover exposé of my wine-making process, sales of Chateau Petit Mont Blanc have been declining at an annual rate of 12%. Using

the preceding exercise as a guide, write down a model for the deflated sales $s(t)$ of Chateau Petit Mont Blanc t years from now.

57. ⊤ **Air Travel: Domestic** The following table shows total domestic air travel on U.S. air carriers in specified months from January 2006 to July 2008 ($t = 0$ represents January 2006):[15]

t	0	3	6	9	12	15	18	21	24	27	30
Revenue Passenger Miles (billions)	43	49	55	47	44	50	57	49	45	48	55

 a. Plot the data and *roughly* estimate the period P and the parameters C, A, and β for a cosine model.

 b. Find the best-fit cosine curve approximating the given data. [You may have to use your estimates from part (a) as initial guesses if you are using Solver.] Plot the given data together with the regression curve (round coefficients to three decimal places).

 c. Complete the following: Based on the regression model, domestic air travel on U.S. air carriers showed a pattern that repeats itself every ___ months, from a low of ___ to a high of ___ billion revenue passenger miles. (Round answers to the nearest whole number.) HINT [See Example 5.]

58. ⊤ **Air Travel: International** The following table shows total international travel on U.S. air carriers in specified months from January 2006 to July 2008 ($t = 0$ represents January 2006):[16]

t	0	3	6	9	12	15	18	21	24	27	30
Revenue Passenger Miles (billions)	18	19	23	19	19	20	24	20	25	20	25

 a. Plot the data and *roughly* estimate the period P and the parameters C, A, and β for a cosine model.

 b. Find the best-fit cosine curve approximating the given data. [You may have to use your estimates from part (a) as initial guesses if you are using Solver.] Plot the given data together with the regression curve (round coefficients to three decimal places).

 c. Complete the following: Based on the regression model, international travel on U.S. air carriers showed a pattern that repeats itself every ___ months, from a low of ___ to a high of ___ billion revenue passenger miles. (Round answers to the nearest whole number.) HINT [See Example 5.]

---

[13]Source: Company reports (www.ge.com/investors).

[14]Based on www.Marsnext.com, a now extinct virtual society.

[15]Source: Bureau of Transportation Statistics (www.bts.gov).

[16]*Ibid.*

Music *Musical sounds exhibit the same kind of periodic behavior as the trigonometric functions. High-pitched notes have short periods (less than 1/1,000 second) while the lowest audible notes have periods of about 1/100 second. Some electronic synthesizers work by superimposing (adding) sinusoidal functions of different frequencies to create different textures. Exercises 59–62 show some examples of how superposition can be used to create interesting periodic functions.*

59. [T] ▼ *Sawtooth Wave*

a. Graph the following functions in a window with $-7 \leq x \leq 7$ and $-1.5 \leq y \leq 1.5$.

$$y_1 = \frac{2}{\pi} \cos x$$

$$y_3 = \frac{2}{\pi} \cos x + \frac{2}{3\pi} \cos 3x$$

$$y_5 = \frac{2}{\pi} \cos x + \frac{2}{3\pi} \cos 3x + \frac{2}{5\pi} \cos 5x$$

b. Following the pattern established above, give a formula for y_{11} and graph it.

c. How would you modify y_{11} to approximate a sawtooth wave with an amplitude of 3 and a period of 4π?

60. [T] ▼ *Square Wave* Repeat the preceding exercise using sine functions in place of cosine functions (which results in an approximation of a square wave).

61. [T] ▼ *Harmony* If we add two sinusoidal functions with frequencies that are simple ratios of each other, the result is a pleasing sound. The following function models two notes an octave apart together with the intermediate fifth:

$$y = \cos(x) + \cos(1.5x) + \cos(2x).$$

Graph this function in the window $0 \leq x \leq 20$ and $-3 \leq y \leq 3$ and estimate the period of the resulting wave.

62. [T] ▼ *Discord* If we add two sinusoidal functions with similar, but unequal, frequencies, the result is a function that "pulsates," or exhibits "beats." (Piano tuners and guitar players use this phenomenon to help them tune an instrument.) Graph the function

$$y = \cos(x) + \cos(0.9x)$$

in the window $-50 \leq x \leq 50$ and $-2 \leq y \leq 2$ and estimate the period of the resulting wave.

COMMUNICATION AND REASONING EXERCISES

63. What are the highs and lows for sales of a commodity modeled by a function of the form $s(t) = A \sin(2\pi t) + B$ (A, B constants)?

64. Your friend has come up with the following model for choral society Tupperware stock inventory: $r(t) = 4 \sin(2\pi(t-2)/3) + 2.3$, where t is time in weeks and $r(t)$ is the number of items in stock. Why is the model not realistic?

65. Your friend is telling everybody that all six trigonometric functions can be obtained from the single function $\sin x$. Is he correct? Explain your answer.

66. Another friend claims that all six trigonometric functions can be obtained from the single function $\cos x$. Is she correct? Explain your answer.

67. If weekly sales of sodas at a movie theater are given by $s(t) = A + B\cos(\omega t)$, what is the largest B can be? Explain your answer.

68. Complete the following: If the cost of an item is given by $c(t) = A + B\cos[\omega(t - \alpha)]$, then the cost fluctuates by ____ with a period of ____ about a base of ____, peaking at time $t =$ ____.

| **16.2** | # Derivatives of Trigonometric Functions and Applications |

We start with the derivatives of the sine and cosine functions.

> ### Theorem Derivatives of the Sine and Cosine Functions
>
> The sine and cosine functions are differentiable with
>
> $$\frac{d}{dx} \sin x = \cos x$$
>
> $$\frac{d}{dx} \cos x = -\sin x \qquad \text{Notice the sign change.}$$

Indefinite Integrals of Some Trig Functions

$$\int \cos x \, dx = \sin x + C \qquad \text{Because } \frac{d}{dx}(\sin x) = \cos x$$

$$\int \sin x \, dx = -\cos x + C \qquad \text{Because } \frac{d}{dx}(-\cos x) = \sin x$$

$$\int \sec^2 x \, dx = \tan x + C \qquad \text{Because } \frac{d}{dx}(\tan x) = \sec^2 x$$

Quick Examples

1. $\int (\sin x + \cos x) \, dx = -\cos x + \sin x + C$ Integral of sum = Sum of integrals
2. $\int (4 \sin x - \cos x) \, dx = -4 \cos x - \sin x + C$ Integral of constant multiple
3. $\int (e^x - \sin x + \cos x) \, dx = e^x + \cos x + \sin x + C$

EXAMPLE 1 Substitution

Evaluate $\int (x + 3) \sin (x^2 + 6x) \, dx$.

Solution There are two parenthetical expressions that we might replace with u. Notice, however, that the derivative of the expression $(x^2 + 6x)$ is $2x + 6$, which is twice the term $(x + 3)$ in front of the sine. Recall that we would like the derivative of u to appear as a factor. Thus, let us take $u = x^2 + 6x$.

$$u = x^2 + 6x$$
$$\frac{du}{dx} = 2x + 6 = 2(x + 3)$$
$$dx = \frac{1}{2(x + 3)} \, du$$

Substituting into the integral, we get

$$\int (x + 3) \sin (x^2 + 6x) \, dx = \int (x + 3) \sin u \left(\frac{1}{2(x + 3)} \right) du = \int \frac{1}{2} \sin u \, du$$

$$= -\frac{1}{2} \cos u + C = -\frac{1}{2} \cos (x^2 + 6x) + C.$$

EXAMPLE 2 Definite Integrals

Compute the following.

a. $\displaystyle\int_0^\pi \sin x \, dx$ **b.** $\displaystyle\int_0^\pi x \sin (x^2) \, dx$

Solution

a. $\displaystyle\int_0^\pi \sin x \, dx = \left[-\cos x \right]_0^\pi = (-\cos \pi) - (-\cos 0) = -(-1) - (-1) = 2$

Thus, the area under one "arch" of the sine curve is exactly two square units!

b. $\displaystyle\int_0^{\pi} x \, \sin(x^2)\, dx = \int_0^{\pi^2} \frac{1}{2}\sin u \, du$ After substituting $u = x^2$

$$= \left[-\frac{1}{2}\cos u \right]_0^{\pi^2}$$

$$= \left[-\frac{1}{2}\cos(\pi^2) \right] - \left[-\frac{1}{2}\cos(0) \right]$$

$$= -\frac{1}{2}\cos(\pi^2) + \frac{1}{2}$$ $\cos(0) = 1$

We can approximate $\frac{1}{2}\cos(\pi^2)$ by a decimal or leave it in the above form, depending on what we want to do with the answer.

Antiderivatives of the Six Trigonometric Functions

The following table gives the indefinite integrals of the six trigonometric functions. (The first two we have already seen.)

Integrals of the Trigonometric Functions

$$\int \sin x \, dx = -\cos x + C$$

$$\int \cos x \, dx = \sin x + C$$

$$\int \tan x \, dx = -\ln|\cos x| + C \qquad \text{Shown below.}$$

$$\int \cot x \, dx = \ln|\sin x| + C \qquad \text{See the Exercise Set.}$$

$$\int \sec x \, dx = \ln|\sec x + \tan x| + C \qquad \text{Shown below.}$$

$$\int \csc x \, dx = -\ln|\csc x + \cot x| + C \qquad \text{See the Exercise Set.}$$

Derivations of Formulas for Antiderivatives of Trigonometic Functions

To show that $\int \tan x \, dx = -\ln|\cos x| + C$, we first write $\tan x$ as $\dfrac{\sin x}{\cos x}$ and put $u = \cos x$ in the integral:

$$\int \tan x \, dx = \int \frac{\sin x}{\cos x}\, dx$$

$$= -\int \frac{\sin x}{u} \frac{du}{\sin x}$$

$$= -\int \frac{du}{u}$$

$$= -\ln|u| + C$$

$$= -\ln|\cos x| + C.$$

$$\boxed{\begin{aligned} u &= \cos x \\[4pt] \frac{du}{dx} &= -\sin x \\[4pt] dx &= -\frac{du}{\sin x} \end{aligned}}$$

To show that $\int \sec x \, dx = \ln|\sec x + \tan x| + C$, first use a little "trick": Write $\sec x$ as $\sec x \left(\dfrac{\sec x + \tan x}{\sec x + \tan x} \right)$ and put u equal to the denominator:

$$\int \sec x \, dx = \int \sec x \left(\frac{\sec x + \tan x}{\sec x + \tan x} \right) dx$$

$$= \int \sec x \frac{\sec x + \tan x}{u} \cdot \frac{du}{\sec x (\tan x + \sec x)}$$

$$= \int \frac{du}{u}$$

$$= \ln|u| + C$$

$$= \ln|\sec x + \tan x| + C.$$

$$u = \sec x + \tan x$$

$$\frac{du}{dx} = \sec x \tan x + \sec^2 x$$

$$= \sec x \, (\tan x + \sec x)$$

$$dx = \frac{du}{\sec x \, (\tan x + \sec x)}$$

Shortcuts

If a and b are constants with $a \neq 0$, then we have the following formulas. (All of them can be obtained using the substitution $u = ax + b$. They will appear in the exercises.)

Shortcuts: Integrals of Expressions Involving ($ax + b$)

Rule	Quick Example
$\int \sin(ax+b)\,dx$ $= -\dfrac{1}{a}\cos(ax+b) + C$	$\int \sin(-4x)\,dx = \dfrac{1}{4}\cos(-4x) + C$
$\int \cos(ax+b)\,dx$ $= \dfrac{1}{a}\sin(ax+b) + C$	$\int \cos(x+1)\,dx = \sin(x+1) + C$
$\int \tan(ax+b)\,dx$ $= -\dfrac{1}{a}\ln|\cos(ax+b)| + C$	$\int \tan(-2x)\,dx = \dfrac{1}{2}\ln|\cos(-2x)| + C$
$\int \cot(ax+b)\,dx$ $= \dfrac{1}{a}\ln|\sin(ax+b)| + C$	$\int \cot(3x-1)\,dx$ $= \dfrac{1}{3}\ln|\sin(3x-1)| + C$
$\int \sec(ax+b)\,dx$ $= \dfrac{1}{a}\ln|\sec(ax+b) + \tan(ax+b)| + C$	$\int \sec(9x)\,dx$ $= \dfrac{1}{9}\ln|\sec(9x) + \tan(9x)| + C$
$\int \csc(ax+b)\,dx$ $= -\dfrac{1}{a}\ln|\csc(ax+b) + \cot(ax+b)| + C$	$\int \csc(x+7)\,dx$ $= -\ln|\csc(x+7) + \cot(x+7)| + C$

EXAMPLE 3 **Sales**

The rate of sales of cypods (one-bedroom living units) in the city-state of Utarek, Mars[21] can be modeled by

$$s(t) = 7.5 \cos(\pi t/6) + 87.5 \text{ units per month},$$

where t is time in months since January 1. How many cypods are sold in a calendar year?

Solution Total sales over one calendar year are given by

$$\int_0^{12} s(t)\, dt = \int_0^{12} [7.5 \cos(\pi t/6) + 87.5]\, dt$$

$$= \left[7.5 \frac{6}{\pi} \sin(\pi t/6) + 87.5t \right]_0^{12} \qquad \text{We used a shortcut on the first term.}$$

$$= \left[7.5 \frac{6}{\pi} \sin(2\pi) + 87.5(12) \right] - \left[7.5 \frac{6}{\pi} \sin(0) + 87.5(0) \right]$$

$$= 87.5(12) \qquad\qquad \sin(2\pi) = \sin(0) = 0.$$

$$= 1{,}050 \text{ cypods}.$$

➡ **Before we go on...** Would it have made any difference in Example 3 if we had computed total sales over the period [12, 24], [6, 18], or any interval of the form $[a, a + 12]$? ∎

Using Integration by Parts with Trig Functions

EXAMPLE 4 **Integrating a Polynomial Times Sine or Cosine**

Calculate $\int (x^2 + 1) \sin(x + 1)\, dx$.

Solution We use the column method of integration by parts described in Section 14.1. Because differentiating $x^2 + 1$ makes it simpler, we put it in the D column and get the following table:

	D	I
$+$	$x^2 + 1$	$\sin(x + 1)$
$-$	$2x$	$-\cos(x + 1)$
$+$	2	$-\sin(x + 1)$
$-\int$	0	$\cos(x + 1)$

---

[21]Based on www.Marsnext.com, a now extinct virtual society.

[Notice that we used the shortcut formulas to repeatedly integrate $\sin(x+1)$.] We can now read the answer from the table:

$$
\begin{aligned}
\int (x^2+1)\sin(x+1)\,dx &= (x^2+1)[-\cos(x+1)] - 2x[-\sin(x+1)] \\
&\quad + 2[\cos(x+1)] + C \\
&= (-x^2-1+2)\cos(x+1) + 2x\sin(x+1) + C \\
&= (-x^2+1)\cos(x+1) + 2x\sin(x+1) + C.
\end{aligned}
$$

EXAMPLE 5 Integrating an Exponential Times Sine or Cosine

Calculate $\int e^x \sin x\,dx$.

Solution The integrand is the product of e^x and $\sin x$, so we put one in the D column and the other in the I column. For this example, it doesn't matter much which we put where.

	D	I
$+$	$\sin x$	e^x
$-$	$\cos x$	e^x
$+\int$	$-\sin x \longrightarrow e^x$	

It looks like we're just spinning our wheels. Let's stop and see what we have:

$$
\int e^x \sin x\,dx = e^x \sin x - e^x \cos x - \int e^x \sin x\,dx.
$$

At first glance, it appears that we are back where we started, still having to evaluate $\int e^x \sin x\,dx$. However, if we add this integral to both sides of the equation above, we can solve for it:

$$
2\int e^x \sin x\,dx = e^x \sin x - e^x \cos x + C.
$$

(Why $+C$?) So,

$$
\int e^x \sin x\,dx = \frac{1}{2}e^x \sin x - \frac{1}{2}e^x \cos x + \frac{C}{2}.
$$

Because $C/2$ is just as arbitrary as C, we write C instead of $C/2$, and obtain

$$
\int e^x \sin x\,dx = \frac{1}{2}e^x \sin x - \frac{1}{2}e^x \cos x + C.
$$

16.3 EXERCISES

Evaluate the integrals in Exercises 1–28. HINT [See Quick Examples page 1173.]

1. $\displaystyle\int (\sin x - 2\cos x)\,dx$

2. $\displaystyle\int (\cos x - \sin x)\,dx$

3. $\displaystyle\int (2\cos x - 4.3\sin x - 9.33)\,dx$

4. $\displaystyle\int (4.1\sin x + \cos x - 9.33/x)\,dx$

5. $\displaystyle\int \left(3.4\sec^2 x + \frac{\cos x}{1.3} - 3.2e^x\right) dx$

6. $\displaystyle\int \left(\frac{3\sec^2 x}{2} + 1.3\sin x - \frac{e^x}{3.2}\right) dx$

7. $\displaystyle\int 7.6\cos(3x - 4)\,dx$ **8.** $\displaystyle\int 4.4\sin(-3x + 4)\,dx$
  HINT [See Example 1.]   HINT [See Example 1.]

9. $\displaystyle\int x\sin(3x^2 - 4)\,dx$ **10.** $\displaystyle\int x\cos(-3x^2 + 4)\,dx$

11. $\displaystyle\int (4x + 2)\sin(x^2 + x)\,dx$

12. $\displaystyle\int (x + 1)[\cos(x^2 + 2x) + (x^2 + 2x)]\,dx$

13. $\displaystyle\int (x + x^2)\sec^2(3x^2 + 2x^3)\,dx$

14. $\displaystyle\int (4x + 2)\sec^2(x^2 + x)\,dx$

15. $\displaystyle\int (x^2)\tan(2x^3)\,dx$ **16.** $\displaystyle\int (4x)\tan(x^2)\,dx$

17. $\displaystyle\int 6\sec(2x - 4)\,dx$ **18.** $\displaystyle\int 3\csc(3x)\,dx$

19. $\displaystyle\int e^{2x}\cos(e^{2x} + 1)\,dx$ **20.** $\displaystyle\int e^{-x}\sin(e^{-x})\,dx$

21. $\displaystyle\int_{-\pi}^{0} \sin x\,dx$ **22.** $\displaystyle\int_{\pi/2}^{\pi} \cos x\,dx$
  HINT [See Example 2.]   HINT [See Example 2.]

23. $\displaystyle\int_{0}^{\pi/3} \tan x\,dx$ **24.** $\displaystyle\int_{\pi/6}^{\pi/2} \cot x\,dx$

25. $\displaystyle\int_{1}^{\sqrt{\pi+1}} x\cos(x^2 - 1)\,dx$ **26.** $\displaystyle\int_{0.5}^{(\pi+1)/2} \sin(2x - 1)\,dx$

27. ▼ $\displaystyle\int_{1/\pi}^{2/\pi} \frac{\sin(1/x)}{x^2}\,dx$ **28.** ▼ $\displaystyle\int_{0}^{\pi/3} \frac{\sin x}{\cos^2 x}\,dx$

In Exercises 29–32, derive each equation, where a and b are constants with $a \neq 0$.

29. ▼ $\displaystyle\int \cos(ax + b)\,dx = \frac{1}{a}\sin(ax + b) + C$

30. ▼ $\displaystyle\int \sin(ax + b)\,dx = -\frac{1}{a}\cos(ax + b) + C$

31. ▼ $\displaystyle\int \cot x\,dx = \ln|\sin x| + C$

32. ▼ $\displaystyle\int \csc x\,dx = -\ln|\csc x + \cot x| + C$

Use the shortcut formulas on page 1175 to calculate the integrals in Exercises 33–40 mentally.

33. $\displaystyle\int \sin(4x)\,dx$ **34.** $\displaystyle\int \cos(5x)\,dx$

35. $\displaystyle\int \cos(-x + 1)\,dx$ **36.** $\displaystyle\int \sin\left(\frac{1}{2}x\right) dx$

37. $\displaystyle\int \sin(-1.1x - 1)\,dx$ **38.** $\displaystyle\int \cos(4.2x - 1)\,dx$

39. $\displaystyle\int \cot(-4x)\,dx$ **40.** $\displaystyle\int \tan(6x)\,dx$

Use geometry (not antiderivatives) to compute the integrals in Exercises 41–44. HINT [First draw the graph.]

41. $\displaystyle\int_{-\pi/2}^{\pi/2} \sin x\,dx$ **42.** $\displaystyle\int_{0}^{\pi} \cos x\,dx$

43. ▼ $\displaystyle\int_{0}^{2\pi} (1 + \sin x)\,dx$ **44.** ▼ $\displaystyle\int_{0}^{2\pi} (1 + \cos x)\,dx$

Use integration by parts to evaluate the integrals in Exercises 45–52. HINT [See Example 4.]

45. $\displaystyle\int x\sin x\,dx$ **46.** $\displaystyle\int x^2\cos x\,dx$

47. $\displaystyle\int x^2\cos(2x)\,dx$ **48.** $\displaystyle\int (2x + 1)\sin(2x - 1)\,dx$

49. ▼ $\displaystyle\int e^{-x}\sin x\,dx$ **50.** ▼ $\displaystyle\int e^{2x}\cos x\,dx$

51. ▼ $\displaystyle\int_{0}^{\pi} x^2\sin x\,dx$ **52.** ▼ $\displaystyle\int_{0}^{\pi/2} x\cos x\,dx$

Recall from Section 14.3 that the average of a function $f(x)$ on an interval $[a, b]$ is

$$\bar{f} = \frac{1}{b - a}\int_{a}^{b} f(x)\,dx.$$

Find the averages of the functions in Exercises 53 and 54 over the given intervals. Plot each function and its average on the same graph.

53. ▼ $f(x) = \sin x$ over $[0, \pi]$

54. ▼ $f(x) = \cos(2x)$ over $[0, \pi/4]$

Decide whether each integral in Exercises 55–58 converges. (See Section 14.5.) If the integral converges, compute its value.

55. $\displaystyle\int_0^{+\infty} \sin x \, dx$ **56.** $\displaystyle\int_0^{+\infty} \cos x \, dx$

57. ▼ $\displaystyle\int_0^{+\infty} e^{-x} \cos x \, dx$ **58.** ▼ $\displaystyle\int_0^{+\infty} e^{-x} \sin x \, dx$

APPLICATIONS

59. *Varying Cost* The cost of producing a bottle of suntan lotion is changing at a rate of $0.04 - 0.1 \sin\left[\dfrac{\pi}{26}(t - 25)\right]$ dollars per week, t weeks after January 1. If it cost \$1.50 to produce a bottle 12 weeks into the year, find the cost $C(t)$ at time t.

60. *Varying Cost* The cost of producing a box of holiday tree decorations is changing at a rate of $0.05 + 0.4 \cos\left[\dfrac{\pi}{6}(t - 11)\right]$ dollars per month, t months after January 1. If it cost \$5 to produce a box on June 1, find the cost $C(t)$ at time t.

61. ▼ *Pets* My dog Miranda is running back and forth along a 12-foot stretch of garden in such a way that her velocity t seconds after she began is

$$v(t) = 3\pi \cos\left[\frac{\pi}{2}(t - 1)\right] \text{ feet per second.}$$

How far is she from where she began 10 seconds after starting the run? HINT [See Example 3.]

62. ▼ *Pets* My cat, Prince Sadar, is pacing back and forth along his favorite window ledge in such a way that his velocity t seconds after he began is

$$v(t) = -\frac{\pi}{2} \sin\left[\frac{\pi}{4}(t - 2)\right] \text{ feet per second.}$$

How far is he from where he began 10 seconds after starting to pace? HINT [See Example 3.]

For Exercises 63–68, recall from Section 14.3 that the average of a function $f(x)$ on an interval $[a, b]$ is

$$\bar{f} = \frac{1}{b - a} \int_a^b f(x) \, dx.$$

63. *Sunspot Activity* The activity of the sun (sunspots, solar flares, and coronal mass ejection) fluctuates in cycles of around 10–11 years. Sunspot activity can be modeled by the following function:[22]

$$N(t) = 57.7 \sin[0.602(t - 1.43)] + 58.8,$$

where t is the number of years since January 1, 1997, and $N(t)$ is the number of sunspots observed at time t. Estimate the average number of sunspots visible over the 2-year period beginning January 1, 2002. (Round your answer to the nearest whole number.)

64. *Solar Emissions* The following model gives the flux of radio emission from the sun:

$$F(t) = 49.6 \sin[0.602(t - 1.48)] + 111,$$

where t is the number of years since January 1, 1997, and $F(t)$ is the flux of solar emissions of a specified wavelength at time t.[23] Estimate the average flux of radio emission over the 5-year period beginning January 1, 2001. (Round your answer to the nearest whole number.)

65. ▼ *Biology* Sigatoka leaf spot is a plant disease that affects bananas. In an infected plant, the percentage of leaf area affected varies from a low of around 5% at the start of each year to a high of around 20% at the middle of each year.[24] Use a sine function model of the percentage of leaf area affected by Sigatoka leaf spot t weeks since the start of a year to estimate, to the nearest 0.1%, the average percentage of leaf area affected in the first quarter (13 weeks) of a year.

66. ▼ *Biology* Apple powdery mildew is an epidemic that affects apple shoots. In a new infection, the percentage of apple shoots infected varies from a low of around 10% at the start of May to a high of around 60% 6 months later.[25] Use a sine function model of the percentage of apple shoots affected by apple powdery mildew t months since the start of a year to estimate, to the nearest 0.1%, the average percentage of apple shoots affected in the first 2 months of a year.

67. ▣ ▼ *Electrical Current* The typical voltage V supplied by an electrical outlet in the United States is given by

$$V(t) = 165 \cos(120\pi t),$$

where t is time in seconds.

---

[22]The model is based on a regression obtained from predicted data for 1997–2006 and the mean historical period of sunspot activity from 1755 to 1995. Source: NASA Science Directorate; Marshall Space Flight Center, August, 2002 (www.science.nasa.gov/ssl/pad/solar/predict.htm).

[23]*Ibid.* Flux measured at a wavelength of 10.7 cm.

[24]Based on graphical data. Source: American Phytopathological Society (www.apsnet.org/education/AdvancedPlantPath/Topics/Epidemiology/CyclicalNature.htm).

[25]*Ibid.*

a. Find the average voltage over the interval $[0, 1/6]$. How many times does the voltage reach a maximum in one second? (This is referred to as the number of **cycles per second**.)

b. Plot the function $S(t) = (V(t))^2$ over the interval $[0, 1/6]$.

c. The **root mean square (RMS)** voltage is given by the formula

$$V_{rms} = \sqrt{\bar{S}},$$

where $\bar{S}$ is the average value of $S(t)$ over one cycle. Estimate V_{rms}.

68. ▼ **Tides** The depth of water at my favorite surfing spot varies from 5 to 15 feet, depending on the time. Last Sunday, high tide occurred at 5:00 am and the next high tide occurred at 6:30 pm. Use the cosine function to model the depth of water as a function of time t in hours since midnight on Sunday morning. What was the average depth of the water between 10:00 am and 2:00 pm?

Income Streams *Recall from Section 14.4 that the total income received from time $t = a$ to time $t = b$ from a continuous income stream of $R(t)$ dollars per year is*

$$\text{Total value} = TV = \int_a^b R(t)\, dt.$$

In Exercises 69 and 70, find the total value of the given income stream over the given period.

69. $R(t) = 50{,}000 + 2{,}000\pi \sin(2\pi t), 0 \le t \le 1$

70. $R(t) = 100{,}000 - 2{,}000\pi \sin(\pi t), 0 \le t \le 1.5$

COMMUNICATION AND REASONING EXERCISES

71. What can you say about the definite integral of a sine or cosine function over a whole number of periods?

72. How are the derivative and antiderivative of $\sin x$ related?

73. ▼ What is the average value of $1 + 2\cos x$ over a large interval?

74. ▼ What is the average value of $3 - \cos x$ over a large interval?

75. ▼ The acceleration of an object is given by $a = K \sin(\omega t - \alpha)$. What can you say about its displacement at time t?

76. ▼ Write down a function whose derivative is -2 times its antiderivative.

CHAPTER 16 REVIEW

KEY CONCEPTS

 Website www.WanerMath.com

Go to the Website at www.WanerMath .com to find a comprehensive and interactive Web-based summary of Chapter 16.

16.1 Trigonometric Functions, Models, and Regression

The **sine** of a real number *p. 1151*

Plotting the graphs of functions based on $\sin x$ *p. 1152*

The general sine function:

$$f(x) = A \sin[\omega(x - \alpha)] + C$$

A is the **amplitude**.

C is the **vertical offset** or height of the **baseline**.

ω is the **angular frequency**.

$P = 2\pi/\omega$ is the **period** or **wavelength**.

α is the **phase shift**. *p. 1153*

Modeling with the general sine function *p. 1154*

The **cosine** of a real number *p. 1155*

Fundamental trigonometric identities:

$$\sin^2 t + \cos^2 t = 1$$
$$\cos t = \sin(t + \pi/2)$$
$$\cos t = \sin(\pi/2 - t)$$
$$\sin t = \cos(t - \pi/2)$$
$$\sin t = \cos(\pi/2 - t) \quad p.\ 1156$$

The general cosine function:

$$f(x) = A \cos[\omega(x - \beta)] + C \quad p.\ 1156$$

Modeling with the general cosine function *p. 1157*

Other trig functions:

$$\tan x = \frac{\sin x}{\cos x}$$
$$\cot x = \cotan x = \frac{\cos x}{\sin x} = \frac{1}{\tan x}$$
$$\sec x = \frac{1}{\cos x}$$
$$\csc x = \cosec x = \frac{1}{\sin x} \quad p.\ 1158$$

16.2 Derivatives of Trigonometric Functions and Applications

Derivatives of sine and cosine:

$$\frac{d}{dx} \sin x = \cos x$$
$$\frac{d}{dx} \cos x = -\sin x \quad p.\ 1164$$

Some trigonometric limits:

$$\lim_{h \to 0} \frac{\sin h}{h} = 1$$
$$\lim_{h \to 0} \frac{\cos h - 1}{h} = 0 \quad p.\ 1165$$

Derivatives of sines and cosines of functions:

$$\frac{d}{dx} \sin u = \cos u \frac{du}{dx}$$
$$\frac{d}{dx} \cos u = -\sin u \frac{du}{dx} \quad p.\ 1166$$

Derivatives of the other trigonometric functions:

$$\frac{d}{dx} \tan x = \sec^2 x$$
$$\frac{d}{dx} \cot x = -\csc^2 x$$

$$\frac{d}{dx} \sec x = \sec x \tan x$$
$$\frac{d}{dx} \csc x = -\csc x \cot x \quad p.\ 1168$$

16.3 Integrals of Trigonometric Functions and Applications

$$\int \cos x \, dx = \sin x + C$$
$$\int \sin x \, dx = -\cos x + C$$
$$\int \sec^2 x \, dx = \tan x + C \quad p.\ 1173$$

Substitution in integrals involving trig functions *p. 1173*

Definite integrals involving trig functions *p. 1173*

Antiderivatives of the other trigonometric functions:

$$\int \tan x \, dx = -\ln|\cos x| + C$$
$$\int \cot x \, dx = \ln|\sin x| + C$$
$$\int \sec x \, dx = \ln|\sec x + \tan x| + C$$
$$\int \csc x \, dx = -\ln|\csc x + \cot x| + C$$
p. 1174

Shortcuts: Integrals of expressions involving $(ax + b)$ *p. 1175*

Using integration by parts with trig functions *p. 1176*

REVIEW EXERCISES

In Exercises 1–4, model the given curve with a sine function. (The scales on the two axes may not be the same.)

1.
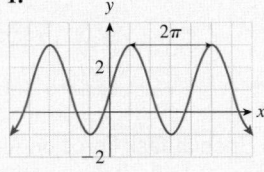

2.
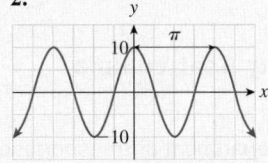

3.
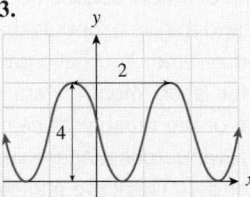

4.
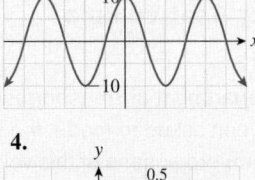

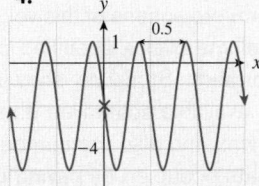

In Exercises 5–8, model the curves in Exercises 1–4 with cosine functions.

5. The curve in Exercise 1   **6.** The curve in Exercise 2

7. The curve in Exercise 3   **8.** The curve in Exercise 4

In Exercises 9–14, find the derivative of the given function.

9. $f(x) = \cos(x^2 - 1)$

10. $f(x) = \sin(x^2 + 1)\cos(x^2 - 1)$

11. $f(x) = \tan(2e^x - 1)$ **12.** $f(x) = \sec\sqrt{x^2 - x}$

13. $f(x) = \sin^2(x^2)$ **14.** $f(x) = \cos^2[1 - \sin(2x)]$

In Exercises 15–22, evaluate the given integral.

15. $\displaystyle\int 4\cos(2x - 1)\, dx$

16. $\displaystyle\int (x - 1)\sin(x^2 - 2x + 1)\, dx$

1181

17. $\int 4x \sec^2(2x^2 - 1)\,dx$ 18. $\int \dfrac{\cos\left(\dfrac{1}{x}\right)}{x^2 \sin\left(\dfrac{1}{x}\right)}\,dx$

19. $\int x\,\tan(x^2 + 1)\,dx$ 20. $\int_0^\pi \cos(x + \pi/2)\,dx$

21. $\int_{\ln(\pi/2)}^{\ln(\pi)} e^x \sin(e^x)\,dx$ 22. $\int_\pi^{2\pi} \tan(x/6)\,dx$

Use integration by parts to evaluate the integrals in Exercises 23 and 24.

23. $\int x^2 \sin x\,dx$ 24. $\int e^x \sin 2x\,dx$

APPLICATIONS: OHaganBooks.com

25. **Sales** After several years in the business, OHaganBooks.com noticed that its sales showed seasonal fluctuations, so that weekly sales oscillated in a sine wave from a low of 9,000 books per week to a high of 12,000 books per week, with the high point of the year being three quarters of the way through the year, in October. Model OHaganBooks.com's weekly sales as a generalized sine function of t, the number of weeks into the year.

26. **Mood Swings** The shipping personnel at OHaganBooks.com are under considerable pressure to cope with the large volume of orders, and periodic emotional outbursts are commonplace. The human resources department has been logging these outbursts over the course of several years, and has noticed a peak of 50 outbursts a week during the holiday season each December and a low point of 15 per week each June (probably attributable to the mild June weather). Model the weekly number of outbursts as a generalized cosine function of t, the number of months into the year ($t = 1$ represents January).

27. **Precalculus for Geniuses** The "For Geniuses" series of books has really been taking off since Duffin House first gained exclusive rights to the series 6 months ago, and revenues from *Precalculus for Geniuses* are expected to follow the curve

$$R(t) = 100,000 + 20,000e^{-0.05t} \sin\left[\frac{\pi}{6}(t - 2)\right] \text{ dollars}$$
$$(0 \le t \le 72),$$

where t is time in months from now and $R(t)$ is the monthly revenue. How fast, to the nearest dollar, will the revenue be changing 20 months from now?

28. **Elvish for Dummies** The sales department at OHaganBooks.com predicts that the revenue from sales of the latest blockbuster *Elvish for Dummies* will vary in accordance with annual releases of episodes of the movie series "Lord of the Rings Episodes 9–12." It has come up with the following model (which includes the effect of diminishing sales):

$$R(t) = 20,000 + 15,000e^{-0.12t} \cos\left[\frac{\pi}{6}(t - 4)\right] \text{ dollars}$$
$$(0 \le t \le 72),$$

where t is time in months from now and $R(t)$ is the monthly revenue. How fast, to the nearest dollar, will the revenue be changing 10 months from now?

29. **Revenue** Refer back to Question 27. Use technology or integration by parts to estimate, to the nearest $100, the total revenue from sales of *Precalculus for Geniuses* over the next 20 months.

30. **Revenue** Refer back to Question 28. Use technology or integration by parts to estimate, to the nearest $100, the total revenue from sales of *Elvish for Dummies* over the next 10 months.

31. **Mars Missions** Having completed his doctorate in biophysics, Billy-Sean O'Hagan will be accompanying the first manned mission to Mars. For reasons too complicated to explain (but having to do with the continuation of his doctoral research project and the timing of messages from his fiancée), during the voyage he will be consuming protein at a rate of

$$P(t) = 150 + 50 \sin\left[\frac{\pi}{2}(t - 1)\right] \text{ grams per day}$$

t days into the voyage. Find the total amount of protein he will consume as a function of time t.

32. **Utilities** Expenditure for utilities at OHaganBooks.com fluctuated from a high of $9,500 in October ($t = 0$) to a low of $8,000 in April ($t = 6$). Construct a sinusoidal model for the monthly expenditure on utilities and use your model to estimate the total annual cost.

Case Study   Predicting Airline Empty Seat Volume

You are a consultant to the Department of Transportation's Special Task Force on Air Traffic Congestion and have been asked to model the volume of empty seats on U.S. airline flights, to make short-term projections of this volume, and to give a formula that estimates the accumulated volume over a specified period of time. You have data from the Bureau of Transportation Statistics showing, for each month starting January 2002, the number of available seat-miles (the total of the number of seats times the number of miles flown), and also the number of revenue passenger-miles (the total of the number of seats occupied by paying passengers times the number

Logic

Introduction

Logic is the underpinning of all reasoned argument. The ancient Greeks recognized its role in mathematics and philosophy, and studied it extensively. Aristotle, in his *Organon*, wrote the first systematic treatise on logic. His work had a heavy influence on philosophy, science and religion through the Middle Ages.

But Aristotle's logic was expressed in ordinary language, so was subject to the ambiguities of ordinary language. Philosophers came to want to express logic more formally and symbolically, more like the way that mathematics is written (Leibniz, in the 17th century, was probably the first to envision and call for such a formalism). It was with the publication in 1847 of G. Boole's *The Mathematical Analysis of Logic* and A. DeMorgan's *Formal Logic* that **symbolic logic** came into being, and logic became recognized as part of mathematics. Since Boole and DeMorgan, logic and mathematics have been inextricably intertwined. Logic is part of mathematics, but at the same time it is the language of mathematics.

The study of symbolic logic is usually broken into several parts. The first and most fundamental is the **propositional logic**. Built on top of this is the **predicate logic**, which is the language of mathematics. In this appendix we give an introduction to propositional logic.

A.1 Statements and Logical Operators

Propositional logic is the study of *propositions*. A **statement**, or **proposition**, is any declarative sentence which is either true (T) or false (F). We refer to T or F as the **truth value** of the statement.

EXAMPLE 1 Statements

a. "$2 + 2 = 4$" is a statement because it can be either true or false.[1] Because it happens to be a true statement, its truth value is T.

b. "$1 = 0$" is also a statement, but its truth value is F.

c. "It will rain tomorrow" is a statement. To determine its truth value, we shall have to wait for tomorrow.

d. "Solve the following equation for x" is not a statement, because it cannot be assigned any truth value whatsoever. It is an imperative, or command, rather than a declarative sentence.

e. "The number 5" is not a statement, because it is not even a complete sentence.

For a much more extensive interactive treatment of logic, including discussion of proofs, rules of inference, and an introduction to the predicate calculus, go online and follow:

Online Text
→ On Line Topics in Finite Mathematics
  → Introduction to Logic

---

[1] Is "$2 + 2 = 4$" a sentence? Read it aloud: "Two plus two equals four," is a perfectly respectable English sentence.

f. "This statement is false" gets us into a bind: If it were true, then, because it is declaring itself to be false, it must be false. On the other hand, if it were false, then its declaring itself false is a lie, so it is true! In other words, if it is true, then it is false, and if it is false, then it is true, and we go around in circles. We get out of this bind by saying that because the sentence cannot be either true or false, we refuse to call it a statement. An equivalent pseudo-statement is: "I am lying," so this sentence is known as **the liar's paradox**.

Note Sentences that refer to themselves, or *self-referential sentences*, as illustrated in Example 1(f), are not permitted to be statements. This eliminates the liar's paradox and several similar problems. ■

We shall use letters like p, q, r, and so on to stand for statements. Thus, for example, we might decide that p should stand for the statement "the moon is round." We write

p: "the moon is round" p is the statement that the moon is round

to express this.

We can form new statements from old ones in several different ways. For example, starting with p: "I am an Anchovian," we can form the **negation** of p: "It is not the case that I am an Anchovian" or simply "I am not an Anchovian."

Negation of a Statement

If p is a statement, then its **negation** is the statement "not p" and is denoted by $\sim p$. We mean by this that, if p is true, then $\sim p$ is false, and vice versa.

Quick Examples

1. If p: "$2 + 2 = 4$," then $\sim p$: "It is not the case that $2 + 2 = 4$," or, more simply, $\sim p$: "$2 + 2 \neq 4$."

2. If q: "$1 = 0$," then $\sim q$: "$1 \neq 0$."

3. If r: "Diamonds are a pearl's best friend," then $\sim r$: "Diamonds are not a pearl's best friend."

4. If s: "All politicians are crooks," then $\sim s$: "Not all politicians are crooks."

5. **Double Negation:** If p is any statement, then the negation of $\sim p$ is $\sim(\sim p)$: "not (not p)," or, in other words, p. Thus $\sim(\sim p)$ has the same meaning as p.

Notes

1. Notice in Quick Example 1 above that $\sim p$ is false, because p is true. However, in Quick Example 2, $\sim q$ is true, because q is false. A statement of the form $\sim q$ can very well be true; it is a common mistake to think it must be false.

2. Saying that not all politicians are crooks is not the same as saying that no politicians are crooks, but is the same as saying that some (meaning one or more) politicians are not crooks.

3. The symbol $\sim$ is our first example of a **logical operator**.

4. When we say in Quick Example 5 above that $\sim(\sim p)$ has the same meaning as p, we mean that they are *logically equivalent*—a notion we will make precise below. ■

Here is another way we can form a new statement from old ones. Starting with p: "I am wise," and q: "I am strong," we can form the statement "I am wise and I am strong." We denote this new statement by $p \wedge q$, read "*p and q*." In order for $p \wedge q$ to be true, *both p and q* must be true. Thus, for example, if I am wise but not strong, then $p \wedge q$ is false. The symbol $\wedge$ is another logical operator. The statement $p \wedge q$ is called the **conjunction** of p and q.

Conjunction

The **conjunction** of p and q is the statement $p \wedge q$, which we read "*p and q*." It can also be said in a number of different ways, such as "*p even though q*." The statement $p \wedge q$ is true when both p and q are true and false otherwise.

Quick Examples

1. If p: "This galaxy will ultimately disappear into a black hole" and q: "$2 + 2 = 4$," then $p \wedge q$ is the statement "Not only will this galaxy ultimately disappear into a black hole, but $2 + 2 = 4$!"

2. If p: "$2 + 2 = 4$" and q: "$1 = 0$," then $p \wedge q$: "$2 + 2 = 4$ and $1 = 0$." Its truth value is F because q is F.

3. With p and q as in Quick Example 1, the statement $p \wedge (\sim q)$ says: "This galaxy will ultimately disappear into a black hole and $2 + 2 \neq 4$," or, more colorfully, as "Contrary to your hopes, this galaxy is doomed to disappear into a black hole; moreover, two plus two is decidedly *not* equal to four!"

Notes

1. We sometimes use the word "but" as an emphatic form of "and." For instance, if p: "It is hot," and q: "It is not humid," then we can read $p \wedge q$ as "It is hot but not humid." There are always many ways of saying essentially the same thing in a natural language; one of the purposes of symbolic logic is to strip away the verbiage and record the underlying logical structure of a statement.

2. A **compound statement** is a statement formed from simpler statements via the use of logical operators. Examples are $\sim p$, $(\sim p) \wedge (q \wedge r)$ and $p \wedge (\sim p)$. A statement that cannot be expressed as a compound statement is called an **atomic statement**.[2] For example, "I am clever" is an atomic statement. In a compound statement such as $(\sim p) \wedge (q \wedge r)$, we refer to p, q, and r as the **variables** of the statement. Thus, for example, $\sim p$ is a compound statement in the single variable p. ∎

Before discussing other logical operators, we pause for a moment to talk about **truth tables**, which give a convenient way to analyze compound statements.

Truth Table

The **truth table** for a compound statement shows, for each combination of possible truth values of its variables, the corresponding truth value of the statement.

---

[2]"Atomic" comes from the Greek for "not divisible." Atoms were originally thought to be the indivisible components of matter, but the march of science proved that wrong. The name stuck, though.

> **Quick Examples**
>
> **1.** The truth table for negation, that is, for $\sim p$, is:
>
> | p | $\sim p$ |
> |---|---|
> | T | F |
> | F | T |
>
> Each row shows a possible truth value for p and the corresponding value of $\sim p$.
>
> **2.** The truth table for conjunction, that is, for $p \wedge q$, is:
>
> | p | q | $p \wedge q$ |
> |---|---|---|
> | T | T | T |
> | T | F | F |
> | F | T | F |
> | F | F | F |
>
> Each row shows a possible combination of truth values of p and q and the corresponding value of $p \wedge q$.

EXAMPLE 2 Construction of Truth Tables

Construct truth tables for the following compound statements.

a. $\sim(p \wedge q)$ **b.** $(\sim p) \wedge q$

Solution

a. Whenever we encounter a complex statement, we work from the inside out, just as we might do if we had to evaluate an algebraic expression like $-(a + b)$. Thus, we start with the p and q columns, then construct the $p \wedge q$ column, and finally, the $\sim(p \wedge q)$ column.

p	q	$p \wedge q$	$\sim(p \wedge q)$
T	T	T	F
T	F	F	T
F	T	F	T
F	F	F	T

Notice how we get the $\sim(p \wedge q)$ column from the $p \wedge q$ column: we reverse all the truth values.

b. Because there are two variables, p and q, we again start with the p and q columns. We then evaluate $\sim p$, and finally take the conjunction of the result with q.

p	q	$\sim p$	$(\sim p) \wedge q$
T	T	F	F
T	F	F	F
F	T	T	T
F	F	T	F

Because we are "and-ing" $\sim p$ with q, we look at the values in the $\sim p$ and q columns and combine these according to the instructions for "and." Thus, for example, in the first row we have $F \wedge T = F$ and in the third row we have $T \wedge T = T$.

Here is a third logical operator. Starting with p: "You are over 18" and q: "You are accompanied by an adult," we can form the statement "You are over 18 or are accompanied by an adult," which we write symbolically as $p \vee q$, read "p or q." Now in English the word "or" has several possible meanings, so we have to agree on which one we want here. Mathematicians have settled on the **inclusive or:** $p \vee q$ means p is true or q is true *or both are true*.[3] With p and q as above, $p \vee q$ stands for "You are over 18 or are accompanied by an adult, or both." We shall sometimes include the phrase "or both" for emphasis, but even if we leave it off we still interpret "or" as inclusive.

Disjunction

The **disjunction** of p and q is the statement $p \vee q$, which we read "p or q." Its truth value is defined by the following truth table.

p	q	$p \vee q$
T	T	T
T	F	T
F	T	T
F	F	F

This is the **inclusive** or, so $p \vee q$ is true when p is true or q is true *or both* are true.

Quick Examples

1. Let p: "The butler did it" and let q: "The cook did it." Then $p \vee q$: "Either the butler or the cook did it."

2. Let p: "The butler did it," and let q: "The cook did it," and let r: "The lawyer did it." Then $(p \vee q) \wedge (\sim r)$: "Either the butler or the cook did it, but not the lawyer."

Note The only way for $p \vee q$ to be false is for *both* p and q to be false. For this reason, we can say that $p \vee q$ also means "p and q are not both false." ∎

To introduce our next logical operator, we ask you to consider the following statement: "If you earn an A in logic, then I'll buy you a new car." It seems to be made up out of two simpler statements,

> p: "You earn an A in logic," and
> q: "I will buy you a new car."

---

[3]There is also the **exclusive or:** "p or q *but not both*." This can be expressed as $(p \vee q) \wedge \sim (p \wedge q)$. Do you see why?

The original statement says: *if p is true, then q is true*, or, more simply, **if** *p*, **then** *q*. We can also phrase this as *p* **implies** *q*, and we write the statement symbolically as $p \rightarrow q$.

   Now let us suppose for the sake of argument that the original statement: "If you earn an A in logic, then I'll buy you a new car," is true. This does *not* mean that you *will* earn an A in logic. All it says is that *if* you do so, then I will buy you that car. Thinking of this as a promise, the only way that it can be broken is if you *do* earn an A and I do *not* buy you a new car. With this in mind, we define the logical statement $p \rightarrow q$ as follows.

Conditional

The **conditional** $p \rightarrow q$, which we read "if *p*, then *q*" or "*p* implies *q*," is defined by the following truth table:

p	q	$p \rightarrow q$
T	T	T
T	F	F
F	T	T
F	F	T

The arrow "$\rightarrow$" is the **conditional** operator, and in $p \rightarrow q$ the statement *p* is called the **antecedent,** or **hypothesis,** and *q* is called the **consequent,** or **conclusion.** A statement of the form $p \rightarrow q$ is also called an **implication.**

1. "If $1 + 1 = 2$ then the sun rises in the east" has the form $p \rightarrow q$ where *p*: "$1 + 1 = 2$" is true and *q*: "the sun rises in the east." is also true. Therefore the statement is true.

2. "If the moon is made of green cheese, then I am Arnold Schwartzenegger" has the form $p \rightarrow q$ where *p* is false. From the truth table, we see that $p \rightarrow q$ is therefore true, regardless of whether or not I am Arnold Schwartzenegger.

3. "If $1 + 1 = 2$ then $0 = 1$" has the form $p \rightarrow q$ where this time *p* is true but *q* is false. Therefore, by the truth table, the given statement is false.

Notes

1. The only way that $p \rightarrow q$ can be false is if *p* is true and *q* is false—this is the case of the "broken promise" in the car example above.

2. If you look at the last two rows of the truth table, you see that we say that "$p \rightarrow q$" is true when *p* is false, *no matter what the truth value of q*. Think again about the promise—if you don't get that A, then whether or not I buy you a new car, I have not broken my promise. It may seem strange at first to say that F $\rightarrow$ T is T and F $\rightarrow$ F is also T, but, as they did in choosing to say that "or" is always inclusive, mathematicians agreed that the truth table above gives the most useful definition of the conditional. ∎

It is usually misleading to think of "if *p* then *q*" as meaning that *p causes q*. For instance, tropical weather conditions cause hurricanes, but one cannot claim that if there are tropical weather conditions, then there are (always) hurricanes. Here is a list of some English phrases that *do* have the same meaning as $p \rightarrow q$.

Some Phrasings of the Conditional

We interpret each of the following as equivalent to the conditional $p \to q$.

If p then q.	p implies q.
q follows from p.	Not p unless q.
q if p.	p only if q.
Whenever p, q.	q whenever p.
p is sufficient for q.	q is necessary for p.
p is a sufficient condition for q.	q is a necessary condition for p.

Quick Example

"If it's Tuesday, this must be Belgium" can be rephrased in several ways as follows:

"Its being Tuesday implies that this is Belgium."
"This is Belgium if it's Tuesday."
"It's Tuesday only if this is Belgium."
"It can't be Tuesday unless this is Belgium."
"Its being Tuesday is sufficient for this to be Belgium."
"That this is Belgium is a necessary condition for its being Tuesday."

Notice the difference between "if" and "only if." We say that "p only if q" means $p \to q$ because, assuming that $p \to q$ is true, p can be true only if q is also. In other words, the only line of the truth table that has $p \to q$ true and p true also has q true. The phrasing "p is a sufficient condition for q" says that it suffices to know that p is true to be able to conclude that q is true. For example, it is sufficient that you get an A in logic for me to buy you a new car. Other things might induce me to buy you the car, but an A in logic would suffice. The phrasing "q is necessary for p" says that for p to be true, q must be true (just as we said for "p only if q").

Q: *Does the commutative law hold for the conditional? In other words, is $p \to q$ the same as $q \to p$?*

A: No, as we can see in the following truth table:

p	q	$p \to q$	$q \to p$
T	T	T	T
T	F	F	T
F	T	T	F
F	F	T	T

not the same

Converse and Contrapositive

The statement $q \to p$ is called the **converse** of the statement $p \to q$. A conditional and its converse are *not* the same.

The statement $\sim q \to \sim p$ is the **contrapositive** of the statement $p \to q$. A conditional and its contrapositive are logically equivalent in the sense we define below: they have the same truth value for all possible values of p and q.

EXAMPLE 3 **Converse and Contrapositive**

Give the converse and contrapositive of the statement "If you earn an A in logic, then I'll buy you a new car."

Solution This statement has the form $p \rightarrow q$ where p: "you earn an A" and q: "I'll buy you a new car." The converse is $q \rightarrow p$. In words, this is "If I buy you a new car then you earned an A in logic."

The contrapositive is $(\sim q) \rightarrow (\sim p)$. In words, this is "If I don't buy you a new car, then you didn't earn an A in logic."

Assuming that the original statement is true, notice that the converse is not necessarily true. There is nothing in the original promise that prevents me from buying you a new car if you do not earn the A. On the other hand, the contrapositive is true. If I don't buy you a new car, it must be that you didn't earn an A; otherwise I would be breaking my promise.

It sometimes happens that we do want both a conditional and its converse to be true. The conjunction of a conditional and its converse is called a **biconditional**.

Biconditional

The **biconditional**, written $p \leftrightarrow q$, is defined to be the statement $(p \rightarrow q) \wedge (q \rightarrow p)$. Its truth table is the following:

p	q	$p \leftrightarrow q$
T	T	T
T	F	F
F	T	F
F	F	T

Phrasings of the Biconditional

We interpret each of the following as equivalent to $p \leftrightarrow q$.

p if and only if q.
p is necessary and sufficient for q.
p is equivalent to q.

Quick Example

"I teach math if and only if I am paid a large sum of money" can be rephrased in several ways as follows:

"I am paid a large sum of money if and only if I teach math."
"My teaching math is necessary and sufficient for me to be paid a large sum of money."
"For me to teach math, it is necessary and sufficient that I be paid a large sum of money."

A.2 Logical Equivalence

We mentioned above that we say that two statements are **logically equivalent** if for all possible truth values of the variables involved the two statements always have the same truth values. If s and t are equivalent, we write $s \equiv t$. This is *not* another logical statement. It is simply the claim that the two statements s and t are logically equivalent. Here are some examples.

EXAMPLE 4 Logical Equivalence

Use truth tables to show the following:

a. $p \equiv \sim(\sim p)$. This is called **double negation**.

b. $\sim(p \wedge q) \equiv (\sim p) \vee (\sim q)$. This is one of **DeMorgan's Laws**.

Solution

a. To demonstrate the logical equivalence of these two statements, we construct a truth table with columns for both p and $\sim(\sim p)$.

same

p	$\sim p$	$\sim(\sim p)$
T	F	T
F	T	F

Because the p and $\sim(\sim p)$ columns contain the same truth values in all rows, the two statements are logically equivalent.

b. We construct a truth table showing both $\sim(p \wedge q)$ and $(\sim p) \vee (\sim q)$.

same

p	q	$p \wedge q$	$\sim(p \wedge q)$	$\sim p$	$\sim q$	$(\sim p) \vee (\sim q)$
T	T	T	F	F	F	F
T	F	F	T	F	T	T
F	T	F	T	T	F	T
F	F	F	T	T	T	T

Because the $\sim(p \wedge q)$ column and $(\sim p) \vee (\sim q)$ column agree, the two statements are equivalent.

➡ **Before we go on...** The statement $\sim(p \wedge q)$ can be read as "It is not the case that both p and q are true" or "p and q are not both true." We have just shown that this is equivalent to "Either p is false or q is false." ∎

Here are the two equivalences known as DeMorgan's Laws.

DeMorgan's Laws

If p and q are statements, then

$$\sim(p \wedge q) \equiv (\sim p) \vee (\sim q)$$
$$\sim(p \vee q) \equiv (\sim p) \wedge (\sim q)$$

> ### Quick Example
>
> Let p: "the President is a Democrat," and q: "the President is a Republican." Then the following two statements say the same thing:
>
> $\sim(p \wedge q)$: "the President is not both a Democrat and a Republican."
>
> $(\sim p) \vee (\sim q)$: "either the President is not a Democrat, or he is not a Republican (or he is neither)."

Here is a list of some important logical equivalences, some of which we have already encountered. All of them can be verified using truth tables as in Example 4. (The verifications of some of these are in the exercise set.)

Important Logical Equivalences

$\sim(\sim p) \equiv p$	the Double Negative Law
$p \wedge q \equiv q \wedge p$	the Commutative Law for Conjunction
$p \vee q \equiv q \vee p$	the Commutative Law for Disjunction
$(p \wedge q) \wedge r \equiv p \wedge (q \wedge r)$	the Associative Law for Conjunction
$(p \vee q) \vee r \equiv p \vee (q \vee r)$	the Associative Law for Disjunction
$\sim(p \vee q) \equiv (\sim p) \wedge (\sim q)$	DeMorgan's Laws
$\sim(p \wedge q) \equiv (\sim p) \vee (\sim q)$	
$p \wedge (q \vee r) \equiv (p \wedge q) \vee (p \wedge r)$	the Distributive Laws
$p \vee (q \wedge r) \equiv (p \vee q) \wedge (p \vee r)$	
$p \wedge p \equiv p$	Absorption Laws
$p \vee p \equiv p$	
$p \rightarrow q \equiv (\sim q) \rightarrow (\sim p)$	Contrapositive Law

Note that these logical equivalences apply to *any* statement. The ps, qs, and rs can stand for atomic statements or compound statements, as we see in the next example.

EXAMPLE 5 Applying Logical Equivalences

a. Apply DeMorgan's law (once) to the statement $\sim([p \wedge (\sim q)] \wedge r)$.

b. Apply the distributive law to the statement $(\sim p) \wedge [q \vee (\sim r)]$.

c. Consider: "You will get an A if either you are clever and the sun shines, or you are clever and it rains." Rephrase the condition more simply using the distributive law.

Solution

a. We can analyze the given statement from the outside in. It is first of all a negation, but further, it is the negation $\sim(A \wedge B)$, where A is the compound statement $[p \wedge (\sim q)]$ and B is r:

$$\sim(\quad \underline{A} \quad \wedge \quad B)$$
$$\sim(\; \overbrace{[p \wedge (\sim q)]} \wedge \quad r)$$

Now one of DeMorgan's laws is

$$\sim(A \wedge B) \equiv (\sim A) \vee (\sim B)$$

Applying this equivalence gives

$$\sim([p \wedge (\sim q)] \wedge r) \equiv (\sim[p \wedge (\sim q)]) \vee (\sim r)$$

b. The given statement has the form $A \wedge [B \vee C]$, where $A = (\sim p)$, $B = q$, and $C = (\sim r)$. So, we apply the distributive law $A \wedge [B \vee C], \equiv [A \wedge B] \vee [A \wedge C]$:

$$(\sim p) \wedge [q \vee (\sim r)] \equiv [(\sim p) \wedge q] \vee [(\sim p) \wedge (\sim r)]$$

(We need not stop here: The second expression on the right is just begging for an application of DeMorgan's law. . .)

c. The condition is "either you are clever and the sun shines, or you are clever and it rains." Let's analyze this symbolically: Let p: "You are clever," q: "The sun shines," and r: "It rains." The condition is then $(p \wedge q) \vee (p \wedge r)$. We can "factor out" the p using one of the distributive laws in reverse, getting

$$(p \wedge q) \vee (p \wedge r) \equiv p \wedge (q \vee r)$$

We are taking advantage of the fact that the logical equivalences we listed can be read from right to left as well as from left to right. Putting $p \wedge (q \vee r)$ back into English, we can rephrase the sentence as "You will get an A if you are clever and either the sun shines or it rains."

➡ **Before we go on...** In part (a) of Example 5 we could, if we wanted, apply DeMorgan's law again, this time to the statement $\sim[p \wedge (\sim q)]$ that is part of the answer. Doing so gives

$$\sim[p \wedge (\sim q)] \equiv (\sim p) \vee \sim(\sim q) \equiv (\sim p) \vee q$$

Notice that we've also used the double negative law. Therefore, the original expression can be simplified as follows:

$$\sim([p \wedge (\sim q)] \wedge r) \equiv (\sim[p \wedge (\sim q)]) \vee (\sim r) \equiv ((\sim p) \vee q) \vee (\sim r)$$

which we can write as

$$(\sim p) \vee q \vee (\sim r)$$

because the associative law tells us that it does not matter which two expressions we "or" first. ∎

A.3 Tautologies, Contradictions, and Arguments

Tautologies and Contradictions

A compound statement is a **tautology** if its truth value is always T, regardless of the truth values of its variables. It is a **contradiction** if its truth value is always F, regardless of the truth values of its variables.

1. $p \vee (\sim p)$ has truth table

p	$\sim p$	$p \vee (\sim p)$
T	F	T
F	T	T

— all T's

and is therefore a tautology.

2. $p \wedge (\sim p)$ has truth table

p	$\sim p$	$p \wedge (\sim p)$
T	F	F
F	T	F

and is therefore a contradiction.

When a statement is a tautology, we also say that the statement is **tautological**. In common usage this sometimes means simply that the statement is self-evident. In logic it means something stronger: that the statement is always true, under all circumstances. In contrast, a contradiction, or **contradictory** statement, is *never* true, under any circumstances.

Some of the most important tautologies are the **tautological implications**, tautologies that have the form of implications. We look at two of them: Direct Reasoning, and Indirect Reasoning:

Modus Ponens or Direct Reasoning

The following tautology is called *modus ponens* or **direct reasoning:**

$$[(p \rightarrow q) \wedge p] \rightarrow q$$

In Words
If an implication and its antecedent (p) are both true, then so is its consequent (q).

If my loving math implies that I will pass this course, and if I do love math, then I will pass this course.

Note You can check that the statement $[(p \rightarrow q) \wedge p] \rightarrow q$ is a tautology by drawing its truth table. ∎

Tautological implications are useful mainly because they allow us to check the validity of **arguments**.

Argument

An **argument** is a list of statements called **premises** followed by a statement called the **conclusion.** If the premises are $P_1, P_2, \ldots, P_n$ and the conclusion is C, then we say that the argument is **valid** if the statement $(P_1 \wedge P_2 \wedge \ldots \wedge P_n) \to C$ is a tautology. In other words, an argument is valid if the truth of all its premises logically implies the truth of its conclusion.

Quick Examples

1. The following is a valid argument:

$$p \to q$$
$$\underline{p}$$
$$\therefore \quad q$$

(This is the traditional way of writing an argument: We list the premises above a line and then put the conclusion below; the symbol "$\therefore$" stands for the word "therefore.") This argument is valid because the statement $[(p \to q) \wedge p] \to q$ is a tautology, namely *modus ponens*.

2. The following is an invalid argument:

$$p \to q$$
$$\underline{q}$$
$$\therefore \quad p$$

The argument is invalid because the statement $[(p \to q) \wedge q] \to p$ is not a tautology. In fact, if p is F and q is T, then the whole statement is F.

The argument in Quick Example 2 above is known as the *fallacy of affirming the consequent.* It is a common invalid argument and not always obviously flawed at first sight, so is often exploited by advertisers. For example, consider the following claim: All Olympic athletes drink Boors, so you should too. The suggestion is that, if you drink Boors, you will be an Olympic athlete:

If you are an Olympic Athlete you drink Boors.	Premise (Let's pretend this is True)
You drink Boors.	Premise (True)
$\therefore$ You are an Olympic Athlete.	Conclusion (May be false!)

This is an error that Boors hopes you will make!

There is, however, a correct argument in which we *deny* the consequent:

Modus Tollens or Indirect Reasoning

The following tautology is called *modus tollens* or **indirect reasoning:**

$$[(p \to q) \wedge (\sim q)] \to (\sim p)$$

In Words

If an implication is true but its consequent (q) is false, then its antecedent (p) is false.

In Argument Form

$$p \to q$$
$$\frac{\sim q}{\therefore \; \sim p}$$

Quick Example

If my loving math implies that I will pass this course, and if I do not pass the course, then it must be the case that I do not love math.
In argument form:

> If I love math, then I will pass this course.
>
> I will not pass the course.
>
> Therefore, I do not love math.

Note This argument is not as direct as *modus ponens;* it contains a little twist: "If I loved math I would pass this course. However, I will not pass this course. Therefore, it must be that I don't love math (else I *would* pass this course)." Hence the name "indirect reasoning."

Note that, again, there is a similar, but fallacious argument to avoid, for instance: "If I were an Olympic athlete then I would drink Boors ($p \to q$). However, I am not an Olympic athlete ($\sim p$). Therefore, I won't drink Boors. ($\sim q$)." This is a mistake Boors certainly hopes you do *not* make! ■

There are other interesting tautologies that we can use to justify arguments. We mention one more and refer the interested reader to the website for more examples and further study.

For an extensive list of tautologies go online and follow:

Chapter L Logic
→ List of Tautologies and Tautological Implications

Disjunctive Syllogism or "One or the Other"

The following tautologies are both known as the **disjunctive syllogism** or **one-or-the-other:**

$$[(p \lor q) \land (\sim p)] \to q \qquad [(p \lor q) \land (\sim q)] \to p$$

In Words
If one or the other of two statements is true, but one is known to be false, then the other must be true.

In Argument Form

$$p \lor q \qquad\qquad p \lor q$$
$$\frac{\sim p}{\therefore \; q} \qquad\qquad \frac{\sim q}{\therefore \; p}$$

Quick Example

The butler or the cook did it. The butler didn't do it. Therefore, the cook did it.
In argument form:

> The butler or the cook did it.
>
> The butler did not do it.
>
> Therefore, the cook did it.

A   EXERCISES

Which of Exercises 1–10 are statements? Comment on the truth values of all the statements you encounter. If a sentence fails to be a statement, explain why. HINT [See Example 1.]

1. All swans are white.   **2.** The fat cat sat on the mat.

3. Look in thy glass and tell whose face thou viewest.[4]

4. My glass shall not persuade me I am old.[5]

5. There is no largest number.

6. 1,000,000,000 is the largest number.

7. Intelligent life abounds in the universe.

8. There may or may not be a largest number.

9. This is exercise number 9.   **10.** This sentence no verb.[6]

Let p: "Our mayor is trustworthy," q: "Our mayor is a good speller," and r = "Our mayor is a patriot." Express each of the statements in Exercises 11–16 in logical form: HINT [See Quick Examples on pages A2, A3, A5.]

11. Although our mayor is not trustworthy, he is a good speller.

12. Either our mayor is trustworthy, or he is a good speller.

13. Our mayor is a trustworthy patriot who spells well.

14. While our mayor is both trustworthy and patriotic, he is not a good speller.

15. It may or may not be the case that our mayor is trustworthy.

16. Our mayor is either not trustworthy or not a patriot, yet he is an excellent speller.

Let p: "Willis is a good teacher," q: "Carla is a good teacher," r: "Willis' students hate math," s: "Carla's students hate math." Express the statements in Exercises 17–24 in words.

17. $p \wedge (\sim r)$ **18.** $(\sim p) \wedge (\sim q)$

19. $q \vee (\sim q)$ **20.** $((\sim p) \wedge (\sim s)) \vee q$

21. $r \wedge (\sim r)$ **22.** $(\sim s) \vee (\sim r)$

23. $\sim (q \vee s)$ **24.** $\sim (p \wedge r)$

Assume that it is true that "Polly sings well," it is false that "Quentin writes well," and it is true that "Rita is good at math." Determine the truth of each of the statements in Exercises 25–32.

25. Polly sings well and Quentin writes well.

26. Polly sings well or Quentin writes well.

27. Polly sings poorly and Quentin writes well.

28. Polly sings poorly or Quentin writes poorly.

29. Either Polly sings well and Quentin writes poorly, or Rita is good at math.

30. Either Polly sings well and Quentin writes poorly, or Rita is not good at math.

31. Either Polly sings well or Quentin writes well, or Rita is good at math.

32. Either Polly sings well and Quentin writes well, or Rita is bad at math.

Find the truth value of each of the statements in Exercises 33–48. HINT [See Quick Examples on page A6.]

33. "If $1 = 1$, then $2 = 2$."   **34.** "If $1 = 1$, then $2 = 3$."

35. "If $1 \neq 0$, then $2 \neq 2$."   **36.** "If $1 = 0$, then $1 = 1$."

37. "A sufficient condition for 1 to equal 2 is $1 = 3$."

38. "$1 = 1$ is a sufficient condition for 1 to equal 0."

39. "$1 = 0$ is a necessary condition for 1 to equal 1."

40. "$1 = 1$ is a necessary condition for 1 to equal 2."

41. "If I pay homage to the great Den, then the sun will rise in the east."

42. "If I fail to pay homage to the great Den, then the sun will still rise in the east."

43. "In order for the sun to rise in the east, it is necessary that it sets in the west."

44. "In order for the sun to rise in the east, it is sufficient that it sets in the west."

45. "The sun rises in the west only if it sets in the west."

46. "The sun rises in the east only if it sets in the east."

47. "In order for the sun to rise in the east, it is necessary and sufficient that it sets in the west."

48. "In order for the sun to rise in the west, it is necessary and sufficient that it sets in the east."

Construct the truth tables for the statements in Exercises 49–62. HINT [See Example 2.]

49. $p \wedge (\sim q)$ **50.** $p \vee (\sim q)$

51. $\sim (\sim p) \vee p$ **52.** $p \wedge (\sim p)$

53. $(\sim p) \wedge (\sim q)$ **54.** $(\sim p) \vee (\sim q)$

55. $(p \wedge q) \wedge r$ **56.** $p \wedge (q \wedge r)$

57. $p \wedge (q \vee r)$ **58.** $(p \wedge q) \vee (p \wedge r)$

59. $p \rightarrow (q \vee p)$ **60.** $(p \vee q) \rightarrow \sim p$

61. $p \leftrightarrow (p \vee q)$ **62.** $(p \wedge q) \leftrightarrow \sim p$

Use truth tables to verify the logical equivalences given in Exercises 63–72.

63. $p \wedge p \equiv p$ **64.** $p \vee p \equiv p$

65. $p \vee q \equiv q \vee p$ **66.** $p \wedge q \equiv q \wedge p$
(Commutative law for   (Commutative law for
disjunction)   conjunction)

---

[4] William Shakespeare Sonnet 3.

[5] *Ibid.*, Sonnet 22.

[6] From *Metamagical Themas: Questing for the Essence of Mind and Pattern* by Douglas R. Hofstadter (Bantam Books, New York 1986).

67. $\sim(p \vee q) \equiv (\sim p) \wedge (\sim q)$

68. $\sim(p \wedge (\sim q)) \equiv (\sim p) \vee q$

69. $(p \wedge q) \wedge r \equiv p \wedge (q \wedge r)$
(Associative law for conjunction)

70. $(p \vee q) \vee r \equiv p \vee (q \vee r)$
(Associative law for disjunction)

71. $p \rightarrow q \equiv (\sim q) \rightarrow (\sim p)$ **72.** $\sim(p \rightarrow q) \equiv p \wedge (\sim q)$

In Exercises 73–78, use truth tables to check whether the given statement is a tautology, a contradiction, or neither. HINT [See Quick Examples on page A11.]

73. $p \wedge (\sim p)$ **74.** $p \wedge p$

75. $p \wedge \sim(p \vee q)$ **76.** $p \vee \sim(p \vee q)$

77. $p \vee \sim(p \wedge q)$ **78.** $q \vee \sim(p \wedge (\sim p))$

Apply the stated logical equivalence to the given statement in Exercises 79–84. HINT [See Example 5a, b.]

79. $p \vee (\sim p)$; the commutative law

80. $p \wedge (\sim q)$; the commutative law

81. $\sim(p \wedge (\sim q))$; DeMorgan's law

82. $\sim(q \vee (\sim q))$; DeMorgan's law

83. $p \vee ((\sim p) \wedge q)$; the distributive law

84. $(\sim q) \wedge ((\sim p) \vee q)$; the distributive law

In Exercises 85–88, use the given logical equivalence to rewrite the given sentence. HINT [See Example 5c.]

85. It is not true that both I am Julius Caesar and you are a fool. DeMorgan's law.

86. It is not true that either I am Julius Caesar or you are a fool. DeMorgan's law.

87. Either it is raining and I have forgotten my umbrella, or it is raining and I have forgotten my hat. The distributive law.

88. I forgot my hat or my umbrella, and I forgot my hat or my glasses. The distributive law.

Give the contrapositive and converse of each of the statements in Exercises 89 and 90, phrasing your answers in words.

89. "If I think, then I am."

90. "If these birds are of a feather, then they flock together."

Exercises 91 and 92 are multiple choice. Indicate which statement is equivalent to the given statement, and say why that statement is equivalent to the given one.

91. "In order for you to worship Den, it is necessary for you to sacrifice beasts of burden."
   (A) "If you are not sacrificing beasts of burden, then you are not worshiping Den."
   (B) "If you are sacrificing beasts of burden, then you are worshiping Den."
   (C) "If you are not worshiping Den, then you are not sacrificing beasts of burden."

92. "In order to read the Tarot, it is necessary for you to consult the Oracle."
   (A) "In order to consult the Oracle, it is necessary to read the Tarot."
   (B) "In order not to consult the Oracle, it is necessary not to read the Tarot."
   (C) "In order not to read the Tarot, it is necessary not to read the Oracle."

In Exercises 93–102, write the given argument in symbolic form (use the underlined letters to represent the statements containing them), then decide whether it is valid or not, If it is valid, name the validating tautology. HINT [See Quick Examples on pages A12, A13, A14.]

93. If I am <u>h</u>ungry I am also <u>t</u>hirsty. I am hungry. Therefore, I am thirsty.

94. If I am not <u>h</u>ungry, then I certainly am not <u>t</u>hirsty either. I am not thirsty, and so I cannot be hungry.

95. For me to bring my <u>u</u>mbrella, it's sufficient that it <u>r</u>ain. It is not raining. Therefore, I will not bring my umbrella.

96. For me to bring my <u>u</u>mbrella, it's necessary that it <u>r</u>ain. But it is not raining. Therefore, I will not bring my umbrella.

97. For me to pass <u>m</u>ath, it is sufficient that I have a <u>g</u>ood teacher. I will not pass math. Therefore, I have a bad teacher.

98. For me to pass <u>m</u>ath, it is necessary that I have a <u>g</u>ood teacher. I will pass math. Therefore, I have a good teacher.

99. I will either pass <u>m</u>ath or I have a <u>b</u>ad teacher. I have a good teacher. Therefore, I will pass math.

100. Either <u>r</u>oses are not red or <u>v</u>iolets are not blue. But roses are red. Therefore, violets are not blue.

101. I am either <u>s</u>mart or <u>a</u>thletic, and I am athletic. So I must not be smart.

102. The president is either <u>w</u>ise or <u>s</u>trong. She is strong. Therefore, she is not wise.

In Exercises 103–108, use the stated tautology to complete the argument.

103. If John is a swan, it is necessary that he is green. John is indeed a swan. Therefore, _____. (*Modus ponens.*)

104. If Jill had been born in Texas, then she would be able to ride horses. But Jill cannot ride horses. Therefore, _____. (*Modus tollens.*)

105. If John is a swan, it is necessary that he is green. But John is not green. Therefore, _____. (*Modus tollens.*)

106. If Jill had been born in Texas, then she would be able to ride horses. Jill was born in Texas. Therefore, _____ (*Modus ponens.*)

107. Peter is either a scholar or a gentleman. He is not, however, a scholar. Therefore, _____. (*Disjunctive syllogism.*)

108. Pam is either a plumber or an electrician. She is not, however, an electrician. Therefore, _____ (*Disjunctive syllogism.*)

4.0 billion metric tons of petrochemicals domestically. $s(5) = 4.1$. In 2010, Pemex sold 4.1 billion metric tons of petrochemicals domestically. **b.** [0, 5] **c.** Graph:

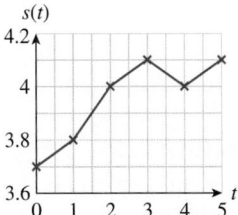

$s(1.5) \approx 3.9$. Pemex sold about 3.9 billion metric tons of petrochemicals domestically in the year ending June 30, 2007. **43.** $f(4) \approx 1,600$, $f(5) \approx 1,700$, $f(7.5) \approx 800$. There were 1.6 million housing starts in 2004, 1.7 million housing starts in 2005, and 800,000 housing starts in the year beginning July 2007.
45. $f(8) - f(0)$. The change in the number of housing starts from 2000 to 2008 was larger in magnitude than the change from 2000 to 2005.
47. a. $P(0) \approx -100$; $P(4) \approx -400$; $P(1.5) \approx 400$. Continental Airlines lost about $100 million in 2005, lost about $400 million in 2009, and made a profit of about $400 million in the year starting July 1, 2006.
b. 0; Continental's net income was increasing most rapidly in 2005. **c.** 2.5; Continental's net income was decreasing most rapidly midway through 2007.
49. a. [0, 10]. $t \geq 0$ is not an appropriate domain because it would predict U.S. trade with China into the indefinite future with no basis. **b.** $280 billion; U.S. trade with China in 2004 was valued at approximately $280 billion. **51. a.** `100*(1-12200/t^4.48)`
b. Graph:

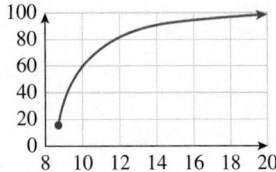

c. Table:

t	9	10	11	12	13	14
$p(t)$	35.2	59.6	73.6	82.2	87.5	91.1
t	15	16	17	18	19	20
$p(t)$	93.4	95.1	96.3	97.1	97.7	98.2

d. 82.2% **e.** 14 months
53. a. $v(10) \approx 58$, $v(16) = 200$, $v(28) = 3,800$. Processor speeds were about 58 MHz in 1990, 200 MHz in 1996, and 3,800 MHz in 2008.
b. `(8*(1.22)^x)*(x<16)+(400*x-6200)*`
`(x>=16)*(x<25)+3800*(x>=25)`

c. Graph:

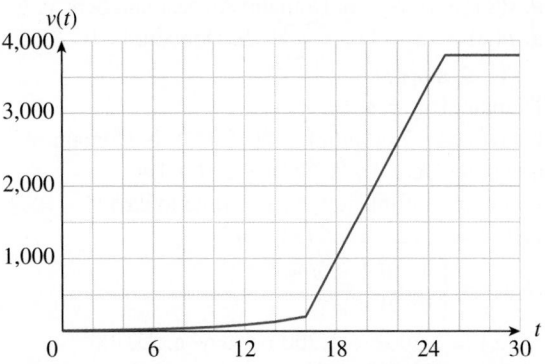

Table:

t	0	2	4	6	8	10
$v(t)$	8.0	12	18	26	39	58
t	12	14	16	18	20	22
$v(t)$	87	130	200	1,000	1,800	2,600
t	24	26	28	30		
$v(t)$	3,400	3,800	3,800	3,800		

d. 2003
55. a.
$T(x) =$

$$\begin{cases} 0.10x & \text{if } 0 < x \leq 8{,}375 \\ 837.50 + 0.15(x - 8{,}375) & \text{if } 8{,}375 < x \leq 34{,}000 \\ 4{,}663.25 + 0.25(x - 34{,}000) & \text{if } 34{,}000 < x \leq 82{,}400 \\ 16{,}763.25 + 0.28(x - 82{,}400) & \text{if } 82{,}400 < x \leq 171{,}850 \\ 41{,}809.25 + 0.33(x - 171{,}850) & \text{if } 171{,}850 < x \leq 373{,}650 \\ 108{,}403.25 + 0.35(x - 373{,}650) & \text{if } 373{,}650 < x \end{cases}$$

b. $5,163.25 **57.** $t; m$ **59.** $y(x) = 4x^2 - 2$ (or $f(x) = 4x^2 - 2$) **61.** False. A graph usually gives infinitely many values of the function while a numerical table will give only a finite number of values.
63. False. In a numerically specified function, only certain values of the function are specified so we cannot know its value on every real number in [0, 10], whereas an algebraically specified function would give values for every real number in [0, 10]. **65.** False: Functions with infinitely many points in their domain (such as $f(x) = x^2$) cannot be specified numerically. **67.** As the text reminds us: to evaluate f of a quantity (such as $x + h$) replace x everywhere by the *whole quantity* $x + h$, getting $f(x + h) = (x + h)^2 - 1$.
69. They are different portions of the graph of the associated equation $y = f(x)$. **71.** The graph of g is the same as the graph of f, but shifted 5 units to the right.

Section 1.2

1. a. $s(x) = x^2 + x$ **b.** Domain: All real numbers **c.** 6
3. a. $p(x) = (x - 1)\sqrt{x + 10}$ **b.** Domain: $[-10, 0)$
c. -14 **5. a.** $q(x) = \frac{\sqrt{10 - x}}{x - 1}$
b. Domain: $0 \le x \le 10; x \ne 1$
c. Undefined **7. a.** $m(x) = 5(x^2 + 1)$ **b.** Domain: All
real numbers **c.** 10 **9.** $N(t) = 200 + 10t$
(N = number of music files, t = time in days)
11. $A(x) = x^2/2$ **13.** $C(x) = 12x$

15. $h(n) = \begin{cases} 4 & \text{if } 1 \le n \le 5 \\ 0 & \text{if } n > 5 \end{cases}$

17. $C(x) = 1,500x + 1,200$ per day **a.** $5,700
b. $1,500 **c.** $1,500 **d.** Variable cost $= $1,500x$; Fixed
cost $= $1,200$; Marginal cost $= $1,500$ per piano
e. Graph:

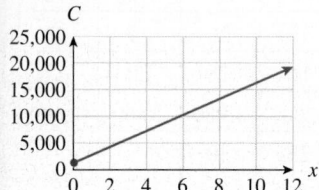

19. a.
$C(x) = 0.4x + 70$, $R(x) = 0.5x$, $P(x) = 0.1x - 70$
b. $P(500) = -20$; a loss of $20 **c.** 700 copies
21. $R(x) = 100x$, $P(x) = -2,000 + 90x - 0.2x^2$; at
least 24 jerseys. **23.** $P(x) = -1.7 - 0.02x + 0.0001x^2$;
approximately 264 thousand square feet
25. $P(x) = 100x - 5,132$, with domain $[0, 405]$.
For profit, $x \ge 52$ **27.** 5,000 units
29. $FC/(SP - VC)$ **31.** $P(x) = 579.7x - 20,000$,
with domain $x \ge 0$; $x = 34.50$ g per day for breakeven
33. a. Graph:

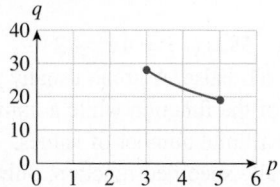

b. Ridership drops by about 3,070 rides per day.
35. a. Graph:

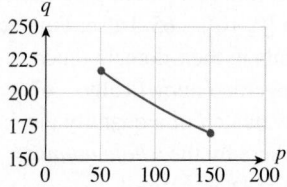

b. 210 million units **c.** 233 million units
d. $R(p) = 0.0009p^3 - 0.63p^2 + 245p$ million dollars/
year; $13 billion/year **37.** $240 per skateboard.
39. a. $110 per phone. **b.** Shortage of 25 million phones

41. a. $3.50 per ride.
Graph:

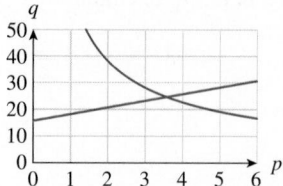

b. A surplus of around 9,170 rides. **43. a.** $12,000
b. $N(q) = 2,000 + 100q^2 - 500q$; This is the cost of
removing q pounds of PCPs per day after the subsidy is
taken into account. **c.** $N(20) = $32,000$; The net cost
of removing 20 pounds of PCPs per day is $32,000.
45. a. (B) b. $36.8 billion **47. a. (C) b.** $20.80 per
shirt if the team buys 70 shirts
Graph:

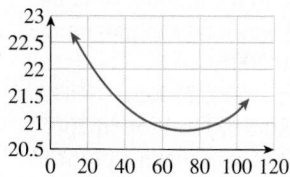

49. a. Graph:

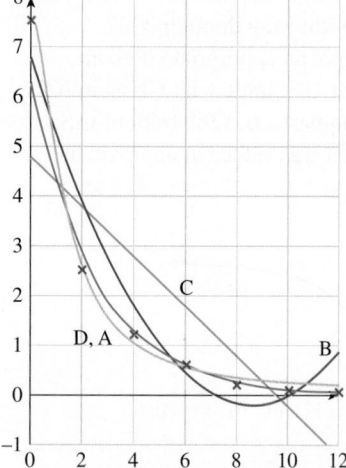

(A), (D) **b.** Model (A); Approximately $0.0021
51. A linear model (A) is the best choice; a plot of the
given points gives a straight line. **53.** Model (D) is
the best choice; the exponential models predict either
perpetually increasing or perpetually decreasing sales,
and Model (C) would give a concave-up parabola.
55. $A(t) = 5,000(1 + 0.003/12)^{12t}$; $5,106 **57.** 2024
59. 31.0 grams, 9.25 grams, 2.76 grams **61.** 20,000
years **63. a.** 1,000 years: 65%, 2,000 years: 42%,
3,000 years: 27% **b.** 1,600 years **65.** 30
67. Curve fitting. The model is based on fitting a
curve to a given set of observed data.

69. The cost of downloading a movie was $4 in January and is decreasing by 20¢ per month.   **71.** Variable; marginal.   **73.** Yes, as long as the supply is going up at a faster rate, as illustrated by the following graph:

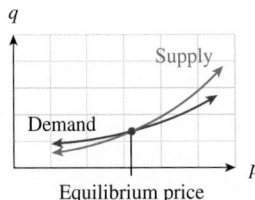

75. Extrapolate both models and choose the one that gives the most reasonable predictions.
77. They are ≥ 0.   **79.** Books per person

Section 1.3

1. 11; $m = 3$ **3.** -4; $m = -1$ **5.** 7; $m = 3/2$
7. $f(x) = -x/2 - 2$ **9.** $f(0) = -5$, $f(x) = -x - 5$
11. f is linear: $f(x) = 4x + 6$ **13.** g is linear:
$g(x) = 2x - 1$ **15.** $-3/2$ **17.** $1/6$ **19.** Undefined
21. 0   **23.** $-4/3$

25.

27.

29.

31.

33.

35.

37.

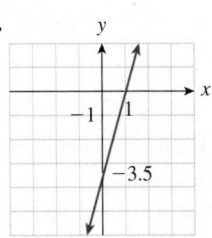

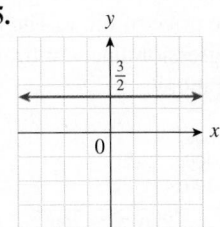

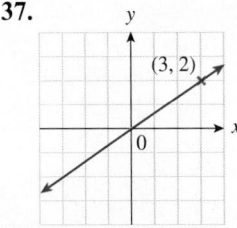

39. 2   **41.** 2   **43.** -2 **45.** Undefined   **47.** 1.5
49. -0.09 **51.** 1/2   **53.** $(d - b)/(c - a)$
55. Undefined   **57.** $-b/a$ **59. a.** 1 **b.** 1/2 **c.** 0
d. 3 **e.** $-1/3$ **f.** -1 **g.** Undefined **h.** $-1/4$ **i.** -2
61. $y = 3x$ **63.** $y = \dfrac{1}{4}x - 1$
65. $y = 10x - 203.5$ **67.** $y = -5x + 6$
69. $y = -3x + 2.25$ **71.** $y = -x + 12$
73. $y = 2x + 4$ **75.** $y = \dfrac{q}{p}x$ **77.** $y = q$
79. Fixed cost = $8,000, marginal cost = $25 per bicycle   **81.** $C = 188x + 20$; $188 per iPhone; $7,540
83. $q = -40p + 2,000$ **85. a.** $q = -2p + 273$; 123 million phones **b.** $1; 2 million
87. a. $q = -4,500p + 41,500$ **b.** Rides/day per $1 increase in the fare; ridership decreases by 4,500 rides per day for every $1 increase in the fare.
c. 14,500 rides/day   **89. a.** $y = 40t + 290$ million pounds of pasta **b.** 890 million pounds
91. a. $N = 5t - 20$ **b.** Billions of dollars per year; Amazon's net income grew at a rate of $5 billion per year. **c.** $20 billion   **93. a.** 2.5 ft/sec **b.** 20 feet along the track **c.** after 6 seconds   **95. a.** 130 miles per hour **b.** $s = 130t - 1,300$ **97. a.** $L = 55n + 475$
b. Pages per edition; *Applied Calculus* is growing at a rate of 55 pages per edition. **c.** 19th edition
99. $F = 1.8C + 32$; 86°F; 72°F; 14°F; 7°F
101. a. $A = 2.4C - 730$ **b.** $100 million lower than the actual $450 million net income American earned in 2007 **c.** Millions of dollars of American's net income per million dollars of Continental's net income; American earned an additional net income of $2.40 per $1 additional net income earned by Continental.
103. $I(N) = 0.05N + 50,000$; $N = $1,000,000$; marginal income is $m = 5$¢ per dollar of net profit
105. Increasing at 400 MHz per year
107. a. $y = 31.1t + 78$ **b.** $y = 90t - 1,100$

c. $y = \begin{cases} 31.1t + 78 & \text{if } 0 \leq t < 20 \\ 90t - 1,100 & \text{if } 20 \leq t \leq 40 \end{cases}$ or

$y = \begin{cases} 31.1t + 78 & \text{if } 0 \leq t \leq 20 \\ 90t - 1,100 & \text{if } 20 < t \leq 40 \end{cases}$

d. $1,960,000, considerably lower than the actual shown on the graph. The actual cost is not linear in the range 1990–2010.

109. $N = \begin{cases} 0.22t + 3 & \text{if } 0 \leq t \leq 5 \\ -0.15t + 4.85 & \text{if } 5 < t \leq 9 \end{cases}$

3.8 million jobs

111. Compute the corresponding successive changes Δx in x and Δy in y, and compute the ratios $\Delta y/\Delta x$. If the answer is always the same number, then the values in the table come from a linear function.

113. $f(x) = -\dfrac{a}{b}x + \dfrac{c}{b}$. If $b = 0$, then $\dfrac{a}{b}$ is undefined, and y cannot be specified as a function of x. (The graph of the resulting equation would be a vertical line.) **115.** slope, 3.   **117.** If m is positive, then y will increase as x increases; if m is negative, then y will decrease as x increases; if m is zero, then y will not change as x changes.   **119.** The slope increases, because an increase in the y-coordinate of the second point increases Δy while leaving Δx fixed. **121.** Bootlags per zonar; bootlags   **123.** It must increase by 10 units each day, including the third. **125.** (B)   **127.** It is linear with slope $m + n$. **129.** Answers may vary. For example, $f(x) = x^{1/3}$, $g(x) = x^{2/3}$ **131.** Increasing the number of items from the break-even number results in a profit: Because the slope of the revenue graph is larger than the slope of the cost graph, it is higher than the cost graph to the right of the point of intersection, and hence corresponds to a profit.

Section 1.4

1. 6   **3.** 86   **5. a.** 0.5 (better fit) **b.** 0.75
7. a. 27.42 **b.** 27.16 (better fit)
9. $y = 1.5x - 0.6667$
Graph:

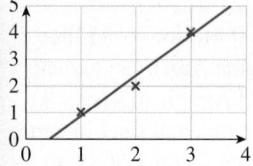

11. $y = 0.7x + 0.85$
Graph:

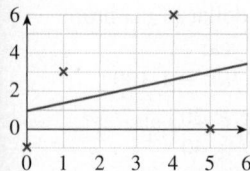

13. a. $r = 0.9959$ (best, not perfect) **b.** $r = 0.9538$
c. $r = 0.3273$ (worst)
15.

x	y	xy	x^2
2	0	0	4
6	70	420	36
10	900	9,000	100
18	970	9,420	140

$y = 112.5x - 351.7$; 998.3 million

17. $y = 3.4t + 5$; $25.4 billion   **19.** $y = 0.135x + 0.15$; 6.9 million jobs   **21. a.** $I = 44S - 220$ **b.** Amazon.com earned $44 million in net income per billion dollars in net sales. **c.** $28 billion
d. Graph:

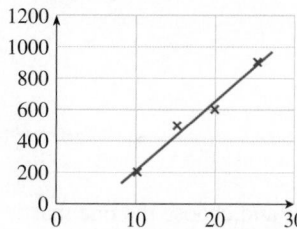

The graph shows a close fit, so the linear model seems reasonable.   **23. a.** $L = 52.70n + 486.30$
Graph:

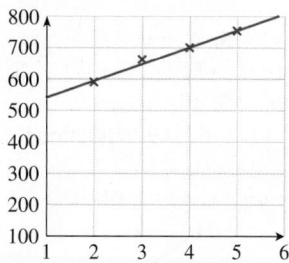

b. *Applied Calculus* is growing at a rate of 52.7 pages per edition.   **25. a.** $y = 1.62x - 23.87$.
Graph:

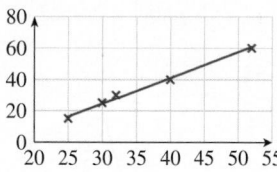

b. Each acre of cultivated land produces about 1.62 tons of soybeans.
27. a. $y = -11.85x + 797.71$; $r \approx -0.414$
b. Continental's net income is not correlated to the price of oil. **c.** The points are nowhere near the regression line, confirming the conclusion in (b).
Graph:

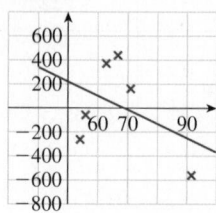

29. a. Regression line: $y = 0.582x + 35.4$ Graph:

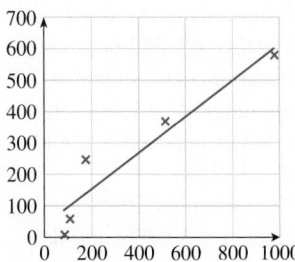

b. There are about 582 additional doctorates in engineering per additional 1,000 doctorates in the natural sciences. **c.** $r \approx 0.949$; a strong correlation. **d.** No, the data points suggest a concave down curve rather than a straight line. **31. a.** $y = 44.0t - 67.8$; $r \approx 0.912$ Graph:

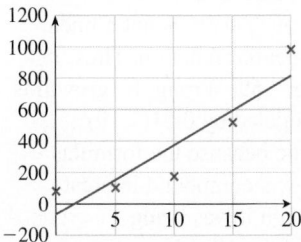

b. The number of natural science doctorates has been increasing at a rate of about 44 per year. **c.** Faster and faster rate; the slopes of successive pairs of points increase as we go from left to right. **d.** No; If r had been equal to 1, then the points would lie exactly on the regression line, which would indicate that the number of doctorates is growing at a constant rate.
33. a. $p = 0.13t + 0.22$; $r \approx 0.97$
Graph:

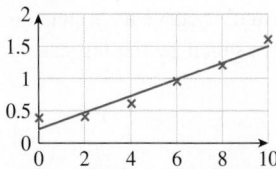

b. Yes; the first and last points lie above the regression line, while the central points lie below it, suggesting a curve.
c.

	A	B	C	D
1	t	p (Observed)	p (predicted)	Residual
2	0	0.38	0.22	0.16
3	2	0.4	0.48	-0.08
4	4	0.6	0.74	-0.14
5	6	0.95	1	-0.05
6	8	1.2	1.26	-0.06
7	10	1.6	1.52	0.08
8				

Notice that the residuals are positive at first, then become negative, and then become positive, confirming the impression from the graph.

35. The line that passes through (a, b) and (c, d) gives a sum-of-squares error SSE $= 0$, which is the smallest value possible.   **37.** The regression line is the line passing through the given points.   **39.** 0   **41.** No. The regression line through $(-1, 1)$, $(0, 0)$, and $(1, 1)$ passes through none of these points.   **43.** (Answers may vary.) The data in Exercise 33 give $r \approx 0.97$, yet the plotted points suggest a curve, not a straight line.

Chapter 1 Review

1. a. 1 **b.** -2 **c.** 0 **d.** -1 **3. a.** 1 **b.** 0 **c.** 0 **d.** -1
5.

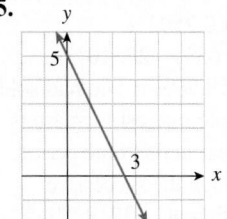

7.

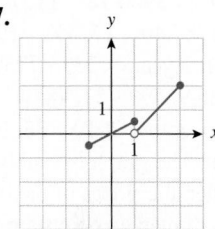

9. Absolute value   **11.** Linear   **13.** Quadratic
15. $y = -3x + 11$ **17.** $y = 1.25x - 4.25$
19. $y = (1/2)x + 3/2$ **21.** $y = 4x - 12$
23. $y = -x/4 + 1$ **25.** $y = -0.214x + 1.14$,
$r \approx -0.33$
27. a. Exponential. Graph:

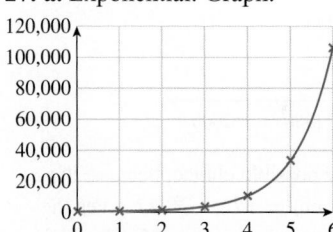

b. The ratios (rounded to 1 decimal place) are:

$V(1)/V(0)$	$V(2)/V(1)$	$V(3)/V(2)$	$V(4)/V(3)$	$V(5)/V(4)$	$V(6)/V(5)$
3	3.3	3.3	3.2	3.2	3.2

They are close to 3.2. **c.** About 343,700 visits/day
29. a. 2.3; 3.5; 6 **b.** For Web site traffic of up to 50,000 visits per day, the number of crashes is increasing by 0.03 per additional thousand visits.
c. 140,000   **31. a.** (A) **b.** (A) Leveling off
(B) Rising (C) Rising; begins to fall after 7 months
(D) Rising   **33. a.** The number of visits would increase by 30 per day. **b.** No; it would increase at a slower and slower rate and then begin to decrease.
c. Probably not. This model predicts that Web site popularity will start to decrease as advertising increases beyond $8,500 per month, and then drop toward zero.   **35. a.** $v = 0.05c + 1,800$
b. $2,150$ new visits per day **c.** $14,000 per month
37. $d = 0.95w + 8$; 86 kg   **39. a.** Cost:
$C = 5.5x + 500$; Revenue: $R = 9.5x$; Profit
$P = 4x - 500$ **b.** More than 125 albums per week
c. More than 200 albums per week

41. a. $q = -80p + 1,060$ **b.** 100 albums per week
c. $9.50, for a weekly profit of $700
43. a. $q = -74p + 1,015.5$ **b.** 239 albums per week

Chapter 2

Section 2.1

1. $INT = \$120, FV = \$2,120$
3. $INT = \$505, FV = \$20,705$
5. $INT = \$250, FV = \$10,250$ **7.** $PV = \$9,090.91$
9. $PV = \$966.18$ **11.** $PV = \$14,457.83$ **13.** $5,200
15. $787.40 **17.** 5% **19.** About 0.2503%
21. In 2 years **23.** 3.775% **25.** 65% **27.** 10%
29. 86.957% **31.** 48.04% **33.** 58.96% if you had sold
in November 2010 **35.** No. Simple interest increase is
linear. The graph is visibly not linear in that time period.
Further, the slopes of the lines through the successive
pairs of marked points are quite different. **37.** 9.2%
39. 3,260,000 **41.** $P = 500 + 46t$ thousand
(t = time in years since 1950)

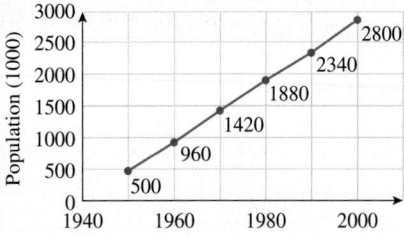

43. Graph (A) is the only possible choice, because the
equation $FV = PV(1 + rt) = PV + PVrt$ gives the
future value as a linear function of time. **45.** Wrong. In
simple interest growth, the change each year is a fixed percentage of the *starting* value, and not the preceding
year's value. (Also see Exercise 46.)
47. Simple interest is always calculated on a constant
amount, PV. If interest is paid into your account, then
the amount on which interest is calculated does not
remain constant.

Section 2.2

1. $13,439.16 **3.** $11,327.08 **5.** $19,154.30
7. $12,709.44 **9.** $613.91 **11.** $810.65
13. $1,227.74 **15.** 5.09% **17.** 10.47% **19.** 10.52%
21. $268.99 **23.** $728.91 **25.** $2,927.15
27. $21,161.79 **29.** $163,414.56
31. $55,526.45 per year **33.** $174,110 **35.** $750.00
37. $27,171.92 **39.** $111,678.96 **41.** $1,039.21

43. The one earning 11.9% compounded monthly
45. Yes. The investment will have grown to about
$394,020 million. **47.** 136 reals **49.** 245 bolivianos
51. 223 bolivars **53.** The Nicaragua investment is
better: It is worth about 1.006 units of currency (in
constant units) per unit invested as opposed to about
1.003 units for Mexico. **55.** 53.81%
57. 65.99% if you had sold in November 2010
59. No. Compound interest increase is exponential,
and exponential curves either increase continually (in
the case of appreciation) or decrease continually (in
the case of depreciation). The graph of the stock price
has both increases and decreases during the given
period, so the curve cannot model compound interest
change. **61.** 31 years; about $26,100 **63.** 2.3 years
65. a. $1,510.31 **b.** $54,701.29 **c.** 23.51%
67. The function $y = P(1 + r/m)^{mx}$ is not a linear
function of x, but an exponential function. Thus, its
graph is not a straight line. **69.** Wrong. Its growth is
exponential and can be modeled by $0.01(1.10)^t$.
71. The graphs are the same because the formulas
give the same function of x; a compound interest
investment behaves as though it was being
compounded once a year at the effective rate.
73. The effective rate exceeds the nominal rate when
the interest is compounded more than once a year
because then interest is being paid on interest accumulated during each year, resulting in a larger effective
rate. Conversely, if the interest is compounded less
often than once a year, the effective rate is less than
the nominal rate. **75.** Compare their future
values in constant dollars. The investment with the
larger future value is the better investment. **77.** The
graphs are approaching a particular curve as m gets
larger, approximately the curve given by the largest
two values of m.

Section 2.3

1. $15,528.23 **3.** $171,793.82 **5.** $23,763.28
7. $147.05 **9.** $491.12 **11.** $105.38 **13.** $90,155.46
15. $69,610.99 **17.** $95,647.68 **19.** $554.60
21. $1,366.41 **23.** $524.14 **25.** $248.85
27. $1,984.65 **29.** $999.61 **31.** $998.47
33. $917.45 **35.** 3.617% **37.** 3.059% **39.** 6.038%
41. You should take the loan from Solid Savings & Loan:
It will have payments of $248.85 per month. The
payments on the other loan would be more than
$300 per month. **43.** $973.54 **45.** $7,451.49

47. Answers using correctly rounded intermediate results:

Year	Interest	Payment on Principal
1	$3,934.98	$1,798.98
2	$3,785.69	$1,948.27
3	$3,623.97	$2,109.99
4	$3,448.84	$2,285.12
5	$3,259.19	$2,474.77
6	$3,053.77	$2,680.19
7	$2,831.32	$2,902.64
8	$2,590.39	$3,143.57
9	$2,329.48	$3,404.48
10	$2,046.91	$3,687.05
11	$1,740.88	$3,993.08
12	$1,409.47	$4,324.49
13	$1,050.54	$4,683.42
14	$661.81	$5,072.15
15	$240.84	$5,491.80

49. 1st 5 years: $402.62/month; last 25 years: $601.73
51. Original monthly payments were $824.79. The new monthly payments will be $613.46. You will save $36,481.77 in interest.   **53.** 10.81%   **55.** 13 years
57. 4.5 years   **59.** 24 years   **61.** He is wrong because his estimate ignores the interest that will be earned by your annuity—both while it is increasing and while it is decreasing. Your payments will be considerably smaller (depending on the interest earned).   **63.** He is not correct. For instance, the payments on a $100,000 10-year mortgage at 12% are $1,434.71, while for a 20-year mortgage at the same rate, they are $1,101.09, which is a lot more than half the 10-year mortgage payment.   **65.** $PV = FV(1 + i)^{-n} =$

$$PMT\frac{(1 + i)^n - 1}{i}(1 + i)^{-n} = PMT\frac{1 - (1 + i)^{-n}}{i}$$

Chapter 2 Review

1. $7,425.00   **3.** $7,604.88   **5.** $6,757.41
7. $4,848.48   **9.** $4,733.80   **11.** $5,331.37
13. $177.58   **15.** $112.54   **17.** $187.57
19. $9,584.17   **21.** 5.346%   **23.** 14.0 years
25. 10.8 years   **27.** 7.0 years   **29.** 168.85%
31. 85.28% if she sold in February 2010.   **33.** No. Simple interest increase is linear. We can compare slopes between successive points to see if the slope remained roughly constant: From December 2002 to August 2004 the slope was $(16.31 - 3.28)/(20/12) = 7.818$ while from August 2004 to March 2005 the slope was $(33.95 - 16.31)/(7/12) = 30.24$. These slopes are quite different.   **35.** 2003

Year	2000	2001	2002	2003	2004
Revenue	$180,000	$216,000	$259,200	$311,040	$373,248

37. At least 52,515 shares   **39.** $3,234.94
41. $231,844   **43.** 7.75%   **45.** $420,275
47. $140,778   **49.** $1,453.06   **51.** $2,239.90 per month
53. $53,055.66   **55.** 5.99%

Chapter 3

Section 3.1

1. (2, 2)   **3.** (3, 1)   **5.** (6, 6)   **7.** (5/3, −4/3)
9. (0, −2)   **11.** $(x, (1 − 2x)/3)$ or $((1 − 3y)/2, y)$
13. No solution   **15.** (5, 0)   **17.** (0.3, −1.1)
19. (116.6, −69.7)   **21.** (3.3, 1.8)   **23.** (3.4, 1.9)
25. 200 quarts of vanilla and 100 quarts of mocha
27. 2 servings of Mixed Cereal and 1 serving of Mango Tropical Fruit   **29. a.** 4 servings of beans and 5 slices of bread   **b.** No. One of the variables in the solution of the system has a negative value.
31. Mix 12 servings of Designer Whey and 2 servings of Muscle Milk for a cost of $9.20.   **33.** 65 g
35. 20 AAPL, 30 GOOG   **37.** 200 BOH, 400 JPM
39. 242 in favor and 193 against   **41.** 5 soccer games and 7 football games   **43.** 7   **45.** $1.50 each
47. 55 widgets   **49.** Demand: $q = −4p + 47$; supply: $q = 4p − 29$; equilibrium price: $9.50
51. 33 pairs of dirty socks and 11 T-shirts   **53.** $1,200
55. The three lines in a plane must intersect in a single point for there to be a unique solution. This can happen in two ways: (1) The three lines intersect in a single point, or (2) two of the lines are the same, and the third line intersects it in a single point.
57. Yes. Even if two lines have negative slope, they will still intersect if the slopes differ.   **59.** You cannot round both of them up, since there will not be sufficient eggs and cream. Rounding both answers down will ensure that you will not run out of ingredients. It may be possible to round one answer down and the other up, and this should be tried.   **61.** (B)   **63.** (B)
65. Answers will vary.   **67.** It is very likely. Two randomly chosen straight lines are unlikely to be parallel.

Section 3.2

1. (3, 1)   **3.** (6, 6)   **5.** $(\frac{1}{2}(1 − 3y), y)$; y arbitrary
7. No solution   **9.** (1/4, 3/4)   **11.** No solution
13. (10/3, 1/3)   **15.** (4, 4, 4)   **17.** $(−1, −3, \frac{1}{2})$
19. (z, z, z); z arbitrary   **21.** No solution
23. (−1, 1, 1)   **25.** $(1, z − 2, z)$; z arbitrary
27. $(4 + y, y, −1)$; y arbitrary

29. $(4 - y/3 + z/3, y, z)$; y arbitrary, z arbitrary
31. $(-17, 20, -2)$ **33.** $\left(-\frac{3}{2}, 0, 1/2, 0\right)$
35. $(-3z, 1 - 2z, z, 0)$; z arbitrary   **37.** $(7/5 - 17z/5 + 8w/5, 1/5 - 6z/5 - 6w/5, z, w)$; z, w arbitrary
39. $(1, 2, 3, 4, 5)$ **41.** $(-2, -2 + z - u, z, u, 0)$;
z, u arbitrary   **43.** $(16, 12/7, -162/7, -88/7)$
45. $(-8/15, 7/15, 7/15, 7/15, 7/15)$ **47.** $(1.0, 1.4, 0.2)$
49. $(-5.5, -0.9, -7.4, -6.6)$ **51.** A pivot is an entry in a matrix that is selected to "clear a column"; that is, use the row operations of a certain type to obtain zeros everywhere above and below it. "Pivoting" is the procedure of clearing a column using a designated pivot.   **53.** $2R_1 + 5R_4$, or $6R_1 + 15R_4$ (which is less desirable)   **55.** It will include a row of zeros.
57. The claim is wrong. If there are more equations than unknowns, there can be a unique solution as well as row(s) of zeros in the reduced matrix, as in Example 6.   **59.** Two   **61.** The number of pivots must equal the number of variables, since no variable will be used as a parameter.   **63.** A simple example is: $x = 1$; $y - z = 1$; $x + y - z = 2$.
65. It has to be the zero solution (each unknown is equal to zero): Putting each unknown equal to zero causes each equation to be satisfied because the right-hand sides are zero. Thus, the zero solution is in fact a solution. Because the solution is unique, this solution is the *only* solution.
67. No: As pointed out in Exercise 65, every homogeneous system has at least one solution (namely, the zero solution), and hence cannot be inconsistent.

Section 3.3

1. 100 batches of vanilla, 50 batches of mocha, and 100 batches of strawberry   **3.** 3 sections of Finite Math, 2 sections of Applied Calculus, and 1 section of Computer Methods   **5.** $25 million for regional music, $16 million for pop/rock music, and $5 million for tropical music   **7.** 4 Airbus A330-300s, 4 Boeing 767-200ERs, and 8 Dreamliners   **9.** 22 tons from Cheesy Cream, 56 tons from Super Smooth & Sons, and 22 tons from Bagel's Best Friend   **11.** 10 evil sorcerers, 50 trolls, and 500 orcs   **13.** $600 to each of the MPBF and the SCN, and $1,200 to the Jets
15. Continental: 120; American: 40; Southwest: 50
17. $5,000 in SHPIX, $2,000 in RYURX, $2,000 in RYIHX   **19.** 100 shares of KT, 50 shares of IFF, 50 shares of QSII   **21.** Microsoft: 88 million, Time Warner: 79 million, Yahoo: 75 million, Google: 42 million
23. The third equation is $x + y + z + w = 100$.
General Solution: $x = 50.5 - 0.5w$, $y = 33.5 - 0.3w$,

$z = 16 - 0.2w$, w arbitrary. State Farm is most impacted by other companies.
25. a. Brooklyn to Long Island: 500 books; Brooklyn to Manhattan: 500 books; Queens to Long Island: 1,000 books; Queens to Manhattan: 1,000 books.
b. Brooklyn to Long Island: none; Brooklyn to Manhattan: 1,000 books; Queens to Long Island: 1,500 books; Queens to Manhattan: 500 books, giving a total cost of $8,000   **27. a.** The associated system of equations has infinitely many solutions.   **b.** No; the associated system of equations still has infinitely many solutions.   **c.** Yes; North America to Australia: 440,000, North America to South Africa: 190,000, Europe to Australia: 950,000, Europe to South Africa: 950,000.   **29. a.** $x + y = 14,000$; $z + w = 95,000$; $x + z = 63,550$, $y + w = 45,450$. The system does not have a unique solution, indicating that the given data are insufficient to obtain the missing data.
b. $(x, y, z, w) = (5,600, 8,400, 57,950, 37,050)$
31. a. No; The general solution is: Eastward Blvd.: $S + 200$; Northwest La.: $S + 50$; Southwest La.: S, where S is arbitrary. Thus it would suffice to know the traffic along Southwest La.   **b.** Yes, as it leads to the solution Eastward Blvd.: 260; Northwest La.: 110; Southwest La.: 60.   **c.** 50 vehicles per day
33. a. With $x =$ traffic on middle section of Bree, $y =$ traffic on middle section of Jeppe, $z =$ traffic on middle section of Simmons, $w =$ traffic on middle section of Harrison, the general solution is $x = w - 100$, $y = w$, $z = w$, w arbitrary.
b. No; there are infinitely many possible values for y.
c. 500 cars/minute **d.** 100 cars/minute **e.** No; a large number of cars can circulate around the middle block without affecting the numbers shown.   **35. a.** No; the corresponding system of equations is underdetermined. The net flow of traffic along any of the three stretches of Broadway would suffice.   **b.** West
37. $10 billion   **39.** $x =$ water, $y =$ gray matter, $z =$ tumor   **41.** $x =$ water, $y =$ bone, $z =$ tumor, $u =$ air   **43.** tumor   **45.** 200 Democrats, 20 Republicans, 13 of other parties   **47.** Yes; $20m in Company X, $5m in Company Y, $10m in Company Z, and $30m in Company W   **49.** It is not realistic to expect to use exactly all of the ingredients. Solutions of the associated system may involve negative numbers or not exist. Only solutions with nonnegative values for all the unknowns correspond to being able to use up all of the ingredients.   **51.** Yes; $x = 100$
53. Yes; $0.3x - 0.7y + 0.3z = 0$ is one form of the equation.   **55.** No; represented by an inequality rather than an equation.   **57.** Answers will vary.

Chapter 3 Review

1. One solution

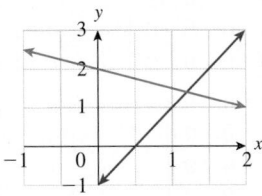

3. Infinitely many solutions

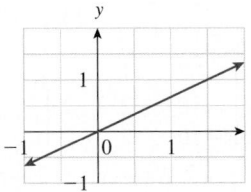

5. One solution

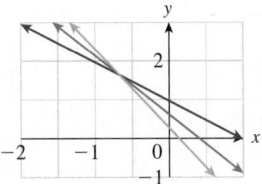

7. $(6/5, 7/5)$ **9.** $(3y/2, y)$; y arbitrary
11. $(-0.7, 1.7)$ **13.** $(-1, -1, -1)$
15. $(z - 2, 4(z - 1), z)$; z arbitrary   **17.** No solution
19. $-40°$ **21.** It is impossible; setting $F = 1.8C$
leads to an inconsistent system of equations.
23. $x + y + z + w = 10$; Linear   **25.** $w = 0$; Linear
27. $-1.3y + z = 0$ or $1.3y - z = 0$; Linear
29. 550 packages from Duffin House, 350 from
Higgins Press   **31.** 1,200 packages from Duffin
House, 200 from Higgins Press   **33.** $40
35. 7 of each   **37.** 5,000 hits per day at
OHaganBooks.com, 1,250 at JungleBooks.com,
3,750 at FarmerBooks.com   **39.** 100 shares of HAL,
20 shares of POM, and 80 shares of WELL
41. Billy-Sean is forced to take exactly the following
combination: Liberal Arts: 52 credits, Sciences:
12 credits, Fine Arts: 12 credits, Mathematics: 48
credits.   **43. a.** $x = 100$, $y = 100 + w$, $z = 300 - w$,
w arbitrary **b.** 100 book orders per day **c.** 300 book
orders per day **d.** $x = 100$, $y = 400$, $z = 0$, $w = 300$
e. 100 book orders per day
45. Yes; New York to OHaganBooks.com: 450 packages, New York to FantasyBooks.com: 50 packages,
Illinois to OHaganBooks.com: 150 packages, Illinois
to FantasyBooks.com: 150 packages

Chapter 4

Section 4.1

1. 1×4; 0 **3.** 4×1; $5/2$ **5.** $p \times q$; e_{22} **7.** 2×2; 3
9. $1 \times n$; d_r **11.** $x = 1$, $y = 2$, $z = 3$, $w = 4$

13. $\begin{bmatrix} 0.25 & -2 \\ 1 & 0.5 \\ -2 & 5 \end{bmatrix}$ **15.** $\begin{bmatrix} -0.75 & -1 \\ 0 & -0.5 \\ -1 & 6 \end{bmatrix}$

17. $\begin{bmatrix} -1 & -1 \\ 1 & -1 \\ -1 & 5 \end{bmatrix}$ **19.** $\begin{bmatrix} 0 & 2 & -2 \\ -2 & 0 & 4 \end{bmatrix}$

21. $\begin{bmatrix} 4 & -1 & -1 \\ 5 & 1 & 0 \end{bmatrix}$ **23.** $\begin{bmatrix} -2 + x & 0 & 1 + w \\ -5 + z & 3 + r & 2 \end{bmatrix}$

25. $\begin{bmatrix} -1 & -2 & 1 \\ -5 & 5 & -3 \end{bmatrix}$ **27.** $\begin{bmatrix} 9 & 15 \\ 0 & -3 \\ -3 & 3 \end{bmatrix}$

29. $\begin{bmatrix} -8.5 & -22.35 & -24.4 \\ 54.2 & 20 & 42.2 \end{bmatrix}$

31. $\begin{bmatrix} 1.54 & 8.58 \\ 5.94 & 0 \\ 6.16 & 7.26 \end{bmatrix}$ **33.** $\begin{bmatrix} 7.38 & 76.96 \\ 20.33 & 0 \\ 29.12 & 39.92 \end{bmatrix}$

35. $\begin{bmatrix} -19.85 & 115.82 \\ -50.935 & 46 \\ -57.24 & 94.62 \end{bmatrix}$

37. Q1 2010: $[3.3 \quad 21 \quad 8.7]$;   Q1 2011:
$[4.1 \quad 19.4 \quad 16.2]$

39. Sales $= \begin{bmatrix} 700 & 1,300 & 2,000 \\ 400 & 300 & 500 \end{bmatrix}$;

Inventory $-$ Sales $= \begin{bmatrix} 300 & 700 & 3,000 \\ 600 & 4,700 & 1,500 \end{bmatrix}$

41. Profit $=$ Revenue $-$ Cost $= \begin{bmatrix} 8,000 & 7,200 & 8,800 \\ 5,600 & 5,760 & 7,040 \\ 2,800 & 3,500 & 4,000 \end{bmatrix}$

43. 2000 distribution $= A = [53.6 \quad 64.4 \quad 100.2 \quad 63.2]$;
2010 distribution $= B = [55.3 \quad 66.9 \quad 114.6 \quad 71.9]$;
Net change 2000 to 2010 $= D = B - A =$
$[1.7 \quad 2.5 \quad 14.4 \quad 8.7]$; 2020 Distribution $= B + D =$
$[57.0 \quad 69.4 \quad 129.0 \quad 80.6]$
45. Total Foreclosures $=$ Foreclosures in California $+$
Foreclosures in Florida $+$ Foreclosures in Texas $=$
$[55,900 \quad 51,900 \quad 54,100 \quad 56,200 \quad 59,400] +$
$[19,600 \quad 19,200 \quad 23,800 \quad 22,400 \quad 23,600] +$
$[8,800 \quad 9,100 \quad 9,300 \quad 10,600 \quad 10,100] =$
$[84,300 \quad 80,200 \quad 87,200 \quad 89,200 \quad 93,100]$
47. Difference $=$ Foreclosures in California $-$
Foreclosures in Florida $= [55,900 \quad 51,900 \quad 54,100$
$56,200 \quad 59,400] - [19,600 \quad 19,200 \quad 23,800 \quad 22,400$
$23,600] = [36,300 \quad 32,700 \quad 30,300 \quad 33,800 \quad 35,800]$
The difference was greatest in April.

49. a. Use =

	Proc	Mem	Tubes
Pom II	2	16	20
Pom Classic	1	4	40

$$\begin{bmatrix} 2 & 16 & 20 \\ 1 & 4 & 40 \end{bmatrix};$$

$$\text{Inventory} = \begin{bmatrix} 500 & 5{,}000 & 10{,}000 \\ 200 & 2{,}000 & 20{,}000 \end{bmatrix};$$

$$\text{Inventory} - 100 \times \text{Use} = \begin{bmatrix} 300 & 3{,}400 & 8{,}000 \\ 100 & 1{,}600 & 16{,}000 \end{bmatrix}$$

b. After 4 months

51. a. $A = \begin{bmatrix} 440 & 190 \\ 950 & 950 \\ 1{,}790 & 200 \end{bmatrix}$ $D = \begin{bmatrix} -20 & 40 \\ 50 & 50 \\ 0 & 100 \end{bmatrix}$

$$2008 \text{ Tourism} = A + D = \begin{bmatrix} 420 & 230 \\ 1{,}000 & 1{,}000 \\ 1{,}790 & 300 \end{bmatrix}$$

b. $\frac{1}{2}(A+B);$ $\begin{bmatrix} 430 & 210 \\ 975 & 975 \\ 1{,}790 & 250 \end{bmatrix}$

53. No; for two matrices to be equal, they must have the same dimension. **55.** The ijth entry of the sum $A + B$ is obtained by adding the ijth entries of A and B.
57. It would have zeros down the main diagonal:

$$A = \begin{bmatrix} 0 & \# & \# & \# & \# \\ \# & 0 & \# & \# & \# \\ \# & \# & 0 & \# & \# \\ \# & \# & \# & 0 & \# \\ \# & \# & \# & \# & 0 \end{bmatrix}$$

The symbols # indicate arbitrary numbers.
59. $(A^T)_{ij} = A_{ji}$ **61.** Answers will vary.

a. $\begin{bmatrix} 0 & -4 \\ 4 & 0 \end{bmatrix}$ **b.** $\begin{bmatrix} 0 & -4 & 5 \\ 4 & 0 & 1 \\ -5 & -1 & 0 \end{bmatrix}$

63. The associativity of matrix addition is a consequence of the associativity of addition of numbers, since we add matrices by adding the corresponding entries (which are real numbers).
65. Answers will vary.

Section 4.2

1. [13]   **3.** [5/6]   **5.** $[-2y + z]$ **7.** Undefined
9. [3 0 −6 −2]   **11.** [−6 37 7]

13. $\begin{bmatrix} -4 & -7 & -1 \\ 9 & 17 & 0 \end{bmatrix}$ **15.** $\begin{bmatrix} 0 & 1 \\ 0 & 0 \end{bmatrix}$ **17.** $\begin{bmatrix} 1 & -1 \\ 1 & -1 \end{bmatrix}$

19. $\begin{bmatrix} 0 & 0 \\ 0 & 0 \end{bmatrix}$ **21.** Undefined **23.** $\begin{bmatrix} 1 & -5 & 3 \\ 0 & 0 & 9 \\ 0 & 4 & 1 \end{bmatrix}$

25. $\begin{bmatrix} 3 \\ -4 \\ 0 \\ 3 \end{bmatrix}$ **27.** $\begin{bmatrix} 0.23 & 5.36 & -21.65 \\ -13.18 & -5.82 & -16.62 \\ -11.21 & -9.9 & 0.99 \\ -2.1 & 2.34 & 2.46 \end{bmatrix}$

29. $A^2 = \begin{bmatrix} 0 & 0 & 1 & 2 \\ 0 & 0 & 0 & 1 \\ 0 & 0 & 0 & 0 \\ 0 & 0 & 0 & 0 \end{bmatrix}; \quad A^3 = \begin{bmatrix} 0 & 0 & 0 & 1 \\ 0 & 0 & 0 & 0 \\ 0 & 0 & 0 & 0 \\ 0 & 0 & 0 & 0 \end{bmatrix};$

$$A^4 = \begin{bmatrix} 0 & 0 & 0 & 0 \\ 0 & 0 & 0 & 0 \\ 0 & 0 & 0 & 0 \\ 0 & 0 & 0 & 0 \end{bmatrix}; \ldots; A^{100} = \begin{bmatrix} 0 & 0 & 0 & 0 \\ 0 & 0 & 0 & 0 \\ 0 & 0 & 0 & 0 \\ 0 & 0 & 0 & 0 \end{bmatrix};$$

31. $\begin{bmatrix} 4 & -1 \\ -1 & -7 \end{bmatrix}$ **33.** $\begin{bmatrix} 4 & -1 \\ -12 & 2 \end{bmatrix}$

35. $\begin{bmatrix} -2 & 1 & -2 \\ 10 & -2 & 2 \\ -10 & 2 & -2 \end{bmatrix}$

37. $\begin{bmatrix} -2+x-z & 2-r & -6+w \\ 10+2z & -2+2r & 10 \\ -10-2z & 2-2r & -10 \end{bmatrix}$

39. a.–d. $P^2 = P^4 = P^8 = P^{1{,}000} = \begin{bmatrix} 0.2 & 0.8 \\ 0.2 & 0.8 \end{bmatrix}$

41. a. $P^2 = \begin{bmatrix} 0.01 & 0.99 \\ 0 & 1 \end{bmatrix}$

b. $P^4 = \begin{bmatrix} 0.0001 & 0.9999 \\ 0 & 1 \end{bmatrix}$

c and d. $P^8 \approx P^{1{,}000} \approx \begin{bmatrix} 0 & 1 \\ 0 & 1 \end{bmatrix}$

43. a.–d. $P^2 = P^4 = P^8 = P^{1{,}000} = \begin{bmatrix} 0.3 & 0.3 & 0.4 \\ 0.3 & 0.3 & 0.4 \\ 0.3 & 0.3 & 0.4 \end{bmatrix}$

The rows of P are the same, and the entries in each row add up to 1. If P is any square matrix with identical rows such that the entries in each row add up to 1, then $P \cdot P = P$.
45. $2x - y + 4z = 3;\ -4x + \frac{3}{4}y + \frac{1}{3}z = -1;\ -3x = 0$
47. $x - y + w = -1;\ x + y + 2z + 4w = 2$

49. $\begin{bmatrix} 1 & -1 \\ 2 & -1 \end{bmatrix}\begin{bmatrix} x \\ y \end{bmatrix} = \begin{bmatrix} 4 \\ 0 \end{bmatrix}$

51. $\begin{bmatrix} 1 & 1 & -1 \\ 2 & 1 & 1 \\ \frac{3}{4} & 0 & \frac{1}{2} \end{bmatrix}\begin{bmatrix} x \\ y \\ z \end{bmatrix} = \begin{bmatrix} 8 \\ 4 \\ 1 \end{bmatrix}$

53. Revenue = Price × Quantity =
$$[15 \quad 10 \quad 12]\begin{bmatrix} 50 \\ 40 \\ 30 \end{bmatrix} = [1{,}510]$$

55. $32.5 million
57. Revenue = Quantity × Price =
$$\begin{bmatrix} 700 & 1{,}300 & 2{,}000 \\ 400 & 300 & 500 \end{bmatrix}\begin{bmatrix} 30 \\ 10 \\ 15 \end{bmatrix} = \begin{bmatrix} 64{,}000 \\ 22{,}500 \end{bmatrix}$$

59. $5,100 billion (or $5.1 trillion)

61. $D = N(F - M)$, where N is the income per person, and F and M are, respectively, the female and male populations in 2020; $400 billion

63. [1.2   1.0], which represents the amount, in billions of pounds, by which cheese production in north central states exceeded that in western states.

65. Number of foreclosures filings handled by firm = Percentage handled by firm × Total number = [8,460   8,860   9,140]

67. The number of foreclosures in California and Florida combined in each of the months shown.

69. $[1 \quad -1 \quad 1] \begin{bmatrix} 54{,}100 & 56{,}200 & 59{,}400 \\ 23{,}800 & 22{,}400 & 23{,}600 \\ 9{,}300 & 10{,}600 & 10{,}100 \end{bmatrix} \begin{bmatrix} 1 \\ 1 \\ 1 \end{bmatrix}$

$= [129{,}900]$

71. $\begin{bmatrix} 2 & 16 & 20 \\ 1 & 4 & 40 \end{bmatrix} \begin{bmatrix} 100 & 150 \\ 50 & 40 \\ 10 & 15 \end{bmatrix} = \begin{bmatrix} 1{,}200 & 1{,}240 \\ 700 & 910 \end{bmatrix}$

73. $AB = \begin{bmatrix} 29.6 \\ 85.5 \\ 97.5 \end{bmatrix} \quad AC = \begin{bmatrix} 22 & 7.6 \\ 47.5 & 38 \\ 89.5 & 8 \end{bmatrix}$

The entries of AB give the number of people from each of the three regions who settle in Australia or South Africa, while the entries in AC break those figures down further into settlers in South Africa and settlers in Australia.   **75.** [54.6   66.0   111.8   70.6]

77. Answers will vary. One example:

$A = [1 \quad 2], B = \begin{bmatrix} 1 & 2 & 3 \\ 4 & 5 & 6 \end{bmatrix}.$

Another example: $A = [1], B = [1 \quad 2]$

79. We finnd that the addition and multiplication of 1×1 matrices is identical to the addition and multiplication of numbers.   **81.** The claim is correct. Every matrix equation represents the equality of two matrices. When two matrices are equal, each of their corresponding entries must be equal. Equating the corresponding entries gives a system of equations.

83. Here is a possible scenario: costs of items A, B, and C in 2013 = [10   20   30], percentage increases in these costs in 2014 = [0.5   0.1   0.20], actual increases in costs = [10 × 0.5   20 × 0.1   30 × 0.20]

85. It produces a matrix whose ij entry is the product of the ij entries of the two matrices.

Section 4.3

1. Yes   **3.** Yes   **5.** No   **7.** $\begin{bmatrix} -1 & 1 \\ 2 & -1 \end{bmatrix}$ **9.** $\begin{bmatrix} 0 & 1 \\ 1 & 0 \end{bmatrix}$

11. $\begin{bmatrix} 1 & -1 \\ -1 & 2 \end{bmatrix}$ **13.** Singular   **15.** $\begin{bmatrix} 1 & -1 & 0 \\ 0 & 1 & -1 \\ 0 & 0 & 1 \end{bmatrix}$

17. $\begin{bmatrix} 1 & -1 & 1 \\ \frac{1}{2} & 0 & -\frac{1}{2} \\ -\frac{1}{2} & 1 & -\frac{1}{2} \end{bmatrix}$

19. $\begin{bmatrix} 1 & \frac{1}{3} & -\frac{1}{3} \\ 1 & -\frac{2}{3} & -\frac{1}{3} \\ -1 & \frac{1}{3} & \frac{2}{3} \end{bmatrix}$

21. Singular   **23.** $\begin{bmatrix} 0 & 1 & -2 & 1 \\ 0 & 1 & -1 & 0 \\ 1 & -1 & 2 & -1 \\ 0 & 1 & -1 & 1 \end{bmatrix}$

25. $\begin{bmatrix} 1 & -2 & 1 & 0 \\ 0 & 1 & -2 & 1 \\ 0 & 0 & 1 & -2 \\ 0 & 0 & 0 & 1 \end{bmatrix}$ **27.** $-2; \begin{bmatrix} \frac{1}{2} & \frac{1}{2} \\ \frac{1}{2} & -\frac{1}{2} \end{bmatrix}$

29. $-2; \begin{bmatrix} -2 & 1 \\ \frac{3}{2} & -\frac{1}{2} \end{bmatrix}$ **31.** $1/36; \begin{bmatrix} 6 & 6 \\ 0 & 6 \end{bmatrix}$

33. 0; Singular

35. $\begin{bmatrix} 0.38 & 0.45 \\ 0.49 & -0.41 \end{bmatrix}$ **37.** $\begin{bmatrix} 0.00 & -0.99 \\ 0.81 & 2.87 \end{bmatrix}$

39. Singular

41. $\begin{bmatrix} 91.35 & -8.65 & 0 & -71.30 \\ -0.07 & -0.07 & 0 & 2.49 \\ 2.60 & 2.60 & -4.35 & 1.37 \\ 2.69 & 2.69 & 0 & -2.10 \end{bmatrix}$

43. $(5/2, 3/2)$

45. $(6, -4)$ **47.** $(6, 6, 6)$ **49. a.** $(10, -5, -3)$

b. $(6, 1, 5)$ **c.** $(0, 0, 0)$ **51. a.** 10/3 servings of beans, and 5/6 slices of bread

b. $\begin{bmatrix} -1/2 & 1/6 \\ 7/8 & -5/24 \end{bmatrix} \begin{bmatrix} A \\ B \end{bmatrix} = \begin{bmatrix} -A/2 + B/6 \\ 7A/8 - 5B/24 \end{bmatrix}$; that is,

$-A/2 + B/6$ servings of beans and $7A/8 - 5B/24$ slices of bread   **53. a.** 100 batches of vanilla, 50 batches of mocha, 100 batches of strawberry   **b.** 100 batches of vanilla, no mocha, 200 batches of strawberry

c. $\begin{bmatrix} 1 & -1/3 & -1/3 \\ -1 & 0 & 1 \\ 0 & 2/3 & -1/3 \end{bmatrix} \begin{bmatrix} A \\ B \\ C \end{bmatrix}$, or

$A - B/3 - C/3$ batches of vanilla, $-A + C$ batches of mocha, and $2B/3 - C/3$ batches of strawberry

55. $5,000 in SHPIX, $2,000 in RYURX, $2,000 in RYIHX   **57.** 100 shares of KT, 50 shares of IFF, 50 shares of QSII   **59.** [54.1   65.7   112.9   70.3]

61. a. $(-0.7071, 3.5355)$ **b.** R^2, R^3 **c.** R^{-1}

63. [37 81 40 80 15 45 40 96 29 59 4 8]

65. CORRECT ANSWER   **67.** (A)   **69.** The inverse does not exist; the matrix is singular.   (If two rows of a matrix are the same, then row reducing it will lead to a row of zeros, and so it cannot be reduced to the identity.)

71. Calculation (See the Student Solutions Manual.)
73. When one or more of the d_i are zero. If that is the case, then the matrix $[D \mid I]$ easily reduces to a matrix that has a row of zeros on the left-hand portion, so that D is singular. Conversely, if none of the d_i are zero, then $[D \mid I]$ easily reduces to a matrix of the form $[I \mid E]$, showing that D is invertible.
75. $(AB)(B^{-1}A^{-1}) = A(BB^{-1})A^{-1} = AIA^{-1} = AA^{-1} = I$ **77.** If A has an inverse, then every system of equations $AX = B$ has a unique solution, namely $X = A^{-1}B$. But if A reduces to a matrix with a row of zeros, then such a system has either infinitely many solutions or no solution at all.

Section 4.4

1. -1 **3.** -0.25 **5.** $[0 \quad 0 \quad 1 \quad 0]$; $e = 2.25$
7. $[1 \quad 0 \quad 0]^T$ or $[0 \quad 1 \quad 0]^T$; $e = 1/4$

9. $\begin{array}{c} \\ a \\ b \end{array}\begin{array}{cc} p & r \\ \begin{bmatrix} 1 & 10 \\ 2 & -4 \end{bmatrix} \end{array}$ **11.** $\begin{array}{c} c \\ 3[-1] \end{array}$ **13.** $\begin{array}{c} b \\ q\,[0] \end{array}$

15. Strictly determined. The row player's optimal strategy is a; the column player's optimal strategy is q; Value: 1   **17.** Not strictly determined Circle the row minima and box the row maxima:

$$\begin{array}{c} \\ a \\ b \end{array}\begin{array}{ccc} p & q & r \\ \begin{bmatrix} 2 & 0 & \boxed{-2} \\ \boxed{-1} & 3 & 0 \end{bmatrix} \end{array} \qquad \begin{array}{c} \\ a \\ b \end{array}\begin{array}{ccc} p & q & r \\ \begin{bmatrix} \boxed{2} & 0 & -2 \\ -1 & \boxed{3} & \boxed{0} \end{bmatrix} \end{array}.$$

Since no entry is both circled and boxed, there are no saddle points, and so the game is not strictly determined.

19. Not strictly determined
Circle the row minima and box the row maxima:

$$\begin{array}{c} \\ P \\ Q \\ R \\ S \end{array}\begin{array}{ccc} a & b & c \\ \begin{bmatrix} 1 & -1 & \boxed{-5} \\ 4 & \boxed{-4} & 2 \\ 3 & -3 & \boxed{-10} \\ 5 & \boxed{-5} & -4 \end{bmatrix} \end{array} \quad \begin{array}{c} \\ P \\ Q \\ R \\ S \end{array}\begin{array}{ccc} a & b & c \\ \begin{bmatrix} 1 & \boxed{-1} & -5 \\ 4 & -4 & \boxed{2} \\ 3 & -3 & -10 \\ \boxed{5} & -5 & -4 \end{bmatrix} \end{array}.$$

Since no entry is both circled and boxed, there are no saddle points, and so the game is not strictly determined.

21. $R = [1/4 \quad 3/4]$, $C = [3/4 \quad 1/4]^T$, $e = -1/4$
23. $R = [3/4 \quad 1/4]$, $C = [3/4 \quad 1/4]^T$, $e = -5/4$
25. Row player: You; Column player: Your friend;

$$\begin{array}{c} \\ H \\ T \end{array}\begin{array}{cc} H & T \\ \begin{bmatrix} -1 & 1 \\ 1 & -1 \end{bmatrix} \end{array}$$

27. $F =$ France; $S =$ Sweden; $N =$ Norway

Your Opponent Defends

$$\text{You Invade} \quad \begin{array}{c} \\ F \\ S \\ N \end{array}\begin{array}{ccc} F & S & N \\ \begin{bmatrix} -1 & 1 & 1 \\ 1 & -1 & 1 \\ 1 & 1 & -1 \end{bmatrix} \end{array}$$

29. Row player: You, Column player: Your opponent; B = Brakpan, N = Nigel, S = Springs;

Your Opponent

$$\text{You} \quad \begin{array}{c} \\ B \\ N \\ S \end{array}\begin{array}{ccc} B & N & S \\ \begin{bmatrix} 0 & 0 & 1{,}000 \\ 0 & 0 & 1{,}000 \\ -1{,}000 & -1{,}000 & 0 \end{bmatrix} \end{array}$$

31. P = PleasantTap; T = Thunder Rumble; S = Strike the Gold, N = None;

Winner

$$\text{You Bet} \quad \begin{array}{c} \\ P \\ T \\ S \end{array}\begin{array}{cccc} P & T & S & N \\ \begin{bmatrix} 25 & -10 & -10 & -10 \\ -10 & 35 & -10 & -10 \\ -10 & -10 & 40 & -10 \end{bmatrix} \end{array}$$

33. You can expect to lose 39 customers.
35. Option 2: Move to the suburbs.   **37. a.** About 66%
b. Yes; spend the whole night studying game theory; 75%
c. Game theory; 57.5%   **39. a.** Lay off 10 workers; cost: $40,000 **b.** 60 inches of snow, costing $350,000
c. Lay off 15 workers.   **41. a.** CE should charge $1,000, and GCS should charge $900; 15% gain in market share for CE. **b.** CE should charge $1,200 (the more CE can charge for the same market, the better!).
43. Pablo vs. Noto; evenly matched   **45.** Both commanders should use the northern route; 2 days.
47. Confess

49. a. $\begin{array}{c} \\ F \\ O \end{array}\begin{array}{cc} F & O \\ \begin{bmatrix} 24 & 21 \\ 25 & 24 \end{bmatrix} \end{array}$

b. Both candidates should visit Ohio, leaving Romney with a 24% chance of winning the election.
51. Allocate 1/7 of the budget to WISH and the rest (6/7) to WASH. Softex will lose approximately $2,860.
53. Like a saddle point in a payoff matrix, the center of a saddle is a low point (minimum height) in one direction and a high point (maximum) in a perpendicular direction.   **55.** Although there is a saddle point in the (2, 4) position, you would be wrong to use saddle points (based on the minimax criterion) to reach the conclusion that row strategy 2 is best. One reason is that the entries in the matrix do not represent payoffs, since high numbers of employees in an area do not necessarily represent benefit to the row player. Another reason for this is that there is no opponent deciding what your job will be in such a way as to force you into the least populated job.

9. $p = 260/3$; $x = 50/3$, $y = 0$, $z = 70/3$, $w = 0$
11. $c = 80$; $x = 20/3$, $y = 20/3$
13. $c = 100$; $x = 0$, $y = 100$, $z = 0$
15. $c = 111$; $x = 1$, $y = 1$, $z = 1$
17. $c = 200$; $x = 200$, $y = 0$, $z = 0$, $w = 0$
19. $p = 136.75$; $x = 0$, $y = 25.25$, $z = 0$, $w = 15.25$
21. $c = 66.67$; $x = 0$, $y = 66.67$, $z = 0$
23. $c = -250$; $x = 0$, $y = 500$, $z = 500$, $w = 1,500$ **25.** Plant 100 acres of tomatoes and no other crops. This will give you a profit of $200,000. (You will be using all 100 acres of your farm.)
27. 10 mailings to the East Coast, none to the Midwest, 10 to the West Coast. Cost: $900. Another solution resulting in the same cost is no mailings to the East Coast, 15 to the Midwest, none to the West Coast.   **29.** 10,000 quarts of orange juice and 2,000 quarts of orange concentrate   **31.** Sell 25,000 regional music albums, 10,000 pop/rock music albums, and 5,000 tropical music albums per day for a maximum revenue of $195,000.   **33.** One serving of cereal, one serving of juice, and no dessert!   **35.** 15 bundles from Nadir, 5 from Sonny, and none from Blunt. Cost: $70,000. Another solution resulting in the same cost is 10 bundles from Nadir, none from Sonny, and 10 from Blunt.   **37.** Mix 6 servings of Riboforce HP and 10 servings of Creatine Transport for a cost of $15.60.   **39. a.** Build 1 convention-style hotel, 4 vacation-style hotels, and 2 small motels. The total cost will amount to $188 million.   **b.** Because 20% of this is $37.6 million, you will still be covered by the subsidy.   **41.** Tucson to Honolulu: 500 boards/week; Tucson to Venice Beach: 120 boards/week; Toronto to Honolulu: 0 boards/week; Toronto to Venice Beach: 410 boards/week. Minimum weekly cost is $9,700.
43. $2,500 from Congressional Integrity Bank, $0 from Citizens' Trust, $7,500 from Checks R Us.
45. Fly 5 people from Chicago to LA, 15 from Chicago to New York, 5 from Denver to LA, none from Denver to New York at a total cost of $4,000.
47. Hire no more cardiologists, 12 rehabilitation specialists, and 5 infectious disease specialists.
49. The solution $x = 0$, $y = 0, \ldots$, represented by the initial tableau may not be feasible. In Phase I we use pivoting to arrive at a basic solution that is feasible.   **51.** The basic solution corresponding to the initial tableau has all the unknowns equal to zero, and this is not a feasible solution because it does not satisfy the given inequality.   **53. (C)**   **55.** Answers may vary. Examples are Exercises 1 and 2.
57. Answers may vary. A simple example is: Maximize $p = x + y$ subject to $x + y \le 10$, $x + y \ge 20$, $x \ge 0$, $y \ge 0$.

Section 5.5

1. Minimize $c = 6s + 2t$ subject to $s - t \ge 2$, $2s + t \ge 1$, $s \ge 0$, $t \ge 0$
3. Maximize $p = 100x + 50y$ subject to $x + 2y \le 2$, $x + y \le 1$, $x \le 3$, $x \ge 0$, $y \ge 0$.   **5.** Minimize $c = 3s + 4t + 5u + 6v$ subject to $s + u + v \ge 1$, $s + t + v \ge 1$, $s + t + u \ge 1$, $t + u + v \ge 1$, $s \ge 0$, $t \ge 0$, $u \ge 0$, $v \ge 0$.   **7.** Maximize $p = 1,000x + 2,000y + 500z$ subject to $5x + z \le 1$, $-x + z \le 3$, $y \le 1$, $x - y \le 0$, $x \ge 0$, $y \ge 0$, $z \ge 0$.
9. $c = 4$; $s = 2$, $t = 2$ **11.** $c = 80$; $s = 20/3$, $t = 20/3$ **13.** $c = 1.8$; $s = 6$, $t = 2$
15. $c = 25$; $s = 5$, $t = 15$ **17.** $c = 30$; $s = 30$, $t = 0$, $u = 0$ **19.** $c = 100$; $s = 0$, $t = 100$, $u = 0$
21. $c = 30$; $s = 10$, $t = 10$, $u = 10$
23. $R = [3/5 \quad 2/5]$, $C = [2/5 \quad 3/5 \quad 0]^T$, $e = 1/5$
25. $R = [1/4 \quad 0 \quad 3/4]$, $C = [1/2 \quad 0 \quad 1/2]^T$, $e = 1/2$ **27.** $R = [0 \quad 3/11 \quad 3/11 \quad 5/11]$, $C = [8/11 \quad 0 \quad 2/11 \quad 1/11]^T$, $e = 9/11$
29. 4 ounces each of fish and cornmeal, for a total cost of 40¢ per can; 5/12¢ per gram of protein, 5/12¢ per gram of fat.   **31.** 100 oz of grain and no chicken, for a total cost of $1; 1/2¢ per gram of protein, 0¢ per gram of fat.   **33.** One serving of cereal, one serving of juice, and no dessert! for a total cost of 37¢; 1/6¢ per calorie and 17/120¢ per % U.S. RDA of vitamin C.
35. 10 mailings to the East coast, none to the Midwest, 10 to the West Coast. Cost: $900; 20¢ per Democrat and 40¢ per Republican. OR 15 mailings to the Midwest and no mailing to the coasts. Cost: $900; 20¢ per Democrat and 40¢ per Republican.
37. Gillian should use 480 sleep spells and 160 shock spells, costing 360,000 pico-shirleys of energy OR 2,880/7 sleep spells and 1,440/7 shock spells.
39. T. N. Spend should spend about 73% of the days in Littleville, 27% in Metropolis, and skip Urbantown. T. L. Down should spend about 91% of the days in Littleville, 9% in Metropolis, and skip Urbantown. The expected outcome is that T. L. Down will lose about 227 votes per day of campaigning.   **41.** Each player should show one finger with probability 1/2, two fingers with probability 1/3, and three fingers with probability 1/6. The expected outcome is that player A will win 2/3 point per round, on average.   **43.** Write moves as (x, y), where x represents the number of regiments sent to the first location and y represents the number sent to the second location. Colonel Blotto should play $(0, 4)$ with probability 4/9, $(2, 2)$ with probability 1/9, and $(4, 0)$ with probability 4/9. Captain Kije has several optimal strategies, one of which is to play $(0, 3)$ with probability 1/30, $(1, 2)$ with probability 8/15, $(2, 1)$ with probability 16/45,

and (3, 0) with probability 7/90. The expected outcome is that Colonel Blotto will win 14/9 points on average.   **45.** The dual of a standard minimization problem satisfying the nonnegative objective condition is a standard maximization problem, which can be solved using the standard simplex algorithm, thus avoiding the need to do Phase I.   **47.** Answers will vary. An example is: Minimize $c = x - y$ subject to $x - y \geq 100$, $x + y \geq 200$, $x \geq 0$, $y \geq 0$. This problem can be solved using the techniques in Section 5.4.
49. If the given problem is a standard minimization problem satisfying the nonnegative objective condition, its dual is a standard maximization problem, and so can be solved using a single-phase simplex method. Otherwise, dualizing may not save any labor, since the dual will not be a standard maximization problem.

Chapter 5 Review

1.
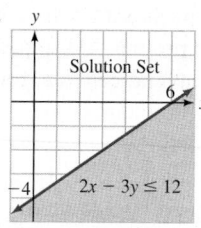
Unbounded

3.
$3x + 2y = 30$
$x + 2y = 20$
Solution Set
Bounded; corner points: (0, 0), (0, 10), (5, 15/2), (10, 0)

5. $p = 21; x = 9, y = 3$ **7.** $c = 22; x = 8, y = 6$
9. $p = 45; \; x = 0, y = 15, z = 15$
11. $p = 220; x = 20, y = 20, z = 60$
13. $c = 30; x = 30, y = 0, z = 0$
15. No solution; feasible region unbounded
17. $c = 50; x = 20, y = 10, z = 0, w = 20$, OR
$x = 30, y = 0, z = 0, w = 20$
19. $c = 60; x = 24, y = 12$ OR $x = 0, y = 60$
21. $c = 20; x = 0, y = 20$
23. $R = [1/2 \quad 1/2 \quad 0], C = [0 \quad 1/3 \quad 2/3]^T, e = 0$
25. $R = [1/27 \quad 7/9 \quad 5/27]$,
$C = [8/27 \quad 5/27 \quad 14/27]^T, \; e = 8/27$
27. (A)   29. 35   **31.** 400 packages from each for a minimum cost of $52,000   **33. a.** (B), (D) **b.** 450 packages from Duffin House, 375 from Higgins Press for a minimum cost of $52,500   **35.** 220 shares of EEE and 20 shares of RRR. The minimum total risk index is 500.   **37.** 240 Sprinkles, 120 Storms, and no Hurricanes   **39.** Order 600 packages from Higgins and none from the others, for a total cost of $90,000.
41. a. Let $x = $ # science credits, $y = $ # fine arts credits, $z = $ # liberal arts credits, and $w = $ # math credits.

Minimize $C = 300x + 300y + 200z + 200w$ subject to: $x + y + z + w \geq 120$; $x - y \geq 0$; $-2x + w \leq 0$; $-y + 3z - 3w \leq 0$; $x \geq 0, y \geq 0, z \geq 0, w \geq 0$.
b. Billy-Sean should take the following combination: Sciences—24 credits, Fine Arts—no credits, Liberal Arts—48 credits, Mathematics—48 credits, for a total cost of $26,400.   **43.** Smallest cost is $20,000; New York to OHaganBooks.com: 600 packages, New York to FantasyBooks.com: 0 packages, Illinois to OHaganbooks.com: 0 packages, Illinois to FantasyBooks.com: 200 packages.
45. FantasyBooks.com should choose between "2 for 1" and "3 for 2" with probabilities 20% and 80%, respectively. OHaganBooks.com should choose between "3 for 1" and "Finite Math" with probabilities 60% and 40%, respectively. OHaganBooks.com expects to gain 12,000 customers from FantasyBooks.com.

Chapter 6

Section 6.1

1. $F = \{\text{spring, summer, fall, winter}\}$
3. $I = \{1, 2, 3, 4, 5, 6\}$ **5.** $A = \{1, 2, 3\}$
7. $B = \{2, 4, 6, 8\}$ **9. a.** $S = \{(H, H), (H, T), (T, H), (T, T)\}$ **b.** $S = \{(H, H), (H, T), (T, T)\}$
11. $S = \{(1, 5), (2, 4), (3, 3), (4, 2), (5, 1)\}$
13. $S = \{(1, 5), (2, 4), (3, 3)\}$ **15.** $S = \emptyset$

17.
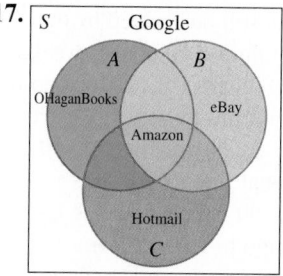

19.
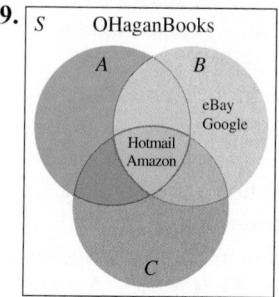

21. A **23.** A **25.** {June, Janet, Jill, Justin, Jeffrey, Jello, Sally, Solly, Molly, Jolly}   **27.** {Jello}   **29.** $\emptyset$
31. {Jello}   **33.** {Janet, Justin, Jello, Sally, Solly,

Molly, Jolly} **35.** {(small, triangle), (small, square), (medium, triangle), (medium, square), (large, triangle), (large, square)} **37.** {(small, blue), (small, green), (medium, blue), (medium, green), (large, blue), (large, green)}

39.

	A	B	C
1		Triangle	Square
2	Blue	Blue Triangle	Blue Square
3	Green	Green Triangle	Green Square

41.

	A	B	C
1		Blue	Green
2	Small	Small Blue	Small Green
3	Medium	Medium Blue	Medium Green
4	Large	Large Blue	Large Green

43. $B \times A = $ {1H, 1T, 2H, 2T, 3H, 3T, 4H, 4T, 5H, 5T, 6H, 6T}
45. $A \times A \times A = $ {HHH, HHT, HTH, HTT, THH, THT, TTH, TTT}
47. {(1,1), (1,3), (1,5), (3,1), (3,3), (3,5), (5,1), (5,3), (5,5)} **49.** ∅ **51.** {(1,1), (1,3), (1,5), (3,1), (3,3), (3,5), (5,1), (5,3), (5,5), (2,2), (2,4), (2,6), (4,2), (4,4), (4,6), (6,2), (6,4), (6,6)} **61.** $A \cap B = $ {Acme, Crafts}
63. $B \cup C = $ {Acme, Brothers, Crafts, Dion, Effigy, Global, Hilbert} **65.** $A' \cap C = $ {Dion, Hilbert}
67. $A \cap B' \cap C' = \emptyset$
69. {2003, 2004, 2005, 2006} × {Sail Boats, Motor Boats, Yachts}

	A	B	C	D
1		**Sail Boats**	**Motor Boats**	**Yachts**
2	**2003**	(2003 Sail Boats)	(2003 Motor Boats)	(2003 Yachts)
3	**2004**	(2004 Sail Boats)	(2004 Motor Boats)	(2004 Yachts)
4	**2005**	(2005 Sail Boats)	(2005 Motor Boats)	(2005 Yachts)
5	**2006**	(2006 Sail Boats)	(2006 Motor Boats)	(2006 Yachts)

71. $I \cup J$ **73.** (B) **75.** Answers may vary. Let $A = $ {1}, $B = $ {2}, and $C = $ {1, 2}. Then $(A \cap B) \cup C = $ {1, 2} but $A \cap (B \cup C) = $ {1}. In general, $A \cap (B \cup C)$ must be a subset of A, but $(A \cap B) \cup C$ need not be; also, $(A \cap B) \cup C$ must contain C as a subset, but $A \cap (B \cup C)$ need not.
77. A universal set is a set containing all "things" currently under consideration. When discussing sets of positive integers, the universe might be the set of all positive integers, or the set of all integers (positive, negative, and 0), or any other set containing the set of all positive integers. **79.** A is the set of suppliers who deliver components on time, B is the set of suppliers

whose components are known to be of high quality, and C is the set of suppliers who do not promptly replace defective components. **81.** Let $A = $ {movies that are violent}, $B = $ {movies that are shorter than two hours}, $C = $ {movies that have a tragic ending}, and $D = $ {movies that have an unexpected ending}. The given sentence can be rewritten as "She prefers movies in $A' \cap B \cap (C \cup D)'$." It can also be rewritten as "She prefers movies in $A' \cap B \cap C' \cap D'$."
83. Removing the comma would cause the statement to be ambiguous, as it could then correspond to either WWII ∪ (Comix ∩ Aliens') or to (WWII ∪ Comix) ∩ Aliens'. (See Exercise 76.)

Section 6.2

1. 9 **3.** 7 **5.** 4 **7.** $n(A \cup B) = 7$, $n(A) + n(B) - n(A \cap B) = 4 + 5 - 2 = 7$
9. 4 **11.** 18 **13.** 72 **15.** 60 **17.** 20 **19.** 6 **21.** 9
23. 4 **25.** $n((A \cap B)') = 9$
$n(A') + n(B') - n((A \cup B)') = 6 + 7 - 4 = 9$

27.
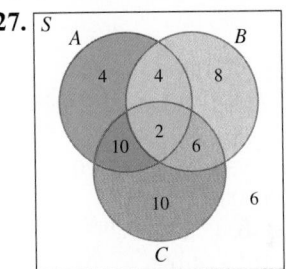

29.
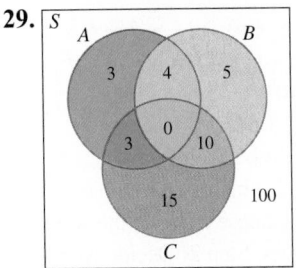

31. 115.4 million **33.** 2 **35.** $C \cap N$ is the set of authors who are both successful and new. $C \cup N$ is the set of authors who are either successful or new (or both). $n(C) = 30$; $n(N) = 20$; $n(C \cup N) = 5$; $n(C \cup N) = 45$; $45 = 30 + 20 - 5$ **37.** $C \cap N'$ is the set of authors who are successful but not new. $n(C \cap N') = 25$ **39.** 31.25%; 83.33%
41. $M \cap C$; $n(M \cap C) = 1.1$ million
43. $B \cap T'$; $n(B \cap T') = 3.1$ million
45. $B \cap (W \cup M)$; $n(B \cap (W \cup M)) = 2.3$ million
47. $V \cap I'$; $n(V \cap I') = 15$ **49.** 80; The number of stocks that were either not pharmaceutical stocks, or were unchanged in value after a year (or both).

51. 3/8; the fraction of Internet stocks that increased in value   **53. a.** 931 **b.** 382

55. a.

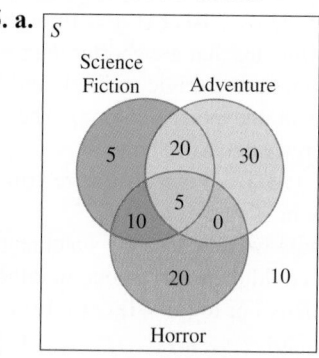

b. 37.5%   **57.** 17   **59.** $n(A) < n(B)$ **61.** The number of elements in the Cartesian product of two finite sets is the product of the number of elements in the two sets. **65.** When $A \cap B \neq \emptyset$ **67.** When $B \subseteq A$
69. $n(A \cup B \cup C) = n(A) + n(B) + n(C) - n(A \cap B) - n(B \cap C) - n(A \cap C) + n(A \cap B \cap C)$

Section 6.3

1. 10   **3.** 30   **5.** 6 outcomes   **7.** 15 outcomes
9. 13 outcomes   **11.** 25 outcomes   **13.** 4   **15.** 93
17. 16   **19.** 30   **21.** 13   **23.** 18   **25.** 25,600
27. 3,381   **29. a.** 288 **b.** 288   **31.** 256   **33.** 10
35. 286   **37.** 4   **39. a.** 8,000,000 **b.** 30,000
c. 4,251,528   **41. a.** $4^3 = 64$ **b.** 4^n **c.** $4^{2.1 \times 10^{10}}$
43. a. $16^6 = 16,777,216$ **b.** $16^3 = 4,096$
c. $16^2 = 256$ **d.** 766
45. $(10 \times 9 \times 8 \times 7 \times 6 \times 5 \times 4) \times (8 \times 7 \times 6 \times 5)$
$= 1,016,064,000$ possible casts   **47. a.** $26^3 \times 10^3 =$
17,576,000 **b.** $26^2 \times 23 \times 10^3 = 15,548,000$
c. $15,548,000 - 3 \times 10^3 = 15,545,000$
49. a. 4   **b.** 4   **c.** There would be an infinite number of routes.   **51. a.** 72   **b.** 36   **53.** 96
55. a. 36 **b.** 37   **57.** Step 1: Choose a day of the week on which Jan. 1 will fall: 7 choices. Step 2: Decide whether or not it is a leap year; 2 choices. Total: $7 \times 2 = 14$ possible calendars.   **59.** 1,900   **61.** Step 1: Choose a position in the left-right direction; m choices. Step 2: Choose a position in the front-back direction; n choices. Step 3: Choose a position in the up-down direction; r choices. Hence, there are $m \cdot n \cdot r$ possible outcomes.
63. 4   **65.** Cartesian product   **67.** The decision algorithm produces every pair of shirts twice, first in one order and then in the other.   **69.** Think of placing the five squares in a row of five empty slots. Step 1: Choose a slot for the blue square; five choices. Step 2: Choose a slot for the green square; four choices. Step 3: Choose the remaining three slots for the yellow squares; one choice. Hence, there are 20 possible five-square sequences.

Section 6.4

1. 720   **3.** 56   **5.** 360   **7.** 15   **9.** 3   **11.** 45   **13.** 20
15. 4,950   **17.** 360   **19.** 35   **21.** 120   **23.** 120
25. 20   **27.** 60   **29.** 210   **31.** 7   **33.** 35   **35.** 24
37. 126   **39.** 196   **41.** 105

43. $\dfrac{C(30, 5) \times 5^{25}}{6^{30}} \approx 0.192$

45. $\dfrac{C(30, 15) \times 3^{15} \times 3^{15}}{6^{30}} \approx 0.144$

47. $C(11, 1)C(10, 4)C(6, 4)C(2, 2)$
49. $C(11, 2)C(9, 1)C(8, 1)C(7, 3)C(4, 1)$
$C(3, 1)C(2, 1)C(1, 1)$
51. $C(10, 2)C(8, 4)C(4, 1)C(3, 1)C(2, 1)C(1, 1)$
53. 24   **55.** $C(13, 2)C(4, 2)C(4, 2) \times 44 = 123,552$
57. $13 \times C(4, 2)C(12, 3) \times 4 \times 4 \times 4 = 1,098,240$
59. 10,200   **61. a.** 252 **b.** 20 **c.** 26   **63. a.** 300 **b.** 3
c. 1 in 100 or .01   **65. a.** 210 **b.** 91 **c.** No   **67. a.** 23!
b. 18! **c.** $19 \times 18!$ **69.** (A)   **71.** (D)   **73. a.** 9,880
b. 1,560 **c.** 11,480
75. a. $C(20, 2) = 190$ **b.** $C(n, 2)$
77. The multiplication principle; it can be used to solve all problems that use the formulas for permutations.
79. (A), (D)   **81.** A permutation. Changing the order in a list of driving instructions can result in a different outcome; for instance, "1. Turn left. 2. Drive one mile." and "1. Drive one mile. 2. Turn left." will take you to different locations.   **83.** Urge your friend not to focus on formulas, but instead learn to formulate decision algorithms and use the principles of counting.
85. It is ambiguous on the following point: Are the three students to play different characters, or are they to play a group of three, such as "three guards"? This should be made clear in the exercise.

Chapter 6 Review

1. $N = \{-3, -2, -1\}$ **3.** $S = \{(1, 2), (1, 3), (1, 4),$
$(1, 5), (1, 6), (2, 1), (2, 3), (2, 4), (2, 5), (2, 6), (3, 1),$
$(3, 2), (3, 4), (3, 5), (3, 6), (4, 1), (4, 2), (4, 3), (4, 5),$
$(4, 6), (5, 1), (5, 2), (5, 3), (5, 4), (5, 6), (6, 1), (6, 2),$
$(6, 3), (6, 4), (6, 5)\}$ **5.** $A \cup B' = \{a, b, d\},$
$A \times B' = \{(a, a), (a, d), (b, a), (b, d)\}$ **7.** $A \times B$
9. $(P \cap E' \cap Q)'$ or $P' \cup E \cup Q'$ **11.** $n(A \cup B) = n(A) + n(B) - n(A \cap B), n(C') = n(S) - n(C)$; 100
13. $n(A \times B) = n(A)n(B), n(A \cup B) = n(A) + n(B) - n(A \cap B), n(A') = n(S) - n(A)$; 21
15. $C(12, 1)C(4, 2)C(11, 3)C(4, 1)C(4, 1)C(4, 1)$
17. $C(4, 1)C(10, 1)$ **19.** 6   **21.** $C(4, 4)C(8, 1) = 8$
23. $C(3, 2)C(9, 3) + C(3, 3)C(9, 2) = 288$
25. The set of books that are either sci-fi or stored

in Texas (or both); $n(S \cup T) = 112{,}000$ **27.** The set of books that are either stored in California or not sci-fi; $n(C \cup S') = 175{,}000$ **29.** The romance books that are also horror books or stored in Texas; $n(R \cap (T \cup H)) = 20{,}000$ **31.** 1,000
33. FarmerBooks.com; 1,800   **35.** FarmerBooks.com; 5,400   **37.** $26 \times 26 \times 26 = 17{,}576$
39. $26 \times 25 \times 9 \times 10 = 58{,}500$ **41.** 60,000
43. 19,600

Chapter 7

Section 7.1

1. $S = \{HH, HT, TH, TT\}$; $E = \{HH, HT, TH\}$
3. $S = \{HHH, HHT, HTH, HTT, THH, THT, TTH, TTT\}$; $E = \{HTT, THT, TTH, TTT\}$

$$5.\ S = \left\{ \begin{array}{l} (1,1), (1,2), (1,3), (1,4), (1,5), (1,6), \\ (2,1), (2,2), (2,3), (2,4), (2,5), (2,6), \\ (3,1), (3,2), (3,3), (3,4), (3,5), (3,6), \\ (4,1), (4,2), (4,3), (4,4), (4,5), (4,6), \\ (5,1), (5,2), (5,3), (5,4), (5,5), (5,6), \\ (6,1), (6,2), (6,3), (6,4), (6,5), (6,6) \end{array} \right\};$$

$E = \{(1,4), (2,3), (3,2), (4,1)\}$

$$7.\ S = \left\{ \begin{array}{l} (1,1), (1,2), (1,3), (1,4), (1,5), (1,6), \\ (2,2), (2,3), (2,4), (2,5), (2,6), \\ (3,3), (3,4), (3,5), (3,6), \\ (4,4), (4,5), (4,6), \\ (5,5), (5,6), \\ (6,6) \end{array} \right\};$$

$E = \{(1,3), (2,2)\}$
9. S as in Exercise 7; $E = \{(2,2), (2,3), (2,5), (3,3), (3,5), (5,5)\}$ **11.** $S = \{m, o, z, a, r, t\}$: $E = \{o, a\}$
13. $S = \{(s, o), (s, r), (s, e), (o, s), (o, r), (o, e), (r, s), (r, o), (r, e), (e, s), (e, o), (e, r)\}$; $E = \{(o, s), (o, r), (o, e), (e, s), (e, o), (e, r)\}$ **15.** $S = \{01, 02, 03, 04, 10, 12, 13, 14, 20, 21, 23, 24, 30, 31, 32, 34, 40, 41, 42, 43\}$; $E = \{10, 20, 21, 30, 31, 32, 40, 41, 42, 43\}$
17. $S = \{$domestic car, imported car, van, antique car, antique truck$\}$; $E = \{$van, antique truck$\}$
19. a. All sets of 4 gummy candies chosen from the packet of 12 **b.** All sets of four gummy candies in which two are strawberry and two are black currant
21. a. All lists of 15 people chosen from 20
b. All lists of 15 people chosen from 20, in which Hillary Clinton occupies the eleventh position
23. $A \cap B$; $n(A \cap B) = 1$ **25.** B'; $n(B') = 33$
27. $B' \cap D'$; $n(B' \cap D') = 2$
29. $C \cup B$; $n(C \cup B) = 12$
31. $W \cap I$ **33.** $E \cup I'$ **35.** $I \cup (W \cap E')$
37. $(I \cup W) \cap E'$ **39.** 56; 4
41. $C(4,1) C(2,1) C(2,1) = 16$

43. $E = \{$Pacific, Mountain, South Atlantic$\}$
45. $E \cup F$ is the event that you choose a region that saw a decrease in housing prices of 7% or more or is on the east coast. $E \cup F = \{$Pacific, Mountain, New England, Middle Atlantic, South Atlantic$\}$. $E \cap F$ is the event that you choose a region that saw a decrease in housing prices of 7% or more and is on the east coast. $E \cap F = \{$South Atlantic$\}$.   **47. a.** Not mutually exclusive **b.** Mutually exclusive   **49.** $S \cap N$ is the event that an author is successful and new. $S \cup N$ is the event that an author is either successful or new; $n(S \cap N) = 5$; $n(S \cup N) = 45$
51. N and E **53.** $S \cap N'$ is the event that an author is successful but not a new author. $n(S \cap N') = 25$
55. 31.25%; 83.33%   **57.** $V \cap I'$; $n(V \cap I') = 15$
59. 80; the number of stocks that were either not pharmaceutical stocks, or were unchanged in value after a year (or both)   **61.** P and E, P and I, E and I, N and E, V and N, V and D, N and D **63.** 3/8; the fraction of Internet stocks that increased in value
65. a. $E' \cap H$ **b.** $E \cup H$ **c.** $(E \cup G)' = E' \cap G'$
67. a. $\{9\}$ **b.** $\{6\}$ **69. a.** The dog's fight drive is weakest. **b.** The dog's fight and flight drives are either both strongest or both weakest. **c.** Either the dog's fight drive is strongest, or its flight drive is strongest.   **71.** $C(6,4) = 15$; $C(1,1)C(5,3) = 10$
73. a. $n(S) = P(7,3) = 210$ **b.** $E \cap F$ is the event that Celera wins and Electoral College is in second or third place. In other words, it is the set of all lists of three horses in which Celera is first and Electoral College is second or third. $n(E \cap F) = 10$ **75.** Subset of the sample space   **77.** E and F do not both occur
79. (B)   **81.** True; consider the following experiment: Select an element of the set S at random.
83. Answers may vary. Cast a die and record the remainder when the number facing up is divided by 2.
85. Yes. For instance, $E = \{(2,5), (5,1)\}$ and $F = \{(4,3)\}$ are two such events.

Section 7.2

1. .4   **3.** .8   **5.** .6
7.

Outcome	HH	HT	TH	TT
Rel. Frequency	.275	.2375	.3	.1875

9. .575   **11.** The second coin *seems* slightly biased in favor of heads, as heads comes up approximately 58% of the time. On the other hand, it is conceivable that the coin is fair and that heads came up 58% of the time purely by chance. Deciding which conclusion is more reasonable requires some knowledge of inferential statistics.   **13.** Yes   **15.** No; relative frequencies cannot be negative.   **17.** Yes   **19.** Missing value: .3 **a.** .6
b. .4   **21.** Answers will vary.   **23.** Answers will vary.
25. a. .5 **b.** .17 **c.** .94

27. a.

Mortgage Status	Current	Past Due	In Foreclosure	Repossessed
Rel. Frequency	.67	.26	.045	.025

b. .33

29. a.

Age	0–14	15–29	30–64	>64
Rel. Frequency	.30	.27	.37	.06

b. .36

31. a.

Test Rating	3	2	1	0
Rel. Frequency	.1	.4	.4	.1

b. .5   **33.** Dial-up: 1256, Cable Modem: 412, DSL: 302, Other: 30

35.

Outcome	Surge	Plunge	Steady
Rel. Frequency	.2	.3	.5

37. .25   **39.** .2   **41.** .7   **43.** 5/6   **45.** 5/16

47.

Outcome	U	C	R
Rel. Frequency	.2	.64	.16

49.

Conventional	No pesticide	Single pesticide	Multiple pesticide
Probability	.27	.13	.60
Organic	No pesticide	Single pesticide	Multiple pesticide
Probability	.77	.13	.10

51. $P(\text{false negative}) = 10/400 = .025$, $P(\text{false positive}) = 10/200 = .05$ **53.** Answers will vary.   **55.** The fraction of times E occurs   **57.** 101; $fr(E)$ can be any number between 0 and 100 inclusive, so the possible answers are $0/100 = 0$, $1/100 = .01$, $2/100 = .02, \ldots, 99/100 = .99$, $100/100 = 1$. **59.** Wrong. For a pair of fair dice, the probability of a pair of matching numbers is 1/6, as Ruth says. However, it is quite possible, although not very likely, that if you cast a pair of fair dice 20 times, you will never obtain a matching pair (in fact, there is approximately a 2.6% chance that this will happen). In general, a nontrivial claim about probability can never be absolutely validated or refuted experimentally. All we can say is that the evidence suggests that the dice are not fair.   **61.** For a (large) number of days, record the temperature prediction for the next day and then check the actual high temperature the next day. Record whether the prediction was accurate (within, say, 2°F of the actual temperature). The fraction of times the prediction was accurate is the relative frequency.

Section 7.3

1. $P(e) = .2$ **a.** .9 **b.** .95 **c.** .1 **d.** .8
3. $P(E) = 1/4$ **5.** $P(E) = 1$ **7.** $P(E) = 3/4$
9. $P(E) = 3/4$ **11.** $P(E) = 1/2$ **13.** $P(E) = 1/9$
15. $P(E) = 0$ **17.** $P(E) = 1/4$
19. $1/12$; $\{(4, 4), (2, 3), (3, 2)\}$
21.

Outcome	1	2	3	4	5	6
Probability	1/9	2/9	1/9	2/9	1/9	2/9

$P(\{1, 2, 3\}) = 4/9$

23.

Outcome	1	2	3	4
Probability	8/15	4/15	2/15	1/15

25. .65   **27.** .1   **29.** .7   **31.** .4   **33.** .25   **35.** 1.0
37. .3   **39.** 1.0   **41.** No; $P(A \cup B)$ should be $\leq$ $P(A) + P(B)$.   **43.** Yes   **45.** No; $P(A \cup B)$ should be $\geq P(A)$.   **47. a.** .93 **b.** .33
49.

Outcome	Hispanic or Latino	White (not Hispanic)	African American	Asian	Other
Probability	0.48	0.29	0.08	0.08	0.07

$P(\text{Neither White nor Asian}) = .63$
51. a. $S = \{$Stock market success, Sold to other concern, Fail$\}$
b.

Outcome	Stock Market Success	Sold to Other Concern	Fail
Probability	.2	.3	.5

c. .5
53.

Outcome	SUV	Pickup	Passenger Car	Minivan
Probability	.25	.15	.50	.10

55. $P(1) = 0$, $P(6) = 0$; $P(2) = P(3) = P(4) = P(5) = 1/4 = .25$; $P(\text{odd}) = .5$ **57.** $P(1) = P(6) = 1/10$; $P(2) = P(3) = P(4) = P(5) = 1/5$, $P(\text{odd}) = 1/2$ **59.** $P(1, 1) = P(2, 2) = \ldots = P(6, 6) = 1/66$; $P(1, 2) = \ldots = P(6, 5) = 1/33$, $P(\text{odd sum}) = 6/11$
61. $P(2) = 15/38$; $P(4) = 3/38$, $P(1) = P(3) = P(5) = P(6) = 5/38$, $P(\text{odd}) = 15/38$ **63.** 5/6
65. .39   **67.** .25   **69.** .25   **71.** .00   **73.** .75   **75.** .45
77. .55   **79.** .01   **81.** .53   **83.** .44   **85.** 22%; 43%
87. All of them   **89.** 88.4%   **91.** Here is one possible experiment: Roll a die and observe which of the following outcomes occurs. Outcome A: 1 or 2 facing up; $P(A) = 1/3$, Outcome B: 3 or 4 facing up; $P(B) = 1/3$, Outcome C: 5 or 6 facing up; $P(C) = 1/3$.   **93.** He is wrong. It is possible to have a run of losses of any length. Tony may have grounds to *suspect* that the game is rigged, but no proof.
95. They are mutually exclusive.   **97.** Wrong. For example, the modeled probability of winning a state

lotto is small but nonzero. However, the vast majority of people who play lotto every day of their lives never win no matter how frequently they play, so the relative frequency is zero for these people. **99.** When $A \cap B = \emptyset$ we have $P(A \cap B) = P(\emptyset) = 0$, so $P(A \cup B) = P(A) + P(B) - P(A \cap B) = P(A) + P(B) - 0 = P(A) + P(B)$. **101.** Zero. According to the assumption, no matter how many thunderstorms occur, lightning cannot strike a given spot more than once, and so, after n trials the relative frequency will never exceed $1/n$, and so will approach zero as the number of trials gets large. Since the modeled probability models the limit of relative frequencies as the number of trials gets large, it must therefore be zero. **103.** $P(A \cup B \cup C) = P(A) + P(B) + P(C) - P(A \cap B) - P(A \cap C) - P(B \cap C) + P(A \cap B \cap C)$

Section 7.4

1. $1/42$ **3.** $7/9$ **5.** $1/7$ **7.** $1/2$ **9.** $41/42$
11. $1/15$ **13.** $4/15$ **15.** $1/5$ **17.** $1/(2^8 \times 5^5 \times 5!)$
19. $.4226$ **21.** $.0475$ **23.** $.0020$ **25.** $1/27^{39}$
27. $1/7$ **29.** Probability of being a Big Winner $= 1/2,118,760 \approx .000000472$. Probability of being a Small-Fry Winner $= 225/2,118,760 \approx .000106194$. Probability of being either a Big Winner or a Small-Fry Winner $= 226/2,118,760 \approx .000106666$.
31. a. $C(600,300)/C(700,400)$ **b.** $C(699,399)/C(700,400)$ or $400/700$ **33.** $P(10, 3)/10^3 = 18/25 = .72$ **35.** $8!/8^8$ **37.** $1/8$ **39.** $1/8$
41. $37/10,000$ **43. a.** $90,720$ **b.** $25,200$
c. $25,200/90,720 = 25/90 \approx .28$ **45.** The four outcomes listed are not equally likely; for example, (red, blue) can occur in four ways. The methods of this section yield a probability for (red, blue) of $C(2, 2)/C(4, 2) = 1/6$ **47.** No. If we do not pay attention to order, the probability is $C(5, 2)/C(9, 2) = 10/36 = 5/18$. If we do pay attention to order, the probability is $P(5, 2)/P(9, 2) = 20/72 = 5/18$ again. The difference between permutations and combinations cancels when we compute the probability.
49. Answers will vary.

Section 7.5

1. $.4$ **3.** $.08$ **5.** $.75$ **7.** $.2$ **9.** $.5$
11.

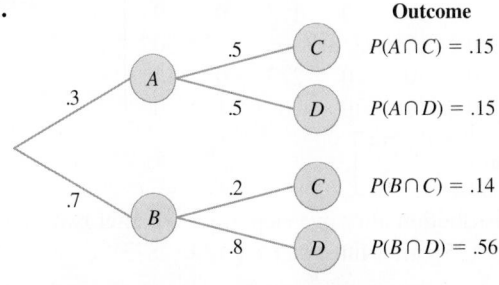

13.

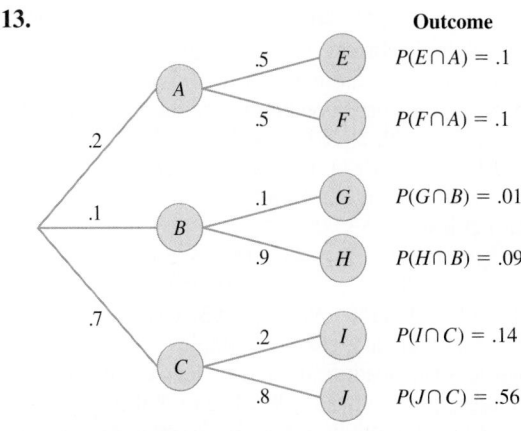

15. $1/10$ **17.** $1/5$ **19.** $2/9$ **21.** $1/84$
23. $5/21$ **25.** $24/175$ **27.** Mutually exclusive
29. Neither **31.** $\dfrac{1}{2} \cdot \dfrac{1}{2} = \dfrac{1}{4}$ Independent
33. $\dfrac{5}{18} \cdot \dfrac{1}{2} \neq \dfrac{1}{9}$ Dependent **35.** $\dfrac{25}{36} \cdot \dfrac{5}{18} \neq \dfrac{2}{9}$
Dependent **37.** $(1/2)^{11} = 1/2,048$
39. $P(D) = .10$; $P(D|M) = .30$ **41.** $P(A|L) = .30$; $P(L|A) = .10$ **43.** $P(E|M) = .55$; $P(E|M') = .05$
45. $.8$ **47. a.** $.24$ **b.** $.059$ **49.** $.34$ **51.** Not independent; $P(\text{giving up} \mid \text{used Brand X}) = .1$ is larger than $P(\text{giving up}) = .05$. **53.** $.00015$
55. $5/6$ **57.** $3/4$ **59.** $11/16$ **61.** $11/14$
63.

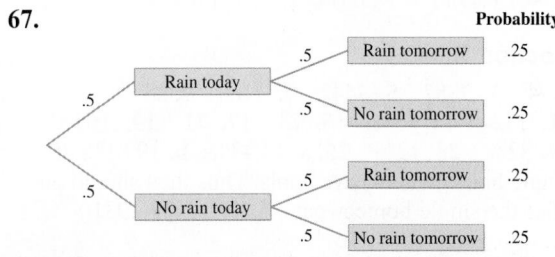

65.

67.

69. .77   **71.** .37   **73.** .40   **75.** .96
77. The claim is correct. The probability that an unemployed person has a high school diploma only is .35, while the corresponding figure for an employed person is .27.   **79.** $P(K \mid D) = 1.31 P(K \mid D')$
81. a. $P(I \mid T) > P(I)$ **b.** It was ineffective.
83. a. .59   **b.** $35,000 or more:
$P(\text{Internet user} \mid < \$35,000) \approx$
$.27 < P(\text{Internet user} \mid \geq \$35,000) \approx .59$.
85. $P(R \mid J)$ **87.** (D)   **89. a.** .000057
b. .015043   **91.** 11%   **93.** 106   **95.** .631
97. Answers will vary. Here is a simple one: E: the first toss is a head; F: the second toss is a head; G: the third toss is a head.   **99.** The probability you seek is $P(E \mid F)$, or should be. If, for example, you were going to place a wager on whether E occurs or not, it is crucial to know that the sample space has been reduced to F (you know that F did occur). If you base your wager on $P(E)$ rather than $P(E \mid F)$ you will misjudge your likelihood of winning.
101. You might explain that the conditional probability of E is not the *a priori* probability of E, but it is the probability of E in a hypothetical world in which the outcomes are restricted to be what is given. In the example she is citing, yes, the probability of throwing a double-six is $1/36$, in the absence of any other knowledge. However by the "conditional probability" of throwing a double-six given that the sum is larger than 7, we might mean the probability of a double-six in a situation in which the dice have already been thrown, but all we know is that the sum is greater than 7. Because there are only 15 ways in which that can happen, the conditional probability is $1/15$. For a more extreme case, consider the conditional probability of throwing a double-six given that the sum is 12.
103. If $A \subseteq B$ then $A \cap B = A$, so $P(A \cap B) = P(A)$ and $P(A \mid B) = P(A \cap B)/P(B) = P(A)/P(B)$.
105. Your friend is correct. If A and B are mutually exclusive, then $P(A \cap B) = 0$. On the other hand, if A and B are independent, then $P(A \cap B) = P(A)P(B)$. Thus, $P(A)P(B) = 0$. If a product is 0, then one of the factors must be 0, so either $P(A) = 0$ or $P(B) = 0$. Thus, it cannot be true that A and B are mutually exclusive, have nonzero probabilities, and are independent all at the same time.
107. $P(A' \cap B') = 1 - P(A \cup B) =$
$1 - [P(A) + P(B) - P(A \cap B)] =$
$1 - [P(A) + P(B) - P(A)P(B)] =$
$(1 - P(A))(1 - P(B)) = P(A')P(B')$

Section 7.6

1. .4   **3.** .7887   **5.** .7442   **7.** .1163   **9.** 26.8%
11. .1724   **13.** 85%   **15.** .73   **17.** .71   **19.** .165
21. 82%   **23.** 12%   **25. a.** 14.43%;   **b.** 19.81% of single homeowners have pools. Thus they should go after the single homeowners.   **27.** 9   **29.** .9310

31. 1.76%   **33.** .20   **35.** K: child killed;
D: Airbag deployed; $P(K \mid D) = 1.31 P(K \mid D')$;
$P(D \mid K) = 1.31(.25)/[1.31(.25) + .75] = .30$
37. Show him an example like Example 1 of this section, where $P(T \mid A) = .95$ but $P(A \mid T) \approx .64$.
39. Suppose that the steroid test gives 10% false negatives and that only 0.1% of the tested population uses steroids. Then the probability that an athlete uses steroids, given that he or she has tested positive, is
$$\frac{(.9)(.001)}{(.9)(.001) + (.01)(.999)} \approx .083.$$ **41.** Draw a tree in which the first branching shows which of R_1, R_2, or R_3 occurred, and the second branching shows which of T or T' then occurred. There are three final outcomes in which T occurs: $P(R_1 \cap T) = P(T \mid R_1)P(R_1)$, $P(R_2 \cap T) = P(T \mid R_2)P(R_2)$, and $P(R_3 \cap T) = P(T \mid R_3)P(R_3)$. In only one of these, the first, does R_1 occur. Thus,

$$P(R_1 \mid T) = \frac{P(R_1 \cap T)}{P(T)}$$
$$= \frac{P(T \mid R_1)P(R_1)}{P(T \mid R_1)P(R_1) + P(T \mid R_2)P(R_2) + P(T \mid R_3)P(R_3)}$$

43. The reasoning is flawed. Let A be the event that a Democrat agrees with Safire's column, and let F and M be the events that a Democrat reader is female and male, respectively. Then A. D. makes the following argument:
$P(M \mid A) = 0.9$, $P(F \mid A') = 0.9$. Therefore, $P(A \mid M) = 0.9$.
According to Bayes' theorem, we cannot conclude anything about $P(A \mid M)$ unless we know $P(A)$, the percentage of all Democrats who agreed with Safire's column. This was not given.

Section 7.7

1. $\begin{bmatrix} 1/4 & 3/4 \\ 1/2 & 1/2 \end{bmatrix}$ **3.** $\begin{bmatrix} 0 & 1 \\ 1/6 & 5/6 \end{bmatrix}$

5. $\begin{bmatrix} 0 & .8 & .2 \\ .9 & 0 & .1 \\ 0 & 0 & 1 \end{bmatrix}$ **7.** $\begin{bmatrix} 1 & 0 & 0 \\ 0 & 1 & 0 \\ 0 & 0 & 1 \end{bmatrix}$

9. $\begin{bmatrix} 1 & 0 & 0 & 0 & 0 & 0 \\ 2/3 & 0 & 1/3 & 0 & 0 & 0 \\ 0 & 2/3 & 0 & 1/3 & 0 & 0 \\ 0 & 0 & 2/3 & 0 & 1/3 & 0 \\ 0 & 0 & 0 & 2/3 & 0 & 1/3 \\ 0 & 0 & 0 & 0 & 1 & 0 \end{bmatrix}$

11. a. $\begin{bmatrix} .25 & .75 \\ 0 & 1 \end{bmatrix}$

b. distribution after one step: [.5   .5]; after two steps: [.25   .75]; after three steps: [.125   .875]

Positive direction; extrapolating in the negative direction eventually leads to negative values, which do not model reality.
65. a. About 4.467×10^{23} ergs **b.** About 2.24%
c. $E = 10^{1.5R+11.8}$ **d.** Proof **e.** 1,000 **67. a.** 75 dB, 69 dB, 61 dB **b.** $D = 95 - 20\log r$ **c.** 57,000 feet
69. Graph:

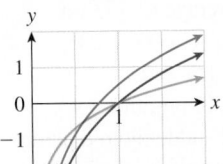

The green curve is $y = \ln x$. The blue curve is $y = 2\ln x$, and the red curve is $y = 2\ln x + 0.5$. Multiplying by A stretches the graph in the y-direction by a factor of A. Adding C moves the graph C units vertically up. **71.** The logarithm of a negative number, were it defined, would be the power to which a base must be raised to give that negative number. But raising a base to a power never results in a negative number, so there can be no such number as the logarithm of a negative number. **73.** Any logarithmic curve $y = \log_b t + C$ will eventually surpass 100%, and hence not be suitable as a long-term predictor of market share. **75.** $\log_4 y$ **77.** 8 **79.** x **81.** Time is increasing logarithmically with population; solving $P = Ab^t$ for t gives $t = \log_b(P/A) = \log_b P - \log_b A$, which is of the form $t = \log_b P + C$. **83.** (Proof)

Section 9.4

1. $N = 7, A = 6, b = 2$;   **3.** $N = 10, A = 4, b = 0.3$;
7/(1+6*2^-x)   10/(1+4*0.3^-x)

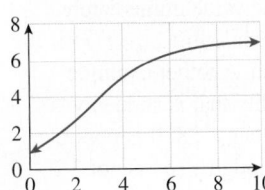   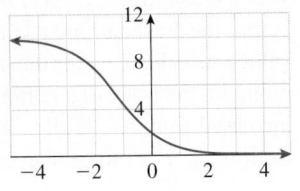

5. $N = 4, A = 7, b = 1.5$;
4/(1+7*1.5^-x)

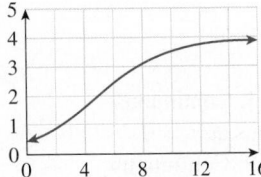

7. $f(x) = \dfrac{200}{1 + 19(2^{-x})}$ **9.** $f(x) = \dfrac{6}{1 + 2^{-x}}$

11. (B)   **13.** (B)   **15.** (C)

17. $y = \dfrac{7.2}{1 + 2.4(1.04)^{-x}}$ **19.** $y = \dfrac{97}{1 + 2.2(0.942)^{-x}}$

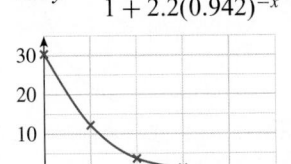

21. a. (A) **b.** 2003   **23. a.** (A) **b.** 20% per year
25. a. 95.0% **b.** $P(x) \approx 25.13(1.064)^x$ **c.** \$18,000

27. $N(t) = \dfrac{10,000}{1 + 9(1.25)^{-t}}$; $N(7) \approx 3,463$ cases

29. $N(t) = \dfrac{3,000}{1 + 29(2^{1/5})^{-t}}$; $t = 16$ days

31. a. $A(t) = \dfrac{6.3}{1 + 4.8(1.2)^{-t}}$; 6,300 articles
b. 5,200 articles   **33. a.** $B(t) = \dfrac{1,090}{1 + 0.410(1.09)^{-t}}$;
1,090 teams **b.** $t \approx -10.3$. According to the model, the number of teams was rising fastest about 10.3 years *prior* to 1990; that is, sometime during 1979. **c.** The number of men's basketball teams was growing by about 9% per year in the past, well before 1979.
35. $y = \dfrac{4,500}{1 + 1.1466(1.0357)^{-t}}$; 2013   **37.** Just as diseases are communicated via the spread of a pathogen (such as a virus), new technology is communicated via the spread of information (such as advertising and publicity). Further, just as the spread of a disease is ultimately limited by the number of susceptible individuals, so the spread of a new technology is ultimately limited by the size of the potential market.
39. It can be used to predict where the sales of a new commodity might level off. **41.** The curve is still a logistic curve, but decreases when $b > 1$ and increases when $b < 1$. **43.** (Proof)

Chapter 9 Review

1.

3. $f: f(x) = 5(1/2)^x$, or $5(2^{-x})$

5.   **7.**

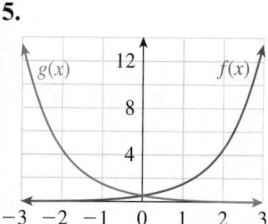 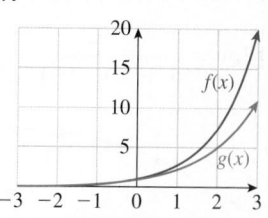

9. $3,484.85   **11.** $3,705.48   **13.** $3,485.50

15. $f(x) = 4.5(9^x)$ **17.** $f(x) = \frac{2}{3}3^x$ **19.** $-\frac{1}{2}\log_3 4$

21. $\frac{1}{3}\log 1.05$

23.

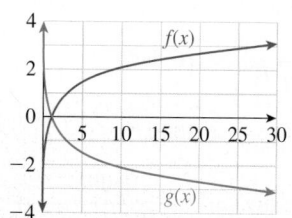

25. $Q = 5e^{-0.00693t}$ **27.** $Q = 2.5e^{0.347t}$
29. 10.2 years   **31.** 10.8 years
33. $f(x) = \dfrac{900}{1 + 8(1.5)^{-x}}$

35. $f(x) = \dfrac{20}{1 + 3(0.8)^{-x}}$ **37. a.** $8,500 per month;
an average of approximately 2,100 hits per day
b. $29,049 per month   **c.** The fact that -0.000005,
the coefficient of c^2, is negative.
39. a. $R = -60p^2 + 950p$; $p = $7.92 per novel,
Monthly revenue $= $3,760.42
b. $P = -60p^2 + 1,190p - 4,700$; $p = $9.92 per
novel, Monthly profit $= $1,200.42   **41. a.** 9.1, 19
b. About 310,000 pounds   **43.** 2016
45. 1.12 million pounds   **47.** $n(t) = 9.6(0.80^t)$ million
pounds of lobster   **49.** (C)

Chapter 10

Section 10.1

1. 0   **3.** 4   **5.** Does not exist   **7.** 1.5   **9.** 0.5
11. Diverges to $+\infty$ **13.** 0   **15.** 1   **17.** 0   **19.** 0
21. a. -2 **b.** -1 **23. a.** 2 **b.** 1 **c.** 0 **d.** $+\infty$
25. a. 0 **b.** 2 **c.** -1 **d.** Does not exist **e.** 2 **f.** $+\infty$
27. a. 1 **b.** 1 **c.** 2 **d.** Does not exist **e.** 1 **f.** 2
29. a. 1 **b.** Does not exist **c.** Does not exist **d.** 1
31. a. 1 **b.** $+\infty$ **c.** $+\infty$ **d.** $+\infty$ **e.** not defined **f.** -1
33. a. -1 **b.** $+\infty$ **c.** $-\infty$ **d.** Does not exist **e.** 2 **f.** 1
35. 210 trillion pesos per month. In the long term, the

model predicts that the value of sold goods in Mexico
will approach 210 trillion pesos per month.   **37.** $+\infty$;
In the long term, Amazon's revenue will grow
without bound.   **39.** 7.0; In the long term, the number
of research articles in *Physical Review* written by
researchers in Europe approaches 7,000 per year.
41. 573. This suggests that students with an exceptionally
large household income earn an average of 573 on
the math SAT test.   **43. a.** $\lim_{t\to14.75^-} r(t) = 21$,
$\lim_{t\to14.75^+} r(t) = 21$, $\lim_{t\to14.75} r(t) = 21$,
$r(14.75) = 0.01$ **b.** Just prior to 2:45 pm, the stock
was approaching $21, but then fell suddenly to a penny
($0.01) at 2:45 exactly, after which time it jumped back
to values close to $21.   **45.** 100; In the long term, the
home price index will level off at 100 points.
47. $\lim_{t\to1^-} C(t) = 0.06$, $\lim_{t\to1^+} C(t) = 0.08$, so
$\lim_{t\to1} C(t)$ does not exist.   **49.** $\lim_{t\to+\infty} I(t) = +\infty$,
$\lim_{t\to+\infty}(I(t)/E(t)) \approx 2.5$. In the long term, U.S.
imports from China will rise without bound and be 2.5
times U.S. exports to China. In the real world, imports
and exports cannot rise without bound. Thus, the given
models should not be extrapolated far into the future.
51. To approximate $\lim_{x\to a} f(x)$ numerically, choose
values of x closer and closer to, and on either side of
$x = a$, and evaluate $f(x)$ for each of them. The limit
(if it exists) is then the number that these values of $f(x)$
approach. A disadvantage of this method is that it may
never give the exact value of the limit, but only an
approximation. (However, we can make this as accurate
as we like.)   **53.** Any situation in which there is a
sudden change can be modeled by a function in which
$\lim_{t\to a^+} f(t)$ is not the same as $\lim_{t\to a^-} f(t)$. One
example is the value of a stock market index before and
after a crash: $\lim_{t\to a^-} f(t)$ is the value immediately
before the crash at time $t = a$, while $\lim_{t\to a^+} f(t)$ is the
value immediately after the crash. Another example
might be the price of a commodity that is suddenly
increased from one level to another.   **55.** It is possible
for $\lim_{x\to a} f(x)$ to exist even though $f(a)$ is not
defined. An example is $\lim_{x\to1} \dfrac{x^2 - 3x + 2}{x - 1}$.
57. An example is $f(x) = (x - 1)(x - 2)$.
59. These limits are all 0.

Section 10.2

1. Continuous on its domain   **3.** Continuous
on its domain   **5.** Discontinuous at $x = 0$
7. Discontinuous at $x = -1$ **9.** Continuous
on its domain   **11.** Continuous on its domain
13. Discontinuous at $x = -1$ and 0   **15.** (A), (B),
(D), (E)   **17.** 0   **19.** -1 **21.** No value possible
23. -1 **25.** Continuous on its domain

27. Continuous on its domain   **29.** Discontinuity at $x = 0$ **31.** Discontinuity at $x = 0$ **33.** Continuous on its domain   **35.** Not unless the domain of the function consists of all real numbers. (It is impossible for a function to be continuous at points not in its domain.) For example, $f(x) = 1/x$ is continuous on its domain—the set of nonzero real numbers—but not at $x = 0$.   **37.** True. If the graph of a function has a break in its graph at any point a, then it cannot be continuous at the point a.   **39.** Answers may vary. $f(x) = 1/[(x-1)(x-2)(x-3)]$ is such a function; it is undefined at $x = 1, 2, 3$ and so its graph consists of three distinct curves.   **41.** Answers may vary.

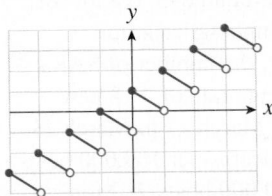

43. Answers may vary. The price of OHaganBooks.com stocks suddenly drops by \$10 as news spreads of a government investigation. Let $f(x)$ = Price of OHaganBooks.com stocks.

Section 10.3

1. $x = 1$ **3.** 2   **5.** Determinate; diverges to $+\infty$
7. Determinate; does not exist   **9.** Determinate; diverges to $-\infty$ **11.** Determinate; 0
13. Indeterminate; $-1/3$ **15.** Indeterminate; 0
17. Determinate; 0   **19.** Determinate; -60 **21.** 1
23. 2   **25.** 0   **27.** 6   **29.** 4   **31.** 2   **33.** 0   **35.** 0
37. 12   **39.** $+\infty$ **41.** Does not exist; left and right (infinite) limits differ.   **43.** $-\infty$ **45.** Does not exist
47. $+\infty$ **49.** 0   **51.** 1/6   **53.** 3/2   **55.** 1/2
57. $+\infty$ **59.** 3/2   **61.** 1/2   **63.** $-\infty$ **65.** 0   **67.** 12
69. 0   **71.** $+\infty$ **73.** 0   **75.** Discontinuity at $x = 0$
77. Continuous everywhere   **79.** Discontinuity at $x = 0$ **81.** Discontinuity at $x = 0$
83. a. $\lim_{t \to 15^-} v(t) = 3{,}800$, $\lim_{t \to 15^+} v(t) = 3{,}800$;
Shortly before 2005 the speed of Intel processors was approaching 3,800 MHz. Shortly after 2005 the speed of Intel processors was close to 3,800 MHz
b. Continuous at $t = 15$; No abrupt changes.
85. a. 0.49, 1.16. Shortly before 1999 annual advertising expenditures were close to \$0.49 billion. Shortly after 1999 annual advertising expenditures were close to \$1.16 billion.   **b.** Not continuous; movie advertising expenditures jumped suddenly in 1999.
87. 1.59; if the trend continued indefinitely, the annual spending on police would be 1.59 times the annual spending on courts in the long run.   **89.** 573. This suggests that students with an exceptionally large

household income earn an average of 573 on the math SAT test.   **91.** $\lim_{t \to +\infty} I(t) = +\infty$, $\lim_{t \to +\infty}(I(t)/E(t)) = 2.5$.   In the long term, U.S. imports from China will rise without bound and be 2.5 times U.S. exports to China. In the real world, imports and exports cannot rise without bound. Thus, the given models should not be extrapolated far into the future.   **93.** $\lim_{t \to +\infty} p(t) = 100$. The percentage of children who learn to speak approaches 100% as their age increases.   **95.** To evaluate $\lim_{x \to a} f(x)$ algebraically, first check whether $f(x)$ is a closed-form function. Then check whether $x = a$ is in its domain. If so, the limit is just $f(a)$; that is, it is obtained by substituting $x = a$. If not, then try to first simplify $f(x)$ in such a way as to transform it into a new function such that $x = a$ is in its domain, and then substitute. A disadvantage of this method is that it is sometimes extremely difficult to evaluate limits algebraically, and rather sophisticated methods are often needed.   **97.** $x = 3$ is not in the domain of the given function f, so, yes, the *function* is undefined at $x = 3$. However, the *limit* may well be defined. In this case, it leads to the indeterminate form $0/0$, telling us that we need to try to simplify, and that leads us to the correct limit of 27.   **99.** She is wrong. Closed-form functions are continuous only at points in their domains, and $x = 2$ is not in the domain of the closed-form function $f(x) = 1/(x-2)^2$.
101. Answers may vary. (1) See Example 1(b): $\lim_{x \to 2} \dfrac{x^3 - 8}{x - 2}$, which leads to the indeterminate form $0/0$ but the limit is 12. (2) $\lim_{x \to +\infty} \dfrac{60x}{2x}$, which leads to the indeterminate form ∞/∞, but where the limit exists and equals 30.   **103.** $\pm\infty/\infty$; The limits are zero. This suggests that limits resulting in $\dfrac{p(\infty)}{e^\infty}$ are zero.
105. The statement may not be true, for instance, if $f(x) = \begin{cases} x + 2 & \text{if } x < 0 \\ 2x - 1 & \text{if } x \geq 0 \end{cases}$, then $f(0)$ is defined and equals -1, and yet $\lim_{x \to 0} f(x)$ does not exist. The statement can be corrected by requiring that f be a closed-form function: "If f is a closed-form function, and $f(a)$ is defined, then $\lim_{x \to a} f(x)$ exists and equals $f(a)$."   **107.** Answers may vary, for example
$$f(x) = \begin{cases} 0 & \text{if } x \text{ is any number other than 1 or 2} \\ 1 & \text{if } x = 1 \text{ or } 2 \end{cases}$$
109. Answers may vary.
(1) $\lim_{x \to +\infty} [(x + 5) - x] = \lim_{x \to +\infty} 5 = 5$
(2) $\lim_{x \to +\infty} [x^2 - x] = \lim_{x \to +\infty} x(x - 1) = +\infty$
(3) $\lim_{x \to +\infty} [(x - 5) - x] = \lim_{x \to +\infty} -5 = -5$

Section 10.4

1. -3 **3.** 0.3 **5.** $-\$25{,}000$ per month **7.** -200 items per dollar **9.** $\$1.33$ per month **11.** 0.75 percentage point increase in unemployment per 1 percentage point increase in the deficit **13.** 4 **15.** 2 **17.** $7/3$

19.

h	Ave. Rate of Change
1	2
0.1	0.2
0.01	0.02
0.001	0.002
0.0001	0.0002

21.

h	Ave. Rate of Change
1	-0.1667
0.1	-0.2381
0.01	-0.2488
0.001	-0.2499
0.0001	-0.24999

23.

h	Ave. Rate of Change
1	9
0.1	8.1
0.01	8.01
0.001	8.001
0.0001	8.0001

25. a. $\$60$ billion per year; World military expenditure increased at an average rate of $\$60$ billion per year during 2005–2010. **b.** $\$50$ billion per year; World military expenditure increased at an average rate of $\$50$ billion per year during 2000–2010. **27. a.** $-20{,}000$ barrels/year; during 2002–2007, daily oil production by Pemex was decreasing at an average rate of 20,000 barrels of oil per year. **b.** (C) **29. a.** 1.7; the percentage of mortgages classified as subprime was increasing at an average rate of around 1.7 percentage points per year between 2000 and 2006. **b.** 2004–2006 **31. a.** 2007–2009; During 2007–2009 immigration to Ireland was decreasing at an average rate of 22,500 people per year. **b.** 2008–2010; During 2008–2010 immigration to Ireland was decreasing at an average rate of 7,500 people per year. **33. a.** $[3, 5]$; -0.25 thousand articles per year. During the period 1993–1995, the number of articles authored by U.S. researchers decreased at an average rate of 250 articles per year. **b.** Percentage rate ≈ -0.1765; Average rate $= -0.09$ thousand articles/year. Over the period 1993–2003, the number of articles authored by U.S. researchers decreased at an average rate of 90 per year, representing a 17.65% decrease over that period. **35. a.** 12 teams per year **b.** Decreased **37. a.** (A) **b.** (C) **c.** (B) **d.** Approximately $-\$0.088$ per year (if we round to two significant digits). This is less than the slope of the regression line, about $-\$0.063$ per year.

39. Answers may vary. Graph:

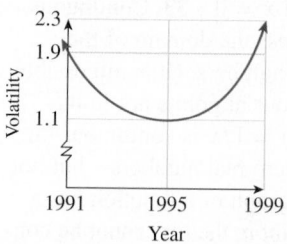

41. The index was increasing at an average rate of 300 points per day. **43. a.** $\$0.15$ per year **b.** No; according to the model, during that 25-year period the price of oil went down from around $\$93$ to a low of around $\$25$ in 1993 before climbing back up. **45. a.** 47.3 new cases per day; the number of SARS cases was growing at an average rate of 47.3 new cases per day over the period March 17 to March 23. **b.** (A) **47. a.** 8.85 manatee deaths per 100,000 boats; 23.05 manatee deaths per 100,000 boats **b.** More boats result in more manatee deaths per additional boat. **49. a.** The average rates of change are shown in the following table

Interval	[0, 40]	[40, 80]	[80, 120]	[120, 160]	[160, 200]
Average Rate of Change of S	1.35	0.80	0.48	0.28	0.17

b. For household incomes between $\$40{,}000$ and $\$80{,}000$, a student's math SAT increases at an average rate of 0.80 points per $\$1{,}000$ of additional income. **c.** (A) **d.** (B) **51.** The average rate of change of f over an interval $[a, b]$ can be determined numerically, using a table of values, graphically, by measuring the slope of the corresponding line segment through two points on the graph, or algebraically, using an algebraic formula for the function. Of these, the least precise is the graphical method, because it relies on reading coordinates of points on a graph. **53.** No, the formula for the average rate of a function f over $[a, b]$ depends only on $f(a)$ and $f(b)$, and not on any values of f between a and b. **55.** Answers will vary. Graph:

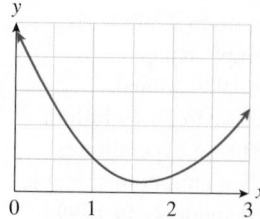

57. 6 units of quantity A per unit of quantity C **59.** (A)

61. Yes. Here is an example:

Year	2000	2001	2002	2003
Revenue ($ billion)	$10	$20	$30	$5

63. (A)

Section 10.5

1. 6   **3.** -5.5

5.

h	1	0.1	0.01
Ave. rate	39	39.9	39.99

Instantaneous rate = $40 per day

7.

h	1	0.1	0.01
Ave. Rate	140	66.2	60.602

Instantaneous rate = $60 per day

9.

h	10	1
C_{ave}	4.799	4.7999

11.

h	10	1
C_{ave}	99.91	99.90

$C'(1,000) = \$4.80$ per item $C'(100) = \$99.90$ per item

13. $1/2$ **15.** 0   **17. a.** R **b.** P **19. a.** P **b.** R
21. a. Q **b.** P **23. a.** Q **b.** R **c.** P **25. a.** R
b. Q **c.** P **27. a.** $(1, 0)$ **b.** None **c.** $(-2, 1)$
29. a. $(-2, 0.3)$, $(0, 0)$, $(2, -0.3)$ **b.** None **c.** None
31. $(a, f(a))$; $f'(a)$ **33.** (B)   **35. a.** (A) **b.** (C)
c. (B) **d.** (B) **e.** (C)   **37.** -2 **39.** -1.5 **41.** -5
43. 16   **45.** 0   **47.** -0.0025

49. a. 3 **b.** $y = 3x + 2$ **51. a.** $\dfrac{3}{4}$ **b.** $y = \dfrac{3}{4}x + 1$

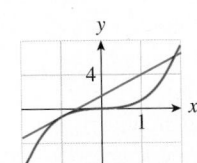 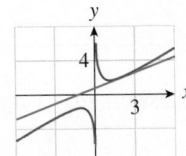

53. a. $\dfrac{1}{4}$ **b.** $y = \dfrac{1}{4}x + 1$

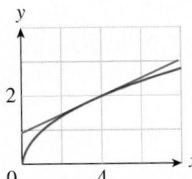

55. 1.000   **57.** 1.000   **59.** (C)   **61.** (A)   **63.** (F)
65. Increasing for $x < 0$; decreasing for $x > 0$.
67. Increasing for $x < -1$ and $x > 1$; decreasing for
$-1 < x < 1$ **69.** Increasing for $x > 1$; decreasing
for $x < 1$.   **71.** Increasing for $x < 0$; decreasing
for $x > 0$.

73. $x = -1.5, x = 0$
Graph:

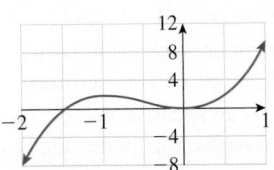

75. Note: Answers depend on the form of technology
used. Excel ($h = 0.1$):

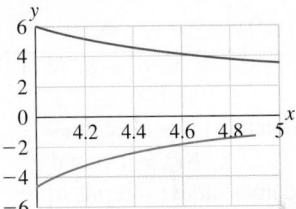

Graphs:

77. $q(100) = 50,000$, $q'(100) = -500$. A total of
50,000 pairs of sneakers can be sold at a price of $100,
but the demand is decreasing at a rate of 500 pairs per
$1 increase in the price.   **79. a.** -0.05; daily oil imports
from Mexico in 2005 were 1.6 million barrels and
declining at a rate of 0.05 million barrels (or 50,000
barrels) per year.   **b.** Decreasing   **81. a.** (B) **b.** (B)
c. (A) **d.** 2004 **e.** 0.033; in 2004 the U.S. prison
population was increasing at a rate of 0.033 million
prisoners (33,000 prisoners) per year.   **83. a.** -96 ft/sec
b. -128 ft/sec   **85. a.** $0.60 per year; the price per
barrel of crude oil in constant 2008 dollars was
growing at an average rate of about 60¢ per year over
the 28-year period beginning at the start of 1980.
b. $-\$12$ per year; the price per barrel of crude oil in
constant 2008 dollars was dropping at an instantaneous
rate of about $12 per year at the start of 1980.
c. The price of oil was decreasing in January 1980, but
eventually began to increase (making the average rate
of change in part (a) positive).   **87. a.** 144.7 new
cases per day; the number of SARS cases was growing
at a rate of about 144.7 new cases per day on March 27.
b. (A)

89. $S(5) \approx 109$, $\left.\dfrac{dS}{dt}\right|_{t=5} \approx 9.1$. After 5 weeks, sales
are 109 pairs of sneakers per week, and sales are
increasing at a rate of 9.1 pairs per week each week.
91. $A(0) = 4.5$ million; $A'(0) = 60{,}000$ subscribers/
week **93. a.** 60% of children can speak at the age of
10 months. At the age of 10 months, this percentage is
increasing by 18.2 percentage points per month.
b. As t increases, p approaches 100 percentage points
(all children eventually learn to speak), and dp/dt
approaches zero because the percentage stops increas-
ing. **95. a.** $A(6) \approx 12.0$; $A'(6) \approx 1.4$; at the start of
2006, about 12% of U.S. mortgages were subprime,
and this percentage was increasing at a rate of about
1.4 percentage points per year
b. Graphs:

Graph of A: Graph of A':

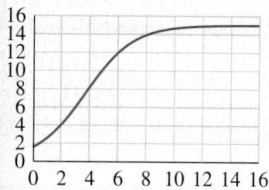   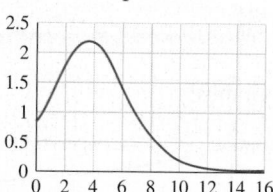

From the graphs, $A(t)$ approaches 15 as t becomes
large (in terms of limits, $\lim_{x \to +\infty} A(t) = 15$) and
$A'(t)$ approaches 0 as t becomes large (in terms of
limits, $\lim_{x \to +\infty} A'(t) = 0$). Interpretation: If the trend
modeled by the function A had continued indefinitely,
in the long term 15% of U.S. mortgages would have
been subprime, and this percentage would not be
changing. **97. a.** (D) **b.** 33 days after the egg was laid
c. 50 days after the egg was laid. Graph:

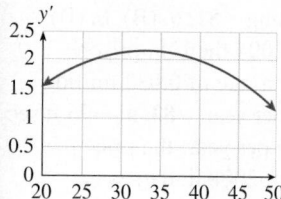

99. $L(.95) \approx 31.2$ meters and $L'(.95) \approx -304.2$ meters/
warp. Thus, at a speed of warp 0.95, the spaceship has
an observed length of 31.2 meters and its length is
decreasing at a rate of 304.2 meters per unit warp,
or 3.042 meters per increase in speed of 0.01 warp.
101. The difference quotient is not defined when
$h = 0$ because there is no such number as 0/0.
103. (D) **105.** The derivative is positive and decreasing toward zero. **107.** Company B. Although the
company is currently losing money, the derivative is
positive, showing that the profit is increasing.
Company A, on the other hand, has profits that are
declining.   **109.** (C) is the only graph in which the
instantaneous rate of change on January 1 is greater
than the one-month average rate of change.   **111.** The
tangent to the graph is horizontal at that point, and so
the graph is almost horizontal near that point.
113. Answers may vary.

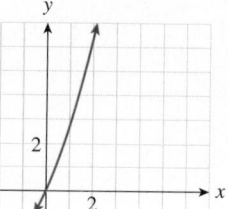

115. If $f(x) = mx + b$, then its average rate of
change over any interval $[x, x + h]$ is
$$\frac{m(x + h) + b - (mx + b)}{h} = m.$$ Because this does not
depend on h, the instantaneous rate is also equal to m.
117. Increasing because the average rate of change
appears to be rising as we get closer to 5 from the left.
(See the bottom row.)
119. Answers may vary.

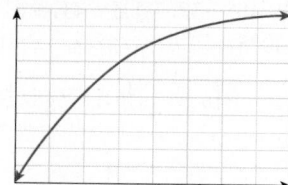

121. Answers may vary.

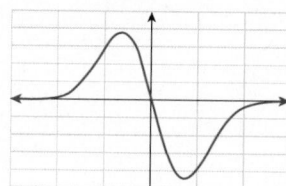

123. (B)   **125.** Answers will vary.

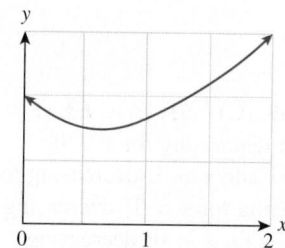

Section 10.6

1. 4 **3.** 3 **5.** 7 **7.** 4 **9.** 14 **11.** 1 **13.** m **15.** $2x$
17. 3 **19.** $6x + 1$ **21.** $2 - 2x$ **23.** $3x^2 + 2$
25. $1/x^2$ **27.** m **29.** -1.2 **31.** 30.6 **33.** -7.1
35. 4.25 **37.** -0.6 **39.** $y = 4x - 7$
41. $y = -2x - 4$ **43.** $y = -3x - 1$
45. $s'(t) = -32t$; $s'(4) = -128$ ft/sec
47. $dI/dt = -0.030t + 0.1$; daily oil imports were
decreasing at a rate of 0.11 million barrels per year.
49. $R'(t) = -90t + 900$; Increasing at a rate of
450 million gallons per year **51.** $f'(8) = 26.6$
manatee deaths per 100,000 boats. At a level of
800,000 boats, the number of manatee deaths
is increasing at a rate of 26.6 manatees per
100,000 additional boats. **53.** Yes; $\lim_{t \to 20^-} C(t) =$
$\lim_{t \to 20^+} C(t) = 700 = C(20)$. **b.** No; $\lim_{t \to 20^-} C'(t) =$
31.1 while $\lim_{t \to 20^+} C'(t) = 90$. Until 1990, the cost
of a Super Bowl ad was increasing at a rate of
$31,100 per year. Immediately thereafter, it was
increasing at a rate of $90,000 per year. **55.** The
algebraic method because it gives the exact value of
the derivative. The other two approaches give only
approximate values (except in some special cases).
57. The error is in the second line: $f(x + h)$ is *not*
equal to $f(x) + h$. For instance, if $f(x) = x^2$, then
$f(x + h) = (x + h)^2$, whereas $f(x) + h = x^2 + h$.
59. The error is in the second line: One could only
cancel the h if it were a *factor* of both the numerator
and denominator; it is not a factor of the numerator.
61. Because the algebraic computation of $f'(a)$ is
exact and not an approximation, it makes no difference
whether one uses the balanced difference quotient or
the ordinary difference quotient in the algebraic
computation. **63.** The computation results in a limit
that cannot be evaluated.

Chapter 10 Review

1. 5 **3.** Does not exist **5. a.** -1 **b.** 3 **c.** Does not
exist **7.** $-4/5$ **9.** -1 **11.** -1 **13.** Does not exist
15. $10/7$ **17.** Does not exist **19.** $+\infty$ **21.** 0
23. Diverges to $-\infty$ **25.** 0 **27.** $2/5$ **29.** 1

31.

h	1	0.01	0.001
Ave. Rate of Change	-0.5	-0.9901	-0.9990

Slope ≈ -1

33.

h	1	0.01	0.001
Avg. Rate of Change	6.3891	2.0201	2.0020

Slope ≈ 2
35. a. P **b.** Q **c.** R **d.** S **37. a.** Q **b.** None
c. None **d.** None **39. a.** (B) **b.** (B) **c.** (B) **d.** (A)
e. (C) **41.** $2x + 1$ **43.** $2/x^2$

45.

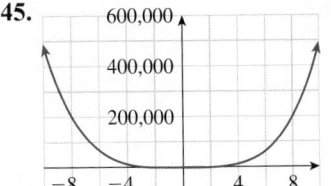

47.

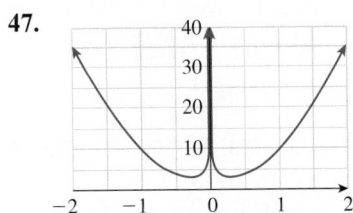

49. a. $P(3) = 25$: O'Hagan purchased the stock
at $25. $\lim_{t \to 3^-} P(t) = 25$: The value of the stock had
been approaching $25 up to the time he bought it.
$\lim_{t \to 3^+} P(t) = 10$: The value of the stock dropped to
$10 immediately after he bought it. **b.** Continuous but
not differentiable. Interpretation: the stock price
changed continuously but suddenly reversed direction
(and started to go up) the instant O'Hagan sold it.
51. a. $\lim_{t \to 3} p(t) \approx 40$; $\lim_{t \to +\infty} p(t) = +\infty$. Close
to 2007 ($t = 3$), the home price index was about 40.
In the long term, the home price index will rise without bound. **b.** 10 (The slope of the linear portion of
the curve is 10.) In the long term, the home price
index will rise about 10 points per year.
53. a. 500 books per week **b.** [3, 4], [4, 5] **c.** [3, 5];
650 books per week **55. a.** 3 percentage points per
year **b.** 0 percentage points per year **c.** (D)
57. a. $72t + 250$ **b.** 322 books per week
c. 754 books per week.

Chapter 11

Section 11.1

1. $5x^4$ **3.** $-4x^{-3}$ **5.** $-0.25x^{-0.75}$ **7.** $8x^3 + 9x^2$
9. $-1 - 1/x^2$ **11.** $\dfrac{dy}{dx} = 10(0) = 0$ (constant multiple
and power rule) **13.** $\dfrac{dy}{dx} = \dfrac{d}{dx}(x^2) + \dfrac{d}{dx}(x)$
(sum rule) $= 2x + 1$ (power rule)
15. $\dfrac{dy}{dx} = \dfrac{d}{dx}(4x^3) + \dfrac{d}{dx}(2x) - \dfrac{d}{dx}(1)$ (sum and
difference) $= 4\dfrac{d}{dx}(x^3) + 2\dfrac{d}{dx}(x) - \dfrac{d}{dx}(1)$
(constant multiples) $= 12x^2 + 2$ (power rule)
17. $f'(x) = 2x - 3$ **19.** $f'(x) = 1 + 0.5x^{-0.5}$
21. $g'(x) = -2x^{-3} + 3x^{-2}$ **23.** $g'(x) = -\dfrac{1}{x^2} + \dfrac{2}{x^3}$
25. $h'(x) = -\dfrac{0.8}{x^{1.4}}$ **27.** $h'(x) = -\dfrac{2}{x^3} - \dfrac{6}{x^4}$

29. $r'(x) = -\dfrac{2}{3x^2} + \dfrac{0.1}{2x^{1.1}}$

31. $r'(x) = \dfrac{2}{3} - \dfrac{0.1}{2x^{0.9}} - \dfrac{4.4}{3x^{2.1}}$

33. $t'(x) = |x|/x - 1/x^2$ **35.** $s'(x) = \dfrac{1}{2\sqrt{x}} - \dfrac{1}{2x\sqrt{x}}$

37. $s'(x) = 3x^2$ **39.** $t'(x) = 1 - 4x$

41. $2.6x^{0.3} + 1.2x^{-2.2}$ **43.** $1.2(1 - |x|/x)$

45. $3at^2 - 4a$ **47.** $5.15x^{9.3} - 99x^{-2}$

49. $-\dfrac{2.31}{t^{2.1}} - \dfrac{0.3}{t^{0.4}}$ **51.** $4\pi r^2$ **53.** 3   **55.** -2 **57.** -5

59. $y = 3x + 2$

61. $y = \dfrac{3}{4}x + 1$

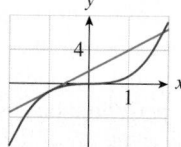
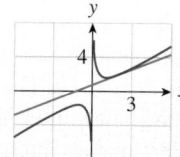

63. $y = \dfrac{1}{4}x + 1$

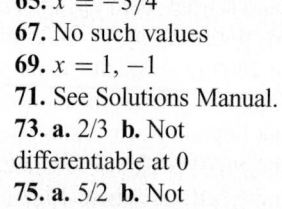

65. $x = -3/4$
67. No such values
69. $x = 1, -1$
71. See Solutions Manual.
73. a. 2/3 **b.** Not differentiable at 0
75. a. 5/2 **b.** Not differentiable at 0

77. Yes; 0   **79.** Yes; 12   **81.** No; 3   **83.** Yes; 3/2
85. Yes; diverges to $-\infty$ **87.** Yes; diverges to $-\infty$
89. $P'(t) = 0.9t - 12$; $P'(20) = 6$; the price of a
barrel of crude oil was increasing at a rate of $6 per
year in 2000.   **91.** 0.55   **93. a.** $s'(t) = -32t$;
$0, -32, -64, -96, -128$ ft/s **b.** 5 seconds;
downward at 160 ft/s   **95. a.** $P'(t) = -4.0t + 6.6$;
Decreasing at a rate of $5.4 billion per year **b.** (B)
97. a. $f'(x) = 7.1x - 30.2$; $f'(8) = 26.6$ manatees
per 100,000 boats; At a level of 800,000 boats, manatee deaths are increasing at a rate of 26.6 deaths each
year per 100,000 additional boats. **b.** Increasing; the
number of manatees killed per additional 100,000
boats increases as the number of boats increases.
99. a. $I - A$ measures the amount by which the
iPhone market share exceeds the Android market
share. $(I - A)'$ measures the rate at which this difference is changing. **b.** (B) **c.** $-0.2t + 0.5$ percentage
points per month; (C); The vertical distance between
the graphs at first increases, and then decreases; the
iPhone increases its advantage over the Android share
at first, but then the Android begins to catch up.
d. -0.3 percentage points per month; In May 2009,

the iPhone's advantage over Android was decreasing at
a rate of 0.3 percentage points per month.
101. After graphing the curve $y = 3x^2$, draw the line
passing through $(-1, 3)$ with slope -6.   **103.** The
slope of the tangent line of g is twice the slope of the
tangent line of f.   **105.** $g'(x) = -f'(x)$ **107.** The
left-hand side is not equal to the right-hand side. The
derivative of the left-hand side is equal to the right-
hand side, so your friend should have written

$$\dfrac{d}{dx}(3x^4 + 11x^5) = 12x^3 + 55x^4.$$ **109.** $\dfrac{1}{2x}$ is not

equal to $2x^{-1}$. Your friend should have written

$$y = \dfrac{1}{2x} = \dfrac{1}{2}x^{-1}, \text{ so } \dfrac{dy}{dx} = -\dfrac{1}{2}x^{-2}.$$ **111.** The

derivative of a constant times a function is the
constant times the derivative of the function, so that
$f'(x) = (2)(2x) = 4x$. Your enemy mistakenly computed the *derivative* of the constant times the derivative of the function. (The derivative of a product of
two functions is not the product of the derivative of the
two functions. The rule for taking the derivative of a
product is discussed later in the chapter.).
113. For a general function f, the derivative of f is

defined to be $f'(x) = \lim\limits_{h \to 0} \dfrac{f(x + h) - f(x)}{h}$. One

then finds by calculation that the derivative of the
specific function x^n is nx^{n-1}. In short, nx^{n-1} is the
derivative of a specific function: $f(x) = x^n$, it is not
the *definition* of the derivative of a general function or
even the definition of the derivative of the function
$f(x) = x^n$.   **115.** Answers may vary.

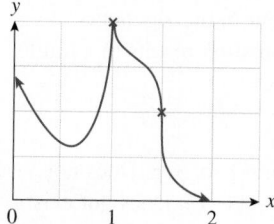

Section 11.2

1. $C'(1,000) = \$4.80$ per item   **3.** $C'(100) = \$99.90$
per item   **5.** $C'(x) = 4$; $R'(x) = 8 - 0.002$;
$P'(x) = 4 - 0.002x$; $P'(x) = 0$ when $x = 2,000$.
Thus, at a production level of 2,000, the profit is
stationary (neither increasing nor decreasing) with
respect to the production level. This may indicate a
maximum profit at a production level of 2,000.
7. a. (B) **b.** (C) **c.** (C)   **9. a.** $C'(x) = 2,500 - 0.04x$;
The cost is going up at a rate of $2,499,840 per televi-
sion commercial. The exact cost of airing the fifth

television commercial is $2,499,820. **b.** $\overline{C}(x) = 150/x + 2,500 - 0.02x$; $\overline{C}(4) = 2,537.42$ thousand dollars. The average cost of airing the first four television commercials is $2,537,420. **11. a.** $R'(x) = 0.90$, $P'(x) = 0.80 - 0.002x$ **b.** Revenue: $450, Profit: $80, Marginal revenue: $0.90, Marginal profit: $-\$0.20$. The total revenue from the sale of 500 copies is $450. The profit from the production and sale of 500 copies is $80. Approximate revenue from the sale of the 501st copy is 90¢. Approximate loss from the sale of the 501st copy is 20¢. **c.** $x = 400$. The profit is a maximum when you produce and sell 400 copies. **13.** The profit on the sale of 1,000 DVDs is $3,000, and is decreasing at a rate of $3 per additional DVD sold. **15.** Profit ≈ $257.07; Marginal profit ≈ 5.07. Your current profit is $257.07 per month, and this would increase at a rate of $5.07 per additional magazine sold. **17. a.** $2.50 per pound **b.** $R(q) = 20,000/q^{0.5}$ **c.** $R(400) = \$1,000$. This is the monthly revenue that will result from setting the price at $2.50 per pound. $R'(400) = -\$1.25$ per pound of tuna. Thus, at a demand level of 400 pounds per month, the revenue is decreasing at a rate of $1.25 per pound. **d.** The fishery should raise the price (to reduce the demand). **19.** $P'(50) = \$350$. This means that, at an employment level of 50 workers, the firm's daily profit will increase at a rate of $350 per additional worker it hires. **21. a.** (B) **b.** (B) **c.** (C)
23. a. $C(x) = 500,000 + 1,600,000x - 100,000\sqrt{x}$;

$$C'(x) = 1,600,000 - \frac{50,000}{\sqrt{x}};$$

$$\overline{C}(x) = \frac{500,000}{x} + 1,600,000 - \frac{100,000}{\sqrt{x}}$$

b. $C'(3) \approx \$1,570,000$ per spot, $\overline{C}(3) \approx \$1,710,000$ per spot. The average cost will decrease as x increases. **25. a.** $2,000 per one-pound reduction in emissions. **b.** 2.5 pounds per day reduction. **c.** $N(q) = 100q^2 - 500q + 4,000$; 2.5 pounds per day reduction. The value of q is the same as that for part (b). The net cost to the firm is minimized at the reduction level for which the cost of controlling emissions begins to increase faster than the subsidy. This is why we get the answer by setting these two rates of increase equal to each other. **27.** $M'(10) \approx 0.0002557$ mpg/mph. This means that, at a speed of 10 mph, the fuel economy is increasing at a rate of 0.0002557 miles per gallon per 1-mph increase in speed. $M'(60) = 0$ mpg/mph. This means that, at a speed of 60 mph, the fuel economy is neither increasing nor decreasing with increasing speed.

$M'(70) \approx -0.00001799$. This means that, at 70 mph, the fuel economy is decreasing at a rate of 0.00001799 miles per gallon per 1-mph increase in speed. Thus 60 mph is the most fuel-efficient speed for the car. **29.** (C) **31.** (D) **33.** (B) **35.** Cost is often measured as a function of the number of items x. Thus, $C(x)$ is the cost of producing (or purchasing, as the case may be) x items. **a.** The average cost function $\overline{C}(x)$ is given by $\overline{C}(x) = C(x)/x$. The marginal cost function is the derivative, $C'(x)$, of the cost function. **b.** The average cost $\overline{C}(r)$ is the slope of the line through the origin and the point on the graph where $x = r$. The marginal cost of the rth unit is the slope of the tangent to the graph of the cost function at the point where $x = r$. **c.** The average cost function $\overline{C}(x)$ gives the average cost of producing the first x items. The marginal cost function $C'(x)$ is the rate at which cost is changing with respect to the number of items x, or the incremental cost per item, and approximates the cost of producing the $(x + 1)$st item. **37.** Answers may vary. An example is $C(x) = 300x$. **39.** The marginal cost **41.** Not necessarily. For example, it may be the case that the marginal cost of the 101st item is larger than the average cost of the first 100 items (even though the marginal cost is decreasing). Thus, adding this additional item will *raise* the average cost. **43.** The circumstances described suggest that the average cost function is at a relatively low point at the current production level, and so it would be appropriate to advise the company to maintain current production levels; raising or lowering the production level will result in increasing average costs.

Section 11.3

1. 3 **3.** $3x^2$ **5.** $2x + 3$ **7.** $210x^{1.1}$
9. $-2/x^2$ **11.** $2x/3$ **13.** $36x^2 - 3$ **15.** $3x^2 - 5x^4$
17. $8x + 12$ **19.** $-8/(5x - 2)^2$ **21.** $-14/(3x - 1)^2$
23. 0 **25.** $-|x|/x^3$ **27.** $3\sqrt{x}/2$
29. $(x^2 - 1) + 2x(x + 1) = (x + 1)(3x - 1)$
31. $(x^{-0.5} + 4)(x - x^{-1}) + (2x^{0.5} + 4x - 5)(1 + x^{-2})$
33. $8(2x^2 - 4x + 1)(x - 1)$
35. $(1/3.2 - 3.2/x^2)(x^2 + 1) + 2x(x/3.2 + 3.2/x)$
37. $2x(2x + 3)(7x + 2) + 2x^2(7x + 2) + 7x^2(2x + 3)$
39. $5.3(1 - x^{2.1})(x^{-2.3} - 3.4) - 2.1x^{1.1}(5.3x - 1)(x^{-2.3} - 3.4) - 2.3x - 3.3(5.3x - 1)(1 - x^{2.1})$

41. $\dfrac{1}{2\sqrt{x}}\left(\sqrt{x} + \dfrac{1}{x^2}\right) + (\sqrt{x} + 1)\left(\dfrac{1}{2\sqrt{x}} - \dfrac{2}{x^3}\right)$

43. $\dfrac{(4x+4)(3x-1)-3(2x^2+4x+1)}{(3x-1)^2}=(6x^2-4x-7)/(3x-1)^2$

45. $\dfrac{(2x-4)(x^2+x+1)-(x^2-4x+1)(2x+1)}{(x^2+x+1)^2}=(5x^2-5)/(x^2+x+1)^2$

47. $\dfrac{(0.23x^{-0.77}-5.7)(1-x^{-2.9})-2.9x^{-3.9}(x^{0.23}-5.7x)}{(1-x^{-2.9})^2}$

49. $\dfrac{\frac{1}{2}x^{-1/2}(x^{1/2}-1)-\frac{1}{2}x^{-1/2}(x^{1/2}+1)}{(x^{1/2}-1)^2}=\dfrac{-1}{\sqrt{x}\left(\sqrt{x}-1\right)^2}$ **51.** $-3/x^4$

53. $\dfrac{[(x+1)+(x+3)](3x-1)-3(x+3)(x+1)}{(3x-1)^2}=(3x^2-2x-13)/(3x-1)^2$

55. $\dfrac{[(x+1)(x+2)+(x+3)(x+2)+(x+3)(x+1)](3x-1)-3(x+3)(x+1)(x+2)}{(3x-1)^2}$

57. $4x^3-2x$ **59.** 64   **61.** 3   **63.** Difference; $4x^3-12x^2+2x-480$ **65.** Sum; $1+2/(x+1)^2$

67. Product; $\left[\dfrac{x}{x+1}\right]+(x+2)\dfrac{1}{(x+1)^2}$

69. Difference; $2x-1-2/(x+1)^2$ **71.** $y=12x-8$
73. $y=x/4+1/2$ **75.** $y=-2$ **77.** $q'(5)=1{,}000$ units/month (sales are increasing at a rate of 1,000 units per month); $p'(5)=-\$10$/month (the price of a sound system is dropping at a rate of \$10 per month); $R'(5)=900{,}000$ (revenue is increasing at a rate of \$900,000 per month).   **79.** \$703 million; increasing at a rate of \$67 million per year   **81.** Decreasing at a rate of \$1 per day   **83.** Decreasing at a rate of approximately \$0.10 per month   **85.** $M'(x)=\dfrac{3{,}000(3{,}600x^{-2}-1)}{\left(x+3{,}600x^{-1}\right)^2}$; $M'(10)\approx0.7670$ mpg/mph. This means that, at a speed of 10 mph, the fuel economy is increasing at a rate of 0.7670 miles per gallon per one mph increase in speed. $M'(60)=0$ mpg/mph. This means that, at a speed of 60 mph, the fuel economy is neither increasing nor decreasing with increasing speed. $M'(70)\approx-0.0540$. This means that, at 70 mph, the fuel economy is decreasing at a rate of 0.0540 miles per gallon per one mph increase in speed. 60 mph is the most fuel-efficient speed for the car. (In the next chapter we shall discuss how to locate largest values in general.)
87. a. $P(t)-I(t)$ represents the daily production of oil in Mexico that was not exported to the United States. $I(t)/P(t)$ represents U.S. imports of oil from Mexico as a fraction of the total produced there. **b.** -0.023 per year; at the start of 2008, the fraction of oil produced in

Mexico that was imported by the United States was decreasing at a rate of 0.023 (or 2.3 percentage points) per year.   **89.** Increasing at a rate of about \$3,420 million per year.   **91.** $R'(p)=-\dfrac{5.625}{(1+0.125p)^2}$; $R'(4)=-2.5$ thousand organisms per hour, per 1,000 organisms. This means that the reproduction rate of organisms in a culture containing 4,000 organisms is declining at a rate of 2,500 organisms per hour, per 1,000 additional organisms.   **93.** Oxygen consumption is decreasing at a rate of 1,600 milliliters per day. This is due to the fact that the number of eggs is decreasing, because $C'(25)$ is positive.   **95.** 20; 33
97. 5/4; $-17/16$ **99.** The analysis is suspect, as it seems to be asserting that the annual increase in revenue, which we can think of as dR/dt, is the product of the annual increases, dp/dt in price, and dq/dt in sales. However, because $R=pq$, the product rule implies that dR/dt is not the product of dp/dt and dq/dt, but is instead $\dfrac{dR}{dt}=\dfrac{dp}{dt}\cdot q+p\cdot\dfrac{dq}{dt}$.   **101.** Answers will vary $q=-p+1{,}000$ is one example.   **103.** Mine; it is increasing twice as fast as yours. The rate of change of revenue is given by $R'(t)=p'(t)q(t)$ because $q'(t)=0$. Thus, $R'(t)$ does not depend on the selling price $p(t)$.   **105.** (A)

Section 11.4

1. $4(2x+1)$ **3.** $-(x-1)^{-2}$ **5.** $2(2-x)^{-3}$
7. $(2x+1)^{-0.5}$ **9.** $-3/(3x-1)^2$

11. $4(x^2 + 2x)^3(2x + 2)$ **13.** $-4x(2x^2 - 2)^{-2}$

15. $-5(2x - 3)(x^2 - 3x - 1)^{-6}$ **17.** $-6x/(x^2 + 1)^4$

19. $1.5(0.2x - 4.2)(0.1x^2 - 4.2x + 9.5)^{0.5}$

21. $4(2s - 0.5s^{-0.5})(s^2 - s^{0.5})^3$ **23.** $-x/\sqrt{1 - x^2}$

25. $\dfrac{3|3x - 6|}{3x - 6}$ **27.** $\dfrac{(-3x^2 + 5)|-x^3 + 5x|}{-x^3 + 5x}$

29. $-[(x + 1)(x^2 - 1)]^{-3/2}(3x - 1)(x + 1)$

31. $6.2(3.1x - 2) + 6.2/(3.1x - 2)^3$

33. $2[(6.4x - 1)^2 + (5.4x - 2)^3] \times$
$[12.8(6.4x - 1) + 16.2(5.4x - 2)^2]$

35. $-2(x^2 - 3x)^{-3}(2x - 3)(1 - x^2)^{0.5}$
$-x(x^2 - 3x)^{-2}(1 - x^2)^{-0.5}$

37. $-56(x + 2)/(3x - 1)^3$ **39.** $3z^2(1 - z^2)/(1 + z^2)^4$

41. $3[(1 + 2x)^4 - (1 - x)^2]^2[8(1 + 2x)^3 + 2(1 - x)]$

43. $6|3x - 1|$ **45.** $\dfrac{|x - (2x - 3)^{1/2}|}{x - (2x - 3)^{1/2}}[1 - (2x - 3)^{-1/2}]$

47. $-\dfrac{\left(\dfrac{1}{\sqrt{2x + 1}} - 2x\right)}{\left(\sqrt{2x + 1} - x^2\right)^2}$

49. $54(1 + 2x)^2(1 + (1 + 2x)^3)^2(1 + (1 + (1 + 2x)^3)^3)^2$

51. $(100x^{99} - 99x^{-2}) \, dx/dt$

53. $(-3r^{-4} + 0.5r^{-0.5}) \, dr/dt$ **55.** $4\pi r^2 dr/dt$

57. $-47/4$ **59.** $1/3$ **61.** $-5/3$ **63.** $1/4$

65. $\left.\dfrac{dP}{dn}\right|_{n=10} = 146{,}454.9$. At an employment

level of 10 engineers, Paramount will increase its profit at a rate of \$146,454.90 per additional engineer hired. **67.** $y = 35(7 + 0.2t)^{-0.25}$; -0.11 percentage points per month.   **69.** $-\$30$ per additional ruby sold. The revenue is decreasing at a rate of \$30 per additional ruby sold.

71. $\dfrac{dy}{dt} = \dfrac{dy}{dx}\dfrac{dx}{dt} = (1.5)(-2) = -3$ murders per 100,000 residents/yr each year.   **73.** $5/6 \approx 0.833$; relative to the 2003 levels, home sales were changing at a rate of about 0.833 percentage points per percentage point change in price. (Equivalently, home sales in 2008 were dropping at a rate of about 0.833 percentage points per percentage point drop in price.)   **75.** 12π mi^2/h   **77.** \$200,000π/week $\approx$ \$628,000/week   **79. a.** $q'(4) \approx 333$ units per month **b.** $dR/dq = \$800$/unit **c.** $dR/dt \approx \$267{,}000$ per month **81.** 3% per year   **83.** 8% per year   **85.** The glob squared, times the derivative of the glob.   **87.** The derivative of a quantity cubed is three times the *original quantity* squared, times the derivative of the quantity, not three times the derivative of the quantity squared. Thus, the correct answer is $3(3x^3 - x)^2(9x^2 - 1)$.

89. First, the derivative of a quantity cubed is three times the *original quantity* squared times the derivative of the quantity, not three times the derivative of the quantity squared. Second, the derivative of a quotient is not the quotient of the derivatives; the quotient rule needs to be used in calculating the derivative of $\dfrac{3x^2 - 1}{2x - 2}$. Thus, the correct result (before simplifying) is

$$3\left(\frac{3x^2 - 1}{2x - 2}\right)^2 \left(\frac{6x(2x - 2) - (3x^2 - 1)(2)}{(2x - 2)^2}\right).$$

91. Following the calculation thought experiment, pretend that you were evaluating the function at a specific value of x. If the last operation you would perform is addition or subtraction, look at each summand separately. If the last operation is multiplication, use the product rule first; if it is division, use the quotient rule first; if it is any other operation (such as raising a quantity to a power or taking a radical of a quantity) then use the chain rule first.   **93.** An

example is $f(x) = \sqrt{x + \sqrt{x + \sqrt{x + \sqrt{x + \sqrt{x + 1}}}}}$.

Section 11.5

1. $1/(x - 1)$ **3.** $1/(x \ln 2)$ **5.** $2x/(x^2 + 3)$

7. e^{x+3} **9.** $-e^{-x}$ **11.** $4^x \ln 4$ **13.** $2^{x^2-1}2x \ln 2$

15. $1 + \ln x$ **17.** $2x \ln x + (x^2 + 1)/x$

19. $10x(x^2 + 1)^4 \ln x + (x^2 + 1)^5/x$

21. $3/(3x - 1)$ **23.** $4x/(2x^2 + 1)$

25. $(2x - 0.63x^{-0.7})/(x^2 - 2.1x^{0.3})$

27. $-2/(-2x + 1) + 1/(x + 1)$

29. $3/(3x + 1) - 4/(4x - 2)$

31. $1/(x + 1) + 1/(x - 3) - 2/(2x + 9)$

33. $5.2/(4x - 2)$

35. $2/(x + 1) - 9/(3x - 4) - 1/(x - 9)$

37. $\dfrac{1}{(x + 1) \ln 2}$ **39.** $\dfrac{1 - 1/t^2}{(t + 1/t) \ln 3}$ **41.** $\dfrac{2 \ln |x|}{x}$

43. $\dfrac{2}{x} - \dfrac{2 \ln(x - 1)}{x - 1}$ **45.** $e^x(1 + x)$

47. $1/(x + 1) + 3e^x(x^3 + 3x^2)$ **49.** $e^x(\ln |x| + 1/x)$

51. $2e^{2x+1}$ **53.** $(2x - 1)e^{x^2-x+1}$ **55.** $2xe^{2x-1}(1 + x)$

57. $4(e^{2x-1})^2$ **59.** $2 \cdot 3^{2x-4} \ln 3$

61. $2 \cdot 3^{2x+1} \ln 3 + 3e^{3x+1}$

63. $\dfrac{2x3^{x^2}[(x^2 + 1)\ln 3 - 1]}{(x^2 + 1)^2}$

65. $-4/(e^x - e^{-x})^2$

67. $5e^{5x-3}$ **69.** $-\dfrac{\ln x + 1}{(x \ln x)^2}$ **71.** $2(x - 1)$ **73.** $\dfrac{1}{x \ln x}$

75. $\dfrac{1}{2x \ln x}$ **77.** $y = (e/\ln 2)(x - 1) \approx 3.92(x - 1)$

79. $y = x$ **81.** $y = -[1/(2e)](x - 1) + e$ **83.** \$163 billion and increasing at a rate of \$5.75 billion per year **85.** \$163 billion and increasing at a rate of \$5.75 billion per year. **87.** $-1{,}653$ years per gram; the age of the specimen is decreasing at a rate of about 1,653 years per additional one gram of carbon 14 present in the sample. (Equivalently, the age of the specimen is increasing at a rate of about 1,653 years per additional one gram less of carbon 14 in the sample.) **89.** Average price: \$1.4 million; increasing at a rate of about \$220,000 per year. **91. a.** $N(15) \approx 1{,}762 \approx 1{,}800$ (rounded to 2 significant digits) wiretap orders; $N'(15) \approx 89.87 \approx 90$ wiretap orders per year (rounded to 2 significant digits). The constants in the model are specified to 2 significant digits, so we cannot expect the answer to be accurate to more than 2 digits. **b.** In 2005, the number of people whose communications were intercepted was about 180,000 and increasing at a rate of about 9,000 people per year. **c.** (C) **93.** \$451.00 per year **95.** \$446.02 per year **97.** $A(t) = 167(1.18)^t$; 280 new cases per day **99. a.** (A) **b.** The math SAT increases by approximately 0.97 points. **c.** $S'(x)$ decreases with increasing x, so that as parental income increases, the effect on math SAT scores decreases. **101. a.** -6.25 years/child; when the fertility rate is 2 children per woman, the average age of a population is dropping at a rate of 6.25 years per one-child increase in the fertility rate. **b.** 0.160 **103.** 3,300,000 cases/week; 11,000,000 cases/week; 640,000 cases/week **105.** 2.1 percentage points per year; the rate of change is the slope of the tangent at $t = 3$. This is also approximately the average rate of change over $[2, 4]$, which is about $4/2 = 2$, in approximate agreement with the answer. **107. a.** 2.1 percentage points per year **b.** $\lim\limits_{t \to +\infty} A(t) = 15$; Had the trend continued indefinitely, the percentage of mortgages that were subprime would have approached 15% in the long term. $\lim\limits_{t \to +\infty} A'(t) = 0$; Had the trend continued indefinitely, the rate of change of the percentage of mortgages that were subprime would have approached 0 percentage points per year in the long term. **109.** 277,000 people per year **111.** 0.000283 g/yr **113.** $R(t) = 350e^{-0.1t}(39t + 68)$ million dollars; $R(2) \approx \$42$ billion; $R'(2) \approx \$7$ billion per year

115. e raised to the glob, times the derivative of the glob. **117.** 2 raised to the glob, times the derivative of the glob, times the natural logarithm of 2.

119. The derivative of $\ln |u|$ is not $\dfrac{1}{|u|}\dfrac{du}{dx}$; it is $\dfrac{1}{u}\dfrac{du}{dx}$. Thus, the correct derivative is $\dfrac{3}{3x + 1}$.

121. The power rule does not apply when the exponent is not constant. The derivative of 3 raised to a quantity is 3 raised to the quantity, times the derivative of the quantity, times $\ln 3$. Thus, the correct answer is $3^{2x} 2 \ln 3$. **123.** No. If $N(t)$ is exponential, so is its derivative. **125.** If $f(x) = e^{kx}$, then the fractional rate of change is $\dfrac{f'(x)}{f(x)} = \dfrac{ke^{kx}}{e^{kx}} = k$, the fractional growth rate. **127.** If $A(t)$ is growing exponentially, then $A(t) = A_0 e^{kt}$ for constants A_0 and k. Its percentage rate of change is then $\dfrac{A'(t)}{A(t)} = \dfrac{kA_0 e^{kt}}{A_0 e^{kt}} = k$, a constant.

Section 11.6

1. $-2/3$ **3.** x **5.** $(y - 2)/(3 - x)$ **7.** $-y$ **9.** $-\dfrac{y}{x(1 + \ln x)}$ **11.** $-x/y$ **13.** $-2xy/(x^2 - 2y)$ **15.** $-(6 + 9x^2 y)/(9x^3 - x^2)$ **17.** $3y/x$ **19.** $(p + 10p^2 q)/(2p - q - 10pq^2)$ **21.** $(ye^x - e^y)/(xe^y - e^x)$ **23.** $se^{st}/(2s - te^{st})$ **25.** $ye^x/(2e^x + y^3 e^y)$ **27.** $(y - y^2)/(-1 + 3y - y^2)$ **29.** $-y/(x + 2y - xye^y - y^2 e^y)$ **31. a.** 1 **b.** $y = x - 3$ **33. a.** -2 **b.** $y = -2x$ **35. a.** -1 **b.** $y = -x + 1$ **37. a.** $-2{,}000$ **b.** $y = -2{,}000x + 6{,}000$ **39. a.** 0 **b.** $y = 1$ **41. a.** -0.1898 **b.** $y = -0.1898x + 1.4721$

43. $\dfrac{2x + 1}{4x - 2}\left[\dfrac{2}{2x + 1} - \dfrac{4}{4x - 2}\right]$

45. $\dfrac{(3x + 1)^2}{4x(2x - 1)^3}\left[\dfrac{6}{3x + 1} - \dfrac{1}{x} - \dfrac{6}{2x - 1}\right]$

47. $(8x - 1)^{1/3}(x - 1)\left[\dfrac{8}{3(8x - 1)} + \dfrac{1}{x - 1}\right]$

49. $(x^3 + x)\sqrt{x^3 + 2}\left[\dfrac{3x^2 + 1}{x^3 + x} + \dfrac{1}{2}\dfrac{3x^2}{x^3 + 2}\right]$

51. $x^x(1 + \ln x)$ **53.** $-\$3{,}000$ per worker. The monthly budget to maintain production at the fixed level P is decreasing by approximately \$3,000 per additional worker at an employment level of 100 workers and a monthly operating budget of \$200,000.

55. −125 T-shirts per dollar; when the price is set at $5, the demand is dropping by 125 T-shirts per $1 increase in price. **57.** $\dfrac{dk}{de}\Big|_{e=15} = -0.307$ carpenters per electrician. This means that, for a $200,000 house whose construction employs 15 electricians, adding one more electrician would cost as much as approximately 0.307 additional carpenters. In other words, one electrician is worth approximately 0.307 carpenters. **59. a.** 22.93 hours. (The other root is rejected because it is larger than 30.) **b.** $\dfrac{dt}{dx} = \dfrac{4t - 20x}{0.4t - 4x}$; $\dfrac{dt}{dx}\Big|_{x=3.0} \approx$ −11.2 hours per grade point. This means that, for a 3.0 student who scores 80 on the examination, 1 grade point is worth approximately 11.2 hours. **61.** $\dfrac{dr}{dy} = 2\dfrac{r}{y}$, so $\dfrac{dr}{dt} = 2\dfrac{r}{y}\dfrac{dy}{dt}$ by the chain rule. **63.** x, y, y, x **65.** Let $y = f(x)g(x)$. Then $\ln y = \ln f(x) + \ln g(x)$, and

$$\frac{1}{y}\frac{dy}{dx} = \frac{f'(x)}{f(x)} + \frac{g'(x)}{g(x)},$$

so $\dfrac{dy}{dx} = y\left(\dfrac{f'(x)}{f(x)} + \dfrac{g'(x)}{g(x)}\right) =$

$$f(x)g(x)\left(\frac{f'(x)}{f(x)} + \frac{g'(x)}{g(x)}\right) = f'(x)g(x) + f(x)g'(x).$$

67. Writing $y = f(x)$ specifies y as an explicit function of x. This can be regarded as an equation giving y as an *implicit* function of x. The procedure of finding dy/dx by implicit differentiation is then the same as finding the derivative of y as an explicit function of x: We take d/dx of both sides. **69.** Differentiate both sides of the equation $y = f(x)$ with respect to y to get $1 = f'(x) \cdot \dfrac{dx}{dy}$, giving $\dfrac{dx}{dy} = \dfrac{1}{f'(x)} = \dfrac{1}{dy/dx}$.

Chapter 11 Review

1. $50x^4 + 2x^3 - 1$ **3.** $9x^2 + x^{-2/3}$ **5.** $1 - 2/x^3$
7. $-\dfrac{4}{3x^2} + \dfrac{0.2}{x^{1.1}} + \dfrac{1.1x^{0.1}}{3.2}$ **9.** $e^x(x^2 + 2x - 1)$
11. $(-3x|x| + |x|/x - 6x)/(3x^2 + 1)^2$
13. $-4(4x - 1)^{-2}$ **15.** $20x(x^2 - 1)^9$
17. $-0.43(x + 1)^{-1.1}[2 + (x + 1)^{-0.1}]^{3.3}$
19. $e^x(x^2 + 1)^9(x^2 + 20x + 1)$
21. $3^x[(x - 1)\ln 3 - 1]/(x - 1)^2$ **23.** $2xe^{x^2-1}$
25. $2x/(x^2 - 1)$ **27.** $x = 7/6$ **29.** $x = \pm 2$
31. $x = (1 - \ln 2)/2$ **33.** None **35.** $\dfrac{2x - 1}{2y}$

37. $-y/x$ **39.** $\dfrac{(2x - 1)^4(3x + 4)}{(x + 1)(3x - 1)^3} \times$
$\left[\dfrac{8}{2x - 1} + \dfrac{3}{3x + 4} - \dfrac{1}{x + 1} - \dfrac{9}{3x - 1}\right]$
41. $y = -x/4 + 1/2$ **43.** $y = -3ex - 2e$
45. $y = x + 2$ **47. a.** 274 books per week **b.** 636 books per week **c.** The function w begins to decrease more and more rapidly after $t = 14$ Graph:

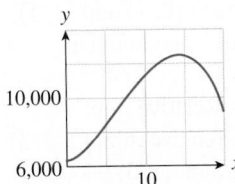

d. Because the data suggest an upward curving parabola, the long-term prediction of sales for a quadratic model would be that sales will increase without bound, in sharp contrast to (c) **49. a.** $2.88 per book **b.** $3.715 per book **c.** Approximately −$0.000104 per book, per additional book sold. **d.** At a sales level of 8,000 books per week, the cost is increasing at a rate of $2.88 per book (so that the 8,001st book costs approximately $2.88 to sell), and it costs an average of $3.715 per book to sell the first 8,000 books. Moreover, the average cost is decreasing at a rate of $0.000104 per book, per additional book sold.
51. a. $3,000 per week (rising) **b.** 300 books per week **53.** $R = pq$ gives $R' = p'q + pq'$. Thus, $R'/R = R'/(pq) = (p'q + pq')/pq = p'/p + q'/q$
55. $110 per year **57. a.** $s'(t) = \dfrac{2,475e^{-0.55(t-4.8)}}{(1 + e^{-0.55(t-4.8)})^2}$; 556 books per week **b.** 0; In the long term, the rate of increase of weekly sales slows to zero. **59.** 616.8 hits per day per week. **61. a.** −17.24 copies per $1. The demand for the gift edition of *The Complete Larry Potter* is dropping at a rate of about 17.24 copies per $1 increase in the price. **b.** $138 per dollar is positive, so the price should be raised.

Chapter 12

Section 12.1

1. Absolute min: $(-3, -1)$, relative max: $(-1, 1)$, relative min: $(1, 0)$, absolute max: $(3, 2)$ **3.** Absolute min: $(3, -1)$ and $(-3, -1)$, absolute max: $(1, 2)$
5. Absolute min: $(-3, 0)$ and $(1, 0)$, absolute max: $(-1, 2)$ and $(3, 2)$ **7.** Relative min: $(-1, 1)$
9. Absolute min: $(-3, -1)$, relative max: $(-2, 2)$, relative min: $(1, 0)$, absolute max: $(3, 3)$

11. Relative max: $(-3, 0)$, absolute min: $(-2, -1)$, stationary nonextreme point: $(1, 1)$ **13.** Absolute max: $(0, 1)$, absolute min: $(2, -3)$, relative max: $(3, -2)$ **15.** Absolute min: $(-4, -16)$, absolute max: $(-2, 16)$, absolute min: $(2, -16)$, absolute max: $(4, 16)$ **17.** Absolute min: $(-2, -10)$, absolute max: $(2, 10)$ **19.** Absolute min: $(-2, -4)$, relative max: $(-1, 1)$, relative min: $(0, 0)$ **21.** Relative max: $(-1, 5)$, absolute min: $(3, -27)$ **23.** Absolute min: $(0, 0)$ **25.** Absolute maxima at $(0, 1)$ and $(2, 1)$, absolute min at $(1, 0)$ **27.** Relative maximum at $(-2, -1/3)$, relative minimum at $(-1, -2/3)$, absolute maximum at $(0, 1)$ **29.** Relative min: $(-2, 5/3)$, relative max: $(0, -1)$, relative min: $(2, 5/3)$ **31.** Relative max: $(0, 0)$; absolute min: $(1/3, -2\sqrt{3}/9)$ **33.** Relative max: $(0, 0)$, absolute min: $(1, -3)$ **35.** No relative extrema   **37.** Absolute min: $(1, 1)$ **39.** Relative max: $(-1, 1 + 1/e)$, absolute min: $(0, 1)$, absolute max: $(1, e - 1)$ **41.** Relative max: $(-6, -24)$, relative min: $(-2, -8)$ **43.** Absolute max $(1/\sqrt{2}, \sqrt{e/2})$, absolute min: $(-1/\sqrt{2}, -\sqrt{e/2})$ **45.** Relative min at $(0.15, -0.52)$ and $(2.45, 8.22)$, relative max at $(1.40, 0.29)$ **47.** Absolute max at $(-5, 700)$, relative max at $(3.10, 28.19)$ and $(6, 40)$, absolute min at $(-2.10, -392.69)$ and relative min at $(5, 0)$.   **49.** Stationary minimum at $x = -1$ **51.** Stationary minima at $x = -2$ and $x = 2$, stationary maximum at $x = 0$ **53.** Singular minimum at $x = 0$, stationary nonextreme point at $x = 1$ **55.** Stationary minimum at $x = -2$, singular nonextreme points at $x = -1$ and $x = 1$, stationary maximum at $x = 2$
57. Answers will vary.   **59.** Answers will vary.

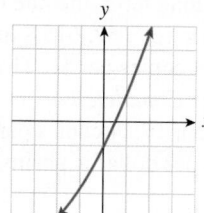   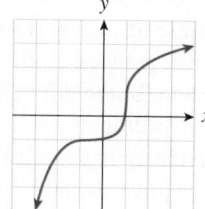

61. Not necessarily; it could be neither a relative maximum nor a relative minimum, as in the graph of $y = x^3$ at the origin.
63. Answers will vary.

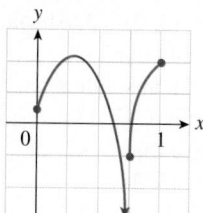

65. The graph oscillates faster and faster above and below zero as it approaches the end-point at 0, so 0 cannot be either a relative minimum or maximum.

Section 12.2

1. $x = y = 5$; $P = 25$ **3.** $x = y = 3$; $S = 6$
5. $x = 2$, $y = 4$; $F = 20$ **7.** $x = 20$, $y = 10$, $z = 20$; $P = 4,000$ **9.** 5×5 **11.** 1,500 per day for an average cost of $130 per iPod   **13.** $\sqrt{40} \approx 6.32$ pounds of pollutant per day, for an average cost of about $1,265 per pound   **15.** 2.5 lb   **17.** 5×10
19. 50×10 for an area of 250 sq. ft.   **21.** 11×22
23. $10   **25.** $78 for a quarterly revenue of $6,084 million, or $6.084 billion   **27.** $4.61 for a daily revenue of $95,680.55   **29. a.** $1.41 per pound
b. 5,000 pounds **c.** $7,071.07 per month
31. 34.5¢ per pound, for an annual (per capita) revenue of $5.95   **33.** $98 for an annual profit of $3,364 million, or $3.364 billion   **35. a.** 656 headsets, for a profit of $28,120 **b.** $143 per headset
37. Height $=$ Radius of base ≈ 20.48 cm.
39. Height ≈ 10.84 cm; Radius ≈ 2.71 cm; Height/Radius $= 4$ **41.** $13\frac{1}{3}$ in $\times\, 3\frac{1}{3}$ in $\times\, 1\frac{1}{3}$ in for a volume of $1,600/27 \approx 59$ cubic inches
43. $5 \times 5 \times 5$ cm   **45.** $l = w = h \approx 20.67$ in, volume $\approx 8,827$ in^3 **47.** $l = 30$ in, $w = 15$ in, $h = 30$ in   **49.** $l = 36$ in, $w = h = 18$ in, $V = 11,664$ in^3 **51. a.** 1.6 years, or year 2001.6;
b. $R_{max} = \$28,241$ million   **53.** $t = 2.5$ or midway through 1972; $D(2.5)/S(2.5) \approx 4.09$. The number of new (approved) drugs per $1 billion of spending on research and development reached a high of around four approved drugs per $1 billion midway through 1972.   **55.** 30 years from now   **57.** 55 days
59. 1,600 copies. At this value of x, average profit equals marginal profit; beyond this the marginal profit is smaller than the average.   **61.** 40 laborers and 250 robots   **63.** 71 employees   **65.** Increasing most rapidly in 1990; increasing least rapidly in 2007
67. Maximum when $t = 17$ days. This means that the embryo's oxygen consumption is increasing most rapidly 17 days after the egg is laid.
69. Graph of derivative:

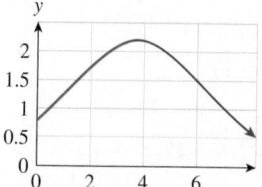

The absolute maximum occurs at approximately (3.7, 2.2) during the year 2003. The percentage of mortgages that were subprime was increasing most rapidly during 2003, when it increased at a rate of around 2.2 percentage points per year.   **71.** You should sell them in 17 years' time, when they will be worth approximately $3,960.   **73.** 25 additional trees   **75.** (D)   **77.** (A); (B)   **79.** The problem is uninteresting because the company can accomplish the objective by cutting away the entire sheet of cardboard, resulting in a box with surface area zero.   **81.** Not all absolute extrema occur at stationary points; some may occur at an endpoint or singular point of the domain, as in Exercises 29, 30, 65, and 66.   **83.** The minimum of dq/dp is the fastest that the demand is dropping in response to increasing price.

Section 12.3

1. 6   **3.** $4/x^3$ **5.** $-0.96x^{-1.6}$ **7.** $e^{-(x-1)}$
9. $2/x^3 + 1/x^2$ **11. a.** $a = -32$ ft/sec^2
b. $a = -32$ ft/sec^2 **13. a.** $a = 2/t^3 + 6/t^4$ ft/sec^2
b. $a = 8$ ft/sec^2 **15. a.** $a = -1/(4t^{3/2}) + 2$ ft/sec^2
b. $a = 63/32$ ft/sec^2 **17.** (1, 0)   **19.** (1, 0)   **21.** None
23. $(-1, 0), (1, 1)$ **25.** Points of inflection at $x = -1$ and $x = 1$ **27.** One point of inflection, at $x = -2$
29. Points of inflection at $x = -2, x = 0, x = 2$
31. Points of inflection at $x = -2$ and $x = 2$
33. $x = 2$; minimum   **35.** Maximum at $x = -2$, minimum at $x = 2$ **37.** Maximum at $t = -1/\sqrt{3}$, minimum at $t = 1/\sqrt{3}$ **39.** Nonextreme stationary point at $x = 0$ minimum at $x = 3$ **41.** Maximum at $x = 0$ **43.** Minimum at $x = -1/\sqrt{2}$; maximum at $x = 1/\sqrt{2}$ **45.** $f'(x) = 8x - 1$; $f''(x) = 8$; $f'''(x) = f^{(4)}(x) = \ldots = f^{(n)}(x) = 0$
47. $f'(x) = -4x^3 + 6x$; $f''(x) = -12x^2 + 6$; $f'''(x) = -24x$; $f^{(4)}(x) = -24$; $f^{(5)}(x) = f^{(6)}(x) = \ldots = f^{(n)}(x) = 0$
49. $f'(x) = 8(2x + 1)^3$; $f''(x) = 48(2x + 1)^2$; $f'''(x) = 192(2x + 1)$; $f^{(4)}(x) = 384$; $f^{(5)}(x) = f^{(6)}(x) = \ldots = f^{(n)}(x) = 0$
51. $f'(x) = -e^{-x}$; $f''(x) = e^{-x}$; $f'''(x) = -e^{-x}$; $f^{(4)}(x) = e^{-x}$; $f^{(n)}(x) = (-1)^n e^{-x}$
53. $f'(x) = 3e^{3x-1}$; $f''(x) = 9e^{3x-1}$; $f'''(x) = 27e^{3x-1}$; $f^{(4)}(x) = 81e^{3x-1}$; $f^{(n)}(x) = 3^n e^{3x-1}$ **55.** -3.8 m/s^2
57. $6t - 2$ ft/s^2; increasing   **59.** Decelerating by 90 million gals/yr^2 **61. a.** 400 ml   **b.** 36 ml/day
c. -1 ml/day^2 **63. a.** 0.6%   **b.** Speeding up
c. Speeding up for $t < 3.33$ (prior to 1/3 of the way through March) and slowing for $t > 3.33$ (after that time)   **65. a.** December 2005: -0.202% (deflation rate of 0.202%) February 2006: 0.363%   **b.** Speeding up

c. Speeding up for $t > 4.44$ (after mid-November) and decreasing for $t < 4.44$ (prior to that time).
67. Concave up for $8 < t < 20$, concave down for $0 < t < 8$, point of inflection around $t = 8$. The percentage of articles written by researchers in the United States was decreasing most rapidly at around $t = 8$ (1991).   **69. a.** (B)   **b.** (B)   **c.** (A)
71. Graphs:
$A(t)$:

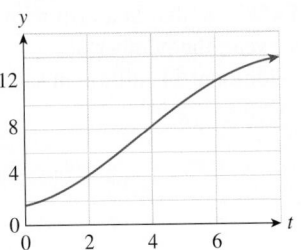

$A'(t)$:

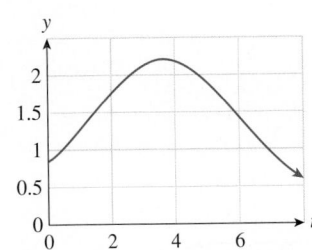

$A''(t)$:

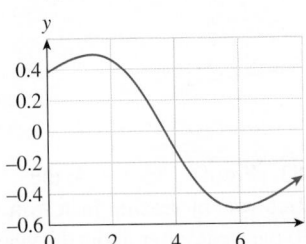

Concave up when $t < 4$; concave down when $t > 4$; point of inflection when $t \approx 4$. The percentage of U.S. mortgages that were subprime was increasing fastest at the beginning of 2004.   **73. a.** 2 years into the epidemic   **b.** 2 years into the epidemic   **75. a.** 2024   **b.** 2026   **c.** 2022; (A)   **77. a.** There are no points of inflection in the graph of S.   **b.** Because the graph is concave up, the derivative of S is increasing, and so the rate of *decrease* of SAT scores with increasing numbers of prisoners was diminishing. In other words, the apparent effect of more prisoners on SAT scores was diminishing.   **79. a.** $\left.\dfrac{d^2n}{ds^2}\right|_{s=3} = -21.494$. Thus, for a firm with annual sales of $3 million, the rate at which new patents are produced decreases with increasing firm size. This means that the returns (as measured in the number of new patents per increase of $1 million in sales) are diminishing as the firm size increases. **b.** $\dfrac{d^2n}{ds^2}\Big|_{s=7} = 13.474$. Thus, for a firm with annual sales of $7 million, the rate at which new patents are produced increases with increasing firm size by 13.474 new patents per $1 million increase in annual sales. **c.** There is a point of inflection when $s \approx 5.4587$, so that in a firm with sales of $5,458,700 per year, the number of new patents produced per additional $1 million in sales is a minimum.

81. Graphs:

$I(t)/P(t)$:

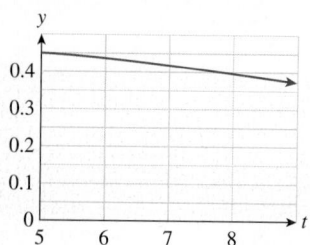

$[I(t)/P(t)]'$:

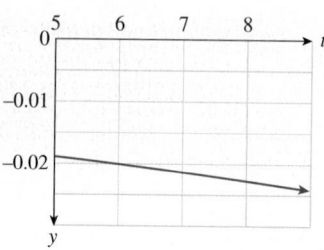

Concave down; (D)   **83.** (Proof)   **85.** $t \approx 4$; the population of Puerto Rico was increasing fastest in 1954.   **87.** About $570 per year, after about 12 years   **89.** Increasing most rapidly in 17.64 years, decreasing most rapidly now (at $t = 0$)   **91.** Non-negative   **93.** Daily sales were decreasing most rapidly in June 2002.

95.

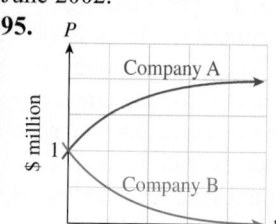

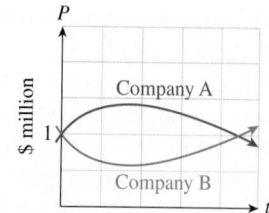

97. At a point of inflection, the graph of a function changes either from concave up to concave down, or vice versa. If it changes from concave up to concave down, then the derivative changes from increasing to decreasing, and hence has a relative maximum. Similarly, if it changes from concave down to concave up, the derivative has a relative minimum.

Section 12.4

1. a. x-intercept: -1; y-intercept: 1 **b.** Absolute min at $(-1, 0)$ **c.** None **d.** None **e.** $y \to +\infty$ as $x \to \pm\infty$
Graph:

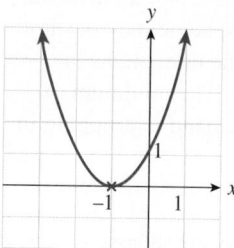

3. a. x-intercepts: $-\sqrt{12}, 0, \sqrt{12}$; y-intercept: 0 **b.** Absolute min at $(-4, -16)$ and $(2, -16)$, absolute max at $(-2, 16)$ and $(4, 16)$ **c.** $(0, 0)$ **d.** None **e.** None
Graph:

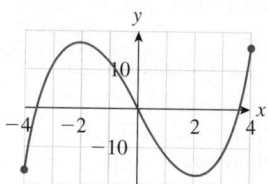

5. a. x-intercepts: $-3.6, 0, 5.1$; y-intercept: 0 **b.** Relative max at $(-2, 44)$, relative min at $(3, -81)$ **c.** $(0.5, -18.5)$ **d.** None **e.** $y \to -\infty$ as $x \to -\infty$; $y \to +\infty$ as $x \to +\infty$
Graph:

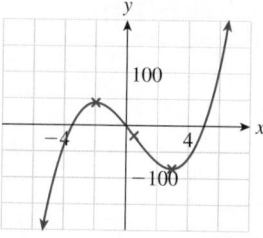

7. a. x-intercepts: $-3.3, 0.1, 1.8$; y-intercept: 1 **b.** Relative max at $(-2, 21)$, relative min at $(1, -6)$ **c.** $(-1/2, 15/2)$ **d.** None **e.** $y \to -\infty$ as $x \to -\infty$; $y \to +\infty$ as $x \to +\infty$

Graph:

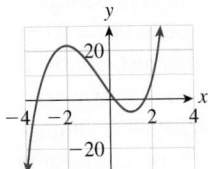

9. a. x-intercepts: -2.9, 4.2; y-intercept: 10
b. Relative max at $(-2, 74)$, relative min at $(0, 10)$, absolute max at $(3, 199)$ **c.** $(-1.12, 44.8)$, $(1.79, 117.3)$
d. None **e.** $y \to -\infty$ as $x \to -\infty$; $y \to -\infty$ as $x \to +\infty$
Graph:

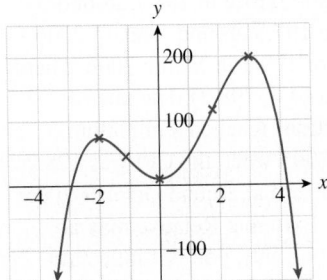

11. a. t-intercepts: $t = 0$; y-intercept: 0 **b.** Absolute min at $(0, 0)$ **c.** $(1/3, 11/324)$ and $(1, 1/12)$ **d.** None
e. $y \to +\infty$ as $t \to -\infty$; $y \to +\infty$ as $t \to +\infty$
Graph:

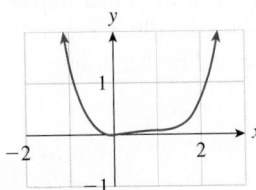

13. a. x-intercepts: None; y-intercept: None
b. Relative min at $(1, 2)$, relative max at $(-1, -2)$
c. None **d.** $y \to -\infty$ as $x \to 0^-$; $y \to +\infty$ as $x \to 0^+$, so there is a vertical asymptote at $x = 0$. **e.** $y \to -\infty$ as $x \to -\infty$; $y \to +\infty$ as $x \to +\infty$
Graph:

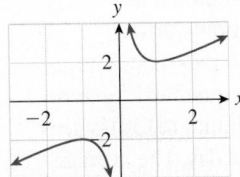

15. a. x-intercept: 0; y-intercept: 0 **b.** None **c.** $(0, 0)$, $(-3, -9/4)$, and $(3, 9/4)$ **d.** None **e.** $y \to -\infty$ as $x \to -\infty$; $y \to +\infty$ as $x \to +\infty$

Graph:

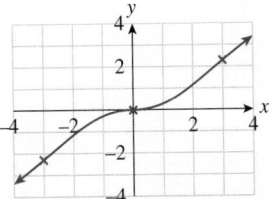

17. a. t-intercepts: None; y-intercept: -1 **b.** Relative min at $(-2, 5/3)$ and $(2, 5/3)$, relative max at $(0, -1)$ **c.** None **d.** $y \to +\infty$ as $t \to -1^-$; $y \to -\infty$ as $t \to -1^+$; $y \to -\infty$ as $t \to 1^-$; $y \to +\infty$ as $t \to 1^+$; so there are vertical asymptotes at $t = \pm 1$. **e.** None
Graph:

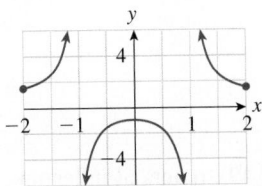

19. a. x-intercepts: -0.6; y-intercept: 1 **b.** Relative maximum at $(-2, -1/3)$, relative minimum at $(-1, -2/3)$ **c.** None **d.** None. **e.** $y \to -\infty$ as $x \to -\infty$; $y \to +\infty$ as $x \to +\infty$
Graph:

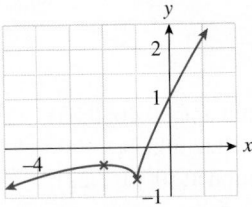

21. a. x-intercepts: None; y-intercept: None
b. Absolute min at $(1, 1)$ **c.** None **d.** Vertical asymptote at $x = 0$ **e.** $y \to +\infty$ as $x \to +\infty$
Graph:

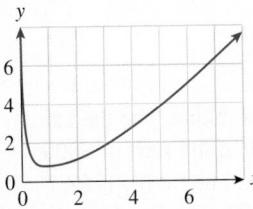

23. a. x-intercepts: ± 0.8; x-intercept: None **b.** None
c. $(1, 1)$ and $(-1, 1)$ **d.** $y \to -\infty$ as $x \to 0$; vertical asymptote at $x = 0$ **e.** $y \to +\infty$ as $x \to \pm\infty$

Graph:

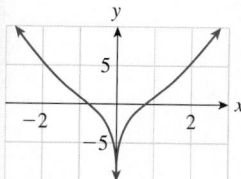

25. a. *t*-intercepts: None; *y*-intercept: 1 **b.** Absolute
min at (0, 1). Absolute max at $(1, e - 1)$, relative max
at $(-1, e^{-1} + 1)$. **c.** None **d.** None **e.** None
Graph:

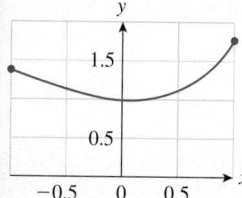

27. Absolute min at $(1.40, -1.49)$; points of inflection:
$(0.21, 0.61)$, $(0.79, -0.55)$
Graph:

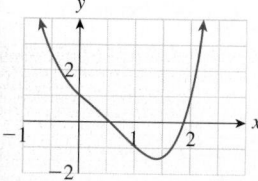

29. $f(x) = e^x - x^3$. Relative min at $(-0.46, 0.73)$,
relative max at $(0.91, 1.73)$, absolute min at
$(3.73, -10.22)$; points of inflection at $(0.20, 1.22)$
and $(2.83, -5.74)$
Graph:

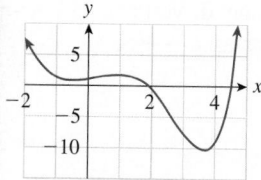

31. *y*-intercept: 125; *t*-intercepts: None. The median
home price was about $125,000 at the start of 2000
$(t = 0)$. Extrema: Absolute minimum at (0, 125);
absolute maximum at (6, 225). The median home price
was lowest in 2000 $(t = 0)$ when it stood at $125,000;
the median home price was at its highest at the start of

2006 $(t = 6)$ at $225,000. Points of inflection at
(3, 175) and (10, 200). The median home price was
increasing most rapidly at the start of 2003 $(t = 3)$
when it was $175,000, and decreasing most rapidly at
the start of 2010 when it was $200,000. Points where
the function is not defined: None. Behavior at infinity:
As $t \to +\infty$, $y \to 175$. Assuming the trend shown in
the graph continues indefinitely, the median home
price will approach a value of $175,000 in the long
term.   **33. a.** Intercepts: No *t*-intercept; *y*-intercept at
$I(0) = 195$. The CPI was never zero during the given
period; in July 2005 the CPI was 195. Absolute min at
(0, 195), absolute max at (8, 199.3), relative max at
(2.9, 198.7), relative min at (6.0, 197.8). The CPI was
at a low of 195 in July 2005, rose to 198.7 around
October 2005, dipped to 197.8 around January 2006,
and then rose to a high of 199.3 in March 2006. There
is a point of inflection at (4.4, 198.2). The rate of
change of the CPI (inflation) reached a minimum
around mid-November 2005 when the CPI was 198.2.
b. The inflation rate was zero at around October 2005
and January 2006.   **35.** Extrema: Relative max at
(0, 100), absolute min at (1, 99); point of inflection
(0.5, 99.5); $s \to +\infty$ as $t \to +\infty$. At time $t = 0$
seconds, the UFO is 100 ft away from the observer,
and begins to move closer. At time $t = 0.5$ seconds,
when the UFO is 99.5 feet away, its distance is
decreasing most rapidly (it is moving toward the
observer most rapidly). It then slows down to a stop at
$t = 1$ sec when it is at its closest point (99 ft away)
and then begins to move further and further away.
Graph:

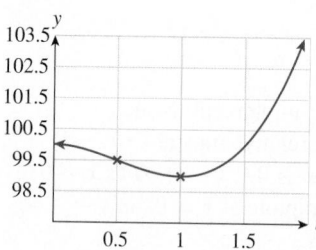

37. Intercepts: None; Absolute minimum at
(1,500, 130); No points of inflection; vertical
asymptote at $x = 0$. As $x \to +\infty$, $y \to +\infty$. The
average cost is never zero, nor is it defined for zero
iPods. The average cost is a minimum ($130) when
1,500 iPods are manufactured per day. The average
cost becomes extremely large for very small or very
large numbers of iPods.

Graph:

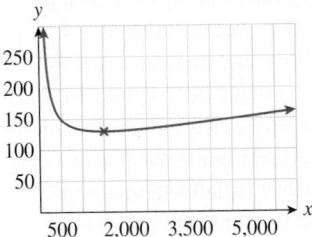

39. Graph of derivative:

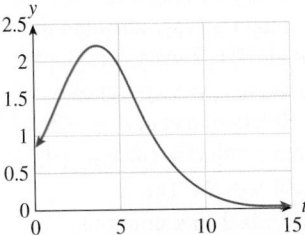

The absolute maximum occurs at approximately (3.7, 2.2); during the year 2003. The percentage of mortgages that were subprime was increasing most rapidly during 2003, when it increased at a rate of around 2.2 percentage points per year. As $t \to +\infty$, $A'(t) \to 0$; In the long term, assuming the trend shown in the model continues, the rate of change of the percentage of mortgages that were subprime approaches zero; that is, the percentage of mortgages that were subprime approaches a constant value. **41.** No; yes. Near a vertical asymptote the value of y increases without bound, and so the graph could not be included between two horizontal lines; hence, no vertical asymptotes are possible. Horizontal asymptotes are possible, as for instance in the graph in Exercise 31.   **43.** It too has a vertical asymptote at $x = a$; the magnitude of the derivative increases without bound as $x \to a$.   **45.** No. If the leftmost critical point is a relative maximum, the function will decrease from there until it reaches the rightmost critical point, so can't have a relative maximum there. **47.** Between every pair of zeros of $f(x)$ there must be a local extremum, which must be a stationary point of $f(x)$, hence a zero of $f'(x)$.

Section 12.5

1. $P = 10,000$; $\dfrac{dP}{dt} = 1,000$ **3.** Let R be the annual revenue of my company, and let q be annual sales. $R = 7,000$ and $\dfrac{dR}{dt} = -700$. Find $\dfrac{dq}{dt}$.   **5.** Let p be the price of a pair of shoes, and let q be the demand

for shoes. $\dfrac{dp}{dt} = 5$. Find $\dfrac{dq}{dt}$.   **7.** Let T be the average global temperature, and let q be the number of Bermuda shorts sold per year. $T = 60$ and $\dfrac{dT}{dt} = 0.1$. Find $\dfrac{dq}{dt}$.   **9. a.** $6/(100\pi) \approx 0.019$ km/sec   **b.** $6/(8\sqrt{\pi}) \approx 0.4231$ km/sec   **11.** $3/(4\pi) \approx 0.24$ ft/min   **13.** 7.5 ft/sec   **15.** Decreasing at a rate of \$1.66 per player per week   **17.** Monthly sales will drop at a rate of 26 T-shirts per month.   **19.** Raise the price by 3¢ per week.   **21.** Increasing at a rate of \$233.68 million per year   **23.** 1 laborer per month **25.** Dropping at a rate of \$2.40 per year.   **27.** The price is decreasing at a rate of approximately 31¢ per pound per month.   **29.** $2,300/\sqrt{4,100} \approx 36$ miles/hour. **31.** About 10.7 ft/sec   **33.** The y coordinate is decreasing at a rate of 16 units per second. **35.** \$534 per year   **37.** Their prior experience must increase at a rate of approximately 0.97 years every year.   **39.** $\dfrac{2,500}{9\pi}\left(\dfrac{3}{5,000}\right)^{2/3} \approx 0.63$ m/sec

41. $\dfrac{\sqrt{1 + 128\pi}}{4\pi} \approx 1.6$ cm/sec   **43.** 0.5137 computers per household, and increasing at a rate of 0.0230 computers per household per year.   **45.** The average SAT score was 904.71 and decreasing at a rate of 0.11 per year.   **47.** Decreasing by 2 percentage points per year **49.** The section is called "related rates" because the goal is to compute the rate of change of a quantity based on a knowledge of the rate of change of a related quantity.   **51.** Answers may vary: A rectangular solid has dimensions 2 cm × 5 cm × 10 cm, and each side is expanding at a rate of 3 cm/second. How fast is the volume increasing?   **53.** (Proof)   **55.** Linear **57.** Let $x =$ my grades and $y =$ your grades. If $dx/dt = 2\, dy/dt$, then $dy/dt = (1/2)\, dx/dt$.

Section 12.6

1. $E = 1.5$; the demand is going down 1.5% per 1% increase in price at that price level; revenue is maximized when $p = \$25$; weekly revenue at that price is \$12,500.   **3. a.** $E = 6/7$; the demand is going down 6% per 7% increase in price at that price level; thus, a price increase is in order.   **b.** Revenue is maximized when $p = 100/3 \approx \$33.33$ **c.** 4,444 cases per week **5. a.** $E = (4p - 33)/(-2p + 33)$ **b.** 0.54; the demand for $E = mc^2$ T-shirts is going down by about 0.54% per 1% increase in the price.   **c.** \$11 per shirt for a daily revenue of \$1,331   **7. a.** $E = 1.81$. Thus, the demand is elastic at the given tuition level, showing that a decrease in tuition will result in an increase in revenue. **b.** They should charge an average of $2,250 per student, and this will result in an enrollment of about 4,950 students, giving a revenue of about $11,137,500. **9. a.** $E = 51$; the demand is going down 51% per 1% increase in price at that price level; thus, a large price decrease is advised. **b.** ¥50　**c.** About 78 paint-by-number sets per month

11. a. $E = -\dfrac{mp}{mp + b}$　**b.** $p = -\dfrac{b}{2m}$　**13. a.** $E = r$

b. E is independent of p. **c.** If $r = 1$, then the revenue is not affected by the price. If $r > 1$, then the revenue is always elastic, while if $r < 1$, the revenue is always inelastic. This is an unrealistic model because there should always be a price at which the revenue is a maximum.　**15. a.** $q = -1,500p + 6,000$. **b.** $2 per hamburger, giving a total weekly revenue of $6,000　**17.** $E \approx 0.77$. At a family income level of $20,000, the fraction of children attending a live theatrical performance is increasing by 0.77% per 1% increase in household income.　**19. a.** $E = \dfrac{1.554xe^{-0.021x}}{-74e^{-0.021x} + 92}$;

$E(100) \approx 0.23$: At a household income level of $100,000, the percentage of people using broadband in 2010 was increasing by 0.23% per 1% increase in household income. **b.** The model predicts elasticity approaching zero for households with large incomes.　**21. a.** $E \approx 0.46$. The demand for computers is increasing by 0.46% per 1% increase in household income. **b.** E decreases as income increases. **c.** Unreliable; it predicts a likelihood greater than 1 at incomes of $123,000 and above. In a more appropriate model, we would expect the curve to level off at or below 1. **d.** $E \approx 0$

23. The income elasticity of demand is
$$\dfrac{dQ}{dY} \cdot \dfrac{Y}{Q} = a\beta P^{\alpha} Y^{\beta-1} \dfrac{Y}{aP^{\alpha}Y^{\beta}} = \beta.$$

25. a. $q = 1,000e^{-0.30p}$　**b.** At $p = \$3$, $E = 0.9$; at $p = \$4$, $E = 1.2$; at $p = \$5$, $E = 1.5$　**c.** $p = \$3.33$ **d.** $p = \$5.36$. Selling at a lower price would increase demand, but you cannot sell more than 200 pounds anyway. You should charge as much as you can and still be able to sell all 200 pounds.　**27.** The price is lowered.　**29.** Start with $R = pq$, and differentiate

with respect to p to obtain $\dfrac{dR}{dp} = q + p\dfrac{dq}{dp}$. For a

stationary point, $dR/dp = 0$, and so $q + p\dfrac{dq}{dp} = 0$.

Rearranging this result gives $p\dfrac{dq}{dp} = -q$, and hence

$-\dfrac{dq}{dp} \cdot \dfrac{p}{q} = 1$, or $E = 1$, showing that stationary

points of R correspond to points of unit elasticity.
31. The distinction is best illustrated by an example. Suppose that q is measured in weekly sales and p is the unit price in dollars. Then the quantity $-dq/dp$ measures the drop in weekly sales per $1 increase in price. The elasticity of demand E, on the other hand, measures the *percentage* drop in sales per 1% increase in price. Thus, $-dq/dp$ measures absolute change, while E measures fractional, or percentage, change.

Chapter 12 Review

1. Relative max: $(-1, 5)$, absolute min: $(-2, -3)$ and $(1, -3)$　**3.** Absolute max: $(-1, 5)$, absolute min: $(1, -3)$　**5.** Absolute min: $(1, 0)$　**7.** Absolute min: $(-2, -1/4)$　**9.** Relative max at $x = 1$, point of inflection at $x = -1$　**11.** Relative max at $x = -2$, relative min at $x = 1$, point of inflection at $x = -1$ **13.** One point of inflection, at $x = 0$

15. a. $a = 4/t^4 - 2/t^3$ m/sec² **b.** 2 m/sec²
17. Relative max: $(-2, 16)$; absolute min: $(2, -16)$; point of inflection: $(0, 0)$; no horizontal or vertical asymptotes

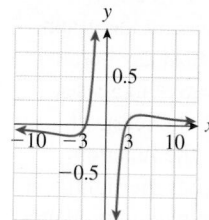

19. Relative min: $(-3, -2/9)$; relative max: $(3, 2/9)$; inflection: $(-3\sqrt{2}, -5\sqrt{2}/36)$, $(3\sqrt{2}, 5\sqrt{2}/36)$; vertical asymptote: $x = 0$; horizontal asymptote: $y = 0$

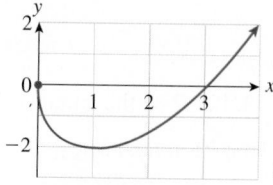

21. Relative max at $(0, 0)$, absolute min at $(1, -2)$, no asymptotes

23. $22.14 per book　**25. a.** Profit $= -p^3 + 42p^2 - 288p - 181$　**b.** $24 per copy; $3,275　**c.** For

maximum revenue, the company should charge $22.14 per copy. At this price, the cost per book is decreasing with increasing price, while the revenue is not decreasing (its derivative is zero). Thus, the profit is increasing with increasing price, suggesting that the maximum profit will occur at a higher price.
27. 12 in × 12 in × 6 in, for a volume of 864 in^3

29. a. $E = \dfrac{2p^2 - 33p}{-p^2 + 33p + 9}$ **b.** 0.52, 2.03; when the

price is $20, demand is dropping at a rate of 0.52% per 1% increase in the price; when the price is $25, demand is dropping at a rate of 2.03% per 1% increase in the price. **c.** $22.14 per book **31. a.** $E = 2p^2 - p$
b. 6; the demand is dropping at a rate of 6% per 1% increase in the price. **c.** $1.00, for a monthly revenue of $1,000 **33. a.** Week 5 **b.** Point of inflection on the graph of s; maximum on the graph of s', t-intercept in the graph of s''. **c.** 10,500; if weekly sales continue as predicted by the model, they will level off at around 10,500 books per week in the long term. **d.** 0; if weekly sales continue as predicted by the model, the rate of change of sales approaches zero in the long term. **35. a.–d.** $10/\sqrt{2}$ ft/sec

Chapter 13

Section 13.1

1. $x^6/6 + C$ **3.** $6x + C$ **5.** $x^2/2 + C$
7. $x^3/3 - x^2/2 + C$ **9.** $x + x^2/2 + C$
11. $-x^{-4}/4 + C$
13. $x^{3.3}/3.3 - x^{-0.3}/0.3 + C$ **15.** $u^3/3 - \ln|u| + C$
17. $\dfrac{4x^{5/4}}{5} + C$ **19.** $3x^5/5 + 2x^{-1} - x^{-4}/4 + 4x + C$
21. $2\ln|u| + u^2/8 + C$ **23.** $\ln|x| - \dfrac{2}{x} + \dfrac{1}{2x^2} + C$
25. $3x^{1.1}/1.1 - x^{5.3}/5.3 - 4.1x + C$
27. $\dfrac{x^{0.9}}{0.3} + \dfrac{40}{x^{0.1}} + C$
29. $2.55t^2 - 1.2\ln|t| - \dfrac{15}{t^{0.2}} + C$
31. $2e^x + 5x|x|/2 + x/4 + C$
33. $12.2x^{0.5} + x^{1.5}/9 - e^x + C$ **35.** $\dfrac{2^x}{\ln 2} - \dfrac{3^x}{\ln 3} + C$
37. $\dfrac{100(1.1^x)}{\ln(1.1)} - \dfrac{x|x|}{3} + C$ **39.** $x^3/3 + 3x^{-5}/10 + C$
41. $1/x - 1/x^2 + C$ **43.** $f(x) = x^2/2 + 1$
45. $f(x) = e^x - x - 1$
47. $C(x) = 5x - x^2/20,000 + 20,000$
49. $C(x) = 5x + x^2 + \ln x + 994$

51. a. $M(t) = 0.8t^4 - 6t^2 + 10t + 1$
b. 83 million members
53. $I(t) = 42,000 + 1200t$; $48,000
55. $S(t) = -15t^3 + 450t^2 + 4,200t$; 54,720 million gallons **57. a.** $H'(t) = 3.5t + 65$ billion dollars per year **b.** $H(t) = 1.75t^2 + 65t + 700$ billion dollars
59. a. $-0.4t + 1$ percentage points per year
b. $-0.2t^2 + t + 13$; 13.8 **61. a.** $s = t^3/3 + t + C$
b. $C = 1$; $s = t^3/3 + t + 1$ **63.** 320 ft/sec
downward **65. a.** $v(t) = -32t + 16$ **b.** $s(t) =$
$-16t^2 + 16t + 185$; zenith at $t = 0.5$ sec $s = 189$ feet, 4 feet above the top of the tower.
67. a. $s = 525t + 25t^2$ **b.** 3 hours **c.** The negative solution indicates that, at time $t = -24$, the tail wind would have been large and negative, causing the plane to be moving backward through that position 24 hours prior to departure and arrive at the starting point of the flight at time 0! **69.** (Proof) **71.** $(1,280)^{1/2} \approx$
35.78 ft/sec **73. a.** 80 ft/sec **b.** 60 ft/sec
c. 1.25 seconds **75.** $\sqrt{2} \approx 1.414$ times as fast
77. The term *indefinite* refers to the arbitrary constant term in the indefinite integral; we do not obtain a definite value for C and hence the integral is "not definite." **79.** Constant; because the derivative of a linear function is constant, linear functions are antiderivatives of constant functions. **81.** No; there are infinitely many antiderivatives of a given function, each pair of them differing by a constant. Knowing the value of the function at a specific point suffices.
83. They differ by a constant, $G(x) - F(x) =$
Constant **85.** Antiderivative, marginal **87.** Up to a constant, $\int f(x)\,dx$ represents the total cost of manufacturing x items. The units of $\int f(x)\,dx$ are the product of the units of $f(x)$ and the units of x.
89. The indefinite integral of the constant 1 is not zero; it is x (+ constant). Correct answer: $3x^2/2 + x + C$
91. There should be no integral sign ($\int$) in the answer. Correct answer: $2x^6 - 2x^2 + C$ **93.** The integral of a constant times a function is the constant times the integral of the function; not the *integral* of the constant times the integral of the function. (In general, the integral of a product is *not* the product of the integrals.) Correct answer: $4(e^x - x^2) + C$ **95.** It is the *integral* of $1/x$ that equals $\ln|x| + C$, and not $1/x$ itself. Correct answer: $\int \frac{1}{x}dx = \ln|x| + C$
97. $\int (f(x) + g(x))\,dx$ is, by definition, an antiderivative of $f(x) + g(x)$. Let $F(x)$ be an antiderivative of $f(x)$ and let $G(x)$ be an antiderivative of $g(x)$. Then, because the derivative of $F(x) + G(x)$ is $f(x) + g(x)$ (by the rule for sums of derivatives), this means that $F(x) + G(x)$ is an antiderivative of $f(x) + g(x)$. In

symbols, $\int (f(x) + g(x))\,dx = F(x) + G(x) + C =$ $\int f(x)\,dx + \int g(x)\,dx$, the sum of the indefinite integrals.   **99.** Answers will vary. $\int x \cdot 1\,dx =$ $\int x\,dx = x^2/2 + C$, whereas $\int x\,dx \cdot \int 1\,dx =$ $(x^2/2 + D) \cdot (x + E)$, which is not the same as $x^2/2 + C$, no matter what values we choose for the constants C, D, and E.   **101.** Derivative; indefinite integral; indefinite integral; derivative

Section 13.2

1. $(3x - 5)^4/12 + C$ **3.** $(3x - 5)^4/12 + C$

5. $-e^{-x} + C$ **7.** $-e^{-x} + C$ **9.** $\dfrac{1}{2}e^{(x+1)^2} + C$

11. $(3x + 1)^6/18 + C$ **13.** $1.6(3x - 4)^{3/2} + C$

15. $2e^{(0.6x+2)} + C$ **17.** $(3x^2 + 3)^4/24 + C$

19. $2(3x^2 - 1)^{3/2}/9 + C$ **21.** $-(x^2 + 1)^{-0.3}/0.6 + C$

23. $(4x^2 - 1)|4x^2 - 1|/16 + C$ **25.** $x + 3e^{3.1x-2} + C$

27. $-(1/2)e^{-x^2+1} + C$ **29.** $-(1/2)e^{-(x^2+2x)} + C$

31. $(x^2 + x + 1)^{-2}/2 + C$

33. $(2x^3 + x^6 - 5)^{1/2}/3 + C$

35. $(x - 2)^7/7 + (x - 2)^6/3 + C$

37. $4[(x + 1)^{5/2}/5 - (x + 1)^{3/2}/3] + C$

39. $20\ln|1 - e^{-0.05x}| + C$ **41.** $3e^{-1/x} + C$

43. $-\dfrac{(4 + 1/x^2)^4}{8} + C$ **45.** $(e^x - e^{-x})/2 + C$

47. $\ln(e^x + e^{-x}) + C$

49. $(1 - e^{3x-1})|1 - e^{3x-1}|/6 + C$

51. $(e^{2x^2-2x} + e^{x^2})/2 + C$ **53.** Derivation

55. Derivation   **57.** $-e^{-x} + C$ **59.** $(1/2)e^{2x-1} + C$

61. $(2x + 4)^3/6 + C$ **63.** $(1/5)\ln|5x - 1| + C$

65. $(1.5x)^4/6 + C$ **67.** $\dfrac{1.5^{3x}}{3\ln(1.5)} + C$

69. $\dfrac{1}{4}(2x + 4)|2x + 4| + C$ **71.** $\dfrac{2^{3x+4} - 2^{-3x+4}}{3\ln 2} + C$

73. $f(x) = (x^2 + 1)^4/8 - 1/8$

75. $f(x) = (1/2)e^{x^2-1}$ **77.** $(5x^2 - 3)^7/70 + C$

79. $2(3e^x - 1)^{1/2}/3 + C$ **81.** $e^{(x^4-8)}/4 + C$

83. $\dfrac{1}{6}\ln|1 + 2e^{3x}| + C$

85. $210t + 1{,}240e^{-0.05t} - 1{,}240$

87. $8e^{0.25t} - 35.854$; \$62 billion

89. $C(x) = 5x - 1/(x + 1) + 995.5$

91. a. $N(t) = 35\ln(5 + e^{0.2t}) - 63$ **b.** 80,000 articles

93. $S(t) = 3{,}600[\ln(3 + e^{0.25t}) - \ln 4]$; 6,310 sets

95. a. $s = (t^2 + 1)^5/10 + t^2/2 + C$ **b.** $C = 9/10$; $s = (t^2 + 1)^5/10 + t^2/2 + 9/10$

97. a. $S(t) = -15(t - 2{,}000)^3 + 450(t - 2{,}000)^2 + 4{,}200(t - 2{,}000) - 16{,}245$ million gallons **b.** 38,475 million gallons   **99.** None; the substitution $u = x$ simply replaces the letter x throughout by the letter u,

and thus does not change the integral at all. For instance, the integral $\int x(3x^2 + 1)\,dx$ becomes $\int u(3u^2 + 1)\,du$ if we substitute $u = x$.   **101.** It may mean that, but it may not; see Example 4.   **103.** (D); to compute the integral, first break it up into a sum of two integrals: $\int \dfrac{x}{x^2 - 1}\,dx + \int \dfrac{3x}{x^2 + 1}\,dx$, and then compute the first using $u = x^2 - 1$ and the second using $u = x^2 + 1$.   **105.** There are several errors: First, the term "dx" is missing in the integral, and this affects all the subsequent steps (since we must substitute for dx when changing to the variable u). Second, when there is a noncanceling x in the integrand, we cannot treat it as a constant. Correct answer: $3(x^2 - 1)^2/4 + C$ **107.** In the fourth step, u was substituted back for x before the integral was taken and du was just changed to dx; they're not equal. Correct answer: $(x^3 - 1)^2/6 + C$ **109.** Proof

Section 13.3

1. 4   **3.** 6   **5.** 0.7456   **7.** 2.3129   **9.** 2.5048   **11.** 30   **13.** 22   **15.** -2 **17.** 0   **19.** 1   **21.** 1/2   **23.** 1/4   **25.** 2   **27.** 0   **29.** 6   **31.** 0   **33.** 0.5   **35.** 3.3045, 3.1604, 3.1436   **37.** 0.0275, 0.0258, 0.0256   **39.** 15.76 liters   **41.** \$99.95   **43.** 41 billion gallons   **45.** \$19 billion   **47. a.** Left sum: about 46,000 articles, right sum: about 55,000 articles   **b.** 50.5; a total of about 50,500 articles in *Physical Review* were written by researchers in Europe in the 16-year period beginning 1983.   **49.** 118,750 degrees   **51.** 54,000 students   **53.** Left sum: 76; right sum: 71; Left sum, as it is the sum of the annual net incomes for the given years.   **55. a.** Left sum $= 15.6$; right sum $= 15.9$ **b.** Pemex sold a total of 15.9 billion tons of petrochemicals in the period 2006 through 2009.   **57.** -84.8; after 4 seconds, the stone is about 84.8 ft below where it started.   **59.** 91.2 ft   **61.** 394; A total of about 394 million members joined Facebook from the start of 2005 to the start of 2010.   **63.** 18,400; a total of 18,400 wiretaps were authorized by U.S. state and federal courts during the 15-year period starting January 1990.   **65.** Yes. The Riemann sum gives an estimated area of 420 square feet.

67. a. Graph:

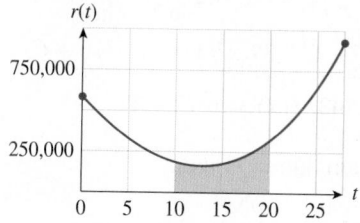

Total expenditure on oil in the United States from 1990 to 2000. **b.** 2,010,000; a total of $2,010,000 million, or $2.01 trillion, was spent on oil in the U.S. from 1990 to 2000.   **69. a.** 99.4% **b.** 0 (to at least 15 decimal places)   **71.** Stays the same   **73.** Increases   **75.** The area under the curve and above the x axis equals the area above the curve and below the x axis.   **77.** Answers will vary. One example: Let $r(t)$ be the rate of change of net income at time t. If $r(t)$ is negative, then the net income is decreasing, so the change in net income, represented by the definite integral of $r(t)$, is negative.
79. Answers may vary.

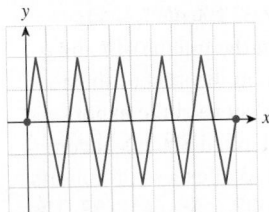

81. The total cost is $c(1) + c(2) + \cdots + c(60)$, which is represented by the Riemann sum approximation of $\int_1^{61} c(t)\,dt$ with $n = 60$.   **83.** $[f(x_1) + f(x_2) + \cdots + f(x_n)]\Delta x = \sum_{k=1}^{n} f(x_k)\Delta x$

85. Answers may vary:

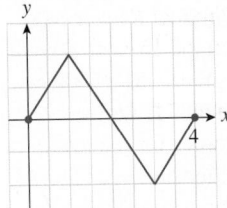

Section 13.4

1. 14/3   **3.** 5   **5.** 0   **7.** 40/3   **9.** −0.9045
11. $2(e - 1)$ **13.** 2/3   **15.** $1/\ln 2$
17. $4^6 - 1 = 4{,}095$ **19.** $(e^1 - e^{-3})/2$ **21.** $3/(2\ln 2)$
23. 40/3   **25.** 5/4   **27.** 0   **29.** $(5/2)(e^3 - e^2)$

31. $(1/3)[\ln 26 - \ln 7]$ **33.** $\dfrac{0.1}{2.2\ln(1.1)}$

35. $e - e^{1/2}$ **37.** 0.2221   **39.** $2 - \ln 3$ **41.** −4/21
43. $3^{5/2}/10 - 3^{3/2}/6 + 1/15$ **45.** 1/2   **47.** 16/3
49. 9/2   **51.** 56/3   **53.** 1/2   **55.** $783
57. 296 miles   **59.** 260 ft   **61.** 54,720 million gallons
63. 391.2 million members   **65.** 18,470; a total of about 18,470 wiretaps were authorized by U.S. state and federal courts during the 15-year period starting January 1990.   **67.** $210t + 1{,}240e^{-0.05t} - 1{,}240$
69. 9 gallons   **71.** 907 T-shirts   **73.** 68 milliliters

75. 20.48; The total value of vehicle transactions on eBay during 2008 and 2009 was about $20.48 billion.
77. Change in cost $= \int_0^x m(t)\,dt = C(x) - C(0)$ by the FTC, so $C(x) = C(0) + \int_0^x m(t)\,dt$. $C(0)$ is the *fixed cost.*   **79.** 80,000 articles   **81. a., b.** (Proofs)
c. 2,401 thousand students   **85.** They are related by the Fundamental Theorem of Calculus, which (briefly) states that the definite integral of a suitable function can be calculated by evaluating the indefinite integral at the two endpoints and subtracting.
87. Computing its definite integral from a to b.
89. Calculate definite integrals by using antiderivatives.   **91.** An example is $v(t) = t - 5$.
93. An example is $f(x) = e^{-x}$.   **95.** By the FTC, $\int_a^x f(t)\,dt = G(x) - G(a)$, where G is an antiderivative of f. Hence, $F(x) = G(x) - G(a)$. Taking derivatives of both sides, $F'(x) = G'(x) + 0 = f(x)$, as required. The result gives us a formula, in terms of area, for an antiderivative of any continuous function.

Chapter 13 Review

1. $x^3/3 - 5x^2 + 2x + C$ **3.** $4x^3/15 + 4/(5x) + C$
5. $(1/2)\ln|2x| + C$ or $(1/2)\ln|x| + C$

7. $-e^{-2x+11}/2 + C$ **9.** $(x^2 + 1)^{2.3}/4.6 + C$

11. $2\ln|x^2 - 7| + C$ **13.** $\dfrac{1}{6}(x^4 - 4x + 1)^{3/2} + C$

15. $-e^{x^2/2} + C$ **17.** $(x + 2) - \ln|x + 2| + C$ or $x - \ln|x + 2| + C$ **19.** 1   **21.** 2.75
23. −0.24   **25.** 0.7778, 0.7500, 0.7471
27. 0   **29.** 1   **31.** 2/7   **33.** $50(e^{-1} - e^{-2})$
35. 52/9   **37.** $[\ln 5 - \ln 2]/8 = \ln(2.5)/8$
39. 32/3   **41.** $(1 - e^{-25})/2$
43. a. $N(t) = 196t + t^2/2 - 0.16t^6/6$ **b.** 4 books
45. a. At a height of $-16t^2 + 100t$ ft   **b.** 156.25 ft
c. After 6.25 seconds   **47.** 25,000 copies.   **49.** 0 books
51. 39,200 hits   **53.** $8,200   **55.** About 86,000 books
57. About 35,800 books

Chapter 14

Section 14.1

1. $2e^x(x - 1) + C$ **3.** $-e^{-x}(2 + 3x) + C$
5. $e^{2x}(2x^2 - 2x - 1)/4 + C$
7. $-e^{-2x+4}(2x^2 + 2x + 3)/4 + C$
9. $2^x[(2 - x)/\ln 2 + 1/(\ln 2)^2] + C$
11. $-3^{-x}[(x^2 - 1)/\ln 3 + 2x/(\ln 3)^2 + 2/(\ln 3)^3] + C$
13. $-e^{-x}(x^2 + x + 1) + C$

15. $\dfrac{1}{7}x(x + 2)^7 - \dfrac{1}{56}(x + 2)^8 + C$

17. $-\dfrac{x}{2(x-2)^2} - \dfrac{1}{2(x-2)} + C$

19. $(x^4 \ln x)/4 - x^4/16 + C$

21. $(t^3/3 + t)\ln(2t) - t^3/9 - t + C$

23. $(3/4)t^{4/3}(\ln t - 3/4) + C$

25. $x \log_3 x - x/\ln 3 + C$

27. $e^{2x}(x/2 - 1/4) - 4e^{3x}/3 + C$

29. $e^x(x^2 - 2x + 2) - e^{x^2}/2 + C$

31. $\dfrac{1}{3}(3x - 4)(2x - 1)^{3/2} - \dfrac{1}{5}(2x - 1)^{5/2} + C$

33. e **35.** $38{,}229/286$

37. $(7/2)\ln 2 - 3/4$ **39.** $1/4$

41. $1 - 11e^{-10}$ **43.** $4\ln 2 - 7/4$

45. $\dfrac{1}{2}x(x-3)|x-3| - \dfrac{1}{6}(x-3)^2|x-3| + C$

47. $2x|x-3| - (x-3)|x-3| + C$

49. $-x^2(-x+4)|-x+4| - \dfrac{2}{3}x(-x+4)^2|-x+4| - \dfrac{1}{6}(-x+4)^3|-x+4| + C$

51. $\dfrac{1}{2}(x^2 - 2x + 3)(x-4)|x-4| -$

$\dfrac{1}{3}(x-1)(x-4)^2|x-4| + \dfrac{1}{12}(x-4)^3|x-4| + C$

53. $28{,}800{,}000(1 - 2e^{-1})$ ft

55. $5{,}001 + 10x - 1/(x+1) - [\ln(x+1)]/(x+1)$

57. $\$4{,}252$ billion   **59.** 19 billion square feet

61. 20,800 million gallons

63. $e^{0.1x}(250[-8x^2 + 70x + 1{,}000] - 2{,}500[-16x + 70] - 400{,}000) + 325{,}000$ **65.** 33,598

67. a. $r(t) = -0.075t + 2.75 + [0.025t - 0.25]\dfrac{|t - 10|}{t - 10}$

b. 42.5 million people   **69.** Answers will vary. Examples are xe^{x^2} and $e^{x^2} = 1 \cdot e^{x^2}$.   **71.** Answers will vary. Examples are Exercises 31 and 32, or, more simply, integrals like $\int x(x+1)^5\,dx$.

73. Substitution   **75.** Parts   **77.** Substitution

79. Parts   **81.** $n + 1$ times   **83.** Proof.

Section 14.2

1. $16/3$ **3.** 9.75 **5.** 2 **7.** $31/3$ **9.** $8/3$ **11.** 4

13. $1/3$ **15.** 1 **17.** 2 **19.** 39 **21.** $e - 3/2$ **23.** $2/3$

25. $3/10$ **27.** $1/20$ **29.** $4/15$ **31.** $1/3$ **33.** 32

35. $2\ln 2 - 1$ **37.** $8\ln 4 + 2e - 16$ **39.** 0.9138

41. 0.3222 **43.** 112.5. This represents your total profit for the week, \$112.50.   **45.** 608; there were approximately 608,000 housing starts from the start of 2002 to the start of 2006 not for sale purposes.

47. a. Graph: (The upper curve is Myspace.)

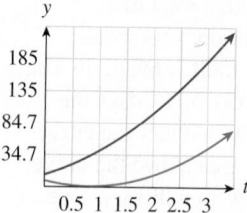

Myspace; about 93 million members   **b.** The area between the curves $y = f(t)$ and $y = m(t)$ over $[0.5, 2]$ **49.** $16{,}078(e^{0.051t} - 1) - 7{,}333.3(e^{0.06t} - 1)$; the total number of wiretaps authorized by federal courts from the start of 1990 up to time t was about $16{,}078(e^{0.051t} - 1) - 7{,}333.3(e^{0.06t} - 1)$.

51. Wrong: It could mean that the graphs of f and g cross, as shown in the caution at the start of this topic in the textbook.   **53.** The area between the export and import curves represents the United States's accumulated trade deficit (that is, the total excess of imports over exports) from 1960 to 2007.   **55.** (A)   **57.** The claim is wrong because the area under a curve can only represent income if the curve is a graph of income *per unit time*. The value of a stock price is not income per unit time—the income can be realized only when the stock is sold, and it amounts to the current market price. The total net income (per share) from the given investment would be the stock price on the date of sale minus the purchase price of \$50.

Section 14.3

1. Average $= 2$ **3.** Average $= 1$

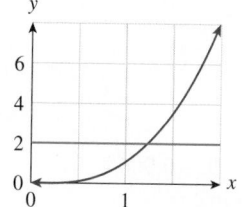   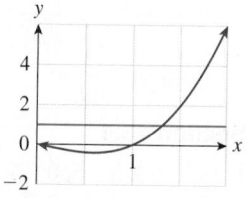

5. Average $= (1 - e^{-2})/2$ **7.** Average $= 17/8$

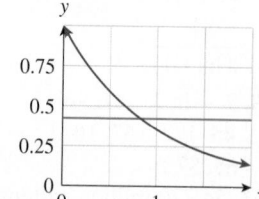   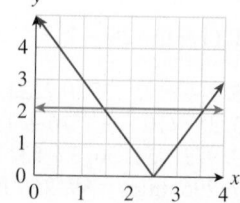

9.

x	0	1	2	3	4	5	6	7
$r(x)$	3	5	10	3	2	5	6	7
$\bar{r}(x)$			6	6	5	10/3	13/3	6

11.

x	0	1	2	3	4	5	6	7
$r(x)$	1	2	6	7	11	15	10	2
$\bar{r}(x)$			3	5	8	11	12	9

13. Moving average:

$$\bar{f}(x) = x^3 - (15/2)x^2 + 25x - 125/4$$

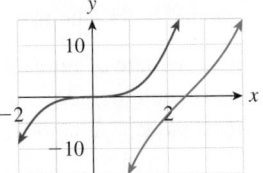

15. Moving average:

$$\bar{f}(x) = (3/25)[x^{5/3} - (x - 5)^{5/3}]$$

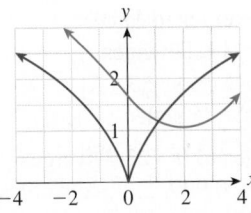

17. $\bar{f}(x) = \dfrac{2}{5}(e^{0.5x} - e^{0.5(x-5)})$

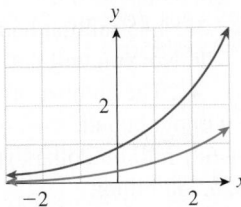

19. $\bar{f}(x) = \dfrac{2}{15}(x^{3/2} - (x - 5)^{3/2})$

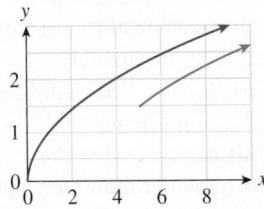

21. $\bar{f}(x) = \dfrac{1}{5}\left[5 - \dfrac{1}{2}|2x - 1| + \dfrac{1}{2}|2x - 11|\right]$

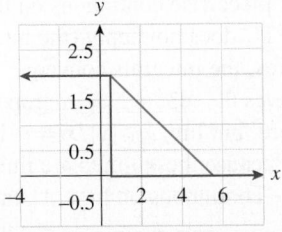

23. $\bar{f}(x) = \dfrac{1}{5}\left[2x - \dfrac{1}{2}(x + 1)|x + 1| + \dfrac{1}{2}x|x| - \right.$

$$\left. 2(x - 5) + \dfrac{1}{2}(x - 4)|x - 4| - \dfrac{1}{2}(x - 5)|x - 5|\right]$$

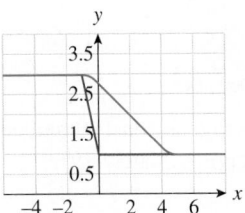

25.

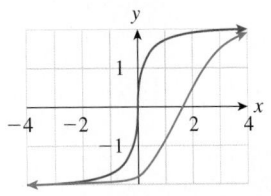

27.

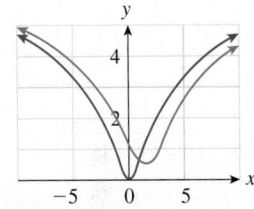

29.

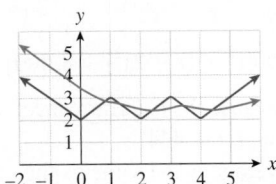

31.

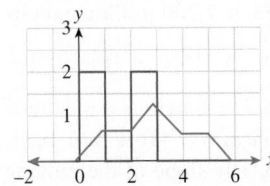

33.

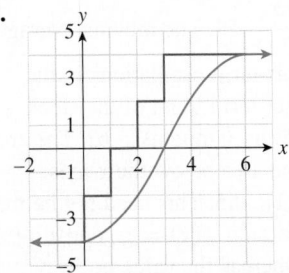

35. $1.8 million   **37.** 16 million members per year
39. 277 million tons   **41.** $10,410.88   **43.** $1,500
45. $2.5 billion per quarter
47.

Year t	2001	2002	2003	2004	2005	2006	2007	2008	2009	2010
Stock Price	39	35	41	51	56	77	94	80	68	73
Moving Average (rounded)				42	46	56	70	77	80	79

The moving average continued to rise, at a lower rate, until it began to fall in 2010.
49. a. To obtain the moving averages from January to June, use the fact that the data repeats every 12 months.

Graph:

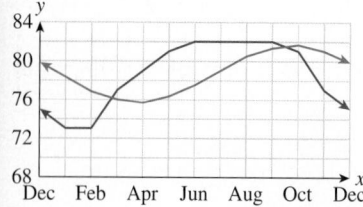

b. The 12-month moving average is constant and equal to the year-long average of approximately 79°.
51. a. Graph:

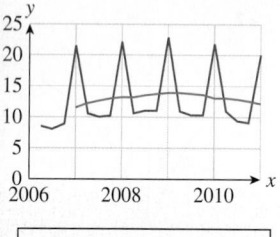

— Sales — Moving average

b. Approximately −0.2 million iPods per quarter
53. a. 7,200 million gallons per year
b. $\frac{1}{2}[-15[t^3 - (t-2)^3] + 450[t^2 - (t-2)^2] + 8,400]$
c. The function is quadratic because the t^3 terms cancel.
55. a. $s = 41t + 526$ **b.** $\bar{s}(t) = 41t + 444$
c. The slope of the moving average is the same as the slope of the original function.
57. $\bar{f}(x) = mx + b - \dfrac{ma}{2}$ **59.** The moving average "blurs" the effects of short-term oscillations in the price and shows the longer-term trend of the stock price. **61.** They repeat every 6 months. **63.** The area above the x-axis equals the area below the x-axis. Example: $y = x$ on $[-1,1]$ **65.** This need not be the case; for instance, the function $f(x) = x^2$ on $[0, 1]$ has average value $1/3$, whereas the value midway between the maximum and minimum is $1/2$.
67. (C) A shorter-term moving average most closely approximates the original function because it averages the function over a shorter period, and continuous functions change by only a small amount over a small period.

Section 14.4

1. $6.25 **3.** $512 **5.** $119.53 **7.** $900 **9.** $416.67
11. $326.27 **13.** $25 **15.** $0.50 **17.** $386.29
19. $225 **21.** $25.50 **23.** $12,684.63
25. $TV = \$300,000, FV = \$434,465.45$
27. $TV = \$350,000, FV = \$498,496.61$
29. $TV = \$389,232.76, FV = \$547,547.16$

31. $TV = \$100,000, PV = \$82,419.99$
33. $TV = \$112,500, PV = \$92,037.48$
35. $TV = \$107,889.50, PV = \$88,479.69$
37. $\bar{p} = \$5,000, \bar{q} = 10,000, CS = \$25$ million, $PS = \$100$ million. The total social gain is $125 million. **39.** €200 billion **41.** $1,190 billion
43. €220 billion **45.** $1,080 billion **47.** $1,943,162.44
49. $3,086,245.73 **51.** $73,802.63 **53.** $1,792,723.35
55. total **57.** $CS = \dfrac{1}{2m}(b - m\bar{p})^2$ **59.** She is correct, provided there is a positive rate of return, in which case the future value (which includes interest) is greater than the total value (which does not). **61.** $PV < TV < FV$

Section 14.5

1. Diverges **3.** Converges to $2e$ **5.** Converges to e^2
7. Converges to ½ **9.** Converges to $1/108$
11. Converges to $3 \times 5^{2/3}$ **13.** Diverges
15. Diverges **17.** Converges to $\dfrac{5}{4}(3^{4/5} - 1)$
19. Diverges **21.** Converges to 0 **23.** Diverges
25. Diverges **27.** 0.9, 0.99, 0.999, . . . ; converges to 1. **29.** 7.602, 95.38, 993.1, . . . ; diverges.
31. 1.368, 1.800, 1.937, 1.980, 1.994, 1.998, 1.999, 2.000, . . . ; converges to 2. **33.** 9.000, 99.00, 999.0, . . . ; diverges to $+\infty$. **35.** 4.45 million homes
37. 13,600 billion cigarettes **39.** No; you will not sell more than 2,000 of them. **41.** The number of graduates each year will rise without bound.
43. a. $R(t) = 350e^{-0.1t}(39t + 68)$ million dollars/yr;
b. $1,603,000 million **45.** $20,700 billion
47. $\int_0^{+\infty} N(t)\, dt$ diverges, indicating that there is no bound to the expected future total online sales of mousse. $\int_{-\infty}^0 N(t)\, dt$ converges to approximately 2.006, indicating that total online sales of mousse prior to the current year amounted to approximately 2 million gallons. **49.** 1 **51.** 0.1587 **53.** $70,833
55. a. 2.468 meteors on average **b.** The integral diverges. We can interpret this as saying that the number of impacts by meteors smaller than 1 megaton is very large. (This makes sense because, for example, this number includes meteors no larger than a grain of dust.) **57. a.** $\Gamma(1) = 1; \Gamma(2) = 1$ **59.** The integrand is neither continuous nor piecewise continuous on the interval $[-1, 1]$, so the FTC does not apply (the integral is improper). **61.** Yes; the integrals converge to 0, and the FTC also gives 0. **63. a.** Not improper. $|x|/x$ is not defined at zero, but $\lim_{x\to 0^-}|x|/x = -1$ and $\lim_{x\to 0^+}|x|/x = 1$. Because these limits are finite, the integrand is piecewise continuous on $[-1, 1]$ and

so the integral is not improper. **b.** Improper, because $x^{-1/3}$ has infinite left and right limits at 0. **c.** Improper, since $(x-2)/(x^2-4x+4)=1/(x-2)$, which has an infinite left limit at 2.   **65.** In all cases, you need to rewrite the improper integral as a limit and use technology to evaluate the integral of which you are taking the limit. Evaluate for several values of the endpoint approaching the limit. In the case of an integral in which one of the limits of integration is infinite, you may have to instruct the calculator or computer to use more subdivisions as you approach $+\infty$.
67. Answers will vary.

Section 14.6

1. $y = \dfrac{x^3}{3} + \dfrac{2x^{3/2}}{3} + C$ **3.** $\dfrac{y^2}{2} = \dfrac{x^2}{2} + C$

5. $y = Ae^{x^2/2}$ **7.** $y = -\dfrac{2}{(x+1)^2 + C}$

9. $y = \pm\sqrt{(\ln x)^2 + C}$ **11.** $y = \dfrac{x^4}{4} - x^2 + 1$

13. $y = (x^3 + 8)^{1/3}$ **15.** $y = 2x$ **17.** $y = e^{x^2/2} - 1$

19. $y = -\dfrac{2}{\ln(x^2+1)+2}$ **21.** With $s(t) = $ monthly

sales after t months, $\dfrac{ds}{dt} = -0.05s$; $s = 1{,}000$ when $t = 0$. Solution: $s = 1{,}000e^{-0.05t}$ quarts per month
23. a. $75 + 125e^{-0.05t}$ **b.** 64.4 minutes
25. $k \approx 0.04274$; $H(t) = 75 + 115e^{-0.04274t}$ degrees Fahrenheit after t minutes   **27.** With $S(t) = $ total

sales after t months, $\dfrac{ds}{dt} = 0.1(100{,}000 - S)$;

$S(0) = 0$. Solution: $S = 100{,}000(1 - e^{-0.1t})$ monitors after t months.   **29.** $q = 0.6078e^{-0.05p}p^{1.5}$
31. $y = e^{-t}(t + 1)$

33. $y = e^{t/2}\left[-2te^{-t/2} - 4e^{-t/2} + 5\right]$

35. $i = 5e^{-t}(e^t - e)\left[1 + \dfrac{|t - 1|}{t - 1}\right]$

Graph:

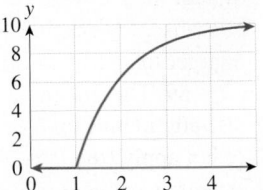

37. a. $\dfrac{dp}{dt} = k(D(p) - S(p)) = k(20{,}000 - 1{,}000p)$

b. $p = 20 - Ae^{-kt}$ **c.** $p = 20 - 10e^{-0.2231t}$ dollars after t months   **39.** Verification

41. $S = \dfrac{2/1{,}999}{e^{-0.5t} + 1/1{,}999}$

Graph:

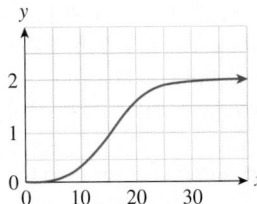

It will take about 27 months to saturate the market.
43. a. $y = be^{Ae^{-at}}$, $A = $ constant   **b.** $y = 10e^{-0.69315e^{-t}}$
Graph:

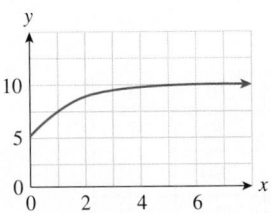

45. A general solution gives all possible solutions to the equation, using at least one arbitrary constant. A particular solution is one specific function that satisfies the equation. We obtain a particular solution by substituting specific values for any arbitrary constants in the general solution.   **47.** Example: $y'' = x$ has general solution $y = \frac{1}{6}x^3 + Cx + D$ (integrate twice).   **49.** $y' = -4e^{-x} + 3$

Chapter 14 Review

1. $(x^2 - 2x + 4)e^x + C$ **3.** $(1/3)x^3 \ln 2x - x^3/9 + C$

5. $\dfrac{1}{2}x(2x + 1)|2x + 1| - \dfrac{1}{12}(2x + 1)^2|2x + 1| + C$

7. $-5x|-x + 3| - \dfrac{5}{2}(-x + 3)|-x + 3| + C$

9. $-e^2 - 39/e^2$ **11.** $\dfrac{3}{2 \cdot 2^{1/3}} - \dfrac{1}{2}$ **13.** $\dfrac{2\sqrt{2}}{3}$ **15.** -1

17. $e - 2$ **19.** $3x - 2$ **21.** $\dfrac{3}{14}[x^{7/3} - (x-2)^{7/3}]$

23. \$1,600   **25.** \$2,500   **27.** 1/4   **29.** Diverges   **31.** 1

33. $y = -\dfrac{3}{x^3 + C}$ **35.** $y = \sqrt{2\ln|x| + 1}$

37. \$18,200   **39. a.** \$2,260
b. $50{,}000e^{0.01t}(1 - e^{-0.04}) \approx 1{,}960.53e^{0.01t}$
41. a. $\bar{p} = 20$, $\bar{q} = 40{,}000$ **b.** $CS = \$240{,}000$, $PS = \$30{,}000$ **43.** Approximately \$910,000
45. a. \$5,549,000   **b.** Principal: \$5,280,000, interest: \$269,000   **47.** \$51 million **49.** The amount in the account would be given by $y = 10{,}000/(1 - t)$, where t is time in years, so would approach infinity 1 year after the deposit.

Chapter 15

Section 15.1

1. a. 1 **b.** 1 **c.** 2 **d.** $a^2 - a + 5$ **e.** $y^2 + x^2 - y + 1$
f. $(x + h)^2 + (y + k)^2 - (x + h) + 1$ **3. a.** 0 **b.** 0.2
c. -0.1 **d.** $0.18a + 0.2$ **e.** $0.1x + 0.2y - 0.01xy$
f. $0.2(x + h) + 0.1(y + k) - 0.01(x + h)(y + k)$
5. a. 1 **b.** e **c.** e **d.** e^{x+y+z} **e.** $e^{x+h+y+k+z+l}$
7. a. Does not exist **b.** 0 **c.** 0 **d.** $xyz/(x^2 + y^2 + z^2)$
e. $(x + h)(y + k)(z + l)/[(x + h)^2 + (y + k)^2 + (z + l)^2]$ **9. a.** Increases; 2.3 **b.** Decreases; 1.4
c. Decreases; 1 unit increase in z **11.** Neither
13. Linear **15.** Linear **17.** Interaction
19. a. 107 **b.** -14 **c.** -113

21.

		$x \rightarrow$		
	10	**20**	**30**	**40**
y **10**	52	107	162	217
↓ **20**	94	194	294	394
30	136	281	426	571
40	178	368	558	748

25. 18, 4, 0.0965, 47,040 **27.** 6.9078, 1.5193, 5.4366, 0
29. Let $z =$ annual sales of Z (in millions of dollars),
$x =$ annual sales of X, and $y =$ annual sales of Y.
The model is $z = -2.1x + 0.4y + 16.2$.

31.

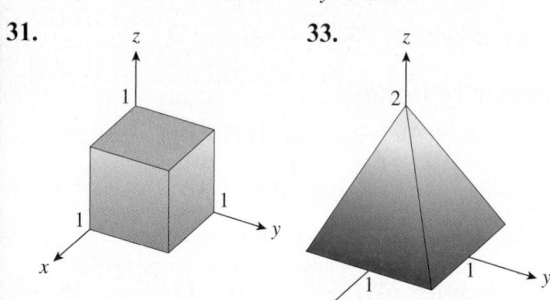

33.

35.

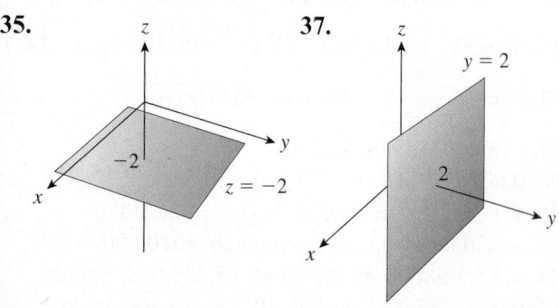

37.

39.

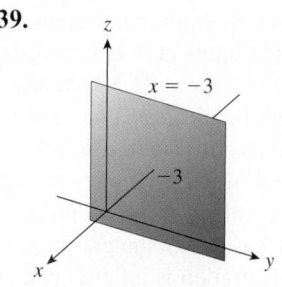

41. (H) **43.** (B) **45.** (F) **47.** (C)

49.

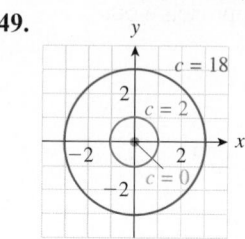

51.

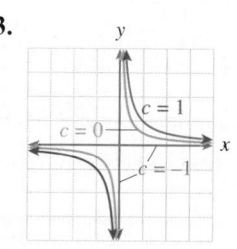

53.

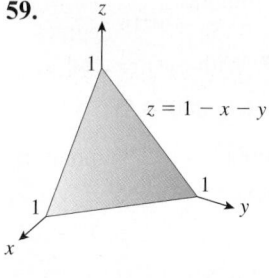

55. a. 4 **b.** 5 **c.** -1
57. $(2, 2)$ and $(-2, -2)$

59.

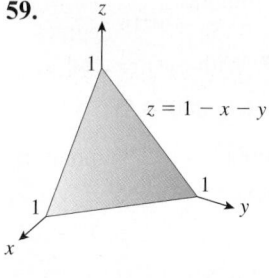

61.

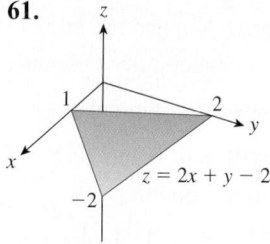

63.

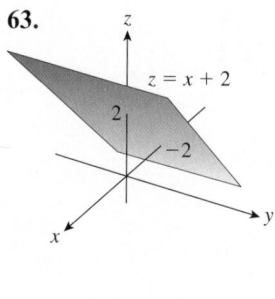

65.

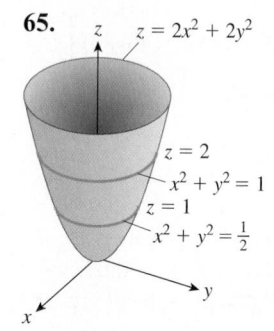

67.

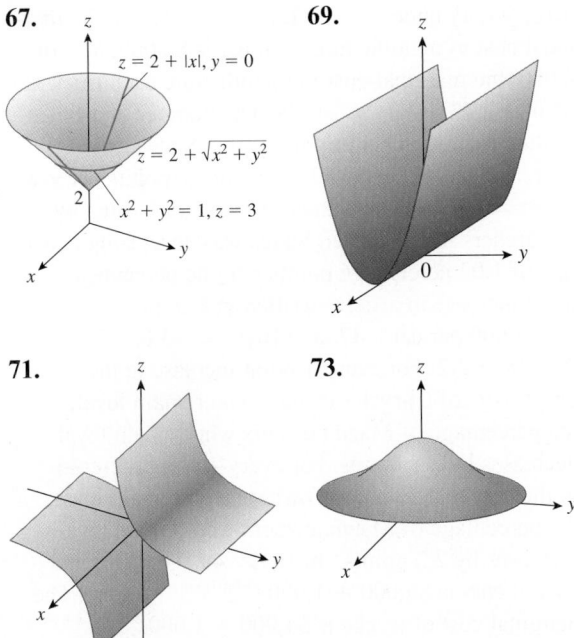

$z = 2 + |x|, y = 0$

$z = 2 + \sqrt{x^2 + y^2}$

$x^2 + y^2 = 1, z = 3$

69.

71.

73.

75. a. The marginal cost of cars is $6,000 per car. The marginal cost of trucks is $4,000 per truck. **b.** The graph is a plane with x-intercept -40, y-intercept -60, and z-intercept 240,000. **c.** The slice $x = 10$ is the straight line with equation $z = 300,000 + 4,000y$. It describes the cost function for the manufacture of trucks if car production is held fixed at 10 cars per week. **d.** The level curve $z = 480,000$ is the straight line $6,000x + 4,000y = 240,000$. It describes the number of cars and trucks you can manufacture to maintain weekly costs at $480,000.
77. $C(x, y) = 10 + 0.03x + 0.04y$, where C is the cost in dollars, $x = \#$ video clips sold per month, $y = \#$ audio clips sold per month   **79. a.** 28% **b.** 21%
c. Percentage points per year   **81.** The graph is a plane with x_1-intercept 0.3, x_2-intercept 33, and x_3-intercept 0.66. The slices by $x_1 = $ constant are straight lines that are parallel to each other. Thus, the rate of change of General Motors' share as a function of Ford's share does not depend on Chrysler's share. Specifically, GM's share decreases by 0.02 percentage points per one percentage-point increase in Ford's market share, regardless of Chrysler's share.   **83. a.** 189 thousand prisoners **b.** $y = 300$: $N = -0.02x + 51$; $y = 500$: $N = 0.02x + 67$; when there are 300,000 prisoners in local jails, the number in federal prisons decreases by 20 per 1,000 additional prisoners in state prisons. When there are 500,000 prisoners in local

jails, the number in federal prisons increases by 20 per 1,000 additional prisoners in state prisons.
85. a. The slices $x = $ constant and $y = $ constant are straight lines. **b.** No. Even though the slices $x = $ constant and $y = $ constant are straight lines, the level curves are not, and so the surface is not a plane. **c.** The slice $x = 10$ has a slope of 3,800. The slice $x = 20$ has a slope of 3,600. Manufacturing more cars lowers the marginal cost of manufacturing trucks.   **87. a.** $9,980 **b.** $R(z) = 9,850 + 0.04z$
89. $f(x, t) = 2.5x + 17.5t + 145$; $311 billion
91. $U(11, 10) - U(10, 10) \approx 5.75$. This means that, if your company now has 10 copies of Macro Publish and 10 copies of Turbo Publish, then the purchase of one additional copy of Macro Publish will result in a productivity increase of approximately 5.75 pages per day.   **93. a.** Answers will vary. $(a, b, c) = (3, 1/4, 1/\pi); (a, b, c) = (1/\pi, 3, 1/4)$.
b. $a = \left(\frac{3}{4\pi}\right)^{1/3}$. The resulting ellipsoid is a sphere with radius a. **95.** 7,000,000
97. a. $100 = K(1,000)^a(1,000,000)^{1-a}$; $10 = K(1,000)^a(10,000)^{1-a}$ **b.** $\log K - 3a = -4$; $\log K - a = -3$ **c.** $a = 0.5$, $K \approx 0.003162$ **d.** $P = 71$ pianos (to the nearest piano)   **99. a.** 4×10^{-3} gram per square meter **b.** The total weight of sulfates in the Earth's atmosphere   **101. a.** The value of N would be doubled. **b.** $N(R, f_p, n_e, f_l, f_i, L) = R f_p n_e f_l f_i L$, where L is the average lifetime of an intelligent civilization **c.** Take the logarithm of both sides, since this would yield the linear function $\ln(N) = \ln(R) + \ln(f_p) + \ln(n_e) + \ln(f_l) + \ln(f_i) + \ln(f_c) + \ln(L)$.
103. They are reciprocals of each other.   **105.** For example, $f(x, y) = x^2 + y^2$.   **107.** For example, $f(x, y, z) = xyz$.   **109.** For example, take $f(x, y) = x + y$. Then setting $y = 3$ gives $f(x, 3) = x + 3$. This can be viewed as a function of the single variable x. Choosing other values for y gives other functions of x.   **111.** The slope is independent of the choice of $y = k$.   **113.** That CDs cost more than cassettes   **115.** plane
117. (B) Traveling in the direction B results in the shortest trip to nearby isotherms, and hence the fastest rate of increase in temperature.   **119.** Agree: Any slice through a plane is a straight line.   **121.** The graph of a function of three or more variables lives in four-dimensional (or higher) space, which makes it difficult to draw and visualize.   **123.** We need one dimension for each of the variables plus one dimension for the value of the function.

Section 15.2

1. $f_x(x, y) = -40$; $f_y(x, y) = 20$; $f_x(1, -1) = -40$; $f_y(1, -1) = 20$ **3.** $f_x(x, y) = 6x + 1$; $f_y(x, y) = -3y^2$; $f_x(1, -1) = 7$; $f_y(1, -1) = -3$
5. $f_x(x, y) = -40 + 10y$; $f_y(x, y) = 20 + 10x$; $f_x(1, -1) = -50$; $f_y(1, -1) = 30$
7. $f_x(x, y) = 6xy$; $f_y(x, y) = 3x^2$; $f_x(1, -1) = -6$; $f_y(1, -1) = 3$ **9.** $f_x(x, y) = 2xy^3 - 3x^2y^2 - y$; $f_y(x, y) = 3x^2y^2 - 2x^3y - x$; $f_x(1, -1) = -4$; $f_y(1, -1) = 4$
11. $f_x(x, y) = 6y(2xy + 1)^2$; $f_y(x, y) = 6x(2xy + 1)^2$; $f_x(1, -1) = -6$; $f_y(1, -1) = 6$ **13.** $f_x(x, y) = e^{x+y}$; $f_y(x, y) = e^{x+y}$; $f_x(1, -1) = 1$; $f_y(1, -1) = 1$
15. $f_x(x, y) = 3x^{-0.4}y^{0.4}$; $f_y(x, y) = 2x^{0.6}y^{-0.6}$; $f_x(1, -1)$ undefined; $f_y(1, -1)$ undefined
17. $f_x(x, y) = 0.2ye^{0.2xy}$; $f_y(x, y) = 0.2xe^{0.2xy}$; $f_x(1, -1) = -0.2e^{-0.2}$; $f_y(1, -1) = 0.2e^{-0.2}$
19. $f_{xx}(x, y) = 0$; $f_{yy}(x, y) = 0$; $f_{xy}(x, y) = f_{yx}(x, y) = 0$; $f_{xx}(1, -1) = 0$; $f_{yy}(1, -1) = 0$; $f_{xy}(1, -1) = f_{yx}(1, -1) = 0$
21. $f_{xx}(x, y) = 0$; $f_{yy}(x, y) = 0$; $f_{xy}(x, y) = f_{yx}(x, y) = 10$; $f_{xx}(1, -1) = 0$; $f_{yy}(1, -1) = 0$; $f_{xy}(1, -1) = f_{yx}(1, -1) = 10$ **23.** $f_{xx}(x, y) = 6y$; $f_{yy}(x, y) = 0$; $f_{xy}(x, y) = f_{yx}(x, y) = 6x$; $f_{xx}(1, -1) = -6$; $f_{yy}(1, -1) = 0$; $f_{xy}(1, -1) = f_{yx}(1, -1) = 6$ **25.** $f_{xx}(x, y) = e^{x+y}$; $f_{yy}(x, y) = e^{x+y}$; $f_{xy}(x, y) = f_{yx}(x, y) = e^{x+y}$; $f_{xx}(1, -1) = 1$; $f_{yy}(1, -1) = 1$; $f_{xy}(1, -1) = f_{yx}(1, -1) = 1$ **27.** $f_{xx}(x, y) = -1.2x^{-1.4}y^{0.4}$; $f_{yy}(x, y) = -1.2x^{0.6}y^{-1.6}$; $f_{xy}(x, y) = f_{yx}(x, y) = 1.2x^{-0.4}y^{-0.6}$; $f_{xx}(1, -1)$ undefined; $f_{yy}(1, -1)$ undefined; $f_{xy}(1, -1)$ & $f_{yx}(1, -1)$ undefined
29. $f_x(x, y, z) = yz$; $f_y(x, y, z) = xz$; $f_z(x, y, z) = xy$; $f_x(0, -1, 1) = -1$; $f_y(0, -1, 1) = 0$; $f_z(0, -1, 1) = 0$
31. $f_x(x, y, z) = 4/(x + y + z^2)^2$; $f_y(x, y, z) = 4/(x + y + z^2)^2$; $f_z(x, y, z) = 8z/(x + y + z^2)^2$; $f_x(0, -1, 1)$ undefined; $f_y(0, -1, 1)$ undefined; $f_z(0, -1, 1)$ undefined   **33.** $f_x(x, y, z) = e^{yz} + yze^{xz}$; $f_y(x, y, z) = xze^{yz} + e^{xz}$; $f_z(x, y, z) = xy(e^{yz} + e^{xz})$; $f_x(0, -1, 1) = e^{-1} - 1$; $f_y(0, -1, 1) = 1$; $f_z(0, -1, 1) = 0$ **35.** $f_x(x, y, z) = 0.1x^{-0.9}y^{0.4}z^{0.5}$; $f_y(x, y, z) = 0.4x^{0.1}y^{-0.6}z^{0.5}$; $f_z(x, y, z) = 0.5x^{0.1}y^{0.4}z^{-0.5}$; $f_x(0, -1, 1)$ undefined; $f_y(0, -1, 1)$ undefined, $f_z(0, -1, 1)$ undefined
37. $f_x(x, y, z) = yze^{xyz}$, $f_y(x, y, z) = xze^{xyz}$, $f_z(x, y, z) = xye^{xyz}$; $f_x(0, -1, 1) = -1$; $f_y(0, -1, 1) = f_z(0, -1, 1) = 0$ **39.** $f_x(x, y, z) = 0$; $f_y(x, y, z) = -\dfrac{600z}{y^{0.7}(1 + y^{0.3})^2}$; $f_z(x, y, z) = \dfrac{2{,}000}{1 + y^{0.3}}$; $f_x(0, -1, 1)$ undefined; $f_y(0, -1, 1)$ undefined;

$f_z(0, -1, 1)$ undefined   **41.** $\partial C/\partial x = 6{,}000$, the marginal cost to manufacture each car is \$6,000. $\partial C/\partial y = 4{,}000$, the marginal cost to manufacture each truck is \$4,000.   **43.** $\partial y/\partial t = -0.78$. The number of articles written by researchers in the United States was decreasing at a rate of 0.78 percentage points per year. $\partial y/\partial x = -1.02$. The number of articles written by researchers in the United States was decreasing at a rate of 1.02 percentage points per one percentage-point increase in articles written in Europe.
45. \$5,600 per car   **47. a.** $\partial M/\partial c = -3.8$, $\partial M/\partial f = 2.2$. For every 1 point increase in the percentage of Chrysler owners who remain loyal, the percentage of Mazda owners who remain loyal decreases by 3.8 points. For every 1 point increase in the percentage of Ford owners who remain loyal, the percentage of Mazda owners who remain loyal increases by 2.2 points.   **b.** 16%   **49.** The marginal cost of cars is $\$6{,}000 + 1{,}000e^{-0.01(x+y)}$ per car. The marginal cost of trucks is $\$4{,}000 + 1{,}000e^{-0.01(x+y)}$ per truck. Both marginal costs decrease as production rises.   **51. a.** \$36,600; \$52,900   **b.** \$270 per year; \$410 per year   **c.** Widening   **d.** The rate at which the income gap is widening
53. $\bar{C}(x, y) = \dfrac{200{,}000 + 6{,}000x + 4{,}000y - 100{,}000e^{-0.01(x+y)}}{x + y}$; $\bar{C}_x(50, 50) = -\$2.64$ per car. This means that at a production level of 50 cars and 50 trucks per week, the average cost per vehicle is decreasing by \$2.64 for each additional car manufactured. $\bar{C}_y(50, 50) = -\$22.64$ per truck. This means that at a production level of 50 cars and 50 trucks per week, the average cost per vehicle is decreasing by \$22.64 for each additional truck manufactured.
55. No; your marginal revenue from the sale of cars is $\$15{,}000 - \dfrac{2{,}500}{\sqrt{x + y}}$ per car and $\$10{,}000 - \dfrac{2{,}500}{\sqrt{x + y}}$ per truck from the sale of trucks. These increase with increasing x and y. In other words, you will earn more revenue per vehicle with increasing sales, and so the rental company will pay more for each additional vehicle it buys.
57. $P_z(10, 100{,}000, 1{,}000{,}000) \approx 0.0001010$ papers/\$
59. a. $U_x(10, 5) = 5.18$, $U_y(10, 5) = 2.09$. This means that if 10 copies of Macro Publish and 5 copies of Turbo Publish are purchased, the company's daily productivity is increasing at a rate of 5.18 pages per day for each additional copy of Macro purchased and by 2.09 pages per day for each additional copy of Turbo purchased.   **b.** $\dfrac{U_x(10, 5)}{U_y(10, 5)} \approx 2.48$ is the ratio

of the usefulness of one additional copy of Macro to one of Turbo. Thus, with 10 copies of Macro and 5 copies of Turbo, the company can expect approximately 2.48 times the productivity per additional copy of Macro compared to Turbo.   **61.** 6×10^9 N/sec
63. a. $A_P(100, 0.1, 10) = 2.59$; $A_r(100, 0.1, 10) = 2{,}357.95$; $A_t(100, 0.1, 10) = 24.72$. Thus, for a \$100 investment at 10% interest, after 10 years the accumulated amount is increasing at a rate of \$2.59 per \$1 of principal, at a rate of \$2,357.95 per increase of 1 in r (note that this would correspond to an increase in the interest rate of 100%), and at a rate of \$24.72 per year. **b.** $A_P(100, 0.1, t)$ tells you the rate at which the accumulated amount in an account bearing 10% interest with a principal of \$100 is growing per \$1 increase in the principal, t years after the investment.

65. a. $P_x = Ka\left(\dfrac{y}{x}\right)^b$ and $P_y = Kb\left(\dfrac{x}{y}\right)^a$. They are equal precisely when $\dfrac{a}{b} = \left(\dfrac{x}{y}\right)^b \left(\dfrac{x}{y}\right)^a$. Substituting $b = 1 - a$ now gives $\dfrac{a}{b} = \dfrac{x}{y}$. **b.** The given information implies that $P_x(100, 200) = P_y(100, 200)$. By part (a), this occurs precisely when $a/b = x/y = 100/200 = 1/2$. But $b = 1 - a$, so $a/(1 - a) = 1/2$, giving $a = 1/3$ and $b = 2/3$.   **67.** Decreasing at 0.0075 parts of nutrient per part of water/sec   **69.** f is increasing at a rate of $\underline{s}$ units per unit of x, f is increasing at a rate of $\underline{t}$ units per unit of y, and the value of f is $\underline{r}$ when $x = \underline{a}$ and $y = \underline{b}$ **71.** the marginal cost of building an additional orbicus; zonars per unit.
73. Answers will vary. One example is $f(x, y) = -2x + 3y$. Others are $f(x, y) = -2x + 3y + 9$ and $f(x, y) = xy - 3x + 2y + 10$.   **75. a.** b is the z-intercept of the plane. m is the slope of the intersection of the plane with the xz-plane. n is the slope of the intersection of the plane with the yz-plane. **b.** Write $z = b + rx + sy$. We are told that $\partial z/\partial x = m$, so $r = m$. Similarly, $s = n$. Thus, $z = b + mx + ny$. We are also told that the plane passes through (h, k, l). Substituting gives $l = b + mh + nk$. This gives b as $l - mh - nk$. Substituting in the equation for z therefore gives $z = l - mh - nk + mx + ny = l + m(x - h) + n(y - k)$, as required.

Section 15.3

1. P: relative minimum; Q: none of the above; R: relative maximum   **3.** P: saddle point; Q: relative maximum; R: none of the above

5. Relative minimum   **7.** Neither
9. Saddle point   **11.** Relative minimum at $(0, 0, 1)$
13. Relative maximum at $(-1/2, 1/2, 3/2)$
15. Saddle point at $(0, 0, 0)$ **17.** Minimum at $(4, -3, -10)$ **19.** Maximum at $(-7/4, 1/4, 19/8)$
21. Relative maximum at $(0, 0, 0)$, saddle points at $(\pm 4, 2, -16)$ **23.** Relative minimum at $(0, 0, 0)$, saddle points at $(-1, \pm 1, 1)$ **25.** Relative minimum at $(0, 0, 1)$ **27.** Relative minimum at $(-2, \pm 2, -16)$, $(0, 0)$ a critical point that is not a relative extremum
29. Saddle point at $(0, 0, -1)$ **31.** Relative maximum at $(-1, 0, e)$ **33.** Relative minimum at $(2^{1/3}, 2^{1/3}, 3(2^{2/3}))$ **35.** Relative minimum at $(1, 1, 4)$ and $(-1, -1, 4)$ **37.** Absolute minimum at $(0, 0, 1)$
39. None; the relative maximum at $(0, 0, 0)$ is not absolute.   **41.** Minimum of $1/3$ at $(c, f) = (2/3, 2/3)$. Thus, at least $1/3$ of all Mazda owners would choose another new Mazda, and this lowest loyalty occurs when $2/3$ of Chrysler and Ford owners remain loyal to their brands.   **43.** It should remove 2.5 pounds of sulfur and 1 pound of lead per day.
45. You should charge \$580.81 for the Ultra Mini and \$808.08 for the Big Stack.   **47.** $l = w = h \approx 20.67$ in, volume $\approx 8{,}827$ cubic inches   **49.** 18 in $\times$ 18 in $\times$ 36 in, volume $= 11{,}664$ cubic inches

51.

53. Continues up indefinitely

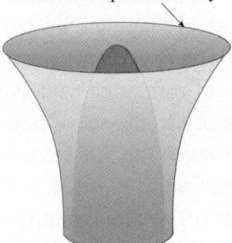

Continues down indefinitely
Function not defined on circle

55. H must be positive.   **57.** No. In order for there to be a relative maximum at (a, b), *all* vertical planes through (a, b) should yield a curve with a relative maximum at (a, b). It could happen that a slice by another vertical plane through (a, b) (such as $x - a = y - b$) does not yield a curve with a relative maximum at (a, b). [An example is $f(x, y) = x^2 + y^2 - \sqrt{xy}$, at the point $(0, 0)$. Look at the slices through $x = 0$, $y = 0$ and $y = x$.]

59. $\bar{C}_x = \dfrac{\partial}{\partial x}\left(\dfrac{C}{x+y}\right) = \dfrac{(x+y)C_x - C}{(x+y)^2}$. If this is

zero, then $(x+y)C_x = C$, or $C_x = \dfrac{C}{x+y} = \bar{C}$.

Similarly, if $\bar{C}_y = 0$, then $C_y = \bar{C}$. This is reasonable because if the average cost is decreasing with increasing x, then the average cost is greater than the marginal cost C_x. Similarly, if the average cost is increasing with increasing x, then the average cost is less than the marginal cost C_x. Thus, if the average cost is stationary with increasing x, then the average cost equals the marginal cost C_x. (The situation is similar for the case of increasing y.)   **61.** The equation of the tangent plane at the point (a, b) is $z = f(a, b) + f_x(a, b)(x - a) + f_y(a, b)(y - b)$. If f has a relative extremum at (a, b), then $f_x(a, b) = 0 = f_y(a, b)$. Substituting these into the equation of the tangent plane gives $z = f(a, b)$, a constant. But the graph of $z = constant$ is a plane parallel to the xy-plane.

Section 15.4

1. 1; $(0, 0, 0)$ **3.** 1.35; $(1/10, 3/10, 1/2)$ **5.** Minimum value $= 6$ at $(1, 2, 1/2)$ **7.** 200; $(20, 10)$ **9.** 16; $(2, 2)$ and $(-2, -2)$ **11.** 20; $(2, 4)$ **13.** 1; $(0, 0, 0)$ **15.** 1.35; $(1/10, 3/10, 1/2)$ **17.** Minimum value $= 6$ at $(1, 2, 1/2)$ **19. a.** $f(x, y, z)$ is the square of the distance from the point (x, y, z) to $(3, 0, 0)$, and the constraint tells us that (x, y, z) must lie on the paraboloid $z = x^2 + y^2$. Because there must be such a point (or points) on the paraboloid closest to $(3, 0, 0)$, the given problem must have at least one solution. Solution: $(x, y, z) = (1, 0, 1)$ for a minimum value of 5. **b.** Same solution as part (a) **c.** There are no critical points using this method. **d.** The constraint equation $y^2 = z - x^2$ tells us that $z - x^2$ cannot be negative, and thus restricts the domain of f to the set of points (x, y, z) with $z - x^2 \geq 0$. However, this information is lost when $z - x^2$ is substituted in the expression for f, and so the substitution in part (c) results in a different optimization problem; one in which there is no requirement that $z - x^2$ be ≥ 0. If we pay attention to this constraint we can see that the minimum will occur when $z = x^2$, which will then lead us to the correct solution.   **21.** $5 \times 10 = 50$ sq. ft.   **23.** $10
25. $(1/\sqrt{3}, 1/\sqrt{3}, 1/\sqrt{3})$, $(-1/\sqrt{3}, -1/\sqrt{3}, 1/\sqrt{3})$, $(1/\sqrt{3}, -1/\sqrt{3}, -1/\sqrt{3})$, $(-1/\sqrt{3}, 1/\sqrt{3}, -1/\sqrt{3})$
27. $(0, 1/2, -1/2)$ **29.** $(-5/9, 5/9, 25/9)$
31. $l \times w \times h = 1 \times 1 \times 2$ **33.** 18 in $\times$ 18 in $\times$ 36 in, volume $= 11{,}664$ cubic inches   **35.** $(2l/h)^{1/3} \times (2l/h)^{1/3} \times 2^{1/3}(h/l)^{2/3}$, where $l = $ cost of

lightweight cardboard and $h = $ cost of heavy-duty cardboard per square foot   **37.** $1 \times 1 \times 1/2$
39. 6 laborers, 10 robots for a productivity of 368 pairs of socks per day   **41.** Method 1: Solve $g(x, y, z) = 0$ for one of the variables and substitute in $f(x, y, z)$. Then find the maximum value of the resulting function of two variables. Advantage (answers may vary): We can use the second derivative test to check whether the resulting critical points are maxima, minima, saddle points, or none of these. Disadvantage (answers may vary): We may not be able to solve $g(x, y, z) = 0$ for one of the variables. Method 2: Use the method of Lagrange multipliers. Advantage (answers may vary): We do not need to solve the constraint equation for one of the variables. Disadvantage (answers may vary): The method does not tell us whether the critical points obtained are maxima, minima, points of inflection, or none of these.   **43.** If the only constraint is an equality constraint, and if it is impossible to eliminate one of the variables in the objective function by substitution (solving the constraint equation for a variable or some other method).   **45.** Answers may vary: Maximize $f(x, y) = 1 - x^2 - y^2$ subject to $x = y$.   **47.** Yes. There may be relative extrema at points on the boundary of the domain. The partial derivatives of the function need not be 0 at such points.   **49.** In a linear programming problem, the objective function is linear, and so the partial derivatives can never all be zero. (We are ignoring the simple case in which the objective function is constant.) It follows that the extrema cannot occur in the interior of the domain (since the partial derivatives must be zero at such points).

Section 15.5

1. $-1/2$ **3.** $e^2/2 - 7/2$ **5.** $(e^3 - 1)(e^2 - 1)$
7. 7/6   **9.** $[e^3 - e - e^{-1} + e^{-3}]/2$ **11.** 1/2
13. $(e - 1)/2$ **15.** 45/2   **17.** 8/3   **19.** 4/3   **21.** 0
23. 2/3   **25.** 2/3   **27.** $2(e - 2)$ **29.** 1/3

31. $\displaystyle\int_0^1 \int_0^{1-x} f(x, y)\, dy\, dx$

33. $\displaystyle\int_0^1 \int_0^{1-x} f(x, y)\, dy\, dx$

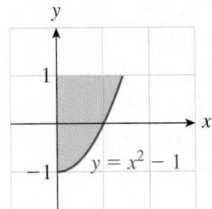

$y = x^2 - 1$

35. $\displaystyle\int_1^4 \int_1^{2/\sqrt{y}} f(x, y)\, dx\, dy$

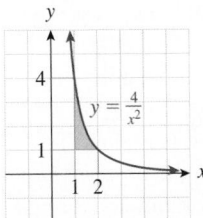

$y = \dfrac{4}{x^2}$

37. 4/3   **39.** 1/6   **41.** 162,000 gadgets
43. \$312,750   **45.** \$17,500   **47.** 8,216   **49.** 1 degree
51. The area between the curves $y = r(x)$ and
$y = s(x)$ and the vertical lines $x = a$ and $x = b$ is
given by $\int_a^b \int_{r(x)}^{s(x)} dy\, dx$ assuming that $r(x) \le s(x)$ for
$a \le x \le b$.   **53.** The first step in calculating an
integral of the form $\int_a^b \int_{r(x)}^{s(x)} f(x, y)\, dy\, dx$ is to evaluate
the integral $\int_{r(x)}^{s(x)} f(x, y)\, dy$, obtained by holding x
constant and integrating with respect to y.
55. Paintings per picasso per dali   **57.** Left-hand side
is $\int_a^b \int_c^d f(x)\, g(y)\, dx\, dy = \int_a^b \left(g(y) \int_c^d f(x)\, dx \right) dy$
(because $g(y)$ is treated as a constant in the inner
integral) $= \left(\int_c^d f(x)\, dx \right) \int_a^b g(y)\, dy$ (because
$\int_c^d f(x)\, dx$ is a constant and can therefore be taken
outside the integral). $\displaystyle\int_0^1 \int_1^2 y e^x\, dx\, dy = \frac{1}{2}(e^2 - e)$
no matter how we compute it.

Chapter 15 Review

1. 0; 14/3; 1/2; $\dfrac{1}{1+z} + z^3$; $\dfrac{x+h}{y+k+(x+h)(z+l)} +$
$(x+h)^2(y+k)$ **3.** Decreases by 0.32 units; increases
by 12.5 units   **5.** Reading left to right, starting
at the top: 4, 0, 0, 3, 0, 1, 2, 0, 2   **7.** Answers may
vary; two examples are $f(x, y) = 3(x - y)/2$ and
$f(x, y) = 3(x - y)^3/8$.

9.

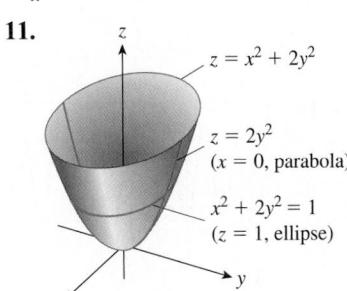

$z = x + y$

11.

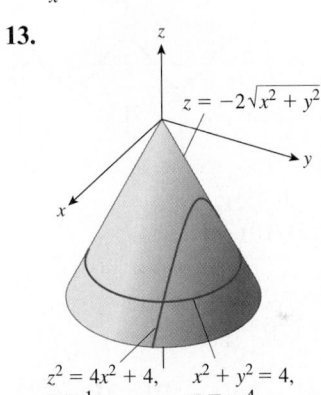

$z = x^2 + 2y^2$

$z = 2y^2$
$(x = 0,\ \text{parabola})$

$x^2 + 2y^2 = 1$
$(z = 1,\ \text{ellipse})$

13.

$z = -2\sqrt{x^2 + y^2}$

$z^2 = 4x^2 + 4,$ $x^2 + y^2 = 4,$
$y = 1$ $z = -4$

15. $f_x = 2x + y,\ f_y = x,\ f_{yy} = 0$ **17.** 0
19. $\dfrac{\partial f}{\partial x} = \dfrac{-x^2 + y^2 + z^2}{(x^2 + y^2 + z^2)^2},\ \dfrac{\partial f}{\partial y} = \dfrac{2xy}{(x^2 + y^2 + z^2)^2},$
$\dfrac{\partial f}{\partial z} = -\dfrac{2xz}{(x^2 + y^2 + z^2)^2},\ \dfrac{\partial f}{\partial x}\Big|_{(0,1,0)} = 1$
21. Absolute minimum at $(1, 3/2)$ **23.** Maximum at
$(0, 0)$, saddle points at $(\pm\sqrt{2}, 1)$ **25.** Saddle point at
$(0, 0)$ **27.** 1/27 at $(1/3, 1/3, 1/3)$ **29.** $(0, 2, \sqrt{2})$
31. 4; $(\sqrt{2}, \sqrt{2})$ and $(-\sqrt{2}, -\sqrt{2})$ **33.** $(0, 2, \sqrt{2})$
35. 2   **37.** ln 5   **39.** 1   **41. a.** $h(x, y) = 5{,}000 -$
$0.8x - 0.6y$ hits per day (x = number of new
customers at JungleBooks.com, y = number of
new customers at FarmerBooks.com) **b.** 250
c. $h(x, y, z) = 5{,}000 - 0.8x - 0.6y + 0.0001z$
(z = number of new Internet shoppers) **d.** 1.4 million
43. a. 2,320 hits per day **b.** $0.08 + 0.00003x$ hits
(daily) per dollar spent on television advertising per
month; increases with increasing x **c.** \$4,000 per
month **45.** (A)   **47. a.** About 15,800 additional
orders per day **b.** 11   **49.** \$23,050

Chapter 16

Section 16.1

1.
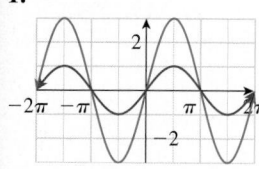

3.
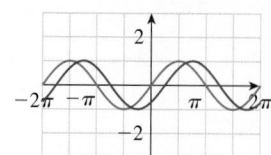

5.
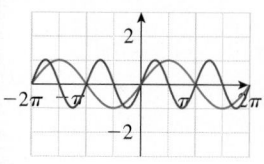

7.
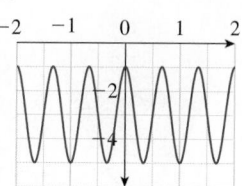

9.
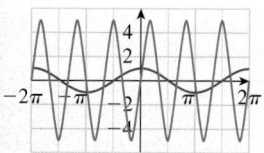

11.
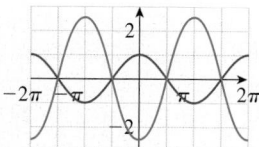

13. $f(x) = \sin(2\pi x) + 1$
15. $f(x) = 1.5 \sin(4\pi(x - 0.25))$
17. $f(x) = 50 \sin(\pi(x - 5)/10) - 50$
19. $f(x) = \cos(2\pi x)$
21. $f(x) = 1.5 \cos(4\pi(x - 0.375))$
23. $f(x) = 40 \cos(\pi(x - 10)/10) + 40$
25. $f(t) = 4.2 \sin(\pi/2 - 2\pi t) + 3$
27. $g(x) = 4 - 1.3 \sin[\pi/2 - 2.3(x - 4)]$ **31.** $\sqrt{3}/2$
37. $\tan(x + \pi) = \tan(x)$ **39. a.** $2\pi/0.602 \approx 10.4$
years. **b.** Maximum: $58.8 + 57.7 = 116.5 \approx 117$;
minimum: $58.8 - 57.7 = 1.1 \approx 1$
c. $1.43 + P/4 + 2P = 1.43 + 23.48 \approx 25$ years, or
about the beginning of 2022   **41. a.** Sales were highest when $t \approx 0, 4, 8,$ and 12, which correspond to the
end of the last quarter or beginning of the first quarter
of each year. Sales were lowest when $t \approx 2, 6,$ and 10,
which correspond to the beginning of the third quarter
of each year. **b.** The maximum quarterly sales were
approximately 21 million iPods per quarter; minimum
quarterly sales were approximately 9 million iPods
per quarter. **c.** Maximum: $15 + 6 = 21$; minimum:
$15 - 6 = 9$ **43.** Amplitude $= 6.00$, vertical
offset $= 15$, phase shift $= -1.85/1.51 \approx -1.23$,
angular frequency $= 1.51$, period ≈ 4.16. From 2008
through 2010, Apple's sales of iPods fluctuated in
cycles of 4.16 quarters about a baseline of 15 million
iPods per quarter. Every cycle, sales peaked at
$15 + 6 = 21$ million iPods per quarter and dipped to a
low of $15 - 6 = 9$ million iPods per quarter. Sales first

peaked at $t = -1.23 + (5/4) \times 4.16 = 3.97$, the end
of 2008.   **45.** $P(t) = 7.5 \sin[\pi(t - 13)/26] + 12.5$
47. $T(t) = 3.5 \cos[\pi(t - 7)/6] + 78.5$
49. $n(t) = 1.25 \sin[\pi(t - 1)/2] + 5.75$
51. $n(t) = 1.25 \cos[\pi(t - 2)/2] + 5.75$
53. $d(t) = 5 \sin(2\pi(t - 1.625)/13.5) + 10$
55. a. $u(t) = 2.5 \sin(2\pi(t - 0.75)) + 7.5$
b. $c(t) = 1.04^t[2.5 \sin(2\pi(t - 0.75)) + 7.5]$
57. a. $C \approx 50, A \approx 8, P \approx 12, \beta \approx 6$

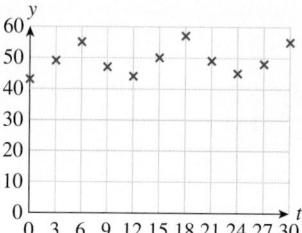

b. $f(t) = 5.882 \cos[2\pi(t - 5.696)/12.263] + 49.238$

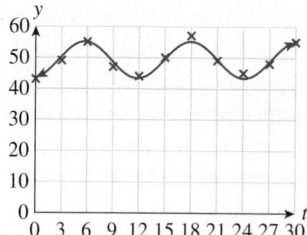

c. 12; 43; 55

59. a.

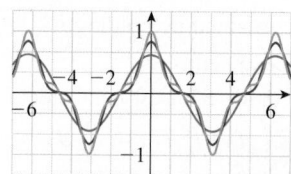

b. $y_{11} = \dfrac{2}{\pi} \cos x + \dfrac{2}{3\pi} \cos 3x + \dfrac{2}{5\pi} \cos 5x$

$\quad + \dfrac{2}{7\pi} \cos 7x + \dfrac{2}{9\pi} \cos 9x + \dfrac{2}{11\pi} \cos 11x$

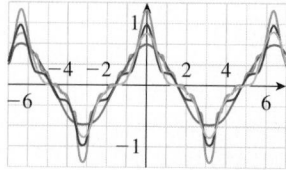

c. Multiply the amplitudes by 3 and change ω to $1/2$:

$$y_{11} = \dfrac{6}{\pi} \cos \dfrac{x}{2} + \dfrac{6}{3\pi} \cos \dfrac{3x}{2} + \dfrac{6}{5\pi} \cos \dfrac{5x}{2}$$

$$\quad + \dfrac{6}{7\pi} \cos \dfrac{7x}{2} + \dfrac{6}{9\pi} \cos \dfrac{9x}{2} + \dfrac{6}{11\pi} \cos \dfrac{11x}{2}.$$

61. The period is approximately 12.6 units.

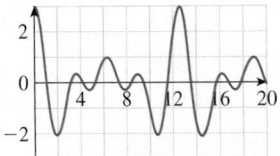

63. Lows: $B - A$; highs: $B + A$.   **65.** He is correct. The other trig functions can be obtained from the sine function by first using the formula $\cos x = \sin(x + \pi/2)$ to obtain cosine, and then using the formulas

$$\tan x = \frac{\sin x}{\cos x}, \quad \cot x = \frac{\cos x}{\sin x}, \quad \sec x = \frac{1}{\cos x},$$

$\csc x = \dfrac{1}{\sin x}$ to obtain the rest   **67.** The largest B

can be is A. Otherwise, if B is larger than A, the low figure for sales would have the negative value of $A - B$.

Section 16.2

1. $\cos x + \sin x$ **3.** $(\cos x)(\tan x) + (\sin x)(\sec^2 x)$
5. $-2 \csc x \cot x - \sec x \tan x + 3$
7. $\cos x - x \sin x + 2x$
9. $(2x - 1)\tan x + (x^2 - x + 1)\sec^2 x$
11. $-[\csc^2 x(1 + \sec x) + \cot x \sec x \tan x]/(1 + \sec x)^2$
13. $-2 \cos x \sin x$ **15.** $2 \sec^2 x \tan x$
17. $3 \cos(3x - 5)$ **19.** $2 \sin(-2x + 5)$

21. $\pi \cos\left[\dfrac{\pi}{5}(x - 4)\right]$ **23.** $-(2x - 1)\sin(x^2 - x)$

25. $(2.2x^{1.2} + 1.2)\sec(x^{2.2} + 1.2x - 1)$
$\tan(x^{2.2} + 1.2x - 1)$ **27.** $\sec x \tan x \tan(x^2 - 1) +$
$2x \sec x \sec^2(x^2 - 1)$ **29.** $e^x[-\sin(e^x) + \cos x - \sin x]$
31. $\sec x$ **37.** $e^{-2x}[-2 \sin(3\pi x) + 3\pi \cos(3\pi x)]$
39. $1.5[\sin(3x)]^{-0.5} \cos(3x)$

41. $\dfrac{x^4 - 3x^2}{(x^2 - 1)^2} \sec\left(\dfrac{x^3}{x^2 - 1}\right) \tan\left(\dfrac{x^3}{x^2 - 1}\right)$

43. $\dfrac{\cot(2x - 1)}{x} - 2 \ln|x| \csc^2(2x - 1)$

45. a. Not differentiable at 0   **b.** $f'(1) \approx 0.5403$
47. 0   **49.** 2   **51.** Does not exist   **53.** $1/\sec^2 y$
55. $-[1 + y \cos(xy)]/[1 + x \cos(xy)]$
57. $c'(t) = 7\pi \cos[2\pi(t - 0.75)]$; $c'(0.75) \approx \$21.99$
per *year* $\approx \$0.42$ per week   **59.** $N'(6) \approx -32.12$ On
January 1, 2003, the number of sunspots was
decreasing at a rate of 32.12 sunspots per year.
61. $c'(t) = 1.035^t \times [\ln(1.035)(0.8 \sin(2\pi t) + 10.2) +$

$1.6\pi \cos(2\pi t)]$; $c'(1) = 1.035[10.2 \ln|1.035| + 1.6\pi] \approx$
$\$5.57$ per year, or $\$0.11$ per week.
63. a. $d(t) = 5 \cos(2\pi t/13.5) + 10$ **b.** $d'(t) =$
$-(10\pi/13.5)\sin(2\pi t/13.5)$; $d'(7) \approx 0.270$. At noon,
the tide was rising at a rate of 0.270 feet per hour.

65. $\dfrac{dV}{dt} = 11{,}000\pi \dfrac{|\sin(100\pi t)|}{\sin(100\pi t)} \cos(100\pi t)$
Graph:

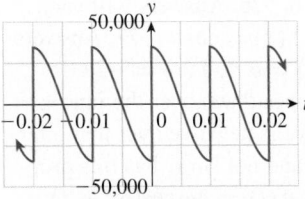

The sudden jumps in the graph are due to the
nondifferentiability of V at the times $0, \pm 0.01,$
$\pm 0.02, \ldots$. The derivative is negative immediately
to the left and positive immediately to the right of
these points.   **67. a.** 1.2 cm above the rest position
b. 0 cm/sec; not moving; moving downward at
18.85 cm/sec   **c.** 2.5 cycles per second   **69. a.** Moving
downward at 0.12 cm/sec; moving downward at
18.66 cm/sec   **b.** 0.1 sec
Graphs: of p:

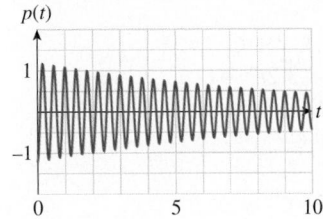

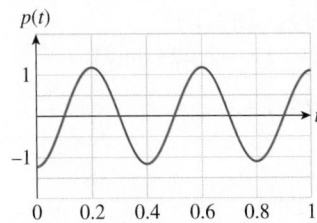

Graphs of p':

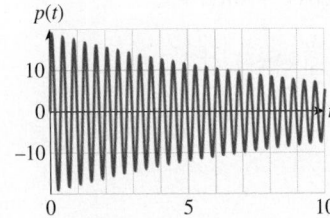

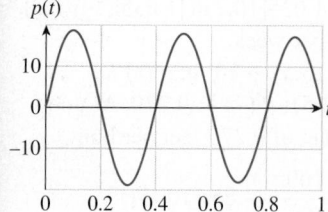

71. a. (III) b. Increasing at a rate of 0.157 degrees per thousand years   **73.** −6; 6   **75.** Answers will vary. Examples: $f(x) = \sin x$; $f(x) = \cos x$ **77.** Answers will vary. Examples: $f(x) = e^{-x}$; $f(x) = -2e^{-x}$
79. The graph of $\cos x$ slopes down over the interval $(0, \pi)$, so that its derivative is negative over that interval. The function $-\sin x$, and not $\sin x$, has this property.   **81.** The velocity is $p'(t) = A\omega \cos(\omega t + d)$, which is a maximum when its derivative, $p''(t) = -A\omega^2 \sin(\omega t + d)$, is zero. But this occurs when $\sin(\omega t + d) = 0$, so that $p(t)$ is zero as well, meaning that the stock is at yesterday's close.
83. The derivative of $\sin x$ is $\cos x$. When $x = 0$, this is $\cos(0) = 1$. Thus, the tangent to the graph of $\sin x$ at the point (0, 0) has slope 1, which means it slopes upward at 45°.

Section 16.3

1. $-\cos x - 2 \sin x + C$
3. $2 \sin x + 4.3 \cos x - 9.33x + C$
5. $3.4 \tan x + (\sin x)/1.3 - 3.2e^x + C$
7. $(7.6/3) \sin(3x - 4) + C$
9. $-(1/6) \cos(3x^2 - 4) + C$
11. $-2 \cos(x^2 + x) + C$
13. $(1/6) \tan(3x^2 + 2x^3) + C$
15. $-(1/6) \ln|\cos(2x^3)| + C$
17. $3 \ln|\sec(2x - 4) + \tan(2x - 4)| + C$
19. $(1/2) \sin(e^{2x} + 1) + C$ **21.** −2   **23.** $\ln(2)$
25. 0   **27.** 1   **33.** $-\dfrac{1}{4} \cos(4x) + C$
35. $-\sin(-x + 1) + C$
37. $[\cos(-1.1x - 1)]/1.1 + C$
39. $-\dfrac{1}{4} \ln|\sin(-4x)| + C$
41. 0   **43.** 2π **45.** $-x \cos x + \sin x + C$
47. $\left[\dfrac{x^2}{2} - \dfrac{1}{4}\right] \sin(2x) + \dfrac{x}{2} \cos(2x) + C$
49. $-\dfrac{1}{2} e^{-x} \cos x - \dfrac{1}{2} e^{-x} \sin x + C$ **51.** $\pi^2 - 4$

53. Average $= 2/\pi$

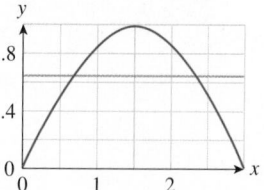

55. Diverges   **57.** Converges to 1/2
59. $C(t) = 0.04t + \dfrac{2.6}{\pi} \cos\left[\dfrac{\pi}{26}(t - 25)\right] + 1.02$
61. 12 feet   **63.** 79 sunspots
65. $P(t) = 7.5 \sin[(\pi/26)(t - 13)] + 12.5$; 7.7%
67. a. Average voltage over [0, 1/6] is zero; 60 cycles per second.
b.

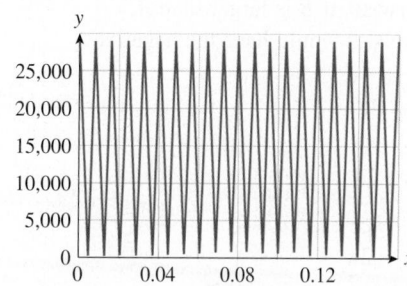

c. 116.673 volts   **69.** $50,000   **71.** It is always zero.
73. 1   **75.** $s = -\dfrac{K}{\omega^2} \sin(\omega t - \alpha) + Lt + M$ for constants L and M

Chapter 16 Review

1. $f(x) = 1 + 2 \sin x$
3. $f(x) = 2 + 2 \sin[\pi(x - 1)] = 2 + 2 \sin[\pi(x + 1)]$
5. $f(x) = 1 + 2 \cos(x - \pi/2)$
7. $f(x) = 2 + 2 \cos[\pi(x + 1/2)] =$ $2 + 2 \cos[\pi(x - 3/2)]$ **9.** $-2x \sin(x^2 - 1)$
11. $2e^x \sec^2(2e^x - 1)$ **13.** $4x \sin(x^2) \cos(x^2)$
15. $2 \sin(2x - 1) + C$ **17.** $\tan(2x^2 - 1) + C$
19. $-\dfrac{1}{2} \ln|(\cos(x^2 + 1)| + C$ **21.** 1
23. $-x^2 \cos x + 2x \sin x + 2 \cos x + C$
25. $s(t) = 10{,}500 + 1{,}500 \sin[(2\pi/52)t - \pi] =$ $10{,}500 + 1{,}500 \sin(0.12083t - 3.14159)$
27. Decreasing at a rate of $3,852 per month
29. $2,029,700
31. $150t - \dfrac{100}{\pi} \cos\left[\dfrac{\pi}{2}(t - 1)\right]$ grams

Index